Springer Collected Works in Mathematics

T0171913

For further volumes:
http://www.springer.com/series/11104

Goro Shimura

Collected Papers IV

1989 – 2001

Reprint of the 2003 Edition

Springer

Goro Shimura
Mathematics Department
Princeton University
Princeton, NJ 08544-1000
USA

ISSN 2194-9875
ISBN 978-1-4939-1837-9 (Softcover)
978-3-540-18087-6 (Hardcover)
DOI 10.1007/978-1-4939-1838-6
Springer New York Heidelberg Dordrecht London

Library of Congress Control Number: 2012954381

Printed on acid-free paper

Springer is part of Springer Science+Business Media (www.springer.com)

Contents

List of Articles

The numbers in parentheses are the articles not included in this collection.

Volume I

[(52)] On a certain ideal of the center of a Frobeniusean algebra, Scientific Papers of the College of General Education, University of Tokyo, 2 (1952), 117–124.

[54] A note on the normalization-theorem of an integral domain, Scientific Papers of the College of General Education, University of Tokyo, 4 (1954), 1–8.

[55] Reduction of algebraic varieties with respect to a discrete valuation of the basic field, American Journal of Mathematics, 77 (1955), 134–176.

[56] On complex multiplications, Proceedings of the International Symposium on Algebraic Number Theory, Tokyo-Nikko, 1955, Science Council of Japan, Tokyo (1956), 23–30.

[57a] La fonction ζ du corps des fonctions modulaires elliptiques, Comptes rendus des séances de l'Académie des Sciences, 244 (1957), 2127–2130.

[(57b)] Kindai-teki Seisu-ron (Modern Number Theory, in Japanese, with Yutaka Taniyama), Kyoritsu Shuppan, 1957.

[58a] Correspondances modulaires et les fonctions zeta de courbes algébriques, Journal of the Mathematical Society of Japan, 10 (1958), 1–28.

[58b] Modules des variétés abéliennes polarisées et fonctions modulaires, Séminaire Henri Cartan, École Normale Supérieure, 1957/58, *Fonctions Automorphes*, Exposé 18–20 (1958).

[(58c)] Fonctions automorphes et variétés abéliennes, Séminaire Bourbaki, 1957/58, Exposé 167 (1958).

[59a] Fonctions automorphes et correspondances modulaires, Proceedings of the International Congress of Mathematicians, Edinburgh, 1958 (1959), 330–338.

[59b] On the theory of automorphic functions, Annals of Mathematics, 70 (1959), 101–144.

[59c] Sur les intégrales attachées aux formes automorphes, Journal of the Mathematical Society of Japan, 11 (1959), 291–311.

[59d] On specializations of abelian varieties (with Shoji Koizumi), Scientific Papers of the College of General Education, University of Tokyo, 9 (1959), 187–211.

[60a] On vector differential forms attached to automorphic forms (with Michio Kuga), Journal of the Mathematical Society of Japan, 12 (1960), 258–270.

[(60b)] Automorphic functions and number theory: I (in Japanese), Sugaku, 11 (1960), 193–205.

[(61a)] Complex multiplication of abelian varieties and its applications to number theory (with Yutaka Taniyama), Publications of the Mathematical Society of Japan, No. 6 (1961).

[61b] On the zeta functions of the algebraic curves uniformized by certain automorphic functions, Journal of the Mathematical Society of Japan, 13 (1961), 275–331.

[(61c)] Automorphic functions and number theory: II (in Japanese), Sugaku, 13 (1961), 65–80.

[62a] On Dirichlet series and abelian varieties attached to automorphic forms, Annals of Mathematics, 76 (1962), 237–294.

[62b] On the class-fields obtained by complex multiplication of abelian varieties, Osaka Mathematical Journal, 14 (1962), 33-44.

[63a] Arithmetic of alternating forms and quaternion hermitian forms, Journal of the Mathematical Society of Japan, 15 (1963), 33–65.

[63b] On analytic families of polarized abelian varieties and automorphic functions, Annals of Mathematics, 78 (1963), 149–192.

[63c] On the cohomology groups attached to certain vector valued differential forms on the product of the upper half planes (with Yozo Matsushima), Annals of Mathematics, 78 (1963), 417–449.

[63d] On modular correspondences for $Sp(n, \mathbf{Z})$ and their congruence relations, Proceedings of the National Academy of Sciences, 49 (1963), 824–828.

[63e] On the fields of definition for fields of automorphic functions, Proceedings of the 1963 Number Theory Conference, Boulder, Colorado, August 5–24, 1963, 26–32.

[64a] Arithmetic of unitary groups, Annals of Mathematics, 79 (1964), 369–409.

[64b] On the field of definition for a field of automorphic functions, Annals of Mathematics, 80 (1964), 160–189.

[64c] Class-fields and automorphic functions, Annals of Mathematics, 80 (1964), 444–463.

[64d] On purely transcendental fields of automorphic functions of several variables, Osaka Journal of Mathematics, 1 (1964), 1–14.

[64e] The zeta function of an algebraic variety and automorphic functions, Summer Research Institute on Algebraic Geometry, Woods Hole, Massachusetts, July 6–31, 1964 (mimeographed notes of lectures).

[65a] On the field of definition for a field of automorphic functions: II, Annals of Mathematics, 81 (1965), 124–165.

[65b] On the zeta function of a fibre variety whose fibres are abelian varieties (with Michio Kuga), Annals of Mathematics, 82 (1965), 478–539.

[66a] A reciprocity law in non-solvable extensions, Journal für die reine und angewandte Mathematik, 221 (1966), 209–220.

[66b] Moduli and fibre systems of abelian varieties, Annals of Mathematics, 83 (1966), 294–338.

[66c] On the field of definition for a field of automorphic functions: III, Annals of Mathematics, 83 (1966), 377–385.

[66d] Moduli of abelian varieties and number theory, Proceedings of the Symposium on Pure Mathematics, 9, *Algebraic groups and discontinuous subgroups*, 1965, American Mathematical Society, (1966), 312–332.

Volume II

[67a] Discontinuous groups and abelian varieties, Mathematische Annalen, 168 (1967), 171–199.

[67b] Construction of class fields and zeta functions of algebraic curves, Annals of Mathematics, 85 (1967), 58–159.

[67c] Number fields and zeta functions associated with discontinuous groups and algebraic varieties, Proceedings of the International Congress of Mathematicians, Moscow, 1966, 100–107 (1967).

[67d] Algebraic number fields and symplectic discontinuous groups, Annals of Mathematics, 86 (1967), 503–592.

[68a] Algebraic varieties without deformation and the Chow variety, Journal of the Mathematical Society of Japan, 20 (1968), 336–341.

[(68b)] Automorphic functions and number theory, Lecture notes in mathematics 54, Springer, 1968.

[68c] An ℓ-adic method in the theory of automorphic forms, the text of a lecture at the conference *Automorphic functions for arithmetically defined groups*, Oberwolfach, Germany, July 28–August 3, 1968.

[69] Local representations of Galois groups, Annals of Mathematics, 89 (1969), 99–124.

[70a] On canonical models of arithmetic quotients of bounded symmetric domains, Annals of Mathematics, 91 (1970), 144–222.

[70b] On canonical models of arithmetic quotients of bounded symmetric domains: II, Annals of Mathematics, 92 (1970), 528–549.

[71a] On arithmetic automorphic functions, Proceedings of the International Congress of Mathematicians, Nice, 1970, vol. 2, 343–348 (1971).

[71b] On the zeta-function of an abelian variety with complex multiplication, Annals of Mathematics, 94 (1971), 504–533.

[71c] Class fields over real quadratic fields in the theory of modular functions, in *Several Complex Variables* II, Maryland 1970, Lecture notes in mathematics 185 (1971), 169–188.

[(71d)] Introduction to the arithmetic theory of automorphic functions, Publications of the Mathematical Society of Japan, No. 11, Iwanami Shoten and Princeton University Press, 1971.

[71e] On elliptic curves with complex multiplication as factors of the Jacobians of modular function fields, Nagoya Mathematical Journal, 43 (1971), 199–208.

[72a] On the field of rationality for an abelian variety, Nagoya Mathematical Journal, 45 (1972), 167–178.

[72b] Class fields over real quadratic fields and Hecke operators, Annals of Mathematics, 95 (1972), 130–190.

[73a] On modular forms of half integral weight, Annals of Mathematics, 97 (1973), 440–481.

[(73b)] Complex multiplication, Proceedings of the International Summer School of Modular Functions of One Variable, Antwerp, 1972, Lecure notes in mathematics, 320 (1973), 37–56.

[(73c)] Modular forms of half integral weight, Proceedings of the International Summer School of Modular Functions of One Variable, Antwerp, 1972, Lecure notes in mathematics, 320 (1973), 57–74.

[73d] On the factors of the jacobian variety of a modular function field, Journal of the Mathematical Society of Japan, 25 (1973), 523–544.

[74] On the trace formula for Hecke operators, Acta mathematica, 132 (1974), 245–281.

[75a] On the holomorphy of certain Dirichlet series, Proceedings of the London Mathematical Society, 3rd ser. 31 (1975), 79–98.

[75b] On the real points of an arithmetic quotient of a bounded symmetric domain, Mathematische Annalen, 215 (1975), 135–164.

Volume III

Volume IV

[90a] Invariant differential operators on hermitian symmetric spaces, Annals of Mathematics, 132 (1990), 237–272.

[90b] On the fundamental periods of automorphic forms of arithmetic type, Inventiones mathematicae, 102 (1990), 399–428.

[(90c)] Some old and recent arithmetical results concerning modular forms and related zeta functions, University of Istanbul, Faculty of Science, Journal of Mathematics, 49 (1990), 45–56.

[91] The critical values of certain Dirichlet series attached to Hilbert modular forms, Duke Mathematical Journal, 63 (1991), 557–613.

[(93a)] Arithmeticity of the special values of various zeta functions and the periods of abelian integrals (in Japanese), Sugaku, 45 (1993), 111–127; English translation by Toshitsune Miyake, Sugaku expositions, 8 (1995), 17–38.

[93b] On the transformation formulas of theta series, American Journal of Mathematics, 115 (1993), 1011–1052.

[93c] On the Fourier coefficients of Hilbert modular forms of half-integral weight, Duke Mathematical Journal, 71 (1993), 501–557.

[94a] Fractional and trigonometric expressions for matrices, American Mathematical Monthly, 101 (1994), 744–758.

[94b] Euler products and Fourier coefficients of automorphic forms on symplectic groups, Inventiones mathematicae, 116 (1994), 531–576.

[94c] Differential operators, holomorphic projection, and singular forms, Duke Mathematical Journal, 76 (1994), 141–173.

[95a] Eisenstein series and zeta functions on symplectic groups, Inventiones mathematicae, 119 (1995), 539–584.

[95b] Zeta functions and Eisenstein series on metaplectic groups, Inventiones mathematicae, 121 (1995), 21–60.

[96a] Convergence of zeta functions on symplectic and metaplectic groups, Duke Mathematical Journal, 82 (1996), 327–347.

[96b] Response, Notices of the American Mathematical Society, vol. 43, No. 11 (November 1996), 1344–1347.

[(97a)] Euler Products and Eisenstein series, CBMS Regional Conference Series in Mathematics, No. 93, American Mathematical Society, 1997.

[97b] Zeta functions and Eisenstein series on classical groups, Proceedings of the National Academy of Sciences, 94, 11133–11137 (1997).

[(98)] Abelian varieties with complex multiplication and modular functions, Princeton University Press, 1998.

[99a] An exact mass formula for orthogonal groups, Duke Mathematical Journal, 97 (1999), 1–66.

[99b] The number of representations of an integer by a quadratic form, Duke Mathematical Journal, 100 (1999), 59–92.

[99c] Generalized Bessel functions on symmetric spaces, Journal für die reine und angewandte Mathematik, 509 (1999), 35–66.

[99d] Some exact formulas on quaternion unitary groups, Journal für die reine und angewandte Mathematik, 509 (1999), 67–102.

[99e] André Weil as I knew him, Notices of the American Mathematical Society, vol. 46, No. 4 (April 1999), 428–433

[(00)] Arithmeticity in the theory of automorphic forms, Mathematical Surveys and Monographs, vol. 82, American Mathematical Society, 2000.

[(01a)] Letter to the editor, Notices of the American Mathematical Society, vol. 48, No. 7 (August 2001), 678.

[01b] Arithmeticity of Dirichlet series and automorphic forms on unitary groups, unpublished.

[01c] The relative regulator of an algebraic number field, unpublished.

89a

Yutaka Taniyama and his time

Very personal recollections

Bulletin of the London Mathematical Society, 21 (1989), 186-196

To write about Taniyama's time, first I have to emphasize that it was the mid to late 1950s, and the situation was totally different from that of Japan today, to say nothing of the comparison with the United States or Europe, now or then. Pollution was not a household word in those days, and in fine weather one could see, from the center of Tokyo, Mt Fuji on the western horizon 70 miles away, with its snowed crown in the morning and silhouetted in the evening. The destruction and deprivation of wartime and the succeeding period were things of the past, but not forgotten. We were no longer hungry. The whole country was aspiring and hopeful, but still very poor. This was so in a collective sense and also on the individual level. Taniyama and his peers were no exception, though it may be said, in any country at any time, one is usually both ambitious and poor at the beginning of one's career.

He was not particularly poorer than others, and I think he never had any great financial difficulties, yet his life was anything but comfortable, as it was for most of us at that time. At least, he enjoyed a fair portion of the universal poverty of the period. For example, he lived in a one-room apartment which consisted of 81 square feet of living space, a sink, and a tiny unfloored part behind the door. Running water, gas and electricity were provided separately in each room, but there was only one toilet on each floor of the two-storey building, shared by all the occupants of the dozen or so rooms of the floor. I remember that his was No. 20 on the second floor, close to the last. Thus it was more like a dormitory than an apartment, but it was more or less typical of the time. To take a bath, he had to go to a public bathhouse, a few minutes' walk from his apartment. The building, a rather shabby wooden structure, was named poetically 'Villa Tranquil Mountains', but that expressed merely an unfulfilled desire, because it stood on a narrow but lively street closely lined by small retail shops, and besides, trains passed on the nearby railroad every few minutes. There was no central heating system; air conditioning was inconceivable at such a place. However, many of the innumerable coffee shops in Tokyo offered a certain luxury of coolness whenever needed, as well as a place for endless mathematical and nonmathematical conversation, with the price of 50 yen a cup of coffee. It was the time when one dollar was 360 yen, and his monthly salary as a lecturer at the University of Tokyo was less than ¥15,000.

Speaking about domestic affairs, he was a lazy type, and so he rarely cooked, and he ate out often at small restaurants. One of his favorite dishes available at several

---

Editors' note. This personal tribute to Y. Taniyama by his friend is published some thirty years after his death in recognition of his outstanding qualities as a man and because of his influence on the development of number theory and algebraic geometry. He was not a member of the London Mathematical Society, but perhaps might have become one in due time. The notes on Taniyama's problems, which have had a profound influence on work on elliptic curves and automorphic functions, were specially translated from the article in *Sûgaku* by G. Shimura, who also added the comments on them for this Notice.

western style restaurants was tongue stew, ¥250 à la carte, another of the few items of modest luxury in which he could indulge occasionally. He almost always wore—he exclusively wore, I am tempted to say—except in the summer, a blue-green suit with a strange metallic sheen. Once he explained to me how he got hold of the suit. His father bought the material very cheap from a peddler. Because of the luster, however, nobody in his family dared to wear it, and finally he volunteered to have it tailored for himself, as he didn't care much about how he looked. His shoelaces were always loose, and he often dragged them on the ground; since he was incapable of keeping them securely tied all the time, he decided not to concern himself about tying them again when they got loose.

Such was the mathematician who departed from this life so quickly, leaving an everlasting source of inspiration to his generation and also to the coming ones.

Yutaka Taniyama was born November 12, 1927, the third son and sixth child of Sahei and Kaku Taniyama. He had three brothers and four sisters. Both his parents lived beyond ninety. His first name, expressed by a single Chinese character, can be pronounced 'Toyo', and he once told me, if my recollection is correct, that was the originally intended pronunciation. But as he grew up, most people, especially those outside his family, read the character 'Yutaka', and so he accepted it at some point, and he was Taniyama Yutaka since then. At least all his papers were written under that name, if in reverse order. I have no information about his childhood, nor his pre-college years, except that he suffered from tuberculosis while attending a senior high school, and had to be out of school for two years. As far as I can remember, he was coughing every ten to fifteen minutes.

His father was a locally well known country doctor practising pediatrics, and often medicine in general, as this type of profession ordinarily demanded. I met him only once. He was a vigorous man in his early eighties, and seemed to belong to the so-called self-made type. Shortly after our encounter, he sent a letter to one of my colleagues at the University of Tokyo, who met him at the same time. Somehow the old man formed an idea that the young scholar was not doing well academically, and advised him to take food rich in vitamin B (or C or possibly calcium) in order to make his brain work more efficiently. Since this was after Yutaka's passing away, I was never able to check whether the father gave a similar piece of advice to the son.

Taniyama graduated from the University of Tokyo in March 1953, I in 1952, though he was older than I; the delay was caused by his illness. I knew him since 1950, but our first serious mathematical contact was in early 1954, when I wrote him a letter requesting that he return the copy of *Mathematische Annalen*, Vol. 124, which contained the paper by Deuring on his algebraic theory of complex multiplication. Taniyama had checked it out from the library some weeks ago. In December of the previous year, I had sent my manuscript on reduction modulo *p* of algebraic varieties to André Weil, who was at Chicago, and I was intending to apply the theory to abelian varieties, in particular to elliptic curves. In his reply to me, Taniyama wrote that he had the same intention, and politely asked me to explain my theory to him some time. In retrospect, I think he was, with a wider knowledge and a better perspective, more mature than I mathematically, though I didn't know it at that time.

I still keep the postcard, postmarked January 23, 1954. After more than thirty years, it naturally aged, but clearly shows his handwriting. It bears the address of his parents' home, where he was temporarily staying. It was in a small undistinguished town called Kisai, one of those half rural and half urban types, about 30 miles north of the University. Incidentally, he was born and raised there. Only God knew, a mere five and a half years later, I would be standing in front of his grave in the cemetery behind a temple in the town.

At the time of our correspondence, he was a 'special research student' and I an assistant, but there was not much difference in substance. If there were, that would have been like the distinction of stipend from salary. He was at the Department of Mathematics where professors taught juniors and seniors, while I belonged to another department, which took care of calculus courses for freshmen and sophomores, and which was located in a different campus called College of General Education. This separation was the main reason why I had little contact with him before the above exchange; a secondary one may have been the shyness on either side. Eventually both of us became lecturers in the latter department. He was an associate professor when he died.

Whatever positions we held, our real status in 1954–55 was, in all practical senses, that of a graduate student with no advisor, but with a certain teaching load, which was, at least in my case, equivalent to two undergraduate courses in an American University. This observation applies to nearly all Japanese mathematicians of my generation. The only significant point was that most of us had tenure even as assistants! At any rate, no senior professors were capable of advising students. Even so some of them occasionally offered unsolicited pieces of advice. Once one of us met accidentally on a train a professor who was about fifty at that time. The latter asked the young man what his research interest was. Hearing that he was studying Siegel's theory of quadratic forms, the old man said, 'Ah, quadratic forms. You, as young as you are, may not know it, but it's Minkowski's work that matters.' My colleague complained of this to me afterward. Mocking the professor's pompous air, he said, 'Of course I knew that Minkowski mattered, but did he add anything to Siegel's work?' I too received similar useless pieces of advice or comments.

I always wondered if those professors were trying to imitate their elders, in particular a very worshipped figure among them, who must have made many such comments, most of which, I am inclined to think, were similarly meaningless. Or maybe they were trying to be useful in their style, without realizing that the younger generation, Taniyama for example, had already leapfrogged them, evidence of which the reader will find below. I should note here that Taniyama himself never made pretentious comments; his advice to his juniors was always practical and professional.

Anyway we ignored such almost comical sayings, but accepted them as a kind of reminder that we could not rely on anybody but ourselves. There were of course some Japanese mathematicians in the generation in between who were already prominent or on the way to becoming so. But practically all of them were either abroad or to leave in a short time. For instance, Kodaira and Iwasawa were in the States, soon to be followed by Igusa and Matsusaka.

Around 1950, Hilbert's fifth problem was a topic much mentioned, and the arithmetization of class field theory, or even lattice theory, was being talked about. As none of these was attractive, not a few chose algebraic geometry. At that time, perhaps Chevalley's *Theory of Lie groups* and Weil's *Foundations of algebraic geometry* were the two most widely read books, the former usually read till the end, and the latter given up after the first twenty pages or so in most cases.

In his undergraduate years, Taniyama read both, as well as the two succeeding books by Weil on algebraic curves and abelian varieties. He once wrote that he was led to number theory under the influence of Masao Sugawara, whose course on algebra he had taken. Sugawara was a senior professor in my department, and published some papers on complex multiplication and also on discontinuous groups in the higher-dimensional spaces. However, I have been puzzled by this acknowledgment by Taniyama, as I found Sugawara uninspiring, though I liked him and even respected him as a man. Speaking for myself, during this period and on a personal level, I was influenced exclusively by the people of my generation, above all by Taniyama, and by none of those above the age of thirty. I think this applies in essence to him too.

Indeed, his training ground was many seminars organized by the students themselves. He was the driving force of such activities, and furiously acquiring mathematical knowledge as much as possible. He must have studied Hecke's papers Nos. 33, 35, 36, and 38 on Dirichlet series and modular forms at some point, perhaps somewhat later. While in the same department, he would kindly lend me his notes on this topic when I was unable to secure the library copy of the journal in question.

His first nonelementary work is titled 'On n-division of abelian function fields', and may be termed his senior thesis, though such was not required. As I have no intention of intermingling detailed expositions of his mathematical works with these personal recollections of mine that are my main objective here, let me just say that this paper gave a proof of the Mordell–Weil theorem, based on an idea of Hasse and the results of Weil's paper (*Ann. of Math.* 1951) and that in 1953 he was the only person in Japan who had any working knowledge of this topic. I remember vividly his presentation of this work in a few lectures in the seminar Chevalley held at the University of Tokyo in the spring of 1954.

As explained earlier, he had been interested in complex multiplication of abelian varieties for some time. He took up first the case of the jacobian variety of a hyperelliptic curve, and eventually that of more general abelian varieties. Since not much had been known in this field, the task was a 'hard fighting' against difficulties and a 'bitter struggle' of trial and error. He used to express any substantial undertaking of a mathematician in those four words (strictly speaking, in four corresponding Chinese characters). 'Effortless' was a word alien to his mathematics at least from his viewpoint, though it may have looked differently to others, and he must have found immense delight in such 'fighting and struggle.' He presented his results at the symposium on algebraic number theory held in Tokyo–Nikko, September, 1955. He met Weil there, and incorporating some of Weil's ideas into his, he later published an improved version of his theory on the relationship between abelian varieties and certain Hecke L-functions, a supreme achievement of the time ('L-functions of number fields and zeta functions of abelian varieties', [3]).

As for the part not included in this paper, the collaboration with me was then planned, as I had some results on the subject on my own. We set out in this task in a fashion which may be called leisurely according to today's standard, as we were living in a relaxed, perhaps too relaxed, atmosphere of no competition, which might be envied by the young mathematicians of the 1980s. I have to thank Yasuo Akizuki for hastening our project by inducing us to write a volume in the series of mathematical monographs of which he was an editor.

During this period of collaboration, I often visited his 'Villa' to discuss the matter, as the place was appreciably closer to our school than my home. He always worked late at night. My diary of 1957 tells that, on the afternoon of Thursday, April 4, 2:20 p.m. to be exact, I visited his apartment where he was still asleep. He said he went to bed at 6:00 a.m. On another occasion, perhaps in the late morning, as my knock at the door was not answered, I went to our department office, half an hour train ride from his residence. Finding him there, I said to him, 'I've just stopped by your place before coming here', to which he answered: 'Hum, was I there then?' Immediately realizing his blunder and much embarrassed, he defended his position: 'But you know I am often asleep at that time of the day'.

I discovered that he was different from me in many ways. For one thing, I was, and still am, an early riser. At that time, I thought he was more rational and I more whimsical, but I might have been wrong. We had something in common however: each of us was a late child in a big family; I am the fifth and last child. I say this because I used to resent the egocentricity of those first sons in Japanese families. Though he was by no means a sloppy type, he was gifted with the special capability of making many mistakes, mostly in the right direction. I envied him for this, and tried in vain to imitate him, but found it quite difficult to make good mistakes.

Our joint work in Japanese entitled *Modern number theory* was published in July, 1957. Our next task was of course to make an English version, desirably in a better form, but somehow we lost enthusiasm. The first obvious reason was that we were relieved by the fact that at least it had been written, if in Japanese. There was another practical side of the matter: I was going to leave for France in the fall of that year, which made me restless in a certain way. However, a more fundamental reason can be given by quoting a passage from the preface of the book:

'We find it difficult to claim that the theory is presented in a completely satisfactory form. In any case, it may be said, we are allowed in the course of progress to climb to a certain height in order to look back at our tracks, and then to take a view of our destination'.

Said more prosaically, it was necessary to search for a better formulation and to refine the results. In that year, we were already thinking about the adelization of the whole theory, and perhaps we should have pursued that direction, but didn't. Also, as a matter of psychological reaction, once one proves something, one is more interested in obtaining more new theorems than polishing old ones. Indeed, both of us were interested in modular forms of various types, and that course looked more exciting. Thus our correspondence between Tokyo and Paris was always on that subject. In the spring of 1958, as he informed me of the news, Tokyo greeted Siegel and Eichler, who gave series of lectures, the former on reduction theory of quadratic forms and the latter on his recent results, in particular his trace formula. Meanwhile, in Paris, the topics of the Cartan Seminar were centering around Siegel modular forms.

I wrote more often than he, who wrote to me only twice in this period. In his second letter dated September 22, 1958, which is the very last of all his letters in existence, he mentions that the correspondence of Hecke's type between Hilbert modular forms and certain Dirichlet series could be formulated on the adele group of GL (2). However, as indicated in the tone of the letter, his enthusiasm was rather restrained. He knew that the mere feasibility of such a formulation was not enough, and a real breakthrough was lacking. Obviously more work was necessary; in fact he wrote: 'Because of the heat, I've laid aside the work for one month, but will start thinking about it soon.' Given a sufficient time of concentration, he would have succeeded, but he had to leave the work unfinished forever, since he was destined to die within two months, the remotest thing both sender and recipient of the letter could imagine at that point.

As for our collaborative work, the situation was changed completely by his death, which I will describe later. Having been left alone, I felt it was my duty to finish it as quickly as possible, though I was not completely satisfied with the formulation I had. Eventually 'Complex multiplication of abelian varieties and its applications to number theory' was published in the spring of 1961. The title had been suggested by him in one of his letters. It took me ten more years to be able to set things in a better perspective, and still another five years or so in order to formulate the theory in terms of theta functions as he might have wished, but alas, the man who would have been pleased by these had been long gone.

To write about the private aspect of his life and his last days, first I have to go back a few years to 1955. Though he and I had been members of the same seminar for some time, our relationship became closer after his joining my department in December that year, which naturally made us engage in the same activities of various kinds. For instance, as our official duties would require, we would be confined together in a department office to grade examination papers for admission to the University, more than 5000 sheets for each of us. Fortunately for us but unfortunately for the examinees, many of them were blank.

On a more agreeable note, we would enjoy, together with some other friends, relaxed times in those coffee shops, and spend a Saturday afternoon at a botanical garden in the city, or at a park in the outskirts. In the evenings, we would eat at a restaurant specializing in whale meat, not a particular delicacy in those days but perhaps unthinkable nowadays. We would also take a long walk after a day of work at our school, visiting a Shinto shrine, where we would purchase 'oracles' printed on small pieces of papers to amuse ourselves; they were supposed to tell our fortunes.

While riding in a train together, he asked me the name of the next station, to which I answered, 'The next stop will be *Station*; and then the next stop will be *Next Station*'. This entertained him very much, as he heard it for the first time, and I had to explain to him that I merely mimicked a line of a comedian who was popular on the radio at that time. Immediately afterward, he bought a radio set, and eventually acquired a record player and a sizeable record collection. In his last letter mentioned above, he wrote, 'Lately I am listening repeatedly to Beethoven's No. 8'. I would think these and movie-going were practically all the entertainments he had alone. One of the movies he enjoyed was 'The King and I'. I don't think he ever played any musical instrument. He was no athlete. He never drank, nor smoked, and had no hobby whatsoever. He was not fond of travelling; rather, it seemed to me, he avoided it whenever he could, perhaps on account of his delicate health. I would think Kyoto was the farthest place he reached in his whole life. As an educated man, he must have read standard classical literature, but I am inclined to think he was not an avid reader of novels by modern writers, Japanese or foreign. Nor was he much interested in history, except in that of mathematics.

There is one thing, however, on which he spent a considerable amount of time and energy throughout his prime years, that is, a kind of journalistic writing on academic matters. The topics are varied: how researchers should be trained, how a new institute for mathematical sciences should be organized, criticism on previous articles by others, book reviews, etc. He wrote these articles rather quickly, and made few revisions after finishing them. Probably he was organizing his thoughts by expressing them in writing. He was an articulate writer, more so than in his speech. Incidentally, he was more cheerful in his letters than in his conversation. To tell the truth, I found this 'hobby' of his regrettable, as I thought he was simply wasting much of his precious time, and the cause in each case was not important enough to justify his expending much effort, though I never told him so explicitly. But on one occasion, a few days after hearing my opinion of laissez-faire, he showed me his rough draft on the issue, in which he caricatured my manner of speaking. Naturally I protested and he dropped my part.

He was always kind to his colleagues, especially to his juniors, and he genuinely cared about their welfare. However, in retrospect, I can surmise, without fearing blame for cynicism, that the cause aside he derived much enjoyment from his writing activities. If so, perhaps there was not much point in my regretting.

I wish to close this rather discursive description of his life by telling about his last months. In those days, we were naturally full of youthful ardor and yearning; this could be said on all matters, academic or nonacademic. Speaking now in the latter sense, I can express the predominant mood at that time by a single sentence: nobody believed in arranged marriages—well, almost nobody. Maybe some of us thought, if jokingly, the institution was for the bourgeoisie, and we the proletariat should denounce it as an evil practice, but of course that was an exaggeration. In fact, when I, together with some of my friends, made a call of condolence at his family's home on a hot summer day of 1959 some eight months after his death, his eldest brother, or it may have been his father, suggested to me the daughter of a well known painter as a possible mate. Embarrassed, I asked a female companion in the party how I should respond. She said that the etiquette book would advise me to say such and such. So I repeated the suggested words mechanically, which caused laughter. And that was that.

I used to fancy the idea that the girl might have originally been picked as a candidate for Yutaka. If that were so—my wife would tease me some years later—I should have married her for that reason. Whatever wishes his family may have had, he chose someone on his own, with the eventual consent of the parents of both sides. Her name was Misako Suzuki. He refers to her as M. S. in his will, to which I will now turn. But first the background.

I think he met her as a friend of a friend of a friend in a small and loosely defined social circle around us. I clearly remember the dinner party she gave at her home, with the help of her mother, for Taniyama, K. Yamazaki (one of the colleagues at our school), his fiancée, and myself, shortly before my departure for France in November 1957. The gathering, nominally a farewell party for me, was rather a quiet one, unlike those on other occasions. Indeed, I remember that she was making fun of his reticence during the meal. The same five had spent an evening together in April of that year, which could have been approximately the first encounter of the two. There were many such evenings in those days, with varied members depending on circumstances.

Misako was, relatively speaking, a newcomer to my social circle, and I never got to know her well, but she appeared to be a typically pleasant girl from a typically upper middle class family, and spoke the standard Tokyo dialect with few inhibitions. She was an only child and about five years younger than he. When informed of their engagement, I was somewhat surprised, since I had vaguely thought she was not his type, but I felt no misgivings.

I was told afterward that they had signed a lease for an apartment, apparently a better one, for their new home, had bought some kitchenware together, and had been preparing for their wedding. Everything looked promising for them and their friends. Then the catastrophe befell them. On the morning of Monday, November 17, 1958, the superintendent of his apartment (the one mentioned earlier) found him dead in his room with a note left on a desk. It was written on three pages of a notebook of the type he had been using for his scholastic work; its first paragraph read like this:

'Until yesterday, I had no definite intention of killing myself. But more than a few must have noticed that lately I have been tired both physically and mentally. As to the cause of my suicide, I don't quite understand it myself, but it is not the result of a particular incident, nor of a specific matter. Merely may I say, I am in the frame of mind that I lost confidence in my future. There may be someone to whom my suicide will be troubling or a blow to a certain degree. I sincerely hope that this incident will cast no dark shadow over the future of that person. At any rate, I cannot deny that this is a kind of betrayal, but please excuse it as my last act in my own way, as I have been doing my own way all my life'.

He went on to describe, quite methodically, his wish of how his belongings should be disposed of, and which books and records were the ones he had borrowed from the library or from his friends, and so on. Specifically he says: 'I would like to leave the records and the player to M. S. provided she will not be upset [by my leaving them to her]'. Also he explains how far he reached in the undergraduate courses on calculus and linear algebra he was teaching, and concludes the note with an apology to his colleagues for all the inconveniences this act would cause.

Thus one of the most brilliant and pioneering minds of the time ended his life by his own will. He had attained the age of thirty-one only five days earlier.

There was an inevitable turmoil, and then a funeral, followed by a gathering of his friends and colleagues in his memory. They were utterly perplexed. Naturally they asked among themselves why he had to kill himself, but no convincing answer was available. According to his fiancée, he was to see her within a few days of that fatal Monday morning. It was as if God designed him to be a virgin mathematician and not a family man. I eventually reconciled myself to that view, but that was much later.

Anyway, after a few weeks, people seemed to have recovered somewhat, if slowly, from the shock and sorrow, and things were going back to routine. Then, on a chilly day of early December, Misako killed herself in the apartment which had been intended for their new home. She left a note, which was never made public. I only heard that it contained a passage to the following effect: 'We promised each other that no matter where we went, we would never be separated. Now that he is gone, I must go too in order to join him.'

When these misfortunes occurred, I was staying in Princeton as a member of the Institute for Advanced Study, and so the details of all the events were told to me by Kuga and Yamazaki after my return to Tokyo in the spring of 1959. Taniyama himself was supposed to be at the Institute in the fall of that year, and I could have spent my second year there, but decided to leave.

By the time I came home, the cherry trees had shed their blossoms, and dark green leaves were already dominating the view. Though a hackneyed expression, spring was swiftly passing. After my absence of one year and a half, the streets of Tokyo with their vitality and vulgarity showed little change. But people changed. So did I. A further period of transformation was still ahead of me, but in those late spring days, I could not but keep reflecting vainly on a simple fact: the type of party we had only two years ago was no longer possible. The years of turbulence had passed.

To conclude this article, I may ask somewhat rhetorically: Who was Yutaka Taniyama? This is not asked about his stature in the history of mathematics. My concern here is what his existence meant to his generation and especially to me. What

I have written may naturally be viewed as a lengthy answer to that question, but to sum up, I should state more clearly one point to which my writing so far has only vaguely alluded: that he was the moral support of many of those who came into mathematical contact with him, including of course myself. Probably he was never conscious of this role he was playing. But I feel his noble generosity in this respect even more strongly now than when he was alive. And yet nobody was able to give him any support when he desperately needed it. Reflecting on this, I am overwhelmed by the bitterest grief.

ACKNOWLEDGMENT. I wish to express my sincere gratitude to Dr Seiji Taniyama and Professor Mitsuo Sugiura for providing me with essential biographical data. Also I wish to thank heartily some of my friends who read the first draft and contributed valuable suggestions, which have been incorporated into this article.

Taniyama's problems

At the International Symposium on Algebraic Number Theory held in Tokyo and Nikko in September, 1955, mimeographed copies of a collection of 36 mathematical problems were distributed to the participants. Two of the problems posed by Taniyama concern the possibility that a certain elliptic curve is a factor of the Jacobian of an automorphic function field. These may be viewed as the origin of the now widely accepted conjecture: every elliptic curve defined over the rational number field is a factor of the Jacobian of a modular function field. Instead of reproducing the problems in their original forms, we present here English translations of their Japanese versions published in *Sûgaku*, Vol. 7 (1956), p. 269. This choice was made for minor technical reasons. At any rate, there is little difference in substance between the English and Japanese versions.

PROBLEM 12. Let C be an elliptic curve defined over an algebraic number field k, and $L_C(s)$ the L-function of C over k in the sense that

$$\zeta_C(s) = \zeta_k(s)\zeta_k(s-1)/L_C(s)$$

is the zeta function of C over k. If Hasse's conjecture is true for $\zeta_C(s)$, then the Fourier series obtained from $L_C(s)$ by the inverse Mellin transformation must be an automorphic form of dimension -2 of a special type (see Hecke). If so, it is very plausible that this form is an elliptic differential of the field of associated automorphic functions. Now, going through these observations backward, is it possible to prove Hasse's conjecture by finding a suitable automorphic form from which $L_C(s)$ can be obtained?

PROBLEM 13. In connection with Problem 12, the following may be set as a problem: to characterize the field of elliptic modular functions of 'Stufe' N, and especially to decompose the Jacobian variety J of this function field into simple factors up to isogeny. Also, it is well known that if $N = q$, a prime, and $q \equiv 3 \pmod 4$, then J contains elliptic curves with complex multiplication. What can one say for general N?

A few comments may be in order as to exactly what Taniyama meant or what he was thinking. An insight into this matter can be gained from a passage in the record of an informal discussion session on September 12, 1955, attended by many of the participants of the Symposium mentioned above. The notes were taken by Taniyama, and published in Japanese in the same issue of *Sûgaku*, p. 228. The relevant part may be translated as follows.

Weil asks Taniyama: Do you think all elliptic functions are uniformized by modular functions?

Taniyama: Modular functions alone will not be enough. I think other special types of automorphic functions are necessary.

In these notes, he refers to Hecke's papers ⟨**3, 4, 5**⟩, and we can safely assume that the reference to Hecke in Problem 12 was to ⟨**4**⟩ which involves not only congruence subgroups of $SL_2(\mathbb{Z})$ but also some Fuchsian groups not commensurable with it. That explains his answer in the above exchange. The 'well known' fact in Problem 13 obviously refers to ⟨**3**⟩.

I should add that he was familiar with the results of ⟨**1**, I; **2**⟩, and ⟨**6**⟩, which were the only papers concerning this topic in 1955.

It should also be noted that, strictly speaking, Problem 12 makes sense only when $k = \mathbb{Q}$, because ⟨**4**⟩ deals with the functional equation with a single $\Gamma(s)$ and so doesn't apply to the case where $k \neq \mathbb{Q}$. However, the reason he formulated the problem for a general number field may be given in the following way. If C has complex multiplication, then, with a suitable choice of k, L_C is the L-function of a certain Hecke character of k, or the product of two such L-functions, according to the choice of k, as proved by Deuring (see ⟨**1**⟩). If $k \neq \mathbb{Q}$, this type of Hecke's L-function itself is not the Mellin transform of a modular form, but can be the product of several such transforms in certain cases. As can be seen from his second problem, he was conscious of this connection of modular forms with the case of complex multiplication, and it is my guess that he therefore considered C over a general k so that this case could be included.

References

⟨**1**⟩. M. Deuring, 'Die Zetafunktion einer algebraischen Kurve vom Geschlechte Eins, I, II, III, IV', *Nachr. Akad. Wiss. Göttingen*, (1953) 85–94, (1955) 13–42, (1956) 37–76, (1957) 55–80.

⟨**2**⟩. M. Eichler, 'Quaternäre quadratische Formen und die Riemannsche Vermutung für die Kongruenzzetafunktion', *Arch. Math.* 5 (1954) 355–366.

⟨**3**⟩. E. Hecke, 'Bestimmung der Perioden gewisser Integrale durch die Theorie der Klassenkörper', *Math. Z.* 28 (1928) 708–727. (*Werke*, No. 27.)

⟨**4**⟩. E. Hecke, 'Über die Bestimmung Dirichletscher Reihen durch ihre Funktionalgleichung', *Math. Ann.* 112 (1936) 664–699. (*Werke*, No. 33.)

⟨**5**⟩. E. Hecke, 'Über Modulfunktionen und die Dirichletschen Reihen mit Eulerscher Produktentwicklung, I, II', *Math. Ann.* 114 (1937) 1–28, 316–351. (*Werke*, Nos. 35 and 36.)

⟨**6**⟩. A. Weil, 'Jacobi sums as 'Grössencharaktere', *Trans. Amer. Math. Soc.* 73 (1952) 487–495.

Bibliography of Y. Taniyama

1. 'Jacobian varieties and number fields', *Proceedings of the International Symposium on algebraic number theory, Tokyo-Nikko* 1955 (Science Council of Japan, Tokyo, 1956), pp. 31–45.

2. 'Jacobian varieties and number fields' (Japanese), *Sûgaku* 7 (1956) 218–220.

3. '*L*-functions of number fields and zeta functions of abelian varieties', *J. Math. Soc. Japan* 9 (1957) 330–366.

4. (With G. Shimura) *Modern number theory* (Japanese; Kyoritsu Publishing Co., 1957), 224 pages.

5. 'Distribution of positive *O*-cycles in absolute classes of an algebraic variety with finite constant field', *Sci. Papers Coll. Gen. Ed. Univ. Tokyo* 8 (1958) 123–137.

6. (With G. SHIMURA) *Complex multiplication of abelian varieties and its applications to number theory*, Publ. Math. Soc. Japan 6 (Math. Soc. Japan, Tokyo 1961), 159 pages.
7. *The complete works of Yutaka Taniyama*. Publication by subscription, ed. by Seiji Taniyama *et al.*, 1962. This volume contains unpublished manuscripts, nonmathematical articles, letters, the last note, a chronological list of events in his life, as well as the above articles except Nos. 4 and 6.

Department of Mathematics
Princeton University
Fine Hall
Princeton
New Jersey 08544, USA

L-functions and eigenvalue problems

Algebraic analysis, geometry, and number theory, Proceedings of the JAMI Conference 1988, Supplement to the American Journal of Mathematics, 1989, 341-396

Introduction. The eigenvalue problems we consider in this paper concern an inhomogeneous equation

(1) $$(L_m - \lambda)f(z) = p(z).$$

Here f and p are functions, possibly with singularities, on the complex upper half plane

(2) $$H = \{z \in \mathbf{C} \mid \mathrm{Im}(z) > 0\},$$

and L_m is the differential operator given by

$$L_m = -y^{-2}(\partial^2/\partial x^2 + \partial^2/\partial y^2) + miy\partial/\partial x,$$

where $z = x + iy$ as usual and m is an integer. We fix a congruence subgroup Γ of $\mathrm{SL}_2(\mathbf{Z})$, and assume that both f and p are Γ-automorphic forms of weight m in the sense that they are invariant under

(3) $$f \mapsto f(\gamma z)(cz + d)^{-m}|cz + d|^m$$

for every $\gamma = \begin{pmatrix} * & * \\ c & d \end{pmatrix} \in \Gamma$. It should be noted that L_m commutes with the map of (3). The homogeneous equation

(4) $$(L_m - \lambda)\varphi = 0$$

has been extensively studied. One of the most noteworthy facts about such λ and φ is the existence of a certain zeta function, defined by Selberg, whose set of zeros coincides essentially with the set $S(\Gamma, m)$ of complex numbers s such that $\lambda = s(1 - s)$ occurs as an eigenvalue of L_m as in (4)

Manuscript received September 1, 1988.

with a cusp form φ. However, no connection of such λ or s with the zeros of classical zeta functions has been discovered.

Now the fundamental principle governing the present paper can be condensed as follows:

A zero of a zeta function is a complex number s such that $s(1-s)$ *occurs as a value* λ *in the inhomogeneous equation* (1) *with a fixed p determined by the zeta function and a rapidly decreasing f.*

We shall present two types of zeta functions and the corresponding f and p. To describe the first type, we let $\Gamma = SL_2(\mathbf{Z})$ and take two holomorphic cusp forms $g(z) = \sum_{n=1}^{\infty} a_n e^{2\pi i n z}$ of weight k and $h(z) = \sum_{n=1}^{\infty} b_n e^{2\pi i n z}$ of weight l with respect to Γ, and put $\kappa = (k+l)/2$, $m = l - k$, and $R(s) = R_0(s + \kappa - 1)$, where

$$R_0(s) = (4\pi)^{-s}\Gamma(s) \sum_{n=1}^{\infty} \bar{a}_n b_n n^{-s}.$$

It is well known that $s(1-s)\pi^{-s}\Gamma(s+|m/2|)\zeta(2s)R(s)$ can be continued to an entire function invariant under $s \mapsto 1 - s$. Now we shall show that there is a function $f(z, s)$ meromorphic on the whole s-plane with the following properties (Theorem 3.1):

(I-0) *$f(z, s)$ is finite at s if $s \notin \{1/2\} \cup S(\Gamma, m)$ and $\Gamma(s+|m/2|)\zeta(2s) \neq 0$. Put $f_s(z) = f(z, s)$ for such an s.*

(I-1) *f_s is a C^∞ Γ-automorphic form of weight m.*

(I-2) $[L_m - s(1-s)]f_s = y^\kappa \overline{g(z)} h(z)$.

(I-3) *$f_s(z) = (2s-1)^{-1}R(s)y^{1-s} + O(e^{-\pi y})$ as $y \to \infty$.*

The significance of the last fact is that f_s is rapidly decreasing at every cusp of Γ if and only if $R(s) = 0$, which is an example of the above principle.

Such a function f can be obtained as an integral

$$f(z, s) = -\int_{\Gamma\backslash H} G_s(z, w)\overline{g(w)}h(w)\operatorname{Im}(w)^\kappa d_H w, \tag{5}$$

where $d_H w$ is the standard invariant measure on H and G_s is the resolvent kernel of equation (1), which was introduced by Roelcke [9] and subsequently studied by several authors. (See Fadeev [3], Neunhöffer [8], Elstrodt [1], Fay [4], Hejhal [7]. In this paper, we employ the notation and results of [7], which is the most comprehensive treatment of the subject.)

As its name suggests, relation (I-2) is an immediate consequence of (5). Likewise, (I-0) and (I-1) follow from the corresponding properties of G_s. Thus the essential point is (I-3), which is also closely connected with the following fact: for a fixed w, $G_s(z, w)$ as a function of z has a Fourier expansion for sufficiently large $\mathrm{Im}(z)$, and the expansion shows that

$$G_s(z, w) = (1 - 2s)^{-1}y^{1-s}E_{-m}(w, s) + O(e^{-\pi y}) \quad \text{as} \quad y \to \infty$$

with a certain Eisenstein series E_{-m} of weight $-m$. Now the above R can be obtained as an integral

$$R(s) = \int_{\Gamma\backslash H} E_{-m}(w, s)\mathrm{Im}(w)^{\kappa}\overline{g(w)}h(w)d_H w, \tag{6}$$

and therefore termwise integration, if valid, would give (I-3), establishing $(2s - 1)^{-1}R(s)y^{1-s}$ as the constant term of $f(z, s)$. This is not so, however. Our proof of (I-3), which is rather involved, will show that the supposed constant term is actually the dominant part of the true constant term of f, the remaining part being rapidly decreasing.

Let us next note a few consequences of this fact. First of all, the self-adjointness of L_m combined with (I-2) and (I-3) shows that if $R(s) = 0$, then

(I-4)

$$\mathrm{Im}[s(1 - s)]\langle f_s, f_s\rangle = \mathrm{Im}\langle p, f_s\rangle$$

$$= -\mathrm{Im}\left[\int_{\Gamma\backslash H}\int_{\Gamma\backslash H} G_s(z, w)\overline{p(z)}p(w)d_H z d_H w\right],$$

where $p(z) = y^{\kappa}\overline{g(z)}h(z)$ and $\langle\ ,\ \rangle$ denotes the Petersson inner product. Thus for such a zero s of R, one has $s(1 - s) \in \mathbf{R}$ if and only if the last double integral is real. We shall also prove (Theorems 3.1 and 3.4):

(I-5) $4t \cdot \mathrm{Im}\langle f_s, p\rangle = |R(s)|^2$ *if* $s = (1/2) + it$ *with* $0 \neq t \in \mathbf{R}$.

(I-6) $(1 - 2s)\langle E_m(z, \bar{s}), f_s\rangle = dR/ds$ *if* $R(s) = 0$.

There is a significant aspect of these relations beyond the fact that they hold: namely, all of them have their counterparts in the second case, in which R is replaced by the L-function $L(s, \chi)$ of a Hecke character χ of an imaginary quadratic field K. We take the archimedean factor of χ is of the type $a \mapsto a^m|a|^{-m}$, and assume for simplicity that the restriction of χ to $\mathbf{Q}$ is trivial. In this case f of (1) is given by

$$g(z, s) = \mu(A/\pi)^s\Gamma(s + |m/2|)\zeta(2s) \sum_{t\in T} \chi(\mathbf{Z}\tau + \mathbf{Z})G_s(z, \tau).$$

Here T is a finite subset of $K \cap H$ such that the lattices $\mathbf{Z}\tau + \mathbf{Z}$ for $\tau \in T$ represent certain ideal classes in K; μ, A, and T are chosen so that

$$(7) \quad \mathfrak{R}(s, \chi) = \mu(A/\pi)^s\Gamma(s + |m/2|)\zeta(2s) \sum_{\tau\in T} \chi(\mathbf{Z}\tau + \mathbf{Z})E_{-m}(\tau, s),$$

where $\mathfrak{R}(s, \chi)$ is $L(s, \chi)$ times its standard gamma factors. Then we have (Theorem 5.2):

(II-0) *If $z \notin \Gamma T$, $g(z, s)$ is meromorphic in s with poles only in $\{1/2\} \cup S(\Gamma, m)$; put $g_s(z) = g(z, s)$ for $s \notin \{1/2\} \cup S(\Gamma, m)$.*

(II-1) *g_s is a Γ-automorphic form of weight m with singularities in ΓT.*

(II-2) $[L_m - s(1 - s)]g_s = 0$ *outside* ΓT.

(II-3) $g_s(z) = (1 - 2s)^{-1}\mathfrak{R}(s, \chi)y^{1-s} + O(e^{-\pi y})$ *as* $y \to \infty$.

We again observe that g_s is rapidly decreasing if and only if $\mathfrak{R}(s, \chi) = 0$. This time (II-2) shows that g_s is an eigenfunction of L_m, but in order to gain a better comprehension of g_s, we have to look more closely at its singularities. Since g_s is a linear combination $\sum_{\tau\in T} c_s(\tau)G_s(z, \tau)$ with complex numbers $c_s(\tau)$, it can be obtained as an integral

$$(8) \quad \int_{\Gamma\backslash H} G_s(z, w) \sum_{\tau\in T} c_s(\tau)\delta_\tau(w)d_H w$$

with Dirac's point-distribution δ_τ. Thus g_s should be viewed as a solution of (1) with $\sum_\tau c_s(\tau)\delta_\tau$ as p. In fact, for each $\tau \in T$, we have

$$g_s(z) = (2\pi)^{-1}e_\tau c_s(\tau)\log|z - \tau| + q_{\tau,s}(z)$$

in a neighborhood of τ, where $q_{\tau,s}$ is continuous at τ and e_τ is a certain integer. Also, (7) may be written in the form

$$\mathcal{R}(s, \chi) = \int_{\Gamma\backslash H} E_{-m}(w, s) \sum_{\tau} c_s(\tau)\delta_\tau(w) d_H w. \tag{9}$$

Now the analogues of (I-4, 5, 6) can be given as follows (Theorem 5.5):

(II-4) $\operatorname{Im}[s(1-s)]\langle g_s, g_s\rangle = -\Sigma_\tau \operatorname{Im}[c_s(\tau)q_{\tau,s}(\tau)]$ *if* $\mathcal{R}(s, \chi) = 0$.

(II-5) $4t\, \Sigma_\tau \operatorname{Im}[c_s(\tau)q_{\tau,s}(\tau)] = |\mathcal{R}(s, \chi)|^2$ *if* $s = (1/2) + it$ *with* $0 \neq t \in \mathbf{R}$.

(II-6) $(2s-1)\langle E_m(z, \bar{s}), g_s\rangle = d\mathcal{R}(s, \chi)/ds$ *if* $\mathcal{R}(s, \chi) = 0$.

Thus there is an unmistakable parallelism between the two cases. It is this parallelism that sustains our interpretation of the zeros of a zeta function in terms of the inhomogeneous equation (1). Though we treat only these two types, it is expected that many more cases can be comprehended according to this principle. Such probabilities are enhanced by the fact or expectations that many other kinds of zeta functions are given by integrals similar to (6) or (9). The reader is referred to Section 6 for more comments on this. It should be noted that the cases with congruence subgroups present some nontrivial difficulties because of their plural inequivalent cusps. We touch little on this, except that our treatment of the second type include groups of prime power level.

Another related problem concerns a *canonical* expression of a Hecke L-function as a linear combination of special values of Eisenstein series, of which (7) is an example, and which is far from obvious for an arbitrary Hecke character of K. Since this is a subject of independent interest, we devote two full sections (4 and 7) to the study of such an expression for a Hecke L-function of an arbitrary CM-field, though we shall not consider equation (1) in the higher-dimensional case.

The introduction would be utterly defective if a work [6] of Hejhal were not mentioned. Indeed, he observed in the paper that the constant term of $G_s(z, \omega)$ for $\omega = e^{2\pi i/3}$ is essentially $\zeta(2s)^{-1}$ times the zeta function of $\mathbf{Q}(\omega)$ and therefore $G_s(z, \omega)$ is rapidly decreasing if the zeta function vanishes at s. The author wishes to acknowledge his indebtedness to this and also to thank heartily Dennis Hejhal for clarifying some technical aspects of G_s on several occasions and communicating some of his observations, though they are not incorporated in the present paper.

1. Automorphic forms with logarithmic singularities. Throughout the paper, we denote by H the upper half plane as in (2) of the introduction. For every 2×2-matrix α, we denote by a_α, b_α, c_α, and d_α the entries of α in the standard order. We let every α of $SL_2(\mathbf{R})$ act on H as usual by $\alpha(z) = \alpha z = (a_\alpha z + b_\alpha)/(c_\alpha z + d_\alpha)$, and define two factors of automorphy $j_\alpha(z)$ and $J_\alpha(z)$ by

$$j_\alpha(z) = j(\alpha, z) = c_\alpha z + d_\alpha, \tag{1.1a}$$

$$J_\alpha(z) = J(\alpha, z) = j_\alpha(z)/|j_\alpha(z)| \qquad (\alpha \in SL_2(\mathbf{R}),\ z \in H). \tag{1.1b}$$

Given an integer m, a function f on H, and $\alpha \in SL_2(\mathbf{R})$, we define a function $f\|_m\alpha$ on H by

$$(f\|_m\alpha)(z) = J_\alpha(z)^{-m} f(\alpha z) \qquad (z \in H). \tag{1.2}$$

We also define differential operators δ_m, ϵ_m, and L_m acting on C^∞ functions f on H or on its open subsets by

$$\delta_m f = y^{1-(m/2)}(\partial/\partial z)(y^{m/2} f) = y\partial f/\partial z - (mi/4)f, \tag{1.3a}$$

$$\epsilon_m f = -y^{1+(m/2)}(\partial/\partial \bar{z})(y^{-m/2} f) = -y\partial f/\partial \bar{z} + (mi/4)f, \tag{1.3b}$$

$$\begin{aligned} L_m &= -y^2(\partial^2/\partial x^2 + \partial^2/\partial y^2) + miy\partial/\partial x \\ &= 4\delta_{m-2}\epsilon_m - m(m-2)/4 = 4\epsilon_{m+2}\delta_m - m(m+2)/4, \end{aligned} \tag{1.3c}$$

where $z = x + iy$. As explained in the introduction, our main object of study is an automorphic form f on H satisfying an inhomogeneous equation $(L_m - \lambda)f = p$ with $\lambda \in \mathbf{C}$ and a function p on H. In this section, we prove some basic formulas on such an f when p is a function of Dirac's type, with no reference to L-functions. The connection with L-functions will be made in Section 5.

We fix a discrete subgroup Γ of $SL_2(\mathbf{R})$ containing -1 such that $\Gamma\backslash H$ has a finite measure. For a point w that is either a cusp of Γ or a point of H we put

$$\Gamma_w = \{\gamma \in \Gamma \mid \gamma(w) = w\}, \qquad e_w = e(w) = [\Gamma_w : \{\pm 1\}], \tag{1.4}$$

e_w being defined only when $w \in H$. We take a subset Ξ of $SL_2(\mathbf{R})$ so that

(1.5a) $\{\xi(\infty) \mid \xi \in \Xi\}$ *is a complete set of* Γ*-inequivalent cusps;*

(1.5b) $$\xi\begin{pmatrix}1 & 1\\ 0 & 1\end{pmatrix}\xi^{-1} \text{ generates } \Gamma_{\xi(\infty)}/\{\pm 1\}.$$

Though Ξ can be empty, we are primarily interested in the case of non-empty Ξ.

We now consider a $\mathbf{C}$-valued measurable function f on H such that

(1.6) $$f\|_m\gamma = \theta(\gamma)f \text{ for every } \gamma \in \Gamma,$$

where θ is a character of Γ with values in $\{z \in \mathbf{C} \mid |z| = 1\}$ such that $\theta(-1) = (-1)^m$. If g is another such function, we put

(1.7a) $$\langle f, g\rangle_\Gamma = \int_{\Gamma\backslash H} \overline{f(z)}g(z)d_H z \qquad (d_H z = y^{-2}dxdy),$$

(1.7b) $$\|f\|_\Gamma = (\langle f, f\rangle_\Gamma)^{1/2},$$

whenever the integral is convergent. The subscript Γ will often be suppressed.

We now specify a set of singularities $\mathcal{Z}$ such that $\mathcal{Z} = \cup_{\nu=1}^{p} \Gamma\zeta_\nu$ with Γ-inequivalent points $\zeta_1, \ldots, \zeta_p$ of H, and impose the following conditions (1.8a,b,c) on f:

(1.8a) *f is C^∞ outside of $\mathcal{Z}$ and $L_m f = \lambda f$ with $\lambda \in \mathbf{C}$ there.*

(1.8b) *For every $\zeta \in \mathcal{Z}$ there is a neighborhood U of ζ, a complex number c, and a continuous function r on U such that:* (i) $f(z) = c\log|z - \zeta| + r(z)$ *for* $\zeta \neq z \in U$, (ii) *r is C^∞ on $U - \{\zeta\}$, and* (iii) $\lim_{z\to\zeta}(z - \zeta)\log|z - \zeta|(|\partial r/\partial z| + |\partial r/\partial \bar{z}|) = 0$.

(1.8c) *For every $\xi \in \Xi$, $(f\|_m\xi)(x + iy) = 0(y^\sigma)$ as $y \to \infty$ uniformly in x with a real number $\sigma < 1/2$.*

If f and g satisfy (1.6) and (1.8a,b,c), then $\langle f, g\rangle$ is meaningful; in particular $|f|$ belongs to $L^2(\Gamma\backslash H)$. We also consider a stronger condition:

(1.9) *For every $\xi \in \Xi$, $(f\|_m\xi)(x + iy) = O(e^{-ay})$ as $y \to \infty$ uniformly in x with $a > 0$.*

We say that f is *rapidly decreasing* (at every cusp of Γ) if (1.9) is satisfied.

Proposition 1.1. *Let f and g be two functions satisfying* (1.6) *and* (1.8a,b,c) *with eigenvalues λ and μ respectively and with the same Ξ and m; put*

$$f(z) = c_\nu \log|z - \zeta_\nu| + r_\nu(z), \qquad g(z) = b_\nu \log|z - \zeta_\nu| + q_\nu(z)$$

with constants c_ν, b_ν and functions r_ν, q_ν as in (1.8b). *Then*

$$(\bar{\lambda} - \mu)\langle f, g\rangle_\Gamma = 2\pi \sum_{\nu=1}^{p} [\bar{c}_\nu q_\nu(\zeta_\nu) - b_\nu \overline{r_\nu(\zeta_\nu)}]/e(\zeta_\nu),$$

where e is defined by (1.4). *In particular, we have*

$$-\operatorname{Im}(\lambda)\langle f, f\rangle_\Gamma = 2\pi \sum_{\nu=1}^{p} \operatorname{Im}[\bar{c}_\nu r_\nu(\zeta_\nu)]/e(\zeta_\nu).$$

Proof. Changing Γ for its suitable subgroup, we may assume that Γ has no elliptic elements. Let $L'_m = 4\delta_{m-2}\epsilon_m$. A simple calculation shows that

$$d(\bar{f}\epsilon_m g \cdot y^{-1} d\bar{z}) = 2i(\overline{\epsilon_m f} \cdot \epsilon_m g - (1/4)\bar{f} L'_m g) d_H z.$$

Take a sufficiently small positive number ρ so that the sets

$$\xi(\{x + iy \mid 0 \leqq x < 1, y > 1/\rho\}) \qquad (\xi \in \Xi),$$

$$K_\nu = \{z \in H \mid |z - \zeta_\nu| < \rho\} \qquad (1 \leqq \nu \leqq p)$$

are mapped into $\Gamma\backslash H$ with no overlapping. Let D be the complement of the union of these sets on the Riemann surface $\Gamma\backslash H$. Then the above equality shows that

$$2i \int_D \overline{\epsilon_m f} \cdot \epsilon_m g d_H z - (i/2) \int_D \bar{f} \cdot L'_m g d_H z = \int_{\partial D} \bar{f} \epsilon_m g \cdot y^{-1} d\bar{z}.$$

Exchange f and g, and take the complex conjugate of the resultant equality; add it to the original one. Since $L_m - L'_m = m(2 - m)/4$, we thus obtain

(1.10)

$$(i/2)\int_D (\overline{L_m f}\cdot g - \bar{f}L_m g)d_H z = \int_{\partial D} (\bar{f}\epsilon_m g\cdot y^{-1}d\bar{z} + g\overline{\epsilon_m f}\cdot y^{-1}dz).$$

Let S denote the line segment connecting $i\rho$ to $1 + i\rho$. Then $-\partial D = \sum_{\nu=1}^{p} \partial K_\nu + \sum_{\xi\in\Xi} \xi(S)$. Let σ be a real number as in (1.8c), which we can assume to be common to both f and g and to all $\xi \in \Xi$. Now we have

(1.11) $$(\epsilon_m f)\|_{m-2}\xi = O(y^\sigma) \textit{ as } y \to \infty \textit{ uniformly in } x.$$

This will be shown at the end of the proof. Therefore, since $\sigma < 1/2$, the integral over $\xi(S)$ tends to 0 as ρ tends to 0. To compute

(*) $$\int_{\partial K_\nu} g\overline{\epsilon_m f}\cdot y^{-1}dz$$

we observe that

$$\overline{\epsilon_m f} = -(\bar{c}_\nu y/2)(z - \zeta_\nu)^{-1} - (\bar{c}_\nu mi/4)\log\rho - y\partial\bar{r}_\nu/\partial z - (mi/4)\overline{r_\nu(z)}$$

on ∂K_ν. Then it is easy to see that the integral of (*) is of the form

$$-b_\nu\bar{c}_\nu\pi i\cdot\log\rho - \bar{c}_\nu\pi i q_\nu(\zeta_\nu) + o(1) \qquad (\rho \to 0).$$

The integral of $\bar{f}\epsilon_m g\cdot y^{-1}d\bar{z}$ over ∂K_ν can be computed in a similar way. Since the left-hand side of (1.10) tends to $(i/2)(\bar{\lambda} - \mu)\langle f, g\rangle$, we obtain the desired formula.

It remains to prove (1.11). Fix an element ξ of Ξ; let $\theta\left(\xi\begin{pmatrix}1&1\\0&1\end{pmatrix}\xi^{-1}\right) = e^{2\pi ip}$ with $p \in \mathbf{R}$. Then $f\|_m\xi$ is C^∞ on the half plane

$$H_M = \{z \in H \mid \mathrm{Im}(z) > M\}$$

for some $M > 0$. A well known argument combining the equation $L_m(f\|_m\xi) = \lambda f\|_m\xi$ with (1.8c) shows that $(f\|_m\xi)(z) = \sum_{r-p\in\mathbf{Z}} \alpha_r(y)e^{2\pi(irx-|r|y)}$ on H_M with functions α_r of the forms

(1.12a) $$\alpha_0(y) = \begin{cases} by^s + b'y^{1-s} & \text{if } \lambda \neq 1/4, \\ by^{1/2} + b'y^{1/2}\log y & \text{if } \lambda = 1/4, \end{cases}$$

(1.12b) $\alpha_r(y) = b_r y^s \Psi(s - (m/2)\operatorname{sgn}(r), 2s; 4\pi|r|y)$ if $r \neq 0$,

where b, b', and b_r are constants, s is a complex number such that $s(1-s) = \lambda$ and $\operatorname{Re}(s) \leqq 1/2$, and $\Psi(a, c; z)$ is a holomorphic function on $\mathbf{C}^2 \times (-i)H$ such that

$$\Psi(a, c; z) = \Gamma(a)^{-1} \int_0^\infty e^{-tz} t^{a-1}(1+t)^{c-a-1}\,dt \tag{1.13}$$

if $ia \in H$ and $iz \in H$.

We now need

LEMMA 1.2. *Let V be a compact subset of $\mathbf{C}$ and $h(z, s)$ a function on $H_M \times V$. Suppose that h is C^∞ in z and each derivative $\partial^{a+b}h/\partial x^a \partial y^b$ is continuous on $H_M \times V$, and h has an expansion of the form*

$$h(z, s) = \sum_{0 \neq r \equiv p (\mathrm{mod}\ \mathbf{Z})} b_r(s) y^s e^{2\pi(irx - |r|y)} \Psi(s_r, 2s; 4\pi|r|y),$$

where $p \in \mathbf{R}$, $s_r = s - (m/2)\operatorname{sgn}(r)$, and the b_r are continuous functions on V. Then $h(z, s) = O(e^{-cy})$ as $y \to \infty$ uniformly in (x, s) with any $c > 0$ such that $c < 2\pi|r|$ for all r appearing in the sum.

Proof. It is well known that $\lim_{t\to\infty} t^a \Psi(a, c; t) = 1$ uniformly for (a, c) in any compact subset of $\mathbf{C}^2$ (see [2, p. 278], [14, (10.8)], for example). Therefore we can find $\eta > M$ such that

$$1/2 < |t^{s_r} \Psi(s_r, 2s; t)| < 2 \quad \text{if} \quad t > \eta \quad \text{and} \quad s \in V.$$

It is also well known that the Fourier expansion of a C^∞ function with continuous parameters is absolutely and locally uniformly convergent. Hence we see that

$$\sum_r e^{-(2\pi|r| - c)y} |b_r(s) y^s (4\pi|r|y)^{-s_r}| \tag{1.14}$$

is convergent for $y > \eta$ uniformly in s with any c as in our lemma. Now we can find $\tau > \eta$ such that each term of (1.14) is decreasing as y is increasing and $y \geqq \tau$. Therefore, if $y \geqq \tau$, we have

$$\sum_r e^{-2\pi|r|y}|b_r(s)y^s\Psi(s_r, 2s; 4\pi|r|y)|$$
$$\leqq 2e^{-cy}\sum_r e^{-(2\pi|r|-c)\tau}|b_r(s)\tau^s(4\pi|r|\tau)^{-s_r}|.$$

Since the last sum is bounded for $s \in V$, we obtain our assertion.

Applying this lemma to $f\|_m\xi$, we see from condition (1.8c) that $\alpha_0(y) = by^s$ with $\mathrm{Re}(s) < 1/2$. This combined with the relation $L_{m-2}\epsilon_m f = \lambda\epsilon_m f$ shows that

$$(\epsilon_m f)\|_{m-2}\xi = cy^s + \sum_{r\neq 0}\beta_r(y)e^{2\pi(irx-|r|y)}$$

on H_M with β_r of type (1.12b) with $m - 2$ instead of m. Therefore (1.11) follows from the above lemma.

Proposition 1.3. *In the setting of Proposition 1.1, suppose $\lambda = \bar{\mu} = s(1-s)$ with $s \in \mathbf{C}$; suppose f and g satisfy (1.6), (1.8a,b), and, instead of (1.8c), the following condition:*

(1.15) *For every $\xi \in \Xi$, there exist constants b_ξ, c_ξ, and a positive number a such that*

$$(f\|_m\xi)(z) = c_\xi y^{1-s} + O(e^{-ay})$$
$$(y \to \infty).$$
$$(g\|_m\xi)(z) = b_\xi y^{\bar{s}} + O(e^{-ay})$$

Then we have

$$(1-2\bar{s})\sum_{\xi\in\Xi} b_\xi\bar{c}_\xi = 2\pi\sum_{\nu=1}^{p}[\bar{c}_\nu q_\nu(\zeta_\nu) - b_\nu\overline{r_\nu(\zeta_\nu)}]/e(\zeta_\nu).$$

Proof. We repeat the proof of Proposition 1.1. This time the left-hand side of (1.10) is 0. The limits of the integrals over $\xi(S)$, however, may be nonzero. In fact,

$$(\epsilon_m f)\|_{m-2}\xi = (ic_\xi/2)(s - 1 + m/2)y^{1-s} + O(e^{-cy})$$

for some $c > 0$, and hence we find that

$$\int_{\xi(S)}\overline{\epsilon_m f}\cdot gy^{-1}dz \to (ib_\xi\bar{c}_\xi/2)(1 - \bar{s} - m/2) \qquad (\rho \to 0).$$

Computing the other integral over $\xi(S)$ in the same way, we obtain the desired formula.

Let us conclude this section by recalling the definition of a cusp form. If f satisfies (1.6) and (1.8a,c) with $\mathcal{Z} = \varnothing$ and $\sigma < 0$, then f is called a *cusp form*. The following facts are well known: (i) every cusp form is rapidly decreasing; (ii) a function satisfying (1.6) and (1.8a) with $\mathcal{Z} = \varnothing$ belongs to $L^2(\Gamma\backslash H)$ if and only if it is either a cusp form or a constant; the constant is precluded if $m \neq 0$ or $\theta \neq 1$; (iii) if f is a nonzero cusp form and $L_m f = \lambda f$, then $|m|(2 - |m|)/4 \leqq \lambda \in \mathbf{R}$.

Given (Γ, θ, m), we take a maximal orthonormal set $\Phi = \Phi(\Gamma, \theta, m) = \{\varphi\}$ of cusp forms of type (Γ, θ, m), adding a constant function if $m = 0$ and $\theta = 1$. For each $\varphi \in \Phi$, we put $L_m\varphi = \lambda_\varphi \varphi$. Let $\Lambda(\Gamma, \theta, m)$ denote the set of such eigenvalues λ_φ for all $\varphi \in \Phi$. Then we let $S(\Gamma, \theta, m)$ denote the set of all complex numbers s such that $s(1 - s) \in \Lambda(\Gamma, \theta, m)$. Thus $S(\Gamma, \theta, m)$ contains 0 and 1 if $m = 0$ and $\theta = 1$.

2. The kernel function $G_s(z, w)$. Let Γ, θ, and m be the same as in Section 1. We define functions $k_s(z, w)$ and $G_s(z, w; \theta, m)$ for $s \in \mathbf{C}$ and $(z, w) \in H^2$ by

$$G_s(z, w; \theta, m) = (1/2) \sum_{\gamma\in\Gamma} \theta(\gamma) J_\gamma(w)^m k_s(z, \gamma w), \tag{2.1a}$$

(2.1b) $k_s(z, w) =$

$$-(4\pi)^{-1} i^m |z - \bar{w}|^m (z - \bar{w})^{-m} (1 - t)^s \tilde{F}(s + m/2, s - m/2, 2s; 1 - t),$$

where $t = |z - w|^2/|z - \bar{w}|^2$ and

$$\tilde{F}(a, b, c; z) = \Gamma(a)\Gamma(b)\Gamma(c)^{-1} F(a, b, c; z) \tag{2.1c}$$

with the standard hypergeometric function F. As mentioned in the introduction, the function G_s was introduced by Roelcke [9], and later investigated by several authors [1, 3, 4, 7, 8]. Let us now recall some of its basic properties given in these papers, in particular in Hejhal [7], following the formulation of that article (with a few notational changes). First of all, we have

(2.2) $k_s(\alpha z, \alpha w) = J_\alpha(z)^m J_\alpha(w)^{-m} k_s(z, w)$ *for every* $\alpha \in \mathrm{SL}_2(\mathbf{R})$,

(2.3a) $$G_s(z, w) = (1/2) \sum_{\gamma\in\Gamma} \theta(\gamma)^{-1} J_\gamma(z)^{-m} k_s(\gamma z, w),$$

(2.3b) $G_s(z, w) = \theta(\gamma)^{-1} J_\gamma(z)^{-m} G_s(\gamma z, w) = \theta(\gamma) J_\gamma(w)^m G_s(z, \gamma w)$ *for every* $\gamma \in \Gamma$.

Here and henceforth we simply write $G_s(z, w)$ for our function if there is no fear of confusion.

The series for G_s in (2.1a) is meaningful and convergent for $\mathrm{Re}(s) > 1$, $\Gamma(s \pm m/2) \neq \infty$, and for $z \notin \Gamma w$. Now $G_s(z, w)$ when $z \notin \Gamma w$ can be continued to the whole plane as a meromorphic function of s, with poles only in the following set:

(2.4) $$P(\Gamma, \theta, m) = S(\Gamma, \theta, m) \cup \{1/2\} \cup P'(\Gamma, \theta, m),$$

where P' is the set of poles of certain Eisenstein series which we will describe below. More precisely, for every fixed $s_0 \in \mathbf{C}$, there exists a nonnegative integer α and a neighborhood U of s_0 such that $(s - s_0)^\alpha G_s(z, w)$ is C^∞ on $U \times \{(z, w) \in H^2 \mid z \notin \Gamma w\}$ and holomorphic in s; we can take $\alpha = 0$ if $s_0 \notin P(\Gamma, \theta, m)$. When G_s is finite, it satisfies

(2.5) $$L_m G_s(z, w) = s(1 - s) G_s(z, w),$$

where z is the variable of differentiation.

Fix an element ξ of Ξ: put $\theta\left(\xi\left(\begin{smallmatrix}1 & 1\\ 0 & 1\end{smallmatrix}\right)\xi^{-1}\right) = e^{2\pi i p}$ with $p \in \mathbf{R}$. Then G_s has the following Fourier expansion at $\xi(\infty)$:

(2.6) $$J_\xi(z)^{-m} G_s(\xi z, w) = (1 - 2s)^{-1} y^{1-s} E_\xi(w, s; \theta^{-1}, -m)$$
$$\sum_{0\neq r\equiv p \,(\mathrm{mod}\ \mathbf{Z})} e^{2\pi(irx - |r|y)} F_{\xi r}(w, s) y^s \Psi(s_r, 2s; 4\pi|r|y).$$

Here the symbols are as follows: $z = x + iy$, s_r and Ψ are as in Lemma 1.2 and (1.13): $F_{\xi r}$ is a certain explicitly defined function; E_ξ is given by

(2.7) $$E_\xi(w, s; \theta, m) = \begin{cases} \sum_{\alpha\in\Gamma_t\backslash\Gamma} \theta(\alpha)^{-1} \mathrm{Im}(w)^s \|_m \xi^{-1}\alpha & \text{if} \quad p \in \mathbf{Z}, \\ 0 & \text{if} \quad p \notin \mathbf{Z}, \end{cases}$$

where $t = \xi(\infty)$. The expansion of (2.6) is valid for sufficiently large y

when w stays in a compact set and $s \notin P(\Gamma, \theta, m)$. It can be shown that E_ξ of (2.7) can be continued as a meromorphic function of s to the whole $\mathbf{C}$. Now $P'(\Gamma, \theta, m)$ of (2.4) is the set of poles of the E_ξ for all $\xi \in \Xi$. It is known that $s = 1$ or $\mathrm{Re}(s) < 1/2$ for every $s \in P'(\Gamma, \theta, m)$. We also note here

$$(2.8)\quad G_s(z, w) - G_{1-s}(z, w)$$
$$= (1 - 2s)^{-1} \sum_{\xi \in \Xi} E_\xi(z, s\colon \theta, m) E_\xi(w, 1 - s; \bar{\theta}, -m)$$

(see [7, p. 319, p. 414]). Fix a real number $a > \mathrm{Max}(1, |m|/2)$. Then

$$(2.9)\quad G_s(z, w) - G_a(z, w)$$
$$= Q_s(z, w) + \sum_{\varphi \in \Phi} \{[s(1 - s) - \lambda_\varphi]^{-1} - [a(1 - a) - \lambda_\varphi]^{-1}\} \varphi(z) \overline{\varphi(w)}$$

with a function $Q_s(z, w)$ which is meromorphic on the whole s-plane with poles only in $\{1/2\} \cup P'(\Gamma, \theta, m)$ (see [7, p. 250, p. 318, p. 414]).

Let us now study the nature of $G_s(z, w)$ as a function of z when w is fixed, by modifying it as follows: for each $s \in \mathbf{C}$ not belonging to $\{1/2\} \cup P'(\Gamma, \theta, m)$, we put

$$(2.10)$$
$$f_s(z, w) = f_{s,w}(z) = \lim_{S \to s} \{G_S(z, w) - \sum [S(1 - S) - \lambda_\varphi]^{-1} \varphi(z) \overline{\varphi(w)}\},$$

where the sum is taken over all φ in Φ such that $s(1 - s) = \lambda_\varphi$. Obviously $f_s = G_s$ if $s \notin S(\Gamma, \theta, m)$. We easily see that $f_{s,w}$ is well defined for $z \notin \Gamma w$.

Proposition 2.1. *For a fixed w and $s \notin \{1/2\} \cup P'(\Gamma, \theta, m)$, $f_{s,w}$ as a function of z satisfies*

$$(2.11)\qquad [L_m - s(1 - s)] f_{s,w} = \sum_{s(1-s) = \lambda_\varphi} \overline{\varphi(w)} \varphi;$$

$f_{s,w}$ is not identically equal to 0 if and only if $\theta(\alpha) J_\alpha(w)^m = 1$ for every $\alpha \in \Gamma_w$. If it is not 0, we have

$$(2.12)\qquad f_{s,w}(z) = (2\pi)^{-1} e_w \log|z - w| + r_{s,w}(z)$$

for z in a sufficiently small neighborhood U of w with a continuous function $r_{s,w}$ on U. Moreover

$$r_{s,w}(z) = p_{s,w}(z) + q_{s,w}(z)\log|z - w| \tag{2.13}$$

with C^∞ functions $p_{s,w}$ and $q_{s,w}$ on U such that $q_{s,w}(w) = 0$. Furthermore, for every $\xi \in \Xi$, we have

$$(f_{s,w}\|_m\xi)(z) - (1 - 2s)^{-1}y^{1-s}E_\xi(w, s; \bar{\theta}, -m) = O(e^{-cy})$$

with some $c > 0$ uniformly in x.

Proof. Put $P_S(z, w) = G_S(z, w) - \Sigma' [S(1 - S) - \lambda_\varphi]^{-1}\varphi(z)\overline{\varphi(w)}$ with φ running over Φ under the condition $\lambda_\varphi = s(1 - s)$. Then P is holomorphic in S on a small disc $|S - s| < \epsilon$ and $L_mP_S - S(1 - S)P_S = \Sigma'$ $\varphi(z)\overline{\varphi(w)}$ from which we obtain (2.11). Now we see from (2.6) and Lemma 1.2 that for every $\xi \in \Xi$ and for $|S - s| = \epsilon/2$,

$$J_\xi(z)^{-m}P_S(\xi z, w) - (1 - 2S)^{-1}y^{1-S}E_\xi(w, S; \bar{\theta}, -m) = O(e^{-cy})$$

with some $c > 0$. Applying the maximum principle to the left-hand side, we obtain the last assertion. (The result also follows from (2.15) below.) Next we observe that f_s satisfies formula (2.3b), from which the "only-if"-part of the second assertion follows immediately. To prove the remaining part, we employ a formula (see [2, p. 74, (2)])

$$\tilde{F}(a, b, a + b; 1 - t) = A(t) - B(t)\log(t) \qquad (0 < t < 1)$$

with convergent power series A and B on $|t| < 1$ such that $B(0) = 1$, provided Γ is finite at a, b, and $a + b$. From this we can easily derive that

(2.14)

$$k_s(z, w) = (2\pi)^{-1}\log|z - w| + A_s(z, w) + B_s(z, w)\log|z - w|$$

with real analytic A_s and B_s on H^2 such that $B_s(z, z) = 0$. Take a neighborhood U of w so that $\gamma U \cap U = \varnothing$ for every $\gamma \in \Gamma$, $\notin \Gamma_w$. Suppose $\theta(\alpha)J_\alpha(w)^m = 1$ for every $\alpha \in \Gamma_w$. If $\mathrm{Re}(s) > 1$ and $\Gamma(s \pm m/2) \neq \infty$, the existence of $r_{s,w}$, $p_{s,w}$, and $q_{s,w}$ immediately follows from (2.14) and (2.1a).

If $\mathrm{Re}(s) \leqq 1$ or $\Gamma(s \pm m/2) = \infty$, it follows from (2.9) and (2.10). Once we have (2.12), it shows that $f_{s,w}$ is a nonzero function.

Now for a fixed (s, w) as above, $f_{s,w}$ satisfies (1.6) and (1.8b) with $Z = \Gamma w$. Moreover the above proposition shows that $f_{s,w}\|_m \xi = O(y^{1-s})$ for every $\xi \in \Xi$. Therefore $f_{s,w}$ satisfies (1.8c) if $\mathrm{Re}(s) > 1/2$. In any case, $r_{s,w}(w)$ is meaningful; for simplicity, we put $r_s(w) = r_{s,w}(w)$. As for (1.8a), however, it is satisfied by $f_{s,w}$ only if $s \notin S(\Gamma, \theta, m)$ or if $\varphi(w) = 0$ for every φ such that $\lambda_\varphi = s(1-s)$. In such a case, the Fourier expansion of $f_{s,w}$ is a mere specialization of (2.6). If $s \in S(\Gamma, \theta, m)$, however, it can be shown that $f_{s,w}$ for a fixed (s, w) has an expansion of the form

$$(2.15)\quad f_{s,w}\|_m \xi = (1-2s)^{-1} y^{1-s} E_\xi(\tau, s; \bar{\theta}, -m) + \sum_{0 \neq r \equiv p(Z)} \{p_{\xi r} V_r(y, s) + q_{\xi r} \partial V_r/\partial s(y, s)\} e^{2\pi(irx - |r|y)}$$

for every $\xi \in \Xi$ and sufficiently large y, where $V_r(y, s) = y^s \Psi(s_r, 2s; 4\pi|r|y)$ and $p_{\xi r}$ and $q_{\xi r}$ are constants.

Lemma 2.2. *If $s \notin \{1/2\} \cup P'(\Gamma, \theta, m)$, we have $\overline{f_s(z, w)} = f_{\bar{s}}(w, z)$; moreover $\overline{r_s(w)} = r_{\bar{s}}(w)$ if $\theta(\alpha) J_\alpha(w)^m = 1$ for every $\alpha \in \Gamma_w$.*

Proof. We first observe that $\overline{G_s(z, w)} = G_{\bar{s}}(w, z)$ and $\overline{Q_s(z, w)} = Q_{\bar{s}}(w, z)$. Therefore the first equality follows immediately from (2.10). Suppose $\mathrm{Re}(s) > 1$ and $\Gamma(s \pm m/2) \neq \infty$; put

$$C_s(z, w) = (1/2) \sum_{\alpha \notin \Gamma_w} \theta(\alpha) J_\alpha(w)^m k_s(z, \alpha w).$$

Then $r_{s,w}(z) = C_s(z, w) + e_w A_s(z, w) + e_w B_s(z, w) \log|z - w|$ with A_s and B_s of (2.14). Since $\overline{k_s(z, w)} = k_{\bar{s}}(w, z)$, we see that $A_{\bar{s}}(z, w) = \overline{A_s(w, z)}$. By (2.2), we have $\overline{C_s(z, w)} = (1/2) \sum_{\beta \notin \Gamma_w} \theta(\beta) J_\beta(z)^m k_{\bar{s}}(w, \beta z)$, and hence $\overline{C_s(w, w)} = C_{\bar{s}}(w, w)$. Thus we obtain $\overline{r_s(w)} = r_{\bar{s}}(w)$ when $\mathrm{Re}(s) > 1$ and $\Gamma(s \pm m/2) \neq \infty$. In the general case, we have

$$r_s(w) = Q_s(w, w) + r_a(w) + \Sigma' \{[s(1-s) - \lambda_\varphi]^{-1} - [a(1-a) - \lambda_\varphi]^{-1}\} |\varphi(w)|^2 - \Sigma'' [a(1-a) - \lambda_\varphi]^{-1} |\varphi(w)|^2,$$

where a is as in (2.9), Σ' resp. Σ'' is the sum over all $\varphi \in \Phi$ such that $\lambda_\varphi \neq s(1-s)$ resp. $\lambda_\varphi = s(1-s)$. Therefore we obtain $\overline{r_s(w)} = r_{\bar{s}}(w)$.

We now fix a finite set of points T of Γ-inequivalent points of H. Fixing an s not contained in $\{1/2\} \cup P'(\Gamma, \theta, m)$, put

$$f = \sum_{\tau \in T} c(\tau) f_{s,\tau}, \quad g = \sum_{\tau \in T} b(\tau) f_{s,\tau}, \tag{2.16a}$$

$$\psi_c = \sum_{\tau \in T} \sum_{\varphi} c(\tau)\overline{\varphi(\tau)}\varphi, \quad \psi_b = \sum_{\tau \in T} \sum_{\varphi} b(\tau)\overline{\varphi(\tau)}\varphi \tag{2.16b}$$

with complex numbers $c(\tau)$ and $b(\tau)$, and with φ running over Φ under the condition $\lambda_\varphi = s(1-s)$. Then for each $\tau \in T$ we have

$$f(z) = c(\tau)(2\pi)^{-1} e_\tau \log|z - \tau| + r_{f,\tau}(z), \tag{2.17a}$$

$$g(z) = b(\tau)(2\pi)^{-1} e_\tau \log|z - \tau| + r_{g,\tau}(z) \tag{2.17b}$$

in a neighborhood of τ, where

$$r_{f,\tau} = c(\tau) r_{s,\tau} + \sum_{\tau' \neq \tau} c(\tau') f_{s,\tau'}, \tag{2.18a}$$

$$r_{g,\tau} = b(\tau) r_{s,\tau} + \sum_{\tau' \neq \tau} b(\tau') f_{s,\tau'}. \tag{2.18b}$$

Proposition 2.3. *The notation being as above, the following assertions hold:*

(1) *If* $\mathrm{Re}(s) > 1/2$ *or both* f *and* g *are rapidly decreasing, then*

$$2i \cdot \mathrm{Im}[s(1-s)]\langle f, g\rangle + \langle f, \psi_b\rangle - \langle \psi_c, g\rangle$$

$$= \sum_{\tau \in T} [b(\tau)\overline{r_{f,\tau}(\tau)} - \overline{c(\tau)} r_{g,\tau}(\tau)]$$

$$= \sum_{\tau \in T} b(\tau)\overline{c(\tau)}[\overline{r_s(\tau)} - r_s(\tau)] + \sum_{\tau \neq \tau'} b(\tau')\overline{c(\tau)}[\overline{f_s(\tau', \tau)} - f_s(\tau, \tau')].$$

(2) *If* $s = (1/2) + it$ *with* $0 \neq t \in \mathbf{R}$, *we have* $\langle \psi_c, g\rangle = \langle f, \psi_b\rangle$ *and*

$$\sum_{\xi \in \Xi} \Big\{ \sum_{\tau \in T} b(\tau) E_\xi(\tau, \bar{s}; \bar{\theta}, -m)\Big\}\Big\{ \sum_{\tau \in T} \overline{c(\tau)} E_\xi(\tau, s; \theta, m)\Big\}$$

$$= 2it \sum_{\tau\in T} [b(\tau)\overline{r_{f,\tau}(\tau)} - \overline{c(\tau)}r_{g,\tau}(\tau)]$$

$$= 4t \sum_{\tau\in T} b(\tau)\overline{c(\tau)}\mathrm{Im}[r_s(\tau)] + 2it \sum_{\tau\neq\tau'} b(\tau')\overline{c(\tau)}[\overline{f_s(\tau',\tau)} - f_s(\tau,\tau')].$$

We note here special cases of these formulas:

$$\text{(2.19a)} \qquad \mathrm{Im}[r_s(\tau)] = \mathrm{Im}[\langle \Sigma\, \overline{\varphi(\tau)}\varphi, f_{s,\tau}\rangle - s(1-s)\|f_{s,\tau}\|^2]$$

if $\mathrm{Re}(s) > 1/2$ *or* $f_{s,\tau}$ *is rapidly decreasing.*

$$\text{(2.19b)} \qquad \langle \Sigma\, \overline{\varphi(\tau)}\varphi, f_{s,\tau}\rangle \in \mathbf{R} \quad \text{if} \quad \mathrm{Re}(s) = 1/2.$$

In these two formulas, φ runs over Φ under the condition $\lambda_\varphi = s(1-s)$.

$$\text{(2.19c)} \quad 4t\cdot \mathrm{Im}[r_s(\tau)] = \sum_{\xi\in\Xi} |E_\xi(\tau, s; \theta, m)|^2 \qquad \text{if} \quad s = (1/2) + it \quad \text{with} \quad 0 \neq t \in \mathbf{R}.$$

$$\text{(2.19d)} \quad 2it[\overline{f_s(\tau',\tau)} - f_s(\tau,\tau')] = \sum_{\xi\in\Xi} E_\xi(\tau, s; \theta, m)E(\tau', \bar{s}; \bar{\theta}, -m)$$

if $\tau \notin \Gamma\tau'$ *and* $s = (1/2) + it$ *with* $0 \neq t \in \mathbf{R}$.

Proof. If $s \notin S(\Gamma, \theta, m)$, the formula of (1) is an immediate consequence of Proposition 1.1. Suppose $s \in S(\Gamma, \theta, m)$; let P_S be as in the proof of Proposition 2.1. For every $\xi \in \Xi$ we see from Lemma 1.2 and its proof that $\epsilon_m(P_S(z,\tau)\|_m\xi)$ is of the form $Ay^{1-S} + O(e^{-by})$ with $b > 0$ uniformly for $|S - s| = \rho$ with a sufficiently small $\rho > 0$. Then the maximum principle tells that $\epsilon_m(f_{s,\tau}\|_m\xi)$ is of the form $ay^{1-s} + O(e^{-by})$. Repeating the proof of Proposition 1.1, we find that the left-hand side of (1.10) tends to $\mathrm{Im}[s(1-s)]\langle f, g\rangle + (i/2)\{\langle\psi_c, g\rangle - \langle f, \psi_b\rangle\}$ and therefore we obtain (1). To prove (2), we first observe that

$$(1-2s)[f_s(z,w) - f_{1-s}(z,w)] = \sum_{\xi\in\Xi} E_\xi(z, s; \theta, m)E_\xi(w, 1-s; \bar{\theta}, -m).$$

If $s = (1/2) + it \neq 1/2$, this together with Lemma 2.2 yields (2.19c,d). Taking a linear combination, we obtain the second part of (2). Define c_ξ

and b_ξ as in (1.15) for the present f and g. Then we have

$$-2itc_\xi = \sum_{\tau\in T} c(\tau)E_\xi(\tau, s; \bar{\theta}, -m),$$

$$-2itb_\xi = \sum_{\tau\in T} b(\tau)E_\xi(\tau, s; \bar{\theta}, -m),$$

and hence

$$4t^2 \sum_\xi b_\xi \bar{c}_\xi = \sum_\xi \Big\{ \sum_{\tau\in T} b(\tau)E_\xi(\tau, s; \bar{\theta}, -m)\Big\}\Big\{ \sum_{\tau\in T} \overline{c(\tau)E_\xi(\tau, \bar{s}; \theta, m)}\Big\}$$

$$= \sum_{\tau,\tau'} b(\tau')\overline{c(\tau)} \sum_\xi E_\xi(\tau', s; \bar{\theta}, -m)E_\xi(\tau, 1-s; \theta, m).$$

By (2.8), we see that the last sum over ξ is invariant under $s \mapsto 1-s$. Hence $4t^2 \sum_\xi b_\xi \bar{c}_\xi$ is equal to the last sum of (2). Repeat the proofs of Propositions 1.1 and 1.3 with the present f and g. Then the left-hand side of (1.10) tends to $(i/2)\{\langle \psi_c, g\rangle - \langle f, \psi_b\rangle\}$ and the right-hand side to

$$(2\bar{s}-1)(i/2)\sum_\xi b_\xi \bar{c}_\xi + (i/2)\sum_{\tau\in T}\big[\overline{c(\tau)}r_{g,\tau}(\tau) - b(\tau)\overline{r_{f,\tau}(\tau)}\big]$$

as shown in the proof of Proposition 1.3. This sum must be 0, because of the last formula of (2). Thus we obtain $\langle \psi_c, g\rangle = \langle f, \psi_b\rangle$, which completes the proof.

It should be noted that a formula for $\|f_{s,\tau}\|$ of the same type as (1) of Proposition 2.3 was given by Elstrodt [1, II, Satz 7.2] when $\mathrm{Re}(s) > 1$ and $\Gamma(s \pm m/2) \neq \infty$.

We now specialize our group Γ to the congruence subgroup

$$(2.20) \qquad \Gamma_0(W) = \{\gamma \in SL_2(\mathbf{Z}) \mid c_\gamma \equiv 0 \pmod{W}\}.$$

Here W is a positive integer. We take a Dirichlet character θ modulo W such that $\theta(-1) = (-1)^m$ and put $\theta(\gamma) = \theta(d_\gamma)$ for $\gamma \in \Gamma_0(W)$; we understand that $\theta(d_\gamma) = 1$ for every γ if $W = 1$. Recall that every cusp of $\Gamma_0(W)$ is $\Gamma_0(W)$-equivalent to $t = u/v$ with a positive divisor v of W and an integer u prime to v such that $0 < u \leqq (v, W/v)$. Now we have

LEMMA 2.4. *Suppose W is the conductor of θ; let $t = u/v$ with u and*

v as above. Then Γ_t contains an element β such that $\mathrm{Tr}(\beta) = 2$ *and* $\theta(\beta) \neq 1$ *if and only if v is not prime to W/v.*

Proof. Take integers x and y so that $uy - vx = 1$; let $\epsilon = \begin{pmatrix} u & x \\ v & y \end{pmatrix}$, $\alpha = \begin{pmatrix} 1 & q \\ 0 & 1 \end{pmatrix}$ with $q \in \mathbf{Q}$, and $\beta = \epsilon\alpha\epsilon^{-1}$. Then $\epsilon(\infty) = t$ and

$$(*) \qquad \beta = \begin{pmatrix} 1 - quv & qu^2 \\ -qv^2 & 1 + quv \end{pmatrix}.$$

This is contained in Γ_t if and only if $\beta \in \mathrm{SL}_2(\mathbf{Z})$ and W divides qv^2; moreover Γ_t consists of the elements $\pm\beta$ with all such q's. Suppose $(v, W/v) = 1$ and $\beta \in \Gamma_t$. Then $q = kW/v$ with $k \in \mathbf{Z}$, so that $quv \in W\mathbf{Z}$. Thus $\theta(d_\beta) = 1$, which proves the "only-if"-part. Suppose $(v, W/v) \neq 1$. Fix a prime p dividing $(v, W/v)$, and put $W = p^e M$ and $v = p^f N$ with integers M and N prime to p. Then $0 < f < e$. Put $r = \mathrm{Max}(0, e - 2f)$ and take $p^r M/N$ as q in (*). Then $\beta \in \Gamma_t$ and $1 + quv = 1 + p^{r+f}Mu$. Considering $(\mathbf{Z}_p/W\mathbf{Z}_p)^\times$ as a direct product factor of $(\mathbf{Z}/W\mathbf{Z})^\times$ in a natural way, denote by θ_p the restriction of θ to $(\mathbf{Z}_p/W\mathbf{Z}_p)^\times$. Then $\theta(1 + quv) = \theta_p(1 + p^{r+f}Mu)$. Let $G_i = \{a \in \mathbf{Z}_p^\times \mid a - 1 \in p^i\mathbf{Z}_p\}$. Since $p \nmid Mu$ and $e \leqq 2r + 2f < 2e$, we see that $1 + p^{r+f}Mu$ generates G_{r+f}/G_e. Therefore $\theta_p(1 + p^{r+f}Mu) \neq 1$, and hence $\theta(d_\beta) \neq 1$, which completes the proof.

Choose Ξ for $\Gamma_0(W)$ and define G_s and E_ξ with respect to $\Gamma_0(W)$ and a primitive θ. Then the above lemma shows that $E_\xi = 0$ unless $\xi(\infty)$ is equivalent to $1/v$ with a positive divisor v of W such that v is prime to W/v. In particular, if W is a prime power > 1, then E_ξ is a nonzero function only if $\xi(\infty)$ is equivalent to 0 or ∞.

3. Main theorems in the first case. In this section we confine ourselves to the case $\Gamma = \mathrm{SL}_2(\mathbf{Z})$, though many, if not all, of our results can be generalized to the case of congruence subgroups. We consider G_s and E_ξ of Section 2 with $\Gamma = \mathrm{SL}_2(\mathbf{Z})$ by taking θ to be the trivial character; thus we write $G_s(z, w; m)$ or $G_s(z, w)$; naturally m must be even and Ξ consists of the identity element. Thus the function of (2.7) simply becomes

$$(3.1\mathrm{a}) \qquad E(z, s; m) = \sum_{\alpha \in \Gamma_\infty \backslash \Gamma} \mathrm{Im}(z)^s \|_m \alpha.$$

We put then

$$(3.1\mathrm{b}) \qquad E^*(z, s; m) = \pi^{-s}\Gamma(s + |m|/2)\zeta(2s)E(z, s; m),$$

where ζ is Riemann's zeta function. It is well known that $s(1-s)E^*$ can be continued to a real analytic function on $H \times \mathbf{C}$ that is entire in s and invariant under $s \mapsto 1-s$ (cf. Section 5).

We are going to consider the inhomogeneous equation $(L_m - \lambda)f = p$ with a function p on H satisfying the following condition:

(3.2a) *p is C^∞ and satisfies $p\|_m \gamma = p$ for every $\gamma \in \Gamma$.*

We then have a Fourier expansion

$$p(z) = \sum_{n \in \mathbf{Z}} e^{2\pi inx} \psi_n(y) \tag{3.2b}$$

with C^∞ functions ψ_n. We impose the following condition on them:

(3.2c) *For every $\eta > 0$, there exist positive constants A, B, and c such that $|\psi_n(y)| \leqq A(|n|+1)^B e^{-(2\pi|n|+c)y}$ for every n and $y > \eta$.*

We easily see that $e^{cy}p(z)$ as well as $e^{cy}\psi_n(y)$ is bounded; thus p is rapidly decreasing. We now put

$$R(s) = \int_{\Gamma \backslash H} p(z)E(z, s; -m)d_H z, \tag{3.3a}$$

$$R^*(s) = \pi^{-s}\Gamma(s + |m/2|)\zeta(2s)R(s). \tag{3.3b}$$

The integral of (3.3a) is convergent if E is finite at s. Moreover we see that $s(1-s)R^*(s)$ is an entire function and $R^*(s) = R^*(1-s)$. The purpose of this section is to study the function

$$f(z, s) = -\int_{\Gamma \backslash H} G_s(z, w; m)p(w)d_H w \tag{3.4}$$

in connection with R. The properties of G_s recalled in Section 2 imply that for every $s_0 \in \mathbf{C}$, there is a nonnegative integer k such that $(s-s_0)^k\Gamma(s + |m/2|)\zeta(2s)G_s(z, w)$ as a function of w is slowly increasing uniformly on a neighborhood of s_0; we can take $k = 0$ if $s \notin \{1/2\} \cup S(\Gamma, 1, m)$ as will be seen in Lemma 5.1 below. Therefore $f(z, s)$ is well-defined as a meromorphic function of s as stated in (i) of the following

Theorem 3.1. *The function $f(z, s)$ has the following properties:*

(i) *$\Gamma(s + |m/2|)\zeta(2s)f(z, s)$ is meromorphic on the whole s-plane with poles only in $\{1/2\} \cup S(\Gamma, 1, m)$.*

(ii) *$f(z, s)$ is a C^∞ function in (z, s) if $s \notin P(\Gamma, 1, m)$; moreover, for every $s_0 \in \mathbf{C}$, there exists a neighborhood U of s_0 and a nonnegative integer n such that $(s - s_0)^n f(z, s)$ is C^∞ on $H \times U$.*

(iii) *$f(\gamma z, s) = J_\gamma(z)^m f(z, s)$ for every $\gamma \in \Gamma$.*

(iv) *$[L_m - s(1 - s)]f(z, s) = p(z)$ if $s \notin P(\Gamma, 1, m)$.*

(v)
$$(2s - 1)[f(z, s) - f(z, 1 - s)] = R(s)E(z, 1 - s; m) = R(1 - s)E(z, s; m).$$

(vi) *If K is a compact subset of $\mathbf{C}$ disjoint with $P(\Gamma, 1, m)$, then $f(z, s) - (2s - 1)^{-1}y^{1-s}R(s) = O(e^{-by})$ as $y \to \infty$ uniformly for $(x, s) \in \mathbf{R} \times K$, where $b = \mathrm{Min}(c, \pi)$.*

Assertion (iii) follows immediately from (2.3b). If $\mathrm{Re}(s) > 1$ and $\Gamma(s \pm m/2) \neq \infty$, (ii) and (iv) follow from a well known principle. In fact, if f is defined by (3.4) with any bounded C^∞ function p and such an s, then f is C^∞ and satisfies (iv) (see [9], [3], and [7, p. 645]). To prove (ii) for an arbitrary s, take any $a > \mathrm{Max}(1, |m/2|)$, and put $Z_s = G_a - G_s$. Then

$$f(z, s) - f(z, a) = \int_{\Gamma\backslash H} Z_s(z, w)p(w)d_H w.$$

Thus our problem is the differentiability of the right-hand side. Take a point $(z_0, s_0) \in H \times \mathbf{C}$; let U denote a neighborhood of (z_0, s_0) which will be made smaller according to our requirements. Since $G_s(z, w; m) = G_s(w, z; -m)$, (2.6) implies that for a suitable nonnegative integer α, G_s as a function of w has an expansion

$$(3.5) \quad (s - s_0)^\alpha G_s(z, u + iv) = r_0(z, s)v^{1-s} + \sum_{0 \neq n \in \mathbf{Z}} e^{2\pi(inu - |n|v)} r_n(z, s)v^s \Psi(s_n', 2s; 4\pi|n|v)$$

for $v > M_U$ and $(z, s) \in U$, where $s_n' = s + (m/2)\mathrm{sgn}(n)$, r_0 and r_n are C^∞ on U, and M_U is a constant depending on U. Since the left-hand side is C^∞ in (z, w, s), we see that

$$(3.6) \quad (\partial^{a+b+c}/\partial x^a \partial y^b \partial s^c)[(s - s_0)^\alpha G_s(z, u + iv)]$$

for every nonnegative a, b, c has a similar expansion. By Lemma 1.2, we see that (3.6) is $O(v^{1-s})$ uniformly on a suitable U. The same is true for G_a with a in place of s. Hence if a is sufficiently large, the partial derivatives of $(s - s_0)^\alpha Z_s$ with respect to (x, y, s) are all $O(v^{1-s})$ uniformly on U. This proves (ii). By analytic continuation, we then obtain (iv) for any $s \notin P(\Gamma, 1, m)$. Now (2.8) specialized to the present case shows

$$(1 - 2s)[G_s(z, w) - G_{1-s}(z, w)] = E(z, 1 - s; m)E(w, s; -m),$$

which together with (3.3a) proves (v).

The principal feature of our theorem is in (vi), whose proof is not so simple. We first observe that the automorphy property of f together with its differentiability shows that

$$f(z, s) = \sum_{n \in \mathbf{Z}} q_n(y, s)e^{2\pi inx} \tag{3.7}$$

with functions q_n which are holomorphic or meromorphic in s according as f itself is holomorphic or meromorphic. To compute q_n, put

$$-G_s^*(z, w) = \sum_{n \in \mathbf{Z}} k_s(z + n, w) = \sum_{n \in \mathbf{Z}} k_s(z, w + n). \tag{3.8}$$

This is meaningful for $z - w \notin \mathbf{Z}$, $\mathrm{Re}(s) > 1$, and $\Gamma(s \pm m/2) \neq \infty$. Obviously

$$f(z, s) = \int_T G_s^*(z, w)p(w)d_H w \qquad (T = \Gamma_\infty \backslash H). \tag{3.9}$$

Now G_s^* has a Fourier expansion of the form

$$G_s^*(z, w) = \sum_{n \in \mathbf{Z}} e^{2\pi in(x-u)-2\pi|n|(y+v)}g_n(y, v, s) \tag{3.10}$$

with functions g_n given as follows (see [7, p. 351], [8, Lemma 4.3]):

$$g_0(y, v; s) = (2s - 1)^{-1} \cdot \begin{cases} y^s v^{1-s} & (y < v), \\ y^{1-s}v^s & (y > v), \end{cases} \tag{3.11a}$$

$$(3.11\text{b})\quad g_n(y, v; s) = |4\pi n|^{2s-1}\Gamma(s_n)y^s v^s \cdot \begin{cases} \Omega(s_n, 2s; 4\pi|n|y)\Psi(s_n, 2s; 4\pi|n|v) & (n \neq 0, y < v), \\ \Psi(s_n, 2s; 4\pi|n|y)\Omega(s_n, 2s; 4\pi|n|v) & (n \neq 0, y > v). \end{cases}$$

Here $s_n = s - (m/2)\mathrm{sgn}(n)$, $z = x + iy$, $w = u + iv$, Ψ is the function of (1.13), and $\Omega(a, c; z)$ is a holomorphic function on $\mathbf{C}^3$ with an integral representation

$$(3.12)\qquad \Omega(a, c; z) = \Gamma(a)^{-1}\Gamma(c-a)^{-1}\int_0^1 e^{zu}u^{a-1}(1-u)^{c-a-1}du,$$

valid for $\mathrm{Re}(c) > \mathrm{Re}(a) > 0$. The expansion of (3.10) is valid if $y \neq v$, $\mathrm{Re}(s) > 1$, and $\Gamma(s \pm m/2) \neq \infty$.

LEMMA 3.2. *Given any compact subset X of $\mathbf{C}^2$ and a positive number η, there exist positive constants A and B depending only on X and η such that if $(a, c) \in X$, one has*

$$|\Psi(a, c; y)| < Ay^B \quad \text{for} \quad y \geqq \eta,$$

$$|\Psi(a, c; y)| < Ay^{-B} \quad \text{for} \quad 0 < y \leqq \eta,$$

$$|\Omega(a, c\, y)| < Ay^B e^y \quad \text{for} \quad y \geqq \eta,$$

$$|\Omega(a, c; y)| < A \quad \text{for} \quad 0 < y \leqq \eta.$$

Moreover $\Psi(a, c; y)y^a \to 1$ as $y \to \infty$ uniformly for $(a, c) \in X$.

These are well known; see [2, p. 262, p. 278]. For the function Ψ, see also [12, Theorem 3.1] and [14, Section 10]. The method in [14] is also applicable to Ω. We need an estimate of the incomplete gamma function

$$\gamma(a, b; y) = \int_y^\infty e^{-bt}t^{a-1}dt \qquad (a \in \mathbf{C}, b > 0, y > 0).$$

We easily find that $\gamma(a, b; y) = y^a e^{-by}\Psi(1, a+1; by)$. Hence Lemma 3.2 shows that

$$(3.13)\qquad |\gamma(a, b; y)| \leqq Ae^{-by}(by)^B|y^a| \quad \text{for} \quad by > \eta > 0$$

when a stays in a compact subset L of $\mathbf{C}$, where A and B are positive constants depending only on L and η.

From (3.3a), (3.1a), and (3.2b), we obtain

$$(3.14) \qquad R(s) = \int_T p(u+iv)v^s d_H w \qquad (T = \Gamma_\infty \backslash H)$$

$$= \int_0^\infty \psi_0(v)v^{s-2}dv$$

for $\mathrm{Re}(s) > 1$. Now (3.9) shows that

$$f(z, s) = \int_0^\infty \int_0^1 G_s^*(z, u+iv)p(u+iv)du \cdot v^{-2}dv.$$

From (3.10) and (3.2b) we obtain

$$\int_0^1 G_s^*(z, u+iv)p(u+iv)du = \sum_{n\in\mathbf{Z}} e^{2\pi(inx-|n|y-n|v)}g_n(y, v, s)\psi_n(v).$$

Therefore we have formally

$$(3.15a) \qquad f(z, s) = \sum_{n\in\mathbf{Z}} e^{2\pi(inx-|n|y)}f_n(y, s),$$

$$(3.15b) \qquad f_n(y, s) = \int_0^\infty e^{-2\pi|n|v}\psi_n(v)g_n(y, v, s)v^{-2}dv.$$

This formal calculation can be justified if

$$(3.16) \qquad \sum_{n\neq 0} e^{-2\pi|n|y} \int_0^\infty e^{-2\pi|n|v}|\psi_n(v)g_n(y, v, s)|v^{-2}dv$$

is convergent. To see this convergence, we take a positive number η and a compact subset K of $\{s \in \mathbf{C} \mid \Gamma(s \pm m/2) \neq \infty\}$ and make estimates of integrals when $s \in K$ and $y > \eta$. Let us first consider f_0. By (3.11a) and (3.14) we have

$$(2s-1)f_0(y,s) = y^{1-s}\int_0^y \psi_0(v)v^{s-2}dv + y^s\int_y^\infty \psi_0(v)v^{-s-1}dv$$

$$= R(s)y^{1-s} - y^{1-s}\int_y^\infty \psi_0(v)v^{s-2}dv + y^s\int_y^\infty \psi_0(v)v^{-1-s}dv$$

if $\mathrm{Re}(s) > 1$. From (3.13) and (3.2c), we obtain

$$|(2s-1)f_0(y,s) - R(s)y^{1-s}| \leqq A_1e^{-cy}y^{B_1}$$

for $y > \eta$ and $s \in K$. Here and in the following $A_i, B_i, C_i, \ldots$ are positive constants which depend only on K, η, and φ and are independent of n.

Next suppose $n \neq 0$. Then

$$f_n(y,s) = |4\pi n|^{2s-1}\Gamma(s_n)y^s[U(y,s) + V(y,s)],$$

$$U(y,s) = \Psi(s_n, 2s; 4\pi|n|y)\int_0^y e^{-2\pi|n|v}\psi_n(v)\Omega(s_n, 2s; 4\pi|n|v)v^{s-2}dv,$$

$$V(y,s) = \Omega(s_n, 2s; 4\pi|n|y)\int_y^\infty e^{-2\pi|n|v}\psi_n(v)\Psi(s_n, 2s; 4\pi|n|v)v^{s-2}dv.$$

Let $U^*(y,s)$ and $V^*(y,s)$ denote the functions obtained from the above expressions for U and V by replacing Ψ, Ω, and $\psi_n(v)v^{s-2}$ by their absolute values. By (3.2c), Lemma 3.2, and (3.13) we have

$$V^*(y,s) \leqq A_2|n|^{B_2}y^{C_2}e^{4\pi|n|y}\int_y^\infty e^{-cv-4\pi|n|v}v^{D_2}dv$$
$$\leqq A_3|n|^{B_3}y^{C_3}e^{-cy}$$

if $s \in K$ and $y > \eta$. As for $U(y,s)$, we first put

(3.17a) $$b_n(s) = |4\pi n|^{2s-1}\Gamma(s_n)c_n(s),$$

(3.17b) $$c_n(s) = \int_0^\infty e^{-2\pi|n|v}\psi_n(v)\Omega(s_n, 2s, 4\pi|n|v)v^{s-2}dv,$$

and let $c_n^*(s)$ denote the integral whose integrand is the absolute value of that of c_n. By (3.2c) and Lemma 3.2, we see that c_n is holomorphic in s for $\mathrm{Re}(s) > M$ with a sufficiently large M and moreover $c_n^*(s) \leqq A_4|n|^{B_4}$ if $s \in K$ and $\mathrm{Re}(s) > M$. Now

$$\begin{aligned} c_n^*(s)|\Psi(s_n, 2s; 4\pi|n|y)| &- U^*(y, s) \\ &= |\Psi(s_n, 2s; 4\pi|n|y)| \int_y^\infty e^{-2\pi|n|v}|\psi_n(v)\Omega(s_n, 2s; 4\pi|n|v)v^{s-2}|dv \\ &\leqq A_5|n|^{B_5}y^{C_5}e^{-cy} \end{aligned}$$

by (3.2c), Lemma 3.2, and (3.13), if $s \in K$ and $y > \eta$. Combining these estimates, we obtain

$$f_0(y, s) = (2s-1)^{-1}y^{1-s}R(s) + h_0(y, s), \tag{3.18a}$$

$$f_n(y, s) = b_n(s)y^s\Psi(s_n, 2s; 4\pi|n|y) + h_n(y, s) \qquad (n \neq 0) \tag{3.18b}$$

with functions $h_n(y, s)$ such that

$$|h_n(y, s)| \leqq A_6(|n|+1)^{B_6}y^{C_6}e^{-cy} \quad \text{for} \quad s \in K \quad \text{and} \quad y > \eta. \tag{3.19}$$

At the same time we see that (3.16) is majorized by

$$A_7 \sum_{n \neq 0} e^{-2\pi|n|y}|n|^{B_7}y^{C_7}$$

if $s \in K$, $\mathrm{Re}(s) > M$, and $y > \eta$. Since the last series is convergent, we have thus established (3.15b) for $\mathrm{Re}(s) > M$ with some M. Since h_n is essentially a sum of several integrals over $[y, \infty)$ which are absolutely and locally uniformly convergent on $\mathbf{C}$, we easily see that $\Gamma(s_n)^{-1}h_n$ is an entire function of s. Moreover it can be easily seen that $\Gamma(s_n)^{-1}h_n$ is C^∞ in (y, s) and in particular $\Gamma(s_n)^{-1}\partial h_n/\partial y$ is holomorphic in s. Since f_n is well-defined by (3.15a) as a function with the differentiability as stated in (ii), (3.18b) shows that b_n can be continued to a meromorphic function on the whole $\mathbf{C}$ independent of y. Therefore we have

$$\begin{aligned} f(z, s) - (2s-1)^{-1}y^{1-s}R(s) &- \sum_{n \in \mathbf{Z}} e^{2\pi(inx-|n|y)}h_n(y, s) \\ &= \sum_{0 \neq n \in \mathbf{Z}} e^{2\pi(inx-|n|y)}b_n(s)y^s\Psi(s_n, 2s; 4\pi|n|y). \end{aligned} \tag{3.20}$$

Now the left-hand side of (3.20) is holomorphic on the domain

$$\{s \in \mathbf{C} \mid s \notin P(\Gamma, 1, m), \Gamma(s \pm m/2) \neq \infty\}. \tag{3.21}$$

Since $\Psi(a, b; y) \neq 0$ for sufficiently large y, b_n must be holomorphic on this domain. Applying Lemma 1.2 to (3.20), we find that it is $O(e^{-\pi y})$ locally uniformly on (3.21). Combining this with (3.19), we obtain (vi) of our theorem for K contained in (3.21). Suppose now s_0 is a pole of $\Gamma(s + m/2)$ or $\Gamma(s - m/2)$, but $s_0 \notin P(\Gamma, 1, m)$. Put

$$A(z, s) = e^{by}\{f(z, s) - (2s - 1)^{-1}y^{1-s}R(s)\}.$$

Then the above argument shows that $(s - s_0)A(z, s)$ is bounded when $y > \eta$ and $|s - s_0| \leqq \epsilon$ for some η and ϵ. Now $A(z, s)$ is holomorphic at s_0 and therefore the maximum principle shows that $A(z, s)$ itself is bounded for $y > \eta$ and $|s - s_0| \leqq \epsilon$. This completes the proof.

THEOREM 3.3. *Define b_n and c_n by* (3.17a,b) *with ψ_n of* (3.2b), *Ω of* (3.12), *and $s_n = s - (m/2)\operatorname{sgn}(n)$. Then the integral for c_n is convergent for sufficiently large* $\operatorname{Re}(s)$, *and c_n can be continued to a meromorphic function on the whole plane such that $\Gamma(s + |m/2|)\zeta(2s)c_n(s)$ has poles only in $\{1/2\} \cup S(\Gamma, 1, m)$. Moreover, put*

$$g(z, s) = (2s - 1)^{-1}y^{1-s}R(s) + \sum_{0 \neq n \in \mathbf{Z}} e^{2\pi(inx - |n|y)}b_n(s)y^s\Psi(s_n, 2s; 4\pi|n|y).$$

Then g is a C^∞ function for $z \in H$ and s in the set of (3.21), *meromorphic on the whole s-plane; moreover we have $L_m g(z, s) = s(1 - s)g(z, s)$,*

$$(2s - 1)[g(z, s) - g(z, 1 - s)] = R(s)E(z, 1 - s; m) = R(1 - s)E(z, s; m),$$

$$(2s - 1)[|4\pi n|^{-s}b_n(s) - |4\pi n|^{s-1}b_n(1 - s)] = R(s)\omega_n(1 - s) = R(1 - s)\omega_n(s),$$

where ω_n is defined by (3.23b) *below.*

Proof. We have already proved these assertions except the last few equalities. That g is an eigenfunction follows immediately from its expression. Put $W_n(y, s) = |4\pi ny|^s \Psi(s_n, 2s, 4\pi|n|y)$ for $n \neq 0$ and $y > 0$. A well known functional equation for Ψ (see [2, p. 257, (6)], [14, (10.10)]) implies that $W_n(y, s) = W_n(y, 1 - s)$. Now we have

$$g(z, s) = (2s - 1)^{-1}y^{1-s}R(s) + \sum_{n \neq 0} e^{2\pi(inx - |n|y)}|4\pi n|^{-s}b_n(s)W_n(y, s).$$

It is well known that $E(z, s; m)$ has the following expansion (see [14, p. 17], for example):

$$(3.22)\quad E(z, s; m) = y^s + \omega_0(s)y^{1-s} + \sum_{0 \neq n \in \mathbf{Z}} e^{2\pi(inx - |n|y)}\omega_n(s)W_n(y, s),$$

$$(3.23a)\quad \omega_0(s) = \pi i^{-m}2^{2-2s}\Gamma(2s - 1)\zeta(2s - 1)/[\Gamma(s + m/2)\Gamma(s - m/2)\zeta(2s)],$$

$$(3.23b)\quad \omega_n(s) = i^{-m}\pi^s|n|^{-s}[\Gamma(s + (m/2)\operatorname{sgn}(n))\zeta(2s)]^{-1}\sum_{0<d|n} d^{2s-1}.$$

Comparing the Fourier coefficients of both sides of (v) of Theorem 3.1 and those of (3.20), we find for $n \neq 0$ that

$$(3.24)\quad h_n(y, s) - h_n(y, 1 - s) = W_n(y, s)\{(2s - 1)^{-1}R(s)\omega_n(1 - s) - |4\pi n|^{-s}b_n(s) + |4\pi n|^{s-1}b_n(1 - s)\}.$$

For a generic s, $|4\pi ny|^k W_n(y, s)$ tends to 1 as $y \to \infty$, where $k = -\operatorname{sgn}(n)m/2$. Therefore (3.19) shows that (3.24) must vanish. A similar vanishing holds also for $n = 0$. This proves the desired equalities.

It should be noted that $g(z, s)$ is not an automorphic form. If it were (for a generic s), it must be an Eisenstein series, which is a contradiction, as it has no term for y^s.

THEOREM 3.4. *The notation being the same as in Theorem* 3.1, *put* $f_s(z) = f(z, s)$ *for* $s \notin P(\Gamma, 1, m)$. *Then the following assertions hold*:

(1) f_s *is not identically equal to* 0 *on* H;

(2) f_s *is rapidly decreasing if and only if* $R(s) = 0$;

(3) $\mathrm{Im}[\langle p, f_s\rangle] = \mathrm{Im}[s(1-s)]\|f_s\|^2$ *if* $R(s) = 0$ *or* $\mathrm{Re}(s) > 1/2$;
(4) $\mathrm{Im}[\langle f_s, p\rangle] = (4t)^{-1}|R(s)|^2$ *if* $s = (1/2) + it$ *with* $0 \neq t \in \mathbf{R}$;
(5) $(1-2s)\langle E(z,\bar{s}; m), f_s\rangle = dR(s)/ds$ *if* $R(s) = 0$.

Proof. The first two assertions follow directly from Theorem 3.1, (iv) and (vi). To prove the remaining ones, let us write s_0 for a fixed point outside $P(\Gamma, 1, m)$ and put $f_* = f(z, s_0)$ and $\lambda_0 = s_0(1-s_0)$. We observe that

$$(3.25)\quad \epsilon_m f_* = (i/2)(s_0 - 1 + m/2)(2s_0 - 1)^{-1} y^{1-s_0} R(s_0) + O(e^{-by}) \qquad (y \to \infty).$$

In fact, Lemma 3.2 combined with the formulas

$$\partial\Omega(a, c; y)/\partial y = a\Omega(a+1, c+1; y),$$

$$\partial\Psi(a, c; y)/\partial y = -a\Psi(a+1, c+1; y)$$

gives an estimate of $\partial h_n/\partial y$ similar to (3.19), and also that of the effect of ϵ_m on the right-hand side of (3.20) in view of Lemma 1.2. If $R(s_0) = 0$, $\langle f_*, f_*\rangle$ is meaningful, and $\epsilon_m f_*$ is rapidly decreasing. Since $(L_m - \lambda_0) f_* = p$, we have $\langle L_m f_*, f_*\rangle - \bar{\lambda}_0\langle f_*, f_*\rangle = \langle p, f_*\rangle$. Moreover $\langle L_m f_*, f_*\rangle = \langle f_*, L_m f_*\rangle$ as can be seen from the proof of Proposition 1.1. Therefore we obtain (3). We can easily verify that the same conclusion holds if $\mathrm{Re}(s) > 1/2$. To prove (5), put $\lambda = s(1-s)$ and $E(z, s) = E(z, s; m)$. That f_* and $\epsilon_m f_*$ are rapidly decreasing guarantees the equality

$$\langle E(z, \bar{s}), L_m f_*\rangle = \langle L_m E(z, \bar{s}), f_*\rangle,$$

from which we can easily derive

$$(\lambda - \lambda_0)\langle E(z, \bar{s}), f_*\rangle = \langle E(z, \bar{s}), \varphi\rangle = R(s).$$

Taking the limit when s tends to s_0, we obtain (5). Finally as for (4), a modification of Proposition 1.2 as in the proof of (2) of Proposition 2.3 gives the desired formula. (See also Remark (C) below.)

So far we have treated $f(z, s)$ with p only under conditions (3.2a,c). To make it more specific, take

$$p(z) = y^{(k+l)/2}\overline{g(z)}h(z) \tag{3.26}$$

with holomorphic cusp forms g and h of weight k and l, respectively, with respect to Γ, with expansions

$$g(z) = \sum_{n=1}^{\infty} a_n e^{2\pi i n z} \quad \text{and} \quad h(z) = \sum_{n=1}^{\infty} b_n e^{2\pi i n z}. \tag{3.27}$$

For simplicity, put

$$m = l - k, \qquad \kappa = (k + l)/2. \tag{3.28}$$

Then p satisfies (3.2a) with this m. Moreover, if we define ψ_n by (3.2b), then

$$\psi_n(y) = y^{\kappa} e^{-2\pi|n|y} \sum_{r=1}^{\infty} e^{-4\pi r y} \cdot \begin{cases} \bar{a}_r b_{n+r} & \text{if} \quad n \geqq 0, \\ \bar{a}_{r-n} b_r & \text{if} \quad n < 0. \end{cases} \tag{3.29}$$

Since $a_n = O(n^{\alpha})$ and $b_n = O(n^{\beta})$ with some α and β, we can easily verify (3.2c) with any $c < 4\pi$. Put

$$D(s) = \sum_{n=1}^{\infty} \bar{a}_n b_n n^{-s}, \tag{3.30a}$$

$$R_0(s) = (4\pi)^{-s}\Gamma(s)D(s). \tag{3.30b}$$

Define $R(s)$ and $f(z, s)$ for the present p. Then (3.14) shows that $R(s) = R_0(s + \kappa - 1)$. Thus the assertions of the above theorems for p of (3.26) concern the series of (3.30a). In this case $c_n(s)$ can be expressed as an explicit infinite series. In fact we have

Corollary 3.5. *For cusp forms g and h as in (3.27) and $0 \neq n \in \mathbf{Z}$, put*

$$A_n(s) = |n|^{s-1}\Gamma(2s)^{-1}\Gamma(s + \kappa - 1)\Gamma(s_n)$$
$$\cdot \begin{cases} \displaystyle\sum_{r=1}^{\infty} \bar{a}_r b_{r+n}(r + n)^{1-\kappa-s} F(s_n, s + \kappa - 1, 2s; n/(r + n)) & \text{if} \quad n > 0, \\ \displaystyle\sum_{r=1}^{\infty} \bar{a}_{r-n} b_r (r - n)^{1-\kappa-s} F(s_n, s + \kappa - 1, 2s; n/(r - n)) & \text{if} \quad n < 0, \end{cases}$$

where $s_n = s - (m/2)\operatorname{sgn}(n)$. *Then these series are absolutely convergent for sufficiently large* $\operatorname{Re}(s)$. *Moreover* A_n *can be continued to a meromorphic function on the whole plane that satisfies*

$$(2s-1)(4\pi)^{-\kappa}[A_n(s) - A_n(1-s)]$$
$$= \omega_n(s)R_0(\kappa - s) = \omega_n(1-s)R_0(s+\kappa-1)$$

with ω_n *of* (3.23b), R_0 *of* (3.30b), *and* $\kappa = (k+l)/2$. *Moreover* $\Gamma(s_n)^{-1}\Gamma(s+|m/2|)\zeta(2s)A_n(s)$ *has poles only in* $\{1/2\} \cup S(\Gamma, 1, m)$.

A result of the same type was obtained by Goldfeld in [5] in the case where g and h are of the same weight.

Proof. Employing a well known formula

$$F(a, b, c; z) = \frac{\Gamma(c)}{\Gamma(a)\Gamma(c-a)} \int_0^1 t^{a-1}(1-t)^{c-a-1}(1-tz)^{-b}dt,$$

valid for $\operatorname{Re}(c) > \operatorname{Re}(a) > 0$ and $|\arg(1-z)| < \pi$, we see from (3.12) that

$$\int_0^\infty e^{-\beta y}\Omega(a, c; \gamma y)y^{b-1}dy = \Gamma(b)\Gamma(c)^{-1}\beta^{-b}F(a, b, c; \gamma/\beta)$$

if $\beta > \gamma > 0$. Therefore b_n of (3.17a) is equal to $(4\pi)^{s-\kappa}|n|^s A_n(s)$. Thus our assertions immediately follow from Theorem 3.3.

Let us conclude this section by a few remarks.

(A) If we take p of (3.26) assuming that $g = h$ and that g is a common eigenfunction of all Hecke operators, then

$$\pi^{-s}\Gamma(s)\zeta(2s)R(s)/[\pi^{-s/2}\Gamma(s/2)\zeta(s)]$$

is an entire function as shown in [11]. Therefore, if $\zeta(s) = 0$, $\operatorname{Re}(s) > 0$, and $\zeta(2s) \neq 0$ (which is the case if $\operatorname{Re}(s) \geq 1/2$), then $R(s) = 0$. Thus Theorem 3.4 is applicable to such zeros of the Riemann zeta function.

(B) Coming back to the general case in which p is a function satisfying (3.2a,b,c), we see from (vi) of Theorem 3.1 that if $\operatorname{Re}(s) > 1/2$, $s \notin P(\Gamma, 1, m)$, and $R(s) \neq 0$, then f_s is L^2, but not rapidly decreasing, a fact which makes a notable contrast to the case of cusp forms.

(C) As to the quantity $\langle p, f_s \rangle$ appearing in Theorem 3.4, it can also be written $[p, p]_s$, if we put

$$(3.31) \qquad [p, q]_s = -\int_{\Gamma\backslash H}\int_{\Gamma\backslash H} G_s(z, w)\overline{p(z)}q(w)d_Hzd_Hw$$

for two rapidly decreasing functions p and q satisfying (3.2a). Therefore the Riemann hypothesis for R concerns whether $[p, p]_s \in \mathbf{R}$ when $R(s) = 0$. It should also be noted that (4) of the theorem can be derived by combining (2.8) with (3.3a); also (4) can be proved under milder conditions than (3.2c).

(D) Next we consider the function $F(z, s) = q(s)f(z, s)$ with $q(s) = \pi^{-s}\Gamma(s + |m/2|)\zeta(2s)$. Then $F(z, s)$ is finite if $s \notin \{1/2\} \cup S(\Gamma, 1, m)$, and

$$[L_m - s(1 - s)]F(z, s) = q(s)p(z),$$

$$F(z, s) - (2s - 1)^{-1}y^{1-s}R^*(s) = O(e^{-by}) \qquad (y \to \infty)$$

with R^* of (3.3b). Now suppose $q(s_0) = 0$, but still $s_0 \notin \{1/2\} \cup S(\Gamma, 1, m)$. Then $F(z, s_0)$ is an eigenform of L_m. If $R^*(s_0) = 0$, $F(z, s_0)$ must be a cusp form. However, the Riemann hypothesis for ζ and R^* would imply the impossibility of the simultaneous vanishing of q and R^* for $0 < \mathrm{Re}(s_0) < 1$. (If $F(z, s_0)$ is not identically equal to 0, which is the case if s_0 is a pole of $f(z, s)$ and s_0 is a simple zero of q, then that it is a cusp form contradicts the assumption $s_0 \notin S(\Gamma, 1, m)$.) Thus it is natural to assume $R^*(s_0) \neq 0$ when $q(s_0) = 0$. Then $F(z, s_0)$ is a cyclopean form in the sense of [14, Section 9] and it must be an Eisenstein series. In other words, if s_0 is a simple zero of q, the residue of $f(z, s)$ at $s = s_0$ is such an Eisenstein series.

(E) Finally we note that (3.2c) is a rather strong condition. For instance, a nonzero holomorphic cusp form does not satisfy it. Even when it is satisfied, the corresponding f may not necessarily be a natural object. As such an example, we can take $p(z) = y^{1/2}|\eta(z)|^2$ with $\eta(z) = q^{1/24}\prod_{n=1}^{\infty}(1 - q^n)$, $q = e^{2\pi iz}$. In this case R^* of (3.3b) is essentially $\zeta(2s)\zeta(2s - 1)$ times gamma factors and therefore $\mathrm{Re}(s) = 1/2$ is not the line of its zeros. If $p(z) = y^2|\eta(z)|^8$, however, R^* is essentially $\zeta(s)^2L(s, \chi)$ times gamma factors, where χ is the Hecke (ideal) character of $\mathbf{Q}(e^{2\pi i/3})$ such that $\chi((\alpha)) = \alpha^2|\alpha|^{-2}$ if $\alpha \equiv 1 \pmod 2$ (cf. [11, (5.6)]). Therefore $\mathrm{Re}(s) = 1/2$ is the critical line in this case.

4. L-functions of a CM-field as special values of Eisenstein series. This and the following sections concern the second case in which a

Hecke L-function of an imaginary quadratic field gives the dominant term of a certain automorphic form. Though we shall eventually consider only quadratic fields, in this section we treat more generally CM-fields of arbitrary degree. Our aim is to find a *canonical expression* for a Hecke L-function of such a field as a special value of an Eisenstein series. This necessitates the introduction of nonmaximal orders and their ideals.

For an algebraic number field K of finite degree, we denote by $K_{\mathbf{A}}^{\times}$, $K_{\mathbf{a}}^{\times}$, $K_{\mathbf{f}}^{\times}$, and O_K the idele group of K, its archimedean and nonarchimedean factors, and the maximal order of K. Throughout this section, we fix a subfield F of K such that $[K : F] = 2$, and denote by ρ the nontrivial automorphism of K over F. A finite prime of F resp. K will be denoted by p resp. v; F_p and K_v then denote their completions; also we put $K_p = K \otimes_F F_p$. For $x \in K_{\mathbf{A}}^{\times}$, we denote by x_v and x_p its v-component and p-component. By an O_F-*lattice* in K, we understand a finitely generated O_F-submodule of K that spans K over F. If A is an O_F-lattice, we denote by A_p its p-closure in K_p. For $x \in K_{\mathbf{A}}^{\times}$, we denote by xA the O_F-lattice such that $(xA)_p = x_pA_p$ for every finite prime p of F. For simplicity, we put $O_p = (O_F)_p$ and similarly denote by O_v the closure of O_K in K_v. By an O_F-*order* in K, we understand an O_F-lattice in K which is a subring of K containing O_F. (Every O_F-order R in K is of the form $R = O_F + BO_K$ with an integral ideal B in F; moreover B is unique for R. We shall not need these facts, however.) If R is an O_F-order, we understand by an R-*ideal* an O_F-lattice of the form xR with $x \in K_{\mathbf{f}}^{\times}$. We call an R-ideal A *integral* if $A \subset R$. Given an integral ideal C in K, we denote by I_R^C the group of all the R-ideals of the form xR with an element x of $K_{\mathbf{f}}^{\times}$ such that $x_v = 1$ for every v dividing C. If $C = BO_K$ with an ideal B in F, we write I_R^C also I_R^B. Notice that if $\alpha \in K^{\times}$ and $\alpha R \in I_R^C$, then α is a v-unit for every $v|C$, but the converse is not necessarily true.

Let us now take a *Hecke character* χ of K, by which we understand a continuous homomorphism of $K_{\mathbf{A}}^{\times}$ into $\{z \in \mathbf{C} \mid |z| = 1\}$ that is trivial on $K^{\times}$. We denote by χ_v and $\chi_{\mathbf{a}}$ the restrictions of χ to $K_v^{\times}$ and $K_{\mathbf{a}}^{\times}$; if Y is an integral ideal in F or K, we put $\chi_Y = \Pi_{v|Y}\,\chi_v$; we view χ_v and $\chi_{\mathbf{a}}$ as characters of $K_{\mathbf{A}}^{\times}$. Further we denote by χ^* the ideal character attached to χ. Let ψ be the Hecke character of F obtained by composing χ with the injection of $F_{\mathbf{A}}^{\times}$ into $K_{\mathbf{A}}^{\times}$. Let Z resp. W be the integral ideal in K resp. F which represents the finite part of the conductor of χ resp. ψ. We can decompose Z into the product $Z = Z_0Z_1$ with two integral ideals Z_0 and Z_1 in K such that Z_0 and Z_0^{ρ} are divisible by the same prime ideals in K and that $Z^{\rho} + Z_1 = O_K$. Such a decomposition is obviously unique. Let $S = F \cap Z$, $S_0 =$

$F \cap Z_0$, and $Q = F \cap Z_1$. Then $QO_K = Z_1Z_1^\rho$ and $S = S_0Q$. Now $Q_p = W_p$ for every $p|Q$. In fact, let v be the prime of K dividing p and Z_1. Then the other prime of K dividing p does not divide Z, and hence $\psi_p(a) = \psi_v(a)$ for every p-unit a. This shows that $W_p = Z_v \cap F_p = Q_p$. Thus $W = W_0Q$ with an integral ideal W_0 in F dividing S_0. We then put $C = W_0Z_1$ and consider an O_F-order R of the form $R = O_F + BO_K$ with an ideal B in F satisfying condition (4.1e) below. To facilitate our later reference, we list here the properties of these ideals and order:

(4.1a) *Z is the conductor of χ; $Z = Z_0Z_1$; Z_0 and Z_0^ρ have the same prime factors in K; $Z_1 + Z^\rho = O_K$.*

(4.1b) $$S = Z \cap F = S_0Q, \qquad S_0 = Z_0 \cap F, \qquad Q = Z_1 \cap F.$$

(4.1c) *W is the conductor of ψ; $W = W_0Q$, $W_0 \supset S_0$.*

(4.1d) $$C = W_0Z_1.$$

(4.1e) $$S_0 \subset B \subset W_0^{-1}S_0.$$

(4.1f) $$R = O_F + BO_K.$$

Now we denote by $\mathfrak{M}_R^C$ the subgroup of $K^\times$ consisting of the elements α satisfying the following set of conditions:

(4.2) *$\alpha R \in I_R^C$; there is an element r of F prime to W_0 such that $\alpha - r \in W_0R_p$ for every $p|W_0$.*

We understand that $\mathfrak{M}_R^C = K^\times$ if $C = O_K$.

Lemma 4.1. *If $xR = R$, $x \in K_{\mathfrak{f}}^\times$, and $v \nmid C$, then $\chi_v(x) = 1$.*

Proof. This is obvious if $v \nmid Z$ since $xR = R$ implies that x_v is a v-unit. Suppose $v|Z$ and $v \nmid C$. Then v divides a prime p of F such that $p|S_0$ and $p \nmid W_0$. Since $x_p \in R_p^\times$ and $R_p \subset O_p + Z_p$, we have $x_p - s \in Z_p$ with $s \in O_p^\times$. Then $\chi_v(x) = \chi_v(s) = \psi_p(s) = 1$ since $p \nmid W$, which proves our lemma.

We define a character χ^R of I_R^C by

(4.3) $$\chi^R(xR) = \prod_{v \nmid C} \chi_v(x) \quad for \quad xR \in I_R^C, \qquad x \in K_{\mathfrak{f}}^\times.$$

This is well-defined in view of the above lemma. We have obviously

(4.4a) $$\chi^R(A) = \chi^*(AO_K) \quad \textit{if} \quad A \in I_R^Z,$$

(4.4b) $$\chi^R(\alpha R) = \chi_C(\alpha)^{-1}\chi_{\mathbf{a}}(\alpha)^{-1} \quad \textit{if} \quad \alpha \in K \quad \textit{and} \quad \alpha R \in I_R^C.$$

Let us now assume that F is totally real and K totally imaginary. For $x \in F_{\mathbf{A}}^\times$ (in particular for $x \in F^\times$) we write $x \gg 0$ if all the archimedean components of x are positive. We define groups G, G^1, and G_+ by

(4.5)
$$G = \mathrm{GL}_2(F), \qquad G^1 = \mathrm{SL}_2(F), \qquad G_+ = \{\alpha \in G \mid \det(\alpha) \gg 0\}.$$

We consider their parabolic subgroups by putting

(4.6) $$P = \{\alpha \in G \mid c_\alpha = 0\}, \qquad P^1 = P \cap G^1, \qquad P_+ = P \cap G_+.$$

We define the adelizations $G_{\mathbf{A}}$, $G_{\mathbf{A}}^1$, $P_{\mathbf{A}}$ etc, as well as the symbols $G_{\mathbf{a}}$, $G_{\mathbf{f}}$, G_p, etc. as usual. Hereafter we use the symbols $\mathbf{a}$ and $\mathbf{f}$ to denote the sets of archimedean and nonarchimedean primes of F. Given two fractional ideals X and X' in F such that XX' is integral, we put

(4.7a) $$S[X', X] = G_{\mathbf{a}} \prod_{p \in \mathbf{f}} S_p[X', X],$$

(4.7b) $$S_p[X', X] = \{y \in G_p \mid \det(y) \in O_p^\times,\ a_y \in O_p,$$
$$b_y \in X_p',\ c_y \in X_p,\ d_y \in O_p\},$$

(4.7c) $$S^1[X', X] = G_{\mathbf{A}}^1 \cap S[X', X],$$

(4.7d) $$\Gamma[X', X] = G_+ \cap S[X', X], \qquad \Gamma^1[X', X] = G^1 \cap S[X', X].$$

Let $H^{\mathbf{a}}$ denote the product of $\mathbf{a}$ copies of the upper half plane H, which consists of the indexed elements $(z_u)_{u \in \mathbf{a}}$ with $z_u \in H$. We let G_+ act on $H^{\mathbf{a}}$ as usual by componentwise action.

We fix a *CM-type* Φ of K, by which we mean a set of embeddings of K into $\mathbf{C}$ such that for a generic element α of K, the elements α^φ and $\alpha^{\rho\varphi}$ for $\varphi \in \Phi$ are exactly the conjugates of α over $\mathbf{Q}$. We then denote by K_Φ the subset of K consisting of the elements τ of K such that $\mathrm{Im}(\tau^{\varphi}) > 0$ for every $\varphi \in \Phi$. Identifying Φ with $\mathbf{a}$ in a natural way, we can consider K_Φ a subset of $H^{\mathbf{a}}$ through the map $\tau \mapsto (\tau^{\varphi})_{\varphi\in\Phi}$. Obviously K_Φ is stable under G_+.

Take an element ζ of K_Φ such that $\zeta^\rho = -\zeta$. For an O_F-lattice A in K, denote by $\lambda_\zeta(A)$ the O_F-linear span of $\mathrm{Tr}_{K/F}(\zeta\alpha\beta^\rho)$ for all α and β in A, and by $\lambda(A)$ the ideal class of $\lambda_\zeta(A)$ modulo $\mathbf{a}$ in F. This is independent of the choice of ζ. Given a fractional ideal X in F, we put

$$I_R^C(X) = \{A \in I_R^C \mid X \in \lambda(A)\}, \tag{4.8a}$$

$$H_X^C = \{\tau \in K_\Phi \mid XQ\tau + O_F \in I_R^C,\ X\tau \subset O_v \text{ for every } v|Z_1\}. \tag{4.8b}$$

LEMMA 4.2.

(i) *H_X^C is stable under $\Gamma[X^{-1}, WX]$.*

(ii) *For every fractional ideal T in F prime to W, the map $\tau \mapsto TXQ\tau + T$ gives a bijection of $\Gamma[X^{-1}, WX]\backslash H_X^C$ onto $I_R^C(T^2XQ)/M_R^C$, where $M_R^C = \{\alpha R \mid \alpha \in \mathfrak{N}_R^C\}$.*

The proof will be given in Section 7.

LEMMA 4.3. *$G_{\mathbf{A}} = P_{\mathbf{A}}S[X', X]$ if $X'X = O_F$. If $XX' \neq O_F$, $P_{\mathbf{A}}S[X', X]$ consists of the elements y of $G_{\mathbf{A}}$ such that $(d_y)_p \neq 0$ and $(d_y)_p^{-1}(c_y)_p \in X_p$ for every $p|XX'$.*

This is an easy exercise. At least the first assertion is well known.

Thus for any fixed X, we have $G_{\mathbf{A}} = P_{\mathbf{A}}S[X^{-1}, X]$ and $G_{\mathbf{A}}^1 = P_{\mathbf{A}}^1S^1[X^{-1}, X]$. To every $y \in G_{\mathbf{A}}$, we can assign a fractional ideal $\mathrm{il}_X(y)$ as follows: let $y = gh$ with $g \in P_{\mathbf{A}}$ and $h \in S[X^{-1}, X]$; then we put $\mathrm{il}_X(y) = d_gO_F$. It is also given by

$$\mathrm{il}_X(y)_p = (c_y)_pX_p^{-1} + (d_y)_pO_p \text{ for every } p \in \mathbf{f}. \tag{4.9}$$

We can easily verify that

$$\mathrm{il}_{stX}(y) = t^{-1}\,\mathrm{il}_X(y\cdot \mathrm{diag}[s^{-1}, t]) \text{ for every } s \text{ and } t \in F_{\mathbf{A}}^\times. \tag{4.10}$$

The ideal class of $\mathrm{il}_X(y)$ depends only on the coset $PyS[X^{-1}, X]$. In addition to X, we fix an integral ideal W in F.

LEMMA 4.4.

(i) $G \cap P_{\mathbf{A}}S[X^{-1}, XW] = P(G^1 \cap P_{\mathbf{A}}^1S^1[X^{-1}, XW])$.

(ii) *There is a finite subset* $\mathfrak{B}$ *of* G^1 *such that* $G^1 \cap P_{\mathbf{A}}^1 S^1[X^{-1}, XW]$ *is a disjoint union of* $P^1\beta\Gamma^1[X^{-1}, XW]$ *for* $\beta \in \mathfrak{B}$.

(iii) $P\alpha\Gamma[X^{-1}, XW] = P\alpha\Gamma^1[X^{-1}, XW]$ *for every* $\alpha \in G$.

(iv) *If* $\mathfrak{B}$ *is a set as in* (ii), *then the ideals* $\mathrm{il}_X(\beta)$ *for* β *in* $\mathfrak{B}$ *form a complete set of representatives for the ideal classes in F.*

(v) *If* $\mathfrak{B}$ *is a set as in* (ii), *then* $G \cap P_{\mathbf{A}}S[X^{-1}, XW]$ *is a disjoint union of* $P\beta\Gamma[X^{-1}, XW]$ *for all* $\beta \in \mathfrak{B}$.

Assertions (ii) and (iv) were given in [13, Lemma 1.6] when $X = O_F$. The general case can be proved in a similar way. The remaining assertions either follow from these or can be verified in a straightforward way.

Before proceeding further, we introduce a notational convention: for two elements $x = (x_v)_{v\in\mathbf{a}}$ and $y = (y_v)_{v\in\mathbf{a}}$ in $\mathbf{C}^{\mathbf{a}}$, we put

$$x^y = \prod_{v\in\mathbf{a}} (x_v)^{y_v} \tag{4.11a}$$

whenever the factors are meaningful. Denoting by u the identity element of the ring $\mathbf{C}^{\mathbf{a}}$, we have the following special case of (4.11a):

$$x^{su} = \Big(\prod_{v\in\mathbf{a}} x_v\Big)^s \qquad (s \in \mathbf{C}). \tag{4.11b}$$

Now, for $\alpha \in G_+$ and $z \in H^{\mathbf{a}}$ we define a factor of automorphy $j(\alpha, z)$ to be the element of $\mathbf{C}^{\mathbf{a}}$ given by

$$j(\alpha, z) = (j_v(\alpha, z))_{v\in\mathbf{a}}, \qquad j_v(\alpha, z) = \det(\alpha_v)^{-1/2}(c_v z_v + d_v), \tag{4.12}$$

where α_v is the image of α under v and (c_v, d_v) is the lower half of α_v.

Given an integral ideal W in F as above, we now consider a Hecke character φ of F such that

$$\varphi(x) = x^m |x|^{i\mu - m} \quad \text{for} \quad x \in F_{\mathbf{a}}^{\times} = (\mathbf{R}^{\times})^{\mathbf{a}}, \tag{4.13a}$$

$$\varphi(O_p^{\times}) = 1 \quad \text{for every} \quad p|W. \tag{4.13b}$$

Here $m \in \mathbf{Z}^{\mathbf{a}}$, $\mu \in \mathbf{R}^{\mathbf{a}}$, and the right-hand side of (4.13a) should be understood in the sense of (4.11a). We assume that $\Sigma_{v\in\mathbf{a}}\, \mu_v = 0$. For $\alpha \in G_+$ and a function f on $H^{\mathbf{a}}$ we define a function $f\|_{m,\mu}\alpha$ by

$$(f\|_{m,\mu}\alpha)(z) = |j(\alpha, z)|^{m - i\mu} j(\alpha, z)^{-m} f(\alpha z). \tag{4.14}$$

Taking a complete set of representatives $\mathfrak{A}$ for $P_+\backslash(G_+ \cap P_{\mathbf{A}}S[X^{-1}, XW])$, we define an Eisenstein series $E(z, s)$, depending on X, W, φ, and m, by

$$(4.15)\quad E(z, s) = E(z, s; X, W; \varphi, m) = \sum_{\alpha \in A} N(\det(\alpha)^{-1}\, \mathrm{il}_X(\alpha)^2)^s \varphi_{\mathbf{a}}(\det(\alpha)^{-1/2} d_\alpha) \cdot \varphi^*(d_\alpha\, \mathrm{il}_X(\alpha)^{-1}) \mathrm{Im}(z)^{su} \|_{m,\mu}\alpha.$$

Here $z \in H^{\mathbf{a}}$, $s \in \mathbf{C}$, and we understand that the factor $\varphi_{\mathbf{a}}(\det(\alpha)^{-1/2} d_\alpha)\varphi^*(d_\alpha\, \mathrm{il}_X(\alpha)^{-1})$ means $\varphi_{\mathbf{a}}(\det(\alpha)^{-1/2})\varphi^*(\mathrm{il}_X(\alpha)^{-1})$ if $W = O_F$. Notice that if $W \neq O_F$ and $\alpha \in G \cap P_{\mathbf{A}}S[X^{-1}, XW]$, then $a_\alpha d_\alpha \neq 0$ and $d_\alpha\, \mathrm{il}_X(\alpha)^{-1}$ is prime to W and that E is independent of the choice of $\mathfrak{A}$. It can easily be seen that

$$(4.16)\quad E\|_{m,\mu}\gamma = \det(\gamma)^{i\mu/2}\varphi_W(d_\gamma)E \textit{ for every } \gamma \in \Gamma[X^{-1}, XW].$$

For simplicity, let us put $\Gamma^1 = \Gamma^1[X^{-1}, XW]$. Take a set $\mathfrak{B}$ as in Lemma 4.4; let A_β be a complete set of representatives for $(P^1 \cap \beta\Gamma^1\beta^{-1})\backslash\beta\Gamma^1$. Then that lemma shows that $\bigcup_{\beta \in \mathfrak{B}} A_\beta$ can be taken as $\mathfrak{A}$, and therefore

$$(4.17\text{a})\quad E(z, s) = \sum_{\beta \in \mathfrak{B}} E_\beta(z, s),$$

$$(4.17\text{b})\quad E_\beta(z, s) = N(\mathrm{il}_X(\beta))^{2s} \sum_{\alpha \in A_\beta} \varphi_{\mathbf{a}}(d_\alpha)\varphi^*(d_\alpha\, \mathrm{il}_X(\beta)^{-1}) \mathrm{Im}(z)^{su} \|_{m,\mu}\alpha.$$

If $X = O_F$, this is essentially the same as the series of [13] and [14]. The series with an arbitrary X is not much different from that special case in its nature. We need to introduce another type of series

$$(4.18)\quad E_Y(z, s) = E_Y(z, s; X, W; \varphi, m) = N(Y)^{2s}\, \mathrm{Im}(z)^{su} \sum_{(c,d)} \varphi_{\mathbf{a}}(d)\varphi^*(dY^{-1})(cz + d)^{-m}|cz + d|^{m - i\mu - 2su}.$$

Here Y is a fractional ideal in F and (c, d) runs over $T[X, Y; W]/O_F^\times$, where

$$T[X, Y; W] = \{(c, d) \in WXY \times Y \mid (c, d) \neq (0, 0), dY^{-1} + W = O_F\}.$$

If $W = O_F$, we understand that $\varphi_{\mathbf{a}}(d)\varphi^*(dY^{-1}) = \varphi^*(Y^{-1})$ even when $d = 0$. We can easily verify that E_Y depends only on the ideal class of Y.

Proposition 4.5. *Let $L_W(s, \varphi) = \Sigma\, \varphi^*(Z)N(Z)^{-s}$, where Z runs over all the integral ideals in F prime to W, and let $L_W(s, \varphi; Y) = \Sigma_{Z\sim Y}$ $\varphi^*(Z)N(Z)^{-s}$, where $Z \sim Y$ means the additional condition that Z belongs to the ideal class of Y. Then, for a fixed X, we have*

$$E_Y(z, s) = \sum_{\beta\in\mathfrak{B}} L_W(2s, \varphi; \mathrm{il}_X(\beta)Y^{-1})E_\beta(z, s),$$

$$\sum_Y E_Y(z, s) = L_W(2s, \varphi)E(z, s),$$

where Y runs over a complete set of representatives for the ideal classes in F.

Proof. A similar result was given in [15, Lemma 6.5]. The present case is somewhat different, but the proof can be given in the same way.

We now take a Hecke character χ of K and define ψ, Z, W, Q, C, and R as before with a choice of B as in (4.1e). Define the L-function $L(s, \chi)$ of χ as usual; define also χ^R by (4.3). Then the main result of this section can be given as follows:

Theorem 4.6. *Let h be the class number of F and let $\mathfrak{X}$ be a complete set of representatives for the ideal classes modulo $\mathbf{a}$ in F. Suppose*

$$\chi(t) = t^m|t|^{i\mu-m} \quad \textit{for} \quad t \in K_{\mathbf{a}}^\times \tag{4.19}$$

with $m \in \mathbf{Z}^{\mathbf{a}}$ and $\mu \in \mathbf{R}^{\mathbf{a}}$, $\Sigma_{v\in\mathbf{a}}\, \mu_v = 0$, where $|t| = ((x_v{}^2 + y_v{}^2)^{1/2})_{v\in\mathbf{a}}$ if $t_v = x_v + iy_v$ with real x_v and y_v. Then

$$h[U : O_F^\times]L(s, \chi) = N(d(K/F))^{-s/2}N(2B^{-1}Q)^sL(2s, \psi)$$
$$\cdot \sum_{X\in\mathfrak{X}} N(X)^s\, \Sigma\, \chi^R(XQ\tau + O_F)^{-1}E(\tau, s; X, W; \psi, m),$$

where $U = R^\times \cap \mathfrak{N}_R^C$, $d(K/F)$ is the different of K relative to F, and τ runs over $\Gamma[X^{-1}, XW]\backslash H_X^C$ with H_X^C defined by (4.8b).

The proof will be given in Section 7.

The reader may have wondered if the nonmaximal order R is really necessary. Indeed, if we merely seek *some* expression for $L(s, \chi)$, then that can be obtained in a rather straightforward fashion. For instance, with S as in (4.1b), define $L_S(s, \chi)$ in the same manner as for $L_W(s, \varphi)$ in Proposition 4.5. Then $L_S(s, \chi)$ has an expression which is similar to that of the above theorem and which involves only ordinary ideals and χ^* instead of χ^R, but $E(\ldots; X, S; \ldots)$ instead of $E(\ldots; X, W; \ldots)$. The main trouble with this expression is that the functional equations for L_S and $E(\ldots; X, S; \ldots)$ are not clear-cut, which presents serious technical difficulties in our later application. That is why nonmaximal orders and χ^R have been considered.

5. Main theorems in the second case. Let us first specialize the material of Section 4 to the case $F = \mathbf{Q}$. We naturally take $X = \mathbf{Z}$ and eliminate the symbol X. We also understand that the letters W, B, Q, etc. which previously meant ideals in F now denote the positive rational numbers that generate them. For instance, $\Gamma[X^{-1}, XW]$ is nothing else but the group $\Gamma_0(W)$ of (2.21). The Eisenstein series of (4.15) takes the form

$$E(z, s; W, \varphi, m) = \sum_{\alpha \in V} \varphi_W(d_\alpha)^{-1} \mathrm{Im}(z)^s \|_m \alpha, \tag{5.1}$$

where φ is a Hecke character of $\mathbf{Q}$ whose conductor divides W, m is a fixed integer, and $V = [P \cap \Gamma_0(W)] \backslash \Gamma_0(W)$. Condition (4.13a) becomes simply

$$\varphi_{\mathbf{a}}(-1) = (-1)^m. \tag{5.2}$$

Let us now put

(5.3)

$$E^*(z, s; W, \varphi, m) = \pi^{-s}\Gamma(s + |m/2|)L_W(2s, \varphi)E(z, s; W, \varphi, m).$$

Then $s(1 - s)E^*$ can be continued to a real analytic function on $H \times \mathbf{C}$ that is holomorphic in s; the factor $s(1 - s)$ is unnecessary if $\varphi \neq 1$ or $m \neq 0$. These facts are well known. A more general result holds for the function of (4.15) (see [14, Theorem 4.1]). It should be remembered that $L_W(s, \varphi) = \sum_{(n, W)=1} \varphi^*(n\mathbf{Z})n^{-s} = \sum_{(n, W)=1} \varphi_W(n)^{-1} n^{-s}$. If W is the conductor of φ, then

(5.4)

$$E^*(z, 1 - s; W, \varphi, m) = \mathrm{sgn}(m)^m q(\bar{\varphi}) W^{3s-2} z^{-m} |z|^m E^*(\delta z, s; W, \bar{\varphi}, m),$$

where $q(\overline{\varphi}) = \Sigma_t\, \varphi_W(t)^{-1} e^{2\pi i t/W}$ with t running over $(\mathbf{Z}/W\mathbf{Z})^\times$, and

$$(5.5) \qquad \delta = W^{-1/2}\begin{pmatrix} 0 & -1 \\ W & 0 \end{pmatrix}.$$

Naturally $q(\overline{\varphi}) = 1$ if $W = 1$. Formula (5.4) is also well known and easily follows from [10, Lemma 3.3], for example.

We now consider the function G_s of Section 2 with $\Gamma = \Gamma_0(W)$ and $\theta = \varphi_W$. We denote $S(\Gamma, \theta, m)$, $P(\Gamma, \theta, m)$, and $P'(\Gamma, \theta, m)$ by $S(W, \varphi, m)$, $P(W, \varphi, m)$, and $P'(W, \varphi, m)$. Transforming G_s, E_ξ, and cusp forms by δ, we easily see that these sets S, P, and P' are invariant under the change of φ for $\overline{\varphi}$. The symbols being as in (2.7), we have

$$(5.6) \qquad E_1(z, s; \theta, m) = E(z, s; W, \varphi, m).$$

Lemma 5.1. *Suppose that W is 1 or a power of a prime and that W is the conductor of φ. Then the following assertions hold*:

(i) $(1-2s)[G_s(z, w; \varphi_W, m) - G_{1-s}(z, w; \varphi_W, m)]$

$$= \begin{cases} E(z, s; 1, 1, m)E(w, 1-s; 1, 1, -m) & \text{if} \quad W = 1, \\ E(z, s; W, \varphi, m)E(w, 1-s; W, \overline{\varphi}, -m) & \\ \qquad + E(z, 1-s; W, \varphi, m)E(w, s; W, \overline{\varphi}, -m) & \text{if} \quad W > 1. \end{cases}$$

(ii) *Let $\beta_m(s, \varphi) = \pi^{-s}\Gamma(s + |m/2|)L(2s, \varphi)$. Suppose $w \notin \Gamma_0(W)z$. If $W > 1$ or $m \neq 0$, $\beta_m(s, \varphi)\beta_m(s, \overline{\varphi})G_s(z, w; \varphi_W, m)$ has poles only in $\{1/2\} \cup S(W, \varphi, m)$; if $\varphi = \overline{\varphi}$, the factor $\beta_m(s, \overline{\varphi})$ is unnecessary. If $W = 1$ and $m = 0$, $\beta_m(s, 1)G_s(z, w; 1, 0)$ has poles only in $\{1/2\} \cup S(W, 1, 0)$.*

Proof. If $W = 1$, (i) follows immediately from (2.8). If $W > 1$, the right-hand side of (2.8) has only two nonvanishing terms corresponding to the cusps 0 and ∞ for the reason explained at the end of Section 2. The term corresponding to ∞ presents no problem. The other term has the form

$$(5.7) \qquad J_\delta(z)^{-m}E(\delta z, s; W, \overline{\varphi}, m)J_\delta(w)^m E(\delta w, 1-s; W, \varphi, -m)$$

with δ of (5.5). By (5.3) and (5.4) we have

$$J_\delta(z)^{-m}E(\delta z, s; W, \varphi, m) = \eta_m x(s, \varphi)E(z, 1-s; W, \bar{\varphi}, m)$$

with $\eta_m = \operatorname{sgn}(m)^m$ and a function $x(s, \varphi)$ depending only on s, φ, and $|m|$. Substituting δz for z, we easily see that $x(s, \varphi)x(1-s, \bar{\varphi}) = (-1)^m$, and hence (5.7) is equal to the last term of the formula of (i). To prove (ii), assume $w \notin \Gamma_0(W)z$. Recall that G_s for $\mathrm{Re}(s) \geqq 1/2$ has poles only in $\{1/2\} \cup S(W, \varphi, m)$. Therefore the formula of (i) implies that if $\mathrm{Re}(s) < 1/2$ and $s \notin S(W, \varphi, m)$, then the poles of G_s are those of the right-hand side of (i). Suppose $m \neq 0$ or $W > 1$. As already mentioned, $\beta_m(s, \varphi)E(z, s; W, \varphi, m)$ is finite everywhere, and therefore we obtain assertion (ii) in this case. Suppose $m = 0$ and $W = 1$. Then $E^*(z, s; 1, 1, 0)$ is holomorphic in s except at simple poles at 0 and 1, from which we obtain the desired result.

Now take an imaginary quadratic field K embedded in $\mathbf{C}$ and a Hecke character χ of K such that

$$\chi(t) = t^m|t|^{-m} \quad \text{for} \quad t \in K_{\mathbf{a}}^\times. \tag{5.8}$$

We then consider its L-function $L(s, \chi)$ and put

$$\mathcal{R}(s, \chi) = |D_K N(Z)|^{s/2}(2\pi)^{-s}\Gamma(s + |m/2|)L(s, \chi), \tag{5.9}$$

where D_K denotes the discriminant of K and Z the conductor of χ. The formula of Theorem 4.6 in the present case yields

$$\mathcal{R}(s, \chi) = 2[U:1]^{-1}(QB^{-1}N(Z)^{1/2})^s \cdot \sum_{\tau \in T} \chi^R(\mathbf{Z}Q\tau + \mathbf{Z})^{-1}E^*(\tau, s; W, \psi, m), \tag{5.10}$$

where ψ is the restriction of χ to $\mathbf{Q}$ (see Section 4), W is the conductor of ψ, Q is defined by (4.1b), and $T = \Gamma_0(W)\backslash H_{\mathbf{Z}}^C$ with $H_{\mathbf{Z}}^C$ defined by

$$H_{\mathbf{Z}}^C = \{\tau \in K \cap H \mid \mathbf{Z}Q\tau + \mathbf{Z} \in I_R^C,\ \tau \in O_v \text{ for every } v|Z_1\}. \tag{5.11}$$

We now state the main result in the present case:

Theorem 5.2. *The notation being as above, suppose that W is either 1 or a power of a prime; if $W > 1$, suppose further that $\bar{\chi} = \chi$ or $\psi \circ N_{K/\mathbf{Q}} = 1$ or B of (4.1e) can be chosen so that $N(Z_0) \neq W_0^3B^2$. Put $S =$*

$S(W, \psi, m)$ for simplicity. Then there exists a function $f(z, s)$ on $H \times \mathbf{C}$, with singularities, satisfying the following conditions (i–vi):

(i) *$\beta_m(s, \bar{\psi}) f$ is C^∞ in (z, s) if $s \notin S \cup \{1/2\}$ and z does not belong to the set $\mathfrak{Z} = \delta(H_{\mathbf{Z}}^{\mathbf{C}}) \cup \omega(H_{\mathbf{Z}}^{\mathbf{C}})$, where β_m is as in Lemma* 5.1 *and $\omega(z) = -\bar{z}$.*

(ii) *$f(\gamma z, s) = \psi_W(d_\gamma)^{-1} J_\gamma(z)^m f(z, s)$ for every $\gamma \in \Gamma_0(W)$.*

(iii) *If $z \notin \mathfrak{Z}$, $\beta_m(s, \bar{\psi}) f(z, s)$ is meromorphic in s with poles only in $S \cup \{1/2\}$.*

(iv) *If $s \notin S \cup \{1/2\}$, then $\beta_m(s, \bar{\psi}) f(z, s)$ as a function of z satisfies* (1.8a,b) *with $\lambda = s(1 - s)$ and $\mathfrak{Z}$ of* (i).

(v) *Suppose $s_0 \notin S \cup \{1/2\}$. Then $(1 - N^{1-2s})\beta_m(s, \bar{\psi})\mathfrak{R}(s, \chi)$ vanishes at s_0 if and only if $[\beta_m(s, \bar{\psi}) f(z, s)]_{s=s_0}$ is rapidly decreasing, where $N = W_0^3 B^2 / N(Z_0)$.*

(vi) *If $\psi = \bar{\psi}$, the factor $\beta_m(s, \psi)$ is unnecessary in the above assertions. If $\chi = \bar{\chi}$ or $\psi \circ N_{K/\mathbf{Q}} = 1$, then the factor $(1 - N^{1-2s})\beta_m(s, \bar{\psi})$ in* (v) *is unnecessary and we can take $\mathfrak{Z}$ to be any of the four sets $H_{\mathbf{Z}}^{\mathbf{C}}$, $\omega(H_{\mathbf{Z}}^{\mathbf{C}})$, $\delta(H_{\mathbf{Z}}^{\mathbf{C}})$, and $\delta\omega(H_{\mathbf{Z}}^{\mathbf{C}})$.*

Proof. We first observe that

$$E^*(z, s; W, \varphi, m) = E^*(-\bar{z}, s; W, \varphi, -m), \tag{5.12}$$

and therefore (5.10) can be transformed into

$$\mathfrak{R}(s, \chi) = \mu A^s \sum_{\tau \in T} \chi^R(\mathbf{Z}Q\tau + \mathbf{Z})^{-1} E^*(-\bar{\tau}, s; W, \psi, -m), \tag{5.13}$$

where $\mu = 2/[U : 1]$ and $A = QB^{-1}N(Z)^{1/2}$. Put

$$g(z, s) = \mu A^s \pi^{-s} \Gamma(s + |m/2|) L(2s, \psi) \cdot \sum_{\tau \in T} \chi^R(\mathbf{Z}Q\tau + \mathbf{Z})^{-1} G_s(z, -\bar{\tau}; \bar{\psi}_W, m). \tag{5.14}$$

For a fixed s, g is a function of type (2.16a), except that here it is not subject to "modification" of (2.10). Now (5.13) combined with (2.6) and (5.6) shows that the constant term of the Fourier expansion of g is exactly $y^{1-s}(1 - 2s)^{-1}\mathfrak{R}(s, \chi)$. If $W = 1$, we can take this g as the desired f. In fact, the required properties (i–vi) (with $\mathfrak{Z} = \omega(H_{\mathbf{Z}}^{\mathbf{C}})$) follow from the corresponding properties of G_s, or rather, from Proposition 2.1, in view of Lemma 5.1. If $W > 1$, g does not necessarily satisfy (v), since $\Gamma_0(W)$ has more than one cusp. Therefore we need to study $g\|_m\delta$ with δ of (5.5). We observe that

$$J_\delta(z)^{-m}G_s(\delta z, w; \theta, m) = (-J_\delta(w))^m G_s(z, \delta w; \overline{\theta}, m) \tag{5.15}$$

and recall that

$$\mathcal{R}(1 - s, \chi) = r(\chi)\mathcal{R}(s, \overline{\chi}) \tag{5.16}$$

with a constant $r(\chi)$ such that $\overline{r(\chi)} = r(\overline{\chi}) = r(\chi)^{-1}$. By (2.6), (5.4), (5.6), and (5.15), the constant term of the expansion of $J_\delta(z)^{-m}g(\delta z, s)$ is

$$\begin{aligned} &y^{1-s}(1 - 2s)^{-1}\mu\epsilon q(\psi)W^{1-3s}A^s M(s, \psi) \\ &\qquad \cdot \sum_{\tau\in T} \chi^R(\mathbf{Z}Q\tau + \mathbf{Z})^{-1}E^*(-\overline{\tau}, 1 - s; W, \psi, -m), \end{aligned} \tag{5.17}$$

where $\epsilon = \operatorname{sgn}(-m)^m$ and $M(s, \psi) = L(2s, \psi)/L(2s, \overline{\psi})$. By (5.13) and (5.16), we see that (5.17) is equal to $y^{1-s}(1 - 2s)^{-1}A_1A_2^s M(s, \psi)\mathcal{R}(s, \overline{\chi})$ with $A_1 = \epsilon r(\chi)q(\psi)WA^{-1}$ and $A_2 = W^{-3}A^2$. We need to consider another function

$$\begin{aligned} h(z, s) = &\mu A^s\pi^{-s}\Gamma(s + |m/2|)L(2s, \overline{\psi}) \\ &\qquad \cdot \sum_{\tau\in T} \chi^R(\mathbf{Z}Q\tau + \mathbf{Z})G_s(z, \tau; \psi_W, m). \end{aligned} \tag{5.18}$$

This satisfies (i–vi) with the same m, but with ψ instead of $\overline{\psi}$ and $\mathcal{Z} = H_{\mathbf{Z}}^C$. Moreover, by (5.10), its constant term is $y^{1-s}(1 - 2s)^{-1}\mathcal{R}(s, \overline{\chi})$; similarly the constant term of $J_\delta(z)^{-m}h(\delta z, s)$ is

$$(-1)^m y^{1-s}(1 - 2s)^{-1}\overline{A}_1A_2^s M(s, \psi)^{-1}\mathcal{R}(s, \chi).$$

As observed at the end of Section 2, G_s as a function of z vanishes at the cusps inequivalent to 0 and ∞; hence the same is true for g and h. Suppose now $\chi = \overline{\chi}$ or $\psi \circ N_{K/\mathbf{Q}} = 1$. Then we have $\mathcal{R}(s, \chi) = \mathcal{R}(s, \overline{\chi})$ and $\psi = \overline{\psi}$ in both cases and observe that any of the four functions g, h, $g\|_m\delta$ and $h\|_m\delta$ has the required properties (i–vi); $\mathcal{Z}$ is $\omega(H_{\mathbf{Z}}^C)$, $H_{\mathbf{Z}}^C$, $\delta\omega(H_{\mathbf{Z}}^C)$, and $\delta(H_{\mathbf{Z}}^C)$, respectively.

In a more general case, we consider f defined by

$$f(z, s) = g(z, s) - (-1)^m A_1 A_2^s M(s, \psi) J_\delta(z)^{-m} h(\delta z, s). \tag{5.19}$$

Its constant term is

$$y^{1-s}(1 - 2s)^{-1}[1 - (W^3 A^{-2})^{1-2s}]\mathcal{R}(s, \chi), \tag{5.20}$$

while the constant term of $f\|_m\delta$ is 0. In view of (4.1a–e), we have $W^3A^{-2} = W_0^3B^2N(Z_0)^{-1}$. This is different from 1 because of our assumption, and hence (5.20) is nonvanishing. Thus the only remaining point is the location of poles of f. Lemma 5.1 shows that $\beta_m(s, \bar{\psi})g$ and $\beta_m(s, \psi)h$ are finite if $z \notin \omega(H_{\mathbf{Z}}^C) \cup \delta(H_{\mathbf{Z}}^C)$ and $s \notin S \cup \{1/2\}$. If $\psi = \bar{\psi}$, we see that g and h without the factor β_m have the same property. This completes the proof.

Proposition 5.3. *The assumptions being the same as in Theorem* 5.2, *define g, h, and f by* (5.14), (5.18), *and* (5.19). *Let $s_0 \notin S \cup \{1/2\}$; suppose $\beta_m(s_0, \psi)\beta_m(s_0, \bar{\psi}) \neq 0$ with β_m of Lemma* 5.1. *Then $g(z, s_0)$ and $h(z, s_0)$ are nonzero functions. Furthermore, if* $\mathrm{Re}(s_0) \neq 1/2$ *or* $W_0 \neq 1$, *$f(z, s_0)$ is a nonzero function.*

Proof. For each $\tau \in \omega(H_{\mathbf{Z}}^C)$ and $s \notin P(W, \psi, m)$ we have

(5.21a)

$$g(z, s) = \mu A^s \beta_m(s, \psi)\chi^R(\mathbf{Z}Q\bar{\tau} + \mathbf{Z})^{-1}\cdot[(2\pi)^{-1}e_\tau \log|z - \tau| + r_{\tau,s}^g(z)]$$

in a neighborhood of τ with a function $r_{\tau,s}^g$ continuous at τ. This is a special case of (2.17a). We have to verify that this is so even when $e_\tau > 1$, which will be seen from Lemma 5.4 below combined with Proposition 2.1. Similarly, for each $\tau \in H_{\mathbf{Z}}^C$ we have

(5.21b)

$$h(z, s) = \mu A^s \beta_m(s, \bar{\psi})\chi^R(\mathbf{Z}Q\tau + \mathbf{Z})\cdot[(2\pi)^{-1}e_\tau \log|z - \tau| + r_{\tau,s}^h(z)]$$

with a function $r_{\tau,s}^h$ continuous at τ. Therefore $g(z, s_0)$ and $h(z, s_0)$ are nonzero functions. As for $f(z, s_0)$, it is sufficient to show that the singularities of g cannot be cancelled by those of $h\|_m\delta$. For that purpose, let $\tau \in H_{\mathbf{Z}}^C$ and $\sigma \in H_{\mathbf{Z}}^C$; suppose $-\bar{\tau} = \delta(\sigma)$. For every $p|W_0$, we have $\mathbf{Z}_p + (BO_K)_p = R_p = \mathbf{Z}_p\sigma + \mathbf{Z}_p = \mathbf{Z}_p\tau + \mathbf{Z}_p$, and hence $W\sigma R_p = W\sigma(\mathbf{Z}_p\bar{\tau} + \mathbf{Z}_p) = \mathbf{Z}_pW\sigma + \mathbf{Z}_p = \mathbf{Z}_p + WR_p = \mathbf{Z}_p + (WBO_K)_p$. This is a contradiction, since the last module is not an R_p-ideal. Thus we cannot have $-\bar{\tau} = \delta(\sigma)$ if

$W_0 \neq 1$. Suppose $-\bar{\tau} = \delta(\sigma)$ and $W_0 = 1$. Then, in a neighborhood of $-\bar{\tau}$, we have

$$f(z, s) = \mu A^s \beta_m(s, \psi)(2\pi)^{-1} e(-\bar{\tau}) \log|z + \bar{\tau}| \cdot [u_1 + u_2(W^{-3}A^2)^{s-1/2}] + \text{(finite part)}$$

with $|u_1| = |u_2| = 1$. Since $W^{-3}A^2 \neq 1$, the singularity vanishes only when $\mathrm{Re}(s) = 1/2$, which completes the proof.

LEMMA 5.4

(i) *H_Z^C contains an elliptic point of $\Gamma_0(W)$ only if $C = Z = Z_1$ and $W = Q = N(Z)$.*

(ii) *Let σ be a point of $K \cap H$ fixed by an elliptic element γ of $\Gamma_0(W)$. Then $J_\gamma(\sigma)^m = \psi_W(d_\gamma)$ if $-\bar{\sigma} \in H_Z^C$ and $J_\gamma(\sigma)^m = \psi_W(d_\gamma)^{-1}$ if $\sigma \in H_Z^C$.*

Proof. Let σ and γ be as in (ii); put $\zeta = c_\gamma \sigma + d_\gamma$. Then ζ is a root of unity other than ± 1, and $\sigma = \alpha(\zeta)$ with $\alpha \in \mathrm{GL}_2(\mathbf{Z})$. Hence $(c_\alpha \zeta + d_\alpha)(\mathbf{Z}\sigma + \mathbf{Z}) = \mathbf{Z}\zeta + \mathbf{Z} = O_K$. Suppose $\sigma \in H_Z^C$. Then $R = O_K$, so that $B = 1$. By (4.1e), this is possible only when $W_0 = S_0$. For every p dividing W_0, we have $\mathbf{Z}_p \sigma + \mathbf{Z}_p = R_p$, and hence $R_p = \mathbf{Z}_p \zeta + \mathbf{Z}_p = \mathbf{Z}_p c_\gamma \sigma + \mathbf{Z}_p = c_\gamma R_p + \mathbf{Z}_p$. Since $p|c_\gamma$, this happens only when $W_0 = 1$, in which case we have $W = Q$ and $C = Z = Z_1$, in view of (4.1a–d). By (5.11), we have $\zeta - d_\gamma = c_\gamma \sigma \in Z_v$ for every $v|Z$. Thus $\psi_W(d_\gamma) = \chi_Z(d_\gamma) = \chi_Z(\zeta) = \chi_{\mathbf{a}}(\zeta)^{-1} = \zeta^{-m} = J_\gamma(\sigma)^{-m}$, which proves the last fact of (ii). The case $-\bar{\sigma} \in H_Z^C$ can be proved by observing that $J_\gamma(\sigma)^{-1} = J_{\omega\gamma\omega}(\omega\sigma)$.

The main point of Theorem 5.2 is in (v), which is parallel to Theorem 3.4, (2). As explained in the introduction, we have the analogues of Theorem 3.4, (3), (4), and (5) in the present case. For simplicity, we state them only for the function g of (5.14) as follows:

THEOREM 5.5. *Let the notation be the same as in Theorem* 5.2, (5.10), (5.13), (5.14), *and* (5.21a); *let β_m be as in Lemma* 5.1.

(i) *If $\mathcal{R}(s, \chi) = \mathcal{R}(s, \bar{\chi}) = 0$, $\beta_m(s, \psi) \neq 0$, and $s \notin \{1/2\} \cup S$, we have $-\mathrm{Im}[s(1-s)] \|g(z, s)\|^2 = |\mu A^s \beta_m(s, \psi)|^2 \,\mathrm{Im}[\Sigma_{\tau \in \omega(T)} r^g_{\tau,s}(\tau)]$.*

(ii) *For $s = (1/2) + it$ with $0 \neq t \in \mathbf{R}$, we have*

$$4t|\mu A^s \beta_m(s, \psi)|^2 \,\mathrm{Im}\Big[\sum_{\tau \in \omega(T)} r^g_{\tau,s}(\tau)\Big] = \begin{cases} |\mathcal{R}(s, \chi)|^2 & \text{if } W = 1, \\ |\mathcal{R}(s, \chi)|^2 + |\mathcal{R}(s, \bar{\chi})|^2 & \text{if } W > 1. \end{cases}$$

(iii) *Put* $E(z, s) = E(z, s; W, \bar{\psi}, m)$. *For* s *as in* (i), *we have*

$$d\mathcal{R}(s, \chi)/ds = (2s - 1)\langle E(z, \bar{s}), g(z, s)\rangle.$$

Proof. The first two assertions are special cases of Proposition 2.3. In (ii), we need Lemma 5.1 and (5.14). To prove (iii), let s_0 denote a particular s as in (i), and use s for the variable. Put $\lambda_0 = s_0(1 - s_0)$, $\lambda = s(1 - s)$, $F_s(z) = E(z, \bar{s})$, $g_0(z) = g(z, s_0)$, and $\alpha(s) = \kappa A^s \beta_m(s, \psi)$. Observe that the conclusion of Proposition 1.1 is applicable to F_s and g_0, even if F_s may not satisfy (1.8c), since g_0 is rapidly decreasing. Thus the formula of the proposition with $c_\nu = 0$ gives

$$(\lambda - \lambda_0)\langle F_s, g_0\rangle = -2\pi \sum_{\nu=1}^{p} b_\nu \overline{F_s(\zeta_\nu)}/e(\zeta_\nu).$$

By (5.21a) and (5.13), the right-hand side is $-\mathcal{R}(s, \chi)\alpha(s_0)/\alpha(s)$. Taking the limit when s tends to s_0, we obtain (iii).

Remark 5.6. (1) If $Z_1 = O_K$, then $Q = 1$, $C = W_0 O_K$, and H_Z^C is stable under the map $\tau \mapsto -\bar{\tau}$. Therefore, replacing $-\bar{\tau}$ by τ in (5.14), we obtain

(5.22) $g(z, s)$

$$= \mu A^s \pi^{-s}\Gamma(s + |m/2|)L(2s, \psi) \cdot \sum_{\tau \in T} \chi^K(\mathbf{Z}\tau + \mathbf{Z})^{-1} G_s(z, \tau; \bar{\psi}_W, m).$$

In particular, if $\psi \circ N_{K/\mathbf{Q}} = 1$, then $\psi = \bar{\psi}$ and $g = h$.

(2) In Theorem 5.2, we have associated an eigenform of weight m to a character χ satisfying (5.8). We can also associate to χ a form of weight $-m$ satisfying (i–vi) with the change of m for $-m$. In fact, let g^* and h^* denote the functions defined by (5.14) and (5.18) with $\bar{\chi}$, $\bar{\psi}$, and $-m$ instead of χ, ψ, and m. If $\psi \circ N_{K/\mathbf{Q}} = 1$, we can take both g^* and h^* as the desired function. In the general case, we take

$$h^*(z, s) - A_1 A_2^s M(s, \psi) J_\delta(z)^m g^*(\delta z, s)$$

with A_1, A_2, and $M(s, \psi)$ as in the proof of Theorem 5.2.

6. A few more comments. Though we have treated only two types of zeta functions, many more cases are likely to be comprehended in the same fashion. Let us first examine possible analogues of $\mathcal{R}(s, \chi)$ when we take $SL_2(\mathbf{C})$ in place of $SL_2(\mathbf{R})$. Naturally we are led to the special values $\mathcal{E}(\sigma, s)$ of an Eisenstein series $\mathcal{E}$ of $SL_2(M)$ with an imaginary quadratic field M, where σ is a nontrivial fixed point of an element of $SL_2(M)$. It can easily be seen that σ determines a definite quaternion algebra A over $\mathbf{Q}$. In the simplest case where $\mathcal{E}$ is of level 1 and M has class number 1, $\zeta_M(2s)\mathcal{E}(\sigma, s)$ is of the form $\Sigma_{0 \neq x \in X} N_{A/\mathbf{Q}}(x)^{-2s}$ with a lattice X in A. In a more general case, we have to deal with linear combinations of such sums under some congruence conditions on x. Now, by virtue of the results of Eichler, Shimizu, Jacquet, and Langlands, we know that there are many elliptic cusp forms $g(z) = \Sigma_{n=1}^{\infty} a_n e^{2\pi i n z}$ such that the series $L(s, g) = \Sigma_n a_n n^{-s}$ can be obtained as such a linear combination. Therefore it is plausible that $L(s, g)$ takes the place of $L(s, \chi)$ when we consider our problem for $SL_2(M)$ instead of $SL_2(\mathbf{Q})$.

Coming back to $SL_2(\mathbf{Q})$, there is another possibility that concerns a real quadratic field, say F. It was shown by Siegel in [16] that the zeta function of F or more generally an L-function of a Hecke character of F can be obtained as a linear combination of integrals of the type

$$\int_1^{\lambda} E(\gamma_t(i), s; 0) t^{-1} dt,$$

where $t \mapsto \gamma_t$ is a homomorphism of $\mathbf{R}^{\times}$ into $SL_2(\mathbf{R})$ and λ is a unit of F. This suggests that we may be able to obtain a function similar to $f(z, s)$ in Theorem 5.2 for the L-functions of F.

More generally, let $E(z, s)$ denote an Eisenstein series of a reductive algebraic group G over a number field, where z is the variable on $G_{\mathbf{R}}$ and s is a complex parameter. Suppose that there is a homomorphic embedding h of another such group G' into G. Then, with a suitable function p on $G'_{\mathbf{R}}$ and a suitable quotient Φ of $G'_{\mathbf{R}}$, one may obtain a zeta function $R(s)$ by an integral

$$R(s) = \int_{\Phi} p(w) E(h(w), s) dw.$$

This is a general principle of producing zeta functions, and those mentioned in the present paper belong to this type. On the other hand, no general theory of the resolvent kernel has been developed in the higher-dimensional case, and it may be too simple-minded to expect the existence of a function which generalizes G_s and whose constant term is $E(z, s)$. Still, there is no reason to believe that the functions investigated in this paper are phenomena found only in the low-dimensional cases. It may fairly be said that further investigations of the inhomogeneous equations similar to $(L_m - \lambda)f = p$ in the higher-dimensional case are very much worth undertaking.

7. Proofs of the results of Section 4. Define K, F, χ, ψ, R, and other symbols as in Section 4, in particular as in (4.1a–f); define also $\mathfrak{M}_R^C$ by (4.2).

LEMMA 7.1. *An element α of $K^\times$ belongs to $\mathfrak{M}_R^C$ if and only if there is an element t of F prime to C such that $\alpha - t \in Z_v$ for $v|Z_1$ and $\alpha - t \in WR_p$ for $p|W_0$.*

Proof. Suppose $\alpha \in \mathfrak{M}_R^C$. Then $\alpha R \in I_R^C$, so that α is a v-unit for $v|Z_1$. Since K_v is isomorphic to F_q if such a v divides q, there is an element s of F such that $\alpha - s \in Z_v$ for $v|Z_1$. With r as in (4.2), we can take $t \in F$ so that $t - r \in WR_p$ for $p|W_0$ and $t - s \in Q_p$ for $p|Q$. This proves the "only if"-part. The "if"-part can be seen by observing that $R_p = (O_K)_p$ if $p|Q$ and that the existence of t implies that $\alpha \in R_p^\times$ for $p|W_0$ and $\alpha \in O_v^\times$ for $v|Z_1$.

LEMMA 7.2. *Let E be an ideal of F divisible by C. Then every R-ideal in I_R^C can be written in the form αA with $\alpha \in \mathfrak{M}_R^C$ and $A \in I_R^E$.*

Proof. Take an element x of $K_{\mathbf{f}}^\times$ such that $x_v = 1$ for $v|C$. We can find an element α of K such that $\alpha - x_p \in x_pER_p$ for every $p|E$. Then $\alpha x_p^{-1}R_p = R_p$ for such p. Therefore $\alpha^{-1}xR \in I_R^E$ and hence $\alpha R \in I_R^C$. Since $\alpha - 1 \in W_0R_p$ for $p|W_0$, we have $\alpha \in \mathfrak{M}_R^C$, which proves our assertion.

For $\alpha \in \mathfrak{M}_R^C$, denote by t_α any element t as in Lemma 7.1, or rather its class modulo W. Then $\alpha \mapsto t_\alpha$ defines a homomorphism of $\mathfrak{M}_R^C$ into $(O_F/W)^\times$.

LEMMA 7.3.
(i) $\chi_C(\alpha) = \psi_W(t_\alpha)$ *if* $\alpha \in \mathfrak{M}_R^C$.
(ii) $\chi_Z(\alpha) = \chi_C(\alpha)$ *if* $\alpha \in K$ *and* $\alpha R \in I_R^Z$.

Proof. Let $\alpha \in \mathfrak{M}_R^C$. Since $\alpha - t_\alpha \in (W_0 + S_0O_K)_p$ for $p|W_0$, there is an element r of W such that $\alpha - t_\alpha - r \in (S_0O_K)_p$ for such p. Then $\alpha - t_\alpha - r \in Z_v$ for $v|C$. Hence $\chi_C(\alpha) = \chi_C(t_\alpha + r) = \psi_W(t_\alpha)$, which proves (i). If $\alpha R \in I_R^Z$, we see that $\alpha \in R_p^\times$ for $p|S_0$, and hence $\chi_v(\alpha) = 1$ for $v|Z_0$, $v \nmid W_0$ by Lemma 4.1. Therefore we obtain (ii).

Define χ^R by (4.3); let $M_R^C = \{\alpha R \mid \alpha \in \mathfrak{N}_R^C\}$. Since $I_R^C = M_R^C I_R^Z$ by Lemma 7.2, we see from (4.4a,b) that χ^R can also be defined by

$$\text{(7.1)} \qquad \chi^R(\alpha A) = \chi_C(\alpha)^{-1}\chi_{\mathbf{a}}(\alpha)^{-1}\chi^*(AO_K)\ \textit{for}\ \alpha \in \mathfrak{N}_R^C\ \textit{and}\ A \in I_R^Z.$$

For an R-ideal $A = xR$ with $x \in K_{\mathbf{f}}^\times$, we define a fractional ideal $N_{K/F}(A)$ in F and a positive rational number $N(A)$ by $N_{K/F}(A) = N_{K/F}(x)O_F$ and $N(A) = \Pi_{v \in \mathbf{f}} |x_v|_v^{-1}$. Obviously $N_{K/F}(A) = N_{K/F}(AO_K)$ and $N(A) = N(AO_K)$.

Define the L-function $L(s, \chi)$ of χ as usual and another function $L(s, \chi^R)$ by

$$\text{(7.2)} \qquad L(s, \chi^R) = \sum_A \chi^R(A)N(A)^{-s},$$

where A runs over all the integral R-ideals in I_R^C. Since $(O_K)_p = R_p$ if $p \nmid S_0$, we can easily verify that $A \mapsto AO_K$ gives an isomorphism of I_R^Z onto I^Z and that $A \subset R$ if and only if $AO_K \subset O_K$.

Proposition 7.4. $L(s, \chi^R) = L(s, \chi)$.

Proof. The nontrivial point of our assertion is that $L(s, \chi^R)$ involves R-ideals which are not necessarily prime to Z. For every finite prime p not dividing W, put $L_p(s) = \Sigma\, \chi_p(x)|x|^s$, where x runs over the nonzero elements of R_p modulo $R_p^\times$. Then we easily see that

$$L(s, \chi^R) = L(s, \chi) \prod_{p|S,\, p \nmid W} L_p(s).$$

Thus our task is to show that $L_p(s) = 1$ if $p|S$ and $p \nmid W$. For such a p, fix a prime element π of $F_{\mathfrak{p}}$ and put $P = \pi O_p$. Then $S_p = P^e$ with $0 < e \in \mathbf{Z}$. We first treat the case where p splits in K. Then $(O_K)_p$ can be identified with $O_p \times O_p$ and R_p with $\{(a, b) \in O_p \times O_p \mid a - b \in P^e\}$. Let $x = (\pi^m c, \pi^n d) \in R_p$ with p-units c and d. Then $x \in R_p$ only in the following two cases: (i) $m \geqq e$ and $n \geqq e$; (ii) $m = n < e$ and $c - d \in P^{e-m}$. Now we have

$$\text{(7.3)} \qquad L_p(s) = \sum_{(m,n),(c,d)} \chi_p(\pi^m c, \pi^n d)|\pi|^{(m+n)s}$$

with all possible $(\pi^m c, \pi^n d)$ modulo $R_p^\times$. Since $p \nmid W$, we have $\chi_p(c, c) = 1$ for every p-unit c. This shows that $Z_p = P^e(O_K)_p$. Fix (m, n) in Case (i).

The terms of L_p with this (m, n) produce

$$\chi_p(\pi^m, \pi^n)|\pi|^{(m+n)s} \sum \chi_p(1, d), \tag{7.4}$$

where d runs over U_0/U_e, $U_k = \{y \in O_p^\times \mid y - 1 \in P^k\}$. Since both prime divisors of p in K divide Z, the sum must be 0. If $m = n < e$, the terms of (7.3) with this (m, n) produce (7.4) with d running over U_{e-m}/U_e. This sum is not 0 only if $m = 0$, in which case the sum gives 1. Next suppose p is ramified; let γ be a prime element of K_p; let $Y = (\gamma O_K)_p$, $Z_p = Y^e$, and $f = [(e+1)/2]$. Then $S_p = P^f$, $R_p = O_p + Y^{2f} = O_p + \gamma P^f$, and

$$L_p(s) = \sum_{m,u} \chi_p(\gamma^m u)|\gamma|^{ms},$$

where (m, u) runs over $\mathbf{Z} \times (O_K)_p^\times / R_p^\times$ under the condition $\gamma^m u \in R_p$. Put

$$U_k = O_p^\times \cdot \{x \in (O_K)_p^\times \mid x - 1 \in Y^k\}.$$

If $m \geqq 2f$, the terms with this fixed m produce $\chi_p(\gamma^m)|\gamma|^{ms} \sum \chi_p(u)$ with u running over U_0/U_{2f}. Obviously the sum is 0. Suppose $m = 2n < 2f$ with $n \in \mathbf{Z}$. Then $\pi^n u \in R_p$ with $u \in (O_K)_p^\times$ if and only if $u \in U_{2f-2n}$. Thus the terms of L_p with this m can be written in the form $\chi_p(\pi^n)|\gamma|^{2ns} \sum \chi_p(u)$ with u running over U_{2f-2n}/U_{2f}. The sum is nonzero only if $f = n$, in which case it is 1. Suppose $m = 2n + 1 < 2f$ and $\gamma\pi^n u \in R_p$ with $u \in (O_K)_p^\times$. Then $\gamma u \in F_p + \gamma P^{f-n}$, so that $\gamma u = a + \gamma\pi b$ with $a \in F_p$ and $b \in O_p$. Then a must be divisible by π, which is impossible, since $\pi \nmid \gamma u$. Thus (7.3) has no terms with $m = 2n + 1 < 2f$. It remains to consider the case where p remains prime in K. This case can be handled in a similar fashion and in fact it is simpler than the ramified case, and so we omit the details.

Let $\mathfrak{I}$ be a complete set of representatives for I_R^C/M_R^C and let $U = R^\times \cap \mathfrak{N}_R^C$. By Lemma 7.4, we have

$$L(s, \chi) = \sum_{A \in \mathfrak{I}} \sum_{\alpha} \chi^R(\alpha A^{-1}) N(\alpha A^{-1})^{-s}, \tag{7.5}$$

where α runs over $(A \cap \mathfrak{N}_R^C)/U$. By (4.4b) and Lemma 7.3, this can also be written

$$L(s, \chi) = \sum_{A \in \mathfrak{I}} \chi^R(A)^{-1} N(A)^s \sum_{\alpha} \psi_W(t_\alpha)^{-1} \chi_{\mathbf{a}}(\alpha)^{-1} N_{K/\mathbf{Q}}(\alpha)^{-s}. \tag{7.6}$$

By Lemma 7.2, we may take $\mathcal{I}$ from $\mathcal{I}_R^Z$. If we do so, $\chi^R(A)^{-1}$ in (7.6) can be written $\chi^*(AO_K)^{-1}$.

LEMMA 7.5. *Let A be an O_F-lattice in K and let $X = A \cap F$. Then $A = X + Y\xi$ with a fractional ideal Y in F and $\xi \in K$.*

Proof. Since A/X is a torsion-free O_F-module, it is O_F-isomorphic to a fractional ideal Y in F. Then $Y^{-1}A/Y^{-1}X$ is isomorphic to O_F. Let ξ be an element of $Y^{-1}A$ which corresponds to 1 by this isomorphism. Then $Y^{-1}A = Y^{-1}X + O_F\xi$, which proves our lemma.

Fix an element ζ of $K^\times$ such that $\zeta^\rho = -\zeta$. For an O_F-lattice A in K, denote by $\lambda_\zeta(A)$ the O_F-linear span of $\mathrm{Tr}_{K/F}(\zeta\alpha\beta^\rho)$ for all α and β in A.

LEMMA 7.6. *Let A be an O_F-lattice in K of the form $A = X\xi + Y\eta$ with ξ, $\eta \in K$ and fractional ideals X and Y in F. Then $\lambda_\zeta(A) = \zeta(\xi\eta^\rho - \xi^\rho\eta)XY$.*

This is an easy exercise.

Let $d(K/F)$ denote the different of K relative to F. Given ζ as above, let D_ζ denote the fractional ideal in F such that $D_\zeta O_K = \zeta d(K/F)$.

LEMMA 7.7. *Let $R = O_F + BO_K$ with an integral ideal B in F. Then $\lambda_\zeta(A) = N_{K/F}(A)BD_\zeta$ for every R-ideal A. Moreover, if $A = X\xi + Y\eta$ as in Lemma 7.6, then $N_{K/F}(A) = \zeta(\alpha - \alpha^\rho)XYB^{-1}D_\zeta^{-1}$ and $N(A) = N(B^{-1}XY)N((\alpha - \alpha^\rho)d(K/F)^{-1})^{1/2}$, where $\alpha = \xi\eta^\rho$.*

Proof. We can easily verify that $\lambda_\zeta(A) = N_{K/F}(A)\lambda_\zeta(R)$, $\lambda_\zeta(R) = B\lambda_\zeta(O_K)$, and $\lambda_\zeta(O_K) = D_\zeta$. Combining these and Lemma 7.6, we obtain our formulas.

We now assume that F is totally real and K totally imaginary, take ζ from K_Φ, and define $\lambda(A)$ as in Section 4. The notation being as in Lemma 7.6, we see that $XY \in \lambda(A)$ if $\xi/\eta \in K_\Phi$.

LEMMA 7.8. *Let $A \in I_R^C(X)$ (see (4.8a)) with a fractional ideal X in F. Then there exists an element τ of K_Φ and μ in $\mathfrak{M}_R^C$ such that $\mu A = X\tau + O_F$.*

Proof. By Lemma 7.2, we may assume that $A \in I_R^W$. Let $Y = A \cap F$. We easily see that Y is prime to W. There is an element g of F such that gY is integral and prime to W. Replacing A by gA, we may assume that Y is integral. By Lemma 7.5, we have $A = Y + E\xi$ with a fractional ideal E and $\xi \in K$. Replacing E and ξ by hE and $h^{-1}\xi$ with a suitable h in F, we may assume that $E + Y = O_F$. Take $r \in Y$ and $s \in EW$ so that $r - s = 1$. Let $\alpha = r + s\xi$ and $\beta = 1 + \xi$. Then we easily see that $Y + E\xi = O_F\alpha +$

$YE\beta$ and $\alpha - 1 = s(1 + \xi) \in WR_p$ for $p|W$, and hence $\alpha \in \mathfrak{N}_R^C$. Thus $\alpha^{-1}A = O_F + YE\alpha^{-1}\beta$. Take $\gamma \in F$ so that $\gamma\alpha^{-1}\beta \in K_\Phi$. By Lemma 7.6 we have $\lambda(A) = \lambda(\alpha^{-1}A) \ni \gamma YE$. Therefore $\gamma YE = \delta X$ with a totally positive δ in F. Put $\tau = \delta\gamma^{-1}\alpha^{-1}\beta$. Then $\tau \in K_\Phi$ and $\alpha^{-1}A = O_F + X\tau$ as desired.

LEMMA 7.9. *The element τ of Lemma 7.8 can be chosen so that $X\tau \subset Z_v$ for every v dividing Z_1.*

Proof. If $v|Z_1$, we see that $X\tau \subset O_v$. Take a totally positive a in F so that aX is integral and prime to Z_1. Then $a^{-1}\tau \in O_v$. Since $O_K = O_F + Z_1$, we can find an element b of O_F such that $a^{-1}\tau - b \in Z_v$ for $v|Z_1$. Putting $\tau' = \tau - ab$, we obtain $\mu A = O_F + X\tau'$ and $X\tau' \subset Z_v$ for every such v.

Proof of Lemma 4.2. Let $\gamma = \begin{pmatrix} a & b \\ c & d \end{pmatrix} \in \Gamma[X^{-1}, WX]$ and $A_\tau = TXQ\tau + T$. Then $A_\tau = (c\tau + d)A_{\gamma(\tau)}$. Suppose $\tau \in H_X^C$. Observe that $X(a\tau + b) \subset O_v$ and $c\tau \in Z_v$ for $v|Z_1$, and $c\tau \in W_0R_p$ for $p|W_0$. Since d is prime to W, we see that $c\tau + d \in \mathfrak{N}_R^C$ by Lemma 7.1 and $X\gamma(\tau) \subset O_v$ for $v|Z_1$, and hence $\gamma(\tau) \in H_X^C$. This proves (i) and that the map of (ii) is well-defined. If $B \in I_R^C(T^2XQ)$, then $T^{-1}B \in I_R^C(XQ)$. Then Lemma 7.9 proves the surjectivity of the map of (ii). Suppose $A_\tau = \alpha A_\sigma$ for $\sigma, \tau \in H_X^C$ and $\alpha \in \mathfrak{N}_R^C$. We can put $\alpha\sigma = a\tau + b$ and $\alpha = c\tau + d$ with a, b, c, d in F. Put $\gamma = \begin{pmatrix} a & b \\ c & d \end{pmatrix}$. Then $\sigma = \gamma\tau$, and we easily see that $\gamma \in \Gamma[Q^{-1}X^{-1}, QX]$. Since $\alpha \in \mathfrak{N}_R^C$, we have $c\tau + d - t_\alpha \in W_0R_p = W_0(X_p\tau + O_p)$ for $p|W_0$. Thus $c \in W_0X_p$ for every such p. Since $c \in QX$, this shows that $c \in WX$. Now $X\tau \subset O_v$ and $(a\tau + b)X = \alpha\sigma X \subset O_v$ for $v|Z_1$, so that $Xb \subset O_p$ for $p|Q$. Since $b \in Q^{-1}X^{-1}$, we have $b \in X^{-1}$, so that $\gamma \in \Gamma[X^{-1}, WX]$. This completes the proof.

LEMMA 7.10. *Let $A = XTQ\tau + T$ with X and T as in Lemma 4.2 and $\tau \in H_X^C$; let $\alpha = a\tau + b$ with $a \in XTQ$ and $b \in T$. Then α belongs to $\mathfrak{N}_R^C$ if and only if $\alpha \neq 0$, $a \in XTW$, and b is prime to W, in which case we have $\chi_C(\alpha) = \psi_W(t_\alpha) = \psi_W(b)$.*

Proof. If $\tau \in H_X^C$, $a \in XTW$, and b is prime to W, then we easily see from Lemma 7.1 that $a\tau + b \in \mathfrak{N}_R^C$. Conversely, suppose $a\tau + b - t \in W_0R_p$ for $p|W_0$ and $a\tau + b - t \in Z_v$ for $v|Z_1$, with an element t of F prime to W. Then $a\tau + b - t \in W_0(XTQ\tau + T)_p$, and hence $a \in W_0X_p$ and $b - t \in W_{0p}$ for every such p. Since $a \in XTQ$, we have $a \in XTW$, and so $a\tau \in Z_v$ for $v|Z_1$. Therefore $b - t \in Z_v$ for every such v, and hence b is prime to W. Since we can take b as t_α, we obtain the last assertion from Lemma 7.3.

Proof of Theorem 4.6. Let χ and $\mathfrak{X}$ be as in the theorem. For any fixed ideal Y in F prime to W, we can take $\bigcup_{X \in \mathfrak{X}} I_R^C(QXY^2)/M_R^C$ as $\mathfrak{I}$ of (7.6). By Lemma 4.2, (ii), we can take the R-ideals $YXQ\tau + Y$ with $\tau \in \Gamma[X^{-1}, WX]\backslash H_X^C$ as a complete set of representatives for $I_R^C(QXY^2)/M_R^C$. Therefore by (7.6) we have

$$(7.7)\quad L(s, \chi) = \sum_{X \in \mathfrak{X}} \sum_{\tau} \chi^R(YXQ\tau + Y)^{-1} \cdot N(YXQ\tau + Y)^s \sum_{\alpha} \psi_W(t_\alpha)^{-1}\chi_{\mathbf{a}}(\alpha)^{-1}N_{K/\mathbf{Q}}(\alpha)^{-s},$$

where τ runs over $\Gamma[X^{-1}, WX]\backslash H_X^C$ and α over $[\mathfrak{N}_R^C \cap (YXQ\tau + Y)]/U$, $U = \mathfrak{N}_R^C \cap R^\times$. By Lemma 7.10, the last sum over α can be written in the form

$$[U : O_F^\times]^{-1} \sum_{a,b} \psi_W(b)^{-1}\chi_{\mathbf{a}}(a\tau + b)^{-1}N_{K/\mathbf{Q}}(a\tau + b)^{-s},$$

where (a, b) runs over all the nonzero elements of $(YXW \times Y)/O_F^\times$ under the condition that $bY^{-1} + W = O_F$. Applying Lemma 7.7 to $YXQ\tau + Y$ and using E_Y of (4.18), we obtain

$$[U : O_F^\times]L(s, \chi) = N(d(K/F))^{-s/2}N(2B^{-1}Q)^s \cdot \sum_{X \in \mathfrak{X}} N(X)^s \sum_{\tau} \chi^R(XQ\tau + O_F)^{-1}E_Y(\tau, s; X, W; \psi, m).$$

Taking the sum over all Y belonging to a complete set of representatives for the ideal classes in F, we obtain the formula of Theorem 4.6 in view of Proposition 4.5.

PRINCETON UNIVERSITY

REFERENCES

[1] J. Elstrodt, Die Resolvente zum Eigenwertproblem der automorphen Formen in der hyperbolischen Ebene I, II, III, *Math. Ann.* **203** (1973), 295–330, *Math. Z.* **132** (1973), 99–134, *Math. Ann.* **208** (1974), 99–132.

[2] A. Erdélyi et al., *Higher Transcendental Functions*, Vol. 1, McGraw-Hill, 1953.

[3] L. D. Fadeev, Expansion in eigenfunctions of the Laplace operator on the fundamental domain of a discrete group on the Lobacevskii plane, *Trans. Moscow Math. Soc.* 17 (1967), 357–386.

[4] J. D. Fay, Fourier coefficients of the resolvent for a Fuchsian group, *J. Reine Angew. Math.* **294** (1977), 143–203.

[5] D. Goldfeld, Analytic and arithmetic theory of Poincaré series, *Astérisque* 61 (1979), 95–107.

[6] D. Hejhal, Some observations concerning eigenvalues of the Laplacian and Dirichlet L-series, in *Recent Progress in Analytic Number Theory*, Vol. 2. Academic Press, 1981, 95–110.

[7] ———, The Selberg trace formula for *PSL*(2, **R**), Vol. 2, *Springer Lecture Notes in Math.* No. 1001, 1983.

[8] H. Neunhöffer, Über die analytische Fortsetzung von Poincaréreihen, *Sitz.-Ber. Heidelberg Akad. Wiss. Math.* 1973, 33–90.

[9] W. Roelcke, Das Eigenwertproblem der automorphen Formen in der hyperbolischen Ebene I, II, *Math. Ann.* **167** (1966), 292–337, **168** (1967), 261–324.

[10] G. Shimura, On modular forms of half integral weight, *Ann. of Math.* 97 (1973), 440–481.

[11] ———, On the holomorphy of certain Dirichlet series, *Proc. London Math. Soc.* 31 (1975), 79–98.

[12] ———, Confluent hypergeometric functions on tube domains, *Math. Ann.* **260** (1982), 269–302.

[13] ———, On Eisenstein series, *Duke Math. J.* **50** (1983), 417–476.

[14] ———, On the Eisenstein series of Hilbert modular groups, *Revista Mat. Iberoamer.* 1, No. 3 (1985), 1–42.

[15] ———, On Hilbert modular forms of half-integral weight, *Duke Math. J.* 55 (1987), 765–838.

[16] C. L. Siegel, *Lectures on Advanced Analytic Number Theory*, Tata Institute of Fundamental Research, 1961.

90a

Invariant differential operators on hermitian symmetric spaces

Annals of Mathematics, 132 (1990), 237-272

Introduction

Our object of study is the ring of invariant differential operators on a hermitian symmetric space G/K of classical and noncompact type. Here, as usual, G is a connected noncompact semisimple Lie group with finite center and K is a maximal compact subgroup of G. It is well-known that the ring, denoted by $\mathscr{D}(G/K)$, is isomorphic to a polynomial ring of l variables, l being the rank of G/K. This is true even in the nonhermitian case. If $l = 1$, it is generated by the Laplace-Beltrami operator $\mathscr{L}$ which is essentially self-adjoint; moreover $-\mathscr{L}$ is nonnegative. In the general case, an easily posable problem is to find an explicitly defined set of generators. We can go one step further by focusing our attention on the nonnegativity, which necessarily limits the range of eigenvalues under a suitable integrability condition on functions. It is thus natural to ask whether there exist some canonically defined operators which generate $\mathscr{D}(G/K)$ and have the property of nonnegativity. The main purpose of the present paper is to give an affirmative answer to this question. In fact, our answer applies to a somewhat more general type of rings of differential operators which includes $\mathscr{D}(G/K)$ as a special case.

To be explicit, we take an irreducible representation $\rho: K \to GL(V)$ with a complex vector space V of finite dimension, and consider the set $C^\infty(\rho)$ of all V-valued C^∞ functions f on G such that $f(xk^{-1}) = \rho(k)f(x)$ for every $k \in K$. We then denote by $\mathscr{D}(\rho)$ the ring of left-invariant differential operators on G which map $C^\infty(\rho)$ into itself. Now the complexification $\mathfrak{g}$ of the Lie algebra of G has abelian subalgebras $\mathfrak{p}_+$ and $\mathfrak{p}_-$ which can be identified with the spaces of holomorphic and antiholomorphic tangent vectors on G/K at the origin. For any complex vector space W let $S_r(W)$ denote the ring of all complex-valued homogeneous polynomial functions on W of degree r, and let $S(W) = \sum_{r=0}^{\infty} S_r(W)$. Through the adjoint representation of G on $\mathfrak{g}$, K acts naturally on $S_r(\mathfrak{p}_+)$. It is a known fact, due to Hua and Schmid, that each irreducible constituent of this representation of K has multiplicity one. Now our principal result can be stated as follows (Theorem 3.6):

Suppose that G is classical and ρ is one-dimensional. Then, to each irreducible constituent Z of $S_r(\mathfrak{p}_+)$, we can assign an element $\mathscr{L}_Z$ of $\mathscr{D}(\rho)$ such that $(-1)^r\mathscr{L}_Z$ is symmetric and nonnegative with respect to a hermitian inner product (defined at least for functions of compact support or for automorphic forms in cocompact cases, for example). The $\mathscr{L}_Z$ for all such Z and all nonnegative integers r form a **C***-linear basis of $\mathscr{D}(\rho)$. Moreover there is a canonically defined set $\{Z_1, \ldots, Z_l\}$ such that $\mathscr{D}(\rho) = \mathbf{C}[\mathscr{L}_{Z_1}, \ldots, \mathscr{L}_{Z_l}]$ and that $\mathscr{L}_{Z_1}, \ldots, \mathscr{L}_{Z_l}$ are algebraically independent.*

This result leads us to another natural question. *Suppose $(\lambda_1, \ldots, \lambda_l)$ is a set of eigenvalues of $\mathscr{L}_{Z_1}, \ldots, \mathscr{L}_{Z_l}$ on a common eigenfunction in a space on which the $\mathscr{L}_Z$ are nonnegative. Then what condition on $(\lambda_1, \ldots, \lambda_l)$ does the nonnegativity of each $\mathscr{L}_Z$ impose?* We have no complete answer to this question, but we shall derive various types of inequalities for the λ_i from the nonnegativity. In particular, we shall show that *if G is simple, then $\lambda_m \le g_m(\lambda_1)$ for each $m \le l$ with a real polynomial g_m of degree m whose leading coefficient is* 1 (Theorem 6.2).

Though these are our main results on $\mathscr{D}(\rho)$, another objective of the present paper is to study the ring in connection with the differential operators which map $C^\infty(\rho)$ into $C^\infty(\rho \otimes \tau)$ with an irreducible constituent τ of $S_r(\mathfrak{p}_+)$ or $S_r(\mathfrak{p}_-)$. In fact, $(-1)^r\mathscr{L}_Z$ is the product of such an operator and its adjoint, which immediately implies the nonnegativity. This fact may be independently emphasized instead of viewing it as a technique of the proof.

Let us now summarize the whole paper explaining some of our ideas. In Section 1, we define a **C**-linear bijection ω of $S(\mathfrak{g})$ onto the universal enveloping algebra $U(\mathfrak{g})$ of $\mathfrak{g}$. Since $S(\mathfrak{g})$ is isomorphic to the symmetric algebra $\mathfrak{S}(\mathfrak{g})$ over $\mathfrak{g}$, ω is merely a paraphrase of the standard map ψ of $\mathfrak{S}(\mathfrak{g})$ onto $U(\mathfrak{g})$, which we also employ. However, we often find ω more convenient than ψ, at least in the hermitian case. Section 2 concerns a symmetric space G/K which is not necessarily hermitian. We shall prove some elementary propositions on $\mathscr{D}(\rho)$ in this general case. In particular, we shall show that *if ρ is one-dimensional, then $\mathscr{D}(\rho)$ is commutative and given by $\omega(I(\mathfrak{p}))$, where $\mathfrak{p} = \mathfrak{p}_+ + \mathfrak{p}_-$ and $I(\mathfrak{p})$ denotes the set of K-invariant elements of* $S(\mathfrak{p})$ (Proposition 2.3). We shall also prove that *if G is classical and ρ is one-dimensional, $\mathscr{D}(\rho)$ is the image of the center of* $U(\mathfrak{g})$ (Theorem 2.4), generalizing a result of Helgason [H1] which concerns the case of trivial ρ.

In Section 3, after introducing a pairing on $\mathfrak{p}$ and observing that $S_r(\mathfrak{p}_+)$ and $S_r(\mathfrak{p}_-)$ are dual to each other, we shall show that $I(\mathfrak{p})$ is ring-isomorphic to the ring J^H of H-invariant elements of $S(\mathfrak{p}_+)$, where H is the isotropy subgroup of the complexification of G at a "generic" point of $\mathfrak{p}_-$. Now, in [S4] we studied the structure of J^H in connection with the decomposition of $S_r(\mathfrak{p}_+)$ into irreducible subspaces. This connection is crucial in the proof of our main theorem. We shall in fact give two different types of generators for $\mathscr{D}(\rho)$, one the straightforward images of the basis elements of $I(\mathfrak{p})$ under ω (Theorem 3.4), and the other their "variations" which give the above $\mathscr{L}_Z$. The symmetricity and nonnegativity will be shown in Section 4, where we shall study operators which map $C^\infty(\rho)$ to $C^\infty(\rho \otimes \tau)$ as mentioned above. Our principal results of this section are the explicit description of the adjoint of each operator of this type (Theorem 4.3) and the decomposition of $(-1)^r\mathscr{L}_Z$ into the product of such an operator and its adjoint (Proposition 4.1).

In Section 5, we shall explain how to choose the representations $Z_1, \dots, Z_l$ for each classical and irreducible G/K by showing that they correspond to the set of generators of J^H given in [S4]. The first part of Section 6 concerns the conditions on the eigenvalues of the $\mathscr{L}_Z$ imposed by their nonnegativity. We shall express some $\mathscr{L}_Z$ as polynomials of the $\mathscr{L}_{Z_i}$, which will produce nontrivial conditions on the eigenvalues (Theorem 6.2, Propositions 6.3 and 6.4). Some observations on $\mathscr{D}(\rho)$ for an arbitrary irreducible ρ as well as a few open problems will be given in the last part of this section. The final section is of expository nature. Representing G/K as a domain of complex matrices in the standard fashion, we shall show that the operators of Sections 3 and 4 are practically the same as those we treated in our previous papers [S1–5], mainly for the purposes of investigating arithmeticity problems.

1. Preliminaries on symmetric algebras

Let V and W be finite-dimensional vector spaces over $\mathbf{C}$ and r a positive integer. We denote by $\mathrm{Ml}_r(W, V)$ the vector space of all $\mathbf{C}$-multilinear maps of $W \times \cdots \times W$ (r copies) into V, and by $S_r(W, V)$ the vector space of all homogeneous polynomial maps of W into V of degree r. In particular, $S_1(W, V) = \mathrm{Hom}(W, V)$. An element Q of $\mathrm{Ml}_r(W, V)$ is called *symmetric* if $Q(x_1, \dots, x_r) = Q(x_{\pi(1)}, \dots, x_{\pi(r)})$ for ever permutation π of $\{1, \dots, r\}$. Given $P \in S_r(W, V)$, there is a unique symmetric element P_* of $\mathrm{Ml}_r(W, V)$ such that

$$P(x) = P_*(x, \dots, x). \tag{1.1}$$

Moreover the map $P \mapsto P_*$ is a $\mathbf{C}$-linear bijection of $S_r(W, V)$ onto the set of all symmetric elements of $\mathrm{Ml}_r(W, V)$.

Taking especially $V = \mathbf{C}$, we put

$$S_r(W) = S_r(W, \mathbf{C}),$$

$$S(W) = \bigcup_{p=0}^{\infty} S^p(W), \qquad S^p(W) = \sum_{r=0}^{p} S_r(W).$$

Thus $S(W)$ is the set of all polynomial functions on W. We also define the symmetric algebra $\mathfrak{S}(W)$ over W as usual, denote by $\mathfrak{S}_r(W)$ its subspace consisting of all homogeneous elements of degree r, and put $\mathfrak{S}^p(W) = \sum_{r=0}^{p} \mathfrak{S}_r(W)$. We note that $S_r(W)$ and $\mathfrak{S}_r(W)$ are dual to each other with respect to the pairing

$$(1.2) \qquad \langle \alpha, x_1 \cdots x_r \rangle = \alpha_*(x_1, \ldots, x_r) \qquad (x_i \in W, \alpha \in S_r(W)).$$

Let $\mathfrak{p}$ be the dual space of W; that is, $\mathfrak{p} = S_1(W)$. (For the moment $\mathfrak{p}$ is merely a vector space.) Then the identity map of $\mathfrak{p}$ to $S_1(W)$ can be extended in an obvious way to an algebra-isomorphism of $\mathfrak{S}_r(\mathfrak{p})$ to $S_r(W)$. Let $\{b_1, \ldots, b_m\}$ be a basis of W and $\{X_1, \ldots, X_m\}$ be the basis of $\mathfrak{p}$ dual to it. Then $S_r(W)$ and $S_r(\mathfrak{p})$ are dual to each other with respect to the pairing

$$(1.3) \qquad \langle \alpha, \beta \rangle = \sum \alpha_*(b_{i_1}, \ldots, b_{i_r}) \beta_*(X_{i_1}, \ldots, X_{i_r})$$

for $\alpha \in S_r(W)$ and $\beta \in S_r(\mathfrak{p})$, where $(i_1, \ldots, i_r)$ runs over $\{1, \ldots, m\}^r$. This is the same as formula (1.2) combined with the identification of $S_r(\mathfrak{p})$ with $\mathfrak{S}_r(W)$. Given $c_1, \ldots, c_r \in W$, we may view $c_1 \cdots c_r$ as an element of $S_r(\mathfrak{p})$ by putting $(c_1 \cdots c_r)(X) = c_1(X) \cdots c_r(X)$ for $X \in \mathfrak{p}$. Similarly, if $Y_1, \ldots, Y_r \in \mathfrak{p}$, we may view $Y_1 \cdots Y_r$ as an element of $S_r(W)$.

LEMMA 1.1. *The symbols b_i and X_i being as above,* $\langle b_{i_1} \cdots b_{i_r}, X_{j_1} \cdots X_{j_r} \rangle \neq 0$ *only when* $(j_1, \ldots, j_r) = (i_{\pi(1)}, \ldots, i_{\pi(r)})$ *for some permutation π of* $\{1, \ldots, r\}$, *in which case*

$$\langle b_{i_1} \cdots b_{i_r}, X_{j_1} \cdots X_{j_r} \rangle = N(i_1, \ldots, i_r)/r!,$$

where $N(i_1, \ldots, i_r)$ is the number of permutations π of $\{1, \ldots, r\}$ such that $(i_{\pi(1)}, \ldots, i_{\pi(r)}) = (i_1, \ldots, i_r)$.

The proof is completely elementary and so may be left to the reader. This lemma will be needed only in Section 6.

Let us now consider a Lie algebra $\mathfrak{g}$ over $\mathbf{C}$. Let $U(\mathfrak{g})$ be its universal enveloping algebra and $U^p(\mathfrak{g})$ its subspace spanned by the elements of the form $Z_1 \cdots Z_s$ with $Z_i \in \mathfrak{g}$, $s \leq p$. We recall that there is a $\mathbf{C}$-linear bijection ψ of $\mathfrak{S}(\mathfrak{g})$ onto $U(\mathfrak{g})$ which is characterized by the property that $\psi(X^r) = X^r$ for every $X \in \mathfrak{g}$. It is often more convenient to take an algebra of type $S(W)$ instead of $\mathfrak{S}(\mathfrak{g})$. Given a complex vector subspace $\mathfrak{p}$ of $\mathfrak{g}$, we take $W = S_1(\mathfrak{p})$

as above, and define an element $\omega(\alpha)$ of $U(\mathfrak{g})$ for each $\alpha \in S_r(W)$ by

$$(1.4) \qquad \omega(\alpha) = \sum \alpha_*(b_{i_1}, \ldots, b_{i_r}) X_{i_1} \cdots X_{i_r},$$

where b_i, X_i, and the summation are the same as in (1.3). If $Y \in \mathfrak{p}$, then Y^r as an element of $S_r(W)$ is defined by $Y^r(u) = Y(u)^r$ for $u \in W$, and hence $(Y^r)_*(u_1, \ldots, u_r) = Y(u_1) \cdots Y(u_r)$. Therefore (1.4) shows that $\omega(Y^r) = (\sum Y(b_i)X_i)^r = Y^r$ (as an element of $U(\mathfrak{g})$). This also shows that if $\alpha(\sum_i t_i b_i) = P(t_1, \ldots, t_m)$ for $t_i \in \mathbf{C}$ with a polynomial P, then $\omega(\alpha) = \psi(P(X_1, \ldots, X_m))$. Thus ω is merely another expression for the restriction of ψ to $\mathfrak{S}(\mathfrak{p})$ in terms of $S(W)$, so that ω is a $\mathbf{C}$-linear injection of $S(W)$ into $U(\mathfrak{g})$ independent of the choice of a basis. Naturally we have $\omega(S_r(W)) = \psi(\mathfrak{S}_r(\mathfrak{p}))$. We note here a well-known fact on ψ by expressing it in terms of ω:

$$(1.5) \quad \omega(\alpha_1 \cdots \alpha_n) - \omega(\alpha_1) \cdots \omega(\alpha_n) \in U^{r-1}(\mathfrak{g}) \quad \text{if } \alpha_1 \cdots \alpha_n \in S^r(W).$$

Let $W = V \oplus V'$ with complex subspaces V and V'. Given $f \in S_q(V)$ and $f' \in S_r(V')$, define $f \otimes f' \in S_{q+r}(W)$ by $(f \otimes f')(x + y) = f(x)f'(y)$ for $x \in V$ and $y \in V'$. Then we see that

$$(1.6) \quad (f \otimes f')_*(x_1 + y_1, \ldots, x_{q+r} + y_{q+r}) = [(q+r)!]^{-1} \sum_\pi f_*(x_{\pi(1)}, \ldots, x_{\pi(q)}) f'_*(y_{\pi(q+1)}, \ldots, y_{\pi(q+r)}),$$

for $x_i \in V$ and $y_i \in V'$, where π runs over all permutations of $\{1, \ldots, q + r\}$. In particular, if f' is the constant 1, then identifying f with $f \otimes 1$, we can embed $S_r(V)$ into $S_r(W)$. Of course this depends on the choice of V'. In our later applications, however, direct-sum decomposition of a vector space will always be canonical and so the above embedding will be canonical too.

2. Invariant differential operators acting on vector-valued functions

Though our main interest is in the hermitian case, we consider in this section more generally a symmetric space given as G/K with a connected noncompact semisimple Lie group G with finite center and a maximal compact subgroup K of G. Let $\mathfrak{g}_0$ and $\mathfrak{k}_0$ be the Lie algebra of G and its subalgebra corresponding to K. Then we have a Cartan decomposition $\mathfrak{g}_0 = \mathfrak{k}_0 + \mathfrak{p}_0$ as usual with a subspace $\mathfrak{p}_0$ of $\mathfrak{g}_0$. We denote by $\mathfrak{g}$, $\mathfrak{k}$, and $\mathfrak{p}$ the complexifications of $\mathfrak{g}_0$, $\mathfrak{k}_0$, and $\mathfrak{p}_0$, respectively. Take a connected Lie group G^c whose Lie algebra is $\mathfrak{g}$ and its connected subgroup K^c corresponding to $\mathfrak{k}$. Let Ad: $G^c \to \mathrm{Aut}(\mathfrak{g})$ denote the adjoint representation of G^c. For each $g \in G^c$ the action of $\mathrm{Ad}(g)$ can be extended to automorphisms of $\mathfrak{S}(\mathfrak{g})$ and $U(\mathfrak{g})$, which we denote also by $\mathrm{Ad}(g)$. Then we have $\psi(\mathrm{Ad}(g)Y) = \mathrm{Ad}(g)\psi(Y)$ for $Y \in \mathfrak{S}(\mathfrak{g})$.

In particular, we consider $\mathrm{Ad}(k)$ for each $k \in K^c$, and its restriction to $\mathfrak{p}$ and also to $\mathfrak{S}(\mathfrak{p})$. We can also define the action of k on $S(\mathfrak{p})$ by $(kf)(x) = f(\mathrm{Ad}(k)^{-1}x)$ for $x \in \mathfrak{p}$ and $f \in S(\mathfrak{p})$. If $\mathfrak{X}$ is any of these spaces (or their subspaces such as $S_r(\mathfrak{p})$ and $U^r(\mathfrak{g})$) on which K^c naturally acts, we denote by $\mathfrak{X}^K$ the subspace consisting of all its K^c-invariant elements. It is well-known that the K^c-invariance is equivalent to the K-invariance. In fact, G^c and K^c are unnecessary in this section.

Now we take a continuous representation $\rho\colon K \to GL(V)$ with a finite-dimensional complex vector space V. For any C^∞-manifold M we denote by $C^\infty(M, V)$ the set of all C^∞ maps of M into V. In particular, when $M = G$, we denote by $C^\infty(\rho)$ the subset of $C^\infty(G, V)$ consisting of the elements f such that $f(xk^{-1}) = \rho(k)f(x)$ for every $k \in K$. We can let every element of $U(\mathfrak{g})$ act on $C^\infty(G, V)$ as a left-invariant differential operator. We then denote by $D(\rho)$ the set of elements of $U(\mathfrak{g})$ which map $C^\infty(\rho)$ into itself, and by $\mathscr{D}(\rho)$ the ring of operators on $C^\infty(\rho)$ given by the elements of $D(\rho)$. In particular, if $V = \mathbf{C}$ and ρ is trivial, $\mathscr{D}(\rho)$ can be identified with the ring of invariant differential operators on G/K, which we denote by $\mathscr{D}(G/K)$. The purpose of this section is to prove elementary results on the structure of $\mathscr{D}(\rho)$. (If ρ is trivial, they are naturally the standard facts on $\mathscr{D}(G/K)$ that can be found in Helgason's book [H2]. I owe Proposition 2.1 below to Harish-Chandra who communicated it to me in June, 1983.)

First we have

$$(2.1)\qquad [(\mathrm{Ad}(k)B)f](x) = \rho(k)(Bf)(xk)$$

$$\text{if } f \in C^\infty(\rho),\ B \in U(\mathfrak{g}),\ k \in K,\ x \in G.$$

This shows that $U(\mathfrak{g})^K \subset D(\rho)$. More generally we can let $\mathrm{End}(V) \otimes U(\mathfrak{g})$ act on $f \in C^\infty(G, V)$ by the rule $(E \otimes B)f = EBf$ for $E \in \mathrm{End}(V)$, $B \in U(\mathfrak{g})$, and then take the operators which map $C^\infty(\rho)$ into itself. However, if ρ is irreducible, such operators can always be obtained from $D(\rho)$ for the following reason. We first observe that $Xf = -d\rho(X)f$ if $X \in \mathfrak{k}$ and $f \in C^\infty(\rho)$. Extending $-d\rho$ to an antihomomorphism ρ_u of $U(\mathfrak{k})$ into $\mathrm{End}(V)$, we obtain

$$(2.2)\qquad Bf = \rho_u(B)f \quad \text{if } B \in U(\mathfrak{k}) \text{ and } f \in C^\infty(\rho).$$

For every $X \in \mathfrak{k}$ and a nonnegative integer m, $\rho_u(U(\mathfrak{k}))$ contains $d\rho(X)^m$, so that it contains $\rho(\exp(X))$. Therefore if ρ is irreducible, ρ_u is surjective; hence we can find for each $E \in \mathrm{End}(V)$ an element ε of $U(\mathfrak{k})$ such that $Ef = \varepsilon f$ for all $f \in C^\infty(\rho)$. Consequently the action of any element of $\mathrm{End}(V) \otimes U(\mathfrak{g})$ on $C^\infty(\rho)$ can be obtained from $U(\mathfrak{g})$. Thus we lose practically nothing by restricting ourselves to $D(\rho)$.

PROPOSITION 2.1. (1) *An element* T *of* $U(\mathfrak{g})$ *annihilates* $C^\infty(\rho)$ *if and only if* $T \in U(\mathfrak{g})\mathfrak{N}_\rho$, *where* $\mathfrak{N}_\rho = \mathrm{Ker}(\rho_u)$.

(2) $D(\rho) = U(\mathfrak{g})^K + U(\mathfrak{g})\mathfrak{N}_\rho$.

(3) *The natural map of* $D(\rho)$ *onto* $\mathscr{D}(\rho)$ *gives an isomorphism of* $U(\mathfrak{g})^K/[U(\mathfrak{g})^K \cap U(\mathfrak{g})\mathfrak{N}_\rho]$ *onto* $\mathscr{D}(\rho)$.

Proof. Obviously every element of $U(\mathfrak{g})\mathfrak{N}_\rho$ annihilates $C^\infty(\rho)$. To prove its converse, recall that $G = \exp(\mathfrak{p}_0)K$, and define for $h \in C^\infty(\mathbf{R}^n, V)$ an element f_h of $C^\infty(\rho)$ by $f_h(\exp(\Sigma_{i=1}^n t_i X_i)k) = \rho(k^{-1})h(t_1, \ldots, t_n)$ for $k \in K$, where $\{X_1, \ldots, X_n\}$ is a basis of $\mathfrak{p}_0$ over $\mathbf{R}$. For $B = P(X_1, \ldots, X_n) \in \mathfrak{S}(\mathfrak{p})$ with a polynomial $P(t_1, \ldots, t_n)$, put $B' = P(\partial/\partial t_1, \ldots, \partial/\partial t_n)$. Then we see that $(\psi(B)f_h)(1) = (B'h)(0)$. Now we have

$$U^r(\mathfrak{g}) = \bigoplus_{s+t \le r} \psi(\mathfrak{S}_s(\mathfrak{p}))\psi(\mathfrak{S}_t(\mathfrak{k})). \tag{2.3}$$

Let $T \in U^r(\mathfrak{g})$. Take a basis $\{B_i\}$ of $\mathfrak{S}^r(\mathfrak{p})$ over $\mathbf{C}$ (consisting of the monomials of the X_i, for example) and write $T = \Sigma_i \psi(B_i)C_i$ with $C_i \in U(\mathfrak{k})$. Suppose $TC^\infty(\rho) = \{0\}$. Then $0 = (Tf_h)(1) = \Sigma_i \rho_u(C_i)(B_i'h)(0)$. Since the B_i' are linearly independent, we see, by choosing a suitable h, that $\rho_u(C_i) = 0$ for ever i, and hence $T \in U(\mathfrak{g})\mathfrak{N}_\rho$, which proves (1). As for (2), we already observed that $U(\mathfrak{g})^K \subset D(\rho)$. In view of (1), our task is therefore to show that $D(\rho) \subset U(\mathfrak{g})^K + U(\mathfrak{g})\mathfrak{N}_\rho$. Let $B \in D(\rho)$. Then (2.1) shows that $B - \mathrm{Ad}(k)B \in U(\mathfrak{g})\mathfrak{N}_\rho$ for every $k \in K$. Put $C = \int_K[\mathrm{Ad}(k)B]\,dk$. Taking the total measure of K to be 1, we see that $C \in U(\mathfrak{g})^K$ and $B - C \in U(\mathfrak{g})\mathfrak{N}_\rho$, which proves (2). Assertion (3) follows immediately from (1) and (2).

We can represent $\mathscr{D}(\rho)$ in a different way. First, for every $\Sigma_i E_i \otimes B_i$ in $\mathrm{End}(V) \otimes \mathfrak{S}(\mathfrak{g})$ with E_i in $\mathrm{End}(V)$ and B_i in $\mathfrak{S}(\mathfrak{g})$, we define $\Psi(\Sigma_i E_i \otimes B_i)$ to be the operator on $C^\infty(G, V)$ given by $\Sigma_i E_i \psi(B_i)$.

PROPOSITION 2.2. *Suppose* ρ *is irreducible. Define the action of* K *on* $\mathrm{End}(V) \otimes \mathfrak{S}(\mathfrak{p})$ *by*

$$k(E \otimes B) = \rho(k)E\rho(k)^{-1} \otimes \mathrm{Ad}(k)B \quad \textit{for } k \in K,\ E \in \mathrm{End}(V), \textit{ and } B \in \mathfrak{S}(\mathfrak{p}).$$

Let $I(\rho, \mathfrak{p})$ *denote the set of all* K*-invariant elements of* $\mathrm{End}(V) \otimes \mathfrak{S}(\mathfrak{p})$. *Then* Ψ *gives a* $\mathbf{C}$*-linear bijection of* $I(\rho, \mathfrak{p})$ *onto* $\mathscr{D}(\rho)$.

Proof. Identifying $U(\mathfrak{g})$ with $\psi(\mathfrak{S}(\mathfrak{p})) \otimes U(\mathfrak{k})$, define a map μ of $U(\mathfrak{g})$ into $\mathrm{End}(V) \otimes \mathfrak{S}(\mathfrak{p})$ by $\mu(\psi(B)C) = \rho_u(C) \otimes B$ for $B \in \mathfrak{S}(\mathfrak{p})$ and $C \in U(\mathfrak{k})$. We see that $\mu \circ \mathrm{Ad}(k) = k \circ \mu$ for every $k \in K$, since $\rho_u(\mathrm{Ad}(k)C) = \rho(k)\rho_u(C)\rho(k)^{-1}$. Now we easily see that every element of $\Psi(I(\rho, \mathfrak{p}))$ maps $C^\infty(\rho)$ into itself. Since ρ is irreducible, every such operator is contained in

$\mathscr{D}(\rho)$ as observed above. By Proposition 2.1, every element of $\mathscr{D}(\rho)$ is represented by an element, say T, of $U^r(\mathfrak{g})^K$ for some r. Then we see that $\mu(T) \in I(\rho, \mathfrak{p})$, and $\Psi(\mu(T))$ gives the same action as T on $C^\infty(\rho)$. This proves the surjectivity. Next take f_h as in the above proof. Let $V = \mathbf{C}^m$ so that $\mathrm{End}(V) = M_m(\mathbf{C})$, and let $\{E_{ij}\}$ be the standard matrix units. Suppose $\sum_{i,j} E_{ij}\psi(B_{ij})f_h = 0$ with $B_{ij} \in \mathfrak{S}(\mathfrak{p})$. Then $\sum_{i,j} E_{ij}(B'_{ij}h)(0) = 0$ with B'_{ij} defined as in the proof of Proposition 2.1. Taking a suitable h, we see that $(B'_{ij}g)(0) = 0$ for every (i, j) and every $g \in C^\infty(\mathbf{R}^n)$, and so $B_{ij} = 0$. This proves the injectivity.

We can now generalize some well-known facts on $\mathscr{D}(G/K)$ as follows:

PROPOSITION 2.3. *If ρ is one-dimensional, the following assertions hold:*
(1) *$\mathscr{D}(\rho)$ is commutative.*
(2) *ψ gives a* C*-linear bijection of $\mathfrak{S}(\mathfrak{p})^K$ onto $\mathscr{D}(\rho)$.*
(3) *$U^r(\mathfrak{g})^K$ and $\psi(\mathfrak{S}^r(\mathfrak{p})^K)$ have the same image in $\mathscr{D}(\rho)$.*

Proof. The first assertion is implicit in the work of Lepowsky. In fact he proved in [L, Theorem 1.3] that $U(\mathfrak{g})^K/[U(\mathfrak{g})^K \cap U(\mathfrak{g})\mathfrak{N}_\rho]$ is anti-isomorphic to a subalgebra of $U(\mathfrak{a}) \otimes [U(\mathfrak{k})/\mathfrak{N}_\rho]$, where $\mathfrak{a}$ is an abelian subalgebra of $\mathfrak{g}$. This combined with Proposition 2.1 proves (1). Next, if ρ is one-dimensional, we have $I(\rho, \mathfrak{p}) = \mathfrak{S}(\mathfrak{p})^K$, and so (2) follows immediately from Proposition 2.2, and (3) from its proof, since $\mu(T) \in \mathfrak{S}^r(\mathfrak{p})^K$ in view of (2.3).

Let $C(\mathfrak{g})$ denote the center of $U(\mathfrak{g})$. In [H1], Helgason proved that the natural map of $C(\mathfrak{g})$ into $\mathscr{D}(G/K)$ is surjective except when G has a factor belonging to certain exceptional types. A close examination of his proof shows that he proved in fact, at least for classical groups, the following relation:

$$\psi(\mathfrak{I}^r(\mathfrak{g})) + U(\mathfrak{g})\mathfrak{k} \supset \psi\big(\mathfrak{S}^r(\mathfrak{p})^K\big) \quad \text{for every } r, \tag{2.4}$$

where $\mathfrak{I}^r(\mathfrak{g})$ is the set of $\mathrm{Ad}(G)$-invariant elements of $\mathfrak{S}^r(\mathfrak{g})$. We now prove such a surjectivity holds for $\mathscr{D}(\rho)$ if ρ is one-dimensional and G is classical.

THEOREM 2.4. *Suppose* (2.4) *holds for G (which is the case at least for all classical G). Then the natural map of $C(\mathfrak{g})$ into $\mathscr{D}(\rho)$ is surjective for every one-dimensional ρ.*

Proof. In view of Proposition 2.3, it is enough to prove by induction on r that the action of $\psi(\mathfrak{S}^r(\mathfrak{p})^K)$ on $C^\infty(\rho)$ can be obtained from $\psi(\mathfrak{I}^r(\mathfrak{g}))$. Let $B \in \psi(\mathfrak{S}^r(\mathfrak{p})^K)$. By (2.4), there exists an element T of $\psi(\mathfrak{I}^r(\mathfrak{g}))$ such that $T - B \in U(\mathfrak{g})\mathfrak{k}$. Now we have, by [H2, Ch. II, Lemma 4.7],

$$U^r(\mathfrak{g}) = U^{r-1}(\mathfrak{g})\mathfrak{k} \oplus \psi(\mathfrak{S}^r(\mathfrak{p})). \tag{2.5}$$

(This is somewhat different from the statement given in [H2], but it can easily be seen that (2.5) holds.) Therefore we see that $T - B \in U^{r-1}(\mathfrak{g})\mathfrak{k}$. Applying (2.3) to $U^{r-1}(\mathfrak{g})$, we find that $T - B = \sum_i \psi(P_i)Q_i$ with finitely many $P_i \in \mathfrak{S}^{r-1}(\mathfrak{p})$ and $Q_i \in U(\mathfrak{k})$. Put $R = \sum_i \rho_u(Q_i)P_i$ with ρ_u of (2.2). Then $R \in \mathfrak{S}^{r-1}(\mathfrak{p})$ and $T - B$ has the same effect as $\psi(R)$ on $C^\infty(\rho)$. Therefore $\psi(R)$ maps $C^\infty(\rho)$ into itself, so that by (2.1) and Proposition 2.1, $\psi(\mathrm{Ad}(k)R - R) \in U(\mathfrak{g})\mathfrak{N}_\rho$ for every $k \in K$. Put $W = \int_K \mathrm{Ad}(k)R\,dk$. Taking the total measure of K to be 1, we see that $W \in \mathfrak{S}^{r-1}(\mathfrak{p})^K$ and $\psi(W - R) \in U(\mathfrak{g})\mathfrak{N}_\rho$, and hence B has the same effect as $T - \psi(W)$ on $C^\infty(\rho)$. Applying induction to $\psi(W)$, we can complete the proof.

It should be noted that when G is simple, K has a nontrivial one-dimensional representation if and only if G/K is hermitian. In this sense the new points of the above theorem and Proposition 2.3 concern essentially the hermitian case.

3. Invariant differential operators in the hermitian case

Let us now assume that G/K of Section 2 is a hermitian symmetric space of noncompact type. The complex structure of G/K determines the decomposition $\mathfrak{p} = \mathfrak{p}_+ \oplus \mathfrak{p}_-$ with the properties

(3.1) $[\mathfrak{k}, \mathfrak{p}_\pm] \subset \mathfrak{p}_\pm, \qquad [\mathfrak{p}_+, \mathfrak{p}_+] = [\mathfrak{p}_-, \mathfrak{p}_-] = \{0\}, \qquad [\mathfrak{p}_+, \mathfrak{p}_-] = \mathfrak{k}.$

Let $\Phi(X, Y)$ be the Killing form on $\mathfrak{g}$. With an arbitrarily fixed positive number s we put

(3.2) $$\langle X, Y \rangle = s^{-1}\Phi(X, Y) \quad \text{for } X, Y \in \mathfrak{p}.$$

Then we have:

(3.3) *If σ is the complex conjugation of $\mathfrak{g}$ with respect to $\mathfrak{g}_0$, then $\langle X^\sigma, Y \rangle$ for $(X, Y) \in \mathfrak{p} \times \mathfrak{p}$ is a positive definite hermitian form.*

(3.4) $$\langle \mathrm{Ad}(k)X, \mathrm{Ad}(k)Y \rangle = \langle X, Y \rangle \quad \textit{for every } k \in K^c.$$

(3.5) $$\langle \mathfrak{p}_+, \mathfrak{p}_+ \rangle = \langle \mathfrak{p}_-, \mathfrak{p}_- \rangle = \{0\}.$$

These are well-known (or can easily be verified). (If G is simple, the symmetric form $\langle X, Y \rangle$ is characterized by the above three properties up to constant factors.) In Section 5 we shall see that $\langle X, Y \rangle$ of (3.2) has a simple expression in terms of a matrix representation of $\mathfrak{p}$ in each classical case.

From now on we view $\mathfrak{p}$ as its own dual by (3.2); then (3.5) shows that $\mathfrak{p}_+$ and $\mathfrak{p}_-$ are canonically dual to each other. We let K^c act on $S(\mathfrak{p})$ by

(3.6a) $$(kf)(x) = f\left(\mathrm{Ad}(k)^{-1}x\right) \qquad (f \in S(\mathfrak{p}), k \in K^c).$$

Then we easily see that $\omega(kf) = \mathrm{Ad}(k)\omega(f)$ for every $k \in K^c$, where ω is the map of $S(\mathfrak{p})$ into $U(\mathfrak{g})$ defined in Section 1. We denote by τ_r^+ and τ_r^- the restrictions of (3.6a) to $S(\mathfrak{p}_+)$ and $S(\mathfrak{p}_-)$; that is,

$$[\tau_r^{\pm}(k)f](x) = f(\mathrm{Ad}(k^{-1})x) \qquad (f \in S_r(\mathfrak{p}_{\pm}),\ k \in K^c). \tag{3.6b}$$

We now assume that G is classical, and fix an element ε of $\mathfrak{p}_-$, whose explicit description in each case will be given in Section 5, with the property that

$$\mathrm{Ad}(K^c)\varepsilon \text{ is dense in } \mathfrak{p}_-. \tag{3.7}$$

We put then

$$H = \{k \in K^c \mid \mathrm{Ad}(k)\varepsilon = \varepsilon\}, \tag{3.8}$$

$$J_r^H = \{f \in S_r(\mathfrak{p}_+) \mid \tau_r^+(h)f = f \text{ for all } h \in H\}, \tag{3.9a}$$

$$J^H = \sum_{r=0}^{\infty} J_r^H. \tag{3.9b}$$

Proposition 3.1. *If G is classical, the following assertions hold:*

(1) *The multiplicity of each irreducible constituent of τ_r^+ is 1.*

(2) *If Z is a K^c-irreducible subspace of $S_r(\mathfrak{p}_+)$, then $Z \cap J_r^H$ is one-dimensional.*

(3) *If $S_r(\mathfrak{p}_+)$ is the direct sum of K^c-irreducible subspaces $\{Z_\lambda\}_{\lambda=1}^{\mu}$, then J_r^H is the direct sum of $\{J_r^H \cap Z_\lambda\}_{\lambda=1}^{\mu}$. In particular, the number of irreducible constituents of τ_r^+ is equal to $\dim(J_r^H)$.*

(4) *There exist explicitly given algebraically independent elements $g_1, \ldots, g_l$ such that $J^H = \mathbf{C}[g_1, \ldots, g_l]$, where l is the rank of G/K. Moreover g_i is of degree i if G is simple.*

The first assertion is due to Hua [Hu] (for the classical groups) and Schmid [Sc] (in the general case including exceptional ones). The second and third ones are restatements of [S4, Proposition 1.2, (ii) and (iv)]. (J_r^H was written $\mathscr{I}_r$ there.) In fact, we gave in [S4] a proof of (1) for the classical groups by reducing the problem to the structure of J_r^H. The last assertion was also given in each case in the proofs of Theorems 2.A, 2.B, 2.C, and 2.D of [S4]. In Section 5 we shall show that the objects ε, H, and $\tau_r^{\pm}$ are consistent with those defined in [S4], and clarify the last point by presenting explicit forms of the g_i. We treated in [S4] only simple groups, but the case of semisimple groups can easily be reduced to that of simple ones.

Let us now put for simplicity

$$I(\mathfrak{p}) = S(\mathfrak{p})^K, \qquad I_r(\mathfrak{p}) = S_r(\mathfrak{p})^K, \qquad I^r(\mathfrak{p}) = S^r(\mathfrak{p})^K. \tag{3.10}$$

Since $\mathfrak{p}$ is self-dual with respect to (3.2), there is a canonical isomorphism of

$\mathfrak{S}(\mathfrak{p})$ onto $S(\mathfrak{p})$, which commutes with the action of K^c and hence sends $\mathfrak{S}(\mathfrak{p})^K$ onto $I(\mathfrak{p})$. Therefore, by Proposition 2.3, $\mathscr{D}(\rho)$ is given by $\omega(I(\mathfrak{p}))$ if ρ is one-dimensional.

As the spaces $\mathfrak{p}_+$ and $\mathfrak{p}_-$ are dual to each other, so are the spaces $S_r(\mathfrak{p}_+)$ and $S_r(\mathfrak{p}_-)$; hence we have a canonical pairing

$$\langle\ ,\ \rangle : S_r(\mathfrak{p}_+) \times S_r(\mathfrak{p}_-) \to \mathbf{C}, \tag{3.11a}$$

which is given by

$$\langle f, g\rangle = \sum_{(i)} f_*(X_{i_1}, \ldots, X_{i_r}) g_*(Y_{i_1}, \ldots, Y_{i_r}) \qquad (f \in S_r(\mathfrak{p}_+),\ g \in S_r(\mathfrak{p}_-)), \tag{3.11b}$$

where $\{X_i\}$ and $\{Y_i\}$ are dual bases of $\mathfrak{p}_+$ and $\mathfrak{p}_-$, and $(i_1, \ldots, i_r)$ runs over $\{1, \ldots, \lambda\}^r$, $\lambda = \dim(\mathfrak{p}_+)$. This is a special case of (1.3). Also as a special case of (1.4) we note

$$\omega(\xi) = \sum_{(i)} \xi_*(Y_{i_1}, \ldots, Y_{i_r}) X_{i_1} \cdots X_{i_r} \quad \text{for } \xi \in S_r(\mathfrak{p}_-). \tag{3.11c}$$

From (3.11b) and (3.4) we can easily derive

$$\langle \tau_r^+(k) f, \tau_r^-(k) g\rangle = \langle f, g\rangle \qquad (k \in K^c). \tag{3.12}$$

Given a K^c-irreducible subspace Z of $S_r(\mathfrak{p}_+)$, we can find a unique K^c-irreducible subspace W of $S_r(\mathfrak{p}_-)$ such that $S_r(\mathfrak{p}_-)$ is the direct sum of W and the annihilator of Z. Then Z and W are dual to each other with respect to (3.11a). Take bases $\{\zeta_1, \ldots, \zeta_\kappa\}$ of Z and $\{\xi_1, \ldots, \xi_\kappa\}$ of W that are dual to each other and put

$$f_Z(x, y) = \sum_{\nu=1}^{\kappa} \zeta_\nu(x) \xi_\nu(y) \qquad (x \in \mathfrak{p}_+,\ y \in \mathfrak{p}_-), \tag{3.13a}$$

$$g_Z(x) = f_Z(x, \varepsilon) \qquad (x \in \mathfrak{p}_+). \tag{3.13b}$$

We can easily see that f_Z belongs to $I_{2r}(\mathfrak{p})$ and is independent of the choice of dual bases $\{\zeta_\nu\}$ and $\{\xi_\nu\}$.

LEMMA 3.2. (1) $\mathbf{C} g_Z = Z \cap J_r^H$.

(2) *The* g_Z *for all* K^c*-irreducible* $Z \subset S_r(\mathfrak{p}_+)$ *form a basis of* J_r^H.

(3) *For* $f \in S(\mathfrak{p})$, *define* $f^+ \in S(\mathfrak{p}_+)$ *by* $f^+(x) = f(x, \varepsilon)$ *for* $x \in \mathfrak{p}_+$. *Then* $f \mapsto f^+$ *gives a ring-isomorphism of* $I(\mathfrak{p})$ *onto* J^H.

Proof. That $f^+ \in J^H$ if $f \in I(\mathfrak{p})$ can be easily seen. For $k \in K^c$ and $f \in I(\mathfrak{p})$, we have $f(x, k^{-1}\varepsilon) = f(kx, \varepsilon) = f^+(kx)$, where we write for simplicity kx for $\mathrm{Ad}(k)x$. Therefore (3.7) shows that f is determined by f^+; hence our map of (3) is injective. In particular we see that $0 \neq g_Z = (f_Z)^+ \in J^H$. Since $g_Z =$

$\sum_\nu \xi_\nu(\varepsilon)\zeta_\nu \in Z$, we obtain (1) in view of Proposition 3.1(2). Then (2) follows from Proposition 3.1(3). Since $J^H = \sum_{r=0}^\infty J_r^H$ and $g_Z = (f_Z)^+$, the map of (3) is surjective.

We can now determine $I(\mathfrak{p})$ as follows:

THEOREM 3.3. *If G is classical, the following assertions hold:*

(1) $I(\mathfrak{p}) = \sum_{r=0}^\infty I_{2r}(\mathfrak{p})$; *that is,* $I_t(\mathfrak{p}) = 0$ *for odd t.*

(2) *The* f_Z *for all* K^c*-irreducible subspaces* Z *of* $S_r(\mathfrak{p}_+)$ *form a* C*-linear basis of* $I_{2r}(\mathfrak{p})$.

(3) *There exists a set* $\{Z_1, \ldots, Z_l\}$ *with a* K^c*-irreducible subspace* Z_i *of* $S(\mathfrak{p}_+)$ *such that* $I(\mathfrak{p}) = \mathbf{C}[f_{Z_1}, \ldots, f_{Z_l}]$, *where* $l = \mathrm{rank}(G/K)$. *Moreover* $f_{Z_1}, \ldots, f_{Z_l}$ *are algebraically independent. If G is simple,* $Z_i \subset S_i(\mathfrak{p}_+)$.

Proof. Let $f \in I(\mathfrak{p})$. By Lemma 3.2 and Proposition 3.1, f^+ is a finite linear combination of the g_Z. Since $f \mapsto f^+$ is an isomorphism, we see that f is a finite linear combination of the f_Z. Therefore the f_Z form a basis of $I(\mathfrak{p})$. This proves (1) and (2), since $f_Z \in I_{2r}(\mathfrak{p})$ if $Z \subset S_r(\mathfrak{p}_+)$. Take the g_i as in Proposition 3.1(4). As will be shown in Section 5, g_i can be given as g_{Z_i} with $Z_i \subset S_i(\mathfrak{p}_+)$ when G is simple. Then (3) follows from Lemma 3.2(3) and an obvious technique of decomposing $\mathfrak{g}$ into simple factors.

We now consider the set $C^\infty(\rho)$ of vector-valued functions and the ring of operators $\mathscr{D}(\rho)$ defined in Section 2 for a representation ρ of K.

THEOREM 3.4. *For every* $f \in I(\mathfrak{p})$, *let* $\Omega(f)$ *denote the element of* $\mathscr{D}(\rho)$ *represented by* $\omega(f)$. *Suppose that* ρ *is one-dimensional. Then the* $\Omega(f_Z)$ *for all* K^c*-irreducible* Z *as above form a* C*-linear basis of* $\mathscr{D}(\rho)$; *each* $\Omega(f_Z)$ *is a symmetric operator (with respect to a certain hermitian inner product). Moreover* $\mathscr{D}(\rho)$ *is the polynomial ring* $\mathbf{C}[\Omega(f_{Z_1}), \ldots, \Omega(f_{Z_l})]$ *generated by l algebraically independent elements* $\Omega(f_{Z_i})$ *with* Z_i *as in Theorem* 3.3(3).

Proof. The first assertion is an immediate consequence of Proposition 2.3 and Theorem 3.3 since $\omega(I(\mathfrak{p})) = \psi(\mathfrak{S}(\mathfrak{p})^K)$. By Proposition 2.3, we know that $\mathscr{D}(\rho)$ is commutative for one-dimensional ρ. Let us now prove by induction on r that $\Omega(f) \in \mathbf{C}[\Omega(f_{Z_1}), \ldots, \Omega(f_{Z_l})]$ for every $f \in I^r(\mathfrak{p})$. We may assume that $f \notin I^{r-1}(\mathfrak{p})$. By Theorem 3.3(3), $f = P(f_{Z_1}, \ldots, f_{Z_l})$ with a polynomial P. Since the f_{Z_i} are algebraically independent, no terms of degree greater than r can occur on the right-hand side. Therefore from (1.5) we obtain $\omega(f) - P(\omega(f_{Z_1}), \ldots, \omega(f_{Z_l})) \in U^{r-1}(\mathfrak{g})$. Since the difference is K-invariant, its action on $C^\infty(\rho)$ can be given by $\omega(h)$ with some $h \in I^{r-1}(\mathfrak{p})$ by virtue of Proposition 2.3(3). Applying induction to h, we obtain the desired result on $\Omega(f)$. To prove the algebraic independence, suppose $P(\omega(f_{Z_1}), \ldots, \omega(f_{Z_l})) = 0$ with a nonzero polynomial P. Put $f = P(f_{Z_1}, \ldots, f_{Z_l})$. By Theorem 3.3(3), $f \neq 0$. Let r be the degree of f. Then the above reasoning shows that $\Omega(f) = \Omega(h)$ with $h \in I^{r-1}(\mathfrak{p})$, which is a contradiction since $f \mapsto \Omega(f)$ is injective by Proposition 2.3. The symmetricity will be shown in Corollary 4.4.

We note here an explicit form of $\omega(f_Z)$.

PROPOSITION 3.5. *Let $\{X_1, \dots, X_\lambda\}$ be a basis of $\mathfrak{p}_+$ and $\{Y_1, \dots, Y_\lambda\}$ the basis of $\mathfrak{p}_-$ dual to it. Put $X_{\lambda+i} = Y_i$ and $Y_{\lambda+i} = X_i$ for $i = 1, \dots, \lambda$. Let Z and W be K^c-irreducible dual subspaces of $S_r(\mathfrak{p}_+)$ and $S_r(\mathfrak{p}_-)$ in the sense described above, and let $\{\zeta_\nu\}_{\nu=1}^{\kappa}$ and $\{\xi_\nu\}_{\nu=1}^{\kappa}$ be their dual bases. Then*

$$(3.14)\quad \omega(f_Z) = [1/(2r)!] \sum_{\nu=1}^{\kappa} \sum_{(i)} \sum_{\pi} (\zeta_\nu)_*(X_{i_1}, \dots, X_{i_r})(\xi_\nu)_*(X_{i_{r+1}}, \dots, X_{i_{2r}}) Y_{i_{\pi(1)}} \cdots Y_{i_{\pi(2r)}},$$

where π runs over all permutations of $\{1, \dots, 2r\}$, $(i_1, \dots, i_r)$ over $\{1, \dots, \lambda\}^r$, and $(i_{r+1}, \dots, i_{2r})$ over $\{\lambda + 1, \dots, 2\lambda\}^r$.

Proof. Let $g \in S_r(\mathfrak{p}_+)$ and $h \in S_r(\mathfrak{p}_-)$. By (1.4) we have

$$\omega(g \otimes h) = \sum (g \otimes h)_*(X_{i_1}, \dots, X_{i_{2r}}) Y_{i_1} \cdots Y_{i_{2r}},$$

where $\{i_1, \dots, i_{2r}\}$ runs over $\{1, \dots, 2\lambda\}^{2r}$. Applying this to $\zeta_\nu \otimes \xi_\nu$ and employing (1.6), we obtain our formula.

We can now state the principal result of this paper as follows:

THEOREM 3.6. *Suppose G is classical. The notation being as in Proposition 3.5, define two elements L_Z and M_Z of $U(\mathfrak{g})$ by*

$$(3.15)\qquad L_Z = \sum_{\nu=1}^{\kappa} \omega(\zeta_\nu)\omega(\xi_\nu), \qquad M_Z = \sum_{\nu=1}^{\kappa} \omega(\xi_\nu)\omega(\zeta_\nu).$$

Then both L_Z and M_Z belong to $U(\mathfrak{g})^K$. Furthermore let $\mathscr{L}_Z$ and $\mathscr{M}_Z$ denote the elements of $\mathscr{D}(\rho)$ represented by L_Z and M_Z, respectively. Then

(1) $(-1)^r\mathscr{L}_Z$ and $(-1)^r\mathscr{M}_Z$ are nonnegative symmetric operators (with respect to a certain hermitian inner product).

Moreover, if ρ is one-dimensional, the following assertions hold:

(2) The $\mathscr{L}_Z$ for all K^c-irreducible Z as above form a **C***-linear basis of $\mathscr{D}(\rho)$.*

(3) $\mathscr{D}(\rho) = \mathbf{C}[\mathscr{L}_{Z_1}, \dots, \mathscr{L}_{Z_l}]$ with $Z_1, \dots, Z_l$ of Theorem 3.3(3); moreover $\mathscr{L}_{Z_1}, \dots, \mathscr{L}_{Z_l}$ are algebraically independent.

(4) Assertions (2) and (3) hold with $\mathscr{M}$ instead of $\mathscr{L}$.

(5) $\mathscr{L}_Z = \mathscr{M}_Z$ for every Z if ρ is trivial.

Proof. We first note, as an immediate consequence of (3.11c),

$$L_Z = \sum_{\nu=1}^{\kappa} \sum_{(i,j)} (\zeta_\nu)_*(X_{i_1},\ldots,X_{i_r})(\xi_\nu)_*(Y_{j_1},\ldots,Y_{j_r})Y_{i_1}\cdots Y_{i_r}X_{j_1}\cdots X_{j_r},$$

$$M_Z = \sum_{\nu=1}^{\kappa} \sum_{(i,j)} (\zeta_\nu)_*(X_{i_1},\ldots,X_{i_r})(\xi_\nu)_*(Y_{j_1},\ldots,Y_{j_r})X_{j_1}\cdots X_{j_r}Y_{i_1}\cdots Y_{i_r},$$

where $(i_1,\ldots,i_r,j_1,\ldots,j_r)$ runs over $\{1,\ldots,\lambda\}^{2r}$. We prove only the assertions concerning L_Z and $\mathscr{L}_Z$, since M_Z and $\mathscr{M}_Z$ can be similarly treated. The K^c-invariance of L_Z can be verified in a straightforward way by means of (3.12). Now each term of (3.14) can be written in the form

$$(\zeta_\nu)_*(X_{i_1},\ldots,X_{i_r})(\xi_\nu)_*(Y_{j_1},\ldots,Y_{j_r})W_1\cdots W_{2r},$$

where $(i_1,\ldots,i_r,j_1,\ldots,j_r)\in\{1,\ldots,\lambda\}^{2r}$ and $(W_1,\ldots,W_{2r})$ is a permutation of $(Y_{i_1},\ldots,Y_{i_r},X_{j_1},\ldots,X_{j_r})$. We easily see that

$$W_1\cdots W_{2r} - Y_{i_1}\cdots Y_{i_r}X_{j_1}\cdots X_{j_r}\in U^{2r-1}(\mathfrak{g}),$$

and hence $\omega(f_Z)-L_Z\in U^{2r-1}(\mathfrak{g})^K$. By Proposition 2.3(3) this shows that $\Omega(f_Z)-\mathscr{L}_Z=\Omega(h)$ with $h\in I^{2r-1}(\mathfrak{p})$. By induction on r, we see that the $\mathscr{L}_Z$ for all Z form a $\mathbf{C}$-basis of $\mathscr{D}(\rho)$. To prove (3), we denote by $\mathscr{D}_t$ (resp. $\mathscr{D}_t'$) for $0\le t\in\mathbf{Z}$ the ring generated by the $\Omega(f_{Z_i})$ (resp. the $\mathscr{L}_{Z_i}$) for all Z_i $(1\le i\le l)$ contained in $S^t(\mathfrak{p}_+)$. Repeating the proof of Theorem 3.4, we can prove by induction on t that $\Omega(f)\in\mathscr{D}_t$ if $f\in I^{2t}(\mathfrak{p})$. Taking f to be the above h, we see that

$$\Omega(f_Z)-\mathscr{L}_Z\in\mathscr{D}_{r-1}\quad\text{if } Z\subset S_r(\mathfrak{p}_+). \tag{3.16}$$

By induction on r we find that $\mathscr{D}_r=\mathscr{D}_r'$, which together with Theorem 3.4 proves (3). Assertions (1) and (5) will be proved in Section 4.

The expressions of L_Z and M_Z in (3.15) suggest some analogues of Propositions 2.2 and 2.3. We first decompose $S(\mathfrak{p})$ into the tensor product $S(\mathfrak{p}_+)\otimes S(\mathfrak{p}_-)$ and define two linear maps $\omega_\pm$: $S(\mathfrak{p})\to U(\mathfrak{g})$ by $\omega_+(\alpha\beta)=\omega(\alpha)\omega(\beta)$ and $\omega_-(\alpha\beta)=\omega(\beta)\omega(\alpha)$ for $\alpha\in S(\mathfrak{p}_+)$ and $\beta\in S(\mathfrak{p}_-)$. Clearly $\omega_\pm$ commutes with $\mathrm{Ad}(k)$ for every $k\in K^c$. Next, for $F=\sum_i(E_i\otimes B_i)$ in $\mathrm{End}(V)\otimes S(\mathfrak{p})$ define $\Psi_\pm(F)$ to be the operators on $C^\infty(G,V)$ represented by $\sum_i E_i\omega_\pm(B_i)$. In this setting we have:

Proposition 3.7. *The two maps $\Psi_\pm$ are $\mathbf{C}$-linear bijections of $I(\rho,\mathfrak{p})$ onto $\mathscr{D}(\rho)$ for every irreducible ρ. If ρ is one-dimensional, $\omega_\pm$ give $\mathbf{C}$-linear bijections of $I(\mathfrak{p})$ onto $\mathscr{D}(\rho)$, and moreover $U^r(\mathfrak{g})^K$, $\omega_+(I^r(\mathfrak{p}))$, and $\omega_-(I^r(\mathfrak{p}))$ have the same image in $\mathscr{D}(\rho)$.*

Proof. That $I(\rho, \mathfrak{p})$ is mapped into $\mathscr{D}(\rho)$ by $\Psi_{\pm}$ can be verified in a straightforward way. Let $I^r(\rho, \mathfrak{p})$ denote the set of elements $\sum_i E_i \otimes B_i$ of $I(\rho, \mathfrak{p})$ with $E_i \in \mathrm{End}(V)$ and $B_i \in S^r(\mathfrak{p})$. To show that $\Psi_+(I^r(\rho, \mathfrak{p})) = \Psi(I^r(\rho, \mathfrak{p}))$ with Ψ of Proposition 2.2, take $A \in I^r(\rho, \mathfrak{p})$. Then $A = \sum_i E_i \otimes B_i$ with $E_i \in \mathrm{End}(V)$ and $B_i = Y_i^m X_i^n$ with $X_i \in \mathfrak{p}_+$, $Y_i \in \mathfrak{p}_-$, $m + n \leq r$. Identify $\mathfrak{p}_{\pm}$ with $S_1(\mathfrak{p}_{\mp})$. Then

$$\Psi(A) - \Psi_+(A) = \sum E_i[\psi(Y_i^m X_i^n) - Y_i^m X_i^n] \in \mathrm{End}(V)U^{r-2}(\mathfrak{g})U(\mathfrak{k}).$$

Since $\mathrm{End}(V) = \rho_u(U(\mathfrak{k}))$, there is an element D of $\omega(S^{r-2}(\mathfrak{p}))U(\mathfrak{k})$ such that D has the same effect as $\Psi(A) - \Psi_+(A)$ on $C^\infty(\rho)$. By Proposition 2.1, there is an element D' of $U(\mathfrak{g})^K$ such that $D - D' \in U(\mathfrak{g})\mathfrak{N}_\rho$. Let $F = \int_K \mathrm{Ad}(k)D\,dk$. Then F belongs to $U(\mathfrak{g})^K \cap \omega(S^{r-2}(\mathfrak{p}))U(\mathfrak{k})$ and has the same effect as $\Psi(A) - \Psi_+(A)$ on $C^\infty(\rho)$. Let $H = \mu(F)$ with the map μ in the proof of Proposition 2.2. Then $H \in I^{r-2}(\rho, \mathfrak{p})$, and $\Psi(H)$ gives the effect of F. Thus $\Psi(A) - \Psi_+(A) \in \Psi(I^{r-2}(\rho, \mathfrak{p}))$. By induction on r, we obtain the surjectivity of Ψ_+ and similarly that of Ψ_-. If $A \neq 0$, taking the smallest r, we see that $\Psi_+(A) \neq 0$, which proves the injectivity. If ρ is one-dimensional, we have $I^r(\rho, \mathfrak{p}) = I^r(\mathfrak{p})$. Therefore we obtain the desired results from Proposition 2.3(3) and what we have just proved.

We have $L_Z = \omega_+(f_Z)$ and $M_Z = \omega_-(f_Z)$, and therefore an alternative proof of Theorem 3.6(2), (3), (4) can be given by repeating the proof of Theorem 3.4 with $\omega_{\pm}$ in place of ω without using formula (3.14). The maps $\omega_{\pm}$ seem more natural than ω or ψ when G/K is hermitian.

Remark 3.8. The pairing (3.2) depends on the choice of s. Once s is fixed, the objects f_Z, L_Z, and M_Z do not depend on the choice of dual bases of Z and W. The change of s multiplies them only by positive real numbers. If G is not simple, obviously there are more choices for the pairing.

Remark 3.9. Two easy facts may be worth noting when G is simple. First, we have $Z_1 = S_1(\mathfrak{p}_+)$, which follows from Proposition 3.1. Next, if C denotes the Casimir element of $U(\mathfrak{g})$, then for every irreducible ρ, both $\mathscr{L}_{Z_1}$ and $\mathscr{M}_{Z_1}$ are given by $aC + b$ with constants a and b. To see this, take $s = 1$ in (3.2). From the definition of C we see that $C = L_{Z_1} + M_{Z_1} + N$ with an element N of $U^2(\mathfrak{k})$, and (3.15) implies that $L_{Z_1} - M_{Z_1} = N'$ with an element N' of $\mathfrak{k}$. Clearly both N and N' belong to $U(\mathfrak{k})^K$; hence their images under ρ_u commute with every element of $\rho(K)$, and so must be constants. This proves the expected fact. As a consequence, we note that $\mathscr{L}_{Z_1}$ and $\mathscr{M}_{Z_1}$ belong to the center of $\mathscr{D}(\rho)$. In particular, if ρ is trivial, they are constant multiples of the Laplace-Beltrami operator, since N and N' give the zero operator.

Remark 3.10. Theorem 3.3(1) can be generalized and proved more directly. In fact, for every irreducible ρ we have

$$I(\rho, \mathfrak{p}) \subset \mathrm{End}(V) \otimes \left[\sum_{r=0}^{\infty} S_r(\mathfrak{p}_+) S_r(\mathfrak{p}_-)\right]. \tag{3.17}$$

This can easily be shown by considering the action of the center of K on $S_r(\mathfrak{p}_+)$ and $S_r(\mathfrak{p}_-)$.

4. The operators D^Z, E^W, and their adjoints

Given G as in Section 3 and a complex vector space V of finite dimension, for $0 \le r \in \mathbf{Z}$ and $f \in C^\infty(G, V)$ we define an $S_r(\mathfrak{p}_+, V)$-valued function $D^r f$ and an $S_r(\mathfrak{p}_-, V)$-valued function $E^r f$ by

$$(D^r f)(X) = X^r f \qquad (X \in \mathfrak{p}_+), \tag{4.1a}$$

$$(E^r f)(Y) = Y^r f \qquad (Y \in \mathfrak{p}_-). \tag{4.1b}$$

These are equivalent to

$$(D^r f)_*(A_1, \ldots, A_r) = A_1 \cdots A_r f \qquad (A_i \in \mathfrak{p}_+), \tag{4.2a}$$

$$(E^r f)_*(B_1, \ldots, B_r) = B_1 \cdots B_r f \qquad (B_i \in \mathfrak{p}_-), \tag{4.2b}$$

where F_* is the symmetric multilinear function attached to F by (1.1).

If V' is another finite-dimensional complex vector space, we shall identify $S_r(V', V)$ with $S_r(V') \otimes V$ by putting

$$(h \otimes v)(u) = h(u)v \qquad (h \in S_r(V'), v \in V, u \in V'). \tag{4.3}$$

Then we may view $D^r f$ and $E^r f$ as functions with values in $S_r(\mathfrak{p}_+) \otimes V$ and $S_r(\mathfrak{p}_-) \otimes V$.

Let Z be a K^c-irreducible subspace of $S_r(\mathfrak{p}_+)$, and W the subspace of $S_r(\mathfrak{p}_-)$ dual to Z with respect to (3.11a) as considered in Section 3. We have $S_r(\mathfrak{p}_+) = Z \oplus M$ with a K^c-stable subspace M which is the annihilator of W, so that there is a canonical projection map φ_Z of $S_r(\mathfrak{p}_+)$ into Z with kernel M. Similarly φ_W: $S_r(\mathfrak{p}_-) \to W$ can be defined. We use the same symbols φ_Z and φ_W for the projection maps of $S_r(\mathfrak{p}_\pm) \otimes V$ to $Z \otimes V$ and $W \otimes V$ which are the identity map on V. We then define a $(Z \otimes V)$-valued function $D^Z f$ and a $(W \otimes V)$-valued function $E^W f$ by

$$D^Z f = \varphi_Z D^r f, \qquad E^W f = \varphi_W E^r f. \tag{4.4}$$

Now $Z \otimes V$ can be identified with $S_1(W, V)$ by the rule

$$(\zeta \otimes v)(\xi) = \langle \zeta, \xi \rangle v \qquad (\zeta \in Z, \xi \in W, v \in V), \tag{4.5}$$

where $\langle\ ,\ \rangle$ is the pairing of (3.11a). This is a special case of (4.3). Similarly

$W \otimes V$ can be identified with $S_1(Z, V)$. Then $D^Z f$ and $E^W f$ have values in $S_1(W, V)$ and $S_1(Z, V)$. With dual bases $\{X_i\}_{i=1}^{\lambda}$ and $\{Y_i\}_{i=1}^{\lambda}$ of $\mathfrak{p}_+$ and $\mathfrak{p}_-$, these functions are given by

$$(4.6a)\qquad (D^Z f)(\xi) = \sum_{(i)} \xi_*(Y_{i_1}, \ldots, Y_{i_r}) X_{i_1} \cdots X_{i_r} f \qquad (\xi \in W),$$

$$(4.6b)\qquad (E^W f)(\zeta) = \sum_{(i)} \zeta_*(X_{i_1}, \ldots, X_{i_r}) Y_{i_1} \cdots Y_{i_r} f \qquad (\zeta \in Z),$$

where $(i_1, \ldots, i_r)$ runs over $\{1, \ldots, \lambda\}^r$. To prove these, it is sufficient to consider the case $V = \mathbf{C}$, since the general case is merely the arrangement of several copies of this special case. If $V = \mathbf{C}$, a value of the function $D^r f$ as an element of the dual of $S_r(\mathfrak{p}_-)$, when restricted to W, gives a value of $D^Z f$ as an element of $S_1(W)$. Thus for $\xi \in W$ we have

$$(D^Z f)(\xi) = \langle D^r f, \xi \rangle = \sum (D^r f)_*(X_{i_1}, \ldots, X_{i_r}) \xi_*(Y_{i_1}, \ldots, Y_{i_r})$$

by (3.11b). This combined with (4.2a) proves (4.6a). The other formula can be proved in the same way. In view of (3.11c), we can express (4.6a, b) in a simpler form:

$$(4.7)\quad (D^Z f)(\xi) = \omega(\xi) f, \qquad (E^W f)(\zeta) = \omega(\zeta) f \qquad (\xi \in W, \zeta \in Z).$$

We now define a contraction operator θ as follows:

$$(4.8)\qquad \theta\colon S_1(Z, S_1(W, V)) \to V,$$

$$\theta p = \sum_{\nu=1}^{\kappa} p(\zeta_\nu, \xi_\nu) \qquad (p \in S_1(Z, S_1(W, V))).$$

Here $\{\zeta_\nu\}$ and $\{\xi_\nu\}$ are dual bases of Z and W. Clearly θ is independent of the choice of such bases. We use the same letter θ for the map of $S_1(W, S_1(Z, V))$ into V which is defined in the same way. Then for a function h on G with values in $S_1(Z, V)$, resp. $S_1(W, V)$, we can define $\theta D^Z h$, resp. $\theta E^W h$, to be V-valued functions. From (4.7) we obtain their explicit forms as follows:

$$(4.9)\qquad \theta D^Z h = \sum_{\nu=1}^{\kappa} \omega(\xi_\nu)[h(\zeta_\nu)], \qquad \theta E^W h = \sum_{\nu=1}^{\kappa} \omega(\zeta_\nu)[h(\xi_\nu)].$$

Now for a V-valued f, $D^Z f$ has values in $S_1(W, V)$; hence $\theta E^W D^Z f$ is meaningful and so is $\theta D^Z E^W f$. From (4.7) and (4.9) we obtain:

Proposition 4.1. *Let L_Z and M_Z be defined by* (3.15). *Then $L_Z = \theta E^W D^Z$ and $M_Z = \theta D^Z E^W$.*

Let us now assume that V is equipped with an inner product $[u, v]_V$, which we take to be $\mathbf{C}$-antilinear in u and $\mathbf{C}$-linear in v. (In order to distinguish a hermitian form from a C-bilinear form, we denote the former by square brackets.) We denote by $C_c^\infty(G, V)$ the subset of $C^\infty(G, V)$ consisting of the functions of compact support. For f and g in $C_c^\infty(G, V)$ we put

$$[f, g]^V = \int_G [f(x), g(x)]_V \, dx, \tag{4.10}$$

where dx is a fixed Haar measure on G. We are going to define canonical inner products on $Z \otimes V$ and $W \otimes V$ (in particular on Z and W). First, let $X \mapsto X^\sigma$ denote the complex conjugation of $\mathfrak{g}$ with respect to $\mathfrak{g}_0$. This interchanges $\mathfrak{p}_+$ and $\mathfrak{p}_-$, and commutes with $\mathrm{Ad}(k)$ for every $k \in K$. For $h \in S_r(\mathfrak{p})$, define $h' \in S_r(\mathfrak{p})$ by

$$h'(X) = \overline{h(X^\sigma)} \qquad (X \in \mathfrak{p}). \tag{4.11}$$

Then $h \mapsto h'$ interchanges $S_r(\mathfrak{p}_+)$ and $S_r(\mathfrak{p}_-)$; moreover we have

$$(\tau_r^{\pm}(k)h)' = \tau_r^{\mp}(k)h' \qquad (k \in K). \tag{4.12}$$

Now define an $\mathbf{R}$-bilinear form $[\ ,\ ]_r$ on $S_r(\mathfrak{p})$ by

$$[\alpha, \beta]_r = \langle \alpha', \beta \rangle \qquad (\alpha, \beta \in S_r(\mathfrak{p})). \tag{4.13}$$

This is K-invariant, hermitian, and positive definite. The K-invariance is obvious. To see the latter properties, take a basis $\{A_i\}$ of $\mathfrak{p}$ over $\mathbf{C}$ so that $\langle A_i^\sigma, A_j \rangle = \delta_{ij}$, which is possible because of (3.3). Then $\{A_i^\sigma\}$ and $\{A_i\}$ are dual bases of $S_r(\mathfrak{p})$. Therefore

$$\langle \alpha', \beta \rangle = \sum_{(i)} \overline{\alpha_*(A_{i_1}, \ldots, A_{i_r})}\beta_*(A_{i_1}, \ldots, A_{i_r}),$$

from which we obtain the expected properties of our bilinear form.

LEMMA 4.2. *Let $\{\zeta_\nu\}_{\nu=1}^\kappa$ and $\{\xi_\nu\}_{\nu=1}^\kappa$ be dual bases of Z and W. Then the following assertions hold:*

(1) *The map $h \mapsto h'$ interchanges Z and W.*

(2) *$\{\xi_\nu'\}$ and $\{\zeta_\nu'\}$ are dual bases of Z and W.*

(3) *Define $\mathbf{R}$-bilinear forms on $Z \otimes V$ and $W \otimes V$ by*

$$[p, q]_{Z \otimes V} = \sum_{\nu=1}^{\kappa} [p(\xi_\nu), q(\zeta_\nu')]_V \quad \textit{for } p, q \in Z \otimes V,$$

$$[u, v]_{W \otimes V} = \sum_{\nu=1}^{\kappa} [u(\zeta_\nu), v(\xi_\nu')]_V \quad \textit{for } u, v \in W \otimes V.$$

Then these are hermitian, positive definite, and independent of the choice of $\{\zeta_\nu\}$ and $\{\xi_\nu\}$.

Proof. Let $S_r(\mathfrak{p}_+) = Z \oplus M$ with the annihilator M of W. The orthogonal complement of Z in $S_r(\mathfrak{p}_+)$ with respect to (4.13), being K-stable, must coincide with M. This means that $\langle \alpha', \beta \rangle = 0$ for every $\alpha \in Z$ and $\beta \in M$, from which we obtain (1). If $\{X_i\}$ and $\{Y_i\}$ are dual bases of $\mathfrak{p}_+$ and $\mathfrak{p}_-$, then so are $\{Y_i^\sigma\}$ and $\{X_i^\sigma\}$. This fact combined with (3.11b) proves (2). To prove (3), we may assume that $V = \mathbf{C}$, since $Z \otimes V$ is essentially the product of copies of Z. Thus

$$[p, q]_Z = \sum_{\nu=1}^{\kappa} \overline{\langle p, \xi_\nu \rangle} \langle q, \zeta_\nu' \rangle .$$

Since $p = \Sigma_\nu \langle p, \xi_\nu \rangle \zeta_\nu$ and $q = \Sigma_\nu \langle q, \zeta_\nu' \rangle \xi_\nu'$, we have $p' = \Sigma_\nu \overline{\langle p, \xi_\nu' \rangle} \zeta_\nu'$ and so $[p, q]_Z = \langle p', q \rangle$ in view of (2). Therefore $[p, q]_Z$ is the restriction of (4.13) to Z. This proves (3), since the case of $[u, v]_W$ can similarly be treated.

THEOREM 4.3. *The sets* $\{(-1)^r \theta E^W, D^Z\}$ *and* $\{(-1)^r \theta D^Z, E^W\}$ *are pairs of mutually adjoint operators in the sense that*

$$[D^Z f, g]^{Z \otimes V} = [f, (-1)^r \theta E^W g]^V$$

$$\textit{for } f \in C_c^\infty(G, V) \textit{ and } g \in C_c^\infty(G, Z \otimes V),$$

$$[E^W f, g]^{W \otimes V} = [f, (-1)^r \theta D^Z g]^V$$

$$\textit{for } f \in C_c^\infty(G, V) \textit{ and } g \in C_c^\infty(G, W \otimes V).$$

Proof. We first recall a well-known fact that

(4.14a) $$[Xp, q]^V = [p, -X^\sigma q]^V$$

for $X \in \mathfrak{g}$ and $p, q \in C_c^\infty(G, V)$, where σ is the same as in (4.11). If $\{A_i\}$ is a self-dual basis of $\mathfrak{p}$, so is $\{A_i^\sigma\}$. From this and (1.4) we can easily derive

(4.14b) $$[\omega(\alpha)p, q]^V = [p, (-1)^r \omega(\alpha')q]^V$$

for every $\alpha \in S_r(\mathfrak{p})$. Therefore by Lemma 4.2 and (4.7) we have

$$[D^Z f, g]^{Z \otimes V} = \sum_{\nu=1}^{\kappa} [(D^Z f)(\xi_\nu), g(\zeta_\nu')]^V$$

$$= \sum_\nu [\omega(\xi_\nu) f, g(\zeta_\nu')]^V = \left[f, (-1)^r \sum_\nu \omega(\xi_\nu')[g(\zeta_\nu')] \right]^V .$$

By (4.9) the last sum is $\theta E^W g$, since $\{\xi_\nu'\}$ and $\{\zeta_\nu'\}$ are dual bases by Lemma

4.2(1). This proves the first equality of our theorem. The same type of argument applies to the second one.

Now (1) of Theorem 3.6 is included in the following result:

COROLLARY 4.4. *Let* f_Z, L_Z, *and* M_Z *be as in* (3.13a) *and Theorem* 3.6. *Then* $\omega(f_Z)$, $(-1)^r L_Z$, *and* $(-1)^r M_Z$ *represent symmetric operators on* $C_c^\infty(G, V)$. *Moreover the latter two are nonnegative in the sense that*

$$\left[f, (-1)^r L_Z f\right]^V \geq 0 \quad \text{and} \quad \left[f, (-1)^r M_Z f\right]^V \geq 0$$

for every $f \in C_c^\infty(G, V)$.

Proof. Combining Proposition 4.1 with the above theorem, we obtain

$$\text{(4.15)} \quad \left[f, (-1)^r L_Z h\right]^V = \left[f, (-1)^r \theta E^W D^Z h\right]^V = \left[D^Z f, D^Z h\right]^{Z \otimes V}$$

for $f, h \in C_c^\infty(G, V)$. Hence L_Z gives a symmetric operator. Taking $f = h$, we find that it is nonnegative. The case of M_Z can be handled in the same way. The assertion concerning f_Z follows from (4.14b), since we can easily derive from Lemma 4.2(2) that $(f_Z)' = f_Z$.

Proof of Theorem 3.6(5). Extend the map $X \mapsto X^\sigma$ to an **R**-linear automorphism of $U(\mathfrak{g})$ and denote it also by σ. From (1.4) we easily see that $\omega(\xi)^\sigma = \omega(\xi')$ for every $\xi \in S(\mathfrak{p}_\pm)$, and so $(L_Z)^\sigma = M_Z$ by Lemma 4.2(2). Assuming ρ to be trivial and fixing $Z \subset S_r(\mathfrak{p}_+)$, let us prove by induction on r that $\mathscr{L}_Z = \mathscr{M}_Z$. We first observe that $\mathscr{L}_Z - \mathscr{M}_Z = P(\mathscr{L}_{Z_1}, \ldots, \mathscr{L}_{Z_l})$ with a polynomial P, where we arrange the Z_i and take $t \leq l$ so that $Z_i \subset S^{r-1}(\mathfrak{p}_+)$ if and only if $i \leq t$. This follows from (3.16) and its analogue with $\mathscr{M}_Z$ in place of $\mathscr{L}_Z$, since $\mathscr{D}_i = \mathscr{D}_i'$ as shown in the paragraph containing (3.16). In particular $\mathscr{L}_{Z_1} = \mathscr{M}_{Z_1}$ as observed in Remark 3.9. The symmetricity implies that P is a real polynomial. Now we have

$$L_Z - M_Z = P(L_{Z_1}, \ldots, L_{Z_t}) + N$$

with $N \in U(\mathfrak{g})\mathfrak{k}$, where we understand by $P(L_{Z_1}, \ldots, L_{Z_t})$ any polynomial expression obtained by formally substituting L_{Z_i} for $\mathscr{L}_{Z_i}$. (Since the L_{Z_i} are not necessarily commutative, the expression may not be unique.) Applying σ, we obtain

$$M_Z - L_Z = P(M_{Z_1}, \ldots, M_{Z_t}) + N^\sigma.$$

By our induction assumption, we have $\mathscr{M}_{Z_i} = \mathscr{L}_{Z_i}$ for $i \leq t$, and hence $\mathscr{M}_Z - \mathscr{L}_Z = \mathscr{L}_Z - \mathscr{M}_Z$, which proves our assertion.

Let us now assume that there is a representation ρ: $K \to \mathrm{GL}(V)$ as in Section 2 such that the metric $[u, v]_V$ is $\rho(K)$-invariant. Then the spaces

$S_r(\mathfrak{p}_\pm, V)$, identified with $S_r(\mathfrak{p}_\pm) \otimes V$ by (4.3), are the representation spaces of $\tau_r^\pm \otimes \rho$:

$$(4.16) \qquad \left[(\tau_r^\pm \otimes \rho)(k)p\right](u) = \rho(k)p\left(\mathrm{Ad}(k)^{-1}u\right)$$
$$(p \in S_r(\mathfrak{p}_\pm, V),\, u \in \mathfrak{p}_\pm,\, k \in K).$$

We see easily that $D^r f \in C^\infty(\tau_r^+ \otimes \rho)$ and $E^r f \in C^\infty(\tau_r^- \otimes \rho)$ if $f \in C^\infty(\rho)$. Denote by τ_Z, resp. τ_W, the restriction of τ_r^+, resp. τ_r^-, to Z resp. W. Then $D^Z f \in C^\infty(\tau_Z \otimes \rho)$, and $E^W f \in C^\infty(\tau_W \otimes \rho)$. Moreover the inner products on $Z \otimes V$ and $W \otimes V$ introduced in Lemma 4.2 are $(\tau_Z \otimes \rho)(K)$-invariant and $(\tau_W \otimes \rho)(K)$-invariant, respectively.

Given a discrete subgroup Γ of G, let $\mathscr{A}(\Gamma, \rho)$ denote the set of elements f of $C^\infty(\rho)$ such that $f(\gamma x) = f(x)$ for every $\gamma \in \Gamma$. Since our differential operators are left-invariant, L_Z and M_Z send $\mathscr{A}(\Gamma, \rho)$ into itself; D^Z and E^W send $\mathscr{A}(\Gamma, \rho)$ into $\mathscr{A}(\Gamma, \tau_Z \otimes \rho)$ and $\mathscr{A}(\Gamma, \tau_W \otimes \rho)$, respectively. For $f, g \in \mathscr{A}(\Gamma, \rho)$ we can define their inner product by

$$(4.17) \qquad [f, g] = \int_{\Gamma \backslash G} [f(x), g(x)]_V \, dx$$

provided the integral is convergent. Then the assertions of Theorem 4.3 and Corollary 4.4 have obvious analogues for such inner products at least when the functions are C^∞ vectors in the space $L^2(\Gamma \backslash G, V)$ of square-integrable V-valued functions on $\Gamma \backslash G$, because (4.14a, b) have analogues in the same sense. It should be noted that both L_Z and M_Z defined on the space of C^∞ vectors represent essentially self-adjoint operators in the sense that the closure of each operator is its adjoint. This follows immediately from a result of Nelson and Stinespring [NS, p. 551, lines 7–9 from bottom] (see also Warner [W, p. 269, Example (3)]).

As will be seen in Section 7, the operators D_Z and E_W as well as θ of (4.8) are essentially the same as those introduced in our previous papers [S1–5]. The content of Theorem 4.3 formulated in terms of inner products of type (4.17) was already given in [S3, Theorem 11.5] for the groups of Types A and C.

5. The explicit forms of groups, Lie algebras, and irreducible subspaces

We now proceed according to the classification of irreducible hermitian symmetric spaces of noncompact and classical type. In each case, we take a vector space T of certain complex matrices and give $\mathbf{C}$-linear isomorphisms $\iota_\pm$ of T to $\mathfrak{p}_\pm$; we also take a linear group $\mathscr{K}$ and fix an isomorphism φ of $\mathscr{K}$ to

K^c. Then we shall find that the action of K^c on $\mathfrak{p}_-$ can be transformed, through ι_- and φ, to that of $\mathscr{K}$ on T given in [S4, p. 466] and also that the representation $\{\tau_r^+ \circ \varphi, S_r(\mathfrak{p}_+)\}$ is the same as $\{\tau_r, S_r(T)\}$ on the same page through ι_+; that is, $\tau_r(k)(h \circ \iota_+) = [\tau_r^+(\varphi(k))h] \circ \iota_+$ for $k \in \mathscr{K}$ and $h \in S_r(\mathfrak{p}_+)$. We note here several formulas which hold for all four types:

$$K^c = \varphi(\mathscr{K}), \tag{5.1}$$

$$\mathfrak{p}_+ = \iota_+(T), \qquad \mathfrak{p}_- = \iota_-(T), \tag{5.2}$$

$$\mathfrak{p}_0 = \{\iota_+(z) + \iota_-(\bar{z}) \mid z \in T\}, \tag{5.3}$$

$$\langle \iota_+(z), \iota_-(w) \rangle = \mathrm{tr}({}^t z w) \quad \text{for } z, w \in T, \tag{5.4}$$

$$\iota_\perp(z)^\sigma = \iota_\top(\bar{z}) \quad \text{for } z \in T, \tag{5.5}$$

where σ is the complex conjugation of $\mathfrak{g}$ with respect to $\mathfrak{g}_0$. In each case we shall specify an element ε of T so that $\iota_-(\varepsilon)$ plays the role of ε of (3.7). Then H of (3.8) coincides with $\varphi(H)$ with H of [S4, p. 466]. We shall then present the g_i of Proposition 3.1(4) as elements of $S_i(T)$ and the Z_i of Theorem 3.3 as subspaces of $S_i(T)$ (when G is simple).

Before discussing individual types, let us first define, in addition to the standard symbols GL_n and SL_n, some classical linear groups as follows:

$$\mathrm{O}(Q, F) = \{\alpha \in \mathrm{GL}_n(F) \mid {}^t\alpha Q \alpha = Q\}, \qquad \mathrm{SO}(Q, F) = \mathrm{O}(Q, F) \cap \mathrm{SL}_n(F),$$

$$\mathrm{O}(n, F) = \mathrm{O}(1_n, F), \qquad \mathrm{SO}(n, F) = \mathrm{SO}(1_n, F), \qquad \mathrm{SO}(n) = \mathrm{SO}(n, \mathbf{R}),$$

$$\mathrm{U}(n, m) = \{\alpha \in \mathrm{GL}_{n+m}(\mathbf{C}) \mid \alpha^* I_{n,m} \alpha = I_{n,m}\}, \qquad I_{n,m} = \mathrm{diag}[1_n, -1_m],$$

$$\mathrm{SU}(n, m) = \mathrm{U}(n, m) \cap \mathrm{SL}_{n+m}(\mathbf{C}), \qquad \mathrm{SO}(n, m) = \mathrm{SO}(I_{n,m}, \mathbf{R}),$$

$$\mathrm{U}(n) = \mathrm{U}(n, 0), \qquad \mathrm{SU}(n) = \mathrm{SU}(n, 0),$$

$$\mathrm{Sp}(n, F) = \{\alpha \in \mathrm{GL}_{2n}(F) \mid {}^t\alpha J_n \alpha = J_n\}, \qquad J_n = \begin{bmatrix} 0 & -1_n \\ 1_n & 0 \end{bmatrix}.$$

Here F is a field and ${}^tQ = Q \in \mathrm{GL}_n(F)$; 1_n denotes the identity matrix of degree n, and $\mathrm{diag}[u, \ldots, z]$ the diagonal arrangement of square matrices $u, \ldots, z$; $\alpha^* = {}^t\bar{\alpha}$ for a complex matrix α. We also denote by F_m^n the set of all $(n \times m)$-matrices with entries in F and put $M_n(F) = F_n^n$.

Type A. $G^c = SL_{n+m}(\mathbf{C})$, $G = SU(n, m)$, $K = G \cap SU(n+m)$,

$$\mathscr{K} = \{(u, v) \in GL_n(\mathbf{C}) \times GL_m(\mathbf{C}) \mid \det(u) = \det(v)\},$$

$$\varphi(u, v) = \operatorname{diag}[{}^t u^{-1}, v] \quad \text{for } (u, v) \in \mathscr{K},$$

$$\mathfrak{g} = \{X \in M_{n+m}(\mathbf{C}) \mid \operatorname{tr}(X) = 0\}, \qquad T = \mathbf{C}_m^n,$$

$$\iota_+(z) = \begin{pmatrix} 0 & z \\ 0 & 0 \end{pmatrix}, \qquad \iota_-(z) = \begin{pmatrix} 0 & 0 \\ {}^t z & 0 \end{pmatrix} \quad \text{for } z \in T,$$

$$\operatorname{Ad}(\varphi(u, v))\iota_+(z) = \iota_+({}^t u^{-1} z v^{-1}), \qquad \operatorname{Ad}(\varphi(u, v))\iota_-(z) = \iota_-(uz \cdot {}^t v),$$

$$[\tau_r^+(\varphi(u, v))h](\iota_+(z)) = h(\iota_+({}^t uzv)) \quad \text{for } h \in S_r(\mathfrak{p}_+),$$

$$\varepsilon = \begin{bmatrix} 1_m \\ 0 \end{bmatrix} \quad \text{if } n \geq m, \qquad \varepsilon = [1_n \quad 0] \quad \text{if } n \leq m.$$

Put $l = \operatorname{Min}(n, m)$ and define l functions λ_i on $M_l(\mathbf{C})$ by

$$\text{(5.6)} \quad \det(t1_l - y) = t^l + \sum_{i=1}^{l} (-1)^i \lambda_i(y) t^{l-i} \qquad (t \in \mathbf{C},\ y \in M_l(\mathbf{C})).$$

Further define l elements $g_1, \ldots, g_l$ of $S(T)$ by $g_i(x) = \lambda_i(x')$ where x' denotes the upper left submatrix of x of size l. Then $J^H = \mathbf{C}[g_1, \ldots, g_m]$ as shown in the proof of [S4, Theorem 2.A]. For each fixed positive integer $i \leq l$, let ρ_i denote the representation

$$GL_n(\mathbf{C}) \times GL_m(\mathbf{C}) \to GL\left(\bigwedge^i \mathbf{C}^n\right) \otimes GL\left(\bigwedge^i \mathbf{C}^m\right).$$

Then the restriction of ρ_i to $\mathscr{K}$ occurs in $(\tau_i, S_i(T))$. Let Z_i be the corresponding subspace of $S_i(T)$. Then Z_i is spanned by all the subdeterminants of $x \in T$ of degree i and g_{Z_i} is a constant multiple of g_i. (In fact, our definition of g_i shows that $g_i(x)$ is a linear combination of those subdeterminants, and hence $g_i \in Z_i \cap J^H = \mathbf{C}g_{Z_i}$.) Thus these are the Z_i we take in Theorem 3.3.

Type B. $\quad G^c = SO(R, \mathbf{C}), \qquad R = \operatorname{diag}[1_n, I_{1,1}J_1],$

$$G = \text{the identity component of } G^c \cap U(n, 2),$$

$$K = G \cap SU(n+2), \qquad \mathscr{K} = SO(n, \mathbf{C}) \times GL_1(\mathbf{C}),$$

$$\varphi(u, v) = \operatorname{diag}[u, v^{-1}, v] \quad \text{for } (u, v) \in \mathscr{K},$$

$$\mathfrak{g} = \{X \in M_{n+2}(\mathbf{C}) \mid {}^t XR = -RX\}, \qquad T = \mathbf{C}_1^n,$$

$$\iota_+(z) = \begin{bmatrix} 0 & 0 & z \\ {}^t z & 0 & 0 \\ 0 & 0 & 0 \end{bmatrix}, \qquad \iota_-(z) = \begin{bmatrix} 0 & z & 0 \\ 0 & 0 & 0 \\ {}^t z & 0 & 0 \end{bmatrix} \quad \text{for } z \in T,$$

where the upper left blocks are of size n,

$$\mathrm{Ad}(\varphi(u,v))\iota_+(z) = \iota_+(uv^{-1}z), \qquad \mathrm{Ad}(\varphi(u,v))\iota_-(z) = \iota_-(uvz),$$

$$[\tau_r^+(\varphi(u,v))h](\iota_+(z)) = h(\iota_+(u^{-1}vz)) \quad \text{for } h \in S_r(\mathfrak{p}_+),$$

$$\varepsilon = {}^t(1,0,\ldots,0).$$

Assuming $n > 2$, define two elements g_1 and g_2 of $S(T)$ by $g_1(z) = {}^t\varepsilon z$ and $g_2(z) = {}^t zz$ for $z \in T$. In the proof of [S4, Lemma 2.4], we showed that $J^H = \mathbf{C}[g_1, g_2]$. Let $Z_1 = S_1(T)$ and $Z_2 = \mathbf{C}g_2$. Since $Z_i \cap J^H = \mathbf{C}g_i$, we can take Z_1 and Z_2 as those of Theorem 3.3. The result holds for $n = 1$ if we ignore Z_2. If $n = 2$, G is not simple and $J^H = S(T) = \mathbf{C}[g_+, g_-]$ with $g_\pm(z) = z_1 \pm iz_2$; so we take $Z_\pm = \mathbf{C}g_\pm$ in place of $\{Z_i\}$.

Type C.

$$G^c = \mathrm{Sp}(n,\mathbf{C}), \qquad G = \mathrm{Sp}(n,\mathbf{R}), \qquad K = G \cap \mathrm{SO}(2n), \qquad \mathscr{K} = \mathrm{GL}_n(\mathbf{C}),$$

$$\varphi(u) = \beta \cdot \mathrm{diag}[{}^t u^{-1}, u]\beta^{-1} \quad \text{for } u \in \mathscr{K}, \qquad \beta = \begin{bmatrix} -i1_n & i1_n \\ 1_n & 1_n \end{bmatrix},$$

$$\mathfrak{g} = \left\{X \in M_{2n}(\mathbf{C}) \,\middle|\, {}^t XJ_n = -J_n X\right\},$$

$$T = \left\{z \in M_n(\mathbf{C}) \,\middle|\, {}^t z = z\right\},$$

$$\iota_\pm(z) = \frac{1}{4}\begin{bmatrix} \mp iz & z \\ z & \pm iz \end{bmatrix} \quad \text{for } z \in T,$$

$$\mathrm{Ad}(\varphi(u))\iota_+(z) = \iota_+({}^t u^{-1} z u^{-1}), \qquad \mathrm{Ad}(\varphi(u))\iota_-(z) = \iota_-(uz \cdot {}^t u),$$

$$[\tau_r^+(\varphi(u))h](\iota_+(z)) = h(\iota_+({}^t uzu)) \quad \text{for } h \in S_r(\mathfrak{p}_+),$$

$$\varepsilon = 1_n.$$

Define $g_i \in S_i(T)$ for $1 \le i \le n$ by $g_i(z) = \lambda_i(z)$ with λ_i of (5.6), taking n to be l. Then $J^H = \mathbf{C}[g_1, \ldots, g_n]$. (See the proof of [S4, Theorem 2.C]. No detailed proof for this is given there because it is similar to and far easier than [S4, Lemma 2.2]. Indeed, the fact follows easily from a simple observation that any element h of J^H is determined by its values at diagonal matrices.) Let Z_i be the subspace of $S_i(T)$ spanned by the functions $\det_i({}^t aza)$ for all $a \in \mathrm{GL}_n(\mathbf{C})$, where $\det_i(x)$ denotes the determinant of the upper left i^2 entries of x. Then $g_i \in Z_i \cap J^H$, so that these Z_i for $1 \le i \le n$ are those of Theorem 3.3.

Type D.

$$G^c = SO(Q, \mathbf{C}), \qquad Q = J_n I_{n,n}, \qquad G = G^c \cap SU(n,n) \qquad (n > 1),$$

$$K = G \cap SU(2n), \qquad \mathscr{K} = GL_n(\mathbf{C}), \qquad \varphi(u) = \mathrm{diag}[{}^t u^{-1}, u] \quad \text{for } u \in \mathscr{K},$$

$$\mathfrak{g} = \{X \in M_{2n}(\mathbf{C}) \mid {}^t X Q = -QX\}, \qquad T = \{z \in M_n(\mathbf{C}) \mid {}^t z = -z\},$$

$$\iota_+(z) = \begin{pmatrix} 0 & z \\ 0 & 0 \end{pmatrix}, \qquad \iota_-(z) = \begin{pmatrix} 0 & 0 \\ -z & 0 \end{pmatrix} \quad \text{for } z \in T,$$

$$\mathrm{Ad}(\varphi(u))\iota_+(z) = \iota_+({}^t u^{-1} z u^{-1}), \qquad \mathrm{Ad}(\varphi(u))\iota_-(z) = \iota_-(uz \cdot {}^t u),$$

$$[\tau_r^+(\varphi(u))h](\iota_+(z)) = h(\iota_+({}^t uzu)) \quad \text{for } h \in S_r(\mathfrak{p}_+),$$

$$\varepsilon = \begin{cases} J_m & \text{if } n = 2m, \\ \mathrm{diag}[J_m, 0] & \text{if } n = 2m+1. \end{cases}$$

For $z \in T$ let $\mathrm{Pf}_i(z)$ denote the Pfaffian of the upper left $(2i)^2$ entries of z. Put $m = [n/2]$ and define $g_i \in S_i(T)$ for $0 \le i \le m$ by

$$(5.7) \qquad \mathrm{Pf}_m(t\varepsilon - z) = \sum_{i=0}^{m} (-1)^i g_i(z) t^{m-i} \qquad (t \in \mathbf{C}).$$

As shown in [S4, pp. 470–471], we have $J^H = \mathbf{C}[g_1, \ldots, g_m]$. Let Z_i be the subspace of $S_i(T)$ spanned by $\mathrm{Pf}_i({}^t aza)$ for all $a \in GL_n(\mathbf{C})$. Then the Z_i for $1 \le i \le m$ are the desired ones in Theorem 3.3, since we can show that $g_i \in Z_i$ as follows:

LEMMA 5.1. *As a representation space of $\mathscr{K} = GL_n(\mathbf{C})$, Z_i is isomorphic to $\wedge^{2i}\mathbf{C}^n$. Moreover it contains g_i.*

Proof. By virtue of [S4, Theorem 2.D], we can easily show that the highest weight of Z_i coincides with that of $\wedge^{2i}\mathbf{C}^n$, which proves the first assertion. For every $z \in T$ define $[z] \in \wedge^2\mathbf{C}^n$ by $[z] = \sum_{p<q} z_{pq} e_p \wedge e_q$, where the e_p are the standard basis vectors of $\mathbf{C}^n$. For $v \in \wedge^2\mathbf{C}^n$ write $v^i = \wedge^i v$ for simplicity. Put $E = e_1 \wedge \cdots \wedge e_n$ and assume that n is odd. Then $[z]^m \wedge e_n = m!\mathrm{Pf}_m(z)E$. Applying this to $t\varepsilon - z$, we find that

$$(5.8) \qquad [z]^i \wedge [\varepsilon]^{m-i} \wedge e_n = i!(m-i)!g_i(z)E.$$

For every $x \in \wedge^{n-2i}\mathbf{C}^n$, define an element $q(x)$ of $S_i(T)$ by $q(x)(z)E = [z]^i \wedge x$. Let $GL_n(\mathbf{C})$ act on $\wedge^k\mathbf{C}^n$ as usual. Since $[az \cdot {}^t a] = a[z]$ for $a \in GL_n(\mathbf{C})$, we have $q(ax)(ax \cdot {}^t a) = \det(a)q(x)(z)$; that is, $\tau_r(a)q(x) = \det(a)q({}^t a^{-1}x)$. Therefore we see that $q(\wedge^{n-2i}\mathbf{C}^n) = Z_i$ in view of our first assertion. Then (5.8) shows that

$$i!(m-i)!g_i = q([\varepsilon]^{m-i} \wedge e_n) \in Z_i.$$

If n is even, we obtain (5.8) and the same conclusion without e_n, which completes the proof.

6. The domain of nonnegative eigenvalues and open problems

The notation being the same as in Theorem 3.6 with a fixed irreducible ρ, let us put

$$\Lambda_Z = (-1)^r \mathscr{L}_Z \tag{6.1}$$

for every K-irreducible $Z \subset S_r(\mathfrak{p}_+)$, and write L_i, $\mathscr{L}_i$, and Λ_i for L_{Z_i}, $\mathscr{L}_{Z_i}$, and Λ_{Z_i}. If ρ is one-dimensional, for any K-irreducible Z we have $\Lambda_Z = h_Z(\Lambda_1, \ldots, \Lambda_l)$ with a polynomial h_Z depending only on Z and ρ. The Λ_i and Λ_Z being symmetric, h_Z has real coefficients. Now, when we let these operators act on a space of functions on which they are nonnegative (such as the subspace of "cusp forms" belonging to $\mathscr{A}(\Gamma, \rho)$ at the end of Section 4), we are naturally interested in the set of eigenvalues $(\lambda_1, \ldots, \lambda_l)$ of $\Lambda_1, \ldots, \Lambda_l$ for a common eigenfunction in the space. Obviously it belongs to the set

$$E_\rho = \{(\lambda_i) \in \mathbf{R}^l \,|\, h_Z(\lambda_1, \ldots, \lambda_l) \geq 0 \text{ for all } Z\}. \tag{6.2}$$

It can happen for some Z that $h_Z(\lambda_1, \ldots, \lambda_l) \geq 0$ whenever $\lambda_i \geq 0$ for every i, but this is not always the case. In general, even for trivial ρ, E_ρ can be a proper subset of the set $\{(\lambda_i) \in \mathbf{R}^l | \lambda_i \geq 0 \text{ for every } i\}$. Thus the study of the nature of E_ρ is a very interesting question. We are going to show that if G is simple and $\operatorname{rank}(G/K) > 1$, there always exists an irreducible subspace Z for which the condition $h_Z(\lambda_1, \ldots, \lambda_l) \geq 0$ does not follow from the nonnegativity of the λ_i.

As preliminaries, we first take a maximal abelian subalgebra $\mathfrak{h}_0$ of $\mathfrak{k}_0$ and its complexification $\mathfrak{h}$, which is a Cartan subalgebra of $\mathfrak{g}$. Then we have

$$\mathfrak{k} = \mathfrak{h} + \sum_{\gamma \in Q} \mathfrak{g}^\gamma, \qquad \mathfrak{p}_+ = \sum_{\alpha \in P} \mathfrak{g}^\alpha, \qquad \mathfrak{p}_- = \sum_{-\alpha \in P} \mathfrak{g}^\alpha,$$

where P and Q are certain sets of roots, and $\mathfrak{g}^\lambda$ is the eigensubspace defined as usual for each $\lambda \in S_1(\mathfrak{h})$ (cf. [H3, pp. 383–384]). Denoting by Φ the Killing form of $\mathfrak{g}$, we take $H_\alpha \in \mathfrak{h}$ for each $\alpha \in P$ such that $\Phi(H_\alpha, H) = \alpha(H)$ for every $H \in \mathfrak{h}$. It is well-known that

$$[X, Y] = \Phi(X, Y) H_\alpha \neq 0 \quad \text{if } 0 \neq X \in \mathfrak{g}^\alpha \text{ and } 0 \neq Y \in \mathfrak{g}^{-\alpha}. \tag{6.3}$$

Therefore we can find $A_\alpha \in \mathfrak{g}^\alpha$ and $B_\alpha \in \mathfrak{g}^{-\alpha}$ for each $\alpha \in P$ so that $\langle A_\alpha, B_\alpha \rangle = 1$. Then we have

$$\langle A_\alpha, B_\beta \rangle = \delta_{\alpha\beta} \qquad (\alpha, \beta \in P), \tag{6.4}$$

since $\Phi(\mathfrak{g}^\alpha, \mathfrak{g}^\beta) = 0$ if $\alpha + \beta \neq 0$. (We can choose A_α so that $B_\alpha = A_\alpha^\sigma$, but this is unnecessary.) Put

$$C_{\alpha\beta} = [A_\alpha, B_\beta] \qquad (\alpha, \beta \in P). \tag{6.5}$$

Then $C_{\alpha\beta} \in \mathfrak{k}$, and

$$C_{\alpha\alpha} = sH_\alpha \quad \text{with } s \text{ of (3.2)}, \tag{6.6}$$

$$[C_{\alpha\beta}, A_\beta] = s\alpha(H_\beta)A_\alpha. \tag{6.7}$$

In fact, (6.6) follows from (6.3). As for (6.7), since $[A_\alpha, A_\beta] = 0$, we have

$$[C_{\alpha\beta}, A_\beta] = [[A_\alpha, B_\beta], A_\beta] = [[A_\beta, B_\beta], A_\alpha] = [sH_\beta, A_\alpha] = s\alpha(H_\beta)A_\alpha,$$

which is (6.7).

Lemma 6.1. *Let $\{\alpha_\nu\}$ and $\{\beta_\nu\}$ be dual bases of $S_r(\mathfrak{p}_+)$ and $S_r(\mathfrak{p}_-)$. Then*

$$\sum_\nu \omega(\alpha_\nu)\omega(\beta_\nu) = \sum_Z L_Z,$$

where Z runs over all the K-irreducible subspaces of $S_r(\mathfrak{p}_+)$.

This is because the left-hand side is independent of the choice of dual bases.

Theorem 6.2. *Suppose that G is simple and ρ is one-dimensional; let (λ_Z, λ_i) be a set of eigenvalues of the Λ_Z and in particular the Λ_i for a common eigenfunction in a space on which the Λ_Z are nonnegative. Then for $Z \subset S_r(\mathfrak{p}_+)$, we have $\lambda_Z \leq \lambda_1^r + q_Z(\lambda_1)$ with a real polynomial q_Z of degree less than r. In particular, for each $m \leq l$ we have $\lambda_m \leq \lambda_1^m + q_m(\lambda_1)$ with a real polynomial q_m of degree less than m.*

Proof. We first define the weight of a polynomial $h(t_1, \ldots, t_l)$ in l indeterminates t_i to be the degree of the polynomial $h(t_1, t_2^2, \ldots, t_l^l)$. Then we have:

(6.8) *For every $f \in I^{2r}(\mathfrak{p})$, the effect of $\omega_+(f)$ on $\mathscr{D}(\rho)$ can be given by $q(\mathscr{L}_1, \ldots, \mathscr{L}_l)$ with a polynomial q of weight at most r.*

In fact, we have $f = Q(f_{Z_1}, \ldots, f_{Z_l})$ with a polynomial Q, which must be of weight $\leq r$ because of the algebraic independence of the f_{Z_i}. By (1.5) we have $\omega_+(f) - Q(L_1, \ldots, L_l) \in U^{2r-1}(\mathfrak{g})^K$, and by Proposition 3.7, the difference has the same effect as $\omega_+(g)$ for some $g \in I^{2r-2}(\mathfrak{p})$. Applying induction to g, we obtain (6.8). Now we can view $\{A_\alpha\}$ and $\{B_\alpha\}$ as dual bases of $S_1(\mathfrak{p}_-)$ and $S_1(\mathfrak{p}_+)$. Since $Z_1 = S_1(\mathfrak{p}_+)$, we have $L_1 = \sum_{\alpha \in P} B_\alpha A_\alpha$. The spaces $S_r(\mathfrak{p}_-)$ and $S_r(\mathfrak{p}_+)$ are spanned by $\{A_{\alpha_1} \cdots A_{\alpha_r}\}$ and $\{B_{\alpha_1} \cdots B_{\alpha_r}\}$. By Lemma 1.1 and Lemma

6.1, we have

$$\sum B_{\alpha_1} \cdots B_{\alpha_r} A_{\alpha_1} \cdots A_{\alpha_r} = \sum_Z L_Z, \tag{6.9}$$

where the right-hand side is the same as in Lemma 6.1 and $(\alpha_1, \ldots, \alpha_r)$ runs over P^r. Comparing the left-hand side with L_1^r and employing (1.5), we obtain $L_1^r - \Sigma_Z L_Z \in U^{2r-1}(\mathfrak{g})^K$. By Proposition 3.7 and (6.8) we have $\mathscr{L}_1^r - \Sigma_Z \mathscr{L}_Z = Q(\mathscr{L}_1, \ldots, \mathscr{L}_l)$ with a polynomial Q of weight less than r. Fixing our attention on one particular Z, we can write

$$\mathscr{L}_Z = \mathscr{L}_1^r - \sum_U \mathscr{L}_U - Q(\mathscr{L}_1, \ldots, \mathscr{L}_l),$$

where U runs over the irreducible subspaces of $S_r(\mathfrak{p}_+)$ other than Z. Obviously Q involves only those $\mathscr{L}_i$ with $i < r$. Thus we obtain

$$\Lambda_Z = \Lambda_1^r - \sum_U \Lambda_U + F(\Lambda_1, \ldots, \Lambda_t),$$

with $t = \mathrm{Min}(r-1, l)$ and a polynomial F of weight $< r$, which must have real coefficients. Since the Λ_U are nonnegative, we obtain $\lambda_Z \leq \lambda_1^r + F(\lambda_1, \ldots, \lambda_t)$. In particular, this holds with λ_r in place of λ_Z for each $r \leq l$. By induction on r we obtain the last assertion of our theorem. Substituting these estimates for λ_i into the inequality about λ_Z, we can complete the proof.

The inequalities of Theorem 6.2 are derived from the nonnegativity of $\Sigma_U \Lambda_U$, and therefore are weaker than those resulting from the nonnegativity of each Λ_U. If $m = 2$, we can prove a more explicit relation as follows:

PROPOSITION 6.3. *Suppose G is simple and classical. Then the following assertions hold*:

(1) $\Sigma_{\beta \in P} \alpha(H_\beta) = p$ *for every* $\alpha \in P$ *with a positive real number p depending only on* $\mathfrak{g}$.

(2) *If* $\mathrm{rank}(G/K) > 1$, *then* $S_2(\mathfrak{p}_+) = Z_2 \oplus U$ *with a K-irreducible subspace U.*

(3) *For every irreducible* ρ, *the equality*

$$L_1^2 = L_2 + L_U + sp\, L_1 + \sum_{(\alpha,\beta) \in P \times P} B_\alpha A_\beta C_{\alpha\beta}$$

holds in $U(\mathfrak{g})$ *with U of* (2), *p of* (1), *and s of* (3.2); *moreover* $\Sigma_{\alpha,\beta} \rho_u(C_{\alpha\beta}) \otimes B_\alpha A_\beta$ *belongs to the set* $I(\rho, \mathfrak{p})$ *of Proposition* 2.2. *Here and in* (4) *it is understood that* $L_U = 0$, $\Lambda_U = 0$ *and* $Z_2 = S_2(\mathfrak{p}_+)$ *if* $\mathrm{rank}(G/K) = 1$.

(4) *If* ρ *is one-dimensional,*

$$\Lambda_U = \Lambda_1^2 + s(p-q)\Lambda_1 - \Lambda_2$$

with a real number q which is equal to $d\rho(H_\alpha)$ *for every* $\alpha \in P$.

Proof. By Proposition 3.1(4), we see that $\dim(J_2^H) = 1$ or 2 according as $\operatorname{rank}(G/K) = 1$ or > 1. Therefore by Proposition 3.1(3) we have $S_2(\mathfrak{p}_+) = Z_2 \oplus U$ with an irreducible U if $\operatorname{rank}(G/K) > 1$, and hence (6.9) can be written in the form

$$L_2 + L_U = \sum_{(\alpha,\beta)\in P\times P} B_\alpha B_\beta A_\alpha A_\beta.$$

Employing (6.5) and (6.7), we easily see that

$$(6.10)\quad L_1^2 = \sum_{(\alpha,\beta)\in P\times P} B_\alpha A_\alpha B_\beta A_\beta$$

$$= \sum_{\alpha,\beta} B_\alpha B_\beta A_\alpha A_\beta + s\sum_\alpha\left[\sum_\beta \alpha(H_\beta)\right] B_\alpha A_\alpha + \sum_{\alpha,\beta} B_\alpha A_\beta C_{\alpha\beta}.$$

Recall that

$$U(\mathfrak{g}) = \omega_+(S(\mathfrak{p}))U(\mathfrak{k}) = \bigoplus_{t,u} [\omega(S_t(\mathfrak{p}_+))\omega(S_u(\mathfrak{p}_-))U(\mathfrak{k})]$$

and observe that each term of the last direct sum is K-stable. Therefore each of the three terms of the right-hand side of (6.10) is K-invariant. Thus the second term belongs to $\omega_+(S(\mathfrak{p}))$ with ω_+ of Proposition 3.7, and hence must be in $\omega_+(I_2(\mathfrak{p}))$, since ω_+ is injective and commutes with K. Therefore it is L_1 times a constant, say p, which must be equal to $\sum_{\beta\in P}\alpha(H_\beta)$ for all α. Let F denote the matrix $(\Phi(H_\alpha, H_\beta))_{(\alpha,\beta)\in P\times P}$ of size P, which is known to be positive definite. Since $\alpha(H_\beta) = \Phi(H_\alpha, H_\beta)$, we see that $Fv = pv$ with the vector v in $\mathbf{R}^P$ whose components are all equal to 1. Thus p is an eigenvalue of F and hence positive. This proves (1). As for the last term of (6.10), we first define a map μ_+ of $U(\mathfrak{g})$ into $\operatorname{End}(V)\otimes S(\mathfrak{p})$ by $\mu_+(\omega_+(D)C) = \rho_u(C)\otimes D$ for $C\in U(\mathfrak{k})$ and $D\in S(\mathfrak{p})$ with ρ_u of (2.2). Then $\mu_+\circ \operatorname{Ad}(k) = k\circ\mu_+$ for every $k\in K$, where the action of k on $\operatorname{End}(V)\otimes S(\mathfrak{p})$ is defined as in Proposition 2.2. Therefore $\sum_{\alpha,\beta}\rho_u(C_{\alpha\beta})\otimes B_\alpha A_\beta$, being the image of an element of $U(\mathfrak{g})^K$ under μ_+, must belong to the set $I(\rho,\mathfrak{p})$ defined in the same proposition. This proves (3). If ρ is one-dimensional, it belongs to $I(\mathfrak{p})$ and so $\sum\rho_u(C_{\alpha\beta})B_\alpha A_\beta = q'L_1$ with a constant q', which is equal to $\rho_u(C_{\alpha\alpha}) = -s\,d\rho(H_\alpha)$. Since $\mathscr{L}_1^2$ is symmetric, q' is real. Thus we obtain $\mathscr{L}_1^2 = \mathscr{L}_2 + \mathscr{L}_U + s(p-q)\mathscr{L}_1$, which proves (4).

From (4) of the above proposition we see that

$$(6.11)\qquad \lambda_1^2 + s(p-q)\lambda_1 \geq \lambda_2,$$

which is an explicit form of the inequality of Theorem 6.2 for $m = 2$.

To obtain further examples, let us confine ourselves to G of Type B of Section 5. Assume $n > 2$. In this case it is more convenient to work with the bases $\{a_i\}$ and $\{b_i\}$ of $\mathfrak{p}_+$ and $\mathfrak{p}_-$ given by $a_i = \iota_+(e_i)$ and $b_i = \iota_-(e_i)$ with the standard basis elements e_i of $\mathbf{C}_1^n = T$. We take the pairing of (5.4), which means that $s = 2n$ for s of (3.2). We put $c_{ij} = [a_i, b_j]$ and $c_0 = c_{ii}$ (this is independent of i). Put also $\alpha = \sum_{i=1}^n a_i^2$ and $\beta = \sum_{i=1}^n b_i^2$. Then $Z_2 = \mathbf{C}\beta$. We have $L_1 = \sum_{i=1}^n b_i a_i$ and $L_2 = n^{-1}\beta\alpha$ since $\langle \alpha, \beta \rangle = n$. We see that $S_3(\mathfrak{p}_+) = Q \oplus R$ with two irreducible subspaces Q and R, and $Q = \beta S_1(\mathfrak{p}_+)$. Then $L_Q = 3(n+2)^{-1}\sum_{i=1}^n \beta b_i \alpha a_i$ since Lemma 1.1 shows that $\langle \alpha a_i, \beta b_i \rangle = (n+2)/3$. By Lemma 6.1 and Lemma 1.1 we have

$$L_Q + L_R = \sum_{i,j,k} b_i b_j b_k a_i a_j a_k, \tag{6.12}$$

where (i, j, k) runs over $\{1, \ldots, n\}^3$.

Proposition 6.4. *The notation being as above, the following equalities hold in* $U(\mathfrak{g})$:

$$L_U = L_1^2 - nL_1 - L_2 - L_1 c_0 - \gamma_1,$$

$$L_Q = 3n(n+2)^{-1} L_2(L_1 - n - 2 - 2c_0),$$

$$L_R = (L_1 - 2n - 2 - 2c_0)(L_1^2 - nL_1 - L_1 c_0 - \gamma_1) + nL_2 - L_Q - 2\gamma_2,$$

$$n(L_2\gamma_1 - \gamma_1 L_2) = 2\sum_{i \gtrless j} (b_i^2 \alpha - \beta a_i^2) c_{ij}^2 + 2\sum_{h,i,j} (\beta a_h a_i - b_h b_i \alpha) c_{hj} c_{ji},$$

$$\alpha\beta - \beta\alpha = 4L_1 c_0 + 2nc_0^2 + n(n-2)c_0 - 4\gamma_1 + 2\sum_{i \gtrless j} c_{ij}^2.$$

Here γ_1 *and* γ_2 *are given by*

$$\gamma_1 = \sum_{i \gtrless j} b_i a_j c_{ij},$$

$$\gamma_2 = \sum_{i \gtrless j} b_i b_j a_i^2 c_{ji} + \sum_{i \gtrless j} b_i^2 a_i a_j c_{ij} + \sum_{h,i,j} b_h b_i a_j a_i c_{hj},$$

and (h, i, j) *runs over all the ordered sets of three different letters.*

The first equality about L_U is essentially a special case of (6.10). All the formulas can be obtained by the same technique as in the proof of Proposition 6.3(3). We omit the details which are straightforward and tedious.

Let us now consider for simplicity the case of trivial ρ. From the above formulas we obtain

$$\Lambda_U = \Lambda_1^2 + n\Lambda_1 - \Lambda_2, \tag{6.13a}$$

$$\Lambda_Q = 3n(n+2)^{-1}\Lambda_2(\Lambda_1 + n + 2), \tag{6.13b}$$

$$\Lambda_R = (\Lambda_1 + 2n + 2)(\Lambda_1^2 + n\Lambda_1) - n\Lambda_2\{4 + 3(n+2)^{-1}\Lambda_1\}. \tag{6.13c}$$

Therefore, if (λ_1, λ_2) is a set of eigenvalues of Λ_1 and Λ_2 on a common eigenfunction for which the nonnegativity holds, we have

$$\lambda_1^2 + n\lambda_1 \geq \lambda_2, \tag{6.14a}$$

$$\lambda_2(\lambda_1 + n + 2) \geq 0, \tag{6.14b}$$

$$\lambda_1(\lambda_1 + n)(\lambda_1 + 2n + 2) \geq n\lambda_2\{4 + 3(n+2)^{-1}\lambda_1\}. \tag{6.14c}$$

Clearly (6.14b) gives no new restriction on nonnegative λ_1 and λ_2. Both (6.14a) and (6.14c) are nontrivial conditions on them. However, the latter implies the former, as can easily be seen.

There is another type of inequality which results from the nonnegativity of $\mathcal{M}_Z$ for a nontrivial one-dimensional ρ. For instance, taking $\rho(\mathrm{diag}[u, v^{-1}, v]) = v^k$ with $k \in \mathbf{Z}$ for $u \in SO(n, \mathbf{C})$ and $v \in GL_1(\mathbf{C})$, we obtain

$$-\mathcal{M}_{Z_1} = \Lambda_1 - nk, \tag{6.15a}$$

$$n(\mathcal{M}_{Z_2} - \Lambda_2) = -4k\Lambda_1 + 2nk^2 + n(n-2)k. \tag{6.15b}$$

The first equality is easy; the second one follows from the formula for $\alpha\beta - \beta\alpha$ in Proposition 6.4. Thus the nonnegativity of $(-1)^r\mathcal{M}_Z$ gives

$$\lambda_1 \geq nk, \tag{6.16a}$$

$$n\lambda_2 \geq 4k\lambda_1 - 2nk^2 - n(n-2)k. \tag{6.16b}$$

Apart from the question about the set E_ρ, there is another natural problem of determining $\mathcal{D}(\rho)$ for an arbitrary ρ. As shown by Theorems 3.6 and 2.4, if ρ is one-dimensional, the operators obtained from L_Z (resp. M_Z) exhaust the ring $\mathcal{D}(\rho)$, which coincides with the image of the center $C(\mathfrak{g})$ of $U(\mathfrak{g})$. In general, $\mathcal{D}(\rho)$ is larger for a higher-dimensional ρ. To see this clearly, take Ψ_+ of Proposition 3.7. Then the $\Psi_+(L_Z)$ for all Z form a $\mathbf{C}$-linear basis of $\Psi_+(I(\mathfrak{p}))$, but $\Psi_+(I(\mathfrak{p}))$ is not a ring for every higher-dimensional ρ, even when $\mathrm{rank}(G/K) = 1$. In fact, the sum $\sum_{\alpha,\beta}\rho_u(C_{\alpha\beta}) \otimes B_\alpha A_\beta$ of Proposition 6.2(3) belongs to $I(\rho, \mathfrak{p})$ but not to $I(\mathfrak{p})$ if $\rho_u(C_{\alpha\beta})$ is not a scalar for some (α, β). Such a pair (α, β) always exists if ρ is not one-dimensional, since the $C_{\alpha\beta}$ span $\mathfrak{k}$.

As shown in Remark 3.9, $\mathcal{L}_1$ belongs to the center of $\mathcal{D}(\rho)$. This is no longer true for $\mathcal{L}_2$ as can be seen from the fourth formula of Proposition 6.4. In fact, if we take for example the restriction of ρ to the semisimple part of K to be the identity map, then we easily see that $L_2\gamma_1 - \gamma_1 L_2 \neq 0$. Thus $\mathcal{D}(\rho)$ is not commutative in general. However, it can be commutative for some G and some higher-dimensional ρ regardless of $\mathrm{rank}(G/K)$. In particular, if $G = SU(n, 1)$, it is always commutative, but not necessarily generated by $\mathscr{L}_1$. Therefore we can ask the following questions for an arbitrary irreducible ρ:

(I) *Is there any, desirably more informative, description of* $\mathscr{D}(\rho)$ *different from Propositions* 2.1, 2.2, *and* 3.7?

(II) *How can one characterize the image of* $C(\mathfrak{g})$ *in* $\mathscr{D}(\rho)$? *Is it always the center of* $\mathscr{D}(\rho)$?

(III) *In what status are the* $\mathscr{L}_Z$ *and* $\mathscr{M}_Z$ *among the whole* $\mathscr{D}(\rho)$?

Let us conclude this section by giving another example of an element of $I(\rho, \mathfrak{p})$ that does not belong to $I(\mathfrak{p})$. Fix Z, $\{\zeta_\nu\}_{\nu=1}^{\kappa}$, and $\{\xi_\nu\}_{\nu=1}^{\kappa}$ as in Theorem 3.6, and define a matrix representation β: $K \to GL(\mathbf{C}^\kappa)$ by $\tau_r^+(k)\zeta_\nu = \sum_{\mu=1}^{\kappa}\beta(k)_{\mu\nu}\zeta_\mu$. Let $\rho = \alpha\beta$ with an arbitrary one-dimensional representation α of K. Put

$$A_Z = \sum_{\mu,\nu=1}^{\kappa} e_{\mu\nu} \otimes \zeta_\nu\xi_\mu, \tag{6.17}$$

where the $e_{\mu\nu}$ are the standard matrix units. Then we can easily verify that A_Z belongs to $I(\rho, \mathfrak{p})$. Obviously $A_Z \notin I(\mathfrak{p})$ if ρ is not one-dimensional.

7. Operators on symmetric domains

Let us now represent G/K for each G of Section 5 as a domain $\mathscr{H}$ contained in the complex vector space T defined there as follows:

Type A: $\mathscr{H} = \{z \in T \mid zz^* < 1_n\}$,

Type B: $\mathscr{H} = \left\{z \in T \,\middle|\, z^*z < 1 + (1/4)|{}^tzz|^2 < 2\right\}$,

Type C: $\mathscr{H} = \{z \in T \mid i(\bar{z} - z) > 0\}$,

Type D: $\mathscr{H} = \{z \in T \mid zz^* < 1_n\}$.

here $z^* = {}^t\bar{z}$; we write $a > b$ for hermitian matrices a and b of the same size if $a - b$ is positive definite. We are going to study the operators on $\mathscr{H}$ corresponding to those of Sections 3 and 4. We let $\mathscr{K}$ act on T so that $\iota_+(\alpha z) = \mathrm{Ad}(\varphi(\alpha))\iota_+(z)$ for $\alpha \in \mathscr{K}$ and $z \in T$. Then G acts on $\mathscr{H}$ as a group of transformations and there is a $\mathscr{K}$-valued (canonical) holomorphic factor of automorphy $\Lambda(g, z)$ defined for $g \in G$ and $z \in \mathscr{H}$ such that

$$d(gz) = \Lambda(g, z)\, dz. \tag{7.1}$$

Moreover there is a point $\mathbf{o}$ of $\mathscr{H}$ such that

$$K = \{g \in G \mid g\mathbf{o} = \mathbf{o}\}, \tag{7.2a}$$

$$\varphi^{-1}(k) = \Lambda(k, \mathbf{o}) \quad \text{for every } k \in K. \tag{7.2b}$$

The explicit forms of gz and $\Lambda(g,z)$ for Type A are as follows: for $g = \begin{pmatrix} a & b \\ c & d \end{pmatrix} \in G$ with a of size n,

$$(7.3a)\qquad gz = (az+b)(cz+d)^{-1},$$

$$(7.3b)\qquad \Lambda(g,z) = (\lambda(g,z), \mu(g,z)) \in \mathscr{K},$$

$$(7.3c)\qquad \lambda(g,z) = \bar{a} + \bar{b}\cdot {}^t z, \qquad \mu(g,z) = cz+d.$$

These formulas can be given by a single equality

$$(7.3d)\qquad g\begin{bmatrix} 1_n & z \\ z^* & 1_m \end{bmatrix} = \begin{bmatrix} 1_n & gz \\ (gz)^* & 1_m \end{bmatrix}\begin{bmatrix} \overline{\lambda(g,z)} & 0 \\ 0 & \mu(g,z) \end{bmatrix}.$$

Formula (7.1) can be written $d(gz) = {}^t\lambda(g,z)^{-1}\,dz\,\mu(g,z)^{-1}$. The origin $\mathbf{o}$ in this case is the zero matrix of T. For these facts and also for the other three types, the reader is referred to [S1, §2], [S2, §§1, 8], [S5, §1].

Given a finite-dimensional complex vector space V and $f \in C^\infty(\mathscr{H}, V)$, we define Df and $\overline{D}f$ as elements of $C^\infty(\mathscr{H}, S_1(T,V))$ as follows. For Type A, they are defined by

$$(7.4)\quad Df(u) = \sum_{p=1}^{n}\sum_{q=1}^{m} u_{pq}\,\partial f/\partial z_{pq}, \qquad \overline{D}f(u) = \sum_{p=1}^{n}\sum_{q=1}^{m} u_{pq}\,\partial f/\partial \bar{z}_{pq},$$

for $u = (u_{pq}) \in T$. Symbolically these can be expressed in the form

$$(7.5)\qquad Df(dz) = \partial f, \qquad \overline{D}f(d\bar{z}) = \bar{\partial} f,$$

which can be taken as the definitions of Df and $\overline{D}f$ for all types. From (7.1) we obtain

$$(7.6a)\qquad D(f\circ\alpha)(u) = [(Df)\circ\alpha](\Lambda(\alpha,z)u) \qquad (\alpha \in G,\ u \in T),$$

$$(7.6b)\qquad \overline{D}(f\circ\alpha)(u) = [(\overline{D}f)\circ\alpha]\big(\overline{\Lambda(\alpha,z)u}\big) \qquad (\alpha \in G,\ u \in T).$$

LEMMA 7.1. *For f as above define $F \in C^\infty(G,V)$ by $F(g) = f(g\mathbf{o})$. Then for every $u \in T$,*

$$[\iota_+(u)F](g) = (Df)(g\mathbf{o})[\Lambda(g,\mathbf{o})u],$$

$$[\iota_-(u)F](g) = (\overline{D}f)(g\mathbf{o})\big[\overline{\Lambda(g,\mathbf{o})u}\big].$$

Proof. Put $X = \iota_+(u) + \iota_-(\bar{u})$ and $Y = i[\iota_+(u) - \iota_-(\bar{u})]$. Then $2\iota_+(u) = X - iY$, $2\iota_-(\bar{u}) = X + iY$, and $X, Y \in \mathfrak{p}_0$. Put $z_A(t) = (\exp tA)(\mathbf{o})$ for $A \in \mathfrak{g}_0$, and $z'_A(t) = dz_A(t)/dt$. By (7.6a, b) we have

$$(AF)(g) = (d/dt)_{t=0} f(gz_A(t))$$

$$= [(Df)(g\mathbf{o})](\Lambda(g,\mathbf{o})z'_A(0)) + [(\overline{D}f)(g\mathbf{o})]\big(\overline{\Lambda(g,\mathbf{o})z'_A(0)}\big).$$

Employing the explicit forms of $\iota_{\pm}(u)$ given in Section 5, we easily find that $z'_X(0) = u$ and $z'_Y(0) = iu$ for all four types, from which we obtain the desired formulas. In fact, the factor $1/4$ in the definition of $\iota_{\pm}$ for Type C was chosen in order to make our formulas valid with no scalar factors.

We now take a holomorphic representation $\rho\colon \mathscr{K} \to GL(V)$, and consider the factor of automorphy $\rho(\Lambda(g, z))$.

LEMMA 7.2. *$\iota_+(u)\rho(\overline{\Lambda(g, o)}) = 0$ and $\iota_-(u)\rho(\Lambda(g, o)) = 0$ for every $u \in T$.*

Proof. Put $P(g) = \iota_-(u)\rho(\Lambda(g, o))$. Since $\iota_-(u)$ is left-invariant, for every $a \in G$ we have

$$P(ag) = \iota_-(u)[\rho(\Lambda(a, go)\rho(\Lambda(g, o))].$$

Now $\Lambda(a, z)$ is holomorphic in z, so that $\iota_-(u)\rho(\Lambda(a, go)) = 0$ by Lemma 7.1, and hence $P(ag) = \rho(\Lambda(a, go))P(g)$. Putting $g = 1$, we obtain $P(a) = \rho(\Lambda(a, o))P(1)$. If $\varphi\colon \mathbf{R} \to \mathscr{K}$ is a smooth map such that $\varphi(0) = 1$, then $\varphi'(0)$ is meaningful as an element of the Lie algebra of $\mathscr{K}$ and $(d/dt)_{t=0}\rho(\varphi(t)) = d\rho(\varphi'(0))$. Applying this to $\varphi(t) = \rho(\Lambda(\exp tA, o))$ with $A \in \mathfrak{g}_0$, we find that

$$(d/dt)_{t=0}\rho(\Lambda(\exp tA, o)) = d\rho(\Lambda(A)),$$

where we put $\Lambda(A) = (d/dt)_{t=0}\Lambda(\exp tA, o)$. Since $\Lambda(g, o)$ has a simple expression in the matrix entries of g, so does $\Lambda(A)$. For example, if G is of Type A, $\Lambda\begin{pmatrix} a & b \\ c & d \end{pmatrix} = (\bar{a}, d)$. With X and Y as in the proof of Lemma 7.1, we thus obtain

$$2P(1) = d\rho(\Lambda(X)) + i\,d\rho(\Lambda(Y)).$$

For Type A, this is obviously 0, which is also the case for the remaining three cases as can easily be seen from the explicit forms of X and Y. (Type B is slightly nontrivial.) This proves the second formula of our lemma. The same type of reasoning applies to the first one.

Let us now put, for $z = go \in \mathscr{H}$ with $g \in G$,

$$\Xi(z) = [\Lambda(g, o)\Lambda(g, o)^*]^{-1} (\in \mathscr{K}), \tag{7.7}$$

where we understand that $(z, w)^* = (z^*, w^*)$ for Types A and B. It can easily be seen that $\Xi(z)$ is well-defined and

$$\Lambda(\alpha, z)^*\Xi(\alpha z)\Lambda(\alpha, z) = \Xi(z) \qquad (\alpha \in G, z \in \mathscr{H}). \tag{7.8}$$

Obviously Ξ is uniquely determined by this under an additional condition $\Xi(o) = 1$. For Type A we can show that

$$\Xi(z) = \left(1_n - \bar{z}\cdot{}^t z, 1_m - z^*z\right).$$

For this and other types see [S1, 2]. Now for $f \in C^\infty(\mathscr{H}, V)$, we define $D_\rho f$ and

Ef as elements of $C^\infty(\mathscr{H}, S_1(T, V))$ by

(7.9) $$(D_\rho f)(u) = \rho(\Xi)^{-1} D[\rho(\Xi)f](u), \qquad (Ef)(u) = \overline{D}f({}^t\Xi^{-1}u)$$

for $u \in T$, where ${}^t(a, b) = ({}^ta, {}^tb)$ for Types A and B.

Proposition 7.3. *For $f \in C^\infty(\mathscr{H}, V)$ define $f^\rho \in C^\infty(G, V)$ by $f^\rho(g) = \rho(\Lambda(g, \mathrm{o}))^{-1} f(g\mathrm{o})$. Then $f^\rho \in C^\infty(\rho \circ \varphi^{-1})$ (see (7.2b)) and*

$$\iota_+(u)f^\rho = (D_\rho f)^{\rho\otimes\tau}(u), \qquad \iota_-(u)f^\rho = (Ef)^{\rho\otimes\sigma}(u)$$

for every $u \in T$, where $\rho \otimes \tau$ and $\rho \otimes \sigma$ are representations of $\mathscr{K}$ on $S_1(T, V)$ defined by

$$[(\rho \otimes \tau)(\alpha)h](u) = \rho(\alpha)h(\alpha^{-1}u), \qquad [(\rho \otimes \sigma)(\alpha)h](u) = \rho(\alpha)h({}^t\alpha u)$$
$$(\alpha \in \mathscr{K}, h \in S_1(T, V), u \in T).$$

Proof. According to our definition, for $u \in T$ and $z = g\mathrm{o}$ we have

$$\begin{aligned}(D_\rho f)^{\rho\otimes\tau}(u) &= \rho\big(\Lambda(g, \mathrm{o})^{-1}\big)(D_\rho f)(z)[\Lambda(g, \mathrm{o})u] \\ &= \rho\big(\Lambda(g, \mathrm{o})^{-1}\big)\rho\big(\Xi(z)^{-1}\big)D[\rho(\Xi(z))f][\Lambda(g, \mathrm{o})u] \\ &= \rho(\Lambda(g, \mathrm{o})^*)\iota_+(u)[\rho(\Xi(g\mathrm{o}))f(g\mathrm{o})]\end{aligned}$$

by (7.7) and Lemma 7.1. By Lemma 7.2 this is equal to

$$\begin{aligned}&\iota_+(u)\{\rho[\Lambda(g, \mathrm{o})^*\Xi(g\mathrm{o})]f(g\mathrm{o})\} \\ &\quad = \iota_+(u)\big\{\rho\big[\Lambda(g, \mathrm{o})^{-1}\big]f(g\mathrm{o})\big\} = \iota_+(u)f^\rho,\end{aligned}$$

which proves the first desired equality. The second one can be proved in a similar and simpler fashion.

Once the above proposition is established, it is easy to see that the operators introduced in [S1–5] are exactly those on $\mathscr{H}$ corresponding to D^Z, E^W, L_Z, and M_Z. The verification, being straightforward, may be left to the reader. Some explicit formulas for those corresponding to L_Z and M_Z for certain Z are given in [S3, (11.20a, b), (11.21a, b)].

Princeton University, Princeton, New Jersey

References

[L] J. Lepowsky, Algebraic results on representations of semisimple Lie groups, Trans. A.M.S. 176 (1973), 1–44.

[H1] S. Helgason, Fundamental solutions of invariant differential operators on symmetric spaces, Amer. J. Math. 86 (1964), 565–601.

[H2] ______, *Groups and Geometric Analysis*, Academic Press, 1984.

[H3] S. Helgason, *Differential Geometry, Lie Groups, and Symmetric Spaces*, Academic Press, 1978.

[Hu] L. K. Hua, *Harmonic Analysis of Functions of Several Complex Variables in the Classical Domains*, Translations of mathematical monographs, Vol. 6, A.M.S., 1963.

[NS] E. Nelson and W. F. Stinespring, Representation of elliptic operators in an enveloping algebra, Amer. J. Math. 81 (1959), 547–560.

[Sc] W. Schmid, Die Randwerte holomorpher Funktionen auf hermitesch symmetrischen Räumen, Invent. Math. 9 (1969), 61–80.

[S1] G. Shimura, The arithmetic of certain zeta functions and automorphic forms on orthogonal groups, Ann. of Math. 111 (1980), 313–375.

[S2] ______, Arithmetic of differential operators on symmetric domains, Duke Math. J. 48 (1981), 813–843.

[S3] ______, Differential operators and the singular values of Eisenstein series, Duke Math. J. 51 (1984), 261–329.

[S4] ______, On differential operators attached to certain representations of classical groups, Invent. Math. 77 (1984), 463–488.

[S5] ______, On a class of nearly holomorphic automorphic forms, Ann. of Math. 123 (1986), 347–406.

[W] G. Warner, *Harmonic Analysis on Semisimple Lie Groups* I, Springer, 1972.

(Received January 17, 1989)

90b

On the fundamental periods of automorphic forms of arithmetic type

Inventiones mathematicae, 102 (1990), 399-428

Introduction

The automorphic forms to be investigated are defined with respect to a congruence subgroup Γ of an algebraic group G which consists of the invertible elements of a quaternion algebra B over a totally real algebraic number field E. We have $G_{\mathbb{R}}=GL_2(\mathbb{R})^{\delta}\times(\mathbb{H}^{\times})^{\delta'}$, where δ resp. δ' is the set of archimedean primes of E unramified resp. ramified in B, and $\mathbb{H}$ denotes the Hamilton quaternions. Such a B is said to be of signature (δ, δ'). We consider a holomorphic map h of H^{δ} into a complex vector space V such that

$$h(\gamma z)=\rho'(\gamma)J(\gamma, z)h(z) \quad \text{for every } \gamma\in\Gamma, \tag{1}$$

where H is the standard upper half complex plane, J is a holomorphic factor of automorphy of a certain weight, and ρ' is an irreducible representation: $(\mathbb{H}^{\times})^{\delta'}\to GL(V)$. We assume that h is $\bar{\mathbb{Q}}$-rational and $h|T_v=\chi(v)h$, that is, h is a common eigenform of the Hecke operators T_v for all finite primes v. In order to speak of such an eigenform, in general we have to deal with an automorphic form defined on the adelization $G_{\mathbb{A}}$ of G, which amounts to several forms h of the above type with several different Γ's. In the introduction, however, let us assume for simplicity that such a common eigenform h is meaningful with a single Γ, which is the case if a certain ideal class group in E is trivial. Assume also that $\{\chi, h\}$ is primitive in the sense that χ does not occur with a group of lower level than Γ. In our previous papers [7] and [9] we investigated the plausibility of assigning some nonzero complex numbers $P(\chi, \delta, \varepsilon; B)$ determined modulo algebraic numbers to each χ, δ, and each subset ε of δ. These constants P are closely connected with the special values of various zeta functions such as $\sum\chi(\mathfrak{a})N(\mathfrak{a})^{-s}$ and also conjecturally satisfy some marked relations. Assuming the weight of χ to be even, let us recall here the following six of them:

(I) *$P(\chi, \delta, \varepsilon; B)$ is independent of B up to algebraic factors.*

(II) *The periods of h, when suitably defined, are linear combinations of $\pi^n P(\chi, \delta, \varepsilon; B)$ for all $\varepsilon\subset\delta$.*

(III) $\overline{P(\chi, \delta, \varepsilon; B)}\sim P(\bar{\chi}, \delta, \varepsilon; B)$.

(IV) $\pi^n P(\chi, \delta, \varepsilon; B)\, P(\bar{\chi}, \delta, \delta-\varepsilon; B) \sim \langle h, h\rangle$.

(V) *If f is a Hilbert modular form with respect to $SL_2(E)$ given by $f(z) = \langle \theta(z, w), h(w)\rangle$ with a certain theta function θ on $H^\iota \times H^\delta$, then every Fourier coefficient of f is an algebraic number times $P(\chi, \delta, \zeta; B)$ with ζ determined by θ.*

(VI) *If B' is of signature (δ', δ) and χ occurs with B', then*

$$P(\chi, \iota, \zeta; M_2(E)) \sim P(\chi, \delta, \delta \cap \zeta; B)\, P(\chi, \delta' \cap \zeta; B')$$

for every $\zeta \subset \iota$.

Here we write $a \sim b$ when a/b is algebraic; $\langle\ ,\ \rangle$ is the inner product defined as usual (see (1.16) below); n is the number of elements of δ; $\iota = \delta \cup \delta'$, that is, ι represents the set of all archimedean primes of E.

In [9] we proved the existence of such P satisfying (I, III–V) or (II–V) according as $\delta' = \emptyset$ or δ consists of a single element. The purpose of the present paper is to establish the invariants P satisfying (II–V) for an arbitrary nonempty δ when B is a division algebra. As for (I) and (VI), which predict algebraic relations between the periods of forms belonging to different algebraic groups (even on spaces of different dimensions), we cannot claim any definitive result. However, we shall at least show that (I) and (VI) follow from our results, (V) in particular, combined with those in [9] under the assumption that certain inner products of Hilbert modular forms do not vanish.

Let us now briefly explain our basic ideas by describing the contents of each section. For a weight k and a subset ζ of δ we consider the space $\mathscr{S}_k^\zeta(\Gamma)$ of all maps $g\colon H^\delta \to V$ which are anti-holomorphic on H^ζ and holomorphic on $H^{\delta-\zeta}$, and which satisfy (1) with J modified in an obvious way. Section 1 is a collection of elementary facts on such $\mathscr{S}_k^\zeta$ and its adelized versions. In Sect. 2 we discuss three types of operators on $\mathscr{S}_k^\zeta$, namely, Hecke operators, complex conjugation, and an operator defined for each $\varepsilon \subset \delta$ which sends an element g of $\mathscr{S}_k^\zeta$ to an element g^ε of $\mathscr{S}_k^{\zeta+\varepsilon}$. Now with such a g we can associate a differential form $[g]$ on H^δ with values in a complex vector space U_k such that $[g] \circ \gamma = \rho_k(\gamma)[g]$ for every $\gamma \in \Gamma$, where ρ_k is a $GL(U_k)$-valued representation of $G_{\mathbf{R}}$ which is the tensor product of ρ' of (1) with a representation of $\prod_{v\in\delta} G_v$. We give several easy formulas about $[g]$ in Sect. 3.

We then consider in Sect. 4 the cohomology groups $H_k^*(\Gamma, S)$ obtained from the complex of all U_k-valued singular cochains φ such that $\varphi(\gamma c) = \rho_k(\gamma)\,\varphi(c)$ for every singular chain c on H^δ. From a theorem in [3] we know that the map $g \mapsto [g]$ defines an injection of $\prod_{\zeta\subset\delta} \mathscr{S}_k^\zeta(\Gamma)$ into $H_k^n(\Gamma, S)$. Then the notion of the periods of h required in (II) can be given as the values of the cochain $[h]$ at equivariant cycles in Eilenberg's sense. Since U has a natural $\mathbf{Q}$-structure, we have $H_k^*(\Gamma, S) = H_k^*(\Gamma, S, \mathbf{Q}) \otimes_{\mathbf{Q}} \mathbf{C}$ with the cohomology groups $H_k^*(\Gamma, S, \mathbf{Q})$ defined over $\mathbf{Q}$. For χ and h as above we can define a subspace K_χ in $H_k^n(\Gamma, S, \mathbf{Q})$ such that $K_\chi \otimes_{\mathbf{Q}} \mathbf{C} = \sum_{\zeta\subset\delta} \mathbf{C}[h^\zeta]$. Transferring the map $[g] \mapsto [g^\varepsilon]$ to K_χ, we can find an element y_ε of K_χ and a complex number p_ε for each $\varepsilon \subset \delta$ which satisfy

$$[h^\zeta] = \sum_{\varepsilon\subset\delta} (-1)^{a(\zeta,\varepsilon)} p_\varepsilon y_\varepsilon \tag{2}$$

for every $\zeta \subset \delta$, where $a(\zeta, \varepsilon)$ is the number of elements in $\zeta \cap \varepsilon$. Then we shall show that the constants $\pi^{-n} p_\varepsilon$ have properties (II), (III), and (IV) (Proposition 4.2 and Theorem 4.4). These results are true even when k is not even.

That the same constants satisfy (V) will be proved in Sect. 5. We already showed in [6] that each Fourier coefficient of $f(z)$ as in (V) is the integral of a certain holomorphic n-form closely related to $[h]$ over a singular n-cube. We shall see that this integral is a period of $[h]$ in the above sense, which is the principal reason why we have chosen singular cohomology in our formulation. The desired result (Theorem 5.1) can then be derived by modifying the technique of [9, Sect. 10]. The final section includes the observations about how (I) and (VI) can be proved under the assumption of a certain nonvanishing (Proposition 6.1).

1. Holomorphic and antiholomorphic automorphic forms

Throughout the paper we use the following notation: For an associative ring R with identity element we denote by $R^\times$ the group of all its invertible elements and by $M_n(R)$ the ring of $n \times n$-matrices with entries in R; we put then $GL_n(R) = GL(R^n) = M_n(R)^\times$. To indicate that a union $X = \bigcup_{i \in I} Y_i$ is disjoint, we write $X = \coprod_{i \in I} Y_i$. The algebraic closure of the rational number field $\mathbb{Q}$ in the complex number field $\mathbb{C}$ is denoted by $\bar{\mathbb{Q}}$.

As explained in the introduction, we consider automorphic forms with respect to a quaternion algebra B over a totally real algebraic number field E. We shall eventually assume that B is a division algebra, but in the first three sections our treatment includes the case $B = M_2(E)$. We let $\mathbf{a}$ and $\mathbf{f}$ denote the sets of archimedean and nonarchimedean primes of E, respectively, $\mathfrak{g}$ the maximal order in E, and $\mathfrak{d}$ the different of E over $\mathbb{Q}$. We put $G = B^\times$ and define the adelizations $E_\mathbf{A}$, $B_\mathbf{A}$, and $G_\mathbf{A} (= B_\mathbf{A}^\times)$ of E, B, and G, and also their archimedean and nonarchimedean factors $E_\mathbf{a}$, $B_\mathbf{a}$, $G_\mathbf{a}$, $E_\mathbf{f}$, $B_\mathbf{f}$, and $G_\mathbf{f}$ as usual. Given a subset ε of $\mathbf{a}$ and a set X which is often a subset of $\mathbb{C}$ such as $\mathbb{Z}$ or $\mathbb{R}$, we denote by X^ε the product of ε copies of X, that is, the set of all indexed elements $(x_v)_{v \in \varepsilon}$ with x_v in X. For x and y in $\mathbb{C}^\mathbf{a}$ we put

$$\|y\| = \sum_{v \in \mathbf{a}} y_v, \tag{1.1a}$$

$$x^y = \prod_{v \in \mathbf{a}} x_v^{y_v}, \tag{1.1b}$$

whenever the factors $x_v^{y_v}$ are well defined according to the context. We use the notation x^y also for $x \in E_\mathbf{A}^\times$ when the product is meaningful. For $x \in \mathbb{R}^\mathbf{a}$ or $x \in E_\mathbf{A}$ we write $x \gg 0$ if $x_v > 0$ for every $v \in \mathbf{a}$.

We denote by $\mathfrak{d}_B$ the product of all the prime ideals of E ramified in B, and by δ resp. δ' the set of all archimedean primes which are unramified resp. ramified in B. Then there is an isomorphism

$$B_\mathbf{a} \to M_2(\mathbb{R})^\delta \times \mathbb{H}^{\delta'}, \tag{1.2}$$

where $\mathbb{H}$ denotes the Hamilton quaternions. For each $v \in \mathbf{a} \cup \mathbf{f}$, we can define the localizations E_v, B_v, G_v, etc., which we consider subsets of $E_\mathbb{A}$, $B_\mathbb{A}$ or $G_\mathbb{A}$. If $x \in B_\mathbb{A}$, its v-component, $\mathbf{a}$-component, and $\mathbf{f}$-component can naturally be defined and are denoted by x_v, $x_\mathbf{a}$, and $x_\mathbf{f}$. We denote by x^* the image of $x \in B$ under the main involution of B, and put $N(x) = xx^*$ and $\mathrm{Tr}(x) = x + x^*$. The maps $x \mapsto x^*$, $x \mapsto N(x)$, and $x \mapsto \mathrm{Tr}(x)$ can be extended to maps of B_v or $B_\mathbb{A}$ into B_v, $B_\mathbb{A}$, E_v, or $E_\mathbb{A}$. We put $x^{-*} = (x^*)^{-1}$.

Throughout the paper we assume that $\delta \neq \emptyset$, put

(1.3) $$n = \text{the number of elements in } \delta,$$

and consider automorphic forms on H^δ, where

$$H = \{z \in \mathbb{C} \mid \mathrm{Im}(z) > 0\}.$$

To define them, we first take an $\mathbb{R}$-rational irreducible polynomial representation $\sigma_m \colon \mathbb{H}^\times \to GL(\mathbb{C}^{m+1})$ of degree m. Changing it for an equivalent representation and identifying $B_\mathbf{a}$ with $M_2(\mathbb{R})^\delta \times \mathbb{H}^{\delta'}$ by a suitably fixed isomorphism (1.2), we may assume:

(1.4a) $$x_v \in M_2(\mathbb{Q}) \quad \textit{for every } v \in \delta \textit{ and every } x \in B;$$

(1.4b) $$\sigma_m(x_v) \in M_{m+1}(\mathbb{Q}) \quad \textit{for every } v \in \delta' \textit{ and every } x \in B;$$

(1.4c) $$\sigma_m(x^*) = {}^t\overline{\sigma_m(x)} \quad \textit{for every } x \in \mathbb{H}.$$

We then identify $G_\mathbf{a}$ with $GL_2(\mathbb{R})^\delta \times (\mathbb{H}^\times)^{\delta'}$. For $\xi \in G_\mathbb{A}$ and $z \in \mathbb{C}^\delta$ with $\mathrm{Im}(z) \in (\mathbb{R}^\times)^\delta$ we define $\xi(z)$ and $j(\xi, z)$ as elements of $\mathbb{C}^\delta$ by

(1.5a) $$\xi z = \xi(z) = ((a_v z_v + b_v)(c_v z_v + d_v)^{-1})_{v \in \delta},$$

(1.5b) $$j(\xi, z) = (|N(\xi_v)|^{-1/2}(c_v z_v + d_v))_{v \in \delta},$$

where a_v, b_v, c_v, d_v are the entries of ξ_v in the standard order. Now we fix a "weight", which is an element k of $\mathbb{Z}^\mathbf{a}$ such that $k_v > 0$ for $v \in \delta$ and $k_v > 1$ for $v \in \delta'$, and define a representation ρ'_k of $G_\mathbb{A}$ by

(1.6a) $$\rho'_k(x) = \bigotimes_{v \in \delta'} \sigma_{k_v - 2}(N(x_v)^{-1/2} x_v) \quad (x \in G_\mathbb{A}),$$

and denote by V_k the representation space of ρ'_k, which is $\bigotimes_{v \in \delta'} \mathbb{C}^{k_v - 1}$. From (1.4c) we obtain

(1.6b) $$\rho'_k(x^*) = \rho'_k(x)^{-1} = {}^t\overline{\rho'_k(x)} \quad (x \in G_\mathbb{A}).$$

Notice that (1.4b) means that V_k has a $\mathbb{Q}$-rational structure in the sense that $V_k = V_k(\mathbb{Q}) \otimes_\mathbb{Q} \mathbb{C}$ with a vector space $V_k(\mathbb{Q})$ over $\mathbb{Q}$ stable under $\rho'_k(x)$ for every $x \in G$.

For each subset ε of δ we put

(1.7) $$G_\varepsilon = \{x \in G_\mathbb{A} \mid N(x_v) < 0 \textit{ for } v \in \varepsilon \textit{ and } N(x_v) > 0 \textit{ for } v \in \delta - \varepsilon\},$$

and for $\beta \in G_\varepsilon$ we define a map β_* of H^δ onto itself by

(1.8) $$(\beta_* z)_v = \begin{cases} \beta_v \bar{z}_v & \text{if } v \in \varepsilon, \\ \beta_v z_v & \text{if } v \in \delta - \varepsilon. \end{cases}$$

We shall often view ε as an element of $\mathbb{Z}^\delta$ by putting $\varepsilon_v = 1$ or 0 according as $v \in \varepsilon$ or $v \notin \varepsilon$. Then $k\varepsilon$ is meaningful as an element of the ring $\mathbb{Z}^\delta$. We shall also view ε as an element of $(\mathbb{Z}/2\mathbb{Z})^\delta$ in a similar way. Thus for two subsets ε and ζ of δ, their sum and difference can be defined as elements of $\mathbb{Z}^\delta$ or $(\mathbb{Z}/2\mathbb{Z})^\delta$, but the distinction will be clear from the context. For instance, if we speak of $G_{\varepsilon+\zeta}$, then naturally $\varepsilon + \zeta \in (\mathbb{Z}/2\mathbb{Z})^\delta$. Notice that $\alpha\beta \in G_{\varepsilon+\zeta}$ if $\alpha \in G_\varepsilon$ and $\beta \in G_\zeta$; moreover $(\alpha\beta)_* = \alpha_* \beta_*$. The zero element 0 of $\mathbb{Z}^\delta$ or $(\mathbb{Z}/2\mathbb{Z})^\delta$ represents $\emptyset$, and therefore we use 0 for $\emptyset$ as an element of either ring; for instance we have

(1.9) $$G_0 = \{x \in G_\mathbb{A} \mid N(x) \gg 0\},$$

which is a special case of (1.7).

Given k and ε as above, we define a (semi-) factor of automorphy $j_k^\varepsilon(\xi, z)$ for $\xi \in G_\mathbb{A}$ and $z \in H^\delta$ by

(1.10a) $$j_k^\varepsilon(\xi, z) = j(\xi, \bar{z})^{k\varepsilon} j(\xi, z)^{k\delta - k\varepsilon} \rho_k'(\xi),$$

where the first two factors on the right-hand side should be understood in the sense of (1.1b). We can easily verify that

(1.10b) $$j_k^\varepsilon(\xi\eta, z) = j_k^\zeta(\xi, \eta_* z) j_k^\varepsilon(\eta, z) \quad \text{if } \eta \in G_{\varepsilon+\zeta},$$

(1.10c) $$j_k^\varepsilon(c, z) = \operatorname{sgn}(c)^k \quad \text{if } c \in E_\mathbb{A}^\times.$$

For any $\xi \in G_\mathbb{A}$ and a V_k-valued function f on H^δ we define a function $f \|_k^\varepsilon \xi$ of the same type by

(1.11) $$(f \|_k^\varepsilon \xi)(z) = j_k^\varepsilon(\xi, z)^{-1} f(\xi_* z) \quad (z \in H^\delta).$$

From (1.10b) we obtain

(1.12) $$f \|_k^\varepsilon (\xi\eta) = (f \|_k^\zeta \xi) \|_k^\varepsilon \eta \quad \text{if } \eta \in G_{\varepsilon+\zeta}.$$

It should be noted that our notation is somewhat different from that of [9, Sects. 3, 6]; k and κ there correspond to $k\delta$ and $k\delta' - 2\delta'$ here; similarly $f \|_{k,\kappa} \xi$ there is now $f \|_k^0 \xi$.

Now given a congruence subgroup Γ of G or G^u defined by

(1.13) $$G^u = \{x \in G \mid N(x) = 1\},$$

we denote by $\mathscr{S}_k^\varepsilon(\Gamma)$ the set of all V_k-valued functions f on H^δ satisfying the following conditions:

(1.14a) $f \|_k^\varepsilon \gamma = f$ *for every* $\gamma \in \Gamma$;

(1.14b) $f(z)$ *is holomorphic in* z_v *for every* $v \in \delta - \varepsilon$ *and antiholomorphic in* z_v *for every* $v \in \varepsilon$;

(1.14c) *If* $B = M_2(E)$, *then* f *is rapidly decreasing at every cusp in the sense that* $f\|_k^0\beta$ *is a holomorphic Hilbert cusp form for some (and hence every)* $\beta \in G \cap G_\varepsilon$.

From (1.12) we easily obtain

(1.15) $$\mathscr{S}_k^\varepsilon(\Gamma)\|_k^\zeta\eta = \mathscr{S}_k^\zeta(\eta^{-1}\Gamma\eta) \quad \textit{if } \eta \in G \cap G_{\varepsilon+\zeta}.$$

We denote by $\mathscr{S}_k^\varepsilon(B)$ the union of $\mathscr{S}_k^\varepsilon(\Gamma)$ for all congruence subgroups Γ of G^u. Given f and g in $\mathscr{S}_k^\varepsilon(B)$, we take Γ so that both f and g belong to $\mathscr{S}_k^\varepsilon(\Gamma)$ and put

(1.16) $$\langle f, g\rangle = \mathrm{vol}(D)^{-1} \int_D {}^t\overline{f(z)}\, g(z)\, \mathrm{Im}(z)^{k\delta}\, d_H^\delta z,$$

where $D = \Gamma\backslash H^\delta$, $\mathrm{vol}(D) = \int_D d_H^\delta z$, and

(1.16a) $$d_H^\delta z = (2i)^{-n} \prod_{v\in\delta} \mathrm{Im}(z_v)^{-2}\, d\bar{z}_v \wedge dz_v.$$

It can easily be seen that

(1.17) $\langle f\|_k^\zeta\beta, g\|_k^\zeta\beta\rangle = \langle f, g\rangle$ *for every* $\zeta \subset \delta$ *and every* $\beta \in G \cap G_{\varepsilon+\zeta}$.

To define automorphic forms on $G_{\mathbf{A}}$, we take an integral ideal $\mathfrak{m}$ in E, fix a maximal order $\mathfrak{o}$ in B, and define an order $\mathfrak{o}_1$ in B by

(1.18) $$\mathfrak{o}_{1v} = \begin{cases} \mathfrak{g}_v + \mathfrak{m}_v\mathfrak{o}_v & \textit{if } v \mid \mathfrak{d}_B, \\ \mu_v^{-1}(\{x \in M_2(F_v) \mid a_x \in \mathfrak{g}_v,\, b_x \in \mathfrak{d}_v^{-1},\, c_x \in \mathfrak{m}\mathfrak{d}_v,\, d_x \in \mathfrak{g}_v\}) & \textit{if } v \nmid \mathfrak{d}_B, \end{cases}$$

where μ_v, for each $v \in \mathbf{f}$ prime to $\mathfrak{d}_B$, is a fixed isomorphism of B_v onto $M_2(E_v)$ such that $\mu_v(\mathfrak{o}_v) = M_2(\mathfrak{g}_v)$. Then we define subgroups $W_\mathfrak{m}$ and $W_\mathfrak{m}^1$ of $G_{\mathbf{A}}$, written simply W and W^1, by

(1.19a) $$W = W_\mathfrak{m} = (G_{\mathbf{a}} \cap G_0) \prod_{v\in\mathbf{f}} \mathfrak{o}_{1v}^\times,$$

(1.19b) $$W^1 = W_\mathfrak{m}^1 = \{x \in W \mid a_v(x) - 1 \in \mathfrak{m}_v \textit{ if } v \mid \mathfrak{m} \textit{ and } v \nmid \mathfrak{d}_B\},$$

where $a_v(x)$ is the a-entry of $\mu_v(x_v)$. Since $N(W) = N(W^1)$, we can easily show by virtue of strong approximation that

(1.20) $$GxW = GxW^1 = \{y \in G_{\mathbf{A}} \mid N(y) \in E^\times N(xW)\}.$$

Take a coset decomposition $E_{\mathbf{A}}^\times = \coprod_{\lambda\in\Lambda} E^\times t_\lambda N(W)$ with elements $t_\lambda \in E_\mathbf{f}^\times$ and then take $x_\lambda \in G_\mathbf{f}$ so that $N(x_\lambda) = t_\lambda$ and $\mu_v(x_\lambda) = \mathrm{diag}[1, t_{\lambda v}]$ for $v \mid \mathfrak{m}$, $v \nmid \mathfrak{d}_B$. Then we have

(1.21) $$G_{\mathbf{A}} = \coprod_{\lambda\in\Lambda} Gx_\lambda W = \coprod_{\lambda\in\Lambda} Gx_\lambda W^1.$$

Fixing such x_λ, we put

(1.22a) $$W_\lambda = x_\lambda W x_\lambda^{-1}, \quad W_\lambda^1 = x_\lambda W^1 x_\lambda^{-1},$$

(1.22b) $$\Gamma_\lambda = G \cap W_\lambda, \quad \Gamma_\lambda^1 = G \cap W_\lambda^1.$$

Take a Hecke character Φ of E which is by definition a continuous homomorphism of $E_{\mathbf{A}}^\times$ into $\mathbf{C}^\times$ such that $\Phi(E^\times)=1$ and $|\Phi|=1$. We assume:

(1.23) *The conductor of* Φ *is prime to* $\mathfrak{d}_B$ *and divides* $\mathfrak{m}$; $\Phi(x)=\operatorname{sgn}(x)^k |x|^{2i\kappa}$ *for every* $x \in E_{\mathbf{a}}^\times$,

where κ is a fixed element of $\mathbf{R}^{\mathbf{a}}$ such that $\|\kappa\|=0$. We then define $\mathscr{S}_k^e(\mathfrak{m}, \Phi_{\mathfrak{m}}, \kappa)$ to be the set of all V_k-valued functions $\mathbf{g}$ on $G_{\mathbf{A}}$ satisfying the following two conditions:

(1.24a) $\mathbf{g}(\alpha x u) = \Phi_{\mathfrak{m}}(d_u)\, \mathbf{g}(x)$ *for every* $\alpha \in G$ *and every* $u \in W_{\mathfrak{f}}$;

(1.24b) *For each* $x \in G_{\mathfrak{f}}$, *there is an element* g_x *of* $\mathscr{S}_k^e(B)$ *such that* $\mathbf{g}(xy) = N(y)^{i\kappa}(g_x \|_k^e y)(\mathbf{i})$ *for all* $y \in G_0 \cap G_{\mathbf{a}}$,

where $\mathbf{i}$ is the point $(i, \ldots, i)$ of H^δ and $\Phi_{\mathfrak{m}} = \prod_{v|\mathfrak{m}} \Phi_v$; d_u is the element of $E_{\mathfrak{f}}^\times$ whose v-component is 1 except when $v|\mathfrak{m}$ and $v \nmid \mathfrak{d}_B$, in which case the v-component is the d-entry of $\mu_v(u_v)$. We define a_u similarly with a-entry instead of d-entry. (The symbol a_u or d_u will appear only in the form $\Phi_{\mathfrak{m}}(a_u)$ or $\Phi_{\mathfrak{m}}(d_u)$.) We denote by $\mathscr{S}_k^e(\mathfrak{m}, \Phi)$ the subset of $\mathscr{S}_k^e(\mathfrak{m}, \Phi_{\mathfrak{m}}, \kappa)$ consisting of all $\mathbf{g}$ satisfying

(1.24c) $$\mathbf{g}(sx) = \Phi(s)\, \mathbf{g}(x) \quad \textit{for every } s \in E_{\mathbf{A}}^\times.$$

With Γ_λ as in (1.22b), we denote by $\mathscr{S}_k^e(\Gamma_\lambda, \Phi_{\mathfrak{m}}, \kappa)$ the subset of $\mathscr{S}_k^e(B)$ consisting of the elements f such that

(1.25) $$f \|_k^e \gamma = \Phi_{\mathfrak{m}}(a_\gamma) N(\gamma)^{i\kappa} f \quad \textit{for every } \gamma \in \Gamma_\lambda.$$

Given $(f_\lambda)_{\lambda \in \Lambda} \in \prod_{\lambda \in \Lambda} \mathscr{S}_k^e(\Gamma_\lambda, \Phi_{\mathfrak{m}}, \kappa)$, define a V_k-valued function $\mathbf{g}$ on $G_{\mathbf{A}}$ by

(1.26) $$\mathbf{g}(\alpha x_\lambda^{-*} u) = \Phi_{\mathfrak{m}}(d_u) N(u)^{i\kappa} (f_\lambda \|_k^e u)(\mathbf{i}) \quad (\alpha \in G, \lambda \in \Lambda, u \in W).$$

This is well-defined and $\mathbf{g} \in \mathscr{S}_k^e(\mathfrak{m}, \Phi_{\mathfrak{m}}, \kappa)$. (Notice that $G_{\mathbf{A}} = \coprod_{\lambda \in \Lambda} G x_\lambda^{-*} W$.) Conversely every $\mathbf{g}$ in $\mathscr{S}_k^e(\mathfrak{m}, \Phi_{\mathfrak{m}}, \kappa)$ can be obtained from $(f_\lambda)_{\lambda \in \Lambda}$ in this fashion. Hereafter we shall identify $\mathbf{g}$ with $(f_\lambda)_{\lambda \in \Lambda}$ and write simply $\mathbf{g} = (f_\lambda)_{\lambda \in \Lambda}$ when (1.26) holds, and identify $\mathscr{S}_k^e(\mathfrak{m}, \Phi_{\mathfrak{m}}, \kappa)$ with $\prod_{\lambda \in \Lambda} \mathscr{S}_k^e(\Gamma_\lambda, \Phi_{\mathfrak{m}}, \kappa)$. For $\mathbf{g} = (f_\lambda)_{\lambda \in \Lambda}$ and $\mathbf{g}' = (f'_\lambda)_{\lambda \in \Lambda}$ we put

(1.27) $$\langle \mathbf{g}, \mathbf{g}' \rangle = |\Lambda|^{-1} \sum_{\lambda \in \Lambda} \langle f_\lambda, f'_\lambda \rangle,$$

where $|\Lambda|$ is the number of elements in Λ. If we replace $\mathfrak{m}$ by its multiple or change x_λ's, then the groups Γ_λ as well as $|\Lambda|$ may need to be changed. However it can easily be verified that $\langle \mathbf{g}, \mathbf{g}' \rangle$ of (1.27) is determined independently of the choice of $\mathfrak{m}$ and x_λ. Also, from (1.24a, b, c) we easily see that $\mathscr{S}_k^e(\mathfrak{m}, \Phi_{\mathfrak{m}}, \kappa)$

is the direct sum of the spaces $\mathcal{S}_k^\varepsilon(\mathfrak{m}, \Phi\psi)$ for all the Hecke characters ψ of $E_{\mathbf{A}}^\times/(E^\times E_{\mathbf{a}}^\times \prod_{v\in\mathbf{f}} \mathfrak{g}_v^\times)$. If $\mathbf{g}\in\mathcal{S}_k^\varepsilon(\mathfrak{m}, \Phi)$, (1.26) takes the form

$$(1.28)\qquad \mathbf{g}(\alpha x_\lambda u)=\Phi(t_\lambda)\Phi_{\mathfrak{m}}(d_u)N(u)^{i\kappa}(f_\lambda\|_k^\varepsilon u)(\mathbf{i}) \quad (\alpha\in G, \lambda\in\Lambda, u\in W).$$

2. Operators on automorphic forms

We are going to introduce the following three types of operators on $\mathcal{S}_k^\varepsilon(\mathfrak{m}, \Phi_{\mathfrak{m}}, \kappa)$: (i) Hecke operators; (ii) complex conjugation; (iii) an operator which sends $\mathcal{S}_k^\zeta(\mathfrak{m}, \Phi_{\mathfrak{m}}, \kappa)$ onto $\mathcal{S}_k^{\varepsilon+\zeta}(\mathfrak{m}, \Phi_{\mathfrak{m}}, \kappa)$. Let us begin with the last one.

Lemma 2.1. *Given $\varepsilon\subset\delta$, let e be the element of $G_{\mathbf{A}}$ such that $e_v=\mathrm{diag}[-1, 1]$ for $v\in\varepsilon$ and $e_v=1$ for all other v. For $\mathbf{g}\in\mathcal{S}_k^\zeta(\mathfrak{m}, \Phi_{\mathfrak{m}}, \kappa)$ define a V_k-valued function $\mathbf{g}^\varepsilon$ on $G_{\mathbf{A}}$ by $\mathbf{g}^\varepsilon(x)=\mathbf{g}(xe)$. Then $\mathbf{g}^\varepsilon\in\mathcal{S}_k^{\varepsilon+\zeta}(\mathfrak{m}, \Phi_{\mathfrak{m}}, \kappa)$; moreover $\mathbf{g}^\varepsilon\in\mathcal{S}_k^{\varepsilon+\zeta}(\mathfrak{m}, \Phi)$ if $\mathbf{g}\in\mathcal{S}_k^\zeta(\mathfrak{m}, \Phi)$. Furthermore if $\mathbf{g}=(f_\lambda)_{\lambda\in\Lambda}$ and $\mathbf{g}^\varepsilon=(f'_\lambda)_{\lambda\in\Lambda}$ with $f_\lambda\in\mathcal{S}_k^\zeta(\Gamma_\lambda, \Phi_{\mathfrak{m}}, \kappa)$ and $f'_\lambda\in\mathcal{S}_k^{\varepsilon+\zeta}(\Gamma_\lambda, \Phi_{\mathfrak{m}}, \kappa)$, then*

$$(2.1)\qquad f'_\mu=(-1)^{\|k\varepsilon\|}|N(\beta)|^{-i\kappa}f_\lambda\|_k^{\varepsilon+\zeta}\beta,$$

where μ is determined by λ and e by the condition $x_\lambda e\in G x_\mu W$, and β is an element of G such that $x_\lambda e\in\beta x_\mu W^1$. Any such β is contained in G_ε and satisfies $\beta\Gamma_\mu=\Gamma_\lambda\beta=G\cap e x_\lambda W x_\mu^{-1}$ and $\beta\Gamma_\mu^1=\Gamma_\lambda^1\beta=G\cap e x_\lambda W^1 x_\mu^{-1}$.

Proof. That $\mathbf{g}^\varepsilon$ satisfies (1.24a) is clear. By (1.21) we can find μ and β as specified. Clearly $\beta\in G_\varepsilon$. Take $w\in W^1$ so that $x_\lambda e=\beta x_\mu w$. Then $\beta=x_\lambda e w^{-1}x_\mu^{-1}$, and hence we easily see that $\beta W_\mu=W_\lambda\beta=e x_\lambda W x_\mu^{-1}$ and $\beta W_\mu^1=W_\lambda^1\beta=e x_\lambda W^1 x_\mu^{-1}$. Taking the intersection of these with G, we obtain the last equalities of our lemma. For $\alpha\in G$ and $u\in W$ we have $\mathbf{g}^\varepsilon(\alpha x_\mu^{-*}u)=\mathbf{g}(x_\mu^{-*}ue)=\mathbf{g}(\beta^* x_\lambda^{-*}e^{-*}w^* ue)$ $=\Phi_{\mathfrak{m}}(d_u)|N(u)|^{i\kappa}|N(\beta)|^{-i\kappa}[f_\lambda\|_k^\zeta(\beta^{-*}ue)](\mathbf{i})$ since $N(e)=N(\beta w)_{\mathbf{a}}$ and $(e^{-*}w^*)_{\mathbf{a}}$ $=\beta_{\mathbf{a}}^{-*}$. Now $f_\lambda\|_k^\zeta(\beta^{-*}ue)=f_\lambda\|_k^\zeta(N(\beta)^{-1}\beta ue)=(-1)^{\|k\varepsilon\|}f_\lambda\|_k^\zeta(\beta ue)$ by (1.10c) and (1.12). Since $j_k^\zeta(e, z)=1$ and $e_*(\mathbf{i})=\mathbf{i}$, we have $[f_\lambda\|_k^\zeta(\beta ue)](\mathbf{i})=[(f_\lambda\|_k^{\varepsilon+\zeta}\beta)\|_k^{\varepsilon+\zeta}u](\mathbf{i})$ by (1.12). Combining these, we obtain $\mathbf{g}^\varepsilon(\alpha x_\mu^{-*}u)=\Phi_{\mathfrak{m}}(d_u)|N(u)|^{i\kappa}(f'_\mu\|_k^{\varepsilon+\zeta}u)(\mathbf{i})$ with f'_μ as in (2.1). From (1.15) we easily see that $f'_\mu\in\mathcal{S}_k^{\varepsilon+\zeta}(\Gamma_\mu, \Phi_{\mathfrak{m}}, \kappa)$, and hence $\mathbf{g}^\varepsilon\in\mathcal{S}_k^{\varepsilon+\zeta}(\mathfrak{m}, \Phi_{\mathfrak{m}}, \kappa)$. If $\mathbf{g}$ satisfies (1.24c), so does $\mathbf{g}^\varepsilon$. This completes the proof.

From the definition of $\mathbf{g}^\varepsilon$ we see that

$$(2.2)\qquad (\mathbf{g}^\varepsilon)^\zeta=\mathbf{g}^{\varepsilon+\zeta} \quad \textit{for every } \varepsilon, \zeta\subset\delta.$$

To consider Hecke operators, we let $\mathcal{Y}$ denote the set of all y in G_0 such that $y_v\in\mathfrak{o}_{1v}$ for all $v|\mathfrak{m}$, $y_v\in\mathfrak{o}_{1v}^\times$ if $v|\mathfrak{m}+\mathfrak{d}_B$, and the a-entry of $\mu_v(y_v)$ belongs to $\mathfrak{g}_v^\times$ if $v|\mathfrak{m}$ and $v\nmid\mathfrak{d}_B$. We let each double coset Wy_0W with $y_0\in\mathcal{Y}$ act on $\mathcal{S}_k^\zeta(\mathfrak{m}, \Phi_{\mathfrak{m}}, \kappa)$ in the following way. Taking a finite subset Y of $G_{\mathbf{f}}$ such that $Wy_0W=\coprod_{y\in Y} Wy$, for $\mathbf{g}\in\mathcal{S}_k^\zeta(\mathfrak{m}, \Phi_{\mathfrak{m}}, \kappa)$ we define $\mathbf{g}|Wy_0W$ by

$$(2.3)\qquad (\mathbf{g}|Wy_0W)(x)=\sum_{y\in Y}\Phi_{\mathfrak{m}}(a_y)^{-1}\mathbf{g}(xy^*) \quad (x\in G_{\mathbf{A}}).$$

It can easily be seen that $\mathbf{g}|Wy_0W\in\mathcal{S}_k^\zeta(\mathfrak{m}, \Phi_{\mathfrak{m}}, \kappa)$.

We now define Hecke operators T_v and S_v (of level $\mathfrak{m}$) for each $v \in \mathbf{f}$ as follows. We first put $T_v = 0$ for $v | (\mathfrak{m} + \mathfrak{d}_B)$ and $S_v = 0$ for $v | \mathfrak{m}$. Excluding these special cases, we take a prime element π_v of E_v and consider an element y_0 of G_v, viewed as an element of $G_{\mathbf{A}}$, of the following two types:

(2.4a) $N(y_0) = \pi_v$, $v \nmid \mathfrak{m} + \mathfrak{d}_B$; *moreover* $\mu_v(y_0) = \mathrm{diag}[1, \pi_v]$ *if* $v \nmid \mathfrak{d}_B$.

(2.4b) $$y_0 = \pi_v, v \nmid \mathfrak{m}.$$

Notice that $y_0 \in \mathfrak{o}_{1v}$ for y_0 of (2.4a) even when $v | \mathfrak{d}_B$. In any case $W y_0 W$ is determined by v and independent of the choice of π_v or y_0. We then denote the operator $W y_0 W$ by T_v for y_0 of type (2.4a) and by S_v for that of type (2.4b). Clearly $\mathscr{S}_k^\zeta(\mathfrak{m}, \Phi)$ is stable under $W y_0 W$ for every $y_0 \in \mathscr{Y}$, and

(2.5) $$(\mathbf{g} | W y_0 W)^\varepsilon = \mathbf{g}^\varepsilon | W y_0 W \quad \textit{for every } \varepsilon \subset \delta.$$

The right-hand side of (2.3) for the operator S_v is simply $\mathbf{g}(\pi_v x)$. Therefore we easily see that

(2.6) $$\mathscr{S}_k^\zeta(\mathfrak{m}, \Phi) = \{\mathbf{g} \in \mathscr{S}_k^\zeta(\mathfrak{m}, \Phi_\mathfrak{m}, \kappa) \mid \mathbf{g} | S_v = \Phi(\pi_v)\mathbf{g} \ \textit{for every } v \in \mathbf{f}, v \nmid \mathfrak{m}\}.$$

Lemma 2.2. *Given an index $\lambda \in \Lambda$ and $y_0 \in \mathscr{Y}$, there is a unique index $\nu \in \Lambda$ and an element α_0 of G such that $x_\lambda y_0 \in \alpha_0 x_\nu W$. For such α_0 and ν we have $W_\lambda \alpha_0 W_\nu = W_\lambda \alpha_0 \Gamma_\nu^1$ and $\Gamma_\lambda \alpha_0 \Gamma_\nu = \Gamma_\lambda \alpha_0 \Gamma_\nu^1$. If $\Gamma_\lambda \alpha_0 \Gamma_\nu = \coprod_{\alpha \in A} \Gamma_\lambda \alpha$, then $W y_0 W = \coprod_{\alpha \in A} W x_\lambda^{-1} \alpha x_\nu$. Moreover, if $\Gamma_\lambda^1 \alpha_0 \Gamma_\nu^1 = \coprod_{\gamma \in C} \Gamma_\lambda^1 \gamma$, then $\Gamma_\lambda \alpha_0 \Gamma_\nu = \coprod_{\gamma \in C} \Gamma_\lambda \gamma$.*

Proof. We can find ν and α_0 as specified by virtue of (1.21). Put $y = x_\lambda^{-1} \alpha_0 x_\nu$. Then $y_0 \in y W$ so that $W y_0 W = W y W$. Put $y_0 = y w$ and $U = y^{-1} W y \cap W^1$. Then $U = w y_0^{-1} W y_0 w^{-1} \cap W^1 = w(y_0^{-1} W y_0 \cap W^1) w^{-1}$. From our definition of $\mathscr{Y}$ we easily see that $N(y_0^{-1} W y_0 \cap W^1) = N(W)$, and hence $N(U) = N(W)$. Now $x_\nu U x_\nu^{-1} = \alpha_0^{-1} W_\lambda \alpha_0 \cap W_\nu^1$ and so $N(\alpha_0^{-1} W_\lambda \alpha_0 \cap W_\nu^1) = N(U) = N(W) = N(W_\nu^1)$. Therefore by [5, II, Lemma 1.1] we have $W_\lambda \alpha_0 W_\nu^1 = W_\lambda \alpha_0 \Gamma_\nu^1$. Observing that W is generated by W^1 and the elements x of $\mathfrak{o}_v$ such that $\mu_v(x) = \mathrm{diag}[a_v, 1]$ with $a_v \in \mathfrak{g}_v^\times$ for all $v | \mathfrak{m}$, $v \nmid \mathfrak{d}_B$, we easily see that $W y_0 W = W y_0 W^1$. Then $W y W^1 = W y_0 w^{-1} W^1 = W y_0 W^1 w^{-1} = W y_0 W = W y W$. Therefore $W_\lambda \alpha_0 W_\nu = x_\lambda W y W x_\nu^{-1} = x_\lambda W y W^1 x_\nu^{-1} = W_\lambda \alpha_0 W_\nu^1 = W_\lambda \alpha_0 \Gamma_\nu^1$, and hence $\Gamma_\lambda \alpha_0 \Gamma_\nu \subset G \cap W_\lambda \alpha_0 W_\nu = G \cap W_\lambda \alpha_0 \Gamma_\nu^1 = \Gamma_\lambda \alpha_0 \Gamma_\nu^1$. Let $\Gamma_\lambda \alpha_0 \Gamma_\nu = \coprod_{\alpha \in A} \Gamma_\lambda \alpha$. Then $W_\lambda \alpha_0 W_\nu = W_\lambda \alpha_0 \Gamma_\nu = \bigcup_{\alpha \in A} W_\lambda \alpha$, and clearly this is a disjoint union. Then $W y_0 W = W y W = x_\lambda^{-1} W_\lambda \alpha_0 W x_\nu = \coprod_{\alpha \in A} x_\lambda^{-1} W_\lambda \alpha x_\nu = \coprod_{\alpha \in A} W x_\lambda^{-1} \alpha x_\nu$. Finally let $\Gamma_\lambda^1 \alpha_0 \Gamma_\nu^1 = \coprod_{\gamma \in C} \Gamma_\lambda^1 \gamma$. Then $\Gamma_\lambda \alpha_0 \Gamma_\nu = \Gamma_\lambda \alpha_0 \Gamma_\nu^1 = \bigcup_{\gamma \in C} \Gamma_\lambda \gamma$. To prove that this is a disjoint union, it is sufficient to do so for a special choice of C. We can take C from $\alpha_0 \Gamma_\nu^1$. Then for γ and γ' in C we see that $\gamma' \gamma^{-1} \in \alpha_0 \Gamma_\nu^1 \alpha_0^{-1}$. From our definition of $\mathscr{Y}$ we easily see that $W \cap y_0 W^1 y_0^{-1} \subset W^1$. Therefore $W_\lambda \cap \alpha_0 W_\nu^1 \alpha_0^{-1} = W_\lambda \cap x_\lambda y_0 w^{-1} W^1 w y_0^{-1} x_\lambda^{-1} = x_\lambda (W \cap y_0 W^1 y_0^{-1}) x_\lambda^{-1} \subset x_\lambda W^1 x_\lambda^{-1} = W_\lambda^1$. Suppose $\Gamma_\lambda \gamma = \Gamma_\lambda \gamma'$. Then $\gamma' \gamma^{-1} \in \Gamma_\lambda \cap \alpha_0 \Gamma_\nu^1 \alpha_0^{-1} \subset G \cap W_\lambda \cap \alpha_0 W_\nu^1 \alpha_0^{-1} \subset G \cap W_\lambda^1 = \Gamma_\lambda^1$, which completes the proof.

Lemma 2.3. *Let* $\mathbf{g}=(f_\lambda)_{\lambda\in\Lambda}\in\mathcal{S}_k^\zeta(\mathfrak{m}, \Phi_{\mathfrak{m}}, \kappa)$ *and* $\mathbf{g}|Wy_0W=(f'_\lambda)_{\lambda\in\Lambda}$. *Given* λ, *take* ν, α_0, *and a set A as in Lemma 2.2. Then*

$$f'_\nu=\sum_{\alpha\in A}\Phi_{\mathfrak{m}}(a_\alpha)^{-1}N(\alpha)^{-i\kappa}f_\lambda\|_k^\zeta\alpha. \tag{2.7}$$

Proof. Notice that $\Phi_{\mathfrak{m}}(\mathbf{a}_\alpha)$ is meaningful since $x_\lambda^{-1}\alpha x_\nu\in Wy_0W$. By Lemma 2.2 we can take Y of (2.3) to be $\{(x_\lambda^{-1}\alpha x_\nu)_{\mathbf{f}}\,|\,\alpha\in A\}$. Then for $u\in G_{\mathbf{a}}\cap G_0$ we have, by (1.26),

$$(f'_\nu\|_k^\zeta u)(\mathbf{i})=N(u)^{-i\kappa}(\mathbf{g}|Wy_0W)(x_\nu^{-*}u)$$
$$=N(u)^{-i\kappa}\sum_{\alpha\in A}\Phi_{\mathfrak{m}}(a_\alpha)^{-1}\mathbf{g}(\alpha_{\mathbf{f}}^*x_\lambda^{-*}u).$$

Now $\mathbf{g}(\alpha_{\mathbf{f}}^*x_\lambda^{-*}u)=\mathbf{g}(\alpha^*\alpha_{\mathbf{a}}^{-*}x_\lambda^{-*}u)=\mathbf{g}(x_\lambda^{-*}\alpha_{\mathbf{a}}^{-*}u)=N(\alpha^{-1}u)^{i\kappa}(f_\lambda\|_k^\zeta\alpha u)(\mathbf{i})=N(\alpha^{-1}u)^{i\kappa}((f_\lambda\|_k^\zeta\alpha)\|_k^\zeta u)(\mathbf{i})$, and hence we obtain (2.7).

By (1.21) we can choose α_0 of Lemma 2.2 so that $x_\lambda y_0\in\alpha_0x_\nu W^1$, and take A so that $\Gamma_\lambda^1\alpha_0\Gamma_\nu^1=\coprod_{\alpha\in A}\Gamma_\lambda^1\alpha$. With this choice of A for y_0 of type (2.4a) or (2.4b), we have

$$f'_\nu=\sum_{\alpha\in A}N(\alpha)^{-i\kappa}f_\lambda\|_k^\zeta\alpha, \tag{2.8}$$

since $\Phi_{\mathfrak{m}}(a_\alpha)=1$ for α in such an A.

To study the effect of complex conjugation, we first observe that there is a $\mathbb{Q}$-rational element A_k of $GL(V_k)$, unique up to scalar factors, such that

$$A_k\overline{\rho'_k(x)}=\rho'_k(x)A_k \quad \textit{for every } x\in G_{\mathbf{A}}. \tag{2.9}$$

Fixing such an A_k, we easily see that

(2.10a) $\quad A_k\overline{f\|_k^\zeta\xi}=(A_k\bar f)\|_k^{\delta-\zeta}\xi \quad$ *for every* $\xi\in G_{\mathbf{A}}$,

(2.10b) $\quad f\in\mathcal{S}_k^\zeta(\Gamma)\Rightarrow A_k\bar f\in\mathcal{S}_k^{\delta-\zeta}(\Gamma)$,

(2.10c) $\quad f\in\mathcal{S}_k^\zeta(\Gamma_\lambda, \Phi_{\mathfrak{m}}, \kappa)\Rightarrow A_k\bar f\in\mathcal{S}_k^{\delta-\zeta}(\Gamma_\lambda, \bar\Phi_{\mathfrak{m}}, -\kappa)$,

(2.10d) $\quad \mathbf{g}=(f_\lambda)_{\lambda\in\Lambda}\in\mathcal{S}_k^\zeta(\mathfrak{m}, \Phi_{\mathfrak{m}}, \kappa)$
$\Rightarrow A_k\bar{\mathbf{g}}=(A_k\bar f_\lambda)_{\lambda\in\Lambda}\in\mathcal{S}_k^{\delta-\zeta}(\mathfrak{m}, \bar\Phi_{\mathfrak{m}}, -\kappa)$ *and* $A_k\overline{(\mathbf{g}^\varepsilon)}=(A_k\bar{\mathbf{g}})^\varepsilon$,

(2.10e) $\quad A_k\overline{\mathbf{g}|T_v}=(A_k\bar{\mathbf{g}})|T_v, \quad A_k\overline{\mathbf{g}|S_v}=(A_k\bar{\mathbf{g}})|S_v$,

(2.10f) $\quad \mathbf{g}\in\mathcal{S}_k^\zeta(\mathfrak{m}, \Phi)\Rightarrow A_k\bar{\mathbf{g}}\in\mathcal{S}_k^{\delta-\zeta}(\mathfrak{m}, \bar\Phi)$.

Let us now consider a common eigenfunction $\mathbf{h}$ in $\mathcal{S}_k^0(\mathfrak{m}, \Phi)$ of the T_v for all $v\in\mathbf{f}$. Writing $\chi(v)$ for the eigenvalue, we have $\mathbf{h}|T_v=\chi(v)\mathbf{h}$. We say that $\{\chi, \mathfrak{m}\}$ is *primitive* if for every divisor $\mathfrak{n}$ of $\mathfrak{m}$ different from $\mathfrak{m}$ and divisible by the conductor of Φ the space $\mathcal{S}_k^0(\mathfrak{n}, \Phi)$ contains no nonzero $\mathbf{g}$ such that $\mathbf{g}|T_v=\chi(v)\mathbf{g}$ for almost all v. We call then $\mathbf{h}$ a *primitive form* (cf. [5, II, pp. 584–585]). For a primitive $\{\chi, \mathfrak{m}\}$ the set of all such $\mathbf{h}$ in $\mathcal{S}_k^0(\mathfrak{m}, \Phi)$ is one-dimensional, provided $\mathfrak{m}$ is prime to $\mathfrak{d}_B$. In the next section we consider the periods of $\mathbf{h}$ under this condition on $\mathfrak{m}$. The restriction is made merely for simplicity. In the general case, in order to obtain one-dimensionality, we have to decompose $\mathscr{S}_k^0(\mathfrak{m}, \Phi)$ according to the representations of $\prod_{v|\mathfrak{d}_B} G_v$, which, if somewhat cumbersome, present no essential difficulties.

3. Vector-valued differential forms

For an integer $m \geqq 0$ and $\begin{pmatrix} u \\ v \end{pmatrix} \in \mathbf{C}^2$, we denote by $\begin{pmatrix} u \\ v \end{pmatrix}^m$ the vector in $\mathbf{C}^{m+1}$ whose components are $u^{m-\nu} v^\nu$ for $0 \leqq \nu \leqq m$; in particular $\begin{pmatrix} u \\ v \end{pmatrix}^0 = 1$. We can then define a representation τ_m: $GL(\mathbf{C}^2) \to GL(\mathbf{C}^{m+1})$ by

$$\tau_m(\alpha)\begin{pmatrix} u \\ v \end{pmatrix}^m = \left(\alpha\begin{pmatrix} u \\ v \end{pmatrix}\right)^m, \tag{3.1}$$

and also an element P_m of $GL(\mathbf{C}^{m+1})$ by

$${}^t\begin{pmatrix} u \\ v \end{pmatrix}^m P_m \begin{pmatrix} x \\ y \end{pmatrix}^m = \det\begin{pmatrix} u & x \\ v & y \end{pmatrix}^m. \tag{3.2}$$

We easily see that ${}^tP_m = (-1)^m P_m$ and

$${}^t\tau_m(\alpha) P_m \tau_m(\alpha) = \det(\alpha)^m P_m \quad \textit{for every } \alpha \in GL(\mathbf{C}^2). \tag{3.3}$$

We now take a weight $k \in \mathbf{Z}^{\mathbf{a}}$ as in Sect. 1 and assume throughout the rest of the paper that $k_v \geqq 2$ for every $v \in \delta$. We then define a representation ρ_k of $G_{\mathbf{A}}$ by

$$\rho_k(x) = [\bigotimes_{v\in\delta} \tau_{k_v - 2}(|N(x_v)|^{-1/2} x_v)] \otimes \rho'_k(x) \quad (x \in G_{\mathbf{A}}) \tag{3.4}$$

with ρ'_k of (1.6a) and denote by U_k the representation space of ρ_k, which can be written $\bigotimes_{v\in\mathbf{a}} \mathbf{C}^{k_v - 1}$.

For every congruence subgroup Γ of G or G^u, let $D_k^r(\Gamma)$ denote the complex vector space of all U_k-valued C^∞ r-forms ω on H^δ such that

$$\omega \circ \gamma = \rho_k(\gamma)\omega \quad \textit{for every } \gamma \in \Gamma, \tag{3.5}$$

where $\omega \circ \gamma$ denotes the transform of ω under the action of γ on H^δ. We then denote by $H_k^r(\Gamma)$ the r-th cohomology group of the complex $\sum_r D_k^r(\Gamma)$ defined as usual by means of exterior differentiation. We are going to associate an element of $D_k^n(\Gamma)$ to each element of $\mathscr{S}_k^\varepsilon(\Gamma)$, where $n = \|\delta\|$ as in (1.3). First, for $z \in \mathbf{C}^\delta$ and $\varepsilon \subset \delta$ we define an element $[z]_k^\varepsilon$ of $\bigotimes_{v\in\delta} \mathbf{C}^{k_v - 1}$ and an n-form $d_\varepsilon z$

on H^δ by

$$[z]_k^\varepsilon = \bigotimes_{v\in\delta} \begin{pmatrix} z'_v \\ 1 \end{pmatrix}^{k_v-2}, \tag{3.6}$$

$$d_\varepsilon z = \bigwedge_{v\in\delta} dz'_v, \tag{3.7}$$

where z'_v denotes $\bar{z}_v$ or z_v according as $v\in\varepsilon$ or $v\in\delta-\varepsilon$. Here we fix an order among the elements of δ once and for all, and define the exterior product of (3.7) according to that order; for instance if the elements of δ are t, u, v, w in our fixed order and $\varepsilon=\{u, w\}$, then $d_\varepsilon z = dz_t \wedge d\bar{z}_u \wedge dz_v \wedge d\bar{z}_w$.

Given $f\in\mathcal{S}_k^\varepsilon(B)$ we define a U_k-valued n-form $[f]$ by

$$[f]=[z]_k^\varepsilon \otimes f d_\varepsilon z. \tag{3.8}$$

Here are some formulas which can easily be verified:

(3.9a) $[f\|_k^{\varepsilon+\zeta}\beta]=(-1)^{\|\varepsilon\|}\rho_k(\beta)^{-1}[f]\circ\beta_*$ *if* $f\in\mathcal{S}_k^\zeta(B)$ *and* $\beta\in G_\varepsilon\cap G$;

(3.9b) $f\in\mathcal{S}_k^\zeta(\Gamma)\Rightarrow[f]\in D_k^n(\Gamma)$;

(3.9c) $\overline{d_\varepsilon z}=d_{\delta-\varepsilon}z$;

(3.9d) $\overline{d_\varepsilon z}\wedge d_\varepsilon z=(-1)^{a(\varepsilon)}(2i)^n\,\mathrm{Im}(z)^{2\delta}\,d_H^\delta z$ *with* $a(\varepsilon)=\|\varepsilon\|+n(n-1)/2$ *and* $d_H^\delta z$ *of* (1.16a);

(3.9e) $\omega\in D_k^r(\Gamma)\to(1\otimes A_k)\bar{\omega}\in D_k^r(\Gamma)$;

(3.9f) $[A_k\bar{f}]=(1\otimes A_k)\overline{[f]}$ *for every* $f\in\mathcal{S}_k^\zeta(B)$.

Here $1\otimes A_k$ is an element of $GL(U_k)$ which is the tensor product of the identity map on $\bigotimes_{v\in\delta}\mathbf{C}^{k_v-1}$ and the map A_k on V_k of (2.9). Define an element Θ_k of $GL(U_k)$ by

$$\Theta_k=\Big(\bigotimes_{v\in\delta}P_{k_v-2}\Big)\otimes{}^tA_k^{-1}. \tag{3.10}$$

From (3.3), (2.9), and (1.6b) we obtain

$$^t\rho_k(x)\,\Theta_k\,\rho_k(x)=\mathrm{sgn}[N(x)^{k\delta}]\,\Theta_k \quad \textit{for every } x\in G_{\mathbf{A}}. \tag{3.11}$$

Let ι denote the identity element of $\mathbf{Z}^{\mathbf{a}}$. Then $\rho_{2\iota}$ is the trivial representation, so that $D_{2\iota}^r(\Gamma)$ consists of Γ-invariant C^∞ scalar-valued r-forms. Given $\omega\in D_k^r(\Gamma)$ and $\omega'\in D_k^s(\Gamma)$, we observe that ${}^t\omega\wedge\Theta_k\omega'$ is meaningful as an element of $D_{2\iota}^{r+s}(\Gamma)$. In particular ${}^t[f]\wedge\Theta_k[g]$ is meaningful as an element of $D_{2\iota}^{2n}(\Gamma)$ for $f\in\mathcal{S}_k^\varepsilon(\Gamma)$ and $g\in\mathcal{S}_k^\zeta(\Gamma)$. Since $d_\varepsilon z\wedge d_\zeta z=0$ if $\varepsilon+\zeta\neq\delta$, we have

$$^t[f]\wedge\Theta_k[g]=0 \quad \textit{if } \varepsilon+\zeta\neq\delta. \tag{3.12a}$$

In the nonvanishing case, from (3.2) and (3.9d) we can derive

(3.12b) $$^t[A_k \bar{f}] \wedge \Theta_k[g] = (-1)^{b(\varepsilon)}(2i)^{\|k\delta\|-n} \cdot {}^t\bar{f} g \operatorname{Im}(z)^{k\delta} d_H^\delta z$$
if $f, g \in \mathscr{S}_k^\varepsilon(B)$, *where* $b(\varepsilon) = \|k\delta - k\varepsilon + \varepsilon\| + n(n-1)/2$.

4. The fundamental periods

Let $S(H^\delta) = \sum_r S_r(H^\delta)$ denote the complex of singular chains on H^δ. A congruence subgroup Γ of $G \cap G_0$ or G^u acts naturally on $S(H^\delta)$. Assuming the condition

(4.1) $$\rho_k(\Gamma \cap E) = 1,$$

we denote by $C_k^r(\Gamma, S)$ the set of all U_k-valued Γ-equivariant r-cochains, by which we mean $\mathbb{Z}$-linear maps $\varphi\colon S_r(H^\delta) \to U_k$ such that $\varphi(\gamma c) = \rho_k(\gamma)\varphi(c)$ for every $\gamma \in \Gamma$. Then the coboundary operator δ is defined as usual by $\delta\varphi(b) = \varphi(\partial b)$, and we obtain the cohomology groups of this complex $\sum_r C_k^r(\Gamma, S)$, which are denoted by $H_k^r(\Gamma, S)$. A well known generalization of de Rham's theorem asserts that $H_k^r(\Gamma, S)$ is isomorphic to $H_k^r(\Gamma)$ in a natural way. It should be noted that the isomorphism is valid even when Γ has nontrivial fixed points. These two types of cohomology groups are isomorphic to a group of the third type which is defined purely algebraically by Γ and ρ_k with no reference to H^δ, a well known fact due to Eilenberg [2]. We shall not use this third type in the present paper, however. Strictly speaking, Γ should be replaced by $\Gamma/(\Gamma \cap E)$, but we use Γ instead of $\Gamma/(\Gamma \cap E)$, since no difficulties arise as long as condition (4.1) is imposed.

Hereafter we assume that B is a division algebra. Then each class of $H_k^n(\Gamma)$ is represented by a unique harmonic form. This fact is a special case of a general result on the cohomology groups of a discrete group acting on a hermitian symmetric space (see the paper by Matsushima and Murakami quoted in [3]). In [3] it was shown that $[f]$ defined by (3.8) for $f \in \mathscr{S}_k^\varepsilon(\Gamma)$ is a harmonic form of type $(n - \|\varepsilon\|, \|\varepsilon\|)$ and that these forms exhaust all harmonic n-forms except when n is even and $k = 2\iota$, in which case we need the forms $\Xi_\zeta = \bigwedge_{v \in \zeta} \operatorname{Im}(z_v)^{-2} d\bar{z}_v \wedge dz_v$ for all $\zeta \subset \delta$ such that $2\|\zeta\| = n$. Thus, assigning to each $f \in \mathscr{S}_k^\varepsilon(\Gamma)$ the class of the cocycle

(4.2) $$[f](c) = \int_c [f] \qquad (c \in S_n(H^\delta)),$$

we obtain a $\mathbb{C}$-linear isomorphism

(4.3) $$\prod_{\varepsilon \subset \delta} \mathscr{S}_k^\varepsilon(\Gamma) \oplus \Big[\sum_{2\|\zeta\| = n} \mathbb{C}\Xi_\zeta\Big] \to H_k^n(\Gamma, S),$$

where $\Sigma\,\mathbb{C}\Xi_\zeta$ is necessary only when n is even and $k = 2\iota$. (For details see [3, §4]. In particular the connection of harmonic forms with $\mathscr{S}_k^\varepsilon(\Gamma)$ is treated in [3, p. 434]. The problem can be reduced to the holomorphic case as explained in the last paragraph on that page. However, the formula about a differential form ω there is not completely correct. For a holomorphic ω, the vector $e_{0\dots 0}$ in that formula should be $[z]_k^0$ of (3.6); $e_{0\dots 0}$ represents essentially, if not exactly, the vector-valued function on $G_{\mathbf{a}}$ corresponding to $[z]_k^0$.)

We now take Γ to be Γ_λ^1 and assume that κ of (1.23) is 0. Thus (1.23) becomes

(4.4) *The conductor of* Φ *is prime to* $\mathfrak{d}_B$ *and divides* $\mathfrak{m}$;
$\Phi(x)=\operatorname{sgn}(x)^k$ *for every* $x\in E_{\mathbf{a}}^\times$.

Notice that if $a\in\Gamma_\lambda^1\cap E$, then $\Phi_{\mathfrak{f}}(a)=1$ since a belongs to the set of (1.19b), and hence $\operatorname{sgn}(a)^k=\Phi_{\mathbf{a}}(a)=1$. Thus Γ_λ^1 satisfies (4.1). Hence we obtain a canonical injection

$$\prod_{\lambda\in\Lambda}\prod_{\varepsilon\subset\delta}\mathscr{S}_k^\varepsilon(\Gamma_\lambda^1)\to\prod_{\lambda\in\Lambda}H_k^n(\Gamma_\lambda^1,S). \tag{4.5}$$

Now $\mathscr{S}_k^\varepsilon(\mathfrak{m},\Phi_{\mathfrak{m}},0)$ is embedded in the product on the left-hand side. If $\mathbf{g}=(f_\lambda)_{\lambda\in\Lambda}\in\mathscr{S}_k^\varepsilon(\mathfrak{m},\Phi_{\mathfrak{m}},0)$, then the corresponding element of $\prod_{\lambda\in\Lambda}H_k^n(\Gamma_\lambda^1,S)$ is $([f_\lambda])_{\lambda\in\Lambda}$, where we identify $[f_\lambda]$ with the cohomology class of the cocycle defined by (4.3). We put $[\mathbf{g}]=([f_\lambda])_{\lambda\in\Lambda}$ and view this as an element of $\prod_{\lambda\in\Lambda}H_k^n(\Gamma_\lambda^1,S)$.

To define Hecke operators T_v and S_v on $\prod_{\lambda\in\Lambda}H_k^n(\Gamma_\lambda^1,S)$, take y_0 of (2.4a, b) and an element α_0 of G so that $x_\lambda y_0\in\alpha_0 x_v W^1$, and consider a coset decomposition $\Gamma_\lambda^1\alpha_0\Gamma_v^1=\coprod_{\alpha\in A}\Gamma_\lambda^1\alpha$. For $\varphi\in C_k^r(\Gamma_\lambda^1,S)$ define $\varphi|T_v$ or $\varphi|S_v$ to be a U_k-valued cochain given by

$$(\varphi|X_v)(c)=\sum_{\alpha\in A}\rho_k(\alpha)^{-1}\varphi(\alpha c)\qquad(c\in S_r(H^\delta)), \tag{4.6a}$$

where X_v is T_v or S_v according as y_0 is of type (2.4a) or (2.4b). We easily see that $\varphi|X_v$ is well defined and belongs to $C_k^r(\Gamma_v^1,S)$; moreover $(\delta\varphi)|X_v=\delta(\varphi|X_v)$. Thus X_v defines a $\mathbb{C}$-linear map of $H_k^r(\Gamma_\lambda^1,S)$ into $H_k^r(\Gamma_v^1,S)$, which we denote again by T_v or S_v. Similarly for $\omega\in D_k^r(\Gamma_\lambda^1)$ we can define $\omega|X_v\in D_k^r(\Gamma_v^1)$ by

$$\omega|X_v=\sum_{\alpha\in A}\rho_k(\alpha)^{-1}\,\omega\circ\alpha, \tag{4.6b}$$

and this naturally defines a map of $H_k^r(\Gamma_\lambda^1)$ into $H_k^r(\Gamma_v^1)$, which is consistent with the map obtained from (4.6a) under the de Rham isomorphism. Since the map $\lambda\mapsto v$ is a permutation of Λ, in this way X_v acts on both $\prod_{\lambda\in\Lambda}H_k^r(\Gamma_\lambda^1)$ and $\prod_{\lambda\in\Lambda}H_k^r(\Gamma_\lambda^1,S)$, and satisfies

$$[\mathbf{g}|X_v]=[\mathbf{g}]|X_v\quad\text{ for }\mathbf{g}\in\mathscr{S}_k^\zeta(\mathfrak{m},\Phi_{\mathfrak{m}},0). \tag{4.7}$$

This follows easily from (2.8) and (3.9a) with $\varepsilon=0$.

We now define an operator $R(\varepsilon)$ on our cohomology groups corresponding to the map $\mathbf{g} \mapsto \mathbf{g}^{\varepsilon}$. Take β, λ, and μ as in Lemma 2.1. For $\varphi \in C_k^r(\Gamma_\lambda^1, S)$ and $\omega \in D_k^r(\Gamma_\lambda^1)$ define $\varphi | R(\varepsilon)$ and $\omega | R(\varepsilon)$ by

$$(4.8\,\mathrm{a}) \qquad (\varphi | R(\varepsilon))(c) = (-1)^{\|k\varepsilon\| + \|\varepsilon\|} \rho_k(\beta)^{-1} \varphi(\beta_* c) \qquad (c \in S_r(H^\delta)),$$

$$(4.8\,\mathrm{b}) \qquad \omega | R(\varepsilon) = (-1)^{\|k\varepsilon\| + \|\varepsilon\|} \rho_k(\beta)^{-1} \omega \circ \beta_*.$$

The last equalities of Lemma 2.1 show that $\varphi | R(\varepsilon) \in C_k^r(\Gamma_\mu^1, S)$, $\omega | R(\varepsilon) \in D_k^r(\Gamma_\mu^1)$, and these are independent of the choice of β. In this way we obtain endomorphisms of $\prod_{\lambda \in \Lambda} H_k^r(\Gamma_\lambda^1, S)$ and $\prod_{\lambda \in \Lambda} H_k^r(\Gamma_\lambda^1)$, which we also denote by $R(\varepsilon)$. From Lemma 2.1 and (3.9a) we immediately see that

$$(4.9) \qquad [\mathbf{g}^{\varepsilon}] = [\mathbf{g}] | R(\varepsilon) \qquad \textit{for every } \mathbf{g} \in \mathscr{S}_k^{\zeta}(\mathfrak{m}, \Phi_{\mathfrak{m}}, 0).$$

We now fix a primitive $\{\chi, \mathfrak{m}\}$ as considered at the end of Sect. 2 with the assumption that

$$(4.10) \qquad \mathfrak{m} \textit{ is prime to } \mathfrak{d}_B.$$

Then we can find a $\mathbb{Q}$-rational nonzero element $\mathbf{h}$ of $\mathscr{S}_k^0(\mathfrak{m}, \Phi)$ such that $\mathbf{h} | T_v = \chi(v)\mathbf{h}$ for all $v \in \mathfrak{f}$. (For the $\mathbb{Q}$-rationality see [5, II, Sect. 2].) Such an $\mathbf{h}$ is uniquely determined up to algebraic factors as a $\mathbb{Q}$-rational element of $\mathscr{S}_k^0(\mathfrak{m}, \Phi_{\mathfrak{m}}, 0)$ such that $\mathbf{h} | T_v = \chi(v)\mathbf{h}$ and $\mathbf{h} | S_v = \Phi(\pi_v)\mathbf{h}$ for $v \in \mathfrak{f}$, $v \nmid \mathfrak{m}$. Therefore from (2.5) we obtain, for every $\varepsilon \subset \delta$,

$$(4.11) \qquad \mathbb{C}\mathbf{h}^{\varepsilon} = \{\mathbf{g} \in \mathscr{S}_k^{\varepsilon}(\mathfrak{m}, \Phi_{\mathfrak{m}}, 0) \mid \mathbf{g} | T_v = \chi(v)\mathbf{g} \textit{ and } \mathbf{g} | S_v = \Phi(\pi_v)\mathbf{g} \quad \textit{for every } v \in \mathfrak{f}, v \nmid \mathfrak{m}\}.$$

Proposition 4.1. *The* $[\mathbf{h}^{\varepsilon}]$ *for all* $\varepsilon \subset \delta$, *as elements of* $\prod_{\lambda \in \Lambda} H_k^n(\Gamma_\lambda^1, S)$, *form a* $\mathbb{C}$-*basis of the space*

$$\{q = (q_\lambda) \in \prod_{\lambda \in \Lambda} H_k^n(\Gamma_\lambda^1, S) \mid \rho_k(\gamma)^{-1} q_\lambda \circ \gamma = \Phi_{\mathfrak{m}}(a_\gamma) q_\lambda \quad \textit{for every } \gamma \in \Gamma_\lambda,$$
$$q | T_v = \chi(v) q \textit{ and } q | S_v = \Phi(\pi_v) q \quad \textit{for every } v \in \mathfrak{f}, v \nmid \mathfrak{m}\},$$

where $\rho_k(\gamma)^{-1} q_\lambda \circ \gamma$ *denotes the cohomology class of the cocycle* $\rho_k(\gamma)^{-1} \varphi \circ \gamma$ *for a cocycle* φ *in the cohomology class of* q_λ.

This follows immediately from (4.11) and the injectivity of the map of (4.5). Notice that the eigenvalues $\chi(v)$ don't occur in the space spanned by the forms Ξ_ε of (4.2).

Now U_k has a natural $\mathbb{Q}$-rational structure consistent with the action of G; namely $U_k = U_k(\mathbb{Q}) \otimes_{\mathbb{Q}} \mathbb{C}$ with a vector space $U_k(\mathbb{Q})$ over $\mathbb{Q}$ stable under $\rho_k(G)$. By considering $U_k(\mathbb{Q})$-valued elements of $C_k^r(\Gamma, S)$, we obtain cohomology groups $H_k^r(\Gamma, S, \mathbb{Q})$ in an obvious way so that $H_k^r(\Gamma, S) = H_k^r(\Gamma, S, \mathbb{Q}) \otimes_{\mathbb{Q}} \mathbb{C}$. From (4.6a) we see that $\prod_{\lambda\in\Lambda} H_k^r(\Gamma_\lambda^1, S, \mathbb{Q})$ is stable under T_v, S_v, and also under $(q_\lambda)\mapsto(\rho_k(\gamma_\lambda)^{-1} q_\lambda\circ\gamma_\lambda)$ with $\gamma_\lambda\in\Gamma_\lambda$ as in Proposition 4.1. Therefore if we put

$$K_\chi=\left(\sum_{\varepsilon\subset\delta}\mathbb{C}[\mathbf{h}^\varepsilon]\right)\cap\prod_{\lambda\in\Lambda} H_k^r(\Gamma_\lambda^1, S, \mathbb{Q}), \tag{4.12}$$

then the same proposition shows that $\sum_{\varepsilon\subset\delta}\mathbb{C}[\mathbf{h}^\varepsilon]=K_\chi\otimes_{\mathbb{Q}}\mathbb{C}$. Now (4.9) implies that $\zeta\mapsto R(\zeta)$ defines a regular representation of the additive group $(\mathbb{Z}/2\mathbb{Z})^\delta$ on the space $\sum_\varepsilon\mathbb{C}[\mathbf{h}^\varepsilon]$. Moreover by (4.8a, b) K_χ is stable under $R(\zeta)$. Put

$$\langle\varepsilon,\zeta\rangle=(-1)^{\|\varepsilon\zeta\|}\quad(\varepsilon,\zeta\subset\delta), \tag{4.13}$$

and observe that this defines a nondegenerate pairing of $(\mathbb{Z}/2\mathbb{Z})^\delta$ with itself. Then we can find a basis $\{y_\varepsilon\}_{\varepsilon\subset\delta}$ of K_χ over $\mathbb{Q}$ such that

$$y_\varepsilon|R(\zeta)=\langle\varepsilon,\zeta\rangle\, y_\varepsilon\quad(\varepsilon,\zeta\subset\delta). \tag{4.14}$$

These y_ε are uniquely determined up to algebraic factors, and form a basis of $\sum_\varepsilon\mathbb{C}[\mathbf{h}^\varepsilon]$. Therefore we can put

$$[\mathbf{h}]=\sum_{\varepsilon\subset\delta} p(\chi,\delta,\varepsilon;B)\, y_\varepsilon \tag{4.15}$$

with complex numbers $p(\chi, \delta, \varepsilon; B)$. By (4.9) and (4.14) we have

$$[\mathbf{h}^\zeta]=\sum_{\varepsilon\subset\delta}\langle\varepsilon,\zeta\rangle\, p(\chi,\delta,\varepsilon;B)\, y_\varepsilon\quad\textit{for every }\zeta\subset\delta. \tag{4.16}$$

Since the $[\mathbf{h}^\zeta]$ and the y_ε span the same space over $\mathbb{C}$, we see that $p(\chi, \delta, \varepsilon; B)\neq 0$ for every ε.

Now $\mathbf{h}$ is unique for χ up to algebraic factors, and so are the y_ε. Therefore, once the x_λ in (1.21) are fixed, the class of $p(\chi, \delta, \varepsilon; B)$ in $\mathbb{C}^\times/\mathbb{Q}^\times$ is determined by χ, δ, ε, and B. If we choose different x_λ's, the groups Γ_λ^1 and the identification of $\mathbf{h}$ with $(h_\lambda)_{\lambda\in\Lambda}$ are changed, and we have to consider $\prod_{\lambda\in\Lambda} H_k^n(\Gamma_\lambda^1, S)$ with different Γ_λ^1. But it can easily be seen by an argument similar to and easier than Lemma 2.1 that this change of x_λ sends $[\mathbf{h}^\zeta]$ and y_ε consistently to the elements in a new cohomology group by a $\mathbb{Q}$-rational isomorphism. Thus we still obtain the same constants p as elements of $\mathbb{C}^\times/\mathbb{Q}^\times$. Since we fix B, we denote the constants simply by $p(\chi, \delta, \varepsilon)$. In fact, it is conjecturable that $p(\chi, \delta, \varepsilon; B)$ modulo $\mathbb{Q}^\times$ is independent of B, a point which will be studied in Sect. 6.

In order to speak of the periods of a cocycle in $C_k^n(\Gamma, S)$, we have to introduce certain U_k-valued cycles which were called *equivariant cycles* by Eilenberg [2]. Namely, by an *equivariant r-cycle of type* (ρ_k, Γ), we understand an element u of $U_k\otimes_{\mathbb{Z}} S_r(H^\delta)$ such that ∂u is a finite sum of the form

$$\partial u=\sum[v\otimes\gamma(c)-{}^t\rho_k(\gamma)v\otimes c] \tag{4.17}$$

with $v \in U_k$, $\gamma \in \Gamma$, and $c \in S_{r-1}(H^\delta)$, where ∂ acts trivially on U_k. Such a u is called $\bar{\mathbb{Q}}$-rational if it belongs to $U_k(\bar{\mathbb{Q}}) \otimes_{\mathbb{Z}} S_r(H^\delta)$. For $\varphi \in C_k^r(\Gamma, S)$ and $w = \sum v \otimes c \in U_k \otimes_{\mathbb{Z}} S_r(H^\delta)$ we put $\varphi(w) = \sum {}^t v \varphi(c)$. If φ is a cocycle and u is an equivariant r-cycle of type (ρ_k, Γ), then it can easily be seen that $\varphi(u)$ depends only on the cohomology class of φ. In particular, if ω is a closed form in $D_k^r(\Gamma)$ viewed as a cochain, then $\omega(u)$ may be called a *period* of ω.

Proposition 4.2. *Let $\mathbf{h}^\zeta = (h_{\zeta\lambda})_{\lambda \in \Lambda}$ for each $\zeta \subset \delta$. Then for every $\bar{\mathbb{Q}}$-rational equivariant n-cycle u of type $(\rho_k, \Gamma_\lambda^1)$ the period $[h_{\zeta\lambda}](u)$ is a $\bar{\mathbb{Q}}$-linear combination of the $p(\chi, \delta, \varepsilon)$ for all $\varepsilon \subset \delta$.*

This is an immediate consequence of (4.16).

Lemma 4.3. *The measure of $\Gamma_\lambda^1 \backslash H^\delta$ is π^n times an algebraic number independent of λ. The same assertion is true with $\Gamma_\lambda^1 \cap G^u$ in place of Γ_λ^1.*

Proof. Let $\mathfrak{o}_\lambda$ be the maximal order in B such that $\mathfrak{o}_{\lambda v} = x_{\lambda v} \mathfrak{o}_v x_{\lambda v}^{-1}$ for every $v \in \mathbf{f}$ with the order $\mathfrak{o}$ fixed in Sect. 1; put $\Gamma_\lambda^0 = \mathfrak{o}_\lambda^\times$ and $\Gamma_\lambda^+ = \{\gamma \in \Gamma_\lambda^0 \mid N(\gamma) \gg 0\}$. Let φ_λ be the map of $\mathfrak{o}_\lambda$ into $\prod_{v \mid \mathfrak{m}} \mathfrak{o}_v / \mathfrak{m} \mathfrak{o}_v$ obtained by combining the natural map of $\mathfrak{o}_\lambda$ into $\mathfrak{o}_\lambda / \mathfrak{m} \mathfrak{o}_\lambda$ with the map $(a_v)_v \mapsto (x_{\lambda v}^{-1} a_v x_{\lambda v})_v$. Then $\varphi_\lambda(\Gamma_\lambda^+)$ and $\varphi_\lambda(\Gamma_\lambda^1 \mathfrak{g}^\times)$ are independent of λ; $N(\Gamma_\lambda^0)$ consists of the elements u of $\mathfrak{g}^\times$ such that $u_v > 0$ for every $v \in \delta'$. These facts follow easily from Eichler's strong approximation theorem [1, Satz 5]. Consequently the indices $[\Gamma_\lambda^0 : \Gamma_\lambda^+]$ and $[\Gamma_\lambda^+ : \Gamma_\lambda^1 \mathfrak{g}^\times]$ are independent of λ. Now Shimizu proved in [4, p. 192] a result concerning Γ_λ^0 which implies that $2^n [\Gamma_\lambda^0 : \Gamma_\lambda^+]^{-1} \operatorname{vol}(\Gamma_\lambda^+ \backslash H^\delta)$ is π^n times an algebraic number independent of λ. Therefore we obtain our first assertion. The second one can be shown in a similar way.

Theorem 4.4. *Let $\mathbf{h}$ be an eigenform in $\mathscr{S}_k^0(\mathfrak{m}, \Phi)$ belonging to a primitive $\{\chi, \mathfrak{m}\}$ under assumptions* (4.4) *and* (4.10). *Then the following assertions hold for every $\varepsilon \subset \delta$:*

(1) $$\overline{p(\chi, \delta, \varepsilon)} \sim p(\bar{\chi}, \delta, \varepsilon),$$

(2) $$p(\chi, \delta, \varepsilon)\, p(\bar{\chi}, \delta, k\delta + \delta + \varepsilon) \sim \pi^n \langle \mathbf{h}, \mathbf{h} \rangle,$$

where $k\delta + \delta + \varepsilon$ should be considered modulo $2\mathbb{Z}^\delta$ and we write $a \sim b$ if $a/b \in \bar{\mathbb{Q}}$.

Proof. Formulas (2.5) and (2.10d–f) show that $A_k \overline{\mathbf{h}^\delta}$ is a form in $\mathscr{S}_k^0(\mathfrak{m}, \bar{\Phi})$ belonging to $\bar{\chi}$. On the other hand, for $\varphi \in C_k^r(\Gamma, S)$ we see that $(1 \otimes A_k)\bar{\varphi}$ is meaningful as an element of $C_k^r(\Gamma, S)$ and $\varphi \mapsto (1 \otimes A_k)\bar{\varphi}$ defines an $\mathbb{R}$-linear automorphism of $\prod_\lambda H_k^r(\Gamma_\lambda^1, S)$; call it $\mathscr{A}$. By (2.10d) and (3.9f) we have $[(A_k \overline{\mathbf{h}^\delta})^\varepsilon] = [A_k \overline{\mathbf{h}^{\delta+\varepsilon}}] = \mathscr{A}[\mathbf{h}^{\delta+\varepsilon}]$, and hence $\mathscr{A} K_\chi = K_{\bar{\chi}}$. From (4.8a) we see that $(\mathscr{A} y) \mid R(\varepsilon) = \mathscr{A}(y \mid R(\varepsilon))$ for $y \in \prod_\lambda H_k^n(\Gamma_\lambda^1, S)$. Therefore $\mathscr{A} y_\varepsilon$ plays the role of y_ε when χ is replaced by $\bar{\chi}$. Applying $\mathscr{A}$ to (4.16) with δ as ζ, we obtain assertion (1). Next, to prove (2), we consider a pairing $U_k \times U_k \to \mathbb{C}$ given by $(a, b) \mapsto {}^t a \Theta_k b$ with Θ_k of (3.10). This is $\bar{\mathbb{Q}}$-valued on $U_k(\bar{\mathbb{Q}}) \times U_k(\bar{\mathbb{Q}})$. In view of (3.11), we can define for $x \in C_k^r(\Gamma, S)$ and $y \in C_k^s(\Gamma, S)$ their cup product $x \cup y \in C_{2i}^{r+s}(\Gamma, S)$. (Here recall that $\rho_{2\iota}$ is the trivial representation.) From this we obtain a bilinear map

$$H_k^r(\Gamma, S) \times H_k^s(\Gamma, S) \to H_{2\iota}^{r+s}(\Gamma, S).$$

(See Eilenberg [2, p. 415].) By a well known principle, for closed forms $\omega \in D_k^r(\Gamma)$ and $\omega' \in D_k^s(\Gamma)$ the exterior product ${}^t\omega \wedge \Theta_k \omega'$ represents the cohomology class of $[\omega] \cup [\omega']$. Now $H_{2\iota}^{2n}(\Gamma) = \mathbb{C} d_H^\delta z$ for every Γ. (This is obvious if Γ is fixed-point-free; the general case can be reduced to that special case.) Therefore a closed form ω in $D_{2\iota}^{2n}(\Gamma_\lambda^1)$ represents a class in $H_{2\iota}^{2n}(\Gamma_\lambda^1, S, \overline{\mathbb{Q}})$ if and only if $\int_{F_\lambda} \omega \in \overline{\mathbb{Q}}$, where F_λ is a fundamental domain for $\Gamma_\lambda^1 \backslash H^\delta$. Notice that the integral depends only on the cohomology class of ω. Put $\mathbf{h}^\eta = (h_{\eta\lambda})_{\lambda \in A}$ for each $\eta \in (\mathbb{Z}/2\mathbb{Z})^\delta$. By (3.12b) we have

$$(-1)^{b(\eta)}(2i)^{\|k\delta\|-n} \operatorname{vol}(F_\lambda)\langle h_{\eta\lambda}, h_{\eta\lambda}\rangle = ([A_k \overline{h_{\eta\lambda}}] \cup [h_{\eta\lambda}])(F_\lambda),$$

where F_λ is viewed as a singular $2n$-chain. From Lemma 2.1, (1.17), and (1.27) we see that $\langle \mathbf{h}, \mathbf{h}\rangle = \langle \mathbf{h}^\eta, \mathbf{h}^\eta\rangle$, and so

$$|A|\, 2^n \langle \mathbf{h}, \mathbf{h}\rangle = |A| \sum_{\eta \subset \delta} \langle \mathbf{h}^\eta, \mathbf{h}^\eta\rangle = \sum_{\eta \subset \delta,\, \lambda \in A} \langle h_{\eta\lambda}, h_{\eta\lambda}\rangle.$$

By Lemma 4.3 we can put

$$|A|(-1)^{\|k\delta\| + n(n-1)/2}\, i^{\|k\delta\|-n}\, 2^{\|k\delta\|} \operatorname{vol}(F_\lambda) = \pi^n M$$

with an algebraic number M independent of λ. Then

$$\pi^n M \langle \mathbf{h}, \mathbf{h}\rangle = \sum_{\eta \subset \delta,\, \lambda \in A} \langle \eta, k\delta + \delta\rangle ([A_k \overline{h_{\eta\lambda}}] \cup [h_{\eta\lambda}])(F_\lambda). \tag{4.18}$$

For simplicity put $p_\varepsilon = p(\chi, \delta, \varepsilon)$; let $y_{\varepsilon\lambda}$ and $y'_{\varepsilon\lambda}$ be $U_k(\overline{\mathbb{Q}})$-valued cocycles which represent the λ-components of y_ε and $\mathscr{A} y_\varepsilon$. By (4.16) and (3.9f) we have

$$[h_{\eta\lambda}] \equiv \sum_{\varepsilon \subset \delta} \langle \varepsilon, \eta\rangle p_\varepsilon y_{\varepsilon\lambda}, \tag{4.19a}$$

$$[A_k \overline{h_{\eta\lambda}}] \equiv \sum_{\zeta \subset \delta} \langle \zeta, \eta\rangle \bar{p}_\zeta y'_{\zeta\lambda}, \tag{4.19b}$$

where we write $a \equiv b$ when $a - b$ is a coboundary. Substituting these into (4.18), we obtain

$$\begin{aligned} \pi^n M \langle \mathbf{h}, \mathbf{h}\rangle &= \sum_{\lambda \in A} \sum_{\varepsilon, \zeta} p_\varepsilon \bar{p}_\zeta (y'_{\zeta\lambda} \cup y_{\varepsilon\lambda})(F_\lambda) \sum_\eta \langle \eta, k\delta + \delta + \varepsilon + \zeta\rangle \\ &= \sum_{\zeta \subset \delta} p_{k\delta+\delta+\zeta} \bar{p}_\zeta X_\zeta, \end{aligned} \tag{4.20}$$

where $X_\zeta = 2^n \sum_{\lambda\in\Lambda} (y'_{\zeta\lambda} \cup y_{k\delta+\delta+\zeta,\lambda})(F_\lambda)$. For $0 \neq \xi \subset \delta$ we have ${}^t[A_k \bar{h}_{\eta\lambda}] \wedge \Theta_k[h_{\xi+\eta,\lambda}] = 0$ by (3.12a), so that

$$0 = \sum_{\eta\subset\delta,\,\lambda\in\Lambda} \langle \eta, k\delta+\delta \rangle ([A_k \bar{h}_{\eta\lambda}] \cup [h_{\xi+\eta,\lambda}])(F_\lambda).$$

Therefore, repeating the above computation with $h_{\xi+\eta,\lambda}$ in place of $h_{\eta\lambda}$ of (4.19a), we obtain

$$\begin{aligned} 0 &= \sum_{\lambda\in\Lambda} \sum_{\varepsilon,\zeta} p_\varepsilon \bar{p}_\zeta (y'_{\zeta\lambda} \cup y_{\varepsilon\lambda})(F_\lambda) \langle \varepsilon, \xi \rangle \sum_\eta \langle \eta, k\delta+\delta+\varepsilon+\zeta \rangle \\ &= \sum_\zeta \langle k\delta+\delta+\zeta, \xi \rangle p_{k\delta+\delta+\zeta} \bar{p}_\zeta X_\zeta. \end{aligned}$$

Multiply this by $\langle \alpha, \xi \rangle$ with a fixed $\alpha \subset \delta$ and take the sum over all nonzero ξ; add it to (4.20). Then we find that

$$\pi^n M \langle \mathbf{h}, \mathbf{h} \rangle = 2^n \bar{p}_{k\delta+\delta+\alpha} p_\alpha X_{k\delta+\delta+\alpha}.$$

Since the X_ζ are algebraic, this together with (1) proves (2).

Remark 4.5. In [9, (10.6)] we attached the period symbols $\mathfrak{p}_\pm(\mathbf{h})$ to each $\mathbf{h}$ of the above type when $n=1$, that is, when δ consists of a single element. These are special cases of the present $p(\chi, \delta, \varepsilon)$; namely

(4.21) $$\mathfrak{p}_+(\mathbf{h}) = p(\chi, \delta, 0), \qquad \mathfrak{p}_-(\mathbf{h}) = p(\chi, \delta, \delta).$$

This can be seen by comparing (4.14) and (4.15) with [9, Lemma 10.2, (10.6)] since $\mathbf{h}^e$ of [9, (10.2)] is $\bar{\mathbf{h}}^\delta$. In this sense Theorem 4.4 includes [9, (10.8) and Proposition 10.3] as special cases.

5. The fundamental periods as factors of Fourier coefficients

In this section we study $p(\chi, \delta, \varepsilon)$ when the weight k is even, that is, when $k \in 2\mathbf{Z}^{\mathbf{a}}$. Our aim is to show that if f is a Hilbert modular form of weight $(k+\iota)/2$ given as $f(z) = \langle \theta(z, w), \mathbf{h}(w) \rangle$ with a certain theta function θ and $\mathbf{h} \in \mathcal{S}_k^0(\mathfrak{m}, \Phi)$ as in Sect. 4, then $\pi^n p(\chi, \delta, \varepsilon)^{-1} f$ has algebraic Fourier coefficients with ε determined by θ. We hereafter put $k = 2m$ with $m \in \mathbf{Z}^{\mathbf{a}}$ assuming that $m_v > 0$ for every v, and also put

(5.1) $$X = \{x \in B \mid \mathrm{Tr}(x) = 0\}.$$

For each $v \in \mathbf{a}$ let N_v denote the real quadratic form $x \mapsto N(x)$ on the real vector space X_v; let $\mathcal{P}_v^m$ denote the real vector space consisting of all $\mathbf{R}$-valued N_v-harmonic homogeneous polynomial functions on X_v of degree $m_v - 1$ in the sense of [6, p. 609]. Then we put

(5.2) $$\mathcal{P}_\delta^m = \bigotimes_{v\in\delta} \mathcal{P}_v^m, \qquad \mathcal{P}_{\delta'}^m = \bigotimes_{v\in\delta'} \mathcal{P}_v^m.$$

We view the elements of these spaces as functions on X through diagonal embeddings of X. As shown in [6, p. 609], if $v\in\delta'$, the representation of G_v on $\mathscr{P}_v^m$ given by $t\mapsto t(\alpha\xi\alpha^*)$ for $t\in\mathscr{P}_v^m$, $\xi\in X_v$, and $\alpha\in G_v$ is equivalent to σ_{k_v-2} with σ of (1.4b, c). Clearly this is a real representation, and hence in view of (1.4b, c), we can make it real orthogonal on G_v^u by changing it for an equivalent one, still retaining condition (1.4b). This means that we can find a V_k-valued function q on $\prod_{v\in\delta'} X_v$ such that

(5.3a) *The components of* q *form an* $\mathbb{R}$*-basis of* $\mathscr{P}_{\delta'}^m$ *and are* $\mathbb{Q}$*-valued on* X;

(5.3b) $q(\alpha\xi\alpha^{-1})=\rho_k'(\alpha)\,q(\xi)$ *with* ρ_k' *of* (1.6a) *for every* $\alpha\in G$ *and every* $\xi\in X$.

Moreover we can now assume

(5.3c) $\rho_k'(x)$ *is real orthogonal for every* $x\in G_{\mathbf{A}}$,

(5.3d) $$A_k=1.$$

For $w\in\mathbb{C}^\delta$ and $\xi\in X$ we put

$$(5.4)\qquad [\xi, w]=\left((-1, w_v)\,\xi_v\begin{pmatrix} w_v\\ 1\end{pmatrix}\right)_{v\in\delta} =(c_v w_v^2-2a_v w_v-b_v)_{v\in\delta}\,(\in\mathbb{C}^\delta),$$

where $\xi_v=\begin{pmatrix} a_v & b_v\\ c_v & -a_v\end{pmatrix}$. An easy calculation shows that

$$(5.5)\qquad [\varphi\xi\varphi^{-1}, \varphi w]=\operatorname{sgn} N(\varphi)\, j(\varphi, w)^{-2}[\xi, w]\qquad (\varphi\in G_{\mathbf{a}}, \xi\in X).$$

Put $U_{k\delta}=\bigotimes_{v\in\delta}\mathbb{C}^{k_v-1}$. Then $U_k=U_{k\delta}\otimes V_k$. We can find a $U_{k\delta}$-valued function r on X such that

$$(5.6)\qquad {}^t r(\xi)[w]_k^0=[\xi, w]^{m\delta-\delta}\qquad (\xi\in X, w\in\mathbb{C}^\delta)$$

with the symbol of (3.6). In fact, the right-hand side, being a polynomial in w_v of degree k_v-2 for each $v\in\delta$, can be expressed as a linear combination of the components of $[w]_k^0$, whose coefficients form the components of $r(\xi)$. From (5.5), (5.3b), (3.1), and (3.4), we easily see that

$$(5.7)\qquad (r\otimes q)(\alpha\xi\alpha^{-1})=\operatorname{sgn} N(\alpha)^{m\delta-\delta}\cdot {}^t\rho_k(\alpha)^{-1}(r\otimes q)(\xi)\qquad (\alpha\in G, \xi\in X).$$

It can easily be seen that the components of $r(\xi)$ form an $\mathbb{R}$-basis of $\mathscr{P}_\delta^m$, but this fact will not be employed.

Given $f\in\mathscr{S}_k^\varepsilon(B)$, consider $[f]$ of (3.8) and observe that

$$(5.8)\qquad {}^t(r\otimes q)(\xi)[f]=[\xi, \bar w]^{m\varepsilon-\varepsilon}[\xi, w]^{m\varepsilon'-\varepsilon'}\cdot {}^t q(\xi)\, f d_\varepsilon w,$$

where $\varepsilon'=\delta-\varepsilon$ and we use w as the variable on H^δ instead of z in (3.8).

To define our theta function, we denote by $\mathscr{L}(X)$ the set of all $\mathbb{C}$-valued locally constant functions on X, that is, the restrictions of the Schwartz-Bruhat functions on $X_{\mathfrak{f}}$ to X. Now our theta function, introduced in [6, p. 612] and [9, (6.10)] has the form

$$(5.9)\qquad \theta(z, w; \eta, \tau) = \mathrm{Im}(z)^{\delta/2}\,\mathrm{Im}(w)^{-k\delta} \cdot \sum_{\alpha\in X} \eta(\alpha)\, q(\alpha)\,[\alpha, \bar{w}]^{m\delta}\, \mathbf{e}_E((\tau/2)\, R'[\alpha; z, w]).$$

Here $z \in H^{\mathbf{a}}$, $w \in H^{\delta}$, and $\eta \in \mathscr{L}(X)$; $\mathbf{e}_E$ is defined by

$$(5.9\,\mathrm{a})\qquad \mathbf{e}_E(u) = \exp(2\pi i \sum_{v\in\mathbf{a}} u_v) \quad \textit{for } u \in \mathbb{C}^{\mathbf{a}};$$

τ is an element of E such that

$$(5.9\,\mathrm{b})\qquad \tau_v > 0 \quad \textit{for } v \in \delta \quad \textit{and} \quad \tau_v < 0 \quad \textit{for } v \in \delta';$$

$R' = (R'_v)_{v\in\mathbf{a}}$ with a certain function R'_v in α_v, z_v, and w_v, whose explicit form is given in [9, (6.10b)] but will not be needed in the present paper. This theta function is an automorphic form in both variables z and w [9, (6.11), Lemmas 6.1, 6.2, and 6.8]. In particular, for a sufficiently small congruence subgroup Γ of G^u it is invariant under $\|_k^0 \gamma$ for every $\gamma \in \Gamma$, and hence the inner product $\langle \theta(z, w; \eta, \tau), h(w)\rangle$ is meaningful in the sense of (1.16) for every $h \in \mathscr{S}_k^0(B)$. Moreover, the inner product is a holomorphic Hilbert cusp form of weight $(k+\iota)/2$ (see [6] and [9, Proposition 6.3]).

We extend this to forms on $G_{\mathbf{A}}$ by putting

$$(5.10)\qquad I(z; \eta, \tau; \mathbf{h}) = \sum_{\lambda\in\Lambda} \langle \theta(z, w; \eta_\lambda, \tau), h_\lambda(w)\rangle$$

for $\mathbf{h} = (h_\lambda)_{\lambda\in\Lambda} \in \mathscr{S}_k^0(\mathfrak{m}, \psi_1^2)$. Here ψ_1 is a Hecke character of E such that

(5.11) $\psi_{1\mathbf{a}}(x) = \mathrm{sgn}(x)^t$ *with an element* $t \in \mathbf{Z}^{\mathbf{a}}$ *such that* $\|t\| \equiv \|m - \delta'\|$ (mod 2); *the conductor of* ψ_1 *is prime to* $\mathfrak{d}_B$ *and divides* $\mathfrak{m}$ *(*$\mathfrak{m}$ *is not necessarily prime to* $\mathfrak{d}_B$*);*

η_λ is given by

$$(5.12)\qquad \eta_\lambda(y) = \psi_1(t_\lambda)^{-1}\, \eta(x_\lambda^{-1} y x_\lambda)$$

with an element η of $\mathscr{L}(X)$ satisfying

$$(5.13\,\mathrm{a})\qquad \eta(\alpha) = 0 \quad \textit{if } \tau N(\alpha) \notin \mathfrak{g};$$

$$(5.13\,\mathrm{b})\qquad \eta(sx) = \psi_1(s)\,\eta(x) \quad \textit{for every } s \in \prod_{v\in\mathbf{f}} \mathfrak{g}_v^\times;$$

$$(5.13\,\mathrm{c})\qquad \eta(uxu^{-1}) = \psi_{1\mathfrak{m}}(a_u^2/N(u))\,\eta(x) \quad \textit{for every } u \in W_{\mathfrak{m}}.$$

For an example of η satisfying (5.13a–c), see [9, pp. 280–281]. For the reason explained above, the function of (5.10) is a Hilbert cusp form of weight $(k+\iota)/2$. Moreover, in [9, Theorem 6.7] we showed that the correspondence $\mathbf{h} \mapsto I(z; \eta, \tau; \mathbf{h})$ commutes with Hecke operators for almost all primes of E. Now our second main theorem is

Theorem 5.1. *Let τ, ψ_1, and $\mathfrak{m}$ be as in* (5.9b) *and* (5.11), *and let* $f(z)=I(z;\eta,\tau;\mathbf{h})$ *with a $\bar{\mathbb{Q}}$-valued η satisfying* (5.13a–c) *and* $\mathbf{h}\in\mathscr{S}^0_{2m}(\mathfrak{b},\psi_1^2)$, *where $\mathfrak{b}$ is a divisor of $\mathfrak{m}$ prime to $\mathfrak{d}_B$. Suppose that $\mathbf{h}$ is $\bar{\mathbb{Q}}$-rational,* $\mathbf{h}|T_v=\chi(v)\mathbf{h}$ *for every $v\in\mathfrak{f}$, and $\{\chi,\mathfrak{b}\}$ is primitive in the sense of Sect. 2. Then all Fourier coefficients of* $\pi^n p(\chi,\delta,t\delta+m\delta)^{-1}f$ *are algebraic, where t is determined by* (5.11) *and $t\delta+m\delta$ is considered modulo $2\mathbb{Z}^\delta$.*

A technical clarification is necessary, as we are dealing with two "levels" $\mathfrak{m}$ and $\mathfrak{b}$. To define the inner products in (5.10), we view $\mathbf{h}$ as an element of $\mathscr{S}^0_{2m}(\mathfrak{m},\psi_1^2)$ and take the elements $(h_\lambda)_{\lambda\in\Lambda}$ in $\prod_{\lambda\in\Lambda}\mathscr{S}^0_{2m}(\Gamma_\lambda,\psi_1^2,0)$. On the other hand we have defined the period symbol p by (4.15) at the exact level of $\mathbf{h}$, which is $\mathfrak{b}$ in the present case. Namely we take a decomposition

$$G_{\mathbf{A}}=\coprod_{v\in N} G r_v W_{\mathfrak{b}}=\coprod_{v\in N} G r_v W_{\mathfrak{b}}^1 \tag{5.14}$$

and put $\Gamma_{\mathfrak{b}v}=G\cap r_v W_{\mathfrak{b}} r_v^{-1}$ and $\Gamma^1_{\mathfrak{b}v}=G\cap r_v W^1_{\mathfrak{b}} r_v^{-1}$. Let $\mathbf{h}^\zeta=(h'_{\zeta v})_{v\in N}$ with $h'_{\zeta v}\in\mathscr{S}^\zeta_k(\Gamma_{\mathfrak{b}v},\Phi_{\mathfrak{b}},0)$. Then (4.16) in the present case means

$$[h'_{\zeta v}]=\sum_{\varepsilon\subset\delta}\langle\varepsilon,\zeta\rangle p_\varepsilon y_{\varepsilon v} \tag{5.15}$$

with $y_{\varepsilon v}\in H^n_k(\Gamma^1_{\mathfrak{b}v},S)$ and $p_\varepsilon=p(\chi,\delta,\varepsilon)$ for simplicity. Now changing the r_v if necessary, we may assume that $x_\lambda\in r_v W^1_{\mathfrak{b}}$, where v is uniquely determined by λ. Let $\mathbf{h}^\zeta=(h_{\zeta\lambda})_{\lambda\in\Lambda}$ with $h_{\zeta\lambda}\in\mathscr{S}^\zeta_k(\Gamma_\lambda,\psi^2_{1m},0)$. Then we easily see that $\Gamma^1_\lambda\subset\Gamma^1_{\mathfrak{b}v}$ and $h_{\zeta\lambda}=h'_{\zeta v}$. The element $y_{\varepsilon v}$ naturally defines an element $y_{\varepsilon\lambda}$ of $H^n_k(\Gamma^1_\lambda,S)$. Then clearly

$$[h_{\zeta\lambda}]=\sum_{\varepsilon\subset\delta}\langle\varepsilon,\zeta\rangle p_\varepsilon y_{\varepsilon\lambda} \tag{5.16}$$

for every $\lambda\in\Lambda$. We shall employ this relation later. (We could have defined the inner product $I(z;\eta,\tau;\mathbf{h})$ independently of the level by putting the factor $|\Lambda|^{-1}$ as we did in (1.27), or by defining it as an integral over a quotient of $G_{\mathbf{A}}$.)

The proof of our theorem requires two more facts.

Lemma 5.2. *Let ε, β, λ, and μ be as in Lemma 2.1 and let t be as in* (5.11). *Then we have* $\eta_\lambda(\beta\alpha\beta^{-1})=(-1)^{\|t\varepsilon\|}\eta_\mu(\alpha)$ *for every $\alpha\in X$.*

Proof. Take e as in Lemma 2.1 and $u\in W^1_{\mathfrak{m}}$ so that $x_\lambda e=\beta x_\mu u$. By (5.12) we have $\eta_\lambda(\beta\alpha\beta^{-1})=\psi_1(t_\lambda)^{-1}\eta(x_\lambda^{-1}\beta\alpha\beta^{-1}x_\lambda)=\psi_1(t_\lambda)^{-1}\eta(eu^{-1}x_\mu^{-1}\alpha x_\mu ue^{-1})$. We can ignore e since $e\in G_{\mathbf{a}}$. Therefore by (5.12) and (5.13c) we have $\eta_\lambda(\beta\alpha\beta^{-1})=\psi_1(N(u)/t_\lambda)\eta(x_\mu^{-1}\alpha x_\mu)=\psi_1(N(u)t_\mu/t_\lambda)\eta_\mu(\alpha)$. Now $t_\lambda N(e)=N(\beta)t_\mu N(u)$, and hence $\psi_1(N(u)t_\mu/t_\lambda)=\psi_1(N(e))=(-1)^{\|t\varepsilon\|}$, which proves our formula.

Lemma 5.3. *Let* $R=\{\alpha\in L\,|\,N(\alpha)=c\}$ *with a $\mathbb{Z}$-latice L in X and a fixed nonzero element c of E. Let Γ be a congruence subgroup of G^u such that $\gamma L\gamma^{-1}=L$ for every $\gamma\in\Gamma$. Then there exists a finite subset S of R such that* $R=\bigcup_{\gamma\in\Gamma}\gamma S\gamma^{-1}$.

Proof. Let $T=\{\alpha\in X_{\mathbf{a}}\,|\,N(\alpha)=c\}$ and $\bar{G}=\{g\in\prod_{v\in\mathbf{a}}SL(X_v)\,|\,N(g(x))=N(x)\}$. In this proof we view E, X, and G as subsets of $E_{\mathbf{a}}$, $X_{\mathbf{a}}$, and $G_{\mathbf{a}}$ through natural embeddings. Each $\gamma \in G^u$ defines an element g_γ of $\bar{G}$ by $g_\gamma(x) = \gamma x \gamma^{-1}$. Then G^u is the spin covering of the identity component of $\bar{G}$ by the map $\gamma \mapsto g_\gamma$. Put $\Gamma' = \{g_\gamma | \gamma \in \Gamma\}$. Since B is a division algebra, $\bar{G} = \Gamma' K$ with a compact subset K. We easily see that $\bar{G}$ acts transitively on T (if T is nonempty), so that $T = \bar{G}\xi = \Gamma' K \xi$ with any $\xi \in T$. Therefore we have $R = \Gamma' S$ with $S = L \cap K\xi$, which is finite because L is a discrete lattice in X_a and $K\xi$ is compact.

Proof of Theorem 5.1. We first put $\Delta_\lambda = G^u \cap \Gamma_\lambda^1$ with Γ_λ^1 defined at level m, $f(z) = \sum_{0 \ll b \in E} l_f(b) e_E(bz/2)$ with $l_f(b) \in \mathbb{C}$, and $R(b) = \{\alpha \in X | -\tau N(\alpha) = b\}$. From (5.12) and (5.13c) we see that $\eta_\lambda(\gamma y \gamma^{-1}) = \eta_\lambda(y)$ for every $\gamma \in \Delta_\lambda$ and hence $j_k^0(\gamma, w)^{-1} \theta(z, \gamma w; \eta_\lambda, \tau) = \theta(z, w; \eta_\lambda, \tau)$ for such γ by [9, (6.11)]. Therefore we can apply [6, Proposition 2.3] (see also Remark 5.4 below and [9, Proposition 6.3]) to the present situation with Δ_λ as Γ there to obtain

$$(5.17) \qquad M\, l_f(b) = \kappa (2b)^{-\delta/2} \sum_{\mu \in \Lambda} \sum_{\alpha \in R(b)/\Delta_\mu} \overline{\eta_\mu(\alpha)}\, P(h_\mu, \alpha, \Delta_\mu).$$

Here $M = \mathrm{vol}(\Delta_\mu \backslash H^\delta)$ which is independent of μ by Lemma 4.3; κ is the order of $\Delta_\mu \cap \{\pm 1\}$ which is clearly equal to the order of $W_m^1 \cap \{\pm 1\}$ and hence independent of μ; $R(b)/\Delta_\mu$ is a complete set of representatives for the set $R(b)$ modulo the transformations $\alpha \mapsto \gamma \alpha \gamma^{-1}$ for all $\gamma \in \Delta_\mu$; this is a finite set by Lemma 5.3; P is the symbol of [6, (2.5)], whose explicit form, with the subscript μ suppressed, is

$$P(h, \alpha, \Delta) = \int_{J^\alpha/\Delta^\alpha} [\alpha, g w^0]^{m\delta} \cdot {}^t q(\alpha)\, h(g w^0)\, d(g\Delta^\alpha).$$

The symbols are as follows: w^0 is an arbitrarily fixed point of H^δ;

$$\Delta^\alpha = \Delta \cap J^\alpha, \qquad J^\alpha = \{g \in SL_2(\mathbb{R})^\delta | g\alpha = \alpha g\}.$$

(In this proof we work within the group $\prod_{v \in \delta} G_v = GL_2(\mathbb{R})^\delta$, and therefore view every element of G as an element of this group through an obvious projection.) Fix an element α of $R(b)/\Delta$. To make the measure $d(g\Delta^\alpha)$ explicit and also to change Δ^α for a more manageable group, take $\varphi \in SL_2(\mathbb{R})^\delta$ so that $(\varphi \alpha \varphi^{-1})_v = \mathrm{diag}[c_v, -c_v]$ with $0 < c_v \in \mathbb{R}$, and put $a(t) = \mathrm{diag}[t, t^{-1}]$ for $t \neq 0$,

$$A = \{a(t) | 0 \neq t \in \mathbb{R}\}, \qquad A_+ = \{a(t) | 0 < t \in \mathbb{R}\}.$$

Then $\tau_v c_v^2 = b_v$ and $\varphi J^\alpha \varphi^{-1} = A^\delta$. Let $J_+ = \varphi^{-1} A_+^\delta \varphi$ and let D denote the subgroup $\{\pm 1\}^\delta$ of A^δ (and of J^α). Then $J^\alpha = D \times J_+$. As explained in [6, p. 615], $\{\pm 1\}\Delta^\alpha/\{\pm 1\}$ is isomorphic to $\mathbb{Z}^n$. Let Δ_+ denote the isomorphic image of Δ^α under the projection of J^α to J_+. Then

$$\kappa P(h, \alpha, \Delta) = 2^n \int_{J_+/\Delta_+} [\alpha, g w^0]^{m\delta} \cdot {}^t q(\alpha)\, h(g w^0)\, d(g\Delta_+)$$

and the measure on J_+ is given by $d^\times t = \prod_{v\in\delta} t_v^{-1}\, dt_v$ if we parametrize J_+ by the map $t \mapsto g(t) = \varphi^{-1} a(t)\varphi$ with $0 \ll t \in \mathbb{R}^{a}$ (see [6, p. 610]). Thus

$$\kappa P(h, \alpha, \Delta) = 2^n \int_F [\alpha, g(t)w^0]^{m\delta} \cdot {}^t q(\alpha)\, h(g(t)w^0)\, d^\times t,$$

where $F = \mathbb{R}_+^{a}/g^{-1}(\Delta_+)$. Let $\{\gamma_1, \ldots, \gamma_n\}$ be a set of generators for Δ_+. Then we can put $\gamma_i = g((\exp(b_{iv}))_{v\in\delta})$ with $b_i \in \mathbb{R}^\delta$.

We now consider an n-dimensional cube

$$K = \{s \in \mathbb{R}^n \mid 0 \leqq s_i \leqq 1 \ \textit{for all } i\}$$

and define a map $\sigma: K \to H^\delta$ by

$$\sigma(s) = g\left(\exp\left(\sum_{i=1}^n b_{iv} s_i\right)_{v\in\delta}\right) w^0 \quad (s \in K).$$

Changing the order of the γ_i if necessary, we may assume that $\det(b_{iv}) > 0$. (Recall that we have a fixed order in the set δ.) Decomposing K into simplexes and mapping them by σ, we obtain a singular n-chain on H^δ, which we also denote by σ. Then the above integral over F is equal to

$$(5.18) \qquad (-c)^\delta \int_\sigma [\alpha, w]^{m\delta-\delta} \cdot {}^t q(\alpha)\, h(w)\, d_0 w.$$

The proof, which was essentially given in [6, pp. 615–616], is as follows. We first observe that $(\varphi\sigma(s))_v = \exp(2\sum_i b_{iv} s_i)\,(\varphi w^0)_v$ and

$$j(\varphi, \sigma(s))_v^{-2}\, d\sigma_v = \sum_{i=1}^n 2 b_{iv} (\varphi\sigma(s))_v\, ds_i.$$

By (5.5) we have $[\alpha, \sigma(s)] = -2c_v\, j(\varphi, \sigma(s))_v^2 (\varphi\sigma(s))_v$. Therefore

$$-c_v\, d\sigma_v = [\alpha, \sigma(s)]_v \sum_{i=1}^n b_{iv}\, ds_i.$$

Since $d^\times t = \det(b_{iv})\, ds_1 \ldots ds_n$, we obtain (5.18). Combining this with (5.8) and restoring the subscript μ, we thus have

$$\kappa P(h_\mu, \alpha, \Delta_\mu) = (-2)^n |b/\tau|^{\delta/2} \cdot {}^t(r\otimes q)(\alpha) \int_\sigma [h_\mu].$$

We now fix any $\varepsilon \subset \delta$ and take λ, μ, and β as in Lemma 2.1. Put $\alpha' = \beta\alpha\beta^{-1}$ and $\varphi_v' = \mathrm{diag}[N(\beta)_v, 1]\varphi_v\beta_v^{-1}$ for each $v \in \delta$. By Lemma 2.1 we have $\beta\Delta_\mu\beta^{-1} = \Delta_\lambda$. Taking φ', $\beta_* w^0$, α', and $\beta_*\sigma(s)$ in place of φ, w^0, α, and $\sigma(s)$, we find that

$$\kappa P(h_\lambda, \alpha', \Delta_\lambda) = (-2)^n |b/\tau|^{\delta/2} \int_{\beta_*\sigma} {}^t(r\otimes q)(\alpha')[h_\lambda].$$

By (3.9a), (5.7), and Lemma 2.1 the last integral is equal to

$$(-1)^{\|m\varepsilon\|}\cdot{}^t(r\otimes q)(\alpha)\int_\sigma [h_{\varepsilon\mu}],$$

where we put $\mathbf{h}^\varepsilon=(h_{\varepsilon\mu})_{\mu\in\Lambda}$ with $h_{\varepsilon\mu}\in\mathscr{S}_k^\varepsilon(\Gamma_\mu^1, \psi_{1\,m}^2, 0)$. Now, for each μ and $\alpha\in R(b)/\Delta_\mu$ we construct a chain σ in the above fashion and denote it by $\sigma_{\mu\alpha}$. Then

$$\kappa P(h_\lambda, \beta\alpha\beta^{-1}, \Delta_\lambda)=(-2)^n|b/\tau|^{\delta/2}(-1)^{\|m\varepsilon\|}\cdot{}^t(r\otimes q)(\alpha)\int_{\sigma_{\mu\alpha}}[h_{\varepsilon\mu}].$$

By Lemma 5.2 we have $\eta_\lambda(\beta\alpha\beta^{-1})=\langle\delta t, \varepsilon\rangle\,\eta_\mu(\alpha)$. Since $\alpha\mapsto\beta\alpha\beta^{-1}$ gives a bijection of $R(b)/\Delta_\mu$ onto $R(b)/\Delta_\lambda$ and $\lambda\mapsto\mu$ is a permutation of Λ, the last formula combined with (5.17) yields, for every fixed ε,

$$\begin{aligned} Ml_f(b)&=\kappa(2b)^{-\delta/2}\sum_{\lambda\in\Lambda}\sum_{\alpha'\in R(b)/\Delta_\lambda}\overline{\eta_\lambda(\alpha')}\,P(h_\lambda,\alpha',\Delta_\lambda)\\ &=\kappa(2b)^{-\delta/2}\sum_{\mu\in\Lambda}\sum_{\alpha\in R(b)/\Delta_\lambda}\overline{\eta_\lambda(\beta\alpha\beta^{-1})}\,P(h_\lambda,\beta\alpha\beta^{-1},\Delta_\lambda)\\ &=(-1)^n|2/\tau|^{\delta/2}\langle t\delta+m\delta,\varepsilon\rangle\sum_{\mu\in\Lambda}\sum_{\alpha\in R(b)/\Delta_\mu}\overline{\eta_\mu(\alpha)}\cdot{}^t(r\otimes q)(\alpha)\int_{\sigma_{\mu\alpha}}[h_{\varepsilon\mu}]. \end{aligned}$$

At this stage we need an alternation operator θ on $S_n(H^\delta)$ which is defined as follows: Let s: $\mathscr{E}\to H^\delta$ be a singular r-simplex, where $\mathscr{E}=\langle e_0\ldots e_r\rangle$ is a standard Euclidean simplex with vertices e_i. Each permutation π of $\{0, \ldots, r\}$ defines a barycentric map φ_π of $\mathscr{E}$ onto itself such that $\varphi_\pi(e_i)=e_{\pi(i)}$. Then $s\circ\varphi_\pi$ is a singular r-simplex, which we denote by $s\circ\pi$. Put

$$\theta_r(s)=\theta(s)=\sum_\pi \operatorname{sgn}(\pi)\,s\circ\pi,$$

where π runs over all such permutations. We can extend θ additively to the whole $S_r(H^\delta)$, and easily verify that $\partial\theta_r=(r+1)\theta_{r-1}\partial$, $\theta\circ\gamma=\gamma\circ\theta$ for every $\gamma\in G_{\mathbf{a}}^u$, and

(5.19) $$\int_{\theta(c)}\omega=(r+1)!\int_c\omega$$

for every r-form ω and every $c\in S_r(H^\delta)$. Let us now show that

(5.20) $$\partial\theta(\sigma)=\sum_{\nu=1}^n(\gamma_\nu u_\nu-u_\nu)$$

with $u_\nu\in S_{n-1}(H^\delta)$. In fact, if we view K as an *alternating* chain which is a sum of *oriented* simplexes, then obviously $\partial K=\sum_{\nu=1}^n(t_\nu d_\nu-d_\nu)$ with the sides d_ν

of K and the Euclidean translations t_ν corresponding to the γ_ν. This means that

$$(5.21)\qquad \partial\sigma=\sum_{\nu=1}^{n}\sum_{s,\tau}(\gamma_\nu s-\operatorname{sgn}(\tau)s\circ\tau)+\sum_{s',\tau'}(s'-\operatorname{sgn}(\tau')s'\circ\tau')$$

with some singular $(n-1)$-simplexes s, s' and some permutations τ, τ' of $\{0, 1, \dots, n-1\}$. Then (5.20) follows immediately from (5.21). (Equality (5.20) may be false with $\partial\sigma$ in place of $\partial\theta(\sigma)$ if $n>1$, which is why the operator θ is necessary.)

Putting $\xi_{\alpha\mu}=\theta(\sigma_{\alpha\mu})$ and applying (5.20) to it, we have

$$\partial\xi_{\alpha\mu}=\sum_{\nu=1}^{n}(\gamma'_\nu u_\nu-u_\nu)$$

with some $u_\nu\in S_{n-1}(H^\delta)$ for each fixed α and μ, where γ'_ν is the element of Δ_μ corresponding to γ_ν. This together with (5.7) shows that $(r\otimes q)(\alpha)\otimes\xi_{\alpha\mu}$ is an equivariant n-cycle of type (ρ_k, Γ^1_μ) since α commutes with the γ'_ν. Now we invoke relation (5.16). Choose a cocycle in the class $y_{\varepsilon\lambda}$ of (5.16) and denote it again by $y_{\varepsilon\lambda}$. Recalling our choice of y_ε in (4.14), we can take $y_{\varepsilon\lambda}$ to be $U_k(\bar{\mathbb{Q}})$-valued. Since the value of a cocycle φ at an equivariant cycle of type (ρ_k, Γ^1_μ) depends only on the cohomology class of φ, we obtain, in view of (5.19),

$$(n+1)!\cdot{}^t(r\otimes q)(\alpha)\int_{\sigma_{\alpha\mu}}[h_{\varepsilon\mu}]=\sum_{\zeta}\langle\zeta,\varepsilon\rangle p_\zeta\cdot{}^t(r\otimes q)(\alpha)\,y_{\zeta\mu}(\zeta_{\alpha\mu}).$$

This result combined with the above formula for $l_f(b)$ produces

$$Ml_f(b)=\sum_{\mu\in A}\sum_{\alpha\in R(b)/\Delta_\mu}\sum_{\zeta}\langle\zeta+t\delta+m\delta,\varepsilon\rangle p_\zeta\,Y(\alpha,\zeta,\mu),$$

where we put

$$Y(\alpha,\zeta,\mu)=(-1)^n|2/\tau|^{\delta/2}(1/(n+1)!)\overline{\eta_\mu(\alpha)}\cdot{}^t(r\otimes q)(\alpha)\,y_{\zeta\mu}(\xi_{\alpha\mu}).$$

This quantity is algebraic since $y_{\zeta\mu}$ is $U_k(\bar{\mathbb{Q}})$-valued and $(r\otimes q)(\alpha)\in U_k(\bar{\mathbb{Q}})$ by virtue of (1.4a) and (5.3a). Taking the sum over all $\varepsilon\subset\delta$, we find that

$$Ml_f(b)=p_{t\delta+m\delta}\sum_{\mu\in A}\sum_{\alpha\in R(b)/\Delta_\mu}Y(\alpha,t\delta+m\delta,\mu).$$

Since $M\pi^{-n}$ is algebraic by Lemma 4.3, this shows that $\pi^n l_f(b)/p_{t\delta+m\delta}$ is algebraic, which completes the proof.

Remark 5.4. Some formulas and statements in [6] and [9] require corrections when the subgroup Γ of $SL_2(\mathbb{R})^\delta$ in question contains -1. First of all, $2^r q^{\xi/2}$ in [6, Proposition 2.3] should be $[\Gamma\cap\{\pm1\}:1]^{-1}2^r q^{\xi/2}$. This is why κ is necessary in (5.17). The factor $[\Gamma\cap\{\pm1\}:1]$ must be applied to the quantities in [6, p. 613, lines 15–17; p. 614, line 5]. The same is necessary on the right-hand sides of the formulas in [9, p. 280, line 21; p. 282, line 2]. A simpler solution is to assume always that $-1 \notin \Gamma$, which should have been done in the proof of [9, p. 283, Theorem 6.7].

We insert here some more corrections to [9].

Page 253, line 6: f should be (f).

Page 255, line 8 from the bottom: Delete "at most".

Page 271, line 15: $aF^{(v)}$ should be "a to $F^{(v)}$".

Page 283, line 11: $2\mathfrak{b}\mathfrak{d}_B$ should be $\mathfrak{b}\prod_{v \nmid \mathfrak{d}_B}(2\mathfrak{g})_v$.

Page 280, line 2: $(z, w; \eta)$ should be $\theta(z, w; \eta)$.

Page 287, (6.4): $\mathfrak{o}_{1v}$ should be defined as in (1.18) of the present paper.

Page 296, line 9 from the bottom: $q(w)$ should be $q(\xi)$.

Page 300: The proof on this page is notationally somewhat confused. It should be organized as in the proof of Theorem 5.1 above. Anyway the result there is included in the present theorem as a special case.

6. Algebraic relations between the periods

So far we have been treating the objects with a fixed B. We now have to consider various B's, and so write $\mathcal{S}_k^\varepsilon(\mathfrak{m}, \Phi; B)$ for $\mathcal{S}_k^\varepsilon(\mathfrak{m}, \Phi)$. We say that B has signature (δ, δ') if δ and δ' are as in (1.2). We also recall that any primitive χ occurring in $\mathcal{S}_k^0(\mathfrak{m}, \Phi; B)$ can occur in $\mathcal{S}_k^0(\mathfrak{n}, \Phi; M_2(E))$ for some $\mathfrak{n}$ with the same k and Φ. Thus k may be called the *weight* of χ. Throughout the rest of the paper we put

$$[E:\mathbf{Q}] = d. \tag{6.1}$$

In our previous papers [7] and [9] we stated conjectures concerning the possibility of associating with each primitive χ some constants $P(\chi, \delta, \varepsilon)$ which are elements of $\mathbf{C}^\times/\overline{\mathbf{Q}}^\times$ defined for every $\delta \subset \mathbf{a}$ and every $\varepsilon \subset \delta$. To recall their properties, we remind the reader of the following convention: these δ and ε are viewed as elements of the ring $(\mathbf{Z}/2\mathbf{Z})^{\mathbf{a}}$; ι denotes the identity element of this ring; the zero element of the ring corresponds to the empty set. Now the supposed properties of P when $k \in 2\mathbf{Z}^{\mathbf{a}}$ are as follows:

(P1) $P(\chi, 0, 0) = 1$.

(P2) $P(\bar{\chi}, \delta, \varepsilon) = \overline{P(\chi, \delta, \varepsilon)}$.

(P3) *If* $\mathbf{h}$ *is a* $\overline{\mathbf{Q}}$*-rational primitive form in* $\mathcal{S}_k^0(\mathfrak{m}, \Phi; B)$ *belonging to* χ *with* B *of signature* (δ, δ'), *then the "periods" of* $\mathbf{h}$ *are* $\overline{\mathbf{Q}}$*-rational combinations of* $\pi^{\|\delta\|} P(\chi, \delta, \varepsilon)$ *for all* $\varepsilon \subset \delta$.

(P4) *If* $\mathbf{h}$ *is as in* (P3), *then* $\langle \mathbf{h}, \mathbf{h}\rangle \overline{\mathbf{Q}}^\times = \pi^{\|\delta\|} P(\chi, \delta, \varepsilon) P(\bar{\chi}, \delta, \delta - \varepsilon)$ *for every* $\varepsilon \subset \delta$.

(P5) *The notation being the same as in Theorem* 5.1, *all the Fourier coefficients of* $P(\chi, \delta, t\delta)^{-1} f$ *are algebraic.*

(P6) $P(\chi, \iota, \zeta) = P(\chi, \delta, \zeta\delta) P(\bar{\chi}, \iota - \delta, \zeta - \zeta\delta)$ *for every* $\zeta \subset \mathbf{a}$.

(P7) $P(\chi, \iota, \zeta) = \pi^{-\|m\| - d} V(\chi, \zeta)$, *where* $m = k/2$ *and* V *is the element of* $\mathbf{C}^\times/\overline{\mathbf{Q}}^\times$ *determined by* [9, (8.2a)].

We should also note here that (i) $V(\chi, \zeta)$ *appears as the essential factor of the critical values of a series of the type* $\sum \varphi(\mathfrak{a}) \chi(\mathfrak{a}) N(\mathfrak{a})^{-s}$, *where* φ *is a Hecke* *ideal character of E and $\mathfrak{a}$ runs over the integral ideals of E*; (ii) *$P(\chi, \delta, \varepsilon)$ appears as the essential factor of the critical values of some Dirichlet series associated with two Hilbert modular forms of integral or half-integral weight*. For precise statements the reader is referred to [9, (8.2a, b), (8.3), Theorem 8.3, Conjecture 9.3] and [8, Theorem 10.2].

The above properties of P were given in [9, Conjecture 9.3] except (P4) which was mentioned in [9, p. 294] and [7, p. 281, (P2)]. Replacing $P(\chi, \delta, \varepsilon)$ with another symbol $P(\chi, \delta, \varepsilon; B)$ that possibly depends on B, we can state weaker versions of (P1–5), which we call (P*1–5). (As for (P6), its weaker version will imply the independence of P from B, as will be seen below.)

Now our results in Sects. 4 and 5 establish (P*2–5) for division algebras B. In fact, if we put

$$P(\chi, \delta, \varepsilon; B)=\pi^{-\|\delta\|}\, p(\chi, \delta, \varepsilon+m\delta; B) \tag{6.2}$$

with $m=k/2$, then Proposition 4.2, Theorem 4.4 (1), (2), and Theorem 5.1 are exactly (P*3), (P*2), (P*4), and (P*5). The results of Sect. 4 are valid even in the case where k is not necessarily even. Probably (P1–7), excluding (P5), if suitably modified, will hold even when $k\notin 2\mathbb{Z}^{\mathbf{a}}$ so long as k_v (mod 2) is independent of v. As for (P7), we should note that $\pi^{-\|m\|-d}\,V(\chi, \zeta)$ has the properties of $P(\chi, \iota, \zeta)$ as stipulated in (P2), (P4), and (P5) for every B with $\delta=\iota$ as shown by [9, Lemma 8.1, (8.2b), and Theorem 9.4]. However, that this is consistent with (6.2) with $\delta=\iota$ is an open problem. If we could prove that $I(z; \eta, \tau; \mathbf{h})\neq 0$ for suitable η and τ in the case $\delta=\iota$, then Theorem 4.4 combined with [9, Theorem 9.4] would imply that we indeed have

$$\pi^{-\|m\|-d}\,V(\chi, \zeta)=\pi^{-d}\,p(\chi, \iota, \zeta+m; B) \tag{6.3}$$

when B is a division algebra, and this quantity satisfies (P2–6) with $\delta=\iota$.

Let us now show that the stronger versions for $\delta\neq\iota$ including (P6) can be proved if we assume the nonvanishing of certain inner products. Let B and B' be of signature (δ, δ') and (δ', δ) respectively with nonempty δ and δ'. Let $\mathbf{h}$ be an element of $\mathscr{S}_k^0(\mathfrak{b}, \psi_1^2; B)$ as in Theorem 5.1. Similarly let $\mathbf{h}'$ be a $\bar{\mathbb{Q}}$-rational element of $\mathscr{S}_k^0(\mathfrak{b}', \psi_1'^2; B')$ belonging to the same χ and satisfying the same conditions as those satisfied by $\mathbf{h}$ with (B, δ, δ') replaced by (B', δ', δ); let $\psi_{1\mathbf{a}}(x)=\operatorname{sgn}(x)^t$ and $\psi'_{1\mathbf{a}}(x)=\operatorname{sgn}(x)^{t'}$ with $t, t'\in\mathbb{Z}^{\mathbf{a}}$. Naturally $\psi_1^2=\psi_1'^2$. Put $f(z)=I(z; \eta, \tau; \mathbf{h})$ with τ and η as in Theorem 5.1 and $f'(z)=I(z; \eta', \tau'; \mathbf{h}')$ with τ' and η' satisfying the corresponding conditions in an obvious sense. From [9, Lemma 6.4] we know that $f\in\mathscr{S}_l(\mathfrak{c}, \varphi\psi_1)$ and $f'\in\mathscr{S}_l(\mathfrak{c}, \varphi'\psi'_1)$ with some integral ideal $\mathfrak{c}$ in E, where $l=m+(\iota/2)$, $\mathscr{S}_l(*, *)$ is defined in [9, p. 277], and φ resp. φ' is the Hecke character of E corresponding to $E(\tau^{1/2})$ resp. $E(\tau'^{1/2})$. We choose τ, τ', ψ_1, and ψ'_1 so that $\varphi\psi_1=\varphi'\psi'_1$. By [9, Theorem 6.7] we have $f|\mathscr{T}_v=N(v)^{-1}\chi(v)f$ and $f'|\mathscr{T}_v=N(v)^{-1}\chi(v)f'$ for almost all $v\in\mathbf{f}$, where $\mathscr{T}_v$ is the Hecke operator at v acting on the forms of weight l defined in [8]. Therefore it is not too optimistic to expect that $\langle f, f'\rangle\neq 0$ for suitable choices of τ, τ', η, and η'. In this setting we have

Proposition 6.1. *If $\langle f, f'\rangle\neq 0$ for some τ, τ', η, and η', then*

$$\pi^{-\|m\|-d}\,V(\chi, t+\delta')=P(\bar{\chi}, \delta, t\delta; B)\,P(\chi, \delta', t'\delta'; B'). \tag{6.4}$$

Proof. Put

$$p=P(\chi,\delta,t\delta;B),\qquad q=P(\chi,\delta',t'\delta';B'),$$

$$f(z)=\sum_{\xi\in E}a(\xi)\,\mathbf{e}_E(\xi z/2),\qquad f'(z)=\sum_{\xi\in E}a'(\xi)\,\mathbf{e}_E(\xi z/2),$$

$$D(s)=\sum_{\xi\in E/U}\overline{a(\xi)}\,a'(\xi)\,\xi^{-1/2}N_{E/\mathbb{Q}}(\xi)^{1/2-s},$$

where U is a sufficiently small subgroup of $\mathfrak{g}^\times$ of finite index such that each term of D depends only on $U\xi$. Put $f_\rho(z)=\overline{f(-\bar{z})}$. Then $f_\rho(z)=\sum_\xi\overline{a(\xi)}\,\mathbf{e}_E(\xi z/2)$; moreover $f_\rho(z)=(-1)^{\|m\delta+t\delta\|}I(z;\bar{\eta},\tau;\overline{\mathbf{h}^\delta})$ as will be shown in Lemma 6.2 below, and so $f_\rho\in\mathcal{S}_l(\mathfrak{c}',\varphi\psi_1)$ with some $\mathfrak{c}'$ by [9, Lemma 6.4]. (In fact this is so with $\mathfrak{c}'=\mathfrak{c}$, but this fact is unnecessary.) By Theorem 5.1 and (6.2) we know that $p^{-1}f$, $\bar{p}^{-1}f_\rho$, and $q^{-1}f'$ have algebraic Fourier coefficients. Therefore we can apply [9, Theorem 8.3(II)] to $(p\bar{q})^{-1}D$ to find that the residue of D at $s=1/2$ is an algebraic number times $\bar{p}q\pi^{-d}R_E V(\chi,\iota-t-\delta')$, since $(\varphi\overline{\psi_1})_{\mathbf{a}}(x)=\operatorname{sgn}(x)^{t+\delta'}$, where R_E is the regulator of E. Now the proof of this fact [9, p. 290, in particular, lines 16–17] shows that the residue is a nonzero constant times $\langle f,f'\rangle$. On the other hand we can also apply [9, Theorem 9.1] to D with f_ρ and f' as f and f' there to find that the residue of D at $s=1/2$ is an algebraic number times $\pi^{\|m\|-d}R_E Q(\chi,\iota)$ with Q of [9, Sect. 7]. Thus we have an equality

$$\bar{p}q\pi^{-d}R_E V(\chi,\iota-t-\delta')=\pi^{\|m\|-d}R_E Q(\chi,\iota)$$

in $\mathbb{C}^\times/\bar{\mathbb{Q}}^\times$. By [9, (8.2b)] we have

$$\pi^{\|2m\|+d}Q(\chi,\iota)=V(\chi,\iota-t-\delta')\,V(\chi,t+\delta').$$

Combining these two equalities, we obtain (6.4).

Now the left-hand side of (6.4) is independent of B and B', and the equality holds for any pair (B,B') of prescribed signature as above, provided χ occurs in both $\mathcal{S}_k^0(B)$ and $\mathcal{S}_k^0(B')$. Notice that if $\|\delta\|$ is even, we can always find B' of signature (δ',δ) such that χ occurs in $\mathcal{S}_k^0(B')$. In this way, on the premise that we can find such a B' and also η,η',τ,τ' such that $\langle f,f'\rangle\neq 0$, we know from (6.4) that the symbol P is independent of B.

There is one more significant consequence of this argument. Since $\varphi\psi_1=\varphi'\psi'_1$, we have $t+\delta'=t'+\delta$. Put $\zeta=t+\delta'$. Observing that $t\delta=\zeta\delta$ and $t'\delta'=\zeta-\zeta\delta$, and suppressing B and B', we can write (6.4) in the form

$$\pi^{-\|m\|-d}V(\chi,\zeta)=P(\bar{\chi},\delta,\zeta\delta)\,P(\chi,\iota-\delta,\zeta-\zeta\delta). \tag{6.5}$$

As already explained, the left-hand side is the incontestable candidate for $P(\chi,\iota,\zeta)$ and therefore once we accept it, (6.5) yields (P6).

It remains to prove a fact employed in the proof of Proposition 6.1.

Lemma 6.2. *Let* $f(z)=I(z;\eta,\tau;\mathbf{h})$ *with the symbols of* (5.11), (5.13a–c), *and Theorem 5.1. Then* $\overline{f(-\bar{z})}=(-1)^{\|m\delta+t\delta\|}I(z;\bar{\eta},\tau;\overline{\mathbf{h}^\delta})$.

Proof. Fix λ in Λ and take e, μ, and β as in Lemma 2.1 with $\varepsilon=\delta$. Let $\mathbf{h}=(h_\lambda)_{\lambda\in\Lambda}$ and $\overline{\mathbf{h}^\delta}=(h^*_\lambda)_{\lambda\in\Lambda}$. Then the same lemma shows that $h^*_\mu=\overline{h_\lambda\|^\delta_k\beta}$. On the other hand, employing the explicit form of $\theta(z, w; \eta, \tau)$ given in [9, (6.10)], we can verify that

$$\overline{\theta(-\bar{z}, \beta\bar{w}; \eta', \tau)} = (-1)^{\|m\delta\|} j(\beta, w)^{k\delta} \rho'_k(\beta)\, \theta(z, w; \eta^*, \tau)$$

with $\eta^*(\xi) = \overline{\eta'(\beta\xi\beta^{-1})}$. This combined with Lemma 5.2 shows that

$$\overline{\theta(-\bar{z}, \beta\bar{w}; \eta_\lambda, \tau)} = (-1)^{\|m\delta + t\delta\|} j(\beta, w)^{k\delta} \rho'_k(\beta)\, \theta(z, w; \overline{\eta_\mu}, \tau).$$

Therefore our assertion can be shown by changing w for $\beta\bar{w}$ in each term of the right-hand side of (5.10).

References

1. Eichler, M.: Allgemeine Kongruenzklasseneinteilungen der Ideale einfacher Algebren über algebraischen Zahlkörpern und ihre *L*-Reihen. J. Reine Angew. Math. **179**, 227–251 (1938)
2. Eilenberg, S.: Homology of spaces with operators. I. Trans. Am. Math. Soc. **61**, 378–417 (1947)
3. Matsushima, Y., Shimura, G.: On the cohomology groups attached to certain vector-valued differential forms on the product of the upper half planes. Ann. Math. **78**, 417–449 (1963)
4. Shimizu, H.: On zeta functions of quaternion algebras. Ann. Math. **81**, 166–193 (1965)
5. Shimura, G.: On certain zeta functions attached to two Hilbert modular forms I, II. Ann. Math. **114**, 127–164, 569–607 (1981)
6. Shimura, G.: The periods of certain automorphic forms of arithmetic type. J. Fac. Sci. Univ. Tokyo (Sect. IA) **28**, 605–632 (1982)
7. Shimura, G.: Algebraic relations between critical values of zeta functions and inner products. Am. J. Math. **104**, 253–285 (1983)
8. Shimura, G.: On Hilbert modular forms of half-integral weight. Duke Math. J. **55**, 765–838 (1987)
9. Shimura, G.: On the critical values of certain Dirichlet series and the periods of automorphic forms. Invent. math. **94**, 245–305 (1988)

The critical values of certain Dirichlet series attached to Hilbert modular forms

Duke Mathematical Journal, 63 (1991), 557-613

Introduction. To define our Dirichlet series, we start with two Hilbert modular forms

(1) $$f(z) = \sum_{\xi \in F} \mu_f(\xi)\mathbf{e}_{\mathbf{a}}(\xi z), \qquad g(z) = \sum_{\xi \in F} \mu_g(\xi)\mathbf{e}_{\mathbf{a}}(\xi z)$$

with respect to a congruence subgroup of $SL_2(F)$, where F is a totally real algebraic number field, $\mathbf{e}_{\mathbf{a}}(\xi z) = \exp(2\pi i \sum_v \xi_v z_v)$ with v running over the set $\mathbf{a}$ of archimedean primes of F, and $z = (z_v)_{v \in \mathbf{a}}$ with variables z_v on the upper half plane H. Let k and l be the weights of f and g, respectively, which are integral or half-integral. Then one of the series to be considered is given in a special case by

(2a) $$D(s, f, g) = \sum_{\xi U} \mu_f(\xi)\mu_g(\xi)\xi^{-(k+l)/2} N_{F/\mathbf{Q}}(\xi)^{-s},$$

where ξ runs over all the totally positive elements of F modulo a group of units U which is chosen so that each term depends only on ξU. In general, this has no Euler product. We investigate also two types of series with Euler products defined by

(2b) $$\mathscr{D}(s, \chi_1, \chi_2) = L'(2s - 2, \Psi_1\Psi_2) \sum_{\mathfrak{a}} \chi_1(\mathfrak{a})\chi_2(\mathfrak{a})N(\mathfrak{a})^{-s},$$

(2c) $$\mathscr{D}(s, \chi_1^2, \eta) = L'(2s - 2, \eta^2\Psi_1^2) \sum_{\mathfrak{a}} \eta^*(\mathfrak{a})\chi_1(\mathfrak{a}^2)N(\mathfrak{a})^{-s}.$$

Here χ_1 and χ_2 are systems of eigenvalues in the spaces of forms of weight k and l, respectively, Ψ_μ is the central (Hecke) character of χ_μ, η is a Hecke (idele) character of F, η^* is the ideal character attached to it, $\mathfrak{a}$ runs over all integral ideals, and L' denotes the L-function defined with some finitely many Euler factors removed. Now our main problem concerns the arithmetic nature of the values and the residues of these series (2a, b, c) at some integers or half-integers belonging to a certain interval.

In the present paper we restrict our investigation to the following case:

(3) $$f \textit{ is a cusp form and } k_v \geqslant l_v \textit{ for every } v \in \mathbf{a}.$$

When both k and l are integral, we obtained in [4, 5, 6] some results which may be stated roughly as follows:

Received 2 July 1990.

There exists an invariant $I(\chi_1)$ such that

$$\text{(4a)}\qquad \left[\frac{\mathscr{D}(m/2, \chi_1, \chi_2)}{(\pi i)^{md}\gamma(\Psi_1\Psi_2)I(\chi_1)}\right]^\sigma = \frac{\mathscr{D}(m/2, \chi_1^\sigma, \chi_2^\sigma)}{(\pi i)^{md}\gamma(\Psi_1^\sigma\Psi_2^\sigma)I(\chi_1^\sigma)}$$

for certain integers m and every $\sigma \in Aut(\mathbf{C})$, where $d = [F: \mathbf{Q}]$, χ_μ^σ is a well-defined transform of χ_μ under σ, and $\gamma(\Phi)$ is the Gauss sum of Φ.

If f is a Hecke eigenform, (4a) for some restricted m may be written in the form

$$\text{(4b)}\qquad [D(m/2, f, g)/I(f)]^\sigma = D(m/2, f^\sigma, g^\sigma)/I(f^\sigma)$$

with an invariant $I(f)$, where $f^\sigma = \sum_\xi \mu_f(\xi)^\sigma \mathbf{e_a}(\xi z)$ and $g^\sigma = \sum_\xi \mu_g(\xi)^\sigma \mathbf{e_a}(\xi z)$. In fact, formulas of type (4b) are true even when f or g is half-integral. Strictly speaking, to obtain such formulas, one has to take a suitable linear combination of several series of type (2a). In the introduction, however, let us assume for simplicity that the class number of F is 1, in which case a single f and a single D are sufficient for (4b).

The previous investigations published after [6] are as follows: the case in which $F = \mathbf{Q}$, $k \in \mathbf{Z}$, $l \notin \mathbf{Z}$ by Sturm [15, 16]; the case in which $F = \mathbf{Q}$ and $k \notin \mathbf{Z}$ by the author [8]; the case in which $k \notin \mathbf{Z}^\mathbf{a}$ and $l \notin \mathbf{Z}^\mathbf{a}$ by the author [13]; the case in which $k + l \notin \mathbf{Z}^\mathbf{a}$ by Im [1]; the special values of (2c) by Sturm [15, 17] for $F = \mathbf{Q}$, and by Im [1] for an arbitrary F. Though these exhaust all possible combinations of k and l under (3), the results were obtained under several conditions, some of which may not absolutely be necessary. To make this point more explicit, let us begin with an essential condition without which results of type (4a, b) cannot be expected, namely,

(5) $\qquad k_v - l_v \pmod 2$ *is independent of v.*

It is also natural to assume that

(6a) $\qquad f\|_k\gamma = \varphi(a_\gamma)f$ *and* $g\|_l\gamma = \varphi'(a_\gamma)g$ *for every $\gamma \in \Gamma(\mathfrak{c})$,*

(6b) $\qquad \Gamma(\mathfrak{c}) = \{\gamma \in SL_2(F) \mid a_\gamma \in \mathfrak{g}, b_\gamma \in \mathfrak{d}^{-1}, c_\gamma \in \mathfrak{cd}, d_\gamma \in \mathfrak{g}\}$,

where $\|_k\gamma$ denotes the action of γ of weight k (see (1.4) below), $\mathfrak{g}$ the maximal order of F, $\mathfrak{d}$ the different of F over $\mathbf{Q}$, $\mathfrak{c}$ an integral ideal in F, φ and φ' are characters of $(\mathfrak{g}/\mathfrak{c})^\times$, and $a_\gamma, b_\gamma, c_\gamma, d_\gamma$ the entries of γ in the standard order. In addition, in all the above-mentioned papers, one had to assume:

(7a) *There exists a Hecke character ψ of finite order such that φ is essentially the restriction of ψ to $\prod_{v|\mathfrak{c}} \mathfrak{g}_v^\times$ and that $\psi_\mathbf{a}(x) = \operatorname{sgn}(x)^{[k]}$, where $[k] = ([k_v])_{v\in\mathbf{a}}$.*

Furthermore, if $k \in \mathbf{Z}^\mathbf{a}$, it was necessary to assume:

(7b) *The equalities of* (6a) *hold for every* γ *in a larger group obtained by changing* SL_2 *in* (6b) *for* GL_2 *and by adding that* $\det(\gamma)$ *is a totally positive unit.*

(8) k_v (mod 2) *is independent of* v.

It should be observed that (7a, b) are rather restrictive, since there are essentially more forms f of type (6a) than those of type (7b), and ψ of (7a) does not necessarily exist for a given φ. Also (8) is indispensable for the existence of χ_1^σ; in other words, f^σ may not be a Hecke eigenform unless (8) is assumed. Now the main purpose of the present paper is to prove (4b) and a suitable reformulation of (4a) without conditions (7a, b) and (8), or rather, under much weaker conditions, which, we believe, cannot essentially be weakened further. The most essential improvement is the removal of (8).

Let us now explain our methods and results, first for the series of (2a) with integral k. The key ideas are to formulate everything within $SL_2(F)$, and to avoid Hecke characters whenever feasible. To be more specific, we consider the Hecke algebra of double cosets $\Gamma(\mathfrak{c})\alpha\Gamma(\mathfrak{c})$ with α restricted to $SL_2(F)$ and let it act on the forms of type (6a). Then we can show (Proposition 3.6 and Theorem 3.7) that *for each eigenform* f *on* $SL_2(F)$ *in this sense there exists a primitive system* χ *of Hecke eigenvalues on* $GL_2(F)$ *such that for every squarefree totally positive element* τ *of* $\mathfrak{g}$ *one has*

$$(9)\qquad \sum_{\mathfrak{a}}{}' N(\mathfrak{a})^{1-s} \sum_{\mathfrak{a}}{}' \varphi(\alpha)\alpha^{-k}\mu_f(\tau\alpha^2)N(\mathfrak{a})^{2-s}$$

$$= \mu_f(\tau) \prod_{\mathfrak{p}\nmid\mathfrak{c},\, \mathfrak{p}|\tau} (1 + N(\mathfrak{p})^{1-s})^{-1} \sum_{\mathfrak{a}}{}' |\chi(\mathfrak{a})|^2 N(\mathfrak{a})^{-s},$$

where $\sum_{\mathfrak{a}}'$ *denotes the sum over all the integral ideals* $\mathfrak{a}$ *prime to* $\mathfrak{c}$, α *an element of* $\mathfrak{g}$ *such that* $\mathfrak{a} = \alpha\mathfrak{g}$, *and* $\mathfrak{p}$ *a prime ideal in* F. It is noteworthy that f determines only $|\chi(\mathfrak{p})|^2$ for almost all $\mathfrak{p}$, but not χ. Now, given $\sigma \in \mathrm{Aut}(\mathbf{C})$, in general the transform of χ under σ cannot be defined without condition (8). However, we can show (Proposition 4.3) that:

(IA) f^σ *is always an eigenform on* $SL_2(F)$ *(but not necessarily on* $GL_2(F)$*) which determines* $|\chi'|^2$ *with some system* χ' *such that* $|\chi'(\mathfrak{p})^2| = |\chi(\mathfrak{p})^2|^\sigma$ *for every* $\mathfrak{p} \nmid \mathfrak{c}$.

Then one of our main results (Theorem 7.2), specialized to those m belonging to the right half of the critical strip in the case of class number 1, asserts:

(IB) *Suppose that* $(\varphi\varphi')(e) = sgn(e)^{k+[l]}$ *for every* $e \in \mathfrak{g}^\times$. *Then* (4b) *holds for certain* m *with* $I(f)$ *defined to be a certain constant times the inner product* $\langle p, p\rangle$, *where* p *is a suitably chosen eigenform belonging to the same eigenvalues as* f.

This result leads us to the notion of a class $|\chi^2|$ which consists of all χ_1 such that $|\chi_1(\mathfrak{p})|^2 = |\chi(\mathfrak{p})|^2$ for almost all $\mathfrak{p}$. Then we can always speak of the transform of the class $|\chi^2|$ under σ, even when the transform of χ under σ is meaningless. Thus the invariant $I(f)$ may also be written $I(|\chi^2|)$.

The case of half-integral k, treated in Section 9, is similar and in fact simpler. In [13] we studied Hecke eigenforms of half-integral weight. The formulation there required a choice of a Hecke character, which could be of infinite order, and therefore could destroy algebraicity. In the present paper we define eigenforms so that no Hecke character is necessary. Then we obtain a formula of type (4b) under a condition on φ and φ' similar to the case of integral k (Theorem 9.3).

To treat (2b) without (8), we first prove a preliminary theorem concerning n systems of eigenvalues $\chi_1, \ldots, \chi_n$ on $GL_2(F)$ of weight $k_1, \ldots, k_n$ and central characters $\Psi_1, \ldots, \Psi_n$ as follows: Let $\sum_{v=0}^{2^n} C_{\mathfrak{p}v}(\bigotimes_{\mu=1}^n \chi_\mu)N(\mathfrak{p})^{-vs}$ denote the Euler $\mathfrak{p}$-factor of the L-function $\mathscr{D}(s, \bigotimes_{\mu=1}^n \chi_\mu)$ (defined as usual so that it gives $\mathscr{D}(s, \chi_1, \chi_2)$ if $n = 2$). Then we obtain (Theorems 11.1 and 11.2):

(IIA) *Suppose that $\Psi_1 \ldots \Psi_n$ is of finite order and $\sum_{\mu=1}^n k_{\mu v} = \kappa \pmod 2$ for every v with an integer κ. Then for every $\sigma \in \mathrm{Aut}(\mathbf{C})$ there exist n systems of eigenvalues $\chi'_1, \ldots, \chi'_n$ with central characters $\Psi'_1, \ldots, \Psi'_n$ such that $(\Psi_1 \ldots \Psi_n)^\sigma = \Psi'_1 \ldots \Psi'_n$, $|\chi_\mu^2|^\sigma = |\chi_\mu'^2|$ in the above sense for every μ, and*

$$\left[N(\mathfrak{p})^{\kappa v/2} C_{\mathfrak{p}v}\left(\bigotimes_{\mu=1}^n \chi_\mu\right)\right]^\sigma = N(\mathfrak{p})^{\kappa v/2} C_{\mathfrak{p}v}\left(\bigotimes_{\mu=1}^n \chi'_\mu\right)$$

for every $\mathfrak{p}$ and v. Moreover the numbers $N(\mathfrak{a})^{\kappa/2} \prod_{\mu=1}^n \chi_\mu(\mathfrak{a})$ for all $\mathfrak{a}$ prime to $\mathfrak{e}$ and $N(\mathfrak{p})^{\kappa v/2} C_{\mathfrak{p}v}(\bigotimes_{\mu=1}^n \chi_\mu)$ for all v and $\mathfrak{p} \nmid \mathfrak{e}$ generate a finite extension of $\mathbf{Q}$ which is totally real or a CM-field, where $\mathfrak{e}$ is the product of the levels for the χ_μ.

Then the reformulation of (4a) is (Theorem 11.3):

(IIB) *Under the same assumptions as in (IIA) with $n = 2$, we have*

$$\left[\frac{\mathscr{D}(m/2, \chi_1, \chi_2)}{(\pi i)^{md}\gamma(\Psi_1\Psi_2)I(|\chi_1^2|)}\right]^\sigma = \frac{\mathscr{D}(m/2, \chi'_1, \chi'_2)}{(\pi i)^{md}\gamma(\Psi'_1\Psi'_2)I(|\chi_1'^2|)}$$

for certain m and every $\sigma \in \mathrm{Aut}(\mathbf{C})$, where $I(|\chi_1^2|)$ is $\langle p, p\rangle$ of (IB) times a certain power of π.

The same type of results hold for (2c) (Proposition 10.2 and Theorem 10.3 for m in the right half of the critical strip):

(IIIA) *Suppose that $\eta\Psi_1$ is of finite order. Then for every $\sigma \in \mathrm{Aut}(\mathbf{C})$ there exists a Hecke character η' and a system χ' with central character Ψ' such that $(\eta\Psi_1)^\sigma = \eta'\Psi'$, and $[\eta^*(\mathfrak{a})\chi_1(\mathfrak{a}^2)]^\sigma = \eta'^*(\mathfrak{a})\chi'(\mathfrak{a}^2)$ for every integral $\mathfrak{a}$. Moreover the numbers $\eta^*(\mathfrak{a})\chi_1(\mathfrak{a}^2)$ for all $\mathfrak{a}$ prime to the level of χ generate a finite extension of $\mathbf{Q}$ which is totally real or a CM-field.*

(IIIB) *The notation and assumption being as in* (IIIA), *we have*

$$\left[\frac{\mathscr{D}(m, \chi_1^2, \eta)}{\pi^{2md}\gamma(\eta^2\Psi_1^2)I(|\chi_1^2|)}\right]^\sigma = \frac{\mathscr{D}(m, \chi'^2, \eta')}{\pi^{2md}\gamma(\eta'^2\Psi'^2)I(|\chi'^2|)}$$

for certain m and every $\sigma \in \mathrm{Aut}(\mathbf{C})$.

The main new point common to (IIA, B) and (IIIA, B) is that they hold even when χ_μ^σ is meaningless. Apart from difference in formulation, our theorems in Sections 7, 9, 10, and 11 include all previous results mentioned above as special cases.

The attentive reader may have noticed a similarity between (IA), (IIA), and (IIIA). In fact, they can be proved by the same methods. In (IA) and (IIIA) the relationship between SL_2 and GL_2 is essential, while in (IIA) that between the group

$$\{\alpha \in GL_2(F)^n | \det(\alpha) \in F\}$$

and $GL_2(F)^n$ is a key point.

Our method of proof of (IB) is an adaptation of that in our previous papers [8, 13], and somewhat different from [4, 6]. Forms of half-integral weight are indispensable even when both k and l are integral. Also, the nonvanishing of certain zeta functions is one of the key points. For integral k this means the nonvanishing of the right-hand side of (9). We shall not go into details here, since our ideas on this have been adopted by a few researchers in their recent investigations, and so are fairly well understood. Once (IIA) and (IIIA) are established, (IIB) and (IIIB) follow easily from (IB), since $\mathscr{D}(s, \chi_1, \chi_2)$ and $\mathscr{D}(s, \chi_1^2, \eta)$ are linear combinations of finitely many series of type (2a).

This paper has been written not only for the purpose of proving the theorems in the best possible forms, but also with the following intention: to highlight some principles, one of which concerns the idea of attaching an invariant to the class $|\chi^2|$ and not to χ, and others expressed by (IA), (IIA), and (IIIA), which have not previously been noticed. These seem applicable to many other types of automorphic forms and zeta functions. The author hopes to treat some cases in the future.

1. Hilbert modular forms. The following notation will be used throughout the paper: For an associative ring R with identity element we denote by $R^\times$ the group of all its invertible elements and by $M_n(R)$ the ring of $n \times n$-matrices with entries in R; we put then $GL_n(R) = M_n(R)^\times$. To indicate that a union $X = \bigcup_{i \in I} Y_i$ is disjoint, we write $X = \bigsqcup_{i \in I} Y_i$. The algebraic closure of the rational number field $\mathbf{Q}$ in the complex number field $\mathbf{C}$ is denoted by $\bar{\mathbf{Q}}$. We put

$$H = \{z \in \mathbf{C} | \mathrm{Im}(z) > 0\},$$

$$\mathbf{T} = \{z \in \mathbf{C} \,|\, |z| = 1\}.$$

For $x = \begin{pmatrix} a & b \\ c & d \end{pmatrix}$ we put $a = a_x$, $b = b_x$, $c = c_x$, and $d = d_x$ if there is no fear of confusion; these are also written $a(x)$, $b(x)$, $c(x)$, and $d(x)$ when x involves more than one letter.

We fix a totally real algebraic number field F of finite degree and denote by $\mathbf{a}$, $\mathbf{h}$, $\mathfrak{g}$, and $\mathfrak{d}$ the set of archimedean primes of F, the set of nonarchimedean (henselian) primes of F, the maximal order of F, and the different of F relative to $\mathbf{Q}$. Given an algebraic group $\mathscr{G}$ defined over F, we define $\mathscr{G}_v$ for each $v \in \mathbf{a} \cup \mathbf{h}$ and the adelization $\mathscr{G}_{\mathbf{A}}$ as usual, and view $\mathscr{G}$ as a subgroup of $\mathscr{G}_{\mathbf{A}}$. We then denote by $\mathscr{G}_{\mathbf{a}}$ and $\mathscr{G}_{\mathbf{h}}$ the archimedean and nonarchimedean factors of $\mathscr{G}_{\mathbf{A}}$, respectively. For an element x of $\mathscr{G}_{\mathbf{A}}$ its $\mathbf{a}$-component, $\mathbf{h}$-component, and v-component are denoted by $x_{\mathbf{a}}$, $x_{\mathbf{h}}$, and x_v. If $v \in \mathbf{h}$, we denote by π_v any prime element of F_v. Given a set X which is often a subset of $\mathbf{C}$ such as $\mathbf{Z}$ or $\mathbf{R}$, we denote by $X^{\mathbf{a}}$ the product of $\mathbf{a}$ copies of X, that is, the set of all indexed elements $(x_v)_{v \in \mathbf{a}}$ with x_v in X. For x and y in $\mathbf{C}^{\mathbf{a}}$ we put

$$\mathbf{e}_{\mathbf{a}}(x) = \exp\left(2\pi i \sum_{v \in \mathbf{a}} x_v\right), \tag{1.1a}$$

$$\|y\| = \sum_{v \in \mathbf{a}} y_v, \tag{1.1b}$$

$$x^y = \prod_{v \in \mathbf{a}} x_v^{y_v}; \tag{1.1c}$$

the last one is employed whenever the factors $x_v^{y_v}$ are well defined according to the context. We use the notation x^y also for $x \in F_{\mathbf{A}}^{\times}$ when the product is meaningful. The identity element of the ring $\mathbf{C}^{\mathbf{a}}$, which will, for the most part, appear as an exponent, is denoted by u. Thus $\|u\| = [F:\mathbf{Q}]$ and $c^u = N_{F/\mathbf{Q}}(c)$ for $c \in F^{\times}$, for example. For $x, y \in \mathbf{R}^{\mathbf{a}}$ or $\in F_{\mathbf{A}}$ we write $x \geqslant y$ if $x_v \geqslant y_v$ for every $v \in \mathbf{a}$ and $x \gg 0$ if $x_v > 0$ for every $v \in \mathbf{a}$. For $a \in F_{\mathbf{A}}^{\times}$ we put

$$\mathrm{sgn}(a) = (a_v/|a_v|)_{v \in \mathbf{a}}. \tag{1.2}$$

For a fractional ideal $\mathfrak{x}$ in F and $t \in F_{\mathbf{A}}^{\times}$ we denote by $N(\mathfrak{x})$ the norm of $\mathfrak{x}$ and by $t\mathfrak{x}$ the fractional ideal in F such that $(t\mathfrak{x})_v = t_v \mathfrak{x}_v$ for every $v \in \mathbf{h}$. We put $N_v = N(\pi_v \mathfrak{g})$. By a *Hecke character* of F, we mean a continuous $\mathbf{T}$-valued character ψ of $F_{\mathbf{A}}^{\times}$ of finite or infinite order such that $\psi(F^{\times}) = 1$. Given such a ψ, we denote by ψ^* the ideal character such that $\psi^*(t\mathfrak{g}) = \psi(t)$ if $t \in F_v^{\times}$ and $\psi(\mathfrak{g}_v^{\times}) = 1$; we put $\psi^*(\mathfrak{a}) = 0$ for every fractional ideal $\mathfrak{a}$ that is not prime to the conductor of ψ.

We now put

$$G = SL_2(F), \qquad \tilde{G} = GL_2(F), \qquad \tilde{G}_{\mathbf{a}+} = \{x \in \tilde{G}_{\mathbf{a}} \mid \det(x) \gg 0\},$$

$$\tilde{G}_{\mathbf{A}+} = \tilde{G}_{\mathbf{a}+}\tilde{G}_{\mathbf{h}}, \qquad \tilde{G}_+ = \tilde{G}_{\mathbf{A}+} \cap \tilde{G},$$

and for two fractional ideals $\mathfrak{x}$ and $\mathfrak{y}$ in F such that $\mathfrak{x}\mathfrak{y} \subset \mathfrak{g}$ we put

$$\tilde{D}[\mathfrak{x}, \mathfrak{y}] = \tilde{G}_{\mathbf{a}+} \prod_{v \in \mathbf{h}} \tilde{D}_v[\mathfrak{x}, \mathfrak{y}], \qquad \tilde{D}_v[\mathfrak{x}, \mathfrak{y}] = \mathfrak{o}[\mathfrak{x}, \mathfrak{y}]_v^{\times},$$

$$\mathfrak{o}[\mathfrak{x}, \mathfrak{y}] = \{x \in M_2(F) \mid a_x \in \mathfrak{g}, b_x \in \mathfrak{x}, c_x \in \mathfrak{y}, d_x \in \mathfrak{g}\},$$

$$D[\mathfrak{x}, \mathfrak{y}] = G_{\mathbf{A}} \cap \tilde{D}[\mathfrak{x}, \mathfrak{y}], \qquad D_v[\mathfrak{x}, \mathfrak{y}] = G_v \cap \tilde{D}_v[\mathfrak{x}, \mathfrak{y}],$$

$$\tilde{\Gamma}[\mathfrak{x}, \mathfrak{y}] = \tilde{G} \cap \tilde{D}[\mathfrak{x}, \mathfrak{y}], \qquad \Gamma[\mathfrak{x}, \mathfrak{y}] = G \cap D[\mathfrak{x}, \mathfrak{y}].$$

By a *half-integral weight* we mean an element k of $\mathbf{Q}^{\mathbf{a}}$ such that $k - u/2 \in \mathbf{Z}^{\mathbf{a}}$; naturally an *integral weight* is an element of $\mathbf{Z}^{\mathbf{a}}$. Given an integral weight k, we define a factor of automorphy $J_k(\tau, z)$ for $\tau \in G_{\mathbf{A}+}$ and $z \in H^{\mathbf{a}}$ by

$$\text{(1.3a)} \qquad J_k(\tau, z) = j(\tau, z)^k, \qquad j(\tau, z) = (j(\tau_v, z_v))_{v \in \mathbf{a}},$$

$$j(\alpha, w) = \det(\alpha)^{-1/2}(c_\alpha w + d_\alpha) \ (\alpha \in GL_2(\mathbf{R}), \det(\alpha) > 0, w \in \mathbf{C}).$$

To define J_k for half-integral k, we need the metaplectic covering $M_{\mathbf{A}}$ of $G_{\mathbf{A}}$ and a factor of automorphy $h(\tau, z)$ of weight $u/2$ defined for $\tau \in \mathrm{pr}^{-1}(P_{\mathbf{A}}C'')$ and $z \in H^{\mathbf{a}}$, where pr is the projection map of $M_{\mathbf{A}}$ to $G_{\mathbf{A}}$, P is the standard parabolic subgroup of G, and C'' is a certain compact subgroup of $G_{\mathbf{A}}$. (For the precise definitions, see [11, Prop. 3.2] or [13, p. 770, (2.11), and Prop. 2.3]. In general the notation and convention in the present paper are the same as those in [13] with the following exceptions: the upper half plane, the maximal order of F, and the set of finite primes of F are denoted by H, $\mathfrak{g}$, and $\mathbf{h}$ here instead of $\mathscr{H}$, $\mathfrak{o}$, and $\mathbf{f}$ there; here $D[\mathfrak{x}, \mathfrak{y}]$ and $\tilde{D}[\mathfrak{x}, \mathfrak{y}]$ contain $G_{\mathbf{a}}$ and $\tilde{G}_{\mathbf{a}+}$, but in [13] their archimedean factors are compact.) Now for a half-integral k we put

$$\text{(1.3b)} \qquad J_k(\tau, z) = h(\tau, z)j(\tau, z)^{k-u/2},$$

$$j(\tau, z) = j(\mathrm{pr}(\tau), z) \qquad (\tau \in \mathrm{pr}^{-1}(P_{\mathbf{A}}C'')).$$

We let $\tilde{G}_{\mathbf{A}+}$ and $M_{\mathbf{A}}$ act on $H^{\mathbf{a}}$ by

$$\xi z = \xi(z) = (\xi_v(z_v))_{v \in \mathbf{a}} \qquad (\xi \in \tilde{G}_{\mathbf{A}+}, z \in H^{\mathbf{a}}),$$

$$\alpha w = \alpha(w) = (a_\alpha w + b_\alpha)/(c_\alpha w + d_\alpha) \qquad (\alpha \in GL_2(\mathbf{R}), \det(\alpha) > 0, w \in H),$$

$$\tau z = \tau(z) = \mathrm{pr}(\tau)(z) \qquad (\tau \in M_{\mathbf{A}}, z \in H^{\mathbf{a}}).$$

For a $\mathbf{C}$-valued function f on $H^{\mathbf{a}}$ and $\tau \in \tilde{G}_{\mathbf{A}+}$ or $\tau \in \mathrm{pr}^{-1}(P_{\mathbf{A}}C'')$ we define a function $f\|_k\tau$ on $H^{\mathbf{a}}$ by

$$\text{(1.4)} \qquad (f\|_k\tau)(z) = J_k(\tau, z)^{-1}f(\tau z) \qquad (z \in H^{\mathbf{a}}).$$

Given two functions f and g on $H^{\mathbf{a}}$ invariant under the operation $\|_k\gamma$ for every γ in a congruence subgroup Γ in G, we define their inner product $\langle f, g\rangle$ by

$$(1.5)\qquad \langle f, g\rangle = \mathrm{vol}(\Phi)^{-1}\int_{\Phi}\overline{f(z)}g(z)\,\mathrm{Im}(z)^k\,d_Hz \qquad (\Phi = \Gamma\backslash H^{\mathbf{a}}),$$

where $d_H(x+iy) = \prod_{v\in\mathbf{a}} y_v^{-2}\,dx_v\,dy_v$ and $\mathrm{vol}(\Phi) = \int_\Phi d_Hz$.

Given a congruence subgroup Γ of G or $\tilde{G}_+$, we denote by $\mathcal{M}_k(\Gamma)$ the vector space of all holomorphic functions f on $H^{\mathbf{a}}$ which satisfy $f\|_k\gamma = f$ for every $\gamma\in\Gamma$ and also the cusp condition when $F = \mathbf{Q}$; we assume $\Gamma\subset C''G_{\mathbf{a}}$ if k is half-integral. We let $\mathcal{M}_k$ denote the union of $\mathcal{M}_k(\Gamma)$ for all such Γ's and $\mathcal{S}_k$ the subset of $\mathcal{M}_k$ consisting of the cusp forms; we put then $\mathcal{S}_k(\Gamma) = \mathcal{S}_k\cap\mathcal{M}_k(\Gamma)$.

For $f\in\mathcal{M}_k$ we can define a function $f_{\mathbf{A}}$ on $G_{\mathbf{A}}$ or $M_{\mathbf{A}}$ according as k is integral or half-integral by

$$(1.6)\qquad f_{\mathbf{A}}(\alpha x) = (f\|_k x)(\mathbf{i}) \qquad (\alpha\in G,\ x\in B),$$

where $\mathbf{i} = iu$ $(\in H^{\mathbf{a}})$ and B is an open subgroup of $G_{\mathbf{A}}$ or $\mathrm{pr}^{-1}(P_{\mathbf{A}}C'')$ such that $f\|_k\gamma = f$ for every $\gamma\in G\cap B$. (It should be remembered that G can be viewed as a subgroup of $M_{\mathbf{A}}$.)

To consider the forms of given level and character, we take an integral ideal $\mathfrak{c}$ in F and a character φ of $(\mathfrak{g}/\mathfrak{c})^\times$. We extend φ to a multiplicative function on

$$\{x\in F_{\mathbf{A}}\,|\,x_v\in\mathfrak{g}_v^\times \text{ for every } v|\mathfrak{c}\}$$

by putting $\varphi(x) = \varphi(a \bmod \mathfrak{c})$ with $a\in\mathfrak{g}$ such that $a - x_v\in\mathfrak{c}_v$ for every $v|\mathfrak{c}$. To make our later formulas applicable to both integral and half-integral weights, we introduce the following symbols:

$$(1.7\text{a})\qquad [k] = \begin{cases} k & \textit{if } k \textit{ is integral},\\ k - u/2 & \textit{if } k \textit{ is half-integral},\end{cases}$$

$$(1.7\text{b})\qquad \nu = \nu_k = \begin{cases} 1 & \textit{if } k \textit{ is integral},\\ 2 & \textit{if } k \textit{ is half-integral},\end{cases}$$

$$(1.7\text{c})\qquad \mathfrak{b} = \begin{cases} \textit{an arbitrarily fixed ideal in } F & \textit{if } k \textit{ is integral},\\ 2^{-1}\mathfrak{d} & \textit{if } k \textit{ is half-integral}.\end{cases}$$

We then denote by $\mathcal{M}_k(\mathfrak{b}, \mathfrak{c}, \varphi)$ the set of all f in $\mathcal{M}_k$ such that

$$(1.8)\qquad f\|_k\gamma = \varphi(a_\gamma)f \quad \textit{for every} \quad \gamma\in\Gamma[\mathfrak{b}^{-1}, \mathfrak{b}\mathfrak{c}]$$

and put $\mathscr{S}_k(\mathfrak{b}, \mathfrak{c}, \varphi) = \mathscr{S}_k \cap \mathscr{M}_k(\mathfrak{b}, \mathfrak{c}, \varphi)$. We naturally assume

(1.9a) $$\mathfrak{c} \subset 4\mathfrak{g} \quad \textit{if } k \textit{ is half-integral},$$

(1.9b) $$\varphi(-1) = (-1)^{\|[k]\|}.$$

PROPOSITION 1.1. *Given $f \in \mathscr{M}_k(\mathfrak{b}, \mathfrak{c}, \varphi)$, there is a complex number $\mu(\xi, t; f)$ determined for $\xi \in F$ and $t \in F_{\mathbf{A}}^{\times}$ such that*

$$f_{\mathbf{A}}\left(r_P\begin{pmatrix} t & s \\ 0 & t^{-1}\end{pmatrix}\right) = t_{\mathbf{a}}^{[k]}|t|_{\mathbf{A}}^{(\nu-1)/2} \sum_{\xi \in F} \mu(\xi, t; f)\mathbf{e}_{\mathbf{a}}(\mathbf{i}t^2\xi/\nu)\mathbf{e}_{\mathbf{A}}(ts\xi/\nu)$$

for every $s \in F_{\mathbf{A}}$, where $\mathbf{e}_{\mathbf{A}}$ is defined in [13, p. 771] and r_P is the identity map of $P_{\mathbf{A}}$ onto itself or the map of $P_{\mathbf{A}}$ into $M_{\mathbf{A}}$ defined in [13, (1.8)] according as k is integral or half-integral. Moreover $\mu(\xi, t; f)$ has the following properties:

(1.10a) $\mu(\xi, t; f) \neq 0$ *only if* $\xi \in \nu t^{-2}\mathfrak{b}\mathfrak{d}^{-1}$ *and* ξ *is totally nonnegative;*

(1.10b) $\mu(\xi b^2, t; f) = b^{[k]}\mu(\xi, bt; f)$ *for every* $b \in F^{\times}$;

(1.10c) $\mu(\xi, t; f) = \mu(\xi, t_{\mathbf{h}}; f)$;

(1.10d) $\varphi(c)\mu(\xi, ct; f) = \mu(\xi, t; f)$ *for every* $c \in \prod_{v \in \mathbf{h}} \mathfrak{g}_v^{\times}$.

Furthermore, if $\beta \in G \cap \mathrm{diag}[r, r^{-1}]D[\mathfrak{b}^{-1}, \mathfrak{b}\mathfrak{c}]$ with $r \in F_{\mathbf{A}}^{\times}$, then

(1.10e) $$J_k(\beta, \beta^{-1}z)f(\beta^{-1}z) = \varphi(d_\beta r)|r_{\mathbf{h}}|^{(\nu-1)/2} \sum_{\xi \in F} \mu(\xi, r; f)\mathbf{e}_{\mathbf{a}}(\xi z/\nu).$$

If $k \notin \mathbf{Z}^{\mathbf{a}}$, this was essentially given in [13, Prop. 3.1], where the result was formulated with a Hecke character ψ of $F_{\mathbf{A}}^{\times}$ such that $\psi = \varphi$ on $\prod_{v \in \mathbf{h}} \mathfrak{g}_v^{\times}$. Putting

(1.11) $$\mu(\xi, t; f) = \psi_{\mathbf{h}}(t)^{-1}\lambda(\xi, t\mathfrak{g}; f, \psi)$$

with the symbol λ there, we obtain our assertions. The point of this new formulation is that we have eliminated the character ψ which is not necessarily unique for f. The case of integral weight can be proved in the same manner. It should be noted that if k is integral and $f \in \mathscr{M}_k(\mathfrak{b}, \mathfrak{c}, \varphi)$, then

(1.12) $$f_{\mathbf{A}}(\alpha x w) = \varphi(a_w)^{-1}J_k(w, \mathbf{i})^{-1}f_{\mathbf{A}}(x)$$

$$\text{for } \alpha \in G,\ x \in G_{\mathbf{A}},\ w \in D[\mathfrak{b}^{-1}, \mathfrak{b}\mathfrak{c}],\ w(\mathbf{i}) = \mathbf{i}.$$

To shorten our formulas we put

$$\mu_f(\xi, t) = \mu(\xi, t; f). \tag{1.13}$$

PROPOSITION 1.2. *Let $0 \ll \tau \in F, q \in F_{\mathbf{h}}^{\times}$, and $f \in \mathcal{M}_k(\mathfrak{b}, \mathfrak{c}, \varphi)$; suppose that $\tau \in q^2\mathfrak{g}$ if $k \notin \mathbf{Z}^{\mathbf{a}}$. Then there exists an element g of $\mathcal{M}_k(\tau q^{-2}\mathfrak{b}, \mathfrak{c}, \varphi)$ or $\mathcal{M}_k(2^{-1}\mathfrak{d}, \tau q^{-2}\mathfrak{c}, \varphi\varepsilon_{\mathbf{h}})$ according as k is integral or half-integral such that $\mu_g(\xi, t) = \varepsilon_{\mathbf{h}}(t)^{\nu-1}\mu_f(\xi/\tau, qt)$, where ε is the Hecke character of F corresponding to $F(\tau^{1/2})/F$.*

If k is half-integral, this is a reformulation of [13, Prop. 3.2]. The case of integral k can be proved by a similar (and, in fact, simpler) argument.

PROPOSITION 1.3. *Let $\mathfrak{f}$ be an integral ideal, and ω a character of $(\mathfrak{g}/\mathfrak{f})^{\times}$ which is primitive in the sense that ω cannot be defined modulo any proper divisor of $\mathfrak{f}$; let n be an element of $\mathbf{Z}^{\mathbf{a}}$ such that $0 \leqslant n \leqslant u$ and $\omega(-1) = (-1)^{\|n\|}$. Put*

$$\theta_{n,\omega}(z) = \sum_{\zeta\mathfrak{g}+\mathfrak{f}=\mathfrak{g}} \omega(\zeta)^{-1}\zeta^n \mathbf{e}_{\mathbf{a}}(\zeta^2 z/2),$$

where the sum is over all $\zeta \in \mathfrak{g}$ prime to $\mathfrak{f}$; we understand that if $\mathfrak{f} = \mathfrak{g}$, the sum includes the term for $\zeta = 0$ and ω denotes the constant 1. Then $\theta_{n,\omega} \in \mathcal{M}_l(2^{-1}\mathfrak{d}, 4\mathfrak{f}^2, \omega)$ with $l = u/2 + n$ and

$$\mu(\xi, t; \theta_{n,\omega}) = \begin{cases} 2\omega((\zeta t)_{\mathfrak{f}})^{-1}\zeta^n & \text{if } 0 \neq \xi = \zeta^2 \text{ and } \zeta t\mathfrak{g} + \mathfrak{f} = \mathfrak{g}, \\ 1 & \text{if } \xi = 0, \mathfrak{f} = \mathfrak{g}, \text{ and } n = 0, \\ 0 & \text{otherwise}, \end{cases}$$

where ω is extended to a character of $\prod_{v|\mathfrak{f}} \mathfrak{g}_v^{\times}$ in an obvious fashion.

Taking a Hecke character χ of F of conductor $\mathfrak{f}$ which coincides with ω on $\prod_{v|\mathfrak{f}} \mathfrak{g}_v^{\times}$, we obtain our assertions immediately from [13, Lemma 4.3 and (4.12)].

PROPOSITION 1.4. *Let ω and $\mathfrak{f}$ be as in Proposition 1.3; let $f \in \mathcal{M}_k(\mathfrak{b}, \mathfrak{c}, \varphi)$ and $\mathfrak{a} = \mathfrak{c} \cap \mathfrak{f}^2 \cap \mathfrak{f}\mathfrak{c}_0$ with the conductor $\mathfrak{c}_0$ of φ. Further let r be an element of $F_{\mathbf{h}}^{\times}$ such that $\nu r\mathfrak{b} = \mathfrak{d}$. Then there exists an element g of $\mathcal{M}_k(\mathfrak{b}, \mathfrak{a}, \varphi\omega^2)$ such that $\mu_g(\xi, t) = \omega(rt^2\xi)^{-1}\mu_f(\xi, t)$ or 0 according as $rt^2\xi\mathfrak{g}$ is prime to $\mathfrak{f}$ or not.*

Proof. Take a Hecke character η of F such that $\eta = \omega^{-1}$ on $\prod_{v\in\mathbf{h}} \mathfrak{g}_v^{\times}$. Define $\mathbf{g}$ on $G_{\mathbf{A}}$ or $M_{\mathbf{A}}$ according as k is integral or half-integral by

$$\mathbf{g}(x) = \sum_p \eta^*(p\mathfrak{f}\mathfrak{d})\eta_{\mathbf{a}}(p) f_{\mathbf{A}}\left(x \cdot r_P\left(\begin{pmatrix} 1 & -\nu r p \\ 0 & 1 \end{pmatrix}_{\mathfrak{f}}\right)\right),$$

where p runs over $\mathfrak{f}^{-1}\mathfrak{d}^{-1}/\mathfrak{d}^{-1}$. Then we can easily verify that $\mathbf{g} = h_{\mathbf{A}}$ with $h \in \mathcal{M}_k(\mathfrak{b}, \mathfrak{a}, \varphi\omega^2)$ and the desired g is h divided by the Gauss sum of η.

2. Hecke operators on G and $\tilde{G}$. In this section k is always integral. Given $\mathfrak{c}$ and φ as in Section 1, we can find an element λ of $\mathbf{R}^{\mathbf{a}}$ such that $\|\lambda\| = 0$ and

$$\varphi(e) = \operatorname{sgn}(e)^k |e|^{-2i\lambda} \quad \textit{for every} \quad e \in \mathfrak{g}^\times. \tag{2.1}$$

For simplicity let us put $\tilde{D}_\mathfrak{c} = \tilde{D}[\mathfrak{d}^{-1}, \mathfrak{c}\mathfrak{d}]$, $\tilde{\Gamma}_\mathfrak{c} = \tilde{\Gamma}[\mathfrak{d}^{-1}, \mathfrak{c}\mathfrak{d}]$, and

$$\tilde{\Gamma}_p = \tilde{G} \cap p\tilde{D}_\mathfrak{c} p^{-1} \quad (p \in \tilde{G}_\mathbf{A}). \tag{2.2}$$

Now λ of (2.1) is not unique for φ, but choosing one, we put

$$\mathscr{S}_k(\tilde{\Gamma}_p, \varphi, \lambda) = \{f \in \mathscr{S}_k | f \|_k \gamma = \varphi(a(p^{-1}\gamma p)) \det(\gamma)^{i\lambda} f \textit{ for every } \gamma \in \tilde{\Gamma}_p\}. \tag{2.3}$$

LEMMA 2.1. *Let $p = \operatorname{diag}[1, e]$ with an element e of $F_\mathbf{h}^\times$ such that $e\mathfrak{d} = \mathfrak{b}$. Then $\mathscr{S}_k(\mathfrak{b}, \mathfrak{c}, \varphi)$ is the direct sum of $\mathscr{S}_k(\tilde{\Gamma}_p, \varphi, \lambda)$ for several λ's satisfying (2.1).*

Proof. Observe that $\Gamma[\mathfrak{b}^{-1}, \mathfrak{b}\mathfrak{c}] = G \cap \tilde{\Gamma}_p$. Choosing a λ as in (2.1), put

$$\rho(\gamma)f = \varphi(a_\gamma) \det(\gamma)^{i\lambda} f \|_k \gamma^{-1}$$

for $\gamma \in \tilde{\Gamma}_p$ and $f \in \mathscr{S}_k(\mathfrak{b}, \mathfrak{c}, \varphi)$. Then ρ defines a representation of $\tilde{\Gamma}_p/\mathfrak{g}^\times\Gamma[\mathfrak{b}^{-1}, \mathfrak{b}\mathfrak{c}]$, and hence we can decompose $\mathscr{S}_k(\mathfrak{b}, \mathfrak{c}, \varphi)$ into the direct sum of subspaces V_θ, where

$$V_\theta = \{f \in \mathscr{S}_k | \rho(\gamma)f = \theta(\det(\gamma))f \textit{ for every } \gamma \in \tilde{\Gamma}_p\}$$

with a character θ of $\mathfrak{g}_+^\times = \{d \in \mathfrak{g}^\times | d \gg 0\}$ such that $\theta(e^2) = 1$ for every $e \in \mathfrak{g}^\times$. Take $\varepsilon \in \mathbf{R}^{\mathbf{a}}$ so that $\|\varepsilon\| = 0$ and $\theta(d) = |d|^{i\varepsilon}$ for every $d \in \mathfrak{g}_+^\times$. Then we see that $V_\theta = \mathscr{S}_k(\tilde{\Gamma}_p, \varphi, \lambda - \varepsilon)$, which proves our lemma, since (2.1) holds with $\lambda - \varepsilon$ in place of λ.

Let us now consider a Hecke character Ψ of F such that

$$\Psi_\mathbf{a}(x) = \operatorname{sgn}(x)^k |x|^{2i\lambda} \textit{ with } \lambda \in \mathbf{R}^{\mathbf{a}}, \|\lambda\| = 0; \textit{the conductor of } \Psi \textit{ divides } \mathfrak{c}. \tag{2.4}$$

We then denote by $\mathscr{S}_k((\mathfrak{c}, \Psi))$ the set of all $\mathbf{C}$-valued functions $\mathbf{f}$ on $\tilde{G}_\mathbf{A}$ satisfying the following three conditions:

(2.5a) $\mathbf{f}(sx) = \Psi(s)\mathbf{f}(x)$ *for every* $s \in F_\mathbf{A}^\times$.

(2.5b) $\mathbf{f}(\alpha x w) = \Psi((d_w)_\mathfrak{c})\mathbf{f}(x)$ *for every* $\alpha \in \tilde{G}$, $x \in \tilde{G}_\mathbf{A}$, *and* $w \in \tilde{D}_\mathfrak{c}$, $w_\mathbf{a} = 1$,

where $d_\mathfrak{c} = (d_v)_{v|\mathfrak{c}}$ *for* $d \in F_\mathbf{A}$.

(2.5c) *For every* $p \in \tilde{G}_\mathbf{h}$ *there is an element* f_p *of* $\mathscr{S}_k$ *such that*

$\mathbf{f}(\det(p)^{-1}py) = \det(y)^{i\lambda}(f_p\|_k y)(\mathbf{i})$ *for every* $y \in \tilde{G}_{\mathbf{a}+}$.

Take a finite subset Q of $\tilde{G}_{\mathbf{h}}$ so that

$$(2.6)\qquad \tilde{G}_{\mathbf{A}} = \bigsqcup_{q\in Q} \tilde{G}q\tilde{D}_{c}.$$

Given φ as in (2.1), we can always find a Hecke character Ψ satisfying (2.4) and such that $\Psi = \varphi$ on $\prod_{v\in\mathbf{h}} \mathfrak{g}_v^{\times}$. ($\Psi$ is not necessarily unique for φ.) Conversely, if Ψ is given and φ is the character of $(\mathfrak{g}/\mathfrak{c})^{\times}$ obtained from $\Psi_{\mathfrak{c}}$, then (2.1) holds with λ of (2.4), and we can view $\mathscr{S}_k((\mathfrak{c}, \Psi))$ as a subspace of $\prod_{q\in Q} \mathscr{S}_k(\tilde{\Gamma}_q, \varphi, \lambda)$ through the map $\mathbf{f} \mapsto (f_q)_{q\in Q}$ with f_q defined by (2.5c). In fact, the map becomes surjective if we drop condition (2.5a). Namely, denoting by $\mathscr{S}_k((\mathfrak{c}, \varphi, \lambda))$ the space of all $\mathbf{f}$ satisfying (2.5b, c), we find that the map $\mathbf{f} \mapsto (f_q)_{q\in Q}$ gives an isomorphism of $\mathscr{S}_k((\mathfrak{c}, \varphi, \lambda))$ onto $\prod_{q\in Q} \mathscr{S}_k(\tilde{\Gamma}_q, \varphi, \lambda)$. Identifying these spaces, we shall often write $\mathbf{f} = (f_q)_{q\in Q}$. With another element $\mathbf{g} = (g_q)_{q\in Q}$ in the same space we put

$$(2.7)\qquad \langle \mathbf{f}, \mathbf{g}\rangle = \sum_{q\in Q} \langle f_q, g_q\rangle.$$

This is independent of the choice of Q. Replacing $\mathscr{S}_k$ in (2.5c) by $\mathscr{M}_k$, we can define the set $\mathscr{M}_k((\mathfrak{c}, \Psi))$ and similarly $\mathscr{M}_k((\mathfrak{c}, \varphi, \lambda))$ and $\mathscr{M}_k(\tilde{\Gamma}_q, \varphi, \lambda)$.

In order to define Hecke operators, we consider the set Y of all y in $\tilde{G}_{\mathbf{A}}$ such that $y_v \in \mathfrak{o}[\mathfrak{d}^{-1}, \mathfrak{c}\mathfrak{d}]_v$ and $a(y_v) \in \mathfrak{g}_v^{\times}$ for every $v|\mathfrak{c}$. For a fixed $y_0 \in Y$ take a coset decomposition $\tilde{D}_{c}y_0\tilde{D}_{c} = \bigsqcup_{w\in W} \tilde{D}_{c}w$ with a finite subset W of $\tilde{G}_{\mathbf{h}}$; then for $\mathbf{f} = (f_q)_{q\in Q} \in \mathscr{M}_k((\mathfrak{c}, \varphi, \lambda))$ we put

$$(2.8)\qquad (\mathbf{f}|\tilde{D}_{c}y_0\tilde{D}_{c})(x) = \sum_{w\in W} \varphi(a_w)^{-1}\mathbf{f}(xw^*) \qquad (x \in \tilde{G}_{\mathbf{A}}),$$

where $w^* = \det(w)w^{-1}$. It can easily be seen that the right-hand side satisfies (2.5b, c), so that we can put $\mathbf{f}|\tilde{D}_{c}y_0\tilde{D}_{c} = (f'_q)_{q\in Q}$ with $f'_q \in \mathscr{M}_k(\tilde{\Gamma}_q, \varphi, \lambda)$. Given $q \in Q$, we can find a unique p in Q and an element α_0 of $\tilde{G}$ such that $qy_0 \in \alpha_0 p\tilde{D}_{c}$. For such an α_0 let $\tilde{\Gamma}_q\alpha_0\tilde{\Gamma}_p = \bigsqcup_{\alpha\in A} \tilde{\Gamma}_q\alpha$. Then we can easily verify that

$$(2.9)\qquad f'_p = \sum_{\alpha\in A} \varphi(a(q^{-1}\alpha p))^{-1} \det(\alpha)^{-i\lambda} f_q\|_k\alpha.$$

We denote by $\mathfrak{T}_v$ the operator $\tilde{D}_{c}y_0\tilde{D}_{c}$ with $y_0 = \mathrm{diag}[1, \pi_v]$ for each $v \in \mathbf{h}$ and by $\mathfrak{S}_v$ the operator $\tilde{D}_{c}\pi_v\tilde{D}_{c}$ or the zero operator according as $v \nmid \mathfrak{c}$ or $v|\mathfrak{c}$. Clearly $\mathscr{M}_k((\mathfrak{c}, \Psi))$ is stable under $\tilde{D}_{c}y_0\tilde{D}_{c}$ for all $y_0 \in Y$ and

$$\mathscr{M}_k((\mathfrak{c}, \Psi)) = \{\mathbf{f} \in \mathscr{M}_k((\mathfrak{c}, \varphi, \lambda)) \,|\, \mathbf{f}|\mathfrak{S}_v = \Psi(\pi_v)\mathbf{f} \text{ for every } v \nmid \mathfrak{c}\}.$$

We now consider $\mathscr{M}_k(\mathfrak{b}, \mathfrak{c}, \varphi)$ and take an element y from $G_{\mathbf{A}} \cap F_{\mathbf{A}}^{\times} Y$. For simplicity put $\Gamma = \Gamma[\mathfrak{b}^{-1}, \mathfrak{b}\mathfrak{c}]$ and $D = D[\mathfrak{b}^{-1}, \mathfrak{b}\mathfrak{c}]$; take p as in Lemma 2.1. Then we can easily find a finite subset W' of $G_{\mathbf{h}}$ such that $\tilde{D}_{c}y\tilde{D}_{c} = \bigsqcup_{w\in W'} \tilde{D}_{c}p^{-1}wp$ and

$Dpyp^{-1}D = \bigsqcup_{w \in W'} Dw$ as well. We can also find an element α_0 and a finite subset B of G such that $Dpyp^{-1}D = D\alpha_0 D = \bigsqcup_{\beta \in B} D\beta$ and $\Gamma\alpha_0\Gamma = \bigsqcup_{\beta \in B} \Gamma\beta$.

Take especially y to be $y_v = \mathrm{diag}[\pi_v^{-1}, \pi_v]$, $v \in \mathbf{h}$. With B and W' as above for this y, we put, for $f \in \mathscr{M}_k(\mathfrak{b}, \mathfrak{c}, \varphi)$ and a function g on $G_{\mathbf{A}}$ of type (1.12),

$$f|R_v = \sum_{\beta \in B} \varphi(a_\beta \pi_v)^{-1} f\|_k \beta. \tag{2.10a}$$

$$(g|R_v)(x) = \sum_{w \in W'} \varphi(a_w \pi_v)^{-1} g(xw^{-1}) \qquad (x \in G_{\mathbf{A}}). \tag{2.10b}$$

Then we can easily verify that these are well-defined, $f|R_v \in \mathscr{M}_k(\mathfrak{b}, \mathfrak{c}, \varphi)$, and $(f|R_v)_{\mathbf{A}} = f_{\mathbf{A}}|R_v$. If $v \nmid \mathfrak{c}$, R_v is independent of the choice of π_v, since π_v can be ignored in (2.10a, b). If $v|\mathfrak{c}$, however, R_v depends on π_v, and therefore, to emphasize this dependence, we denote the operator by $R(\pi_v)$.

Proposition 2.2. *Let $f \in \mathscr{M}_k(\mathfrak{b}, \mathfrak{c}, \varphi)$ and $\xi \in t^{-2}\mathfrak{b}\mathfrak{d}^{-1}$. If $v|\mathfrak{c}$, we have $\mu(\xi, t; f|R(\pi_v)) = N_v^2 \mu_f(\xi, t\pi_v)$, and if $v \nmid \mathfrak{c}$, we have*

$$\mu(\xi, t; f|R_v) = N_v^2 \mu_f(\xi, t\pi_v) - \mu_f(\xi, t)[1 - N_v \varepsilon(\pi_v^{-1}\xi t^2 \mathfrak{b}^{-1}\mathfrak{d})] + \mu_f(\xi, \pi_v^{-1} t),$$

where $N_v = N(\pi_v \mathfrak{g})$ and $\varepsilon(\mathfrak{a})$ for a fractional ideal $\mathfrak{a}$ denotes 1 or 0 according as $\mathfrak{a}$ is integral or not; we understand that $\varepsilon(0) = 1$, which applies to the case $\xi = 0$.

Proof. If $v \nmid \mathfrak{c}$, the following $N_v^2 + N_v$ elements of G_v form a complete set of representatives for $D \backslash D y_v D$:

$$\alpha_j = \begin{pmatrix} 1/\pi & j/(\pi b) \\ 0 & \pi \end{pmatrix} \qquad (j \in \mathfrak{g}_v/\pi^2\mathfrak{g}_v),$$

$$\beta_h = \begin{pmatrix} 1 & h/(\pi b) \\ 0 & 1 \end{pmatrix} \qquad (h \in (\mathfrak{g}_v/\pi\mathfrak{g}_v)^\times),$$

$$\gamma = \begin{pmatrix} \pi & 0 \\ 0 & 1/\pi \end{pmatrix}.$$

Here $\pi = \pi_v$ and b is an element of F_v such that $b\mathfrak{g}_v = \mathfrak{b}_v$. If $v|\mathfrak{c}$, we need only the α_j. Taking these to be the elements w of (2.10b), we can compute the Fourier expansion of $(f|R_v)_{\mathbf{A}}$ in the sense of Proposition 1.1 to obtain the desired result. The procedure is similar to and in fact much simpler than the proof of [13, Proposition 5.4], which deals with the case of half-integral k.

Take a Hecke character Ψ of F such that $\Psi = \varphi$ on $\prod_{v \in \mathbf{h}} \mathfrak{g}_v^\times$. For $f \in \mathscr{M}_k(\mathfrak{b}, \mathfrak{c}, \varphi)$ and a fractional ideal $\mathfrak{m}$ we put

$$A_f(\xi, \mathfrak{m}; \Psi) = N(\mathfrak{m})\Psi_{\mathfrak{c}}(t)\mu_f(\xi, t) \tag{2.11}$$

with an element t of $F_{\mathbf{A}}^{\times}$ such that $\mathfrak{m} = t\mathfrak{g}$. This is well defined in view of (1.10d); moreover it is independent of the choice of Ψ if $\mathfrak{m}$ is prime to $\mathfrak{c}$. We consider the ring of formal Dirichlet series $\sum_{\mathfrak{m}\subset\mathfrak{g}} c(\mathfrak{m})M(\mathfrak{m})$ with $c(\mathfrak{m}) \in \mathbf{C}$ and a system of formal multiplicative symbols $M(\mathfrak{m})$ for the fractional ideals $\mathfrak{m}$ in F as follows: the $M(\mathfrak{p})$ for the prime ideals $\mathfrak{p}$ are independent indeterminates; $M(\mathfrak{g}) = 1$ and $M(\mathfrak{m}\mathfrak{n}) = M(\mathfrak{m})M(\mathfrak{n})$. For each $v \in \mathbf{h}$ we put $M_v = M(\pi_v\mathfrak{g})$ with a prime element π_v of F_v.

PROPOSITION 2.3. *Let $f \in \mathcal{M}_k(\mathfrak{b}, \mathfrak{c}, \varphi)$ and let $\mathfrak{t}$ be an integral ideal; let $0 \ll \tau \in \mathfrak{b}\mathfrak{d}^{-1}$ and $\tau\mathfrak{b}^{-1}\mathfrak{d} = \mathfrak{q}^2\mathfrak{r}$ with integral ideals $\mathfrak{q}$ and $\mathfrak{r}$. Suppose that $\mathfrak{r}$ has no square factors prime to $\mathfrak{c}\mathfrak{t}$ and $f|R_v = \omega_v f$ with $\omega_v \in \mathbf{C}$ for every $v \in \mathbf{h}$ prime to $\mathfrak{t}$. Then*

$$\sum_{\mathfrak{m}+\mathfrak{t}=\mathfrak{q}} A_f(\tau, \mathfrak{q}^{-1}\mathfrak{m}; \Psi)M(\mathfrak{m})$$

$$= A_f(\tau, \mathfrak{q}^{-1}; \Psi) \prod_{v \nmid \mathfrak{c}\mathfrak{r}\mathfrak{t}} (1 + M_v) \prod_{v|\mathfrak{c},\, v\nmid\mathfrak{t}} (1 - c_v M_v)^{-1} \prod_{v\nmid\mathfrak{c}\mathfrak{t}} (1 - c_v M_v + M_v^2)^{-1},$$

where $c_v = N_v^{-1}\omega_v + N_v^{-1} - 1$ if $v \nmid \mathfrak{c}$ and $c_v = N_v^{-1}\Psi(\pi_v)\omega_v$ if $v|\mathfrak{c}$.

Proof. Fixing τ and v, put $\mathfrak{p} = \pi_v\mathfrak{g}$, $A(\mathfrak{m}) = A_f(\tau, \mathfrak{m}; \Psi)$, $\omega = \omega_v$, and $N = N_v$. Then for an integral $\mathfrak{m}$ and v prime to $\mathfrak{c}\mathfrak{t}\mathfrak{m}$ Proposition 2.2 shows that

$$N^{-1}\omega A(\mathfrak{q}^{-1}\mathfrak{m}) = A(\mathfrak{q}^{-1}\mathfrak{p}\mathfrak{m}) - [N^{-1} - \varepsilon(\mathfrak{p}^{-1}\mathfrak{r})]A(\mathfrak{q}^{-1}\mathfrak{m}),$$

$$N^{-1}\omega A(\mathfrak{q}^{-1}\mathfrak{p}^n\mathfrak{m}) = A(\mathfrak{q}^{-1}\mathfrak{p}^{n+1}\mathfrak{m}) - (N^{-1} - 1)A(\mathfrak{q}^{-1}\mathfrak{p}^n\mathfrak{m}) + A(\mathfrak{q}^{-1}\mathfrak{p}^{n-1}\mathfrak{m}),$$

$(0 < n \in \mathbf{Z})$. Put $E = \sum_{n=0}^{\infty} A(\mathfrak{q}^{-1}\mathfrak{p}^n\mathfrak{m})x^n$ with an indeterminate x. Then from the above equalities we obtain

$$N^{-1}\omega E = x^{-1}[E - A(\mathfrak{q}^{-1}\mathfrak{m})] - (N^{-1} - 1 - x)E + [\varepsilon(\mathfrak{p}^{-1}\mathfrak{r}) - 1]A(\mathfrak{q}^{-1}\mathfrak{m}),$$

and hence $E = A(\mathfrak{q}^{-1}\mathfrak{m})[1 + (1 - \varepsilon(\mathfrak{p}^{-1}\mathfrak{r}))x](1 - c_v x + x^2)^{-1}$. Therefore we easily see that

$$\sum_{\mathfrak{m}+\mathfrak{t}=\mathfrak{g}} A(\mathfrak{q}^{-1}\mathfrak{m})M(\mathfrak{m})$$

$$= \sum_{\mathfrak{m}+\mathfrak{p}\mathfrak{t}=\mathfrak{g}} A(\mathfrak{q}^{-1}\mathfrak{m})M(\mathfrak{m})\cdot[1 + (1 - \varepsilon(\mathfrak{p}^{-1}\mathfrak{r}))M_v][1 - c_v M_v + M_v^2]^{-1}.$$

If $v|\mathfrak{c}$ and $v \nmid \mathfrak{t}$, the last two factors must be replaced by $(1 - c_v M_v)^{-1}$. Our formula follows easily from these facts.

PROPOSITION 2.4. *If $v \nmid \mathfrak{c}$, the operator R_v is hermitian with respect to the inner product of* (1.5). *Moreover the R_v for all $v \in \mathbf{h}$ form a commutative ring.*

These assertions can be proved by the standard argument. It should be noted that $\Gamma\alpha_0\Gamma = \Gamma\alpha_0^{-1}\Gamma$ for the element α_0 of (2.10a). It follows from this proposition that $\mathcal{S}_k(\mathfrak{b}, \mathfrak{c}, \varphi)$ is spanned by common eigenforms of the R_v for all $v \nmid \mathfrak{c}$.

3. The connection between eigenforms on G and those on $\tilde{G}$. In this section we assume k to be integral. For $\mathbf{f} \in \mathscr{M}_k((\mathfrak{c}, \Psi))$ and an integral ideal $\mathfrak{a}$ we define a complex number $c(\mathfrak{a}, \mathbf{f})$ as follows. For each $p \in G_{\mathbf{h}}$ define f_p by (2.5c) and put

$$f_p(z) = \sum_{\xi \in F} \lambda_p(\xi)\mathbf{e}_{\mathbf{a}}(\xi z). \tag{3.1}$$

Take $r \in F_{\mathbf{h}}^{\times}$ and $\xi \in F$, $\gg 0$, so that $\mathfrak{a} = \xi r^{-1}\mathfrak{g}$, and put

$$c(\mathfrak{a}, \mathbf{f}) = \lambda_p(\xi)\xi^{-k/2-i\lambda} \tag{3.2}$$

with $p = \mathrm{diag}[1, r]$ and λ of (2.4). This number can also be defined as a coefficient of the global Fourier expansion of $\mathbf{f}$ independently of the choice of r and ξ (see [6, (2.17), (2.18)] and [7, (9.27), (9.28)]).

Now suppose that $\mathbf{f}$ is an eigenform of $\mathfrak{T}_v$ for every $v \in \mathbf{h}$; put $\mathbf{f}|\mathfrak{T}_v = \chi(v)\mathbf{f}$. Then we call χ a *system of eigenvalues*, and say that χ is *primitive* if $\mathbf{f}$ is primitive in the sense of [6, p. 652]. For an integral ideal $\mathfrak{a}$ let $\mathfrak{T}(\mathfrak{a})$ denote the sum of all different $\tilde{D}_{\mathfrak{c}} y \tilde{D}_{\mathfrak{c}}$ with $y \in Y$ such that $\det(y)\mathfrak{g} = \mathfrak{a}$. Then $\mathbf{f}$ is an eigenform of $\mathfrak{T}(\mathfrak{a})$ as well. Putting $\mathbf{f}|\mathfrak{T}(\mathfrak{a}) = \chi(\mathfrak{a})\mathbf{f}$, we have

$$\chi(\mathfrak{a})c(\mathfrak{g}, \mathbf{f}) = N(\mathfrak{a})c(\mathfrak{a}, \mathbf{f}), \tag{3.3a}$$

$$\sum_{\mathfrak{a}} \chi(\mathfrak{a})M(\mathfrak{a}) = \prod_{v \in \mathbf{h}} [1 - \chi(v)M_v + \Psi_*(v)N_v M_v^2]^{-1}. \tag{3.3b}$$

Here and throughout the rest of the paper $\sum_{\mathfrak{a}}$ denotes the sum over all the integral ideals $\mathfrak{a}$ in F, the symbol M is as in Proposition 2.3, and $\Psi_*(v)$ denotes $\Psi(\pi_v)$ or 0 according as $v \nmid \mathfrak{c}$ or $v|\mathfrak{c}$. (See [6, (2.13), (2.20)], [7, (9.29)].) We call $\mathbf{f}$ *normalized* if $c(\mathfrak{g}, \mathbf{f}) = 1$. We recall a simple fact

$$\Psi^*(\mathfrak{a})\bar{\chi}(\mathfrak{a}) = \chi(\mathfrak{a}) \quad \textit{if} \quad \mathfrak{a} \quad \textit{is prime to} \quad \mathfrak{c}, \tag{3.4}$$

which was proved in [6, Prop. 2.5] under the assumption that Ψ is of finite order. The same proof is applicable to the general case; alternatively, the general case can be reduced to the special case by virtue of [7, Prop. 9.7].

PROPOSITION 3.1. *Let $\mathbf{f} \in \mathscr{M}_k((\mathfrak{c}, \Psi))$ with Ψ as in (2.4) and let φ be the character of $(\mathfrak{g}/\mathfrak{c})^{\times}$ obtained from the restriction of Ψ to $\prod_{v \in \mathbf{h}} \mathfrak{g}_v^{\times}$. Define f_p for $p \in \tilde{G}_{\mathbf{h}}$ by (2.5c). If $q = \mathrm{diag}[1, t]$ with an element t of $F_{\mathbf{h}}^{\times}$, then $f_q \in \mathscr{M}_k(t\mathfrak{d}, \mathfrak{c}, \varphi)$. Moreover, if $s = \mathrm{diag}[1, tr^{-2}]$ with $r \in F_{\mathbf{h}}^{\times}$, then $f_s(z) = \Psi(r)\sum_{\xi \in F} \mu(\xi, r; f_q)\mathbf{e}_{\mathbf{a}}(\xi z)$. Furthermore, suppose that $\mathbf{f}|\mathfrak{T}(\mathfrak{a}) = \chi(\mathfrak{a})\mathbf{f}$ for every integral ideal $\mathfrak{a}$. Then for $\mathfrak{a} = \xi t^{-1}\mathfrak{g}$ with $0 \ll \xi \in F$ we have $\chi(\mathfrak{a})\lambda_1(1) = N(\mathfrak{a})\xi^{-k/2-i\lambda}\lambda_q(\xi)$ with $\lambda_p(\xi)$ of (3.1), and also $\chi(\mathfrak{m})\mu(1, 1; f_1) = N(\mathfrak{m})\xi^{-k/2-i\lambda} \cdot \Psi(r)\mu(\xi, r; f_q)$ if $\mathfrak{m} = \xi r^2 t^{-1}\mathfrak{g}$.*

Proof. Clearly $f_q \in \mathscr{M}_k(t\mathfrak{d}, \mathfrak{c}, \varphi)$. Let $\mathrm{diag}[r, r^{-1}] = \beta w$ with $\beta \in G$ and $w \in D[t^{-1}\mathfrak{d}^{-1}, t\mathfrak{c}\mathfrak{d}]$. Then for $y \in G_{\mathbf{a}}$ we have $\Psi(r)^{-1}(f_s\|_k y)(\mathbf{i}) = \Psi(r)^{-1}\mathbf{f}(t^{-1}r^2 sy) =$

$\mathbf{f}(t^{-1}\beta wqy) = \varphi(d_w)\mathbf{f}(t^{-1}qw_{\mathbf{a}}y) = \varphi(rd_\beta)^{-1}(f_q\|\beta^{-1}y)(\mathbf{i})$. This combined with (1.10e) proves the formula for f_s. From (3.2) and (3.3a) we see that $\chi(\mathfrak{a})\lambda_1(1) = N(\mathfrak{a})\xi^{-k/2-i\lambda}\lambda_q(\xi)$ if $t\mathfrak{a} = \xi\mathfrak{g}$. Taking q to be s, we obtain the last assertion.

Given two systems of eigenvalues χ and χ' occurring in $\mathscr{S}_k((\mathfrak{c}, \Psi))$ and $\mathscr{M}_{k'}((\mathfrak{c}', \Psi'))$, we put

(3.5a) $$D(s, \chi, \chi') = \sum_{\mathfrak{a}} \chi(\mathfrak{a})\chi'(\mathfrak{a})N(\mathfrak{a})^{-s},$$

(3.5b) $$\mathscr{D}(s, \chi, \chi') = L_{\mathfrak{c}\mathfrak{c}'}(2s-2, \Psi\Psi')D(s, \chi, \chi'),$$

where for an integral ideal $\mathfrak{e}$ and a Hecke character θ of F of conductor $\mathfrak{f}$ we put

(3.6) $$L_{\mathfrak{e}}(s, \theta) = \prod_{v \nmid \mathfrak{e}\mathfrak{f}} [1 - \theta(\pi_v)N_v^{-s}]^{-1}.$$

To obtain the Euler product for $\mathscr{D}(s, \chi, \chi')$, put

$$1 - \chi(v)X + \Psi_*(v)N_vX^2 = (1 - \alpha_vX)(1 - \beta_vX),$$
$$1 - \chi'(v)X + \Psi'_*(v)N_vX^2 = (1 - \alpha'_vX)(1 - \beta'_vX)$$

with an indeterminate X. Then we have

(3.7) $\mathscr{D}(s, \chi, \chi')$
$$= \prod_{v \in \mathbf{h}} [(1 - \alpha_v\alpha'_vN_v^{-s})(1 - \alpha_v\beta'_vN_v^{-s})(1 - \beta_v\alpha'_vN_v^{-s})(1 - \beta_v\beta'_vN_v^{-s})]^{-1}.$$

PROPOSITION 3.2. *Let κ be the element of $\mathbf{R}^{\mathbf{a}}$ such that $\|\kappa\| = 0$ and $(\Psi/\Psi')(x) = |x|^{i\kappa}$ for $0 \ll x \in F_{\mathbf{a}}^\times$. Then the product*

$$\mathscr{D}(s, \chi, \overline{\chi'}) \prod_{v \in \mathbf{a}} \Gamma(s - 2 + (k_v + k'_v + i\kappa_v)/2)\Gamma(s - 1 + (|k_v - k'_v| + i\kappa_v)/2)$$

can be continued to the whole s-plane as a meromorphic function, which is holomorphic except for possible simple poles at 1 and 2. The pole at $s = 1$ occurs only when $k = k'$, $\Psi = \Psi'$, and $\mathfrak{c} = \mathfrak{c}' = \mathfrak{g}$. The pole at $s = 2$ occurs only when $k = k'$ and $\Psi = \Psi'$, in which case $D(s, \chi, \overline{\chi'})$ at $s = 2$ has residue

$$2^{\|2k+u\|-1}\pi^{\|k\|}[\mathfrak{g}_+^\times : \mathfrak{g}^{\times 2}]^{-1}R_F \prod_{v \in \mathbf{a}} \Gamma(k_v)^{-1}\langle \mathbf{f}', \mathbf{f}\rangle,$$

where $\mathbf{f}$ and $\mathbf{f}'$ are normalized eigenforms belonging to χ and χ', respectively, $\mathfrak{g}_+^\times = \{a \in \mathfrak{g} | a \gg 0\}$, $\mathfrak{g}^{\times 2} = \{a^2 | a \in \mathfrak{g}^\times\}$, and R_F is the regulator of F.

Our assertions were proved in [6, Prop. 4.13] and [9, Prop. 5.1] when both Ψ and Ψ' are of finite order and $k_v - k'_v \pmod 2$ is independent of v. We shall give a proof in the general case in Section 6.

PROPOSITION 3.3. *Suppose that both χ and χ' occur in the space of cusp forms. For an integral ideal $\mathfrak{f}$ let $\mathscr{D}^{\mathfrak{f}}(s, \chi, \chi')$ denote the Euler product obtained from* (3.5b) *by eliminating the v-factors for all $v|\mathfrak{f}$. Then $\mathscr{D}^{\mathfrak{f}}(s, \chi, \chi') \neq 0$ for* $\mathrm{Re}(s) \geqslant 2$.

Proof. With $t \in \mathbf{R}$ put

$$A(s) = \mathscr{D}^{\mathfrak{f}}(s+it, \chi, \chi')\mathscr{D}^{\mathfrak{f}}(s-it, \bar{\chi}, \overline{\chi'})\mathscr{D}^{\mathfrak{f}}(s, \chi, \bar{\chi})\mathscr{D}^{\mathfrak{f}}(s, \chi', \overline{\chi'}).$$

Then it can easily be verified that

$$\log A(s) = \sum_{m=1}^{\infty} \sum_{v \in \mathbf{h}} m^{-1} N_v^{-ms} |\alpha_v^m + \beta_v^m + \gamma_v^m + \delta_v^m|^2, \tag{3.8}$$

where $\gamma_v = \overline{\alpha'_v} N_v^{it}$ and $\delta_v = \overline{\beta'_v} N_v^{it}$. By Proposition 3.2 $A(s)$ is holomorphic on the whole plane except for possible poles on $\mathrm{Re}(s) = 2$. (Notice that the pole at $s = 1$ mentioned there comes from a gamma factor.) Let $\mathrm{Re}(s) = \sigma_0$ be the line of convergence of the right-hand side of (3.8). Since the gamma factors of $\mathscr{D}$ given in Proposition 3.2 produce many zeros of $A(s)$, we see that $\sigma_0 \neq -\infty$. Suppose $\sigma_0 > 2$. Then, for real $s > \sigma_0$ we have $\log A(s) \geqslant 0$ and hence $A(s) \geqslant 1$, so that $A(\sigma_0) \geqslant 1$. Since A is holomorphic at σ_0, this means that $\log A(s)$ can be holomorphically continued to a neighborhood of σ_0, which contradicts the well-known fact that a Dirichlet series with nonnegative coefficients is not holomorphic at the real point on the line of convergence. Therefore $\sigma_0 \leqslant 2$. This implies in particular that $A(s) = \exp[\log A(s)] \neq 0$ for $\mathrm{Re}(s) > 2$, and hence $\mathscr{D}^{\mathfrak{f}}(s, \chi, \chi') \neq 0$ for $\mathrm{Re}(s) > 2$. Suppose $\mathscr{D}^{\mathfrak{f}}(2+it, \chi, \chi') = 0$; then $\mathscr{D}^{\mathfrak{f}}(2-it, \bar{\chi}, \overline{\chi'}) = 0$, and hence A is holomorphic at $s = 2$. Consequently A is holomorphic at any real point, in particular at σ_0. Repeating the above argument, we find that $A(\sigma_0) \geqslant 1$ and $\log A(s)$ is holomorphic at σ_0, which is a contradiction. This completes the proof.

LEMMA 3.4. *Let $\tilde{R}_v = \tilde{D}_{\mathfrak{c}} \cdot \mathrm{diag}[\pi_v^{-1}, \pi_v]\tilde{D}_{\mathfrak{c}}$. Then $\mathfrak{T}_v^2 = \mathfrak{S}_v(\tilde{R}_v + N_v + 1)$ on $\mathscr{M}_k((\mathfrak{c}, \varphi, \lambda))$ for every $v \nmid \mathfrak{c}$.*

This is an elementary result which concerns essentially the Hecke algebra of $GL_2(F_v)$ relative to $\tilde{D}_v[\mathfrak{d}^{-1}, \mathfrak{d}]$.

PROPOSITION 3.5. *Given a system of eigenvalues χ occurring in $\mathscr{S}_k((\mathfrak{c}, \Psi))$, there exists an element f of $\mathscr{S}_k(\mathfrak{d}, \mathfrak{c}, \varphi)$ such that $\mu_f(1, 1) = 1$ and $f|R_v = \omega_v f$ for every $v \in \mathbf{h}$, where φ is as in Proposition 3.1, and*

$$\omega_v = \begin{cases} |\chi(v)|^2 - N_v - 1 & \text{if } v \nmid \mathfrak{c}, \\ \Psi(\pi_v)^{-1}\chi(v)^2 & \text{if } v | \mathfrak{c}. \end{cases} \tag{3.9}$$

Proof. Take Q of (2.6) so that the element p of Lemma 2.1 belongs to Q. Let $0 \neq \mathbf{f} = (f_q)_{q \in Q} \in \mathscr{S}_k((\mathfrak{c}, \Psi))$. We easily see that $(f_p)_{\mathbf{A}}(x) = \mathbf{f}(xp^{-*})$ for $x \in G_{\mathbf{A}}$. Moreover we have

$$(3.10a) \qquad (f_p|R_v)_{\mathbf{A}}(x) = (\mathbf{f}|\tilde{R}_v)(xp^{-*}) \qquad \text{for } x \in G_{\mathbf{A}} \text{ if } v \nmid \mathfrak{c},$$

$$(3.10b) \qquad \Psi(\pi_v)(f_p|R_v(\pi_v))_{\mathbf{A}}(x) = (\mathbf{f}|\mathfrak{T}_v^2)(xp^{-*}) \qquad \text{for } x \in G_{\mathbf{A}} \text{ if } v|\mathfrak{c}.$$

Indeed, as already noted, for $y_0 = y_v = \mathrm{diag}[\pi_v^{-1}, \pi_v]$, $v \nmid \mathfrak{c}$ we can take W of (2.8) to be $p^{-1}W'p$ with W' of (2.10b). Then (3.10a) can be proved by a direct calculation, and (3.10b) in a similar way. Now $f_1 \in \mathscr{S}_k(\mathfrak{d}, \mathfrak{c}, \varphi)$, and $\mu_f(1, 1) = \lambda_1(1) = 1$ if $\mathbf{f}$ is a normalized eigenform. Therefore formulas (3.10a, b) and (3.4) together with Lemma 3.4 prove our assertion.

PROPOSITION 3.6. *Let $0 \neq f \in \mathscr{S}_k(\mathfrak{b}, \mathfrak{c}, \varphi)$ and suppose that $f|R_v = \eta_v f$ with $\eta_v \in \mathbf{C}$ for every $v \nmid \mathfrak{t}$ with an integral ideal $\mathfrak{t}$. Then there exists a Hecke character Ψ of F of type* (2.4) *such that $\Psi = \varphi$ on $\prod_{v \in \mathbf{h}} \mathfrak{g}_v^\times$ and also a system of eigenvalues χ which occurs in $\mathscr{S}_k((\mathfrak{c}, \Psi))$ such that η_v coincides with ω_v of* (3.9) *for every $v \nmid \mathfrak{t}$.*

Proof. Take Q and p as in the above proof. By Lemma 2.1 f is a finite sum $\sum_\lambda g_\lambda$ with $g_\lambda \in \mathscr{S}_k(\Gamma_p, \varphi, \lambda)$, where each λ satisfies (2.1). Fixing one λ, define $\mathbf{g}_\lambda = (g_q)_{q \in Q}$ to be an element of $\mathscr{S}_k((\mathfrak{c}, \varphi, \lambda))$ such that $g_{\lambda p} = g_\lambda$ and $g_{\lambda q} = 0$ for $p \neq q \in Q$. Now $\mathscr{S}_k((\mathfrak{c}, \varphi, \lambda))$ is the direct sum of $\mathscr{S}_k((\mathfrak{c}, \Psi))$ with some Hecke characters Ψ satisfying (2.4) and such that $\Psi = \varphi$ on $\prod_{v \in \mathbf{h}} \mathfrak{g}_v^\times$. Observe that all such $\mathscr{S}_k((\mathfrak{c}, \varphi, \lambda))$, as sets of functions on $\tilde{G}_{\mathbf{A}}$, form a direct sum, which we denote by S. Putting $\mathbf{g} = \sum_\lambda \mathbf{g}_\lambda$, we see that $\mathbf{g} \in S$, and $f_{\mathbf{A}}(x) = \sum_\lambda (g_{\lambda p})_{\mathbf{A}}(x) = \sum_\lambda \mathbf{g}_\lambda(xp^{-*}) = \mathbf{g}(xp^{-*})$ for $x \in G_{\mathbf{A}}$. For $\mathbf{f} = (f_q)_{q \in Q} \in S$ put $U(\mathbf{f}) = f_p$. Then $f = U(\mathbf{g}) \in U(S)$. Let $\tilde{R}_v$ be defined for $v \nmid \mathfrak{c}$ as in Lemma 3.4. For $v|\mathfrak{c}$ let $\tilde{R}_v$ be the operator that coincides with $\Psi(\pi_v)^{-1}\mathfrak{T}_v^2$ on $\mathscr{S}_k((\mathfrak{c}, \Psi))$. Let A (resp. A') denote the ring of operators on S (resp. $U(S)$) generated by $\mathbf{C}$ and the $\tilde{R}_v$ (resp. the R_v) for $v \nmid \mathfrak{t}$. Then (3.10a, b) show that $U \circ \alpha = h(\alpha) \circ U$ for every $\alpha \in A$ with a homomorphism h of A onto A'. Now every nontrivial $\mathbf{C}$-linear homomorphism of A (resp. A') into $\mathbf{C}$ corresponds to a common eigenform in S (resp. $U(S)$). Let $\lambda: A' \to \mathbf{C}$ be defined by $\lambda(\beta)f = f|\beta$ for $\beta \in A'$. Then we find a nonzero element $\mathbf{f}_0$ in S such that $\mathbf{f}_0|\alpha = \lambda(h(\alpha))\mathbf{f}_0$ for every $\alpha \in A$. Now each $\mathscr{S}_k((\mathfrak{c}, \Psi))$ is a direct sum of the subspaces $V(\Psi, \Phi)$ consisting of all its elements $\mathbf{f}$ such that $\mathbf{f}|\mathfrak{T}_v = \Phi(v)\mathbf{f}$ for every $v \nmid \mathfrak{c}$, where Φ is the restriction of a system of eigenvalues restricted to the v's prime to $\mathfrak{c}$. Then $\mathbf{f}_0 = \sum_{\Psi, \Phi} \mathbf{f}_{\Psi, \Phi}$ with $\mathbf{f}_{\Psi, \Phi} \in V(\Psi, \Phi)$. Since $V(\Psi, \Phi)$ is stable under A, we see that $\mathbf{f}_{\Psi, \Phi}|\alpha = \lambda(h(\alpha))\mathbf{f}_{\Psi, \Phi}$ for every (Ψ, Φ). We can pick a (Ψ, Φ) such that $\mathbf{f}_{\Psi, \Phi} \neq 0$. Let

$$W = \{\mathbf{j} \in V(\Psi, \Phi) | \lambda(h(\alpha))\mathbf{j} = \mathbf{j}|\alpha \text{ for every } \alpha \in A\}.$$

This space W, being stable under the $\mathfrak{T}_v$ for all $v|\mathfrak{c}$, has a common eigenfunction of all such $\mathfrak{T}_v$, which determines a system of eigenvalues with the required property.

The M_v and $M(\mathfrak{a})$ being the multiplicative symbols of Section 2, put

$$(3.11)\qquad Z_{\mathfrak{c}}(\{M_v\}) = \prod_{v \nmid \mathfrak{c}} (1 - M_v)^{-1} = \sum_{\mathfrak{a}+\mathfrak{c}=\mathfrak{g}} M(\mathfrak{a}).$$

THEOREM 3.7. *Let f, τ, $\mathfrak{q}$, $\mathfrak{r}$, $\mathfrak{t}$, and $A_f(\tau, \mathfrak{m}; \Psi)$ be the same as in Proposition 2.3. Suppose that f is a cusp form and (3.9) holds for every $v \nmid \mathfrak{t}$ with a system of eigenvalues χ on $\mathcal{S}_k((\mathfrak{c}, \Psi))$. Then*

$$Z_{\mathfrak{ct}}(\{N_v M_v\}) \sum_{\mathfrak{m}+\mathfrak{t}=\mathfrak{g}} A_f(\tau, \mathfrak{q}^{-1}\mathfrak{m}; \Psi) N(\mathfrak{m}) M(\mathfrak{m})$$

$$= A_f(\tau, \mathfrak{q}^{-1}; \Psi) \sum_{\mathfrak{a}+\mathfrak{ct}=\mathfrak{g}} |\chi(\mathfrak{a})|^2 M(\mathfrak{a}) \prod_{v|\mathfrak{c},\, v \nmid \mathfrak{t}} (1 - \chi_v^2 M_v)^{-1} \prod_{v|\mathfrak{r},\, v \nmid \mathfrak{ct}} (1 + N_v M_v)^{-1},$$

where we put $\chi_v = \chi(v)$ for simplicity.

Proof. From the Euler product of (3.7) with $\chi' = \bar{\chi}$ we can easily derive that

$$(3.12)\qquad Z_{\mathfrak{c}}(\{N_v^2 M_v^2\}) \sum_{\mathfrak{a}} |\chi(\mathfrak{a})|^2 M(\mathfrak{a}) = Z_{\mathfrak{c}}(\{N_v M_v\})^2 \cdot \prod_{v|\mathfrak{c}} (1 - |\chi_v|^2 M_v)^{-1}$$

$$\cdot \prod_{v \nmid \mathfrak{c}} [1 - (|\chi_v|^2 - 2N_v) M_v + N_v^2 M_v^2]^{-1}.$$

Let c_v be as in Proposition 2.3. By (3.9) we have $N_v c_v = |\chi_v|^2 - 2N_v$ for $v \nmid \mathfrak{c}$, and hence our assertion follows from Proposition 2.3 combined with (3.12).

We note that if every prime factor of $\mathfrak{c}$ divides $\mathfrak{t}$, the equality of Theorem 3.7 takes a simpler form:

$$(3.13)\qquad Z_{\mathfrak{t}}(\{N_v M_v\}) \sum_{\mathfrak{m}+\mathfrak{t}=\mathfrak{g}} A_f(\tau, \mathfrak{q}^{-1}\mathfrak{m}; \Psi) N(\mathfrak{m}) M(\mathfrak{m})$$

$$= A_f(\tau, \mathfrak{q}^{-1}; \Psi) \prod_{v|\mathfrak{r},\, v \nmid \mathfrak{t}} (1 + N_v M_v)^{-1} \sum_{\mathfrak{a}+\mathfrak{t}=\mathfrak{g}} |\chi(\mathfrak{a})|^2 M(\mathfrak{a}).$$

In [13, Theorems 5.5, 6.1, and 6.2] we showed that given an eigenform g in $\mathcal{S}_l(2^{-1}\mathfrak{d}, \mathfrak{c}, \varphi)$ with half-integral l and a Hecke character ψ such that $\psi = \varphi$ on $\prod_{v \in \mathbf{h}} \mathfrak{g}_v^\times$, there exists a system χ occurring in $\mathcal{S}_{2l-u}((2^{-1}\mathfrak{c}, \psi^2))$ such that $\sum_{\mathfrak{a}} \chi(\mathfrak{a}) M(\mathfrak{a})$ is determined by the Fourier coefficients of g (cf. Proposition 9.1 and Lemma 9.6 below). The above theorem is an analogue of this result. Namely, an eigenform f in $\mathcal{S}_k(\mathfrak{b}, \mathfrak{c}, \varphi)$ with integral k determines $\sum_{\mathfrak{a}} |\chi(\mathfrak{a})|^2 M(\mathfrak{a})$ in a similar fashion. It should be noted that only $|\chi(v)|^2$ for almost all v, but not χ itself, can be determined by f.

4. Rationality of automorphic forms and the class $|\chi|^2$. Let k be an integral or a half-integral weight. We let every automorphism σ of $\mathbf{C}$ act on $\mathcal{M}_k$ by means of its action on the Fourier coefficients of each form. More generally we can let σ act on the space $\mathcal{N}_k$ of all nearly holomorphic forms of weight k. For details the reader is referred to [13, p. 811]. We recall that σ maps $\mathcal{M}_k$, $\mathcal{N}_k$, and $\mathcal{S}_k$ onto $\mathcal{M}_{k\sigma}$, $\mathcal{N}_{k\sigma}$, and $\mathcal{S}_{k\sigma}$, where $k\sigma = (k'_v)_{v\in\mathbf{a}}$ with $k_v = k'_{v\sigma}$ when $v\sigma$ is defined by $x_{v\sigma} = (x_v)^\sigma$ for $x \in F$. By [13, Prop. 8.1], we have

$$\mathcal{M}_k(\mathfrak{b}, \mathfrak{c}, \varphi)^\sigma = \mathcal{M}_{k\sigma}(\mathfrak{b}, \mathfrak{c}, \varphi^\sigma), \qquad \mathcal{S}_k(\mathfrak{b}, \mathfrak{c}, \varphi)^\sigma = \mathcal{S}_{k\sigma}(\mathfrak{b}, \mathfrak{c}, \varphi^\sigma). \tag{4.1}$$

We now ask the following question:

Given a system of eigenvalues χ occurring in $\mathcal{S}_k((\mathfrak{c}, \Psi))$ and $\sigma \in \mathrm{Aut}(\mathbf{C})$, can one define the transform χ^σ of χ under σ?

If k_v (mod 2) is independent of v and Ψ is of finite order, we can indeed define χ^σ which occurs in $\mathcal{S}_{k\sigma}((\mathfrak{c}, \Psi^\sigma))$ as proved in [6, Prop. 2.6]. However, if k_v (mod 2) is not independent of v or Ψ is of infinite order, there is no clear-cut candidate for χ^σ; for one thing, Ψ^σ is meaningless if Ψ is of infinite order. Thus, in the most general case, the answer is negative. Still, we shall give some positive results by introducing two types of classes of χ, one denoted by $[\chi]$ and the other by $|\chi^2|$, and by showing that the transforms of these classes under σ are well defined. In this section we consider the class $|\chi^2|$ which is related to the eigenforms on G; the class $[\chi]$, which is finer than $|\chi^2|$, will be treated in Section 8. We begin with

LEMMA 4.1. *For $f \in \mathcal{M}_k(\mathfrak{b}, \mathfrak{c}, \varphi)$ and $\sigma \in \mathrm{Aut}(\mathbf{C})$ we have $\mu(\xi, t; f)^\sigma = \mu(\xi, t; f^\sigma)$.*

Proof. This was essentially proved in [13, Prop. 8.6]. In fact, let $\alpha \in G \cap \mathrm{diag}[e, e^{-1}]D[\mathfrak{b}^{-1}, \mathfrak{bc}]$ with $e \in F_{\mathbf{h}}^\times$. Then the first six lines of the proof of [13, Prop. 8.6] show that given σ and a (meromorphic) form q of integral weight m, there exists an element β of $G \cap \mathrm{diag}[e, e^{-1}]D[\mathfrak{b}^{-1}, \mathfrak{bc}]$ such that $\varphi(ed_\beta) = \varphi(ed_\alpha)$ and $q^\sigma\|_{m\sigma}\alpha^{-1} = (q\|_m\beta^{-1})^\sigma$. The desired formula for integral k follows from this and (1.10e). If k is half-integral, we put $g = \theta^{-1}f$ and $m = [k]$ with $\theta(z) = \sum_{\zeta\in\mathfrak{g}} \mathbf{e}_{\mathbf{a}}(\zeta^2 z/2)$. Then $g^\sigma\|_{m\sigma}\alpha^{-1} = (g\|_m\beta^{-1})^\sigma$ and

$$|e|^{-1/2}h(\alpha, \alpha^{-1}z)\theta(\alpha^{-1}z) = |e|^{-1/2}h(\beta, \beta^{-1}z)\theta(\beta^{-1}z) = \sum_{\zeta\in\mathfrak{a}} \mathbf{e}_{\mathbf{a}}(\zeta^2 z/2),$$

where $\mathfrak{a} = e^{-1}\mathfrak{g}$, as noted in the same proof. Taking the product of the equalities, we obtain the desired formula for half-integral k.

We call two systems χ and χ' of eigenvalues equivalent if $|\chi(v)|^2 = |\chi'(v)|^2$ for $v \notin S$ with a finite subset S of $\mathbf{h}$. In view of (3.12), this means that $\sum_{\mathfrak{a}} |\chi(\mathfrak{a})|^2 M(\mathfrak{a})$ and $\sum_{\mathfrak{a}} |\chi'(\mathfrak{a})|^2 M(\mathfrak{a})$ have the same Euler v-factor for $v \notin S$. Then we denote by $|\chi^2|$ the equivalence class containing χ, and by $\mathcal{S}_k(\mathfrak{b}, \mathfrak{c}, \varphi, |\chi^2|)$ the set of all forms f in $\mathcal{S}_k(\mathfrak{b}, \mathfrak{c}, \varphi)$ such that $f|R_v = (|\chi(v)|^2 - N_v - 1)f$ for almost all $v \nmid \mathfrak{c}$. From Propositions 3.6 and 2.4 we can easily derive that $\mathcal{S}_k(\mathfrak{b}, \mathfrak{c}, \varphi)$ is the direct sum of $\mathcal{S}_k(\mathfrak{b}, \mathfrak{c}, \varphi, |\chi^2|)$ for some different $|\chi^2|$'s and that these subspaces are mutually orthogonal with respect to the inner product of (1.5).

LEMMA 4.2. *For $f \in \mathcal{M}_k(\mathfrak{b}, \mathfrak{c}, \varphi)$, $v \in \mathbf{h}$, and $\sigma \in \mathrm{Aut}(\mathbf{C})$ we have $(f|R_v)^\sigma = f^\sigma|R_v$.*

This follows immediately from Proposition 2.2 and Lemma 4.1.

Proposition 4.3. *Given a system of eigenvalues χ on $\mathscr{S}_k((\mathfrak{c}, \Psi))$ and $\sigma \in \mathrm{Aut}(\mathbf{C})$, there exists a Hecke character Ψ' and a system χ' of eigenvalues on $\mathscr{S}_{k\sigma}((\mathfrak{c}, \Psi'))$ such that $\Psi' = \Psi^\sigma$ on $\prod_{v \in \mathbf{h}} \mathfrak{g}_v^\times$ and*

$$\Psi'(\pi_v)^{-1}\chi'(v)^2 = [\Psi(\pi_v)^{-1}\chi(v)^2]^\sigma \quad \textit{for every} \quad v \in \mathbf{h}. \tag{4.2}$$

Proof. Take f as in Proposition 3.5. By Lemma 4.2 we can apply Proposition 3.6 to f^σ to obtain the desired Ψ' and χ'.

From (3.4) we see that (4.2) implies $|\chi'(v)^2| = |\chi(v)^2|^\sigma$ for every $v \nmid \mathfrak{c}$, which together with (3.12) shows that

$$|\chi'(\mathfrak{a})^2| = |\chi(\mathfrak{a})^2|^\sigma \quad \text{for every } \mathfrak{a} \text{ prime to } \mathfrak{c}. \tag{4.3}$$

Since $\chi(\mathfrak{p}^m) = \chi(\mathfrak{p})^m$ for every prime factor $\mathfrak{p}$ of $\mathfrak{c}$, we can thus derive from (4.2) that

$$\Psi'(r)^{-1}\chi'(r\mathfrak{g})^2 = [\Psi(r)^{-1}\chi(r\mathfrak{g})^2]^\sigma \quad \text{for every} \quad r \in F_{\mathbf{h}}^\times \quad \text{such that} \quad r\mathfrak{g} \subset \mathfrak{g}. \tag{4.4}$$

Hereafter we denote by $|\chi^2|^\sigma$ the class $|\chi'^2|$ with χ' of this proposition. From Lemma 4.2 we see that

$$\mathscr{S}_k(\mathfrak{b}, \mathfrak{c}, \varphi, |\chi^2|)^\sigma = \mathscr{S}_{k\sigma}(\mathfrak{b}, \mathfrak{c}, \varphi^\sigma, |\chi^2|^\sigma). \tag{4.5}$$

Lemma 4.4. *Let W be a subspace of $\mathscr{M}_k$ which is finite-dimensional over $\mathbf{C}$ and spanned by $\bar{\mathbf{Q}}$-rational elements. Let K be the subfield of $\mathbf{C}$ consisting of the elements x invariant under all $\sigma \in \mathrm{Aut}(\mathbf{C})$ such that $W^\sigma = W$. Then $[K:\mathbf{Q}] < \infty$ and W is spanned by K-rational elements.*

Proof. Let Φ be a finite set of $\bar{\mathbf{Q}}$-rational elements which spans W. For every $f \in \Phi$ let K_f denote the field generated over $\mathbf{Q}$ by the Fourier coefficients of f at ∞. Then $[K_f : \mathbf{Q}] < \infty$ as shown in [6, Prop. 1.3] and [13, page 812, lines 20–23]. Then we see that K is contained in the composite K' of K_f for all $f \in \Phi$, and hence $[K:\mathbf{Q}] < \infty$. Moreover, as a subfield of $\bar{\mathbf{Q}}$, K corresponds to $\{\sigma \in \mathrm{Gal}(\bar{\mathbf{Q}}/\mathbf{Q}) \mid f^\sigma \in W \text{ for all } f \in \Phi\}$. Let L be the Galois closure of K' over K. For $f \in \Phi$ and $a \in L$ let $g_a = \sum_\sigma (af)^\sigma$ where σ runs over $\mathrm{Gal}(L/K)$. Then g_a is K-rational and belongs to W. Since f is an L-linear combination of such g_a for some a's, we obtain the last assertion.

Proposition 4.5. *Let $\tilde{F}$ be the Galois closure of F over $\mathbf{Q}$, and F_k the subfield of $\tilde{F}$ such that for $\sigma \in \mathrm{Gal}(\tilde{F}/F)$ one has $k\sigma = k$ if and only if σ is the identity map on F_k. Let $F_k(\varphi)$ denote the extension of F_k generated by the values of φ, and $F_k(\varphi, |\chi^2|)$ the smallest extension of $F_k(\varphi)$ which contains $|\chi(v)|^2$ for almost all v. Then $\mathscr{S}_k(\mathfrak{b}, \mathfrak{c}, \varphi)$ (for any weight k) and $\mathscr{S}_k(\mathfrak{b}, \mathfrak{c}, \varphi, |\chi^2|)$ are spanned by $F_k(\varphi)$-rational elements and $F_k(\varphi, |\chi^2|)$-rational elements, respectively. Moreover $F_k(\varphi)$ and $F_k(\varphi, |\chi^2|)$ are finite over $\mathbf{Q}$, and are totally real or CM-fields.*

Proof. Clearly $F_k(\varphi)$ is totally real or a CM-field. By [13, Prop. 8.1(5)] $\mathscr{S}_k(\mathfrak{b}, \mathfrak{c}, \varphi)$ is spanned by $\bar{\mathbf{Q}}$-rational elements, and hence (4.1) combined with Lemma 4.4 shows that $\mathscr{S}_k(\mathfrak{b}, \mathfrak{c}, \varphi)$ is spanned by $F_k(\varphi)$-rational elements. Now Lemma 4.2 shows that such elements (for integral k) are stable under the R_v for all $v \nmid \mathfrak{c}$. Since all such R_v form a commutative semisimple ring, their eigenvalues generate a finite algebraic extension of $\mathbf{Q}$, which must be totally real in view of Proposition 2.4 and Lemma 4.2. Consequently $F_k(\varphi, |\chi^2|)$ is totally real or a CM-field, and finite over $\mathbf{Q}$. The remaining part of our proposition now follows from Lemma 4.4 and Proposition 4.3.

Our proof together with Proposition 3.6 shows that if χ occurs at level $\mathfrak{c}$, then $|\chi(v)|^2$ is algebraic for every $v \nmid \mathfrak{c}$. In fact, the algebraicity is true even for $v|\mathfrak{c}$ as will be seen in Theorem 8.5.

Fix $\mathfrak{b}$, $\mathfrak{c}$, φ, and $|\chi^2|$ as above. For $g \in \mathscr{N}_k$ let $r_k(g)$ denote the element of $\mathscr{S}_k(\mathfrak{b}, \mathfrak{c}, \varphi, |\chi^2|)$ with the property that

$$\langle r_k(g), h\rangle = \langle g, h\rangle \quad \textit{for every} \quad h \in \mathscr{S}_k(\mathfrak{b}, \mathfrak{c}, \varphi, |\chi^2|). \tag{4.6}$$

Then r_k is a $\mathbf{C}$-linear map of $\mathscr{N}_k$ into $\mathscr{S}_k(\mathfrak{b}, \mathfrak{c}, \varphi, |\chi^2|)$.

LEMMA 4.6. *For every $\sigma \in \mathrm{Aut}(\mathbf{C})$ we have $r_k(g)^\sigma = r_{k\sigma}(g^\sigma)$, where $r_{k\sigma}$ is the map of $\mathscr{N}_{k\sigma}$ into $\mathscr{S}_{k\sigma}(\mathfrak{b}, \mathfrak{c}, \varphi^\sigma, |\chi^2|^\sigma)$.*

Proof. Take $\chi_1, \ldots, \chi_t$ so that $\chi = \chi_1$ and $\mathscr{S}_k(\mathfrak{b}, \mathfrak{c}, \varphi)$ is the direct sum of $\mathscr{S}_k(\mathfrak{b}, \mathfrak{c}, \varphi, |\chi_i^2|)$ for $1 \leqslant i \leqslant t$. Let $\mathscr{T}_k(\mathfrak{b}, \mathfrak{c}, \varphi)$ be the orthogonal complement of $\mathscr{S}_k(\mathfrak{b}, \mathfrak{c}, \varphi)$ in $\mathscr{S}_k$. Then $\mathscr{T}_k(\mathfrak{b}, \mathfrak{c}, \varphi) + \sum_{i>1} \mathscr{S}_k(\mathfrak{b}, \mathfrak{c}, \varphi, |\chi_i^2|)$ is the orthogonal complement of $\mathscr{S}_k(\mathfrak{b}, \mathfrak{c}, \varphi, |\chi^2|)$ in $\mathscr{S}_k$. Let q_k denote the orthogonal projection map of $\mathscr{S}_k$ into $\mathscr{S}_k(\mathfrak{b}, \mathfrak{c}, \varphi, |\chi^2|)$. Since [13, Lemma 8.3] has an obvious analogue for integral k, we can apply the argument of [13, p. 817, lines 19–28] to the present case to find that $\mathscr{T}_k(\mathfrak{b}, \mathfrak{c}, \varphi)^\sigma = \mathscr{T}_{k\sigma}(\mathfrak{b}, \mathfrak{c}, \varphi^\sigma)$. This combined with Proposition 4.4 shows that $q_k(f)^\sigma = q_{k\sigma}(f^\sigma)$. In [13, Prop. 9.4] we obtained a $\mathbf{C}$-linear map p_k of $\mathscr{N}_k$ into $\mathscr{S}_k$ such that $p_k(g)^\sigma = p_{k\sigma}(g^\sigma)$ and $\langle p_k(g), h\rangle = \langle g, h\rangle$ for every $g \in \mathscr{N}_k$ and $h \in \mathscr{S}_k$. Then $\langle q_k(p_k(g)), h\rangle = \langle g, h\rangle$ for every $h \in \mathscr{S}_k(\mathfrak{b}, \mathfrak{c}, \varphi, |\chi^2|)$, and hence $q_k \circ p_k = r_k$. Therefore our assertion follows from the "commutativity" of p_k and q_k with σ.

5. Eisenstein series. Throughout the rest of the paper we put

$$d = [F : \mathbf{Q}]. \tag{5.0}$$

Given a weight q and an integral ideal $\mathfrak{c}$, we take a Hecke character ψ and $\kappa \in \mathbf{R}^{\mathbf{a}}$ such that

$$\psi_{\mathbf{a}}(x) = \mathrm{sgn}(x)^{[q]}|x_{\mathbf{a}}|^{i\kappa}, \quad \|\kappa\| = 0, \quad \textit{and the conductor of } \psi \textit{ divides } \mathfrak{c}. \tag{5.1a}$$

We also assume that $\mathfrak{c} \subset 4\mathfrak{g}$ if q is half-integral. Throughout this section we put $D = D[\mathfrak{e}^{-1}, \mathfrak{e}\mathfrak{c}]$ and $\Gamma = \Gamma[\mathfrak{e}^{-1}, \mathfrak{e}\mathfrak{c}]$ with a fractional ideal $\mathfrak{e}$ in F. We assume

(5.1b) $\quad e^{-1} \subset 2\mathfrak{d}^{-1}$ *and* $e\mathfrak{c} \subset 2\mathfrak{d}$ *if* $q \notin \mathbf{Z}^{\mathbf{a}}$; e *is arbitrary if* $q \in \mathbf{Z}^{\mathbf{a}}$.

This makes $J_q(\tau, z)$ of (1.3b) meaningful for $\tau \in \mathrm{pr}^{-1}(P_{\mathbf{A}}D)$.

For $\beta \in G$ we define an ideal $\mathfrak{a}_\beta$ in F by $\mathfrak{a}_\beta = d_p\mathfrak{g}$ with an element p of $P_{\mathbf{A}}$ such that $\beta \in pD$. This is well defined independently of the choice of p; it depends on $\mathfrak{c}$ however. Let B be a complete set of representatives for $P\backslash(G \cap P_{\mathbf{A}}D)/\Gamma$. We then put

$$(5.2) \qquad E(z, s; q, \psi, \Gamma) = \sum_{\beta \in B} E_\beta(z, s; q, \psi, \Gamma),$$

$$E_\beta(z, s; q, \psi, \Gamma) = N(\mathfrak{a}_\beta)^{2s} \sum_{\alpha \in R_\beta} \psi_{\mathbf{a}}(d_\alpha)\psi^*(d_\alpha \mathfrak{a}_\beta^{-1}) y^{su+(i\kappa-q)/2} \|_q \alpha.$$

Here $z \in H^{\mathbf{a}}$, $s \in \mathbf{C}$, $y = \mathrm{Im}(z)$, and $R_\beta = (P \cap \beta\Gamma\beta^{-1})\backslash\beta\Gamma$; we understand that $\psi_{\mathbf{a}}(d_\alpha)\psi^*(d_\alpha\mathfrak{a}_\beta^{-1}) = \psi^*(\mathfrak{a}_\beta^{-1})$ if $\mathfrak{c} = \mathfrak{g}$. We can easily verify that

$$(5.3) \quad E_\beta(z, s; q, \psi, \Gamma) = \psi_{\mathbf{a}}(d_\beta)\psi^*(d_\beta\mathfrak{a}_\beta^{-1})N(\mathfrak{a}_\beta)^{2s} \cdot \sum_{\alpha \in R_\beta} \psi(a(\beta^{-1}\alpha)_{\mathfrak{c}}) y^{su+(i\kappa-q)/2} \|_q \alpha.$$

These series were investigated in our previous papers. To recall some of their properties necessary for our present purposes, let us put

$$(5.4) \qquad C(z, s; q, \psi, \Gamma) = \begin{cases} L_{\mathfrak{c}}(2s, \psi)E(z, s; q, \psi, \Gamma) & \text{if } q \in \mathbf{Z}^{\mathbf{a}}, \\ L_{\mathfrak{c}}(4s-1, \psi^2)E(z, s; q, \psi, \Gamma) & \text{if } q \notin \mathbf{Z}^{\mathbf{a}}, \end{cases}$$

$$(5.5) \qquad \Gamma(s; q, \kappa) = \begin{cases} \prod_{v \in \mathbf{a}} \Gamma(s + (|q_v| + i\kappa_v)/2) & \text{if } q \in \mathbf{Z}^{\mathbf{a}}, \\ \prod_{v \in \mathbf{a}} \Gamma_v(s + (i\kappa_v/2)) & \text{if } q \notin \mathbf{Z}^{\mathbf{a}}, \end{cases}$$

$$(5.6) \qquad \Gamma_v(s) = \Gamma(s + (2\theta_v - 1)/4) \cdot \begin{cases} \Gamma(s + q_v/2) & \text{if } 2q_v \geqslant -1, \\ \Gamma(s - q_v/2) & \text{if } 2q_v < -1, \end{cases}$$

where θ_v is 0 or 1, determined by the condition that it is congruent modulo 2 to $q_v - 1/2$ or $q_v + 1/2$ according as $2q_v \geqslant -1$ or $2q_v < -1$.

PROPOSITION 5.1. *There exists a real analytic function on* $H^{\mathbf{a}} \times \mathbf{C}$ *which coincides with* $s(s-1)D(z, s)$ *or* $(s - 3/4)D(z, s)$ *for* $\mathrm{Re}(s) > 1$ *according as* q *is integral or half-integral, where*

$$D(z, s) = \Gamma(s; q, \kappa)C(z, s; q, \psi, \Gamma).$$

Moreover, if q *is integral,* D *has a pole at* $s = 0$ *if and only if* $\mathfrak{c} = \mathfrak{g}$, $\psi = 1$, *and* $q = 0$; D *has a pole at* $s = 1$ *if and only if* $\psi = 1$ *and* $q = 0$, *in which case the residue of* C *at* $s = 1$ *is* $2^{d-2}h_F R_F D_F^{-1}\pi^d N(e\mathfrak{c})^{-1} \prod_{v|\mathfrak{c}} [1 - N_v^{-1}]$, *where* h_F *denotes the class number of* F, R_F *the regulator of* F, *and* D_F *the discriminant of* F. *If* q *is half-integral,* D *has a simple pole at* $s = 3/4$ *if and only if* $\psi^2 = 1$ *and* $|q_v| - 1/2 \in 2\mathbf{Z}$ *for every* $v \in \mathbf{a}$.

These were given in, or can easily be derived from, [6, Prop. 3.3], [11 Corollary 6.2], [12, Theorem 4.1, Prop. 4.3], and [13, Prop. 9.5]. The group Γ here is more general than those in these papers, but can be treated by the same methods. As for the residue of D for half-integral q, see Proposition 5.2 below.

We define the Gauss sum $\gamma(\psi)$ of ψ by

$$\gamma(\psi) = \sum_{x \in R} \psi_{\mathbf{a}}(x)\psi^*(x\mathfrak{d}\mathfrak{f})\mathbf{e}_{\mathbf{a}}(x), \tag{5.7}$$

where $\mathfrak{f}$ is the conductor of ψ and $R = \mathfrak{f}^{-1}\mathfrak{d}^{-1}/\mathfrak{d}^{-1}$. We understand that $\gamma(\psi) = \psi^*(\mathfrak{d})$ if $\mathfrak{f} = \mathfrak{g}$.

PROPOSITION 5.2. *Suppose $\psi_{\mathbf{a}}(x) = \operatorname{sgn}(x)^{[q]}$; put*

$$\Lambda_q = \{\lambda \in \mathbf{Z} \mid -q_v < \lambda \leqslant q_v,\ \lambda \equiv q_v \ (\mathrm{mod}\ 2) \textit{ for every } v \in \mathbf{a}\} \qquad \textit{if } q \in \mathbf{Z}^{\mathbf{a}},$$

$$\Lambda_q = \{\lambda \in (1/2)\mathbf{Z} \mid 3/2 \leqslant \lambda \leqslant q_v,\ \lambda \equiv q_v \ (\mathrm{mod}\ 2) \textit{ for every } v \in \mathbf{a}\}$$

$$\cup \{\lambda \in (1/2)\mathbf{Z} \mid -q_v < \lambda \leqslant 1/2,\ \lambda \equiv -q_v \ (\mathrm{mod}\ 2) \textit{ for every } v \in \mathbf{a}\}$$

if $q \notin \mathbf{Z}^{\mathbf{a}}$.

For $\lambda \in \Lambda_q$ put $C^*(z, \lambda/2; q, \psi, \Gamma) = \alpha_{\lambda,q,\psi} C(z, \lambda/2; q, \psi, \Gamma)$ with

$$\alpha_{\lambda,q,\psi} = \begin{cases} \gamma(\psi)^{-1} D_F^{1/2} i^{\|q\|} \pi^{-\|q+\lambda u\|/2} & \textit{if } q \in \mathbf{Z}^{\mathbf{a}}, \\ \gamma(\psi^2)^{-1} D_F^{1/2} \pi^{d-\|3\lambda u+q\|/2} & \textit{if } q \notin \mathbf{Z}^{\mathbf{a}},\ \lambda \geqslant 3/2, \\ \gamma(\psi)^{-1} 2^{d/2} i^{\|[q]\|} \pi^{-\|q+\lambda u\|/2} & \textit{if } q \notin \mathbf{Z}^{\mathbf{a}},\ \lambda \leqslant 1/2. \end{cases}$$

Exclude the case in which $\lambda = 3/2$, $F = \mathbf{Q}$, and $\psi^2 = 1$ simultaneously. Then C^* as a function of z is a $\overline{\mathbf{Q}}$-rational element of $\mathcal{N}_q$, and

$$C^*(z, \lambda/2; q, \psi, \Gamma)^{\sigma} = C^*(z, \lambda/2; q\sigma, \psi^{\sigma}, \Gamma)$$

for every $\sigma \in (\overline{\mathbf{Q}}/\mathbf{Q})$. Moreover, suppose that $0 \leqslant q - u/2 \in 2\mathbf{Z}^{\mathbf{a}}$ and $\psi^2 = 1$. Let $R(z, q, \psi, \Gamma)$ be the residue of $\beta C(z, s; q, \psi, \Gamma)$ at $s = 3/4$ with

$$\beta = \gamma(\psi)^{-1} 2^{d/2} D_F^{1/2} R_F^{-1} \pi^{-d-\|[q]/2\|}.$$

Then R as a function of z is a $\overline{\mathbf{Q}}$-rational element of $\mathcal{N}_q$, and

$$R(z, q, \psi, \Gamma)^{\sigma} = R(z, q\sigma, \psi, \Gamma)$$

for every $\sigma \in \operatorname{Gal}(\overline{\mathbf{Q}}/\mathbf{Q})$.

If $q \in \mathbf{Z}^{\mathbf{a}}$ and $\Gamma = \Gamma[2\mathfrak{d}^{-1}, 2^{-1}\mathfrak{c}\mathfrak{d}]$, our assertion was given in [13, Proposition 9.7]. The remaining cases can be derived by the same methods from the results of our previous papers. The necessary formulas are [10, (7.11), (7.12), (7.13), (7.14)], [11, (6.1), (6.2), Theorem 6.1, Propositions 6.3 and 6.4], [12, (4.17), (4.18a), (4.20), the last two lines of p. 17, the first two lines of p. 18], [13, (8.6), (8.8)], [14, (4.12)], and [6, (3.10), Lemma 4.12]. In fact, the detailed proof in the case of half-integral weight was given by Im in [1, Section 1], though his formulation is somewhat different from ours. The class of groups Γ treated in all these papers is more special than the present one, but there is no difficulty in extending the results to the general case.

6. The series $D(s; f, g; \psi)$ and its analytic properties. Given f in $\mathscr{M}_k(\mathfrak{b}, \mathfrak{c}, \varphi)$ and g in $\mathscr{M}_l(\mathfrak{b}', \mathfrak{c}', \varphi')$ with integral or half-integral k and l, we take a Hecke character ψ of F such that

(6.1a) $$\psi(a) = (\varphi/\varphi')(a) \qquad \text{for every} \quad a \in \prod_{v \in \mathbf{h}} \mathfrak{g}_v^\times,$$

(6.1b) $$\psi_{\mathbf{a}}(x) = \operatorname{sgn}(x)^{[k]-[l]}|x|^{i\kappa}, \qquad \kappa \in \mathbf{R}^{\mathbf{a}}, \qquad \|\kappa\| = 0.$$

We then put

(6.2) $$\mathfrak{e} = \mathfrak{b} + \mathfrak{b}', \qquad \mathfrak{h} = \mathfrak{e}^{-1}(\mathfrak{b}\mathfrak{c} \cap \mathfrak{b}'\mathfrak{c}'),$$

(6.3a) $$\mathscr{D}_{\mathfrak{h}}(s; f, g; \psi) = \begin{cases} L_{\mathfrak{h}}(2s, \psi)D(s; f, g; \psi) & \text{if } k - l \in \mathbf{Z}^{\mathbf{a}}, \\ L_{\mathfrak{h}}(4s - 1, \psi^2)D(s; f, g; \psi) & \text{if } k - l \notin \mathbf{Z}^{\mathbf{a}}, \end{cases}$$

(6.3b) $$D(s; f, g; \psi) = \sum_{\xi, t} \psi_{\mathbf{h}}(t)\mu_f(\xi, t)\overline{\mu_g(\xi, t)}\xi^{-([k]+[l]+i\kappa)/2}N(\xi t^2\mathfrak{g})^{c-s},$$

where $L_{\mathfrak{h}}$ is as in (3.6), $s \in \mathbf{C}$, and (ξ, t) runs over $(F^\times \times F_{\mathbf{A}}^\times)/Z$ with

$$Z = \left\{(b^2, b^{-1}a) \,\middle|\, a \in F_{\mathbf{a}}^\times \prod_{v \in \mathbf{h}} \mathfrak{g}_v^\times, b \in F^\times\right\};$$

c is a rational number determined by

(6.3c) $$(2 - 2c)u = k - [k] + l - [l].$$

We easily see that the sum is formally well defined. The weights k and l should be nonzero, since our series becomes trivially 0 otherwise. Let $\mathscr{A}$ be a complete set of representatives for the ideal classes in F. For $\mathfrak{a} = t\mathfrak{g}$ with $t \in F_{\mathbf{h}}^\times$ we can put

(6.4) $$\lambda(\xi, \mathfrak{a}) = \psi_{\mathbf{h}}(t)\mu_f(\xi, t)\overline{\mu_g(\xi, t)},$$

since the right-hand side depends only on $t\mathfrak{g}$. Then we have

(6.5) $$D(s; f, g; \psi) = \sum_{\mathfrak{a}\in\mathscr{A}} \sum_{\xi} \lambda(\xi, \mathfrak{a})\xi^{-(\|k\|+\|l\|+i\kappa)/2} N(\xi\mathfrak{a}^2)^{c-s},$$

where ξ runs over $\{\xi \in \mathfrak{e}\mathfrak{d}^{-1}\mathfrak{a}^{-2} | \xi \gg 0\}/\{a^2 | a \in \mathfrak{g}^\times\}$. Since the $\mu_f(\xi, t)$ and $\mu_g(\xi, t)$ are Fourier coefficients of some elements of $\mathscr{M}_k$ and $\mathscr{M}_l$ as shown in (1.10e), we can easily show that our series is convergent for sufficiently large Re(s).

THEOREM 6.1. *Suppose that either f or g is a cusp form. Then the product*

$$\mathscr{D}_\mathfrak{h}(s; f, g; \psi)\Gamma(s; k - l, \kappa) \prod_{v\in\mathbf{a}} \Gamma(s - 1 + (k_v + l_v + i\kappa_v)/2)$$

with $\Gamma(s; q, \kappa)$ of (5.5) can be continued to the whole s-plane as a meromorphic function, which is holomorphic except for possible simple poles at the following points: $s = 0$ only if $k = l \in \mathbf{Z}^\mathbf{a}$, $\mathfrak{b} = \mathfrak{b}'$, $\mathfrak{c} = \mathfrak{c}' = \mathfrak{g}$, and $\psi = 1$; $s = 1$ only if $k = l$ and $\psi = 1$; $s = 3/4$ only if $\psi^2 = 1$ and $|k_v - l_v| - 1/2 \in 2\mathbf{Z}$ for every $v \in \mathbf{a}$. The residue of $D(s; f, g; \psi)$ at $s = 1$ is

$$2^{d-1}(4\pi/v_k)^{\|k\|} \prod_{v\in\mathbf{a}} \Gamma(k_v)^{-1} h_F R_F \langle g, f\rangle,$$

where h_F and R_F are as in Proposition 5.1, and v_k is defined by (1.7b).

To prove this, we first put

(6.6) $$\Gamma = \Gamma[\mathfrak{e}', \mathfrak{h}\mathfrak{e}], \qquad D = D[\mathfrak{e}', \mathfrak{h}\mathfrak{e}],$$

where $\mathfrak{e}' = \mathfrak{c}^{-1}$ or $2\mathfrak{c}^{-1}$ according as $k + l \in \mathbf{Z}^\mathbf{a}$ or $k + l \notin \mathbf{Z}^\mathbf{a}$; $\mathfrak{e}$ and $\mathfrak{h}$ are as in (6.2). For $\mathfrak{a} \in \mathscr{A}$ we take $r \in F_\mathbf{h}^\times$ so that $\mathfrak{a} = r\mathfrak{g}$ and also take β in $G \cap \mathrm{diag}[r, r^{-1}]D$. Then $\mathfrak{a}^{-1}$ coincides with the ideal $\mathfrak{a}_\beta$ defined in Section 5. Fixing a β for each $\mathfrak{a}$, let B be the set of such β's for all $\mathfrak{a} \in \mathscr{A}$. Then B represents $P\backslash(G \cap P_A D)/\Gamma$. This was given in [10, Lemma 1.6] for a special type of D; the general case can be proved in the same way. Put

$$f_\beta(z) = J_k(\beta, \beta^{-1}z)f(\beta^{-1}z) = \sum_{\xi\in F} \lambda_\beta(\xi)\mathbf{e}_\mathbf{a}(\xi z/v_k),$$

$$g_\beta(z) = J_l(\beta, \beta^{-1}z)g(\beta^{-1}z) = \sum_{\xi\in F} \lambda'_\beta(\xi)\mathbf{e}_\mathbf{a}(\xi z/v_l).$$

From (1.10e) we see that $\lambda_\beta(\xi)\overline{\lambda'_\beta(\xi)}$ is $\psi_\mathbf{a}(d_\beta)^{-1}\psi^*(d_\beta^{-1}\mathfrak{a}_\beta)N(\mathfrak{a}_\beta)^{2-2c}$ times $\lambda(\xi, \mathfrak{a}_\beta^{-1})$ of (6.4), and hence

(6.7) $D(s; f, g; \psi)$

$$= \sum_{\beta\in B} N(\mathfrak{a}_\beta)^{2s-2}\psi_\mathbf{a}(d_\beta)\psi^*(d_\beta\mathfrak{a}_\beta^{-1})\cdot\sum_{\xi} \lambda_\beta(\xi)\overline{\lambda'_\beta(\xi)}\xi^{-(k+l+i\kappa)/2}N(\xi)^{1-s},$$

where ξ runs over $\{\xi \in \mathfrak{e}\mathfrak{d}^{-1}\mathfrak{a}_\beta^2 | \xi \gg 0\}/\{a^2 | a \in \mathfrak{g}^\times\}$. Put

$$K_\beta(z) = y^{su+(k+l+i\kappa)/2} F_\beta(z), \qquad y = \mathrm{Im}(z),$$

$$F_\beta(z) = \begin{cases} f_\beta(z)\overline{g_\beta(z)} & \text{if } k + l \in \mathbf{Z}^{\mathbf{a}}, \\ f_\beta(z)\overline{g_\beta(z/2)} & \text{if } k \notin \mathbf{Z}^{\mathbf{a}}, l \in \mathbf{Z}^{\mathbf{a}}, \\ f_\beta(z/2)\overline{g_\beta(z)} & \text{if } k \in \mathbf{Z}^{\mathbf{a}}, l \notin \mathbf{Z}^{\mathbf{a}}. \end{cases}$$

Then we have

$$K_\beta(\alpha z) = \varepsilon_\beta \psi_0(a(\beta^{-1}\alpha)_{\mathfrak{h}})\overline{J_{k-l}(\alpha, z)}^{-1} \cdot |j(\alpha, z)|^{-2su+k-l-i\kappa} K_1(z) \tag{6.8}$$

for every $\alpha \in \beta\Gamma$, where $\psi_0 = \psi$ and $\varepsilon_\beta = 1$ except when $k \in \mathbf{Z}^{\mathbf{a}}$ and $l \notin \mathbf{Z}^{\mathbf{a}}$, in which case $\psi_0 = \theta\psi$ with the Hecke character θ corresponding to $F(\sqrt{-1})$ and $\varepsilon_\beta = \theta_{\mathbf{a}}(d_\beta)\theta^*(d_\beta\mathfrak{a}_\beta^{-1})$. This is easy if both k and l are integral. The case involving a half-integral weight is not so trivial, since $h(\gamma, z)$ is only a "partial" factor of automorphy. However, by [13, Pro. 2.3 (iii)] we have

$$h(\gamma\gamma', z) = h(\gamma, \gamma' z)h(\gamma', z) \tag{6.9}$$

if $\gamma \in G \cap P_A C''$ and $\gamma' \in G \cap C'' G_{\mathbf{a}}$. Furthermore, our choice of β and D shows that if $\alpha_0 = \varepsilon\alpha\varepsilon^{-1}$ with $\varepsilon = \mathrm{diag}[1, 2]$, then $\alpha_0 \in \beta\Gamma[\mathfrak{b}^{-1}, \mathfrak{b}\mathfrak{c}]$, from which we can easily derive $f_\beta((\alpha z)/2) = \varphi(a(\beta^{-1}\alpha))J_k(\alpha, z)f(z/2)$ for $\alpha \in \beta\Gamma$ if $k \in \mathbf{Z}^{\mathbf{a}}$. This together with (6.9) proves (6.8) when $k \in \mathbf{Z}^{\mathbf{a}}$ and $l \notin \mathbf{Z}^{\mathbf{a}}$; θ appears because of the difference of $h(\alpha, z)^2$ from $j(\alpha, z)^u$ (see [11, Lemma 3.5]). The remaining cases can be handled in the same fashion.

Let $\Phi = \Gamma \backslash H^{\mathbf{a}}$, $\Psi = (P \cap \beta\Gamma\beta^{-1})\backslash H^{\mathbf{a}}$, and $R_\beta = (P \cap \beta\Gamma\beta^{-1})\backslash \beta\Gamma$. Observe that $\beta\Gamma\beta^{-1} = \Gamma[\mathfrak{e}'\mathfrak{a}_\beta^{-2}, \mathfrak{e}\mathfrak{h}\mathfrak{a}_\beta^2]$ and

$$P \cap \beta\Gamma\beta^{-1} = \left\{ \begin{pmatrix} a & b \\ 0 & a^{-1} \end{pmatrix} \middle| a \in \mathfrak{g}^\times, b \in \mathfrak{e}'\mathfrak{a}_\beta^{-2} \right\},$$

and hence Ψ is represented by $(\mathbf{R}^{\mathbf{a}}/\mathfrak{e}'\mathfrak{a}_\beta^{-2}) \times Y$ with

$$Y = \{y \in \mathbf{R}^{\mathbf{a}} | y \gg 0\}/\{a^2 | a \in \mathfrak{g}^\times\}.$$

Applying (1.10a, b, d, e) to f and g, we easily see that K_β is invariant under $P \cap \beta\Gamma\beta^{-1}$. Therefore, putting $r = (k + l + i\kappa)/2$, we have

$$\int_\Psi K_\beta(z)\, d_H z = D_F^{1/2} N(\mathfrak{e}'\mathfrak{a}_\beta^{-2}) \cdot \int_Y \sum_{\xi \in F} \lambda_\beta(\xi)\overline{\lambda'_\beta(\xi)} \mathbf{e}_{\mathbf{a}}(2i\xi y/v_0) y^{(s-2)u+r}\, dy^u, \tag{6.10}$$

where $\nu_0 = 1$ if both k and l are integral and $\nu_0 = 2$ otherwise. The last integral over Y is

$$(4\pi/\nu_0)^{\|u-su-r\|} \prod_{v \in \mathbf{a}} \Gamma(s - 1 + r_v) \sum_{\xi} \lambda_\beta(\xi)\overline{\lambda'_\beta(\xi)}\xi^{-r}N(\xi)^{1-s},$$

where $\sum_\xi$ is the same as in (6.7). On the other hand, Ψ is given by $\bigcup_{\alpha \in R_\beta} \alpha\Phi$, and hence (6.8) together with (5.3) shows that

$$\int_\Psi K_\beta(z)\, d_H z = A_\beta^{-1} \int_\Phi F_1(z)\overline{E_\beta(z, \bar{s}; k - l, \bar{\psi}_0, \Gamma)}y^k\, d_H z$$

with $A_\beta = \psi_{\mathbf{a}}(d_\beta)\psi^*(d_\beta \mathfrak{a}_\beta^{-1})N(\mathfrak{a}_\beta)^{2s}$. Combining this result with (6.7) and (6.10), we obtain

$$\text{(6.11)} \quad \mathscr{D}_{\mathfrak{h}}(s; f, g; \psi)(4\pi/\nu_0)^{\|u-su-r\|} \prod_{v \in \mathbf{a}} \Gamma(s - 1 + r_v)$$
$$= D_F^{-1/2}N(\mathfrak{e}')^{-1} \int_\Phi F_1(z)\overline{C(z, \bar{s}; k - l, \bar{\psi}_0, \Gamma)}y^k\, d_H z.$$

Theorem 6.1 now follows from this expression and Proposition 5.1, since C times an appropriate factor is slowly increasing at every cusp as guaranteed by [12, Prop. 4.4]. As for the residue of D at $s = 1$, we need

$$\text{(6.12)} \quad \operatorname{vol}(\Gamma[\mathfrak{e}^{-1}, \mathfrak{e}\mathfrak{h}]\backslash H^{\mathbf{a}}) = 2\pi^{-d}D_F^{3/2}\zeta_F(2)N(\mathfrak{h}) \prod_{v|\mathfrak{h}} [1 + N_v^{-1}],$$

which follows easily from [12, (4.28)].

Proof of Proposition 3.2. Let $\mathbf{f} = (f_q)_{q \in Q}$ and $\mathbf{f}' = (f'_q)_{q \in Q}$ be the normalized eigenforms in $\mathscr{S}_k((\mathfrak{c}, \Psi))$ and $\mathscr{M}_{k'}((\mathfrak{c}', \Psi'))$ with eigenvalues $\chi(\mathfrak{v})$ and $\chi'(\mathfrak{v})$, respectively. Let $\mathscr{I}$ be the ideal group of F; let $\mathscr{P} = \{\xi\mathfrak{g} \mid \xi \in F^\times\}$ and $\mathscr{P}_+ = \{\xi\mathfrak{g} \mid 0 \ll \xi \in F^\times\}$. Take complete sets of representatives $\mathscr{A}$ and $\mathscr{B}$ for $\mathscr{I}/\mathscr{P}$ and $\mathscr{I}/\mathscr{P}_+$, respectively. Then

$$\text{(6.13)} \quad [\mathscr{I} : \mathscr{P}] \sum_{\mathfrak{a}} \chi(\mathfrak{a})\overline{\chi'(\mathfrak{a})}N(\mathfrak{a})^{-s} = \sum_{\mathfrak{m} \in \mathscr{A}} \sum_{\mathfrak{t} \in \mathscr{B}} \sum_{\xi} \chi(\xi\mathfrak{m}^2\mathfrak{t}^{-1})\overline{\chi'(\xi\mathfrak{m}^2\mathfrak{t}^{-1})}N(\xi\mathfrak{m}^2\mathfrak{t}^{-1})^{-s},$$

where ξ runs over $\{\xi \in \mathfrak{m}^{-2}\mathfrak{t} \mid \xi \gg 0\}/\mathfrak{g}_+^\times$. For each $\mathfrak{m}$ and $\mathfrak{t}$ take elements t and r of $F_{\mathbf{h}}^\times$ so that $\mathfrak{t} = t\mathfrak{g}$ and $\mathfrak{m} = r\mathfrak{g}$. By Proposition 3.1, for $\mathfrak{a} = \xi\mathfrak{m}^2\mathfrak{t}^{-1}$ we have

$$\text{(6.14)} \quad \chi(\mathfrak{a})\overline{\chi'(\mathfrak{a})} = N(\mathfrak{a})^2(\Psi/\Psi')(r)\mu(\xi, r; f_q)\overline{\mu(\xi, r; f'_q)}\xi^{-(k+k'+i\kappa)/2},$$

where $q = \operatorname{diag}[1, t]$ and κ is as in Proposition 3.2. Therefore, by (6.5), we obtain

$$\text{(6.15)} \quad [\mathscr{I} : \mathscr{P}][\mathfrak{g}_+^\times : U] \sum_{\mathfrak{a}} \chi(\mathfrak{a})\overline{\chi'(\mathfrak{a})}N(\mathfrak{a})^{-s} = \sum_{\mathfrak{t} \in \mathscr{B}} N(\mathfrak{t})^{s-2}D(s - 1; f_q, f'_q; \Psi/\Psi')$$

where $U = \{e^2 | e \in \mathfrak{g}^\times\}$, and consequently

$$(6.16) \qquad [\mathscr{I} : \mathscr{P}][\mathfrak{g}_+^\times : U]\mathscr{D}(s, \chi, \overline{\chi'}) \sum_{\mathfrak{t} \in \mathscr{B}} N(\mathfrak{t})^{s-2}\mathscr{D}_{\mathfrak{h}}(s-1; f_q, f'_q; \Psi/\Psi'),$$

where $\mathfrak{h} = \mathfrak{c} \cap \mathfrak{c}'$. This together with Theorem 6.1 proves Proposition 3.2.

7. Main theorems for integral k. We are going to study the arithmeticity of the values and the residues of the series $D_{\mathfrak{h}}(s; f, g; \psi)$ of (6.3a) for certain rational numbers s belonging to a critical strip. The essential conditions we need are:

(i) *f is a cusp eigenform of certain Hecke operators;*
(ii) *$k_v \geqslant l_v$ for every $v \in \mathbf{a}$,*

where k and l are the weights of f and g, respectively. We treat the case of integral k in this section and that of half-integral k in Section 9. In either case, l can be integral or half-integral, and so there are four essentially different cases. Thus, throughout this section, we assume k to be integral, and denote complex conjugation by ρ. We also denote by $\mathscr{S}_k(\varphi, |\chi^2|)$ the union of the sets $\mathscr{S}_k(\mathfrak{b}, \mathfrak{c}, \varphi, |\chi^2|)$ for all $(\mathfrak{b}, \mathfrak{c})$ with which $\mathscr{S}_k(\mathfrak{b}, \mathfrak{c}, \varphi, |\chi^2|)$ is meaningful.

THEOREM 7.1. *Let K be a totally real algebraic number field or a* CM*-field; suppose that K contains $F(\varphi, |\chi^2|)$ of Proposition 4.5. Let g, h, and p be elements of $\mathscr{S}_k(\varphi, |\chi^2|)$. Suppose that $k \geqslant 2u$. If p is nonzero and K-rational, then*

$$(7.1) \qquad (\langle g, h\rangle/\langle p, p\rangle)^\sigma = \langle g^{\rho\sigma\rho}, h^\sigma\rangle/\langle p^\sigma, p^\sigma\rangle$$

for every $\sigma \in \mathrm{Aut}(\mathbf{C})$.

THEOREM 7.2. *Let $f \in \mathscr{S}_k(\mathfrak{b}, \mathfrak{c}, \varphi, |\chi^2|)$ and $g \in \mathscr{M}_l(\mathfrak{b}', \mathfrak{c}', \varphi')$ with integral or half-integral l. Let ψ be a Hecke character of F such that $\psi(x) = \mathrm{sgn}(x)^{k-[l]}(\varphi/\varphi')(x_{\mathbf{h}})$ for every $x \in F_{\mathbf{a}}^\times \prod_{v \in \mathbf{h}} \mathfrak{g}_v^\times$. Define Λ_q as in Proposition 5.2 and put*

$$A(m/2; f, g; \psi) = \beta_{m,\psi}\mathscr{D}_{\mathfrak{h}}(m/2; f, g; \psi)$$

for every $m \in \Lambda_{k-l}$ with $\mathfrak{h}$ of (6.2) and

$$\beta_{m,\psi} = \begin{cases} \gamma(\psi)^{-1} i^{\|k-l\|}\pi^{-\|k\|-md} & \text{if } l \in \mathbf{Z}^{\mathbf{a}}, \\ \gamma(\psi^2)^{-1}\pi^{d-2md-\|k\|} & \text{if } l \notin \mathbf{Z}^{\mathbf{a}},\ m \geqslant 3/2, \\ \gamma(\psi)^{-1} i^{\|k-[l]\|} D_F^{1/2}\pi^{(d/2)-md-\|k\|} & \text{if } l \notin \mathbf{Z}^{\mathbf{a}},\ m \leqslant 1/2, \end{cases}$$

where $d = [F : \mathbf{Q}]$ and $\gamma(\ \ )$ is defined by (5.7). Exclude the case in which $F = \mathbf{Q}$, $\psi^2 = 1$, and $m = 3/2$ simultaneously. Then

$$[A(m/2; f, g; \psi)/\langle p, p\rangle]^\sigma = A(m/2; f^\sigma, g^{\rho\sigma\rho}; \psi^\sigma)/\langle p^\sigma, p^\sigma\rangle$$

for every $\sigma \in \mathrm{Aut}(\mathbf{C})$, where p is a function as in Theorem 7.1.

THEOREM 7.3. *Let $R(f, g, \psi)$ be the residue of $D(s; f, g; \psi)$ at $s = s_0$ with f, g, and ψ as in Theorem 7.2 in the following two cases:*

(1) $k = l$, $\varphi = \varphi'$, $\psi = 1$, *and* $s_0 = 1$;
(2) $0 \leqslant k - l - u/2 \in 2\mathbf{Z}^{\mathbf{a}}$, $\psi^2 = 1$, *and* $s_0 = 3/4$.

Put $R^*(f, g, \psi) = \varepsilon R(f, g, \psi)$ *with*

$$\varepsilon = \begin{cases} R_F^{-1}\pi^{-\|k\|} & \text{in Case (1)}, \\ \gamma(\psi)^{-1} i^d D_F^{1/2} R_F^{-1} \pi^{d-\|k\|} & \text{in Case (2)}. \end{cases}$$

Then

$$[R^*(f, g, \psi)/\langle p, p\rangle]^\sigma = R^*(f^\sigma, g^{\rho\sigma\rho}, \psi)/\langle p^\sigma, p^\sigma\rangle$$

for every $\sigma \in \mathrm{Aut}(\mathbf{C})$, where p is a function as in Theorem 7.1.

As mentioned in the introduction, the above type of results were previously obtained in [4, 5, 6] for integral l, and in Sturm [15, 16] and Im [1] for half-integral l. Our theorems include these in essence as special cases.

Remark 7.4. Suppose $l \in \mathbf{Z}^{\mathbf{a}}$. Then $\Lambda_{k-l} \subset \mathbf{Z}$; moreover Λ_{k-l} is nonempty if and only if $k_v - l_v \pmod 2$ is independent of v and $k_v > l_v$ for all $v \in \mathbf{a}$. If $l \notin \mathbf{Z}^{\mathbf{a}}$, then Λ_{k-l} is nonempty if and only if $k_v - l_v \pmod 2$ is independent of v and $k_v - l_v \geqslant 3/2$ for all $v \in \mathbf{a}$. In either case, $k \geqslant 2u$ if Λ_{k-l} is nonempty. A Hecke character ψ as in Theorem 7.2 exists if and only if

$$(\varphi/\varphi')(e) = \mathrm{sgn}(e)^{k-[l]} \quad \text{for every} \quad e \in \mathfrak{g}^\times. \tag{7.2}$$

Thus Theorem 7.2 is truly meaningful only when these conditions on k, l, φ, and φ' are satisfied.

The remaining part of this section is devoted to the proof of the above theorems. We begin with

LEMMA 7.5. *Given $r \in \mathbf{Z}^{\mathbf{a}}$, two positive integers l, m, and $\lambda \in \mathbf{Z}^{\mathbf{a}}$, $\|\lambda\| = 0$, suppose that $|a|^{i\lambda}$ is a root of unity for every $a \in \mathfrak{g}^\times$. Then there exists a Hecke character ω of F satisfying the following three conditions*: (i) $\omega_{\mathbf{a}}(x) = \mathrm{sgn}(x)^r|x|^{i\lambda}$; (ii) *$\omega$ and ω^m are ramified exactly at the same prime ideals of F*; (iii) *l divides the conductor of ω^m.*

Proof. Since $\{a \in \mathfrak{g}^\times \mid \mathrm{sgn}(a)^r|a|^{i\lambda} = 1\}$ is a subgroup of $\mathfrak{g}^\times$ of finite index, we can find, by Chevalley's theorem, an integral ideal $\mathfrak{a}$ such that $\mathrm{sgn}(a)^r|a|^{i\lambda} = 1$ if $a \in \mathfrak{g}^\times$ and $a - 1 \in \mathfrak{a}$. Let

$$X = F_{\mathbf{a}}^\times \cdot \left\{ x \in \prod_{v \in \mathbf{h}} \mathfrak{g}_v^\times \mid x_v - 1 \in \mathfrak{a}_v \text{ for every } v \mid \mathfrak{a} \right\}.$$

Define ψ on X by $\psi(x) = \mathrm{sgn}(x)^r|x|^{i\lambda}$. We can extend ψ to $F^\times X$ so that $\psi(F^\times) = 1$. Since $[F_{\mathbf{A}}^\times : F^\times X]$ is finite, ψ can be extended to a Hecke character of F, denoted again by ψ. We are going to obtain the desired ω as $\psi\eta$ with a Hecke character η such that $\eta_{\mathfrak{a}} = 1$. Now, for each prime number p, we can find a chain

$$\mathbf{Q} = M(p^0) \subset M(p) \subset \cdots \subset M(p^n) \subset \cdots$$

with a totally real cyclic extension $M(p^n)$ of $\mathbf{Q}$ of degree p^n in which p is completely ramified and all other primes are unramified. Let F' be the Galois closure of F over $\mathbf{Q}$ and $\mathfrak{q}$ the product of all prime factors of p in F'. Then the conductor of $M(p^n)F'$ over F' is of the form $\mathfrak{q}^{\nu(n)}$ with an integer $\nu(n)$. Clearly $\lim_{n\to\infty} \nu(n) = \infty$. Therefore if we denote by $q_p(n)$ the largest power of p dividing the conductor of $M(p^n)F$ over F, then $\lim_{n\to\infty} q_p(n) = \infty$. Now, given l and m, let P be the set of all prime numbers which divide $lmN(\mathfrak{a})$. For each $p \in P$ let L_p resp. m_p the largest power of p which divides $lN(\mathfrak{a})^2$ resp. m. Take a sufficiently large integer n_p so that $[M(p^{n_p}m_p)F : M(p^{n_p})F] = m_p$ and $q_p(n_p) \geqslant L_p$. Let η_p be a Hecke character of F whose kernel corresponds to $M(p^{n_p}m_p)F$. Then the conductor of $\eta_p^{m_p t}$ is divisible by L_p for every integer t prime to p. Putting $\omega = \psi \prod_{p|lN(\mathfrak{a})} \eta_p$, we obtain the desired character.

We insert here a simple fact:

LEMMA 7.6. *Let χ be a system of eigenvalues occurring in $\mathscr{S}_k((\mathfrak{c}, \Psi))$, and η a Hecke character of F of conductor $\mathfrak{q}$. Let $\mathfrak{c}_0$ be the conductor of Ψ and let $\mathfrak{m} = \mathfrak{c} \cap \mathfrak{q}^2 \cap \mathfrak{q}\mathfrak{c}_0$. Then there exists a system of eigenvalues χ_1 occurring in $\mathscr{S}_k((\mathfrak{m}, \eta^2\Psi))$ such that $\chi_1(\mathfrak{a}) = \eta^*(\mathfrak{a})\chi(\mathfrak{a})$ for every integral ideal $\mathfrak{a}$.*

For the proof see [6, Proposition 4.4 and 4.5] and [7, Prop. 9.7].

To prove Theorem 7.1, we may assume, choosing suitable $\mathfrak{b}$ and $\mathfrak{c}$, that g, h, and p are contained in $\mathscr{S}_k(\mathfrak{b}, \mathfrak{c}, \varphi, |\chi^2|)$ and that $2^{-1}\mathfrak{d} \subset \mathfrak{b}$. By Proposition 3.6 we may assume that χ occurs in $\mathscr{S}_k((\mathfrak{c}, \Psi))$ with a Hecke character Ψ which coincides with φ on $\prod_{v\in\mathbf{h}} \mathfrak{g}_v^\times$. We can find a multiple $\mathfrak{t}$ of $\mathfrak{c}$ such that $f|R_v = (|\chi(v)|^2 - N_v - 1)f$ for every $f \in \mathscr{S}_k(\mathfrak{b}, \mathfrak{c}, \varphi, |\chi|^2)$ and every $v \nmid \mathfrak{t}$. By Lemma 7.5 we can find a Hecke character η of F such that $\eta^2 \neq 1$, $\eta_{\mathfrak{a}} = 1$, and $\mathfrak{t}^2$ divides the conductor of η, which we denote by $\mathfrak{f}$. Define a character ω on $\prod_{v\in\mathbf{h}} \mathfrak{g}_v^\times$ by $\omega = \varphi/\eta$. Since φ is defined modulo $\mathfrak{c}$, we see that ω defines a primitive character of $(\mathfrak{g}/\mathfrak{f})^\times$. Take $n \in \mathbf{Z}^{\mathbf{a}}$ so that $0 \leqslant n \leqslant u$ and $n - k \in 2\mathbf{Z}^{\mathbf{a}}$, and also $\theta_{n,\omega}$ as in Proposition 1.3.

Take an arbitrary $\tau \in F$ such that $0 \ll \tau \in \mathfrak{g}$. Since $\mathfrak{d} \subset \mathfrak{b}$, we have a decomposition $\tau\mathfrak{b}^{-1}\mathfrak{d} = \mathfrak{q}^2\mathfrak{r}$ with integral ideals $\mathfrak{q}$ and $\mathfrak{r}$ such that $\mathfrak{q}$ is prime to $\mathfrak{f}$ and $\mathfrak{r}$ has no square factors prime to $\mathfrak{f}$. By Proposition 1.2 with $(1/2)\theta_{n,\omega}$ as f there, we obtain an element g of $\mathscr{M}_l(2^{-1}\mathfrak{d}, 4\tau\mathfrak{q}^{-2}\mathfrak{f}^2, \omega\varepsilon_{\mathbf{h}})$ with $l = n + u/2$ such that $\xi \neq 0$ and $\mu_g(\xi, t) \neq 0$ only when there is an element ζ of $F^\times$ such that $\xi = \tau\zeta^2$ and $\zeta t\mathfrak{q} + \mathfrak{f} = \mathfrak{g}$, in which case

$$\mu_g(\xi, t) = \varepsilon_{\mathbf{h}}(t)\zeta^n\omega((\zeta t)_{\mathfrak{f}})^{-1}, \tag{7.3}$$

where ε is as in Proposition 1.2.

Observing that $\varphi/(\omega\varepsilon) = \eta\varepsilon$ on $\prod_{v\in\mathbf{h}} \mathfrak{g}_v^\times$, we can consider $D(s; f, g; \eta\varepsilon)$ with any fixed f in $\mathscr{S}_k(\mathfrak{b}, \mathfrak{c}, \varphi, |\chi|^2)$. This series can be written

$$\sum_{\zeta, t} (\eta\varepsilon)_{\mathbf{h}}(t)\mu_f(\tau\zeta^2, t)\varepsilon_{\mathbf{h}}(t)\zeta^n\omega((\zeta t)_{\mathfrak{f}})(\tau\zeta^2)^{-(k+n)/2}N(\tau\zeta^2 t^2\mathfrak{g})^{3/4-s},$$

where the sum is taken over all different integral ideals $\zeta t\mathfrak{q}$ prime to $\mathfrak{f}$. Writing $A_f(\xi, \mathfrak{m})$ for $A_f(\xi, \mathfrak{m}; \Psi)$ of (2.11) and putting $\mathfrak{m} = \zeta t\mathfrak{q}$, we find that

$$D(s; f, g; \eta\varepsilon) = N(\tau)^{3/4-s}\tau^{-(k+n)/2} \cdot \sum_{\mathfrak{m}} \eta^*(\mathfrak{q}^{-1}\mathfrak{m})A_f(\tau, \mathfrak{q}^{-1}\mathfrak{m})N(\mathfrak{q}^{-1}\mathfrak{m})^{1/2-2s},$$

where $\mathfrak{m}$ runs over all the integral ideals prime to $\mathfrak{f}$. This together with (3.13) shows that

$$\begin{aligned} L_{\mathfrak{f}}(2s - 1/2, \eta)&D(s; f, g; \eta\varepsilon) \\ &= N(\tau)^{3/4-s}\tau^{-(k+n)/2}\eta^*(\mathfrak{q}^{-1})N(\mathfrak{q})^{2s-1/2} \\ &\quad \cdot A_f(\tau, \mathfrak{q}^{-1}) \prod_{v|\tau, v\nmid\mathfrak{f}} [1 + \eta(\pi_v)N_v^{1/2-2s}]^{-1} \sum_{\mathfrak{a}+\mathfrak{f}=\mathfrak{g}} |\chi(\mathfrak{a})|^2\eta^*(\mathfrak{a})N(\mathfrak{a})^{-2s-1/2}. \end{aligned}$$

By Lemma 7.6 there exists a system χ' occurring in $\mathscr{S}_k(\mathfrak{f}^2, \eta^2\Psi))$ such that $\chi'(\mathfrak{a}) = \eta^*(\mathfrak{a})\chi(\mathfrak{a})$ for every $\mathfrak{a}$. Then the last sum over $\mathfrak{a}$ is $L_{\mathfrak{f}}(4s - 1, \eta^2)^{-1} \cdot \mathscr{D}^{\mathfrak{f}}(2s + 1/2, \bar{\chi}, \chi')$ with the function $\mathscr{D}^{\mathfrak{f}}$ of Proposition 3.3. Observe that $\mathfrak{e}$, $\mathfrak{h}$ of (6.2) and Γ of (6.6) in the present setting are $\mathfrak{b}$, $2\mathfrak{r}\mathfrak{f}^2$, and $\Gamma[2\mathfrak{b}^{-1}, 2\mathfrak{b}\mathfrak{r}\mathfrak{f}^2]$, respectively. Call this group Γ_τ. Then, combining the last equality with (6.11), we obtain

$$\begin{aligned} \text{(7.4a)} \quad &\int_\Phi f(z/2)\overline{g(z)C(z, \bar{s}; k - l, \bar{\eta}\varepsilon\theta, \Gamma_\tau)}y^k\, d_H z \\ &\qquad = A_f(\tau, \mathfrak{q}^{-1})B(s, \tau, \eta)L_{\mathfrak{f}}(2s - 1/2, \eta)^{-1}\mathscr{D}^{\mathfrak{f}}(2s + 1/2; \bar{\chi}, \chi'), \end{aligned}$$

where θ is the Hecke character corresponding to $F(\sqrt{-1})$, and

$$\begin{aligned} \text{(7.4b)} \quad B(s, \tau, \eta) &= N(\tau)^{3/4-s}\tau^{-(k+n)/2}\eta^*(\mathfrak{q}^{-1})N(\mathfrak{q})^{2s-1/2}N(2\mathfrak{b}^{-1}) \\ &\quad \cdot D_F^{1/2}(2\pi)^{\|u-su-(k+l)/2\|} \prod_{v\in\mathbf{a}} \Gamma(s - 1 + (k_v + l_v)/2) \\ &\quad \cdot \prod_{v|\mathfrak{h}, v\nmid\mathfrak{f}} [1 - \eta(\pi_v)^2N_v^{1-4s}] \prod_{v|\tau, v\nmid\mathfrak{f}} [1 + \eta(\pi_v)N_v^{1/2-2s}]^{-1}. \end{aligned}$$

We evaluate (7.4a) at $s = 3/4$ and apply Proposition 5.2 to the present C. Since $2u \leqslant k - n \in 2\mathbf{Z}^{\mathbf{a}}$, we can take $\lambda = 3/2$. Writing simply α for $\alpha_{\lambda, q, \psi}$ there, we have

$$(7.5)\qquad \int_\Phi \overline{f(z/2)g(z)C^*(z, 3/4; k-l, \bar{\eta}\varepsilon\theta, \Gamma_\tau)}y^k\, d_H z$$

$$= A_f(\tau, \mathfrak{q}^{-1})\bar{\alpha}B(3/4, \tau, \eta)L_\mathfrak{f}(1, \eta)^{-1}\mathscr{D}^\mathfrak{f}(2; \bar{\chi}, \chi').$$

The last factor $\mathscr{D}^\mathfrak{f}(2; \bar{\chi}, \chi')$ is nonzero by Proposition 3.3, and $L_\mathfrak{f}(1, \eta)^{-1} \neq 0$ since $\eta \neq 1$; clearly $B(3/4, \tau, \eta) \neq 0$. Hence the right-hand side of (7.5) is $A_f(\tau, \mathfrak{q}^{-1})$ times a nonzero number. Put

$$h_\tau(z) = r_k(g(2z)C^*(2z, 3/4; k-l, \bar{\eta}\varepsilon\theta, \Gamma_\tau))$$

with r_k of (4.6). Observe that

$$\langle q(z/2), f(z/2)\rangle = 2^{\|k\|}\langle q, f\rangle$$

for any $q \in \mathscr{N}_k$. Take $y \in F_\mathbf{h}^\times$ so that $y\mathfrak{g} = \mathfrak{q}^{-1}$. Then from (2.11), (4.6), and (7.5) we obtain

$$(7.6)\qquad \langle h_\tau, f\rangle = \mu_f(\tau, y)\beta(\tau, y, \varphi, \eta)$$

with a nonzero number $\beta(\tau, y, \varphi, \eta)$ given by

$$(7.7)\qquad \beta(\tau, y, \varphi, \eta) = 2^{-\|k\|}\mathrm{vol}(\Gamma_\tau\backslash H^\mathbf{a})^{-1}\varphi(y)N(\mathfrak{q})^{-1} \cdot \bar{\alpha}B(3/4, \tau, \eta)L_\mathfrak{f}(1, \eta)^{-1}\mathscr{D}^\mathfrak{f}(2; \bar{\chi}, \chi').$$

Equality (7.6) holds for every $f \in \mathscr{S}_k(\mathfrak{b}, \mathfrak{c}, \varphi, |\chi^2|)$. (The character η is fixed.)

Suppose $f \neq 0$. Take t in $F_\mathbf{h}^\times$ so that $t^{-2}\mathfrak{b}\mathfrak{d}^{-1} \subset \mathfrak{g}$. As can be seen from (1.10a, e), we have $\mu_f(\tau, t) \neq 0$ for some τ such that $0 \ll \tau \in t^{-2}\mathfrak{b}\mathfrak{d}^{-1}$. If we consider the decomposition $\tau\mathfrak{b}^{-1}\mathfrak{d} = \mathfrak{q}^2\mathfrak{r}$ as before with this τ and put $\mathfrak{m} = t\mathfrak{q}$, we can easily verify that $\mathfrak{m}$ is integral and prime to $\mathfrak{f}$. We have then $A_f(\tau, \mathfrak{q}^{-1}\mathfrak{m}) \neq 0$, and hence $A_f(\tau, \mathfrak{q}^{-1}) \neq 0$ in view of Proposition 2.3. By (7.6) we have $\langle h_\tau, f\rangle \neq 0$. This shows that the h_τ for all τ such that $0 \ll \tau \in \mathfrak{g}$ span $\mathscr{S}_k(\mathfrak{b}, \mathfrak{c}, \varphi, |\chi^2|)$.

Now take another f' from $\mathscr{S}_k(\mathfrak{b}, \mathfrak{c}, \varphi, |\chi^2|)$ and another $\tau' \in \mathfrak{g}, \gg 0$. Assuming that $\langle h_\tau, f\rangle \neq 0$, from (7.6) we obtain

$$(7.8)\qquad \langle h_{\tau'}, f'\rangle/\langle h_\tau, f\rangle = [\mu(\tau', y'; f')/\mu(\tau, y; f)]\beta(\tau', y', \varphi, \eta)/\beta(\tau, y, \varphi, \eta),$$

where y' is defined for τ' in the same manner as y for τ. Let $\sigma \in \mathrm{Aut}(\bar{\mathbf{Q}})$. By (4.5) σ sends $\mathscr{S}_k(\mathfrak{b}, \mathfrak{c}, \varphi, |\chi^2|)$ onto $\mathscr{S}_{k\sigma}(\mathfrak{b}, \mathfrak{c}, \varphi^\sigma, |\chi^2|^\sigma)$. We now repeat the above computation with $f^\sigma, g^\sigma, \eta^\sigma, k\sigma$, and $l\sigma$ in place of f, g, η, k, and l. (The ideal $\mathfrak{f}$ stays the same.) Let h_τ^* be the function corresponding to h_τ defined for these new "transforms". By Lemma 4.6 and Proposition 5.2 we easily see that $h_\tau^* = (h_\tau)^\sigma$, and hence by (7.6) we have $\langle (h_\tau)^\sigma, f^\sigma\rangle = \langle h_\tau^*, f^\sigma\rangle = \mu(\tau, y; f^\sigma)\beta(\tau, y, \varphi^\sigma, \eta^\sigma)$, which is nonzero in view of

Lemma 4.1. Now from (7.7) and (7.4b) we can easily derive

$$[\beta(\tau', y', \varphi, \eta)/\beta(\tau, y, \varphi, \eta)]^\sigma = \beta(\tau', y', \varphi^\sigma, \eta^\sigma)/\beta(\tau, y, \varphi^\sigma, \eta^\sigma),$$

which combined with (7.8) and Lemma 4.1 shows that

$$(7.9)\qquad [\langle h_{\tau'}, f'\rangle/\langle h_\tau, f\rangle]^\sigma = \langle (h_{\tau'})^\sigma, f'^\sigma\rangle/\langle (h_\tau)^\sigma, f^\sigma\rangle.$$

Let ρ denote complex conjugation. Observe that $(h_\tau)^{\rho\sigma} = (h_\tau)^{\sigma\rho}$, since $\rho\sigma = \sigma\rho$ on any totally real field and any cyclotomic field.

Now take an arbitrary element j of $\mathcal{S}_k(\mathfrak{b}, \mathfrak{c}, \varphi, |\chi^2|)$. Then $j = \sum_{\xi \in X} a_\xi h_\xi$ with $a_\xi \in \mathbf{C}$ and a finite subset X of $\mathfrak{g}$ consisting of totally positive elements. Thus $\langle j, f'\rangle = \sum_\xi a_\xi^\rho \langle h_\xi, f'\rangle$, and so by (7.9) we have

$$(\langle j, f'\rangle/\langle h_\tau, f\rangle)^\sigma = \sum_\xi a_\xi^{\rho\sigma}\langle (h_\xi)^\sigma, f'^\sigma\rangle/\langle (h_\tau)^\sigma, f^\sigma\rangle$$

$$= \langle j^{\rho\sigma\rho}, f'^\sigma\rangle/\langle (h_\tau)^\sigma, f^\sigma\rangle.$$

In particular, take f to be h_τ. Then we obtain

$$(7.10)\qquad (\langle j, f'\rangle/\langle h_\tau, h_\tau\rangle)^\sigma = \langle j^{\rho\sigma\rho}, f'^\sigma\rangle/\langle (h_\tau)^\sigma, (h_\tau)^\sigma\rangle.$$

Take both j and f' to be a K-rational element p with K of Theorem 7.1. Then $p^{\rho\sigma} = p^{\sigma\rho}$, and hence

$$(\langle p, p\rangle/\langle h_\tau, h_\tau\rangle)^\sigma = \langle p^\sigma, p^\sigma\rangle/\langle (h_\tau)^\sigma, (h_\tau)^\sigma\rangle.$$

Combining this with (7.10), we obtain

$$(\langle j, f'\rangle/\langle p, p\rangle)^\sigma = \langle j^{\rho\sigma\rho}, f'^\sigma\rangle/\langle p^\sigma, p^\sigma\rangle,$$

which implies (7.1) and completes the proof of Theorem 7.1.

Let us now prove Theorem 7.2. As explained in Remark 7.4, we assume that $k_v - l_v \pmod 2$ is independent of v. We apply (6.11) to the functions f and g of Theorem 7.2. With the notation of (6.11) put

$$(7.11)\qquad w(z) = r_k(g(v_l z)C^*(v_l z, \lambda/2; k - l, \overline{\psi_0}, \Gamma)),$$

where $\psi_0 = \psi$ or $\psi\theta$ according as $l \in \mathbf{Z}^\mathbf{a}$ or $l \notin \mathbf{Z}^\mathbf{a}$. Fix any m in Λ_{k-l}. Write simply α for the constant $\alpha_{m,k-l,\eta}$ with $\eta = \overline{\psi_0}$. Evaluating (6.11) at $s = m/2$, we obtain

$$\bar{\alpha}\mathscr{D}_\mathfrak{b}(m/2; f, g; \psi)(4\pi/v_l)^{(2d-md-\|k+l\|)/2}\prod_{v\in\mathbf{a}}\Gamma((m-2+k_v+l_v)/2)$$

$$= D_F^{-1/2}N(v_l^{-1}\mathfrak{e})\mathrm{vol}(\Gamma\backslash H^\mathbf{a})v_l^{\|k\|}\langle w, f\rangle.$$

If we define $A(m/2;\dots)$ as in Theorem 7.2, then we see that

$$A(m/2; f, g; \psi) = B(k, l, m, \mathrm{e}, \psi, \Gamma)\langle w, f\rangle \tag{7.12}$$

with a quantity $B(\cdots)$ such that

$$B(k, l, m, \mathrm{e}, \psi, \Gamma)^{\sigma} = B(k\sigma, l\sigma, m, \mathrm{e}, \psi^{\sigma}, \Gamma) \tag{7.13}$$

for every $\sigma \in \mathrm{Gal}(\bar{\mathbf{Q}}/\mathbf{Q})$. This is straightforward except when $l \notin \mathbf{Z}^{\mathbf{a}}$ and $m \leqslant 1/2$, in which case we have to deal with the difference between $\gamma(\psi\theta)$ and $\gamma(\psi)$. However, we know that $\gamma(\theta) \in i^{d}\mathbf{Z}$ as proved in [13, Lemma 9.3], and thus $[i^{d}\gamma(\psi\theta)/\gamma(\psi)]^{\sigma} = i^{d}\gamma(\psi^{\sigma}\theta)/\gamma(\psi^{\sigma})$. Therefore, with $\beta_{m,\psi}$ defined as in Theorem 7.2, we obtain (7.13) in all cases. Replacing f, g, ψ by $f^{\sigma}, g^{\sigma}, \psi^{\sigma}$ in (7.12), we obtain

$$A(m/2; f^{\sigma}, g^{\sigma}; \psi^{\sigma}) = B(k\sigma, l\sigma, m, \mathrm{e}, \psi^{\sigma}, \Gamma)\langle w', f^{\sigma}\rangle, \tag{7.14}$$

where w' is defined by (7.11) with $k\sigma, l\sigma, g^{\sigma}, \psi_0^{\sigma}$ in place of k, l, g, ψ_0. By Lemma 4.6 and Proposition 5.2 we have $w' = w^{\sigma}$. The formula of Theorem 7.2 now follows from this combined with (7.1), (7.12), (7.13), and (7.14).

Theorem 7.3 in Case (1) follows immediately from Theorem 6.1 and (7.1). In Case (2), we take the residue of (6.11) at $s = 3/4$ and employ Proposition 5.2.

8. The critical values of $D(s, \chi, \eta)$ and the class $[\chi]$. The purpose of this section is to reformulate some results of [6] and also to introduce the class $[\chi]$ mentioned in Section 4. Given a system of eigenvalues χ occurring in $\mathscr{S}_k((\mathfrak{c}, \Psi))$ and a Hecke character η of F, we put

$$k_0 = \mathrm{Max}\{k_v \mid v \in \mathbf{a}\}, \qquad k^0 = \mathrm{Min}\{k_v \mid v \in \mathbf{a}\}, \tag{8.1a}$$

$$\kappa = \begin{cases} 0 & \text{if } k_0 \text{ is even}, \\ 1/2 & \text{if } k_0 \text{ is odd}, \end{cases} \tag{8.1b}$$

$$D(s, \chi, \eta) = \sum_{\mathfrak{a}} \eta^{*}(\mathfrak{a})\chi(\mathfrak{a})N(\mathfrak{a})^{-s-1}, \tag{8.2a}$$

$$A(t, \chi, \eta) = (2\pi i)^{-(t+\kappa)d}\gamma(\eta)^{-1}D(t, \chi, \eta), \tag{8.2b}$$

$$E(\chi) = (2\pi i)^{(1-2\kappa)d}\pi^{\|k\|}\gamma(\Psi)\langle \mathbf{f}, \mathbf{f}\rangle, \tag{8.2c}$$

where $\mathbf{f}$ is the normalized eigenform belonging to χ and t is an element of $2^{-1}\mathbf{Z}$ satisfying

$$2t \equiv k_v \pmod 2 \quad \text{and} \quad |2t| < k_v \quad \text{for every} \quad v \in \mathbf{a}. \tag{8.3}$$

Notice that the choice of the exponent $-s-1$ in (8.2a) makes 0 the center of the critical strip for the series D. A number t satisfying (8.3) exists if and only if $k^0 \geqslant 2$ and the following condition is satisfied:

(8.4) $\qquad k_v \pmod 2$ *is independent of* v.

Now, under the assumtion that Ψ is of finite order, we can show that for every $\sigma \in \mathrm{Gal}(\bar{\mathbf{Q}}/\mathbf{Q})$ *satisfying the condition*

(8.5) $$[\xi^{(k_0u-k)/2}]^\sigma = \xi^{(k_0u-k\sigma)/2} \quad \text{for all totally positive } \xi \text{ in } F,$$

there exists a system χ_σ *occurring in* $\mathscr{S}_{k\sigma}((\mathfrak{c}, \Psi^\sigma))$ *such that*

(8.6) $$[N_v^{k_0/2}\chi(v)]^\sigma = N_v^{k_0/2}\chi_\sigma(v)$$

for every $v \in \mathbf{h}$; *moreover* χ_σ *is primitive if and only if* χ *is primitive.* This was given in [6, Prop. 2.6]. If k_0 is even, we can write χ^σ for χ_σ for an obvious reason. We shall prove in Theorem 8.5 below a result which includes this fact as a special case. Notice that (8.4) implies (8.5) for every $\sigma \in \mathrm{Gal}(\bar{\mathbf{Q}}/\mathbf{Q})$, and hence χ_σ is always meaningful under (8.4).

THEOREM 8.1. *Let* χ *be a primitive system occurring in* $\mathscr{S}_k((\mathfrak{c}, \Psi))$. *Suppose that (8.4) is satisfied,* $k^0 \geqslant 2$, *and* Ψ *is of finite order. Then there exists a complex number* $V(\chi, r)$, *defined for each* $r \in \mathbf{Z}^{\mathbf{a}}/2\mathbf{Z}^{\mathbf{a}}$, *with the following properties:*

(I) *If* t *is as in (8.3) and* η *is a Hecke character of* F *such that* $\eta_{\mathbf{a}}(x) = \mathrm{sgn}(x)^{r+(t+\kappa)u}$, *then* $A(t, \chi, \eta)/V(\chi, r) \in \bar{\mathbf{Q}}$ *and*

$$[A(t, \chi, \eta)/V(\chi, r)]^\sigma = A(t, \chi_\sigma, \eta^\sigma)/V(\chi_\sigma, r)$$

for every $\sigma \in \mathrm{Gal}(\bar{\mathbf{Q}}/\mathbf{Q})$.

(II) *If* $q, r \in \mathbf{Z}^{\mathbf{a}}/2\mathbf{Z}^{\mathbf{a}}$ *and* $q + r \equiv u \pmod{2\mathbf{Z}^{\mathbf{a}}}$, *then* $V(\chi, q)V(\chi, r)/E(\chi) \in \bar{\mathbf{Q}}$ *and*

$$[V(\chi, q)V(\chi, r)/E(\chi)]^\sigma = V(\chi_\sigma, q)V(\chi_\sigma, r)/E(\chi_\sigma)$$

for every $\sigma \in \mathrm{Gal}(\bar{\mathbf{Q}}/\mathbf{Q})$.

If $k^0 \geqslant 3$ or $F = \mathbf{Q}$, this theorem is merely a reformulation of [5, Theorem 1] and [6, Theorem 4.3]. In fact, $D(s - k_0/2, \chi, \eta)$ coincides with $D(s, \mathbf{f}, \eta)$ of [6, (4.1b)]. Therefore putting $V(\chi_\sigma, r) = (2\pi i)^{((k_0/2)-\kappa)d}u(r + \kappa u - (k_0/2)u, \mathbf{f}^\sigma)$ with the symbol $u(*, *)$ of [6, Theorem 4.3] and the normalized eigenform $\mathbf{f}^\sigma$ belonging to χ_σ defined in [6, Prop. 2.6], we immediately obtain our theorem when $k^0 \geqslant 3$ or $F = \mathbf{Q}$. Now, to treat the case in which $k^0 = 2$ and $F \neq \mathbf{Q}$, we need a recent result of Rohrlich [2]. Let us now state a fact which easily follows from [2]:

LEMMA 8.2. *Given* r, l, *and* λ *as in Lemma 7.5, a primitive system of eigenvalues* χ, *and* $s_0 \in \mathbf{C}$, *there exists a Hecke character* ψ *such that* $D(s_0, \chi, \psi) \neq 0$, $\psi_{\mathbf{a}}(x) = \mathrm{sgn}(x)^r|x|^{i\lambda}$, *and* l *divides the conductor of* ψ.

Proof. Suppose that χ occurs in $\mathscr{S}_k((\mathfrak{c}, \Psi))$. By Lemma 7.5 we can find a Hecke character ω such that ω and ω^2 are ramified at the same finite places, $\omega_{\mathbf{a}}(x) = \operatorname{sgn}(x)^r|x|^{i\lambda}$, and $l\mathfrak{c}^2$ divides the conductor of ω^2. By Lemma 7.6 there is a system of eigenvalues χ_1 in $\mathscr{S}_k((\mathfrak{c}_1, \omega^2\Psi))$ such that $D(s, x, \omega) = D(s, \chi_1, \iota)$, where ι is the trivial character and $\mathfrak{c}_1$ is a multiple of $\mathfrak{c}$ whose prime factors divide the conductor of ω. Clearly $\omega^2\Psi$ has the same conductor as ω^2, and hence χ_1 is primitive. Applying Rohrlich's result in [2] to χ_1, we find a Hecke character θ such that $\theta_{\mathbf{a}} = 1$, the conductor of θ is prime to the conductor of ω, and $D(s_0, \chi_1, \theta) \neq 0$. Take $\psi = \theta\omega$. Then $D(s_0, \chi, \psi) = D(s_0, \chi_1, \theta) \neq 0$.

Coming back to Theorem 8.1, suppose that $k^0 = 2$. Then 0 is the only t satisfying (8.3). In [6, (4.16)] we proved the following fact: *if η and θ are Hecke characters of F such that $\eta_{\mathbf{a}}(x) = \operatorname{sgn}(x)^r$ and $\theta_{\mathbf{a}}(x) = \operatorname{sgn}(x)^{u-r}$, then*

$$(8.7) \qquad [A(0, \chi, \eta)A(0, \chi, \theta)/E(\chi)]^\sigma = A(0, \chi_\sigma, \eta^\sigma)A(0, \chi_\sigma, \theta^\sigma)/E(\chi_\sigma)$$

for every $\sigma \in$ Aut(**C**). Now, given any $r \in \mathbf{Z}^{\mathbf{a}}$, we can find, by virtue of Lemma 8.2, the above θ and η so that $D(0, \chi, \theta)D(0, \chi, \eta) \neq 0$. Then (8.7) shows that $D(0, \chi_\sigma, \theta^\sigma) \neq 0$. We now put $V(\chi_\sigma, r) = E(\chi_\sigma)/A(0, \chi_\sigma, \theta^\sigma)$ with such a θ. Then we obtain (I, II) of Theorem 8.1 for $k^0 = 2$ directly from (8.7).

Remark 8.3. Let χ, η, and χ_1 be as in Lemma 7.6. Clearly $D(s, \chi, \eta) = D(s, \chi_1, \iota)$. However, χ_1 may not be primitive even if χ is primitive. Let χ_2 be the primitive system attached to χ_1. Then

$$(8.8) \qquad D(s, \chi, \eta) = D(s, \chi_1, \iota) = D(s, \chi_2, \iota)P(s)$$

with a finite linear combination P of $N(\mathfrak{q})^{-s}$ with some integral ideals $\mathfrak{q}$. Indeed, let $\mathbf{f}_i$ be the normalized eigenform belonging to χ_i. Then $\mathbf{f}_1$ is a finite sum $\sum_{\mathfrak{q}} a_{\mathfrak{q}}\mathbf{f}_{\mathfrak{q}}$ with $a_{\mathfrak{q}} \in \mathbf{C}$ and forms $\mathbf{f}_{\mathfrak{q}}$ such that $c(\mathfrak{a}, \mathbf{f}_{\mathfrak{q}}) = c(\mathfrak{q}^{-1}\mathfrak{a}, \mathbf{f}_2)$ (see [6, Prop. 2.3 and p. 652]). Then $D(s, \chi_1, \iota) = \sum_{\mathfrak{q}} a_{\mathfrak{q}} N(\mathfrak{q})^{-s-k_0/2} D(s, \chi_2, \iota)$ as expected.

Proposition 8.4. *Let χ and Ψ be as in Theorem* 8.1. *Let χ' be another primitive system such that $\chi'(v) = \eta(\pi_v)\chi(v)$ for almost all $v \in \mathbf{h}$ with a Hecke character η such that $\eta_{\mathbf{a}}(x) = \operatorname{sgn}(x)^q$, $q \in \mathbf{Z}^{\mathbf{a}}$. Then*

$$[\gamma(\eta)V(\chi, r)/V(\chi', q + r)]^\sigma = \gamma(\eta^\sigma)V(\chi_\sigma, r)/V(\chi'_\sigma, q + r)$$

for every $\sigma \in$ Gal($\bar{\mathbf{Q}}/\mathbf{Q}$).

Proof. Let $\mathfrak{b}$ be the product of the conductors of χ and η. Take any t satisfying (8.3). By Lemma 8.2, for each $r \in \mathbf{Z}^{\mathbf{a}}/2\mathbf{Z}^{\mathbf{a}}$ there exists a Hecke character ψ such that $\psi_{\mathbf{a}}(x) = \operatorname{sgn}(x)^{q+r+(t+\kappa)u}$, $\mathfrak{b}^2$ divides the conductor of ψ, and $D(t, \chi', \psi) \neq 0$. Observe that $D(t, \chi, \eta\psi) = D(t, \chi', \psi)$ and $D(t, \chi_\sigma, \eta^\sigma\psi^\sigma) = D(t, \chi'_\sigma, \psi^\sigma)$. Thus we have

$$\frac{\gamma(\eta)V(\chi, r)}{V(\chi', q + r)} = \frac{A(t, \chi', \psi)}{V(\chi', q + r)} \cdot \frac{V(\chi, r)}{A(t, \chi, \eta\psi)} \cdot \frac{\gamma(\eta)\gamma(\psi)}{\gamma(\eta\psi)}.$$

Applying σ to this equality and invoking [6, Lemma 4.12], we obtain the desired formula.

Let us now turn to the problem of defining a certain class $[\chi]$ for χ occurring in $\mathscr{S}_k((\mathfrak{c}, \Psi))$ with Ψ of finite or infinite order. Take λ as in (2.4). For $a \in \mathfrak{g}^\times$ we have $|a|^{2i\lambda} = \operatorname{sgn}(a)^k \Psi_{\mathbf{h}}(a)^{-1}$, and hence $|a|^{i\lambda}$ is a root of unity. Now, let σ be an element of $\operatorname{Gal}(\bar{\mathbf{Q}}/\mathbf{Q})$ satisfying (8.5). As noted in [6, Remark 2.7], we have

$$k \equiv k\sigma \pmod{2\mathbf{Z}^{\mathbf{a}}}. \tag{8.9}$$

Given $r, s \in \mathbf{Z}^{\mathbf{a}}$ such that $\|r\| = \|s\|$, we take two Hecke characters ψ and ψ' of F such that

(8.10)

$$\psi_{\mathbf{a}}(x) = \operatorname{sgn}(x)^r |x|^{i\lambda}, \qquad \psi'_{\mathbf{a}}(x) = \operatorname{sgn}(x)^s |x|^{i\lambda'} \textit{ with some } \lambda' \in \mathbf{R}^{\mathbf{a}}, \ \|\lambda'\| = 0, \qquad \textit{and}$$

$$\psi' = \psi^\sigma \text{ on } \prod_{v \in \mathbf{h}} \mathfrak{g}_v^\times .$$

Lemma 7.5 guarantees the existence of ψ such that $\psi_{\mathbf{a}}(x) = \operatorname{sgn}(x)^r |x|^{i\lambda}$. Once ψ is chosen, take λ' so that $\operatorname{sgn}(a)^s \psi_{\mathbf{h}}(a)^\sigma = |a|^{-i\lambda'}$ for every $a \in \mathfrak{g}^\times$. Then define ψ' on $F_{\mathbf{a}}^\times \prod_{v \in \mathbf{h}} \mathfrak{g}_v^\times$ by $\psi'(x) = \operatorname{sgn}(x)^s |x|^{i\lambda'} \psi_{\mathbf{h}}(x)^\sigma$ and extend it to a Hecke character of F. Now $\psi^{-2}\Psi$ is of finite order, and hence $(\psi^{-2}\Psi)^\sigma$ is meaningful as a Hecke character of F. Put

$$\Psi' = \psi'^2 (\psi^{-2}\Psi)^\sigma. \tag{8.11}$$

Then Ψ' is a Hecke character of F and satisfies (2.4) with $k\sigma$ and λ' in place of k and λ; $\Psi' = \Psi^\sigma$ on $\prod_{v \in \mathbf{h}} \mathfrak{g}_v^\times$; thus $\mathfrak{c}$ remains the same.

Theorem 8.5. *Let χ be a system of eigenvalues occurring in $\mathscr{S}_k((\mathfrak{c}, \Psi))$ with Ψ as in (2.4), and ψ a Hecke character of F such that $\psi^{-2}\Psi$ is of finite order. Then $\psi(\pi_v)^{-1}\chi(v)$ is algebraic for every $v \in \mathbf{h}$. Moreover let σ be an element of* $\operatorname{Gal}(\bar{\mathbf{Q}}/\mathbf{Q})$ *satisfying (8.5). Then, with ψ' and Ψ' as in (8.10) and (8.11), there exists a system of eigenvalues χ' occurring in $\mathscr{S}_{k\sigma}((\mathfrak{c}, \Psi'))$ such that*

$$[\psi(\pi_v)^{-1} N_v^{k_0/2} \chi(v)]^\sigma = \psi'(\pi_v)^{-1} N_v^{k_0/2} \chi'(v) \tag{8.12}$$

for every $v \in \mathbf{h}$. Furthermore, χ' is primitive if and only if χ is primitive.

Proof. We modify the proof of [6, Prop. 2.6] which concerns the case where Ψ is of finite order. We take Q of (2.6) so that it consists of elements of the form $\operatorname{diag}[1, t]$ with $t \in F_{\mathbf{h}}^\times$. Extend σ to an element of $\operatorname{Aut}(\mathbf{C})$ and denote it also by σ. Let $\mathfrak{f} = (f)_{q \in Q} \in \mathscr{S}_k((\mathfrak{c}, \varphi, \lambda))$ and let $\psi_{\mathbf{h}}(\det(y_0)^{-1})\mathfrak{f}|\tilde{D}_{\mathfrak{c}} y_0 \tilde{D}_{\mathfrak{c}} = (f'_q)_{q \in Q}$ with a fixed y_0 in the set Y of Section 2. Put $h_q = \psi_{\mathbf{h}}(\det(q)) f_q$ and $h'_q = \psi_{\mathbf{h}}(\det(q)) f'_q$. Fix an element

q of Q and take p, α_0, and A as in (4.9). Then

$$h'_p = \sum_{\alpha \in A} C_\alpha h_q \|_k \alpha \tag{8.13}$$

with $C_\alpha = \varphi(a_\alpha)^{-1}\psi_{\mathbf{h}}(\det(y_0^{-1}q^{-1}\alpha p))$. Since $\det(y_0^{-1}q^{-1}\alpha p) \in \det(\tilde{D}_c)$, we see that C_α is a root of unity. Now $\det(\gamma)^{i\lambda}$ in (2.3) is a root of unity, so that $\mathscr{S}_k(\bar{\Gamma}_p, \varphi, \lambda)$ is spanned by $\bar{\mathbf{Q}}$-rational elements. Let $\mathscr{U}$ be the subspace of $\mathscr{S}_k((\mathfrak{c}, \varphi, \lambda)) = \prod_{q \in Q} \mathscr{S}_k(\bar{\Gamma}_q, \varphi, \lambda)$ consisting of the elements $(f_q)_{q \in Q}$ such that $\psi_{\mathbf{h}}(\det(q))f_q$ is $\bar{\mathbf{Q}}$-rational for every $q \in Q$. Then $\mathscr{U} \otimes_{\bar{\mathbf{Q}}} \mathbf{C} = \mathscr{S}_k((\mathfrak{c}, \varphi, \lambda))$, and moreover, $\mathscr{U}$ is stable under $\psi_{\mathbf{h}}(\det(y_0))^{-1}\tilde{D}_c y_0 \tilde{D}_c$ because of relation (8.13). This proves that $\psi(\pi_v)^{-1}\chi(v)$ is algebraic for every $v \in \mathbf{h}$. Now we have

$$h_q \|_k \gamma = \varphi(a_\gamma)\psi_{\mathbf{h}}(\det(\gamma))^{-1}h_q \qquad \textit{for every} \qquad \gamma \in \bar{\Gamma}_q.$$

By [6, Prop. 1.6] we have

$$h_q^\sigma \|_{k\sigma} \gamma = \det(\gamma)^{k\sigma/2}[\det(\gamma)^{-k/2}]^\sigma \varphi^\sigma(a_\gamma)\psi'_{\mathbf{h}}(\det(\gamma))^{-1}h_q^\sigma$$

for such a γ. Now (8.5) implies that for every $e \in \mathfrak{g}_+^\times$ we have $e^{k\sigma/2} = (e^{k/2})^\sigma$, and hence $h_q^\sigma \in \mathscr{S}_{k\sigma}(\bar{\Gamma}_q, \varphi^\sigma, \lambda')$. We now apply σ to (8.13). To make our computation simpler, we first observe that A in (8.13) can be chosen so that $\det(\alpha) = \det(\alpha_0)$ for every $\alpha \in A$. Moreover, the reasoning of [6, p. 653, lines 7–14] shows that there is a permutation $\alpha \mapsto \beta_\alpha$ of A and an element γ_α of $\bar{\Gamma}_q \cap G$ for each $\alpha \in A$ such that

$$[\det(\alpha)^{-k/2}h_q\|_k\alpha]^\sigma = \det(\gamma_\alpha\beta_\alpha)^{-k\sigma/2}h_q^\sigma\|_{k\sigma}\gamma_\alpha\beta_\alpha$$

and $a_\alpha \equiv a(\gamma_\alpha\beta_\alpha) \pmod{\mathfrak{c}}$. Thus we have

$$\begin{aligned}
[\det(\alpha_0)^{-k_0u/2}h'_p]^\sigma &= \sum_{\alpha \in A} [\det(\alpha)^{(k-k_0u)/2}]^\sigma C_\alpha^\sigma[\det(\alpha)^{-k/2}h_q\|_k\alpha]^\sigma \\
&= \sum_{\alpha \in A} \det(\alpha)^{(k\sigma-k_0u)/2}C_\alpha^\sigma \det(\gamma_\alpha\beta_\alpha)^{-k\sigma/2}h_q^\sigma\|_{k\sigma}\gamma_\alpha\beta_\alpha \\
&= \sum_{\alpha \in A} \det(\beta_\alpha)^{-k_0u/2}\varphi^\sigma(a_\alpha)^{-1}\varphi^\sigma(a(\gamma_\alpha))\psi'_{\mathbf{h}}(\det(y_0^{-1}q^{-1}\beta_\alpha p))h_q^\sigma\|_{k\sigma}\beta_\alpha \\
&= \det(\alpha_0)^{-k_0u/2} \sum_{\alpha \in A} \varphi^\sigma(a(\beta_\alpha))\psi'_{\mathbf{h}}(\det(y_0^{-1}q^{-1}\beta_\alpha p))h_q^\sigma\|_{k\sigma}\beta_\alpha.
\end{aligned}$$

Let $N_q = N(\det(q)\mathfrak{g})$, $M = N(\det(y_0)\mathfrak{g})$, $g_q = (N_q^{k_0/2})^\sigma N_q^{-k_0/2}h_q^\sigma$, and $\mathbf{g} = (\psi'_{\mathbf{h}}(\det(q)^{-1}g_q)_{q \in Q}$; define similarly $\mathbf{g}'$ with $h_q'^\sigma$ in place of h_q^σ. Observing that $\det(\alpha_0)^u N_p = MN_q$, we obtain

$$\mathbf{g}' = (M^{k_0/2})^\sigma M^{-k_0/2}\psi'_{\mathbf{h}}(\det(y_0))^{-1}\mathbf{g}|\tilde{D}_c y_0 \tilde{D}_c.$$

Suppose $f|\tilde{D}_c y_0 \tilde{D}_c = \omega f$. Then $h_q'^{\sigma} = [\psi_{\mathbf{h}}(\det(y_0))^{-1}\omega]^{\sigma} h_q^{\sigma}$, and hence $\mathbf{g}|\tilde{D}_c y_0 \tilde{D}_c = \omega' \mathbf{g}$ with

$$\omega' = \psi_{\mathbf{h}}(\det(y_0)^{-1})^{\sigma}\psi_{\mathbf{h}}'(\det(y_0))M^{k_0/2}(M^{-k_0/2})^{\sigma}\omega^{\sigma}.$$

Taking $y_0 = \pi_v$ for $v \nmid c$, we see that

$$\mathbf{g}|\mathfrak{S}_v = \psi(\pi_v^{-2})^{\sigma}\psi'(\pi_v)^2\Psi(\pi_v)^{\sigma}\mathbf{g} = \Psi'(\pi_v)\mathbf{g},$$

and hence $\mathbf{g} \in \mathscr{S}_{k\sigma}((c, \Psi'))$. Similarly, taking $y_0 = \mathrm{diag}[1, \pi_v]$, we obtain $\mathbf{g}|\mathfrak{T}_v = \chi'(v)\mathbf{g}$ with $\chi'(v)$ as in (8.12). This proves our theorem except the last assertion, which can easily be seen since we can go back from χ' to χ.

In the above theorem, Ψ' and χ' depend on the choice of ψ and ψ'. (If Ψ is of finite order, we can take $\psi = \psi' = 1$. Then χ' is uniquely determined by χ and σ, and occurs in $\mathscr{S}_{k\sigma}((c, \Psi^{\sigma}))$.) To deal with this ambiguity of Ψ' and χ', we consider the set of all χ_1 such that $\chi_1(\mathfrak{a}) = \theta^*(\mathfrak{a})\chi(\mathfrak{a})$ for every integral $\mathfrak{a}$ with a *completely unramified* Hecke character θ of F, by which we mean a character such that $\theta_{\mathbf{a}}(x) = |x|^{i\kappa}$ with some $\kappa \in \mathbf{R}^{\mathbf{a}}$, $\|\kappa\| = 0$, and $\theta(\mathfrak{g}_v^{\times}) = 1$ for every $v \in \mathbf{h}$. We denote this set by $[\chi]$. Now, if we fix r and s in (8.10) and Theorem 8.5, then we can easily verify that $[\chi']$ with χ' of Theorem 8.5 is completely determined by $[\chi]$, r, s, and σ, and so we denote it by $[\chi]^{(\sigma,r,s)}$. It should be remembered that σ must satisfy condition (8.5), which is the case if (8.4) is satisfied.

Remark 8.6. Coming back to the setting of Theorem 8.1, let us say that χ is of *finite type* if Ψ is of finite order and k satisfies (8.4). From Proposition 8.4 we see that if $\chi \in [\chi']$ and both χ and χ' are of finite type, then

$$[V(\chi, r)/V(\chi', r)]^{\sigma} = V(\chi_{\sigma}, r)/V(\chi'_{\sigma}, r). \tag{8.14}$$

Also, we have $[\chi']^{(\sigma,r,r)} = [\chi'_{\sigma}]$. Therefore, if we denote $V(\chi', r)$ and $E(\chi')$ by $V(C, r)$ and $E(C)$ for $C = [\chi']$, then Theorem 8.1 can be restated, with these new symbols $V(C, r)$ and $E(C)$, as two assertions valid for every χ of finite type in C.

9. Main theorems for half-integral k. In this section k is half-integral. We consider $\mathscr{M}_k(2^{-1}\mathfrak{d}, \mathfrak{c}, \varphi)$ with an integral ideal $\mathfrak{c}$ divisible by 4 and a character φ of $(\mathfrak{g}/\mathfrak{c})^{\times}$ such that $\varphi(-1) = (-1)^{\|[k]\|}$. Given any $r \in \mathbf{Z}^{\mathbf{a}}$ such that $\|r\| \equiv \|[k]\| \pmod 2$, we can find an element κ of $\mathbf{R}^{\mathbf{a}}$ such that $\|\kappa\| = 0$ and

$$\varphi(e) = \mathrm{sgn}(e)^r|e|^{-i\kappa} \quad \textit{for every} \quad e \in \mathfrak{g}^{\times}. \tag{9.1}$$

Then there exists a Hecke character ψ of F such that $\psi_{\mathbf{a}}(x) = \mathrm{sgn}(x)^r|x|^{i\kappa}$ and $\psi = \varphi$ on $\prod_{v \in \mathbf{h}} \mathfrak{g}_v^{\times}$ when φ is extended as in Section 1. With any choice of such a ψ, $\mathscr{M}_k(2^{-1}\mathfrak{d}, \mathfrak{c}, \varphi)$ coincides with $\mathscr{M}_k(\mathfrak{c}, \psi)$ of [13]. Put $D = D[2\mathfrak{d}^{-1}, 2^{-1}\mathfrak{d}\mathfrak{c}]$, $\Gamma = G \cap D$, and $U = \mathrm{pr}^{-1}(D)$, where pr is the projection map of the metaplectic group M_A onto

G_A. Fix a prime v in $\mathbf{h}$ and put $\sigma = r_P(\mathrm{diag}[\pi_v^{-1}, \pi_v])$ with the lifting $r_P\colon P_A \to M_A$ of [13, (1.9)]. Taking a coset decomposition $G \cap U\sigma U = \bigsqcup_{\alpha\in A} \Gamma\alpha$, we define, as in [13, (5.5)], a Hecke operator T_v on $\mathcal{M}_k(\mathfrak{c}, \psi)$ by

$$(9.2)\qquad (f\,|\,T_v)(z) = N_v^{-3/2}(\psi/\psi_{\mathfrak{c}})(\pi_v) \sum_{\alpha\in A} \psi_{\mathfrak{c}}(a_\alpha)^{-1} J_{\Xi}(\alpha, z)^{-1} f(\alpha z) \qquad (z \in H^{\mathbf{a}})$$

for $f \in \mathcal{M}_k(\mathfrak{c}, \psi)$, where J_Ξ is defined by [13, (5.3)] (independently of ψ). This is well defined even when $v\,|\,\mathfrak{c}$. Suppose that $v \nmid \mathfrak{c}$; put $Q_v = \psi(\pi_v)^{-1} T_v$. Then

$$(9.3)\qquad (f\,|\,Q_v)(z) = N_v^{-3/2} \sum_{\alpha\in A} \varphi(a_\alpha)^{-1} J_{\Xi}(\alpha, z)^{-1} f(\alpha z) \qquad (z \in H^{\mathbf{a}}),$$

and hence Q_v is an operator on $\mathcal{M}_k(2^{-1}\mathfrak{d}, \mathfrak{c}, \varphi)$ defined independently of the choice of ψ. From [13, Prop. 5.3] we obtain

$$(9.4)\qquad \langle f\,|\,Q_v, g\rangle = \langle f, g\,|\,Q_v\rangle$$

for every $v \nmid \mathfrak{c}$, $f \in \mathcal{M}_k(2^{-1}\mathfrak{d}, \mathfrak{c}, \varphi)$, and $g \in \mathcal{S}_k(2^{-1}\mathfrak{d}, \mathfrak{c}, \varphi)$. We now reformulate a main result of [13] as follows:

Proposition 9.1. *Let $f \in \mathcal{S}_k(2^{-1}\mathfrak{d}, \mathfrak{c}, \varphi)$ and let $\mathfrak{a} = 2^{-1}\mathfrak{c} \cap \mathfrak{c}_1^2$ with the conductor $\mathfrak{c}_1$ of φ; let $\mathfrak{t}$ be a multiple of $\mathfrak{c}$. Suppose that $k \geqslant 3u/2$, $f\,|\,Q_v = \omega_v f$ with $\omega_v \in \mathbf{C}$ for every $v \nmid \mathfrak{t}$, and the following condition is satsfied:*

(9.5) *$k_v > 3/2$ for some $v \in \mathbf{a}$, or $k = 3u/2$ and f is orthogonal to the theta series of the type described in [13, Theorem 6.1(III)].*

Then there exists a primitive system of eigenvalues χ in $\mathcal{S}_{2k-u}((\mathfrak{a}', \iota))$ with a divisor $\mathfrak{a}'$ of $\mathfrak{a}$ such that $\chi(v) = N_v\omega_v$ for every $v \nmid \mathfrak{t}$, where ι denotes the trivial Hecke character of F. Moreover $f\,|\,Q_v = N_v^{-1}\chi(v) f$ for every $v \nmid \mathfrak{c}$.

Proof. Take any Hecke character ψ as above. By [13, Lemma 8.8, Theorem 6.1(III)] there exists a system of eigenvalues χ_0 in $\mathcal{S}_{2k-u}((2^{-1}\mathfrak{c}, \psi^2))$ such that $\chi_0(v) = \psi(\pi_v)N_v\omega_v$ for $v \nmid \mathfrak{t}$. By [13, Prop. 8.9(1)] we have $f\,|\,Q_v = N_v^{-1}\psi(\pi_v)^{-1}\cdot\chi_0(v)$ for every $v \nmid \mathfrak{c}$. Taking ψ^{-1} to be η of Lemma 7.6, we find a system χ_1 in $\mathcal{S}_{2k-u}((\mathfrak{a}, \iota))$ such that $\chi_1(v) = \psi(\pi_v)^{-1}\chi_0(v)$ for every $v \nmid \mathfrak{c}$. Replacing χ_1 by a primitive χ associated with it, we obtain our proposition.

The difference of this new formulation from the old one in [13] is that χ is completely determined by a given eigenform f without any choice of a Hecke character ψ.

The notation being as in the above proposition, we denote by $\mathcal{S}_k(\mathfrak{c}, \varphi, \chi)$ the set of all f in $\mathcal{S}_k(2^{-1}\mathfrak{d}, \mathfrak{c}, \varphi)$ such that $f\,|\,Q_v = N_v^{-1}\chi(v) f$ for every $v \nmid \mathfrak{c}$, which is the case if $f\,|\,Q_v = N_v^{-1}\chi(v) f$ for almost all v as proved in the proposition. We can now state our main theorems for half-integral k:

THEOREM 9.2. *For $f \in \mathscr{S}_k(\mathfrak{c}, \varphi, \chi)$ and $q \in \mathscr{S}_k$ put*

$$I(q, f) = 2^{d/2} i^d \pi^{\|k\|+d/2} \langle q, f \rangle.$$

Then for every $\sigma \in \mathrm{Aut}(\mathbf{C})$ we have

$$[I(q, f)/V(\chi, u)]^\sigma = I(q^{\rho\sigma\rho}, f^\sigma)/V(\chi^\sigma, u),$$

where V is the constant of Theorem 8.1.

We shall see in Proposition 9.9 below that f^σ belongs to $\mathscr{S}_{k\sigma}(\mathfrak{c}, \varphi, \chi^\sigma)$, and hence $I(q^{\rho\sigma\rho}, f^\sigma)$ is meaningful.

THEOREM 9.3. *Let $f \in \mathscr{S}_k(\mathfrak{c}, \varphi, \chi)$ and $g \in \mathscr{M}_l(\mathfrak{b}', \mathfrak{c}', \varphi')$ with integral or half-integral l. Let ψ be a Hecke character of F such that $\psi(x) = \mathrm{sgn}(x)^{[k-l]} \cdot (\varphi/\varphi')(x_{\mathfrak{h}})$ for every $x \in F_{\mathbf{a}}^\times \prod_{v \in \mathbf{h}} \mathfrak{g}_v^\times$. Define Λ_q as in Proposition 5.2 and put*

$$A(m/2; f, g; \psi) = \beta_{m,\psi} \mathscr{D}_{\mathfrak{h}}(m/2; f, g; \psi)$$

for every $m \in \Lambda_{k-l}$ with $\mathfrak{h}$ of (6.2) and

$$\beta_{m,\psi} = \begin{cases} \gamma(\psi)^{-1} i^{d+\|k-l\|} \pi^{d-md} & \text{if } l \notin \mathbf{Z}^{\mathbf{a}}, \\ \gamma(\psi^2)^{-1} i^d \pi^{2d(1-m)} & \text{if } l \in \mathbf{Z}^{\mathbf{a}}, m \geqslant 3/2, \\ \gamma(\psi)^{-1} i^{\|k-l\|+d/2} D_F^{1/2} \pi^{(d/2)-md} & \text{if } l \in \mathbf{Z}^{\mathbf{a}}, m \leqslant 1/2, \end{cases}$$

where $d = [F : \mathbf{Q}]$ and $\gamma(\ )$ is defined by (5.7). Exclude the case in which $F = \mathbf{Q}$, $\psi^2 = 1$, and $m - 3/2$ simultaneously. Then

$$[A(m/2; f, g; \psi)/V(\chi, u)]^\sigma = A(m/2; f^\sigma, g^{\rho\sigma\rho}; \psi^\sigma)/V(\chi^\sigma, u)$$

for every $\sigma \in \mathrm{Aut}(\mathbf{C})$.

THEOREM 9.4. *Let $R(f, g, \psi)$ be the residue of $D(s; f, g; \psi)$ at $s = s_0$ with f, g, and ψ as in Theorem 9.3 in the following two cases:*

(1) $k = l$, $\varphi = \varphi'$, $\psi = 1$, *and* $s_0 = 1$;
(2) $0 \leqslant k - l - u/2 \in 2\mathbf{Z}^{\mathbf{a}}$, $\psi^2 = 1$, *and* $s_0 = 3/4$.

Put $R^*(f, g, \psi) = \varepsilon R(f, g, \psi)$ with

$$\varepsilon = \begin{cases} R_F^{-1} i^d \pi^d & \text{in Case (1)}, \\ \gamma(\psi)^{-1} i^d D_F^{1/2} R_F^{-1} \pi^d & \text{in Case (2)}. \end{cases}$$

Then

$$[R^*(f, g, \psi)/V(\chi, u)]^\sigma = R^*(f^\sigma, g^{\rho\sigma\rho}, \psi)/V(\chi^\sigma, u)$$

for every $\sigma \in \mathrm{Aut}(\mathbf{C})$.

The above theorems give improvements on the previous results in [8, 13] and Im [1]. In all these papers, it was assumed that φ coincides with a Hecke character of finite order on $\prod_{v \in \mathbf{h}} \mathfrak{g}_v^\times$ and the same is true for φ'. Here we have replaced that condition by a weaker one on φ/φ', which seems absolutely necessary.

We begin with some auxiliary lemmas.

LEMMA 9.5. *Let* $f \in \mathcal{M}_k(2^{-1}\mathfrak{d}, \mathfrak{c}, \varphi)$. *Then for every* $v \nmid \mathfrak{c}$, $t \in F_\mathbf{h}^\times$, *and* $\xi \in t^{-2}\mathfrak{g}$ *we have*

$$\mu(\xi, t; f|Q_v) = \mu_f(\xi, \pi_v t) + N_v^{-1}\left(\frac{\xi t_v^2}{v}\right)\mu_f(\xi, t) + N_v^{-1}\mu_f(\xi, t/\pi_v),$$

where (d/v) *denotes the quadratic residue symbol, that is, the number of solutions of* $x^2 \equiv d \pmod{\pi_v \mathfrak{g}_v}$ *minus* 1.

This is merely a reformulation of [13, Prop. 5.4].

LEMMA 9.6. *Let* $f \in \mathcal{M}_k(2^{-1}\mathfrak{d}, \mathfrak{c}, \varphi)$ *and let* $\mathfrak{t}$ *be a multiple of* $\mathfrak{c}$. *Suppose that* $f|Q_v = \omega_v f$ *with* $\omega_v \in \mathbf{C}$ *for every* $v \nmid \mathfrak{t}$. *Put* $A_f(\xi, \mathfrak{m}) = \varphi(t)\mu_f(\xi, t)$ *for every ideal* $\mathfrak{m} = t\mathfrak{g}$ *prime to* $\mathfrak{c}$ *with* $t \in F_\mathbf{h}^\times$. *Then this is well defined. Moreover, let* $0 \ll \tau \in \mathfrak{g}$ *and* $\tau\mathfrak{g} = \mathfrak{q}^2\mathfrak{r}$ *with integral ideals* $\mathfrak{q}$ *and* $\mathfrak{r}$ *such that* $\mathfrak{q}$ *is prime to* $\mathfrak{t}$ *and* $\mathfrak{r}$ *has no square factor prime to* $\mathfrak{t}$. *Then*

$$Z_\mathfrak{t}(\{\varepsilon_\tau(\pi_v)N_v^{-1}M_v\}) \sum_{\mathfrak{m}+\mathfrak{t}=\mathfrak{g}} A_f(\tau, \mathfrak{q}^{-1}\mathfrak{m})M(\mathfrak{m})$$
$$= A_f(\tau, \mathfrak{q}^{-1}) \prod_{v \nmid \mathfrak{t}} [1 - \omega_v M_v + N_v^{-1}M_v^2]^{-1},$$

where ε_τ *is the Hecke character of* F *corresponding to* $F(\tau^{1/2})$.

This is similar to Proposition 2.3 and follows from Lemma 9.5; in fact, it is a reformulation of [13, Theorem 5.5].

LEMMA 9.7. *For* $f \in \mathcal{M}_k(2^{-1}\mathfrak{d}, \mathfrak{c}, \varphi)$, $v \nmid \mathfrak{c}$, *and* $\sigma \in \mathrm{Aut}(\mathbf{C})$ *we have* $(f|\mathbf{Q}_v)^\sigma = f^\sigma|Q_v$.

This follows immediately from Lemmas 4.1 and 9.5 (cf. formula (4.1)).

LEMMA 9.8. *Let* $f \in \mathcal{M}_k(2^{-1}\mathfrak{d}, \mathfrak{c}, \varphi)$ *and let* g *be the element of* $\mathcal{M}_k(2^{-1}\mathfrak{d}, \tau q^{-2}\mathfrak{c}, \varphi\varepsilon_\mathbf{h})$ *as in Proposition* 1.2. *If* $f|Q_v = \omega_v f$ *for a prime* v *not dividing* $\tau q^{-2}\mathfrak{c}$, *then* $g|Q_v = \varepsilon(\pi_v)\omega_v g$.

This follows easily from Lemma 9.5.

PROPOSITION 9.9. (i) *$\mathscr{S}_k(\mathfrak{c}, \varphi, \chi)^\sigma = \mathscr{S}_{k\sigma}(\mathfrak{c}, \varphi^\sigma, \chi^\sigma)$ for every $\sigma \in \mathrm{Aut}(\mathbf{C})$.*

(ii) *Define $F_k(\varphi)$ as in Proposition 4.5; let $F_k(\varphi, \chi)$ denote the subfield of $\bar{\mathbf{Q}}$ such that an element σ of $\mathrm{Gal}(\bar{\mathbf{Q}}/F_k(\varphi))$ is the identity on $F_k(\varphi, \chi)$ if and only if $\chi^\sigma = \chi$. Then $F_k(\varphi, \chi)$ coincides with the smallest extension of $F_k(\varphi)$ containing the $\chi(v)$ for almost all v.*

(iii) *$F_k(\varphi, \chi)$ is finite over $\mathbf{Q}$, and is totally real or a CM-field.*

(iv) *$\mathscr{S}_k(\mathfrak{c}, \varphi, \chi)$ is spanned by $F_k(\varphi, \chi)$-rational elements.*

Proof. The first assertion follows immediately from (4.1) and Lemma 9.7. The remaining part can be proved in the same manner as in Proposition 4.5, in view of (9.4), Lemmas 9.5 and 4.4, and the commutativity of the Q_v (see [13, Prop. 5.2]).

PROPOSITION 9.10. *Given $\mathfrak{c}$, φ, and χ, there exists a $\mathbf{C}$-linear map r_k of $\mathscr{N}_k$ into $\mathscr{S}_k(\mathfrak{c}, \varphi, \chi)$ with the following properties:*

(9.6) $\langle r_k(f), g\rangle = \langle f, g\rangle$ *for* $f \in \mathscr{N}_k$ *and* $g \in \mathscr{S}_k(\mathfrak{c}, \varphi, \chi)$,

(9.7) $r_k(f)^\sigma = r_{k\sigma}(f^\sigma)$ *for every* $\sigma \in \mathrm{Aut}(\mathbf{C})$,

where $r_{k\sigma}$ is the projection map of $\mathscr{N}_{k\sigma}$ into $\mathscr{S}_{k\sigma}(\mathfrak{c}, \varphi^\sigma, \chi^\sigma)$.

Proof. This was proved in [13, (9.5), (9.7)] when ψ of (9.2) is of finite order. The same proof essentially applies to the general case. In fact, r_k can be obtained as the composite $q_k \circ p_k$ with the map p_k of $\mathscr{N}_k$ into $\mathscr{S}_k$ obtained in [13, Proposition 9.4] and the orthogonal projection q_k of $\mathscr{S}_k$ into $\mathscr{S}_k(\mathfrak{c}, \varphi, \chi)$. The only point to be shown is $q_k(f)^\sigma = q_{k\sigma}(f^\sigma)$. We first observe that if $k \neq 3u/2$, $\mathscr{S}_k(2^{-1}\mathfrak{d}, \mathfrak{c}, \varphi)$ is the direct sum of the subspaces $\mathscr{S}_k(\mathfrak{c}, \varphi, \chi)$ with finitely many different χ's and these subspaces are mutually orthogonal. This follows from (9.4), since the Q_v are mutually commutative. If $k = 3u/2$, we need, for the reason explained in [13, Theorem 6.1(III)], the subspace of $\mathscr{S}_k(2^{-1}\mathfrak{d}, \mathfrak{c}, \varphi)$ spanned by theta series, whose orthogonal complement gives the sum of $\mathscr{S}_k(\mathfrak{c}, \varphi, \chi)$. In any case, if we denote by $\mathscr{R}_k(\mathfrak{c}, \varphi, \chi)$ the orthogonal complement of $\mathscr{S}_k(\mathfrak{c}, \varphi, \chi)$ in $\mathscr{S}_k(2^{-1}\mathfrak{d}, \mathfrak{c}, \varphi)$, then $\mathscr{R}_k(\mathfrak{c}, \varphi, \chi)^\sigma = \mathscr{R}_{k\sigma}(\mathfrak{c}, \varphi^\sigma, \chi^\sigma)$. Now let $W_k(\mathfrak{c}, \varphi)$ denote the orthogonal complement of $\mathscr{S}_k(2^{-1}\mathfrak{d}, \mathfrak{c}, \varphi)$ in $\mathscr{S}_k$. Then $W_k(\mathfrak{c}, \varphi)^\sigma = W_{k\sigma}(\mathfrak{c}, \varphi^\sigma)$. This is proved in [13, p. 817], where ψ is assumed to be of finite order, but the same proof is applicable to the present $W_k(\mathfrak{c}, \varphi)$, since $\psi_\mathfrak{c}$ in the definition of $P(f)$ there can be replaced by φ. Put $\mathscr{T}_k(\mathfrak{c}, \varphi, \chi) = \mathscr{R}_k(\mathfrak{c}, \varphi, \chi) + W_k(\mathfrak{c}, \varphi)$. Then this sum is the orthogonal complement of $\mathscr{S}_k(\mathfrak{c}, \varphi, \chi)$ in $\mathscr{S}_k$ and $\mathscr{T}_k(\mathfrak{c}, \varphi, \chi)^\sigma = \mathscr{T}_{k\sigma}(\mathfrak{c}, \varphi^\sigma, \chi^\sigma)$, from which the desired formula $q_k(f)^\sigma = q_{k\sigma}(f)^\sigma$ follows immediately.

To prove Theorem 9.2 we modify the proof in [13, pp. 827–829], which is similar to the proof of Theorem 7.1. We start from $\mathscr{S}_k(\mathfrak{c}, \varphi, \chi)$ with χ as in Proposition 9.1. By Lemma 7.5 we can find a Hecke character η such that $\eta_\mathbf{a}(x) = \mathrm{sgn}(x)^u$ and the conductor of η is divisible by $\mathfrak{c}^2$. Define a character ω on $\prod_{v \in \mathbf{h}} \mathfrak{g}_v^\times$ by $\omega = \varphi/\eta$. Let $\mathfrak{f}$ be the conductor of ω, which is clearly the conductor of η and $\mathfrak{f} \subset \mathfrak{c}^2$. Take any $\tau \in \mathfrak{g}$, $\gg 0$; let $\tau\mathfrak{g} = \mathfrak{q}^2\mathfrak{r}$ with integral ideals $\mathfrak{q}$ and $\mathfrak{r}$ such that $\mathfrak{q}$ is prime to $\mathfrak{f}$ and $\mathfrak{r}$

has no square factors prime to $\mathfrak{f}$. Take $n \in \mathbf{Z}^{\mathbf{a}}$ so that $0 \leqslant n \leqslant u$ and $[k] - n - u \in 2\mathbf{Z}^{\mathbf{a}}$; let $l = n + u/2$. Then there is an element g of $\mathcal{M}_l(2^{-1}\mathfrak{d}, 4\mathfrak{r}\mathfrak{f}^2, \omega\varepsilon_{\mathbf{h}})$ with the properties described in the paragraph including (7.3). We can then consider $D(s; f, g; \eta\varepsilon)$ for any fixed f in $\mathcal{S}_k(\mathfrak{c}, \varphi, \chi)$. A computation similar to that in Section 7 shows that

$$D(s; f, g; \eta\varepsilon) = N(\tau)^{1/2-s}\tau^{-([k]+n)/2} \cdot \sum_{\mathfrak{m}+\mathfrak{f}=\mathfrak{g}} \eta^*(\mathfrak{q}^{-1}\mathfrak{m})A_f(\tau, \mathfrak{q}^{-1}\mathfrak{m})N(\mathfrak{q}^{-1}\mathfrak{m})^{1-2s},$$

which together with Lemma 9.6 yields

$$L_{\mathfrak{f}}(2s, \eta\varepsilon)D(s; f, g; \eta\varepsilon) = N(\tau)^{1/2-s}\tau^{-([k]+n)/2}\eta^*(\mathfrak{q}^{-1})N(\mathfrak{q})^{2s-1} \cdot A_f(\tau, \mathfrak{q}^{-1})\prod_{v\nmid\mathfrak{f}}[1 - \omega_v\eta(\pi_v)N_n^{1-2s} + \eta(\pi_v)^2N_v^{1-4s}]^{-1}.$$

The last product over $v \nmid \mathfrak{f}$ can be written $D(2s - 1, \chi, \eta)$. Combining this with (6.11), we obtain

$$\int_{\Phi} f(z)\overline{g(z)C(z, \bar{s},; k - l, \bar{\eta}\varepsilon, \Gamma_\tau)}y^k\, d_Hz = A_f(\tau, \mathfrak{q}^{-1})B(s, \tau, \eta)D(2s - 1, \chi, \eta), \tag{9.8}$$

where $\Gamma_\tau = \Gamma(2\mathfrak{d}^{-1}, 2\mathfrak{d}\mathfrak{r}\mathfrak{f}^2]$ (that is, $\mathfrak{h}$ of Section 6 is now $4\mathfrak{r}\mathfrak{f}^2$) and

$$B(s, \tau, \eta) = D_F^{1/2}N(2\mathfrak{d}^{-1})(2\pi)^{\|u-su-(k+l)/2\|}\prod_{v\in\mathbf{a}}\Gamma(s - 1 + (k_v + l_v)/2) \cdot N(\tau)^{1/2-s}\tau^{-([k]+n)/2}\eta^*(\mathfrak{q}^{-1})N(\mathfrak{q})^{2s-1}\prod_{v|\mathfrak{r},\, v\nmid\mathfrak{f}}[1 - (\eta\varepsilon)(\pi_v)N_v^{-2s}].$$

We evaluate our functions at $s = 1/2$. By Lemma 8.2 we can choose η (independently of f) so that $D(0, \chi, \eta) \neq 0$. Let α be $\alpha_{\lambda,q,\psi}$ of Proposition 5.2 with $\lambda = 1$, $q = k - l$, and $\psi = \bar{\eta}\varepsilon$; put

$$h_\tau = r_k(g(z)C^*(z, 1/2; k - l, \bar{\eta}\varepsilon, \Gamma_\tau))$$

with C^* in the same proposition and r_k of Proposition 9.10. Take $y \in F_{\mathbf{h}}^\times$ so that $y\mathfrak{g} = \mathfrak{q}^{-1}$. Then from (9.6) and (9.8) we obtain

$$\langle h_\tau, f\rangle = \mu_f(\tau, y)\beta(\tau, y, \varphi, \eta)A(0, \chi, \eta) \tag{9.9}$$

with A of (8.2b) and with a nonzero number $\beta(\tau, y, \varphi, \eta)$ given by

$$\beta(\tau, y, \varphi, \eta) = \mathrm{vol}(\Gamma_\tau\backslash H^{\mathbf{a}})^{-1}\varphi(y)\bar{\alpha}B(1/2, \tau, \eta)\gamma(\eta).$$

From this point on we can proceed in the same fashion as in [13, pp. 828–829] or as in Section 7. We then find that the h_τ for all $\tau \in \mathfrak{g}$, $\gg 0$, span $\mathcal{S}_k(\mathfrak{c}, \varphi, \chi)$. Repeating the argument of [13, p. 829, lines 9–18] and applying Theorem 8.1 to (9.9) divided by $V(\chi, u)$, we obtain Theorem 9.2. Theorems 9.3 and 9.4 can be proved by the same technique as in the proof of Theorems 7.2 and 7.3. We simply evaluate (6.11) at $s = m/2$ or take its residue at s_0 and apply (9.6), (9.7), Propositions 5.1 and 5.2 to the present setting.

10. Critical values of $\mathscr{D}(s, \{\chi^2, \eta\})$. The series in question is defined by

$$\mathscr{D}(s, \{\chi^2, \eta\}) = L_{\mathfrak{c}}(2s-2, \eta^2\Psi^2) \sum_{\mathfrak{a}} \eta^*(\mathfrak{a})\chi(\mathfrak{a}^2)N(\mathfrak{a})^{-s} \tag{10.1}$$

with a system of eigenvalues χ on $\mathscr{S}_k((\mathfrak{c}, \Psi))$ and a Hecke character η. Let $\mathfrak{f}$ be the conductor of η. It can easily be verified that

$$Z_{\mathfrak{c}}(\{(\eta\Psi)(\pi_v)N_v M_v\}) \sum_{\mathfrak{a}} \eta^*(\mathfrak{a})\chi(\mathfrak{a}^2)M(\mathfrak{a}) = \sum_{\mathfrak{a}} \eta^*(\mathfrak{a})\chi(\mathfrak{a})^2 M(\mathfrak{a}), \tag{10.2}$$

(10.3)
$$\mathscr{D}(s, \{\chi^2, \eta\}) = \prod_{v \nmid \mathfrak{f}} [(1-\alpha_v^2\eta(\pi_v)N_v^{-s})(1-\alpha_v\beta_v\eta(\pi_v)N_v^{-s})(1-\beta_v^2\eta(\pi_v)N_v^{-s})]^{-1},$$

where α_v and β_v are as in (3.7). Denote χ_1 of Lemma 7.6 by $\eta\chi$. Then (3.5b) and (10.2) show that

$$\mathscr{D}(s, \eta\chi, \chi) = L_{\mathfrak{c}}(s-1, \eta\Psi)\mathscr{D}(s, \{\chi^2, \eta\}). \tag{10.4}$$

PROPOSITION 10.1. *Suppose that $(\eta\Psi)_{\mathbf{a}}(x) = \operatorname{sgn}(x)^{k+n}|x|^{i\kappa}$ with $n \in \mathbf{Z}^{\mathbf{a}}, 0 \leqslant n \leqslant u$ and $\kappa \in \mathbf{R}^{\mathbf{a}}$, $\|\kappa\| = 0$. Let $\tau_v = 0$ or 1 according as $k_v + n_v$ is even or odd. Then the product*

$$\mathscr{D}(s, \{\chi^2, \eta\}) \prod_{v \mid 2, v \nmid \mathfrak{c}\mathfrak{f}} [1-(\eta\Psi)(\pi_v)^2 N_v^{2-2s}]$$

$$\cdot \prod_{v \in \mathbf{a}} \Gamma((s-\tau_v+i\kappa_v)/2)\Gamma((s-1+k_v-n_v+i\kappa_v)/2)\Gamma((s-2+k_v+n_v+i\kappa_v)/2)$$

can be continued to the whole s-plane as a mermorphic function, which is holomorphic except for a possible simple pole at $s = 2$. The pole at $s = 2$ occurs if and only if $\eta\Psi$ is a nontrivial character of order 2 and $\eta(\pi_v)\chi(v) = \overline{\chi(v)}$ for almost all v, in which case $|k_v - n_v - 1/2| - 1/2 \in 2\mathbf{Z}$ for every $v \in \mathbf{a}$.

This generalizes Theorem 1 of [3] which concerns the case $F = \mathbf{Q}$. We shall give some results on the possible poles of $\mathscr{D}$ on the line $\operatorname{Re}(s) = 1$ in Proposition 10.4 below. Postponing the proof, we state a key fact necessary for the main theorem:

PROPOSITION 10.2. *Suppose that $\eta\Psi$ is of finite order. Then for every $\sigma \in \operatorname{Aut}(\mathbf{C})$ there exist Hecke characters η', Ψ' and a system of eigenvalues χ' in $\mathscr{S}_{k\sigma}((\mathfrak{c}, \Psi'))$ such that $\eta' = \eta^\sigma$ on $\prod_{v \in \mathbf{h}} \mathfrak{g}_v^\times$, $\eta'\Psi' = (\eta\Psi)^\sigma$, and $\eta'^*(\mathfrak{a})\chi'(\mathfrak{a}^2) = [\eta^*(\mathfrak{a})\chi(\mathfrak{a}^2)]^\sigma$ for every $\mathfrak{a}$.*

Moreover the numbers $\eta^*(\mathfrak{a})\chi(\mathfrak{a}^2)$ *for all* $\mathfrak{a}$ *prime to* $\mathfrak{c}$ *generate a finite extension of* $\mathbf{Q}$ *which is totally real or a CM-field.*

Proof. Take χ' and Ψ' as in Proposition 4.3, and put $\eta' = (\eta\Psi)^\sigma\Psi'^{-1}$. Then $\eta' = \eta^\sigma$ on $\prod_{v\in\mathbf{h}}\mathfrak{g}_v^\times$ and we obtain the first assertion from (4.4) and (10.2). The remaining part will be proved in Section 11.

The symbols being as in the above proposition, we put $\{\chi'^2, \eta'\} = \{\chi^2, \eta\}^\sigma$. It should be noted that we then have $|\chi^2|^\sigma = |\chi'^2|$ in view of (3.4) and (10.2).

THEOREM 10.3. *Suppose that* $(\eta\Psi)_\mathbf{a}(x) = \operatorname{sgn}(x)^{k+n}$ *with* $n \in \mathbf{Z}^\mathbf{a}$, $0 \leqslant n \leqslant u$. *Put*

$$A(m, \{\chi^2, \eta\}) = \alpha_m \cdot \left[\mathscr{D}(s, \{\chi^2, \eta\}) \prod_{v|2, v\nmid\mathfrak{f}} [1 - (\eta\Psi)(\pi_v)^2 N_v^{2-2s}]\right]_{s=m}$$

$$(m \in M_1 \cup M_2),$$

$$M_1 = \{m \in \mathbf{Z} \,|\, 2 \leqslant m \leqslant k_v - n_v, \quad m \equiv k_v - n_v \ (\mathrm{mod}\ 2) \textit{ for every } v\},$$

$$M_2 = \{m \in \mathbf{Z} \,|\, n_v - k_v + 1 < m \leqslant 1, \quad m \equiv k_v - n_v + 1 \ (\mathrm{mod}\ 2) \textit{ for every } v\},$$

$$\alpha_m = \begin{cases} \gamma(\eta^2\Psi^2)^{-1}\pi^{2d-2md-\|k\|} & \textit{if } m \in M_1, \\ \gamma(\eta\Psi)^{-1} i^{\|k-n\|} D_F^{1/2} \pi^{d-md-\|k\|} & \textit{if } m \in M_2. \end{cases}$$

Then $A(m, \{\chi^2, \eta\})/\langle p, p\rangle \in \bar{\mathbf{Q}}$ *and*

$$[A(m, \{\chi^2, \eta\})/\langle p, p\rangle]^\sigma = A(m, \{\chi^2, \eta\}^\sigma)/\langle p^\sigma, p^\sigma\rangle \tag{10.5}$$

for every $\sigma \in \mathrm{Gal}(\bar{\mathbf{Q}}/\mathbf{Q})$, *where* p *is an element of* $\mathscr{S}_k(\varphi, |\chi^2|)$ *as in Theorem* 7.1.

We have $M_1 \cup M_2 \neq \varnothing$ if and only if $k_v - n_v$ (mod 2) is independent of v and $k \geqslant n + 2u$. Such an n exists if and only if $k \geqslant 2u$. Notice that the quantity A is finite at 2 when $2 \in M$ by virtue of the last part of Proposition 10.1.

The above result for $F = \mathbf{Q}$ was proved by Sturm in [15, 17]. The general case was treated by Im [1] under the assumptions that both Ψ and η are of finite order and σ satisfies (8.5), which was necessary for the existence of χ_σ.

Proof of Proposition 10.1. Let $l = n + u/2$. Taking $\eta_\mathfrak{f}^{-1}$ to be ω in Proposition 1.3, we obtain an element g of $\mathscr{M}_l(2^{-1}\mathfrak{d}, 4\mathfrak{f}^2, \eta_\mathfrak{f}^{-1})$ such that for $\xi \in F^\times$ one has $\mu_g(\xi, t) \neq 0$ only if $\xi = \zeta^2$ with an element ζ of $F^\times$ such that $\zeta t\mathfrak{g} + \mathfrak{f} = \mathfrak{g}$, in which case $\mu_g(\xi, t) = \eta_\mathfrak{f}(\zeta t)\zeta^n$. Take f as in Proposition 3.5. Then $D(s; f, g; \eta\Psi)$ is meaningful and equals

$$\sum (\eta\Psi)_\mathbf{h}(t)\mu_f(\zeta^2, t)\zeta^n\eta_\mathfrak{f}(\zeta t)^{-1}|\zeta|^{-k-n-i\kappa}N(\zeta t\mathfrak{g})^{3/2-2s},$$

where the sum is taken over all different integral ideals $\zeta t\mathfrak{g}$ prime to $\mathfrak{f}$. Writing $A(\mathfrak{m})$ for $A_f(1, \mathfrak{m}; \Psi)$ of (2.11) and putting $\mathfrak{m} = \zeta t\mathfrak{g}$, we find that this is equal to

$$\sum_{\mathfrak{m}+\mathfrak{f}=\mathfrak{g}} \eta^*(\mathfrak{m})\Psi^*(\mathfrak{m}')A(\mathfrak{m})N(\mathfrak{m})^{1/2-2s},$$

where $\mathfrak{m}' = \prod_{v \nmid \mathfrak{c}} \mathfrak{m}_v$. Combining this with Theorem 3.7, (3.4), and (10.2), we obtain

$$D(s; f, g; \eta\Psi) = \sum_{\mathfrak{a}} \eta^*(\mathfrak{a})\chi(\mathfrak{a}^2)N(\mathfrak{a})^{-2s-1/2},$$

and hence, with $\mathfrak{h} = 2\mathfrak{c} \cap 4\mathfrak{f}^2$, we have

$$(10.6) \quad \mathscr{D}_{\mathfrak{h}}(s; f, g; \eta\Psi) = L_{\mathfrak{h}}(4s-1, \eta^2\Psi^2)D(s; f, g; \eta\Psi)$$
$$= \mathscr{D}(2s+1/2, \{\chi^2, \eta\}) \prod_{v|2, v\nmid \mathfrak{c}\mathfrak{f}} [1 - (\eta\Psi)(\pi_v)^2 N_v^{1-4s}].$$

Thus the first assertion of Proposition 10.1 follows immediately from Theorem 6.1. The assertion concerning the pole at $s = 2$ follows from Proposition 3.2, (10.4), and Theorem 6.1.

Proof of Theorem 10.3. From our assumption and Theorem 8.5 we easily see that $\eta^*(\mathfrak{a})\chi(\mathfrak{a}^2) \in \bar{\mathbf{Q}}$ for every $\mathfrak{a}$, and hence the algebraicity follows if (10.5) holds for every $\sigma \in \mathrm{Aut}(\mathbf{C})$. Take f as above. Given $\sigma \in \mathrm{Aut}(\mathbf{C})$, take η', Ψ', and χ' as in Proposition 10.2. Since the series $L(s, \eta'^2\Psi'^2)$ and $\mathscr{D}(s, \{\chi^2, \eta\}^\sigma)$ depend only on the values $\eta'^*(\mathfrak{a})\chi'^*(\mathfrak{a}^2)$ and $\eta'\Psi'$, we may assume that Ψ' and χ' are those of the proof of Proposition 4.3. By Lemma 4.2 we see that f^σ has the properties of f in Proposition 3.5 with φ, χ, and Ψ replaced by φ^σ, χ', and Ψ'. We assume that $M_1 \cup M_2 \neq \varnothing$, and hence $k_v - n_v \pmod 2$ is independent of v. Then we have $(\eta'\Psi')_{\mathbf{a}}(x) = \mathrm{sgn}(x)^{k\sigma+n\sigma}$, and (10.6) holds with f^σ, g^σ, η', Ψ', and χ' in place of f, g, η, Ψ, and χ. Therefore Theorem 10.3 follows immediately from Theorem 7.2, unless $F = \mathbf{Q}$, $\eta^2\Psi^2 = 1$, and $m = 2$ simultaneously. This exceptional case was proved by Sturm in [17]. Here we give a different and less technical proof.

Thus we assume that the basic field is $\mathbf{Q}$, $\eta^2\Psi^2 = 1$, and $m = 2 \equiv k - n \pmod 2$, and take a system χ in $\mathscr{S}_k((c\mathbf{Z}, \Psi))$. Here k, n, and c are integers; η and Ψ are Hecke characters of $\mathbf{Q}$. We choose a totally real cyclic extension F of $\mathbf{Q}$ of degree 3 whose discriminant is prime to c and the conductor of η. Let ω be a Hecke character of $\mathbf{Q}$ of degree 3 whose kernel corresponds to F. A well known fact on base change, due to Saito, Shintani, and Langlands, shows that there is a system of eigenvalues χ_0 in $\mathscr{S}_{ku}((\mathfrak{c}, \Psi \circ N_{F/\mathbf{Q}}))$ with some $\mathfrak{c}$ such that

$$\sum_{\mathfrak{a}} \chi_0(\mathfrak{a})N(\mathfrak{a})^{-s} = \prod_{i=0}^{2}\left[\sum_{a=1}^{\infty} \omega^*(a\mathbf{Z})^i\chi(a\mathbf{Z})a^{-s}\right].$$

Now an easy calculation shows that

$$(10.7) \quad \mathscr{D}(s, \{\chi_0^2, \eta \circ N_{F/\mathbf{Q}}\}) = \prod_{i=0}^{2} \mathscr{D}(s, \{(\omega^i\chi)^2, \eta\}),$$

where $\omega^i\chi$ is a system occurring in $\mathscr{S}_k((c'\mathbf{Z}, \omega^{2i}\Psi))$ with some c' such that $(\omega^i\chi)(a\mathbf{Z}) = \omega^*(a\mathbf{Z})^i\chi(a\mathbf{Z})$. We evaluate (10.7) at $s = 2$. If $i = 1$ or 2, we have

$\eta^2(\omega^{2i}\Psi)^2 \neq 1$, and hence our result in the nonexceptional case applies to $\mathscr{D}(2, \{(\omega^i\chi)^2, \eta\})$. We can also apply the same result to the left-hand side of (10.7), which is not zero by virtue of (10.4) and Proposition 3.2. Therefore the desired result follows from the fact

$$\left[\langle q, q\rangle^{-1} \prod_{i=0}^{2} \langle p_i, p_i\rangle\right]^\sigma = \langle q^\sigma, q^\sigma\rangle^{-1} \prod_{i=0}^{2} \langle p_i^\sigma, p_i^\sigma\rangle. \tag{10.8}$$

Here p_i is an element of $\mathscr{S}_k(\omega^i\varphi, |\chi^2|)$ as in Theorem 7.1 with $\mathbf{Q}$ as the base field, and q is an element of $\mathscr{S}_{ku}(\Phi, |\chi_0^2|)$ with similar properties, where Φ is the restriction of $\Psi \circ N_{F/\mathbf{Q}}$ to $\prod_{v|\mathfrak{c}} \mathfrak{g}_v^\times$. To prove (10.8), we change η for a character η' such that its conductor is prime to cD_F, $\eta'^6\Psi^6 \neq 1$, and $(\eta'\Psi)_\mathbf{a}(x) = \operatorname{sgn}(x)^{k+n}$. Then (10.7) holds with η' in place of η, and our result in the nonexceptional case applies to both sides of (10.7) which are nonzero. On the other hand we have

$$\left[\gamma(\xi \circ N_{F/\mathbf{Q}})^{-1} \prod_{i=0}^{2} \gamma(\omega^i\xi)\right]^\sigma = \gamma(\xi^\sigma \circ N_{F/\mathbf{Q}})^{-1} \prod_{i=0}^{2} \gamma(\omega^i\xi^\sigma) \tag{10.9}$$

for every Hecke character ξ of $\mathbf{Q}$, which follows easily from [6, (3.10)]. Therefore we obtain (10.8), which completes the proof in the exceptional case.

PROPOSITION 10.4. *Let $\mathscr{R}(s)$ denote $\mathscr{D}(s, \{\chi^2, \eta\})$ times the product of gamma functions over $v \in \mathbf{a}$ in Proposition* 10.1, *and let $\mathfrak{e}$ be the conductor of $\eta\Psi$. Then $\mathscr{R}$ has a pole at $s = 1$ only in the following two cases*: (i) $\mathfrak{c} = \mathfrak{f} = \mathfrak{g}$, $\eta^2\Psi^2 = 1$, *and* $\eta\Psi \neq 1$; (ii) $(\eta\Psi)(\pi_w) = 1$ *for some* $w|\mathfrak{c}$, $w \nmid \mathfrak{e}$ *and* $(\eta\Psi)(\pi_v)^2 = 1$ *for some* $v|2$, $v \nmid \mathfrak{c}\mathfrak{f}$. *Moreover, if $\eta\Psi$ is of finite order and every prime factor of $2\mathfrak{g} + \mathfrak{c}$ divides $\mathfrak{e}$, then $\mathscr{R}$ is finite for* $\operatorname{Re}(s) = 1$, $s \neq 1$.

This can be proved in the same manner as in [3, Theorem 2, p. 97].

11. Critical values of $\mathscr{D}(s, \chi, \chi')$. Given n elements $A_1, \ldots, A_n$ *of* $M_2(\mathbf{C})$, we put

$$\det(1 - tA_\mu) = 1 - b_\mu t + c_\mu t^2 \qquad (1 \leqslant \mu \leqslant n), \tag{11.1a}$$

$$\det\left[1 - t \bigotimes_{\mu=1}^{n} A_\mu\right] = \sum_{r=0}^{2^n} h_r(b_1, \ldots, b_n, c_1, \ldots, c_n)t^r, \tag{11.1b}$$

where t is an indeterminate. For $1 \leqslant \mu \leqslant n$ let χ_μ be a system of eigenvalues occurring in $\mathscr{S}_{k_\mu}((\mathfrak{c}_\mu, \Psi_\mu))$. Then for $b_\mu = \chi_\mu(v)$, $c_\mu = N_v\Psi_{\mu*}(v)$, and $t = N_v^{-s}$ the polynomial of (11.1b) gives the Euler v-factor of the L-function of $\bigotimes_{\mu=1}^n \chi_\mu$, where $\Psi_{\mu*}(v)$ denotes $\Psi_\mu(\pi_v)$ or 0 according as $v \nmid \mathfrak{c}_\mu$ or $v|\mathfrak{c}_\mu$. We denote by $\mathscr{E}_v(t, \bigotimes_{\mu=1}^n \chi_\mu)$ the polynomial of (11.1b) in this case.

THEOREM 11.1. *The notation being as above, suppose that $\Psi_1 \ldots \Psi_n$ is of finite order and $\sum_{\mu=1}^n k_{\mu v}$* (mod 2) *is independent of v. Then for every $\sigma \in$* Aut(**C**) *there exist*

Hecke characters $\Psi_1', \ldots, \Psi_n'$ *of* F *and a system of eigenvalues* χ_μ' *in* $\mathscr{S}_{k_\mu\sigma}((\mathfrak{c}_\mu, \Psi_\mu'))$ *for each* μ *such that* $\Psi_\mu' = \Psi_\mu^\sigma$ *on* $\prod_{v\in\mathbf{h}} \mathfrak{g}_v^\times$, $\Psi_1' \ldots \Psi_n' = (\Psi_1 \ldots \Psi_n)^\sigma$, $|\chi_\mu^2|^\sigma = |\chi_\mu'^2|$ *for every* μ, *and*

$$\text{(11.2a)} \qquad \left[N_v^{rk_0/2} \prod_{\mu=1}^n \chi_\mu(v)^{d_\mu} \Psi_{\mu*}(v)^{e_\mu} \right]^\sigma = N_v^{rk_0/2} \prod_{\mu=1}^n \chi_\mu'(v)^{d_\mu} \Psi_{\mu*}'(v)^{e_\mu}$$

for every $v \in \mathbf{h}$ *and every* $(d_1, \ldots, d_n, e_1, \ldots, e_n) \in \mathbf{Z}^{2n}$ *such that* $d_\mu \geqslant 0$, $e_\mu \geqslant 0$, *and* $d_1 + 2e_1 = \cdots = d_n + 2e_n = r$, *where* $k_0 = \operatorname{Max}\{\sum_{\mu=1}^n k_{\mu v} | v \in \mathbf{a}\}$.

In the setting of this theorem we write $\{\chi_1', \ldots, \chi_n'\} = \{\chi_1, \ldots, \chi_n\}^\sigma$. We note here two equalities which can be derived from (11.2a):

$$\text{(11.2b)} \qquad \left[N(\mathfrak{a})^{k_0/2} \prod_{\mu=1}^n \chi_\mu(\mathfrak{a}) \right]^\sigma = N(\mathfrak{a})^{k_0/2} \prod_{\mu=1}^n \chi_\mu'(\mathfrak{a}) \text{ for every integral ideal } \mathfrak{a},$$

$$\text{(11.2c)} \qquad \mathscr{E}_v\left(N_v^{k_0/2} t, \bigotimes_{\mu=1}^n \chi_\mu \right)^\sigma = \mathscr{E}_v\left(N_v^{k_0/2} t, \bigotimes_{\mu=1}^n \chi_\mu' \right) \quad \text{for every} \quad v \in \mathbf{h}.$$

Here the action of σ on a polynomial in t is defined by its action on the coefficients. In fact, if $\mathfrak{a}$ is a prime power $\mathfrak{p}^r$, then $\chi_\mu(\mathfrak{p}^r)$ is a $\mathbf{Z}$-linear combination of $\chi_\mu(v)^d \Psi_{\mu*}(v)^e$ such that $d + 2e = r$, and hence (11.2b) can be derived from (11.2a) by decomposing $\mathfrak{a}$ into prime powers. As for (11.2c), it is enough to observe that

$$h_r(s_1 b_1, \ldots, s_n b_n, s_1^2 c_1, \ldots, s_n^2 c_n) = (s_1 \ldots s_n)^r h_r(b_1, \ldots, b_n, c_1, \ldots, c_n),$$

and hence $h_r(b_1, \ldots, b_n, c_1, \ldots, c_n)$ is a $\mathbf{Z}$-linear combination of monomials $b_1^{d_1} \ldots b_n^{d_n} c_1^{e_1} \ldots c_n^{e_n}$ such that $d_1 + 2e_1 = \cdots = d_n + 2e_n = r$.

THEOREM 11.2. *The notation and assumptions being as in Theorem* 11.1, *put* $\Psi = \Psi_1 \ldots \Psi_n$ *and* $\mathfrak{e} = \mathfrak{c}_1 \ldots \mathfrak{c}_n$. *Let* L *(resp.* L_0*) be the field generated over* $\mathbf{Q}$ *by* $N_v^{rk_0/2} \prod_{\mu=1}^n \chi_\mu(v)^{d_\mu} \Psi_{\mu*}(v)^{e_\mu}$ *for all* $v \in \mathbf{h}$ *(resp. all* $v \nmid \mathfrak{e}$*) and all* (d_μ, e_μ) *as in that theorem. Further let* L' *(resp.* L_0'*) be the field generated over* $\mathbf{Q}$ *by the values of* Ψ, $|\chi_\mu(v)^2|$ *for all* $v \nmid \mathfrak{e}$ *and all* μ, *and* $N(\mathfrak{a})^{k_0/2} \prod_{\mu=1}^n \chi_\mu(\mathfrak{a})$ *for all* $\mathfrak{a}$ *(resp.* $\mathfrak{a}$ *prime to* $\mathfrak{e}$*). Then* $L_0 = L_0' \subset L' \subset L$, $[L : \mathbf{Q}] < \infty$, *and* L_0 *is totally real or a* CM*-field.*

Proof. Clearly $\Psi(\pi_v) = \prod_{\mu=1}^n \Psi_{\mu*}(v) \in L_0$ *for* $v \nmid \mathfrak{e}$. Dividing one quantity of type (11.2a) by another, we easily see that $|\chi_\mu(v)|^2 = \Psi_{\mu*}(v)^{-1} \chi_\mu(v)^2 \in L_0$ for $v \nmid \mathfrak{e}$. In deriving (11.2b) from (11.2a) we have seen that $N(\mathfrak{a})^{k_0/2} \prod_{\mu=1}^n \chi_\mu(\mathfrak{a})$ belongs to L, and to L_0 if $\mathfrak{a}$ is prime to $\mathfrak{e}$. Combining these, we obtain $L' \subset L$ and $L_0' \subset L_0$. Conversely, given (d_μ, e_μ) and r, let e be the largest e_μ and let $d = r - 2e$. The quantity of (11.2a) is 0 if $e_\mu > 0$ and $v | \mathfrak{c}_\mu$ for some μ. Assuming that $e_\mu = 0$ whenever $v | \mathfrak{c}_\mu$, we have

$$N_v^{rk_0/2} \prod_{\mu=1}^n \chi_\mu(v)^{d_\mu} \Psi_{\mu*}(v)^{e_\mu} = N_v^{ek_0} \Psi(\pi_v)^e \cdot \left[N_v^{k_0/2} \prod_{\mu=1}^n \chi_\mu(v) \right]^d \prod_{\mu=1}^n [\Psi_\mu(\pi_v)^{-1} \chi_\mu(v)^2]^{e-e_\mu}.$$

This shows that L is generated over L' by $\Psi_\mu(\pi_v)^{-1}\chi_\mu(v)^2$ for $v|\mathfrak{e}$, $1 \leqslant \mu \leqslant n$ and also that $L_0 \subset L_0'$. Therefore, to prove that $[L:\mathbf{Q}] < \infty$, it is sufficient to show that $N(\mathfrak{a})^{k_0/2}\prod_{\mu=1}^n \chi_\mu(\mathfrak{a})$ for all $\mathfrak{a}$ generate a finite extension, since Ψ is of finite order and the $|\chi_\mu(v)|^2$ generate a finite extension as shown by Proposition 4.5. Let $\mathbf{f}_\mu$ be the normalized eigenform belonging to χ_μ. Put $\mathbf{f}_\mu = (f_{\mu q})_{q\in Q}$ and $f_{\mu q}(z) = \sum_{\xi\in F} c_{\mu q}(\xi)\mathbf{e}_{\mathbf{a}}(\xi z)$. Take q in the form $q = \mathrm{diag}[1, t]$ with $t \in F_{\mathbf{h}}^\times$. By Proposition 3.1, for $\mathfrak{a} = \xi t^{-1}\mathfrak{g}$ with $0 \ll \xi \in F$ we have $\chi_\mu(\mathfrak{a}) = N(\mathfrak{a})\xi^{-k_\mu/2-i\lambda_\mu}c_{\mu q}(\xi)$, where λ_μ is determined by $\Psi_{\mu\mathbf{a}}(x) = \mathrm{sgn}(x)^{k_\mu}|x|^{2i\lambda_\mu}$. Then we have

$$N(\mathfrak{a})^{k_0/2}\prod_{\mu=1}^n \chi_\mu(\mathfrak{a}) = N(t\mathfrak{g})^{-k_0/2}N(\mathfrak{a})^n\xi^{(k_0u-l)/2}\prod_{\mu=1}^n c_{\mu q}(\xi),$$

where $l = \sum_{\mu=1}^n k_\mu$. Now the $c_{\mu q}(\xi)$ for all ξ, μ, and all $q \in Q$ generate a finite extension of $\mathbf{Q}$ as shown in [6, Prop. 1.3] and [13, page 812, lines 20–23]. Therefore the quantity of the last equality is contained in a finite extension of $\mathbf{Q}$ as desired. To prove the last assertion of our theorem, let ρ denote complex conjugation, and b the quantity inside the brackets of (11.2a). From (3.4) we obtain $\Psi(\pi_v)^r b^\rho = b$ if $v \nmid \mathfrak{e}$. Now (11.2a) shows that b^σ is the same type of quantity as b with (χ_μ', Ψ_μ') in place of (χ_μ, Ψ_μ), and hence $\Psi^\sigma(\pi_v)^r b^{\sigma\rho} = b^\sigma$. Therefore we obtain $b^{\rho\sigma} = b^{\sigma\rho}$, so that $\rho\sigma = \sigma\rho$ on L_0 for every $\sigma \in \mathrm{Aut}(\mathbf{C})$. This completes the proof.

Proof of the last assertion of Proposition 10.2. Taking $\chi_1 = \chi$ and $\chi_2 = \eta\chi$, we obtain the desired result from Theorem 11.2 and (10.2).

THEOREM 11.3. *Let the notation and assumptions be the same as in Theorem* 11.1 *with* $n = 2$. *Put*

$$A(m/2, \{\chi_1, \chi_2\}) = \alpha_m \mathscr{D}(m/2, \chi_1, \chi_2) \qquad (m \in M),$$

$$\alpha_m = \gamma(\Psi_1\Psi_2)^{-1} i^{\|k_1-k_2\|}\pi^{2d-md-\|k_1\|},$$

$$M = \{m \in \mathbf{Z} \mid k_{2v} - k_{1v} < m - 2 \leqslant k_{1v} - k_{2v} \quad \textit{and}$$

$$m \equiv k_{1v} - k_{2v} \ (\mathrm{mod}\ 2) \textit{ for every } v \in \mathbf{a}\}.$$

Let φ_1 *be the restriction of* Ψ_1 *to* $\prod_{v\in\mathbf{h}}\mathfrak{g}_v^\times$, *and* p *an element as in Theorem* 7.1 *with* k_1, φ_1, *and* $|\chi_1^2|$ *as* k, φ, *and* $|\chi^2|$ *there. Then* $A(m/2, \{\chi_1, \chi_2\})/\langle p, p\rangle \in \bar{\mathbf{Q}}$ *and*

$$[A(m/2, \{\chi_1, \chi_2\})/\langle p, p\rangle]^\sigma = A(m/2, \{\chi_1, \chi_2\}^\sigma)/\langle p^\sigma, p^\sigma\rangle \tag{11.3}$$

for every $\sigma \in \mathrm{Gal}(\bar{\mathbf{Q}}/\mathbf{Q})$.

To prove Theorems 11.1 and 11.3, we let K resp. $\mathfrak{o}$ denote the direct sum of n copies of F resp. $\mathfrak{g}$. We view F as a subring of K as usual, and put

$$\mathscr{G} = \{\alpha \in GL_2(K) \mid \det(\alpha) \in F\},$$

where $\det((x_\mu)_{\mu=1}^n) = (\det(x_\mu))_{\mu=1}^n$ for $(x_\mu)_{\mu=1}^n \in GL_2(K) = GL_2(F)^n$. We fix an ideal $\mathfrak{c} = \mathfrak{c}_1 \oplus \cdots \oplus \mathfrak{c}_n$ of $\mathfrak{o}$ with n integral ideals $\mathfrak{c}_\mu$ in F and put

$$E = E_\mathfrak{c} = \mathscr{G}_\mathbf{A} \cap \prod_{\mu=1}^{n} \tilde{D}[\mathfrak{d}^{-1}, \mathfrak{c}_\mu \mathfrak{d}],$$

$$\Delta_p = \mathscr{G} \cap pEp^{-1} \qquad (p \in \mathscr{G}_\mathbf{h}).$$

For $1 \leqslant \mu \leqslant n$ let V_μ be a finite-dimensional vector space over $\mathbf{C}$ of $\mathbf{C}$-valued functions on a set A_μ. To $\bigotimes_{\mu=1}^n h_\mu \in \bigotimes_{\mu=1}^n V_\mu$ we assign a function on $\prod_{\mu=1}^n A_\mu$ whose value at $(x_\mu)_{\mu=1}^n$ is $\prod_{\mu=1}^n h_\mu(x_\mu)$. By [9, Lemma 1.1] this gives an isomorphism of $\bigotimes_{\mu=1}^n V_\mu$ onto the vector space of all $\mathbf{C}$-valued functions f on $\prod_{\mu=1}^n A_\mu$ such that f as a function on A_μ belongs to V_μ. Therefore we denote this latter space by $\bigotimes_{\mu=1}^n V_\mu$, and view $\bigotimes_{\mu=1}^n h_\mu$ as a function on $\prod_{\mu=1}^n A_\mu$. We apply this principle to the cases in which $A_\mu = F_\mathbf{A}^\times$, $\tilde{G}_\mathbf{A}$ or $H^\mathbf{a}$. For instance, for $k = (k_1, \ldots, k_n) \in (\mathbf{Z}^\mathbf{a})^n$ we denote by $\mathscr{S}_k$ the set of functions on $(H^\mathbf{a})^n$ which is the union of $\bigotimes_{\mu=1}^n \mathscr{S}_{k_\mu}(\Gamma_\mu)$ with all choices of congruence subgroups $\Gamma_1, \ldots, \Gamma_n$ of G.

Let $\Psi_1, \ldots, \Psi_n$ be Hecke characters of F. Then $\bigotimes_{\mu=1}^n \Psi_\mu$ defines a character of $K_\mathbf{A}^\times$. Let ψ denote its restriction to $F_\mathbf{A}^\times \prod_{v \in \mathbf{h}} \mathfrak{o}_v^\times$. Observe that $\psi_\mathbf{a}(x) = \operatorname{sgn}(x)^r |x|^{2i\lambda}$ with $r \in \mathbf{Z}^\mathbf{a}$ and $\lambda \in \mathbf{R}^\mathbf{a}$, $\|\lambda\| = 0$. Then with $k = (k_1, \ldots, k_n) \in (\mathbf{Z}^\mathbf{a})^n$ such that $r - \sum_{\mu=1}^n k_\mu \in 2\mathbf{Z}^\mathbf{a}$ we denote by $\mathscr{S}_k((\mathfrak{c}, \psi))$ the set of all $\mathbf{C}$-valued functions $\mathbf{f}$ on $\mathscr{G}_\mathbf{A}$ satisfying the following three conditions:

(11.4a) $\mathbf{f}(sx) = \psi(s)\mathbf{f}(x)$ *for every* $s \in F_\mathbf{A}^\times$.

(11.4b) $\mathbf{f}(\alpha x w) = \psi_\mathfrak{c}(d_w)\mathbf{f}(x)$ *for every* $\alpha \in \mathscr{G}, x \in \mathscr{G}_\mathbf{A}$, *and* $w \in E_\mathfrak{c}, w_\mathbf{a} = 1$.

(11.4c) *For every* $p \in \mathscr{G}_\mathbf{h}$ *there is an element* f_p *of* $\mathscr{S}_k$ *such that* $\mathbf{f}(\det(p)^{-1}py) = \det(y)^{i\lambda}(f_p\|_k y)(\mathbf{i})$ *for every* $y \in \mathscr{G}_\mathbf{a}$, $\det(y) \gg 0$.

Here for a function g on $(H^\mathbf{a})^n$ and $\alpha = (\alpha_\mu)_{\mu=1}^n \in (\tilde{G}_{\mathbf{A}+})^n$ we put $(g\|_k\alpha)(z) = g(\alpha(z))\prod_{\mu=1}^n j(\alpha_\mu, z_\mu)^{-k_\mu}$.

Let φ be the character of $(\mathfrak{o}/\mathfrak{c})^\times$ obtained by restricting ψ to $\prod_{v \in \mathbf{h}} \mathfrak{o}_v^\times$. We denote by $\mathscr{S}_k(\Delta_p, \varphi, \lambda)$ the set of all g in $\mathscr{S}_k$ such that $g\|_k\gamma = \varphi(a(p^{-1}\gamma p))\det(\gamma)^{i\lambda} g$ for every $\gamma \in \Delta_p$, and $\mathscr{S}_k((\mathfrak{c}, \varphi, \lambda))$ the set of all $\mathbf{f}$ satisfying (11.4b, c). Take a coset decomposition

$$\mathscr{G}_\mathbf{A} = \bigsqcup_{p \in P} \mathscr{G}pE \tag{11.5}$$

with a subset P of $\mathscr{G}_\mathbf{h}$. For simplicity we choose P so that it consists of diagonal elements. Then $\mathbf{f} \mapsto (f_p)_{p \in P}$ gives an isomorphism of $\mathscr{S}_k((\mathfrak{c}, \varphi, \lambda))$ onto $\prod_{p \in P} \mathscr{S}_k(\Delta_p, \varphi, \lambda)$. We write then $\mathbf{f} = (f_p)_{p \in P}$.

To define Hecke operators, we recall that the operator $\tilde{D}_\mathfrak{c} y_0 \tilde{D}_\mathfrak{c}$ was defined in Section 2 for the elements y_0 in a certain set Y, which we now call $Y(\mathfrak{c})$. Then we put $\mathscr{Y} = \mathscr{G}_{\mathbf{A}} \cap \prod_{\mu=1}^{n} Y(\mathfrak{c}_\mu)$. With any y in $\mathscr{Y}$, let $EyE = \bigsqcup_{w \in W} Ew$ with $W \subset \mathscr{G}_{\mathbf{h}}$. For $\mathbf{f} \in \mathscr{S}_k((\mathfrak{c}, \varphi, \lambda))$ we define $\mathbf{f}|EyE$ to be the element in the same set given by

$$(\mathbf{f}|EyE)(x) = \sum_{w \in W} \varphi((a_w)_{\mathfrak{c}})^{-1}\mathbf{f}(xw^*) \qquad (x \in \mathscr{G}_{\mathbf{A}}).$$

Given $q \in P$, we can find a unique p in P and an element α_0 of $\mathscr{G}$ such that $qy \in \alpha_0 pE$. Let $\Delta_q\alpha_0\Delta_p = \bigsqcup_{\alpha \in A} \Delta_q\alpha$. If $\mathbf{f} = (f_p)_{p \in P}$ and $\mathbf{f}|EyE = (f_p')_{p \in P}$, then we easily see that (2.9) holds with the present symbols.

Suppose now that $(\Psi_\mu)_{\mathbf{a}}(x) = \operatorname{sgn}(x)^{k_\mu}|x|^{2i\lambda_\mu}$ with $\lambda_\mu \in \mathbf{R}^{\mathbf{a}}$, $\|\lambda_\mu\| = 0$. Then $\lambda = \sum_{\mu=1}^{n} \lambda_\mu$. Define k_0 as in Theorem 11.1. We now assume

$$(11.6) \quad \lambda = 0 \quad \textit{and} \quad \sum_{\mu=1}^{n} k_{\mu v} \equiv k_0 \pmod 2 \quad \textit{for every} \quad v \in \mathbf{a}.$$

Then (2.9) can be written

$$(11.7) \qquad f_p' = \sum_{\alpha \in A} \varphi(a_\alpha)^{-1} f_q\|_k\alpha.$$

We let an element σ of Aut(**C**) act on $\mathscr{S}_k$ by the rule $(\bigotimes_{\mu=1}^{n} f_\mu)^\sigma = \bigotimes_{\mu=1}^{n} f_\mu^\sigma$ for $f_\mu \in \mathscr{S}_{k_\mu}(\Gamma_\mu)$ as above. This is well defined, and

$$(11.8) \qquad \mathscr{S}_k(\Delta_p, \varphi, 0)^\sigma = \mathscr{S}_{k\sigma}(\Delta_p, \varphi^\sigma, 0),$$

where $k\sigma = (k_\mu\sigma)_{\mu=1}^{n}$. This follows easily from [6, Prop. 1.6], or rather, from its proof, combined with the fact that $\Delta_p = \mathscr{G} \cap \prod_{\mu=1}^{n} \tilde{\Gamma}_{p_\mu}$ if $p = (p_\mu)_{\mu=1}^{n}$ and $\tilde{\Gamma}_{p_\mu}$ is defined by (2.2) with $\mathfrak{c} = \mathfrak{c}_\mu$.

PROPOSITION 11.4. *For* $\mathbf{f} = (f_p)_{p \in P} \in \mathscr{S}_k((\mathfrak{c}, \varphi, 0))$ *and* $\sigma \in$ Aut(**C**) *define* $\mathbf{f}^\sigma = (\varepsilon(p)f_p^\sigma)_{p \in P}$, *where* $\varepsilon(x) = N(\det(x)\mathfrak{g})^{-k_0/2}[N(\det(x)\mathfrak{g})^{k_0/2}]^\sigma$ *for* $x \in \mathscr{G}_{\mathbf{A}}$. *Then under* (11.6) *we have* $\mathbf{f}^\sigma \in \mathscr{S}_{k\sigma}((\mathfrak{c}, \varphi^\sigma, 0))$ *and* $(\mathbf{f}|EyE)^\sigma = \varepsilon(y)\mathbf{f}^\sigma|EyE$.

Proof. The first assertion follows immediately from (11.8). To prove the second one, we take A of (11.7) so that $\det(\alpha) = \det(\alpha_0)$ for every $\alpha \in A$. Then a suitable modification of the argument of [6, p. 653, lines 7–14] shows that there is a permutation $\alpha \mapsto \beta_\alpha$ of A and an element γ_α of $\Delta_q \cap SL_2(K)$ such that $a_\alpha \equiv a(\gamma_\alpha\beta_\alpha)$ (mod $\mathfrak{c}$) and

$$[\det(\alpha)^{-m/2} f_q\|_k\alpha]^\sigma = \det(\gamma_\alpha\beta_\alpha)^{-m\sigma/2} f_q^\sigma\|_{k\sigma}\gamma_\alpha\beta_\alpha,$$

where $m = \sum_{\mu=1}^{n} k_\mu$. Put $\mathbf{f}|EyE = (f_p')_{p \in E}$. Since $\varepsilon(qy) = \varepsilon(\alpha_0 p)$, we have

$$\begin{aligned} \varepsilon(p)f_p'^\sigma &= \varepsilon(qy)\det(\alpha_0)^{k_0u/2}[\det(\alpha_0)^{-k_0u/2}f_p']^\sigma \\ &= \varepsilon(qy)\det(\alpha_0)^{k_0u/2}\sum_{\alpha \in A}[\det(\alpha)^{m-k_0u/2}]^\sigma\varphi^\sigma(a_\alpha)^{-1}[\det(\alpha)^{-m/2}f_q\|_k\alpha]^\sigma \end{aligned}$$

$$= \varepsilon(qy)\det(\alpha_0)^{k\sigma u/2} \sum_{\alpha\in A} \det(\beta_\alpha)^{-k\sigma u/2}\varphi^\sigma(a_\alpha)^{-1} f_q^\sigma \|_k \gamma_\alpha\beta_\alpha$$

$$= \varepsilon(y) \sum_{\alpha\in A} \varphi^\sigma(a(\beta_\alpha))^{-1}\varepsilon(q) f_q^\sigma \|_{k\sigma}\beta_\alpha,$$

which proves the second assertion.

Let Y_v denote EyE with $y = \pi_v$ for $v \nmid \mathfrak{c}_1 \ldots \mathfrak{c}_n$. Then the above proposition shows that $\mathbf{f}^\sigma | Y_v = \psi^\sigma(\pi_v)\mathbf{f}^\sigma$ if $\mathbf{f} \in \mathscr{S}_k((\mathfrak{c}, \psi))$. From this fact we can easily see that ψ^σ and $\mathbf{f}^\sigma$ satisfy (11.4a), so that under (11.6) we have

$$\mathscr{S}_k((\mathfrak{c}, \psi))^\sigma = \mathscr{S}_{k\sigma}((\mathfrak{c}, \psi^\sigma)). \tag{11.9}$$

That ψ^σ is the restriction of $\bigotimes_{\mu=1}^n \Psi'_\mu$ with some Hecke characters Ψ'_μ such that $(\Psi'_\mu)_\mathbf{a}(x) = \operatorname{sgn}(x)^{k_\mu\sigma}|x|^{2i\kappa_\mu}$, $\kappa_\mu \in \mathbf{R}^\mathbf{a}$, $\|\kappa_\mu\| = 0$ can be seen as follows. First we can find a Hecke character Ψ'_μ for each $\mu > 1$ such that $(\Psi'_\mu)_\mathbf{a}$ is of that form and $\Psi'_\mu = \Psi_\mu^\sigma$ on $\prod_{v\in\mathbf{h}} \mathfrak{g}_v^\times$ (cf. the paragraph below (8.10)). Then $(\Psi'_2 \ldots \Psi'_n)^{-1}$ times the restriction of ψ^σ to $F_\mathbf{A}^\times$ gives the desired Ψ'_1.

We now consider χ_μ in $\mathscr{S}_{k_\mu}((\mathfrak{c}_\mu, \Psi_\mu))$ for $1 \leqslant \mu \leqslant n$ without assuming (11.6). Let $\mathbf{f}_\mu$ be the normalized eigenform in $\mathscr{S}_{k_\mu}((\mathfrak{c}_\mu, \Psi_\mu))$ belonging to χ_μ, and $\mathbf{f}$ the restriction of $\bigotimes_{\mu=1}^n \mathbf{f}_\mu$ to $\mathscr{G}_\mathbf{A}$. Clearly $0 \neq \mathbf{f} \in \mathscr{S}_k((\mathfrak{c}, \psi))$. Given (d_μ, e_μ) as in Theorem 11.1, let $X_{v,d,e}$ denote the operator EyE with $y = (\operatorname{diag}[\pi_v^{e_\mu}, \pi_v^{d_\mu+e_\mu}])_{\mu=1}^n$, under the assumption that $e_\mu = 0$ whenever $v | \mathfrak{c}_\mu$. If $e_\mu > 0$ and $v | \mathfrak{c}_\mu$ for some μ, then we put $X_{v,d,e} = 0$. Further for $v \nmid \mathfrak{c}_\mu$ let Z_v^μ denote the operator EyE with $y = (1, \ldots, 1\ \operatorname{diag}[\pi_v^{-1}, \pi_v], 1, \ldots, 1)$, where the nontrivial component is the μ-th one. Then we easily see that

$$\mathbf{f} | X_{v,d,e} = \left(\prod_{\mu=1}^n \chi_\mu(v)^{d_\mu}\Psi_{\mu*}(v)^{e_\mu}\right)\mathbf{f} \qquad (v \in \mathbf{h}), \tag{11.10a}$$

$$\mathbf{f} | Z_v^\mu = (|\chi_\mu(v)^2| - N_v - 1)\mathbf{f} \qquad (v \nmid \mathfrak{c}_\mu). \tag{11.10b}$$

Proposition 11.5. *Let $\mathbf{g}$ be a nonzero element of $\mathscr{S}_k((\mathfrak{c}, \psi))$ such that $\mathbf{g} | X_{v,d,e} = b_{v,d,e}\mathbf{g}$ for every (v, d, e) and $\mathbf{g} | Z_v^\mu = c_v^\mu \mathbf{g}$ for every $v \nmid \mathfrak{c}_\mu$ with $b_{v,d,e}$ and c_v^μ in $\mathbf{C}$. Then there exist Hecke characters $\Psi'_1, \ldots, \Psi'_n$ of F and a system of eigenvalues χ'_μ in $\mathscr{S}_{k_\mu}((\mathfrak{c}_\mu, \Psi'_\mu))$ for each μ such that $\psi(x) = \Psi'_1(x_1)\ldots\Psi'_n(x_n)$ for $x = (x_\mu)_{\mu=1}^n \in F_\mathbf{A}^\times \prod_{v\in\mathbf{h}} \mathfrak{o}_v^\times$, $b_{v,d,e} = \prod_{\mu=1}^n \chi'_\mu(v)^{d_\mu}\Psi'_{\mu*}(v)^{e_\mu}$ for every (v, d, e) and $c_v^\mu = |\chi'_\mu(v)^2| - N_v - 1$ for every $v \nmid \mathfrak{c}_\mu$.*

Proof. Let φ_μ be the restriction of φ to the μ-th factor $(\mathfrak{g}/\mathfrak{c}_\mu)^\times$ of $(\mathfrak{o}/\mathfrak{c})^\times$. By the same argument as in the proof of Lemma 2.1, we can easily show that $\mathscr{S}_k(\Delta_p, \varphi, \lambda)$ is a finite direct sum of $\bigotimes_{\mu=1}^n \mathscr{S}_{k_\mu}(\tilde{\Gamma}_{p_\mu}, \varphi_\mu, \lambda_\mu)$ with $\{\lambda_\mu\}_{\mu=1}^n$ such that $\lambda = \sum_{\mu=1}^n \lambda_\mu$. For a function $\mathbf{q}$ on $\tilde{G}_\mathbf{A}^n$ let $\mathbf{q}_0$ denote its restriction to $\mathscr{G}_\mathbf{A}$. From the above fact we can derive that $\mathbf{g}$ is a finite sum $\mathbf{g} = \sum \mathbf{j}_0$, where each $\mathbf{j}$ belongs to $\bigotimes_{\mu=1}^n \mathscr{S}_{k_\mu}((\mathfrak{c}_\mu, \varphi_\mu, \lambda_\mu))$. Let S be the sum of all such tensor products necessary for the expression $\mathbf{g} = \sum \mathbf{j}_0$. Now let $\tilde{X}_{v,d,e}$ denote the operator $\bigotimes_{\mu=1}^n (\mathfrak{T}_v)^{d_\mu}(\mathfrak{S}_v)^{e_\mu}$ on S, and $\tilde{Z}_v^\mu$ the operators $1 \otimes \cdots \otimes 1 \otimes \tilde{R}_v \otimes 1 \otimes \cdots \otimes 1$ on S with $\tilde{R}_v$, $v \nmid \mathfrak{c}_\mu$, of Lemma 3.4 as the μ-th factor. (In each case the μ-th factor should be defined with level $\mathfrak{c}_\mu$.) Let A (resp. A_0) be the ring of operators on S (resp. S_0) generated by $\mathbf{C}$, $\tilde{X}_{v,d,e}$, and $\tilde{Z}_v^\mu$ (resp. $\mathbf{C}$, $X_{v,d,e}$, and Z_v^μ). Then we see that there is a homomorphism h of A onto A_0 such that $h(\tilde{X}_{v,d,e}) = X_{v,d,e}$, $h(\tilde{Z}_v^\mu) = Z_v^\mu$, and $(\mathbf{q}|\alpha)_0 = \mathbf{q}_0|h(\alpha)$ for every $\mathbf{q} \in S$ and $\alpha \in A$. Our assumption on $\mathbf{g}$ implies that there is a homomorphism $\xi\colon h_0 \to \mathbf{C}$ such that $\xi(X_{v,d,e}) = b_{v,d,e}$ and $\xi(Z_v^\mu) = c_v^\mu$. Then $\xi \circ h$ defines a homomorphism of A into $\mathbf{C}$, so that we can find an element $\mathbf{y}$ of S such that $\mathbf{y}|\alpha = \xi(h(\alpha))\mathbf{y}$ for every $\alpha \in A$. Now $\mathscr{S}_{k_\mu}((\mathfrak{c}_\mu, \varphi_\mu, \lambda_\mu))$ is a direct sum of $\mathscr{S}_{k_\mu}((\mathfrak{c}_\mu, \Psi'_\mu))$ with Ψ'_μ such that $(\Psi'_\mu)_\mathbf{a}(x) = \operatorname{sgn}(x)^{k_\mu}|x|^{i\lambda_\mu}$ and $\Psi'_\mu = \varphi_\mu$ on $\prod_{v|\mathfrak{c}_\mu} \mathfrak{g}_v^\times$, and therefore S is a direct sum of the spaces

$$V(\{\Psi'_\mu\}_{\mu=1}^n, \{\Phi_\mu\}_{\mu=1}^n) = \left\{\xi \in \bigotimes_{\mu=1}^n \mathscr{S}_{k_\mu}((\mathfrak{c}_\mu, \Psi'_\mu))|\mathbf{x}|(\mathfrak{T}_{v_1} \otimes \cdots \otimes \mathfrak{T}_{v_n}) \right.$$

$$\left. = \Phi_1(v_1)\ldots\Phi_n(v_n)\mathbf{x} \quad \text{if} \quad v_\mu \nmid \mathfrak{c}_\mu \right\},$$

where Φ_μ is a system of eigenvalues restricted to the primes not dividing $\mathfrak{c}_\mu$. Let U be one of the spaces $V(\{\Psi'_\mu\}_{\mu=1}^n, \{\Phi_\mu\}_{\mu=1}^n)$ such that $\mathbf{y}$ has a nonzero projection $\mathbf{y}'$ on it. Since U is stable under A, we see that $\mathbf{y}'|\alpha = \xi(h(\alpha))\mathbf{y}'$ for every $\alpha \in A$. Put

$$W = \{\mathbf{z} \in U | \xi(h(\alpha))\mathbf{z} = \mathbf{z}|\alpha \quad \text{for every} \quad \alpha \in A\}.$$

Then $W \neq 0$ and stable under $1 \otimes \cdots \otimes 1 \otimes \mathfrak{T}_v \otimes 1 \otimes \cdots \otimes 1$ for every $v \in \mathbf{h}$ with $\mathfrak{T}_v$ on any factor, and therefore has a common eigenform $\mathbf{f}$ of all such operators. Then $\mathbf{f}$ as a function on the μ-th factor $\tilde{G}_\mathbf{A}$ defines a system χ'_μ, and clearly these χ'_μ have the desired properties. From our construction we see that $\bigotimes_{\mu=1}^n \Psi'_\mu = \psi$ on $F^\times \prod_{v\in\mathbf{h}} \mathfrak{o}_v^\times$. Taking $d_\mu = 0$ and $e_\mu = 1$, we obtain $(\prod_{\mu=1}^n \Psi'_\mu)(\pi_v) = b_{v,d,e} = \psi(\pi_v)$ for $v \nmid \mathfrak{c}_1 \ldots \mathfrak{c}_n$, and hence we have $\bigotimes_{\mu=1}^n \Psi'_\mu = \psi$ on $F_\mathbf{A}^\times \prod_{v\in\mathbf{h}} \mathfrak{o}_v^\times$ as desired.

Proof of Theorem 11.1. We use the same notation as in the above proof. Let $\mathbf{f} = (\bigotimes_{\mu=1}^n \mathbf{f}_\mu)_0$ with the normalized eigenform $\mathbf{f}_\mu$ belonging to χ_μ. Then $\mathbf{f} \in \mathscr{S}_k((\mathfrak{c}, \psi))$, $\mathbf{f}|X_{v,d,e} = b_{v,d,e}\mathbf{f}$ and $\mathbf{f}|Z_v^\mu = c_v^\mu\mathbf{f}$ with $b_{v,d,e} = \prod_{\mu=1}^n \chi_\mu(v)^{d_\mu}\Psi_{\mu*}(v)^{e_\mu}$ and $c_v^\mu = |\chi_\mu(v)^2| - N_v - 1$ as observed in (11.10a, b). By (11.9) and Proposition 11.4 we have $\mathbf{f}^\sigma \in \mathscr{S}_{k\sigma}((\mathfrak{c}, \psi^\sigma))$, $\mathbf{f}^\sigma|X_{v,d,e} = \varepsilon_v^r(b_{v,d,e})^\sigma\mathbf{f}^\sigma$, and $\mathbf{f}^\sigma|Z_v^\mu = (c_v^\mu)^\sigma\mathbf{f}^\sigma$, where $\varepsilon_v^r = N_v^{-rk_0/2}(N_v^{rk_0/2})^\sigma$ and $r = d_\mu + 2e_\mu$. Applying Proposition 11.5 to $\mathbf{f}^\sigma$, we obtain Hecke characters $\Psi'_1, \ldots, \Psi'_n$ of F and a system χ'_μ in $\mathscr{S}_{k_\mu\sigma}((\mathfrak{c}_\mu, \Psi'_\mu))$ for each μ such that $\psi^\sigma = \bigotimes_{\mu=1}^n \Psi'_\mu$ on $F_\mathbf{A}^\times \prod_{v\in\mathbf{h}} \mathfrak{o}_v^\times$, $\varepsilon_v^r(b_{v,d,e})^\sigma = \prod_{\mu=1}^n \chi'_\mu(v)^{d_\mu}\Psi'_{\mu*}(v)^{e_\mu}$ and $(c_v^\mu)^\sigma = |\chi'_\mu(v)^2| - N_v - 1$ as desired.

Proof of Theorem 11.3. From Theorem 8.5 we easily see that $\chi_1(\mathfrak{a})\chi_2(\mathfrak{a}) \in \bar{\mathbf{Q}}$ for every $\mathfrak{a}$, and hence the algebraicity follows if (11.3) holds for every $\sigma \in \operatorname{Aut}(\mathbf{C})$. To prove this, we first observe that we can take P of (11.5) to be $\{p1_K | p \in Q\}$ with Q of

(2.6). For simplicity let us denote $p1_K$ also by p. The functions $\mathbf{f}_\mu$ and $\mathbf{f}$ being as in the above proof, put $\mathbf{f}_\mu = (f_{\mu q})_{q\in Q}$ and $\mathbf{f} = (f_p)_{p\in P}$. Then $f_p = f_{1p} \otimes f_{2p}$. Taking χ_1 and $\overline{\chi_2}$ to be χ and χ' of (6.16), we obtain

$$(11.11a)\quad [\mathscr{I}:\mathscr{P}][\mathfrak{g}_+^\times : U]\mathscr{D}(s, \chi_1, \chi_2) = \sum_{\mathfrak{t}\in\mathscr{B}} N(\mathfrak{t})^{s-2}\mathscr{D}_\mathfrak{h}(s-1; f_{1q}, f_{2q}^\rho; \Psi_1\Psi_2),$$

where ρ denotes complex conjugation. Given $\sigma \in \mathrm{Aut}(\mathbf{C})$, put $\varepsilon(\mathfrak{a}) = N(\mathfrak{a})^{-k_0/2}\cdot [N(\mathfrak{a})^{k_0/2}]^\sigma$. Since $k_{1v} + k_{2v} \equiv k_0 \pmod 2$, we have $\varepsilon(\xi\mathfrak{g})\xi^{(k_1\sigma+k_2\sigma)/2} = [\xi^{(k_1+k_2)/2}]^\sigma$. Let Ψ'_μ and χ'_μ be determined for σ as in Theorem 11.1. Applying σ to (6.14), we obtain

$$\begin{aligned}\chi'_1(\mathfrak{a})\chi'_2(\mathfrak{a}) &= \varepsilon(\mathfrak{a})[\chi_1(\mathfrak{a})\chi_2(\mathfrak{a})]^\sigma \\ &= \varepsilon(q)(\Psi'_1\Psi'_2)(r)\mu(\xi, r; f_{1q}^\sigma)\overline{\mu(\xi, r; f_{2q}^{\sigma\rho})}\xi^{-(k_1+k_2)\sigma/2}.\end{aligned}$$

Therefore if we replace f_{1q} and f_{2q}^ρ by $\varepsilon(q)f_{1q}^\sigma$ and $f_{2q}^{\sigma\rho}$, the computation of (6.13) through (6.16) shows that

(11.11b)

$$[\mathscr{I}:\mathscr{P}][\mathfrak{g}_+^\times : U]\mathscr{D}(s, \chi'_1, \chi'_2) = \sum_{\mathfrak{t}\in\mathscr{B}} \varepsilon(\mathfrak{t})N(\mathfrak{t})^{s-2}\mathscr{D}_\mathfrak{h}(s-1; f_{1q}^\sigma, f_{2q}^{\sigma\rho}; \Psi'_1\Psi'_2).$$

We now evaluate (11.11a, b) at $m/2$ with m in the set M of our theorem. Then our assertion follows immediately from Theorem 7.2, which is indeed applicable to f_{1q} in view of (3.10a).

References

1. J. Im, *Special values of Dirichlet series attached to Hilbert modular forms*, to appear in Amer. J. Math.
2. D. Rohrlich, *Nonvanishing of L-functions for* GL(2), Inv. Math. **97** (1989), 381–403.
3. G. Shimura, *On the holomorphy of certain Dirichlet series*, Proc. London Math. Soc., 3rd ser. **31** (1975), 79–98.
4. ———, *The special values of the zeta functions associated with cusp forms*, Comm. pure appl. Math. **29** (1976), 783–804.
5. ———, *On the periods of modular forms*, Math. Ann. **229** (1977), 211–221.
6. ———, *The special values of the zeta functions associated with Hilbert modular forms*, Duke Math. J. **45** (1978), 637–679.
7. ———, *The arithmetic of certain zeta functions and automorphic forms on orthogonal groups*, Ann. of Math. **111** (1980), 313–375.
8. ———, *The critical values of certain zeta functions associated with modular forms of half-integral weight*, J. Math. Soc. Japan **33** (1981), 649–672.
9. ———, *Alegbraic relations between critical values of zeta functions and inner products*, Amer. J. Math. **104** (1983), 253–285.
10. ———, *On Eisenstein series*, Duke Math. J. **50** (1983), 417–476.
11. ———, *On Eisenstein series of half-integral weight*, Duke Math. J. **52** (1985), 281–314.
12. ———, *On the Eisenstein series of Hilbert modular groups*, Revista Mat. Iberoamer. **1**, 3 (1985), 1–42.
13. ———, *On Hilbert modular forms of half-integral weight*, Duke Math. J. **55** (1987), 765–838.
14. ———, *Nearly holomorphic functions on hermitian symmetric spaces*, Math. Ann. **278** (1987), 1–28.

15. J. STURM, *Special values of zeta functions and Eisenstein series of half integral weight*, Amer. J. Math. **102** (1980), 219–240.
16. ———, *Addendum to special values of zeta functions*, ibid. 781–783.
17. ———, *Evaluation of the symmetric square at the near center point*, Amer. J. Math. **111** (1989), 585–598.

DEPARTMENT OF MATHEMATICS, PRINCETON UNIVERSITY, PRINCETON, NEW JERSEY 08544-1000

On the transformation formulas of theta series

American Journal of Mathematics, 115 (1993), 1011-1052

Introduction. One type of the series we will investigate can be given, in the simplest case in which the basic number field is **Q**, as follows:

(1) $g(\mathfrak{z}, u; \lambda)$

$$= \sum_{\xi} \lambda(\xi) \mathbf{e}\left(2^{-1}\mathrm{tr}\left[(2iy)^{-1}u \cdot {}^t u + x \cdot {}^t \xi S \xi + iy \cdot {}^t(A\xi)A\xi + 2uA\xi\right]\right).$$

Here S is a nondegenerate rational symmetric matrix of signature $(q - s, s)$, $\mathfrak{z} = x + iy$ is a variable in the Siegel half space H_n of degree n, u is a variable in the space $\mathbf{C}_q^n$ of all complex $(n \times q)$-matrices, ξ runs over the set $\mathbf{Q}_n^q$ of all rational $(q \times n)$-matrices, λ is a **C**-valued function on $\mathbf{Q}_n^q$ that is a finite linear combination of characteristic functions of cosets of $\mathbf{Q}_n^q$ modulo a **Z**-lattice, A is a real matrix such that $S = {}^t A \cdot \mathrm{diag}[1_{q-s}, -1_s]A$, and $\mathbf{e}(c) = \exp(2\pi \mathbf{i} c)$ for $c \in \mathbf{C}$.

If $q = 1$, $S = A = 1$, and λ is suitably chosen, the series g becomes a classical theta function of Jacobi-Riemann. If S is positive definite, $n = 1$, $u = 0$, and q is even, g is an elliptic modular form of weight $q/2$ studied by Hecke. Also, Siegel introduced the series, with $u = 0$, in his theory of quadratic forms in [Si].

Now g satisfies a transformation formula of the following type: for every

$$\alpha = \begin{pmatrix} a & b \\ c & d \end{pmatrix} \in Sp(n, \mathbf{Q})$$

with blocks a, b, c, d of size n one has

(2) $$J(\alpha, \mathfrak{z})^{-1} g\big(\alpha(\mathfrak{z}), M_{q,s}(\alpha, \mathfrak{z})u; \lambda\big) = g(\mathfrak{z}, u; \lambda'),$$

where λ' is another function of λ's type,

(3) $$J(\alpha, \mathfrak{z}) = \det(c\mathfrak{z} + d)^{q/2-s} |\det(c\mathfrak{z} + d)|^s,$$

Manuscript received October 7, 1992.

and $M_{q,S}(\alpha, \mathfrak{z})$ is a factor of automorphy acting on $\mathbf{C}_q^n$ which involves both $c\mathfrak{z}+d$ and $c\bar{\mathfrak{z}}+d$. In particular, if α belongs to a congruence subgroup that depends on S and λ, then we have $\lambda' = \lambda$. That such a formula holds is well known, but the explicit description of λ' for given λ and α is a complicated problem. Besides, if q is odd, there is the question of how one should specify the branch of $\det(c\mathfrak{z}+d)^{1/2}$, on which clearly λ' depends. These points become more nontrivial when we consider such a series over an arbitrary number field.

In order to set up a framework in which one can deal with these difficulties in a conceptual way, Weil introduced in [W] the metaplectic cover, which we write $M_{\mathbf{A}}^{(n)}$ in the introduction, of the group $Sp(n, \mathbf{A})$ for the ring of adeles $\mathbf{A}$. There is a lifting injection of $Sp(n, \mathbf{Q})$ into this group $M_{\mathbf{A}}^{(n)}$, which is defined as a group of unitary operators on $L^2(\mathbf{A}^n)$. For $x \in M_{\mathbf{A}}^{(n)}$ and a suitable $\Phi \in L^2(\mathbf{A}^n)$ put $\Theta(x, \Phi) = \sum_{\xi\in\mathbf{Q}^n} (x\Phi)(\xi)$. Then Weil showed

$$\Theta(\alpha x, \Phi) = \Theta(x, \Phi) \quad \textit{for every } \alpha \in Sp(n, \mathbf{Q}). \tag{4}$$

This is theoretically satisfactory, but in various applications, one is confronted by the following two types of problems:

(I) To obtain (2) from (4) when $q = 1$, we take Φ to be the product of its archimedean and nonarchimedean parts, the former being the exponential function on the right-hand side of (1) depending on $\mathfrak{z}$ and u, and the latter being essentially λ. Call this $\Phi[\mathfrak{z}, u, \lambda]$. Then for $x \in M_{\mathbf{A}}^{(n)}$ we have

$$x\Phi[\mathfrak{z}, u, \lambda] = J(x, \mathfrak{z})^{-1}\Phi[\mathfrak{z}', u', \lambda'] \tag{5}$$

with certain $J(x, \mathfrak{z})$, $\mathfrak{z}'$, and u', which are $\det(c\mathfrak{z}+d)^{1/2}$, $\alpha(\mathfrak{z})$, and ${}^t(c\mathfrak{z}+d)^{-1}u$ if $x = \alpha \in Sp(n, \mathbf{Q})$. This together with (4) gives (2). However, (5) holds only with the product $J(x, \mathfrak{z})^{-1}\lambda'$ determined by x and λ. In general there is no way of determining $J(x, \mathfrak{z})$ and λ' separately, because the "global" action of an element of $M_{\mathbf{A}}^{(n)}$ cannot be canonically decomposed into the product of all "local actions."

(II) If $q = 1$, (4) is essentially equivalent with (2) in the sense described in (I), but to treat g of (1) as $\Theta(x, \Phi)$ when $q > 1$, one has to consider $M_{\mathbf{A}}^{(nq)}$ in addition to $M_{\mathbf{A}}^{(n)}$. In fact, Weil's version of (2) in the general case is formula (4) with nq instead of n, and x in a certain subgroup $Mp(S)$ of $M_{\mathbf{A}}^{(nq)}$ depending on S, and α in $Mp(S) \cap Sp(nq, \mathbf{Q})$. However, formula (2) is formulated, at least in its crudest form, only with $Sp(n, \mathbf{Q})$ and H_n, without $Sp(nq, \mathbf{Q})$. Such is often necessary in applications.

The purpose of this paper is to present several formulas in which these points are settled in explicit forms. More specifically we put $G = Sp(n, F)$ with an arbitrary number field F of finite degree, consider its adelization $G_{\mathbf{A}}$, and take the metaplectic cover $M_{\mathbf{A}}$ of $G_{\mathbf{A}}$; we denote by pr the projection map of $M_{\mathbf{A}}$ onto $G_{\mathbf{A}}$. The archimedean factor $G_{\mathbf{a}}$ of $G_{\mathbf{A}}$ modulo its maximal compact subgroup is a symmetric space $\mathcal{H}$ of which H_n is a special case. Then for a nondegenerate symmetric matrix S of size q with entries in F we can define the generalization of $g(\mathfrak{z}, u; \lambda)$ with $\mathfrak{z} \in \mathcal{H}$, u in a certain complex vector space $\mathcal{U}$, and λ as a function on the set F_n^q of all $(q \times n)$-matrices over F. Let $\theta(\mathfrak{z}, u; \lambda)$ denote this g when $q = 1$ and $S = 1$ and let $\Phi[\mathfrak{z}, u, \lambda]$ be the function of (I) in this case. Our first basic formula (Theorem 1.2) is a refined version of (5) which has the form

$$(6) \qquad \sigma\Phi[\mathfrak{z}, u, \lambda] = h(\sigma, \mathfrak{z})^{-1}\Phi[\mathfrak{z}', u', {}^{\sigma}\lambda] \qquad (\mathfrak{z} \in \mathcal{H},\ u \in \mathcal{U}),$$

which holds for every σ in a subset $\mathfrak{M}$ of $M_{\mathbf{A}}$. Here $\mathfrak{M}$ is the inverse image of a subset of $G_{\mathbf{A}}$ which contains $G_{\mathbf{a}}$, a large enough open compact subgroup of the nonarchimedean factor of $G_{\mathbf{A}}$, and $P_{\mathbf{A}}$, where P is the standard parabolic subgroup of G. There is an action of $\mathfrak{M}$ on the vector space of all λ, written $\lambda \mapsto {}^{\sigma}\lambda$; $h(\sigma, \mathfrak{z})$ is a factor of automorphy such that if $F = \mathbf{Q}$ and the archimedean part of $\mathrm{pr}(\sigma)$ is $\begin{pmatrix} a & b \\ c & d \end{pmatrix}$, then $h(\sigma, \mathfrak{z})$ is $\det(cz + d)^{1/2}$ times a root of unity.

Now ${}^{\sigma}\lambda$ depends only on λ and the nonarchimedean part of $\mathrm{pr}(\sigma)$, and therefore (6) means, to some extent, the decomposition of σ into its archimedean and nonarchimedean parts. Anyway (6) together with (4) implies immediately (Proposition 1.3)

$$(7) \qquad \theta\big(\alpha(\mathfrak{z}), M_{1,0}(\alpha, \mathfrak{z})u; {}^{\alpha}\lambda\big) = h(\alpha, \mathfrak{z})\theta(\mathfrak{z}, u; \lambda) \ \ \textit{for every} \ \ \alpha \in G \cap \mathfrak{M}.$$

However, $h(\sigma, \mathfrak{z})$ is a factor of automorphy only in a weak sense, and the action of σ on λ is only partially associative (Theorem 1.2, (4)). Despite these drawbacks our formulation has the advantage that ${}^{\sigma}\lambda$ is computable only with the information on the nonarchimedean part of $\mathrm{pr}(\sigma)$, and that $h(\sigma, \mathfrak{z})$ is also computable for σ in almost all cases necessary for applications.

Our second formula (Theorem 3.2) concerns g with an arbitrary S of size $q \geq 1$, and can be given as

$$(8) \quad J(\alpha, \mathfrak{z})^{-1} g\big(\alpha(\mathfrak{z}), M_{q,s}(\alpha, \mathfrak{z})u; {}^{\alpha}\lambda\big) = g(\mathfrak{z}, u; \lambda) \ \ \textit{for every} \ \ \alpha \in G \cap \mathfrak{M}.$$

Here $\mathfrak{z}$ is still in $\mathcal{H}$, but u is in $\mathcal{U}^q$, and λ is a function on F_n^q; $J(\alpha, \mathfrak{z})$ is such that for $F = \mathbf{Q}$,

$$J(\alpha, \mathfrak{z}) = \begin{cases} \det(c\mathfrak{z} + d)^{q/2-s} |\det(c\mathfrak{z} + d)|^s & \textit{if } q \textit{ is even}, \\ h(\alpha, \mathfrak{z})^q \det(c\mathfrak{z} + d)^{-s} |\det(c\mathfrak{z} + d)|^s & \textit{if } q \textit{ is odd}. \end{cases}$$

If q is odd, all the comments on h and ${}^{\sigma}\lambda$ in (7) apply to the present $J(\alpha, \mathfrak{z})$ and ${}^{\alpha}\lambda$. If q is even, however, our result is more clear-cut: (8) holds for *all* $\alpha \in G$; besides $J(\alpha, \mathfrak{z})$ is actually a factor of automorphy, and the action $\lambda \mapsto {}^{\alpha}\lambda$ is associative. In both cases the action $\lambda \mapsto {}^{\alpha}\lambda$ depends on S, but $J(\alpha, \mathfrak{z})$ depends only on the signature of S. Thus (8) for $F = \mathbf{Q}$ is formulated within $Sp(n, \mathbf{Q})$ and H_n, without $Sp(nq, \mathbf{Q})$ as desired.

Weaker versions of (7) and (8) for totally real F were obtained in our previous papers [S3] and [S4, §11], in which some explicit formulas for $h(\alpha, \mathfrak{z})$ were also given without proofs. In Section 2 we give such formulas for an arbitrary F with detailed proofs.

As applications of (7) and (8) we investigate three more types of series. The first of these is defined by putting $u = 0$ in (1) and multiplying each term of (1) by $p(\xi)$ with a harmonic polynomial p of the most general type. Then a formula similar to (8) holds with J modified suitably according to the choice of p (Theorem 3.5 and Proposition 3.6). The second type, when $F = \mathbf{Q}$, is given as a vector-valued series

$$\theta_m(\mathfrak{z}, \lambda) = \sum_{\xi} \lambda(\xi)\rho_m(y^{-1/2})K_m(\xi y^{1/2})\mathbf{e}(2^{-1}\xi\mathfrak{z}\cdot{}^t\xi),$$

where K_m is a vector whose components are Hermite polynomials on $\mathbf{R}_n^1$ of a fixed degree m and ρ_m is a representation of $GL_n(\mathbf{C})$ on the space of all homogeneous polynomial functions of degree m on $\mathbf{C}^n$. Then we have (Theorem 4.3)

$$\theta_m(\alpha(\mathfrak{z}), {}^{\alpha}\lambda) = h(\alpha, \mathfrak{z})\rho_m(c\mathfrak{z} + d)\theta_m(\mathfrak{z}, \lambda) \quad \textit{for every} \quad \alpha \in G \cap \mathfrak{M}.$$

The third type is a special case of the first one. We take S to be the norm form of pure quaternions in an arbitrary quaternion algebra B over F. In general the series g is viewed as a function on $M_{\mathbf{A}} \times \mathcal{G}_{\mathbf{A}}$, where $\mathcal{G}$ is the orthogonal group of S. In the present paper we don't treat the behavior of g under $\mathcal{G}$, which is almost always straightforward. However, we investigate this series defined with B under the action of $M_{\mathbf{A}} \times B_{\mathbf{A}}^{\times}$, since this case is important in the following application.

In our recent paper [S5] we have been able to prove that a Fourier coefficient of a holomorphic Hilbert modular form of half-integral weight gives essentially the central value of the zeta function of the corresponding form of integral weight, which generalizes a previous result of Waldspurger in the elliptic modular case. Now our method can be extended to automorphic forms on the covering $M_{\mathbf{A}}$ of $Sp(1, F)$ for an arbitrary F, but such a generalization requires an explicit transformation formula of the theta series defined for $B = M_2(F)$, which is one of the reasons why such series are treated in the present paper.

There are two more applications of our formulas. One concerns the generalization of Siegel's product formula for an inhomogeneous quadratic form in [Si] to the higher-dimensional case over an arbitrary number field. In his thesis [F], employing a weaker version of Theorem 3.2, Fractman obtained such a generalization for totally real F. Our present Theorem 3.2, being given for an arbitrary F, certainly provides the necessary formula for the investigation in the most general case. The other application is to the arithmetic theory of automorphic forms on

$G_{\mathbf{A}}$ and $M_{\mathbf{A}}$, in particular, Eisenstein series of half-integral weight defined with respect to parabolic subgroups different from P, which we hope to treat in the future. Some of our results of Sections 2 and 3 are given specifically for that purpose.

Though the present paper is restricted to the case of quadratic forms, clearly the theta series of a hermitian form over an involutorial algebra can be handled by our methods. We don't treat such series here, since we believe that the expected formulas for them can be derived from our Theorems 1.2 and 3.2 by the same technique as in the proof of the latter theorem.

1. Metaplectic groups and factors of automorphy. The following notation will be used throughout the paper: For an associative ring R with identity element we denote by $R^\times$ the group of all its invertible elements and by R_n^m the module of all $m \times n$-matrices with entries in R; we put $R^m = R_1^m$ for simplicity. For $x = \begin{pmatrix} a & b \\ c & d \end{pmatrix} \in R_{2n}^{2n}$ with a, b, c, and d in R_n^n, we put $a = a_x$, $b = b_x$, $c = c_x$, and $d = d_x$ if there is no fear of confusion. For a complex hermitian matrix x we write $x > 0$ if x is positive definite.

We fix an algebraic number field F of finite degree and denote by $\mathbf{a}$, $\mathbf{r}$, $\mathbf{c}$, and $\mathbf{h}$, the sets of archimedean primes, real archimedean primes, imaginary archimedean primes, and nonarchimedean primes of F; further we denote by $\mathfrak{g}$ and $\mathfrak{d}$ the maximal order of F and the different of F relative to $\mathbf{Q}$.

Given an algebraic group $\mathcal{G}$ defined over F, we define $\mathcal{G}_v$ for each $v \in \mathbf{a} \cup \mathbf{h}$ and the adelization $\mathcal{G}_{\mathbf{A}}$ as usual, and view $\mathcal{G}$ as a subgroup of $\mathcal{G}_{\mathbf{A}}$. We then denote by $\mathcal{G}_{\mathbf{a}}$ and $\mathcal{G}_{\mathbf{h}}$ the archimedean and nonarchimedean factors of $\mathcal{G}_{\mathbf{A}}$, respectively. For $x \in \mathcal{G}_{\mathbf{A}}$ we denote by x_v, $x_{\mathbf{a}}$, and $x_{\mathbf{h}}$ its projections to $\mathcal{G}_v$, $\mathcal{G}_{\mathbf{a}}$, and $\mathcal{G}_{\mathbf{h}}$. If $v \in \mathbf{r}$ (resp. $v \in \mathbf{c}$), we view v as an injection of F into $\mathbf{R}$ (resp. $\mathbf{C}$), and identify F_v with $\mathbf{R}$ (resp. $\mathbf{C}$); this means that we fix one of the two injections for each $v \in \mathbf{c}$. For a fractional ideal $\mathfrak{x}$ in F and $t \in F_{\mathbf{A}}^\times$ we denote by $N(\mathfrak{x})$ the norm of $\mathfrak{x}$ and by $t\mathfrak{x}$ the fractional ideal in F such that $(t\mathfrak{x})_v = t_v\mathfrak{x}_v$ for every $v \in \mathbf{h}$; we put $|t|_v = |t_v|_v$ with the normalized valuation $|\ \ |_v$ at v and $|t|_{\mathbf{A}} = \prod_{v \in \mathbf{a} \cup \mathbf{h}} |t|_v$. Given two sets S and $\mathbf{b}$, we denote by $S^{\mathbf{b}}$ the product of $\mathbf{b}$ copies of S, that is, the set of all indexed elements $(x_v)_{v \in \mathbf{b}}$ with x_v in S. For the most part $\mathbf{b}$ is a subset of $\mathbf{a}$. We put

$$\text{(1.1a)} \qquad \mathbf{T} = \{\zeta \in \mathbf{C} \mid |\zeta| = 1\},$$

$$\text{(1.1b)} \qquad \mathbf{e}(z) = e^{2\pi i z} \quad (z \in \mathbf{C}),$$

and define a character $\mathbf{e}_v : F_v \to \mathbf{T}$ for each v by $\mathbf{e}_v(x) = \mathbf{e}(x)$ for $v \in \mathbf{r}$, $\mathbf{e}_v(x) = \mathbf{e}(x+\bar{x})$ for $v \in \mathbf{c}$, and $\mathbf{e}_v(x) = \mathbf{e}(-\{\textit{fractional part of } \mathrm{Tr}_{F_v/\mathbf{Q}_p}(x)\})$ if $v \in \mathbf{h}$ and p is the rational prime divisible by v. We then put $\mathbf{e}_{\mathbf{A}}(x) = \prod_{v \in \mathbf{a} \cup \mathbf{h}} \mathbf{e}_v(x_v)$ for $x \in F_{\mathbf{A}}$.

With a positive integer n we now put

$$X = F_n^1, \quad L = \mathfrak{g}_n^1, \quad L^* = \mathfrak{d}^{-1}L,$$

$$G = Sp(n, F) = \{\alpha \in GL_{2n}(F) \mid {}^t\alpha\iota\alpha = \iota\}, \qquad \iota = \iota_n = \begin{pmatrix} 0 & -1_n \\ 1_n & 0 \end{pmatrix},$$

$$P = \{\alpha \in G \mid c_\alpha = 0\},$$

$$\Omega_{\mathbf{A}} = \left\{x \in G_{\mathbf{A}} \mid \det(c_x) \in F_{\mathbf{A}}^\times\right\}, \qquad \Omega_v = \{x \in G_v \mid \det(c_x) \neq 0\}.$$

We let G act on $X \times X = F_{2n}^1$ by right multiplication. Then we can define the metaplectic groups $Mp(X_{\mathbf{A}})$ and $Mp(X_v)$ in the sense of [W] for each $v \in \mathbf{a} \cup \mathbf{h}$ with respect to the alternating form $(x, y) \mapsto x\iota_n \cdot {}^t y$ on $F_{2n}^1 \times F_{2n}^1$. Recall that these groups, written $M_{\mathbf{A}}$ and M_v for simplicity, are groups of unitary transformations on $L^2(X_{\mathbf{A}})$ and on $L^2(X_v)$; there is an exact sequence

$$1 \longrightarrow \mathbf{T} \longrightarrow M_{\mathbf{A}} \longrightarrow G_{\mathbf{A}} \longrightarrow 1$$

and also a similar exact sequence for each v with M_v and G_v in place of $M_{\mathbf{A}}$ and $G_{\mathbf{A}}$. We denote by pr the projection maps of $M_{\mathbf{A}}$ and M_v to $G_{\mathbf{A}}$ and G_v. There is a natural lift $r : G \to M_{\mathbf{A}}$ by which we can consider G a subgroup of $M_{\mathbf{A}}$. There are also two types of lifts

(1.2) $$r_P : P_{\mathbf{A}} \to M_{\mathbf{A}}, \qquad r_\Omega : \Omega_{\mathbf{A}} \to M_{\mathbf{A}},$$

which satisfy the formulas

(1.3a) $$[r_P(\alpha)f](x) = |\det(a_\alpha)|_{\mathbf{A}}^{1/2} \mathbf{e}_{\mathbf{A}}(xa_\alpha \cdot {}^t b_\alpha \cdot {}^t x/2) f(xa_\alpha) \ \ if\ \alpha \in P_{\mathbf{A}},$$

(1.3b) $$r_\Omega(\alpha\beta\gamma) = r_P(\alpha) r_\Omega(\beta) r_P(\gamma) \ \ if\ \alpha, \gamma \in P_{\mathbf{A}},\ and\ \beta \in \Omega_{\mathbf{A}},$$

(1.3c) $$[r_\Omega(\beta)f](x) = |\det(c_\beta)|_{\mathbf{A}}^{1/2} \int_{X_{\mathbf{A}}} f(xa_\beta + yc_\beta)\mathbf{e}_{\mathbf{A}}(q_\beta(x, y))dy \ \ if\ \beta \in \Omega_{\mathbf{A}},$$

$$q_\beta(x, y) = (1/2)xa_\beta \cdot {}^t b_\beta \cdot {}^t x + (1/2)yc_\beta \cdot {}^t d_\beta \cdot {}^t y + xb_\beta \cdot {}^t c_\beta \cdot {}^t y.$$

Moreover $r_P = r$ on P and $r_\Omega = r$ on $G \cap \Omega_{\mathbf{A}}$. There are also similar lifts of P_v and Ω_v into M_v given by the same formulas with the subscript $\mathbf{A}$ replaced by v. We denote these lifts also by r_P and r_Ω, since the distinction will be clear from the context. Here the measure on $X_{\mathbf{A}}$ is the n-fold product of the measure $\prod_{v \in \mathbf{a} \cup \mathbf{h}} d_v x$ on $F_{\mathbf{A}}$, $\int_{\mathfrak{g}_v} d_v x = N(\mathfrak{d}_v)^{-1/2}$ for $v \in \mathbf{h}$, $\int_0^1 d_v x = 1$ for $v \in \mathbf{r}$, $d_v(x + iy) = 2dxdy$ for $v \in \mathbf{c}$.

If $v \in \mathbf{c}$, G_v is $Sp(n, \mathbf{C})$. To express its quotient by a maximal compact subgroup in an explicit form, we take the Hamilton quaternion algebra $\mathbf{H} = \mathbf{R}+i\mathbf{R}+j\mathbf{R}+k\mathbf{R}$ with the quaternion units i, j, and k as usual, and identify $\mathbf{R}+i\mathbf{R}$ with $\mathbf{C}$. For $a \in \mathbf{H}$ we denote by $\bar{a}$ the quaternion conjugate of a. Further we define a ring-injection $\kappa : \mathbf{H}_n^n \to \mathbf{C}_{2n}^{2n}$ by

$$\text{(1.4a)} \qquad \kappa(x+iy) = \begin{pmatrix} x & -y \\ y & x \end{pmatrix} \quad \textit{for } x, y \in \mathbf{R}_n^n$$

$$\text{(1.4b)} \qquad \kappa(z+jw) = \kappa(z) + iE \cdot \kappa(w) \textit{ for } z, w \in \mathbf{C}_n^n, \quad E = \mathrm{diag}[1_n, -1_n].$$

We note that $\det(\kappa(\xi)) > 0$ for every $\xi \in GL_n(\mathbf{H})$. Hereafter, whenever we speak of a point $x + iy$ of $\mathbf{C}_n^n$ or $z + jw$ of $\mathbf{H}_n^n$, the letters x, y, z, and w should be understood in the sense of (1.4a, b). We now put

$$H_n = \{ z \in \mathbf{C}_n^n \mid {}^t z = z,\ \mathrm{Im}(z) > 0 \},$$
$$H_n' = \{ z + jw \in \mathbf{H}_n^n \mid {}^t z = z \in \mathbf{C}_n^n,\ 0 < {}^t\overline{w} = w \in \mathbf{C}_n^n \}.$$

We can let $Sp(n, \mathbf{R})$ act on H_n and $Sp(n, \mathbf{C})$ on H_n' respectively, and define factors of automorphy $\mu_0(\alpha, \mathfrak{z})$ for α in each group and $\mathfrak{z}$ in each space as follows:

$$\text{(1.5a)} \qquad \alpha\mathfrak{z} = \alpha(\mathfrak{z}) = (a_\alpha\mathfrak{z} + b_\alpha)(c_\alpha\mathfrak{z} + d_\alpha)^{-1}, \qquad \mu_0(\alpha, \mathfrak{z}) = c_\alpha\mathfrak{z} + d_\alpha.$$

These can be given by a single equality:

$$\text{(1.5b)} \qquad \alpha \begin{pmatrix} \mathfrak{z} \\ 1 \end{pmatrix} = \begin{pmatrix} \alpha\mathfrak{z} \\ 1 \end{pmatrix} \mu_0(\alpha, \mathfrak{z}).$$

Here $\mu_0(\alpha, \mathfrak{z}) \in GL_n(\mathbf{C})$ for $\alpha \in Sp(n, \mathbf{R})$ and $\mathfrak{z} \in H_n$, and $\mu_0(\alpha, \mathfrak{z}) \in GL_n(\mathbf{H})$ for $\alpha \in Sp(n, \mathbf{C})$ and $\mathfrak{z} \in H_n'$. (For the reader's convenience, in Section 6 we give an easy exposition of these facts for $Sp(n, \mathbf{C})$ and H_n'.) We now define a space $\mathcal{H}$ and a vector space $\mathcal{U}$ by

$$\text{(1.6)} \qquad \mathcal{H} = \prod_{v \in \mathbf{a}} H_v, \qquad \mathcal{U} = \prod_{v \in \mathbf{a}} U_v,$$

$$H_v = H_n \textit{ and } U_v = \mathbf{C}^n \textit{ if } v \in \mathbf{r},$$
$$H_v = H_n' \textit{ and } U_v = \mathbf{C}^{2n} \textit{ if } v \in \mathbf{c}.$$

For $\sigma \in M_{\mathbf{A}}$, $\alpha = \mathrm{pr}(\sigma) \in G_{\mathbf{A}}$, and $\mathfrak{z} \in \mathcal{H}$ we put

$$\mu(\sigma, \mathfrak{z}) = \mu(\alpha, \mathfrak{z}) = (\mu(\alpha_v, \mathfrak{z}_v))_{v\in\mathbf{a}}, \tag{1.7a}$$

$$\mu(\alpha_v, \mathfrak{z}_v) = \begin{cases} \mu_0(\alpha_v, \mathfrak{z}_v) & \text{if } v \in \mathbf{r}, \\ \kappa(\mu_0(\alpha_v, \mathfrak{z}_v)) & \text{if } v \in \mathbf{c}, \end{cases} \tag{1.7b}$$

$$m(\sigma, \mathfrak{z}) = m(\alpha, \mathfrak{z}) = (m(\alpha_v, \mathfrak{z}_v))_{v\in\mathbf{a}}, \quad m(\alpha_v, \mathfrak{z}_v) = \det(\mu(\alpha_v, \mathfrak{z}_v)), \tag{1.7c}$$

$$m(\sigma, \mathfrak{z})^{\mathbf{1}} = m(\alpha, \mathfrak{z})^{\mathbf{1}} = \prod_{v\in\mathbf{a}} m(\alpha_v, \mathfrak{z}_v), \tag{1.7d}$$

$$Y(\mathfrak{z}) = (Y_v(\mathfrak{z}_v))_{v\in\mathbf{a}}, \tag{1.7e}$$

$$Y_v(x+iy) = \det(y) \quad \text{if } v \in \mathbf{r} \text{ and } x+iy \in H_n, \tag{1.7f}$$

$$Y_v(z+jw) = \det(w)^2 \quad \text{if } v \in \mathbf{c} \text{ and } z+jw \in H'_n. \tag{1.7g}$$

Notice that $0 < m(\alpha_v, \mathfrak{z}_v) \in \mathbf{R}$ if $v \in \mathbf{c}$. From (1.5b) we can easily derive

$$Y(\alpha\mathfrak{z}) = \left(|m(\alpha_v, \mathfrak{z}_v)^{-2}|Y_v(\mathfrak{z}_v)\right)_{v\in\mathbf{a}} \qquad (\alpha \in G_{\mathbf{A}},\ \mathfrak{z} \in \mathcal{H}). \tag{1.7h}$$

We let G_v act on $H_v \times U_v$ by

$$\alpha(\mathfrak{z}, u) = (\alpha\mathfrak{z},\ {}^t\mu(\alpha, \mathfrak{z})^{-1}u),$$

and let $G_{\mathbf{A}}$ act on $\mathcal{H}$ and $\mathcal{H} \times \mathcal{U}$ by

$$\alpha(\mathfrak{z}) = (\alpha_v\mathfrak{z}_v)_{v\in\mathbf{a}}, \qquad \alpha(\mathfrak{z}, u) = (\alpha_v\mathfrak{z}_v,\ {}^t\mu(\alpha_v, \mathfrak{z}_v)^{-1}u_v)_{v\in\mathbf{a}}$$

for $\mathfrak{z} \in \mathcal{H}$ and $\alpha \in G_{\mathbf{A}}$, ignoring $\alpha_{\mathbf{h}}$. We define the action of an element σ of $M_{\mathbf{A}}$ (resp. M_v) on $\mathcal{H}$ and $\mathcal{H} \times \mathcal{U}$ (resp. H_v and $H_v \times U_v$) to be the same as that of $\mathrm{pr}(\sigma)$.

Let us now define a **C**-valued function $\varphi(x; \mathfrak{z}, u)$ for $x \in X_{\mathbf{a}}$, $\mathfrak{z} \in \mathcal{H}$, and $u \in \mathcal{U}$ by

$$\text{(1.8a)} \qquad \varphi(x; \mathfrak{z}, u) = \prod_{v\in\mathbf{a}} \varphi_v(x_v; \mathfrak{z}_v, u_v),$$

$$\text{(1.8b)} \qquad \varphi_v(x_v; \mathfrak{z}, u_v) = \mathbf{e}\big((1/2)^t u(\mathfrak{z} - \bar{\mathfrak{z}})^{-1}u + (1/2)x\mathfrak{z} \cdot {}^t x + xu\big) \quad \textit{if} \ v \in \mathbf{r},$$

$$\text{(1.8c)} \qquad \varphi_v(x_v; z + jw, u_v) = \mathbf{e}\big({}^t u(2i \cdot \kappa(w))^{-1}u + \mathrm{Re}(xz \cdot {}^t x) + i\mathrm{Re}(\bar{x}w \cdot {}^t x) + 2(\mathrm{Re}(x), \mathrm{Im}(x))u\big) \quad \textit{if} \ \ v \in \mathbf{c},$$

where the subscript v is suppressed on the right-hand sides of the last two formulas.

LEMMA 1.1R. *If $v \in \mathbf{r}$, M_v can be identified with the group formed by all couples (α, g) with $\alpha \in G_v$ and a holomorphic function g on H such that $g(\mathfrak{z})^2/m(\alpha, \mathfrak{z})$ is a constant belonging to* **T**, *the law of composition being $(\alpha, g)(\alpha', g') = \big(\alpha\alpha', g(\alpha'(\mathfrak{z}))g'(\mathfrak{z})\big)$. Moreover if $\xi = (\alpha, g) \in M_v$, then $(\xi\varphi_v)(x; \mathfrak{z}, u) = g(\mathfrak{z})^{-1}\varphi_v(x; \xi(\mathfrak{z}, u))$. In particular $r_\Omega(\iota_n) = (\iota_n, \det(-i\mathfrak{z})^{1/2})$.*

This was given in [S3, Prop.3.1] and the paragraph following its proof. If $\sigma = r_\Omega(\iota_n)$, formula (1.3c) together with an easy calculation shows that

$$\text{(1.9a)} \qquad \sigma\varphi_v(x; \mathfrak{z}, u) = \int_{\mathbf{R}_n^1} \varphi_v(y; \mathfrak{z}, u)\mathbf{e}(-x \cdot {}^t y)dy = \det(-i\mathfrak{z})^{-1/2}\varphi_v(x; \sigma(\mathfrak{z}, u)),$$

which proves the last assertion.

LEMMA 1.1C. *If $v \in \mathbf{c}$, M_v can be identified with $G_v \times \mathbf{T}$ whose action on $L^2(X_v)$ is such that $((\alpha, t)\varphi_v)(x; \mathfrak{z}, u) = t \cdot m(\alpha, \mathfrak{z})^{-1/2}\varphi_v(x; \alpha(\mathfrak{z}, u))$ for $\alpha \in G_v$ and $t \in \mathbf{T}$.*

Proof. Define $\rho : \mathbf{C}_n^1 \to \mathbf{R}_{2n}^1$, $\varepsilon : \mathbf{H}_n^n \to \mathbf{C}_{2n}^{2n}$, and $\kappa' : \mathbf{C}_{2n}^{2n} \to \mathbf{R}_{4n}^{4n}$ by

$$\rho(x) = (\mathrm{Re}(x), \mathrm{Im}(x)), \quad \varepsilon(\mathfrak{z}) = E \cdot \kappa(\mathfrak{z}), \quad \kappa'\left(\begin{pmatrix} a & b \\ c & d \end{pmatrix}\right) = \begin{pmatrix} E\kappa(a)E & E\kappa(b) \\ \kappa(c)E & \kappa(d) \end{pmatrix}.$$

Then we easily see that $\varepsilon(H_n') \subset H_{2n}$, $\kappa'(Sp(n, \mathbf{C})) \subset Sp(2n, \mathbf{R})$, $\varepsilon(\alpha\mathfrak{z}) = \kappa'(\alpha) \cdot \varepsilon(\mathfrak{z})$, and $\mu(\alpha, \mathfrak{z}) = \mu_0\big(\kappa'(\alpha), \varepsilon(\mathfrak{z})\big)$ for $\alpha \in Sp(n, \mathbf{C})$ and $\mathfrak{z} \in H_n'$. Let $\varphi^{(n)}(x; z, u)$ denote the function of (1.8b) for $x \in \mathbf{R}_n^1$, $z \in H_n$, and $u \in \mathbf{C}^n$. Then for $v \in \mathbf{c}$ it can be easily verified that $\varphi_v(x; \mathfrak{z}, u) = \varphi^{(2n)}(\rho(x); 2\varepsilon(\mathfrak{z}), 2u)$. Therefore from

(1.9a) and (1.3c) we easily obtain

$$(1.9b) \qquad r_\Omega(\iota_n)\varphi_v(x;\,\mathfrak{z},\,u) = \det(\kappa(\mathfrak{z}))^{-1/2}\varphi_v(x;\,-\mathfrak{z}^{-1},\,{}^t\kappa(\mathfrak{z})^{-1}u).$$

Also from (1.3a) we obtain $r_P(\alpha)\varphi_v(x;\,\mathfrak{z},\,u) = |\det(a_\alpha)|_v^{1/2}\varphi_v(x;\,\alpha(\mathfrak{z},\,u))$. Combining these two formulas, we know that for each $\gamma \in G_v$ there is an element $\tilde{\gamma}$ of M_v such that

$$\tilde{\gamma}\varphi_v(x;\,\mathfrak{z},\,u) = m(\gamma,\,\mathfrak{z})^{-1/2}\varphi_v(x;\,\gamma(\mathfrak{z},\,u)).$$

Our Lemma is merely a paraphrase of this fact. □

For $\xi = (\alpha,\,g) \in M_v$, $v \in \mathbf{r}$ as in Lemma 1.1R we put $g(\mathfrak{z}) = g(\xi,\,\mathfrak{z})$; for $\xi = (\alpha,\,t) \in M_v$, $v \in \mathbf{c}$ as in Lemma 1.1C we put $g(\xi,\,\mathfrak{z}) = t^{-1}m(\alpha,\,\mathfrak{z})^{1/2}$. Then clearly

$$(1.10) \qquad \xi\varphi_v(x;\,\mathfrak{z},\,u) = g(\xi,\,\mathfrak{z})^{-1}\varphi_v(x;\,\xi(\mathfrak{z},\,u))$$

and $g(\xi\xi',\,\mathfrak{z}) = g(\xi,\,\xi'\mathfrak{z})g(\xi',\,\mathfrak{z})$ for $\xi,\,\xi' \in M_v$ in both cases. We note an easy fact:

$$(1.11) \qquad g(r_P(\alpha),\,\mathfrak{z}) = |\det(d_\alpha)|_v^{1/2} \quad \textit{if} \quad \alpha \in P_v,\ v \in \mathbf{a}.$$

For two fractional ideals $\mathfrak{x}$ and $\mathfrak{y}$ in F such that $\mathfrak{x}\mathfrak{y} \subset \mathfrak{g}$ we put

$$(1.12a) \quad D[\mathfrak{x},\,\mathfrak{y}] = G_{\mathbf{a}}\prod_{v\in\mathbf{h}} D_v[\mathfrak{x},\,\mathfrak{y}] \ (\subset G_{\mathbf{A}}), \qquad \Gamma[\mathfrak{x},\,\mathfrak{y}] = G \cap D[\mathfrak{x},\,\mathfrak{y}],$$

$$(1.12b) \quad D_v[\mathfrak{x},\,\mathfrak{y}] = \{x \in G_v \mid a_x \in (\mathfrak{g}_v)_n^n,\ b_x \in (\mathfrak{x}_v)_n^n,\ c_x \in (\mathfrak{y}_v)_n^n,\ d_x \in (\mathfrak{g}_v)_n^n\}.$$

We define a subgroup C^θ of $G_{\mathbf{A}}$, which may be called the "theta-subgroup," by

$$(1.13a) \qquad C^\theta = G_{\mathbf{a}}\prod_{v\in\mathbf{h}} C_v^\theta,$$

$$(1.13b) \qquad C_v^\theta = \{\xi \in D_v[\mathfrak{d}^{-1},\,\mathfrak{d}] \mid \chi_v((x,\,y)\xi) = \chi_v(x,\,y) \ \textit{for every} \ x \in L_v \ \textit{and} \ y \in L_v^*\} \text{ if } v \in \mathbf{h},$$

where $\chi_v(x,\,y) = \mathbf{e}_v(x \cdot {}^ty/2)$ for $x,\,y \in (F_v)_n^1$. Then it can be shown that

$$(1.14) \qquad C_v^\theta = \{\xi \in D_v[\mathfrak{d}^{-1},\mathfrak{d}] \mid (a_\xi \cdot {}^tb_\xi)_{ii} \in 2\mathfrak{d}_v^{-1} \ \textit{and} \ (c_\xi \cdot {}^td_\xi)_{ii} \in 2\mathfrak{d}_v \ \textit{for} \ 1 \leqq i \leqq n\},$$

where α_{ii} denotes the (i, i)-entry of α. We easily see that

$$C_v^\theta \supset D_v[2\mathfrak{d}^{-1}, 2\mathfrak{d}] \cup D_v[2\mathfrak{d}^{-1}, 2\mathfrak{d}]\varepsilon_v \tag{1.15}$$

with $\varepsilon \in G_\mathbf{A}$ given by

$$\varepsilon_\mathbf{a} = 1_{2n}, \qquad \varepsilon_v = \begin{pmatrix} 0 & -\delta_v^{-1}1_n \\ \delta_v 1_n & 0 \end{pmatrix} \quad for \quad v \in \mathbf{h}, \tag{1.16}$$

where δ is an arbitrarily fixed element of $F_\mathbf{h}^\times$ such that $\mathfrak{d} = \delta\mathfrak{g}$. Since C_v^θ coincides with $Ps(X_v, L_v)$ of [W, n°36], we obtain a lift

$$r_v : C_v^\theta \to M_v \tag{1.17}$$

which is written $\mathbf{r}'_L$ there.

We are going to define a factor of automorphy of half-integral weight $h(\sigma, \mathfrak{z})$ for $\mathfrak{z} \in \mathcal{H}$ and σ in the set

$$\mathfrak{M} = \left\{ \sigma \in M_\mathbf{A} \mid \mathrm{pr}(\sigma) \in P_\mathbf{A} C^\theta \right\}. \tag{1.18}$$

We denote by $\mathcal{S}(X_\mathbf{h})$ and $\mathcal{S}(X_\mathbf{A})$ the Schwartz-Bruhat spaces of $X_\mathbf{h}$ and $X_\mathbf{A}$. We shall often view an element ℓ of $\mathcal{S}(X_\mathbf{h})$ as a function on X by restricting ℓ to the image of X in $X_\mathbf{h}$. Given $\ell \in \mathcal{S}(X_\mathbf{h})$, we put

$$\ell_\mathbf{A}(x; \mathfrak{z}, u) = \ell(x_\mathbf{h})\varphi(x_\mathbf{a}; \mathfrak{z}, u) \quad for \ x \in X_\mathbf{A}, \mathfrak{z} \in \mathcal{H}, u \in \mathcal{U}. \tag{1.19}$$

For fixed $\mathfrak{z}$ and u we view $\ell_\mathbf{A}$ as an element of $\mathcal{S}(X_\mathbf{A})$, so that $\sigma\ell_\mathbf{A}$ for $\sigma \in M_\mathbf{A}$ is meaningful. Now there is another action of $\mathfrak{M}$ on $\mathcal{S}(X_\mathbf{h})$ as follows:

THEOREM 1.2. *For every $\sigma \in \mathfrak{M}$ its action on $\mathcal{S}(X_\mathbf{h})$ which is a $\mathbf{C}$-linear automorphism, written $\ell \mapsto {}^\sigma\ell$ for $\ell \in \mathcal{S}(X_\mathbf{h})$, and a function $h(\sigma, \mathfrak{z})$ of $\mathfrak{z} \in \mathcal{H}$ can be defined by the formula*

$$(\sigma\ell_\mathbf{A})(x; \mathfrak{z}, u) = h(\sigma, \mathfrak{z})^{-1}({}^\sigma\ell)_\mathbf{A}(x; \sigma(\mathfrak{z}, u)). \tag{0}$$

Moreover, this action and h have the following properties:

(1) $h(\sigma, \mathfrak{z})\prod_{v\in\mathbf{c}} m(\sigma, \mathfrak{z})_v^{-1/2}$ for a fixed σ depends only on $\mathfrak{z}_\mathbf{r} = (\mathfrak{z}_v)_{v\in\mathbf{r}}$ and is holomorphic in $\mathfrak{z}_\mathbf{r}$;

(2) $h(\sigma, \mathfrak{z})^2 = \zeta m(\sigma, \mathfrak{z})^1$ with $\zeta \in \mathbf{T}$;

(3) $h(t \cdot r_P(\gamma), \mathfrak{z}) = t^{-1}|\det(d_\gamma)_\mathbf{a}|_\mathbf{A}^{1/2}$ if $t \in \mathbf{T}$ and $\gamma \in P_\mathbf{A}$;

(4) $h(\rho\sigma\tau, \mathfrak{z}) = h(\rho, \mathfrak{z})h(\sigma, \tau\mathfrak{z})h(\tau, \mathfrak{z})$ and ${}^{(\rho\sigma\tau)}\ell = {}^\rho({}^\sigma({}^\tau\ell))$ if $\mathrm{pr}(\rho) \in P_\mathbf{A}$ and $\mathrm{pr}(\tau) \in C^\theta$;

(5) $^{\sigma}\ell$ *depends only on* ℓ *and* $\mathrm{pr}(\sigma)_{\mathbf{h}}$;

(6) $\{\sigma \in \mathfrak{M} \mid {}^{\sigma}\ell = \ell\}$ *contains an open subgroup of* $M_{\mathbf{A}}$ *for every* $\ell \in \mathcal{S}(X_{\mathbf{h}})$;

(7) *If* $\ell(x) = \prod_{v\in\mathbf{h}} \ell_v(x_v)$ *with* $\ell_v \in \mathcal{S}(X_v)$, ℓ_v *is the characteristic function of* L_v
for almost all v, *and* $\sigma = r_{\Omega}(\eta)$, $\eta \in P_{\mathbf{A}}C^*$ *with* $C^* = \{\alpha \in C^{\theta} \mid L_v^*(c_\alpha)_v = L_v$ *for every* $v \in \mathbf{h}\}$, *then* ${}^{\sigma}\ell(x) = \prod_{v\in\mathbf{h}} [r_{\Omega}(\eta_v)\ell_v](x_v)$;

(8) *If* ℓ *is as in (7) and* $\mathrm{pr}(\sigma) = \beta\alpha$ *with* $\beta \in P_{\mathbf{A}}$ *and* $\alpha \in C^{\theta}$, *then* ${}^{\sigma}\ell(x) = \prod_{v\in\mathbf{h}} [r_P(\beta_v)r_v(\alpha_v)\ell_v](x_v)$, *where* r_v *is the lift of (1.17).*

Proof. Let $\tau \in M_{\mathbf{A}}$ and $\alpha = \mathrm{pr}(\tau)$; suppose $\alpha \in C^{\theta}$. Take $\xi_v \in M_v$ for each $v \in \mathbf{a}$ so that $\mathrm{pr}(\xi_v) = \alpha_v$. Then we can define an element γ of $M_{\mathbf{A}}$ such that $\mathrm{pr}(\gamma) = \alpha$ and

$$(1.20) \qquad (\gamma\ell_{\mathbf{A}})(x;\, \mathfrak{z},\, u) = \prod_{v\in\mathbf{h}} [r_v(\alpha_v)\ell_v](x_v) \prod_{v\in\mathbf{a}} (\xi_v\varphi_v)(x_v;\, \mathfrak{z}_v,\, u_v)$$

for $\ell = \prod_{v\in\mathbf{h}} \ell_v$ as in (7) (see [W, n°38]). Then $\tau = \zeta\gamma$ with $\zeta \in \mathbf{T}$. Now every element σ of $\mathfrak{M}$ can be written $\sigma = r_P(\beta)\tau$ with such a τ and $\beta \in P_{\mathbf{A}}$. Applying $r_P(\beta)$ to (1.20), we obtain

$$(1.21) \qquad (\sigma\ell_{\mathbf{A}})(x;\, \mathfrak{z},\, u) = \zeta\ell'(x_{\mathbf{h}}) \prod_{v\in\mathbf{a}} [r_P(\beta_v)\xi_v\varphi_v](x_v;\, \mathfrak{z}_v,\, u_v),$$

where $\ell' = \prod_{v\in\mathbf{h}} r_P(\beta_v)r_v(\alpha_v)\ell_v$. By (1.10) we can write (1.21) in the form

$$(1.22) \qquad (\sigma\ell_{\mathbf{A}})(x;\, \mathfrak{z},\, u) = h(\mathfrak{z})^{-1}\ell'_{\mathbf{A}}(x;\, \sigma(\mathfrak{z},\, u))$$

with $h(\mathfrak{z}) = \zeta^{-1}\prod_{v\in\mathbf{a}} g\big(r_P(\beta_v)\xi_v,\, \mathfrak{z}_v\big)$. We have assumed that $\ell = \prod_{v\in\mathbf{h}} \ell_v$, but clearly $\ell \mapsto \ell'$ can be extended to a $\mathbf{C}$-linear automorphism of $\mathcal{S}(X_{\mathbf{h}})$, which we write again $\ell \mapsto \ell'$. Then (1.22) holds for every $\ell \in \mathcal{S}(X_{\mathbf{h}})$ with the same $h(\mathfrak{z})$. We put $h(\sigma, \mathfrak{z}) = h(\mathfrak{z})$ and ${}^{\sigma}\ell = \ell'$. To show that these are independent of the choice of β and τ, take ℓ to be the characteristic function λ of $\prod_{v\in\mathbf{h}} L_v$, and recall that $r_v(C_v^{\theta})\lambda_v = \lambda_v$ (see [W, n°21]). Now (1.3a) shows that $(r_P(\beta_v)\lambda_v)(0) = |\det(d_\beta)|_v^{-1/2}$. Therefore, putting $x = 0$ and $u = 0$ in (1.22), we obtain

$$(1.23) \qquad (\sigma\lambda_{\mathbf{A}})(0;\, \mathfrak{z},\, 0) = |\det(d_\beta)_{\mathbf{h}}|_{\mathbf{A}}^{-1/2} \cdot h(\mathfrak{z})^{-1}.$$

It can easily be seen that $|\det(d_\beta)_{\mathbf{h}}|_{\mathbf{A}}$ depends only on $\mathrm{pr}(\sigma)$. (In fact the number is $N(\mathrm{il}(\mathrm{pr}(\sigma)))^{-1}$ with the symbol il of (2.1) below.) Thus $h(\mathfrak{z})$ is determined by σ, and consequently formula (0) is established with ${}^{\sigma}\ell$ well defined. Clearly (1), (2), (4), (5), and (8) follow easily from our definition of $h(\sigma, \mathfrak{z})$ and ${}^{\sigma}\ell$; (3) follows from (1.11) if we take $\xi_v = 1$ and $\alpha = 1$; (6) can be derived from the fact that $\{\gamma \in C_v^\theta \mid r_v(\gamma)\ell_v = \ell_v\}$ is open for every $\ell_v \in \mathcal{S}(X_v)$ (see [W, n°21, n°36]). Finally let $\sigma = r_\Omega(\eta)$ with $\eta = \beta\alpha$ with $\beta \in P_{\mathbf{A}}$ and $\alpha \in C^*$. Then $r_\Omega(\eta_v) = r_P(\beta_v)r_\Omega(\alpha_v)$; moreover $r_\Omega(\alpha_v) = r_v(\alpha_v)$ by [W, p.168, last line]. Therefore (7) follows from (8). □

Remark. As (1) shows, $(\mathfrak{z}_v)_{v\in\mathbf{c}}$ is not so essential in $h(\sigma, \mathfrak{z})$. In fact, if we replace φ by $\prod_{v\in\mathbf{c}} Y_v(\mathfrak{z}_v)^{1/4}\varphi(x; \mathfrak{z}, u)$, then h is replaced by a function of $\mathfrak{z}_{\mathbf{r}}$ as in (1). But this may be rather artificial.

Given $\ell \in \mathcal{S}(X_{\mathbf{h}})$, we define a theta function $\theta(\mathfrak{z}, u; \ell)$ for $(\mathfrak{z}, u) \in \mathcal{H} \times \mathcal{U}$ by

$$\theta(\mathfrak{z}, u; \ell) = \sum_{\xi\in X} \ell_{\mathbf{A}}(\xi; \mathfrak{z}, u). \tag{1.24}$$

PROPOSITION 1.3. *For every $\alpha \in G \cap \mathfrak{M}$ we have*

$$\theta\big(\alpha(\mathfrak{z}, u); {}^{\alpha}\ell\big) = h(\alpha, \mathfrak{z})\theta(\mathfrak{z}, u; \ell).$$

Moreover, for every $\beta \in G$ and $\xi \in \prod_{v\in\mathbf{a}} M_v$ such that $\mathrm{pr}(\xi) = \beta_{\mathbf{a}}$, *there is a* $\mathbf{C}$*-linear automorphism $\ell \mapsto \ell'$ of $\mathcal{S}(X_{\mathbf{h}})$ such that*

$$\theta\big(\beta(\mathfrak{z}, u); \ell'\big) = \prod_{v\in\mathbf{a}} g(\xi_v, \mathfrak{z}_v)\theta(\mathfrak{z}, u; \ell).$$

Proof. By [W, Theorem 4 or 6] we have $\sum_{\xi\in X} (\alpha\ell_{\mathbf{A}})(\xi; \mathfrak{z}, u) = \sum_{\xi\in X} \ell_{\mathbf{A}}(\xi; \mathfrak{z}, u)$ for every $\alpha \in G$, which combined with (0) of Theorem 1.2 proves the first assertion. Now G is generated by $G \cap \mathfrak{M}$. Therefore the second assertion follows from the first one. □

PROPOSITION 1.4. *With β and ξ of Proposition 1.3 put $g(\mathfrak{z}) = \prod_{v\in\mathbf{a}} g(\xi_v, \mathfrak{z}_v)$. Then there is a congruence subgroup Δ of G such that*

$$h(\beta\gamma\beta^{-1}, \beta\mathfrak{z}) = g(\gamma\mathfrak{z})h(\gamma, \mathfrak{z})g(\mathfrak{z})^{-1}$$

fo every $\gamma \in \Delta$.

Proof. Take any nonzero ℓ in $\mathcal{S}(X_{\mathbf{h}})$ and define ℓ' as in Proposition 1.3. By Theorem 1.2(6) we can find a congruence subgroup Γ of G such that ${}^{\gamma}\ell = \ell$ and ${}^{\gamma}\ell' = \ell'$ for every $\gamma \in \Gamma$. Let $\gamma \in \beta^{-1}\Gamma\beta \cap \Gamma$. By Proposition 1.3 we then have

$$\begin{aligned} h(\beta\gamma\beta^{-1}, \beta\mathfrak{z})\theta(\beta(\mathfrak{z}, u); \ell') &= \theta(\beta\gamma\beta^{-1}\beta(\mathfrak{z}, u); \ell') = g(\gamma\mathfrak{z})\theta(\gamma(\mathfrak{z}, u); \ell) \\ &= g(\gamma\mathfrak{z})h(\gamma, \mathfrak{z})\theta(\mathfrak{z}, u; \ell) \\ &= g(\gamma\mathfrak{z})h(\gamma, \mathfrak{z})g(\mathfrak{z})^{-1}\theta(\beta(\mathfrak{z}, u); \ell'). \end{aligned}$$

Since $\theta(\mathfrak{z}, u; \ell')$ is a nonzero real analytic function, we obtain our assertion. □

The significance of this proposition can be explained in the following way: Suppose a function f on $\mathcal{H}$ satisfies $f(\gamma\mathfrak{z}) = h(\gamma, \mathfrak{z})f(\mathfrak{z})$ for every γ in a congruence subgroup of G. Put $f'(\mathfrak{z}) = g(\mathfrak{z})^{-1}f(\beta\mathfrak{z})$ with β and g as above. Then our proposition shows that $f'(\gamma\mathfrak{z}) = h(\gamma, \mathfrak{z})f'(\mathfrak{z})$ for every γ in a (sufficiently small) congruence subgroup of G. This is far more nontrivial than the corresponding fact for functions of integral weight.

2. Some formulas concerning $h(\sigma, \mathfrak{z})$**.** Since $G_{\mathbf{A}} = P_{\mathbf{A}}D[\mathfrak{d}^{-1}, \mathfrak{d}]$, we can write every element x of $G_{\mathbf{A}}$ in the form $x = pw$ with $p \in P_{\mathbf{A}}$ and $w \in D[\mathfrak{d}^{-1}, \mathfrak{d}]$. Then we define a fractional ideal $\mathrm{il}(x)$ by

$$\mathrm{il}(x) = \det(d_p)\mathfrak{g}, \tag{2.1}$$

which can easily be shown to be well defined. For ${}^ts = s \in (F^n_n)_{\mathbf{A}}$ we put

$$\nu_0(s) = \mathrm{il}\left(\begin{pmatrix} 1_n & 0 \\ s & 1_n \end{pmatrix}\right)^{-1}, \qquad \nu(s) = N(\nu_0(s)), \tag{2.2}$$

$$\gamma(s) = \prod_{v\in\mathbf{h}} \gamma_v(s), \tag{2.3a}$$

$$\gamma_v(s) = N(\mathfrak{d}_v)^{n/2}\int_{L_v} \mathbf{e}_v(xs_v \cdot {}^tx/2)d_vx, \tag{2.3b}$$

$$\omega_v(s) = \nu(\delta_v^2 s_v)^{1/2}\gamma_v(s), \tag{2.3c}$$

where δ_v is an element of $F_v^\times$ as in (1.16). We define a quadratic ideal character θ of the ideal group of F by

$$\theta(\mathfrak{a}) = \left(\frac{F(\sqrt{-1})/F}{\mathfrak{a}}\right) \tag{2.4}$$

for every ideal $\mathfrak{a}$ prime to 2. We understand that θ is the trivial character if -1 is a square in F.

LEMMA 2.1. *Suppose $v \in \mathbf{h}$ and $v \nmid 2$. Then $\omega_v(s)$ depends only on s_v modulo $\{ b \in (\mathfrak{d}_v^{-1})^n_n \mid {}^tb = b \}$. Moreover, $\omega_v(s) = \gamma_v(s)/|\gamma_v(s)|$, $\omega_v(s)^2 = \theta(\nu_0(\delta_v^2 s_v))$, and*

$$\gamma_v(cs) = \gamma_v(s)\left(\frac{c}{\nu_0(\delta_v^2 s_v)}\right)$$

for every $c \in \mathfrak{g}_v^\times$, *where the last fractional expression is the quadratic residue symbol.*

This was given in [S2, Lemmas 4.2, 4.4] and [S3, Lemma 3.3] except the equality $\omega_v(s) = \gamma_v(s)/|\gamma_v(s)|$, which follows from (2.3c), since $|\omega_v(s)| = 1$.

LEMMA 2.2. *If* $\sigma \in \mathfrak{M}$, *and* σ *or* $\sigma\iota^{-1}$ *belongs to* $r_\Omega(\Omega_\mathbf{A} \cap P_\mathbf{A} C^\theta)$, *then* $h(\sigma, \mathfrak{z})$ *is completely determined by Theorem 1.2, (1), (2), and the following formulas:*

$$\lim_{r\to\infty} h(\sigma, r\mathbf{i})/|h(\sigma, r\mathbf{i})| = \gamma(-c_\alpha^{-1}d_\alpha)/|\gamma(-c_\alpha^{-1}d_\alpha)|$$

$$\text{if} \quad \sigma = r_\Omega(\alpha) \quad \text{with} \quad \alpha \in \Omega_\mathbf{A} \cap P_\mathbf{A} C^\theta,$$

$$\lim_{r\to 0} h(\sigma, r\mathbf{i})/|h(\sigma, r\mathbf{i})| = \gamma(\delta^{-2}d_\alpha^{-1}c_\alpha)/|\gamma(\delta^{-2}d_\alpha^{-1}c_\alpha)|$$

$$\text{if} \quad \sigma = r_\Omega(\alpha\iota^{-1})\iota \quad \text{with} \quad \alpha \in \Omega_\mathbf{A}\iota \cap P_\mathbf{A} C^\theta,$$

where $\mathbf{i}$ *denotes the point of* $\mathcal{H}$ *such that* $\mathbf{i}_v = i1_n$ *for* $v \in \mathbf{r}$ *and* $\mathbf{i}_v = j1_n$ *for* $v \in \mathbf{c}$. *In particular*

$$h(\iota, \mathfrak{z}) = \prod_{v\in\mathbf{r}} \det(-i\mathfrak{z}_v)^{1/2} \prod_{v\in\mathbf{c}} \det(\kappa(\mathfrak{z}_v))^{1/2}, \tag{2.5}$$

where $\det(-i\mathfrak{z}_v)^{1/2}$ *is chosen so that it is positive when* $\mathrm{Re}(\mathfrak{z}_v) = 0$.

Proof. Let us simply write a, b, c, d for $a_\alpha, b_\alpha, c_\alpha, d_\alpha$. Let $\sigma = r_\Omega(\alpha)$ with $\alpha \in \Omega_\mathbf{A} \cap P_\mathbf{A} C^\theta$. By (1.23), (2.1), and (1.3c) we have

$$\begin{aligned}
N(\mathrm{il}(\alpha))^{1/2} h(\sigma, \mathfrak{z})^{-1} &= (\sigma\lambda_\mathbf{A})(0; \mathfrak{z}, 0) \\
&= |\det(c)|_\mathbf{A}^{1/2} \int_{X_\mathbf{A}} \lambda_\mathbf{A}(yc; \mathfrak{z}, 0)\mathbf{e}_\mathbf{A}(yc \cdot {}^t d \cdot {}^t y/2)dy \\
&= |\det(c)|_\mathbf{A}^{-1/2} \int_{X_\mathbf{A}} \lambda_\mathbf{A}(x; \mathfrak{z}, 0)\mathbf{e}_\mathbf{A}(xc^{-1}d \cdot {}^t x/2)dx \\
&= |\det(c)|_\mathbf{A}^{-1/2} \prod_{v\in\mathbf{a}} \int_{X_v} \varphi_v(x; (\mathfrak{z} + c^{-1}d)_v, 0)d_v x \\
&\quad \times \prod_{v\in\mathbf{h}} \int_{L_v} \mathbf{e}_v(xc^{-1}d \cdot {}^t x/2)d_v x.
\end{aligned}$$

From (1.9a, b) we see that the integral over X_v is equal to $\det(-i(\mathfrak{z}+c^{-1}d)_v)^{-1/2}$ if $v \in \mathbf{r}$ and to $\det(\kappa(\mathfrak{z} + c^{-1}d)_v)^{-1/2}$ if $v \in \mathbf{c}$. The integral over L_v is $N(\mathfrak{d}_v)^{-n/2}\gamma_v(c^{-1}d)$. Since $\overline{\gamma_v(s)} = \gamma_v(-s)$, we obtain the first formula. Observe that $\iota_{\mathbf{h}} = (\mathrm{diag}[\delta 1_n, \delta^{-1}1_n]\varepsilon)_{\mathbf{h}}$ with ε of (1.16), and hence $\iota \in P_{\mathbf{A}}C^{\theta}$. Specializing the above calculation, we find our assertion concerning $h(\iota, \mathfrak{z})$. Now if $\rho = \mathrm{diag}[\delta 1_n, \delta^{-1}1_n]$, then (8) of Theorem 1.2 together with (1.3a) shows that ${}^{\iota}\lambda = \prod_{v\in\mathbf{h}} r_P(\rho_v)\lambda_v = N(\mathfrak{d})^{-n/2}\lambda'$, where $\lambda'(x) = \lambda(\delta x)$. Now let $\sigma = r_{\Omega}(\alpha\iota^{-1})\iota$ with $\alpha \in \Omega_{\mathbf{A}}\iota \cap P_{\mathbf{A}}C^{\theta}$. By (0) of Theorem 1.2 we have

$$\begin{aligned}(\sigma\lambda_{\mathbf{A}})(x; \mathfrak{z}, 0) &= \big(r_{\Omega}(\alpha\iota^{-1})\iota\lambda_{\mathbf{A}}\big)(x; \mathfrak{z}, 0)\\ &= N(\mathfrak{d})^{-n/2}h(\iota, \mathfrak{z})^{-1}\big(r_{\Omega}(\alpha\iota^{-1})\lambda'_{\mathbf{A}}\big)(x; \iota(\mathfrak{z}), 0).\end{aligned}$$

Since $\alpha\iota^{-1} = \begin{pmatrix} -b & a \\ -d & c \end{pmatrix}$, a calculation similar to the above one shows that

$$\begin{aligned}(r_{\Omega}(\alpha\iota^{-1})\lambda'_{\mathbf{A}})(0; \iota(\mathfrak{z}), 0) = {}& |\det(d)|_{\mathbf{A}}^{-1/2}N(\mathfrak{d})^{n/2} \cdot \gamma(-\delta^{-2}d^{-1}c)\\ &\times \prod_{v\in\mathbf{r}} \det(i(\mathfrak{z}^{-1}+d^{-1}c)_v)^{-1/2}\\ &\times \prod_{v\in\mathbf{c}} \det(\kappa(\mathfrak{z}^{-1}+d^{-1}c)_v)^{-1/2}.\end{aligned}$$

Putting $\mathfrak{z} = r\mathbf{i}$ and taking the limit as $r \to 0$, we obtain the second formula. □

LEMMA 2.3. *If* $w \in D[\mathfrak{d}^{-1}, \mathfrak{d}]$ *and* $d_w \in GL_n(F_{\mathbf{A}})$, *then* $\nu_0(d_w^{-1}c_w) = \det(d_w)\mathfrak{g}$.

Proof. This is because

$$\begin{pmatrix} 1 & 0 \\ d^{-1}c & 1 \end{pmatrix} = \begin{pmatrix} {}^t d & -{}^t b \\ 0 & d^{-1} \end{pmatrix}\begin{pmatrix} a & b \\ c & d \end{pmatrix},$$

where we suppress the subscript w for simplicity.

PROPOSITION 2.4. *Suppose* $\alpha \in G \cap P_{\mathbf{A}}D[2\mathfrak{d}^{-1}, 2\mathfrak{d}]$. *Then* d_α *is invertible,* $\nu_0(d_\alpha^{-1}c_\alpha) = \det(d_\alpha)\mathrm{il}(\alpha)^{-1}$, $\det(d_\alpha)\mathrm{il}(\alpha)^{-1}$ *is prime to 2, and*

$$h(\alpha, \mathfrak{z})^2 = \mathrm{sgn}\big(N_{F/\mathbf{Q}}(\det(d_\alpha))\big)\theta\big(\det(d_\alpha)\mathrm{il}(\alpha)^{-1}\big)m(\alpha, \mathfrak{z})^{\mathbf{1}}.$$

Proof. Let $\alpha = pw$ with $p \in P_{\mathbf{A}}$ and $w \in D[2\mathfrak{d}^{-1}, 2\mathfrak{d}]$. Then $c_\alpha = d_p c_w$ and $d_\alpha = d_p d_w$. If $v|2$, we have $1 = {}^t a_w d_w - {}^t c_w b_w \equiv {}^t a_w d_w \pmod 4$, and hence $(d_w)_v \in GL_n(\mathfrak{g}_v)$. Thus $\det(d_\alpha) \neq 0$; moreover $(d_\alpha^{-1}c_\alpha)_v = (d_w^{-1}c_w)_v \equiv 0 \pmod{2\mathfrak{d}_v}$, so that $\gamma_v(\delta^{-2}d_\alpha^{-1}c_\alpha) = 1$ for $v|2$. By Lemma 2.3 $\nu_0(d_\alpha^{-1}c_\alpha) = \nu_0(d_w^{-1}c_w) = \det(d_w)\mathfrak{g} = \det(d_\alpha)\det(d_p)^{-1}\mathfrak{g} = \det(d_\alpha)\mathrm{il}(\alpha)^{-1}$ and this is prime to 2. Since $\alpha = \alpha\iota^{-1}\iota = r_{\Omega}(\alpha\iota^{-1})r_{\Omega}(\iota)$, Lemma 2.2 shows that

$$\lim_{\mathfrak{z}\to 0} h(\alpha, \mathfrak{z})^2/|h(\alpha, \mathfrak{z})|^2 = \gamma(\delta^{-2}d_\alpha^{-1}c_\alpha)^2/|\gamma(\delta^{-2}d_\alpha^{-1}c_\alpha)|^2.$$

By Lemma 2.1 $\prod_{v\nmid 2}\omega_v(\delta^{-2}d_\alpha^{-1}c_\alpha)^2 = \theta(\nu_0(d_\alpha^{-1}c_\alpha))$. This combined with Theorem 1.2 (2) proves our last assertion, since $\lim_{\mathfrak{z}\to 0} m(\alpha,\mathfrak{z})^1 = N_{F/\mathbf{Q}}(\det(d_\alpha))$. □

Proposition 2.5. *If $\alpha \in \Gamma[2\mathfrak{d}^{-1}, 2\mathfrak{d}]$, then $\det(d_\alpha)\mathfrak{g}$ is prime to 2, $\gamma_v(\delta^{-2}d_\alpha^{-1}c_\alpha) = 1$ for $v|2$, and*

$$\lim_{\mathfrak{z}\to 0} h(\alpha,\mathfrak{z}) = |N_{F/\mathbf{Q}}(\det(d_\alpha))|\gamma(\delta^{-2}d_\alpha^{-1}c_\alpha)$$

$$= \sum_{x\in L/Ld_\alpha} \mathbf{e}_\mathfrak{h}(\delta^{-2}xd_\alpha^{-1}c_\alpha\cdot{}^t x/2),$$

where $\mathbf{e}_\mathfrak{h}(y) = \mathbf{e}_\mathbf{A}(y_\mathfrak{h})$.

Proof. Taking $\alpha = w$ in the above proof, we obtain the first two assertions, and find that $\nu_0(d_\alpha^{-1}c_\alpha) = \det(d_\alpha)\mathfrak{g}$. By Lemma 2.1 $|\gamma_v(\delta^{-2}d_\alpha^{-1}c_\alpha)| = \nu((d_\alpha^{-1}c_\alpha)_v)^{-1/2}$ for $v \nmid 2$, and hence $|\gamma(\delta^{-2}d_\alpha^{-1}c_\alpha)| = N(\det(d_\alpha)\mathfrak{g})^{-1/2} = \lim_{\mathfrak{z}\to 0}|h(\alpha,\mathfrak{z})|^{-1}$. This combined with Lemma 2.2 proves our first equality. Employing (2.3b), we can easily express the quantity in question as a sum over L/Ld_α as stated. □

Let us now study the behavior of $h(\sigma,\mathfrak{z})$ and the action of $\mathfrak{M}$ on $\mathcal{S}(V_\mathfrak{h})$ under the reflection $\mathfrak{z} \mapsto -\mathfrak{z}^\rho$, where x^ρ is defined for $x \in (\mathbf{C}_m^m)^\mathbf{r} \times (\mathbf{H}_m^m)^\mathbf{c}$ by $x_v^\rho = \bar{x}_v$ if $v \in \mathbf{r}$ and $x_v^\rho = \iota x_v \iota^{-1}$ if $v \in \mathbf{c}$. Observe that this reflection maps $\mathcal{H}$ onto itself. Putting $\alpha^* = E\alpha E^{-1}$ for $\alpha \in G_\mathbf{A}$ with $E = \mathrm{diag}[1_n, -1_n]$, we see that $(C^\theta)^* = C^\theta$, $\alpha^*(-\mathfrak{z}^\rho) = -\alpha(\mathfrak{z})^\rho$ for $\mathfrak{z} \in \mathcal{H}$, $\mu_0(\alpha,\mathfrak{z})^\rho = \mu_0(\alpha^*, -\mathfrak{z}^\rho)$, and $\overline{\mu(\alpha,\mathfrak{z})} = \mu(\alpha^*, -\mathfrak{z}^\rho)$.

Proposition 2.6. *There is an automorphism of $M_\mathbf{A}$ which is written $\sigma \mapsto \sigma^*$, consistent with $\alpha \mapsto E\alpha E^{-1}$ for $\alpha \in G$, and determined by the relation $(\sigma f)^* = \sigma^* f^*$ for $f \in L^2(X_\mathbf{A})$, where $f^*(x) = \overline{f(-x)}$. Moreover, $\mathrm{pr}(\sigma)^* = \mathrm{pr}(\sigma^*)$, $r_P(\alpha)^* = r_P(\alpha^*)$ for $\alpha \in P_\mathbf{A}$, $r_\Omega(\beta)^* = r_\Omega(\beta^*)$ for $\beta \in \Omega_\mathbf{A}$, $t^* = t^{-1}$ for $t \in \mathbf{T}$, $\mathfrak{M}^* = \mathfrak{M}$, $\overline{h(\sigma,\mathfrak{z})} = h(\sigma^*, -\mathfrak{z}^\rho)$ and ${}^{\sigma^*}(\ell^*) = ({}^\sigma\ell)^*$ for every $\sigma \in \mathfrak{M}$ and $\ell \in \mathcal{S}(X_\mathfrak{h})$, where ℓ^* is defined by $\ell^*(x) = \overline{\ell(-x)}$.*

Proof. From (1.3a, c) we easily see that $[r_P(\alpha)f]^* = r_P(\alpha^*)f^*$ for $\alpha \in P_\mathbf{A}$ and $[r_\Omega(\beta)f]^* = r_\Omega(\beta^*)f^*$ for $\beta \in \Omega_\mathbf{A}$. Since $M_\mathbf{A}$ is generated by $r_P(P_\mathbf{A})$, $r_\Omega(\Omega_\mathbf{A})$, and $\mathbf{T}$, these equalities prove our assertions except the last two, which follow from Theorem 1.2(0) and (1.23) combined with the relation $(\ell_\mathbf{A})^*(x;\mathfrak{z},u) = (\ell^*)_\mathbf{A}(x; -\mathfrak{z}^\rho, \bar{u})$ that can easily be verified. □

We conclude this section by investigating $h(\alpha,\mathfrak{z})$ for α in the subgroup $Sp(r,F)\times Sp(s,F)$ of $Sp(r+s,F)$. To emphasize the dimension, let us denote the symbols G, P, X, $\mathcal{H}$, $\mathcal{U}$, $M_{\mathbf{A}}$, and $\mathfrak{M}$ of Section 1 by $G^{(n)}$, $P^{(n)}$, $X^{(n)}$, $\mathcal{H}^{(n)}$, $\mathcal{U}^{(n)}$, $M_{\mathbf{A}}^{(n)}$, and $\mathfrak{M}^{(n)}$. Let $n = r + s$ with positive integers r and s. For $\alpha \in G_{\mathbf{A}}^{(r)}$ and $\beta \in G_{\mathbf{A}}^{(s)}$ we define an element $[\alpha, \beta]$ of $G_{\mathbf{A}}^{(n)}$ by

$$[\alpha, \beta] = \begin{pmatrix} a_\alpha & 0 & b_\alpha & 0 \\ 0 & a_\beta & 0 & b_\beta \\ c_\alpha & 0 & d_\alpha & 0 \\ 0 & c_\beta & 0 & d_\beta \end{pmatrix}. \tag{2.6}$$

Let us now study how this injection $(\alpha, \beta) \mapsto [\alpha, \beta]$ of $G_{\mathbf{A}}^{(r)} \times G_{\mathbf{A}}^{(s)}$ into $G_{\mathbf{A}}^{(n)}$ can be extended to their metaplectic coverings. First for $f \in L^2(X_{\mathbf{A}}^{(r)})$ and $f' \in L^2(X_{\mathbf{A}}^{(s)})$ we define $f \otimes f' \in L^2(X_{\mathbf{A}}^{(n)})$ by $(f \otimes f')(x, x') = f(x)f'(x')$ for $x \in X_{\mathbf{A}}^{(r)}$ and $x' \in X_{\mathbf{A}}^{(s)}$. Given $\sigma \in M_{\mathbf{A}}^{(r)}$ and $\sigma' \in M_{\mathbf{A}}^{(s)}$, we have a unique unitary operator $\langle \sigma, \sigma' \rangle$ on $L^2(X_{\mathbf{A}}^{(n)})$ such that $\langle \sigma, \sigma' \rangle (f \otimes f') = \sigma f \otimes \sigma' f'$ for all such f and f'. Then, checking the formula of [W, (15)], we find that $\langle \sigma, \sigma' \rangle \in M_{\mathbf{A}}^{(n)}$ and $\mathrm{pr}(\langle \sigma, \sigma' \rangle) = [\mathrm{pr}(\sigma), \mathrm{pr}(\sigma')]$. (Notice that $(\sigma, \sigma') \mapsto \langle \sigma, \sigma' \rangle$ is not injective.) For $\mathfrak{z} \in \mathcal{H}^{(r)}$ and $\mathfrak{z}' \in \mathcal{H}^{(s)}$ define $[\mathfrak{z}, \mathfrak{z}'] \in \mathcal{H}^{(n)}$ by $[\mathfrak{z}, \mathfrak{z}']_v = \mathrm{diag}[\mathfrak{z}_v, \mathfrak{z}'_v]$; similarly for $u \in \mathcal{U}^{(r)}$ and $u' \in \mathcal{U}^{(s)}$ define $[u, u'] \in \mathcal{U}^{(n)}$ by $[u, u']_v = {}^t({}^t u_v, {}^t u'_v)$. Clearly

$$\varphi(x, x'; [\mathfrak{z}, \mathfrak{z}'], [u, u']) = \varphi(x; \mathfrak{z}, u)\varphi(x'; \mathfrak{z}', u'). \tag{2.7}$$

Observe that $\langle \sigma, \sigma' \rangle \in \mathfrak{M}^{(n)}$ if $\sigma \in \mathfrak{M}^{(r)}$ and $\sigma' \in \mathfrak{M}^{(s)}$.

PROPOSITION 2.7. *If $\sigma \in \mathfrak{M}^{(r)}$ and $\sigma' \in \mathfrak{M}^{(s)}$, we have*

$$h(\langle \sigma, \sigma' \rangle, [\mathfrak{z}, \mathfrak{z}']) = h(\sigma, \mathfrak{z})h(\sigma', \mathfrak{z}') \quad \text{and} \quad {}^{\langle \sigma, \sigma' \rangle}(\ell \otimes \ell') = ({}^{\sigma}\ell) \otimes ({}^{\sigma'}\ell')$$

for $\ell \in \mathcal{S}(X_{\mathbf{h}}^{(r)})$ and $\ell' \in \mathcal{S}(X_{\mathbf{h}}^{(s)})$, where $(\ell \otimes \ell')(x, x') = \ell(x)\ell'(x')$. Moreover, if $\alpha \in G^{(r)}$ and $\alpha' \in G^{(s)}$, then $[\alpha, \alpha']$ as an element of $M_{\mathbf{A}}^{(n)}$ coincides with $\langle \alpha, \alpha' \rangle$, and hence

$$h([\alpha, \alpha'], [\mathfrak{z}, \mathfrak{z}']) = h(\alpha, \mathfrak{z})h(\alpha', \mathfrak{z}')$$

if $\alpha \in G^{(r)} \cap \mathfrak{M}^{(r)}$ and $\alpha' \in G^{(s)} \cap \mathfrak{M}^{(s)}$.

Proof. Formula (2.7) together with (1.23) proves the first equality, which together with Theorem 1.2(0) proves the second one. Let $\alpha \in G^{(r)}$ and $\alpha' \in G^{(s)}$. If $\alpha \in P^{(r)}$ and $\alpha' \in P^{(s)}$, formula (1.3a) gives $[\alpha, \alpha'] = \langle \alpha, \alpha' \rangle$. Since G is generated by P and ι, our proof would be complete if we could show $[\iota_r, 1] = \langle \iota_r, 1 \rangle$ and $[1, \iota_s] = \langle 1, \iota_s \rangle$. Now, from Lemma 2.2 we easily see that

$$h([\alpha, 1], [\mathfrak{z}, \mathfrak{z}']) = h(\alpha, \mathfrak{z}) \ \text{ if } \alpha \in G^{(r)} \text{ and } \det(d_\alpha) \neq 0. \tag{2.8}$$

Observe that $\iota_r = \beta\gamma$ with

$$\beta = \begin{pmatrix} 0 & -1_r \\ 1_r & 2\cdot 1_r \end{pmatrix}, \qquad \gamma = \begin{pmatrix} 1_r & -2\cdot 1_r \\ 0 & 1_r \end{pmatrix}.$$

Clearly $\det(d_\beta)\det(d_\gamma) \neq 0$, and hence the equality of (2.8) is true with β and γ in place of α. Since $\gamma \in C^\theta$, Theorem 1.2(4) shows that

$$h(\iota_r, \mathfrak{z}) = h(\beta, \gamma\mathfrak{z})h(\gamma, \mathfrak{z}) = h([\beta, 1], [\gamma\mathfrak{z}, \mathfrak{z}'])h([\gamma, 1], [\mathfrak{z}, \mathfrak{z}']) = h([\iota_r, 1], [\mathfrak{z}, \mathfrak{z}']).$$

On the other hand we already know that $h(\langle \iota_r, 1\rangle, [\mathfrak{z}, \mathfrak{z}']) = h(\iota_r, \mathfrak{z})$. Observe that if $h(\sigma, \mathfrak{w}) = h(\tau, \mathfrak{w})$ for some $\mathfrak{w}$ and $\mathrm{pr}(\sigma) = \mathrm{pr}(\tau)$, then $\sigma = \tau$. Therefore we have $\langle \iota_r, 1\rangle = [\iota_r, 1]$. Similarly $\langle 1, \iota_s\rangle = [1, \iota_s]$, which completes the proof. □

3. Transformation formulas of general theta series. For $X = (x_{ij}) \in \mathbf{C}_q^q$ and $Y \in \mathbf{C}_n^n$ we define $X \otimes Y \in \mathbf{C}_{nq}^{nq}$ by

$$X \otimes Y = \begin{bmatrix} x_{11}Y & \cdots & x_{1q}Y \\ \cdots & \cdots & \cdots \\ x_{q1}Y & \cdots & x_{qq}Y \end{bmatrix}.$$

Let V be a q-dimensional vector space over F and $S : V \times V \to F$ a nondegenerate F-bilinear symmetric form. For each $v \in \mathbf{a} \cup \mathbf{h}$ we have an F_v-bilinear symmetric form $S_v : V \times V \to F_v$. Let $I_v = \mathrm{diag}[1_{r_v}, -1_{s_v}]$ with the signature (r_v, s_v) of S_v if $v \in \mathbf{r}$; let $I_v = 1_q$ if $v \in \mathbf{c}$. For each $v \in \mathbf{a}$ we take and fix an F_v-linear bijection $A_v : V_v \to (F^q)_v$ so that $S_v(x, y) = {}^t(A_v x)I_v(A_v y)$ and put $T_v(x, y) = {}^t\overline{(A_v x)}(A_v y)$. For $p = (p_1, \ldots, p_n) \in V_v^n$ with $p_i \in V_v$ we define elements $S_v[p]$ and $T_v\{p\}$ of $(F_v)_n^n$ by $S_v[p] = (S_v(p_i, p_j))_{i,j=1}^n$, $T_v\{p\} = (T_v(p_i, p_j))_{i,j=1}^n$. Notice that $S_v[p]$ is meaningful also for $v \in \mathbf{h}$. A straightforward calculation shows:

LEMMA 3.1. *$T_v(x, x) = \mathrm{Re}[S_v(x, x)] + 2Q_v[A_v x]$ for every $x \in V_v$, where*

$$Q_v[y] = \begin{cases} \sum_{j=r_v+1}^{q} y_j^2 & \text{if } v \in \mathbf{r}, \\ {}^t\mathrm{Im}(y)\mathrm{Im}(y) & \text{if } v \in \mathbf{c}, \end{cases}$$

for $y \in F_v^q$.

Let us again emphasize the dimension as we did in Section 2, by using the symbols $G^{(n)}$, $\mathcal{H}^{(n)}$, $\mathcal{U}^{(n)}$; in addition we use $D^{(n)}[\mathfrak{x}, \mathfrak{y}]$ and $\theta^{(n)}$ for $D[\mathfrak{x}, \mathfrak{y}]$ and θ.

We now define a theta function $g(\mathfrak{z}, u; \lambda)$ for $\mathfrak{z} \in \mathcal{H}^{(n)}$, $u \in \mathcal{U}^{(nq)}$, and $\lambda \in \mathcal{S}(V^n_{\mathbf{h}})$ by

$$g(\mathfrak{z}, u; \lambda) = \sum_{\xi \in V^n} \lambda(\xi_{\mathbf{h}})\Phi(\xi; \mathfrak{z}, u), \tag{3.0a}$$

$$\Phi(p; \mathfrak{z}, u) = \prod_{v \in \mathbf{a}} \Phi_v(p_v; \mathfrak{z}_v, u_v) \qquad (p \in V^n_{\mathbf{A}}), \tag{3.0b}$$

$$\Phi_v(p; x+iy, u) = \mathbf{e}\Big({}^t u(1_q \otimes 4iy)^{-1}u + 2^{-1}\mathrm{tr}\big(xS_v[p] + iyT_v\{p\}\big) + \mathrm{tr}\big(u'(A_v p)\big)\Big)$$
$$\textit{for } p \in V^n_v,\ x+iy \in H_n,\ \textit{and } u \in \mathbf{C}^{nq} \textit{ if } v \in \mathbf{r},$$

$$\Phi_v(p; z+jw, u) = \mathbf{e}\Big({}^t u\big(2i\kappa(1_q \otimes w)\big)^{-1}u + \mathrm{tr}[\mathrm{Re}(zS_v[p]) + i\mathrm{Re}(\overline{w}T_v\{p\})]$$
$$+ 2\mathrm{tr}[u'(\mathrm{Re}(A_v p), \mathrm{Im}(A_v p))]\Big)$$
$$\textit{for } p \in V^n_v,\ z+jw \in H'_n,\ \textit{and } u \in \mathbf{C}^{2nq} \textit{ if } v \in \mathbf{c},$$

where $A_v p = (A_v p_1 \ \dots \ A_v p_n) \in (F_v)^q_n$ for $p = (p_1, \dots, p_n) \in V^n_v$, $u' = (u_1 \ \dots \ u_q) \in \mathbf{C}^n_q$ for ${}^t u = ({}^t u_1 \ \dots \ {}^t u_q) \in \mathbf{C}^1_{nq}$ with $u_i \in \mathbf{C}^n$ if $v \in \mathbf{r}$, and

$$u' = \begin{pmatrix} a_1 & \cdots & a_q \\ b_1 & \cdots & b_q \end{pmatrix} \in \mathbf{C}^{2n}_q$$

for ${}^t u = ({}^t a_1 \ \dots \ {}^t a_q \ {}^t b_1 \ \dots \ {}^t b_q) \in \mathbf{C}^1_{2nq}$ with $a_i, b_i \in \mathbf{C}^n$ if $v \in \mathbf{c}$. If $q = 1$, $V = F$, $S(x, x) = x^2$, and $A_v = 1$ for all v, then we see that $g(\mathfrak{z}, u; \lambda)$ coincides with $\theta^{(n)}(\mathfrak{z}, u; \lambda)$. In the general case g can be obtained as a "pullback" of $\theta^{(nq)}$ as will be shown below. It should be noted that $g(\mathfrak{z}, u; \lambda) = 0$ for every $(\mathfrak{z}, u)$ only if $\lambda = 0$.

We define the action of $G_{\mathbf{A}}$ on $\mathcal{H}^{(n)} \times \mathcal{U}^{(nq)}$ by $\alpha(\mathfrak{z}, u) = (\alpha\mathfrak{z}, w)$ for $\alpha \in G_{\mathbf{A}}$ with

$$w_v = \mathrm{diag}[1_{r_v} \otimes {}^t\mu(\alpha, \mathfrak{z})_v^{-1},\ 1_{s_v} \otimes {}^t\overline{\mu(\alpha, \mathfrak{z})}_v^{-1}]u_v \quad \textit{if } v \in \mathbf{r}, \tag{3.1a}$$

$$w'_v = {}^t\mu(\alpha, \mathfrak{z})_v^{-1}\, u'_v \qquad \textit{if } v \in \mathbf{c}. \tag{3.1b}$$

We now put $\mathfrak{M}_q = \mathfrak{M}$ if q is odd and $\mathfrak{M}_q = G_{\mathbf{A}}$ if q is even, and define a factor of automorphy $J^S(\alpha, \mathfrak{z})$ for $\alpha \in \mathfrak{M}_q$ by

$$J^S(\alpha, \mathfrak{z}) = \begin{cases} \prod_{v \in \mathbf{r}} m(\alpha, \mathfrak{z})_v^{(q/2)-s_v} |m(\alpha, \mathfrak{z})_v|^{s_v} \prod_{v \in \mathbf{c}} m(\alpha, \mathfrak{z})_v^{q/2} & \textit{if } q \textit{ is even}, \\ h(\alpha, \mathfrak{z})^q \prod_{v \in \mathbf{r}} m(\alpha, \mathfrak{z})_v^{-s_v} |m(\alpha, \mathfrak{z})_v|^{s_v} & \textit{if } q \textit{ is odd}. \end{cases} \tag{3.2a}$$

If q is even, J^S is a factor of automorphy; if q is odd, however, Theorem 1.2(4) implies the following weaker property:

$$J^S(\alpha\beta\gamma, \mathfrak{z}) = J^S(\alpha, \mathfrak{z})J^S(\beta, \gamma\mathfrak{z})J^S(\gamma, \mathfrak{z}) \quad \text{if } \operatorname{pr}(\alpha) \in P_{\mathbf{A}},\ \beta \in \mathfrak{M},\ \text{and } \operatorname{pr}(\gamma) \in C^{\theta}. \tag{3.2b}$$

THEOREM 3.2. *Let χ be the Hecke character of $F_{\mathbf{A}}^{\times}$ corresponding to the extension $F(\det(S)^{1/2})/F$ or $F((-1)^{q/4}\det(S)^{1/2})$ according as q is odd or even; let* pr *denote the identity map of $G_{\mathbf{A}}$ onto itself if q is even. Then every $\sigma \in \mathfrak{M}_q$ gives a* **C**-*linear automorphism $\lambda \mapsto {}^{\sigma}\lambda$ of $\mathcal{S}(V_{\mathbf{h}}^n)$ with the following properties:*

(0) $J^S(\alpha, \mathfrak{z})^{-1}g(\alpha(\mathfrak{z}, u); {}^{\alpha}\lambda) = g(\mathfrak{z}, u; \lambda)$ *if* $\alpha \in G \cap \mathfrak{M}_q$.

(1) The map $\lambda \mapsto {}^{\sigma}\lambda$ does not depend on $\{A_v\}_{v\in\mathbf{a}}$ (though it depends on S).

(2) ${}^{(\sigma\tau)}\lambda = {}^{\sigma}({}^{\tau}\lambda)$ *for every σ and τ in $G_{\mathbf{A}}$ if q is even;* ${}^{(\rho\sigma\tau)}\lambda = {}^{\rho}({}^{\sigma}({}^{\tau}\lambda))$ *whenever* $\operatorname{pr}(\rho) \in P_{\mathbf{A}}$ *and* $\operatorname{pr}(\tau) \in C^{\theta}$ *if q is odd.*

(3) ${}^{\sigma}\lambda$ *depends only on λ and* $\operatorname{pr}(\sigma)_{\mathbf{h}}$.

(4) $\operatorname{pr}(\{\sigma \in \mathfrak{M}_q \mid {}^{\sigma}\lambda = \lambda\})$ *contains an open subgroup of $G_{\mathbf{A}}$ for every* $\lambda \in \mathcal{S}(V_{\mathbf{h}}^n)$.

(5) For $\operatorname{pr}(\sigma) = \tau \in P_{\mathbf{h}}$ *we have*

$$({}^{\sigma}\lambda)(x) = |\det(a_\tau)_{\mathbf{h}}|_{\mathbf{A}}^{q/2}\chi_{\mathbf{h}}(\det(a_\tau))\mathbf{e}_{\mathbf{h}}\Big(\operatorname{tr}(S[x]a_\tau \cdot {}^t b_\tau)/2\Big)\lambda(xa_\tau),$$

where $S[x] = (S_v[x_v])_{v\in\mathbf{h}}$, $\mathbf{e}_{\mathbf{h}}(y) = \mathbf{e}_{\mathbf{A}}(y_{\mathbf{h}})$, *and* $(x_1, \ldots, x_n)a = (\sum_{i=1}^n x_i a_{ij})_{j=1}^n$ *for* $(x_1, \ldots, x_n) \in V_{\mathbf{h}}^n$ *and* $a \in (F_{\mathbf{h}})_n^n$.

(6) If $\iota = \begin{pmatrix} 0 & -1_n \\ 1_n & 0 \end{pmatrix}$, *we have*

$$({}^{\iota}\lambda)(x) = i^r \int_Y \lambda(y)\mathbf{e}_{\mathbf{h}}\left(-\sum_{i=1}^n S(x_i, y_i)\right) d^S y,$$

where $Y = V_{\mathbf{h}}^n$, $d^S y$ *is the Haar measure on Y such that the measure of* $(\sum_{i=1}^q \mathfrak{g}_v e_i)^n$ *for each* $v \in \mathbf{h}$ *with a basis* $\{e_i\}_{i=1}^q$ *of V over F is* $N(\mathfrak{d}_v)^{-qn/2} \cdot |\det(S(e_i, e_j))|_v^{n/2}$, *and* $r = (n/2)\sum_{v\in\mathbf{r}}(r_v - s_v)$ *or* $-n\sum_{v\in\mathbf{r}} s_v$ *according as q is even or odd.*

Since V has no fixed coordinate system, $\det(S)$ means the coset of $\det(S(e_i, e_j))$ modulo $\{a^2 \mid a \in F^{\times}\}$ with $\{e_i\}_{i=1}^n$ as in (6) above. Clearly χ is well defined. Thus our theorem is "coordinate-free" (as far as V is concerned). In the proof, however, it is convenient to use a matrix representation, and so hereafter we assume that $V = F^q$ and $S(x, y) = {}^t xSy$ with ${}^t S = S \in GL_q(F)$. Then $A_v \in GL_q(F_v)$, $S_v = {}^t A_v I_v A_v$, and $T_v = {}^t \overline{A}_v A_v$ for each $v \in \mathbf{a}$. In this setting, the formula of (6) can be written

$$(3.3)\qquad ({}^{\iota}\lambda)(x) = i^r |N_{F/\mathbf{Q}}(\det(S)|^{-n/2} \int_Y \lambda(y) \mathbf{e}_{\mathbf{h}}(-\mathrm{tr}({}^t xSy))dy,$$

where $Y = (F_n^q)_{\mathbf{h}}$, and the measure of $(\mathfrak{g}_v)_n^q$ is $N(\mathfrak{d}_v)^{-nq/2}$ for each $v \in \mathbf{h}$.

The proof of our theorem requires some preliminaries. We first define an embedding $\psi : \mathcal{H}^{(n)} \to \mathcal{H}^{(nq)}$ and an injective homomorphism $\alpha \mapsto \alpha_S$ of $G_{\mathbf{A}}^{(n)}$ into $G_{\mathbf{A}}^{(nq)}$ by $\psi(\mathfrak{z}) = (\psi_v(\mathfrak{z}_v))_{v \in \mathbf{a}}$,

$$\psi_v(x + iy) = S_v \otimes x + iT_v \otimes y \qquad \text{if } v \in \mathbf{r},$$

$$\psi_v(z + jw) = S_v \otimes z + j(T_v \otimes w) \qquad \text{if } v \in \mathbf{c},$$

$$\alpha_S = \begin{bmatrix} 1_q \otimes a_\alpha & S \otimes b_\alpha \\ S^{-1} \otimes c_\alpha & 1_q \otimes d_\alpha \end{bmatrix}.$$

It can easily be verified that $\psi(\alpha(\mathfrak{z})) = \alpha_S(\psi(\mathfrak{z}))$ and

$$\mu(\alpha_S, \psi(\mathfrak{z}))_v = (A_v \otimes 1_n)^{-1} \mathrm{diag}[1_{r_v} \otimes \mu(\alpha, \mathfrak{z})_v, 1_{s_v} \otimes \overline{\mu(\alpha, \mathfrak{z})_v}](A_v \otimes 1_n) \quad \text{if } v \in \mathbf{r},$$

$$\mu_0(\alpha_S, \psi(\mathfrak{z}))_v = (A_v \otimes 1_n)^{-1} [\mu_0(\alpha, \mathfrak{z})_v]_q (A_v \otimes 1_n) \quad \text{if } v \in \mathbf{c},$$

where $[z + jw]_q = 1_q \otimes z + j(1_q \otimes w)$ for $z + jw \in \mathbf{H}_n^n$. From these we obtain immediately

$$(3.4)\qquad m(\alpha_S, \psi(\mathfrak{z}))^{\mathbf{1}} = (m(\alpha, \mathfrak{z})^{\mathbf{1}})^q \prod_{v \in \mathbf{r}} m(\alpha, \mathfrak{z})_v^{-s_v} \overline{m(\alpha, \mathfrak{z})}_v^{s_v}.$$

To find a relationship between $J^S(\alpha, \mathfrak{z})$ and $h(\alpha_S, \psi(\mathfrak{z}))$, we take integral ideals $\mathfrak{b}$ and $\mathfrak{c}$ such that $\alpha_S \in D^{(nq)}[2\mathfrak{d}^{-1}, 2\mathfrak{d}]$ for every $\alpha \in D^{(n)}[2\mathfrak{b}\mathfrak{d}^{-1}, 2\mathfrak{c}\mathfrak{d}]$. Then $S \in GL_q(\mathfrak{g}_v)$ for every $v \nmid \mathfrak{b}\mathfrak{c}$.

LEMMA 3.3. *For $\alpha \in G \cap P_{\mathbf{A}}D[2\mathfrak{b}\mathfrak{d}^{-1}, 2\mathfrak{c}\mathfrak{d}]$ we have*

$$h(\alpha_S, \psi(\mathfrak{z})) = \chi_{\mathbf{a}}(\det(d_\alpha))\chi^*(\det(d_\alpha)\mathrm{il}(\alpha)^{-1})J^S(\alpha, \mathfrak{z}),$$

where χ^* is the ideal character associated with the Hecke character χ of Theorem 3.2.

Proof. When F is totally real, this was essentially given in [S4, Prop.11.3], whose proof is, in essence, applicable to the present situation. For the reader's convenience, we reproduce the proof with necessary modifications. From Theorem 1.2 (1), (2), and (3.4) we easily see that

$$(3.5)\qquad h(\alpha_S, \psi(\mathfrak{z})) = tJ^S(\alpha, \mathfrak{z}) \quad \textit{with} \quad t \in \mathbf{T}.$$

By Lemma 2.2 we have

$$\lim_{\mathfrak{z}\to 0} h(\alpha, \mathfrak{z})/|h(\alpha, \mathfrak{z})| = \gamma/|\gamma|, \quad \lim_{\mathfrak{z}\to 0} h(\alpha_S, \psi(\mathfrak{z}))/|h(\alpha_S, \psi(\mathfrak{z}))| = \gamma'/|\gamma'|$$

with $\gamma = \gamma(\delta^{-2}d_\alpha^{-1}c_\alpha)$ and $\gamma' = \gamma(\delta^{-2}S^{-1} \otimes d_\alpha^{-1}c_\alpha)$. Let $\alpha \in P_{\mathbf{A}}\tau$ with $\tau \in D[2\mathfrak{b}\mathfrak{d}^{-1}, 2\mathfrak{c}\mathfrak{d}]$; write simply c and d for c_τ and d_τ. Clearly $d_\alpha^{-1}c_\alpha = d^{-1}c$. From (3.2a) and (3.4) we see that $\gamma'/|\gamma'| = t\chi_{\mathbf{a}}(\det(d_\alpha))\xi$, where $\xi = 1$ or $\xi = \gamma^q/|\gamma^q|$ according as q is even or odd. If $v \nmid 2\mathfrak{b}\mathfrak{c}$, then $S^{-1} = {}^tT_v \cdot \mathrm{diag}[s_1, \ldots, s_q]T_v$ with $T_v \in GL_q(\mathfrak{g}_v)$ and $s_i \in \mathfrak{g}_v^\times$. By Lemma 2.1 we see that

$$\gamma_v(\delta^{-2}S^{-1} \otimes d^{-1}c) = \prod_{i=1}^{q} \gamma_v(\delta^{-2}s_i d^{-1}c) = \gamma_v(\delta^{-2}d^{-1}c)^q \left(\frac{\det(S)}{\nu_0(d^{-1}c)_v}\right).$$

By Lemma 2.3 and Proposition 2.4 we have

$$\nu_0(d^{-1}c) = \nu_0(d_\alpha^{-1}c_\alpha) = \det(d_\alpha)\mathrm{il}(\alpha)^{-1}.$$

If $v|2\mathfrak{b}\mathfrak{c}$, then $|\det(d)|_v = 1$, and hence both $(2^{-1}\delta^{-1}d^{-1}c)_v$ and $(2^{-1}\delta^{-1}S^{-1} \otimes d^{-1}c)_v$ are v-integral. Therefore (2.3b) shows that $\gamma_v = \gamma'_v = 1$. Notice that $\prod_{v\nmid 2} \gamma_v^q/|\gamma_v|^q = \theta(\nu_0(d^{-1}c))^{q/2}$ by Lemma 2.1, if q is even. Combining all these, we obtain our lemma. □

Proof of Theorem 3.2. We are identifying V with F^q and V^n with F_n^q, and hence $S_v[p] = {}^tpS_vp$ and $T_v\{p\} = {}^t\overline{p}T_vp$ for $p \in (F_n^q)_v$. Define $\mathcal{A} : \mathcal{U}^{(nq)} \to \mathcal{U}^{(nq)}$ by $\mathcal{A}(u)_v = ({}^tA_v \otimes 1_n)u_v$ or ${}^t\kappa(A_v \otimes 1_n)u_v$ according as $v \in \mathbf{r}$ or $v \in \mathbf{c}$, and also ω :

$F^1_{nq} \to F^q_n$ by $\omega(x_1 \ \ldots \ x_q) = {}^t({}^tx_1 \ \ldots \ {}^tx_q)$ for $x_i \in F^1_n$. Then a straightforward calculation shows that

$$g(\mathfrak{z}, u; \lambda) = \theta^{(nq)}\big(\psi(\mathfrak{z}), \mathcal{A}(u); \lambda \circ \omega\big), \tag{3.6a}$$

$$\alpha_S\big(\psi(\mathfrak{z}), \mathcal{A}(u)\big) = \big(\psi(\alpha\mathfrak{z}), \mathcal{A}(w)\big) \ \ \textit{if} \ \alpha(\mathfrak{z}, u) = (\alpha\mathfrak{z}, w). \tag{3.6b}$$

For $\alpha \in G \cap \mathfrak{M}_q$ we can find an element $\xi \in \prod_{v\in\mathbf{a}} Mp((F^1_{nq})_v)$ such that $\mathrm{pr}(\xi) = (\alpha_S)_\mathbf{a}$ and $\prod_{v\in\mathbf{a}} g(\xi_v, \psi(\mathfrak{z})_v) = J^S(\alpha, \mathfrak{z})$. Then, for $\ell = \lambda \circ \omega$ Proposition 1.3 together with (3.6a, b) shows that

$$J^S(\alpha, \mathfrak{z})g(\mathfrak{z}, u; \lambda) = \theta^{(nq)}\big(\alpha_S(\psi(\mathfrak{z}), \mathcal{A}(u)); \ell'\big) = g\big(\alpha(\mathfrak{z}, u); \lambda'\big)$$

with $\lambda' = \ell' \circ \omega^{-1}$. Putting $\lambda' = {}^\alpha\lambda$, we obtain (0). Let $\mathfrak{f}$ be the conductor of χ; let $E = D[2\mathfrak{b}\mathfrak{d}^{-1}, 2\mathfrak{c}\mathfrak{d}]$ and $E' = \{\,\alpha \in E \mid \chi_\mathfrak{f}(\det(d_\alpha)) = 1\,\}$, where $\chi_\mathfrak{f} = \prod_{v|\mathfrak{f}} \chi_v$. If $\alpha \in G \cap P_\mathbf{A}E$, then Lemma 3.3 and Proposition 1.3, combined with the above argument, show that

$$ {}^\alpha\lambda \circ \omega = \chi_\mathbf{a}(\det(d_\alpha))\chi^*(\det(d_\alpha)\mathrm{il}(\alpha)^{-1}) \cdot {}^\beta(\lambda \circ \omega) \tag{*}$$

with $\beta = \alpha_S$. In particular, ${}^\alpha\lambda \circ \omega = {}^\beta(\lambda \circ \omega)$ if $\alpha \in G \cap E'$. This together with Theorem 1.2, (6) shows that ${}^\alpha\lambda = \lambda$ for every α in a congruence subgroup Γ of G depending on λ. Take an open subgroup D of E' so that $\Gamma \supset G \cap D$. We take $D \subset C^\theta$ if q is odd. By strong approximation, given $\sigma \in \mathfrak{M}_q$, we can take $\alpha \in G$ so that $\mathrm{pr}(\sigma) \in \alpha D$. Then $\alpha \in \mathfrak{M}_q$. Define ${}^\sigma\lambda$ to be ${}^\alpha\lambda$. It is then easy to verify that this is well defined and has property (2) for even q, and also properties (3) and (4). To prove (5), let $\mathrm{pr}(\sigma) = \tau \in P_\mathbf{h}$. Given $\lambda \in \mathcal{S}((F^q_n)_\mathbf{h})$, take $\alpha \in G$ so that $\tau \in \alpha E'$ and that ${}^\sigma\lambda = {}^\alpha\lambda$. Then we easily see that

$$\chi_\mathbf{a}(\det(d_\alpha))\chi^*(\det(d_\alpha)\mathrm{il}(\alpha)^{-1}) = \chi(\det(a_\tau)). \tag{**}$$

Let $\ell = \lambda \circ \omega$ and $\beta = \alpha_S$. From (*) and (**) we obtain ${}^\beta\ell = \chi(\det(a_\tau))({}^\alpha\lambda) \circ \omega$. Changing E' for a smaller group if necessary, we may assume that τ_S and α_S have the same effect on ℓ. Then we obtain

$$\chi(\det(a_\tau))({}^\alpha\lambda) \circ \omega = {}^\varphi\ell \ \ \textit{with} \ \ \mathrm{pr}(\varphi) = \tau_S. \tag{***}$$

Taking φ to be $r_P(\tau_S)$, from Theorem 1.2(8) and (1.3a) we obtain

$${}^\varphi\ell(y) = |\det(1_q \otimes a_\tau)|^{1/2}_\mathbf{A}\, \mathbf{e_h}\big(y(S \otimes a_\tau \cdot {}^tb_\tau) \cdot {}^ty/2\big)\, \ell(y(1_q \otimes a_\tau)),$$

which combined with (***) proves (5). To prove (6), or rather (3.3), we first observe that (3.5) is valid for $\alpha = \iota$. Now the first formula of Lemma 2.2 shows that both $h(\iota, \mathfrak{z})$ and $h(\iota_S, \psi(\mathfrak{z}))$ are positive if $\mathfrak{z} \in \mathbf{R}\mathbf{i}$ with $\mathbf{i}$ of that lemma, and hence $i^r \cdot h(\iota_S, \psi(\mathfrak{z})) = J^S(\iota, \mathfrak{z})$ with r as in (6). Then from (3.6a, b) we obtain ${}^{\iota}\lambda \circ \omega = i^r \cdot {}^{\gamma}(\lambda \circ \omega)$ with $\gamma = \iota_S$. Observe that γ belongs to the set $P_{\mathbf{A}}C^*$ of Theorem 1.2(7) (of degree nq). Therefore, by (1.3c) we have

$$^{\gamma}\ell(x) = |\det(c_\gamma)_{\mathbf{h}}|_{\mathbf{A}}^{1/2} \int_Y \ell(yc_\gamma)\mathbf{e}_{\mathbf{h}}(xb_\gamma \cdot {}^t c_\gamma \cdot {}^t y)dy \qquad (Y = (F^1_{nq})_{\mathbf{h}}),$$

which can easily be transformed to (3.3). To prove (2) when q is odd, first let $\sigma \in \mathfrak{M}$ and $\mathrm{pr}(\rho) \in C^\theta$. With an open normal subgroup D of C^θ, take $\alpha \in G \cap \mathrm{pr}(\sigma)D$ and $\beta \in G \cap \mathrm{pr}(\rho)D$. Then $\alpha\beta \in G \cap \mathrm{pr}(\sigma\rho)D$. Take D so small that ${}^{\sigma}({}^{\beta}\lambda) = {}^{\alpha}({}^{\beta}\lambda)$, ${}^{\beta}\lambda = {}^{\rho}\lambda$, and ${}^{(\sigma\rho)}\lambda = {}^{(\alpha\beta)}\lambda$. Now $\beta \in G \cap C^\theta$, and hence (3.2b) shows that $J^S(\alpha\beta, \mathfrak{z}) = J^S(\alpha, \beta\mathfrak{z})J^S(\beta, \mathfrak{z})$, and hence we obtain ${}^{(\alpha\beta)}\lambda = {}^{\alpha}({}^{\beta}\lambda)$ from (0). Thus ${}^{(\sigma\rho)}\lambda = {}^{(\alpha\beta)}\lambda = {}^{\alpha}({}^{\beta}\lambda) = {}^{\sigma}({}^{\beta}\lambda) = {}^{\sigma}({}^{\rho}\lambda)$. Next let $\mathrm{pr}(\tau) \in P_{\mathbf{A}}$ and $\sigma \in \mathfrak{M}$. Then $\sigma = \pi\rho$ with $\mathrm{pr}(\pi) \in P_{\mathbf{A}}$ and $\mathrm{pr}(\rho) \in C^\theta$. From (***) we see that ${}^{(\tau\pi)}\zeta = {}^{\tau}({}^{\pi}\zeta)$ for every $\zeta \in \mathcal{S}((F^q_m)_{\mathbf{h}})$. Therefore ${}^{(\tau\sigma)}\lambda = {}^{(\tau\pi\rho)}\lambda = {}^{(\tau\pi)}({}^{\rho}\lambda) = {}^{\tau}({}^{\pi}({}^{\rho}\lambda)) = {}^{\tau}({}^{(\pi\rho)}\lambda) = {}^{\tau}({}^{\sigma}\lambda)$. This proves (2). Finally, to prove (1), it is sufficient to show that ${}^{\alpha}\lambda$ for $\alpha \in G \cap \mathfrak{M}_q$ is independent of $\{A_v\}$. If q is even this follows from (5) and (6), since P and ι generate G. Suppose q is odd. By (2) and (5), it is sufficient to show that ${}^{\alpha}\lambda$ for $\alpha \in G \cap C^\theta$ is independent of $\{A_v\}$. Now $\alpha^m \in D[2\mathfrak{b}\mathfrak{d}^{-1}, 2\mathfrak{c}\mathfrak{d}]$ for some positive integer m. Therefore Lemma 3.3 shows that $h(\alpha_S, \psi(\mathfrak{z})) = tJ^S(\alpha, \mathfrak{z})$ with a root of unity t. Then ${}^{\alpha}\lambda \circ \omega = t^{-1} \cdot {}^{\gamma}(\lambda \circ \omega)$ with $\gamma = \alpha_S$. Now the set of all $\{A_v\}$ is not necessarily connected, but it can easily be shown that the set of all $\{T_v\}$ is connected. Since $h(\alpha_S, \psi(\mathfrak{z}))$ is continuous in $\{T_v\}$, we see that t does not depend on $\{A_v\}$. This completes the proof. □

We are going to introduce a series involving harmonic polynomials on V^n. Given a finite-dimensional complex vector space W and $0 \le a \in \mathbf{Z}$, we denote by $\mathfrak{S}_a(W)$ the vector space of all $\mathbf{C}$-valued homogeneous polynomial functions on W of degree a. We then denote by $\mathcal{P}_a(\mathbf{C}^q_m)$ the vector subspace of $\mathfrak{S}_a(\mathbf{C}^q_m)$ spanned by the functions p satisfying the condition

$$\text{(3.7a)} \qquad \sum_{i=1}^{q} \partial^2 p/\partial x_{ih}\partial x_{ik} = 0 \quad \textit{for every } h \textit{ and } k,$$

where $x = (x_{ih})$ is a variable on $\mathbf{C}^q_m$. For instance, we can take $p(x) = \varphi({}^t\rho x)$ with $\varphi \in \mathfrak{S}_a(\mathbf{C}^m_m)$ and $\rho \in \mathbf{C}^q_m$ satisfying the condition

$$\text{(3.7b)} \qquad {}^t\rho_h\rho_k = 0 \quad \textit{whenever} \quad \partial^2\varphi/\partial y_{hi}\partial y_{kj} \neq 0 \textit{ for some } i \textit{ and } j,$$

where ρ_h denotes the h-th column of ρ, and $y = (y_{hi})$ is a variable on $\mathbf{C}^m_m$.

LEMMA 3.4. *Let* $\omega(x) = \exp\left(\sum_{h,k=1}^{m}\sum_{i=1}^{q} c_{hk}x_{ih}x_{ik}\right)$ *for* $x \in \mathbf{R}_m^q$ *with* $c_{hk} = c_{kh} \in \mathbf{C}$. *Then* $[p(D)(\omega\psi)](0) = [p(D)\psi](0)$ *for every* $p \in \mathcal{P}_a(\mathbf{C}_m^q)$ *and every* C^∞ *function* ψ *in* x, *where* D *is the* $(q \times m)$*-matrix whose* (i, h)*-entry is* $\partial/\partial x_{ih}$.

Proof. We first observe that if α is a polynomial in n variables $y_1, \ldots, y_n$ and $\alpha_i = \partial\alpha/\partial y_i$, then

$$(*)\qquad [\alpha(\partial/\partial y_1, \ldots, \partial/\partial y_n)(y_i\beta)](0) = [\alpha_i(\partial/\partial y_1, \ldots, \partial/\partial y_n)\beta](0)$$

for every i and every C^∞ function β. This is completely elementary. Now our lemma is trivial if $a = 0$. Assume that it is true for degree $< a$ and that $a > 0$. We have $ap(x) = \sum_{i,h} x_{ih}p_{ih}(x)$ with $p_{ih} = \partial p/\partial x_{ih}$, and hence

$$\begin{aligned}[ap(D)(\omega\psi)](0) &= \sum_{i,h}[p_{ih}(D)\partial/\partial x_{ih}(\omega\psi)](0)\\ &= \sum_{i,h}[p_{ih}(D)(\omega\cdot\partial\psi/\partial x_{ih})](0) + \sum_{i,h,k}2c_{hk}[p_{ih}(D)(x_{ik}\omega\psi)](0).\end{aligned}$$

Since $p_{ih} \in \mathcal{P}_{a-1}(\mathbf{C}_m^q)$, by our induction assumption the first sum on the last line is $\sum_{i,h}[p_{ih}(D)\partial\psi/\partial x_{ih}](0)$, which is $[ap(D)\psi](0)$. By (*) the second sum equals $\sum_{i,h,k} 2c_{hk}[(\partial p_{ih}/\partial x_{ik})(D)(\omega\psi)](0)$, which is 0 by (3.7a). This completes the proof. □

Coming back to the space V and the form S, for each $v \in \mathbf{r}$ put

$$(3.8a)\qquad X_v^+ = \{x \in V_v \mid (A_v x)_i = 0 \ \ for \ \ i > r_v\},$$

$$(3.8b)\qquad X_v^- = \{x \in V_v \mid (A_v x)_i = 0 \ \ for \ \ i \le r_v\},$$

where y_i for $y \in \mathbf{R}^q$ means the i-th component of y. Clearly $V_v = X_v^+ \oplus X_v^-$. For $y \in V_v$, we denote by y^+ and y^- the projections of y to X_v^+ and X_v^-.

Given $m, m' \in \mathbf{Z}^{\mathbf{r}}$ and $t \in \mathbf{Z}^{\mathbf{c}}$ whose components are all nonnegative, we denote by $\mathcal{P}_{m,m',t}(V^n)$ the vector space over $\mathbf{C}$ spanned by all functions p on $V_{\mathbf{a}}^n = \prod_{v\in\mathbf{a}} V_v^n$ of the form

$$(3.9)\quad p(x) = \prod_{v\in\mathbf{r}} p_v(A_v x_v^+)p'_v(A_v x_v^-)\prod_{v\in\mathbf{c}} p''_v(\mathrm{Re}(A_v x_v), \mathrm{Im}(A_v x_v)) \qquad (x \in V_{\mathbf{a}}^n)$$

with $p_v \in \mathcal{P}_{m_v}(\mathbf{C}_n^{r_v})$, $p'_v \in \mathcal{P}_{m'_v}(\mathbf{C}_n^{s_v})$ if $v \in \mathbf{r}$, and $p''_v \in \mathcal{P}_{t_v}(\mathbf{C}_{2n}^q)$ if $v \in \mathbf{c}$, where $A_v x_v^\pm = (A_v x_{v1}^\pm, \ldots, A_v x_{vn}^\pm)$ for $x_v = (x_{v1}, \ldots, x_{vn})$ with $x_{vi} \in V_v$. Write p of

(3.9) as $p = (p_v, p'_v, p''_v)$. We let every element μ of $GL_n(\mathbf{C})^{\mathbf{r}} \times GL_{2n}(\mathbf{C})^{\mathbf{c}}$ act $\mathbf{C}$-linearly on $\mathcal{P}_{m,m',t}(V^n)$ by defining $\mu p = (\mu_v p_v, \overline{\mu}_v p'_v, \mu_v p''_v)$ with $(\nu s)(y) = s(y\nu)$ for $s = p_v, p'_v, p''_v$ and $\nu = \mu_v$ or $\overline{\mu}_v$. Notice that this action is compatible with both (3.7a) and (3.7b).

Now for $\mathfrak{z} \in \mathcal{H}$, $\lambda \in \mathcal{S}(V^n_{\mathbf{h}})$, and $p \in \mathcal{P}_{m,m',t}(V^n)$ we consider a series

$$f(\mathfrak{z}; \lambda, p) = \sum_{\xi \in V^n} \lambda(\xi_{\mathbf{h}}) p(\xi_{\mathbf{a}}) \Phi(\xi; \mathfrak{z}, 0). \tag{3.10}$$

THEOREM 3.5. *The notation ${}^{\alpha}\lambda$ being as in Theorem 3.2, we have*

$$J^S(\alpha, \mathfrak{z})^{-1} f(\alpha(\mathfrak{z}); {}^{\alpha}\lambda, {}^{t}\mu(\alpha, \mathfrak{z})^{-1} p) = f(\mathfrak{z}; \lambda, p)$$

for every $\alpha \in G \cap \mathfrak{M}_q$.

Proof. Write the variable u in the form ${}^t u_v = (u^1_{v1}, \ldots, u^n_{v1}, \ \ldots \ , u^1_{vq}, \ldots, u^n_{vq})$ if $v \in \mathbf{r}$ and ${}^t u_v = (a^1_{v1}, \ldots, a^n_{v1}, \ \ldots \ , a^1_{vq}, \ldots, a^n_{vq}, b^1_{v1}, \ldots, b^n_{v1}, \ \ldots \ , b^1_{vq}, \ldots, b^n_{vq})$ if $v \in \mathbf{c}$. We can then define a differential operator

$$B = \prod_{v \in \mathbf{r}} p_v(D_v) p'_v(D'_v) \prod_{v \in \mathbf{c}} p''_v(E_v, E'_v)$$

on $\mathcal{U}$, where $D_v = (\partial u^j_{vi})$ with $1 \le i \le r_v$, $1 \le j \le n$, $D'_v = (\partial u^j_{vi})$ with $r_v < i \le q$, $1 \le j \le n$, $E_v = (\partial / \partial a^j_{vi})$, $E'_v = (\partial / \partial b^j_{vi})$ with $1 \le i \le q$, $1 \le j \le n$. Employing Lemma 3.4 we can easily verify that $[Bg(\mathfrak{z}, u; \lambda)]_{u=0} = 2^M (2\pi i)^{M+N} f(\mathfrak{z}; \lambda, p)$, where $N = \sum_{v \in \mathbf{r}} (m_v + m'_v)$ and $M = \sum_{v \in \mathbf{c}} t_v$. Therefore we obtain our assertion by applying B to the equality of Theorem 3.2, (0). □

We can associate with the above f a function $f_{\mathbf{A}}(x; \lambda, p)$ with a variable x on $G_{\mathbf{A}}$ or $M_{\mathbf{A}}$, according as q is even or odd, by

$$f_{\mathbf{A}}(\alpha w; \lambda, p) = J^S(w, \mathbf{i})^{-1} f(w(\mathbf{i}); {}^{w}\lambda, {}^{t}\mu(w, \mathbf{i})^{-1} p) \tag{3.11}$$

for $\alpha \in G$ and $w \in \mathfrak{M}_q$, where $\mathbf{i}$ is as in Lemma 2.2; we take $\mathrm{pr}(w) \in C^{\theta}$ if q is odd. This is well defined by virtue of Theorem 3.5. Now we have

PROPOSITION 3.6. *For every $\alpha \in G$ and $y \in \mathfrak{M}_q$ such that $y(\mathbf{i}) = \mathbf{i}$ and that $\mathrm{pr}(y) \in C^{\theta}$ if q is odd, we have*

$$f_{\mathbf{A}}(\alpha x y; \lambda, p) = J^S(y, \mathbf{i})^{-1} f_{\mathbf{A}}(x; {}^{y}\lambda, {}^{t}\mu(y, \mathbf{i})^{-1} p).$$

Proof. Given x, take $\beta \in G$ and $w \in \mathfrak{M}_q$ so that $x = \beta w$ and $\mathrm{pr}(w) \in C^\theta$. Then

$$\begin{aligned} f_{\mathbf{A}}(\alpha xy, \lambda, p) &= f_{\mathbf{A}}(wy, \lambda, p) \\ &= J^S(wy, \mathbf{i})^{-1} f\big(w(\mathbf{i}), {}^{(wy)}\lambda, {}^t\mu(wy, \mathbf{i})^{-1} p\big) \\ &= J^S(y, \mathbf{i})^{-1} J^S(w, \mathbf{i})^{-1} f\big(w(\mathbf{i}), {}^w({}^y\lambda), {}^t\mu(w, \mathbf{i})^{-1} \cdot {}^t\mu(y, \mathbf{i})^{-1} p\big) \\ &= J^S(y, \mathbf{i})^{-1} f_{\mathbf{A}}(x, {}^y\lambda, {}^t\mu(y, \mathbf{i})^{-1} p). \qquad \square \end{aligned}$$

PROPOSITION 3.7. *The notation being as in Proposition 2.6, we have* $J^S(\sigma^*, -\mathfrak{z}^\rho) = \overline{J^S(\sigma, \mathfrak{z})}$ *and* ${}^{\sigma^*}(\lambda^*) = ({}^\sigma\lambda)^*$ *for every* $\sigma \in \mathfrak{M}_q$ *and* $\lambda \in \mathcal{S}(V_{\mathbf{h}}^n)$, *where* λ^* *is defined by* $\lambda^*(x) = \overline{\lambda(-x)}$.

Proof. The first equality follows immediately from Proposition 2.6 and (3.2a). By virtue of strong approximation and (4) of Theorem 3.2, it is sufficient to prove the second assertion when $\sigma \in G \cap \mathfrak{M}_q$, in which case the desired fact follows from (0) of Theorem 3.2, since $g(-\mathfrak{z}^\rho, \bar{u}; \lambda^*) = \overline{g(\mathfrak{z}, u; \lambda)}$. $\square$

Remarks. (I) Define an algebraic group $\mathfrak{G}$ by

$$\mathfrak{G} = \{\, \alpha \in GL(V) \mid S(\alpha x, \alpha x) = S(x, x) \,\}.$$

Fixing $(A_v)_{v \in \mathbf{a}}$ as above, put $A_v^\alpha = A_v \alpha_v$ for every $\alpha \in \mathfrak{G}_{\mathbf{a}}$. Then we can define our series g and f with A_v^α in place of A_v. Thus g and f are essentially parametrized by $\mathfrak{G}_{\mathbf{a}}$. In Section 5 we shall discuss these series for $q = 3$ as functions on the quotient of $\mathfrak{G}_{\mathbf{a}}$ by its maximal compact subgroup.

(II) For every $p'' \in \mathcal{P}_{t_v}(\mathbf{C}_{2n}^q)$ we can find an element q of $\mathcal{P}_{t_v}(\mathbf{C}_{2n}^q)$ such that $p''\big(\mathrm{Re}(u), \mathrm{Im}(u)\big) = q(u, \bar{u})$ for $u \in \mathbf{C}_n^q$. Therefore, in (3.9), instead of $p_v''\big(\mathrm{Re}(A_v x_v), \mathrm{Im}(A_v x_v)\big)$ we can also take $q_v(A_v x_v, \overline{A_v x_v})$ with $q_v \in \mathcal{P}_{t_v}(\mathbf{C}_{2n}^q)$. Theorem 3.4 is still valid if we define $(\mu_v q_v)(y) = q_v(y \tau \mu_v \tau^{-1})$ with $\tau = \begin{pmatrix} 1_n & -i1_n \\ 1_n & i1_n \end{pmatrix}$.

(III) If $r \le \mathrm{Min}(q, m)$, we easily see that the subdeterminants of $x \in \mathbf{C}_m^q$ of degree r define elements of $\mathcal{P}_r(\mathbf{C}_m^q)$. In particular, if $r_v = q = n$ and $v \in \mathbf{r}$, we can take $p_v(A_v x_v^+) = \det(A_v x_v)$ in (3.9). In this case $\mu_v p_v = \det(\mu_v) p_v$. If $q = n$ and $v \in \mathbf{c}$, we can take $p_v''\big(\mathrm{Re}(A_v x_v), \mathrm{Im}(A_v x_v)\big)$ to be any subdeterminant of $\big(\mathrm{Re}(A_v x_v), \mathrm{Im}(A_v x_v)\big)$ of degree n.

3.b. Some more explicit transformation formulas. Let us now derive from Theorem 3.2 more explicit formulas for ${}^\alpha\lambda$ in two forms convenient in applications. For simplicity, we put ${}^\alpha\lambda = {}^\beta\lambda$ if $\alpha = \mathrm{pr}(\beta)$ with $\beta \in \mathfrak{M}_q$, q odd. This is meaningful in view of Theorem 3.2(3). Now our first formula is:

LEMMA 3b.1. *Let $\mathfrak{f}$ be the conductor of χ of Theorem 3.2, and χ^* the ideal character associated with χ. For $\lambda \in S(V_{\mathbf{h}}^n)$ put*

$$U_\lambda = \Big\{ \sigma \in D[2\mathfrak{d}^{-1}, 2\mathfrak{f}\mathfrak{d}] \mid {}^\sigma\lambda = \lambda \text{ and } \det(d_\sigma)_v \equiv 1 \pmod{\mathfrak{f}_v} \text{ for every } v|\mathfrak{f} \Big\}.$$

Then, for every $\alpha \in \mathrm{diag}[p, {}^tp^{-1}]U_\lambda$ with $p \in GL_n(F_{\mathbf{h}})$ we have

$$({}^\alpha\lambda)(x) = |\det(p)_{\mathbf{h}}|_{\mathbf{A}}^{q/2} \chi_{\mathbf{h}}(\det(p))\lambda(xp),$$

where xp is as in Theorem3.2(5). In particular, if such an α belongs to G, then

$$({}^\alpha\lambda)(x) = \chi_{\mathbf{a}}\big(\det(d_\alpha)\big)\chi^*\big(\det(d_\alpha)\mathrm{il}(\alpha)^{-1}\big)N(\mathrm{il}(\alpha))^{q/2}\lambda(xp).$$

Proof. Let $\alpha = \tau\sigma$ with $\tau = \mathrm{diag}[p, {}^tp^{-1}]$ and $\sigma \in U_\lambda$. By Theorem 3.2(2, 5) we have $({}^\alpha\lambda)(x) = ({}^\tau\lambda)(x) = |\det(p)_{\mathbf{h}}|_{\mathbf{A}}^{q/2}\chi_{\mathbf{h}}(\det(p))\lambda(xp)$. Suppose $\alpha \in G$. Since ${}^tpd_\alpha = d_\sigma$, we have $\chi_{\mathbf{a}}\big(\det(d_\alpha)\big)\chi^*\big(\det(d_\alpha)\mathrm{il}(\alpha)^{-1}\big) = \chi_{\mathbf{h}}(\det(p))\chi_{\mathfrak{f}}(\det(d_\sigma^{-1})) = \chi_{\mathbf{h}}(\det(p))$, which completes the proof.

PROPOSITION 3b.2. *Given $\lambda \in S(V_{\mathbf{h}}^n)$, let M be a $\mathfrak{g}$-lattice in V^n such that $\lambda(x+u) = \lambda(x)$ for every $u \in M$. Further let $\mathfrak{x}$, $\mathfrak{y}$, and $\mathfrak{z}$ be fractional ideals of F with the following properties:*

(i) $2(1+\delta_{ij})^{-1}S(x_i, x_j) \in \mathfrak{x}$ *for every i, j and every $x \in V^n$ such that $\lambda(x) \neq 0$.*

(ii) $2(1+\delta_{ij})^{-1}S(y_i, y_j) \in \mathfrak{y}$ *for every i, j and every $y \in M'$, where*

$$M' = \left\{ y \in V^n \,\middle|\, \sum_{i=1}^n S(x_i, y_i) \in \mathfrak{d}^{-1} \text{ for every } x \in M \right\}.$$

(iii) $\lambda(xa) = \lambda(x)$ *for every $a \in \prod_{v\in\mathbf{h}} GL_n(\mathfrak{g}_v)$ such that $a_v - 1 \in (\mathfrak{z}_v)_n^n$ for every $v \in \mathbf{h}$, where xa is as in Theorem 3.2(5).*
Then with χ as in Theorem 3.2 we have

$$({}^\gamma\lambda)(x) = \chi_{\mathfrak{f}}\big(\det(a_\gamma)\big)\lambda\big(x(a_\gamma)_{\mathfrak{z}}\big)$$

for every $\gamma \in B$, where $(a_\gamma)_{\mathfrak{z}}$ is the projection of a_γ to $\prod_{v|\mathfrak{z}} GL_n(F_v)$,

$$B = D[2\mathfrak{d}^{-1}\mathfrak{x}^{-1}, 2\mathfrak{d}^{-1}\mathfrak{y}^{-1} \cap \{2^{-1}\mathfrak{d}\mathfrak{x}(\mathfrak{f} \cap \mathfrak{z})\}]$$

if q is even, and $B = D[2\mathfrak{d}^{-1}\mathfrak{a}, 2^{-1}\mathfrak{d}\mathfrak{a}^{-1}\mathfrak{b}]$, $\mathfrak{a} = \mathfrak{x}^{-1} \cap \mathfrak{g}$, $\mathfrak{b} = \mathfrak{f} \cap \mathfrak{z} \cap 4\mathfrak{a} \cap 4\mathfrak{d}^{-2}\mathfrak{a}\mathfrak{y}^{-1}$ if q is odd.

This was essentially given in [S4, Proposition 11.7], except that the factor of automorphy there for even q is different from the present one. The point of this new formulation for even q is that the result can be given in an improved form. Anyway we are going to give a somewhat simpler proof. We start with

LEMMA 3b.3. *Let $\mathfrak{b}$ and $\mathfrak{c}$ be fractional ideals in F such that $\mathfrak{bc}$ is integral, and $\mathfrak{a}$ an integral ideal such that $\mathfrak{a} \subset \mathfrak{b} \cap \mathfrak{c}$. Further let $E(\mathfrak{b}) = G_{\mathbf{a}} \prod_{v \in \mathbf{h}} E_v(\mathfrak{b})$ and $E'(\mathfrak{c}) = G_{\mathbf{a}} \prod_{v \in \mathbf{h}} E'_v(\mathfrak{c})$, where $E_v(\mathfrak{b})$ resp. $E'_v(\mathfrak{c})$ denotes the set of all elements of G_v of the form $\begin{pmatrix} 1_n & b \\ 0 & 1_n \end{pmatrix}$ resp. $\begin{pmatrix} 1_n & 0 \\ c & 1_n \end{pmatrix}$ with $b \in (\mathfrak{b}_v)^n_n$ resp. $c \in (\mathfrak{c}_v)^n_n$. Then $D[\mathfrak{b}, \mathfrak{c}]$ is generated by $E(\mathfrak{b})$, $E'(\mathfrak{c})$, and $D[\mathfrak{a}, \mathfrak{a}]$.*

Proof. Since $D_v[\mathfrak{a}, \mathfrak{a}] = D_v[\mathfrak{b}, \mathfrak{c}]$ if $v \nmid \mathfrak{a}$, it is sufficient to show that $D_v[\mathfrak{b}, \mathfrak{c}]$ is generated by $E_v(\mathfrak{b})$, $E'_v(\mathfrak{c})$, and $D_v[\mathfrak{a}, \mathfrak{a}]$. If $v \nmid \mathfrak{bc}$, then $D_v[\mathfrak{b}, \mathfrak{c}]$ is conjugate to $D_v[\mathfrak{g}, \mathfrak{g}] = Sp(n, \mathfrak{g}_v)$, and hence our assertion follows from a well known fact that $Sp(n, \mathfrak{g}_v)$ is generated by $E_v(\mathfrak{g})$, $E'_v(\mathfrak{g})$, and $\mathrm{diag}[a, {}^t a^{-1}]$ with $a \in GL_n(\mathfrak{g}_v)$. If $v | \mathfrak{bc}$ and $\begin{pmatrix} a & b \\ c & d \end{pmatrix} \in D_v[\mathfrak{b}, \mathfrak{c}]$, then ${}^t ad \equiv 1 \pmod{(\mathfrak{bc})_v}$, and hence $a \in GL_n(\mathfrak{g}_v)$, and

$$\begin{pmatrix} a & b \\ c & d \end{pmatrix} = \begin{pmatrix} 1 & 0 \\ ca^{-1} & 1 \end{pmatrix} \begin{pmatrix} a & 0 \\ 0 & {}^t a^{-1} \end{pmatrix} \begin{pmatrix} 1 & a^{-1}b \\ 0 & 1 \end{pmatrix},$$

which proves our lemma. □

Proof of Proposition 3b.2. We use the matrix representation as in the proof of Theorem 3.2. Let $\alpha \in P_{\mathbf{A}}$ with $a_\alpha = 1_n$. Then Theorem 3.2(5) shows that ${}^\alpha\lambda(x) = \lambda(x)\mathbf{e_h}(\mathrm{tr}({}^t xSxb_\alpha)/2)$. Therefore ${}^\alpha\lambda = \lambda$ if $\alpha \in E(2\mathfrak{d}^{-1}\mathfrak{r}^{-1})$. Put $\beta = \iota^{-1}\alpha\iota$ and $\lambda' = {}^\iota\lambda$. Substituting $y+z$ for y in (3.3), we find that $\lambda'(x) = \mathbf{e_h}(\mathrm{tr}({}^t xSz))\lambda'(x)$ for every $z \in M$, and hence $\lambda'(x) \neq 0$ only if $x \in M'$. By (ii) this means that $\lambda'(x) \neq 0$ only if ${}^t xSx$ has entries in $\mathfrak{y}$. Therefore ${}^\alpha\lambda' = \lambda'$ if $\alpha \in E(2\mathfrak{d}^{-1}\mathfrak{y}^{-1})$. Suppose $\beta \in E'(2\mathfrak{d} \cap 2\mathfrak{d}^{-1}\mathfrak{y}^{-1})$. Since $\beta \in C^\theta$ and $\alpha \in P_{\mathbf{A}}$, we have ${}^\iota({}^\beta\lambda) = {}^{(\iota\beta)}\lambda = {}^{(\alpha\iota)}\lambda = {}^\alpha({}^\iota\lambda) = {}^\iota\lambda$, and hence ${}^\beta\lambda = \lambda$. By Lemma 3b.1 the expected formula for ${}^\gamma\lambda$ is true for $\gamma \in D[\mathfrak{e}, \mathfrak{e}]$ with a suitable ideal $\mathfrak{e}$. We have seen that it is also true for $\gamma \in E(2\mathfrak{d}^{-1}\mathfrak{r}^{-1}) \cup E'(2\mathfrak{d} \cap 2\mathfrak{d}^{-1}\mathfrak{y}^{-1})$, which together with Lemma 3b.3 proves our proposition for odd q, since ${}^\delta({}^\varepsilon\lambda) = {}^{\delta\varepsilon}\lambda$ at least for $\delta, \varepsilon \in C^\theta$. If q is even, the associativity is true for all $\delta, \varepsilon \in G_{\mathbf{A}}$, and so ${}^\beta\lambda = \lambda$ for $\beta \in E(2\mathfrak{d}^{-1}\mathfrak{y}^{-1})$. Therefore we can take B in the form stated in our proposition. □

In the proof of [S4, Proposition 11.7] we employed the following lemma for which we only mentioned some ideas of the proof; also its special case was stated as Lemma 3.4 in the same paper. The fact, though unnecessary in the present paper, is of independent interest, and so we give here a proof.

LEMMA 3b.4. *The notation being as in Lemma 3b.3, let* $T(\mathfrak{b}) = G \cap E(\mathfrak{b})$ *and* $T'(\mathfrak{c}) = G \cap E'(\mathfrak{c})$. *Then* $\Gamma[\mathfrak{b}, \mathfrak{c}]$ *is generated by* $T(\mathfrak{b})$, $T'(\mathfrak{c})$, *and* $\Gamma[\mathfrak{a}, \mathfrak{a}]$.

Proof. Let X be an open normal subgroup of $D[\mathfrak{b}, \mathfrak{c}]$ contained in $D[\mathfrak{a}, \mathfrak{a}]$. Given $\alpha \in \Gamma[\mathfrak{b}, \mathfrak{c}]$, Lemma 3b.3 allows us to take $u_1, \dots, u_m$ in $E(\mathfrak{b}) \cup E'(\mathfrak{c}) \cup D[\mathfrak{a}, \mathfrak{a}]$ so that $\alpha = u_1 \cdots u_m$. By strong approximation, $u_i \in \beta_i X$ with some $\beta_i \in G$. If $u_i \in D[\mathfrak{a}, \mathfrak{a}]$, then $\beta_i \in G \cap D[\mathfrak{a}, \mathfrak{a}] = \Gamma[\mathfrak{a}, \mathfrak{a}]$. If $u_i \in E(\mathfrak{b})$, we can take β_i from $T(\beta)$, and similarly if $u_i \in E'(\mathfrak{c})$, we can take β_i from $T'(\mathfrak{c})$. Then $\alpha = u_1 \cdots u_m \in \beta_1 \cdots \beta_m X$, and hence $\alpha = \beta_1 \cdots \beta_m \gamma$ with $\gamma \in G \cap X \subset \Gamma[\mathfrak{a}, \mathfrak{a}]$, which completes the proof. □

4. Theta series with Hermite polynomials. Coming back to the setting of Section 1, we consider certain series involving Hermite polynomials

$$h_m(x) = (-1)^m \exp(x^2/2)(d/dx)^m \exp(-x^2/2) \qquad (0 \le m \in \mathbf{Z}). \tag{4.1}$$

Given $0 \le p \in \mathbf{Z}$, put $N(p) = N(p, n) = \{\nu \in \mathbf{Z}^n \mid \nu_i \ge 0 \text{ for every } i, \sum_{i=1}^n \nu_i = p\}$. For each $\nu \in N(p)$ and z in $\mathbf{C}^n$ or $\mathbf{C}_n^1$ we put $z^\nu = \prod_{i=1}^n z_i^{\nu_i}$ and

$$h_\nu(x) = \prod_{i=1}^n h_{\nu_i}(x_i) \qquad (x \in \mathbf{R}_n^1). \tag{4.2}$$

Define a representation

$$\rho_p = \rho_p^{(n)} : GL_n(\mathbf{C}) \longrightarrow GL(\mathbf{C}^{N(p)}) \tag{4.3}$$

by $(az)^\mu = \sum_{\nu \in N(p)} \rho_p(a)_{\mu\nu} z^\nu$ for $\mu \in N(p)$, $z \in \mathbf{C}^n$, and $a \in GL_n(\mathbf{C})$. Then we can easily verify that

$$h_\mu(x) = \sum_{\nu \in N(p)} \rho_p(s)_{\mu\nu} h_\nu(xs) \quad \text{if } x \in \mathbf{R}_n^1 \text{ and } s \in GL_n(\mathbf{R}),\ {}^t ss = 1_n. \tag{4.4}$$

We define $K_p^{(n)} : \mathbf{R}_n^1 \longrightarrow \mathbf{R}^{N(p)}$ by $K_p^{(n)}(x) = \left(h_\nu(2\pi^{1/2}x)\right)_{\nu \in N(p, n)}$, and put $L(x, y) = \rho_p(y^{-1/2}) K_p^{(n)}(xy^{1/2})$ for $x \in \mathbf{R}_n^1$ and $0 < y = {}^t y \in \mathbf{R}_n^n$. Then we have

$$L(xa, y) = \rho_p({}^t a) L(x, ay \cdot {}^t a) \quad \text{for every } a \in GL_n(\mathbf{R}). \tag{4.5}$$

This can be derived from (4.4) by observing that $(ay \cdot {}^t a)^{-1/2} a y^{1/2}$ is orthogonal.

Let us now investigate the action of M_v for $v \in \mathbf{a}$ on the functions

$$\omega_p(x, \mathfrak{z}) = \rho_p^{(n)}(y^{-1/2})K_p^{(n)}(xy^{1/2})\varphi_v(x;\, \mathfrak{z},\, 0) \qquad (v \in \mathbf{r},\, x \in X_v,\, \mathfrak{z} \in H_n,\, y = \mathrm{Im}(\mathfrak{z})), \tag{4.6a}$$

$$\omega_p(x, \mathfrak{z}) = \rho_p^{(2n)}(\kappa(w)^{-1/2})K_p^{(2n)}\big((\mathrm{Re}(x), \mathrm{Im}(x))\kappa(2w)^{1/2}\big)\varphi_v(x;\, \mathfrak{z},\, 0) \qquad (v \in \mathbf{c},\, x \in X_v,\, \mathfrak{z} = z + jw \in H'_n), \tag{4.6b}$$

where the φ_v are the functions of (1.8b, c).

LEMMA 4.1. *Let $v \in \mathbf{a}$ and $\sigma \in M_v$ (cf. Lemmas 1.1R, C, and (1.10)). Then*

$$\sigma\omega_p(x, \mathfrak{z}) = g(\sigma, \mathfrak{z})^{-1}\rho_p\big(\mu(\sigma, \mathfrak{z})^{-1}\big)\omega_p(x, \sigma(\mathfrak{z})),$$

where $\rho_p = \rho_p^{(n)}$ if $v \in \mathbf{r}$ and $\rho_p = \rho_p^{(2n)}$ if $v \in \mathbf{c}$.

Proof. We first assume that $v \in \mathbf{r}$ and prove the formula in the special case:

$$\sigma\omega_p(x, \mathbf{i}_v) = g(\sigma, \mathbf{i}_v)^{-1}\rho_p\big(\mu(\sigma, \mathbf{i}_v)^{-1}\big)\omega_p(x, \mathbf{i}_v) \quad \text{if } \mathrm{pr}(\sigma) \in C_v,\ v \in \mathbf{r}, \tag{4.7}$$

where $\mathbf{i}$ is as in Lemma 2.2 and

$$C_v = \{\, \alpha \in G_v \mid \alpha(\mathbf{i}_v) = \mathbf{i}_v \,\} \qquad (v \in \mathbf{a}). \tag{4.8}$$

For $u, u' \in \mathbf{R}_n^1$ and $\zeta \in \mathbf{T}$ define a unitary operator $U(u, u'; \zeta)$ on $L^2(\mathbf{R}_n^1)$ by

$$[U(u, u'; \zeta)f](x) = \zeta\mathbf{e}(x \cdot {}^t u')f(x + u) \qquad (x \in \mathbf{R}_n^1,\, f \in L^2(\mathbf{R}_n^1))$$

(cf. [W, p.149]) and define a function Bf on $\mathbf{C}_n^1$ by

$$(Bf)(z) = \exp(\pi z \cdot {}^t z/2)\int_{\mathbf{R}_n^1} \exp(-\pi x \cdot {}^t x)\mathbf{e}(x \cdot {}^t z)f(x)dx \qquad (z \in \mathbf{C}_n^1). \tag{4.9}$$

It was shown by Bargman [B] that B gives a unitary isomorphism of $L^2(\mathbf{R}_n^1)$ onto the Hilbert space $\mathfrak{H}(\mathbf{C}_n^1)$ of all holomorphic functions k on $\mathbf{C}_n^1$ such that

$$\int_{\mathbf{C}_n^1} \exp(-\pi z \cdot {}^t\bar{z})|k(z)|^2|dzd\bar{z}| < \infty.$$

Observing that $s \mapsto \mu(s, \mathbf{i}_v)$ gives an isomorphism of C_v onto the unitary group of degree n, we define unitary operators $T(s)$ on $L^2(\mathbf{R}_n^1)$ and $T'(s)$ on $\mathfrak{H}(\mathbf{C}_n^1)$ for $s \in C_v$ by $T(s) = B^{-1}T'(s)B$ and $[T'(s)k](z) = k\big(z\,\overline{\mu(s, \mathbf{i}_v)}\big)$. Then a direct calculation (cf. [I, p.35, Th.7]) shows that

$$[BU(u, u'; \zeta)B^{-1}k](z) = \zeta \cdot e^{\pi X(z;\, u, u')}k(z + u' - iu) \quad \text{for every } k \in \mathfrak{H}(\mathbf{C}_n^1)$$

with $X(z; u, u') = -(u' + iu) \cdot {}^t z - 2^{-1}(u \cdot {}^t u + u' \cdot {}^t u') - iu \cdot {}^t u'$. Put $iw + w' = (iu + u')\mu(s, \mathbf{i}_v)$ and $k' = T'(s)k$. Then $(w, w') = (u, u')s$ and

$$\begin{aligned} BT(s)^{-1}U(u, u'; \zeta)T(s)B^{-1}k &= T'(s)^{-1}BU(u, u'; \zeta)B^{-1}k' \\ &= T'(s)^{-1}[\zeta \cdot e^{\pi X(z; u, u')}k'(z + u' - iu)] \\ &= \zeta \mathbf{e}(2^{-1}(w \cdot {}^t w' - u \cdot {}^t u'))e^{\pi X(z; w, w')}k(z + w' - iw) \\ &= BU\big(w, w'; \zeta \mathbf{e}(q_s(u, u'))\big)B^{-1}k \end{aligned}$$

with q_s of (1.3c). This shows that

$$T(s)^{-1}U(u, u'; \zeta)T(s) = U\big((u, u')s; \zeta \mathbf{e}(q_s(u, u'))\big).$$

In view of the principle of [W, p.157, p.183], this proves that $T(s) \in M_v$ and $\mathrm{pr}(T(s)) = s$. Put $\psi(x) = \omega_p(x, \mathbf{i}_v)$. Since

$$\sqrt{2}\int_{\mathbf{R}} h_m(2\sqrt{\pi}\, t)\mathbf{e}(it^2)\mathbf{e}(tx)dt = (i\sqrt{\pi}\, x)^n \mathbf{e}(ix^2/4),$$

a direct calculation of (4.9) with ψ as f shows that $(B\psi)(z) = b \cdot (z^\nu)_{\nu \in N(p)}$ with $b = 2^{-n/2} i^p \pi^{p/2}$. Clearly $T'(s)$ maps $B\psi$ to $\rho_p\big(\mu(s, \mathbf{i}_v)^{-1}\big)B\psi$. Now any element $\sigma \in M_v$ with $\mathrm{pr}(\sigma) = s \in C_v$ is of the form $\sigma = \xi T(s)$ with $\xi \in \mathbf{T}$. Then $\sigma\psi = \xi B^{-1}T'(s)B\psi = \xi\rho_p\big(\mu(s, \mathbf{i}_v)\big)^{-1}\psi$. If $p = 0$, then $\psi(x) = \varphi_v(x; \mathfrak{z}, 0)$, and hence Lemma 1.1R shows that $\xi = g(\sigma, \mathbf{i}_v)^{-1}$, which proves (4.7). Now it is sufficient to prove our lemma when $\mathrm{pr}(\sigma) \in P_v$ and $\sigma = r_\Omega(\iota)$, since such elements together with $\mathbf{T}$ generate M_v. The desired formula can easily be verified if $\mathrm{pr}(\sigma) \in P_v$, by virtue of (1.3a) and (4.5). To prove it for $\sigma = r_\Omega(\iota)$, given $\mathfrak{z} \in H_n$, take $\alpha, \beta \in P_v$ so that $\mathfrak{z} = \alpha(\mathbf{i}_v)$ and $\iota(\mathfrak{z}) = \beta(\mathbf{i}_v)$. Let $\tau = r_P(\beta)^{-1}r_\Omega(\iota)r_P(\alpha)$. Then $\mathrm{pr}(\tau) \in C_v$, and by (4.7) we have

$$\begin{aligned} r_\Omega(\iota)\omega_p(x, \mathfrak{z}) &= r_\Omega(\iota)[g(\alpha, \mathbf{i}_v)\rho_p(\mu(\alpha, \mathbf{i}_v))r_P(\alpha)\omega_p(x, \mathbf{i}_v)] \\ &= g(\alpha, \mathbf{i}_v)\rho_p(\mu(\alpha, \mathbf{i}_v))r_P(\beta)\tau\omega_p(x, \mathbf{i}_v) \\ &= g(\alpha, \mathbf{i}_v)\rho_p(\mu(\alpha, \mathbf{i}_v))r_P(\beta)g(\tau, \mathbf{i}_v)^{-1}\rho_p(\mu(\tau, \mathbf{i}_v))^{-1}\omega_p(x, \mathbf{i}_v) \\ &= g(\alpha, \mathbf{i}_v)g(\tau, \mathbf{i}_v)^{-1}g(\beta, \mathbf{i}_v)^{-1}\rho_p\big(\mu(\alpha, \mathbf{i}_v)\mu(\tau, \mathbf{i}_v)^{-1}\mu(\beta, \mathbf{i}_v)^{-1}\big) \\ &\quad \times \omega_p(x, \iota(\mathfrak{z})) \\ &= g(\iota, \mathfrak{z})^{-1}\rho_p\big(\mu(\iota, \mathfrak{z})^{-1}\big)\omega_p(x, \iota(\mathfrak{z})). \end{aligned}$$

This proves our lemma for $v \in \mathbf{r}$. If $v \in \mathbf{c}$, it is again sufficient to prove the cases $\mathrm{pr}(\sigma) \in P_v$ and $\sigma = r_\Omega(\iota)$. The former case follows directly from (1.3a) and

(4.5). As for $r_\Omega(\iota)$, (1.3c) gives the desired formula in the form

$$\text{(4.10)} \quad \int_{X_v} \omega_p(y, \mathfrak{z})\mathbf{e}_v(-x \cdot {}^t y)dy = \det(\kappa(\mathfrak{z}))^{-1/2}\rho_p^{(2n)}(\kappa(\mathfrak{z})^{-1})\omega_p(x, \iota(\mathfrak{z})).$$

Observing that $\omega_p(x, \mathfrak{z})$ is a constant times the pullback of ω_p on H_{2n} in the same sense as in the proof of Lemma 1.1C, we can easily derive (4.10) from the corresponding formula for $r_\Omega(\iota)$ in the case $v \in \mathbf{r}$, which we have established in the above proof. This completes the proof. □

For $\ell \in \mathcal{S}(X_\mathbf{h})$, $x \in X_\mathbf{A}$, $\mathfrak{z} \in \mathcal{H}$, and $m \in \mathbf{Z}^\mathbf{a}$ with $m_v \geq 0$ for all v, put

$$\text{(4.11)} \quad \omega_m(x; \mathfrak{z}, \ell) = \ell(x_\mathbf{h}) \bigotimes_{v\in\mathbf{a}} \omega_{m_v}(x_v; \mathfrak{z}_v).$$

This has values in $\bigotimes_{v\in\mathbf{a}} \mathbf{C}^{N(m_v, n_v)}$, where $n_v = n$ if $v \in \mathbf{r}$ and $n_v = 2n$ if $v \in \mathbf{c}$. Calling this vector space W_m, we can define a representation

$$\text{(4.12)} \quad \rho_m = \bigotimes_{v\in\mathbf{a}} \rho_{m_v}^{(n_v)} : \prod_{v\in\mathbf{a}} GL_{n_v}(\mathbf{C}) \longrightarrow GL(W_m).$$

LEMMA 4.2. *For every* $\sigma \in \mathfrak{M}$ *we have*

$$\sigma\omega_m(x; \mathfrak{z}, \ell) = h(\sigma, \mathfrak{z})^{-1}\rho_m(\mu(\sigma, \mathfrak{z})^{-1})\omega_m(x; \sigma(\mathfrak{z}), {}^\sigma\ell).$$

Proof. As in the proof of Theorem 1.2, we can take σ in the form $\sigma = \zeta r_P(\beta)\gamma$ with $\zeta \in \mathbf{T}$, $\beta \in P_\mathbf{A}$, and $\gamma \in M_\mathbf{A}$, $\mathrm{pr}(\gamma) \in C^\theta$. Take also $\xi_v \in M_v$ as in (1.20). For the same reason as in (1.21), we have

$$\sigma\omega_m(x; \mathfrak{z}, \ell) = \zeta \cdot {}^\sigma\ell(x_\mathbf{h}) \bigotimes_{v\in\mathbf{a}} [r_P(\beta_v)\xi_v\omega_{m_v}](x_v, \mathfrak{z}_v),$$

since l' of (1.21) is ${}^\sigma\ell$. By Lemma 4.1, the right-hand side equals

$$\zeta \prod_{v\in\mathbf{a}} g\big(r_P(\beta_v)\xi_v, \mathfrak{z}_v\big)^{-1}\rho_m\big(\mu(\sigma, \mathfrak{z})^{-1}\big)\omega_m(x; \sigma(\mathfrak{z}), {}^\sigma\ell).$$

This proves our lemma, because of our definition of $h(\sigma, \mathfrak{z})$ in the proof of Theorem 1.2. □

We now define functions $\theta_m(\mathfrak{z}, \ell)$ for $\mathfrak{z} \in \mathcal{H}$ and $\theta_{m,\mathbf{A}}(\sigma, \ell)$ for $\sigma \in M_{\mathbf{A}}$ by

$$\theta_m(\mathfrak{z}, \ell) = \sum_{\xi \in X} \omega_m(\xi; \mathfrak{z}, \ell), \tag{4.13a}$$

$$\theta_{m,\mathbf{A}}(\sigma, \ell) = \sum_{\xi \in X} (\sigma\omega_m)(\xi; \mathbf{i}, \ell). \tag{4.13b}$$

Notice that $\theta_0(\mathfrak{z}, \ell)$ coincides with $\theta(\mathfrak{z}, 0; \ell)$ with θ of (1.24).

THEOREM 4.3. *The following formulas hold:*

(1) $\theta_m(\alpha\mathfrak{z}, {}^{\alpha}\ell) = h(\alpha, \mathfrak{z})\rho_m(\mu(\alpha, \mathfrak{z}))\theta_m(\mathfrak{z}, \ell)$ *for every* $\alpha \in G \cap \mathfrak{M}$.

(2) $\theta_{m,\mathbf{A}}(\alpha\sigma\tau, \ell) = h(\tau, \mathbf{i})^{-1}\rho_m(\mu(\tau, \mathbf{i})^{-1})\theta_{m,\mathbf{A}}(\sigma, {}^{\tau}\ell)$ *for every* $\alpha \subset G$ *and every* $\tau \in \mathfrak{M}$ *such that* $\tau(\mathbf{i}) = \mathbf{i}$.

(3) $\theta_m(\sigma(\mathbf{i}), \ell) = h(\sigma, \mathbf{i})\rho_m(\mu(\sigma, \mathbf{i}))\theta_{m,\mathbf{A}}(\sigma, {}^{\sigma}\ell)$ *if* $\sigma \in \mathfrak{M}$.

Proof. By [W, Theorem 4 or 6] we have $\theta_{m,\mathbf{A}}(\alpha\sigma, \ell) = \theta_{m,\mathbf{A}}(\sigma, \ell)$ for every $\alpha \in G$. This together with Lemma 4.2 proves (2). By Lemma 4.2 and (4.11b) we have $\theta_{m,\mathbf{A}}(\sigma, \ell) = h(\sigma, \mathbf{i})^{-1}\rho_m(\mu(\sigma, \mathbf{i})^{-1})\theta_m(\sigma(\mathbf{i}), {}^{\sigma}\ell)$ if $\sigma \in \mathfrak{M}$, which proves (3). Take σ so that $\mathrm{pr}(\sigma) \in G_{\mathbf{a}}$. Then ${}^{\sigma}\ell = \ell$, and for $\alpha \in G \cap \mathfrak{M}$ we have $\theta_{m,\mathbf{A}}(\sigma, \ell) = \theta_{m,\mathbf{A}}(\alpha\sigma, \ell) = h(\alpha\sigma, \mathbf{i})^{-1}\rho_m(\mu(\alpha\sigma, \mathbf{i})^{-1})\theta_m(\alpha\sigma(\mathbf{i}), {}^{\alpha}\ell)$, which proves (1). □

Remark. For every $\varphi \in \mathfrak{S}_p(\mathbf{C}_n^1)$ the functions $\varphi(u)$ and $\varphi(\bar{u})$ with $u \in \mathbf{C}_n^1$ are $\mathbf{C}$-linear combinations of $h_\nu(\mathrm{Re}(u), \mathrm{Im}(u))$ for $\nu \in N(p, 2n)$. This is because

$$(x \pm iy)^m = \sum_{k=0}^{m} \binom{m}{k} (\pm i)^k h_k(y) h_{m-k}(x). \tag{4.14}$$

In fact, this formula is a special case of

$$(x\sigma)^p = \sum_{\nu \in N(p)} p!(\nu!)^{-1} \sigma^\nu h_\nu(x) \tag{4.15}$$

$$(x \in \mathbf{R}_n^1,\ \sigma \in \mathbf{C}^n,\ {}^t\sigma\sigma = 0 \text{ if } p > 1),$$

where $\nu! = \prod_{k=1}^{n} \nu_k!$. This follows from an easy relation

$$(x\sigma)^p \exp(-x \cdot {}^t x/2) = \left(-\sum_{k=1}^{n} \sigma_k \partial/\partial x_k\right)^p \exp(-x \cdot {}^t x/2).$$

From (4.15) it follows that $q({}^t x)$ with $q \in \mathcal{P}_p(\mathbf{C}_1^n)$ is a $\mathbf{C}$-linear combination of $h_\nu(x)$ for $\nu \in N(p, n)$.

5. A three-dimensional case. In this section we take $n = 1$ (and hence $G = SL_2(F)$ and $\mathcal{H} = (H_1)^{\mathbf{r}} \times (H_1')^{\mathbf{c}})$, and consider the case in which V and S of Section 3 are given by

$$V = \{\alpha \in B \mid \mathrm{tr}(\alpha) = 0\},$$
$$S(\alpha, \beta) = (-\sigma/2)\mathrm{tr}(\alpha\beta) \quad for \quad \alpha, \beta \in V.$$

Here B is a quaternion algebra over F, $\sigma \in F^\times$, and tr is the standard trace map of B into F. Thus $q = 3$ and $S(\alpha, \alpha) = \sigma N(\alpha)$ with the standard norm map N of B into F. Let $\mathbf{s}$ resp. $\mathbf{t}$ denote the set of all $v \in \mathbf{r}$ unramified resp. ramified in B. We assume

$$\sigma_v > 0 \quad for \ v \in \mathbf{s} \ and \quad \sigma_v < 0 \quad for \ v \in \mathbf{t}. \tag{5.1}$$

For each $v \in \mathbf{s} \cup \mathbf{c}$ (resp. $v \in \mathbf{t}$) we fix an identification of B_v with $(F_v)_2^2$ (resp. $\mathbf{H}$), so that $B_{\mathbf{a}} = (\mathbf{R}_2^2)^{\mathbf{s}} \times \mathbf{H}^{\mathbf{t}} \times (\mathbf{C}_2^2)^{\mathbf{c}}$. We put

$$\mathcal{G} = B^\times, \qquad \mathcal{G}_{\mathbf{A}+} = \{x \in \mathcal{G}_{\mathbf{A}} \mid N(x)_v > 0 \ for \ every \ v \in \mathbf{s}\}, \tag{5.2a}$$

$$\mathcal{G}_+ = \mathcal{G} \cap \mathcal{G}_{\mathbf{A}+}, \quad \mathcal{G}^1 = \{\alpha \in B \mid N(\alpha) = 1\}, \quad \mathcal{H}_B = H_1^{\mathbf{s}} \times (H_1')^{\mathbf{c}} \tag{5.2b}$$

Then we can let $\mathcal{G}_{\mathbf{a}}^1$ act on $\mathcal{H}_B$ in an obvious way, ignoring $\prod_{v \in \mathbf{t}} \mathcal{G}_v^1$. Now S_v has signature (1, 2) for every $v \in \mathbf{s}$ and (0, 3) for every $v \in \mathbf{t}$. Therefore $I_v = \mathrm{diag}[1, -1_2]$ if $v \in \mathbf{s}$, $I_v = -1_3$ if $v \in \mathbf{t}$, and $I_v = 1_3$ if $v \in \mathbf{c}$. We put

$$H_v^- = \begin{cases} \{x + iy \in \mathbf{C} \mid y < 0\} & if \ v \in \mathbf{r}, \\ \{z + jw \in \mathbf{H} \mid w < 0\} & if \ v \in \mathbf{c}. \end{cases}$$

We can let G_v act on H_v^- and define $\mu_0(\alpha, \mathfrak{z})$ for $\alpha \in G_v$ and $\mathfrak{z} \in H_v^-$ by (1.5a, b); we put then $m(\alpha, \mathfrak{z}) = \mu_0(\alpha, \mathfrak{z})$ if $v \in \mathbf{r}$ and $m(\alpha, \mathfrak{z}) = |\mu_0(\alpha, \mathfrak{z})|^2$ if $v \in \mathbf{c}$. These are also meaningful for $\alpha \in \mathcal{G}_v^1$, $v \in \mathbf{s} \cup \mathbf{c}$. If we define an automorphism ρ of $\mathbf{H}$ by $x^\rho = ixi^{-1}$ for $x \in \mathbf{H}$, then $\alpha(\mathfrak{z})^\rho = \alpha(\mathfrak{z}^\rho)$ and $\mu_0(\alpha, \mathfrak{z})^\rho = \mu_0(\alpha, \mathfrak{z}^\rho)$ for every $\alpha \in G_v$ and $\mathfrak{z} \in H_v \cup H_v^-$, $v \in \mathbf{c}$. Furthermore, for $\alpha \in GL_2(F_v)$ ($\det(\alpha) > 0$ if $v \in \mathbf{r}$) we define $\alpha(\mathfrak{z})$ and $m(\alpha, \mathfrak{z})$ by $\alpha(\mathfrak{z}) = \tilde{\alpha}(\mathfrak{z})$ and $m(\alpha, \mathfrak{z}) = m(\tilde{\alpha}, \mathfrak{z})$, $\tilde{\alpha} = \det(\alpha)^{-1/2}\alpha$, with any choice of $\det(\alpha)^{-1/2}$ if $v \in \mathbf{c}$.

We are going to define the series f of (3.10) as a function on $\mathcal{H} \times \mathcal{H}_B$. We first define $A_v^0 : V_v \to F_v^3$ for $v \in \mathbf{a}$ by

$$A_v^0 \begin{pmatrix} a & b \\ c & -a \end{pmatrix} = \begin{cases} {}^t((1/2)(c-b), (1/2)(c+b), a) & if \ v \in \mathbf{s}, \\ {}^t((1/2)(c-b), (-i/2)(c+b), ia) & if \ v \in \mathbf{c}, \end{cases} \tag{5.3a}$$

$$A_v^0(ia + jb + kc) = {}^t(a, b, c) \quad if \ v \in \mathbf{t}. \tag{5.3b}$$

Then $S_v(\alpha, \beta) = \mathrm{sgn}(\sigma_v)\sigma_v{}^t(A_v^0\alpha)I_v(A_v^0\beta)$, where we understand that $\mathrm{sgn}(\sigma_v) = 1$ if $v \in \mathbf{c}$. For each $\gamma \in \mathcal{G}_v^1$, $v \in \mathbf{a}$, define $A_v^\gamma : V_v \to F_v^3$ by $A_v^\gamma(\xi) = A_v^0(\gamma^{-1}\xi\gamma)$ and put $T_v^\gamma\{\alpha\} = {}^t(A_v^\gamma\alpha)(A_v^\gamma\alpha)$ for $\alpha \in V_v$. Then we see that $T_v^\gamma\{\alpha\} = N(\alpha)$ if $v \in \mathbf{t}$ and $T_v^\gamma\{\alpha\} = (1/2)\mathrm{tr}({}^t\overline{\alpha}(\gamma \cdot {}^t\overline{\gamma})^{-1}\alpha\gamma \cdot {}^t\overline{\gamma})$ if $v \in \mathbf{s} \cup \mathbf{c}$, and hence T_v^γ depends only on $\gamma(\mathbf{i}_v)$, where $\mathbf{i}_v = i$ for $v \in \mathbf{r}$ and $\mathbf{i}_v = j$ for $v \in \mathbf{c}$. Thus, given $\mathfrak{w} \in \mathcal{H}_B$, we put $T_v^\gamma = T_v^{\mathfrak{w}}$ when $\mathfrak{w}_v = \gamma(\mathbf{i}_v)$.

To express the $T_v^{\mathfrak{w}}$ in explicit forms, we put

$$(5.4)\quad [\xi, \mathfrak{w}] = [\xi; \mathfrak{w}, \mathfrak{w}], \qquad [\xi; \mathfrak{w}, \mathfrak{w}'] = (-1 \ \ \mathfrak{w})\xi\begin{pmatrix}\mathfrak{w}'\\ 1\end{pmatrix} \quad (\in \mathbf{H}),$$
$$\text{for } \xi \in \mathbf{C}_2^2 \text{ and } \mathfrak{w}, \mathfrak{w}' \in \mathbf{C} + j\mathbf{R}.$$

If $v \in \mathbf{s}$, we consider $[\xi; \mathfrak{w}, \mathfrak{w}']$ only for $\mathfrak{w}, \mathfrak{w}' \in \mathbf{C}$ and $\xi \in \mathbf{R}_2^2$; then of course $[\xi; \mathfrak{w}, \mathfrak{w}'] \in \mathbf{C}$. Now, for $v \in \mathbf{s} \cup \mathbf{c}$, $\alpha, \beta \in \mathcal{G}_v^1$, $\mathfrak{w}, \mathfrak{w}' \in H_v \cup H_v^-$, and $\xi \in (F_v)_2^2$ we can easily verify that

$$(5.5)\ [\alpha^{-1}\xi\beta; \mathfrak{w}, \mathfrak{w}'] = \begin{cases} \mu_0(\alpha, \mathfrak{w})\mu_0(\beta, \mathfrak{w}')[\xi; \alpha\mathfrak{w}, \beta\mathfrak{w}'] & \text{if } v \in \mathbf{s}, \\ k\overline{\mu_0(\alpha, \mathfrak{w})}k^{-1}[\xi; \alpha\mathfrak{w}, \beta\mathfrak{w}']\mu_0(\beta, \mathfrak{w}') & \text{if } v \in \mathbf{c}. \end{cases}$$

For $\mathfrak{w} \in \mathcal{H}_B$ and $v \in \mathbf{s} \cup \mathbf{c}$ we have

$$(5.6\text{a})\qquad T_v^{\mathfrak{w}}\{\xi\} = \mathrm{Re}(N(\xi)) + 2^{-1}|[\xi, \mathfrak{w}_v]/\eta_v(\mathfrak{w}_v)|^2 \quad \text{for} \quad \xi \in V_v,$$

$$(5.6\text{b})\quad 2^{-1}|[\xi + d1_2, \mathfrak{w}_v]/\eta_v(\mathfrak{w}_v)|^2 = T_v^{\mathfrak{w}}\{\xi\} + |d|^2$$
$$\text{if } \xi \in V_v,\ d \in F_v, \text{ and } N(\xi) = -d^2,$$

where $\eta_v(x + iy) = y$ if $v \in \mathbf{s}$ and $\eta_v(z + jw) = w$ if $v \in \mathbf{c}$. These can be easily verified by reducing the problem to the case $\mathfrak{w}_v = \mathbf{i}_v$ by (5.5) and invoking Lemma 3.1.

Let $E = i\mathbf{R} + j\mathbf{R} + k\mathbf{R}$. For $0 \leq m \in \mathbf{Z}$ we denote by $\mathcal{P}_m(E)$ resp. $\mathcal{P}_m(E^2)$ the vector space of all $\mathbf{C}$-valued polynomial functions $p(x)$ on E resp. $p(x, y)$ on $E \times E$ of the form $p(x) = q(A^0x)$ resp. $p(x, y) = q(A^0x, A^0y)$ with $q \in \mathcal{P}_m(\mathbf{C}^3)$ resp. $q \in \mathcal{P}_m(\mathbf{C}_2^3)$, where $\mathcal{P}_m$ is the set defined in Section 3 and A^0 is the injection $E \to \mathbf{C}^3$ given by $A^0(ia + jb + kc) = {}^t(a, b, c)$. We view $\mathcal{P}_m(E)$ resp. $\mathcal{P}_m(E^2)$ as a representation space of $\mathbf{H}^\times/\{\pm 1\}$ resp. $GL_2(\mathbf{C}) \times (\mathbf{H}^\times/\{\pm 1\})$ by

$$(5.7\text{a})\qquad (\mu p)(x) = p(\overline{\mu}x\mu) \quad \text{for } p \in \mathcal{P}_m(E) \text{ and } \mu \in \mathbf{H}^\times,$$

$$(5.7\text{b})\quad ((\varepsilon, \mu)p)(x, y) = q\big((A^0(\overline{\mu}x\mu), A^0(\overline{\mu}y\mu))\varepsilon\big)$$
$$\text{for } p(x, y) = q(A^0x, A^0y), \quad \mu \in \mathbf{H}^\times, \text{ and } \varepsilon \in GL_2(\mathbf{C}).$$

Now, given $n \in \mathbf{Z}^{\mathbf{a}}$ with $n_v \geq 0$ for every $v \in \mathbf{a}$, we put $\mathcal{Q} = \prod_{v \in \mathbf{t}} \mathcal{P}_{n_v}(E) \times \prod_{v \in \mathbf{c}} \mathcal{P}_{n_v}(E^2)$ and consider a function

$$(5.8)\ \psi(\xi, \mathfrak{w}, p) = \prod_{v \in \mathbf{s}} [\xi_v, \overline{\mathfrak{w}_v}]^{n_v} \prod_{v \in \mathbf{t}} p_v(\xi_v) \prod_{v \in \mathbf{c}} p_v\big(k[\xi_v, \mathfrak{w}_v, \mathfrak{w}_v^\rho]i,\ k[\xi_v, \mathfrak{w}_v]\big)$$

for $\xi \in V_{\mathbf{A}}$, $\mathfrak{w} \in \mathcal{H}_B$ and $p = (p_v)_{v \in \mathbf{t} \cup \mathbf{c}} \in \mathcal{Q}$. Notice that $\xi_v \in E$ for $v \in \mathbf{t}$, and both $k[\xi_v, \mathfrak{w}_v]$ and $k[\xi_v, \mathfrak{w}_v, \mathfrak{w}_v^\rho]i$ belong to E. Combining (5.7a, b) for all $v \in \mathbf{t} \cup \mathbf{c}$, we obtain a representation

$$(5.9) \qquad \rho : GL_2(\mathbf{C})^{\mathbf{c}} \times (\mathbf{H}^\times/\{\pm\})^{\mathbf{t} \cup \mathbf{c}} \longrightarrow GL(\mathcal{Q}).$$

From (5.5) we easily see that

$$(5.10) \qquad \psi(\beta^{-1}\xi\beta, \mathfrak{w}, p) = \prod_{v \in \mathbf{s}} \overline{m(\beta, \mathfrak{w})_v^{2n_v}} \prod_{v \in \mathbf{t}} N(\beta)_v^{-n_v} \cdot \psi\big(\xi, \beta(\mathfrak{w}), \rho(1, \mu_0(\beta, \mathfrak{w}))p\big) \text{ for every } \beta \in \mathcal{G}_+,$$

where $\mu_0(\beta, \mathfrak{w})_v = \beta_v$ for $v \in \mathbf{t}$ and $\mu_0(\beta, \mathfrak{w})_v = \pm\mu_0(\det(\beta_v)^{-1/2}\beta_v, \mathfrak{w}_v)$ if $v \in \mathbf{c}$.

We now define a series Θ by

$$(5.11)\ \Theta(\mathfrak{z}, \mathfrak{w}; \lambda, p) = \prod_{v \in \mathbf{s}} \eta_v(\mathfrak{z}_v)^{1/2}\eta_v(\mathfrak{w}_v)^{-2n_v} \sum_{\xi \in V} \lambda(\xi_{\mathbf{h}})\psi(\xi, \mathfrak{w}, p)\Psi(\xi; \mathfrak{z}, \mathfrak{w})$$

for $(\mathfrak{z}, \mathfrak{w}) \in \mathcal{H} \times \mathcal{H}_B$, $\lambda \in \mathcal{S}(V_{\mathbf{h}})$, and p as in (5.8), where

$$\begin{aligned}\Psi(\xi; \mathfrak{z}, \mathfrak{w}) &= \mathbf{e}\Bigg(\sum_{v \in \mathbf{t}} 2^{-1}\sigma_v N(\xi)_v \bar{\mathfrak{z}}_v + \sum_{v \in \mathbf{s}} \Big\{2^{-1}\sigma_v N(\xi)_v \mathfrak{z}_v + 4^{-1} i y_v \sigma_v |[\xi_v, \mathfrak{w}_v]/\eta_v(\mathfrak{w}_v)|^2\Big\} \\ &+ \sum_{v \in \mathbf{c}} \Big\{\mathrm{Re}\big(\sigma_v N(\xi)_v z_v\big) + i w_v |\sigma_v| \big(\mathrm{Re}(N(\xi)_v) + 2^{-1}|[\xi_v, \mathfrak{w}_v]/\eta_v(\mathfrak{w}_v)|^2\big)\Big\}\Bigg),\end{aligned}$$

$y_v = \eta_v(\mathfrak{z}_v)$ for $v \in \mathbf{s}$ and $\mathfrak{z}_v = z_v + j w_v$ for $v \in \mathbf{c}$.

Let us now show that Θ is a special case of f of (3.10). From (5.6a) we see that Ψ is $\Phi(\xi; \mathfrak{z}, 0)$ of (3.10). Thus our task is to show that each v-factor of ψ is a v-factor of (3.9). This is clear for $v \in \mathbf{t}$. If $v \in \mathbf{s} \cup \mathbf{c}$, (5.10) reduces the problem to the case $\mathfrak{w}_v = \mathbf{i}_v$. For $v \in \mathbf{s}$, we have $[\xi, i] = \tau A_v^0 \xi^-$ with $\tau = (0, -2, -2i)$.

Clearly $\tau \cdot {}^t\tau = 0$. If $v \in \mathbf{c}$ and $\xi = \begin{pmatrix} a & b \\ c & -a \end{pmatrix} \in V_v$, we have

$$2\mathrm{Re}(A_v^0\xi) = {}^t(\mathrm{Re}(c-b), \mathrm{Im}(b+c), -2\mathrm{Im}(a)),$$
$$2\mathrm{Im}(A_v^0\xi) = {}^t(\mathrm{Im}(c-b), -\mathrm{Re}(b+c), 2\mathrm{Re}(a)),$$
$$k[\xi, j, -j]i = -2i \cdot \mathrm{Im}(a) + j \cdot \mathrm{Re}(c-b) + k \cdot \mathrm{Im}(b+c),$$
$$k[\xi, j] = 2i \cdot \mathrm{Re}(a) + j \cdot \mathrm{Im}(c-b) - k \cdot \mathrm{Re}(b+c).$$

Thus the last factor of (5.8) fits in that of (3.9).

PROPOSITION 5.1. *The formula of Theorem 3.5 in the present case can be written*

$$\Theta(\alpha\mathfrak{z}, \mathfrak{w}; {}^\alpha\lambda, \rho({}^t\mu(\alpha, \mathfrak{z})^{-1}, 1)p) = J(\alpha, \mathfrak{z})\Theta(\mathfrak{z}, \mathfrak{w}; \lambda, p),$$
$$J(\alpha, \mathfrak{z}) = h(\alpha, \mathfrak{z})^{-3} \textstyle\prod_{v\in\mathbf{s}} m(\alpha, \mathfrak{z})_v^{n_v+2} \prod_{v\in\mathbf{t}} m(\alpha, \mathfrak{z})_v^{n_v+3} \prod_{v\in\mathbf{c}} m(\alpha, \mathfrak{z})_v^3$$

for every $\alpha \in G \cap \mathfrak{M}$. The character χ of Theorem 3.2 in the present case corresponds to $F(\sigma^{1/2})/F$. Moreover, for every $\beta \in \mathcal{G}_+$ we have

$$\Theta(\mathfrak{z}, \beta\mathfrak{w}; \lambda, \rho(1, \mu_0(\beta, \mathfrak{w}))p) = \prod_{v\in\mathbf{s}} m(\beta, \mathfrak{w})_v^{2n_v} \prod_{v\in\mathbf{t}} N(\beta)_v^{n_v} \cdot \Theta(\mathfrak{z}, \mathfrak{w}; \lambda^\beta, p),$$

where $\lambda^\beta(\xi) = \lambda(\beta\xi\beta^{-1})$.

Proof. The first assertion follows immediately from Theorem 3.5. Notice that $s_v = 2$ for $v \in \mathbf{s}$ and $s_v = 3$ for $v \in \mathbf{t}$. It is easy to see that $\det(S)/\sigma$ is a square in F, which proves the second assertion. The last formula follows directly from (5.10). □

For $0 \le m \in \mathbf{Z}$ let $s_m : GL_2(\mathbf{C}) \to GL_{m+1}(\mathbf{C})$ be an irreducible polynomial representation, homogeneous of degree m, which is unique up to equivalence. Define $r_m : \mathbf{H}^\times \to GL_{m+1}(\mathbf{C})$ by $r_m = s_m \circ \kappa$ with the injection $\kappa : \mathbf{H} \to \mathbf{C}_2^2$ of Section 1. Let $\mathcal{P}'_m(\mathbf{C}_r^q)$ denote the subspace of $\mathfrak{S}_m(\mathbf{C}_r^q)$ spanned by all functions of the form $f(x) = \varphi({}^t\sigma x)$ with $\varphi \in \mathfrak{S}_m(\mathbf{C}_r^1)$ and $\sigma \in \mathbf{C}^q$ satisfying the condition ${}^t\sigma\sigma = 0$, which is dropped if $a \le 1$. Clearly $\mathcal{P}'_m(\mathbf{C}_r^q) \subset \mathcal{P}_m(\mathbf{C}_r^q)$. Let $\mathcal{P}'_m(E^2)$ denote the subspace of $\mathcal{P}_m(E^2)$ corresponding to $\mathcal{P}'_m(\mathbf{C}_2^3)$. Then $\mathcal{P}'_m(E^2)$ is stable under $GL_2(\mathbf{C}) \times \mathbf{H}^\times$.

LEMMA 5.2. *The representation of $\mathbf{H}^\times$ on $\mathcal{P}_m(E)$ defined by (5.7a) is equivalent to r_{2m}. The representation of $GL_2(\mathbf{C}) \times \mathbf{H}^\times$ on $\mathcal{P}'_m(E^2)$ defined by (5.7b) is equivalent to $s_m \otimes r_{2m}$.*

Proof. The first assertion was proved in [S1, Lemma 1.1]. To prove the second one, let $\mathcal{F}$ denote the group of all $\gamma \in GL_3(\mathbf{C})$ such that ${}^t\gamma U\gamma = cU$ with $c \in \mathbf{C}$, where $U = \mathrm{diag}\left[\begin{pmatrix} 0 & 1 \\ 1 & 0 \end{pmatrix}, 1\right]$. Take $R \in GL_3(\mathbf{R})$ so that $U = {}^tRR$. Then it is sufficient to prove the corresponding fact on the representation ρ of $GL_2(\mathbf{C}) \times \mathcal{F}$ on $\mathcal{P}'_m(\mathbf{C}^3_2)$ defined by $[\rho(\alpha, \beta)q](x) = q({}^t(R\beta R^{-1})x\alpha)$ for $q \in \mathcal{P}'_m(\mathbf{C}^3_2)$. Define an element k of $\mathfrak{S}_m(\mathbf{C}^1_2)$ by $k(y) = y_1^m$, and put $h(x) = k({}^t(Re_1)x)$ for $x \in \mathbf{C}^3_2$ with ${}^te_1 = (1, 0, 0)$. Then $h \in \mathcal{P}'_m(\mathbf{C}^3_2)$. Now $\mathcal{P}'_m(\mathbf{C}^3_2)$ is spanned by all functions p of the form $p(x) = \varphi({}^twx)$ with $\varphi \in \mathfrak{S}_m(\mathbf{C}^1_2)$ and $0 \neq w \in \mathbf{C}^3$, ${}^tww = 0$. (This is so even if $m \leq 1$.) Since $\mathfrak{S}_m(\mathbf{C}^1_2)$ is an irreducible representation space of $GL_2(\mathbf{C})$ and $w = R\beta e_1$ with $\beta \in \mathcal{F}$, we easily see that p is a finite linear combination of functions of the form $\rho(\alpha, \beta)h$ with $(\alpha, \beta) \in GL_2(\mathbf{C}) \times \mathcal{F}$. Also it can easily be seen that h is a highest weight vector in an obvious sense. Thus ρ is irreducible. From the weight of h we immediately see that it corresponds to $s_m \otimes r_{2m}$, which proves our lemma. □

6. Appendix: The action of $Sp(n, \mathbf{C})$ on H'_n. It is well known that $Sp(n, \mathbf{R})$ acts on H_n by the rule (1.5a) and H_n is the quotient of $Sp(n, \mathbf{R})$ by its maximal compact subgroup. We present here an easy exposition of the corresponding statements for $Sp(n, \mathbf{C})$ and H'_n, as well as several related facts, since these, if elementary, are perhaps new to most readers. As in Section 1, we consider $\mathbf{C}$ as a subring of $\mathbf{H}$. For $\alpha \in \mathbf{H}^n_n$ we put $\alpha^* = {}^t\bar{\alpha}$ and $\alpha^\rho = i\alpha i^{-1}$. If $\alpha^* = \alpha$ and $\xi^*\alpha\xi > 0$ for every $\xi \in \mathbf{H}^n_1, \neq 0$, then we write $\alpha > 0$.

LEMMA 6.1. *$Sp(n, \mathbf{C}) = \{\alpha \in GL_{2n}(\mathbf{H}) \mid \alpha^* j\iota_n\alpha = j\iota_n,\ \alpha^* k\iota_n\alpha = k\iota_n\}$.*

Proof. If α belongs to the right-hand side, then $j^{-1}\alpha^* j = k^{-1}\alpha^* k$, and hence $\alpha i = i\alpha$. Thus $\alpha \in \mathbf{C}^{2n}_{2n}$, and ${}^t\alpha\iota_n\alpha = j^{-1}\alpha^* j\iota_n\alpha = \iota_n$, which shows that $\alpha \in Sp(n, \mathbf{C})$. The converse can be proved by reversing the order of our arguments. □

We now put

$$\text{(6.1)} \qquad \mathfrak{S}_n = \{z + jw \in \mathbf{H}^n_n \mid {}^tz = z \in \mathbf{C}^n_n,\ w^* = w \in \mathbf{C}^n_n\},$$

$$\text{(6.2)} \qquad \eta(z + jw) = w \quad \textit{for} \quad z + jw \in \mathfrak{S}_n,$$

and observe that if $\mathfrak{z} \in \mathbf{H}^n_n$, then

$$\mathfrak{z} \in \mathfrak{S}_n \Longleftrightarrow j\mathfrak{z}^\rho = \mathfrak{z}^* j \Longleftrightarrow k\mathfrak{z} = \mathfrak{z}^* k.$$

Moreover, for $\mathfrak{z}$ and $\mathfrak{w}$ in $\mathfrak{S}_n$ we have

$$\text{(6.3a)} \qquad (\mathfrak{w}^* \quad 1)j\iota_n\begin{pmatrix} \mathfrak{z} \\ 1_n \end{pmatrix} = j(\mathfrak{z} - \mathfrak{w}^\rho),$$

(6.3b) $$(\mathfrak{w}^* \quad 1)k\iota_n \begin{pmatrix} \mathfrak{z} \\ 1_n \end{pmatrix} = k(\mathfrak{z} - \mathfrak{w}).$$

Lemma 6.2. *Put* $\mathfrak{X} = \{\, \xi \in \mathbf{H}_n^{2n} \mid \xi^* j\iota_n \xi < 0,\ \xi^* k\iota_n \xi = 0 \,\}$. *Then* $H_n' \times GL_n(\mathbf{H})$ *can be bijectively mapped onto* $\mathfrak{X}$ *by*

$$(\mathfrak{z}, \mu) \mapsto \begin{pmatrix} \mathfrak{z}\mu \\ \mu \end{pmatrix}$$

for $\mathfrak{z} \in H_n'$ *and* $\mu \in GL_n(\mathbf{H})$.

Proof. That the image $\begin{pmatrix} \mathfrak{z}\mu \\ \mu \end{pmatrix}$ belongs to $\mathfrak{X}$ can be seen from (6.3a, b). Let $\begin{pmatrix} x \\ y \end{pmatrix} \in \mathfrak{X}$ with $x, y \in \mathbf{H}_n^n$. Then $y^*kx = x^*ky$ and $y^*jx - x^*jy < 0$. If $yu = 0$ with $u \in \mathbf{H}_n^n$, then $u^*(y^*jx - x^*jy)u = 0$, and hence $u = 0$, which shows that y is invertible. Put $\mathfrak{z} = xy^{-1}$. Then $k\mathfrak{z} = \mathfrak{z}^*k$ and $j\mathfrak{z} - \mathfrak{z}^*j < 0$, and hence $\mathfrak{z} \in H_n'$. Then the desired bijectivity is clear. □

If $\alpha \in Sp(n, \mathbf{C})$ and $\xi \in \mathfrak{X}$, then $\alpha\xi \in \mathfrak{X}$, as can be seen from Lemma 6.1. In particular $\alpha \begin{pmatrix} \mathfrak{z} \\ 1 \end{pmatrix} \in \mathfrak{X}$ if $\mathfrak{z} \in H_n'$. By Lemma 6.2 we have

$$\alpha \begin{pmatrix} \mathfrak{z} \\ 1 \end{pmatrix} = \begin{pmatrix} \mathfrak{w} \\ 1 \end{pmatrix} \mu$$

with $\mathfrak{w} \in H_n'$ and $\mu \in GL_n(\mathbf{H})$. We then define $\alpha\mathfrak{z}$ to be $\mathfrak{w}$ and put $\mu = \mu_0(\alpha, \mathfrak{z})$. Clearly $\mathfrak{w}\mu = a_\alpha\mathfrak{z} + b_\alpha$ and $\mu = c_\alpha\mathfrak{z} + d_\alpha$. Thus we obtain (1.5a, b). From (1.5b) we easily see that $(\alpha\beta)\mathfrak{z} = \alpha(\beta\mathfrak{z})$ and $\mu_0(\alpha\beta, \mathfrak{z}) = \mu_0(\alpha, \beta\mathfrak{z})\mu_0(\beta, \mathfrak{z})$. It is also easy to verify that

(6.4) $$Sp(n, \mathbf{C}) \cap U(2n) = \{\, \alpha \in Sp(n, \mathbf{C}) \mid \alpha(j1_n) = j1_n \,\}.$$

Taking $\mathfrak{z} = \mathfrak{w}$ in (6.3a) and substituting $\alpha\mathfrak{z}$ for $\mathfrak{z}$, we easily obtain

(6.5) $$\eta(\mathfrak{z}) = \mu_0(\alpha, \mathfrak{z})^* \eta(\alpha\mathfrak{z}) \mu_0(\alpha, \mathfrak{z}) \quad \text{for every } \alpha \in Sp(n, \mathbf{C}).$$

Similarly, from (6.3b) we obtain

$$k(\mathfrak{z} - \mathfrak{w}) = \mu_0(\alpha, \mathfrak{w})^* k(\alpha\mathfrak{z} - \alpha\mathfrak{w}) \mu_0(\alpha, \mathfrak{z}),$$

and therefore

$$(6.6)\qquad d(\alpha\mathfrak{z}) = k^{-1}(\mu_0(\alpha,\mathfrak{z})^*)^{-1}k \cdot d\mathfrak{z} \cdot \mu_0(\alpha,\mathfrak{z})^{-1}.$$

From this we easily obtain

PROPOSITION 6.3. *The jacobian of the map* $\mathfrak{z} \mapsto \alpha\mathfrak{z}$ *for* $\alpha \in Sp(n, \mathbf{C})$ *is* $m(\alpha, \mathfrak{z})^{-(2n+1)}$. *The standard (Lebesgue) measure on the Euclidean space* $\mathfrak{S}_n$ *times* $\det(\eta(\mathfrak{z}))^{-(2n+1)}$ *gives an* $Sp(n, \mathbf{C})$*-invariant measure on* H'_n.

Let $P = \{\alpha \in Sp(n, \mathbf{C}) \mid c_\alpha = 0\}$. Given $\mathfrak{z} = z + jw \in H'_n$, if we put

$$\beta = \begin{pmatrix} {}^t\bar{a} & za^{-1} \\ 0 & a^{-1} \end{pmatrix}$$

with $a = w^{1/2}$, then $\beta \in P$ and $\mathfrak{z} = \beta(j1_n)$. Thus $Sp(n, \mathbf{C})$ acts on H'_n transitively, and consequently H'_n is isomorphic to $Sp(n, \mathbf{C})/K$, and $Sp(n, \mathbf{C}) = PK$ if we denote by K the group of (6.4).

DEPARTMENT OF MATHEMATICS, PRINCETON UNIVERSITY, PRINCETON, NJ 08544

REFERENCES

[B] V. Bargman, On a Hilbert space of analytic functions and an associated integral transform: Part I, *Comm. Pure Appl. Math.*, **14** (1961), 187–214.

[F] G. Fractman, On the product formula for quadratic forms, thesis, Princeton University, 1991.

[I] J. Igusa, *Theta Functions*, Springer-Verlag, New York, 1972.

[S1] G. Shimura, The periods of certain automorphic forms of arithmetic type, *J. Fac. Sci. Univ. Tokyo Sect. IA Math.*, **28** (1982), 605–632.

[S2] ———, On Eisenstein series, *Duke Math. J.*, **50** (1983), 417–476.

[S3] ———, On Eisenstein series of half-integral weight, *Duke Math. J.*, **52** (1985), 281–314.

[S4] ———, On Hilbert modular forms of half-integral weight, *Duke Math. J.* **55** (1987), 765–838.

[S5] ———, On the Fourier coefficients of Hilbert modular forms of half-integral weight, *Duke Math. J.*, **71** (1993), 501–557.

[Si] C. L. Siegel, Indefinite quadratische Formen und Funktionentheorie I, II, *Math. Ann.*, **124** (1952), 17–54, 364–387.

[W] A. Weil, Sur certain groupes d'opérateurs unitaires, *Acta Math.*, **111** (1964), 143–211.

On the Fourier coefficients of Hilbert modular forms of half-integral weight

Duke Mathematical Journal, 71 (1993), 501-557

Introduction. To explain our main problem, let us first recall the correspondence between an elliptic modular cusp form $f(z) = \sum_{n=1}^{\infty} a(n)e^{2\pi i n z}$ of half-integral weight $\kappa/2$ and a form of weight $\kappa - 1$, where κ is an odd integer $\geqslant 3$. Suppose that f is of level N with a multiple N of 4 and of character ψ, and that f is an eigenform of Hecke operators. Then there is a modular form $g(z) = \sum_{n=1}^{\infty} A_n e^{2\pi i n z}$ of weight $\kappa - 1$ belonging to the same eigenvalues; moreover, if we normalize g so that $A_1 = 1$ and put $D(s, g) = \sum_{n=1}^{\infty} A_n n^{-s}$, then we have

$$(1) \qquad L(s + 1 - m, \varphi) \sum_{n=1}^{\infty} a(rn^2)n^{-s} = a(r)D(s, g)$$

for every squarefree $r > 0$, where $L(s, \varphi)$ is the Dirichlet L-function of φ, $\varphi(n) = \psi(n)(\frac{-1}{n})^m(\frac{r}{n})$, and $m = (\kappa - 1)/2$. In the paper [73] in which these results were obtained, we asked the question whether there was a connection between these Fourier coefficients $a(r)$ and $D(s, q)$.

An extremely interesting and provocative answer was given by Waldspurger in [Wa]. Namely, he proved that $r^{1-\kappa/2}\overline{\psi}(r)a(r)^2/D(m, g, \overline{\varphi})$ depends only on f and $\{r\mathbf{Q}_p^{\times 2}\}_{p|N}$, where $D(s, g, \overline{\varphi}) = \sum_{n=1}^{\infty} \overline{\varphi}(n)A_n n^{-s}$. A formula of a more exact nature which gives the quotient in an explicit form was proved by Kohnen and Zagier in the case where ψ is trivial and the level of g is odd and squarefree, with some conditions on r when the level is not 1 (see [KZ] and [K]). A similar but somewhat different type of result was independently obtained by Niwa in [N] under the assumptions that $N = 4$, r is a prime, and $r \equiv (-1)^{m+1} \pmod 4$.

The purpoose of the present paper is to prove several formulas of such exact nature for the Hilbert modular forms of an arbitrary level and an arbitrary character. To describe our results, let us call F the basic number field which is totally real and denote by $\mathbf{a}$ the set of all archimedean primes of F. Then our Hilbert modular forms are functions on $H^{\mathbf{a}}$, where H is the standard upper half plane. We start with a cusp form f of half-integral weight k and (idele) character ψ, where $k = (m_v + 1/2)_{v\in\mathbf{a}}$ with $m = (m_v)_{v\in\mathbf{a}} \in \mathbf{Z}^{\mathbf{a}}$. For each fractional ideal $\mathfrak{a}$ in F the Fourier expansion of f at a cusp corresponding to $\mathfrak{a}$ has the form

$$\sum_{\xi\in F} \lambda(\xi, \mathfrak{a}; f) \exp\left(\pi i \sum_{v\in\mathbf{a}} \xi_v z_v\right)$$

Received 15 December 1992.

with $\lambda(\xi, \mathfrak{a}; f) \in \mathbf{C}$, where $z = (z_v)_{v \in \mathbf{a}}$ is a variable on $H^{\mathbf{a}}$. We showed in [87] that if f is an eigenform of Hecke operators, then it determines a system of Hecke eigenvalues χ in the space of cusp forms of weight $2m$ and character ψ^2. Instead of a squarefree positive integer r, we take an arbitrary totally positive element τ of F and put $(\tau) = \mathfrak{q}^2\mathfrak{r}$ with a fractional ideal $\mathfrak{q}$ and a squarefree integral ideal $\mathfrak{r}$. In [87] we obtained a generalization of (1) in which $\lambda(\tau, \mathfrak{q}^{-1}; f)$ takes the place of $a(r)$. Since $\lambda(b^2\xi, \mathfrak{a}; f) = b^m\psi_{\mathbf{a}}(b)\lambda(\xi, b\mathfrak{a}; f)$ for every $b \in F^\times$, $\lambda(\tau, \mathfrak{q}^{-1}; f)$ may be symbolically called an $\mathfrak{r}$-th Fourier coefficient of f. Now one of our main results (Theorem 3.4) states that under some conditions we have

$$(2)\quad \overline{\lambda(\tau, \mathfrak{q}^{-1}; f)}\lambda(\tau, \mathfrak{q}^{-1}; f^{\#})\langle \mathbf{g}, \mathbf{g}\rangle/\langle f, f\rangle$$
$$= S\cdot\bar{\gamma}\tau^m\bar{\psi}_{\mathbf{a}}(\tau)N(\mathfrak{h})^{-1}\prod_{\mathfrak{p}}[\bar{\chi}(\mathfrak{p}) - \bar{\varphi}^*(\mathfrak{p})]\left[\sum_{\mathfrak{m}}\bar{\chi}(\mathfrak{m})\varphi^*(\mathfrak{m})N(\mathfrak{m})^{-s}\right]_{s=1}.$$

Here $\mathbf{g}$ is a normalized eigenform belonging to χ, $f^{\#}$ is a certain transform of f, $\varphi^*(\mathfrak{a}) = (\frac{\tau}{\mathfrak{a}})\psi^*(\mathfrak{a})$ with the ideal character $\psi^*(\mathfrak{a})$ associated with ψ, γ is the Gauss sum of φ^*, $\mathfrak{h}$ is the conductor of φ^*, $\mathfrak{m}$ runs over all the integral ideals in F, $\mathfrak{p}$ over all the prime factors of the level of f prime to $\mathfrak{h}$, and S is an explicitly given constant. (The absence of $(\frac{-1}{n})$ is due to our new formulation.) In many cases the first two factors of (2) can be reduced to a square times a simple constant. The product over $\mathfrak{p}$ in (2) is one of the new features previously unnoticed; it often leads to the vanishing of $a(r)$ under a congruence condition on r (see Examples (I)–(III) in Section 3B).

Clearly a formula of type (2) cannot be given with the information on χ alone without specifying f. In fact, we obtain (2) under a certain uniqueness assumption on f relative to τ (see (3.21) of the text). However, without such an assumption, we can prove a formula which gives the sum $\sum_i \overline{\lambda(\tau, \mathfrak{q}^{-1}; f_i')}\lambda(\tau, \mathfrak{q}^{-1}; f_i^{\#})$ for dual bases $\{f_i\}$ and $\{f_i'\}$ of the space of the forms of a fixed level belonging to the same eigenvalues for almost all primes (Theorem 3.2). In addition to these results, we shall prove some more different types of formulas under various conditions (Theorems 3.6, 9.1, and 9.2). For the reader's convenience almost all these results specialized to the case $F = \mathbf{Q}$ are stated in Section 3B translated into the terminology of [73]. Our formulas in this case include Niwa's, but not those of Kohnen and Zagier, who chose f for a given primitive g so that $a(n) \neq 0$ only if $(-1)^m n \equiv 0, 1 \pmod 4$. In the case they treated, we shall give a different type of result for the forms f which they did not consider (see Example (III), Theorem 3B.6 in Section 3B, and Theorem 9.2). Anyway, our results cover the case of arbitrary (N, ψ) which was not treated by these three researchers.

Let us now describe our methods, taking $F = \mathbf{Q}$ for simplicity. We first introduce a theta function $\Theta_r(z, w)$ depending on r, and for f and g as in the beginning prove two integral formulas (Propositions 5.6 and 5.8)

$$(3)\qquad Ca(r)g(w) = \langle \overline{f(rz)}, \Theta_r(z, w)\rangle,$$

$$(4)\qquad Af(rz) = \langle \Theta_r(z, w), g(w)\rangle,$$

where A and C are constants. That each right-hand side of (3) and (4) is a form of the expected weight belonging to a congruence subgroup is relatively easy, but that it is really given as in the formulas is far more nontrivial. The first of these, (3), was essentially given in [87] with C explicitly determined. The latter one is more subtle. It was shown in [88] that the right-hand side of (4) is a Hecke eigenform of weight $\kappa/2$ and of level Nr. However, even if we assume that f is uniquely determined at level N by its eigenvalues, the result in [88] alone cannot justify (4). We overcome this difficulty by showing that the inner product in question is invariant under $z \mapsto z + (1/r)$, which requires several technical lemmas (5.4, 5.5, and 5.7). Once (4) is established, from (3) and (4) we obtain

$$(5) \qquad A = C'\overline{a(r)}\langle g, g\rangle/\langle f, f\rangle$$

with an explicit constant C'.

Next we take the theta function $\theta_0(z) = \sum_{n\in\mathbf{Z}} \exp(2\pi i n^2 z)$ and observe that

$$(6) \qquad \langle\theta_0(z)E(z, s), f(rz)\rangle = \xi(s) \sum_{n=1}^{\infty} a(rn^2)n^{-2s},$$

where E is a standard Eisenstein series of weight m and ξ is a certain gamma factor. (This was one of the main points in [73].) This together with (1) and (4) shows that

$$(7) \qquad Aa(r)\xi(s)D(2s, g) = L(2s + 1 - m, \varphi)\langle\langle\overline{\Theta_r(z, w)}, \theta_0(z)E(z, s)\rangle, g(w)\rangle.$$

By somewhat complicated calculations (Sections 6 and 7) we can prove that $\langle\overline{\Theta}_r, \theta_0 E\rangle$ is essentially the square of an Eisenstein series E_1. Now a principle employed in [76] shows that $\langle E_1^2, g\rangle$ is essentially $D(m, g, \overline{\varphi})D(2s, g)$ (Section 8). Combining this with (5) and (7), we obtain the desired relation between $|a(r)|^2$ and $D(m, g, \overline{\varphi})$.

The whole series of argument is valid only under some strong conditions. For instance, in general $\langle\overline{\Theta}_r, \theta_0 E\rangle$ is essentially E_1^2 only if we modify Θ_r suitably. For this reason, under weaker conditions, we have to transform f and g by $z \mapsto -(Nz)^{-1}$ and $w \mapsto -(Nw/2)^{-1}$, which is why we have $f^{\#}$ instead of f in (2). These transformations, as well as the formation of $f(rz)$ from $f(z)$, become nontrivial for F of an arbitrary degree (Sections 1 and 2). This is one of the factors which make the paper considerably longer than what it would have been if it were restricted to the case $F = \mathbf{Q}$.

Though the present investigation is restricted to the holomorphic modular forms, our methods are applicable to the nonholomorphic forms. In fact, the case of eigenforms of Maass's type on $SL_2(F)$ and its metaplectic cover is being successfully investigated by Kamal Khuri-Makdisi. There is no doubt that the most general case of an arbitrary F can be handled in the same fashion. It may be pointed out also that we empolyed neither trace formula nor automorphic representations, simply because we obtain the formulas without them. While their dispensability makes our treatment require less calculations and so is an advantage, we cannot deny the possibility that more refined results may be obtained by combining our methods with them. Another point worth commenting on is about the basic formula (1). It should be emphasized that the correspondence $f \mapsto g$ involves *infinitely many equalities (1) valid for all squarefree r*. Though the correspondence has been generalized in various directions in many papers, there are few investigations in terms of Fourier coefficients in the sense of (1) for general r, which leaves much to be desired. We hope that the present paper will arouse the interest of future researchers on this subject.

1. Hilbert modular forms. The following notation will be used throughout the paper: For an associative ring R with identity element we denote by $R^\times$ the group of all its invertible elements and by $M_n(R)$ the ring of $n \times n$ matrices with entries in R. To indicate that a union $X = \bigcup_{i \in I} Y_i$ is disjoint, we write $X = \bigsqcup_{i \in I} Y_i$. For $x = \begin{pmatrix} a & b \\ c & d \end{pmatrix}$ we put $a = a_x$, $b = b_x$, $c = c_x$, and $d = d_x$ if there is no fear of confusion; these are also written $a(x)$, $b(x)$, $c(x)$, and $d(x)$ when x involves more than one letter; further we put $\hat{x} = \begin{pmatrix} d & -b \\ -c & a \end{pmatrix}$.

We fix a totally real algebraic number field F of degree d and denote by $\mathbf{a}$, $\mathbf{h}$, $\mathfrak{g}$, D_F, and $\mathfrak{d}$ the set of archimedean primes of F, the set of nonarchimedean primes of F, the maximal order of F, the discriminant of F, and the different of F relative to $\mathbf{Q}$. Given an algebraic group $\mathscr{G}$ defined over F, we define $\mathscr{G}_v$ for each $v \in \mathbf{a} \cup \mathbf{h}$ and the adelization $\mathscr{G}_\mathbf{A}$ as usual and view $\mathscr{G}$ as a subgroup of $\mathscr{G}_\mathbf{A}$. We then denote by $\mathscr{G}_\mathbf{a}$ and $\mathscr{G}_\mathbf{h}$ the archimedean and nonarchimedean factors of $\mathscr{G}_\mathbf{A}$, respectively. For an element x of $\mathscr{G}_\mathbf{A}$ its $\mathbf{a}$-component, $\mathbf{h}$-component, and v-component are denoted by $x_\mathbf{a}$, $x_\mathbf{h}$, and x_v. Given a set X, we denote by $X^\mathbf{a}$ the product of $\mathbf{a}$ copies of X, that is, the set of all indexed elements $(x_v)_{v \in \mathbf{a}}$ with x_v in X. For x and y in $\mathbf{C}^\mathbf{a}$ we put

(1.1a) $$\mathbf{e}_\mathbf{a}(x) = \exp\left(2\pi i \sum_{v \in \mathbf{a}} x_v\right),$$

(1.1b) $$\|y\| = \sum_{v \in \mathbf{a}} y_v,$$

(1.1c) $$x^y = \prod_{v \in \mathbf{a}} x_v^{y_v},$$

where $x_v^{y_v}$ should be understood according to the context. We use the notation $\mathbf{e}_\mathbf{a}(x)$ and x^y also for $x \in F_\mathbf{A}^\times$. The identity element of the ring $\mathbf{C}^\mathbf{a}$, which will, for the most part, appear as an exponent, is denoted by u. Thus $\|u\| = d$ and $c^u = N_{F/\mathbf{Q}}(c)$ for $c \in F^\times$. For $x, y \in \mathbf{R}^\mathbf{a}$ or $\in F_\mathbf{A}$ we write $x \geqslant y$ if $x_v \geqslant y_v$ for every $v \in \mathbf{a}$, and $x \gg 0$ if $x_v > 0$ for every $v \in \mathbf{a}$. For $a \in F_\mathbf{A}^\times$ we put

(1.2) $$\mathrm{sgn}(a) = (a_v/|a_v|)_{v \in \mathbf{a}}.$$

For a fractional ideal $\mathfrak{x}$ in F and $t \in F_{\mathbf{A}}^{\times}$ we denote by $N(\mathfrak{x})$ the norm of $\mathfrak{x}$ and by $t\mathfrak{x}$ the fractional ideal in F such that $(t\mathfrak{x})_v = t_v \mathfrak{x}_v$ for every $v \in \mathbf{h}$. For $v \in \mathbf{h}$ we denote by π_v any prime element of F_v and put $N_v = N(\pi_v \mathfrak{g})$. By a *Hecke character* of F, we mean a continuous unitary character ψ of $F_{\mathbf{A}}^{\times}$ of finite or infinite order such that $\psi(F^{\times}) = 1$. Given such a ψ, we denote by ψ^* the ideal character such that $\psi^*(t\mathfrak{g}) = \psi(t)$ if $t \in F_v^{\times}$ and $\psi(\mathfrak{g}_v^{\times}) = 1$; we put $\psi^*(\mathfrak{a}) = 0$ for every fractional ideal $\mathfrak{a}$ that is not prime to the conductor of ψ.

We now put

$$H = \{z \in \mathbf{C} \mid \operatorname{Im}(z) > 0\}, \qquad \mathbf{T} = \{t \in \mathbf{C} \mid |t| = 1\},$$

$$G = SL_2(F), \qquad \tilde{G} = GL_2(F), \qquad \tilde{G}_{\mathbf{a}+} = \{x \in \tilde{G}_{\mathbf{a}} \mid \det(x) \gg 0\},$$

$$\tilde{G}_{\mathbf{A}+} = \tilde{G}_{\mathbf{a}+}\tilde{G}_{\mathbf{h}}, \qquad \tilde{G}_{+} = \tilde{G}_{\mathbf{A}+} \cap \tilde{G},$$

and for two fractional ideals $\mathfrak{x}$ and $\mathfrak{y}$ in F such that $\mathfrak{x}\mathfrak{y} \subset \mathfrak{g}$ we put

$$\tilde{D}[\mathfrak{x}, \mathfrak{y}] = \tilde{G}_{\mathbf{a}+} \prod_{v \in \mathbf{h}} \tilde{D}_v[\mathfrak{x}, \mathfrak{y}] \ (\subset \tilde{G}_{\mathbf{A}}), \qquad \tilde{D}_v[\mathfrak{x}, \mathfrak{y}] = \mathfrak{o}[\mathfrak{x}, \mathfrak{y}]_v^{\times},$$

$$\mathfrak{o}[\mathfrak{x}, \mathfrak{y}] = \{x \in M_2(F) \mid a_x \in \mathfrak{g}, b_x \in \mathfrak{x}, c_x \in \mathfrak{y}, d_x \in \mathfrak{g}\},$$

$$D[\mathfrak{x}, \mathfrak{y}] = G_{\mathbf{A}} \cap \tilde{D}[\mathfrak{x}, \mathfrak{y}], \qquad D_v[\mathfrak{x}, \mathfrak{y}] = G_v \cap \tilde{D}_v[\mathfrak{x}, \mathfrak{y}],$$

$$\tilde{\Gamma}[\mathfrak{x}, \mathfrak{y}] = \tilde{G} \cap \tilde{D}[\mathfrak{x}, \mathfrak{y}], \qquad \Gamma[\mathfrak{x}, \mathfrak{y}] = G \cap D[\mathfrak{x}, \mathfrak{y}].$$

In general, the notation and conventions in the present paper are the same as those in [87] with the following exceptions: the upper half plane, the maximal order of F, and the set of finite primes of F are denoted by H, $\mathfrak{g}$, and $\mathbf{h}$ here instead of $\mathscr{H}$, $\mathfrak{o}$, and $\mathfrak{f}$ there; here $D[\mathfrak{x}, \mathfrak{y}]$ and $\tilde{D}[\mathfrak{x}, \mathfrak{y}]$ contain $G_{\mathbf{a}}$ and $\tilde{G}_{\mathbf{a}+}$, but in [87] their archimedean factors are compact.

We now take the metaplectic group $Mp(F_{\mathbf{A}})$ of Weil [We]. We denote by $M_{\mathbf{A}}$ this group, which is an extension of $G_{\mathbf{A}}$ with kernel $\mathbf{T}$, and by pr the projection map of $M_{\mathbf{A}}$ to $G_{\mathbf{A}}$. For $y = \operatorname{pr}(x)$ with $x \in M_{\mathbf{A}}$ we put $a_x = a_y$, $b_x = b_y$, $c_x = c_y$, and $d_x = d_y$. By a *half-integral weight* we mean an element k of $\mathbf{Q}^{\mathbf{a}}$ such that $k - u/2 \in \mathbf{Z}^{\mathbf{a}}$; naturally, an *integral weight* is an element of $\mathbf{Z}^{\mathbf{a}}$. Given a weight n, we define a factor of automorphy $J_n(\tau, z)$ for $z \in H^{\mathbf{a}}$ by

$$J_n(\tau, z) = \begin{cases} j(\tau, z)^n & \text{if } n \in \mathbf{Z}^{\mathbf{a}} \text{ and } \tau \in \tilde{G}_{\mathbf{A}+}, \\ h(\tau, z) j(\tau, z)^{n-u/2} & \text{if } n \notin \mathbf{Z}^{\mathbf{a}}, \tau \in M_{\mathbf{A}}, \text{ and } \operatorname{pr}(\tau) \in P_{\mathbf{A}} C'', \end{cases}$$

$$j(\xi, z) = j(\operatorname{pr}(\xi), z) \qquad (\xi \in M_{\mathbf{A}}),$$

$$j(\tau, z) = (j(\tau_v, z_v))_{v \in \mathbf{a}} \qquad (\tau \in \tilde{G}_{\mathbf{A}+}),$$

$$j(\alpha, w) = \det(\alpha)^{-1/2}(c_\alpha w + d_\alpha) \qquad (\alpha \in GL_2(\mathbf{R}), \det(\alpha) > 0, w \in \mathbf{C}).$$

Here $h(\tau, z)$ is a quasi factor of automorphy of weight $u/2$ defined for $\tau \in \mathrm{pr}^{-1}(P_{\mathbf{A}}C'')$ and $z \in H^{\mathbf{a}}$ (see [85a, Prop. 3.2] or [87, Prop. 2.3]), and

$$\text{(1.3a)} \qquad P = \{\alpha \in G \mid c_\alpha = 0\},$$

$$\text{(1.3b)} \qquad C' = D[2\mathfrak{d}^{-1}, 2\mathfrak{d}], \qquad C'' = C' \cup C'\varepsilon,$$

$$\text{(1.3c)} \qquad \varepsilon \in G_{\mathbf{A}}, \qquad \varepsilon_{\mathbf{a}} = 1, \qquad \varepsilon_v = \begin{pmatrix} 0 & -\delta_v^{-1} \\ \delta_v & 0 \end{pmatrix} \textit{ for } v \in \mathbf{h},$$

where δ is an arbitrarily fixed element of $F_{\mathbf{h}}^\times$ such that $\mathfrak{d} = \delta\mathfrak{g}$. We view G as a subgroup of $M_{\mathbf{A}}$ through the canonical lifting $G \to M_{\mathbf{A}}$. Thus $h(\tau, z)$ is meaningful for $\tau \in G \cap P_{\mathbf{A}}C''$. We let $\tilde{G}_{\mathbf{a}+}$ act on $H^{\mathbf{a}}$ as usual and define the action of an element τ of $\tilde{G}_{\mathbf{A}+}$ or $M_{\mathbf{A}}$ on $H^{\mathbf{a}}$ to be the same as that of $\tau_{\mathbf{a}}$ or $\mathrm{pr}(\tau)_{\mathbf{a}}$. We note here (see [85a, (3.10b, c), (3.13)] and [87, (1.13), Lemma 2.5])

$$\text{(1.4a)} \qquad xy = yx \textit{ if } x, y \in M_{\mathbf{A}}, \qquad \mathrm{pr}(x) \in G_{\mathbf{a}}, \qquad \textit{and } \mathrm{pr}(y) \in G_{\mathbf{h}},$$

$$\text{(1.4b)} \qquad h(\xi, z)^2 = \zeta \cdot j(\xi, z)^u \qquad \textit{with } \zeta \in \mathbf{T} \textit{ if } \mathrm{pr}(\xi) \in P_{\mathbf{A}}C'',$$

$$\text{(1.4c)} \qquad h(\alpha\beta\gamma, z) = h(\alpha, z)h(\beta, \gamma(z))h(\gamma, z)$$

$$\textit{if } \mathrm{pr}(\alpha) \in P_{\mathbf{A}}, \qquad \mathrm{pr}(\beta) \in P_{\mathbf{A}}C'', \qquad \textit{and } \mathrm{pr}(\gamma) \in C'',$$

$$\text{(1.4d)} \qquad h(t \cdot r_P(\pi), z) = t^{-1}|d_\pi|_{\mathbf{a}}^{1/2} \qquad \textit{if } t \in \mathbf{T} \textit{ and } \pi \in P_{\mathbf{A}},$$

$$\text{(1.4e)} \quad h(x, z) = h(y, z) \quad \textit{if both } \mathrm{pr}(x) \textit{ and } \mathrm{pr}(y) \textit{ belong to } C' \textit{ and } x = \sigma y \sigma^{-1}$$
$$\textit{with } \sigma = r_P(\mathrm{diag}[r, r^{-1}]), \; r \in F_{\mathbf{h}}^\times,$$

where r_P is the map of $P_{\mathbf{A}}$ into $M_{\mathbf{A}}$ defined in [87, (1.8)].

Given a $\mathbf{C}$-valued function f on $H^{\mathbf{a}}$ and τ in $M_{\mathbf{A}}$ or $\tilde{G}_{\mathbf{A}+}$, we define a function $f\|_n\tau$ on $H^{\mathbf{a}}$ by

$$(f\|_n\tau)(z) = J_n(\tau, z)^{-1}f(\tau(z)) \qquad (z \in H^{\mathbf{a}}).$$

We then denote by $\mathscr{M}_n$ the set of all holomorphic Hilbert modular forms of weight n with respect to congruence subgroups of G, and by $\mathscr{S}_n$ the subset of $\mathscr{M}_n$ consisting of the cusp forms (see [87, p. 778]). Given two functions f and g on $H^{\mathbf{a}}$ invariant under the operation $\|_n\gamma$ for every γ in a congruence subgroup Γ of G, we define their inner product $\langle f, g\rangle$ by

$$\text{(1.5a)} \qquad \langle f, g\rangle = \mathrm{vol}(\Phi)^{-1}\int_\Phi \overline{f(z)}g(z)\mathrm{Im}(z)^n \, d_H z \qquad (\Phi = \Gamma\backslash H^{\mathbf{a}}),$$

where $d_H(x+iy) = \prod_{v\in\mathbf{a}} y_v^{-2}\, dx_v\, dy_v$ and $\mathrm{vol}(\Phi) = \int_\Phi d_H z$. We note here

$$\mathrm{vol}(\Gamma[\mathfrak{x}, \mathfrak{y}]\backslash H^{\mathbf{a}}) = 2\pi^{-d} D_F^{3/2} \zeta_F(2) N(\mathfrak{x}\mathfrak{y}) \prod_{v|\mathfrak{x}\mathfrak{y}} [1 + N_v^{-1}], \tag{1.5b}$$

where ζ_F denotes the zeta function of F.

Throughout the rest of the paper, we fix a half-integral weight k and put

$$m = k - u/2. \tag{1.6}$$

For $f \in \mathcal{M}_k$ we can define a function $f_{\mathbf{A}}$ on $M_{\mathbf{A}}$ by

$$f_{\mathbf{A}}(\alpha x) = (f\|_k x)(\mathbf{i}) \qquad \text{for } \alpha \in G \text{ and } \mathrm{pr}(x) \in B, \tag{1.7}$$

where $\mathbf{i}$ is the "origin" iu of $H^{\mathbf{a}}$ and B is an open subgroup of C'' such that $f\|_k\gamma = f$ for every $\gamma \in G \cap B$. This is well defined independently of the choice of B.

Given two integral ideals $\mathfrak{b}$ and $\mathfrak{b}'$ and a Hecke character ψ of F whose conductor divides $4\mathfrak{b}\mathfrak{b}'$, we denote by $\mathcal{M}_k(\mathfrak{b}, \mathfrak{b}'; \psi)$ the set of all f in $\mathcal{M}_k$ such that

$$f\|_k\gamma = \psi_{\mathfrak{f}}(a_\gamma) f \qquad \text{for every } \gamma \in \Gamma[2\mathfrak{b}\mathfrak{d}^{-1}, 2\mathfrak{b}'\mathfrak{d}], \tag{1.8}$$

where $\mathfrak{f}$ is the conductor of ψ and $\psi_{\mathfrak{a}} = \prod_{v|\mathfrak{a}} \psi_v$ for any integral ideal $\mathfrak{a}$; we put $\mathcal{S}_k(\mathfrak{b}, \mathfrak{b}'; \psi) = \mathcal{S}_k \cap \mathcal{M}_k(\mathfrak{b}, \mathfrak{b}'; \psi)$. The ideal $4\mathfrak{b}\mathfrak{b}'$ may be called the "level" of our group. We always assume

$$\psi_{\mathbf{a}}(-1) = (-1)^{\|m\|}. \tag{1.9}$$

If $f \in \mathcal{M}_k(\mathfrak{b}, \mathfrak{b}'; \psi)$, we have

$$f_{\mathbf{A}}(\alpha x w) = \psi_{\mathfrak{f}}(a_w)^{-1} J_k(w, \mathbf{i})^{-1} f_{\mathbf{A}}(x) \tag{1.10}$$

$$\text{for } \alpha \in G \text{ and } \mathrm{pr}(w) \in D[2\mathfrak{b}\mathfrak{d}^{-1}, 2\mathfrak{b}'\mathfrak{d}],\ w(\mathbf{i}) = \mathbf{i}.$$

PROPOSITION 1.1. *Given $f \in \mathcal{M}_k(\mathfrak{b}, \mathfrak{b}'; \psi)$, there is a complex number $\lambda(\xi, \mathfrak{m}; f, \psi)$, determined for $\xi \in F$ and a fractional ideal $\mathfrak{m}$, such that*

$$f_{\mathbf{A}}\left(r_P\begin{pmatrix} t & s \\ 0 & t^{-1}\end{pmatrix}\right) = \psi_{\mathbf{h}}(t)^{-1} t_{\mathbf{a}}^{m} |t|_{\mathbf{A}}^{1/2} \sum_{\xi\in F} \lambda(\xi, t\mathfrak{g}; f, \psi) \mathbf{e}_{\mathbf{a}}(it^2\xi/2) \mathbf{e}_{\mathbf{A}}(ts\xi/2)$$

for every $t \in F_{\mathbf{A}}^\times$ and $s \in F_{\mathbf{A}}$, where $\mathbf{e}_{\mathbf{A}}$ is the basic character $F_{\mathbf{A}} \to \mathbf{T}$ of [87, p. 771]. Moreover, $\lambda(\xi, \mathfrak{m}; f, \psi)$ has the following properties:

$$\lambda(\xi, \mathfrak{m}; f, \psi) \neq 0 \qquad \textit{only if } \xi \in \mathfrak{b}^{-1}\mathfrak{m}^{-2} \textit{ and } \xi \textit{ is totally nonnegative}; \tag{1.11a}$$

$$\lambda(\xi b^2, \mathfrak{m}; f, \psi) = b^m \psi_{\mathbf{a}}(b) \lambda(\xi, b\mathfrak{m}; f, \psi) \qquad \textit{for every } b \in F^\times. \tag{1.11b}$$

Furthermore, if $\beta \in G \cap \mathrm{diag}[r, r^{-1}]D[2\mathfrak{b}\mathfrak{d}^{-1}, 2\mathfrak{b}'\mathfrak{d}]$ *with* $r \in F_{\mathbf{A}}^{\times}$, *then*

$$(1.11c)\quad J_k(\beta, \beta^{-1}z)f(\beta^{-1}z) = \psi_{\mathbf{h}}(r)^{-1}\psi_{\mathfrak{f}}(d_\beta r)|r_{\mathbf{h}}|^{1/2}\sum_{\xi\in F}\lambda(\xi, r\mathfrak{g}; f, \psi)\mathbf{e}_{\mathbf{a}}(\xi z/2).$$

If $\mathfrak{b} = \mathfrak{g}$, this is a restatement of [87, Prop. 3.1]. The general case can be proved in a similar way. Alternatively, it can be reduced to the special case, though this reduction is not completely trivial as it requires Lemma 1.3 below as well as (1.4c, d, e).

It should be noted that ψ is not unique for f. In fact, $f \in \mathcal{M}_k(\mathfrak{b}, \mathfrak{b}'; \omega\psi)$ for any Hecke character ω such that $\omega(\mathfrak{g}_v^{\times}) = 1$ for all $v \in \mathbf{h}$. (See also [87, p. 779, lines 10–14 and Remark 6.3].) When ψ is fixed, we simply write $\lambda_f(\xi, \mathfrak{m})$ or $\lambda(\xi, \mathfrak{m}; f)$ for $\lambda(\xi, \mathfrak{m}; f, \psi)$. Now we have

$$(1.12)\qquad \psi_{\mathbf{a}}(x) = \mathrm{sgn}(x)^r|x|^{i\mu}$$

with $r \in \mathbf{Z}^{\mathbf{a}}$ and $\mu \in \mathbf{R}^{\mathbf{a}}$. For simplicity we shall always assume $\|\mu\| = 0$, though this is not absolutely necessary. From (1.9) we see that $\|r\| \equiv \|m\| \pmod 2$. We say that ψ is *normalized for* f if (1.12) holds with $r = m$. Given $f \in \mathcal{M}_k(\mathfrak{b}, \mathfrak{b}'; \psi)$, we can always replace ψ by a character that is normalized for f, as can be seen from the following:

LEMMA 1.2. *Let* $\mathfrak{a}$ *be an integral ideal,* α *a character of* $(\mathfrak{g}/\mathfrak{a})^{\times}$, *viewed also as a character of* $\prod_{v|\mathfrak{a}}\mathfrak{g}_v^{\times}$ *in an obvious way, and* r *an element of* $\mathbf{Z}^{\mathbf{a}}$ *such that* $\alpha(-1) = (-1)^{\|r\|}$. *Then there exist a Hecke character* ω *of* F *and an element* κ *of* $\mathbf{R}^{\mathbf{a}}$ *such that* $\omega_{\mathbf{a}}(x) = \mathrm{sgn}(x)^r|x|^{i\kappa}$, $\|\kappa\| = 0$, $\omega = \alpha$ *on* $\prod_{v|\mathfrak{a}}\mathfrak{g}_v^{\times}$, *and* ω *is unramified at every* $v \in \mathbf{h}$ *prime to* $\mathfrak{a}$.

Proof. Assigning $\mathrm{sgn}(e)^r\alpha(e)$ to $e \in \mathfrak{g}^{\times}$, we obtain a character of $\mathfrak{g}^{\times}/\{\pm 1\}$. Therefore we can find an element κ of $\mathbf{R}^{\mathbf{a}}$ such that $\|\kappa\| = 0$ and $\mathrm{sgn}(e)^r\alpha(e) = |e|^{-i\kappa}$ for every $e \in \mathfrak{g}^{\times}$. Define $\omega(x)$ for $x \in F_{\mathbf{a}}^{\times}\prod_{v\in\mathbf{h}}\mathfrak{g}_v^{\times}$ by $\omega(x) = \mathrm{sgn}(x)^r|x|^{i\kappa}\alpha((x_v)_{v|\mathfrak{a}})$. Since $\omega(\mathfrak{g}^{\times}) = 1$, we can extend ω to a Hecke character of F with the desired properties.

LEMMA 1.3. *Let* $\rho = r_P(\zeta)$, $\zeta = \mathrm{diag}[\sigma, \sigma^{-1}]$ *with* $\sigma \in F_{\mathbf{A}}^{\times}$ *and* $\eta = \rho\xi\rho^{-1}$ *with* $\xi \in M_{\mathbf{A}}$. *If both* ξ *and* η *belong to* $\mathrm{pr}^{-1}(C')$, *then* $J_k(\eta, \sigma_{\mathbf{a}}^2 z) = J_k(\xi, z)$.

Proof. Let $\alpha = r_P(\zeta_{\mathbf{h}})$ and $\beta = r_P(\zeta_{\mathbf{a}})$. Then $\eta = \alpha\beta\xi\beta^{-1}\alpha^{-1}$. By (1.4e) we have $J_k(\eta, z) = J_k(\beta\xi\beta^{-1}, z)$. Since $\mathrm{pr}(\beta) \in P_{\mathbf{a}}$, by (1.4a, c) we have $J_k(\beta\xi\beta^{-1}, \sigma_a^2 z) = J_k(\beta, z)J_k(\xi, z)J_k(\beta^{-1}, z) = J_k(\xi, z)$ as expected.

PROPOSITION 1.4. *Let* $\mathfrak{y}$ *be a fractional ideal; let* $0 \ll \tau \in F$ *and* $f \in \mathcal{M}_k(\mathfrak{b}, \mathfrak{b}'; \psi)$. *Then there exists an element* h *of* $\mathcal{M}_k(\tau\mathfrak{y}^{-2}\mathfrak{b} \cap \mathfrak{g}, \tau^{-1}\mathfrak{y}^2\mathfrak{b}' \cap \mathfrak{g}; \psi\varepsilon_\tau)$ *such that*

$$(1.13)\qquad \lambda(\xi, \mathfrak{m}; h, \psi\varepsilon_\tau) = \lambda(\tau\xi, \mathfrak{y}^{-1}\mathfrak{m}; f, \psi)$$

for every $(\xi, \mathfrak{m})$, *where* ε_τ *is the Hecke character of* F *corresponding to the extension* $F(\tau^{1/2})/F$. *Moreover, let* $p(z) = f(z/\tau)$ *and* $\rho = \mathrm{diag}[y^{-1}, y]$ *with any* $y \in F_{\mathbf{h}}^{\times}$ *such that* $y\mathfrak{g} = \mathfrak{y}$. *Then* $h_{\mathbf{A}}(x) = (\psi\varepsilon_\tau)(y)^{-1}N(\mathfrak{y})^{-1/2}p_{\mathbf{A}}(x\cdot r_P(\rho))$. *(This implies that* $h(z) = f(z/\tau)$ *if* $\mathfrak{y} = \mathfrak{g}$.*) Furthermore, if* f *is a cusp form, so is* h, *and* $\langle f, f\rangle = \tau^{-k}N(\mathfrak{y})\langle h, h\rangle$.

Proof. Put $g'(x) = f_{\mathbf{A}}(xr_P(\rho))$ for $x \in M_{\mathbf{A}}$. By (1.10) and (1.4e), $g'(\alpha x w) = \overline{\psi}_{\mathfrak{f}}(a_w)J_k(w, \mathbf{i})^{-1}g'(x)$ if $\alpha \in G$ and $\mathrm{pr}(w) \in D[2(y^{-2}\mathfrak{b} \cap \mathfrak{g})\mathfrak{d}^{-1}, 2(y^2\mathfrak{b}' \cap \mathfrak{g})\mathfrak{d}]$, $w(\mathbf{i}) = \mathbf{i}$. Let g be the function of (1.11c) with y^{-1} as r. Since $g \in \mathscr{M}_k$, $g_{\mathbf{A}}(\alpha x w) = J_k(w, \mathbf{i})^{-1}g_{\mathbf{A}}(x)$ if $\alpha \in G$ and $\mathrm{pr}(w) \in B$, $w(\mathbf{i}) = \mathbf{i}$ with an open subgroup B of C'. Let us now show that $g_{\mathbf{A}} = \psi_{\mathfrak{f}}(d_\beta r)g'$. Since $M_{\mathbf{A}} = G \cdot r_P(P_{\mathbf{a}})\{w \in \mathrm{pr}^{-1}(B) | w(\mathbf{i}) = \mathbf{i}\}$ for any such B, it is sufficient to prove the equality on $r_P(P_{\mathbf{a}})$. Thus let $x = r_P(\pi)$, $\pi = \begin{pmatrix} t & s \\ 0 & t^{-1} \end{pmatrix} \in P_{\mathbf{a}}$, $t \gg 0$, and $z = \pi(\mathbf{i})$. Then $g_{\mathbf{A}}(x) = (g\|_k x)(\mathbf{i}) = t^k g(z)$ and $g'(x) = f_{\mathbf{A}}(r_P(\pi\rho)) = \psi(y)t^k|y|_{\mathbf{A}}^{-1/2}\sum_\xi \lambda_f(\xi, t\mathfrak{y}^{-1})\mathbf{e}_{\mathbf{a}}(\xi z/2)$. Comparing this with (1.11c), we obtain the desired equality and find that $g \in \mathscr{M}_k(y^{-2}\mathfrak{b} \cap \mathfrak{g}, y^2\mathfrak{b}' \cap \mathfrak{g}; \psi)$. Taking π more generally in $P_{\mathbf{A}}$ in this comparison, we obtain $\lambda_{cg}(\xi, \mathfrak{m}) = \lambda_f(\xi, \mathfrak{y}^{-1}\mathfrak{m})$ with $c = \psi(y)^{-1}N(\mathfrak{y})^{-1/2}\overline{\psi}_{\mathfrak{f}}(d_\beta r)$. This proves the first two assertions in the case $\tau = 1$. To prove them in the general case, we first observe that if $y = \sigma_{\mathbf{h}}$ with $\sigma \in F^\times$, then our result shows

$$(f\|_k\gamma)_{\mathbf{A}}(x) = f_{\mathbf{A}}(xr_P(\gamma_{\mathbf{h}}^{-1})) \qquad \textit{if } \gamma = \mathrm{diag}[\sigma, \sigma^{-1}], \sigma \in F^\times, \tag{1.14}$$

since we can take γ^{-1} as β in (1.11c). Therefore we see that changing τ and $\mathfrak{y}$ for $\sigma^2\tau$ and $\sigma\mathfrak{y}$ with a suitable $\sigma \in F^\times$, we may assume that $1/\tau \in \mathfrak{g}$. Put $p(z) = f(z/\tau)$. Then the argument of [87, p. 782] (with $1/\tau$ in place of τ) shows that $p \in \mathscr{M}_k(\mathfrak{b}, \tau^{-1}\mathfrak{b}'; \psi\varepsilon_\tau)$ and $\lambda_p(\xi, \mathfrak{m}) = \lambda_f(\tau\xi, \mathfrak{m})$. Applying the above result to p, we obtain an element h of $\mathscr{M}_k(y^{-2}\mathfrak{b} \cap \mathfrak{g}, \tau^{-1}y^2\mathfrak{b}' \cap \mathfrak{g}; \psi\varepsilon_\tau)$ such that $\lambda_h(\xi, \mathfrak{m}) = \lambda_p(\xi, \mathfrak{y}^{-1}\mathfrak{m})$ and $h_{\mathbf{A}}(x) = (\psi\varepsilon_\tau)(y)^{-1}N(\mathfrak{y})^{-1/2}p_{\mathbf{A}}(xr_P(\rho))$. Then $\lambda_h(\xi, \mathfrak{m}) = \lambda_f(\tau\xi, \mathfrak{y}^{-1}\mathfrak{m})$, and in particular $\lambda_h(\xi, \mathfrak{g}) = \lambda_f(\tau\xi, \mathfrak{y}^{-1})$. This is nonvanishing only if $\tau\xi \in \mathfrak{b}^{-1}\mathfrak{y}^2$, and hence $h(z + b) = h(z)$ for every $b \in 2\tau\mathfrak{b}\mathfrak{y}^{-2}\mathfrak{d}^{-1}$. By [87, Lemma 3.4] this shows that $h \in \mathscr{M}_k(\tau y^{-2}\mathfrak{b} \cap \mathfrak{g}, \tau^{-1}y^2\mathfrak{b}' \cap \mathfrak{g}; \psi\varepsilon_\tau)$ since ε_τ is defined modulo $4(\tau\mathfrak{y}^{-2} \cap \tau^{-1}\mathfrak{y}^2)$. As for the last two assertions, the above reasoning shows that $h(z) = N(\mathfrak{y})^{-1/2}t \cdot J_k(\beta, \beta^{-1}z)p(\beta^{-1}z)$ with $t \in \mathbf{T}$ and $\beta \in G$, which yields the desired facts.

The function h being as in Proposition 1.4, from (1.11a) and (1.13) we obtain

$$\lambda_h(\xi, \mathfrak{m}) \neq 0 \Rightarrow \tau\xi \in \mathfrak{b}^{-1}\mathfrak{y}^2\mathfrak{m}^{-2}, \tag{1.15}$$

and in particular

$$\lambda_h(\xi, \mathfrak{y}) \neq 0 \Rightarrow \tau\xi \in \mathfrak{b}^{-1}. \tag{1.16}$$

PROPOSITION 1.5. *The notation being as in Proposition 1.4, suppose that* $\mathfrak{b}'\mathfrak{y}^2 \subset \tau\mathfrak{g}$. *Then the map* $f \mapsto h$ *determined by (1.13) gives a bijection of* $\mathscr{M}_k(\mathfrak{b}, \mathfrak{b}'; \psi)$ *onto the set of all* h *in* $\mathscr{M}_k(\tau\mathfrak{y}^{-2}\mathfrak{b} \cap \mathfrak{g}, \tau^{-1}\mathfrak{y}^2\mathfrak{b}'; \psi\varepsilon_\tau)$ *satisfying (1.16).*

Proof. Clearly, $h \in \mathscr{M}_k(\tau\mathfrak{y}^{-2}\mathfrak{b} \cap \mathfrak{g}, \tau^{-1}\mathfrak{y}^2\mathfrak{b}'; \psi\varepsilon_\tau)$ if $\mathfrak{b}'\mathfrak{y}^2 \subset \tau\mathfrak{g}$. To prove the surjectivity, let h be an element of the set satisfying (1.16). Applying Proposition 1.4 to h with τ^{-1} and $\mathfrak{y}^{-1}$ in place of τ and $\mathfrak{y}$, we obtain an element g of $\mathscr{M}_k(\mathfrak{b} \cap \tau^{-1}\mathfrak{y}^2, \mathfrak{b}'; \psi)$ such that $\lambda_g(\xi, \mathfrak{m}) = \lambda_h(\xi/\tau, \mathfrak{y}\mathfrak{m})$. Then $\lambda_g(\xi, \mathfrak{g}) \neq 0$ only if $\xi \in \mathfrak{b}^{-1}$. This implies that $g(z+b) = g(z)$ for every $b \in 2\mathfrak{b}\mathfrak{d}^{-1}$, and hence $g \in \mathcal{M}_k(\mathfrak{b}, \mathfrak{b}'; \psi)$ by virtue of the principle of [87, Lemma 3.4]. This proves our proposition.

LEMMA 1.6. *Let $f \in \mathcal{M}_k(\mathfrak{a}\mathfrak{b}, \mathfrak{b}'; \psi)$ with an integral ideal $\mathfrak{a}$ such that $\mathfrak{a}_v = \mathfrak{g}_v$ for $v \nmid 2\mathfrak{b}\mathfrak{b}'$. Then there exists a unique element g of $\mathcal{M}_k(\mathfrak{b}, \mathfrak{b}'; \psi)$ such that $\lambda_g(\xi, \mathfrak{m}) = \lambda_f(\xi, \mathfrak{m})$ if $\xi\mathfrak{b} \subset \mathfrak{m}^{-2}$.*

Proof. Let $\Gamma[2\mathfrak{b}\mathfrak{d}^{-1}, 2\mathfrak{b}'\mathfrak{d}] = \bigsqcup_{\alpha\in A} \Gamma[2\mathfrak{a}\mathfrak{b}\mathfrak{d}^{-1}, 2\mathfrak{b}'\mathfrak{d}]\alpha$ with $A \subset G$ and

$$D[2\mathfrak{b}\mathfrak{d}^{-1}, 2\mathfrak{b}'\mathfrak{d}] = \bigsqcup_{w\in W} D[2\mathfrak{a}\mathfrak{b}\mathfrak{d}^{-1}, 2\mathfrak{b}'\mathfrak{d}]\,\mathrm{pr}(w)$$

with $W \subset \mathrm{pr}^{-1}(G_{\mathbf{h}})$. Put $g = N(\mathfrak{a})^{-1}\sum_{\alpha\in A}\psi_{\mathfrak{f}}(a_\alpha)^{-1} f\|_k\alpha$. We easily see that $g \in \mathcal{M}_k(\mathfrak{b}, \mathfrak{b}'; \psi)$ and $g_{\mathbf{A}}(x) = N(\mathfrak{a})^{-1}\sum_{w\in W}\psi_{\mathfrak{f}}(a_w)^{-1} f_{\mathbf{A}}(xw^{-1})$. By our assumption on $\mathfrak{a}$, we can take A to be the set of elements $\begin{pmatrix}1 & b\\ 0 & 1\end{pmatrix}$ with $b \in 2\mathfrak{b}\mathfrak{d}^{-1}/2\mathfrak{a}\mathfrak{b}\mathfrak{d}^{-1}$. Taking x in $r_P(P_{\mathbf{A}})$, we can verify that $\lambda_g(\xi, \mathfrak{m}) = \lambda_f(\xi, \mathfrak{m})$ if $\xi\mathfrak{b} \subset \mathfrak{m}^{-2}$. Since $\lambda_g(\xi, \mathfrak{m}) = 0$ otherwise, g is uniquely determined.

2. Hecke operators and inverters. To define Hecke operators, let us first put

$$U_0 = \mathrm{pr}^{-1}(C'), \tag{2.1a}$$

$$\Xi = \{w_1 r_P(\mathrm{diag}[\zeta^{-1}, \zeta])w_2 \mid \zeta \in F_{\mathbf{h}}^\times,\ \zeta\mathfrak{g} \subset \mathfrak{g},\ w_1, w_2 \in U_0\}. \tag{2.1b}$$

For w_1, w_2, and $\eta = r_P(\mathrm{diag}[\zeta^{-1}, \zeta])$ as in (2.1b), we put

$$J_\Xi(w_1\eta w_2, z) = J_k(w_1 w_2, z). \tag{2.2}$$

As shown in [87, p. 786], this is well defined and satisfies

$$J_\Xi(y_1 x y_2, z) = J_k(y_1, xy_2(z))J_\Xi(x, y_2(z))J_k(y_2, z) \qquad \text{if } y_1, y_2 \in U_0. \tag{2.3}$$

For each $v \in \mathbf{h}$ we define a Hecke operator T_v acting on $\mathcal{M}_k(\mathfrak{b}, \mathfrak{b}'; \psi)$ as follows: Put $U = \mathrm{pr}^{-1}(D[2\mathfrak{b}\mathfrak{d}^{-1}, 2\mathfrak{b}'\mathfrak{d}])$; with $\sigma = r_P(\mathrm{diag}[\pi_v^{-1}, \pi_v])$ take a finite subset W of $M_{\mathbf{A}}$ so that $\mathrm{pr}(W) \subset G_{\mathbf{h}}$ and $U\sigma U = \bigsqcup_{w\in W} Uw$. Then for $f \in \mathcal{M}_k(\mathfrak{b}, \mathfrak{b}'; \psi)$ we can define an element $f|T_v$ of the same space by

$$(f|T_v)_{\mathbf{A}}(x) = N_v^{-3/2}\psi(\pi_v)\sum_{w\in W}\psi((\pi_v a_w)_{\mathfrak{c}})^{-1} J_\Xi(w, \mathbf{i})^{-1} f_{\mathbf{A}}(xw^{-1}), \tag{2.4}$$

where $\mathfrak{c} = 4\mathfrak{b}\mathfrak{b}'$. This is a generalization of [87, (5.6)].

PROPOSITION 2.1. *For $f \in \mathscr{M}_k(\mathfrak{b}, \mathfrak{b}'; \psi)$ and $\mathfrak{p} = \pi_v \mathfrak{g}$, $v \in \mathbf{h}$, we have*

$$\lambda(\xi, \mathfrak{m}; f|T_v) = \lambda_f(\xi, \mathfrak{p}\mathfrak{m}) \qquad \text{if } \mathfrak{p}|2\mathfrak{b}\mathfrak{b}',$$

$$\lambda(\xi, \mathfrak{m}; f|T_v) = \lambda_f(\xi, \mathfrak{p}\mathfrak{m}) + \psi^*(\mathfrak{p})N(\mathfrak{p})^{-1}\left(\frac{\xi c^2}{\mathfrak{p}}\right)\lambda_f(\xi, \mathfrak{m})$$

$$+ \psi^*(\mathfrak{p})^2 N(\mathfrak{p})^{-1}\lambda_f(\xi, \mathfrak{m}\mathfrak{p}^{-1}) \qquad \text{if } \mathfrak{p} \nmid 2\mathfrak{b}\mathfrak{b}',$$

provided $(\xi\mathfrak{b}\mathfrak{m}^2)_v \subset \mathfrak{g}_v$, where c is an element of F_v such that $\mathfrak{m}_v = c\mathfrak{g}_v$ and $(\frac{d}{\mathfrak{p}})$ is the number of solutions of $x^2 \equiv d \pmod{\mathfrak{p}}$ minus 1.

Proof. If $\xi \notin \mathfrak{b}^{-1}\mathfrak{m}^{-2}$ and $(\xi\mathfrak{b}\mathfrak{m}^2)_v \subset \mathfrak{g}_v$, then $\xi \notin (\mathfrak{b}^{-1}\mathfrak{m}^{-2})_w$ for some $w \in \mathbf{h}$, $\neq v$, and hence $\xi \notin \mathfrak{b}^{-1}\mathfrak{p}^n\mathfrak{m}^{-2}$ for every $n \in \mathbf{Z}$. In this case both sides of our equalities are 0. Therefore we may assume that $\xi \in \mathfrak{b}^{-1}\mathfrak{m}^{-2}$. Under this condition, our proposition was proved in [87, Prop. 5.4] when $\mathfrak{b} = \mathfrak{g}$. The same proof applies to the general case with obvious modifications.

In order to describe algebraic properties of various Dirichlet series, we introduce a system of multiplicative symbols $M(\mathfrak{x})$ defined for the fractional ideals $\mathfrak{x}$ in F with the following properties: $M(\mathfrak{g}) = 1$ and $M(\mathfrak{x}\mathfrak{y}) = M(\mathfrak{x})M(\mathfrak{y})$; the $M(\mathfrak{p})$ for the prime ideals $\mathfrak{p}$ are independent indeterminates. For each $v \in \mathbf{h}$ we put $M_v = M(\pi_v\mathfrak{g})$, $N_v = N(\pi_v\mathfrak{g})$. Throughout the rest of the paper we denote by $\sum_{\mathfrak{m}}$ (unless stated otherwise) the sum over all the integral ideals $\mathfrak{m}$ in F.

PROPOSITION 2.2. *Let $0 \neq f \in \mathscr{M}_k(\mathfrak{b}, \mathfrak{b}'; \psi)$, $0 \ll \tau \in F$, and $\tau\mathfrak{b} = \mathfrak{q}^2\mathfrak{r}$ with a fractional ideal $\mathfrak{q}$ and a squarefree integral ideal $\mathfrak{r}$. Suppose that $f|T_v = \omega_v f$ with $\omega_v \in \mathbf{C}$ for every $v \in \mathbf{h}$. Then*

$$\sum_{\mathfrak{m}} \lambda_f(\tau, \mathfrak{q}^{-1}\mathfrak{m})M(\mathfrak{m})$$

$$= \lambda_f(\tau, \mathfrak{q}^{-1}) \prod_{v \in \mathbf{h}} [1 - (\psi'\varepsilon_\tau^*)(\pi_v\mathfrak{g})N_v^{-1}M_v][1 - \omega_v M_v + \psi'(\pi_v\mathfrak{g})^2 N_v^{-1} M_v^2]^{-1},$$

where ε_τ is as in Proposition 1.4 and $\psi'(\pi_v\mathfrak{g}) = \psi(\pi_v)$ or 0 according as $v \nmid 2\mathfrak{b}\mathfrak{b}'$ or $v|2\mathfrak{b}\mathfrak{b}'$.

This can be derived from Proposition 2.1 in exactly the same fashion as in [87, Theorem 5.5], which concerns the special case $\mathfrak{b} = \mathfrak{g}$.

We are going to define an operator, which may be called an *inverter*, and corresponds to $z \mapsto -1/z$ in the one-dimensional case. We begin with:

LEMMA 2.3. *There exists a unique element $\tilde{\varepsilon}$ of $M_{\mathbf{A}}$ such that $h(\tilde{\varepsilon}, z) = J_k(\tilde{\varepsilon}, z) = 1$ and $\mathrm{pr}(\tilde{\varepsilon})$ equals ε of (1.3c). Moreover, given $f \in \mathscr{M}_k$, let B be an open subgroup of C'' such that $f_{\mathbf{A}}(x) = (f\|_k x)(\mathbf{i})$ for $x \in \mathrm{pr}^{-1}(B)$, and ω an element of G such that $\omega\varepsilon \in B$. Then $(f\|_k\omega)_{\mathbf{A}}(x) = f_{\mathbf{A}}(x\tilde{\varepsilon})$.*

Proof. The existence of $\tilde{\varepsilon}$ follows easily from the fact that $h(y,z)\in \mathbf{T}$ and $h(ty,z)=t^{-1}h(y,z)$ for $t\in\mathbf{T}$ if $\mathrm{pr}(y)\in C''\cap G_{\mathbf{h}}$. Let $h(x)=f_{\mathbf{A}}(x\tilde{\varepsilon})$. From (1.4c) we see that $h(\alpha xw)=J_k(w,\mathbf{i})^{-1}h(x)$ for $\alpha\in G$ and $\mathrm{pr}(w)\in\varepsilon B\varepsilon^{-1}$, $w(\mathbf{i})=\mathbf{i}$, and hence we can define a function g on $H^{\mathbf{a}}$ by $g(x(\mathbf{i}))=h(x)J_k(x,\mathbf{i})$ for $\mathrm{pr}(x)\in\varepsilon B\varepsilon^{-1}$. Then for $\mathrm{pr}(y)\in G_{\mathbf{a}}$ we have $g(y(\mathbf{i}))J_k(y,\mathbf{i})^{-1}=f_{\mathbf{A}}(y\tilde{\varepsilon})=f_{\mathbf{A}}(\omega y\tilde{\varepsilon})=(f\|_k\omega y\tilde{\varepsilon})(\mathbf{i})$ since $\mathrm{pr}(\omega y\tilde{\varepsilon})\in B$. Now ω, y, and $\tilde{\varepsilon}$ belong to $\mathrm{pr}^{-1}(C'')$, and hence $J_k(\omega y\tilde{\varepsilon},\mathbf{i})=J_k(\omega,y(\mathbf{i}))J_k(y,\mathbf{i})$ since $J_k(\tilde{\varepsilon},\mathbf{i})=1$. Thus, for $z=y(\mathbf{i})$ we have $g(z)=J_k(\omega,z)^{-1}f(\omega z)$, which completes the proof.

For $f\in\mathcal{M}_k(\mathfrak{b},\mathfrak{b}';\psi)$ we define an element f^* of $\mathcal{M}_k$ by

$$(f^*)_{\mathbf{A}}(x)=\psi_{\mathbf{h}}(\delta)f_{\mathbf{A}}(x\tilde{\varepsilon}), \tag{2.5a}$$

which amounts to $f^*=\psi_{\mathbf{h}}(\delta)f\|_k\omega$, with $\tilde{\varepsilon}$ and ω of Lemma 2.3 and δ of (1.3c). Then we easily see that f^* is well defined independently of the choice of δ, and $f\mapsto f^*$ gives a $\mathbf{C}$-linear bijection of $\mathcal{M}_k(\mathfrak{b},\mathfrak{b}';\psi)$ onto $\mathcal{M}_k(\mathfrak{b}',\mathfrak{b};\psi^{-1})$, as can easily be seen by verifying (1.10) for $f^*_{\mathbf{A}}$; moreover $(f^*)^*=\psi_{\mathbf{a}}(-1)f$. Since $h(\tilde{\varepsilon}^{-1},z)=h(\tilde{\varepsilon},z)^{-1}=1$, we can take $\tilde{\varepsilon}^{-1}$ and $(-1)_{\mathbf{h}}\delta$ in place of $\tilde{\varepsilon}$ and δ, and find that

$$(f^*)_{\mathbf{A}}(x)=\psi_{\mathbf{h}}(-\delta)f_{\mathbf{A}}(x\tilde{\varepsilon}^{-1}). \tag{2.5b}$$

If $\xi\in G\cap\varepsilon^{-1}B$, then $f^*=\psi_{\mathbf{h}}(-\delta)f\|_k\xi^{-1}$.

Lemma 2.4. *Let $0\ll\tau\in\mathfrak{g}$, $\alpha\in G\cap\varepsilon D[2\mathfrak{d}^{-1},2\tau\mathfrak{d}]$, and $\beta=\eta\alpha\eta^{-1}$ with $\eta=\mathrm{diag}[\tau,1]$. Then $\beta\in P_{\mathbf{A}}C''$, $c_\alpha\neq 0$, $c_\alpha\mathfrak{d}^{-1}$ is prime to 2τ, and $J_k(\beta,\tau z)=\varepsilon_\tau^*(c_\alpha\mathfrak{d}^{-1})J_k(\alpha,z)$ with ε_τ of Proposition 1.4. Moreover, if $\alpha\in\varepsilon E\cup\varepsilon^{-1}E$, where*

$$E=\{x\in D[2\mathfrak{d}^{-1},2\tau\mathfrak{d}]\,|\,a(x_v)-1\in 4\tau\mathfrak{g}_v \text{ for every } v|2\tau\}, \tag{2.6}$$

then $J_k(\beta,\tau z)=\varepsilon_\tau(\delta)J_k(\alpha,z)$.

Proof. Let $\alpha=\varepsilon\zeta$ with $\zeta\in D[2\mathfrak{d}^{-1},2\tau\mathfrak{d}]$. Then $\beta=\eta\varepsilon\eta^{-1}\cdot\eta\zeta\eta^{-1}\in\mathrm{diag}[\tau,\tau^{-1}]\varepsilon\cdot D[2\tau\mathfrak{d}^{-1},2\mathfrak{d}]\subset P_{\mathbf{A}}C''$. Put $c=c_\alpha$ and $d=d_\alpha$ for simplicity. Then $c_{\mathbf{h}}=(\delta a_\zeta)_{\mathbf{h}}$ and $d_{\mathbf{h}}=(\delta b_\zeta)_{\mathbf{h}}$. Since $(a_\zeta)_v$ is a v-unit for $v|2\tau$, we see that $c\neq 0$ and $c\mathfrak{d}^{-1}$ is prime to 2τ. Clearly, $c_\beta=c/\tau$ and $j(\beta,\tau z)=j(\alpha,z)$. From [87, (2.18a)] we obtain $h(\beta,\tau z)/h(\alpha,z)=|\gamma/\gamma'|\gamma'/\gamma$ with $\gamma=\gamma(-d/c)$ and $\gamma'=\gamma(-\tau d/c)$, where $\gamma(*)$ is the Gauss sum defined in [87, (2.17a, b)]. If $v|2\tau$, we easily see that $d/c\in 2\mathfrak{d}_v^{-1}$ and hence $\gamma_v(-d/c)=\gamma_v(-\tau d/c)=1$. Suppose $v\nmid 2\tau$; then $\gamma_v(-\tau d/c)/\gamma_v(-d/c)=\left(\frac{\tau}{\mathfrak{x}_v}\right)$ by [85a, Lemma 3.3], where $\mathfrak{x}_v^{-1}=c^{-1}d\mathfrak{d}_v^2\mathfrak{d}_v^{-1}+\mathfrak{g}_v$. Since $\zeta\in D[2\mathfrak{d}^{-1},2\tau\mathfrak{d}]$, we have $\mathfrak{x}_v^{-1}=a_\zeta^{-1}b_\zeta\mathfrak{d}_v+\mathfrak{g}_v=a_\zeta^{-1}\mathfrak{g}_v=c^{-1}\mathfrak{d}_v$. Therefore $h(\beta,\tau z)=\varepsilon_\tau^*(c\mathfrak{d}^{-1})h(\alpha,z)$, which proves the first assertion. If $\alpha\in\varepsilon E$, we see that $(c\delta^{-1})_v-1\in 4\tau\mathfrak{g}_v$ for every $v|2\tau$, and hence $\varepsilon_\tau^*(c\mathfrak{d}^{-1})=(\varepsilon_\tau)_{\mathbf{h}}(c\delta^{-1})=\varepsilon_\tau(\delta)$, from which we obtain the second assertion since ε can be replaced by ε^{-1}.

Lemma 2.5. *Let $\sigma=r_P(\eta)$, $\eta=\mathrm{diag}[r^{-1},r]$ with an element r of $F_{\mathbf{h}}^\times$ such that $r\mathfrak{g}\subset\mathfrak{g}$. Then $\tilde{\varepsilon}\sigma\tilde{\varepsilon}^{-1}=\sigma^{-1}$. Moreover, if $r\mathfrak{g}$ is prime to 2, then $J_{\Xi}(x,z)=J_{\Xi}(\tilde{\varepsilon}x\tilde{\varepsilon}^{-1},z)$ for every $x\in U_0\sigma U_0$.*

Proof. Let $x = \alpha_1\sigma\alpha_2$ with α_i in U_0. Then $J_\Xi(x, z) = J_k(\alpha_1\alpha_2, z)$. Put $\tilde{\varepsilon}\alpha_i\tilde{\varepsilon}^{-1} = \beta_i$ and $\zeta = \varepsilon\iota_\mathbf{a}$ with $\iota = \begin{pmatrix} 0 & -1 \\ 1 & 0 \end{pmatrix}$. Then $\beta_i \in U_0$, $\zeta\eta = \eta^{-1}\zeta$, and $\tilde{\varepsilon}x\tilde{\varepsilon}^{-1} = \beta_1\tilde{\varepsilon}\sigma\tilde{\varepsilon}^{-1}\beta_2$. Now take the subset $\Omega_\mathbf{A}$ of $G_\mathbf{A}$ and the map r_Ω of $\Omega_\mathbf{A}$ into $M_\mathbf{A}$ of [85a, (3.1c)]. From [85a, (3.2) and (3.3)] we can easily derive that $r_\Omega(\zeta)r_P(\eta) = r_P(\eta^{-1})r_\Omega(\zeta)$. Put $r_\Omega(\zeta)^{-1}\tilde{\varepsilon} = \gamma$. Then $\mathrm{pr}(\gamma) \in G_\mathbf{a}$, and hence $\gamma\sigma = \sigma\gamma$ by (1.4a). Therefore $\tilde{\varepsilon}\sigma\tilde{\varepsilon}^{-1} = r_\Omega(\zeta)\sigma r_\Omega(\zeta)^{-1} = \sigma^{-1}$, which proves the first assertion. Thus $\tilde{\varepsilon}x\tilde{\varepsilon}^{-1} = \beta_1\sigma^{-1}\beta_2$. By (2.3) we have $J_\Xi(\beta_1\sigma^{-1}\beta_2, z) = J_k(\beta_1, \beta_2 z)J_\Xi(\sigma^{-1}, \beta_2 z)J_k(\beta_2, z)$. Now [87, Lemma 5.1] shows that if $r\mathfrak{g}$ is prime to 2, then $J_\Xi(\sigma^{-1}, z) = 1$, and hence $J_\Xi(\tilde{\varepsilon}x\tilde{\varepsilon}^{-1}, z) = J_k(\beta_1\beta_2, z) = J_k(\tilde{\varepsilon}\alpha_1\alpha_2\tilde{\varepsilon}^{-1}, z) = J_k(\alpha_1\alpha_2, z)$ since $J_k(\tilde{\varepsilon}, z) = J_k(\tilde{\varepsilon}^{-1}, z) = 1$. This completes the proof.

LEMMA 2.6. *Let $\mathfrak{y}$ be a fractional ideal and τ a totally positive element of F such that $\mathfrak{y}^2 \subset \tau\mathfrak{b}$. Given $f \in \mathcal{M}_k(\mathfrak{b}, \mathfrak{b}'; \psi)$, let h be the element of $\mathcal{M}_k(\mathfrak{g}, \tau^{-1}\mathfrak{y}^2\mathfrak{b}'; \psi\varepsilon_\tau)$ such that $\lambda_h(\xi, \mathfrak{m}) = \lambda_f(\tau\xi, \mathfrak{y}^{-1}\mathfrak{m})$ as given in Proposition 1.4. Then $\lambda(\xi, \mathfrak{m}; h^*) = \tau^k N(\mathfrak{y})^{-1}\lambda(\xi/\tau, \mathfrak{y}\mathfrak{m}; f^*)$.*

Proof. Replacing $\mathfrak{y}$ and τ by $b\mathfrak{y}$ and $b^2\tau$ with a suitable $b \in F$, we may assume that $\tau^{-1} \in \mathfrak{g}$ and $\mathfrak{g} \subset \mathfrak{y}$. Take $y \in F_\mathbf{h}^\times$ so that $y\mathfrak{g} = \mathfrak{y}$. Put $p(z) = f(z/\tau)$. By Proposition 1.4, $h_\mathbf{A}(x) = (\overline{\psi}\varepsilon_\tau)(y)N(\mathfrak{y})^{-1/2}p_\mathbf{A}(x\sigma)$ with $\sigma = r_P(\mathrm{diag}[y^{-1}, y])$. Put $\eta = \mathrm{diag}[\tau^{-1}, 1]$. Given an open subgroup E' of the group E of (2.6) with τ^{-1} in place of τ, put $B = E' \cap \varepsilon E'\varepsilon^{-1} \cap \eta^{-1}\varepsilon E'\varepsilon^{-1}\eta$. Let $\omega \in G \cap \varepsilon^{-1}B$, $\varepsilon\omega = \beta$, $\zeta = \eta\omega\eta^{-1}$, and $\pi = \mathrm{diag}[\tau, \tau^{-1}]$. Then $(\zeta\omega^{-1})_\mathbf{h} = (\eta\varepsilon^{-1}\beta\eta^{-1}\beta^{-1}\varepsilon)_\mathbf{h} = (\pi^{-1}\varepsilon^{-1}\eta\beta\eta^{-1}\beta^{-1}\varepsilon)_\mathbf{h}$, and $\varepsilon^{-1}\eta\beta\eta^{-1}\beta^{-1}\varepsilon \in E'$. Now take E' so that the last formula of Lemma 2.3 is applicable to f, p, h, and $q = f\|_k\pi^{-1}$, and $q\|_k\gamma = q$ for every $\gamma \in G \cap E'$. Then by (1.4c) $f\|_k\zeta = (f\|_k\pi^{-1})\|_k\omega$, and hence $(f\|_k\zeta)_\mathbf{A}(x) = (f\|_k\pi^{-1})_\mathbf{A}(x\tilde{\varepsilon}) = f_\mathbf{A}(x\tilde{\varepsilon}r_P(\pi_\mathbf{h}))$ by (1.14). By Lemma 2.5, (2.5a), and (1.14), $f_\mathbf{A}(x\tilde{\varepsilon}r_P(\pi_\mathbf{h})) = f_\mathbf{A}(xr_P(\pi_\mathbf{h}^{-1})\tilde{\varepsilon}) = \overline{\psi}(\delta)f_\mathbf{A}^*(xr_P(\pi_\mathbf{h}^{-1})) = \overline{\psi}(\delta)(f^*\|_k\pi)_\mathbf{A}(x)$. Thus $(f\|_k\zeta)(z) = \overline{\psi}(\delta)\tau^k f^*(\tau^2 z)$. By Lemma 2.4 we have $(f\|_k\zeta)(z/\tau) = J_k(\zeta, z/\tau)^{-1}f(\zeta(z/\tau)) = \varepsilon_\tau(\delta)J_k(\omega, z)^{-1}f(\omega(z)/\tau) = \varepsilon_\tau(\delta)\cdot(p\|_k\omega)(z) = \overline{\psi}(\delta)p^*(z)$. Hence $p^*(z) = \tau^k f^*(\tau z)$, which implies

$$\lambda(\tau\xi, \mathfrak{m}; p^*) = \tau^k\lambda(\xi, \mathfrak{m}; f^*). \tag{2.7}$$

Now $\sigma^{-1}\tilde{\varepsilon} = \tilde{\varepsilon}\sigma$ by Lemma 2.5, and hence $p_\mathbf{A}^*(x\sigma^{-1}) = (\psi\varepsilon_\tau)(\delta)p_\mathbf{A}(x\sigma^{-1}\tilde{\varepsilon}) = (\psi\varepsilon_\tau)(\delta)p_\mathbf{A}(x\tilde{\varepsilon}\sigma) = (\psi\varepsilon_\tau)(\delta y)N(\mathfrak{y})^{1/2}h_\mathbf{A}(x\tilde{\varepsilon}) = (\psi\varepsilon_\tau)(y)N(\mathfrak{y})^{1/2}h_\mathbf{A}^*(x)$. By Proposition 1.4 this implies $\lambda(\xi, \mathfrak{m}; h^*) = N(\mathfrak{y})^{-1}\lambda(\xi, \mathfrak{y}\mathfrak{m}; p^*)$, which combined with (2.7) proves the desired formula.

LEMMA 2.7. (i) *Let $f \mapsto h$ denote the map determined by (1.13) with fixed τ and $\mathfrak{y}$. Then $f|T_v \mapsto h|T_v$ for every prime v such that $\mathfrak{y}_v^2 = \tau\mathfrak{g}_v$. If in particular $\mathfrak{y}^2 \subset \tau\mathfrak{b}$, then $f|T_v \mapsto h|T_v$ for every prime v not dividing $(\tau\mathfrak{b})^{-1}\mathfrak{y}^2$.*

(ii) *If $f \in \mathcal{M}_k(\mathfrak{b}, \mathfrak{b}'; \psi)$ and $v \nmid 2\mathfrak{b}\mathfrak{b}'$, then $(f|T_v)^* = \psi(\pi_v)^2 f^*|T_v$.*

Proof. The first assertion can easily be verified by means of Proposition 2.1. As for (ii), let $U = \mathrm{pr}^{-1}(D[2\mathfrak{b}\mathfrak{d}^{-1}, 2\mathfrak{b}'\mathfrak{d}])$ and $U' = \mathrm{pr}^{-1}(D[2\mathfrak{b}'\mathfrak{d}^{-1}, 2\mathfrak{b}\mathfrak{d}])$. Then $\tilde{\varepsilon}U\tilde{\varepsilon}^{-1} = U'$. Let $U\sigma U = \bigsqcup_{w \in W} Uw$ with σ as in Lemma 2.5 and $W \subset \mathrm{pr}^{-1}(G_\mathbf{h})$. Modifying the proof of [87, Lemma 5.1], we easily find that $U\sigma U = U\sigma^{-1}U$. By Lemma 2.5 we have $U'\sigma U' = \tilde{\varepsilon}U\sigma U\tilde{\varepsilon}^{-1} = \bigsqcup_{w\in W} U'\tilde{\varepsilon}w\tilde{\varepsilon}^{-1}$ and $J_{\Xi}(\tilde{\varepsilon}w\tilde{\varepsilon}^{-1}, z) = J_{\Xi}(w, z)$. Our assertion now follows from this fact, (2.4), and (2.5a) in a straightforward way.

LEMMA 2.8. (i) *The operators T_v for all $v \in \mathbf{h}$ are mutually commutative.*
(ii) *If $f \in \mathcal{M}_k(\mathfrak{b}, \mathfrak{b}'; \psi)$, $g \in \mathcal{S}_k(\mathfrak{b}, \mathfrak{b}'; \psi)$, and $v \nmid 2\mathfrak{b}\mathfrak{b}'$, then $\langle f, g|T_v\rangle = \psi(\pi_v)^2 \cdot \langle f|T_v, g\rangle$.*

If $\mathfrak{b} = \mathfrak{g}$, these were given in [87, Propositions 5.2 and 5.3]. The same proofs, if suitably modified, apply to the general case.

Remark 2.9. Suppose that a nonzero element f of $\mathcal{S}_k(\mathfrak{b}, \mathfrak{b}'; \psi)$ satisfies $f|T_v = \omega_v f$ with $\omega_v \in \mathbf{C}$ for every $v \in \mathbf{h}$. Then, from Lemma 2.7(ii) and Lemma 2.8(ii) we obtain

$$(2.8) \qquad \omega_v = \psi(\pi_v)^2\overline{\omega}_v \quad \textit{and} \quad f^*|T_v = \overline{\omega}_v f^* \quad \textit{for every } v \nmid 2\mathfrak{b}\mathfrak{b}'.$$

Suppose further that f is characterized by the ω_v for $v \nmid 2\mathfrak{b}\mathfrak{b}'$ in the sense that

$$(2.9) \qquad \mathbf{C}f = \{g \in \mathcal{S}_k(\mathfrak{b}, \mathfrak{b}'; \psi) \mid g|T_v = \omega_v g \textit{ for every } v \nmid 2\mathfrak{b}\mathfrak{b}'\}.$$

Then Lemma 2.7(ii) shows that

$$(2.10) \qquad \mathbf{C}f^* = \{h \in \mathcal{S}_k(\mathfrak{b}', \mathfrak{b}; \overline{\psi}) \mid h|T_v = \overline{\omega}_v h \textit{ for every } v \nmid 2\mathfrak{b}\mathfrak{b}'\}.$$

By Lemma 2.8(i) this space is stable under T_v for every $v \in \mathbf{h}$. Therefore, under (2.9), f^* is an eigenform of the T_v for all $v \in \mathbf{h}$.

PROPOSITION 2.10. *Given $f \in \mathcal{M}_k(\mathfrak{b}, \mathfrak{b}'; \psi)$, there exists an element $f^{\#}$ of $\mathcal{M}_k(\mathfrak{b}, \mathfrak{b}'; \overline{\psi})$, unique for f and $\mathfrak{b}$, satisfying*

$$(2.11) \quad \lambda(\xi, \mathfrak{m}; f^{\#}) = \lambda(\xi, \mathfrak{b}\mathfrak{m}; f^*) \qquad \textit{for every } (\xi, \mathfrak{m}) \textit{ such that } \xi\mathfrak{b} \subset \mathfrak{m}^{-2}.$$

Moreover, $(f|T_v)^{\#} = \psi(\pi_v)^2 f^{\#}|T_v$ for every $v \nmid 2\mathfrak{b}\mathfrak{b}'$.

Proof. Taking $(f^*, \mathfrak{b}', \mathfrak{b}, 1, \mathfrak{b}^{-1})$ as $(f, \mathfrak{b}, \mathfrak{b}', \tau, \mathfrak{y})$ in Proposition 1.4, we obtain an element h of $\mathcal{M}_k(\mathfrak{b}^2\mathfrak{b}', \mathfrak{g}; \overline{\psi})$ such that $\lambda_h(\xi, \mathfrak{m}) = \lambda(\xi, \mathfrak{b}\mathfrak{m}; f^*)$ for every $(\xi, \mathfrak{m})$. Clearly $h \in \mathcal{M}_k(\mathfrak{b}^2\mathfrak{b}', \mathfrak{b}'; \overline{\psi})$. Therefore, by Lemma 1.6, we obtain a unique element $f^{\#}$ of $\mathcal{M}_k(\mathfrak{b}, \mathfrak{b}'; \overline{\psi})$ such that $\lambda(\xi, \mathfrak{m}; f^{\#}) = \lambda_h(\xi, \mathfrak{m})$ if $\xi\mathfrak{b} \subset \mathfrak{m}^{-2}$. The last assertion follows easily from Proposition 2.1 and Lemma 2.7(ii).

3. Statement of the main theorems. Let us first recall several basic facts on Hilbert modular forms of integral weight. Let $\mathfrak{z}$ be an integral ideal in F and let $0 \leqslant n \in \mathbf{Z}^{\mathbf{a}}$. We then consider a Hecke character Ψ of F such that

$$(3.1) \quad \Psi_{\mathbf{a}}(x) = \operatorname{sgn}(x)^n |x|^{2i\mu} \textit{ with } \mu \in \mathbf{R}^{\mathbf{a}}, \|\mu\| = 0; \textit{ the conductor of } \Psi \textit{ divides } \mathfrak{z}.$$

For simplicity put

$$\tilde{D}_{\mathfrak{z}} = \tilde{D}[\mathfrak{d}^{-1}, \mathfrak{z}\mathfrak{d}]. \tag{3.2}$$

We then denote by $\mathscr{S}_n((\mathfrak{z}, \Psi))$ (resp. $\mathscr{M}_n((\mathfrak{z}, \Psi))$) the set of all **C**-valued functions **f** on $\tilde{G}_{\mathbf{A}}$ satisfying the following three conditions:

(3.3a) $\mathbf{f}(sx) = \Psi(s)\mathbf{f}(x)$ *for every* $s \in F_{\mathbf{A}}^{\times}$.

(3.3b) $\mathbf{f}(\alpha x w) = \Psi((d_w)_{\mathfrak{z}})\mathbf{f}(x)$ *for every* $\alpha \in \tilde{G}$, $x \in \tilde{G}_{\mathbf{A}}$, *and* $w \in \tilde{D}_{\mathfrak{z}}$, $w_{\mathbf{a}} = 1$, *where* $d_{\mathfrak{z}} = (d_v)_{v|\mathfrak{z}}$ *for* $d \in F_{\mathbf{A}}$.

(3.3c) *For every* $p \in \tilde{G}_{\mathbf{h}}$ *there is an element* f_p *of* $\mathscr{S}_n$ *(resp.* $\mathscr{M}_n$*) such that* $\mathbf{f}(\det(p)^{-1}py) = \det(y)^{i\mu}(f_p\|_n y)(\mathbf{i})$ *for every* $y \in \tilde{G}_{\mathbf{a}+}$.

Take a finite coset decomposition of $\tilde{G}_{\mathbf{A}}$ of the form

$$\tilde{G}_{\mathbf{A}} = \bigsqcup_{\lambda=1}^{\kappa} \tilde{G}x_{\lambda}\tilde{D}_{\mathfrak{z}}, \qquad x_{\lambda} = \mathrm{diag}[1, t_{\lambda}] \tag{3.4}$$

with $t_{\lambda} \in F_{\mathbf{h}}^{\times}$. By (3.3c) for each λ we obtain an element f_{λ} of $\mathscr{M}_n$ such that

$$\mathbf{f}(t_{\lambda}^{-1}x_{\lambda}y) = \det(y)^{i\mu}(f_{\lambda}\|_n y)(\mathbf{i}) \qquad \textit{for every } y \in \tilde{G}_{\mathbf{a}}. \tag{3.5}$$

Then f_{λ} satisfies

$$f_{\lambda}\|_n \gamma = \Psi_{\mathfrak{z}}(a_{\gamma}) \det(\gamma)^{i\mu} f_{\lambda} \qquad \textit{for every } \gamma \in \tilde{\Gamma}[(t_{\lambda}\mathfrak{d})^{-1}, t_{\lambda}\mathfrak{z}\mathfrak{d}]. \tag{3.6}$$

Since **f** is completely determined by $(f_1, \ldots, f_{\kappa})$, we shall write $\mathbf{f} = (f_1, \ldots, f_{\kappa})$, and for $\mathbf{g} = (g_1, \ldots, g_{\kappa}) \in \mathscr{S}_n((\mathfrak{z}, \Psi))$ we put

$$\langle \mathbf{f}, \mathbf{g} \rangle = \sum_{\lambda=1}^{\kappa} \langle f_{\lambda}, g_{\lambda} \rangle. \tag{3.7}$$

This is independent of the choice of $\{t_{\lambda}\}$. Putting

$$f_{\lambda}(z) = \sum_{\xi \in F} c_{\lambda}(\xi)\mathbf{e}_{\mathbf{a}}(\xi z), \tag{3.8a}$$

we define $c(\mathfrak{m}, \mathbf{f})$ for a fractional ideal $\mathfrak{m}$ by

$$c(\mathfrak{m}, \mathbf{f}) = c_{\lambda}(\xi)\xi^{-n/2-i\mu} \qquad \textit{if } \mathfrak{m} = \xi t_{\lambda}^{-1}\mathfrak{g}. \tag{3.8b}$$

If $\mathbf{f} \in \mathscr{S}_n((\mathfrak{z}, \Psi))$, this can also be defined by the Fourier expansion

$$\mathbf{f}\left(\begin{pmatrix} y & x \\ 0 & 1 \end{pmatrix}\right) = \sum_{0 \ll \zeta \in F} c(\zeta y\mathfrak{g}, \mathbf{f})(\zeta y)^{n/2+i\mu}\mathbf{e}_{\mathbf{a}}(i\zeta y)\mathbf{e}_{\mathbf{A}}(\zeta x), \tag{3.9}$$

which holds for $0 \ll y \in F_{\mathbf{A}}^\times$ and $x \in F_{\mathbf{A}}$. For every integral ideal $\mathfrak{n}$ in F we can define a $\mathbf{C}$-linear endomorphism $\mathfrak{T}(\mathfrak{n})$ of $\mathscr{M}_n(\mathfrak{z}, \Psi))$ such that

$$c(\mathfrak{m}, \mathbf{f}|\mathfrak{T}(\mathfrak{n})) = \sum_{\mathfrak{a} \supset \mathfrak{m}+\mathfrak{n}} \Psi_*(\mathfrak{a})N(\mathfrak{a}^{-1}\mathfrak{n})c(\mathfrak{a}^{-2}\mathfrak{m}\mathfrak{n}, \mathbf{f}), \tag{3.10}$$

where $\Psi_*(\mathfrak{a})$ denotes $\Psi^*(\mathfrak{a})$ or 0 according as $\mathfrak{a}$ is prime to $\mathfrak{z}$ or not. (For these see [78] and [80, pp. 352–3].)

Let $\mathbf{f}$ be a common eigenform of $\mathfrak{T}(\mathfrak{n})$ for all $\mathfrak{n}$; put $\mathbf{f}|\mathfrak{T}(\mathfrak{n}) = \chi(\mathfrak{n})\mathbf{f}$ and $\chi(v) = \chi(\pi_v\mathfrak{g})$ for each $v \in \mathbf{h}$. Then we call χ a *system of eigenvalues*. We have

$$c(\mathfrak{m}, \mathbf{f}) = N(\mathfrak{m})^{-1}\chi(\mathfrak{m})c(\mathfrak{g}, \mathbf{f}), \tag{3.11}$$

$$\sum_{\mathfrak{m}} \chi(\mathfrak{m})M(\mathfrak{m}) = \prod_{v \in \mathbf{h}} [1 - \chi(v)M_v + \Psi_*(\pi_v\mathfrak{g})N_vM_v^2]^{-1}, \tag{3.12}$$

$$\chi(\mathfrak{m}) = \Psi^*(\mathfrak{m})\bar{\chi}(\mathfrak{m}) \qquad \text{if } \mathfrak{m} \text{ is prime to } \mathfrak{z}. \tag{3.13}$$

We call such an eigenform $\mathbf{f}$ *normalized* if $c(\mathfrak{g}, \mathbf{f}) = 1$. We put

$$D(s, \chi) = \sum_{\mathfrak{m}} \chi(\mathfrak{m})N(\mathfrak{m})^{-s-1}, \tag{3.14a}$$

$$D(s, \mathbf{g}) = \sum_{\mathfrak{m}} c(\mathfrak{m}, \mathbf{g})N(\mathfrak{m})^{-s} \qquad (\mathbf{g} \in \mathscr{S}_n((\mathfrak{z}, \Psi))). \tag{3.14b}$$

More generally, for a Hecke character ω of F and an integral ideal $\mathfrak{x}$ we put

$$D(s, \chi, \omega) = \sum_{\mathfrak{m}} \omega^*(\mathfrak{m})\chi(\mathfrak{m})N(\mathfrak{m})^{-s-1}, \tag{3.15a}$$

$$D(s, \mathbf{g}, \omega, \mathfrak{x}) = \sum_{\mathfrak{m}} \omega^*(\mathfrak{m})c(\mathfrak{x}\mathfrak{m}, \mathbf{g})N(\mathfrak{m})^{-s}. \tag{3.15b}$$

Notice that the choice of exponent $-s-1$ makes 0 the center of the critical strip.

We define an *inverter* $J_{\mathfrak{z}}$ acting on $\mathscr{S}_n((\mathfrak{z}, \Psi))$ by

$$(\mathbf{f}|J_{\mathfrak{z}})(x) = \Psi(\det(x)^{-1})\mathbf{f}(x\pi) \qquad (x \in \tilde{G}_{\mathbf{A}}), \tag{3.16}$$

where $\pi = \begin{pmatrix} 0 & -1 \\ \delta^2 s & 0 \end{pmatrix} \in \tilde{G}_{\mathbf{h}}$ with an element s of $F_{\mathbf{h}}^\times$ such that $s\mathfrak{g} = \mathfrak{z}$. This is independent of the choice of s and δ; moreover $\mathbf{f}|J_{\mathfrak{z}} \in \mathscr{S}_n((\mathfrak{z}, \Psi^{-1}))$. This inverter depends on $\mathfrak{z}$, whereas that in Section 2 is independent of the level. We have

$$(\mathbf{f}|\mathfrak{T}(\mathfrak{m}))|J_{\mathfrak{z}} = \Psi^*(\mathfrak{m})(\mathbf{f}|J_{\mathfrak{z}})|\mathfrak{T}(\mathfrak{m}) \qquad \text{if } \mathfrak{m} \text{ is prime to } \mathfrak{z}. \tag{3.17}$$

(See [78, (2.49), (2.50)].) This combined with (3.13) shows that, if $\mathbf{f}|\mathfrak{T}(\mathfrak{m}) = \chi(\mathfrak{m})\mathbf{f}$ and $\mathfrak{m}$ is prime to $\mathfrak{z}$, then $(\mathbf{f}|J_{\mathfrak{z}})\mathfrak{T}(\mathfrak{m}) = \bar{\chi}(\mathfrak{m})\mathbf{f}|J_{\mathfrak{z}}$.

PROPOSITION 3.1. (I) *Let* $f \in \mathscr{S}_k(\mathfrak{b}, \mathfrak{b}'; \psi)$ *with* $k \geqslant 3u/2$; *let* $0 \ll \tau \in F$ *and* $\tau\mathfrak{b} = \mathfrak{q}^2\mathfrak{r}$ *with a fractional ideal* $\mathfrak{q}$ *and a squarefree integral ideal* $\mathfrak{r}$. *Then there exists an element* $\mathbf{g}_\mathfrak{r}$ *of* $\mathscr{M}_{2m}((2\mathfrak{b}\mathfrak{b}', \psi^2))$ *such that*

$$(3.18a)\quad \sum_{\mathfrak{m}} c(\mathfrak{m}, \mathbf{g}_\mathfrak{r})M(\mathfrak{m}) = \sum_{\mathfrak{m}} \lambda_f(\tau, \mathfrak{q}^{-1}\mathfrak{m})M(\mathfrak{m}) \prod_{v \nmid 2\mathfrak{r}\mathfrak{b}\mathfrak{b}'} [1 - (\psi\varepsilon_\tau)(\pi_v)N_v^{-1}M_v]^{-1},$$

where ε_τ *is the Hecke character of F corresponding to* $F(\tau^{1/2})/F$.

(II) *Suppose further that* f *is a nonzero common eigenfunction of* T_v *for every* $v \in \mathbf{h}$; *let* $f | T_v = \omega_v f$. *Then there exists a system of eigenvalues* χ *occurring in* $\mathscr{M}_{2m}((2\mathfrak{b}\mathfrak{b}', \psi^2))$ *such that* $\chi(v) = N_v\omega_v$ *for every* $v \in \mathbf{h}$, *and* $\mathbf{g}_\mathfrak{r}$ *of (I) is* $\lambda_f(\tau, \mathfrak{q}^{-1})$ *times the normalized eigenform belonging to* χ.

(III) *The function* $\mathbf{g}_\mathfrak{r}$ *of (I) belongs to* $\mathscr{S}_{2m}((2\mathfrak{b}\mathfrak{b}', \psi^2))$ *if* $k \neq 3u/2$, *or if* $k = 3u/2$ *and* f *is orthogonal to the theta series of [87, Theorem 6.1 (III)].*

This was given in [87, Theorems 6.1 and 6.2] when $\mathfrak{b} = \mathfrak{g}$. The same proof with easy modifications is applicable to the general case, as will be explained in the proof of Proposition 5.6 below.

To state our main theorems, we start with f of $\mathscr{S}_k(\mathfrak{b}, \mathfrak{b}'; \psi)$ and χ as in (II) of the above proposition, assuming the orthogonality condition of (III) if $k = 3u/2$. For simplicity we put $\mathfrak{c} = 4\mathfrak{b}\mathfrak{b}'$. We fix an element $\tau \in F$, $\gg 0$, and consider a unique decomposition $\tau\mathfrak{b} = \mathfrak{q}^2\mathfrak{r}$ with a fractional ideal $\mathfrak{q}$ and a square free integral ideal $\mathfrak{r}$. Put $\varphi = \psi\varepsilon_\tau$ with ε_τ as in the above proposition; let $\mathfrak{h}$ denote the conductor of φ. Clearly $4\mathfrak{r} \cap \mathfrak{c} \subset \mathfrak{h}$; moreover, $\mathfrak{r}\mathfrak{c}$ and $\mathfrak{h}\mathfrak{c}$ have the same prime factors. By Proposition 2.2 and (3.12) we have

$$(3.18b)\quad \sum_{\mathfrak{m}} \lambda_f(\tau, \mathfrak{q}^{-1}\mathfrak{m})M(\mathfrak{m}) \prod_{v \nmid \mathfrak{r}\mathfrak{c}} [1 - \varphi(\pi_v)N_v^{-1}M_v]^{-1}$$
$$= \lambda_f(\tau, \mathfrak{q}^{-1}) \sum_{\mathfrak{m}} \chi(\mathfrak{m})N(\mathfrak{m})^{-1}M(\mathfrak{m}).$$

Put

$$(3.19a)\quad X = \{l \in \mathscr{S}_k(\mathfrak{b}, \mathfrak{b}'; \psi) | N_v^{-1}\chi(v)l = l | T_v \textit{ for almost all } v \in \mathbf{h}\},$$

$$(3.19b)\quad Y = \{\mathbf{q} \in \mathscr{S}_{2m}((2\mathfrak{b}\mathfrak{b}', \psi^2)) | \chi(v)\mathbf{q} = \mathbf{q} | \mathfrak{T}(\pi_v \mathfrak{g}) \textit{ for almost all } v \in \mathbf{h}\}.$$

These spaces can be defined somewhat differently; see Lemma 3.11 below.

We now consider the following three conditions:

(3.20a) ψ *is normalized for* f; *that is,* $\psi_\mathbf{a}(x) = \operatorname{sgn}(x)^m |x|^{i\mu}$ *with* $\mu \in \mathbf{R}^\mathbf{a}$, $\|\mu\| = 0$.

(3.20b) $\mathfrak{r}$ *divides* $\mathfrak{h}$.

(3.20c) *If* v *is a common prime factor of 2 and* $\mathfrak{r}$, *then* φ_v *satisfies either*
(i) $(\mathfrak{r}\mathfrak{c})_v = \mathfrak{h}_v = 4\mathfrak{r}_v$ *and* $\varphi_v(1 + 4x) = \varphi_v(1 + 4x^2)$ *for every* $x \in \mathfrak{g}_v$; *or*
(ii) $(\mathfrak{r}\mathfrak{c})_v \neq \mathfrak{h}_v \subset 4\mathfrak{r}_v$.

Notice that (i) means that the restriction of φ_v to $1 + 4\mathfrak{g}_v$ is invariant under the automorphism $1 + 4x \mapsto 1 + 4x^2$ (modulo $1 + \mathfrak{h}_v$), which is the case if $N_v = 2$.

Now our first main result is:

THEOREM 3.2. *Let $\{f_1, \ldots, f_\alpha\}$ and $\{\mathbf{g}_1, \ldots, \mathbf{g}_\beta\}$ be bases of X and Y over $\mathbf{C}$, respectively; let $\{f'_1, \ldots, f'_\alpha\}$ and $\{\mathbf{g}'_1, \ldots, \mathbf{g}'_\beta\}$ be their dual bases in the sense that $\langle f_i, f'_j\rangle = \delta_{ij}$ and $\langle \mathbf{g}_p, \mathbf{g}'_q\rangle = \delta_{pq}$. Further, let $f_i^{\#}$ be the element of $\mathscr{S}_k(\mathfrak{b}, \mathfrak{b}'; \overline{\psi})$ determined by (2.11) with f_i^* in place of f^* and let $\mathbf{g}_p^* = \mathbf{g}_p | J_{2\mathfrak{b}\mathfrak{b}'}$. Then, under (3.20a, b, c) we have*

$$\sum_{i=1}^{\alpha} \lambda(\tau, \mathfrak{q}^{-1}; f_i^{\#}, \overline{\psi})\overline{\lambda(\tau, \mathfrak{q}^{-1}; f'_i; \psi)}$$

$$= Q \sum_{\mathfrak{t} \supset \mathfrak{j}} \mu(\mathfrak{t})\overline{\varphi}^*(\mathfrak{t})N(\mathfrak{t})^{-1} \sum_{p=1}^{\beta} \overline{c(\mathfrak{g}, \mathbf{g}'_p)}D(0, \mathbf{g}_p^*, \varphi, \mathfrak{t}^{-1}\mathfrak{h}^{-1}\mathfrak{r}\mathfrak{c})$$

with

$$Q = \overline{\gamma(\varphi)}2^{d-1-\|k\|}\tau^m\overline{\psi}_{\mathbf{a}}(\tau)\pi^{-\|m\|}h_F[\mathfrak{g}_+^\times : \mathfrak{g}^{\times 2}]N(\mathfrak{h})^{-1}\prod_{v\in\mathbf{a}}(m_v - 1)!,$$

where $\mathfrak{j}$ is the product of all the prime factors of $\mathfrak{c}$ prime to $\mathfrak{h}$, μ the Moebius function on the ideals in F, $\gamma(\varphi)$ the Gauss sum of φ (see (4.1) below), h_F the class number of F, $\mathfrak{g}_+^\times = \{a \in \mathfrak{g}^\times | a \gg 0\}$, and $\mathfrak{g}^{\times 2} = \{a^2 \in \mathfrak{g}^\times | a \in \mathfrak{g}^\times\}$.

The proof will be completed in Section 9.

Remark 3.3. (A) It can be easily be seen that the quantity on either side of the first equality is independent of the choice of dual bases. As a typical basis of Y, we can take the one defined as follows: $\mathbf{g}_1$ is the normalized primitive eigenform and $\mathbf{g}_p$ is such that $c(\mathfrak{m}, \mathbf{g}_p) = c(\mathfrak{a}_p^{-1}\mathfrak{m}, \mathbf{g}_1)$ with an integral ideal $\mathfrak{a}_p$. Let $\mathfrak{z}$ be the exact level of $\mathbf{g}_1$. Then $\mathfrak{z} = (2\mathfrak{x})^{-1}\mathfrak{c}$ with an integral ideal $\mathfrak{x}$. Put $\mathbf{g}_0 = \mathbf{g}_1 | J_{\mathfrak{z}}$. By Lemma 3.10 below we have $c(\mathfrak{m}, \mathbf{g}_p^*) = c(\mathfrak{a}_p\mathfrak{x}^{-1}\mathfrak{m}, \mathbf{g}_0)$, and hence

$$D(s, \mathbf{g}_p^*, \varphi, \mathfrak{t}^{-1}\mathfrak{h}^{-1}\mathfrak{r}\mathfrak{c}) = D(s, \mathbf{g}_0, \varphi, (\mathfrak{x}\mathfrak{t}\mathfrak{h})^{-1}\mathfrak{r}\mathfrak{c}\mathfrak{a}_p).$$

Now $\langle \mathbf{g}_p, \mathbf{g}_q\rangle / \langle \mathbf{g}_1, \mathbf{g}_1\rangle$ has a simple rational expression in terms of $c(\mathfrak{p}, \mathbf{g}_1)$ for the prime factors $\mathfrak{p}$ of $\mathfrak{a}_p\mathfrak{a}_q$, as shown in [76, (3.2)] and [78, Prop. 4.14]. Since $\mathbf{g}_q = \sum_{p=1}^{\beta} \langle \mathbf{g}_p, \mathbf{g}_q\rangle \mathbf{g}'_p$, we obtain $c(\mathfrak{g}, \mathbf{g}'_p)$ by solving the system of linear equations $\delta_{1q} = \sum_{p=1}^{\beta} \langle \mathbf{g}_p, \mathbf{g}_q\rangle c\langle \mathfrak{g}, \mathbf{g}'_p)$. By (3.23a) below we can show that $D(s, \mathbf{g}_0, \varphi, \mathfrak{y})$ for any $\mathfrak{y}$ is $D(s, \mathbf{g}_0, \varphi, \mathfrak{g})$ times an easily computable rational expression in $N(\mathfrak{p})^{-s}$. Thus the sum $\sum_{p=1}^{\beta}$ on the right-hand side of the first equality of Theorem 3.2 is $\langle \mathbf{g}_1, \mathbf{g}_1\rangle^{-1}D(s, \mathbf{g}_0, \varphi, \mathfrak{g})$ times such an expression involving $c(\mathfrak{p}, \mathbf{g}_1)$ and $c(\mathfrak{p}, \mathbf{g}_0)$. The same type of remark applies to Theorem 3.4 below.

(B) The elements f'_i can be obtained from f_i by solving $f_j = \sum_{i=1}^{\alpha} \langle f_i, f_j\rangle f'_i$. If the f_i have algebraic Fourier coefficients, then $\langle f_i, f_j\rangle$ is an algebraic number times a period of $\mathbf{g}_1$, as shown in [82a, Theorem 1] (when $F = \mathbf{Q}$), [87, Theorem 10.5], and [91, Theorem 9.2] (in the general case).

The above theorem concerns the "average" of the quantities defined for all the elements of X and Y. Under some conditions we can in fact obtain an equality for each individual form. We shall give one type of results as Theorems 9.1 and 9.2 in Section 9, and another as Theorem 3.4 below by imposing a condition:

(3.21) *If $f' \in \mathscr{S}_k(\mathfrak{b}, \mathfrak{b}'; \psi)$ and $f'|T_v = N_v^{-1}\chi(v)f'$ for every $v \nmid \mathfrak{h}^{-1}\mathfrak{r}^2\mathfrak{c}$, then f' is a constant times f.*

Condition (3.20b) implies that $\mathfrak{h} = \mathfrak{h}_0\mathfrak{r}$ with a divisor $\mathfrak{h}_0$ of $\mathfrak{c}$. Then the ideal $\mathfrak{h}^{-1}\mathfrak{r}^2\mathfrak{c}$ in (3.21) can be written $\mathfrak{r}\mathfrak{h}_0^{-1}\mathfrak{c}$.

THEOREM 3.4. *With f, χ, τ, $\mathfrak{q}$, and $\mathfrak{r}$ as above, define f^* by (2.11); let $\mathbf{g}$ be the normalized eigenform in $\mathscr{S}_{2m}((2\mathfrak{b}\mathfrak{b}', \psi^2))$ belonging to χ, and let $\mathbf{g}^* = \mathbf{g}|J_{2\mathfrak{b}\mathfrak{b}'}$. Then, under conditions (3.20a, b, c) and (3.21) we have*

$$\overline{\lambda(\tau, \mathfrak{q}^{-1}; f, \psi)}\lambda(\tau, \mathfrak{q}^{-1}; f^*, \overline{\psi})\langle \mathbf{g}, \mathbf{g}\rangle/\langle f, f\rangle$$

$$= Q \sum_{\mathfrak{t} \supset \mathfrak{j}} \mu(\mathfrak{t})\overline{\varphi}^*(\mathfrak{t})N(\mathfrak{t})^{-1} D(0, \mathbf{g}^*, \varphi, \mathfrak{t}^{-1}\mathfrak{h}^{-1}\mathfrak{r}\mathfrak{c})$$

with Q and other symbols defined as in Theorem 3.2. Moreover, if $\mathbf{g}^*|\mathfrak{T}(\mathfrak{n}) = \chi'(\mathfrak{n})\mathbf{g}^*$ for every $\mathfrak{n}$ with some χ', then the above sum over $\mathfrak{t}$ can be written

$$N(\mathfrak{h}\mathfrak{r}^{-1}\mathfrak{c}^{-1})\chi'(\mathfrak{j}^{-1}\mathfrak{h}^{-1}\mathfrak{r}\mathfrak{c}) \prod_{v|\mathfrak{j}} [\chi'(v) - \overline{\varphi}(\pi_v)] \cdot c(\mathfrak{g}, \mathbf{g}^*)D(0, \chi', \varphi).$$

Since $\mathbf{g}^*|\mathfrak{T}(\mathfrak{n}) = \overline{\chi}(\mathfrak{n})\mathbf{g}^*$ for $\mathfrak{n}$ prime to $2\mathfrak{b}\mathfrak{b}'$, the assumption on $\mathbf{g}^*$ in the latter part of the theorem amounts to the condition that $\mathbf{g}^*$ is an eigenform of $\mathfrak{T}(\pi_v\mathfrak{g})$ for every $v|2\mathfrak{b}\mathfrak{b}'$. This is satisfied if χ is primitive and $2\mathfrak{b}\mathfrak{b}'$ is exactly its level, in which case $\chi'(\mathfrak{n}) = \overline{\chi}(\mathfrak{n})$ for *every* $\mathfrak{n}$.

Under certain conditions the first two factors of the left-hand side of the first formula of Theorem 3.4 becomes essentially the square of a Fourier coefficient:

THEOREM 3.5. *The notation and assumptions being as in the first half of Theorem 3.4, suppose that (2.9) holds and $f^* \neq 0$. Then there exists a nonzero element f_1 in the space $\mathbf{C}f$, unique up to real constant factors, such that*

$$\lambda(\xi, \mathfrak{m}; f_1^*, \overline{\psi}) = \overline{\lambda(\xi, \mathfrak{m}; f_1, \psi)}[\langle f_1^*, f_1^*\rangle/\langle f_1, f_1\rangle]^{1/2}$$

for every $(\xi, \mathfrak{m})$. For such an f_1 we have

$$\overline{\lambda(\tau, \mathfrak{q}^{-1}; f_1, \psi)}\lambda(\tau, \mathfrak{q}^{-1}; f_1^*, \overline{\psi}) = \overline{\lambda(\tau, \mathfrak{q}^{-1}; f_1, \psi)^2}[\langle f_1^*, f_1^*\rangle/\langle f_1, f_1\rangle]^{1/2}.$$

It should be noted that (2.9) holds if and only if X of (3.19a) is one-dimensional, because of Lemma 3.11 below.

Proof. There is an element f^ρ of $\mathscr{S}_k(\mathfrak{b}, \mathfrak{b}'; \bar{\psi})$ such that $\lambda(\xi, \mathfrak{m}; f^\rho, \bar{\psi}) = \overline{\lambda_f(\xi, \mathfrak{m})}$ for every $(\xi, \mathfrak{m})$ and $f^\rho | T_v = N_v^{-1}\bar{\chi}(v) f^\rho$ for every $v \in \mathbf{h}$. This was proved in [87, Prop. 8.6] when $\mathfrak{b} = \mathfrak{g}$ and ψ is of finite order, but the proof is applicable to the general case. By Proposition 2.10 and (2.9) we have $f^\# = \alpha(f) f^\rho$ with $\alpha(f) \in \mathbf{C}$. Clearly $|\alpha(f)|^2 = \langle f^\#, f^\# \rangle / \langle f, f \rangle$. Since $\alpha(cf) = \alpha(f)c/\bar{c}$ for $c \in \mathbf{C}^\times$, we can find $c \in \mathbf{C}^\times$ so that $0 < \alpha(cf) \in \mathbf{R}$. Then cf is the desired f_1. Clearly, f_1 is unique up to real factors.

We can obtain formulas without $\mathbf{g}^*$ and $f^\#$ under stronger conditions. Namely, we impose

(3.22a) *If* $0 \neq f' \in \mathscr{S}_k(\mathfrak{b}, \mathfrak{b}''; \psi)$ *with a divisor* $\mathfrak{b}''$ *of* $\mathfrak{b}'$ *and* $f' | T_v = N_v^{-1}\chi(v) f'$ *for every* $v \nmid \mathfrak{h}^{-1}\mathfrak{r}^2\mathfrak{c}$, *then* $\mathfrak{b}'' = \mathfrak{b}'$ *and* f' *is a constant times* f.

(3.22b) $4\mathfrak{r}\mathfrak{b} \supset \mathfrak{h} \cap 4\mathfrak{g}$; $\mathfrak{h}^{-1}\mathfrak{r}\mathfrak{c}$ *is prime to* $\mathfrak{r}$; $\mathfrak{h}_v = \mathfrak{c}_v$ or $\mathfrak{c}_v \neq 4\mathfrak{g}_v$ *if* $v | 2$ *and* $v \nmid \mathfrak{r}$.

THEOREM 3.6. *Let f, χ, $\mathbf{g}$, and $\mathfrak{j}$ be as in Theorem 3.4. Then, under (3.20a, b, c) and (3.22a, b) we have*

$$\mu(\mathfrak{h}^{-1}\mathfrak{r}\mathfrak{c})\varphi^*(\mathfrak{h}^{-1}\mathfrak{r}\mathfrak{c})|\lambda_f(\tau, \mathfrak{q}^{-1})|^2 = R \cdot D(0, \chi, \bar{\varphi})\langle f, f\rangle / \langle \mathbf{g}, \mathbf{g}\rangle$$

with

$$R = 2^{-1-\|k\|}\tau^k \pi^{-\|m\|} h_F [\mathfrak{g}_+^\times : \mathfrak{g}^{\times 2}] \prod_{v \in \mathbf{a}} (m_v - 1)!$$

$$\cdot N(\mathfrak{q}\mathfrak{r})^{-1}\chi(\mathfrak{j}^{-1}\mathfrak{h}^{-1}\mathfrak{r}\mathfrak{c}) \prod_{v | \mathfrak{j}} [\chi(v) - \varphi(\pi_v)].$$

The proof of Theorems 3.4 and 3.6 will be completed in Section 8.

The left-hand side of the first equality of Theorem 3.6 vanishes if $\mathfrak{h}^{-1}\mathfrak{r}\mathfrak{c}$ is not prime to $\mathfrak{h}$ or it has a nontrivial square factor. It would be interesting to know whether this gives the vanishing of $D(0, \chi, \bar{\varphi})$ under suitable conditions.

COROLLARY 3.7. *The notation and assumptions being as in Theorems 3.5 and 3.6, suppose that* $\mathbf{g}^* | \mathfrak{T}(\pi_v \mathfrak{g}) = \bar{\chi}(v)\mathbf{g}^*$ *for every* $v | \mathfrak{c}$. *Then*

$$\overline{\lambda(\tau, \mathfrak{q}^{-1}; f_1)} = \lambda(\tau, \mathfrak{q}^{-1}; f_1)\mu(\mathfrak{h}^{-1}\mathfrak{r}\mathfrak{c})\bar{\varphi}^*(\mathfrak{h}^{-1}\mathfrak{r}\mathfrak{c})\bar{\psi}_{\mathbf{a}}(\tau)c(\mathbf{g}, \mathbf{g}^*)|\gamma(\varphi)|/\gamma(\varphi),$$

and moreover $\langle f_1, f_1\rangle / \langle f_1^\#, f_1^\#\rangle = N(\mathfrak{h}^{-1}\mathfrak{r}\mathfrak{c}\mathfrak{b}')$ *with* $\mathfrak{h}$ *defined with any such* τ *for which* $\lambda(\tau, \mathfrak{q}^{-1}; f_1) \neq 0$.

Proof. Put $K = \langle f_1, f_1\rangle / \langle f_1^\#, f_1^\#\rangle$. Comparing the formula of Theorem 3.4 with that of Theorem 3.6, we obtain

$$\overline{\lambda(\tau, \mathfrak{q}^{-1}; f_1)}^2 = A|\lambda(\tau, \mathfrak{q}^{-1}; f_1)|^2 \mu(\mathfrak{h}^{-1}\mathfrak{r}\mathfrak{c})\bar{\varphi}^*(\mathfrak{h}^{-1}\mathfrak{r}\mathfrak{c})$$

with $A = K^{1/2}\tau^{-u/2}N(2\mathfrak{q}\mathfrak{c}^{-1})\overline{\psi}_{\mathbf{a}}(\tau)\overline{\gamma(\varphi)}c(\mathfrak{g}, \mathbf{g}^*)$. Since $|c(\mathfrak{g}, \mathbf{g}^*)| = 1$, $\tau^{u/2} = N(\mathfrak{b}^{-1}\mathfrak{q}^2\mathfrak{r})$, and $|\gamma(\varphi)|^2 = N(\mathfrak{h})$, we obtain our assertions.

In the rest of this section we prove four relatively easy lemmas.

LEMMA 3.8. *Let* $\mathbf{f} = (f_\lambda)_{\lambda=1}^{\kappa} \in \mathscr{S}_n((\mathfrak{z}, \Psi))$ *and* $\alpha \in G \cap \rho x_\lambda \tilde{D}_{\mathfrak{z}} x_\lambda^{-1}$ *with a fixed* λ *and* $\rho = \mathrm{diag}[r^{-1}, r]$, $r \in F_{\mathbf{h}}^{\times}$. *Then*

$$(f_\lambda \|_n \alpha^{-1})(z) = \Psi(r)\Psi_{\mathfrak{z}}(r^{-1}d_\alpha) \sum_{\xi \in F} c(\xi r^{-2} t_\lambda^{-1}\mathfrak{g}, \mathbf{f})\xi^{n/2+i\mu}\mathbf{e}_{\mathbf{a}}(\xi z).$$

Proof. Define f_p by (3.3c) with $p = r\rho x_\lambda$ and put $\alpha = \rho x_\lambda w x_\lambda^{-1}$. Then for $y \in \tilde{G}_{\mathbf{a}+}$ we have $\det(y)^{i\mu}(f_p\|_n y)(\mathbf{i}) = \mathbf{f}(r^{-1}t_\lambda^{-1}\rho x_\lambda y) = \mathbf{f}(r^{-1}t_\lambda^{-1}\alpha x_\lambda w^{-1} y) = \Psi(r)^{-1}\Psi_{\mathfrak{z}}(d_w)^{-1}\cdot \det(y)^{i\mu}(f_\lambda\|_n w^{-1}y)(\mathbf{i})$. Since $w_{\mathbf{a}} = \alpha_{\mathbf{a}}$, we obtain $f_\lambda\|_n\alpha^{-1} = \Psi(r)\Psi_{\mathfrak{z}}(r^{-1}d_\alpha)f_p$. Now $c(\mathfrak{m}, \mathbf{f})$ is given by (3.8a, b) independently of the choice of t_λ, and hence $f_p(z) = \sum_{\xi\in F} c(\xi r^{-2}t_\lambda^{-1}\mathfrak{g}, \mathbf{f})\xi^{n/2+i\mu}\mathbf{e}_{\mathbf{a}}(\xi z)$, which proves our lemma.

LEMMA 3.9. *The series of (3.14a, b) and (3.15a, b) can be continued as holomorphic functions in s to the whole plane. Furthermore, for* $(s, t) \in \mathbf{C}^2$, *a Hecke character* ω *of F, an integral ideal* $\mathfrak{x}$, *and* $\mathbf{f} \in \mathscr{S}_n((\mathfrak{z}, \Psi))$ *put*

$$X(s, t) = \sum_{\mathfrak{m}, \mathfrak{n}} \omega^*(\mathfrak{m})c(\mathfrak{x}\mathfrak{m}\mathfrak{n}, \mathbf{f})N(\mathfrak{m})^{-s}N(\mathfrak{n})^{-t},$$

where $\mathfrak{m}$ *and* $\mathfrak{n}$ *run independently over the integral ideals in F. Then this is convergent for sufficiently large* Re(*s*) *and* Re(*t*), *and* $L_{\mathfrak{x}\mathfrak{z}}(s + t + 1, \omega\Psi)X(s, t)$ *can be continued to a holomorphic function on* $\mathbf{C}^2$. *Moreover,* $\lim_{t\to+\infty} X(s, t) = D(s, \mathbf{f}, \omega, \mathfrak{x})$ *for every* $s \in \mathbf{C}$.

Proof. Take a basis $\{\mathbf{f}_1, \ldots, \mathbf{f}_r\}$ of $\mathscr{S}_n((\mathfrak{z}, \Psi))$ and define an $r \times r$ matrix $\Xi(\mathfrak{n})$ by $\mathfrak{F}|\mathfrak{T}(\mathfrak{n}) = \Xi(\mathfrak{n})\mathfrak{F}$, where $\mathfrak{F} = {}^t(\mathbf{f}_1, \ldots, \mathbf{f}_r)$. Let $C(\mathfrak{n}) = {}^t(c(\mathfrak{n}, \mathbf{f}_1), \ldots, c(\mathfrak{n}, \mathbf{f}_r))$. From (3.10) we obtain

$$\Xi(\mathfrak{n})C(\mathfrak{m}) = \sum_{\mathfrak{a} \supset \mathfrak{m}+\mathfrak{n}} \Psi_*(\mathfrak{a})N(\mathfrak{a}^{-1}\mathfrak{n})C(\mathfrak{a}^{-2}\mathfrak{m}\mathfrak{n}),$$

and in particular $\Xi(\mathfrak{n})C(\mathfrak{g}) = N(\mathfrak{n})C(\mathfrak{n})$. Now we have two formal identities, of which the first one follows from the second one:

$$\text{(3.23a)} \qquad \sum_{\mathfrak{m}} \Xi(\mathfrak{x}\mathfrak{m})M(\mathfrak{m}) = \prod_{v|\mathfrak{x}} E_v(M_v, 0) \sum_{\mathfrak{m}} \Xi(\mathfrak{m})M(\mathfrak{m}),$$

$$\text{(3.23b)} \quad \sum_{\mathfrak{m},\mathfrak{n}} \Xi(\mathfrak{x}\mathfrak{m}\mathfrak{n})M(\mathfrak{m})M'(\mathfrak{n}) = \prod_{v\in\mathbf{h}} E_v(M_v, M'_v) \sum_{\mathfrak{m}} \Xi(\mathfrak{m})M(\mathfrak{m}) \sum_{\mathfrak{n}} \Xi(\mathfrak{n})M'(\mathfrak{n}),$$

$$E_v = \begin{cases} 1 - \psi_v N_v M_v M'_v & \text{if } v \nmid \mathfrak{x}, \\ \Xi(\pi_v\mathfrak{g}) - \psi_v N_v(M_v + M'_v) & \text{if } \mathfrak{x}_v = \pi_v\mathfrak{g}_v, \\ \Xi(\mathfrak{x}_v) - \psi_v N_v \Xi(\pi_v^{-1}\mathfrak{x}_v)(M_v + M'_v) + \psi_v^2 N_v^2 \Xi(\pi_v^{-2}\mathfrak{x}_v)M_v M'_v & \text{if } \mathfrak{x} \subset \pi_v^2\mathfrak{g}, \end{cases}$$

where M' is another system of multiplicative symbols and $\psi_v = \Psi_*(\pi_v \mathfrak{g})$. The proof, being elementary, is left to the reader. Our first assertion is well known except when the series is of type (3.15b) with $\mathfrak{x} \neq \mathfrak{g}$, since each series is a linear combination of the Mellin transforms of functions $f(iy)$ with some $f \in \mathscr{S}_n$. If $\mathfrak{x} \neq \mathfrak{g}$, we apply the operator of (3.23a) to the vector $C(\mathfrak{g})$ to obtain

$$(3.24) \qquad N(\mathfrak{x}) \sum_{\mathfrak{m}} \omega^*(\mathfrak{m}) C(\mathfrak{x}\mathfrak{m}) N(\mathfrak{m})^{-s}$$
$$= \prod_{v|\mathfrak{x}} E_v(\omega^*(\pi_v \mathfrak{g}) N_v^{-1-s}, 0) \sum_{\mathfrak{m}} \omega^*(\mathfrak{m}) C(\mathfrak{m}) N(\mathfrak{m})^{-s},$$

which proves the desired holomorphy. Now we can find r integral ideals $\mathfrak{y}_1, \ldots, \mathfrak{y}_r$ such that $C(\mathfrak{y}_1), \ldots, C(\mathfrak{y}_r)$ are linearly independent. For each j we have

$$\sum_{\mathfrak{m}} \omega^*(\mathfrak{m}) \Xi(\mathfrak{m}) N(\mathfrak{m})^{-s} C(\mathfrak{y}_j)$$
$$= \sum_{\mathfrak{a} \supset \mathfrak{y}_j} (\omega^* \Psi_*)(\mathfrak{a}) N(\mathfrak{a}^{-1}\mathfrak{y}_j) N(\mathfrak{a})^{-s} (D(s, \mathfrak{f}_i, \omega, \mathfrak{a}^{-1}\mathfrak{y}_j))_{i=1}^r .$$

This proves that $\sum_{\mathfrak{m}} \omega^*(\mathfrak{m}) \Xi(\mathfrak{m}) N(\mathfrak{m})^{-s}$ is entire. From (3.23a, b) we obtain

$$N(\mathfrak{x}) \sum_{\mathfrak{m}, \mathfrak{y}} \omega^*(\mathfrak{m}) C(\mathfrak{x}\mathfrak{m}\mathfrak{n}) N(\mathfrak{m})^{-s} N(\mathfrak{y})^{-t} L_{\mathfrak{x}\mathfrak{z}}(s + t + 1, \omega\Psi)$$
$$= \sum_{\mathfrak{y}} \Xi(\mathfrak{y}) N(\mathfrak{y})^{-1-t} \prod_{v|\mathfrak{x}} E_v(\omega^*(\pi_v \mathfrak{g}) N_v^{-1-s}, N_v^{-1-t}) \sum_{\mathfrak{m}} \omega^*(\mathfrak{m}) C(\mathfrak{m}) N(\mathfrak{m})^{-s}.$$

Our assertion on X now follows from this equality and (3.24).

LEMMA 3.10. *For* $\mathbf{f} \in \mathscr{S}_n((\mathfrak{z}, \Psi))$ *and an integral ideal* $\mathfrak{y}$ *there exists an element* $\mathbf{f}|\mathfrak{y}$ *of* $\mathscr{S}_n((\mathfrak{y}\mathfrak{z}, \Psi))$ *such that* $c(\mathfrak{m}, \mathbf{f}|\mathfrak{y}) = c(\mathfrak{m}\mathfrak{y}^{-1}, \mathbf{f})$ *for every fractional ideal* $\mathfrak{m}$. *Moreover,* $(\mathbf{f}|\mathfrak{y})|J_{\mathfrak{x}\mathfrak{y}\mathfrak{z}} = (\mathbf{f}|J_{\mathfrak{z}})|\mathfrak{x}$ *for every integral ideal* $\mathfrak{x}$.

Proof. Put $\mathbf{g}(x) = \mathbf{f}(x \cdot \mathrm{diag}[q^{-1}, 1])$ with an element q of $F_{\mathbf{h}}^\times$ such that $\mathfrak{y} = q\mathfrak{g}$. We can then verify that $c(\mathfrak{m}, \mathbf{g}) = c(\mathfrak{m}\mathfrak{y}^{-1}, \mathbf{f})$ by means of (3.9). The last assertion follows easily from (3.16).

LEMMA 3.11. *Define* X *and* Y *by* (3.19a, b). *Then* $l|T_v = N_v^{-1}\chi(v)l$ *and* $\mathbf{q}|\mathfrak{T}(\pi_v\mathfrak{g}) = \chi(v)\mathbf{q}$ *for every* $v \nmid 2\mathfrak{b}\mathfrak{b}'$, *every* $l \in X$, *and every* $\mathbf{q} \in Y$.

This was shown in [87, Prop. 8.9(i)] for X with $\mathfrak{b} = \mathfrak{g}$, but the proof (or rather its idea) applies to both X and Y in the general case.

3B. The case $F = \mathbf{Q}$. Our present formulation for $F = \mathbf{Q}$ is somewhat different from that of [73]. Let us now translate the main theorems into the terminology of [73]. We put $\Gamma_0(N) = \Gamma[\mathbf{Z}, N\mathbf{Z}]$ for $0 < N \in \mathbf{Z}$, $\mathbf{e}(w) = \exp(2\pi i w)$ for $w \in \mathbf{C}$, and

$$\theta_0(z) = \sum_{n \in \mathbf{Z}} \mathbf{e}(n^2 z) \quad (z \in H), \tag{3B.1}$$

$$\theta_0(\gamma z) = h_0(\gamma, z)\theta_0(z) \quad (\gamma \in \Gamma_0(4)). \tag{3B.2}$$

Then $\theta_0(z) = \theta(2z)$ with θ of (6.7), and hence $h_0(\gamma, z) = h(\gamma', 2z)$ for every $\gamma \in \Gamma_0(4)$, where $\gamma' = \sigma\gamma\sigma^{-1}$ with $\sigma = \operatorname{diag}[2, 1]$. For $f \in \mathscr{S}_l$ with integral or half-integral l we put

$$f(z) = \sum_{n \in \mathbf{Q}} a(n, f)\mathbf{e}(nz). \tag{3B.3}$$

Let $k = \kappa/2$ with an odd integer $\kappa \geqslant 3$. For a multiple N of 4 and a Dirichlet character ψ_0 modulo N such that $\psi_0(-1) = 1$, we denote by $S_k(N, \psi_0)$ the space of all cusp forms f on H such that $f(\gamma z) = \psi_0(d_\gamma)h_0(\gamma, z)^\kappa f(z)$ for every $\gamma \in \Gamma_0(N)$. Put $m = (\kappa - 1)/2$; let ψ be the Hecke character of $\mathbf{Q}$ such that

(3B.4) $\prod_{p|N} \psi_p(a) = \left(\dfrac{-1}{a}\right)^m \psi_0(a)^{-1}$ *for* $a \in (\mathbf{Z}/N\mathbf{Z})^\times$, $\psi_p(\mathbf{Z}_p^\times) = 1$ *for* $p \nmid N$, *and* $\psi_{\mathbf{a}}(x) = \operatorname{sgn}(x)^m$.

Then we easily see that $t(z) \mapsto t(2z)$ maps $\mathscr{S}_k(\mathbf{Z}, (N/4)\mathbf{Z}; \psi)$ onto $S_k(N, \psi_0)$, and $a(n, f) = \lambda_t(n, \mathbf{Z})$ if $f(z) = t(2z)$. Moreover $T^N_{\kappa, \psi_0}(p^2)$ of [73], which we write $T(p^2)$ for simplicity, corresponds to $p^m T_p$ through this map. Let $\mathscr{T}_n$ for $0 < n \in \mathbf{Z}$ denote the Hecke operator as defined by Hecke in [H]. Then $\mathscr{T}_n = n^{m-1}\mathfrak{T}(n\mathbf{Z})$ with $\mathfrak{T}$ of Section 3 for the forms of weight $2m$. We now take $f \in S_k(N, \psi_0)$ which is an eigenform of all $T(p^2)$, put $f | T(p^2) = \omega(p)f$ for each prime p, and assume that if $\kappa = 3$, f is orthogonal to $\sum_{a \in M} a\mathbf{e}(ba^2 z)$ for every $b \in \mathbf{Q}, > 0$, and every coset M of $\mathbf{Q}/\mathbf{Z}$. We then have an eigenform g such that $g | \mathscr{T}_p = \omega(p)g$ for every p in the space $S_{2m}((N/2, \psi_0^2))$ consisting of all cusp forms q such that $q \|_{2m} \gamma = \psi_0^2(d_\gamma)q$ for every $\gamma \in \Gamma_0(N/2)$. This is a special case of Proposition 3.1.

Given $q \in S_{2m}((N/2, \psi_0^2))$, a Dirichlet character η, and $0 < t \in \mathbf{Z}$, we put

$$\mathscr{D}(s, q, \eta, t) = \sum_{n=1}^{\infty} \eta(n)a(tn, q)n^{-s}, \qquad \mathscr{D}(s, q, \eta) = \mathscr{D}(s, q, \eta, 1). \tag{3B.5}$$

If $q | \mathscr{T}_n = \omega(n)q$ for every n, then $\omega(n)a(1, q) = a(n, q)$. We then put

$$\mathscr{D}(s, \omega, \eta) = \sum_{n=1}^{\infty} \eta(n)\omega(n)n^{-s}, \qquad \mathscr{D}(s, \omega) = \sum_{n=1}^{\infty} \omega(n)n^{-s}. \tag{3B.6}$$

For q as above and $f \in S_k(N, \psi_0)$ we put

(3B.7a) $$\hat{f}(z) = (-iN^{1/2}z)^{-k}f(-1/(Nz)),$$

(3B.7b) $$\tilde{f}(z) = \sum_{n=1}^{\infty} a(nN/4, \hat{f})\mathbf{e}(nz),$$

(3B.7c) $$q^*(z) = (q|J_{N/2})(z) = (2/N)^m z^{-2m} q(-2/(Nz)).$$

Then $q^* \in S_{2m}((N/2, \psi_0^{-2}))$ and $\hat{f} \in \mathscr{S}_k(N, \hat{\psi}_0)$, where $\hat{\psi}_0(n) = \overline{\psi}_0(n)(\frac{N}{n})$ (see [73, Prop. 1.4]). Moreover $\tilde{f} \in \mathscr{S}_k(N, \overline{\psi}_0)$. This is a special case of Proposition 2.10; the fact also follows from [73, Props. 1.4 and 1.5]. We also put

(3B.8a) $$X = \{l \in S_k(N, \psi_0) | \omega(p)l = l|T(p^2) \textit{ for almost all } p\},$$

(3B.8b) $$Y = \{g \in S_{2m}((N/2, \psi_0^2)) | \omega(p)g = g|\mathscr{T}_p \textit{ for almost all } p\}.$$

We take a squarefree integer $r \geqslant 1$ and denote by φ_0 the primitive Dirichlet character such that $\varphi_0(n) = \psi_0(n)(\frac{-1}{n})^m(\frac{r}{n})$ for n prime to Nr. Let h be the conductor of φ_0. Then we assume

(3B.9a) *r divides h;*
(3B.9b) *if r is even, h is divisible by* 8.

We can now rephrase the main theorems of Section 3 in the present terminology.

THEOREM 3B.1. *Let $\{f_1, \ldots, f_\alpha\}$ and $\{g_1, \ldots, g_\beta\}$ be bases of X and Y over* **C**, *respectively; let $\{f_1', \ldots, f_\alpha'\}$ and $\{g_1', \ldots, g_\beta'\}$ be their dual bases in the sense that $\langle f_i, f_j'\rangle = \delta_{ij}$ and $\langle g_p, g_q'\rangle = \delta_{pq}$. Then, under (3B.9a, b) we have*

$$\sum_{i=1}^{\alpha} a(r, \tilde{f}_i)\overline{a(r, f_i')} = P \sum_{t|j} \mu(t)\overline{\varphi}_0(t)t^{m-1} \sum_{p=1}^{\beta} \overline{a(1, g_p')}\mathscr{D}(m, g_p^*, \varphi_0, t^{-1}h^{-1}rN)$$

with

$$P = (-i)^m 2^{-k} h^{m-1} N^{(1-k)/2}\overline{\gamma(\varphi_0)}\pi^{-m}(m-1)!,$$

where j is the product of all the prime factors of N prime to h, μ the Moebius function, and $\gamma(\varphi_0) = \sum_{n=1}^{h} \varphi_0(n)\mathbf{e}(n/h)$.

The translation of Remark 3.3 into the terminology of the present setting is straightforward and therefore may be left to the reader. Next we impose:

(3B.10) *If $f' \in S_k(N, \psi_0)$ and $f'|T(p)^2 = \omega(p)f'$ for every $p \nmid h^{-1}r^2N$, then f' is a constant times f.*

THEOREM 3B.2. *Let f be a nonzero element of $S_k(N, \psi_0)$ such that $f|T(p^2) = \omega(p)f$ for every p, and g the normalized eigenform in $S_{2m}((N/2, \psi_0^2))$ such that $g|\mathcal{T}_p = \omega(p)g$ for every p. Then, under (3B.9a, b) and (3B.10) we have*

$$\overline{a(r, f)}a(r, \tilde{f})\langle g, g\rangle/\langle f, f\rangle = P\sum_{t|j} \mu(t)\overline{\varphi}_0(t)t^{m-1}\mathscr{D}(m, g^*, \varphi_0, t^{-1}h^{-1}rN)$$

with P and j as in Theorem 3B.1. Moreover, if $g^*|\mathcal{T}_m = \omega'(n)g^*$ for every n, then the above sum over t can be written

$$\omega'(j^{-1}h^{-1}rN)\prod_{p|j}[\omega'(p) - \overline{\varphi}_0(p)p^{m-1}]a(1, g^*)\mathscr{D}(m, \omega', \varphi_0).$$

THEOREM 3B.3. *The notation and assumptions being as in Theorem 3B.2, suppose that $\tilde{f} \neq 0$ and*

(3B.11) $$\mathbf{C}f = \{l \in S_k(N, \psi_0) \mid l|T(p^2) = \omega(p)l \text{ for every } p \nmid N\}.$$

Then there exists a nonzero element f_1 in the space $\mathbf{C}f$, unique up to real constant factors, such that $a(n, \tilde{f}_1) = \overline{a(n, f_1)}[\langle \tilde{f}_1, \tilde{f}_1\rangle/\langle f_1, f_1\rangle]^{1/2}$ for every n. For such an f_1 we have

$$\overline{a(r, f_1)}a(r, \tilde{f}_1) = \overline{a(r, f_1)}^2[\langle \tilde{f}_1, \tilde{f}_1\rangle/\langle f_1, f_1\rangle]^{1/2}.$$

We now impose the following conditions:

(3B.12a) *If $0 \neq f' \in S_k(4K, \psi_0)$ with a divisor K of $N/4$ and $f'|T(p^2) = \omega(p)f'$ for every $p \nmid h^{-1}r^2N$, then $4K = N$ and f' is a constant times f.*

(3B.12b) *$4r$ divides the least common multiple of h and 4; $h^{-1}rN$ is prime to r; $2 \nmid h^{-1}rN$ or $8|N$ if r is odd.*

THEOREM 3B.4. *Let f, ω, g, and j be as in Theorem 3B.2. Then, under (3B.9a, b) and (3B.12a, b) we have*

$$\mu(h^{-1}rN)\varphi_0(h^{-1}rN)|a(r, f)|^2 = S\cdot\mathscr{D}(m, \omega, \overline{\varphi}_0)\langle f, f\rangle/\langle g, g\rangle$$

with

$$S = 2^{-1}h^{m-1}N^{1-m}r^{1/2}\pi^{-m}(m-1)!\,\omega(j^{-1}h^{-1}rN)\prod_{p|j}[\omega(p) - \varphi_0(p)p^{m-1}].$$

COROLLARY 3B.5. *The notation and assumptions being as in Theorems 3B.3 and 3B.4, suppose that $g^*|\mathcal{T}_p = \overline{\omega}(p)g^*$ for every $p|N$. Then*

$$\overline{a(r, f_1)} = a(r, f_1)\mu(h^{-1}rN)\overline{\varphi}_0(h^{-1}rN)(-i)^m a(1, g^*)|\gamma(\varphi_0)|/\gamma(\varphi_0),$$

and moreover $\langle f_1, f_1\rangle/\langle\tilde{f}_1, \tilde{f}_1\rangle = 2^{2m-1}N^{2-k}r/h$ *with h defined with any such r for which* $a(r, f_1) \neq 0$.

Examples. (I) First take $N = 4, \psi_0 = 1$, and $k = 9/2$. Then $S_{9/2}(4, 1)$ is generated by $f(z) = \theta_0(z)^{-3}\eta(2z)^{12}$, where $\eta(z) = \mathbf{e}(z/24)\prod_{n=1}^{\infty}(1 - \mathbf{e}(nz))$. Then Theorems 3B.2 and 3B.3 are applicable to all squarefree $r \geqslant 1$. We have $g^* = g = \eta(z)^8\eta(2z)^8$, $\hat{f} = \tilde{f} = f$, and $\varphi_0(n) = (\frac{r}{n})$. Observing that $\omega(2) = -8$, we obtain a factor $-8(1 + (\frac{r}{2}))$ as $\prod_{p|j}$ in Theorem 3B.2 when $r \equiv 1 \pmod 4$. Therefore $a(r, f) = 0$ if $r \equiv 5 \pmod 8$. Here r is squarefree, but by Proposition 2.2 (or by [73, Theorem 1.9])we easily see that $a(n, f) = 0$ for every positive integer n such that $n \equiv 5 \pmod 8$. In fact, the functional equation for $\mathcal{D}(s, g, \varphi_0)$ shows that $\mathcal{D}(4, g, \varphi_0) = 0$ for $r \equiv 5 \pmod 8$. Thus the vanishing of $a(r, f)$ follows from this fact with no knowledge of $\prod_{p|j}$. Anyway, the formula of Theorem 3B.2 gives $\mathcal{D}(4, g, \varphi_0)$ effecitvely for all other r's.

(II) Let $f(z) = \theta_0(3z)^{-1}[\eta(2z)\eta(6z)]^3$ and $g(z) = [\eta(z)\eta(2z)\eta(3z)\eta(6z)]^2$. These generate $S_{5/2}(12, 1)$ and $S_4((6, 1))$, respectively. We have $g^* = g$, $\varphi_0(n) = (\frac{r}{n})$, $\hat{f}(z) = 3^{1/4}\theta_0(z)^{-1}[\eta(2z)\eta(6z)]^3$, and $\tilde{f} = 3^{1/4}f$. In this case the vanishing of $\prod_{p|j}$ implies that $a(n, f) = 0$ if $n \equiv 2 \pmod 3$ or $n \equiv 5 \pmod 8$. The functional equation of $\mathcal{D}$ gives $\mathcal{D}(2, g, \varphi_0) = 0$ for $r \equiv 13$ or $17 \pmod{24}$. If $r \not\equiv 2 \pmod 3$ or $r \not\equiv 5 \pmod 8$, then $\mathcal{D}(2, g, \varphi_0)$ can be obtained from the formula of Theorem 3B.2. However, in the remaining cases (that is, if $r \equiv 2, 5, 11, 14, 22$, or $23 \pmod{24}$) it gives no information about $\mathcal{D}(2, g, \varphi_0)$.

(III) Let $N = 4$, $q_1(z) = \theta(z)^{-3}\eta(z)^8\eta(2z)^8$, and $q_2(z) = 16\theta(z)^{-3}\eta(2z)^8\eta(4z)^8$. Then $S_{13/2}(4, 1) = \mathbf{C}q_1 + \mathbf{C}q_2$, $\hat{q}_1 = q_2$, $\hat{q}_2 = q_1$, and $q_1 + q_2 = \theta(z)\eta(2z)^{12}$. This space corresponds to $S_{12}((2, 1)) = \mathbf{C}\Delta(z) + \mathbf{C}\Delta(2z) = \mathbf{C}g_1 + \mathbf{C}g_2$, where $\Delta = \eta^{24}$, $g_i = \Delta - \alpha_i\Delta(2z), \alpha_1 = -12 + 4\sqrt{-119}, \alpha_2 = \bar{\alpha}_1$; each g_i is a normalized eigenform of all Hecke operators of level 2, and corresponds to the eigenforms f_1 and f_2 in $S_{13/2}(4, 1)$ given by $f_i = 8(\alpha_i + 56)q_1 + (7\alpha_i + 2^9)q_2$. Now Theorems 3B.2 and 3B.4 are applicable to (f_i, g_i) for every squarefree r such that $r \equiv 3 \pmod 4$. Notice that $\mathcal{D}(s, \Delta, \varphi_0) = \mathcal{D}(s, g_i, \varphi_0)$ if h is even. These theorems do not apply to other r's. Now a result in [KZ] specialized to this case (see also [K]) shows a formula between the Fourier coefficients of $8q_1 + 7q_2$ and $\mathcal{D}(6, \Delta, \sigma)$ for every even quadratic Dirichlet character σ. (The form $8q_1 + 7q_2$ is characterized, up to constant factors, by the property that $a(r, 8q_1 + 7q_2) = 0$ for $r \equiv 2, 3 \pmod 4$.) Though this is satisfactory as a result on $\mathcal{D}(6, \Delta, \sigma)$, it does not answer the natural question about the nature of $a(r, f_i)$. Our Theorems 3B.2 and 3B.4 give answers to that question if $r \equiv 3 \pmod 4$. The answer applicable to all squarefree r will be given in Theorem 3B.6 below.

Let us now consider the following situation which generalizes that of (III) above:

(3B.13a) *$N = 4M$ with an odd integer M; there is a normalized primitive eigenform $g_0(z) = \sum_{n=1}^{\infty}\omega_0(n)\mathbf{e}(nz)$ in $S_{2m}((M, \psi_0^2))$ such that $\omega_0(p) = \omega(p)$ for $p \neq 2$; M is the exact level of g_0.*

Then $Y = \mathbf{C}g_0(z) + \mathbf{C}g_0(2z)$. Put $g_i(z) = g_0(z) - \alpha_i g_0(2z)$, where α_1 and α_2 are the roots of $x^2 - \omega_0(2)x + \psi_0^2(2)2^{2m-1} = 0$. Each g_i is a normalized eigenform of level $2M$ and $g_i(z) = \sum_{n=1}^{\infty} \omega_i(n)e(nz)$ with $\omega_i(p) = \omega_0(p)$ for $p \neq 2$ and $\omega_i(2) = \omega_0(2) - \alpha_i$. Let us now assume

(3B.13b) *$\alpha_1 \neq \alpha_2$ and $X = \mathbf{C}f_1 + \mathbf{C}f_2$ with two eigenforms f_1 and f_2 such that $f_i | T(p^2) = \omega_i(p)f_i$ for every prime p.*

Put $u = \langle g_1, g_1\rangle/\langle g_0, g_0\rangle$, $t_1 = \langle g_1, g_2\rangle/\langle g_0, g_0\rangle$, and $t_2 = \bar{t}_1$. Then the principle of [76, (3.2)] shows that

(3B.14) $$u = (9 - 2^{2-2m}|\omega_0(2)|^2)/6, \qquad t_1 = (1 - 2^{-2m}\bar{\alpha}_1\alpha_2)/3.$$

We then define two constants d_i and e_i by

(3B.15) $$d_i = 2^{-m}(\bar{t}_i\alpha_{i+1} - u\alpha_i)/(u^2 - t_1t_2), \qquad e_i = 2^m(u - \bar{t}_i)/(u^2 - t_1t_2),$$

where we understand that $\alpha_3 = \alpha_1$.

THEOREM 3B.6. *The notation being as above and as in Theorem 3B.1, let f_1' and f_2' be the elements of $\mathbf{C}f_1 + \mathbf{C}f_2$ such that $\langle f_i, f_j'\rangle = \delta_{ij}$, and let $g^0(z) = M^{-m}z^{-2m}g_0(-1/(Mz))$. Further, let l be the product of all the prime factors of M prime to h. Put $A_i = a(r, (f_i')^{\sim})\overline{a(r, f_i)}$ and*

$$B = \mathscr{D}(m, g^0, \varphi_0)\prod_{p|l}[\bar{\omega}_0(p) - \bar{\varphi}_0(p)p^{m-1}].$$

Then, under (3B.9a, b) and (3B.13a, b), A_i can be given as follows:

Case I: $2|h$. $\langle g_0, g_0\rangle A_i = PB\bar{\omega}_0(rN/(lh))d_i$.
Case II: $2\nmid h$. $\langle g_0, g_0\rangle A_i = PB\bar{\omega}_0(rM/(lh))$
$\cdot[d_i\{\bar{\omega}_0(4) - 2m\bar{\varphi}_0(2) + 2^{2m-2}\bar{\varphi}_0(4)\} + e_i\{\bar{\omega}_0(2) - 2^m\bar{\varphi}_0(2)\}]$.

Moreover, we have

$$\langle g_0, g_0\rangle[(2^{2m} - \bar{\alpha}_1\alpha_2)A_1 + (2^{2m} - \bar{\alpha}_2\alpha_1)A_2]$$

$$= \begin{cases} 0 & \text{if } 2|h, \\ 2^{3m-1}3PB\bar{\omega}_0(rM/(lh))[\bar{\omega}_0(2) - 2^m\bar{\varphi}_0(2)] & \text{if } 2\nmid h. \end{cases}$$

This is a special case of Theorem 9.2, translated into the present language. Notice that h is even if and only if rN/h is odd, because of (3B.9a, b). Also, a special case of Remark 9.3 can be given as:

PROPOSITION 3B.7. *Under (3B.13a, b), suppose that $a(r, f_1) \neq 0$ for some squarefree r satisfying (3B.9a, b) for which r^2N/h is odd. Then, changing f_2 for its suitable constant multiple, we can have $a(r, f_1) = a(r, f_2)$ for all such r's, $\langle f_1, f_1\rangle = \langle f_2, f_2\rangle$, and $\langle f_1, f_2\rangle/\langle f_1, f_1\rangle = \langle g_1, g_2\rangle/\langle g_1, g_1\rangle$.*

Coming back to Example (III), we note that f_1 and f_2 have been chosen so that the relations of Proposition 3B.7 hold for them.

4. Preliminaries on Gauss sums and Eisenstein series. Given a Hecke character ω of F of conductor $\mathfrak{f}$, we define the Gauss sum $\gamma(\omega)$ by

$$\gamma(\omega) = \sum_{t \in \mathfrak{f}^{-1}\mathfrak{d}^{-1}/\mathfrak{d}^{-1}} \omega_{\mathbf{a}}(t)\omega^*(t\mathfrak{f}\mathfrak{d})e_{\mathbf{a}}(t). \tag{4.1}$$

We understand that $\gamma(\omega) = \omega^*(\mathfrak{d})$ if $\mathfrak{f} = \mathfrak{g}$ and that the symbol $\omega_{\mathbf{a}}(t)\omega^*(t\mathfrak{x})$ for $t = 0$ denotes $\omega^*(\mathfrak{x})$ or 0 according as $\mathfrak{f} = \mathfrak{g}$ or $\mathfrak{f} \neq \mathfrak{g}$. Notice that if $t \in F^\times$, $\mathfrak{x} = x\mathfrak{g}$ with $x \in F_{\mathbf{h}}^\times$, and $t\mathfrak{x}$ is prime to $\mathfrak{f}$, then $\omega_{\mathbf{a}}(t)\omega^*(t\mathfrak{x}) = \omega(x)\omega_{\mathfrak{f}}(tx)^{-1}$.

LEMMA 4.1. *Let $\mathfrak{x}$ be a fractional ideal and $\mathfrak{z}$ an integral ideal divisible by $\mathfrak{f}$. Then for $b \in (\mathfrak{d}\mathfrak{x}\mathfrak{z})^{-1}$ we have*

$$\sum_{t \in \mathfrak{x}/\mathfrak{z}\mathfrak{x}} \omega_{\mathbf{a}}(t)\omega^*(t\mathfrak{x}^{-1})e_{\mathbf{a}}(bt) = \begin{cases} \gamma(\omega)N(\mathfrak{f}^{-1}\mathfrak{z})\overline{\omega}_{\mathbf{a}}(b)\overline{\omega}^*(b\mathfrak{f}\mathfrak{d}\mathfrak{x}) & \text{if } b \in (\mathfrak{f}\mathfrak{d}\mathfrak{x})^{-1}, \\ 0 & \text{if } b \notin (\mathfrak{f}\mathfrak{d}\mathfrak{x})^{-1}. \end{cases}$$

Moreover, let μ denote the Moebius function on the ideals in F. Then

$$\sum_{s \in \mathfrak{x}/\mathfrak{z}\mathfrak{x}}{}' \omega_{\mathbf{a}}(s)\omega^*(s\mathfrak{x}^{-1})e_{\mathbf{a}}(bs) = \gamma(\omega)\sum_{\mathfrak{a}} \mu(\mathfrak{a}^{-1}\mathfrak{f}^{-1}\mathfrak{z})N(\mathfrak{a})\omega^*(\mathfrak{a}^{-1}\mathfrak{f}^{-1}\mathfrak{z})\overline{\omega}_{\mathbf{a}}(b)\overline{\omega}^*(b\mathfrak{a}^{-1}\mathfrak{z}\mathfrak{d}\mathfrak{x}),$$

where s runs under the condition that $s\mathfrak{x}^{-1} + \mathfrak{z} = \mathfrak{g}$ and $\mathfrak{a}$ runs over the integral ideals dividing $\mathfrak{f}^{-1}\mathfrak{z} + b\mathfrak{z}\mathfrak{d}\mathfrak{x}$.

Proof. The first formula for $\mathfrak{z} = \mathfrak{f}$, if formulated somewhat differently, is a standard fact on Gauss sums. The general case can easily be reduced to that special case. The left-hand side of the second formula is equal to

$$\sum_{s \in \mathfrak{x}/\mathfrak{z}\mathfrak{x}} \left\{ \sum_{\mathfrak{b} \supset s\mathfrak{x}^{-1} + \mathfrak{z}} \mu(\mathfrak{b}) \right\} \omega_{\mathbf{a}}(s)\omega^*(s\mathfrak{x}^{-1})e_{\mathbf{a}}(bs)$$

with no extra condition on s. If $\mathfrak{b} \supset \mathfrak{z}$ and $\mathfrak{b} \not\supset \mathfrak{f}^{-1}\mathfrak{z}$, then $\omega^*(\mathfrak{b}) = 0$, and hence the sum can be written

$$\sum_{\mathfrak{b} \supset \mathfrak{f}^{-1}\mathfrak{z}} \mu(\mathfrak{b})\omega^*(\mathfrak{b}) \sum_{s \in \mathfrak{b}\mathfrak{x}/\mathfrak{z}\mathfrak{x}} \omega_{\mathbf{a}}(s)\omega^*(s\mathfrak{b}^{-1}\mathfrak{x}^{-1})e_{\mathbf{a}}(bs).$$

Applying the first formula to the last sum over $\mathfrak{b}\mathfrak{x}/\mathfrak{z}\mathfrak{x}$ and putting $\mathfrak{f}^{-1}\mathfrak{z} = \mathfrak{a}\mathfrak{b}$, we obtain the desired result.

Given ω as above, take $n \in \mathbf{Z}^{\mathbf{a}}$ and $\lambda \in \mathbf{R}^{\mathbf{a}}$ such that $\|\lambda\| = 0$ and

$$\omega_{\mathbf{a}}(x) = \operatorname{sgn}(x)^n |x|^{i\lambda}. \tag{4.2}$$

Taking also an integral ideal $\mathfrak{z}$ divisible by $\mathfrak{f}$ and fractional ideals $\mathfrak{x}$ and $\mathfrak{y}$, we consider a series

$$(4.3)\quad E(z, s; n, \omega, \mathfrak{z}; \mathfrak{x}, \mathfrak{y})$$
$$= N(\mathfrak{y})^{2s} y^{su+(i\lambda-n)/2} \sum_{(c,d)\in R} \omega_{\mathbf{a}}(d)\omega^*(d\mathfrak{y}^{-1})(cz+d)^{-n}|cz+d|^{n-i\lambda-2su},$$

where $z \in H^{\mathbf{a}}$, $y = \mathrm{Im}(z)$, $s \in \mathbf{C}$, and $R = T/\mathfrak{g}^{\times}$ with

$$(4.4)\qquad T = \{(c, d) \in \mathfrak{x}\mathfrak{y}\mathfrak{z} \times \mathfrak{y} \,|\, d\mathfrak{y}^{-1} + \mathfrak{z} = \mathfrak{g}, (c, d) \neq (0, 0)\}.$$

This is similar to the series of [87, (6.14)]. We have

$$(4.5)\quad E(z, s; n, \omega, \mathfrak{z}; \mathfrak{x}, \mathfrak{y})$$
$$= \sum_{\mathfrak{a}\supset\mathfrak{f}^{-1}\mathfrak{z}} \mu(\mathfrak{a})\omega^*(\mathfrak{a})N(\mathfrak{a})^{-2s}E(z, s; n, \omega, \mathfrak{f}; \mathfrak{a}^{-1}\mathfrak{f}^{-1}\mathfrak{z}\mathfrak{x}, \mathfrak{a}\mathfrak{y}).$$

This can be seen by inserting $\sum_{\mathfrak{a}\supset\mathfrak{z}+d\mathfrak{y}^{-1}} \mu(\mathfrak{a})$ in the sum of the right-hand side of (4.3) instead of the condition $d\mathfrak{y}^{-1} + \mathfrak{z} = \mathfrak{g}$, and observing that $\omega^*(d\mathfrak{y}^{-1}) = \omega^*(\mathfrak{a})\omega^*(d\mathfrak{a}^{-1}\mathfrak{y}^{-1})$ and this is nonvanishing only if $\mathfrak{a} \supset \mathfrak{f}^{-1}\mathfrak{z}$.

Since $G_{\mathbf{A}} = P_{\mathbf{A}}D[\mathfrak{x}^{-1}, \mathfrak{x}]$, every y in $G_{\mathbf{A}}$ can be written $y = pw$ with $p \in P_{\mathbf{A}}$ and $w \in D[\mathfrak{x}^{-1}, \mathfrak{x}]$. We can then define a fractional ideal $i_{\mathfrak{x}}(y)$ by $i_{\mathfrak{x}}(y) = d_p\mathfrak{g}$, which depends only on y and $\mathfrak{x}$. Throughout the rest of this section we fix $\mathfrak{x}$ and put $i_{\mathfrak{x}}(y) = \mathfrak{a}_y$. Notice that $\mathfrak{a}_y = c_y\mathfrak{x}^{-1} + d_y\mathfrak{g}$.

With ω and $\mathfrak{z}$ as above, put $D_0 = D[\mathfrak{x}^{-1}, \mathfrak{x}\mathfrak{z}]$ and $\Gamma_0 = \Gamma[\mathfrak{x}^{-1}, \mathfrak{x}\mathfrak{z}]$. Assigning the ideal class of $\mathfrak{a}_y$ to $Py_0\Gamma_0$, we obtain a bijection of $P\backslash(G \cap P_{\mathbf{A}}D_0)/\Gamma_0$ onto the ideal class group of F. This can be proved in the same manner as in [83, Lemmas 1.4, 1.5, and 1.6], which include the special case $\mathfrak{x} = \mathfrak{g}$. More generally, take an arbitrary open subgroup D^* of D_0 and put $\Gamma^* = G \cap D^*$. Then $G \cap P_{\mathbf{A}}D^* = \bigsqcup_{\beta\in B} P\beta\Gamma^*$ with a finite subset B of G. We now put

$$(4.6)\qquad E(z, s; n, \omega, \Gamma^*) = \sum_{\alpha\in R} \omega_{\mathbf{a}}(d_\alpha)\omega^*(d_\alpha\mathfrak{a}_\alpha^{-1})N(\mathfrak{a}_\alpha)^{2s}y^{su+(i\lambda-n)/2}\|_n\alpha,$$

$$(4.7)\qquad E_\beta(z, s; n, \omega, \Gamma^*) = N(\mathfrak{a}_\beta)^{2s} \sum_{\alpha\in R_\beta} \omega_{\mathbf{a}}(d_\alpha)\omega^*(d_\alpha\mathfrak{a}_\beta^{-1})y^{su+(i\lambda-n)/2}\|_n\alpha.$$

where $R = P\backslash(G \cap P_{\mathbf{A}}D^*)$ and $R_\beta = (P \cap \beta\Gamma^*\beta^{-1})\backslash\beta\Gamma^*$. Since R can be given by $\bigsqcup_{\beta\in B} R_\beta$, we obtain

$$(4.8)\qquad E(z, s; n, \omega, \Gamma^*) = \sum_{\beta\in B} E_\beta(z, s; n, \omega, \Gamma^*).$$

These were studied in [83, 85b, 87, 91]. We can easily verify that

$$(4.9)\qquad J_n(\gamma, z)^{-1}E_\beta(\gamma z, s; X) = \omega_{\mathfrak{z}}(a_\gamma)^{-1}E_\beta(z, s; X) \qquad \textit{for every } \gamma \in \Gamma^*,$$

where X means (n, ω, Γ^*). For any integral ideal $\mathfrak{c}$ we define "modified and partial" L-series of ω by putting

$$L_{\mathfrak{c}}(s, \omega) = \sum_{\mathfrak{m}} \omega^*(\mathfrak{m})N(\mathfrak{m})^{-s}, \tag{4.10a}$$

$$L_{\mathfrak{c}}(s, \omega, \mathfrak{y}) = \sum_{\mathfrak{m} \sim \mathfrak{y}} \omega^*(\mathfrak{m})N(\mathfrak{m})^{-s}, \tag{4.10b}$$

where $\mathfrak{m}$ in (4.10a) runs over all the integral ideals prime to $\mathfrak{c}$, and with the additional condition that $\mathfrak{m} = a\mathfrak{y}$ with $a \in F^\times$ in (4.10b). Further, we put

$$C(z, s; n, \omega, \Gamma_0) = L_{\mathfrak{z}}(2s, \omega)E(z, s; n, \omega, \Gamma_0). \tag{4.11}$$

Then we have

$$E(z, s; n, \omega, \mathfrak{z}; \mathfrak{x}, \mathfrak{y}) = \sum_{\beta \in B} L_{\mathfrak{z}}(2s, \omega, \mathfrak{y}^{-1}\mathfrak{a}_\beta)E_\beta(z, s; n, \omega, \Gamma_0), \tag{4.12a}$$

$$C(z, s; n, \omega, \Gamma_0) = \sum_{\mathfrak{y} \in \mathscr{A}} E(z, s; n, \omega, \mathfrak{z}; \mathfrak{x}, \mathfrak{y}), \tag{4.12b}$$

where $\mathscr{A}$ is a complete set of representatives for the ideal classes in F. These are similar to [87, Lemma 6.5] and can be proved in the same manner.

LEMMA 4.2. *For every* $\gamma \in G \cap \mathrm{diag}[t^{-1}, t]D_0$ *with* $t \in F_{\mathbf{h}}^\times$ *we have*

$$J_n(\gamma^{-1}, z)^{-1}E(\gamma^{-1}z, s; n, \omega, \mathfrak{z}; \mathfrak{x}, \mathfrak{y})$$
$$= N(\mathfrak{a}_\gamma)^{2s}\omega_{\mathbf{a}}(d_\gamma)\omega^*(d_\gamma\mathfrak{a}_\gamma^{-1})E(z, s; n, \omega, \mathfrak{z}; \mathfrak{a}_\gamma^2\mathfrak{x}, \mathfrak{a}_\gamma^{-1}\mathfrak{y}).$$

This can be verified easily by observing that $T_{\mathfrak{z}}[\mathfrak{x}, \mathfrak{y}] = T_{\mathfrak{z}}[\mathfrak{a}_\gamma^2\mathfrak{x}, \mathfrak{a}_\gamma^{-1}\mathfrak{y}]\gamma$, where $T_{\mathfrak{z}}[\mathfrak{x}, \mathfrak{y}]$ denotes the set of (4.4).

To study the Fourier expansion of (4.3), we recall a formula

$$D_F^{1/2}N(\mathfrak{m}) \sum_{a \in \mathfrak{m}} (z + a)^{-\alpha}(\bar{z} + b)^{-\beta} = \sum_{h \in (\mathfrak{b}\mathfrak{m})^{-1}} \mathbf{e}_{\mathbf{a}}(hx)\xi(y, h; \alpha, \beta). \tag{4.13}$$

Here $z \in H^{\mathbf{a}}$, $\alpha \in \mathbf{C}^{\mathbf{a}}$, $\beta \in \mathbf{C}^{\mathbf{a}}$, $\mathrm{Re}(\alpha_v + \beta_v) > 1$ for every $v \in \mathbf{a}$, $\mathfrak{m}$ is a fractional ideal in F, and $\xi(y, h; \alpha, \beta) = \prod_{v \in \mathbf{a}} \xi(y_v, h_v; \alpha_v, \beta_v)$ with certain confluent hypergeometric functions ξ. We note here

$$\xi(y, h; \alpha, \beta) = i^{\beta-\alpha}(2\pi)^{\alpha+\beta}\Gamma(\alpha)^{-1}\Gamma(\beta)^{-1}h^{\alpha+\beta-1}e^{-2\pi hy} \tag{4.14}$$
$$\cdot \int_0^\infty e^{-4\pi hyt}(t + 1)^{\alpha-1}t^{\beta-1}\,dt \qquad (y > 0,\ h > 0,\ (\alpha, \beta) \in \mathbf{C}^2,\ \mathrm{Re}(\beta) > 0).$$

For these, see [75, Lemma 1], [82b, (1.32), (4.34K)], and [85b, Lemma 5.1]. To apply (4.13) to the series of (4.3), we first observe

$$N(\mathfrak{y})^{-2s}y^{-su-(i\lambda-n)/2}E(z, s; n, \omega, \mathfrak{z}; \mathfrak{x}, \mathfrak{y}) = \sum_{0\neq d\in\mathfrak{y}/\mathfrak{g}^\times} \omega^*(d\mathfrak{y}^{-1})|d|^{-2su}$$

$$+ \sum_{0\neq c\in\mathfrak{x}\mathfrak{y}\mathfrak{z}/\mathfrak{g}^\times}\sum_{b\in\mathfrak{y}/\mathfrak{y}\mathfrak{z}} \omega_{\mathbf{a}}(b)\omega^*(b\mathfrak{y}^{-1})\sum_{a\in\mathfrak{y}\mathfrak{z}}(cz+b+a)^{-n}|cz+b+a|^{n-i\lambda-2su},$$

where $d\mathfrak{y}^{-1}$ and $b\mathfrak{y}^{-1}$ for $b\neq 0$ must be prime to $\mathfrak{z}$. Thus we have

$$(4.15)\quad y^{-su-(i\lambda-n)/2}E(z, s; n, \omega, \mathfrak{z}; \mathfrak{x}, \mathfrak{y})$$

$$= L_{\mathfrak{z}}(2s, \omega, \mathfrak{y}^{-1}) + D_F^{-1/2}N(\mathfrak{z})^{-1}N(\mathfrak{y})^{2s-1}\sum_{b,c}\omega_{\mathbf{a}}(b)\omega^*(b\mathfrak{y}^{-1})\overline{\omega}_{\mathbf{a}}(c)|c|^{u-2su}$$

$$\sum_{h\in(\mathfrak{d}\mathfrak{y}\mathfrak{z})^{-1}} \mathbf{e}_{\mathbf{a}}(chx+hb)\xi(y, ch; su+(n+i\lambda)/2, su+(i\lambda-n)/2),$$

where (b, c) runs over the same range as above. With n, ω, $\mathfrak{z}$, and $\mathfrak{x}$ fixed and with γ as in Lemma 4.2, put

$$(4.16)\qquad G_\gamma(z, s) = J_n(\gamma^{-1}, z)^{-1}C(\gamma^{-1}z, s; n, \omega, \Gamma_0).$$

Then Lemma 4.2 together with (4.5) shows that

$$\overline{\omega}_{\mathbf{a}}(d_\gamma)\overline{\omega}^*(d_\gamma\mathfrak{a}_\gamma^{-1})N(\mathfrak{a}_\gamma)^{-2s}G_\gamma(z, s)$$

$$= \sum_{\mathfrak{y}\in\mathscr{A}}\sum_{\mathfrak{t}\supset\mathfrak{f}^{-1}\mathfrak{z}} \mu(\mathfrak{t})\omega^*(\mathfrak{t})N(\mathfrak{t})^{-2s}E(z, s; n, \omega, \mathfrak{f}; (\mathfrak{t}\mathfrak{f})^{-1}\mathfrak{z}\mathfrak{a}_\gamma^2\mathfrak{x}, \mathfrak{y})$$

since $\mathfrak{y}$ in (4.12b) can be replaced by $\mathfrak{c}\mathfrak{y}$ with any fractional ideal $\mathfrak{c}$. Thus by (4.15) and Lemma 4.1 we obtain

$$(4.17)\quad \overline{\omega}_{\mathbf{a}}(d_\gamma)\overline{\omega}^*(d_\gamma\mathfrak{a}_\gamma^{-1})N(\mathfrak{a}_\gamma)^{-2s}y^{-su-(i\lambda-n)/2}G_\gamma(z, s)$$

$$= L_{\mathfrak{z}}(2s, \omega) + D_F^{-1/2}\gamma(\omega)N(\mathfrak{f})^{-1}\sum_{\mathfrak{t}\supset\mathfrak{f}^{-1}\mathfrak{z}}\mu(\mathfrak{t})\omega^*(\mathfrak{t})N(\mathfrak{t})^{-2s}\sum_{\mathfrak{y}\in\mathscr{A}}N(\mathfrak{y})^{2s-1}$$

$$\cdot\sum_{h,c}\overline{\omega}_{\mathbf{a}}(c)|c|^{u-2su}\overline{\omega}_{\mathbf{a}}(h)\overline{\omega}^*(h\mathfrak{f}\mathfrak{d}\mathfrak{y})\mathbf{e}_{\mathbf{a}}(chx)$$

$$\cdot\,\xi(y, ch; su+(n+i\lambda)/2, su+(i\lambda-n)/2),$$

where h runs over $(\mathfrak{d}\mathfrak{f}\mathfrak{y})^{-1}$ and c over $(F^\times\cap\mathfrak{t}^{-1}\mathfrak{a}_\gamma^2\mathfrak{x}\mathfrak{y}\mathfrak{z})/\mathfrak{g}^\times$.

We conclude this section by noting that

$$(4.18)\qquad \int_0^\infty e^{-2\pi hy}\xi(y, h; \alpha, \beta)y^{p-1}\,dy$$
$$= i^{\beta-\alpha}2^{-p}(2\pi)^{\alpha+\beta-p}h^{\alpha+\beta-p-1}\frac{\Gamma(p)\Gamma(p-\alpha-\beta+1)}{\Gamma(\alpha)\Gamma(p-\alpha+1)}$$

if $h > 0$, $\mathrm{Re}(\beta) > 0$, $\mathrm{Re}(p) > 0$, and $\mathrm{Re}(p-\alpha-\beta+1) > 0$. This follows from (4.14) combined with a well-known formula for the beta function.

5. Key propositions on theta integrals. For $w, w' \in \mathbf{C}^{\mathbf{a}}$ and $\xi \in M_2(F_{\mathbf{A}})$ we put

$$(5.1a)\qquad [\xi, w, w'] = \left((-1, w_v)\xi_v\begin{pmatrix} w'_v \\ 1 \end{pmatrix}\right)_{v\in\mathbf{a}}$$
$$= (c_v w_v w'_v + d_v w_v - a_v w'_v - b_v)_{v\in\mathbf{a}}\ (\in \mathbf{C}^{\mathbf{a}}),$$

$$(5.1b)\qquad [\xi, w] = [\xi, w, w],$$

where $\xi_v = \begin{pmatrix} a_v & b_v \\ c_v & d_v \end{pmatrix}$. An easy calculation shows that

$$(5.2)\qquad [\hat{\beta}\xi\gamma, w, w'] = \det(\beta\gamma)_{\mathbf{a}}^{1/2} j(\beta, w) j(\gamma, w')[\xi, \beta w, \gamma w']$$
$$(\beta, \gamma \in \tilde{G}_{\mathbf{A}+},\ \xi \in M_2(F_{\mathbf{A}})).$$

Let us now put

$$(5.3)\qquad V = \{\xi \in M_2(F) \mid \mathrm{tr}(\xi) = 0\}$$

and denote by $\mathscr{S}(V_{\mathbf{h}})$ and $\mathscr{S}(V_v)$ for $v \in \mathbf{h}$ the Schwartz-Bruhat spaces of $V_{\mathbf{h}}$ and V_v. We also denote by $\mathscr{L}(V)$ the set of all locally constant functions on V in the sense of [88, Notation]. By restricting each element of $\mathscr{S}(V_{\mathbf{h}})$ to the image of V in $V_{\mathbf{h}}$, we obtain an isomorphism of $\mathscr{S}(V_{\mathbf{h}})$ onto $\mathscr{L}(V)$, and so we shall identify these two spaces.

Given $\eta \in \mathscr{L}(V)$, we define a theta function Θ on $H^{\mathbf{a}} \times H^{\mathbf{a}}$ by

$$(5.4a)\qquad \Theta(z, w; \eta) = y^{u/2}\,\mathrm{Im}(w)^{-2m}\sum_{\xi\in V}\eta(\xi)[\xi, \overline{w}]^m \mathbf{e}_{\mathbf{a}}(2^{-1}R[\xi, z, w]),$$

where $z \in H^{\mathbf{a}}$, $w \in H^{\mathbf{a}}$, $y = \mathrm{Im}(z)$, and $R[\xi, z, w]$ is an element of $\mathbf{C}^{\mathbf{a}}$ whose v-component for each $v \in \mathbf{a}$ is defined by

$$(5.4b)\qquad R[\xi, z, w]_v = \det(\xi_v)z_v + (iy_v/2)\,\mathrm{Im}(w_v)^{-2}\,|[\xi, w]_v|^2.$$

We can easily verify that

(5.5)
$$J_{2m}(\alpha, w)^{-1}\Theta(z, \alpha w; \eta) = \Theta(z, w; \eta^{\alpha}) \quad \textit{for every } \alpha \in \tilde{G}_+, \textit{ where } \eta^{\alpha}(\xi) = \eta(\alpha\xi\alpha^{-1}).$$

The present Θ times $y^{-u/2}$ is a special case of [87, (11.14)], and therefore, for every $\eta \in \mathscr{L}(V)$ and every $\gamma \in G \cap P_{\mathbf{A}}C''$ we have

(5.6)
$$\overline{J_k(\gamma, z)}^{-1}\Theta(\gamma z, w; {}^{\gamma}\eta) = \Theta(z, w; \eta)$$

with some ${}^{\gamma}\eta \in \mathscr{L}(V)$, as noted in [87, Prop. 11.8] and [88, Lemma 6.1]. (In fact, if $\gamma \in G \cap P_{\mathbf{A}}C'$, we have $K(\gamma, z) = |j(\gamma, z)^u| \overline{J_k(\gamma, z)}$ for K of [87, Prop. 11.8]. The factor $y^{u/2}$ in (5.4a) has the effect of eliminating $|j(\gamma, z)^u|$. Thus ${}^{\gamma}\eta$ coincides with that of [87, Prop. 11.8] for such a γ. For a more general type of element, however, ${}^{\gamma}\eta$ defined by (5.6) may differ from that of [87, Prop. 11.8] by a constant factor. For example, for $\iota = \begin{pmatrix} 0 & -1 \\ 1 & 0 \end{pmatrix}$, we have $K(\iota, z) = (-1)^d |j(\iota, z)^u| \overline{J_k(\iota, z)}$, and hence that constant is $(-1)^d$.)

Lemma 5.1. *Let $\mathfrak{a}$ and $\mathfrak{b}$ be integral ideals and $\mathfrak{y}$ a fractional ideal in F; let ξ be a character of $(\mathfrak{g}/\mathfrak{a})^{\times}$ $(= \prod_{v|\mathfrak{a}} (\mathfrak{g}_v/\mathfrak{a}_v)^{\times})$ and ω a $\mathbf{C}$-valued function on $\mathfrak{b}\mathfrak{y}/\mathfrak{a}\mathfrak{b}\mathfrak{y}$ such that $\omega(ab) = \xi(a)\omega(b)$ for $a \in (\mathfrak{g}/\mathfrak{a})^{\times}$ and $b \in \mathfrak{b}\mathfrak{y}/\mathfrak{a}\mathfrak{b}\mathfrak{y}$. Further, let ζ be the element of $\mathscr{S}(V_{\mathbf{h}})$ such that $\zeta(x) \neq 0$ for $x \in V$ only if $x \in \mathfrak{o}[\mathfrak{b}\mathfrak{y}, \mathfrak{y}^{-1}]$, in which case $\zeta(x) = \omega(b_x \pmod{\mathfrak{a}\mathfrak{b}\mathfrak{y}})$. Define $\zeta' \in \mathscr{S}(V_{\mathbf{h}})$ by $\zeta'(x) = \zeta(px)$ with an element p of $F_{\mathbf{h}}^{\times}$ such that $p^{-1}\mathfrak{g} \subset \mathfrak{g}$. Then $\zeta(wxw^{-1}) = \xi((a_w^2/\det(w))_{\mathfrak{a}} \pmod{\mathfrak{a}})\zeta(x)$ for every $w \in \tilde{D}[(2^{-1}\mathfrak{a}\mathfrak{b} \cap 2\mathfrak{g})\mathfrak{y}, \mathfrak{y}^{-1}]$ and*

$$\overline{J_k(\gamma, z)}^{-1}\Theta(\gamma z, w; \zeta') = \xi(a_{\gamma} \pmod{\mathfrak{a}})^{-1}\Theta(z, w; \zeta')$$

for every $\gamma \in \Gamma[2\mathfrak{d}^{-1}, 2^{-1}p^{-2}(\mathfrak{a}\mathfrak{b} \cap 4\mathfrak{g})\mathfrak{d}]$.

Proof. The first assertion can easily be verified by writing the matrix entries of wxw^{-1} in terms of those of w and x. The second assertion follows directly from (5.6) and [87, Prop. 11.7]. In fact, the ideals $\mathfrak{x}$, $\mathfrak{y}$, and $\mathfrak{z}$ in that proposition are given in the present setting by $\mathfrak{g}$, $p^2(\mathfrak{g} + 4(\mathfrak{a}\mathfrak{b})^{-1})\mathfrak{d}^{-2}$, and $\mathfrak{a}$, respectively.

Lemma 5.2. *Put $r(z) = \langle \Theta(z, w; \eta), g(w) \rangle$ with $g \in \mathscr{S}_{2m}$. Then r belongs to $\mathscr{S}_k$ and has a Fourier expansion of the form $r(z) = \sum_{\alpha \in W} \overline{\eta(\alpha)} q_{g\alpha} \mathbf{e}_{\mathbf{a}}(-\det(\alpha)z/2)$ with a certain subset W of V and some $q_{g\alpha} \in \mathbf{C}$ independent of η.*

This is included in [82a, Th. 2.2, Prop. 2.3]. (See also [88, Prop. 6.3] and corrections in [90, p. 425].) Notice that the inner product is meaningful because of (5.5).

Lemma 5.3. *For $\eta \in \mathscr{L}(V)$ there exists an open subgroup U of $D[2\mathfrak{d}^{-1}, 2\mathfrak{d}]$ such that, if $\gamma \in G \cap \mathrm{diag}[p, 1/p]U$ with $p \in F_{\mathbf{h}}^{\times}$, then ${}^{\gamma}\eta(x) = |p|^{3/2}\eta(px)$.*

This is contained in [88, Lemma 6.2], which is a special case of [87, Prop. 11.5].

LEMMA 5.4. *Let L be a finite field of odd characteristic, and φ the character of $L^\times$ of order 2. Extend φ to L by putting $\varphi(0) = 0$. Then for a, b, and c in L such that $a^2 - bc \neq 0$, we have $\sum_{x \in L} \varphi(cx^2 + 2ax + b) = -\varphi(c)$.*

Proof. This is obvious if $c = 0$. Suppose $c \neq 0$; then the sum is $\varphi(c) \sum_{x \in L} \varphi((cx + a)^2 + bc - a^2)$. Thus it is sufficient to prove that $\sum_{x \in L} \varphi(x^2 + y) = -1$ if $y \neq 0$. For that purpose, put $\gamma = \sum_{t \in L} \varphi(t)h(t)$ with a nontrivial character $h: L \to \mathbf{T}$. It is well known that $\sum_{x \in L} h(ax^2) = \varphi(a)\gamma$ if $a \in L^\times$. Therefore with $y \neq 0$ we have $\gamma \sum_{x \in L} \varphi(x^2 + y) = \sum_{x \in L} \sum_{t \in L} \varphi(t)h(t(x^2 + y)) = \sum_{t \in L} \varphi(t)h(ty) \sum_{x \in L} h(tx^2) = \sum_{t \in L} \varphi(t)^2 h(ty)\gamma = \sum_{t \in L^\times} h(ty)\gamma = -\gamma$ as expected.

LEMMA 5.5. *Let $v \in \mathbf{h}$ and let η_v be an element of $\mathcal{S}(V_v)$ such that $\eta_v(x) \neq 0$ only if $x \in \mathfrak{o}[f^{-1}\mathfrak{g}, e\mathfrak{g}]_v$, in which case $\eta_v(x) = \varphi(fb_x)$, where e and f are fixed elements of $F_v^\times$ such that $e \in f\mathfrak{g}_v$, and φ is a continuous nontrivial character of $\mathfrak{g}_v^\times$ extended to $\mathfrak{g}_v$ so that $\varphi(a) = 0$ for $a \notin \mathfrak{g}_v^\times$. Let $\mathfrak{h}$ be the conductor of φ and π a prime element of F_v. Suppose that $\varphi^2 = 1$ on $\mathfrak{g}_v^\times$ if $e/f \in \mathfrak{g}_v^\times$ and $v \nmid 2$; if $v | 2$, we assume either of the following two sets of conditions: (i) $e/f \in \mathfrak{g}_v^\times$, $\mathfrak{h} = 4\pi\mathfrak{g}_v$, and $\varphi(1 + 4x) = \varphi(1 + 4x^2)$ for every $x \in \mathfrak{g}_v$; (ii) $e/f \notin \mathfrak{g}_v^\times$ and $\mathfrak{h} \subset 4\pi\mathfrak{g}_v$. Let Y be a complete set of representatives for $D_v[(2f)^{-1}\mathfrak{h}, e\mathfrak{g}] \backslash D_v[(2\pi f)^{-1}\mathfrak{h}, e\mathfrak{g}]$. Then the sum $\sum_{y \in Y} s_y \eta_v(yxy^{-1})$ is independent of Y, where $s_y = 1$ if $e/f \in \mathfrak{g}_v^\times$ and $v \nmid 2$, and $s_y = \varphi(a_y)^{-2}$ otherwise. Moreover, the sum is 0 if $\pi \nmid \det(x)$.*

Proof. We easily see that $\eta_v(wxw^{-1}) = \varphi(a_w)^2 \eta_v(x)$ for $w \in D_v[(2f)^{-1}\mathfrak{h}, e\mathfrak{g}]$, from which the first assertion follows. To prove the second one, we may assume that $x \in \mathfrak{o}[f^{-1}\mathfrak{g}, e\mathfrak{g}]_v$. Suppose that $e/f \in \mathfrak{g}_v^\times$ and $v \nmid 2$. Then we can take Y to be the set consisting of $\begin{pmatrix} 0 & -f^{-1} \\ f & 0 \end{pmatrix}$ and $\begin{pmatrix} 1 & t/f \\ 0 & 1 \end{pmatrix}$ for $t \in \mathfrak{g}/\pi\mathfrak{g}$. Then for $x = \begin{pmatrix} a & b/f \\ ce & -a \end{pmatrix}$ we have $\sum_{y \in Y} \eta_v(yxy^{-1}) = \varphi(-ce/f) + \sum_t \varphi(b - 2at - ct^2 e/f)$. This combined with Lemma 5.4 proves the desired fact in this case. If $e/f \notin \mathfrak{g}_v^\times$ and $v \nmid 2$, we need only $\begin{pmatrix} 1 & t/f \\ 0 & 1 \end{pmatrix}$ for $t \in \pi^{-1}\mathfrak{h}/\mathfrak{h}$, and obtain $\sum_{y \in Y} \varphi(a_y)^{-2} \eta_v(yxy^{-1}) = \sum_t \varphi(b - 2at - ct^2 e/f)$. Since $t^2 e/f \in \mathfrak{h}$, the last sum equals $\sum_t \varphi(b - 2at)$. There are four cases depending on whether or not b is a v-unit, and on whether or not $\mathfrak{h} = \pi\mathfrak{g}_v$. In all cases the sum is 0 if $a \in \mathfrak{g}_v^\times$. Since $\det(x) \equiv -a^2 \pmod{\pi}$, this proves the last assertion. If $v | 2$, we obtain $\sum_t \varphi(b - 2at - ct^2 e/f)$ with $t \in (2\pi)^{-1}\mathfrak{h}/2^{-1}\mathfrak{h}$ as the sum in question. Suppose $e/f \notin \mathfrak{g}_v^\times$. Then $t^2 e/f \in \mathfrak{h}$ and the desired conclusion follows for the same reason as in the case $v \nmid 2$. If $e/f \in \mathfrak{g}_v^\times$, we may assume $e = f$ and put $t = 2s$ with $s \in \mathfrak{g}_v/\pi\mathfrak{g}_v$. Then the sum becomes $\sum_s \varphi(b - 4as - 4cs^2)$. This is 0 if $b \notin \mathfrak{g}_v^\times$. Suppose $b \in \mathfrak{g}_v^\times$ and $\pi \nmid \det(x)$; put $d = b^{-1}$. Then the sum times $\varphi(d)$ is $\sum_s \varphi(1 - 4ads - 4cds^2)$. By our assumption on φ this equals $\sum_s \varphi(1 - 4(a^2d^2 + cd)s^2)$. Now $a^2d^2 + cd = -d^2 \cdot \det(x)$, and hence the sum equals $\sum_r \varphi(r)$ with $r \in (1 + \pi^{-1}\mathfrak{h})/(1 + \mathfrak{h})$, which is 0. This completes the proof.

We now take $f \in \mathscr{S}_k(\mathfrak{b}, \mathfrak{b}'; \psi)$ and χ as in Proposition 3.1(II). With a fixed $\tau \in F$, $\gg 0$, we define $\mathfrak{c}$, φ, $\mathfrak{q}$, $\mathfrak{r}$, and $\mathfrak{h}$ as in Section 3. Conditions (3.20a, b, c) and (3.21) are unnecessary for the moment. Take t_λ and x_λ as in (3.4). To make our notation simpler, we put

(5.7) $$\mathfrak{e} = 2^{-1}\mathfrak{c}\mathfrak{d}, \qquad \mathfrak{e}_\lambda = 2^{-1}t_\lambda\mathfrak{c}\mathfrak{d} \qquad (\lambda = 1, \ldots, \kappa).$$

Define an element η of $\mathscr{S}(V_\mathbf{h})$ as follows:

(5.8) *$\eta(x) \neq 0$ for $x \in V$ only when $x \in \mathfrak{o}[\mathfrak{e}^{-1}, \mathfrak{e}]$, in which case $\eta(x) = \sum_t \varphi_\mathbf{a}(t)\varphi^*(2t\mathfrak{r})\mathbf{e}_\mathbf{a}(-b_x t)$, where t runs over $(2\mathfrak{r})^{-1}/2^{-1}\mathfrak{c}$ under the condition that $2t\mathfrak{r} + \mathfrak{r}\mathfrak{c} = \mathfrak{g}$.*

Given $\xi \in \mathscr{S}(V_\mathbf{h})$, we define elements $\xi_\lambda \in \mathscr{S}(V_\mathbf{h})$ by

(5.9) $$\xi_\lambda(y) = \varphi(t_\lambda)^{-1}\xi(x_\lambda^{-1} y x_\lambda).$$

We easily see that

(5.10a) *$\eta_\lambda(x) \neq 0$ for $x \in V$ only when $x \in \mathfrak{o}[\mathfrak{e}_\lambda^{-1}, \mathfrak{e}_\lambda]$, in which case $\eta_\lambda(x) = \sum_t \varphi_\mathbf{a}(t)\varphi^*(2tt_\lambda^{-1}\mathfrak{r})\mathbf{e}_\mathbf{a}(-b_x t)$, where t runs over $t_\lambda(2\mathfrak{r})^{-1}/t_\lambda 2^{-1}\mathfrak{c}$ under the condition that $2tt_\lambda^{-1}\mathfrak{r} + \mathfrak{r}\mathfrak{c} = \mathfrak{g}$,*

(5.10b) $$\eta_\lambda(sx) = \varphi(s)\eta_\lambda(x) \qquad \textit{for every } s \in \prod_{v \in \mathbf{h}} \mathfrak{g}_v^\times.$$

Moreover, by Lemma 5.1 we have

(5.11) $$\eta_\lambda(wyw^{-1}) = \varphi_\mathfrak{h}(a_w^2/\det(w))\eta_\lambda(y) \qquad \textit{for every } w \in \tilde{D}[\mathfrak{r}(t_\lambda\mathfrak{d})^{-1}, \mathfrak{e}_\lambda],$$

(5.12) $$\overline{J_k(\gamma, z)}^{-1}\Theta(\gamma z, w; \eta_\lambda) = \varphi_\mathfrak{h}(a_\gamma)^{-1}\Theta(z, w; \eta_\lambda) \qquad \textit{for every } \gamma \in \Gamma[2\mathfrak{d}^{-1}, 2^{-1}\mathfrak{r}\mathfrak{c}\mathfrak{d}].$$

Taking $\mathfrak{q}\mathfrak{r}$ as $\mathfrak{y}$ in Proposition 1.4, we obtain an element h of $\mathscr{S}_k(\mathfrak{g}, \mathfrak{r}\mathfrak{b}\mathfrak{b}', \varphi)$ such that

(5.13) $$\lambda_h(\xi, \mathfrak{m}) = \lambda_f(\tau\xi, \mathfrak{q}^{-1}\mathfrak{r}^{-1}\mathfrak{m}) \qquad \textit{for every } (\xi, \mathfrak{m}).$$

Then we define a function $g_{\tau\lambda}(w)$ on $H^\mathbf{a}$ by

(5.14a) $$C \cdot g_{\tau\lambda}(w) = \int_\Phi h(z)\Theta(z, w; \eta_\lambda)y^k d_H z, \qquad \Phi = \Gamma_{\mathfrak{r}\mathfrak{c}} \backslash H^\mathbf{a},$$

(5.14b) $$C = 2^{1+d+\|m\|} i^{\|m\|} N(\mathfrak{r}\mathfrak{c}),$$

where $\Gamma_{\mathfrak{r}\mathfrak{c}} = \Gamma[2\mathfrak{d}^{-1}, 2^{-1}\mathfrak{r}\mathfrak{c}\mathfrak{d}]$. The integral is meaningful in view of (5.12).

PROPOSITION 5.6. *Let* $\mathbf{g}$ *be the normalized eigenform in* $\mathscr{S}_{2m}((2^{-1}\mathfrak{c}, \psi^2))$ *belonging to* χ. *Then* $\lambda_f(\tau, \mathfrak{q}^{-1})\mathbf{g} = (g_{\tau 1}, \ldots, g_{\tau\kappa})$ *with* $g_{\tau\lambda}$ *of (5.14a).*

Proof. Let us first consider $g_{\tau\lambda}$ without assuming f to be an eigenform. By [87, Prop. 7.1] we know that $g_{\tau\lambda} \in \mathscr{M}_{2m}$, and moreover $g_{\tau\lambda} \in \mathscr{S}_{2m}$ if $k \neq 3u/2$, or if $k = 3u/2$ and f is orthogonal to the theta functions described there. Moreover, (5.5) combined with (5.11) shows that $g_{\tau\lambda}\|_{2m}\gamma = \varphi_{\mathfrak{h}}(a_\gamma^2/\det(\gamma))g_{\tau\lambda}$ for every $\gamma \in \tilde{\Gamma}[\mathfrak{r}(t_\lambda\mathfrak{d})^{-1}, \mathfrak{e}_\lambda]$. Put $g_{\tau\lambda}(w) = \sum_{0 \ll \xi \in F} c_\lambda(\xi)\mathbf{e}_{\mathbf{a}}(\xi w)$. Now η_λ is exactly $(\varphi_{\mathfrak{r}\mathfrak{c}}/\varphi)(t_\lambda)D_F^{1/2}N(2^{-1}t_\lambda\mathfrak{c})$ times the function of [87, (7.7), (7.8)] with $\mathfrak{r}\mathfrak{c}$, $t_\lambda(2\mathfrak{r})^{-1}$, and φ as $\mathfrak{c}$, $\mathfrak{x}$, and ψ there. Thus $g_{\tau\lambda}$ is a constant times the function g studied in [87, pp. 798–805] with f there replaced by h here. Now $\lambda_h(1, \mathfrak{m}) = \lambda_f(\tau, (\mathfrak{q}\mathfrak{r})^{-1}\mathfrak{m})$, and this is nonzero only if $\mathfrak{m} \subset \mathfrak{r}$ because of (1.11a). Therefore the reasoning of [87, p. 804, line 10 through p. 805, line 5] shows that

$$(5.15)\qquad \sum_{0 \ll \xi \in F/\mathfrak{g}_+^\times} c_\lambda(\xi)\xi^{-m-i\mu}M(\xi t_\lambda^{-1}\mathfrak{g}) = \sum_{\mathfrak{m}} \lambda_f(\tau, \mathfrak{q}^{-1}\mathfrak{m})M(\mathfrak{m}) \sum_{\mathfrak{n}} \varphi^*(\mathfrak{n})N(\mathfrak{n})^{-1}M(\mathfrak{n}),$$

where μ is as in (1.12) and $\mathfrak{n}$ runs over all the integral ideals of F prime to $\mathfrak{r}\mathfrak{c}$ and equivalent to $(t_\lambda\mathfrak{m})^{-1}$ modulo $\mathbf{a}$. (The constant C was chosen so that the equality holds with no extra factor.) From this equality we see that $c_\lambda(\xi) \neq 0$ only if $\xi \in t_\lambda\mathfrak{g}$, and hence $g_{\tau\lambda}$ satisfies (3.6) with ψ^2 and $2\mathfrak{b}\mathfrak{b}'$ as Ψ and $\mathfrak{z}$ there. For $(z, x) \in H^{\mathbf{a}} \times \tilde{G}_{\mathbf{A}}$ put

$$\Theta(z, x) = \mathrm{Im}(z)^{u/2} \sum_{\xi \in V} \eta((x^{-1}\xi x)_{\mathbf{h}})\varphi(\det(x))[(x^{-1}\xi x)_{\mathbf{a}}, -\mathbf{i}]^m\mathbf{e}_{\mathbf{a}}(2^{-1}R[\xi, z, x(\mathbf{i})]).$$

Then $\Theta(z, sx) = \psi(s)^2\Theta(z, x)$ for $s \in F_{\mathbf{h}}^\times$, $\Theta(z, \alpha xw) = \varphi_{\mathfrak{h}}(d_w)^2\Theta(z, x)$ for $\alpha \in \tilde{G}$ and $w \in \tilde{G}_{\mathbf{h}} \cap \tilde{D}[\mathfrak{r}\mathfrak{d}^{-1}, \mathfrak{e}]$, and $\Theta(z, t_\lambda^{-1}x_\lambda y) = \det(y)^{i\mu}j(y, \mathbf{i})^{-2m}\Theta(z, y(\mathbf{i}); \eta_\lambda)$ for $y \in \tilde{G}_{\mathbf{a}+}$. Therefore if we put $C\mathbf{g}_\tau(x) = \int_\Phi h(z)\Theta(z, x)y^k d_H z$, then $\mathbf{g}_\tau(sx) = \psi(s)^2\mathbf{g}_\tau$ for $s \in F_{\mathbf{h}}^\times$ and $\mathbf{g}_\tau = (g_{\tau\lambda})_{\lambda=1}^\kappa$ in the sense of (3.5). Thus $\mathbf{g}_\tau \in \mathscr{S}_{2m}((2\mathfrak{b}\mathfrak{b}', \psi^2))$. Moreover, if we define $c(\mathfrak{m}, \mathbf{g}_\tau)$ by (3.8b) with $n = 2m$ and take the sum of (5.15) for $\lambda = 1, \ldots, \kappa$, then we obtain (3.18a), and consequently (I) and (III) of Proposition 3.1. Suppose $f|T_v = \omega_v f$ for every $v \in \mathbf{h}$. Then (3.18a) together with Proposition 2.2 shows that

$$(5.16)\qquad \sum_{\mathfrak{m}} c(\mathfrak{m}, \mathbf{g}_\tau)M(\mathfrak{m}) = \lambda_f(\tau, \mathfrak{q}^{-1}) \prod_{v \in \mathbf{h}} [1 - \omega_v M_v + \psi'(\pi_v\mathfrak{g})^2 N_v^{-1} M_v^2]^{-1}.$$

By (1.11c) we have $\lambda_f(\tau, \mathfrak{g}) \neq 0$ for some $\tau \gg 0$. For such a τ the series of Proposition 2.2, as well as (5.16), does not vanish, and hence $\mathbf{g}_\tau \neq 0$. Then (5.16) shows that $\mathbf{g}_\tau$ is a common eigenfunction of $\mathfrak{T}(\mathfrak{n})$ for all $\mathfrak{n}$, and $\mathbf{g}_\tau|\mathfrak{T}(\pi_v\mathfrak{g}) = N_v\omega_v\mathbf{g}_\tau$. This establishes Proposition 3.1, and at the same time proves the present proposition.

For an integral ideal $\mathfrak{a}$ we define elements $\zeta^{\mathfrak{a}}$ and $\zeta_{\mathfrak{a}}$ of $\mathscr{S}(V_{\mathbf{h}})$ by

(5.17) $\zeta^{\mathfrak{a}}(x) \neq 0$ *for* $x \in V$ *only when* $x \in \mathfrak{o}[\mathfrak{a}\mathfrak{e}^{-1}, \mathfrak{e}]$, *in which case* $\zeta^{\mathfrak{a}}(x) = \overline{\varphi}_{\mathbf{a}}(b_x)\overline{\varphi}^*(b_x\mathfrak{a}^{-1}\mathfrak{e})$,

(5.18) $\zeta_{\mathfrak{a}}(x) \neq 0$ *for* $x \in V$ *only when* $x \in \mathfrak{o}[\mathfrak{a}\mathfrak{e}^{-1}, \mathfrak{e}]$ *and* $b_x\mathfrak{a}^{-1}\mathfrak{e}$ *is prime to* $\mathfrak{r}\mathfrak{c}$, *in which case* $\zeta_{\mathfrak{a}}(x) = \overline{\varphi}_{\mathbf{a}}(b_x)\overline{\varphi}^*(b_x\mathfrak{a}^{-1}\mathfrak{e})$.

If we define ζ_λ^a and $\zeta_{a\lambda}$ by (5.9), they satisfy (5.17) and (5.18) with $\mathfrak{e}_\lambda$ in place of $\mathfrak{e}$, and

$$(5.19)\qquad \zeta_\lambda^a(sx) = \varphi(s)\zeta_\lambda^a(x) \text{ for every } s \in \prod_{v \in \mathbf{h}} \mathfrak{g}_v^\times .$$

Lemma 5.7. *Suppose that (3.20b, c) and (3.21) are satisfied. Let $\mathfrak{a}$ be an integral ideal such that $\mathfrak{a}\mathfrak{h} \supset \mathfrak{r}\mathfrak{c}$ and $(\mathfrak{a}\mathfrak{h})_v = (\mathfrak{r}\mathfrak{c})_v$ for every $v|\mathfrak{r}$. Put*

$$l(z) = \sum_{\lambda=1}^{\kappa} \langle \Theta(z, w; \zeta_\lambda^a), g_\lambda(w) \rangle .$$

Then $l = M_a h$ with a constant M_a, which is 0 if $4\mathfrak{r}\mathfrak{b} \supset \mathfrak{a}\mathfrak{h} \cap 4\mathfrak{g} \neq \mathfrak{r}\mathfrak{c}$ and (3.22a) is assumed.

Proof. Fix $\mathfrak{a}$ and write simply ζ for ζ^a. For $p \in F_{\mathbf{h}}^\times$ define an element ζ_λ^p of $\mathscr{S}(V_{\mathbf{h}})$ by $\zeta_\lambda^p(y) = \varphi(p)^{-1}|p|\zeta_\lambda(py)$. Now we can put $\mathfrak{a}\mathfrak{h} \cap 4\mathfrak{g} = 4\mathfrak{r}\mathfrak{t}$ with a divisor $\mathfrak{t}$ of $\mathfrak{b}\mathfrak{b}'$. Then $2\mathfrak{r}\mathfrak{t}\mathfrak{e}_\lambda^{-1} \supset \mathfrak{r}(t_\lambda\mathfrak{d})^{-1}$. By Lemma 5.1 we have

$$(5.20)\qquad \zeta_\lambda^p(wyw^{-1}) = \varphi_{\mathfrak{h}}(a_w^2/\det(w))\zeta_\lambda^p(y) \qquad \text{for every } w \in \tilde{D}[2\mathfrak{r}\mathfrak{t}\mathfrak{e}_\lambda^{-1}, \mathfrak{e}_\lambda],$$

$$(5.21)\qquad \overline{J_k(\gamma, z)}^{-1}\Theta(\gamma z, w; \zeta_\lambda^p) = \varphi_{\mathfrak{h}}(a_\gamma)^{-1}\Theta(z, w; \zeta_\lambda^p)$$

for every $\gamma \in \Gamma[2\mathfrak{d}^{-1}, 2p^{-2}\mathfrak{r}\mathfrak{t}\mathfrak{d}]$, if $p^{-1}\mathfrak{g} \subset \mathfrak{g}$. By [88, Lemma 6.4] $l \in \mathscr{S}_k(\mathfrak{g}, \mathfrak{f}; \varphi)$ with some $\mathfrak{f}$, and for $\mathfrak{m} = p\mathfrak{g}$ we have

$$(5.22)\qquad \sum_{\xi \in F} \lambda_l(\xi, \mathfrak{m})\mathbf{e}_{\mathbf{a}}(\xi z/2) = \sum_{\lambda=1}^{\kappa} \langle \Theta(z, w; \zeta_\lambda^p), g_\lambda(w) \rangle .$$

(Here are some corrections to [88] in addition to those made in [90, p. 425]: ψ on the second and third lines of [88, Lemma 6.4] should be ψ_1. Note also that τ there is 1 in the present setting.) By (5.21) we see that the function of (5.22) belongs to $\mathscr{S}_k(\mathfrak{g}, p^{-2}\mathfrak{r}\mathfrak{t}; \varphi)$ if $p^{-1}\mathfrak{g} \subset \mathfrak{g}$. In particular, $l \in \mathscr{S}_k(\mathfrak{g}, \mathfrak{r}\mathfrak{t}; \varphi)$. Notice that $\zeta = \prod_{v \in \mathbf{h}} \zeta_v$ with $\zeta_v \in \mathscr{S}(V_v)$, and ζ_v for $v \nmid \mathfrak{h}^{-1}\mathfrak{r}^2\mathfrak{c}$ is a constant times a function of type [88, (6.14a, b)] with $2^{-1}\mathfrak{c}$ and φ as $\mathfrak{b}$ and ψ_1 there. Hence [88, Th. 6.7] (or rather its proof) shows that $l|T_v = N_v^{-1}\chi(v)l$ for $v \nmid \mathfrak{h}^{-1}\mathfrak{r}^2\mathfrak{c}$. (Strictly speaking, if $v|2$ and $v|\mathfrak{h}$, the condition that $\mathfrak{h}_v$ divides $2^{-1}\mathfrak{c}_v$ in [88, p. 281, line 1] is not satisfied, but the proof in [88, pp. 284–5] is applicable even to this case.) Let us now show that

$$(5.23)\qquad \lambda_l(\xi, \mathfrak{m}) \neq 0 \Rightarrow \xi \in \mathfrak{r}\mathfrak{m}^{-2}.$$

In view of (1.11b) we may assume that $\mathfrak{m} = p\mathfrak{g} \supset \mathfrak{g}$ and $p_v = 1$ for every $v|\mathfrak{r}$. If $\mathfrak{r} = \mathfrak{g}$, (5.23) follows from (1.11a). Assume $\mathfrak{r} \neq \mathfrak{g}$ and fix a prime v dividing $\mathfrak{r}$. By (3.20b) φ_v is nontrivial, and our condition on $\mathfrak{a}$ implies that $(2\mathfrak{r}\mathfrak{t}\mathfrak{e}_\lambda^{-1})_v = (\mathfrak{h}\mathfrak{a}(2\mathfrak{e}_\lambda)^{-1})_v = (\mathfrak{r}(t_\lambda\mathfrak{d})^{-1})_v$. Let Y_λ be a complete set of representatives for

$$D_v[\mathfrak{r}(t_\lambda\mathfrak{d})^{-1}, \mathfrak{e}_\lambda]\backslash D_v[(t_\lambda\mathfrak{d})^{-1}, \mathfrak{e}_\lambda].$$

For each $y \in Y_\lambda$ take $\alpha_{\lambda y} \in G$ so that $y\alpha_{\lambda y}^{-1} \in D[\mathfrak{r}(t_\lambda \mathfrak{d})^{-1}, \mathfrak{e}_\lambda]$ and

(5.24) $$a(y\alpha_{\lambda y}^{-1}) - 1 \in (\mathfrak{r}\mathfrak{c})_{v'} \quad \textit{for every } v'|\mathfrak{r}\mathfrak{c}.$$

Then $\alpha_{\lambda y} \in \Gamma[(t_\lambda \mathfrak{d})^{-1}, \mathfrak{e}_\lambda]$ and $g_\lambda \|_{2m} \alpha_{\lambda y} = s_y^{-1} g_\lambda$, where $s_y = 1$ or $\varphi_v(a_y)^{-2}$ according as $v \nmid \mathfrak{c}$ or $v|\mathfrak{c}$. Let $|Y|$ denote the number of elements in Y_λ, which is the same for all λ. Then the function of (5.22) is equal to

(5.25) $$|Y|^{-1} \sum_{\lambda=1}^{\kappa} \sum_{y \in Y_\lambda} \langle \Theta(z, w; \zeta_\lambda^{\mathfrak{p}})\|_{2m}\alpha_{\lambda y}, g_\lambda \|_{2m}\alpha_{\lambda y}\rangle$$
$$= \sum_{\lambda=1}^{\kappa} \left\langle |Y|^{-1} \sum_{y \in Y} s_y \Theta(z, w; \zeta_\lambda^{\mathfrak{p}})\|_{2m}\alpha_{\lambda y}, g_\lambda \right\rangle$$
$$= \sum_{\lambda=1}^{\kappa} \langle \Theta(z, w; \varepsilon_\lambda), g_\lambda \rangle,$$

where $\varepsilon_\lambda(x) = |Y|^{-1} \sum_{y \in Y_\lambda} s_y \zeta_\lambda^{\mathfrak{p}}(\alpha_{\lambda y} x \alpha_{\lambda y}^{-1})$ in view of (5.5). Our definition of $\zeta_\lambda^{\mathfrak{p}}$ guarantees the decomposition $\zeta_\lambda^{\mathfrak{p}}(x) = \zeta_{\lambda v}(x_v) \prod_{v \neq u \in \mathbf{h}} (\zeta_\lambda^{\mathfrak{p}})_u(x_u)$ with $(\zeta_\lambda^{\mathfrak{p}})_u \in \mathscr{S}(V_u)$ and $\zeta_{\lambda v}$ of the type described in Lemma 5.5. (Notice that $\mathfrak{a}_v = (e/f)\mathfrak{g}_v$ with e and f of the lemma. The conditions in the lemma can be verified by recalling that $\mathfrak{h} \supset 4\mathfrak{r} \cap \mathfrak{c}$, $\varphi = \psi\varepsilon_\mathfrak{r}$, and employing (3.20c).) From (5.20) and (5.24) we see that

$$\sum_{y \in Y_\lambda} s_y \zeta_\lambda^{\mathfrak{p}}(\alpha_{\lambda y} x \alpha_{\lambda y}^{-1}) = \sum_{y \in Y_\lambda} s_y \zeta_\lambda^{\mathfrak{p}}(yxy^{-1}) = \prod_{v \neq u \in \mathbf{h}} (\zeta_\lambda^{\mathfrak{p}})_u(x_u) \sum_{y \in Y_\lambda} s_y \zeta_{\lambda v}(yx_v y^{-1}).$$

By Lemma 5.5 this is nonvanishing only if $\det(x_v) \in \mathfrak{r}_v$. Thus $\varepsilon_\lambda(x) \neq 0$ only if $\det(x) \in \mathfrak{r}_v$. By Lemma 5.2 we see that $\lambda_l(\xi, \mathfrak{m}) \neq 0$ only if $\xi \in \mathfrak{r}_v$. This is so for every $v|\mathfrak{r}$. Now $\lambda_l(\xi, \mathfrak{m}) \neq 0$ only if $\xi \in \mathfrak{m}^{-2}$. Since $\mathfrak{m}$ is prime to $\mathfrak{r}$, this proves (5.23). In particular, (1.16) holds with l and $\mathfrak{q}\mathfrak{r}$ as h and $\mathfrak{y}$. Therefore we can apply Proposition 1.5 to l to find an element f' of $\mathscr{S}_k(\mathfrak{b}, \mathfrak{b}'; \psi)$ such that $f' \mapsto l$ in the sense of (1.13) with $\mathfrak{y} = \mathfrak{q}\mathfrak{r}$. By Lemma 2.7(i), $f'|T_v = N_v^{-1}\chi(v)f'$ for $v \nmid \mathfrak{h}^{-1}\mathfrak{r}^2\mathfrak{c}$. Then (3.21) implies that $f' = Mf$ with some constant M, and hence $l = Mh$. Suppose $4\mathfrak{r}\mathfrak{b} \supset 4\mathfrak{r}\mathfrak{t} \neq \mathfrak{r}\mathfrak{c}$. Then $\mathfrak{g} \supset \mathfrak{b}^{-1}\mathfrak{t} \supsetneqq \mathfrak{b}'$ and f' can be taken from $\mathscr{S}_k(\mathfrak{b}, \mathfrak{b}^{-1}\mathfrak{t}; \psi)$, and hence must be 0 under (3.22a), which completes the proof.

PROPOSITION 5.8. *Let f, h, and $\mathbf{g}$ be as above; let η_λ and ζ_a be as in (5.8) and (5.18); put $\mathbf{g} = (g_1, \ldots, g_\kappa)$. Then, under (3.20b, c) and (3.21) we have*

(5.26a) $$Ah(z) = \sum_{\lambda=1}^{\kappa} \langle \Theta(z, w; \eta_\lambda), g_\lambda(w) \rangle$$

with

(5.26b) $$A = \overline{\lambda_f(\tau, \mathfrak{q}^{-1})} \frac{2^{1+d+\|m\|} i^{-\|m\|} \tau^{-k} N(\mathfrak{q}\mathfrak{r}^2\mathfrak{c}) \langle \mathbf{g}, \mathbf{g} \rangle}{\mathrm{vol}(\Gamma[2\mathfrak{d}^{-1}, 2^{-1}\mathfrak{r}\mathfrak{c}\mathfrak{d}] \backslash H^{\mathbf{a}}) \langle f, f \rangle}.$$

Moreover, if in addition, (3.22a, b) are assumed, then

(5.27a) $$KAh(z) = \sum_{\lambda=1}^{\kappa} \langle \Theta(z, w; \zeta_{\mathfrak{g}\lambda}), g_\lambda(w) \rangle$$

with

(5.27b) $$K = \varphi_{\mathbf{a}}(-1)\gamma(\varphi)\mu(\mathfrak{h}^{-1}\mathfrak{rc})\varphi^*(\mathfrak{h}^{-1}\mathfrak{rc})N(\mathfrak{rc})^{-1}.$$

Proof. For $\xi \in \mathscr{L}(V)$ put $I(\xi) = \sum_{\lambda=1}^{\kappa} \langle \Theta(z, w; \xi_\lambda), g_\lambda(w) \rangle$ and $I(\mathfrak{a}) = I(\zeta^{\mathfrak{a}})$. We observe that if $x \in V \cap \mathfrak{o}[\mathfrak{e}^{-1}, \mathfrak{e}]$, then

$$\zeta_{\mathfrak{g}}(x) = \sum_{\mathfrak{a}} \mu(\mathfrak{a})\overline{\varphi}^*(\mathfrak{a})\overline{\varphi}_{\mathbf{a}}(b_x)\overline{\varphi}^*(b_x\mathfrak{a}^{-1}\mathfrak{e}),$$

where $\mathfrak{a}$ runs over all the divisors of $b_x\mathfrak{e} + \mathfrak{h}^{-1}\mathfrak{rc}$. Therefore we have $\zeta_{\mathfrak{g}} = \sum_{\mathfrak{a} \supset \mathfrak{h}^{-1}\mathfrak{rc}} \mu(\mathfrak{a})\overline{\varphi}^*(\mathfrak{a})\zeta^{\mathfrak{a}}$, and hence $I(\zeta_{\mathfrak{g}}) = \sum_{\mathfrak{a}} \mu(\mathfrak{a})\varphi^*(\mathfrak{a})I(\mathfrak{a})$. Here $\mathfrak{a}$ may be assumed squarefree and prime to $\mathfrak{h}$, as $\mu(\mathfrak{a})\overline{\varphi}^*(\mathfrak{a}) = 0$ otherwise. Under (3.22a, b) Lemma 5.7 shows that $I(\mathfrak{a}) = 0$ unless $\mathfrak{ah} = \mathfrak{rc}$, and hence

(5.28) $$I(\zeta_{\mathfrak{g}}) = \mu(\mathfrak{h}^{-1}\mathfrak{rc})\varphi^*(\mathfrak{h}^{-1}\mathfrak{rc})I(\mathfrak{h}^{-1}\mathfrak{rc}).$$

Similarly Lemma 4.1 shows that

$$\eta = \varphi_{\mathbf{a}}(-1)\gamma(\varphi) \sum_{\mathfrak{a} \supset \mathfrak{h}^{-1}\mathfrak{rc}} \mu(\mathfrak{a}^{-1}\mathfrak{h}^{-1}\mathfrak{rc})\varphi^*(\mathfrak{a}^{-1}\mathfrak{h}^{-1}\mathfrak{rc})N(\mathfrak{a})\zeta^{\mathfrak{a}}.$$

Put $\mathfrak{t} = \mathfrak{a}^{-1}\mathfrak{h}^{-1}\mathfrak{rc}$. Then $\mathfrak{t}$ may be assumed squarefree and prime to $\mathfrak{h}$. Thus we obtain

$$I(\eta) = \varphi_{\mathbf{a}}(-1)\overline{\gamma(\varphi)} \sum_{\mathfrak{t}} \mu(\mathfrak{t})\overline{\varphi}^*(\mathfrak{t})N((\mathfrak{th})^{-1}\mathfrak{rc})I((\mathfrak{th})^{-1}\mathfrak{rc}).$$

From this and Lemma 5.7 we see that $I(\eta) = Ah$ with a constant A. Combining this result with Proposition 5.6, we obtain

$$A \cdot \mathrm{vol}(\Gamma \backslash H^{\mathbf{a}})\langle h, h \rangle = \overline{C}\langle \mathbf{g}_\tau, \mathbf{g} \rangle = \overline{C}\overline{\lambda_f(\tau, \mathfrak{q}^{-1})}\langle \mathbf{g}, \mathbf{g} \rangle,$$

where $\Gamma = \Gamma[2\mathfrak{d}^{-1}, 2^{-1}\mathfrak{rcd}]$. This together with the last assertion of Proposition 1.4 proves (5.26b). If (3.22a, b) are assumed, we find, for the same reason as above, that

$$I(\eta) = \varphi_{\mathbf{a}}(-1)\overline{\gamma(\varphi)}N(\mathfrak{h}^{-1}\mathfrak{rc})I(\mathfrak{h}^{-1}\mathfrak{rc}).$$

This together with (5.28) and the fact that $|\gamma(\varphi)|^2 = N(\mathfrak{h})$ shows that $I(\zeta_{\mathfrak{g}}) = KI(\eta)$ with K as in (5.27b).

6. $\lambda_f(\tau, \mathfrak{q}^{-1})D(2s, \chi)$ as a sum of inner products. The purpose of this section is to obtain formulas for $\lambda_f(\tau, \mathfrak{q}^{-1})D(2s, \chi)$ and $\sum_{\mathfrak{m}} \lambda(\tau, \mathfrak{q}^{-1}\mathfrak{b}\mathfrak{m}; f^*)N(\mathfrak{m})^{-2s}$ first as integrals, and eventually as sums of certain inner products. We begin with an observation about the inverter of (3.16). For $\mathbf{f} = (f_\lambda)_{\lambda=1}^{\kappa} \in \mathscr{S}_{2m}((\mathfrak{z}, \Psi))$ as in Section 3, put $\mathbf{f}|J_{\mathfrak{z}} = (f'_\lambda)_{\lambda=1}^{\kappa} \in \mathscr{S}_{2m}((\mathfrak{z}, \Psi^{-1}))$. Then $f'_\nu = \det(q_\lambda)^{-i\mu} f_\lambda \|_{2m} q_\lambda^{-1}$ with a permutation $\lambda \mapsto \nu$ and $q_\nu \in \tilde{G}_{\mathbf{a}}$ given as follows: for each λ we can find a unique ν and a totally positive element r_λ of F such that $t_\lambda t_\nu \mathfrak{z}\mathfrak{d}^2 = r_\lambda \mathfrak{g}$; then $q_\lambda = \begin{pmatrix} 0 & -1 \\ r_\lambda & 0 \end{pmatrix}$. This can easily be verified by observing that $(q_\lambda)_{\mathbf{h}} x_\lambda = t_\lambda x_\nu \pi \cdot \mathrm{diag}[a, 1]$ with $a \in \prod_{v \in \mathbf{h}} \mathfrak{g}_v^\times$ (cf. [78, p. 655]).

We apply this to $\mathbf{g} = (g_\lambda)_{\lambda=1}^{\kappa}$ of Section 5. Putting $\mathbf{g}^* = \mathbf{g}|J_{2\mathfrak{b}\mathfrak{b}'} = (g'_\lambda)_{\lambda=1}^{\kappa}$, from (5.26a) and (5.5) we obtain

$$\text{(6.1)} \qquad Ah(z) = \sum_{\lambda=1}^{\kappa} \langle r_\lambda^{-i\mu} \Theta(z, w; \eta_\lambda) \|_{2m} q_\lambda^{-1}, g'_\nu(w) \rangle$$

$$= \sum_{\nu=1}^{\kappa} \langle \Theta(z, w; \rho_\nu), g'_\nu(w) \rangle,$$

where $\rho_\nu(y) = r_\lambda^{-i\mu} \eta_\lambda(q_\lambda^{-1} y q_\lambda)$. Taking an element s of $F_{\mathbf{h}}^\times$ so that $s\mathfrak{g} = 2\mathfrak{b}\mathfrak{b}'$ and π as in (3.16) with this s, put $\rho(y) = \varphi(\delta^2 s)\eta(\pi^{-1} y \pi)$. Then we easily see that $\rho_\nu(y) = \varphi(t_\nu)\rho(x_\nu^{-1} y x_\nu)$.

We are going to apply the inverter ω of Lemma 2.3 to (6.1). With ${}^\gamma\eta$ defined by (5.6), we note that (1.4c) implies ${}^\alpha({}^\beta({}^\gamma\eta)) = {}^{\alpha\beta\gamma}\eta$ for α, β, and γ there. Take $\omega \in G \cap B\varepsilon^{-1}$ with a small open subgroup B of C'. Put $\alpha = \iota\omega^{-1}$ with $\iota = \begin{pmatrix} 0 & -1 \\ 1 & 0 \end{pmatrix}$. Then $\alpha \in \mathrm{diag}[-\delta, -\delta^{-1}]B$ and ${}^\iota\rho_\nu = {}^\alpha({}^\omega\rho_\nu)$ since $\omega \in C''$. Put $\sigma_\nu = \overline{\varphi}(-\delta) \cdot {}^\omega\rho_\nu$. Now Lemma 5.3 shows that ${}^\alpha\sigma_\nu(y) = |\delta|^{3/2} \sigma_\nu(-\delta y)$ if B is sufficiently small. Thus

$$\text{(6.2)} \qquad \sigma_\nu(y) = \overline{\varphi}(-\delta)|\delta|^{-3/2}({}^\iota\rho_\nu)(-\delta^{-1} y).$$

Since $\omega \in C''$, we have $J_k(\omega^{-1}, z) = J_k(\omega, \omega^{-1} z)^{-1}$ by (1.4c), and hence by (5.6)

$$\overline{J_k(\omega^{-1}, z)^{-1} \Theta(\omega^{-1} z, w; \rho_\nu)} = \varphi(-\delta) \Theta(z, w; \sigma_\nu).$$

Applying this to (6.1), by (2.5b) we find that

$$\text{(6.3)} \qquad Ah^*(z) = \sum_{\lambda=1}^{\kappa} \langle \Theta(z, w; \sigma_\lambda), g'_\lambda \rangle.$$

By [87, Lemma 11.6] for every $\xi \in \mathscr{S}(V_{\mathbf{h}})$ we have

$$ {}^\iota\xi(x) = 2^d \int_{V_{\mathbf{h}}} \xi(y) \mathbf{e}_{\mathbf{h}}(-tr(\hat{x} y)/2)\, dy,$$

where $\mathbf{e}_{\mathbf{h}}(a) = \mathbf{e}_{\mathbf{A}}(a_{\mathbf{h}})$ for $a \in F_{\mathbf{A}}$ (see [87, p. 771]) and $x \mapsto \hat{x}$ denotes the main involution of $M_2(F)$. (As remarked in the paragraph after formula (5.6), $'\xi$ differs from that of [87, Lemma 11.6] by the factor $(-1)^d$, which cancels i^{-p} in the formula given in the lemma.) We easily see that

$$'(\rho_\lambda)(x) = \varphi(t_\lambda)('\rho)(x_\lambda^{-1} x x_\lambda). \tag{6.4}$$

Since $\rho(y) = \varphi(\delta^2 s)\eta(\pi^{-1} y \pi)$ and η is defined by (5.8), we can put $\rho\begin{pmatrix} n & p \\ q & -n \end{pmatrix} =$ $\mu_1(n)\mu_2(p)\mu_3(q)$ with $\mu_i \in \mathscr{S}(F_{\mathbf{h}})$. Then, for $x = \begin{pmatrix} a & b \\ c & -a \end{pmatrix}$ we have

$$'\rho(x) = 2^d \int_{F_{\mathbf{h}}} \mathbf{e}_{\mathbf{h}}(an)\mu_1(n)\,dn \int_{F_{\mathbf{h}}} \mathbf{e}_{\mathbf{h}}(cp/2)\mu_2(p)\,dp \int_{F_{\mathbf{h}}} \mathbf{e}_{\mathbf{h}}(bq/2)\mu_3(q)\,dq.$$

We see that μ_1 is the characteristic function of $\prod_{v \in \mathbf{h}} \mathfrak{g}_v$, $\mu_2(p) = \mu_1(\delta p)$, and $\mu_3(q) \neq 0$ only if $q \in \prod_{v \in \mathbf{h}} \mathfrak{d}_v$, in which case $\mu_3(q) = \varphi_{\mathbf{h}}(2r\delta^2 s)\sum_t \overline{\varphi}_{\mathfrak{h}}(2tr)\mathbf{e}_{\mathbf{h}}(-qt/(\delta^2 s))$, where $r \in F_{\mathbf{h}}^\times$, $r\mathfrak{g} = \mathfrak{r}$, and t runs over $(2\mathfrak{r})^{-1}/2\mathfrak{b}\mathfrak{b}'$ under the condition that $2t\mathfrak{r}$ is prime to $\mathfrak{r}\mathfrak{c}$. Thus we find that

$$'\rho(x) = 2^d |\delta|^{3/2} \mu_1(\delta a)\lambda_2(b)\mu_1(c/2),$$

where $\lambda_2(b) \neq 0$ only if $b \in 2(\mathfrak{r}\mathfrak{c}\mathfrak{d}^2)^{-1}$ and $2^{-1} b\mathfrak{r}\mathfrak{c}\mathfrak{d}^2$ is prime to $\mathfrak{r}\mathfrak{c}$, in which case $\lambda_2(b) = \varphi_{\mathbf{h}}(r\delta^2 s)\overline{\varphi}_{\mathfrak{h}}(br\delta^2 s)$. Combining this result with (6.2) and (6.4), we find that

(6.5) *$\sigma_\lambda(x) \neq 0$ for $x \in V$ only if $a_x \in \mathfrak{g}$, $b_x \in 2(t_\lambda \mathfrak{r}\mathfrak{c}\mathfrak{d})^{-1}$, $c_x \in 2t_\lambda \mathfrak{d}$, and $2^{-1} b_x \cdot t_\lambda \mathfrak{r}\mathfrak{c}\mathfrak{d}$ is prime to $\mathfrak{r}\mathfrak{c}$, in which case $\sigma_\lambda(x) = 2^d \varphi_{\mathbf{a}}(b_x)\varphi^*(2^{-1} b_x t_\lambda \mathfrak{r}\mathfrak{c}\mathfrak{d})$.*

By (5.6) and [87, Prop. 11.7] we have

$$\overline{J_k(\gamma, z)}^{-1}\Theta(\gamma z, w; \sigma_\lambda) = \varphi_{\mathfrak{h}}(a_\gamma)\Theta(z, w; \sigma_\lambda) \tag{6.6}$$

for every $\gamma \in \Gamma[(2\mathfrak{d})^{-1}\mathfrak{r}\mathfrak{c}, 2\mathfrak{d}]$.

We now assume condition (3.20a). Then $\varphi_{\mathbf{a}}(x) = \operatorname{sgn}(x)^m |x|^{i\mu}$. Noting that $h^* \in \mathscr{S}_k(4^{-1}\mathfrak{r}\mathfrak{c}, \mathfrak{g}; \overline{\varphi})$ and the standard theta function

$$\theta(z) = \sum_{b \in \mathfrak{g}} \mathbf{e}_{\mathbf{a}}(b^2 z/2) \tag{6.7}$$

belongs to $\mathscr{M}_{u/2}(\mathfrak{g}, \mathfrak{g}; \varepsilon_1)$, where ε_1 denotes the trivial Hecke character (see [87, Lemma 4.3]), we consider an integral

$$\int_\Phi h^*(z)\theta(z)\overline{C(z, \bar{s} + (1/2); m, \varphi, \Gamma)}\, y^k\, d_H z \qquad (\Phi = \Gamma\backslash H^{\mathbf{a}}), \tag{6.8}$$

where $\Gamma = \Gamma[(2\mathfrak{d})^{-1}\mathfrak{r}\mathfrak{c}, 2\mathfrak{d}]$ and C is the function of (4.11). (The symbols ω, $\mathfrak{x}$, $\mathfrak{z}$, Γ_0, and D_0 there are now φ, $(2\mathfrak{d})^{-1}\mathfrak{r}\mathfrak{c}$, $\mathfrak{r}\mathfrak{c}$, Γ, and $D[(2\mathfrak{d})^{-1}\mathfrak{r}\mathfrak{c}, 2\mathfrak{d}]$; we take $\Gamma^* = \Gamma_0 = \Gamma$ in (4.6–11).)

PROPOSITION 6.1. *The integral of (6.8) equals*

$$D_F^{-1/2}2^{1-d}\tau^{u/2+i\mu}N(\mathfrak{q}^{-1}\mathfrak{c})(2\pi)^{\|-su-m/2\|}\prod_{v\in\mathbf{a}}\Gamma(s+q_v)$$

$$\cdot L_{\mathfrak{c}}(2s+1,\overline{\varphi})\sum_{\mathfrak{m}}\lambda(\tau,\mathfrak{q}^{-1}\mathfrak{b}\mathfrak{m};f^*)N(\mathfrak{m})^{-2s},$$

where $q = (m - i\mu)/2$. Similarly, if we take h, $\overline{\varphi}$, and $\Gamma[2\mathfrak{d}^{-1}, 2^{-1}\mathfrak{r}\mathfrak{c}\mathfrak{d}]$ in place of h^*, φ, and Γ, then the integral gives

$$\lambda_f(\tau,\mathfrak{q}^{-1})2^{1+d}N(\mathfrak{r})^{-2s}D_F^{-1/2}(2\pi)^{\|-su-m/2\|}D(2s,\chi)\prod_{v\in\mathbf{a}}\Gamma(s+\overline{q}_v).$$

Proof. Let $G \cap P_{\mathbf{A}}D_0 = \bigsqcup_{\beta\in B}P\beta\Gamma$ and $R_\beta = (P\cap\beta\Gamma\beta^{-1})\backslash\beta\Gamma$ as in (4.6) and (4.7). Then (6.8) equals $L_{\mathfrak{r}\mathfrak{c}}(2s+1,\overline{\varphi})\sum_{\beta\in B}X_\beta$ with

$$X_\beta = \int_\Phi h^*(z)\overline{\theta(z)E_\beta(z,\overline{s}+(1/2);m,\varphi,\Gamma)}y^k d_H z. \tag{6.9}$$

We may assume that $\beta\in\mathrm{diag}[p,p^{-1}]D_0$ with $p\in F_{\mathbf{h}}^\times$. For each $\beta\in B$ put

$$l_\beta(z) = J_k(\beta,\beta^{-1}z)h^*(\beta^{-1}z),\qquad \theta_\beta(z) = J_{u/2}(\beta,\beta^{-1}z)\theta(\beta^{-1}z).$$

Then for $\alpha\in\beta\Gamma$ we can easily verify that

$$(l_\beta\overline{\theta}_\beta)(\alpha z) = \overline{\varphi}_{\mathfrak{h}}(a(\beta^{-1}\alpha))\overline{J_m(\alpha,z)}^{-1}|j(\alpha,z)|^{2k}(h^*\overline{\theta})(z) \tag{6.10}$$

and also that $(l_\beta\overline{\theta}_\beta)(z)y^q$ is $(P\cap\beta\Gamma\beta^{-1})$-invariant. Let $\Psi = (P\cap\beta\Gamma\beta^{-1})\backslash H^{\mathbf{a}}$ with a fixed β. Since Ψ is represented by $\bigsqcup_{\alpha\in R_\beta}\alpha\Phi$, from (6.9, 10) we obtain

$$\varphi_{\mathbf{a}}(d_\beta)\varphi^*(d_\beta\mathfrak{a}_\beta^{-1})N(\mathfrak{a}_\beta)^{-2s-1}X_\beta = \int_\Psi(l_\beta\overline{\theta}_\beta)(z)y^{su+u+q}d_H z. \tag{6.11}$$

Observe that $\beta\Gamma\beta^{-1} = \Gamma[(2\mathfrak{d}\mathfrak{a}_\beta^2)^{-1}\mathfrak{r}\mathfrak{c}, 2\mathfrak{d}\mathfrak{a}_\beta^2]$, and hence Ψ can be given by $(\mathbf{R}^{\mathbf{a}}/(2\mathfrak{d}\mathfrak{a}_\beta^2)^{-1}\mathfrak{r}\mathfrak{c})\times Y$ with

$$Y = \{y\in\mathbf{R}^{\mathbf{a}}\,|\,y\gg 0\}/\{a^2\,|\,a\in\mathfrak{g}^\times\}. \tag{6.12}$$

Using this and (1.11c) for l_β and θ_β, we find that

$$X_\beta = N(\mathfrak{a}_\beta)^{2s}D_F^{-1/2}N(2^{-1}\mathfrak{r}\mathfrak{c})\int_Y\sum_{\xi\in F}\lambda(\xi,\mathfrak{a}_\beta^{-1};h^*)\overline{\lambda_\theta(\xi,\mathfrak{a}_\beta^{-1})}\mathbf{e}_{\mathbf{a}}(\mathbf{i}\xi y)y^{su-u+q}\,dy.$$

The last integral over Y equals

$$(2\pi)^{\|-q-su\|}\prod_{v\in\mathbf{a}}\Gamma(s+q_v)\sum_{\xi}\lambda(\xi,\mathfrak{a}_\beta^{-1};h^*)\overline{\lambda_\theta(\xi,\mathfrak{a}_\beta^{-1})}\xi^{-q-su},$$

where ξ runs over $F^\times/\{a^2 \mid a\in\mathfrak{g}^\times\}$. By [87, (4.12)] $\lambda_\theta(\xi,\mathfrak{m})$ for $\xi\in F^\times$ is nonzero only if $\xi=b^2$ with $b\in\mathfrak{m}^{-1}$, in which case $\lambda_\theta(\xi,\mathfrak{m})=2$. This fact together with (1.11b) and (3.20a) shows that the above sum over ξ equals $2\sum_b\lambda(1,b\mathfrak{a}_\beta^{-1};h^*)\cdot|b|^{-2su}$, where b runs over $(F^\times\cap\mathfrak{a}_\beta)/\mathfrak{g}^\times$. Combining all these, we finally find that the integral of (6.8) equals

$$D_F^{-1/2}N(\mathfrak{r}\mathfrak{c})2^{1-d}(2\pi)^{\|-m/2-su\|}\prod_{v\in\mathbf{a}}\Gamma(s+q_v)L_{\mathfrak{c}}(2s+1,\overline{\varphi})\sum_{\mathfrak{m}}\lambda(1,\mathfrak{m};h^*)N(\mathfrak{m})^{-2s}.$$

By Lemma 2.6 and (1.11b) we have $\lambda(1,\mathfrak{m};h^*)=\tau^kN(\mathfrak{q}\mathfrak{r})^{-1}\lambda(1/\tau,\mathfrak{q}\mathfrak{r}\mathfrak{m};f^*)=\tau^{u/2+i\mu}N(\mathfrak{q}\mathfrak{r})^{-1}\lambda(\tau,\mathfrak{q}^{-1}\mathfrak{b}\mathfrak{m};f^*)$. Therefore we obtain the first assertion of our proposition. The case of h instead of h^* can be proved in a similar fashion. In this case $\sum_{\mathfrak{m}}\lambda_h(1,\mathfrak{m})N(\mathfrak{m})^{-2s}$ appears as the essential factor. Since $\lambda_h(1,\mathfrak{m})=\lambda_f(\tau,(\mathfrak{q}\mathfrak{r})^{-1}\mathfrak{m})$ and this is nonzero only if $\mathfrak{m}\subset\mathfrak{r}$, we eventually obtain, employing (3.18b), the desired result.

Let $I(h^*)$ denote the integral of (6.8). Then (6.3) shows that

$$A\cdot I(h^*)=\sum_{\lambda=1}^{\kappa}\langle M_\lambda(w,\bar{s}),g'_\lambda(w)\rangle$$

with M_λ given by

$$M_\lambda(w,s)=\int_\Phi\theta(z)\Theta(z,w;\sigma_\lambda)C(z,s+1/2)y^kd_Hz,$$

where C is as in (6.8); we suppress the symbols m, φ, and Γ for simplicity. Take B as in the above proof. Then

$$M_\lambda(w,s)=L_{\mathfrak{c}}(2s+1,\varphi)\sum_{\beta\in B}I_\beta, \tag{6.13}$$

$$I_\beta=\int_\Phi\theta(z)\Theta(z,w;\sigma_\lambda)E_\beta(z,s+(1/2))y^kd_Hz. \tag{6.14}$$

For each $\beta\in B$ put

$$\Theta_\beta(z,w;\sigma_\lambda)=\varphi_{\mathbf{a}}(d_\beta)\varphi^*(d_\beta\mathfrak{a}_\beta^{-1})\overline{J_k(\beta,\beta^{-1}z)}\Theta(\beta^{-1}z,w;\sigma_\lambda).$$

We can take β from $\mathrm{diag}[p,p^{-1}]U$ with $p\in F_{\mathbf{h}}^\times$ such that $(p_v)_{v|\mathfrak{b}}=1$ and with any small open subgroup U of $G_{\mathbf{A}}$. If U is chosen sufficiently small, we have $\varphi_{\mathfrak{b}}(d_\beta)=1$

and

$$\Theta_\beta(z, w; \sigma_\lambda) = e_\beta |p|^{3/2} \Theta(z, w; \sigma_{p\lambda}) \tag{6.15}$$

with $e_\beta = \overline{\varphi}^*(\mathfrak{a}_\beta)$ and $\sigma_{p\lambda}(x) = \sigma_\lambda(px)$ by virtue of Lemma 5.3. From (6.6) we can derive

$$\theta_\beta(\alpha z)\Theta_\beta(\alpha z; w, \sigma_\lambda) = \varphi_{\mathbf{a}}(d_\alpha)\varphi^*(d_\alpha \mathfrak{a}_\beta^{-1})|j(\alpha, z)|^{2k} J_m(\alpha, z)^{-1}\theta(z)\Theta(z, w; \sigma_\lambda) \tag{6.16}$$

for every $\alpha \in \beta\Gamma$. The same reasoning as in (6.11) shows that

$$I_\beta = e_\beta N(\mathfrak{a}_\beta)^{2s+5/2} \int_\Psi \theta_\beta(z)\Theta(z, w; \sigma_{p\lambda}) y^{su+u+(m+i\mu)/2} d_H z.$$

Employing expressions (5.4a, b) and making the same type of computation as we did on the right-hand side of (6.11), we obtain

$$I_\beta = e_\beta D_F^{-1/2} 2^{1-d+\|t\|} \pi^{-\|t\|} N(\mathfrak{r}\mathfrak{c}) N(\mathfrak{a}_\beta)^{2s+1} \prod_{v \in \mathbf{a}} \Gamma(t_v) S_{\beta\lambda}(w, s), \tag{6.17}$$

$$S_{\beta\lambda}(w, s) = \sum_{\alpha, b} \sigma_{p\lambda}(\alpha)\mu_\beta(b)[\alpha, w]^{-m}|[\alpha, w]/\mathrm{Im}(w)|^{2m-2t}, \tag{6.18}$$

where μ_β is the characteristic function of $\mathfrak{a}_\beta$, $t = su + (u + m + i\mu)/2$, and (α, b) runs over $X/\mathfrak{g}^\times$ with

$$X = \{(\alpha, b) \in V \times F \mid \alpha \neq 0, -\det(\alpha) = b^2\}. \tag{6.19}$$

Combining this result with Proposition 6.1, we finally obtain

$$\begin{aligned} &A \cdot \Delta(s) \sum_{\mathfrak{m}} \lambda(\tau, \mathfrak{q}^{-1}\mathfrak{b}\mathfrak{m}; f^*) N(\mathfrak{m})^{-2s} \prod_{v \in \mathbf{a}} \Gamma(s + q_v)\Gamma(s + q_v + (1/2))^{-1} \\ &\quad = \sum_{\lambda=1}^{\kappa} \left\langle \sum_{\beta \in B} \overline{\varphi}^*(\mathfrak{a}_\beta) N(\mathfrak{a}_\beta)^{2\bar{s}+1} S_{\beta\lambda}(w, \bar{s}), g'_\lambda(w) \right\rangle, \end{aligned} \tag{6.20}$$

with

$$\Delta(s) = 2^{-(d/2)-2sd-\|m\|} \pi^{d/2} N(\mathfrak{q}\mathfrak{r})^{-1} \tau^{u/2+i\mu}. \tag{6.21}$$

We can make the same type of calculation by taking (5.27a) and the second integral of Proposition 6.1 in place of (6.3) and (6.8). In this case we obtain

$$\begin{aligned} &KA \cdot \Delta'(s)\lambda_f(\tau, \mathfrak{q}^{-1}) D(2s, \chi) \prod_{v \in \mathbf{a}} \Gamma(s + \bar{q}_v)\Gamma(s + \bar{q}_v + (1/2))^{-1} \\ &\quad = L_{\mathfrak{c}}(2s + 1, \varphi) \sum_{\lambda=1}^{\kappa} \left\langle \sum_{\beta \in B} \varphi^*(\mathfrak{a}_\beta) N(\mathfrak{a}_\beta)^{2\bar{s}+1} S'_{\beta\lambda}(w, \bar{s}), g_\lambda(w) \right\rangle, \end{aligned} \tag{6.22}$$

with

$$S'_{\beta\lambda}(w, s) = \sum_{\alpha, b} \zeta_{p\lambda}(\alpha)\mu_\beta(b)[\alpha, w]^{-m}|[\alpha, w]/\mathrm{Im}(w)|^{2m-2t'}, \tag{6.23}$$

$$\Delta'(s) = 2^{-(d/2)-2sd-\|m\|}\pi^{d/2}N(\mathfrak{r})^{-2s}, \tag{6.24}$$

where $\zeta_{p\lambda}(x) = \zeta_{\mathfrak{g}\lambda}(px)$, $t' = su + (u + m - i\mu)/2$, and $\sum_{\alpha,b}$ is the same as in (6.18).

7. The analysis of $S_{\beta\lambda}$ and $S'_{\beta\lambda}$. We first observe that $M_2(F) = V + F1_2$ and the map $(\alpha, s) \mapsto \alpha + s1_2$ gives a bijection of the set $\{(\alpha, s) \in V \times F \mid \alpha \neq 0, -\det(\alpha) = s^2\}$ onto

$$W = \{\xi \in M_2(F) \mid \mathrm{rank}(\xi) = 1\} \tag{7.1}$$

and that $[\alpha + s1_2, w] = [\alpha, w]$ for $\alpha \in V$.

Let the notation be the same as in (6.23). Fixing β, λ, and $p \in F_{\mathbf{h}}^\times$ such that $\mathfrak{a}_\beta = p^{-1}\mathfrak{g}$, define $Y_{p\lambda} \in \mathcal{S}(M_2(F)_{\mathbf{h}})$, viewed as a function on $M_2(F)$, by

$$Y_{p\lambda}(\alpha + s1_2) = \zeta_{p\lambda}(\alpha)\mu_\beta(s) \qquad \textit{for } \alpha \in V \textit{ and } s \in F, \tag{7.2}$$

where $\zeta_{p\lambda}(x) = \zeta_{\mathfrak{g}\lambda}(px)$ with $\zeta_{\mathfrak{g}\lambda}$ of (5.18). Then we see that

$$S'_{\beta\lambda}(w, s) = \sum_{\xi \in W/\mathfrak{g}^\times} Y_{p\lambda}(\xi)[\xi, w]^{-m}|[\xi, w]/\mathrm{Im}(w)|^{2m-2t'}, \tag{7.3}$$

(7.4) *$Y_{p\lambda}(x) \neq 0$ only when $px \in \prod_{v\in\mathbf{h}} o[\mathfrak{e}_\lambda^{-1}, \mathfrak{e}_\lambda]_v$, $p(a_x - d_x) \in \prod_{v\in\mathbf{h}} 2\mathfrak{g}_v$, and $pb_x\mathfrak{e}_\lambda + \mathfrak{r}\mathfrak{c} = \mathfrak{g}$, in which case $Y_{p\lambda}(x) = \varphi(e_\lambda)^{-1}\varphi_{\mathfrak{h}}(e_\lambda pb_x)$,*

where e_λ is an element of $F_{\mathbf{h}}^\times$ such that $e_\lambda\mathfrak{g}$ is $\mathfrak{e}_\lambda$ of (5.7).

For simplicity let us put

$$D_\lambda = D[\mathfrak{r}\mathfrak{c}\mathfrak{e}_\lambda^{-1}, 2\mathfrak{e}_\lambda], \qquad D'_\lambda = D[\mathfrak{r}\mathfrak{c}\mathfrak{e}_\lambda^{-1}, \mathfrak{e}_\lambda],$$

$$D^*_\lambda = \{x \in D_\lambda \mid (a_x - d_x)_v \in 2\mathfrak{g}_v \textit{ for every } v|2\},$$

$$\Gamma_\lambda = G \cap D_\lambda, \qquad \Gamma'_\lambda = G \cap D'_\lambda, \qquad \Gamma^*_\lambda = G \cap D^*_\lambda.$$

We can easily verify that

$$Y_{p\lambda}(\hat{\beta}\xi\gamma) = \varphi_{\mathfrak{h}}(d_\beta d_\gamma)Y_{p\lambda}(\xi) \qquad \textit{for every } \beta, \gamma \in \Gamma^*_\lambda. \tag{7.5}$$

We consider a special element v of W given by

$$v = \begin{pmatrix} 0 & 1 \\ 0 & 0 \end{pmatrix}. \tag{7.6}$$

From (5.2) we obtain

$$-[\hat{\beta}v\gamma, w] = j(\beta, w)j(\gamma, w) \qquad (\beta, \gamma \in G). \tag{7.7}$$

LEMMA 7.1. *The set* $P_{\mathbf{A}}D[\mathfrak{x}, \mathfrak{y}]$ *consists of the elements* $\begin{pmatrix} a & b \\ c & d \end{pmatrix}$ *of* $G_{\mathbf{A}}$ *such that* $d_v \neq 0$ *and* $d_v^{-1}c_v \in \mathfrak{y}_v$ *for every* $v|\mathfrak{x}\mathfrak{y}$. *In particular,* $P_{\mathbf{A}}D[\mathfrak{x}, \mathfrak{y}] = G_{\mathbf{A}}$ *if* $\mathfrak{x}\mathfrak{y} = \mathfrak{g}$.

This follows immediately from our defiition of $D[\mathfrak{x}, \mathfrak{y}]$, since $P_v D_v[\mathfrak{x}, \mathfrak{y}] = G_v$ if $\mathfrak{x}_v\mathfrak{y}_v = \mathfrak{g}_v$.

LEMMA 7.2. (i) $P \times P = \{(\beta, \gamma) \in G \times G | \hat{\beta}v\gamma \in Fv\}$.

(ii) *Let S be a complete set of representatives for* $P\backslash G$. *Then* $(g, \beta, \gamma) \mapsto g\hat{\beta}v\gamma$ *gives a bijection of* $F^\times \times S \times S$ *onto W.*

These assertions can be proved in an elementary way.

LEMMA 7.3. $\{\xi \in W | Y_{p\lambda}(\xi) \neq 0\} \subset \{\hat{\beta}v\gamma | \beta, \gamma \in G \cap P_{\mathbf{A}}D'_\lambda\}$.

Proof. Let $\xi = \begin{pmatrix} a & b \\ c & d \end{pmatrix} \in W$ with $Y_{p\lambda}(\xi) \neq 0$. From (7.4) we see that both a and d belong to $p^{-1}\mathfrak{g}$ and $b \neq 0$. Now we have $\xi = \hat{\beta}v\gamma$ with $\beta = \begin{pmatrix} 1 & 0 \\ -d/b & 1 \end{pmatrix}$ and $\gamma = \begin{pmatrix} 1/b & 0 \\ a & b \end{pmatrix}$. Again from (7.4) we see that both a/b and d/b belong to $(\mathfrak{e}_\lambda)_v$ for every $v|\mathfrak{r}\mathfrak{c}$. Hence $\beta, \gamma \in P_{\mathbf{A}}D'_\lambda$ by Lemma 7.1. This proves our lemma.

Fixing λ and p, let us drop (if not consistently) the subscripts λ and p from the symbols D_λ, D^*_λ, $Y_{p\lambda}$, etc. Put

$$U = \left\{a \in \prod_{v \in \mathbf{h}} \mathfrak{g}_v^\times \,|\, a_v^2 \equiv 1 \ (\mathrm{mod}\ 2\mathfrak{g}_v)\ \textit{for every}\ v|2\right\}.$$

Take a complete set of representatives R for $F_{\mathbf{A}}^\times/F^\times F_{\mathbf{a}}^\times U$ so that $r_v = 1$ for every $r \in R$ and every $v|\mathfrak{r}\mathfrak{c}$. Put $\delta_r = \mathrm{diag}[r^{-1}, r]$. By strong approximation, for each $r \in R$, we can find an element ε_r of $G \cap \delta_r D^*$ such that $a(\varepsilon_r) - 1 \in (\mathfrak{r}\mathfrak{c})_v$ for every $v|\mathfrak{r}\mathfrak{c}$. Let Q be a complete set of representatives for $\mathfrak{e}_\lambda/2\mathfrak{e}_\lambda$. Put $\tau_q = \begin{pmatrix} 1 & 0 \\ q & 1 \end{pmatrix}$ for $q \in Q$.

LEMMA 7.4. *The notation being as above, we have* $G \cap P_{\mathbf{A}}D = G \cap P_{\mathbf{A}}D^* = \bigsqcup_{r \in R} P\varepsilon_r\Gamma^*$ *and* $G \cap P_{\mathbf{A}}D' = \bigsqcup_{r \in R, q \in Q} P\varepsilon_r\Gamma^*\tau_q$.

Proof. By the same reasoning as in [83, Lemma 1.3] we can show that $P_{\mathbf{A}} = \bigsqcup_{r \in R} P\delta_r(P_{\mathbf{A}} \cap D^*)$, and hence $P_{\mathbf{A}}D^* = \bigsqcup_{r \in R} P\delta_r D^*$. Then $G \cap P_{\mathbf{A}}D^* = \bigsqcup_{r \in R} P\varepsilon_r\Gamma^*$. Clearly $P_{\mathbf{A}}D = P_{\mathbf{A}}D^*$. By Lemma 7.1 we easily see that $P_{\mathbf{A}}D' = \bigsqcup_{q \in Q} P_{\mathbf{A}}D\tau_q$, and hence $G \cap P_{\mathbf{A}}D' = \bigsqcup_{q \in Q}(G \cap P_{\mathbf{A}}D^*)\tau_q = \bigsqcup_{q,r} P\varepsilon_r\Gamma^*\tau_q$ as expected.

Let T_r be a complete set of representatives for $(P \cap \varepsilon_r \Gamma^* \varepsilon_r^{-1}) \backslash \varepsilon_r \Gamma^*$. Then $\bigsqcup_{r \in R} \bigsqcup_{q \in Q} T_r \tau_q$ is a complete set of representatives for $P \backslash (G \cap P_{\mathbf{A}} D')$. Now, by Lemmas 7.2 and 7.3 we have

$$S'_{\beta\lambda}(w, s) = \sum_{g \in F^\times/\mathfrak{g}^\times} \sum_{\beta, \gamma} Y(g\hat{\beta}\nu\gamma)[g\hat{\beta}\nu\gamma, w]^{-m} |[g\hat{\beta}\nu\gamma, w]/\mathrm{Im}(w)|^{2m-2t'},$$

where (β, γ) runs over $[P \backslash (G \cap P_{\mathbf{A}} D')]^2$. Writing $\beta = \alpha\tau_q$ and $\gamma = \alpha'\tau_{q'}$ with $\alpha \in T_r$ and $\alpha' \in T_{r'}$, we obtain, employing (7.7),

$$(-1)^{\|m\|} S'_{\beta\lambda}(w, s)$$

$$= \sum_{g \in F^\times/\mathfrak{g}^\times} \sum_{r, r' \in R} \sum_{\alpha \in T_r, \alpha' \in T_{r'}} \sum_{q, q' \in Q} Y(g\hat{t}_q \hat{\alpha}\nu\alpha'\tau_{q'})$$

$$\cdot g^{-m}|g|^{2m-2t'} j(\alpha\tau_q, w)^{-m} j(\alpha'\tau_{q'}, w)^{-m} |j(\alpha\tau_q, w) j(\alpha'\tau_{q'}, w)/\mathrm{Im}(w)|^{2m-2t'}.$$

Now we have

(7.8) *$Y(g\hat{t}_q\hat{\alpha}\nu\alpha'\tau_{q'}) \neq 0$ only if $q = q'$, in which case $Y(g\hat{t}_q\hat{\alpha}\nu\alpha'\tau_{q'}) = Y(g\hat{\alpha}\nu\alpha')$.*

To show this, put $\hat{\alpha}\nu\alpha' = \begin{pmatrix} a & b \\ c & d \end{pmatrix}$. Since $r_v = r'_v = 1$ for every $v | \mathfrak{r}\mathfrak{c}$, we see that $b \in \mathfrak{g}_v^\times$, $a \in (2\mathfrak{e}_\lambda)_v$, $d \in (2\mathfrak{e}_\lambda)_v$, and $c \in (2\mathfrak{e}_\lambda)_v^2$ for every such v. Put $u = \hat{t}_q\hat{\alpha}\nu\alpha'\tau_{q'} = \begin{pmatrix} x & y \\ z & w \end{pmatrix}$ and suppose $Y(gu) \neq 0$. Then (7.4) shows that $(pgb\mathfrak{e}_\lambda)_v \in \mathfrak{g}_v^\times$ and $[pg(a + bq' - d + bq)]_v = [pg(x - w)]_v \in 2\mathfrak{g}_v$ for every $v | 2$, and hence $q - q' \in 2e_\lambda \mathfrak{g}_v$ for every such v. Thus we have $q = q'$, which is the first part of (7.8). Then the second part can easily be verified since $y = b$.

Let $\alpha = \varepsilon_r\gamma$ and $\alpha' = \varepsilon_{r'}\gamma'$ with $\gamma, \gamma' \in \Gamma^*$. Then by (7.5) $Y(g\hat{\alpha}\nu\alpha') = \varphi_{\mathfrak{h}}(d_\gamma d_{\gamma'}) Y(g\hat{\varepsilon}_r\nu\varepsilon_{r'})$. By (7.4) we have $Y(g\hat{\varepsilon}_r\nu\varepsilon_{r'}) \neq 0$ only if $pgrr'\mathfrak{e}_\lambda + \mathfrak{r}\mathfrak{c} = \mathfrak{g}$, in which case $Y(g\hat{\varepsilon}_r\nu\varepsilon_{r'}) = \varphi(e_\lambda)^{-1}\varphi_{\mathfrak{h}}(pgrr'e_\lambda)$ with e_λ as in (7.4). Since $\varphi_{\mathfrak{h}}(p) = \varphi_{\mathfrak{h}}(rr') = 1$, we thus obtain

$$(-1)^{\|m\|} S'_{\beta\lambda}(w, s)$$

$$= \sum_{q \in Q} \sum_{r, r' \in R} \sum_{g} \overline{\varphi}_{\mathbf{a}}(g)\overline{\varphi}^*(g\mathfrak{e}_\lambda) g^{-m}|g|^{2m-2t'}$$

$$\cdot \sum_{\alpha, \alpha'} \varphi_{\mathfrak{h}}(d_\alpha d_{\alpha'}) j(\alpha\tau_q, w)^{-m} j(\alpha'\tau_q, w)^{-m} |j(\alpha\tau_q, w) j(\alpha'\tau_q, w)/\mathrm{Im}(w)|^{2m-2t'},$$

where g runs over $(F^\times \cap (prr'\mathfrak{e}_\lambda)^{-1})/\mathfrak{g}^\times$ under the condition that $gprr'\mathfrak{e}_\lambda$ is prime to

$\mathfrak{r}\mathfrak{c}$. Let us now put

$$E_r^*(w, s; \xi) = E_{\varepsilon_r}(w, s; m, \xi, \Gamma_\lambda^*), \qquad E_r(w, s; \xi) = E_{\varepsilon_r}(w, s; m, \xi, \Gamma_\lambda),$$

$$E_\lambda(w, s; \xi) = E(w, s; m, \xi, \Gamma_\lambda), \qquad C_\lambda(w, s; \xi) = C(w, s; m, \xi, \Gamma_\lambda),$$

with the symbols of (4.6, 4.7, 4.11), where ξ is any Hecke character whose conductor divides $2\mathfrak{r}\mathfrak{c}$. Then

$$(7.9a) \qquad (-1)^{\|m\|} S'_{\beta\lambda}(w, s) = \sum_{q \in Q} T'_{\beta\lambda}(w, s) \|_{2m} \tau_q$$

with

$$(7.9b) \qquad T'_{\beta\lambda}(w, s) = \varphi^*(pg) N(p\mathfrak{e}_\lambda)^{2s+1}$$

$$\cdot \sum_{r, r'} L_{\mathfrak{r}\mathfrak{c}}(2s + 1, \overline{\varphi}, prr'\mathfrak{e}_\lambda) E_r^*(w, s + 1/2; \overline{\varphi}) E_{r'}^*(w, s + 1/2; \overline{\varphi}).$$

We can take a subset R' of R so that $G \cap P_{\mathbf{A}} D_\lambda = \bigsqcup_{r \in R'} P\varepsilon_r \Gamma_\lambda$. Then $E_\lambda(w, s; \xi) = \sum_{r \in R'} E_r(w, s; \xi)$. Since $P_{\mathbf{A}} D_\lambda = P_{\mathbf{A}} D_\lambda^*$, we have $E_\lambda(w, s; \xi) = E(w, s; m, \xi, \Gamma_\lambda^*)$. Also $P\varepsilon_r \Gamma_\lambda$ is a union of $P\varepsilon_t \Gamma_\lambda^*$ for finitely many t's, and hence E_r is the sum of E_t^* for such t's. Observing that (4.10b) depends only on the ideal class of $\mathfrak{y}$, we obtain

$$(7.10) \qquad \sum_{r, r' \in R} L_{\mathfrak{r}\mathfrak{c}}(2s, \overline{\varphi}, prr'\mathfrak{e}_\lambda) E_r^*(w, s; \overline{\varphi}) E_{r'}^*(w, s; \overline{\varphi})$$

$$= \sum_{r, r' \in R'} L_{\mathfrak{r}\mathfrak{c}}(2s, \overline{\varphi}, prr'\mathfrak{e}_\lambda) E_r(w, s; \overline{\varphi}) E_{r'}(w, s; \overline{\varphi}).$$

If we put

$$\mathscr{F}(w, s; \xi, \mathfrak{y}) = \sum_{r \in R'} L_{\mathfrak{r}\mathfrak{c}}(2s, \xi, r\mathfrak{y}) E_r(w, s; \xi),$$

then (7.10) equals

$$(7.11) \qquad \sum_{r \in R'} \mathscr{F}(w, s; \overline{\varphi}, pr\mathfrak{e}_\lambda) E_r(w, s; \overline{\varphi}).$$

We also see from (4.12a) that

$$\mathscr{F}(w, s; \xi, \mathfrak{y}) = E(w, s; m, \xi, 2\mathfrak{r}\mathfrak{c}; (\mathfrak{r}\mathfrak{c})^{-1}\mathfrak{e}_\lambda, \mathfrak{y}^{-1}),$$

and hence (4.12b) shows that

$$(7.12) \qquad C_\lambda(w, s; \xi) = \sum_{\mathfrak{y} \in \mathscr{A}} \mathscr{F}(w, s; \xi, \mathfrak{y}).$$

Let us now consider the set B chosen in Section 6. Recalling that $\mathfrak{a}_\beta = p^{-1}\mathfrak{g}$ and combining (7.9–12) and (4.8), we obtain

$$(7.13)\qquad \sum_{\beta\in B} \varphi^*(\mathfrak{a}_\beta)N(\mathfrak{a}_\beta)^{2s}T'_{\beta\lambda}(w, s-1/2)$$

$$= N(\mathfrak{e}_\lambda)^{2s} \sum_{r\in R'}\sum_{\beta\in B} \mathscr{F}(w, s; \overline{\varphi}, \mathfrak{a}_\beta^{-1}r\mathfrak{e}_\lambda)E_r(w, s; \overline{\varphi})$$

$$= N(\mathfrak{e}_\lambda)^{2s}C_\lambda(w, s; \overline{\varphi}) \sum_{r\in R'} E_r(w, s; \overline{\varphi})$$

$$= N(\mathfrak{e}_\lambda)^{2s}C_\lambda(w, s; \overline{\varphi})E_\lambda(w, s; \overline{\varphi}).$$

This finishes the analysis of $S'_{\beta\lambda}$. Taking $\sigma_{p\lambda}$ in place of $\zeta_{p\lambda}$, we can similarly obtain a formula for $S_{\beta\lambda}$. In this case, the groups D_λ, D'_λ, and D^*_λ must be replaced by $D[\mathfrak{c}\mathfrak{e}_\lambda^{-1}, 2\mathfrak{r}\mathfrak{e}_\lambda]$, $D[\mathfrak{c}\mathfrak{e}_\lambda^{-1}, \mathfrak{r}\mathfrak{e}_\lambda]$, and

$$\{x \in D[\mathfrak{c}\mathfrak{e}_\lambda^{-1}, 2\mathfrak{r}\mathfrak{e}_\lambda] \,|\, (a_x - d_x)_v \in 2\mathfrak{g}_v \textit{ for every } v|2\}.$$

Repeating the whole argument, we find that

$$(7.14a)\qquad S_{\beta\lambda}(w, s) = (-1)^{\|m\|}2^d \sum_q T_{\beta\lambda}(w, s)\|_{2m}\tau_q,$$

where q runs over $\mathfrak{r}\mathfrak{e}_\lambda/2\mathfrak{r}\mathfrak{e}_\lambda$, and

$$(7.14b)\qquad \sum_{\beta\in B} \overline{\varphi}^*(\mathfrak{a}_\beta)N(\mathfrak{a}_\beta)^{2s}T_{\beta\lambda}(w, s-1/2)$$

$$= N(\mathfrak{r}\mathfrak{e}_\lambda)^{2s}C(w, s; m, \varphi, \Gamma^\lambda)E(w, s; m, \varphi, \Gamma^\lambda),$$

where $\Gamma^\lambda = \Gamma[\mathfrak{c}\mathfrak{e}_\lambda^{-1}, 2\mathfrak{r}\mathfrak{e}_\lambda]$.

8. The final calculations. The final stage of our proof consists of explicit calculations of the right-hand sides of (6.20) and (6.22) by means of expressions (7.14b) and (7.13). From now on, let us use z instead of w. To avoid difficulties about convergence, we introduce a complex variable t in addition to s, and put

$$(8.1)\qquad \mathscr{E}_\lambda(z, s, t) = N(\mathfrak{r}\mathfrak{e}_\lambda)^{s+t+1}C(z, s+1/2; m, \varphi, \Gamma^\lambda)E(z, t+1/2; m, \varphi, \Gamma^\lambda).$$

For τ_q as in (7.14a), we have $g'_\lambda\|_{2m}\tau_q = g'_\lambda$. Therefore, from (7.14a, b) we see that the right-hand side of (6.20) equals the value at $s = t$ of the function

$$(8.2)\qquad (-1)^{\|m\|}2^{2d} \sum_{\lambda=1}^{\kappa} \langle \mathscr{E}_\lambda(z, \bar{s}, \bar{t}), g'_\lambda(z)\rangle.$$

Fixing one λ, let $G \cap P_A D[\mathfrak{c}\mathfrak{e}_\lambda^{-1}, 2\mathfrak{r}\mathfrak{e}_\lambda] = \bigsqcup_{\alpha \in A} P\alpha\Gamma^\lambda$, $\Phi = \Gamma^\lambda \backslash H^\mathbf{a}$, and $\Psi_\alpha = (P \cap \alpha\Gamma^\lambda\alpha^{-1}) \backslash H^\mathbf{a}$. We then consider

$$(8.3) \qquad \int_\Phi g'_\lambda(z)\overline{C_\lambda(z, \bar{s} + 1/2)E_\lambda(z, \bar{t} + 1/2)}y^{2m}d_H z,$$

where C_λ and E_λ are the functions C and E of (8.1) with m, φ, and Γ^λ suppressed. This is equal to $\sum_{\alpha \in A} I_\alpha$ with

$$I_\alpha = \int_\Phi g'_\lambda(z)\overline{C_\lambda(z, \bar{s} + 1/2)E_{\lambda\alpha}(z, \bar{t} + 1/2)}y^{2m}d_H z,$$

where $E_{\lambda\alpha}$ is the function of (4.7) defined for the present E_λ. For each $\alpha \in A$ put

$$g_\lambda^\alpha = g'_\lambda \|_{2m} \alpha^{-1}, \qquad C_\lambda^\alpha(z) = C_\lambda(z, \bar{s} + 1/2)\|_m \alpha^{-1}.$$

We take α from $\mathrm{diag}[r^{-1}, r]U$ with $r \in F_\mathbf{h}^\times$ and a sufficiently small subgroup U of C'. Then the standard technique employed in Section 6 shows that

$$I_\alpha = \overline{\varphi}_\mathbf{a}(d_\alpha)\overline{\varphi}^*(d_\alpha \mathfrak{a}_\alpha^{-1})N(\mathfrak{a}_\alpha)^{2t+1}\int_{\Psi_\alpha} g_\lambda^\alpha(z)\overline{C_\lambda^\alpha(z)}y^{tu+(u+3m-i\mu)/2}d_H z.$$

Observe that Ψ_α can be given by $(\mathbf{R}^\mathbf{a}/2\mathfrak{a}_\alpha^{-2}(t_\lambda\mathfrak{d})^{-1}) \times Y$ with Y of (6.12). By Lemma 3.8 g_λ^α has the following expansion:

$$g_\lambda^\alpha(z) = [\psi_\mathbf{a}(d_\alpha)\psi^*(d_\alpha\mathfrak{a}_\alpha^{-1})]^2 \sum_{0 \ll \xi \in F} c(\xi t_\lambda^{-1}\mathfrak{a}_\alpha^{-2}, \mathbf{g}^*)\xi^{m-i\mu}\mathbf{e}_\mathbf{a}(\xi z).$$

On the other hand, C_λ^α is exactly $G_\gamma(z, \bar{s} + 1/2)$ of (4.16) with α, m, φ, μ, $\mathfrak{h}$, $2\mathfrak{r}\mathfrak{c}$, and Γ^λ as γ, n, ω, λ, $\mathfrak{f}$, $\mathfrak{z}$, and Γ_0 there, and hence its Fourier expansion is given by substituting $\bar{s} + 1/2$ for s in (4.17) and changing the symbols accordingly. Thus we have

$$\varphi_\mathbf{a}(d_\alpha)\varphi^*(d_\alpha\mathfrak{a}_\alpha^{-1})N(\mathfrak{a}_\alpha)^{-2s-1}y^{-su-(u-i\mu-m)/2}\overline{C_\lambda^\alpha(z)}$$

$$= L_\mathfrak{c}(2s + 1, \overline{\varphi}) + D_F^{-1/2}\overline{\gamma(\varphi)}N(\mathfrak{h})^{-1}\sum_\mathfrak{t} \mu(\mathfrak{t})\overline{\varphi}^*(\mathfrak{t})N(\mathfrak{t})^{-2s-1}\sum_{\mathfrak{y} \in \mathscr{A}} N(\mathfrak{y})^{2s}\sum_{h,b} \varphi_\mathbf{a}(b)|b|^{-2su}$$

$$\cdot \overline{\varphi_\mathbf{a}(h)\varphi^*(h\mathfrak{h}\mathfrak{d}\mathfrak{y})\mathbf{e}_\mathbf{a}(-bhx)\xi(y, bh; \bar{s}u + (u + m + i\mu)/2, \bar{s}u + (u - m + i\mu)/2)},$$

where h runs over $(\mathfrak{d}\mathfrak{h}\mathfrak{y})^{-1}$, b over $(F^\times \cap \mathfrak{t}^{-1}\mathfrak{a}_\alpha^2 t_\lambda \mathfrak{r}\mathfrak{c}\mathfrak{d}\mathfrak{y})/\mathfrak{g}^\times$, and $\mathfrak{t}$ over the integral ideals dividing $2\mathfrak{h}^{-1}\mathfrak{r}\mathfrak{c}$. Employing (4.18), or rather its variation with $\bar{\xi}$ in place of ξ, we can compute the integral over Ψ_α in a straightforward way (in the same manner as we did for (6.11)) to find that

$$I_\alpha = M\overline{\gamma(\varphi)}N(\mathfrak{a}_\alpha)^{2s+2t}N(2(\mathfrak{t}_\lambda\mathfrak{d}\mathfrak{h})^{-1})\sum_{\mathfrak{t},\mathfrak{y}}\mu(\mathfrak{t})\overline{\varphi}^*(\mathfrak{t})N(\mathfrak{t})^{-2s-1}N(\mathfrak{y})^{2s}$$

$$\cdot\sum_{0\ll a\in F/U}\sum_{bh=a}c(at_\lambda^{-1}\mathfrak{a}_\alpha^{-2},\mathfrak{g}^*)\varphi^*(h\mathfrak{h}\mathfrak{d}\mathfrak{y})|b|^{-tu-su}|h|^{su-tu},$$

where $\mathfrak{t}$, $\mathfrak{y}$, b, and h run over the same sets as before under the restriction $bh = a$, $U = \{e^2 | e \in \mathfrak{g}^\times\}$, and

$$(8.4)\quad M = i^{\|m\|}2^{d-2td-2\|m\|}\pi^{(s-t+1)d-\|m\|}$$

$$\cdot\prod_{v\in\mathbf{a}}\Gamma(t-s+m_v)\Gamma(s+t+2q_v)\Gamma(s+q_v+1/2)^{-1}\Gamma(t+q_v+1/2)^{-1}$$

with $q = (m - i\mu)/2$. This is valid for sufficiently large $\mathrm{Re}(s)$ and $\mathrm{Re}(t-s)$. Putting $\mathfrak{m} = h\mathfrak{h}\mathfrak{d}\mathfrak{y}$ and $\mathfrak{n} = b\mathfrak{t}(\mathfrak{a}_\alpha^2 t_\lambda \mathfrak{r}\mathfrak{c}\mathfrak{d}\mathfrak{y})^{-1}$, we find that the quantity of (8.2) equals

$$(8.5)\quad M\overline{\gamma(\varphi)}(-1)^{\|m\|}2^{d(2-s-t)}N(\mathfrak{h}^{-1}\mathfrak{r}\mathfrak{c})N(\mathfrak{h}\mathfrak{d})^{t-s}h_F[\mathfrak{g}_+^\times : U]\mathrm{vol}(\Gamma^1\backslash H^{\mathbf{a}})^{-1}$$

$$\cdot\sum_{\mathfrak{t}\supset 2\mathfrak{h}^{-1}\mathfrak{r}\mathfrak{c}}\mu(\mathfrak{t})\overline{\varphi}^*(\mathfrak{t})N(\mathfrak{t})^{t-s-1}\sum_{\mathfrak{m},\mathfrak{n}}\varphi^*(\mathfrak{m})c(\mathfrak{m}\mathfrak{n}(\mathfrak{t}\mathfrak{h})^{-1}\mathfrak{r}\mathfrak{c},\mathfrak{g}^*)N(\mathfrak{m})^{s-t}N(\mathfrak{n})^{-s-t},$$

where $\mathfrak{m}$ and $\mathfrak{n}$ run independently over all the integral ideals, $\mathfrak{g}_+^\times = \{e \in \mathfrak{g}^\times | e \gg 0\}$, and h_F is the class number of F. By (3.20b) $\mathfrak{h} = \mathfrak{r}\mathfrak{h}_0$ with a divisor $\mathfrak{h}_0$ of $\mathfrak{c}$, and hence $\mathfrak{h}^{-1}\mathfrak{r}\mathfrak{c} = \mathfrak{h}_0^{-1}\mathfrak{c}$. Since $\mathfrak{t}$ may be assumed to be squarefree and prime to $\mathfrak{h}$, the condition $\mathfrak{t} \supset 2\mathfrak{h}^{-1}\mathfrak{r}\mathfrak{c}$ can be replaced by $\mathfrak{t} \supset \mathfrak{h}_0^{-1}\mathfrak{c}$, and then by $\mathfrak{t} \supset \mathfrak{j}$ with $\mathfrak{j}$ defined as in Theorem 3.2.

Let $Y_\mathfrak{t}(s, t)$ denote the last sum over $\mathfrak{m}$, $\mathfrak{n}$ in (8.5). By Lemma 3.9 $Y_\mathfrak{t}(s, s)$ is meaningful for $\mathrm{Re}(s) > 0$. Observe that if $s = t$, then

$$M = i^{\|m\|}2^{-\|m\|}\pi^{d/2-\|m\|}\prod_{v\in\mathbf{a}}\Gamma(m_v)\Gamma(s+q_v)\Gamma(s+q_v+1/2)^{-1}.$$

Therefore, putting $s = t$ in (8.5) and comparing the result with the left-hand side of (6.20), we obtain

$$\overline{\lambda_f(\tau,\mathfrak{q}^{-1})}\sum_{\mathfrak{m}}\lambda(\tau,\mathfrak{q}^{-1}\mathfrak{b}\mathfrak{m};f^*)N(\mathfrak{m})^{-2s}\langle \mathbf{g},\mathbf{g}\rangle/\langle f,f\rangle$$

$$= Q\sum_{\mathfrak{t}\supset\mathfrak{j}}\mu(\mathfrak{t})\overline{\varphi}^*(\mathfrak{t})N(\mathfrak{t})^{-1}Y_\mathfrak{t}(s,s)$$

with Q as given in Theorem 3.2. Take the limit of this equality when $s \to +\infty$. Then Lemma 3.9 gives the first formula of Theorem 3.4.

Suppose that $\mathbf{g}^*|\mathfrak{T}(\mathfrak{n}) = \chi'(\mathfrak{n})\mathbf{g}^*$ for every $\mathfrak{n}$. Then (3.10) shows that $c(\mathfrak{x}\mathfrak{m}, \mathbf{g}^*) = N(\mathfrak{x})^{-1}\chi'(\mathfrak{x})c(\mathfrak{m}, \mathbf{g}^*)$ if $\mathfrak{x} \supset \mathfrak{c}$. Therefore we have

$$(8.6a) \qquad \sum_{\mathfrak{t}\supset\mathfrak{j}} \mu(\mathfrak{t})\bar{\varphi}^*(\mathfrak{t})N(\mathfrak{t})^{-1}D(0, \mathbf{g}^*, \varphi, \mathfrak{t}^{-1}\mathfrak{h}^{-1}\mathfrak{r}\mathfrak{c})$$
$$= D(0, \mathbf{g}^*, \varphi, \mathfrak{g})N(\mathfrak{h}\mathfrak{r}^{-1}\mathfrak{c}^{-1}) \sum_{\mathfrak{t}\supset\mathfrak{j}} \mu(\mathfrak{t})\bar{\varphi}^*(\mathfrak{t})\chi'(\mathfrak{t}^{-1}\mathfrak{h}^{-1}\mathfrak{r}\mathfrak{c}).$$

This proves the last formula of Theorem 3.4, since we can easily verify that

$$(8.6b) \qquad \sum_{\mathfrak{t}\supset\mathfrak{j}} \mu(\mathfrak{t})\bar{\varphi}^*(\mathfrak{t})\chi'(\mathfrak{t}^{-1}\mathfrak{h}^{-1}\mathfrak{r}\mathfrak{c}) = \chi'(\mathfrak{j}^{-1}\mathfrak{h}^{-1}\mathfrak{r}\mathfrak{c}) \prod_{v|\mathfrak{j}} [\chi'(v) - \bar{\varphi}(\pi_v)].$$

To prove Theorem 3.6, we use (6.22), (7.9a), and (7.13). We consider an integral of type (8.3) with g_λ in place of g'_λ; E_λ and C_λ should be those of (7.13). Then we find that the sum over λ in (6.22) equals the value at $s = t$ of the function

$$(8.7) \quad M'\gamma(\varphi)2^{d(1-s-t)}N(\mathfrak{r})N(\mathfrak{h}\mathfrak{d})^{t-s}h_F[\mathfrak{g}_+^\times : U]\mathrm{vol}(\Gamma\backslash H^n)^{-1}$$
$$\cdot \sum_{\mathfrak{t}\supset\mathfrak{h}^{-1}\mathfrak{r}\mathfrak{c}} \mu(\mathfrak{t})\varphi^*(\mathfrak{t})N(\mathfrak{t})^{t-s} \sum_{\mathfrak{m},\mathfrak{n}} \chi(\mathfrak{m}\mathfrak{n}\mathfrak{t}^{-1}\mathfrak{h}^{-1}\mathfrak{c})\bar{\varphi}^*(\mathfrak{m})N(\mathfrak{m})^{s-t-1}N(\mathfrak{n})^{-s-t-1},$$

where $\Gamma = \Gamma[2\mathfrak{r}\mathfrak{d}^{-1}, \mathfrak{c}\mathfrak{d}]$ and M' is defined by (8.4) with $\bar{q}$ in place of q. Let $Z_\mathfrak{t}(s, t)$ denote the last sum over $\mathfrak{m}$ and $\mathfrak{n}$. Put $\mathfrak{a} = \mathfrak{t}^{-1}\mathfrak{h}^{-1}\mathfrak{r}\mathfrak{c}$. Then $\mathfrak{h}_0^{-1}\mathfrak{c} \subset \mathfrak{a} \subset \mathfrak{g}$ and $\mathfrak{a}$ is prime to $\mathfrak{r}$ by (3.22b). Now $\bar{\varphi}^*(\mathfrak{m})\chi(\mathfrak{m}\mathfrak{n}\mathfrak{h}^{-1}\mathfrak{t}^{-1}\mathfrak{c}) \neq 0$ only if $\mathfrak{m}\mathfrak{n}\mathfrak{h}^{-1}\mathfrak{t}^{-1}\mathfrak{c} \subset \mathfrak{g}$ and $\mathfrak{m}$ is prime to $\mathfrak{h}$. Since $\mathfrak{m}\mathfrak{n}\mathfrak{h}^{-1}\mathfrak{t}^{-1}\mathfrak{c} = \mathfrak{m}\mathfrak{n}\mathfrak{a}\mathfrak{r}^{-1}$, these two conditions are satisfied only if $\mathfrak{n} \subset \mathfrak{r}$. Putting $\mathfrak{n} = \mathfrak{r}\mathfrak{x}$, we have

$$(8.8) \qquad Z_\mathfrak{t}(s, t) = \sum_{\mathfrak{m},\mathfrak{x}} \chi(\mathfrak{a}\mathfrak{m}\mathfrak{x})\bar{\varphi}^*(\mathfrak{m})N(\mathfrak{m})^{s-t-1}N(\mathfrak{r}\mathfrak{x})^{-s-t-1}.$$

where $\mathfrak{m}$ and $\mathfrak{x}$ run independently over the integral ideals. Since $\mathfrak{a} \supset \mathfrak{c}$ and χ is defined at level $2\mathfrak{b}\mathfrak{b}'$, we have $\chi(\mathfrak{a}\mathfrak{m}\mathfrak{x}) = \chi(\mathfrak{a})\chi(\mathfrak{m}\mathfrak{x})$. Therefore, by (3.23b), we see that

$$(8.9) \qquad Z_\mathfrak{t}(s, s) = \chi(\mathfrak{a})N(\mathfrak{r})^{-2s-1}L_\mathfrak{c}(2s + 1, \varphi)^{-1}D(2s, \chi)D(0, \chi, \bar{\varphi}).$$

Putting $s = t$ in (8.7) and comparing the result with the left-hand side of (6.22), we obtain the formula of Theorem 3.6. This time the quantity of type (8.6b) appears with χ and φ in place of χ' and $\bar{\varphi}$.

9. Proof of Theorem 3.2 and supplementary results. We take an eigenform f in $\mathscr{S}_k(\mathfrak{b}, \mathfrak{b}'; \psi)$ as in Proposition 3.1 and also the corresponding system of eigenvalues χ in $\mathscr{S}_{2m}((2\mathfrak{b}\mathfrak{b}', \psi^2))$. With X, Y, $\{f_1, \ldots, f_\alpha\}$, $\{\mathbf{g}_1, \ldots, \mathbf{g}_\beta\}$, τ, $\mathfrak{q}$, $\mathfrak{r}$, $\mathfrak{h}$, and φ as in Theorem 3.2, put $\mathbf{g}_j = (g_{j\lambda})_{\lambda=1}^{\kappa}$ and define elements h_i of $\mathscr{S}_k(\mathfrak{g}, \mathfrak{r}\mathfrak{b}\mathfrak{b}'; \varphi)$ by

$$(9.1) \qquad \lambda(\xi, \mathfrak{m}; h_i) = \lambda(\tau\xi, \mathfrak{q}^{-1}\mathfrak{r}^{-1}\mathfrak{m}; f_i) \qquad \textit{for every } (\xi, \mathfrak{m}).$$

For a fixed i, define $g_{\tau\lambda}$ by (5.14a) with h_i in place of h. Then the proof of Proposition 5.6 shows that $\mathbf{g}_\tau = (g_{\tau\lambda})_{\lambda=1}^{\kappa}$ belongs to $\mathscr{S}_{2m}((2\mathfrak{b}\mathfrak{b}', \psi^2))$, and (3.18a) holds with f_i in place of f, and hence, in particular, $c(\mathfrak{g}, \mathbf{g}_\tau) = \lambda(\tau, \mathfrak{q}^{-1}; f_i)$. From this fact and Proposition 2.1 we can easily derive that $\mathbf{g}_\tau \in Y$. Thus we can put

$$C \sum_{j=1}^{\beta} b_{ij} g_{j\lambda}(w) = \int_{\Phi} h_i(z)\Theta(z, w; \eta_\lambda) y^k d_H z \tag{9.2}$$

with $b_{ij} \in \mathbf{C}$, where C, Φ, and η_λ are as in (5.14a, b). We also have

$$\sum_{j=1}^{\beta} b_{ij} c(\mathfrak{g}, \mathbf{g}_j) = \lambda(\tau, \mathfrak{q}^{-1}; f_i). \tag{9.3}$$

In the opposite direction, we consider

$$l(z) = \sum_{\lambda=1}^{\kappa} \langle \Theta(z, w; \eta_\lambda), g_{j\lambda}(w) \rangle,$$

assuming (3.20b, c) but not (3.21). The proofs of Lemma 5.7 and Proposition 5.8 show that $l \in \sum_{i=1}^{\alpha} \mathbf{C} h_i$, and hence

$$\sum_{i=1}^{\alpha} a_{ij} h_i(z) = \sum_{\lambda=1}^{\kappa} \langle \Theta(z, w; \eta_\lambda), g_{j\lambda}(w) \rangle \tag{9.4}$$

with $a_{ij} \in \mathbf{C}$. This together with (9.2) gives

$$\mathrm{vol}(\Phi) \sum_{i=1}^{\alpha} a_{ij} \langle h_p, h_i \rangle = \bar{C} \sum_{q=1}^{\beta} \bar{b}_{pq} \langle \mathbf{g}_q, \mathbf{g}_j \rangle.$$

Applying the last part of Proposition 1.4 to $\langle h_p, h_i \rangle$, we obtain

$$\mathrm{vol}(\Phi)\tau^k N(\mathfrak{q}\mathfrak{r})^{-1} \sum_{i=1}^{\alpha} a_{ij} \langle f_p, f_i \rangle = \bar{C} \sum_{q=1}^{\beta} \bar{b}_{pq} \langle \mathbf{g}_q, \mathbf{g}_j \rangle. \tag{9.5}$$

Starting from (9.4), we can now repeat the calculations of Sections 6, 7, and 8 with $\sum_{i=1}^{\alpha} a_{ij} h_i$ and $\mathbf{g}_j$ in place of Ah and $\mathbf{g}$, and without (3.21). Then we find that the quantity

$$\Delta(s) \sum_{i=1}^{\alpha} a_{ij} \sum_{\mathfrak{m}} \lambda(\tau, \mathfrak{q}^{-1}\mathfrak{b}\mathfrak{m}; f_i^*) N(\mathfrak{m})^{-2s} \prod_{v \in \mathbf{a}} \Gamma(s + q_v)\Gamma(s + q_v + (1/2))^{-1}$$

with Δ of (6.21) is equal to the value of (8.5) at $s = t$ with $\mathbf{g}_j^*$ in place of $\mathbf{g}^*$. This gives

$$(9.6)\qquad \overline{C}^{-1}\operatorname{vol}(\Phi)\tau^k N(\mathfrak{q}\mathfrak{r})^{-1}\sum_{i=1}^{\alpha} a_{ij}\lambda(\tau, \mathfrak{q}^{-1}\mathfrak{b}; f_i^*)$$

$$= Q\sum_{\mathfrak{t}\supset\mathfrak{j}} \mu(\mathfrak{t})\overline{\varphi}^*(\mathfrak{t})N(\mathfrak{t})^{-1}D(0, \mathbf{g}_j^*, \varphi, (\mathfrak{t}\mathfrak{h})^{-1}\mathfrak{r}\mathfrak{c})$$

with Q of Theorem 3.2. With f_i' and $\mathbf{g}_j'$ as in Theorem 3.2 we have $f_i = \sum_{p=1}^{\alpha}\langle f_p, f_i\rangle f_p'$ and $\mathbf{g}_j = \sum_{q=1}^{\beta}\langle \mathbf{g}_q, \mathbf{g}_j\rangle \mathbf{g}_q'$; also $\lambda(\tau, \mathfrak{q}^{-1}\mathfrak{b}; f_p'^*) = \lambda(\tau, \mathfrak{q}^{-1}; f_p'^{\#})$. Therefore from (9.5) and (9.6) we obtain

$$(9.7)\qquad \sum_{p=1}^{\alpha}\overline{b}_{pq}\lambda(\tau, \mathfrak{q}^{-1}; f_p'^{\#}) = Q\sum_{\mathfrak{t}\supset\mathfrak{j}}\mu(\mathfrak{t})\overline{\varphi}^*(\mathfrak{t})N(\mathfrak{t})^{-1}D(0, \mathbf{g}_q'^*, \varphi, (\mathfrak{t}\mathfrak{h})^{-1}\mathfrak{r}\mathfrak{c}).$$

This together with (9.3) proves the formula of Theorem 3.2 in which the members of the dual bases are interchanged.

THEOREM 9.1. *The notation and assumptions being as in Theorem 3.2, suppose that each f_i is an eigenform of the T_v for all $v \in \mathbf{h}$. Let $\mathbf{g}_i$ be the normalized eigenform in $\mathcal{S}_{2m}((2\mathfrak{b}\mathfrak{b}', \psi^2))$ corresponding to f_i in the sense of Proposition 3.1(II). Suppose further that $\mathbf{g}_1, \ldots, \mathbf{g}_\alpha$ are linearly independent over* $\mathbf{C}$. *Then*

$$\lambda(\tau, \mathfrak{q}^{-1}; f_p'^{\#})\overline{\lambda(\tau, \mathfrak{q}^{-1}; f_p)} = Q\sum_{\mathfrak{t}\supset\mathfrak{j}}\mu(\mathfrak{t})\overline{\varphi}^*(\mathfrak{t})N(\mathfrak{t})^{-1}D(0, \mathbf{g}_p'^*, \varphi, (\mathfrak{t}\mathfrak{h})^{-1}\mathfrak{r}\mathfrak{c})$$

for every $p \leqslant \alpha$ with Q as in Theorem 3.2, where $\mathbf{g}_p'$ is the element of $\sum_{i=1}^{\alpha}\mathbf{C}\mathbf{g}_i$ such that $\langle \mathbf{g}_i, \mathbf{g}_j'\rangle = \delta_{ij}$ and $\mathbf{g}_p'^* = \mathbf{g}_p'|J_{2\mathfrak{b}\mathfrak{b}'}$ with J of (3.16).

Proof. We take $\sum_{i>\alpha}\mathbf{C}\mathbf{g}_i$ to be the orthogonal complement of $\sum_{i=1}^{\alpha}\mathbf{C}\mathbf{g}_i$ in Y. By Proposition 5.6 the left-hand side of (9.2) is $C\lambda(\tau, \mathfrak{q}^{-1}; f_i)g_{i\lambda}$. Therefore we have $b_{pq} = \delta_{pq}\lambda(\tau, \mathfrak{q}^{-1}; f_p)$. This together with (9.7) proves the desired formula.

Let us now discuss an exemplary situation to which the above theorem is applicable. With an eigenform f and a system of eigenvalues χ as before, let χ_0 be the primitive system of eigenvalues associated with χ, and $\mathfrak{z}$ its exact level. Let $\mathbf{f}$ be the normalized eigenform in $\mathcal{S}_{2m}((\mathfrak{z}, \psi^2))$ belonging to χ_0. We assume that

(9.8a) $\mathfrak{p}\mathfrak{z} = 2\mathfrak{b}\mathfrak{b}'$ *with a prime factor* $\mathfrak{p}$ *of 2 which is prime to* $\mathfrak{z}$ *and unramified over* $\mathbf{Q}$.

Then $Y = \mathbf{C}\mathbf{f} + \mathbf{C}\mathbf{f}|\mathfrak{p}$, where $\mathbf{f}|\mathfrak{p}$ is the form such that $\mathbf{C}(\mathfrak{m}, \mathbf{f}|\mathfrak{p}) = c(\mathfrak{p}^{-1}\mathfrak{m}, \mathbf{f})$. Let α_1 and α_2 be the roots of $x^2 - \chi_0(\mathfrak{p})x + \psi^*(\mathfrak{p})^2N(\mathfrak{p}) = 0$, and let $\mathbf{g}_i = \mathbf{f} - N(\mathfrak{p})^{-1}\alpha_i\mathbf{f}|\mathfrak{p}$. Each $\mathbf{g}_i$ is a normalized eigenform belonging to χ_i with χ_i such that $\chi_i(v) = \chi_0(v)$ for $\pi_v\mathfrak{g} \neq \mathfrak{p}$ and $\chi_i(\mathfrak{p}) = \chi_0(\mathfrak{p}) - \alpha_i$. In addition to (9.8a) we assume

(9.8b) $\alpha_1 \neq \alpha_2$ *and* $X = \mathbf{C}f_1 + \mathbf{C}f_2$ *with two eigenforms f_i such that $f_i|T_v = N_v^{-1}\chi_i(v)$ for all $v \in \mathbf{h}$.*

Then Theorem 9.1 is applicable to these f_i and $\mathbf{g}_i$. By the principle of [76, (3.2)] and [78, Prop. 4.14] we obtain

$$\text{(9.9a)}\qquad \langle \mathbf{f}|\mathfrak{p}, \mathbf{f}\rangle = \langle \mathbf{f}, \mathbf{f}\rangle\chi(\mathfrak{p})/(N+1), \qquad \langle \mathbf{f}|\mathfrak{p}, \mathbf{f}|\mathfrak{p}\rangle = \langle \mathbf{f}, \mathbf{f}\rangle,$$

$$\text{(9.9b)}\qquad \langle \mathbf{g}_1, \mathbf{g}_1\rangle = \langle \mathbf{g}_2, \mathbf{g}_2\rangle = \langle \mathbf{f}, \mathbf{f}\rangle[(1+N)^2 - |\chi(\mathfrak{p})|^2]/(N^2+N),$$

$$\text{(9.9c)}\qquad \langle \mathbf{g}_1, \mathbf{g}_2\rangle = \langle \mathbf{f}, \mathbf{f}\rangle(1 - N^{-2}\bar{\alpha}_1\alpha_2)(N-1)/(N+1),$$

where $N = N(\mathfrak{p})$. Put $r = \langle \mathbf{g}_1, \mathbf{g}_1\rangle/\langle \mathbf{f}, \mathbf{f}\rangle$, $t_1 = \langle \mathbf{g}_1, \mathbf{g}_2\rangle/\langle \mathbf{f}, \mathbf{f}\rangle$, $t_2 = \bar{t}_1$, $\hat{\mathbf{f}} = \mathbf{f}|J_3$ and $\mathbf{f}^* = \mathbf{f}|J_{\mathfrak{p}3}$. By Lemma 3.10 we have $\mathbf{f}^* = \hat{\mathbf{f}}|\mathfrak{p}$ and $(\mathbf{f}|\mathfrak{p})^* = \hat{\mathbf{f}}$. Since $\mathbf{g}_1 = \langle \mathbf{f}, \mathbf{f}\rangle(r\mathbf{g}_1' + t_2\mathbf{g}_2')$ and $\mathbf{g}_2 = \langle \mathbf{f}, \mathbf{f}\rangle(r\mathbf{g}_2' + t_1\mathbf{g}_1')$, we have $\langle \mathbf{f}, \mathbf{f}\rangle g_i'^* = d_i\hat{\mathbf{f}} + e_i\hat{\mathbf{f}}|\mathfrak{p}$ for $i = 1, 2$ with

$$\text{(9.10)}\qquad d_i = N^{-1}(\alpha_{i+1}\bar{t}_i - \alpha_i r)/(r^2 - t_1t_2), \qquad e_i = (r - \bar{t}_i)/(r^2 - t_1t_2),$$

where we understand that $\alpha_3 = \alpha_1$. We see that e_i and d_i are rational expressions in α_i and $\bar{\alpha}_i$.

THEOREM 9.2. *The notation being as above and as in Theorem 3.2, let $\mathfrak{k} = \mathfrak{j}$ or $\mathfrak{p}^{-1}\mathfrak{j}$ according as $\mathfrak{p}\nmid\mathfrak{j}$ or $\mathfrak{p}|\mathfrak{j}$; put $\hat{\mathbf{f}} = \mathbf{f}|J_3$ and*

$$E = c(\mathfrak{g}, \hat{\mathbf{f}})D(0, \bar{\chi}_0, \varphi)N(\mathfrak{h}(\mathfrak{rc})^{-1})\prod_{v|\mathfrak{k}}[\bar{\chi}_0(v) - \bar{\varphi}(\pi_v)],$$

$$\Lambda_i = \lambda(\tau, \mathfrak{q}^{-1}; f_i'^*)\overline{\lambda(\tau, \mathfrak{q}^{-1}; f_i)} \qquad (i = 1, 2).$$

Then, under (3.20a, b, c) and (9.8a, b), Λ_i can be given as follows:

Case I: $\mathfrak{p}\nmid\mathfrak{h}^{-1}\mathfrak{rc}$. $\langle \mathbf{f}, \mathbf{f}\rangle\Lambda_i = QE\bar{\chi}_0((\mathfrak{k}\mathfrak{h})^{-1}\mathfrak{rc})d_i$.

Case II: $\mathfrak{p}|\mathfrak{h}^{-1}\mathfrak{rc}$, $\mathfrak{p}|\mathfrak{h}$. $\langle \mathbf{f}, \mathbf{f}\rangle\Lambda_i = QE\bar{\chi}_0((\mathfrak{p}\mathfrak{k}\mathfrak{h})^{-1}\mathfrak{rc}))[d_i\bar{\chi}_0(\mathfrak{p}) + e_iN(\mathfrak{p})]$.

Case III: $\mathfrak{p}|\mathfrak{h}^{-1}\mathfrak{rc}$, $\mathfrak{p}\nmid\mathfrak{h}$. $\langle \mathbf{f}, \mathbf{f}\rangle\Lambda_i = QE\bar{\chi}_0((\mathfrak{p}\mathfrak{j}\mathfrak{h})^{-1}\mathfrak{rc}))$
$\cdot[d_i\{\bar{\chi}_0(\mathfrak{p}^2) - 2\bar{\varphi}^*(\mathfrak{p})\bar{\chi}_0(\mathfrak{p}) + \bar{\varphi}^*(\mathfrak{p}^2)\} + e_iN(\mathfrak{p})\{\bar{\chi}_0(\mathfrak{p}) - 2\bar{\varphi}^*(\mathfrak{p})\}]$.

Moreover, we have

$$[(N(\mathfrak{p})^2 - \bar{\alpha}_1\alpha_2)\Lambda_1 + (N(\mathfrak{p})^2 - \bar{\alpha}_2\alpha_1)\Lambda_2]\langle \mathbf{f}, \mathbf{f}\rangle/(N(\mathfrak{p})^2 + N(\mathfrak{p}))$$

$$= QEN(\mathfrak{p})\bar{\chi}_0((\mathfrak{p}\mathfrak{j}\mathfrak{h})^{-1}\mathfrak{rc}))\cdot\begin{cases} 0 & (\textit{Case I}),\\ 1 & (\textit{Case II}),\\ \bar{\chi}_0(\mathfrak{p}) - 2\bar{\varphi}^*(\mathfrak{p}) & (\textit{Case III}).\end{cases}$$

Proof. For $\mathbf{x} \in \mathscr{S}_{2m}((2\mathfrak{b}\mathfrak{b}', \psi^2))$ put $S(\mathbf{x}) = \sum_{\mathfrak{t}\supset\mathfrak{j}}\mu(\mathfrak{t})\bar{\varphi}^*(\mathfrak{t})N(\mathfrak{t})^{-1}D(0, \mathbf{x}, \varphi, (\mathfrak{t}\mathfrak{h})^{-1}\mathfrak{rc})$. By Theorem 9.1 we have $\langle \mathbf{f}, \mathbf{f}\rangle\Lambda_i = Q[d_iS(\hat{\mathbf{f}}) + e_iS(\hat{\mathbf{f}}|\mathfrak{p})]$. By (9.8a) $\mathfrak{p}^2$ is the highest power of $\mathfrak{p}$ dividing $\mathfrak{c}$, and hence $(\mathfrak{t}\mathfrak{h})^{-1}\mathfrak{rc} = \mathfrak{a}\mathfrak{p}^b$ with $0 \leqslant b \leqslant 2$ and a factor $\mathfrak{a}$ of $\mathfrak{z}^2$. Thereofre (3.23a) reduces $D(s, \mathbf{x}, \varphi, (\mathfrak{t}\mathfrak{h})^{-1}\mathfrak{r}\mathfrak{c})$ for $\mathbf{x} = \hat{\mathbf{f}}$ and $\hat{\mathbf{f}}|\mathfrak{p}$ to $D(s, \hat{\mathbf{f}}, \varphi, \mathfrak{g})$ times certain explicit factors. Thus some straightforward calculations together with (8.6b) (with $\bar{\chi}$ and $\mathfrak{f}$ in place of χ' and $\mathfrak{j}$) prove our formulas for $\langle \mathbf{f}, \mathbf{f} \rangle \Lambda_i$. Next, since $\mathbf{f} = \sum_{i=1}^{2} \langle \mathbf{g}_i, \mathbf{f} \rangle \mathbf{g}_i'$ and $\langle \mathbf{g}_1, \mathbf{f} \rangle = \langle \mathbf{f}, \mathbf{g}_2 \rangle = \langle \mathbf{g}_1, \mathbf{g}_2 \rangle N/(N-1)$ with $N = N(\mathfrak{p})$, from Theorem 9.1 we obtain

$$(N^2 + N)QS(\hat{\mathbf{f}}|\mathfrak{p}) = \langle \mathbf{f}, \mathbf{f} \rangle [(N^2 - \bar{\alpha}_1\alpha_2)\Lambda_1 + (N^2 - \bar{\alpha}_2\alpha_1)\Lambda_2].$$

Combining this with the formula for $S(\hat{\mathbf{f}}|\mathfrak{p})$ which we already have, we obtain the last formula of our theorem.

Remark 9.3. To obtain f_1' and f_2' from f_1 and f_2, we need the information about $\langle f_i, f_j \rangle$. Let us now show that, under some conditions, we can normalize f_1 and f_2 so that the quantities $\langle f_i, f_j \rangle$ are proportional to the quantities $\langle \mathbf{g}_i, \mathbf{g}_j \rangle$. For this purpose we assume (9.8a, b) and consider an element τ of F, $\gg 0$, such that (3.20b, c) are satisfied and that

$$(9.11) \qquad \lambda(\tau, \mathfrak{q}^{-1}; f_1) \neq 0 \quad \textit{and} \quad \mathfrak{p} \nmid \mathfrak{h}^{-1}\mathfrak{r}^2\mathfrak{c}.$$

Then $\mathfrak{p}|\mathfrak{h}$. Moreover, both f_1 and f_2 satisfy (3.21) since $\chi_1(\mathfrak{p}) \neq \chi_2(\mathfrak{p})$. Therefore by Proposition 5.8 we can put $(a_{ij}) = \mathrm{diag}[a_1, a_2]$, and hence (9.5) takes the form

$$(9.12) \qquad Ba_i\langle f_p, f_i \rangle = \overline{\lambda(\tau, \mathfrak{q}^{-1}; f_p)} \langle \mathbf{g}_p, \mathbf{g}_i \rangle,$$

where $B = \overline{C}^{-1} \operatorname{vol}(\Phi)\tau^k N(\mathfrak{q}\mathfrak{r})^{-1}$. From (9.11) and (9.12) we see that $\lambda(\tau, \mathfrak{q}^{-1}; f_2) \neq 0$ and

$$(9.13) \qquad \frac{\lambda(\tau, \mathfrak{q}^{-1}; f_2)}{\lambda(\tau, \mathfrak{q}^{-1}; f_1)} = \frac{\langle f_1, f_2 \rangle}{\langle f_1, f_1 \rangle} \cdot \frac{\langle \mathbf{g}_1, \mathbf{g}_1 \rangle}{\langle \mathbf{g}_1, \mathbf{g}_2 \rangle} = \frac{\langle f_2, f_2 \rangle}{\langle f_2, f_1 \rangle} \cdot \frac{\langle \mathbf{g}_2, \mathbf{g}_1 \rangle}{\langle \mathbf{g}_2, \mathbf{g}_2 \rangle}.$$

Thus the quotient $\lambda(\tau, \mathfrak{q}^{-1}; f_2)/\lambda(\tau, \mathfrak{q}^{-1}; f_1)$ for any τ satisfying (3.20b, c) and (9.11) is a constant independent of τ. Changing f_2 for its suitable constant multiple, we may assume that this quotient is 1. Then

$$(9.14a) \qquad \langle f_1, f_1 \rangle = \langle f_2, f_2 \rangle,$$

$$(9.14b) \qquad \langle f_i, f_j \rangle / \langle f_1, f_1 \rangle = \langle \mathbf{g}_i, \mathbf{g}_j \rangle / \langle \mathbf{g}_1, \mathbf{g}_1 \rangle.$$

References

[H] E. Hecke, *Über Modulfunktionen und die Dirichletschen Reihen mit Eulerscher Produktentwicklung I*, Math. Ann. **114** (1937), 1–28; *II*, Math. Ann. **114** (1937), 316–351. Also in *Mathematische Werke*, Vanderhoeck & Ruprecht, Göttingen, 1959, # #35 and 36.

[K] W. Kohnen, *Fourier coefficients of modular forms of half-integral weight*, Math. Ann. **271** (1985), 237–268.

[KZ] W. Kohnen and D. Zagier, *Values of L-series of modular forms at the center of the critical strip*, Invent. Math. **64** (1981), 175–198.

[N] S. Niwa, *On Fourier coefficients and certain "periods" of modular forms*, Proc. Japan Acad. Ser. A Math. Sci. **58**:2 (1982), 90–92.

[73] G. Shimura, *On modular forms of half integral weight*, Ann. of Math. **97** (1973), 440–481.

[75] ———, *On the holomorphy of certain Dirichlet series*, Proc. London Math. Soc. (3) **31** (1975), 79–98.

[76] ———, *The special values of the zeta functions associated with cusp forms*, Comm. Pure Appl. Math. **29** (1976), 783–804.

[78] ———, *The special values of the zeta functions associated with Hilbert modular forms*, Duke Math. J. **45** (1978), 637–679; *Corrections*, Duke Math. J. **48** (1981), 697.

[80] ———, *The arithmetic of certain zeta functions and automorphic forms on orthogonal groups*, Ann. of Math. **111** (1980), 313–375.

[81] ———, *The critical values of certain zeta functions associated with modular forms of half-integral weight*, J. Math. Soc. Japan **33** (1981), 649–672.

[82a] ———, *The periods of certain automorphic forms of arithmetic type*, J. Fac. Sci. Univ. Tokyo Sect. IA Math. **28** (1982), 605–632.

[82b] ———, *Confluent hypergeometric functions on tube domains*, Math. Ann. **260** (1982), 269–302.

[83] ———, *On Eisenstein series*, Duke Math. J. **50** (1983), 417–476.

[85a] ———, *On Eisenstein series of half-integral weight*, Duke Math. J. **52** (1985), 281–314.

[85b] ———, *On the Eisenstein series of Hilbert modular groups*, Rev. Mat. Iberoamericana **1**:3 (1985), 1–42.

[87] ———, *On Hilbert modular forms of half-integral weight*, Duke Math. J. **55** (1987), 765–838.

[88] ———, *On the critical values of certain Dirichlet series and the periods of automorphic forms*, Invent. Math. **94** (1988), 245–305.

[90] ———, *On the fundamental periods of automorphic forms of arithmetic type*, Invent. Math. **102** (1990), 399–428.

[91] ———, *The critical values of certain Dirichlet series attached to Hilbert modular forms*, Duke Math. J. **63** (1991), 557–613.

[Wa] J.-L. Waldspurger, *Sur les coefficients de Fourier des formes modularies de poids demi-entier*, J. Math. Pures Appl. (9) **60** (1981), 375–484.

[We] A. Weil, *Sur certain groupes d'opérateurs unitaires*, Acta Math. **111** (1964), 143–211.

Department of Mathematics, Princeton University, Princeton, New Jersey 08544, USA

Fractional and trigonometric expressions for matrices

American Mathematical Monthly, 101 (1994), 744-758

1. INTRODUCTION. We start with two elementary facts:

(I) *Every rational number x can be expressed as a quotient c/d with two integers c and d which have no common divisor greater than* 1. *Moreover, if $x = c'/d'$ is another such expression, then $(c', d') = \pm(c, d)$.*

(II) *Every real number x can be expressed as a quotient c/d with two real numbers c and d such that $c^2 + d^2 = 1$. Moreover, if $x = c'/d'$ is another such expression, then $(c', d') = \pm(c, d)$.*

One can add some geometric flavor to the latter statement by using $\sin\theta$ and $\cos\theta$ instead of c and d.

There is a certain parallelism between these two statements, but I guess most readers will wonder if it is really essential or merely rhetorical. This point aside, we can pose the question of whether the same types of statements can be made for matrices, which is the meaning of the title of this article. Considering only the first type for the moment, we can ask:

(A) *Given a matrix x with rational entries, can one find matrices c and d with integral entries so that $x = d^{-1}c$ and that this is an expression "reduced to the lowest terms"? If the answer is affirmative, to what extent is the pair (c, d) unique?*

Here we don't have to assume that x is square; if x is an $m \times n$-matrix, then c must be of the same shape and d invertible and $m \times m$. Similarly we can ask for an expression $x = cd^{-1}$ with c and d of appropriate shapes. There is of course a trivial answer: take d to be the scalar which is the least common denominator of the entries of x. But look at the equality

$$\begin{pmatrix} 2 & 0 \\ 0 & 3 \end{pmatrix}^{-1} \begin{pmatrix} 7 & 1 \\ 0 & 5 \end{pmatrix} = \begin{pmatrix} 6 & 0 \\ 0 & 6 \end{pmatrix}^{-1} \begin{pmatrix} 21 & 3 \\ 0 & 10 \end{pmatrix}.$$

Which side is more appropriately called the quotient reduced to "the lowest terms"? Clearly the left-hand side looks more "reduced."

This example will tell that perhaps the answer to (A) should be formulated in terms of *elementary divisors* of integral matrices, and in fact (A) can be answered easily that way, as will be shown below. However, our aim is not only to give an answer to (A), which is quite elementary, but also to supply a certain conceptual background for this type of question. At the same time, we shall show that (II), as well as its matrix version, can be understood in the same framework. In the last two sections we shall give higher-dimensional analogues of $\cos\theta$ and $\sin\theta$, as well as their hyperbolic counterparts, in the context of statement (II), which are closely related to the decomposition of the type $G = K \cdot \exp(\mathfrak{p})$ for *compact and noncompact* Lie groups G. Our theorems or lemmas in the other sections can hardly be called new, but it seems that some of them have never been stated in the forms as we present them in this article.

2. FRACTIONAL EXPRESSIONS FOR RATIONAL AND REAL MATRICES. Let us first make some notational conventions. For an associative ring R with identity element, $M_n^m(R)$ denotes the module of all $m \times n$-matrices with entries in R and $GL_n(R)$ the group of all invertible elements of the ring $M_n^n(R)$. The zero element of $M_n^m(R)$ is denoted by 0_n^m or simply by 0, the identity element of $M_n^n(R)$ by 1_n or simply by 1, and the transpose of a matrix X by ${}^tX$. For square matrices $A_1, \ldots, A_r$ we denote by $\mathrm{diag}[A_1, \ldots, A_r]$ the square matrix with the A_i in the diagonal blocks and 0 everywhere else. As usual, **Z**, **Q**, and **R** denote the ring of integers, the field of rational numbers, and the field of real numbers.

We now recall

Lemma 1. *Given $X \in M_n^m(\mathbf{Q})$ of rank r, there exist $A \in GL_m(\mathbf{Z})$, $B \in GL_n(\mathbf{Z})$ and positive rational numbers $e_1, \ldots, e_r$ such that $e_{i+1}/e_i \in \mathbf{Z}$ for all $i < r$ and*

$$AXB = \begin{pmatrix} E & 0_{n-r}^r \\ 0_r^{m-r} & 0_{n-r}^{m-r} \end{pmatrix}, \qquad E = \mathrm{diag}[e_1, \ldots, e_r]. \tag{2.1}$$

Moreover, the e_i are uniquely determined by X.

This is usually stated only for $X \in M_n^m(\mathbf{Z})$, but for $X \in M_n^m(\mathbf{Q})$ if we take $g \in Z$, > 0, so that $gX \in M_n^m(\mathbf{Z})$ and apply the theorem to gX, then we immediately obtain the result in the above form. The numbers $e_1, \ldots, e_r$ are called the *elementary divisors* of X.

We call an element X of $M_n^m(\mathbf{Z})$ *primitive* if $\mathrm{rank}(X) = \mathrm{Min}(m, n)$ and the elementary divisors of X are all equal to 1. If $m = n$, clearly X is primitive if and only if $X \in GL_n(\mathbf{Z})$.

Lemma 2. *Let $X \in M_n^m(\mathbf{Z})$ and $m \le n$. Then the following conditions on X are mutually equivalent:*

(1) *X is primitive.*

(2) *There exists an element Y of $M_n^{n-m}(\mathbf{Z})$ such that $\binom{X}{Y} \in GL_n(\mathbf{Z})$.*

(3) *There exists an element W of $M_m^n(\mathbf{Z})$ such that $XW = 1_m$.*

(4) *If $C \in M_m^1(\mathbf{Q})$ and $CX \in M_n^1(\mathbf{Z})$, then $C \in M_m^1(\mathbf{Z})$.*

(5) *For every $l \in \mathbf{Z}$, > 0, if $C \in M_m^l(\mathbf{Q})$ and $CX \in M_n^l(\mathbf{Z})$, then $C \in M_m^l(\mathbf{Z})$.*

Proof: Put $s = n - m$. If X is primitive, our definition implies that $AXB = (1_m \quad 0_s^m)$ with $A \in GL_m(\mathbf{Z})$ and $B \in GL_n(\mathbf{Z})$. Let $Y = (0_m^s \quad 1_s)B^{-1}$. Then

$$\begin{pmatrix} X \\ Y \end{pmatrix} = \begin{pmatrix} A^{-1} & 0 \\ 0 & 1_s \end{pmatrix} B^{-1} \in GL_n(\mathbf{Z}),$$

which proves that (1) $\Rightarrow$ (2). With Y as in (2) put $\binom{X}{Y}^{-1} = (W \quad Z)$ with $W \in M_m^n(\mathbf{Z})$ and $Z \in M_s^n(\mathbf{Z})$. Then $XW = 1_m$, and hence (2) $\Rightarrow$ (3). With W as in (3) let $C \in M_m^1(\mathbf{Q})$ and $CX \in M_n^1(\mathbf{Z})$. Then $C = CXW \in M_m^1(\mathbf{Z})$, and hence (3) $\Rightarrow$ (4). Decomposing C of (5) into row vectors, we easily see that (4) $\Leftrightarrow$ (5). Finally, assume (4). If $\mathrm{rank}(X) < m$, then there is a vector $v \in M_m^1(\mathbf{Q})$, $\neq 0$, such that $vX = 0$. Changing v for its suitable rational multiple if necessary, we may assume that $v \notin M_m^1(\mathbf{Z})$. This cannot happen under (4). Thus $\mathrm{rank}(X) = m$. Now take A,

B, E, and $e_1, \dots, e_r$ as in Lemma 1. Then $E^{-1}AX = (1_m \quad 0)B^{-1} \in M_n^m(\mathbf{Z})$. Then (5) implies that $E^{-1}A \in M_m^m(\mathbf{Z})$, and hence $E^{-1} \in M_m^m(\mathbf{Z})$. Therefore $e_i = 1$ for every i, which proves that (4) ⇒ (1) and our proof is complete.

Now an answer to (A) can be given as follows:

Theorem 1. *Given $x \in M_n^m(\mathbf{Q})$, the following assertions hold:*

(1) *There exist $c \in M_n^m(\mathbf{Z})$ and $d \in M_m^m(\mathbf{Z}) \cap GL_m(\mathbf{Q})$ such that $(c \quad d)$ is primitive and $x = d^{-1}c$.*

(2) *If c and d are as in (1) and if $x = d'^{-1}c'$ with $c' \in M_n^m(\mathbf{Z})$ and $d' \in M_m^m(\mathbf{Z}) \cap GL_m(\mathbf{Q})$, then $c' = hd$ and $d' = hd$ with $h \in M_m^m(\mathbf{Z})$. In addition if $(c' \quad d')$ is also primitive, then $h \in GL_m(\mathbf{Z})$.*

(3) *Let $e_1, \dots, e_r$ be the elementary divisors of x and for each i let $e_i = f_i/g_i$ with relatively prime positive integers f_i and g_i. If c and d are as in (1), then the elementary divisors of c are $f_1, \dots, f_r$ and the elementary divisors of d are $1, \dots, 1, g_r, \dots, g_1$ with 1 repeated $m - r$ times.*

Proof: Assuming the existence of c and d as in (1), suppose $x = d'^{-1}c'$ with $c' \in M_n^m(\mathbf{Z})$ and $d' \in M_m^m(\mathbf{Z}) \cap GL_m(\mathbf{Q})$; put $h = d'd^{-1}$. Then $d' = hd$, $c' = d'x = hc$, and so $h(c \quad d) = (c' \quad d') \in M_n^m(\mathbf{Z})$. By (5) of Lemma 2 we have $h \in M_m^m(\mathbf{Z})$. If $(c' \quad d')$ is also primitive, then exchanging $(c \quad d)$ for $(c' \quad d')$, we find that $h^{-1} \in M_m^m(\mathbf{Z})$, and hence $h \in GL_m(\mathbf{Z})$. This proves (2). To prove (1) and (3), we apply Lemma 1 to x to find an expression $axb = \begin{pmatrix} e & 0 \\ 0 & 0 \end{pmatrix}$ with $a \in GL_m(\mathbf{Z})$, $b \in GL_n(\mathbf{Z})$, $e = \mathrm{diag}[e_1, \dots, e_r]$, $0 < e_i \in \mathbf{Q}$, $e_{i+1}/e_i \in \mathbf{Z}$. Put $e_i = f_i/g_i$ as in (3) and $d = \mathrm{diag}[g_1, \dots, g_r, 1_{m-r}]a$, $f = \mathrm{diag}[f_1, \dots, f_r]$, $q = \begin{pmatrix} f & 0 \\ 0 & 0 \end{pmatrix}$, and $c = qb^{-1}$, where the zeros in q are arranged so that $q \in M_n^m(\mathbf{Z})$. Then $x = d^{-1}c$ and

$$(c \quad d)\begin{pmatrix} b & 0 \\ 0 & a^{-1} \end{pmatrix} = \begin{pmatrix} f_1 & & 0 & 0 & g_1 & & 0 & 0 \\ & \ddots & & & & \ddots & & \\ 0 & & f_r & 0 & 0 & & g_r & 0 \\ 0 & \cdots & 0 & 0 & 0 & \cdots & 0 & 1_{m-r} \end{pmatrix}.$$

We easily see that the right-hand side is primitive (by checking (4) of Lemma 2, for example), and hence $(c \quad d)$ is primitive since $\mathrm{diag}[b, a^{-1}] \in GL_{m+n}(\mathbf{Z})$. This proves (1). Put $k_i = e_{i+1}/e_i$. Then $k_i f_i g_{i+1} = g_i f_{i+1}$. Since f_i and g_i are relatively prime, we see that $f_i | f_{i+1}$, and similarly $g_{i+1} | g_i$. Therefore we obtain (3).

Remarks. (R1) If $(c \quad d)$ is as in Theorem 1, then (3) of Lemma 2 guarantees elements $a \in M_m^n(\mathbf{Z})$ and $b \in M_m^m(\mathbf{Z})$ such that $ca + db = 1_m$. Thus, as in number theory, $(c \quad d)$ may be called *relatively prime*. It should be noted that as can be seen from the example $x = \mathrm{diag}[2, 1]^{-1}\mathrm{diag}[1, 2]$, the maximum elementary divisors of c and d may not be relatively prime.

(R2) The expression $x = cd^{-1}$ can be treated in the same fashion by making obvious modifications. For example, in this case we assume that $\begin{pmatrix} c \\ d \end{pmatrix}$ is primitive.

(R3) Clearly all the above results can be extended to the case of matrices with entries in the field of quotients of a principal ideal domain, or in a central division algebra over a $\mathfrak{p}$-adic field. This comment applies also to (1) of Theorems 3, 4, 5 and Lemma 6 below.

Let us now turn to real matrices. In fact, we shall treat not only real matrices but also complex and quaternion matrices, since all these can be handled easily by the same technique. To make our exposition uniform, we let **K** denote any one of the following three objects: the real number field **R**, the complex number field **C**, and the division ring of Hamilton quaternions **H**. For $X = (x_{ij}) \in M_n^m(\mathbf{K})$ we define $X^* \in M_m^n(\mathbf{K})$ by $X^* = (y_{ij})$ with $y_{ij} = \bar{x}_{ji}$, where $\bar{x}$ is the image of x under complex conjugation or quaternion conjugation. We put also

$$U_n(\mathbf{K}) = \{\alpha \in GL_n(\mathbf{K}) | \alpha\alpha^* = 1_n\},$$

$$S_n(\mathbf{K}) = \{X \in M_n^n(\mathbf{K}) | X^* = X\},$$

and call an element X of $S_n(\mathbf{K})$ *positive definite* (resp. *nonnegative*) if $y^*Xy > 0$ (resp. $y^*Xy \geq 0$) for every $y \in M_1^n(\mathbf{K})$, $\neq 0$. If $\mathbf{K} = \mathbf{R}$, then $X^* = {}^tX$ and $U_n(\mathbf{K})$ is the orthogonal group of degree n. For $X, Y \in S_n(\mathbf{K})$ we write $X > Y$ and $Y < X$ if $X - Y$ is positive definite.

Lemma 3. *If $y \in M_n^m(\mathbf{K})$, then yy^* is nonnegative. Conversely, every positive definite element s of $S_n(\mathbf{K})$ can be written in the form $s = r^2$ with a positive definite element r of $S_n(\mathbf{K})$ and also in the form $s = aa^*$ with an upper triangular matrix a in $M_n^n(\mathbf{K})$ whose diagonal elements are positive real numbers. Both r and a are unique for s.*

This is well known. For $0 < s \in S_n(\mathbf{K})$ we write $s^{1/2} = r$ with the above element r; we then put $s^{-1/2} = (s^{-1})^{1/2}$.

Now the higher-dimensional analogue of (II) of the introduction is given by

Theorem 2. *Given $x \in M_n^m(\mathbf{K})$, there exist $c \in M_n^m(\mathbf{K})$ and $d \in GL_m(\mathbf{K})$ such that $x = d^{-1}c$ and $dd^* + cc^* = 1_m$. If $x = d'^{-1}c'$ is another such expression, then $c' = hc$ and $d' = hd$ with $h \in U_m(\mathbf{K})$. Moreover there is a unique choice of such $(c \quad d)$ for x with the property that $0 < d \in S_m(\mathbf{K})$.*

Proof: Observing that $1_m + xx^*$ is positive definite, we find an element e of $GL_m(\mathbf{K})$ such that $1_m + xx^* = ee^*$. Putting $d = e^{-1}$ and $c = dx$, we obtain the first assertion. Given another such pair (c', d'), put $h = d'e$. Then $d' = hd$, $c' = hc$, and $hh^* = d'ee^*d'^* = d'(1_m + xx^*)d'^* = d'd'^* + c'c'^* = 1_m$, and hence $h \in U_m(\mathbf{K})$. By Lemma 3 we can put $e = (1_m + xx^*)^{1/2}$. Then $0 < d \in S_m(\mathbf{K})$. If $0 < d' \in S_m(\mathbf{K})$, then $d'^2 = d^*h^*hd = d^2$, and hence $d' = d$ by Lemma 3. This completes the proof.

Remarks. (R4) In the above theorem, we can take d to be an upper triangular matrix with positive diagonal elements. When d is so chosen, the pair (c, d) is unique for x. This can be shown as in the above proof by means of Lemma 3. Can we prove a similar uniqueness in Theorem 1? We shall answer this question in (R7) below.

(R5) We are tempted to call an element X of $M_n^m(\mathbf{K})$, $m \leq n$, *primitive* if $XX^* = 1_m$, though that may not be good terminology. If we do use it, then $(c \quad d)$ with c, d as in Theorem 2 is primitive. Moreover, the analogue of (1) $\Leftrightarrow$ (2) of Lemma 2 can be stated as follows:

An element X of $M_n^m(\mathbf{K})$, $m \leq n$, is primitive if and only if there is an element Y of $M_n^{n-m}(\mathbf{K})$ such that $\binom{X}{Y} \in U_n(\mathbf{K})$.

Now the last part of Theorem 2 and its proof can be stated in the following way.

Lemma 4. *Put $\varphi(x) = (1_m + xx^*)^{-1/2}x$ for $x \in M_n^m(\mathbf{K})$. Then φ gives a one-to-one map of $M_n^m(\mathbf{K})$ onto $\{y \in M_n^m(\mathbf{K}) \mid 1_m > yy^*\}$, and the inverse map ψ of φ is given by $\psi(y) = (1_m - yy^*)^{-1/2}y$.*

This is an easy exercise (cf. [**2**, Lemma 2.3]).

3. A GROUP-THEORETICAL INTERPRETATION. Let us now show that there are some group-theoretical facts which give more substance to the parallelism between Theorems 1 and 2. For $F = \mathbf{Q}$ or $\mathbf{K}$ (or for any ring F) we define subgroups $P_n(F)$ of $GL_n(F)$ and $P_{n,m}(F)$ of $GL_{m+n}(F)$ by

$$P_{n,m}(F) = \left\{ \begin{pmatrix} a & b \\ 0 & d \end{pmatrix} \in GL_{m+n}(F) \,\middle|\, a \in GL_n(F),\, b \in M_m^n(F),\, d \in GL_m(F) \right\},$$

$$P_n(F) = \bigcap_{r=0}^{n} P_{r,n-r}(F),$$

where we understand that $P_{n,0}(F) = P_{0,n}(F) = GL_n(F)$. Clearly $P_n(F)$ is the group of all upper triangular matrices of $GL_n(F)$.

Theorem 3. (1) $GL_n(\mathbf{Q}) = P_n(\mathbf{Q})GL_n(\mathbf{Z}) = P_{r,n-r}(\mathbf{Q})GL_n(\mathbf{Z})$ $(0 \leq r \leq n)$.
(2) $GL_n(\mathbf{K}) = P_n(\mathbf{K})U_n(\mathbf{K}) = P_{r,n-r}(\mathbf{K})U_n(\mathbf{K})$ $(0 \leq r \leq n)$.

Proof: Though these are well known, we give a proof here for the reader's convenience. Since $P_{r,n-r}(F) \subset P_n(F)$, it is sufficient to prove the first equality in each case. We first prove Case (1) by induction on n. It is trivial if $n = 1$. Given $\xi \in GL_n(\mathbf{Q})$, $n > 1$, let x be the last row of ξ. Then $x = qy$ with $0 \neq q \in \mathbf{Q}$ and a primitive element y of $M_n^1(\mathbf{Z})$. Take an element α of $GL_n(\mathbf{Z})$ whose last row is y. Then $y = (0_{n-1}^1 \quad 1)\alpha$, so that $x\alpha^{-1} = (0_{n-1}^1 \quad q)$. Thus we can put $\xi\alpha^{-1} = \begin{pmatrix} r & s \\ 0 & q \end{pmatrix}$ with $r \in M_{n-1}^{n-1}(\mathbf{Q})$ and $s \in M_1^{n-1}(\mathbf{Q})$. By induction we find $\tau \in P_{n-1}(\mathbf{Q})$ and $\sigma \in GL_{n-1}(\mathbf{Z})$ such that $r = \tau\sigma$. Then $\xi\alpha^{-1} \cdot \mathrm{diag}[\sigma^{-1}, 1] \in P_n(\mathbf{Q})$, which gives (1). To prove (2), let $\xi \in GL_n(\mathbf{K})$. By Lemma 3 we can find $\eta \in P_n(\mathbf{K})$ such that $\xi\xi^* = \eta\eta^*$. Then $\eta^{-1}\xi \in U_n(\mathbf{K})$ and $\xi = \eta \cdot \eta^{-1}\xi$, which proves (2).

The groups $P_{n,m}$ and P_n are examples of *parabolic subgroups* of algebraic groups, and the decompositions of the above types are well known for classical groups. We shall give some more examples in Theorem 5 below.

Let us now derive the essential part of Theorems 1 and 2 from Theorem 3. Given $x \in M_n^m(\mathbf{Q})$, we consider an element

$$\begin{pmatrix} 1_n & 0_m^n \\ x & 1_m \end{pmatrix} \text{ of } GL_{m+n}(\mathbf{Q}).$$

By (1) of Theorem 3 we have

$$\begin{pmatrix} 1_n & 0_m^n \\ x & 1_m \end{pmatrix} = \begin{pmatrix} p & q \\ 0 & s \end{pmatrix}\begin{pmatrix} a & b \\ c & d \end{pmatrix} \tag{3.1}$$

with

$$\begin{pmatrix} p & q \\ 0 & s \end{pmatrix} \in P_{n,m}(\mathbf{Q}) \quad \text{and} \quad \begin{pmatrix} a & b \\ c & d \end{pmatrix} \in GL_{n+m}(\mathbf{Z}).$$

Then $(c \quad d)$ is primitive and $(x \quad 1) = (sc \quad sd)$. Therefore d is invertible and $x = d^{-1}c$.

Similarly, for $x \in M_n^m(\mathbf{K})$ we have (3.1) with

$$\begin{pmatrix} p & q \\ 0 & s \end{pmatrix} \in P_{n,m}(\mathbf{K}) \quad \text{and} \quad \begin{pmatrix} a & b \\ c & d \end{pmatrix} \in U_{m+n}(\mathbf{K}).$$

Again d is invertible and $x = d^{-1}c$. Since $\begin{pmatrix} a & b \\ c & d \end{pmatrix} \in U_{m+n}(\mathbf{K})$, we have $cc^* + dd^* = 1$ (or $(c \quad d)$ is primitive in the sense of (R5)).

Thus the parallelism between Theorems 1 and 2 can be interpreted as the parallelism between (1) and (2) of Theorem 3, or rather, between the decompositions given by (3.1) in both cases.

Theorem 3 positions $U_n(\mathbf{K})$ as the counterpart of $GL_n(\mathbf{Z})$. As the counterpart of Lemma 1 in this sense we obtain

Lemma 5. *Given $X \in M_n^m(\mathbf{K})$ of rank r, there exist $A \in U_m(\mathbf{K})$, $B \in U_n(\mathbf{K})$ and positive real numbers $e_1, \ldots, e_r$ such that $e_i \le e_{i+1}$ for all $i < r$ and* (2.1) *holds. Moreover, the e_i are uniquely determined by X.*

The statement for X belonging to the semisimple part of $GL_n(\mathbf{K})$ is a special case of the fact in the theory of Lie groups which is usually stated as $G = K\overline{A_+}K$ (cf. [1, p. 402]).

Proof: The uniqueness follows from the easy fact that the e_i^2 are the nonzero eigenvalues of XX^*. To obtain the desired expression, we view (left matrix multiplication by) X as a (right-) $\mathbf{K}$-linear map of $M_1^n(\mathbf{K})$ into $M_1^m(\mathbf{K})$. Let V be the orthogonal complement of $\mathrm{Ker}(X)$ in $M_1^n(\mathbf{K})$. Then X is essentially a map of V onto XV. This reduces our problem to the case in which $m = n$ and $X \in GL_n(\mathbf{K})$. In this case we can find an element A of $U_n(\mathbf{K})$ so that $AXX^*A^* = \mathrm{diag}[f_1, \ldots, f_n]$, $0 < f_1 < \cdots < f_n$. Let $e_i = f_i^{1/2}$ and $B = X^{-1}A^{-1}\,\mathrm{diag}[e_1, \ldots, e_n]$. Then $BB^* = 1_n$ and we obtain the desired expression.

Remarks. (R6) As mentioned in (R3), we can prove the same types of results with any $\mathfrak{p}$-adic field F and its ring A of $\mathfrak{p}$-adic integers in place of $\mathbf{Q}$ and $\mathbf{Z}$. In this case, $GL_n(A)$ is a maximal compact subgroup of $GL_n(F)$, as $U_n(\mathbf{K})$ is in $GL_n(\mathbf{K})$. (The fact follows easily from the $\mathfrak{p}$-adic version of Lemma 1 and Lemma 5.) This kind of analogy belongs to the standard philosophy in number theory today, but we shall not make any $\mathfrak{p}$-adic discussion in this article.

(R7) In Theorem 1, for a given x the element d can be changed for any element in the coset $GL_m(\mathbf{Z})d$. This means that if we fix a complete set of representatives D for $GL_m(\mathbf{Z}) \setminus GL_m(\mathbf{Q})$, we can choose d from D. Now by Theorem 3(1) we have $GL_m(\mathbf{Q}) = GL_m(\mathbf{Z})P_m(\mathbf{Q})$. Therefore we can take D to be a subset of $P_m(\mathbf{Q})$. More precisely, if we take a complete set of representatives D_0 for $[GL_m(\mathbf{Z}) \cap P_m(\mathbf{Q})] \setminus P_m(\mathbf{Q})$, then, *given x, there exists a unique $(c \quad d)$ as in Theorem* 1(1) *with $d \in D_0$.*

4. FRACTIONAL EXPRESSIONS FOR HERMITIAN AND SKEWHERMITIAN MATRICES. By restricting our matrices to symmetric, alternating, hermitian, or skewhermitian ones, we can still make clear-cut statements on their fractional expressions. Though the results follow immediately from Theorems 1 and 2, they reflect the decompositions of symplectic, orthogonal, and unitary groups in the same sense as in Theorem 3, and therefore, may make the parallelism more cogent.

With A denoting **Z**, **Q**, or **K** and the symbol $\varepsilon = (\varepsilon_1, \varepsilon_2)$ with $\varepsilon_\nu = +$ or $-$ we put

$$S^\varepsilon(A) = \{x \in M_n^n(A) | x^\rho = -\varepsilon_1 x\},$$

$$G^\varepsilon(A) = \{X \in GL_{2n}(A) | XJ_\varepsilon X^\rho = J_\varepsilon\}, \qquad J_\varepsilon = \begin{pmatrix} 0 & \varepsilon_1 1_n \\ 1_n & 0 \end{pmatrix},$$

$$C^\varepsilon(\mathbf{K}) = G^\varepsilon(\mathbf{K}) \cap U_{2n}(\mathbf{K}),$$

$$V^\varepsilon(A) = \{w \in M_{2n}^n(A) | wJ_\varepsilon w^\rho = 0, \operatorname{rank}(w) = n\}.$$

Here $X^\rho = X^*$ if $\varepsilon_2 = +$ and $X^\rho = {}^tX$ if $\varepsilon_2 = -$; we consider $\varepsilon_2 = -$ only when $A = \mathbf{K} = \mathbf{C}$. If $A = \mathbf{H}$, the condition $\operatorname{rank}(w) = n$ means that $wM_1^{2n}(\mathbf{H}) = M_1^n(\mathbf{H})$. Clearly $S^{(-,+)}(\mathbf{K}) = S_n(\mathbf{K})$. If $w = (c \quad d) \in M_{2n}^n(A)$ with $c, d \in M_n^n(A)$ and $\operatorname{rank}(w) = n$, then

$$w \in V^\varepsilon(A) \Leftrightarrow cd^\rho = -\varepsilon_1 dc^\rho \Leftrightarrow cd^\rho \in S^\varepsilon(A). \tag{4.1}$$

If A is **Z**, **Q**, or **R**, the group $G^{(-,+)}(A)$ is usually denoted by $Sp(n, A)$. If $w \in V^\varepsilon(A)$, $\alpha \in GL_n(A)$, and $\beta \in G^\varepsilon(A)$, then $\alpha w \beta \in V^\varepsilon(A)$. Since $(0 \quad 1_n) \in V^\varepsilon(A)$, we have $(0 \quad 1_n)\beta \in V^\varepsilon(A)$. Clearly $(0 \quad 1_n)\beta$ is the lower half of β. We also put

$$W^\varepsilon(\mathbf{Z}) = \{w \in V^\varepsilon(\mathbf{Z}) | w \text{ is primitive}\}, \tag{4.2}$$

$$W^\varepsilon(\mathbf{K}) = \{w \in V^\varepsilon(\mathbf{K}) | ww^* = 1_n\}. \tag{4.3}$$

Clearly an element $(c \quad d)$ of $V^\varepsilon(\mathbf{K})$ belongs to $W^\varepsilon(\mathbf{K})$ if and only if $cc^* + dd^* = 1_n$. Now our fractional expressions for the matrices in $S^\varepsilon(A)$ are given by

Theorem 4. (1) *Given $x \in S^\varepsilon(\mathbf{Q})$, there exists $(c \quad d) \in W^\varepsilon(\mathbf{Z})$ with an invertible d such that $x = d^{-1}c$.*

(2) *Given $x \in S^\varepsilon(\mathbf{K})$, there exists $(c \quad d) \in W^\varepsilon(\mathbf{K})$ with an invertible d such that $x = d^{-1}c$.*

Proof: For $x \in S^\varepsilon(\mathbf{Q})$ take c and d as in (1) of Theorem 1. Then ${}^tc \cdot {}^td^{-1} = {}^t(d^{-1}c) = -\varepsilon_1 d^{-1}c$, and hence $(c \quad d) \in W^\varepsilon(\mathbf{Z})$, which proves (1). Assertion (2) can be derived from Theorem 2 in the same fashion.

As the analogue of Lemma 2(2) and the statement in (R5) we obtain:

Lemma 6. (1) $W^\varepsilon(\mathbf{Z}) = (0 \quad 1_n)G^\varepsilon(\mathbf{Z})$.

(2) $W^\varepsilon(\mathbf{K}) = (0 \quad 1_n)C^\varepsilon(\mathbf{K})$. *More precisely, $\alpha \mapsto (0 \quad 1_n)\alpha$ gives a one-to-one map of $C^\varepsilon(\mathbf{K})$ onto $W^\varepsilon(\mathbf{K})$.*

Proof: To prove (1), we first observe that $(0 \quad 1_n)G^\varepsilon(\mathbf{Z}) \subset W^\varepsilon(\mathbf{Z})$ since $(0 \quad 1_n)$ is primitive and $G^\varepsilon(\mathbf{Z}) \subset GL_{2n}(\mathbf{Z})$. Conversely, take $x \in W^\varepsilon(\mathbf{Z})$. By (2) of Lemma 2, we can find some $y \in \mathbf{Z}_{2n}^n$ such that $\binom{y}{x} \in GL_{2n}(\mathbf{Z})$. Put $\alpha = \binom{y}{x}$. Then

$$\alpha J_\varepsilon \cdot {}^t\alpha = \begin{pmatrix} y \\ x \end{pmatrix} \begin{pmatrix} J_\varepsilon \cdot {}^t y & J_\varepsilon \cdot {}^t x \end{pmatrix} = \begin{pmatrix} u & v \\ \varepsilon_1 \cdot {}^t v & 0 \end{pmatrix}$$

with $u, v \in M_n^n(\mathbf{Z})$. Since $\alpha J_\varepsilon \cdot {}^t\alpha \in GL_{2n}(\mathbf{Z})$, we see that $v \in GL_n(\mathbf{Z})$. Put $\beta = \operatorname{diag}[\varepsilon_1 v^{-1}, 1_n]$. Then

$$\beta\alpha J_\varepsilon \cdot {}^t\alpha \cdot {}^t\beta = \begin{pmatrix} z & \varepsilon_1 1_n \\ 1_n & 0 \end{pmatrix}$$

with $z \in M_n^n(\mathbf{Z})$. If $(a \quad b)$ is the upper half of $\beta\alpha$, then $z = \varepsilon_1 a \cdot {}^t b + b \cdot {}^t a$. Put

$$\gamma = \begin{pmatrix} 1_n & -b \cdot {}^t a \\ 0 & 1_n \end{pmatrix}.$$

Then $\gamma\beta\alpha J_\varepsilon \cdot {}^t\alpha \cdot {}^t\beta \cdot {}^t\gamma = J_\varepsilon$, and so $\gamma\beta\alpha \in G^\varepsilon(\mathbf{Z})$. Now we see that $(0 \quad 1_n)\gamma\beta = (0 \quad 1_n)$, and hence $(0 \quad 1_n)\gamma\beta\alpha = (0 \quad 1_n)\alpha = x$, which proves (1). Assertion (2) follows from

$$C^\varepsilon(\mathbf{K}) = \left\{ \begin{pmatrix} d' & \varepsilon_1 c' \\ c & d \end{pmatrix} \middle| (c \quad d) \in W^\varepsilon(\mathbf{K}) \right\}, \tag{4.4}$$

where $(d' \quad c') = (\bar{d} \quad \bar{c})$ if $\mathbf{K} = \mathbf{C}$ and $\varepsilon_2 = -$; $(d' \quad c') = (d \quad c)$ otherwise. In fact, it is easy to verify that any matrix on the right-hand side is contained in $C^\varepsilon(\mathbf{K})$. Conversely, let

$$\alpha \in C^\varepsilon(\mathbf{K}), \quad (0 \quad 1_n)\alpha = (c \quad d), \quad \text{and} \quad \beta = \begin{pmatrix} d' & \varepsilon_1 c' \\ c & d \end{pmatrix}.$$

Then $\beta \in C^\varepsilon(\mathbf{K})$ and $(0 \quad 1_n)\alpha = (0 \quad 1_n)\beta$, and hence $\alpha\beta^{-1} = \begin{pmatrix} a & b \\ 0 & 1 \end{pmatrix}$ with $a, b \in M_n^n(\mathbf{K})$. Since $\alpha\beta^{-1} \in C^\varepsilon(\mathbf{K})$, we see that $a = 1$ and $b = 0$, which proves (4.4).

To obtain the analogue of Theorem 3 in the present case, put

$$P^\varepsilon(\mathbf{Q}) = G^\varepsilon(\mathbf{Q}) \cap P_{n,n}(\mathbf{Q}), \qquad P^\varepsilon(\mathbf{K}) = G^\varepsilon(\mathbf{K}) \cap P_{n,n}(\mathbf{K}).$$

Theorem 5. (1) $G^\varepsilon(\mathbf{Q}) = P^\varepsilon(\mathbf{Q})G^\varepsilon(\mathbf{Z})$.
(2) $G^\varepsilon(\mathbf{K}) = P^\varepsilon(\mathbf{K})C^\varepsilon(\mathbf{K})$.

Proof: To prove (1), let $\xi \in G^\varepsilon(\mathbf{Q})$. By Theorem 3(1) we have $\xi = \eta\alpha$ with $\eta \in P_{n,n}(\mathbf{Q})$ and $\alpha \in GL_{2n}(\mathbf{Z})$. Put

$$\eta = \begin{pmatrix} a & b \\ 0 & d \end{pmatrix} \quad \text{and} \quad \xi = \begin{pmatrix} p & q \\ r & s \end{pmatrix}$$

with a, b, etc. of size n. Then $d^{-1}(r \quad s) = (0 \quad 1_n)\alpha$. Since $\xi \in G^\varepsilon(\mathbf{Q})$, the left-hand side of the last equality, $d^{-1}(r \quad s)$, belongs to $V^\varepsilon(\mathbf{Q})$. The right-hand side $(0 \quad 1_n)\alpha$ is primitive. Thus $(0 \quad 1_n)\alpha \in W^\varepsilon(\mathbf{Z})$. By Lemma 6(1) we have $(0 \quad 1_n)\alpha = (0 \quad 1_n)\beta$ with $\beta \in G^\varepsilon(\mathbf{Z})$. Put $\gamma = \alpha\beta^{-1}$. Then $(0 \quad 1_n)\gamma = (0 \quad 1_n)$, and hence $\gamma \in P_{n,n}(\mathbf{Q})$. Now $\xi = \eta\alpha = \eta\gamma\beta$, which proves (1) if we can show that $\eta\gamma \in P^\varepsilon(\mathbf{Q})$, but this is indeed the case, since $\eta\gamma \in P_{n,n}(\mathbf{Q})$ and $\eta\gamma = \xi\beta^{-1} \in G^\varepsilon(\mathbf{Q})$. Similarly let $\xi \in G^\varepsilon(\mathbf{K})$. By Theorem 3(2) we have $\xi = \eta\alpha$ with $\eta \in P_{n,n}(\mathbf{K})$ and $\alpha \in U_{2n}(\mathbf{K})$. With d, r, and s as above, $(0 \quad 1_n)\alpha = d^{-1}(r \quad s) \in W^\varepsilon(\mathbf{K})$. By Lemma 6(2) we have $(0 \quad 1_n)\alpha = (0 \quad 1_n)\beta$ with $\beta \in C^\varepsilon(\mathbf{K})$. Repeating the above argument with $\mathbf{K}$ in place of $\mathbf{Q}$, we obtain (2).

The above theorem enables us to explain Theorem 4 again by means of the decomposition of (3.1). Indeed, if $x \in S^\varepsilon(A)$, then $\begin{pmatrix} 1 & 0 \\ x & 1 \end{pmatrix} \in G^\varepsilon(A)$. By Theorem 5 we have (3.1) with $\begin{pmatrix} p & q \\ 0 & s \end{pmatrix} \in P^\varepsilon(A)$ and $\begin{pmatrix} a & b \\ c & d \end{pmatrix} \in C$, where C denotes $C^\varepsilon(\mathbf{K})$ or $G^\varepsilon(\mathbf{Z})$ according as $A = \mathbf{K}$ or $\mathbf{Q}$. Then we obtain $x = d^{-1}c$ with $(c \quad d) \in W^\varepsilon(\mathbf{K})$ or $W^\varepsilon(\mathbf{Z})$.

We conclude this section by adding some historical notes. The expression $x = d^{-1}c$ of a symmetric rational matrix x by means of $(c \quad d)$ belonging to $W^{(-,+)}(\mathbf{Z})$ was first considered by Siegel in [3, p. 653]. In particular, he observes that $|\det(d)|$ is the product of the reduced denominators of the elementary divisors of x (as shown in Theorem 1(3)), and puts $\nu(x) = |\det(d)|$. He employed this function ν in his investigations of Eisenstein series in [3] and of indefinite quadratic forms in [4]. It should also be noted that Lemma 6(1) for $G^\varepsilon(\mathbf{Q}) = Sp(n, \mathbf{Q})$ was first given by Siegel. The decomposition of an algebraic group over a local field as in Theorems 3 and 5, as well as the generalization of Siegel's function ν, has been employed in recent papers on automorphic forms.

5. MATRIX-VALUED TRIGONOMETRIC FUNCTIONS. As mentioned at the beginning, c and d in the one-dimensional case can be given as $\cos\theta$ and $\sin\theta$. Let us now give higher-dimensional analogues of these trigonometric functions which play similar roles. With fixed m and n we put

$$W(\mathbf{K}) = \{(c \quad d) \in M_n^m(\mathbf{K}) \times M_m^m(\mathbf{K}) \,|\, cc^* + dd^* = 1_m\},$$
$$W'(\mathbf{K}) = \{(c \quad d) \in W(\mathbf{K}) \,|\, d = d^*\}.$$

For $X \in M_n^n(\mathbf{K})$ we put $\exp(X) = \sum_{\nu=0}^{\infty} X^\nu/\nu!$ as usual, and we define an $M_n^m(\mathbf{K})$-valued function $\mathbf{c}$, an $M_m^m(\mathbf{K})$-valued function $\mathbf{d}$, and a $GL_{m+n}(\mathbf{K})$-valued function E on $M_n^m(\mathbf{K})$ by

$$\mathbf{c}(y) = \sum_{\nu=0}^{\infty} (-yy^*)^\nu y/(2\nu+1)!, \qquad \mathbf{d}(y) = \sum_{\nu=0}^{\infty} (-yy^*)^\nu/(2\nu)!,$$

$$E(y) = \exp\left(\begin{pmatrix} 0_n^n & -y^* \\ y & 0_m^m \end{pmatrix}\right) \qquad (y \in M_n^m(\mathbf{K})).$$

We have clearly

$$\mathbf{c}(y)^* = \mathbf{c}(y^*), \quad \mathbf{d}(y)^* = \mathbf{d}(y), \quad {}^t\mathbf{c}(y) = \mathbf{c}({}^ty), \quad {}^t\mathbf{d}(y) = \mathbf{d}(\bar{y}),$$
$$E(y) = \begin{pmatrix} \mathbf{d}(y^*) & -\mathbf{c}(y)^* \\ \mathbf{c}(y) & \mathbf{d}(y) \end{pmatrix}.$$

Since $\exp(X) \in U_N(\mathbf{K})$ if $X^* = -X \in M_N^N(\mathbf{K})$, the matrix $E(y)$ belongs to $U_{m+n}(\mathbf{K})$, and hence

$$\mathbf{c}(y)\mathbf{c}(y)^* + \mathbf{d}(y)^2 = 1_m, \qquad \mathbf{c}(y)^*\mathbf{c}(y) + \mathbf{d}(y^*)^2 = 1_n,$$
$$\mathbf{c}(y)\mathbf{d}(y^*) = \mathbf{d}(y)\mathbf{c}(y).$$

Theorem 6. *Every element x of $M_n^m(\mathbf{K})$ can be written $x = \mathbf{d}(y)^{-1}\mathbf{c}(y)$ with an element $y \in M_n^m(\mathbf{K})$ such that $\mathbf{d}(y)$ is invertible. Moreover, we have*

$$W(\mathbf{K}) = U_m(\mathbf{K})W'(\mathbf{K}) = U_m(\mathbf{K})\{(\mathbf{c}(y) \quad \mathbf{d}(y)) \,|\, y \in M_n^m(\mathbf{K})\}, \tag{5.1}$$
$$W'(\mathbf{K}) = \{(\mathbf{c}(y) \quad \mathbf{d}(y)) \,|\, y \in M_n^m(\mathbf{K})\} \quad \text{if } m \le n, \tag{5.2}$$
$$U_{m+n}(\mathbf{K}) = (U_n(\mathbf{K}) \times U_m(\mathbf{K}))E(M_n^m(\mathbf{K})), \tag{5.3}$$

where $U_n(\mathbf{K}) \times U_m(\mathbf{K}) = \{\mathrm{diag}[a, b] \,|\, a \in U_n(\mathbf{K}),\ b \in U_m(\mathbf{K})\}$.

Proof: The first assertion follows from Theorem 2 and (5.1). Let $(c\ \ d) \in W(\mathbf{K})$. By Lemma 5 we have $d = uev$ with $u, v \in U_m(\mathbf{K})$ and a real diagonal matrix e. Put $w = v^{-1}u^{-1}$ and $(e\ \ f) = w(c\ \ d)$. Then $(e\ \ f) \in W(\mathbf{K})$ and $f = v^{-1}ev$. Thus $f^* = f$ and so $(e\ \ f) \in W'(\mathbf{K})$, which proves the first equality of (5.1). Assuming now $m \leq n$, let us prove (5.2). Clearly the right-hand side is contained in the left-hand side. To prove the opposite inclusion, let $(c\ \ d) \in W'(\mathbf{K})$. Take $p \in U_m(\mathbf{K})$ so that $pdp^{-1} = \mathrm{diag}[a, -1_s, 1_t]$, $m = r + s + t$, with a real diagonal matrix a of size r whose diagonal entries are different from ± 1. Then $pc(pc)^* = 1_r - pd^2p^{-1} = \mathrm{diag}[1_r - a^2, 0^{m-r}_{m-r}]$. From this we can conclude that

$$pc = \begin{pmatrix} z \\ 0^{m-r}_n \end{pmatrix} \quad \text{with} \quad z \in M^r_n(\mathbf{K}).$$

We can then find a real diagonal matrix b of size r such that $b^2 = 1_r - a^2$, and also a diagonal matrix σ of size r such that $\exp(i\sigma) = ib + a$. By Lemma 5 we can find $q \in U_n(\mathbf{K})$ such that $z = (v\ \ 0^r_{n-r})q$ with $v \in GL_r(\mathbf{K})$. Then $vv^* = zz^* = b^2$. Put

$$y = p^{-1}\begin{pmatrix} \sigma b^{-1}v & 0^r_s & 0^r_u \\ 0^s_r & \pi 1_s & 0^s_u \\ 0^t_r & 0^t_s & 0^t_u \end{pmatrix} q,$$

where $u = n - r - s$ and π is a real number such that $e^{i\pi} = -1$. Then $yy^* = p^{-1}\,\mathrm{diag}[\sigma^2, \pi^2 1_s, 0]p$, and a straightforward calculation shows that $\mathbf{c}(y) = c$ and $\mathbf{d}(y) = d$, which proves (5.2), and the second equality of (5.1) as well, under the condition $m \leq n$. If $m > n$, this reasoning fails since $r + s$ may be greater than n. To avoid this difficulty, after taking p, a, b, and σ for a given $(c\ \ d) \in W'(\mathbf{K})$, we put $u = p^{-1}\,\mathrm{diag}[1_r, -1_s, 1_t]p$ and $(e\ \ f) = u(c\ \ d)$. Then $pfp^{-1} = \mathrm{diag}[a, 1_{s+t}]$, $(pe)(pe)^* = \mathrm{diag}[b^2, 0]$, and

$$pe = \begin{pmatrix} z \\ 0^{m-r}_n \end{pmatrix} \quad \text{with} \quad z \in M^r_n(\mathbf{K}).$$

Taking v and q as above, put

$$y = p^{-1}\begin{pmatrix} \sigma b^{-1}v & 0^r_{n-r} \\ 0^{m-r}_r & 0^{m-r}_{n-r} \end{pmatrix} q.$$

Then $yy^* = p^{-1}\,\mathrm{diag}[\sigma^2, 0]p$, $\mathbf{c}(y) = e$, and $\mathbf{d}(y) = f$, which proves (5.1) in general. To prove (5.3), let $\alpha = \begin{pmatrix} a & b \\ c & d \end{pmatrix} \in U_{m+n}(\mathbf{K})$. Since $(c\ \ d) \in W(\mathbf{K})$, we have $(c\ \ d) = u(\mathbf{c}(y)\ \ \mathbf{d}(y))$ with $u \in U_m(\mathbf{K})$ and $y \in M^m_n(\mathbf{K})$ by (5.1). Put $\beta = E(y)$. Then $(0\ \ 1_m)\alpha = (0\ \ u)\beta$, and hence we can put $\mathrm{diag}[1_n, u^{-1}]\alpha\beta^{-1} = \begin{pmatrix} g & h \\ 0 & 1_n \end{pmatrix}$ with $g, h \in M^n_n(\mathbf{K})$. Since the left-hand side belongs to $U_{m+n}(\mathbf{K})$, we see that $g \in U_n(\mathbf{K})$ and $h = 0$. This proves (5.3), since the opposite inclusion is obvious.

It should be noted that the equality of (5.2) is false if $m > n$. In fact, take $(c\ \ d) = (0\ \ -1_m)$. Suppose that $m > n$ and $d = \mathbf{d}(y)$ with $y \in M^m_n(\mathbf{K})$. We can find $p \in U_m(\mathbf{K})$ and a real diagonal matrix r such that $yy^* = pr^2p^{-1}$. Then $-1_m = p^{-1}\mathbf{d}(y)p = \mathbf{d}(r)$, and hence $r = \mathrm{diag}[a_1\pi, \ldots, a_m\pi]$ with odd integers a_ν. This is a contradiction, since $\mathrm{rank}(yy^*) \leq n < m$.

To consider the case of matrices in $S^\varepsilon(\mathbf{K})$, we put

$$Z^\varepsilon(\mathbf{K}) = \{(c\ \ d) \in W^\varepsilon(\mathbf{K}) \mid c^\rho = -\varepsilon_1 c,\ d^* = d\}$$

and restrict the functions $\mathbf{c}$ and $\mathbf{d}$ to $S^\varepsilon(\mathbf{K})$. Clearly, for $s \in S^\varepsilon(\mathbf{K})$ we have

$$\mathbf{c}(s)^\rho = -\varepsilon_1\mathbf{c}(s), \qquad \mathbf{d}(s)^* = \mathbf{d}(s), \qquad \mathbf{d}(s)^\rho = \mathbf{d}(s^*),$$

$$\mathbf{c}(s)\mathbf{d}(s)^\rho = \mathbf{d}(s)\mathbf{c}(s), \qquad \mathbf{c}(s)\mathbf{c}(s)^* + \mathbf{d}(s)^2 = 1_n.$$

From these and (4.4) we see that

$$\{(\mathbf{c}(s) \quad \mathbf{d}(s)) | s \in S^\varepsilon(\mathbf{K})\} \subset Z^\varepsilon(\mathbf{K})$$
$$= \{(c \quad d) \in M^n_{2n}(\mathbf{K}) | c^\rho = -\varepsilon_1 c,\ d^* = d,\ cd^\rho = dc,\ cc^* + dd^* = 1_n\}, \tag{5.4}$$

$$E(S^\varepsilon(\mathbf{K})) \subset \{\alpha \in C^\varepsilon(\mathbf{K}) | \alpha^*\lambda = \lambda\alpha\}, \qquad \lambda = \operatorname{diag}[1_n, -1_n]. \tag{5.5}$$

Theorem 7. *Every element x of $S^\varepsilon(\mathbf{K})$ can be written $x = \mathbf{d}(s)^{-1}\mathbf{c}(s)$ with an element $s \in S^\varepsilon(\mathbf{K})$ such that $\mathbf{d}(s)$ is invertible. Moreover, excluding the two cases in which $(\varepsilon, \mathbf{K}) = ((+,+), \mathbf{R})$ or $((+,-), \mathbf{C})$, we have*

$$W^\varepsilon(\mathbf{K}) = U_n(\mathbf{K})Z^\varepsilon(\mathbf{K}), \tag{5.6}$$

$$Z^\varepsilon(\mathbf{K}) = \{(\mathbf{c}(s) \quad \mathbf{d}(s)) | s \in S^\varepsilon(\mathbf{K})\}, \tag{5.7}$$

$$E(S^\varepsilon(\mathbf{K})) = \{\alpha \in C^\varepsilon(\mathbf{K}) | \alpha^*\lambda = \lambda\alpha\}, \tag{5.8}$$

$$C^\varepsilon(\mathbf{K}) = \{\operatorname{diag}[u', u] | u \in U_n(\mathbf{K})\}E(S^\varepsilon(\mathbf{K})), \tag{5.9}$$

where $u' = \bar{u}$ if $(\varepsilon, \mathbf{K}) = ((-,-), \mathbf{C})$ and $u' = u$ otherwise.

Proof: The first assertion follows from Theorem 4(2), (5.6), and (5.7) except in the two exceptional cases, which will be treated at the end of the proof. To prove (5.7), let $(c \quad d) \in Z^\varepsilon(\mathbf{K})$. We repeat the proof of Theorem 6 with some modifications. Taking p and a as in that proof, we put $g = pcp^\rho$ and $h = pdp^{-1}$. Then $g^\rho = -\varepsilon_1 g$, $gg^* = \operatorname{diag}[1 - a^2, 0]$, and $gh^\rho = hg$. From this we can conclude that $g = \operatorname{diag}[v, 0]$ with $v = -\varepsilon_1 v^\rho \in M_r^r(\mathbf{K})$, $va = av$. We may assume that $a = \operatorname{diag}[a_1 1_{r_1}, \ldots, a_k 1_{r_k}]$ with $r = r_1 + \cdots + r_k$ and the a_ν which are all different. Then $v = \operatorname{diag}[v_1, \ldots, v_k]$ with $v_\nu = -\varepsilon_1 v_\nu^\rho \in M_{r_\nu}^{r_\nu}(\mathbf{K})$. Put $b = \operatorname{diag}[b_1 1_{r_1}, \ldots, b_k 1_{r_k}]$ and $\sigma = \operatorname{diag}[\sigma_1 1_{r_1}, \ldots, \sigma_k 1_{r_k}]$ with real numbers b_ν and σ_ν such that $b_\nu^2 = 1 - a_\nu^2$ and $\exp(i\sigma_\nu) = ib_\nu + a_\nu$. If $\varepsilon_1 = -$, put $y = p^{-1}\operatorname{diag}[\sigma b^{-1}v, \pi 1_s, 0](p^\rho)^{-1}$. Then $y \in S^\varepsilon(\mathbf{K})$, $c = \mathbf{c}(y)$, and $d = \mathbf{d}(y)$. If $\varepsilon = (+,+)$ and $\mathbf{K} \neq \mathbf{R}$, then we get the same conclusion with $y = p^{-1}\operatorname{diag}[\sigma b^{-1}v, i\pi 1_s, 0]p$. This combined with (5.4) proves (5.7). Then (5.8) follows immediately from (5.7) and (4.4). To prove (5.9), given $\alpha \in C^\varepsilon(\mathbf{K})$, put $\beta = \lambda\alpha^*\lambda\alpha$. Then $\beta \in C^\varepsilon(\mathbf{K})$ and $\beta^* = \lambda\beta\lambda$, so that $\beta = E(y)$ with $y \in S^\varepsilon(\mathbf{K})$ by (5.7). Put $\gamma = E(-y/2)$. Then $\lambda\alpha^{-1}\lambda\alpha = \beta = \gamma^2 = \lambda\gamma^{-1}\lambda\gamma$, and hence $\lambda\gamma\alpha^{-1}\lambda = \gamma\alpha^{-1}$. This shows that $\gamma\alpha^{-1}$ commutes with λ, which implies that $\gamma\alpha^{-1} = \operatorname{diag}[w, z]$ with $w, z \in U_n(\mathbf{K})$. Since $\gamma\alpha^{-1} \in C^\varepsilon(\mathbf{K})$, we see that $w = z'$. This proves (5.9). Finally, (5.6) follows from Lemma 6(2), (5.9), and (5.4).

In the exceptional cases we proceed as follows: Put $u = p^{-1}\operatorname{diag}[1_r, -1_s, 1_t]p$ and $(e \quad f) = u(c \quad d)$. Then $pep^\rho = g$, $pfp^{-1} = \operatorname{diag}[a, 1_{s+t}]$. Putting $y = p^{-1}\operatorname{diag}[\sigma b^{-1}v, 0](p^\rho)^{-1}$, we obtain $y \in S^\varepsilon(\mathbf{K})$, $e = \mathbf{c}(y)$, and $f = \mathbf{d}(y)$. Thus, in the exceptional cases we have

$$Z^\varepsilon(\mathbf{K}) \subset U_n(\mathbf{K})\{(\mathbf{c}(s) \quad \mathbf{d}(s)) | s \in S^\varepsilon(\mathbf{K})\}. \tag{5.10}$$

Let us now prove the first assertion in those cases. Given $x \in S^\varepsilon(\mathbf{K})$, let $d = (1 + xx^*)^{-1/2}$ and $c = dx$. As noted in the proof of Theorem 2 (cf. also Lemma 4),

$(c \quad d) \in W(\mathbf{K})$, $d^* = d$, and $x = d^{-1}{}^tc$. Since ${}^tx = -x$, we have $\bar{d} = (1 + x^*x)^{-1/2}$. If we assume that $dx = x\bar{d}$, then ${}^tc = -c$ and $c \cdot {}^td = dc$, and hence $(c \quad d) \in Z^\varepsilon(\mathbf{K})$. By (5.10) we have $(c \quad d) = a(\mathbf{c}(s) \quad \mathbf{d}(s))$ with $a \in U_n(\mathbf{K})$ and $s \in S^\varepsilon(\mathbf{K})$, which proves the desired assertion. The assumed equality $dx = x\bar{d}$ can be proved as follows. Put $f = d^{-1}$ and $g = \bar{d}^{-1}$. Then $f^2x = (1 + xx^*)x = x(1 + x^*x) = xg^2$. Suppose $gv = \lambda v$ with $0 \neq v \in M_1^n(\mathbf{K})$ with $\lambda \in \mathbf{R}$. Then $\lambda > 0$ and $f^2xv = \lambda^2xu$, and hence $fxv = \lambda xv = xgv$. Since $M_1^n(\mathbf{K})$ is spanned by all such v's, we obtain $fx = xg$, and so $dx = x\bar{d}$. This completes the proof.

Let us note that (5.3) and (5.9) are special cases of a general principle

$$G = K \cdot \exp(\mathfrak{p}). \tag{5.11}$$

Here G is a connected Lie group, K is a compact subgroup of G, and $\mathfrak{p}$ is the subset of the Lie algebra $\mathfrak{g}$ of G defined by

$$\mathfrak{p} = \{X \in \mathfrak{g} \mid d\theta X = -X\},$$

where θ is an analytic automorphism of G of order 2 such that K is a subgroup of $K_\theta = \{g \in G \mid \theta g = g\}$ containing the identity component of K_θ. The explicit form of θ in our cases is: $\theta g = \lambda g \lambda^{-1}$, $\lambda = \mathrm{diag}[1_n, -1_m]$ with $m = n$ for (5.9). Relation (5.11) is usually stated only for noncompact groups, but actually it is true in general for the following reason. Since G/K is a complete Riemannian manifold, any point on it can be connected to the origin by a geodesic. Now any geodesic passing through the origin is the image of $\{\exp(tX) \mid t \in \mathbf{R}\}$, $X \in \mathfrak{p}$, under the natural map $G \to G/K$ (see [**1**, pp. 208–9]), which proves (5.11).

In the exceptional cases $C^\varepsilon(\mathbf{K})$ is not connected. So let us denote by $C_0^\varepsilon(\mathbf{K})$ the identity component of $C^\varepsilon(\mathbf{K})$, and put

$$W_0^\varepsilon(\mathbf{K}) = (0 \quad 1_n)C_0^\varepsilon(\mathbf{K}), \qquad Z_0^\varepsilon(\mathbf{K}) = \{(\mathbf{c}(s) \quad \mathbf{d}(s)) \mid s \in S^\varepsilon(\mathbf{K})\}. \tag{5.12}$$

By Lemma 6(2) $W_0^\varepsilon(\mathbf{K})$ is one of the connected components of $W^\varepsilon(\mathbf{K})$.

To see the nature of our group $C^\varepsilon(\mathbf{K})$ more clearly, let us use the traditional notation by putting

$$O(n) = U_n(\mathbf{R}), \qquad SO(n) = \{\alpha \in O(n) \mid \det(\alpha) = 1\},$$
$$U(n) = U_n(\mathbf{C}), \qquad SU(n) = \{\alpha \in U(n) \mid \det(\alpha) = 1\}.$$

If $(\varepsilon, \mathbf{K}) = ((+,-), \mathbf{C})$, it is easy to see that the map $\alpha \mapsto \zeta\alpha\zeta^{-1}$ with

$$\zeta = 2^{-1/2}\begin{pmatrix} 1_n & i1_n \\ i1_n & 1_n \end{pmatrix}$$

gives an isomorphism of $C^{(+,-)}(\mathbf{C})$ onto $O(2n)$. Therefore $C_0^{(+,-)}(\mathbf{C}) = SU(2n) \cap C^{(+,-)}(\mathbf{C}) = \zeta^{-1}SO(2n)\zeta$.

In the case $(\varepsilon, \mathbf{K}) = ((+,+), \mathbf{R})$, equality (4.4) together with a simple relation

$$\frac{1}{2}\begin{pmatrix} 1 & 1 \\ -1 & 1 \end{pmatrix}\begin{pmatrix} d & c \\ c & d \end{pmatrix}\begin{pmatrix} 1 & -1 \\ 1 & 1 \end{pmatrix} = \begin{pmatrix} d+c & 0 \\ 0 & d-c \end{pmatrix}$$

shows that

$$\begin{pmatrix} d & c \\ c & d \end{pmatrix} \mapsto (d+c, d-c)$$

gives an isomorphism of $C^{(+,+)}(\mathbf{R})$ onto $O(n) \times O(n)$, and $\det\begin{pmatrix} d & c \\ c & d \end{pmatrix} = \det[(d+c)(d-c)]$. Therefore

$$C_0^{(+,+)}(\mathbf{R}) = \left\{\begin{pmatrix} d & c \\ c & d \end{pmatrix} \in C^{(+,+)}(\mathbf{R}) \,\middle|\, \det(d+c) = \det(d-c) = 1\right\}.$$

Theorem 8. *Suppose that* $(\varepsilon, \mathbf{K}) = ((+,+), \mathbf{R})$ *or* $((+,-), \mathbf{C})$. *Then*

$$W_0^\varepsilon(\mathbf{K}) = U_n'(\mathbf{K}) Z_0^\varepsilon(\mathbf{K}), \tag{5.13}$$

$$C_0^\varepsilon(\mathbf{K}) = \{\operatorname{diag}[\bar{a}, a] \mid a \in U_n'(\mathbf{K})\} E(S^\varepsilon(\mathbf{K})), \tag{5.14}$$

$$Z_0^\varepsilon(\mathbf{K}) = \{(c \quad d) \in Z^\varepsilon(\mathbf{K}) \mid d \in S_n'\}, \tag{5.15}$$

where $U_n'(\mathbf{K})$ *denotes the identity component of* $U_n(\mathbf{K})$, *and* S_n' *the subset of* $S_n(\mathbf{C})$ *consisting of the elements with* -1 *as an eigenvalue with even multiplicity (including the case of zero multiplicity). Moreover,* $Z^\varepsilon(\mathbf{K})$ *has two connected components, and the component containing* $(0 \quad 1_n)$ *is* $Z_0^\varepsilon(\mathbf{K})$.

Proof: We have $S^\varepsilon(\mathbf{K}) = \{-{}^t y = y \in M_n^n(\mathbf{K})\}$ in the present cases. From our definition of $C^\varepsilon(\mathbf{K})$ we easily see that its Lie algebra is

$$\left\{ \begin{pmatrix} \bar{x} & -y^* \\ y & x \end{pmatrix} \,\middle|\, -x^* = x \in M_n^n(\mathbf{K}),\ y \in S^\varepsilon(\mathbf{K}) \right\}.$$

Therefore (5.14) follows immediately from (5.11). Then (5.13) follows from (5.12) and (5.14). To prove (5.15), let $w \in S^\varepsilon(\mathbf{K})$. Take $q \in U_n(\mathbf{K})$ so that $qww^*q^* = \operatorname{diag}[a_1^2, \dots, a_n^2]$ with $0 \le a_\nu \in \mathbf{R}$. Then $q\mathbf{d}(w)q^* = \operatorname{diag}[\cos a_1, \dots, \cos a_n]$. By Lemma 7 below we easily see that $\mathbf{d}(w) \in S_n'$, and therefore the left-hand side of (5.15) is contained in the right-hand side. Conversely, suppose $(c \quad d) \in Z^\varepsilon(\mathbf{K})$ and $d \in S_n'$. Take $p \in U_n(\mathbf{K})$ so that $pdp^{-1} = \operatorname{diag}[a, -1_s, 1_t]$ as in the proof of Theorem 6. Then s is even. Put

$$h = p^{-1} \operatorname{diag}\left[0_r^r, \pi \begin{pmatrix} 0 & -1_{s/2} \\ 1_{s/2} & 0 \end{pmatrix}, 0_t^t\right] \bar{p}$$

and $\omega = \operatorname{diag}[\bar{p}, p]$. Then $\omega E(h) \omega^{-1} = \operatorname{diag}[\delta, \delta]$ with $\delta = \operatorname{diag}[1_r, -1_s, 1_t]$. Take $y = p^{-1} \operatorname{diag}[\sigma b^{-1} v, 0] \bar{p}$ as in the last part of the proof of Theorem 7. Since $\begin{pmatrix} 0 & -h^* \\ h & 0 \end{pmatrix}$ commutes with $\begin{pmatrix} 0 & -y^* \\ y & 0 \end{pmatrix}$, we have $E(h+y) = E(h)E(y)$. Now $E(h) = \operatorname{diag}[\bar{u}, u]$ with $u = p^{-1}\delta p$, and hence $(\mathbf{c}(h+y), \mathbf{d}(h+y)) = (u\mathbf{c}(y), u\mathbf{d}(y)) = (c, d)$. This completes the proof of (5.15). Now $Z_0^\varepsilon(\mathbf{K})$, being a continuous image of a connected set $S^\varepsilon(\mathbf{K})$, is connected. Suppose $(c \quad d) \in Z^\varepsilon(\mathbf{K})$ and $d \notin S_n'$. Take $x \in U_n'(\mathbf{K})$ so that $xdx^{-1} = \operatorname{diag}[-1_s, w]$ with a real diagonal matrix w whose eigenvalues are different from -1. Then s is odd and $xc \cdot {}^t x = \operatorname{diag}[0_s^s, z]$ with $w \in M_{n-s}^{n-s}(\mathbf{K})$ for the same reason as in the proof of Theorems 6 and 7. This shows that $(c \quad d)$ belongs to the set

$$\{(x^{-1} \operatorname{diag}[0, e]\bar{x}, x^{-1} \operatorname{diag}[-1, f]x) \mid x \in U_n'(\mathbf{K}), (e \quad f) \in Z_{n-1}\}, \tag{5.16}$$

where Z_{n-1} is the set $Z_0^\varepsilon(\mathbf{K})$ defined with $n-1$ in place of n. Since both $U_n'(\mathbf{K})$ and Z_{n-1} are connected, (5.16) is connected. Thus $Z^\varepsilon(\mathbf{K})$ has at most two connected components. To show that it is not connected, take any $(c \quad d) \in Z^\varepsilon(\mathbf{K})$. In the proof of Theorem 7, we obtained an expression $pdp^{-1} = \operatorname{diag}[a_1 1_{r_1}, \dots, a_k 1_{r_k}, -1_s, 1_t]$, and also $v_\nu = -\varepsilon_1 v_\nu^\rho \in M_{r_\nu}^{r_\nu}(\mathbf{K})$. Since v_ν is invertible, r_ν must be even. From this we can conclude that the parity of the number of negative eigenvalues of d is a continuous function on $Z^\varepsilon(\mathbf{K})$. This proves the last assertion of our theorem, since both $(0, 1_n)$ and $(0, \operatorname{diag}[-1, 1_{n-1}])$ belong to $Z^\varepsilon(\mathbf{K})$.

Lemma 7. *If* $w = -{}^t w \in M_n^n(\mathbf{C})$, *then the multiplicity of every nonzero eigenvalue of* ww^* *is even.*

Proof: Take $u \in U_n(\mathbf{C})$ so that $d = uww^*u^*$ is diagonal. Put $z = uw \cdot {}^t u = x + iy$ with $x, y \in M_n^n(\mathbf{R})$. Then ${}^t z = -z$, ${}^t x = -x$, ${}^t y = -y$, and $z\bar{z} = -d$. Since d is real, we have $xy = yx$. Thus x and y are commuting normal matrices, and so simultaneously diagonalizable by an element of $U_n(\mathbf{C})$. Therefore we can find an element v of $U_n(\mathbf{C})$ such that $vzv^* = \mathrm{diag}[c_1, \ldots, c_n]$ with $c_n \in \mathbf{C}$. Then $vdv^* = \mathrm{diag}[|c_1|^2, \ldots, |c_n|^2]$. Since ${}^t z = -z$, we see that $\{-c_1, \ldots, -c_n\}$ coincides with $\{c_1, \ldots, c_n\}$ as a whole. Therefore we obtain our lemma.

The decomposition for the group $C^{(-,+)}(\mathbf{K})$ in (5.9) can be expressed in a somewhat different way. We first define a ring-injection κ: $M_n^n(\mathbf{H}) \to M_{2n}^{2n}(\mathbf{C})$ by

$$\kappa(u + vj) = \begin{pmatrix} u & -v \\ \bar{v} & \bar{u} \end{pmatrix} \quad \text{for } u, v \in M_n^n(\mathbf{C}),$$

where j is one of the standard quaternion units. Now it can easily be verified that

$$\begin{pmatrix} d & -c \\ c & d \end{pmatrix} \mapsto \begin{cases} ic + d & (\mathbf{K} = \mathbf{R}), \\ (ic + d, -ic + d) & (\mathbf{K} = \mathbf{C}), \\ i\kappa(c) + \kappa(d) & (\mathbf{K} = \mathbf{H}) \end{cases}$$

gives an isomorphism of $C^{(-,+)}(\mathbf{R})$ onto $U(n)$, $C^{(-,+)}(\mathbf{C})$ onto $U(n) \times U(n)$, and $C^{(-,+)}(\mathbf{H})$ onto $U(2n)$. Then (5.9) can be given in the forms

$$U(n) = O(n)\exp(i \cdot S_n(\mathbf{R})),$$
$$U(n) \times U(n) = \{(u, u) | u \in U(n)\}\{(u, u^{-1}) | u \in U(n)\},$$
$$U(2n) = \kappa(U_n(\mathbf{H}))\exp(i \cdot \kappa(S_n(\mathbf{H}))).$$

6. THINGS THAT ARE HYPERBOLIC. As a natural variation on our theme, we can take the group

$$H_{n,m}(\mathbf{K}) = \{X \in GL_{n+m}(\mathbf{K}) | XI_{n,m}X^* = I_{n,m}\}, \qquad I_{n,m} = \mathrm{diag}[1_n, -1_m],$$

in place of $U_{n+m}(\mathbf{K})$, and can still prove corresponding theorems. In this case the matrices whose fractional expression is the question must be restricted to the ball

$$B_n^m(\mathbf{K}) = \{y \in M_n^m(\mathbf{K}) | yy^* < 1_m\}.$$

We put also

$$X(\mathbf{K}) = X_{m,n}(\mathbf{K}) = \{(f \quad g) \in M_n^m(\mathbf{K}) \times M_m^m(\mathbf{K}) | gg^* - ff^* = 1_m\},$$
$$X'(\mathbf{K}) = \{(f \quad g) \in X(\mathbf{K}) | g^* = g > 0\},$$
$$X^\varepsilon(\mathbf{K}) = X_{n,n}(\mathbf{K}) \cap V^\varepsilon(\mathbf{K}), \qquad H^\varepsilon(\mathbf{K}) = G^\varepsilon(\mathbf{K}) \cap H_{n,n}(\mathbf{K}),$$
$$Y^\varepsilon(\mathbf{K}) = \{(f \quad g) \in X^\varepsilon(\mathbf{K}) | f^\rho = -\varepsilon_1 f, g* = g > 0\},$$
$$\Omega_{n,m}(\mathbf{K}) = \left\{\begin{pmatrix} a & b \\ c & d \end{pmatrix} \in GL_{m+n}(\mathbf{K}) \middle| cc^* < dd^*\right\},$$

where $c \in M_n^m(\mathbf{K})$. As an analogue of (4.4) we have

$$H^\varepsilon(\mathbf{K}) = \left\{\begin{pmatrix} d' & -\varepsilon_1 c' \\ c & d \end{pmatrix} \middle| (c \quad d) \in X^\varepsilon(\mathbf{K})\right\},$$

where $(d' \quad c')$ is the same as in (4.4).

Theorem 9. (1) $\Omega_{n,m}(\mathbf{K}) = P_{n,m}(\mathbf{K})H_{n,m}(\mathbf{K})$.

(2) $\Omega_{n,m}(\mathbf{K}) \cap G^\varepsilon(\mathbf{K}) = P^\varepsilon(\mathbf{K})H^\varepsilon(\mathbf{K})$.

(3) $(f \quad g) \mapsto g^{-1}f$ *gives a one-to-one map of* $U_m(\mathbf{K}) \backslash X(\mathbf{K})$ *onto* $B_n^m(\mathbf{K})$.

These assertions correspond to Theorem 3(2), Theorem 5(2), and Theorem 2, and can be proved in a straightforward way.

Let us now define $\mathfrak{f}(x)$, $\mathfrak{g}(x)$, and $F(x)$ for $x \in M_n^m(\mathbf{K})$ by

$$\mathfrak{f}(x) = \sum_{\nu=0}^{\infty} (xx^*)^{\nu} x/(2\nu+1)!, \qquad \mathfrak{g}(x) = \sum_{\nu=0}^{\infty} (xx^*)^{\nu}/(2\nu)!,$$

$$F(x) = \exp\left(\begin{pmatrix} 0_n^n & x^* \\ x & 0_m^m \end{pmatrix}\right) \qquad (x \in M_n^m(\mathbf{K})).$$

Then we have

$$\mathfrak{f}(x)^* = \mathfrak{f}(x^*), \mathfrak{g}(x)^* = \mathfrak{g}(x), \quad \text{and} \quad F(x) = \begin{pmatrix} \mathfrak{g}(x^*) & \mathfrak{f}(x^*) \\ \mathfrak{f}(x) & \mathfrak{g}(x) \end{pmatrix}.$$

Theorem 10. *The map $x \mapsto \mathfrak{g}(x)^{-1}\mathfrak{f}(x)$ gives a one-to-one map of $M_n^m(\mathbf{K})$ onto $B_n^m(\mathbf{K})$. Moreover we have*

$$X(\mathbf{K}) = U_m(\mathbf{K})X'(\mathbf{K}), \qquad X'(\mathbf{K}) = \{(\mathfrak{f}(x) \quad \mathfrak{g}(x)) | x \in M_n^m(\mathbf{K})\},$$

$$H_{n,m}(\mathbf{K}) = (U_n(\mathbf{K}) \times U_m(\mathbf{K}))F(M_n^m(\mathbf{K})).$$

Theorem 11. *The map $s \mapsto \mathfrak{g}(s)^{-1}\mathfrak{f}(s)$ gives a one-to-one map of $S^{\varepsilon}(\mathbf{K})$ onto $B_n^n(\mathbf{K}) \cap S^{\varepsilon}(\mathbf{K})$. Moreover we have*

$$X^{\varepsilon}(\mathbf{K}) = U_n(\mathbf{K})Y^{\varepsilon}(\mathbf{K}), \qquad Y^{\varepsilon}(\mathbf{K}) = \{(\mathfrak{f}(s) \quad \mathfrak{g}(s)) | s \in S^{\varepsilon}(\mathbf{K})\},$$

$$F(S^{\varepsilon}(\mathbf{K})) = \{\alpha \in H^{\varepsilon}(\mathbf{K}) | \alpha^* = \alpha\},$$

$$H^{\varepsilon}(\mathbf{K}) = \{\mathrm{diag}[u', u] | u \in U_n(\mathbf{K})\}F(S^{\varepsilon}(\mathbf{K})),$$

where $u' = \bar{u}$ if $(\varepsilon, \mathbf{K}) = ((\pm, -), \mathbf{C})$ and $u' = u$ otherwise.

All these, except the injectivity of the maps in both theorems, can be proved by the same technique as for the corresponding facts in Section 5. In fact the present case is much easier. To see the injectivity, let $y = g^{-1}f = d^{-1}c$ with $(f \quad g)$ and $(c \quad d)$ in $X'(\mathbf{K})$. Then $d = hg$ with $h \in U_m(\mathbf{K})$, $d^2 = g^2$, and hence $d = g$ and $h = 1$. Suppose $f = \mathfrak{f}(x)$ and $g = \mathfrak{g}(x)$ with $x \in M_n^m(\mathbf{K})$. Put $T = \begin{pmatrix} 0_n^n & x^* \\ x & 0_m^m \end{pmatrix}$. Then $T^* = T$ and

$$F(x) = \exp(T) = \begin{pmatrix} k & f^* \\ f & g \end{pmatrix} \quad \text{with } k = (1_n + f^*f)^{1/2}.$$

Now it is well known (and easy to prove) that $z \mapsto \exp(z)$ gives a bijection of $S_N(\mathbf{K})$ onto $\{y \in S_N(\mathbf{K}) | y > 0\}$. Therefore x is unique for $(f \quad g)$, which proves the desired injectivity.

REFERENCES

1. S. Helgason, *Differential geometry, Lie groups, and symmetric spaces*, Academic Press, 1978.
2. G. Shimura, Confluent hypergeometric functions on tube domains, *Math. Ann.* 260 (1982), 269–302.
3. C. L. Siegel, Einführung in die Theorie der Modulfunktionen n-ten Grades, *Math. Ann.* 116 (1939), 617–657 (= Gesammelte Abhandlungen II, No. 32).
4. C. L. Siegel, On the theory of indefinite quadratic forms, *Ann. of Math.* 45 (1944), 576–622 (= Gesammelte Abhandlungen II, No. 45).

Department of Mathematics
Princeton University
Princeton, New Jersey 08544-1000

94b

Euler products and Fourier coefficients of automorphic forms on symplectic groups

Inventiones mathematicae, 116 (1994), 531-576

Introduction

Our automorphic forms are defined on the group

$$G=\mathrm{Sp}(n,F)=\{\alpha\in \mathrm{GL}_{2n}(F)\,|\,{}^t\alpha\,\iota\,\alpha=\iota\},\qquad \iota=\begin{pmatrix}0 & -1_n\\ 1_n & 0\end{pmatrix},$$

with a totally real algebraic number field F. To describe our problems and results, let us take $F=\mathbf{Q}$ for simplicity and denote by Γ the congruence subgroup of $\mathrm{Sp}(n,\mathbf{Z})$ consisting of α such that all the entries of c_α are divisible by a positive integer N, where we denote by a_α, b_α, c_α, d_α the standard a-, b-, c-, d-blocks of α. Then we consider a holomorphic modular form $f(z)$ with a variable z in the Siegel half-space H such that

$$f(\gamma(z))=\varphi(\det(d_\gamma))\det(c_\gamma z+d_\gamma)^k f(z)\qquad \text{for every } \gamma\in\Gamma,$$

where k is an integer and φ is a Dirichlet character modulo N such that $\varphi(-1)^n=(-1)^{kn}$. Then f has a Fourier expansion of the form

$$f(z)=\sum_\sigma \mu(\sigma)\exp(2\pi i\cdot \mathrm{tr}(\sigma z))$$

with complex numbers $\mu(\sigma)$, where σ runs over all half-integral symmetric matrices. For a fixed half-integral positive definite symmetric matrix τ and an arbitrary primitive Dirichlet character η we consider

$$(1)\qquad D(s;\tau,f,\eta)=\sum_\xi \mu({}^t\xi\tau\xi)\,\eta(\det(\xi))\det(\xi)^m|\det(\xi)|^{-m-s},$$

where ξ runs over all integral nonsingular matrices modulo right multiplication by the elements of $\mathrm{GL}_n(\mathbf{Z})$ and m is an integer such that $(\varphi\eta)(-1)=(-1)^{k+m}$. Now our first main purpose of this paper is to show that if f is an eigenform of certain Hecke operators, then D is essentially an Euler product obtained from the eigenvalues. To be explicit, we associate with such an f an Euler product of the form

$$Z(s, f, \eta) = \prod_p W_p(\eta(p)\, p^{-s})^{-1}, \tag{2}$$

where p runs over all prime numbers and W_p is a polynomial with constant term 1 and of degree $2n+1$ or $\leqq n$ according as $p \nmid N$ or $p | N$. Then we shall prove

$$D(s; \tau, f, \eta)\, g(s)\, \Lambda(s) = Z(s, f, \eta)\, h(s) \tag{3}$$

with a certain product of Hecke L-functions Λ and certain finite Dirichlet series $g(s) = \sum_n g_n n^{-s}$ and $h(s) = \sum_n h_n n^{-s}$ with n prime to N and such that $g_1 = 1$ and $h_1 = \mu(\tau)$; moreover τ can be chosen so that $h(s) = \mu(\tau) \neq 0$ (Corollaries 5.2 and 5.3). In fact we shall prove a more general formula for the series defined with an arbitrary nonnegative τ (including the case $\tau = 0$) without assuming f to be an eigenform (Theorem 5.1). In the general case in which F is not necessarily $\mathbf{Q}$, our series D must be defined with the Fourier coefficients of a function on $G_{\mathbf{A}}$, which amounts to the Fourier coefficients of the Fourier expansions of f in the classical sense at several "essential cusps."

If $n = 1$, equality (3) can be derived easily from the classical Hecke theory, and in fact its analytic property was investigated in [S4]. The first result in the higher-dimensional case is due to Andrianov. In [A] he proved a formula of type (3) for $G = Sp(n, \mathbf{Q})$, but his series is the partial sum of (1) consisting of the terms for ξ such that $\det(\xi)$ is prime to N, and his Euler product is defined only over the primes p prime to N. Besides, his proof required lengthy calculations. A simpler approach was introduced by Böcherer in [B2], but he still excluded the primes dividing N, and he did not completely prove Andrianov's equality. In our treatment, we define Hecke operators and Euler factors for all primes, and then, combining an adelic variation of an idea of [B2] with some new technique, we shall give a simpler proof in a more general setting.

Our second aim is naturally meromorphic continuation of $Z(s, f, \eta)$ for a cusp form f to the whole s-plane, which follows from an inner product expression

$$\mathscr{G}(s)\, Z(s, f, \eta) = P(s) \langle f(z), \theta(z)\, \mathscr{E}(z, s) \rangle, \tag{4}$$

where $\mathscr{G}$ is an explicitly given product of gamma functions, P is a product of elementary factors, and θ is a certain theta function, and $\mathscr{E}$ is an Eisenstein series of standard type, which is normalized so that its poles can be located in a precise way (Theorem 8.1 and (8.11)). This equality is the n-dimensional version of what we showed in [S4]. For n even, an equality of this type was employed by Andrianov and Kalinin in [AK] to prove meromorphic continuation in the case $G = Sp(n, \mathbf{Q})$. Now (4) follows easily from (3), but that is so because (the generalization of) (1) is defined properly with no artificial exclusion of terms. Also, if n is odd, θ and $\mathscr{E}$ are functions of half-integral weight, which require special care. For these reasons, the result of [AK] was restricted to the case when n is even and there is a τ such that $\mu(\tau) \neq 0$ and the conductor of η is divisible by all the prime factors of $N \cdot \det(2\tau)$. No such conditions are necessary for our theorem.

Meromorphic continuation of Z was also treated by Piatetski-Shapiro and Rallis in [PR1], [PR2], and [PR3]. In the last paper they proved that Z, with some special η and with sufficiently many finite number of Euler factors eliminated, has an inner product expression similar to (4). In [PR1] and [PR2] they introduced another kind of inner product expression for Z with trivial η involving a certain pullback of an Eisenstein series on $Sp(2n)$. Though they presented general principles applicable to a wide class of groups, their results leave ample room for elaboration. Besides, it is not clear whether the above Z with an arbitrary η can be handled by that method. An explicit calculation along the line of ideas of [PR1] was done by Böcherer in [B1], but still in the case of modular forms with respect to $Sp(n, \mathbf{Z})$. In order to examine the analytic properties of Z in the general case, it is essential to define a *nonvanishing* inner product expression with all Euler factors properly defined, but this is a highly nontrivial question, to which no definite answer seems to have been given, and which we are trying to answer in an explicit way in the present paper.

Though our treatment is restricted to holomorphic modular forms on a symplectic group with a relatively simple factor of automorphy, practically all our methods are applicable to a more general class of forms (even on other types of algebraic groups), provided the nature of analyticity at archimedean primes is understood. It may be added as a final remark that analytic continuation is not our sole main purpose and (3) should not be viewed merely as a preliminary step. In fact, the series D is defined for every τ, and that we have a relation of type (3) for each choice of τ is a significant fact by itself, irrespective of its connection with the question of analytic continuation.

1. Modular forms and their Fourier expansions

The following notation will be used throughout the paper: For an associative ring R with identity element and an R-module M we denote by $R^\times$ the group of all its invertible elements and by M_n^m the R-module of all $m \times n$-matrices with entries in M; we put $M^m = M_1^m$ for simplicity. Sometimes an object with a superscript such as S^n of (1.20a) below is used with a different meaning, but the distinction will be clear from the context. The identity and zero elements of the ring R_n^n are denoted by 1_n and 0_n (when n needs to be stressed). The transpose of a matrix x is denoted by ${}^t x$. We put $\tilde{x} = {}^t x^{-1}$ if x is square and invertible. If $x_1, \dots, x_r$ are square matrices, $\mathrm{diag}[x_1, \dots, x_r]$ denotes the matrix with $x_1, \dots, x_r$ in the diagonal blocks and 0 in all other blocks. For $x = \begin{pmatrix} a & b \\ c & d \end{pmatrix} \in R_{2n}^{2n}$ with a, b, c, and d in R_n^n, we put $a = a_x$, $b = b_x$, $c = c_x$, and $d = d_x$ if there is no fear of confusion. For complex hermitian matrices x and y we write $x > y$ if $x - y$ is positive definite, and $x \geqq y$ if $x - y$ is nonnegative. To indicate that a union $X = \bigcup_{i \in I} Y_i$ is disjoint, we write $X = \coprod_{i \in I} Y_i$. We understand that $\prod_{i=\alpha}^{\beta} = 1$ if $\alpha > \beta$.

We fix a totally real algebraic number field F of finite degree and denote by $\mathbf{a}$ and $\mathbf{h}$ the sets of archimedean primes and nonarchimedean primes of

F; further we denote by $\mathfrak{g}$, $\mathfrak{d}$, and D_F the maximal order of F, the different of F relative to $\mathbf{Q}$, and the discriminant of F.

Given an algebraic group $\mathfrak{G}$ defined over F, we define $\mathfrak{G}_v$ for each $v \in \mathbf{a} \cup \mathbf{h}$ and the adelization $\mathfrak{G}_{\mathbf{A}}$ and view $\mathfrak{G}$ and $\mathfrak{G}_v$ as subgroups of $\mathfrak{G}_{\mathbf{A}}$ as usual. We then denote by $\mathfrak{G}_{\mathbf{a}}$ and $\mathfrak{G}_{\mathbf{h}}$ the archimedean and nonarchimedean factors of $\mathfrak{G}_{\mathbf{A}}$, respectively. For $x \in \mathfrak{G}_{\mathbf{A}}$ we denote by x_v, $x_{\mathbf{a}}$, and $x_{\mathbf{h}}$ its projections to $\mathfrak{G}_v$, $\mathfrak{G}_{\mathbf{a}}$, and $\mathfrak{G}_{\mathbf{h}}$. If $v \in \mathbf{a}$, we view v as an injection of F into $\mathbf{R}$, and identify F_v with $\mathbf{R}$. For a fractional ideal $\mathfrak{x}$ in F and $t \in F_{\mathbf{A}}^{\times}$ we denote by $N(\mathfrak{x})$ the norm of $\mathfrak{x}$ and by $t\mathfrak{x}$ the fractional ideal in F such that $(t\mathfrak{x})_v = t_v \mathfrak{x}_v$ for every $v \in \mathbf{h}$; we put $|t_v| = |t_v|_v$ with the normalized valuation $|\ |_v$ at v and $|t|_{\mathbf{A}} = \prod_{v \in \mathbf{a} \cup \mathbf{h}} |t_v|$. For $v \in \mathbf{h}$ we denote by π_v an unspecified prime element of F_v. Given a set X, we denote by $X^{\mathbf{a}}$ the product of $\mathbf{a}$ copies of X, that is, the set of all indexed elements $(x_v)_{v \in \mathbf{a}}$ with x_v in X. For example, for $t \in F_{\mathbf{A}}^{\times}$ we define an element $\operatorname{sgn}(t)$ of $\mathbf{R}^{\mathbf{a}}$ by $\operatorname{sgn}(t) = (t_v/|t_v|)_{v \in \mathbf{a}}$. We put

$$\mathbf{T} = \{\zeta \in \mathbf{C} \mid |\zeta| = 1\}, \tag{1.1a}$$

$$\mathbf{e}(z) = e^{2\pi i z} \quad (z \in \mathbf{C}), \tag{1.1b}$$

and define a character $\mathbf{e}_v : F_v \to \mathbf{T}$ for each v by $\mathbf{e}_v(x) = \mathbf{e}(x)$ for $v \in \mathbf{a}$, and $\mathbf{e}_v(x) = \mathbf{e}(-\{\textit{fractional part of } \operatorname{Tr}_{F_v/\mathbf{Q}_p}(x)\})$ if $v \in \mathbf{h}$ and p is the rational prime divisible by v. We then put $\mathbf{e}_{\mathbf{A}}(x) = \prod_{v \in \mathbf{a} \cup \mathbf{h}} \mathbf{e}_v(x_v)$ and $\mathbf{e}_{\mathbf{h}}(x) = \mathbf{e}_{\mathbf{A}}(x_{\mathbf{h}})$ for $x \in F_{\mathbf{A}}$, and $\mathbf{e}_{\mathbf{a}}(x) = \mathbf{e}(\sum_{v \in \mathbf{a}} x_v)$ for $x \in F_{\mathbf{A}}$ or $x \in \mathbf{C}^{\mathbf{a}}$. By a *Hecke character* of F we understand a continuous homomorphism ψ of $F_{\mathbf{A}}^{\times}$ into $\mathbf{T}$ such that $\psi(F^{\times}) = 1$. For such a ψ we denote by ψ_v, $\psi_{\mathbf{a}}$, and $\psi_{\mathbf{h}}$ its restrictions to $F_v^{\times}$, $F_{\mathbf{a}}^{\times}$, and $F_{\mathbf{h}}^{\times}$, respectively, and by ψ^* the ideal character associated with ψ, understanding that $\psi^*(\mathfrak{a}) = 0$ if $\mathfrak{a}$ is not prime to the conductor of ψ. We put also $\psi_{\mathfrak{c}} = \prod_{v|\mathfrak{c}} \psi_v$ for every integral ideal $\mathfrak{c}$. We have $\psi_v(x) = \operatorname{sgn}(x)^{t_v} |x|^{i\kappa_v}$ for $x \in F_v^{\times}$, $v \in \mathbf{a}$, with $t_v \in \mathbf{Z}$ and $\kappa_v \in \mathbf{R}$. We always assume that ψ is *normalized* in the sense that $\sum_{v \in \mathbf{a}} \kappa_v = 0$.

We now define an algebraic group G, its subgroup P, and a space H by

$$G = \{\alpha \in GL_{2n}(F) \mid {}^t\alpha \iota \alpha = \iota\}, \quad \iota = \begin{pmatrix} 0 & -1_n \\ 1_n & 0 \end{pmatrix}, \tag{1.2a}$$

$$P = \{\alpha \in G \mid c_\alpha = 0\}, \tag{1.2b}$$

$$H = \{z \in \mathbf{C}_n^n \mid {}^t z = z, \operatorname{Im}(z) > 0\}. \tag{1.2c}$$

For $\alpha \in G_v$, $v \in \mathbf{a}$, and $z \in H$ we define $\alpha z = \alpha(z) \in H$, $\Delta(z) \in \mathbf{R}$, and factors of automorphy $\mu(\alpha, z)$ and $j(\alpha, z)$ by

$$\alpha z = \alpha(z) = (a_\alpha z + b_\alpha)(c_\alpha z + d_\alpha)^{-1}, \quad \Delta(z) = \det(\operatorname{Im}(z)), \tag{1.3a}$$

$$\mu(\alpha, z) = c_\alpha z + d_\alpha, \quad j(\alpha, z) = \det[\mu(\alpha, z)]. \tag{1.3b}$$

For $\alpha \in G_{\mathbf{A}}$ and $z=(z_v)_{v\in\mathbf{a}}\in H^{\mathbf{a}}$ we put

(1.4a) $$\alpha z=(\alpha_v z_v)_{v\in\mathbf{a}}, \qquad \Delta(z)=(\Delta(z_v))_{v\in\mathbf{a}},$$

(1.4b) $$\mu(\alpha, z)=(\mu(\alpha_v, z_v))_{v\in\mathbf{a}}, \qquad j_\alpha(z)=j(\alpha, z)=(j(\alpha_v, z_v))_{v\in\mathbf{a}}.$$

All these are elements of $(\mathbf{C}_n^n)^{\mathbf{a}}$ or $\mathbf{C}^{\mathbf{a}}$. For two fractional ideals $\mathfrak{x}$ and $\mathfrak{y}$ in F such that $\mathfrak{x}\mathfrak{y}\subset\mathfrak{g}$ we put

(1.5a) $$D[\mathfrak{x}, \mathfrak{y}]=G_{\mathbf{a}}\prod_{v\in\mathbf{h}} D_v[\mathfrak{x}, \mathfrak{y}](\subset G_{\mathbf{A}}), \qquad \Gamma[\mathfrak{x}, \mathfrak{y}]=G\cap D[\mathfrak{x}, \mathfrak{y}],$$

(1.5b) $$D_v[\mathfrak{x}, \mathfrak{y}]=\{x\in G_v \mid a_x\in(\mathfrak{g}_v)_n^n,\ b_x\in(\mathfrak{x}_v)_n^n,\ c_x\in(\mathfrak{y}_v)_n^n,\ d_x\in(\mathfrak{g}_v)_n^n\},$$

(1.5c) $$D_0[\mathfrak{x}, \mathfrak{y}]=\{x\in D[\mathfrak{x}, \mathfrak{y}] \mid x_{\mathbf{a}}(\mathbf{i})=\mathbf{i}\},$$

where $\mathbf{i}$ is the "origin" of $H^{\mathbf{a}}$ given by

(1.6) $$\mathbf{i}=(i1_n, \ldots, i1_n).$$

To simplify our notation, for x and y in $\mathbf{C}^{\mathbf{a}}$ we put

(1.7) $$\|x\|=\sum_{v\in\mathbf{a}} x_v, \qquad x^y=\prod_{v\in\mathbf{a}} x_v^{y_v},$$

where $x_v^{y_v}$ should be understood according to the context. For $w\in F_{\mathbf{A}}^{\times}$, in particular for $w\in F^{\times}$, we shall often put $w^y=(w_{\mathbf{a}})^y$ for simplicity. The identity element of the ring $\mathbf{C}^{\mathbf{a}}$, which will, for the most part, appear as an exponent, is denoted by u. For example, $\|u\|=[F:\mathbf{Q}]$, $c^u=N_{F/\mathbf{Q}}(c)$ for $c\in F^{\times}$, and

(1.8) $$j_\alpha(z)^u=\prod_{v\in\mathbf{a}} j(\alpha_v, z_v) \qquad (\alpha\in G_{\mathbf{A}}, z\in H^{\mathbf{a}}).$$

We now put $X=F_n^1$ and let G act on $X\times X=F_{2n}^1$ by right multiplication. Then we can define the metaplectic group $Mp(X_{\mathbf{A}})$ in the sense of [W] with respect to the alternating form $(x, y)\mapsto x\iota\cdot{}^t y$ on $F_{2n}^1\times F_{2n}^1$. Recall that this group, written $M_{\mathbf{A}}$ for simplicity, is a group of unitary transformations on $L^2(X_{\mathbf{A}})$ with an exact sequence

$$1\to\mathbf{T}\to M_{\mathbf{A}}\to G_{\mathbf{A}}\to 1.$$

We denote by pr the projection map of $M_{\mathbf{A}}$ to $G_{\mathbf{A}}$. There is a natural lift r: $G\to M_{\mathbf{A}}$ by which we can consider G a subgroup of $M_{\mathbf{A}}$. There is also a lift

(1.9) $$r_P\colon P_{\mathbf{A}}\to M_{\mathbf{A}}.$$

For $\tau\in M_{\mathbf{A}}$, $\alpha=\mathrm{pr}(\tau)$ and $z\in H_n^{\mathbf{a}}$ we put $a_\tau=a_\alpha$, $b_\tau=b_\alpha$, $c_\tau=c_\alpha$, $d_\tau=d_\alpha$ and $\tau(z)=\tau z=\alpha z$.

We define a subset $\mathfrak{M}$ of $M_{\mathbf{A}}$, a subgroup Γ^θ of G, and a subgroup C^θ of $G_{\mathbf{A}}$ by

(1.10a) $$\mathfrak{M}=\{\sigma\in M_{\mathbf{A}}\,|\,\mathrm{pr}(\sigma)\in P_{\mathbf{A}}\,C^\theta\},$$

(1.10b) $$\Gamma^\theta=G\cap C^\theta,\qquad C^\theta=G_{\mathbf{a}}\prod_{v\in\mathbf{h}}C^\theta_v,$$

(1.10c) $$C^\theta_v=\{\xi\in D_v[\mathfrak{d}^{-1},\mathfrak{d}]\,|\,\chi_v((x,y)\xi)=\chi_v(x,y)\ \text{for every } x\in(\mathfrak{g}_v)^1_n \text{ and } y\in(\mathfrak{d}_v^{-1})^1_n\}\quad \text{if } v\in\mathbf{h},$$

where $\chi_v(x,y)=\mathbf{e}_v(x\cdot{}^t y/2)$ for $x,y\in(F_v)^1_n$. We note that

(1.11) $$D[2\mathfrak{d}^{-1},2\mathfrak{d}]\subset C^\theta.$$

For each $\sigma\in\mathfrak{M}$ we can define a holomorphic function $h_\sigma(z)=h(\sigma,z)$ of $z\in H^{\mathbf{a}}$ which has the following properties (see [S7, pp. 294–5], [S11, Th. 1.2]):

(1.12a) If $\alpha=\mathrm{pr}(\sigma)$, then $h_\sigma(z)^2/j_\alpha(z)^u$ is a constant belonging to $\mathbf{T}$; in particular, h_σ is a constant belonging to $\mathbf{T}$ if $\mathrm{pr}(\sigma)\in G_{\mathbf{h}}$;

(1.12b) $$h(t\cdot r_P(\gamma),z)=t^{-1}|\det(d_\gamma)_{\mathbf{a}}|^{1/2}_{\mathbf{A}}\quad \text{if } t\in\mathbf{T} \text{ and } \gamma\in P_{\mathbf{A}};$$

(1.12c) $$h(\rho\sigma\tau,z)=h(\rho,z)\,h(\sigma,\tau z)\,h(\tau,z)\quad \text{if } \mathrm{pr}(\rho)\in P_{\mathbf{A}},\ \sigma\in\mathfrak{M}, \text{ and } \mathrm{pr}(\tau)\in C^\theta.$$

By a *half-integral weight* we mean an element k of $\mathbf{Q}^{\mathbf{a}}$ such that $k-u/2\in\mathbf{Z}^{\mathbf{a}}$; naturally an *integral weight* means an element of $\mathbf{Z}^{\mathbf{a}}$. We put $[k]=k$ if $k\in\mathbf{Z}^{\mathbf{a}}$ and $[k]=k-u/2$ if $k\notin\mathbf{Z}^{\mathbf{a}}$. For an integral or a half-integral weight k we define a factor of automorphy j^k by

(1.13a) $$j^k_\alpha(z)=j^k(\alpha,z)=j_\alpha(z)^k\qquad (k\in\mathbf{Z}^{\mathbf{a}},\ \alpha\in G_{\mathbf{A}},\ z\in H^{\mathbf{a}}),$$

(1.13b) $$j^k_\sigma(z)=j^k(\sigma,z)=h_\sigma(z)\,j_\alpha(z)^{[k]}\qquad (k\notin\mathbf{Z}^{\mathbf{a}},\ \sigma\in\mathfrak{M},\ \alpha=\mathrm{pr}(\sigma),\ z\in H^{\mathbf{a}}).$$

Then for a function f on $H^{\mathbf{a}}$ and $\xi\in G_{\mathbf{A}}$ or $\in\mathfrak{M}$ we define a function $f\|_k\xi$ on $H^{\mathbf{a}}$ by

(1.14) $$(f\|_k\xi)(z)=j^k_\xi(z)^{-1}f(\xi z)\qquad (z\in H^{\mathbf{a}}).$$

Given a congruence subgroup Γ of G, we denote by $\mathcal{M}_k(\Gamma)$ the vector space of all holomorphic functions f on $H^{\mathbf{a}}$ which satisfy $f\|_k\gamma=f$ for every $\gamma\in\Gamma$ and also the cusp condition if $n=1$ and $F=\mathbf{Q}$. Here we assume that $\Gamma\subset\Gamma[2\mathfrak{d}^{-1},2\mathfrak{d}]$ if $k\notin\mathbf{Z}^{\mathbf{a}}$. We then denote by $\mathcal{S}_k(\Gamma)$ the subspace of $\mathcal{M}_k(\Gamma)$ consisting of all cusp forms. Further we denote by $\mathcal{M}_k$ resp. $\mathcal{S}_k$ the union of $\mathcal{M}_k(\Gamma)$ resp. $\mathcal{S}_k(\Gamma)$ for all congruence subgroups Γ of G. To treat both integral and half-integral weights uniformly, we understand that pr means the identity map of G onto itself when k is integral.

To consider a more specific type of modular forms, we take a weight k, a fractional ideal $\mathfrak{b}$ and an integral ideal $\mathfrak{c}$, and a character φ of $(\mathfrak{g}/\mathfrak{c})^\times$, assuming that

(1.15a) $$\varphi(-1)^n=(-1)^{n\|[k]\|},$$

(1.15b) $$\mathfrak{b}^{-1}\subset 2\mathfrak{d}^{-1} \text{ and } \mathfrak{b}\mathfrak{c}\subset 2\mathfrak{d} \quad \text{if } k\notin \mathbf{Z}^{\mathbf{a}}.$$

The latter condition implies that j_α^k is always meaningful for $\alpha\in\Gamma[\mathfrak{b}^{-1},\mathfrak{b}\mathfrak{c}]$ and that $\mathfrak{c}\subset 4\mathfrak{g}$ if $k\notin\mathbf{Z}^{\mathbf{a}}$. Then we denote by $C_k(\mathfrak{b},\mathfrak{c},\varphi)$ the set of all $\mathbf{C}$-valued functions g on $H^{\mathbf{a}}$ such that

(1.16) $$g\|_k\gamma=\varphi(\det(a_\gamma))\,g \quad \text{for every } \gamma\in\Gamma[\mathfrak{b}^{-1},\mathfrak{b}\mathfrak{c}].$$

Given such a g, we can define a function $g_{\mathbf{A}}$ on $G_{\mathbf{A}}$ or on $M_{\mathbf{A}}$, according as $k\in\mathbf{Z}^{\mathbf{a}}$ or $k\notin\mathbf{Z}^{\mathbf{a}}$, by

(1.17) $$g_{\mathbf{A}}(\alpha w)=\varphi(\det(d_w)_{\mathfrak{c}})(g\|_k w)(\mathbf{i}) \quad \text{for } \alpha\in G \text{ and } w\in \mathrm{pr}^{-1}(D[\mathfrak{b}^{-1},\mathfrak{b}\mathfrak{c}]),$$

where $x_{\mathfrak{c}}=(x_v)_{v|\mathfrak{c}}\in\prod_{v|\mathfrak{c}}\mathfrak{g}_v^\times$, and φ is viewed as a character of $\prod_{v|\mathfrak{c}}\mathfrak{g}_v^\times$ in a natural way. Clearly

(1.18) $$g_{\mathbf{A}}(\alpha x w)=\varphi(\det(d_w)_{\mathfrak{c}})\,j^k(w,\mathbf{i})^{-1}\,g_{\mathbf{A}}(x)$$
$$\text{for every } \alpha\in G \text{ and } w\in\mathrm{pr}^{-1}(D_0[\mathfrak{b}^{-1},\mathfrak{b}\mathfrak{c}]).$$

Conversely a function on $G_{\mathbf{A}}$ or on $M_{\mathbf{A}}$ satisfying such a transformation formula can always be obtained as $g_{\mathbf{A}}$ with $g\in C_k(\mathfrak{b},\mathfrak{c},\varphi)$. We put

(1.19) $$\mathscr{M}_k(\mathfrak{b},\mathfrak{c},\varphi)=\mathscr{M}_k\cap C_k(\mathfrak{b},\mathfrak{c},\varphi), \qquad \mathscr{S}_k(\mathfrak{b},\mathfrak{c},\varphi)=\mathscr{S}_k\cap C_k(\mathfrak{b},\mathfrak{c},\varphi).$$

Let us now put

(1.20a) $$S=S^n=\{\xi\in F_n^n \mid {}^t\xi=\xi\},$$

(1.20b) $$S_{v+}=\{x\in S_v \mid x\geqq 0\} \qquad (v\in\mathbf{a}),$$

(1.20c) $$S_+=\{\xi\in S \mid \xi\in S_{v+} \text{ for every } v\in\mathbf{a}\},$$

(1.20d) $$S(\mathfrak{x})=S\cap\mathfrak{x}_n^n, \qquad S_{\mathbf{h}}(\mathfrak{x})=\prod_{v\in\mathbf{h}}S(\mathfrak{x})_v,$$

where $\mathfrak{x}$ is any fractional ideal in F. The superscript n in S^n is dropped for the moment, but we shall later reinstate it when the dimension needs to be stressed. We note also that every element $f\in\mathscr{M}_k$ has an expansion of the form

$$f(z)=\sum_{\tau\in S_+}c(\tau)\,\mathbf{e}_{\mathbf{a}}(\mathrm{tr}(\tau z))$$

with $c(\tau)\in\mathbf{C}$. The adele version of such an expansion is given by

Proposition 1.1 *Given $f \in \mathscr{M}_k(\mathfrak{b}, \mathfrak{c}, \varphi)$, there is a complex number $\mu(\tau, q; f)$, written also $\mu_f(\tau, q)$, determined for $\tau \in S_+$ and $q \in GL_n(F)_{\mathbf{A}}$, such that*

$$f_{\mathbf{A}}\left(r_P\begin{pmatrix} q & s\tilde{q} \\ 0 & \tilde{q}\end{pmatrix}\right) = \det(q)_{\mathbf{a}}^{[k]} |\det(q)_{\mathbf{a}}|^{k-[k]} \cdot \sum_{\tau \in S_+} \mu(\tau, q; f)\, \mathbf{e}_{\mathbf{a}}(\mathrm{tr}(\mathbf{i} \cdot {}^t q \tau q))\, \mathbf{e}_{\mathbf{A}}(\mathrm{tr}(\tau s)),$$

for every $s \in S_{\mathbf{A}}$, where $\tilde{q} = {}^t q^{-1}$ and r_P stands for the identity map if $k \in \mathbf{Z}^{\mathbf{a}}$. Moreover $\mu_f(\tau, q)$ has the following properties:

(1.21a) $\quad \mu_f(\tau, q) \neq 0 \quad$ *only if* $\mathbf{e}_{\mathbf{h}}(\mathrm{tr}({}^t q \tau q s)) = 1 \quad$ *for every* $s \in S_{\mathbf{h}}(\mathfrak{b}^{-1})$;

(1.21b) $\quad \mu_f(\tau, q) = \mu_f(\tau, q_{\mathbf{h}})$;

(1.21c) $\quad \mu_f({}^t b \tau b, q) = \det(b)^{[k]} |\det(b)|^{k-[k]} \mu_f(\tau, bq) \quad$ *for every* $b \in GL_n(F)$;

(1.21d) $\quad \varphi(\det(a)_{\mathfrak{c}})\, \mu_f(\tau, qa) = \mu_f(\tau, q) \quad$ *for every* $a \in \prod_{v \in \mathbf{h}} GL_n(\mathfrak{g}_v)$.

Furthermore, if $\beta \in G \cap \mathrm{diag}[r, \tilde{r}]\, D[\mathfrak{b}^{-1}, \mathfrak{b}\mathfrak{c}]$ with $r \in GL_n(F)_{\mathbf{A}}$, then

$$j^k(\beta, \beta^{-1}z)\, f(\beta^{-1}z) = \varphi(\det(d_\beta r)_{\mathfrak{c}}) \sum_{\tau \in S_+} \mu_f(\tau, r)\, \mathbf{e}_{\mathbf{a}}(\mathrm{tr}(\tau z)). \tag{1.22}$$

These were given in [S9, Prop. 3.1] and [S10, Prop. 1.1] when $n = 1$. The higher-dimensional case can be proved easily in the same fashion. Formula (1.22) shows that the $\mu_f(\tau, r)$ for $\tau \in S_+$ may be called the Fourier coefficients of f at the cusp corresponding to r; in particular, the $\mu_f(\tau, 1)$ are exactly the Fourier coefficients of f in the ordinary sense.

Given φ as above, we can always find a Hecke character ψ of F, unramified at every $v \in \mathbf{h}$, $\nmid \mathfrak{c}$, such that $\psi = \varphi$ on $\prod_{v|\mathfrak{c}} \mathfrak{g}_v^\times$. More precisely, given any $t \in \mathbf{Z}^{\mathbf{a}}$ such that $\varphi(-1) = (-1)^{\|t\|}$, we can find an element v of $\mathbf{R}^{\mathbf{a}}$ such that $\|v\| = 0$ and

(1.23a) $\quad \varphi(e) = \mathrm{sgn}(e)^t |e|^{-iv} \quad$ for every $e \in \mathfrak{g}^\times$.

Then there exists a Hecke character ψ of F such that

(1.23b) $\quad \psi(x) = \mathrm{sgn}(x)^t |x_{\mathbf{a}}|^{iv} \varphi(x_{\mathfrak{c}}) \quad$ for every $x \in F_{\mathbf{a}}^\times \prod_{v \in \mathbf{h}} \mathfrak{g}_v^\times$.

Such a ψ is not necessarily unique for φ. Condition (1.15a) implies

$$\psi_{\mathbf{h}}(-1)^n = (-1)^{n\|[k]\|}. \tag{1.24}$$

We shall denote $C_k(\mathfrak{b}, \mathfrak{c}, \varphi)$, $\mathscr{M}_k(\mathfrak{b}, \mathfrak{c}, \varphi)$, and $\mathscr{S}_k(\mathfrak{b}, \mathfrak{c}, \varphi)$ also by $C_k(\mathfrak{b}, \mathfrak{c}, \psi)$, $\mathscr{M}_k(\mathfrak{b}, \mathfrak{c}, \psi)$, and $\mathscr{S}_k(\mathfrak{b}, \mathfrak{c}, \psi)$.

2. Formal Hecke operators and formal Euler products

We fix ideals $\mathfrak{b}$ and $\mathfrak{c}$ as before, and put $D=D[\mathfrak{b}^{-1},\mathfrak{bc}]$ and $D_v=D_v[\mathfrak{b}^{-1},\mathfrak{bc}]$ for simplicity. Put also

(2.1 a) $$E^*=GL_n(F)_{\mathbf{a}}E, \quad E=\prod_{v\in\mathbf{h}}E_v, \quad E_v=GL_n(\mathfrak{g}_v),$$

(2.1 b) $$R^*=\{x\in(F_n^n)_{\mathbf{A}}\,|\,x_{\mathbf{h}}\in R\}, \quad R=\prod_{v\in\mathbf{h}}R_v, \quad R_v=(\mathfrak{g}_v)_n^n,$$

(2.1 c) $$B=R\cap GL_n(F)_{\mathbf{h}}, \quad B_v=GL_n(F_v)\cap R_v.$$

and define subsets $\mathscr{Z}$, $\mathscr{Z}_0$, Z, and Z_0 of $G_{\mathbf{A}}$ by

(2.2 a) $$\mathscr{Z}=DZD, \quad \mathscr{Z}_0=DZ_0D, \quad \mathscr{Z}_{0v}=D_vZ_{0v}D_v,$$

(2.2 b) $$Z=\{\operatorname{diag}[\tilde{e},e]\,|\,e\in GL_n(F)_{\mathbf{h}}\}, \quad \tilde{e}={}^te^{-1},$$

(2.2 c) $$Z_0=\{\operatorname{diag}[\tilde{e},e]\,|\,e\in B\}, \quad Z_{0v}=\{\operatorname{diag}[\tilde{q},q]\,|\,q\in B_v\} \quad (v\in\mathbf{h}).$$

Lemma 2.1 $\mathscr{Z}_0\subset DP_{\mathbf{h}}$.

Proof. It is sufficient to show that $\sigma D_v\subset D_vP_v$ for every $\sigma\in Z_{0v}$, $v\in\mathbf{h}$. Put $C_v=D_v[\mathfrak{b}^{-1},\mathfrak{b}]$. Let $\sigma=\operatorname{diag}[\tilde{q},q]\in Z_{0v}$ and $\alpha\in D_v$. Since $G_v=C_vP_v$, we have $\sigma\alpha=\beta^{-1}\pi$ with $\beta\in C_v$ and $\pi\in P_v$. If $v\nmid\mathfrak{c}$, then $C_v=D_v$, which proves the desired fact for such a v. If $v|\mathfrak{c}$, both a_α and d_α belong to $GL_n(\mathfrak{g}_v)$. Since $0=c_\pi=c_\beta\tilde{q}a_\alpha+d_\beta qc_\alpha$, we have $c_\beta=-d_\beta qc_\alpha a_\alpha^{-1}\cdot{}^tq\in(\mathfrak{b}_v\mathfrak{c}_v)_n^n$, which implies that $\beta\in D_v$ as desired.

Lemma 2.2 *If $v|\mathfrak{c}$, we have $D_v\sigma D_v\cap P_v=K_v\sigma K_v$ and $D_v\sigma D_v=D_v\sigma K_v$ for every $\sigma\in Z_{0v}$, where $K_v=D_v\cap P_v$.*

Proof. Given $\alpha=\begin{pmatrix}a&b\\c&d\end{pmatrix}\in D_v$, $v|\mathfrak{c}$, put $\beta=\begin{pmatrix}{}^td&-{}^tb\\0&d^{-1}\end{pmatrix}$. Then $\beta\in K_v$ and $\beta\alpha=\begin{pmatrix}1&0\\s&1\end{pmatrix}$ with $s\in S_v$. Now let $\sigma=\operatorname{diag}[\tilde{q},q]\in Z_{0v}$ and $\pi=\alpha_1\sigma\alpha_2^{-1}\in P_v$ with $\alpha_i\in D_v$. Applying the result just proved to α_i, we obtain elements $\beta_i\in K_v$ such that $\beta_i\alpha_i=\begin{pmatrix}1&0\\s_i&1\end{pmatrix}$ with $s_i\in S_v$. Then $\beta_1\pi\beta_2^{-1}=\begin{pmatrix}\tilde{q}&0\\e&q\end{pmatrix}$ with some $e\in(F_v)_n^n$. Since $\beta_1\pi\beta_2^{-1}\in P_v$, we see that $e=0$, which shows that $\pi\in K_v\sigma K_v$, and hence $D_v\sigma D_v\cap P_v=K_v\sigma K_v$. Next, take any γ, $\varepsilon\in D_v$. By Lemma 2.1, $\gamma\sigma\varepsilon=\xi\zeta$ with $\xi\in D_v$ and $\zeta\in P_v$. Then $\zeta\in D_v\sigma D_v\cap P_v=K_v\sigma K_v$, and hence $\gamma\sigma\varepsilon\in\xi K_v\sigma K_v\subset D_v\sigma K_v$, from which we obtain $D_v\sigma D_v=D_v\sigma K_v$.

Lemma 2.3 *Let $\sigma=\operatorname{diag}[\tilde{q},q]\in Z_{0v}$. If $v|\mathfrak{c}$, then $D_v\sigma D_v=\coprod_{d,b}D_v\begin{pmatrix}\tilde{d}&\tilde{d}b\\0&d\end{pmatrix}$ with $d\in E_v\backslash E_vqE_v$ and $b\in S(\mathfrak{b}^{-1})_v/{}^tdS(\mathfrak{b}^{-1})_vd$ with $S(*)$ of* (1.20d).

Proof. Since $D_v\sigma D_v=D_v\sigma K_v$, each coset of $D_v\backslash D_v\sigma D_v$ has a representative in σK_v. Then our assertion can be verified in a straightforward way.

Lemma 2.4 *The set $\mathscr{Z}_0$ is closed under multiplication.*

Proof. Since $G_v=\mathscr{X}_{0v}$ if $v\nmid\mathfrak{c}$, it is sufficient to prove that $\mathscr{X}_{0v}$ is closed under multiplication for $v|\mathfrak{c}$. Let $D_v\sigma D_v=\coprod_{d,b} D_v\begin{pmatrix}\tilde{d} & \tilde{d}b\\ 0 & d\end{pmatrix}$, $\sigma\in Z_{0v}$, $v|\mathfrak{c}$ as in Lemma 2.3, and also let $D_v\tau D_v=\coprod_{e,g} D_v\begin{pmatrix}\tilde{e} & \tilde{e}g\\ 0 & e\end{pmatrix}$ be a decomposition of the same type for $\tau\in Z_{0v}$. Then $D_v\sigma D_v\tau D_v=\coprod_{d,b,e,g} D_v\begin{pmatrix}\tilde{d} & \tilde{d}b\\ 0 & d\end{pmatrix}\begin{pmatrix}\tilde{e} & \tilde{e}g\\ 0 & e\end{pmatrix}=\coprod_{d,b,e,g} D_v\cdot\mathrm{diag}[\tilde{d}\tilde{e}, de]$ $\begin{pmatrix}1 & x\\ 0 & 1\end{pmatrix}$, where $x=g+{}^{t}ebe$. Since $\begin{pmatrix}1 & x\\ 0 & 1\end{pmatrix}\in D_v$ and $\mathrm{diag}[\tilde{d}\tilde{e}, de]\in Z_{0v}$, we obtain the desired result.

Lemma 2.5 *Given* $x\in(F_n^n)_{\mathbf{h}}$, *there exists a pair* $(c,d)\in R\times B$ *such that* $x=d^{-1}c$ *and* $cL+dL=L$, *where* $L=\mathfrak{g}_1^n$ *and* $cL+dL$ *is the* $\mathfrak{g}$*-lattice in* F_1^n *such that* $(cL+dL)_v=c_vL_v+d_vL_v$ *for every* $v\in\mathbf{h}$. *If* $x=d'^{-1}c'$ *with another such pair* (c',d'), *then* $(c',d')\in E(c,d)$. *Moreover, for each* $v\in\mathbf{h}$ $\det(d_v)^{-1}\mathfrak{g}_v$ *is the product of all the elementary divisors of* x_v *that are not integral.*

This follows easily from the theory of elementary divisors for matrices over F_v, $v\in\mathbf{h}$ (cf. [S6, Lemma 4.1]). The notation being as in the above lemma, we put

$$\nu_0(x)=\det(d)\,\mathfrak{g},\qquad \nu(x)=N(\nu_0(x))\qquad (x\in(F_n^n)_{\mathbf{h}}). \tag{2.3}$$

Notice that $\nu_0(x)$ depends only on x mod R.

Lemma 2.6 *With L as in Lemma* 2.5, *put*

$$W=\{(g,h)\in B\times B\,|\,gL+hL=L,\ h_v\in E_v \text{ for every } v|\mathfrak{c}\},$$
$$S'=\{\sigma\in S_{\mathbf{h}}\,|\,\sigma_v\in S(\mathfrak{b}^{-1})_v \text{ for every } v|\mathfrak{c}\}.$$

Then a complete set of representatives for $D\backslash\mathscr{X}_0$ *is given by*

$$\left\{\begin{pmatrix}g^{-1}h & g^{-1}\sigma\cdot{}^{t}h^{-1}\\ 0 & {}^{t}g\cdot{}^{t}h^{-1}\end{pmatrix}\middle|\,(g,h)\in E\backslash W/(E\times 1),\ \sigma\in S'/gS_{\mathbf{h}}(\mathfrak{b}^{-1})\cdot{}^{t}g\right\}.$$

Proof. This is essentially the local problem of finding $D_v\backslash\mathscr{X}_{0v}$. If $v|\mathfrak{c}$, the answer is given by Lemma 2.3. If $v\nmid\mathfrak{c}$, then the question is about $D_v\backslash G_v$. Since $G_v=D_vP_v$, we can take representatives from $(D_v\cap P_v)\backslash P_v$. Let $\begin{pmatrix}a & b\\ 0 & \tilde{a}\end{pmatrix}\in P_v$. By Lemma 2.5 we can put $a=g^{-1}h$ with $(g,h)\in B_v\times B_v$ such that $gL_v+hL_v=L_v$. Then we can find an element $\sigma\in S_v$ such that $\begin{pmatrix}a & b\\ 0 & \tilde{a}\end{pmatrix}=\begin{pmatrix}g^{-1}h & g^{-1}\sigma\cdot{}^{t}h^{-1}\\ 0 & {}^{t}g\cdot{}^{t}h^{-1}\end{pmatrix}$. Since $E_v(g,h)$ is determined by a and $D_v\cap P_v=\left\{\begin{pmatrix}e & s\tilde{e}\\ 0 & \tilde{e}\end{pmatrix}\middle|\,e\in E_v,\ s\in S(\mathfrak{b}^{-1})_v\right\}$, left multiplication by the elements of $D_v\cap P_v$ reduces the representatives to those in the desired set.

Lemma 2.7 *If* $\alpha=\begin{pmatrix}g^{-1}h & g^{-1}\sigma\cdot{}^{t}h^{-1}\\ 0 & {}^{t}g\cdot{}^{t}h^{-1}\end{pmatrix}$ *with* $(g,h)\in W$ *and* $\sigma\in S_{\mathbf{h}}$, *then* $\nu_0(\alpha)=\det(gh)\,\nu_0(\sigma)$. *Moreover, if* $\alpha\in D\cdot\mathrm{diag}[\tilde{e},e]\,D$ *with* $e\in B$, *then* $\det(e)\,\mathfrak{g}=\det(gh)\,\nu_0(b\sigma)$, *where b is an element of* $F_{\mathbf{h}}^{\times}$ *such that* $b\mathfrak{g}=\mathfrak{b}$.

Proof. Let $\alpha \in D \cdot \operatorname{diag}[\tilde{e}, e] D$ and $\beta = \operatorname{diag}[1_n, b 1_n]$. Then $\beta^{-1} D \beta = D[\mathfrak{g}, \mathfrak{c}] \subset D[\mathfrak{g}, \mathfrak{g}]$ and hence $\det(e)\mathfrak{g} = \nu_0(\beta^{-1}\alpha\beta)$. Therefore, applying the first assertion to $\beta^{-1}\alpha\beta$, we obtain the second one. Now the first assertion is essentially due to Böcherer [B2, Prop. 2]. His proof, given in terms of rational matrices, applies to the elements of G_v with $v \in \mathbf{h}$, but requires the following fact:

If an element $\sigma = (s_{ij})$ of S_v has $\{\pi_v^{e_1}, \ldots, \pi_v^{e_t}\}$ as its set of nonzero elementary divisors with $e_1 \leqq \ldots \leqq e_r < 0$, $1 \leqq r \leqq t \leqq n$ and $e_i \geqq 0$ for $i > r$, then there exists a subset α of $\{1, \ldots, n\}$ consisting of r numerals such that $\det(s_{ij})_{i,j \in \alpha} \in \pi^{e_1 + \cdots + e_r}\mathfrak{g}_v^\times$.

Since this is nontrivial and is not proved in [B2], we give here a proof, which proceeds by induction on n. Let us first assume that $s_{ii} \in \pi^{e_1}\mathfrak{g}_v^\times$ for some i, say $i=1$. Put $a = s_{11}$, $\sigma = \begin{pmatrix} a & b \\ {}^t b & d \end{pmatrix}$, and $\tau = \begin{pmatrix} 1 & -a^{-1}b \\ 0 & 1_{n-1} \end{pmatrix}$. Then ${}^t\tau\sigma\tau = \operatorname{diag}[a, e]$ with $e = d - {}^t b a^{-1} b$. By induction, after permuting the numerals $2, \ldots, n$, we can put $e = \begin{pmatrix} f & * \\ * & * \end{pmatrix}$ with f of size $r-1$ such that $\det(f) \in \pi^{e_2 + \cdots + e_r}\mathfrak{g}_v^\times$. Let σ' resp. τ' be the upper left submatrix of size r of the rearrangement of σ resp. τ according to the chosen permutation. Then we easily see that ${}^t\tau'\sigma'\tau' = \operatorname{diag}[a, f]$. Since $\det(\tau') = 1$, σ' gives the desired principal submatrix of size r. Next assume that $s_{ii}\mathfrak{g}_v \neq \pi^{e_1}\mathfrak{g}_v$ for every i and $s_{12} \in \pi^{e_1}\mathfrak{g}_v^\times$. Let $a = \begin{pmatrix} s_{11} & s_{12} \\ s_{21} & s_{22} \end{pmatrix}$. Then $\det(a) \in \pi^{2e_1}\mathfrak{g}_v^\times$, and hence $e_2 = e_1$. If $r > 2$, we repeat the above argument with a as the present matrix of size 2, and with $b \in (F_v)^2_{n-2}$. Since $a^{-1}b \in (\mathfrak{g}_v)^2_{n-2}$, our reasoning is valid in the present case, and induction completes the proof.

We are going to state some of our results in terms of formal Dirichlet series of the form $\sum_{\mathfrak{a} \subset \mathfrak{g}} c(\mathfrak{a})[\mathfrak{a}]$. Here $c(\mathfrak{a}) \in \mathbf{C}$ and $\{[\mathfrak{a}]\}$ is a system of formal multiplicative symbols defined for the fractional ideals $\mathfrak{a}$ in F as follows: the $[\mathfrak{p}]$ for the prime ideals $\mathfrak{p}$ are independent indeterminates; $[\mathfrak{g}] = 1$ and $[\mathfrak{a}\mathfrak{b}] = [\mathfrak{a}][\mathfrak{b}]$. For each $v \in \mathbf{h}$ we put $[v] = [\pi_v \mathfrak{g}]$ with a prime element π_v of F_v. Now put

(2.4) $$S^* = \prod_{v \in \mathbf{h}} S_v^*, \quad S_v^* = \{x \in S_v \mid \mathbf{e}_v(\operatorname{tr}(x \cdot S(\mathfrak{g})_v)) = 1\}$$

and let δ be an element of $F_{\mathbf{h}}^\times$ such that $\delta\mathfrak{g} = \mathfrak{d}$. For $\zeta \in S_{\mathbf{A}}$ such that $\zeta_{\mathbf{h}} \in \delta S^*$ we define a formal Dirichlet series $\alpha(\zeta)$ in the above sense by

(2.5a) $$\alpha(\zeta) = \sum_{\sigma \in S_{\mathbf{h}}/S_{\mathbf{h}}(\mathfrak{g})} \mathbf{e}_{\mathbf{h}}(-\delta^{-1}\operatorname{tr}(\zeta\sigma))[\nu_0(\sigma)].$$

Clearly $S_{\mathbf{h}}/S_{\mathbf{h}}(\mathfrak{g})$ can be replaced by $S/S(\mathfrak{g})$; also we have

(2.5b) $$\alpha(\zeta) = \prod_{v \in \mathbf{h}} \alpha_v(\zeta_v), \quad \alpha_v(\zeta_v) = \sum_{\sigma \in S_v/S(\mathfrak{g})_v} \mathbf{e}_v(-\delta_v^{-1}\operatorname{tr}(\zeta_v\sigma))[\nu_0(\sigma)].$$

We studied these types of series in [S6] (with $\nu(\sigma)^{-s}$ in place of $[\nu_0(\sigma)]$) and obtained their expressions as Euler products. In the next section we shall give more complete results on them. Notice that α and α_v are independent of the choice of δ, and

$$(2.6)\qquad \alpha(c\cdot {}^t u\zeta u)=\alpha(\zeta) \quad \text{if } c\in\prod_{v\in\mathbf{h}}\mathfrak{g}_v^\times \text{ and } u\in E,$$

since $v_0(x)$ depends only on ExE.

Lemma 2.8 *Let S' and $\mathfrak{b}$ be as in Lemmas 2.6 and 2.7. Let $\zeta\in S_\mathbf{h}$ and $g\in B$; assume ${}^t g\zeta g\in\mathfrak{b}\delta S^*$. Then*

$$\sum_{\sigma\in S'/gS_\mathbf{h}(\mathfrak{b}^{-1})\cdot{}^t g} \mathbf{e}_\mathbf{h}(-\delta^{-1}\mathrm{tr}(\zeta\sigma))[v_0(\mathfrak{b}\sigma)]\neq 0$$

only if $\zeta\in\mathfrak{b}\delta S^*$, in which case the sum equals $|\det(g)|_\mathbf{A}^{-n-1}\prod_{v\nmid \mathfrak{c}}\alpha_v(b_v^{-1}\zeta_v)$.

This is because $v_0(\sigma)$ depends only on σ mod $S_\mathbf{h}(\mathfrak{g})$ and $[S_\mathbf{h}(\mathfrak{g})\colon gS_\mathbf{h}(\mathfrak{g})\cdot{}^t g] =|\det(g)|_\mathbf{A}^{-n-1}$.

We now consider the Hecke algebra $\mathfrak{R}(D,\mathscr{Z})$ consisting of all formal finite sums $\sum c_\sigma D\sigma D$ with $c_\sigma\in\mathbf{Q}$ and $\sigma\in\mathscr{Z}$, with the law of multiplication defined as in [S1], [S2], and [S3, Ch. 3], and similarly the algebra $\mathfrak{R}(D_v,\mathscr{Z}_{0v})$. Fixing $v\in\mathbf{h}$ and taking n indeterminates $t_1,\dots,t_n$, we define a $\mathbf{Q}$-linear map

$$(2.7\text{a})\qquad \omega_v\colon \mathfrak{R}(D_v,\mathscr{Z}_{0v})\to\mathbf{Q}[t_1,\dots,t_n,t_1^{-1},\dots,t_n^{-1}]$$

as follows: First we observe that every coset $E_v y$ with $y\in GL_n(F_v)$ contains an upper triangular matrix whose diagonal elements are powers $\pi_v^{a_1},\dots,\pi_v^{a_n}$ of a prime element π_v of F_v. Then we put $\omega_0(E_v y)=\prod_{i=1}^n(|\pi_v|^t t_i)^{a_i}$. Now, given $\sigma\in\mathscr{Z}_{0v}$, we can take a coset decomposition $D_v\sigma D_v=\coprod_{d\in X}\coprod_{y\in Y_d}D_v\begin{pmatrix}\tilde d & \tilde d y\\ 0 & d\end{pmatrix}$ with $X\subset GL_n(F_v)$ and $Y_d\subset S_v$. Then we put

$$(2.7\text{b})\qquad \omega_v(D_v\sigma D_v)=\sum_{d\in X}\#(Y_d)\,\omega_0(E_v d),$$

where $\#(Y_d)$ denotes the number of elements of Y_d. It is well known that ω_v is a ring-injection if $v\nmid\mathfrak{c}$. (Since $D_v[\mathfrak{b}^{-1},\mathfrak{b}]=\beta D_v[\mathfrak{g},\mathfrak{g}]\beta^{-1}$ with $\beta =\mathrm{diag}[1_n, b_v 1_n]$, $b_v\mathfrak{g}_v=\mathfrak{b}_v$, the problem can be reduced to the case $\mathfrak{b}=\mathfrak{g}$.)

Before treating the case $v|\mathfrak{c}$, we define $\omega'\colon \mathfrak{R}(E_v,B_v)\to\mathbf{Q}[t_1,\dots,t_n]$ by $\omega'(E_v qE_v)=\sum_d\omega_0(E_v d)$ for $E_v qE_v=\coprod_d E_v d$. It is well known that ω' gives a ring-injection of $\mathfrak{R}(E_v,B_v)$ into $\mathbf{Q}[t_1,\dots,t_n]$. Now, if $v|\mathfrak{c}$, we can define a $\mathbf{Q}$-linear map

$$\iota\colon \mathfrak{R}(D_v,\mathscr{Z}_{0v})\to\mathfrak{R}(E_v,B_v)$$

by $\iota(D_v\sigma D_v)=|\det(q)|^{-n-1}E_v qE_v$ for σ and q as in Lemma 2.3. The proof of Lemma 2.4 shows that this is a ring-isomorphism, and also that $\omega_v=\omega'\circ\iota$. Therefore ω_v is a ring-injection even if $v|\mathfrak{c}$. Consequently $\mathfrak{R}(D_v,\mathscr{Z}_{0v})$ is commutative for every $v\in\mathbf{h}$.

We can view $\mathfrak{R}(D_v,\mathscr{Z}_{0v})$ as a subalgebra of $\mathfrak{R}(D,\mathscr{Z})$ by identifying $D_v\tau D_v$ with $D\tau D$ for every $\tau\in\mathscr{Z}_{0v}$. The algebra $\mathfrak{R}(D,\mathscr{Z}_0)$, being generated by $\mathfrak{R}(D_v,\mathscr{Z}_{0v})$

for all $v \in \mathbf{h}$, is commutative, since clearly $D\sigma D \cdot D\tau D = D\sigma\tau D = D\tau\sigma D = D\tau D \cdot D\sigma D$ if $\sigma \in \mathscr{X}_{0v}$, $\tau \in \mathscr{X}_{0v'}$, and $v \neq v'$.

For $e \in GL_n(F)_{\mathbf{h}}$ let T_e denote the element of $\mathfrak{R}(D, \mathscr{X})$ defined by $T_e = D \cdot \mathrm{diag}[\tilde{e}, e]\, D$. This is an element of $\mathfrak{R}(D_v, \mathscr{X}_{0v})$ if $e \in B_v$. Clearly T_e depends only on EeE. We now consider formal Dirichlet series

$$\mathfrak{T} = \sum_{e \in E \backslash B/E} [\det(e)\,\mathfrak{g}]\, T_e, \qquad \mathfrak{T}_v = \sum_{q \in E_v \backslash B_v/E_v} [\det(q)\,\mathfrak{g}]\, T_q,$$

$$\omega_v(\mathfrak{T}_v) = \sum_{q \in E_v \backslash B_v/E_v} [\det(q)\,\mathfrak{g}]\, \omega_v(T_q)$$

with coefficients in $\mathfrak{R}(D, \mathscr{X})$ and $\mathbf{Q}[t_1, \ldots, t_n, t_1^{-1}, \ldots, t_n^{-1}]$. Clearly $\mathfrak{T} = \prod_{v \in \mathbf{h}} \mathfrak{T}_v$.

Theorem 2.9

$$\omega_v(\mathfrak{T}_v) = \prod_{i=1}^{n} (1 - |\pi_v|^{-n} t_i [v])^{-1} \qquad \text{if } v \mid \mathfrak{c},$$

$$\omega_v(\mathfrak{T}_v) = \frac{1-[v]}{1-|\pi_v|^{-n}[v]} \prod_{i=1}^{n} \frac{1-|\pi_v|^{-2i}[v]^2}{(1-|\pi_v|^{-n} t_i [v])(1-|\pi_v|^{-n} t_i^{-1} [v])} \qquad \text{if } v \nmid \mathfrak{c}.$$

Proof. It is well known (and in fact easy to prove) that

$$(2.8) \qquad \sum_{q \in E_v \backslash B_v/E_b} \omega'(E_v q E_v)[\det(q)\,\mathfrak{g}]$$

$$= \sum_{r \in E_v \backslash B_v} \omega_0(E_v r)[\det(r)\,\mathfrak{g}] = \prod_{i=1}^{n} (1 - |\pi_v| t_i [v])^{-1}.$$

For σ and q of Lemma 2.3 we have $\omega_v(D_v \sigma D_v) = |\det(q)|^{-n-1} \omega'(E_v q E_v)$, and hence we obtain our formula for $v \mid \mathfrak{c}$ directly from (2.8). Suppose $v \nmid \mathfrak{c}$. By the local version of Lemmas 2.6 and 2.7 we have

$$\omega_v(\mathfrak{T}_v) = \sum_{g,h} \sum_{\sigma} \omega_0(E_v \cdot {}^t g \cdot {}^t h^{-1})[\det(gh)\, \nu_0(b_v \sigma)],$$

where $(g, h) \in E_v \backslash W_v/(E_v \times 1)$, $W_v = \{(g, h) \in B_v \times B_v \mid g L_v + h L_v = L_v\}$, $L_v = (\mathfrak{g}_v)_1^n$, $\sigma \in S_v / g S(\mathfrak{b}^{-1})_v \, {}^t g$, and b is as in Lemma 2.7. By Lemma 2.8 for each fixed g we have

$$\sum_{\sigma} [\nu_0(b_v \sigma)] = |\det(g)|^{-n-1} \alpha_v(0),$$

where $\alpha_v(0)$ is the series of (2.5b) with $\zeta_v = 0$. Hence

$$\omega_v(\mathfrak{T}_v) = \alpha_v(0) \sum_{g,h} |\det(g)|^{-n-1} \omega_0(E_v \cdot {}^t g \cdot {}^t h^{-1})[\det(gh)\,\mathfrak{g}].$$

Put $Z = E_v \backslash (B_v \times B_v)/(E_v \times 1)$ and

$$\Omega = \sum_{(x,y) \in Z} |\det(x)|^{-n-1} \omega_0(E_v \cdot {}^t x \cdot {}^t y^{-1})[\det(xy)\,\mathfrak{g}].$$

Since every element (x, y) of $B_v \times B_v$ can be written $(x, y) = (cy, ch)$ with $c \in B_v$ and $(g, h) \in W_v$, we easily see that

$$\alpha_v(0)\,\Omega = \omega_v(\mathfrak{T}_v) \sum_{c \in E_v \backslash B_v} |\det(c)|^{-n-1} [\det(c)\,\mathfrak{g}]^2.$$

The last sum equals $\prod_{i=1}^{n} (1 - |\pi_v|^{-n-i}[v]^2)^{-1}$, since it is practically the zeta function of $M_n(F_v)$. (This can also be seen by putting $t_i = |\pi_v|^{-n-1-i}$ in (2.8).) Let X resp. Y be a complete set of representatives for B_v/E_v resp. $E_v \backslash B_v$. Since $X \times Y$ represents Z, we can replace Z in the sum Ω by $X \times Y$. Taking ${}^t x$ and ${}^t y$ to be upper triangular matrices, we find that

$$\Omega = \sum_{w \in E_v \backslash B_v} |\det(w)|^{-n-1} \omega_0(E_v w)[\det(w)\,\mathfrak{g}] \sum_{z \in B_v/E_v} \omega_0(E_v z^{-1})[\det(z)\,\mathfrak{g}]$$

$$= \prod_{i=1}^{n} (1 - |\pi_v|^{-n} t_i [v])^{-1} (1 - |\pi_v|^{-n} t_i^{-1} [v])^{-1}$$

by (2.8) and its variation. Now in [S6, Prop. 5.1] we showed (see also Theorem 3.2 below) that

$$\alpha_v(0) = \frac{1 - [v]}{1 - |\pi_v|^{-n}[v]} \prod_{i=1}^{n} \frac{1 - |\pi_v|^{-2i}[v]^2}{1 - |\pi_v|^{-n-i}[v]^2}.$$

Combining all these formulas, we obtain the desired result for $\omega_v(\mathfrak{T}_v)$, $v \nmid \mathfrak{c}$.

The above expression for $\omega_v(\mathfrak{T}_v)$ for $v \nmid \mathfrak{c}$ was proved by Böcherer in [B2]. We adapted his proof to our setting.

Let us now consider the Hecke algebra $\mathfrak{R}(\Gamma, G \cap \mathscr{X}_0)$ with $\Gamma = G \cap D = \Gamma[\mathfrak{b}^{-1}, \mathfrak{b}\mathfrak{c}]$. From strong approximation in $G_{\mathbf{A}}$ we easily obtain:

(2.9a) $\quad D x D = D x \Gamma = \Gamma x D \quad$ for every $x \in G_{\mathbf{A}}$;

(2.9b) $\quad \Gamma \alpha \Gamma = G \cap D \alpha D \quad$ for every $\alpha \in G$;

(2.9c) $\quad \Gamma \alpha \Gamma = \coprod_{\beta \in B} \Gamma \beta \Rightarrow D \alpha D = \coprod_{\beta \in B} D \beta.$

Lemma 2.10 *The map $\Gamma \alpha \Gamma \mapsto D \alpha D$ gives a $\mathbf{Q}$-linear isomorphism of $\mathfrak{R}(\Gamma, G \cap \mathscr{X}_0)$ onto $\mathfrak{R}(D, \mathscr{X}_0)$, and consequently $\mathfrak{R}(\Gamma, G \cap \mathscr{X}_0)$ is commutative.*

Proof. Since $G_{\mathbf{A}} = GD$, every x in $\mathscr{X}_0$ belongs to αD for some $\alpha \in G$. Then $\alpha \in G \cap \mathscr{X}_0$ and $D \alpha D = D x D$, which proves the surjectivity of the map. The injectivity follows from (2.9b). That it is a homomorphism is an easy consequence of (2.9c).

3. The series $\alpha(\zeta)$

Let us fix a prime $v \in \mathbf{h}$, take a prime element π_v of F_v, and take also an element δ_v of F_v so that $\delta_v \mathfrak{g}_v = \mathfrak{d}_v$; put $\mathfrak{p} = \pi_v \mathfrak{g}$, $q = N(\mathfrak{p})$, and

$$T_v = T_v^n = \{x \in S_v^n \mid 2x_{ij} \in (1+\delta_{ij})\,\mathfrak{g}_v \text{ for every } i \text{ and } j\}, \tag{3.1}$$

$$X_m(l \times n) = (\mathfrak{g}_v)_n^l / \pi_v^m (\mathfrak{g}_v)_n^l, \qquad S_m(n) = \pi_v^{-m} S(\mathfrak{g})_v / S(\mathfrak{g})_v. \tag{3.2}$$

Clearly $T_v^n = \delta_v S_v^*$ with S_v^* of (2.4). We can easily prove

Lemma 3.1 (1) *Given $\sigma \in S_v$, there exists an element $\alpha \in GL_n(\mathfrak{g}_v)$ such that ${}^t\alpha\sigma\alpha = \operatorname{diag}[\sigma_1, \ldots, \sigma_r]$ with σ_i of size 1 and 2; each σ_i of size 2 is necessary only if $v|2$, and is of the form $\sigma_i = \begin{pmatrix} a & b \\ b & d \end{pmatrix}$, $b\mathfrak{g}_v \supsetneqq a\mathfrak{g}_v + d\mathfrak{g}_v$.*

(2) *If $\tau \in T_v^l$ and $x \in (\mathfrak{g}_v)_n^l$, then ${}^t x \tau x \in T_v^n$.*

(3) *If $\sigma \in T_v^n$, then $\det(2\sigma) \in \mathfrak{g}_v$ if n is even, and $\det(2\sigma) \in 2\mathfrak{g}_v$ if n is odd.*

For $\sigma \in S_v$ let $k(\sigma)$ be the nonnegative integer such that $\nu(\sigma) = q^{k(\sigma)}$. We then define, for $\zeta \in T_v^n$, formal power-series $\alpha_\zeta(t)$ and $\alpha_\zeta^m(t)$ in an indeterminate t by

$$\alpha_\zeta(t) = \sum_{\sigma \in S_v/S(\mathfrak{g})_v} \mathbf{e}_v(-\delta_v^{-1} \operatorname{tr}(\zeta\sigma))\, t^{k(\sigma)}, \tag{3.3a}$$

$$\alpha_\zeta^m(t) = \sum_{\sigma \in S_m(n)} \mathbf{e}_v(-\delta_v^{-1} \operatorname{tr}(\zeta\sigma))\, t^{k(\sigma)}. \tag{3.3b}$$

This means that putting $t = [v]$, we obtain $\alpha_\zeta([v]) = \alpha_v(\zeta)$ with α_v of (2.5b), and $\alpha_\zeta^m([v])$ is its partial sum. Clearly $\alpha_\zeta^m(t)$ is a polynomial in t of degree at most mn, and $\alpha_\zeta(t) = \lim_{m \to \infty} \alpha_\zeta^m(t)$ if α_ζ is convergent at t. The coefficients of α_ζ and α_ζ^m are in $\mathbf{Z}$, since they are sums of roots of unity invariant under the Galois action. Since $\alpha_\zeta = \alpha_\eta$ if $\eta = {}^t\beta\zeta\beta$ with $\beta \in GL_n(\mathfrak{g}_v)$, Lemma 3.1(1) reduces the problem of determining α_ζ to the case where $\zeta = \operatorname{diag}[\xi, 0]$ with nonsingular ξ.

We call an element ψ of T_v^n *regular* if n is even and $\det(2\psi) \in \mathfrak{g}_v^\times$, of if n is odd and $\det(2\psi) \in 2\mathfrak{g}_v^\times$. Given $\tau \in T_v^n \cap GL_n(F_v)$, n even, put $\rho = (-1)^{n/2} \det(\tau)$. We then define $\varepsilon(\tau)$ as follows: $\varepsilon(\tau) = 1$ if $F_v(\rho^{1/2}) = F_v$, $\varepsilon(\tau) = -1$ if $F_v(\rho^{1/2})$ is an unramified quadratic extension of F_v, $\varepsilon(\tau) = 0$ if v is ramified in $F_v(\rho^{1/2})$. Notice that $\varepsilon(\tau) \neq 0$ if τ is regular (even when $v|2$). Now our main result of this section is

Theorem 3.2 *If $\zeta = \operatorname{diag}[\xi, 0] \in T_v^n$ with $\xi \in T_v^r \cap GL_r(F_v)$, then $\alpha_\zeta = f_\zeta g_\zeta$ with a polynomial f_ζ with coefficients in $\mathbf{Z}$ whose constant term is 1 and a rational function g_ζ given as follows:*

$$g_\zeta(t) = \frac{(1-t)\prod_{i=1}^{[n/2]}(1-q^{2i}t^2)}{(1-\varepsilon q^{(2n-r)/2}t)\prod_{i=1}^{[(n-r)/2]}(1-q^{2n-2i-r+1}t^2)} \qquad \textit{if } r \textit{ is even,}$$

$$g_\zeta(t) = \frac{(1-t)\prod_{i=1}^{[n/2]}(1-q^{2i}t^2)}{\prod_{i=1}^{[(n-r+1)/2]}(1-q^{2n-2i-r+2}t^2)} \qquad \textit{if } r \textit{ is odd,}$$

where $\varepsilon=\varepsilon(\xi)$. *In particular,* $\alpha_\zeta=g_\zeta$ *if* ξ *is regular or* $\zeta=0$ *with the understanding that* $r=0$ *and* $\varepsilon=1$ *for* $\zeta=0$.

These results were proved in [S6] at least in the following cases: (i) $\zeta=0$; (ii) $\det(\delta\xi)\in\mathfrak{g}_v^\times$ and $v\nmid 2$. In [S6] we also introduced two more types of series similar to α in connection with modular forms of half-integral weight and also those on $SU(n,n)$, and proved results of the same type for them. The case with arbitrary ξ in the case $G=Sp(n,\mathbf{Q})$ was treated by Kitaoka in [K], in which he invoked the result of [S6] for $\zeta=0$. The final results in [K] are stated for all primes p including 2, but the proof for $p=2$ is hardly convincing. More general results for the three types of series in [S6] (as well as some other cases, but still under the condition $v\nmid 2$ in the present case) were obtained in [F1] by Feit, who later treated the case $v|2$ in [F2]. Here we present a much simpler argument, which, together with the result in [S6] for $\zeta=0$ and a basic result in [F1], gives a proof for all primes v including those dividing 2. Hereafter till the end of this section we fix v and suppress the subscript v for simplicity. Thus F denotes the local field, $\mathfrak{g}$ its valuation ring, and $\mathfrak{p}$ the maximal ideal of $\mathfrak{g}$.

For $0<m\in\mathbf{Z}$, $\varphi\in T^n$, and $\psi\in T^l$ we denote by $N_m(\psi,\varphi)$ the number of elements x in $X_m(l\times n)$ such that ${}^t x\psi x\equiv\varphi \pmod{\pi^m T^n}$.

Lemma 3.3 *Let* $\varphi\in T^n\cap GL_n(F)$ *and* $\psi\in T^l\cap GL_l(F)$ *with* $l\geqq n$; *let* e *be the integer determined by* $\det(2\varphi)\in\pi^e\mathfrak{g}^\times$. *Then* $q^{mn(n+1)/2-mnl}N_m(\psi,\varphi)$ *is independent of* m *if* $m>2e$.

This is due to Siegel [Si, No. 20, Hilfssatz 13; No. 26, Hilfssatz 58]. Strictly speaking, his results for $v|2$ are formulated with $S(\mathfrak{g})_v$ instead of T_v^n, but his proof given in [No. 20, Hilfssatz 13], with obvious modifications, applies to the present formulation.

Lemma 3.4 *If* $0\leqq m\in\mathbf{Z}$, $\sigma\in S(\mathfrak{p}^{-m})$, *and* Φ *is a regular element of* T^2, *then*

$$\sum_{x\in X_m(2\times n)} \mathbf{e}_v(\delta^{-1}\operatorname{tr}({}^t x\Phi x\sigma))=\varepsilon(\Phi)^{k(\sigma)}q^{2mn}\nu(\sigma)^{-1},$$

where $k(\sigma)$ *is as in* (3.3a).

Proof. Lemma 3.1(1) reduces our problem to the case of σ_i of size 1 or 2 in that lemma. Let s be the smallest integer such that $\pi^s\sigma$ is v-integral. Denote by Z_m the sum in question. We easily see that $Z_m=q^{2n(m-s)}Z_s$, and so it is sufficient to treat the case $m=s$. We prove this case by induction on s. The case $s=0$ is trivial. Suppose $s=1$ and $n=1$; put $\varepsilon=\varepsilon(\Phi)$. Then for $c\in\mathfrak{g}$ the number of $x\in X_1(2\times 1)$ such that ${}^t x\Phi x\equiv c \pmod{\mathfrak{p}}$ is $q+\varepsilon(q-1)$ if $c\in\mathfrak{p}$ and $q-\varepsilon$ if $c\notin\mathfrak{p}$. Hence we easily see that $Z_1=\varepsilon q$. Next suppose $s=1$, $n=2$, and $v|2$; we can put $\pi\sigma=\begin{pmatrix} a & b\\ b & d\end{pmatrix}$ with $b\in\mathfrak{g}_v^\times$ and a, $d\in\mathfrak{p}$. Then $Z_1 = \sum_{x,y\in X_m(2\times 1)} \mathbf{e}_v(\delta^{-1}\pi^{-1}\cdot{}^t x\cdot 2b\Phi y)$. Since $2b\Phi\in GL_2(\mathfrak{g})$, the sum over y is non-zero only when $x=0$, in which case it is q^2. Thus $Z_1=q^2$. Assume now $s\geqq 2$ and $\sigma=\pi^{-s}\tau$, $\tau\in\mathfrak{g}^\times$ if $n=1$ and $\tau=\begin{pmatrix} a & b\\ b & d\end{pmatrix}$ as above if $n=2$. Then

$$Z_s=\sum_{w\in X_s(2\times n)} \mathbf{e}_v(\delta^{-1}\pi^{-s}\operatorname{tr}({}^t w\Phi w\tau)).$$

Putting $w=u+\pi^{s-1}z$, we obtain

$$Z_s=\sum_{u\in X_{s-1}(2\times n)} \mathbf{e}_v(\delta^{-1}\pi^{-s}\operatorname{tr}({}^t u\Phi u\tau)) \sum_{z\in X_1(2\times n)} \mathbf{e}_v(\delta^{-1}\pi^{-1}\operatorname{tr}({}^t u\cdot 2\Phi z\tau)).$$

Since both 2Φ and τ are invertible modulo $\mathfrak{p}$, the last sum over z is nonzero only if u is divisible by π, and hence

$$Z_s=q^{2n}\sum_{x\in X_{s-2}(2\times n)} \mathbf{e}_v(\delta^{-1}\pi^{2-s}\operatorname{tr}({}^t x\Phi x\tau)).$$

Therefore induction settles our problem.

Lemma 3.5 *Let Ψ be a regular element of T^{2h}, $0<h\in\mathbf{Z}$, and let $\zeta\in T^n$. Then $\alpha_\zeta^m(\varepsilon(\Phi)\,q^{-h})=q^{mn(n+1)/2-2mnh}N_m(\Psi,\zeta)$ for every $m\in\mathbf{Z}$, >0.*

Proof. By Lemma 3.1(1) we can put $\Psi=\operatorname{diag}[\Phi_1,\ldots,\Phi_h]$ with Φ_i of size 2. By Lemma 3.4 we have

$$\sum_{w\in X_m(2h\times n)} \mathbf{e}_v(\delta^{-1}\operatorname{tr}({}^t w\Psi w\sigma))=(\varepsilon q^{-h})^{k(\sigma)}q^{2mnh}$$

with $\varepsilon=\varepsilon(\Psi)$, and hence

$$\begin{aligned}\alpha_\zeta^m(\varepsilon q^{-h})&=q^{-2mnh}\sum_{\sigma\in S_m(n)} \mathbf{e}_v(-\delta^{-1}\operatorname{tr}(\zeta\sigma)) \sum_{w\in X_m(2h\times n)} \mathbf{e}_v(\delta^{-1}\operatorname{tr}({}^t w\Psi w\sigma))\\ &=q^{-2mnh}\sum_{w\in X_m(2h\times n)}\sum_{\sigma\in S_m(n)} \mathbf{e}_v(\delta^{-1}\operatorname{tr}[({}^t w\Psi w-\zeta)\sigma]).\end{aligned}$$

This proves our lemma, since the right-hand side is $q^{mn(n+1)/2-2mnh}N_m(\Psi,\zeta)$.

By Lemma 3.3 the last number is independent of m if $2h\geq n$ and $m>2e$, where e is determined as in that lemma with ζ as φ. In particular, take $\Phi_i=\begin{pmatrix}0 & 1/2\\ 1/2 & 0\end{pmatrix}$ for every i. Then $\varepsilon=1$, and therefore, if the series α_ζ is convergent at $t=q^{-h}$, then $\alpha_\zeta(q^{-h})=\alpha_\zeta^m(q^{-h})$ for $m>2e$. Fix one such m. Then the power-series $\alpha_\zeta(t)$ coincides with the polynomial $\alpha_\zeta^m(t)$ at infinitely many points $t=q^{-h}$, and therefore we obtain the first part of

Proposition 3.6 *Given $\zeta\in T^n\cap \mathrm{GL}_n(F)$, let e be the integer such that $\det(2\zeta)\,\mathfrak{g}_v=\pi^e\mathfrak{g}_v$. Then α_ζ is a polynomial in t with coefficients in $\mathbf{Z}$ of degree at most $(2e+1)\,n$, which is divisible by the polynomial f given as follows:*

$$f(t)=(1-t)(1-\varepsilon(\zeta)\,q^{n/2}t)^{-1}\prod_{i=1}^{n/2}(1-q^{2i}t^2)\quad \textit{if } n \textit{ is even},$$

$$f(t)=(1-t)\prod_{i=1}^{(n-1)/2}(1-q^{2i}t^2)\quad \textit{if } n \textit{ is odd}.$$

Moreover $\alpha_\zeta=f$ if ζ is regular.

Proof. For sufficiently large m we have

$$\alpha_\zeta(1)=\alpha_\zeta^m(1)=\sum_{\sigma\in S_m(n)} \mathbf{e}_v(-\delta^{-1}\operatorname{tr}(\zeta\sigma))=0,$$

and hence α_ζ is divisible by $1-t$. If $2h<n$, we see that $N_m(\Psi,\zeta)=0$ for Ψ of Lemma 3.5 and sufficiently large m, and hence $\alpha_\zeta^m(\varepsilon(\Psi)q^{-h})=0$ by Lemma 3.5. This means that α_ζ is divisible by both $1-q^h t$ and $1+q^h t$, since $\varepsilon(\Psi)$ can be both 1 and -1 with a suitable choice of Ψ. Next suppose that $n=2h$. If ${}^t x\Psi x-\zeta\in\pi^m T^n$, then $\det(x)^2\det(2\Psi)\equiv\det(2\zeta)(\mathrm{mod}\ \pi^m)$, which, for a sufficiently large m, implies that $\varepsilon(\zeta)=\varepsilon(\Psi)$. Thus $N_m(\Psi,\zeta)=0$ if $\varepsilon(\zeta)\neq\varepsilon(\Psi)$. Therefore for $\eta=\pm 1$ we have $\alpha_\zeta(\eta q^{-h})=0$ if $\eta\neq\varepsilon(\zeta)$, that is, α_ζ is divisible by $(1-q^n t^2)(1-\varepsilon(\zeta)q^h t)^{-1}$. Combining all these, we obtain the divisibility of α_ζ by f. If n is even and $\det(2\zeta)\in\mathfrak{g}^\times$, then α_ζ has degree at most n. Since its constant term is 1, we have $\alpha_\zeta=f$. It remains to prove the case when n is odd and $\det(2\zeta)\in 2\mathfrak{g}^\times$. For $n=1$ our assumption implies that $\zeta\in\mathfrak{g}^\times$. Then, by (3.3a) we have

$$\alpha_\zeta(t)=1+\sum_{k=1}^{\infty}t^k(M_k-M_{k-1})\quad\text{with } M_k=\sum_{\sigma\in\mathfrak{p}^{-k}/\mathfrak{g}}\mathbf{e}_v(-\delta^{-1}\zeta\sigma),$$

and hence $\alpha_\zeta(t)=1-t$. Let us now prove that $\alpha_\zeta=f$ by induction on n, assuming $n>2$. With $\Psi\in T_v^{2h}$, $2h>n$, as in Lemma 3.5, we compute $N_m(\Psi,\zeta)$. By Lemma 3.1 we may assume that $\zeta=\mathrm{diag}[\xi,\eta]$ with $\xi\in T^2$ and $\eta\in T^{n-2}$, both regular. Let w be an element of $\mathfrak{g}_n^{2h}$ such that ${}^t w\Psi w-\zeta\in\pi^m T^n$. Put $w=(x\ y)$ with $x\in\mathfrak{g}_2^{2h}$ and $y\in\mathfrak{g}_{n-2}^{2h}$, and $a={}^t x\Psi x$. Then $a-\xi\in\pi^m T^2$, and hence x is primitive in the sense that there is an element z of $(\mathfrak{g})_{2h-2}^{2h}$ such that $g=(x\ z)\in GL_{2h}(\mathfrak{g})$. Put ${}^t g\Psi g=\begin{pmatrix}a & b\\ {}^t b & d\end{pmatrix}$ and $f=g\begin{pmatrix}1 & -a^{-1}b\\ 0 & 1\end{pmatrix}$. Since $2a\in GL_2(\mathfrak{g})$ and $2b\in\mathfrak{g}_{2h-2}^2$, we see that $f\in GL_{2h}(\mathfrak{g})$ and ${}^t f\Psi f=\mathrm{diag}[a,e]$ with $e\in T^{2h-2}$. Then we have $f^{-1}w=\begin{pmatrix}1 & k\\ 0 & l\end{pmatrix}$ with $k\in(\mathfrak{g}_v)_{n-2}^2$ and $l\in(\mathfrak{g})_{n-2}^{2h-2}$, and

$$\begin{pmatrix}1 & 0\\ {}^t k & {}^t l\end{pmatrix}\begin{pmatrix}a & 0\\ 0 & e\end{pmatrix}\begin{pmatrix}1 & k\\ 0 & l\end{pmatrix}-\zeta={}^t w\Psi w-\zeta\in\pi^m T^n,$$

and hence $2ak\in(\mathfrak{p}^m)_{n-2}^2$ and ${}^t kak+{}^t lel-\eta\in\pi^m T^{n-2}$. Therefore $w\equiv f\cdot\begin{pmatrix}1 & 0\\ 0 & l\end{pmatrix}(\mathrm{mod}\ \pi^m)$ and ${}^t lel-\eta\in\pi^m T^{n-2}$. From this we can conclude that $N_m(\Psi,\zeta)=\sum_x N_m(e,\eta)$, where $x\in X_m(2h\times 2)$, ${}^t x\Psi x-\xi\in\pi^m T^2$, and e is chosen for each x by means of the above procedure. If m is sufficiently large, we see that $\varepsilon(a)=\varepsilon(\xi)$, and hence $\varepsilon(e)=\varepsilon(\Psi)\varepsilon(\xi)$, which is independent of x. Therefore Lemma 3.5 shows that $N_m(e,\eta)$ is independent of x, and thus $N_m(\Psi,\zeta)=N_m(\Psi,\xi)N_m(e,\eta)$. By Lemma 3.5 this gives

$$\alpha_\zeta^m(\varepsilon(\Psi)q^{-h})=\alpha_\xi^m(\varepsilon(\Psi)q^{-h})\,\alpha_\eta^m(\varepsilon(e)q^{1-h}).$$

Taking Ψ so that $\varepsilon(\Psi)=1$, we obtain $\alpha_\zeta(t)=\alpha_\xi(t)\,\alpha_\eta(\varepsilon(\xi)qt)$, which together with induction proves the desired equality.

Let us now complete the proof of Theorem 3.2. The case $\zeta=0$ was already proved in [S6, Proposition 5.1]. If $\det(\zeta)\neq 0$, the result is exactly the above proposition. Now let $\zeta=\mathrm{diag}[\xi,0]\in T^n$ with an arbitrary $\xi\in T^r$, $0<r<n$. Then [F1, Theorem 4.1] shows that $\alpha_\zeta(t)/\alpha_\xi(q^{n-r}t)$ depends only on n, r, and F_v. Call

this quotient $h_v^{n,r}$. Taking $\xi=0$ and employing the result for $\zeta=0$, we obtain the explicit form of $h_v^{n,r}$. In fact

(3.4) $$\frac{\alpha_\zeta(t)}{\alpha_\zeta(q^{n-r}t)}=h_v^{n,r}(t)=\frac{1-t}{1-q^{n-r}t}\prod_{i=1}^{n-r}\frac{1-q^{2i}t^2}{1-q^{n+i}t^2}.$$

Assuming now that $\det(\xi)\neq 0$, we can apply Proposition 3.6 to α_ξ, and so obtain α_ζ as stated in our theorem.

4. The Möbius function for torsion g-modules

Our formulation of the main results in the next section requires a generalized Möbius function μ defined as follows:

Lemma 4.1 *To every finitely generated torsion* g*-module A one can uniquely assign an integer* $\mu(A)$ *so that*

(4.1a) $$\sum_{B\subset A}\mu(B)=\begin{cases}1 & \text{if } A=\{0\},\\ 0 & \text{if } A\neq\{0\}.\end{cases}$$

Moreover μ *has the following properties:*

(4.1b) $$\sum_{B\subset A}\mu(A/B)=\begin{cases}1 & \text{if } A=\{0\},\\ 0 & \text{if } A\neq\{0\}.\end{cases}$$

(4.1c) $\mu(A\oplus B)=\mu(A)\,\mu(B)$ *if* $\mathfrak{a}A=\mathfrak{b}B=\{0\}$ *with relatively prime integral ideals* $\mathfrak{a}$ *and* $\mathfrak{b}$.

(4.1d) $\mu((\mathfrak{g}/\mathfrak{p})^r)=(-1)^r N(\mathfrak{p})^{r(r-1)/2}$ *if* $0\leqq r\in\mathbf{Z}$ *and* $\mathfrak{p}$ *is a prime ideal of F.*

(4.1e) $\mu(A)\neq 0$ *if and only if A is annihilated by a squarefree integral ideal.*

Proof. We can define $\mu(A)$ inductively by $\mu(A)=-\sum_{B\subsetneqq A}\mu(B)$, starting from $\mu(\{0\})=1$, which shows also the uniqueness. To prove (4.1b), we may assume that $A=L/N$ with two g-lattices L and N in F^n. For every g-lattice X in F^n put $X'=\{y\in F^n\,|\,{}^t yX\subset\mathfrak{g}\}$. Given a g-submodule B of A, take a g-lattice M so that $N\subset M\subset L$ and $B=M/N$. Put $\psi(B)=\varphi(M'/L')$ with any fixed g-isomorphism φ of N'/L' onto A. Then ψ gives a one-to-one map of the set of all g-submodules of A onto itself, and $\psi(B)\cong A/B$ and $A/\psi(B)\cong B$. Therefore $\sum_{B\subset A}\mu(A/B)$ $=\sum_{B\subset A}\mu(\psi(B))=\sum_{C\subset A}\mu(C)$, which combined with (4.1a) gives (4.1b). Next, if A and B are as in (4.1c), then every g-submodule of $A\oplus B$ is of the form $A'\oplus B'$ with g-submodules A' of A and B' of B. Then (4.1c) can be derived from the relation $\mu(A)=-\sum_{C\subsetneqq A}\mu(C)$ by induction. The formula of (4.1d) follows from the equality $\sum_{r=0}^{n}(-1)^r N(\mathfrak{p})^{r(r-1)/2} c_r^n = 0$ if $n>0$ and c_r^n denotes the number of $\mathfrak{g}$-submodules of $(\mathfrak{g}/\mathfrak{p})^n$ isomorphic to $(\mathfrak{g}/\mathfrak{p})^r$ (see [S3, Lemma 3.23]). To prove (4.1e), we first observe that $\mu(\mathfrak{g}/\mathfrak{p}^2)=0$ for every prime ideal $\mathfrak{p}$. Given A, let C be the maximum $\mathfrak{g}$-submodule of A that is annihilated by a squarefree integral ideal. Suppose $A \neq C$; then $C \neq \{0\}$ and $-\mu(A) = \sum_{D \subset C} \mu(D) + \sum_{B \not\subset C, B \subsetneqq A} \mu(B)$. The first sum on the right-hand side is 0. Therefore we obtain $\mu(A)=0$ by induction. The converse part follows from (4.1c, d).

Let $\mathscr{L}$ denote the set of all $\mathfrak{g}$-lattices in F^n. For $L \in \mathscr{L}$ and $y \in GL_n(F)_{\mathbf{A}}$ we denote by yL the $\mathfrak{g}$-lattice in F^n such that $(yL)_v = y_v L_v$ for every $v \in \mathbf{h}$. For L and M in $\mathscr{L}$ we define a fractional ideal $\{L/M\}$ and a multiplicative symbol $[L/M]$ (in the sense of Sect. 2) by

$$\{L/M\} = \det(y)\,\mathfrak{g}, \qquad [L/M] = [\{L/M\}] = [\det(y)\,\mathfrak{g}] \tag{4.2}$$

with any $y \in GL_n(F)_{\mathbf{A}}$ such that $M = yL$. These are well defined. Clearly we have $[L/M][M/N] = [L/N]$. If $L, M \in \mathscr{L}$ and $M \subset L$, we can speak of $\mu(L/M)$. Moreover, for each $v \in \mathbf{h}$ we can speak of $\mu(L_v/M_v)$ either by viewing L_v/M_v as a $\mathfrak{g}$-module, or by defining μ for $\mathfrak{g}_v$-modules, which makes no difference. From (4.1c) we easily obtain

$$\mu(L/M) = \prod_{v \in \mathbf{h}} \mu(L_v/M_v). \tag{4.3}$$

We now take a subset Λ of $\mathscr{L}$ satisfying the following condition: *if $K \subset L \subset M$, $K, M \in \Lambda$, and $L \in \mathscr{L}$, then $L \in \Lambda$.* We also take a nonempty class $\mathscr{F}$ of finitely generated torsion $\mathfrak{g}$-modules with the following property: *if $B \subset A$, then A belongs to $\mathscr{F}$ if and only if both B and A/B belong to $\mathscr{F}$.* (For example, with a fixed integral ideal $\mathfrak{c}$ we can take $\mathscr{F}$ to be the class consisting of the $\mathfrak{g}$-modules annihilated by integral ideals prime to $\mathfrak{c}$.) We write $L < M$ and $M > L$ if $L \subset M$ and M/L belongs to $\mathscr{F}$. Then the following lemma can easily be verified:

Lemma 4.2 *For two functions α and β defined on Λ with values in a $\mathbf{Z}$-module, we have*

$$\alpha(L) = \sum_{L<M\in\Lambda} \beta(M) \quad \textit{for every } L \in \Lambda$$
$$\Leftrightarrow \quad \beta(L) = \sum_{L<M\in\Lambda} \mu(M/L)\,\alpha(M) \quad \textit{for every } L \in \Lambda,$$
$$\alpha(L) = \sum_{L>M\in\Lambda} \beta(M) \quad \textit{for every } L \in \Lambda$$
$$\Leftrightarrow \quad \beta(L) = \sum_{L>M\in\Lambda} \mu(L/M)\,\alpha(M) \quad \textit{for every } L \in \Lambda.$$

Here each sum may be an infinite sum, and therefore we have to assume that it is convergent in a suitable sense, or it is a formal sum. (This comment applies also to Lemma 4.4 below.) For example, with a fixed $K \in \mathscr{L}$ let $\Xi(K)$ denote the set of all functions ξ on Λ such that for every $L \in \Lambda$ the value $\xi(L)$ is a formal Dirichlet series (in the sense of Sect. 2) of the form $\xi(L)$

$=\sum_{\mathfrak{m}\subset\{L/K\}} c_{\mathfrak{m}}[\mathfrak{m}]$. If $\alpha(L)=\sum_{L<M\in\Lambda}\xi(M)$ and $\beta(L)=\sum_{L<M\in\Lambda}\mu(M/L)\xi(M)$ with $\xi\in\Xi(K)$, then both α and β are meaningful and belong to $\Xi(K)$.

Notice that if $\mathfrak{g}=\mathbf{Z}$, then $n\mapsto\mu(\mathbf{Z}/n\mathbf{Z})$ is the classical Möbius function, and the first half of Lemma 4.2 is exactly the classical Möbius inversion formula if we take $\Lambda=\{n\mathbf{Z}\,|\,0<n\in\mathbf{Z}\}$ and consider $\alpha(n\mathbf{Z})$ a function of n.

Lemma 4.3 *For any fixed $L\in\mathscr{L}$ we have*

$$\sum_{L\subset M\in\mathscr{L}}[M/L]=\sum_{L\supset M\in\mathscr{L}}[L/M]=\sum_{x\in B/E}[\det(x)\mathfrak{g}]=\prod_{i=1}^{n}\prod_{v\in\mathbf{h}}(1-|\pi_v|^{1-i}[v])^{-1},$$

$$\sum_{L\subset M\in\mathscr{L}}\mu(M/L)[M/L]=\sum_{L\supset M\in\mathscr{L}}\mu(L/M)[L/M]=\prod_{i=1}^{n}\prod_{v\in\mathbf{h}}(1-|\pi_v|^{1-i}[v]).$$

These relations follow immediately from (2.8) and Lemma 4.2.

Lemma 4.4 *Let α and γ be functions defined on Λ with values in a* **Z**-*module Y, and let δ be a function on Λ with values in* End(Y). *If*

$$\alpha(L)=\sum_{L<N\in\Lambda}\delta(N)\sum_{\substack{K\in\Lambda\\ L+K=N}}\gamma(K)$$

for every $L\in\Lambda$, then

$$\sum_{L<M\in\Lambda}\mu(M/L)\,\alpha(M)=\sum_{L<M\in\Lambda}\mu(M/L)\,\delta(M)\sum_{L\supset K\in\Lambda}\lambda(K)$$

for every $L\in\Lambda$.

Proof. For fixed L and N in Λ we have

$$\sum_{\substack{K\in\Lambda\\ L+K=N}}\gamma(K)=\sum_{N\supset K\in\Lambda}\Big(\sum_{L+K\subset M\subset N}\mu(N/M)\Big)\gamma(K)=\sum_{L\subset M\subset N}\mu(N/M)\,\varepsilon(M),$$

where $\varepsilon(M)=\sum_{M\supset K\in\Lambda}\gamma(K)$, and hence

$$\alpha(L)=\sum_{L<N\in\Lambda}\delta(N)\sum_{L\subset M\subset N}\mu(N/M)\,\varepsilon(M)$$
$$=\sum_{L<M\in\Lambda}\sum_{M<N\in\Lambda}\mu(N/M)\,\delta(N)\,\varepsilon(M)=\sum_{L<M\in\Lambda}\beta(M),$$

where $\beta(M)=\sum_{M<N\in\Lambda}\mu(N/M)\,\delta(N)\,\varepsilon(M)$. Therefore Lemma 4.2 gives the desired conclusion.

5. The Dirichlet series obtained from Fourier coefficients

We now consider $\mathscr{M}_k(\mathfrak{b},\mathfrak{c},\varphi)$ as in Sect. 1 with an integral weight k and a character φ of $(\mathfrak{g}/\mathfrak{c})^\times$ satisfying (1.15a). To define Hecke operators on this space, however, we have to take a Hecke character ψ such that $\psi(x)=\varphi(x_{\mathfrak{c}})$ for every

$x \in \prod_{v \in \mathbf{h}} \mathfrak{g}_v^\times$ (cf. (1.23b) and (1.24)). We put then $\mathcal{M}_k(\mathfrak{b}, \mathfrak{c}, \psi) = \mathcal{M}_k(\mathfrak{b}, \mathfrak{c}, \varphi)$ as we did at the end of Sect. 1. For every $x \in M_n(F)_\mathbf{A}$ such that $x_\mathfrak{c}$ is invertible we put $\Psi(x) = \psi_\mathfrak{c}(\det(x)_\mathfrak{c})$. As in Sect. 2 we put $D = D[\mathfrak{b}^{-1}, \mathfrak{b}\mathfrak{c}]$ and $\Gamma = G \cap D$. Taking an element $\sigma = \operatorname{diag}[\tilde{e}, e]$ of Z_0 (see Sect. 2), put $G \cap D\sigma D = \Gamma\alpha\Gamma = \coprod_{\gamma \in C} \Gamma\gamma$ with $\alpha \in G$ and $C \subset G$. For x, $y \in D$ and $z = x\sigma y$ we easily see that $(a_x^{-1} a_z a_y^{-1} \cdot {}^t e)_v \equiv 1 \pmod{\mathfrak{c}_v}$ for every $v | \mathfrak{c}$, and hence $(a_z)_\mathfrak{c}$ is invertible for every $z \in D\sigma D$ and $\Psi(a_{uzw}) = \Psi(a_u)\,\Psi(a_z)\,\Psi(a_w)$ for $u, w \in D$. Given $f \in \mathcal{M}_k(\mathfrak{b}, \mathfrak{c}, \psi)$, we define a function $f | T_{e,\psi}$ on $H^\mathbf{a}$ by

$$(5.1\text{a}) \qquad f | T_{e,\psi} = \sum_{\gamma \in C} \Psi(a_\gamma)^{-1} f \|_k \gamma.$$

It can easily be verified that this is well defined independently of the choice of C, and belongs to $\mathcal{M}_k(\mathfrak{b}, \mathfrak{c}, \psi)$. Moreover if $D\sigma D = \coprod_{y \in Y} Dy$ with $Y \subset G_\mathbf{h}$, then

$$(5.1\text{b}) \qquad (f | T_{e,\psi})_\mathbf{A}(x) = \sum_{y \in Y} \Psi(a_y)^{-1} f_\mathbf{A}(xy^{-1}) \quad (x \in G_\mathbf{A}).$$

Clearly $T_{e,\psi}$ depends only on EeE and ψ. It must be remembered, however, that it depends on the choice of ψ, as ψ is not unique for φ. We easily see that $T_e \mapsto T_{e,\psi}$ defines a ring-homomorphism of $\mathfrak{R}(D, \mathscr{Z}_0)$ into $\operatorname{End}(\mathcal{M}_k(\mathfrak{b}, \mathfrak{c}, \psi))$. Let us now define a formal Dirichlet series $f | \mathfrak{T}_\psi$ whose coefficients are modular forms by

$$(5.2) \qquad f | \mathfrak{T}_\psi = \sum_{e \in E \backslash B / E} [\det(e)\,\mathfrak{g}]\, f | T_{e,\psi}.$$

For an integral ideal $\mathfrak{a}$ let $T_\psi(\mathfrak{a})$ denote the sum of $T_{e,\psi}$ for all EeE such that $e \in B$ and $\det(e)\,\mathfrak{g} = \mathfrak{a}$. Then $f | \mathfrak{T}_\psi = \sum_\mathfrak{a} f | T_\psi(\mathfrak{a})[\mathfrak{a}]$. Here and henceforth $\sum_\mathfrak{a}$ denotes the sum over all the integral ideals $\mathfrak{a}$ in F.

As in Sect. 4 let $\mathscr{L}$ denote the set of all $\mathfrak{g}$-lattices in F_1^n. We put $L_0 = \mathfrak{g}_1^n$ and we shall often express an element L of $\mathscr{L}$ in the form $L = yL_0$ with $y \in GL_n(F)_\mathbf{h}$. For $\tau \in S$ put

$$(5.3) \qquad \mathscr{L}_\tau = \{L \in \mathscr{L} \mid {}^t l \tau l \in \mathfrak{b}\mathfrak{d}^{-1} \text{ for every } l \in L\}.$$

We easily see that $\mathscr{L}_\tau$ consists of all the $\mathfrak{g}$-lattices yL_0 with $y \in GL_n(F)_\mathbf{h}$ such that ${}^t y \tau_\mathbf{h} y \in \mathfrak{b}S^*$ with S^* of (2.4). Notice that if $L \in \mathscr{L}$ and $L \subset K \in \mathscr{L}_\tau$, then $L \in \mathscr{L}_\tau$; moreover, if $\det(\tau) \neq 0$, the set $\{M \in \mathscr{L}_\tau \mid L \subset M\}$ is finite. For L and M in $\mathscr{L}$ let us write $L < M$ if $L \subset M$ and $L_v = M_v$ for every $v | \mathfrak{c}$.

We consider the Fourier expansion of Proposition 1.1 and investigate their relationship with the formal series $\mathfrak{T}_\psi$. By (1.21a), for $y \in GL_n(F)_\mathbf{h}$ we have

$$(5.4) \qquad \mu_f(\tau, y) \neq 0 \;\Rightarrow\; yL_0 \in \mathscr{L}_\tau \;\Leftrightarrow\; {}^t y \tau_\mathbf{h} y \in \mathfrak{b}S^*.$$

Now our first main result can be stated as follows:

Theorem 5.1 *For $f \in \mathcal{M}_k(\mathfrak{b}, \mathfrak{c}, \psi)$, $\tau \in S_+$, and $L = pL_0 \in \mathcal{L}_\tau$ with $p \in GL_n(F)_\mathbf{h}$ define formal Dirichlet series $D(\tau, L; f)$, $a(\tau, L)$, and $A(\tau, L)$ by*

$$D(\tau, L; f) = \sum_{x \in B/E} \psi_\mathfrak{c}(\det(px)) |\det(x)|_\mathbf{A}^{-n-1} \mu(\tau, px; f)[\det(x)\mathfrak{g}],$$

$$A(\tau, L) = |\det(p)|_\mathbf{A}^{-n-1} [\det(p)\mathfrak{g}]^2 \sum_{L < M \in \mathcal{L}_\tau} \mu(M/L)\, a(\tau, M),$$

$$a(\tau, L) = |\det(p)|_\mathbf{A}^{n+1} [\det(p)^{-2}\mathfrak{g}]\, \alpha'(\theta \cdot {}^t p \tau p),$$

$$\alpha'(\zeta) = \prod_{v \nmid \mathfrak{c}} \alpha_v(\zeta_v),$$

where $\mu(M/L)$ is the Möbius function introduced in Sect. 4, θ is an element of $F_\mathbf{h}^\times$ such that $\theta \mathfrak{g} = \mathfrak{b}^{-1}\mathfrak{d}$, and α_v is the series of (2.5b). *Then*

$$A(\tau, L)\, D(\tau, L; f) = \sum_{L < M \in \mathcal{L}_\tau} \mu(M/L)\, \psi_\mathfrak{c}(\det(pw))[\det(w)^{-1}\mathfrak{g}]\, \mu(\tau, pw; f|\mathfrak{T}_\psi),$$

$$\psi_\mathfrak{c}(\det(p))\, \mu(\tau, p; f|\mathfrak{T}_\psi) = \sum_{L < M \in \mathcal{L}_\tau} [\det(w)^{-1}\mathfrak{g}]\, A(\tau, M)\, D(\tau, M; f),$$

where w in the last two sums is an element of $GL_n(F)_\mathbf{h}$ chosen for each M so that $M = pwL_0$. In particular, if $f|T_\psi(\mathfrak{a}) = \lambda(\mathfrak{a}) f$ with $\lambda(\mathfrak{a}) \in \mathbf{C}$ for every integral ideal $\mathfrak{a}$, then

$$A(\tau, L)\, D(\tau, L; f) = \sum_\mathfrak{a} \lambda(\mathfrak{a})[\mathfrak{a}] \sum_{L < M \in \mathcal{L}_\tau} \mu(M/L)\, \psi_\mathfrak{c}(\det(pw))[\det(w)^{-1}\mathfrak{g}]\, \mu_f(\tau, pw),$$

$$\psi_\mathfrak{c}(\det(p))\, \mu(\tau, p; f) \sum_\mathfrak{a} \lambda(\mathfrak{a})[\mathfrak{a}] = \sum_{L < M \in \mathcal{L}_\tau} [\det(w)^{-1}\mathfrak{g}]\, A(\tau, M)\, D(\tau, M; f).$$

Remark. The series $D(\tau, L; f)$, $a(\tau, L)$, and $A(\tau, L)$ are defined independently of the choice of p, as can be seen from (1.21d) and (2.6). The Fourier coefficients $\mu(\tau, x; f)$ are defined without ψ, but $D(\tau, L; f)$ depends on the choice of ψ. The sum $\sum_{L < M \in \mathcal{L}_\tau}$ is a finite sum if $\det(\tau) \neq 0$. In general it may be an infinite sum. We shall show, however, in Proposition 5.4 below that in all cases $A(\tau, L)$ can be expressed as an easy Euler product times a finite sum which is essentially a lower-dimensional version of $A(\tau, L)$. Substituting $(\psi/\psi_\mathfrak{c})(t)[t\mathfrak{g}]$ for $[t\mathfrak{g}]$, $t \in F_\mathbf{h}^\times$, in $D(\tau, L; f)$ and multiplying by $(\psi/\psi_\mathfrak{c})(\det(p))$, we obtain

$$D'(\tau, L; f) = \sum_{x \in B/E} \psi(\det(px)) |\det(x)|_\mathbf{A}^{-n-1} \mu(\tau, px; f)[\det(x)\mathfrak{g}]. \tag{5.5}$$

In our later analytic study, we shall see that D' is a more natural series than D. However, since the Euler product expression for D requires fewer symbols than (and is "equivalent" to) D', we investigate D in this section.

Proof. By Lemmas 2.6, 2.7, and (5.1b), for $x \in G_\mathbf{A}$ we have

$$(f|\mathfrak{T}_\psi)_\mathbf{A}(x) = \sum \psi_\mathfrak{c}(\det(gh^{-1}))\, f_\mathbf{A}(xy^{-1})[\det(gh)\, v_0(b\sigma)],$$

where $y=\begin{pmatrix} g^{-1}h & g^{-1}\sigma\cdot{}^th^{-1} \\ 0 & {}^tg\cdot{}^th^{-1}\end{pmatrix}$ with g, h, and σ as in Lemma 2.6, and b as in Lemma 2.7. Substituting $\begin{pmatrix} q & s\tilde{q} \\ 0 & \tilde{q}\end{pmatrix}$ for x and making a straightforward calculation, we find, for every $q\in GL_n(F)_{\mathbf{h}}$, that

$$\mu(\tau, q; f|\mathfrak{T}_\psi)=\sum_{g,h,\sigma}\psi_{\mathfrak{c}}(\det(h^{-1}g))\,\mu(\tau, qh^{-1}g; f)$$
$$\cdot\, \mathbf{e}_{\mathbf{h}}(-\mathrm{tr}({}^th^{-1}\cdot{}^tq\tau qh^{-1}\cdot\sigma))[\det(gh)\,v_0(b\sigma)].$$

By (5.4) we may assume that ${}^tg\cdot{}^th^{-1}\cdot{}^tq\tau qh^{-1}g\in bS^*$. Therefore, by Lemma 2.8 we have

$$\mu(\tau, q; f|\mathfrak{T}_\psi)=\sum_{g,h}\psi_{\mathfrak{c}}(\det(h^{-1}g))|\det(g)|_{\mathbf{A}}^{-n-1}[\det(gh)\,\mathfrak{g}]$$
$$\cdot\,\mu(\tau, qh^{-1}g; f)\,\alpha'(\theta\cdot{}^th^{-1}\cdot{}^tq\tau qh^{-1}), \tag{5.6}$$

where (g, h) runs over $E\backslash W/(E\times 1)$ under the condition that ${}^th^{-1}\cdot{}^tq\tau qh^{-1}\in bS^*$. Then for $L=qL_0\in\mathscr{L}_\tau$ with a fixed $q\in GL_n(F)_{\mathbf{h}}$ it can easily be seen that $(g, h)\mapsto(qh^{-1}L_0,\, qh^{-1}gL_0)$ gives a one-to-one map of the set of all such (g, h) onto the set of all (N, K) in $\mathscr{L}_\tau\times\mathscr{L}_\tau$ such that $L+K=N$ and $L<N$. For a fixed τ and $M=yL_0\in\mathscr{L}$ with $y\in GL_n(F)_{\mathbf{h}}$ put

$$c(M)=\psi_{\mathfrak{c}}(\det(y))\,\mu(\tau, y; f)|\det(y)|_{\mathbf{A}}^{-n-1}[\det(y)\,\mathfrak{g}],$$
$$c'(M)=\psi_{\mathfrak{c}}(\det(y))\,\mu(\tau, y; f|\mathfrak{T}_\psi)[\det(y)^{-1}\mathfrak{g}].$$

These are well defined because of (1.21d). Therefore (5.6) can be written

$$c'(L)=\sum_{L<N\in\mathscr{L}_\tau} a(\tau, N)\sum_{L+K=N} c(K).$$

By Lemma 4.4 we obtain

$$\sum_{L<M\in\mathscr{L}_\tau}\mu(M/L)\,c'(M)=\sum_{L<M\in\mathscr{L}_\tau}\mu(M/L)\,a(\tau, M)\sum_{H\subset L} c(H)$$

for every $L\in\mathscr{L}_\tau$. This gives the first equality of our theorem. The second equality follows immediately from this and Lemma 4.2. The last two equalities are immediate consequences of the first two.

Suppose now $f|T_\psi(\mathfrak{a})=\lambda(\mathfrak{a})\,f$ for every $\mathfrak{a}$ as in the last part of the above theorem. By Theorem 2.9 for each $v\in\mathbf{h}$ we can determine n complex numbers $\lambda_{v,i}$ by the relation

$$\sum_{m=0}^{\infty}\lambda(\pi_v^m\,\mathfrak{g})\,t^m=\begin{cases}\displaystyle\prod_{i=1}^{n}(1-|\pi_v|^{-n}\lambda_{v,i}t)^{-1} & \text{if } v|\mathfrak{c},\\[2ex] \displaystyle\frac{1-t}{1-|\pi_v|^{-n}t}\prod_{i=1}^{n}\frac{1-|\pi_v|^{-2i}t^2}{(1-|\pi_v|^{-n}\lambda_{v,i}t)(1-|\pi_v|^{-n}\lambda_{v,i}^{-1}t)} & \text{if } v\nmid\mathfrak{c}.\end{cases}$$

We then put

$$Z_v(t)=\begin{cases}\prod_{i=1}^{n}(1-|\pi_v|^{-n}\lambda_{v,i}t)^{-1} & \text{if } v\mid \mathfrak{c},\\ \left[(1-|\pi_v|^{-n}t)\prod_{i=1}^{n}(1-|\pi_v|^{-n}\lambda_{v,i}t)(1-|\pi_v|^{-n}\lambda_{v,i}^{-1}t)\right]^{-1} & \text{if } v\nmid \mathfrak{c}.\end{cases}$$

Corollary 5.2 *Let f and Z_v be as above and let $\tau\in S_+\cap \mathrm{GL}_n(F)$. Then for $L=pL_0\in\mathscr{L}_\tau$ with $p\in \mathrm{GL}_n(F)_{\mathbf{h}}$ we have*

$$D(\tau,L;f)\cdot\prod_{v\in\mathfrak{b}}g_v([v])\cdot\prod_{v\nmid\mathfrak{c}}\left\{h_v([v])^{-1}\prod_{i=1}^{[(n+1)/2]}(1-|\pi_v|^{2i-2-2n}[v]^2)^{-1}\right\}$$
$$=\prod_{v\in\mathbf{h}}Z_v([v])\cdot\sum_{L<M\in\mathscr{L}_\tau}\mu(M/L)\,\psi_{\mathfrak{c}}(\det(pw))[\det(w^{-1}\mathfrak{g}]\,\mu_f(\tau,pw),$$

where $M=pwL_0$ as in Theorem 5.1, $\mathfrak{b}$ is the (finite) set of the primes $v\nmid\mathfrak{c}$ such that $(\theta\cdot{}^tp\tau p)_v$ is not regular, the g_v are polynomials with constant term 1, $h_v=1$ if n is odd, and $h_v(t)=1-\rho_\tau^*(\pi_v\mathfrak{g}_v)|\pi_v|^{-n/2}t$ with the Hecke character ρ_τ^* of F corresponding to $F(c^{1/2})/F$, $c=(-1)^{n/2}\det(\tau)$, if n is even.

Proof. Since every M in $\mathscr{L}$ containing L can be expressed as $M=p\tilde{x}L_0$ with $x\in B$, we have $A(\tau,L)=\prod_{v\nmid\mathfrak{c}}A_v(\tau,L)$ with

(5.7) $$A_v(\tau,L)=\sum_x\mu(L_0/xL_0)|\det(x)|^{-n-1}[\det(x)^2\mathfrak{g}]\,\alpha_v(x^{-1}(\theta\cdot{}^tp\tau p)_v\tilde{x}),$$

where x runs over B_v/E_v under the condition that $x^{-1}(\theta\cdot{}^tp\tau p)_v\tilde{x}\in T_v$. Therefore by Theorem 3.2 we can put

(5.8) $$A(\tau,L)=\prod_{v\in\mathfrak{b}}g_v([v])\prod_{v\nmid\mathfrak{c}}h_v([v])^{-1}(1-[v])\prod_{i=1}^{[n/2]}(1-|\pi_v|^{-2i}[v]^2)$$

with $\mathfrak{b}$, g_v, and h_v as stated above, and hence the desired equality follows from Theorem 5.1.

Given $f\in\mathscr{M}_k(\mathfrak{b},\mathfrak{c},\psi)$ and $\tau\in S_+\cap\mathrm{GL}_n(F)$, let $\mathscr{L}_{\tau,f}$ denote the set of all $L=hL_0\in\mathscr{L}_\tau$ with $h\in\mathrm{GL}_n(F)_{\mathbf{h}}$ such that $\mu(\tau,h;f\,|\,T_\psi(\mathfrak{a}))\neq 0$ for some $\mathfrak{a}$. If f is an eigenfunction of $T_\psi(\mathfrak{a})$ for all $\mathfrak{a}$, then $\mathscr{L}_{\tau,f}$ is the set of all $L=hL_0\in\mathscr{L}_\tau$ such that $\mu(\tau,h;f)\neq 0$.

Corollary 5.3 *For $L\in\mathscr{L}_\tau$ with $\tau\in S_+\cap\mathrm{GL}_n(F)$, the following two conditions are equivalent:*

(i) *$L\in\mathscr{L}_{\tau,f}$ and there is no element M of $\mathscr{L}_{\tau,f}$, other than L itself, such that $L<M$.*

(ii) *$D(\tau,L;f)\neq 0$ and there is no element M of $\mathscr{L}_\tau$, other than L itself, such that $L<M$ and $D(\tau,M;f)\neq 0$.*

Moreover, if L satisfies these conditions, then

(5.9 a) $$\psi_{\mathfrak{c}}(\det(p))\,\mu(\tau,p;f\,|\,\mathfrak{T}_\psi)=A(\tau,L)\,D(\tau,L;f)$$

for $p \in GL_n(F)_{\mathbf{h}}$ *such that* $L = pL_0$. *Suppose in particular that* $f|T_\psi(\mathfrak{a}) = \lambda(\mathfrak{a})f$ *as above. Then for such L and p we have*

$$\text{(5.9 b)}\quad \psi_{\mathfrak{c}}(\det(p))\,\mu_f(\tau, p)\prod_{v\in\mathbf{h}} Z_v([v])$$
$$= D(\tau, L; f)\cdot \prod_{v\in\mathbf{b}} g_v([v]) \prod_{v\nmid\mathfrak{c}} \left\{ h_v([v])^{-1} \prod_{i=1}^{[(n+1)/2]} (1 - |\pi_v|^{2i-2-2n}[v]^2)^{-1} \right\}$$

with the symbols as in Corollary 5.2.

Clearly, if f is a nonzero cusp form, there exists a lattice L satisfying (i) and (ii) for some $\tau \in S_+ \cap GL_n(F)$.

Proof. If L satisfies (i) (resp. (ii)), then the first (resp. second) equality of Theorem 5.1 proves (5.9a). Now from (5.7) we see that $A(\tau, L)$ is a nonvanishing formal Dirichlet series, since its leading coefficient is 1. Therefore (5.9a) shows that if L satisfies (i), then $D(\tau, L; f) \neq 0$; if L satisfies (ii), then $\mu(\tau, p; f|\mathfrak{T}_\psi) \neq 0$. Consequently (i) and (ii) are equivalent. The last part of our corollary is an immediate consequence of Corollary 5.2.

Remark. If $n = 1$ and $\psi(\pi_v)[v]$ is substituted for $[v]$ for every $v \nmid \mathfrak{c}$, then (5.9b) is essentially the expression of $\sum_{m=1}^{\infty} \mu_f(\tau m^2, p)\, m^{-s}$ as an Euler product of degree 3 studied in [S4] and [S10]. In [A] Andrianov treated a series of type $D(\tau, L; f)$ for $G = Sp(n, \mathbf{Q})$ assuming τ to be nonsingular, and obtained an equality similar to that of Corollary 5.2. His formula involves a factor X which is a Fourier coefficient of the transform of f by a certain generalized (and somewhat involved) Hecke operator. In contrast with his result, our Theorem 5.1 is applicable to any $\tau \in S_+$, even to $\tau = 0$, and X is replaced by the last sum over M in Corollary 5.2, which is simpler. It can be simplified further as in (5.9b) if τ and L are suitably chosen. A more essential difference of our formulation from his is that all our equalities are stated for Dirichlet series involving all the integral ideals, while his series involves only the ideals (or rather, the integers) prime to the level. This point will become crucial when we study analytic continuation of our Euler products in Sect. 8. It should also be noted that once Theorem 5.1 is established, we can immediately derive the equalities for the partial series involving only the ideals divisible by the primes in any specified set.

Coming back to the case of an arbitrary τ, we easily see that

$$D({}^t g\tau g, L; f) = \det(g)^k D(\tau, gL; f) \quad\text{and}\quad A({}^t g\tau g, L) = A(\tau, gL)$$

for every $g \in GL_n(F)$. If $\operatorname{rank}(\tau) = r < n$, we can find g so that ${}^t g\tau g = \operatorname{diag}[\sigma, 0]$ with $\sigma \in S^r \cap GL_r(F)$. Thus, in the study of $D(\tau, L; f)$ and $A(\tau, L)$ we may assume that τ is of the form $\tau = \operatorname{diag}[\sigma, 0]$ with such a σ.

Proposition 5.4 *Let* $\tau = \operatorname{diag}[\sigma, 0]$ *with* $\sigma \in S^r \cap GL_r(F)$ *and let* L' *be the image of* L *under the projection map* $(x_i)_{i=1}^n \mapsto (x_i)_{i=1}^r$ *of* F^n *into* F^r. *Then*

$$A(\tau, L) = \tilde{A}(\sigma, L') \prod_{v\nmid\mathfrak{c}} \left\{ (1 - [v])(1 - |\pi_v|^{r-n}[v])^{-1} \prod_{i=1}^{n-r} (1 - |\pi_v|^{-2i}[v]^2) \right\},$$

where $\tilde{A}(\sigma, L')$ is the formal Dirichlet series obtained from $A(\sigma, L')$ by substituting $N(\mathfrak{a})^{n-r}[\mathfrak{a}]$ for $[\mathfrak{a}]$.

Proof. Take any $p \in GL_n(F)_{\mathbf{h}}$ so that $L = pL_0$. Define an algebraic subgroup W of $GL_n(F)$ by

$$\text{(5.10)} \qquad W = \left\{ \begin{pmatrix} a & b \\ 0 & d \end{pmatrix} \middle| a \in GL_r(F), b \in F^r_{n-r}, d \in GL_{n-r}(F) \right\}.$$

Since $GL_n(F_v) = W_v E_v$, we may assume that ${}^t p \in W_{\mathbf{h}}$, and we can take $(B_v \cap W_v)/(E_v \cap W_v)$ in place of B_v/E_v in formula (5.7), which is valid even when $\det(\tau) = 0$. Put $p = \begin{pmatrix} e & 0 \\ * & * \end{pmatrix}$ with $e \in GL_r(F)_{\mathbf{h}}$ and $\zeta = \theta \cdot {}^t e \sigma_{\mathbf{h}} e$. Then $\theta \cdot {}^t p \tau_{\mathbf{h}} p = \mathrm{diag}[\zeta, 0]$. Fix a prime $v \nmid \mathfrak{c}$. For $x = \begin{pmatrix} a & b \\ 0 & d \end{pmatrix} \in B_v \cap W_v$ put $|\det(a)| = |\pi_v|^g$ and $|\det(d)| = |\pi_v|^h$. By (4.1e) we have $\mu(L_0/xL_0) \neq 0$ only if L_0/xL_0 is isomorphic to $(\mathfrak{g}/\mathfrak{p})^{g+h}$. Put $K = (\mathfrak{g}_v)^r$ and $K' = (\mathfrak{g}_v)^{n-r}$. By (4.1d) we obtain

$$\mu(L_0/xL_0) = \mu(K/aK)\,\mu(K'/dK')|\pi_v|^{-gh}.$$

We have $x^{-1}(\theta \cdot {}^t p \tau p)_v \tilde{x} = \mathrm{diag}[a^{-1}\zeta_v \tilde{a}, 0]$, and thus

$$A_v(\tau, L) = \sum_{a,b,d} \mu(K/aK)\,\mu(K'/dK') \cdot |\pi_v|^{-gh-(g+h)(n+1)}[v]^{2g+2h} \alpha_v(\mathrm{diag}[a^{-1}\zeta_v \tilde{a}, 0]),$$

where $x = \begin{pmatrix} a & b \\ 0 & d \end{pmatrix}$ runs over $(B_v \cap W_v)/(E_v \cap W_v)$ under the conditions that $\mu(L_0/xL_0) \neq 0$ and $a^{-1}\zeta_v \tilde{a} \in T^r_v$. For a fixed (a, d) the number of possible b in this set of representatives is $|\pi_v|^{g(h+r-n)}$. Therefore the above sum over a, b, d can be written

$$\sum_{a \in B^r_v/E^r_v} \mu(K/aK)|\det(a)|^{r-2n-1}[\det(a)\mathfrak{g}]^2 \alpha_v(\mathrm{diag}[a^{-1}\zeta_v \tilde{a}, 0]) \cdot \sum_{d \in B^{n-r}_v/E^{n-r}_v} \mu(K'/dK')|\det(d)|^{-n-1}[\det(d)\mathfrak{g}]^2,$$

where B^r and E^r are the r-dimensional versions of B and E. By Lemma 4.3 the last sum over d equals $\prod_{i=1}^{n-r}(1 - |\pi_v|^{-n-i}[v]^2)$. By (3.4) we can express $\alpha_v(\mathrm{diag}[a^{-1}\zeta_v \tilde{a}, 0])$ as the product of $h^{n,r}_v([v])$ and $\alpha_v(a^{-1}\zeta_v \tilde{a})$ with $[v]$ replaced by $|\pi_v|^{r-n}[v]$. Then we see that the sum over a gives the v-factor of $\tilde{A}(\sigma, L')$ times $h^{n,r}_v([v])$. This completes the proof.

We conclude this section by noting a few formulas which will become necessary in our later study of the analytic nature of our Dirichlet series. With a basis $\{f_1, \ldots, f_\kappa\}$ of $\mathscr{S}_k(\mathfrak{b}, \mathfrak{c}, \varphi)$ over $\mathbf{C}$ we consider a column vector $\mathbf{f} = (f_i)^\kappa_{i=1}$ and define an element $X(\mathfrak{a})$ of $\mathbf{C}^\kappa_\kappa$ by

$$\text{(5.11)} \qquad \mathbf{f} | T_\psi(\mathfrak{a}) = X(\mathfrak{a})\mathbf{f}.$$

Then we have $f|\mathfrak{T}_\psi = \sum_a [a]\, X(a)\, f$. From Theorem 5.1 we obtain

$$\sum_a [a]\, X(a)(\psi_c(\det(p))\, \mu(\tau, p; f_i))_{i=1}^{\kappa}$$
$$= \Big(\sum_{L<M\in\mathscr{L}_\tau} [\det(w)^{-1} g]\, A(\tau, M)\, D(\tau, M; f_i)\Big)_{i=1}^{\kappa},$$

where $L = pL_0$ and $M = pwL_0$ as in that theorem. Since the f_i are linearly independent, (1.22) shows that for any fixed $p\in GL_n(F)_{\mathbf{h}}$ we can find $\tau_1, \ldots, \tau_\kappa$ so that the matrix $(\psi_c(\det(p))\, \mu(\tau_h, p; f_i))_{h,i=1}^{\kappa}$ is invertible. Call this matrix Ξ and put

$$(5.12)\qquad D_h(f_i) = \sum_{L<M\in\mathscr{L}_{\tau_h}} [\det(w)^{-1} g]\, A(\tau_h, M)\, D(\tau_h, M; f_i).$$

Then we obtain

$$(5.13)\qquad \sum_a [a]\, X(a)\, \Xi = (D_h(f_i))_{h,i=1}^{\kappa}.$$

Take a disjoint decomposition

$$(5.14)\qquad GL_n(F)_{\mathbf{A}} = \coprod_{q\in Q} GL_n(F)\, qE^*$$

with a subset Q of $GL_n(F)_{\mathbf{h}}$. Then pB/E for any fixed $p\in GL_n(F)_{\mathbf{h}}$ is given by $\coprod_{q\in Q} \{\xi_{\mathbf{h}}\, q \mid \xi\in B_{p,q}/U_q\}$, where

$$(5.15)\qquad B_{p,q} = \{\xi\in GL_n(F) \mid \xi_{\mathbf{h}}\in pBq^{-1}\}, \qquad U_q = GL_n(F)\cap qE^*q^{-1}.$$

Therefore from (5.5) we obtain

$$(5.16)\qquad D'(\tau, pL_0; f) = \sum_{q\in Q} \psi(\det(q))\, |\det(q^{-1})|_{\mathbf{A}}^{n+1}$$
$$\cdot \sum_{\xi\in B_{p,q}/U_q} \psi_{\mathbf{h}}(\det(\xi))\, |\det(\xi)|^{(n+1)u}\, \mu_f(\tau, \xi q)[\det(p^{-1}\xi q)\, g].$$

6. Theta functions

Putting $\mathscr{V} = F_n^n$, we let $\mathscr{S}(\mathscr{V}_{\mathbf{h}})$ denote the Schwartz-Bruhat space of $\mathscr{V}_{\mathbf{h}}$. For $z\in H^{\mathbf{a}}$, $\lambda\in\mathscr{S}(\mathscr{V}_{\mathbf{h}})$, a totally positive element τ of S, and $\mu\in\mathbf{Z}^{\mathbf{a}}$ such that $0\leqq\mu\leqq u$ put $l = \mu + (nu/2)$ and

$$(6.1)\qquad \theta(z, \lambda) = \sum_{\xi\in\mathscr{V}} \lambda(\xi_{\mathbf{h}})\, \det(\xi)^\mu\, \mathbf{e}_{\mathbf{a}}(\operatorname{tr}({}^t\xi\tau\xi z)).$$

This is a special case of the function defined in [S11, Sect. 3]. (In fact, take $V = F_1^n$, $q = n$, $S(x, y) = 2\cdot{}^t x\tau y$ for $x, y\in V$, identify V^n with $\mathscr{V} = F_n^n$, and take $p(\xi) = \det(\xi)^\mu$ for $\xi\in\mathscr{V}$ as explained in Remark (III) at the end of Sect. 3 of that paper. Then θ of (6.1) can be obtained as $f(z; \lambda, p)$ of [S11, (3.10)].) We now put $\mathfrak{M}_n=\mathfrak{M}$ with $\mathfrak{M}$ of (1.10a) if n is odd and $\mathfrak{M}_n=G_{\mathbf{A}}$ if n is even, and define a factor of automorphy $J(\alpha, z)$ for $\alpha\in\mathfrak{M}_n$ by

$$(6.2)\qquad J(\alpha, z)=\begin{cases} j^l(\alpha, z) & \text{if } n \text{ is even},\\ h(\alpha, z)^n j(\alpha, z)^\mu & \text{if } n \text{ is odd}.\end{cases}$$

In [S11, Theorems 3.2 and 3.5] we established an action of $\mathfrak{M}_n$ on $\mathcal{S}(\mathcal{V}_{\mathbf{h}})$, written $(x, \lambda)\mapsto {}^x\lambda$ for $x\in\mathfrak{M}_n$ and $\lambda\in\mathcal{S}(\mathcal{V}_{\mathbf{h}})$, with the property that

$$(6.3)\qquad \theta(\alpha z, {}^\alpha\lambda)=J(\alpha, z)\,\theta(z, \lambda)\quad \text{for every } \alpha\in G\cap\mathfrak{M}_n.$$

Moreover, for each λ, our function $\theta(z, \lambda)$ is an element of $\mathcal{M}_l$. Now, as explained in [S11, (3.11)], we can associate with the above θ a function $\theta'_{\mathbf{A}}(x, \lambda)$ with a variable x on $G_{\mathbf{A}}$ or $M_{\mathbf{A}}$, according as n is even or odd, by

$$(6.4)\qquad \theta'_{\mathbf{A}}(x, \lambda)=J(w, \mathbf{i})^{-1}\theta(w(\mathbf{i}), {}^w\lambda)$$

for $x=\alpha w$ with $\alpha\in G$ and $w\in\mathfrak{M}_n$, where $\mathbf{i}$ is as in (1.6); we take $\mathrm{pr}(w)\in C^\theta$ if n is odd. This is well defined. It should be noted that $\theta'_{\mathbf{A}}$ is the function associated with θ by the principle of (1.17) only if $J=j^l$, which is not necessarily true if n is odd.

Proposition 6.1 *Let χ be the Hecke character of F corresponding to the extension $F(\det(2\tau)^{1/2})/F$ or $F((-1)^{n/4}\det(\tau)^{1/2})/F$ according as n is odd or even. Then*

$$(6.5)\qquad \theta'_{\mathbf{A}}\left(r_P\begin{pmatrix} q & s\tilde{q}\\ 0 & \tilde{q}\end{pmatrix}, \lambda\right)=\chi(\det(q))\det(q)_{\mathbf{a}}^{\mu}|\det(q)|_{\mathbf{A}}^{n/2}\cdot\sum_{\xi\in\mathcal{V}}\lambda(\xi_{\mathbf{h}}\,q)\det(\xi)^\mu\,\mathbf{e}_{\mathbf{a}}(\mathrm{tr}(\mathbf{i}\cdot{}^t q\cdot{}^t\xi\tau\xi q))\,\mathbf{e}_{\mathbf{A}}(\mathrm{tr}({}^t\xi\tau\xi s))$$

for every $q\in\mathrm{GL}_n(F)_{\mathbf{A}}$ and $s\in S_{\mathbf{A}}$. Moreover if $\beta=r_P(\mathrm{diag}[r, \tilde{r}])\,w$ with $\beta\in G$, $r\in\mathrm{GL}_n(F)_{\mathbf{h}}$, and $w\in\mathfrak{M}_n$, $\mathrm{pr}(w)\in D[2\mathfrak{d}^{-1}, 2\mathfrak{d}]$, then

$$(6.6)\qquad J(\beta, \beta^{-1}z)\,\theta(\beta^{-1}z, \lambda)=\chi_{\mathbf{h}}(\det(r))|\det(r)|_{\mathbf{A}}^{n/2}\cdot\sum_{\xi\in\mathcal{V}}({}^w\lambda)(\xi_{\mathbf{h}}\,r)\det(\xi)^\mu\,\mathbf{e}_{\mathbf{a}}(\mathrm{tr}({}^t\xi\tau\xi z)).$$

Proof. Given $x\in\mathfrak{M}_n$, take $\alpha\in G$ and w as in (6.4). Then $\alpha\in\mathfrak{M}_n$. Put $z=w(\mathbf{i})$. By (6.3) and (6.4) we have $\theta'_{\mathbf{A}}(x, \lambda)=J(w, \mathbf{i})^{-1}\theta(z, {}^w\lambda)=J(w, \mathbf{i})^{-1}J(\alpha, z)^{-1}\cdot\theta(\alpha z, {}^\alpha({}^w\lambda))$. By [S11, (3.2b), Th. 3.2(2)] we have $J(\alpha w, \mathbf{i})=J(\alpha, z)\,J(w, \mathbf{i})$ and ${}^\alpha({}^w\lambda)={}^x\lambda$ since $\mathrm{pr}(w)\in C^\theta$. Therefore

$$(6.7)\qquad \theta'_{\mathbf{A}}(x, \lambda)=J(x, \mathbf{i})^{-1}\theta(x(\mathbf{i}), {}^x\lambda)\quad \text{if } x\in\mathfrak{M}_n.$$

Take $x=r_P\begin{pmatrix} q & s\tilde{q}\\ 0 & \tilde{q}\end{pmatrix}$. By [S11, Th. 3.2(5)] we have

$$({}^x\lambda)(y)=|\det(q)_{\mathbf{h}}|_{\mathbf{A}}^{n/2}\,\chi_{\mathbf{h}}(\det(q))\,\mathbf{e}_{\mathbf{h}}(\mathrm{tr}({}^t y\tau ys))\,\lambda(yq)\quad (y\in\mathcal{V}_{\mathbf{h}}).$$

This combined with (6.1) and (6.7) gives (6.5). To prove (6.6), take the element x so that $\mathrm{pr}(x)\in P_{\mathbf{a}}$ and $x(\mathbf{i})=z$. Since $\beta^{-1}z=w^{-1}x(\mathbf{i})$ with β as in (6.6) and $G_{\mathbf{a}}$ acts trivially on $\mathcal{S}(\mathcal{V}_{\mathbf{h}})$, formula (6.4) shows that $J(\beta, \beta^{-1}z)\,\theta(\beta^{-1}z, \lambda)=J(\beta, \beta^{-1}z)$

$\cdot J(w^{-1}x, \mathbf{i})\,\theta'_{\mathbf{A}}(w^{-1}x, {}^{w}\lambda) = J(\beta w^{-1}x, \mathbf{i})\,\theta'_{\mathbf{A}}(\beta w^{-1}x, {}^{w}\lambda) = J(g, \mathbf{i})\theta'_{\mathbf{A}}(g, {}^{w}\lambda)$, where $g = r_P\begin{pmatrix} rq & rs\tilde{q} \\ 0 & \tilde{r}\tilde{q} \end{pmatrix}$. Thus we obtain (6.6) from (6.5).

We now take a Hecke character ω of F such that

$$\omega_{\mathbf{a}}(-1)^n = (-1)^{n\|\mu\|}, \tag{6.8}$$

denote the conductor of ω by $\mathfrak{f}$, and take an element p of $GL_n(F)_{\mathbf{h}}$. We then define a series $\theta_{\omega,p}$ by

$$\theta_{\omega,p}(z) = \sum_{\xi \in \mathscr{V} \cap pR^*} \omega_{\mathbf{a}}(\det(\xi))\,\omega^*(\det(p^{-1}\xi)\,\mathfrak{g}\,\det(\xi)^{\mu}\,\mathbf{e}_{\mathbf{a}}(\mathrm{tr}({}^t\xi\tau\xi z)), \tag{6.9}$$

where R^* is defined by (2.1b) and it is understood that $\omega_{\mathbf{a}}(b)\,\omega^*(b\mathfrak{a})$ for $b=0$ and a fractional ideal $\mathfrak{a}$ denotes $\omega^*(\mathfrak{a})$ or 0 according as $\mathfrak{f}=\mathfrak{g}$ or $\mathfrak{f}\neq\mathfrak{g}$. To see that $\theta_{\omega,p}$ is a special case of (6.1), define $\lambda \in \mathscr{S}(\mathscr{V}_{\mathbf{h}})$ by $\lambda(x) = \omega_{\mathbf{h}}(\det(p))^{-1}\lambda_0(p^{-1}x)$, $\lambda_0 = \prod_{v \in \mathbf{h}} \lambda_{0v}$ with $\lambda_{0v} \in \mathscr{S}(\mathscr{V}_v)$ given as follows: if $v \nmid \mathfrak{f}$, λ_{0v} is the characteristic function of R_v; if $v|\mathfrak{f}$, then $\lambda_{0v}(x) = 0$ for $x \notin E_v$ and $\lambda_{0v}(x) = \omega_v(\det(x))^{-1}$ for $x \in E_v$. Then we easily see that (6.1) gives (6.9).

Proposition 6.2 *Let ρ_τ be the Hecke character of F corresponding to the extension $F(c^{1/2})/F$ with $c = (-1)^{[n/2]}\det(2\tau)$ and let $\omega' = \omega\rho_\tau$. Then there exist a fractional ideal $\mathfrak{b}$ and an integral ideal $\mathfrak{c}$, satisfying the inclusion relations of* (1.20b) *if n is odd, and such that the conductor of ω' divides $\mathfrak{c}$, $\theta_{\omega,p} \in \mathscr{M}_l(\mathfrak{b}, \mathfrak{c}, \omega')$, and moreover, if $\beta \in G \cap \mathrm{diag}[r, \tilde{r}]\,D[\mathfrak{b}^{-1}, \mathfrak{b}\mathfrak{c}]$ with $r \in GL_n(F)_{\mathbf{h}}$, then*

$$j^l(\beta, \beta^{-1}z)\,\theta_{\omega,p}(\beta^{-1}z) = \omega'(\det(r))^{-1}\,\omega'_{\mathfrak{c}}(\det(d_\beta r))|\det(r)|_{\mathbf{A}}^{n/2} \cdot \sum_{\xi \in \mathscr{V} \cap pR^*r^{-1}} \omega_{\mathbf{a}}(\det(\xi))\,\omega^*(\det(\xi p^{-1}r)\,\mathfrak{g})\,\det(\xi)^{\mu}\,\mathbf{e}_{\mathbf{a}}(\mathrm{tr}({}^t\xi\tau\xi z)). \tag{6.10}$$

In particular, suppose that ${}^tg\cdot 2\tau g \in \mathfrak{x}$ for every $g \in pL_0$ and ${}^th(2\tau)^{-1}h \in 4\mathfrak{t}^{-1}$ for every $h \in \tilde{p}L_0$ with fractional ideals $\mathfrak{x}$ and $\mathfrak{t}$, where $L_0 = \mathfrak{g}_1^n$; let $\mathfrak{e}$ be the conductor of ρ_τ. Then we can take $(\mathfrak{b}, \mathfrak{c}) = (2^{-1}\mathfrak{d}\mathfrak{x}, \mathfrak{e}\cap\mathfrak{f}\cap\mathfrak{x}^{-1}\mathfrak{f}^2\mathfrak{t})$ if n is even and $(\mathfrak{b}, \mathfrak{c}) = (2^{-1}\mathfrak{d}\mathfrak{a}^{-1}, \mathfrak{e}\cap\mathfrak{f}\cap 4\mathfrak{a}\cap\mathfrak{a}\mathfrak{f}^2\mathfrak{t})$ if n is odd, where $\mathfrak{a} = \mathfrak{x}^{-1}\cap\mathfrak{g}$.

Proof. By [S11, Prop. 3b.2] there exist a fractional ideal $\mathfrak{b}$ and an integral ideal $\mathfrak{c}$, satisfying the inclusion relations of (1.15b) if n is odd, and such that the conductors of χ and ω divide $\mathfrak{c}$, and

$${}^{w}\lambda = (\chi\bar{\omega})_{\mathfrak{c}}(\det(a_w))\,\lambda \quad \text{for every } w \in D[\mathfrak{b}^{-1}, \mathfrak{b}\mathfrak{c}]. \tag{6.11}$$

Now [S11, Prop. 2.4] shows that $J(\alpha, z) = [\varepsilon_{\mathbf{a}}(\det(d_\alpha))\,\varepsilon^*(\det(d_\alpha)\,\mathfrak{a}_\alpha^{-1})]^{(n-1)/2} \cdot j^l(\alpha, z)$ for every $\alpha \in G \cap D[2\mathfrak{d}^{-1}, 2\mathfrak{d}]$ if n is odd, where ε is the Hecke character of F corresponding to the extension $F(\sqrt{-1})/F$ and $\mathfrak{a}_\alpha$ is defined by (7.2) below. This combined with (6.3), (6.6), and (6.11) proves our assertions up to formula (6.10). Let $\mathfrak{e}'$ be the conductor of χ. Then $\mathfrak{e}' = \mathfrak{e}$ if n is even and $\mathfrak{e}\cap 4\mathfrak{g} = \mathfrak{e}'\cap 4\mathfrak{g}$ if n is odd. Therefore [S11, Prop. 3b.2] gives our assertion on $(\mathfrak{b}, \mathfrak{c})$ as stated above. (In fact, the ideal $\mathfrak{y}$ of that proposition is $4(\mathfrak{d}^2\mathfrak{f}^2\mathfrak{t})^{-1}$ in the present case.)

7. Eisenstein series

To define our Eisenstein series, we take a set of data $\{k, \chi, \mathfrak{b}, \mathfrak{c}\}$ which consists of an integral or a half-integral weight k, a Hecke character χ, a fractional ideal $\mathfrak{b}$, and an integral ideal $\mathfrak{c}$, satisfying (1.15b) as well as the following conditions:

(7.1) $$\chi_{\mathbf{a}}(x) = \operatorname{sgn}(x)^{[k]}|x_{\mathbf{a}}|^{i\kappa} \quad \text{with } \kappa \in \mathbf{R}^{\mathbf{a}}, \ \|\kappa\| = 0,$$

(7.2) $$\chi_v(a) = 1 \quad \text{if } v \in \mathbf{h},\ a \in \mathfrak{g}_v^\times \text{ and } a - 1 \in \mathfrak{c}_v.$$

Put again $D = D[\mathfrak{b}^{-1}, \mathfrak{b}\mathfrak{c}]$ and $\Gamma = G \cap D$. To simplify our notation, for $\alpha \in G \cap P_{\mathbf{A}} D$ we put

(7.3) $$\mathfrak{a}_\alpha = \det(d_p)\,\mathfrak{g}, \quad \chi[\alpha] = \chi_{\mathbf{h}}(\det(d_p)^{-1})\,\chi_{\mathfrak{c}}(\det(d_w)^{-1})$$
$$\text{if } \alpha = pw \text{ with } p \in P_{\mathbf{A}} \text{ and } w \in D.$$

These symbols depend on $\mathfrak{b}$. In the following treatment we consider them with a fixed $\mathfrak{b}$. We now put

(7.4) $$E(z, s; k, \chi, \Gamma) = \sum_{\alpha \in R} N(\mathfrak{a}_\alpha)^{2s} \chi[\alpha] (\Delta^{su - (k - i\kappa)/2} \|_k \alpha)(z)$$

for $z \in H^{\mathbf{a}}$ and $s \in \mathbf{C}$ with $R = P \backslash (G \cap P_{\mathbf{A}} D)$ and Δ of (1.4b). Define an element η of $G_{\mathbf{A}}$ by

(7.5) $$\eta_{\mathbf{a}} = 1_{2n}, \quad \eta_v = \begin{pmatrix} 0 & -\delta_v^{-1} 1_n \\ \delta_v 1_n & 0 \end{pmatrix} \quad \text{for } v \in \mathbf{h},$$

and take an element η_0 of $G \cap D'\eta^{-1}$, where δ is an arbitrarily fixed element of $F_{\mathbf{h}}^\times$ such that $\mathfrak{d} = \delta\mathfrak{g}$ and $D' = \{x \in D \mid a(x)_v - 1 \in (\mathfrak{c}_v)_n^n$ *for every* $v \in \mathbf{h}\}$. We then put

(7.6) $$E^*(z, s) = \chi(\delta)^{-n} j^k(\eta_0, z)^{-1} E(\eta_0(z), s; k, \chi, \Gamma).$$

We easily see that this does not depend on the choice of δ and η_0; moreover, E as a function of z belongs to $C_k(\mathfrak{b}, \mathfrak{c}, \chi^{-1})$ of Sect. 1 and E^* to $C_k(\mathfrak{d}^2(\mathfrak{b}\mathfrak{c})^{-1}, \mathfrak{c}; \chi)$. Therefore, by the principle of (1.17) we can define a function $E_{\mathbf{A}}^*$ on $G_{\mathbf{A}}$ or $M_{\mathbf{A}}$, according as k is integral or half-integral, corresponding to E^*. (For a more direct adelic definition of $E_{\mathbf{A}}^*$ or $E_{\mathbf{A}}$, see [S6, (2.17a), (2.19)] and [S7, (4.7a), (4.9a)].) Then we have a Fourier expansion

$$E_{\mathbf{A}}^*\left(r_P\begin{pmatrix} q & \sigma\tilde{q} \\ 0 & \tilde{q} \end{pmatrix}\right) = \sum_{\tau \in S} c(\tau, q, s)\, \mathbf{e}_{\mathbf{A}}(\operatorname{tr}(\tau\sigma)) \quad (q \in GL_n(F_{\mathbf{A}}), \sigma \in S_{\mathbf{A}})$$

with $c(\tau, q, s) \in \mathbf{C}$. Since $E^*_{\mathbf{A}}(\xi) = (E^* \|_k \xi)(\mathbf{i})$ for $\mathrm{pr}(\xi) \in G_{\mathbf{a}}$, we easily see that

$$E^*(x+iy, s) = \det(y)^{-k/2} \sum_{\tau \in S} c(\tau, (1, y^{1/2}), s)\, \mathbf{e}_{\mathbf{a}}(\mathrm{tr}(\tau x)), \tag{7.7}$$

where $(1, y^{1/2})$ denotes the element q of $S_{\mathbf{A}}$ such that $q_{\mathbf{h}} = 1$ and $q_v = y_v^{1/2}$ for every $v \in \mathbf{a}$. To obtain the explicit form of $c(\tau, q, s)$, we first put

$$\xi(g, h; s, s') = \int_{S_v} \mathbf{e}_v(\mathrm{tr}(-hx)) \det(x+ig)^{-s} \det(x-ig)^{-s'} dx$$

$$(s, s' \in \mathbf{C},\ 0 < g \in S_v,\ h \in S_v,\ v \in \mathbf{a}),$$

$$\Xi(y, w; t, t') = \prod_{v \in \mathbf{a}} \xi(y_v, w_v; t_v, t'_v) \qquad (t, t' \in \mathbf{C}^{\mathbf{a}},\ y \in S_{\mathbf{a}},\ y_v > 0,\ w \in S_{\mathbf{a}}).$$

The function ξ was investigated in [S5]. Also, for $\zeta \in S_{\mathbf{A}}$ such that $\zeta_{\mathbf{h}} \in \delta S^*$ with S^* of (2.4) and an integral ideal $\mathfrak{a}$ divisible by the conductor of χ we put

$$\alpha^0_{\mathfrak{a}}(\zeta, s, \chi) = \prod_{v \nmid \mathfrak{a}} \sum_{\sigma \in S_v/S(\mathfrak{g})_v} \mathbf{e}_v(-\delta_v^{-1}\mathrm{tr}(\zeta\sigma))\, \chi^*(\nu_0(\sigma))\, \nu(\sigma)^{-s},$$

$$\alpha^1_{\mathfrak{a}}(\zeta, s, \chi) = \prod_{v \nmid \mathfrak{a}} \sum_{\sigma \in S_v/S(\mathfrak{g})_v} \mathbf{e}_v(-\delta_v^{-1}\mathrm{tr}(\zeta\sigma))\, \chi^*(\nu_0(\sigma))\, \omega_v(\delta_v^{-1}\sigma)\, \nu(\sigma)^{-s},$$

$$\omega_v(\sigma) = \nu(\delta_v \sigma)^{1/2} \int_{L_v} \mathbf{e}_v({}^t x \sigma x/2)\, dx \qquad (\sigma \in S_v,\ L_v = (\mathfrak{g}_v)^n_1),$$

where δ is as in (7.5) and the measure of L_v is taken to be 1. We always assume that $\mathfrak{a}$ is divisible by 2 for $\alpha^1_{\mathfrak{a}}$. As shown in [S6, Lemma 4.4], $\omega_v(\sigma)$ depends only on σ modulo $S(\mathfrak{d}^{-1})_v$.

Proposition 7.1 *Suppose that* $\mathfrak{c} \neq \mathfrak{g}$ *and* $\det(q_v) > 0$ *for every* $v \in \mathbf{a}$; *let* $y = {}^t q_{\mathbf{a}} q_{\mathbf{a}}$. *Then* $c(\tau, q, s) \neq 0$ *only if* $({}^t q \tau q)_v \in (\mathfrak{d}\mathfrak{b}^{-1}\mathfrak{c}^{-1})_v T_v$ *with* T_v *of* (3.1) *for every* $v \in \mathbf{h}$, *in which case*

$$c(\tau, q, s) = C \cdot \chi_{\mathbf{h}}(\det(-q))^{-1} |\det(q)_{\mathbf{h}}|_{\mathbf{A}}^{n+1-2s} |D_F|^{-2ns+3n(n+1)/4} N(\mathfrak{b}\mathfrak{c})^{-n(n+1)/2}$$

$$\cdot \det(y)^{su+i\kappa/2}\, \Xi(y, \tau; su+(k+i\kappa)/2, su-(k-i\kappa)/2)$$

$$\cdot \alpha^{\lambda}_{\mathfrak{c}}(\theta^{-1} \cdot {}^t q \tau q, 2s, \chi),$$

where $C = 1$ *and* $\lambda = 0$ *if* $k \in \mathbf{Z}^{\mathbf{a}}$, *and* $C = \mathbf{e}(n[F:\mathbf{Q}]/8)$ *and* $\lambda = 1$ *if* $k \notin \mathbf{Z}^{\mathbf{a}}$; θ *is an element of* $F^\times_{\mathbf{h}}$ *such that* $\theta\mathfrak{g} = \mathfrak{b}^{-1}\mathfrak{d}$ *if* $k \in \mathbf{Z}^{\mathbf{a}}$, *and* $\theta = 1$ *if* $k \notin \mathbf{Z}^{\mathbf{a}}$.

If $k \notin \mathbf{Z}^{\mathbf{a}}$, our assumption (1.15b) implies that $\mathfrak{b}_v = \mathfrak{d}_v$ for $v \nmid \mathfrak{c}$. Therefore $(\theta^{-1} \cdot {}^t q \tau q)_v \in T_v$ if $v \nmid \mathfrak{c}$ for both integral and half-integral k.

Our proposition was given in [S6] and [S7] under the assumption that $q_{\mathbf{h}} = 1$ and $\mathfrak{b}$ is either $\mathfrak{g}$ or $2^{-1}\mathfrak{d}$ according as $k \in \mathbf{Z}^{\mathbf{a}}$ or $k \notin \mathbf{Z}^{\mathbf{a}}$. The general case can be proved in the same manner by making obvious modifications.

Now, as Langlands showed, the series E and E^* can be continued to the whole s-plane as meromorphic functions. Therefore we can study their singularities by examining the analytic nature of $c(\tau, q, s)$. For this purpose we first put

$$\Gamma_0(s)=1, \quad \Gamma_n(s)=\pi^{n(n-1)/4}\prod_{\nu=0}^{n-1}\Gamma(s-\nu/2) \quad (s\in\mathbf{C}, 0<n\in\mathbf{Z}), \tag{7.8}$$

$$L_{\mathfrak{e}}(s,\chi)=\prod_{v\in\mathbf{h}, v\nmid\mathfrak{e}\mathfrak{f}}(1-\chi_v(\pi_v)|\pi_v|^s)^{-1}, \tag{7.9}$$

where $\mathfrak{e}$ is any integral ideal and $\mathfrak{f}$ is the conductor of χ.

Proposition 7.2 *With τ, q, and θ such that $c(\tau, q, s)\neq 0$ as in Theorem 7.1, put $r=\operatorname{rank}(\tau)$ and ${}^t g\tau g=\operatorname{diag}[\tau', 0]$ with $g\in GL_n(F)$ and $\tau'\in S^r$. Let ρ_τ be the Hecke character corresponding to $F(c^{1/2})/F$, where $c=(-1)^{[r/2]}\det(2\tau')$, if $r>0$; let $\rho_\tau=1$ if $r=0$. Then*

$$\alpha_{\mathfrak{c}}^1(\theta^{-1}\cdot{}^t q\tau q, 2s, \chi)=\Lambda_{\mathfrak{c}}(s)^{-1}\Lambda_\tau(s)\prod_{v\in\mathbf{c}}f_{\tau,q,v}(\chi(\pi_v)|\pi_v|^{2s})$$

with a finite subset $\mathbf{c}$ of $\mathbf{h}$, polynomials $f_{\tau,q,v}$ with coefficients in $\mathbf{Z}$ independent of χ, and functions $\Lambda_{\mathfrak{c}}$ and Λ_τ given as follows:

$$\Lambda_{\mathfrak{c}}(s)=\begin{cases} L_{\mathfrak{c}}(2s,\chi)\prod_{i=1}^{[n/2]}L_{\mathfrak{c}}(4s-2i,\chi^2) & \text{if } k\in\mathbf{Z}^{\mathbf{a}}, \\ \prod_{i=1}^{[(n+1)/2]}L_{\mathfrak{c}}(4s-2i+1,\chi^2) & \text{if } k\notin\mathbf{Z}^{\mathbf{a}}, \end{cases} \tag{7.10}$$

$$\Lambda_\tau(s)=L_{\mathfrak{c}}(2s-n+r/2,\chi\rho_\tau)\prod_{i=1}^{[(n-r)/2]}L_{\mathfrak{c}}(4s-2n+r+2i-1,\chi^2) \quad \text{if } 2k_v+r\in 2\mathbf{Z},$$

$$\Lambda_\tau(s)=\prod_{i=1}^{[(n-r+1)/2]}L_{\mathfrak{c}}(4s-2n+r+2i-2,\chi^2) \quad \text{if } 2k_v+r\notin 2\mathbf{Z}.$$

The set $\mathbf{c}$ is determined as follows: $\mathbf{c}=\emptyset$ if $r=0$. If $r>0$, let L' be the image of $g^{-1}qL_0$ in F^r under the projection map of Proposition 5.4, and let $L'=a\mathfrak{g}_1^r$ with $a\in GL_r(F)_{\mathbf{h}}$. Then $\mathbf{c}$ consists of all the primes v not dividing $\mathfrak{c}$ such that $(\theta^{-1}\cdot{}^t a\tau' a)_v$ is not regular in the sense of Sect. 3.

Proof. As in the proof of Proposition 5.4, we can find an element e of E such that ${}^t(g^{-1}qe)$ belongs to $W_{\mathbf{h}}$ with W of (5.7). Put $g^{-1}qe=\begin{pmatrix} p & 0 \\ * & * \end{pmatrix}$ with $p\in GL_r(F)_{\mathbf{h}}$. Since $\theta^{-1}\cdot{}^t e\cdot{}^t q\tau qe=\operatorname{diag}[\theta^{-1}\cdot{}^t p\tau' p, 0]$, we can replace $\theta^{-1}\cdot{}^t q\tau q$ with $\operatorname{diag}[\theta^{-1}\cdot{}^t p\tau' p, 0]$. We also see that $L'=p\mathfrak{g}_1^r$. Therefore we obtain our expression for $\alpha_{\mathfrak{c}}^0$ immediately from Theorem 3.2. Similar results for the factors of $\alpha_{\mathfrak{c}}^1(\zeta,\ldots)$ were first obtained in [S6] under some conditions on ζ, which were later removed by Feit in [F1] (see in particular [S6, Prop. 6.2], [F1, Theorem 6,2], and [S7, Prop. 5.1]). Applying the results to the present case, we obtain the desired expression. If $q_{\mathbf{h}}=1$, our proposition was essentially given in [S6] and [S7], in which $\mathfrak{c}$ was assumed to be prime to 2. Thus, the new points of the present formulation are: we can remove that assumption when $k\in\mathbf{Z}^{\mathbf{a}}$; $q_{\mathbf{h}}$ is arbitrary. (If $k\notin\mathbf{Z}^{\mathbf{a}}$, assumption (1.15b) implies that $\mathfrak{c}$ is divisible by 4.)

Theorem 7.3 *Let $\mathcal{P}(s)=\mathcal{G}(s)\,\Lambda_{\mathfrak{c}}(s)\,E(z,s;k,\chi,\Gamma)$ with $\mathcal{G}(s)$ defined as follows:*

$$\mathcal{G}(s)=\mathcal{G}_{k,\kappa}(s)=\prod_{v\in\mathbf{a}}\gamma(s+i\kappa_v/2,|k_v|),$$

$$\gamma(s,h)=\begin{cases}\Gamma\left(s+\dfrac{h}{2}-\left[\dfrac{2h+n}{4}\right]\right)\Gamma_n\left(s+\dfrac{h}{2}\right) & (n/2<h\in\mathbf{Z},\ n\ \text{even}),\\ \Gamma_n\left(s+\dfrac{h}{2}\right) & (n/2<h\in\mathbf{Z},\ n\ \text{odd}),\\ \Gamma_{2h+1}\left(s+\dfrac{h}{2}\right)\displaystyle\prod_{i=h+1}^{[n/2]}\Gamma(2s-i) & (0\leqq h\leqq n/2,\ h\in\mathbf{Z}),\end{cases}$$

$$\gamma(s,h)=\begin{cases}\Gamma\left(s+\dfrac{h-1}{2}-\left[\dfrac{2h+n-2}{4}\right]\right)\Gamma_n\left(s+\dfrac{h}{2}\right) & (n/2<h\notin\mathbf{Z},\ n\ \text{odd}),\\ \Gamma_n\left(s+\dfrac{h}{2}\right) & (n/2<h\notin\mathbf{Z},\ n\ \text{even}),\\ \Gamma_{2h+1}\left(s+\dfrac{h}{2}\right)\displaystyle\prod_{i=[h]+1}^{[(n-1)/2]}\Gamma\left(2s-\dfrac{1}{2}-i\right) & (0<h\leqq n/2,\ h\notin\mathbf{Z}).\end{cases}$$

Then $\mathcal{P}(s)$ is a meromorphic function in s on the whole $\mathbf{C}$ with only finitely many poles, and each pole is simple. In particular, $\mathcal{P}$ is an entire function of s if $\chi^2\neq 1$. If $\chi^2=1$ and $\mathfrak{c}\neq\mathfrak{g}$, the poles are determined as follows: Let $m=\operatorname{Max}_{v\in\mathbf{a}}|k_v|$.

If $m>n/2$, $\mathcal{P}$ has no pole except for a possible pole at $s=(n+2)/4$ which occurs only if $2|k_v|-n\in 4\mathbf{Z}$ for every v such that $2|k_v|>n$. If $m\leqq n/2$, $\mathcal{P}$ has possible poles only in the set

(i) $\{j/2\,|\,j\in\mathbf{Z},\ [(n+3)/2]\leqq j\leqq n+1-m\}$ *if* $k\in\mathbf{Z}^{\mathbf{a}}$,
(ii) $\{(2j+1)/4\,|\,j\in\mathbf{Z},\ 1+[n/2]\leqq j\leqq n+(1/2)-m\}$ *if* $k\notin\mathbf{Z}^{\mathbf{a}}$.

If $\chi^2=1$, $\mathfrak{c}=\mathfrak{g}$, and $k\in\mathbf{Z}^{\mathbf{a}}$, each pole belongs to the first set or to

(iii) $\{j/2\,|\,j\in\mathbf{Z},\ 0\leqq j\leqq[n/2]\}$,

where $j=0$ is unnecessary if $\chi\neq 1$.

If $k\in\mathbf{Z}^{\mathbf{a}}$, the "best" gamma factors, as well as the location of poles were essentially determined by Feit in [F1, F2, F3]. Our present formulation covers more general cases with "mixed" and "negative" weights. However, our result in the case in which $k\in\mathbf{Z}^{\mathbf{a}}$, $\mathfrak{c}=\mathfrak{g}$, and $\chi^2=1$ is not as strong as that in [F3]. In this case we content ourselves with a somewhat weaker statement which can be proved with no additional computation. The results for $k\notin\mathbf{Z}^{\mathbf{a}}$ are new, except that the one-dimensional case was treated in [S8].

The remaining part of this section is devoted to the proof of this theorem. For two meromorphic functions f and g on $\mathbf{C}$ let us write $f \succ g$ if f/g is entire, and $f \sim g$ if $f \succ g$ and $g \succ f$. We easily see that

(7.11 a) $$\Gamma(s+m) \succ \Gamma(s) \quad \text{if } 0 \leqq m \in \mathbf{Z},$$

(7.11 b) $$\Gamma_a(s+m/2) \succ \Gamma_a(s) \quad \text{if } 0 \leqq a \in \mathbf{Z},\quad 0 \leqq m \in \mathbf{Z}, \text{ and } ma \in 2\mathbf{Z},$$

(7.11 c) $$\prod_{i=1}^{m} \Gamma(2s-i) \sim \Gamma_m(s)\, \Gamma_m(s-1/2),$$

(7.11 d) $$\Gamma_{m+n}(s) \sim \Gamma_m(s)\, \Gamma_n(s-m/2), \quad \Gamma_{2m}(s) \sim \prod_{i=1}^{m} \Gamma(2s+1-2i).$$

Given τ and $v \in \mathbf{a}$, let $r = \operatorname{rank}(\tau)$, $t = n-r$; let a_v resp. b_v be the number of positive resp. negative eigenvalues of τ_v. Further determine $\varepsilon_v = \pm 1$ by $\varepsilon_v = [k_v] + [r/2] + b_v \pmod 2$. Writing simply a, b, and ε for a_v, b_v, and ε_v, put

$$g_{\tau,v}(s) = g^1_{\tau,v}(s)/g^2_{\tau,v}(s),$$

$$g^1_{\tau,v}(s) = \frac{\Gamma_t(2s-(n+1)/2)}{\Gamma_{a+t}(s+k_v/2)\,\Gamma_{b+t}(s-k_v/2)},$$

$$g^2_{\tau,v}(s) = \Gamma\left(s - \frac{n+t}{4} + \frac{\varepsilon}{2}\right) \prod_{i=1}^{[t/2]} \Gamma\left(2s+i-\frac{n+t+1}{2}\right) \quad \text{if } 2k_v + r \text{ is even},$$

$$g^2_{\tau,v}(s) = \prod_{i=1}^{[(t+1)/2]} \Gamma\left(2s+i-\frac{n+t+2}{2}\right) \quad \text{if } 2k_v + r \text{ is odd}.$$

Now a well known fact on L-functions says that the product

(7.12) $$\Lambda_\tau(s) \prod_{v \in \mathbf{a}} g^2_{\tau,v}(s+i\kappa_v/2)$$

is meromorphic on the whole s-plane. It is entire if $\chi^2 \neq 1$; if $\chi^2 = 1$ and $\mathfrak{c} \neq \mathfrak{g}$, then (7.12) times $p_\tau(s)$ is entire, where p_τ is determined by

$$p_\tau(s) = \begin{cases} \displaystyle\prod_{i=\alpha}^{[t/2]} \left(2s - \frac{n+t}{2} - 1 + i\right) & (2k_v + r \in 2\mathbf{Z}), \\ \displaystyle\prod_{i=1}^{[(t+1)/2]} \left(2s - \frac{n+t+3}{2} + i\right) & (2k_v + r \notin 2\mathbf{Z}), \end{cases}$$

with $\alpha = \operatorname{Max}_{v \in \mathbf{a}} \{\varepsilon_v\}$. In [S5, Theorem 4.2, (4.34.K)] we showed that $\xi(y, \tau_v; s, s')$ is a holomorphic function in (s, s') on the whole $\mathbf{C} \times \mathbf{C}$ times $\Gamma_t(s+s'-(n+1)/2)\,\Gamma_{a+t}(s)^{-1}\,\Gamma_{b+t}(s')^{-1}$. Therefore $\Xi(y, \tau; su+(k+i\kappa)/2, su-(k-i\kappa)/2)$ is $\prod_{v \in \mathbf{a}} g^1_{\tau,v}(s+i\kappa_v/2)$ times a holomorphic function in s. By Proposition 7.2 $\Lambda_\tau(s)\, c(\tau, q, s)$ is the product of (7.12) and $\prod_{v \in \mathbf{a}} g_{\tau,v}(s+i\kappa_v/2)$ times a finite factor.

Thus, in view of (7.6) and (7.7), our task is to show that $\mathscr{G}(s)\prod_{v\in\mathbf{a}} g_{\tau,v}(s+i\kappa_v/2)$ times (7.12) has poles only at the points described in our theorem. To make it clearer, put

$$q_m(s)=\begin{cases}\prod_{i=1}^{[n/2]+1-m}\left(2s-\left[\frac{n+1}{2}\right]-i\right) & (m\in\mathbf{Z}),\\ \prod_{i=1}^{[(n+1)/2]-[m]}\left(2s-\left[\frac{n}{2}\right]-\frac{1}{2}-i\right) & (m\notin\mathbf{Z}).\end{cases}$$

Now the set (i) or (ii) of our theorem consists of the zeros of $q_m(s)$. Observe that the reduced denominator of q_m/p_τ can be written

$$h_{m,\tau}(s)=\begin{cases}\prod_{i=\beta}^{[t/2]+1-\alpha}\left(2s-\left[\frac{n+1}{2}\right]-i\right) & (m\in\mathbf{Z},\ r\in 2\mathbf{Z}),\\ \prod_{i=\beta}^{[(t+1)/2]}\left(2s-\left[\frac{n+1}{2}\right]-i\right) & (m\in\mathbf{Z},\ r\notin 2\mathbf{Z}),\end{cases}$$

$$h_{m,\tau}(s)=\begin{cases}\prod_{i=\beta}^{[t/2]+1-\alpha}\left(2s-\left[\frac{n}{2}\right]-\frac{1}{2}-i\right) & (m\notin\mathbf{Z},\ r\notin 2\mathbf{Z}),\\ \prod_{i=\beta}^{[(t+1)/2]}\left(2s-\left[\frac{n}{2}\right]-\frac{1}{2}-i\right) & (m\notin\mathbf{Z},\ r\in 2\mathbf{Z}),\end{cases}$$

where $\beta=\mathrm{Max}\{1,[n/2]+2-m\}$ if $m\in\mathbf{Z}$ and $\beta=\mathrm{Max}\{1,[(n+3)/2]-[m]\}$ if $m\notin\mathbf{Z}$.

To prove our theorem, let us first assume that $\mathfrak{c}\neq\mathfrak{g}$ and $\chi^2=1$. Then our assertions would follow if we could show that

$$(4s-n-2)\,h_{m,\tau}(s)^{-1}\prod_{v\in\mathbf{a}}\gamma(s+i\kappa_v/2,|k_v|)\,g_{\tau,v}(s+i\kappa_v/2)\succ 1,$$

where the first factor $4s-n-2$ is necessary only when $m>n/2$ and $2|k_v|-n\in 4\mathbf{Z}$ for every v such that $2|k_v|>n$. To do this, we pick one $v\in\mathbf{a}$ and prove

$$(4s-n-2)^{\delta}h_{\tau,v}(s)^{-1}\gamma(s,|k_v|)\,g_{\tau,v}(s)\succ 1, \tag{7.13a}$$

$$\gamma(s,|k_w|)\,g_{\tau,w}(s)\succ 1 \quad \text{for every } w\neq v \tag{7.13b}$$

for every τ, where $h_{\tau,v}$ is the polynomial $h_{m,\tau}$ with $|k_v|$ and ε_v in place of m and α. (Notice that $h_{m,\tau}$ divides $h_{|k_v|,\tau}$ if $|k_v|>n/2$.) Here we put $\delta=0$ and choose a v such that $|k_v|>n/2$ and $2|k_v|-n\notin 4\mathbf{Z}$ if such a v exists; otherwise we take an arbitrary v such that $m=|k_v|$ and put $\delta=0$ or 1 according as $m\leqq n/2$ or $m>n/2$. If $\chi^2\neq 1$, our task is simpler, since we only have to show (7.13b). Therefore we shall only prove (7.13a), which essentially includes (7.13b).

For simplicity, let us fix τ and v, and write k, ε, g^i, and h_k for k_v, ε_v, $g^i_{\tau,v}$, and $h_{\tau,v}$. Now we easily see that $\Gamma_t(2s-(n+1)/2)/g^2(s)\sim f(s)$ with

$$f(s)=\Gamma\left(s-\frac{n+t-2\varepsilon}{4}\right)^{-1}\prod_{i=1}^{[(t+1)/2]}\Gamma\left(2s-\frac{n-\eta}{2}-i\right)\qquad(2k+r\text{ even}),$$

$$f(s)=\prod_{i=1}^{[t/2]}\Gamma\left(2s-\frac{n-\eta}{2}-i\right)\qquad(2k+r\text{ odd}),$$

(7.14) $\quad\varepsilon=\pm1,\quad \varepsilon\equiv[k]+[r/2]+b\ (\mathrm{mod}\ 2),\quad \eta=(1-(-1)^{2k+n})/2.$

Then our problem is to show that

(7.15) $$\frac{(4s-n-2)^{\delta}\gamma(s,|k|)f(s)}{h_k(s)\,\Gamma_{a+t}(s+k/2)\,\Gamma_{b+t}(s-k/2)}\succ 1$$

for every nonnegative integers a, b, and t such that $a+b+t=n$. We may assume that $k\geqq 0$, since we can exchange a and b without changing ε. (This is so even when $k\notin\mathbf{Z}$.) Fixing a set of data $\{n, a, b, t, k\}$, we shall denote by $F_k(s)$ the left-hand side of (7.15) without the factors $4s-n-2$ and h_k. To simplify our formulas, we shall also put $\sigma=s+k/2$.

Case IA. $k\in\mathbf{Z}$, $0<k\leqq n/2$. In this case we find that

$$F_k(s)\sim\frac{\Gamma_{2k+1}(\sigma)\prod_{i=k+1}^{[(n+t)/2]}\Gamma(2s-i)}{\Gamma_{a+t}(\sigma)\,\Gamma_{b+t}(\sigma-k)\,\Gamma(s-(n+t-2\varepsilon)/4)},$$

where $\Gamma(s-(n+t-2\varepsilon)/4)$ should be ignored if r is odd. Let us first assume $k<(a-b)/2$. Then $h_k=1$ and $a+t\geqq 2k+1$, and hence, assuming r to be even, by (7.11 c, d) we have

(7.16) $$\begin{aligned}&\Gamma_{a+t}(\sigma)\,\Gamma_{b+t}(\sigma-k)/\Gamma_{2k+1}(\sigma)\\ &\sim\Gamma_{a+t-2k-1}(\sigma-k-1/2)\,\Gamma_{b+t}(\sigma-k)\\ &\sim\Gamma_{2c-1}(\sigma-d-1/2)\,\Gamma_{b+t}(\sigma-k-1/2)\,\Gamma_{b+t}(\sigma-k)\\ &\sim\Gamma(\sigma-d)^{-1}\,\Gamma_{2c}(\sigma-d)\prod_{i=1}^{b+t}\Gamma(2\sigma-2k-i)\\ &\sim\Gamma(\sigma-d)^{-1}\prod_{i=1}^{c}\Gamma(2\sigma-2d+1-2i)\prod_{i=1}^{b+t}\Gamma(2\sigma-2k-i),\end{aligned}$$

where $c=(a-b-2k)/2$ and $d=(2k+b+t)/2$. Thus

$$F_k(s)\sim\frac{\Gamma(\sigma-d)\prod_{i=1}^{c}\Gamma(2\sigma-2d-i)}{\Gamma(\sigma-(2k+n+t-2\varepsilon)/4)\prod_{i=1}^{c}\Gamma(2\sigma-2d+1-2i)}.$$

Since $(2k+n+t-2\varepsilon)/4-d=(a-b-2k-2\varepsilon)/4\geqq 0$, $\in \mathbf{Z}$, we see that $F_k \succ 1$ as desired. If r is odd, (7.16) becomes

$$\Gamma_{a+t}(\sigma)\,\Gamma_{b+t}(\sigma-k)/\Gamma_{2k+1}(\sigma)\sim\Gamma_{2e}(\sigma-f)\prod_{i=1}^{b+t}\Gamma(2\sigma-2k-i)$$

with $e=(a-b-2k-1)/2$ and $f=(2k+b+t+1)/2$, and hence, applying (7.11d) to Γ_{2e}, we obtain

$$F_k(s)\sim\prod_{i=1}^{e}\Gamma(2s-k-b-t-i)/\Gamma(2s-k-b-t-2i)\succ 1$$

as expected.

Next suppose $k\geqq(a-b)/2$. We consider F_{k-1} with the same a, b, and t. Notice that the constant ε for this function is $1-\varepsilon$. Assuming r to be even, we see that $F_k(s-k/2)/F_{k-1}(s-(k-1)/2)\sim AB$ with

$$A=\frac{\Gamma_{2k+1}(s)\,\Gamma_{b+t}(s-k+1)}{\Gamma_{2k-1}(s)\,\Gamma_{b+t}(s-k)},$$

$$B=\frac{\Gamma(2s-k-(n+t)/2)\,\Gamma(s+1-(2k+n+t+2\varepsilon)/4)}{\Gamma(2s-2k)\,\Gamma(2s-2k+1)\,\Gamma(s-(2k+n+t-2\varepsilon)/4)}.$$

We easily see that $A\sim\Gamma(2s-2k)\prod_{i=1}^{b+t}(2s-2k+1-i)$ and

$$B\sim\Gamma(2s-2k)^{-1}(s-(2k+n+t)/4)^{1-\varepsilon}\prod_{i=1}^{d}(2s-2k+1-i)^{-1},$$

where $d=(n+t+2-2k)/2$. We have also $h_k(s-k/2)/h_{k-1}(s-(k-1)/2)=C$ with

$$C=\begin{cases}(2s-2k-d)(2s-2k+1-d) & \text{if } \varepsilon=0 \text{ and } k>1+r/2,\\ 2s-2k-d & \text{if } \varepsilon=0 \text{ and } k=1+r/2,\end{cases}$$

and $C=1$ otherwise. Then we can easily verify that $AB/C\succ 1$ in all cases. The same conclusion can be obtained for odd r by a similar and simpler reasoning. Thus, considering $k-1$, $k-2$, ..., we can reduce the problem $F_k/h_k\succ 1$ to $F_0\succ 1$ or to the case $k<(a-b)/2$, which has been settled.

Case IB. $k=0$. Observe that F_0 is symmetric in a and b and that the above proof in the case $k<(a-b)/2$ is valid for $k=0$. Therefore it is sufficient to prove the case $a=b$. Put $c=a+t$ in this case. Then $c=(n+t)/2$ and $F_0(s)\sim A/B$ with $A=\Gamma(s)\prod_{i=1}^{c}\Gamma(2s-i)$ and $B=\Gamma_c(s)^2\,\Gamma(s-c/2)$. By (7.11c, d) we have $A\sim\Gamma(s)\,\Gamma_c(s-1/2)\,\Gamma_c(s)\sim\Gamma_{c+1}(s)\,\Gamma_c(s)\sim B$, and hence $F_0\sim 1$ as desired.

Case IC. $n/2<k\in\mathbf{Z}$. This case can be handled by the same technique as in Case IIC below, and so we omit the details.

Case IIA. $0<k\leqq n/2$, $k\notin \mathbf{Z}$. The case $2k<a-b$ can be proved by the same technique as in Case IA. If $2k\geqq a-b$, the problem can be reduced to the case $k=1/2$ by considering the quotient of type AB/C in Case IA.

Case IIB. $k=1/2$. For the above reason we may assume that $a-b\leqq 1$. We have also $h_{1/2}=1$. If r is odd, we have

$$F_k(s)\sim\frac{\prod_{i=1}^{(n+t+1)/2}\Gamma(2\sigma-i)}{\Gamma_{a+t}(\sigma)\,\Gamma_{b+t}(\sigma-1/2)\,\Gamma(\sigma-(1+n+t-2\varepsilon)/4)}$$

with $\sigma=s+1/4$. Put $c=(b-a+1)/2$. Then $0\leqq c\in\mathbf{Z}$, and

$$\begin{aligned}\Gamma_{a+t}(\sigma)\,\Gamma_{b+t}(\sigma-1/2)\,\Gamma(\sigma-(b+t+1)/2)&\\ \sim\Gamma_{a+t}(\sigma)\,\Gamma_{b+t+1}(\sigma-1/2)&\\ \sim\Gamma_{a+t}(\sigma)\,\Gamma_{a+t}(\sigma-1/2)\,\Gamma_{2c}(\sigma-(a+t+1)/2)&\\ \sim\prod_{i=1}^{a+t}\Gamma(2\sigma-i)\prod_{i=1}^{c}\Gamma(2\sigma-a-t-2i)&\end{aligned}$$

by (7.11c, d), and hence

$$F_k(s)\sim\frac{\Gamma(\sigma-(b+t+1)/2)\prod_{i=1}^{c}\Gamma(2\sigma-a-t-i)}{\Gamma(\sigma-(1+n+t-2\varepsilon)/4)\prod_{i=1}^{c}\Gamma(2\sigma-a-t-2i)}.$$

From (7.14) we obtain $c\equiv\varepsilon \pmod 2$. Thus, if $c=0$, we have $2(b+t+1)=1+n+t-2\varepsilon$, and hence $F_k\sim 1$. If $c>0$, we compare the factors for $i>1$ in the numerator with the factors for $i<c$ in the denominator and find that

$$F_k(s)\succ\frac{\Gamma(\sigma-(b+t+1)/2)\,\Gamma(2\sigma-a-t-1)}{\Gamma(\sigma-(1+n+t-2\varepsilon)/4)\,\Gamma(2\sigma-a-t-2c)}.$$

Now it is easy to verify that the last quotient is a polynomial in σ as expected. If r is even, we have $b\geqq a$, and

$$F_k(s)\sim\frac{\prod_{i=1}^{(n+t)/2}\Gamma(2\sigma-i)}{\Gamma_{a+t}(\sigma)\,\Gamma_{b+t}(\sigma-1/2)}.$$

A computation similar to the above one for odd r shows that

$$\Gamma_{a+t}(\sigma)\,\Gamma_{b+t}(\sigma-1/2)\sim\prod_{i=1}^{a+t}\Gamma(2\sigma-i)\prod_{i=1}^{c}\Gamma(2\sigma-a-t-2i),$$

with $c=(b-a)/2$. Therefore we have

$$F_k(s)\sim\prod_{i=1}^{c}(\Gamma(2\sigma-a-t-i)/\Gamma(2\sigma-a-t-2i))\succ 1,$$

which completes the proof for $k=1/2$.

Case IIC. $n/2<k\notin\mathbf{Z}$. For r odd, (7.14) in this case implies that

$$2k+n+t-2\varepsilon\equiv 2b+2t+2 \pmod 4. \tag{7.17}$$

Let us first assume that both t and r are odd. Then

$$F_k(s)=\frac{\Gamma_n(\sigma)\prod_{i=1}^{(t+1)/2}\Gamma(2s-(n-1)/2-i)}{\Gamma_{a+t}(\sigma)\,\Gamma_{b+t}(\sigma-k)\,\Gamma(s-(n+t-2\varepsilon)/4)},$$

$$h_k(s)=\prod_{i=1}^{[t/2]+1-\varepsilon}\left(2s-\frac{n+1}{2}-i\right).$$

By (7.11 d) we have $\Gamma_n(\sigma)/\Gamma_{a+t}(\sigma)\sim\Gamma_b(\sigma-(a+t)/2)$ and

$$\begin{aligned}\Gamma_{b+t}(\sigma-k)&\sim\Gamma_{b+1}(\sigma-k)\,\Gamma_{t-1}(\sigma-k-(b+1)/2)\\&\sim\Gamma_b(\sigma-k)\,\Gamma(\sigma-k-b/2)\prod_{i=1}^{(t-1)/2}\Gamma(2s-k-b-2i),\end{aligned} \tag{7.18}$$

and therefore, putting $c=n+t-2\varepsilon$, we obtain

$$\frac{F_k(s)}{h_k(s)}\sim\frac{\Gamma_b(\sigma-(a+t)/2)(2s-(n+t+2)/2)^{\varepsilon}\prod_{i=1}^{(t+1)/2}\Gamma(2s-i-(n+1)/2)}{\Gamma_b(\sigma-k)\,\Gamma(s-(k+b)/2)\,\Gamma(s-c/4)\prod_{i=1}^{(t-1)/2}\Gamma(2s-k-b-2i)}.$$

By (7.11 b) we have $\Gamma_b(\sigma-(a+t)/2)\succ\Gamma_b(\sigma-k)$, and hence

$$\begin{aligned}\frac{F_k(s)}{h_k(s)}&\succ\frac{(2s-(n+t+2)/2)^{\varepsilon}\,\Gamma(2s-(n+3)/2)}{\Gamma(s-(k+b)/2)\,\Gamma(s-c/4)}\\&\sim\frac{\Gamma(s-(n+3)/4)\,\Gamma(s-(n+1)/4)}{\Gamma(s-(k+b)/2)\,\Gamma(s-(n+t+2\varepsilon)/4)}.\end{aligned}$$

In view of (7.17) we can easily verify that the last fractional expression is always $\succ 1$. The case in which both r and t are even can be handled in a similar and much simpler way.

Let us next assume that r is odd and t even. Using a formula similar to (7.18), we see that $(4s-n-2)^{\delta}F_k(s)/h_k(s)\sim AB$ with $A=(4s-n-2)^{\delta-1}$ and

$$B=\frac{\Gamma_b(\sigma-(a+t)/2)\,\Gamma(\sigma-(1/2)-[(2k+n-2)/4])}{\Gamma_b(\sigma-k)\,\Gamma(s-(n+t+2\varepsilon)/4)}.$$

If b is even, then a is odd, and both Γ_b can be eliminated. By (7.17) we have $2k+n+t-2\varepsilon\equiv 2 \pmod 4$. There are four combinations of (ε, δ); in each case we can easily verify that $AB \succ 1$. Next suppose b is odd; then a is even. In this case we express B in the form

$$\frac{\Gamma_{b-1}(\sigma-(a+t)/2)\,\Gamma(\sigma-(n-1)/2)\,\Gamma(\sigma-(1/2)-[(2k+n-2)/4])}{\Gamma_{b-1}(\sigma-k)\,\Gamma(\sigma-k-(b-1)/2)\,\Gamma(s-(n+t+2\varepsilon)/4)}.$$

We can eliminate both Γ_{b-1} and observe that $2k+n+t-2\varepsilon\in 4\mathbf{Z}$. Then the desired conclusion can easily be verified. The remaining case in which r is even and t odd can be proved in a similar and simpler way.

Let us finally treat the case $\mathfrak{c}=\mathfrak{g}$. Since $\mathfrak{c}\subset 4\mathfrak{g}$ if k is half-integral, we consider only integral k here. Let $\Gamma=\Gamma[\mathfrak{b}^{-1}, \mathfrak{b}]$ and $\Gamma'=\Gamma[\mathfrak{b}^{-1}, \mathfrak{b}\mathfrak{p}]$ with a prime ideal $\mathfrak{p}$. In [S6, Prop. 2.4(ii)], we showed that $E(z, s; k, \chi, \Gamma)$ is a linear combination of $E(z, s; k, \chi, \Gamma')\|_k \alpha$ with $\alpha\in\Gamma'\backslash\Gamma$. (In [S6] it was assumed that χ is of finite order and $k\in\mathbf{Z}u$, but the proof there is valid in the present case.) Therefore we can at least claim that $\mathscr{G}(s)\,\Lambda_{\mathfrak{p}}(s)\,E(z, s; k, \chi, \Gamma)$ has poles only at the points described in our theorem for $\mathfrak{c}\neq\mathfrak{g}$. Now

$$\Lambda_{\mathfrak{p}}(s)/\Lambda_{\mathfrak{g}}(s)=[1-\chi(\mathfrak{p})\,N(\mathfrak{p})^{-2s}]\prod_{t=1}^{[n/2]}[1-\chi(\mathfrak{p})^2\,N(\mathfrak{p})^{2t-4s}].$$

Thus, it is sufficient to show that the functions $\Lambda_{\mathfrak{p}}/\Lambda_{\mathfrak{g}}$ for all $\mathfrak{p}$ have no common zero outside the set given in (iii) of our theorem. Let s be a common zero of the $\Lambda_{\mathfrak{p}}/\Lambda_{\mathfrak{g}}$ for all $\mathfrak{p}$. Suppose $\chi(\mathfrak{p})^2 N(\mathfrak{p})^{2t-4s}=1$ for some $\mathfrak{p}$ and some t. Then $\mathrm{Re}(2t-4s)=0$, and hence $\chi(\mathfrak{p})^2 N(\mathfrak{p})^{2t-4s}=1$ for all $\mathfrak{p}$ with the same t. Consequently $\chi(\mathfrak{a})^2=N(\mathfrak{a})^{4s-2t}$ for every ideal $\mathfrak{a}$. This implies that $|\alpha|^{2i\kappa}=|\alpha|^{(2t-4s)u}$ for every $\alpha\in F^\times$ with κ of (7.1), and hence $4s-2t=0$ since $\|\kappa\|=0$. Thus $\chi^2=1$ and $s=t/2$, $1\leq t\leq[n/2]$. If $\chi(\mathfrak{p})\,N(\mathfrak{p})^{-2s}=1$ at the beginning of this reasoning, we arrive at the conclusion that $\chi=1$ and $s=0$. This completes our proof.

8. Analytic continuation of Euler products and the series $D(s; f, g; \chi)$

Given f in $\mathscr{M}_k(\mathfrak{b}, \mathfrak{c}, \varphi)$ and g in $\mathfrak{M}_l(\mathfrak{b}', \mathfrak{c}', \varphi')$ with integral or half-integral k and l, we take a Hecke character χ of F such that

(8.1 a) $$\chi(a)=(\varphi/\varphi')(a_{\mathfrak{c}\mathfrak{c}'})\quad\text{for every } a\in\prod_{v\in\mathbf{h}}\mathfrak{g}_v^\times,$$

(8.1 b) $$\chi_{\mathbf{a}}(x)=\mathrm{sgn}(x)^{[k]-[l]}|x_{\mathbf{a}}|^{i\kappa},\quad \kappa\in\mathbf{R}^{\mathbf{a}},\quad \|\kappa\|=0.$$

As noted in Sect. 1, such χ and κ can always be found for given φ and φ'. Let $S^+=S_+\cap GL_n(F)$. For $(\sigma, q)\in S^+\times GL_n(F)_{\mathbf{A}}$ we put

(8.2) $$\iota_{\sigma,q}=[U_{\sigma,q}:\{\pm 1\}]^{-1},\quad U_{\sigma,q}=\{a\in U_q\,|\,{}^t a\sigma a=\sigma\}$$

with U_q of (5.15). The number $\iota_{\sigma,q}$ is always 1 if $n=1$, but it can be nontrivial in the higher-dimensional case. We then put

$$(8.3)\quad D(s;f,g;\chi)=\sum_{\sigma,q}\chi_{\mathbf{h}}(\det(q))\,\iota_{\sigma,q}\,\mu_f(\sigma,q)\,\overline{\mu_g(\sigma,q)}\,\det(\sigma)^{-\lambda}|\det(\sigma q^2)_{\mathbf{h}}|_{\mathbf{A}}^{s},$$

where $s\in\mathbf{C}$, $\lambda=(k+l+i\kappa)/2$, and (σ,q) runs over $S^+\times \mathrm{GL}_n(F)_{\mathbf{A}}$ modulo the equivalence relation defined as follows: (σ,q) and (σ',q') are equivalent if $\sigma'={}^t b\sigma b$ and $q'=b^{-1}qa$ with some $b\in\mathrm{GL}_n(F)$ and $a\in E^*$. We easily see that the sum is formally well defined. The weights k and l should be nonzero, since our series becomes trivially 0 otherwise.

Taking a subset Q of $\mathrm{GL}_n(F)_{\mathbf{h}}$ as in (5.14), we easily see that

$$(8.4)\quad D(s;f,g;\chi)=\sum_{q\in Q}\chi_{\mathbf{h}}(\det(q))|\det(q)|_{\mathbf{A}}^{2s}\cdot\sum_{\sigma\in S^+/U_q}\iota_{\sigma,q}\,\mu_f(\sigma,q)\,\overline{\mu_g(\sigma,q)}\,\det(\sigma)^{-\lambda-su},$$

where S^+/U_q is a complete set of representatives for S^+ modulo the equivalence relation $\sim$ in S^+ defined by: $\sigma\sim\sigma'$ if and only if $\sigma'={}^t a\sigma a$ with $a\in U_q$. Since the $\mu_f(\sigma,q)$ and $\mu_g(\sigma,q)$ are Fourier coefficients of some elements of $\mathcal{M}_k$ and $\mathcal{M}_l$ as shown in (1.22), we can easily see that our series is convergent for sufficiently large $\mathrm{Re}(s)$.

To obtain an integral expression for our series, put $\mathfrak{e}=\mathfrak{b}+\mathfrak{b}'$ and $\mathfrak{h}=\mathfrak{e}^{-1}(\mathfrak{b}\mathfrak{c}\cap\mathfrak{b}'\mathfrak{c}')$. Then $D[\mathfrak{e}^{-1},\ \mathfrak{e}\mathfrak{h}]=D[\mathfrak{b}^{-1},\ \mathfrak{b}\mathfrak{c}]\cap D[\mathfrak{b}'^{-1},\mathfrak{b}'\mathfrak{c}']$. Define a $G_{\mathbf{a}}$-invariant measure $d_H z$ on $H^{\mathbf{a}}$ by

$$d_H z=\Delta(z)^{-(n+1)u}\bigwedge_{v\in\mathbf{a}}\bigwedge_{p\leq q}(dx_{pq}^v\wedge dy_{pq}^v),$$

where $z_v=(x_{pq}^v+iy_{pq}^v)$. Now put $\Gamma=\Gamma[\mathfrak{e}^{-1},\mathfrak{e}\mathfrak{h}]$ and $\Phi=\Gamma\backslash H^{\mathbf{a}}$. Assuming that fg is a cusp form and $\mathrm{Re}(s)$ is sufficiently large, we have

$$(8.5)\quad \int_{\Phi} f(z)\,\overline{g(z)\,E(z,\bar{s}+(n+1)/2;k-l,\varepsilon\bar{\chi},\Gamma)}\,\Delta(z)^k d_H z$$
$$=[D_F^{1/2}N(\mathfrak{e})^{-1}]^{n(n+1)/2}(4\pi)^{-n\|su+\lambda\|}\prod_{v\in\mathbf{a}}\Gamma_n(s+\lambda_v)\,D(s;f,g;\chi),$$

where Γ_n is as in (7.8) and ε is the identity character except when $k\in\mathbf{Z}^{\mathbf{a}}$ and $l\notin\mathbf{Z}^{\mathbf{a}}$, in which case ε is the Hecke character corresponding to $F(\sqrt{-1})/F$. If $n=1$, this was given in [S10]. The general case can be proved by making obvious modifications in the argument of [S10, pp. 582–584]. Combining (8.5) with Theorem 7.3, we see that our series of (8.3) times suitable gamma factors and a function of type (7.10) can be continued to a meromorphic function on the whole s-plane with only finitely many poles, and each pole is simple.

Let us now consider the special case in which $f\in\mathcal{S}_k(\mathfrak{b},\mathfrak{c},\psi)$ with $k\in\mathbf{Z}^{\mathbf{a}}$ as in Sect. 5. With a Hecke character η of F and $\tau\in S^+$ we define a Dirichlet series $D(s;\tau,L;f,\eta)$ for $L=pL_0\in\mathcal{L}_\tau$, $p\in\mathrm{GL}_n(F)_{\mathbf{h}}$, by

$$(8.6)\quad D(s;\tau,L;f,\eta)=\sum_{x\in B/E}\psi(\det(px))\,\mu_f(\tau,px)\,\eta^*(\det(x)\,\mathfrak{g})|\det(x)|_{\mathbf{A}}^{s-n-1}.$$

This is obtained from $D'(\tau, L; f)$ of (5.5) by substituting $\eta^*(\mathfrak{a})N(\mathfrak{a})^{-s}$ for $[\mathfrak{a}]$. Take $\mu \in \mathbf{Z}^{\mathbf{a}}$ and $\kappa \in \mathbf{R}^{\mathbf{a}}$ so that $0 \leqq \mu \leqq u$, $\|\kappa\| = 0$, and

$$(\psi\eta)_{\mathbf{a}}(x) = \operatorname{sgn}(x)^{k+\mu}|x_{\mathbf{a}}|^{i\kappa}. \tag{8.7}$$

Define $\theta_{\omega,p}$ as in Sect. 6 with the present τ, p, and with $\omega = \eta^{-1}$. Let ρ_τ be as in Proposition 6.2 and let $l = \mu + nu/2$. We see that $(\psi\eta\rho_\tau)_{\mathbf{a}}(x) = \operatorname{sgn}(x)^{k-[l]}|x_{\mathbf{a}}|^{i\kappa}$, and $\theta_{\omega,p} \in \mathcal{M}_l(\mathfrak{b}', \mathfrak{c}', \rho_\tau/\eta)$ with some $\mathfrak{b}'$ and $\mathfrak{c}'$ by Proposition 6.2, and hence $D(s; f, \theta_{\omega,p}; \psi\eta\rho_\tau)$ is meaningful. By (1.22), (5.16), (6.10), and (8.4) we can easily verify that

$$\begin{aligned} 2 \cdot D(2s + (3n/2) + 1; \tau, L; f, \eta) \\ = |\det(p)|_{\mathbf{A}}^{-2s-n/2} \det(\tau)^{\lambda + su} D(s; f, \theta_{\omega,p}; \psi\eta\rho_\tau), \end{aligned} \tag{8.8}$$

where $\lambda = (k + \mu + (nu/2) + i\kappa)/2$. Now our second main result of this paper is

Theorem 8.1 *Define a matrix representation $T_\psi(\mathfrak{a}) \to X(\mathfrak{a})$ of Hecke operators on $\mathcal{S}_k(\mathfrak{b}, \mathfrak{c}, \psi)$ by (5.11); for a Hecke character η of F put*

$$\mathcal{Z}(s, \eta) = \sum_{\mathfrak{a}} \eta^*(\mathfrak{a})\,\psi^*\Big(\prod_{v \nmid \mathfrak{c}} \mathfrak{a}_v\Big)\, X(\mathfrak{a})\, N(\mathfrak{a})^{-s}, \tag{8.9}$$

where $\mathfrak{a}$ runs over all the integral ideals in F. Then $\mathcal{Z}$ can be continued to a meromorphic function on the whole s-plane. Moreover, for an element f of $\mathcal{S}_k(\mathfrak{b}, \mathfrak{c}, \psi)$ that is an eigenfunction of $T_\psi(\mathfrak{a})$ for every $\mathfrak{a}$, put

$$Z(s, f, \eta) = \prod_{v \in \mathbf{h}} Z_v\big((\psi/\psi_{\mathfrak{c}})(\pi_v)\,\eta^*(\pi_v\,\mathfrak{g})|\pi_v|^s\big), \tag{8.10}$$

with $Z_v(t)$ defined for f as in Sect. 5, where π_v is a prime element of F_v. Then Z can be continued to a meromorphic function on the whole s-plane.

Remark. The factor $\psi^*(\prod_{v \nmid \mathfrak{c}} \mathfrak{a}_v)$ in (8.9) or $(\psi/\psi_{\mathfrak{c}})(\pi_v)$ in (8.10) is natural and essential. To see this, let us assume for simplicity that the class number of F is 1 and $\psi_{\mathbf{a}}(x) = \operatorname{sgn}(x)^k$. Then

$$\begin{aligned} D(s; \tau, L; f, \eta) = |\det(p)|_{\mathbf{A}}^{n+1-s} \\ \cdot \sum_{\xi \in B_{p,1}/U_1} \mu_f({}^t\xi\tau\xi, 1)\,\eta^*(\det(p^{-1}\xi)\,\mathfrak{g})|\det(\xi)|^{(n+1-s)u-k}. \end{aligned}$$

If $n = 1$, $0 < \tau \in \mathbf{Z}$, and $f(z) = \sum_{m=1}^{\infty} \mu(m)\,\mathbf{e}(mz)$, then the above series for $p = 1$ is $\sum_{m=1}^{\infty} \mu(\tau m^2)\,\eta'(m)\,m^{2-k-s}$ with a Dirichlet character η', and $Z(s, f, \eta)$ in this case is the Euler product of degree 3 investigated in [S4] and [S10].

Proof. By (5.12) and (5.13) the analytic nature of $\mathcal{Z}(s, \eta)$ can be reduced to that of the series obtained from $A(\tau_h, M)$ and $D(\tau_h, M; f_i)$ by substituting $\psi^*(\prod_{v \nmid \mathfrak{c}} \mathfrak{a}_v)\,\eta^*(\mathfrak{a})\,N(\mathfrak{a})^{-s}$ for $[\mathfrak{a}]$. The latter series produces $D(s; \tau_h, M; f_i, \eta)$, which is meromorphic on the whole $\mathbf{C}$ as observed above. As for $A(\tau_h, M)$, (5.8) shows that it is essentially a product of L-functions, and hence we obtain the meromorphy of $\mathscr{Z}(s,\eta)$ on $\mathbf{C}$. If f is an eigenform, we take τ and p so that $L=pL_0$ satisfies conditions (i) and (ii) of Corollary 5.3. (Such τ and p exist, since f is a cusp form.) Then $\mu_f(\tau,p)\neq 0$ and the desired meromorphy of $Z(s,f,\eta)$ follows from (5.9b).

The last point can be made more explicit. Namely, combining (8.5) and (8.8) with (5.9b), for τ and $L=pL_0$ as above we obtain

$$(8.11)\qquad Z(s,f,\eta)\,\mathscr{G}_{k-l,\kappa}((2s-n)/4)\prod_{v\in\mathbf{a}}\Gamma_n(2^{-1}(s-n-1+k_v+\mu_v+i\kappa_v))$$
$$=[2\psi_c(\det(p))\,\mu_f(\tau,p)]^{-1}[D_F^{-1/2}N(\mathfrak{e})]^{n(n+1)/2}$$
$$\cdot(4\pi)^{n\|s'u+\lambda\|}\det(\tau)^{s'u+\lambda}|\det(p)|_{\mathbf{A}}^{n+1-s}$$
$$\cdot\prod_{v\in\mathbf{b}}g_v((\psi/\psi_c)(\pi_v)\,\eta^*(\pi_v\mathfrak{g})|\pi_v|^s)(\Lambda_c/\Lambda_{\mathfrak{h}})((2s-n)/4)$$
$$\cdot\int_{\Phi}f(z)\,\overline{\theta_{\omega,p}(z)}\,\mathscr{E}(z,(2s-n)/4)\,\Delta(z)^k d_H z,$$

where $s'=(2s-3n-2)/4$, $\mathscr{G}_{k-l,\kappa}$ is as in Theorem 7.3,

$$(8.12\text{a})\qquad \Lambda_{\mathfrak{a}}(s)=\begin{cases}L_{\mathfrak{a}}(2s,\rho_\tau\psi\eta)\prod_{i=1}^{n/2}L_{\mathfrak{a}}(4s-2i,\psi^2\eta^2) & \text{if } n \text{ is even,}\\ \prod_{i=1}^{(n+1)/2}L_{\mathfrak{a}}(4s-2i+1,\psi^2\eta^2) & \text{if } n \text{ is odd,}\end{cases}$$

$$(8.12\text{b})\qquad \mathscr{E}(z,s)=\mathscr{G}_{k-l,\kappa}(s)\,\Lambda_{\mathfrak{h}}(s)\,\overline{E(z,\bar{s};k-l,\varepsilon\rho_\tau\overline{\psi\eta},\Gamma)};$$

$\mathbf{b}$ and g_v are determined by p and τ as in Corollary 5.2 (see also (5.7) and (5.8)); $\mathfrak{e}$, $\mathfrak{h}$, and $\Gamma=\Gamma[\mathfrak{e}^{-1},\mathfrak{e}\mathfrak{h}]$ are determined for $(\mathfrak{b},\mathfrak{c},\mathfrak{b}',\mathfrak{c}')$ as before; ε is the Hecke character corresponding to $F((-1)^{n/2})/F$.

Now formula (8.11) tells about possible poles of $Z(s,f,\eta)$, or rather, the left-hand side of (8.11). Clearly the last integral has poles only at the poles of $\mathscr{E}$, which are described in Theorem 7.3. Another type of poles may be produced by the factor $(\Lambda_c/\Lambda_{\mathfrak{h}})((2s-n)/4)$, which is a finite product over the primes v such that $v\nmid\mathfrak{c}$ and $v|\mathfrak{h}$. Probably they are not poles of (8.11) except possibly for those on the real line.

To obtain a condition under which this type of poles do not occur, let us impose the following condition on $\mathfrak{b}$:

$$(8.13)\qquad \mathfrak{d}_v\subset\mathfrak{b}_v\quad\text{if } n \text{ is odd and } v\nmid 2.$$

We lose practically nothing by this assumption since we can change $\mathfrak{b}$ for an ideal satisfying the condition by taking a suitable inner automorphism of G. Now let $\mathfrak{e}_\tau$, $\mathfrak{f}$, and $\mathfrak{f}'$ denote the conductors of ρ_τ, η, and η^2, respectively.

Lemma 8.2 *The notation being as in Corollary 5.2, let* $\mathbf{d}$ *be the set of primes* v *such that* $(\theta\cdot{}^t p\tau p)_v$ *is not regular. Then* $\theta_{\omega,p}\in\mathscr{M}_l(\mathfrak{b}',\mathfrak{c}',\rho_\tau/\eta)$ *with ideals* $\mathfrak{b}'$, $\mathfrak{c}'$ *with the following properties:* $\mathfrak{c}'$ *is divisible only by the members of* $\mathbf{d}$ *and the prime factors of* $2\mathfrak{f}\cap\mathfrak{b}^{-1}\mathfrak{d}$, *where* $2\mathfrak{f}\cap\mathfrak{b}^{-1}\mathfrak{d}$ *can be replaced by* $\mathfrak{f}$ *if* n *is even;* $\mathfrak{b}'_v=\mathfrak{b}_v$ *if* n *is even or* $v\nmid 2$.

Proof. Define $\mathfrak{x}$ and $\mathfrak{t}$ as in Proposition 6.2. By (5.4) we can take $\mathfrak{x}=2\mathfrak{d}^{-1}\mathfrak{b}$. By Proposition 6.2 we have $\theta_{\omega,p}\in\mathcal{M}_l(\mathfrak{b}',\mathfrak{c}',\rho_\tau/\eta)$ with $(\mathfrak{b}',\mathfrak{c}')=(\mathfrak{b},\mathfrak{e}_\tau\cap\mathfrak{f}\cap\mathfrak{x}^{-1}\mathfrak{f}^2\mathfrak{t})$ if n is even and $(\mathfrak{b}',\mathfrak{c}')=(\mathfrak{b}+2^{-1}\mathfrak{d},\mathfrak{e}_\tau\cap\mathfrak{f}\cap 4\mathfrak{g}\cap 4\mathfrak{x}^{-1}\cap\mathfrak{f}^2\mathfrak{t}\cap\mathfrak{x}^{-1}\mathfrak{f}^2\mathfrak{t})$ if n is odd. Observe that if α is a regular element of T_v, then so is $(4\alpha)^{-1}$, except when $v|2$ and n is odd. Therefore, assuming that $v\not|2$ or n is even, we can take $\mathfrak{t}_v=\mathfrak{x}_v$ if $v\notin\mathbf{d}$, and clearly $v\not|\mathfrak{e}_\tau$ if $v\notin\mathbf{d}$. Therefore we obtain our assertions.

From this lemma we can immediately derive

Proposition 8.3 (i) *n even: If $(\theta\cdot{}^t p\tau p)_v$ is regular and $v\not|\mathfrak{f}$ for every $v\not|\mathfrak{c}\mathfrak{f}'$, then* $\Lambda_\mathfrak{c}=\Lambda_\mathfrak{h}$.

(ii) *n odd: If $(\theta\cdot{}^t p\tau p)_v$ is regular and $v\not|2\mathfrak{f}\cap\mathfrak{b}^{-1}\mathfrak{d}$ for every $v\not|\mathfrak{c}\mathfrak{f}'$, then $\Lambda_\mathfrak{c}=\Lambda_\mathfrak{h}$.*

If τ and p can be chosen so that $\Lambda_\mathfrak{c}=\Lambda_\mathfrak{h}$, then, as can be seen from Theorem 7.3, the product of (8.11) has only finitely many poles and each pole is simple; it is an entire function at least in the following two cases: (i) $\psi^2\eta^2\neq 1$; (ii) $\psi^2\eta^2=1$, $\mathfrak{c}\neq\mathfrak{g}$, and $0<k_v-\mu_v-n\notin 2\mathbf{Z}$ for at least one $v\in\mathbf{a}$.

References

[A] Andrianov, A.N.: The multiplicative arithmetic of Siegel modular forms. Usp. Mat. Nauk, **34**, 67–135 (1979); English transl. Russian Math. Surv. **34**, 75–148 (1979)

[AK] Andrianov, A.N., Kalinin, V.L.: On the analytic properties of standard zeta functions of Siegel modular forms. Mat. Sb. **106**(148), 323–339 (1978); English transl. Math. USSR Sb. **35**, 1–17 (1979)

[B1] Böcherer, S.: Über die Funktionalgleichung automorpher *L*-Funktionen zur Siegelschen Modulgruppe. J. R. Ang. Math. **362**, 146–168 (1985)

[B2] Böcherer, S.: Ein Rationalitätssatz für formale Heckereihen zur Siegelschen Modulgruppe. Abh. Math. Sem. Univ. Hamburg **56**, 35–47 (1986)

[F1] Feit, P.: Poles and residues of Eisenstein series for symplectic and unitary groups. Memoirs. Am. Math. Soc. **61**, No. 346 (1986)

[F2] Feit, P.: Explicit formulas for local factors in the Euler products for Eisenstein series. Nagoya Math. J. **113**, 37–87 (1989)

[F3] Feit, P.: Locating the poles of Eisenstein series of level 1 on SL_n, Sp_n and $SU(n,n;*)$. Preprint 1987

[K] Kitaoka, Y.: Dirichlet series in the theory of Siegel modular forms. Nagoya Math. J. **95**, 73–84 (1984)

[PR1] Piatetski-Shapiro, I., Rallis, S.: *L*-functions for classical groups. Lecture notes, Institute for Advanced Study, 1984

[PR2] Piatetski-Shapiro, I., Rallis, S.: ε-factors of representations of classical groups. Proc. Nat. Acad. Sci. **83**, 4589–4593 (1986)

[PR3] Piatetski-Shapiro, I., Rallis, S.: A new way to get Euler products. J. R. Ang. Math. **392**, 110–124 (1988)

[S1] Shimura, G.: On the zeta functions of the algebraic curves uniformized by certain automorphic functions. J. Math. Soc. Japan **13**, 275–331 (1961)

[S2] Shimura, G.: On modular correspondences for $Sp(n,\mathbf{Z})$ and their congruence relations. Proc. Nat. Acad. Sci. **49**, 824–828 (1963)

[S3] Shimura, G.: Introduction to the arithmetic theory of automorphic functions. Publ. Math. Soc. Japan, No. 11, Iwanami Shoten and Princeton Univ. Press, 1971

[S4] Shimura, G.: On the holomorphy of certain Dirichlet series. Proc. London Math. Soc. 3rd ser. **31**, 79–98 (1975)

[S5] Shimura, G.: Confluent hypergeometric functions on tube domains. Math. Ann. **260**, 269–302 (1982)

[S6] Shimura, G.: On Eisenstein series. Duke Math. J. **50**, 417–476 (1983)

[S7] Shimura, G.: On Eisenstein series of half-integral weight. Duke Math. J. **52**, 281–314 (1985)

[S8] Shimura, G.: On the Eisenstein series of Hilbert modular groups. Revista Mat. Iberoamericana **1**, 1–42 (1985)

[S9] Shimura, G.: On Hilbert modular forms of half-integral weight. Duke Math. J. **55**, 765–838 (1987)

[S10] Shimura, G.: The critical values of certain Dirichlet series attached to Hilbert modular forms. Duke Math. J. **63**, 557–613 (1991)

[S11] Shimura, G.: On the transformation formulas of theta series. Am. J. M. (to appear)

[Si] Siegel, C.L.: Gesammelte Abhandlungen I–IV. Berlin Heidelberg New York: Springer (1966/1979)

[W] Weil, A.: Sur certain groupes d'opérateurs unitaires. Acta Math. **111**, 143–211 (1964)

Differential operators, holomorphic projection, and singluar forms

Duke Mathematical Journal, 76 (1994), 141-173

Introduction. In this paper we present two types of results as applications of the theory of differential operators on hermitian symmetric spaces which we developed in our previous papers. The first subject concerns a projection map which associates a holomorphic modular form h to a nearly holomorphic form f so that $\langle \varphi, f \rangle = \langle \varphi, h \rangle$ holds for every holomorphic cusp form φ, where $\langle\ ,\ \rangle$ is an inner product. In our previous investigations, [S6] for example, we employed the map $f \mapsto h$ in the Hilbert modular case as an indispensable tool for the algebraicity of the critical values of certain zeta functions. In the present paper we extend it to the case of an arbitrary classical group G acting on a hermitian symmetric space $\mathscr{H}$ of noncompact type with future applications to the algebraicity problems for such a G in view.

The second topic is a generalization of the following fact:

(1) *A Siegel modular form is singular in the sense that it is annihilated by certain differential operators if and only if it is of singular weight.*

This was proved by Resnikoff, Maass, and Freitag for scalar-valued forms and later by Howe for vector-valued forms with L^2-integrability. See [R1], [R2], [M], [F1], and [H]. All these concern the forms on $Sp(n, \mathbf{Q})$ except that Resnikoff treated some forms of a certain special type on tube domains. Also [F2] may be mentioned as a recent article for $Sp(n, \mathbf{Q})$ which lists practically all relevant papers. In the present paper we give a uniform treatment applicable to both tube and nontube classical domains and even to the cocompact case.

The main idea behind these two types of results is the relations

(2) $$\langle D_\rho^Z f, g \rangle = (-1)^p \langle f, \theta E^Z g \rangle, \qquad \langle E^Z f, h \rangle = (-1)^p \langle f, \theta D_{\rho\otimes\sigma_Z}^Z h \rangle.$$

Here Z is an irreducible subspace of the pth symmetric product of the holomorphic tangent space of $\mathscr{H}$, ρ is a representation of the maximal compact subgroup of G, D_ρ^Z, $D_{\rho\otimes\sigma_Z}^Z$, and E^Z are certain differential operators on $\mathscr{H}$ depending on Z and ρ, and θ is a contraction operator. In the simplest case in which $\mathscr{H}$ is the upper half plane and $p = 1$, the operators are $y^{-k}(\partial/\partial z)y^k$ and $y^2\partial/\partial\bar{z}$, where $y = \mathrm{Im}(z)$. It may be noted that both $\theta E^Z D_\rho^Z$ and $\theta D_{\rho\otimes\sigma_Z}^Z E^Z$ are essentially self-adjoint G-covariant operators, which generalize the Casimir operator.

Received 13 December 1993. Revision received 16 March 1994.

The above relations were obtained in [S3] and [S9], but the convergence condition was not given in the most desirable form. In the first two sections we make a survey of the basic properties of these operators more systematically than in our previous papers, which is actually a secondary aim of this paper. We then state a sufficient convergence condition for (2) so that the formulas can be employed in the proof of our main theorems (Theorem 2.1, (2.23)). We establish the map $f \mapsto h$ for (essentially) scalar-valued f in Section 3 by expressing h in the form $h = P(L_1, \ldots, L_m)f$ with a polynomial P in several generalized Laplacians $L_1, \ldots, L_m$ of the form $\theta D^Z_{\rho\otimes\sigma_Z} E^Z$ with simplest Z's (Theorem 3.3). In Section 4 we give explicit forms of D^Z_ρ and E^Z for certain "basic" Z, which may be of independent interest (Theorem 4.7). We then prove generalizations of (1) in Section 5. We first show that the forms of singular weight are annihilated by certain differential operators (Theorem 5.2). This is applicable to almost any G and $\mathscr{H}$ of classical type, even to the cocompact case. However, we prove the converse only when $\mathscr{H}$ is a tube and G has sufficiently many translations so that an automorphic form has a Fourier expansion (Theorem 5.5). As far as scalar-valued forms for such G and $\mathscr{H}$ are concerned, our results of type (1) may be called complete (Theorem 5.6, Corollary 5.7), but we do not try to give similarly complete results for vector-valued forms; in fact, for such forms, we content ourselves only with presenting the basic principles, on which future researchers will be able to make further elaboration according to their purposes.

Notation. For C^∞ manifolds V and W we denote by $C^\infty(V, W)$ the set of all C^∞ maps of V into W. For an associative ring R with identity element we denote by $R^\times$ the group of all its invertible elements and by R^m_n the set of all $m \times n$-matrices with entries in R; we put $R^m = R^m_1$ for simplicity. We denote by 1_n the identity element of R^n_n and put

$$I_{n,m} = \begin{pmatrix} 1_n & 0 \\ 0 & -1_m \end{pmatrix}, \qquad J_n = \begin{pmatrix} 0 & -1_n \\ 1_n & 0 \end{pmatrix}.$$

For $\alpha \in \mathbf{C}^m_n$ we put $\alpha^* = {}^t\bar{\alpha}$ and denote by $\det_i(\alpha)$ the determinant of the upper left i^2 entries of α. If ${}^t\alpha = -\alpha$, we denote by $\mathrm{Pf}_i(\alpha)$ the Pfaffian of the upper left $(2i)^2$ entries of α. For two complex hermitian matrices x and y of the same size we write $x > y$ if $x - y$ is positive definite.

1. Groups, domains, and basic functions. We are going to treat four classical types of irreducible hermitian symmetric spaces H of noncompact type, called Types A, B, C, and D. For each type of H we have a semisimple Lie group G, its maximal compact subgroup K such that G/K can be identified with H, and a complex vector space T which can be identified with the holomorphic (or non-holomorphic) tangent space of H, on which the complexification K^c of K acts. The domain H has a natural Kähler structure, and so there is a basic nearly holomorphic map $r: H \to T$, whose significance will be explained in Section 3.

There are also positive definite hermitian matrices $\xi(z)$ and $\eta(z)$ defined for $z \in H$, which are closely connected with r, the Kähler structure of H, and the canonical factor of automorphy for the elements of G. Let us now give explicit forms of these objects (cf. [S4, p. 466], [S9, § 5], [S5, pp. 357, 367]).

Type A. $G = SU(n, m) = \{\alpha \in SL_{n+m}(\mathbf{C}) | \alpha^* I_{n,m} \alpha = I_{n,m}\}$,

$$K^c = \{(a, b) \in GL_n(\mathbf{C}) \times GL_m(\mathbf{C}) | \det(a) = \det(b)\},$$

$$T = \mathbf{C}_m^n, \qquad H = \{z \in T | zz^* < 1_n\},$$

$$\xi(z) = 1_n - \bar{z} \cdot {}^t z, \qquad \eta(z) = 1_m - z^* z, \qquad r(z) = -\xi(z)^{-1} \bar{z} = -\bar{z} \cdot {}^t \eta(z)^{-1}.$$

Type B. $G =$ the identity component of $\tilde{G}$,

$$\tilde{G} = \{\alpha \in SU(n, 2) | {}^t \alpha R \alpha = R\}, \qquad R = \operatorname{diag}\left[1_n, -\begin{pmatrix} 0 & 1 \\ 1 & 0 \end{pmatrix}\right] \qquad (n > 2),$$

$$K^c = SO(n, \mathbf{C}) \times GL_1(\mathbf{C}) = \{(a, b) \in SL_n(\mathbf{C}) \times GL_1(\mathbf{C}) | {}^t aa = 1_n\},$$

$$T = \mathbf{C}^n, \qquad H = \{z \in T | z^* z < 1 + |{}^t zz/2|^2 < 2\},$$

$$\xi(z) = 1_n - \bar{z} \cdot {}^t z + \eta(z)^{-1} (w\bar{z} - z) \cdot {}^t (\bar{w} z - \bar{z}), \qquad w = {}^t zz/2,$$

$$\eta(z) = 1 + |w|^2 - z^* z, \qquad r(z) = \eta(z)^{-1} (\bar{w} z - \bar{z}).$$

Type C. $G = Sp(n, \mathbf{C}) \cap SU(n, n) = \{\alpha \in SU(n, n) | {}^t \alpha J_n \alpha = J_n\}$,

$$K^c = GL_n(\mathbf{C}), \qquad T = \{z \in \mathbf{C}_n^n | {}^t z = z\}, \qquad H = \{z \in T | z^* z < 1_n\},$$

$$\xi(z) = \eta(z) = 1_n - z^* z, \qquad r(z) = -\xi(z)^{-1} \bar{z}.$$

Type D. $G = \{\alpha \in SU(n, n) | \alpha J_n = J_n \bar{\alpha}\} \qquad (n > 1)$,

$$K^c = GL_n(\mathbf{C}), \qquad T = \{z \in \mathbf{C}_n^n | {}^t z = -z\}, \qquad H = \{z \in T | z^* z < 1_n\},$$

$$\xi(z) = \eta(z) = 1_n - z^* z, \qquad r(z) = -\xi(z)^{-1} \bar{z}.$$

Here H for each type is given as a bounded domain. For simplicity we assume $n > 2$ for Type B, though the cases $n \leqslant 2$ can be included in our treatment with easy modifications. The group G is simple except when it is of Type D and $n = 2$.

Now H is isomorphic to a tube domain if and only if

(1.1) *$n = m$ for Type A, $n \in 2\mathbf{Z}$ for Type D, and n is arbitrary for Types B and C.*

We present here the explicit forms of the objects associated to H of tube form under (1.1), and also that of a real vector subspace U of T such that $T = U \otimes_{\mathbf{R}} \mathbf{C}$ and that $H = U + iP$ with a domain of positivity P in U (cf. [S4, § 3]).

Type A. $G = \{\alpha \in SL_{2n}(\mathbf{C}) | \alpha^* J_n \alpha = J_n\}$,

$$K^c = \{(a, b) \in GL_n(\mathbf{C}) \times GL_n(\mathbf{C}) | \det(a) = \det(b)\},$$

$$T = \mathbf{C}_n^n, \qquad U = \{x \in \mathbf{C}_n^n | x^* = x\}, \qquad H = \{z \in T | i(z^* - z) > 0\},$$

$${}^t\xi(z) = \eta(z) = i(z^* - z), \qquad r(z) = ({}^tz - \bar{z})^{-1}.$$

Type B. $G =$ the identity component of $\tilde{G}$,

$$\tilde{G} = \{\alpha \in SL_{n+2}(\mathbf{R}) | {}^t\alpha \tilde{Q} \alpha = \tilde{Q}\}, \qquad \tilde{Q} = \operatorname{diag}\left[Q, -\begin{pmatrix} 0 & 1 \\ 1 & 0 \end{pmatrix}\right] \qquad (n > 2),$$

$$K^c = \{(a, b) \in SL_n(\mathbf{C}) \times GL_1(\mathbf{C}) | {}^taQa = 1_n\}, \qquad T = \mathbf{C}^n,$$

$$U = \mathbf{R}^n, \qquad H = \{z \in T | {}^t(z - \bar{z})Q(z - \bar{z}) > 0, i \cdot {}^t(z - \bar{z})Qe > 0\},$$

$$\xi(z) = Q - \eta(z)^{-1}(z - \bar{z}) \cdot {}^t(z - \bar{z}), \qquad \eta(z) = (1/2){}^t(z - \bar{z})Q(z - \bar{z}),$$

$$r(z) = \eta(z)^{-1}Q(z - \bar{z}).$$

Here Q is a symmetric element of $GL_n(\mathbf{R})$ of signature $(n - 1, 1)$ such that $Q^2 = 1_n$; e is a fixed element of $\mathbf{R}^n$ such that ${}^teQe < 0$.

Type C. $G = Sp(n, \mathbf{R}) = \{\alpha \in GL_{2n}(\mathbf{R}) | {}^t\alpha J_n \alpha = J_n\}$,

$$K^c = GL_n(\mathbf{C}), \qquad T = \{z \in \mathbf{C}_n^n | {}^tz = z\}, \qquad U = \{x \in \mathbf{R}_n^n | {}^tx = x\},$$

$$H = \{z \in T | i(\bar{z} - z) > 0\}, \qquad \xi(z) = \eta(z) = i(\bar{z} - z), \qquad r(z) = (z - \bar{z})^{-1}.$$

Type D. $G = \{\alpha \in SL_{2n}(\mathbf{C}) | {}^t\alpha L \alpha = L, \bar{\alpha} J_n' = J_n' \alpha\}, 2 \leqslant n = 2\ell \in 2\mathbf{Z}$,

$$L = \begin{pmatrix} 0 & 1_n \\ 1_n & 0 \end{pmatrix}, \qquad J_n' = \operatorname{diag}[J_1, \ldots, J_1] (\in \mathbf{C}_{2n}^{2n}), \qquad T = \{z \in \mathbf{C}_n^n | {}^tz = -z\},$$

$$U = \{x \in T | (J_\ell' x)^* = J_\ell' x\}, \qquad H = \{z \in T | i[(J_\ell' z)^* - J_\ell' z] > 0\},$$

$$K^c = GL_n(\mathbf{C}), \quad \xi(z) = \eta(z) = i[(J_\ell' z)^* - J_\ell' z], \qquad r(z) = (J_\ell' z^* J_\ell' - z)^{-1}.$$

For $\alpha \in G$ and $z \in H$ (with H of bounded or tube form) we write $\alpha(z)$ or simply αz for the image of z under α. For Types A, C, and D, if we write $\alpha = \begin{pmatrix} a & b \\ c & d \end{pmatrix}$ with

a of size n, then $\alpha z = (az + b)(cz + d)^{-1}$. For Type B, αz is still rational in α and z; see [S1, pp. 320–321] or (5.11) below for its explicit form.

We view K^c $(=GL_n(\mathbf{C}))$ for Types C and D as a subgroup of $GL_n(\mathbf{C}) \times GL_n(\mathbf{C})$ by the embedding $a \mapsto (a, a)$. We let m denote 1, n, and n for Types B, C, and D, respectively. Then, in all cases K^c is a subgroup of $GL_n(\mathbf{C}) \times GL_m(\mathbf{C})$. Hereafter we shall write an element of K^c in the form $(a, b) \in GL_n(\mathbf{C}) \times GL_m(\mathbf{C})$ with the understanding that $a = b$ for Types C and D. For example, $(\xi(z), \eta(z)) \in K^c$ for all types. We let K^c act on T by $(a, b)u = au \cdot {}^t b$ for $(a, b) \in K^c$ and $u \in T$.

In each case there is a holomorphic factor of automorphy $(\lambda(\alpha, z), \mu(\alpha, z))$ with values in K^c for $\alpha \in G$ and $z \in H$ with the property that

$$d(\alpha z) = {}^t\lambda(\alpha, z)^{-1}\, dz\, \mu(\alpha, z)^{-1}. \tag{1.2}$$

Thus $\lambda(\alpha, z) \in GL_n(\mathbf{C})$ and $\mu(\alpha, z) \in GL_m(\mathbf{C})$; $\lambda = \mu$ for Types C and D. (The explicit forms of λ and μ can be found in [S1]–[S5].) We then put

$$\delta(z) = \det(\eta(z)), \qquad j(\alpha, z) = \det(\mu(\alpha, z)). \tag{1.3}$$

Now we have

$$\lambda(\alpha, z)^*\xi(\alpha z)\lambda(\alpha, z) = \xi(z), \qquad \mu(\alpha, z)^*\eta(\alpha z)\mu(\alpha, z) = \eta(z), \tag{1.4a}$$

$$|j(\alpha, z)|^2\delta(\alpha z) = \delta(z). \tag{1.4b}$$

2. Differential operators. Throughout the rest of the paper we denote by G an algebraic group over $\mathbf{Q}$, define $G_{\mathbf{Q}}$ and $G_{\mathbf{R}}$ as usual, denote by $G_{\mathbf{R}}^0$ the identity component of $G_{\mathbf{R}}$, and put $G^0 = G_{\mathbf{R}}^0 \cap G_{\mathbf{Q}}$. We assume that

$$G_{\mathbf{R}}^0 = \prod_{s \in \mathbf{s}} G_s \times \mathfrak{R}' \tag{2.1}$$

with a finite set of indices $\mathbf{s}$, groups G_s belonging to the four types of Section 1, and a compact group $\mathfrak{R}'$. (Typically, G may have a structure of algebraic group over a totally real algebraic number field F, $\mathbf{s}$ is a set of archimedean primes of F, G_s is the identity component of the localization of G at s, and $\mathfrak{R}'$ is the product of all the localizations of G at the archimedean primes not belonging to $\mathbf{s}$. Naturally we have to assume that $\mathfrak{R}'$ is compact and G_s belongs to the four types.)

For each $s \in \mathbf{s}$ we take the objects H, T, K^c, etc. associated with G_s, and denote them by H_s, T_s, K_s^c, etc. We then put

$$\mathscr{H} = \prod_{s \in \mathbf{s}} H_s, \qquad \mathfrak{R} = \prod_{s \in \mathbf{s}} K_s^c \times \mathfrak{R}', \tag{2.2a}$$

$$\alpha z = \alpha(z) = (\alpha_s z_s)_{s \in \mathbf{s}}, \qquad \Xi(z) = ((\xi_s(z_s), \eta_s(z_s))_{s \in \mathbf{s}}, 1)(\in \mathfrak{R}), \tag{2.2b}$$

$$r(z) = (r_s(z_s))_{s \in \mathbf{s}}, \qquad \Lambda_\alpha(z) = ((\lambda(\alpha_s, z_s), \mu(\alpha_s, z_s))_{s \in \mathbf{s}}, \alpha')(\in \mathfrak{R}). \tag{2.2c}$$

for $z = (z_s)_{s \in \mathbf{s}} \in \mathscr{H}$ and $\alpha = ((\alpha_s)_{s \in \mathbf{s}}, \alpha') \in G_{\mathbf{R}}^0$ with $\alpha_s \in G_s$ and $\alpha' \in \mathfrak{R}$.

We now consider a representation $\{\rho, X\}$ of $\mathfrak{K}$ such that: X is a finite-dimensional complex vector space and ρ is a homomorphism $\mathfrak{K} \to GL(X)$ which is equivalent to a direct sum $\bigoplus_\mu (\rho_\mu \otimes \sigma_\mu)$ with rational (that is, complex analytic) representations ρ_μ of $\prod_{s \in \mathbf{s}} K_s^c$ and continuous representations σ_μ of $\mathfrak{K}'$. Given such a $\{\rho, X\}$, $f \in C^\infty(\mathscr{H}, X)$, and $\alpha \in G_{\mathbf{R}}^0$, we define $f \|_\rho \alpha \in C^\infty(\mathscr{H}, X)$ by

$$(f \|_\rho \alpha)(z) = \rho(\Lambda_\alpha(z))^{-1} f(\alpha z) \qquad (z \in \mathscr{H}). \tag{2.3}$$

For a congruence subgroup Γ of G^0 we denote by $C_\rho(\Gamma)$ the set of all $f \in C^\infty(\mathscr{H}, X)$ such that $f \|_\rho \gamma = f$ for every $\gamma \in \Gamma$, and by C_ρ the union of $C_\rho(\Gamma)$ for all such Γ. We then denote by $\mathscr{M}_\rho$ and $\mathscr{M}_\rho(\Gamma)$ the subsets of C_ρ and $C_\rho(\Gamma)$ consisting of the holomorphic functions satisfying the cusp condition, which is required only if G has a factor isogenous to $SL_2(\mathbf{Q})$. Suppose that X has an inner product $\langle x, y \rangle_X$ which is $\mathbf{C}$-linear in y and $\mathbf{C}$-antilinear in x and which satisfies

$$\langle x, \rho(a, b, c) y \rangle_X = \langle \rho(a^*, b^*, c^{-1}) x, y \rangle_X \tag{2.4}$$

for $(a, b, c) \in \mathfrak{K}$, where $(a, b) = (a_s, b_s)_{s \in \mathbf{s}} \in \prod_{s \in \mathbf{s}} K_s^c$, $c \in \mathfrak{K}'$, and $(a^*, b^*) = (a_s^*, b_s^*)_{s \in \mathbf{s}}$. Then, for $f, g \in C_\rho$ we define their inner product $\langle f, g \rangle$ by

$$\langle f, g \rangle = \mu(\Phi)^{-1} \int_\Phi \langle f(z), \rho(\Xi(z)) g(z) \rangle_X \, d\mu(z) \tag{2.5}$$

whenever the integral is convergent, where $d\mu(z)$ is a fixed $G_{\mathbf{R}}^0$-invariant measure on $\mathscr{H}$, $\mu(\Phi) = \int_\Phi d\mu(z)$, and $\Phi = \Gamma \backslash \mathscr{H}$ with Γ such that $f, g \in C_\rho(\Gamma)$.

Given a positive integer p and finite-dimensional complex vector spaces X and Y, we denote by $Ml_p(Y, X)$ the vector space of all $\mathbf{C}$-multilinear maps of $Y \times \cdots \times Y$ (p copies) into X, and by $S_p(Y, X)$ the vector space of all homogeneous polynomial maps of Y into X of degree p. Thus $S_1(Y, X) = Ml_1(Y, X)$, and it is the vector space of all $\mathbf{C}$-linear maps of Y into X. We put $S_0(Y, X) = Ml_0(Y, X) = X$ and $S_p(Y) = S_p(Y, \mathbf{C})$. We call an element g of $Ml_p(Y, X)$ *symmetric* if $g(y_{\pi(1)}, \ldots, y_{\pi(p)}) = g(y_1, \ldots, y_p)$ for every permutation π of $\{1, \ldots, p\}$. Given $h \in S_p(Y, X)$, there is a unique symmetric element of $Ml_p(Y, X)$, which we write h_*, such that $h(y) = h_*(y, \ldots, y)$.

We are going to define differential operators on $\mathscr{H}$ with respect to each component z_s of the variable point $z = (z_s)_{s \in \mathbf{s}}$ on $\mathscr{H}$. Given $s \in \mathbf{s}$, $0 \leqslant p \in \mathbf{Z}$, and a representation $\{\rho, X\}$ of $\mathfrak{K}$ as above, we define representations $\{\rho \otimes \tau_s^p, Ml_p(T_s, X)\}$ and $\{\rho \otimes \sigma_s^p, Ml_p(T_s, X)\}$by

$$[(\rho \otimes \tau_s^p)(a, b, c) h](u_1, \ldots, u_p) = \rho(a, b, c) h({}^t a_s u_1 b_s, \ldots, {}^t a_s u_p b_s), \tag{2.6a}$$

$$[(\rho \otimes \sigma_s^p)(a, b, c) h](u_1, \ldots, u_p) = \rho(a, b, c) h(a_s^{-1} u_1 {}^t b_s^{-1}, \ldots, a_s^{-1} u_p {}^t b_s^{-1}), \tag{2.6b}$$

for $(a, b, c) \in \mathfrak{K}$ with $(a, b) \in \prod_{s\in\mathbf{s}} K_s^c$ and $c \in \mathfrak{K}'$, $h \in Ml_p(T_s, X)$, and $u_i \in T_s$. We use the same symbols $\rho \otimes \tau_s^\rho$ and $\rho \otimes \sigma_s^\rho$ for their restrictions to $S_p(T_s, X)$, and write them simply τ_s^ρ and σ_s^ρ if $X = \mathbf{C}$ and ρ is trivial.

We view T_s as its own dual over $\mathbf{C}$ by the pairing $(u, v) \mapsto \mathrm{tr}({}^t uv)$. Then, for $g \in S_p(T_s)$ and $h \in S_p(T_s, X)$ we put

$$(2.7)\qquad [g, h] = \sum g_*(a_{v_1}, \ldots, a_{v_p})h_*(b_{v_1}, \ldots, b_{v_p}),$$

where $\{a_v\}_{v\in N}$ and $\{b_v\}_{v\in N}$ are dual bases of T_s, and $(v_1, \ldots, v_p)$ runs over N^p. Then $[g, h]$ is an element of X determined independently of the choice of dual bases. In particular, taking $X = \mathbf{C}$, we can view $S_p(T_s)$ as its own dual by the pairing $(g, h) \mapsto [g, h]$.

Since T_s has a natural $\mathbf{R}$-structure, we can speak of an $\mathbf{R}$-rational basis of T_s over $\mathbf{C}$. Take any such basis $\{\varepsilon_v\}_{v\in N}$, and for $u \in T_s$ define $u_v \in \mathbf{C}$ by $u = \sum_{v\in N} u_v\varepsilon_v$. We also put $z_s = \sum_{v\in N} z_{sv}\varepsilon_v$ with $z_{sv} \in \mathbf{C}$ for the variable z_s on H_s. Then, for $f \in C^\infty(\mathscr{H}, X)$ we define $D_s f, \bar{D}_s f, C_s f, E_s f \in C^\infty(\mathscr{H}, S_1(T_s, X))$ by

$$(2.8a)\qquad (D_s f)(u) = \sum_{v\in N} u_v \partial f/\partial z_{sv}, \qquad (\bar{D}_s f)(u) = \sum_{v\in N} u_v \partial f/\partial \bar{z}_{sv},$$

$$(2.8b)\qquad (C_s f)(u) = (D_s f)({}^t\xi_s u\eta_s), \qquad (E_s f)(u) = (\bar{D}_s f)(\xi_s u \cdot {}^t\eta_s)$$

for $u \in T_s$. These are indpendent of the choice of $\{\varepsilon_v\}_{v\in N}$. The last two formulas can be written $C_s f = \tau_s^1(\Xi)D_s f$ and $E_s f = \sigma_s^1(\Xi^{-1})\bar{D}_s f$ with Ξ of (2.2b), and the first two are equivalent to the expression

$$(2.8c)\qquad df = \sum_{s\in\mathbf{s}} (D_s f)(dz_s) + \sum_{s\in\mathbf{s}} (\bar{D}_s f)(d\bar{z}_s).$$

Notice that $E_s f = 0$ if and only if f is holomorphic in z_s. We also define $D_s^p f, \bar{D}_s^p f, C_s^p f$, and $E_s^p f$ by

$$(2.9a)\qquad D_s^p f = D_s D_s^{p-1} f, \qquad \bar{D}_s^p f = \bar{D}_s \bar{D}_s^{p-1} f, \qquad D_s^0 f = \bar{D}_s^0 f = f,$$

$$(2.9b)\qquad C_s^p f = C_s C_s^{p-1} f, \qquad E_s^p f = E_s E_s^{p-1} f, \qquad C_s^0 f = E_s^0 f = f.$$

These have values in $Ml_p(T_s, X)$, but it can be shown that they are symmetric. (See [S5]; the fact follows also from Proposition 2.2 below.) Therefore we can view them as elements of $C^\infty(\mathscr{H}, S_p(T_s, X))$. We then define $D_{\rho,s}^p f \in C^\infty(\mathscr{H}, S_p(T_s, X))$ by

$$(2.10)\qquad D_{\rho,s}^p f = (\rho \otimes \tau_s^p)(\Xi)^{-1} C_s^p[\rho(\Xi)f].$$

In particular, writing $D_{\rho,s} = D_{\rho,s}^1$, we have

$$(2.11)\qquad (D_{\rho,s} f)(u) = \rho(\Xi)^{-1} D_s[\rho(\Xi)f](u) \qquad (u \in T_s).$$

These operators, as well as D_ρ^Z and E^Z below, were introduced in our previous papers [S1]–[S5], [S8]. We recall here some of their basic properties. First of all, for every $\alpha \in G_{\mathbf{R}}^0$ we have

$$(2.12)\qquad D_{\rho,s}^p(f\|_\rho\alpha) = (D_{\rho,s}^p f)\|_{\rho\otimes\tau_s^p}\alpha, \qquad E_s^p(f\|_\rho\alpha) = (E_s^p f)\|_{\rho\otimes\sigma_s^p}\alpha.$$

The representation τ_s^p or σ_s^p of $\mathfrak{R}$ on $S_p(T_s)$ is essentially that of K_s^c, and it is well known that it is the direct sum of irreducible representations, and each irreducible constituent has multiplicity one. (Since the τ_s^p-irreducibility is the same as the σ_s^p-irreducibility, we shall simply speak of an irreducible subspace of $S_p(T_s)$.) Thus, for each irreducible subspace Z of $S_p(T_s)$, we can define the projection map φ_Z of $S_p(T_s)$ onto Z. Now we can identify $S_p(T_s, X)$ with $S_p(T_s)\otimes X$ by the rule

$$(2.13)\qquad (h\otimes x)(u) = h(u)x \qquad \text{for } h\in S_p(T_s),\ x\in X, \text{ and } u\in T_s.$$

(This justifies the notation $\rho\otimes\tau_s^p$ and $\rho\otimes\sigma_s^p$.) Using the same symbol φ_Z for the map $h\otimes x\mapsto(\varphi_Z h)\otimes x$ of $S_p(T_s)\otimes X$ to $Z\otimes X$, we define $D_\rho^Z f$, $E^Z f\in C^\infty(\mathscr{H}, Z\otimes X)$ by

$$(2.14)\qquad D_\rho^Z f = \varphi_Z D_{\rho,s}^p f, \qquad E^Z f = \varphi_Z E_s^p f.$$

Then, for every $\alpha\in G_{\mathbf{R}}^0$ we have

$$(2.15)\qquad D_\rho^Z(f\|_\rho\alpha) = (D_\rho^Z f)\|_{\rho\otimes\tau_Z}\alpha, \qquad E^Z(f\|_\rho\alpha) = (E^Z f)\|_{\rho\otimes\sigma_Z}\alpha,$$

where τ_Z and σ_Z are the restrictions of τ_s^p and σ_s^p to Z. It can easily be verified that $S_p(T_s)$ is the direct sum of Z and its annihilator with respect to (2.7), which is $\mathfrak{R}$-stable, and hence Z is its own dual. Therefore we can identify $Z\otimes X$ with $S_1(Z, X)$ by the rule

$$(2.16)\qquad (\omega\otimes x)(\zeta) = [\zeta, \omega]x \qquad \text{for } \omega, \zeta\in Z \text{ and } x\in X.$$

Then φ_Z as a map $S_p(T_s, X)\to S_1(Z, X)$ can be given by $(\varphi_Z g)(\zeta) = [\zeta, g]$ for $g\in S_p(T_s, X)$ and $\zeta\in Z$, and hence $D_\rho^Z f$ and $E^Z f$ as $S_1(Z, X)$-valued functions can be given by

$$(2.17)\qquad (D_\rho^Z f)(\zeta) = [\zeta, D_{\rho,s}^p f], \qquad (E^Z f)(\zeta) = [\zeta, E_s^p f] \qquad (\zeta\in Z).$$

The symbols $\rho\otimes\tau_Z$ and $\rho\otimes\sigma_Z$ as representations on the space $S_1(Z, X)$ can be given by

$$(2.18a)\qquad [(\rho\otimes\tau_Z)(a, b, c)h](\zeta) = \rho(a, b, c)h(\tau_Z({}^t a_s, {}^t b_s)\zeta),$$

$$(2.18b)\qquad [(\rho\otimes\sigma_Z)(a, b, c)h](\zeta) = \rho(a, b, c)h(\sigma_Z({}^t a_s, {}^t b_s)\zeta)$$

for $h\in S_1(Z, X)$, $\zeta\in Z$, and $(a, b, c)\in\mathfrak{R}$ as in (2.6a, b).

We now define a contraction operator $\theta\colon Z \otimes Z \otimes X \to X$ by $\theta(\zeta \otimes \omega \otimes x) = [\zeta, \omega]x$. This as a map $S_1(Z, S_1(Z, X)) \to X$ can be given by

$$\theta h = \sum_{\mu} h(\zeta_\mu, \omega_\mu) \qquad \text{for } h \in S_1(Z, S_1(Z, X)) \tag{2.19}$$

with bases $\{\zeta_\mu\}$ and $\{\omega_\mu\}$ of Z such that $[\zeta_\mu, \omega_\nu] = \delta_{\mu\nu}$. If $g \in C^\infty(\mathscr{H}, S_1(Z, X))$, then $D^Z_{\rho\otimes\sigma_Z} g$ and $E^Z g$ have values in $S_1(Z, S_1(Z, X))$, so that $\theta D^Z_{\rho\otimes\sigma_Z} g$ and $\theta E^Z g$ are meaningful as X-valued functions. In particular, for $f \in C^\infty(\mathscr{H}, X)$ the symbols $\theta D^Z_{\rho\otimes\sigma_Z} E^Z f$ and $\theta E^Z D^Z_\rho f$ are elements of $C^\infty(\mathscr{H}, X)$. We then put

$$L^Z_\rho f = (-1)^p \theta D^Z_{\rho\otimes\sigma_Z} E^Z f, \qquad M^Z_\rho f = (-1)^p \theta E^Z D^Z_\rho f. \tag{2.20}$$

Then for every $\alpha \in G^0_{\mathbf{R}}$ we have

$$L^Z_\rho(f \|_\rho \alpha) = (L^Z_\rho f)\|_\rho \alpha, \qquad M^Z_\rho(f \|_\rho \alpha) = (M^Z_\rho f)\|_\rho \alpha. \tag{2.21}$$

If $Z = S_1(T_s)$ and this is identified with T_s, then the map θ can be viewed as a map $S_1(T_s, S_1(T_s, X)) \to X$, and is given by

$$\theta h = \sum_{\nu \in N} h(a_\nu, b_\nu) \qquad \text{for } h \in S_1(T_s, S_1(T_s, X)) \tag{2.22}$$

with dual bases $\{a_\nu\}$ and $\{b_\nu\}$ of T_s.

THEOREM 2.1. *Let Z be an irreducible subspace of $S_p(T_s)$. Then, for $f, f' \in C_\rho$, $g \in C_{\rho\otimes\tau_Z}$, and $h \in C_{\rho\otimes\sigma_Z}$ we have*

$$\langle D^Z_\rho f, g\rangle = (-1)^p \langle f, \theta E^Z g\rangle, \qquad \langle E^Z f, h\rangle = (-1)^p \langle f, \theta D^Z_{\rho\otimes\sigma_Z} h\rangle,$$

$$\langle L^Z_\rho f, f'\rangle = \langle f, L^Z_\rho f'\rangle, \qquad \langle M^Z_\rho f, f'\rangle = \langle f, M^Z_\rho f'\rangle,$$

$$\langle L^Z_\rho f, f\rangle \geqslant 0, \quad \langle M^Z_\rho f, f\rangle \geqslant 0$$

under suitable convergence conditions (see below).

This was obtained in [S3, Theorem 11.5, Corollary 11.8] and [S4, p. 486]. An equivalent result formulated on G_s was given in [S9, Theorem 4.3, Corollary 4.4]. All these were formulated for functions on $\Gamma\backslash\mathscr{H}$ or on $\Gamma\backslash G^0_{\mathbf{R}}$ of compact support. To state a sufficient convergence condition in a more general case, we first note that given $f \in C_\rho(\Gamma)$, the function $\tilde{f}$ on $G^0_{\mathbf{R}}$ defined by $\tilde{f}(g) = \rho(\Lambda(g, \mathbf{o})^{-1}) f(g\mathbf{o})$ for $g \in G^0_{\mathbf{R}}$ is left Γ-invariant, where $\mathbf{o} = (\mathbf{o}_s)$ is a point of $\mathscr{H}$ such that $\{\alpha \in G_s \mid \alpha\mathbf{o}_s = \mathbf{o}_s\}$ is the standard maximal compact subgroup of G_s with which the results of [S9] are formulated. Let $\mathfrak{g}_s$ be the Lie algebra of G_s and $\mathfrak{g}_s = \mathfrak{k}_s + \mathfrak{p}_s$ its Cartan decomposition. Given $\{\rho, X\}$, $\{\rho', X'\}$, $f \in C_\rho(\Gamma)$, $h \in C_{\rho'}(\Gamma)$, and a positive integer p, we say that (f, h) *is an integrable pair of type* (p, s) if $\psi(Y_1 \cdots Y_\mu \tilde{f})\psi'(Y'_1 \cdots Y'_\nu \tilde{h})$

belongs to $L^1(\Gamma\backslash G^0_{\mathbf{R}})$ for every $\psi \in S_1(X)$, $\psi' \in S_1(X')$, and every Y_i, $Y'_j \in \mathfrak{p}_s$ with $\mu \geqslant 0$ and $\nu \geqslant 0$ such that $\mu + \nu = p$ or $\mu + \nu = p - 1$. Now the first (resp. second) formula of Theorem 2.1 is valid if

$$(2.23) \qquad (f, g) \text{ (resp. } (f, h)) \text{ is an integrable pair of type } (p, s).$$

The reason for this will be explained in the appendix. As noted in [S9, p. 257], the formulas are valid if $\tilde{f}$, $\tilde{g}$, $\tilde{h}$ are C^∞ vectors in $L^2(\Gamma\backslash G^0_{\mathbf{R}})$. Another sufficient condition is that all the holomorphic and antiholomorphic derivatives of f are rapidly decreasing and g, h are slowly increasing at the cusps of G^0. Sufficient conditions for the last four formulas of Theorem 2.1 can be given in a similar manner or in the style of (2.23), since they are straightforward consequences of the first two formulas.

The relationship between the formulation on H_s and that on G_s is explained in [S9, § 7]. We supplement it by stating a stronger form of the last proposition of that paper. Let $\{\rho, X\}$ be a rational representation of K^c_s. For $f \in C^\infty(H_s, X)$ define $f^\rho \in C^\infty(G_s, X)$ by

$$(2.24) \qquad f^\rho(g) = \rho(\lambda(g, \mathbf{o}_s), \mu(g, \mathbf{o}_s))^{-1} f(g\mathbf{o}_s) \qquad \text{for } g \in G_s$$

with $\mathbf{o}_s$ as above. Consider also $\iota_+ \colon T_s \to \mathfrak{p}_+$ and $\iota_- \colon T_s \to \mathfrak{p}_-$ as in [S9, § 5]. Then, we have the following proposition as a generalization of [S9, Proposition 7.3].

Proposition 2.2. *For an irreducible subspace Z of $S_p(T_s)$, let V and W be the corresponding subspaces of $S_p(\mathfrak{p}_+)$ and $S_p(\mathfrak{p}_-)$ under ι_+ and ι_-. Define $D^p f^\rho$, $E^p f^\rho$, $D^V f^\rho$, and $E^W f^\rho$ by [S9, (4.1a,b), (4.7)]. Then*

$$(D^p_{\rho,s} f)^{\rho\otimes\tau^p} = (D^p f^\rho) \circ \iota_+, \qquad (E^p_s f)^{\rho\otimes\sigma^p} = (E^p f^\rho) \circ \iota_-,$$

$$(D^Z_\rho f)^{\rho\otimes\tau_Z} \circ j_- = D^V f^\rho, \qquad (E^Z f)^{\rho\otimes\sigma_Z} \circ j_+ = E^W f^\rho,$$

where $j_\pm$ are the isomorphisms $V \to Z$ and $W \to Z$ defined by $j_+(\zeta) = \zeta \circ \iota_+$ and $j_-(\omega) = \omega \circ \iota_-$.

Proof. We prove the first two formulas by induction on p. The case $p = 1$, with the subscript s suppressed, can be written

$$\iota_+(u) f^\rho = (D_\rho f)^{\rho\otimes\tau}(u), \qquad \iota_-(u) f^\rho = (Ef)^{\rho\otimes\sigma}(u) \qquad (u \in T_s),$$

which is exactly [S9, Proposition 7.3]. Now assuming the case for p and noting $D^{p+1}_\rho = D_{\rho\otimes\tau^p} D^p_\rho$, for $u_i, v \in T_s$, we obtain

$$\begin{aligned}(D^{p+1}_\rho f)(u_1, \ldots, u_p)(v) &= \iota_+(v)(D^p_\rho f)^{\rho\otimes\tau^p}(u_1, \ldots, u_p) \\ &= \iota_+(v)(D^p f^\rho)(\iota_+(u_1), \ldots, \iota_+(u_p))\end{aligned}$$

$$= \iota_+(v)\iota_+(u_1)\ldots\iota_+(u_p)f^\rho$$
$$= (D^{p+1}f^\rho)(\iota_+(u_1), \ldots, \iota_+(u_p), \iota_+(v)),$$

which proves the case for $p + 1$. From [S9, (5.4)] we can easily derive that $(\varphi_Z h)^{\rho\otimes\tau_Z} \circ j_- = \varphi_V(h^{\rho\otimes\tau^p} \circ \iota_+^{-1})$ for every $h \in C^\infty(H_s, S_p(T_s, X))$ with φ_V of [S9, (4.4)], where $\varphi_Z h$ is viewed as an element of $C^\infty(H_s, S_1(Z, X))$. Therefore $(D_\rho^Z f)^{\rho\otimes\tau_Z} \circ j_- = (\varphi_Z D_\rho^p f)^{\rho\otimes\tau_Z} \circ j_- = \varphi_V((D_\rho^p f)^{\rho\otimes\tau^p} \circ \iota_+^{-1}) = \varphi_V D^p f^\rho = D^V f^\rho$ as expected. The formulas for E^p and E^Z can be proved in the same manner.

We conclude this section by making some comments on automorphic forms of half-integral weight. To include them in our formulation, we consider a pair $\{\rho, X\}$ such that

$$(2.25) \qquad \rho(a, b, c) = \prod_{s\in\mathbf{s}} \det(b_s)^{e_s}\rho_0(a, b, c)$$

for $(a, b, c) \in \mathfrak{K}$ with a representation $\{\rho_0, X\}$ of $\mathfrak{K}$ as above and $e_s = 0$ or $1/2$, that is, $e = (e_s)_{s\in\mathbf{s}} \in \{0, 1/2\}^{\mathbf{s}}$. We call such $\{\rho, X\}$ a *quasi representation* of $\mathfrak{K}$, and say that it is *irreducible* if ρ_0 is irreducible. Then, for $f \in C^\infty(\mathscr{H}, X)$ and $\alpha \in G_{\mathbf{R}}^0$ the symbol $f\|_\rho\alpha$ can be defined by (2.3), but this depends on the choice of a branch of $\prod_{s\in\mathbf{s}} \det(\mu(\alpha_s, z_s))^{e_s}$. However, formulas (2.12), (2.15), and (2.21) are all meaningful, since the choice of $\rho(\Lambda_\alpha(z))$ makes the choice of $(\rho\otimes\pi)(\Lambda_\alpha(z))$ for $\pi = \tau_s^p, \sigma_s^p, \tau_Z, \sigma_Z$ obvious. There is no problem for $\rho(\Xi)$, since $\det(\eta_s(z_s))$ is positive and we only have to take its positive square root. Thus there is no ambiguity about the operators $D_{\rho,s}^p$, D_ρ^Z, L_ρ^Z, and M_ρ^Z. To define $C_\rho(\Gamma)$ and $\mathscr{M}_\rho(\Gamma)$, we have to assume that there is a consistent choice of $\rho(\Lambda_\gamma(z))$ for $\gamma \in \Gamma$. The existence of such a Γ is not guaranteed for an arbitrarily given G, but we develop our theory formally without caring about this point. We only impose the condition that

$$(2.26) \qquad f\otimes f \in C_\iota \qquad \text{with } \tau(a, b, c) = \prod_{s\in\mathbf{s}} \det(b_s)^{2e_s}(\rho_0\otimes\rho_0)(a, b, c).$$

As for Proposition 2.2, it is still valid if we define f^ρ on a suitable covering of $G_{\mathbf{R}}^0$.

3. Near holomorphy and holomorphic projection. So far the function r has played no role in our treatment. To explain its signficance, let us first recall basic facts on nearly holomorphic functions on a complex Kähler manifold, the notion of which was introduced in [S5], [S7]. Given such a manifold W of complex dimension N with a fundamental 2-form Ω, we can put $\Omega = i\sum_{p,q=1}^N \partial^2\varphi/\partial z_p\partial\bar{z}_q \cdot dz_p \wedge d\bar{z}_q$ with a real-valued function φ and local complex coordinate functions $\{z_1, \ldots, z_N\}$ in a coordinate neighborhood U. Define N functions r_p on U by $r_p = \partial\varphi/\partial z_p$. Then we call a function on W *nearly holomorphic* if it is a polynomial in the r_p with holomorphic functions as coefficients in each such neighborhood U. The r_p are algebraically independent over the field of all meromorphic functions on U, as can easily be seen from [S5, Lemma 2.1].

Before applying this principle to our space $\mathscr{H}$, let us first state some basic formulas concerning the derivatives of ξ, η, and r. We fix one $s \in \mathbf{s}$ and consider the behavior of a function on $\mathscr{H}$ only with respect to z_s. For simplicity let us drop temporarily the subscript s from the objects T_s, D_s, $\bar{D}_s$, ξ_s, etc. Then for $u \in T$ we have

$$(3.1a)\qquad \eta^{-1}(D\eta)(u) = {}^t ru \qquad \text{(All types)},$$

$$(3.1b)\qquad \xi^{-1}(D\xi)(u) = \begin{cases} r \cdot {}^t u & \text{(Types A, C, and D)}, \\ r \cdot {}^t u - Qu \cdot {}^t rQ & \text{(Type B)}, \end{cases}$$

$$(3.1c)\qquad (Dr)(u) = \begin{cases} -r \cdot {}^t ur & \text{(Types A, C, and D)}, \\ ({}^t rQr/2)Qu - r \cdot {}^t ur & \text{(Type B)}, \end{cases}$$

$$(3.1d)\qquad (\bar{D}r)(u) = -\xi^{-1}u \cdot {}^t\eta^{-1} \qquad \text{(All types)}.$$

Here we understand that $Q = 1_n$ if H is bounded and is of Type B. These were given in [S5, Lemma 2.2 and p. 367] for most types. All the remaining cases can be verified in the same fashion. From (3.1a) and (3.1d) we obtain, for $u, v \in T$,

$$(3.2a)\qquad (D \log \delta)(u) = \operatorname{tr}({}^t ru),$$

$$(3.2b)\qquad (\bar{D}D \log \delta)(u, v) = -\operatorname{tr}({}^t u\xi^{-1}v \cdot {}^t\eta^{-1}).$$

The last formula can be written $\bar{\partial}\partial \log \delta = -\operatorname{tr}({}^t dz \cdot \xi^{-1} d\bar{z} \cdot {}^t\eta^{-1})$. Since ξ and η are hermitian and positive definite, we see that H is a Kähler manifold with $i\bar{\partial}\partial \log \delta$ as its fundamental 2-form. For the product space $\mathscr{H} = \prod_{s \in \mathbf{s}} H_s$, we have to take $i \sum_{s \in \mathbf{s}} \bar{\partial}\partial \log \delta_s(z_s)$. Then (3.2a) shows that the entries of the functions $(r_s)_{s \in \mathbf{s}}$ are exactly the r_p discussed at the beginning.

Taking an $\mathbf{R}$-rational basis $\{\varepsilon_\nu\}_{\nu \in N}$ of T_s, put $z_s = \sum_{\nu \in N} z_{s\nu}\varepsilon_\nu$ and $r_s = \sum_{\nu \in N} r_{s\nu}\varepsilon_\nu$. Define vector fields $\partial/\partial r_{s\nu}$ and $\partial/\partial \bar{r}_{s\nu}$ by

$$(3.3)\qquad \partial/\partial z_{s\mu} = \sum_{\nu \in N} (\partial \bar{r}_{s\nu}/\partial z_{s\mu})\partial/\partial \bar{r}_{s\nu}, \qquad \partial/\partial \bar{z}_{s\mu} = \sum_{\nu \in N} (\partial r_{s\nu}/\partial \bar{z}_{s\mu})\partial/\partial r_{s\nu}.$$

These are well defined in view of (3.1d). Then, from (3.1d) and (2.8b) we can easily derive, for $u = \sum_{\nu \in N} u_{s\nu}\varepsilon_\nu \in T_s$,

$$(3.4)\qquad (C_s f)(u) = -\sum_{\nu \in N} u_{s\nu}\partial f/\partial \bar{r}_{s\nu}, \qquad (E_s f)(u) = -\sum_{\nu \in N} u_{s\nu}\partial f/\partial r_{s\nu}.$$

LEMMA 3.1. *The difference* $r_s(\alpha_s z_s) - \lambda(\alpha_s, z_s) r_s(z_s) \cdot {}^t\mu(\alpha_s, z_s)$ *is holomorphic in* z *for every* $\alpha \in G^0_{\mathbf{R}}$.

Proof. Dropping the subscript s for simplicity and denoting by 1_T the identity map $T \to T$, we see from (3.4) that $Er = -1_T$. Define $\rho: K^c \to GL(T)$ by $\rho(a, b)u = au \cdot {}^t b$. Then $r\|_\rho \alpha = \lambda(\alpha, z)^{-1} r(\alpha z) \cdot {}^t\mu(\alpha, z)^{-1}$. Clearly $[(\rho \otimes \sigma^1)(a, b)(-1_T)] = -1_T$, and hence by (2.12) we have $E(r\|_\rho \alpha) = (Er)\|_{\rho\otimes\sigma^1}\alpha = Er$. Thus $E(r\|_\rho \alpha - r) = 0$, so that $r\|_\rho \alpha - r$ is holomorphic as expected.

Given $p = (p_s)_{s \in \mathbf{s}} \in \mathbf{Z}^{\mathbf{s}}$ with $p_s \geqslant 0$ for every s and a quasi representation $\{\rho, X\}$ of $\mathfrak{R}$, we denote by $\mathcal{N}^p(X)$ the set of all $f \in C^\infty(\mathcal{H}, X)$ which are polynomials in the components of $r = (r_s)_{s \in \mathbf{s}}$, of degree $\leqslant p_s$ in r_s, with holomorphic maps of $\mathcal{H}$ into X as coefficients. Then, (3.4) together with [S5, Lemma 2.1] shows that $\mathcal{N}^p(X)$ consists of all $f \in C^\infty(\mathcal{H}, X)$ such that $E_s^{p_s+1} f = 0$ for every $s \in \mathbf{s}$. Moreover the components of r_s for all s are algebraically independent over the field of all meromorphic functions on $\mathcal{H}$. For example, if we view an element f of $\mathcal{N}^p(X)$ as a function of z_s and suppress other variables $z_{s'}$ for $s' \in \mathbf{s}$, $\neq s$, then

$$f(z) = \sum_{i=0}^{p_s} f_i(z_s, r_s(z_s)) \tag{3.5}$$

with a holomorphic map $f_i: H_s \to S_i(T_s, X)$ for each i, where $f_i(z_s, u)$ means the value $f_i(z_s) \in S_i(T_s, X)$ evaluated at $u \in T_s$. From (2.12) we see that $\mathcal{N}^p(X)$ is stable under the maps $f \mapsto f \circ \alpha$ and $f \mapsto f\|_\rho \alpha$ for every $\alpha \in G^0_{\mathbf{R}}$. If $f \in \mathcal{N}^p(X)$, then clearly $E_s f$ is of degree $\leqslant p_s - 1$ in r_s; moreover, $D_{\rho,s} f$ is of degree $\leqslant p_s + 1$ in r_s. The proof given in [S5, Lemma 3.3] for Types A and B is valid in all cases. The elements of $\bigcup_p \mathcal{N}^p(X)$ are called (X-valued) *nearly holomorphic* functions on $\mathcal{H}$, as defined at the beginning of this section. Given a congruence subgroup Γ of G^0 we denote by $\mathcal{N}^p_\rho(\Gamma)$ the subset of $\mathcal{N}^p(X) \cap C_\rho(\Gamma)$ consisting of the functions satisfying the cusp condition, which is required only when G has a simple factor isogenous to $SL_2(\mathbf{Q})$. (See [S7, (5.4b)] for the precise statement of the cusp condition.) We then denote by $\mathcal{N}^p_\rho$ the union of $\mathcal{N}^p_\rho(\Gamma)$ for all Γ. Clearly $\mathcal{N}^p_\rho\|_\rho \alpha = \mathcal{N}^p_\rho$ for every $\alpha \in G^0$.

Now the purpose of this section is to find a certain projection map $\mathcal{N}^p_\rho \to \mathcal{M}_\rho$. It is necessary to consider $\theta h = \sum_{\nu \in N} h(a_\nu, b_\nu)$ defined by (2.22) with any pair of dual bases $\{a_\nu\}$ and $\{b_\nu\}$ of T for several specific $h \in Ml_2(T, X) = S_1(T, S_1(T, X))$. (We again fix one $s \in \mathbf{s}$, and drop the subscript s from the objects T_s, τ^p_s, etc.) For example, for $h(x, y) = {}^t xy$ we can easily verify that

$$\theta h = \sum_{\nu \in N} {}^t a_\nu b_\nu = \lambda(T) 1_m, \tag{3.6}$$

where $\lambda(T) = n$, n, $(n+1)/2$, and $(n-1)/2$ for Types A, B, C, and D with the convention that $m = 1$, n, and n for Types B, C, and D, respectively.

We now define a $\mathbf{C}$-linear endomorphism ψ of $S_p(T)$ by

$$\psi = 0 \quad \text{if} \quad p = 1, \tag{3.7a}$$

(3.7b) $$(\psi h)(x) = \sum_{v \in N} h_*(a_v, x \cdot {}^t b_v x, x, \ldots, x) \qquad \text{(Types A, C, D; } p > 1\text{)},$$

(3.7c) $$(\psi h)(x) = \sum_{v \in N} h_*(a_v, x \cdot {}^t b_v x - ({}^t x Q x/2) Q b_v, x, \ldots, x) \qquad \text{(Types B; } p > 1\text{)}$$

for $h \in S_p(T)$ and $x \in T$. We can easily verify that $\psi \tau^p(\alpha) = \tau^p(\alpha)\psi$ for every $\alpha \in K^c$, and hence, for each irreducible subspace Z of $S_p(T)$ there is a constant c_Z such that $\psi h = c_Z h$ for every $h \in Z$. Thus $c_Z = 0$ if $p = 1$.

LEMMA 3.2. *The constant c_Z is a rational number such that $-1 \leqslant c_Z \leqslant 1$ for Type A, $(2-n)/2 \leqslant c_Z \leqslant 1$ for Type B, $-1/2 \leqslant c_Z \leqslant 1$ for Type C, and $-1 \leqslant c_Z \leqslant 1/2$ for Type D. Moreover c_Z attains either end value for Z in $S_2(T_s)$.*

Proof. Type A. Let $\ell = \mathrm{Min}(m, n)$ and let $\omega = \sum_{i=1}^{\ell} e_{ii}$ $(\in T)$ with the standard matrix units e_{ij}. We can take a highest-weight vector of Z in the form $h(x) = \prod_{i=1}^{\ell} \det_i(x)^{c_i}$ with $0 \leqslant c_i \in \mathbf{Z}$ (see [S4, Theorem 2.A]). Take $k \in Ml_p(T, \mathbf{C})$ so that $h(x) = k(x, \ldots, x)$ by the rule which can be illustrated by the following example: if $p = 9, c_1 = c_2 = 2, c_3 = 1$, then

$$k(r, s, t, u, v, w, x, y, z) = r_{11}s_{11}\begin{vmatrix} t_{11} & u_{12} \\ t_{21} & u_{22} \end{vmatrix} \cdot \begin{vmatrix} v_{11} & w_{12} \\ v_{21} & w_{22} \end{vmatrix} \cdot \begin{vmatrix} x_{11} & y_{12} & z_{13} \\ x_{21} & y_{22} & z_{23} \\ x_{31} & y_{32} & z_{33} \end{vmatrix}.$$

Then $p!h_* = \sum_\pi k_\pi$ with $k_\pi(x_1, \ldots, x_p) = k(x_{\pi(1)}, \ldots, x_{\pi(p)})$. Since $h(\omega) = 1$, we have $c_Z = (\psi h)(\omega) = (1/p!)\sum_\pi \sum_v k_\pi(a_v, \omega \cdot {}^t b_v \omega, \omega, \ldots, \omega)$. Each $k_\pi(x, y, \omega, \ldots, \omega)$ belongs to the following three types of functions: $x_{ii}y_{ii}$, $x_{ii}y_{jj}$ $(i \neq j)$, and $x_{ii}y_{jj} - x_{ji}y_{ij}$ $(i \neq j)$. The value $\sum_v k_\pi(a_v, \omega \cdot {}^t b_v \omega, \omega, \ldots, \omega)$ is 1, 0, and -1, respectively. Therefore we obtain $-1 \leqslant c_Z \leqslant 1$. Now $S_2(T_s)$ has two irreducible subspaces for which h is $\det_1(x)^2$ and $\det_2(x)$. Then we easily see that $c_Z = 1$ and -1. Since this part of the proof for the other three types is similar, we shall not indicate it explicitly in each case.

Type B. The space Z is spanned by the functions h of the form $h(x) = Q[x]^e \varphi(x)^f$, where e and f are nonnegative integers depending on Z such that $2e + f = p$, $Q[x] = {}^t x Q x$, and $\varphi(x) = {}^t u Q x$ with $u \in T$ such that $Q[u] = 0$ (see [S4, Theorem 2.B]). Define $k \in Ml_p(T, \mathbf{C})$ by

$$k(x_1, \ldots, x_p) = \prod_{i=1}^{e} {}^t x_{2i-1} Q x_{2i} \prod_{j=1}^{f} \varphi(x_{2e+j}).$$

Then $p!h_* = \sum_\pi k_\pi$ with k_π defined as in the proof for Type A. Put $c_v = x \cdot {}^t e_v x - (Q[x]/2) Q e_v$ with the standard basis vectors e_v of $\mathbf{C}^n$. Then $k_\pi(e_v, c_v, x, \ldots, x)$ belongs to the following five types of functions:

$${}^t e_v Q c_v Q[x]^{e-1} \varphi(x)^f, \quad \varphi(e_v)\varphi(c_v) Q[x]^e \varphi(x)^{f-2}, \quad {}^t x Q c_v \varphi(e_v) Q[x]^{e-1} \varphi(x)^{f-1},$$

$${}^t x Q e_v \varphi(c_v) Q[x]^{e-1} \varphi(x)^{f-1}, \quad {}^t x Q c_v \cdot {}^t x Q e_v Q[x]^{e-2} \varphi(x)^f.$$

Then we find that $\sum_{\nu=1}^{n} k_\pi(e_\nu, c_\nu, x, \ldots, x)$ equals $(1 - n/2)h(x)$ for the first type, $h(x)$ for the second type, and $(1/2)h(x)$ for the remaining three types. Therefore $1 - n/2 \leqslant c_Z \leqslant 1$.

Type C. We can employ the same technique as for Type A with 1_n in place of ω, $h(x) = \prod_{i=1}^{n} \det_i(x)^{c_i}$, and the same k. Then $p!c_Z = \sum_\pi \sum_\nu k_\pi(a_\nu, b_\nu, 1, \ldots, 1)$. Each $k_\pi(x, y, 1, \ldots, 1)$ belongs to the following three types of functions: $x_{ii}y_{ii}$, $x_{ii}y_{jj}$ $(i \neq j)$, and $x_{ii}y_{jj} - x_{ij}y_{ij}$ $(i \neq j)$. The value $\sum_\nu k_\pi(a_\nu, b_\nu, 1, \ldots, 1)$ is 1, 0, and $-1/2$, respectively, and hence $-1/2 \leqslant c_Z \leqslant 1$.

Type D. For $x \in T$ define $[x] = \bigwedge^2 \mathbf{C}^n$ by $[x] = \sum_{a<b} x_{ab} e_a \wedge e_b$ with e_a as above. Then $x \mapsto [x]$ gives a $\mathbf{C}$-linear isomorphism of T onto $\bigwedge^2 \mathbf{C}^n$, and $r!\,\mathrm{Pf}_r(x)\varepsilon = [x]^r \wedge e_{2r+1} \wedge \cdots \wedge e_n$ for $r \leqslant n/2$, where $\varepsilon = e_1 \wedge e_2 \wedge \cdots \wedge e_n$, and Pf_r is normalized so that $\mathrm{Pf}_r(\omega) = 1$ for $\omega = \mathrm{diag}[\iota, \ldots, \iota, 0]$, $\iota = \begin{pmatrix} 0 & 1 \\ -1 & 0 \end{pmatrix}$, with the last 0 in ω ignored if n is even. Let g_r be the element of $Ml_r(T, \mathbf{C})$ defined by

$$r!g_r(u_1, \ldots, u_r)\varepsilon = [u_1] \wedge \cdots \wedge [u_r] \wedge e_{2r+1} \wedge \cdots \wedge e_n.$$

Then g_r is symmetric and $g_r = (\mathrm{Pf}_r)_*$. Given Z, we can take a highest-weight vector h of Z in the form $h(x) = \prod_{i=1}^{[n/2]} \mathrm{Pf}_i(x)^{c_i}$ with $0 \leqslant c_i \in \mathbf{Z}$ (see [S4, Theorem 2.D]). Then $p!h_* = \sum_\pi k_\pi$ as in the above three cases with k defined by the rule illustrated by the following example:

$$k(r, s, t, u, v, w, x, y, z) = g_1(r)g_1(s)g_2(t, u)g_2(v, w)g_3(x, y, z)$$

if $p = 9$, $c_1 = c_2 = 2$, and $c_3 = 1$. Then $k_\pi(x, y, \omega, \ldots, \omega)$ equals either $g_r(x, \omega, \ldots, \omega)g_s(y, \omega, \ldots, \omega)$ with $r + s \leqslant p$ or $g_r(x, y, \omega, \ldots, \omega)$ with $2 \leqslant r \leqslant p$. We can easily verify that $rg_r(x, \omega, \ldots, \omega) = \sum_{i=1}^{r} x_{2i-1,2i}$ and

$$r(r-1)g_r(x, y, \omega, \ldots, \omega) = \sum (x_{ab}y_{cd} - x_{ac}y_{bd} + x_{ad}y_{bc} + x_{cd}y_{ab} - x_{bd}y_{ac} + x_{bc}y_{ad}),$$

where $(a, b, c, d) = (2i-1, 2i, 2j-1, 2j)$, $1 \leqslant i < j \leqslant r$. Now $\{\varepsilon_{ij}\}$ and $\{\varepsilon_{ij}/2\}$ with $\varepsilon_{ij} = e_{ij} - e_{ji}$, $i < j$, form dual bases of T. Straightforward (and tedious) calculations show that $\sum_{i<j} k_\pi(\varepsilon_{ij}, \omega({}^t\varepsilon_{ij}/2)\omega, \omega, \ldots, \omega)$ is $[2\,\mathrm{Max}(r, s)]^{-1}$ for k_π of the first type, and -1 for the second type, and hence we obtain the desired result.

Given a quasi representation $\{\rho, X\}$ of $\mathfrak{R}$, we define an operator $L_{\rho,s}$ acting on $C^\infty(\mathscr{H}, X)$ by

$$L_{\rho,s} = -\theta D_{\rho\otimes\sigma_1^1,s}E_s. \tag{3.8}$$

This is a special case of (2.20), and $L_{\rho,s}(f\,\|_\rho\,\alpha) = (L_{\rho,s}f)\|_\rho\,\alpha$ for every $\alpha \in G^0_{\mathbf{R}}$.

THEOREM 3.3. *Let $0 \neq p = (p_s)_{s\in\mathbf{s}} \in \mathbf{Z}^{\mathbf{s}}$ with $p_s \geqslant 0$ for every $s \in \mathbf{s}$ and*

$$\rho(a, b, c) = \prod_{s\in\mathbf{s}} \det(b_s)^{k_s} \cdot \rho'(c)$$

for $(a, b) = (a_s, b_s)_{s\in\mathbf{s}} \in \prod_{s\in\mathbf{s}} K_s^c$ *and* $c \in \mathfrak{K}'$ *with a representation* $\{\rho', X\}$ *of* $\mathfrak{K}'$ *and* $k = (k_s)_{s\in\mathbf{s}} \in ((1/2)\mathbf{Z})^{\mathbf{s}}$. *For an irreducible subspace* Z *of* $S_i(T_s)$, *put*

$$\alpha_Z = i\{k_s - \lambda_s + (1-i)c_Z\},$$

where $\lambda_s = m+n,\ n,\ n+1$, *or* $n-1$, *according as* G_s *is of Type A, B, C, or D. Suppose that for each* s *such that* $p_s > 0$ *the number* k_s *satisfies the following inequalities:*

$$k_s > m+n+p_s-1 \quad \text{or} \quad k_s < m+n+1-p_s \quad \text{if } G_s \text{ is of Type A},$$

$$k_s > n+p_s-1 \quad \text{or} \quad k_s < n-(n-2)(p_s-1)/2 \quad \text{if } G_s \text{ is of Type B},$$

$$k_s > n+p_s \quad \text{or} \quad k_s < n+(3-p_s)/2 \quad \text{if } G_s \text{ is of Type C},$$

$$k_s > n+(p_s-3)/2 \quad \text{or} \quad k_s < n-p_s \quad \text{if } G_s \text{ is of Type D}.$$

Put $A_s^i = \prod_Z (1-\alpha_Z^{-1}L_{\rho,s})$ *for* $0 < i \leqslant p_s$, *where* Z *runs over all the irreducible subspaces of* $S_i(T_s)$. *(Notice that the estimate of* c_Z *given in Lemma 3.2 shows that* $\alpha_Z \neq 0$.*) Given* $f \in \mathcal{N}_\rho^p$, *put* $h = (\prod_{s\in\mathbf{s}'}\prod_{i=1}^{p_s} A_s^i)f$, *where* $\mathbf{s}' = \{s \in \mathbf{s} \mid p_s > 0\}$. *Then* $h \in \mathcal{M}_\rho$ *and* $f = h + \sum_{s\in\mathbf{s}'} L_{\rho,s}t_s$ *with* $t_s \in \mathcal{N}_\rho^p$.

Proof. We fix one s and consider f as a function of z_s, suppressing the remaining variables. By (3.5) we have $f = \sum_{i=1}^q f_i$, $f_i(z_s) = g_i(r_s(z_s))$, $q = p_s$, with a holomorphic map $g_i\colon H_s \to S_i(T_s)$. Then $E_s f = \sum_{i=1}^q E_s f_i$, and $E_s f_i$ is of degree $i-1$ in r_s. To study the highest term $E_s f_q$, write simply g for g_q. Then g_* is a holomorphic map of H_s into $Ml_q(T_s, X)$. Let us now write simply T, r, k, D, E, etc. for T_s, r_s, k_s, D_s, E_s, etc. By (3.4), for $u \in T$ we have

$$(Ef_q)(u) = -\sum_{v\in N} u_v(\partial/\partial r_v)g_*(r, \ldots, r) = -qg_*(u, r, \ldots, r).$$

By (2.6b) and (2.11) we see that $(D_{\rho\otimes\sigma^1}Ef_q)(u, v) = \det(\eta)^{-k}Y(\xi u\cdot{}^t\eta, v)$ with $Y(u, v) = D\{\det(\eta)^k(Ef_q)(\xi^{-1}u\cdot{}^t\eta^{-1})\}(v)$ for $u, v \in T$. Then by (3.1a,b) we have $D(\det(\eta)^k)(v) = k\cdot\det(\eta)^k\operatorname{tr}({}^t rv)$, and

$$D(\xi^{-1}u\cdot{}^t\eta^{-1})(v) = \begin{cases} -r\cdot{}^t v\xi^{-1}u\cdot{}^t\eta^{-1} - \xi^{-1}u\cdot{}^t\eta^{-1}\cdot{}^t vr & \text{(Types A, C, D)},\\ ((Qv\cdot{}^t rQ - r\cdot{}^t v)\xi^{-1}u\eta^{-1} - \xi^{-1}u\eta^{-1}\cdot{}^t vr & \text{(Type B)}.\end{cases}$$

Therefore

$$\begin{aligned} -q^{-1}(D_{\rho\otimes\sigma^1}Ef_q)(u, v) = k\cdot\operatorname{tr}({}^t rv)g_*(u, r, \ldots, r) - g_*(w, r, \ldots, r)\\ + (q-1)g_*(u, (Dr)(v), r, \ldots, r)\\ + \sum_{v\in N} v_v(\partial g_*/\partial z_v)(u, r, \ldots, r), \end{aligned}$$

where $w = r \cdot {}^t vu - Qv \cdot {}^t rQu + u \cdot {}^t vr$ for Type B and $w = r \cdot {}^t vu + u \cdot {}^t vr$ for the other three types. Applying θ to this equality, we obtain $kg(r)$ from the first term on the right-hand side and $-\lambda_s g(r)$ from the second term with λ_s given as in our theorem. Now $(Dr)(v)$ is given by (3.1c), and therefore θ times the third term is $(1-q)(\psi g)(r)$. The last sum $\sum_v$ is of degree at most $q-1$ in r_s. Thus we obtain

$$L_{\rho,s} f \equiv L_{\rho,s} f_q \equiv p_s\{k_s - \lambda_s + (1-p_s)\psi\} g(r) \qquad (\mathrm{mod}\ \mathcal{N}^{p'}),$$

where $p'_s = p_s - 1$ and $p'_t = p_t$ for $s \neq t \in \mathbf{s}$. (This is true even if $p_s = 1$.) Let φ_Z be the projection map $S_q(T_s) \to Z$. Then $p_s\{k_s - \lambda_s + (1-p_s)\psi\} = \sum_Z \alpha_Z \varphi_Z$. Now $L_{\rho,s}$ maps $\mathcal{N}_\rho^p$ into itself, and hence we easily see that $A_s^q f \in \mathcal{N}_\rho^{p'}$ with A_s^q defined in our theorem. Therefore if $h = (\prod_{s\in\mathbf{s}'} \prod_{i=1}^{p_s} A_s^i) f$, then $h \in \mathcal{N}_\rho^0 = \mathcal{M}_\rho$. Since $\prod_i A_s^i$ is a polynomial in $L_{\rho,s}$ whose constant term is 1, we obtain the desired expression for f.

Corollary 3.4. *The notation being as in Theorem 3.3, we have* $\langle \varphi, f \rangle = \langle \varphi, h \rangle$ *for* $\varphi \in \mathcal{M}_\rho$, *if either* φ *is a cusp form, or* $\Gamma\backslash\mathcal{H}$ *is compact for a congruence subgroup* Γ.

Proof. We have $f = h + \sum_{s\in\mathbf{s}'} L_{\rho,s} t_s$ with $t_s \in \mathcal{N}_\rho^p$. By Theorem 2.1

$$\langle \varphi, L_{\rho,s} t_s \rangle = \langle \varphi, -\theta D_{\rho\otimes\sigma_s^1,s} E_s t_s \rangle = \langle E_s \varphi, E_s t_s \rangle = 0,$$

since $E_s \varphi = 0$ because of the holomorphy of φ. This proves the desired equality. We need a suitable convergence condition, which is certainly satisfied for φ or Γ as above.

Remark 3.5. If $G_\mathbf{Q}$ is the semisimple part of $B^\times$ with a quaternion algebra B over a totally real number field, then we have a complete description of $\mathcal{N}^p$ (see [S7, Theorems 5.2, 5.4]). Therefore, in this case we can establish a holomorphic form h with the property of Corollary 3.4 in a different way, without any condition on k, and even with the additional property that the map $f \mapsto h$ keeps the rationality over a number field (see [S6, Proposition 9.4], [S8, Lemma 4.1]). A similar but less effective formula was obtained for G of a general type in [S5, Proposition 3.4] and [S7, Proposition 3.3]. The results of these types are crucial in the proof of the algebraicity of critical values of certain zeta functions. As we said in the introduction, it is our intention to employ the results in this section in our future investigation of algebraicity problems.

4. Explicit formulas for D_ρ^Z and E^Z. Fixing $s \in \mathbf{s}$ and taking $\mathbf{R}$-rational dual bases $\{\varepsilon_v\}_{v\in N}$ and $\{\varepsilon'_v\}_{v\in N}$ of T_s, put $z_s = \sum_{v\in N} z_{sv}\varepsilon_v$ as in Section 2, and

$$(4.1) \qquad \mathcal{D}_s = \sum_{v\in N} \varepsilon'_v \partial/\partial z_{sv}, \qquad \bar{\mathcal{D}}_s = \sum_{v\in N} \varepsilon'_v \partial/\partial \bar{z}_{sv}.$$

These are independent of the choice of bases, and $\mathcal{D}_s$ is practically the same as the matrix differential operator D of [S4, § 4] except when T_s is of Type D, in which case $\mathscr{D}_s = (1/2)D$. Given $g \in S_p(T_s)$, we define $g(\mathscr{D}_s)$ by

$$(4.2) \qquad g(\mathscr{D}_s) = g_*(\mathscr{D}_s, \ldots, \mathscr{D}_s) = \sum g_*(\varepsilon'_{v_1}, \ldots, \varepsilon'_{v_p})\partial^p/\partial z_{sv_1} \cdots \partial z_{sv_p},$$

where $(v_1, \ldots, v_p)$ runs over N^p, and we define $g(\overline{\mathscr{D}}_s)$ similarly. Then we can easily verify that

$$(4.3a) \qquad g(\mathscr{D}_s)f = [g, D_s^p f] \qquad \text{for every } f \in C^\infty(\mathscr{H}, X),$$

$$(4.3b) \qquad g(\mathscr{D}_s)h = p![g, h] \qquad \text{for every } h \in S_p(T_s)$$

with $[g, h]$ of (2.7). In the first formula we view $D_s^p f$ as $S_p(T_s, X)$-valued.

LEMMA 4.1. *Let $\delta_s(z_s) = \det(\eta_s(z_s))$. For each irreducible subspace Z of $S_p(T_s)$ there is a polynomial ψ_Z such that*

$$(4.4) \qquad \zeta(\mathscr{D}_s)\delta_s(z_s)^t = \psi_Z(t)\delta_s(z_s)^t\zeta(r_s(z_s))$$

for every $t \in \mathbf{C}$ and for every $\zeta \in Z$. Moreover ψ_Z can be given explicitly in terms of the highest weight of Z.

Proof. This is a special case of [S4, Theorem 4.3]. In fact, $\eta(z)$ can be given as $cz + d$ or $cz + (d/2)\cdot {}^t zQ^{-1}z + e$ as in that theorem (with c, d, and e involving $\bar{z}$), where the subscript s is suppressed for simplicity. Therefore $\zeta(\mathscr{D}_s)\delta^t = \psi_Z(t)\delta^t\zeta(\alpha)$ with $\alpha = {}^t c \cdot {}^t(cz + d)^{-1}$ or $\alpha = \mathscr{D}(\log[cz + (d/2)\cdot {}^t zQ^{-1}z + e])$ and a polynomial ψ_Z whose explicit form is given in [S4, Theorem 4.1] with the convention of [S4, Theorem 4.3]. (Notice that $\psi_Z(t)$ corresponds to $\beta_\sigma(s)$ of those theorems. For Type D, $\beta_\sigma(2s)$ appears in [S4, Theorem 4.3]. Since $\mathscr{D}_s = (1/2)D$, we have $\psi_Z(t) = 2^{-p}\beta_\sigma(2t)$.) Take in particular $Z = S_1(T_s)$ and $\zeta_u(x) = \mathrm{tr}({}^t ux)$ with $u \in T$. Then $\zeta_u(\mathscr{D}_s)\delta^t = (D\delta^t)(u) = t\delta^t\,\mathrm{tr}({}^t ur) = t\delta^t\zeta_u(r)$, and hence $\alpha = r$, since $\psi_Z(t) = t$ in this case. This proves our lemma.

We now define three invariants κ, ι, and ℓ, and an element q_s of $S_\ell(T_s)$ for each type as follows:

Type A. $\quad \kappa = n$, $\iota = 1$, $\ell = \mathrm{Min}(n, m)$, $q_s(u) = \det_\ell(u)$.
Type B. $\quad \kappa = n/2$, $\iota = 1$, $\ell = 2$, $q_s(u) = {}^t uQu$.
Type C. $\quad \kappa = (n + 1)/2$, $\iota = 1$, $\ell = n$, $q_s(u) = \det(u)$.
Type D. $\quad \kappa = (n - 1)/2$, $\iota = 1/2$, $\ell = [n/2]$, $q_s(u) = \mathrm{Pf}_\ell(u)$.

The numbers κ, ι, ℓ, and n depend on s, and so κ_s, ι_s, ℓ_s, and n_s may be clearer; for simplicity, however, we fix s in this section and use the letters without subscript. These are defined in all cases, but we use κ, ι, and q_s only when H_s is isomorphic to a tube domain. For such an H_s, $0 < k \in \mathbf{Z}$, and $t \in \mathbf{C}$, we define differential operators Δ_s^k and $\Delta_{s,t}$ on $\mathscr{H}$ by

(4.5a) $$\Delta_s^k = q_s^k(\mathscr{D}_s), \quad \overline{\Delta}_s^k = q_s^k(\overline{\mathscr{D}}_s), \quad \Delta_s = \Delta_s^1, \quad \overline{\Delta}_s = \overline{\Delta}_s^1,$$

(4.5b) $$\Delta_{s,t} f = \delta_s(z_s)^{\kappa-\iota-t}\Delta_s(\delta_s(z_s)^{t+\iota-\kappa} f).$$

Lemma 4.2. *Assuming (1.1) for H_s, for every $\alpha \in G_{\mathbf{R}}^0$ and a fixed $s \in \mathbf{s}$ we have*

(4.6a) $$\Delta_s^k(f \|_{\kappa-\iota k}\alpha) = (\Delta_s^k f)\|_{\kappa+\iota k}\alpha,$$

(4.6b) $$\Delta_{s,t}(f \|_t \alpha) = (\Delta_{s,t} f)\|_{t+2\iota}\alpha,$$

where $(f \|_t \alpha)(z) = j(\alpha_s, z_s)^{-t} f(\alpha z)$ with any choice of a branch of $j(\alpha_s, z_s)^{-t}$ with the convention that $j(\alpha_s, z_s)^{-t-\mu} = j(\alpha_s, z_s)^{-t} j(\alpha_s, z_s)^{-\mu}$ for $\mu \in \mathbf{Z}$.

Proof. The formula for $\Delta_s^k f$ for each type was given in [S2, Lemma 7.1, (8.6), (8.7), and (8.8)]. The formula for $\Delta_{s,t}$ can be derived from that for $\Delta_s f$ by substituting $\delta_s(z_s)^{t+\iota-\kappa} f$ for f.

Assuming (1.1) for H_s, put $Z = \mathbf{C}q_s^k$, $0 \leqslant k \in \mathbf{Z}$ with q_s as above. Then Z is an irreducible subspace of $S_{k\ell}(T_s)$ and so D_ρ^Z is meaningful for every quasi representation $\{\rho, X\}$ of $\mathfrak{R}$; moreover, $D_\rho^Z f$, as an $S_1(T_s, X)$-valued function, is completely determined by $(D_\rho^Z f)(q_s^k)$.

Lemma 4.3 *With $Z = \mathbf{C}q_s^k$ we have*

$$(D_\rho^Z f)(q_s^k) = \delta_s^{\kappa-\iota k}\rho(\Xi)^{-1}\Delta_s^k(\delta_s^{\iota k-\kappa}\rho(\Xi)f),$$

$$(E^Z f)(q_s^k) = \delta_s^{\kappa+\iota k}\overline{\Delta}_s^k(\delta_s^{\iota k-\kappa} f).$$

Proof. If H_s is of Type C, we observe that $\Phi_k(h)$ of [S2, (2.13)] equals $[q_s^k, h]$, and hence [S2, (2.15)] shows that $\Delta_\rho^{(k)} f = [q_s^k, D_\rho^{(kn)} f] = (D_\rho^Z f)(q_s^k)$. Therefore our formula for Type C is merely a reformulation of [S2, Proposition 7.2]. The formulas for D_ρ^Z in the other cases are given in the first paragraph of [S2, p. 842]. Those for $E^Z f$ are given in [S3, (2.14)] for Types A and C. The idea of proof explained in [S3, p. 278, lines 11–13] applies to the other cases.

Now we define the basic irreducible subspaces $Z_h \subset S_h(T_s)$ for $0 \leqslant h \leqslant \ell$ according to the type of T_s as follows:

Type A. $\det_h \in Z_h$ $(1 \leqslant h \leqslant \mathrm{Min}(n, m))$.

Type B. $Z_1 = S_1(T_s)$, $Z_2 = \mathbf{C}q_s$.

Type C. $\det_h \in Z_h$ $(1 \leqslant h \leqslant n)$.

Type D. $\mathrm{Pf}_h \in Z_h$ $(1 \leqslant h \leqslant [n/2])$.

We take $Z_0 = \mathbf{C} = S_0(T_s)$ in all cases. These Z_h for $h > 0$ are the same as those given in [S9, § 5].

LEMMA 4.4. *With Z_h as above, the following assertions hold.*

(1) *Let W (resp. Z) be irreducible subspaces of $S_{h-i}(T_s)$ (resp. $S_i(T_s)$), $0 \leqslant i \leqslant h \leqslant \ell$. Then $[Z_h, WZ] \neq 0$ if and only if $W = Z_{h-i}$ and $Z = Z_i$.*

(2) *Under (1.1) the map $(\varphi, \psi) \mapsto [q_s, \varphi\psi]$ for $(\varphi, \psi) \in Z_{\ell-i} \times Z_i$ defines a nondegenerate pairing.*

Proof. Since these assertions are trivial for Type B, we prove them for the other three types. Suppose $W \neq Z_{h-i}$; then the highest weight of W is higher than that of Z_h. Let φ be a highest-weight vector of W, and f (resp. g) an eigenvector of the diagonal elements of K_s^c in Z_h (resp. $S_i(T_s)$). Since $[\tau_s^h(a, b)f, \varphi g] = [f, \tau_s^h({}^ta, {}^tb)(\varphi g)]$ for $(a, b) \in K_s^c$, we see that $[f, \varphi g] = 0$, and hence $[Z_h, \varphi S_i(T_s)] = 0$. Now $\tau_s^h(K_s^c)[\varphi S_i(T_s)]$ spans $WS_i(T_s)$, and hence we obtain $[Z_h, WS_i(T_s)] = 0$. Similarly $[Z_h, S_{h-i}(T_s)Z] = 0$ if $Z \neq Z_i$. Since $[Z_h, S_h(T_s)] \neq 0$, we must have $[Z_h, Z_{h-i}Z_i] \neq 0$, which proves (1). If $h = \ell$, we have $Z_\ell = \mathbf{C}q_s$, and hence $[q_s, Z_{\ell-i}Z_i] \neq 0$. Then $(\varphi, \psi) \mapsto [q_s, \varphi\psi]$ must be nondegenerate, because of the irreducibility of $Z_{\ell-i}$ and Z_i.

LEMMA 4.5. *If $0 \leqslant i \leqslant h \leqslant \ell$, every element of Z_h is a finite sum $\sum_\mu f_\mu g_\mu$ with $f_\mu \in Z_{h-i}$ and $g_\mu \in Z_i$. In particular, q_s is such a finite sum with $f_\mu \in Z_{\ell-i}$ and $g_\mu \in Z_i$.*

Proof. Let Y be the set of such sums with $f_\mu \in Z_{h-i}$ and $g_\mu \in Z_i$. Then Y is a K_s^c-stable subspace of $S_h(T_s)$, and $[Z_h, Y] \neq 0$ by Lemma 4.4 (1), and hence $Z_h \subset Y$ as desired.

As a consequence of this lemma, for $0 \leqslant i \leqslant h \leqslant \ell$ we have

(4.7) $$\zeta(\mathcal{D}_s)f = 0 \quad \text{for every } \zeta \in Z_i$$
$$\Rightarrow \quad \zeta(\mathcal{D}_s)f = 0 \quad \text{for every } \zeta \in Z_h \quad \Rightarrow \quad \Delta_s f = 0.$$

Assuming (1.1) for H_s, let $\{\omega_\nu\}$ (resp. $\{\zeta_\nu\}$) be a basis of $Z_{\ell-i}$ (resp. Z_i), $0 \leqslant i \leqslant \ell$. We call them a *canonical pair of bases of $Z_{\ell-i}$ and Z_i* if $[q_s, \omega'_\mu\zeta'_\nu] = \delta_{\mu\nu}$ holds for the bases $\{\omega'_\nu\}$ and $\{\zeta'_\nu\}$ of $Z_{\ell-i}$ and Z_i which are dual to $\{\omega_\nu\}$ and $\{\zeta_\nu\}$, respectively. Given such a pair, we have

(4.8) $$[q_s, gh] = \sum_\nu [g, \omega_\nu][h, \zeta_\nu]$$

for every $g \in S_{\ell-i}(T_s)$ and $h \in S_i(T_s)$. In fact, if $W = Z_{\ell-i}$ and $Z = Z_i$, the above lemma shows that $[q_s, gh] = [q_s, \varphi_W g \cdot \varphi_Z h]$. Now $\varphi_W g = \sum_\nu [g, \omega_\nu]\omega'_\nu$ and $\varphi_Z h = \sum_\nu [h, \zeta_\nu]\zeta'_\nu$, and hence we obtain (4.8).

LEMMA 4.6. *For $0 \leqslant i \leqslant \ell$ let $\{\omega_\nu^i\}$ and $\{\zeta_\nu^i\}$ be a canonical pair of bases of $Z_{\ell-i}$ and Z_i. Then for $f, g \in C^\infty(\mathcal{H}, \mathbf{C})$ we have*

$$\Delta_s(fg) = \sum_{i=0}^{\ell} \binom{\ell}{i} \sum_\nu \omega_\nu^i(\mathcal{D}_s)f \cdot \zeta_\nu^i(\mathcal{D}_s)g.$$

Proof. For $u \in T_s$ we have $D_s^\ell(fg)(u) = \sum_{i=0}^{\ell} \binom{\ell}{i} (D_s^{\ell-i}f)(u)(D_s^i g)(u)$, and hence, by (4.3a) and (4.8)

$$\Delta_s(fg) = [q_s, D_s^\ell(fg)] = \sum_{i=0}^{\ell} \binom{\ell}{i} [q_s, (D_s^{\ell-i}f)(D_s^i g)]$$

$$= \sum_{i=0}^{\ell} \binom{\ell}{i} \sum_{v} [\omega_v^i, D_s^{\ell-i}f][\zeta_v^i, D_s^i g]$$

$$= \sum_{i=0}^{\ell} \binom{\ell}{i} \sum_{v} \omega_v^i(\mathscr{D}_s)f \cdot \zeta_v^i(\mathscr{D}_s)g, \qquad \text{Q.E.D.}$$

Condition (1.1) is necessary for this lemma, but unnecessary for the following result.

THEOREM 4.7. *Let $\{\rho, X\}$ be a quasi representation of $\mathfrak{K}$ and let $Z = Z_h \subset S_h(T_s)$, $0 < h \leqslant \ell$, with a fixed $s \in \mathbf{s}$. Then, for $f \in C^\infty(\mathscr{H}, X)$, $\zeta \in Z$, and $\alpha \in G_{\mathbf{R}}^0$ we have*

(4.9a) $$[\zeta(\mathscr{D}_s)f] \|_\lambda \alpha = \zeta'(\mathscr{D}_s)(f \|_\lambda \alpha),$$

(4.9b) $$[\delta_s^\lambda \zeta(\bar{\mathscr{D}}_s)(\delta_s^{-\lambda}f)] \circ \alpha = \delta_s^\lambda \zeta''(\bar{\mathscr{D}}_s)[\delta_s^{-\lambda}(f \circ \alpha)],$$

(4.9c) $$(D_\rho^Z f)(\zeta) = \delta_s^\lambda \rho(\Xi)^{-1} \zeta(\mathscr{D}_s)[\delta_s^{-\lambda}\rho(\Xi)f],$$

(4.9d) $$(E^Z f)(\zeta) = \delta_s^\lambda \zeta^*(\bar{\mathscr{D}}_s)(\delta_s^{-\lambda}f).$$

Here $\zeta' = \sigma_Z(\Lambda_\alpha(z)^{-1})\zeta$, $\zeta'' = \sigma_Z(\overline{\Lambda_\alpha(z)}^{-1})\zeta$, and $\zeta^* = \tau_Z(\Xi)\zeta$; $\lambda = h - 1$ for Types A and D, $\lambda = (h-1)(n-2)/2$ for Type B, and $\lambda = (h-1)/2$ for Type C.

Proof. These formulas were given in [S3, (2.9a,b), (2.11)] for Types A (with $m = n$) and C. Here we give a proof in all cases by the same technique. We first assume (1.1) for H_s. Taking fg of Lemma 4.6 to be $\delta_s^{t+\iota-\kappa}f$ with $t \in \mathbf{R}$, we have

$$\Delta_{s,t}f = \delta_s^{\kappa-\iota-t} \sum_{i=0}^{\ell} \binom{\ell}{i} \sum_{v} \omega_v^i(\mathscr{D}_s)\delta_s^{t+\iota-\kappa} \cdot \zeta_v^i(\mathscr{D}_s)f$$

$$= \sum_{i=0}^{\ell} \binom{\ell}{i} \psi_{\ell-i}(t+\iota-\kappa) \sum_{v} \omega_v^i(r_s)\zeta_v^i(\mathscr{D}_s)f$$

by Lemma 4.1, where $\psi_i = \psi_{Z_i}$. The explicit form of ψ_i given in [S4, Theorems 4.1 and 4.3] (see also the proof of Lemma 4.1) is

$$\text{(4.10a)} \qquad \psi_i(t) = \begin{cases} \prod_{a=1}^{i} (t + a - 1) & \text{(Type A)}, \\ \prod_{a=1}^{i} (t + (a - 1)/2) & \text{(Type C)}, \\ \prod_{a=1}^{i} (t + a - 1) & \text{(Type D)}, \end{cases}$$

$$\text{(4.10b)} \quad \psi_0(t) = 1, \quad \psi_1(t) = t, \quad \psi_2(t) = t(t - 1 + n/2) \quad \text{(Type B)}.$$

With λ determined for a fixed h, $0 < h \leqslant \ell$, as above, we see that $\psi_{\ell-i}(\lambda + \iota - \kappa) = 0$ for $i < h$ and $\psi_{\ell-h}(\lambda + \iota - \kappa) \neq 0$, and hence

$$\text{(4.11)} \qquad \Delta_{s,\lambda} f = \sum_{i=h}^{\ell} \binom{\ell}{i} \psi_{\ell-i}(\lambda + \iota - \kappa) \sum_{\nu} \omega_\nu^i(r_s) \zeta_\nu^i(\mathscr{D}_s) f.$$

Since $\zeta(\mathscr{D}_s)$ and $\zeta'(\mathscr{D}_s)$ of (4.9a) are holomorphic differential operators, it is sufficient to prove (4.9a) for holomorphic f. By (4.6b) we have $\Delta_{s,\lambda}(f \|_\lambda \alpha) = (\Delta_{s,\lambda} f)\|_{\lambda+2\iota}\alpha$. If f is holomorphic, (4.11) shows that both sides of the last equality are polynomials in r_s with holomorphic functions as coefficients. Now $\sigma_s^\ell(a, b) q_s = \det(b)^{-2\iota} q_s$ for every $(a, b) \in K_s^c$. Fix $\alpha \in G_{\mathbf{R}}^0$ and write $\zeta' = \sigma_s^p(\Lambda_\alpha(z)^{-1})\zeta$ for $\zeta \in S_p(T_s)$, $0 \leqslant p \leqslant \ell$. Then we see that $\{(\omega_\nu^i)'\}$ and $\{j(\alpha_s, z_s)^{-2\iota}(\zeta_\nu^i)'\}$ form a canonical pair of bases of $Z_{\ell-i}$ and Z_i. Therefore, changing $\{\omega_\nu^i\}$ and $\{\zeta_\nu^i\}$ for these bases, from (4.11) we obtain

$$\text{(4.12)} \qquad j(\alpha_s, z_s)^{2\iota} \Delta_{s,\lambda} f = \sum_{i=h}^{\ell} \binom{\ell}{i} \psi_{\ell-i}(\lambda + \iota - \kappa) \sum_{\nu} (\omega_\nu^i)'(r_s)(\zeta_\nu^i)'(\mathscr{D}_s) f.$$

On the other hand,

$$\text{(4.13)} \quad j(\alpha_s, z_s)^{2\iota} \cdot (\Delta_{s,\lambda} f)\|_{\lambda+2\iota}\alpha = \sum_{i=h}^{\ell} \binom{\ell}{i} \psi_{\ell-i}(\lambda + \iota - \kappa) \sum_{\nu} \omega_\nu^i(r_s \circ \alpha)[\zeta_\nu^i(\mathscr{D}_s) f]\|_\lambda \alpha.$$

Substitute $f\|_\lambda \alpha$ for f in (4.12) and compare the result with (4.13). By Lemma 3.1 the highest terms in r_s produce the equality $(\zeta_\nu^h)'(\mathscr{D}_s)(f\|_\lambda \alpha) = [\zeta_\nu^h(\mathscr{D}_s) f]\|_\lambda \alpha$ for every ν. This proves (4.9a) under (1.1).

Let us now prove (4.9a) for Type A with $n \neq m$. Clearly we may assume $n > m$. We consider the objects for Type A with $G = SU(n, n)$ and denote them by G°, H°, $\mathscr{D}^\circ$, etc. (Since our formulas concern essentially the functions on H_s, we drop the subscript s for simplicity.) Given $f \in C^\infty(H, X)$, define $f^\circ \in C^\infty(H^\circ, X)$ by $f^\circ(z\ w) = f(z)$ for $(z\ w) \in H^\circ$ with $z \in H$. We can similarly define $\zeta^\circ \in Z_h^\circ$ for $\zeta \in Z_h$. For $\alpha \in G$ and $z \in H$ put $\alpha^\circ = \operatorname{diag}[\alpha, 1_{n-m}]$ and $z^\circ = (z\ 0)(\in H^\circ)$. Then we have $(f\|_\lambda\alpha)^\circ = f^\circ\|_\lambda\alpha^\circ$ and $[\zeta(\mathscr{D})f]^\circ = \zeta^\circ(\mathscr{D}^\circ)f^\circ$. Therefore we can derive (4.9a) by evaluating the corresponding formula on H° at z°. As for Type D with $n = 2\ell + 1$, we define G°, H°, $\mathscr{D}^\circ$, etc. to be objects of degree $n+1$ and embed G and H into G° and H° in a natural way. Then we obtain (4.9a) from the corresponding formula on H°.

To prove (4.9c), define $R: C^\infty(\mathscr{H}, X) \to C^\infty(\mathscr{H}, S_1(Z, X))$ by

$$(4.14) \qquad (Rf)(\zeta) = \delta_s^\lambda \rho(\Xi)^{-1}\zeta(\mathscr{D}_s)[\delta_s^{-\lambda}\rho(\Xi)f] \qquad (\zeta \in Z, f \in C^\infty(\mathscr{H}, X)).$$

From (4.9a), (1.4a,b), and (2.18a) we can easily derive $R(f\|_\rho\alpha) = (Rf)\|_{\rho\otimes\tau_Z}\alpha$ for every $\alpha \in G_{\mathbf{R}}^0$, that is, R satisfies the same transformation formula as D_ρ^Z. Assuming that H_s is of bounded form, evaluate (4.14) at $w \in \mathscr{H}$ with $w_s = 0$. From (3.1a,b,c) we see that $D_s^i\xi_s = 0$ and $D_s^i\eta_s = 0$ for $i > 0$ at w, and hence $D_s^h[\delta_s^{-\lambda}\rho(\Xi)f] = D_s^h f$ at w. Therefore $(Rf)(\zeta)(w) = [\zeta(\mathscr{D}_s)f](w)$ (cf. [S2, Lemma 2.2]). Similarly $(D_\rho^Z f)(\zeta)(w) = [\zeta(\mathscr{D}_s)f](w)$ by (2.17), and hence $(Rf)(\zeta)(w) = (D_\rho^Z f)(\zeta)(w)$. Given $z \in \mathscr{H}$, we can find $\alpha \in G_{\mathbf{R}}^0$ so that $\alpha(w) = z$. Then, with $\zeta' = \sigma_Z(\Lambda_\alpha(w)^{-1})\zeta$ we have

$$\begin{aligned}
\rho(\Lambda_\alpha(w)^{-1})(Rf)(\zeta)(\alpha w) &= ((Rf)\|_{\rho\otimes\tau_Z}\alpha)(\zeta')(w) \\
&= R(f\|_\rho\alpha)(\zeta')(w) \\
&= D_\rho^Z(f\|_\rho\alpha)(\zeta')(w) \\
&= ((D_\rho^Z f)\|_{\rho\otimes\tau_Z}\alpha)(\zeta')(w) \\
&= \rho(\Lambda_\alpha(w)^{-1})(D_\rho^Z f)(\zeta)(\alpha w).
\end{aligned}$$

This proves that $D_\rho^Z f = Rf$, which together with (4.14) proves (4.9c) for H_s of bounded form. If H_s is of tube form, we take a map β of the bounded domain to H_s so that βw is a given point on H_s. Then we can apply the same technique to obtain the desired formula. This requires various formulas for such a β. For Types A and C, necessary formulas are given and this procedure is explained in detail in [S2] and [S3, p. 278]. Types B and D can be handled in the same manner (see the last paragraph of [S2, p. 842]). We thus obtain (4.9c) for all types. Now (4.9b) can be obtained by taking complex conjugation of (4.9a). Employing (4.9b), we can show that the right-hand side of (4.9d) satisfies the same transformation formula as E^Z. Then we obtain (4.9d) by the same type of argument as in the proof of (4.9c).

Remark 4.8. The significance of Z_h can be partly explained by the fact that the operators L_ρ^Z of (2.20) for $Z = Z_{h,s}$, $1 \leqslant h \leqslant \ell$, generate the ring of all G_s-invariant operators on C_ρ, provided that the restriction of ρ to K_s^c is scalar-valued. The same is true with M_ρ^Z in place of L_ρ^Z. For details see [S9]. These operators can be expressed explicitly by means of Theorem 4.7.

5. Singular forms and singular weights. The purpose of this section is to study the elements f of $\mathcal{M}_\rho$ satisfying $\zeta(\mathcal{D}_s)f = 0$ for every $\zeta \in Z_i$ with a fixed i such that $0 < i \leqslant \ell$. We call such an f *singular*. This notion depends on the choice of s, but in certain cases it is independent of s, as will be shown later.

LEMMA 5.1. *Let $\pi\colon K_s^c \to GL(X)$ be such that $\pi(a, b) = \det(b)^e\pi_0(a, b)$ with $e = 0$ or $1/2$ and a rational representation $\{\pi_0, X\}$ of K_s^c for some fixed $s \in \mathbf{s}$. Then for every $p \in \mathbf{Z}$, $\geqslant 0$, the following assertions hold:*

(1) *$[D_s^p\pi(\xi_s, \eta_s)](u) = \pi(\xi_s, \eta_s)\psi(u, r_s)$ for $u \in T_s$ with some $\psi \in S_p(T_s, S_p(T_s, \operatorname{End}(X)))$.*

(2) *For every $\zeta \in S_p(T_s)$ there exists an element ω of $S_p(T_s, \operatorname{End}(X))$ such that $\pi(\xi_s, \eta_s)^{-1}\zeta(\mathcal{D}_s)\pi(\xi_s, \eta_s) = \omega(r_s)$.*

Proof. The first formula can be written, with the subscript s dropped,

$$[D^p\pi(\xi, \eta)](u_1, \ldots, u_p) = \pi(\xi, \eta)\varphi(u_1, \ldots, u_p, r, \ldots, r) \tag{5.1}$$

for $u_i \in T$ with some $\varphi \in Ml_{2p}(T, \operatorname{End}(X))$. We prove this by induction on p. First we have

$$[D\pi(\xi, \eta)](u) = \pi(\xi, \eta)\, d\pi(\xi^{-1}(D\xi)(u), \eta^{-1}(D\eta)(u)) \qquad (u \in T),$$

where $d\pi(x, y) = d\pi_0(x, y) + e \cdot \operatorname{tr}(y)1_X$ for (x, y) in the Lie algebra of K_s^c. If $e = 0$, this was given in [S5, p. 361, line 15]; the case $e = 1/2$ can be easily derived from that result. Now $\xi^{-1}(D\xi)(u)$ and $\eta^{-1}(D\eta)(u)$ are bilinear in (r, u) as given in (3.1a,b). This proves the case $p = 1$. Assuming (5.1) for a p, apply D to it and observe that $(Dr)(u) = \varphi(u, r)$ with $\varphi \in S_1(T, S_2(T))$ as shown in (3.1c). Then we obtain (5.1) for degree $p + 1$, which proves (1). Employing (1), for $\zeta \in S_p(T)$ we have

$$\zeta(\mathcal{D}_s)\pi(\xi, \eta) = [\zeta, D^p\pi(\xi, \eta)] = \pi(\xi, \eta)[\zeta(*), \psi(*, r)],$$

which proves (2).

For each integer h such that $0 < h \leqslant \ell$ define λ_h by

$$\lambda_h = \begin{cases} h - 1 & \text{(Types A and D)}, \\ (h - 1)/2 & \text{(Type C)}, \end{cases} \tag{5.2a}$$

$$\lambda_1 = 0, \qquad \lambda_2 = (n - 2)/2 \qquad \text{(Type B)}. \tag{5.2b}$$

This λ_h is the number λ in Theorem 4.7. To emphasize the dependence on $s \in \mathbf{s}$, we denote the symbols ℓ, λ_h, and Z_h by ℓ_s, $\lambda_{h,s}$, and $Z_{h,s}$.

THEOREM 5.2. *Let $\{\rho, X\}$ be a quasi representation of $\mathfrak{K}$, and ρ_s for a fixed $s \in \mathbf{s}$ the restriction of ρ to K_s^c; let h be an integer such that $0 < h \leqslant \ell_s$. Suppose that $\rho_s(a, b) = \det(b)^\lambda\pi(a, b)$ with $\lambda = \lambda_{h,s}$ and $\pi(a, b) = \det(b)^e\pi_0(a, b)$ with π and π_0 as in Lemma 5.1. Let $Z = Z_{h,s}$ and $f \in \mathcal{M}_\rho$. Suppose that*

(1) $\zeta(\mathscr{D}_s)\pi(\xi_s, \eta_s) = 0$ *for every* $\zeta \in Z$.
(2) $(D_\rho^Z f, f)$ *is an integrable pair of type* (h, s) *in the sense of (2.23).*
Then $D_\rho^Z f = 0$ *and* $\zeta(\mathscr{D}_s)f = 0$ *for every* $\zeta \in Z$.

Proof. By (4.9c), for every $\zeta \in Z$ we have

$$(D_\rho^Z f)(\zeta) = \pi(\xi_s, \eta_s)^{-1}\zeta(\mathscr{D}_s)[\pi(\xi_s, \eta_s)f].$$

The right-hand side is a finite sum $\pi(\xi_s, \eta_s)^{-1}\sum \alpha(\mathscr{D}_s)\pi(\xi_s, \eta_s)\cdot\beta(\mathscr{D}_s)f$ with $\alpha \in S_i(T_s)$ and $\beta \in S_{h-i}(T_s)$, $0 \leqslant i < h$. The terms for $i = h$ do not appear because of our assumption (1). By Lemma 5.1(2) this can be written $\sum \gamma(r_s)\beta(\mathscr{D}_s)f$ with $\gamma \in S_i(T_s, \mathrm{End}(X))$, $0 \leqslant i < h$. Now E^h annihilates such a sum, and hence $E^Z D_\rho^Z f = 0$. Then, by Theorem 2.1

$$\langle D_\rho^Z f, D_\rho^Z f\rangle = (-1)^h\langle f, \theta E^Z D_\rho^Z f\rangle = 0,$$

and hence $D_\rho^Z f = 0$. We need assumption (2) here. Now $D_\rho^Z f$, as a polynomial in r_s, has $\zeta(\mathscr{D}_s)f$ as its constant term, and hence $\zeta(\mathscr{D}_s)f = 0$.

It should be noted that condition (1) is trivially satisfied if π is trivial, that is, if $\rho_s(a, b) = \det(b)^\lambda 1_X$, where 1_X is the identity map of X onto itself. Condition (2) is satisfied if $\Gamma\backslash\mathscr{H}$ is compact for a congruence subgroup Γ of G^0. In the noncompact case, (2) is a nontrivial condition, but we shall see later that it is always satisfied under certain conditions on G and ρ. Nontrivial singular forms in the cocompact case can be obtained as pullbacks of singular forms on $Sp(n, \mathbf{Q})$.

The above theorem is applicable to both tube and nontube domains. We now assume that $\mathscr{H}$ is a tube and G belongs to the following types of groups:

Type A: $G = \{\alpha \in SL_{2n}(F_1) \mid {}^t\alpha' J_n \alpha = J_n\}$.

Type B: $G = \{\alpha \in SL_{n+2}(F) \mid {}^t\alpha\tilde{P}\alpha = \tilde{P}\}$, $\quad \tilde{P} = \mathrm{diag}\left[P, -\begin{pmatrix}0 & 1\\ 1 & 0\end{pmatrix}\right]$.

Type C: $G = \{\alpha \in GL_{2n}(F) \mid {}^t\alpha J_n \alpha = J_n\}$.

Type D: $G = \{\alpha \in GL_n(B) \mid {}^t\alpha' J_\ell \alpha = J_\ell\}$, $\quad n = 2\ell$.

In each case we take a basic totally real algebraic number field F of finite degree and denote by $\mathbf{s}$ the set of all archimedean primes of F. In addition, for Type A we take a totally imaginary quadratic extension F_1 of F, and for Type D take a totally definite quaternion algebra B over F; we then denote by α' the image of α under the Galois involution of F_1/F or the main involution of B/F. For Type B we take $P = {}^tP \in GL_n(F)$ whose signature is $(n-1, 1)$ at every $s \in \mathbf{s}$.

In fact, there are more groups acting on tube domains to which our theory is applicable. For example, a quaternion unitary group of even degree over a totally indefinite quaternion algebra over F is of Type C. For simplicity, however, we restrict our treatment to the above types.

Now we can take G_s for $s \in \mathbf{s}$ to be the identity component of the localization of G at s. We view each $s \in \mathbf{s}$ as an embedding of F, F_1, and B into $\mathbf{R}$, $\mathbf{C}$, and the Hamilton quaternions $\mathbf{H}$. Then G_s is of the type described in Section 1 associated with H_s of tube type. This is clear for Types A, B, and C. In fact, for Type B, we take $A_s \in GL_n(\mathbf{R})$ so that $P^s = {}^tA_sQA_s$, put $\tilde{A}_s = \mathrm{diag}[A_s, 1_2]$, and observe that $\alpha \mapsto \tilde{A}_s\alpha^s\tilde{A}_s^{-1}$ embeds G into the group of Type B of tube form given in Section 1, where α^s is the image of α under s. As for Type D, we fix an embedding of $\mathbf{C}$ into $\mathbf{H}$, as well as an element $\mathbf{j}$ of $\mathbf{H}$, so that $\mathbf{H} = \mathbf{C} + \mathbf{j}\mathbf{C}$ and $\mathbf{j}c = \bar{c}\mathbf{j}$ for $c \in \mathbf{C}$. We then define embeddings φ_n, ψ_n: $\mathbf{H}_n^n \to \mathbf{C}_{2n}^{2n}$ by

$$\varphi_n((a_{ik} + b_{ik}\mathbf{j})) = \begin{bmatrix} c_{11} & \cdots & c_{1n} \\ \cdots & \cdots & \cdots \\ c_{n1} & \cdots & c_{nn} \end{bmatrix}, \qquad c_{ik} = \begin{pmatrix} a_{ik} & -b_{ik} \\ \bar{b}_{ik} & \bar{a}_{ik} \end{pmatrix} \quad (a_{ik}, b_{ik} \in \mathbf{C}),$$

$$\psi_n(x) = \mathrm{diag}[J'_\ell, 1_n]^{-1}\varphi_n(x)\,\mathrm{diag}[J'_\ell, 1_n] \qquad \text{for } x \in \mathbf{H}_n^n,$$

where $J'_\ell = \mathrm{diag}[J_1, \ldots, J_1]$ $(\in \mathbf{C}_{2\ell}^{2\ell})$. Now each s embeds G into $\{\alpha \in GL_n(\mathbf{H}) | {}^t\bar{\alpha}J_\ell\alpha = J_\ell\}$, which is mapped under ψ_n onto the group

$$\{\alpha \in SL_{2n}(\mathbf{C}) | {}^t\alpha L\alpha = L, \bar{\alpha}J'_n = J'_n\alpha\}$$

of Type D of tube form given in Section 1. Therefore we see that the present G of Type D fits in our formulation. Let us now put

$$T_{\mathbf{Q}} = \{b \in (F_1)_n^n | {}^tb' = b\} \qquad \text{(Type A)},$$

$$T_{\mathbf{Q}} = F^n \qquad \text{(Type B)},$$

$$T_{\mathbf{Q}} = \{b \in F_n^n | {}^tb = b\} \qquad \text{(Type C)},$$

$$T_{\mathbf{Q}} = \{b \in B_\ell^\ell | {}^tb' = b\} \qquad \text{(Type D)}.$$

Then the map

$$b \mapsto (b_s)_{s \in \mathbf{a}} \quad \text{with} \quad b_s = \begin{cases} b^s & \text{(Types A and C)}, \\ A_sb^s & \text{(Type B)}, \\ J'_\ell\varphi_\ell(b^s) & \text{(Type D)} \end{cases}$$

embeds $T_{\mathbf{Q}}$ into $\prod_{s \in \mathbf{a}} U_s$, where U_s is the real vector subspace of T_s given in Section 1. Hereafter we identify $T_{\mathbf{Q}}$ with its image under this map. Then $\prod_{s \in \mathbf{a}} U_s = T_{\mathbf{Q}} \otimes_{\mathbf{Q}} \mathbf{R}$. Define a pairing $\{\ ,\ \}$: $\prod_{s \in \mathbf{a}} T_s \times \prod_{s \in \mathbf{a}} T_s \to \mathbf{C}$ by

$$\{u, v\} = \begin{cases} \sum_{s \in \mathbf{s}} \operatorname{tr}({}^t u_s v_s) & \text{(Types A and C)}, \\ -\sum_{s \in \mathbf{s}} {}^t u_s Q v_s & \text{(Type B)}, \\ \sum_{s \in \mathbf{s}} \operatorname{tr}(J'_\ell u_s J'_\ell v_s) & \text{(Type D)}, \end{cases} \tag{5.3}$$

and put

$$\mathbf{e}(c) = e^{2\pi i c} \quad \text{for } c \in \mathbf{C}, \qquad \mathbf{e}(u, v) = \mathbf{e}(\{u, v\}) \quad \text{for } u, v \in \prod_{s \in \mathbf{s}} T_s. \tag{5.4}$$

Then $\{u, v\} \in \mathbf{Q}$ for $u, v \in T_{\mathbf{Q}}$. Now G^0 has a subgroup $R_{\mathbf{Q}}$ whose action on $\mathscr{H}$ consists of the translations $(z_s)_{s \in \mathbf{s}} \mapsto (z_s + c_s)_{s \in \mathbf{s}}$ with $(c_s)_{s \in \mathbf{s}} \in T_{\mathbf{Q}}$. Therefore, given a quasi representation $\{\rho, X\}$ as before, every element f of $\mathscr{M}_\rho$ has an expansion of the form

$$f(z) = \sum_{b \in L} c(b)\mathbf{e}(b, z) \tag{5.5}$$

with $c(b) \in X$ and a $\mathbf{Z}$-lattice L in $T_{\mathbf{Q}}$. This is so even if the weight is "half-integral," because of (2.26). For $\zeta \in S_p(T_s)$ we easily see that

$$\zeta(\mathscr{D}_s)f(z) = (2\pi i)^p \sum_{b \in L} \zeta_1(b_s)c(b)\mathbf{e}(b, z), \tag{5.6}$$

where $\zeta_1 = \zeta$ for Types A and C, $\zeta_1(u) = \zeta(-Qu)$ for Type B, and $\zeta_1(u) = \zeta(-J'_\ell u J'_\ell)$ for Type D.

LEMMA 5.3. *If ρ is irreducible, then, for $0 \neq f \in \mathscr{M}_\rho$, the $c(b)$ of (5.5) for all $b \in L$ generate X over* $\mathbf{C}$.

Proof. Let $K_{\mathbf{Q}}$ be the subgroup of G consisting of the elements of the forms

$$\operatorname{diag}[({}^t\beta')^{-1}, \beta] \quad \text{with } \beta \in GL_\ell(M) \qquad \text{(Types A, C, D)}, \tag{5.7a}$$

$$\operatorname{diag}[\alpha, \beta^{-1}, \beta] \quad \text{with } \beta \in F^\times, \alpha \in SL_n(F), {}^t\alpha P \alpha = P \qquad \text{(Type B)}, \tag{5.7b}$$

where $\beta' = \beta$ for Type C, and M denotes F_1, F, and B for Types A, C, and D, respectively; for Type A we assume $\det(\beta) \in F$. Let $K_{\mathbf{R}}$ be the subgroup of $G_{\mathbf{R}}$ defined in a similar fashion with $F \otimes_{\mathbf{Q}} \mathbf{R}$ in place of F, and $K^0_{\mathbf{R}}$ its identity component. If γ is as in (5.7a,b), then $\Lambda_\gamma(z) = (\overline{\beta^s}, \beta^s)_{s \in \mathbf{s}}$ for Type A, $\Lambda_\gamma(z) = (\beta^s, \beta^s)_{s \in \mathbf{s}}$ for Types C and D, and $\Lambda_\gamma(z) = ({}^t(A_s \alpha^s A_s^{-1})^{-1}, \beta^s)_{s \in \mathbf{s}}$ for Type B. (Strictly speaking, we have to assume that $\gamma \in G^0_{\mathbf{R}}$ for Types A and B.) Write Λ_γ for $\Lambda_\gamma(z)$ as it does not involve z. Put $\mathscr{G} = \{\Lambda_\gamma | \gamma \in K^0_{\mathbf{R}}\}$. Since the Lie algebra of $\mathscr{G}$ generates that of $\mathfrak{K}$ over $\mathbf{C}$, ρ is irreducible on $\mathscr{G}$. Let K^1 (resp. $\mathscr{G}^1$) denote the semisimple part of $K^0_{\mathbf{R}}$ (resp. $\mathscr{G}$). Take Γ so that $f \in \mathscr{M}_\rho(\Gamma)$. Then $\mathscr{G}^1/\{\Lambda_\gamma | \gamma \in \Gamma \cap K_{\mathbf{Q}} \cap K^1\}$ is of finite volume, and hence ρ must be irreducible on $\{\Lambda_\gamma | \gamma \in \Gamma \cap K_{\mathbf{Q}} \cap K^1\}$ by virtue of Borel's result in [B]. Put $V = \{h \in S_1(X) | h \circ f = 0\}$; suppose $0 \neq h \in V$. If $h'(x) = h(\rho(\Lambda_\gamma^{-1})x)$ with $\gamma \in \Gamma \cap K_{\mathbf{Q}} \cap K^1$, then $h'(f(\gamma z)) = h(f \|_\rho \gamma) = 0$, and hence $h' \in V$. Then $V = S_1(X)$ because of the above irreducibility, which implies that $f = 0$, a contradiction. Therefore $V = \{0\}$, which proves our lemma.

LEMMA 5.4. *Suppose that $\sum_{b \in L} \varphi_b(z)\mathbf{e}(b, z) = 0$ on the whole $\mathscr{H}$, where L is a $\mathbf{Z}$-lattice in $T_{\mathbf{Q}}$ and each φ_b is a polynomial in the entries of $(z_s)_{s \in \mathbf{s}}$ whose total degree is $\leqslant p$ with an integer p independent of b. Then $\varphi_b = 0$ for all $b \in L$.*

Proof. We prove this by induction on p. The case $p = 0$ is obvious. Let $M = \{c \in T_{\mathbf{Q}} | \{L, c\} \subset \mathbf{Z}\}$. Then we have $\sum_{b \in L} [\varphi_b(z + c) - \varphi_b(z)]\mathbf{e}(b, z) = 0$ for every $c \in M$. By induction we have $\varphi_b(z + c) = \varphi_b(z)$ for every $c \in M$. Clearly such a polynomial φ_b must be a constant, and hence is 0.

Given $\{\pi, X\}$ as in Lemma 5.1 with G of the present type, we define $R_\pi(z_s)$ $(\in GL(X))$ for $z_s \in H_s$ by

$$(5.8)\qquad R_\pi(z_s) = \begin{cases} \pi({}^t z_s, z_s) & \text{(Types A and C)}, \\ \pi(Q - w^{-1} z_s \cdot {}^t z_s, -w), \quad w = {}^t z_s Q z_s/2 & \text{(Type B)}, \\ \pi(J'_\ell z_s, J'_\ell z_s) & \text{(Type D)}. \end{cases}$$

Each right-hand side is well defined, since it can be written $\pi(\lambda(\alpha_s, z_s), \mu(\alpha_s, z_s))$ with some $\alpha \in G^0$ as will be shown in the proof of Theorem 5.5 below. Now for $\zeta \in S_t(T_s)$ we easily see that

$$(5.9)\qquad \zeta(\mathscr{D}_s)\pi(\xi_s, \eta_s) = 0 \iff \zeta(\mathscr{D}_s)R_\pi(z_s) = 0.$$

Therefore condition (1) of Theorem 5.2 is equivalent to

$$(5.10)\qquad \zeta(\mathscr{D}_s)R_\pi(z_s) = 0 \qquad \text{for every } \zeta \in Z_h.$$

THEOREM 5.5. *Let the symbols Z, λ, ρ_s, and π be as in Theorem 5.2 with G of the present type. If ρ is irreducible and $\zeta(\mathscr{D}_s)f = 0$ for every $\zeta \in Z$ with some $f \in \mathscr{M}_\rho$, $\neq 0$, then $\zeta(\mathscr{D}_s)R_\pi(z_s) = 0$ for every $\zeta \in Z$.*

Proof. We first take an element α of G^0 such that $R_\pi(z_s) = \pi(\lambda(\alpha_s, z_s), \mu(\alpha_s, z_s))$. For Types A, C, and D this can be achieved by $\alpha = J_\ell$. As for Type B, we note that for $\beta_s \in G_s$ and $z_s \in H_s$, the point $z' = \beta_s z_s$ and $\mu(\beta_s, z_s)$ are determined by

$$(5.11)\qquad \beta_s \begin{bmatrix} z_s \\ w \\ 1 \end{bmatrix} = \begin{bmatrix} z' \\ w' \\ 1 \end{bmatrix} \mu(\beta_s, z_s), \qquad w = {}^t z_s Q z_s/2,\ w' = {}^t z' Q z'/2.$$

Without losing generality we may assume that Q, P, and A_s are diagonal. Consider the element $\alpha = \mathrm{diag}\left[Q, -\begin{pmatrix}0 & 1\\ 1 & 0\end{pmatrix}\right]$ of $SL_{n+2}(F)$. Then $\alpha \in G$ and this defines an element of G^0 whose s-component is $\alpha_s = \mathrm{diag}\left[Q, -\begin{pmatrix}0 & 1\\ 1 & 0\end{pmatrix}\right]$ for every $s \in \mathbf{s}$. Computing λ and μ by means of (5.11) and (1.2), we find that α has the desired property. Now, given f as in our theorem, put $g = f\|_\rho \alpha$. By (4.9a) $\zeta(\mathscr{D}_s)(f\|_\lambda \alpha) = 0$ for every $\zeta \in Z$, and hence $\zeta(\mathscr{D}_s)[R_\pi(z_s)g] = 0$ for every $\zeta \in Z$. Since ρ is irreducible, π is also irreducible, and so $R_\pi(z_s) = q_s^\nu(z_s)\psi(z_s)$ with some $\nu \in (1/2)\mathbf{Z}$ and some $\psi \in S_d(T_s, \mathrm{End}(X))$, $0 \leqslant d \in \mathbf{Z}$. For $\omega \in S_i(T_s)$ we see that $\omega(\mathscr{D}_s)R_\pi = q_s^{\nu-i}\gamma$ with $\gamma \in S_p(T_s, \mathrm{End}(X))$, $p = d - i + i\ell$. Putting $\zeta(\mathscr{D}_s)R_\pi = q_s^{\nu-h}\beta(z_s)$, we have

$$0 = \zeta(\mathscr{D}_s)[R_\pi g] = q_s(z_s)^{\nu-h}\left[\beta(z_s)g(z_s) + \sum_{0 \leqslant i < h} q_s(z_s)^{h-i}\sum_\gamma \gamma(z_s)g_\gamma(z_s)\right]$$

with finitely many $\gamma \in S_p(T_s, \mathrm{End}(X))$ and certain derivatives g_γ of g with respect to z_s. Observing that $\deg(q_s^{h-i}\gamma) > \deg(\beta)$ and applying Lemma 5.4 to the present situation, we see that $[\zeta(\mathscr{D}_s)R_\pi]g(z_s) = 0$, which together with Lemmas 5.3 and 5.4 proves that $\zeta(\mathscr{D}_s)R_\pi = 0$ as expected.

We now define the rank of an element b of $T_\mathbf{Q}$, written $\mathrm{rank}(b)$, as follows. For Types A and C, it is the rank of the matrix in the ordinary sense. For Type D, it is the rank (over B) of the B-module generated by the columns of b. For Type B, $\mathrm{rank}(b) = 2$ if ${}^t bPb \neq 0$, $\mathrm{rank}(b) = 1$ if ${}^t bPb = 0$ and $b \neq 0$, and $\mathrm{rank}(0) = 0$. We easily see that $\mathrm{rank}(b) < i$ with $i \leqslant \ell$ if and only if $\zeta(b_s) = 0$ for every $\zeta \in Z_{i,s}$, where s is any fixed element of $\mathbf{s}$. Given f, $\neq 0$, as in (5.5), we put

$$\text{(5.12)} \qquad \mathrm{rank}(f) = \mathrm{Max}\{\mathrm{rank}(b) \mid c(b) \neq 0\}.$$

Then (5.6) shows that

$$\text{(5.13)} \qquad \mathrm{rank}(f) < i \Leftrightarrow \zeta(\mathscr{D}_s)f = 0 \quad \text{for every } \zeta \in Z_{i,s} \text{ with some } s \in \mathbf{s}$$
$$\Leftrightarrow \zeta(\mathscr{D}_s)f = 0 \quad \text{for every } \zeta \in Z_{i,s} \text{ and every } s \in \mathbf{s}.$$

Denoting $c(b)$ of (5.5) by $c(b, f)$, we call f a *cusp form* if $c(b, f\|_\rho \alpha) = 0$ for every $\alpha \in G^0$ and every $b \in T_\mathbf{Q}$ of rank $< \ell$. From (4.9a) and (5.13) we obtain

$$\text{(5.14)} \qquad \mathrm{rank}(f\|_\rho \alpha) = \mathrm{rank}(f) \quad \text{for every } \alpha \in G^0.$$

We now state Theorems 5.2 and 5.5 in a stronger form when the restriction of ρ to K_s^c for some $s \in \mathbf{s}$ is essentially scalar-valued.

THEOREM 5.6. *Let $0 \neq f \in \mathcal{M}_\rho$ with an irreducible quasi representation $\{\rho, X\}$ of $\mathfrak{K}$. Suppose that $\rho(a_s, b_s) = \det(b_s)^e 1_X$ for $(a_s, b_s) \in K_s^c$ with $e \in (1/2)\mathbf{Z}$ for some $s \in \mathbf{s}$, where 1_X denotes the identity map of X onto itself. Then $e = \lambda_h$ with h such that $0 < h \leqslant \ell$ if and only if* $\operatorname{rank}(f) = h - 1$.

Proof. We first prove (with any fixed $s \in \mathbf{s}$)

$$(*) \qquad \operatorname{rank}(f) < h \qquad \text{if } e = \lambda_h,\ 0 < h \leqslant \ell.$$

Assuming that $e = \lambda_h$, $0 < h \leqslant \ell$, put $Z = Z_{h,s}$ and $g = f\|_\rho\alpha$ with an arbitrary $\alpha \in G^0$. By (2.15) and (4.9c) $(D_\rho^Z f)\|_{\rho\otimes\tau_Z}\alpha = D_\rho^Z g$ and $(D_\rho^Z g)(\zeta) = \zeta(\mathscr{D}_s)g$ for every $\zeta \in Z$. If $h = \ell$, then $Z = \mathbf{C}q_s$. From (5.6) we see that $D_\rho^Z g$ has nonzero Fourier coefficients only for the elements of $T_\mathbf{Q}$ of rank ℓ, which means that $D_\rho^Z f$ is a cusp form. Thus condition (2) of Theorem 5.2 is satisfied. Since π is trivial in the present setting, condition (1) is also satisfied, and hence that theorem implies that $\Delta_s f = 0$, that is, $\operatorname{rank}(f) < \ell$, which proves (*) when $h = \ell$. To prove (*) in general let us first assume that our objects are of Type A, C, or D. Our proof proceeds by induction on ℓ. To emphasize ℓ, let us write $T_\mathbf{Q}^\ell$, $\mathscr{H}^\ell$, and $\mathscr{M}_\rho^\ell$ for $T_\mathbf{Q}$, $\mathscr{H}$, and $\mathscr{M}_\rho$. For $f \in \mathscr{M}_\rho^\ell$ we define a function Φf on $\mathscr{H}^{\ell-1}$ by

$$(5.15) \qquad (\Phi f)(z) = \lim_{r\to\infty} f((\operatorname{diag}[z_s, ru])_{s\in\mathbf{s}}) \qquad \text{for } z = (z_s)_{s\in\mathbf{s}} \in \mathscr{H}^{\ell-1},$$

where $u = i$ for Types A and C, and $u = -iJ_1$ for Type D. Then we easily see that $\Phi f \in \mathscr{M}_\tau^{\ell-1}$ with a certain representation $\{\tau, X\}$ such that $\tau(a_s, b_s) = \det(b_s)^e 1_X$. Moreover, if f is given by (5.5), then

$$(5.16) \qquad (\Phi f)(z) = \sum_\sigma c(\operatorname{diag}[\sigma, 0])\mathbf{e}(\sigma, z),$$

where σ runs over $T_\mathbf{Q}^{\ell-1}$. Suppose that $e = \lambda_h$, $0 < h < \ell$. We want to show that $D_\rho^Z f$ is a cusp form. Take $g = f\|_\rho\alpha$ as above and put $g = \sum_b c(b, g)\mathbf{e}(b, z)$. By (5.6) we have $(D_\rho^Z g)(\zeta) = \zeta(\mathscr{D}_s)g = (2\pi i)^h \sum_b \zeta_1(b_s)c(b, g)\mathbf{e}(b, z)$. Suppose $c(b_0, g) \neq 0$ and $\operatorname{rank}(b_0) < \ell$ for some $b_0 \in T_\mathbf{Q}$. Then we can find an element $\gamma = \operatorname{diag}[({}^t\beta')^{-1}, \beta]$ as in (5.7a) such that $\beta b_0 \cdot {}^t\beta' = \operatorname{diag}[d, 0]$ with $d \in T_\mathbf{Q}^{\ell-1}$. Then $c(\operatorname{diag}[d, 0], g\|_\rho\gamma^{-1}) = \rho(\Lambda_\gamma)c(b_0, g) \neq 0$. Now $\Phi(g\|_\rho\gamma^{-1}) \in \mathscr{M}_\tau^{\ell-1}$ and our induction implies that $\operatorname{rank}(d) < h$, and hence $\operatorname{rank}(b_0) < h$. Therefore the coefficient $\zeta_1(b_s)c(b, g)$ is 0 if $\operatorname{rank}(b) < \ell$. This means that $D_\rho^Z f$ is a cusp form. Therefore, by Theorem 5.2 and (5.13) we have $\operatorname{rank}(f) < h$.

Postponing the proof of (*) for Type B, let us now derive our theorem from (*) for all types. Suppose $\operatorname{rank}(f) = i - 1$ with $i \leqslant \ell$; let $\nu = e - \lambda_i$. By (5.13) and Theorem 5.5 we have $\zeta(\mathscr{D}_s)R_\kappa(z_s) = 0$ for every $\zeta \in Z_{i,s}$ with $\pi(a, b) = \det(b)^\nu 1_X$. By (5.9) and Lemma 4.1 we have $\psi_i(\nu) = 0$ with ψ_i of (4.10a,b). Then we see that $e = \lambda_p$ with $1 \leqslant p \leqslant i$, and (*) shows that $\operatorname{rank}(f) < p$, and hence $p = i$. Thus $e = \lambda_i$, which proves the "if" part of our theorem. Since the "if" part together with (*) implies the "only if" part, it only remains to prove (*) for Type B when $h = 1$.

Our task is to show that $\mathrm{rank}(f) < 1$ if $e = \lambda_1 = 0$. Let us first assume that ${}^t bPb \neq 0$ for every nonzero $b \in F^n$. Then, for $(D^Z_\rho g)(\zeta) = \zeta(\mathscr{D}_s)g$ with $g = f\|_\rho \alpha$ as above, we see that $\zeta(\mathscr{D}_s)g$ has no terms for $\mathrm{rank}(b) < 2$. This shows that $D^Z_\rho f$ is a cusp form, which together with Theorem 5.2 shows that $\mathrm{rank}(f) < 1$. Let us next assume that ${}^t bPb = 0$ for some nonzero $b \in F^n$. Then we may assume that $P = \mathrm{diag}\left[R, -\begin{pmatrix} 0 & 1 \\ 1 & 0 \end{pmatrix}\right]$, $Q = \mathrm{diag}\left[1_{n-2}, -\begin{pmatrix} 0 & 1 \\ 1 & 0 \end{pmatrix}\right]$, and $A_s = \mathrm{diag}[B_s, 1_2]$ with $R = {}^t R \in GL_{n-2}(F)$ and $B_s \in GL_{n-2}(\mathbf{R})$ such that $R^s = {}^t B_s B_s$, and

$$H_s = \left\{ z \in \mathbf{C}^n \,\middle|\, \sum_{v=1}^{n-2} y_v^2 < 2y_{n-1}y_n,\ y_{n-1} + y_n > 0 \right\},$$

where $y = \mathrm{Im}(z)$. For $f \in \mathscr{M}_\rho$ we define a function Φf on $H_1^{\mathbf{s}}$, with $H_1 = \{z \in \mathbf{C} \mid \mathrm{Im}(z) > 0\}$, by

(5.17) $$(\Phi f)(z) = \lim_{r\to\infty} f({}^t(0, \ldots, 0, z_s, ir)_{s\in\mathbf{s}}) \quad \text{for } z = (z_s)_{s\in\mathbf{s}} \in H_1^{\mathbf{s}}.$$

For $\gamma = \begin{pmatrix} a & b \\ c & d \end{pmatrix} \in SL_2(F)$ and $z, w \in H_1^{\mathbf{s}}$ put

(5.18a) $$\tilde{\gamma} = \mathrm{diag}[1_{n-2}, \omega(\gamma)], \qquad \omega(\gamma) = \begin{bmatrix} a & 0 & 0 & b \\ 0 & d & -c & 0 \\ 0 & -b & a & 0 \\ c & 0 & 0 & d \end{bmatrix},$$

(5.18b) $$[z, w] = {}^t(0, \ldots, 0, z, w) (\in \mathscr{H}).$$

Then we can easily verify that $\tilde{\gamma} \in G^0$, $\tilde{\gamma}([z, w]) = [\gamma z, w]$, and $\mu(\tilde{\gamma}_s, [z, w]_s) = (c^s z_s + d^s)_{s\in\mathbf{s}}$, and hence Φf is an automorphic form on $H_1^{\mathbf{s}}$ of Type C. Moreover

(5.19) $$(\Phi f)(z) = \sum_{a\in F} c({}^t(0, \ldots, 0, a))\mathbf{e}\left(\sum_{s\in\mathbf{s}} a^s z_s\right).$$

Since $e = 0$, our result for Type C implies that $\Phi(f\|_\rho \alpha)$ is a constant for every $\alpha \in G^0$. Then the same technique as for the other three types (using (5.7b)) shows that $D^Z_\rho f$ is a cusp form, which together with Theorem 5.2 shows that $\mathrm{rank}(f) < 1$. This completes the proof.

COROLLARY 5.7. *Let $f \in \mathscr{M}_\rho$ with an irreducible quasi representation $\{\rho, X\}$ of $\mathfrak{K}$. Suppose that $\rho(a_s, b_s) = \det(b_s)^e 1_X$ for $(a_s, b_s) \in K_s^c$ with $e \in (1/2)\mathbf{Z}$ and some $s \in \mathbf{s}$. Then the following assertions hold:*

(1) *f is a constant if $e = 0$.*

(2) *$f = 0$ if $e < 0$.*

(3) *If* $f \neq 0$ *and* $e = \lambda_h$ *with* $0 < h \leqslant \ell$, *then* f *is not a cusp form.*

(4) *If* $f \neq 0$, $e = \lambda_h$ *with* $0 < h \leqslant \ell$, *and* $\rho(a_t, b_t) = \det(b_t)^{e'} 1_X$ *for* $(a_t, b_t) \in K_t^c$ *with* $e' \in (1/2)\mathbf{Z}$ *for any other* $t \in \mathbf{s}$, *then* $e' = \lambda_h$.

Proof. Assertions (3) and (4) follow trivially from Theorem 5.6. The same is true for (1) since $\lambda_1 = 0$ and f is a constant if $\operatorname{rank}(f) = 0$. Suppose now $e < 0$. We can find a positive integer k and a nonzero element g of $\mathcal{M}_\sigma$ with $\sigma(a, b) = \prod_{s \in \mathbf{s}} \det(b_s)^{-ke}$ such that g has zeros on $\mathcal{H}$. By (1) the product $g \cdot (f \otimes \cdots \otimes f)$, with f repeated k times, must be a constant vector, and hence $f = 0$, since $f \otimes \cdots \otimes f$ is not holomorphic otherwise.

COROLLARY 5.8. *Suppose that* G *is of Type B and* ${}^t bPb \neq 0$ *for every nonzero* $b \in F^n$. *If* ρ *is as in Theorem 5.6 with* $e = (n-2)/2$ *for some* $s \in \mathbf{s}$, *then* $\mathcal{M}_\rho = \{0\}$.

Proof. This is because there is no nonconstant singular form in such a case.

APPENDIX

Our purpose is to explain why (2.23) is sufficient for the validity of the first two formulas of Theorem 2.1. Let G be a unimodular Lie group, L its Lie algebra, and Γ a closed unimodular subgroup of G. Then $\Gamma\backslash G$ has a G-invariant measure μ.

PROPOSITION A.1. *Let* f *and* h *be* $\mathbf{C}$*-valued* Γ*-invariant* C^1 *functions on* G *and let* $X \in L$. *Then*

$$\int_{\Gamma\backslash G} Xf \cdot h \, d\mu = -\int_{\Gamma\backslash G} f \cdot Xh \, d\mu,$$

provided fh, $Xf \cdot h$, *and* $f \cdot Xh$ *all belong to* $L^1(\Gamma\backslash G)$.

This, as well as the lemma below, is usually (or rather, always) stated only for C^∞ functions on $\Gamma\backslash G$ of compact support or for C^∞ vectors in $L^2(\Gamma\backslash G)$, which is why we give here the proof for the statement in the above form. Anyway, the assertion can be obtained by taking fh to be φ in the following.

LEMMA A.2. *Let* $X \in L$. *If* φ *is a* $\mathbf{C}$*-valued* Γ*-invariant* C^1 *function on* G *such that both* φ *and* $X\varphi$ *belong to* $L^1(\Gamma\backslash G)$, *then* $\int_{\Gamma\backslash G} X\varphi \, d\mu = 0$.

Proof. Put $c = \int_{\Gamma\backslash G} X\varphi \, d\mu$ and $F(g, t) = (X\varphi)(g \cdot \exp(tX))$ for $g \in G$ and $t \in \mathbf{R}$. Then $\int_{\Gamma\backslash G} F(g, t) \, d\mu(g) = c$ for every t. Similarly $\int_{\Gamma\backslash G} |F(g, t)| \, d\mu(g) = \int_{\Gamma\backslash G} |X\varphi| \, d\mu$, and hence $F(g, t)$ is integrable on $[0, 1] \times (\Gamma\backslash G)$. Therefore

$$\begin{aligned} c &= \int_0^1 \int_{\Gamma\backslash G} F(g, t) \, d\mu(g) \, dt \\ &= \int_{\Gamma\backslash G} \int_0^1 F(g, t) \, dt \, d\mu(g) \\ &= \int_{\Gamma\backslash G} [\varphi(g \cdot \exp(X)) - \varphi(g)] \, d\mu(g) = 0, \end{aligned}$$

which proves our lemma. Here we employ the fact that $F(g, t) = (d/dt)\varphi(g \cdot \exp(tX))$, which holds only under the assumption that φ is C^1.

Now, Proposition 2.2 translates Theorem 2.1 into [S9, Theorem 4.3], whose proof is valid as long as formulas (4.14a,b) of [S9] are applicable to the functions on $G^0_{\mathbf{R}}$ corresponding to f, g, and h. By virtue of Proposition A.1, (2.23) guarantees those formulas, and hence gives a sufficient condition for the first two formulas of Theorem 2.1.

REFERENCES

[B] A. Borel, *Density properties for certain subgroups of semi-simple groups without compact components*, Ann. of Math. (2) **72** (1960), 179–188.

[F1] E. Freitag, *Holomorphe Differential formen zu Kongruenzgruppen der Siegelschen Modulgruppe*, Invent. Math. **30** (1975), 181–196.

[F2] ———, *The classification of singular weights*, preprint, 1990.

[H] R. Howe, "Automorphic forms of low rank" in *Non-commutative Harmonic Analysis and Lie Groups*, Lecture Notes in Math. **880**, Springer, Berlin, 1981, 211–248.

[M] H. Maass, *Siegel's Modular Forms and Dirichlet Series*, Lecture Notes in Math. **216**, Springer, Berlin, 1971.

[R1] H. L. Resnikoff, *On a class of linear differential equations for automorphic forms in several complex variables*, Amer. J. Math. **95** (1973), 321–332.

[R2] ———, *Automorphic forms of singular weight are singular forms*, Math. Ann. **215** (1975), 172–193.

[S1] G. Shimura, *The arithmetic of certain zeta functions and automorphic forms on orthogonal groups*, Ann. of Math. (2) **111** (1980), 313–375.

[S2] ———, *Arithmetic of differential operators on symmetric domains*, Duke Math. J. **48** (1981), 813–843.

[S3] ———, *Differential operators and the singular values of Eisenstein series*, Duke Math. J. **51** (1984), 261–329.

[S4] ———, *On differential operators attached to certain representations of classical groups*, Invent. Math. **77** (1984), 463–488.

[S5] ———, *On a class of nearly holomorphic automorphic forms*, Ann. of Math. (2) **123** (1986), 347–406.

[S6] ———, *On Hilbert modular forms of half-integral weight*, Duke Math. J. **55** (1987), 765–838.

[S7] ———, *Nearly holomorphic functions on hermitian symmetric spaces*, Math. Ann. **278** (1987), 1–28.

[S8] ———, *On the critical values of certain Dirichlet series and the periods of automorphic forms*, Invent. Math. **93** (1988), 1–61.

[S9] ———, *Invariant differential operators on hermitian symmetric spaces*, Ann. of Math. (2) **132** (1990), 237–272.

Department of Mathematics, Princeton University, Princeton, New Jersey 08544, USA

Eisenstein series and zeta functions on symplectic groups

Inventiones mathematicae, 119 (1995), 539-584

Introduction

There are three main problems to be investigated in the present paper. Two of them concern certain Eisenstein series on $Sp(n, F)$ with a totally real algebraic number field F, and the other is on the zeta functions associated with cusp forms on the same type of group, which appear naturally as normalizing factors of the Eisenstein series. To describe our problems and results, let us take for simplicity $F = \mathbf{Q}$, put $G^n = Sp(n, \mathbf{Q})$ and call $P^{n,r}$ for $0 \leqq r < n$ the parabolic subgroup consisting of the elements of G^n whose lower left $(n-r) \times (n+r)$-block is 0. The factor of automorphy is $\det(cz+d)^k$ as usual with a positive integer k. Then we denote by $\mathscr{M}_k^n$ the space of all holomorphic modular forms of weight k with respect to all congruence subgroups of G^n, and by $\mathscr{S}_k^n$ its subspace consisting of all cusp forms. Given a congruence subgroup Γ of G^n and $f \in \mathscr{S}_k^r$ of a suitable type, we define an Eisenstein series $E_k^{n,r}(z, s; f, \Gamma)$ of weight k by

$$E_k^{n,r}(z, s; f, \Gamma) = \sum_{\alpha \in A} \{f(\omega_r(z))[\delta(z)/\delta(\omega_r(z))]^{s-k/2}\}\|_k\alpha$$

where z is a variable on the Siegel half-space H_n of degree n, $s \in \mathbf{C}$, $A = (\Gamma \cap P^{n,r})\backslash\Gamma, \delta(z) = \det(\mathrm{Im}(z)$, $\omega_r(z)$ is the upper left $(r \times r)$-block of z, and $\|_k\alpha$ is the standard weight k action of α. Our first problem is the nature of E at some critical points s. This problem is similar and also closely related to the study of the critical values of various zeta functions. One expects that given a reasonable E, there is a finite set of critical points X such that $E(z, s)$ is "arithmetic" for each $s \in X$. This requires of course the notion of arithmeticity, which is preceded by still another notion of "near holomorphy," since $E(z, s)$ for $s \in X$ is not holomorphic in z in general. These notions were introduced in our previous papers. In particular, we showed that if $r = 0$, then there is indeed a natural X such that $E_k^{n,0}(z, s)$ for $s \in X$ is nearly holomorphic and arithmetic. In the present paper we prove a similar result of near holomorphy for an arbitrary r, leaving the question of arithmeticity to a subsequent paper.

Our second problem on E is to show that the space of all holomorphic modular forms on G^n is spanned by cusp forms and Eisenstein series. To be precise, we show that $\mathscr{M}_k^n = \bigoplus_{r=0}^n \mathscr{E}_k^{n,r}$, where $\mathscr{E}_k^{n,r}$ denotes the space spanned by $E_k^{n,r}(z, k/2; f, \Gamma)\|_k\alpha$ for all possible (f, Γ) and all $\alpha \in G^n$ with the convention that $\mathscr{E}_k^{n,n} = \mathscr{S}_k^n$. In fact, Klingen [K] proved a precise result of this type for $Sp(n, \mathbf{Z})$, and a similar, if not so precise, result was obtained by Harris [H] for more general types of groups. However, these concern the cases in which the Eisenstein series are convergent, which means that $k > 2n$ for our series. We will show that the decomposition is true even when the series are divergent, provided k is "not too small."

To prove these, we have to introduce another type of series $E(z, s; f, \chi, C)$, where χ is a Hecke character of F and C is an open subgroup of $G_{\mathbf{A}}^n$ with compact nonarchimedean part. If $F = \mathbf{Q}$, it can be given as

$$(1)\quad \sum_{\alpha \in A} \psi(\det(d_\alpha))\{f(\omega_r(z))[\delta(z)/\delta(\omega_r(z))]^{s-k/2}\}\|_k\alpha, \ A = (\Gamma \cap P^{n,r})\backslash\Gamma,$$

with a Dirichlet character ψ and a suitable Γ. In addition to this, we have to introduce the zeta function attached to a Hecke eigenform f in $\mathscr{S}_k^r$, which is actually our third main topic, and which is defined by

$$Z(s, f, \eta) = \prod_p W_p(\eta(p)p^{-s})^{-1},$$

where η is an arbitrary primitive Dirichlet character, p runs over all prime numbers prime to the level of f, and W_p is a polynomial with constant term 1 and of degree $2r + 1$. There are two essential facts:

(2) $E_k^{n,r}(z, s; f, \Gamma)$ *is always a finite linear combination of the transformations of the series of type* $E(z, s; f, \chi, C)$;

(3) $\Lambda(s)Z(2s, f, \eta)E(z, s; f, \chi, C) = \langle f'(w), E_k^{n+r,0}(\mathrm{diag}[z, w], s)\rangle$.

Here Λ is a product of Hecke L-functions and gamma functions, f' is an easy transform of f, $E_k^{n+r,0}$ is a series of type (1) on G^{n+r} with a constant 1 as f, transformed by a suitable element of G^{n+r}, $(z, w) \in H_n \times H_r$, and χ is chosen suitably. In principle, (3) is a known relation. In fact, Garrett gave a formula for $E_k^{n+r,0}(\mathrm{diag}[z, w])$ (without the parameter s) on $Sp(n, \mathbf{Z})$, from which one could derive (3) with trivial η at $s = k/2$ in that special case [Ga1], and that derivation was done explicitly by Böcherer [B]. If $n = r$, the function E on the left-hand side of (3) is merely f. The equality in this case with trivial η was employed by Piatetski-Shapiro and Rallis [PR] in their investigation of Euler products of type Z, but their result, involving some unknown factors, is not so clear-cut as (3). However, our problems require a nonvanishing exact formula in the general case with no ambiguous bad factors. We also need precise information about the poles of $E_k^{n+r,0}$ times an explicit normalizing factor. To achieve these, we choose $E_k^{n+r,0}$ on the right-hand side carefully, so that its pullback on $H_n \times H_r$ can be employed effectively. In fact, a seemingly natural choice of $E_k^{n+r,0}$ often leads to a vanishing inner product. Another technical problem we have to solve is to define $E(z,s;f,\chi,C)$ so that (2) holds for an arbitrary pair (f,Γ) and also that it appears in (3). Strictly speaking, (3) requires the condition that $\eta(-1)=(-1)^k$, and the handling of Z for general η is more involved.

Anyway, as a consequence of (3) and its variation we can prove the following result, which is stated here, for simplicity, only when $F = \mathbf{Q}$ in a form weaker than what is actually proved:

If f is a Hecke eigenform belonging to a principal congruence subgroup, then $Z(s, f, \eta)$ times explicitly defined gamma factors can be continued to a meromorphic function on the whole plane with finitely many poles, which are all simple. In particular the product is entire if $\eta^2 \neq 1$.

In our previous paper [94a] we proved a similar result for f of a somewhat special type with a full Euler product with no exclusion of Euler factors. Our present result (Theorem 6.1) covers the whole $\mathscr{S}_k^r$, but requires the elimination of some Euler factors. We can also give explicitly a certain finite set which contains the possible poles of Z. This is better than what we obtained in [94a].

Once (2) and (3) are established, our results on near holomorphy (Theorems 7.3, 7.4, and 7.5) follow easily from the corresponding results in the case $r=0$ obtained in [83b, 85a, 87]. The decomposition of $\mathscr{M}_k^n$ (Theorems 8.12, 8.13, and 8.14) is more complicated; we have to analyze the nature of Fourier coefficients by means of generalized confluent hypergeometric functions we investigated in [82].

Although our main results concern the case of integral weight, we have included in our treatment the case of half-integral weight to the extent that $E_k^{n,r}(z,s;f,\Gamma)$ and $E(z,s;f,\chi,C)$ are defined so that (2) holds, and that $\mathscr{M}_k^n = \bigoplus_{r=0}^n \mathscr{E}_k^{n,r}$ can be proved in the convergent case. It may also be added that the three problems in the present paper are investigated not only for their own sake, but also for the purpose of future applications to the arithmeticity problems, which we hope to treat in subsequent papers.

1. Parabolic subgroups and Eisenstein series

We use the same notation and terminology as in our previous paper [94a]; we recall here some of the basic symbols and conventions. For an associative ring R with identity element and an R-module M we denote by $R^\times$ the group of all its invertible elements and by M_n^m the R-module of all $m \times n$-matrices with entries in M; we put $M^m = M_1^m$ for simplicity. Sometimes an object with a superscript such as G^n of (1.2a) below is used with a different meaning, but the distinction will be clear from the context. We put $\tilde{x} = {}^t x^{-1}$ if a matrix x is square and invertible. However, $\tilde{X}$ for an object X other than matrices will be defined differently. For complex hermitian matrices x and y we write $x > y$ if

$x-y$ is positive definite, and $x \geqq y$ if $x-y$ is nonnegative. To indicate that a union $X = \bigcup_{i\in I} Y_i$ is disjoint, we write $X = \bigsqcup_{i\in I} Y_i$.

We fix a totally real algebraic number field F of finite degree and denote by $\mathbf{a}$, $\mathbf{h}$, $\mathfrak{g}$, and $\mathfrak{d}$ the sets of archimedean primes and nonarchimedean primes of F, the maximal order of F, and the different of F relative to $\mathbf{Q}$. Given an algebraic group $\mathfrak{G}$ defined over F, we define $\mathfrak{G}_v$ for each $v \in \mathbf{a} \cup \mathbf{h}$ and the adelization $\mathfrak{G}_\mathbf{A}$ and view $\mathfrak{G}$ and $\mathfrak{G}_v$ as subgroups of $\mathfrak{G}_\mathbf{A}$ as usual; we then put $\mathfrak{G}_\mathbf{a} = \prod_{v\in\mathbf{a}} \mathfrak{G}_v$ and $\mathfrak{G}_\mathbf{h} = \mathfrak{G}_\mathbf{A} \cap \prod_{v\in\mathbf{h}} \mathfrak{G}_v$. For $x \in \mathfrak{G}_\mathbf{A}$ we denote by $x_v, x_\mathbf{a}$, and $x_\mathbf{h}$ its projections to $\mathfrak{G}_v, \mathfrak{G}_\mathbf{a}$, and $\mathfrak{G}_\mathbf{h}$. For a fractional ideal $\mathfrak{x}$ in F and $t \in F_\mathbf{A}^\times$ we denote by $t\mathfrak{x}$ the fractional ideal in F such that $(t\mathfrak{x})_v = t_v\mathfrak{x}_v$ for every $v \in \mathbf{h}$; we put $|t_v| = |t_v|_v$ with the normalized valuation $|\ |_v$ at v and $|t|_\mathbf{A} = \prod_{v\in\mathbf{a}\cup\mathbf{h}} |t_v|$. Given a set X, we denote by $X^\mathbf{a}$ the product of $\mathbf{a}$ copies of X, that is, the set of all indexed elements $(x_v)_{v\in\mathbf{a}}$ with x_v in X. For example, for $t \in F_\mathbf{A}^\times$ we define an element $\operatorname{sgn}(t)$ of $\mathbf{R}^\mathbf{a}$ by $\operatorname{sgn}(t) = (t_v/|t_v|)_{v\in\mathbf{a}}$. We put

(1.1a) $$\mathbf{T} = \{\zeta \in \mathbf{C} \mid |\zeta| = 1\},$$

(1.1b) $$\mathbf{e}(z) = e^{2\pi iz} \quad (z \in \mathbf{C}),$$

and define the standard characters $\mathbf{e}_\mathbf{A} : F_\mathbf{A} \to \mathbf{T}$ and $\mathbf{e}_v : F_v \to \mathbf{T}$ for each v as usual so that $\mathbf{e}_\mathbf{A}(x) = \prod_{v\in\mathbf{a}\cup\mathbf{h}} \mathbf{e}_v(x_v)$, $\mathbf{e}_v(x_v) = \mathbf{e}(x_v)$ for $v \in \mathbf{a}$, and $\mathbf{e}_\mathbf{A}(F) = 1$. We then put $\mathbf{e}_\mathbf{h}(x) = \mathbf{e}_\mathbf{A}(x_\mathbf{h})$ for $x \in F_\mathbf{A}$ and $\mathbf{e}_\mathbf{a}(x) = \mathbf{e}(\sum_{v\in\mathbf{a}} x_v)$ for $x \in F_\mathbf{A}$ or $x \in \mathbf{C}^\mathbf{a}$. We now define an algebraic group G and spaces H and $\mathscr{H}$ by

(1.2a) $$G = G^n = \{\alpha \in GL_{2n}(F) \mid {}^t\alpha\iota\alpha = \iota\}, \quad \iota = \iota_n = \begin{pmatrix} 0 & -1_n \\ 1_n & 0 \end{pmatrix},$$

(1.2b) $$H = H_n = \{z \in \mathbf{C}_n^n \mid {}^tz = z, \operatorname{Im}(z) > 0\},$$

(1.2c) $$\mathscr{H} = \mathscr{H}_n = H_n^\mathbf{a}.$$

We drop the superscript n or the subscript n, as we do for the moment, whenever the dimensionality is clear from the context, but reinstate it when it needs to be stressed. This applies to other symbols $M_\mathbf{A}^n$, $\mathscr{G}^n$, $\mathscr{M}_k^n$, etc. defined below. For $\alpha \in G_v, v \in \mathbf{a}$, and $z \in H$ we define $\alpha z = \alpha(z) \in H$, $\delta(z) \in \mathbf{R}$, and factors of automorphy $\mu(\alpha, z)$ and $j(\alpha, z)$ by

(1.3a) $$\alpha z = \alpha(z) = (a_\alpha z + b_\alpha)(c_\alpha z + d_\alpha)^{-1}, \quad \delta(z) = \det(\operatorname{Im}(z)),$$

(1.3b) $$\mu(\alpha, z) = c_\alpha z + d_\alpha, \quad j(\alpha, z) = \det[\mu(\alpha, z)].$$

Here and throughout the paper, $a_\alpha, b_\alpha, c_\alpha$, and d_α denote the standard a-, b-, c-, and d-blocks of α. For $\alpha \in G_\mathbf{A}$ and $z = (z_v)_{v\in\mathbf{a}} \in \mathscr{H}$ we put

(1.4a) $$\alpha z = (\alpha_v z_v)_{v\in\mathbf{a}}, \quad \delta(z) = (\delta(z_v))_{v\in\mathbf{a}},$$

(1.4b) $$\mu(\alpha, z) = (\mu(\alpha_v, z_v))_{v\in\mathbf{a}}, \quad j_\alpha(z) = j(\alpha, z) = (j(\alpha_v, z_v))_{v\in\mathbf{a}}.$$

All these are elements of $(\mathbf{C}_n^n)^{\mathbf{a}}$ or $\mathbf{C}^{\mathbf{a}}$. For two fractional ideals $\mathfrak{x}$ and $\mathfrak{y}$ in F such that $\mathfrak{x}\mathfrak{y} \subset \mathfrak{g}$ we put

(1.5a) $D[\mathfrak{x}, \mathfrak{y}] = G_{\mathbf{a}} \prod_{v \in \mathbf{h}} D_v[\mathfrak{x}, \mathfrak{y}] (\subset G_{\mathbf{A}}), \quad \Gamma[\mathfrak{x}, \mathfrak{y}] = G \cap D[\mathfrak{x}, \mathfrak{y}],$

(1.5b) $D_v[\mathfrak{x}, \mathfrak{y}] = \{x \in G_v | a_x \in (\mathfrak{g}_v)_n^n, b_x \in (\mathfrak{x}_v)_n^n, c_x \in (\mathfrak{y}_v)_n^n, d_x \in (\mathfrak{g}_v)_n^n\},$

(1.5c) $D_0[\mathfrak{x}, \mathfrak{y}] = \{x \in D[\mathfrak{x}, \mathfrak{y}] | x_{\mathbf{a}}(\mathbf{i}) = \mathbf{i}\},$

where $\mathbf{i}$ is the "origin" of $\mathscr{H}$ given by

(1.6) $$\mathbf{i} = \mathbf{i}_n = (i1_n, \dots, i1_n).$$

To simplify our notation, for x and y in $\mathbf{C}^{\mathbf{a}}$ we put

(1.7) $$x^y = \prod_{v \in \mathbf{a}} x_v^{y_v},$$

where $x_v^{y_v}$ should be understood according to the context. For $w \in F_{\mathbf{A}}^{\times}$, in particular for $w \in F^{\times}$, we shall often put $w^y = (w_{\mathbf{a}})^y$ for simplicity. The identity element $(1, \dots, 1)$ of the ring $\mathbf{C}^{\mathbf{a}}$ is denoted by u. For example we have

(1.8) $$j_\alpha(z)^u = \prod_{v \in \mathbf{a}} j(\alpha_v, z_v) \quad (\alpha \in G_{\mathbf{A}}, z \in \mathbf{H}).$$

A *weight* is either half-integral or integral; by a *half-integral weight* we mean an element k of $\mathbf{Q}^{\mathbf{a}}$ such that $k - u/2 \in \mathbf{Z}^{\mathbf{a}}$; by an *integral weight* we mean an element of $\mathbf{Z}^{\mathbf{a}}$. For a weight k we put

(1.9) $$[k] = k \text{ if } k \in \mathbf{Z}^{\mathbf{a}}; \quad [k] = k - u/2 \text{ if } k \notin \mathbf{Z}^{\mathbf{a}}.$$

To define forms of half-integral weight, we take the metaplectic covering $M_{\mathbf{A}} = M_{\mathbf{A}}^n$ of $G_{\mathbf{A}}$ as in [93] and [94a], and denote by pr the projection map of $M_{\mathbf{A}}$ onto $G_{\mathbf{A}}$. We always view G as a subgroup of $M_{\mathbf{A}}$ by the canonical lifting. For $\alpha = \mathrm{pr}(\sigma)$ with $\sigma \in M_{\mathbf{A}}$ we put $a_\sigma = a_\alpha, b_\sigma = b_\alpha, c_\sigma = c_\alpha, d_\sigma = d_\alpha$, and $\sigma(z) = \sigma z = \alpha z$. For a weight k and $z \in \mathscr{H}$ we define a factor of automorphy $j^k(\alpha, z)$ by

(1.10a) $$j_\alpha^k(z) = j^k(\alpha, z) = j_\alpha(z)^k \quad (k \in \mathbf{Z}^{\mathbf{a}}, \alpha \in G_{\mathbf{A}}),$$

(1.10b) $$j_\sigma^k(z) = j^k(\sigma, z) = h_\sigma(z) j_\alpha(z)^{[k]} \quad (k \notin \mathbf{Z}^{\mathbf{a}}, \sigma \in \mathfrak{M}, \alpha = \mathrm{pr}(\sigma)).$$

Here h is a basic factor of automorphy of weight $u/2$, and $\mathfrak{M}$ is the subset of $M_{\mathbf{A}}$ defined by $\mathfrak{M} = \mathfrak{M}^n = \mathrm{pr}^{-1}(P_{\mathbf{A}}^{n,0} C^\theta)$, where $P^{n,0}$ is defined by (1.21b) below and C^θ is a certain subgroup of $G_{\mathbf{A}}$ containing $D[2\mathfrak{d}^{-1}, 2\mathfrak{d}]$ (see [93, (1.13a, b)]). Then for a function f on $\mathscr{H}$ and $\xi \in G_{\mathbf{A}}$ or $\in \mathfrak{M}$ we define a function $f\|_k \xi$ on $\mathscr{H}$ by

(1.11) $$(f\|_k \xi)(z) = j_\xi^k(z)^{-1} f(\xi z) \quad (z \in \mathscr{H}).$$

Once k is fixed, we shall often write $f\|\xi$ for $f\|_k\xi$. (In principle, we keep the subscript k in the theorems, lemmas, and some formulas, but drop it in the proof.)

Let G' denote the image of G in $G_{\mathbf{a}}$ under the map $x \mapsto x_{\mathbf{a}}$, and let

$$\mathscr{G} = \mathscr{G}^n = \{\sigma \in M_{\mathbf{A}} | \mathrm{pr}(\sigma) \in G'\}. \tag{1.12}$$

Since $G_{\mathbf{a}} \subset \mathrm{pr}(\mathfrak{M})$, we see that $\mathscr{G}$ is a subgroup of $M_{\mathbf{A}}$ contained in $\mathfrak{M}$, and $\mathrm{pr}(\mathscr{G}) = G'$. Combining pr with the isomorphism $G' \to G$, we obtain a surjection $\mathscr{G} \to G$, which gives an exact sequence

$$1 \to \mathbf{T} \to \mathscr{G} \to G \to 1. \tag{1.13}$$

If $\sigma \in \mathscr{G}$ and σ is mapped onto $\alpha \in G$, (that is, if $\mathrm{pr}(\sigma) = \alpha_{\mathbf{a}}$), then $h_\sigma^2/j_\alpha^u \in \mathbf{T}$, and σ is completely determined by α and the function h_σ. On the other hand, consider the group consisting of all couples (α, p) formed by $\alpha \in G$ and a holomorphic function p on $\mathscr{H}$ such that $p(z)^2/j_\alpha^u$ is a constant belonging to $\mathbf{T}$, the group-law being defined by $(\alpha, p)(\alpha', p') = (\alpha\alpha', p(\alpha' z)p'(z))$. For an element (α, p) in this group, there is a unique element σ of $\mathscr{G}$ such that $\mathrm{pr}(\sigma) = \alpha_{\mathbf{a}}$ and $p = h_\sigma$. Therefore this group can be identified with $\mathscr{G}$.

Let $\beta \in G \cap P_{\mathbf{A}}^{n,0} C^\theta$. Then, through the natural injection of G into $M_{\mathbf{A}}$, β may be viewed as an element of $\mathfrak{M}$, so that h_β is meaningful. Let $\hat{\beta}$ denote the element of $\mathscr{G}$ represented by (β, h_β). From [93, Th.1.2(4)] we obtain

$$\widehat{\alpha\beta\gamma} = \hat{\alpha}\hat{\beta}\hat{\gamma} \quad \text{if } \alpha \in P^{n,0} \text{ and } \gamma \in \Gamma^\theta, \tag{1.14}$$

where $\Gamma^\theta = G \cap C^\theta$. We call a subgroup Δ of $\mathscr{G}$ a *congruence subgroup* of $\mathscr{G}$ if the projection map of $\mathscr{G}$ onto G of (1.12) gives an isomorphism of Δ onto a congruence subgroup Γ of G, and the inverse of this isomorphism coincides with the map $\alpha \mapsto \hat{\alpha}$ on a congruence subgroup of $\Gamma^\theta \cap \Gamma$. Any conjugate $\xi\Delta\xi^{-1}$ of such a Δ with $\xi \in \mathscr{G}$ is also a congruence subgroup. For the proof see [85a, Prop.1.3] or [93, Prop.1.4].

To treat modular forms of integral or half-integral weight uniformly, we fix a weight k and make the following conventions: *The symbol $\mathscr{G}$ denotes the group defined above if $k \neq [k]$; it denotes G and* pr *means the identity map if* $k = [k]$. Given a congruence subgroup Γ of $\mathscr{G}$, we denote by $\mathscr{M}_k^n(\Gamma)$ the vector space of all holomorphic functions f on $\mathscr{H}$ which satisfy $f\|_k\gamma = f$ for every $\gamma \in \Gamma$ and also the cusp condition if $n = 1$ and $F = \mathbf{Q}$. We then denote by $\mathscr{S}_k^n(\Gamma)$ the subspace of $\mathscr{M}_k^n(\Gamma)$ consisting of all cusp forms. Further we denote by $\mathscr{M}_k^n$ resp. $\mathscr{S}_k^n$ the union of $\mathscr{M}_k^n(\Gamma)$ resp. $\mathscr{S}_k^n(\Gamma)$ for all congruence subgroups Γ of $\mathscr{G}^n$. We make another convention that $\mathscr{M}_k^0 = \mathscr{S}_k^0 = \mathscr{M}_k^0(\Gamma) = \mathscr{S}_k^0(\Gamma) = \mathbf{C}$. If $k \neq [k]$ and Γ is a congruence subgroup of Γ^θ, we shall identify Γ with $\hat{\Gamma}$; thus $\mathscr{M}_k^n(\hat{\Gamma}) = \mathscr{M}_k^n(\Gamma)$ and $\mathscr{S}_k^n(\hat{\Gamma}) = \mathscr{S}_k^n(\Gamma)$. If K is an open subgroup of $G_{\mathbf{A}}$ such that $K \cap G_{\mathbf{h}}$ is compact and $\Gamma = K \cap G$, then we put

$$\mathscr{M}_k^n(K) = \mathscr{M}_k^n(\Gamma), \quad \mathscr{S}_k^n(K) = \mathscr{S}_k^n(\Gamma). \tag{1.15}$$

Here we assume that $K \subset C^\theta$ if $k \neq [k]$.

For a Hecke character ψ of F (that is, a continuous homomorphism of $F_{\mathbf{A}}^{\times}$ into $\mathbf{T}$ such that $\psi(F^{\times})=1$) we denote by $\psi_v, \psi_{\mathbf{a}}$, and $\psi_{\mathbf{h}}$ its restrictions to $F_v^{\times}, F_{\mathbf{a}}^{\times}$, and $F_{\mathbf{h}}^{\times}$, respectively, and by ψ^* the ideal character associated with ψ, putting $\psi^*(\mathfrak{a})=0$ if $\mathfrak{a}$ is not prime to the conductor of ψ. We put also $\psi_{\mathfrak{c}}=\prod_{v|\mathfrak{c}}\psi_v$ for every integral ideal $\mathfrak{c}$. We have $\psi_v(x)=\operatorname{sgn}(x)^{t_v}|x|^{i\kappa_v}$ for $x\in F_v^{\times}, v\in\mathbf{a}$, with $t_v\in\mathbf{Z}$ and $\kappa_v\in\mathbf{R}$. We always assume that ψ is *normalized* in the sense that $\sum_{v\in\mathbf{a}}\kappa_v=0$, and say that ψ is *k-normalized* if we can take $t=[k]$.

Take a weight k, a fractional ideal $\mathfrak{b}$, an integral ideal $\mathfrak{c}$, and a Hecke character ψ (which is normalized but not necessarily k-normalized) such that

$$\psi_v(a)=1 \text{ if } v\in\mathbf{h}, a\in\mathfrak{g}_v^{\times}, \text{ and } a-1\in\mathfrak{c}_v. \tag{1.16}$$

Let K be an open subgroup of $D[\mathfrak{b}^{-1},\mathfrak{bc}]$ and let $\Gamma=G\cap K$. Assume that $K\subset C^{\theta}$ if $k\neq[k]$. Then we consider a C^{∞} function g on $\mathscr{H}$ satisfying

$$g\|_k\gamma=\psi_{\mathfrak{c}}(\det(a_\gamma))g \text{ for every } \gamma\in\Gamma. \tag{1.17}$$

Given such a g, we can define a function $g_{\mathbf{A}}$ on $G_{\mathbf{A}}$ or on $M_{\mathbf{A}}$ according as $k=[k]$ or $k\neq[k]$ by

$$g_{\mathbf{A}}(\alpha w)=\psi_{\mathfrak{c}}(\det(d_w))(g\|_k w)(\mathbf{i}) \text{ for } \alpha\in G \text{ and } w\in \operatorname{pr}^{-1}(K). \tag{1.18}$$

Clearly

$$g_{\mathbf{A}}(\alpha x w)=\psi_{\mathfrak{c}}(\det(d_w))j^k(w,\mathbf{i})^{-1}g_{\mathbf{A}}(x) \tag{1.19}$$
$$\text{for every } \alpha\in G \text{ and } w\in\operatorname{pr}^{-1}(K\cap D_0[\mathfrak{b}^{-1},\mathfrak{bc}]).$$

Conversely, a function $g_{\mathbf{A}}$ on $G_{\mathbf{A}}$ or on $M_{\mathbf{A}}$ satisfying (1.19) can always be obtained from a function g on $\mathscr{H}$ satisfying (1.17) by means of (1.18). We denote by $\mathscr{M}_k(K,\psi)$ and also by $\mathscr{M}_k(\Gamma,\psi)$ the set of all $g\in\mathscr{M}_k$ satisfying (1.17) and put $\mathscr{S}_k(K,\psi)=\mathscr{S}_k(\Gamma,\psi)=\mathscr{S}_k\cap\mathscr{M}_k(K,\psi)$. Obviously $\mathscr{M}_k(K,\psi)=\{0\}$ if $-1_{2n}\in K$ and $\psi_{\mathbf{h}}(-1)^n\neq(-1)^{n\|[k]\|}$.

Fixing an integer r such that $0\leqq r\leqq n$, we write each element α of $(F_{\mathbf{A}})_{2n}^{2n}$ in the form

$$\alpha=\begin{pmatrix} a_1 & a_2 & b_1 & b_2\\ a_3 & a_4 & b_3 & b_4\\ c_1 & c_2 & d_1 & d_2\\ c_3 & c_4 & d_3 & d_4\end{pmatrix}. \tag{1.20}$$

where x_1 is of size r, and x_4 is of size $n-r$. Then we write $x_i=x_i(\alpha)$ and $x_\alpha=x(\alpha)=\begin{pmatrix} x_1(\alpha) & x_2(\alpha)\\ x_3(\alpha) & x_4(\alpha)\end{pmatrix}$ for $x=a,b,c,d$, and $i=1,2,3,4$. We understand that $x(\alpha)=x_1(\alpha)$ if $r=n$, and $x(\alpha)=x_4(\alpha)$ if $r=0$. We define a parabolic subgroup $P^{n,r}$ of G^n by

(1.21a)

$$P^{n,r}=\{\alpha\in G^n | a_2(\alpha)=c_2(\alpha)=0, c_3(\alpha)=d_3(\alpha)=0, c_4(\alpha)=0\} (0<r<n),$$

$$P^{n,0}=\{\alpha\in G^n | c(\alpha)=0\},\quad P^{n,n}=G^n, \tag{1.21b}$$

and define also maps $\pi_r : (F_{\mathbf{A}})^{2n}_{2n} \to (F_{\mathbf{A}})^{2r}_{2r}$ and $\lambda_r : (F_{\mathbf{A}})^{2n}_{2n} \to F_{\mathbf{A}}$ by

(1.22) $$\pi_r(\alpha) = \begin{pmatrix} a_1(\alpha) & b_1(\alpha) \\ c_1(\alpha) & d_1(\alpha) \end{pmatrix}, \quad \lambda_r(\alpha) = \det(d_4(\alpha)).$$

These define homomorphisms $P^{n,r}_{\mathbf{A}} \to G^r_{\mathbf{A}}$ and $P^{n,r}_{\mathbf{A}} \to F^{\times}_{\mathbf{A}}$. We understand that $G^0_{\mathbf{A}} = \pi_0(P^{n,0}_{\mathbf{A}}) = 1$, $\pi_n(\alpha) = \alpha$, and $\lambda_n(\alpha) = 1$ for $\alpha \in G^n_{\mathbf{A}}$. We note that

(1.23a) $$P^{n,t} \cap P^{n,r} = \{\alpha \in P^{n,t} \mid \pi_t(\alpha) \in P^{t,r}\},$$

(1.23b) $$\pi_t(P^{n,t} \cap P^{n,r}) = P^{t,r} \text{ if } t \geqq r.$$

Assuming $r > 0$, for $z \in \mathbf{C}^n_n$ we let $\omega_r(z)$ denote the upper left $(r \times r)$-block of z, and use the same letter ω_r for the map $(\mathbf{C}^n_n)^{\mathbf{a}} \to (\mathbf{C}^r_r)^{\mathbf{a}}$ defined by $\omega_r((z_v)_{v \subset \mathbf{a}}) = (\omega_r(z_v))_{v \in \mathbf{a}}$. Then for $\alpha \in P^{n,r}_{\mathbf{A}}$ and $z \in \mathscr{H}_n$ we have

(1.24) $$\omega_r(\alpha z) = \pi_r(\alpha)\omega_r(z), \quad j(\alpha, z) = \lambda_r(\alpha_{\mathbf{a}})j(\pi_r(\alpha), \omega_r(z)).$$

For $\beta \in G^r_{\mathbf{A}}$ and $\gamma \in G^{n-r}_{\mathbf{A}}$ we define an element $\beta \times \gamma$ of $G^n_{\mathbf{A}}$ by

(1.25) $$\beta \times \gamma = \begin{pmatrix} a_\beta & 0 & b_\beta & 0 \\ 0 & a_\gamma & 0 & b_\gamma \\ c_\beta & 0 & d_\beta & 0 \\ 0 & c_\gamma & 0 & d_\gamma \end{pmatrix}.$$

With $\mathscr{G}^n$ defined as above, we put

(1.26) $$\mathscr{P}^{n,r} = \{(\alpha, p) \in \mathscr{G}^n \mid \alpha \in P^{n,r}\}.$$

and define homomorphisms $\pi_r : \mathscr{P}^{n,r} \to \mathscr{G}^r$ and $\lambda_r : \mathscr{P}^{n,r} \to F_{\mathbf{a}}\times$ by

(1.27a) $$\pi_r((\alpha, p)) = (\pi_r(\alpha), |\lambda_r(\alpha)|^{-u/2} p'), \quad p'(z) = p\begin{pmatrix} z & w \\ {}^t w & z' \end{pmatrix},$$

(1.27b) $$\lambda_r((\alpha, p)) = \lambda_r(\alpha_{\mathbf{a}}),$$

where it should be observed that $p'(z)$ does not depend on the choice of w and z'. We make the convention that $\mathscr{P}^{n,r} = P^{n,r}$ if $k = [k]$. Clearly

(1.28) $$j^k(\xi, z) = j^k(\pi_r(\xi), \omega_r(z))\lambda_r(\xi)^{[k]}|\lambda_r(\xi)|^{k-[k]} \text{ if } \xi \in \mathscr{P}^{n,r}.$$

For $0 \leqq r \leqq n$, a congruence subgroup Γ of $\mathscr{G}^n$, and $X = \mathscr{M}$ or $\mathscr{S}$, we put

(1.29)
$$X^r_k(\Gamma \cap \mathscr{P}^{n,r}) = \{f \in X^r_k \mid f\|_k \pi_r(\gamma) = \operatorname{sgn}[\lambda_r(\gamma)]^{[k]} f \text{ for every } \gamma \in \Gamma \cap \mathscr{P}^{n,r}\}.$$

This space is $\{0\}$ unless the following condition is satisfied:

(1.30) *There is a homomorphism φ of $\pi_r(\Gamma \cap \mathscr{P}^{n,r})$ into $\{\pm 1\}$ such that*

$$\varphi(\pi_r(\gamma)) = \operatorname{sgn}[\lambda_r(\gamma)]^{[k]} \textit{ for every } \gamma \in \Gamma \cap \mathscr{P}^{n,r}.$$

Under (1.30) we have

(1.31)

$$X_k^r(\Gamma \cap \mathscr{P}^{n,r}) = \{f \in X_k^r \| f\|_k\varepsilon = \varphi(\varepsilon)f \ \textit{for every}\ \varepsilon \in \pi_r(\Gamma \cap \mathscr{P}^{n,r})\}.$$

For $f \in \mathscr{S}_k^r(\Gamma \cap \mathscr{P}^{n,r}), z \in \mathscr{H}_n$, and $s \in \mathbf{C}$ we put

$$\delta(z,s;f) = f(\omega_r(z))[\delta(z)/\delta(\omega_r(z))]^{su-k/2}, \tag{1.32}$$

and note that

$$\delta(z,s;f)\|_k\beta = \delta(z,s;f\|_k\pi_r(\beta))\lambda_r(\beta)^{-[k]}|\lambda_r(\beta)|^{[k]-2su} \ \textit{if}\ \beta \in \mathscr{P}^{n,r}. \tag{1.33}$$

We now define an Eisenstein series $E_k^{n,r}(z,s;f,\Gamma)$ by

$$E_k^{n,r}(z,s;f,\Gamma) = \sum_{\gamma\in A}\delta(z,s;f)\|_k\gamma, \quad A = (\Gamma \cap \mathscr{P}^{n,r})\backslash\Gamma. \tag{1.34}$$

The sum is formally well defined, since $f \in \mathscr{S}_k^r(\Gamma \cap \mathscr{P}^{n,r})$. It is convergent for $\mathrm{Re}(2s) > n + r + 1$. The series is defined even when $r = 0$ or n. If $r = n$, we have $\mathscr{S}_k^n(\Gamma \cap \mathscr{P}^{n,n}) = \mathscr{S}_k^n(\Gamma)$ and

$$E_k^{n,n}(z,s;f,\Gamma) = f. \tag{1.35}$$

If $r = 0$, we have $\delta(z,s;c) = c\delta(z)^{su-k/2}$ for a constant $c \in \mathbf{C}$, and

(1.36)

$$\mathscr{S}_k^0(\Gamma \cap \mathscr{P}^{n,0}) = \begin{cases} \mathbf{C} & \text{if } \mathrm{sgn}[\lambda_0(\gamma)]^{[k]} = p_\gamma \text{ for every } \gamma \in \Gamma \cap \mathscr{P}^{n,0}, \\ \{0\} & \text{otherwise,} \end{cases}$$

where we understand that $\gamma = (\mathrm{pr}(\gamma), p_\gamma)$ if $k \neq [k]$ and $p_\gamma = 1$ if $k = [k]$. Thus

$$E_k^{n,0}(z,s;c,\Gamma) = c\sum_{\gamma\in A}\delta^{su-k/2}\|_k\gamma, \quad A = (\Gamma \cap \mathscr{P}^{n,0})\backslash\Gamma. \tag{1.37}$$

2. Some elementary facts and technical lemmas

Let us now put

$$S = S^n = \{\xi \in F_n^n | {}^t\xi = \xi\}, \tag{2.1a}$$

$$S_v^+ = \{x \in S_v | x > 0\} \quad (v \in \mathbf{a}), \tag{2.1b}$$

$$S_+ = \{\xi \in S | \xi_v \geqq 0 \text{ for every } v \in \mathbf{a}\}. \tag{2.1c}$$

Every element $f \in \mathscr{M}_k$ has an expansion

$$f(z) = \sum_{h\in L\cap S_+} c(h)\mathbf{e_a}(\mathrm{tr}(hz)). \tag{2.2}$$

with $c(h) \in \mathbf{C}$ and a $\mathbf{Z}$-latice L in S.

Proposition 2.1. (1) *$\mathcal{M}_k \neq \mathcal{S}_k$ only if $k = k_0 u$ with $k_0 \in (1/2)\mathbf{Z}$.*
(2) *$\mathcal{S}_k \neq \{0\}$ only if $k_v \geqq n/2$ for every $v \in \mathbf{a}$.*
(3) *If $f \in \mathcal{M}_k$, $\neq 0$, has expansion (2.2) and $k_v \geqq n/2$ for some $v \in \mathbf{a}$, then $c(h) \neq 0$ for some $h \in S \cap GL_n(F)$.*

Assertion (1) follows easily from the fact that $f({}^t aza) = \det(a)^{[-k]} f(z)$ for every $f \in \mathcal{M}_k$ and every a in a congruence subgroup of $GL_n(\mathfrak{g})$. If $F = \mathbf{Q}$, assertions (2) and (3) are due to Resnikoff and Freitag. For the proof in the general case, see [94b, Theorem 5.6 and Corollary 5.7(3)].

For two complex-valued functions f and g on $\mathcal{H}$ such that $f\|_k\gamma = f$ and $g\|_k\gamma = g$ for every γ in a congruence subgroup Γ of $\mathcal{G}$, we put

$$\langle f, g\rangle = \operatorname{vol}(\Gamma\backslash\mathcal{H})^{-1} \int_{\Gamma\backslash\mathcal{H}} \overline{f(z)} g(z)\delta(z)^k d_H z, \tag{2.3a}$$

whenever the integral is convergent, with $\operatorname{vol}(\Gamma\backslash\mathcal{H})$ and $d_H z$ given by

$$\operatorname{vol}(\Gamma\backslash\mathcal{H}) = \int_{\Gamma\backslash\mathcal{H}} d_H z, \quad d_H z = \delta(z)^{-(n+1)u} \bigwedge_{v\in\mathbf{a}} \bigwedge_{p\leqq q} (dx^v_{pq} \wedge dy^v_{pq}), \tag{2.3b}$$

where $z_v = (x^v_{pq} + iy^v_{pq})$.

Lemma 2.2. *Let Γ be a congruence subgroup of $\mathcal{G}^n$ and R a complete set of representatives for $(\mathcal{P}^{n,r} \cap \tau\Gamma\tau^{-1})\backslash\tau\Gamma$ with a fixed $\tau \in \mathcal{G}^n$ and $r < n$; let $g(z) = h(\omega_r(z), \operatorname{Im}(z))$ for $z \in \mathcal{H}_n$ with a function h on $\mathcal{H}_r \times \prod_{v\in\mathbf{a}} S_v^+$. Suppose that $g\|_k\alpha = g$ for every $\alpha \in \mathcal{P}^{n,r} \cap \tau\Gamma\tau^{-1}$. Then $\langle \sum_{\gamma\in R} g\|_k\gamma, f\rangle = 0$ for every $f \in \mathcal{S}^n_k$, under a suitable condition on the convergence.*

Proof. Clearly $\langle \sum_{\gamma\in R} g\|\gamma, f\rangle = \langle \sum_{\gamma\in R} g\|(\gamma\tau^{-1}), f\|\tau^{-1}\rangle$. Since $R\tau^{-1}$ gives $(\mathcal{P}^{n,r} \cap \tau\Gamma\tau^{-1})\backslash\tau\Gamma\tau^{-1}$, the question can be reduced to the case $\tau = 1$. If $f \in \mathcal{S}^n_k(\Gamma')$ with $\Gamma' \subset \Gamma$, then f can be replaced by $[\Gamma : \Gamma']^{-1} \sum_{\alpha\in\Gamma'\backslash\Gamma} f\|\alpha$, and so we may assume that $f \in \mathcal{S}^n_k(\Gamma)$. Put $\Phi = \Gamma\backslash\mathcal{H}_n$ and $\Psi = (\mathcal{P}^{n,r} \cap \Gamma)\backslash\mathcal{H}_n$. Then

$$\int_\Phi \overline{\sum_{\gamma\in R}(g\|\gamma)(z)} f(z)\delta(z)^k d_H(z) = \int_\Psi \overline{g(z)} f(z)\delta(z)^k d_H(z)$$
$$= \int_\Psi \overline{h(z_1, \operatorname{Im}(z))} f\begin{pmatrix} z_1 & z_2 \\ {}^t z_2 & z_4 \end{pmatrix} \delta(z)^k d_H(z).$$

Writing $z_v = x_v + iy_v$ (and changing Γ for a suitable subgroup if necessary), we can express the last integral (times a nonzero constant) in the form

$$\int_Z \overline{h(z_1, \operatorname{Im}(z))} \left\{ \int_X f\begin{pmatrix} z_1 & z_2 \\ {}^t z_2 & z_4 \end{pmatrix} dx_2 dx_4 \right\} \delta(z)^{k-(n+1)u} d(z_1, y_2, y_4),$$

where Z is a certain domain to which (z_1, y_2, y_4) belongs, and X is the (x_2, x_4)-space modulo a lattice. Since f is a cusp form, the integral over X vanishes. In this proof we have to assume a convergence condition which guarantees $\int_\Phi = \int_\Psi = \int_Z \int_X$, but this is always satisfied in the case we treat later.

Lemma 2.3. *For a* $\mathbf{Z}$*-lattice* L *in* S, $z \in \mathscr{H}_n$, *and* $(\alpha, \beta) \in \mathbf{C}^{\mathbf{a}} \times \mathbf{C}^{\mathbf{a}}$ *put*

$$X(z, L; \alpha, \beta) = \sum_{a \in L} \det(z + a)^{-\alpha} \det(\bar{z} + a)^{-\beta},$$

where $\det(w)^s$ *and* $\det(\overline{w})^s$ *for* $s \in \mathbf{C}$ *and* $w \in H_n$ *are defined so that* $\det(i1_n)^s = e^{ns\pi i/2}$ *and* $\det(-i1_n)^s = e^{-ns\pi i/2}$. *Then this is convergent if* $\mathrm{Re}(\alpha_v + \beta_v) > n$ *for every* $v \in \mathbf{a}$, *and* $\prod_{v \in \mathbf{a}} \Gamma_n(\alpha_v + \beta_v - (n+1)/2)^{-1} X(z, L; \alpha, \beta)$ *can be continued as a holomorphic function in* (α, β) *to the whole* $\mathbf{C}^{\mathbf{a}} \times \mathbf{C}^{\mathbf{a}}$, *where*

$$\Gamma_n(s) = \pi^{n(n-1)/4} \prod_{v=0}^{n-1} \Gamma(s - v/2). \tag{2.4}$$

Moreover X *has a Fourier expansion*

$$X(x + iy, L; \alpha, \beta) = \mathrm{vol}(S_{\mathbf{a}}/L)^{-1} \sum_{h \in L'} \mathbf{e}_{\mathbf{a}}(\mathrm{tr}(hx)) \Xi^n(y, h; \alpha, \beta),$$

$$\Xi^n(y, h; \alpha, \beta) = \prod_{v \in \mathbf{a}} \xi^n(y_v, h_v; \alpha_v, \beta_v),$$

where $L' = \{b \in S \mid Tr_{F/\mathbf{Q}}(\mathrm{tr}(bL)) \subset \mathbf{Z}\}$ *and* ξ^n *is the function* ξ *of* [82, (1.25), (4.34.K)].

This was given in [82, Theorem 6.1] when $F = \mathbf{Q}$. Our lemma is its straightforward generalization. It was also shown in [82] that $\xi^n(y, h; \alpha, \beta)$ for each fixed (g, h) is meromorphic in (α, β) on the whole $\mathbf{C}^2$.

Lemma 2.4. (1) $0 < i^{n(\alpha-\beta)} \xi^n(y, h; \alpha, \beta) \in \mathbf{R}$ *if* $(n-1)/2 < \alpha \in \mathbf{R}$, $(n-1)/2 < \beta \in \mathbf{R}$, *and* $\alpha + \beta > n + 1$.

(2) *Suppose* $0 < g \in S_v, h \in S_v, v \in \mathbf{a}$, $p = \omega_r({}^t aga)$ *and* $a^{-1} h {}^t a^{-1} = diag[q, 0_t], t = n - r$ *with* $a \in GL_n(\mathbf{R})$ *and* $q \in S_v^r$. *Then*

$$\xi^n(g, h; \alpha, \beta) = \xi^r(p, q; \alpha - t/2, \beta - t/2) \Gamma_t(\alpha + \beta - \kappa(n)) \Gamma_t(\alpha)^{-1} \Gamma_t(\beta)^{-1} \pi^{-rt/2}$$
$$\cdot\, i^{t(\beta-\alpha)} (2\pi)^{n\kappa(n) - r\kappa(r)} 2^{t(1-\alpha-\beta-r)} \det(g)^{\kappa(n)-\alpha-\beta} \det(p)^{\alpha+\beta-t-\kappa(r)},$$

where $\kappa(n) = (n+1)/2$ *and* $i^s = \mathbf{e}(s/4)$.

The first assertion can easily be derived from [82, (4.34.K), (4.25), (4.19), (4.7.K), (3.6), (3.2)]. The second one follows immediately from [82, (1.29), (4.3), Prop.4.1]. It should be observed that [82, Prop.4.1] is true even if $\det(k) = 0$, since that case can easily be reduced to the case $\det(k) \neq 0$.

Lemma 2.5. *Let* $\zeta_0 = 1_{2n}$ *and* $\zeta_r = \iota_r \times 1_{2n-2r}$ *for* $r > 0$ *with the notation of* (1.2a) *and* (1.25). *Then* $G^n = \bigsqcup_{r=0}^n P^{n,0} \zeta_r P^{n,0}$ *and* $P^{n,0} \zeta_r P^{n,0} = \{\alpha \in G^n \mid \mathrm{rank}(c_\alpha) = r\}$.

This is well known.

We now consider $E_k^{n,0}(z, s; 1, \Gamma)$ with $k = k_0 u, k_0 \in (1/2)\mathbf{Z}$, and for any fixed $\beta \in \mathscr{G}^n$ we take the Fourier expansion

(2.5) $\delta(z)^{k/2-su} \cdot [E_k^{n,0}(z,s;1,\Gamma)\|_k\beta] = \sum_{h\in L} \mathbf{e_a}(\mathrm{tr}(hx))c(y,h,s), z = x+iy$,

with a $\mathbf{Z}$-lattice L in S and $c(y,h,s)$ which is meromorphic in s. The expansion is valid if the left-hand side is finite at s; if it has a pole of order m at $s = s_0$, then the equality is still valid if both sides are multiplied by $(s-s_0)^m$.

Lemma 2.6. *If $h \in S^{n-1} \cap GL_{n-1}(F)$, then*

$$\text{(2.6a)}\quad \begin{aligned} c(y,\mathrm{diag}[h,0],s) &= a(h,s)\Xi^{n-1}(\omega_{n-1}(y),h;s+k_0/2,s-k_0/2)\\ &\quad + b(h,s)\Xi^n(y,\mathrm{diag}[h,0];s+k_0/2,s-k_0/2) \text{ if } n>1, \end{aligned}$$

$$\text{(2.6b)}\quad c(y,0,s) = a(s) + b(s)\Xi^1(y,0;s+k_0/2,s-k_0/2) \text{ if } n=1$$

with functions $a(h,s), b(h,s)$, $a(s)$, and $b(s)$ meromorphic (at least) for $\mathrm{Re}(2s) > n+1$, *where we write $\Xi^*(y,h;s,s')$ for $\Xi^*(y,h;su,s'u)$.*

Proof. Observing that $\zeta_r \in \mathfrak{M}$, write simply ζ_r for $\hat{\zeta}_r$, $\mathscr{P}$ for $\mathscr{P}^{n,0}$, and put $R = \{(\alpha,p)\in\mathscr{P} \mid a_\alpha = 1, p=1\}$, $\Delta = \mathscr{P}\cap\Gamma$, and $\Delta^1 = R\cap\beta^{-1}\Gamma\beta$. By Lemma 2.5 we have $E_k^{n,0}(z,s;\Gamma,1)\|\beta = \sum_{r=0}^n E_r$ with $E_r = \sum_{\gamma\in\Delta\backslash T_r}\delta^{su-k/2}\|\gamma$, $T_r = \mathscr{P}\zeta_r\mathscr{P}\cap\Gamma\beta$. Take a subset A of T_r so that $T_r = \bigsqcup_{\alpha\in A}\Delta\alpha\Delta^1$. Then $\Delta\backslash T_r$ is given by $\bigsqcup_{\alpha\in A}\alpha X_\alpha$ with $X_\alpha = (\alpha^{-1}\Delta\alpha\cap\Delta^1)\backslash\Delta^1$, and hence $E_r = \sum_{\alpha\in A}E_\alpha^r$ with $E_\alpha^r = \sum_{\gamma\in X_r}\delta^{su-k/2}\|\alpha\gamma$. Let us now fix r and α and examine the Fourier expansion of E_α^r. Take $\psi,\omega\in\mathscr{P}$ so that $\alpha = \psi\zeta_r\omega$. Then $X_\alpha = \omega^{-1}X'\omega$ with $X' = (\zeta_r^{-1}\psi^{-1}\Delta\psi\zeta_r\cap\omega\Delta^1\omega^{-1})\backslash\omega\Delta^1\omega^{-1}$. For an even positive integer N put

$$\Delta_N = \{\alpha\in\mathscr{P}\cap\Gamma^0 \mid \mathrm{pr}(\alpha) - 1_{2n} \in N\mathfrak{g}_{2n}^{2n}\},\quad \Delta_N^1 = R\cap\Delta_N,$$

and take N so that $\Delta_N \subset \psi^{-1}\Delta\psi$ and $\Delta_N^1 \subset \omega\Delta^1\omega^{-1}$. Then $E_\alpha^r = \sum_{\tau\in Z}E_{\alpha,\tau}^r$, $E_{\alpha,\tau}^r = p^{-1}\sum_{\sigma\in Y}\delta^{su-k/2}\|\alpha\omega^{-1}\sigma\tau\omega$ with

$$p = [\zeta_r^{-1}\psi^{-1}\Delta\psi\zeta_r\cap\omega\Delta^1\omega^{-1} : \zeta_r^{-1}\Delta_N\zeta_r\cap\Delta_N^1],$$

$Y = (\zeta_r^{-1}\Delta_N\zeta_r\cap\Delta_N^1)\backslash\Delta_N^1$, and $Z = \Delta_N^1\backslash\omega\Delta^1\omega^{-1}$. Observing that

$$\zeta_r^{-1}\Delta_N\zeta_r\cap\Delta_N^1 = \{\alpha\in\Delta_N^1 \mid b_1(\alpha) = 0\},$$

we can take Y to be the set of all $\sigma\in\Delta_N^1$ such that $\mathrm{pr}(\sigma) = \begin{pmatrix}1_r & b\\ 0 & 1_r\end{pmatrix}\times 1_{2n-2r}$ with ${}^tb = b \in N\mathfrak{g}_r^r$. Put $\sigma = \sigma_b$. Then, with a constant $A_{\omega,\psi}$ we have

$$E_{\alpha,\tau}^r(\omega^{-1}\tau^{-1}z) = p^{-1}A_{\omega,\psi}|\lambda_0(\psi)|^{k-2su}\sum_b \delta(z)^{su-k/2}\|\zeta_r\sigma_b.$$

Since $j(\zeta_r,\sigma_b z) = \det[\omega_r(z)+b]$, by Lemma 2.3 the last sum is a constant times

$$\delta(z)^{su-k/2}\sum_{q\in M}\mathbf{e_a}(\mathrm{tr}(q\omega_r(x)))\Xi^r(\omega_r(y),q;s+k_0/2,s-k_0/2)$$

with a $\mathbf{Z}$-lattice M in S^r. Since $\tau\omega\in\mathscr{P}$, we can thus put

(2.7) $\delta(z)^{k/2-su}E^r_{\alpha,\tau}(z)$

$$= \sum_{q \in M} c_q(s)\mathbf{e}_{\mathbf{a}}(\mathrm{tr}(\mathrm{diag}[q,0]ax \cdot {}^t a))\Xi^r(\omega_r(ay \cdot {}^t a), q; s + k_0/2, s - k_0/2)$$

with some $a \in GL_n(F)$ and a simple function $c_q(s)$. The series $\sum_{r=0}^n \sum_\alpha \sum_{\gamma \in X_\tau} \delta^{su-k/2}\|\alpha\gamma$ is convergent for $\mathrm{Re}(2s) > n+1$, and hence (2.5) can be obtained by summing up (2.7) for all r, α, and τ. Given $h \in S^{n-1} \cap GL_{n-1}(F), n > 1$, the coefficient $c(y, \mathrm{diag}[h,0], s)$ can be obtained from the series of (2.7) for $r = n$ and $r = n-1$. For $r = n$ we get the last term of (2.6a) because of the equality

$$\xi^n({}^t aga, h; \alpha, \beta) = |\det(a)|^{n+1-2\alpha-2\beta}\xi^n(g, ah \cdot {}^t a; \alpha, \beta),$$

given in [82, (1.29), (3.1.K)], which holds for every $a \in GL_n(\mathbf{R})$. Now suppose ${}^t a \cdot \mathrm{diag}[q,0]a = \mathrm{diag}[h,0]$ with $q \in S^{n-1}$. Then we easily see that $h = {}^t a' q a'$ and $\omega_{n-1}(ay \cdot {}^t a) = a'\omega_{n-1}(y) \cdot {}^t a'$. with $a' = \omega_{n-1}(a)$. Therefore, from (2.7) with $r = n-1$ we obtain the remaining term of (2.6a). Each term is a Fourier coefficient of the partial sum of our absolutely convergent series, and so is meromorphic in s. Since Ξ^r is a nonvanishing meromorphic function as can be seen from Lemma 2.4(1), both $a(h,s)$ and $b(h,s)$ must be meromorphic for $\mathrm{Re}(2s) > n+1$. The case $n = 1$ is similar and simpler.

3. The adelized definition of Eisenstein series

Lemma 3.1. *Let $P' = \bigcap_{r=0}^{n-1} P^{n,r}$. Then $G_{\mathbf{A}} = P'_{\mathbf{A}} D_0[\mathfrak{b}^{-1}, \mathfrak{b}]$ for every fractional ideal $\mathfrak{b}$. Moreover, for every integral ideal $\mathfrak{c}$ we have*

(3.1) $P^{n,0}_{\mathbf{A}} D[\mathfrak{b}^{-1}, \mathfrak{b}\mathfrak{c}] = P'_{\mathbf{A}} D[\mathfrak{b}^{-1}, \mathfrak{b}\mathfrak{c}]$

$$= \{x \in G_{\mathbf{A}} \mid \det(d_x)_v \neq 0 \text{ and } (d_x^{-1} c_x)_v \in (\mathfrak{b}_v\mathfrak{c}_v)^n_n \text{ for every } v|\mathfrak{c}\}.$$

The first assertion is well known. It is an easy exercise to prove (3.1).

Take a fractional ideal $\mathfrak{b}$ and fix an integer $r < n$. The above lemma shows that $G_{\mathbf{A}} = P^{n,r}_{\mathbf{A}} D_0[\mathfrak{b}^{-1}, \mathfrak{b}]$. Therefore we can define a real number $\varepsilon_r(x)$ and an ideal $\mathfrak{a}_r(x)$ for every $x \in G_{\mathbf{A}}$ or $x \in M_{\mathbf{A}}$ by

(3.2)

$$\mathfrak{a}_r(x) = \lambda_r(p)\mathfrak{g}, \quad \varepsilon_r(x) = |\lambda_r(p)|_{\mathbf{A}} \text{ if } \mathrm{pr}(x) \in pD_0[\mathfrak{b}^{-1}, \mathfrak{b}] \text{ with } p \in P^{n,r}_{\mathbf{A}}.$$

Then we easily see that these are well defined, and

(3.3) $\varepsilon_r(x_{\mathbf{h}}) = N(\mathfrak{a}_r(x))^{-1}, \quad \varepsilon_r(x_{\mathbf{a}})^2 = [\delta(\omega_r(x(\mathbf{i})))/\delta(x(\mathbf{i}))]^u.$

if $x \in G_{\mathbf{A}}$. The symbols ε_r and $\mathfrak{a}_r$ depend not only on r, but also on $\mathfrak{b}$. In the following treatment we consider these with a fixed $\mathfrak{b}$.

Lemma 3.2. *Let $U = F_{\mathbf{a}}^{\times}\prod_{v\in\mathbf{h}}\mathfrak{g}_v^{\times}$ and let C be an open subgroup of $D[\mathfrak{b}^{-1},\mathfrak{b}]$ such that $\lambda_r(P_{\mathbf{A}}^{n,r}\cap C)=U$. Then the map $x\mapsto\mathfrak{a}_r(x)$ gives a bijection of $P^{n,r}\backslash P_{\mathbf{A}}^{n,r}C/C$ onto the ideal class group of F, and $P_{\mathbf{A}}^{n,r}C=\bigsqcup_{\xi\in X}P^{n,r}\xi C$ for any subset X of $P_{\mathbf{A}}^{n,r}$ such that $F_{\mathbf{A}}^{\times}=\bigsqcup_{\xi\in X}F^{\times}U\lambda_r(\xi)$.*

This can be proved by making obvious modifications in the proofs of [83b, Lemmas 1.3, 1.4, and 1.6], which concern the case $r=0$.

For example, take a subset T of $F_{\mathbf{h}}^{\times}$ so that $F_{\mathbf{A}}^{\times}=\bigsqcup_{t\in T}F^{\times}Ut$, and put $\tau_t=\mathrm{diag}[1_{n-1},t^{-1},1_{n-1},t]$. Then $P_{\mathbf{A}}^{n,r}C=\bigsqcup_{t\in T}P^{n,r}\tau_tC$. By strong approximation we have $\tau_tC\cap G\neq\varnothing$. Therefore we obtain

Lemma 3.3. *With C as in Lemma 3.2 put*

$$\tilde{C}=\bigcup_{t\in F_{\mathbf{h}}^{\times}}\mathrm{diag}[1_{n-1},t^{-1},1_{n-1},t]C. \tag{3.4}$$

Then there exists a subset Z of $G\cap\tilde{C}$ such that $P_{\mathbf{A}}^{n,r}C=\bigsqcup_{\zeta\in Z}P^{n,r}\zeta C$ for every r. For such a Z we have $G\cap P_{\mathbf{A}}^{n,r}C=\bigsqcup_{\zeta\in Z}P^{n,r}\zeta(G\cap C)$.

Notice that $\mathfrak{a}_r(\xi)=\mathfrak{a}_0(\xi)$ for every $\xi\in\tilde{C}$.

As shown in [93, Section 2], there is a homomorphism of $M_{\mathbf{A}}^r\times M_{\mathbf{A}}^{n-r}$ into $M_{\mathbf{A}}^n$, written $(\sigma,\sigma')\mapsto\langle\sigma,\sigma'\rangle$, such that $\mathrm{pr}(\langle\sigma,\sigma'\rangle)=\mathrm{pr}(\sigma)\times\mathrm{pr}(\sigma')$ and $\langle t\sigma,\sigma'\rangle=\langle\sigma,t\sigma'\rangle=t\langle\sigma,\sigma'\rangle$ for $t\in\mathbf{T}$. Moreover, [93, Prop. 2.7] shows that if $\alpha\in G^r$ and $\beta\in G^{n-r}$, then $\alpha\times\beta$ as an element of $M_{\mathbf{A}}^n$ coincides with $\langle\alpha,\beta\rangle$, and that if $\sigma\in\mathfrak{M}^r$ and $\sigma'\in\mathfrak{M}^{n-r}$, then $\langle\sigma,\sigma'\rangle\in\mathfrak{M}^n$ and

$$h(\langle\sigma,\sigma'\rangle,\mathrm{diag}[z,z'])=h(\sigma,z)h(\sigma',z')\quad(z\in H_r^{\mathbf{a}},z'\in H_{n-r}^{\mathbf{a}}). \tag{3.5a}$$

We define $\lambda_r:\mathrm{pr}^{-1}(P_{\mathbf{A}}^{n,r})\to F_{\mathbf{A}}^{\times}$ by

$$\lambda_r(\xi)=\lambda_r(\mathrm{pr}(\xi))\quad(\xi\in\mathrm{pr}^{-1}(P_{\mathbf{A}}^{n,r})). \tag{3.5b}$$

Lemma 3.4. *There exists a homomorphism $\pi_r:\mathrm{pr}^{-1}(P_{\mathbf{A}}^{n,r})\to M_{\mathbf{A}}^r$ with the following properties:*

(1) $\mathrm{pr}\circ\pi_r=\pi_r\circ\mathrm{pr}$.

(2) *If $\xi\in\mathrm{pr}^{-1}(P_{\mathbf{A}}^{n,r})$ and $\pi_r(\xi)\in\mathfrak{M}^r$, then $\xi\in\mathfrak{M}^n$ and $j^k(\xi,z)=\lambda_r(\xi)^{[k]}\cdot|\lambda_r(\xi)|^{k-[k]}j^k(\pi_r(\xi),\omega_r(z))$.*

(3) *The restrictions of π_r and λ_r to $\mathcal{P}^{n,r}$ coincide with the maps of (1.27a, b).*

(4) *$\ell_r\circ\pi_r=\pi_r\circ\ell_n$, where ℓ_r denotes the canonical lift $G^r\to M_{\mathbf{A}}^r$.*

Proof. Let $W=\{\alpha\in P^{n,r}\,|\,\pi_r(\alpha)=1\}$. Then $W\subset P^{n,0}$ and $P^{n,r}$ is the semi-direct product of W and $G^r\times 1$, and hence we easily see that $\mathrm{pr}^{-1}(P_{\mathbf{A}}^{n,r})$ is the semi-direct product of $r_P(W_{\mathbf{A}})$ and $\langle M_{\mathbf{A}}^r,1\rangle$ by virtue of Lemma 3.5 below, where r_P denotes the canonical lift $P_{\mathbf{A}}^{n,0}\to M_{\mathbf{A}}$. Therefore we can define π_r by $\pi_r(r_P(\sigma)\langle\alpha,1\rangle)=\alpha$ for $\sigma\in W_{\mathbf{A}}$ and $\alpha\in M_{\mathbf{A}}^r$. Then (4) can be verified easily. If $\alpha\in\mathfrak{M}^r$, $\eta=r_P(\sigma)$, and $\xi=\eta\langle\alpha,1\rangle$, then $\xi\in r_P(P_{\mathbf{A}}^{n,0})\mathfrak{M}^n=\mathfrak{M}^n$, and by [93, Theorem 1.2(3, 4) and Prop.2.7] we have

$$h(\xi,z) = h(\eta,z)h(\langle\alpha,1\rangle,z) = |\lambda_r(\xi)|^{u/2}h(\alpha,\omega_r(z))$$

at least when $z = \mathrm{diag}[\omega_r(z),z']$. In view of (1.24) the equality is true for every $z \in \mathscr{H}_n$. Then (2) and (3) can be verified easily.

Lemma 3.5. *$r_P(W_{\mathbf{A}})$ is a normal subgroup of* $\mathrm{pr}^{-1}(P_{\mathbf{A}}^{n,r})$.

Proof. It is sufficient to show that $\langle\alpha,1\rangle r_P(\sigma)\langle\alpha,1\rangle^{-1} \in r_P(W_{\mathbf{A}})$ for every $\sigma \in W_{\mathbf{A}}$. Let $\sigma' = (\beta\times 1)\sigma(\beta\times 1)^{-1}$ with $\beta = \mathrm{pr}(\alpha)$. Then $\langle\alpha,1\rangle r_P(\sigma)\langle\alpha,1\rangle^{-1} = t\cdot r_P(\sigma')$ with some $t \in \mathbf{T}$. Thus our task is to show that $t = 1$. Now $\sigma = \tau\rho$ with elements τ and ρ of $W_{\mathbf{A}}$ such that $a_4(\tau) = 1_{n-r}$ and $\rho = 1_{2r}\times\mathrm{diag}[\tilde{e},e]$ with $e \in GL_{n-r}(F_{\mathbf{A}})$. Our problem is easy if $\sigma = \rho$, and hence we may assume that $\sigma = \tau$. To prove that $t = 1$, it is sufficient to show that $\langle\alpha,1\rangle r_P(\sigma)$ and $r_P(\sigma')\langle\alpha,1\rangle$ as unitary operators have the same action on some nonzero function $(f\otimes f')(u,u') = f(u)f'(u')$, where u and u' are variables on $(F_r^1)_{\mathbf{A}}$ and $(F_{n-r}^1)_{\mathbf{A}}$. This is clearly so if $\alpha \in r_P(P_{\mathbf{A}}^{r,0})$. Since $P_{\mathbf{A}}^{r,0}$ and $\imath_r$ of (1.2a) generate a dense subgroup of $G_{\mathbf{A}}^r$, it is sufficient to show that $\langle\gamma,1\rangle r_P(\sigma)$ and $r_P(\sigma')\langle\gamma,1\rangle$ have the same action on $f\otimes f'$ when $\gamma = r_\Omega(\imath_r)$, where $r_\Omega(\xi)$ is the canonical lift of $\xi \in G_{\mathbf{A}}^r$ such that $c_\xi \in GL_r(F_{\mathbf{A}})$. Now the action of each operator can be given explicitly (see [93, (1.3a, c)]) and the desired fact can be verified by several lines of straightforward calculation.

To define our Eisenstein series, we take a set of data $(k,\mathfrak{b},\mathfrak{c},\mathfrak{e},\chi)$ (in addition to fixed n and r) which consists of a weight k, a fractional ideal $\mathfrak{b}$, integral ideals $\mathfrak{c},\mathfrak{e}$, and a Hecke character χ satisfying the following conditions:

(3.6a) $$\chi_{\mathbf{a}}(x) = \mathrm{sgn}(x)^{[k]}|x_{\mathbf{a}}|^{i\kappa} \text{ with } \kappa \in \mathbf{R}^{\mathbf{a}}, \sum_{v\in\mathbf{a}}\kappa_v = 0\,;$$

(3.6b) $$\chi_v(a) = 1 \text{ if } v \in \mathbf{h},\ a \in \mathfrak{g}_v^\times, \text{ and } a-1 \in \mathfrak{c}_v\,;$$

(3.6c) $$\mathfrak{c} \subset \mathfrak{e} \text{ and } \mathfrak{e} \text{ is prime to } \mathfrak{e}^{-1}\mathfrak{c}\,;$$

(3.6d) $$\mathfrak{b}^{-1} \subset 2\mathfrak{d}^{-1} \text{ and } \mathfrak{b}\mathfrak{c} \subset 2\mathfrak{d} \text{ if } k \neq [k]\,.$$

The last condition implies that j_α^k is always meaningful if $\mathrm{pr}(\alpha) \in D[\mathfrak{b}^{-1},\mathfrak{b}\mathfrak{c}]$ and that $\mathfrak{c} \subset 4\mathfrak{g}$ if $k \neq [k]$.

For simplicity let us put $D = D[\mathfrak{b}^{-1},\mathfrak{b}\mathfrak{c}]$, $D_0 = D_0[\mathfrak{b}^{-1},\mathfrak{b}\mathfrak{c}]$, and

(3.7a) $$C = C^{n,r} = C_0^{n,r}G_{\mathbf{a}},\quad C_0 = C_0^{n,r} = \{x \in D_0 \mid (a_1(x)-1)_v \in (\mathfrak{e}_v)_r^r,$$
$$a_2(x)_v \in (\mathfrak{e}_v)_{n-r}^r, b_1(x)_v \in (\mathfrak{b}_v^{-1}\mathfrak{e}_v)_r^r \text{ for every } v|\mathfrak{e}\}\,,$$

(3.7b) $$K = K^r = \{x \in D^r[\mathfrak{b}^{-1}\mathfrak{e}\,,\mathfrak{b}\mathfrak{c}] \mid (a_x-1)_v \in (\mathfrak{e}_v)_r^r \text{ for every } v|\mathfrak{e}\}\,.$$

Taking an element f of $\mathscr{S}_k^r(K,\chi^{-1})$, we define a function μ on $G_{\mathbf{A}}^n$ or on $M_{\mathbf{A}}$ according as k is integral or half-integral as follows: $\mu(x) = 0$ if $\mathrm{pr}(x) \notin P_{\mathbf{A}}^{n,r}C$; if $x = pw$ with $\mathrm{pr}(p) \in P_{\mathbf{A}}^{n,r}$ and $\mathrm{pr}(w) \in C_0$, then

(3.8) $$\mu(x) = \chi(\lambda_r(p))^{-1}\chi_{\mathfrak{c}}(\det(d_w))^{-1}j^k(w,\mathbf{i})^{-1}f_{\mathbf{A}}(\pi_r(p))\,.$$

This is well-defined; moreover, we have

$$\mu(\alpha x w) = \chi_{\mathfrak{c}}(\det(d_w))^{-1} j^k(w,\mathbf{i})^{-1}\mu(x) \tag{3.9}$$
$$\textit{for every } \alpha \in P^{n,r} \textit{ and } w \in \mathrm{pr}^{-1}(C_0).$$

Then we define a function $E_{\mathbf{A}}(x,s;f,\chi,C)$ for $s \in \mathbf{C}$ and $x \in G_{\mathbf{A}}$ or $x \in M_{\mathbf{A}}$ according as k is integral or half-integral by

$$E_{\mathbf{A}}(x,s) = E_{\mathbf{A}}(x,s;f,\chi,C) = \sum_{\alpha\in A} \mu(\alpha x)\varepsilon_r(\alpha x)^{-2s}, \quad A = P^{n,r}\backslash G^n. \tag{3.10}$$

For each s where this is convergent, it defines a function satisfying a formula of type (1.19), and so it corresponds (via a formula of type (1.18)) to a function E on $\mathscr{H}_n$ such that $E\|_k\gamma = \chi_{\mathfrak{c}}(\det(a_\gamma))^{-1}E$ for every $\gamma \in G^n \cap C$. We denote this E by $E(z,s;\ f,\chi,C)$ or simply by $E(z,s)$. To simplify our notation, for $\alpha \in G \cap P_{\mathbf{A}}^{n,0}D$ we put $\mathfrak{a}_\alpha = \mathfrak{a}_0(\alpha)$ and

$$\chi[\alpha] = \begin{cases} \chi_{\mathbf{a}}(\det(d_\alpha))\chi^*(\det(d_\alpha)\mathfrak{a}_\alpha^{-1}) & \textit{if } \mathfrak{c} \neq \mathfrak{g}, \\ \chi^*(\mathfrak{a}_\alpha)^{-1} & \textit{if } \mathfrak{c} = \mathfrak{g}. \end{cases} \tag{3.11}$$

Notice that $\det(d_\alpha) \neq 0$ and $\det(d_\alpha)\mathfrak{a}_\alpha^{-1}$ is prime to $\mathfrak{c}$ if $\mathfrak{c} \neq \mathfrak{g}$. We easily see that

$$\chi[\alpha] = \chi_{\mathbf{h}}(t)^{-1}\chi_{\mathfrak{c}}(t/\det(d_\alpha)) \text{ if } \mathfrak{a}_\alpha = t\mathfrak{g}, t \in F_{\mathbf{A}}^{\times}, \tag{3.12a}$$

$$\chi[\beta\alpha\gamma] = \chi[\beta]\chi[\alpha]\chi[\gamma] \text{ if } \beta \in P^{n,0} \text{ and } \gamma \in \Gamma[\mathfrak{b}^{-1},\mathfrak{bc}]. \tag{3.12b}$$

We also put

$$\delta(s,f) = \delta(z,s;f,\kappa) = f(\omega_r(z))[\delta(z)/\delta(\omega_r(z))]^{s u-(k-i\kappa)/2}. \tag{3.13}$$

If $\kappa = 0$, this is the same as the function of (1.32).

Lemma 3.6. *Let Z be as in Lemma 3.3 with the present C; let $\Gamma_0 = G \cap C$ and $R_\zeta = (P^{n,r} \cap \zeta\Gamma_0\zeta^{-1})\backslash\zeta\Gamma_0$. Then*

$$E(z,s;f,\chi,C) = \sum_{\zeta\in Z} N(\mathfrak{a}_\zeta)^{2s} \sum_{\alpha\in R_\zeta} \chi[\alpha]\delta(s,f)\|_k\alpha.$$

Proof. Put $R = \bigcup_{\zeta\in Z} R_\zeta$. We easily see that R gives $P^{n,r}\backslash(G \cap P_{\mathbf{A}}^{n,r}C)$. Therefore, for $z = x(\mathbf{i})$ with $\mathrm{pr}(x) \in G_{\mathbf{a}}$ we have

$$E(z,s) = \sum_{\alpha\in R} \mu(\alpha x)\varepsilon_r(\alpha x)^{-2s} j^k(x,\mathbf{i}). \tag{3.14}$$

Let us now determine $\mu(\alpha x)$ for $\alpha \in R_\zeta$. Since R is contained in the set $\tilde{C}$ of (3.4), we can put $\alpha x = \sigma w$ with $\mathrm{pr}(w) \in C_0$, $\sigma = r_P(\tau)$, $\tau_{\mathbf{h}} = \mathrm{diag}[1_{n-1}, t^{-1}, 1_{n-1}, t]$, $t \in F_{\mathbf{h}}^{\times}$, and $\tau_{\mathbf{a}} \in P'_{\mathbf{a}}$ with P' of Lemma 3.1. Then

$$\mu(\alpha x) = \chi_{\mathbf{h}}(t)^{-1}\chi_{\mathbf{a}}(\lambda_r(\tau))^{-1}\chi_{\mathfrak{c}}(\det(d_w))^{-1} j^k(w,\mathbf{i})^{-1} f_{\mathbf{A}}(\pi_r(\sigma)).$$

Since $\mathfrak{a}_\alpha = t\mathfrak{g}$ and $\det(d_\alpha)_\mathbf{h} = t \cdot \det(d_w)_\mathbf{h}$, we have $\chi_\mathbf{h}(t)\chi_\mathfrak{c}(\det(d_w)) = \chi_\mathbf{h}(t) \cdot \chi_\mathfrak{c}(\det(d_\alpha)/t) = \chi[\alpha]^{-1}$ by (3.12a). Now $\omega_r(\alpha z) = \omega_r(\sigma(\mathbf{i})) = \pi_r(\sigma)(\mathbf{i})$ and $\mathrm{pr}(\pi_r(\sigma)) \in G^r_\mathbf{a}$, and so by (1.18) and Lemma 3.4(2) $f_\mathbf{A}(\pi_r(\sigma)) = j^k(\pi_r(\sigma), \mathbf{i})^{-1} \cdot f(\omega_r(\alpha z)) = \lambda_r(\tau_\mathbf{a})^{[k]}|\lambda_r(\tau_\mathbf{a})|^{k-[k]}j^k(\sigma, \mathbf{i})^{-1}f(\omega_r(\alpha z))$. Thus

$$\begin{aligned}\mu(\alpha x) &= \chi[\alpha]\chi_\mathbf{a}(\lambda_r(\tau))^{-1}\lambda_r(\tau_\mathbf{a})^{[k]}|\lambda_r(\tau_\mathbf{a})|^{k-[k]}j^k(\sigma w, \mathbf{i})^{-1}f(\omega_r(\alpha z))\\ &= \chi[\alpha]|\lambda_r(\tau_\mathbf{a})|^{k-i\kappa}j^k(\alpha, z)^{-1}j^k(x, \mathbf{i})^{-1}f(\omega_r(\alpha z)).\end{aligned}$$

By (1.24) we have $|\lambda_r(\tau_\mathbf{a})|^2 = \delta(\omega_r(\sigma(\mathbf{i}))/\delta(\sigma(\mathbf{i})) = \delta(\omega_r(\alpha z))/\delta(\alpha z)$ and $\varepsilon_r(\alpha x) = \varepsilon_r(\tau_\mathbf{h})\varepsilon_r(\tau_\mathbf{a}) = N(\mathfrak{a}_\zeta)^{-1}|\lambda_r(\tau_\mathbf{a})|^u$ by (3.3). Substituting these into (3.14), we obtain our lemma.

Put $\Gamma_0 = G \cap C^{n,r}$ and

(3.15a) $\qquad \Gamma_u = \{\gamma \in \Gamma_0 | \det(d_\gamma) - e \in \mathfrak{c}$ with some $e \in \mathfrak{g}^\times\}$,

(3.15b) $\qquad \Gamma_1 = \{\gamma \in \Gamma_0 | a_\gamma - 1 \in \mathfrak{c}^n_n, b_\gamma \in (\mathfrak{b}^{-1}\mathfrak{c})^n_n\}$.

Then we have

(3.16) $\qquad \Gamma_u = (\Gamma_u \cap P^{n,r})\Gamma_1$.

To see this, let $\alpha \in \Gamma_u$ with α written as in (1.20). Then $\det(d_4) \equiv \det(d_\alpha) \equiv e \pmod{\mathfrak{c}}$ with $e \in \mathfrak{g}^\times$. By strong approximation we can find $g \in GL_{n-r}(\mathfrak{g})$ so that $g - d_4 \in \mathfrak{c}^{n-r}_{n-r}$. Put $h = \begin{pmatrix} 1_r & -d_2g^{-1} \\ 0 & g^{-1}\end{pmatrix}$ and $\beta = \begin{pmatrix} \tilde{h} & -{}^tdbh \\ 0 & h\end{pmatrix}$. Then $h \in GL_n(\mathfrak{g})$, $hd_\alpha - 1_n \in \mathfrak{c}^n_n$, $\beta \in \Gamma_u \cap P^{n,r}$, and $\beta\alpha \in \Gamma_1$, which proves (3.16).

Lemma 3.7. *Suppose that* $\mathfrak{e} = \mathfrak{c}$; *let* Φ *be the set of all Hecke characters* χ *satisfying* (3.6a,b) *with* $\kappa = 0$ *and a fixed* k. *Then* $\Phi \neq \emptyset$ *if and only if* $\mathrm{sgn}(a)^{[k]} = 1$ *for every* $a \in \mathfrak{g}^\times$ *such that* $a - 1 \in \mathfrak{c}$, *in which case we have*

$$[\mathcal{P}^{n,r} \cap \Gamma : \mathcal{P}^{n,r} \cap \Gamma_1]\#(\Phi)E^{n,r}_k(z, s; f, \Gamma) = \sum_{\tau \in \Gamma_1\backslash\Gamma}\sum_{\chi\in\Phi} E(z, s; f, \chi, C)\|_k\tau$$

for every congruence subgroup Γ *of* $\mathcal{G}$ *containing* Γ_1 *and every* $f \in \mathcal{S}^r_k(\Gamma \cap \mathcal{P}^{n,r})$.

This is a generalization of [83b, Prop. 2.4, (i)] and can be proved in the same manner.

4. Integral expressions of certain Eisenstein series

In this section we put $P^n = P^{n,0}$ for simplicity. We are going to consider the pullback of an Eisenstein series on G^{m+n} to $G^m \times G^n$, where $G^m \times G^n$ is identified with the subgroup of G^{m+n} consisting of all the elements of the form (1.25) with (m, n) in place of $(r, n - r)$. Put $N = m + n$ and

$$W = \{w \in F^N_{2N} | w\iota_N \cdot {}^tw = 0, \mathrm{rank}(w) = N\},$$

and let $GL_N(F)$ and G^N act on W on the left and right.

Lemma 4.1. *Each orbit of $GL_N(F)\backslash W/(G^m \times G^n)$ has a representative whose last N columns form an invertible matrix.*

Proof. Let $w=(a\,b\,a'\,b')\in W$ with $a,a'\in F_m^N$ and $b,b'\in F_n^N$. Put $x=(a\,a')$ $(\in F_{2m}^N)$ and $y=(b\,b')$ $(\in F_{2n}^N)$. Then $x\imath_m\cdot{}^t x=-y\imath_n\cdot{}^t y$. Changing w for ξw with a suitable ξ of $GL_N(F)$, we may assume that $x\imath_m\cdot{}^t x=-y\imath_n\cdot{}^t y=\mathrm{diag}[\imath_r,0]$ with $0\leqq r\leqq \mathrm{Min}(m,n)$. Let $\{e_i\}_{i=1}^{2m}$ be the standard basis of F_{2m}^1, and x_i the i-th row of x. Then $x_1,\ldots,x_{2r}$ are linearly independent. Assuming $r>0$, put

$$Z=\{z\in F_{2m}^1 \mid x_i\imath_m^t z=0\ (1\leqq i\leqq 2r)\}.$$

Then $F_{2m}^1=\sum_{i=1}^{2r}Fx_i\oplus Z$ and $x_j\in Z$ for $j>2r$. Since $x_j\imath_m\cdot{}^t x_k=0$ for $j,k>2r$, we can find $\alpha\in G^m$ so that $x_i\alpha=e_i$ and $x_{r+i}\alpha=e_{m+i}$ for $i\leqq r$, and $x_j\alpha\in\sum_{k>r}Fe_{m+k}$ for $j>2r$. If $r=0$, we can take $\alpha\in G^m$ so that $x_j\in\sum_{k>0}Fe_{m+k}$ for every j. Thus for $r\geqq 0$ we obtain

$$x\alpha=\begin{pmatrix}1_r & 0 & 0 & 0\\ 0 & 0 & 1_r & 0\\ 0 & 0 & 0 & u\end{pmatrix}$$

with some $u\in F_{m-r}^{N-2r}$. Similarly we can find $\beta\in G^n$ so that

$$y\beta=\begin{pmatrix}0 & 0 & 1_r & 0\\ 1_r & 0 & 0 & 0\\ 0 & 0 & 0 & v\end{pmatrix}$$

with some $v\in F_{n-r}^{N-2r}$. If $r=0$, we have $x\alpha=(0\ u)$ and $y\beta=(0\ v)$. Thus

$$w(\alpha\times\beta)=\begin{pmatrix}1_r & 0 & 0 & 0 & 0 & 0 & 1_r & 0\\ 0 & 0 & 1_r & 0 & 1_r & 0 & 0 & 0\\ 0 & 0 & 0 & 0 & 0 & u & 0 & v\end{pmatrix}.$$

Since $\mathrm{rank}(w)=N$, we have $\det(u\ v)\neq 0$, which proves our lemma.

Lemma 4.2. *For $0\leqq r\leqq \mathrm{Min}(m,n)$ let τ_r denote the element of G^N given by*

$$\tau_r=\begin{pmatrix}1_N & 0\\ \rho_r & 1_N\end{pmatrix},\quad \rho_r=\begin{pmatrix}0_m & e_r\\ {}^t e_r & 0_n\end{pmatrix},\quad e_r=\begin{pmatrix}1_r & 0\\ 0 & 0\end{pmatrix}\in F_n^m.$$

Then the τ_r form a complete set of representatives for $P^N\backslash G^N/(G^m\times G^n)$.

Proof. Put $w_0=(0\,1_N)$. Then $\alpha\mapsto w_0\alpha$ for $\alpha\in G^N$ gives a bijection of $P^N\backslash G^N$ onto $GL_N(F)\backslash W$, and also a bijection of $P^N\backslash G^N/(G^m\times G^n)$ onto $GL_N(F)\backslash W/(G^m\times G^n)$. Thus our task is to find suitable representatives for $GL_N(F)\backslash W/(G^m\times G^n)$. By Lemma 4.1 each coset has a representative $w=(c\ d)$ with $\det(d)\neq 0$. Put $d^{-1}c=\begin{pmatrix}f_1 & f_2\\ {}^t f_2 & f_4\end{pmatrix}$, $\alpha_1=\begin{pmatrix}1_m & 0\\ f_1 & 1_m\end{pmatrix}$, $\beta_1=\begin{pmatrix}1_n & 0\\ f_4 & 1_n\end{pmatrix}$, $r=\mathrm{rank}(f_2)$, $\alpha_2=\mathrm{diag}[{}^t p,p^{-1}]$, and $\beta_2=\mathrm{diag}[{}^t q,q^{-1}]$ with $p\in GL_m(F)$ and $q\in GL_n(F)$ such that $pf_2\cdot{}^t q=e_r$ with e_r as in our lemma. Then

$$\mathrm{diag}[p,q]d^{-1}w(\alpha_1^{-1}\alpha_2\times\beta_1^{-1}\beta_2)=(\rho_r\ 1_N).$$

This shows that $G^N = \bigcup_r P^N\tau_r(G^m \times G^n)$. Now the rank of the matrix x in the proof of Lemma 4.1 depends only on $GL_N(F)w(G^m \times G^n)$. For $w = (\rho_r\ 1_N)$ the rank is $r+m$, and hence our union must be disjoint.

Lemma 4.3. *Put* $U_r = (G^m \times G^n) \cap \tau_r^{-1}P^N\tau_r$ *with the notation as in Lemma 4.2. Then* $U_0 = P^m \times P^n$,

$$U_r = \{\beta \times \gamma \in P^{m,r} \times P^{n,r} | \pi_r(\beta)\kappa_r = \kappa_r\pi_r(\gamma)\}$$

$$\text{with } \kappa_r = \begin{pmatrix} 0 & 1_r \\ 1_r & 0 \end{pmatrix} \text{ for } r > 0,$$

$$P^{m,r} \times P^{n,r} = \bigsqcup_{\xi\in G^r} U_r((\xi \times 1_{2m-2r}) \times 1_{2n}) = \bigsqcup_{\xi\in G^r} U_r(1_{2m} \times (\xi \times 1_{2n-2r})),$$

$$P^N\tau_r(G^m \times G^n) = \bigsqcup P^N\tau_r((\xi \times 1_{2m-2r})\beta \times \gamma) = \bigsqcup P^N\tau_r((\beta \times (\xi \times 1_{2n-2r})\gamma),$$

where ξ *runs over* G^r, β *over* $P^{m,r}\backslash G^m$, *and* γ *over* $P^{n,r}\backslash G^n$.

Proof. The first two equalities are merely computational. These together with Lemma 4.2 prove the last equality. We ignore ξ for $r = 0$.

For $z \in H_m, w \in H_n$, and ξ, β, γ as in the last equality of Lemma 4.3 we can easily verify that

$$\begin{aligned} j(\tau_r((\xi \times 1_{2m-2r})\beta \times \gamma), \mathrm{diag}[z, w]) \\ = j(\beta, z)j(\gamma, w)j(\eta, \omega_r(\beta z))\det[\eta\omega_r(\beta z) + \omega_r(\gamma w)] \\ = j(\beta, z)j(\gamma, w)j(\zeta, \omega_r(\gamma w))\det[\omega_r(\beta z) + \zeta\omega_r(\gamma w)], \end{aligned} \tag{4.1}$$

where $\eta = \iota_r^{-1}\xi$ and $\zeta = \varepsilon\xi^{-1}\iota_r\varepsilon$ with $\varepsilon = \mathrm{diag}[1_r, -1_r]$. If $r = n$, we can clearly take $\gamma = 1$. In this special case we have

Lemma 4.4. *Let* $\mathfrak{e}$ *be an integral ideal as in* (3.6c), *and* σ *an element of* $G_{\mathbf{h}}^N$ *given by*

$$\sigma_v = \begin{cases} \mathrm{diag}[1_m, \theta_v^{-1}1_n, 1_m, \theta_v 1_n]\tau_n & \textit{if } v|\mathfrak{e}, \\ \mathrm{diag}[1_m, \theta_v^{-1}1_n, 1_m, \theta_v 1_n] & \textit{if } v \nmid \mathfrak{e}, \end{cases}$$

where θ *is an element of* $F_{\mathbf{h}}^\times$ *such that* $\theta\mathfrak{g} = \mathfrak{b}$. *Suppose that* $m \geqq n$. *Then*

$$(P^N\tau_n(G^m \times G^n)) \cap (P_{\mathbf{A}}^N D^N[\mathfrak{b}^{-1}, \mathfrak{b}\mathfrak{c}]\sigma) = \bigsqcup_{\xi\in X, \beta\in B} P^N\tau_n((\xi \times 1_{2m'})\beta \times 1_{2n}),$$

where $m' = m - n$, B *is a subset of* $G^m \cap \tilde{C}^{m,n}$ *which represents* $P^{m,n}\backslash(G^m \cap P_{\mathbf{A}}^{m,n}C^{m,n})$ *completely with* $C^{m,n}$ *and* $\tilde{C}$ *of* (3.4), *and* $X = G^n \cap G_{\mathbf{a}}^n \prod_{v\in\mathbf{h}} X_v$ *with*

$$X_v = \begin{cases} \{x \in D_v^n[\mathfrak{b}^{-1}\mathfrak{c}, \mathfrak{b}\mathfrak{c}] | a_x - 1 \in (\mathfrak{e}_v)_n^n\} & \textit{if } v|\mathfrak{e}, \\ D_v^n[\mathfrak{b}^{-1}\mathfrak{c}, \mathfrak{b}]W_v D_v^n[\mathfrak{b}^{-1}, \mathfrak{b}\mathfrak{c}] & \textit{if } v|\mathfrak{e}^{-1}\mathfrak{c}, \\ G_v^n & \textit{if } v \nmid \mathfrak{c}, \end{cases}$$

$$W_v = \{\mathrm{diag}[q, \tilde{q}] | q \in GL_n(F_v) \cap (\mathfrak{c}_v)_n^n\};$$

we take $B = \{1_{2m}\}$ *if* $m = n$.

Proof. For simplicity put $D'' = D^n[\mathfrak{b}^{-1}, \mathfrak{b}\mathfrak{c}]$. Let $\alpha = (\xi \times 1_{2m'})\beta$ with $\xi \in G^n$ and $\beta \in P^{m,n}\backslash G^m$. By Lemma 4.3 $P^N\tau_n(G^m \times G^n)$ is a disjoint union of $P^N\tau_n(\alpha \times 1_{2n})$ with such α's. Thus our task is to determine $P^N\tau_n(\alpha \times 1_{2n})$ contained in $P^N_{\mathbf{A}}D^N\sigma$. Write α in the form (1.20) with (m,n) instead of (n,r) and let $\varepsilon = (\tau_n(\alpha \times 1_{2n})\sigma^{-1})_v$ with a fixed $v \in \mathbf{h}$. Then (with the subscript v suppressed)

$$(c_\varepsilon\ d_\varepsilon) = \begin{pmatrix} c_1 & c_2 & 0-\theta d_1 & d_1 & d_2 & 0 \\ c_3 & c_4 & -\theta d_3 & d_3 & d_4 & 0 \\ a_1 - 1_n & a_2 & -\theta b_1 & b_1 & b_2 & \theta^{-1}1_n \end{pmatrix} \quad if\ v|\mathfrak{e}\ ,$$

$$(c_\varepsilon\ d_\varepsilon) = \begin{pmatrix} c_1 & c_2 & \theta 1_n & d_1 & d_2 & 0 \\ c_3 & c_4 & 0 & d_3 & d_4 & 0 \\ a_1 & a_2 & 0 & b_1 & b_2 & \theta^{-1}1_n \end{pmatrix} \quad if\ v \nmid \mathfrak{e}\,.$$

From (3.1) and the above explicit forms of c_ε and d_ε we see that $\varepsilon \in P^N_v D^N_v$ only if $d \in GL_m(F_v)$ and $d^{-1}c \in (\mathfrak{b}_v\mathfrak{c}_v)^m_m$ whenever $v|\mathfrak{c}$, and hence $\alpha \in P^m_{\mathbf{A}}D^m \subset P^{m,n}_{\mathbf{A}}D^m$ by (3.1). Then $\beta \in P^{m,n}_{\mathbf{A}}D^m$, since $\xi \times 1 \in P^{m,n}$. Thus we may restrict β to $P^{m,n}\backslash(G^m \cap P^{m,n}_{\mathbf{A}}D^m)$. By Lemma 3.3 we may assume that $\beta = hw$ with $w \in D^m$ and $h = \mathrm{diag}[1_{m-1}, t^{-1}, 1_{m-1}, t]$, $t \in F^\times_{\mathbf{h}}$. If $m = n$, we can take $\beta = 1$, and so can take $h = w = 1$. Assuming $m > n$, we have $\xi \times 1 \in P^m_{\mathbf{A}}D^m$ since $h(\xi \times 1)w = \alpha \in P^m_{\mathbf{A}}D^m$. Consequently $d_\xi \in GL_n(F_v)$ and $d_\xi^{-1}c_\xi \in (\mathfrak{b}_v\mathfrak{c}_v)^n_n$ for every $v|\mathfrak{c}$ by (3.1). Write w in the form (1.20) with $p_i = a(w)_i$, $q_i = b(w)_i$, $r_i = c(w)_i$, $s_i = d(w)_i$; put $y = \mathrm{diag}[1_{m-n-1}, t]$ and $e = d_\xi^{-1}c_\xi$. Then

$$\begin{pmatrix} c_1 & c_2 & d_1 & d_2 \\ c_3 & c_4 & d_3 & d_4 \end{pmatrix} = \begin{pmatrix} d_\xi & 0 \\ 0 & y \end{pmatrix} \begin{pmatrix} ep_1 + r_1 & ep_2 + r_2 & eq_1 + s_1 & eq_2 + s_2 \\ r_3 & r_4 & s_3 & s_4 \end{pmatrix}.$$

Compute c_ε and d_ε with these c_i and d_i; then from the condition that $d_\varepsilon \in GL_N(F_v)$ and $d_\varepsilon^{-1}c_\varepsilon \in (\mathfrak{b}_v\mathfrak{c}_v)^N_N$ whenever $v|\mathfrak{c}$, we can easily derive that $d_\xi \in GL_n(\mathfrak{g}_v)$, $(s_1 - d_\xi^{-1})_v \in (\mathfrak{e}_v)^n_n$ and $(s_3)_v \in (\mathfrak{e}_v)^{m-n}_n$ for every $v|\mathfrak{e}$. Moreover we can find $f \in S^n_{\mathbf{h}}$ such that $(\tilde{s}_1 f - q_1)_v \in (\mathfrak{b}_v^{-1}\mathfrak{e}_v)^n_n$ for $v|\mathfrak{e}$. By strong approximation we can find an element $\zeta \in G^n \cap D^n$ such that $\zeta\begin{pmatrix} \tilde{s}_1 & \tilde{s}_1 f \\ 0 & s_1 \end{pmatrix}_v \in K^n_v$ for $v|\mathfrak{e}$ with K of (3.7b). Then $(\zeta \times 1)w \in C^{m,n}$. Changing ξ, β, and w for $\xi\zeta^{-1}$, $(\zeta \times 1)\beta$, and $(\zeta \times 1)w$, we may now assume that $w \in C^{m,n}$. Then we can easily verify that $\sigma(w \times 1)\sigma^{-1} \in D^N$, and hence $\tau_n(h(\xi \times 1) \times 1)\sigma^{-1} \in P^N_{\mathbf{A}}D^N$, which is true also when $m = n$, since $h = w = 1$. Computing again $d_\varepsilon^{-1}c_\varepsilon$ with $h(\xi \times 1)$ as α, we find that $\tau_n(h(\xi \times 1) \times 1)\sigma^{-1} \in P^N_{\mathbf{A}}D^N$ if and only if $\xi \in X_v$ for $v|\mathfrak{e}$ and $\xi \in X'_v$ for $v|\mathfrak{e}^{-1}\mathfrak{c}$, where

$$X'_v = \{x \in G^n_v | d_x \in GL_n(F_v),\ d_x^{-1} \equiv \theta_v b_x d_x^{-1} \equiv \theta_v^{-1} d_x^{-1} c_x \equiv 0 \pmod{(\mathfrak{c}_v)^n_n}\}\,.$$

Since $P^{m,n}\backslash(G^m \cap P^{m,n}_{\mathbf{A}}C^{m,n})$ can be represented by the elements $\beta = hw \in G^m \cap hC^{m,n}$ as above, we obtain our assertion because of the following result.

Lemma 4.5. *If* $v|\mathfrak{e}^{-1}\mathfrak{c}$, *we have* $X'_v = X_v = D^n_v[\mathfrak{b}^{-1}\mathfrak{c}, \mathfrak{b}\mathfrak{c}^\mu]W_v D^n_v[\mathfrak{b}^{-1}\mathfrak{c}^\nu, \mathfrak{b}\mathfrak{c}]$ *with any nonnegative integers* μ *and* ν.

Proof. Let $Z_v = D_v^n[\mathfrak{b}^{-1}\mathfrak{c}, \mathfrak{b}\mathfrak{c}^{\mu}]W_v D_v^n[\mathfrak{b}^{-1}\mathfrak{c}^{\nu}, \mathfrak{b}\mathfrak{c}]$. Clearly $Z_v \subset X_v$. For $\begin{pmatrix} a & b \\ c & d \end{pmatrix} \in G_v^n$ with d invertible, we have

$$(4.2) \qquad \begin{pmatrix} a & b \\ c & d \end{pmatrix} = \begin{pmatrix} 1 & bd^{-1} \\ 0 & 1 \end{pmatrix} \begin{pmatrix} \tilde{d} & 0 \\ 0 & d \end{pmatrix} \begin{pmatrix} 1 & 0 \\ d^{-1}c & 1 \end{pmatrix},$$

which shows that $X_v' \subset Z_v$. We easily see that $D_v^n[\mathfrak{b}^{-1}\mathfrak{c}, \mathfrak{b}]W_v \subset X_v'$. Now the above proof of Lemma 4.4 with $m = n$ shows that $\xi \in X_v'$ if and only if $\tau_n(\xi \times 1_{2n})\sigma_v^{-1} \in P_v^N D_v^N$. From this we easily see that $X_v' D_v^n[\mathfrak{b}^{-1}, \mathfrak{b}\mathfrak{c}] \subset X_v'$, and hence $X_v \subset X_v'$, which completes the proof.

We now consider $E_{\mathbf{A}}^{\bullet}$ of (3.10) with N, 0, $D^N[\mathfrak{b}^{-1}, \mathfrak{b}\mathfrak{c}]$ in place of n, r, C and the constant 1 as f. We assume k to be integral and put $\mathfrak{a}_x = \mathfrak{a}_0(x)$ for $x \in G_{\mathbf{A}}^N$. In this case $\mu(x) = 0$ for $x \notin P_{\mathbf{A}}^N D^N[\mathfrak{b}^{-1}, \mathfrak{b}\mathfrak{c}]$; if $x = pw$ with $p \in P_{\mathbf{A}}^N$ and $w \in D_0^N[\mathfrak{b}^{-1}, \mathfrak{b}\mathfrak{c}]$, then

$$(4.3a) \qquad \mu(x) = \chi(\det(d_p))^{-1}\chi_{\mathfrak{c}}(\det(d_w))^{-1} j^k(w, \mathbf{i})^{-1}$$
$$= \chi_{\mathbf{h}}(\det(d_p))^{-1}\chi_{\mathfrak{c}}(\det(d_w))^{-1} j^k(x, \mathbf{i})^{-1} |j(x, \mathbf{i})|^{k-i\kappa},$$

$$(4.3b) \qquad \varepsilon_N(x_{\mathbf{h}}) = N(\mathfrak{a}_x)^{-1}, \qquad \varepsilon_N(x_{\mathbf{a}}) = |j(x, \mathbf{i})|^u$$

Let $E_{\mathbf{A}}^N(x)$ and $E^N(z)$ denote the functions $E_{\mathbf{A}}(x, s)$ and $E(z, s)$ in this case. Given σ as in Lemma 4.4, strong approximation guarantees an element ρ of $G^N \cap D^N[\mathfrak{b}^{-1}, \mathfrak{b}\mathfrak{c}]\sigma$ such that $a(\sigma\rho^{-1})_v - 1_N \in (\mathfrak{c}_v)_N^N$ for every $v|\mathfrak{c}$. Then we easily see that $E^N\|\rho$ corresponds to $E_{\mathbf{A}}^N(x\sigma^{-1})$. Therefore, taking $y \in G_{\mathbf{a}}^N$ so that $z = y(\mathbf{i})$ and putting $p_\alpha(z) = \mu(\alpha y\sigma^{-1})\varepsilon(\alpha y\sigma^{-1})^{-2s} j^k(y, \mathbf{i})$, we easily find that

$$(4.4) \qquad (E^N\|_k\rho)(z) = \sum_{\alpha\in A} p_\alpha(z)$$
$$= \sum_{\alpha\in A} N(\mathfrak{a}_0(\alpha\sigma^{-1}))^{2s}\mu\big((\alpha\sigma^{-1})_{\mathbf{h}}\big) j^k(\alpha, z)^{-1}\delta(\alpha z)^{su-(k-i\kappa)/2},$$

where $A = P^N\backslash G^N$. Every element x of $G_{\mathbf{A}}^n$ can be written $x = \rho \cdot \mathrm{diag}[q, \tilde{q}]\rho'$ with $\rho, \rho' \in D^n[\mathfrak{b}^{-1}, \mathfrak{b}]$ and $q \in GL_n(F)_{\mathbf{h}} \cap \prod_{v\in\mathbf{h}}(\mathfrak{g}_v)_n^n$. We then put

$$(4.5) \qquad \ell_0(x) = \det(q)\mathfrak{g}, \quad \ell_1(x) = \prod_{v\nmid\mathfrak{c}} \ell_0(x)_v, \quad \ell(x) = N\big(\ell_0(x)\big).$$

Lemma 4.6. *Let $g = \alpha\sigma^{-1}$ with $\alpha = \tau_n\big((\xi \times 1_{2m'})\beta \times 1_{2n}\big)$, $\xi \in X$, $\beta \in B$ and σ as in Lemma* 4.4. *Then $\mathfrak{a}_g = \mathfrak{b}^{-n}\mathfrak{a}_\beta\ell_0(\xi)^{-1}$ and*

$$\mu(g_{\mathbf{h}}) = \chi(\theta^n)\chi[\beta]\chi^*\big(\ell_1(\xi)\big)\chi_{\mathfrak{c}}\big(\det(d_\xi)\big)^{-1}.$$

Proof. Put $\beta = hw$ and $f = \tau_n\big(h(\xi \times 1_{2m'}) \times 1_{2n}\big)\sigma^{-1}$ with h and w as in the proof of Lemma 4.4. Since $g = f\sigma(w \times 1_{2n})\sigma^{-1}$ and $\sigma(w \times 1)\sigma^{-1} \in D^N$, we have $\mu(g_{\mathbf{h}}) = \mu(f_{\mathbf{h}})\chi_{\mathfrak{c}}\big(\det(d_w)\big)^{-1}$ and $\mathfrak{a}_g = \mathfrak{a}_f$. Now, if $e^{-1}(c_f\ d_f)$ is the lower half of an element of $D^N[\mathfrak{b}^{-1}, \mathfrak{b}\mathfrak{c}]$ with $e \in GL_n(F)_{\mathbf{A}}$, then $\mu(f_{\mathbf{h}}) = (\chi_{\mathfrak{c}}/\chi_{\mathbf{h}})\big(\det(e)\big)\chi_{\mathfrak{c}}\big(\det(d_f)\big)^{-1}$ and $\mathfrak{a}_f = \det(e)\mathfrak{g}$. The computation in the proof of Lemma 4.4 shows that $\det(d_f) = t\theta^{-n}\det(d_\xi)$. Thus our task is to find a suitable e. If $v|\mathfrak{e}$, we can take $e_v = \mathrm{diag}[1_{m-1}, t, \theta^{-1}1_n]_v$. If $v \nmid \mathfrak{e}$, we have

$$(c_f\ d_f)_v = \begin{pmatrix} c_\xi & 0 & \theta 1_n & d_\xi & 0 & 0 \\ 0 & 0 & 0 & 0 & y & 0 \\ a_\xi & 0 & 0 & b_\xi & 0 & \theta^{-1}1_n \end{pmatrix}_v$$

with $y = \mathrm{diag}[1_{m-n-1}, t]$. By our definition of X we can find $\rho \in D_v^n[\mathfrak{b}^{-1}\mathfrak{c}, \mathfrak{b}]$, $\rho' \in D_v^n[\mathfrak{b}^{-1}, \mathfrak{b}\mathfrak{c}]$, and $q \in (\mathfrak{c}_v)_n^n$ so that $\rho\xi = \mathrm{diag}[q_v, \tilde{q}_v]\rho'$. Put

$$\eta = \begin{pmatrix} 0 & \theta^{-1}1_n \\ \theta 1_n & 0 \end{pmatrix} \rho \begin{pmatrix} 0 & \theta^{-1}1_n \\ \theta 1_n & 0 \end{pmatrix}.$$

Then $\eta \in D_v^n[\mathfrak{b}^{-1}, \mathfrak{b}\mathfrak{c}]$ and, with the subscripts v and ξ suppressed, we have

$$\begin{pmatrix} 0 & {}^tq \\ \theta 1_n & 0 \end{pmatrix} \rho \begin{pmatrix} 0 & 1_n \\ 1_n & 0 \end{pmatrix} \begin{pmatrix} c & \theta 1_n & d & 0 \\ a & 0 & b & \theta^{-1}1_n \end{pmatrix}$$

$$= \begin{pmatrix} 0 & \theta\cdot{}^tq & 1_n & 0 \\ \theta q & 0 & 0 & 1_n \end{pmatrix} (\rho' \times \eta).$$

Since the right-hand side is the lower half of an element of $D_v^{2n}[\mathfrak{b}^{-1}, \mathfrak{b}\mathfrak{c}]$, we can take e_v so that $\det(e_v) = t_v\theta_v^{-n}\det(q_v)^{-1}$. If we put $q = (q_v)_{v\in\mathbf{h}}$ with $q_v = 1$ for $v|\mathfrak{e}$, then $\mathfrak{a}_g = \mathfrak{a}_f = \det(e)\mathfrak{g} = t\theta^{-n}\det(q)^{-1}\mathfrak{g} = \mathfrak{a}_\beta\mathfrak{b}^{-n}\ell_0(\xi)^{-1}$ and

$$\begin{aligned}\mu(f_{\mathbf{h}}) &= (\chi/\chi_{\mathfrak{c}})(t^{-1}\theta^n\det(q))\chi_{\mathfrak{c}}\left(t^{-1}\theta^n\det(d_\xi)^{-1}\right) \\ &= \chi(t^{-1}\theta^n)\chi^*(\ell_1(\xi))\chi_{\mathfrak{c}}\left(\det(d_\xi)\right)^{-1}.\end{aligned}$$

Since $\chi[\beta]^{-1} = \chi(t)\chi_{\mathfrak{c}}\left(\det(d_w)\right)$, we obtain the last formula of our lemma. In this proof we tacitly assumed $m > n$, but the case $m = n$ can be proved in the same manner by ignoring y.

Lemma 4.7. *If f is a holomorphic function on $\mathcal{H}_n$, $k \in \mathbf{R}^{\mathbf{a}}$, $\mathbf{s} = (s_v)_{v\in\mathbf{a}} \in \mathbf{C}^{\mathbf{a}}$, $\mathrm{Re}(s_v) \geqq 0$ and $\mathrm{Re}(s_v) + k_v > 2n$ for every $v \in \mathbf{a}$, and $\delta(z)^k f(z)^2$ is bounded, then*

$$\int_{\mathcal{H}} \det(z - \overline{w})^{-k}|\det(z - \overline{w})|^{-2\mathbf{s}}\delta(w)^{\mathbf{s}+k}f(w)d_Hw = c_{n,k}(\mathbf{s})\delta(z)^{-\mathbf{s}}f(z),$$

where

$$c_{n,k}(\mathbf{s}) = \prod_{v\in\mathbf{a}}(2i)^{-nk_v}\left(2^{n+3-4s_v}\pi^{n+1}\right)^{n/2}\Gamma_n\left(s_v + k_v - (n+1)/2\right)\Gamma_n(s_v + k_v)^{-1}$$

with $\Gamma_n(s)$ of (2.4)

The basic principles for the formulas of this type are well known and explained in [Go, exposés 6, 10]. The constant corresponding to $c_{n,k}(\mathbf{s})$ is given erroneously in [Go, exp.6], but it is not difficult to find the correct value. Under our assumptions $|\delta(z)^{\alpha/2}\det(z - \overline{z}_1)^{-\mathbf{s}}f(z)|$ with $\alpha = k + \mathbf{s}$ is bounded on $\mathscr{H}$ for any fixed $z_1 \in \mathscr{H}$, and so the problem can be reduced to the case $s = 0$.

Let us now study the value of (4.4) at $\mathrm{diag}[z, w]$ with $(z, w) \in \mathscr{H}_m \times \mathscr{H}_n$, assuming $m \geqq n$. By (4.4) and Lemma 4.2 we have $E^N \|_k \rho = \sum_{r=0}^{n} \mathscr{E}_r$ with $\mathscr{E}_r = \sum_{\alpha \in A_r} p_\alpha$, $A_r = P^N \backslash P^N \tau_r(G^m \times G^n)$. Fixing r, take a complete set of representatives R^m for $P^{m,r} \backslash G^m$, and put $p_{\xi,\beta,\gamma} = p_\alpha$ for $\alpha = \tau_r\big((\xi \times 1_{2m-2r})\beta \times \gamma\big)$. Then, by Lemma 4.3 $\mathscr{E}_r = \sum_{\xi \in G^r} \sum_{\beta \in R^m} \sum_{\gamma \in R^n} p_{\xi,\beta,\gamma}$. Put

$$D' = \left\{ x \in D^N[\mathfrak{b}^{-1}, \mathfrak{b}\mathfrak{c}] \mid \det(d_x)_v - 1 \in \mathfrak{c}_v \text{ for every } v | \mathfrak{c} \right\}.$$

Then we easily see that $p_\alpha \| \alpha' = p_{\alpha\alpha'}$ for $\alpha' \in G^N \cap \sigma^{-1} D' \sigma$. Let Γ be a congruence subgroup of G^n such that $1_{2m} \times \Gamma \subset \sigma^{-1} D' \sigma$. Take $T \subset G^n$ so that $G^n = \bigsqcup_{\tau \in T} P^{n,r} \tau \Gamma$. Then we can take $R^n = \bigsqcup_{\tau \in T} R_\tau$ with $R_\tau = (P^{n,r} \cap \tau \Gamma \tau^{-1}) \backslash \tau \Gamma$. Put $g_\tau = \sum_{\xi \in G^r} \sum_{\beta \in R^m} p_{\xi,\beta,\tau} \| (1 \times \tau^{-1})$. Then $\mathscr{E}_r = \sum_{\tau \in T} \sum_{\gamma \in R_\tau} g_\tau \| (1 \times \gamma)$. Now, given $\eta \in P^{n,r} \cap \tau \Gamma \tau^{-1}$, put $\zeta = \kappa_r \pi_r(\eta)^{-1} \kappa_r$ with κ_r of Lemma 4.3. Then by that lemma we have

$$\tau_r\big((\xi \times 1_{2m-2r})\beta \times \tau\big)(1_{2m} \times \tau^{-1} \eta \tau) \in P_N \tau_r\big((\zeta\xi \times 1_{2m-2r})\beta \times \tau\big),$$

and hence $p_{\xi,\beta,\tau} \| (1 \times \tau^{-1} \eta \tau) = p_{\zeta\xi,\beta,\tau}$, since p_α depends only on $P^N \alpha$. This shows that $g_\tau \| (1 \times \eta) = g_\tau$ for such an η. From (4.1) we easily see that $g_\tau\big(\mathrm{diag}[z, w]\big)$ depends only on z, $\omega_r(w)$, and $\delta(w)$. Therefore, by Lemma 2.2

$$\langle f(w), \mathscr{E}_r(\mathrm{diag}[z, w]) \rangle = 0 \tag{4.6}$$

for every $f \in \mathscr{S}_k^n$ if $n > r$, at least for sufficiently large $\mathrm{Re}(s)$.

Next suppose $r = n$. Observing that $\varepsilon X \varepsilon = X$ with $\varepsilon = \mathrm{diag}[1_n, -1_n]$, we obtain, by (4.1), (4.4), Lemmas 4.4 and 4.6,

$$\chi(\theta)^{-n} N(\mathfrak{b})^{2ns} \mathscr{E}_n(z, w) = \sum_{\beta \in B} \sum_{\xi \in X} \chi[\beta] N(\mathfrak{a}_\beta)^{2s} \ell(\xi)^{-2s} \chi^*(\ell_1(\xi))$$
$$\cdot \chi_\mathfrak{c}(\det(d_\xi))^{-1} j(\beta, z)^{-k} j(\xi^{-1}\iota^{-1}, w)^{-k} M(\beta z, \xi^{-1}\iota^{-1} w),$$

where $\mathscr{E}_n(z, w) = \mathscr{E}_n(\mathrm{diag}[z, w])$ and

$$M(z, w) = \det[\omega_n(z) + w]^{-k} |\delta(z)^{-1/2} \delta(w)^{-1/2} \det[\omega_n(z) + w]|^{k - i\kappa - 2su}.$$

Let $Y = \iota X \iota^{-1}$. Then $Y = G^n \cap G^n_{\mathbf{a}} \prod_{v \in h} Y_v$ with

$$Y_v = \begin{cases} \{ y \in D^n_v[\mathfrak{b}\mathfrak{c}, \mathfrak{b}^{-1}\mathfrak{c}] | a_y - 1 \in (\mathfrak{e}_v)^n_n \} & \text{if } v | \mathfrak{e}, \\ D^n_v[\mathfrak{b}, \mathfrak{b}^{-1}\mathfrak{c}] Z_v D^n_v[\mathfrak{b}\mathfrak{c}, \mathfrak{b}^{-1}] & \text{if } v | \mathfrak{e}^{-1}\mathfrak{c}, \\ G_v & \text{if } v \nmid \mathfrak{c}, \end{cases} \tag{4.7a}$$

$$Z_v = \{ \mathrm{diag}[\tilde{q}, q] | q \in GL_n(F_v) \cap (\mathfrak{c}_v)^n_n \}. \tag{4.7b}$$

Let $f \in \mathscr{S}_k^n(\Delta, \chi^{-1})$, $\Delta = G \cap K_1$, with an open subgroup K_1 of $D^n[\mathfrak{b}, \mathfrak{b}^{-1}\mathfrak{c}]$ such that $\Delta Y \subset Y$. Put $\ell'(\eta) = \ell(\iota^{-1}\eta\iota)$ and $\ell_1'(\eta) = \ell_1(\iota^{-1}\eta\iota)$ for $\eta \in Y$. Then

$$\chi(\theta)^{-n} N(\mathfrak{b})^{2ns} \mathrm{vol}(\Delta \backslash \mathscr{H}) \langle f(w), \mathscr{E}_n(z, w) \rangle \tag{4.8}$$

$$= \nu_\Delta \sum_{\beta\in B} \sum_{\eta\in\Delta\backslash Y} \chi[\beta] N(\mathfrak{a}_\beta)^{2s} \chi^*(\ell_1'(\eta))\ell'(\eta)^{-2s} \chi_{\mathfrak{c}}(\det(a_\eta))^{-1}$$

$$\cdot\, j(\beta,z)^{-k} \int_{\mathscr{H}} M(\beta z, w)\overline{(f\|_k \eta\iota)(w)}\delta(w)^k d_H w,$$

where $\nu_\Delta = [\Delta \cap \{\pm 1\} : 1]$. By Lemma 4.7, the last line of (4.8) equals

$$c_{n,k}(\mathbf{s})\delta(z,s;(f\|_k\eta\iota)^c,\kappa)\|_k\beta, \tag{4.9}$$

where $\mathbf{s} = su - (k - i\kappa)/2$, $f^c(w) = \overline{f(-\bar{w})}$, and $\delta(z,s;\ldots)$ is as in (3.13). To simplify our sum expression, let us put

$$\mathscr{D}(s,g) = \sum_{\eta\in\Delta\backslash Y} \chi_{\mathfrak{c}}(\det(a_\eta))^{-1}\chi^*(\ell_1'(\eta))\ell'(\eta)^{-s} g\|_k\eta. \tag{4.10}$$

Observe that $\varepsilon Y\iota\varepsilon = Y\iota^{-1}$ and $(f\|\alpha)^c = f^c\|(\varepsilon\alpha\varepsilon)$. Thus, assuming that $m > n$ and $\varepsilon\Delta\varepsilon = \Delta$, and combining (4.6), (4.8), (4.9), and Lemma 3.6 together, we obtain

$$\chi(\theta)^{-n} N(\mathfrak{b})^{2ns}\mathrm{vol}(\Delta\backslash\mathscr{H})\langle f(w), (E^N\|_k\rho)(\mathrm{diag}[z,w])\rangle \tag{4.11}$$

$$= c_{n,k}(\mathbf{s})\nu_\Delta \sum_{\beta\in B}\chi[\beta]N(\mathfrak{a}_\beta)^{2s}\delta(s,\mathscr{D}(2s,f^c)\|_k\iota^{-1})\|_k\beta$$

$$= c_{n,k}(\mathbf{s})\nu_\Delta \cdot E(z,s;\mathscr{D}(2s,f^c)\|_k\iota^{-1},\chi,C)$$

with $E(\ldots,C)$ of that lemma, at least for sufficiently large $\mathrm{Re}(s)$. Notice that $\mathscr{D}(2s,f^c)\|_k\iota^{-1}$ for a fixed s belongs to $\mathscr{S}_k^n(K^n,\chi^{-1})$ with K of (3.7b). If $m = n$, we have $B = \{1\}$, and therefore (4.11) is true even in that case, if we understand that $E(z,s;g,\chi,C) = g$. Thus our next task is to investigate $\mathscr{D}(s,f^c)$. However, (4.11) is not completely satisfactory and we have to replace $E^N\|_k\rho$ by its certain derivative, which we will do in the next section.

5. Application of differential operators

In this section we work with $\mathscr{H}_{2n}$ and G^{2n}. Put $T_v^n = S_v^n \otimes_{\mathbf{R}} \mathbf{C}$ and $T_{\mathbf{a}}^n = \prod_{v\in\mathbf{a}} T_v^n$. For a variable W on $T_{\mathbf{a}}^{2n}$ we put

$$W = \begin{pmatrix} z & q \\ {}^t q & w \end{pmatrix} \text{ with } z, w \in T_{\mathbf{a}}^n \text{ and } q \in (\mathbf{C}_n^n)^{\mathbf{a}}. \tag{5.1}$$

Fixing a $v \in \mathbf{a}$, we denote by $\mathfrak{S}$ the vector space of all $\mathbf{C}$-valued homogeneous polynomial functions on T_v^{2n} of degree n. Temporarily let us simply write $W = \begin{pmatrix} z & q \\ {}^t q & w \end{pmatrix}$ for the variable W_v on T_v^{2n}; thus z, $w \in T_v^n$, and $q \in \mathbf{C}_n^n$. We define $\tau : GL_{2n}(\mathbf{C}) \to GL(\mathfrak{S})$ by $(\tau(a)\varphi)(W) = \varphi({}^t aWa)$, and define also elements σ and η of $\mathfrak{S}$ by $\sigma(W) = \det(z)$ and $\eta(W) = \det(q)$ with z and q of (5.1). Let Z denote the subspace of $\mathfrak{S}$ spanned by $\tau(a)\sigma$ for all $a \in GL_{2n}(\mathbf{C})$. We have then

(5.2) $\eta \in Z$ and $\tau(\mathrm{diag}[b, c])\eta = \det(bc)\eta$ for every $b, c \in GL_n(\mathbf{C})$.

The latter formula is easy. To see that $\eta \in Z$, denote by Z' the subspace of $\mathfrak{S}$ spanned by $\tau(a)\eta$ for all $a \in GL_{2n}(\mathbf{C})$. Let Y be a τ-irreducible subspace of Z'. The highest weight vector φ of Y is of the form $\varphi(W) = \prod_{i=1}^{n} \det(\omega_i(W))^{c_i}$ with nonnegative integers c_i such that $\sum_{i=1}^{n} ic_i = n$ (see [84b, Th.2.C], for example). If $\varphi \neq \sigma$, then φ has degree $\geqq 2$ in z_{11}. Now we easily see that $[\tau(a)\eta](W)$ is of degree $\leqq 1$ in z_{11} for every $a \in GL_{2n}(\mathbf{C})$, and hence $\varphi \in Z'$ only when $\varphi = \sigma$. This proves that $Z' = Z$, and so $\eta \in Z$.

We now define a matrix differential operator ∂_v on $\mathscr{H}_{2n}$ by

(5.3) $$\partial_v = (\partial_{vij})_{i,j=1}^{2n}, \quad \partial_{vij} = 2^{-1}(1 + \delta_{ij})\partial/\partial W_{vij}.$$

with the variable $W_v = (W_{vij})$ on the v-th factor H_{2n} of $\mathscr{H}_{2n}$. Then for each $\zeta \in \mathfrak{S}$ we can define a differential operator $\zeta(\partial_v)$ of degree n. Given $k \in \mathbf{Z}^{\mathbf{a}}$, $\zeta \in Z$, a subset ε of $\mathbf{a}$, and a C^∞ function g on $\mathscr{H}_{2n}$, we define functions $B_\eta^\varepsilon g$ and $B_{k,v,\zeta} g$ on $\mathscr{H}_{2n}$ and $A_k^\varepsilon g$ on $\mathscr{H}_n \times \mathscr{H}_n$ by

(5.4a) $$B_\eta^\varepsilon = \prod_{v\in\varepsilon} B_{k,v,\eta}, \quad (B_{k,v,\zeta} g)(W) = \delta(W)^{\lambda u - k}\zeta(\partial_v)\left[\delta(W)^{k-\lambda u} g\right],$$

(5.4b) $$(A_k^\varepsilon g)(z, w) = \left(\prod_{v\in\varepsilon} \eta(\partial_v) g\right)(\mathrm{diag}[z, w])$$

where $\lambda = (n-1)/2$. For a function h on $\mathscr{H}_n \times \mathscr{H}_n$ and $\beta \times \gamma \in G^n \times G^n$ we define a function $h\|_k(\beta \times \gamma)$ on $\mathscr{H}_n \times \mathscr{H}_n$ by

(5.5) $$[h\|_k(\beta \times \gamma)](z, w) = j_\beta(z)^{-k} j_\gamma(w)^{-k} h(\beta z, \gamma w).$$

Lemma 5.1. *The notation being as above, one has*

$$(A_k^\varepsilon g)\|_{k+\varepsilon}(\beta \times \gamma) = A_k^\varepsilon\big(g\|_k(\beta \times \gamma)\big)$$

for every $\beta \times \gamma \in G^n \times G^n$, where ε is viewed as the element of $\mathbf{Z}^{\mathbf{a}}$ whose v-component is 1 or 0 according as $v \in \varepsilon$ or $v \notin \varepsilon$.

Proof. We have

$$\delta(W)^{k-\lambda u} B_\eta^\varepsilon g = \left(\prod_{v\in\varepsilon} \eta(\partial_v)\right)\left[\delta(W)^{k-\lambda u} g\right]$$

$$= \delta(W)^{k-\lambda u}\left(\prod_{v\in\varepsilon} \eta(\partial_v)\right) g + \sum [\varphi(\partial)\delta(W)^{k-\lambda u}][\varphi'(\partial) g],$$

where $\partial = (\partial_v)_{v\in\varepsilon}$ and φ, φ' are polynomials involving only $(q_v)_{v\in\varepsilon}$. By [84b, Theorem 4.3] or [94b, Lemma 4.1] we have $\zeta(\partial_v)\delta(W_v)^s = c(s)\delta(W_v)^s \zeta\big(\mathrm{Im}(W_v)^{-1}\big)$ for every homogeneous polynomial ζ with a polynomial c. If ζ is nonconstant and involves only q_v, then $\zeta\big(\mathrm{Im}(W_v)^{-1}\big) = 0$ for $W_v = \mathrm{diag}[z_v, w_v]$. This proves

(5.6) $$(A_k^{\varepsilon}g)(z, w) = (B_{\eta}^{\varepsilon}g)(\mathrm{diag}[z, w]).$$

Define $\rho : GL_{2n}(\mathbf{C})^{\mathbf{a}} \to \mathbf{C}^{\times}$ by $\rho(x) = \det(x)^k$. Then the operator D_ρ^Z can be defined as in [86, (1.24)] or [94b, (2.14)]; moreover, $B_{k,v,\zeta}g = (D_\rho^Z g)(\zeta)$ by [94b, (4.9c)] or [84a, (2.9a)], and $D_\rho^Z(g\|_\rho \alpha) = (D_\rho^Z g)\|_{\rho\otimes\tau}\alpha$ for every $\alpha \in G^{2n}$ by [86, (1.25)] or [94b, (2.15)]. This means that

(5.7) $$B_{k,v,\zeta}(g\|_k \alpha) = (B_{k,v,\xi}g)\|_k \alpha \text{ if } \xi \in Z \text{ and } \zeta = \tau({}^t\mu(\alpha_v, W_v))\xi$$

with μ of (1.4b). A similar formula holds for $\prod_{v\in\varepsilon} B_{k,v,\xi}$. Evaluating it at $W = \mathrm{diag}[z, w]$ with $\xi = \eta$ and $\alpha = \beta \times \gamma$ and combining the result with (5.6), we obtain our lemma.

We now recall a formula ([84b, Th.4.3], [94b, Lemma 4.1, (4.10a)])

$$\zeta(\partial_v)\big[\delta(W_v)^p\big] = \psi(p)(2i)^{-n}\zeta\big(\mathrm{Im}(W_v)^{-1}\big)\delta(W_v)^p,$$

$$\psi(p) = \prod_{h=1}^{n}\big(p + (h-1)/2\big), \quad (p \in \mathbf{C}).$$

Taking $g = \delta(W)^t$ with $t \in \mathbf{C}^{\mathbf{a}}$ in (5.7), we obtain

$$\begin{aligned} B_{k,v,\zeta}(\delta^t\|_k \alpha) &= \big\{\delta^{\lambda u-k}\big[\xi(\partial_v)(\delta^{t+k-\lambda u})\big]\big\}\|_k \alpha \\ &= (2i)^{-n}\psi(t_v + k_v - \lambda)\big[\xi\big(\mathrm{Im}(W_v)^{-1}\big)\delta^t\big]\|_k \alpha \\ &= (2i)^{-n}\psi(t_v + k_v - \lambda)\xi\big(\mathrm{Im}(\alpha_v W_v)^{-1}\big)\cdot\big[\delta^t\|_k \alpha\big]. \end{aligned}$$

This for $\zeta = \eta$ together with (5.6) yields

$$A_k^{\varepsilon}(\delta^t\|_k \alpha) = (\delta^t\|_k \alpha)(W)\prod_{v\in\varepsilon}(2i)^{-n}\psi(t_v + k_v - \lambda)\eta\big(\mu_v^{-1}\overline{\mu_v}\,\mathrm{Im}(W_v)^{-1}\big)$$

with $W_v = \mathrm{diag}[z_v, w_v]$ and $\mu_v = \mu(\alpha_v, W_v)$. An easy calculation shows that if α is τ_r of Lemma 4.2, then the last factor $\eta(\cdots)$ equals 0 or $(-2i)^n j(\tau_n, W_v)^{-1}$ according as $r < n$ or $r = n$. Thus $A_k^{\varepsilon}(\delta^t\|_k \tau_r) = 0$ for $r < n$ and

(5.8) $$A_k^{\varepsilon}(\delta^t\|_k \tau_n) = (\delta^t\|_{k'} \tau_n)\big(\mathrm{diag}[z, w]\big)\prod_{v\in\varepsilon}(-1)^n\psi(t_v + k_v - \lambda),$$

where $k' = k + \varepsilon$. We now apply A_k^{ε} to $E^N\|_k\rho = \sum_{r=0}^{n}\mathscr{E}_r$ of Section 4. The notation being as in the paragraph containing (4.6), for $m = n$ and $r < n$ we have, by Lemma 5.1, $A_k^{\varepsilon}\big(\delta^t\|_k \tau_r(\beta \times \gamma)\big) = A_k^{\varepsilon}(\delta^t\|_k \tau_r)\|_{k'}(\beta \times \gamma) = 0$, and hence $A_k^{\varepsilon}(\mathscr{E}_r) = 0$. If $r = n$, then $B = \{1\}$, and for $t = su - (k - i\kappa)/2$ we have

$$\begin{aligned} (A_k^{\varepsilon}\mathscr{E}_n)(z, w) &= L\sum_{\xi\in X}\ell(\xi)^{-2s}\chi^*\big(\ell_1(\xi)\big)\chi_c\,\big(\det(d_\xi)\big)^{-1}A_k^{\varepsilon}\big(\delta^t\|_k\tau_n(\xi \times 1)\big) \\ &= c_0 L\sum_{\xi\in X}\ell(\xi)^{-2s}\chi^*\big(\ell_1(\xi)\big)\chi_c\,\big(\det(d_\xi)\big)^{-1} \\ &\quad\cdot j(\xi^{-1}\iota^{-1}, w)^{-k'}M'(z, \xi^{-1}\iota^{-1}w), \end{aligned}$$

where $L = \chi(\theta)^n N(\mathfrak{b})^{-2ns}$, $c_0 = \prod_{v\in\varepsilon}(-1)^n\psi(s-\lambda+(k_v+i\kappa_v)/2)$, and

$$M'(z,w) = \det(z+w)^{-k'}|\delta(z)^{-1/2}\delta(w)^{-1/2}\det(z+w)|^{k-i\kappa-2su}.$$

Take $f \in \mathscr{S}^n_{k'}(\Delta, \chi^{-1})$ with Δ as in (4.8). Then a calculation similar to that led to (4.11) gives

$$\text{(5.9)}\qquad \operatorname{vol}(\Delta\backslash\mathscr{H})\langle f(w), (A^\varepsilon_k(E^{2n}\|_k\rho))(\operatorname{diag}[z,w])\rangle$$
$$= \nu_\Delta c_0\chi(\theta)^n N(\mathfrak{b})^{-2ns}c_{n,k'}(su-(k-i\kappa)/2)\mathscr{D}(2s, f^c)\|f\|^{-1}_{k'}.$$

The operator of type A^ε_k was employed by Böcherer for $G = Sp(n, \mathbf{Q})$ in [B]. Here we have treated it in a more general framework of [84a, b] and [94b], which is more transparent. In fact, the results can be generalized to the case of vector-valued forms, which we hope to treat in a future paper.

6. Hecke operators and zeta functions

In this and next sections k is always integral. Since we work with G^n for the most part, we often suppress the superscript n. We take $\mathfrak{b}, \mathfrak{c}, \mathfrak{e}$ as in Section 3 and ψ as in (1.16). We don't assume ψ to be k-normalized. We put

(6.1a)
$$K = \{x \in D[\mathfrak{b}^{-1}\mathfrak{e}, \mathfrak{bc}] \mid (a_x - 1)_v \in (\mathfrak{e}_v)^n_n \text{ for every } v|\mathfrak{e}\},\quad \Gamma = G\cap K,$$

$$\text{(6.1b)}\quad Q(\mathfrak{e}) = \{r \in GL_n(F)_\mathbf{h} \cap \textstyle\prod_{v\in\mathbf{h}}(\mathfrak{g}_v)^n_n \mid r_v = 1 \text{ for every } v|\mathfrak{e}\},$$

$$\text{(6.1c)}\qquad R(\mathfrak{e}) = \{\operatorname{diag}[\tilde{r}, r] \mid r\in Q(\mathfrak{e})\},$$

$$\text{(6.1d)}\qquad E = \textstyle\prod_{v\in\mathbf{h}} E_v,\qquad E_v = GL_n(\mathfrak{g}_v).$$

Given $\sigma = \operatorname{diag}[\tilde{r}, r] \in R(\mathfrak{e})$, take $A \subset G$ so that $G\cap K\sigma K = \bigsqcup_{\alpha\in A}\Gamma\alpha$. Then for $f \in \mathscr{M}_k(K,\psi)$ we define an element $f|T_{r,\psi}$ of $\mathscr{M}_k(K,\psi)$ by

$$\text{(6.2)}\qquad f|T_{r,\psi} = \sum_{\alpha\in A}\psi_\mathfrak{c}(\det(a_\alpha)_\mathfrak{c})^{-1}f\|_k\alpha.$$

Notice that $(a_\alpha)_\mathfrak{c}$ is invertible. We put $T_{r,\psi} = T_{c,\psi}$ if $r = c1_n$ with $c \in F^\times_\mathbf{h}$. Strictly speaking, $T_{r,\psi}$ depends on K, but we use it for different K's with the same $\mathfrak{c}$ and $\mathfrak{e}$, but with different $\mathfrak{b}$'s, as there will be no fear of confusion.

For an integral ideal $\mathfrak{a}$ we denote by $T_\psi(\mathfrak{a})$ the sum of $T_{r,\psi}$ for all different ErE with $r \in Q(\mathfrak{e})$ such that $\det(r)\mathfrak{g} = \mathfrak{a}$. Clearly this is nonzero only if $\mathfrak{a}$ is prime to $\mathfrak{e}$. Let us now take a nonzero element f of $\mathscr{S}^n_k(K,\psi)$ that is an eigenfunction of $T_\psi(\mathfrak{a})$ for every $\mathfrak{a}$. We then put $f|T_\psi(\mathfrak{a}) = \lambda(\mathfrak{a})f$ and

(6.3a) $$Z_\psi(s,\lambda) = \Lambda_{\mathfrak{c}}^{2n}(s,\psi)\sum_{\mathfrak{a}}\lambda(\mathfrak{a})\psi^*(\mathfrak{a}')N(\mathfrak{a})^{-s},$$

(6.3b) $$\Lambda_{\mathfrak{c}}^{m}(s,\psi) = L_{\mathfrak{c}}(s,\psi)\prod_{i=1}^{[m/2]} L_{\mathfrak{c}}(2s-2i,\psi^2),$$

(6.3c) $$L_{\mathfrak{c}}(s,\psi) = \prod_{v\in\mathbf{h},\, v\nmid\mathfrak{c}} \left(1-\psi(\pi_v)|\pi_v|^s\right)^{-1},$$

where π_v is a prime element of F_v, $\mathfrak{a}$ runs over all the integral ideals of F prime to $\mathfrak{e}$, and $\mathfrak{a}' = \prod_{v\nmid\mathfrak{c}} \mathfrak{a}_v$. Since our ring of Hecke operators is a homomorphic image of the abstract ring of [94a, §2], clearly [94a, Theorem 2.9] shows that Z_ψ has an Euler product

(6.4) $$Z_\psi(s,\lambda) = \prod_{v\in\mathbf{h},\, v\nmid\mathfrak{e}} W_v\left((\psi/\psi_{\mathfrak{c}})(\pi_v)|\pi_v|^s\right)^{-1},$$

where W_v is a polynomial with constant term 1 and of degree $2n+1$ or $\leqq n$ according as $v\nmid\mathfrak{c}$ or $v|\mathfrak{c}$. For a Hecke character φ of F, an integral ideal $\mathfrak{x}$, and $\varepsilon\in\mathbf{Z}^{\mathbf{a}}$ with $0\leqq\varepsilon_v\leqq 1$, we define $Z_{\psi,\mathfrak{x}}(s,\lambda,\varphi)$ and a gamma-factor $\Gamma_{k,\kappa,\varepsilon}^n$ by

(6.5) $$Z_{\psi,\mathfrak{x}}(s,\lambda,\varphi) = \prod_{v\in\mathbf{h},\, v\nmid\mathfrak{c}\mathfrak{x}} W_v\left((\psi/\psi_{\mathfrak{c}})(\pi_v)\varphi^*(\pi_v\mathfrak{g})\,|\pi_v|^s\right)^{-1},$$

(6.6a)
$$\Gamma_{k,\kappa,\varepsilon}^n(s) = \prod_{v\in\mathbf{a}}\Gamma_n\left(s+(k_v+\varepsilon_v-n-1+i\kappa_v)/2\right)g^n\left(s+(i\kappa_v/2),\,k_v-\varepsilon_v\right),$$

(6.6b) $g^n(s,h)$
$$=\begin{cases}\Gamma_n\left(s+\frac{h-n}{2}\right)\Gamma\left(s+\frac{h}{2}-\left[\frac{h+n}{2}\right]\right) & \text{if } h\geqq n,\\ \Gamma_{2h+2-n}\left(s+\frac{h-n}{2}\right)\Gamma\left(s-\frac{h}{2}\right)\prod_{a=h+2}^{n}\Gamma(2s-a) & \text{if } (n-2)/2\leqq h<n.\end{cases}$$

By Proposition 2.1(2), we may assume that $k_v\geqq n/2$ for every $v\in\mathbf{a}$, which is why $g^n(s,h)$ is defined only for $h\geqq(n-2)/2$. We write $Z_\psi(s,\lambda,\varphi)$ for $Z_{\psi,\mathfrak{g}}(s,\lambda,\varphi)$.

Theorem 6.1. *Let φ be a Hecke character of F with conductor $\mathfrak{f}$; suppose $(\varphi\psi)_{\mathbf{a}}(x) = \mathrm{sgn}(x_{\mathbf{a}})^{k+\varepsilon}|x_{\mathbf{a}}|^{i\kappa}$ with $\varepsilon\in\mathbf{Z}^{\mathbf{a}}$, $0\leqq\varepsilon_v\leqq 1$, and $\kappa\in\mathbf{R}^{\mathbf{a}}$, $\sum_{v\in\mathbf{a}}\kappa_v = 0$. Let f be an eigenform in $\mathcal{S}_k(K,\psi)$ with eigenvalues $\lambda(\mathfrak{a})$ as above, and $\mathfrak{z}$ the product of all prime factors v of $\mathfrak{e}^{-1}\mathfrak{c}$ such that $f|T_{\pi_v,\psi}=0$. Then the product $\mathcal{Z} = \Gamma_{k,\kappa,\varepsilon}^n(s/2)Z_{\psi,\mathfrak{z}}(s,\lambda,\varphi)$ can be continued meromorphically to the whole s-plane with finitely many poles. Moreover each pole is simple. In particular, $\mathcal{Z}$ is an entire function of s if $\varphi^2\psi^2\neq 1$. To describe the set of possible poles of $\mathcal{Z}$, put $\ell = \mathrm{Max}_{v\in\mathbf{a}}\{k_v-\varepsilon_v\}$.*

(1) *Suppose $\varphi^2\psi^2 = 1$ and $\mathfrak{c}\mathfrak{f}\neq\mathfrak{g}$. If $\ell>n$, $\mathcal{Z}$ has no pole except for a possible pole at $s=n+1$ which occurs only if $k_v-\varepsilon_v-n\in 2\mathbf{Z}$ for every v such that $k_v-\varepsilon_v>n$. If $\ell\leqq n$, $\mathcal{Z}$ has possible poles only in the set*

$$\{j\in\mathbf{Z}\mid n+1\leqq j\leqq 2n+1-\ell\}.$$

(2) *If* $\varphi^2\psi^2 = 1$ *and* $\mathfrak{cf} = \mathfrak{g}$, *each pole belongs to the set given in* (1) *or to*

$$\{ j \in \mathbf{Z} \mid 0 \leqq j \leqq n \},$$

where $j = 0$ *is unnecessary if* $\varphi\psi \neq 1$. (*See Theorem* 6.4 *below for the case* $n = 1$.)

To prove this, we first introduce some auxiliary objects given by

$$K' = \{ x \in D[\mathfrak{b}^{-1}\mathfrak{c}, \mathfrak{be}] \mid (a_x - 1)_v \in (\mathfrak{e}_v)_n^n \text{ for every } v|\mathfrak{e} \}, \quad \Gamma' = G \cap K',$$

$$R(\mathfrak{e}, \mathfrak{c}) = \{ \mathrm{diag}[\tilde{q}, q] \in R(\mathfrak{e}) \mid q_v \in (\mathfrak{c}_v)_n^n \text{ for every } v|\mathfrak{e}^{-1}\mathfrak{c} \}.$$

Let $\tau = \mathrm{diag}[\tilde{q}, q] \in R(\mathfrak{e}, \mathfrak{c})$. Take $B \subset G$ so that $G \cap K\tau K' = \bigsqcup_{\beta \in B} \Gamma\beta$. For $f \in \mathscr{M}_k(K, \psi)$ we define a function $f|U_{q,\psi}$ on $\mathscr{H}$ by

$$f|U_{q,\psi} = \sum_{\beta \in B} \psi_{\mathfrak{c}}\left(\det(a_\beta)_{\mathfrak{c}}\right)^{-1} f\|_k\beta. \tag{6.7}$$

This is well defined and belongs to $\mathscr{M}_k(K', \psi)$. Notice again that $(a_\beta)_{\mathfrak{c}}$ (as well as $(a_y)_{\mathfrak{c}}$ of (6.8) below) is invertible.

Lemma 6.2. *Let* h *be an element of* $F_{\mathbf{h}}^\times$ *such that* $h\mathfrak{g} = \mathfrak{e}^{-1}\mathfrak{c}$ *and* $h_v = 1$ *for* $v|\mathfrak{c}$. *Then* $U_{hr,\psi} = T_{r,\psi}U_{h,\psi}$ *for every* $r \in Q(\mathfrak{e})$, *where* $U_{h,\psi}$ *denotes* $U_{q,\psi}$ *with* $q = h1_n$. *Moreover, for* $f \in \mathscr{M}_k(K, \psi)$ *we have* $f|T_{h,\psi} \neq 0$ *only if* $f|U_{h,\psi} \neq 0$.

Proof. With q and τ as above, let $K\tau K' = \bigsqcup_{y \in Y} Ky$ with $Y \subset G_{\mathbf{h}}$. Then

$$(f|U_{q,\psi})_{\mathbf{A}}(x) = \sum_{y \in Y} \psi_{\mathfrak{c}}\left(\det(a_y)_{\mathfrak{c}}\right)^{-1} f_{\mathbf{A}}(xy^{-1}) \quad (x \in G_{\mathbf{A}}) \tag{6.8}$$

and a similar formula holds for $f|T_{r,\psi}$ (see [94a, (5.1b)]). Therefore, to prove the first assertion, it is sufficient to show that $K\sigma\eta K' = K\sigma K \cdot K\eta K'$ for $\sigma = \mathrm{diag}[\tilde{r}, r]$ and $\eta = \mathrm{diag}[h^{-1}1_n, h1_n]$, where the product of double cosets should be understood in the sense of [71, Section 3]. Clearly this is a local problem, and so we consider $(K\sigma\eta K')_v$ for $v|\mathfrak{e}^{-1}\mathfrak{c}$. As shown in [94a, Lemma 2.3], $(K\sigma K)_v = \bigsqcup_{d,b} K_v\begin{pmatrix} \tilde{d} & \tilde{d}b \\ 0 & d \end{pmatrix}$ with $d \in E_v \backslash E_v r_v E_v$ and $b \in S(\mathfrak{b}^{-1})_v / {}^t d S(\mathfrak{b}^{-1})_v d$, where

$$S(\mathfrak{x}) = \{ \alpha \in \mathfrak{x}_n^n \mid \alpha = {}^t\alpha \}. \tag{6.9}$$

Similarly (using the inverse of (4.2), for example) we have $(K\tau K')_v = \bigsqcup_{f,g} K_v \cdot \begin{pmatrix} \tilde{f} & \tilde{f}g \\ 0 & f \end{pmatrix}$ with $f \in E_v \backslash E_v q_v E_v$ and $g \in S(\mathfrak{b}^{-1}\mathfrak{c})_v / {}^t f S(\mathfrak{b}^{-1})_v f$. Taking especially q to be $h1_n$, we easily see that the matrices $\begin{pmatrix} \tilde{d} & \tilde{d}b \\ 0 & d \end{pmatrix}\begin{pmatrix} \tilde{f} & \tilde{f}g \\ 0 & f \end{pmatrix}$ represent $K_v \backslash (K\sigma\eta K')_v$, which implies that $(K\sigma\eta K')_v = (K\sigma K)_v \cdot (K\eta K')_v$ as expected.

To prove the second assertion, we use the Fourier expansion

$$(6.10)\ f_{\mathbf{A}}\left(\begin{pmatrix} q & s\tilde{q} \\ 0 & \tilde{q} \end{pmatrix}\right) = \det(q)_{\mathbf{a}}^{k} \sum_{\tau \in S_+} \mu(\tau, q; f)\mathbf{e}_{\mathbf{a}}\big(\mathrm{tr}(\mathrm{i} \cdot {}^t q\tau q)\big)\mathbf{e}_{\mathbf{A}}\big(\mathrm{tr}(\tau s)\big),$$

which holds for every $f \in \mathscr{M}_k(\Gamma, \psi)$, $s \in S_{\mathbf{A}}$, and $q \in GL_n(F)_{\mathbf{A}}$. The Fourier coefficient $\mu(\tau, q; f)$, written also $\mu_f(\tau, q)$, is a complex number, and has the following property:

$$(6.11)\quad \mu_f(\tau, q) \neq 0 \textit{ only if } \mathbf{e}_{\mathbf{h}}\big(\mathrm{tr}({}^t q\tau q s)\big) = 1 \textit{ for every } s \in S(\mathfrak{b}^{-1}\mathfrak{c}).$$

Expansion (6.10) is somewhat more general than that of [94a, Prop.1.1], but the present case can be proved similarly. Using the representatives for $K\backslash K\eta K$ and those for $K\backslash K\eta K'$ given above and putting $M = N(\mathfrak{e}^{-1}\mathfrak{c})$, we can easily verify that

$$\mu(\tau, q; f|T_{h,\psi}) = \begin{cases} \psi(h)^n M^{n(n+1)}\mu_f(\tau, hq) & \textit{if } \mathbf{e}_{\mathbf{h}}\big(\mathrm{tr}\big({}^t q\tau q S(\mathfrak{b}^{-1}\mathfrak{e})\big)\big) = 1, \\ 0 & \textit{otherwise}, \end{cases}$$

$$\mu(\tau, q; f|U_{h,\psi}) = \begin{cases} \psi(h)^n M^{n(n+1)/2}\mu_f(\tau, hq) & \textit{if } \mathbf{e}_{\mathbf{h}}\big(\mathrm{tr}\big({}^t q\tau q S(\mathfrak{b}^{-1}\mathfrak{c})\big)\big) = 1, \\ 0 & \textit{otherwise}. \end{cases}$$

The last assertion of Lemma 6.2 now follows from these relations.

Given a fractional ideal $\mathfrak{t} = t\mathfrak{g}$ with $t \in F_{\mathbf{h}}^{\times}$, put $\tau = \mathrm{diag}[t^{-1}1_n, t1_n]$. We can then define an isomorphism $I_{\mathfrak{t}}$ of $\mathscr{M}_k(K, \psi)$ onto $\mathscr{M}_k(\tau^{-1}K\tau, \psi)$ by

$$(6.12)\qquad (f|I_{\mathfrak{t}})_{\mathbf{A}}(x) = \psi(t^n) f_{\mathbf{A}}(x\tau^{-1}) \quad (x \in G_{\mathbf{A}}).$$

We easily see that $I_{\mathfrak{t}}$ is independent of the choice of t and commutes with the operators $T_{r,\psi}$ (defined with respect to $\mathfrak{b}$ and $\mathfrak{t}^{-2}\mathfrak{b}$).

Let us now take χ as in (3.6a, b), and assume that $\chi = \psi$ on $\prod_{v|\mathfrak{c}} \mathfrak{g}_v^{\times}$. Then $\mathscr{S}_k(K, \psi) = \mathscr{S}_k(K, \chi)$, $(\chi/\psi)^*(\mathfrak{a})\psi^*(\mathfrak{a}')T_\psi(\mathfrak{a}) = \chi^*(\mathfrak{a}')T_\chi(\mathfrak{a})$, where $\mathfrak{a}' = \prod_{v|\mathfrak{c}} \mathfrak{a}_v$, and $U_{h,\chi} = (\chi/\psi)^*(\mathfrak{e}^{-1}\mathfrak{c})^n U_{h,\psi}$; also $f^c \in \mathscr{S}_k(K, \overline{\chi})$ if $f \in \mathscr{S}_k(K, \chi)$, $(f|I_{\mathfrak{t}})^c = f^c|I_{\mathfrak{t}}$, and $\big(f|T_{r,\chi}\big)^c = f^c|T_{r,\overline{\chi}}$. Put $\tau = \mathrm{diag}[\theta^{-1}1_n, \theta 1_n]$ with θ of Lemma 4.4. Then we see that Y_v of (4.7a) coincides with $\big(\tau^{-1}KR(\mathfrak{e}, \mathfrak{c})K'\tau\big)_v$ for every v. Let $f|T_\psi(\mathfrak{a}) = \lambda(\mathfrak{a})f$ with $f \in \mathscr{S}_k(K, \psi)$ as before. Define $\mathscr{D}$ by (4.10) with $G \cap \tau^{-1}K\tau$ as Δ there. Then, by Lemma 6.2 we have

$$(6.13)\ \mathscr{D}(s, f|I_{\mathfrak{b}}) = N(\mathfrak{e}^{-1}\mathfrak{c})^{-ns}\sum_{\mathfrak{a}} \chi^*(\mathfrak{a}')N(\mathfrak{a})^{-s} f|T_\chi(\mathfrak{a})U_{h,\chi}I_{\mathfrak{b}}$$

$$= N(\mathfrak{e}^{-1}\mathfrak{c})^{-ns}\sum_{\mathfrak{a}} \lambda(\mathfrak{a})\psi^*(\mathfrak{a}')(\chi/\psi)^*(\mathfrak{a})N(\mathfrak{a})^{-s} f|U_{h,\chi}I_{\mathfrak{b}}.$$

Let us now put

$$(6.14)\qquad \mathscr{P}_k^N(z, s) = \Lambda_{\mathfrak{c}}^N(2s, \chi)\mathscr{G}_{k,\kappa}^N(s)E^N(z, s; 1, \chi, C)$$

with E^N which we wrote $E^N(z, s)$ in Section 4 and a product of gamma functions $\mathscr{G}_{k,\kappa}^N$ whose explicit form will be given in (6.18a, b) below. Since Δ is conjugate to $\Gamma = G \cap K$ in G, (6.13) together with (4.11) yields

$$\text{(6.15a)}\ \chi(\theta)^{-n}N(\mathfrak{e}^{-1}\mathfrak{c}\mathfrak{b})^{2ns}\mathrm{vol}(\Gamma\backslash\mathscr{H})\left\langle (f^c|I_\mathfrak{b})(w),\ (\mathscr{P}_k^N\|\rho)(\mathrm{diag}[z, w], s)\right\rangle$$

$$= \nu_\Gamma\cdot c_{n,k}\big(su-(k-i\kappa)/2\big)\mathscr{G}_{k,\kappa}^N(s)E\big(z, s;\ (f|U_{h,\chi}I_\mathfrak{b})\|\iota^{-1}, \chi, C\big)$$

$$\cdot\ Z_\psi(2s, \lambda, \chi/\psi)\prod_{i=n+1}^{[(n+m)/2]} L_\mathfrak{c}(4s-2i, \chi^2).$$

In particular, taking $m = n$ and changing k for $k-\varepsilon$ in (5.9), we obtain

(6.15b)

$$\chi(\theta)^{-n}N(\mathfrak{e}^{-1}\mathfrak{c}\mathfrak{b})^{2ns}\mathrm{vol}(\Gamma\backslash\mathscr{H})\left\langle (f^c|I_\mathfrak{b})(w), \big(A_{k-\varepsilon}^{\varepsilon}(\mathscr{P}_{k-\varepsilon}^{2n}\|\rho)\big)(\mathrm{diag}[z, w], s)\right\rangle$$

$$= \nu_\Gamma c' c_{n,k}\big(su-(k-\varepsilon-i\kappa)/2\big)\mathscr{G}_{k-\varepsilon,\kappa}^{2n}(s)Z_\psi(2s, \lambda, \chi/\psi)(f|U_{h,\chi}I_\mathfrak{b})\|\iota^{-1},$$

where $c' = \prod_{v\in\varepsilon}(-1)^n\prod_{a=1}^n\big(s+(k_v+i\kappa_v-a)/2\big)$. If we change $\mathfrak{e}$ for its multiple satisfying (3.6c), K is changed for a smaller group, and so f is still an eigenform with respect to the new group, but this may eliminate some Euler factors from (6.5). However, this change is effective in our investigation of $E\big(z, s; g, \chi, C\big)$; in fact we may simply assume that $\mathfrak{e} = \mathfrak{c}$. In this special case we have $U_{h,\chi} = 1$ in (6.13) and (6.15a, b), but there is another expression for $\mathscr{D}$. Thus, if $\mathfrak{e} = \mathfrak{c}$, we see that Y_v of (4.7a) coincides with $(\iota KR(\mathfrak{c})K\iota^{-1})_v$ for every v, and $f\mapsto f\|\iota$ gives an isomorphism of $\mathscr{S}_k(K)$ onto $\mathscr{S}_k(\iota K\iota^{-1})$. Moreover, the group $\iota K\iota^{-1}$ is obtained from K by changing $\mathfrak{b}$ for $\mathfrak{b}^{-1}$, and $(f\|\iota)|T_r = (f|T_r)\|\iota$, as can easily be verified, where we write simply T_r for $T_{r,\chi}$ and $T_{r,\bar\chi}$, since $\chi\big(\det(a_x)_\mathfrak{c}\big) = 1$ for every $x\in KR(\mathfrak{c})K$ and $x\in\iota KR(\mathfrak{c})K\iota^{-1}$. Thus

$$\text{(6.16)}\qquad \mathscr{D}\big(s, f\|\iota\big)\|\iota^{-1} = \sum_\mathfrak{a}\lambda(\mathfrak{a})\psi^*(\mathfrak{a}')(\chi/\psi)^*(\mathfrak{a})N(\mathfrak{a})^{-s}f.$$

Combining this with (4.11), for $\mathfrak{e} = \mathfrak{c}$ we obtain

$$\text{(6.17)}\qquad \chi(\theta)^{-n}N(\mathfrak{b})^{2ns}\mathrm{vol}(\Gamma\backslash\mathscr{H})\left\langle (f\|\iota)^c(w), (\mathscr{P}_k^N\|\rho)(\mathrm{diag}[z, w], s)\right\rangle$$

$$= \nu_\Gamma\cdot c_{n,k}(\mathbf{s})\mathscr{G}_{k,\kappa}^N(s)E\big(z, s;\ f, \chi, C\big)Z_\psi(2s, \lambda, \chi/\psi)\prod_{i=n+1}^{[(n+m)/2]} L_\mathfrak{c}(4s-2i, \chi^2),$$

where $\mathbf{s} = su-(k-i\kappa)/2$. Here is the explicit form of $\mathscr{G}_{k,\kappa}^N$:

$$\text{(6.18a)}\qquad \mathscr{G}_{k,\kappa}^N(s) = \prod_{v\in\mathbf{a}}\gamma^N\big(s+i\kappa_v/2, |k_v|\big),$$

$$\text{(6.18b)}\ \gamma^N(s, h) = \begin{cases}\Gamma\left(s+\frac{h}{2}-\left[\frac{2h+N}{4}\right]\right)\Gamma_N\left(s+\frac{h}{2}\right) & (h \geqq N/2\in\mathbf{Z}),\\ \Gamma_N\left(s+\frac{h}{2}\right) & (h > N/2\notin\mathbf{Z}),\\ \Gamma_{2h+1}\left(s+\frac{h}{2}\right)\prod_{a=h+1}^{[N/2]}\Gamma(2s-a) & (0\leqq h < N/2).\end{cases}$$

We are now ready to complete the proof of Theorem 6.1. We first observe that $\mathscr{G}^{2n}_{k-\varepsilon,\kappa}(s)c'c_{n,k}\big(su-(k-\varepsilon-i\kappa)/2\big)$ in (6.15b) coincides with $A\cdot B^s\Gamma^n_{k,\kappa,\varepsilon}(s)$ with some positive numbers A and B. Given φ as in our theorem, let us first assume that $\varphi=1$ on $\prod_{v\nmid\mathfrak{e}}\mathfrak{g}_v^\times$ and $\mathfrak{f}\supset\mathfrak{c}$. Then (6.15b) is valid for $\chi=\varphi\psi$. Defining $\mathfrak{z}$ as in our theorem, change $\mathfrak{e}$ for $\mathfrak{c}\prod_{r|\mathfrak{z}}\mathfrak{c}_r$. This changes $Z_\psi(s,\lambda,\varphi)$ for $Z_{\psi\mathfrak{z}}(s,\lambda,\varphi)$. If $h\mathfrak{g}=\prod_{v\nmid\mathfrak{z}\mathfrak{c}}\mathfrak{c}_v$ with $h\in F_\mathbf{h}^\times$, then $f|T_{h,\psi}\neq 0$, and hence $f|U_{h,\psi}\neq 0$ by Lemma 6.2. Consequently $f|U_{h,\chi}\neq 0$. Thus (6.15b) reduces the analytic nature of $\Gamma^n_{k,\kappa,\varepsilon}(s)Z_{\psi\mathfrak{z}}(2s,\lambda,\varphi)$ to that of $\mathscr{P}^{2n}_{k-\varepsilon}$. Now $\mathscr{P}^{2n}_{k-\varepsilon}$ has a meromorphic continuation to the whole s-plane whose poles are described in [94a, Theorem 7.3]. (See *Correction* at the end of the present paper.) Therefore we obtain our theorem in this case.

Let us next take a more general φ and let $\mathfrak{c}^0=\mathfrak{c}\cap\mathfrak{f}$. Decompose $\mathfrak{c}^0$ into the product $\mathfrak{c}^0=\mathfrak{e}^0\mathfrak{e}^1$ of relatively prime integral ideals $\mathfrak{e}^0$ and $\mathfrak{e}^1$ so that $\mathfrak{e}^0_v=\mathfrak{c}^0_v$ for every $v|\mathfrak{e}\mathfrak{z}\mathfrak{f}$ and $\mathfrak{e}^0_v=\mathfrak{g}_v$ for $v\nmid\mathfrak{e}\mathfrak{z}\mathfrak{f}$. Let K^0 denote the group K defined with $(\mathfrak{b},\mathfrak{c}^0,\mathfrak{e}^0)$ in place of $(\mathfrak{b},\mathfrak{c},\mathfrak{e})$ and let $\chi=\varphi\psi$. Then $f\in\mathscr{S}_k(K^0,\chi)=\mathscr{S}_k(K^0,\psi)$. Therefore, applying our result in the above special case to the present case, we can complete our proof.

Remark 6.3. (I) Since $\mathscr{S}_k$ is the union of $\mathscr{S}_k(K)$ for all K of type (6.1a) with $\mathfrak{e}=\mathfrak{c}$, Theorem 6.1 covers practically the whole $\mathscr{S}_k$. We are eliminating (or rather, not defining) the Euler v-factors for $v|\mathfrak{e}$. If we were interested only in "almost all primes," we could take $\mathfrak{e}=\mathfrak{c}$ at the beginning, which would eliminate W_v^{-1} for all $v|\mathfrak{c}$. The case $\mathfrak{e}=\mathfrak{g}$ is most transparent, and the function Z in this case was investigated in [94a]. If $n=1$, we can take $\mathfrak{e}=\mathfrak{g}$ and $K=D[\mathfrak{b}^{-1},\mathfrak{b}\mathfrak{c}]$ with no fear of losing generality.

(II) The polynomial W_v of (6.4), if $v|\mathfrak{e}^{-1}\mathfrak{c}$, is of degree $\leqq n$, and it can easily be seen that its coefficient of degree n is a nonzero constant times the eigenvalue of $T_{\pi_v,\psi}$ on f. Therefore $Z_\psi/Z_{\psi,\mathfrak{z}}$ is the product of W_v^{-1} for all $v|\mathfrak{e}^{-1}\mathfrak{c}$ such that W_v is of degree $<n$. In particular, we have $Z_{\psi\mathfrak{z}}=Z_\psi$ if $n=1$. Thus, if $n=1$, taking $\mathfrak{e}=\mathfrak{g}$, we have a natural Euler product in Theorem 6.1 with no artificial exclusion of Euler factors.

(III) The meromorphic continuation of $Z_\psi(s,\lambda,\varphi)$ was obtained in [94a, Theorem 8.1] in the case $\mathfrak{e}=\mathfrak{g}$. (As to comparisons with the previous results by Andrianov and Kalinin [AK] and also by Piatetzki-Shapiro and Rallis [PR], the reader is referred to the introduction of [94a].) The gamma factor for Z_ψ obtained in the first line of [94a, (8.11)] coincides with $\Gamma^n_{k,\kappa}(s/2)$ given in Theorem 6.1. This is noteworthy, since the Eisenstein series employed in [94a] is of weight $k-\varepsilon-nu/2$ on G^n, while the present proof employs that of weight $k-\varepsilon$ on G^{2n}. This coincidence seems to indicate that the normalizing gamma factors for the Eisenstein series obtained in [94a, Theorem 7.3] are natural and perhaps best possible. Even the information about the possible poles given in Theorem 6.1 is the same as that obtained from [94a, (8.11)] except that: (i) If n is odd, $k=(n+1)u/2$, and $\varepsilon=u$, then the poles of $\mathscr{E}(z,(2s-n)/4)$ in [94a, (8.11)] described in [94a, Theorem 7.3] form a smaller set than the set of possible poles given in Theorem 6.1; (ii) the same can be said if $\varphi^2\psi^2=1$ and

$\mathfrak{c}\mathfrak{f} = \mathfrak{g}$; (iii) the expression [94a, (8.11)] involves a factor $\Lambda_{\mathfrak{c}}/\Lambda_{\mathfrak{b}}$, which makes the result on poles derived from [94a, (8.11)] less clear-cut than Theorem 6.1.

(IV) In [KR] Kudla and Rallis proved that the possible poles of zeta functions of type (6.4), *without gamma factors*, for trivial $\varphi\psi$ are simple and belong to an explicitly given set of $2[(n+2)/2]$ points. Our theorem, though formulated only for scalar-valued holomorphic forms, concerns the product of such a zeta function and some gamma factors which may produce many poles. Therefore, we obtain a much smaller set of possible poles for $Z_{\psi,\mathfrak{z}}$ than that of [KR]. In fact, Theorem 6.1 shows (at least) that it is either entire or has at most one simple pole at $n+1$ except in the following two cases:

Case a: $\ell > n$, $\varphi^2\psi^2 = 1$, and $\mathfrak{c}\mathfrak{f} = \mathfrak{g}$. Each pole of $Z_{\psi,\mathfrak{z}}$ other than $n+1$ belongs to

$$\{ j \in \mathbf{Z} \mid \mathrm{Max}\{0, 2n-\ell\} \leqq j \leqq n,\ j \equiv \ell + 1 \pmod 2 \}.$$

Case b: $\ell < n$ and $\varphi^2\psi^2 = 1$. Each pole of $Z_{\psi\mathfrak{z}}$ belongs to

$$\{ j \in \mathbf{Z} \mid \mathrm{Max}_{v\in\mathbf{a}}\{2n+1-k_v-\varepsilon_v\} \leqq j \leqq 2n+1-\ell \}.$$

Here the symbols are the same as in Theorem 6.1. Once k and ε are explicitly given, we can eliminate some more poles. Moreover Theorem 6.1 guarantees many zeros of $Z_{\psi,\mathfrak{z}}$.

We conclude this section by examining the case $n = 1$ more closely. If $f \in \mathscr{S}^1_k(D[\mathfrak{b}^{-1}, \mathfrak{b}\mathfrak{c}], \psi)$ and f is an eigenform as above, then by [91, Prop. 3.6] there exists a k-normalized Hecke character ψ' and a system of eigenvalues χ in $\mathscr{S}_k((\mathfrak{c}, \psi'))$ in the sense of that paper such that $\psi' = \psi$ on $\prod_{v\in\mathbf{h}} \mathfrak{g}_v^\times$, $\lambda(\mathfrak{p}) = |\chi(\mathfrak{p})|^2 - N(\mathfrak{p}) - 1$ for $\mathfrak{p} \nmid \mathfrak{c}$, and $\lambda(\mathfrak{p}) = (\psi/\psi')(\pi_v)\chi(\mathfrak{p})^2$ for $\mathfrak{p}|\mathfrak{c}$, where $\mathfrak{p} = \pi_v\mathfrak{g}$. Then we easily see that $Z_\psi(s, \lambda, \varphi)$ coincides with $\mathscr{D}(s, \{\chi^2, \varphi\psi/\psi'\})$ of [91, (10.3)]. (As mentioned in Remark 6.3(II), no Euler factor is eliminated.) Conversely, if we start from χ and ψ', then by [91, Prop. 3.5] we can find $f \in \mathscr{S}^1_k(D[\mathfrak{b}^{-1}, \mathfrak{b}\mathfrak{c}], \psi')$ so that $\mathscr{D}(s, \{\chi^2, \varphi\}) = Z_{\psi'}(s, \lambda, \varphi)$.

Theorem 6.4. *The notation being as above, the product* $\Gamma^1_{k,\kappa,\varepsilon}(s/2)Z_\psi(s, \lambda, \varphi)$ *has possible poles only at* $s = 1$ *and* $s = 2$. *Each pole is simple. The pole at* $s = 1$ *occurs only if* $\mathfrak{c}\mathfrak{f} = \mathfrak{g}$, $\varphi^2\psi^2 = 1$, *and* $\varphi\psi \neq 1$. *The pole at* $s = 2$ *occurs if and only if* $\varphi^2\psi^2 = 1$, $\varphi\psi \neq 1$, *and* $\eta^*(\mathfrak{p})\chi(\mathfrak{p}) = \overline{\chi(\mathfrak{p})}$ *for almost all primes* $\mathfrak{p}$, *where* $\eta = \varphi\psi/\psi'$, *in which case* $|k_v - \varepsilon_v - 1/2| - 1/2 \in 2\mathbf{Z}$ *for every* $v \in \mathbf{a}$.

Proof. The gamma factor of $\mathscr{D}$ given in [91, Prop. 10.1] coincides with $\Gamma^1_{k,\kappa,\varepsilon}(s/2)$. Therefore, combining Theorem 6.1 with [91, Propositions 10.1 and 10.4], we obtain our result.

7. The series $E_k^{m,n}$ and its critical values

By a *nearly holomorphic function* on $\mathscr{H}$ we understand a $\mathbf{C}$-valued function on $\mathscr{H}$ that is a polynomial in the entries of $\left(\mathrm{Im}(z_v)^{-1}\right)_{v\in\mathbf{a}}$ with holomorphic functions as coefficients. (For basic facts on such functions, see [86, 87, 94b]).

Given a congruence subgroup Γ of G^n we denote by $\mathcal{N}_k^n(\Gamma)$ the set of all nearly holomorphic functions g on $\mathcal{H}$ such that $g\|_k\gamma = g$ for every $\gamma \in \Gamma$, and that for every $\alpha \in G^n$ we have an expansion of the form

$$(7.1) \qquad (g\|_k\alpha)(z) = \sum_{\sigma\in S_+} p_\sigma\left(\left(\mathrm{Im}(z_v)^{-1}\right)_{v\in\mathbf{a}}\right)\mathbf{e}_\mathbf{a}(\mathrm{tr}(\sigma z)) \quad (z \in \mathcal{H}),$$

where p_σ is a polynomial function on $S_\mathbf{a}$. In fact, this expansion follows automatically from the invariance property $g\|_k\Gamma = g$ if $F \neq \mathbf{Q}$. We then denote by $\mathcal{N}_k^n$ the union of $\mathcal{N}_k^n(\Gamma)$ for all Γ, and by $\mathcal{N}_{k,d}^n$ for $d \in \mathbf{Z}^\mathbf{a}$, $d_v \geqq 0$, the set of all g in $\mathcal{N}_k^n$ whose degree as a polynomial in the entries of $\mathrm{Im}(z_v)^{-1}$ is of degree $\leqq d_v$ for each $v \in \mathbf{a}$. Clearly $\mathcal{N}_{k,0}^n = \mathcal{M}_k^n$.

We now consider C and K of (3.7a, b) with $\mathfrak{e} = \mathfrak{c}$. Take an integral weight k, a character χ satisfying (3.6a, b), and an eigenform f in $\mathcal{S}_k^n(K)$ with respect to $T_\tau(\mathfrak{a})$, and put $f|T_\tau(\mathfrak{a}) = \lambda(\mathfrak{a})f$, where τ stands for the trivial Hecke character. Then $Z_\tau(s, \lambda, \chi)$ can be defined by (6.5) with $\mathfrak{c}$ and τ as $\mathfrak{e}$ and ψ there. Let us now put, for $m > n > 0$,

(7.2a)

$$\mathcal{F}^{m,n}(z, s;\ f, \chi, C) = E(z, s;\ f, \chi, C)Z_\tau(2s, \lambda, \chi)\prod_{i=n+1}^{[(n+m)/2]} L_\mathfrak{c}(4s - 2i, \chi^2).$$

If $n = 0$ and $f = 1$, we take $D = D^m[\mathfrak{b}^{-1}, \mathfrak{bc}]$ and put

$$(7.2b) \qquad \mathcal{F}^m(z, s) = \mathcal{F}^{m,0}(z, s;\ 1, \chi, D) = E(z, s;\ 1, \chi, D)\Lambda_\mathfrak{c}^m(2s, \chi).$$

Then (7.2a) is valid even for $n = 0$ if we understand that $C = D$ and

$$(7.3) \qquad Z_\tau(s, \lambda, \chi) = L_\mathfrak{c}(s, \chi) \quad \text{if } n = 0.$$

From (6.17), we obtain

$$(7.4) \qquad v_\Gamma \cdot c_{n,k}(su - (k - i\kappa)/2)\mathcal{F}^{m,n}(z, s;\ f, \chi, C)$$
$$= \chi(\theta)^{-n}N(\mathfrak{b})^{2ns}\mathrm{vol}(\Gamma\backslash\mathcal{H})\left\langle (f\|\iota)^c(w), (\mathcal{F}^{m+n}\|\rho)(\mathrm{diag}[z, w], s)\right\rangle.$$

Theorem 7.1. *The notation being as above, the product*

$$(7.5) \qquad c_{n,k}(su - (k - i\kappa)/2)\mathcal{G}_{k,\kappa}^{m+n}(s)\mathcal{F}^{m,n}(z, s;\ f, \chi, C)$$

can be continued to a meromorphic function on the whole s-plane with only finitely many poles, which are all simple, where we understand that $c_{0,k} = 1$.

This follows immediately from (7.4) and [94a, Theorem 7.3], which tells about the analytic properties of $\mathcal{P}^N$ of (6.14). The location of all possible poles can be determined by that theorem. For instance, (7.5) is an entire function of s if $\chi^2 \neq 1$.

Lemma 7.2. $Z_\tau(s, \lambda, \chi) \neq 0$ *at least for* $\mathrm{Re}(s) \in \Lambda(n)$, *where*

$$\Lambda(n) = \begin{cases} \{x \in \mathbf{R} \mid x \geqq 1\} & \text{if } n = 0, \\ \{x \in \mathbf{R} \mid x \geqq 2\} & \text{if } n = 1, \\ \{x \in \mathbf{R} \mid x > (3n/2) + 1\} & \text{if } n = 2^r \text{ with } 0 < r \in \mathbf{Z}, \\ \{x \in \mathbf{R} \mid x > (5n/3) + 1\} & \text{otherwise.} \end{cases}$$

Moreover, $Z_\tau(s, \lambda, \chi)$ *for such an* s *is finite except when* $s = n + 1 \leqq 2$, *in which case* Z_τ *is finite or has a simple pole at* s.

This result for $n > 1$ follows from Duke, Howe, and Li [DHL, Prop. 5.6]. Strictly speaking [DHL] concerns only the case $F = \mathbf{Q}$, but clearly the method applies to the present general case. The case $n = 0$ is classical; the case $n = 1$ follows easily from [91, Prop. 3.3 and (10.4)].

Let us now consider the case in which $k = k_0 u$ with $k_0 \in \mathbf{Z}$ and $\kappa = 0$. If $k_0 > m + n + 1$, then $E_k^{m,n}$ of (1.34) and $E(z, s; f, \chi, C)$ are convergent at $s = k_0/2$, and their values are holomorphic in z, and hence belong to $\mathscr{M}_k^m$. For a smaller k_0 we have:

Theorem 7.3. *Suppose that* $m > n \geqq 0$ *and* $k = k_0 u$ *with* $(m + n + 1)/2 \leqq k_0 \in \mathbf{Z}$. *Let* f *and* χ *be as above with* $\kappa = 0$ *in (3.6a); let* $g \in \mathscr{S}_k^n(\Gamma')$ *with an arbitrary congruence subgroup* Γ' *of* G^n.

(I) *Suppose that* $k_0 \in \Lambda(n)$ *and that* $F \neq \mathbf{Q}$ *or* $2k_0 \notin \{m + n + 2, m + n + 3\}$. *Then* $E_k^{m,n}(z, k_0/2; g, \Gamma')$ *and* $E(z, k_0/2; f, \chi, C)$ *belong to* $\mathscr{M}_k^m$. *Moreover,* $E(z, k_0/2; f, \chi, C)$ *belongs to* $\mathscr{M}_k^m$ *even if* $F = \mathbf{Q}$ *and* $2k_0 \in \{m + n + 2, m + n + 3\}$, *provided* $k_0 \in \Lambda(n)$ *and* $\chi^2 \neq 1$.

(II) *If* $F = \mathbf{Q}$ *and* $k_0 = (m + n + 3)/2 \in \Lambda(n)$, *then both* $E_k^{m,n}(z, k_0/2; g, \Gamma')$ *and* $E(z, k_0/2; f, \chi, C)$ *belong to* $\mathscr{N}_{k,m}^m$.

(III) *Let* $\mu = m + n + 1 - k_0$. *Then* $\mathscr{F}^{m,n}(z, \mu/2; f, \chi, C)$ *belongs to* $\mathscr{M}_k^m$ *except in the following two cases:*

(A) $\mu = 0$, $\mathfrak{c} = \mathfrak{g}$, and $\chi = 1$; (B) $0 < \mu \leqq (m + n)/2$, $\mathfrak{c} = \mathfrak{g}$, and $\chi^2 = 1$.

Proof. These results for $n = 0$ are contained in [87, Prop.4.1] (see also [83b], [85a]). By Lemma 3.7 the results for $E_k^{m,n}(z, k_0/2; g, \Gamma')$, $n > 0$, follow from those for $E(z, k_0/2; f, \chi, C)$, since Γ' contains Γ_1 of (3.15b) for a suitable $\mathfrak{c}$, and $\mathscr{S}_k^n(K)$ is spanned by eigenforms. To investigate $E(z, k_0/2; f, \chi, C)$, we evaluate (7.4) at $s = k_0/2$. We easily see that $c_{n,k}(\mathbf{s})$ is finite and nonzero for this value of s, and $\Lambda_\mathfrak{c}^{2n}(k_0, \chi) \neq 0$. Thus by Lemma 7.2 $E(z, k_0/2; f, \chi, C)$ is a constant times

$$\langle (f \| \iota)^c(w), (E^{m+n} \| \rho)(\mathrm{diag}[z, w], k_0/2) \rangle . \tag{7.6}$$

Now, by [87, Prop.4.1] $E^{m+n}(Z, k_0/2)$ is an element of $\mathscr{M}_k^{m+n}$ (at least) for k_0 as in our theorem, and hence we can write

$$(E^{m+n} \| \rho)(\mathrm{diag}[z, w], k_0/2) = \sum_{i \in I} p_i(z) q_i(w) \tag{7.7}$$

with $p_i \in \mathcal{M}_k^m$, $q_i \in \mathcal{M}_k^n$, and a finite set I. (This is elementary; see [83a, Lemma 1.1] for a general principle.) Therefore (7.6) equals $\sum_{i\in I} \langle (f\|\iota)^c, q_i \rangle p_i(z)$, from which we obtain (I). If $F = \mathbf{Q}$ and $k_0 = (m+n+3)/2$, [83b, Prop. 10.2] shows (as noted in [85a, p.291]) that

$$(7.8)\qquad (E^{m+n}\|\rho)(Z, k_0/2) = \sum_{r=0}^{m+n} \mathrm{tr}\left[\rho_r(Z - \bar{Z})^{-1} h_r(Z)\right],$$

where ρ_r is a representation of $GL_{m+n}(\mathbf{C})$ on $\bigwedge^r \mathbf{C}^{m+n}$ and h_r is a holomorphic function on $\mathcal{H}_{m+n}$ with values in $\mathrm{End}\left(\bigwedge^r \mathbf{C}^{m+n}\right)$. Then (7.8) for $Z = \mathrm{diag}[z, w]$ is nearly holomorphic in z of degree $\leqq m$. Hence (7.7) holds with $p_i \in \mathcal{N}_{k,m}^m$ and $q_i \in \mathcal{N}_{k,n}^n$, which proves (II). Assertion (III) can be derived from [87, Prop.4.1(v)] in the same fashion, by simply evaluating (7.4) at $\mu/2$.

Theorem 7.4. *Suppose that $m > n > 0$ and $k = k_0 u$ with $0 < k_0 < (m+n+1)/2$; put $s_0 = (m+n+1-k_0)/2$. Then $\mathcal{F}^{m,n}(z, s; f, \chi, C)$ has at most a simple pole at s_0, which occurs only when $\chi^2 = 1$. Moreover, the residue is an element of $\mathcal{M}_k^m$.*

This can be derived from (7.4) and [85a, Theorem 2.7] by the same type of reasoning as in the above proof. Notice that both $c_{n,k}(su - k/2)$ and $\Lambda_{\mathfrak{c}}^{m+n}(2s, \chi)$ are finite and nonzero for $s = s_0$.

As for a more general weight and more general critical values, we have

Theorem 7.5. *Suppose that $k_v \geqq (m+n+1)/2$ for every $v \in \mathbf{a}$ and $k_v \pmod 2$ is independent of v; let χ be as in (3.6a, b) with $\kappa = 0$.*

(I) *Let μ be an integer such that $\mu \in \Lambda(n)$, $(m+n+1)/2 \leqq \mu \leqq k_v$, and $\mu \equiv k_v \pmod 2$ for every $v \in \mathbf{a}$. Then $E_k^{m,n}(z, \mu/2; g, \Gamma')$ belongs to $\mathcal{N}_{k,d}^m$, where*

$$d = \begin{cases} (m+n)(k-\mu+2)/2 & \text{if } \mu = (m+n+3)/2 \text{ and } F = \mathbf{Q}, \\ (m+n)(k-\mu u)/2 & \text{otherwise,} \end{cases}$$

except when $F = \mathbf{Q}$ and $\mu = (m+n+2)/2$.

(II) *Let μ be an integer such that $m+n+1-k_v \leqq \mu \leqq k_v$ for every $v \in \mathbf{a}$ and that*

$$\mu \equiv \begin{cases} k_v + 1 \pmod 2 & \text{if } \mu < (m+n+1)/2 \notin \mathbf{Z}, \\ k_v \pmod 2 & \text{otherwise.} \end{cases}$$

Then $\mathcal{F}^{m,n}(z, \mu/2; f, \chi, C)$ belongs to $\mathcal{N}_{k,d}^m$, where

$$d = \begin{cases} (m+n)(k-\mu+2)/2 \text{ if } \mu = (m+n+3)/2, F = \mathbf{Q}, & \text{and } \chi^2 = 1, \\ (m+n)\{k - |\mu-\ell|u - \ell u\}/2,\ \ell = (m+n+1)/2, & \text{otherwise,} \end{cases}$$

except in the three cases which are (A), (B) of Theorem 7.3 and

(C) $\qquad \mu = (m+n+2)/2,\ F = \mathbf{Q}, \text{and } \chi^2 = 1.$

These results can be derived from [87, Theorem 4.2] by evaluating (7.4) at $\mu/2$ and employing the same argument as in the above proof.

Remark 7.6. It is conjecturable that Lemma 7.2 holds with $\Lambda(n) = \{x \in \mathbf{R} \mid x \geqq n+1\}$ at least if $k_v \geqq n+1$ for every $v \in \mathbf{a}$. If this is so, we can eliminate the conditions $k_0 \in \Lambda(n)$ in Theorem 7.3 and $\mu \in \Lambda(n)$ in Theorem 7.5.

8. The operator Φ and the space of Eisenstein series

Throughout this section $k = k_0 u$ with $k_0 \in (1/2)\mathbf{Z}$, Γ denotes an unspecified congruence subgroup of $\mathscr{G}^n$, and ρ a real variable on $(0, \infty)$. Given a function $f : \mathscr{H}_n \to \mathbf{C}$, we define $\Phi f : \mathscr{H}_{n-1} \to \mathbf{C}$ by

$$(8.1) \qquad (\Phi f)(w) = \lim_{\rho\to\infty} f\bigl(\mathrm{diag}[w, \rho\mathbf{i}_1]\bigr) \qquad (w \in \mathscr{H}_{n-1})$$

with $\mathbf{i}_n$ of (1.6), whenever the limit exists. If $n = 1$, we ignore w, and so Φf is a constant. We then define the ν–th power Φ^ν of Φ for $\nu \leqq n$ inductively by $\Phi^\nu = \Phi\Phi^{\nu-1}$, with the identity map as Φ^0. This sends f to a function on $\mathscr{H}_{n-\nu}$. If f has a Fourier expansion of the form $f(z) = \sum_{\tau\in S_+} c(\tau)\mathbf{e}\bigl(\mathrm{tr}(\tau z)\bigr)$ with $c(\tau) \in \mathbf{C}$, then it can easily be shown that $\Phi^\nu f$ is meaningful, and

$$(8.2) \quad (\Phi^\nu f)(w) = \lim_{\rho\to\infty} f\bigl(\mathrm{diag}[w, \rho\mathbf{i}_\nu]\bigr) = \sum_{\sigma\in S_+^{n-\nu}} c\bigl(\mathrm{diag}[\sigma, 0]\bigr)\mathbf{e}\bigl(\mathrm{tr}(\sigma w)\bigr)$$

for $w \in \mathscr{H}_{n-\nu}$ if $n > \nu$, and $\Phi^n f = c(0)$.

Lemma 8.1. *If $f \in \mathscr{M}_k(\Gamma)$, then $\Phi^{n-r} f$ belongs to $\mathscr{M}_k^r(\Gamma \cap \mathscr{P}^{n,r})$ of (1.29) and $\Phi^{n-r}(f\|_k\alpha) = \lambda_r(\alpha)^{-[k]}|\lambda_r(\alpha)|^{[k]-k}(\Phi^{n-r} f)\|_k\pi_r(\alpha)$ for every $\alpha \in \mathscr{P}^{n,r}$.*

Proof. The first assertion follows from the second one, which can be verified by a straightforward calculation.

Lemma 8.2. *Let X be a complete set of representatives for $\mathscr{P}^{n,r}\backslash\mathscr{G}^n/\Gamma$ and let $f \in \mathscr{M}_k^n(\Gamma)$. Then*

(1) $\Phi^{n-r}(f\|_k\alpha) = 0$ for every $\alpha \in \mathscr{G}^n$ if and only if $\Phi^{n-r}(f\|_k\xi^{-1}) = 0$ for every $\xi \in X$, in which case $\Phi^{n-r-1}(f\|_k\alpha)$ is a cusp form for every $\alpha \in \mathscr{G}^n$.

(2) f is a cusp form if and only if $\Phi(f\|_k\alpha) = 0$ for every $\alpha \in \mathscr{G}^n$.

Proof. Assertion (1) follows from Lemma 8.1 and assertion (2), which is practically the definition of cusp form.

Lemma 8.3. *Let $\alpha \in G_v^n$, $v \in \mathbf{a}$, and $z = \mathrm{diag}[z_1, i\rho]$ with $z_1 = x_1 + iy_1 \in H_{n-1}$ and $\rho > 0$. Put $c_\alpha = \begin{pmatrix} * & * \\ e & e' \end{pmatrix}$ and $d_\alpha = \begin{pmatrix} * & * \\ f & f' \end{pmatrix}$ with $e, f \in \mathbf{R}_{n-1}^{n-r}$ and $e', f' \in \mathbf{R}^{n-r}$. Then $\delta\bigl(\omega_r(\alpha z)\bigr)|j_\alpha(z)|^2 = \rho \cdot \det(y_1)\det[A + \rho e' \cdot {}^t e' + \rho^{-1} f' \cdot {}^t f']$ with $A = (ex_1 + f)y_1^{-1} \cdot {}^t(ex_1 + f) + ey_1 \cdot {}^t e$.*

Proof. Put $y = \mathrm{Im}(z)$, $Y = \mathrm{Im}(\alpha z) = \begin{pmatrix} g & * \\ * & * \end{pmatrix} = \begin{pmatrix} * & * \\ * & h \end{pmatrix}^{-1}$ with $g \in S_r^r$ and $h \in S_v^{n-r}$. Then $\delta(\omega_r(\alpha z)) = \det(g) = \det(Y)\det(h)$ by [83b, Lemma 6.1] and $|j_\alpha(z)|^2 \det(Y) = \rho \det(y_1)$. Now $Y^{-1} = (c_\alpha z + d_\alpha)y^{-1} \cdot {}^t(c_\alpha \bar{z} + d_\alpha) = (cx+d)y^{-1} \cdot {}^t(cx+d) + cy \cdot {}^t c$, and hence we easily find that $h = A + \rho e' \cdot {}^t e' + \rho^{-1} f' \cdot {}^t f'$, which proves our lemma.

Lemma 8.4. *For $r < n$ and $\alpha \in G^n$ one has $\Phi((\delta \circ \omega_r)^{-1} |j_\alpha|^{-2}) \neq 0$ if and only if $\alpha \in P^{n,r} P^{n,n-1}$.*

Proof. For simplicity let us assume $F = \mathbf{Q}$; the general case can be proved in the same manner. We use the technique of [K]. The notation being as in Lemma 8.3, we easily see that $\rho \cdot \det[A + \rho e' \cdot {}^t e' + \rho^{-1} f' \cdot {}^t f']$ is a polynomial in ρ. Suppose $\Phi((\delta \circ \omega_r)^{-1} |j_\alpha|^{-2}) \neq 0$; then this polynomial is a constant. Since A is nonnegative, we have $\rho \cdot \det[A + \rho e' \cdot {}^t e' + \rho^{-1} f' \cdot {}^t f'] \geqq \rho \cdot \det(A)$, and hence $\det(A) = 0$. We can thus find $v \in \mathbf{R}^1_{n-r}$, $\neq 0$ such that $vA \cdot {}^t v = 0$. Then $v(ex_1 + f) = 0$ and $ve = 0$, so that $v(e\ f) = 0$. This shows that rank $(e\ f) < n - r$. Therefore we can find $g \in SL_{n-r}(F)$ so that the last row of $g(e\ f)$ is 0. Put $\beta = \mathrm{diag}[1_r, {}^t g^{-1}, 1_r, g]$ and change α for $\beta\alpha$. Then $\delta(\omega_r(\alpha z)) |j_\alpha(z)|^2$ doesn't change. Thus we may assume that the bottom rows of e and f are 0. Let p and q be the last elements of e' and f'. Then $pf' = qe'$. Suppose $p \neq 0$. Then $\rho \cdot \det[A + \rho e' \cdot {}^t e' + \rho^{-1} f' \cdot {}^t f']$ becomes 0 for $\rho = iq/p$. On the other hand, the equality of Lemma 8.3 shows that the quantity is positive for $0 < \rho \in \mathbf{R}$, and hence our polynomial is not a constant. Therefore $p = 0$, that is, the present α belongs to $P^{n,n-1}$, which means that the original α belongs to $P^{n,r} P^{n,n-1}$ as expected. Conversely, if $\alpha = \gamma\varepsilon$ with $\gamma \in P^{n,r}$ and $\varepsilon \in P^{n,n-1}$, by (1.24) we have $\delta(\omega_r(\alpha z)) |j_\alpha(z)|^2 = |\lambda_r(\gamma)\lambda_{n-1}(\varepsilon) j_\xi(z_1)|^2 \delta(\omega_r(\xi z_1))$ with $\xi = \pi_{n-1}(\varepsilon)$ for $z = \mathrm{diag}[z_1, i\rho]$, which proves the 'if'-part.

Lemma 8.5. *Let $\delta(z, s; f)$ be as in* (1.32) *with $f \in \mathscr{S}_k^r$ and let $\alpha \in \mathscr{G}^n$. Then for* $\mathrm{Re}(s) > 0$ *and $t \geqq r$ we have $\Phi^{n-t}\{\delta(z)^{k/2-su}[\delta(z, s; f) \|_k \alpha]\} \neq 0$ only if $\alpha = \beta\gamma$ with $\beta \in \mathscr{P}^{n,r}$ and $\gamma \in \mathscr{P}^{n,t}$, in which case for $z = \mathrm{diag}[w, w']$ with $(w, w') \in \mathscr{H}_t \times \mathscr{H}_{n-t}$ one has*

$$\begin{aligned} (8.3) \qquad & \delta(z)^{k/2-su}[\delta(z, s; f) \|_k \alpha] \\ &= \delta(w)^{k/2-su} |\lambda_r(\beta)\lambda_t(\gamma)|^{[k]-2su} (\lambda_r(\beta)\lambda_t(\gamma))^{-[k]} \delta(w, s; f \|_k \pi_r(\beta)) \|_k \pi_t(\gamma). \end{aligned}$$

Proof. From (1.33) and (1.28) we can easily derive (8.3). Now it is well known that $|f(z')| \leqq C\delta(z')^{-k/2}$ for every $z' \in \mathscr{H}_r$ with a constant C. Therefore, for $\alpha \in \mathscr{G}$ we have

$$\begin{aligned} |\delta(z)^{k/2-su}[\delta(z, s; f) \| \alpha]| &= |f(\omega_r(\alpha z)) \delta(\omega_r(\alpha z))^{k/2-su} j_\alpha(z)^{-2su}| \\ &\leqq C |\delta(\omega_r(\alpha z)) j_\alpha(z)^2|^{-su}. \end{aligned}$$

By Lemma 8.4 the last quantity, with $z = \mathrm{diag}[w, \rho i_1]$, $w \in \mathscr{H}_{n-1}$, and $\mathrm{Re}(s) > 0$, tends to 0 as $\rho \to \infty$ if $\alpha \notin \mathscr{P}^{n,r} \mathscr{P}^{n,n-1}$. This proves the case $t = n-1$.

Suppose that our lemma is true for Φ^{n-t}, $t > r$, and $\Phi^{n-t+1}\{\delta(z)^{k/2-su}[\delta(z, s; f)\|\alpha]\} \neq 0$. Then $\alpha = \beta\gamma$ with $\beta \in \mathscr{P}^{n,r}$ and $\gamma \in \mathscr{P}^{n,t}$, and by (8.3), $\Phi\{\delta(w)^{k/2-su}[\delta(w, s; f\|\pi_r(\beta))\|\pi_t(\gamma)]\} \neq 0$, where $w \in \mathscr{H}_t$. Hence $\pi_t(\gamma) = \xi\eta$ with $\xi \in \mathscr{P}^{t,r}$ and $\eta \in \mathscr{P}^{t,t-1}$. By (1.23b) we can put $\xi = \pi_t(\xi')$ and $\eta = \pi_t(\eta')$ with $\xi' \in \mathscr{P}^{n,t} \cap \mathscr{P}^{n,r}$ and $\eta' \in \mathscr{P}^{n,t} \cap \mathscr{P}^{n,t-1}$. Then $\alpha = \beta'\gamma'$ with $\beta' = \beta\xi'$ and $\gamma' = \xi'^{-1}\gamma$. Clearly $\beta' \in \mathscr{P}^{n,r}$. Since $\gamma' \in \mathscr{P}^{n,t}$ and $\pi_t(\gamma') = \eta \in \mathscr{P}^{t,t-1}$, we have $\gamma' \in \mathscr{P}^{n,t-1}$ by (1.23a). Therefore induction proves our lemma.

Lemma 8.6. *For each ξ in the set X of Lemma 8.2 let Z_ξ be a complete set of representatives for $(P \cap Q)\backslash Q/(\xi\Gamma\xi^{-1} \cap Q)$, where $P = \mathscr{P}^{n,t}$ and $Q = \mathscr{P}^{n,r}$, $r \leqq t$. For every $\zeta \in Z_\xi$ such that $\zeta\xi\Gamma \cap P \neq \varnothing$ choose and fix an element $\eta \in \zeta\xi\Gamma \cap P$. Let Y denote the set of all such η's. Then Y is a finite set, $P = \bigsqcup_{\eta\in Y}(P \cap Q)\eta(\Gamma \cap P)$, and $\mathscr{G}^t = \bigsqcup_{\eta\in Y}\mathscr{P}^{t,r}\pi_t(\eta)\pi_t(\Gamma \cap P)$. Moreover, if R_η is a complete set of representatives for $(\eta\Gamma\eta^{-1} \cap Q \cap P)\backslash(\eta\Gamma \cap P)$, then $\bigsqcup_{\eta\in Y}\zeta^{-1}R_\eta$ gives $\bigsqcup_{\xi\in X}(\xi\Gamma\xi^{-1} \cap Q)\backslash(\xi\Gamma \cap QP)$, where ζ is taken for each η so that $\zeta \in Z_\xi$ and $\eta \in \zeta\xi\Gamma$, and π_t gives a bijection of R_η onto $(\eta_t\Delta\eta_t^{-1} \cap \mathscr{P}^{t,r})\backslash\eta_t\Delta$, where $\Delta = \pi_t(\Gamma \cap P)$ and $\eta_t = \pi_t(\eta)$.*

Proof. Clearly $Q\xi\Gamma = \bigsqcup_{\zeta\in Z_\xi}(P \cap Q)\zeta\xi\Gamma$ for every $\xi \in X$. Thus $\mathscr{G} = \bigsqcup_{\xi\in X} Q\xi\Gamma = \bigsqcup_{\xi,\zeta}(P \cap Q)\zeta\xi\Gamma$, and hence $P = \bigsqcup_{\xi,\zeta}(P \cap Q)(\zeta\xi\Gamma \cap P) = \bigsqcup_{\eta\in Y}(P \cap Q)\eta(\Gamma \cap P)$. Applying π_t to this equality, we obtain $\mathscr{G}^t = \bigcup_{\eta\in Y}\mathscr{P}^{t,r}\pi_t(\eta)\pi_t(\Gamma \cap P)$ by (1.23b). That this union is disjoint can easily be derived from the fact that if $\alpha \in P$ and $\pi_t(\alpha) = 1$, then $\alpha \in P \cap Q$, which follows from (1.23a). Since $\mathscr{P}^{t,r}\backslash\mathscr{G}^t/\Gamma'$ is finite for any congruence subgroup Γ' of $\mathscr{G}^t$, Y must be finite. Now $Q = Q^{-1} = \bigsqcup_{\zeta\in Z_\xi}(\xi\Gamma\xi^{-1} \cap Q)\zeta^{-1}(P \cap Q)$, and hence $QP = \bigsqcup_{\zeta\in Z_\xi}(\xi\Gamma\xi^{-1} \cap Q)\zeta^{-1}P$. Therefore $(\xi\Gamma\xi^{-1} \cap Q)\backslash(\xi\Gamma \cap QP)$ is represented by the disjoint union of $(\xi\Gamma\xi^{-1} \cap Q)\backslash[\xi\Gamma \cap (\xi\Gamma\xi^{-1} \cap Q)\zeta^{-1}P]$, which is represented by $(\xi\Gamma\xi^{-1} \cap Q \cap \zeta^{-1}P\zeta)\backslash(\xi\Gamma \cap \zeta^{-1}P)$. With $\eta \in \zeta\xi\Gamma \cap P$, this is clearly represented by $\zeta^{-1}R_\eta$. Finally we have $\pi_t(\eta\Gamma \cap P) = \eta_t\Delta$, and our last assertion can easily be verified by means of (1.23a, b).

Lemma 8.7. *The notation being as in Lemma 8.6, suppose that $\lambda_t(\gamma)^{[k]} = 1$ for every $\gamma \in \Gamma \cap P$; let $p_\xi \in \mathscr{S}_k^r(\xi\Gamma\xi^{-1} \cap Q)$ for each $\xi \in X$. Then*

$$\Phi^{n-t}\left\{\delta(z)^{k/2-su}\sum_{\xi\in X}E_k^{n,r}(z, s;\, p_\xi,\, \xi\Gamma\xi^{-1})\|_k\xi\right\} \tag{8.4}$$

$$= \delta(w)^{k/2-su}\sum_{\eta\in Y}|c_\eta|^{[k]-2su}E_k^{t,r}\left(w, s;\, q_\eta,\, \eta_t\Delta\eta_t^{-1}\right)\|_k\,\eta_t$$

at least for $\mathrm{Re}(2s) > n + r + 1$, *where $q_\eta = \lambda_r(\zeta)^{[k]}\lambda_t(\eta)^{-[k]}p_\xi\|_k\pi_r(\zeta^{-1})$ and $c_\eta = \lambda_r(\zeta^{-1})\lambda_t(\eta)$ with $\zeta \in Z_\xi$ such that $\eta \in \zeta\xi\Gamma \cap P$.*

Proof. Let us first show (8.4) by formally applying Φ^{n-t} termwise. Let $T_\xi = (\xi\Gamma\xi^{-1} \cap Q)\backslash(\xi\Gamma \cap QP)$. By Lemma 8.5 the left-hand side of (8.4) equals (formally)

$$(8.5)\qquad \sum_{\xi\in X}\sum_{\alpha\in T_\xi} \Phi^{n-t}\left\{\delta(z)^{k/2-su}\left[\delta(z, s; f)\|\alpha\right]\right\}.$$

By Lemma 8.6 $\bigsqcup_{\xi\in X} T_\xi$ can be replaced by $\bigsqcup_{\eta\in Y}\zeta^{-1}R_\eta$. Let $\gamma=\varepsilon\eta\in R_\eta$. Then $\eta^{-1}\varepsilon\eta\in\Gamma\cap P$, and hence $|\lambda_t(\varepsilon)|^u=\lambda_t(\varepsilon)^{[k]}=1$; thus $|\lambda_t(\gamma)|^{[k]}=|\lambda_t(\eta)|^{[k]}$ and $|\lambda_t(\gamma)|^u=|\lambda_t(\eta)|^u$. Therefore, by Lemma 8.5, (8.5) becomes

$$(8.6)\qquad \delta(w)^{k/2-su}\sum_{\eta\in Y}|c_\eta|^{[k]-2su}\sum_{\gamma\in R_\eta}\delta(w, s; q_\eta)\|\pi_t(\gamma)$$

with c_η and q_η as stated in our lemma. Since $\lambda_r(\beta)=\lambda_t(\beta)\lambda_r(\pi_t(\beta))$ for every $\beta\in P\cap Q$, we easily see that $q_\eta\in\mathcal{S}^r_k(\eta_t\Delta\eta_t^{-1}\cap\mathcal{P}^{t,r})$. By Lemma 8.6 $\pi_t(R_\eta)$ gives $(\eta_t\Delta\eta_t^{-1}\cap\mathcal{P}^{t,r})\backslash\eta_t\Delta$ and hence the last sum over γ in (8.6) is $E^{r,t}(w, s; q_\eta, \eta_t\Delta\eta_t^{-1})\|\eta_t$, which proves (8.4) at least in the formal sense. Now the condition $\mathrm{Re}(2s)>n+r+1$ guarantees the absolute and uniform convergence of (1.34) for $E_k^{t,r}$, $r\leq t\leq n$, in a domain of the form

$$(8.7)\qquad D=\{x+iy\in\mathcal{H}_t\,|\,x\in A,\ y_v>p_v \text{ for every } v\in\mathbf{a}\},$$

where A is a compact subset of $S^t_{\mathbf{a}}$ and $0<p_v\in S^t_v$. Therefore each application of Φ can be done termwise, and therefore the above formal proof is meaningful.

Remark 8.8. The condition that $\lambda_t(\gamma)^{[k]}=1$ for every $\gamma\in\Gamma\cap P$ is not absolutely necessary. Without assuming it, we still have (8.4) by replacing Δ by $\pi(\{\gamma\in\Gamma\cap P\,|\,\lambda_t(\gamma)^{[k]}=1\})$, and changing q_η suitably, but the formulation and the proof are somewhat more involved.

Lemma 8.9. (1) *For* $\alpha\in\mathcal{G}^n$ *and* $\mathrm{Re}(2s)>n+r+1$ *we have*

$$\Phi^{n-r}\{\delta(z)^{k/2-su}[E_k^{n,r}(z, s; \Gamma, f)\|_k\alpha]\}=\begin{cases}0 & \text{if } \alpha\notin\Gamma\mathcal{P}^{n,r},\\ \delta^{k/2-su}f & \text{if } \alpha=1.\end{cases}$$

(2) *If* $\mathcal{F}=\delta(z)^{k/2-su}\sum_{\xi\in X}E_k^{n,r}(z, s; p_\xi, \xi\Gamma\xi^{-1})\|_k\xi$ *with* X *and* p_ξ *of Lemma 8.7, then* $\Phi^{n-r}(\mathcal{F}\|_k\eta^{-1})=\delta^{k/2-su}p_\eta$ *for every* $\eta\in X$ *and* $\mathrm{Re}(2s)>n+r+1$.

Proof. Though these are essentially special cases of Lemma 8.7, it is easier to derive them directly from Lemma 8.5. In fact, termwise application of Φ^{n-r} to the function of (1) produces a nonzero quantity only when $\Gamma\alpha\cap\mathcal{P}^{n,r}\neq\emptyset$. If $\alpha=1$, nonvanishshing can appear only from $(\Gamma\cap\mathcal{P}^{n,r})\backslash(\Gamma\cap\mathcal{P}^{n,r})$. Therefore, taking $\beta=\gamma=1$ in Lemma 8.5, we obtain (1). This formal proof can be justified for the same reason as in the proof of Lemma 8.7. Assertion (2) follows immediately from (1), since $\xi\eta^{-1}\in\xi\Gamma\xi^{-1}\mathcal{P}^{n,r}$ only if $\xi=\eta$.

Lemma 8.10. *Suppose that:* (i) $k_0\in\mathbf{Z}$; (ii) $k_0\geqq(n+r+2)/2$ *if* $F\neq\mathbf{Q}$, *and* $k_0>(n+r+3)/2$ *if* $F=\mathbf{Q}$; (iii) k_0 *belongs to the set* $\Lambda(r)$ *of Lemma* 7.2. *Then equality* (8.4) *and the equalities of Lemma* 8.9 *are valid for* $s=k_0/2$.

Proof. Our task is to derive $\Phi^{n-t}\{\mathscr{A}(z, k_0/2)\} = \mathscr{A}'(w, k_0/2)$ from the type of relation $\Phi^{n-t}\{\mathscr{A}(z, s)\} = \mathscr{A}'(w, s)$ established in Lemmas 8.7 and 8.9. Clearly it is sufficient to prove the case $n - t = 1$. We first consider $\Phi\{\mathscr{F}(z, s)\}$ for $\mathscr{F}(z, s) = \delta(z)^{k/2-su}\big[E_k^{n,0}(z, s; 1, \Gamma)\|\alpha\big]$, $\alpha \in G^n$. By Lemma 8.7, for $\mathrm{Re}(2s) > n + 1$ we have $\Phi\{\mathscr{F}(z, s)\} = \mathscr{F}'(z_1, s)$ with a well defined Eisenstein series $\mathscr{F}'$ defined for $z_1 \in \mathscr{H}_{n-1}$. Take Fourier expansions

(8.8a) $$\mathscr{F}(x + iy, s) = \sum_{h \in S} \mathbf{e}_{\mathbf{a}}\big(\mathrm{tr}(hx)\big)c(y, h, s),$$

(8.8b) $$\mathscr{F}'(x_1 + iy_1, s) = \sum_{h_1 \in S^{n-1}} \mathbf{e}_{\mathbf{a}}(\mathrm{tr}(h_1x_1))c(y_1, h_1, s)$$

as in (2.5). Taking a suitable $\mathbf{Z}$-lattice L in S, we can put

(8.9) $$c(y, h, s) = \mathrm{vol}(S_{\mathbf{a}}/L)^{-1} \int_{S_{\mathbf{a}}/L} \mathscr{F}(x + iy, s)\mathbf{e}_{\mathbf{a}}\big(-\mathrm{tr}(hx)\big)dx.$$

Writing $h \in S_{\mathbf{A}}$ in the form $h = \begin{pmatrix} h_1 & h_2 \\ {}^th_2 & h_4 \end{pmatrix}$ with h_1 of size $n - 1$, we may assume that

$$L = \{\, h \in S \mid h_1 \in L_1,\ h_2 \in L_2,\ h_4 \in L_4 \,\}$$

with $\mathbf{Z}$-lattices L_1, L_2, and L_4 in S^{n-1}, F_1^{n-1}, and F. Let $\varepsilon = \begin{pmatrix} 1_n & b \\ 0 & 1_n \end{pmatrix} \in G_{\mathbf{a}}$ with $b = \begin{pmatrix} 0 & b_2 \\ {}^tb_2 & b_4 \end{pmatrix} \in S_{\mathbf{a}}$. Changing α and γ of (8.3) for $\alpha\varepsilon$ and $\gamma\varepsilon$, we easily see that $\Phi(\mathscr{F}\|\varepsilon) = \mathscr{F}'(z_1, s)$ for $\mathrm{Re}(2s) > n + 1$, since $\pi_{n-1}(\varepsilon) = 1$. This shows that $\lim_{\rho\to\infty} \mathscr{F}\big(x + i \cdot \mathrm{diag}[y_1, \rho u], s\big) = \mathscr{F}'(x_1 + iy_1, s)$. Since $\mathscr{F}$ is bounded in D of (8.7), from (8.9) we obtain

$$\lim_{\rho\to\infty} c\big(\mathrm{diag}[y_1, \rho u], h, s\big) = \mathrm{vol}(S_{\mathbf{a}}/L)^{-1} \int_{S_{\mathbf{a}}/L} \mathscr{F}'(x_1 + iy_1, s)\mathbf{e}_{\mathbf{a}}\big(-\mathrm{tr}(hx)\big)dx$$

for $\mathrm{Re}(2s) > n + 1$. This is nonzero only if $h_2 = 0$ and $h_4 = 0$, in which case

(8.10) $$\begin{aligned} &\lim_{\rho\to\infty} c\big(\mathrm{diag}[y_1, \rho u], \mathrm{diag}[h_1, 0], s\big) \\ &= \mathrm{vol}(S_{\mathbf{a}}^{n-1}/L_1)^{-1} \int_{S_{\mathbf{a}}^{n-1}/L_1} \mathscr{F}'(x_1 + iy_1, s)\mathbf{e}_{\mathbf{a}}\big(-\mathrm{tr}(h_1x_1)\big)dx_1 \\ &= c'(y_1, h_1, s), \end{aligned}$$

with a suitable choice of L_1 at the beginning. Suppose now $\det(h_1) \neq 0$. Then, by (2.6a) and Lemma 2.4 we can put

(8.11) $$\begin{aligned} &c\big(\mathrm{diag}[y_1, \rho u], \mathrm{diag}[h_1, 0], s\big) \\ &= a(h, s)\Xi^{n-1}\big(y_1, h_1; s + k_0/2, s - k_0/2\big) + \rho^{d(\kappa(n)-2s)}\eta(y_1, h_1, s), \end{aligned}$$

where $d = [F : \mathbf{Q}]$, $\kappa(n) = (n + 1)/2$, and η is a function defined independently of ρ at least for $\mathrm{Re}(2s) > n + 1$. Then (8.10) shows that

(8.12) $$c'(y_1, h_1, s) = a(h, s)\Xi^{n-1}\big(y_1, h_1; s + k_0/2, s - k_0/2\big)$$

for such s, which implies that $a(h, s)$ has meromorphic continuation to the whole s-plane. Then (8.11) shows that the same is true for $\eta(y_1, h_1, s)$. Now suppose that both $\mathscr{F}(z, k_0/2)$ and $\mathscr{F}'(z_1, k_0/2)$ are holomorphic, which is the case under our assumptions (with $r = 0$) on k_0 by Theorem 7.3(I). Then $c(y, h, k_0/2) = c(h)\mathbf{e}_{\mathbf{a}}\big(i \cdot \mathrm{tr}(hy)\big)$ with a constant $c(h)$, which is nonzero only if $h \in S_+$, and similarly $c'(y_1, h_1, k_0/2) = c'(h_1)\mathbf{e}_{\mathbf{a}}\big(i \cdot \mathrm{tr}(h_1 y_1)\big)$ with a constant $c'(h_1)$. If h_1 is totally positive, then, from (8.11) and (8.12) we see that $\eta(y_1, h_1, s)$ is finite at $s = k_0/2$, and

$$\{c\big(\mathrm{diag}[h_1, 0]\big) - c'(h_1)\}\mathbf{e}_{\mathbf{a}}\big(i \cdot \mathrm{tr}(h_1 y_1)\big) = \rho^{d(\kappa(n)-k_0)}\eta(y_1, h_1, k_0/2).$$

This shows that $c\big(\mathrm{diag}[h_1, 0]\big) = c'(h_1)$ if $k_0 \neq \kappa(n)$ and h_1 is totally positive. Therefore, by Prop. 2.1(3) we obtain $\Phi\{\mathscr{F}(z, k_0/2)\} = \mathscr{F}'(z_1, k_0/2)$, provided $k_0 > (n+1)/2$. This proves the first half of our lemma for $t = n - 1$ and $r = 0$.

Let us now consider the action of Φ on $E_k^{n,r}$ with $r > 0$. Let $\mathscr{A}(z, s)$ denote the function inside the brackets of (8.4) and $\mathscr{A}'(z_1, s)$ the right-hand side of (8.4) with z_1 in place of w. By Lemma 3.7 $\mathscr{A}(z, s)$ is a finite linear combination of functions of the form $\delta(z)^{k/2-su}\big[E(z, s; f, \chi, C)\|_k\alpha\big]$ with $\mathfrak{e} = \mathfrak{c}$ and $\alpha \in G^n$. We may assume that each f is an eigenform as in Section 6, since $\mathscr{S}_k^r(K)$ with K of (3.7b) is spanned by such eigenforms. Changing (m, n) for (n, r) in (7.4) and applying $\|\alpha$, we obtain

$$\nu_\Gamma \cdot c_{r,k}(su - k/2)\big[E\big(z, s; f, \chi, C\big)\|\alpha\big]Z_\tau(2s, \lambda, \chi)\Lambda_{\mathfrak{c}}^{2r}(2s, \chi)^{-1}$$

$$= \chi(\theta)^{-r}N(\mathfrak{b})^{2rs}\mathrm{vol}(\Gamma\backslash\mathscr{H})\langle(f\|\iota)^c(w), \big(E^{n+r}\|\rho(\alpha \times 1_{2r})\big)\big(\mathrm{diag}[z, w], s\big)\rangle.$$

By Lemma 3.6 we have

$$E^{n+r}(Z, s) = \sum_{\eta \in Y} N(\mathfrak{a}_\eta)^{2s}\chi[\eta]E_k^{n+r,0}(Z, s; 1, \eta\Gamma_0\eta^{-1})\|\eta^{-1}$$

with a finite subset Y of G^{n+r} and some Γ_0. Since $\mathrm{diag}[z, w] = \beta\big(\mathrm{diag}[w, z]\big)$ with a suitable $\beta \in G^{n+r}$, we can thus put

(8.13) $$\mathscr{A}(z, s) = \sum_{i \in I} D_i(s)\langle f_i(w), \delta(w)^{su-k/2}\mathscr{F}_i(\mathrm{diag}[w, z], s)\rangle,$$

where I is a finite set of indices, $f_i \in \mathscr{S}_k^r$, D_i is a meromorphic function finite at $s = k_0/2$, and $\mathscr{F}_i$ is a function of type $\mathscr{F}$ as above with $n + r$ in place of n. We can put $\Phi\{\mathscr{F}_i(Z, s)\} = \mathscr{F}_i'(Z_1, s)$ with $\mathscr{F}_i'$ of type $\mathscr{F}'$. Then

(8.14) $$\mathscr{A}'(z_1, s) = \sum_{i \in I} D_i(s)\langle f_i(w), \delta(w)^{su-k/2}\mathscr{F}_i'\big(\mathrm{diag}[w, z_1], s\big)\rangle,$$

for $\mathrm{Re}(2s) > n + r + 1$, since $\mathscr{F}_i$ is bounded on any strip of type (8.7). Analytic continuation validates (8.14) for every s where the functions are finite. Evaluate (8.13) and (8.14) at $s = k_0/2$ and apply Φ to (8.13) at $s = k_0/2$.

Since $\Phi\{\mathscr{F}_i(Z, k_0/2)\} = \mathscr{F}'_i(Z_1, k_0/2)$, we find (using (7.7), for example) that $\Phi\{\mathscr{A}(z, k_0/2)\} = \mathscr{A}'(z_1, k_0/2)$, which proves the case $r > 0$. This completes the proof, as our assertion for the equalities of Lemma 8.9 are merely special cases.

We now put, with $k = k_0 u$, $k_0 \in (1/2)\mathbf{Z}$,

$$E_k^{n,r}(z; f, \Gamma) = E_k^{n,r}(z, k_0/2; f, \Gamma). \tag{8.15}$$

whenever the right-hand side is finite. In general this may not be holomorphic in z. We denote by $\tilde{\mathscr{E}}_k^{n,r}$ the vector space spanned over $\mathbf{C}$ by $E_k^{n,r}(z; f, \Gamma)\|_k\alpha$ for all $\alpha \in \mathscr{G}^n$, all $f \in \mathscr{S}_k^r(\Gamma \cap \mathscr{P}^{n,r})$, and for all congruence subgroups Γ of $\mathscr{G}^n$; we then put $\mathscr{E}_k^{n,r} = \tilde{\mathscr{E}}_k^{n,r} \cap \mathscr{M}_k^n$. Clearly these spaces are stable under the operator $\|_k\xi$ for every $\xi \in \mathscr{G}^n$. From (1.35) we see that $\mathscr{E}_k^{n,n} = \mathscr{S}_k^n$.

Lemma 8.11. *Given $k_0 \in (1/2)\mathbf{Z}$, suppose that $k_0 > 2n$ if $k_0 \notin \mathbf{Z}$ and that k_0, n, and r are as in Lemma 7.10 if $k_0 \in \mathbf{Z}$. Then the following assertions hold:*

(1) $\mathscr{E}_k^{n,r} = \tilde{\mathscr{E}}_k^{n,r}$.
(2) $\Phi^{n-t}(\mathscr{E}_k^{n,r}) \subset \mathscr{E}_k^{t,r}$ *for* $r \leqq t \leqq n$ *and in particular* $\Phi^{n-r}(\mathscr{E}_k^{n,r}) = \mathscr{S}_k^r$.
(3) $\Phi^s(\mathscr{E}_k^{n,r}) = 0$ *if* $s > n - r$.
(4) *If* $g \in \mathscr{E}_k^{n,r}$ *and* $\Phi^{n-r}(g\|_k\alpha) = 0$ *for every* $\alpha \in \mathscr{G}^n$, *then* $g = 0$.

Proof. The first assertion follows immediately from Theorem 6.3(I), and the second one from Lemmas 8.7, 8.9, and 8.10. Then $\Phi^{n-r+1}(\mathscr{E}_k^{n,r}) = \Phi(\mathscr{S}_k^r) = 0$, which gives (3). To prove (4), let $g = \sum_{i \in I} E_k^{n,r}(z; f_i, \Gamma_i)\|\alpha_i$ with a finite set of indices I, $\alpha_i \in \mathscr{G}^n$, $f_i \in \mathscr{S}_k^r(\Gamma_i \cap \mathscr{P}^{n,r})$, and congruence subgroups Γ_i. Take a congruence subgroup Γ so that $\Gamma \subset \bigcap_{i \in I} \alpha_i^{-1}\Gamma_i\alpha_i$ and $\lambda_t(\Gamma \cap \mathscr{P}^{n,t})^{[k]} = 1$ for every $t \geqq r$. For each i we can find a finite set B_i such that $\Gamma_i\alpha_i = \bigsqcup_{\beta \in B_i} \beta\Gamma$. Then $\beta\Gamma\beta^{-1} \subset \Gamma_i$ for every $\beta \in B_i$, and

$$g = \sum_{i \in I} \sum_{\beta \in B_i} E_k^{n,r}(z; c_{i,\beta} f_i, \beta\Gamma\beta^{-1})\|\beta$$

with some constants $c_{i,\beta}$. Thus we may assume at the beginning that $\Gamma_i = \alpha_i\Gamma\alpha_i^{-1}$ for every $i \in I$ with some Γ. Now if $\alpha \in \sigma\xi\Gamma$ with $\sigma \in \mathscr{P}^{n,r}$ and $\xi \in \mathscr{G}$, then

$$E_k^{n,r}(z; f, \alpha\Gamma\alpha^{-1})\|\alpha = \lambda_r(\sigma)^{-[k]} E_k^{n,r}\big(z; f\|\pi_r(\sigma), \xi\Gamma\xi^{-1}\big)\|\xi.$$

Therefore, with $X = \mathscr{P}^{n,r}\backslash\mathscr{G}/\Gamma$ we can put $g = \sum_{\xi \in X} E_k^{n,r}(z; h_\xi, \xi\Gamma\xi^{-1})\|\xi$. Suppose that $\Phi^{n-r}(g\|\alpha) = 0$ for every $\alpha \in \mathscr{G}$. Then by Lemmas 8.9(2) and 8.10 we have $h_\eta = \Phi^{n-r}(g\|\eta^{-1}) = 0$ for every $\eta \in X$, so that $g = 0$, which proves (4).

We are now ready to state our main results on the structure of the spaces of Eisenstein series.

Theorem 8.12. *Let $k = k_0 u$ with $k_0 \in (1/2)\mathbf{Z}$ such that: $k_0 > 2n$ if $k_0 \notin \mathbf{Z}$; if $k_0 \in \mathbf{Z}$, then $k_0 \in \bigcap_{r=1}^{n-1} \Lambda(r)$ with $\Lambda(r)$ as in Lemma 7.2, and moreover $k_0 > n$ if $F \neq \mathbf{Q}$ and $k_0 > n+1$ if $F = \mathbf{Q}$. Then $\mathscr{M}_k^n = \bigoplus_{r=0}^n \mathscr{E}_k^{n,r}$ and $\mathscr{M}_k^n(\Gamma) = \bigoplus_{r=0}^n \big[\mathscr{M}_k^n(\Gamma) \cap \mathscr{E}_k^{n,r}\big]$ for every congruence subgroup Γ of $\mathscr{G}^n$.*

Proof. Suppose that $\sum_{r=0}^{n} p_r = 0$ with $p_r \in \mathscr{E}_k^{n,r}$. By Lemma 8.11(2) we have $\Phi^n(p_0\|\alpha) = -\sum_{r=1}^{n} \Phi^n(p_r\|\alpha) = 0$ for every $\alpha \in \mathscr{G}$, and hence $p_0 = 0$ by Lemma 8.11(3). Similarly we find that $\Phi^{n-1}(p_1\|\alpha) = 0$ for every $\alpha \in \mathscr{G}$, which means that $p_1 = 0$ for the same reason. Repeating this process, we obtain $p_r = 0$ for every r, which proves that the $\mathscr{E}_k^{n,r}$ for $0 \leqq r \leqq n$ form a direct sum. Now given $g \in \mathscr{M}_k^n$, take Γ, so that $g \in \mathscr{M}_k^n(\Gamma)$ and $\lambda_r(\Gamma \cap \mathscr{P}^{n,r})^{[k]} = 1$ for every r; take also $X_r = \mathscr{P}^{n,r}\backslash\mathscr{G}/\Gamma$. Put $c_\xi = \Phi^n(g\|\xi^{-1})$ for each $\xi \in X_0$ and $f_0 = \sum_{\xi\in X_0} E_k^{n,0}(z;\, c_\xi,\, \xi\Gamma\xi^{-1})\|\xi$. Then by Lemma 8.9(2) and Lemma 8.10 $\Phi^n\big((g - f_0)\|\xi^{-1}\big) = 0$ for every $\xi \in X_0$. By Lemma 8.2(1) we have $\Phi^{n-1}\big((g - f_0)\|\alpha\big) \in \mathscr{S}_k^1$ for every $\alpha \in \mathscr{G}$. Put $p_\eta = \Phi^{n-1}\big((g - f_0)\|\eta^{-1}\big)$ for $\eta \in X_1$, and $f_1 = \sum_{\eta\in X_1} E_k^{n,1}\big(z;\, p_\eta,\, \eta\Gamma\eta^{-1}\big)\|\eta$. Then $\Phi^{n-1}\big((g - f_0 - f_1)\|\eta^{-1}\big) = 0$ for every $\eta \in X_1$, and hence $\Phi^{n-2}\big((g - f_0 - f_1)\|\alpha\big) \in \mathscr{S}_k^2$ for every $\alpha \in \mathscr{G}$ by Lemma 8.2(1). Continuing in this fashion, we find $f_r \in \mathscr{E}_k^{n,r}$ for $r \leqq n-1$ so that if we put $f_n = g - \sum_{r=0}^{n-1} f_r$, then $\Phi(f_n\|\alpha) = 0$ for every $\alpha \in X_{n-1}$, which means that $f_n \in \mathscr{S}_k^n = \mathscr{E}_k^{n,n}$. This proves the first equality. If $g \in \mathscr{M}_k^n(\Gamma)$, then $\sum_{r=o}^{n} f_r\|\gamma = \sum_{r=o}^{n} f_r$ for every $\gamma \in \Gamma$, so that $f_r\|\gamma = f_r$, that is, $f_r \in \mathscr{M}_k^n(\Gamma) \cap \mathscr{E}_k^{n,r}$, which completes the proof.

Theorem 8.13. *Let $\mathscr{E}_k^n = \sum_{r=0}^{n-1} \mathscr{E}_k^{n,r}$ with $k = k_0 u$ as in Theorem 8.12. Then*

$$\mathscr{E}_k^n = \{f \in \mathscr{M}_k^n \,|\, \langle f, g\rangle = 0 \text{ for every } g \in \mathscr{S}_k^n\},$$

$$\mathscr{E}_k^{n,r} = \{ f \in \mathscr{E}_k^n \,|\, \Phi(f\|_k\alpha) \in \mathscr{E}_k^{n-1,r} \text{ for every } \alpha \in \mathscr{G}^n\} \text{ if } r < n.$$

Proof. If $f \in \mathscr{E}_k^{n,r}$, $r < n$, and $g \in \mathscr{S}_k$, then $\langle f, g\rangle = 0$ by Lemma 2.2. This together with Theorem 8.12 gives the first equality. Also, by Lemma 8.11(2) $\Phi(f\|\alpha) \in \mathscr{E}_k^{n-1,r}$ for every $\alpha \in \mathscr{G}$. Conversely suppose that $f = \sum_{s=0}^{n-1} g_s$ with $g_s \in \mathscr{E}_k^{n,s}$ and that $\Phi(f\|\alpha) \in \mathscr{E}_k^{n-1,r}$ for every $\alpha \in \mathscr{G}$. Since $\Phi(g_s\|\alpha) \in \mathscr{E}_k^{n-1,s}$ and the $\mathscr{E}_k^{n-1,s}$ for $0 \leqq s < n$ form a direct sum, we obtain $\Phi(g_s\|\alpha) = 0$ for $s \neq r$. By Lemma 8.2(2) $g_s \in \mathscr{E}_k^{n,s} \cap \mathscr{S}_k^n = \{0\}$ for $s \neq r$, and hence $f = g_r \in \mathscr{E}_k^{n,r}$, which proves the second equality.

Theorem 8.14. *Let k be as in Theorem 8.12. Given a congruence subgroup Γ of $\mathscr{G}^n$, let $\mathscr{E}_k^{n,r}(\Gamma) = \mathscr{M}_k(\Gamma) \cap \mathscr{E}_k^{n,r}$ and let $X = \mathscr{P}^{n,r}\backslash\mathscr{G}^n/\Gamma$ with a fixed $r < n$. Then $f \mapsto \big(\Phi^{n-r}(f\|\xi^{-1})\big)_{\xi\in X}$ gives a $\mathbf{C}$-linear isomorphism of $\mathscr{E}_k^{n,r}(\Gamma)$ onto $\prod_{\xi\in X} \mathscr{S}_k^r\big(\xi\Gamma\xi^{-1} \cap \mathscr{P}^{n,r}\big)$. Moreover, if $f \in \mathscr{E}_k^{n,r}(\Gamma)$ and $p_\xi = \Phi^{n-r}(f\|\xi^{-1})$, then $f = \sum_{\xi\in X} E_k^{n,r}(z;\, p_\xi,\, \xi\Gamma\xi^{-1})\|_k\xi$.*

Proof. If $f \in \mathscr{E}_k^{n,r}(\Gamma)$, then $\Phi^{n-r}(f\|\xi^{-1}) \in \mathscr{S}_k^r\big(\xi\Gamma\xi^{-1} \cap \mathscr{P}^{n,r}\big)$ by Lemma 8.1 and Lemma 8.11(2), and so our map is meaningful. The injectivity of the map follows from Lemma 8.2(1) and Lemma 8.11(3). Now, given $p_\xi \in \mathscr{S}_k^r\big(\xi\Gamma\xi^{-1} \cap \mathscr{P}^{n,r}\big)$ for each $\xi \in X$, put $g = \sum_{\xi\in X} E_k^{n,r}(z;\, p_\xi,\, \xi\Gamma\xi^{-1})\|_k\xi$. Then $g \in \mathscr{E}_k^{n,r}(\Gamma)$ and $\Phi^{n-r}(g\|\xi^{-1}) = p_\xi$ for every $\xi \in X$ by Lemma 8.10, which proves the surjectivity.

Remark 8.15. If the conjecture made in Remark 7.6 in the case $k_0 \in \mathbf{Z}$ is true, then we can eliminate the condition $k_0 \in \bigcap_{r=1}^{n-1} \Lambda(r)$ in Theorem 8.12. Our results for $k_0 \notin \mathbf{Z}$ in this section are restricted to the cases in which the Eisenstein series are convergent, but naturally it is expected that they can be extended in the same manner as in the case of integral k_0. If $n = 1$, we obtained far more complete results in [85b] for both integral and half-integral k.

The space $\mathscr{E}_k^{n,r}(\Gamma)$ can occasionally vanish even for large k_0. For example, it can easily be seen that $\mathscr{M}_k^n(\Gamma) = \mathscr{S}_k^n(\Gamma)$ if n is even, k_0 is odd, and Γ is the principal congruence subgroup of $Sp(n, \mathbf{Z})$ of level 2.

Correction to [94a]. Page 564, line 10 from the bottom: "the first set" should read "the set of possible poles described in the case in which $\chi^2 = 1$ and $c \neq g$."

References

[AK] A.N. Andrianov, V.L. Kalinin: On the analytic properties of standard zeta functions of Siegel modular forms, Mat. Sb. 106 (148) (1978) 323-339; English transl. Math. USSR Sb. **35** (1979) 1–17

[B] S. Böcherer: Über die Funktionalgleichung automorpher L-Funktionen zur Siegelschen Modulgruppe, J. Math. **362** (1985) 146–168

[DHL] W. Duke, R. Howe, J.-S. Li: Estimating Hecke Eigenvalues of Siegel modular forms, Duke M.J. **67** (1992) 219–240

[Ga1] P.B. Garrett: Pullbacks of Eisenstein series: Applications, Automorphic Forms of Several Variables, Taniguchi Symposium, Katata, 1983, Birkhäuser, 1984

[Ga2] P.B. Garrett: Integral representations of Eisenstein series and L-functions. Number theory, trace formulas, and discrete groups, Symp. in honor of A. Selberg, 1987, Academic Press, 1989, pp. 241–264

[Go] R. Godement, Fonctions automorphes, Séminaire Cartan 1957-58, exposés 6 and 10

[H] M. Harris: Eisenstein series on Shimura varieties, Ann. Math. **119** (1984) 59–94

[K] H. Klingen: Zum Darstellungssatz für Siegelsche Modulformen. Math. Z. **102** (1967) 30–43; Berichtigung, Math. Z. **105** (1968) 399–400

[KR] S. Kudla, S. Rallis: Poles of Eisenstein series and L-functions, Israel Mathematical Conference Proceedings, vol.3, 1990, pp 81–110

[PR] I. Piatetski-Shapiro, S. Rallis: ε-factors of representations of classical groups. Proc. Nat. Acad. Sci. **83** (1986) 4589–4593

[71] G. Shimura: Introduction to the arithmetic theory of automorphic functions, Publ. Math. Soc. Japan, No.11, Iwanami Shoten and Princeton University Press, 1971

[82] G. Shimura: Confluent hypergeometric functions on tube domains. Math. Ann. **260** (1982) 269–302

[83a] G. Shimura: Algebraic relations between critical values of zeta functions and inner products Am J. Math. **104** (1983) 253–285

[83b] G. Shimura: On Eisenstein series, Duke Math. J. **50** (1983) 417–476

[84a] G. Shimura: Differential operators and the singular values of Eisenstein series. Duke Math. J. **51** (1984) 261–329

[84b] G. Shimura: On differential operators attached to certain representations of classical groups. Invent. math., **77** (1984) 463–488

[85a] G. Shimura: On Eisenstein series of half-integral weight, Duke Math. J. **52** (1985) 281–314

[85b] G. Shimura: On the Eisenstein series of Hilbert modular groups. Rev. Mat. Iberoamericana 1, No. 3, (1985) 1–42

[86] G. Shimura: On a class of nearly holomorphic automorphic forms. Ann. Math. **123** (1986) 347–406

[87] G. Shimura: Nearly holomorphic functions on hermitian symmetric spaces. Math. Ann. **278** (1987) 1–28

[91] G. Shimura: The critical values of certain Dirichlet series attached to Hilbert modular forms, Duke Math. J. **63** (1991) 557–613

[93] G. Shimura: On the transformation formulas of theta series. Am. J. M. **115** (1993) 1011–1052

[94a] G. Shimura: Euler products and Fourier coefficients of automorphic forms on symplectic groups. Invent. math. **116** (1994) 531–576

[94b] G. Shimura: Differential operators, holomorphic projection, and singular forms. Duke Math. J. (to appear)

95b

Zeta functions and Eisenstein series on metaplectic groups

Inventiones mathematicae, 121 (1995), 21-60

Introduction

The main objectives of this paper are: (1) to construct an Euler product from a Hecke eigenform of half-integral weight; (2) to prove its meromorphic continuation to the whole plane; (3) to examine its relationship with the Fourier coefficients of the form; and (4) to apply these results to the investigations of Eisenstein series of half-integral weight. The automorphic forms are holomorphic ones, considered on the metaplectic cover $M^n_{\mathbf{A}}$ of $G^n_{\mathbf{A}}$, where $G^n = Sp(n, F)$ with a totally real algebraic number field F. The first three are generalizations of the corresponding results in the elliptic and Hilbert modular cases obtained in [S2] and [S6]. At the same time all four closely parallel similar results in the case of integral weight in [S9] and [S10].

To describe our results, let us take for simplicity F to be $\mathbf{Q}$, in which case the weight k is half an odd positive integer. We consider a holomorphic form f of weight k, with respect to a principal congruence subgroup of $Sp(n, \mathbf{Z})$ of level N, which is an eigenform in the sense that $f|T(m) = \lambda(m)f$ with a complex number $\lambda(m)$, where N is a multiple of 4 and $T(m)$ is a Hecke operator defined for each positive integer m in a natural way. As our first main result we shall prove (Theorem 4.4) that

$$(1) \qquad \sum_m \lambda(m)m^{-s} = \prod_p W_p(p^{-s})^{-1} \prod_{i=1}^{n} (1 - p^{2i-1-2s}),$$

where m (resp. p) runs over all positive integers (resp. prime numbers) prime to N and W_p is a polynomial with constant term 1 and of degree $2n$. In fact, we can establish the Euler factors W_p even for the prime factors p of N if f is of a certain special type. It should be noted that the corresponding Euler products for integral weight have degree $2n + 1$ instead of $2n$.

Next, given a Dirichlet character η, put

$$(2) \qquad Z(s, \lambda, \eta) = \prod_p W_p(\eta(p)p^{-s})^{-1}.$$

Then we shall show (Theorem 6.1) that *if f is a cusp form, then $Z(s,\lambda,\eta)$ times explicitly defined gamma factors can be continued to a meromorphic function on the whole plane with finitely many poles, which are all simple. In particular the product is entire if $\eta^2 \neq 1$.*

Our third main result (Theorem 5.1, Proposition 5.4) asserts that if $f(z) = \sum_\sigma \mu(\sigma)\exp(2\pi i \cdot \mathrm{tr}(\sigma z))$ is the Fourier expansion of f with σ running over symmetric rational matrices of size n and z in the Siegel half-space H^n, then

$$(3) \qquad g(s)\mathscr{L}(s)\sum_\xi \mu({}^t\xi\tau\xi)\det(\xi)^{-[k]}|\det(\xi)|^{n-s+1/2} = h(s)Z(s,\lambda,1).$$

Here τ is an arbitrarily fixed positive definite rational matrix of size n, ξ runs over all integral nonsingular matrices, congruent to 1 modulo N, modulo right multiplication by the elements $\gamma \in GL_n(\mathbf{Z})$ such that $\gamma \equiv 1_n \pmod{N}$, g and h are finite Dirichlet series, and $\mathscr{L}$ is a product of certain Hecke L-functions.

Our fourth problem concerns Eisenstein series of the following two types:

$$(4) \qquad E_k^{n,r}(z,s;f,\Gamma) = \sum_{\alpha\in A}\left\{f(\omega_r(z))\,[\Delta(z)/\Delta(\omega_r(z))]^{s-k/2}\right\}\|_k\alpha,$$

$$(5) \quad E(z,s;f,\eta,\Gamma) = \sum_{\alpha\in A}\eta(\det(d_\alpha))\left\{f(\omega_r(z))\,[\Delta(z)/\Delta(\omega_r(z))]^{s-k/2}\right\}\|_k\alpha.$$

Here $0 < r < n$, Γ is a congruence subgroup of G^n, f is a cusp form of weight k on $G^n, A = (\Gamma \cap P^{n,r})\backslash\Gamma$ with the parabolic subgroup $P^{n,r}$ consisting of the elements of G^n whose lower left $(n-r)\times(n+r)$-block is 0, $\Delta(z) = \det(\mathrm{Im}(z))$, $\omega_r(z)$ is the upper left $(r\times r)$-block of z, and η is a Dirichlet character such that $\eta(-1) = (-1)^{[k]}$; Γ must be of a special type in (5); f must satisfy certain conditions in each case. We first prove that if f is a Hecke eigenform, then

$$(6) \qquad \Lambda(s)Z(2s,\lambda,\eta)E(z,s;f,\eta,\Gamma) = \langle f'(w), E^{n+r}(\mathrm{diag}[z,w],s)\rangle.$$

Here Λ is a product of Hecke L-functions and gamma functions, f' is an easy transform of $f,(z,w) \in H^n \times H^r$, and E^{n+r} is a series on G^{n+r} of type (5) with $(n+r,0)$ in place of (n,r) transformed by an element of G^{n+r}. This result implies that the left-hand side of (6) times suitable gamma functions has meromorphic continuation on the whole plane with finitely many poles, which are all simple (Theorem 6.3). Then we shall show that the values (or residues) of (4) and (5) for integer or half-integer values of s in a certain interval are holomorphic or nearly holomorphic (Theorems 6.4, 6.5, and 6.6).

The problem of establishing the Euler product of (1) can be reduced to the computation of a certain integral (or an infinite sum) over a lift of $P_{\mathbf{A}}$ into $M^n_{\mathbf{A}}$, where $P = P^{n,0}$. The crucial point is how to choose the lift, which must be compatible with right and left action of the lift of a fixed compact subgroup of $G_{\mathbf{A}}$, and which is not a group homomorphism. It should be observed that the standard lift that is a homomorphism does not serve the purpose. The existence of the desired lift is nontrivial, but not so difficult compared with the explicit description of the lift of each parabolic element, which is essential for the calculation of the Euler factors. In the elliptic modular case, the point is to show that the desired lift of $\begin{pmatrix} 1 & h/p \\ 0 & 1 \end{pmatrix}$ for an odd prime p and $h \in \mathbf{Z}, 0 < h < p$, is specified by a fourth root of unity $p^{-1/2}\sum_{m=1}^{p-1}\exp(2\pi i h m^2/p)$, as shown in [S2]. Thus our task is to determine such a root of unity for each element of $P_{\mathbf{A}}$. The final formula is not complicated, but requires a lengthy proof; in fact, we devote the whole Section 3 to this question.

Once the lift of $P_{\mathbf{A}}$ is established, our objectives can be accomplished in much the same fashion as in the case of integral weight. However, a few non-trivial modifications are necessary. One of them is the analysis of the pullback of an Eisenstein series on $M_{\mathbf{A}}^{n+r}$ to $M_{\mathbf{A}}^{n} \times M_{\mathbf{A}}^{r}$ which leads to equality (6). This is complicated even in the case of integral weight, but more intricate in the present case, because of the subtlety of the factor of automorphy mentioned above. Another is the involvement of an infinite series of type

$$\sum_{\sigma} \mathbf{g}(\sigma)\exp(-2\pi i \cdot \mathrm{tr}(\tau\sigma))\,\nu(\sigma)^{-s},$$

where σ runs over all $\mathbf{Z}_2$-integral rational symmetric matrices modulo integral ones, $\mathbf{g}(\sigma)$ is the signature of the Gauss sum of σ, and $\nu(\sigma)$ is the denominator of σ in Siegel's sense. The crucial fact is that this is a product of certain L-functions. The series without $\mathbf{g}(\sigma)$ was employed when the weight was integral. Both series were investigated in our previous papers [S4, S5] and also by Feit [F]. Fortunately, the results in these papers are completely adequate for our purpose.

In addition to the four problems treated in this paper, there are two more types of results obtained in [S9] and [S10] whose analogues we do not pursue here. We shall make some comments on these at the end of the paper.

1. Factors of automorphy and modular forms

We use the same notation and terminology as in our previous papers [S9] and [S10]; we recall here some of the basic symbols and conventions. For an associative ring R with identity element and an R-module M we denote by $R^{\times}$ the group of all its invertible elements and by M_n^m the R-module of all $m \times n$-matrices with entries in M. We put $\tilde{x} = {}^t x^{-1}$ if a matrix x is square and invertible.

We fix a totally real algebraic number field F of finite degree and denote by $\mathbf{a}, \mathbf{h}, \mathfrak{g}$, and $\mathfrak{d}$ the sets of archimedean primes and nonarchimedean primes of F, the maximal order of F, and the different of F relative to $\mathbf{Q}$. For a fractional ideal $\mathfrak{r}$ in F and $t \in F_{\mathbf{A}}^{\times}$ we denote by $t\mathfrak{r}$ the fractional ideal in F such that $(t\mathfrak{r})_v = t_v\mathfrak{r}_v$ for every $v \in \mathbf{h}$; we put $|t_v| = |t_v|_v$ with the normalized valuation $|\ \ |_v$ at v and $|t|_{\mathbf{A}} = \prod_{v \in \mathbf{a} \cup \mathbf{h}} |t_v|$. Throughout the paper we denote by δ an arbitrarily fixed element of $F_{\mathbf{h}}^{\times}$ such that $\delta\mathfrak{g} = \mathfrak{d}$. Given a set X, we denote by $X^{\mathbf{a}}$ the product of $\mathbf{a}$ copies of X, that is, the set of all indexed elements $(x_v)_{v \in \mathbf{a}}$ with x_v in X. For example, for $t \in F_{\mathbf{A}}^{\times}$ we define an element $\mathrm{sgn}(t)$ of $\mathbf{R}^{\mathbf{a}}$ by $\mathrm{sgn}(t) = (t_v/|t_v|)_{v \in \mathbf{a}}$.

With $\mathbf{T} = \{\zeta \in \mathbf{C} \mid |\zeta| = 1\}$ we define characters $\mathbf{e_A} : F_\mathbf{A} \to \mathbf{T}$ and $\mathbf{e}_v : F_v \to \mathbf{T}$ for each $v \in \mathbf{a} \cup \mathbf{h}$ as in [S9]. We then put $\mathbf{e_h}(x) = \mathbf{e_A}(x_\mathbf{h})$ for $x \in F_\mathbf{A}$ and $\mathbf{e_a}(x) = \mathbf{e}\left(\sum_{v \in \mathbf{a}} x_v\right)$ for $x \in F_\mathbf{A}$ or $x \in \mathbf{C}^\mathbf{a}$, where $\mathbf{e}(c) = \exp(2\pi i c)$ for $c \in \mathbf{C}$.

With a positive integer n we now put

$$G = G^n = \{\alpha \in GL_{2n}(F) \mid {}^t\alpha \iota \alpha = \iota\}, \quad \iota = \iota_n = \begin{pmatrix} 0 & -1_n \\ 1_n & 0 \end{pmatrix},$$

$$P = P^n = \{\alpha \in G \mid c_\alpha = 0\}, \quad \Omega = \Omega^n = \{x \in G_\mathbf{A} \mid \det(c_x) \in F_\mathbf{A}^\times\},$$

$$H = H^n = \{z \in \mathbf{C}_n^n \mid {}^t z = z, \operatorname{Im}(z) > 0\}, \quad \mathscr{H} = \mathscr{H}^n = (H^n)^\mathbf{a}.$$

Here and throughout the paper, $a_\alpha, b_\alpha, c_\alpha$, and d_α denote the standard a-, b-, c-, and d-blocks of α; also we write $x > 0$ (resp. $x \geqq 0$) to indicate that x is positive definite (resp. nonnegative). We drop the superscript n or the subscript n whenever the dimension is clear from the context, but reinstate it when it needs to be stressed. This applies to other symbols $M_\mathbf{A}^n, \mathscr{G}^n, \mathscr{M}_k^n$, etc. defined below. We put $X = F_n^1$ and let G act on $X \times X = F_{2n}^1$ by right multiplication. Then we can define the metaplectic groups $Mp(X_\mathbf{A})$ and $Mp(X_v)$ in Weil's sense for each $v \in \mathbf{a} \cup \mathbf{h}$ with respect to the alternating form $(x, y) \mapsto x\iota_n \cdot {}^t y$ on $F_{2n}^1 \times F_{2n}^1$. Recall that these groups, written $M_\mathbf{A}^n$ and M_v^n for simplicity, are groups of unitary transformations on $L^2(X_\mathbf{A})$ and on $L^2(X_v)$; there is an exact sequence

$$1 \to \mathbf{T} \to M_\mathbf{A}^n \to G_\mathbf{A} \to 1$$

and also a similar exact sequence for each v with M_v and G_v in place of $M_\mathbf{A}$ and $G_\mathbf{A}$. We denote by pr the projection maps of $M_\mathbf{A}$ and M_v to $G_\mathbf{A}$ and G_v. There is a natural lift $r : G \to M_\mathbf{A}$ by which we can consider G a subgroup of $M_\mathbf{A}$. There are also two types of lifts

(1.1) $$r_P : P_\mathbf{A} \to M_\mathbf{A}, \quad r_\Omega : \Omega \to M_\mathbf{A},$$

which have the following properties:

(1.2a) $$r_P = r \text{ on } P; \quad r_\Omega = r \text{ on } G \cap \Omega;$$

(1.2b) $$r_\Omega(\alpha\beta\gamma) = r_P(\alpha) r_\Omega(\beta) r_P(\gamma) \quad \text{if } \alpha, \gamma \in P_\mathbf{A}, \text{ and } \beta \in \Omega.$$

We insert here an easy fact:

(1.3) *If* $x, y \in M_\mathbf{A}$, $\operatorname{pr}(x) \in G_\mathbf{h}$, and $\operatorname{pr}(y) \in G_\mathbf{a}$, then $xy = yx$.

For $\gamma \in G_v, v \in \mathbf{a}$, and $w \in H$ we define $\gamma w = \gamma(w) \in H, \Delta(w) \in \mathbf{R}$, and a factor of automorphy $j(\gamma, w)$ by

(1.4a) $$\gamma w = \gamma(w) = (a_\gamma w + b_\gamma)(c_\gamma w + d_\gamma)^{-1}, \quad \Delta(w) = \det(\operatorname{Im}(w)),$$

(1.4b) $$j(\gamma, w) = \det(c_\gamma w + d_\gamma).$$

For $\alpha \in G_{\mathbf{A}}$ and $z = (z_v)_{v \in \mathbf{a}} \in \mathscr{H}$ we put

(1.5a) $$\alpha z = \alpha(z) = (\alpha_v z_v)_{v \in \mathbf{a}}, \quad \Delta(z) = (\Delta(z_v))_{v \in \mathbf{a}},$$

(1.5b) $$j_\alpha(z) = j(\alpha, z) = (j(\alpha_v, z_v))_{v \in \mathbf{a}}.$$

All these are elements of $(\mathbf{C}_n^n)^{\mathbf{a}}$ or $\mathbf{C}^{\mathbf{a}}$. For two fractional ideals $\mathfrak{x}$ and $\mathfrak{y}$ in F such that $\mathfrak{x}\mathfrak{y} \subset \mathfrak{g}$ we put

(1.6a) $$D[\mathfrak{x}, \mathfrak{y}] = D^n[\mathfrak{x}, \mathfrak{y}] = G_{\mathbf{a}} \prod_{v \in \mathbf{h}} D_v^n[\mathfrak{x}, \mathfrak{y}] \ (\subset G_{\mathbf{A}}),$$

(1.6b) $$D_v^n[\mathfrak{x}, \mathfrak{y}] = \{x \in G_v \mid a_x \in (\mathfrak{g}_v)_n^n, \ b_x \in (\mathfrak{x}_v)_n^n, \ c_x \in (\mathfrak{y}_v)_n^n, \ d_x \in (\mathfrak{g}_v)_n^n\}.$$

By a *half-integral weight* we mean an element k of $\mathbf{Q}^{\mathbf{a}}$ such that $k - u/2 \in \mathbf{Z}^{\mathbf{a}}$, where u denotes the identity element $(1, \ldots, 1)$ of the ring $\mathbf{C}^{\mathbf{a}}$. An *integral weight* is an element of $\mathbf{Z}^{\mathbf{a}}$. For a weight k we put

(1.7) $$[k] = k \text{ if } k \in \mathbf{Z}^{\mathbf{a}}; \quad [k] = k - u/2 \text{ if } k \notin \mathbf{Z}^{\mathbf{a}}.$$

In [S5] and [S7] we defined a factor of automorphy h of weight $u/2$. To recall its properties, let us first put

(1.8a) $$\mathfrak{M} = \mathfrak{M}^n = \{\alpha \in M_{\mathbf{A}} \mid \mathrm{pr}(\sigma) \in P_{\mathbf{A}} C^{\theta}\}, \quad C^{\theta} = C^{n,\theta} = G_{\mathbf{a}} \prod_{v \in \mathbf{h}} C_v^{\theta},$$

(1.8b) $$C_v^{\theta} = \{\xi \in D_v[\mathfrak{d}^{-1}, \mathfrak{d}] \mid (a_\xi \cdot {}^t b_\xi)_{ii} \in 2\mathfrak{d}_v^{-1} \text{ and}$$
$$(c_\xi \cdot {}^t d_\xi)_{ii} \in 2\mathfrak{d}_v \text{ for } 1 \leqq i \leqq n\},$$

where x_{ii} denotes the (i,i)-entry of x. We note that $D[2\mathfrak{d}^{-1}, 2\mathfrak{d}] \subset C^{\theta}$.

For $\alpha = \mathrm{pr}(\sigma)$ with $\sigma \in M_{\mathbf{A}}$ and $z \in \mathscr{H}$ we put $a_\sigma = a_\alpha$, $b_\sigma = b_\alpha$, $c_\sigma = c_\alpha$, $d_\sigma = d_\alpha$, and $\sigma(z) = \sigma z = \alpha z$. Now for each $\sigma \in \mathfrak{M}$ we have a holomorphic function $h_\sigma(z) = h(\sigma, z)$ of $z \in \mathscr{H}$ with the following properties:

(1.9a) $$h(\sigma, z)^2 = \zeta \cdot j(\mathrm{pr}(\sigma), z)^u \text{ with } \zeta \in \mathbf{T}; \quad h(\sigma, z) \in \mathbf{T} \text{ if } \mathrm{pr}(\sigma)_{\mathbf{a}} = 1;$$

(1.9b) $$h(t \cdot r_P(\gamma), z) = t^{-1} |\det(d_\gamma)_{\mathbf{a}}|_{\mathbf{A}}^{1/2} \text{ if } t \in \mathbf{T} \text{ and } \gamma \in P_{\mathbf{A}};$$

(1.9c) $$h(\rho\sigma\tau, z) = h(\rho, z) h(\sigma, \tau z) h(\tau, z) \text{ if } \mathrm{pr}(\rho) \in P_{\mathbf{A}} \text{ and } \mathrm{pr}(\tau) \in C^{\theta}.$$

For x and y in $\mathbf{C}^{\mathbf{a}}$ we put $x^y = \prod_{v \in \mathbf{a}} x_v^{y_v}$ whenever each factor $x_v^{y_v}$ is meaningful (according to the context). For a weight k and $z \in \mathscr{H}$ we define a factor of automorphy $j^k(*, z)$ by

(1.10a) $$j_\alpha^k(z) = j^k(\alpha, z) = j_\alpha(z)^k \quad (k = [k], \ \alpha \in G_{\mathbf{A}}),$$

(1.10b) $$j_\sigma^k(z) = j^k(\sigma, z) = h_\sigma(z) j_\alpha(z)^{[k]} \quad (k \neq [k], \ \sigma \in \mathfrak{M}, \ \alpha = \mathrm{pr}(\sigma)).$$

Then for a function f on $\mathscr{H}$ and $\xi \in G_{\mathbf{A}}$ or $\in \mathfrak{M}$ we define a function $f\|_k \xi$ on $\mathscr{H}$ by

(1.11) $$(f\|_k \xi)(z) = j_\xi^k(z)^{-1} f(\xi z) \quad (z \in \mathscr{H}).$$

Once k is fixed, we shall often write $f\|\xi$ for $f\|_k\xi$. We denote by $\mathscr{M}_k^n$ the set of all holomorphic modular forms of weight k on $\mathscr{H}$ with respect to congruence subgroups of G, and by $\mathscr{S}_k^n$ its subset consisting of all the cusp forms (see [S9, Sect. 1]).

We now take a weight k, a fractional ideal $\mathfrak{b}$, an integral ideal $\mathfrak{c}$, and a Hecke character ψ of F such that

$$(1.12) \qquad \psi_v(a) = 1 \text{ if } v \in \mathbf{h},\ a \in \mathfrak{g}_v^\times, \text{ and } a - 1 \in \mathfrak{c}_v .$$

Whenever we speak of a Hecke character, we always assume that it is $\mathbf{T}$-valued and normalized in the sense that it takes the value 1 for $cu \in F_\mathbf{a}^\times$ with $0 < c \in \mathbf{R}$. Let K be an open subgroup of $D[\mathfrak{b}^{-1}, \mathfrak{b}\mathfrak{c}]$ and let $\Gamma = G \cap K$. Assume that $K \subset C^0$ if $k \neq [k]$. Then we denote by $\mathscr{M}_k^n(\Gamma, \psi)$ the set of all $g \in \mathscr{M}_k^n$ satisfying

$$(1.13) \qquad g\|_k\gamma = \psi_\mathfrak{c}(\det(a_\gamma))g \quad \text{for every} \quad \gamma \in \Gamma ,$$

where $\psi_\mathfrak{c} = \prod_{v|\mathfrak{c}} \psi_v$. Given a function g on $\mathscr{H}$ satisfying (1.13), we can define a function $g_\mathbf{A}$ on $G_\mathbf{A}$ or on $M_\mathbf{A}$ according as $k = [k]$ or $k \neq [k]$ by

$$(1.14) \qquad g_\mathbf{A}(\alpha w) = \psi_\mathfrak{c}(\det(d_w))(g\|_k w)(\mathbf{i}) \text{ for } \alpha \in G \text{ and } w \in \mathrm{pr}^{-1}(K) ,$$

where $\mathbf{i}$ is the "origin" of $\mathscr{H}$ given by

$$(1.15) \qquad \mathbf{i} = \mathbf{i}_n = (i1_n, \ldots, i1_n) ,$$

and we understand, throughout the paper, that pr means the identity map $G_\mathbf{A} \to G_\mathbf{A}$ if $k = [k]$. Clearly

$$(1.16) \qquad g_\mathbf{A}(\alpha x w) = \psi_\mathfrak{c}(\det(d_w))j^k(w, \mathbf{i})^{-1} g_\mathbf{A}(x)$$

$$\text{if } \alpha \in G,\ \mathrm{pr}(w) \in K \text{ and } w(\mathbf{i}) = \mathbf{i} .$$

Conversely, a function $g_\mathbf{A}$ on $G_\mathbf{A}$ or on $M_\mathbf{A}$ satisfying (1.16) can always be obtained from a function g on $\mathscr{H}$ satisfying (1.13) by means of (1.14). We put

$$(1.17a) \qquad \mathscr{M}_k^n(K, \psi) = \{g_\mathbf{A} \mid g \in \mathscr{M}_k^n(\Gamma, \psi)\} ,$$

$$(1.17b) \qquad \mathscr{S}_k^n(K, \psi) = \{g_\mathbf{A} \mid g \in \mathscr{S}_k^n(\Gamma, \psi)\}, \quad \mathscr{S}_k^n(\Gamma, \psi) = \mathscr{S}_k^n \cap \mathscr{M}_k^n(\Gamma, \psi) .$$

(In [S9, S10] the symbols $\mathscr{M}_k(K, \psi)$ and $\mathscr{S}_k(K, \psi)$ were defined to be the same as $\mathscr{M}_k(\Gamma, \psi)$ and $\mathscr{S}_k(\Gamma, \psi)$. In the present paper, however, we use the former for the functions on $M_\mathbf{A}$ and the latter for those on $\mathscr{H}$. It should be noted that K is a subgroup of $G_\mathbf{A}$, not $M_\mathbf{A}$.) If $\psi_\mathfrak{c}(\det(d_w)) = 1$ for every $w \in K$, then we denote these sets by $\mathscr{S}_k^n(K), \mathscr{S}_k^n(\Gamma)$, etc. without the letter ψ.

Let us now put

$$(1.18a) \qquad S = S^n = \{\xi \in F_n^n \mid {}^t\xi = \xi\} ,$$

$$(1.18b) \qquad S_+ = \{\xi \in S \mid \xi_v \geqq 0 \text{ for every } v \in \mathbf{a}\} ,$$

$$(1.18c) \qquad S(\mathfrak{r}) = S \cap \mathfrak{r}_n^n, \quad S_\mathbf{h}(\mathfrak{r}) = \prod_{v \in \mathbf{h}} S(\mathfrak{r})_v ,$$

where $\mathfrak{r}$ is any fractional ideal in F. We now take a divisor $\mathfrak{e}$ of $\mathfrak{c}$ such that $\mathfrak{e}^{-1}\mathfrak{c} + \mathfrak{e} = \mathfrak{g}$, and assume that K has the form

$$(1.19) \qquad K = \{x \in D[\mathfrak{b}^{-1}\mathfrak{e}, \mathfrak{b}\mathfrak{c}] \mid (a_x - 1)_v \in (\mathfrak{e}_v)_n^n \text{ for every } v \in \mathbf{h}\}.$$

Proposition 1.1. *Given $f \in \mathcal{M}_k(\Gamma, \psi)$, there is a complex number $\mu(\tau, q; f)$, written also $\mu_f(\tau, q)$, determined for $\tau \in S_+$ and $q \in GL_n(F)_{\mathbf{A}}$, such that*

$$(1.20) \qquad f_{\mathbf{A}}\left(r_P\begin{pmatrix} q & s\tilde{q} \\ 0 & \tilde{q}\end{pmatrix}\right) = \det(q)_{\mathbf{a}}^{[k]}|\det(q)_{\mathbf{a}}|^{k-[k]} \cdot \sum_{\tau \in S_+} \mu(\tau, q; f)\mathbf{e}_{\mathbf{a}}(\mathrm{tr}(\mathbf{i} \cdot {}^t q\tau q))\mathbf{e}_{\mathbf{A}}(\mathrm{tr}(\tau s)),$$

for every $s \in S_{\mathbf{A}}$, where r_P stands for the identity map if $k \in \mathbf{Z}^{\mathbf{a}}$. Moreover $\mu_f(\tau, q)$ has the following properties:

$$(1.21\text{a}) \qquad \mu_f(\tau, q) \neq 0 \textit{ only if } \mathbf{e}_{\mathbf{h}}(\mathrm{tr}({}^t q\tau q s)) = 1 \textit{ for every } s \in S_{\mathbf{h}}(\mathfrak{b}^{-1}\mathfrak{e});$$

$$(1.21\text{b}) \qquad \mu_f(\tau, q) = \mu_f(\tau, q_{\mathbf{h}});$$

$$(1.21\text{c}) \qquad \mu_f({}^t b\tau b, q) = \det(b)^{[k]}|\det(b)|^{k-[k]}\mu_f(\tau, bq) \textit{ for every } b \in GL_n(F);$$

$$(1.21\text{d}) \qquad \psi_{\mathbf{h}}(\det(a))\mu_f(\tau, qa) = \mu_f(\tau, q) \textit{ for every } \mathrm{diag}[a, \tilde{a}] \in K.$$

Furthermore, if $\beta \in G \cap \mathrm{diag}[r, \tilde{r}]K$ with $r \in GL_n(F)_{\mathbf{A}}$, then

$$(1.22) \qquad j^k(\beta, \beta^{-1}z)f(\beta^{-1}z) = \psi(\det(d_\beta r)_{\mathfrak{c}}) \sum_{\tau \in S_+} \mu_f(\tau, r)\mathbf{e}_{\mathbf{a}}(\mathrm{tr}(\tau z)).$$

Proof. This was given in [S9, Proposition 1.1] when $K = D[\mathfrak{b}^{-1}, \mathfrak{b}\mathfrak{c}]$, but we only indicated that the proof could be given in the same fashion as in the one-dimensional case [S6]. For the reader's convenience we sketch a proof here. Given $w = \begin{pmatrix} q & s\tilde{q} \\ 0 & \tilde{q}\end{pmatrix} \in P_{\mathbf{A}}$, take $\beta \in G \cap wK$, and put $z = w(\mathbf{i}), \sigma = r_P(w_{\mathbf{h}}), \tau = r_P(w_{\mathbf{a}})$, and $\rho = \beta^{-1}\sigma$. Then $\mathrm{pr}(\rho) \in K, \rho\tau(\mathbf{i}) = \beta^{-1}z$, and hence by (1.14) we have $f_{\mathbf{A}}(r_P(w)) = f_{\mathbf{A}}(\beta\rho\tau) = \psi_{\mathfrak{c}}(\det(d_\rho))(f\|\rho\tau)(\mathbf{i}) = \psi_{\mathfrak{c}}(\det(d_\rho))j^k(\rho\tau, \mathbf{i})^{-1}f(\beta^{-1}z)$. By (1.9b, c) $1 = j^k(\sigma, z) = j^k(\beta\rho, z) = j^k(\beta, \beta^{-1}z)j^k(\rho, z)$ and so $j^k(\rho\tau, \mathbf{i}) = j^k(\rho, z)j^k(\tau, \mathbf{i}) = j^k(\beta, \beta^{-1}z)^{-1}\det(q)_{\mathbf{a}}^{-[k]}|\det(q)_{\mathbf{a}}|^{[k]-k}$. We easily see that $(a_\rho)_v = ({}^t d_\beta q)_v$ for every $v|\mathfrak{c}$. Thus we obtain

$$f_{\mathbf{A}}(r_P(w)) = \psi_{\mathfrak{c}}(\det(qd_\beta))^{-1}\det(q)_{\mathbf{a}}^{[k]}\,|\det(q)_{\mathbf{a}}|^{k-[k]}j^k(\beta, \beta^{-1}z)f(\beta^{-1}z).$$

We can put $j^k(\beta, \beta^{-1}z)f(\beta^{-1}z) = \sum_{\tau \in S_+}\lambda(\tau)\mathbf{e}_{\mathbf{a}}(\mathrm{tr}(\tau z))$. Therefore we obtain (1.20) with a certain constant $\lambda(\tau, q, s)$ in place of $\mu(\tau, q; f)$. Since β depends only on $w_{\mathbf{h}}$, we see that $\lambda(\tau, q, s)$ is independent of $q_{\mathbf{a}}$ and $s_{\mathbf{a}}$. Then employing (1.16), we see that λ is independent of s, and so we can put $\lambda(\tau, q, s) = \mu_f(\tau, q)$. Then we can derive (1.21a, b, c, d) from (1.16) and obtain (1.22) by taking $\mathrm{diag}[r, \tilde{r}]_{\mathbf{h}} = w_{\mathbf{h}}$.

For $\beta \in G^r_{\mathbf{A}}$ and $\gamma \in G^{n-r}_{\mathbf{A}}$ we define an element $\beta \times \gamma$ of $G^n_{\mathbf{A}}$ by

$$(1.23) \qquad \beta \times \gamma = \begin{pmatrix} a_\beta & 0 & b_\beta & 0 \\ 0 & a_\gamma & 0 & b_\gamma \\ c_\beta & 0 & d_\beta & 0 \\ 0 & c_\gamma & 0 & d_\gamma \end{pmatrix}.$$

We recall that there is a homomorphism of $M^r_{\mathbf{A}} \times M^{n-r}_{\mathbf{A}}$ into $M^n_{\mathbf{A}}$, written $(\sigma, \sigma') \mapsto \langle \sigma, \sigma' \rangle$, such that $\mathrm{pr}(\langle \sigma, \sigma' \rangle) = \mathrm{pr}(\sigma) \times \mathrm{pr}(\sigma')$ and $\langle t\sigma, \sigma' \rangle = \langle \sigma, t\sigma' \rangle = t\langle \sigma, \sigma' \rangle$ for $t \in \mathbf{T}$. Moreover, if $\alpha \in G^r$ and $\beta \in G^{n-r}$, then $\alpha \times \beta$ as an element of $M^n_{\mathbf{A}}$ coincides with $\langle \alpha, \beta \rangle$; if $\sigma \in \mathfrak{M}^r$ and $\sigma' \in \mathfrak{M}^{n-r}$, then $\langle \sigma, \sigma' \rangle \in \mathfrak{M}^n$ and

$$(1.24) \qquad h(\langle \sigma, \sigma' \rangle, \mathrm{diag}[z, z']) = h(\sigma, z) h(\sigma', z') \quad (z \in \mathscr{H}^r, z' \in \mathscr{H}^{n-r}).$$

For these the reader is referred to [S7, Section 2]. We can also easily show that

$$(1.25) \qquad r_P(\alpha \times \beta) = \langle r_P(\alpha), r_P(\beta) \rangle, \quad r_\Omega(\alpha \times \beta) = \langle r_\Omega(\alpha), r_\Omega(\beta) \rangle.$$

Lemma 1.2. *Given $\sigma \in M_{\mathbf{A}}$ such that $\mathrm{pr}(\sigma)_{\mathbf{a}} = 1$, there exists an open subgroup C of $C^0 \cap \sigma C^0 \sigma^{-1}$ such that $h(\sigma^{-1}\xi\sigma, z) = h(\xi, z)$ for every $\xi \in C$.*

This will be proved in the next section after the proof of Lemma 2.2.

Lemma 1.3. *Given $\sigma \in M_{\mathbf{A}}$ as in Lemma 1.2, let ρ be an element of $K \cdot \mathrm{pr}(\sigma) \cap G$ such that $\psi_{\mathfrak{c}}(\det(d(\rho\sigma^{-1}))) = 1$. Then there is a unique decomposition $\rho = \rho_1\rho_2$ with $\rho_1, \rho_2 \in M_{\mathbf{A}}$ such that $\mathrm{pr}(\rho_2) = \rho_{\mathbf{h}}$ and $h(\rho_2\sigma^{-1}, z) = 1$. Moreover, if g and $g_{\mathbf{A}}$ are as in (1.13) and (1.14), then $g_{\mathbf{A}}(x\sigma^{-1})$ corresponds to $g\|_k\rho_1$ by (a generalization of) the principle of (1.14). In particular, if $\mathrm{pr}(\sigma) \in C^0$ and $h(\sigma, z) = 1$, then $g\|_k\rho_1 = g\|_k\rho$.*

Proof. Take any decomposition $\rho = \rho_1\rho_2$ in $M_{\mathbf{A}}$ so that $\mathrm{pr}(\rho_2) = \rho_{\mathbf{h}}$. Then $\mathrm{pr}(\rho_2\sigma^{-1}) \in G_{\mathbf{h}} \cap K$, and so ρ_2 is uniquely determined by the condition $h(\rho_2\sigma^{-1}, z) = 1$. Put $y = \rho_2\sigma^{-1}$ and $f(x) = g_{\mathbf{A}}(x\sigma^{-1})$. By Lemma 1.2 we easily see that $f(\alpha x w) = j^k(w, \mathbf{i})^{-1} f(x)$ for $\alpha \in G$ and $\mathrm{pr}(w) \in C, w(\mathbf{i}) = \mathbf{i}$, with an open subgroup C of C^0. By (1.16) $g_{\mathbf{A}}(xy) = g_{\mathbf{A}}(x)$, and hence for $\mathrm{pr}(x) \in G_{\mathbf{a}}$ we have $f(x) = g_{\mathbf{A}}(x\rho_2^{-1}) = g_{\mathbf{A}}(\rho\rho_2^{-1}x) = g_{\mathbf{A}}(\rho_1 x) = (g\|(\rho_1 x))(\mathbf{i}) = ((g\|\rho_1)\|x)(\mathbf{i})$. This proves that f corresponds to $g\|\rho_1$. If $\mathrm{pr}(\sigma) \in C^0$ and $h(\sigma, z) = 1$, then $h(\rho_2, z) = 1$ by (1.9c), and hence $j^k(\rho_1, z) = j^k(\rho, z)$, so that $g\|\rho_1 = g\|\rho$.

2. A new factor of automorphy J^k

From now on k will always be half-integral unless specified otherwise. Since j^k, in particular h, has only a weak property of automorphy expressed by (1.9c), it is not satisfactory for the definition of Hecke operators. We are going to introduce another (quasi) factor of automorphy $J^k(\alpha,z)$ defined for $\alpha \in \mathscr{Y}$ and $z \in \mathscr{H}$ satisfying

(2.1a) $$J^k(\xi,z) = j^k(\xi,z) \quad \text{if } \xi \in U,$$

(2.1b) $$J^k(\xi\alpha\eta,z) = J^k(\xi,\alpha\eta z)\, J^k(\alpha,\eta z)\, J^k(\eta,z) \quad \text{if } \alpha \in \mathscr{Y} \text{ and } \xi,\eta \in U,$$

(2.1c) $$J^k(\alpha,z) = j^{[k]}(\mathrm{pr}(\alpha),z)\, J^{u/2}(\alpha,z).$$

Here $\mathscr{Y}, U$, and other symbols are defined as follows:

$$P' = \{\alpha \in G \mid b_\alpha = 0\}, \quad \Omega' = \{\alpha \in G_{\mathbf{A}} \mid d_\alpha \in GL_n(F)_{\mathbf{A}}\},$$

$$E = \prod_{v\in\mathbf{h}} E_v, \quad E_v = GL_n(\mathfrak{g}_v), \quad B = GL_n(F)_{\mathbf{h}} \cap \prod_{v\in\mathbf{h}} B_v, \quad B_v = (\mathfrak{g}_v)_n^n \cap GL_n(F_v),$$

$$D = D[2\mathfrak{d}^{-1}, 2\mathfrak{d}], \quad D_v = D_v[2\mathfrak{d}^{-1}, 2\mathfrak{d}], \quad Z = DZ_0D, \quad Z' = DZ_0'D,$$

$$Z_0 = \{\mathrm{diag}[\tilde{q}, q] \mid q \in B\}, \quad Z_0' = \{\mathrm{diag}[q, \tilde{q}] \mid q \in B\},$$

$$U = \{\alpha \in M_{\mathbf{A}} \mid \mathrm{pr}(\alpha) \in D\},$$

$$\mathscr{Y} = \{\alpha \in M_{\mathbf{A}} \mid \mathrm{pr}(\alpha) \in Z\}, \quad \mathscr{Y}' = \{\alpha \in M_{\mathbf{A}} \mid \mathrm{pr}(\alpha) \in Z'\}.$$

Clearly $P' = \iota^{-1}P\iota$ and $\Omega' = \Omega\iota$. For $\alpha \in \Omega'$ we put

(2.2) $$r'(\alpha) = r_\Omega(\alpha\iota^{-1})\iota.$$

Lemma 2.1. *The map r' has the following properties:*

(1) *If $\alpha \in \Omega'$ and $\pi \in P_{\mathbf{A}}$, then $\pi\alpha \in \Omega'$, $r'(\pi) = r_P(\pi)$, and $r'(\pi\alpha) = r_P(\pi)r'(\alpha)$.*

(2) *If $\alpha \in \Omega'$, $f = r_\Omega(g)$, and $g = \mathrm{diag}[\tilde{p}, p]\iota$ with $p \in GL_n(F)_{\mathbf{A}}$, then $\alpha g^{-1} \in \Omega$, $f^{-1} = r_\Omega(g^{-1})$, and $r'(\alpha) = r_\Omega(\alpha g^{-1})f$.*

(3) *If $\alpha \in \Omega'$ and $\beta \in P'_{\mathbf{A}}$, then $\alpha\beta \in \Omega'$ and $r'(\alpha\beta) = r'(\alpha)r'(\beta)$.*

(4) *$r'(\alpha) = \alpha$ if $\alpha \in G \cap \Omega'$.*

(5) *$r'(\alpha \times \beta) = \langle r'(\alpha), r'(\beta)\rangle$.*

Proof. That $\pi\alpha \in \Omega'$, $\alpha g^{-1} \in \Omega$, and $\alpha\beta \in \Omega'$ can easily be verified. Assertion (1) follows immediately from (1.2a, b). Putting $s = \mathrm{diag}[\tilde{p}, p]$, by (1.2a, b) we have $f = r_P(s)\iota$ and $r_\Omega(\alpha\iota^{-1}s^{-1}) = r_\Omega(\alpha\iota^{-1})r_P(s)^{-1} = r'(\alpha)f^{-1}$, which proves the last equality of (2). Taking $\alpha = 1$, we obtain $r_\Omega(g^{-1}) = f^{-1}$. As for (3), similarly $r'(\alpha\beta) = r_\Omega(\alpha\iota^{-1}\iota\beta\iota^{-1})\iota = r_\Omega(\alpha\iota^{-1})r_P(\iota\beta\iota^{-1})\iota = r'(\alpha)\iota^{-1}r_P(\iota\beta\iota^{-1})\iota = r'(\alpha)r_\Omega(\iota^{-1}\iota\beta\iota^{-1})\iota = r'(\alpha)r'(\beta)$ as expected. Assertion (4) follows directly from (1.2a), and assertion (5) from (1.25).

We define $\gamma(s) \in \mathbf{C}$ and $\mathbf{g}(s) \in \mathbf{T}$ for $s \in S_{\mathbf{A}}$ and also $\mathbf{t}_0(\alpha) \in \mathbf{T}$ for $\alpha \in \Omega'$ by

$$\text{(2.3a)} \qquad \gamma(s) = \prod_{v \in \mathbf{h}} \gamma_v(s), \quad \gamma_v(s) = N(\mathfrak{d}_v)^{n/2} \int_{L_v} \mathbf{e}_v({}^t x s_v x/2) d_v x,$$

$$\text{(2.3b)} \qquad \mathbf{g}(s) = \gamma(s)/|\gamma(s)|, \quad \mathbf{t}_0(\alpha) = \mathbf{g}(\delta^{-2} d_\alpha^{-1} c_\alpha).$$

Here $L = \mathfrak{g}_1^n$ and $d_v x$ is the additive Haar measure such that L_v has measure $N(\mathfrak{d}_v)^{-n/2}$. For $\alpha \in \Omega'$ we define a holomorphic function $p(\alpha, z)$ of $z \in \mathscr{H}$ by

$$\text{(2.4)} \qquad p(\alpha, z) = t \cdot j(\alpha, z)^{u/2} \quad (z \in \mathscr{H})$$

with an element $t \in \mathbf{T}$ satisfying the condition that

$$\text{(2.5)} \qquad \lim_{z \to 0} p(\alpha, z) = |\det(d_\alpha)^{u/2}|.$$

By [S7, Lemma 2.2] we have

$$\text{(2.6)} \qquad h(r'(\alpha), z) = \mathbf{t}_0(\alpha) p(\alpha, z) \quad \text{if } \alpha \in \Omega' \cap P_{\mathbf{A}} C^{\theta}.$$

Lemma 2.2. *Put* $U' = \{x \in M_{\mathbf{A}} \mid \mathrm{pr}(x) \in C^{\theta}\}$. *If* $\xi, \eta \in U'$, $\sigma = r_P(w)$, $w = \mathrm{diag}[\tilde{q}, q]$ *with* $q \in GL_n(F)_{\mathbf{h}}$ *such that* $q_v \in B_v$ *or* $q_v^{-1} \in B_v$ *for each* $v \in \mathbf{h}$, *and* $\xi\sigma = \sigma\eta$, *then* $h(\xi, z) = h(\eta, z)$ *at least in the following three cases:* (i) $\xi, \eta \in U$; (ii) $q_v \in E_v$ *for every* $v|2$; (iii) $\eta \in U$ *and* $q_v \in B_v$ *for every* $v|2$.

Proof. Take ξ_1 and ξ_2 in U' so that $\xi = \xi_1\xi_2$, $\mathrm{pr}(\xi_1) \in G_{\mathbf{h}}$, and $\mathrm{pr}(\xi_2) \in G_{\mathbf{a}}$. By (1.9c) it is sufficient to show that $h(\xi_1, z) = h(\eta\xi_2^{-1}, z)$. By (1.3) we have $\xi_1\sigma = \sigma\eta\xi_2^{-1}$. Thus our problem can be reduced to the case where $\mathrm{pr}(\xi)_{\mathbf{a}} = \mathrm{pr}(\eta)_{\mathbf{a}} = 1$. We can also assume that q is diagonal. We first consider Cases (i) and (ii). Decomposing w into a finite product, we can further reduce the problem to the case where $q_v \notin E_v$ at exactly one v. Call that prime s. Changing σ for σ^{-1} if necessary, we may assume that $q_s \in B_s$. In view of the definition of $h(\xi, z)$ in [S5, (3.10a)] and [S7, (1.23)] we have $h_\xi = h_\eta$ (for $\xi, \eta \in U'$) if and only if $(\xi\varphi)(0) = (\eta\varphi)(0)$ where $\varphi = \varphi_{\mathbf{a}} \prod_{v \in \mathbf{h}} \varphi_v$, $\varphi_{\mathbf{a}}$ is the standard function of [S5, (3.9c)] and [S7, (1.8a)], and φ_v is the characteristic function of $(\mathfrak{g}_v)_n^1$. Put $f = \mathrm{pr}(\xi)$. We can find $t \in \mathbf{T}$ so that $\xi\varphi = t\varphi_{\mathbf{a}} \prod_{v \in \mathbf{h}} r_v(f_v)\varphi_v$, where r_v is the lift $C_v^\theta \to M_v$ of [S7, (1.17)]. Then $\eta\varphi = t\varphi_{\mathbf{a}} \prod_{v \in \mathbf{h}} r_P(w_v)^{-1} r_v(f_v) r_P(w_v)\varphi_v$. Since $r_v(C_v^\theta)\varphi_v = \varphi_v$, we have $\xi\varphi = t\varphi$, and $\eta\varphi = t\varphi_{\mathbf{a}} \cdot r_P(w_s)^{-1} r_s(f_s) r_P(w_s)\varphi_s \prod_{v \neq s} \varphi_v$. We can now define an element ξ' of U' so that $\xi'\varphi = \xi\varphi$, $\mathrm{pr}(\xi')_v = 1$ for $v \neq s$, and $\mathrm{pr}(\xi')_s = f_s$. Put $\eta' = \sigma^{-1}\xi'\sigma$. Then $\eta' \in U'$ and $\eta'\varphi = \eta\varphi$; $\xi', \eta' \in U$ if $\xi, \eta \in U$. Changing ξ and η for ξ' and η', we may thus assume that $f_v = 1$ for $v \neq s$. To prove $h_\xi = h_\eta$ in this setting, we first assume that $s \nmid 2$. We may also assume that $q = \mathrm{diag}[1_m, p]$ with $p \in (\pi_s \mathfrak{g}_s)_{n-m}^{n-m}$, $0 \leqq m < n$, where π_s is a prime element of F_s. Put

$$a_f = \begin{pmatrix} a_1 & a_2 \\ a_3 & a_4 \end{pmatrix}, \quad b_f = \begin{pmatrix} b_1 & b_2 \\ b_3 & b_4 \end{pmatrix}, \quad c_f = \begin{pmatrix} c_1 & c_2 \\ c_3 & c_4 \end{pmatrix}, \quad d_f = \begin{pmatrix} d_1 & d_2 \\ d_3 & d_4 \end{pmatrix}$$

with a_1, b_1, c_1, d_1 of size m. Since $w^{-1}fw \in D[\mathfrak{d}^{-1}, \mathfrak{d}]$, we easily see that the entries of $(\delta^{-1}c_2)_s$, $(\delta^{-1}c_3)_s$, $(\delta^{-1}c_4)_s$, $(d_3)_s$, and $(a_2)_s$ are divisible by π_s. Therefore we can find $v \in D_s^m[\mathfrak{d}^{-1}, \mathfrak{d}]$ (viewed as an element of $D^m[\mathfrak{d}^{-1}, \mathfrak{d}]$) and $\lambda \in D_s[\mathfrak{d}^{-1}, \pi_s\mathfrak{d}]$ (viewed as an element of $D[\mathfrak{d}^{-1}, \mathfrak{d}]$) such that $f = (v \times 1)\lambda$. Then $\xi = \langle \tau, 1\rangle r'(\lambda)$ with an element $\tau \in M_{\mathbf{A}}^m$ such that $\mathrm{pr}(\tau) = v$. Since $w = 1_{2m} \times \mathrm{diag}[\tilde{p}, p]$, we have $\eta = \langle \tau, 1\rangle r_P(w)^{-1} r'(\lambda) r_P(w) = \langle \tau, 1\rangle r'\left(w^{-1}\lambda w\right)$ by Lemma 2.1. Put $\mu = w^{-1}\lambda w$. Then our task is to show that $h(r'(\lambda), z), h(r'(\mu), z)$ which holds if $\mathbf{g}\left(\delta^{-2}d_\lambda^{-1}c_\lambda\right) = \mathbf{g}\left(\delta^{-2}d_\mu^{-1}c_\mu\right)$. Since $\lambda_v = 1$ for $v \neq s$ and $\lambda_s \in D_s[\mathfrak{d}^{-1}, \pi_s\mathfrak{d}]$, we see that $(d_\lambda)_s \in E_s$ and $(c_\lambda)_s \in (\mathfrak{d}_s)_n^n$, and hence $\mathbf{g}\left(\delta^{-2}d_\lambda^{-1}c_\lambda\right) = 1$. Since $\eta \in U'$, we have $\mu_s \in D_s[\mathfrak{d}^{-1}, \mathfrak{d}]$; hence $\left(d_\mu\right)_s \in (\mathfrak{g}_s)_n^n$ and $\left(c_\mu\right)_s \in (\mathfrak{d}_s)_n^n$. Now $\det(d_\mu) = \det\left(q^{-1}d_\lambda q\right) = \det(d_\lambda) \in \mathfrak{g}_s^\times$, and so we obtain $\mathbf{g}\left(\delta^{-2}d_\mu^{-1}c_\mu\right) = 1$ as expected. So far we tacitly assumed $m > 0$, but if $m = 0$, we can take $\lambda = f$ and the same reasoning works. This proves Case (ii). Next assume that $s|2$. In this case we assume that $\xi, \eta \in U$. Put $g = w^{-1}fw$. We have $\xi = t'r'(f)$ with some $t' \in \mathbf{T}$ and $\eta = \sigma^{-1}\xi\sigma = t'r'(g)$ by Lemma 2.1. Since $f \in D_s[2\mathfrak{d}^{-1}, 2\mathfrak{d}]$, we have $d_f \in E_s$ and $c_f \in (2\mathfrak{d}_s)_n^n$, and so $h(r'(f), z) = \mathbf{g}(\delta^{-2}d_f^{-1}c_f) = 1$. Similarly $h(r'(g), z) = 1$. This proves Case (i). Finally suppose $\eta \in U$ and $q_v \in B_v$ for every $v|2$. Put $w = w_1w_2$ with $w_{1v} = 1$ for $v|2$ and $w_{2v} = 1$ for $v \nmid 2$; put $\sigma_i = r_P(w_i)$ and $\zeta = \sigma_1\eta\sigma_1^{-1}$. Then $\zeta = \sigma_2^{-1}\xi\sigma_2 \in U$. Our result in Case (i) gives $h_\eta = h_\zeta$. Thus, changing η for ζ, we may assume that $q_v = 1$ for $v \nmid 2$. Also, our technique in Cases (i, ii) shows that we may assume that $\mathrm{pr}(\xi)_v = \mathrm{pr}(\eta)_v = 1$ for $v \nmid 2$. Put $\mathrm{pr}(\xi) = f$ and $\mathrm{pr}(\eta) = g$. Then $\xi = t'r'(f)$ and $\eta = t'r'(g)$ with some $t' \in \mathbf{T}$ as in Case (i). Since $g \in D$, we see that $d_g \in E$ and $\left(c_g\right)_v \in (2\mathfrak{d}_v)_n^n$ for every $v \in \mathbf{h}$, and hence $h(r'(g), z) = \mathbf{g}\left(\delta^{-2}d_g^{-1}c_g\right) = 1$. Now $d_f \in B$ and $\det(d_f) = \det(d_g)$, so that $d_f \in E$; also $(c_f)_v = (qc_g \cdot {}^t q)_v \in (2\mathfrak{d}_v)_n^n$ for every $v \in \mathbf{h}$. Thus $h(r'(f), z) = 1$. This completes the proof.

Proof of Lemma 1.2. If $\xi = \xi_1\xi_2$ with $\mathrm{pr}(\xi_1) = \mathrm{pr}(\xi)_{\mathbf{h}}$, then $\sigma^{-1}\xi\sigma = \sigma^{-1}\xi_1\sigma\xi_2$ by (1.3), and our problem is to show $h(\sigma^{-1}\xi_1\sigma, z) = h(\xi_1, z)$. Thus we may assume that $\mathrm{pr}(\xi) \in G_{\mathbf{h}}$. We first consider the case $\sigma = r_P(w)$ with $w \in P_{\mathbf{h}}$. Put $\eta = \sigma^{-1}\xi\sigma$ and use the same technique as in the proof of Lemma 2.2. Let $\varphi' = \varphi_{\mathbf{a}}\prod_{v\in\mathbf{h}}\varphi_v'$, $\varphi_v' = r_P(w_v)\varphi_v$, with φ there. Now there is an open subgroup C of $C^0 \cap \sigma C^0\sigma^{-1}$ such that $\varphi_{\mathbf{a}}\prod_{v\in\mathbf{h}} r_v(x_v)\varphi_v' = \varphi'$ for every $x \in C$. Then $\eta\varphi = \xi\varphi$ if $\mathrm{pr}(\xi) \in C \cap G_{\mathbf{h}}$, and hence $h_\xi = h_\eta$. Since $G_{\mathbf{A}} = P_{\mathbf{A}}D[\mathfrak{d}^{-1}, \mathfrak{d}]$, this result reduces our problem to the case where $\sigma \in G_{\mathbf{h}} \cap D[\mathfrak{d}^{-1}, \mathfrak{d}]$. By [S7, Lemma 3b.3] $D[\mathfrak{d}^{-1}, \mathfrak{d}]$ is generated by $P_{\mathbf{h}} \cap D[\mathfrak{d}^{-1}, \mathfrak{d}], P_{\mathbf{h}}' \cap D[\mathfrak{d}^{-1}, \mathfrak{d}]$, and $D[2\mathfrak{d}^{-1}, 2\mathfrak{d}]$, and therefore we easily see that it is sufficient to treat the case $\sigma \in P_{\mathbf{h}}' \cap D[\mathfrak{d}^{-1}, \mathfrak{d}]$. Take $\omega \in M_{\mathbf{A}}$ so that $\mathrm{pr}(\omega) = \mathrm{diag}[\delta^{-1}1_n, \delta 1_n]\iota_{\mathbf{h}}$; put $\tau = \omega\sigma\omega^{-1}$, and $\zeta = \omega\xi\omega^{-1}$. Since $\mathrm{pr}(\omega) \in C^0$, if $\mathrm{pr}(\xi) \in C^0 \cap \sigma C^0\sigma^{-1}$, then $h(\sigma^{-1}\xi\sigma, z) = h(\omega\sigma^{-1}\xi\sigma\omega^{-1}, z) = h(\tau^{-1}\zeta\tau, z)$ and $h(\xi, z) = h(\zeta, z)$. Since $\tau \in P_{\mathbf{h}}$, we obtain $h(\tau^{-1}\zeta\tau, z) = h(\zeta, z)$ if $\mathrm{pr}(\zeta)$ belongs to a sufficiently small open subgroup of C^0. This completes the proof.

Given an element $\alpha = \xi_1\sigma\xi_2 \in \mathscr{Z}$ with $\xi_i \in U$ and $\mathrm{pr}(\sigma) \in Z_0$, we put

(2.7) $$J^k(\alpha, z) = j^k(\xi_1\xi_2, z) \quad (z \in \mathscr{H}).$$

This is well defined. In fact, suppose $\alpha = \eta_1\tau\eta_2$ with $\eta_i \in U$ and $\mathrm{pr}(\tau) \in Z_0$. Then $\tau = \beta\sigma\gamma, \beta = r_P(\mathrm{diag}[\tilde{b}, b]), \gamma = r_P(\mathrm{diag}[\tilde{c}, c])$ with $b, c \in E$ and $\beta^{-1}\eta_1^{-1}\xi_1 \sigma = \sigma\gamma\eta_2\xi_2^{-1}$. Lemma 2.2 shows that $h(\beta^{-1}\eta_1^{-1}\xi_1, z) = h(\gamma\eta_2\xi_2^{-1}, z)$. By (1.9b) we have $h(\beta, z) = h(\gamma, z) = 1$. These together with (1.9c) prove $h(\xi_1\xi_2, z) = h(\eta_1\eta_2, z)$ as desired. From (2.7) we immediately obtain (2.1a, b, c).

Lemma 2.3. *Let $\xi = r_P(\mathrm{diag}[s, \tilde{s}])$ with $s \in GL_n(F)_{\mathbf{h}}$. If $s_v \in E_v$ for every $v|2$, then $\xi \in \mathscr{Z}$, $U\xi U = U\xi^{-1}U$, $J^k(\xi, z) = 1$, and $J^k(\eta^{-1}, z) = J^k(\eta, \eta^{-1}z)^{-1}$ for every $\eta \in U\xi U$.*

This lemma will be proved at the end of the next section.

Given $\alpha \in P_{\mathbf{h}} \cap Z$, we can find $(g, h) \in B \times B$ such that $a_\alpha = g^{-1}h$ and $g_vL_v + h_vL_v = L_v$ for every $v \in \mathbf{h}$ (cf. [S9, Lemma 2.5]). Put $\sigma = gb_\alpha \cdot {}^th$. Then $\sigma \in S_{\mathbf{h}}$ and

(2.8) $$\alpha = \begin{pmatrix} g^{-1}h & g^{-1}\sigma \cdot {}^th^{-1} \\ 0 & {}^tg \cdot {}^th^{-1} \end{pmatrix}.$$

Lemma 2.4. *For this α we have $J^k(r_P(\alpha), z) = \mathbf{g}(-\sigma)$.*

This will also be proved at the end of the next section. As explained in the introduction, Lemma 2.4 is crucial in our whole theory. If $n = 1$, it is essentially the formulas for β_h in [S2, p. 451] and [S6, (5.8a, b)].

Put $\alpha' = \varepsilon\alpha\varepsilon^{-1}$ for $\alpha \in G_{\mathbf{A}}$ with $\varepsilon = \mathrm{diag}[1_n, -1_n]$. In [S7, Prop. 2.6] we established an automorphism $\sigma \to \sigma'$ of $M_{\mathbf{A}}$ with the following properties: it is consistent with the injection $G \to M_{\mathbf{A}}$; $\mathrm{pr}(\sigma') = \mathrm{pr}(\sigma)'$, $t' = t^{-1}$ for $t \in \mathbf{T}$, $r_P(\alpha') = r_P(\alpha)'$ for $\alpha \in P_{\mathbf{A}}$, and $r_\Omega(\beta') = r_\Omega(\beta)'$ for $\beta \in \Omega$; $h(\sigma', -\bar{z}) = \overline{h(\sigma, z)}$ if $\sigma \in \mathfrak{M}$.

Lemma 2.5. *If $\xi \in \mathscr{Z}$, then $\xi' \in \mathscr{Z}$ and $J^k(\xi', -\bar{z}) = \overline{J^k(\xi, z)}$. In particular, $J^k(\xi', z) = J^k(\xi, z)^{-1}$ if $\mathrm{pr}(\xi) \in G_{\mathbf{h}}$.*

Proof. Let $\xi = \alpha\sigma\beta$ with $\alpha, \beta \in U$ and $\sigma = r_P(\mathrm{diag}[\tilde{q}, q])$, $q \in B$. Then $\xi' = \alpha'\sigma'\beta'$, $\sigma' = \sigma$, and α', $\beta' \in U$. Therefore $\xi' \in \mathscr{Z}$, $J^k(\xi, z) = j^k(\alpha\beta, z)$, and $J^k(\xi', z) = j^k(\alpha'\beta', z)$, and hence our first formula follows from the properties of $h(\xi', z)$ stated above. If $\mathrm{pr}(\xi) \in G_{\mathbf{h}}$, then $J^k(\xi, z) \in \mathbf{T}$, so the second formula is merely a special case.

Lemma 2.6. *Suppose that $\alpha \in \mathscr{Z}$, $\beta \in U$,* $\mathrm{pr}(\alpha)\mathrm{pr}(\beta) = \mathrm{pr}(\beta)\mathrm{pr}(\alpha)$, *and either* $\mathrm{pr}(\alpha)_{\mathbf{a}} = 1$ *or* $\mathrm{pr}(\beta)_{\mathbf{a}} = 1$. *Then $\alpha\beta = \beta\alpha$.*

Proof. Clearly $\alpha\beta = t\beta\alpha$ with $t \in \mathbf{T}$. By (2.1b) $J^k(\alpha, \beta z)J^k(\beta, z) = J^k(\alpha\beta, z) = t^{-1}J^k(\beta\alpha, z) = t^{-1}J^k(\beta, \alpha z)J^k(\alpha, z)$. If $\mathrm{pr}(\alpha)_{\mathbf{a}} = 1$, then $J^k(\alpha, z)$ is a constant and $\alpha z = z$, and hence $t = 1$ as expected. The case $\mathrm{pr}(\beta)_{\mathbf{a}} = 1$ is similar.

3. A reciprocal of $J^{u/2}$

We are going to give an alternative definition of $J^{u/2}$ which leads to the proof of Lemmas 2.3 and 2.4.

Lemma 3.1. (1) $Z' \subset DP'_{\mathbf{A}} \cap P_{\mathbf{A}} D$,
(2) $Z' = \{x \in G_{\mathbf{A}} \mid \det(d_x)_v \neq 0,\ (d_x)_v^{-1} \in (\mathfrak{g}_v)_n^n,\ (b_x d_x^{-1})_v \in (2\mathfrak{d}_v^{-1})_n^n,$
$(d_x^{-1} c_x)_v \in (2\mathfrak{d}_v)_n^n$ *for every* $v|2\}$.

Proof. Assertion (1) can be proved in the same manner as in [S9, Lemma 2.1]. Since $G_v \subset Z'$ if $v \nmid 2$, equality (2) concerns only the prime factors of 2. If $x = \begin{pmatrix} a & b \\ c & d \end{pmatrix}$ of G_v, $v|2$, belongs to the set on the right-hand side, then

$$\begin{pmatrix} a & b \\ c & d \end{pmatrix} = \begin{pmatrix} 1 & bd^{-1} \\ 0 & 1 \end{pmatrix} \begin{pmatrix} \tilde{d} & 0 \\ 0 & d \end{pmatrix} \begin{pmatrix} 1 & 0 \\ d^{-1}c & 1 \end{pmatrix},$$

and hence $x \in Z'$. Conversely, we can verify in a straightforward way that $D_v \cdot \operatorname{diag}[q, \tilde{q}] D_v$ is contained in the set on the right-hand side.

For $\xi \in Z' \cap \Omega'$ we define as element $\mathbf{t}(\xi)$ by

$$\mathbf{t}(\xi) = \kappa(\xi)/|\kappa(\xi)|, \quad \kappa(\xi) = \prod_{v \in \mathbf{h}} \kappa(\xi_v), \tag{3.1}$$

$$\kappa(\xi_v) = \int_{L_v \times L_v} \mathbf{e}_v\left((2\delta_v^2)^{-1} \cdot {}^t x d^{-1} c x + 2\delta_v^{-1} \cdot {}^t x d^{-1} y - 2 \cdot {}^t y b d^{-1} y\right) dx\, dy,$$

where $\xi_v = \begin{pmatrix} a & b \\ c & d \end{pmatrix}$. Clearly $\kappa(\xi_v) = 1$ for almost all v. We will see later that $\kappa(\xi) \neq 0$. From the above lemma we see that if $v|2$, then $\kappa(\xi_v)$ is the measure of $L_v \times L_v$, and hence we obtain

$$\mathbf{t}(\xi_v) = 1 \quad \text{if } v|2. \tag{3.2}$$

Define an element τ of G^{2n} and an element f of $G_{\mathbf{h}}^{2n}$ by

$$\tau = \begin{pmatrix} 1_{2n} & 0 \\ \rho_n & 1_{2n} \end{pmatrix}, \quad \rho_n = \begin{pmatrix} 0 & 1_n \\ 1_n & 0 \end{pmatrix}, \quad f = 1_{2n} \times \operatorname{diag}\left[(2\delta)^{-1} 1_n, 2\delta 1_n\right]_{\mathbf{h}}. \tag{3.3}$$

Lemma 3.2. *If* $A = \tau(\xi \times 1) f^{-1}$ *with* $\xi \in Z'$, *then* $A \in P_{\mathbf{A}}^{2n} D^{2n}[2\mathfrak{d}^{-1}, 2\mathfrak{d}]$.

Proof. Put $D^{2n} = D^{2n}[2\mathfrak{d}^{-1}, 2\mathfrak{d}]$ for simplicity. We have

$$(c_A \quad d_A)_{\mathbf{h}} = \begin{pmatrix} c_\xi & 2\delta 1_n & d_\xi & 0 \\ a_\xi & 0 & b_\xi & (2\delta)^{-1} 1_n \end{pmatrix}_{\mathbf{h}}. \tag{3.4}$$

By [S10, (3.1)] $A \in P_{\mathbf{A}}^{2n}D^{2n}$ if and only if $\det(d_A)_v \neq 0$ and $(d_A^{-1}c_A)_v \in (2\mathfrak{d}_v)_{2n}^{2n}$ for every $v|2$. By (3.4) this is so if and only if

$$\text{(3.5)} \quad \det(d_\xi)_v \neq 0, \ (d_\xi)_v^{-1} \in (\mathfrak{g}_v)_n^n, \ (b_\xi d_\xi^{-1})_v \in (2^{-1}\mathfrak{d}_v^{-1})_n^n, \\ \text{and } (d_\xi^{-1}c_\xi)_v \in (2\mathfrak{d}_v)_n^n \quad \text{for every} \quad v|2 .$$

Therefore our assertion follows from Lemma 3.1.

Let $A = \tau(\xi \times 1)f^{-1}$ with $\xi \in Z' \cap \Omega'$. From Lemma 3.2 and (3.4) we see that $d_A \in GL_{2n}(F)_{\mathbf{A}}$ and $A \in P_{\mathbf{A}}^{2n}D^{2n}$, and also that

$$\text{(3.6)} \qquad \gamma(\delta^{-2}d_A^{-1}c_A) = N(\mathfrak{d}\,)^n\kappa(\xi)\,,$$

and hence $\kappa(\xi) \neq 0$, and (2.6) shows that

$$\text{(3.7)} \qquad h\left(r'(A),\ \operatorname{diag}[z,w]\right) = \mathbf{t}(\xi)p(\xi,z)q(\xi(z),w)\ ,$$

where $q(z,w)$ is a function of $(z,w) \in \mathscr{H}^n \times \mathscr{H}^n$ given by

$$\text{(3.8)} \qquad q(z,w) = \prod_{v \in \mathbf{a}} \det(1_n - z_v w_v)^{1/2}\,,$$

with the stipulation that $q > 0$ for pure imaginary z and w. Now put

$$\text{(3.9)} \qquad H_0(\xi,z) = \mathbf{t}(\xi)p(\xi,z) \quad (\xi \in Z' \cap \Omega')\ .$$

Lemma 3.3. *There is a holomorphic function $H(\alpha,z)$ of $z \in \mathscr{H}^n$ defined for each $\alpha \in \mathscr{Z}'$ with the following properties:*

(1) $H(\alpha\zeta,z) = H(\alpha,\zeta z)h(\zeta,z)$ *if* $\zeta \in U$,

(2) $H(\zeta\alpha,z) = h(\zeta,\alpha z)H(\alpha,z)$ *if* $\zeta \in U$,

(3) $H(r'(\beta),z) = H_0(\beta,z)$ *if* $\beta \in P'_{\mathbf{A}} \cap Z'$,

(4) $H(\xi,z) = H_0(\xi,z)$ *if* $\xi \in G \cap Z'$.

Proof. In this proof we repeatedly employ the formulas of Lemma 2.1, (3) and (4) in particular, though we will not mention which one we are using in each instance. Let β, $\beta\eta \in Z' \cap \Omega'$ with $\eta \in D$; take an element $\zeta \in U$ such that $\operatorname{pr}(\zeta) = \eta$. Then $r'(\tau(\beta\eta \times 1)) = r'(\tau(\beta \times 1))\langle t\zeta, 1\rangle$ with $t \in \mathbf{T}$. Changing ζ for $t\zeta$, we may assume that $t = 1$. In this setting we have

$$\text{(3.10)} \qquad H_0(\beta\eta,z) = H_0(\beta,\eta z)h(\zeta,z)\,.$$

To prove this, put $g = r_P(f)$, $A = \tau(\beta\eta \times 1)f^{-1}$, and $B = \tau(\beta \times 1)f^{-1}$. Then

$$r'(A) = r'(\tau(\beta\eta \times 1))g^{-1} = r'(\tau(\beta \times 1))\langle\zeta, 1\rangle g^{-1} = r'(B)g\langle\zeta, 1\rangle g^{-1}.$$

Since $f = 1 \times e$ with $e \in G_{\mathbf{A}}^n$, we have $g\langle\zeta, 1\rangle g^{-1} = \langle\zeta, 1\rangle$, and so $r'(A) = r'(B)\langle\zeta, 1\rangle$. Since $\eta \times 1 \in D^{2n}[2\mathfrak{d}^{-1}, 2\mathfrak{d}]$, by (1.9c) we have

$$h\left(r'(A), \mathfrak{z}\right) = h\left(r'(B), (\eta \times 1)\mathfrak{z}\right) h(\langle\zeta, 1\rangle, \mathfrak{z}) \quad \left(\mathfrak{z} \in \mathcal{H}^{2n}\right). \tag{3.11}$$

For $\mathfrak{z} = \mathrm{diag}[z, w]$ this together with (3.7) and (1.24) gives

$$\mathbf{t}(\beta\eta)\, p(\beta\eta, z)\, q(\beta\eta(z), w) = \mathbf{t}(\beta)\, p(\beta, z)\, q(\beta\eta(z), w)\, h(\zeta, z),$$

from which we obtain (3.10).

Now for $\beta \in Z' \cap \Omega'$ we define an element β^* of $M_{\mathbf{A}}$ by $r'(\tau(\beta \times 1)) = \tau\langle\beta^*, 1\rangle$. Observe that $\beta^* = \beta$ if $\beta \in G$. Given $\alpha \in \mathscr{Z}'$ we can find an element $\beta \in Z' \cap \Omega'$ and $\varphi \in U$ such that $\alpha = \beta^*\varphi$. (In fact, by strong approximation we can find $\xi \in G$ and $\varphi \in U$ such that $\alpha = \xi\varphi$. Clearly $\xi \in Z'$. Since $G \cap Z' \subset G \cap P_{\mathbf{A}}D \subset \Omega'$, we can take ξ as β. However, we consider a more general β.) With such β and φ we put

$$H(\alpha, z) = H_0(\beta, \varphi z)\, h(\varphi, z). \tag{3.12}$$

To show that this is well defined, take $\gamma \in Z' \cap \Omega'$ and $\psi \in U$ so that $\beta^*\varphi = \gamma^*\psi$, and put $\eta = \beta^{-1}\gamma$. Then $\eta = \mathrm{pr}(\varphi\psi^{-1}) \in D$. Define ζ as in (3.10). Then

$$\tau\langle\beta^*\zeta, 1\rangle = r'(\tau(\beta \times 1))\langle\zeta, 1\rangle = r'(\tau(\gamma \times 1)) = \tau\langle\gamma^*, 1\rangle,$$

which shows that $\varphi = \zeta\psi$. Therefore by (3.10)

$$\begin{aligned} H_0(\gamma, \psi z)h(\psi, z) &= H_0(\beta\eta, \psi z)h(\psi, z) \\ &= H_0(\beta, \eta\psi z)h(\zeta, \psi z)h(\psi, z) \\ &= H_0(\beta, \varphi z)h(\varphi, z), \end{aligned}$$

and hence H is well defined. We are going to show that this H has the properties of our lemma. Clearly (1) follows immediately from (3.12). If $\beta \in P'_{\mathbf{A}}$, then we have $r'(\tau(\beta \times 1)) = r'(\tau)r'(\beta \times 1) = \tau\langle r'(\beta), 1\rangle$, and hence $\beta^* = r'(\beta)$, which together with (3.12) proves (3). Similarly, if $\beta \in G \cap Z'$, then $\beta^* = \beta$, which proves (4).

As a preliminary step to (2) put $D_2 = \prod_{v|2} D_v[2\mathfrak{d}^{-1}, 2\mathfrak{d}]$ and $M = D_2(P_{\mathfrak{h}} \cap D)$. Clearly $M \subset Z' \cap \Omega'$. For $\gamma \in M$ we have $\gamma^* \in U$ and $h(\gamma^*, z) = H(\gamma^*, z) = H_0(\gamma, z) = \mathbf{t}(\gamma)$ by (1) and (3.12). Since $\gamma \in D$ and $d_{\gamma} \in E$, we have $\mathbf{t}(\gamma) = 1$. Thus γ^* is the unique element of U such that $\mathrm{pr}(\gamma^*) = \gamma$ and $h(\gamma^*, z) = 1$. Hence $(\gamma_1\gamma_2)^* = \gamma_1^*\gamma_2^*$ for $\gamma_1, \gamma_2 \in M$. In particular, $\pi^* = r_P(\pi)$ if $\pi \in P_{\mathfrak{h}} \cap D$.

To prove (2), express α in the form $\alpha = \beta\varepsilon\gamma$ with $\varepsilon = r_P(\mathrm{diag}[q, \tilde{q}]), q \in B$ and $\beta, \gamma \in U$. Put $Q = U \cap \beta\varepsilon U(\beta\varepsilon)^{-1}$, and let $\zeta \in Q$. Then $\zeta\beta\varepsilon = \beta\varepsilon\xi$ with $\xi \in U$. Now we can put $H(\beta\varepsilon, z) = t \cdot h(\beta, z)$ with $t \in \mathbf{T}$. Then, employing (1), we obtain

$$H(\zeta\alpha, z) = H(\zeta\beta\varepsilon\gamma, z) = H(\beta\varepsilon\xi\gamma, z) = H(\beta\varepsilon, \xi\gamma z)h(\xi\gamma, z) = t \cdot h(\beta\xi\gamma, z).$$

Now $h(\beta^{-1}\zeta\beta,z) = h(\xi,z)$ by Lemma 2.2, and hence $h(\beta\xi\gamma,z) = h(\zeta\beta\gamma,z)$. Therefore

$$\begin{aligned} t\cdot h(\beta\xi\gamma,z) &= t\cdot h(\zeta\beta\gamma,z) \\ &= h(\zeta,\alpha z)t\cdot h(\beta,\gamma z)h(\gamma,z) \\ &= h(\zeta,\alpha z)H(\beta\varepsilon,\gamma z)h(\gamma,z) \\ &= h(\zeta,\alpha z)H(\alpha,z). \end{aligned}$$

Thus (2) is true if $\zeta \in Q$ with Q determined for α as above. Now every element of U is a product $\zeta'\zeta''$ with $\zeta'' \in Q$ and $\zeta' \in U$ such that $\mathrm{pr}(\zeta')_v = 1$ for almost all v. Such a ζ' is an element of $\mathbf{T}$ times a product of elements ζ of the types $\zeta = r_P(\eta)$ with $\eta \in D_v \cap P'_v$, $d_\eta = 1$, or $\zeta = r'(\eta)$ with $\eta \in D_v \cap P_v$. Thus it is sufficient to prove (2) for ζ of such forms. Fix such a ζ. Since $G_v = P'_v D_v$ for $v \nmid 2$, we have $Z' \subset D_2 P'_{\mathbf{h}} D$. Therefore, given $\alpha \in \mathscr{Z}'$, we can find $\beta \in P'_{\mathbf{h}}$ and $\sigma \in D_2$ such that $\mathrm{pr}(\alpha) \in \sigma\beta D$. Put $\xi = \sigma^* r'(\beta)$. Then $\xi \in \mathscr{Z}'$ and $\alpha = \xi\varphi$ with $\varphi \in U$. Suppose (2) is true for such a ζ with this ξ in place of α. Then, by (1)

$$H(\zeta\alpha,z) = H(\zeta\xi,\varphi z)h(\varphi,z) = h(\zeta,\alpha z)H(\xi,\varphi z)h(\varphi,z) = h(\zeta,\alpha z)H(\alpha,z),$$

which proves (2) for the given α. This means that we may assume that $\alpha = \sigma^* r'(\beta)$ with such σ and β. We easily see that $(\sigma\beta)^* = \sigma^* r'(\beta) = \alpha$, and hence $H(\alpha,z) = H_0(\sigma\beta,z) = \mathbf{t}(\sigma\beta) = \mathbf{t}(\beta)$ by (3.12) and (3.2). Replace σ by $\rho\sigma$ with any $\rho \in D_2$ and put $\zeta = \rho^*$. Then $H(\zeta\alpha,z) = \mathbf{t}(\beta) = H(\alpha,z)$. Since $h(\rho^*,z) = 1$, we thus obtain (2) for $\zeta = \rho^*$, $\rho \in D_2$.

Next, if $\zeta = r'(\eta)$ with $\eta \in D_v \cap P'_v$, $d_\eta = 1$, $v \nmid 2$, we have $\zeta\sigma^* = \sigma^*\zeta$ by Lemma 2.6, and hence $\zeta\alpha = \sigma^* r'(\eta\beta)$. Changing β for $\eta\beta$ in the above discussion, we obtain $H(\zeta\alpha,z) = \mathbf{t}(\eta\beta)$. Now we have $\mathbf{t}(\eta\beta) = \mathbf{t}(\beta)$ by Lemma 3.4 below and hence we obtain (2) in this case, since $h(\zeta,z) = \mathbf{t}_0(\eta) = 1$.

Finally let $\zeta = r_P(\eta)$ with $\eta \in P_v \cap D_v$, $v \nmid 2$. We have $\zeta = \eta^*$, and so $\zeta\alpha = \eta^*\sigma^* r'(\beta) = (\eta\sigma)^* r'(\beta) = (\eta\sigma\beta)^*$, and hence $H(\zeta\alpha,z) = H_0(\eta\sigma\beta,z) = \mathbf{t}(\eta\sigma\beta) = \mathbf{t}(\eta\beta)$ by (3.12) and (3.2). Since we easily see that $\mathbf{t}(\eta\beta) = \mathbf{t}(\beta)$ and $h(\zeta,z) = 1$, we obtain the desired result. This completes the proof.

Lemma 3.4. *If* $v \nmid 2$, $\xi = \begin{pmatrix} {}^t g \cdot {}^t h^{-1} & 0 \\ g^{-1}\sigma \cdot {}^t h^{-1} & g^{-1}h \end{pmatrix} \in P'_v$ *with* $\sigma \in S_v$ *and* $(g,h) \in B_v \times B_v$ *such that* $gL_v + hL_v = L_v$, *then* $\kappa(\xi) = |\det(\delta_v h)|\gamma(\delta_v^{-2}\sigma)$. *Moreover* $\kappa(\xi) = \kappa(\eta\xi)$ *for every* $\eta \in D_v \cap P'_v$ *such that* $d_\eta = 1$.

Proof. We have

$$\kappa(\xi) = \int_{L_v \times L_v} \mathbf{e}_v\left(2\delta_v^{-1}\cdot {}^t x h^{-1} g y + (2\delta_v^2)^{-1}\cdot {}^t x h^{-1}\sigma \cdot {}^t h^{-1} x\right) dx\, dy.$$

Now $\int_{L_v} \mathbf{e}_v(2\delta_v^{-1}\cdot {}^t x h^{-1} g y) dy \neq 0$ only if $2\cdot {}^t g \cdot {}^t h^{-1} x \in L_v$. For $x \in L_v$ this is so if and only if $x \in {}^t h L_v$ since $gL_v + hL_v = L_v$. Putting $x = {}^t h z$ with $z \in L_v$, we thus obtain $\kappa(\xi) = |\delta_v|^{n/2}|\det(h)| \int_{L_v} \mathbf{e}_v({}^t z(2\delta_v^2)^{-1}\sigma z) dz = |\det(\delta_v h)|\gamma(\delta_v^{-2}\sigma)$

as expected. Now changing ξ for $\eta\xi$ we obtain $\sigma + gc_\eta \cdot {}^t g$ in place of σ. Therefore the last assertion follows immediately from the first one.

Lemma 3.5. *$J^{u/2}(\alpha, z) = H(\alpha^{-1}, \alpha z)^{-1}$ for every $\alpha \in \mathfrak{Z}$.*

Proof. In view of (2.1a, b) and Lemma 3.3(1, 2) it is sufficient to prove the equality for $\alpha = r_P(\beta)$, $\beta = \mathrm{diag}[\tilde{q}, q]$ with $q \in B$. For such an α we have $J^{u/2}(\alpha, z) = 1$. By Lemma 3.3(3) and (3.9) $H(\alpha^{-1}, z) = H_0(\beta^{-1}, z) = \mathbf{t}(\beta^{-1})$. From the integral expression for $\kappa(\beta_v^{-1})$ we immediately see that $\mathbf{t}(\beta^{-1}) = 1$, which completes the proof.

Lemma 3.6. *If $\xi^* = \varphi^{-1}\xi\varphi$ with $\xi \in \mathfrak{Z}$ and $\varphi = r_\Omega(g)$, $g = \mathrm{diag}[\delta^{-1}1_n, \delta 1_n]\iota$, then $\xi^* \in \mathfrak{Z}'$ and $h(\varphi, \xi^* z)H(\xi^*, z) = J^{u/2}(\xi, \varphi z)h(\varphi, z)$.*

Proof. We have $\xi = \beta\sigma\gamma$ with $\beta, \gamma \in U$ and $\sigma = r_P(w)$, $w = \mathrm{diag}[\tilde{q}, q]$, $q \in B$. Put $w' = \mathrm{diag}[q, \tilde{q}]$ and $\zeta^* = \varphi^{-1}\zeta\varphi$ for every $\zeta \in \mathfrak{Z}$. Then $\xi^* = \beta^*\sigma^*\gamma^*$, $wg = gw'$, and $r_P(w)\varphi = r_\Omega(wg) = r_\Omega(gw') = \varphi r_P(w')$. Thus $\sigma^* = r_P(w')$. Since $g^{-1}Dg = D$, we have $\beta^*, \gamma^* \in U$, and hence $\xi^* \in \mathfrak{Z}'$. By Lemma 3.3(3) and (3.9) we have $H(\sigma^*, z) = 1$, and hence $H(\xi^*, z) = h(\beta^*\gamma^*, z)$ by Lemma 3.3(1, 2). Since φ, β, and γ belong to $\mathrm{pr}^{-1}(C^\theta)$, by (1.9c) we have

$$
\begin{aligned}
h(\varphi, \xi^* z)H(\xi^*, z) &= h(\varphi, \varphi^{-1}\xi\varphi z)h(\varphi^{-1}\beta\gamma\varphi, z) \\
&= h(\beta\gamma\varphi, z) \\
&= h(\beta\gamma, \varphi z)h(\varphi, z) \\
&= J^{u/2}(\xi, \varphi z)h(\varphi, z)
\end{aligned}
$$

as expected.

Lemma 3.7. *If $\beta = g^{-1}\alpha g$ with $g = \mathrm{diag}[\delta^{-1}1_n, \delta 1_n]\iota$ and $\alpha \in P_\mathbf{h} \cap Z$, then $J^k(r_P(\alpha), z) = \mathbf{t}(\beta)$.*

Proof. Put $\varphi = r_\Omega(g)$. By Lemma 2.1(2) we have $r'(\beta)\varphi^{-1} = r_\Omega(\beta g^{-1}) = r_\Omega(g^{-1}\alpha) = r_\Omega(g^{-1})r_P(\alpha) = \varphi^{-1}r_P(\alpha)$. By Lemma 3.6 we have $h(\varphi, z)H(r'(\beta), z) = J^{u/2}(r_P(\alpha), z)h(\varphi, z)$, and hence $J^{u/2}(r_P(\alpha), z) = H(r'(\beta), z)$. This together with Lemma 3.3(3) and (3.9) proves the desired equality.

We conclude this section by proving Lemmas 2.3 and 2.4.

Proof of Lemma 2.3. Since $\mathrm{pr}(\xi) \in Z$ and $\mathrm{pr}(\xi)^{-1} \in D \cdot \mathrm{pr}(\xi)D$, we have $\xi \in \mathfrak{Z}$ and $U\xi^{-1}U = U\xi U$. By the above lemma, we have $J^k(\xi, z) = \mathbf{t}(\tau)$ with $\tau = \mathrm{diag}[\tilde{s}, s]$. Now

$$\kappa(\tau_v) = \int_{L_v \times L_v} \mathbf{e}_v\left(2\delta_v^{-1} \cdot {}^t x s_v^{-1} y\right) dx\, dy,$$

and it can easily be seen that this is a real positive number, and hence $\mathbf{t}(\tau) = 1$, from which we obtain $J^k(\xi, z) = 1$. Changing ξ for ξ^{-1}, we obtain $J^k(\xi^{-1}, z) = 1$. Let $\eta = \beta\xi\gamma$ with $\beta, \gamma \in U$. Since $\eta^{-1} = \gamma^{-1}\xi^{-1}\beta^{-1}$ and

J^k is a factor of automorphy on U by (2.1a) and (1.9c), from (2.1b) we obtain $J^k(\eta^{-1},z) = J^k(\gamma^{-1}\beta^{-1},z) = J^k(\beta\gamma,\gamma^{-1}\beta^{-1}z)^{-1} = J^k(\eta,\eta^{-1}z)^{-1}$, which completes the proof.

Proof of Lemma 2.4. For α as in Lemma 2.4 take β as in Lemma 3.7. Then we see that β_v has the same form as ξ_r of Lemma 3.4 with $-\delta_r^2\sigma_r$ in place of σ. Therefore $J^k(r_P(\alpha),z) = \mathbf{t}(\beta) = \mathbf{g}(-\sigma)$, which proves Lemma 2.4.

4. Hecke operators and formal Euler products

We now consider formal Dirichlet series of the form $\sum_{\mathfrak{a}\subset\mathfrak{g}} c(\mathfrak{a})[\mathfrak{a}]$ with multiplicative symbols $[\mathfrak{a}]$ defined for the fractional ideals $\mathfrak{a}$ in F (see [S9, Sect. 2]). For each $v \in \mathbf{h}$ we take a prime element π_v of F_v and put $[v] = [\pi_v\mathfrak{g}]$. The coefficients $c(\mathfrak{a})$ are in $\mathbf{C}$ for the most part, but they may be in the space $\mathcal{M}_k$ or in the rings we are going to introduce in this section. Now, the notation being as in (1.18a, b, c), put

$$S^* = \{\sigma \in S_{\mathbf{h}} \,|\, \mathbf{e}(\mathrm{tr}(S_{\mathbf{h}}(\mathfrak{g})\sigma)) = 1\}\,. \tag{4.1}$$

For $\zeta \in S_{\mathbf{A}}$ such that $\zeta_{\mathbf{h}} \in \delta S^*$ we define a formal Dirichlet series $\alpha(\zeta)$ by

$$\alpha(\zeta) = \prod_{v\nmid 2}\alpha_v(\zeta_v), \quad \alpha_v(\zeta_v) = \sum_{\sigma\in S_v/S(\mathfrak{g})_v} \mathbf{g}\left(\delta_v^{-1}\sigma\right)\mathbf{e}_v\left(-\delta_v^{-1}\mathrm{tr}(\zeta_v\sigma)\right)[\nu_0(\sigma)]\,, \tag{4.2}$$

where $\nu_0(\sigma)$ is the denominator ideal of σ defined in [S9, (2.3)].

Given $\xi \in S_v^r \cap GL_r(F_v)$ with odd r, put $\rho = \left[(-1)^{(r-1)/2}\det(2\xi)\right]^{1/2}$. We then define $\varepsilon(\xi)$ as follows: $\varepsilon(\xi) = 1$ if $\rho \in F_v$; $\varepsilon(\xi) = -1$ if $\rho \notin F_v$ and v is unramified in $F_v(\rho)$; $\varepsilon(\xi) = 0$ otherwise.

Proposition 4.1. *For $\zeta \in S(\mathfrak{g})_v$, $v\nmid 2$, let $\alpha_\zeta(t)$ denote the formal power series in an indeterminate t such that $\alpha_\zeta([v]) = \alpha_v(\zeta)$. If $\zeta = \mathrm{diag}[\xi,0]$ with $\xi \in GL_r(F_v)$, then $\alpha_\zeta = f_\zeta g_\zeta$ with a polynomial f_ζ with coefficients in $\mathbf{Z}$ whose constant term is 1 and a rational function g_ζ given as follows:*

$$g_\zeta(t) = \frac{\prod_{i=1}^{[(n+1)/2]}(1-q^{2i-1}t^2)}{(1-\varepsilon(\xi)q^{(2n-r)/2}t)\prod_{i=1}^{[(n-r)/2]}(1-q^{2n-2i-r+1}t^2)} \quad \textit{if } r \textit{ is odd}\,,$$

$$g_\zeta(t) = \frac{\prod_{i=1}^{[(n+1)/2]}(1-q^{2i-1}t^2)}{\prod_{i=1}^{[(n-r+1)/2]}(1-q^{2n-2i-r+2}t^2)} \quad \textit{if } r \textit{ is even}\,,$$

where $q = |\pi_v|^{-1}$. In particular, $\alpha_\zeta = g_\zeta$ if $\det(\xi)$ is a v-unit or $\zeta = 0$ with the understanding that $r = 0$ for $\zeta = 0$. Moreover

$$\alpha_\zeta(t)/\alpha_\xi(q^{n-r}t) = \prod_{i=1}^{n-r}(1-q^{2i-1}t)/(1-q^{n+i}t^2)\,.$$

The first part concerning α_ζ was obtained in [S4] when $\zeta = 0$ or $\det(\xi)$ is a v-unit. The general case was proved by Feit [F]. The proof is also given in [S5, Prop. 5.1]. The last formula can be obtained by combining [S5, (5.9)] with [S9, (3.4)]. (An idea of Feit [F] is essential here.)

To define Hecke operators, we take a fractional ideal $\mathfrak{b}$, an integral ideal $\mathfrak{c}$ divisible by 4, and a divisor $\mathfrak{e}$ of $\mathfrak{c}$ such that

$$(4.3)\qquad \mathfrak{b}^{-1} \subset 2\mathfrak{d}^{-1}, \quad \mathfrak{bc} \subset 2\mathfrak{d}, \quad \text{and} \quad \mathfrak{e} + \mathfrak{e}^{-1}\mathfrak{c} = \mathfrak{g}.$$

Then $D\left[\mathfrak{b}^{-1}, \mathfrak{bc}\right] \subset D\left[2\mathfrak{d}^{-1}, 2\mathfrak{d}\right]$. With U, Z_0, and $\mathfrak{X}$ as in Section 2, we put

$$(4.4a)\qquad K = G_{\mathbf{a}} \prod_{v \in \mathbf{h}} K_v, \quad K_v = \left\{x \in D_v\left[\mathfrak{b}^{-1}\mathfrak{e}, \mathfrak{bc}\right] \mid a_x - 1 \in (\mathfrak{e}_v)^n_n\right\},$$

$$(4.4b)\qquad \mathfrak{X}^0 = \{\alpha \in \mathfrak{X} \mid \mathrm{pr}(\alpha) \in G_{\mathbf{h}} \cap KZ_0K, \quad \mathrm{pr}(\alpha)_v \in K_v \text{ for every } v|\mathfrak{e}\},$$

$$(4.4c)\qquad U^0 = \{\alpha \in U \mid \mathrm{pr}(\alpha) \in G_{\mathbf{h}} \cap K\},$$

$$(4.4d)\qquad W = \left\{(\alpha, t) \mid t \in \mathbf{T}, \alpha \in \mathfrak{X}^0\right\}, \quad V = \left\{(\alpha, 1) \in W \mid \alpha \in U^0\right\},$$

$$(4.4e)\qquad J(\alpha) = J^k(\alpha, \mathbf{i}) \quad (\alpha \in \mathfrak{X}^0).$$

By [S9, Lemma 2.4] $\mathfrak{X}^0$ is closed under multiplication. We define a law of multiplication in W by

$$(4.5)\qquad (\alpha, t)(\alpha', t') = \left(\alpha\alpha', tt'J(\alpha)J(\alpha')/J(\alpha\alpha')\right).$$

In view of (2.1a, b) this is associative. By Lemma 2.3 and (2.1b) we have

(4.6a)
$$J(\alpha^{-1}) = J(\alpha)^{-1} \quad \text{and} \quad (\alpha, t)^{-1} = (\alpha^{-1}, t^{-1}) \text{ if } \mathrm{pr}(\alpha)_v \in K_v \text{ for every } v|2,$$

$$(4.6b)\qquad J(\alpha\beta) = J(\alpha)J(\beta) \quad \text{if} \quad \alpha \in U^0 \quad \text{or} \quad \beta \in U^0.$$

Clearly V is a subgroup of W. We thus consider the Hecke ring $\mathfrak{R}(V, W)$ consisting of all formal finite sums $\sum_\sigma c_\sigma V\sigma V$ with $c_\sigma \in \mathbf{C}$ and $\sigma \in W$ with the law of multiplication defined as in [S1, Sect. 3.1].

Proposition 4.2. *The ring $\mathfrak{R}(V, W)$ is commutative.*

Proof. Since $(1, t)$ commutes with every element of W, it is sufficient to show that $AB = BA$ for $A = V(\alpha, 1)V$ and $B = V(\beta, 1)V$. Notice that A as a set consists of all $(\varphi, 1)$ with $\varphi \in U^0\alpha U^0$.

(I) We first assume that $\alpha = r_P(\zeta)$, $\beta = r_P(\omega)$, $\zeta = \mathrm{diag}[\tilde{f}, f]$, $\omega = \mathrm{diag}[\tilde{g}, g]$ with $f \in \prod_{v|\mathfrak{a}} B_v$, $g \in \prod_{v|\mathfrak{b}} B_v$, where $\mathfrak{a}$ and $\mathfrak{b}$ are relatively prime integral ideals. Then $K\zeta K = \bigsqcup_\sigma K\zeta\sigma$, $K\omega K = \bigsqcup_\tau K\omega\tau$ with $\sigma, \tau \in K$ such that $\sigma_v = 1$ for $v \nmid \mathfrak{a}$ and $\tau_v = 1$ for $v \nmid \mathfrak{b}$. Clearly $A = \bigsqcup_\xi V(\alpha\xi, 1)$ and $B = \bigsqcup_\eta V(\beta\eta, 1)$ with $\xi, \eta \in U^0$ such that $\mathrm{pr}(\xi) = \sigma$ and $\mathrm{pr}(\eta) = \tau$. Then $\alpha\xi\beta\eta = \alpha\beta\xi\eta$ by

Lemma 2.6 and $J(\alpha\xi\beta\eta) = J(\xi\eta) = J(\xi)J(\eta) = J(\alpha\xi)J(\beta\eta)$. Thus $(\alpha\xi, 1)(\beta\eta, 1) = (\alpha\beta\xi\eta, 1)$. From this we easily see that AB is $V(\alpha\beta, 1)V$ with multiplicity 1. Since $\alpha\beta = \beta\alpha$, we obtain $AB = BA$.

(II) Using the same notation as in (I), let us now assume that $f \in B_v$, $g \in B_v$ with the same fixed prime v (instead of assuming $\mathfrak{a} + \mathfrak{b} = \mathfrak{g}$). The result of (I) reduces our task to the problem of proving $AB = BA$ in this case. We first assume $v|\mathfrak{c}$. Our problem is trivial if $v|\mathfrak{c}$, and so we assume $v|\mathfrak{e}^{-1}\mathfrak{c}$. As shown in [S9, Lemma 2.3] we can take $\zeta\sigma$ and $\omega\tau$ in the forms $\zeta\sigma = \begin{pmatrix} \tilde{d} & \tilde{d}b \\ 0 & d \end{pmatrix}$, $\omega\tau = \begin{pmatrix} \tilde{e} & \tilde{e}c \\ 0 & e \end{pmatrix}$ with $d \in fE_v$, $e \in gE_v$, and $b, c \in (\mathfrak{b}_v^{-1})_n^n$. Then $\zeta\sigma\omega\tau = \begin{pmatrix} \tilde{d}\tilde{e} & \tilde{d}\tilde{e}x \\ 0 & de \end{pmatrix}$ with $x = c + {}^t ebe$. If γ denotes $\zeta\sigma$, $\omega\tau$, or $\zeta\sigma\omega\tau$, then we have $J(r_P(\gamma)) = 1$ by (2.7) or Lemma 2.4. Thus $(\varphi, 1)(\psi, 1) = (\varphi\psi, 1)$ for $\varphi = r_P(\zeta\sigma)$ and $\psi = r_P(\omega\tau)$. From this we easily see that $U^0\alpha U^0 \mapsto K_v\zeta_v K_v$ is a ring injection. Since $\mathfrak{R}(K_v, KZ_0K \cap G_v)$ is commutative (see [S9, p. 542]), we obtain the desired conclusion.

(III) Next we assume $v \nmid \mathfrak{c}$. By Lemma 2.3 if $\sigma \in \mathscr{X}^0$ and $\mathrm{pr}(\sigma) \in G_v$, then $U^0\sigma U^0 = U^0\sigma^{-1}U^0$, and hence $V(\sigma, t)^{-1}V = V(\sigma, t^{-1})V$ for every $t \in \mathbf{T}$. In particular $A = V(\alpha, 1)^{-1}V$ and $B = V(\beta, 1)^{-1}V$. Take a component $V\zeta V$ of AB; let $c(\zeta)$ be the multiplicity of $V\zeta V$ in AB. Then the proof of [S1, Prop. 3.8] shows that $c(\zeta)$ equals the multiplicity of $V\zeta^{-1}V$ in BA. Therefore we have $AB = BA$ if we can show that $c(\zeta) = c(\zeta^{-1})$. (The matter would be simpler if $V\zeta V = V\zeta^{-1}V$. But that is false in general.) For this purpose put $\zeta = (\gamma, t)$; we may assume that $\gamma \in r_P(G_v \cap Z_0)$. Take subsets X and Y of $\mathscr{X}^0$ such that $A = \bigsqcup_{\xi \in X} V(\xi, 1)$ and $B = \bigsqcup_{\eta \in Y} V(\eta, 1)$. Then $c(\zeta)$ is the number of $(\xi, \eta) \in X \times Y$ such that $(\xi, 1)(\eta, 1) \in V(\gamma, t)$. It can easily be seen that is so if and only if $\xi\eta \in U^0\gamma$ and $t = J(\xi)J(\eta)/J(\xi\eta)$. Consider the automorphism $\alpha \to \alpha'$ of $M_{\mathbf{A}}$ of Lemma 2.5. Then we easily see that $\gamma' = \gamma$, $A = \bigsqcup_{\xi \in X} V(\xi', 1)$, and $B = \bigsqcup_{\eta \in Y} V(\eta', 1)$. If $\xi\eta \in U^0\gamma$ and $t = J(\xi)J(\eta)/J(\xi\eta)$, then $\xi'\eta' \in U^0\gamma$ and $t^{-1} = J(\xi')J(\eta')/J(\xi'\eta')$ by Lemma 2.5, and vice versa. Since $V\zeta^{-1}V = V(\gamma, t^{-1})V$, this shows that $c(\zeta)$ is the multiplicity of $V\zeta^{-1}V$ in AB. This completes the proof.

For $\zeta = (\gamma, t) \in W$ and a $\mathbf{C}$-valued function g on $M_{\mathbf{A}}$ we define $J(\zeta) \in \mathbf{T}$ and a function $g|\zeta$ on $M_{\mathbf{A}}$ by

$$(4.7) \qquad J(\zeta) = J(\gamma)t, \quad (g|\zeta)(x) = J(\zeta)^{-1}g(x\gamma^{-1}) \quad (x \in M_{\mathbf{A}}).$$

Clearly $J(\zeta\zeta') = J(\zeta)J(\zeta')$ and $g|(\zeta\zeta') = (g|\zeta)|\zeta'$. Let us now take a Hecke character ψ as in (1.12). Given $g \in \mathscr{M}_k(K, \psi)$, and $X = V(\sigma, t)V$ with $(\sigma, t) \in W$, we consider a decomposition $X = \bigsqcup_\zeta V\zeta$, and define a function $g|X_\psi$ on $M_{\mathbf{A}}$ by

$$(4.8) \qquad (g|X_\psi)(x) = \sum_\zeta \psi_{\mathfrak{c}}(\det(a_\zeta))^{-1}(g|\zeta)(x \in M_{\mathbf{A}}),$$

where $a_\zeta = a_\gamma$ when $\zeta = (\gamma, t)$. We write $X_\psi = T_{q,\psi}$ if $t = 1$ and $\sigma = r_P(\mathrm{diag}[\tilde{q}, q])$ with $q \in B$. (Since $\sigma \in \mathscr{X}^0$, we have $\mathrm{diag}[\tilde{q}, q]_v \in K_v$ for every $v|\mathfrak{e}$.) It can easily be seen that $g|X_\psi \in \mathscr{M}_k(K, \psi)$ and $X \mapsto X_\psi$ is a ring homomorphism of $\mathfrak{R}(V, W)$ into the ring of operators on $\mathcal{M}_k(K, \psi)$. Put $\Gamma = G \cap K$. Then $G \cap (K \operatorname{diag}[\tilde{q}, q]K) = \Gamma\xi\Gamma = \bigsqcup_\alpha \Gamma\alpha$ with some elements ξ and α of $G \cap Z$. Moreover, if $g = f_{\mathbf{A}}$ with $f \in \mathcal{M}_k(\Gamma, \psi)$, then $g|T_{q,\psi} = (f|T_{q,\psi})_{\mathbf{A}}$ with $f|T_{q,\psi}$ defined on $\mathscr{H}$ by

(4.9) $$\big(f|T_{q,\psi}\big)(z) = \sum_\alpha \psi_{\mathfrak{c}}(\det(a_\alpha))^{-1} J^k(\alpha, z)^{-1} f(\alpha z) \quad (z \in \mathscr{H}).$$

With $\mathfrak{e}$ as in (4.3) and B, E as in Section 2, we put

(4.10a) $$B'_v = \{x \in B_v | x_v - 1 \in (\mathfrak{e}_v)^n_n\}, \quad E'_v = B'_v \cap E_v,$$

(4.10b) $$B' = \{x \in B | x_v \in B'_v \text{ for every } v\}, \quad E' = B' \cap E.$$

Given $\alpha \in D[\mathfrak{b}^{-1}, \mathfrak{b}] \cdot \operatorname{diag}[\tilde{q}, q] D[\mathfrak{b}^{-1}, \mathfrak{b}]$ with $q \in B$, put

(4.11) $$\ell_0(\alpha) = \det(q)\mathfrak{g}, \quad \ell(x) = N(\ell_0(x)).$$

Then, by [S9, Lemma 2.7], for α of (2.8) we have

(4.12) $$\ell_0(\alpha) = \det(gh)\nu_0(b\alpha),$$

where b is an element of $F_{\mathbf{h}}^\times$ such that $\mathfrak{b} = b\mathfrak{g}$.

Let $\mathfrak{R}'$ denote the factor ring of $\mathfrak{R}(V, W)$ modulo its ideal generated by $V(\alpha, 1)V - t \cdot V(\alpha, t)V$ for all $(\alpha, t) \in W$. Since the elements of the ideal act as 0 on $\mathcal{M}_k(K, \psi)$, we can naturally define the action of $\mathfrak{R}'$ on $\mathcal{M}_k(K, \psi)$. Let A_q denote the element of $\mathfrak{R}'$ represented by $V(r_P(\sigma), 1)V$ with $\sigma = \operatorname{diag}[\tilde{q}, q]$, $q \in B'$, and $\mathfrak{R}'_v$ denote the subalgebra of $\mathfrak{R}'$ generated by the A_q for all $q \in B'_v$. Clearly $\mathfrak{R}'_v = \mathbf{C}$ if $v|\mathfrak{e}$. We then define a map

(4.13a) $$\Phi_v : \mathfrak{R}'_v \to \mathfrak{R}(E'_v, GL_n(F_v))$$

as follows. By [S9, Lemma 2.1] if $\sigma = \operatorname{diag}[\tilde{q}, q]$ with $q \in B'_v$, then $K_v \backslash K_v\sigma K_v$ can be taken from P_v. Therefore it can easily be seen that

(4.13b) $$K_v\sigma K_v = \bigsqcup_{x \in X} \bigsqcup_{s \in Y_x} \bigsqcup_{d \in R_x} K_v\xi_{d,s}, \quad \xi_{d,s} = \begin{pmatrix} \tilde{d} & sd \\ 0 & d \end{pmatrix},$$

where $X \subset GL_n(F_v)$, R_x is a subset of xE'_v representing $E'_v \backslash E'_v x E'_v$, and $Y_x \subset S_v$. Then we put

(4.14a) $$\Phi_v(A_q) = \sum_{x \in X} \sum_{s \in Y_x} J(r_P(\xi_{x,s}))^{-1} E'_v x E'_v,$$

where we put $J(\alpha) = J((\alpha, 1))$ for $\alpha \in \mathscr{Z}^0$ for simplicity. Clearly $J(\alpha) = J^k(\alpha, z)$. Since $J(r_P(\xi_{d,s})) = J(r_P(\xi_{x,s}))$ for every $d \in xE'_v$, (4.14a) can be written symbolically

(4.14b) $$\Phi_v(A_q) = \sum_{d,s} J\big(r_P(\xi_{d,s})\big)^{-1} E'_v d.$$

Therefore we easily see that Φ_v is a well defined ring homomorphism.

Lemma 4.3. *The map Φ_v is injective*

Proof. This is trivial if $v|\mathfrak{e}$. Our task for v/e is to show that the $\Phi_v(A_q)$ for q in any finite subset Q of $E_v\backslash B_v/E_v$ are linearly independent over $\mathbf{C}$. Suppose $v \nmid \mathfrak{c}$. Let r be an element of Q such that $|\det(r)| = \mathrm{Min}_{q\in Q}|\det(q)|$. Suppose $E_v r^{-1}E_v$ appears in $\Phi_v(A_q)$ for some $q \in Q$. Then $K_v\mathrm{diag}[\tilde{q}, q]K_v$ contains an element of the form $\begin{pmatrix} {}^t r & sr^{-1} \\ 0 & r^{-1} \end{pmatrix}$. By (4.12) we have $\det(q)\mathfrak{g} = \det(r)\nu_0(b_v s)$, and hence $\nu_0(b_v s) = \mathfrak{g}$, that is, $\begin{pmatrix} 1 & s \\ 0 & 1 \end{pmatrix} \in K_v$. Consequently $E_v q E_v = E_v r E_v$. Thus $E_v r^{-1}E_v$ doesn't appear in $\Phi_v(A_q)$ if $r \neq q \in Q$. Also our argument shows that $E_v r^{-1}E_v$ appears in $\Phi_v(A_r)$ with coefficient 1. This proves the case $v \nmid \mathfrak{c}$. Next suppose $v|\mathfrak{e}^{-1}\mathfrak{c}$. By [S9, Lemma 2.3], if $\sigma = \mathrm{diag}[\tilde{q}, q]$, then $K_v\sigma K_v = \bigsqcup_{d,b}\eta_{d,b}, \eta_{d,b} = \begin{pmatrix} \tilde{d} & \tilde{d}b \\ 0 & d \end{pmatrix}$, with $d \in E_v\backslash E_v q E_v$ and $b \in S(\mathfrak{b}^{-1})_v/{}^t dS(\mathfrak{b}^{-1})_v d$. Then $J(r_P(\eta_{d,b})) = 1$ by (2.7), and hence $A_q = |\det(q)|_{\mathbf{A}}^{-n-1}E_v q E_v$, which proves the desired fact.

Fixing v prime to $\mathfrak{e}$ and taking a prime element π_v of F_v and n indeterminates $t_1, \ldots, t_n$, we now define a ring-injection

$$\text{(4.15)} \qquad \omega_v' : \mathfrak{R}(E_v, GL_n(F_v)) \to \mathbf{C}[t_1, \ldots, t_n, t_n^{-1}, \ldots, t_n^{-1}]$$

(as usual) as follows: If $E_v x E_v = \bigsqcup_{y\in Y} E_v y$ and y is an upper triangular matrix with diagonal elements $\pi_v^{a_1}, \ldots, \pi_v^{a_n}$, then $\omega_v'(E_v x E_v) = \sum_{y\in Y}\omega_v'(E_v y)$ with $\omega_v'(E_v y) = \prod_{i=1}^n (|\pi_v|^i t_i)^{a_i}$. Assuming $v \nmid \mathfrak{e}$ and putting $\omega_v = \omega_v' \circ \Phi_v$, we obtain a ring homomorphism

$$\text{(4.16)} \qquad \omega_v : \mathfrak{R}_v' \to \mathbf{C}[t_1, \ldots, t_n, t_1^{-1}, \ldots, t_n^{-1}] \quad (v \nmid \mathfrak{e}),$$

which is injective by virtue of Lemma 4.3. We then put

$$\text{(4.17a)} \qquad \mathfrak{T}_v = \sum_{q\in E_v'\backslash B_v'/E_v'} [\det(q)\mathfrak{g}]A_q, \quad (v \in \mathbf{h}),$$

$$\text{(4.17b)} \qquad \omega_v(\mathfrak{T}_v) = \sum_{q\in E_v\backslash B_v/E_v} [\det(q)\mathfrak{g}]\omega_v(A_q) \quad (v \nmid \mathfrak{e}).$$

Theorem 4.4. *One has*

$$\mathfrak{T}_v = 1 \text{ if } v|\mathfrak{e}, \quad \omega_v(\mathfrak{T}_v) = \prod_{i=1}^n \left(1 - |\pi_v|^{-n}t_i[v]\right)^{-1} \quad \text{if} \quad v|\mathfrak{e}^{-1}\mathfrak{c},$$

$$\omega_v(\mathfrak{T}_v) = \prod_{i=1}^n \frac{1 - |\pi_v|^{1-2i}[v]^2}{(1 - |\pi_v|^{-n}t_i[v])\left(1 - |\pi_v|^{-n}t_i^{-1}[v]\right)} \quad \text{if} \quad v \nmid \mathfrak{c}.$$

Consequently $\mathbf{C}[t_1, \ldots, t_n, t_1^{-1}, \ldots, t_n^{-1}]$ *(resp.* $\mathbf{C}[t_1, \ldots, t_n]$*) is integral over* $\omega_v(\mathfrak{R}_v')$ *if* $v \nmid \mathfrak{c}$ *(resp.* $v\,|\,\mathfrak{e}^{-1}\mathfrak{c}$*).*

Proof. The case $v|\mathfrak{e}$ is trivial. Suppose $v \nmid \mathfrak{c}$; then $\mathfrak{b}_v = \mathfrak{d}_v$ by (4.3). By [S9, Lemma 2.6] a complete set of representatives for $K_v\backslash(G_v \cap KZ_0K)$ is given by

$$\left\{ \begin{pmatrix} g^{-1}h & g^{-1}\sigma \cdot {}^t h^{-1} \\ 0 & {}^t g \cdot {}^t h^{-1} \end{pmatrix} \middle| (g,h) \in E_v \backslash W_v/(E_v \times 1), \quad \sigma \in S_v/gS(\mathfrak{d}^{-1})_v \cdot {}^t g \right\} .$$

where $W_v = \{(g,h) \in B_v \times B_v | gL_v + hL_v = L_v\}$. Therefore, by Lemma 2.4, (4.14b), and (4.12) we have

$$\omega_v(\mathfrak{T}_v) = \sum_{g,h} \sum_{\sigma} \omega_v'(E_v \cdot {}^t g \cdot {}^t h^{-1}) \mathbf{g}(\sigma) [\det(gh) \nu_0 (\delta_v \sigma)] .$$

For each fixed g we have $\sum_\sigma \mathbf{g}(\sigma)[\nu_0(\delta_v\sigma)] = |\det(g)|^{-n-1}\alpha_v(0)$, where $\alpha_v(0)$ is the series of (4.2) with $\zeta_v = 0$. Hence

$$\omega_v(\mathfrak{T}_v) = \alpha_v(0) \sum_{g,h} |\det(g)|^{-n-1} \omega_v'(E_v \cdot {}^t g \cdot {}^t h^{-1}) [\det(gh)\mathbf{g}] .$$

In the proof of [S9, Theorem 2.9] we showed that the last sum over (g,h) equals

$$\prod_{i=1}^{n} \left(1 - |\pi_v|^{-n-i}[v]^2\right) \left(1 - |\pi_v|^{-n} t_i [v]\right)^{-1} \left(1 - |\pi_v|^{-n} t_i^{-1} [v]\right)^{-1} .$$

Now $\alpha_v(0) = \prod_{i=1}^n \left(1 - |\pi_v|^{-i}[v]^2\right) / \left(1 - |\pi_v|^{-2i}[v]^2\right)$ by Lemma 4.1. Combining these, we obtain the formula for $\omega_v(\mathfrak{T}_v)$ when $v \nmid \mathfrak{c}$. Finally suppose $v | \mathfrak{e}^{-1}\mathfrak{c}$. As shown in the proof of Lemma 4.3, we have $\omega_v(A_q) = |\det(q)|_{\mathbf{A}}^{-n-1} \omega_v'(E_v q E_v)$. Therefore a direct calculation, which is the same as in the proof of [S9, Theorem 2.9], gives the desired expression. Now our formulas tell that the polynomial $\prod_{i=1}^n (X - |\pi_v|^{-n} t_i)(X - |\pi_v|^{-n} t_i^{-1})$ or $\prod_{i=1}^n (X - |\pi_v|^{-n} t_i)$ has coefficients in $\omega_v(\mathfrak{R}_v')$, and hence we obtain the last assertion of our theorem.

In the next section we consider a nonzero common eigenfunction in $\mathscr{S}_k(K,\psi)$ of $T_\psi(\mathfrak{a})$ for all $\mathfrak{a}$, which certainly exists because of the commutativity of our ring. Moreover $\mathscr{S}_k(K,\psi)$ is spanned by the eigenfunctions of $T_\psi(\mathfrak{a})$ for all $\mathfrak{a}$ prime to $\mathfrak{c}$. This follows from the commutativity combined with the following result:

Lemma 4.5. *If $f \in \mathscr{S}_k(\Gamma,\psi)$, $g \in \mathscr{M}_k(\Gamma,\psi)$, $q \in B'$, and $q_v \in E_v$ for every $v|\mathfrak{c}$, then $\langle f | T_{q,\psi}, g\rangle = \langle f, g | T_{q,\psi}\rangle$, where $\langle\,,\rangle$ is the inner product of* [S10, (2.3a)].

Proof. Clearly $K\tau K = K\tau^{-1}K$ for $\tau = \mathrm{diag}[\tilde{q}, q]$. Take $\xi \in G$ so that $\Gamma\xi\Gamma = G \cap K\tau K$. Then $\Gamma\xi\Gamma = \Gamma\xi^{-1}\Gamma$, and hence by [S1, Lemma 3.5] $\Gamma\xi\Gamma = \bigsqcup_{\alpha\in A}\Gamma\alpha = \bigsqcup_{\alpha\in A}\alpha\Gamma$ with a suitable $A \subset \Gamma\xi\Gamma$. Put $(f|\alpha)(z) = J^k(\alpha,z)^{-1}f(\alpha z)$. Since $\Gamma\xi\Gamma = \bigsqcup_{\alpha\in A}\Gamma\alpha^{-1}$, we have $g|T_{q,\psi} = \sum_{\alpha\in A}\psi_{\mathfrak{c}}(\det(a_\alpha))g|\alpha^{-1}$. Clearly $\langle f|\alpha, g|\alpha\rangle = \langle f, g\rangle$, and $(g|\alpha^{-1})|\alpha = g$ by Lemma 2.3, and hence we obtain our lemma.

Clearly the same type of result for integral k can be proved in the same manner, in which the fact corresponding to Lemma 2.3 is trivially true.

5. Hecke eigenvalues and Fourier coefficients

We use the same notation as in Section 4. Let $\mathscr{L}$ denote the set of all $\mathfrak{g}$-lattices in F_1^n. We put $L_0 = \mathfrak{g}_1^n$ and for $M \in \mathscr{L}$ and $y \in GL_n(F)_{\mathbf{h}}$ we define yM to be the element of $\mathscr{L}$ such that $(yM)_v = y_v M_v$ for every $v \in \mathbf{h}$. For $\tau \in S$ put

$$\mathscr{L}_\tau = \left\{ L \in \mathscr{L} \,|\, {}^t l \tau l \in \mathfrak{b}\mathfrak{e}^{-1}\mathfrak{d}^{-1} \quad \text{for every } l \in L \right\}. \tag{5.1}$$

Notice that $\{M \in \mathscr{L}_\tau | L \subset M\}$ is a finite set if $\det(\tau) \neq 0$. For L and M in $\mathscr{L}$ we write $L < M$ if $L \subset M$ and $L_v = M_v$ for every $v | \mathfrak{c}$.

Let B' and E' be defined by (4.10b). For an integral ideal $\mathfrak{a}$ we denote by $T_\psi(\mathfrak{a})$ the sum of $T_{q,\psi}$ for all different $E'qE'$ with $q \in B'$ such that $\det(q)\mathfrak{g} = \mathfrak{a}$. Clearly $T_\psi(\mathfrak{a}) = 0$ if $\mathfrak{a}$ is not prime to $\mathfrak{e}$. Given $f \in \mathscr{M}_k(K, \psi)$ or $\in \mathscr{M}_k(\Gamma, \psi)$, we define $f|T_{q,\psi}$ as in Section 4, and define a formal Dirichlet series $f|\mathfrak{T}_\psi$ whose coefficients are modular forms by

$$f|\mathfrak{T}_\psi = \sum_{\mathfrak{a}} [\mathfrak{a}] \cdot f|T_\psi(\mathfrak{a}), \tag{5.2}$$

where $\mathfrak{a}$ runs over all the integral ideals in F. We now consider the Fourier expansion of Proposition 1.1 and investigate their relationship with the formal series $f|\mathfrak{T}_\psi$. By (1.21a), for $y \in GL_n(F)_{\mathbf{h}}$ and S^* as in (4.1) we have

$$\mu_f(\tau, y) \neq 0 \Longrightarrow {}^t y \tau_{\mathbf{h}} y \in be^{-1}S^* \Longleftrightarrow yL_0 \in \mathscr{L}_\tau, \tag{5.3}$$

where b and e are elements of $F_{\mathbf{h}}^\times$ such that $b\mathfrak{g} = \mathfrak{b}$ and $e\mathfrak{g} = \mathfrak{c}$. Now our main result on the relationship between $\mu_f(\tau, y)$ and Hecke operators can be stated as follows:

Theorem 5.1. *For $f \in \mathscr{M}_k(K, \psi)$ with ψ and K as in* (1.12) *and* (1.19), *$\tau \in S_+$, and $L = pL_0 \in \mathscr{L}_\tau$ with $p \in GL_n(F)_{\mathbf{h}}$ define formal Dirichlet series $D(\tau, p; f)$, $a(\tau, L)$, and $A(\tau, L)$ by*

$$D(\tau, p; f) = \sum_{x \in B'/E'} \psi_{\mathfrak{c}}(\det(px)) |\det(x)|_{\mathbf{A}}^{-n-1} \mu(\tau, px; f)[\det(x)\mathfrak{g}],$$

$$A(\tau, L) = |\det(p)|_{\mathbf{A}}^{-n-1} [\det(p)\mathfrak{g}]^2 \sum_{L < M \in \mathscr{L}_\tau} \mu(M/L) a(\tau, M),$$

$$a(\tau, L) = |\det(p)|_{\mathbf{A}}^{n+1} [\det(p)^{-2}\mathfrak{g}] \alpha'({}^t p \tau p), \quad \alpha'(\zeta) = \prod_{v | \mathfrak{c}} \alpha_v(\zeta_v),$$

where $\mu(M/L)$ is the Möbius function introduced in [S9, Sect. 4] *and α_v is the series of* (4.2). *Then*

$$A(\tau, L) D(\tau, p; f) = \sum_{L < M \in \mathscr{L}_\tau} \mu(M/L) \psi_{\mathfrak{c}}(\det(pw)) [\det(w)^{-1}\mathfrak{g}] \mu(\tau, pw; f|\mathfrak{T}_\psi),$$

$$\psi_{\mathfrak{c}}(\det(p)) \mu(\tau, p; f|\mathfrak{T}_\psi) = \sum_{L < M \in \mathscr{L}_\tau} [\det(w)^{-1}\mathfrak{g}] A(\tau, M) D(\tau, pw; f),$$

*where w in the last two sums is an element of $GL_n(F)_{\mathbf{h}}$ chosen for each M so that $M = pwL_0$ and $w^{-1} \in B'$. In particular, if $f|T_\psi(\mathfrak{a}) = \lambda(\mathfrak{a})f$ with $\lambda(\mathfrak{a}) \in$ **C** for every integral ideal $\mathfrak{a}$, then*

$$A(\tau, L)D(\tau, p; f) = \sum_{\mathfrak{a}} \lambda(\mathfrak{a})[\mathfrak{a}] \sum_{L<M\in\mathscr{L}_\tau} \mu(M/L)\psi_{\mathfrak{c}}(\det(pw))$$

$$\cdot\,[\det(w)^{-1}\mathfrak{g}]\mu_f(\tau, pw)\,,$$

$$\psi_{\mathfrak{c}}(\det(p))\mu(\tau, p; f)\sum_{\mathfrak{a}} \lambda(\mathfrak{a})[\mathfrak{a}] = \sum_{L<M\in\mathscr{L}_\tau} [\det(w)^{-1}\mathfrak{g}]A(\tau, M)D(\tau, pw; f)\,.$$

Remark. This is completely parallel to [S9, Theorem 5.1]. The series $a(\tau, L)$ and $A(\tau, L)$ are defined independently of the choice of p. However, $D(\tau, p; f)$ depends on pE'. The sum $\sum_{L<M\in\mathscr{L}_\tau}$ is a finite sum if $\det(\tau) \neq 0$. In general it may be an infinite sum. As for the nature of $A(\tau, L)$, see (5.5) and Proposition 5.5 below.

Proof. We practically repeat the proof of [S9, Theorem 5.1] with the following modification: the factor $J(*)$ of (4.4e) must be included in our calculation. As already observed, we can take a subset R of $P_{\mathbf{h}}$ that gives a set of representatives for $(G_{\mathbf{h}} \cap K)\backslash \mathrm{pr}(\mathscr{Z}^0)$. Then

$$(f|\mathfrak{T}_\psi)(x) = \sum_{y\in R} J(r_P(y))^{-1}\psi_{\mathfrak{c}}(\det(a_y))^{-1}f(x \cdot r_P(y)^{-1})[\ell_0(y)] \quad (x \in M_{\mathbf{A}})$$

with ℓ_0 of (4.11). Put

$$W = \{(g, h) \in B' \times B' | gL_0 + hL_0 = L_0, h_v \in E'_v \text{ for every } v|\mathfrak{c}\}\,,$$

$$S' = \{\sigma \in S_{\mathbf{h}} | \sigma_v \in S(\mathfrak{b}^{-1}\mathfrak{e})_v \text{ for every } v|\mathfrak{c}\}\,.$$

Then we can take R to be the set of all $y = \begin{pmatrix} g^{-1}h & g^{-1}\sigma \cdot {}^t h^{-1} \\ 0 & {}^t g \cdot {}^t h^{-1} \end{pmatrix}$ with $(g, h) \in E'\backslash W/(E' \times 1)$ and $\sigma \in S'/gS_{\mathbf{h}}(\mathfrak{b}^{-1}\mathfrak{e}) \cdot {}^t g$. This follows directly from [S9, Lemma 2.6]. Then, by (4.12) and Lemma 2.4, we obtain

$$(f|\mathfrak{T}_\psi)(x) = \sum \mathbf{g}(\sigma)\psi_{\mathfrak{c}}(\det(h^{-1}g))f(x \cdot r_P(y)^{-1})[\det(gh)\nu_0(b\sigma)]\,.$$

Substituting $r_P\left(\begin{pmatrix} q & s\tilde{q} \\ 0 & \tilde{q} \end{pmatrix}\right)$ for x and making a straightforward calculation, we find, for every $q \in GL_n(F)_{\mathbf{h}}$, that

$$\mu(\tau, q; f|\mathfrak{T}_\psi) = \sum_{g,h,\sigma} \psi_{\mathfrak{c}}(\det(h^{-1}g))\mu(\tau, qh^{-1}g; f)$$

$$\cdot\,\mathbf{g}(\sigma)\mathbf{e}_{\mathbf{h}}(-\mathrm{tr}({}^t h^{-1} \cdot {}^t q\tau qh^{-1} \cdot \sigma))[\det(gh)\nu_0(b\sigma)]\,.$$

By (5.3) we may assume that ${}^t g \cdot {}^t h^{-1} \cdot {}^t q\tau_{\mathbf{h}} qh^{-1}g \in be^{-1}S^*$. Now, for ${}^t g\zeta g \in be^{-1}S^*$,

$$\sum_{\sigma\in S'/g S_{\mathbf{h}}(\mathfrak{b}^{-1}\mathfrak{e})\cdot {}^t g} \mathbf{g}(\sigma)\mathbf{e}_{\mathbf{h}}(-\mathrm{tr}(\zeta\sigma))[\nu_0(b\sigma)] \neq 0$$

only if $\zeta \in be^{-1}S^*$, in which case the sum equals $|\det(g)|_{\mathbf{A}}^{-n-1}\prod_{v|\mathfrak{c}}\alpha_v(\zeta_v)$, for the same reason as in [S9, Lemma 2.8]. (Recall here that $\mathfrak{b}^{-1} \subset 2\mathfrak{d}^{-1}$ and $\mathfrak{b}_v = \mathfrak{d}_v$ for $v|\mathfrak{c}$.) Therefore we have

$$\mu(\tau, q; f|\mathfrak{T}_\psi) = \sum_{g,h} \psi_{\mathfrak{c}}(\det(h^{-1}g))|\det(g)|_{\mathbf{A}}^{-n-1}[\det(gh)\mathfrak{g}]$$

$$\cdot\, \mu(\tau, qh^{-1}g; f)\alpha'({}^th^{-1}\cdot {}^tq\tau qh^{-1}),$$

where (g, h) runs over $E'\backslash W/(E' \times 1)$ under the condition that ${}^th^{-1}\cdot {}^tq\tau_{\mathbf{h}}qh^{-1} \in be^{-1}S^*$. Now fix $L = pL_0 \in \mathscr{L}_\tau$ as in our theorem; let Λ be the set of all $M \in \mathscr{L}_\tau$ such that $M_v = L_v$ for every $v|\mathfrak{c}$. Given $M \in \Lambda$, we can take $y \in GL_n(F)_{\mathbf{h}}$ so that $M = yL_0$ and $p_v^{-1}y_v \in E'_v$ for every $v|\mathfrak{c}$. Then we put

$$c(M) = \psi_{\mathfrak{c}}(\det(y))\mu(\tau, y; f)|\det(y)|_{\mathbf{A}}^{-n-1}[\det(y)\mathfrak{g}]\,,$$
$$c'(M) = \psi_{\mathfrak{c}}(\det(y))\mu(\tau, y; f|\mathfrak{T}_\psi)[\det(y)\mathfrak{g}]^{-1}\,.$$

Now repeating the argument of [S9, p.554, lines 11–23] with all lattices restricted to those in Λ, we can complete the proof. (See *Correction* at the end of the present paper.)

Suppose now $f|T_\psi(\mathfrak{a}) = \lambda(\mathfrak{a})f$ for every $\mathfrak{a}$ as in the last part of the above theorem. By Theorem 4.4 for each v prime to $\mathfrak{e}$ we can determine n complex numbers $\lambda_{v,i}$ by the relation

$$(5.4\text{a})\qquad \sum_{m=0}^{\infty}\lambda(\pi_v^m\mathfrak{g})t^m = \begin{cases}\prod_{i=1}^{n}(1-|\pi_v|^{-n}\lambda_{v,i}t)^{-1} & \text{if } v|\mathfrak{e}^{-1}\mathfrak{c}\,,\\ \prod_{i=1}^{n}\dfrac{1-|\pi_v|^{1-2i}t^2}{(1-|\pi_v|^{-n}\lambda_{v,i}t)(1-|\pi_v|^{-n}\lambda_{v,i}^{-1}t)} & \text{if } v\nmid\mathfrak{c}\,,\end{cases}$$

where π_v is a prime element of F_v. We then put $W_v(t) = 1$ if $v|\mathfrak{e}$ and

$$(5.4\text{b})\qquad W_v(t) = \begin{cases}\prod_{i=1}^{n}(1-|\pi_v|^{-n}\lambda_{v,i}t) & \text{if } v|\mathfrak{e}^{-1}\mathfrak{c}\,,\\ \prod_{i=1}^{n}(1-|\pi_v|^{-n}\lambda_{v,i}t)(1-|\pi_v|^{-n}\lambda_{v,i}^{-1}t) & \text{if } v\nmid\mathfrak{c}\,.\end{cases}$$

Corollary 5.2. *Let f and W_v be as above and let $\tau \in S_+ \cap GL_n(F)$. Then for $L = pL_0 \in \mathscr{L}_\tau$ with $p \in GL_n(F)_{\mathbf{h}}$ we have*

$$D(\tau, p; f)\cdot\prod_{v\in\mathbf{b}}g_v([v])\cdot\prod_{v\nmid\mathfrak{c}}\left\{h_v([v])^{-1}\prod_{i=1}^{[n/2]}\left(1-|\pi_v|^{2i-1-2n}[v]^2\right)^{-1}\right\}$$

$$= \prod_{v\in\mathbf{h}}W_v([v])^{-1}\cdot\sum_{L<M\in\mathscr{L}_\tau}\mu(M/L)\psi_{\mathfrak{c}}(\det(pw))[\det(w)^{-1}\mathfrak{g}]\mu_f(\tau, pw)\,,$$

where $M = pwL_0$ as in Theorem 5.1, $\mathbf{b}$ is the (finite) set of the primes $v \in \mathbf{h}$ such that $v \nmid \mathfrak{c}$ and $({}^tp\tau p)_v \notin E_v$, the g_v are polynomials with constant term 1, $h_v = 1$ if n is even, and $h_v(t) = 1 - \rho_\tau^*(\pi_v\mathfrak{g}_v)|\pi_v|^{-n/2}t$ with the Hecke ideal *character* ρ_τ^* *of* F *corresponding to* $F(c^{1/2})/F$, $c = (-1)^{(n-1)/2}\det(2\tau)$, *if* n *is odd.*

Proof. We can repeat the proof of [S9, Cor.5.2] by employing Proposition 4.1 instead of [S9, Theorem 3.2]. We note that

$$(5.5)\qquad A(\tau, L) = \prod_{v\in\mathbf{b}} g_v([v]) \prod_{v\nmid\mathfrak{c}} \left\{ h_v([v])^{-1} \prod_{i=1}^{[(n+1)/2]} \left(1 - |\pi_v|^{1-2i}[v]^2\right) \right\}$$

with $\mathbf{b}$, g_v, and h_v as stated above.

Lemma 5.3. *Let* $0 \neq f \in \mathscr{M}_k$ *with integral or half-integral* k. *Then the following four conditions are mutually equivalent:*

(1) $k_v \geqq n/2$ *for some* $v \in \mathbf{a}$.

(2) $k_v \geqq n/2$ *for every* $v \in \mathbf{a}$.

(3) $\mu_f(\tau, p) \neq 0$ *for some* $\tau \in S_+ \cap GL_n(F)$ *and some* $p \in GL_n(F)_\mathbf{h}$.

(4) *For every* $p \in GL_n(F)_\mathbf{h}$ *there is a* $\tau \in S_+ \cap GL_n(F)$ *such that* $\mu_f(\tau, p) \neq 0$.

Moreover, if $0 \neq f \in \mathscr{S}_k$, *then these conditions are satisfied.*

Proof. Given $p \in GL_n(F)_\mathbf{h}$, put $f_p(z) = \sum_{\tau\in S_+} \mu_f(\tau, p)\mathbf{e_a}(\mathrm{tr}(\tau z))$. By (1.22) and [S7, Proposition 1.4] we see that $0 \neq f_p \in \mathscr{M}_k$. Now (1) is equivalent to (2) by [S12, Cor.5.7(4)]; also [S12, Theorem 5.6] says that (1) is equivalent to (4) and also that (3) implies (1). The last assertion follows from [S12, Cor.5.7(3)].

Proposition 5.4. *Let* $0 \neq f \in \mathscr{M}_k(K, \psi)$; *suppose that the conditions of Lemma 5.3 are satisfied. Then there exist* $\tau \in S_+ \cap GL_n(F)$ *and* $p \in GL_n(F)_\mathbf{h}$ *such that*

$$(5.6)\qquad 0 \neq \psi_\mathfrak{c}(\det(p))\mu(\tau, p; f|\mathfrak{T}_\psi) = A(\tau, pL_0)D(\tau, p; f)\,.$$

If in particular $f|T_\psi(\mathfrak{a}) = \lambda(\mathfrak{a})f$ *for every* $\mathfrak{a}$, *then* $\mu_f(\tau, p) \neq 0$ *and*

$$(5.7)\qquad \psi_\mathfrak{c}(\det(p))\mu_f(\tau, p) \prod_{v\in\mathbf{h}} W_v([v])^{-1}$$

$$= D(\tau, p; f) \cdot \prod_{v\in\mathbf{b}} g_v([v]) \cdot \prod_{v\nmid\mathfrak{c}} \left\{ h_v([v])^{-1} \prod_{i=1}^{[n/2]} \left(1 - |\pi_v|^{2i-1-2n}[v]^2\right)^{-1} \right\}$$

with the symbols as in Corollary 5.2.

Proof. We have $\mu_f(\tau, 1) \neq 0$ for some $\tau \in S_+ \cap GL_n(F)$. Then $D(\tau, 1; f) \neq 0$. Let $\mathscr{L}'_\tau$ be the set of all lattices $L \in \mathscr{L}_\tau$ such that $L_0 < L$ and $L = pL_0$ with some $p \in GL_n(F)_\mathbf{h}$ such that $D(\tau, p; f) \neq 0$. Since $\mathscr{L}'_\tau$ is a finite set containing

L_0, it has a maximal element. Writing it pL_0 with p such that $D(\tau, p; f) \neq 0$, we obtain (5.6) and (5.7) from Theorem 5.1 and Corollary 5.2.

Proposition 5.5. *Let* $\tau = \mathrm{diag}[\sigma, 0]$ *with* $\sigma \in S^r \cap GL_r(F)$ *and let* L' *be the image of* L *under the projection map* $(x_i)_{i=1}^n \mapsto (x_i)_{i=1}^r$ *of* F^n *into* F^r. *Then*

$$A(\tau, L) = \tilde{A}(\sigma, L') \prod_{v|\mathfrak{c}} \prod_{i=1}^{n-r} \left(1 - |\pi_v|^{1-2i}[v]^2\right),$$

where $\tilde{A}(\sigma, L')$ *is the formal Dirichlet series obtained from* $A(\sigma, L')$ *by substituting* $N(\mathfrak{a})^{n-r}[\mathfrak{a}]$ *for* $[\mathfrak{a}]$.

Proof. We repeat the proof of [S9, Prop. 5.4] with A and α_v understood in the present sense. At the end of the proof we employ the last formula of Proposition 4.1 instead of the corresponding formula [S9, (3.4)] needed in the previous case. Otherwise the argument is the same.

Remark. (1) If $n = 1$, Proposition 5.4 is essentially the same as [S6, Theorem 5.5] and [S8, Proposition 2.2]. In fact, if we substitute $\psi^*(\prod_{v|\mathfrak{c}} \mathfrak{a}_v)N(\mathfrak{a})^{-3/2} \cdot M(\mathfrak{a})$ for $[\mathfrak{a}]$ in $D(\tau, p, f)$ of Proposition 5.4, then we obtain the series $\sum_{\mathfrak{a}} \lambda_f(2\tau, \mathfrak{q}^{-1}\mathfrak{a}) M(\mathfrak{a})$ of [S8, Proposition 2.2]. If in particular $F = \mathbf{Q}$, the series is essentially $\sum_{m=1}^{\infty} \mu_f(\tau m^2, 1) m^{-s}$, whose nature we investigated in [S3].

(2) In [S9, Section 5] we treated $\mathcal{M}_k(K, \psi)$ with integral k only when $\mathfrak{e} = \mathfrak{g}$, that is, when $K = D[\mathfrak{b}^{-1}, \mathfrak{b}\mathfrak{c}]$. The results there can easily be generalized to the case of K of type (1.19) in the same manner as in the above treatment for half-integral k.

6. Main theorems on zeta functions and Eisenstein series

The notation being as in Sections 4 and 5, we now consider the case $\mathfrak{b} = 2^{-1}\mathfrak{d}$. We take a nonzero element f of $\mathcal{S}_k(K, \psi)$ such that $f | T_\psi(\mathfrak{a}) = \lambda(\mathfrak{a}) f$ for every $\mathfrak{a}$ as in Theorem 5.1, and put

$$(6.1\mathrm{a}) \qquad Z_\psi(s, \lambda) = \Lambda_{\mathfrak{c}}^{2n}(s, \psi) \sum_{\mathfrak{a}} \lambda(\mathfrak{a}) \psi^*(\mathfrak{a}') N(\mathfrak{a})^{-s},$$

$$(6.1\mathrm{b}) \qquad \Lambda_{\mathfrak{c}}^m(s, \psi) = \prod_{i=1}^{[(m+1)/2]} L_{\mathfrak{c}}(2s - 2i + 1, \psi^2),$$

$$(6.1\mathrm{c}) \qquad L_{\mathfrak{c}}(s, \psi) = \prod_{v \in \mathbf{h},\, v \nmid \mathfrak{c}} (1 - \psi(\pi_v)|\pi_v|^s)^{-1},$$

where π_v is a prime element of F_v, $\mathfrak{a}$ runs over all the integral ideals of F, ψ^* is the ideal character associated to ψ, and $\mathfrak{a}' = \prod_{v|\mathfrak{c}} \mathfrak{a}_v$. By (5.4a, b) Z_ψ has an Euler product

$$(6.2) \qquad Z_\psi(s, \lambda) = \prod_{v \in \mathbf{h},\, v \nmid \mathfrak{c}} W_v((\psi/\psi_{\mathfrak{c}})(\pi_v)|\pi_v|^s)^{-1},$$

with W_v of (5.4b). This is the most natural Euler product attached to λ if $\mathfrak{c} = \mathfrak{g}$. In [S10] which concerns the case of integral weight, we tried to keep as many Euler factors as possible. But the technique becomes very complicated in the present setting, and therefore we content ourselves with dealing only with the Euler factors for v prime to $\mathfrak{c}$. We thus take $\mathfrak{e} = \mathfrak{c}$. Then $\mathcal{S}_k(K, \psi) = \mathcal{S}_k(K)$ and $T_\psi(\mathfrak{a})$ is independent of ψ. For a Hecke character φ of F and $\varepsilon \in \mathbf{Z}^{\mathbf{a}}$ with $0 \leqq \varepsilon_v \leqq 1$, we define $Z(s, \lambda, \varphi)$ and a gamma-factor $\Gamma_{k,\kappa,\varepsilon}^{m,n}$ by

$$(6.3) \qquad Z(s, \lambda, \varphi) = \prod_{v \in \mathbf{h},\, v \nmid \mathfrak{c}} W_v(\varphi^*(\pi_v \mathfrak{g})|\pi_v|^s)^{-1},$$

$$(6.4a) \quad \Gamma_{k,\kappa,\varepsilon}^{m,n}(s) = \prod_{v \in \mathbf{a}} \Gamma_n(s + (k_v + \varepsilon_v - n - 1 + i\kappa_v)/2) g(s + (i\kappa_v/2), k_v - \varepsilon_v),$$

$$(6.4b) \quad g(s,h) = \begin{cases} \Gamma_m\left(s + \dfrac{h-n}{2}\right) & \text{if } h > (m+n)/2 \in \mathbf{Z}, \\ \Gamma\left(s + \dfrac{h-1}{2} - \left[\dfrac{2h+m+n-2}{4}\right]\right) \Gamma_m\left(s + \dfrac{h-n}{2}\right) & \\ \qquad \text{if } h > (m+n)/2 \notin \mathbf{Z}, & \\ \Gamma_{2h+1-n}\left(s + \dfrac{h-n}{2}\right) \displaystyle\prod_{a=[h]+1}^{[(m+n-1)/2]} \Gamma\left(2s - a - \frac{1}{2}\right) & \\ \qquad \text{if } (n-2)/2 < h \leqq (m+n)/2, & \\ \Gamma\left(s - \dfrac{n-2}{4}\right) \displaystyle\prod_{a=(n+1)/2}^{n-1} \Gamma\left(2s - a - \frac{1}{2}\right) & \\ \qquad \text{if } h = (n-2)/2, & \end{cases}$$

$$(6.5) \quad \Gamma_0(s) = 1, \quad \Gamma_n(s) = \pi^{n(n-1)/4} \prod_{\nu=0}^{n-1} \Gamma(s - \nu/2) \quad (s \in \mathbf{C}, 0 < n \in \mathbf{Z}).$$

By Lemma 5.3 we may assume that $k_v \geqq n/2$ for every $v \in \mathbf{a}$, which is why $g^n(s,h)$ is defined only for $h \geqq (n-2)/2$.

Theorem 6.1. *Let φ be a Hecke character of F; suppose $\varphi_{\mathbf{a}}(x) = \mathrm{sgn}(x_{\mathbf{a}})^{[k]+\varepsilon} \cdot |x_{\mathbf{a}}|^{i\kappa}$ with $\varepsilon \in \mathbf{Z}^{\mathbf{a}}$, $0 \leqq \varepsilon_v \leqq 1$, and $\kappa \in \mathbf{R}^{\mathbf{a}}, \sum_{v \in \mathbf{a}} \kappa_v = 0$. Let f be an eigenform in $\mathcal{S}_k(K, \psi)$ with eigenvalues $\lambda(\mathfrak{a})$ as above. Then the product $\mathcal{Z}(s) = \Gamma_{k,\kappa,\varepsilon}^{n,n}(s/2) Z(s, \lambda, \varphi)$ can be continued meromorphically to the whole s-plane with finitely many poles. Moreover each pole is simple. In particular, $\mathcal{Z}$ is an entire function of s if $\varphi^2 \neq 1$. If $\varphi^2 = 1$, the poles are determined as follows: Let $\ell = \mathrm{Max}_{v \in \mathbf{a}}\{k_v - \varepsilon_v\}$. If $\ell > n$, $\mathcal{Z}$ has no pole. If $\ell < n$, $\mathcal{Z}$ has possible poles only in the set*

$$\{(2j+1)/2 \mid j \in \mathbf{Z}, n+1 \leqq j \leqq 2n + (1/2) - \ell\}.$$

This theorem, as well as the following ones in this section, will be proved in Section 8.

Proposition 6.2. *$Z(s,\lambda,\varphi)$ is holomorphic and nonzero at least for* $\mathrm{Re}(s) \in \Lambda(n,k)$, *where*

$$\text{(6.6a)}\quad \Lambda(n,k) = \begin{cases} \mathbf{R} & \text{if } n = 0, \\ \{x \in \mathbf{R} \mid x \geqq 2\} & \text{if } n = 1 \text{ and } k_v > 3/2 \text{ for some } v \in \mathbf{a}, \\ \{x \in \mathbf{R} \mid x > (3n/2)+1\} & \text{otherwise}, \end{cases}$$

and we understand that

$$\text{(6.6b)}\qquad Z(s,\lambda,\varphi) = 1 \quad \text{if } n = 0\,.$$

If $n = 1$ and $k_v > 3/2$ for some $v \in \mathbf{a}$, then [S6, Theorem 6.1] and [S8, Proposition 3.1] show that $Z(s,\lambda,\varphi)$ is the twist of a zeta function on $GL_2(F)$ for the reason given at the end of Section 5. Therefore the nonvanishing for $\mathrm{Re}(s) \geqq 2$ follows from [S3, Prop. 4.16] (cf. also [S6, Lemma 10.3]). (In fact, by [S6, Theorem 6.1] the same can be said even if $k = 3u/2$, provided f is orthogonal to certain theta series, and therefore Theorems 6.4 and 6.6 below in such a case can be stated in somewhat better forms.) The remaining part will be proved in a forthcoming paper [S13].

To recall the Eisenstein series introduced in [S10], we take positive integers m and n such that $m \geqq n$ and consider the parabolic subgroup $P^{m,n}$ of G^m consisting of the elements whose lower left $(m-n)\times(m+n)$-block is 0. Taking $\mathfrak{b} = 2^{-1}\mathfrak{d}$ and with $\mathfrak{e}$ as in (4.3), we put

$$\text{(6.7)}\quad C = C^{m,n} = \{x \in D^m[2\mathfrak{d}^{-1}, 2^{-1}\mathfrak{d}\mathfrak{c}] \mid (a_1(x)-1)_v \in (\mathfrak{e}_v)^n_n,$$
$$a_2(x)_v \in (\mathfrak{e}_v)^n_{m-n},\, b_1(x)_v \in (2\mathfrak{d}_v^{-1}\mathfrak{e}_v)^n_n \text{ for every } v|\mathfrak{c}\}\,.$$

We also consider K of (4.4a) and a Hecke character χ such that

$$\text{(6.8a)}\qquad \chi_{\mathbf{a}}(x) = \mathrm{sgn}(x)^{[k]}|x_{\mathbf{a}}|^{i\kappa} \quad \text{with } \kappa \in \mathbf{R}^{\mathbf{a}}, \sum_{v\in\mathbf{a}} \kappa_v = 0\,;$$

$$\text{(6.8b)}\qquad \chi_v(a) = 1 \text{ if } v \in \mathbf{h},\ a \in \mathfrak{g}_v^{\times}, \text{ and } a - 1 \in \mathfrak{c}_v\,.$$

We can then define an Eisenstein series $E_{\mathbf{A}}(x,s;f,\chi,C)$ for $(x,s) \in M^m_{\mathbf{A}} \times \mathbf{C}$ and $f \in \mathscr{S}^n_k(K \cap G^n, \chi^{-1})$ by [S10, (3.10)]. This corresponds to a function $E(z,s;f,\chi,C)$ of $(z,s) \in \mathscr{H}^m \times \mathbf{C}$, which is given by

$$\text{(6.9)}\qquad E(z,s;f,\chi,C) = \sum_{\beta\in B} \chi[\beta]N(\mathfrak{a}_\beta)^{2s}\Delta_{k,\kappa}(z,s,f)\|_k\beta\,,$$

where B is a certain set of representatives for $P^{m,n}\backslash(G \cap P^{n,r}_{\mathbf{A}}C)$, $\chi[\beta]$ is a complex number of [S10, (3.11)], $\mathfrak{a}_\beta$ is a fractional ideal defined by (8.5b) below, and

$$\text{(6.10)}\qquad \Delta_{k,\kappa}(z,s,f) = f(\omega_n(z))\,[\Delta(z)/\Delta(\omega_n(z))]^{su-(k-i\kappa)/2} \quad (z \in \mathscr{H}^m)\,.$$

(For details, see [S10, Lemmas 3.3 and 3.6].)

We again consider the case $\mathfrak{e} = \mathfrak{c}$. (Our final results are stated only in this case. However, we treat the general case at least in the first four pages of

Section 7.) Then we are simply taking f from $\mathscr{S}_k^n(K \cap G^n)$. Put

(6.11)
$$\mathscr{F}^{m,n}(z,s;f,\chi,C) = E(z,s;f,\chi,C)Z(2s,\lambda,\chi) \prod_{i=n+1}^{[(m+n+1)/2]} L_{\mathfrak{c}}(4s-2i+1,\chi^2).$$

If $n = 0$ and $f = 1$, we take $D = D^m[\mathfrak{b}^{-1}, \mathfrak{bc}]$ and put

(6.12)
$$\mathscr{F}^m(z,s) = \mathscr{F}^{m,0}(z,s;1,\chi,D) = E(z,s;1,\chi,D)\Lambda_{\mathfrak{c}}^m(2s,\chi).$$

Then (6.11) is valid even for $n = 0$ under our convention (6.6b).

Theorem 6.3. *The notation being as above, the product*

(6.13)
$$\Gamma_{k,\kappa,0}^{m,n}(s)\mathscr{F}^{m,n}(z,s;f,\chi,C)$$

can be continued to a meromorphic function on the whole s-plane with only finitely many poles, which are all simple. The set of poles of (6.13) *is contained in the set of poles of* $\Gamma_{k,\kappa,0}^{m+n,0}(s)\mathscr{F}^{m+n}$, *which is described in* [S9, *Theorem* 7.3]. *In particular,* (6.13) *is entire in s if* $\chi^2 \neq 1$.

For $g \in \mathscr{S}_k^n$ and a congruence subgroup Γ of $\mathscr{G}^m$ we define an Eisenstein series $E_k^{m,n}(z,s;g,\Gamma)$ by

(6.14)
$$E_k^{m,n}(z,s;g,\Gamma) = \sum_{\gamma\in A} \Delta_{k,0}(z,s;g)\|_k\gamma, \quad A = (\Gamma \cap \mathscr{P}^{m,n})\backslash\Gamma.$$

Here $\mathscr{P}^{m,n}$ is defined by [S10, (1.26)] and g must belong to the set $\mathscr{S}_k^n(\Gamma \cap \mathscr{P}^{m,n})$ defined by [S10, (1.29)]. By Lemma 5.3 the series of (6.9) or (6.14) is nontrivial only when $k_v \geqq n/2$ for every $v \in \mathbf{a}$.

Let us first consider the case in which $k = k_0 u$ with $k_0 \equiv 1/2 \pmod{\mathbf{Z}}$ and $\kappa = 0$. If $k_0 > m+n+1$, then $E_k^{m,n}(z,s;g,\Gamma)$ and $E(z,s;f,\chi,C)$ are convergent at $s = k_0/2$, and their values are holomorphic in z, and hence belong to $\mathscr{M}_k^m$. For a smaller k_0 we have the following result, in which $\mathscr{N}_{k,d}^m$ denotes the set of nearly holomorphic forms on $\mathscr{H}^m$ defined in [S10, Sect. 7].

Theorem 6.4. *Suppose that* $m > n \geqq 0$ *and* $k = k_0 u$ *with* $(m+n+1)/2 \leqq k_0 \equiv 1/2 \pmod{\mathbf{Z}}$. *Let f and* χ *be as above with* $\kappa = 0$ *in* (6.8a); *let* $g \in \mathscr{S}_k^n(\Gamma \cap \mathscr{P}^{m,n})$ *with an arbitrary congruence subgroup* Γ *of* G^m.

(I) *Suppose that* $k_0 \in \Lambda(n, k_0 u)$ *and that* $F \neq \mathbf{Q}$ *or* $2k_0 \notin \{m+n+2, m+n+3\}$. *Then* $E_k^{m,n}(z,k_0/2;g,\Gamma)$ *and* $E(z,k_0/2;f,\chi,C)$ *belong to* $\mathscr{M}_k^m$. *Moreover,* $E(z,k_0/2;f,\chi,C)$ *belongs to* $\mathscr{M}_k^m$ *even if* $F = \mathbf{Q}$ *and* $2k_0 \in \{m+n+2, m+n+3\}$, *provided* $k_0 \in \Lambda(n,k_0 u)$ *and* $\chi^2 \neq 1$.

(II) *If* $F = \mathbf{Q}$ *and* $k_0 = (m+n+3)/2 \in \Lambda(n,k_0 u)$, *then both* $E_k^{m,n}(z,k_0/2;g,\Gamma)$ *and* $E(z,k_0/2;f,\chi,C)$ *belong to* $\mathscr{N}_{k,m}^m$.

(III) *Let* $\mu = m+n+1-k_0$. *Then* $\mathscr{F}^{m,n}(z,\mu/2;f,\chi,C)$ *belongs to* $\mathscr{M}_k^m$.

Theorem 6.5. *Suppose that $m > n \geqq 0$ and $k = k_0 u$ with $n/2 \leqq k_0 \leqq (m+n)/2$; put $s_0 = (m+n+1-k_0)/2$. Then $\mathscr{F}^{m,n}(z,s;f,\chi,C)$ has at most a simple pole at s_0, which occurs only when $\chi^2 = 1$. Moreover, the residue is an element of $\mathscr{M}_k^m$.*

As for a more general weight and more general critical values, we have

Theorem 6.6. *Suppose that $m > n \geqq 0$, $k_v \geqq (m+n+1)/2$ for every $v \in \mathbf{a}$, and $[k]_v$ (mod 2) is independent of v; let χ be as in (6.8a,b) with $\kappa = 0$.*

(I) *Let μ be an element of $(1/2)\mathbf{Z}$ such that $\mu \in \Lambda(n,k)$, $(m+n+1)/2 \leqq \mu \leqq k_v$, and $\mu \equiv k_v$ (mod 2) for every $v \in \mathbf{a}$. Then $E_k^{m,n}(z,\mu/2;g,\Gamma)$ belongs to $\mathscr{N}_{k,d}^m$ except when $F = \mathbf{Q}$ and $2\mu = m+n+2$, where*

$$d = \begin{cases} (m+n)(k-\mu+2)/2 & \text{if } 2\mu = m+n+3 \text{ and } F = \mathbf{Q}, \\ (m+n)(k-\mu u)/2 & \text{otherwise.} \end{cases}$$

(II) *Let μ be an element of $(1/2)\mathbf{Z}$ such that $m+n+1-k_v \leqq \mu \leqq k_v$ for every $v \in \mathbf{a}$ and that*

$$\mu \equiv \begin{cases} k_v + m + n \pmod 2 & \text{if } 2\mu < m+n+1, \\ k_v \pmod 2 & \text{if } 2\mu \geqq m+n+1. \end{cases}$$

Then $\mathscr{F}^{m,n}(z,\mu/2;f,\chi,C)$ belongs to $\mathscr{N}_{k,d}^m$ except when $2\mu = m+n+2, F = \mathbf{Q}$, and $\chi^2 = 1$, where

$$d = \begin{cases} (m+n)(k-\mu+2)/2 & \text{if } 2\mu = m+n+3,\ F = \mathbf{Q}, \text{ and } \chi^2 = 1, \\ (m+n)\{k - |\mu-\ell|u - \ell u\}/2,\ \ell = (m+n+1)/2, & \text{otherwise}, \end{cases}$$

Remark 6.7. It is conjecturable that Proposition 6.2 holds with $\Lambda(n,k) = \{x \in \mathbf{R} \mid x \geqq n+1\}$ if $k_v > n+1$ for every $v \in \mathbf{a}$. If that is so, we can eliminate the conditions $k_0 \in \Lambda(n, k_0 u)$ in Theorem 6.4 and $\mu \in \Lambda(n,k)$ in Theorem 6.6.

7. The pullback of an Eisenstein series

In this section we put $N = m+n$ with two positive integers m and n such that $m \geqq n$. Our aim is to express $E(z,s;f,\ldots)$ of (6.9) as an inner product of f and a certain pullback of an Eisenstein series on $\mathscr{H}^N$ to $\mathscr{H}^m \times \mathscr{H}^n$. We consider K of (4.4a) with $\mathfrak{b} = 2^{-1}\mathfrak{d}$, and put $D^n = D^n[2\mathfrak{d}^{-1}, 2^{-1}\mathfrak{d}\mathfrak{c}]$ for simplicity. With a divisor $\mathfrak{e}$ of $\mathfrak{c}$ as in (4.3), define $\sigma \in G_{\mathbf{h}}^N$ and $\tau_r \in G^N$ for $0 \leqq r \leqq n$ by

$$\text{(7.1a)} \qquad \sigma_v = \begin{cases} \operatorname{diag}[1_m, (2/\delta_v)1_n, 1_m, (\delta_v/2)1_n]\tau_n & \text{if } v|\mathfrak{e}, \\ \operatorname{diag}[1_m, (2/\delta_v)1_n, 1_m, (\delta_v/2)1_n] & \text{if } v\nmid\mathfrak{e}, \end{cases}$$

$$\text{(7.1b)} \qquad \tau_r = \begin{pmatrix} 1_N & 0 \\ \rho_r & 1_N \end{pmatrix}, \quad \rho_r = \begin{pmatrix} 0_m & e_r \\ {}^t e_r & 0_n \end{pmatrix}, \quad e_r = \begin{pmatrix} 1_r & 0 \\ 0 & 0 \end{pmatrix} \in F_n^m.$$

We shall eventually take $\mathfrak{e} = \mathfrak{c}$, but for the moment we treat the general case.

To consider an Eisenstein series on $M_{\mathbf{A}}^N$, we take a Hecke character χ as in (6.8a, b). Then our Eisenstein series $E_{\mathbf{A}}$ is defined by

$$E_{\mathbf{A}}(x) = E_{\mathbf{A}}(x, s) = \sum_{\alpha \in A} \mu(\alpha x)\varepsilon(\alpha x)^{-2s} \quad (x \in M_{\mathbf{A}}^N, A = P^N \backslash G^N), \tag{7.2}$$

where μ and ε are defined so that $\mu(x) = 0$ if $\mathrm{pr}(x) \notin P_{\mathbf{A}}^N D^N$; if $x = pw$ with $\mathrm{pr}(p) \in P_{\mathbf{A}}^N$ and $\mathrm{pr}(w) \in D^N$, $w(\mathbf{i}) = \mathbf{i}$, then

$$\begin{aligned} \mu(x) &= \chi(\det(d_p))^{-1}\chi_{\mathfrak{c}}(\det(d_w))^{-1} j^k(w, \mathbf{i})^{-1} \\ &= \chi_{\mathbf{h}}(\det(d_p))^{-1}\chi_{\mathfrak{c}}(\det(d_w))^{-1} j^k(x, \mathbf{i})^{-1}|j(x, \mathbf{i})|^{k-i\kappa}, \end{aligned} \tag{7.3a}$$

$$\varepsilon(x) = N(\mathfrak{a}_x)^{-1}|j(x, \mathbf{i})|^u, \quad \mathfrak{a}_x = \det(d_p)\mathfrak{g}. \tag{7.3b}$$

We denote by $E^N(z)$ the function on $\mathscr{H}^N$ corresponding to $E_{\mathbf{A}}$. This type of series was investigated in [S5]; it is also a special case of [S10, (3.10)].

Observing that $\sigma \in P'_{\mathbf{A}}$, we put $\hat{\sigma} = r'(\sigma)$ with r' of (2.2). Take $\rho \in G^N \cap D^N\sigma$ such that $a(\rho\sigma^{-1})_v - 1_N \in (\mathfrak{c}_v)_N^N$ for every $v|\mathfrak{c}$. By Lemma 1.3 $E_{\mathbf{A}}(x\hat{\sigma}^{-1})$ corresponds to $E^N\|_k\rho_1$ if we choose $\rho_1 \in M_{\mathbf{A}}$ as in that lemma. Take a variable element y of $M_{\mathbf{A}}^N$ such that $\mathrm{pr}(y) \in G_{\mathbf{a}}^N$ and put $y(\mathbf{i}) = z$. Then

$$\left(E^N\|_k\rho_1\right)(z) = \sum_{\alpha \in A} \mu\left(\alpha\hat{\sigma}^{-1}y\right)\varepsilon\left(\alpha\hat{\sigma}^{-1}y\right)^{-2s} j^k(y, \mathbf{i}). \tag{7.4}$$

By [S10, Lemma 4.2] A is given by $\bigsqcup_{r=0}^n A_r$ with $A_r = P^N\backslash P^N\tau_r(G^m \times G^n)$. Thus $(E^N\|_k\rho_1)(z) = \sum_{r=0}^n \mathscr{E}_r$ with $\mathscr{E}_r = \sum_{\alpha \in A_r} p_\alpha$, where $p_\alpha(z) = \mu(\alpha\hat{\sigma}^{-1}y)\cdot \varepsilon(\alpha\hat{\sigma}^{-1}y)^{-2s}j^k(y, \mathbf{i})$. Clearly p_α depends only on $P^N\alpha$ and

$$\mu(x) = \lambda(\mathrm{pr}(x)_{\mathbf{h}})j^k(x, \mathbf{i})^{-1}|j(x, \mathbf{i})|^{k-i\kappa} \tag{7.5}$$

with a function λ on $G_{\mathbf{h}}^N$ such that $\lambda(\xi q) = \chi_{\mathfrak{c}}\left(\det(d_q)\right)^{-1}\lambda(\xi)$ for every $q \in G_{\mathbf{h}}^N \cap D^N$. Therefore

$$p_\alpha(z) = N\left(\mathfrak{a}(\alpha\hat{\sigma}^{-1})\right)^{2s}\lambda\left(\alpha_{\mathbf{h}}\sigma^{-1}\right) j^k(\alpha\hat{\sigma}^{-1}, z)^{-1}\Delta(\alpha z)^{su-(k-i\kappa)/2}, \tag{7.6}$$

where $\mathfrak{a}(x) = \mathfrak{a}_x$. By Lemma 1.2 there exists an open subgroup C' of D^N such that $j^k(\alpha\zeta\hat{\sigma}^{-1}, z) = j^k(\alpha\hat{\sigma}^{-1}, \zeta z)j^k(\zeta, z)$ for every $\zeta \in C'$, and hence $p_\alpha\|_k\omega = p_{\alpha\omega}$ for every ω in a suitable congruence subgroup Γ_1 of G^N independent of α. Put $p_{\xi,\beta,\gamma} = p_\alpha$ for $\alpha = \tau_r((\xi \times 1_{2m-2r})\beta \times \gamma)$ with $\xi \in G^r$, $\beta \in G^m$, and $\gamma \in G^n$. By [S10, Lemma 4.3] $\mathscr{E}_r = \sum_{\xi \in G^r}\sum_{\beta \in R^m}\sum_{\gamma \in R^n} p_{\xi,\beta,\gamma}$, where $R^n = P^{n,r}\backslash G^n$. Take a congruence subgroup Γ_2 of G^n so that $1 \times \Gamma_2 \subset \Gamma_1$ and take $Y \subset G^n$ so that $G^n = \bigsqcup_{\eta \in Y} P^{n,r}\eta\Gamma_2$. Then $p_{\xi,\beta,\varepsilon}\|(1 \times \gamma) = p_{\xi,\beta,\varepsilon\gamma}$ for $\gamma \in \Gamma_2$ and R^n is given by $\bigsqcup_{\eta \in Y} \eta S_\eta$ with $S_\eta = (\eta^{-1}P^{n,r}\eta \cap \Gamma_2)\backslash\Gamma_2$. Put $g_\eta = \sum_{\xi \in G^r}\sum_{\beta \in R^m} p_{\xi,\beta,\eta}$. Then $\mathscr{E}_r = \sum_{\eta \in Y}\sum_{\gamma \in S_\eta} g_\eta\|(1 \times \gamma)$. Moreover, repeating the argument of several lines preceding [S10, (4.6)], we find that $g_\eta\|(1 \times \gamma) = g_\eta$ if $\gamma \in \eta^{-1}P^{n,r}\eta \cap \Gamma_2$. Therefore, by [S10, Lemma 2.2] we obtain

$$\langle f(w), \mathscr{E}_r(\mathrm{diag}[\mathfrak{z}, w])\rangle = 0 \quad \text{for every } f \in \mathscr{S}_k^n \text{ if } n > r, \tag{7.7}$$

where $(\mathfrak{z}, w) \in \mathscr{H}^m \times \mathscr{H}^n$, for sufficiently large Re(s). (Since η may not belong to $\mathfrak{M}^n$, we have to take an element $\hat{\eta}$ of the group $\mathscr{G}^n$ of [S10, (1.12)] lying above η. Then observing that $\hat{\eta}^{-1}\mathscr{P}^{n,r}\hat{\eta}\cap\Gamma_2$ coincides with $\eta^{-1}P^{n,r}\eta\cap\Gamma_2$ within $\mathscr{G}^n$, we can apply [S10, Lemma 2.2] to the present setting.)

To study the nature of $\mathscr{E}_n$ more closely, let us put $\tau = \tau_n$ for simplicity. In [S10, Lemma 4.4] we showed that

$$(7.8) \qquad (P^N\tau_n(G^m \times G^n)) \cap (P^N_{\mathbf{A}}D^N\sigma) = \bigsqcup_{\xi\in X,\beta\in B} P^N\tau_n((\xi \times 1_{2m'})\beta \times 1_{2n}),$$

where $m' = m - n$, B is a subset of $G^m \cap \tilde{C}^{m,n}$ which represents $P^{m,n}\backslash(G^m\cap P^{m,n}_{\mathbf{A}}C^{m,n})$ completely with $C^{m,n}$ of (6.7) and $\tilde{C}$ of [S10, (3.4)], and $X = G^n \cap G^n_{\mathbf{a}} \prod_{v\in\mathbf{h}} X_v$ with

$$(7.9a) \qquad X_v = \begin{cases} \{x \in D^n_v[2\mathfrak{d}^{-1}\mathfrak{c}, 2^{-1}\mathfrak{d}\mathfrak{c}] \,|\, a_x - 1 \in (\mathfrak{c}_v)^n_n\} & \text{if } v|\mathfrak{c}, \\ D^n_v[2\mathfrak{d}^{-1}\mathfrak{c}, 2^{-1}\mathfrak{d}\mathfrak{c}^{\mu}]W_vD^n_v[2\mathfrak{d}^{-1}\mathfrak{c}^{v}, 2^{-1}\mathfrak{d}\mathfrak{c}] & \text{if } v|\mathfrak{e}^{-1}\mathfrak{c}, \\ G^n_v & \text{if } v \nmid \mathfrak{c}, \end{cases}$$

$$(7.9b) \qquad W_v = \{\mathrm{diag}[q, \tilde{q}] \,|\, q \in GL_n(F_v) \cap (\mathfrak{c}_v)^n_n\};$$

we take $B = \{1_{2m}\}$ if $m = n$. By [S10, Lemma 4.5] we can take μ and v to be any nonnegative integers. Now for $\alpha = \tau((\xi \times 1)\beta \times 1)$ with $\xi \in X$ and $\beta \in B$ we see that $\mathfrak{a}(\alpha\hat{\sigma}^{-1})$ and $\lambda(\alpha_{\mathbf{h}}\sigma^{-1})$ are exactly $\mathfrak{a}_g$ and $\mu(g_{\mathbf{h}})$ of [S10, Lemma 4.6], and hence

$$(7.10) \qquad p_\alpha(z) = \chi_{\mathbf{h}}(\delta/2)^n N\left(2^{-n}\mathfrak{d}^n\mathfrak{a}_\beta^{-1}\ell_0(\xi)\right)^{-2s}\chi[\beta] \cdot \chi^*(\ell_1(\xi))\chi_{\mathfrak{c}}\left(\det(d_\xi)\right)^{-1} j^k(\alpha\hat{\sigma}^{-1}, z)^{-1}\Delta(\alpha z)^{su-(k-i\kappa)/2},$$

where $\chi[\beta]$ is as in [S10, (3.11)], ℓ_0 is as in (4.11), and $\ell_1(x) = \prod_{v\nmid\mathfrak{c}} \ell_0(x)_v$, and hence

$$(7.11) \qquad \mathscr{E}_n = \chi_{\mathbf{h}}(\delta/2)^n N(2\mathfrak{d}^{-1})^{2ns}\sum_{\beta\in B}\chi[\beta]N(\mathfrak{a}_\beta)^{2s} \cdot \sum_{\xi\in X}\chi^*(\ell_1(\xi))\ell(\xi)^{-2s}\chi_{\mathfrak{c}}\left(\det(d_\xi)\right)^{-1} j^k(\alpha\hat{\sigma}^{-1}, z)^{-1}\Delta(\alpha z)^{su-(k-i\kappa)/2}.$$

Since $\beta \in B$, we have $\beta = qw$ with $w \in C^{m,n}$ and $q = \mathrm{diag}\left[1_{m-1}, t^{-1}, 1_{m-1}, t\right]$, $t \in F^\times_{\mathbf{h}}$. If $m = n$, we understand that $\beta = q = w = 1$.

Lemma 7.1. *For $\alpha = \tau((\xi \times 1)\beta \times 1)$, $\xi \in X$, $\beta = qw$, and $\hat{\sigma}$ as above, put $A = \tau((\xi \times 1)q \times 1)\sigma^{-1}$. Then $A \in \Omega' \cap P^N_{\mathbf{A}}D^N[2\mathfrak{d}^{-1}, 2\mathfrak{d}]$, $\mathfrak{t}_0(A) = \mathfrak{t}(\xi)$, and*

$$(7.12) \qquad j^k(\alpha\hat{\sigma}^{-1}, z) = j^k\left(r'(A), (\beta \times 1)z\right)j^k(\beta \times 1, z) \quad (z \in \mathscr{H}^N).$$

Proof. Put $w^* = r_P(q)^{-1}\beta$. Since $r' = r$ on $G \cap \Omega'$, we have, by Lemma 2.1,

$$\begin{aligned}\alpha &= \tau((\xi \times 1) \times 1)(\beta \times 1)\\ &= r'(\tau((\xi \times 1) \times 1))r_P(q \times 1)\langle w^*, 1\rangle\\ &= r'(\tau((\xi \times 1)q \times 1))\langle w^*, 1\rangle\\ &= r'(A)\hat{\sigma}\langle w^*, 1\rangle,\end{aligned}$$

and hence $\alpha\hat{\sigma}^{-1} = r'(A)\eta$ with $\eta = \hat{\sigma}\langle w^*, 1\rangle\hat{\sigma}^{-1}$. Put $p = \mathrm{diag}\left[2^{-1}1_n, 2 \cdot 1_n\right]_{\mathfrak{c}} \times 1_{2m}$, where the subscript $\mathfrak{c}$ means the projection to $\prod_{v|\mathfrak{c}} G_v$. Then $p\sigma^{-1}p^{-1} = fg$ with $f = \mathrm{diag}[1_m, (\delta/2)1_n, 1_m, (2/\delta)1_n]$ and an element g of $C^{N,\theta} \cap \prod_{v|\mathfrak{c}} P'_v$. Let $\pi = r_P(p)$, $\varphi = r_P(f)$, and $\gamma = r'(g)$. Then $\pi\hat{\sigma}^{-1}\pi^{-1} = \varphi\gamma$. Now $\pi = \langle\pi_1, 1\rangle$ with $\pi_1 = r_P\left(\mathrm{diag}\left[2^{-1}1_n, 1_{m-n}, 2 \cdot 1_n, 1_{m-n}\right]_{\mathfrak{c}}\right)$. Put $\omega = \pi_1\omega^*\pi_1^{-1}$ and $\zeta = \gamma^{-1}\langle\omega, 1\rangle\gamma$. Then $\eta = \pi^{-1}\gamma^{-1}\varphi^{-1}\langle\omega, 1\rangle\varphi\gamma\pi = \pi^{-1}\zeta\pi$, since $\varphi = \langle 1, \varphi_1\rangle$ with $\varphi_1 \in M^n_{\mathbf{A}}$. We easily see that $\mathrm{pr}(\omega) \in C^{m,\theta}$, and hence $\mathrm{pr}(\zeta) \in C^{N,\theta}$. On the other hand $\mathrm{pr}(\eta) = \sigma(w \times 1)\sigma^{-1} \in D^N[2\mathfrak{d}^{-1}, 2\mathfrak{d}]$ as can easily be verified. Since $\alpha\sigma^{-1} \in P^N_{\mathbf{A}}D^N$ by (7.8), we see that $A \in P^N_{\mathbf{A}}D^N[2\mathfrak{d}^{-1}, 2\mathfrak{d}]$. By Lemma 2.2(iii) we have $h_\zeta = h_\eta$ and $h(\omega, z) = h(w^*, z)$. Since $\mathrm{pr}(\gamma) \in C^{N,\theta} \cap G^N_{\mathbf{h}}$ and $q \in P^N_{\mathbf{h}}$, by (1.9b, c) we have $h(\zeta, z) = h(\langle\omega, 1\rangle, z)$, and $h(\langle w^*, 1\rangle, z) = h(\beta \times 1, z)$. Combining all these we obtain $j^k(\eta, z) = j^k(\beta \times 1, z)$. Since $\mathrm{pr}(\eta) \in D[2\mathfrak{d}^{-1}, 2\mathfrak{d}]$, by (1.9c) we have $j^k(\alpha\hat{\sigma}^{-1}, z) = j^k(r'(A)\eta, z) = j^k(r'(A), \eta z)\, j^k(\eta, z)$, which proves (7.12). Put $\xi = \left(\begin{smallmatrix} a & b \\ c & d \end{smallmatrix}\right)$. Since $\xi \in X$, we easily see that $\xi \in Z'$, $A \in \Omega'$, and

$$d_A^{-1}c_A = \begin{pmatrix} d^{-1}c & 0 & (\delta/2)e \\ 0 & 0 & 0 \\ (\delta/2)\cdot {}^t e & 0 & -(\delta/2)^2 bd^{-1} \end{pmatrix},$$

$$e_v = d_v^{-1} \text{ if } v \nmid \mathfrak{e}, \quad e_v = d_v^{-1} - 1 \text{ if } v | \mathfrak{e}\ .$$

Therefore $\mathbf{t}_0(A) = \mathbf{g}(\delta^{-2}d_A^{-1}c_A) = \lambda/|\lambda|$, $\lambda = \prod_{v\in\mathbf{h}} \lambda_v$ with

$$\lambda_v = \int_{L_v \times L_v} \mathbf{e}_v\left((2\delta_v^2)^{-1} \cdot {}^t x d_v^{-1} c_v x + (2\delta_v)^{-1} \cdot {}^t x e_v y - 8^{-1} \cdot {}^t y b_v d_v^{-1} y\right) dx\, dy\,.$$

From (7.9a) we easily see that $0 < \lambda_v \in \mathbf{R}$ for $v|\mathfrak{e}$, and also for $v|\mathfrak{c}^{-1}\mathfrak{c}$ by virtue of [S10, Lemma 4.5]. Similarly we can verify that $0 < \kappa(\xi_v) \in \mathbf{R}$ for $v|\mathfrak{c}$. Changing the variable y for $4y$ in the above integral expression, we see that $\lambda_v = \kappa(\xi_v)$ for $v \nmid 2$. Thus $\mathbf{t}_0(A) = \mathbf{t}(\xi)$. This completes the proof of Lemma 7.1.

To avoid confusion, we hereafter denote by $\mathfrak{z}$ a variable on $\mathscr{H}^m$; we use (z, w) for a variable on $\mathscr{H}^n \times \mathscr{H}^n$. For $l \in \mathbf{R}^{\mathbf{a}}$ define $M^l(z, w)$ by

(7.13) $$M^l(z, w) = \prod_{v\in\mathbf{a}} \det(z_v + w_v)^{l_v}$$

with $\det(z)^c$ on H^n determined so that $\det(i1_n)^c = \mathbf{e}(nc/4)$. We can easily verify that

(7.14) $$j^{\nu}(\tau(\xi \times 1_{2m}),\ \mathrm{diag}[\mathfrak{z}, w]) = M^{\nu}(\omega_n(\mathfrak{z}), \zeta w)\, j^{\nu}(\zeta, w)$$

for $v \in \mathbf{Z}^{\mathbf{a}}$, where $\zeta = \varepsilon\xi^{-1}\iota\varepsilon$, $\varepsilon = \mathrm{diag}[1_n, -1_n]$, and $\omega_n(\mathfrak{z})$ is the upper left $(n \times n)$-block of $\mathfrak{z}$. In particular, for A of Lemma 7.1 we have

$$j^{[k]}(A, \mathrm{diag}[\mathfrak{z}, w]) = M^{[k]}(\omega_n(\mathfrak{z}), \zeta w) j^{[k]}(\zeta, w). \tag{7.15}$$

Lemma 7.2. *For* $\alpha = \tau((\xi \times 1)\beta \times 1)$ *and* A *as in Lemma* 7.1, *we have*

$$t \cdot j^k\left(r'(A), \mathrm{diag}[\mathfrak{z}, w]\right) = M^k\left(\omega_n(\mathfrak{z}), \xi_*^{-1}\iota^{-1}w\right) J^k(\xi_*^{-1}, \iota^{-1}w) j^k(\iota^{-1}, w),$$

where $\xi_* = \varepsilon\xi\varepsilon$ *and* $t = \mathbf{e}(n[F:\mathbf{Q}]/8)$.

Proof. We have $\mathbf{t}_0(A) = \mathbf{t}(\xi)$ by Lemma 7.1. Hence by (2.6), (3.9), and Lemma 3.3(4) we have

$$\begin{aligned} h\left(r'(A), \mathrm{diag}[\mathfrak{z}, w]\right) &= \mathbf{t}_0(A)\, p(\tau(\xi \times 1_{2m}), \mathrm{diag}[\mathfrak{z}, w]) \\ &= \mathbf{t}(\xi) q(\xi\omega_n(\mathfrak{z}), w)\, p(\xi, \omega_n(\mathfrak{z})) \\ &= q(\xi\omega_n(\mathfrak{z}), w) H(\xi, \omega_n(\mathfrak{z})). \end{aligned}$$

By (7.15) we can find an element $t \in \mathbf{T}$ such that

$$t \cdot q(\xi z, w) H(\xi, z) = M^{u/2}(z, \zeta w) J^{u/2}(\xi_*^{-1}, \iota^{-1}w) h(\iota^{-1}, w).$$

By Lemma 3.5 this can be written, with ιw in place of w,

$$t \cdot q(\xi z, \iota w) = M^{u/2}\left(z, \xi_*^{-1}w\right) J^{u/2}\left(\xi_*^{-1}, w\right) J^{u/2}\left(\xi^{-1}, \xi z\right) h\left(\iota^{-1}, \iota w\right).$$

Take $w = \xi z$. Then, by Lemma 2.5 we have

$$t \cdot q(w, \iota w) = M^{u/2}\left(\xi^{-1}w, \xi_*^{-1}w\right) \overline{J^{u/2}(\xi^{-1}, -\overline{w})} J^{u/2}\left(\xi^{-1}, w\right) h\left(\iota^{-1}, \iota w\right).$$

By [S7, (2.5)] $h(\iota^{-1}, w) = h(\iota, w) = \det(-iw)^{u/2}$. Take w to be pure imaginary. Then the left-hand side is a positive constant times t, and the right-hand side is $\mathbf{e}(n[F:\mathbf{Q}]/8)$ times a positive number. This proves our formula for $k = u/2$, which together with (7.15) proves the general case.

Let us hereafter put $g(\mathfrak{z}, w) = g(\mathrm{diag}[\mathfrak{z}, w])$ for any function g on $\mathscr{H}^N$. Then for every $f \in \mathscr{S}_k^n$ we have, by (7.7),

$$\begin{aligned} \left\langle \left(f \| \iota^{-1}\right)(w), \left(E^N \|_k \rho_1\right)(\mathfrak{z}, w) \right\rangle &= \left\langle (f \| \iota^{-1})(w), \mathscr{E}_n(\mathfrak{z}, w) \right\rangle \\ &= \left\langle f(w), \mathscr{E}_n(\mathfrak{z}, \iota w) j^k(\iota, w)^{-1} \right\rangle. \end{aligned} \tag{7.16}$$

Since $\varepsilon X \varepsilon = X$, from (7.11), (7.12), and Lemma 7.2 we obtain

(7.17)

$$\begin{aligned} \mathscr{E}_n(\mathfrak{z}, \iota w) j^k(\iota, w)^{-1} &= t \cdot \chi_{\mathbf{h}}(\delta/2)^n N(2\mathfrak{d}^{-1})^{2ns} \sum_{\beta \in B} \chi[\beta] N(\mathfrak{a}_\beta)^{2s} \sum_{\xi \in X} \ell(\xi)^{-2s} \\ &\quad \cdot \chi^*(\ell_1(\xi))\, \chi_{\mathbf{c}}(\det(d_\xi))^{-1} J^k(\xi^{-1}, w)^{-1} \\ &\quad \cdot j^k(\beta, \mathfrak{z})^{-1} \Delta(\beta\mathfrak{z})^{\mathbf{s}} M_k^{\mathbf{s}}(\omega_n(\beta\mathfrak{z}), \xi^{-1}w), \end{aligned}$$

where $\mathbf{s} = su - (k - i\kappa)/2$ and

$$M_k^{\mathbf{s}}(z, w) = M^k(z, w)^{-1} \Delta(w)^{\mathbf{s}} |\det(z + w)|^{-2\mathbf{s}}.$$

For $\alpha \in \mathfrak{Z}$ and a function p on $\mathcal{H}^n$ put

$$(7.18) \qquad (p|\alpha)(w) = J^k(\alpha, w)^{-1} p(\alpha w) \quad (w \in \mathcal{H}^n).$$

This may be different from $p\|_k\alpha$, but $p|\alpha = p\|_k\alpha$ if $\mathrm{pr}(\alpha) \in D^n[2\mathfrak{d}^{-1}, 2\mathfrak{d}]$.

We now assume that $\mathfrak{e} = \mathfrak{c}$. Let $\Gamma = K \cap G$ with K of (4.4a). Then $-1 \notin \Gamma$ and $\chi_{\mathfrak{c}}(\det(d_\xi)) = 1$ for $\xi \in X$; also $J^k(\xi, w)^{-1} = J^k(\xi^{-1}, \xi w)$ for $\xi \in X$ by Lemma 2.3. For $f \in \mathcal{S}_k^n(\Gamma)$, $\xi \in X$, and a function g on $\mathcal{H}$ we easily see that

$$(7.19) \qquad \int_\Phi \overline{f(w)} \sum_{\gamma \in \Gamma} (g|\xi^{-1}\gamma)(w) \Delta(w)^k d_H w = \int_{\mathcal{H}} \overline{(f|\xi)(w)} g(w) \Delta(w)^k d_H w$$

under a suitable convergence condition, where $\Phi = \Gamma \backslash \mathcal{H}$ and $d_H w$ is as in [S10, (2.3b)]. Applying this principle to the function of (7.17), we thus obtain, from (7.16),

$$(7.20) \quad \mathrm{vol}(\Phi) \left\langle \left(f\|\iota^{-1}\right)(w),\ \left(E^N\|\rho_1\right)(\mathfrak{z}, w) \right\rangle$$
$$= t \cdot \chi_{\mathfrak{h}}(\delta/2)^n N(2\mathfrak{d}^{-1})^{2ns} \sum_{\xi \in \Gamma \backslash X} \ell(\xi)^{-2s} \chi^*(\ell_1(\xi)) \sum_{\beta \in B} \chi[\beta] N(\mathfrak{a}_\beta)^{2s}$$
$$\cdot\, j^k(\beta, \mathfrak{z})^{-1} \Delta(\beta\mathfrak{z})^{\mathbf{s}} \int_{\mathcal{H}} \overline{(f|\xi)(w)} M_k^{\mathbf{s}}(\omega_n(\beta\mathfrak{z}), w) \Delta(w)^k d_H w.$$

Substitute $-\overline{w}$ for w and put $f^c(w) = \overline{f(-\overline{w})}$. By [S10, Lemma 4.7] the integral over $\mathcal{H}$ equals $c_{n,k}(\mathbf{s}) \Delta(\omega_n(\beta\mathfrak{z}))^{-\mathbf{s}} (f|\xi)^c(\omega_n(\beta\mathfrak{z}))$, where

$$c_{n,k}(\mathbf{s}) = \prod_{v \in \mathbf{a}} (2i)^{-nk_v} \left(2^{n+3-4s_v} \pi^{n+1}\right)^{n/2} \Gamma_n(s_v + k_v - (n+1)/2) \Gamma_n(s_v + k_v)^{-1}.$$

By Lemma 2.5 we have $(f|\xi)^c = f^c|(\varepsilon\xi\varepsilon)$. Therefore, putting

$$(7.21) \qquad \mathcal{D}(s, f^c) = \sum_{\xi \in \Gamma \backslash X} \chi^*(\ell_1(\xi)) \ell(\xi)^{-s} f^c|\xi,$$

and employing formula (6.9), for sufficiently large $\mathrm{Re}(s)$ we obtain

$$(7.22) \qquad \mathrm{vol}(\Phi) \left\langle \left(f\|\iota^{-1}\right)(w),\ \left(E^N\|\rho_1\right)(\mathfrak{z}, w) \right\rangle$$
$$= t \cdot c_{n,k}(\mathbf{s}) \chi_{\mathfrak{h}}(\delta/2)^n N(2^{-1}\mathfrak{d})^{2ns} E(\mathfrak{z}, s; \mathcal{D}(2s, f^c), \chi, C).$$

8. Proof of the theorems in Section 6

Let us now take $m = n$ in (7.22). We then have $E(\ldots; g, \ldots) = g$. In this special case we employ the differential operator A_k^{ε} of [S10, (5.4b)], which is meaningful for $k \notin \mathbf{Z}^{\mathbf{a}}$ and any fixed subset ε of $\mathbf{a}$, viewed as an element of

$\mathbf{Z}^{\mathbf{a}}$. We first observe that [S10, Lemma 5.1] holds (with the same proof) even when $k \notin \mathbf{Z}^{\mathbf{a}}$ in the sense that

$$(8.1) \qquad (A_k^{\varepsilon} g)\|_{k+\varepsilon}\langle \beta, \gamma\rangle = A_k^{\varepsilon}(g\|_k \langle \beta, \gamma\rangle) \quad \text{for every } \beta \times \gamma \in \mathscr{G}^n \times \mathscr{G}^n,$$

where $\mathscr{G}^n$ is the group of [S10, (1.12)], and $\langle \beta, \gamma\rangle$ is the element of $\mathscr{G}^{2n}$ corresponding naturally to $\beta \times \gamma$. In the paragraph below that lemma we showed that $A_k^{\varepsilon}\mathscr{E}_r = 0$ for $r < n$ if k is integral. The same result holds in the present case of half-integral weight. We don't need any subtle analysis of the factor of automorphy here; we only have to replace elements of G by elements of $\mathscr{G}$, as we do now in the proof of the formula

$$(8.2) \qquad \left[A_k^{\varepsilon}\left(\Delta^{\mathbf{s}}\|_k \alpha\hat{\sigma}^{-1}\right)\right](z,w) = c_0\left(\Delta^{\mathbf{s}}\|_{k+\varepsilon}\alpha\hat{\sigma}^{-1}\right)(z,w)$$

with $c_0 = \prod_{v\in\varepsilon}\prod_{i=0}^{n-1}(-s_v - k_v + i/2)$ if $\alpha = \tau_n(\xi \times 1)$ and $\hat{\sigma}$ are as in (7.10). (We have $\beta = 1$ since $m = n$.) To prove this, let Z denote a variable on $\mathscr{H}^{2n}$; let $\eta = (\tau_n, p(Z))$ be an element of $\mathscr{G}^{2n}$ lying above τ_n. Then $h(\alpha\hat{\sigma}^{-1}, Z) = p((\xi \times 1)Z)q(Z)$ with q such that $q(Z)^{-2}j(\xi \times 1, Z)^u$ is a constant in $\mathbf{T}$. Then $(\xi \times 1, q)$ defines an element of $\mathscr{G}^{2n}$; call it ζ. Then $\Delta^{\mathbf{s}}\|_k\alpha\hat{\sigma}^{-1} = \Delta^{\mathbf{s}}\|_k\eta\zeta$, and $\zeta = \langle \beta, \gamma\rangle$ with some $\beta, \gamma \in \mathscr{G}^n$. Now we have

$$(8.3) \qquad A_k^{\varepsilon}(\Delta^{\mathbf{s}}\|_k\eta) = c_0(\Delta^{\mathbf{s}}\|_{k+\varepsilon}\eta)(z,w).$$

This was given in [S10, (5.8)] for $k \in \mathbf{Z}^{\mathbf{a}}$. The proof given there is valid in the present case. Combining this with (8.1), we obtain (8.2).

Applying A_k^{ε} to (7.11) and employing (8.2), we obtain

$$(A_k^{\varepsilon}\mathscr{E}_n)\|_{k+\varepsilon}(1 \times \iota) = c_0\chi_{\mathbf{h}}(\delta/2)^n N(2^{-1}\mathfrak{d})^{2ns} \cdot \sum_{\xi\in X}\chi^*(\ell_1(\xi))\ell(\xi)^{-2s}\left(\Delta^{\mathbf{s}}\|_{k+\varepsilon}\alpha\hat{\sigma}^{-1}(1\times\iota)\right)(z,w),$$

where $\alpha = \tau_n(\xi \times 1)$. The last factor $(\Delta^{\mathbf{s}}\|_{k+\varepsilon}\ldots)(z,w)$ can be replaced by

$$t \cdot J^{k+\varepsilon}(\xi^{-1}, w)^{-1}\Delta(z)^{\mathbf{s}}M_{k+\varepsilon}^{\mathbf{s}}(z, \xi^{-1}w)$$

for the same reason as in (7.17). Therefore, by integration as in (7.20), for $f \in \mathscr{S}_{k+\varepsilon}^N(\Gamma)$ we thus obtain

$$(8.4) \qquad \operatorname{vol}(\Phi)\left\langle\left(f\|_{k+\varepsilon}\iota^{-1}\right)(w), A_k^{\varepsilon}\left(E^{2n}\|_k\rho_1\right)(z,w)\right\rangle$$
$$= tc_0c_{n,k+\varepsilon}(su - (k - i\kappa)/2)\chi_{\mathbf{h}}(\delta/2)^n N(2^{-1}\mathfrak{d})^{2ns}\mathscr{D}(2s, f^c).$$

We are now ready to prove Theorems 6.1 and 6.2. Given φ, f, and ε as in the theorem, let $\mathfrak{c}' = \mathfrak{c} \cap \mathfrak{f}$ with the conductor $\mathfrak{f}$ of φ. Then we consider (8.4) with $(f^c, k - \varepsilon, \mathfrak{c}', \varphi)$ in place of $(f, k, \mathfrak{c}, \chi)$. By (7.9a) we see that the elements ξ for $\xi \in \Gamma\backslash X$ represent completely $K\backslash\operatorname{pr}(\mathscr{X}^0)G_{\mathbf{a}}$, and hence

$$\mathscr{D}(s, f) = Z(s, \lambda, \varphi)\Lambda_{\mathfrak{c}}^{2n}(s, \varphi)^{-1}$$

by (6.1a). From [S9, Theorem 7.3] we know that $\mathscr{G}(s)\Lambda_{\mathfrak{c}}^{2n}(2s,\chi)E^{2n}$ with a certain $\mathscr{G}$ given there are $\Lambda_{\mathfrak{c}}^{2n}$ of (6.1b) can be continued to a meromorphic function on the whole plane, whose poles belong to an explicitly given finite set. Therefore verifying that the product $c_0 c_{n,k+\varepsilon}(su-(k-i\kappa)/2)\mathscr{G}(s)$ with k replaced by $k-\varepsilon$ equals $A\cdot B^s\Gamma_{k,\kappa,\varepsilon}^{n,n}(s)$ with positive numbers A and B, we obtain Theorem 6.1.

Next, from (6.1a) and (7.22) we obtain

$$(8.5)\quad \mathrm{vol}(\Gamma\backslash\mathscr{H})\langle(f\|\iota)^c(w),\ (\mathscr{F}^{m+n}\|\rho_1)(\mathrm{diag}[\mathfrak{z},w],s)\rangle$$
$$= t\cdot\chi_{\mathbf{h}}(\delta/2)^n N(2^{-1}\mathfrak{d})^{2ns}c_{n,k}(su-(k-i\kappa)/2)\mathscr{F}^{m,n}(\mathfrak{z},s;f,\chi,C),$$

where $\mathscr{F}^{m+n}$ is defined by (6.12). This together with [S9, Theorem 7.3] proves Theorem 6.3.

To prove Theorems 6.4, 6.5, and 6.6, we first observe that the results about $E_k^{m,n}(z,s;g,\Gamma')$ follow from those about $E(z,s;f,\chi,C)$ by virtue of [S10, Lemma 3.7]. Also, as noted in the paragraph preceding Lemma 4.5, $\mathscr{S}_k^n(K)$ is spanned by eigenforms, and so we may assume that f is an eigenform for which $Z(s,\lambda,\chi)$ is meaningful. Then we can derive Theorems 6.4 and 6.5 from [S5, Theorems 2.3, 2.4, and 2.5] by evaluating (7.22) and (8.5) at $s=k_0/2$ and repeating the argument in the proof of [S10, Theorem 7.3]. Here we need the nonvanishing of Z at $s=k_0$, which is guaranteed by Proposition 6.2 under the condition that $k_0\in\Lambda(n,k_0u)$. Similarly Theorem 6.6 can be derived from [S11, Theorem 4.3].

Remark. In [S9] we showed an integral expression

$$(8.6)\qquad \mathscr{Q}(s)Z(s,\lambda,\varphi)=\langle f(z),\theta(z)\mathscr{F}^m(z,s)\rangle$$

when the weight is integral and $\mathfrak{e}=\mathfrak{g}$. Here θ is a theta function and $\mathscr{Q}$ is the product of certain gamma functions and elementary factors. A similar expression can be obtained in the case of half-integral weight. Also, in [S10] we proved that if $k=k_0u$ with $k_0\in(1/2)\mathbf{Z}$, then $\mathscr{M}_k^n=\bigoplus_{r=0}^n\mathscr{E}_k^{n,r}$, where $\mathscr{E}_k^{n,r}$ is the space spanned by $E_k^{n,r}(z,k_0/2;f,\Gamma)\|_k\alpha$ for all possible (f,Γ) and all $\alpha\in G^n$ with the convention that $\mathscr{E}_k^{n,n}=\mathscr{S}_k^n$. For $k_0\notin\mathbf{Z}$ we proved it under the condition $k_0>2n$; if $k_0\in\mathbf{Z}$ we obtained a better result. We can actually show that $\mathscr{M}_k^n=\bigoplus_{r=0}^n\mathscr{E}_k^{n,r}$ for both integral and half-integral k_0 provided: $k_0>(3n-1)/2$ if $n>2$; $k_0>3$ if $n=2$ and $F=\mathbf{Q}$; $k_0>2$ if $n=2$ and $F\neq\mathbf{Q}$. The details of the proof, as well as (8.6) for half-integral k, will be given in [S13].

Corrections to [S9]. Page 551, line 14: Read $\gamma(K)$ for $\lambda(K)$.

Page 556, line 7 from the bottom: The left-hand side of the first equality must be multiplied by $\psi_{\mathfrak{c}}(\det(g))$.

Page 565, line 9 and Page 567, line 5: Read "0 or 1" for ±1.

Corrections to [S10]. Theorem 7.3: Read $\mathscr{S}_k^n(\Gamma'\cap\mathscr{P}^{m,n})$ and G^m for $\mathscr{S}_k^n(\Gamma')$ and G^n.

Lemma 8.11: Read Lemma 8.10 for Lemma 7.10.

References

[F] Feit, P.: Poles and residues of Eisenstein series for symplectic and unitary groups. Memoirs, Amer. Math. Soc. **61**, No. 346 (1986)

[S1] Shimura, G.: Introduction to the arithmetic theory of automorphic functions. Publ. Math. Soc. Japan, No.11, Iwanami Shoten and Princeton Univ. Press, 1971

[S2] Shimura, G.: On modular forms of half integral weight, Ann. of Math. **97** (1973), 440–481

[S3] Shimura, G.: The special values of the zeta functions associated with Hilbert modular forms. Duke M. J. **45** (1978) 637–679

[S4] Shimura, G.: On Eisenstein series. Duke Math. J. **50** (1983), 417–476

[S5] Shimura, G.: On Eisenstein series of half-integral weight. Duke Math. J. **52** (1985) 281–314

[S6] Shimura, G.: On Hilbert modular forms of half-integral weight. Duke M. J. **55** (1987) 765–838

[S7] Shimura, G.: On the transformation formulas of theta series. Amer. J. M. **115** (1993) 1011–1052

[S8] Shimura, G.: On the Fourier coefficients of Hilbert modular forms of half-integral weight. Duke M. J. **71** (1993) 501–557

[S9] Shimura, G.: Euler products and Fourier coefficients of automorphic forms on symplectic groups. Inv. Math. **116** (1994) 531–576

[S10] Shimura, G.: Eisenstein series and zeta functions on symplectic groups. Inv. Math. **119** (1995) 539–584

[S11] Shimura, G.: Nearly holomorphic functions on hermitian symmetric spaces. Math. Ann. **278** (1987) 1–28

[S12] Shimura, G.: Differential operators, holomorphic projection, and singular forms. Duke M. J. **76** (1994) 141–173

[S13] Shimura, G.: Convergence of zeta functions on symplectic and metaplectic groups (to appear)

96a

Convergence of zeta functions on symplectic and metaplectic groups

Duke Mathematical Journal, 82 (1996), 327-347

Introduction. Each of our zeta functions is associated with a holomorphic Hecke eigenform f of integral or half-integral weight with respect to a congruence subgroup of $G^n = Sp(n, F)$, where F is a totally real algebraic number field. The form f can be considered on $G^n_{\mathbf{A}}$ or on the metaplectic cover $M^n_{\mathbf{A}}$ of $G^n_{\mathbf{A}}$ accordingly. The zeta function has the Euler product expression

$$Z(s) = \prod_{\mathfrak{p}} W_{\mathfrak{p}}(N(\mathfrak{p})^{-s})^{-1}, \tag{1}$$

where $\mathfrak{p}$ runs over all the prime ideals of F, and $W_{\mathfrak{p}}$, except finitely many $\mathfrak{p}$'s, is a polynomial of degree $2n+1$ or $2n$ according as the weight is integral or half-integral. It may be noted that such Euler products on $M^n_{\mathbf{A}}$ and their meromorphic continuation have been obtained in our recent paper [S10]. Those on $G^n_{\mathbf{A}}$ are well known (cf. the introduction of [S7]).

Now our first main purpose is to show that the right-hand side of (1) is absolutely convergent, and consequently $Z(s) \neq 0$, for $\mathrm{Re}(s) > (3n/2) + 1$ (Theorem A). Here, for some technical reasons, we take $s = n + 1/2$ to be the center of the critical strip. Duke, Howe, and Li showed in [DHL] that if the form is on $Sp(n, \mathbf{Q})_{\mathbf{A}}$, then the absolute convergence holds for $\mathrm{Re}(s) > (5n/2) + 1$ in general, and in particular for $\mathrm{Re}(s) > (3n/2) + 1$ if $n = 2^r$ with $0 < r \in \mathbf{Z}$. Our present result applies to every n, and even to the Euler products on $M^n_{\mathbf{A}}$.

The bound $(3n/2) + 1$ is best possible, since the right-hand side of (1) does not converge at this point for a certain f. This fact was shown in [DHL] for even n as a consequence of a result of Rallis. We shall prove more generally that given any n, Z has a pole at $(3n/2) + 1$ only if the weight of f is of a "relatively small" restricted type, and it must be integral or half-integral according as n is even or odd, and moreover that such a pole occurs for every n with a certain theta series as f (Theorem C). In [S8] and [S10], we obtained some related results on the location of possible poles of Z. We shall state the results in more refined forms as Theorems B1 and B2. In this and other problems in the present paper, we consider not only Z itself but also its twists by Hecke characters of F.

As an application of Theorem A, we shall show that if the weight is "not too small," the space $\mathscr{M}^n_k(\Gamma)$ of all holomorphic modular forms of weight k with respect to a congruence subgroup Γ of G^n is spanned by cusp forms and Eisenstein

Received 6 January 1995.

series in the sense that

(2) $$\mathcal{M}_k^n(\Gamma) = \bigoplus_{r=0}^{n} \mathcal{E}_k^{n,r}(\Gamma),$$

where $\mathcal{E}_k^{n,n}(\Gamma)$ is the space of all cusp forms in $\mathcal{M}_k^n(\Gamma)$, and $\mathcal{E}_k^{n,r}(\Gamma)$ for $0 \leqslant r < n$ denotes the space spanned by certain Eisenstein series associated with cusp forms on $G_{\mathbf{A}}^r$ or on $M_{\mathbf{A}}^r$, where it is understood that the cusp forms are constants if $r = 0$. In general, the weight k of modular forms is "multiple," but for this kind of problem, we may view it as a single element of $(1/2)\mathbf{Z}$. The decomposition of type (2) was obtained by Klingen in [K] when $\Gamma = Sp(n, \mathbf{Z})$ and $2n < k \in 2\mathbf{Z}$. In our previous paper [S8], we obtained (2) for an arbitrary Γ and every integral $k > b_n$ with some b_n substantially smaller than $2n$ if k is integral, and also for half-integral $k > 2n$. Now, employing the nonvanishing of Z and its twists for $\mathrm{Re}(s) > (3n/2) + 1$, we shall prove (2) for both integral and half-integral k under the following condition: $k \geqslant 3n/2$ if $n > 2$; $k > 3$ if $n = 2$ and $F = \mathbf{Q}$; $k > 2$ if $n = 2$ and $F \neq \mathbf{Q}$ (Theorem 5.2). This is better than what was proved in [S8].

Our proof for the convergence relies on the connection of Z with a certain Dirichlet series $D(s)$, which we defined in [S7] and [S10] by means of the Fourier coefficients of f. We shall prove that D is convergent for $\mathrm{Re}(s) > (3n/2) + 1$, which leads to the desired result. It may be added as a final remark that our method is applicable to nonholomorphic forms and also to some groups other than symplectic groups.

1. Preliminaries and the main theorems. We use the same notation and terminology as in our previous papers [S7], [S8], and [S10]; we recall here some of the basic symbols and conventions. For an associative ring R with identity element and an R-module M, we denote by $R^\times$ the group of all its invertible elements and by M_n^m the R-module of all $m \times n$-matrices with entries in M. We put $\tilde{x} = {}^t x^{-1}$ if a matrix x is square and invertible. We fix a totally real algebraic number field F of finite degree and denote by $\mathbf{a}$, $\mathbf{h}$, $\mathfrak{g}$, and $\mathfrak{d}$ the sets of archimedean primes and nonarchimedean primes of F, the maximal order of F, and the different of F relative to $\mathbf{Q}$. For a fractional ideal $\mathfrak{x}$ in F and $t \in F_{\mathbf{A}}^\times$, we denote by $t\mathfrak{x}$ the fractional ideal in F such that $(t\mathfrak{x})_v = t_v \mathfrak{x}_v$ for every $v \in \mathbf{h}$ and put $|t|_{\mathbf{A}} = \prod_{v \in \mathbf{a} \cup \mathbf{h}} |t_v|$. With $\mathbf{T} = \{\zeta \in \mathbf{C} \,|\, |\zeta| = 1\}$, we define a basic character $\mathbf{e}_{\mathbf{A}}\colon F_{\mathbf{A}} \to \mathbf{T}$ as in [S7], and put $\mathbf{e}_{\mathbf{a}}(x) = \exp(2\pi i \sum_{v \in \mathbf{a}} x_v)$ for $x \in F_{\mathbf{A}}$ and $x \in \mathbf{C}^{\mathbf{a}}$.

With a positive integer n, we now put

$$G = \{\alpha \in GL_{2n}(F) \,|\, {}^t\alpha\iota\alpha = \iota\}, \qquad \iota = \begin{pmatrix} 0 & -1_n \\ 1_n & 0 \end{pmatrix},$$

$$P = \{\alpha \in G \,|\, c_\alpha = 0\},$$

$$H = \{z \in \mathbf{C}_n^n \,|\, {}^t z = z, \mathrm{Im}(z) > 0\}, \qquad \mathcal{H} = H^{\mathbf{a}}.$$

Here and throughout the paper, a_α, b_α, c_α, and d_α denote the standard a-, b-, c-, and d-blocks of α; also we write $x > 0$ (resp., $x \geqslant 0$) to indicate that x is positive definite (resp., nonnegative). We denote by $M_{\mathbf{A}}$ the standard metaplectic cover of $G_{\mathbf{A}}$, for which there is an exact sequence

$$1 \to \mathbf{T} \to M_{\mathbf{A}} \to G_{\mathbf{A}} \to 1.$$

We let pr denote the projection map of $M_{\mathbf{A}}$ to $G_{\mathbf{A}}$. There is a natural lift $r: G \to M_{\mathbf{A}}$ by which we can consider G a subgroup of $M_{\mathbf{A}}$, and also a lift $r_P: P_{\mathbf{A}} \to M_{\mathbf{A}}$. For $\gamma \in G_v$, $v \in \mathbf{a}$, and $w \in H$, we define $\gamma w = \gamma(w) \in H$, $\Delta(w) \in \mathbf{R}$, and a factor of automorphy $j(\gamma, w)$ by

$$\text{(1.1a)} \qquad \gamma w = \gamma(w) = (a_\gamma w + b_\gamma)(c_\gamma w + d_\gamma)^{-1}, \qquad \Delta(w) = \det(\operatorname{Im}(w)),$$

$$\text{(1.1b)} \qquad j(\gamma, w) = \det(c_\gamma w + d_\gamma).$$

For $\alpha \in G_{\mathbf{A}}$ and $z = (z_v)_{v \in \mathbf{a}} \in \mathscr{H}$, we put

$$\text{(1.2a)} \qquad \alpha z = \alpha(z) = (\alpha_v z_v)_{v \in \mathbf{a}}, \qquad \Delta(z) = (\Delta(z_v))_{v \in \mathbf{a}},$$

$$\text{(1.2b)} \qquad j(\alpha, z) = (j(\alpha_v, z_v))_{v \in \mathbf{a}}.$$

These are elements of $(\mathbf{C}_n^n)^{\mathbf{a}}$ or $\mathbf{C}^{\mathbf{a}}$. For two fractional ideals $\mathfrak{x}$ and $\mathfrak{y}$ in F such that $\mathfrak{x}\mathfrak{y} \subset \mathfrak{g}$, we put

$$\text{(1.3a)} \qquad D[\mathfrak{x}, \mathfrak{y}] = G_{\mathbf{a}} \prod_{v \in \mathbf{h}} D_v[\mathfrak{x}, \mathfrak{y}] \ (\subset G_{\mathbf{A}}),$$

$$\text{(1.3b)} \qquad D_v[\mathfrak{x}, \mathfrak{y}] = \{x \in G_v \mid a_x \in (\mathfrak{g}_v)_n^n,\ b_x \in (\mathfrak{x}_v)_n^n,\ c_x \in (\mathfrak{y}_v)_n^n,\ d_x \in (\mathfrak{g}_v)_n^n\}.$$

By a *half-integral weight*, we mean an element k of $\mathbf{Q}^{\mathbf{a}}$ such that $k - u/2 \in \mathbf{Z}^{\mathbf{a}}$, where u denotes the identity element $(1, \ldots, 1)$ of the ring $\mathbf{C}^{\mathbf{a}}$. By an *integral weight*, we mean an element of $\mathbf{Z}^{\mathbf{a}}$. For a weight k, we put

$$\text{(1.4)} \qquad [k] = k \quad \text{if } k \in \mathbf{Z}^{\mathbf{a}}; \qquad [k] = k - u/2 \quad \text{if } k \notin \mathbf{Z}^{\mathbf{a}}.$$

In [S2] and [S6], we defined a factor of automorphy $h_\sigma(z) = h(\sigma, z)$ of weight $u/2$. This is defined for $z \in \mathscr{H}$ and σ in the subset $\mathfrak{M}$ of $M_{\mathbf{A}}$ defined by $\mathfrak{M} = \operatorname{pr}^{-1}(P_{\mathbf{A}} C^\theta)$, with a certain subgroup C^θ of $G_{\mathbf{A}}$ containing $D[2\mathfrak{d}^{-1}, 2\mathfrak{d}]$ (see [S6, (1.13a, b)]). For $\alpha = \operatorname{pr}(\sigma)$ with $\sigma \in M_{\mathbf{A}}$, we put $a_\sigma = a_\alpha$, $b_\sigma = b_\alpha$, $c_\sigma = c_\alpha$, $d_\sigma = d_\alpha$, and $\sigma(z) = \sigma z = \alpha z$. The function $h_\sigma(z)$ is holomorphic in z, and $h_\sigma(z)^2 = \zeta \cdot j(\operatorname{pr}(\sigma), z)^u$ with $\zeta \in \mathbf{T}$. For other properties of h, see [S6] and [S10]. For x and y in $\mathbf{C}^{\mathbf{a}}$, we put $\|x\| = \sum_{v \in \mathbf{a}} x_v$, and also $x^y = \prod_{v \in \mathbf{a}} x_v^{y_v}$ whenever each factor $x_v^{y_v}$ is meaningful (according to the context). For a weight k and $z \in \mathscr{H}$, we define a factor of automorphy $j_*^k(z)$ by

$$j_\alpha^k(z) = j(\alpha, z)^k \qquad (k = [k], \alpha \in G_{\mathbf{A}}), \tag{1.5a}$$

$$j_\sigma^k(z) = h(\sigma, z) j(\alpha, z)^{[k]} \qquad (k \neq [k], \sigma \in \mathfrak{M}, \alpha = \mathrm{pr}(\sigma)). \tag{1.5b}$$

Then for a function f on $\mathscr{H}$ and $\xi \in G_{\mathbf{A}}$ or $\in \mathfrak{M}$, we define a function $f\|_k\xi$ on $\mathscr{H}$ by

$$(f\|_k\xi)(z) = j_\xi^k(z)^{-1} f(\xi z) \qquad (z \in \mathscr{H}). \tag{1.6}$$

We denote by $\mathscr{M}_k$ the set of all holomorphic modular forms of weight k on $\mathscr{H}$ with respect to congruence subgroups of G, and by $\mathscr{S}_k$ its subset consisting of all the cusp forms (see [S7, §1]). For a congruence subgroup Γ, contained in C^θ if $k \neq [k]$, we denote by $\mathscr{M}_k(\Gamma)$ (resp., $\mathscr{S}_k(\Gamma)$) the set of elements f of $\mathscr{M}_k$ (resp., $\mathscr{S}_k$) such that $f\|_k\gamma = f$ for every $\gamma \in \Gamma$.

By a *Hecke character* of F, we mean a continuous homomorphism of $F_{\mathbf{A}}^\times/F^\times$ into $\mathbf{T}$. We always assume that it is normalized in the sense that it takes the value 1 for $cu \in F_{\mathbf{a}}^\times$ with $0 < c \in \mathbf{R}$. We now take a weight k, a fractional ideal $\mathfrak{b}$, an integral ideal $\mathfrak{c}$, and a Hecke character ψ of F such that

$$\psi_v(a) = 1 \ \text{ if } v \in \mathbf{h},\ a \in \mathfrak{g}_v^\times, \text{ and } a - 1 \in \mathfrak{c}_v. \tag{1.7a}$$

Let K be an open subgroup of $D[\mathfrak{b}^{-1}, \mathfrak{bc}]$, and let $\Gamma = G \cap K$. We assume that

$$\mathfrak{b}^{-1} \subset 2\mathfrak{d}^{-1} \quad \text{and} \quad \mathfrak{bc} \subset 2\mathfrak{d} \qquad \text{if } k \neq [k]. \tag{1.7b}$$

This implies that $\mathfrak{c} \subset 4\mathfrak{g}$ and $K \subset C^\theta$ if $k \neq [k]$. Then we denote by $\mathscr{M}_k(K, \psi)$ the set of all $g \in \mathscr{M}_k$ satisfying

$$g\|_k\gamma = \psi_{\mathfrak{c}}(\det(a_\gamma))g \qquad \text{for every } \gamma \in \Gamma, \tag{1.8}$$

where $\psi_{\mathfrak{c}} = \prod_{v|\mathfrak{c}} \psi_v$, and put $\mathscr{S}_k(K, \psi) = \mathscr{S}_k \cap \mathscr{M}_k(K, \psi)$. Given such a g, we can define a function $g_{\mathbf{A}}$ on $G_{\mathbf{A}}$ or on $M_{\mathbf{A}}$, according as $k = [k]$ or $k \neq [k]$, by

$$g_{\mathbf{A}}(\alpha w) = \psi_{\mathfrak{c}}(\det(d_w))(g\|_k w)(\mathbf{i}) \qquad \text{for } \alpha \in G \text{ and } w \in \mathrm{pr}^{-1}(K), \tag{1.9}$$

where $\mathbf{i}$ is the "origin" of $\mathscr{H}$ given by

$$\mathbf{i} = \mathbf{i}_n = (i1_n, \ldots, i1_n), \tag{1.10}$$

and we understand, throughout the paper, that pr means the identity map $G_{\mathbf{A}} \to G_{\mathbf{A}}$ if $k = [k]$. For two complex-valued functions f and g on $\mathscr{H}$ such that

$f\|_k\gamma = f$ and $g\|_k\gamma = g$ for every γ in a congruence subgroup Γ of G, we put

$$(1.11)\qquad \langle f, g\rangle = \mathrm{vol}(\Gamma\backslash\mathscr{H})^{-1}\int_{\Gamma\backslash\mathscr{H}} \overline{f(z)}g(z)\Delta(z)^k\, d_H z,$$

whenever the integral is convergent, with $\mathrm{vol}(\Gamma\backslash\mathscr{H})$ and $d_H z$ given by

$$(1.12)\quad \mathrm{vol}(\Gamma\backslash\mathscr{H}) = \int_{\Gamma\backslash\mathscr{H}} d_H z, \qquad d_H z = \Delta(z)^{-(n+1)u}\bigwedge_{v\in\mathbf{a}}\bigwedge_{p\leqslant q}(dx^v_{pq}\wedge dy^v_{pq}),$$

where $z_v = (x^v_{pq} + iy^v_{pq})$.

Let us now put

$$(1.13a)\qquad S = \{\xi \in F^n_n \mid {}^t\xi = \xi\},$$

$$(1.13b)\qquad S^+ = \{\xi \in S \mid \xi_v > 0 \text{ for every } v \in \mathbf{a}\},$$

$$(1.13c)\qquad S_+ = \{\xi \in S \mid \xi_v \geqslant 0 \text{ for every } v \in \mathbf{a}\},$$

$$(1.13d)\qquad S(\mathfrak{x}) = S \cap \mathfrak{x}^n_n, \qquad S_{\mathbf{h}}(\mathfrak{x}) = \prod_{v\in\mathbf{h}} S(\mathfrak{x})_v,$$

where $\mathfrak{x}$ is any fractional ideal in F. We now take a divisor $\mathfrak{e}$ of $\mathfrak{c}$ such that $\mathfrak{e}^{-1}\mathfrak{c} + \mathfrak{e} = \mathfrak{g}$, and assume in addition to (1.7b) that K has the form

$$(1.14)\qquad K = \{x \in D[\mathfrak{b}^{-1}\mathfrak{e}, \mathfrak{b}\mathfrak{c}] \mid (a_x - 1)_v \in (\mathfrak{e}_v)^n_n \text{ for every } v \in \mathbf{h}\}.$$

We shall mainly be interested in the extreme cases $\mathfrak{e} = \mathfrak{g}$ and $\mathfrak{e} = \mathfrak{c}$. We recall here a basic result [S10, Proposition 1.1].

PROPOSITION 1.1. *Given $f \in \mathscr{M}_k(K, \psi)$, there is a complex number $\mu(\tau, q; f)$, written also $\mu_f(\tau, q)$, determined for $\tau \in S_+$ and $q \in GL_n(F)_{\mathbf{A}}$, such that*

$$(1.15)\qquad f_{\mathbf{A}}\left(r_P\begin{pmatrix} q & s\tilde{q} \\ 0 & \tilde{q}\end{pmatrix}\right) = \det(q)^{[k]}_{\mathbf{a}}|\det(q)_{\mathbf{a}}|^{k-[k]}$$
$$\cdot\sum_{\tau\in S_+}\mu(\tau, q; f)\mathbf{e}_{\mathbf{a}}(\mathrm{tr}(\mathbf{i}\cdot{}^tq\tau q))\mathbf{e}_{\mathbf{A}}(\mathrm{tr}(\tau s))$$

for every $s \in S_{\mathbf{A}}$, where r_P stands for the identity map if $k = [k]$. Moreover, $\mu_f(\tau, q)$ has the following properties:

$$(1.16a)\qquad \mu_f(\tau, q) \neq 0 \quad \textit{only if } \mathbf{e}_{\mathbf{h}}(\mathrm{tr}({}^tq\tau q s)) = 1 \quad \textit{for every } s \in S_{\mathbf{h}}(\mathfrak{b}^{-1}\mathfrak{e});$$

$$(1.16b)\qquad \mu_f(\tau, q) = \mu_f(\tau, q_{\mathbf{h}});$$

(1.16c) $\mu_f({}^t b\tau b, q) = \det(b)^{[k]}|\det(b)|^{k-[k]}\mu_f(\tau, bq)$ *for every* $b \in GL_n(F)$;

(1.16d) $\psi_{\mathbf{h}}(\det(a))\mu_f(\tau, qa) = \mu_f(\tau, q)$ *for every* $\operatorname{diag}[\tilde{a}, a] \in K$.

Furthermore, if $\beta \in G \cap \operatorname{diag}[r, \tilde{r}]K$ *with* $r \in GL_n(F)_{\mathbf{A}}$, *then*

$$(1.17)\qquad j_\beta^k(\beta^{-1}z)f(\beta^{-1}z) = \psi_c(\det(d_\beta r)) \sum_{\tau \in S_+} \mu_f(\tau, r)\mathbf{e}_{\mathbf{a}}(\operatorname{tr}(\tau z)).$$

In our previous papers, we defined Hecke operators $T_\psi(\mathfrak{a})$ acting on $\mathcal{M}_k(K, \psi)$ for every integral ideal $\mathfrak{a}$ in F. We showed that if $0 \neq f \in \mathcal{M}_k(K, \psi)$ and $f|T_\psi(\mathfrak{a}) = \lambda(\mathfrak{a})f$ with a complex number $\lambda(\mathfrak{a})$ for each $\mathfrak{a}$, then for every Hecke character η, we have

$$(1.18)\qquad \Lambda_{\mathfrak{c}}^{2n,k}(s, \eta\psi) \sum_{\mathfrak{a}} \eta^*(\mathfrak{a})\psi^*(\mathfrak{a}')\lambda(\mathfrak{a})N(\mathfrak{a})^{-s} = \prod_{\mathfrak{p}} W_{\mathfrak{p}}(\psi^*(\mathfrak{p}_c)\eta^*(\mathfrak{p})N(\mathfrak{p})^{-s})^{-1}.$$

Here $\mathfrak{a}$ runs over all the integral ideals and $\mathfrak{p}$ over all the prime ideals; η^* denotes the ideal character associated with η; $\mathfrak{a}' = \prod_{v \nmid \mathfrak{c}} \mathfrak{a}_v$; $W_{\mathfrak{p}}(t)$ is a polynomial in t with constant term 1, and its degree $d_{\mathfrak{p}}$ is as follows: $d_{\mathfrak{p}} = 0$, that is, $W_{\mathfrak{p}} = 1$ if $\mathfrak{p}|\mathfrak{e}$; $d_{\mathfrak{p}} \leqslant n$ if $\mathfrak{p}|\mathfrak{e}^{-1}\mathfrak{c}$; $d_{\mathfrak{p}} = 2n+1$ if $\mathfrak{p} \nmid \mathfrak{c}$ and $k = [k]$; $d_{\mathfrak{p}} = 2n$ if $\mathfrak{p} \nmid \mathfrak{c}$ and $k \neq [k]$. Finally, $\Lambda_{\mathfrak{c}}^{*,k}$ is given by

$$(1.19)\qquad \Lambda_{\mathfrak{c}}^{n,k}(s, \chi) = \begin{cases} L_{\mathfrak{c}}(s, \chi) \prod_{i=1}^{[n/2]} L_{\mathfrak{c}}(2s-2i, \chi^2) & \text{if } k = [k], \\ \prod_{i=1}^{[(n+1)/2]} L_{\mathfrak{c}}(2s-2i+1, \chi^2) & \text{if } k \neq [k], \end{cases}$$

$$(1.20)\qquad L_{\mathfrak{c}}(s, \chi) = \prod_{\mathfrak{p} \nmid \mathfrak{c}} (1 - \chi^*(\mathfrak{p})N(\mathfrak{p})^{-s})^{-1}.$$

For these, see [S7, §2], [S8, §6], and [S10, §6]. Strictly speaking, the result in [S7] for integral k concerns only the case $\mathfrak{e} = \mathfrak{g}$. However, since the ring of Hecke operators generated by $T_\psi(\mathfrak{a})$ is a homomorphic image of the abstract ring of [S7, §2], we obtain (1.18) in the general case from the formulas of [S7, Theorem 2.9].

Let us now denote the function of (1.18) by $Z_\psi(s, \lambda, \eta)$. If $\mathfrak{e} = \mathfrak{c}$, we can simplify the matter by taking $\psi = 1$ without losing anything. If $\mathfrak{e} = \mathfrak{g}$, however, the Euler product of (1.18) is most natural. Now our first main result of this paper can be stated as follows.

THEOREM A. *The Euler product on the right-hand side of (1.18) is absolutely convergent, and consequently* $Z_\psi(s, \lambda, \eta) \neq 0$, *at least for* $\operatorname{Re}(s) > (3n/2) + 1$.

This and Theorem C will be proved in Section 3. As immediate consequences of Theorem A, we obtain improved versions of [S8, Theorems 7.3 and 7.5], which concern holomorphy and near holomorphy of certain Eisenstein series at integer or half-integer points in an interval. Namely, the set $\Lambda(n)$ for $n > 1$ in those theorems can be replaced by $\{x \in \mathbf{R} \mid x > (3n/2) + 1\}$ so that the interval can be replaced by a larger one. We shall give another type of improved results in Section 5.

In [S8, Theorem 6.1] and [S10, Theorem 6.1], we showed that $Z_\psi(s, \lambda, \eta)$ times explicitly defined gamma factors is meromorphic on the whole plane and has only finitely many poles. Here we reformulate the results in a somewhat improved form. It should be noted that since our f is a cusp form, we have $k_v \geqslant n/2$ for every $v \in \mathbf{a}$ by virtue of [S9, Corollary 5.7(3)].

THEOREM B1. *Suppose that* $\mathfrak{e} = \mathfrak{c}$ *and* $\psi = 1$. *Let* $\mathfrak{f}$ *be the conductor of* η; *take* $\varepsilon \in \mathbf{Z}^{\mathbf{a}}$, $0 \leqslant \varepsilon_v \leqslant 1$, *and* $\nu \in \mathbf{R}^{\mathbf{a}}$, $\sum_{v \in \mathbf{a}} \nu_v = 0$, *so that* $\eta_{\mathbf{a}}(x) = \mathrm{sgn}(x_{\mathbf{a}})^{[k]+\varepsilon} |x_{\mathbf{a}}|^{i\nu}$; *put* $\ell = \mathrm{Max}_{v \in \mathbf{a}} \{k_v - \varepsilon_v\}$. *Then the product* $\mathscr{Z} = \mathscr{G}(s/2) Z_\psi(s, \lambda, \varphi)$ *can be continued meromorphically to the whole s-plane with finitely many poles, where* $\mathscr{G}$ *is given in [S8, (6.6a–b)] and [S10, (6.4a–b)]. Moreover, each pole is simple. In particular,* $\mathscr{Z}$ *is an entire function of s if* $\eta^2 \neq 1$. *If* $\eta^2 = 1$, *the possible poles of* $\mathscr{Z}$ *are determined as follows.*

(1) $\mathfrak{c}\mathfrak{f} \neq \mathfrak{g}$, $\ell > n$: *$\mathscr{Z}$ has no pole except for a possible pole at* $s = n + 1$, *which occurs only if k is integral and* $k_v - \varepsilon_v - n \in 2\mathbf{Z}$ *for every v such that* $k_v - \varepsilon_v > n$.

(2) $\mathfrak{c}\mathfrak{f} \neq \mathfrak{g}$, $\ell \leqslant n$: *$\mathscr{Z}$ has possible poles only in the set*

$$\{j + \alpha \mid j \in \mathbf{Z},\, n + 1 \leqslant j \leqslant \mathrm{Min}(2n - \ell, 3n/2) + 1 - \alpha\},$$

where α *is 0 or 1/2, according as k is integral or half-integral.*

(3) $\mathfrak{c}\mathfrak{f} = \mathfrak{g}$: *In addition to the possible poles given in (1) or (2) according as* $\ell > n$ *or* $\ell < n$, *$\mathscr{Z}$ may have poles only in the set*

$$\{j \in \mathbf{Z} \mid [(n + 1)/2] \leqslant j \leqslant n\}.$$

Proof. The only new points are cases (2) and (3). Case (3) will be proved in Section 4. As for case (2), in [S8, Theorem 6.1] and [S10, Theorem 6.1], we gave the bound $2n + 1 - \ell - \alpha$ in place of $\mathrm{Min}(2n - \ell, 3n/2) + 1 - \alpha$. Now Z_ψ is holomorphic for $\mathrm{Re}(s) > (3n/2) + 1$ by Theorem A. Verifying that $\mathscr{G}(s/2)$ is finite in that half-plane, we obtain the set of (2) as stated.

THEOREM B2. *Let the notation and the assumptions be the same as in Theorem B1; suppose* $n > 1$. *(See [S8, Theorem 6.4] and Theorem D below for the case* $n = 1$.*) Then* $Z_\psi(s, \lambda, \eta)$ *is a meromorphic function on the whole s-plane with finitely many poles, and each pole is simple. In particular,* Z_ψ *is an entire function of s if* $\eta^2 \neq 1$. *If* $\eta^2 = 1$, *the possible poles of* Z_ψ *are determined as follows.*

(1) $\ell > n$, $\mathfrak{c}\mathfrak{f} \neq \mathfrak{g}$: *$Z_\psi$ has no pole except for a possible pole at* $s = n + 1$, *which occurs only if k is integral,* $k_v \geqslant n$ *for every* $v \in \mathbf{a}$, *and* $k_v - \varepsilon_v - n \in 2\mathbf{Z}$ *for every v such that* $k_v - \varepsilon_v > n$.

(2) $\ell > n$, $\mathfrak{cf} = \mathfrak{g}$: *In addition to the possible poles given in (1), Z_ψ may have poles only in the set*

$$\{j \in \mathbf{Z} \mid [(n+1)/2] \leqslant j \leqslant n,$$
$$2n - k_v < j \equiv k_v - \varepsilon_v + 1 \ (\mathrm{mod}\ 2) \text{ for every } v \in \mathbf{a}\}.$$

(3) $\ell \leqslant n$: *Z_ψ has possible poles only in the set*

$$\{j + \alpha \mid j \in \mathbf{Z},\ n + 1 \leqslant j \leqslant \mathrm{Min}(2n - \ell, 3n/2) + 1 - \alpha,$$
$$j \geqslant 2n + 1 - \alpha - k_v - \varepsilon_v \text{ for every } v \in \mathbf{a}\},$$

where α is 0 or 1/2, according as k is integral or half-integral.

Proof. If $k = [k]$, $\eta^2 = 1$, and $\ell > n > 1$, [S8, (6.6a)] shows that $\mathscr{G}(s/2)$ has the following product as its factor:

$$\prod_{v \in \mathbf{a}} \Gamma((1/2)(s + k_v + \varepsilon_v - 2n))\Gamma((1/2)(s + k_v + \varepsilon_v + 1 - 2n))$$
$$\cdot \Gamma((1/2)(s + k_v - \varepsilon_v + 1 - 2n))\Gamma((1/2)(s + k_v - \varepsilon_v + 2 - 2n)).$$

This has all integers $\leqslant 2n - k_v$ as its poles. Also $\mathscr{G}(s/2)$ has another type of factor

$$\prod_{v \in \mathbf{a}} \Gamma\left(\frac{s + k_v - \varepsilon_v}{2} - \left[\frac{k_v - \varepsilon_v + n}{2}\right]\right),$$

which has poles at the integers s such that $n \geqslant s \equiv k_v - \varepsilon_v \ (\mathrm{mod}\ 2)$. Combining these facts with Theorem B1, we obtain (1) and (2). Case (3) can be verified in a similar way.

The poles of Z_ψ on $G_{\mathbf{A}}$ (without gamma factors and with some Euler $\mathfrak{p}$-factors eliminated) were treated also by Kudla and Rallis in [KR], but their result is of a different nature from ours, since it is "generic" in the sense that it does not refer to the weight (or rather, the type of automorphic representation), nor to the signature of η.

We now consider a weight ℓ of the form $\ell = (nu/2) + \mu$ with $\mu \in \mathbf{Z}^{\mathbf{a}}$ such that $0 \leqslant \mu_v \leqslant 1$ for every $v \in \mathbf{a}$. Let $\mathscr{T}_\ell$ denote the space spanned by the functions g on $\mathscr{H}$ of the form

$$(1.21) \qquad g(z) = \sum_{\xi \in V} \omega(\xi_{\mathbf{h}}) \det(\xi)^{\mu} \mathbf{e}_{\mathbf{a}}(\mathrm{tr}({}^t\xi\tau\xi z)) \qquad (z \in \mathscr{H}).$$

Here $V = F_n^n$, ω is a Schwartz-Bruhat function on $V_{\mathbf{h}}$, and $\tau \in S^+$. As noted in [S7, §6], this is a special case of the functions investigated in [S6, §3], and so $\mathscr{T}_\ell \subset \mathscr{M}_\ell$, and in particular $\mathscr{T}_\ell \subset \mathscr{S}_\ell$ if $\mu \neq 0$.

THEOREM C. (1) *$Z_\psi(s, \lambda, \eta)$ has a pole at $s = (3n/2) + 1$ only if $k = (nu/2) + \mu$ with μ as above.*

(2) *If $\mathfrak{e} = \mathfrak{c}$ and $k = (nu/2) + \mu$ as above with $\mu \neq 0$, then $Z_\psi(s, \lambda, \eta)$ has a pole at $s = (3n/2) + 1$ only if $f \in \mathcal{T}_k$.*

(3) *Given a Hecke character η of F and k as in (2), there exists an eigenform f contained in $\mathcal{T}_k$ such that $Z_\psi(s, \lambda, \eta)$ has a pole at $s = (3n/2) + 1$.*

In fact, the first assertion follows essentially from Theorem B2. There is room for improvement in this theorem. For instance, probably the conditions $\mathfrak{e} = \mathfrak{c}$ and $\mu \neq 0$ in (2) are unnecessary, and also other types of generalizations are possible. In this paper, however, we content ourselves with relatively easy clear-cut statements that require no lengthy discussion.

Coming back to the general setting of Theorem A, we are tempted to conjecture that $Z_\psi(s, \lambda, \eta) \neq 0$ for $\mathrm{Re}(s) \geqslant n + 1$ if $k_v \geqslant n + 1$ for every $v \in \mathbf{a}$. If $n = 1$, we have indeed a better result.

THEOREM D. *If $n = 1$, we have $Z_\psi(s, \lambda, \eta) \neq 0$ for $\mathrm{Re}(s) \geqslant 2$ in the following three cases: (i) $k = [k]$; (ii) $k \neq [k]$ and $k_v > 3/2$ for some $v \in \mathbf{a}$; (iii) $k = 3u/2$and f is orthogonal to the series of $\mathcal{T}_{3u/2}$. Moreover, in these cases $Z_\psi(s, \lambda, \eta)$ is finite for every $s \in \mathbf{C}$ except that it has a possible simple pole at $s = 1$ and $s = 2$ if $k = [k]$.*

This is not new; we state it here for the reader's convenience. If $k = [k]$, $Z_\psi(s, \lambda, \eta)$ coincides with a function of [S5, (10.3)] as explained at the end of [S8, §6], and hence nonvanishing follows from [S5, Proposition 3.3 and (10.4)]. The information concerning possible poles is given in [S8, Theorem 6.4]. As for the case $k \neq [k]$, we refer the reader to [S10, Proposition 6.2] and the succeeding paragraph. In this case, Z_ψ is entire.

2. Theta functions with imprimitive characters. Fixing an integral ideal $\mathfrak{e}$ in F, we put

$$E = \prod_{v \in \mathbf{h}} E_v, \quad E' = \prod_{v \in \mathbf{h}} E'_v, \quad E_v = GL_n(\mathfrak{g}_v), \quad E'_v = \{x \in E_v \mid x_v - 1 \in (\mathfrak{e}_v)_n^n\},$$

$$V = F_n^n, \quad R = \prod_{v \in \mathbf{h}} (\mathfrak{g}_v)_n^n \ (\subset V_\mathbf{h}),$$

$$R' = \{x \in R \mid x_v \in E'_v \text{ for every } v \mid \mathfrak{e}\}, \quad R^* = R' \cdot V_\mathbf{a} \ (\subset V_\mathbf{A}).$$

We take a Hecke character η of F and $\mu \in \mathbf{Z}^\mathbf{a}$ such that $0 \leqslant \mu_v \leqslant 1$ for every $v \in \mathbf{a}$, and denote the conductor of η by $\mathfrak{f}$. Taking also an element τ of S^+ and an element p of $GL_n(F)_\mathbf{h}$, we define a series θ by

$$\theta(z) = \sum_{\xi \in V \cap pR^*} \eta_\mathbf{a}(\det(\xi))\eta^*(\det(p^{-1}\xi)\mathfrak{g}) \det(\xi)^\mu \mathbf{e}_\mathbf{a}(\mathrm{tr}({}^t\xi\tau\xi z)), \tag{2.1}$$

where η^* is the ideal character associated with η, and it is understood that $\eta_{\mathbf{a}}(b)\eta^*(b\mathfrak{a})$ for $b = 0$ and a fractional ideal $\mathfrak{a}$ denotes $\eta^*(\mathfrak{a})$ or 0, according as $\mathfrak{f} = \mathfrak{g}$ or $\mathfrak{f} \neq \mathfrak{g}$. This is a special case of (1.21). As a generalization of [S7, Proposition 6.2] which concerns the case $\mathfrak{e} = \mathfrak{g}$, we can state the following proposition.

Proposition 2.1. *Let ρ_τ be the Hecke character of F corresponding to the extension $F(c^{1/2})/F$ with $c = (-1)^{[n/2]}\det(2\tau)$; let $\mathfrak{f}' = \mathfrak{f} \cap \mathfrak{e}$, $\ell = \mu + (nu/2)$, and $\eta' = \eta\rho_\tau$. Then there exist a fractional ideal $\mathfrak{b}$ and an integral ideal $\mathfrak{c}$ such that the conductor of η' divides $\mathfrak{c}$, $\theta \in \mathcal{M}_\ell(D[\mathfrak{b}^{-1}, \mathfrak{b}\mathfrak{c}], \eta')$, and $D[\mathfrak{b}^{-1}, \mathfrak{b}\mathfrak{c}] \subset D[2\mathfrak{d}^{-1}, 2\mathfrak{d}]$ if n is odd. Moreover, if $\beta \in G \cap \operatorname{diag}[r, \tilde{r}]D[\mathfrak{b}^{-1}, \mathfrak{b}\mathfrak{c}]$ with $r \in GL_n(F)_{\mathbf{h}}$, then*

$$(2.2)\quad j_\beta^\ell(\beta^{-1}z)\theta(\beta^{-1}z) = \eta'(\det(r))^{-1}\eta'_{\mathfrak{c}}(\det(d_\beta r))|\det(r)|_{\mathbf{A}}^{n/2} \cdot \sum_{\xi \in V \cap pR^*r^{-1}} \eta_{\mathbf{a}}(\det(\xi))\eta^*(\det(\xi p^{-1}r)\mathfrak{g})\det(\xi)^\mu \mathbf{e}_{\mathbf{a}}(\operatorname{tr}({}^t\xi\tau\xi z)).$$

In particular, suppose that ${}^tg \cdot 2\tau g \in \mathfrak{x}$ for every $g \in pL_0$ and ${}^th(2\tau)^{-1}h \in 4\mathfrak{t}^{-1}$ for every $h \in \tilde{p}L_0$ with fractional ideals $\mathfrak{x}$ and $\mathfrak{t}$, where $L_0 = \mathfrak{g}_1^n$; let $\mathfrak{h}$ be the conductor of ρ_τ. Then we can take $(\mathfrak{b}, \mathfrak{c}) = (2^{-1}\mathfrak{d}\mathfrak{x}, \mathfrak{h} \cap \mathfrak{f}' \cap \mathfrak{x}^{-1}\mathfrak{f}'^2\mathfrak{t})$ if n is even and $(\mathfrak{b}, \mathfrak{c}) = (2^{-1}\mathfrak{d}\mathfrak{a}^{-1}, \mathfrak{h} \cap \mathfrak{f}' \cap 4\mathfrak{a} \cap \mathfrak{a}\mathfrak{f}'^2\mathfrak{t})$ if n is odd, where $\mathfrak{a} = \mathfrak{x}^{-1} \cap \mathfrak{g}$.

Proof. Using the notation of [S7, §6], define $\lambda \in \mathcal{S}(V_{\mathbf{h}})$ by $\lambda(x) = \eta_{\mathbf{h}}(\det(p)^{-1}) \cdot \lambda'(p^{-1}x)$, $\lambda'(x) = \prod_{v \in \mathbf{h}} \lambda'_v(x_v)$, with

$$\lambda'_v(y) = \begin{cases} 1 & \text{if } y \in (\mathfrak{g}_v)_n^m, \quad v \nmid \mathfrak{f}', \\ \eta_v(\det(y)^{-1}) & \text{if } y \in E'_v, \quad v \mid \mathfrak{f}', \\ 0 & \text{otherwise}. \end{cases}$$

Then the function $\theta(z, \lambda)$ of [S7, (6.1)] coincides with θ of (2.1), and therefore we obtain our result from [S6, Proposition 3b.2] and [S7, Proposition 6.1], modifying the proof of [S7, Proposition 6.2] in an obvious way.

From the above proposition and (1.17), we obtain

$$(2.3)\quad \mu_\theta(\sigma, r) = \sum_\xi \eta'(\det(r))^{-1}|\det(r)|_{\mathbf{A}}^{n/2}\eta_{\mathbf{a}}(\det(\xi))\eta^*(\det(\xi p^{-1}r)\mathfrak{g})\det(\xi)^\mu,$$

where ξ runs over $V \cap pR^*r^{-1}$ under the condition that ${}^t\xi\tau\xi = \sigma$.

3. Proof of Theorems A and C. Take $\mathfrak{b}$, $\mathfrak{c}$, $\mathfrak{e}$, and K as in (1.7b) and (1.14); define E' as in Section 2. Fixing $\mathfrak{e}$ for the moment, for $q \in GL_n(F)_{\mathbf{h}}$ and $\sigma \in S^+$ put

$$U_q = GL_n(F) \cap qE^*q^{-1}, \qquad E^* = GL_n(F)_{\mathbf{a}}E',$$

$$\nu_{\sigma,q} = [U_{\sigma,q} : 1]^{-1}, \qquad U_{\sigma,q} = \{a \in U_q \mid {}^ta\sigma a = \sigma\}.$$

Then we consider an equivalence relation $\sim$ in S^+ defined by: $\sigma \sim \sigma'$ if and only if $\sigma' = {}^t a\sigma a$ with $a \in U_q$. Let $f \in \mathscr{S}_k(K, \psi)$ and $g \in \mathscr{M}_\ell(K', \psi')$ with integral or half-integral k and ℓ. Here K' is a group of the same type with possibly different $\mathfrak{b}$, $\mathfrak{c}$, and $\mathfrak{e}$. We assume that

$$\text{(3.1a)} \qquad \psi'_{\mathfrak{h}}(\det(a))\mu_g(\sigma, qa) = \mu_g(\sigma, q) \qquad \text{for every } a \in E',$$

$$\text{(3.1b)} \qquad (\psi/\psi')_{\mathbf{a}}(x) = \operatorname{sgn}(x_{\mathbf{a}})^{[k]-[\ell]}|x_{\mathbf{a}}|^{i\kappa} \qquad \text{with } \kappa \in \mathbf{R}^{\mathbf{a}}, \quad \|\kappa\| = 0.$$

Notice that (1.16d) implies (3.1a) with f and ψ in place of g and ψ'. We then define, for $s \in \mathbf{C}$,

$$\text{(3.2)} \quad D_{q,\kappa}(s; f, g) = \sum_{\sigma \in S^+/U_q} \nu_{\sigma,q}\mu_f(\sigma, q)\overline{\mu_g(\sigma, q)} \det(\sigma)^{-su-\lambda}, \quad \lambda = (k + \ell + i\kappa)/2.$$

This is well defined in view of (1.16b–d) and (3.1a–b).

Lemma 3.1. *With f and q as above, suppose that for some $\sigma \in S^+$ we have $\mu_f(\sigma, q) \neq 0$ and $U_{\sigma,q}$ contains an element a such that* $\det(a) = -1$. *Then* $\psi_{\mathbf{a}}(-1) = (-1)^{\|[k]\|}$.

Proof. This follows immediately from (1.6c–d).

Lemma 3.2. (1) *The right-hand side of (3.2) is absolutely convergent for* $\operatorname{Re}(s) > 0$ *if* $g \in \mathscr{S}_\ell$.

(2) *$D_{q,\kappa}(s; f, g)$ can be continued to a meromorphic function on the whole plane, which is holomorphic for* $\operatorname{Re}(s) > 0$. *Moreover, it is holomorphic at $s = 0$ if $k \neq \ell$ or $\kappa \neq 0$.*

(3) *If $k = \ell$ and $\kappa = 0$, then $D_{q,0}(s; f, g)$ has at most a simple pole at $s = 0$ whose residue is a positive number times $\langle g, f\rangle$.*

Proof. By strong approximation, we can find an element $\beta \in G \cap \operatorname{diag}[q, \tilde{q}] \cdot (K \cap K')$ such that $\psi(\det(d_\beta q)_{\mathfrak{c}}) = \psi'(\det(d_\beta q)_{\mathfrak{c}'}) = 1$, where $\mathfrak{c}'$ is the ideal determined for K' similarly to $\mathfrak{c}$. Put

$$f_q(z) = \sum_{\sigma \in S_+} \mu_f(\sigma, q)\mathbf{e}_{\mathbf{a}}(\operatorname{tr}(\sigma z)), \qquad g_q(z) = \sum_{\sigma \in S_+} \mu_g(\sigma, q)\mathbf{e}_{\mathbf{a}}(\operatorname{tr}(\sigma z)).$$

By (1.17), $f_q = j_\beta^k(\beta^{-1}z)f(\beta^{-1}z) \in \mathscr{S}_k$ and $g_q = j_\beta^\ell(\beta^{-1}z)g(\beta^{-1}z) \in \mathscr{M}_\ell$. Then the standard argument shows that

$$\text{(3.3)} \qquad \int_{\Gamma\backslash\mathscr{H}} f_q(z)\overline{g_q(z)E(z, \bar{s} + (n+1)/2; k - \ell, \kappa, \Gamma)}\Delta(z)^k\, d_H z$$

$$= c \cdot (4\pi)^{-n\|su+\lambda\|} \prod_{v \in \mathbf{a}} \Gamma_n(s + \lambda_v) D_{q,\kappa}(s; f, g),$$

where Γ is a sufficiently small congruence subgroup of G, c is a positive constant,

(3.4a) $$\Gamma_n(s) = \pi^{n(n-1)/4} \prod_{v=0}^{n-1} \Gamma(s - v/2),$$

and E is the Eisenstein series defined by

(3.4b) $$E(z, s; m, \kappa, \Gamma) = \sum_{\alpha \in A} \Delta(z)^{su-(m+i\kappa)/2} \|_m \alpha, \qquad A = (P \cap \Gamma)\backslash\Gamma.$$

By Lemma 3.3 below, $E(z, s; m, \kappa, \Gamma)$ is holomorphic for $\mathrm{Re}(s) > (n+1)/2$. Moreover, it is well known that if it has a pole of order t at s_0, then $(s - s_0)^t E$ as a function of z is slowly increasing at every cusp. Observe also that $\langle g, f\rangle = \langle g_q, f_q\rangle$. If $f \neq 0$, [S9, Corollary 5.7(3)] shows that $k_v > (n-1)/2$ for every $v \in \mathbf{a}$, and hence $\prod_{v \in \mathbf{a}} \Gamma_n(k_v)^{-1} \neq 0$. Therefore, (2) and (3) follow immediately from (3.3) and Lemma 3.3 below. To prove (1), we observe that for real s the Cauchy-Schwarz inequality gives

$$\sum_{\sigma \in S^+/U} v_{\sigma,q} |\mu_f(\sigma, q)\mu_g(\sigma, q) \det(\sigma)^{-su-\lambda}| \leqslant [D_{q,0}(s; f, f) D_{q,0}(s; g, g)]^{1/2},$$

and therefore assertion (1) in the general case follows from the special case $f = g$. Now we know that $D_{q,0}(s; f, f)$ is holomorphic for $\mathrm{Re}(s) > 0$. Since all its coefficients are nonnegative, the series defining $D_{q,0}(s; f, f)$ must be convergent for $\mathrm{Re}(s) > 0$. This completes the proof.

LEMMA 3.3. *$E(z, s; m, \kappa, \Gamma)$ can be continued to a meromorphic function on the whole s-plane, which is holomorphic for* $\mathrm{Re}(s) > (n+1)/2$. *Moreover, it has a pole at $s = (n+1)/2$ only if $m = \kappa = 0$, in which case it has a simple pole at $s = (n+1)/2$ with a positive real number as its residue.*

Proof. Given Γ, we can find an integral ideal $\mathfrak{h}$ divisible by 4 such that Γ contains the group $\Gamma_1 = \{\gamma \in G \cap D[2\mathfrak{d}^{-1}\mathfrak{h}, 2^{-1}\mathfrak{d}\mathfrak{h}] \mid a_\gamma - 1 \in \mathfrak{h}_n^n\}$. Then we have

$$[P \cap \Gamma : P \cap \Gamma_1] \# (\Phi) E(z, s; m, \kappa, \Gamma) = \sum_{\alpha \in \Gamma_1 \backslash \Gamma} \sum_{\chi \in \Phi} E(z, s; m, \chi, \Gamma_0) \|_k \alpha,$$

where $\Gamma_0 = G \cap D[2\mathfrak{d}^{-1}, 2^{-1}\mathfrak{d}\mathfrak{h}]$, $E(z, s; m, \chi, \Gamma_0)$ is a series defined in [S7, (7.4)], Φ is the set of all Hecke characters χ such that $\chi_{\mathbf{a}}(x) = \mathrm{sgn}(x_{\mathbf{a}})^{[m]} |x_{\mathbf{a}}|^{-i\kappa}$, and the conductor of χ divides $\mathfrak{h}$. If $\kappa = 0$, this is a special case of [S8, Lemma 3.7]. The case $\kappa \neq 0$ can be proved in the same manner as in [S1, Proposition 2.4] and [S2, Lemma 2.2]. Thus, the analytic nature of $E(z, s; m, \kappa, \Gamma)$ can be reduced to that of $E(z, s; m, \chi, \Gamma_0)$, which is given in [S7, Theorem 7.3]. Also, the Fourier expansion of a certain transform of the latter series is given in [S1], [S2], and [S7, (7.7), Propositions 7.1 and 7.2]. From these we easily see that our functions are all holomorphic for $\mathrm{Re}(s) > (n+1)/2$; $E(z, s; m, \chi, \Gamma_0)$ has a pole at $s = (n+1)/2$

only if $m = 0$ and $\chi = 1$, in which case it has a simple pole there with a positive number as its residue. Combining this with the above equality, we obtain our lemma. In fact, the result about the residue can be obtained more directly without χ, but we will not discuss it here.

To make our argument clearer, we introduce the ring of formal Dirichlet series $\sum c(\mathfrak{a})[\mathfrak{a}]$ with $c(\mathfrak{a}) \in \mathbf{C}$, where $[\mathfrak{a}]$ is a formal multiplicative symbol defined for each fractional ideal $\mathfrak{a}$ in F with the properties that $[\mathfrak{a}\mathfrak{b}] = [\mathfrak{a}][\mathfrak{b}]$, $[\mathfrak{g}] = 1$, and the $[\mathfrak{p}]$ for all the prime ideals $\mathfrak{p}$ are independent indeterminates. The series must always be of the form $\sum_{\mathfrak{a} \subset \mathfrak{x}} c(\mathfrak{a})[\mathfrak{a}]$ with some fixed fractional ideal $\mathfrak{x}$.

LEMMA 3.4. *Suppose a Dirichlet series $\sum_{\mathfrak{a} \subset \mathfrak{g}} c_{\mathfrak{a}} N(\mathfrak{a})^{-s}$ with $c_{\mathfrak{a}} \in \mathbf{C}$ is absolutely convergent for $\mathrm{Re}(s) > \alpha$ and can be decomposed formally into an Euler product in the sense that $\sum_{\mathfrak{a}} c_{\mathfrak{a}}[\mathfrak{a}] = \prod_{\mathfrak{p}} V_{\mathfrak{p}}([\mathfrak{p}])^{-1}$ with complex polynomials $V_{\mathfrak{p}}(x)$ such that $V_{\mathfrak{p}}(0) = 1$ defined for all prime ideals $\mathfrak{p}$. Then, for any $\mathbf{T}$-valued ideal character φ, the infinite product $\prod_{\mathfrak{p}} V_{\mathfrak{p}}(\varphi(\mathfrak{p})N(\mathfrak{p})^{-s})^{-1}$ is convergent to the Dirichlet series $\sum_{\mathfrak{a} \subset \mathfrak{g}} c_{\mathfrak{a}} \varphi(\mathfrak{a}) N(\mathfrak{a})^{-s}$ and nonvanishing for $\mathrm{Re}(s) > \alpha$.*

Proof. We can formally put $V_{\mathfrak{p}}(x)^{-1} = 1 + \sum_{n=1}^{\infty} b_{\mathfrak{p},n} x^n$ with $b_{\mathfrak{p},n} \in \mathbf{C}$. Then $\sum_{\mathfrak{p}} \sum_{n=1}^{\infty} b_{\mathfrak{p},n} N(\mathfrak{p})^{-ns}$ is a partial series $\sum_{\mathfrak{a}} c_{\mathfrak{a}} N(\mathfrak{a})^{-s}$, and therefore $\sum_{\mathfrak{p}} \sum_{n=1}^{\infty} |b_{\mathfrak{p},n} \varphi(\mathfrak{p})^n N(\mathfrak{p})^{-ns}| < \infty$ for $\mathrm{Re}(s) > \alpha$. Thus, the infinite product must be convergent. Clearly, each factor $V_{\mathfrak{p}}(\varphi(\mathfrak{p})N(\mathfrak{p})^{-s})^{-1}$ is nonvanishing, and hence we obtain our lemma.

Now, given $f \in \mathscr{S}_k(K, \psi)$, a Hecke character η, $\tau \in S^+$, and $p \in GL_n(F)_{\mathbf{h}}$, we define a formal Dirichlet series $C(f, \eta)$ and an ordinary Dirichlet series $D(s, f, \eta)$ by

$$C(f, \eta) = \sum_{x \in B'/E'} \psi(\det(px))\eta^*(\det(x)\mathfrak{g})\mu_f(\tau, px)|\det(x)|_{\mathbf{A}}^{-n-1}[\det(x)\mathfrak{g}], \tag{3.5a}$$

$$D(s, f, \eta) = \sum_{x \in B'/E'} \psi(\det(px))\eta^*(\det(x)\mathfrak{g})\mu_f(\tau, px)|\det(x)|_{\mathbf{A}}^{s-n-1}. \tag{3.5b}$$

These depend on p and τ, but we fix them in the following treatment. Taking a disjoint decomposition

$$GL_n(F)_{\mathbf{A}} = \bigsqcup_{q \in Q} GL_n(F)qE^* \tag{3.6}$$

with a finite subset Q of $GL_n(F)_{\mathbf{h}}$, put

$$X_q = GL_n(F) \cap pB^*q^{-1}, \qquad X_{\sigma,q} = \{\xi \in X_q | \sigma = {}^t\xi\tau\xi\} \qquad (q \in Q, \sigma \in S^+),$$

$$B^* = B' \cdot GL_n(F)_{\mathbf{a}}, \qquad B' = R' \cap GL_n(F)_{\mathbf{h}}$$

with R' of Section 2. Then B'/E' is given by $\bigsqcup_{q \in Q}\{p^{-1}\xi_{\mathbf{h}}q | \xi \in X_q/U_q\}$, and hence

$$(3.7)\quad C(f,\eta)=\sum_{q\in Q}|\det(p^{-1}q)|_{\mathbf{A}}^{-n-1}$$
$$\cdot\sum_{\xi\in X_q/U_q}\mu_f(\tau,\xi q)\psi_{\mathbf{h}}(\det(\xi q))\eta^*(\det(p^{-1}\xi q)\mathfrak{g})|\det(\xi)|^{(n+1)u}[\det(p^{-1}\xi q)\mathfrak{g}].$$

Take $\mu\in\mathbf{Z}^{\mathbf{a}}$ and $\kappa\in\mathbf{R}^{\mathbf{a}}$ so that $0\leqslant\mu_v\leqslant 1$ for every $v\in\mathbf{a}$, $\|\kappa\|=0$, and

$$(3.8)\qquad (\psi\eta)_{\mathbf{a}}(x)=\operatorname{sgn}(x_{\mathbf{a}})^{[k]-\mu}|x_{\mathbf{a}}|^{i\kappa}.$$

Let g be the series of (2.1) with η^{-1} in place of η. By Proposition 2.1, this belongs to $\mathcal{M}_\ell(D[\mathfrak{x},\mathfrak{y}],\eta^{-1}\rho_\tau)$ with $\ell=\mu+(nu/2)$, ρ_τ as in that proposition, and suitable $\mathfrak{x}$, $\mathfrak{y}$. Let R and $Y_{\sigma,q}$ be complete sets of representatives for S^+/U_q and $X_{\sigma,q}/U_{\sigma,q}$, respectively. Then $\bigsqcup_{\sigma\in R}Y_{\sigma,q}$ gives X_q/U_q. Notice that if there is $\sigma\in S^+$, as in Lemma 3.1, then $\eta_{\mathbf{a}}(-1)=(-1)^{\|\mu\|}$ by (3.8). Let $C_{q,\kappa}(f,g)$ denote the formal Dirichlet series obtained from the right-hand side of (3.2) with $\det(\sigma)^{-\lambda}[\det(\sigma)\mathfrak{g}]$ in place of $\det(\sigma)^{-su-\lambda}$. Then, by (1.16c) and (2.3), we have

$$\det(\tau)^{\lambda}[\det(\tau)\mathfrak{g}]^{-1}|\det(q)|_{\mathbf{A}}^{-n/2}(\eta\rho_\tau)(\det(q))C_{q,\kappa}(f,g)$$
$$=\sum_{\xi\in X_q/U_q}\psi_{\mathbf{h}}(\det(\xi))\eta^*(\det(p^{-1}\xi q)\mathfrak{g})\mu_f(\tau,\xi q)|\det(\xi)|^{-nu/2}[\det(\xi)\mathfrak{g}]^2.$$

Together with (3.7), this gives

$$(3.9\text{a})\quad C(N(\mathfrak{a})^{-1-3n/2}[\mathfrak{a}]^2)$$
$$=\det(\tau)^{\lambda}|\det(p)|_{\mathbf{A}}^{-n/2}\sum_{q\in Q}(\psi\eta\rho_\tau)(\det(q))[\det(p^{-2}q^2\tau^{-1})\mathfrak{g}]C_{q,\kappa}(f,g),$$

$$(3.9\text{b})\quad D(2s+(3n/2)+1,f,\eta)$$
$$=\det(\tau)^{su+\lambda}|\det(p)|_{\mathbf{A}}^{-2s-n/2}\sum_{q\in Q}(\psi\eta\rho_\tau)(\det(q))|\det(q)|_{\mathbf{A}}^{2s}D_{q,\kappa}(s;f,g),$$

where the left-hand side of (3.9a) means the formal series obtained from $C(f,\eta)$ by substituting $N(\mathfrak{a})^{-1-3n/2}[\mathfrak{a}]^2$ for $[\mathfrak{a}]$.

LEMMA 3.5. *Let $\mathfrak{a}$ be an integral ideal such that $2\notin\mathfrak{a}$ and μ an element of $\mathbf{Z}^{\mathbf{a}}$ such that $0\leqslant\mu_v\leqslant 1$ for every $v\in\mathbf{a}$; let ψ be a Hecke character of F. Then there exists a Hecke character η of F of which the conductor divides $\mathfrak{a}$ and which satisfies (3.8) with some κ.*

Proof. Observing that -1 does not belong to the set

$$W(\mathfrak{a})=\left\{x\in F_{\mathbf{a}}^{\times}\prod_{v\in\mathbf{h}}\mathfrak{g}_v^{\times}\,\middle|\,x_v-1\in\mathfrak{a}_v\text{ for every }v\in\mathbf{h}\right\},$$

we can find $\kappa \in R^{\mathbf{a}}$ so that $\|\kappa\| = 0$ and $\psi_{\mathbf{a}}(a)\,\mathrm{sgn}(a)^{[k]-\mu} = |a|^{i\kappa}$ for every $a \in \mathfrak{g}^{\times} \cap W(\mathfrak{a})$. Then we can define a character $\eta\colon F^{\times}W(\mathfrak{a}) \to \mathbf{T}$ by

$$\eta(cx) = \psi_{\mathbf{a}}(x)^{-1}\,\mathrm{sgn}(x_{\mathbf{a}})^{[k]-\mu}|x_{\mathbf{a}}|^{i\kappa} \qquad \text{for } c \in F^{\times} \text{ and } x \in W(\mathfrak{a}).$$

Since $[F_{\mathbf{A}}^{\times}\colon F^{\times}W(\mathfrak{a})]$ is finite, η can be extended to a Hecke character with the required properties.

We are now ready to prove Theorem A. Given an eigenform f, take $\mathfrak{a}$, μ, and η as in Lemma 3.5; we take any $\mu \neq 0$. Then we consider (3.9a–b) for this η. Put simply $D(s) = D(s, f, \eta) = \sum_{\mathfrak{a}} b_{\mathfrak{a}} N(\mathfrak{a})^{-s}$ and $s_0 = (3n/2) + 1$. Now by [S7, Corollary 5.2] and [S10, Corollary 5.2] (see also Remark at the end of [S10, §5]), we have

$$M(s)Z_{\psi}(s, \lambda, \eta) = D(s)P(s)\Lambda_{\mathfrak{c}}^{n,k-\ell}(s - n/2, \psi\eta\rho_{\tau}). \tag{3.10}$$

Here M and P are finite Dirichlet series, and P has constant term 1; $\Lambda_{\mathfrak{c}}^{n,*}$ is as in (1.19) and ρ_{τ} is as in Proposition 2.1. More precisely, we have an equality between formal Dirichlet series which leads to (3.10). Moreover, by [S7, (5.9b)] and [S10, (5.7)] we can choose τ and p so that $M(s)$ is a nonzero constant. Let us call (3.10) in that special case (3.10p). Since $\mu \neq 0$, g is a cusp form, and so Lemma 3.2 is applicable to $D_{q,\kappa}(s; f, g)$ in (3.9b). Then Lemma 3.2, together with (3.9a–b), shows that $\sum_{\mathfrak{a}} |b_{\mathfrak{a}} N(\mathfrak{a})^{-s}| < \infty$ for $\mathrm{Re}(s) > s_0$. (We are not claiming the absolute convergence of the right-hand side of (3.5b), which may be suggested if we consider only (3.9b) without (3.9a).) The same holds for each factor following $D(s)$ in (3.10p), and hence the Dirichlet series expressing Z_{ψ} has the same property. Therefore, by Lemma 3.4, the Euler product $\prod_{\mathfrak{p}\nmid\mathfrak{a}} W_{\mathfrak{p}}(\varphi(\mathfrak{p})N(\mathfrak{p})^{-s})^{-1}$ for any ideal character φ is absolutely convergent and nonvanishing for $\mathrm{Re}(s) > s_0$. Now we can change $(\mathfrak{a}, \eta)$ for $(\mathfrak{a}', \eta')$ with $\mathfrak{a}'$ prime to $\mathfrak{a}$. Then we see that $W_{\mathfrak{p}}(\varphi(\mathfrak{p})N(\mathfrak{p})^{-s}) \neq 0$ even for $\mathfrak{p}|\mathfrak{a}$. This proves Theorem A.

Let us next prove Theorem C.

(I) Suppose first that k cannot be written in the form $k = (nu/2) + \mu$ as in Theorem B. Then, by Lemma 3.2(2), (3.9b), and (3.10p), we see that $Z_{\psi}(s, \lambda, \eta)$ is holomorphic at s_0. This proves (1) of Theorem C.

(II) Let us next consider $\mathcal{T}_k$ with $k = (nu/2) + \mu$, $\mu \neq 0$. Put $\Gamma = K \cap G$ and $\mathcal{U} = \mathcal{T}_k \cap \mathcal{S}_k(\Gamma)$; let $\mathcal{R}$ denote the orthogonal complement of $\mathcal{U}$ in $\mathcal{S}_k(\Gamma)$. Then both $\mathcal{U}$ and $\mathcal{R}$ are stable under the operators $T_{\psi}(\mathfrak{a})$, since we are assuming $\mathfrak{e} = \mathfrak{c}$ here, and hence each $T_{\psi}(\mathfrak{a})$ is hermitian (see [S10, Lemma 4.5]). Let f be an eigenform belonging to $\mathcal{R}$. For the same reason as in (I), $Z_{\psi}(s, \lambda, \eta)$ has a pole at s_0 only when the function g attached to η has the same weight as f, that is, when $g \in \mathcal{T}_k$. Take a subgroup Γ' of Γ so that $g \in \mathcal{M}_k(\Gamma')$; let $A = \Gamma'\backslash\Gamma$. Then $\sum_{\alpha\in A} g\|_k\alpha \in \mathcal{U}$, and hence $0 = \langle f, \sum_{\alpha\in A} g\|_k\alpha\rangle = \sum_{\alpha\in A}\langle f\|_k\alpha^{-1}, g\rangle = \#(A)\langle f, g\rangle$. Thus, $\langle f, g\rangle = 0$. Then we see from Lemma 3.2(3), (3.9b), and (3.10p) that $Z_{\psi}(s, \lambda, \eta)$ has no pole at s_0. This proves (2).

(III) Finally, suppose that $k = (nu/2) + \mu$, $\mu \neq 0$, and $Z_\psi(s, \lambda, \eta)$ has no pole at s_0 for every eigenform $f \in \mathcal{T}_k \cap \mathcal{S}_k(K, \psi)$ for all possible (K, ψ), where η is a fixed Hecke character. Define g to be θ of (2.1) as before with any fixed $\tau \in S^+$, $p = 1_n$, and η^{-1} in place of η. We can find an integral ideal $\mathfrak{a}$ such that if $\mathfrak{a}$ divides $\mathfrak{e}$ of Section 2, then ${}^t\xi\tau\xi = \tau$ with $\xi \in R^* \cap GL_n(F)$ only when $\xi = 1_n$. For such an $\mathfrak{e}$ we have $g \neq 0$. By Proposition 2.1, we can find $\mathfrak{b}$ and $\mathfrak{c}$ such that $g \in \mathcal{M}_k(D[\mathfrak{b}^{-1}, \mathfrak{bc}], \eta^{-1}\rho_\tau)$; we may assume that $\mathfrak{c}$ is a multiple of $\mathfrak{a}$. We now consider K of (1.14) with $\mathfrak{e} = \mathfrak{c}$ and take ψ to be $\eta^{-1}\rho_\tau$. Then $g \in \mathcal{S}_k(K, \psi) \cap \mathcal{T}_k$ and (3.8) is satisfied with $\kappa = 0$. Take any eigenform f in $\mathcal{S}_k(K, \psi) \cap \mathcal{T}_k$. Changing $\mathfrak{c}$ for its multiple, we can eliminate finitely many Euler factors from both sides of (3.10), since (3.10) was derived from the corresponding equality of formal Dirichlet series as noted above. After this change, we may assume that $P = 1$. The new Z_ψ has no pole at s_0. In view of Lemma 3.2(3), (3.9b), and (3.10), we see that $\langle g, f\rangle = 0$, and hence $\langle g, h\rangle = 0$ for every $h \in \mathcal{S}_k(K, \psi) \cap \mathcal{T}_k$, since this space can be spanned by such eigenforms f. But this is a contradiction, since $\langle g, g\rangle \neq 0$. This proves (3).

4. An integral expression for Z. In [S7, (8.11)], we obtained a certain integral expression similar to (3.3) for $Z_\psi(s, f, \eta)$ with integral k and $\mathfrak{e} = \mathfrak{g}$. We can in fact obtain a formula of the same type also for half-integral k. The precise statement, valid for both integral and half-integral k, is as follows. We assume $\mathfrak{e} = \mathfrak{g}$. Let f be a nonzero Hecke eigenform in $\mathcal{S}_k(D[\mathfrak{b}^{-1}, \mathfrak{bc}], \psi)$ and η a Hecke character of F. Take μ and κ as in (3.8). Then there exist a $\tau \in S^+$ and a $p \in GL_n(F)_\mathbf{h}$ with which the following formula holds:

$$(4.1)\quad Z_\psi(s, f, \eta)\mathcal{G}_{k-\ell,\kappa}((2s-n)/4) \prod_{v \in \mathbf{a}} \Gamma_n(2^{-1}(s - n - 1 + k_v + \mu_v + i\kappa_v))$$

$$= [2\psi_\mathfrak{c}(\det(p))\mu_f(\tau, p)]^{-1}[D_F^{-1/2}N(\mathfrak{x})]^{n(n+1)/2}$$

$$\cdot (4\pi)^{n\|s'u+\lambda\|} \det(\tau)^{s'u+\lambda} |\det(p)|_\mathbf{A}^{n+1-s} \prod_{v \in \mathbf{b}} g_v(\psi(\pi_v)\eta^*(\pi_v \mathfrak{g})|\pi_v|^s)$$

$$\cdot (\Lambda_\mathfrak{c}^{n,k-\ell}/\Lambda_\eta^{n,k-\ell})(s - n/2) \int_\Phi f(z)\overline{\theta(z)}\mathcal{E}(z, (2s-n)/4)\Delta(z)^k d_H z.$$

Here $\mathcal{G}_{k-\ell,\kappa}$ is as in [S7, Theorem 7.3]; $\ell = \mu + (nu/2)$, $\lambda = (k + \ell + i\kappa)/2$, $s' = (2s - 3n - 2)/4$; D_F is the discriminant of F; θ is the function of (2.1) defined with $\mathfrak{e} = \mathfrak{g}$ and η^{-1} in place of η; $\mathbf{b}$ is a finite subset of $\mathbf{h}$ and g_v is a polynomial; they are determined by p and τ as in [S7, Corollary 5.2] and [S10, Corollary 5.2]; π_v is a prime element of F_v;

$$(4.2)\qquad \mathcal{E}(z, s) = \mathcal{G}_{k-l,\kappa}(s)\Lambda_\eta^{n,k-\ell}(2s, \chi)\overline{E(z, \bar{s}; k - l, \varepsilon\bar{\chi}, \Gamma)}, \qquad \chi = \psi\eta\rho_\tau;$$

$E(\cdots)$ is defined by [S7, (7.4)]; ε is the identity character except when $k \in \mathbf{Z}^\mathbf{a}$ and $l \notin \mathbf{Z}^\mathbf{a}$, in which case ε is the Hecke character corresponding to $F(\sqrt{-1})/F$; $\Lambda_*^{n,*}$

is as in (1.19) with χ of (4.2); $\Phi = \Gamma\backslash\mathscr{H}$, $\Gamma = \Gamma[\mathfrak{x}^{-1}, \mathfrak{x}\mathfrak{y}]$, $\mathfrak{x} = \mathfrak{b} + \mathfrak{b}'$, $\mathfrak{y} = \mathfrak{x}^{-1}(\mathfrak{b}\mathfrak{c} \cap \mathfrak{b}'\mathfrak{c}')$ with $(\mathfrak{b}', \mathfrak{c}')$ such that $g \in \mathscr{M}_\ell(D[\mathfrak{b}'^{-1}, \mathfrak{b}'\mathfrak{c}'], \eta^{-1}\rho_\tau)$.

Let us briefly explain why (4.1) holds. Since $\mathfrak{e} = \mathfrak{g}$, we have $E' = E$ and $E^* = GL_n(F)_{\mathbf{a}}E$. Then we see that

$$2 \cdot \sum_{q \in Q} \chi(\det(q))|\det(q)|_{\mathbf{A}}^{2s} D_{q,\kappa}(s; f, \theta) \tag{4.3}$$

is exactly $D(s; f, \theta; \chi)$ of [S7, (8.4)]. The factor 2 appears because $\iota_{\sigma,q}$ of [S7, (8.2)] is $2\nu_{\sigma,q}$. This combined with (3.9b) gives

$$2 \cdot D(2s + (3n/2) + 1, f, \eta) = \det(\tau)^{\lambda+su}|\det(p)|_{\mathbf{A}}^{-2s-n/2} D(s; f, \theta; \chi). \tag{4.4}$$

Now in [S7, (8.5)], we showed that for f of any weight $D(s; f, \theta; \chi)$ times suitable gamma factors has an integral expression similar to (3.3). Combining this with [S7, (5.9b)], [S10, (5.7)], and (4.4), we obtain the above formula.

Let us now prove (3) of Theorem B1. This concerns only integral k, since 4 divides $\mathfrak{c}$ if $k \neq [k]$. In [S8, Theorem 6.1] we gave a set $\{j \in \mathbf{Z} | 0 \leqslant j \leqslant n\}$, which is larger than the set given in (3). To make it smaller, we observe that (4.1) is applicable to the present case since $\mathfrak{e} = \mathfrak{c} = \mathfrak{g}$ and also that the gamma factors in (4.1) coincide with those given in [S8, Theorem 6.1] as noted in [S8, Remark 6.3, (III)]. Now [S7, Theorem 7.3] tells about the possible poles of $\mathscr{E}$ of (4.2) (see Corrections at the end of [S8]), and so we obtain the set as stated in (3) of Theorem B1. In fact, if n is odd, $\mathscr{E}$ is of half-integral weight, and so [S7, Theorem 7.3] tells that there is no additional poles. However, $(\Lambda_{\mathfrak{c}}^{n,k-\ell}/\Lambda_{\mathfrak{y}}^{n,k-\ell})(s - n/2)$ may produce poles (for both even and odd n), but they are contained in the set of (3) of Theorem B1. This completes the proof.

5. Holomorphic Eisenstein series. Our aim in this section is to show that the orthogonal complement of $\mathscr{S}_k$ in $\mathscr{M}_k$ can be spanned by explicitly defined Eisenstein series. Since it is easy to see that $\mathscr{S}_k \neq \mathscr{M}_k$ only if $k = k_0 u$ with $k_0 \in (1/2)\mathbf{Z}$ as noted in [S8, Proposition 2.1(1)], we may assume that $k = k_0 u$ as we will do later, but at first we treat a more general k. To emphasize the dimensionality, in this section we often (but not always) write G^n, P^n, $\mathscr{H}^n$, $\mathscr{M}_k^n$, and $\mathscr{S}_k^n$ for the corresponding objects without the superscript n. Fixing an integer r such that $0 \leqslant r < n$, we write each element α of $G_{\mathbf{A}}^n$ in the form

$$\alpha = \begin{bmatrix} a_1 & a_2 & b_1 & b_2 \\ a_3 & a_4 & b_3 & b_4 \\ c_1 & c_2 & d_1 & d_2 \\ c_3 & c_4 & d_3 & d_4 \end{bmatrix}, \tag{5.1}$$

where x_1 is of size r, and x_4 is of size $n - r$. Then we write $x_i = x_i(\alpha)$ and $x_\alpha = x(\alpha) = \begin{pmatrix} x_1(\alpha) & x_2(\alpha) \\ x_3(\alpha) & x_4(\alpha) \end{pmatrix}$ for $x = a, b, c, d$, and $i = 1, 2, 3, 4$. We understand that

$x(\alpha) = x_4(\alpha)$ if $r = 0$. Let $P^{n,r}$ for $0 \leqslant r < n$ denote the parabolic subgroup of G^n consisting of the elements α such that a_2, c_2, c_3, c_4, and d_3 of (5.1) are all 0; we have $P^{n,0} = P^n$.

Let G' denote the image of G in $G_{\mathbf{a}}$ under the map $x \mapsto x_{\mathbf{a}}$, and let

$$\mathscr{G} = \mathscr{G}^n = \{\sigma \in M_{\mathbf{A}} \mid \mathrm{pr}(\sigma) \in G'\}, \tag{5.2}$$

$$\mathscr{P}^{n,r} = \{\xi \in \mathscr{G}^n \mid \mathrm{pr}(\xi) \in P^{n,r}_{\mathbf{a}}\}. \tag{5.3}$$

Since $G_{\mathbf{a}} \subset \mathrm{pr}(\mathfrak{M})$, we see that $\mathscr{G}$ is a subgroup of $M_{\mathbf{A}}$ contained in $\mathfrak{M}$, and $\mathrm{pr}(\mathscr{G}) = G'$. Combining pr with the isomorphism $G' \to G$, we obtain a surjection $\mathrm{pr}_0 \colon \mathscr{G} \to G$, which gives an exact sequence

$$1 \to \mathbf{T} \to \mathscr{G} \to G \to 1. \tag{5.4}$$

As explained in [S8], $\mathscr{G}$ can be identified with the group of all couples (α, p) formed by $\alpha \in G$ and a holomorphic function p on $\mathscr{H}$ such that p^2/j^u_α is a constant belonging to $\mathbf{T}$, the group-law being defined by $(\alpha, p)(\alpha', p') = (\alpha\alpha', p(\alpha' z)p'(z))$. Such a couple (α, p) defines a unique element σ of $\mathscr{G}$ such that $\mathrm{pr}_0(\sigma) = \alpha$ and $p = h_\sigma$.

We call a subgroup Ξ of $\mathscr{G}$ a *congruence subgroup* of $\mathscr{G}$ if pr_0 gives an isomorphism of Ξ onto a congruence subgroup Γ of G, and the inverse of this isomorphism coincides with the map $\alpha \mapsto (\alpha, h_\alpha)$ on a congruence subgroup of $C^\theta \cap \Gamma$. If Γ is a congruence subgroup of G contained in C^θ, we can identify it with its image in $\mathscr{G}$ under the map $\alpha \mapsto (\alpha, h_\alpha)$. To treat modular forms of integral and half-integral weight uniformly, we fix a weight k and make the following conventions: *The symbols $\mathscr{G}$ and $\mathscr{P}^{n,r}$ denote the groups defined above if $k \neq [k]$; they denote G and $P^{n,r}$ if $k = [k]$.* Given a congruence subgroup Γ of $\mathscr{G}$, we denote by $\mathscr{M}^n_k(\Gamma)$ the set of all $f \in \mathscr{M}_k$ such that $f\|_k\gamma = f$ for every $\gamma \in \Gamma$, and put $\mathscr{S}^n_k(\Gamma) = \mathscr{M}^n_k(\Gamma) \cap \mathscr{S}^n_k$. We now define maps $\pi_r \colon \mathscr{P}^{n,r} \to \mathscr{G}^r$ and $\lambda_r \colon \mathscr{P}^{n,r} \to F^\times$ by

$$\pi_r(\alpha) = \begin{pmatrix} a_1(\alpha) & b_1(\alpha) \\ c_1(\alpha) & d_1(\alpha) \end{pmatrix}, \qquad \lambda_r(\alpha) = \det(d_4(\alpha)) \qquad (k = [k]), \tag{5.5a}$$

$$\pi_r((\alpha, p)) = (\pi_r(\alpha), |\lambda_r(\alpha)|^{-u/2} p'), \qquad p'(z) = p\begin{pmatrix} z & w \\ {}^t w & z' \end{pmatrix} \qquad (k \neq [k]), \tag{5.5b}$$

$$\lambda_r((\alpha, p)) = \lambda_r(\alpha) \qquad (k \neq [k]). \tag{5.5c}$$

For a congruence subgroup Γ of $\mathscr{G}^n$ and $X = \mathscr{M}$ or $\mathscr{S}$, we put

$$X^r_k(\Gamma \cap \mathscr{P}^{n,r}) = \{f \in X^r_k \mid f\|_k \pi_r(\gamma) = \mathrm{sgn}[\lambda_r(\gamma)]^{[k]} f \text{ for every } \gamma \in \Gamma \cap \mathscr{P}^{n,r}\}. \tag{5.6}$$

This space is $\{0\}$ unless the following condition is satisfied:

(5.7) There is a homomorphism φ of $\pi_r(\Gamma \cap \mathcal{P}^{n,r})$ into $\{\pm 1\}$ such that $\varphi(\pi_r(\gamma)) = \operatorname{sgn}[\lambda_r(\gamma)]^{[k]}$ for every $\gamma \in \Gamma \cap \mathcal{P}^{n,r}$.

Under (5.7), we have

$$X_k^r(\Gamma \cap \mathcal{P}^{n,r}) = \{f \in X_k^r \mid f\|_k\varepsilon = \varphi(\varepsilon)f \text{ for every } \varepsilon \in \pi_r(\Gamma \cap \mathcal{P}^{n,r})\}. \tag{5.8}$$

For $f \in \mathcal{S}_k^r(\Gamma \cap \mathcal{P}^{n,r})$, $z \in \mathcal{H}^n$, and $s \in \mathbf{C}$, we put

$$\delta(z, s; f) = f(\omega_r(z))[\Delta(z)/\Delta(\omega_r(z))]^{su-k/2}, \tag{5.9}$$

where $\omega_r(z)$ is the upper left $(r \times r)$-submatrix of z.

We now define an Eisenstein series $E_k^{n,r}(z, s; f, \Gamma)$ by

$$E_k^{n,r}(z, s; f, \Gamma) = \sum_{\gamma \in A} \delta(z, s; f)\|_k\gamma, \qquad A = (\Gamma \cap \mathcal{P}^{n,r})\backslash\Gamma. \tag{5.10}$$

The sum is formally well defined and convergent for $\operatorname{Re}(2s) > n + r + 1$. If $r = 0$, we have $\delta(z, s; c) = c\Delta(z)^{su-k/2}$ for a constant $c \in \mathbf{C}$, and

$$\mathcal{S}_k^0(\Gamma \cap \mathcal{P}^{n,0}) = \begin{cases} \mathbf{C} & \text{if } \operatorname{sgn}[\lambda_0(\gamma)]^{[k]} = p_\gamma \quad \text{for every } \gamma \in \Gamma \cap \mathcal{P}^{n,0}, \\ \{0\} & \text{otherwise}, \end{cases} \tag{5.11}$$

where we understand that $\gamma = (\mathrm{pr}_0(\gamma), p_\gamma)$ if $k \neq [k]$ and $p_\gamma = 1$ if $k = [k]$. Thus,

$$E_k^{n,0}(z, s; c, \Gamma) = c \sum_{\gamma \in A} \delta^{su-k/2}\|_k\gamma, \qquad A = (\Gamma \cap \mathcal{P}^{n,0})\backslash\Gamma. \tag{5.12}$$

For the reason explained at the beginning of this section, we now assume that $k = k_0 u$ with $k_0 \in (1/2)\mathbf{Z}$ and put

$$E_k^{n,r}(z; f, \Gamma) = E_k^{n,r}(z, k_0/2; f, \Gamma) \tag{5.13}$$

whenever the right-hand side is finite. In general, this may not be holomorphic in z. For $0 \leqslant r < n$, we denote by $\tilde{\mathcal{E}}_k^{n,r}$ the vector space spanned over $\mathbf{C}$ by $E_k^{n,r}(z; f, \Gamma)\|_k\alpha$ for all $\alpha \in \mathcal{G}^n$, all $f \in \mathcal{S}_k^r(\Gamma \cap \mathcal{P}^{n,r})$, and for all congruence subgroups Γ of $\mathcal{G}^n$; we then put $\mathcal{E}_k^{n,r} = \tilde{\mathcal{E}}_k^{n,r} \cap \mathcal{M}_k^n$. Clearly, these spaces are stable under the operator $\|_k\xi$ for every $\xi \in \mathcal{G}^n$. We also put $\mathcal{E}_k^{n,n} = \mathcal{S}_k^n$. To state our results, we need two more symbols, one of which is a subset $\Lambda(n, k_0)$ of $\mathbf{R}$ defined by

$$\Lambda(n, k_0) = \begin{cases} \{x \in \mathbf{R} \mid x \geqslant 2\} & \text{if } n = 1 \text{ and } k_0 \neq 3/2, \\ \{x \in \mathbf{R} \mid x > (3n/2) + 1\} & \text{if } n > 1 \text{ or } k_0 = 3/2. \end{cases} \tag{5.14}$$

The other is the map Φ defined as follows. Given a function $f: \mathscr{H}^n \to \mathbf{C}$, we define $\Phi f: \mathscr{H}^{n-1} \to \mathbf{C}$ by

$$(5.15) \qquad (\Phi f)(w) = \lim_{\rho \to \infty} f(\mathrm{diag}[w, \rho \mathbf{i}_1]) \qquad (w \in \mathscr{H}^{n-1})$$

whenever the limit exists, where ρ is real and $\mathbf{i}_n$ is as in (1.10). If $n = 1$, we ignore w, and so Φf is a constant. We then define the νth power Φ^ν of Φ for $\nu \leqslant n$ inductively by $\Phi^\nu = \Phi\Phi^{\nu-1}$, with the identity map as Φ^0. This sends f to a function on $\mathscr{H}^{n-\nu}$. It can easily be seen that if $f \in \mathscr{M}_k^n$, then $\Phi^{n-r} f$ is well defined and belongs to $\mathscr{M}_k^r$.

PROPOSITION 5.1. *Given $k_0 \in (1/2)\mathbf{Z}$, suppose that $k_0 \geqslant (n + r + 2)/2$ if $F \neq \mathbf{Q}$, and $k_0 > (n + r + 3)/2$ if $F = \mathbf{Q}$. Suppose also that $k_0 \in \Lambda(r, k_0)$ if $r > 0$. Then the function of (5.13) is meaningful and the following assertions hold:*
(1) $\mathscr{E}_k^{n,r} = \tilde{\mathscr{E}}_k^{n,r}$.
(2) $\Phi^{n-t}(\mathscr{E}_k^{n,r}) \subset \mathscr{E}_k^{t,r}$ *for* $r \leqslant t \leqslant n$ *and in particular* $\Phi^{n-r}(\mathscr{E}_k^{n,r}) = \mathscr{S}_k^r$.
(3) $\Phi^s(\mathscr{E}_k^{n,r}) = 0$ *if* $s > n - r$.
(4) *If* $g \in \mathscr{E}_k^{n,r}$ *and* $\Phi^{n-r}(g\|_k \alpha) = 0$ *for every* $\alpha \in \mathscr{G}^n$, *then* $g = 0$.

Proof. This is an improved version of [S8, Lemma 8.11], and can be proved in the same manner. A crucial fact is [S8, Lemma 8.10], which was stated only for integral k_0. (The results of [S8, §8] up to Lemma 8.9 were given for both integral and half-integral k_0, and neither change nor improvement is necessary.) Now [S10, (8.5)] together with [S10, Theorem 6.4(I)] allows us to repeat its proof even for half-integral k_0. In the proof, we needed the nonvanishing of $Z_\psi(s, \lambda, \eta)$ at $s = k_0$. For this, we employed a weaker result than Theorem A for integral k_0. Now Theorem A gives a stronger result for integral k_0 than what we had in [S8], and also a similar result for half-integral k_0. Therefore, we obtain an improved version of [S8, Lemma 8.10] including the case of half-integral k_0. Then repeating the proof of [S8, Lemma 8.11] with these improvements, we obtain our proposition.

Once this proposition is established, we can prove the following theorems in exactly the same fashion as in the proofs of Theorems 8.12, 8.13, and 8.14 of [S8], by using the above proposition in place of [S8, Lemma 8.11]. For the reason explained above, the present theorems apply to more values of k_0 than in the previous theorems. We assume $n > 1$ in these theorems.

THEOREM 5.2. *Let $k = k_0 u$ with $k_0 \in (1/2)\mathbf{Z}$. Then $\mathscr{M}_k^n = \bigoplus_{r=0}^n \mathscr{E}_k^{n,r}$ and $\mathscr{M}_k^n(\Gamma) = \bigoplus_{r=0}^n [\mathscr{M}_k^n(\Gamma) \cap \mathscr{E}_k^{n,r}]$ for every congruence subgroup Γ of $\mathscr{G}^n$ provided the following condition is satisfied:*

$$(5.16) \qquad k_0 \geqslant 3n/2 \quad \text{if } n > 2; \qquad k_0 > 3 \quad \text{if } n = 2 \text{ and } F = \mathbf{Q};$$
$$k_0 > 2 \quad \text{if } n = 2 \text{ and } F \neq \mathbf{Q}.$$

THEOREM 5.3. *Let $\mathscr{E}_k^n = \sum_{r=0}^{n-1} \mathscr{E}_k^{n,r}$, $k = k_0 u$ with k_0 as in (5.16). Then*

$$\mathscr{E}_k^n = \{f \in \mathscr{M}_k^n \mid \langle f, g \rangle = 0 \text{ for every } g \in \mathscr{S}_k^n\},$$

$$\mathscr{E}_k^{n,r} = \{f \in \mathscr{E}_k^n \mid \Phi(f\|_k\alpha) \in \mathscr{E}_k^{n-1,r} \text{ for every } \alpha \in \mathscr{G}^n\} \qquad \text{if } r < n.$$

THEOREM 5.4. *Let $k = k_0 u$ with k_0 as in (5.16). Given a congruence subgroup Γ of $\mathscr{G}^n$, let $\mathscr{E}_k^{n,r}(\Gamma) = \mathscr{M}_k(\Gamma) \cap \mathscr{E}_k^{n,r}$ and let $X = \mathscr{P}^{n,r}\backslash\mathscr{G}^n/\Gamma$ with a fixed $r < n$. Then $f \mapsto (\Phi^{n-r}(f\|\xi^{-1}))_{\xi \in X}$ gives a* **C***-linear isomorphism of $\mathscr{E}_k^{n,r}(\Gamma)$ onto $\prod_{\xi \in X} \mathscr{S}_k^r(\xi\Gamma\xi^{-1} \cap \mathscr{P}^{n,r})$. Moreover, if $f \in \mathscr{E}_k^{n,r}(\Gamma)$ and $p_\xi = \Phi^{n-r}(f\|\xi^{-1})$, then $f = \sum_{\xi \in X} E_k^{n,r}(z; p_\xi, \xi\Gamma\xi^{-1})\|_k\xi$.*

As a final remark, we note that if $n = 1$, a more complete result applicable to every $k_0 \in (1/2)\mathbf{Z}$ was obtained in [S3].

REFERENCES

[DHL] W. DUKE, R. HOWE, AND J.-S. LI, *Estimating Hecke eigenvalues of Siegel modular forms*, Duke Math. J. **67** (1992), 219–240.

[K] H. KLINGEN, *Zum Darstellungssatz für Siegelsche Modulformen*, Math. Z. **102** (1967), 30–43; *Berichtigung*, Math. Z. **105** (1968), 399–400.

[KR] S. KUDLA AND S. RALLIS, *A regularized Siegel-Weil formula: the first term identity*, Ann. of Math. (2) **140** (1994), 1–80.

[S1] G. SHIMURA, *On Eisenstein series*, Duke Math. J. **50** (1983), 417–476.

[S2] ———, *On Eisenstein series of half-integral weight*, Duke Math. J. **52** (1985), 281–314.

[S3] ———, *On the Eisenstein series of Hilbert modular groups*, Rev. Mat. Iberoamericana 1: 3 (1985), 1–42.

[S4] ———, *On Hilbert modular forms of half-integral weight*, Duke Math. J. **55** (1987), 765–838.

[S5] ———, *The critical values of certain Dirichlet series attached to Hilbert modular forms*, Duke Math. J. **63** (1991), 557–613.

[S6] ———, *On the transformation formulas of theta series*, Amer. J. Math. **115** (1993), 1011–1052.

[S7] ———, *Euler products and Fourier coefficients of automorphic forms on symplectic groups*, Invent. Math. **116** (1994), 531–576.

[S8] ———, *Eisenstein series and zeta functions on symplectic groups*, Invent. Math. **119** (1995), 539–584.

[S9] ———, *Differential operators, holomorphic projection, and singular forms*, Duke Math. J. **76** (1994), 141–173.

[S10] ———, *Zeta functions and Eisenstein series on metaplectic groups*, Invent. Math. **121** (1995), 21–60.

DEPARTMENT OF MATHEMATICS, PRINCETON UNIVERSITY, PRINCETON, NEW JERSEY 08544, USA

Response

Notices of the American Mathematical Society,
vol. 43, No. 11 (November 1996), 1344-1347*

I always thought this prize was for an old person, certainly someone older than I, and so it was a surprise to me, if a pleasant one, to learn that I was chosen as a recipient. Though I am not so young, I am not so old either, and besides, I have been successful in making every newly appointed junior member of my department think that I was also a fellow new appointee. This time I failed, and I should be grateful to the selection committee for discovering that I am a person at least old enough to have his lifetime work spoken of.

There are many prizes conferred by various kinds of institutions, but in the present case, I view it as something from my friends, which makes me really happy. So let me just say thank you, my friends!

* * *

I would like to take this opportunity to give a historical perspective of a topic on which I worked in the 1950s and 1960s, intermingled with some of my personal recollections. It concerns arithmetic Fuchsian groups which can be obtained from an indefinite quaternion algebra B over a totally real algebraic number field F. For such a B one has

$$B \otimes_{\mathbf{Q}} \mathbf{R} = M_2(\mathbf{R})^r \times \mathbf{H}^{d-r},$$

where $d = [F : \mathbf{Q}]$, $0 \leq r \leq d$, $M_2(\mathbf{R})$ is the matrix algebra over $\mathbf{R}$ of size 2, and $\mathbf{H}$ is the Hamilton quaternions. Assuming $r > 0$ and taking a subring R of B that contains $\mathbf{Z}$ and spans B over $\mathbf{Q}$, denote by Γ the group of invertible elements of R whose projection to any factor $M_2(\mathbf{R})$ has determinant 1. Then we can view Γ as a subgroup of $SL_2(\mathbf{R})^r$ through the projection map to $M_2(\mathbf{R})^r$, and so we can let Γ act on the product H^r of r copies of the upper half plane H. In this way we obtain an algebraic variety $\Gamma\backslash H^r$, which is an algebraic curve if $r = 1$. It is known that $\Gamma\backslash H^r$ is compact if and only if B is a division algebra. In particular, we can take B to be the matrix algebra $M_2(F)$ over F of size 2, in which case $r = d$ and the meromorphic functions on $\Gamma\backslash H^d$ are called *Hilbert modular functions.*

If $F = \mathbf{Q}$, the group Γ was first discovered by Poincaré [7], apparently in 1886. He reminisced in his *Science et Méthode* as follows: "One day, while walking on a cliff, it occurred to me, with the same customary characters of shortness, suddenness, and immediate certainty, that arithmetic transformations of indefinite ternary quadratic forms were identical to those of non-Euclidean geometry."

One interesting aspect of this work is that the quotient $\Gamma\backslash H$ is compact if B is a division algebra. Until then the only Fuchsian groups he or anybody else knew

1

were those obtained from hypergeometric series, among which the arithmetically defined ones were the classical modular groups; in all those cases the quotient is not compact. (Uniformization of an arbitrary compact Riemann surface was proved independently by Koebe and Poincaré only in 1907.) Poincaré's group was generalized to the case $1 = r \leq d$ with an arbitrary F by Fricke [3] in 1893. It is also discussed in the last chapter of the thick volume [4] of Fricke and Klein published in 1897. These mathematicians employed an indefinite ternary quadractic form instead of a quaternion algebra. Since $SO(2, 1)$ is covered by $SL_2(\mathbf{R})$, the unit group of the given ternary form produces a discrete subgroup of $SL_2(\mathbf{R})$.

After Fricke's investigations, which showed that the action of the groups on H is properly discontinuous, no significant progress was made in this area for the next fifty years. In 1912, Hecke published his thesis work [5] concerning Hilbert modular functions in the case of $M_2(F)$ with $d = 2$. In its introduction he said that the results of Fricke on the Fuchsian groups of the above type seemed to be "without specific meaning in number theory." Later developments proved that he was wrong. Taking his tender age of twenty-five into consideration, we may forgive him, and may even justify his comment, allowing him a thirty-year warranty, since it could apply to all papers on this subject in that period, one by Heegner [6], for example, which I cite here in order to show that the topic was not forgotten but was being treated without any new ideas. It should also be pointed out that Hecke's own work was critically flawed, though generally speaking he was headed in the right direction, except for that comment.

Eichler may have been the first person who was seriously interested in this group. He wrote his dissertation with Brandt on quaternion algebras, and later worked on more general types of simple algebras. He once told me that Brandt did not think much of non-quaternion algebras, and was unhappy with Eichler's turning to them. In reality, there was no need for him to be unhappy, since the fact that Eichler started with quaternion algebras determined his course thereafter, which was vastly successful. In a lecture he gave in Tokyo, he drew a hexagon on the blackboard, and called its vertices clockwise as follows: automorphic forms, modular forms, quadratic forms, quaternion algebras, Riemann surfaces, and algebraic functions. Anyway in the mid-1950s Eichler was developing the theory of Hecke operators for the Fuchsian groups of Poincaré's type (see [1], for example). He also gave a formula for the genus of $\Gamma\backslash H$ somewhat earlier. However, there were no other number-theoretical investigations on these algebraic curves by that time.

In 1957 while in Paris I became interested in this class of groups. I had just finished my first work on the zeta functions of elliptic modular curves. Though I knew that it needed elaboration, I was more interested in finding other curves whose zeta functions could be determined. I was also trying to formulate the theory of complex multiplication in higher dimension in terms of the values of automorphic functions of several variables, Siegel modular functions, for example. It turned out that these two problems were inseparably connected to each other. Also, nobody else was working on such questions. I can assure the reader that I had no intention of humiliating Hecke posthumously.

So I took up the group of the above type. My aim was to find an algebraic curve C defined over an algebraic number field k that is complex analytically isomorphic to $\Gamma\backslash H$, and to determine the zeta function of C. Such a C is called a *model of* $\Gamma\backslash H$ *over* k. Naturally I started with the simplest case $F = \mathbf{Q}$. Since it was relatively easy to see that $\Gamma\backslash H$ in this case parametrizes a family of certain two-dimensional abelian varieties, I was soon able to prove that the curve had a $\mathbf{Q}$-rational model. The proof required a theory of the field of moduli of a polarized abelian variety, but luckily, I had it at my disposal, since I had been forced to develop such a theory in order to get a better formulation of complex multiplication as mentioned above.

In June 1958 I visited three schools in Germany: Münster, Göttingen, and Marburg. I gave a talk at each place, but remember only that at Göttingen I spoke about the field of moduli of a polarized abelian variety, and its application to the field of definition for the field of automorphic functions. At the end of my talk I mentioned briefly the $\mathbf{Q}$-rationality of the curve $\Gamma\backslash H$ for Poicaré's Γ.

Siegel was among the audience, and pressed on the last point. I began to explain the idea, but he interrupted me and simply wanted to know whether I really had the proof. So I said "Yes," and that was that. Siegel said nothing, but apparently he was not convinced, and expressed his doubts to Klingen, who in 1970 told me about Siegel's skepticism at that time. I can easily guess the rationale behind his disbelief: since $\Gamma\backslash H$ is compact, there is no natural Fourier expansion of an automorphic form, so that there is no way of defining the rationality of automorphic functions, and that was exactly why Hecke made the comment mentioned above. Eventually I determined the zeta function of the curve, and gave a talk on that topic at the ICM, Edinburgh, in September 1958. The full details were published in [8] in 1961.

There was no such incident at Marburg, where I met Eichler. I remember that after dinner at his home, he played a religious piece of music on the phonograph, which I think was by Bach. I am sure it was not by Mozart, as he did not think much of the composer. He was a tall and handsome man, whose look immediately reminded me of the knight in the movie "The Seventh Seal" by Ingmar Bergman, which I had seen in Paris a few months earlier. As for Siegel, who was sixty-one at that time, calling him a big mass of flesh would have been misleading and even derogatory, but that was my first impression. Though he must have looked awesome to many, he assumed no airs, and there was a certain homely atmosphere around him, which made him less intimidating, at least to me.

Coming back to my work, at first I thought that these curves obtained from a division quaternion algebra B over $\mathbf{Q}$ might not be modular, (and strictly speaking, that is true, see the next paragraph), but I realized that *no nonmodular* $\mathbf{Q}$*-rational elliptic curves could be obtained* for the following reason: Eichler had shown, by means of his trace formula, the following fact: the Euler products on

B are already included in those obtained from elliptic modular forms [2]. The Tate conjecture on this was explicitly stated much later, but the idea was known to many people, and so it was natural for me to think that two elliptic curves with the same zeta function are isogenous. This fact concerning B, in addition to the results I had about the zeta functions of modular curves, may have been the strongest reason for my stating the conjecture that every $\mathbf{Q}$-rational elliptic curve is modular.

Let me insert here a remark on the curves obtained from a division algebra B. I showed much later in [11] that the natural models of the curves have *no real points* even when the genus of $\Gamma\backslash H$ is 1, and in that sense they are not modular! They are not *elliptic curves* in the strict sense, though their jacobian varieties are. This point may explain the raison d'être of those curves. I wonder if there is any recent investigation on this phenomenon.

The curves with $F \neq \mathbf{Q}$ were more difficult. After going back to Tokyo in the spring of 1959, I decided to investigate more general families of abelian varieties. By specifying the types of endomorphism algebra and polarization of abelian varieties, one obtains a quotient $\Delta\backslash S$ that parametrizes abelian varieties of a prescribed type, where S is a hermitian symmetric domain of noncompact type, and Δ is an arithmetic subgroup of a certain algebraic group. The above $\Gamma\backslash H$ for Poicaré's Γ is an easiest example of $\Delta\backslash S$; one simply takes B to be the endomorphism algebra. For a certain reason, however, the algera B with $0 < r < d$ never appears as the endomorphism algebra of an abelian variety, which was the main difficulty. Then I realized that choosing an algebra different from B, one obtains $\Delta\backslash S$ that is essentially the same as $\Gamma\backslash H$ for an arbitrary B of the above type. I think that was sometime in the fall of 1960. I knew at that point that the problem was approachable, and even knew that the curves had models over a number field, but did not know how to state the theorems in the best possible forms, not to mention how to prove them.

In a series of papers published in 1963–65 I investigated the number fields over which the varieties $\Delta\backslash S$ can be defined. In many higher-dimensional cases, the results were best possible, but in the one-dimensional case that was the main question, I was not satisfied. So I turned to a higher-dimensional case of a different nature. In a famous paper on symplectic geometry [12] Siegel defined a certain arithmetic subgroup Γ' of $Sp(n, \mathbf{R})$ which was a generalization of Fricke's group, and which was also defined relative to F. If $n > 1$ and $F \neq \mathbf{Q}$, this group does not appear as the above group Δ associated with a family of abelian varieties. But in the summer of 1963, while in Boulder, Colorado, I found that there was an injection $\Gamma' \to \Delta$ with some Δ, which produced a holomorphic embedding $\Gamma'\backslash S' \to \Delta\backslash S$, where S' is the Siegel upper half space of degree n. If $n = 1$, $\Gamma'\backslash S'$ is exactly the algebraic curve $\Gamma\backslash H$ in question, and moreover the embedding is essentially birational over $\mathbf{C}$. Anyway, employing this embedding, I was able to find a number field over which $\Gamma'\backslash S'$ is defined for an arbitray n. When I was asked to contribute a paper to the volume in honor of Siegel's seventieth birthday, I naturally took this as the topic, and sent the manuscript to the editor in the fall of 1965.

Around the same time, perhaps in early September that year, I finally had a definite idea of settling the original question in the one-dimensional case: to employ many different $\Delta\backslash S$ for a given $\Gamma\backslash H$. By means of this idea together with a finer theory of variety of moduli of polarized abelian varieties, by June 1966 I was able to finish the paper [9] in which I determined the zeta function of the curve $\Gamma\backslash H$ with any totally real F. At the same time I determined the class fields generated by the values of automorphic functions, not only in the one-dimensional case but also in the case where B is totally indefinite, including the Hilbert modular case. By doing so, I showed that similar theories could be developed in a parallel way in both Fricke's and Hecke's cases. In fact, those are the two extreme cases of a more general class of arithmetic quotients for which one can do number-theoretical investigations Hecke wished to do in his case, a fact Hecke never realized.

I dedicated the paper to Weil. At some point I said to him jokingly that he became sufficiently old that I could now dedicate a paper to him, to which he replied, "I can't stop it." Meanwhile my paper dedicated to Siegel appeared in the *Mathematische Annalen* [10]; I also sent a reprint of my Annals article to him, as I had been doing regularly with my earlier papers. Here is what he wrote me about these:

Göttingen, 15 May 1967

Dear Professor Shimura:

After a long trip around the world I returned to Göttingen and I found your last paper from the Annals of Mathematics together with the work which you kindly dedicated on the occasion of my 70th birthday.

I am sending you my most cordial thanks for your kindness. I have now begun to study these two papers, and both of them seem to be of great interest, from the arithmetical and the analytical point of view.

During many years I have regretted that Hecke's earlier work on Hilbert's modular function and class field theory had not been continued by later mathematicians. I am glad to see in your last paper how much you have already achieved in this direction.

I was very pleased to see from your other paper that you have obtained decisive results concerning those groups which I introduced in my paper on symplectic geometry.

Best congratulations for the success of your previous work, and best wishes for the future!

Yours sincerely

Carl Ludwig Siegel

I was naturally gratified and even moved, but frankly I was somewhat disappointed by his mentioning only the Hilbert modular case, which was far easier than the case of curves that was the main feature of my paper. Therefore I was not sure whether he perceived the full scope of the work. Perhaps he thought what he said was enough, which is true, and so I should not complain. In fact, reading this letter after almost thirty years, I now think that the letter tells more about the sender than about the recipient.

To clarify this point, we have to know what kind of a man Siegel was. Of course he established himself as one of the giants in the history of mathematics long ago. He was not known, however, for his good-naturedness. Around 1980 I sat next to Natasha Brunswick at a dinner table, when she proclaimed, "Siegel is mean!" I don't remember how our conversation led to that statement, but many of those who knew him would agree with her opinion. Hel Braun, one of his few students, apparently disliked him. He was indisputably original, and even original in his perverseness. Once at a party he played a piano piece and challenged the audience to tell who the composer was. Hearing no answer, he said it was a sonata by Mozart, Köchel number such and such, played backward. On the other hand, he had a certain sense of humor. When Weil asked him which work of his he thought best, he replied, "Oh, I think a few watercolors I made in Greece some years ago are pretty good."

In any case, it would be wrong to presume him to be a mathematician who did what he wanted to do, unconcerned about what other people might think of his work. I believe he was not that aloof. He must have known who he was, but at the same time he must have felt unappreciated by the younger generation. That was Eichler's opinion, and I am inclined to agree with him.

After his retirement, Siegel took a long trip around the world as he mentions in his letter. On coming back to Göttingen, one day he went into his office in the university, and found on his desk a copy of the volume of the *Mathematische Annalen* dedicated to him, which pleased him greatly. And here was a man thirty-four years younger than he, completely outside of his German influence, who took up the topic on which he expended considerable effort many years ago, with genuine appreciation of his work.

Perhaps he was not so crabbed as many people had imagined, and it is possible that he wrote a few more letters like the above one. At any rate, when he wrote that letter, he knew that at least one of his papers was really understood, and at that moment he was capable of appreciating the progress made by the new generation, of which he had often been contemptuous. I am indeed glad to be the recipient of the letter which showed this great mathematician as a warm-hearted man with no trace of ill-temperedness, nor any cynicism.

References

[1] M. Eichler, Modular correspondences and their representations, J. Ind. Math. Soc. **20** (1956), 163–206.

[2] M. Eichler, Quadratische Formen und Modulfunktionen, Acta arith. **4** (1958), 217–239.

[3] R. Fricke, Zur gruppentheoretischen Grundlegung der automorphen Functionen, Math. Ann.**42** (1893), 564–597.

[4] R. Fricke and F. Klein, Vorlesungen über die Theorie der automorphen Funktionen, I. Leipzig, Teubner, 1897.

[5] E. Hecke, Höhere Modulfunktionen und ihre Anwendung auf die Zahlentheorie, Math. Ann. **71** (1912), 1–37 (=Werke, 21–57).

[6] K. Heegner, Transformierbare automorphe Funktionen und quadratishe Formen I, II, Math. Z. **43** (1937), 162–204, 321–352.

[7] H. Poincaré, Les fonctions fuchsiennes et l'arithmétique, J. de Math. 4 ser. 3 (1887), 405–464 (=Oeuvres, vol.2, 463–511).

[8] G. Shimura, On the zeta functions of the algebraic curves uniformized by certain automorphic functions, J. Math. Soc. Japan, **13** (1961), 275–331.

[9] ______, Construction of class fields and zeta functions of algebraic curves, Ann. of Math. **85** (1967), 58–159.

[10] ______, Discontinuous groups and abelian varieties, Math. Ann. **168** (1967), 171–199.

[11] ______, On the real points of an arithmetic quotient of a bounded symmetric domain, Math. Ann. **215** (1975), 135–164.

[12] C. L. Siegel, Symplectic Geometry, Amer. J. Math. **65** (1943), 1–86 (=Gesammelte Abhandlungen, II, 274–359).

Zeta functions and Eisenstein series on classical groups

Proceedings of the National Academy of Sciences, 94, 11133-11137 (1997)*

Abstract. We construct an Euler product from the Hecke eigenvalues of an automorphic form on a classical group, and prove its analytic continuation to the whole complex plane when the group is a unitary group over a CM-field and the eigenform is holomorphic. We also prove analytic continuation of an Eisenstein series on another unitary group containing the group just mentioned defined with such an eigenform. As an application of our methods, we prove an explicit class number formula for a totally definite hermitian form over a CM-field.

1. Introduction. Given a reductive algebraic group G over an algebraic number field, we denote by $G_{\mathbf{A}}$, $G_{\mathbf{a}}$, and $G_{\mathbf{h}}$ its adelization, the archimedean factor of $G_{\mathbf{A}}$, and the nonarchimedean factor of $G_{\mathbf{A}}$. We take an open subgroup D of $G_{\mathbf{A}}$ of the form $D = D_0 G_{\mathbf{a}}$ with a compact subgroup D_0 of $G_{\mathbf{A}}$ such that $D_0 \cap G_{\mathbf{a}}$ is maximal compact in $G_{\mathbf{a}}$. Choosing a specific type of representation of D_0, we can define automorphic forms on $G_{\mathbf{A}}$ as usual. For simplicity we consider here the forms invariant under $D_0 \cap G_{\mathbf{h}}$. Each Hecke operator is given by $D\tau D$ with τ in a subset $\mathfrak{X}$ of $G_{\mathbf{A}}$, which is a semigroup containing D and the localizations of G for almost all nonarchimedean primes. Taking an automorphic form $\mathbf{f}$ such that $\mathbf{f}|D\tau D = \lambda(\tau)\mathbf{f}$ with a comlex number $\lambda(\tau)$ for every $\tau \in \mathfrak{X}$ and a Hecke ideal character χ of F, we put

$$\mathfrak{T}(s, \mathbf{f}, \chi) = \sum_{\tau \in D\backslash \mathfrak{X}/D} \lambda(\tau)\chi\big(\nu_0(\tau)\big)N\big(\nu_0(\tau)\big)^{-s}, \tag{1.1}$$

where $\nu_0(\tau)$ is *the denominator ideal of* τ and $N\big(\nu_0(\tau)\big)$ is its norm. Now our first main result is that if G is symplectic, orthogonal, or unitary, then

$$\Lambda(s, \chi)\mathfrak{T}(s, \mathbf{f}, \chi) = \prod_{\mathfrak{p}} W_{\mathfrak{p}}\big[\chi(\mathfrak{p})N(\mathfrak{p})^{-s}\big]^{-1}, \tag{1.2}$$

where $\Lambda(s, \chi)$ is an explicitly determined product of L-functions depending on χ, $W_{\mathfrak{p}}$ is a polynomial determined for each $v \in \mathbf{h}$ whose constant term is 1, and $\mathfrak{p}$ runs over all the prime ideals of the basic number field. This is a purely algebraic result concerning only nonarchimedean primes.

Let $\mathcal{Z}(s, \mathbf{f}, \chi)$ denote the right-hand side of (1.2). As our second main result we obtain a product $\mathfrak{G}(s)$ of gamma factors such that $\mathfrak{G}\mathcal{Z}$ can be continued to the whole s-plane as a meromorphic function with finitely many poles, when G is a unitary group of an arbitrary signature distribution over a CM-field and $\mathbf{f}$ corresponds to holomorphic forms.

Now these problems are closely connected with the theory of Eisenstein series E on a group G' in which G is embedded. To describe the series, let $\mathfrak{Z}'$ denote the symmetric space on which G' acts. Then the series as a function of $(z, s) \in \mathfrak{Z}' \times \mathbf{C}$ can be given (in the classical style) in the form

$$(1.3)\qquad E(z,\,s;\,\mathbf{f},\,\chi)=\sum_{\alpha\in A}\delta(z,\,s,\,\mathbf{f},\,\chi)\|\alpha,\qquad A=(P\cap\Gamma)\backslash\Gamma,$$

where Γ is a congruence subgroup of G', P is a parabolic subgroup of G' which is a semidirect product of a unipotent group and $G \times GL_m$ with some m. The adelized version of δ will be explicitly described in §5. Now our third main result is that there exist an explicit product $\mathfrak{G}'$ of gamma factors and an explicit product Λ' of L-functions such that $\mathfrak{G}'(s)\Lambda'(s)\mathcal{Z}(s,\,\mathbf{f},\,\chi)E(z,\,s;\,\mathbf{f},\,\chi)$ can be continued to the whole s-plane as a meromorphic function with finitely many poles.

Though the above results concern holomorphic forms, our method is applicable to the unitary group of a totally definite hermitian form over a CM-field. In this case we can give an explicit class number formula for such a hermitian form, which is the fourth main result of this paper.

2. For an associative ring R with identity element we denote by $R^\times$ the group of all its invertible elements and by R^m_n the R-module of all $m \times n$-matrices with entries in R. To indicate that a union $X = \bigcup_{i\in I} Y_i$ is disjoint, we write $X = \bigsqcup_{i\in I} Y_i$.

Let K be an associative ring with identity element and an involution ρ. For a matrix x with entries in K we put $x^* = {}^t x^\rho$, and $\widehat{x} = (x^*)^{-1}$ if x is square and invertible. Given a finitely generated left K-module V, we denote by $GL(V)$ the group of all K-linear automorphisms of V. We let $GL(V)$ act on V on the right; namely we denote by $w\alpha$ the image of $w \in V$ under $\alpha \in GL(V)$. Given $\varepsilon = \pm 1$, by an *ε-hermitian form on V* we understand a biadditive map $\varphi : V \times V \to K$ such that $\varphi(x,\,y)^\rho = \varepsilon\varphi(y,\,x)$ and $\varphi(ax,\,by) = a\varphi(x,\,y)b^\rho$ for every $a,\,b \in K$. Assuming that φ is nondegenerate, we put

$$(2.1)\qquad G^\varphi = G(\varphi) = G(V,\,\varphi) = \{\,\gamma \in GL(V)\,|\,\varphi(x\gamma,\,y\gamma) = \varphi(x,\,y)\,\}\,.$$

Given $(V,\,\varphi)$ and $(W,\,\psi)$, we define an ε-hermitian form $\varphi \oplus \psi$ on $V \oplus W$ by

$$(2.2)\qquad (\varphi\oplus\psi)(x+y,\,x'+y') = \varphi(x,\,x') + \psi(y,\,y') \qquad (x,\,x' \in V;\; y,\,y' \in W).$$

We then write $(V \oplus W,\,\varphi\oplus\psi) = (V,\,\varphi) \oplus (W,\,\psi)$. If both φ and ψ are nondegenerate, we can view $G^\varphi \times G^\psi$ as a subgroup of $G^{\varphi\oplus\psi}$. The element $(\alpha,\,\beta)$ of $G^\varphi \times G^\psi$ viewed as an element of $G^{\varphi\oplus\psi}$ will be denoted by $\alpha \times \beta$ or by $(\alpha,\,\beta)$. Given a positive integer r, we put $H_r = I'_r \oplus I_r$, $I_r = I'_r = K^1_r$ and

$$(2.3)\qquad \eta_r(x+u,\,y+v) = u\cdot {}^t y^\rho + \varepsilon x \cdot {}^t v^\rho \qquad (x,\,y \in I'_r;\; u,\,v \in I_r).$$

We shall always use H_r, I'_r, I_r, η_r in this sense. We understand that $H_0 = \{0\}$ and $\eta_0 = 0$.

Hereafter we fix V and a nondegenerate φ on V, assuming that K is a division ring whose characteristic is different from 2. Let J be a K-submodule of V which is *totally φ-isotropic,* by which we mean that $\varphi(J,\,J) = 0$. Then we can find a decomposition $(V,\,\varphi) = (Z,\,\zeta) \oplus (H,\,\eta)$ and an isomorphism f of $(H,\,\eta)$ onto $(H_r,\,\eta_r)$ such that $f(J) = I_r$. In this setting we define *the parabolic subgroup P^φ_J of G^φ relative to J* by

$$(2.4)\qquad P^\varphi_J = \{\,\pi \in G^\varphi\,|\,J\pi = J\,\},$$

and define homomorphisms $\pi^{\varphi}_{\zeta} : P^{\varphi}_{J} \to G^{\zeta}$ and $\lambda^{\varphi}_{J} : P^{\varphi}_{J} \to GL(J)$ such that $z\alpha - z\pi^{\varphi}_{\zeta}(\alpha) \in J$ and $w\alpha = w\lambda^{\varphi}_{J}(\alpha)$ if $z \in Z$, $w \in J$ and $\alpha \in P^{\varphi}_{J}$.

Taking a fixed nonnegative integer m, we put

$$(2.5)\qquad (W, \psi) = (V, \varphi) \oplus (H_m, \eta_m), \qquad (X, \omega) = (W, \psi) \oplus (V, -\varphi).$$

We can naturally view $G^{\psi} \times G^{\varphi}$ as a subgroup of G^{ω}. Since $W = V \oplus H_m$, we can put $X = V \oplus H_m \oplus V$ with the first summand V in W, and write every element of X in the form (u, h, v) with $(u, h) \in V \oplus H_m = W$ and $v \in V$. Put

$$U = \{ (v, i, v) \mid v \in V, i \in I_m \}.$$

Observing that U is totally ω-isotropic, we can define P^{ω}_{U}.

Proposition 1. *Let $\lambda(\varphi)$ be the maximum dimension of totally φ-isotropic K-submodules of V. Then*

$$(2.6)\qquad P^{\omega}_{U}\backslash G^{\omega}/[G^{\psi} \times G^{\varphi}]$$

has exactly $\lambda(\varphi)$ orbits. Moreover,

$$(2.7)\qquad P^{\omega}_{U}[G^{\psi} \times G^{\varphi}] = \bigsqcup_{\beta, \xi} P^{\omega}_{U}((\xi \times 1_H)\beta, 1_V),$$

with ξ running over G^{φ} and β over $P^{\psi}_{I}\backslash G^{\psi}$, where $H = H_m$ and $I = I_m$.

In fact, we can give an explicit set of representatives $\{\tau_e\}_{e=1}^{\lambda(\varphi)}$ for (2.6) and also an explicit set of representatives for $P^{\omega}_{U}\backslash P^{\omega}_{U}\tau_e[G^{\psi} \times G^{\varphi}]$ in the same manner as in (2.7). This proposition plays an essential role in the analysis of our Eisensten series $E(z, s; \mathbf{f}, \chi)$.

3. In this section K is a locally compact field of characteritic 0 with respect to a discrete valuation. Our aim is to establish the Euler factor $W_{\mathfrak{p}}$ of (1.2). We denote by $\mathfrak{r}$ and $\mathfrak{q}$ the valuation ring and its maximal ideal; we put $q = [\mathfrak{r} : \mathfrak{q}]$ and $|x| = q^{-\nu}$ if $x \in K$ and $x \in \pi^{\nu}\mathfrak{r}^{\times}$ with $\nu \in \mathbf{Z}$. We assume that K has an automorphism ρ such that $\rho^2 = 1$, and put $F = \{ x \in K \mid x^{\rho} = x \}$, $\mathfrak{g} = F \cap \mathfrak{r}$, and $\mathfrak{d}^{-1} = \{ x \in K \mid \mathrm{Tr}_{K/F}(x\mathfrak{r}) \subset \mathfrak{g} \}$ if $K \neq F$. We consider (V, φ) as in §2 with $V = K^1_n$ and φ defined by $\varphi(x, y) = x\varphi y^*$ for $x, y \in V$ with a matrix φ of the form

$$(3.1)\qquad \varphi = \begin{bmatrix} 0 & 0 & \varepsilon\delta^{-\rho}1_r \\ 0 & \theta & 0 \\ \delta^{-1}1_r & 0 & 0 \end{bmatrix}, \quad \theta = \varepsilon\theta^* \in GL_t(K), \quad \delta \in K^{\times},$$

where $t = n - 2r$. We assume that θ is anisotropic and also that:

$$(3.2a)\qquad \varepsilon = \pm 1 \quad \text{and } \delta = 2 \text{ if } K = F,$$

$$(3.2b)\qquad \varepsilon = 1, \quad \delta\mathfrak{r} = \mathfrak{d}, \quad \text{and} \quad \delta^{\rho} = -\delta \quad \text{if } K \neq F.$$

Thus our group G^{φ} is orthogonal, symplectic, or unitary. The element δ of (3.2b) can be obtained by putting $\delta = u - u^{\rho}$ with u such that $\mathfrak{r} = \mathfrak{g}[u]$. We include the case $rt = 0$ in our discussion. If $t = 0$, we simply ignore θ; this is always so if $K = F$ and $\varepsilon = -1$. We have $\varphi = \theta$ if $r = 0$.

Denoting by $\{e_i\}$ the standard basis of K^1_n, we put

$$J=\sum_{i=1}^{r} Ke_{r+t+i},\quad T=\sum_{i=1}^{t} Ke_{r+i},\quad N=\{\,u\in T\,|\,\varphi(u,u)\in\mathfrak{g}\,\},$$
$$M=\sum_{i=1}^{r}(\mathfrak{r}e_i+\mathfrak{r}e_{r+t+i})+N,\quad C=\{\,\gamma\in G^{\varphi}\,|\,M\gamma=M\,\},\quad E=GL_r(\mathfrak{r}).$$

Then $G^{\varphi}=P_J^{\varphi}C$. We choose $\{e_{r+i}\}_{i=1}^{t}$ so that $N=\sum_{i=1}^{t}\mathfrak{r}e_{r+i}$. Then we can find an element λ of $\mathfrak{r}_t^t$ such that

$$\theta=\delta^{-1}\lambda+\varepsilon(\delta^{-1}\lambda)^{*}. \tag{3.3}$$

Put

$$S=S^r=\{\,h\in K_r^r\,|\,h^{*}=-\varepsilon(\delta^{\rho}/\delta)h\,\}. \tag{3.4}$$

We can write every element of P_J^{φ} in the form

$$\xi=\begin{bmatrix} a & b & c\\ 0 & e & f\\ 0 & 0 & d\end{bmatrix},\quad \widehat{a}=d\in GL_r(K),\ \ e\in G^{\theta}, \tag{3.5}$$
$$b\in K_t^r,\ f=-\delta e\theta b^{*}d,\ c=(s-b\lambda b^{*})d,\ s\in S.$$

If $t=0$, we simply ignore b, e, and f, so that $\xi=\begin{bmatrix} a & sd\\ 0 & d\end{bmatrix}$; we have $\xi=e$ if $r=0$.

We consider the Hecke algebra $\mathfrak{R}(E,GL_r(K))$ consisting of all formal finite sums $\sum c_x ExE$ with $c_x\in\mathbf{Q}$ and $x\in GL_r(K)$, with the law of multiplication defined as in [3]. Taking r indeterminates $t_1,\ldots,t_r$, we define a $\mathbf{Q}$-linear map

$$\omega_0:\mathfrak{R}(E,GL_r(K))\to\mathbf{Q}[t_1,\ldots,t_r,t_1^{-1},\ldots,t_r^{-1}] \tag{3.6}$$

as follows: Given ExE with $x\in GL_r(K)$, we can put $ExE=\bigsqcup_y Ey$ with upper triangular y whose diagonal entries are $\pi^{e_1},\ldots,\pi^{e_r}$ with $e_i\in\mathbf{Z}$. Then we put

$$\omega_0(ExE)=\sum_y\omega_0(Ey),\quad \omega_0(Ey)=\prod_{i=1}^{r}(q^{-i}t_i)^{e_i}. \tag{3.7}$$

Next we consider the Hecke algebra $\mathfrak{R}(C,G^{\varphi})$ consisting of all formal finite sums $\sum c_{\tau}C\tau C$ with $c_{\tau}\in\mathbf{Q}$ and $\tau\in G^{\varphi}$. We then define a $\mathbf{Q}$-linear map

$$\omega:\mathfrak{R}(C,G^{\varphi})\to\mathbf{Q}[t_1,\ldots,t_r,t_1^{-1},\ldots,t_r^{-1}] \tag{3.8}$$

as follows: Given $C\tau C$ with $\tau\in G^{\varphi}$, we can put $C\tau C=\bigsqcup_{\xi}C\xi$ with $\xi\in P$ of the form (3.5). We then put

$$\omega(C\tau C)=\sum_{\xi}\omega(C\xi),\qquad \omega(C\xi)=\omega_0(Ed_{\xi}), \tag{3.9}$$

where ω_0 is given by (3.6) and d_{ξ} is the d-block in (3.5). We can prove that this is well defined, and gives a ring-injection.

Given $x\in K_n^m$, we denote by $\nu_0(x)$ the ideal of $\mathfrak{r}$ which is the inverse of the product of all the elementary divisor ideals of x not contained in $\mathfrak{r}$; we put then $\nu(x)=[\mathfrak{r}:\nu_0(x)]$. We call x *primitive* if $\mathrm{rank}(x)=\mathrm{Min}(m,n)$ and all the elementary divisor ideals of x are $\mathfrak{r}$.

Proposition 2. *Given ξ as in (3.5), suppose that both e and $(\delta\theta)^{-1}(e-1)$ have coefficients in $\mathfrak{r}$ if $t>0$. Let $a=g^{-1}h$ with primitive $[g\ \ h]\in\mathfrak{r}_{2r}^r$ and $gb=j^{-1}k$ with primitive $[j\ \ k]\in\mathfrak{r}_{r+t}^r$. Then*

$$\nu_0\big((\delta\varphi)^{-1}(\xi-1)\big) = \det(ghj^2)\nu_0(jgsg^*j^*),$$

where we take $j = 1_r$ *if* $t = 0$.

We now define a formal Dirichlet series $\mathcal{T}$ by

$$\mathcal{T}(s) = \sum_{\tau\in A}\omega(C\tau C)\nu(\tau)^{-s}, \quad A = C\backslash G^{\varphi}/C. \tag{3.10}$$

This is a formal version of the Euler factor of (1.2) at a fixed nonarchimedean prime.

Theorem 1. *Suppose that* $\delta\varphi \in GL_n(\mathfrak{r})$; *put* $p = [\mathfrak{g} : \mathfrak{g}\cap\mathfrak{q}]$. *(Thus* $p = q$ *if* $K = F$.*) Then*

$$\mathcal{T}(s) = \frac{1-p^{-s}}{1-p^{r-s}}\prod_{i=1}^{r}\frac{(1-p^{2i-2s})}{(1-p^{r-s}t_i)(1-p^{r-s}t_i^{-1})} \qquad (K = F,\ \varepsilon = -1),$$

$$\mathcal{T}(s) = \prod_{i=1}^{r}\frac{(1-p^{2i-2-2s})}{(1-p^{r+t-2-s}t_i)(1-p^{r-s}t_i^{-1})} \qquad (K = F,\ \varepsilon = 1),$$

$$\mathcal{T}(s) = \frac{\prod_{i=1}^{2r}(1-\theta^{i-1}p^{i-1-2s})}{\prod_{i=1}^{r}(1-q^{r+t-1-s}t_i)(1-q^{r-s}t_i^{-1})} \qquad (K \neq F).$$

Here $\theta^i = 1$ *if* i *is even; for* i *odd,* θ^i *is* -1 *or* 0 *according as* $\mathfrak{d} = \mathfrak{r}$ *or* $\mathfrak{d} \neq \mathfrak{r}$.

This can be proved in the same manner as in [5] by means of Proposition 2.

Since we are going to take localizations of a global unitary group, we have to consider $G^{\varphi} = G(V, \varphi)$ of (2.1) with $V = K_n^1$, $K = F \times F$, and ρ defined by $(x, y)^{\rho} = (y, x)$, where F is a locally compact field of characteristic 0 with resepct to a discrete valuation. Let $\mathfrak{g}$ and $\mathfrak{p}$ be the valution ring of F and its maximal ideal; put $\mathfrak{r} = \mathfrak{g}\times\mathfrak{g}$ and $p = [\mathfrak{g} : \mathfrak{p}]$. We consider $\mathfrak{R}(C, G^{\varphi})$ with $C = G^{\varphi}\cap GL_n(\mathfrak{r})$. Then the projection map pr of $GL_n(K)$ onto $GL_n(F)$ gives an isomorphism $\eta : \mathfrak{R}(C, G^{\varphi}) \to \mathfrak{R}(E_1, GL_n(F))$, where $E_1 = GL_n(\mathfrak{g})$. To be explicit, we have $\eta(C(x, {}^tx^{-1})C) = E_1xE_1$. Let ω_1 denote the map of (3.6) defined with n, E_1, and F in place of r, E, and K. Putting $\omega = \omega_1\circ\eta$, we obtain a ring-injection

$$\omega : \mathfrak{R}(C, G^{\varphi}) \to \mathbf{Q}[t_1, \ldots, t_n, t_1^{-1}, \ldots, t_n^{-1}]. \tag{3.11}$$

For $z = (x, y) \in K_n^n$ with $x, y \in F_n^n$ put $\nu_1(z) = \nu(x)$ and $\nu_2(z) = \nu(y)$, where ν is defined with respect to $\mathfrak{g}$ instead of $\mathfrak{r}$. We then put

$$\mathcal{T}(s, s') = \sum_{\tau\in R}\omega(C\tau C)\nu_1(\tau)^{-s}\nu_2(\tau)^{-s'}, \quad R = C\backslash G^{\varphi}/C. \tag{3.12}$$

Then we obtain

$$\mathcal{T}(s, s') = \prod_{i=1}^{n}\frac{1-p^{i-1-s-s'}}{(1-p^{n-s}t_i^{-1})(1-p^{-1-s'}t_i)}. \tag{3.13}$$

4. We now take a totally imaginary quadratic extension K of a totally real algebraic number field F of finite degree. We denote by $\mathbf{a}$ (resp. $\mathbf{h}$) the set of archimedean (resp. nonarchimedean) primes of F; further we denote by $\mathfrak{g}$ (resp. $\mathfrak{r}$) the maximal order of F (resp. K). Let V be a vector space over K

of dimension n. We take a K-valued nondegenerate ε-hermitian form φ on V with $\varepsilon = 1$ with respect to the Galois involution of K over F, and define G^φ as in §2. For every $v \in \mathbf{a} \cup \mathbf{h}$ and an object X we denote by X_v its localization at v. For $v \in \mathbf{h}$ not splitting in K and for $v \in \mathbf{a}$ we take a decomposition

$$(V_v, \varphi_v) = (T_v, \theta'_v) \oplus (H_{r_v}, \eta_{r_v}) \tag{4.1}$$

with anisotropic θ'_v and a nonnegative integer r_v. Put $t_v = \dim(T_v)$. Then $n = 2r_v + t_v$. If n is odd, then $t_v = 1$ for every $v \in \mathbf{h}$. If n is even, then $t_v = 0$ for almost all $v \in \mathbf{h}$ and $t_v = 2$ for the remaining $v \in \mathbf{h}$. If n is odd, by replacing φ by $c\varphi$ with a suitable $c \in F$, we may assume that φ is represented by a matrix whose determinant times $(-1)^{(n-1)/2}$ belongs to $N_{K/F}(K)$.

We fix an element κ of K such that $\kappa^\rho = -\kappa$ and $i\kappa_v\varphi_v$ has signature $(r_v + t_v, r_v)$ for every $v \in \mathbf{a}$. Then $G(i\kappa_v\varphi_v)$ modulo a maximal compact subgroup is a hermitian symmetric space which we denote by $\mathfrak{Z}^\varphi_v$. We take a suitable point $\mathbf{i}_v$ of $\mathfrak{Z}^\varphi_v$ which plays the role of "origin" of the space. If $r_v = 0$, we understand that $\mathfrak{Z}^\varphi_v$ consists of a single point $\mathbf{i}_v$. We put $\mathfrak{Z}^\varphi = \prod_{v\in\mathbf{a}} \mathfrak{Z}^\varphi_v$. To simplify our notation, for $x \in K^\times_\mathbf{A}$ or $x \in (\mathbf{C}^\times)^\mathbf{a}$, $a \in \mathbf{Z}^\mathbf{a}$, and $c \in (\mathbf{C}^\times)^\mathbf{a}$ we put

$$x^a = \prod_{v\in\mathbf{a}} x_v^{a_v}, \qquad |x|^c = \prod_{v\in\mathbf{a}} (x_v\overline{x}_v)^{c_v/2}. \tag{4.2}$$

For $\xi \in G^\varphi_v$ and $w \in \mathfrak{Z}^\varphi_v$ we define $\xi w \in \mathfrak{Z}^\varphi_v$ in a natural way and define also a scalar factor of automorphy $j_\xi(w)$ so that $\det(\xi)^{r_v} j_\xi(w)^{-n}$ is the jacobian of ξ. Given $k, \nu \in \mathbf{Z}^\mathbf{a}$, $z \in \mathfrak{Z}^\varphi$, and $\alpha \in G^\varphi_\mathbf{A}$, we put

$$\alpha z = (\alpha_v z_v)_{v\in\mathbf{a}}, \qquad j^{k,\nu}_\alpha(z) = \det(\alpha)^\nu j_\alpha(z)^k. \tag{4.3}$$

Then, for a function $f : \mathfrak{Z}^\varphi \to \mathbf{C}$ we define $f\|_{k,\nu}\alpha : \mathfrak{Z}^\varphi \to \mathbf{C}$ by

$$(f\|_{k,\nu}\alpha)(z) = j^{k,\nu}_\alpha(z)^{-1} f(\alpha z) \qquad (z \in \mathfrak{Z}^\varphi). \tag{4.4}$$

Now, given a congruence subgroup Γ of G^φ, we denote by $\mathcal{M}^\varphi_{k,\nu}(\Gamma)$ the vector space of all holomorphic functions f on $\mathfrak{Z}^\varphi$ which satisfy $f\|_{k,\nu}\gamma = f$ for every $\gamma \in \Gamma$ and also the cusp condition if G^φ is of the elliptic modular type. We then denote by $\mathcal{S}^\varphi_{k,\nu}(\Gamma)$ the set of all cusp forms belonging to $\mathcal{M}^\varphi_{k,\nu}(\Gamma)$. Further we denote by $\mathcal{M}^\varphi_{k,\nu}$ resp. $\mathcal{S}^\varphi_{k,\nu}$ the union of $\mathcal{M}^\varphi_{k,\nu}(\Gamma)$ resp. $\mathcal{S}^\varphi_{k,\nu}(\Gamma)$ for all congruence subgroups Γ of G. If φ is anisotropic, we understand that $\mathcal{S}^\varphi_{0,\nu} = \mathbf{C}$.

Let D be an open subgroup of $G^\varphi_\mathbf{A}$ such that $D \cap G^\varphi_\mathbf{h}$ is compact. We then denote by $\mathcal{S}^\varphi_{k,\nu}(D)$ the set of all functions $\mathbf{f} : G^\varphi_\mathbf{A} \to \mathbf{C}$ satisfying the following conditions:

(4.5) $\mathbf{f}(\alpha x w) = \mathbf{f}(x)$ *if* $\alpha \in G^\varphi$ *and* $w \in D \cap G^\varphi_\mathbf{h}$;

(4.6) *For every* $p \in G^\varphi_\mathbf{h}$ *there exists an element* $f_p \in \mathcal{S}^\varphi_{k,\nu}$ *such that* $\mathbf{f}(py) = (f_p\|_{k,\nu}y)(\mathbf{i}^\varphi)$ *for every* $y \in G^\varphi_\mathbf{a}$, *where* $\mathbf{i}^\varphi = (\mathbf{i}_v)_{v\in\mathbf{a}}$.

We now take D in a special form. We take a maximal $\mathfrak{r}$-lattice M in V whose norm is $\mathfrak{g}$ in the sense of [2, p.375], and put

$$C = \{\alpha \in G^\varphi_\mathbf{A} \mid M_v\alpha_v = M_v \ \text{ for every } \ v \in \mathbf{h}\}, \tag{4.7}$$

(4.8) $$\widetilde{M} = \{ x \in V \mid \varphi(x, M) \subset \mathfrak{d}^{-1} \},$$

(4.9) $$D = D^{\varphi} = \left\{ \gamma \in C \mid \widetilde{M}_v(\gamma_v - 1) \subset \mathfrak{c}_v M_v \ \text{ for every } \ v \in \mathbf{h} \right\},$$

where $\mathfrak{d}$ is the different of K relative to F and $\mathfrak{c}$ is a fixed integral $\mathfrak{g}$-ideal. Clearly $\widetilde{M}$ is an $\mathfrak{r}$-lattice in V containing M, and we easily see that D^{φ} is an open subgroup of $G^{\varphi}_{\mathbf{A}}$. We assume that

(4.10) $$v|\mathfrak{c} \quad \text{if} \quad \widetilde{M}_v \neq M_v.$$

Define a subgroup $\mathfrak{X}$ of $G^{\varphi}_{\mathbf{A}}$ by

(4.11) $$\mathfrak{X} = \{ y \in G^{\varphi}_{\mathbf{A}} \mid y_v \in D \ \text{ for every } \ v|\mathfrak{c} \}.$$

We then consider the algebra $\mathfrak{R}(D, \mathfrak{X})$ consisting of all the finite linear combinations of $D\tau D$ with $\tau \in \mathfrak{X}$, and define its action on $\mathcal{S}^{\varphi}_{k,\nu}(D)$ as follows. Given $\tau \in \mathfrak{X}$ and $\mathbf{f} \in \mathcal{S}^{\varphi}_{k,\nu}(D)$, take a finite subset Y of $G^{\varphi}_{\mathbf{h}}$ so that $D\tau D = \bigsqcup_{\eta \in Y} D\eta$ and define $\mathbf{f}|D\tau D : G^{\varphi}_{\mathbf{A}} \to \mathbf{C}$ by

(4.12) $$(\mathbf{f}|D\tau D)(x) = \sum_{\eta \in Y} \mathbf{f}(x\eta^{-1}) \qquad (x \in G^{\varphi}_{\mathbf{A}}).$$

These operators form a commutative ring of normal operators on $\mathcal{S}^{\varphi}_{k,\nu}(D)$.

For $x \in G^{\varphi}_{\mathbf{A}}$ we define an ideal $\nu_0(x)$ of $\mathfrak{r}$ by

(4.13) $$\nu_0(x) = \prod_{v \in \mathbf{h}} \nu_0(x_v),$$

where $\nu_0(x_v)$ is defined as in §3 with respect to an $\mathfrak{r}_v$-basis of M_v. Clearly $\nu_0(x)$ depends only on CxC.

Let $\mathbf{f}$ be an element of $\mathcal{S}^{\varphi}_{k,\nu}(D)$ that is a common eigenfunction of all the $D\tau D$ with $\tau \in \mathfrak{X}$, and let $\mathbf{f}|D\tau D = \lambda(\tau)\mathbf{f}$ with $\lambda(\tau) \in \mathbf{C}$. Given a Hecke idele character χ of K such that $|\chi| = 1$, define a Dirichlet series $\mathfrak{T}(s, \mathbf{f}, \chi)$ by

(4.14) $$\mathfrak{T}(s, \mathbf{f}, \chi) = \sum_{\tau \in D \backslash \mathfrak{X}/D} \lambda(\tau)\chi^*\big(\nu_0(\tau)\big)N\big(\nu_0(\tau)\big)^{-s},$$

where χ^* is the ideal character associated with χ and $N(\mathfrak{a})$ is the norm of an ideal $\mathfrak{a}$. Denote by χ_1 the restriction of χ to $F^{\times}_{\mathbf{A}}$, and by θ the Hecke character of F corresponding to the quadratic extension K/F. For any Hecke character ξ of F put

(4.15) $$L_{\mathfrak{c}}(s, \xi) = \prod_{\mathfrak{p} \nmid \mathfrak{c}} \left[1 - \xi^*(\mathfrak{p})N(\mathfrak{p})^{-s}\right]^{-1}.$$

From Theorem 1 and (3.13) we see that

(4.16) $$\mathfrak{T}(s, \mathbf{f}, \chi) \prod_{i=1}^{n} L_{\mathfrak{c}}(2s - i + 1, \chi_1\theta^{i-1}) = \prod_{\mathfrak{q} \nmid \mathfrak{c}} W_{\mathfrak{q}}\left[\chi^*(\mathfrak{q})N(\mathfrak{q})^{-s}\right]^{-1}$$

with a polynomial $W_{\mathfrak{q}}$ of degree n whose constant term is 1, where $\mathfrak{q}$ runs over all the prime ideals of K prime to $\mathfrak{c}$. Let $\mathcal{Z}(s, \mathbf{f}, \chi)$ denote the function of (4.16). Put

(4.17) $$\Gamma_m(s) = \pi^{m(m-1)/2} \prod_{k=0}^{m-1} \Gamma(s - k).$$

Theorem 2. *Suppose that* $\chi_{\mathbf{a}}(b) = b^{\mu}|b|^{i\kappa-\mu}$ *with* $\mu \in \mathbf{Z}^{\mathbf{a}}$ *and* $\kappa \in \mathbf{R}^{\mathbf{a}}$ *such that* $\sum_{v\in\mathbf{a}} \kappa_v = 0$. *Put* $m = k + 2\nu - \mu$ *and*

$$\mathcal{R}(s, \mathbf{f}, \chi) = \prod_{v\in\mathbf{a}} \gamma_v(s + (i\kappa_v/2)) \cdot \mathcal{Z}(s, \mathbf{f}, \chi)$$

with γ_v *defined by*

$$\gamma_v(s) = p_v(s)q_v(s)\Gamma_{r_v}\left(s - n + r_v + \frac{k_v + |m_v|}{2}\right)\Gamma_{n-r_v}\left(s - r_v + \frac{|\mu_v - 2\nu_v|}{2}\right),$$

$$p_v(s) = \begin{cases} \Gamma_{r_v}\left(s + \dfrac{|k_v - m_v|}{2}\right)\Gamma_{r_v}\left(s + \dfrac{k_v - m_v}{2}\right)^{-1} & \text{if } m_v \geq 0, \\ \Gamma_{r_v}\left(s - \dfrac{k_v + m_v}{2}\right)\Gamma_{r_v}\left(s - \dfrac{k_v - m_v}{2}\right)^{-1} & \text{if } m_v < 0, \end{cases}$$

$$q_v(s) = \prod_{i=1}^{n-\ell-1} \Gamma\left(s - \frac{\ell}{2} - \left[\frac{i}{2}\right]\right)\Gamma\left(s - \frac{\ell}{2} - i\right)^{-1}, \quad \ell = |\mu_v - 2\nu_v|.$$

Then $\mathcal{R}(s, \mathbf{f}, \chi)$ *can be continued to the whole s-plane as a meromorphic function with finitely many poles, which are all simple. It is entire if* $\chi_1 \neq \theta^{\nu}$ *for* $\nu = 0, 1$.

We can give an explicitly defined finite set of points in which the possible poles of $\mathcal{R}$ belong. Notice that p_v and q_v are polynomials; in particular, $p_v = 1$ if $0 \leq m_v \leq k_v$ and $q_v = 1$ if $|\mu_v - 2\nu_v| \geq n - 1$.

The results of the above type and also of the type of Theorem 3 below were obtained in [5], [6], and [7] for the forms on the symplectic and metaplectic groups over a totally real algebraic number field. The Euler product of type $\mathcal{Z}$, its analytic continuation, and its relationship with the Fourier coefficients of $\mathbf{f}$ have been obtained by Oh [1] for the group G^{φ} as above when $\varphi = \eta_r$.

5. We now put $(W, \psi) = (V, \varphi) \oplus (H_m, \eta_m)$ as in (2.5) with (V, φ) of §4 and $m \geq 0$. Writing simply $I = I_m$, we can consider the parabolic subgroup P_I^{ψ} of G^{ψ}. We put $P^{\psi} = P_I^{\psi}$ for simplicity, $\lambda_0(\alpha) = \det\left(\lambda_I^{\psi}(p)\right)$ for $p \in P^{\psi}$, and

$$L = \textstyle\sum_{i=1}^{m}(\mathfrak{r}\varepsilon_i + \mathfrak{d}^{-1}\varepsilon_{m+n+i}) + M \tag{5.1}$$

with M of §4 and the standard basis $\{\varepsilon_i, \varepsilon_{m+n+i}\}_{i=1}^{m}$ of H_m. We can define the space $\mathfrak{Z}^{\psi}$ and its origin $\mathbf{i}^{\psi}$ in the same manner as for G^{φ}. We then put

$$C^{\psi} = \left\{x \in G_{\mathbf{A}}^{\psi} \mid Lx = L\right\}, \quad C_0^{\psi} = \left\{x \in C^{\psi} \mid x(\mathbf{i}^{\psi}) = \mathbf{i}^{\psi}\right\}, \tag{5.2}$$

$$D^{\psi} = \left\{x \in C^{\psi} \mid \widetilde{M}_v(e_v - 1) \subset \mathfrak{c}_v M_v \text{ for every } v \in \mathbf{h}\right\}. \tag{5.3}$$

Here e_v is the element of $\mathrm{End}(V_v)$ defined for x_v by $wx_v - we_v \in (H_m)_v$ for $w \in V_v$. We define an $\mathbf{R}$-valued function h on $G_{\mathbf{A}}^{\psi}$ by

$$h(x) = |\lambda_0(p)|_{\mathbf{A}} \quad \text{if } x \in pC_0^{\psi} \text{ with } p \in P_{\mathbf{A}}. \tag{5.4}$$

Taking $\mathbf{f} \in \mathcal{S}_{k,\nu}^{\varphi}(D^{\varphi})$ and χ as in §4, we define $\mu : G_{\mathbf{A}}^{\psi} \to \mathbf{C}$ as follows: $\mu(x) = 0$ if $x \notin P_{\mathbf{A}}^{\psi}D^{\psi}$; if $x = pw$ with $p \in P_{\mathbf{A}}^{\psi}$ and $w \in D^{\psi} \cap C_0^{\psi}$, then we put

$$\mu(x) = \chi(\lambda_0(p))^{-1}\chi_{\mathfrak{c}}(\lambda_0(w))^{-1} j_w^{k,\nu}(\mathbf{i}^{\psi})^{-1}\mathbf{f}(\pi_{\varphi}^{\psi}(p)), \tag{5.5}$$

where $\chi_{\mathfrak{c}} = \prod_{v|\mathfrak{c}} \chi_v$. Then we define $E(x, s)$ for $x \in G^{\psi}_{\mathbf{A}}$ and $s \in \mathbf{C}$ by

$$E(x, s) = E(x, s; \mathbf{f}, \chi, D^{\psi}) = \sum_{\alpha \in A} \mu(\alpha x) h(\alpha x)^{-s}, \quad A = P^{\psi}_{I} \backslash G^{\psi}. \tag{5.6}$$

This is meaningful if $\chi_{\mathbf{a}}(b) = b^{k+2\nu} |b|^{i\kappa - k - 2\nu}$ with $\kappa \in \mathbf{R}^{\mathbf{a}}$, $\sum_{v \in \mathbf{a}} \kappa_v = 0$, and the conductor of χ divides $\mathfrak{c}$. We take such a χ in the following theorem. The series of (5.6) is the adelized version of a collection of several series of type (1.3).

Theorem 3. *Define* γ_v *and* q_v *as in Theorem 2 with* $m = 0$. *Put*

$$\gamma'_v(s) = q'(s, |k_v|) \gamma_v(s) q_v(s)^{-1} \Gamma_m (s - n + (k_v/2)),$$

$$q'(s, \ell) = \prod_{i=1}^{m+n-\ell-1} \Gamma\left(s - \frac{\ell}{2} - \left[\frac{i}{2}\right]\right) \Gamma\left(s - \frac{\ell}{2} - i\right)^{-1}.$$

Then the product

$$\prod_{v \in \mathbf{a}} \gamma'_v(s + (i\kappa_v/2)) \prod_{j=n}^{m+n-1} L_{\mathfrak{c}}(2s - j, \chi_1 \theta^j) \cdot \mathcal{Z}(s, \mathbf{f}, \chi) E(x, s; \mathbf{f}, \chi, D^{\psi})$$

can be continued to the whole s*-plane as a meromorphic function with finitely many poles, which are all simple.*

We can give an explicitly defined finite set of points in which the possible poles of the above product belong.

6. Let G be an arbitrary reductive algebraic group over $\mathbf{Q}$. Given an open subgroup U of $G_{\mathbf{A}}$ containing $G_{\mathbf{a}}$ and such that $U \cap G_{\mathbf{h}}$ is compact, we put $U^a = aUa^{-1}$ and $\Gamma^a = G \cap U^a$ for every $a \in G_{\mathbf{A}}$. We assume that $G_{\mathbf{a}}$ acts on a symmetric space $\mathcal{W}$ and we let G act on $\mathcal{W}$ via its projection to $G_{\mathbf{a}}$. We also assume that $\Gamma^a \backslash \mathcal{W}$ has finite measure, written $\mathrm{vol}(\Gamma^a \backslash \mathcal{W})$, with respect to a fixed $G_{\mathbf{a}}$-invariant measure on $\mathcal{W}$. Taking a complete set of representatives $\mathcal{B}$ for $G \backslash G_{\mathbf{A}} / U$, we put

$$\mathfrak{m}(G, U) = \mathfrak{m}(U) = \sum_{a \in \mathcal{B}} [\Gamma^a \cap T : 1]^{-1} \mathrm{vol}(\Gamma^a \backslash \mathcal{W}), \tag{6.1}$$

where T is the set of elements of G which act trivially on $\mathcal{W}$, and we assume that $[\Gamma^a \cap T : 1]$ is finite. Clearly $\mathfrak{m}(U)$ does not depend on the choice of $\mathcal{B}$. We call $\mathfrak{m}(G, U)$ *the mass of G with respect to U*. If $G_{\mathbf{a}}$ is compact, we take $\mathcal{W}$ to be a single point of measure 1 on which $G_{\mathbf{a}}$ acts trivially. Then we have

$$\mathfrak{m}(G, U) = \mathfrak{m}(U) = \sum_{a \in \mathcal{B}} [\Gamma^a : 1]^{-1}. \tag{6.2}$$

We can show that $\mathfrak{m}(U') = [U : U'] \mathfrak{m}(U')$ if U' is a subgroup of U. If strong approximation holds for the semisimple factor of G, then it often happens that both $[\Gamma^a \cap T : 1]$ and $\mathrm{vol}(\Gamma^a \backslash \mathcal{W})$ depend only on U, so that

$$\mathfrak{m}(G, U) = \mathfrak{m}(U) = \#(G \backslash G_{\mathbf{A}} / U) [\Gamma^1 \cap T : 1]^{-1} \mathrm{vol}(\Gamma^1 \backslash \mathcal{W}). \tag{6.3}$$

If $G_{\mathbf{a}}$ is compact and U is sufficiently small, then $\Gamma^a = \{1\}$ for every a, in which case we have $\mathfrak{m}(U) = \#(G \backslash G_{\mathbf{A}} / U)$. If U is the stabilizer of a lattice L in a vector space on which G acts, then $\#(G \backslash G_{\mathbf{A}} / U)$ is the number of classes in the genus of L. Therefore $\mathfrak{m}(U)$ may be viewed as a refined version of the class number in this sense.

Coming back to the unitary group G^φ of §4, we can prove:

Theorem 4. *Suppose that $G^\varphi_{\mathbf{a}}$ is compact. Let M be a $\mathfrak{g}$-maximal lattice in V of norm $\mathfrak{g}$ and let $\mathfrak{d}$ be the different of K relative to F. Define an open subgroup D of $G^\varphi_{\mathbf{A}}$ by (4.9) with an integral ideal $\mathfrak{c}$. If n is odd, assume that φ is represented by a matrix whose determinant times $(-1)^{(n-1)/2}$ belongs to $N_{K/F}(K)$; if n is even, assume that $\mathfrak{c}$ is divisible by the product $\mathfrak{e}$ of all prime ideals for which $t_v = 2$. Then*

$$\mathfrak{m}(G^\varphi, D) = 2 \cdot \left\{ \prod_{k=1}^{n} (n-k)! \right\}^d D_F^{(n^2-n)/2} N(\mathfrak{c})^{n^2} A$$
$$\cdot \prod_{k=1}^{n} \left\{ N(\mathfrak{d})^{k/2} D_F^{1/2} (2\pi)^{-kd} L_{\mathfrak{c}}(k, \theta^k) \right\},$$

where $d = [F : \mathbf{Q}]$, D_F is the discriminant of F, and $A = N(\mathfrak{e})^n N(\mathfrak{d})^{-n/2}$ or $A = 1$ according as n is even or odd.

If n is odd, we can also consider $\mathfrak{m}(D')$ for

$$D' = \left\{ \gamma \in C \mid M_v(\gamma_v - 1) \subset \mathfrak{c}_v M_v \quad \text{for every} \quad v \in \mathbf{h} \right\} \tag{6.4}$$

with an arbitrary integral ideal $\mathfrak{c}$. Then $\mathfrak{m}(D') = 2^{-\tau}\mathfrak{m}(D)$, where τ is the number of primes in F ramified in K.

7. Let us now sketch the proof of the above theorems. The full details will be given in [8]. We first take $\mathcal{B} \subset G^\varphi_{\mathbf{h}}$ so that $G^\varphi_{\mathbf{A}} = \bigsqcup_{b \in \mathcal{B}} G^\varphi b D^\varphi$. Given $E(x, s)$ as in (5.6), for each $q \in G^\psi_{\mathbf{h}}$ we can define a function $E_q(z, s)$ of $(z, s) \in \mathfrak{Z}^\psi \times \mathbf{C}$ by

$$E(qy, s) = E_q(y(\mathbf{i}^\psi), s) j_y^{k,\nu}(\mathbf{i}^\psi)^{-1} \quad \text{for every} \quad y \in G^\psi_{\mathbf{a}}. \tag{7.1}$$

The principle is the same as in (4.6), and so it is sufficient to prove the assertion of Theorem 3 with $E_q(z, s)$ in place of $E(x, s)$. In particular, we can take q to be $q = b \times 1_{2m}$ with $b \in \mathcal{B}$. Define (X, ω) as in (2.5). Then there is an isomorphism of (X, ω) to (H_{m+n}, η_{m+n}) which maps P^ω_U of Proposition 1 to the standard parabolic subgroup P of $G(\eta_{m+n})$. Therefore we can identify $\mathfrak{Z}^\omega$ with the space $\mathfrak{H}^{\mathbf{a}}$ with

$$\mathfrak{H} = \left\{ z \in \mathbf{C}^{m+n}_{m+n} \mid i(z^* - z) \ \text{ is positive definite} \right\}. \tag{7.2}$$

We can also define an Eisenstein series $E'(x, s; \chi)$ for $x \in G^\omega_{\mathbf{A}}$ and $s \in \mathbf{C}$, which is defined by (5.6) with $\left(G(\eta_{m+n})_{\mathbf{A}}, P, 1\right)$ in place of $\left(G^\psi_{\mathbf{A}}, P^\psi, \mathfrak{f}\right)$. Taking E' and $(q, a) \in G^\omega_{\mathbf{h}}$ (with $a \in \mathcal{B}$) in place of $E(x, s)$ and q, we can define a function $E'_{q,a}(\mathfrak{z}, s)$ of $(\mathfrak{z}, s) \in \mathfrak{H}^{\mathbf{a}} \times \mathbf{C}$ in the same manner as in (7.1). There is also an injection ι of $\mathfrak{Z}^\psi \times \mathfrak{Z}^\varphi$ into $\mathfrak{H}^{\mathbf{a}}$ compatible with the embedding $G^\psi \times G^\varphi \to G(\eta_{m+n})$. We put then

$$g^\circ(z, w) = \delta(w, z)^{-k} g(\iota(z, w)) \qquad (z \in \mathfrak{Z}^\psi, \ w \in \mathfrak{Z}^\varphi) \tag{7.3}$$

for every function g on $\mathfrak{H}^{\mathbf{a}}$, where $\delta(w, z)$ is a natural factor of automorphy associated with the embedding ι. Take a Hecke eigenform $\mathbf{f}$ as in §4 and define f_a by the principle of (4.6). Then, employing Proposition 1, we can prove

$$(7.4) \qquad A(s)\mathcal{T}(s, \mathbf{f}, \chi)E_q(z, s) = \sum_{a \in B} \int_{\Phi_a} (E'_{q,a})^\circ(z, w; s) f_a(w)\delta(w)^k dw,$$

where $q = b \times 1_{2m}$, A is a certian gamma factor, and $\Phi_a = \Gamma^a \backslash \mathfrak{Z}^\varphi$. The computation is similar to, but more involved than, that of [6, §4]. Since the analytic nature of $E'_{q,a}$ can be seen from the results of [4], we can derive Theorem 3 from (7.4).

Take $m = 0$. Then $\psi = \varphi$ and $E_q(z, s) = f_b(z)$. Then the analytic nature of $\mathcal{T}(s, \mathbf{f}, \chi)$, and consequently that of $\mathcal{Z}(s, \mathbf{f}, \chi)$, can be derived from (7.4). However, here we have to assume that $\chi_{\mathbf{a}}(b) = b^{k+2\nu}|b|^{i\kappa-k-2\nu}$ with $\kappa \in \mathbf{R}^{\mathbf{a}}$, $\sum_{v \in \mathbf{a}} \kappa_v = 0$, and the conductor of χ divides $\mathfrak{c}$. The latter condition on $\mathfrak{c}$ is a minor matter, but the condition on $\chi_{\mathbf{a}}$ is essential. To obtain $\mathcal{Z}(s, \mathbf{f}, \chi)$ with an arbitrary χ, we have to replace $E'_{q,a}$ by $\mathfrak{D}E''_{q,a}$, where E'' is a series of type E' with $2\nu - \mu$ in place of k and $\mathfrak{D}$ is a certain differential operator on $\mathfrak{H}^{\mathbf{a}}$.

As for Theorem 4, we take again $\psi = \varphi$ and observe that a constant function can be taken as $\mathbf{f}$ if $G^\varphi_{\mathbf{a}}$ is compact. The space $\mathfrak{Z}^\varphi$ consists of a single point. The integral on the right-hand side of (7.4) is merely the value $(E'_{q,a})^\circ(z, w; s)$. We can compute its residue at $s = n$ explicitly. Comparing it with the residue on the left-hand side, we obtain Theorem 4 when $\mathfrak{c}$ satisfies (4.10). If n is odd, we can remove this condition by computing a group index of type $[U : U']$.

References

1. L. Oh, Zeta functions and Fourier coefficients of automorphic forms on unitary groups, Thesis, Princeton University, 1996.
2. G. Shimura, Arithmetic of unitary groups, Ann. of Math. **83** (1964), 369–409.
3. G. Shimura, Introduction to the arithmetic theory of automorphic functions, Publ. Math. Soc. Japan, No.11, Iwanami Shoten and Princeton Univ. Press, 1971.
4. G. Shimura, On Eisenstein series, Duke Math. J. **50** (1983), 417–476.
5. G. Shimura, Euler products and Fourier coefficients of automorphic forms on symplectic groups, Inv. math. **116** (1994), 531–576.
6. G. Shimura, Eisenstein series and zeta functions on symplectic groups, Inv. math. **119** (1995), 539–584.
7. G. Shimura, Zeta functions and Eisenstein series on metaplectic groups, Inv. math. **121** (1995), 21–60.
8. G. Shimura, Euler Products and Eisenstein series, CBMS Reg. Conf. Ser. in Math. No.93, Amer. Math. Soc., 1997, to appear.

An exact mass formula for orthogonal groups

Duke Mathematical Journal, 97 (1999), 1-66

Introduction. Our mass formula concerns the orthogonal group G^φ of a nondegenerate symmetric bilinear form $\varphi : V \times V \to F$, where V is an n-dimensional vector space over an algebraic number field F. We take F to be totally real, and we assume in the introduction that φ is totally definite, though we shall also treat the indefinite case. Given a lattice L in V, we consider its genus Λ, which consists of all the lattices in V that are equivalent to L with respect to the localizations of G^φ at all nonarchimedean primes. It is well known that there exists a finite set of lattices $\{L_i\}_{i=1}^h$ such that Λ is the disjoint union $\bigsqcup_{i=1}^h \{L_i\alpha | \alpha \in G^\varphi\}$. Here, as well as in the text, we let G^φ act on V on the right. Then we define the *mass of the genus* Λ to be the sum

$$\mathfrak{m}(\varphi, \Lambda) = \sum_{i=1}^{h} [\Gamma_i : 1]^{-1}, \quad \Gamma_i = \{\alpha \in G^\varphi | L_i\alpha = L_i\}.$$

Following a pioneer work of Minkowski, Siegel showed that the mass is an infinite product $\prod_v \{2/e_v(\varphi)\}$ with certain representation densities $e_v(\varphi)$ defined at all archimedean and nonarchimedean primes v of F. He calculated these factors explicitly except for finitely many nonarchimedean primes, finding that the product consists essentially of several special values of the Dedekind zeta function and an L-function of F, but he did not give an exact form. In fact he said something to the effect that his formula was the most practicable way of expressing the relationship between the global theory of quadratic forms and the local theory, since the determination of the bad factors was a tiresome and complicated task, which Minkowski had undertaken unsuccessfully some fifty years earlier. The formulation of this problem in terms of the Tamagawa number provided a perspective applicable to a wide class of algebraic groups, but it may fairly be said that it has encouraged the researchers in this field to continue avoiding the issue.

In order to face the issue squarely, we first notice that the expression of the class number of an algebraic number field in terms of the residue of the Dedekind zeta function is given only for the maximal order, and the class number of an arbitrary order must be given relative to that of the maximal order. Therefore, in the case of orthogonal groups, it is natural to expect a clear-cut formula only for a special type of lattice. Now Eichler introduced the notion of a *maximal lattice*, which is maximal among the lattices on which the quadratic form $x \mapsto \varphi(x, x)$ takes integral values. It is with this type of lattice that we shall give our exact mass formula. Eichler himself

Received 2 April 1997.
1991 *Mathematics Subject Classification.* 11E1Z, 11F55.

computed the densities for maximal lattices in a few easy cases, but he did not go much beyond the point Siegel had reached.

To give an idea about the type of formula we shall prove, let us take for simplicity the case in which the basic field is $\mathbf{Q}$ and n is odd; we also assume that $\det(\varphi)$ is a square, a condition always satisfied by a suitable constant multiple of any given φ, if n is odd. Then the mass of the genus Λ of maximal lattices can be given by

$$\mathfrak{m}(\varphi, \Lambda) = \prod_p \left\{2^{-1}(p+1)^{-1}(p^{2m}-1)\right\} \cdot \prod_{k=1}^{m} \left\{(2k-1)!(2\pi)^{-2k}\zeta(2k)\right\}.$$

Here $m = (n-1)/2$, ζ is the Riemann zeta function, and p runs over all the primes for which φ is equivalent to

$$\begin{bmatrix} 0 & 0 & 1_{m-1} \\ 0 & \theta_p & 0 \\ 1_{m-1} & 0 & 0 \end{bmatrix}$$

over $\mathbf{Q}_p$ with an anisotropic θ_p of degree 3. A similar formula involving a special value of an L-function can be proved for even n and also for indefinite φ (Theorem 5.8). In the indefinite case, the group Γ_i is infinite, and our mass formula gives the volume of the fundamental domain for it in an exact form (Theorem 5.10). Strictly speaking, we shall treat the mass with respect to the special orthogonal group, and so the mass given in our theorem is twice the above $\mathfrak{m}$.

We proved a similar formula in the unitary case in [S5] (see also [S6]), in which we defined the mass $\mathfrak{m}(G, D)$ for an algebraic group G of a general type and an arbitrary open subgroup D of $G_{\mathbf{A}}$ whose nonarchimedean factor is compact, in such a way that the above mass is $\mathfrak{m}(G^{\varphi}, D)$ when D is the stabilizer of L in an obvious sense. Our method in the present case is the same as in [S5]. Since we already explained the basic ideas of the proof in the introduction of [S5] and also in [S6], we mention here only the following: Without relying on Siegel's method, we obtain the mass $\mathfrak{m}(G^{\varphi}, D)$ for a sufficiently small D by comparing the residue of a certain zeta function of G^{φ} with that of an Eisenstein series on a split orthogonal group. We then obtain $\mathfrak{m}(G^{\varphi}, C)$ for the stabilizer C of a maximal lattice by computing $[C : D]$.

However, we shall give in Section 8 some formulas that relate the densities $e_v(\varphi)$ with the local factors of $[C : D]$. These, combined with the results of Section 3 and Siegel's formula or the known fact on the Tamagawa number, will give an alternative proof of our formula, though we shall not discuss this method in detail. There are several definite reasons for not taking this route in the present paper. First of all, our methods allow us to obtain the desired result for every $n \geq 2$ uniformly with no induction process. In addition, by treating a zeta function of G^{φ} and an Eisenstein series on a split orthogonal group, we are able to obtain various by-products. For instance, we shall prove meromorphic continuation of a certain Euler product on G^{φ} (Theorem 5.14) similar to the zeta function of a quaternary form that gives rise to an elliptic cusp form of weight 2. In fact, we can discuss an Euler product attached to a more general type of automorphic eigenform on G^φ and, moreover, an Eisenstein series on G^φ relative to its parabolic subgroup of the most general type, in exactly the same manner as we did in [S5] in the unitary case. In order to keep the paper a reasonable length, however, we treat only the simplest type of eigenform and an Eisenstein series relative to a parabolic subgroup of the easiest type in the split case, simply because we need it in our proof.

Notation. For an associative ring R with identity element and an R-module M, we denote by $R^\times$ the group of all its invertible elements and by M_n^m the R-module of all $m \times n$-matrices with entries in M. For $x \in R_n^m$ and an ideal $\mathfrak{a}$ of R or a fractional ideal $\mathfrak{a}$ of a number field, we write $x \prec \mathfrak{a}$ if all the entries of x belong to $\mathfrak{a}$. (There is a variation of this; see §2.1.) If $x_1, \ldots, x_r$ are square matrices, $\mathrm{diag}[x_1, \ldots, x_r]$ denotes the matrix with $x_1, \ldots, x_r$ in the diagonal blocks and 0 in all other blocks. For an invertible matrix x we put $\widehat{x} = {}^t x^{-1}$. We write $h > k$ for complex hermitian matrices h and k if $h - k$ is positive definite. For a finite set X we denote by $\#X$ or $\#(X)$ the number of elements in X. By a Hecke character χ of a number field F we mean a continuous $\mathbf{T}$-valued character of the idèle group of F trivial on $F^\times$, and by χ^* we denote the ideal character associated with χ. Here $\mathbf{T} = \{z \in \mathbf{C} \,|\, |z| = 1\}$.

1. Preliminaries on orthogonal groups

1.1. Given a finite-dimensional vector space V over a field F, we denote by $GL(V, F)$ or simply by $GL(V)$ the group of all F-linear automorphisms of V. We let $GL(V)$ act on V on the right. Given a nondegenerate symmetric F-bilinear form $\varphi : V \times V \to F$, we put

$$G^\varphi = G(\varphi) = G(V, \varphi) = \{\alpha \in GL(V) \,|\, \varphi(x\alpha, y\alpha) = \varphi(x, y)\}, \tag{1.1.1}$$

$$G_+^\varphi = G_+(\varphi) = \{\alpha \in G(\varphi) \,|\, \det(\alpha) = 1\}, \tag{1.1.2}$$

$$\varphi[x] = \varphi(x, x) \qquad (x \in V). \tag{1.1.3}$$

Whenever we speak of (V, φ) with $V = F_n^1$, we understand that φ is given by $\varphi(x, y) = x\varphi_0 \cdot {}^t y$ for $x, y \in F_n^1$ with $\varphi_0 = {}^t\varphi_0 \in GL_n(F)$. We then simply write φ for φ_0. In this case we can put

$$G^\varphi = G(\varphi) = \{\alpha \in GL_n(F) \,|\, \alpha\varphi \cdot {}^t\alpha = \varphi\}. \tag{1.1.4}$$

We denote by $\mathbf{d}(\varphi)$ the element in $F^\times / \{a^2 \,|\, a \in F^\times\}$ represented by $\det(\varphi)$.

In our treatment we often need a quaternion algebra over a field F, by which we mean a central simple algebra B over F such that $[B : F] = 4$. Whenever such a B is given, we denote by ι its main involution and put $\mathrm{Tr}_{B/F}(x) = x + x^\iota$ and $N_{B/F}(x) = xx^\iota$.

1.2. We now consider (V, φ) over a field F of characteristic different from 2; we put $n = \dim(V)$. We denote by $\mathfrak{C} = \mathfrak{C}^\varphi$ the Clifford algebra associated with (V, φ), and by $\mathfrak{C}_+ = \mathfrak{C}_+^\varphi$ its subalgebra consisting of the even elements. We then put

$$(1.2.1) \qquad \mathcal{G} = \mathcal{G}^\varphi = \left\{a \in \mathfrak{C}^\times \middle| aVa^{-1} = V\right\}, \qquad \mathcal{G}_+ = \mathcal{G}_+^\varphi = \mathcal{G}^\varphi \cap \mathfrak{C}_+^\times.$$

For each $a \in \mathcal{G}$ define $\tau_a = \tau(a) \in GL(V)$ by $x\tau_a = a^{-1}xa$ for $x \in V$. It is well known that

$$(1.2.2) \qquad G_+^\varphi = \left\{\tau_a \middle| a \in \mathcal{G}_+\right\}.$$

We define the spinor norm

$$(1.2.3) \qquad \sigma : G_+^\varphi \to F^\times / F^{\times 2}$$

as usual, where $F^{\times 2} = \left\{c^2 \middle| c \in F^\times\right\}$. For $c \in F^\times$ and $\xi \in G_+^\varphi$ we write $\sigma(\xi) \equiv c$ if c represents $\sigma(\xi)$. To study the nature of $\sigma(\xi)$, we first consider low-dimensional cases.

1.3. Lemma. *Suppose $n = 2$; let $K = F((-d)^{1/2})$ with an element d of F representing $\mathbf{d}(\varphi)$ if φ is anisotropic; let $K = F \times F$ if φ is isotropic. In either case, let ρ denote the (unique) nontrivial automorphism of K over F. Then the following assertions hold.*

(1) *K is a quadratic extension of F if φ is anisotropic. Moreover, for both isotropic and anisotropic φ, there exists an element $c \in F^\times$ such that (V, φ) is isomorphic to (K, ψ), where $\psi[x] = cxx^\rho$ for $x \in K$.*

(2) *The algebra K and the coset $c \cdot N_{K/F}(K^\times)$ are determined by the isomorphism class of (V, φ) and vice versa.*

(3) *G_+^ψ consists of the maps $x \mapsto bx$ with $b \in K$ such that $bb^\rho = 1$, and G^ψ is generated by G_+^ψ and ρ.*

(4) *$\mathfrak{C}^\varphi$ is a division algebra if and only if $c \notin N_{K/F}(K^\times)$, in which case φ is anisotropic.*

(5) *The spinor norm of the map of (3) for $b = a/a^\rho$ with $a \in K^\times$ is represented by aa^ρ.*

These can be verified in a straightforward way and may be called well known. Notice that if $a = (c, 1) \in F \times F = K$, then $a/a^\rho = (c, c^{-1})$ and $aa^\rho = (c, c)$.

1.4. Lemma. *If $n = 3$, the following assertions hold.*

(1) *There exists a quaternion algebra B over F such that (V, φ) is isomorphic to (W, ψ), where $W = \left\{x \in B \middle| x^\iota = -x\right\}$ and $\psi[x] = dxx^\iota$ with an element d representing $\mathbf{d}(\varphi)$.*

(2) *For $a \in B^\times$ define $\tau_a \in GL(W)$ by $x\tau_a = a^{-1}xa$ for $x \in W$. Then $G_+^\psi = \left\{\tau_a \middle| a \in B^\times\right\}$, and G^ψ is generated by G_+^ψ and -1_W.*

(3) *B of (1) is a division algebra if and only if φ is anisotropic.*

(4) *$\sigma(\tau_a) \equiv aa^\iota$ for every $a \in B^\times$.*

Proof. Take an F-basis $\{e_i\}_{i=1}^3$ of V such that $\varphi(e_i, e_j) = c_i\delta_{ij}$ with $c_i \in F$. Define a subset W of $\mathfrak{C}_+$ by $W = Fe_1e_2 + Fe_2e_3 + Fe_3e_1$; put $B = F + W$ and $\zeta = e_1e_2e_3$. Clearly $B = \mathfrak{C}_+$ and it is a quaternion algebra over F. Now we have $W = \zeta V$ and $(\zeta x)(\zeta x)^\iota = e\varphi[x]$ for $x \in V$ with $e = c_1c_2c_3$, which proves (1). If $a \in B^\times$, then $aWa^{-1} = W$. Since ζ belongs to the center of $\mathfrak{C}$, for $x \in \mathfrak{C}^\times$ we have $xVx^{-1} = V$ if and only if $xWx^{-1} = W$, so that $\mathcal{G}_+ = B^\times$. The remaining part of our lemma can easily be verified. □

1.5. Lemma. *Let B be a quaternion algebra over F, and let $\varphi[x] = xx^\iota$ for $x \in B$. For $(a, b) \in B \times B$ such that $aa^\iota = bb^\iota \neq 0$, define $\tau_{a,b} \in GL(B)$ by $x\tau_{a,b} = b^{-1}xa$. Then G_+^φ consists of all such $\tau_{a,b}$, and G^φ is generated by G_+^φ and ι. Moreover, $\sigma(\tau_{a,b}) \equiv aa^\iota$.*

Proof. Defining a map $f : B \to B_2^2$ by $f(x) = \left[\begin{smallmatrix} 0 & x \\ x^\iota & 0 \end{smallmatrix}\right]$, we can easily identify B_2^2 with $\mathfrak{C}^\varphi$ with respect to this embedding, and we find that $\mathfrak{C}_+$ consists of $\mathrm{diag}[a, b]$ for all $a, b \in B$. Then our assertions can easily be verified by means of (1.2.2). □

1.6. Suppose that $(V, \varphi) = (H_r, \eta_r) \oplus (Z, \zeta)$ with $H_r = F_{2r}^1$ and

$$\eta = \eta_r = \begin{bmatrix} 0 & 1_r \\ 1_r & 0 \end{bmatrix}. \tag{1.6.1}$$

We put $n = \dim(V)$, $t = \dim(Z)$, and $J' = \sum_{i=1}^r Fe_i$, $J = \sum_{i=1}^r Fe_{r+i}$ with the standard basis $\{e_i\}_{i=1}^{2r}$ of F_{2r}^1. We consider a parabolic subgroup $P = P_J^\varphi$ of G^φ defined by

$$P = P_J^\varphi = \{\alpha \in G^\varphi \,|\, J\alpha = J\}. \tag{1.6.2}$$

We use matrix expression with respect to the decomposition $V = J' \oplus Z \oplus J$. Thus we put

$$\varphi = \begin{bmatrix} 0 & 0 & 1_r \\ 0 & \zeta & 0 \\ 1_r & 0 & 0 \end{bmatrix}, \qquad \zeta = {}^t\zeta \in GL_t(F). \tag{1.6.3}$$

As shown in [S5, Lemma 2.10], P consists of the elements of the form

$$\xi = \begin{bmatrix} a & b & c \\ 0 & e & f \\ 0 & 0 & d \end{bmatrix} \tag{1.6.4}$$

with $\widehat{a} = d \in GL_r(F), e \in G^\zeta, b \in F_t^r, f = -e\zeta \cdot {}^tbd$, and $c = (s - b\mu \cdot {}^tb)d, s \in S^r$, where μ is any fixed element of F_r^r such that $\zeta = \mu + {}^t\mu$ ($\mu = 2^{-1}\zeta$, for example) and

$$S^r = \{h \in F_r^r \,|\, {}^th = -h\}. \tag{1.6.5}$$

We can include the case $(V,\varphi)=(H_r,\eta_r)$ in our discussion. Then we ignore ζ, b, e, and f in (1.6.3) and (1.6.4).

1.7. LEMMA. *If ξ is of the form (1.6.4) and* $\det(e)=1$, *then* $\sigma(\xi)\sigma(e)^{-1}\equiv\det(a)$. *In particular, if $\varphi=\eta_r$ and $\xi=\mathrm{diag}[a,\widehat{a}]$, then* $\sigma(\xi)\equiv\det(a)$.

Proof. Put $\alpha=\mathrm{diag}[2\cdot 1_r, 1_t, 2^{-1}1_r]$ and

$$\beta=\begin{bmatrix}1 & b & -2^{-1}b\zeta\cdot{}^tb\\ 0 & 1 & -\zeta\cdot{}^tb\\ 0 & 0 & 1\end{bmatrix},\qquad \gamma=\begin{bmatrix}1 & 0 & s\\ 0 & 1 & 0\\ 0 & 0 & 1\end{bmatrix}.$$

Then $\alpha\beta\alpha^{-1}=\beta^2$ and $\alpha\gamma\alpha^{-1}=\gamma^4$, so that $\sigma(\beta)=\sigma(\beta)^2\equiv 1$ and $\sigma(\gamma)=\sigma(\gamma)^4\equiv 1$. Thus our task is to show that $\sigma(\xi)\sigma(e)^{-1}\equiv\det(a)$ for $\xi=\mathrm{diag}[a,e,\widehat{a}]$. We first note that $\sigma(\tau\times\rho)=\sigma(\tau)\sigma(\rho)$ if $\tau\times\rho\in G_+(\varphi\oplus\psi)$ with $\tau\in G_+^{\varphi}$ and $\rho\in G_+^{\psi}$. This reduces our question to the case $t=0$ and $\varphi=\eta_r$, that is, the case in which $\xi=\mathrm{diag}[a,\widehat{a}]$. Take such a ξ with $a=\mathrm{diag}[b,1_{n-1}], b\in F^{\times}$. By Lemma 1.3(5), we have $\sigma\big(\mathrm{diag}[b,b^{-1}]\big)\equiv b$, and hence $\sigma(\xi)\equiv b=\det(a)$. It is well known that $SL_n(F)$ is the commutator subgroup of $GL_n(F)$ if the characteristic of F is not 2. Therefore $\sigma(\xi)\equiv 1$ if $\det(a)=1$. Combining these we obtain our lemma. □

2. Orthogonal groups over a number field

2.1. We now assume that F is an algebraic number field. We denote by $\mathbf{a}$ and $\mathbf{h}$ the sets of archimedean primes and nonarchimedean primes of F; we put $\mathbf{v}=\mathbf{a}\cup\mathbf{h}$. Further, we denote by $\mathfrak{g}$ the maximal order of F. Given an algebraic group G over F, we denote by $G_{\mathbf{A}}$ the adelization of G and by G_v for $v\in\mathbf{v}$ the localization of G at v. The archimedean and nonarchimedean factors of $G_{\mathbf{A}}$ are denoted by $G_{\mathbf{a}}$ and $G_{\mathbf{h}}$. If $G\subset GL(V)$ with a vector space V over F, then for $\alpha\in G_{\mathbf{A}}$ and a $\mathfrak{g}$-lattice L in V, we denote by $L\alpha$ the $\mathfrak{g}$-lattice in V determined by $(L\alpha)_v=L_v\alpha_v$ for every $v\in\mathbf{h}$. In particular, for $x\in F_{\mathbf{A}}^{\times}$ we denote by $x\mathfrak{g}$ the fractional ideal such that $(x\mathfrak{g})_v=x_v\mathfrak{g}_v$. Also we put $|x|_{\mathbf{A}}=\prod_{v\in\mathbf{v}}|x_v|_v$, where $|\ |_v$ is the normalized valuation at v. Given $v\in\mathbf{h}$, a matrix x with entries in F_v, and a $\mathfrak{g}_v$-ideal $\mathfrak{a}$, we write $x\prec\mathfrak{a}$ if all the entries of x belong to $\mathfrak{a}$. Similarly, for a matrix y with entries in $F_{\mathbf{A}}$ and a $\mathfrak{g}$-ideal $\mathfrak{b}$, we write $y\prec\mathfrak{b}$ if all the entries of y_v belong to $\mathfrak{b}_v$ for every $v\in\mathbf{h}$.

We now consider (V,φ) and G^{φ} over F. For each $v\in\mathbf{v}$ we naturally denote by φ_v the F_v-bilinear extension of φ to $V_v\times V_v$. For a $\mathfrak{g}$-lattice L in V we denote by $\mu(L)$ the fractional ideal generated by $\varphi[x]$ for all $x\in L$. We call L *maximal* (with respect to φ) if there is no $\mathfrak{g}$-lattice L', other than L itself, that contains L and such that $\mu(L')=\mu(L)$. For a fractional ideal $\mathfrak{a}$ we call L $\mathfrak{a}$-maximal if $\mu(L)\subset\mathfrak{a}$ and there is no $\mathfrak{g}$-lattice L', other than L itself, that contains L and such that $\mu(L')\subset\mathfrak{a}$. Clearly an $\mathfrak{a}$-maximal lattice is maximal; conversely a maximal lattice L is $\mathfrak{a}$-maximal if $\mathfrak{a}=\mu(L)$. Maximal lattices in V_v for $v\in\mathbf{h}$ with respect to φ_v can be defined similarly; for all these see [S5, §4.7].

Hereafter in this section we put $G = G_+^\varphi$ with a fixed φ. Observe that the spinor norm of (1.2.3) can be extended to the map

(2.1.1) $$\sigma : G_{\mathbf{A}} \to F_{\mathbf{A}}^\times / F_{\mathbf{A}}^{\times 2},$$

where $F_{\mathbf{A}}^{\times 2} = \{c^2 | c \in F_{\mathbf{A}}^\times\}$. (Indeed, we first take the spinor norm $(\mathcal{G}_+^\varphi)_{\mathbf{A}} \to F_{\mathbf{A}}^\times$, from which we easily obtain the map of (2.1.1).) To simplify our notation, if H is a subgroup of G (resp., G_v, $G_{\mathbf{A}}$), we denote by $\sigma(H)$ the subgroup X of $F^\times$ (resp., $F_v^\times, F_{\mathbf{A}}^\times$) containing $F^{\times 2}$ (resp., $F_v^{\times 2}, F_{\mathbf{A}}^{\times 2}$) such that $X/F^{\times 2}$ (resp., $X/F_v^{\times 2}, X/F_{\mathbf{A}}^{\times 2}$) is $\sigma(H)$ in the strict sense.

Let $\mathbf{c}$ be the set of all $v \in \mathbf{a}$ such that G_v is compact. Clearly every $v \in \mathbf{c}$ is real. Put

(2.1.2) $$K = F\big([(-1)^{n/2}\det(\varphi)]^{1/2}\big) \quad \text{if } n \text{ is even},$$

(2.1.3) $$W = \begin{cases} N_{K/F}(K_{\mathbf{A}}^\times) & \text{if } n = 2, \\ \{x \in F_{\mathbf{A}}^\times | x_v > 0 \text{ for every } v \in \mathbf{c}\} & \text{if } n > 2, \end{cases}$$

(2.1.4) $$G_{\mathbf{A}}^0 = \{\alpha \in G_{\mathbf{A}} | \sigma(\alpha) \equiv 1\}, \qquad G^0 = \{\alpha \in G | \sigma(\alpha) \equiv 1\}.$$

We are abusing the notation here, since $G_{\mathbf{A}}^0$ is not the adelization of G^0.

2.2. **Lemma.** (1) $\sigma(G_v) = W \cap F_v^\times$ *for every* $v \in \mathbf{v}$.

(2) $\sigma(G_{\mathbf{A}}) = W$.

(3) $\sigma(G) = W \cap F^\times$.

Proof. Since $\sigma(G_v)$ consists of the product $\varphi[x_1]\cdots\varphi[x_{2m}]$ with an even number of elements x_i of V_v such that $\varphi[x_i] \neq 0$, we obtain (1) if $v \in \mathbf{a}$ or φ is isotropic at v. If $n = 2$, all our assertions follow from Lemma 1.3(5). Therefore, to prove (1) for $n > 2$, we may assume that $v \in \mathbf{h}$ and φ is anisotropic at v. If $n = 4$, it is well known that (V, φ) is isomorphic to (B, φ) of Lemma 1.5 with a division quaternion algebra B over F_v. Since every element of F_v can be written bb^ι with $b \in B$, we obtain our assertion. The same conclusion holds for $n = 3$ by virtue of Lemma 1.4(3).

It remains to prove (2) and (3) for $n > 2$. We have $\prod_{v\in\mathbf{h}} \mathfrak{g}_v^\times \subset \sigma(G_{\mathbf{A}})$, as will be shown in Lemma 2.5, which combined with (1) proves (2). Assertion (3) for $n = 3$ follows from Lemma 1.4(4). Suppose $n > 3$; take $\kappa \in F$ so that $\kappa\varphi$ is positive definite for every $v \in \mathbf{c}$. Let $\alpha \in W \cap F^\times$. Then, by the Hasse principle, there exist elements $x, y \in V$ such that $\varphi[x] = \kappa\alpha$ and $\varphi[y] = \kappa$. Then $\varphi[x]\varphi[y] = \kappa^2\alpha$, so that $\alpha \in \sigma(G)$. This completes the proof. □

2.3. **Lemma.** *Let D be an open subgroup of $G_{\mathbf{A}}$ containing $G_{\mathbf{a}}$ such that $D \cap G_{\mathbf{h}}$ is compact. Put $D^q = qDq^{-1}$ with any $q \in G_{\mathbf{A}}$. Then the following assertions hold:*

(1) $[D : D \cap D^q] = [D^q : D \cap D^q]$;

(2) *$GyG^0_{\mathbf{A}}D = \{x \in G_{\mathbf{A}} | \sigma(x) \in F^\times\sigma(y)\sigma(D)\}$ for every $y \in G_{\mathbf{A}}$;*

(3) *$GG^0_{\mathbf{A}}D$ is a normal subgroup of $G_{\mathbf{A}}$, and $G\backslash G_{\mathbf{A}}/G^0_{\mathbf{A}}D$ can be identified with $G_{\mathbf{A}}/GG^0_{\mathbf{A}}D$;*

(4) *the map $x \mapsto \sigma(x)$ defines a bijection of $G\backslash G_{\mathbf{A}}/G^0_{\mathbf{A}}D$ onto $F^\times_{\mathbf{A}}/F^\times\sigma(D)$ or $F^\times W/F^\times\sigma(D)$ according as $n > 2$ or $n = 2$.*

Proof. Assertion (1) was proved in [S5, Proposition 8.13(3)]. Since $G^0_{\mathbf{A}}$ is a normal subgroup of $G_{\mathbf{A}}$, its product with any subgroup of $G_{\mathbf{A}}$ is a subgroup of $G_{\mathbf{A}}$. To prove (2), let $\sigma(xy^{-1}d^{-1}) \equiv a$ with $d \in D$ and $a \in F^\times$. Then $a \in W \cap F^\times$, and hence $\sigma(\alpha) \equiv a$ with some $\alpha \in G$ by Lemma 2.2(3). Then $y^{-1}\alpha^{-1}xd^{-1} \in G^0_{\mathbf{A}}$, so that $x \in GyG^0_{\mathbf{A}}D$. This proves (2), since $\sigma(GyG^0_{\mathbf{A}}D) \subset \sigma(y)F^\times\sigma(D)$. Taking $y = 1$, we obtain (3). Now the image of the map of (4) is $F^\times W/F^\times\sigma(D)$. Clearly $F^\times W = F^\times_{\mathbf{A}}$ if $n > 2$. Therefore (4) follows from (2). □

2.4. Lemma. *The notation being as in Lemma 2.3, suppose $n \geq 3$ and $G_{\mathbf{a}}$ is not compact. Put $\Gamma = G \cap D$, $\Gamma^q = G \cap D^q$, and $\Gamma' = \Gamma \cap \Gamma^q$. Then the following assertions hold:*

(1) *$G^0_{\mathbf{A}} \subset G^0D$;*

(2) *$GyD = \{x \in G_{\mathbf{A}} | \sigma(y^{-1}x) \in F^\times\sigma(D)\}$ for every $y \in G_{\mathbf{A}}$;*

(3) *GD is a normal subgroup of $G_{\mathbf{A}}$, and $G\backslash G_{\mathbf{A}}/D$ can be identified with $G_{\mathbf{A}}/GD$;*

(4) *the map $x \mapsto \sigma(x)$ defines a bijection of $G\backslash G_{\mathbf{A}}/D$ onto $F^\times_{\mathbf{A}}/F^\times\sigma(D)$;*

(5) *$[\Gamma : \Gamma'] = [\Gamma^q : \Gamma']$.*

Proof. To prove (1), we may take D in the form

$$D = \{y \in G_{\mathbf{A}} | Ly = L, L_v(y_v - 1) \subset \mathfrak{a}_v L_v \text{ for every } v \in \mathbf{h}\}$$

with a $\mathfrak{g}$-lattice L in V and an integral ideal $\mathfrak{a}$. Given $x \in G^0_{\mathbf{A}}$, put $\mathbf{p} = \{v \in \mathbf{h} | x_v \notin D_v \text{ or } v|\mathfrak{c}\}$. By virtue of strong approximation due to Eichler and Kneser (see [E1] and [K]), we can find an element $\alpha \in G^0$ such that $L_v\alpha = L_v$ if $v \notin \mathbf{p}$ and $L_v(\alpha - x_v^{-1}) \subset \mathfrak{a}_v L_v x_v^{-1}$ if $v \in \mathbf{p}$. Then $L_v(\alpha x_v - 1) \subset \mathfrak{a}_v L_v$ for $v \in \mathbf{p}$, and so $L_v\alpha x_v = L_v$ since $\det(\alpha x_v) = 1$. Thus $\alpha x \in D$, which proves (1). Taking yDy^{-1} in place of D, we have $GyG^0_{\mathbf{A}}D = GG^0_{\mathbf{A}}yDy^{-1}y = GyDy^{-1}y = GyD$. Therefore (2), (3), and (4) follow immediately from Lemma 2.3. To prove (5), put $D' = D \cap D^q$. Since $D \cap GD' = \Gamma D'$, we have $[D : \Gamma D'] = [GD : GD']$. Also $[\Gamma D' : D'] = [\Gamma : \Gamma']$, and hence

$$[D : D'] = [D : \Gamma D'][\Gamma D' : D'] = [GD : GD'][\Gamma : \Gamma'].$$

If we replace D by D^q, the left-hand side stays the same in view of Lemma 2.3(1). By (2) we have $GD = GD^q$, and therefore we obtain $[\Gamma : \Gamma'] = [\Gamma^q : \Gamma']$. □

2.5. Lemma. *Let $C_v^+ = \{\alpha \in G_v | L_v\alpha = L_v\}$ with a maximal lattice L in V. Then $\mathfrak{g}_v^\times \subset \sigma(C_v^+)$ except when $n = 2$ and φ_v is anisotropic, in which case $\sigma(C_v^+) = N_{K/F}(K_v^\times)$.*

Proof. If φ_v is isotropic, then by [S5, Lemma 5.6] we can find a Witt decomposition $V_v = \sum_{i=1}^{r_v}(F_v e_i + F_v f_i) + T$ such that $L_v = \sum_{i=1}^{r_v}(\mathfrak{g}_v e_i + \mathfrak{a}_v f_i) + N$ with $r_v > 0$, a fractional ideal $\mathfrak{a}$, and a maximal lattice N in T. Therefore C_v^+ contains an element ξ represented by a matrix $\mathrm{diag}[a, 1_{t_v}, \widehat{a}]$ with an arbitrary $a \in GL_{r_v}(\mathfrak{g}_v)$ with respect to the decomposition. By Lemma 1.7 we have $\sigma(\xi) \equiv \det(a)$, and so we obtain our assertion. Suppose φ_v is anisotropic and $t_v \geq 3$; then $C_v^+ = G_v$ and $\sigma(G_v) = F_v^\times$ by Lemmas 1.4 and 1.5. If $t_v = 2$, then we obtain the desired conclusion from Lemma 1.3(5). □

2.6. LEMMA. *The notation being as in §1.6, suppose that $r > 0$ and $n > 2$. Let D be as in Lemma 2.3, $\mathbf{p}$ a finite subset of $\mathbf{h}$, and e an element of $G_+(\zeta)_\mathbf{h}$. Then every coset of $G\backslash G_\mathbf{A}/D$ contains an element q of the form $q = \mathrm{diag}[a, e, \widehat{a}]$ with $a \in GL_r(F)_\mathbf{h}$ such that $a_v = 1$ for every $v \in \mathbf{p}$.*

Proof. Let $x \in G_\mathbf{A}$. Then we can find an element b of $F_\mathbf{h}^\times$ such that $\sigma(x)\sigma(e) \in bF^\times\sigma(D)$ and $b_v = 1$ for every $v \in \mathbf{p}$. Let $q = \mathrm{diag}[a, e, \widehat{a}]$ with an element a of $GL_r(F)_\mathbf{h}$ such that $\det(a) = b$ and $a_v = 1$ for every $v \in \mathbf{p}$. By Lemma 1.7, $\sigma(xq) = \sigma(x)\sigma(e)b \in F^\times\sigma(D)$, and so $x \in GqD$ by Lemma 2.4(2), which proves our lemma. □

2.7. For $x = \left[\begin{smallmatrix} a & b \\ c & d \end{smallmatrix}\right] \in (F_\mathbf{A})_{2n}^{2n}$ with a, b, c, and d of size n, we write $a = a_x, b = b_x, c = c_x$, and $d = d_x$. We then define $\lambda(x) \in F_\mathbf{A}$ by

(2.7.1) $$\lambda(x) = \det(d_x).$$

With fixed $\mathfrak{g}$-ideals $\mathfrak{y}$ and $\mathfrak{z}$ such that $\mathfrak{y}\mathfrak{z} \subset \mathfrak{g}$, we put

(2.7.2) $$D'[\mathfrak{y}, \mathfrak{z}] = \Big\{x \in GL_{2n}(F)_\mathbf{A} \,\Big|\, \det(x)_\mathbf{h} \in \prod_{v \in \mathbf{h}} \mathfrak{g}_v^\times, \quad a_x \prec \mathfrak{g}, b_x \prec \mathfrak{y}, c_x \prec \mathfrak{z}, d_x \prec \mathfrak{g}\Big\}.$$

We put $C'[\mathfrak{z}] = D'[\mathfrak{z}^{-1}, \mathfrak{z}]$ and define a subgroup P' of $GL_{2n}(F)$ by

(2.7.3) $$P' = \{x \in GL_{2n}(F) \mid c_x = 0\}.$$

In [S5, Lemma 9.2] we showed

(2.7.4) $$P'_\mathbf{A} D'[\mathfrak{y}, \mathfrak{z}] = \{x \in GL_{2n}(F)_\mathbf{A} \mid (d_x)_v \in GL_n(F_v) \text{ and } (d_x^{-1}c_x)_v \prec \mathfrak{z}_v \text{ for every } v|\mathfrak{y}\mathfrak{z}\}.$$

In particular $GL_{2n}(F)_\mathbf{A} = P'_\mathbf{A} C'[\mathfrak{z}]$. Therefore every element x of $GL_{2n}(F)_\mathbf{A}$ can be written $x = yz$ with $y \in P'_\mathbf{A}$ and $z \in C'[\mathfrak{z}]$. We then define a $\mathfrak{g}$-ideal $\mathrm{il}_\mathfrak{z}(x)$ by

(2.7.5) $$\mathrm{il}_\mathfrak{z}(x) = \lambda(y)\mathfrak{g}.$$

This is well defined, but it depends on $\mathfrak{z}$. On the other hand, given $s \in (F_{\mathbf{A}})^n_n$, we can define a $\mathfrak{g}$-ideal $\nu_0(s)$ by the condition $\nu_0(s)_v = \nu_0(s_v)$ for every $v \in \mathbf{h}$, where $\nu_0(s_v)$ is the *denominator ideal* of s_v, which is the inverse of the product of all the elementary divisors of s_v not contained in $\mathfrak{g}_v$. (For the basic properties of ν_0, see [S5, Section 3].)

2.8. LEMMA. *For $x \in GL_{2n}(F)_{\mathbf{A}}$ the following assertions hold.*

(1) *If $\tau = \mathrm{diag}[1_n, \kappa 1_n]$ with $\kappa \in F_{\mathbf{A}}^{\times}$, then* $\mathrm{il}_{\kappa\mathfrak{z}}(\tau x \tau^{-1}) = \mathrm{il}_{\mathfrak{z}}(x)$.

(2) *If $d_x \in GL_n(F)_{\mathbf{A}}$ and $\mathfrak{z} = \mu\mathfrak{g}$ with $\mu \in F_{\mathbf{A}}^{\times}$, then* $\nu_0(\mu^{-1}d_x^{-1}c_x) = \lambda(x) \cdot \mathrm{il}_{\mathfrak{z}}(x)^{-1}$.

(3) *If $x \in GL_{2n}(F) \cap P'_{\mathbf{A}}D'[\mathfrak{y}, \mathfrak{z}]$ and $\mathfrak{y}\mathfrak{z} \neq \mathfrak{r}$, then $\lambda(x) \neq 0$ and $\lambda(x)\mathrm{il}_{\mathfrak{z}}(x)^{-1}$ is prime to $\mathfrak{y}\mathfrak{z}$.*

This is included in [S5, Lemma 9.4].

2.9. We now consider $G^{\eta} = G(\eta_n)$ with $\eta = \eta_n$ as in (1.6.1); we assume that $n > 1$ until the end of this section. We put

$$P = P^{\eta} = P' \cap G^{\eta} = \{x \in G^{\eta} | c_x = 0\}, \tag{2.9.1}$$

$$Q = \{x \in P | b_x = 0\}, \qquad R = \{x \in P | a_x = 1\}, \tag{2.9.2}$$

$$D[\mathfrak{y}, \mathfrak{z}] = G^{\eta} \cap D'[\mathfrak{y}, \mathfrak{z}], \qquad C[\mathfrak{z}] = G^{\eta} \cap C'[\mathfrak{z}]. \tag{2.9.3}$$

Then $G^{\eta}_{\mathbf{A}} = P_{\mathbf{A}}C[\mathfrak{z}]$. Hereafter we fix a fractional ideal $\mathfrak{b}$ and an integral ideal $\mathfrak{c}$ and put simply $D = D[\mathfrak{b}^{-1}, \mathfrak{b}\mathfrak{c}]$ in the rest of this section. If $x \in P$, then $a_x \cdot {}^t d_x = 1$, and hence

$$P \subset G^{\eta}_{+}. \tag{2.9.4}$$

2.10. LEMMA. (1) *The map $x \mapsto \mathrm{il}_{\mathfrak{b}}(x)$ gives a bijection of $P\backslash P_{\mathbf{A}}D/D$ onto the ideal class group of F.*

(2) *Given a finite subset $\mathbf{p}$ of $\mathbf{h}$, there exists a finite subset H of $Q_{\mathbf{h}}$ such that $P_{\mathbf{A}}D = \bigsqcup_{h \in H} PhD$ and $h_v = 1$ for every $(h, v) \in H \times \mathbf{p}$.*

(3) *$PgD = gPD$ for every $g \in P_{\mathbf{A}}$.*

(4) *$P_{\mathbf{A}}D = G^{\eta}_{\mathbf{A}} \cap P'_{\mathbf{A}}D'[\mathfrak{b}^{-1}, \mathfrak{b}\mathfrak{c}]$.*

Proof. Put $Y = D \cap P_{\mathbf{A}}$. Then $\lambda(Y) = F_{\mathbf{a}}^{\times} \prod_{v \in \mathbf{h}} \mathfrak{g}_v^{\times}$. In [S5, Lemma 9.6(3)] we showed that the map $x \mapsto \lambda(x)$ gives a bijection of $P\backslash P_{\mathbf{A}}/Y$ onto $F_{\mathbf{A}}^{\times}/F^{\times}\lambda(Y)$. Since there is an obvious bijection of $P\backslash P_{\mathbf{A}}/Y$ onto $P\backslash P_{\mathbf{A}}D/D$, we obtain (1). Given a finite subset $\mathbf{p}$ of $\mathbf{h}$, we can represent $F_{\mathbf{A}}^{\times}/F^{\times}\lambda(Y)$ by some elements of $\prod_{v \notin \mathbf{p}} F_v^{\times}$. Therefore we can easily find $H \subset Q_{\mathbf{h}}$ so that $F_{\mathbf{A}}^{\times} = \bigsqcup_{h \in H} F^{\times}\lambda(hY)$ and $h_v = 1$ for every $(h, v) \in H \times \mathbf{p}$. Then we obtain (2). Given $g \in P_{\mathbf{A}}$, we have $\lambda(gPY) = F^{\times}\lambda(gY)$ and $\lambda(g^{-1}PgY) = F^{\times}\lambda(Y)$, and hence $gPY \subset PgY$ and $g^{-1}PgY \subset PY$. Thus we obtain $PgY = gPY$, which is (3). As for (4), it is included in [S5, Lemma 9.2] as a special case. □

2.11. LEMMA. *Let x be an element of $G^\eta_{\mathbf{A}}$ such that* $\det(x_v) = 1$ *for every $v|\mathfrak{c}$. Then for every $\alpha \in G^\eta$ there exists an element $q \in Q_{\mathbf{h}}$ such that $x\alpha \in G^\eta qD$ and $q_v = 1$ for every $v|\mathfrak{c}$. The same holds for $x\alpha_{\mathbf{h}}$ instead of $x\alpha$.*

Proof. Put $y = \alpha^{-1}x\alpha$. Then $\det(y_v) = 1$ for every $v|\mathfrak{c}$ and $x\alpha = \alpha y$, so that it is sufficient to show that $y \in G^\eta qD$ with q of the above type. Let z be the projection of y to $\prod_{v|\mathfrak{c}} G^\eta_v$; put $y = zw$. Since $\det(z) = 1$, by Lemma 2.6 we can find $\beta \in G^\eta$ such that $\beta z \in Q_{\mathbf{h}}D$. Now $w_v = 1$ for $v|\mathfrak{c}$ and $G^\eta_v = P_vD_v$ for $v \nmid \mathfrak{c}$, and hence we see that $\beta y = \beta zw \in P_{\mathbf{A}}D$. By Lemma 2.10(2) we have $\beta y \in \gamma qD$ with $\gamma \in P$ and $q \in Q_{\mathbf{h}}$ such that $q_v = 1$ for every $v|\mathfrak{c}$. Then $y \in G^\eta qD$ as expected. The last assertion follows from the fact that $\alpha_{\mathbf{a}} \in D$. □

3. Some facts on nonarchimedean localizations

3.1. In this section we consider (V_v, φ_v) for $v \in \mathbf{h}$. To simplify our notation, we fix v and drop the subscript v. Thus F, $\mathfrak{g}$, and $\mathfrak{p}$ denote a finite algebraic extension of $\mathbf{Q}_p$ for some prime number p, its valuation ring, and the maximal ideal. We put $q = [\mathfrak{g} : \mathfrak{p}]$, denote by π an arbitrarily fixed prime element of F, and define the normalized valuation $x \mapsto |x|$ of F as usual so that $|\pi| = q^{-1}$. We assume that φ is *normalized* in the sense that

(3.1.1) $\quad \mathbf{d}(\varphi)$ is represented by an element of $\mathfrak{g}^\times$ if n is odd.

We can put

(3.1.2) $$(V, \varphi) = \left(H_r, 2^{-1}\eta_r\right) \oplus (T, \theta), \qquad t = \dim(T),$$

with η_r as in (1.6.1) and an anisotropic θ. Clearly $n = 2r + t$, and θ is normalized if t is odd. It is well known that $t \le 4$.

More explicitly, using matrices, we can take (V, φ) as follows: $V = F^1_n$ and $\varphi(x, y) = x\varphi \cdot {}^t y$ for $x, y \in V$ with φ of the form

(3.1.3) $$\varphi = \begin{bmatrix} 0 & 0 & 2^{-1}1_r \\ 0 & \theta & 0 \\ 2^{-1}1_r & 0 & 0 \end{bmatrix}, \qquad \theta = {}^t\theta \in GL_t(F).$$

Let $\{e_i\}_{i=1}^n$ be the standard F-basis of V. Then

(3.1.4) $$H_r = \sum_{i=1}^r (Fe_i + Fe_{r+t+i}), \qquad T = \sum_{j=1}^t Fe_{r+j}.$$

To simplify our notation, we write $f_i = e_{r+t+i}$. We now put

$$L = \sum_{i=1}^{r} (\mathfrak{g}e_i + \mathfrak{g}f_i) + M, \quad M = \{x \in T \,|\, \theta[x] \in \mathfrak{g}\}, \tag{3.1.5}$$

$$\widetilde{L} = \{x \in V \,|\, 2\varphi(x, L) \subset \mathfrak{g}\}, \quad \widetilde{M} = \{x \in T \,|\, 2\theta(x, M) \subset \mathfrak{g}\}. \tag{3.1.6}$$

By [S5, Lemma 5.6], L (resp., M) is a $\mathfrak{g}$-maximal lattice in V (resp., T). Choosing $\{e_{r+j}\}_{j=1}^{t}$ suitably, we may assume that

$$M = \sum_{j=1}^{t} \mathfrak{g}e_{r+j}. \tag{3.1.7}$$

We have $2\theta \prec \mathfrak{g}$, since $\theta[z] \in \mathfrak{g}$ for every $z \in M$.

Conversely, given a $\mathfrak{g}$-maximal lattice L in V, we can find a basis $\{e_i\}_{i=1}^{n}$ of V such that (3.1.2)–(3.1.5) and (3.1.7) hold; see [S5, Lemma 5.6]. We easily see that

$$\widetilde{L} = \sum_{i=1}^{r} (\mathfrak{g}e_i + \mathfrak{g}f_i) + \widetilde{M}, \tag{3.1.8}$$

$$\left[\widetilde{L} : L\right] = \left[\widetilde{M} : M\right] = |\det(2\varphi)|^{-1} = |\det(2\theta)|^{-1}. \tag{3.1.9}$$

3.2. Let us now describe T, θ, M, and $\widetilde{M}$ explicitly. When $t \geq 2$, we denote by K the quadratic extension of F in each case described below, and we denote by $\rho, \mathfrak{r}, \mathfrak{q}$, and $\mathfrak{d}$ the Galois involution of K over F, the maximal order of K, its maximal ideal, and the different of K relative to F.

$t = 1$. Our assumption on $\mathbf{d}(\varphi)$ allows us to put $T = F, \theta[x] = cx^2$ with $c \in \mathfrak{g}^\times$ and $M = \mathfrak{g}$. Clearly $\widetilde{M} = 2^{-1}\mathfrak{g}$.

$t = 2$. Let $K = F\big([-\det(\theta)]^{1/2}\big)$. By Lemma 1.3 we can put $T = K$ and $\theta[x] = cxx^\rho$ with $c \in F^\times$. Clearly we may assume that $c \in \mathfrak{g}^\times$ or $c \in \pi\mathfrak{g}^\times$. The latter case is unnecessary if $\mathfrak{d} \neq \mathfrak{r}$. So we always assume that $c \in \mathfrak{g}^\times$ if $\mathfrak{d} \neq \mathfrak{r}$. In any case we have $M = \mathfrak{r}$ and $2\theta(x, y) = c(xy^\rho + yx^\rho)$; $\widetilde{M} = c^{-1}\mathfrak{r}$ if $\mathfrak{d} = \mathfrak{r}$, and $\widetilde{M} = \mathfrak{d}^{-1}$ if $\mathfrak{d} \neq \mathfrak{r}$.

$t = 4$. Let B be a division quaternion algebra over F, and K an unramified quadratic extension of F contained in B. Then $B = K + K\omega$ with an element ω such that $\omega^2 = \pi$ and $\omega a = a^\rho\omega$ for every $a \in K$. Let $\mathfrak{O}$ be the maximal order of B. It is well known that $\mathfrak{O} = \mathfrak{r} + \mathfrak{r}\omega$ (cf. [S5, Proposition A8.7]). It is also well known that an anisotropic θ with $t = 4$ is equivalent to the norm form on B; that is, we can put $T = B$ and $\theta[x] = xx^\iota$ for $x \in B$ (cf. [E2, Satz 7.3]). Then we have $M = \mathfrak{O}$, $2\theta(x, y) = xy^\iota + yx^\iota$, and $\widetilde{M} = \mathfrak{r} + \mathfrak{r}\omega^{-1} = \omega^{-1}\mathfrak{O}$.

$t = 3$. Take B, K, $\mathfrak{r}$, $\mathfrak{O}$, and ω as in the case $t = 4$; put $W = \{x \in B \,|\, x^\iota = -x\}$. Our normalization of φ together with Lemma 1.4 implies that we can put $T = W$ and $\theta[x] = cxx^\iota$ for $x \in W$ with some $c \in \mathfrak{g}^\times$. Then clearly $M = \mathfrak{O} \cap W$. Now take $z \in \mathfrak{r}$ so that $\mathfrak{r} = \mathfrak{g}[z]$ and put $\zeta = z - z^\rho$. Then $\zeta \in \mathfrak{r}^\times$ and $M = \mathfrak{g}\zeta + \mathfrak{r}\omega$. Since $2\theta(x, y) = c(xy^\iota + yx^\iota)$, we can easily verify that $\widetilde{M} = 2^{-1}\mathfrak{g}\zeta + \mathfrak{r}\omega^{-1}$.

From these and (3.1.9) we easily see that if $\widetilde{L} \neq L$, then

$$(3.2.1)\qquad [\widetilde{L}:L] = \begin{cases} [\mathfrak{g}:2\mathfrak{g}] & \text{if } t=1 \text{ and } v|2,\\ q^2 & \text{if } t=2,\ \mathfrak{d}=\mathfrak{r}, \text{ and } \widetilde{M}\neq M,\\ [\mathfrak{r}:\mathfrak{d}] & \text{if } t=2 \text{ and } \mathfrak{d}\neq\mathfrak{r},\\ [\mathfrak{g}:2\mathfrak{g}]q^2 & \text{if } t=3,\\ q^2 & \text{if } t=4. \end{cases}$$

3.3. **Lemma.** *Let χ be a* **T**-*valued character of the additive group F such that $\mathfrak{g} = \{a \in F \mid \chi(a\mathfrak{g}) = 1\}$. Put $I(\varphi) = \sum_{s\in\mathfrak{g}/\mathfrak{p}} B_s(\varphi)$, where $B_s(\varphi) = \sum_{x\in L/\mathfrak{p}L} \chi(s\varphi[x]/\pi)$. Then*

$$I(\varphi) = \begin{cases} q^n - q^r + q^{r+1} & \text{if } t=0,\\ q^n - q^{r+2} + q^{r+1} & \text{if } t=2,\ \mathfrak{d}=\mathfrak{r}, \text{ and } \widetilde{M}=M,\\ q^n - q^{r+2} + q^{r+3} & \text{if } t=2,\ \mathfrak{d}=\mathfrak{r}, \text{ and } \widetilde{M}\neq M,\\ q^n - q^{r+4} + q^{r+3} & \text{if } t=4,\\ q^n & \text{otherwise.} \end{cases}$$

Proof. Clearly $B_s(\varphi) = B_s(2^{-1}\eta_r)B_s(\theta)$ and $B_s(2^{-1}\eta_r) = B_s(2^{-1}\eta_1)^r$. We easily see that $B_s(2^{-1}\eta_1) = q$ if $s \in \mathfrak{g}^\times$. Since $B_0(\varphi) = q^n$, we have

$$\begin{aligned} I(\varphi) &= q^n + \sum_{s\in(\mathfrak{g}/\mathfrak{p})^\times} B_s(\varphi) = q^n + q^r \sum_{s\in(\mathfrak{g}/\mathfrak{p})^\times} B_s(\theta)\\ &= q^n + q^r\big(I(\theta) - q^t\big) = q^n - q^{r+t} + q^r I(\theta). \end{aligned}$$

If $t = 0$, we have $B_s(\theta) = 1$, so that $I(\varphi) = q^n - q^r + q^{r+1}$. If $t = 1$, we can put $T = F$, $M = \mathfrak{g}$, and $\theta[x] = cx^2$ with $c \in \mathfrak{g}^\times$. Then $I(\theta) = \sum_{x\in\mathfrak{g}/\mathfrak{p}} \sum_{s\in\mathfrak{g}/\mathfrak{p}} \chi(scx^2/\pi) = q$, as can easily be seen. Suppose $t = 2$; then we have $T = K$, $\theta[x] = cxx^\rho$ with $c \in F^\times$, and $M = \mathfrak{r}$; $c \in \mathfrak{g}^\times$ or $c \in \pi\mathfrak{g}^\times$. Clearly $I(\theta) = q^3$ if $c \in \mathfrak{p}$. If $c \in \mathfrak{g}^\times$, then $\sum_{s\in\mathfrak{g}/\mathfrak{p}} \chi(scxx^\rho/\pi) = 0$ or q according as $x \notin \mathfrak{q}$ or $x \in \mathfrak{q}$, and hence $I(\theta) = q^2$ or q according as K is ramified or unramified over F.

Next suppose $t = 3$; then $M = \mathfrak{g}\zeta + \mathfrak{r}\omega$ and $\theta[x] = cxx^\iota$ as described above. Then $\theta[a\zeta + b\omega] = c(a^2\zeta\zeta^\rho - \pi bb^\rho)$ for $a \in \mathfrak{g}$ and $b \in \mathfrak{r}$, and hence $I(\theta) = q^2 \sum_{a\in\mathfrak{g}/\mathfrak{p}} \sum_{s\in\mathfrak{g}/\mathfrak{p}} \chi(sca^2\zeta\zeta^\rho/\pi) = q^3$ as can easily be seen.

Finally suppose $t = 4$; then $T = B$ and $M = \mathfrak{O}$ as above. There are exactly q^2 elements $x \in \mathfrak{O}/\mathfrak{p}\mathfrak{O}$ such that $xx^\iota \in \mathfrak{p}$. Therefore we easily find that $I(\theta) = q^3$. Substituting these values of $I(\theta)$ into the above formula for $I(\varphi)$, we obtain our lemma. □

3.4. **Lemma.** *Let $\mathfrak{r}$ denote -1 or 0 according as K is unramified or ramified over F. Put $U = \{a \in \mathfrak{r} \mid aa^\rho = 1\}$, $U_m = U \cap (1 + \mathfrak{p}^m\mathfrak{d})$, and $Y_m = \{a \in \mathfrak{r}^\times \mid a^\rho - a \in \mathfrak{p}^m\mathfrak{d}\}$*

for $0 \le m \in \mathbf{Z}$. *Define a homomorphism* g *of* $\mathfrak{r}^\times$ *into* U *by* $g(x) = x^\rho/x$. *Then* $Y_0 = \mathfrak{r}^\times$, $[U : U_0] = 2 + \tau$, *and* $g(Y_m) = U_m$ *for every* $m \ge 0$; $Y_m = \mathfrak{g}^\times(1 + \mathfrak{p}^m\mathfrak{r})$, $U_m = g(1 + \mathfrak{p}^m\mathfrak{r})$, *and* $[U_0 : U_m] = [\mathfrak{r}^\times : Y_m] = q^m(1 - \tau q^{-1})$ *if* $m > 0$.

3.5. Lemma. *Let* $\mathfrak{P}$ *denote the maximal ideal of* $\mathfrak{O}$. *Then* $N_{B/F}(1 + \mathfrak{p}^{m+1}\mathfrak{O}) = N_{B/F}(1 + \mathfrak{p}^m\mathfrak{P}) = 1 + \mathfrak{p}^{m+1}$ *for* $0 \le m \in \mathbf{Z}$.

These two lemmas were proved in [S5, Lemmas 17.5 and 17.8].

3.6. We define subgroups C_m^φ and D_m^φ of G^φ by

$$C_m = C_m^\varphi = \{\gamma \in G^\varphi \mid L\gamma = L,\, L(\gamma - 1) \subset \mathfrak{p}^m L\}, \tag{3.6.1}$$

$$D_m = D_m^\varphi = \{\gamma \in C_0 \mid \widetilde{L}(\gamma - 1) \subset \mathfrak{p}^m L\}. \tag{3.6.2}$$

We define C_m^θ and D_m^θ in a similar manner with M and $\widetilde{M}$ in place of L and $\widetilde{L}$. For simplicity we put $C^\varphi = C_0^\varphi$ and $C^\theta = C_0^\theta$. We easily see that

$$C_m^\varphi = \{\gamma \in C_0^\varphi \mid \gamma - 1 \prec \mathfrak{p}^m\}, \qquad D_m^\varphi = \left\{\gamma \in C_0^\varphi \mid \varphi^{-1}(\gamma - 1) \prec 2\mathfrak{p}^m\right\}. \tag{3.6.3}$$

3.7. Lemma. (1) $D_1^\theta \subset G_+^\theta$.
(2) *The index* $[C^\theta : D_1^\theta]$ *can be given as follows:*

2	*if* $t = 1$,
$2(q+1)$	*if* $t = 2$, $\mathfrak{d} = \mathfrak{r}$, *and* $\widetilde{M} = M$,
$2q(q+1)$	*if* $t = 2$, $\mathfrak{d} = \mathfrak{r}$, *and* $\widetilde{M} \ne M$,
$4q$	*if* $t = 2$ *and* $\mathfrak{d} \ne \mathfrak{r}$,
$4q^3(q+1)$	*if* $t = 3$,
$4q^5(q+1)^2$	*if* $t = 4$.

Proof. In this proof let us simply write C and D_1 for C^θ and D_1^θ. Since θ is anisotropic, we have $C = G^\theta$ by [S5, Lemma 5.4]. The case $t = 1$ is trivial. Suppose $t = 2$; let the notation be as in Lemma 3.4. Then by Lemma 1.3(3) we see that $C = \{1, \rho\}U$ and $G_+^\theta = U$. To show that $D_1 \subset U$, take $a \in U$ and define $\gamma \in C$ by $x\gamma = ax^\rho$ for $x \in K$. If $\gamma \in D_1$, then $ax^\rho - x \in \mathfrak{p}\mathfrak{r}$ for every $x \in \mathfrak{d}^{-1}$. It is an easy exercise to show that this cannot happen for any $a \in U$. Once we know that $D_1 \subset U$, the definition of D_1 shows that $D_1 = U_1$ if $c \in \mathfrak{g}^\times$ and $D_1 = U_2$ if $c \in \mathfrak{p}$. Since the index $[U : U_m]$ is given in Lemma 3.4, we obtain the indices for $t = 2$.

Next suppose $t = 3$. Put $C' = \left\{\tau_a \mid a \in \mathfrak{O}^\times\right\}$, where the notation is as in §3.2 and $x\tau_a = a^{-1}xa$. In the following we write an element a of $\mathfrak{O}^\times$ always in the form $a = c + d\omega$ with $c, d \in \mathfrak{r}$. Clearly $c \in \mathfrak{r}^\times$. Since $B^\times$ is generated by $\mathfrak{O}^\times$ and ω, we obtain, from Lemma 1.4(2), $C = G^\theta = \{\pm 1, \pm\tau_\omega\}C'$. We first prove that $D_1 \subset C'$, that is,

$$\{-\tau_a, \pm\tau_{\omega a} | a \in \mathfrak{O}^\times\} \cap D_1 = \varnothing. \tag{3.7.1}$$

Suppose $-\tau_{\omega a} \in D_1$ with $a \in \mathfrak{O}^\times$. Then $(\omega a)^{-1}\zeta\omega a + \zeta \in 2\pi M$, so that $a\zeta - \zeta a \in 2\pi\mathfrak{O}$, that is, $d\omega\zeta \in \pi\mathfrak{O}$. Thus $d \in \mathfrak{p}\mathfrak{r}$. Since $\mathfrak{r}\omega^{-1} \subset \widetilde{M}$, for every $h \in \mathfrak{r}$ we have $(\omega a)^{-1}h\omega^{-1}\omega a + h\omega^{-1} \in \pi\mathfrak{O}$. Then $ha\omega + \omega ah \in \pi^2\mathfrak{O}$. Since $d \in \mathfrak{p}\mathfrak{r}$, we see that $hc + c^\rho h^\rho \in \mathfrak{p}\omega\mathfrak{O} \cap F = \mathfrak{p}^2$ for every $h \in \mathfrak{r}$. But $\mathrm{Tr}(c\mathfrak{r}) = \mathfrak{g}$ since $c \in \mathfrak{r}^\times$, a contradiction. Thus $-\tau_{\omega a} \notin D_1$. Next suppose $\tau_{\omega a} \in D_1$ with $a \in \mathfrak{O}^\times$. This time we obtain $a\zeta + \zeta a \in 2\pi\mathfrak{O}$, so that $c\zeta \in \pi\mathfrak{r}$, which is impossible since $c\zeta \in \mathfrak{r}^\times$. Finally suppose $-\tau_a \in D_1$ with $a \in \mathfrak{O}^\times$. Then $\zeta a + a\zeta \in 2\pi\mathfrak{O}$, and again $c\zeta \in \pi\mathfrak{r}$, which cannot happen.

Thus $D_1 \subset C'$. Let $a = c + d\omega \in \mathfrak{O}^\times$; then $\tau_a \in D_1$ if and only if $xa - ax \in \mathfrak{p}\mathfrak{O}$ for every $x \in \widetilde{M}$. Now $\zeta a - a\zeta = 2d\zeta\omega$, and $h\omega^{-1}a - ah\omega^{-1} = h(c^\rho - c)\omega^{-1} + hd^\rho - dh^\rho$ for every $h \in \mathfrak{r}$. By Lemma 3.4, for $c \in \mathfrak{r}^\times$ we have $c^\rho - c \in \mathfrak{p}^2\mathfrak{r}$ if and only if $c \in \mathfrak{g}^\times(1 + \mathfrak{p}^2\mathfrak{r})$. From these we easily see that $\tau_a \in D_1$ if and only if $d \in \mathfrak{p}\mathfrak{r}$ and $c \in \mathfrak{g}^\times(1 + \mathfrak{p}^2\mathfrak{r})$. Since $\mathfrak{P}^3 = \mathfrak{p}\mathfrak{P} = \mathfrak{p}^2\mathfrak{r} + \mathfrak{p}\mathfrak{r}\omega$, we have

$$D_1 = \{\tau_a | a \in \mathfrak{g}^\times(1 + \mathfrak{P}^3)\}. \tag{3.7.2}$$

Now for $a \in \mathfrak{O}^\times$ we have $\tau_a = 1$ if and only if $a \in \mathfrak{g}^\times$. Thus we obtain

$$\begin{aligned}[C : D_1] &= 4[C' : D_1] = 4[\mathfrak{O}^\times : \mathfrak{g}^\times(1 + \mathfrak{P}^3)] \\ &= 4[\mathfrak{O}^\times : 1 + \mathfrak{P}^3]/[\mathfrak{g}^\times : 1 + \mathfrak{p}^2] = 4q^3(q+1).\end{aligned}$$

This proves the case $t = 3$.

Finally suppose $t = 4$. Employing the symbols and results of Lemma 1.5, we have $C = G^\theta = \{1, \iota\}G_+^\theta$ and $G_+^\theta = \{\tau_{a,b} | a, b \in B^\times, aa^\iota = bb^\iota\}$. Suppose $aa^\iota = bb^\iota = s \in F^\times$. Then we can find $r \in F^\times$ so that $r^2\pi^e/s \in \mathfrak{g}^\times$ with $0 \le e \le 1$. Since $\tau_{r,r} = 1$, we see that G_+^θ is generated by $\tau_{a,b}$ and $\tau_{\omega a, \omega b}$ with (a, b) in the set

$$R = \{(a, b) \in \mathfrak{O}^\times \times \mathfrak{O}^\times | aa^\iota = bb^\iota\}.$$

Thus $\{\tau_{a,b} | (a, b) \in R\}$ has index 4 in C. Let us now prove that D_1 is contained in this subgroup. First suppose that $\iota\tau_{a,b} \in D_1$ with some $(a, b) \in R$. Then $b^{-1}x^\iota a - x \in \mathfrak{p}\mathfrak{O}$ for every $x \in \mathfrak{r} + \mathfrak{r}\omega^{-1}$. Put $a = c + d\omega$ and $b = e + f\omega$ with $c, d, e, f \in \mathfrak{r}$. Taking an element y of $\mathfrak{r}^\times$ to be x, we see that $c^{-1}e - y^\rho/y \in \mathfrak{p}\mathfrak{r}$ and, in particular, $c^{-1}e - 1 \in \mathfrak{p}\mathfrak{r}$. By Lemma 3.4 this means that $U \subset 1 + \mathfrak{p}\mathfrak{r}$, a contradiction. Next suppose that $\iota\tau_{\omega a, \omega b} \in D_1$ with some $(a, b) \in R$. Then $b^{-1}\omega^{-1}x^\iota\omega a - x \in \mathfrak{p}\mathfrak{O}$ for every $x \in \mathfrak{r} + \mathfrak{r}\omega^{-1}$. Take $x = \omega^{-1}y$ with $y \in \mathfrak{r}^\times$. Then, with c and e as above, we find that $c^{-1}e^\rho + y^\rho/y \in \mathfrak{p}^2\mathfrak{r}$ for every $y \in \mathfrak{r}^\times$, which leads to a contradiction for the same reason as above. Finally suppose that $\tau_{\omega a, \omega b} \in D_1$ with some $(a, b) \in R$. Taking $x = \omega^{-1}y$ with $y \in \mathfrak{r}^\times$, we obtain $c^{-1}e^\rho - y^\rho/y \in \mathfrak{p}^2\mathfrak{r}$, again a contradiction. This concludes the proof of the inclusion $D_1 \subset \{\tau_{a,b} | (a, b) \in R\}$.

Suppose $\tau_{a,b} \in D_1$; then $b^{-1}xa - x \in \mathfrak{p}\mathfrak{O}$ for every $x \in \mathfrak{r} + \mathfrak{r}\omega^{-1}$. Putting $a = c + d\omega$ and $b = e + f\omega$ with $c, d, e, f \in \mathfrak{r}$, we can easily verify that that is so if and only if $c - e \in \mathfrak{p}\mathfrak{r}$, $c^\rho - e \in \mathfrak{p}^2\mathfrak{r}$, and $dy - fy^\rho \in \mathfrak{p}\mathfrak{r}$, $d^\rho y - fy^\rho \in \mathfrak{p}\mathfrak{r}$ for every $y \in \mathfrak{r}$. In particular, $d - f \in \mathfrak{p}\mathfrak{r}$ and $d^\rho - f \in \mathfrak{p}\mathfrak{r}$. Now we can find $y \in \mathfrak{r}^\times$ such that $\mathfrak{r} = \mathfrak{g}[y]$. Put $u = y^\rho/y$. Since $y^\rho - y \in \mathfrak{r}^\times$, we see that $u - 1 \in \mathfrak{r}^\times$. With this y, we find that $d - fu \in \mathfrak{p}\mathfrak{r}$, so that $f(1-u) \in \mathfrak{p}\mathfrak{r}$. Thus we see that $f \in \mathfrak{p}\mathfrak{r}$ and $d \in \mathfrak{p}\mathfrak{r}$. From these we can conclude that D_1 consists of the $\tau_{a,b}$ with (a, b) in the set

$$S = \{(c + d\omega, e + f\omega) \in R \mid c \in \mathfrak{r}, e \in \mathfrak{r}, d \in \mathfrak{p}\mathfrak{r}, e \in \mathfrak{p}\mathfrak{r}, c - c^\rho \in \mathfrak{p}\mathfrak{r}, c^\rho - e \in \mathfrak{p}^2\mathfrak{r}\}.$$

Since S contains $\tau_{a,a}$ for every $a \in \mathfrak{g}^\times$, we see that $[C : D_1] = 4[R : S]$. To compute $[R : S]$, put

$$\mathfrak{R} = \left\{(a, b) \in (\mathfrak{O}/\mathfrak{p}\mathfrak{P})^\times \times (\mathfrak{O}/\mathfrak{p}\mathfrak{P})^\times \,\middle|\, aa^\iota - bb^\iota \in \mathfrak{p}^2\right\},$$
$$\mathcal{S} = \left\{\left(a, a^\rho\right) \in \mathfrak{R} \,\middle|\, a \in \left(\mathfrak{r}/\mathfrak{p}^2\mathfrak{r}\right)^\times, a - a^\rho \in \mathfrak{p}\mathfrak{r}\right\}.$$

In view of Lemma 3.5 we see that the natural map of R to $\mathfrak{R}$ is surjective. We can also easily verify that S is the inverse image of $\mathcal{S}$ under that map. Thus $[R : S] = [\mathfrak{R} : \mathcal{S}]$. Now $\#(\mathfrak{O}/\mathfrak{p}\mathfrak{P})^\times = q^4(q^2 - 1)$, and for each element $x \in (\mathfrak{g}/\mathfrak{p}^2)^\times$ there exist exactly $q^3(q + 1)$ elements $b \in (\mathfrak{O}/\mathfrak{p}\mathfrak{P})^\times$ such that $bb^\iota - x \in \mathfrak{p}^2$. Therefore $\#\mathfrak{R} = q^7(q + 1)(q^2 - 1)$. On the other hand, using the symbols and the results of Lemma 3.4, we obtain

$$\#\mathcal{S} = \left[Y_1 : 1 + \mathfrak{p}^2\mathfrak{r}\right] = \left[\mathfrak{r}^\times : 1 + \mathfrak{p}^2\mathfrak{r}\right] / \left[\mathfrak{r}^\times : Y_1\right] = q^2(q - 1).$$

Thus $[\mathfrak{R} : \mathcal{S}] = q^5(q + 1)^2$, which gives the desired value of $[C : D_1]$ for $t = 4$. This completes the proof. □

To state the next lemma, we need some symbols defined as follows:

$$X = \{x \in L \mid \varphi[x] \in \mathfrak{p}\}, \qquad X_1 = \{x \in X \mid x \notin \mathfrak{p}L\},$$
$$X_0 = \{x \in X_1 \mid \varphi[x] = 0\}.$$

For $x \in L$ let $\mu(x)$ denote the element of $L/\mathfrak{p}L$ represented by x; put $Y = \mu(X)$, $Y_1 = \mu(X_1)$, $Y_0 = \mu(X_0)$, $A = \#Y$, and $A_0 = \#Y_0$.

3.8. Lemma. *If $r > 0$, then*

(3.8.1) $$\#\{\mu(x, w) \mid x \in X_0, w \in X_0, 2\varphi(x, w) = 1\} = q^{n-2}A_0,$$

and A_0 is given by

$$A_0 = \begin{cases} q^{t-1}(q^{n-t}-1) & \text{if } n \text{ is odd}, \\ (q^m-1)(q^{m-1}+1) & \text{if } t=0, \\ (q^m+1)(q^{m-1}-1) & \text{if } t=2,\ \mathfrak{d}=\mathfrak{r}, \text{ and } \widetilde{M}=M, \\ q^2(q^{m-2}+1)(q^{m-1}-1) & \text{if } t=2,\ \mathfrak{d}=\mathfrak{r}, \text{ and } \widetilde{M}\neq M, \\ q(q^{n-2}-1) & \text{if } t=2 \text{ and } \mathfrak{d}\neq\mathfrak{r}, \\ q^2(q^{m-1}+1)(q^{m-2}-1) & \text{if } t=4, \end{cases}$$

where $m = n/2$.

Proof. For a fixed $x_0 \in X_0$ we want to count the number of $\mu(w)$ with $w \in X_0$ such that $2\varphi(x_0, w) = 1$. By [S5, Lemma 5.14] there exists an element $\gamma \in C$ such that $x_0 = e_1\gamma$. (Take $(1, n-2, 2\varphi, 2\mathfrak{g}, \mathfrak{g})$ as $(r, t, \varphi, \mathfrak{a}, \mathfrak{z})$ there.) Thus we may assume that $x_0 = e_1$. Suppose $w \subset X_0$ and $2\varphi(e_1, w) = 1$; put $w = \sum_{i=1}^{r}(a_i e_i + b_i f_i) + z$ with $a_i, b_i \in \mathfrak{g}$ and $z \in M$. Then $b_1 = 1$, and so $a_1 = -\sum_{i=2}^{r} a_i b_i - \theta[z]$. Then we see that the number of $\mu(w)$ is exactly q^{n-2}. Thus our task is to determine A_0. Put $E = \sum_{i=1}^{r}(\mathfrak{g}e_i + \mathfrak{g}f_i)$. We first prove that

$$(*) \qquad \mu(y+z) \in Y_0 \quad \text{if } y \in E,\ y \notin \mathfrak{p}E,\ z \in M, \text{ and } y+z \in X.$$

In fact, put $y = \sum_{i=1}^{r}(a_i e_i + b_i f_i)$ with $a_i, b_i \in \mathfrak{g}$. Suppose $y+z \in X$ and $y \notin \mathfrak{p}E$; then we may assume that $a_1 \notin \mathfrak{p}$. We have $\sum_{i=1}^{r} a_i b_i + \theta[z] = d$ with $d \in \mathfrak{p}$. Therefore, changing b_1 for $b_1 - a_1^{-1}d$, we obtain an element $y' \in E$ such that $\mu(y+z) = \mu(y'+z)$ and $y'+z \in X_0$. This proves $(*)$. If $t = 0$, we obtain $Y_1 = Y_0$ from $(*)$.

Next suppose $t = 2$; as explained in §3.2, we can put $T = K, M = \mathfrak{r}$, and $\theta[z] = czz^\rho$ with $c \in \mathfrak{g}^\times$ or $c\mathfrak{g} = \mathfrak{p}$. Let $x = y+z \in X_0$ with $y \in E$ and $z \in M$. If $y \in \mathfrak{p}E$, then $2^{-1}\eta[y] \in \mathfrak{p}^2$, so that $czz^\rho \in \mathfrak{p}^2$. Then we easily see that $z \in \mathfrak{p}\mathfrak{r}$, so that $x \in \mathfrak{p}L$, a contradiction. Therefore $y \notin \mathfrak{p}E$, and hence $(*)$ shows that

$$(**) \qquad Y_0 = \{\mu(y+z) \mid y+z \in X, y \in E, \notin \mathfrak{p}E, z \in M\},$$

$$(***) \qquad Y - Y_0 = \{\mu(z) \mid z \in M, \theta[z] \in \mathfrak{p}\}.$$

Clearly the same holds for $t = 1$. Thus we see that $A - A_0 = q$ if K is ramified over F; $A - A_0 = q^2$ if $t = 2$, K is unramified over F, and $c \in \mathfrak{p}$; $A - A_0 = 1$ if $t \le 2$, and those two cases are excluded.

Suppose $t = 3$; let $x = y+z \in X_0$ with $y \in \mathfrak{p}E$ and $z \in M$. Put $z = a\zeta + b\omega$ with $a \in \mathfrak{g}$ and $b \in \mathfrak{r}$. Then $c(a^2\zeta^2 + \pi bb^\rho) = -\theta[z] \in \mathfrak{p}^2$, so that $a \in \mathfrak{p}$ since $\zeta \in \mathfrak{r}^\times$, and consequently $b \in \mathfrak{p}\mathfrak{r}$. Thus $z \in \mathfrak{p}M$, a contradiction. Therefore $(**)$ and $(***)$ hold in this case. We easily see that $Y - Y_0 = \mu(\mathfrak{p}\zeta + \mathfrak{r}\omega)$, so that $A - A_0 = q^2$.

If $t = 4$, the same type of argument validates $(**)$ and $(***)$, and we obtain $Y - Y_0 = \mu(\omega\mathfrak{O})$. Thus $A - A_0 = q^2$.

Now observe that for $x \in L$ the sum $\sum_{s\in\mathfrak{g}/\mathfrak{p}} \chi(s\varphi[x]/\pi)$ is q or 0 according as $x \in X$ or $x \notin X$. Therefore qA coincides with the sum $I(\varphi)$ of Lemma 3.3. Combining the value of $I(\varphi)$ given there with the above result on $A - A_0$, we obtain A_0 as stated in our lemma. □

3.9. PROPOSITION. *Put* $m = [n/2]$ *and* $k = n(n-1)/2$. *Then the number* $2^{-1}q^{-\nu k}[C^\varphi : D_\nu^\varphi]$ *for* $0 < \nu \in \mathbf{Z}$ *is a constant given as follows:*

$$(1-q^{-m})\prod_{i=1}^{m-1}(1-q^{-2i}) \qquad \text{if } t = 0,$$

$$\prod_{i=1}^{m}(1-q^{-2i}) \qquad \text{if } t = 1,$$

$$(1+q^{-m})\prod_{i=1}^{m-1}(1-q^{-2i}) \qquad \text{if } t = 2,\ \mathfrak{d} = \mathfrak{r},\ \text{and } \widetilde{M} = M,$$

$$2(q+1)(1+q^{1-m})^{-1}\prod_{i=1}^{m-1}(1-q^{-2i}) \qquad \text{if } t = 2,\ \mathfrak{d} = \mathfrak{r},\ \text{and } \widetilde{M} \neq M,$$

$$2\prod_{i=1}^{m-1}(1-q^{-2i}) \qquad \text{if } t = 2 \text{ and } \mathfrak{d} \neq \mathfrak{r},$$

$$2(q+1)\prod_{i=1}^{m-1}(1-q^{-2i}) \qquad \text{if } t = 3,$$

$$2(q+1)(1-q^{1-m})^{-1}\prod_{i=1}^{m-1}(1-q^{-2i}) \qquad \text{if } t = 4.$$

Proof. Here we treat only the case $\nu = 1$. Our assertion in the general case can be obtained by combining this special case with the equality $[D_\nu^\varphi : D_{\nu+1}^\varphi] = q^k$, which is proven in Lemma 8.5. Now, if $r = 0$, the above numbers are exactly those given in Lemma 3.7. Therefore we assume $r > 0$. Let $\mu(x, w)$ be an element of the set of (3.8.1). As shown in the proof of [S5, Lemma 5.14] (or by [S5, Lemma 4.9(1)]), there exists an element γ of C such that $x = e_1\gamma$ and $w = f_1\gamma$. Put $C = C_0^\varphi$ and

$$C' = \{\alpha \in C \mid e_1\alpha - e_1 \in \mathfrak{p}L,\ f_1\alpha - f_1 \in \mathfrak{p}L\}.$$

Then $C_1 \subset C' \subset C$, and Lemma 3.8 shows that $[C : C'] = q^{n-2}A_0$. Now take a weak Witt decomposition $V = Fe_1 + Ff_1 + W$ with $W = \sum_{i>1}(Fe_i + Ff_i) + T$. Then we can put $(V, \varphi) = (H_1, 2^{-1}\eta_1) \oplus (W, \sigma)$, where σ is the restriction of φ to W. Write $\alpha \in C'$ in the form

(3.9.1)
$$\alpha = \begin{bmatrix} a & b & c \\ g & e & f \\ h & \ell & d \end{bmatrix}$$

with a, e, and d of size $1, n-2, 1$ with respect to the decomposition $V = Fe_1 \oplus W \oplus Ff_1$. Then we see that the blocks b, c, h, and ℓ have coefficients in $\mathfrak{p}$. In [S5, Lemma 17.2] we proved that under such a condition on b, c, h, and ℓ we have $\sigma^{-1}f \prec 2\mathfrak{p}$ and $\sigma^{-1}g \prec 2\mathfrak{p}$. Since $a-1 \in \mathfrak{p}, d-1 \in \mathfrak{p}$, and $2\sigma \prec \mathfrak{g}$, this means that $\alpha \in C_1$ if and only if $e-1 \prec \mathfrak{p}$. The same lemma guarantees an element $k \in C^\sigma$ such that $\sigma^{-1}(k^{-1}e-1) \prec 2\mathfrak{p}$. Then

$$\sigma^{-1}(e-1) - \sigma^{-1}(k-1) = \sigma^{-1}(e-k) = \widehat{k}\sigma^{-1}\big(k^{-1}e-1\big) \prec 2\mathfrak{p},$$

and so $k \in D_1^\sigma$ if and only if $\sigma^{-1}(e-1) \prec 2\mathfrak{p}$. Therefore [S5, Lemma 17.3(1)] shows that $\alpha \in D_1$ if and only if $k \in D_1^\sigma$. Thus, assigning k to α, we obtain an isomorphism of C'/D_1 onto C^σ/D_1^σ. In this way we find that

(3.9.2)
$$[C : D_1] = q^{n-2}A_0\big[C^\sigma : D_1^\sigma\big].$$

Let $\lambda(n)$ denote the number $2^{-1}q^{-k}[C^\varphi : D_1^\varphi]$ that is our question. Then from (3.9.2) we obtain $\lambda(n) = q^{1-n}A_0\lambda(n-2)$. Employing the value of A_0 given in Lemma 3.8, by induction we obtain $\lambda(n)$ as stated in our proposition, since the case $n = t$ is already proved as remarked at the beginning. □

4. Orthogonal groups over R and symmetric spaces

4.1. In this section we take $\mathbf{R}$ to be our basic field and consider (V, φ) with $V = \mathbf{R}_n^1$. Thus our group G^φ is given with φ of a special matrix form as follows:

(4.1.1)
$$G^\varphi = G(\varphi) = \big\{\alpha \in GL_n(\mathbf{R}) \big| \alpha\varphi \cdot {}^t\alpha = \varphi\big\},$$

(4.1.2)
$$\varphi = \begin{bmatrix} 0 & 0 & -1_r \\ 0 & \theta & 0 \\ -1_r & 0 & 0 \end{bmatrix}, \qquad 0 < \theta = {}^t\theta \in GL_t(\mathbf{R}).$$

We put $q = r + t$. For the moment we assume that $r > 0$, though later we consider the case $r = 0$. We ignore θ if $n = 2r$. Put

(4.1.3)
$$S_+^m = \big\{A \in \mathbf{R}_m^m \big| {}^tA = A > 0\big\},$$

(4.1.4)
$$\mathfrak{Z} = \mathfrak{Z}(r, \theta) = \left\{ \begin{bmatrix} x \\ y \end{bmatrix} \in \mathbf{R}_r^q \,\middle|\, x \in \mathbf{R}_r^r, y \in \mathbf{R}_r^t, {}^tx + x > {}^ty\theta^{-1}y \right\},$$

(4.1.5)
$$\mathfrak{Y} = \big\{Y \in GL_n(\mathbf{R}) \big| {}^tY\varphi^{-1}Y = \mathrm{diag}[A, -B] \text{ with } A \in S_+^q, B \in S_+^r\big\}.$$

For $z = \left[\begin{smallmatrix} x \\ y \end{smallmatrix}\right] \in \mathfrak{X}$ and $w = \left[\begin{smallmatrix} u \\ v \end{smallmatrix}\right] \in \mathfrak{X}$ we put

$$B(z) = \begin{bmatrix} {}^t x & {}^t y & x \\ 0 & \theta & y \\ -1_r & 0 & 1_r \end{bmatrix}, \tag{4.1.6}$$

$$\xi'(w, z) = \begin{bmatrix} u + {}^t x & {}^t y \\ v & \theta \end{bmatrix}, \qquad \xi(w, z) = x + {}^t u - {}^t v \theta^{-1} y, \tag{4.1.7}$$

$$\xi(z) = \xi(z, z), \qquad \xi'(z) = \xi'(z, z), \tag{4.1.8}$$

$$\delta(w, z) = \det\left[2^{-1}\xi(w, z)\right], \qquad \delta(z) = \delta(z, z). \tag{4.1.9}$$

Then we can easily verify that

$$ {}^t B(w)\varphi^{-1} B(z) = \begin{bmatrix} \xi'(w, z) & z - w \\ {}^t w - {}^t z & -\xi(w, z) \end{bmatrix}, \tag{4.1.10}$$

$$\det[B(z)] = 2^r \det(\theta)\delta(z), \tag{4.1.11}$$

$$\det\left[\xi'(z)\right] = \det(\theta)\det[\xi(z)]. \tag{4.1.12}$$

4.2. We can repeat what we did in [S5, Section 6], replacing $\mathfrak{Z}$ and $\mathfrak{Y}$ there by the present $\mathfrak{X}$ and $\mathcal{Y}$. Namely, we first show that the map $(z, \lambda, \mu) \mapsto B(z)\mathrm{diag}[\lambda, \mu]$ gives a bijection of $\mathfrak{X} \times GL_q(\mathbf{R}) \times GL_r(\mathbf{R})$ onto $\mathcal{Y}$ in the same manner as in [S5, Lemma 6.2]. Then for $\alpha \in G^\varphi$ and $z \in \mathfrak{X}$ we can define $\alpha z \in \mathfrak{X}, \kappa_\alpha(z) \in GL_q(\mathbf{R})$, and $\mu_\alpha(z) \in GL_r(\mathbf{R})$ by the relation

$$\alpha B(z) = B(\alpha z)\mathrm{diag}[\kappa_\alpha(z), \mu_\alpha(z)]. \tag{4.2.1}$$

We then define a scalar factor of automorphy j by

$$j(\alpha, z) = j_\alpha(z) = \det\left[\mu_\alpha(z)\right]. \tag{4.2.2}$$

From (4.2.1) we can easily derive that

$${}^t\kappa_\alpha(w)\xi'(\alpha w, \alpha z)\kappa_\alpha(z) = \xi'(w, z), \tag{4.2.3}$$

$${}^t\mu_\alpha(w)\xi(\alpha w, \alpha z)\mu_\alpha(z) = \xi(w, z), \tag{4.2.4}$$

$${}^t\kappa_\alpha(w)(\alpha z - \alpha w)\mu_\alpha(z) = z - w, \tag{4.2.5}$$

$$\delta(\alpha w, \alpha z) j_\alpha(w) j_\alpha(z) = \delta(w, z), \qquad \delta(\alpha w) = j_\alpha(w)^{-2}\delta(w), \tag{4.2.6}$$

$$\det\left[\kappa_\alpha(w)\right] = \det(\alpha) j_\alpha(w). \tag{4.2.7}$$

Then we see that for $z = (z_{ij}) \in \mathfrak{X}$ the form

$$\mathbf{d}z = \delta(z)^{-n/2} \bigwedge_{i=1}^{q} \bigwedge_{j=1}^{r} dz_{ij} \tag{4.2.8}$$

gives a G^φ-invariant measure on $\mathfrak{X}$. We note that

(4.2.9) $$\{\pm 1\} = \{\alpha \in G^\varphi \,|\, \alpha z = z \text{ for every } z \in \mathfrak{X}\}.$$

We define the *origin* **1** of $\mathfrak{X}$ and a subgroup C of G^φ by

(4.2.10) $$C = \{\gamma \in G^\varphi \,|\, \gamma(\mathbf{1}) = \mathbf{1}\}, \qquad \mathbf{1} = \begin{bmatrix} 1_r \\ 0 \end{bmatrix}.$$

Then from (4.2.3) and (4.2.4) we see that $\gamma \mapsto (\kappa_\alpha(\mathbf{1}), \mu_\alpha(\mathbf{1}))$ gives an isomorphism of C onto $G(\mathrm{diag}[1_r, 2\theta^{-1}]) \times G(1_r)$. Consequently $j_\gamma(\mathbf{1}) = \pm 1$ for every $\gamma \in C$. Let us now prove

(4.2.11) $$\delta(w, z) > 0 \quad \text{for every } (w, z) \in \mathfrak{X} \times \mathfrak{X}.$$

Since $\mathfrak{X}$ is connected, it is sufficient to show that $\delta(w, z) \neq 0$ on $\mathfrak{X} \times \mathfrak{X}$. We may also assume that $w = \mathbf{1}$ in view of (4.2.6). Put $p = \xi(\mathbf{1}, z)$ for $z = \left[\begin{smallmatrix} x \\ y \end{smallmatrix}\right]$. Then $p + {}^t p = 2 \cdot 1_r + x + {}^t x > 0$, so that $\det(p) \neq 0$ as expected.

4.3. We now consider the case $t = 0$, in which φ becomes $-\eta_r$ with η_r of (1.6.1). In this special case, we denote the space $\mathfrak{X}$ by $\mathcal{H}_r$. Thus

(4.3.1) $$\mathcal{H}_r = \{w \in \mathbf{R}_r^r \,|\, {}^t w + w > 0\}.$$

Then we can easily verify that

(4.3.2) $$B(w) = \begin{bmatrix} {}^t w & w \\ -1_r & 1_r \end{bmatrix}, \qquad \xi'(w) = \xi(w) = w + {}^t w,$$

(4.3.4) $$\kappa_\alpha(w) = d_\alpha - c_\alpha \cdot {}^t w, \qquad \mu_\alpha(w) = d_\alpha + c_\alpha w,$$

(4.3.5) $$\det(w) > 0 \quad \text{for every } w \in \mathcal{H}_r.$$

4.4. Our space $\mathfrak{X}$ is diffeomorphic to a "ball" $\mathfrak{B}$ on which various objects have simpler expressions. We first put

(4.4.1) $$\xi(w, z) = 1_r - {}^t w z, \qquad \xi'(w, z) = 1_q - w \cdot {}^t z \quad (w, z \in \mathbf{R}_r^q),$$

(4.4.2) $$\delta(w, z) = \det[\xi(w, z)], \qquad \delta(z) = \delta(z, z),$$

(4.4.3) $$\mathfrak{B} = \{z \in \mathbf{R}_r^q \,|\, \xi(z, z) > 0\}.$$

Taking an element τ of $GL_t(\mathbf{R})$ such that $\tau \cdot {}^t\tau = 2\theta$, we can define a map $\mathfrak{t} : \mathfrak{B} \to \mathfrak{X}$ by

(4.4.4) $$\mathfrak{t}\left(\begin{bmatrix} x \\ y \end{bmatrix}\right) = \begin{bmatrix} (1+x)(1-x)^{-1} \\ \tau y (1-x)^{-1} \end{bmatrix} \quad (x \in \mathbf{R}_r^r,\ y \in \mathbf{R}_r^t).$$

Then t is a diffeomorphism of $\mathfrak{B}$ onto $\mathfrak{Z}$. Moreover, if we put $j(z) = \det(1-x)$ for $z = \begin{bmatrix} x \\ y \end{bmatrix}$, then

$$j(w)j(z)\delta(\mathrm{t}w, \mathrm{t}z) = \delta(w, z) \quad \text{for every } w, z \in \mathfrak{B}, \tag{4.4.5}$$

and t sends the measure of (4.2.8) on $\mathfrak{Z}$ back to $2^{rn/2}\det(\theta)^{r/2}$ times the measure on $\mathfrak{B}$ defined by the same formula. These can be proved in the same manner as in [S5, Lemma A2.3].

4.5. Lemma. *For every $w' \in \mathfrak{Z}$ we have*

$$\begin{aligned} &\delta(w')^s \int_{\mathfrak{Z}} \delta(w)^s \delta(w, w')^{-2s} \mathbf{d}w \\ &\quad = 2^{rn/2}\det(\theta)^{r/2}\pi^{qr/2}\Gamma_r(s-(q-1)/2)\Gamma_r(s+(1/2))^{-1} \end{aligned} \tag{4.5.1}$$

if $\mathrm{Re}(s) > (n-2)/2$, *where* $\mathbf{d}w$ *is defined by (4.2.8) and*

$$\Gamma_m(s) = \pi^{m(m-1)/4}\prod_{k=0}^{m-1}\Gamma(s-(k/2)). \tag{4.5.2}$$

Proof. To prove this, we need an integral formula proved in [S5, Lemma A2.7]:

$$\begin{aligned} \int_{\mathfrak{B}} \det(1_q - y\cdot{}^t y)^{s-(n+1)/2} dy &= \int_{\mathbf{R}_r^q} \det(1_q + x\cdot{}^t x)^{-s} dx \\ &= \pi^{qr/2}\Gamma_q(s-r/2)/\Gamma_q(s) \\ &= \pi^{qr/2}\Gamma_r(s-q/2)/\Gamma_r(s) \end{aligned} \tag{4.5.3}$$

if $\mathrm{Re}(s) > (n-1)/2$, where dx and dy are the standard measure on $\mathbf{R}_r^q$. Now take $\gamma \in G^\varphi$ so that $\gamma\mathbf{1} = w'$. Substituting γw for w, we see that the left-hand side of (4.5.1) equals $\int_{\mathfrak{Z}} \delta(w)^s \delta(w, \mathbf{1})^{-2s}\mathbf{d}w$. Putting $w = \mathrm{t}(z)$ with $z \in \mathfrak{B}$ and employing (4.4.5), we find that the last integral over $\mathfrak{Z}$ equals $2^{rn/2}\det(\theta)^{r/2}$ times the first integral of (4.5.3) with $s+1/2$ in place of s. This proves our lemma. □

4.6. To define embeddings $G^\varphi \times G^\varphi \to G(\eta_n)$ and $\mathfrak{Z} \times \mathfrak{Z} \to \mathcal{H}_n$, we first put

$$\omega = \mathrm{diag}[\varphi, -\varphi]. \tag{4.6.1}$$

For two points $z = \begin{bmatrix} x \\ y \end{bmatrix}$ and $w = \begin{bmatrix} u \\ v \end{bmatrix}$ of $\mathfrak{Z}$ we put

$$\iota(z, w) = \begin{bmatrix} x & 0 & 0 \\ y & 2^{-1}\theta & 0 \\ -{}^t v\theta^{-1} y & -{}^t v & {}^t u \end{bmatrix}. \tag{4.6.2}$$

We now need a matrix R of size $2n$ given by

$$(4.6.3)\qquad R=\begin{bmatrix} 1_r & 0 & 0 & 0 & 0 & 0\\ 0 & 2^{-1}1_t & 0 & 0 & 2^{-1}1_t & 0\\ 0 & 0 & 0 & -1_r & 0 & 0\\ 0 & 0 & 1_r & 0 & 0 & 0\\ 0 & -\theta^{-1} & 0 & 0 & \theta^{-1} & 0\\ 0 & 0 & 0 & 0 & 0 & 1_r \end{bmatrix}.$$

We can easily verify that

$$(4.6.4)\qquad R\omega\cdot{}^tR=-\eta_n.$$

For $(\beta,\gamma)\in G^\varphi\times G^\varphi$ we put

$$(4.6.5)\qquad [\beta,\gamma]_R=R\cdot\mathrm{diag}[\beta,\gamma]R^{-1}.$$

From (4.6.4) we see that $[\beta,\gamma]_R\in G^\eta$. Now, by the same technique as in [S5, Proposition 6.11], we can show that $\iota(z,w)\in\mathcal{H}_n$, and

$$(4.6.6)\qquad \iota(\beta z,\gamma w)=[\beta,\gamma]_R\iota(z,w)\quad \text{for every } (\beta,\gamma)\in G^\varphi\times G^\varphi,$$

$$(4.6.7)\qquad j\big([\beta,\gamma]_R,\iota(z,w)\big)=\det(\gamma)j_\gamma(w)j_\beta(z),$$

$$(4.6.8)\qquad \delta\big(\iota(z,w)\big)=\det\big(2^{-1}\theta\big)\delta(z)\delta(w).$$

4.7. If φ is positive definite, we understand that $\mathfrak{Z}$ consists of a single point written as **1**, and we let G^φ act on it trivially. We also put

$$(4.7.1)\qquad B(\mathbf{1})=\xi'(\mathbf{1})=\varphi,\quad \kappa_\alpha(\mathbf{1})=\widehat{\alpha},\quad R=\begin{bmatrix} 2^{-1}1_n & 2^{-1}1_n\\ -\varphi^{-1} & \varphi^{-1}\end{bmatrix},$$

$$(4.7.2)\qquad j_\alpha(\mathbf{1})=\delta(\mathbf{1})=\delta(\mathbf{1},\mathbf{1})=1,\quad \iota(\mathbf{1},\mathbf{1})=2^{-1}\varphi,$$

and ignore ξ and μ_α. The last three formulas of §4.6 are valid in this special case.

5. Main theorems

5.1. We now return to (V,φ) over a number field F and define K by (2.1.2). We naturally assume that $n>1$. For each $v\in\mathbf{v}$ we take a Witt decomposition

$$(5.1.1)\qquad (V_v,\varphi_v)=\big(T_v,\theta'_v\big)\oplus\big(H_{r_v},\eta_{r_v}\big)$$

with anisotropic θ'_v and a nonnegative integer r_v. Here $H_r=(F_v)^1_{2r}$ and η_r is as in (1.6.1). Put $t_v=\dim(T_v)$. Then $n=2r_v+t_v$. It is well known that $t_v\le 4$ if $v\in\mathbf{h}$. We take a $\mathfrak{g}$-maximal lattice L in V and put

$$(5.1.2)\qquad \widetilde{L}=\{x\in V\,|\,2\varphi(x,L)\subset\mathfrak{g}\}.$$

5.2. PROPOSITION. (1) *$t_v > 2$ only for finitely many $v \in \mathbf{h}$.*

(2) *If n is even, $v \in \mathbf{h}$, and $t_v < 4$, then $t_v = 2$ if and only if v does not split in K.*

(3) *If $v \in \mathbf{h}$, $t_v = 2$, and v is unramified in K, then the element c of §3.2 (which depends on v) does not belong to $\mathfrak{g}_v^\times$ only for finitely many v's.*

Proof. If $t_v = 4$ or $t_v = 0$, we see that $\mathbf{d}(\varphi_v)$ is represented by $(-1)^{n/2}$, so that v splits in K. Therefore we obtain (2), and $K \otimes_F F_v$ is the field K of §3.2 if $t_v = 2$. Now $[\widetilde{L}_v : L_v]$ is given by (3.2.1). Since $\widetilde{L}_v = L_v$ for almost all v, we obtain (1) and (3). □

We note that $\widetilde{L}_v \neq L_v$ can happen exactly in the following cases:

$$\begin{aligned} &t_v = 1 \ \text{ and } v|2,\\ &t_v = 2, v \text{ is unramified in } K, \text{ and } c \text{ of §3.2 is}\\ &\qquad\qquad \text{a prime element of } K_v,\\ &t_v = 2 \ \text{ and } v \text{ is ramified in } K,\\ &t_v \geq 3. \end{aligned} \tag{5.2.1}$$

5.3. We now assume that our basic number field F is totally real. We fix an F-basis of V and represent φ and the elements of $GL(V)$ by matrices of F_n^n. For $\alpha \in F_n^n$ and $v \in \mathbf{v}$ we have a well-defined element α_v of $(F_v)_n^n$.

Let us now take $v \in \mathbf{a}$. Since $\theta'_v > 0$ or $\theta'_v < 0$, we see that φ_v has signature $(r_v + t_v, r_v)$ or $(r_v, r_v + t_v)$. We now take and fix an element κ of F such that

$$\kappa_v \varphi_v \text{ has signature } (r_v + t_v, r_v) \text{ for every } v \in \mathbf{a}. \tag{5.3.1}$$

For each $v \in \mathbf{a}$ we fix an element $\sigma_v \in GL_n(\mathbf{R})$ so that

$$\kappa_v \sigma_v \varphi_v \cdot {}^t\sigma_v = \varphi'_v \quad \text{with } \varphi'_v = \begin{bmatrix} 0 & 0 & -1_{r_v} \\ 0 & \theta_v & 0 \\ -1_{r_v} & 0 & 0 \end{bmatrix}, \tag{5.3.2}$$

with $0 < \theta_v = {}^t\theta_v \in GL_{t_v}(\mathbf{R})$. We then put

$$\mathfrak{Z}^\varphi = \prod_{v \in \mathbf{a}} \mathfrak{Z}_v^\varphi, \qquad \mathfrak{Z}_v^\varphi = \mathfrak{Z}(r_v, \theta_v) \tag{5.3.3}$$

with $\mathfrak{Z}(r, \theta)$ of (4.1.4) with the convention of §4.7 if $r_v = 0$. If we put

$$\mathbf{b} = \{v \in \mathbf{a} | r_v > 0\}, \tag{5.3.4}$$

then $\mathfrak{Z}^\varphi$ is essentially $\prod_{v \in \mathbf{b}} \mathfrak{Z}_v^\varphi$. Clearly $\sigma_v G(\varphi_v) \sigma_v^{-1} = G(\varphi'_v)$. Hereafter we identify the v-factor G_v^φ of $G_{\mathbf{A}}^\varphi$ with $G(\varphi'_v)$ by means of the isomorphism $\xi \mapsto \sigma_v \xi \sigma_v^{-1}$. In Section 4 we defined the action of $G(\varphi'_v)$ on $\mathfrak{Z}_v^\varphi$, and we also defined factors of automorphy. Therefore we can let $G_{\mathbf{a}}^\varphi$ act naturally on $\mathfrak{Z}^\varphi$ via the isomorphism $\xi \mapsto (\sigma_v \xi_v \sigma_v^{-1})_{v \in \mathbf{a}}$. More generally, it is convenient to define the action of $G_{\mathbf{A}}^\varphi$ on $\mathfrak{Z}^\varphi$ by ignoring nonarchimedean components. Namely, for $\alpha \in G_{\mathbf{A}}^\varphi$ and $z = (z_v)_{v \in \mathbf{a}} \in \mathfrak{Z}^\varphi$

we define αz, $j(\alpha, z)$, and $\delta(z)$ by

$$\alpha z = \alpha(z) = \left((\sigma_v \alpha_v \sigma_v^{-1})(z_v)\right)_{v \in \mathbf{a}} \quad (\in \mathfrak{Z}^\varphi), \tag{5.3.5}$$

$$j(\alpha, z) = j_\alpha(z) = \left(j(\sigma_v \alpha_v \sigma_v^{-1}, z_v)\right)_{v \in \mathbf{a}} \quad (\in \mathbf{R}^\mathbf{a}), \tag{5.3.6}$$

$$\delta(z) = \left(\delta(z_v)\right)_{v \in \mathbf{a}} \quad (\in \mathbf{R}^\mathbf{a}). \tag{5.3.7}$$

Here the v-components $\xi(z_v)$, $j(\xi, z_v)$, and $\delta(z_v)$ are the symbols defined in Section 4. According to the convention of (4.7.2) we have $j(\alpha_v, z_v) = \delta(z_v) = 1$ if $r_v = 0$. In view of (4.2.9), an element of G^φ gives the identity map on nontrivial $\mathfrak{Z}^\varphi$ if and only if it belongs to $\{\pm 1\}$. We define a point $\mathbf{1} = \mathbf{1}^\varphi$ of $\mathfrak{Z}^\varphi$ and a subgroup $C_\mathbf{a}^\varphi$ of $G_\mathbf{a}^\varphi$ by

$$\mathbf{1} = \mathbf{1}^\varphi = (\mathbf{1}_v)_{v \in \mathbf{a}}, \qquad \mathbf{1}_v = \begin{bmatrix} 1_{r_v} \\ 0 \end{bmatrix} \in \mathfrak{Z}_v^\varphi, \tag{5.3.8}$$

$$C_\mathbf{a}^\varphi = \left\{\alpha \in G_\mathbf{a}^\varphi \,\middle|\, \alpha(\mathbf{1}) = \mathbf{1}\right\}, \tag{5.3.9}$$

where $\mathbf{1}_v$ for $r_v = 0$ is the point that constitutes $\mathfrak{Z}_v^\varphi$ as defined in §4.7. Thus $\mathfrak{Z}^\varphi$ consists of the single point $\mathbf{1}^\varphi$ if $G_\mathbf{a}^\varphi$ is compact.

5.4. To simplify our notation, for $x \in F_\mathbf{A}^\times$ or $x \in \mathbf{C}^\mathbf{a}$, $a \in \mathbf{R}^\mathbf{a}$, and $c \in \mathbf{C}^\mathbf{a}$, we put

$$\|x\| = \sum_{v \in \mathbf{a}} x_v, \qquad x^a = \prod_{v \in \mathbf{a}} x_v^{a_v}, \qquad |x|^c = \prod_{v \in \mathbf{a}} |x_v|^{c_v}, \tag{5.4.1}$$

where $x_v^{a_v}$ should be understood according to the context.

The identity element of the ring $\mathbf{C}^\mathbf{a}$, which will appear as an exponent for the most part, is denoted by u. For example, $\|u\| = [F : \mathbf{Q}]$ and

$$x^u = \prod_{v \in \mathbf{a}} x_v \quad (x \in F_\mathbf{A}^\times), \tag{5.4.2}$$

so that $c^u = N_{F/\mathbf{Q}}(c)$ for $c \in F^\times$. For $\alpha \in G_\mathbf{A}^\varphi$ and $k \in \mathbf{Z}^\mathbf{a}$, we define a factor of automorphy j_α^k by

$$j_\alpha^k(z) = j^k(\alpha, z) = j_\alpha(z)^k \quad (\alpha \in G_\mathbf{A}^\varphi, z \in \mathfrak{Z}^\varphi). \tag{5.4.3}$$

The factor $j_\alpha(z)^k$ depends only on $(k_v)_{v \in \mathbf{b}}$, but for some natural and technical reasons we take k in $\mathbf{Z}^\mathbf{a}$, not in $\mathbf{Z}^\mathbf{b}$. Then, for a function $f : \mathfrak{Z}^\varphi \to \mathbf{C}$ we define $f\|_k\alpha : \mathfrak{Z}^\varphi \to \mathbf{C}$ by

$$(f\|_k\alpha)(z) = j_\alpha^{-k}(z) f(\alpha z) \quad (z \in \mathfrak{Z}^\varphi). \tag{5.4.4}$$

5.5. Let G denote either G^φ or G_+^φ; let D be an open subgroup of $G_\mathbf{A}$ containing $G_\mathbf{a}$ and such that $D \cap G_\mathbf{h}$ is compact. For each $x \in G_\mathbf{A}$ put $\Gamma^x = G \cap xDx^{-1}$. Take

$\mathfrak{B} \subset G_{\mathbf{A}}$ so that $G_{\mathbf{A}} = \bigsqcup_{a \in \mathfrak{B}} GaD$. We then define real numbers $\mathrm{m}(G, D)$ and $\nu(\Gamma^a)$ (cf. [S5, §24.1]) by

$$\mathrm{m}(G, D) = \sum_{a \in \mathfrak{B}} \nu(\Gamma^a), \tag{5.5.1}$$

$$\nu(\Gamma^a) = \begin{cases} [\Gamma^a : 1]^{-1} & \text{if } G_{\mathbf{a}} \text{ is compact,} \\ [\Gamma^a \cap \{\pm 1\} : 1]^{-1} \mathrm{vol}(\Gamma^a \backslash \mathfrak{Z}^\varphi) & \text{otherwise,} \end{cases} \tag{5.5.2}$$

where $\mathrm{vol}(\Gamma^a \backslash \mathfrak{Z}^\varphi)$ is defined with respect to the measure given by the form

$$\mathbf{d}z = \prod_{v \in \mathbf{b}} \mathbf{d}z_v, \qquad \mathbf{d}z_v = \delta(z_v)^{-n/2} \prod_{h=1}^{r_v+t_v} \prod_{k=1}^{r_v} d(z_v)_{hk}. \tag{5.5.3}$$

5.6. Lemma. *Let $D = \{\alpha \in G^\varphi_{\mathbf{A}} \mid H\alpha = H\}$ and $D' = (G^\varphi_+)_{\mathbf{A}} \cap D$ with a $\mathfrak{g}$-lattice H in V. Then the following assertions hold:*

(1) $\mathrm{m}(G^\varphi_+, D') = 2\mathrm{m}(G^\varphi, D)$;

(2) $G^\varphi_{\mathbf{A}} = \bigsqcup_{a \in A} G^\varphi a D$ *with a finite subset A of* $(G^\varphi_+)_{\mathbf{h}}$;

(3) *put* $\widetilde{\Gamma}^x = G^\varphi \cap xDx^{-1}$ *and* $\Gamma^x = G^\varphi_+ \cap xDx^{-1}$ *for* $x \in G^\varphi_{\mathbf{A}}$; *then*

$$2/[\widetilde{\Gamma}^x : \Gamma^x] = 2/\#(\det(\widetilde{\Gamma}^x)) = \#(G^\varphi_+ \backslash G^\varphi x D/D)$$

for every $x \in G^\varphi_{\mathbf{A}}$.

Proof. For simplicity put $G^+ = G^\varphi_+$. Since $-1 \in \det(D \cap G^\varphi_v)$ for every $v \in \mathbf{h}$ by [S5, Lemma 5.11], we see that $G^\varphi_{\mathbf{A}} = G^+_{\mathbf{h}} D$, from which we obtain (2). Next, since $[G^\varphi : G^+] = 2$, we have $\#(G^+ \backslash G^\varphi x D/D) \leq 2$ and $[\widetilde{\Gamma}^x : \Gamma^x] \leq 2$. Now

$$G^\varphi x D = G^+ x D \Longleftrightarrow G^\varphi x D x^{-1} = G^+ x D x^{-1} \Longleftrightarrow G^\varphi \subset G^+ x D x^{-1}$$
$$\Longleftrightarrow G^\varphi \subset G^+ \widetilde{\Gamma}^x \Longleftrightarrow -1 \in \det(\widetilde{\Gamma}^x) \Longleftrightarrow [\widetilde{\Gamma}^x : \Gamma^x] = 2.$$

This proves (3). Now with A as in (2), take B_a so that $G^\varphi a D = \bigsqcup_{b \in B_a} G^+ b D$. We can take B_a in $G^+_{\mathbf{A}}$ for the reason explained above. Then $G^+_{\mathbf{A}} = \bigsqcup_{a \in A} \bigsqcup_{b \in B_a} G^+ b D'$ and so

$$\mathrm{m}(G^+, D') = \sum_{a \in A} \sum_{b \in B_a} \nu(\Gamma^b) = \sum_{a \in A} \sum_{b \in B_a} [\widetilde{\Gamma}^b : \Gamma^b] \nu(\widetilde{\Gamma}^b)$$

by [S5, Lemma 24.2(2)]. By (3) the sum over b equals $2\nu(\widetilde{\Gamma}^a)$. This proves (1). □

5.7. We now impose the condition

(5.7.1) $\mathbf{d}(\varphi)\mathfrak{g}$ is the square of a fractional ideal if n is odd,

which is compatible with our local assumption (3.1.1). This is not a strong assumption, since we can always find $c \in F^\times$ such that $c\varphi$ satisfies (5.7.1) and $G^\varphi = G^{c\varphi}$. We fix a $\mathfrak{g}$-maximal $\mathfrak{g}$-lattice L in V and an integral ideal $\mathfrak{c}$, and we define $\widetilde{L}$ by (5.1.2). We fix an F-basis $\{e_i\}$ of V as we did in §5.3, and we put $L' = \sum_{i=1}^{n} \mathfrak{g}e_i$. We then take and fix an element $\sigma \in GL_n(F)_{\mathbf{h}}$ so that $L'\sigma = L$, and we put

$$\varphi' = \sigma\varphi\cdot {}^t\sigma, \tag{5.7.2}$$

$$C^{\varphi} = \{\alpha \in G^{\varphi}_{\mathbf{A}} \,|\, L\alpha = L\}, \tag{5.7.3}$$

$$D^{\varphi} = \{\gamma \in C^{\varphi} \,|\, \widetilde{L}_v(\gamma_v - 1) \subset \mathfrak{c}_v L_v \text{ for every } v|\mathfrak{c}\}, \tag{5.7.4}$$

$$C^{+} = (G^{\varphi}_{+})_{\mathbf{A}} \cap C^{\varphi}, \qquad D^{+} = (G^{\varphi}_{+})_{\mathbf{A}} \cap D^{\varphi}. \tag{5.7.5}$$

(Hereafter until the end of this section we do not need σ_v and φ'_v of (5.3.2) for $v \in \mathbf{a}$.) For a subgroup X of $GL_n(F)_{\mathbf{A}}$, we denote by X_v its projection to $GL_n(F_v)$. In fact, in all cases in which this notation is used we have $X_v = X \cap GL_n(F_v)$. We observe that $\widetilde{L} = L'(2\varphi')^{-1}\sigma$, so that from (5.7.4) we obtain

$$D^{\varphi}_v = \left\{\gamma \in C^{\varphi}_v \,\middle|\, (2\varphi'_v)^{-1}\left(\sigma\gamma\sigma^{-1} - 1\right)_v \prec \mathfrak{c}_v\right\} \quad \text{if } v|\mathfrak{c}. \tag{5.7.6}$$

Since L is $\mathfrak{g}$-maximal, [S5, Lemma 8.10] shows that L'_v is $\mathfrak{g}_v$-maximal with respect to φ'_v for every $v \in \mathbf{h}$. Taking L'_v to be L of [S5, Lemma 5.6] and changing σ_v suitably we can put

$$\left(\sigma\varphi\cdot{}^t\sigma\right)_v = \varphi'_v = \begin{bmatrix} 0 & 0 & 2^{-1}1_{r_v} \\ 0 & \theta_v & 0 \\ 2^{-1}1_{r_v} & 0 & 0 \end{bmatrix}, \qquad \theta_v = {}^t\theta_v \in \left(2^{-1}\mathfrak{g}_v\right)^{t_v}_{t_v} \tag{5.7.7}$$

with respect to a suitable $\mathfrak{g}_v$-basis of L'_v.

Given any integral ideal $\mathfrak{c}$ and a Hecke character χ of F, we put

$$L_{\mathfrak{c}}(s, \chi) = \prod_{\mathfrak{p}\nmid\mathfrak{c}} \left(1 - \chi^*(\mathfrak{p})N(\mathfrak{p})^{-s}\right)^{-1}, \tag{5.7.8}$$

$$\zeta_{F,\mathfrak{c}}(s) = \prod_{\mathfrak{p}\nmid\mathfrak{c}} \left(1 - N(\mathfrak{p})^{-s}\right)^{-1}, \tag{5.7.9}$$

where $\mathfrak{p}$ runs over all the prime ideals in F prime to $\mathfrak{c}$, and χ^* denotes the ideal character attached to χ.

5.8. THEOREM. *Suppose that φ is anisotropic and (5.7.1) is satisfied. Let $\mathfrak{e}$ be the product of all the prime ideals for which $\widetilde{L}_v \neq L_v$ (see (5.2.1)). Define an open* *subgroup* $D^+ = D(\mathfrak{c})$ *of* $\left(G_+^\varphi\right)_{\mathbf{A}}$ *by (5.7.5) with an integral ideal* $\mathfrak{c}$ *in* F. *Put*

$$\mathfrak{m}_n(\mathfrak{c}) = 2N(\mathfrak{c})^{n\mu} D_F^{[\mu^2]} \prod_{k=1}^{[\mu]} \left\{ D_F^{1/2}\left[(2k-1)!(2\pi)^{-2k}\right]^d \zeta_{F,\mathfrak{c}}(2k) \right\}$$

$$\cdot \begin{cases} 2^{-\mu d} & \text{if } n \text{ is odd,} \\ N(\mathfrak{d})^{1/2} D_F^{1/2}\left[(m-1)!(2\pi)^{-m}\right]^d L_{\mathfrak{c}}(m, \psi) & \text{if } n = 2m \in 2\mathbf{Z}, \end{cases}$$

$$b_\varphi = \prod_{v\in\mathbf{a}} 2^{r_v n/2} \pi^{(n-r_v)r_v/2} \det(\theta_v)^{r_v/2} \Gamma_{r_v}(r_v/2)\Gamma_{r_v}(n/2)^{-1},$$

where $\mu = (n-1)/2$, $d = [F : \mathbf{Q}]$, D_F *is the discriminant of* F, ψ *is the Hecke character of* F *corresponding to the extension* K/F *with* K *of (2.1.2),* $\mathfrak{d}$ *is the different of* K *relative to* F, *and* (r_v, θ_v) *for* $v \in \mathbf{a}$ *is defined by (5.1.1). Then*

$$\mathfrak{m}\left(G_+^\varphi, D(\mathfrak{c})\right) = b_\varphi \mathfrak{m}_n(\mathfrak{c})\left[\widetilde{L} : L\right]^\mu \prod_{v|\mathfrak{e},\, v\nmid\mathfrak{c}} \lambda_v \cdot \begin{cases} 1 & \text{if } n \text{ is odd,} \\ N(\mathfrak{d})^{-1/2} & \text{if } n \text{ is even,} \end{cases}$$

with λ_v *given as follows:*

$$\begin{array}{ll} 1 & \text{if } t_v = 1, \\ 2^{-1}(1+q)^{-1}\left(1+q^{1-m}\right)\left(1+q^{-m}\right) & \text{if } t_v = 2,\ \mathfrak{d}_v = \mathfrak{r}_v, \text{ and } \widetilde{M}_v \neq M_v, \\ 2^{-1} & \text{if } t_v = 2 \text{ and } \mathfrak{d}_v \neq \mathfrak{r}_v, \\ 2^{-1}(1+q)^{-1}\left(1-q^{1-n}\right) & \text{if } t_v = 3, \\ 2^{-1}(1+q)^{-1}\left(1-q^{1-m}\right)\left(1-q^{-m}\right) & \text{if } t_v = 4. \end{array}$$

Here $m = n/2$ *and* q *is the norm of the prime ideal at* v.

The proof will be completed in Section 7. We add several remarks.

(1) If $G_{\mathbf{a}}^\varphi$ is compact, that is, if φ is totally definite, then $r_v = 0$ for every $v \in \mathbf{a}$, so that $b_\varphi = 1$.

(2) It should be noted that the factor $D_F^{1/2}\left[(2k-1)!(2\pi)^{-2k}\right]^d \zeta_{F,\mathfrak{c}}(2k)$ is a rational number, which is why the number $\mathfrak{m}_n(\mathfrak{c})$ is expressed as above. The same is true for $N(\mathfrak{d})^{1/2} D_F^{1/2}\left[(m-1)!(2\pi)^{-m}\right]^d L_{\mathfrak{c}}(m, \psi)$, provided that $\psi_{\mathbf{a}}(x) = \operatorname{sgn}(x)^{mu}$, which is the case if and only if r_v is even for every $v \in \mathbf{a}$. (If $K = F$, we must have $4|n$.) The rationality of this type is well known; see [S5, Section A6], for example. In the present case, however, the rationality can be derived by induction on n from the fact that $\mathfrak{m}(G_+^\varphi, D(\mathfrak{c}))$ is a rational number if $G_{\mathbf{a}}^\varphi$ is compact.

(3) The definition of the space $\mathfrak{X}^\varphi$, as well as its measure, depends on the choice of $(\theta_v)_{v\in\mathbf{a}}$, which is why the factor $\det(\theta_v)^{r_v}$ appears in the definition of b_φ. We can eliminate it if we take $\theta_v = 1_{t_v}$ or if we normalize the measure of (5.5.3) by multiplying by $\prod_{v\in\mathbf{a}} \det(\theta_v)^{-r_v/2}$.

(4) As for the factor $[\widetilde{L}:L]$, we have of course $[\widetilde{L}:L]=\prod_{v|\mathfrak{e}}[\widetilde{L}_v:L_v]$, and $[\widetilde{L}_v:L_v]$ is given by (3.2.1). Then we see that

$$2^{-d\mu}[\widetilde{L}:L]^{\mu}\prod_{v|\mathfrak{e}}\lambda_v=\prod_{\mathfrak{p}}{}'\left\{2^{-1}(N(\mathfrak{p})+1)^{-1}(N(\mathfrak{p})^{n-1}-1)\right\}\quad\text{if } n \text{ is odd},$$

$$N(\mathfrak{d})^{-1/2}[\widetilde{L}:L]^{\mu}=N(\mathfrak{d})^{m-1}\prod_{\mathfrak{p}}{}''N(\mathfrak{p})^{n-1}\quad\text{if } n \text{ is even},$$

where $\prod'_{\mathfrak{p}}$ is the product over all the primes of F for which $t_v=3$ and $\prod''_{\mathfrak{p}}$ is the product over all the primes of F dividing $\mathfrak{e}$ but unramified in K.

(5) Our formula is valid for $n\geq 2$. If $n=2$ and K is totally imaginary, it is reduced to a classically known fact on $L_{\mathfrak{c}}(1,\psi)$. This is similar to what we observed in the unitary case; see [S5, §24.5]. The same can be said even when K has a real archimedean prime, except that our formula gives the transcendental factor of $L(1,\psi)$ in terms of $\mathrm{vol}(\Gamma^a\backslash\mathfrak{Z}^{\varphi})$ instead of the quotient of the regulator of K by that of F.

(6) Let D' be an arbitrary open subgroup of $(G^{\varphi}_{+})_{\mathbf{A}}$ containing $(G^{\varphi}_{+})_{\mathbf{a}}$ and such that $D'\cap(G^{\varphi}_{+})_{\mathbf{h}}$ is compact. Then

$$[C^+:C^+\cap D']\mathfrak{m}(G^{\varphi}_{+},C^+)=[D':C^+\cap D']\mathfrak{m}(G^{\varphi}_{+},D') \tag{5.8.1}$$

by [S5, Lemma 24.2]. Therefore the problem of obtaining $\mathfrak{m}(G^{\varphi}_{+},D')$ can be reduced to that of some group indices, which is nontrivial in general. In Lemma 8.7 we give an interpretation of the quotient of the group indices by means of local representation densities.

5.9. We now consider G^0 and $G^0_{\mathbf{A}}$ of (2.1.4) in the present case. For simplicity, hereafter we write simply G^+ for G^{φ}_{+}. Let D be an open subgroup of $G^+_{\mathbf{A}}$ such that $D\cap G^+_{\mathbf{h}}$ is compact. For each $a\in G^+_{\mathbf{A}}$ we take a decomposition $G^+aG^0_{\mathbf{A}}D=\bigsqcup_{x\in X_a}G^+xD$ with a finite subset X_a of $G^+_{\mathbf{A}}$, and we put

$$\mathfrak{m}(a,D)=\sum_{x\in X_a}\nu(\Gamma^x),\qquad \Gamma^x=G^+\cap xDx^{-1}, \tag{5.9.1}$$

with $\nu(\Gamma^x)$ of (5.5.2). If $\mathcal{A}$ is a complete set of representatives for $G^+_{\mathbf{A}}/G^+G^0_{\mathbf{A}}D$ (see Lemma 2.3(3)), then clearly

$$\mathfrak{m}(G^+,D)=\sum_{a\in\mathcal{A}}\mathfrak{m}(a,D). \tag{5.9.2}$$

If $G^{\varphi}_{\mathbf{a}}$ is not compact and $n\geq 3$, then

$$\mathfrak{m}(a,D)=\nu(\Gamma^a)=[\Gamma^a\cap\{\pm 1\}:1]^{-1}\mathrm{vol}(\Gamma^a\backslash\mathfrak{Z}^{\varphi}), \tag{5.9.3}$$

since $G^+aG^0_{\mathbf{A}}D=G^+aD$ by Lemma 2.4.

To explain the meaning of $\mathfrak{m}(a, D)$, consider the case in which D is given by $D = \{\alpha \in G_{\mathbf{A}}^{+} | \Lambda\alpha = \Lambda\}$ with a $\mathfrak{g}$-lattice Λ in V. Now, by the *spinor genus* of L we mean the set of all lattices of the form Λx with $x \in G_{\mathbf{A}}^{0} G^{+}$. Then the map $x \mapsto \Lambda x^{-1}$ gives a bijection of $G_{\mathbf{A}}^{+}/G^{+}G_{\mathbf{A}}^{0}D$ onto the set of spinor genera contained in the genus of Λ. (In view of the fact mentioned on the first two lines of the proof of Theorem 5.6, the genus of Λ defined with respect to G^{+} is the same as that defined with respect to G^{φ}.) Thus the number of such spinor genera is exactly the group index $[F_{\mathbf{A}}^{\times} : F^{\times}\sigma(D)]$ or $[F^{\times}W : F^{\times}\sigma(D)]$ of Lemma 2.3(4). We may call $\mathfrak{m}(G^{+}, D)$ in this case the *mass of the genus* of Λ, and $\mathfrak{m}(a, D)$ the *mass of the spinor genus* of Λa^{-1}. We can show that the latter does not depend on a, which is included in the first part of the following theorem.

5.10. THEOREM. *The notation being as in Theorem 5.8, let D be an open subgroup of $G_{\mathbf{A}}^{+}$ such that $D \cap G_{\mathbf{h}}^{+}$ is compact. Let σ be the spinor norm map of (1.2.3) and (2.1.1). Suppose that φ is anisotropic. Then the following assertions hold.*

(1) *We have*

$$\mathfrak{m}(a, D) = \mathfrak{m}(G^{+}, D) \cdot \begin{cases} [F_{\mathbf{A}}^{\times} : F^{\times}\sigma(D)]^{-1} & \text{if } n > 2, \\ [F^{\times}W : F^{\times}\sigma(D)]^{-1} & \text{if } n = 2, \end{cases}$$

for every $a \in G_{\mathbf{A}}^{+}$, where W is defined by (2.1.3).

(2) *Suppose that $G_{\mathbf{a}}^{\varphi}$ is not compact and $n \geq 3$. Then*

$$\nu(\Gamma^{a}) = [F_{\mathbf{A}}^{\times} : F^{\times}\sigma(D)]^{-1}\mathfrak{m}(G^{+}, D).$$

Moreover, if n is odd, then $\mathrm{vol}(\Gamma^{a}\backslash\mathfrak{Z}^{\varphi})$ *is a rational number times*

$$\prod_{v \in \mathbf{a}} 2^{r_v/2} \det(\theta_v)^{r_v/2} \pi^{r_v(n-r_v)/2};$$

if n is even, it is a rational number times

$$\pi^{e/2} N(\mathfrak{d})^{1/2} |D_F|^{1/2} (2\pi)^{-md} L(m, \psi) \prod_{v \in \mathbf{a}} \det(\theta_v)^{r_v/2} \pi^{r_v(n-r_v)/2},$$

where e is the number of $v \in \mathbf{a}$ for which r_v is odd.

Proof. By Lemma 2.3(2) we have $G^{+}bG_{\mathbf{A}}^{0}D = G^{+}aG_{\mathbf{A}}^{0}Dc$ if $b = ac$, and hence $\bigsqcup_{x \in X_a} G^{+}xDc = \bigsqcup_{y \in X_b} G^{+}yD$. Applying [S5, Lemma 24.2(3)] to this setting, we obtain

$$[c^{-1}Dc : D \cap c^{-1}Dc]\mathfrak{m}(a, D) = [D : D \cap c^{-1}Dc]\mathfrak{m}(b, D).$$

This together with Lemma 2.3(1) shows that $\mathfrak{m}(a, D) = \mathfrak{m}(b, D)$. From this and Lemma 2.3(4) we obtain our first assertion. The first part of (2) follows from (1) and (5.9.3). As for the second part, it is sufficient to prove it for $D = D(\mathfrak{g})$. Then the desired fact follows from the explicit form of $\mathfrak{m}(G^{+}, D(\mathfrak{c}))$ in Theorem 5.8 in view of Remarks (2) and (4) subsequent to that theorem. □

To define certain Euler products on G^φ, we first prove a lemma.

5.11. LEMMA. *Define a subgroup $\mathfrak{X}$ of $G^\varphi_{\mathbf{A}}$ and a subgroup Ξ of G^+ by*

$$\mathfrak{X} = \{y \in G^\varphi_{\mathbf{A}} \,|\, y_v \in D^\varphi_v \text{ for every } v|\mathfrak{c}\}, \tag{5.11.1}$$

$$\Xi = \mathfrak{X} \cap G^+, \tag{5.11.2}$$

where $D^\varphi_v = D^\varphi \cap G^\varphi_v$ with D^φ of (5.7.4). For $a \in G^+_{\mathbf{A}}$ put $\Gamma^a = G^+ \cap aD^\varphi a^{-1}$; let R^a be a complete set of representatives for $\Gamma^a \backslash \Xi$. Then the following assertions hold.

(1) *There exists a finite subset $\mathfrak{B}$ of $G^+_{\mathbf{h}}$ such that*

$$G^+_{\mathbf{A}} = \bigsqcup_{a \in \mathfrak{B}} G^+ a D^+ \tag{5.11.3}$$

and $a_v = 1$ for every $a \in \mathfrak{B}$ and every $v|\mathfrak{c}$, where $D^+ = G^+_{\mathbf{A}} \cap D^\varphi$.

(2) *For any fixed $b \in \mathfrak{X}$, $D^\varphi \backslash \mathfrak{X}$ can be given by $\bigsqcup_{a \in \mathfrak{B}} a^{-1} R^a b$.*

Proof. Assertion (1) is a special case of [S5, Lemma 8.12]. As for (2), since $-1 \in \det(D^\varphi_v)$ for $v \nmid \mathfrak{c}$, we have $\mathfrak{X} = \mathfrak{X}b = D^\varphi(\mathfrak{X} \cap G^+_{\mathbf{A}})b = \bigcup_{a \in \mathfrak{B}} D^\varphi(\mathfrak{X} \cap D^+ a^{-1} G^+)b = \bigcup_{a \in \mathfrak{B}} D^\varphi a^{-1} \Xi b$. We easily see that the last union is a disjoint one. Then (2) can easily be verified. □

5.12. With σ as in (5.7.2), for $x \in G^\varphi_{\mathbf{A}}$ we define an integral ideal $\nu^\sigma(x)$ by

$$\nu^\sigma(x) = \nu_0(\sigma x \sigma^{-1}), \tag{5.12.1}$$

where ν_0 is defined on $(F_{\mathbf{A}})^n_n$ in §2.7. Clearly $\nu^\sigma(x)$ depends only on $C^\varphi x C^\varphi$.

Let us denote by $\mathcal{M}(D^\varphi)$ the set of all $\mathbf{C}$-valued functions $\mathbf{f}$ on $G^\varphi_{\mathbf{A}}$ such that $\mathbf{f}(\alpha x w) = \mathbf{f}(x)$ for every $\alpha \in G^\varphi$ and $w \in D^\varphi$. (In essence we are considering constant functions on $\mathfrak{T}^\varphi$, but still our zeta functions defined below are highly nontrivial.) Let $\mathfrak{R}(D^\varphi, \mathfrak{X})$ denote the Hecke algebra consisting of all formal finite sums $\sum_\tau c_\tau D^\varphi \tau D^\varphi$ with $c_\tau \in \mathbf{C}$ and $\tau \in \mathfrak{X}$ with the law of multiplication defined as usual. For $\tau \in \mathfrak{X}$ let $D^\varphi \tau D^\varphi = \bigsqcup_{y \in Y} D^\varphi y$ with a finite subset Y of $G^\varphi_{\mathbf{A}}$. Then we can let $D^\varphi \tau D^\varphi$ act on $\mathcal{M}(D^\varphi)$ by the rule

$$(\mathbf{f}|D^\varphi \tau D^\varphi)(x) = \sum_{y \in Y} \mathbf{f}(xy^{-1}). \tag{5.12.2}$$

We extend this to the action of the whole $\mathfrak{R}(D^\varphi, \mathfrak{X})$ on $\mathcal{M}(D^\varphi)$.

Taking a Hecke character χ of F and assuming that $\mathbf{f}|D^\varphi \tau D^\varphi = \varepsilon(\tau)\mathbf{f}$ with $\varepsilon(\tau) \in \mathbf{C}$ for every $\tau \in \mathfrak{X}$, we define a Dirichlet series $\mathfrak{T}(s, \mathbf{f}, \chi)$ and its modification

$Z(s, \mathbf{f}, \chi)$ by

$$\mathfrak{T}(s, \mathbf{f}, \chi) = \sum_{\tau \in D^\varphi \backslash \mathfrak{X} / D^\varphi} \chi^*\big(\nu^\sigma(\tau)\big) N\big(\nu^\sigma(\tau)\big)^{-s} \varepsilon(\tau), \tag{5.12.3}$$

$$Z(s, \mathbf{f}, \chi) = \mathfrak{T}(s, \mathbf{f}, \chi) \prod_{i=1}^{[n/2]} L_\mathfrak{c}(2s+2-2i, \chi^2). \tag{5.12.4}$$

Then for $x \in G_\mathbf{A}^\varphi$ we have clearly

$$\mathfrak{T}(s, \mathbf{f}, \chi)\mathbf{f}(x) = \sum_{y \in D^\varphi \backslash \mathfrak{X}} \chi^*\big(\nu^\sigma(y)\big) N\big(\nu^\sigma(y)\big)^{-s} \mathbf{f}(xy^{-1}). \tag{5.12.5}$$

By Lemma 5.11(2) we can take $\bigsqcup_{a \in \mathfrak{B}} a^{-1} R^a b$ as $D^\varphi \backslash \mathfrak{X}$ with any fixed $b \in \mathfrak{X}$. Then we obtain

$$\mathfrak{T}(s, \mathbf{f}, \chi)\mathbf{f}(b) = \sum_{a \in \mathfrak{B}} \mathbf{f}(a) \sum_{\xi \in R^a} \chi^*\big(\nu^\sigma(a^{-1}\xi b)\big) N\big(\nu^\sigma(a^{-1}\xi b)\big)^{-s}. \tag{5.12.6}$$

5.13. Proposition. *Suppose that $t_v \leq 1$ or $\widetilde{L}_v = L_v$ for every $v \in \mathbf{h}, \nmid \mathfrak{c}$. (Then $t_v \leq 2$ for every $v \in \mathbf{h}, \nmid \mathfrak{c}$.) Then the following assertions hold.*

(1) *For each fixed $\mathbf{f}$ and a prime ideal $\mathfrak{p}$ of F there exists a polynomial $W_\mathfrak{p}$ of degree n whose constant term is 1 such that*

$$Z(s, \mathbf{f}, \chi) = \prod_{\mathfrak{p} \nmid \mathfrak{c}} W_\mathfrak{p}\big(\chi^*(\mathfrak{p}) N(\mathfrak{p})^{-s}\big)^{-1} \tag{5.13.1}$$

for every Hecke character χ of F. Moreover,

$$W_\mathfrak{p}(X) = \big(1 - N(\mathfrak{p})^{2m-2} X^2\big)^{[t_v/2]} \prod_{i=1}^{m-[t_v/2]} (1 - \mu_{v,i} X)(1 - \mu'_{v,i} X) \tag{5.13.2}$$

with complex numbers $\mu_{v,i}$, and $\mu'_{v,i}$ such that $\mu_{v,i}\mu'_{v,i} = N(\mathfrak{p})^{n-2}$, where $m = [n/2]$.

(2) *The series of (5.12.3) and the product of (5.13.1) are absolutely convergent for* $\mathrm{Re}(s) > n$.

Proof. This is similar to [S5, Proposition 20.4] and can be derived from [S5, Theorem 16.16], which gives an explicit rational form of each local Euler factor of the above $\mathfrak{T}$. For example, the theorem combined with [S5, Lemma 16.18] says that if $t_v = 2$, the Euler factor of $\mathfrak{T}$ at v has the form

$$\prod_{i=1}^{m-1} \frac{1 - q^{2i-2-2s}}{(1 - \mu_{v,i} q^{-s})(1 - \mu'_{v,i} q^{-s})} \tag{5.13.3}$$

with $\mu_{v,i}$, and $\mu'_{v,i}$ such that $\mu_{v,i}\mu'_{v,i} = q^{n-2}$, where $q = N(\mathfrak{p})$. Multiplying both numerator and denominator by $1 - q^{2m-2-2s}$, we obtain the desired fact for such a v. The case $t_v \leq 1$ is similar and simpler. The only nontrivial point here is that we can include the prime factors v of 2 if $t_v = 1$. This is because of [S5, Lemma 16.19], which allows us to apply [S5, Theorem 16.16] to such a v. □

5.14. THEOREM. *The notation and assumptions being as in Proposition 5.13, suppose that φ is anisotropic, $\mathfrak{c}$ is divisible by the ideal $\mathfrak{e}$ of Theorem 5.8, and $\chi_{\mathbf{a}}(x) = |x|^{i\kappa}$ with $\kappa \in \mathbf{R}^{\mathbf{a}}$ such that $\|\kappa\| = 0$. Then $\mathcal{M}(D^\varphi)$ is spanned by eigenfunctions $\mathbf{f}$ of the above type, and $Z(s, \mathbf{f}, \chi)$ can be continued to a meromorphic function on the whole s-plane.*

Proof. That $\mathcal{M}(D^\varphi)$ is spanned by eigenfunctions follows from the fact that the action of $\mathfrak{R}(D^\varphi, \mathfrak{X})$ on $\mathcal{M}(D^\varphi)$ defines a commutative ring of normal operators, which can be proved in exactly the same manner as in [S5, Proposition 20.4(1)]. To prove the last part of our theorem, given $\mathbf{f}$, put $\mathbf{g}(x) = \mathbf{f}(x\zeta)$ with a fixed $\zeta \in C^\varphi$. Observe that $\zeta D^\varphi \zeta^{-1} = D^\varphi$, $\zeta \mathfrak{X} \zeta^{-1} = \mathfrak{X}$, and $\zeta D^\varphi \tau D^\varphi \zeta^{-1} = D^\varphi \zeta\tau\zeta^{-1} D^\varphi = D^\varphi \tau D^\varphi$ for every $\tau \in \mathfrak{X}$. Therefore we can easily verify that $\mathbf{g}$ is an eigenfunction with the same eigenvalues as $\mathbf{f}$. Now $G^\varphi_{\mathbf{A}} = G^+_{\mathbf{A}} C^\varphi$, so that $G^\varphi_{\mathbf{A}} \doteq \bigcup_{b \in \mathfrak{B}} G^+ b C^\varphi$. This means that $\mathbf{f}$ is determined by the values $\mathbf{f}(b\zeta)$ for $b \in \mathfrak{B}$ and $\zeta \in C^\varphi$. Thus, changing $\mathbf{f}$ for $\mathbf{g}$ with a suitable $\zeta \in C^\varphi$, we may assume that $\mathbf{f}(b) \neq 0$ for some $b \in \mathfrak{B}$, without changing the series $\mathfrak{T}$ and Z. Using such an $\mathbf{f}$, we shall prove the desired meromorphic continuation in §7.11. □

5.15. PROPOSITION. *In the setting of Proposition 5.13 take $\mathbf{f}$ to be the constant 1 on the whole $G^\varphi_{\mathbf{A}}$. Then $\mathbf{f}$ is an eigenfunction of $D^\varphi \tau D^\varphi$ with eigenvalue $\lambda(\tau) = \#(D^\varphi \backslash D^\varphi \tau D^\varphi)$. Moreover, we have*

(5.15.1)
$$\mathfrak{T}(s, \mathbf{f}, \chi) = \sum_{\tau \in D^\varphi \backslash \mathfrak{X}} \chi^*\big(\nu^\sigma(\tau)\big) N\big(\nu^\sigma(\tau)\big)^{-s},$$

(5.15.2)
$$Z(s, \mathbf{f}, \chi) = \begin{cases} L_{\mathfrak{c}}\big(s+1-(n/2), \chi\psi\big) \prod_{i=1}^{n-1} L_{\mathfrak{c}}(s+1-i, \chi) & \textit{if } n \textit{ is even,} \\ \prod_{i=1}^{n-1} L_{\mathfrak{c}}(s+1-i, \chi) & \textit{if } n \textit{ is odd.} \end{cases}$$

Proof. This is again similar to [S5, Lemma 20.11] and can be proved in the same manner, by taking $\mu_{v,i} = q^{i-1}$ and $\mu'_{v,i} = q^{n-1-i}$ in (5.13.3). Notice that if n is even, we have a factor $1 - q^{m-1-s}$ or $1 + q^{m-1-s}$ according as $t_v = 0$ or $t_v = 2$, which is why the factor $L_{\mathfrak{c}}\big(s+1-(n/2), \chi\psi\big)$ appears. □

5.16. Examples. We consider the case $F = \mathbf{Q}$ and $\varphi = 1_n$; put $m = [n/2]$. To classify the spaces $((\mathbf{Q}_p)^1_n, 1_n)$, take the quaternion algebra B over $\mathbf{Q}$ whose norm form is the sum of four squares. This is ramified exactly at 2 and ∞. Since B_p for any prime number p contains an element of norm -1, we see that 1_4 is equivalent to -1_4 over $\mathbf{Q}_p$. This means that 1_8 is equivalent to η_4 over $\mathbf{Q}_p$; 1_5 is equivalent to $\mathrm{diag}[1, -1_4]$ and, hence, to $\mathrm{diag}[\eta_1, -1_3]$. Similarly 1_6 and 1_7 are equivalent to $\mathrm{diag}[\eta_2, -1_2]$ and $\mathrm{diag}[\eta_3, -1]$, respectively, over $\mathbf{Q}_p$. Therefore we can easily find the values of $[\widetilde{L} : L]$ and other invariants that appear in the formula of Theorem 5.8. Recall the formula $\zeta(2k) = (2\pi)^{2k}|B_{2k}|/[2\cdot(2k)!]$, where B_{2k} denotes the Bernoulli number. We thus obtain the formula for the mass of the genus of maximal lattices as follows:

$$\mathfrak{m}(G_+^\varphi, C^+) = \prod_{k=1}^{m}(4k)^{-1}|B_{2k}|\cdot\begin{cases}2 & \text{if } n\pm 1 \in 8\mathbf{Z},\\ 3^{-1}(2^{2m}-1) & \text{if } n\pm 3\in 8\mathbf{Z},\end{cases}$$

$$\mathfrak{m}(G_+^\varphi, C^+) = (2m)^{-1}|B_m|\prod_{k=1}^{m-1}(4k)^{-1}|B_{2k}| \cdot\begin{cases}2 & \text{if } n \in 8\mathbf{Z},\\ 3^{-1}(2^m-1)(2^{m-1}-1) & \text{if } n-4\in 8\mathbf{Z},\end{cases}$$

$$\mathfrak{m}(G_+^\varphi, C^+) = 2^{n-1}(m-1)!(2\pi)^{-m}L(m,\psi)\prod_{k=1}^{m-1}(4k)^{-1}|B_{2k}| \quad \text{if } n\pm 2\in 8\mathbf{Z}.$$

Here C^+ is defined by (5.7.5) with a maximal lattice L in $\mathbf{Q}_n^1$, $m = [n/2]$, and ψ is the primitive Dirichlet character modulo 4. The formula in the case $n \in 8\mathbf{Z}$ was given by Siegel in [Si, *III*, p. 447]. His formula gives the mass for G^φ instead of G_+^φ, and so it is one-half of our mass (see Lemma 5.6(1)).

Let us now treat the case $n = 9$. Take the standard basis $\{e_i\}_{i=1}^9$ of $\mathbf{Q}_9^1$; put $W = \sum_{i=2}^9 \mathbf{Q}e_i$ and $L = \mathbf{Z}e_1 + M$ with a $\mathbf{Z}$-maximal lattice M in W; let σ denote the restriction of φ to W. Then $\pm e_1$ are the only elements $g \in L$ such that $\varphi[g] = 1$ and $\{x \in L \mid \varphi(g, x) = 0\}$ is a $\mathbf{Z}$-maximal lattice in an 8-dimensional subspace. From this we easily see that for $\gamma \in G^\varphi$ we have $L\gamma = L$ if and only if $e_1\gamma = \pm e_1$ and $M\gamma = M$. Now it is classically known that the order of $\{\alpha \in G^\sigma \mid M\alpha = M\}$ is $2^{14}\cdot 3^5\cdot 5^2\cdot 7$, which follows from the above mass formula for $n = 8$ combined with the fact that the genus of M consists of one class, due to Mordell [M]. Then we observe that this number is the order of $\{\gamma \in G_+^\varphi \mid L\gamma = L\}$. Now from the above formula in the case $n = 9$, we find that the mass is exactly the inverse of the order, and hence the class number of the genus of L is 1.

6. Eisenstein series on G^η

6.1. Assuming that F is totally real, we consider G^η with $\eta = \eta_n$, $n > 1$. We can specialize what we did in §5.3 to this case by taking $\kappa = -1$, $\sigma_v = 1_{2n}$, and $r_v = n$

in (5.3.2). Then $\mathfrak{X}^{\varphi}$ becomes $\mathcal{H}_n^{\mathbf{a}}$ with $\mathcal{H}_n$ of (4.3.1). For simplicity we write $\mathcal{H}$ for $\mathcal{H}_n$. We can then let $G_{\mathbf{A}}^{\eta}$ act on $\mathcal{H}^{\mathbf{a}}$ and define $j_\alpha(z)$ and $\delta(z)$ by (5.3.6) and (5.3.7). Notice that $j_\alpha(z)_v = \det(c_\alpha z + d_\alpha)_v$ for each $v \in \mathbf{a}$.

We define P, Q, and R by (2.9.1) and (2.9.2), and we put

(6.1.1) $$Y = \{g \in F_n^n \mid {}^t g = g\},$$

(6.1.2) $$S = \{h \in F_n^n \mid {}^t h = -h\}, \quad S(\mathfrak{a}) = S \cap \mathfrak{a}_n^n,$$

(6.1.3) $$\tau(b) = \begin{bmatrix} 1_n & b \\ 0 & 1_n \end{bmatrix} \quad (b \in S_{\mathbf{A}}).$$

In (6.1.2), $\mathfrak{a}$ is a fractional ideal in F. The symbols $Y_{\mathbf{A}}, S_{\mathbf{A}}, Y_{\mathbf{h}}, S_{\mathbf{h}}, Y_{\mathbf{a}}$, and $S_{\mathbf{a}}$ are meaningful. Every element w of $\mathcal{H}^{\mathbf{a}}$ can be written in the form $w = x + y$ with $x \in S_{\mathbf{a}}, y \in Y_{\mathbf{a}}, y_v > 0$ for every $v \in \mathbf{a}$. We hereafter understand the expression $x + y$ always in this sense. Thus $\delta(x+y) = \det(y)\ (\in \mathbf{R}^{\mathbf{a}})$.

6.2. We now put

(6.2.1) $$\mathbf{e}(z) = e^{2\pi i z} \quad (z \in \mathbf{C})$$

and define characters $\mathbf{e}_{\mathbf{A}} : F_{\mathbf{A}} \to \mathbf{T}$ and $\mathbf{e}_v : F_v \to \mathbf{T}$ for each $v \in \mathbf{v}$ as usual (see [S5, §18.2]). (A correction: $y \in \mathbf{Q}$ in [S5, p. 147, line 2] should read $y \in \bigcup_{n=0}^{\infty} p^{-n}\mathbf{Z}$.) We then put $\mathbf{e}_{\mathbf{A}}(x) = \prod_{v \in \mathbf{v}} \mathbf{e}_v(x_v)$ and $\mathbf{e}_{\mathbf{h}}(x) = \mathbf{e}_{\mathbf{A}}(x_{\mathbf{h}})$ for $x \in F_{\mathbf{A}}$, and $\mathbf{e}_{\mathbf{a}}(x) = \mathbf{e}\left(\sum_{v \in \mathbf{a}} x_v\right)$ for $x \in F_{\mathbf{A}}$ or $x \in \mathbf{C}^{\mathbf{a}}$. Given a positive integer m and a square matrix X of size m, we put

(6.2.2) $$\mathbf{e}_{\mathbf{s}}^m(X) = \mathbf{e}_{\mathbf{s}}(\mathrm{tr}(X))$$

whenever the right-hand side is meaningful, where $\mathbf{s}$ is one of the symbols $\mathbf{A}, v, \mathbf{a}$, and $\mathbf{h}$. To study the Fourier expansion of a periodic function on $S_{\mathbf{a}}$, we consider a pairing $S_{\mathbf{a}} \times S_{\mathbf{a}} \to \mathbf{T}$ given by $(x, x') \mapsto \mathbf{e}_{\mathbf{a}}^n(-xx')$. Let f be a C^∞ function on $\mathcal{H}^{\mathbf{a}}$ such that $f(z+a) = f(z)$ for every a in a $\mathbf{Z}$-lattice L in S. Then we have a Fourier expansion

(6.2.3) $$f(x+y) = \sum_{h \in M} c(h, y)\mathbf{e}_{\mathbf{a}}^n(-hx) \quad (x + y \in \mathcal{H}^{\mathbf{a}})$$

with $c(h, y)$ determined for $h \in M$, which is a C^∞ function of y, where

(6.2.4) $$M = \{x \in S \mid \mathrm{Tr}_{F/\mathbf{Q}}(\mathrm{tr}(xL)) \subset \mathbf{Z}\}.$$

It is often convenient to write the above expansion in the form

(6.2.5) $$f(x+y) = \sum_{h \in S} c(h, y)\mathbf{e}_{\mathbf{a}}^n(-hx) \quad (x + y \in \mathcal{H}^{\mathbf{a}}),$$

by defining $c(h, y)$ to be 0 for $h \notin M$.

6.3. Given fractional ideals $\mathfrak{y}$ and $\mathfrak{z}$ such that $\mathfrak{y}\mathfrak{z} \subset \mathfrak{g}$, we consider $D[\mathfrak{y}, \mathfrak{z}]$ as in (2.9.3) and put

$$D_0[\mathfrak{y}, \mathfrak{z}] = \{x \in D[\mathfrak{y}, \mathfrak{z}] \mid x(\mathbf{1}) = \mathbf{1}\}, \quad \mathbf{1} = (\mathbf{1}_v)_{v \in \mathbf{a}}, \ \mathbf{1}_v = 1_n. \tag{6.3.1}$$

Hereafter in this section we fix a fractional ideal $\mathfrak{b}$. We have $G^\eta_{\mathbf{A}} = P_{\mathbf{A}} D_0[\mathfrak{b}^{-1}, \mathfrak{b}]$, and so every element x of $G^\eta_{\mathbf{A}}$ belongs to $pD_0[\mathfrak{b}^{-1}, \mathfrak{b}]$ for some $p \in P_{\mathbf{A}}$. With such a p we define a real positive number $\varepsilon(x)$ by

$$\varepsilon(x) = |\lambda(p)|_{\mathbf{A}} \tag{6.3.2}$$

with λ of (2.7.1). This is well defined.

To define our Eisenstein series, we take, in addition to $\mathfrak{b}$, an element k of $Z^{\mathbf{a}}$, an integral ideal $\mathfrak{c}$, and a Hecke character χ of F such that

$$\chi_v(a) = 1 \quad \text{if } v \in \mathbf{h}, a \in \mathfrak{g}_v^\times, \text{and } a - 1 \in \mathfrak{c}_v, \tag{6.3.3}$$

$$\chi_{\mathbf{a}}(b) = b^k |b|^{i\kappa - k}, \tag{6.3.4}$$

where $\kappa \in \mathbf{R}^{\mathbf{a}}$ and $\sum_{v \in \mathbf{a}} \kappa_v = 0$. Hereafter, until the end of this section, we put,

$$D = D^\eta = D[\mathfrak{b}^{-1}, \mathfrak{b}\mathfrak{c}], \qquad D_0 = \{x \in D \mid x(\mathbf{1}) = \mathbf{1}\}. \tag{6.3.5}$$

Next we define a function μ on $G^\eta_{\mathbf{A}}$ as follows:

$$\mu(x) = 0 \quad \text{if } x \notin P_{\mathbf{A}} D, \tag{6.3.6}$$

$$\mu(pw) = \chi(\lambda(p))^{-1} \chi_{\mathfrak{c}}(\lambda(w))^{-1} j_w^k(\mathbf{1})^{-1} \quad (p \in P_{\mathbf{A}}, w \in D_0), \tag{6.3.7}$$

where $\chi_{\mathfrak{c}} = \prod_{v|\mathfrak{c}} \chi_v$. Notice that $\lambda(w)_v \in \mathfrak{g}_v^\times$ for $v|\mathfrak{c}$. We can easily verify (cf. [S5, Lemma 18.5]) that the symbol μ is well defined and

$$\varepsilon(x_{\mathbf{a}}) = |j_x(\mathbf{1})|^u, \qquad \varepsilon(x_{\mathbf{h}}) = N(\mathrm{il}_{\mathfrak{b}}(x))^{-1}, \tag{6.3.8}$$

$$\mu(\alpha x w) = \chi_{\mathfrak{c}}(\lambda(w))^{-1} j_w^k(\mathbf{1})^{-1} \mu(x) \quad \text{if } \alpha \in P \text{ and } w \in D_0, \tag{6.3.9}$$

$$\mu(x_{\mathbf{a}}) = j_x^k(\mathbf{1})^{-1} |j_x(\mathbf{1})|^{k - i\kappa}, \tag{6.3.10}$$

where u is the symbol of (5.4.2) and $N(\mathfrak{a})$ denotes the norm of $\mathfrak{a}$.

6.4. Now our Eisenstein series $E(x, s)$ is defined for $(x, s) \in G^\eta_{\mathbf{A}} \times \mathbf{C}$ by

$$E(x, s) = E(x, s; \chi, D) = \sum_{\alpha \in A} \mu(\alpha x) \varepsilon(\alpha x)^{-2s}, \qquad A = P \backslash G^\eta. \tag{6.4.1}$$

From (6.3.9) we see that the sum is formally well defined, and

$$E(\alpha x w, s) = \chi_{\mathfrak{c}}(\lambda(w))^{-1} j_w^k(\mathbf{1})^{-1} E(x, s) \quad \text{if } \alpha \in G^\eta \text{ and } w \in D_0. \tag{6.4.2}$$

The convergence of (6.4.1) will be discussed after the proof of Lemma 6.5 below. For $x = pw$ as in (6.3.7), we note that

(6.4.3) $$\mu(x)\varepsilon(x)^{-2s} = \varepsilon(x_{\mathbf{h}})^{-2s}\chi_{\mathbf{h}}\big(\lambda(p)\big)^{-1}\chi_{\mathbf{c}}\big(\lambda(w)\big)^{-1}j_x^k(\mathbf{1})^{-1}|j_x(\mathbf{1})|^{k-i\kappa-2su}.$$

This follows from (6.3.7), (6.3.8), and (6.3.10). Since $j_x^k(\mathbf{1}) \in \mathbf{R}$, the value of (6.4.3), as well as our series $E(x,s)$, depends only on k (mod $2\mathbf{Z}^{\mathbf{a}}$). Therefore we lose nothing by assuming that $0 \le k_v \le 1$ for every $v \in \mathbf{a}$.

To study the analytic nature of E, it is more convenient to consider a certain transform of $E_{\mathbf{A}}$ instead of $E_{\mathbf{A}}$ itself. Namely, we put

(6.4.4) $$E^*(x,s) = E\big(x\eta_{\mathbf{h}}, s; \chi, D\big)$$

with $\eta = \eta_n$. From (6.4.2) we can derive in a straightforward way that

(6.4.5) $$E^*(\alpha x w, s) = \chi_{\mathbf{c}}\big(\lambda(w)\big)j_w^k(\mathbf{1})^{-1}E^*(x,s)$$
$$\text{if } \alpha \in G^\eta \text{ and } w \in D_0\big[\mathfrak{bc}, \mathfrak{b}^{-1}\big].$$

Now, for a fixed $q \in GL_n(F)_{\mathbf{A}}$ we have a Fourier expansion

(6.4.6) $$E^*\left(\begin{bmatrix} q & \sigma\widehat{q} \\ 0 & \widehat{q}\end{bmatrix}, s\right) = \sum_{h\in S} c(h,q,s)\mathbf{e}_{\mathbf{A}}^n(-h\sigma) \quad (\sigma \in S_{\mathbf{A}})$$

with $c(h,q,s) \in \mathbf{C}$. Observing that $E^*(xw,s) = E^*(x,s)$ for every $w \in R_{\mathbf{h}} \cap D[\mathfrak{bc}, \mathfrak{b}^{-1}]$, we see that

(6.4.7) $$c(h,q,s) \ne 0 \Longrightarrow \big({}^t qhq\big)_v \prec (2\mathfrak{bc}\mathfrak{d}_F)_v^{-1} \quad \text{for every } v \in \mathbf{h},$$

where $\mathfrak{d}_F$ is the different of F relative to $\mathbf{Q}$. In view of (6.4.2) and (6.4.5), for $r \in G_{\mathbf{h}}^\eta$ we can define functions $E_r(z,s)$ and $E_r^*(z,s)$ of $(z,s) \in \mathcal{H}^{\mathbf{a}} \times \mathbf{C}$ by

(6.4.8) $$E_r\big(\xi(\mathbf{1}),s\big) = j_\xi^k(\mathbf{1})E(r\xi,s),\ E_r^*\big(\xi(\mathbf{1}),s\big) = j_\xi^k(\mathbf{1})E^*(r\xi,s)$$

for $\xi \in G_{\mathbf{a}}^\eta$. For the moment the functions E, E^*, E_r, E_r^*, and $c(h,q,s)$ are defined only for the values of s for which the series of (6.4.1) is convergent, but eventually they become meromorphic functions of s on the whole complex plane.

6.5. Lemma. (1) *If* $r\eta_{\mathbf{h}} = \alpha^{-1}tw$ *with* $r, t \in G_{\mathbf{h}}^\eta, \alpha \in G^\eta$, *and* $w \in D[\mathfrak{bc}, \mathfrak{b}^{-1}]$, *then* $E_r(z,s) = \chi_{\mathbf{c}}(\lambda(w))j_\alpha^k(z)^{-1}E_t^*(\alpha z, s)$.

(2) *For* $t = \mathrm{diag}[q_1, \widehat{q}_1]$ *with* $q_1 \in GL_n(F)_{\mathbf{h}}$ *and* $z = x + y \in \mathcal{H}^{\mathbf{a}}$, *put* $c_t(h,y,s) = \det(y)^{-k/2}c(h,q,s)$ *with* $q \in GL_n(F)_{\mathbf{A}}$ *such that* $q_{\mathbf{h}} = q_1$ *and* $q_{\mathbf{a}} = y^{1/2}$. *Then*

$$E_t^*(z,s) = \sum_{h\in S} c_t(h,y,s)\mathbf{e}_{\mathbf{a}}^n(-hx).$$

(3) *For* $r \in Q_{\mathbf{h}}$ *we have*

$$E_r(z,s) = \chi_{\mathbf{h}}\big(\lambda(r)^{-1}\big)|\lambda(r)|_{\mathbf{A}}^{-2s} \sum_{\alpha \in A'} N\big(\mathfrak{a}_r(\alpha)\big)^{2s} \chi[\alpha]_r \delta(z)^{su-(k-i\kappa)/2} \|_k \alpha,$$

where $A' = P\backslash(G^{\eta} \cap P_{\mathbf{A}} r D r^{-1})$, $\mathfrak{a}_r(\alpha) = \mathrm{il}_{\mathfrak{b}}(r^{-1}\alpha r)$, *and*

$$\text{(6.5.1)} \qquad \chi[\alpha]_r = \begin{cases} \chi_{\mathbf{a}}\big(\lambda(\alpha)\big)\chi^*\big(\lambda(\alpha)\mathfrak{a}_r(\alpha)^{-1}\big) & \text{if } \mathfrak{c} \neq \mathfrak{g}, \\ \chi^*(\mathfrak{a}_r(\alpha))^{-1} & \text{if } \mathfrak{c} = \mathfrak{g}. \end{cases}$$

Proof. If the symbols are as in (1), then $w_{\mathbf{a}} = \alpha_{\mathbf{a}}$ and for $z = \xi(\mathbf{1})$ with $\xi \in G_{\mathbf{a}}^{\eta}$ we have

$$E_r(z,s) = j_{\xi}^k(\mathbf{1})E(r\xi,s) = j_{\xi}^k(\mathbf{1})E^*(r\eta_{\mathbf{h}}\xi,s) = j_{\xi}^k(\mathbf{1})E^*(tw\xi,s).$$

Take $p \in P_{\mathbf{a}}$ and $w' \in D_0[\mathfrak{b}\mathfrak{c}, \mathfrak{b}^{-1}] \cap G_{\mathbf{a}}^{\eta}$ so that $w_{\mathbf{a}}\xi = pw'$. Then $E_r(z,s) = j_{\xi}^k(\mathbf{1}) \cdot E^*(tpw'w_{\mathbf{h}},s) = j_{\xi}^k(\mathbf{1})j^k(w',\mathbf{1})^{-1}\zeta E^*(tp,s)$ with $\zeta = \chi_{\mathfrak{c}}\big(\lambda(w)\big)$ by (6.4.5). Now $\xi w'^{-1} = w_{\mathbf{a}}^{-1}p = \alpha_{\mathbf{a}}^{-1}p$, so that $j_{\xi}^k(\mathbf{1})j^k(w',\mathbf{1})^{-1} = j^k\big(\alpha^{-1},\ p(\mathbf{1})\big)j_p^k(\mathbf{1})$. Since $p(\mathbf{1}) = \alpha z$, we thus obtain

$$E_r(z,s) = \zeta j^k\big(\alpha^{-1}, \alpha z\big)E_t^*\big(p(\mathbf{1}),s\big) = \zeta j_{\alpha}^k(z)^{-1}E_t^*(\alpha z,s),$$

which proves (1). Next, the notation being as in (2), put $p = \tau(x)\mathrm{diag}[q,\widehat{q}]$. Then $p(\mathbf{1}) = z$ and we obtain the formula for E_t^* from (6.4.6) and (6.4.8). To prove (3), we first observe that each term of $\sum_{\alpha \in A'}$ depends only on $P\alpha$. For $\xi \in G_{\mathbf{a}}^{\eta}$ we have $E(r\xi,s) = \sum_{\alpha \in A}\mu(\alpha r\xi)\varepsilon(\alpha r\xi)^{-2s}$. Since $\mu = 0$ outside $P_{\mathbf{A}}D$, the sum is over all the α such that $\alpha r\xi \in P_{\mathbf{A}}D$, which is so if and only if $\alpha \in G^{\eta} \cap P_{\mathbf{A}}rDr^{-1}$. Thus we can replace A by A'. By Lemma 2.10(2), $P_{\mathbf{A}}rDr^{-1} = \bigsqcup_{\tau \in T} P\tau rDr^{-1}$ with a finite subset T of $Q_{\mathbf{h}}$. Then we can take A' in $\bigsqcup_{\tau \in T} \tau rDr^{-1} \cap G^{\eta}$. Let $\alpha \in \tau rDr^{-1} \cap G^{\eta}$. Take $\sigma \in P_{\mathbf{A}}$ so that $\alpha_{\mathbf{a}}\xi \in \sigma_{\mathbf{a}}D_0$ and $\sigma_{\mathbf{h}} = \tau$. Then $\alpha r\xi = \sigma rw$ with $w \in D_0$. Now we have $\mathfrak{a}_r(\alpha) = \lambda(\tau)\mathfrak{g}$, $\varepsilon(\alpha r\xi) = \varepsilon(\sigma rw) = |\lambda(\sigma_{\mathbf{a}}r)|_{\mathbf{A}}N\big(\mathfrak{a}_r(\alpha)\big)^{-1}$, and by (6.3.7) we have

$$\begin{aligned} \mu(\alpha r\xi) = \mu(\sigma rw) &= \chi\big(\lambda(\sigma r)\big)^{-1}\chi_{\mathfrak{c}}\big(\lambda(w)\big)^{-1}j_w^k(\mathbf{1})^{-1} \\ &= \chi_{\mathbf{h}}\big(\lambda(\tau r)\big)^{-1}\chi_{\mathbf{a}}\big(\lambda(\sigma)\big)^{-1}\chi_{\mathfrak{c}}\big(\lambda(w)\big)^{-1}j_w^k(\mathbf{1})^{-1}. \end{aligned}$$

Putting $z = \xi(\mathbf{1})$ and $\mathbf{s} = su - (k - i\kappa)/2$, we have $E_r(z,s) = j_{\xi}^k(\mathbf{1})\sum_{\alpha \in A'}\mu(\alpha r\xi) \cdot \varepsilon(\alpha r\xi)^{-2s}$, $\alpha z = \sigma(\mathbf{1})$, $j_{\alpha}^k(z)j_{\xi}^k(\mathbf{1}) = j_{\sigma}^k(\mathbf{1})j_w^k(\mathbf{1}) = \lambda(\sigma)^k j_w^k(\mathbf{1})$, and $\delta(\alpha z)^{\mathbf{s}} = \delta(\sigma\mathbf{1})^{\mathbf{s}} = |\lambda(\sigma)|^{-2\mathbf{s}}$. Therefore

$$\begin{aligned} |\lambda(\sigma)|^{-2su}j_w^k(\mathbf{1})^{-1} &= |\lambda(\sigma)|^{-2su}\lambda(\sigma)^k j_{\alpha}^k(z)^{-1}j_{\xi}^k(\mathbf{1})^{-1} \\ &= \chi_{\mathbf{a}}\big(\lambda(\sigma)\big)j_{\xi}^k(\mathbf{1})^{-1}\delta(z)^{\mathbf{s}}\|_k\alpha. \end{aligned}$$

On the other hand, $\chi[\alpha]_r = \chi_{\mathbf{a}}(\lambda(\alpha))(\chi_{\mathbf{h}}/\chi_{\mathfrak{c}})(\lambda(\tau^{-1}\alpha)) = \chi_{\mathbf{h}}(\lambda(\tau)^{-1})\chi_{\mathfrak{c}}(\lambda(\alpha^{-1}\tau)) = \chi_{\mathbf{h}}(\lambda(\tau)^{-1})\chi_{\mathfrak{c}}(\lambda(w))^{-1}$. Combining these, we obtain the equality of (3). □

Now for the same reasons as in [S5,§12.11], the convergence of $E(x,s)$ can be reduced to that of $E_r(z,s)$, and to that of $\sum_{\alpha\in B}|\delta^{s-k/2}\|\alpha|$ where $B=(P\cap\Gamma)\backslash\Gamma$ with a congruence subgroup Γ of G^η. In [S5, Section 18] we proved the convergence of Eisenstein series in the unitary and symplectic cases. Modifying the argument there in an obvious way, we find that the last series is convergent locally uniformly in $\mathcal{H}^{\mathbf{a}}$ for $\mathrm{Re}(s)>(n-1)/2$.

6.6. Lemma. *If $\mathfrak{c}\neq\mathfrak{g}$ and $x\in P_{\mathbf{A}}G^\eta_{\mathbf{a}}$, then*

$$E^*(x,s)=\sum_{\alpha\in\eta R}\mu(\alpha x\eta_{\mathbf{h}})\varepsilon(\alpha x\eta_{\mathbf{h}})^{-2s}. \tag{6.6.1}$$

This can be proved in the same manner as in [83, Lemma 2.3] (or [S5, Lemma 18.8]), since $P\eta P=\{\alpha\in G^\eta\,|\,\det(c_\alpha)\neq 0\}$, as can easily be seen (cf. [S5, Lemma 2.12(2)]). Our next task is to determine $c(h,q,s)$.

6.7. For $\sigma\in S_{\mathbf{A}}$ we define a positive integer $\nu[\sigma]$ by

$$\nu[\sigma]=[\mathfrak{g}:\nu_0(\sigma)], \tag{6.7.1}$$

where $\nu_0(\sigma)$ is the ideal defined in §2.7. We specify a Haar measure dx on $S_{\mathbf{A}}$ by taking $\int_{S_{\mathbf{A}}/S}dx=1$. For $v\in\mathbf{h}$, we put $\Lambda_v=S(\mathfrak{g})_v$ and define a Haar measure d_vx on S_v by taking $\int_{\Lambda_v}d_vx=1$. If $v\in\mathbf{a}$, we define d_vx on S_v by $d_vx=\prod_{j<k}dx_{jk}$, where x_{jk} is the (j,k)-entry of x. Then we have (cf. [S5, §A6.6])

$$dx=c(S)\prod_{v\in\mathbf{v}}d_vx_v \tag{6.7.2}$$

with a constant $c(S)$ given by

$$c(S)=|D_F|^{-n(n-1)/4}, \tag{6.7.3}$$

where D_F is the discriminant of F. For $q\in GL_n(F)_{\mathbf{A}}$ we can easily verify

$$d({}^tqxq)=|\det(q)|_{\mathbf{A}}^{n-1}dx. \tag{6.7.4}$$

6.8. Putting $r=\mathrm{diag}[q,\widehat{q}]$ and employing $\tau(\sigma)$ of (6.1.3), from (6.4.6) we obtain

$$c(h,q,s)=\int_{S_{\mathbf{A}}/S}E^*(\tau(\sigma)r,s)\mathbf{e}_{\mathbf{A}}^n(h\sigma)d\sigma. \tag{6.8.1}$$

To simplify our notation, put $g(x)=\mu(x)\varepsilon(x)^{-2s}$. Putting $x=\tau(\sigma)r$ and $\alpha=\eta\tau(b)$ with $b\in S$ in (6.6.1), under the assumption $\mathfrak{c}\neq\mathfrak{g}$ we obtain

$$E^*(\tau(\sigma)r,s)=\sum_{b\in S}g(\eta\tau(b+\sigma)r\eta_{\mathbf{h}}).$$

Substituting this into (6.8.1), we find that

$$c(h,q,s)=\int_{S_{\mathbf{A}}} g\big(\eta\tau(\sigma)r\eta_{\mathbf{h}}\big)\mathbf{e}^n_{\mathbf{A}}(h\sigma)d\sigma. \tag{6.8.2}$$

Our definition of μ and ε shows that we can put $g(x)=\prod_{v\in\mathbf{v}} g(x_v)$. Therefore from (6.7.2) we obtain

$$c(h,q,s)=c(S)\prod_{v\in\mathbf{v}} c_v(h,q_v,s), \tag{6.8.3}$$

$$c_v(h,q_v,s)=\int_{S_v} g\big(\eta_v\tau(\sigma)r_v\zeta_v\big)\mathbf{e}^n_v(h\sigma)d\sigma, \tag{6.8.4}$$

where $\zeta_v=\eta_v$ or 1 according as $v\in\mathbf{h}$ or $v\in\mathbf{a}$.

6.9. For $\sigma\in S_v$ we have

$$\eta_v\tau(\sigma)r_v\eta_v=\begin{bmatrix}\widehat{q} & 0\\ \sigma\widehat{q} & q\end{bmatrix}_v \quad (v\in\mathbf{h}), \tag{6.9.1}$$

$$\eta_v\tau(\sigma)r_v=\begin{bmatrix}0 & \widehat{q}\\ q & \sigma\widehat{q}\end{bmatrix}_v \quad (v\in\mathbf{a}). \tag{6.9.2}$$

Now $\mu(x_{\mathbf{a}})\varepsilon(x_{\mathbf{a}})^{-2s}=j_x^k(\mathbf{1})^{-1}|j_x(\mathbf{1})|^{k-i\kappa-2su}$ by (6.4.3). Therefore, for $v\in\mathbf{a}$, suppressing the subscript v for simplicity, we have

$$\begin{aligned} g\big(\eta\tau(\sigma)r\big)&=\det(q+\sigma\widehat{q})^{-k}|\det(q+\sigma\widehat{q})|^{k-i\kappa-2s}\\ &=\det(q)^k|\det(q)|^{2s+i\kappa-k}\det(q\cdot{}^tq+\sigma)^{-i\kappa-2s}. \end{aligned}$$

Then we define functions ξ and Ξ by

$$\xi(y,h;s)=\int_{S_v}\mathbf{e}^n_v(hx)\det(x+y)^{-2s}dx \tag{6.9.3}$$

$$(s\in\mathbf{C},\ 0<y\in Y_v,\ h\in S_v,\ v\in\mathbf{a}),$$

$$\Xi(y,w;t)=\prod_{v\in\mathbf{a}}\xi(y_v,w_v;t_v) \tag{6.9.4}$$

$$(t\in\mathbf{C}^{\mathbf{a}},\ y\in Y_{\mathbf{A}},\ y_v>0,\ w\in S_{\mathbf{A}}).$$

Then we obtain

$$\prod_{v\in\mathbf{a}} c_v(h,q_v,s)=\chi_{\mathbf{a}}\big(\det(q)\big)^{-1}|\det(q)|^{2su+2i\kappa}\,\Xi\big(q\cdot{}^tq,h;su+(i\kappa/2)\big). \tag{6.9.5}$$

6.10. Lemma. *Put $\mu=(n-1)/2$. Then the integral of (6.9.3) is convergent for* $\mathrm{Re}(s)>\mu-(1/4)$ *and can be continued as a meromorphic function of s to the whole complex plane. For $h=0$, we have*

$$\xi(y,0;s)=2^{n(\mu+1-2s)}\pi^{n^2/2}\det(y)^{\mu-2s}\Gamma_n(2s-\mu)\Gamma_n(s)^{-1}\Gamma_n\big(s+(1/2)\big)^{-1},$$

where $\Gamma_n(s)$ is defined by (4.5.2).

This is proven in Sections 9 and 11.

6.11. LEMMA. *Let $x = \eta_{\mathbf{h}}\tau(\sigma)r_{\mathbf{h}}\eta_{\mathbf{h}}$ with $r = \mathrm{diag}[q, \widehat{q}]$, $\sigma \in S_{\mathbf{h}}$, and $v \in \mathbf{h}$. Then*

$$(6.11.1)\qquad g(x_v) = \chi\left(\det(q)_v\right)^{-1}|\det(q)_v|^{-2s} \cdot \begin{cases} \nu\left[\left(b_0^{-1}q^{-1}\sigma\widehat{q}\right)_v\right]^{-2s}\chi^*\left(\nu_0\left(b_0^{-1}q^{-1}\sigma\widehat{q}\right)_v\right) & \text{if } v\nmid \mathfrak{c}, \\ 1 \quad \text{if } v|\mathfrak{c} \text{ and } (q^{-1}\sigma\widehat{q})_v \prec \mathfrak{b}_v\mathfrak{c}_v, \\ 0 \quad \text{otherwise}, \end{cases}$$

where b_0 is an element of $F_{\mathbf{h}}^{\times}$ such that $\mathfrak{b} = b_0\mathfrak{g}$.

Proof. Put $x = pw$ with $p \in P_{\mathbf{h}}$ and $w \in D$. Then Lemma 2.8(2) shows that

$$(6.11.2)\qquad \lambda(p^{-1}x)\mathfrak{g} = \lambda(x)\mathrm{il}_{\mathfrak{b}}(x)^{-1} = \nu_0(b_0^{-1}d_x^{-1}c_x),$$

since $d_x = q \in GL_n(F)_{\mathbf{h}}$. By (6.9.1) we have $d_x^{-1}c_x = q^{-1}\sigma\widehat{q}$, and hence, by (6.3.8), we have

$$\varepsilon(x) = N\left(\mathrm{il}_{\mathfrak{b}}(x)\right)^{-1} = \nu\left[b_0^{-1}q^{-1}\sigma\widehat{q}\right]|\det(q)|_{\mathbf{A}}.$$

Now $d_x = d_p d_w$, and so by (6.11.2) we obtain

$$\chi_{\mathbf{h}}\left(\lambda(p)\right)^{-1}\chi_{\mathfrak{c}}\left(\lambda(w)\right)^{-1} = \chi_{\mathbf{h}}\left(\lambda(x)\right)^{-1}(\chi_{\mathbf{h}}/\chi_{\mathfrak{c}})\left(\lambda(p^{-1}x)\right)$$
$$= \chi_{\mathbf{h}}\left(\det(q)\right)^{-1}\chi^*\left(\nu_0(b_0^{-1}d_x^{-1}c_x)\right).$$

Here notice that the ideal of (6.11.2) is prime to $\mathfrak{c}$ by Lemma 2.8(3). Combining these, we obtain (6.11.1) for $v\nmid\mathfrak{c}$. If $v|\mathfrak{c}$, our definition of μ together with (2.7.4) tells us that $g(x_v) \neq 0$ only if $(d_x^{-1}c_x)_v \prec \mathfrak{b}_v\mathfrak{c}_v$, that is, only if $(q^{-1}\sigma\widehat{q})_v \prec \mathfrak{b}_v\mathfrak{c}_v$, in which case the v-factor of the ideal of (6.11.2) is $\mathfrak{g}$; therefore, the above result for $v\nmid\mathfrak{c}$ holds also for $v|\mathfrak{c}$ if we ignore the factors involving ν and ν_0. This completes the proof. □

For each $v \in \mathbf{h}$ take an element δ_v of F_v such that $\delta_v\mathfrak{g}_v = (\mathfrak{d}_F)_v$. Given $\zeta \in S_{\mathbf{A}}$ such that $2\zeta_v \prec \mathfrak{g}_v$ for every $v \in \mathbf{h}$, we put

$$(6.11.3)\qquad \alpha_{\mathfrak{c}}(\zeta, s, \chi) = \prod_{v\nmid\mathfrak{c}}\alpha(\zeta_v, s, \chi),$$

$$\alpha(\zeta_v, s, \chi) = \sum_{\sigma\in S_v/\Lambda_v}\mathbf{e}_v^n(\delta_v^{-1}\zeta\sigma)\chi^*(\nu_0(\sigma))\nu[\sigma]^{-s}, \quad \Lambda_v = S(\mathfrak{g})_v.$$

6.12. PROPOSITION. *Suppose that $\mathfrak{c} \neq \mathfrak{g}$. Then $c(h, q, s) \neq 0$ only if $({}^tqhq)_v \prec (2\mathfrak{b}\mathfrak{c}\mathfrak{d}_F)_v^{-1}$ for every $v \in \mathbf{h}$, in which case*

$$c(h, q, s) = c(S)\chi\left(\det(q)\right)^{-1}N(\mathfrak{b}\mathfrak{c})^{-n\mu}|\det(q)_{\mathbf{h}}|_{\mathbf{A}}^{2\mu-2s}|\det(q)_{\mathbf{a}}|^{2su+2i\kappa}$$
$$\cdot\alpha_{\mathfrak{c}}(\omega\cdot{}^tqhq, 2s, \chi)\Xi\left(q\cdot{}^tq, h; su + (i\kappa/2)\right),$$

where $\mu = (n-1)/2$ *and* ω *is an element of* $F_{\mathbf{h}}^{\times}$ *such that* $\omega\mathfrak{g} = \mathfrak{b}\mathfrak{d}_F$.

Proof. Suppose $c(h,q,s) \neq 0$; put $h' = b_0 \cdot {}^t qhq$. By (6.4.7), $h'_v \prec (2\mathfrak{c}\mathfrak{d}_F)_v^{-1}$ for every $v \in \mathbf{h}$. Put $m_v = N(\mathfrak{b}_v\mathfrak{c}_v)^{-n\mu}|\det(q)_v|^{2\mu}$. Substituting $b_0 q\sigma \cdot {}^t q$ for σ in (6.8.4) and employing (6.11.1) and (6.7.4), for $v \nmid \mathfrak{c}$ we find that

$$\begin{aligned}\chi\big(\det(q)_v\big)|\det(q)_v|^{2s} c_v(h,q_v,s) &= m_v \int_{S_v} \nu[\sigma]^{-2s}\chi^*\big(\nu_0(\sigma)_v\big)\mathbf{e}_v^n(h'\sigma)d\sigma \\ &= m_v \sum_{\sigma\in S_v/\Lambda_v} \nu[\sigma]^{-2s}\chi^*\big(\nu_0(\sigma)_v\big)\mathbf{e}_v^n(h'\sigma) \\ &= m_v\alpha\big(\delta_v h'_v, 2s, \chi\big).\end{aligned}$$

If $v|\mathfrak{c}$, we have $c_v(h,q_v,s) = m_v\chi(\det(q)_v)^{-1}|\det(q)_v|^{-2s}$. Taking the product of (6.9.5) and c_v for all $v \in \mathbf{h}$, we obtain the desired formula. □

6.13. **Proposition.** *Suppose that* $\mathfrak{c} \neq \mathfrak{g}$; *let* h, q, *and* ω *be as in Lemma 6.11 and Proposition 6.12. Put* $r = \mathrm{rank}(h)$ *and take a prime element* π_v *of* F_v *for each* $v \in \mathbf{h}$. *Then*

$$\alpha_{\mathfrak{c}}\big(\omega \cdot {}^t qhq, 2s, \chi\big) = \Lambda_{\mathfrak{c}}(s)^{-1}\Lambda_{h,\mathfrak{c}}(s)\prod_{v\in\mathbf{b}} f_{h,q,v}\big(\chi(\pi_v)|\pi_v|^{2s}\big)$$

with a finite subset $\mathbf{b}$ *of* $\mathbf{h}$, *polynomials* $f_{h,q,v}$ *with constant term* 1 *and coefficients in* $\mathbf{Z}$ *independent of* χ, *and functions* $\Lambda_{\mathfrak{c}}$ *and* $\Lambda_{h,\mathfrak{c}}$ *given as follows:*

$$\Lambda_{\mathfrak{c}}(s) = \Lambda_{\mathfrak{c}}^n(s,\chi) = \prod_{i=1}^{[n/2]} L_{\mathfrak{c}}(4s+2-2i,\chi^2), \tag{6.13.1}$$

$$\Lambda_{h,\mathfrak{c}}(s) = \prod_{i=1}^{[(n-r)/2]} L_{\mathfrak{c}}(4s-2n+r+2i+1,\chi^2). \tag{6.13.2}$$

Here $L_{\mathfrak{c}}$ *is defined by (5.7.8). The set* $\mathbf{b}$ *is determined as follows:* $\mathbf{b} = \varnothing$ *if* $r = 0$. *If* $r > 0$, *take* $g_v \in GL_n(\mathfrak{g}_v)$ *for each* $v \nmid \mathfrak{c}$ *so that* $(\omega \cdot {}^t qhq)_v = {}^t g_v \mathrm{diag}[\xi_v, 0]g_v$ *with* $\xi_v \in S_v^r$. *Then* $\mathbf{b}$ *consists of all the* v *prime to* $\mathfrak{c}$ *such that* $\det(2\xi_v) \notin \mathfrak{g}_v^{\times}$.

Proof. The series $\alpha(\zeta_v,s,\chi)$ was investigated in [S5, Sections 13–15]. If $v \nmid \mathfrak{c}$, [S5, Theorem 13.6] shows that $\alpha(\zeta_v,s,\chi) = (p_\zeta f_\zeta)\big(\chi(\pi_v)|\pi_v|^s\big)$ with a polynomial f_ζ and a rational function p_ζ of the form

$$p_\zeta(t) = \prod_{i=1}^{[n/2]}\big(1-q^{2i-2}t^2\big)\prod_{i=1}^{[(n-r)/2]}\big(1-q^{2n-r-2i-1}t^2\big)^{-1},$$

where $q = |\pi_v|^{-1}$; f_ζ has coefficients in $\mathbf{Z}$ and constant term 1. Moreover, $f_\zeta = 1$ if $r = 0$ or if $\zeta_v = {}^t g_v \mathrm{diag}[\xi_v, 0]g_v$ with $g_v \in GL_n(\mathfrak{g}_v)$ and $\det(2\xi_v) \in \mathfrak{g}_v^{\times}$. Therefore we obtain our proposition. □

6.14. THEOREM. *Suppose* $\mathfrak{c} \neq \mathfrak{g}$; *let* $r \in G^{\eta}_{\mathbf{h}} \cap G^{\eta} Q_{\mathbf{h}} D[\mathfrak{b}\mathfrak{c}, \mathfrak{b}^{-1}]\eta_{\mathbf{h}}$. *Then* $E_r(z,s)$ *of (6.4.8) can be continued to a meromorphic function of* s *on the whole plane that is holomorphic for* $\mathrm{Re}(s) > \mu$, *where* $\mu = (n-1)/2$. *Moreover,* E_r *is holomorphic at* $s = \mu$ *except when* $\chi^2 = 1$, *in which case* E_r *has a simple pole at* $s = \mu$. *If* $k = 0$ *and* χ *is trivial, then the residue of* E_r *at* $s = \mu$ *is the number* $R(\mathfrak{c})$ *given by*

$$R(\mathfrak{c}) = 4^{-1} r(\mathfrak{c}) \gamma_n^{-d} |D_F|^{n(1-n)/4} N(\mathfrak{b}\mathfrak{c})^{n(1-n)/2} \Lambda_{\mathfrak{c}}(\mu)^{-1} \prod_{i=2}^{[n/2]} \zeta_{F,\mathfrak{c}}(2i-1),$$

where $\gamma_n = 2^{(n^2-3n+2)/2} \pi^{-n^2/2} \Gamma_n(n/2)$, $d = [F : \mathbf{Q}]$, $\Lambda_{\mathfrak{c}}(s)$ *is the function of Proposition 6.13 defined with trivial* χ, *and* $r(\mathfrak{c})$ *is the residue of* $\zeta_{F,\mathfrak{c}}(s)$ *at* $s = 1$.

Notice that $\Lambda_{\mathfrak{c}}$ is finite and nonzero at $s = \mu$ and also that $0 < R(\mathfrak{c}) \in \mathbf{R}$. We can easily verify that

$$\gamma_n = \begin{cases} 2^{(m-1)^2} \pi^{-m^2} \prod_{i=1}^{m-1} (2i)! & \text{if } n = 2m \in 2\mathbf{Z}, \\ 2^{m^2-m} \pi^{-m^2-m} \prod_{i=1}^{m} (2i-1)! & \text{if } n-1 = 2m \in 2\mathbf{Z}. \end{cases} \tag{6.14.1}$$

Proof. In §11.6 we shall derive meromorphic continuation of E_r to the whole plane from an integral formula, which reduces the problem to the properties of an Eisenstein series on a symplectic group. The holomorphy of $E_r(z,s)$ for $\mathrm{Re}(s) > \mu$ follows from its convergence we already mentioned. Now Lemma 6.5(1), (2) reduces the analytic nature of E_r to that of E_t^* with $t \in Q_{\mathbf{h}}$, which can be determined by the behavior of $c(h,q,s)$. We easily see that $\Lambda_{\mathfrak{c}}(s)^{-1}$ and $\Lambda_{h;\mathfrak{c}}(s)$ for $h \neq 0$ are holomorphic for $\mathrm{Re}(s) > \mu - (1/4)$. Also, by Lemma 6.10, $\Xi(y, h; su + (i\kappa/2))$ is holomorphic for $\mathrm{Re}(s) > \mu - (1/4)$ and for every h. Therefore, by Propositions 6.12 and 6.13, $c(h,q,s)$ with $h \neq 0$ is holomorphic for $\mathrm{Re}(s) > \mu - (1/4)$. Thus our question is reduced to $c(0,q,s)$. Now $\Lambda_{0,\mathfrak{c}}$ is holomorphic at $s = \mu$ unless $\chi^2 = 1$, in which case $\Lambda_{0,\mathfrak{c}}$ has a simple pole at $s = \mu$. Employing the formula for $\xi(y, 0, \mu)$ of (9.6.1), we find that the residue of $c_t(0, y, s)$ at $s = \mu$ is exactly the constant $R(\mathfrak{c})$ given in our theorem. The notation being as in Lemma 6.5(1), we have $E_r(z,s) = E_t^*(\alpha z, s)$ if $k = 0$ and χ is trivial. Therefore we obtain our theorem. □

7. Proof of the main theorems

7.1. With (V, φ) as in §5.3 we put

$$(X, \omega) = (V, \varphi) \oplus (V, -\varphi). \tag{7.1.1}$$

We take a basis of V as in §5.3; we then take the union of this basis and its copy to be the basis of X and consider matrix representation with respect to these bases. In particular, we can put

$$\omega = \mathrm{diag}[\varphi, -\varphi]. \tag{7.1.2}$$

An element (β, γ) of $G^\varphi \times G^\varphi$ can naturally be viewed as an element of G^ω; as a matrix it can be written $\mathrm{diag}[\beta, \gamma]$. Expressing an element of X in the form (u, v) with $u, v \in V$ with respect to the decomposition of (7.1.1), we put

$$U = \{(v, v) \mid v \in V\}, \tag{7.1.3}$$

$$P^\omega = \{\gamma \in G^\omega \mid U\gamma = U\}. \tag{7.1.4}$$

Since U is totally ω-isotropic, P^ω is the parabolic subgroup of G^ω relative to U defined in [S5, (2.1.1)].

Now with κ as in (5.3.2) and L as in §5.1, we can take $\lambda \in F_n^n$ so that

$$\kappa\varphi = \lambda + {}^t\lambda \quad \text{and} \quad 2x\lambda \cdot {}^t y \in \kappa\mathfrak{g} \quad \text{for every } x, y \in L. \tag{7.1.5}$$

Indeed, take fractional ideals $\mathfrak{a}_1, \ldots, \mathfrak{a}_n$ and an F-basis $\{g_i\}$ of V so that $L = \sum_{i=1}^n \mathfrak{a}_i g_i$. Then $\varphi(g_i, g_i) \in \mathfrak{a}_i^{-2}$ and $2\varphi(g_i, g_j) \in (\mathfrak{a}_i\mathfrak{a}_j)^{-1}$. Let g be the $(n \times n)$-matrix whose ith row is g_i. Define $\lambda \in F_n^n$ so that $(g\lambda \cdot {}^t g)_{ii} = \kappa\varphi(g_i, g_i)/2$, $(g\lambda \cdot {}^t g)_{ij} = \kappa\varphi(g_i, g_j)$ for $i < j$, and $(g\lambda \cdot {}^t g)_{ij} = 0$ for $i > j$. Then (7.1.5) is satisfied. With this λ we put

$$\eta = \begin{bmatrix} 0 & 1_n \\ 1_n & 0 \end{bmatrix}, \qquad S = \begin{bmatrix} 1_n & -\lambda \\ -1_n & -{}^t\lambda \end{bmatrix}. \tag{7.1.6}$$

We easily see that

$$S\eta \cdot {}^t S = -\kappa\omega, \qquad US = \left\{ \begin{bmatrix} 0_n^1 & x \end{bmatrix} \in F_{2n}^1 \mid x \in F_n^1 \right\}. \tag{7.1.7}$$

Consider G^η and define the standard parabolic subgroup P^η by (2.9.1). Then

$$S^{-1}G^\omega S = G^\eta, \qquad S^{-1}P^\omega S = P^\eta = \{\pi \in G^\eta \mid US\pi = US\}. \tag{7.1.8}$$

7.2. Lemma. *Define D^η of (6.3.5) with $\mathfrak{b} = 2\kappa^{-1}\mathfrak{g}$ and the integral ideal $\mathfrak{c}$ we took in §5.7; define also C^+, D^+, and Ξ by (5.7.5) and (5.11.2); put $\Sigma = \mathrm{diag}[\sigma, \widehat{\sigma}]$ $(\in G_{\mathbf{h}}^\eta)$ with σ of §5.7. Then the following assertions hold:*

(1) $\Xi = \{\xi \in G^\varphi \mid S^{-1}(\xi, 1)S \in P_{\mathbf{A}}^\eta D^\eta \Sigma\}$ if $2 \notin \mathfrak{c}$;

(2) $D^+ = \{\alpha \in C^+ \mid (\Sigma S^{-1}(\alpha, 1)S\Sigma^{-1})_v \in D_v^\eta$ for every $v \mid \mathfrak{c}\}$;

(3) $D^+ = \{\alpha \in C^+ \mid \Sigma S^{-1}(\alpha, 1)S\Sigma^{-1} \in D^\eta\}$ if $(2\varphi')_v^{-1} \prec \mathfrak{g}_v$ for every $v \nmid \mathfrak{c}$.

Proof. Define P' by (2.7.3) and $D' = D'[\mathfrak{b}^{-1}, \mathfrak{bc}]$ by (2.7.2). Put

$$Y = \begin{bmatrix} 1_n & \sigma\lambda \cdot {}^t\sigma \\ 0 & 1_n \end{bmatrix}. \tag{7.2.1}$$

This is not necessarily an element of $G_{\mathbf{A}}^\varphi$; however, from (7.1.5) we see that $Y \in P_{\mathbf{A}}' \cap D'$. Now, with φ' of (5.7.2) we have

$$Y^{-1}\Sigma S^{-1}(\beta, \gamma)S\Sigma^{-1}Y = \begin{bmatrix} \sigma\beta\sigma^{-1} & 0 \\ (\kappa\varphi')^{-1}\sigma(\gamma - \beta)\sigma^{-1} & \widehat{\sigma}\widehat{\gamma} \cdot {}^t\sigma \end{bmatrix}. \tag{7.2.2}$$

Since $\Sigma \in P^\eta_{\mathbf{A}}$ and $P^\eta_{\mathbf{A}} D^\eta = G^\eta_{\mathbf{A}} \cap P'_{\mathbf{A}} D'$ by Lemma 2.10(4), for $\xi \in G^\varphi_{\mathbf{A}}$ we see that $S^{-1}(\xi, 1)S \in P^\eta_{\mathbf{A}} D^\eta \Sigma$ if and only if $Y^{-1}\Sigma S^{-1}(\xi, 1)S\Sigma^{-1}Y \in P'_{\mathbf{A}} D'$. Let $[g \quad h]$ be the lower half of (7.2.2). By (2.7.4), the matrix of (7.2.2) belongs to $P'_{\mathbf{A}} D'$ if and only if $(h^{-1}g)_v \prec \mathfrak{b}_v \mathfrak{c}_v$ for every $v|\mathfrak{c}$. If $(\beta, \gamma) = (\xi, 1)$, then $h^{-1}g = (\kappa\varphi')^{-1}(1 - \sigma\xi\sigma^{-1})$, so that $(h^{-1}g)_v \prec \mathfrak{b}_v\mathfrak{c}_v$ if and only if $\xi_v \in D^\varphi_v$ for every $v|\mathfrak{c}$ in view of (5.7.6). Now from $S^{-1}(\xi, 1)S \in P^\eta_{\mathbf{A}} D^\eta \Sigma$, we see that $\det(\xi)_v - 1 \in \mathfrak{c}_v$ for every $v|\mathfrak{c}$, and so $\det(\xi) = 1$ if $2 \notin \mathfrak{c}$ and $\xi \in G^\varphi$. Therefore we obtain (1). Next, if $\xi \in C^+$, then $\sigma\xi\sigma^{-1} \prec \mathfrak{g}$, and so from (7.2.2) and (5.7.6) we similarly obtain (2). Finally, if $(2\varphi')^{-1}_v \prec \mathfrak{g}_v$ for every $v \nmid \mathfrak{c}$, then for $\alpha \in C^+$ and such a v, we have $(\sigma\alpha\sigma^{-1})_v \prec \mathfrak{g}_v$ and $(2\varphi')^{-1}_v(\sigma\alpha\sigma^{-1} - 1)_v \prec \mathfrak{g}_v$, so that $\left(Y^{-1}\Sigma S^{-1}(\alpha, 1)S\Sigma^{-1}Y\right)_v \in D'_v$ which implies that $\left(\Sigma S^{-1}(\alpha, 1)S\Sigma^{-1}\right)_v \in D^\varphi_v$. This proves (3). □

7.3. Lemma. *Suppose that $2\varphi'_v \in GL_n(\mathfrak{g}_v)$ for every $v \nmid \mathfrak{c}$. Let $\alpha = (\xi, 1)$ with $\xi \in \Xi$ and $q = S^{-1}(b, a)S$ with $a, b \in G^\varphi_{\mathbf{h}}$ such that $a_v = b_v = 1$ for every $v|\mathfrak{c}$; let $x = S^{-1}\alpha S q \Sigma^{-1}$. Then the following assertions hold:*

(1) *$\nu^\sigma(a^{-1}\xi b)$ is prime to $\mathfrak{c}$;*
(2) *$\mathrm{il}_\mathfrak{b}(x) = \det(\sigma)\nu^\sigma(a^{-1}\xi b)^{-1}$;*
(3) *$(\widehat{\sigma} d_x - 1)_v \prec \mathfrak{c}_v$ for every $v|\mathfrak{c}$, where d_x is the d-block of x in the sense of §2.7.*

Proof. Since $\Sigma x = \Sigma S^{-1}(\xi b, a)S\Sigma^{-1}$, we have, by (7.2.2),

$$Y^{-1}\Sigma x Y = \begin{bmatrix} \sigma\xi b\sigma^{-1} & 0 \\ (\kappa\varphi')^{-1}\sigma(a - \xi b)\sigma^{-1} & \widehat{\sigma}\widehat{a}\cdot{}^t\sigma \end{bmatrix}.$$

Let r and s denote the c-block and d-block of the last matrix. Putting $\mu = 2\kappa^{-1}$, we find that $\mu^{-1}s^{-1}r = (2\varphi')^{-1}\sigma(1 - a^{-1}\xi b)\sigma^{-1}$. Since $Y \in P_{\mathbf{A}} \cap D'$, we have

$$\det(\sigma)^{-1}\mathrm{il}_\mathfrak{b}(x) = \mathrm{il}_\mathfrak{b}(\Sigma x) = \mathrm{il}_\mathfrak{b}\left(Y^{-1}\Sigma x Y\right) = \nu_0\left(\mu^{-1}s^{-1}r\right)^{-1}$$

by Lemma 2.8(2). If $v|\mathfrak{c}$, then $a_v = b_v = 1$ and $\xi \in D^\varphi_v$, so that $(\sigma\xi\sigma^{-1})_v \in \mathfrak{g}_v$ and $(\mu^{-1}s^{-1}r)_v \prec \mathfrak{c}_v$ by (5.7.6). Thus $\nu^\sigma(\xi)_v = \nu_0(\mu^{-1}s^{-1}r)_v = \mathfrak{g}_v$ if $v|\mathfrak{c}$. If $v \nmid \mathfrak{c}$, then $2\varphi'_v \in GL_n(\mathfrak{g}_v)$, so that $\nu_0(\mu^{-1}s^{-1}r) = \nu_0(\sigma a^{-1}\xi b\sigma^{-1}) = \nu^\sigma(a^{-1}\xi b)$. This proves (1) and (2). As for (3), if $v|\mathfrak{c}$ we have

$$(\Sigma x)_v = Y_v \begin{bmatrix} \sigma\xi\sigma^{-1} & 0 \\ (\kappa\varphi')^{-1}\sigma(1 - \xi)\sigma^{-1} & 1 \end{bmatrix}_v \begin{bmatrix} 1 & -\sigma\lambda\cdot{}^t\sigma \\ 0 & 1 \end{bmatrix}_v.$$

Observing that $\left[(\kappa\varphi')^{-1}\sigma(1 - \xi)\sigma^{-1}\right]_v \prec \mathfrak{b}_v\mathfrak{c}_v$ and $(\sigma\lambda\cdot{}^t\sigma)_v \prec \mathfrak{b}_v^{-1}$, we obtain (3). □

7.4. For each $v \in \mathbf{a}$ take φ'_v of (5.3.2) and put $U_v = R_v Q_v Z_v \Sigma_v$ with

$$Q_v = \begin{bmatrix} 1_n & -2^{-1}\varphi'_v \\ -1_n & -2^{-1}\varphi'_v \end{bmatrix}, \qquad Z_v = \begin{bmatrix} 1_n & \tau_v \\ 0 & 1_n \end{bmatrix}, \qquad \Sigma_v = \begin{bmatrix} \sigma & 0 \\ 0 & \sigma_v \end{bmatrix}$$

where $\tau = 2^{-1}\sigma({}^t\lambda - \lambda)\cdot{}^t\sigma$ and where R_v is the matrix of (4.6.3) defined with r_v, t_v as r, t there. We can easily verify that $Q_v\eta\cdot{}^tQ_v = \mathrm{diag}[-\varphi'_v, \varphi'_v]$, which together with (4.6.4) shows that $R_vQ_v \in G^\eta_v$. Clearly $Z_v \in G^\eta_v$, and hence $U_v \in G^\eta_v$.

7.5. Lemma. *Define* $\iota_U : \mathfrak{Z}^\varphi \times \mathfrak{Z}^\varphi \to \mathcal{H}^{\mathbf{a}}_n$ *by*

$$\iota_U(z, w) = \left(U_v^{-1}\iota(z_v, w_v)\right)_{v\in\mathbf{a}} \quad (z, w \in \mathfrak{Z}^\varphi), \tag{7.5.1}$$

where $\iota(z_v, w_v)$ *is defined by (4.6.2) or (4.7.2) with* r_v, θ_v, *and* φ'_v *as* r, θ, *and* φ *there. Also, for* $(\beta, \gamma) \in G^\varphi \times G^\varphi$ *define* $[\beta, \gamma]_S \in G^\eta$ *by*

$$[\beta, \gamma]_S = S^{-1}(\beta, \gamma)S \tag{7.5.2}$$

with S of (7.1.6). Then

$$[\beta, \gamma]_S\iota_U(z, w) = \iota_U(\beta z, \gamma w), \tag{7.5.3}$$

$$j\left([\beta, \gamma]_S, \iota_U(z, w)\right) = \delta(w, z)^{-1}\delta(\gamma w, \beta z)\det(\gamma)j_\gamma(w)j_\beta(z), \tag{7.5.4}$$

$$j\left(U^{-1}, \iota(z, w)\right) = (-1)^{r+t}2^r\det(\sigma)_{\mathbf{a}}\delta(w, z), \tag{7.5.5}$$

$$\delta\left(\iota_U(z, w)\right) = \det\left(2^{-1}\theta\right)\left|2^r\det(\sigma)_{\mathbf{a}}\delta(w, z)\right|^{-2}\delta(z)\delta(w). \tag{7.5.6}$$

Proof. Put $\rho_v = \mathrm{diag}[\sigma_v, \sigma_v]$ with σ_v of (5.3.2). Then $\rho_vS_v = Q_vZ_v\Sigma_v$, and hence, putting $X_v = Z_v\Sigma_v$, $\beta'_v = \sigma_v\beta_v\sigma_v^{-1}$, and $\gamma'_v = \sigma_v\gamma_v\sigma_v^{-1}$, we obtain

$$\begin{aligned}\left(S^{-1}(\beta, \gamma)S\right)_v &= X_v^{-1}Q_v^{-1}\rho_v\,\mathrm{diag}[\beta_v, \gamma_v]\rho_v^{-1}Q_vX_v\\ &= X_v^{-1}Q_v^{-1}\mathrm{diag}\left[\beta'_v, \gamma'_v\right]Q_vX_v\\ &= U_v^{-1}\left[\beta'_v, \gamma'_v\right]_RU_v,\end{aligned}$$

where $[\ ,\ ]_R$ is defined by (4.6.5). As explained in §5.3, the action of γ (resp., β) on $\mathfrak{Z}^\varphi$ is defined by $(\gamma'_v)_{v\in\mathbf{a}}$ (resp., $(\beta'_v)_{v\in\mathbf{a}}$). Therefore (7.5.3) follows from (4.6.6). Put $B_v = R_vQ_v$. Then a direct calculation shows that

$$B_v^{-1} = \begin{bmatrix} 2^{-1}1_r & 0 & 2^{-1}1_r & 0 & 0 & 0\\ 0 & 0 & 0 & 0 & -2^{-1}\theta & 0\\ 0 & 0 & 0 & 2^{-1}1_r & 0 & -2^{-1}1_r\\ 0 & 0 & 0 & 1_r & 0 & 1_r\\ 0 & -2\theta^{-1} & 0 & 0 & 0 & 0\\ 1_r & 0 & -1_r & 0 & 0 & 0 \end{bmatrix}, \tag{7.5.7}$$

where we write simply r and t for r_v and t_v. For $z = \begin{bmatrix} x \\ y \end{bmatrix} \in \mathfrak{Z}(r_v, \theta_v)$ and $w = \begin{bmatrix} u \\ v \end{bmatrix} \in \mathfrak{Z}(r_v, \theta_v)$, we easily find that

$$\mu\left(B_v^{-1}, \iota(z, w)\right) = \begin{bmatrix} 1_r & 0 & 1_r\\ -2\theta^{-1}y & -1_t & 0\\ x + {}^tv\theta^{-1}y & {}^tv & -{}^tu \end{bmatrix}.$$

Taking the determinant, we obtain

$$j\big(B_v^{-1}, \iota(z,w)\big) = (-1)^{r+t}2^r\delta(w,z)$$

with $\delta(w,z)$ of (4.1.9). Dropping the subscript v for simplicity and observing that $j(X^{-1},\mathfrak{z}) = \det(\sigma)$ for every $\mathfrak{z} \in \mathcal{H}_n$, we have thus

$$j\big(U^{-1}, \iota(z,w)\big) = j\big(X^{-1}B^{-1}, \iota(z,w)\big) = (-1)^{r+t}2^r \det(\sigma)\delta(w,z),$$

which is (7.5.5). Combining this result with (4.2.6) and (4.6.8), we obtain (7.5.6). Since $j(U, U^{-1}\mathfrak{z}) = j\big(U^{-1},\mathfrak{z}\big)^{-1}$, we have, by (4.6.7),

$$\begin{aligned} j([\beta,\gamma]_S, \iota_U(z,w)) &= j\big(U^{-1}[\beta',\gamma']_R U, \iota_U(z,w)\big) \\ &= j\big(U^{-1}, [\beta',\gamma']_R\iota(z,w)\big) j\big([\beta',\gamma']_R, \iota(z,w)\big) j\big(U, U^{-1}\iota(z,w)\big) \\ &= \delta(\gamma w, \beta z)\det(\gamma) j_\gamma(w) j_\beta(z)\delta(w,z)^{-1}, \end{aligned}$$

which proves (7.5.4). □

7.6. Let us now assume that φ is anisotropic. Then we have

$$G^\omega = \bigsqcup_{\xi \in G^\varphi} P^\omega(\xi, 1).$$

This is a special case of [S5, (2.7.1)], since l there is 0 in the present case. Consequently

(7.6.1) $$G^\eta = \bigsqcup_{\xi \in G^\varphi} P^\eta S^{-1}(\xi, 1)S.$$

We now consider E of (6.4.1) with the set of data $\{k, \nu, \mathfrak{b}, \mathfrak{c}, \chi\}$ satisfying (6.3.3) and (6.3.4); here we assume that $2 \notin \mathfrak{c} \subset \mathfrak{e}$ with $\mathfrak{e}$ of Theorem 5.8, and we take $\mathfrak{b} = 2\kappa^{-1}\mathfrak{g}$ as in Lemma 7.2. We also take $\mathfrak{B}$ as in Lemma 5.11(1). Pick two elements $a, b \in \mathfrak{B}$. We are interested in $E(xq\Sigma_{\mathbf{h}}^{-1}, s)$ with the elements Σ and $q = S^{-1}(b,a)S$ of $G_{\mathbf{h}}^\eta$ given in Lemmas 7.2 and 7.3. To be explicit, we have

$$E\big(xq\Sigma_{\mathbf{h}}^{-1}, s\big) = \sum_{\alpha \in A} \mu\big(\alpha xq\Sigma_{\mathbf{h}}^{-1}\big)\varepsilon\big(\alpha xq\Sigma_{\mathbf{h}}^{-1}\big)^{-2s} \quad \big(x \in G_{\mathbf{A}}^\eta, A = P^\eta\backslash G^\eta\big).$$

The function μ is given by (6.3.6) and (6.3.7).

Now, from $E(xq\Sigma_{\mathbf{h}}^{-1}, s)$ we obtain a function $\mathscr{E}_{a,b}(\mathfrak{z}, s)$ of $(\mathfrak{z}, s) \in \mathcal{H}_n^{\mathbf{a}} \times \mathbf{C}$ by the same principle as in (6.4.8), that is, $\mathscr{E}_{a,b}\big(y(\mathbf{1}), s\big) = E\big(q\Sigma_{\mathbf{h}}^{-1}y, s\big) j_y^k(\mathbf{1})$ for every $y \in G_{\mathbf{a}}^\eta$. Then putting $\mathfrak{z} = y(\mathbf{1})$, we have $\mathscr{E}_{a,b}(\mathfrak{z}, s) = \sum_{\alpha \in A} p_\alpha(\mathfrak{z})$ with

(7.6.2) $$p_\alpha(\mathfrak{z}) = \mu\big(\alpha q\Sigma_{\mathbf{h}}^{-1}y\big)\varepsilon\big(\alpha q\Sigma_{\mathbf{h}}^{-1}y\big)^{-2s} j_y^k(\mathbf{1}).$$

By (7.6.1) we can take the elements $S^{-1}(\xi, 1)S$ with $\xi \in G^{\varphi}$ to be A. Fix $\xi \in G^{\varphi}$, and put $\alpha = S^{-1}(\xi, 1)S$ and $x = \alpha q \Sigma_{\mathbf{h}}^{-1} y$. Taking this to be x of (6.4.3), we easily see that

$$p_{\alpha}(\mathfrak{z}) = \mu\big(\alpha_{\mathbf{h}} q \Sigma_{\mathbf{h}}^{-1}\big)\varepsilon\big(\alpha_{\mathbf{h}} q \Sigma_{\mathbf{h}}^{-1}\big)^{-2s}\delta(\mathfrak{z})^{su-(k-i\kappa)/2}\|_{k}\alpha, \tag{7.6.3}$$

where $\delta(\mathfrak{z}) = \det\big(2^{-1}(\mathfrak{z} + {}^{t}\mathfrak{z})\big)$. Since $\mu = 0$ outside of $P_{\mathbf{A}}^{\eta} D^{\eta}$, our sum is extended over all $\xi \in G^{\varphi}$ such that $S^{-1}(\xi b, a)S \in P_{\mathbf{A}}^{\eta} D^{\eta} \Sigma$. By Lemma 7.2(1), this is so if and only if $\xi \in \Xi$, since $G_{v}^{\eta} = P_{v}^{\eta} D_{v}^{\eta}$ for $v \nmid \mathfrak{c}$ and $a_v = b_v = 1$ for $v | \mathfrak{c}$. Therefore we hereafter assume $\xi \in \Xi$. To simplify our notation, put $\nu_{a,b}^{\sigma}(\xi) = \nu^{\sigma}(a^{-1}\xi b)$, $N_{a,b}^{\sigma}(\xi) = N\big(\nu^{\sigma}(a^{-1}\xi b)\big)$, and $\mathbf{s} = su - (k - i\kappa)/2$. Since $x \in P_{\mathbf{A}}^{\eta} D^{\eta}$, we can put $x = pw$ with $p \in P_{\mathbf{A}}^{\eta}$ and $w \in D^{\eta}$. Then we have $d_x = d_p d_w$ and $\det(d_p)\mathfrak{g} = \mathrm{il}_{\mathfrak{b}}(x)$, and so by Lemma 7.3,

$$\begin{aligned}
\mu(x_{\mathbf{h}}) &= \chi_{\mathbf{h}}\big(\det(d_p)\big)^{-1}\chi_{\mathfrak{c}}\big(\det(d_w)\big)^{-1} \\
&= (\chi_{\mathbf{h}}/\chi_{\mathfrak{c}})\big(\det(d_p)\big)^{-1}\chi_{\mathfrak{c}}\big(\det(d_x)\big)^{-1} \\
&= \chi^{*}\big(\nu_{a,b}^{\sigma}(\xi)\big)(\chi_{\mathbf{h}}/\chi_{\mathfrak{c}})\big(\det(\sigma)\big)^{-1}\chi_{\mathfrak{c}}\big(\det(\sigma)\big)^{-1} \\
&= \chi_{\mathbf{h}}\big(\det(\sigma)\big)^{-1}\chi^{*}\big(\nu_{a,b}^{\sigma}(\xi)\big).
\end{aligned}$$

We have also, by (6.3.8) and Lemma 7.3(2),

$$\varepsilon\big(\alpha_{\mathbf{h}} q \Sigma_{\mathbf{h}}^{-1}\big) = \varepsilon(x_{\mathbf{h}}) = N\big(\mathrm{il}_{\mathfrak{b}}(x)\big)^{-1} = N_{a,b}^{\sigma}(\xi)|\det(\sigma_{\mathbf{h}})|_{\mathbf{A}}.$$

So far we have been considering an arbitrary weight k, and we can still continue our discussion with it. However, since our theorems concern the case $k = 0$, let us now take k to be 0. Put $H_{a,b}(z, w; s) = \mathscr{E}_{a,b}\big(\iota_U(z, w), s\big)$. From (7.5.3) and (7.5.6) we obtain

$$(\delta^{\mathbf{s}} \circ \alpha)\big(\iota_U(z, w)\big) = \det\big(2^{-1}\theta\big)^{\mathbf{s}}|2^{r}\det(\sigma_{\mathbf{a}})|^{-2\mathbf{s}}|\delta(w)\delta(\xi z)\delta(w, \xi z)^{-2}|^{\mathbf{s}}.$$

Combining these and suppressing s in $H_{a,b}(z, w; s)$, we thus obtain

$$H_{a,b}(z, w) = A(s)\sum_{\xi \in \Xi}\chi^{*}\big(\nu_{a,b}^{\sigma}(\xi)\big)N_{a,b}^{\sigma}(\xi)^{-2s}\left|\delta(\xi z)\delta(w)\delta(w, \xi z)^{-2}\right|^{\mathbf{s}} \tag{7.6.4}$$

with

$$A(s) = \chi_{\mathbf{h}}\big(\det(\sigma)\big)^{-1}|\det(\sigma_{\mathbf{h}})|_{\mathbf{A}}^{-2s}\det\big(2^{-1}\theta\big)^{\mathbf{s}}|2^{r}\det(\sigma_{\mathbf{a}})|^{-2\mathbf{s}}.$$

7.7. Put $\Gamma^{a} = G_{+}^{\varphi} \cap aD^{\varphi}a^{-1}$. Assuming that $\mathfrak{Z}^{\varphi}$ is nontrivial, we consider

$$\int_{\Phi_a} H_{a,b}(z, w)\mathbf{d}w \quad (\Phi_a = \Gamma^{a}\backslash\mathfrak{Z}^{\varphi}), \tag{7.7.1}$$

where $\mathbf{d}w$ is defined by (5.5.3). Take a complete set of representatives R^a for $\Gamma^a\backslash\Xi$, and take an eigenfunction $\mathbf{f} \in \mathcal{M}(D^\varphi)$ as in §5.10. Observe that each term of (7.6.4) is invariant under $(w, \xi) \mapsto (\gamma w, \gamma\xi)$ for every $\gamma \in \Gamma^a$. Let $H'(z, w)$ denote the function defined by the right-hand side of (7.6.4) with $\sum_{\xi \in R^a}$ in place of $\sum_{\xi\in\Xi}$. Then

$$H_{a,b}(z, w) = \sum_{\gamma\in\Gamma^a} H'(z, \gamma w),$$

so that (7.7.1) equals

$$\ell \int_{\mathfrak{Z}^\varphi} H'(z, w)\mathbf{d}w,$$

where $\ell = [\Gamma^a \cap \{\pm 1\} : 1]$. By (4.5.1) this gives

(7.7.2) $$\ell \cdot c(s)A(s) \sum_{\xi\in R^a} \chi^*(\nu^\sigma_{a,b}(\xi))N^\sigma_{a,b}(\xi)^{-2s}$$

with

$$c(s) = \prod_{v\in\mathbf{a}} 2^{r_v n/2} \det(\theta_v)^{r_v/2}\pi^{r_v(n-r_v)/2}\Gamma_{r_v}(s_v - (n - r_v - 1)/2)\Gamma_{r_v}(s_v + 1/2)^{-1},$$

where $s_v = s + (i\kappa_v/2)$.

7.8. Employing (5.12.4) and (5.12.6), from the above calculation we obtain

(7.8.1) $$c(s)A(s)Z(2s, \mathbf{f}, \chi)\mathbf{f}(b) = \ell^{-1}\Lambda_\mathfrak{c}(s)\sum_{a\in\mathfrak{B}} \mathbf{f}(a)\int_{\Phi_a} H_{a,b}(z, w; s)\mathbf{d}w$$

with $\Lambda_\mathfrak{c}$ of (6.13.1).

Let $Z(s)$ denote the function $Z(s, \mathbf{f}, \chi)$ with trivial χ. We are going to take the residue of the above equality in this case at $s = \mu$. Recall that $H_{a,b}(z, w; s) = E(yq\Sigma_\mathbf{h}^{-1}, s)$ with $q = S^{-1}(b, a)S$ and $y \in G^\eta_\mathbf{a}$ such that $y(\mathbf{1}) = \iota_U(z, w)$. Thus the residue of $H_{a,b}(z, w; s)$ at $s = \mu$ is the residue of E_r at μ, where $r = q\Sigma_\mathbf{h}^{-1}$. Since $\det(r_v) = 1$ for every $v \in \mathbf{h}$, Lemma 2.11 shows that $r\eta_\mathbf{h} \in G^\eta Q^\eta_\mathbf{h} D^\eta[\mathfrak{b}\mathfrak{c}, \mathfrak{b}^{-1}]$, and hence Theorem 6.14 is applicable to the present E_r. Thus the residue of $H_{a,b}$ at $s = \mu$ is $R(\mathfrak{c})$ of that theorem. Now $A(s) = |\det(\sigma_\mathbf{h})|_\mathbf{A}^{-2s}|\det(2^{-1}\theta)2^{-2r}\det(\sigma_\mathbf{a})^{-2}|^{su}$. From (5.3.2) and (5.7.7) we obtain $|2^{-n}\kappa_v^n \det(2\varphi_v)\det(\sigma_v)^2| = |\det(\theta_v)|$ for $v \in \mathbf{a}$ and $|\det(\sigma_v)^2\det(2\varphi_v)| = |\det(2\theta_v)|$ for $v \in \mathbf{h}$. Since $\mathfrak{b} = 2\kappa^{-1}\mathfrak{g}$ and $|\det(2\theta)_\mathbf{h}|_\mathbf{A}^{-1} = [\widetilde{L} : L]$ by (3.1.9), we thus obtain

(7.8.2) $$A(\mu) = 2^{-nd\mu}[\widetilde{L} : L]^\mu N(\mathfrak{b})^{-n\mu},$$

where $d = [F : \mathbf{Q}]$.

7.9. In the above computation we assumed that $\mathfrak{Z}^\varphi$ is nontrivial, that is, φ_v is indefinite for some $v \in \mathbf{a}$. If φ is totally definite, then we understand that $\mathfrak{Z}^\varphi$ consists of a single point $\mathbf{1}^\varphi$, and (7.6.4) can be written as

$$H_{a,b}(\mathbf{1}^\varphi, \mathbf{1}^\varphi; s) = A(s)\big[\Gamma^a : 1\big] \sum_{\xi \in R^a} \chi^*\big(\nu^\sigma_{a,b}(\xi)\big) N^\sigma_{a,b}(\xi)^{-2s}.$$

Therefore, instead of (7.8.1), we have

$$A(s)Z(2s, \mathbf{f}, \chi)\mathbf{f}(b) = \Lambda_\mathfrak{c}(s) \sum_{a \in \mathfrak{B}} \big[\Gamma^a : 1\big]^{-1} \mathbf{f}(a) H_{a,b}(\mathbf{1}^\varphi, \mathbf{1}^\varphi; s). \tag{7.9.1}$$

7.10. To prove Theorem 5.8, we first assume that $2 \notin \mathfrak{c} \subset \mathfrak{e}$. Take $\mathbf{f}$ to be the constant 1 on $G^\varphi_\mathbf{A}$. Then Z in this case is given by (5.15.2). Let ρ be its residue at $s = \mu$. Take the residue of (7.8.1) or (7.9.1) at $s = \mu$ and observe that the right-hand side gives $\Lambda_\mathfrak{c}(\mu) R(\mathfrak{c}) \mathfrak{m}\big(G^\varphi_+, D(\mathfrak{c})\big)$. Thus

$$\Lambda_\mathfrak{c}(\mu) R(\mathfrak{c}) \mathfrak{m}\big(G^\varphi_+, D(\mathfrak{c})\big) = A(\mu)\rho \cdot \begin{cases} 1 & \text{if } G^\varphi_\mathbf{a} \text{ is compact,} \\ c(\mu) & \text{otherwise.} \end{cases}$$

Now $R(\mathfrak{c})$ and $A(\mu)$ are given by Theorem 6.14 and (7.8.2); ρ can be obtained from (5.15.2). For example, $\rho = 2^{-1} r(\mathfrak{c}) L_\mathfrak{c}(n/2, \psi) \prod_{i=2}^{n-1} \zeta_{F,\mathfrak{c}}(i)$ with $r(\mathfrak{c})$ of Theorem 6.14 if n is even. Using these explicit values and arranging various factors suitably, we obtain the formula of Theorem 5.8 when $2 \notin \mathfrak{c} \subset \mathfrak{e}$. Notice that $c(\mu) = b_\varphi$.

Next, given an arbitrary $\mathfrak{c}$, take an integral ideal $\mathfrak{a}$ prime to $\mathfrak{c}$ so that $2 \notin \mathfrak{ac} \subset \mathfrak{e}$. Then the above result is applicable to $\mathfrak{m}(G^\varphi_+, D(\mathfrak{ac}))$. By [S5, Lemma 24.2(1)] we have

$$\begin{aligned} \mathfrak{m}\big(G^\varphi_+, D(\mathfrak{c})\big) &= \mathfrak{m}\big(G^\varphi_+, D(\mathfrak{ac})\big)[D(\mathfrak{c}) : D(\mathfrak{ac})]^{-1} \\ &= \mathfrak{m}\big(G^\varphi_+, D(\mathfrak{ac})\big) \prod_{v|\mathfrak{a}} \big[G^\varphi_v \cap C_+ : G^\varphi_v \cap D(\mathfrak{a})\big]^{-1}. \end{aligned}$$

For $v|\mathfrak{a}$ put $D^\varphi_v = \big\{\gamma \in G^\varphi_v \big| L_v\gamma = L_v, \widetilde{L}_v(\gamma - 1) \subset \mathfrak{a}_v L_v\big\}$. We can choose $\mathfrak{a}$ so that $D^\varphi_v \subset \big(G^\varphi_+\big)_v$ for every $v|\mathfrak{a}$. Then $[G^\varphi_v \cap C_+ : G^\varphi_v \cap D(\mathfrak{a})]$ coincides with $2^{-1}[C^\varphi_v : D^\varphi_v]$ with C^φ of (5.7.3), and the last index can be given by Proposition 3.9. Employing the formula for $\mathfrak{m}\big(G^\varphi_+, D(\mathfrak{ac})\big)$ we already have and arranging again various factors suitably, we obtain the formula of Theorem 5.8 for an arbitrary $\mathfrak{c}$.

7.11. To complete the proof of Theorem 5.14, we may assume that $\mathbf{f}(b) \neq 0$, as we already remarked there. Then the desired meromorphic continuation of $Z(2s, \mathbf{f}, \chi)$ follows from (7.8.1) or (7.9.1).

8. Representation densities

8.1. The purpose of this section is to show that the index $[C : D_1]$ of Proposition 3.9 is closely connected with the local representation density at a given prime in Siegel's sense. It is actually a "modified density," whose definition is somewhat different from that of Siegel, though we shall show that our density differs from his only by a power of q.

The setting is local and is the same as in Section 3. We first put

(8.1.1) $$R = \mathfrak{g}_n^n, \quad E = GL_n(\mathfrak{g}),$$

(8.1.2) $$G^\psi = \{\alpha \in GL_n(F) | \alpha\psi \cdot {}^t\alpha = \psi\}, \qquad C^\psi = G^\psi \cap E,$$

(8.1.3) $$T = \{x \in F_n^n | {}^t x = x, x_{ii} \in \mathfrak{g}, 2x_{ij} \in \mathfrak{g} \text{ if } i \neq j\},$$

(8.1.4) $$Z_m^\psi = \left\{z \in R | {}^t z\psi^{-1} z - \psi^{-1} \in 4\mathfrak{p}^m T\right\},$$

(8.1.5) $$Y_m^\psi = \{y \in R | {}^t y\psi y - \psi \prec \mathfrak{p}^m\}.$$

Here and in the following, ψ denotes an element of $T \cap GL_n(F)$ and m a positive integer; we assume moreover that $\psi \in R$ in the definition of Y_m^ψ. We easily see that if $z \in Z_m^\psi$ (resp., $y \in Y_m^\psi$), then $z + 2\psi w \in Z_m^\psi$ (resp., $y + w \in Y_m^\psi$) for every $w \in \mathfrak{p}^m R$. Therefore we can speak of Z_m^ψ modulo $2\pi^m \psi R$ and Y_m^ψ modulo $\pi^m R$, which we denote by $Z_m^\psi / 2\pi^m \psi R$ and $Y_m^\psi / \pi^m R$. We then put

(8.1.6) $$g_m^\psi = \#(Z_m^\psi / 2\pi^m \psi R), \qquad e_m^\psi = \#(Y_m^\psi / \pi^m R),$$

(8.1.7) $$g(\psi) = \lim_{m \to \infty} q^{-km} g_m^\psi, \quad e(\psi) = \lim_{m \to \infty} q^{-km} e_m^\psi,$$

where $k = n(n-1)/2$. The number $e(\psi)$ is exactly the representation density of ψ at a fixed prime defined by Siegel. Since $2T \subset R$, the symbols $Y_m^{2\psi}$, $e_m^{2\psi}$, and $e(2\psi)$ are meaningful for every $\psi \in T$.

8.2. **Lemma.** (1) *If $z \in Z_m^\psi$, then $z \in E$ and $\psi^{-1} z\psi \in E$.*

(2) *If $z \in Z_m^\psi$, then there exists an element ξ of C^ψ such that $\xi - z \in 2\pi^m \psi R$.*

(3) *Let $D_m^\psi = C^\psi \cap (1 + 2\pi^m \psi R)$. Then, assigning αD_m^ψ to the class of α modulo $2\pi^m \psi R$ for $\alpha \in C^\psi$, we obtain a bijection of C^ψ / D_m^ψ onto $Z_m^\psi / 2\pi^m \psi R$.*

Proof. The first two assertions were proved in [S5, Lemma 17.2]; see the two lines preceding [S5, Lemma 17.3]. As for (3), since $\alpha\psi = \psi\widehat{\alpha}$ for every $\alpha \in C^\psi$, we easily see from (3.6.3) that the map is well defined, and it is surjective by virtue of (2). Then it is clearly injective. □

8.3. **Lemma.** (1) *Let τ be an element of $R \cap GL_n(F)$ and ν a positive integer such that $\pi^\nu R \subset 2\tau R\tau$. Then*

$$\#\{x \in R\tau / \mathfrak{p}^\nu R | {}^t\tau x + {}^t x\tau \prec \mathfrak{p}^\nu\} = q^{\nu k} |\det(2\tau)|^{-1}.$$

(2) *Let a be an element of $GL_n(F)$ and λ a positive integer such that $\pi^\lambda R \subset {}^t aRa$. Then*

$$\#\{w \in {}^t aRa / \pi^\lambda R | {}^t w + w \in 2\pi^\lambda T\} = q^{\lambda k} |\det(a)|^{n-1}.$$

Proof. Take $u, v \in E$ so that $u\tau v$ is diagonal. Changing τ for $u\tau v$, we can reduce our problem to the case of diagonal τ. Thus put $\tau = \mathrm{diag}[s_1, \ldots, s_n]$ with s_i such that $s_i\mathfrak{g} \supset s_{i+1}\mathfrak{g}$ and $x = z\tau$ with $z \in R$. Then $x_{ij} = z_{ij}s_j$, so that ${}^t\tau x + {}^t x\tau \prec \mathfrak{p}^\nu$ if and only if $s_i s_j(z_{ij} + z_{ji}) \in \mathfrak{p}^\nu$. Therefore we have to count the number of $z_{ij} \in \mathfrak{g}/s_j^{-1}\mathfrak{p}^\nu$ satisfying that condition. Clearly $z_{ii} \in 2^{-1}s_i^{-2}\mathfrak{p}^\nu/s_i^{-1}\mathfrak{p}^\nu$. If $i < j$, we can choose z_{ji} (mod $s_i^{-1}\mathfrak{p}^\nu$) arbitrarily. Then we have to take z_{ij} (mod $s_j^{-1}\mathfrak{p}^\nu$) so that $z_{ij} = -z_{ji} + w$ with an arbitrary $w \in (s_i s_j)^{-1}\mathfrak{p}^\nu/s_j^{-1}\mathfrak{p}^\nu$. Combining these, we obtain (1). Assertion (2) can be proved in a similar, and in fact, simpler, way. □

8.4. LEMMA. *Let κ be the integer such that $2\mathfrak{g} = \mathfrak{p}^\kappa$, and let $\psi \in T$. Then*
(1) $e(2\psi) = q^{\kappa n}g(\psi)|\det(2\psi)|^{-1}$,
(2) $e(\psi) = q^{-\kappa n(n+1)/2}e(2\psi) = g(\psi)q^{-k\kappa}|\det(2\psi)|^{-1}$ *if* $\psi \in R$.

Proof. To prove (1), put

$$W_m^\psi = \{x \in R \mid {}^t x\psi x - \psi \in 4\mathfrak{p}^m \psi T\psi\}.$$

Then we easily see that the map $z \to \psi^{-1}z\psi$ gives a bijection of Z_m^ψ onto W_m^ψ, and also that W_m^ψ modulo $2\pi^m R\psi$ is meaningful. Take a positive integer ν so that $\pi^\nu R \subset 4\psi R\psi$. Assigning y (mod $2\pi^m R\psi$) to $y \in Y_{m+\nu}^{2\psi}$, we obtain a map of $Y_{m+\nu}^{2\psi}/\pi^{m+\nu}R$ into $W_m^\psi/2\pi^m R\psi$. This is surjective. Indeed, given $x \in W_m^\psi$, by Lemma 8.2(2) we can find an element α of C^ψ such that $\psi^{-1}(\alpha - \psi x\psi^{-1}) \prec 2\mathfrak{p}^m$. Then $\widehat{\alpha} - x \in 2\pi^m R\psi$ and $\widehat{\alpha} \in Y_m^{2\psi}$, which gives the surjectivity. Now two elements y and z of $Y_{m+\nu}^{2\psi}$ go to the same element of $W_m^\psi/2\pi^m R\psi$ if and only if $z = y + 2\pi^m w$ with $w \in R\psi$. By Lemma 8.2(1) we see that $\psi^{-1}\widehat{y}\psi \in E$. Fixing $y \in Y_{m+\nu}^{2\psi}$, put $y' = y + \pi^m\psi^{-1}\widehat{y}\psi u$ with $u \in R(2\psi)$. Then $y' \in Y_{m+\nu}^{2\psi}$ if and only if $2\psi u + {}^t u(2\psi) \prec \mathfrak{p}^\nu$, provided $m \geq \nu$. Thus the number of the elements of $Y_{m+\nu}^{2\psi}/\pi^{m+\nu}R$ that go to an element of W_m^ψ modulo $2\pi^m R\psi$ is exactly the number we determined in Lemma 8.3 if we take 2ψ to be τ there. This shows that

$$e_{m+\nu}^{2\psi} = g_m^\psi q^{\nu k}|\det(4\psi)|^{-1}$$

for $m \geq \nu$, from which we obtain (1). To prove (2), we observe that $Y_m^\psi = Y_{m+\kappa}^{2\psi}$, and hence $e_{m+\kappa}^{2\psi} = q^{\kappa n^2}e_m^\psi$. Thus $e(2\psi) = e(\psi)q^{\kappa n(n+1)/2}$, which combined with (1) gives (2). □

8.5. LEMMA. *Define D_m^ψ as in Lemma 8.2. Then* $\left[D_m^\psi : D_{m+1}^\psi\right] = q^k$.

Proof. Given $\alpha \in D_m^\psi$, put $\alpha = 1 + 2\pi^m\psi x$ with $x \in R$. Then ${}^t x + x = -2\pi^m \cdot {}^t x\psi x \in 2\pi^m T$, so that $x_{ii} \in \mathfrak{p}^m$ and $x_{ij} + x_{ji} \in \mathfrak{p}^m$ for $i \neq j$. Clearly $\alpha \in D_{m+1}^\psi$ if and only if $x \prec \mathfrak{p}$. Assigning $(x_{ij} (\mathrm{mod}\, \mathfrak{p}))_{i<j}$ to α, we obtain an injection of D_m^ψ/D_{m+1}^ψ into $(\mathfrak{g}/\mathfrak{p})^k$. To see that this is surjective, take $(y_{ij})_{i<j} \in \mathfrak{g}^k$. Define $x \in \mathfrak{g}_n^n$ by $x_{ii} = 0$

and $x_{ij} = -x_{ji} = y_{ij}$ for $i < j$, and put $z = 1+2\pi^m\psi x$. Then $z \in Z^{\psi}_{2m}$, so that Lemma 8.2(2) guarantees an element $\alpha \in C^{\psi}$ such that $\alpha - z \in 2\pi^{2m}\psi R$. Since clearly $\alpha \in D^{\psi}_m$, we obtain the desired surjectivity, which completes the proof. □

8.6. THEOREM. *For every $\psi \in T \cap GL_n(F)$, the following assertions hold:*
(1) $g(\psi) = q^{-k}[C^{\psi} : D^{\psi}_1]$;
(2) $e(2\psi) = q^{\kappa n-k}|\det(2\psi)|^{-1}[C^{\psi} : D^{\psi}_1]$ *with κ of Lemma 8.4;*
(3) *in particular, for φ of §3.1 we have $e(2\varphi) = q^{\kappa n-k}[\widetilde{L} : L][C^{\varphi} : D^{\varphi}_1]$ with $[\widetilde{L} : L]$ of (3.2.1).*

Proof. By Lemma 8.2(3) and Lemma 8.5 we have $g^{\psi}_m = [C^{\psi} : D^{\psi}_m] = [C^{\psi} : D^{\psi}_1]q^{k(m-1)}$, from which we obtain (1). This combined with Lemma 8.4(1) proves (2) and (3), since $|\det(2\varphi)|^{-1} = [\widetilde{L} : L]$ as noted in (3.1.9). □

8.7. LEMMA. *If $\psi = a\varphi\cdot{}^t a$ with $a \in GL_n(F)$, then $g(\varphi)[C^{\psi} : C^{\psi} \cap aC^{\varphi}a^{-1}] = |\det(a)|^{n-1}g(\psi)[C^{\varphi} : C^{\varphi} \cap a^{-1}C^{\psi}a]$.*

Proof. We see that $G^{\psi} = aG^{\varphi}a^{-1}$ and $a^{-1}C^{\psi}a = G^{\varphi} \cap a^{-1}Ea$, so that $C^{\varphi} \cap a^{-1}C^{\psi}a = C^{\varphi} \cap a^{-1}Ea = E \cap a^{-1}C^{\psi}a$. For any $b \in GL_n(F)$ put

$$A^{\psi}_m(b) = \left\{z \in E \cap b^{-1}Eb \,\middle|\, {}^t z\psi^{-1}z - \psi^{-1} \in 4\mathfrak{p}^m T\right\},$$
$$B^{\psi}_m(b) = \left\{z \in E \cap b^{-1}Eb \,\middle|\, {}^t z\psi^{-1}z - \psi^{-1} \in 4\mathfrak{p}^m \cdot {}^t bTb\right\}.$$

Take m so that $\pi^m a^{-1}Ra \subset \pi R$. Then we easily see that the map of C^{ψ} to Z^{ψ}_m given in Lemma 8.2(2) gives a surjective map of $C^{\psi} \cap aEa^{-1}$ onto $A^{\psi}_m(a^{-1})/2\pi^m\psi R$, and hence

(*) $$g^{\psi}_m = [C^{\psi} : C^{\psi} \cap aEa^{-1}]\#(A^{\psi}_m(a^{-1})/2\pi^m\psi R).$$

Similarly, if $\pi^m aRa^{-1} \subset \pi R$, then

(**) $$g^{\varphi}_m = [C^{\varphi} : C^{\varphi} \cap a^{-1}Ea]\#(A^{\varphi}_m(a)/2\pi^m\varphi R).$$

Clearly the map $x \mapsto a^{-1}xa$ gives a bijection of $A^{\psi}_m(a^{-1})/2\pi^m\psi R$ onto $B^{\varphi}_m(a)/2\pi^m \varphi\cdot{}^t aRa$. To count the number of elements in the last set, take a positive integer h so that $\pi^h R \subset {}^t aRa$. Take $y \in A^{\varphi}_{m+h}(a)$ and to the class of y modulo $2\pi^{m+h}\varphi R$ assign the class of y modulo $2\pi^m\varphi\cdot{}^t aRa$. Let $z \in B^{\varphi}_m(a)$; then $aza^{-1} \in A^{\psi}_m(a^{-1})$. Take $\beta \in C^{\psi} \cap aEa^{-1}$ so that $\beta - aza^{-1} \in 2\pi^m\psi R$ and put $\gamma = a^{-1}\beta a$. Then $\gamma \in C^{\varphi} \cap a^{-1}Ea$, $\gamma - z \in 2\pi^m\varphi\cdot{}^t aRa$, and $\gamma \in A^{\varphi}_{m+h}(a)$. Thus we obtain a surjective map of $A^{\varphi}_{m+h}(a)/2\pi^{m+h}\varphi R$ onto $B^{\varphi}_m(a)/2\pi^m\varphi\cdot{}^t aRa$. Now let $y' = y+2\pi^m\varphi x$ with $y \in A^{\varphi}_{m+h}(a)$ and $x \in {}^t aRa$. Since $\widehat{y} \in {}^t aE\widehat{a}$, we can put $x = \widehat{y}w$ with $w \in {}^t aRa$. Take a sufficiently large m such that $\pi^m \cdot {}^t aRaT \cdot {}^t aRa \subset \pi^h T$. Then $y' \in A^{\varphi}_{m+h}(a)$

if and only if $w + {}^t w \in 2\pi^h T$. Applying Lemma 8.3(2) to this setting, we find that

$$\#\left(A^{\varphi}_{m+h}(a)/2\pi^{m+h}\varphi R\right) = q^{kh}|\det(a)|^{n-1}\#\left(B^{\varphi}_{m}(a)/2\pi^{m}\varphi\cdot{}^t aRa\right).$$

Combining this with (∗) and (∗∗), we find that

$$g^{\varphi}_{m+h}\left[C^{\psi} : C^{\psi}\cap aC^{\varphi}a^{-1}\right] = q^{kh}|\det(a)|^{n-1}g^{\psi}_{m}\left[C^{\varphi} : C^{\varphi}\cap a^{-1}C^{\psi}a\right],$$

from which we obtain the desired equality. □

Theorem 8.6 may be viewed as an interpretation of each nonarchimedean local factor of (5.8.1) in terms of representation densities.

8.8. Let us now return to the global setting of Section 5. Take $\psi = {}^t\psi \in GL_n(F)$ such that $x\psi\cdot{}^t y \in \mathfrak{g}$ for every $x, y \in \mathfrak{g}^1_n$. Then for each $v \in \mathbf{h}$ we let $g_v(\psi)$ and $e_v(\psi)$ denote the number defined by (8.1.7) at v; $e_v(\psi)$ is meaningful only when $\psi \prec \mathfrak{g}$. Put $\varepsilon(\psi) = \prod_{v\in\mathbf{h}} 2^{-1}e_v(\psi)$. This is the product of representation densities considered by Minkowski and Siegel. From Theorem 8.6(2) and Proposition 3.9 we see that $\varepsilon(2\varphi)$ is essentially the inverse of the product considered in Theorem 5.8, up to a factor which is the product of representation densities at the archimedean primes. In this way we can recover Siegel's formula for 2φ when φ is anisotropic.

Conversely, Theorem 5.8 can be obtained by combining his formula with Proposition 3.9 and Theorem 8.6. This works even for isotropic φ, but we shall not go into details here. However, there is one nontrivial technical point. In our formulation we took a $\mathfrak{g}$-maximal lattice L, but it may not have a free $\mathfrak{g}$-basis, and so we may not be able to represent φ (or 2φ) globally by a matrix with entries in $\mathfrak{g}$. Thus $e_v(2\varphi)$ in our formulation means the density defined with respect to the local matrix representation of φ_v with respect to a $\mathfrak{g}_v$-basis of L_v. This modification is indispensable if one wishes to obtain Theorem 5.8 from Siegel's formula.

The case of an arbitrary $\mathfrak{g}$-lattice M in V can be handled, at least theoretically, by means of (5.8.1) or by Lemma 8.7, though actual calculation is almost always complicated.

9. The function $\xi(y, h; s)$

9.1. In this section we deal only with real and complex matrices. We fix an integer $n > 1$ throughout, and we define X, Y, and Q as follows:

(9.1.1) $$X = \left\{x \in \mathbf{R}^n_n \,\middle|\, {}^t x = -x\right\}, \qquad Y = \left\{y \in \mathbf{R}^n_n \,\middle|\, {}^t y = y\right\},$$

(9.1.2) $$Q = \left\{y \in Y \,\middle|\, y > 0\right\}.$$

We define $\delta : \mathbf{C}^n_n \to \mathbf{R}$ by

(9.1.3) $$\delta(z) = |\det(z)| \quad \text{for } z \in \mathbf{C}^n_n.$$

Hereafter we let U denote either X or Y. For every $p \in Q$ and $a \in GL_n(\mathbf{R})$ we define $\mathbf{R}$-linear endomorphisms T_p^U and R_a^U of U by

$$T_p^U(v) = (1/2)(pv + vp), \qquad R_a^U(v) = av \cdot {}^t a \qquad (v \in U). \tag{9.1.4}$$

9.2. LEMMA. *If $a \in GL_n(\mathbf{R})$, ${}^t aa = 1$, $q \in Q$, $p = aq \cdot {}^t a$, and $w \in U$, then $R_a^U T_q^U = T_p^U R_a^U$ and $\mathrm{tr}[{}^t(R_a^U w)((T_p^U)^{-1} R_a^U w)] = \mathrm{tr}[{}^t w \cdot (T_q^U)^{-1} w]$.*

Proof. The first formula can be verified in a straightforward way. Employing it, we obtain $(T_p^U)^{-1} R_a^U w = R_a^U (T_q^U)^{-1} w$, and hence, the left-hand side of the second equality equals

$$\mathrm{tr}\Big[{}^t(aw{}^t a)a \cdot \big(T_q^U\big)^{-1}(w) \cdot {}^t a\Big] = \mathrm{tr}\Big[a \cdot {}^t w \big(T_q^U\big)^{-1}(w) \cdot {}^t a\Big] = \mathrm{tr}\Big[{}^t w \big(T_q^U\big)^{-1} w\Big]$$

as expected. □

9.3. LEMMA. *Let $\lambda_1, \ldots, \lambda_n$ be the eigenvalues of an element p of Q. Then*

$$\det\big(T_p^X\big) = \delta(p)^{-1} \det\big(T_p^Y\big) = \prod_{i<k} \big\{(\lambda_i + \lambda_k)/2\big\}.$$

Proof. This can easily be verified if p is diagonal. The general case can be reduced to the case of diagonal p by the first equality of Lemma 9.2. □

9.4. LEMMA. *For every $p \in Q$ and every $w \in U$ we have*

$$\int_U \exp\{-\pi \cdot \mathrm{tr}({}^t vpv)\} \mathbf{e}(-\mathrm{tr}({}^t vw)) dv$$
$$= 2^{n(1-n)/4} \det\big(T_p^U\big)^{-1/2} \exp\Big\{-\pi \cdot \mathrm{tr}\Big({}^t w \cdot \big(T_p^U\big)^{-1} w\Big)\Big\}.$$

Proof. By Lemma 9.2 we can reduce the problem to the case of diagonal p. If $p = \mathrm{diag}[\lambda_1, \ldots, \lambda_n]$, $U = X$, $v = (v_{jk})$, and $w = (w_{jk})$, then the integral over U is

$$\prod_{j<k} \int_{\mathbf{R}} \exp(-\pi(\lambda_j + \lambda_k) v_{jk}^2) \exp(-4\pi i v_{jk} w_{jk}) dv_{jk}$$
$$= \prod_{j<k} (\lambda_j + \lambda_k)^{-1/2} \exp(-4\pi(\lambda_j + \lambda_k)^{-1} w_{jk}^2)$$
$$= 2^{n(1-n)/4} \det\big(T_p^X\big)^{-1/2} \exp\big(-\pi \cdot {}^t w \big(T_p^X\big)^{-1} w\big),$$

which gives the desired formula for $U = X$. If $U = Y$, we have extra factors for $j = k$, but the formula can be given in the same manner. □

9.5. We now define our generalized Bessel functions ξ_U by

$$\xi_U(g,h;s)=\int_U \mathbf{e}\big(-\mathrm{tr}({}^t xh)\big)\delta(x+\mathbf{j}g)^{-2s}dx \qquad (9.5.1)$$

$$(g\in Q, h\in U, s\in \mathbf{C}).$$

Here dv is the obvious Euclidean measure on U and $\mathbf{j}=1$ if $U=X$ and $\mathbf{j}=\sqrt{-1}$ if $U=Y$. If $U=X$, this is exactly the function ξ defined in (6.9.3); ξ_Y is a special case of the functions studied in [S1], and in that sense we already know its analytic nature. In particular, the integral for ξ_Y is convergent for $\mathrm{Re}(s)>n/2$; see the last two lines of [S1, p. 274]. The convergence of the integral for ξ_X is treated in §9.6. We need ξ_Y as a tool for investigating ξ_X. Assuming the convergence, we easily see that

$$\xi_U\big(ag\cdot{}^t a,h;s\big)=\delta(a)^{n+\varepsilon-4s}\xi_U\big(g,{}^t aha;s\big) \quad \big(a\in GL_n(\mathbf{R})\big), \qquad (9.5.2)$$

where $\varepsilon=-1$ if $U=X$ and $\varepsilon=1$ if $U=Y$. Recall a well-known formula

$$\int_Q \exp\big(-\mathrm{tr}(pq)\big)\delta(p)^{s-\kappa}dp=\Gamma_n(s)\delta(q)^{-s} \quad (q\in Q, \mathrm{Re}(s)>\mu), \qquad (9.5.3)$$

where $\kappa=(n+1)/2$, $\mu=\kappa-1$, and Γ_n is as in (4.5.2). Putting $u=y^{-1}xy^{-1}$, we have

$$\delta\big(x+\mathbf{j}y^2\big)^2=\delta(y)^4\delta(\mathbf{j}+u)^2=\delta(y)^4\delta\big((\mathbf{j}+u)(\mathbf{j}+u)^*\big)=\delta(y)^4\delta\big(1+u\cdot{}^t u\big),$$

where we write $x^*={}^t\overline{x}$. Therefore, by (9.5.3) we have

$$\pi^{-ns}\Gamma_n(s)\delta(y)^{4s}\delta\big(x+\mathbf{j}y^2\big)^{-2s}=\int_Q \exp\big(-\pi\cdot\mathrm{tr}[p(1+u\cdot{}^t u)]\big)\delta(p)^{s-\kappa}dp$$

if $\mathrm{Re}(s)>\mu$. Thus by (9.5.1) we have at least formally

$$\pi^{-ns}\Gamma_n(s)\delta(y)^{4s}\xi_U\big(y^2,h;s\big)$$
$$=\int_Q \exp\big(-\pi\cdot\mathrm{tr}(p)\big)\int_U \exp\big(-\pi\cdot\mathrm{tr}({}^t upu)\big)\mathbf{e}\big(-\mathrm{tr}({}^t xh)\big)dx\cdot\delta(p)^{s-\kappa}dp.$$

This shows that if $\mathrm{Re}(s)>\mu$, then the integral of (9.5.1) is convergent if and only if the last double integral is convergent, in which case the equality holds. Since $x=yuy$, by Lemma 9.4 we see that the last double integral equals

$$2^{n(1-n)/4}\delta(y)^{n+\varepsilon}\int_Q \exp\Big\{-\pi\cdot\mathrm{tr}\Big(p+{}^t k\cdot\big(T_p^U\big)^{-1}k\Big)\Big\}\delta(p)^{s-\kappa}\det\big(T_p^U\big)^{-1/2}dp,$$

where $k=yhy$. In particular, taking $h=0$, we obtain

$$2^{n(n-1)/4}\pi^{-ns}\delta(y)^{4s-n-\varepsilon}\Gamma_n(s)\xi_U\big(y^2,0;s\big) \qquad (9.5.4)$$
$$=\int_Q \exp\big(-\pi\cdot\mathrm{tr}(p)\big)\delta(p)^{s-\kappa}\det\big(T_p^U\big)^{-1/2}dp.$$

From Lemma 9.3 we see that $\det(T_p^U)^{-1} \le \delta(p)^{-(n+\varepsilon)/2}$, and hence the last integral over Q is convergent at least for $\mathrm{Re}(s) > \mu+(n+\varepsilon)/4$. Thus ξ_X for an arbitrary h is meaningful at least for such values of s.

By Lemma 9.3 the right-hand side of (9.5.4) for $U = X$ is the same as that for $U = Y$ with $s+1/2$ in place of s. Thus

(9.5.5) $$\xi_X(g,0;s) = \pi^{-n/2}\Gamma_n(s+1/2)\Gamma_n(s)^{-1}\xi_Y(g,0;s+1/2).$$

Now observe that $\xi_Y(g,h;s)$ is exactly the function $\xi(g,h;s,s)$ of [S1, (1.25)]. Therefore from [S1, (1.31)] we obtain

(9.5.6) $$\xi_Y(g,0;s) = 2^{n(\kappa+1-2s)}\pi^{n\kappa}\delta(g)^{\kappa-2s}\Gamma_n(2s-\kappa)\Gamma_n(s)^{-2}$$

if $\mathrm{Re}(s) > n/2$. This combined with (9.5.5) shows that

(9.5.7) $$\xi_X(g,0;s) = 2^{n(\kappa-2s)}\pi^{n^2/2}\delta(g)^{\mu-2s}\Gamma_n(s+1/2)^{-1}\Gamma_n(s)^{-1}\Gamma_n(2s-\mu),$$

where $\mu = (n-1)/2$. This proves the last part of Lemma 6.10.

9.6. Let us now prove that the integral for ξ_X in (9.5.1) is convergent for $\mathrm{Re}(s) > \mu-(1/4)$ so that the function $\xi(g,h;s)$ is holomorphic at least for $\mathrm{Re}(s) > \mu-(1/4)$. If $h = 0$, notice that $\Gamma_n(2s-\mu)/\Gamma_n(s)$ takes the value 2^{-1} at $s = \mu$, and hence

(9.6.1) $$\xi_X(g,0;\mu) = 2^{-1}2^{n(3-n)/2}\pi^{n^2/2}\delta(y)^{-\mu}\Gamma_n(n/2)^{-1}.$$

Also the right-hand side of (9.5.7) is holomorphic for $\mathrm{Re}(s) > \mu - 1/4$. Therefore for $h = 0$ and $U = X$ the integral of (9.5.1) must be absolutely convergent for $\mathrm{Re}(s) > \mu - 1/4$ by virtue of a general principle [S5, Lemma A1.5], which is similar to the well-known fact that a Dirichlet series with nonnegative coefficients has a singularity at the real point on the line of convergence. Clearly the convergence holds for an arbitrary h. This proves the first assertion of Lemma 6.10.

10. Linear relations of certain Eisenstein series

10.1. In this section we employ the standard terminology of ideal theory in the matrix algebra F_n^n. For example, given a $\mathfrak{g}$-lattice X in F_n^n, by the right (resp., left) order of X we mean the set of all $\xi \in F_n^n$ such that $X\xi \subset X$ (resp., $\xi X \subset X$). If $\mathfrak{O}'$ is the right (resp., left) order of X, then we call X a right (resp., left) $\mathfrak{O}'$-ideal. We call X integral if $X \subset \mathfrak{O}'$. Now assuming that $\mathfrak{O}'$ is maximal, we denote by $\mathfrak{n}(X)$ the fractional ideal in F generated by $\det(x)$ for all $x \in X$. We put then $N(X) = N(\mathfrak{n}(X))$. Throughout this section we put $\mathfrak{O} = \mathfrak{g}_n^n$. Fixing $\mathfrak{b}$ and $\mathfrak{c}$ as in Section 6, we put $D = D[\mathfrak{b}^{-1}, \mathfrak{b}\mathfrak{c}]$ and $\Gamma = G^\eta \cap D$. Also, for $\alpha \in GL_{2n}(F)$ we put

(10.1.1) $$Y_\alpha = \mathfrak{b}^{-1}c_\alpha\mathfrak{O}+d_\alpha\mathfrak{O}.$$

We can easily verify that $\mathfrak{n}(Y_\alpha) = \mathfrak{il}_\mathfrak{b}(\alpha)$ and $Y_\alpha = Y_\beta$ if $\alpha \in \beta D[\mathfrak{b}^{-1}, \mathfrak{b}]$.

Put

$$W = \{x \in F^n_{2n} \mid x\eta_n \cdot {}^t x = 0, \operatorname{rank}(x) = n\}, \tag{10.1.2}$$

$$x_0 = [0 \quad 1_n]. \tag{10.1.3}$$

Then $W = x_0 G^\eta$. We write an element of F^n_{2n} in the form $[c \quad d]$ or (c, d) with c, d in F^n_n. Notice that $x_0\alpha = [c_\alpha \quad d_\alpha]$ for $\alpha \in G^\eta$.

Now we take and fix an integral right $\mathfrak{O}$-ideal Z such that $\mathfrak{n}(Z)$ is prime to $\mathfrak{c}$.

10.2. Lemma. $\{[c \quad d] \in (\mathfrak{b}\mathfrak{c}Z \times Z) \cap W \mid \det(d)\mathfrak{g} + \mathfrak{c} = \mathfrak{g}\} \subset x_0(G^\eta \cap P_\mathbf{A} D)$.

Proof. Let $[c \quad d] = x_0\alpha$ with $\alpha \in G^\eta$. If $d \in Z, \det(d)\mathfrak{g} + \mathfrak{c} = \mathfrak{g}$, and $c \in \mathfrak{b}\mathfrak{c}Z$, then $d_v \in GL_n(\mathfrak{g}_v)$ and $d_v^{-1}c_v \prec \mathfrak{b}_v\mathfrak{c}_v$ for every $v|\mathfrak{c}$. By (2.7.4) and Lemma 2.10(4) we have $\alpha \in P_\mathbf{A} D$, which proves our lemma. □

10.3. Lemma. *Let Δ be a subgroup of $GL_n(F)$, T a complete set of representatives for $\Delta\backslash GL_n(F)$, and S_β a complete set of representatives for $(P \cap \beta\Gamma\beta^{-1})\backslash\beta\Gamma$ with an arbitrarily fixed $\beta \in G^\eta$. Then the elements $\tau x_0 \sigma$ with $\tau \in T$ and $\sigma \in S_\beta$ form a complete set of representatives for $\Delta\backslash x_0 P\beta\Gamma$.*

This can be verified in a straightforward way, using the simple fact that $P = \{\alpha \in G^\eta \mid x_0\alpha \in GL_n(F)x_0\}$.

10.4. We put $\Delta_Z = \mathfrak{O}_1^\times$, where $\mathfrak{O}_1$ is the left order of Z. Let χ be a Hecke character of F satisfying (6.3.3) and (6.3.4); then we define a function ℓ on F^n_{2n} and a symbol $j(c, d)$ by

$$\ell(c, d) = \begin{cases} \chi_\mathfrak{c}(\det(d))^{-1} & \text{if } [c \quad d] \in \mathfrak{b}\mathfrak{c}Z \times Z, \det(d)\mathfrak{g} + \mathfrak{c} = \mathfrak{g}, \\ 0 & \text{otherwise.} \end{cases} \tag{10.4.1}$$

$$j(c, d) = \det(cw + d)^{-k}|\det(cw + d)|^{-2\mathbf{s}}, \qquad \mathbf{s} = su - (k - i\kappa)/2, \tag{10.4.2}$$

where u, k, κ are as in (5.4.2) and (6.3.4), and w is a variable in $\mathcal{H}^\mathbf{a}$. Notice that $j(c, d)$ depends only on k modulo $2\mathbf{Z}^\mathbf{a}$, since $\det(cw + d)$ is real. We then put

$$I(Z, \chi) = \sum_{c,d} \ell(c, d)j(c, d), \tag{10.4.3}$$

where (c, d) runs over $\Delta_Z\backslash[(\mathfrak{b}\mathfrak{c}Z \times Z) \cap W]$. This is formally well defined.

10.5. We now take $B \subset G^\eta$ so that $G^\eta \cap P_\mathbf{A} D = \bigsqcup_{\beta\in B} P\beta\Gamma$ and $\beta_v \in D_v$ for every $v|\mathfrak{c}$. Such a B exists for the following reason: We first take $H \subset Q_\mathbf{h}$ as in Lemma 2.10(2) so that $h_v = 1$ for every $h \in H$ and every $v|\mathfrak{c}$. Then $G^\eta \cap P_\mathbf{A} D = \bigsqcup_{h\in H}(G^\eta \cap PhD)$, and hence we can take B to be a subset of $G^\eta \cap \bigcup_{h\in H} hD$. Then B has the required properties. Notice that $(Y_\beta)_v = \mathfrak{O}_v$ for every $\beta \in B$ and every $v|\mathfrak{c}$, and $P\backslash(G^\eta \cap P_\mathbf{A} D)$ can be given by $\bigsqcup_{\beta\in B} S_\beta$ with S_β as in Lemma 10.3.

Since $x_0(G^\eta \cap P_{\mathbf{A}}D) = \bigsqcup_{\beta\in B} x_0 P\beta\Gamma$, by Lemmas 10.2 and 10.3 we obtain

(10.5.1) $$I(Z,\chi) = \sum_{\beta\in B}\sum_{\tau\in T}\sum_{\alpha\in S_\beta} \ell(\tau x_0\alpha) j(\tau x_0\alpha),$$

where $T = \Delta_Z\backslash GL_n(F)$. Let $\alpha \in \beta\Gamma$ and $\tau \in T$. Then $\alpha_v \in D_v$ for every $v|\mathfrak{c}$. Suppose $\ell(\tau x_0\alpha) \neq 0$. Then $\tau(c_\alpha, d_\alpha) \in \mathfrak{b}\mathfrak{c}Z \times Z$ and $\det(\tau d_\alpha)\mathfrak{g} + \mathfrak{c} = \mathfrak{g}$, so that $\det(\tau) \in \mathfrak{g}_v^\times$ for every $v|\mathfrak{c}$. Since $Y_\beta = Y_\alpha$, we have $\tau Y_\beta = \tau(\mathfrak{b}^{-1}c_\alpha\mathfrak{O} + d_\alpha\mathfrak{O}) \subset Z$, so that $\tau \in ZY_\beta^{-1}$. Conversely, suppose $\tau \in ZY_\beta^{-1}$ and $\det(\tau) \in \mathfrak{g}_v^\times$ for every $v|\mathfrak{c}$. Then, for $\alpha \in \beta\Gamma$ we have $\tau Y_\alpha \subset Z$, so that $\tau(c_\alpha, d_\alpha) \in \mathfrak{b}Z \times Z$. Fix a prime $v|\mathfrak{c}$. Since $(Y_\beta)_v = \mathfrak{O}_v$, we see that $\tau \in \mathfrak{O}_v = Z_v$, which combined with the assumption $\det(\tau) \in \mathfrak{g}_v^\times$ shows that $\tau \in \mathfrak{O}_v^\times$. Now $c_\alpha \prec \mathfrak{b}_v\mathfrak{c}_v$, and hence $\tau c_\alpha \in \mathfrak{b}\mathfrak{c}Z$. Thus $\tau(c_\alpha, d_\alpha) \in \mathfrak{b}\mathfrak{c}Z \times Z$.

Concluding from these observations, we can restrict τ in (10.5.1) to the elements in the set $T_\beta = \Delta_Z\backslash\big[GL_n(F) \cap ZY_\beta^{-1}\big]$, and

$$\begin{aligned}\ell(\tau x_0\alpha) j(\tau x_0\alpha) &= \chi_\mathfrak{c}\big(\det(\tau)\big)^{-1}\det(\tau)^{-k}|\det(\tau)|^{-2s}\chi_\mathfrak{c}\big(\lambda(\alpha)\big)^{-1} j(x_0\alpha)\\ &= \chi'\big(\det(\tau)\mathfrak{g}\big)|\det(\tau)|^{-2su}\chi_\mathfrak{c}\big(\lambda(\alpha)\big)^{-1} j(x_0\alpha),\end{aligned}$$

where $\chi'(\mathfrak{a}) = \chi^*(\mathfrak{a})$ if $\mathfrak{a}$ is prime to $\mathfrak{c}$ and $\chi'(\mathfrak{a}) = 0$ if $\mathfrak{a}$ is not prime to $\mathfrak{c}$. Since $\delta(w)^s j(x_0\alpha) = \delta(w)^s\|_k\alpha$, we thus obtain, employing $\chi[\alpha]_1$ of (6.5.1),

(10.5.2) $$\begin{aligned}\delta(w)^s I(Z,\chi) = \sum_{\beta\in B}\sum_{\tau\in T_\beta} &\chi'\big(\mathfrak{n}(Y_\beta Z^{-1}\tau)\big)N\big(Y_\beta Z^{-1}\tau\big)^{-2s}\\ &\cdot\chi^*\big(\mathfrak{n}(Z)\big)N(Z)^{-2s}\sum_{\alpha\in S_\beta} N\big(\mathfrak{il}_\mathfrak{b}(\alpha)\big)^{2s}\chi[\alpha]_1\delta(w)^s\|_k\alpha.\end{aligned}$$

We now define an L-function $L_\mathfrak{c}^{(n)}$ of F_n^n and its partial series by

(10.5.3) $$L_\mathfrak{c}^{(n)}(s,\chi) = \sum_X \chi^*\big(\mathfrak{n}(X)\big)N(X)^{-s},$$

(10.5.4) $$L_\mathfrak{c}^{(n)}(s,\chi;\mathfrak{a}) = \sum_{X\sim\mathfrak{a}} \chi^*\big(\mathfrak{n}(X)\big)N(X)^{-s}.$$

Here in (10.5.3) X runs over all the integral left $\mathfrak{O}$-ideals such that $\mathfrak{n}(X)$ is prime to $\mathfrak{c}$, and in (10.5.4) it runs over all such X's for which $\mathfrak{a}^{-1}\mathfrak{n}(X)$ is principal, where $\mathfrak{a}$ is a fixed fractional ideal in F. Notice that these sums do not change when we replace $\mathfrak{O}$ by any maximal order in F_n^n. Recall also that for left $\mathfrak{O}$-ideals X and X', there is an element $\sigma \in GL_n(F)$ such that $X' = X\sigma$ if and only if $\mathfrak{n}(X^{-1}X')$ is principal.

Let E_β denote the last sum over S_β. Then we obtain

$$\chi^*\big(\mathfrak{n}(Z)\big)^{-1}N(Z)^{2s}\delta(w)^s I(Z,\chi) = \sum_{\beta\in B} E_\beta L_\mathfrak{c}^{(n)}\big(2s,\chi;\mathfrak{n}(Y_\beta Z^{-1})\big).$$

Since $\bigsqcup_{\beta\in B} S_\beta$ gives $P\backslash(G^\eta \cap P_{\mathbf{A}}D)$, we see that $\sum_{\beta\in B} E_\beta$ coincides with $E_r(z,s)$ of Lemma 6.5(3) defined with $r=1$.

Now take a complete set of representatives A for the ideal class group of F consisting of ideals prime to $\mathfrak{c}$; for each $\mathfrak{a}\in A$ take an integral right $\mathfrak{O}$-ideal $Z_\mathfrak{a}$ such that $\mathfrak{n}(Z_\mathfrak{a})=\mathfrak{a}$. Then we have

$$\sum_{\mathfrak{a}\in A} \chi^*(\mathfrak{a})^{-1}N(\mathfrak{a})^{2s}\delta(w)^{\mathbf{s}}I(Z_\mathfrak{a},\chi) = L_\mathfrak{c}^{(n)}(2s,\chi)E_1(w,s). \tag{10.5.5}$$

Since each Euler factor of $L_\mathfrak{c}^{(n)}$ is a local L-function of $M_n(\mathfrak{g}_v)$ (cf. [S5, Lemma 3.13]), we have

$$L_\mathfrak{c}^{(n)}(2s,\chi)=\prod_{i=1}^{n} L_\mathfrak{c}(2s+1-i,\chi). \tag{10.5.6}$$

We note here an easy fact that will be needed later.

10.6. LEMMA. *If* $2\notin\mathfrak{c}$, *then* $\det(cw+d)=\det(d-c\cdot{}^t w)$ *for every* $(c,d)\in(\mathfrak{b}\mathfrak{c}Z\times Z)\cap W$.

Proof. By Lemma 10.2 such a $[c\,d]$ is of the form $x_0\alpha$ with $\alpha\in G^\eta\cap P_{\mathbf{A}}D$. Since $\det(P_{\mathbf{A}})=1$ and $\det(D_v)=1$ if $2\notin\mathfrak{c}_v\neq\mathfrak{g}_v$, we have $\det(\alpha)=1$, which combined with (4.2.7) and (4.3.4) proves our lemma. □

11. Theta functions

11.1. In this section we write $y[x]=xy\cdot{}^t x$ for an $(n\times m)$-matrix x and an $(m\times m)$-matrix y. We define symbols X, $\mathfrak{G}$, and $\mathfrak{P}$ by

$$X=F_{2n}^n,\qquad \mathfrak{G}=Sp(n,F),\qquad \mathfrak{P}=\{\alpha\in\mathfrak{G}\,|\,c_\alpha=0\}, \tag{11.1.1}$$

and denote by $\mathcal{S}(X_\mathbf{h})$ the Schwartz-Bruhat space of $X_\mathbf{h}$. Now, for $w\in\mathcal{H}_n$ we put

$$A(w)=\begin{bmatrix}\xi(w)^{1/2} & 0\\ 0 & \xi(w)^{1/2}\end{bmatrix}B(w)^{-1},\qquad T(w)={}^tA(w)A(w) \tag{11.1.2}$$

with $B(w)$ and $\xi(w)$ of (4.3.2). Then we have ${}^tA(w)\,\mathrm{diag}[1_n,-1_n]A(w)=-\eta$. We are going to define a theta function on $\mathfrak{H}_n^\mathbf{a}\times\mathcal{H}^\mathbf{a}$, where

$$\mathfrak{H}=\mathfrak{H}_n=\{z\in\mathbf{C}_n^n\,|\,{}^tz=z,\ i(\bar{z}-z)>0\}. \tag{11.1.3}$$

Take $h,m\in\mathbf{Z}^\mathbf{a}$ so that $0\le h_v\le 1$ and $0\le m_v\le 1$ for every $v\in\mathbf{a}$. For $\sigma=[\sigma_1\ \ \sigma_2]\in X_\mathbf{a}=(\mathbf{R}_{2n}^n)^\mathbf{a}$ with $\sigma_i\in(\mathbf{R}_n^n)^\mathbf{a}$, put

$$p(\sigma)=\det(\sigma_1)^h\det(\sigma_2)^m. \tag{11.1.4}$$

For $z = x + iy \in \mathfrak{H}^{\mathbf{a}}$, $w \in \mathcal{H}^{\mathbf{a}}$, and $\ell \in \mathcal{S}(X_{\mathbf{h}})$, we put

$$f(z, w; \ell) = \sum_{\sigma \in X} \ell(\sigma) p(\sigma \cdot {}^t A_w) \mathbf{e}_{\mathbf{a}}^n (2^{-1}(-x\eta[\sigma] + iyT_w[\sigma])). \tag{11.1.5}$$

Here we understand that $x, y \in (\mathbf{R}_n^n)^{\mathbf{a}}$, $A_w = (A(w_v))_{v \in \mathbf{a}}$, and $T_w = (T(w_v))_{v \in \mathbf{a}}$, and we view each $\sigma \in X$ as an element of $X_{\mathbf{a}} = (\mathbf{R}_{2n}^n)^{\mathbf{a}}$ through the projection map of $X_{\mathbf{A}}$ to $X_{\mathbf{a}}$; $\mathbf{e}_{\mathbf{a}}^n$ is defined by (6.2.2). We put also $\ell(\sigma) = \ell(\sigma_{\mathbf{h}})$ for $\sigma \in X_{\mathbf{A}}$.

Now the above series is a special case of [S3, (3.10)]. The symbols q, n, S, and V^n there correspond to $2n, n, -\eta$, and X here. To simplify our notation, put

$$j_\alpha(z) = \bigl(\det\bigl((c_\alpha)_v z_v + (d_\alpha)_v\bigr)\bigr)_{v \in \mathbf{a}}, \qquad \delta(z) = \bigl(\det\bigl(\mathrm{Im}(z_v)\bigr)\bigr)_{v \in \mathbf{a}}, \tag{11.1.6}$$

$$J(\alpha, z) = j_\alpha(z)^{h-m} |j_\alpha(z)|^{nu+2m} \quad (\alpha \in \mathfrak{G}_{\mathbf{A}}, z \in \mathfrak{H}_n^{\mathbf{a}}). \tag{11.1.7}$$

By [S3, Theorems 3.2 and 3.5] there is an action of $\mathfrak{G}_{\mathbf{A}}$ on $\mathcal{S}(X_{\mathbf{h}})$, written $\ell \mapsto {}^\sigma \ell$ for $\sigma \in \mathfrak{G}_{\mathbf{A}}$ and $\ell \in \mathcal{S}(X_{\mathbf{h}})$, such that

$$f(\alpha z, w; {}^\alpha \ell) = J(\alpha, z) f(z, w; \ell) \quad \text{for every } \alpha \in \mathfrak{G}. \tag{11.1.8}$$

Notice that

$$A(w) = \begin{bmatrix} \xi(w)^{-1/2} & 0 \\ 0 & \xi(w)^{-1/2} \end{bmatrix} \begin{bmatrix} 1_n & -w \\ 1_n & {}^t w \end{bmatrix}, \tag{11.1.9}$$

$$p\bigl([d\,c] \cdot {}^t A_w\bigr) = 2^{\|h+m\| n/2} \delta(w)^{-(h+m)/2} \det(d - c \cdot {}^t w)^h \det(d + cw)^m. \tag{11.1.10}$$

From [S3, Theorem 3.2(5)] we obtain

$$\tau \in \mathfrak{P}_{\mathbf{h}} \Longrightarrow ({}^\tau \ell)(x) = \bigl|\det(a_\tau)\bigr|_{\mathbf{A}}^n \mathbf{e}_{\mathbf{h}}^n\bigl(-2^{-1}\eta[x] a_\tau \cdot {}^t b_\tau\bigr) \ell({}^t a_\tau x). \tag{11.1.11}$$

Suppressing w, we define a function $f_{\mathbf{A}}(y; \ell)$ for $y \in \mathfrak{G}_{\mathbf{A}}$ by

$$f_{\mathbf{A}}(y; \ell) = J(y, \mathbf{i})^{-1} f\bigl(y(\mathbf{i}); {}^y \ell\bigr), \tag{11.1.12}$$

where $\mathbf{i} = (\mathbf{i}_v)_{v \in \mathbf{a}}$, $\mathbf{i}_v = i 1_n$. From (11.1.8) we can easily derive

$$f_{\mathbf{A}}(\alpha y x; \ell) = J(x, \mathbf{i})^{-1} f\bigl(y; {}^x \ell\bigr) \quad \text{if } \alpha \in \mathfrak{G} \text{ and } x, y \in \mathfrak{G}_{\mathbf{A}}, x(\mathbf{i}) = \mathbf{i}. \tag{11.1.13}$$

11.2. Lemma. *For every $q \in GL_n(F)_{\mathbf{A}}$ and $s \in S_{\mathbf{A}}$ we have*

(11.2.1)

$$f_{\mathbf{A}}\left(\begin{bmatrix} q & s\widehat{q} \\ 0 & \widehat{q} \end{bmatrix}; \ell\right) = \det(q)_{\mathbf{a}}^{m+h} |\det(q)|_{\mathbf{A}}^n \cdot \sum_{\sigma \in X} \ell({}^t q \sigma) p(\sigma \cdot {}^t A_w) \mathbf{e}_{\mathbf{a}}^n (2^{-1} i \cdot T_w[{}^t q \sigma]) \mathbf{e}_{\mathbf{A}}^n(-2^{-1}\eta[\sigma] s).$$

Moreover, if $\beta = \mathrm{diag}[\widehat{r}, r]g$ *with* $\beta \in \mathfrak{G}$, $r \in GL_n(F)_{\mathbf{h}}$, *and* $g \in \mathfrak{G}_{\mathbf{A}}$, *then*

$$(11.2.2)\quad J(\beta^{-1}, z)^{-1} f(\beta^{-1}z, w; \ell) = |\det(r)|_{\mathbf{A}}^{-n} \cdot \sum_{\sigma \in X} ({}^{g}\ell)(r^{-1}\sigma) p(\sigma \cdot {}^{t}A_w) \mathbf{e}_{\mathbf{a}}^{n}(2^{-1}(-x\eta[\sigma] + iyT_w[\sigma])).$$

Proof. Equality (11.2.1) can be obtained by taking $\begin{bmatrix} q & s\widehat{q} \\ 0 & \widehat{q} \end{bmatrix}$ as y in (11.1.12) and employing (11.1.11). As for (11.2.2), given such β, r, g, and $z \in \mathfrak{H}_n^{\mathbf{a}}$, let $t = p \cdot \mathrm{diag}[\widehat{r}, r]$ with $p \in \mathfrak{P}_{\mathbf{a}}$ such that $p(\mathbf{i}) = z$. Then, by (11.1.8) and (11.1.12), we have

$$J(\beta^{-1}, z)^{-1} f(\beta^{-1}z, w; \ell) = f(z, w; {}^{\beta}\ell) = f(t(\mathbf{i}), w; {}^{tg}\ell) = J(t, \mathbf{i}) f_{\mathbf{A}}(t; {}^{g}\ell).$$

Applying (11.2.1) to $f_{\mathbf{A}}(t; {}^{g}\ell)$, we obtain the desired result. □

11.3. Put $k = m - h$. Fixing a Hecke character χ satisfying (6.3.3) and (6.3.4) with this k, take $\ell \in \mathcal{S}(X_{\mathbf{h}})$ so that $\ell([d\, c])$ for $[d\, c] \in F_{2n}^{n}$ gives $\ell(c, d)$ of (10.4.1) defined with $\mathfrak{O}$ as Z there. Put $\mathfrak{D}[\mathfrak{x}, \mathfrak{y}] = \mathfrak{G}_{\mathbf{A}} \cap D'[\mathfrak{x}, \mathfrak{y}]$ with D' of (2.7.2). Applying [S3, Proposition 3b.2] to the present f, we obtain

$$(11.3.1)\qquad {}^{\gamma}\ell = \chi_{\mathbf{c}}(\det(a_\gamma))^{-1}\ell \quad \text{if } \gamma \in \mathfrak{D}[\mathfrak{y}^{-1}, \mathfrak{y}\mathfrak{c}]$$

with $\mathfrak{y} = \mathfrak{d}\mathfrak{b}\mathfrak{c}^2$, where $\mathfrak{d}$ is the different of F relative to $\mathbf{Q}$.

Let $\beta = \mathrm{diag}[\widehat{r}, r]g$ with $\beta \in \mathfrak{G}, r \in GL_n(F)_{\mathbf{h}}$, and $g \in \mathfrak{D}[\mathfrak{y}^{-1}, \mathfrak{y}\mathfrak{c}]$. Put $f_\beta(z) = J(\beta^{-1}, z)^{-1} f(\beta^{-1}z, w; \ell)$ and $\Gamma = \mathfrak{G} \cap \mathfrak{D}[\mathfrak{y}^{-1}, \mathfrak{y}\mathfrak{c}]$. By (11.2.2) and (11.3.1) we have

$$(11.3.2)\qquad f_\beta(z) = |\det(r)|_{\mathbf{A}}^{-n} \chi_{\mathbf{c}}(\det(ra_\beta)^{-1}) \cdot \sum_{\sigma \in X} \ell(r^{-1}\sigma) p(\sigma \cdot {}^{t}A_w) \mathbf{e}_{\mathbf{a}}^{n}(2^{-1}(-x\eta[\sigma] + iyT_w[\sigma])).$$

From (11.1.8) and (11.3.1) we obtain

$$(11.3.3)\qquad \alpha \in \beta\Gamma \Longrightarrow f_\beta(\alpha z) = \chi_{\mathbf{c}}(\lambda(\alpha^{-1}\beta)) J(\alpha, z) f_1(z).$$

Now we note a simple but crucial relation:

$$(11.3.4)\qquad \sigma = [d\, c] \quad \text{and} \quad \eta[\sigma] = 0 \Longrightarrow \det(T_w[\sigma]) = \delta(w)^{-1} \det(d + cw)^2.$$

In fact, if $\eta[\sigma] = 0$, then T_w on the right-hand side can be replaced by $T_w + \eta$. Now

$$T_w + \eta = {}^{t}A_w\{1_{2n} + \mathrm{diag}[-1_n, 1_n]\}A_w = {}^{t}A_w\, \mathrm{diag}[0, 2 \cdot 1_n]A_w,$$

which combined with (11.1.9) gives (11.3.4). For $\alpha \in \mathfrak{G}$ we define an ideal $\mathfrak{a}_\alpha$ and a number $\chi[\alpha]$ by

$$(11.3.5)\qquad \mathfrak{a}_\alpha = \mathrm{il}_{\mathfrak{y}}(\alpha),$$

(11.3.6) $$\chi[\alpha] = \begin{cases} \chi_{\mathbf{a}}(\lambda(\alpha))\chi^*(\lambda(\alpha)\mathfrak{a}_\alpha^{-1}) & \text{if } \mathfrak{c} \neq \mathfrak{g}, \\ \chi^*(\mathfrak{a}_\alpha)^{-1} & \text{if } \mathfrak{c} = \mathfrak{g}. \end{cases}$$

Let $\Phi = \Gamma\backslash\mathfrak{H}^{\mathbf{a}}$ and $\Psi_\beta = (\mathfrak{P} \cap \beta\Gamma\beta^{-1})\backslash\mathfrak{H}^{\mathbf{a}}$ with a fixed β as above. Put also

$$\Delta_r = GL_n(F) \cap GL_n(F)_{\mathbf{a}} \prod_{v\in\mathbf{h}} r_v GL_n(\mathfrak{g}_v) r_v^{-1},$$

$$\mathfrak{S} = \{x \in (\mathbf{R}_n^n)^{\mathbf{a}} \mid {}^t x = x\}, \qquad \mathfrak{S}^+ = \{y \in \mathfrak{S} \mid y_v > 0 \text{ for every } v \in \mathbf{a}\}.$$

Then we can take Ψ_β in the form $\Psi_\beta = \mathfrak{X} + i\mathfrak{Y}$ with a subset $\mathfrak{X}$ in $\mathfrak{S}$ such that

(11.3.7) $$\mathrm{vol}(\mathfrak{X}) = \left[D_F^{n/2} N(\mathfrak{a}_\beta^{-2}\mathfrak{y}^{-n})\right]^{(n+1)/2}$$

and $\mathfrak{Y} = \mathfrak{S}^+/\Delta_r$, where we let an element a of Δ_r act on $\mathfrak{S}^+$ by $y \mapsto {}^t aya$ (cf. [S5, Lemma 18.15]).

11.4. We now assume that

(11.4.1) $$2 \notin \mathfrak{c} \quad \text{and} \quad h_v + m_v > 0 \quad \text{for some } v \in \mathbf{a}.$$

The last condition means that $p(\sigma \cdot {}^t A_w) \neq 0$ only when $\mathrm{rank}(\sigma) = n$. Thus, if $\eta[\sigma] = 0$ and $p(\sigma \cdot {}^t A_w) \neq 0$, then σ belongs to the set W of (10.1.2). Therefore from (11.3.2) we obtain

(11.4.2) $$\int_{\mathfrak{X}} f_\beta(x+iy)dx = c \sum_{\sigma\in W} \ell(r^{-1}\sigma)p(\sigma \cdot {}^t A_w)\mathbf{e}_{\mathbf{a}}^n(2^{-1}iyT_w[\sigma]),$$

(11.4.3) $$c = |\det(r)|_{\mathbf{A}}^{-n}\chi_{\mathfrak{c}}(\det(ra_\beta)^{-1})\mathrm{vol}(\mathfrak{X}).$$

Let $\nu = (h+m+i\kappa)/2$ and $\mathbf{s}_1 = su + \nu - (n+1)u/2$; let $M(\sigma, y)$ denote $\delta(z)^{\mathbf{s}_1}$ times the typical term of the last sum over W. Then for $a \in \Delta_r$ we can easily verify that $M(a\sigma, y) = M(\sigma, {}^t aya)$. Therefore, putting $R = \Delta_r\backslash W$, we obtain

$$\sum_{\sigma\in W} M(\sigma, y) = \sum_{\sigma\in R}\sum_{a\in\Delta_r} M(a\sigma, y) = \sum_{\sigma\in R}\sum_{a\in\Delta_r} M(\sigma, {}^t aya),$$

so that

(11.4.4) $$\int_{\mathfrak{Y}}\int_{\mathfrak{X}} f_\beta(x+iy)\delta(z)^{\mathbf{s}_1}dxdy = 2c\int_{\mathfrak{S}^+}\sum_{\sigma\in R} M(\sigma, y)dy.$$

The factor 2 is necessary because -1 gives the identity map on $\mathfrak{S}^+$. To evaluate the last integral, we employ (9.5.3). Thus putting $\mathbf{s}_2 = su + \nu + (n+1)u/2$, we can transform (11.4.4) into the form

(11.4.5) $$\int_{\Psi_\beta} f_\beta(z)\delta(z)^{\mathbf{s}_2}\mathbf{d}z = 2c\Gamma_n(su+\nu)\pi^{-n\|su+\nu\|}$$
$$\cdot \sum_{\sigma\in R} \ell(r^{-1}\sigma)p(\sigma \cdot {}^t A_w)\det(T_w[\sigma])^{-su-\nu},$$

where $\Gamma_n\big((s_v)_{v\in\mathbf{a}}\big) = \prod_{v\in\mathbf{a}} \Gamma_n(s_v)$ and $\mathbf{d}z$ is the standard invariant measure on $\mathfrak{H}^{\mathbf{a}}$ (see [S5, (7.14.5)]). By Lemma 10.6, (11.1.10), and (11.3.4) we see that $p(\sigma\cdot{}^tA_w)\det\big(T_w[\sigma]\big)^{-su-v}$ for $\sigma = [d\,c]$ coincides with $2^{\|h+m\|n/2}\delta(w)^{su+i\kappa/2}j(c,d)$ with $j(c,d)$ of (10.4.2). We now take r so that $r_v = 1$ for every $v|\mathfrak{c}$. Then we see that $\ell(r^{-1}\sigma)$ for $\sigma = [d\,c] \in F_{2n}^n$ is the function $\ell(c,d)$ of (10.4.1) defined with $Z = r\mathfrak{O}$, and Δ_r is Δ_Z there. Thus the last sum over R equals

$$(11.4.6)\qquad 2^{\|h+m\|n/2}\delta(w)^{su+i\kappa/2}I(r\mathfrak{O},\chi).$$

11.5. For simplicity put $\mathfrak{D} = \mathfrak{D}[\mathfrak{y}^{-1},\mathfrak{y}\mathfrak{c}]$. Now we can find a subset B of $\mathfrak{G}$ such that $\mathfrak{G}\cap\mathfrak{P}_{\mathbf{A}}\mathfrak{D} = \bigsqcup_{\beta\in B}\mathfrak{P}\beta\Gamma$ and $\beta \in \mathrm{diag}[\widehat{r}, r]\mathfrak{D}$ with $r \in GL_n(F)_{\mathbf{h}}$; we may assume that $r_v = 1$ for every $v|\mathfrak{c}$. Moreover, the $\mathfrak{a}_\beta$ for $\beta \in B$ form a complete set of representatives for the ideal class group of F. (For these see [S4, §7] or [S5, Lemma 18.7(4), (5)].) Let $\mathcal{A} = \mathfrak{P}\backslash(\mathfrak{G}\cap\mathfrak{P}_{\mathbf{A}}\mathfrak{D})$ and $R_\beta = (\mathfrak{P}\cap\beta\Gamma\beta^{-1})\backslash\beta\Gamma$. Define a function $\mathscr{E}$ on $\mathfrak{H}_n^{\mathbf{a}}\times\mathbf{C}$ by

$$\mathscr{E}(z,s) = \sum_{\alpha\in\mathcal{A}} N(\mathfrak{a}_\alpha)^{2s}\chi[\alpha]\delta(z)^{\mathbf{s}}\|_k\alpha,$$

where $\mathbf{s} = su - (m-h-i\kappa)/2$. This is the function given in [S4, (7.4)] and [S5, Lemma 18.7(3)] that corresponds to an Eisenstein series on $\mathfrak{G}_{\mathbf{A}}$ as in [S5, §18.6] with a set of data $\{\mathfrak{y}, \mathfrak{c}, k, \chi\}$ here in place of $\{\mathfrak{b}, \mathfrak{c}, k, \chi\}$ there. Since $\mathcal{A}$ can be given by $\bigsqcup_{\beta\in B} R_\beta$, we have

$$\mathscr{E}(z,s) = \sum_{\beta\in B} N(\mathfrak{a}_\beta)^{2s}\chi[\beta]\sum_{\alpha\in R_\beta}\chi_{\mathfrak{c}}\big(\lambda(\alpha^{-1}\beta)\big)\delta(z)^{\mathbf{s}}\|_k\alpha.$$

From (11.3.3) we obtain

$$\big(f_\beta\delta^{\mathbf{s}_2-u/2}\big)\circ\alpha = \chi_{\mathfrak{c}}\big(\lambda(\alpha^{-1}\beta)\big)f_1\delta^{m+nu/2}\big(\delta^{\mathbf{s}}\|_k\alpha\big) \quad \text{if } \alpha\in\beta\Gamma.$$

Since Ψ_β is equivalent to $\bigsqcup_{\alpha\in R_\beta}\alpha\Phi$, we see that

$$\begin{aligned}\int_\Phi f_1(z)\mathscr{E}(z,s)\delta(z)^{m+nu/2}\mathbf{d}z &= \sum_{\beta\in B} N(\mathfrak{a}_\beta)^{2s}\chi[\beta]\int_\Phi\sum_{\alpha\in R_\beta}\big(f_\beta\delta^{\mathbf{s}_2-u/2}\big)\circ\alpha\,\mathbf{d}z\\ &= \sum_{\beta\in B} N(\mathfrak{a}_\beta)^{2s}\chi[\beta]\int_{\Psi_\beta} f_\beta\delta^{\mathbf{s}_2-u/2}\mathbf{d}z.\end{aligned}$$

By (11.3.7), (11.4.3), and (11.4.5), we see that the last sum over B with $s+1/2$ in place of s is

$$\begin{aligned}&2\cdot 2^{\|h+m\|n/2}\big[D_F^{n/2}N(\mathfrak{y})^{-n}\big]^{(n+1)/2}\Gamma_n(su+v)\pi^{-n\|su+v\|}\\ &\qquad\cdot\delta(w)^{su+i\kappa/2}\sum_{\beta\in B} N(\mathfrak{a}_\beta)^{2s}\chi[\beta]\chi_{\mathfrak{c}}\Big(\det\big(a_\beta^{-1}\big)\Big)I(r\mathfrak{O},\chi).\end{aligned}$$

Since $\chi[\beta]\chi_{\mathfrak{c}}\big(\det(a_\beta^{-1})\big) = \chi^*(a_\beta)^{-1}$, the last sum is exactly the left-hand side of (10.5.5). Thus we obtain

(11.5.1)

$$\int_\Phi f_1(z)\mathscr{E}(z, s+1/2)\delta(z)^{m+nu/2}\mathbf{d}z = \Gamma_n(su+\nu)\pi^{-n\|su+\nu\|}$$
$$\cdot 2^{1+\|h+m\|n/2}\big[D_F^{n/2}N(\mathfrak{y})^{-n}\big]^{(n+1)/2} L_{\mathfrak{c}}^{(n)}(2s,\chi)\delta(w)^{su+i\kappa/2}E_1(w,s).$$

11.6. By [S4, Theorem 7.3] (cf. [S5, Theorem 19.3]) there is a product $g(s)$ of gamma functions such that

$$g(s)L_{\mathfrak{c}}(2s+1,\chi)\Lambda_{\mathfrak{c}}(s)\mathscr{E}(z, s+1/2) \tag{11.6.1}$$

is a meromorphic function of s on the whole plane; in particular, it is an entire function of s if $\chi^2 \neq 1$. Here $\Lambda_{\mathfrak{c}}$ is coincidentally the function of (6.13.1).

Now, given $\{\mathfrak{b}, \mathfrak{c}, k, \chi\}$ as in Section 6, we may assume that $0 \leq k_v \leq 1$ for every $v \in \mathbf{a}$ as explained in §6.4. Take $m = u$ and $h = u - k$. Then f_1 is rapidly decreasing at every cusp. Therefore from (11.5.1) we can conclude that

$$g(s)\Gamma_n(su+\nu)L_{\mathfrak{c}}(2s+1,\chi)\Lambda_{\mathfrak{c}}(s)L_{\mathfrak{c}}^{(n)}(2s,\chi)E_1(w,s) \tag{11.6.2}$$

can be continued to a meromorphic function of s on the whole plane and, in particular, to an entire function of s if $\chi^2 \neq 1$. The formula of [S4, Theorem 7.3] (or [S5, Theorem 19.3]) shows that $g(s) = \prod_{v\in\mathbf{a}} \gamma(s + i\kappa_v/2, k_v)$ with

$$\gamma(s,e) = \Gamma_{2e+1}\big(s+(e+1)/2\big)\prod_{i=e+1}^{[n/2]}\Gamma(2s+1-i). \tag{11.6.3}$$

In this section we considered the series $E_r(w,s)$ only for $r = 1$, but we can treat the case of an arbitrary r by choosing a suitable Z. We do not go into details here, since it is somewhat involved and inessential in this paper.

Let us now derive from the above result that the function $\xi(y,h;s)$ of (6.9.3) can be continued to a meromorphic function on the whole s-plane. Take $F = \mathbf{Q}$, $\mathfrak{b} = \mathbf{Z}$, and $\mathfrak{c} = c_0\mathbf{Z}$ with an integer $c_0 > 2$. We take $q \in GL_2(\mathbf{Q})_{\mathbf{a}}$ so that $\det(q)_{\mathbf{a}} > 0$; take a Hecke character χ of $\mathbf{Q}$ such that $\chi^2 \neq 1$ and $\chi_{\mathbf{a}}(x) = x/|x|$ for $x \in \mathbf{R}^\times$. We take $k = 1$ and c_0 so that χ satisfies (6.3.3). The case $h = 0$ was treated in Section 9. So assume $h \neq 0$. By virtue of (9.5.2) we may take h in the form $h = \mathrm{diag}[2^{-1}\eta_\tau, 0]$ with $0 < \tau \leq n/2$. Putting $y = (q\cdot{}^tq)_{\mathbf{a}}$, from Propositions 6.12 and 6.13 we obtain

$$c(h,q,s) = c_0^{-n\mu}\det(y)^{2s}\Lambda_{\mathfrak{c}}(s)^{-1}\Lambda_{h,\mathfrak{c}}(s)\xi(y,h;s), \tag{11.6.4}$$

since $\mathbf{b}$ of Proposition 6.13 is empty. Combining this with the above fact on $E_1(z,s)$, we see that

$$g(s)\Gamma_n\big(s+(1/2)\big)\Lambda_{h,\mathfrak{c}}(s)\prod_{i=0}^{n}L_{\mathfrak{c}}(2s+1-i,\chi)\cdot\xi(y,h;s) \tag{11.6.5}$$

can be continued to an entire function of s. This proves meromorphic continuation of $\xi(y, h; s)$ to the whole s-plane for $h \neq 0$.

References

[E1] M. Eichler, *Die Ähnlichkeitsklassen indefiniter Gitter*, Math. Z. **55** (1952), 216–252.

[E2] ———, *Quadratische Formen und orthogonale Gruppen*, 2d ed., Grundlehren Math. Wiss. **63**, Springer-Verlag, Berlin, 1974.

[K] M. Kneser, *Klassenzahlen indefiniter quadratischer Formen in drei oder mehr Veränderlichen*, Arch. Math. (Basel) **7** (1956), 323–332.

[M] L. J. Mordell, *The definite quadratic forms in eight variables with determinant unity*, J. Math. Pures Appl. (9) **17** (1938), 41–46.

[S1] G. Shimura, *Confluent hypergeometric functions on tube domains*, Math. Ann. **260** (1982), 269–302.

[S2] ———, *On Eisenstein series*, Duke Math. J. **50** (1983), 417–476.

[S3] ———, *On the transformation formulas of theta series*, Amer. J. Math. **115** (1993), 1011–1052.

[S4] ———, *Euler products and Fourier coefficients of automorphic forms on symplectic groups*, Invent. Math. **116** (1994), 531–576.

[S5] ———, *Euler Products and Eisenstein Series*, CBMS Regional Conf. Ser. in Math. **93**, Amer. Math. Soc., Providence, 1997.

[S6] ———, "Zeta functions and Eisenstein series on classical groups" in *Elliptic Curves and Modular Forms (Washington, D.C., 1996)*, Proc. Nat. Acad. Sci. U.S.A. **94** (1997), 11133–11137.

[Si] C. L. Siegel, *Gesammelte Abhandlungen, I–III*, Springer-Verlag, Berlin, 1966; *IV*, Springer-Verlag, Berlin, 1979.

Department of Mathematics, Princeton University, Princeton, New Jersey 08544-1000, USA

The number of representations of an integer by a quadratic form

Duke Mathematical Journal, 100 (1999), 59-92

Introduction. Our quadratic form $x \mapsto \varphi[x]$ is defined on a vector space V of dimension n (> 1) over a totally real algebraic number field F. We let G^φ denote the orthogonal group of φ, and $\mathfrak{g}$ the maximal order of F. For a $\mathfrak{g}$-lattice L in V and an element $h \in \mathfrak{g}$, we denote by $N(L,h)$ the number of elements $x \in L$ such that $\varphi[x] = h$. Here we assume that φ is totally positive definite, h is totally positive, and $\varphi[x] \in \mathfrak{g}$ for every $x \in L$. As previous researchers on this topic discovered, in order to obtain a meaningful formula in the most general case, we have to consider a certain average of several numbers of type $N(L,h)$ instead of a single $N(L,h)$ as follows. We first take a set of lattices $\{L_i\}_{i=1}^k$ that are representatives for the classes belonging to the genus of L. Then we put

$$\mathfrak{m}(L) = \sum_{i=1}^{k} [\Gamma_i : 1]^{-1},$$

$$R(L,h) = \sum_{i=1}^{k} [\Gamma_i : 1]^{-1} N(L_i, h),$$

where $\Gamma_i = \{\gamma \in G^\varphi \mid L_i\gamma = L_i\}$. The purpose of this paper is to give an exact formula for $R(L,h)$ when L is *maximal* in the sense that it is maximal among the lattices on which φ takes values in $\mathfrak{g}$. In the previous work [S5], we gave an exact formula for $\mathfrak{m}(L)$. Thus the present paper is its natural continuation.

Before stating the formula, let us first recall the result of Siegel on this topic. For a prime ideal $\mathfrak{p}$ in F and $0 < m \in \mathbf{Z}$, let $A_m(\mathfrak{p})$ denote the number of elements y in $L/\mathfrak{p}^m L$ such that $\varphi[y] - h \in \mathfrak{p}^m$. Then, as Siegel showed, $N(\mathfrak{p})^{m(1-n)} A_m(\mathfrak{p})$ is a constant $d_\mathfrak{p}(h)$ independent of m if m is sufficiently large. Suppose that V (resp. L) is the vector space (resp. the module) of all n-dimensional row vectors with entries in F (resp. $\mathfrak{g}$) and $\varphi[x] = x\varphi_0 \cdot {}^t x$ with a totally positive symmetric matrix φ_0 with entries in $\mathfrak{g}$. Then he proved that

$$\text{(1)} \qquad \frac{R(L,h)}{\mathfrak{m}(L)} = c_n D_F^{(1-n)/2} \pi^{dn/2} \Gamma\left(\frac{n}{2}\right)^{-d} N_{F/\mathbf{Q}}\big(\det(\varphi_0)^{-1} h^{n-2}\big)^{1/2} \prod_{\mathfrak{p}} d_\mathfrak{p}(h),$$

where D_F is the discriminant of F, $\mathfrak{p}$ runs over all the prime ideals in F, $c_n = 1$ if $n > 2$, $c_n = 1/2$ if $n = 2$, and $d = [F : \mathbf{Q}]$; the infinite product $\prod_\mathfrak{p} d_\mathfrak{p}(h)$ must be

Received 4 November 1997. Revision received 19 October 1998.
1991 *Mathematics Subject Classification.* Primary 11E25; Secondary 11E12, 11E45.

interpreted suitably when it is divergent. This is actually a special case of his formula, which concerns the representation $\alpha\varphi_0 \cdot {}^t\alpha = \psi$ for a symmetric matrix ψ of size $\leq n$. He also showed that the product $\prod_{\mathfrak{p}} d_{\mathfrak{p}}(h)^{-1}$ coincides, up to a finite number of bad factors, with a special value of a certain L-function if n is even, and with the quotient of two such values if n is odd. However, he did not give explicit forms for the bad factors. In the present paper, we shall give an exact formula for $R(L,h)/\mathfrak{m}(L)$, as well as that for $R(L,h)$, with no ambiguous factors, when L is maximal.

For simplicity, let us state our formula here only in the case where n is even, though we shall treat the case of odd n too. Thus if $n = 2m \in 2\mathbf{Z}$, for a maximal lattice L we have (see Theorem 1.5)

$$(2) \qquad \begin{aligned} R(L,h) = {} & c_n D_F^{(m-1)^2} \prod_{k=1}^{m-1} \left\{ D_F^{1/2}\{(2k-1)!(2\pi)^{-2k}\}^d \zeta_F(2k) \right\} \\ & \cdot N_{F/\mathbf{Q}}(h)^{m-1} N(\mathfrak{d})^{m-1} \prod_{\mathfrak{p}|\mathfrak{e}} \varepsilon_{\mathfrak{p}} \prod_{\mathfrak{p}|h\mathfrak{e}} \gamma_{\mathfrak{p}}(m). \end{aligned}$$

Here ζ_F is the Dedekind zeta function of F; $\mathfrak{d}$ is the different of $F(\kappa^{1/2})$ relative to F, where $\kappa = (-1)^m \det(\varphi)$; $\mathfrak{e}$ is the product of all prime ideals $\mathfrak{p}$ for which $\widetilde{L}_{\mathfrak{p}} \neq L_{\mathfrak{p}}$, where

$$\widetilde{L} = \{x \in V \mid 2\varphi(x,L) \subset \mathfrak{g}\};$$

$\varepsilon_{\mathfrak{p}}$ is a rational number explicitly determined by the isomorphism class of the localization of φ at $\mathfrak{p}$; and $\gamma_{\mathfrak{p}}(s)$ is a local Dirichlet series explicitly determined by that isomorphism class and h. (See §1.6 for the explicit forms of $\varepsilon_{\mathfrak{p}}$ and $\gamma_{\mathfrak{p}}$.) In the most general case, $\gamma_{\mathfrak{p}}$ has a rather complicated expression, but it can be simplified considerably if h is odd or squarefree. In Section 6, we shall specialize our formulas to the two- and three-dimensional cases. Though there are many previous investigations in such cases, it seems that our formulas in Theorems 6.2 and 6.4 have never been stated in the forms we present there. We shall also add a few examples at the end.

Since we gave an exact formula for $\mathfrak{m}(L)$ in [S5], we obtain the formula for $R(L,h)$ by computing $R(L,h)/\mathfrak{m}(L)$. This can be done in two ways. To explain the ideas, let us take $F = \mathbf{Q}$ for simplicity. We first define a theta series $\theta(g,z)$ for $g \in G_{\mathbf{A}}^{\varphi}$ and z in the upper half-plane in a natural way so that

$$(3) \qquad \int_{G^{\varphi}\backslash G_{\mathbf{A}}^{\varphi}} \theta(g,z)\,dg = \mathfrak{m}(L)^{-1} \sum_{h=0}^{\infty} R(L,h)\exp(2\pi i h z).$$

Then the Siegel-Weil formula shows that

$$(4) \qquad c_n E(x,0) = \int_{G^{\varphi}\backslash G_{\mathbf{A}}^{\varphi}} \theta(g, x+i)\,dg \quad (x \in \mathbf{R})$$

with a certain Eisenstein series $E(x,s)$ defined for $x \in F_{\mathbf{A}}$ and $s \in \mathbf{C}$. This was proven for $n > 4$ in [W2] and for every even n in [KR]. In fact, the equality holds even for

$n = 3$ as we shall show in §5.3. Now we can compute each Fourier coefficient of $E(x, s)$ as a certain Euler product by the methods of our previous papers on Eisenstein series cited in [S4]. Each nonarchimedean Euler factor $c_{\mathfrak{p}}(h, s)$ of a Fourier coefficient can be given as a certain local integral, and has the property that $c_{\mathfrak{p}}(h, 0) = d_{\mathfrak{p}}(h)$. The main technical point is to find an explicit rational expression for $c_{\mathfrak{p}}$. That will be done in Sections 3 and 4. The discussion in this part for the prime factors of 2 is naturally very delicate, but the final formulas are not unbearably complicated. Since $R(L, h)/\mathfrak{m}(L)$ is a Fourier coefficient of the right-hand side of (4), as (3) shows, we obtain the desired exact formula.

We can actually dispense with the Siegel-Weil formula for the following reason. Since $c_{\mathfrak{p}}(h, s)$ can be defined without the series E, we can start with the definition of $c_{\mathfrak{p}}(h, s)$. Then the explicit form of $c_{\mathfrak{p}}$ gives the exact value of $c_{\mathfrak{p}}(h, 0) = d_{\mathfrak{p}}(h)$. This combined with Siegel's formula (1) gives our formula (2), since we already established the exact formula for $\mathfrak{m}(L)$ in [S5], except that this method involves some nontrivial technical points, which will be explained in §5.3. Though this is completely practicable, we employ the series E, because equality (4) is a natural way of expressing formulas of Siegel's type, and it also naturally leads to the integral expression for $c_{\mathfrak{p}}$.

It should be added that Siegel gave a formula even for an indefinite form, and that the Siegel-Weil formula is valid for such a form. In order to obtain an exact formula for the quantity corresponding to $R(L, h)$ in the indefinite case, we only have to compute a certain integral at each archimedean prime, since our results concerning the explicit forms for $c_{\mathfrak{p}}$ are completely local and therefore applicable to the general case. Though the computation of the archimedean integrals is not very difficult, if somewhat involved, in the present paper we restrict ourselves to the definite case, which keeps our exposition a reasonable length.

To avoid any possible misunderstanding, one final remark should also be made. Siegel's original formulas, if generalized as indicated at the end of §5.3, provide probably the best formulation applicable to an *arbitrary* lattice, and our exact formulas in the present and previous papers apply only to the *maximal* lattices; therefore the nature of our formulas is different from, if related to, that of Siegel's formulas.

1. Main theorem

1.1. Our notation is basically the same as in [S4] and [S5]. We note here some of the most frequently used symbols. For a finite set X, we denote by $\#X$ or $\#(X)$ the number of elements in X. For an associative ring R with identity element and a left or right ideal I of R, we denote by $R^{\times}$ the group of all its invertible elements and by I_n^m the R-module of all $m \times n$-matrices with entries in I. For $x = \begin{bmatrix} a & b \\ c & d \end{bmatrix} \in R_{2n}^{2n}$ with a, b, c, and d in R_n^n, we put $a = a_x, b = b_x, c = c_x$, and $d = d_x$ if there is no fear of confusion. For a complex hermitian matrix x, we write $x > 0$ if x is positive definite.

We fix a totally real algebraic number field F of finite degree and denote by $\mathbf{a}$ and $\mathbf{h}$ the sets of archimedean and nonarchimedean primes of F; we put $\mathbf{v} = \mathbf{a} \cup \mathbf{h}$; further, we denote by $\mathfrak{g}$, $\mathfrak{d}_F$, and D_F the maximal order of F, the different of F relative to $\mathbf{Q}$, and the discriminant of F, respectively. For each $v \in \mathbf{h}$, we denote by q_v the norm of the prime ideal at v. Given a set S, we denote by $S^{\mathbf{a}}$ the product of $\mathbf{a}$ copies of S (i.e., the set of all indexed elements $(x_v)_{v\in\mathbf{a}}$ with x_v in S). For a Hecke character χ of F, we denote by $L(s,\chi)$ the L-function of χ, by χ^* the ideal character associated with χ, and by $\zeta_F(s)$ the Dedekind zeta function of F.

Given an algebraic group G defined over F, we define G_v for each $v \in \mathbf{v}$ and the adelization $G_{\mathbf{A}}$ as usual, and view G as a subgroup of $G_{\mathbf{A}}$. We then denote by $G_{\mathbf{a}}$ and $G_{\mathbf{h}}$ the archimedean and nonarchimedean factors of $G_{\mathbf{A}}$, respectively. For $x \in G$, we denote by $x_v, x_{\mathbf{a}}$, and $x_{\mathbf{h}}$ its projections to $G_v, G_{\mathbf{a}}$, and $G_{\mathbf{h}}$. For a fractional ideal $\mathfrak{x}$ in F, we denote by $N(\mathfrak{x})$ the norm of $\mathfrak{x}$. For $t \in F_{\mathbf{A}}^{\times}$, we put $|t|_v = |t_v|_v$ with the normalized valuation $|\ |_v$ at v and $|t|_{\mathbf{A}} = \prod_{v\in\mathbf{v}} |t|_v$. We put

$$\mathbf{T} = \{\zeta \in \mathbf{C} \mid |\zeta| = 1\},$$
$$\mathbf{e}(z) = e^{2\pi i z} \quad (z \in \mathbf{C}),$$

and define characters $\mathbf{e}_{\mathbf{A}} : F_{\mathbf{A}} \to \mathbf{T}$ and $\mathbf{e}_v : F_v \to \mathbf{T}$ for each v as usual (see [S4, §18.2]). We then put $\mathbf{e}_{\mathbf{h}}(x) = \mathbf{e}_{\mathbf{A}}(x_{\mathbf{h}})$ and $\mathbf{e}_{\mathbf{a}}(x) = \mathbf{e}_{\mathbf{A}}(x_{\mathbf{a}})$ for $x \in F_{\mathbf{A}}$. We extend the latter symbol $\mathbf{e}_{\mathbf{a}}$ by putting $\mathbf{e}_{\mathbf{a}}(w) = \mathbf{e}\bigl(\sum_{v\in\mathbf{a}} w_v\bigr)$ for $w \in \mathbf{C}^{\mathbf{a}}$.

Whenever we take a finite-dimensional vector space Y over F, we normalize the Haar measure dx on $Y_{\mathbf{A}}$ so that the measure of $Y_{\mathbf{A}}/Y$ is 1, where the measure on Y is the standard discrete measure. If $Y = F_m^1$, then $dx = \prod_{v\in\mathbf{v}} dx_v$ with dx_v determined as follows. If $v \in \mathbf{a}$, dx_v is the standard measure of $Y_v = (F_v)_m^1 = \mathbf{R}_m^1$; if $v \in \mathbf{h}$, the measure of $(\mathfrak{g}_v)_m^1$ with respect to dx_v is $N(\mathfrak{d}_v)^{-m/2}$. We call these dx_v the normalized measures with respect to $\mathfrak{g}_m^1$. However, we also employ other types of Haar measures on Y_v, which will be specified each time.

We always let $GL(Y,F)$ act on Y on the right, so that the image of $x \in Y$ under $\alpha \in GL(Y,F)$ is written $x\alpha$. Let G be an F-rational algebraic subgroup of $GL(Y,F)$. For $\beta \in G_{\mathbf{A}}$ and a $\mathfrak{g}$-lattice L in Y, we denote by $L\beta$ the $\mathfrak{g}$-lattice in Y determined by $(L\beta)_v = L_v\beta_v$ for every $v \in \mathbf{h}$, where L_v is the closure of L in Y_v. In particular, for a fractional ideal $\mathfrak{x}$ in F and for $t \in F_{\mathbf{A}}^{\times}$, we denote by $t\mathfrak{x}$ the fractional ideal such that $(t\mathfrak{x})_v = t_v\mathfrak{x}_v$ for every $v \in \mathbf{h}$. We denote by δ an arbitrarily fixed element of $F_{\mathbf{h}}^{\times}$ such that $\delta\mathfrak{g} = \mathfrak{d}_F$.

1.2. Given a finite-dimensional vector space V over F and a nondegenerate symmetric F-bilinear form $\varphi : V \times V \to F$, we put

$$G^{\varphi} = G(V,\varphi) = \{\alpha \in GL(V) \mid \varphi(x\alpha, y\alpha) = \varphi(x,y)\},$$
$$G_+^{\varphi} = \{\alpha \in G(\varphi) \mid \det(\alpha) = 1\},$$
$$\varphi[x] = \varphi(x,x) \quad (x \in V).$$

Whenever we speak of (V,φ) with $V = F_n^1$, we understand that φ is given by $\varphi(x,y) = x\varphi_0 \cdot {}^t y$ for $x, y \in F_n^1$ with $\varphi_0 = {}^t\varphi_0 \in GL_n(F)$. We then simply write

φ for φ_0. In this case, we can put

$$G^\varphi = \{\alpha \in GL_n(F) \mid \alpha\varphi \cdot {}^t\alpha = \varphi\}.$$

We denote by $\mathbf{d}(\varphi)$ the element in $F^\times/\{a^2 \mid a \in F^\times\}$ represented by $\det(\varphi)$.

We now fix (V, φ), put $n = \dim(V)$, and assume that φ is totally positive definite; then $G^\varphi_\mathbf{a}$ is compact. Let L be a $\mathfrak{g}$-lattice in V such that $\varphi[x] \in \mathfrak{g}$ for every $x \in L$. Then we put

$$(1.2.1) \qquad C = \{\alpha \in G^\varphi_\mathbf{A} \mid L\alpha = L\}, \qquad C_+ = C \cap (G^\varphi_+)_\mathbf{A},$$

and take finite subsets $\mathfrak{B}$ of $G^\varphi_\mathbf{A}$ and $\mathfrak{B}_+$ of $(G^\varphi_+)_\mathbf{A}$ so that

$$(1.2.2) \qquad G^\varphi_\mathbf{A} = \bigsqcup_{a \in \mathfrak{B}} G^\varphi a C, \qquad (G^\varphi_+)_\mathbf{A} = \bigsqcup_{b \in \mathfrak{B}_+} G^\varphi_+ b C_+.$$

For each $h \in \mathfrak{g}$ and $a \in G^\varphi_\mathbf{A}$, put

$$(1.2.3) \qquad N(L, h) = \#\{\xi \in L \mid \varphi[\xi] = h\},$$

$$(1.2.4) \qquad \Gamma^a = G^\varphi \cap aCa^{-1}, \qquad \Gamma^a_+ = G^\varphi_+ \cap \Gamma^a.$$

Since φ is totally positive definite, $N(L, h)$ is finite, and it is not zero only if h is totally positive or $h = 0$. We define invariants $\mathfrak{m}(G^\varphi, C)$, $R(L, h)$, and $R_+(L, h)$ by

$$(1.2.5) \qquad \mathfrak{m}(G^\varphi, C) = \sum_{a \in \mathfrak{B}} [\Gamma^a : 1]^{-1},$$

$$(1.2.6) \qquad R(L, h) = \sum_{a \in \mathfrak{B}} [\Gamma^a : 1]^{-1} N(La^{-1}, h),$$

$$(1.2.7) \qquad R_+(L, h) = \sum_{b \in \mathfrak{B}_+} [\Gamma^b_+ : 1]^{-1} N(Lb^{-1}, h).$$

These are independent of the choice of $\mathfrak{B}$ and $\mathfrak{B}_+$. Clearly $R(L, 0) = \mathfrak{m}(G^\varphi, C)$. We introduced $\mathfrak{m}(G^\varphi, C)$ in [S4] and called it *the mass of G^φ relative to C*. For simplicity, we put $\mathfrak{m}(L) = \mathfrak{m}(G^\varphi, C)$, which is consistent with our notation in the introduction.

PROPOSITION 1.3. $R_+(L, h) = 2R(L, h)$.

Proof. For each $a \in \mathfrak{B}$, take B_a so that $G^\varphi a C = \bigsqcup_{b \in B_a} G^\varphi_+ b C$. We may assume that $B_a \subset (G^\varphi_+)_\mathbf{h}$ since $G^\varphi_\mathbf{A} = (G^\varphi_+)_\mathbf{h} C$, as shown in the proof of [S5, Lemma 5.6]. Then $(G^\varphi_+)_\mathbf{A} = \bigsqcup_{a \in \mathfrak{B}} \bigsqcup_{b \in B_a} G^\varphi_+ b C_+$ so that

$$R_+(L, h) = \sum_{a \in \mathfrak{B}} \sum_{b \in B_a} [\Gamma^b_+ : 1]^{-1} N(Lb^{-1}, h)$$

$$= \sum_{a \in \mathfrak{B}} \#(B_a) [\Gamma^a_+ : 1]^{-1} N(La^{-1}, h).$$

Now in [S5, Lemma 5.6(3)], we showed that $\#(B_a) = 2[\Gamma^a : \Gamma^a_+]^{-1}$, which combined with the above equality proves our lemma. □

1.4. For each $v \in \mathbf{v}$, we can define the localization of (V, φ) at v as usual, which we denote by (V_v, φ_v). This can be decomposed in the form

$$(V_v, \varphi_v) = (T_v, \theta'_v) \oplus (H_{r_v}, 2^{-1}\eta_{r_v}) \tag{1.4.1}$$

with anisotropic θ'_v and a nonnegative integer r_v. Here $H_r = (F_v)^1_{2r}$ and

$$\eta_r = \begin{bmatrix} 0 & 1_r \\ 1_r & 0 \end{bmatrix}. \tag{1.4.2}$$

Put $t_v = \dim(T_v)$. Then $n = 2r_v + t_v$ and $t_v \leq 4$ if $v \in \mathbf{h}$. Given a $\mathfrak{g}$-lattice L in V, we put

$$\widetilde{L} = \{x \in V \mid 2\varphi(x, L) \subset \mathfrak{g}\}. \tag{1.4.3}$$

We call L $\mathfrak{g}$-*maximal* if $\varphi[x] \in \mathfrak{g}$ for every $x \in L$, and L is maximal among such $\mathfrak{g}$-lattices.

We assume that φ is *normalized* in the sense that

$$\mathbf{d}(\varphi)\mathfrak{g} \quad \text{is the square of a fractional ideal if } n \text{ is odd.} \tag{1.4.4}$$

Now the principal result of this paper can be given as follows.

THEOREM 1.5. *Suppose that φ is totally positive definite and satisfies condition (1.4.4). Let L be a $\mathfrak{g}$-maximal lattice in V and h a totally positive element of $\mathfrak{g}$; let $\mathfrak{e}$ be the product of all primes $v \in \mathbf{h}$ for which $\widetilde{L}_v \neq L_v$. Put $d = [F : \mathbf{Q}]$ and $K = F(\kappa^{1/2})$ with $\kappa = (-1)^{(n-1)/2}\det(\varphi)h$ if n is odd and $\kappa = (-1)^{n/2}\det(\varphi)$ if n is even. Define $\gamma_v(s)$ by the formulas given in §1.6 below.*

(I) *Suppose n is odd; put $n = 2\mu + 1$ and denote by ψ_h the Hecke character of F corresponding to K/F. Let $\mathfrak{f}$ be the product of the prime ideals for which $t_v = 3$. Then we have*

$$\begin{aligned} R(L, h) = D_F^{\mu^2-\mu} \prod_{k=1}^{\mu-1} \left\{ D_F^{1/2} \{(2k-1)!(2\pi)^{-2k}\}^d \zeta_F(2k) \right\} \\ \cdot N_{F/\mathbf{Q}}(h)^{(n-2)/2} D_F^{1/2} \{(\mu-1)!(2\pi)^{-\mu}\}^d L(\mu, \psi_h) \\ \cdot 2^{d(n-2)} \prod_{v|\mathfrak{f}} \{2q_v(q_v+1)\}^{-1} (q_v^{n-1} - 1) \prod_{v|h\mathfrak{e}} \gamma_v\left(\frac{n}{2}\right), \end{aligned}$$

$$\frac{R(L, h)}{\mathfrak{m}(L)} = \frac{2^{d(n-2)} N_{F/\mathbf{Q}}(h)^{(n-2)/2} \{(\mu-1)!(2\pi)^{-\mu}\}^d L(\mu, \psi_h)}{D_F^{\mu} N(\mathfrak{f}) \{(n-2)!(2\pi)^{1-n}\}^d \zeta_F(n-1)} \prod_{v|h\mathfrak{e}} \gamma_v\left(\frac{n}{2}\right).$$

(II) *Suppose n is even; put $n = 2m$; denote by $\mathfrak{r}$ the maximal order of K, by $\mathfrak{d}$ the different of K relative to F, and by ψ the Hecke character of F corresponding to*

K/F. Then

$$R(L,h) = c_n D_F^{(m-1)^2} \prod_{k=1}^{m-1} \left\{ D_F^{1/2}\{(2k-1)!(2\pi)^{-2k}\}^d \zeta_F(2k) \right\}$$
$$\cdot N_{F/\mathbf{Q}}(h)^{m-1} N(\mathfrak{d})^{m-1} \prod_{v|\mathfrak{e}} \varepsilon_v \prod_{v|h\mathfrak{e}} \gamma_v(m),$$

$$\frac{R(L,h)}{\mathfrak{m}(L)} = \frac{c_n N_{F/\mathbf{Q}}(h)^{m-1}}{D_F^{(n-1)/2} N(\mathfrak{d})^{1/2} N(\mathfrak{f})\{(m-1)!(2\pi)^{-m}\}^d L(m,\psi)} \prod_{v|h\mathfrak{e}} \gamma_v(m),$$

where $\mathfrak{f}$ is the product of the prime factors of $\mathfrak{e}$ unramified in K, ε_v is the quantity defined in §1.6 below, $c_n = 1$ if $n > 2$, and $c_2 = 1/2$.

The proof will be completed in Section 5. We add here a few remarks.

First of all, condition (1.4.4) is inessential since, if n is odd, we can always find $c \in F^\times$ such that $\mathbf{d}(c\varphi)$ is represented by any specified number of $F^\times$.

Next, the factor $D_F^{1/2}\{(2k-1)!(2\pi)^{-2k}\}^d \zeta_F(2k)$ is a rational number as already noted in Remark (5) after Theorem 5.8 of [S5]. Now let $\mathfrak{d}_h$ be the different of K relative to F when n is odd. Then it can easily be seen that $[N(\mathfrak{d}_h)/N_{F/\mathbf{Q}}(h)]^{1/2}$ is a rational number, and hence the second line of factors of $R(L,h)$ in (I) is rational for the reason explained in the same remark.

In Section 6, we shall discuss lower-dimensional cases and also give some examples.

1.6. To define ε_v and $\gamma_v(s)$, put $K_v = K \otimes_F F_v$ if n is even. If $t_v = 2$, then, as explained in [S5, §3.2 and Proposition 5.2], (T_v, θ'_v) of (1.4.1) is isomorphic to (K_v, θ_v) with $\theta_v[x] = c_v \cdot N_{K/F}(x)$ for $x \in K_v$. Here c_v is an element of F_v such that $c_v \in \mathfrak{g}_v^\times$ if $\mathfrak{d}_v \neq \mathfrak{r}_v$, and c_v is a v-unit or a prime element of F_v if $\mathfrak{d}_v = \mathfrak{r}_v$.

As shown in [S5, §3.2], we have $\widetilde{L}_v \neq L_v$ exactly in the following cases:

$$\text{(1.6.1)}\quad \begin{aligned} &t_v = 1 \quad \text{and } v|2,\\ &t_v = 2,\quad \mathfrak{d}_v = \mathfrak{r}_v, \text{ and } c_v \text{ is a prime element of } K_v,\\ &t_v = 2 \quad \text{and} \quad \mathfrak{d}_v \neq \mathfrak{r}_v,\\ &t_v \geq 3. \end{aligned}$$

Now ε_v is the number given as follows:

$$\begin{aligned} &\{2q(1+q)\}^{-1}(q^{m-1}+1)(q^m+1) && \text{if } t_v = 2, \mathfrak{d}_v = \mathfrak{r}_v, \text{ and } c_v \notin \mathfrak{g}_v^\times,\\ &2^{-1} && \text{if } t_v = 2 \text{ and } \mathfrak{d}_v \neq \mathfrak{r}_v,\\ &\{2q(1+q)\}^{-1}(q^{m-1}-1)(q^m-1) && \text{if } t_v = 4. \end{aligned}$$

Here and in the formulas for γ_v below, we simply write q for q_v.

To give the formulas for $\gamma_v(s)$, determine $\nu \in \mathbf{Z}$ by $|h|_v^{-1} = q^\nu$ and put $\lambda = [\nu/2]$. If t_v is odd, we put $\psi_h^*(\mathfrak{p}) = \xi$, where $\mathfrak{p}$ is the prime ideal at v; also, we denote by κ the integer such that $2\mathfrak{g}_v = \mathfrak{p}_v^\kappa$ and put $e = 2\kappa + 1$. Thus $\kappa > 0$ if and only if $v \mid 2$. Further, if $v \mid 2$ and $\nu = 2\lambda \in 2\mathbf{Z}$, then we denote by ε the largest integer $\le e$ such that $f\pi^{-\nu}h \in (1+\mathfrak{p}_v^\varepsilon)\mathfrak{g}_v^{\times 2}$, where $\mathfrak{g}_v^{\times 2} = \{a^2 \mid a \in \mathfrak{g}_v^\times\}$ and f is an element of $\mathfrak{g}_v^\times$ such that $(-1)^{(n-1)/2}\det(\varphi)f$ is a square in F_v. (Such an f exists because of (1.4.4).) We put $\varepsilon = 0$ if $\nu \notin 2\mathbf{Z}$ or $v \nmid 2$.

Now $\gamma_v(s)$ can be given as follows:

$$\gamma_v(s) = \frac{1-q^{(\nu+1)(1-s)}}{1-q^{1-s}} \quad \text{if } t_v = 0,$$

$$\gamma_v(s) = q^{\lambda(2-2s)} + \frac{\left(1-\xi q^{(1/2)-s}\right)\left(1-q^{\lambda(2-2s)}\right)}{1-q^{2-2s}}$$
$$\text{if } t_v = 1 \text{ except when } v \mid 2 \text{ and } \nu \in 2\mathbf{Z},$$

$$\gamma_v(s) = \frac{\left(1-\xi q^{(1/2)-s}\right)\left(1-q^{(\nu+e+1)(1-s)}\right)}{1-q^{2-2s}} + \xi q^{(\nu+e)(1-s)-(1/2)}$$
$$\text{if } t_v = 1, v \mid 2, \nu \in 2\mathbf{Z}, \text{ and } \xi \neq 0,$$

$$\gamma_v(s) = \frac{1-q^{(\nu+\varepsilon+1)(1-s)}}{1-q^{2-2s}} \quad \text{if } t_v = 1, v \mid 2, \nu \in 2\mathbf{Z}, \text{ and } \xi = 0,$$

$$\gamma_v(s) = \frac{1-\left(-q^{1-s}\right)^{\nu+1}}{1+q^{1-s}} \quad \text{if } t_v = 2, \mathfrak{d}_v = \mathfrak{r}_v, \text{ and } c_v \in \mathfrak{g}_v^\times,$$

$$\gamma_v(s) = \frac{1-\left(-q^{1-s}\right)^{\nu+2} + q^{2-s}\left(1-\left(-q^{1-s}\right)^\nu\right)}{\left(1+q^{-s}\right)\left(1+q^{1-s}\right)}$$
$$\text{if } t_v = 2, \mathfrak{d}_v = \mathfrak{r}_v, \text{ and } c_v \notin \mathfrak{g}_v^\times,$$

$$\gamma_v(s) = 1 + \psi_v(c_v h)\left(|h|_v^{-1}N(\mathfrak{d}_v)\right)^{1-s} \quad \text{if } t_v = 2 \text{ and } \mathfrak{d}_v \neq \mathfrak{r}_v,$$

$$\gamma_v(s) = \frac{1-q^{3-2s} - q^{(\lambda+\kappa)(2-2s)}\left(1-\xi q^{(3/2)-s}\right)\left(q^{2-2s}+\xi q^{(3/2)-s}\right)}{\left(1+\xi q^{(1/2)-s}\right)\left(1-q^{2-2s}\right)}$$
$$\text{if } t_v = 3 \text{ and } \xi \neq 0,$$

$$\gamma_v(s) = \frac{1-q^{3-2s} + q\cdot q^{(\nu+\varepsilon+1)(1-s)} - q^{(\nu+\varepsilon+3)(1-s)}}{\left(1-q^{1-2s}\right)\left(1-q^{2-2s}\right)} \quad \text{if } t_v = 3 \text{ and } \xi = 0,$$

$$\gamma_v(s) = \frac{1-q^{(\nu+2)(1-s)} - q^{2-s}\left(1-q^{\nu(1-s)}\right)}{\left(1-q^{-s}\right)\left(1-q^{1-s}\right)} \quad \text{if } t_v = 4.$$

When $v \mid 2$ and $\nu = 2\lambda$, the number ε is either $e-1$ or an odd positive integer $\le e$, and any such integer can occur as ε for some h. Moreover, $\varepsilon < e-1$, $\varepsilon = e-1$, or

$\varepsilon = e$ according as $\psi_h^*(\mathfrak{p})$ is 0, -1, or 1. Also, if $\varepsilon < e - 1$, then $\mathfrak{p}^{e-\varepsilon}$ is the exact power of $\mathfrak{p}$ that divides the relative discriminant of K over F. Thus ε is completely determined by $\psi_h^*(\mathfrak{p})$ if $\mathfrak{p}_v = 2\mathfrak{g}_v$, in which case $e = 3$. These points will be explained in §5.4.

1.7. Let us now consider (V_v, φ_v) and L_v for $v \in \mathbf{h}$. To simplify our notation, we fix v and drop the subscript v. Thus, $F, \mathfrak{g}$, and $\mathfrak{p}$ denote a finite algebraic extension of $\mathbf{Q}_p$ for some prime number p, its valuation ring, and the maximal ideal; $q = [\mathfrak{g} : \mathfrak{p}]$; L is a (local) $\mathfrak{g}$-maximal lattice in V. We denote by π an arbitrarily fixed prime element of F, and define the normalized valuation $x \mapsto |x|$ of F as usual so that $|\pi| = q^{-1}$. As explained in [S5, Section 3], by changing the coordinate system, we can take (V, φ) as follows: $V = F_n^1$ and $\varphi(x, y) = x\varphi \cdot {}^t y$ for $x, y \in V$ with a matrix φ of the form

$$\varphi = \begin{bmatrix} 0 & 0 & 2^{-1}1_r \\ 0 & \theta & 0 \\ 2^{-1}1_r & 0 & 0 \end{bmatrix}, \quad \theta = {}^t\theta \in GL_t(F). \tag{1.7.1}$$

Here θ is anisotropic. Thus $n = 2r + t$. Taking the standard F-basis $\{e_i\}_{i=1}^n$ of V, we put

$$H_r = \sum_{i=1}^{r} (Fe_i + Fe_{r+t+i}), \qquad T = \sum_{j=1}^{t} Fe_{r+j}, \tag{1.7.2}$$

$$L = \sum_{i=1}^{r} (\mathfrak{g}e_i + \mathfrak{g}e_{r+t+i}) + M, \qquad M = \{x \in T \mid \theta[x] \in \mathfrak{g}\}. \tag{1.7.3}$$

Choosing $\{e_{r+j}\}_{j=1}^t$ suitably, we may assume that

$$M = \sum_{j=1}^{t} \mathfrak{g}e_{r+j}. \tag{1.7.4}$$

We end this section by noting an easy fact.

Proposition 1.8. *If L and h are as in Theorem 1.5 and $n \geq 4$, then $R(L, h) \neq 0$.*

Proof. If $n \geq 4$, the Hasse principle guarantees an element $x \in V$ such that $\varphi[x] = h$, since we are assuming that h is totally positive. Let $W = \{w \in V \mid \varphi(x, w) = 0\}$, and let J be a $\mathfrak{g}$-lattice in W such that $\varphi[y] \in \mathfrak{g}$ for every $y \in J$. Then we can find a maximal lattice L' containing $\mathfrak{g}x + J$. By [S4, Lemma 5.9], we have $L' = L\alpha$ with a suitable $\alpha \in G_{\mathbf{A}}^{\varphi}$. Since $N(L', h) > 0$, we obtain our assertion. □

2. Eisenstein series

2.1. In addition to G^{φ}, we need to consider a symplectic group over F. Thus, with a positive integer m, we put

$$\mathcal{G} = \mathcal{G}^m = Sp(m, F) = \{\alpha \in GL_{2m}(F) \mid {}^t\alpha\iota\alpha = \iota\}, \qquad \iota = \iota_m = \begin{bmatrix} 0 & -1_m \\ 1_m & 0 \end{bmatrix}.$$

For each $v \in \mathbf{a}$, we let $\mathcal{G}_v$ act on the space

$$\mathfrak{H}_m = \{z \in \mathbf{C}_m^m \mid {}^tz = z, \mathrm{Im}(z) > 0\}$$

as usual, and let $\mathcal{G}_{\mathbf{A}}$ act on $\mathfrak{H}_m^{\mathbf{a}}$ by putting $\alpha(z) = (\alpha_v z_v)_{v \in \mathbf{a}}$ for $\alpha \in \mathcal{G}_{\mathbf{A}}$ and $z \in \mathfrak{H}_m^{\mathbf{a}}$. We then put

$$j(\alpha, z) = \bigl(j(\alpha_v, z_v)\bigr)_{v \in \mathbf{a}} (\in \mathbf{C}^{\mathbf{a}}) \quad (\alpha \in \mathcal{G}_{\mathbf{A}}^m, z \in \mathfrak{H}_m^{\mathbf{a}}),$$
$$j(\xi, w) = \det(c_\xi w + d_\xi) \qquad (\xi \in Sp(n, \mathbf{R}), w \in \mathfrak{H}_m),$$

and denote by $\mathbf{i}$ "the origin" of $\mathfrak{H}_m^{\mathbf{a}}$ given by $\mathbf{i} = (\mathbf{i}_v)_{v \in \mathbf{a}}$, with $\mathbf{i}_v = i\,1_m$.

We also put

$$Y = F_m^1, \qquad \mathcal{P} = \mathcal{P}^m = \{\alpha \in \mathcal{G} \mid c_\alpha = 0\},$$
$$\Omega_{\mathbf{A}} = \{x \in \mathcal{G}_{\mathbf{A}} \mid \det(c_x) \in F_{\mathbf{A}}^\times\}, \qquad \Omega_v = \{x \in \mathcal{G}_v \mid \det(c_x) \neq 0\}.$$

We let $\mathcal{G}$ act on $Y \times Y = F_{2m}^1$ by right multiplication. Then we can define the metaplectic groups $Mp(Y_{\mathbf{A}})$ and $Mp(Y_v)$ in the sense of [W1] for each $v \in \mathbf{v}$ with respect to the alternating form $(x, y) \mapsto x\iota_m \cdot {}^ty$ on $F_{2m}^1 \times F_{2m}^1$. These groups, which we write $M_{\mathbf{A}}$ (or $M_{\mathbf{A}}^n$) and M_v for simplicity, are groups of unitary transformations on $L^2(Y_{\mathbf{A}})$ and on $L^2(Y_v)$. There is an exact sequence

$$1 \longrightarrow \mathbf{T} \longrightarrow M_{\mathbf{A}} \longrightarrow \mathcal{G}_{\mathbf{A}} \longrightarrow 1$$

and also a similar exact sequence for each v with M_v and $\mathcal{G}_v$ in place of $M_{\mathbf{A}}$ and $\mathcal{G}_{\mathbf{A}}$. There is a natural lift $r : \mathcal{G} \to M_{\mathbf{A}}$ by which we consider $\mathcal{G}$ a subgroup of $M_{\mathbf{A}}$. There are also two types of lifts

(2.1.1) $$r_P : \mathcal{P}_{\mathbf{A}} \to M_{\mathbf{A}}, \qquad r_\Omega : \Omega_{\mathbf{A}} \to M_{\mathbf{A}},$$

which satisfy the formulas

(2.1.2) $$[r_P(\alpha) f](x) = |\det(a_\alpha)|_{\mathbf{A}}^{1/2} \mathbf{e}_{\mathbf{A}}(x a_\alpha \cdot {}^tb_\alpha \cdot {}^tx/2) f(x a_\alpha) \quad \text{if } \alpha \in \mathcal{P}_{\mathbf{A}},$$

(2.1.3)

$$[r_\Omega(\beta) f](x) = |\det(c_\beta)|_{\mathbf{A}}^{1/2} \int_{Y_{\mathbf{A}}} f(x a_\beta + y c_\beta) \mathbf{e}_{\mathbf{A}}(q_\beta(x, y))\, dy \quad \text{if } \beta \in \Omega_{\mathbf{A}},$$
$$q_\beta(x, y) = \frac{1}{2} \cdot x a_\beta \cdot {}^tb_\beta \cdot {}^tx + \frac{1}{2} \cdot y c_\beta \cdot {}^td_\beta \cdot {}^ty + x b_\beta \cdot {}^tc_\beta \cdot {}^ty.$$

Moreover, $r_P = r$ on $\mathcal{P}$ and $r_\Omega = r$ on $\mathcal{G} \cap \Omega_{\mathbf{A}}$. There are also similar lifts of $\mathcal{P}_v$ and Ω_v into M_v given by the same formulas with the subscript $\mathbf{A}$ replaced by v. We denote these lifts also by r_P and r_Ω, since the distinction will be clear from the context. Here the measure on Y_v for each v is the normalized one with respect to $\mathfrak{g}_m^1$.

2.2. Given $\xi \in (F_{\mathbf{A}})_m^m$, we define its denominator ideal $\nu_0(\xi)$ as in [S4, §3.7] and denote by $\nu(\xi)$ its norm. With a fixed fractional ideal $\mathfrak{b}$ in F, we define subgroups C' and C'_0 of $\mathcal{G}_{\mathbf{A}}^m$ by

(2.2.1) $$C' = \{x \in \mathcal{G}_{\mathbf{A}}^m \mid a_x \prec \mathfrak{g}, b_x \prec \mathfrak{b}^{-1}, c_x \prec \mathfrak{b}, d_x \prec \mathfrak{g}\},$$

(2.2.2) $$C'_0 = \{x \in C' \mid x(\mathbf{i}) = \mathbf{i}\}.$$

Here we write $\alpha \prec \mathfrak{a}$ for an element α of $(F_{\mathbf{A}})_m^n$ and a fractional ideal $\mathfrak{a}$ if α_v has entries in $\mathfrak{a}_v$ for every $v \in \mathbf{h}$. Noticing that $\mathcal{G}_{\mathbf{A}}^m = \mathcal{P}_{\mathbf{A}} C'_0$, we define $\varepsilon : \mathcal{G}_{\mathbf{A}}^m \to \mathbf{R}$ by

(2.2.3) $$\varepsilon(p\gamma) = \left|\det(d_p)\right|_{\mathbf{A}} \quad \text{for } p \in \mathcal{P}_{\mathbf{A}} \text{ and } \gamma \in C'_0.$$

This depends on $\mathfrak{b}$. We shall later take $\mathfrak{b}$ to be $\mathfrak{d}_F$, but until the end of this section, we consider ε with an arbitrarily fixed $\mathfrak{b}$. Take $\beta \in F_{\mathbf{h}}^{\times}$ so that $\mathfrak{b} = \beta\mathfrak{g}$. Then we have

(2.2.4) $$\varepsilon(\alpha) = |j(\alpha, \mathbf{i})| \quad \text{if } \alpha \in \mathcal{G}_v \text{ and } v \in \mathbf{a},$$

(2.2.5) $$\varepsilon\left(\begin{bmatrix} 0 & -1 \\ 1 & \sigma \end{bmatrix}\right) = |\beta_v|^{-m}\nu(\beta_v\sigma) \quad \text{for every } \sigma = {}^t\sigma \in (F_v)_m^m, v \in \mathbf{h}.$$

The former is easy; see [S4, (18.5.1)]. To prove the latter, given σ, let $\left[\begin{smallmatrix} 0 & -1 \\ 1 & \sigma \end{smallmatrix}\right] = pw$ with $p \in \mathcal{P}_v$ and $w \in C' \cap \mathcal{G}_v$; put $\alpha = \mathrm{diag}[1_m, \beta_v 1_m]$ and $y = \alpha^{-1}w\alpha$. Then $d_p c_y = \beta_v^{-1}1_m$ and $d_p d_y = \sigma$, and hence $\beta_v\sigma = c_y^{-1}d_y$. Since $y \in GL_{2m}(\mathfrak{g}_v)$, we have $\nu(\beta_v\sigma) = |\det(c_y)|_v^{-1} = |\det(\beta_v d_p)|_v$, which proves (2.2.5).

2.3. We now consider (V, φ) of §1.2 with totally positive φ. We identify V with F_n^1 and view φ as a symmetric matrix, so that $\varphi[x] = x\varphi \cdot {}^tx$ for $x \in V = F_n^1$. Let $\mathcal{S}(V_{\mathbf{h}})$ denote the Schwartz-Bruhat space of $V_{\mathbf{h}}$. For a fixed $\ell \in \mathcal{S}(V_{\mathbf{h}})$ and $z \in \mathfrak{H}_1^{\mathbf{a}}$, we define a function f_z on $V_{\mathbf{A}}$ by

(2.3.1) $$f_z(x) = \ell(x_{\mathbf{h}})\mathbf{e}_{\mathbf{a}}(\varphi[x_{\mathbf{a}}]z) \quad (x \in V_{\mathbf{A}}).$$

For $\alpha = \left[\begin{smallmatrix} a & b \\ c & d \end{smallmatrix}\right] \in \mathcal{G}_{\mathbf{A}}^1$ and $\sigma \in F_{\mathbf{A}}$, we put

(2.3.2) $$\alpha_\varphi = \begin{bmatrix} a1_n & 2b\varphi \\ c(2\varphi)^{-1} & d1_n \end{bmatrix}, \qquad \tau(\sigma) = \begin{bmatrix} 1 & \sigma \\ 0 & 1 \end{bmatrix},$$

(2.3.3) $$\rho(\sigma) = r_P(\tau(\sigma)_\varphi).$$

Notice that $\alpha_\varphi \in \mathcal{G}_{\mathbf{A}}^n$ and $\rho(\sigma) \in M_{\mathbf{A}}^n$. Thus we can let $\alpha_\varphi\rho(\sigma)$ act on $L^2(X_{\mathbf{A}})$ if $\alpha \in \mathcal{G}^1$. We then define a $\mathbf{C}$-valued function E on $F_{\mathbf{A}} \times \mathbf{C}$ by

(2.3.4) $$E(\sigma) = E(\sigma, s) = \sum_{\alpha \in A} \varepsilon(\alpha\tau(\sigma))^{-s}\{\alpha_\varphi\rho(\sigma)f_{\mathbf{i}}\}(0) \quad (A = \mathcal{P}^1\backslash\mathcal{G}^1)$$

for $\sigma \in F_{\mathbf{A}}$ and $s \in \mathbf{C}$. This is formally well defined; strictly speaking, however, $E(\sigma)$ is meaningful only for the values of s for which the series is convergent. Since $E(\sigma+b) = E(\sigma)$ for every $b \in F$, we have a Fourier expansion

$$E(\sigma, s) = \sum_{h \in F} c(h, s)\mathbf{e}_{\mathbf{A}}(h\sigma) \tag{2.3.5}$$

with $c(h, s) \in \mathbf{C}$, which can be given by

$$c(h, s) = \int_{F_{\mathbf{A}}/F} E(\sigma, s)\mathbf{e}_{\mathbf{A}}(-h\sigma)\,d\sigma. \tag{2.3.6}$$

Now $\mathcal{G}^1 = \mathcal{P}^1 \sqcup \mathcal{P}^1\iota\mathcal{P}^1$, and $\mathcal{P}^1\backslash\mathcal{G}^1$ can be given by 1 and $\iota\tau(b)$ for all $b \in F$. Therefore

$$E(\sigma) = \varepsilon\big(\tau(\sigma)\big)^{-s}\{\rho(\sigma)f_{\mathbf{i}}\}(0) + E_1(\sigma)$$

with

$$E_1(\sigma) = \sum_{b \in F} \varepsilon\big(\iota\tau(\sigma+b)\big)^{-s}\{\iota_\varphi\rho(\sigma+b)f_{\mathbf{i}}\}(0).$$

Put $\{\rho(\sigma)f_{\mathbf{i}}\}(x) = g_\sigma(x)$; from (2.1.2) we see that $g_\sigma(0) = f_{\mathbf{i}}(0) = \ell(0)$; also, we have $\varepsilon(\tau(\sigma)) = 1$ since $\tau(\sigma) \in \mathcal{P}^1_{\mathbf{A}}$. Therefore $E(\sigma) = \ell(0) + E_1(\sigma)$, and hence

$$c(h, s) = \int_{F_{\mathbf{A}}} \varepsilon\big(\iota\tau(\sigma)\big)^{-s}\{\iota_\varphi g_\sigma\}(0)\mathbf{e}_{\mathbf{A}}(-h\sigma)\,d\sigma + \begin{cases} \ell(0) & \text{if } h = 0, \\ 0 & \text{if } h \neq 0. \end{cases}$$

2.4. To compute $c(h, s)$, we first recall a confluent hypergeometric function ξ and define Ξ as follows:

$$\xi(g, h; s, s') = \int_{\mathbf{R}} \mathbf{e}(-hx)(x+ig)^{-s}(x-ig)^{-s'}\,dx \quad (s, s' \in \mathbf{C}, 0 < g \in \mathbf{R}, h \in \mathbf{R}),$$

$$\Xi(y, w; s, s') = \prod_{v \in \mathbf{a}} \xi(y_v, w_v; s, s') \quad (y \in F_{\mathbf{A}}, y_v > 0, w \in F_{\mathbf{A}}).$$

Here z^{-s} and $\bar{z}^{-s}$ for $z \in \mathfrak{H}_1$ are determined so that their values at $z = i$ are $\mathbf{e}(-s/4)$ and $\mathbf{e}(s/4)$, respectively. Assume that $\ell(x) = \prod_{v \in \mathbf{h}} \ell_v(x_v)$ with $\ell_v \in \mathcal{S}(V_v)$, and ℓ_v is the characteristic function of L_v for almost all $v \in \mathbf{h}$ with some $\mathfrak{g}$-lattice L in V. From (2.1.2) we see that $g_\sigma(y) = \prod_{v \in \mathbf{v}} g_v(y_v, \sigma_v)$ with $g_v(w, \sigma_v) = \mathbf{e}_v\big(\varphi[w]\sigma_v\big)\ell_v(w)$ for $v \in \mathbf{h}$ and $g_v(w, \sigma_v) = \mathbf{e}_v\big(\varphi[w](i+\sigma_v)\big)$ for $v \in \mathbf{a}$. Therefore, by (2.1.3) we have

$$\begin{aligned}(\iota_\varphi g_\sigma)(0) &= \int_{V_{\mathbf{A}}} g_\sigma\big(y(2\varphi)^{-1}\big)\,dy = \int_{V_{\mathbf{A}}} g_\sigma(y)\,dy \\ &= \prod_{v \in \mathbf{v}} \int_{V_v} g_v(y_v, \sigma_v)\,dy_v.\end{aligned}$$

Here the dy_v are normalized with respect to $\mathfrak{g}_n^1$. If $v \in \mathbf{a}$, we have

$$\int_{V_v} g_v(w, \sigma_v)\,dw = \det(2\varphi)_v^{-1/2} \det\left[-i(i+\sigma_v)\right]^{-n/2}.$$

For $v \in \mathbf{a}$, let $d\sigma_v$ be the standard measure of $F_v = \mathbf{R}$; for $v \in \mathbf{h}$, let $d\sigma_v$ be the measure on F_v such that the measure of $\mathfrak{g}_v$ is 1. Then, as noted in [S4, (18.9.3) and §A6.6],

(2.4.1) $$d\sigma = D_F^{-1/2} \prod_{v \in \mathbf{v}} d\sigma_v.$$

Now $\varepsilon(\xi) = \prod_{v\in\mathbf{v}} \varepsilon(\xi_v)$ for $\xi \in \mathscr{G}_{\mathbf{A}}$, and so changing σ for $\delta^{-1}\sigma$, we can put, for $h \neq 0$,

$$c(h,s) = D_F^{1/2} \prod_{v\in\mathbf{v}} c_v(h,s)$$

with

$$c_v(h,s) = \int_{F_v} \mathbf{e}_v(-h\delta_v^{-1}\sigma_v)\varepsilon\left(\begin{bmatrix} 0 & -1 \\ 1 & \delta_v^{-1}\sigma_v \end{bmatrix}\right)^{-s} \int_{V_v} g_v(w, \delta_v^{-1}\sigma_v)\,dw\,d\sigma_v,$$

provided the infinite product $\prod_{v\in\mathbf{v}} c_v(h,s)$ is convergent. Employing (2.2.4) and (2.2.5), we obtain

$$c_v(h,s) = \mathbf{e}\left(\frac{n}{8}\right) \det(2\varphi)_v^{-1/2} \xi\left(1, h_v; \frac{s+n}{2}, \frac{s}{2}\right) \quad (v \in \mathbf{a}),$$

$$c_v(h,s) = |\beta|_v^s \int_{F_v} \nu(\zeta_v\sigma)^{-s} \int_{V_v} \chi_v\big((\varphi[x]-h)\sigma\big)\ell_v(x)\,dx\,d\sigma \quad (v \in \mathbf{h}),$$

where $\zeta_v = \beta_v/\delta_v$ and χ_v is defined by

(2.4.2) $$\chi_v(\xi) = \mathbf{e}_v(\delta_v^{-1}\xi) \quad (\xi \in F_v, v \in \mathbf{h}).$$

We easily see that the convergence of $\prod_{v\in\mathbf{v}} c_v(h,s)$ can be reduced to that of $\sum_{\sigma\in F/\mathfrak{g}} |\nu(\sigma)^{-s}|$, which is guaranteed for sufficiently large $\mathrm{Re}(s)$ as noted in [S4, page 120].

LEMMA 2.5. *With a fixed $v \in \mathbf{h}$ and $0 \neq h \in \mathfrak{g}$, let ℓ_v be the characteristic function of a $\mathfrak{g}_v$-lattice L_v in V_v, and let $W_k = \{x \in L_v \mid \varphi[x] - h \in \mathfrak{p}^k\}$ for $0 < k \in \mathbf{Z}$, where $\mathfrak{p}$ is the maximal ideal of $\mathfrak{g}_v$. Put $q = [\mathfrak{g}_v : \mathfrak{p}]$, and assume that $\varphi[x] \in \mathfrak{g}_v$ for every $x \in L_v$. Let A_k be the number of elements in W_k considered modulo $\mathfrak{p}^k L_v$. Then there exist a rational number d and an integer λ such that $q^{k(1-n)} A_k = d$ for every $k \geq \lambda$. Moreover, $c_v(h,s)$ is a polynomial in q^{-s} and $c_v(h,0) = \mathrm{vol}(L_v)d$.*

Proof. For simplicity, in this proof we suppress the subscript v. Thus the symbols F, $\mathfrak{g}$, and L are local objects. The existence of d and λ is essentially due to Siegel, who assumed that $\varphi(x, y) \in \mathfrak{g}$ for every $x, y \in L$; see [S4, Proposition 14.3] for the proof of the present statement. Put $\mu_k = \mathrm{vol}(W_k)/\mathrm{vol}(L)$. Then clearly $\mu_k = q^{-kn}A_k$. Put $Y = \mathfrak{g} \cap F^\times$; let ω be the characteristic function of $\mathfrak{g}$. By [S4, (3.13.3)], for every $\sigma \in F$, we have

$$(2.5.1) \qquad (1-q^{-s})^{-1}\nu(\sigma)^{-s} = \int_Y \omega(\sigma y)|y|^s\,dy,$$

where dy is the Haar measure of $F^\times$ such that the measure of $\mathfrak{g}^\times$ is 1. Thus, assuming the convergence, we have

$$\begin{aligned}(1-q^{-s})^{-1}|\beta|^{-s}c(h,s) &= \int_Y |y|^s \int_L \int_F \omega(\sigma\zeta y)\chi\bigl(\sigma(\varphi[x]-h)\bigr)\,d\sigma\,dx\,dy \\ &= \sum_{k=0}^{\infty} q^{-ks} \int_L \int_F \omega(\sigma\zeta\pi^k)\chi\bigl(\sigma(\varphi[x]-h)\bigr)\,d\sigma\,dx \\ &= \sum_{k=0}^{\infty} q^{-ks} \int_L \int_{\mathfrak{p}^{-k-e}} \chi\bigl(\sigma(\varphi[x]-h)\bigr)\,d\sigma\,dx,\end{aligned}$$

where e is determined by $\zeta\mathfrak{g} = \mathfrak{p}^e$. The last integral over $\mathfrak{p}^{-e-k}$ is q^{e+k} or zero according as $x \in W_{e+k}$ or $x \notin W_{e+k}$. Therefore, putting $\tau = \mathrm{vol}(L)$, we obtain

$$\begin{aligned}(1-q^{-s})^{-1}|\beta|^{-s}c(h,s) &= \tau \sum_{k=0}^{\infty} q^{e+k-ks}\mu_{e+k} \\ &= \tau \sum_{k=0}^{\nu-1} \mu_{e+k}q^{e+k-ks} + d\tau \sum_{k=\nu}^{\infty} q^{-ks}\end{aligned}$$

for sufficiently large ν, from which we obtain our lemma. □

2.6. Let us now take a $\mathfrak{g}$-maximal lattice L in V, and define $\widetilde{L}$ by (1.4.3). As noted in §1.7, we can find an element ξ of $GL_n(F)_{\mathbf{h}}$ such that $L = (\mathfrak{g}_n^1)\xi$ and

$$(2.6.1) \qquad \xi_v\varphi\cdot{}^t\xi_v = \begin{bmatrix} 0 & 0 & 2^{-1}1_{r_v} \\ 0 & \theta_v & 0 \\ 2^{-1}1_{r_v} & 0 & 0 \end{bmatrix}, \quad \theta_v = {}^t\theta_v \in GL_{t_v}(F_v)$$

for each $v \in \mathbf{h}$, with an anisotropic $\theta_v \in GL_{t_v}(F_v)$. Let μ_v be the measure of L_v with respect to the measure dx_v on V_v normalized with respect to $\mathfrak{g}_n^1$. Then

$$\prod_{v\in\mathbf{h}} \mu_v = |\det(\xi)|_{\mathbf{A}} D_F^{-n/2}.$$

We have $|\det(\xi)|_{\mathbf{A}}^2 \prod_{v\in\mathbf{h}} |\det(2\varphi)|_v = \prod_{v\in\mathbf{h}} |\det(2\theta_v)|_v$, which equals $[\widetilde{L}:L]^{-1}$ by [S5, (3.1.9)]. Thus, we obtain

$$\prod_{v\in\mathbf{h}} \mu_v = \left\{D_F^{-n} N_{F/\mathbf{Q}}\bigl(\det(2\varphi)\bigr)[\widetilde{L}:L]^{-1}\right\}^{1/2}. \tag{2.6.2}$$

For each $v \in \mathbf{h}$, let dx_v^* denote the measure on V_v such that L_v has measure 1. Then $dx_v = \mu_v dx_v^*$, so that

$$N_{F/\mathbf{Q}}\bigl(\det(2\varphi)\bigr)^{-1/2} \prod_{v\in\mathbf{h}} \frac{dx_v}{dx_v^*} = D_F^{-n/2}[\widetilde{L}:L]^{-1/2}.$$

We now consider $E(\sigma, s)$ defined with $\mathfrak{b} = \mathfrak{d}_F$. Taking $\beta_v = \delta_v$, we thus obtain

$$c(h,s) = \mathbf{e}\left(\frac{dn}{8}\right) D_F^{\gamma(s)} [\widetilde{L}:L]^{-1/2}\, \Xi\left(1, h; \frac{s+n}{2}, \frac{s}{2}\right) \prod_{v\in\mathbf{h}} b_v(h,s)$$

with

$$d = [F:\mathbf{Q}], \qquad \gamma(s) = 2^{-1}(1-n) - s,$$

and

$$b_v(h,s) = \int_{F_v} \nu(\sigma)^{-s} \int_{V_v} \chi_v\bigl(\sigma(\varphi[x]-h)\bigr)\ell_v(x)\,dx_v^*\,d\sigma.$$

Take ℓ_v to be the characteristic function of L_v for each $v \in \mathbf{h}$. We recall here an easy fact that if $\eta = 2^{-1}\begin{bmatrix} 0 & 1_r \\ 1_r & 0 \end{bmatrix}$ and $N = (\mathfrak{g}_v)_{2r}^1$, then

$$\int_N \chi_v\bigl(\eta[y]\sigma\bigr)\,dy = \nu(\sigma)^{-r} \quad \text{for every } \sigma \in F_v, \tag{2.6.3}$$

where we take the measure of N to be 1. Indeed, let $\mathfrak{p}$ be the prime ideal at v, and let $q = N(\mathfrak{p})$. Now if $\sigma \in \mathfrak{p}^{-t}$ with $0 < t \in \mathbf{Z}$, then the left-hand side of (2.6.3) is $q^{-2rt}\sum_x \chi_v(\eta[x]\sigma)$, where x runs over $N/\mathfrak{p}_v^t N$. By [S4, Lemma 14.7(1)], the last sum is $q^{2rt}\nu(\sigma)^{-r}$, which proves (2.6.3). Employing this fact, we find that

$$b_v(h,s) = \int_{F_v} \nu(\sigma)^{-s-r_v} \chi_v(-\sigma h) \int_{M_v} \chi_v\bigl(\sigma\theta[x]\bigr)\,dx\,d\sigma, \tag{2.6.4}$$

where M_v is the $\mathfrak{g}_v$-lattice written as M in (1.7.3) and (1.7.4).

3. The nonarchimedean factors of a Fourier coefficient

3.1. We use local symbols $F, \mathfrak{g}, \mathfrak{p}, \pi$, and q as in §1.7, and consider (T, θ) and M defined there. We fix a prime $v \in \mathbf{h}$ and suppress the subscript v. We recall that

$\dim(T) = t$ and $M = \{x \in T \mid \theta[x] \in \mathfrak{g}\}$. We simply write χ for χ_v defined by (2.4.2). This is a $\mathbf{T}$-valued character such that $\mathfrak{g} = \{a \in F \mid \chi(a\mathfrak{g}) = 1\}$. We define a Haar measure dx (resp. $d\sigma$) on the additive group T (resp. F) so that the measure of M (resp. $\mathfrak{g}$) is 1. The purpose of this section is to find an explicit rational expression for

$$\beta(s) = \int_F \nu(\sigma)^{(t/2)-s}\chi(-\sigma h)\int_M \chi\bigl(\sigma\theta[x]\bigr)\,dx\,d\sigma, \tag{3.1.1}$$

where h is a nonzero fixed element of $\mathfrak{g}$, $s \in \mathbf{C}$, and $\nu(\sigma)$ is the norm of the denominator ideal of σ. This is essentially the function b_v of (2.6.4); more precisely,

$$\beta\left(s + \frac{n}{2}\right) = b_v(h, s). \tag{3.1.2}$$

Here $n = \dim(V)$. However, in this section we do not need (V, φ), so that we use the letter n as a parameter in various sums of the form $\sum_{n=0}^{\infty} A_n$. We include in our discussion the case $t = 0$; we then understand that the integral over M is 1.

Employing (2.5.1) and taking $(1, 1, \theta, M)$ in place of $(\beta, \zeta, \varphi, L)$ in the three lines of calculation below (2.5.1), we obtain

$$(1 - q^{-s})^{-1}\beta\left(s + \frac{t}{2}\right) = \sum_{n=0}^{\infty} q^{-ns}\int_M\int_{\mathfrak{p}^{-n}} \chi\bigl(\sigma(\theta[x] - h)\bigr)\,d\sigma\,dx.$$

If $t = 0$, we must ignore M, so that the last double integral becomes the integral of $\chi(-\sigma h)$ over $\mathfrak{p}^{-n}$, which is q^n or zero according as $h \in \mathfrak{p}^n$ or $h \notin \mathfrak{p}^n$. Therefore we easily see that

$$\beta(s) = \frac{(1 - q^{-s})\left(1 - q^{(\nu+1)(1-s)}\right)}{(1 - q^{1-s})} \quad \text{if } t = 0, \tag{3.1.3}$$

where ν is determined by $h\mathfrak{g} = \mathfrak{p}^{\nu}$.

Assuming $t > 0$, put $X_n(h) = \{x \in M \mid \theta[x] - h \in \mathfrak{p}^n\}$ and denote by $\mu_n(h)$ the measure of $X_n(h)$. Then

$$\beta\left(s + \frac{t}{2}\right) = (1 - q^{-s})\sum_{n=0}^{\infty} \mu_n(h) q^{n-ns}. \tag{3.1.4}$$

We are going to calculate $\mu_n(h)$ and then $\beta(s)$ according to the classification of (T, M, θ) given in [S5, §3.2]. In all cases we put $h = \pi^{\nu}k$ with $\nu \in \mathbf{Z}$ and $k \in \mathfrak{g}^{\times}$, and reduce our problem to $\mu_m(k)$. Before stating a basic lemma, we put

$$\mathfrak{g}^{\times 2} = \{a^2 \mid a \in \mathfrak{g}^{\times}\}, \qquad 1 + \mathfrak{p}^m = \{1 + a \mid a \in \mathfrak{p}^m\}. \tag{3.1.5}$$

LEMMA 3.2. (1) *Let* $2\mathfrak{g} = \mathfrak{p}^{\kappa}$ *with* $0 \le \kappa \in \mathbf{Z}$. *Then* $1 + \mathfrak{p}^{2\kappa+1} \subset \mathfrak{g}^{\times 2}$ *and* $1 + \mathfrak{p}^{2\kappa} \not\subset \mathfrak{g}^{\times 2}$.

(2) $1 + 4\mathfrak{p}^m = \{a^2 \mid a \in 1 + 2\mathfrak{p}^m\}$ *for* $0 < m \in \mathbf{Z}$.

(3) *Suppose* $2 \in \mathfrak{p}$ *and* $b \in \mathfrak{g}^{\times}$; *then* $F(\sqrt{b})$ *is unramified over* F *if and only if* $b \in \mathfrak{g}^{\times 2}(1 + 4\mathfrak{g})$.

Proof. Clearly the left-hand side of (2) contains the right-hand side. Conversely, let $c \in \mathfrak{p}^m, m > 0$. By Hensel's lemma, there exists an element $d \in \mathfrak{g}$ such that $d^2-d-c=0$. Changing d for $1-d$ if necessary, we may assume that $d \notin \mathfrak{p}$. Then $(1+2d^{-1}c)^2 = 1+4c$, which proves (2). Taking $m=1$, we obtain $1+\mathfrak{p}^{2\kappa+1} = 1+4\mathfrak{p} \subset \mathfrak{g}^{\times 2}$. If $2 \notin \mathfrak{p}$, then $\mathfrak{g}^\times \neq \mathfrak{g}^{\times 2}$, which proves (1) when $2 \notin \mathfrak{p}$. Suppose $2 \in \mathfrak{p}$, $x \in 1+\mathfrak{p}^{2\kappa}$, and $x = y^2$ with $y \in \mathfrak{g}$. Since $y \mapsto y^2$ is an automorphism of $\mathfrak{g}/\mathfrak{p}$, we have $y \in 1+\mathfrak{p}$, so that we can put $y = 1+\pi z$ with $z \in \mathfrak{g}$. Then $x = 1+\pi^2 z(z+\pi^{-1}2)$, and hence $z(z+\pi^{-1}2) \in \mathfrak{p}^{2\kappa-2}$. We see that $z \in \mathfrak{p}^{\kappa-1}$. Indeed, this is trivial if $\kappa = 1$. If $\kappa > 1$ and $z \notin \mathfrak{p}^{\kappa-1}$, then $z+\pi^{-1}2 \notin \mathfrak{p}^{\kappa-1}$ and so $z(z+\pi^{-1}2) \notin \mathfrak{p}^{2\kappa-2}$, a contradiction. Thus $z \in \mathfrak{p}^{\kappa-1}$. Put $z = \pi^{\kappa-1}w$ and $2 = \pi^\kappa t$. Then $x = 1+\pi^{2\kappa}w(w+t)$. If $1+\mathfrak{p}^{2\kappa} \subset \mathfrak{g}^{\times 2}$, then $w \mapsto w(w+t)$ must give a surjective map of $\mathfrak{g}/\mathfrak{p}$ onto itself. Clearly that is not the case, which proves (1). Finally, suppose $2 \in \mathfrak{p}$ and $a^2 b = 1+4d$ with $a \in \mathfrak{g}^\times$ and $d \in \mathfrak{g}$. Then $(2x+1)^2 - a^2 b = 4(x^2+x-d)$, and hence $F(\sqrt{b})$ is generated by a root of $x^2+x-d=0$. This proves the "if" part of (3). To prove the converse part, take $c \in 1+4\mathfrak{g}$ so that $c \notin \mathfrak{g}^{\times 2}$, as guaranteed by (1). Then $F(\sqrt{c})$ is an unramified quadratic extension of F. If $b \in \mathfrak{g}^\times$ and $F(\sqrt{b})$ is unramified over F, then $\sqrt{b} \in F(\sqrt{c})$, so that $b/c \in \mathfrak{g}^{\times 2}$, which completes the proof. □

3.3. Suppose $t=1$; then we can take $T=F$, $M=\mathfrak{g}$, and $\theta[x] = cx^2$ for $x \in F$ with $c \in \mathfrak{g}^\times$. We define $\xi : F^\times \to \{0, \pm 1\}$ as follows:

$$\xi(x) = \begin{cases} 1 & \text{if } \sqrt{x} \in F, \\ -1 & \text{if } F(\sqrt{x}) \text{ is an unramified quadratic extension of } F, \\ 0 & \text{if } F(\sqrt{x}) \text{ is a ramified quadratic extension of } F. \end{cases} \tag{3.3.1}$$

Let $h = \pi^\nu k$ as in §3.1; then $X_n(h) = \mathfrak{p}^{[(n+1)/2]}$ if $n \le \nu$; for $n > \nu$, we see that $X_n(h) = \pi^{\nu/2} X_{n-\nu}(k)$ or $\varnothing$ according as ν is even or odd. Suppose $2 \notin \mathfrak{p}$; then by Lemma 3.2(1), an element a of $\mathfrak{g}^\times$ is a square modulo $\mathfrak{p}$ if and only if $a \in \mathfrak{g}^{\times 2}$, in which case $\xi(a) = 1$; otherwise $\xi(a) = -1$. Therefore we easily see that

$$\mu_n(h) = \begin{cases} q^{-[(n+1)/2]} & \text{if } \nu \ge n, \\ \big(1+\xi(ch)\big) q^{(\nu/2)-n} & \text{if } n > \nu \in 2\mathbf{Z}, \\ 0 & \text{if } n > \nu \notin 2\mathbf{Z}. \end{cases} \tag{3.3.2}$$

By an elementary calculation, we find that

$$\begin{aligned} \beta(s) &= \frac{1-q^{1-2s}}{1-\xi q^{(1/2)-s}} \left\{ \frac{1-q^{(\lambda+1)(2-2s)}}{1-q^{2-2s}} - \xi q^{(1/2)-s} \frac{1-q^{\lambda(2-2s)}}{1-q^{2-2s}} \right\} \\ &= \left(1-q^{1-2s}\right) \left\{ \frac{q^{\lambda(2-2s)}}{1-\xi q^{(1/2)-s}} + \frac{1-q^{\lambda(2-2s)}}{1-q^{2-2s}} \right\}, \end{aligned} \tag{3.3.3}$$

where $\lambda = [\nu/2]$ and $\xi = \xi(ch)$. This result was already given in [S2, page 308].

3.4. Now suppose $2 \in \mathfrak{p}$, still with $t = 1$; then the value of $\mu_n(h)$ is the same as in (3.3.2) except in the case $n > \nu \in 2\mathbf{Z}$. To study this case, we let e denote the smallest positive integer such that $1+\mathfrak{p}^e \subset \mathfrak{g}^{\times 2}$. By Lemma 3.2(1), $e = 2\kappa+1$ with κ defined there. Put

$$[\mathfrak{g}^\times : (1+\mathfrak{p}^m)\mathfrak{g}^{\times 2}] = \tau(m) \quad (0 < m \in \mathbf{Z}). \tag{3.4.1}$$

Clearly $[\mathfrak{g}^\times : \mathfrak{g}^{\times 2}] = \tau(m) = \tau(e)$ for $m \geq e$. Let $Z_m = \{x \in \mathfrak{g}^\times \mid x^2 - 1 \in \mathfrak{p}^m\}$. Since the map $x \mapsto x^2$ gives an isomorphism of $\mathfrak{g}^\times/Z_m$ onto $(1+\mathfrak{p}^m)\mathfrak{g}^{\times 2}/(1+\mathfrak{p}^m)$, we have

$$\begin{aligned}\tau(m)[\mathfrak{g}^\times : Z_m] &= \tau(m)[(1+\mathfrak{p}^m)\mathfrak{g}^{\times 2} : 1+\mathfrak{p}^m]\\ &= [\mathfrak{g}^\times : 1+\mathfrak{p}^m] = [\mathfrak{g}^\times : Z_m][Z_m : 1+\mathfrak{p}^m],\end{aligned}$$

and hence

$$\tau(m) = [Z_m : 1+\mathfrak{p}^m]. \tag{3.4.2}$$

Given $b \in \mathfrak{g}^\times$, let $\varepsilon(b)$ be the largest integer $\varepsilon \leq e$ such that $b \in (1+\mathfrak{p}^\varepsilon)\mathfrak{g}^{\times 2}$. Then, for $h = \pi^\nu k$ and $n > \nu \in 2\mathbf{Z}$, we have $X_n(h) \neq \varnothing$ if and only if there exists $r \in \mathfrak{g}^\times$ such that $r^2 - k/c \in \mathfrak{p}^{n-\nu}$. That is so if and only if $\varepsilon(k/c) \geq \mathrm{Min}(e, n-\nu)$, in which case $X_n(h) = \pi^{\nu/2} r Z_{n-\nu}$. Therefore we obtain

$$\mu_n(h) = \begin{cases} 0 & \text{if } \varepsilon(k/c) < \mathrm{Min}(e, n-\nu),\\ \tau(n-\nu)q^{(\nu/2)-n} & \text{otherwise.}\end{cases} \tag{3.4.3}$$

LEMMA 3.5. *Suppose $2 \in \mathfrak{p}$ and $2\mathfrak{g} = \mathfrak{p}^\kappa$; put $e = 2\kappa+1$. Then the following assertions hold.*

(1) $\tau(m) = q^{[m/2]}$ *for $m < e$, and $\tau(e) = 2q^\kappa$.*

(2) *For $b \in \mathfrak{g}^\times$, we have $\xi(b) = 1$ if $\varepsilon(b) = e$, $\xi(b) = -1$ if $\varepsilon(b) = e-1$, and $\xi(b) = 0$ if $\varepsilon(b) < e-1$.*

(3) *$\varepsilon(b)$ is odd except when $\varepsilon(b) = e-1$.*

(4) $[(1+\mathfrak{p}^{e-1})\mathfrak{g}^{\times 2} : (1+\mathfrak{p}^e)\mathfrak{g}^{\times 2}] = 2.$

(5) *If $\xi(b) = 0$, then the relative discriminant of $F(\sqrt{b})$ over F is $\mathfrak{p}^{e-\varepsilon(b)}$.*

Proof. We first prove, by induction,

$$Z_{2a-1} = Z_{2a} = 1+\mathfrak{p}^a \quad \text{for } 0 < a \leq \kappa. \tag{3.5.1}$$

Let $x \in Z_1$. Since $z \mapsto z^2$ is an automorphism of $\mathfrak{g}/\mathfrak{p}$, we have $x \in 1+\mathfrak{p}$. Clearly $x^2 \in 1+\mathfrak{p}^2$ if $x \in 1+\mathfrak{p}$. This shows that $Z_1 = Z_2 = 1+\mathfrak{p}$. Next assume that $Z_{2a} = 1+\mathfrak{p}^a$ and $0 < a < \kappa$; let $x \in Z_{2a+1}$. Then $x = 1+\pi^a y$ with $y \in \mathfrak{g}$, and $2\pi^a y + \pi^{2a}y^2 \in \mathfrak{p}^{2a+1}$. Since $a < \kappa$, we must have $y \in \mathfrak{p}$, so that $x \in 1+\mathfrak{p}^{a+1}$. Clearly $x^2 \in Z_{2a+2}$ if $x \in 1+\mathfrak{p}^{a+1}$. This completes the proof of (3.5.1), which combined with (3.4.2) shows that $\tau(m) = q^{[m/2]}$ for $m < e$. Now let $x \in Z_e$ Since $Z_{2\kappa} = 1+\mathfrak{p}^\kappa$, we can put $x = 1+2y$ with $y \in \mathfrak{g}$. Then $x \in Z_e$ if and only if $y^2+y \in \mathfrak{p}$. Thus $1+2\mathfrak{p} \subset Z_e$ and $[Z_e : 1+2\mathfrak{p}] = 2$, and hence $\tau(e) = 2q^\kappa$. As for (2), we have $\xi(b) = 1 \Leftrightarrow b \in \mathfrak{g}^{\times 2} \Leftrightarrow \varepsilon(b) = e$. By Lemma 3.2(3), $\xi(b) = 0 \Leftrightarrow \varepsilon(b) < e-1$. Therefore we obtain (2). Next, from (1) and (3.4.1), we see that $(1+\mathfrak{p}^{m-1})\mathfrak{g}^{\times 2} = (1+\mathfrak{p}^m)\mathfrak{g}^{\times 2} \neq (1+\mathfrak{p}^{m+1})\mathfrak{g}^{\times 2}$ if m is odd and less than e. Therefore we obtain (3). Assertion (4) follows immediately from (3.4.1) and (1). To prove (5), suppose $\xi(b) = 0$; put $\varepsilon = \varepsilon(b)$. By (2) and (3), we can put $\varepsilon = 2c+1$ with $0 \le c < \kappa$. Since $b \in (1+\mathfrak{p}^\varepsilon)\mathfrak{g}^{\times 2}$ and $b \notin (1+\mathfrak{p}^{\varepsilon+1})\mathfrak{g}^{\times 2}$, we can put $a^2 b = 1+\pi^\varepsilon d$ with $a, d \in \mathfrak{g}^\times$. Put $y = \pi^{-c}(1+a\sqrt{b})$. Then $y^2 - 2\pi^{-c}y - \pi d = 0$. Since this is an Eisenstein equation, the different of $F(\sqrt{b})$ relative to F is generated by $2y - 2\pi^{-c}$, and hence it is generated by $2\pi^{-c}$, from which we obtain (5). □

3.6. Suppose $2 \in \mathfrak{p}$ and $t = 1$. If $\nu \notin 2\mathbf{Z}$, then (3.3.3) is valid. Suppose $\nu = 2\lambda \in 2\mathbf{Z}$; put $\varepsilon = \varepsilon(k/c)$. From (3.3.2), we obtain

$$(1-q^{-s})\sum_{n=0}^{\nu}\mu_n(h)q^{n-ns} = \frac{(1-q^{-2s})(1-q^{\lambda(1-2s)})}{1-q^{1-2s}} + q^{\lambda(1-2s)}(1-q^{-s}).$$

Also, from (3.4.3) and Lemma 3.5(1), we see that

$$\sum_{n>\nu}\mu_n(h)q^{n-ns}$$

$$= q^{\lambda(1-2s)}\sum_{m=1}^{\varepsilon}q^{[m/2]-ms} + \begin{cases} q^{\lambda(1-2s)+\kappa-es} + 2q^{\lambda(1-2s)+\kappa}\displaystyle\sum_{m=e+1}^{\infty}q^{-ms} & \text{if } \varepsilon = e, \\ 0 & \text{if } \varepsilon < e. \end{cases}$$

Now, if $\varepsilon = 2f+1$ with $0 \le f \in \mathbf{Z}$ (which is the case if $\varepsilon \neq e-1$, as noted in Lemma 3.5(3)), we have

$$\sum_{m=2}^{\varepsilon}q^{[m/2]-ms} = \frac{(1+q^{-s})(q^{1-2s} - q^{(f+1)(1-2s)})}{1-q^{1-2s}}.$$

Combining these and substituting $s - 1/2$ for s, we obtain

(3.6.1)

$$\beta(s) = \xi\left(1+\xi q^{(1/2)-s}\right)q^{(\nu+e)(1-s)-(1/2)} + \frac{(1-q^{1-2s})(1-q^{(\nu+g+1)(1-s)})}{1-q^{2-2s}},$$

where $g = e$ if $\varepsilon(k/c) \ge e-1$, $g = \varepsilon(k/c)$ if $\varepsilon(k/c) < e-1$, and $\xi = \xi(ch)$.

3.7. If $t = 2$, we can put $T = K$ and $M = \mathfrak{r}$ with a quadratic extension K of F and its valuation ring $\mathfrak{r}$; then we have $\theta[x] = cxx^\rho$ for $x \in K$, where ρ is the nontrivial automorphism of K over F; $c \in \mathfrak{g}^\times$ if K is ramified over F; c is a prime element of F or $c \in \mathfrak{g}^\times$ if K is unramified over F. We denote by $\mathfrak{q}$ the maximal ideal of $\mathfrak{r}$, and by $\mathfrak{d}$ the different of K relative to F. By local class field theory, we have $[F^\times : N_{K/F}(K^\times)] = 2$. Therefore we have a unique character

$$\xi_0 : F^\times \to \{\pm 1\} \tag{3.7.1}$$

such that $\mathrm{Ker}(\xi_0) = N_{K/F}(K^\times)$.

Let us first treat the case where $\mathfrak{d} = \mathfrak{r}$ and $c \in \mathfrak{g}^\times$. Let $h = \pi^\nu k$ as in §3.1; then $X_n(h) = \mathfrak{q}^{[(n+1)/2]}$ if $n \le \nu$; for $n > \nu$, we see that $X_n(h) = \pi^{\nu/2} X_{n-\nu}(k)$ or $\varnothing$ according as ν is even or odd. Since $N_{K/F}(\mathfrak{r}^\times) = \mathfrak{g}^\times$, we easily see that $\mu_m(k) = (q+1)q^{-1-m}$ for $m > 0$, and hence

$$\mu_n(h) = \begin{cases} q^{-n} & \text{if } \nu \ge n \in 2\mathbf{Z}, \\ q^{-n-1} & \text{if } \nu \ge n \notin 2\mathbf{Z}, \\ (q+1)q^{-1-n} & \text{if } n > \nu \in 2\mathbf{Z}, \\ 0 & \text{if } n > \nu \notin 2\mathbf{Z}. \end{cases} \tag{3.7.2}$$

Then an easy calculation shows that

$$\beta(s) = \frac{(1+q^{-s})\left(1-(-q^{1-s})^{\nu+1}\right)}{1+q^{1-s}}. \tag{3.7.3}$$

Next suppose that $\mathfrak{d} = \mathfrak{r}$ and $c\mathfrak{g} = \mathfrak{p}$. By a similar reasoning, we obtain

$$\mu_n(h) = \begin{cases} q^{-n} & \text{if } \nu \ge n \in 2\mathbf{Z}, \\ q^{1-n} & \text{if } \nu \ge n \notin 2\mathbf{Z}, \\ (q+1)q^{-n} & \text{if } n > \nu \notin 2\mathbf{Z}, \\ 0 & \text{if } n > \nu \in 2\mathbf{Z}. \end{cases} \tag{3.7.4}$$

Then we obtain

$$\beta(s) = \frac{1-(-q^{1-s})^{\nu+2} + q^{2-s}\left(1-(-q^{1-s})^{\nu}\right)}{1+q^{1-s}}. \tag{3.7.5}$$

3.8. We now assume that $\mathfrak{d} \ne \mathfrak{r}$ and put $\mathfrak{d} = \mathfrak{q}^e$ with a positive integer e. We take a prime element π_0 of K and take π to be $\pi_0\pi_0^\rho$. From local class field theory we know that $1+\mathfrak{p}^e \subset N_{K/F}(\mathfrak{r}^\times)$ and $[\mathfrak{g}^\times : N_{K/F}(\mathfrak{r}^\times)] = 2$. Then the restriction of ξ to $\mathfrak{g}^\times$ has kernel $N_{K/F}(\mathfrak{r}^\times)$. (If $\mathfrak{d} = \mathfrak{q}$, this is nothing but the quadratic residue symbol modulo $\mathfrak{p}$.) If $0 < m < e$, we have $\mathrm{Tr}_{K/F}(\mathfrak{q}^m) \subset \mathfrak{p}^m$ and $\mathfrak{g}^\times = (1+\mathfrak{p}^m)N_{K/F}(\mathfrak{r}^\times)$, and hence we easily see that the map $x \mapsto xx^\rho$ gives an isomorphism of $(\mathfrak{r}/\mathfrak{q}^m)^\times$ onto $(\mathfrak{g}/\mathfrak{p}^m)^\times$. For $m \ge e$, the map gives a homomorphism of $(\mathfrak{r}/\mathfrak{q}^{2m})^\times$ into $(\mathfrak{g}/\mathfrak{p}^m)^\times$, whose image is $N_{K/F}(\mathfrak{r}^\times)/(1+\mathfrak{p}^m)$. Therefore, for every $k \in \mathfrak{g}^\times$, we have

$$\mu_m(k) = \begin{cases} q^{-m} & \text{if } 0 < m < e, \\ 2q^{-m} & \text{if } m \geq e \text{ and } \xi_0(ck) = 1, \\ 0 & \text{if } m \geq e \text{ and } \xi_0(ck) = -1. \end{cases} \tag{3.8.1}$$

Now, for $h = \pi^\nu k$ as in §3.1, we see that $X_n(h) = \mathfrak{q}^n$ if $n \leq \nu$ and that $X_n(h) = \pi_0^\nu X_{n-\nu}(k)$ for $n > \nu$. Thus

$$\mu_n(h) = \begin{cases} q^{-n} & \text{if } n < \nu+e, \\ 2q^{-n} & \text{if } n \geq \nu+e \text{ and } \xi_0(ch) = 1, \\ 0 & \text{if } n \geq \nu+e \text{ and } \xi_0(ch) = -1. \end{cases} \tag{3.8.2}$$

Therefore we obtain

$$\beta(s) = 1 + \xi_0(ch)q^{(\nu+e)(1-s)}. \tag{3.8.3}$$

3.9. Suppose $t = 4$; then we can put $T = B$ with a division quaternion algebra B over F, $M = \mathfrak{O}$ with the unique maximal order $\mathfrak{O}$ of B, and $\theta[x] = xx^\iota$ for $x \in B$, where ι is the main involution of B. Let $\mathfrak{P}$ denote the maximal ideal of $\mathfrak{O}$, and ω an element such that $\omega^2 = \pi$. Then $X_n(h) = \mathfrak{P}^n$ for $n \leq \nu$, and $X_n(h) = \omega^\nu X_{n-\nu}\big((-1)^\nu k\big)$ for $n > \nu$. For every $m > 0$, the map $x \mapsto xx^\iota$ gives a surjective homomorphism of $\mathfrak{O}^\times$ onto $(\mathfrak{g}/\mathfrak{p}^m)^\times$ whose kernel has measure $(q+1)q^{-1-m}$, which is the measure of $X_m(a)$ for every $a \in \mathfrak{g}^\times$. Thus

$$\mu_n(h) = \begin{cases} q^{-2n} & \text{if } n \leq \nu, \\ (q+1)q^{-1-n-\nu} & \text{if } n > \nu. \end{cases}$$

Using these values of $\mu_n(h)$, we obtain

$$\begin{aligned} \sum_{n=0}^{\infty} \mu_n(h)q^{n-ns} &= \sum_{n=0}^{\nu} q^{-n-ns} + (q+1)q^{-1-\nu}\sum_{m=1}^{\infty} q^{-(m+\nu)s} \\ &= \frac{1-q^{-(\nu+1)(s+1)}}{1-q^{-1-s}} + \frac{(q+1)q^{-(\nu+1)(s+1)}}{1-q^{-s}}, \end{aligned}$$

and hence

$$\beta(s) = \frac{1-q^{(\nu+2)(1-s)} - q^{2-s}\big(1-q^{\nu(1-s)}\big)}{1-q^{1-s}}. \tag{3.9.1}$$

4. The function β in the case $t = 3$

4.1. Let us now discuss the case $t = 3$, which is most complicated. In this case we can put $T = \{x \in B \mid x^\iota = -x\}$, $M = T \cap \mathfrak{O}$, and $\theta[x] = cxx^\iota$ for $x \in T$ with $B, \mathfrak{O}$ as in §3.9 and $c \in \mathfrak{g}^\times$. We put $\mathrm{Tr}(x) = x + x^\iota$ for $x \in B$. Notice that $xy + yx = \mathrm{Tr}(xy)$ if $x, y \in T$. Take an unramified quadratic extension K of F contained in B and an element ω so that $B = K + K\omega$, $\omega^2 = \pi$, and $\omega a = a^\iota \omega$ for every $a \in K$. In [S5, §3.2], we have seen that $M = \mathfrak{g}\zeta + \mathfrak{r}\omega$, where $\mathfrak{r}$ is the maximal order of K and $\zeta = -\zeta^\iota \in \mathfrak{r}^\times$. We note that

$$\mathfrak{P}^{2m} \cap M = \pi^m M = \mathfrak{p}^m\zeta + \mathfrak{p}^m\mathfrak{r}\omega, \qquad \mathfrak{P}^{2m-1} \cap M = \mathfrak{p}^m\zeta + \mathfrak{p}^{m-1}\mathfrak{r}\omega, \tag{4.1.1}$$

$$\{a \in \mathfrak{r} \mid a^\iota = -a\} = \mathfrak{g}\zeta. \tag{4.1.2}$$

Given $h = \pi^\nu k$ as in §3.1, we easily see that $X_n(h) = \mathfrak{P}^n \cap M$ if $n \le \nu$, so that $\mu_n(h) = q^{-[3n/2]}$ if $n \le \nu$.

Define $\xi : F^\times \to \{0, \pm 1\}$ as in (3.3.1), and put

$$Y_m(b) = \{y \in M \mid y^2 - b \in \mathfrak{p}^m\} \quad (b \in \mathfrak{g}). \tag{4.1.3}$$

If $n > \nu$, put $b = -h/(\pi^{2\lambda}c)$ with $\lambda = [\nu/2]$. Then $b \in \mathfrak{g}^\times$ or $b\mathfrak{g} = \mathfrak{p}$, and $X_n(h) = \pi^\lambda Y_{n-2\lambda}(b)$. Thus our task is to find the measure of $Y_m(b)$, which we denote by $\mu'_m(b)$.

LEMMA 4.2. (1) *If $0 \ne b \in \mathfrak{g}$ and $\xi(b) \ne 1$, then there exists an element z of M such that $z^2 = b$, and consequently $Y_m(b) \ne \varnothing$ for every m.*

(2) *Let e be the smallest positive integer such that $1 + \mathfrak{p}^e \subset \mathfrak{g}^{\times 2}$. Suppose $b \in \mathfrak{g}^\times$ and $\xi(b) = 1$; then $Y_m(b) = \varnothing$ for $m \ge e$.*

Proof. If b is as in (1), then $F(\sqrt{b})$, being a quadratic extension of F, can be embedded in B. Therefore B has an element z such that $z^2 = b$. Then $z^\iota = -z$, and so $z \in M$. This proves (1). To prove (2), let $y \in Y_m(b)$ with $b \in \mathfrak{g}^\times$ and $m \ge e$. Then $y^{-2}b \in 1 + \mathfrak{p}^e$, and hence $y^{-2}b = c^2$ with $c \in \mathfrak{g}^\times$; thus $b = (cy)^2$. Since $cy \notin F$, we have $\xi(b) \ne 1$. This proves (2). □

LEMMA 4.3. *If $\xi(b) = -1$, $b \in \mathfrak{g}^\times$, and $\mathfrak{p}^m = 4\mathfrak{p}^a$ with $a > 0$, then $\mu'_m(b) = q \cdot \mu'_{m+1}(b)$.*

Proof. Let $y \in Y_m(b)$. By Lemma 3.2(2), there exists an element $d \in 1 + 2\mathfrak{p}^a$ such that $d^2 = y^{-2}b$. Put $z = dy$. Then $z^2 = b$ and $z - y \in 2\mathfrak{p}^a M$. To prove that $z + 2\pi^a M \subset Y_m(b)$, let $w = r\zeta + s\omega$ with $r \in \mathfrak{g}$ and $s \in \mathfrak{r}$. Then $(z + 2\pi^a w)^2 - b = 2\pi^a\,\mathrm{Tr}(zw) + 4\pi^{2a}w^2$. Now changing K suitably, we may assume that $K = F(z)$ and $\zeta = z$, since $\xi(b) = -1$. Thus $\mathrm{Tr}(zw) = \mathrm{Tr}(rz^2) = 2rb$, and so $z + 2\pi^a w \in Y_m(b)$. Therefore we can find a finite set Z such that $Y_m(b) = \bigsqcup_{z \in Z}(z + 2\pi^a M)$ and $z^2 = b$ for every $z \in Z$. We also see that $z + 2\pi^a w \in Y_{m+1}(b)$ if and only if $r \in \mathfrak{p}$, from which we obtain our assertion. □

LEMMA 4.4. *If $2 \notin \mathfrak{p}$ and $b \in \mathfrak{g}^\times$, then $\mu'_m(b) = (1 - \xi(b))q^{-m}$ for every $m > 0$.*

Proof. If $2 \notin \mathfrak{p}$ and $b \in \mathfrak{g}^\times$, then $\xi(b) \ne 0$, and we can take $e = 1$ in Lemma 4.2(2). Therefore $Y_1(b) \ne \varnothing$ if and only if $\xi(b) = -1$. Assuming $\xi(b) = -1$, from

Lemma 4.3, we obtain $\mu'_m(b) = q \cdot \mu'_{m+1}(b)$ for $m > 0$. Therefore it is sufficient to show that $\mu'_1(b) = 2q^{-1}$. For that purpose take $z \in M$ so that $z^2 = b$ as guaranteed by Lemma 4.2(1); let $y \in Y_1(b)$. Since $\mathfrak{O}/\mathfrak{P}$ is a field, $\pm z$ represent the solutions of the equation $x^2 = b$ in $\mathfrak{O}/\mathfrak{P}$. Therefore $y - z \in \mathfrak{P} \cap M$ or $y + z \in \mathfrak{P} \cap M$. Conversely, if $y \pm z \in \mathfrak{P} \cap M$, then $y^2 - b \in \mathfrak{P} \cap F = \mathfrak{p}$, so that $y \in Y_1(b)$. We have $2z \notin \mathfrak{P}$, since $2 \notin \mathfrak{p}$, and thus $Y_1(b)$ is the disjoint union of $z + \mathfrak{P} \cap M$ and $-z + \mathfrak{P} \cap M$. From (4.1.1) we see that $\mu'_1(b) = 2q^{-1}$ as expected. □

LEMMA 4.5. *Suppose $b \in \mathfrak{g}^\times$ and $2 \in \mathfrak{p}$; define κ, e, and $\varepsilon(b)$ as in §3.4. Then the following assertions hold.*

(1) *If $\varepsilon(b) < e - 1$, then*

$$\mu'_m(b) = \begin{cases} q^{-[3m/2]} & \text{for } m \le \varepsilon(b), \\ (q+1)q^{-m-(\varepsilon+1)/2} & \text{for } m > \varepsilon(b), \end{cases}$$

where $\varepsilon = \varepsilon(b)$.

(2) *If $\varepsilon(b) \ge e - 1$, then*

$$\mu'_m(b) = \begin{cases} q^{-[3m/2]} & \text{for } m < e, \\ (1 - \xi(b))q^{-m-\kappa} & \text{for } m \ge e. \end{cases}$$

Proof. We first prove

$$\varepsilon(\zeta^2) = e - 1. \tag{4.5.1}$$

By Lemma 3.2(1) we can find $a \in 1 + 4\mathfrak{g}$ such that $a \notin \mathfrak{g}^{\times 2}$. By Lemma 3.2(3) $F(\sqrt{a})$ is an unramified quadratic extension of F, and hence is isomorphic to K. Therefore $\zeta^2/a \in \mathfrak{g}^{\times 2}$ by (4.1.2). This proves (4.5.1). Put $d = \zeta^{-2}b$. From Lemma 3.5(4) and the definition of $\varepsilon(b)$, we easily see that $\varepsilon(d) = \varepsilon(b)$ if $\varepsilon(b) < e - 1$, $\varepsilon(d) = e$ if $\varepsilon(b) = e - 1$, and $\varepsilon(d) = e - 1$ if $\varepsilon(b) = e$. Now every element y of M can be written $y = \zeta(r + s\omega)$ with $r \in \mathfrak{g}$ and $s \in \mathfrak{r}$. Then $y \in Y_m(b)$ if and only if

$$r^2 - d - \pi s s' \in \mathfrak{p}^m. \tag{4.5.2}$$

Let σ_n be the measure of the set $\{r \in \mathfrak{g} \mid r^2 - d \in \mathfrak{p}^n\}$, and ρ_ν^m the measure of the set $\{s \in \mathfrak{r} \mid ss' - \pi^\nu u \in \mathfrak{p}^m\}$ for a fixed $u \in \mathfrak{g}^\times$. By (3.7.2), this is independent of u. Then $\sigma_n - \sigma_{n+1}$ is the measure of the set $\{r \in \mathfrak{g} \mid (r^2 - d)\mathfrak{g} = \mathfrak{p}^n\}$, and so

$$\mu'_m(b) = \sum_{n=1}^{\infty} (\sigma_n - \sigma_{n+1})\rho_{n-1}^{m-1}. \tag{4.5.3}$$

The number ρ_{n-1}^{m-1} can be given by (3.7.2) if we take $(m-1, n-1)$ to be (n, ν) there; thus

$$\rho_{n-1}^{m-1} = \begin{cases} q^{1-m} & \text{if } n \geq m \notin 2\mathbf{Z}, \\ q^{-m} & \text{if } n \geq m \in 2\mathbf{Z}, \\ (q+1)q^{-m} & \text{if } m > n \notin 2\mathbf{Z}, \\ 0 & \text{if } m > n \in 2\mathbf{Z}. \end{cases} \tag{4.5.4}$$

As for σ_n, by Lemma 3.5(1) and (3.4.3) we have

$$\sigma_n = \begin{cases} q^{[n/2]-n} & \text{if } \varepsilon(d) \geq n < e, \\ 2q^{\kappa-n} & \text{if } \varepsilon(d) = e \leq n, \\ 0 & \text{if } \varepsilon(d) < \operatorname{Min}(e, n). \end{cases} \tag{4.5.5}$$

Therefore we have

$$\mu'_m(b) = \sum_{n=1}^{\varepsilon(d)} (\sigma_n - \sigma_{n+1}) \rho_{n-1}^{m-1} \quad \text{if } \varepsilon(d) < e. \tag{4.5.6}$$

Also, we see that

$$\sigma_{2a-1} = \sigma_{2a} \quad \text{if } 2a \leq \varepsilon(d). \tag{4.5.7}$$

Suppose $\varepsilon(b) < e-1$; by Lemma 3.5(3) we can put $\varepsilon(b) = 2f+1$ with $0 \leq f \in \mathbf{Z}$. Since $\varepsilon(d) = \varepsilon(b)$, we have

$$\mu'_m(b) = \sum_{a=1}^{f} (\sigma_{2a} - \sigma_{2a+1}) \rho_{2a-1}^{m-1} + \sigma_{2f+1} \rho_{2f}^{m-1}.$$

Put $g = [(m+1)/2]$. By (4.5.4), $\sum_{a=1}^{f}$ can be replaced by $\sum_{a=g}^{f}$. (We understand that $\sum_{a=x}^{y} = 0$ if $x > y$.) Therefore, if $\varepsilon(b) > m = 2g \in 2\mathbf{Z}$, then $g \leq f$, so that by (4.5.4) and (4.5.5),

$$\mu'_m(b) = \sum_{a=g}^{f} (q^{-a} - q^{-a-1}) q^{-m} + q^{-f-1} \cdot q^{-m} = q^{-g-m} = q^{-3m/2}.$$

If $\varepsilon(b) > m = 2g-1$, the factor q^{-m} is replaced by q^{1-m}, and so $\mu'_m(b) = q^{1-g-m} = q^{-[3m/2]}$, which is also valid for $\varepsilon(b) = m$, in which case $\sum_{a=g}^{f} = 0$. If $\varepsilon(b) < m$, then $\mu'_m(b) = \sigma_{2f+1} \rho_{2f}^{m-1} = (q+1) q^{-f-1-m}$, which completes the proof of the formula of (1).

Next suppose $\varepsilon(b) = e-1$; then $\varepsilon(d) = e$, so that by (4.5.3) and (4.5.7) we have

$$\mu'_m(b) = \sum_{a=1}^{\kappa} (\sigma_{2a} - \sigma_{2a+1}) \rho_{2a-1}^{m-1} + \sum_{c=0}^{\infty} (\sigma_{e+c} - \sigma_{e+c+1}) \rho_{2\kappa+c}^{m-1}.$$

Take $m = e$. Then (4.5.4) shows that the first sum on the right-hand side is zero and $\rho_{2\kappa+c}^{m-1} = q^{1-e}$; $\sigma_{e+c} = 2q^{\kappa-e-c}$ by (4.5.5). Thus $\mu_e'(b) = 2q^{-e-\kappa}$. By Lemma 3.5(2), we have $\xi(b) = -1$, and hence, by Lemma 4.3, $\mu_m'(b) = q^{e-m}\mu_e'(b) = 2q^{-m-\kappa}$ for $m \geq e$. Suppose $m < e$. If $m = 2g \in 2\mathbf{Z}$, then by (4.5.4) and (4.5.5) we have

$$\mu_m'(b) = \sum_{a=g}^{\kappa-1}(q^{-a} - q^{-a-1})q^{-m} + (q^{-\kappa} - \sigma_e)q^{-m} + \sigma_e q^{-m}.$$

This is q^{-g-m}. (If $g = \kappa$, we have $\sum_{a=g}^{\kappa-1} = 0$, but the result is the same.) If $m = 2g-1$ with $\kappa \geq g \in \mathbf{Z}$, then similarly $\mu_m'(b) = q^{1-g-m}$. This completes the proof of (2) when $\varepsilon(b) = e-1$.

Finally suppose $\varepsilon(b) = e$. Then $\varepsilon(d) = e-1$, and from (4.5.5), (4.5.6), and (4.5.7) we obtain

$$\mu_m'(b) = \sum_{a=1}^{\kappa-1}(\sigma_{2a} - \sigma_{2a+1})\rho_{2a-1}^{m-1} + \sigma_{2\kappa}\rho_{2\kappa-1}^{m-1}.$$

By (4.5.4) this is zero if $m \geq e$. If $m = 2g$ with $\kappa \geq g \in \mathbf{Z}$, then

$$\mu_m'(b) = \sum_{a=g}^{\kappa-1}(q^{-a} - q^{-a-1})q^{-m} + q^{-\kappa-m} = q^{-g-m}.$$

If $m = 2g-1$ with $\kappa \geq g \in \mathbf{Z}$, then similarly $\mu_m'(b) = q^{1-g-m}$. This completes the proof of our lemma. □

We note here an easy special case:

(4.5.8) $$\mu_1'(b) = q^{-1} \quad \text{if } b \in \mathfrak{g}^\times \text{ and } 2 \in \mathfrak{p}.$$

LEMMA 4.6. *If $b\mathfrak{g} = \mathfrak{p}$, the following assertions hold.*

(1) Let $W_m = \pi^{m-1}\mathfrak{P} \cap M$ and $m > 0$. Then $y + w \in Y_m(b)$ for every $y \in Y_m(b)$ and $w \in W_m$.

(2) If $m > 1$ and $y \in Y_m(b)$, then there exists an element $z \in M$ such that $z - y \in W_m$ and $z^2 = b$.

Proof. The first assertion can easily be verified. To prove (2), let $y \in Y_m(b)$ with $m > 1$. Clearly y^2 is a prime element of F, and so $y^2 = \omega^2 tt^\iota$ with $t \in \mathfrak{r}^\times$. Changing ω for ωt, we may assume that $y^2 = \omega^2$. Then there exists an element $a \in B^\times$ such that $y = a\omega a^{-1}$. Changing K for aKa^{-1}, we may assume that $y = \omega$. Put $x_1 = y + sy$ with $s \in \pi^{m-1}\mathfrak{r}$. Then

$$x_1^2 - b = y^2(1+s)(1+s)^\iota - b = y^2 - b + y^2(s + s^\iota) + y^2 ss^\iota.$$

Since $\mathrm{Tr}(\mathfrak{r}) = \mathfrak{g}$, we can find $s \in \pi^{m-1}\mathfrak{r}$ so that $\mathrm{Tr}(s) = y^{-2}b - 1$. Then $x_1^2 - b \in \mathfrak{p}^{m+1}$ and $x_1 - y \in W_m$. By induction we can find $\{x_\nu\}_{\nu=1}^\infty \subset M$ such that $x_0 = y$, $x_{\nu+1} - x_\nu \in$

$W_{m+\nu}$, and $x_\nu^2 - b \in \mathfrak{p}^{m+\nu}$. Let $z = \lim_{\nu\to\infty} x_\nu$. Then $z^2 = b$ and $z - y \in W_m$ as desired. □

Lemma 4.7. *If $b\mathfrak{g} = \mathfrak{p}$, then $\mu_1'(b) = q^{-1}$ and $\mu_m'(b) = q^{-m-1}(q+1)$ for $m > 1$.*

Proof. We have $W_m = \mathfrak{p}^m\zeta + \mathfrak{p}^{m-1}\mathfrak{r}\omega$ as noted in (4.1.1). We easily see that $Y_1(b) = \mathfrak{P} \cap M = W_1$, so that $\mu_1'(b) = q^{-1}$. To study $Y_2(b)$, put $Z = \{s \in \mathfrak{r} \mid ss^\iota - b/\pi \in \mathfrak{p}\}$. Let $y = r\zeta + s\omega \in Y_2(b)$ with $r \in \mathfrak{g}$ and $s \in \mathfrak{r}$. Then $y^2 - b = r^2\zeta^2 + ss^\iota\pi - b$, and we easily see that $Y_2(b)$ consists of the elements $r\zeta + s\omega$ with $r \in \mathfrak{p}$ and $s \in Z$. Now the map $s \mapsto ss^\iota$ gives a surjective map of $(\mathfrak{r}/\mathfrak{p}\mathfrak{r})^\times$ onto $(\mathfrak{g}/\mathfrak{p})^\times$, and each element of $(\mathfrak{g}/\mathfrak{p})^\times$ has $q+1$ inverse images. Therefore $Z = \bigsqcup_{i=1}^{q+1}(s_i + \mathfrak{p}\mathfrak{r})$ with suitable s_i, and hence $\mu_2'(b) = q^{-3}(q+1)$. Next assume $m > 1$; then by Lemma 4.6 we have $Y_m(b) = \bigsqcup_{y\in A}(y + W_m)$ with a suitable A such that $y^2 = b$ for every $y \in A$. Clearly $Y_{m+1}(b) = \bigsqcup_{y\in A}(y + W_m) \cap Y_{m+1}(b)$. Let $y \in A$ and $w \in W_m$. Since $w^2 \in \mathfrak{p}^{m+1}$, we see that $y + w \in Y_{m+1}(b)$ if and only if $\mathrm{Tr}(yw) \in \mathfrak{p}^{m+1}$. Let $w = r\zeta + s\omega$ with $r \in \mathfrak{p}^m$ and $s \in \mathfrak{p}^{m-1}\mathfrak{r}$. Changing K suitably, we may assume that $\omega = y$. Then $\mathrm{Tr}(yw) \in \mathfrak{p}^{m+1}$ if and only if $\mathrm{Tr}(s) \in \mathfrak{p}^m$. Let W' be the set of all such w's, and $\mathfrak{z}$ the set of all such s's. Then $[W_m : W'] = [\mathfrak{p}^{m-1}\mathfrak{r} : \mathfrak{z}] = q$, so that $\mu_{m+1}'(b) = q^{-1}\mu_m'(b)$. This combined with the result concerning $\mu_2'(b)$ completes the proof. □

4.8. Returning to the original question of $\mu_n(h)$, we have seen that $\mu_n(h) = q^{-[3n/2]}$ if $n \le \nu$; and that if $n > \nu$, then $\mu_n(h) = q^{-3\lambda}\mu_{n-2\lambda}'(b)$ with $b = -h/(\pi^{2\lambda}c)$ as in §4.1, where $\lambda = [\nu/2]$. Therefore, from Lemmas 4.4 and 4.7, we obtain, putting $\xi = \xi(-ch)$ and excluding the case in which $n \ge \nu \in 2\mathbf{Z}$ and $2 \in \mathfrak{p}$,

$$\mu_n(h) = \begin{cases} q^{-3i} & \text{if } n = 2i \le \nu, \\ q^{-3i-1} & \text{if } n = 2i+1 \le \nu, \\ (1-\xi)q^{-\lambda-n} & \text{if } n > \nu \in 2\mathbf{Z}, \\ (1+q^{-1})q^{-\lambda-n} & \text{if } n > \nu \notin 2\mathbf{Z}. \end{cases} \tag{4.8.1}$$

If $2 \in \mathfrak{p}$ and $n \ge \nu = 2\lambda \in 2\mathbf{Z}$, then $\mu_{n-2\lambda}'(b)$ can be given by Lemma 4.5. Thus we obtain the formulas for β, valid for both even and odd $\mathfrak{p}$, as follows:

$$\beta(s) = \frac{1 - q^{3-2s} - q^{(\lambda+\kappa)(2-2s)}\left(1 - \xi q^{(3/2)-s}\right)\left(q^{2-2s} + \xi q^{(3/2)-s}\right)}{1 - q^{2-2s}} \quad \text{if } \xi \ne 0,$$

$$\beta(s) = \frac{1 - q^{3-2s} + q\cdot q^{(\nu+\varepsilon+1)(1-s)} - q^{(\nu+\varepsilon+3)(1-s)}}{1 - q^{2-2s}} \quad \text{if } \xi = 0,$$

where κ is determined by $2\mathfrak{g} = \mathfrak{p}^\kappa$, $\varepsilon = \varepsilon(b)$ if $2 \in \mathfrak{p}$ and ν is even, and $\varepsilon = 0$ otherwise.

These can be obtained from the formulas for $\mu_n(h)$ as follows. First, suppose $2 \in \mathfrak{p}$, $\nu = 2\lambda$, and $\xi = 0$. By Lemma 3.5, parts (2) and (3), we can put $\varepsilon = 2a+1$ with $a \in \mathbf{Z}$.

From Lemma 4.5(1), we see that $\mu_n(h) = q^{-3\lambda}\mu'_{n-\nu}(b) = q^{-[3n/2]}$ for $\nu < n \le \varepsilon+\nu$. Since this is also true for $n \le \nu$, we obtain

$$(\dagger) \quad \sum_{n=0}^{\infty} \mu_n(h) q^{n-ns} = \sum_{n=0}^{2\lambda+2a+1} q^{-[3n/2]+n-ns} + q^{-3\lambda} \sum_{m=\varepsilon+1}^{\infty} \mu'_m(b) q^{(m+2\lambda)(1-s)}.$$

To simplify our notation, put $x = q^{-1-2s}$. Then we easily see that the first sum on the right-hand side equals $(1+q^{-s})(1-x^{\lambda+a+1})/(1-x)$. By Lemma 4.5(1) the second term equals

$$x^{\lambda}(q+1)q^{-a-1} \sum_{m=\varepsilon+1}^{\infty} q^{-ms} = \frac{(q+1)x^{\lambda+a+1}}{1-q^{-s}}.$$

Thus $\beta(s+(3/2)) = (1-q^{-s})\sum_{n=0}^{\infty} \mu_n(h) q^{n-ns} = B/(1-x)$ with

$$\begin{aligned} B &= \left(1-q^{-2s}\right)\left(1-x^{\lambda+a+1}\right) + (q+1)x^{\lambda+a+1}(1-x) \\ &= 1 - qx + qx^{\lambda+a+1} - x^{\lambda+a+2}, \end{aligned}$$

since $q^{-2s} = qx$. Substituting $s-3/2$ for s, or rather, substituting q^{2-2s} for x, we obtain the formula for β when $2 \in \mathfrak{p}$, $\nu = 2\lambda$, and $\xi = 0$.

Next suppose $\xi \ne 0$, still with $v \mid 2$ and $\nu = 2\lambda$. By Lemma 3.5(2), we have $e-1 \le \varepsilon(b) \le e$, and so by Lemma 4.5(2), this time we have

$$(\ddagger) \quad \begin{aligned} \sum_{n=0}^{\infty} \mu_n(h) q^{n-ns} &= \sum_{n=0}^{\nu+e-1} q^{-[3n/2]+n-ns} + q^{-3\lambda} \sum_{m=e}^{\infty} \mu'_m(b) q^{(m+2\lambda)(1-s)} \\ &= \frac{(1+q^{-s})(1-x^{\lambda+\kappa})}{1-x} + x^{\lambda+\kappa} + \frac{(1-\xi)q^{-s}x^{\lambda+\kappa}}{1-q^{-s}}. \end{aligned}$$

Then we obtain the desired formula in this case. If $\nu = 2\lambda+1$, then the calculation of β for both odd and even v is similar to ($\dagger$) with $a = 0$; if $\nu = 2\lambda$ and $2 \notin \mathfrak{p}$, then it is similar to ($\ddagger$) with $\kappa = 0$. Therefore we obtain the above formulas for β in all cases.

5. Proof of the main theorem

5.1. We are going to apply a formula of [W2] to the series of (2.3.4). In order to do this, we have to verify that $\{\alpha_\varphi \mid \alpha \in \mathscr{G}^1\}$ is the group $Ps(V/A)$ of [W1, §49]. We first note that $A = F_n^n$ in the present case, and put $B((x,y),(x_1,y_1)) = y(2\varphi)\cdot {}^t x_1$ for $(x,y),(x_1,y_1) \in V \times V$. This is the form B of [W1, §31]. (We let A act on $V = F_n^1$ by right multiplication, though in [W1] and [W2], V is a left A-module denoted by X.) Put also $\iota(z,z_1) = B(z,z_1) - B(z_1,z)$ for $z, z_1 \in V \times V$. Then $Ps(V/A)$ is the group consisting of the elements γ of $GL(V \times V, F)$ such that $(za)\gamma = (z\gamma)a$ for every $a \in A$ and $\iota(z\gamma, z_1\gamma) = \iota(z,z_1)$ for every $z, z_1 \in V \times V$ (see [W1, §§31 and 49]). Now define $\psi : V \times V \to F_n^1 \times F_n^1$ by $\psi(x,y) = (x, y(2\varphi))$. Then we easily see that

$\iota(z,z_1) = \psi(z)\iota_n \cdot {}^t\psi(z_1)$ and $\psi(z\gamma) = \psi(z)\gamma_\varphi$ for every $z, z_1 \in V \times V$ and $\gamma \in F_2^2$. Therefore we have $Ps(V/A) = \{\alpha_\varphi \mid \alpha \in \mathcal{G}^1\}$. Consequently E of (2.3.4) for $s = 0$ is an Eisenstein series of [W2, (30)].

5.2. For a fixed $\mathfrak{g}$-lattice L in $V = F_n^1$ as in §1.2, let ℓ be an element of $\mathcal{S}(V_\mathbf{h})$ defined by $\ell(x) = \prod_{v\in\mathbf{h}} \ell_v(x_v)$, where ℓ_v is the characteristic function of L_v. Let us now take f_z of (2.3.1) with this ℓ and $z \in \mathfrak{H}_1^\mathbf{a}$. We then put

$$\theta(g,z) = \sum_{\xi\in V} f_z(\xi g) = \sum_{\xi\in V} \ell(\xi g)\mathbf{e}_\mathbf{a}(\varphi[\xi]z) \quad (g \in G_\mathbf{A}^\varphi), \tag{5.2.1}$$

where we put $\ell(x) = \ell(x_\mathbf{h})$ for simplicity. We normalize the Haar measure on $G_\mathbf{A}^\varphi$ so that the measure of $G^\varphi\backslash G_\mathbf{A}^\varphi$ is 1. With C and $\mathfrak{B}$ as in (1.2.1) and (1.2.2), we see that $G^\varphi\backslash G_\mathbf{A}^\varphi$ can be identified with $\bigsqcup_{a\in\mathfrak{B}} (\Gamma^a\backslash aC)$, and hence we have $1 = \sum_{a\in\mathfrak{B}}[\Gamma^a : 1]^{-1}\operatorname{vol}(C)$. Thus

$$\mathfrak{m}(G^\varphi, C) = \operatorname{vol}(C)^{-1}. \tag{5.2.2}$$

Now we have

$$\begin{aligned}\int_{G^\varphi\backslash G_\mathbf{A}^\varphi} \theta(g,z)\,dg &= \sum_{a\in\mathfrak{B}} \int_{\Gamma^a\backslash aC} \sum_{\xi\in V} \ell(\xi g)\mathbf{e}_\mathbf{a}(\varphi[\xi]z)\,dg \\ &= \operatorname{vol}(C)\sum_{a\in\mathfrak{B}}[\Gamma^a : 1]^{-1}\sum_{\xi\in V}\ell(\xi a)\mathbf{e}_\mathbf{a}(\varphi[\xi]z) \\ &= \operatorname{vol}(C)\sum_{a\in\mathfrak{B}}[\Gamma^a : 1]^{-1}\sum_{h\in\mathfrak{g}} N(La^{-1}, h)\mathbf{e}_\mathbf{a}(hz) \\ &= \mathfrak{m}(G^\varphi, C)^{-1}\sum_{h\in\mathfrak{g}} R(L,h)\mathbf{e}_\mathbf{a}(hz).\end{aligned}$$

In other words, $\mathfrak{m}(G^\varphi, C)^{-1}R(L,h)$ is a Fourier coefficient of $\int_{G^\varphi\backslash G_\mathbf{A}^\varphi} \theta(g,z)\,dg$.

5.3. Take $\sigma = x \in F_\mathbf{a} = \mathbf{R}^\mathbf{a}$ in (2.3.4). From (2.1.2) we easily see that $\rho(\sigma)f_\mathbf{i} = f_{x+\mathbf{i}}$. Now the Siegel-Weil formula gives

$$c_n E(x,0) = \int_{G^\varphi\backslash G_\mathbf{A}^\varphi} \theta(g, x+\mathbf{i})\,dg, \tag{5.3.1}$$

where $c_n = 1$ if $n > 2$ and $c_2 = 1/2$. This was proven in [W2] when $n > 4$, in which case the right-hand side of (2.3.4) is convergent at $s = 0$. The divergent case was given in [KR] when n is even. However, to apply the result in the paper to our case, we have to make sure of some nontrivial points. This will be done after stating an immediate consequence of the formula. We shall then add some more remarks.

Now for $0 < y \in \mathbf{R}$, we have (see [S1, (4.34.K) and (4.35.K)])

$$\xi\left(y, h; \frac{n}{2}, 0\right) = \begin{cases} \mathbf{e}\left(\dfrac{-n}{8}\right)(2\pi)^{n/2}\Gamma\left(\dfrac{n}{2}\right)^{-1} h^{(n-2)/2}\mathbf{e}(iyh) & \text{if } h > 0, \\ 0 & \text{if } h < 0. \end{cases}$$

Therefore, comparing the Fourier coefficients on both sides of (5.3.1), we obtain $R(L, h) = c_n \mathfrak{m}(G^\varphi, C)\mathbf{e}_{\mathbf{a}}(-ih)c(h, 0)$ for a totally positive h in F, and hence

$$\begin{aligned} R(L,h) = c_n D_F^{(1-n)/2}\big[\widetilde{L} : L\big]^{-1/2}(2\pi)^{nd/2}\Gamma\left(\frac{n}{2}\right)^{-d} N_{F/\mathbf{Q}}(h)^{(n-2)/2} \\ \cdot\, \mathfrak{m}(G^\varphi, C)\left\{\prod_{v\in\mathbf{h}} b_v(h,s)\right\}_{s=0}. \end{aligned} \tag{5.3.2}$$

Thus our task is to find the value of $\prod_{v\in\mathbf{h}} b_v(h, s)$ at $s = 0$, since we already know the value of $\mathfrak{m}(G^\varphi, C)$ as will be explained in §5.5.

Returning to the question of validity of (5.3.1) in the divergent case, we must note that the series in [KR] is somewhat different from our E in (5.3.1). More precisely, the series in [KR] is of the form $E'(\gamma, s)$ defined for $\gamma \in \mathcal{G}^1_{\mathbf{A}}$, and $E'\big(\tau(\sigma), s\big)$ coincides with the series $E(\sigma, s)$ of (2.3.4) defined with $\mathfrak{b} = \mathfrak{g}$. Now E of (5.3.1) is defined with $\mathfrak{b} = \mathfrak{d}_F$. Still we have $E'\big(\tau(\sigma), 0\big) = E(\sigma, 0)$ since $c_v(h, 0)$ is independent of the choice of $\mathfrak{b}$ as shown in Lemma 2.5, and the change of $\mathfrak{b}$ does not affect $c_v(h, s)$ for almost all v. Thus we can apply the result of [KR] to obtain (5.3.1) for even n.

The only remaining case that cannot be covered by [W2] and [KR] is the case $n = 3$. Now, in [R] it is shown that (5.3.1) holds with a constant c_n, even when $n = 3$, independent of ℓ. This c_n must be 1 for the following reason.

As mentioned in the introduction, equality (5.3.2) is essentially a special case of Siegel's results in [Si]. In fact, by Lemma 2.5, $b_v(h, 0)$ for each $v \in \mathbf{h}$ is the local representation density $d_{\mathfrak{p}}(h)$ he defined, so that (5.3.2) is exactly (1) of the introduction. Therefore (5.3.2) is included in his results, except that he assumed that $L = \mathfrak{g}^1_n$ and $\varphi(x, y) \in \mathfrak{g}$ for $x, y \in L$. The latter assumption can easily be removed, since we easily see that the validity of (5.3.2) (or rather, of his formula) for (φ, h) is equivalent with that for $(2\varphi, 2h)$. However, the removal of the assumption that $L = \mathfrak{g}^1_n$ is not that simple; one has to reformulate everything in terms of an arbitrary lattice from the beginning. That was done by Eichler in [E]. In this way, we can justify (5.3.2) without the Siegel-Weil formula (5.3.1). Anyway, we know (without [E]) that $c_3 = 1$ in (5.3.1).

5.4. Define K and $\mathfrak{e}$ as in Theorem 1.5. Observe that K_v, for $v \in \mathbf{h}$, is a quadratic extension of F_v if and only if $t_v = 2$, in which case K_v is the local field K of §3.6. For every such v, we fix an element c_v of $\mathfrak{g}_v$ as in §1.6.

For each $v \in \mathbf{h}$, define $\beta(s)$ by (3.1.1) at v, and call it $\beta_v(s)$; then put

$$\gamma_v(s) = \beta_v(s) \cdot \begin{cases} \left(1 - \psi^*(\mathfrak{p})q^{-s}\right)^{-1} & \text{if } n \in 2\mathbf{Z}, \\ \dfrac{1 - \psi_h^*(\mathfrak{p})q^{(1/2)-s}}{1 - q^{1-2s}} & \text{if } n \notin 2\mathbf{Z}, \end{cases} \tag{5.4.1}$$

where $\mathfrak{p}$ is the prime ideal at v and $q = N(\mathfrak{p})$. In view of the definition of $\mathfrak{e}$, from (3.1.3), (3.3.3), and (3.7.3) we see that $\gamma_v = 1$ for $v \nmid h\mathfrak{e}$. Therefore, by (3.1.2),

$$\begin{aligned} \prod_{v \in \mathbf{h}} b_v(h, s) &= \prod_{v \in \mathbf{h}} \beta_v\left(s + \frac{n}{2}\right) \\ &= \prod_{v | h\mathfrak{e}} \gamma_v\left(s + \frac{n}{2}\right) \cdot \begin{cases} L\left(s + \dfrac{n}{2}, \psi\right)^{-1} & \text{if } n \in 2\mathbf{Z}, \\ \dfrac{L(s + (n-1)/2, \psi_h)}{\zeta_F(2s + n - 1)} & \text{if } n \notin 2\mathbf{Z}. \end{cases} \end{aligned} \tag{5.4.2}$$

Now in Sections 3 and 4, we determined β_v in each case. Therefore we obtain the formulas for γ_v as stated in §1.6. We need to observe that $\psi_h^*(\mathfrak{p})$ coincides with $\xi(ch)$ of (3.3.2) and (3.6.1) if $t_v = 1$, and with $\xi(-ch)$ of §4.8 if $t_v = 3$. Also, ε of §1.6 coincides with $\varepsilon(k/c)$ of §3.6 if $t_v = 1$, and with $\varepsilon(b)$ of §4.8 if $t_v = 3$. Therefore the statements about ε at the end of §1.6 follow from Lemma 3.5, parts (2), (3), and (5). If n is even, ξ_0 of (3.7.1) coincides with ψ_v. Substituting the value of (5.4.2) at $s = 0$ into (5.3.2), we obtain the formulas of Theorem 1.5 concerning $R(L, h)/\mathfrak{m}(G^\varphi, C)$, since

$$[\widetilde{L} : L] = \begin{cases} 2^d N(\mathfrak{f})^2 & \text{if } n \notin 2\mathbf{Z}, \\ N(\mathfrak{d})N(\mathfrak{f})^2 & \text{if } n \in 2\mathbf{Z}, \end{cases} \tag{5.4.3}$$

where $\mathfrak{f}$ is the product of the prime factors of $\mathfrak{e}$ unramified in K if n is even, and $\mathfrak{f}$ is the product of the prime ideals for which $t_v = 3$ if n is odd. This formula follows immediately from [S5, (3.2.1)].

5.5. To complete the proof of Theorem 1.5, put $\mu = (n-1)/2$. Then we have

$$\begin{aligned} \mathfrak{m}(G^\varphi, C) = D_F^{[\mu^2]} \prod_{k=1}^{[\mu]} \left\{ D_F^{1/2}[(2k-1)!(2\pi)^{-2k}]^d \zeta_F(2k) \right\} \cdot [\widetilde{L} : L]^\mu \prod_{v | \mathfrak{e}} \lambda_v \\ \cdot \begin{cases} 2^{-\mu d} & \text{if } n \notin 2\mathbf{Z}, \\ D_F^{1/2}[(m-1)!(2\pi)^{-m}]^d L(m, \psi) & \text{if } n \in 2\mathbf{Z}, \end{cases} \end{aligned}$$

where λ_v is the number given as follows:

$$\begin{array}{ll} 1 & \text{if } t_v = 1, \\ 2^{-1}(1+q_v)^{-1}(1+q_v^{1-m})(1+q_v^{-m}) & \text{if } t_v = 2, \mathfrak{d}_v = \mathfrak{r}_v, \text{ and } c_v \notin \mathfrak{g}_v^\times, \\ 2^{-1} & \text{if } t_v = 2 \text{ and } \mathfrak{d}_v \neq \mathfrak{r}_v, \\ 2^{-1}(1+q_v)^{-1}(1-q_v^{1-n}) & \text{if } t_v = 3, \\ 2^{-1}(1+q_v)^{-1}(1-q_v^{1-m})(1-q_v^{-m}) & \text{if } t_v = 4. \end{array}$$

In fact, in [S5, Theorem 5.8] we gave a formula for $\mathfrak{m}(G_+^\varphi, D(\mathfrak{c}))$, where $D(\mathfrak{c})$ is a certain subgroup of C of "level $\mathfrak{c}$." Since $C \cap (G_+^\varphi)_\mathbf{A} = D(\mathfrak{g})$ and $\mathfrak{m}(G_+^\varphi, D(\mathfrak{g})) = 2\mathfrak{m}(G^\varphi, C)$ by [S5, Lemma 5.6(1)], we obtain the above formula from the formula of [S5, Theorem 5.8] by putting $\mathfrak{c} = \mathfrak{g}$ and dividing by 2. Substitute this into (5.3.2); rearrange the product by employing (5.4.2); then we obtain the formulas for $R(L, h)$ of Theorem 1.5.

6. The two- and three-dimensional cases and examples

6.1. Let us first consider the case in which $n = 2$, $V = K$ with a quadratic extension K of F, and $\varphi[x] = N_{K/F}(x)$. Then this K is exactly the field K of Theorem 1.5. As noted in [S5, Lemma 1.3], we have

$$G_+^\varphi = \{x \in K \mid xx^\rho = 1\},$$

where ρ is the nontrivial automorphism of K over F; G^φ is generated by G_+^φ and ρ. Since G^φ is totally definite, K must be totally imaginary. The genus of maximal lattices in K consists of the fractional ideals $\mathfrak{a}$ of K such that $\mathfrak{a}\mathfrak{a}^\rho = \mathfrak{r}$. Let Γ be the group of all roots of unity in K. Clearly $\Gamma_+^a = \Gamma$ for every $a \in G_\mathbf{h}^\varphi$. Let C_K be the ideal class group of K, and $C_{K/F}$ the subgroup of C_K consisting of the ideal classes represented by the ideals $\mathfrak{b}$ such that $\mathfrak{b}^\rho = \mathfrak{b}$. Then the ideals $\mathfrak{b}^{-1}\mathfrak{b}^\rho$ for $\mathfrak{b} \in C_K/C_{K/F}$ form a complete set of representatives for the classes of maximal lattices with respect to G_+^φ, which we call A. (The index $[C_K : C_{K/F}]$ can be given by [S4, (24.5.5), (24.5.7)].) Then we have the following theorem.

THEOREM 6.2. *Let the notation be as in Theorem 1.5(II) with $n = 2$, §1.6, and §6.1. Let $\mathfrak{h}$ (resp. $\mathfrak{h}'$) be the product of all the prime factors of h that split (resp. remain prime) in K. Then $\sum_{\mathfrak{a} \in A} N(\mathfrak{a}, h) \neq 0$ only if e_v is even for every $v \mid \mathfrak{h}'$, in which case*

$$\sum_{\mathfrak{a} \in A} N(\mathfrak{a}, h) = [\Gamma : 1] \prod_{v \mid \mathfrak{h}} (e_v + 1) \prod_{v \mid \mathfrak{d}\mathfrak{d}^\rho} \frac{1 + \psi_v(h)}{2},$$

where e_v is the integer such that $|h|_v^{-1} = q_v^{e_v}$.

Observe that $\gamma_v(1) = e_v + 1$ if $v|\mathfrak{h}$, $\gamma_v(1) = (1 + (-1)^{e_v})/2$ if $t_v = 2$ and $\mathfrak{d}_v = \mathfrak{r}_v$, and $\gamma_v(1) = 1 + \psi_v(h)$ if $\mathfrak{d}_v \neq \mathfrak{r}_v$. Therefore our equality follows immediately from Theorem 1.5(II). Here we are looking at $R_+(\mathfrak{g},h)$, which equals $2R(\mathfrak{g},h)$ by Proposition 1.3.

6.3. Next, suppose $n=3$. Then we can take $V=\{x\in B \mid x^\iota=-x\}$ and $\varphi[x]=cxx^\iota$ with a quaternion algebra B over F, where ι is the main involution of B and $c\in F^\times$. To obtain a simpler formulation, we assume that $c\in\mathfrak{g}^\times$, which can be achieved by changing φ for its suitable constant multiple. (This assumption is somewhat stronger than condition (1.4.4).) We have $t_v=3$ if and only if B_v is a division algebra; B must be totally definite, since φ is so. Let $\mathfrak{O}$ be a maximal order of B. Then we can take $L=\mathfrak{O}\cap V$. As noted in [S5, Lemma 1.4], the group G_+^φ consists of the elements $\tau_a\in GL(V)$ for all $a\in B^\times$, defined by $x\tau_a=a^{-1}xa$ for $x\in V$; G^φ is generated by G_+^φ and -1_V. The question of representation $\varphi[x]=h$ is the same as the question of $xx^\iota=h/c$, and therefore it is sufficient to treat the case $\varphi[x]=xx^\iota$.

THEOREM 6.4. *Let the notation be as in Theorem 1.5(I), §1.6, and §6.3; in particular, let h be a totally positive element of $\mathfrak{g}$ and let $K=F((-h)^{1/2})$; let $\mathfrak{f}$ be the product of the prime ideals for which $t_v=3$. Also let $\mathfrak{r}$ be the maximal order of K, and $\mathfrak{d}$ the different of K relative to F. Suppose that $\varphi[x]=xx^\iota$, $h\mathfrak{g}$ is a squarefree ideal, and 2 is unramified in F. Let $\mathfrak{y}$ be the product of all the prime factors of 2 in F, not dividing $h\mathfrak{f}$, that remain prime in K. Let η_K (resp. η_F) be the class number of K (resp. F). Then*

$$\mathfrak{m}(L)=D_F^{3/2}(2\pi)^{-2d}\zeta_F(2)\prod_{v\mid\mathfrak{f}}2^{-1}(q_v-1),$$

$$\begin{aligned}R(L,h)&=2^dN_{F/\mathbf{Q}}(h)^{1/2}D_F^{1/2}(2\pi)^{-d}L(1,\psi_h)\prod_{v\mid\mathfrak{f}}\tau_v\prod_{v\mid\mathfrak{y}}(1+2q_v^{-1})\\&=2^{d-1}N_{F/\mathbf{Q}}(h)^{1/2}N(\mathfrak{d})^{-1/2}[\mathfrak{r}^\times:\mathfrak{g}^\times]^{-1}\eta_F^{-1}\eta_K\prod_{v\mid\mathfrak{f}}\tau_v\prod_{v\mid\mathfrak{y}}(1+2q_v^{-1})\end{aligned}$$

with the number τ_v given by

$$\tau_v=\begin{cases}0 & \text{if } v \text{ splits in } K,\\ q_v^{-1} & \text{if } v \text{ remains prime in } K \text{ and } v\mid 2,\\ 1 & \text{if } v \text{ remains prime in } K \text{ and } v\nmid 2,\\ 2^{-1} & \text{if } v \text{ is ramified in } K.\end{cases}$$

In this theorem, we assume that $h\mathfrak{g}$ is squarefree and 2 is unramified in F, since the formulas become relatively simple under such conditions. Anyway, the result concerning $\mathfrak{m}(L)$ follows immediately from [S5, Theorem 5.8], which is valid without those conditions; the first equality concerning $R(L,h)$ is a special case of Theorem 1.5(I). To obtain the last equality, we need

$$L(1,\psi_h)=2^{d-1}\pi^dD_F^{-1/2}N(\mathfrak{d})^{-1/2}[\mathfrak{r}^\times:\mathfrak{g}^\times]^{-1}\eta_F^{-1}\eta_K, \tag{6.4.1}$$

which follows from the residue formulas for ζ_F and ζ_K.

6.5. The above two theorems, when $F = \mathbf{Q}$, include the classical results concerning the sums of two and three squares. The case of four squares, however, is not included in Theorem 1.5. Indeed, take $F = \mathbf{Q}$, $V = B$ with a definite quaternion algebra B over $\mathbf{Q}$ unramified at every $p \neq 2$, and $\varphi[x] = xx^{\iota}$; let $\mathfrak{O}$ be a maximal order of B. Then $\mathfrak{O}$ is a maximal lattice whose genus consists of a single class, and hence $N(\mathfrak{O}, h) = R(\mathfrak{O}, h)/\mathfrak{m}(\mathfrak{O})$. We can easily find a $\mathbf{Z}$-basis $\{f_i\}_{i=0}^{3}$ of $\mathfrak{O}$ such that

$$\varphi\left[\sum_{i=0}^{3} a_i f_i\right] = \sum_{i=0}^{3} a_i^2 + a_0 \sum_{i=1}^{3} a_i \tag{6.5.1}$$

for $a_i \in \mathbf{Q}$. (This is of course equivalent to the sum of four squares over $\mathbf{Q}$, but not over $\mathbf{Z}$.) Thus Theorem 1.5(II) gives $N(\mathfrak{O}, h)$, that is, the number of representations of h by (6.5.1) with $a_i \in \mathbf{Z}$.

By an easy calculation we obtain, from the second equality of Theorem 1.5(II),

$$N(\mathfrak{O}, h) = 24 \prod_{p \nmid 2} \left(\sum_{i=0}^{e_p} p^i \right) \tag{6.5.2}$$

if $h = \prod_p p^{e_p}$ is the prime decomposition of a given positive integer h.

6.6. Let us now consider the case in which $F = \mathbf{Q}$ and $\varphi = 1_8$. This case is similar to the case of §6.5. We have $t_v = 0$ for every $v \in \mathbf{h}$ (cf. [S5, §5.16]). It is also well known that the genus of maximal lattices consists of a single class, so that we have again $N(L, h) = R(L, h)/\mathfrak{m}(L)$ with any maximal lattice L. Therefore, the second equality of Theorem 1.5(II) shows that

$$N(L, h) = 240 \prod_{p \mid h} \left(\sum_{i=0}^{e_p} p^{3i} \right) \tag{6.6.1}$$

if $h = \prod_p p^{e_p}$.

6.7. Let us next take $F = \mathbf{Q}$, $\varphi = 1_9$, and $h = 1$. Then ψ_h is the trivial character. Employing the known values of $\zeta(4)$ and $\zeta(8)$, we obtain $R(H, 1)/\mathfrak{m}(H) = 2^8 \gamma_2(9/2)$ for any maximal lattice H in V. From the formula of γ_v with $t_v = 1, v \mid 2$, and $\xi = 1$, we obtain $\gamma_2(9/2) = 121/2^7$, and so $R(H, 1)/\mathfrak{m}(H) = 242$. We have seen in [S5, §5.16] that the genus of H consists of a single class. Therefore, we have $N(H, 1) = 242$, which can also be obtained from (6.6.1) combined with the fact that H is the direct sum of the lattice L of §6.5 and an obvious $\mathfrak{g}$-module of rank 1, as shown in [S5, §5.16]. (In other words, we have shown the equality $2 \cdot 121 = 240 + 2$ in a very roundabout way.)

6.8. Let us finally take $F = \mathbf{Q}$ and $\varphi = 1_{10}$. Let $X = \sum_{i=1}^{8} \mathbf{Q}e_i$ and $Y = \mathbf{Q}e_9 + \mathbf{Q}e_{10}$ with the standard basis $\{e_i\}_{i=1}^{10}$ of $V = \mathbf{Q}_{10}^1$; let L (resp. M) be a maximal lattice in X (resp. Y), and let $\Lambda = L + M$. Since $[\widetilde{\Lambda} : \Lambda] = [\widetilde{M} : M] = 4$, we easily see that Λ is maximal. Now ψ is the character of conductor 4, and $L(5, \psi) = 5\pi^5/1536$; so the second equality of Theorem 1.6(II) for $h = 1$ gives $R(\Lambda, 1)/\mathfrak{m}(\Lambda) = 1028/5$, which is not an integer. This shows that the genus of Λ has more than one class.

The last fact can be seen by looking only at $\mathfrak{m}(\Lambda)$. First, we obtain $\mathfrak{m}(\Lambda) = (2^{18} \cdot 3^5 \cdot 5 \cdot 7)^{-1}$ from a formula in [S5, §5.16]. Then we can show that $C \cap G^\varphi$ has order $2^{17} \cdot 3^5 \cdot 5^2 \cdot 7$ by the same type of technique as for 1_9 at the end of the same section, from which we obtain the desired fact.

References

[E] M. Eichler, *Quadratische Formen und orthogonale Gruppen*, 2d ed., Grundlehren Math. Wiss. **63**, Springer-Verlag, Berlin, 1974.

[KR] S. Kudla and S. Rallis, *On the Weil-Siegel formula*, J. Reine Angew. Math. **387** (1988), 1–68.

[R] S. Rallis, *L-functions and the Oscillator Representation*, Lecture Notes in Math. **1245**, Springer-Verlag, Berlin, 1987.

[S1] G. Shimura, *Confluent hypergeometric functions on tube domains*, Math. Ann. **260** (1982), 269–302.

[S2] ———, *On Eisenstein series of half-integral weight*, Duke Math. J. **52** (1985), 281–314.

[S3] ———, *On the transformation formulas of theta series*, Amer. J. Math. **115** (1993), 1011–1052.

[S4] ———, *Euler Products and Eisenstein Series*, CBMS Regional Conf. Ser. in Math. **93**, Amer. Math. Soc., Providence, 1997.

[S5] ———, *An exact mass formula for orthogonal groups*, Duke Math. J. **97** (1999), 1–66.

[Si] C. L. Siegel, *Über die analytische Theorie der quadratischen Formen, I*, Ann. of Math. (2) **36** (1935), 527–606; *II*, **37** (1936), 230–263; *III*, **38** (1937), 212–291.

[W1] A. Weil, *Sur certains groupes d'opérateurs unitaires*, Acta Math. **111** (1964), 143–211.

[W2] ———, *Sur la formule de Siegel dans la théorie des groupes classiques*, Acta Math. **113** (1965), 1–87.

Department of Mathematics, Princeton University, Princeton, New Jersey 08544, USA; goro@math.princeton.edu

Generalized Bessel functions on symmetric spaces

Journal für die reine und angewandte Mathematik, 509 (1999), 35-66*

Introduction

The following notation will be used throughout the paper: For an associative ring R with identity element and an R-module M we denote by $R^\times$ the group of all invertible elements of R and by M_n^m the R-module of all $m \times n$-matrices with entries in M; we put $M^m = M_1^m$ for simplicity. The identity and zero elements of the ring R_n^n are denoted by 1_n and 0_n (when n needs to be stressed). The transpose of a matrix x is denoted by ${}^t x$. If $x_1, \ldots, x_r$ are square matrices, $\mathrm{diag}[x_1, \ldots, x_r]$ denotes the matrix with $x_1, \ldots, x_r$ in the diagonal blocks and 0 in all other blocks. We denote by $\mathbf{H}$ the Hamilton quaternion algebra, and for $z \in \mathbf{H}_n^m$ we put $z^* = {}^t\bar{z}$, where the bar denotes quaternion conjugation. For $y = y^* \in \mathbf{H}_n^n$ we write $y > 0$ if $x^* y x > 0$ for every $x \in \mathbf{H}_1^n, \neq 0$, and $y > z$ if $y - z > 0$.

The functions we investigate in the present paper are defined in connection with five types of symmetric spaces H, which are referred to as Cases I–V. Each H can be obtained as G/K with a semisimple Lie group G and a maximal compact subgroup K of G, and the unipotent radical of a maximal parabolic subgroup of G acts on H as an abelian group of "translations." To give an idea about our functions, let us take Case II, in which G is the quaternion unitary group of signature (n, n). Then $H = X + Q$, with

(1) $$X = \{ x \in \mathbf{H}_n^n \mid x^* = -x \}, \quad Q = \{ y \in \mathbf{H}_n^n \mid y^* = y > 0 \},$$

and our function is defined by

(2) $$\xi_n(g, h; s) = \int_X \exp(\pi i \cdot \tau_0(hv))\delta_0(v+g)^{-s} dv \qquad (g \in Q,\ h \in X,\ s \in \mathbf{C}),$$

where τ_0 (resp. δ_0) is the reduced trace (resp. norm) map of $\mathbf{H}_n^n$ to $\mathbf{R}$. If $n = 1$ and $h \neq 0$, we can easily show that

$$\pi^{-s}\Gamma(s)\xi_1(g, h; s) = |h/g|^{s-(3/2)} K(s - (3/2), \pi|gh|),$$

where $K(s, u)$ is a classical Bessel function

$$K(s, u) = \int_0^\infty \exp(-ut - ut^{-1}) t^{s-1} dt \qquad (s \in \mathbf{C},\ 0 < u \in \mathbf{R}).$$

The function ξ_n appears naturally as an archimedean factor of a Fourier coefficient of an Eisenstein series on G. The purpose of the present paper is to investigate the analytic nature of this type of function in all five cases.

Before describing our results, let us insert here a remark on our choice of H. All hermitian symmetric spaces of tube type belong to the above class of H. We do not include them in our discussion, however, since our generalized Bessel functions in those hermitian cases become special cases of what we already investigated in [S1], if the group in question is classical. Thus our five types of spaces include all irreducible H of the above type which are classical and nonhermitian. The group G is $O(n,\, n;\, \mathbf{R})$, $O(2n,\, \mathbf{C})$, $Sp(n,\, \mathbf{C})$, and $O(n+1,\, 1;\, \mathbf{R})$ in Cases I, III, IV, and V, respectively. In Case I, for example, the spaces X, Q, and our function ξ_n are defined by taking $\mathbf{R}_n^n$ in place of $\mathbf{H}_n^n$ in (1) and (2); τ_0 and δ_0 are just the trace and the determinant of a real matrix.

Let us now describe our main results. We first prove that the integral of (2) is convergent if and only if $\mathrm{Re}(s) > \lambda$ for some explicitly determined λ, and therefore ξ_n is holomorphic at least in that region (Proposition 2.2). In Cases I and II we have $\lambda = (2n-3)/4$ and $\lambda = 2n - (1/2)$, respectively. For $h = 0$ we determine ξ_n explicitly as an elementary product of gamma functions. In Section 4 we give an integral expression for ξ_n by means of ξ_r and ξ_{n-r} (Proposition 4.3). Then we prove that ξ_n can be continued as a meromorphic function of s to the whole complex plane. In Case I we can prove the following precise result (Theorem 4.5): if h has rank $2m$, then

$$\xi_n(g,\, h;\, s) \prod_{i=1}^{n-2m} \Gamma(2s+1-(n+i)/2)^{-1} \prod_{k=1}^{n-m} \Gamma(2s+1-k)$$

is holomorphic in s and real analytic in $(g,\, s)$ on the whole $Q \times \mathbf{C}$. In Cases II–IV, however, we have a similar but much weaker result (Theorem 4.6), for which we employ a result of Wallach in [W] and also a result of Fefferman and Narasimhan in [FN]. In Case V we can find an explicit expression for our function in terms of $K(s,\, u)$, and so our problems in this case can be completely settled (Section 6). In Section 5 we first present H as G/K; we then show that the type of function, which has the form $\delta_0(y)^{s/2} \exp\left(\pi i \cdot \tau_0(hx)\right)\xi_n(y,\, h;\, s)$ in Case II, as a function of $x+y \in H$, for every fixed $h \in X$, when finite, is an eigenfunction of every G-invariant differential operator on H (Theorem 5.3).

The function ξ_n in Case I has been introduced in [S4], and employed in the explicit Fourier expansion of Eisenstein series on an orthogonal group, in the same sense that the functions of [S1] were employed in the expansion of Eisenstein series on symplectic and unitary groups in some papers by the author. Also the explicit expression for $\xi_n(y,\, 0;\, s)$ in terms of gamma functions played an essential role in the determination of the residue of the series at a critical point, which, in turn, was crucial in the proof of the exact mass formula for an orthogonal group of a general type obtained in [S4]. In a subsequent paper [S5] we shall show that our functions can be employed effectively in the investigation of Eisenstein series on a certain quaternion unitary group, and also of the exact mass formula for such a group.

1. Spaces and basic functions

As usual we view $\mathbf{C}$ as a subfield of $\mathbf{H}$. Given $0 < y = y^* \in \mathbf{H}_n^n$, we denote by $y^{1/2}$ a unique element a of $\mathbf{H}_n^n$ such that $a^* = a > 0$ and $a^2 = y$. We put

$$\text{(1.1)} \qquad \mathbf{e}(z) = \exp(2\pi i z) \qquad (z \in \mathbf{C}).$$

Among our five types of spaces, Case V, which concerns a hyperbolic space, is much simpler than the other cases, and so we treat it separately in Section 6. Thus, in Sections 1 through 5 we deal only with the first four cases. The formulas will be cross-referred to, for example, (1.8.III) and (2.4), which are valid in Case III and in all cases, respectively.

In each case we have three vector spaces X, Y, Z over $\mathbf{R}$, a domain of positivity Q in Y, and an open subset H of Z such that

$$\text{(1.2)} \qquad Z = X \oplus Y\mathbf{j}, \qquad H = \{ x + y\mathbf{j} \in Z \mid x \in X,\ y \in Q \},$$

where $\mathbf{j} = 1_n$ in Cases I and II, and $\mathbf{j} = 1_{2n}$ in Case III; in Case IV, $\mathbf{j} = j1_n$ with a fixed element j of $\mathbf{H}$ such that $j^2 = -1$ and $jc = \bar{c}j$ for $c \in \mathbf{C}$. As we said in the introduction and as will be shown in Section 5, H can be given as a quotient G/K. In the first four sections, however, the group G is unnecessary. Now X, Y, Z, and Q are given as follows:

$$\text{(1.3.I)} \qquad Z = \mathbf{R}_n^n, \quad X = \{ x \in Z \mid {}^t x = -x \}, \quad Y = \{ y \in Z \mid {}^t y = y \},$$

$$\text{(1.3.II)} \qquad Z = \mathbf{H}_n^n, \quad X = \{ x \in Z \mid x^* = -x \}, \quad Y = \{ y \in Z \mid y^* = y \},$$

$$\text{(1.3.III)} \qquad Z = \{ z \in \mathbf{R}_{2n}^{2n} \mid {}^t z J_n = J_n z \}, \quad J_n = \begin{bmatrix} 0 & -1_n \\ 1_n & 0 \end{bmatrix},$$

$$X = \{ x \in Z \mid {}^t x = -x \}, \quad Y = \{ y \in Z \mid {}^t y = y \},$$

$$\text{(1.3.IV)} \qquad Z = X + Y\mathbf{j}\ (\subset \mathbf{H}_n^n),$$

$$X = \{ x \in \mathbf{C}_n^n \mid {}^t x = x \}, \quad Y = \{ y \in \mathbf{C}_n^n \mid y^* = y \},$$

$$\text{(1.4)} \qquad Q = \{ y \in Y \mid y > 0 \}.$$

We assume $n > 1$ in Cases I and III.

To clarify the meaning of the objects in Case III, define an injection $\varphi : \mathbf{C}_n^m \to \mathbf{R}_{2n}^{2m}$, and also a matrix E by

$$\text{(1.5.III)} \qquad \varphi(p + iq) = \begin{bmatrix} p & -q \\ q & p \end{bmatrix} \quad (p, q \in \mathbf{R}_n^m), \qquad E = E_n = \mathrm{diag}[1_n, -1_n].$$

We also define an injection $\psi : \mathbf{H}_n^m \to \mathbf{C}_{2n}^{2m}$ by

$$\text{(1.6)} \qquad \psi(c + dj) = \begin{bmatrix} c & d \\ -\bar{d} & \bar{c} \end{bmatrix} \qquad (c, d \in \mathbf{C}_n^m).$$

Then we easily see that $\varphi(x^*) = {}^t\varphi(x) = E_n \varphi({}^t x) E_m$ for $x \in \mathbf{C}_n^m$, $\psi(z^*) = \psi(z)^*$ for $z \in \mathbf{H}_n^m$, and

$$\text{(1.7.III)} \qquad X = \{ \varphi(s)E \mid {}^t s = -s \in \mathbf{C}_n^n \}, \quad Y = \{ \varphi(p) \mid p^* = p \in \mathbf{C}_n^n \}.$$

In Cases I, II, and IV, we speak of *the eigenvalues of* $p \in Q$ as usual. However, in Case III, by *the eigenvalues of* $q = \varphi(p)$ in Q we understand the eigenvalues of p. We define $\delta : Z \to \mathbf{R}$ and $\tau : Z \to \mathbf{R}$ by

$$\text{(1.8.I)} \qquad \delta(z) = |\det(z)|,$$
$$\text{(1.8.II, IV)} \qquad \delta(z) = \det\left(\psi(z)\right)^{1/2} \quad \textit{for } z \in \mathbf{H}_n^n,$$
$$\text{(1.8.III)} \qquad \delta(z) = |\det(z)|^{1/2},$$
$$\text{(1.9.I)} \qquad \tau(z) = \mathrm{tr}(z),$$
$$\text{(1.9.II, IV)} \qquad \tau(z) = (1/2)\mathrm{tr}[\psi(z)] \quad \textit{for } z \in \mathbf{H}_n^n,$$
$$\text{(1.9.III)} \qquad \tau(z) = (1/2)\mathrm{tr}(z).$$

In Section 2 we shall show that $\delta(z) \neq 0$ for every $z \in H$. Since H is connected, this implies that $\det(z) > 0$ for every $z \in H$ in Cases I and III. Notice also that $\det\left(\psi(z)\right) > 0$ for every $z \in GL_n(\mathbf{H})$. In all cases $\delta(cz) = |c|^n\delta(z)$ for $c \in \mathbf{R}$.

We define a group $\mathfrak{g}$ and its subgroup $\mathfrak{g}_1$ in each case by

$$\text{(1.10.I)} \qquad \mathfrak{g} = GL_n(\mathbf{R}),$$
$$\text{(1.10.II)} \qquad \mathfrak{g} = GL_n(\mathbf{H}),$$
$$\text{(1.10.III)} \qquad \mathfrak{g} = \varphi(GL_n(\mathbf{C})),$$
$$\text{(1.10.IV)} \qquad \mathfrak{g} = GL_n(\mathbf{C}),$$
$$\text{(1.11)} \qquad \mathfrak{g}_1 = \{a \in \mathfrak{g} \mid aa^* = 1\}.$$

Hereafter we let U denote either X or Y. For every $p \in Q$ and $a \in \mathfrak{g}$ we define $\mathbf{R}$-linear endomorphisms T_p^U and R_a^U of U by

$$\text{(1.12.IV)} \qquad T_p^X(v) = (1/2)(pv + v \cdot {}^tp), \quad R_a^X(v) = av \cdot {}^ta, \quad \text{(Case IV, } U = X),$$
$$\text{(1.13)} \qquad T_p^U(v) = (1/2)(pv + vp), \quad R_a^U(v) = ava^*, \quad \text{(all other cases)},$$

where $v \in U$.

Lemma 1.1. *Both T_p^U and R_a^U are invertible. Moreover, for $a \in \mathfrak{g}_1$, $q \in Q$, $p = aqa^*$, and $w \in U$ we have $R_a^U T_q^U = T_p^U R_a^U$ and*
$$\tau\left[R_a^U(w)^*(T_p^U)^{-1}\{R_a^U(w)\}\right] = \tau\left[w^*(T_q^U)^{-1}w\right].$$

PROOF. The first equality can be verified in a straightforward way. Clearly R_a^U is invertible. The invertibility of T_p^U can be seen by taking q to be diagonal with a suitable a. For details, see the following lemma and its proof. Now, employing the first equailty, we obtain $(T_p^U)^{-1}\{R_a^U(w)\} = R_a^U\{(T_q^U)^{-1}(w)\}$, and hence, the left-hand side of the second equality for $U = X$ in Case IV equals

$$\tau\left[(aw \cdot {}^ta)^*a \cdot (T_q^U)^{-1}(w) \cdot {}^ta\right] = \tau\left[\bar{a}w^*(T_q^U)^{-1}(w) \cdot {}^ta\right] = \tau\left[w^*(T_q^U)^{-1}w\right]$$

as expected. In all other cases we have a^* in place of ${}^ta$, and the final result is the same.

Lemma 1.2. *Let $\lambda_1, \ldots, \lambda_n$ be the eigenvalues of an element p of Q. Then*
$$\delta(p)^\nu \det(T_p^X) = \det(T_p^Y) = \delta(p) \prod_{i<k}\{(\lambda_i + \lambda_k)/2\}^\iota,$$

$$\prod_{i<k}\{(\lambda_i + \lambda_k)/2\} \geq \delta(p)^{(n-1)/2},$$

where $(\nu, \iota) = (1, 1), (-2, 4), (1, 2),$ *and* $(-1, 2)$ *in Cases I, II, III, and IV, respectively.*

PROOF. The last formula follows from the fact that $(\lambda_i + \lambda_k)/2 \geq (\lambda_i\lambda_k)^{1/2}$. As for the first line of equalities, the first equality of Lemma 1.1 reduces our problem to the case of diagonal p. Take for example $U = X$ in Case III. Then putting $x = \varphi(s)E$ and $p = \varphi(h)$ with $-{}^ts = s \in \mathbf{C}_n^n$ and $h^* = h \in \mathbf{C}_n^n$, we have $T_p^X(x) = (1/2)\varphi(hs + s \cdot {}^th)E$. Taking $h = \mathrm{diag}[\lambda_1, \ldots, \lambda_n]$, we easily find the desired formula. All other cases can be verified in the same way.

We will be considering integrals over a vector space W over $\mathbf{R}$ or on its open subsets. In each case W has an obvious $\mathbf{R}$-basis, and we define a measure dw on each W (if w is a variable on W) by identifying W with $\mathbf{R}^d$, $d = \dim(W)$, with respect to that basis. This applies in particular to U. For $U = Y$ in Case IV, for example, our basis consists of $\varphi(e_{jk})$, $\varphi(ie_{jk})$ for $j < k$, and $\varphi(e_{jj})$ with the standard matrix units e_{jk}.

Lemma 1.3. *For every* $p \in Q$ *and every* $w \in U$ *we have*

$$\int_U \exp\{-\pi \cdot \tau(v^*pv)\}\mathbf{e}(-\tau(v^*w))dv$$
$$= 2^{\iota n(1-n)/4} \det\left(T_p^U\right)^{-1/2} \exp\left\{-\pi \cdot \tau\left(w^*(T_p^U)^{-1}(w)\right)\right\}.$$

PROOF. By Lemma 1.1 we can reduce the problem to the case of diagonal p. If $p = \mathrm{diag}[\lambda_1, \ldots, \lambda_n]$, $U = X$ in Case I, $v = (v_{jk})$, and $w = (w_{jk})$, then the integral over U is

$$\prod_{j<k} \int_{\mathbf{R}} \exp\left(-\pi(\lambda_j + \lambda_k)v_{jk}^2\right) \exp(-4\pi i v_{jk} w_{jk}) dv_{jk}$$
$$= \prod_{j<k} (\lambda_j + \lambda_k)^{-1/2} \exp(-4\pi(\lambda_j + \lambda_k)^{-1} w_{jk}^2)$$
$$= 2^{n(1-n)/4} \det(T_p^X)^{-1/2} \exp\left(-\pi \cdot \mathrm{tr}({}^tw(T_p^X)^{-1}w)\right),$$

which gives the desired formula for $U = X$. If $U = Y$, we have extra factors for $j = k$, and obtain the formula as expected. The other three cases can be proved in the same manner, though Case III is somewhat (but not much) more involved.

2. Basic properties of generalized Bessel functions

We now define our generalized Bessel functions ξ_U by

$$\xi_U(g, h; s) = \int_U \mathbf{e}(-\tau(hv^*))\delta(v + g\mathbf{j}')^{-2s} dv \qquad (g \in Q,\ h \in U,\ s \in \mathbf{C}). \tag{2.1}$$

Here the symbol $\mathbf{j}'$ and $\delta(v + g\mathbf{j}')$ should be understood as follows: $\mathbf{j}'$ is $\mathbf{j}$ as above if $U = X$ in all cases; if $U = Y$, then

$$\delta(v + g\mathbf{j}') = \begin{cases} |\det(v + ig)| & \text{(Cases I, IV)}, \\ \left|\det\left[\psi(v) + i\psi(g)\right]^{1/2}\right| & \text{(Case II)}, \\ |\det(v + gJ_n)^{1/2}| & \text{(Case III)}, \end{cases}$$

Notice that in Case III we have $\delta(v+g\mathbf{j}') = |\det(x+ip)|$ if $v = \varphi(x)$ and $g = \varphi(p)$. That $\delta(v+g\mathbf{j}') > 0$ will be shown in (2.8) below.

The function ξ_X is our main object of study; ξ_Y is a special case of the functions studied in [S1], and we already know its analytic nature. In particular, the integral for ξ_Y is convergent for $\mathrm{Re}(2s) > \iota(n-1)+1$; see the last two lines of [S1], p. 274. The domain of convergence of the integral for ξ_X will be given in Proposition 2.2 below. We need ξ_Y as a tool for investigating ξ_X. Assuming the convergence, we easily see that

$$\xi_U(a^*ga,\, h;\, s) = \delta(a)^{2\mu-4s}\xi_U(g,\, R_a^U(h);\, s) \qquad (a \in \mathfrak{g}), \tag{2.2}$$

where $\mu = \mu(U) = \dim(U)/n$. In other words,

(2.3) $\delta(g)^{2s-\mu}\xi_U(g,\, h;\, s)$ is invariant under $(g,\, h) \mapsto (a^*ga,\, (R_a^U)^{-1}h)$ for every $a \in \mathfrak{g}$.

Since $\delta(v+g\mathbf{j}') = \delta(-v+g\mathbf{j}')$ as will be shown shortly, we obtain

$$\xi_U(g,\, h;\, s) = \xi_U(g,\, -h;\, s). \tag{2.4}$$

To study the nature of this function, we first recall a well-known formula

$$\int_Q \exp\big(-\tau(pq)\big)\delta(p)^{s-\kappa}dp = \Gamma_n^\iota(s)\delta(q)^{-s} \qquad (q \in Q,\ \mathrm{Re}(s) > \kappa - 1). \tag{2.5}$$

Here

$$\kappa = 1 + \iota(n-1)/2, \tag{2.6a}$$

$$\Gamma_n^\iota(s) = \pi^{\iota n(n-1)/4}\prod_{k=0}^{n-1}\Gamma\big(s-(\iota k/2)\big) \tag{2.6b}$$

with ι given as in Lemma 1.2. (We can identify Q in Case III with Q in Case IV through the map φ.) For $x \in U$ and $y \in Q$ put

$$u = (R_y^U)^{-1}(x). \tag{2.7}$$

Then in all cases we have

$$\delta(x+y^2\mathbf{j}')^2 = \delta(y)^4\delta(\mathbf{j}'+u)^2 = \delta(y)^4\delta\big((\mathbf{j}'+u)(\mathbf{j}'+u)^*\big) = \delta(y)^4\delta(1+uu^*). \tag{2.8}$$

From this we easily see that $\delta(-v+g\mathbf{j}') = \delta(v+g\mathbf{j}') > 0$ for every $v \in U$ and every $g \in Q$. Thus, by (2.5) we have

$$\pi^{-ns}\Gamma_n^\iota(s)\delta(y)^{4s}\delta(x+y^2\mathbf{j}')^{-2s} = \int_Q \exp\big(-\pi\cdot\tau[p(1+uu^*)]\big)\,\delta(p)^{s-\kappa}dp$$

if $\mathrm{Re}(s) > \kappa - 1$. Thus by (2.1) we have at least formally

$$\begin{aligned}&\pi^{-ns}\Gamma_n^\iota(s)\delta(y)^{4s}\xi_U(y^2,\, h;\, s)\\ &= \int_Q \exp\big(-\pi\cdot\tau(p)\big)\delta(p)^{s-\kappa}\int_U \exp\big(-\pi\cdot\tau(puu^*)\big)\mathbf{e}\big(-\tau(hx^*)\big)dx\,dp.\end{aligned}$$

Therefore, if $\mathrm{Re}(s) > \kappa - 1$, then the integral of (2.1) is convergent if and only if the last double integral is convergent, in which case the equality holds. Changing the variable x for u, by Lemma 1.3 we see that

$$\begin{aligned}(2.9)\qquad &\pi^{-ns}\Gamma_n^{\iota}(s)\delta(y)^{4s-2\mu}\xi_U(y^2,\,h;\,s)\\ &= 2^{\iota n(1-n)/4}\int_Q \exp\left\{-\pi\cdot\tau\left(p+k^*(T_p^U)^{-1}(k)\right)\right\}\delta(p)^{s-\kappa}\det\left(T_p^U\right)^{-1/2}dp,\end{aligned}$$

where $k = R_y^U(h)$ and $\mu = \mu(U)$. In particular, taking $h = 0$, we obtain

$$\begin{aligned}(2.10)\qquad &2^{\iota n(n-1)/4}\pi^{-ns}\delta(y)^{4s-2\mu}\Gamma_n^{\iota}(s)\xi_U(y^2,\,0;\,s)\\ &= \int_Q \exp\left(-\pi\cdot\tau(p)\right)\delta(p)^{s-\kappa}\det\left(T_p^U\right)^{-1/2}dp.\end{aligned}$$

From Lemma 1.2 we see that $\det\left(T_p^U\right)^{-1/2} \leq \delta(p)^{-a}$ with a positive constant a depending on U, and hence the last integral over Q is convergent at least for $\mathrm{Re}(s) > \kappa + a - 1$. Thus ξ_X for an arbitrary h is meaningful at least for such values of s. Observe that $\mu(Y) - \mu(X) = \nu$. Then by Lemma 1.2 the right-hand side of (2.10) for $U = X$ coincides with that for $U = Y$ with $s + (\nu/2)$ in place of s. Thus

$$(2.11)\qquad \xi_X(g,\,0;\,s) = \pi^{-n\nu/2}\Gamma_n^{\iota}(s+(\nu/2))\Gamma_n^{\iota}(s)^{-1}\xi_Y(g,\,0;\,s+(\nu/2)).$$

Now take $U = Y$ and observe that $\xi_Y(g,\,h;\,s)$ is exactly the function $\xi(g,\,h;\,s,\,s)$ of [S1], (1.25). Therefore from [S1], (1.31), we obtain

$$(2.12)\qquad \xi_Y(g,\,0;\,s) = 2^{n-2ns}(2\pi)^{n\kappa}\delta(g)^{\kappa-2s}\Gamma_n^{\iota}(2s-\kappa)\Gamma_n^{\iota}(s)^{-2}$$

if $\mathrm{Re}(2s) > 2\kappa - 1$. This combined with (2.11) shows that

$$(2.13)\qquad \xi_X(g,\,0;\,s) = 2^n\pi^{n\kappa-(n\nu/2)}\delta(2g)^{\mu-2s}\frac{\Gamma_n^{\iota}(2s-\mu)}{\Gamma_n^{\iota}(s+(\nu/2))\Gamma_n^{\iota}(s)},$$

where $\mu = \mu(X)$. Here we employed the formula

$$(2.14)\qquad \mu(X) = \kappa - \nu,$$

which can easily be verified (see the table below).

Lemma 2.1. *Let μ be a measure on a set X in the standard sense of measure space; let p and q be nonnegative real-valued measurable functions on X such that $q(x) > 0$ for every $x \in X$; further let $f(s)$ be a holomorphic function defined on a domain D containing a half plane $\left\{s \in \mathbf{C} \,\middle|\, \mathrm{Re}(s) > b\right\}$ as well as the point b for some $b \in \mathbf{R}$. Suppose that $p(x)q(x)^{-s}$ as a function of $(x,\,s) \in X \times \left\{s \in \mathbf{C} \,\middle|\, |s - s_0| = r\right\}$ is measurable for every $s_0 \in D$ and $r \geq 0$; suppose also that $|pq^{-s}|$ is integrable on X and*

$$f(s) = \int_X p(x)q(x)^{-s}d\mu(x)$$

for $\mathrm{Re}(s) > b$. Then $|pq^{-s}|$ is integrable for $\mathrm{Re}(s) > a$ with some $a < b$.

PROOF. This was given in [S3], Lemma A1.5, under the assumption that $q(x) \geq t$ for every $x \in X$ with a positive constant t. The present lemma removes that condition. To see this, put $A = \left\{x \in X \,\middle|\, q(x) \geq 1\right\}$ and $B = \left\{x \in X \,\middle|\, q(x) < 1\right\}$. Then our integral over X is the sum of the integrals over A and B. Now $|q(x)^{-s'}| \leq |q(x)^{-s}|$ for $x \in B$ if $\mathrm{Re}(s') < \mathrm{Re}(s)$. Since our integral over X is convergent for $\mathrm{Re}(s) > b$, we see that the integral over B is finite for every $s \in \mathbf{C}$. Therefore our problem can be reduced to the integral over A, to which [S3], Lemma A1.5, is applicable. Therefore we obtain our lemma.

Proposition 2.2. *Let λ be the largest pole of the right-hand side of (2.13). Then the integral for ξ_X in (2.1) is convergent if and only if $\mathrm{Re}(s) > \lambda$. Consequently the function $\xi_X(g, h; s)$ is holomorphic (at least) for $\mathrm{Re}(s) > \lambda$.*

PROOF. We consider the integral of (2.1) for $h = 0$ and $U = X$. We know that the integral is convergent for sufficiently large $\mathrm{Re}(s)$, and can be continued to a holomorphic function of s on the domain $\mathrm{Re}(s) > \lambda$. Then, by Lemma 2.1 the integral must be absolutely convergent for $\mathrm{Re}(s) > \lambda$. Clearly the convergence holds for an arbitrary h.

Here we give a table for the invariants we have introduced so far:

	I	II	III	IV
ι	1	4	2	2
κ	$(n+1)/2$	$2n-1$	n	n
ν	1	-2	1	-1
$\mu(X)$	$(n-1)/2$	$2n+1$	$n-1$	$n+1$
λ	$(2n-3)/4$	$2n-(1/2)$	$n-(3/2)$	n

The value of λ in each case can easily be found. It is a simple pole of (2.13). It should be noted that it coincides with the largest pole of the numerator $\Gamma_n^\iota(2s-\mu)$ of the right-hand side of (2.13) in Cases II and IV. However, in Cases I and III, λ is the second largest pole of $\Gamma_n^\iota(2s-\mu)$. Notice also that $\kappa = \mu(Y)$, so that $\dim(Z) = n(\kappa + \mu(X))$.

Lemma 2.3. *Let $R_c = \{ g \in Q \,|\, g > c1_n \}$ in Case I with $c > 0$; let M be a compact subset of $\mathbf{C}$. Then there exist a neighborhoold N of 0 in $\mathbf{R}_n^n$ and a positive constant p such that $|\det(u+g+iy)^s| \geq p|\det(u+g)^s|$ for every $s \in M$ and every $(u, g, y) \in X \times R_c \times N$, where $\det(u+g+iy)^s$ is defined so that it coincides with $\det(u+g)^s$ for $y = 0$.*

PROOF. Here notice that $X \times R_c \times N$ is simply connected for an easy choice of N, and $\det(u+g) > 0$, since $\det(u+g) \neq 0$ by (2.8). For $x \in \mathbf{R}_n^n$ let $\|x\|$ denote the positive number such that $\|x\|^2$ is the maximum eignevalue of ${}^t xx$. Then $\|xx'\| \leq \|x\| \cdot \|x'\|$. Let $N = \{ y \in \mathbf{R}_n^n \,|\, \|y\| < c/2 \}$. Take $g \in R_c$ and $y \in N$. Put $a = g^{-1/2}$, $x = aua$, $z = aya$, and $w = z(1+x)^{-1}$. Then

$$\det(u+g+iy) = \det(g)\det(1+x+iz)$$
$$= \det(g)\det(1+x)\det(1+iw) = \det(u+g)\prod_{k=1}^{n}(1+i\lambda_k),$$

where $\lambda_1, \ldots, \lambda_n$ are the characteristic roots of w. Since $\|a\| < c^{-1/2}$ and $\|y\| < c/2$, we have $\|z\| < 1/2$, so that ${}^t zz < 4^{-1}1_n$. Thus

$$ {}^t ww = {}^t(1+x)^{-1} \cdot {}^t zz(1+x)^{-1} < 4^{-1}\left[(1+x)(1+{}^t x)\right]^{-1} = 4^{-1}(1+x \cdot {}^t x)^{-1} \leq 4^{-1} 1_n. $$

If $0 \neq v \in \mathbf{C}^n$ and $wv = \lambda_k v$, then $|\lambda_k|^2 v^* v = v^*({}^t ww)v < 4^{-1} v^* v$, so that $|\lambda_k| < 1/2$. Thus $|\det(u+g+iy)| = \det(u+g)\prod_{k=1}^{n} |1+i\lambda_k| \geq 2^{-n}\det(u+g)$. Clearly $\det(u+g+iy)^s$ can be defined for every $s \in \mathbf{C}$ and every $(u, g, y) \in X \times R_c \times N$ as stated in our lemma. Since $|\arg(1+i\lambda_k)| \leq \pi/6$, for $s = \sigma + it$ with $\sigma, t \in \mathbf{R}$ we have

$$ |\det(u+g+iy)^s| \geq e^{-n|t|\pi/6} |\det(u+g+iy)|^{\sigma}, $$

which proves our lemma.

Proposition 2.4. *For each fixed $h \in X$ the function $\xi_X(g, h; s)$ is real analytic in (g, s) on $Q \times \{ s \in \mathbf{C} \mid \mathrm{Re}(s) > \lambda \}$.*

PROOF. To prove this, consider Case IV for example. We can find an $\mathbf{R}$-linear ring-injection α of $\mathbf{H}_n^n$ into $\mathbf{R}_{4n}^{4n}$ such that $\alpha(x^*) = {}^t\alpha(x)$. We now consider, for $(g, w) \in Q \times \alpha(Y)$,

$$ \int_X \mathbf{e}(-\tau(hv^*)) \det\left[\alpha(vj^{-1}+g)+iw\right]^{-s/2} dv. \tag{2.15} $$

Observe that the integral of (2.1) is obtained by putting $w = 0$ in (2.15). Since ${}^t\alpha(vj) = -\alpha(vj)$, ${}^t\alpha(g) = \alpha(g)$, and ${}^t w = w$, given $c \in \mathbf{R}, > 0$, Lemma 2.3 guarantees a neighborhood N of 0 in $\alpha(Y)$ such that the convergence of (2.15) for $g > c1_n$ and $w \in N$ follows from that of (2.1). Thus (2.15) defines a holomorphic function of $(\alpha(g) + iw, s)$ for such (g, w) and $\mathrm{Re}(s) > \lambda$. This proves our proposition.

In Section 4 we shall prove meromorphic continuation of $\xi_X(g, h; s)$ to the whole $\mathbf{C}$.

Remark. (1) In (2.1) we can define ξ_U also with $\tau(hv)$ instead of $\tau(hv^*)$ on the right-hand side except in Case IV, $U = X$. This is trivial if $U = Y$; for $U = X$ the fact can be seen from (2.4).

(2) We can define the spaces X, Y, Z, Q, and H by taking $\mathbf{C}$ in place of $\mathbf{H}$ in (1.3.II). Then $X = iY$, and hence if we define ξ_X in this case by (2.1), then taking $h = ik$ and $v = iw$ with $k, w \in Y$ in (2.1), we obtain

$$ \xi_X(g, -ik; s) = \int_Y \mathbf{e}(-\mathrm{tr}(kw))|\det(g+iw)|^{-2s} dw $$

for $g \in Q$ and $k \in Y$. This is exactly the function $\xi(g, k; s, s)$ of [S1], (1.25), Case II, and its analytic nature was investigated in that paper. It is for this reason that this case is not included in the present paper.

3. Some lemmas and the lowest dimensional cases

To emphasize the dimensionality, we hereafter denote X, Z, Q, H, and ξ_X by X_n, Z_n, Q_n, H_n, and ξ_n. Also we put $\|x\| = \left(\sum_{\nu=1}^{n} |x_\nu|^2\right)^{1/2}$ for $x \in \mathbf{H}^n$.

Lemma 3.1. (1) *Let α and β be two elements of X_n in Case I such that $\alpha\beta = \beta\alpha$. Then there exists an element γ of $\mathfrak{g}_1$ such that $\gamma\alpha \cdot {}^t\gamma = \mathrm{diag}[a_1, \ldots, a_r, 0_s]$ and $\gamma\beta \cdot {}^t\gamma = \mathrm{diag}[b_1, \ldots, b_t, 0_u]$ with a_ν and b_ν in X_2.*

(2) *Let* σ *be an element of* $\mathbf{C}_n^n$ *such that* ${}^t\sigma = \varepsilon\sigma$ *with* $\varepsilon = 1$ *or* -1. *Then there exists a unitary matrix* η *such that* $\eta\sigma \cdot {}^t\eta$ *is a real diagonal matrix with nonnegative diagonal entries if* $\varepsilon = 1$ *and* $\eta\sigma \cdot {}^t\eta = \mathrm{diag}[a_1, \ldots, a_r, 0_s]$ *with* a_ν *of size 2 if* $\varepsilon = -1$.

PROOF. Let α and β be as in (1); suppose $\alpha \neq 0$. Since they are skew-hermitian, we can find a nonzero element x of $\mathbf{C}^n$ such that $\alpha x = i\lambda x$ and $\beta x = i\mu x$ with $\lambda, \mu \in \mathbf{R}$, $\lambda \neq 0$. Put $x = g + ih$ with $g, h \in \mathbf{R}^n$. Then $\alpha g = -\lambda h$, $\alpha h = \lambda g$, $\beta g = -\mu h$, $\beta h = \mu g$. Since $\lambda \neq 0$, we see that $g \neq 0$ and $h \neq 0$; also ${}^t hg = \lambda^{-1} \cdot {}^t h\alpha h = 0$, and $\lambda\|g\|^2 = {}^t g\alpha h = -{}^t h\alpha g = \lambda\|h\|^2$. Therefore, changing x for its suitable constant multiple, we may assume that $\|g\| = \|h\| = 1$. Let V be the orthogonal complement of $\mathbf{R}g + \mathbf{R}h$ in $\mathbf{R}^n$. Then we easily see that V is stable under α and β. Therefore by induction we obtain (1). Given σ as in (2), take a unitary matrix ξ so that $\xi\sigma\sigma^*\xi^*$ is diagonal. Put $\tau = \xi\sigma \cdot {}^t\xi = \alpha + i\beta$ with real matrices α and β. Since $\tau\overline{\tau}$ is real, we see that $\alpha\beta = \beta\alpha$. If $\varepsilon = -1$, take γ as in (1) for the present α and β. Then $\gamma\xi$ gives the desired element η. If $\varepsilon = 1$, we can find a real orthogonal matrix ζ such that both $\zeta\alpha \cdot {}^t\zeta$ and $\zeta\beta \cdot {}^t\zeta$ are diagonal. Thus we can put $\zeta(\alpha + i\beta) \cdot {}^t\zeta = \mathrm{diag}[p_1 u_1^2, \ldots, p_n u_n^2]$ with $0 \le p_\nu \in \mathbf{R}$ and $u_\nu \in \mathbf{C}$, $|u_\nu| = 1$. Then $\mathrm{diag}[u_1, \ldots, u_n]^{-1}\zeta\xi$ gives the desired element η in this case.

Lemma 3.2. *Given* $h \in X_n$, *there exists an element* $a \in \mathfrak{g}_1$ *such that* $R_a^X(h)$ *is of the following form in each case:* $R_a^X(h) = \mathrm{diag}[k_1, \ldots, k_r, 0_s]$ *with* $k_\nu \in X_2$ *in Case I,* $k_\nu \in \mathbf{R}j$ *in Case II, and* $0 \le k_\nu \in \mathbf{R}$ *in Case IV;* $R_a^X(h) = \varphi(\mathrm{diag}[a_1, \ldots, a_r, 0_s])E$ *with* ${}^t a_\nu = -a_\nu \in \mathbf{C}_2^2$ *in Case III.*

PROOF. In Case I (resp. Case IV) this follows from Lemma 3.1 (1) (resp. (2)). In Case II suppose $h \neq 0$. We let h act on $\mathbf{H}^n$ on the left and view $\mathbf{H}^n$ as a right $\mathbf{C}$-module. Then we can find an element e of $\mathbf{H}^n$ such that $\|e\| = 1$ and $he = ec$ with $0 \neq c \in \mathbf{C}$. Then $c = e^*he = -\overline{c}$. Put $V = \{x \in \mathbf{H}^n \mid e^*x = 0\}$. Then we easily see that $hV \subset V$. Thus by induction we can find an element $a \in GL_n(\mathbf{H})$ such that $aa^* = 1$ and $aha^* = id$ with a real diagonal matrix d. Since $j = sis^{-1}$ with an element s of $\mathbf{H}$ such that $|s| = 1$, we obtain the desired result in Case II. In Case III put $h = \varphi(\sigma)E$ with ${}^t\sigma = -\sigma \in \mathbf{C}_n^n$. Take η as in Lemma 3.1 (2) for this σ and put $a = \varphi(\eta)$. Then $a \in \mathfrak{g}_1$ and $R_a^X(h)$ has the desired form.

Lemma 3.3. *Let* $\mathbf{K}$ *denote either* $\mathbf{R}$, $\mathbf{C}$, *or* $\mathbf{H}$; *put* $\mathfrak{B} = \{x \in \mathbf{K}_r^q \mid xx^* < 1_q\}$ *and* $n = q + r$ *with positive integers* q *and* r. *Then for* $\mathrm{Re}(s) > \kappa(n) - 1$ *we have*

$$\int_{\mathfrak{B}} \delta(1_q - xx^*)^{s-\kappa(n)}\, dx = \int_{\mathbf{K}_r^q} \delta(1_q + xx^*)^{-s} dx$$
$$= \pi^{\iota qr/2}\Gamma_q^\iota(s - \iota r/2)/\Gamma_q^\iota(s) = \pi^{\iota qr/2}\Gamma_r^\iota(s - \iota q/2)/\Gamma_r^\iota(s),$$

where dx *is the standard measure on* $\mathbf{K}_r^q$, $\iota = [\mathbf{K} : \mathbf{R}]$, $\kappa(n) = 1 + \iota(n-1)/2$, $\delta(z) = |\det(z)|$ *if* $\mathbf{K}$ *is* $\mathbf{R}$ *or* $\mathbf{C}$, $\delta(z)$ *is defined by (1.8.II,IV) if* $\mathbf{K} = \mathbf{H}$, *and* Γ_m^ι *is defined by (2.6b).*

This was given in [S3], Lemma A2.7 and (A2.7.4), when **K** is **R** or **C**, but the case **K** = **H** can be proved in the same manner as noted in [S3], Page 219. We note here an easy fact:

(3.1) $$\delta(1_q - xy^*) = \delta(1_r - y^*x) \quad \textit{for every } x, y \in \mathbf{K}_r^q.$$

This follows from

$$\begin{bmatrix} 1_q & 0 \\ y^* & 1_r \end{bmatrix}\begin{bmatrix} 1_q & x \\ 0 & 1_r - y^*x \end{bmatrix} = \begin{bmatrix} 1_q & x \\ y^* & 1_r \end{bmatrix} = \begin{bmatrix} 1_q & x \\ 0 & 1_r \end{bmatrix}\begin{bmatrix} 1_q - xy^* & 0 \\ y^* & 1_r \end{bmatrix}.$$

Lemma 3.4. *Let* **K** *be as above. Then* $\delta(1_m + xx^*) \geq 1 + \tau(xx^*)$ *for every* $x \in \mathbf{K}_n^m$.

PROOF. Since $\tau(xx^*) = \tau(x^*x)$ and $\delta(1_m + xx^*) = \delta(1_n + x^*x)$ by (3.1), we may assume that $m \leq n$. Then, by [S1], Lemma 2.4, there exist $a \in GL_m(\mathbf{K})$ and $b \in GL_n(\mathbf{K})$ such that $aa^* = 1_m$, $bb^* = 1_n$, and $axb = [d \ \ 0]$ with a real diagonal d. Then $\delta(1_m + xx^*) = \det(1+d^2) = \prod_{\nu=1}^m (1+d_{\nu\nu}^2) \geq 1 + \sum_{\nu=1}^m d_{\nu\nu}^2 = 1 + \tau(d^2) = 1 + \tau(xx^*)$.

Lemma 3.5. *Given two positive integers* m *and* α, *there exists a positive constant* A *depending only on* m *and* α *such that*

$$\int_{\mathbf{R}^m} \exp\left(-c(1 + {}^t xx)^{1/\alpha}\right)dx \leq Ae^{-c}\left(c^{-m/2} + c^{-m\alpha/2}\right) \quad \textit{for every} \quad c \in \mathbf{R}, > 0.$$

PROOF. In this proof we denote by a_1 and a_2 some positive constants depending only on m and α. Putting ${}^t xx = r^2$ with $0 < r \in \mathbf{R}$, we see that the integral in question equals $a_1 \int_0^\infty \exp\left(-c(1+r^2)^{1/\alpha}\right)r^{m-1}dr$. Then putting $1 + r^2 = (1+t)^\alpha$, we can transform the last integral into

(3.2) $$(\alpha/2)\int_0^\infty e^{-c-ct}(1+t)^{\alpha-1}\left[(1+t)^\alpha - 1\right]^{(m-2)/2} dt.$$

Now $(1+t)^\alpha - 1 = \alpha t\left(1 + \sum_{\nu=1}^{\alpha-1} b_\nu t^\nu\right)$ with $0 < b_\nu \in \mathbf{R}$. Therefore we easily see that the quotient

$$\frac{(1+t)^{\alpha-1}\left[(1+t)^\alpha - 1\right]^{(m-2)/2}}{t^{(m-2)/2}\left(1 + t^{(\alpha-1)m/2}\right)}$$

is bounded for $0 < t \in \mathbf{R}$, since it is bounded as t tends to 0 or ∞. Thus (3.2) is majorized by

$$a_2 e^{-c}\int_0^\infty e^{-ct}\left(t^{(m/2)-1} + t^{(\alpha m/2)-1}\right)dt.$$

The last integral equals $\Gamma(m/2)c^{-m/2} + \Gamma(\alpha m/2)c^{-\alpha m/2}$, which proves our lemma.

Let us now recall the classical Bessel function $K(s, y)$ defined by

(3.3) $$K(s, y) = \int_0^\infty \exp\left(-yt - yt^{-1}\right)t^{s-1}dt \qquad (s \in \mathbf{C},\ 0 < y \in \mathbf{R}).$$

We have $K(s, y) = 2K_s(2y)$ with the traditional symbol K_s. The integral is convergent for every s, and so $K(s, y)$ for each fixed y is an entire function of s; also changing t for t^{-1}, we obtan $K(s, y) = K(-s, y)$. We note an easy formula:

$$\int_0^\infty \exp\big(-a^2t - b^2t^{-1}\big)t^{s-1}dt = |b/a|^s K(s, |ab|) \qquad (a, b \in \mathbf{R}^\times). \tag{3.4}$$

We need two types of estimate:

$$|K(s, y)| \le b(A, y_0)e^{-2y} \quad \textit{if} \quad s \in A \text{ and } y \ge y_0, \tag{3.5}$$

$$|y^s K(s, y)| \le \Gamma\big(\mathrm{Re}(s)\big) \quad \textit{if} \quad \mathrm{Re}(s) > 0. \tag{3.6}$$

Here A is an arbitrary compact subset of $\mathbf{C}$, $0 < y_0 \in \mathbf{R}$, and $b(A, y_0)$ is a positive constant depending only on A and y_0. To prove (3.5), we first note $t + t^{-1} = 2 + (t^{1/2} - t^{-1/2})^2$, and so, for $y \ge y_0$,

$$\begin{aligned}|e^{2y}K(s, y)| &= \left|\int_0^\infty \exp\big[-y(t^{1/2} - t^{-1/2})^2\big]t^{s-1}\,dt\right| \\ &\le \int_0^\infty \exp\big[-y_0(t^{1/2} - t^{-1/2})^2\big]|t^{s-1}|\,dt = e^{2y_0}K(\mathrm{Re}(s), y_0).\end{aligned}$$

This proves (3.5). Next, let $\sigma = \mathrm{Re}(s) > 0$; then, by (3.4),

$$|y^s K(s, y)| \le \int_0^\infty \exp(-t - y^2t^{-1})t^{\sigma-1}dt \le \int_0^\infty e^{-t}t^{\sigma-1}dt = \Gamma(\sigma).$$

This proves (3.6).

Let $\mathbf{K}$ and ι be as in Lemma 3.3. Then for $a \in \mathbf{K}^\times$ we have

$$\pi^{-s}\Gamma(s)\int_{\mathbf{K}} \mathbf{e}(a\bar{x} + x\bar{a})(1 + |x|^2)^{-s}\,dx = |2a|^{s-(\iota/2)}K\big(s - (\iota/2),\, 2\pi|a|\big). \tag{3.7}$$

Indeed. the left-hand side equals

$$\begin{aligned}&\int_{\mathbf{K}} \mathbf{e}(a\bar{x} + x\bar{a})\int_0^\infty \exp\big(-\pi t(1 + |x|^2)\big)t^{s-1}dt\,dx \\ &= \int_0^\infty \exp(-\pi t)t^{s-1}\int_{\mathbf{K}} \mathbf{e}(a\bar{x} + x\bar{a})\exp(-\pi t|x|^2)dx\,dt \\ &= \int_0^\infty \exp(-\pi t - \pi|2a|^2t^{-1})t^{s-(\iota/2)-1}dt,\end{aligned}$$

which combined with (3.4) gives the right-hand side of (3.7).

Let us now show that ξ_X for $n = 1$ in Cases II and IV, and for $n = 2$ in Cases I and III can be expressed by means of $K(s, y)$. In fact if $h \ne 0$, we have

$$\pi^{-2s}\Gamma(2s)\xi_2(g, h; s) = |4\delta(g^{-1}h)|^{s-(1/4)}K\big(2s - (1/2),\, 2\pi\delta(gh)^{1/2}\big), \tag{3.8.I}$$

$$\pi^{-s}\Gamma(s)\xi_1(g, h; s) = |h/g|^{s-(3/2)}K\big(s - (3/2),\, \pi|gh|\big), \tag{3.8.II}$$

$$\pi^{-2s}\Gamma(2s)\xi_2(g, h; s) = |4\delta(g^{-1}h)|^{s-(1/2)}K\big(2s - 1,\, 2\pi\delta(gh)^{1/2}\big), \tag{3.8.III}$$

$$\pi^{-s}\Gamma(s)\xi_1(g, h; s) = |h/g|^{s-1}K\big(s - 1,\, \pi|gh|\big). \tag{3.8.IV}$$

In view of (2.3) it is sufficient to prove the case $g = 1$. In Cases II and IV the desired formula can be obtained from (2.9) by putting $n = 1$. In Case I put $\zeta = \begin{bmatrix} 0 & -1 \\ 1 & 0 \end{bmatrix}$, $h = a\zeta$, and $v = -x\zeta$ with $a, x \in \mathbf{R}$ in formula (2.1). Then

$$\xi_2(1_2, h; s) = \int_{\mathbf{R}} \mathbf{e}(2ax)(1 + x^2)^{-2s}\,dx,$$

and hence we obtain (3.8.I) from (3.7) with $\mathbf{K} = \mathbf{R}$. Similarly, in Case III put $h = \varphi(a\zeta)E$ and $v = -\varphi(x\zeta)E$ with $a, x \in \mathbf{C}$. Then

$$\xi_2(1_4,\, h;\, s) = \int_{\mathbf{C}} \mathbf{e}(a\overline{x} + x\overline{a})(1 + |x|^2)^{-2s}\, dx.$$

Thus we obtain (3.8.III) from (3.7) with $\mathbf{K} = \mathbf{C}$.

Remark. The integral expression for ξ_U in (2.9) is a natural generalization of (3.3). In general, however, unlike (3.3), the integral over Q of (2.9) can be divergent for some s even if h is invertible. In fact, take $n = 2$ and $h \neq 0$ in Cases I and III. Then (2.9) combined with (3.8.I, III) shows that the integrals over Q in those cases are of the forms

(3.9.I) $\qquad ab^s(2s-1)^{-1}K(2s-(1/2),\, c),$

(3.9.III) $\qquad ab^s\Gamma(s-1)\Gamma(s+(1/2))^{-1}K(2s-1,\, c)$

with positive numbers a, b, and c. Therefore the integrals over Q are convergent only for $\mathrm{Re}(s) > 1/2$ and $\mathrm{Re}(s) > 1$ respectively.

4. Analytic continuation of ξ_X

For $x \in GL_n(\mathbf{H})$ we denote by $\rho(x)$ (resp. $\sigma(x)$) the maximum (resp. minimum) eigenvalue of xx^*. Clearly $\rho(x) = \rho(x^*)$ and $\sigma(x) = \sigma(x^*)$.

Lemma 4.1. *If $h \in GL_n(\mathbf{C})$, $g = g^* \in GL_n(\mathbf{C})$, and $g > 1$, then $\sigma(h) \leq \sigma(g^{1/2}h \cdot {}^tg^{1/2})$.*

PROOF. For $x \in \mathbf{C}^n$ we have $\|gx\| \geq \|x\|$ if $g > 1$, and $\|hx\|^2 = x^*h^*hx \geq \sigma(h)\|x\|^2$, and hence $\|\overline{g}h^*ghx\|^2 \geq \|h^*ghx\|^2 \geq \sigma(h)\|ghx\|^2 \geq \sigma(h)\|hx\|^2 \geq \sigma(h)^2 \cdot \|x\|^2$. Therefore we obtain our lemma.

Lemma 4.2. *If $w \in H$, then w is an invertible matrix, and $w^{-1} \in Z$.*

PROOF. For every $w \in H$ we have seen in (2.8) that w is invertible. From the definition of Z we immediately see that $w^{-1} \in Z$ in Cases I–III. In Case IV we can let $Sp(n, \mathbf{C})$ act on H as we shall note in Section 5. The element J_n of $Sp(n, \mathbf{C})$ maps w to $-w^{-1}$, which is in H, and hence $w^{-1} \in Z$.

The size of matrices in X_n is $2n$ in Case III; it is n otherwise. To make our exposition uniform, we put $n' = 2n$ in Case III and $n' = n$ in the other three cases. This symbol n' will appear only as a subscript as in $1_{n'}$ in Propositions 4.3 and 4.4 below.

To study the nature of ξ_X as a function of s, we first observe that by (2.2) it is sufficient to consider the case $g = 1_{n'}$. Then the following proposition reduces the problem to the lower-dimensional cases.

Proposition 4.3. *Let $n = q + r$ with positive integers q and r. Let $h = \varphi(\mathrm{diag}[c,\, d])E_n$, $k = \varphi(c)E_q$, and $\ell = \varphi(d)E_r$ in Case III; let $h = \mathrm{diag}[k,\, \ell]$ with $k \in X_q$ and $\ell \in X_r$ in all other cases. If $\mathrm{Re}(s)$ is greater than the constant λ of Proposition 2.2, then*

$$\xi_n(1_{n'},\, h;\, s) = \int_S \xi_q(1_{q'},\, k + \ell\{y\};\, s - (\iota r/2))\xi_r(1_{r'} + y^*y,\, \ell;\, s)dy.$$

Here $S = \mathbf{R}_r^q$, $\mathbf{H}_r^q$, $\varphi(\mathbf{C}_r^q)E_r$, and $\mathbf{C}_r^q$ in Cases I, II, III, and IV, respectively; $\ell\{y\} = \bar{y}\bar{\ell}y^*$ in Case IV and $\ell\{y\} = y\ell y^*$ otherwise; dy is the measure on S obtained by identifying S with $\mathbf{R}^{\iota qr}$ in an obvious way.

PROOF. We first treat Case IV. In this proof, as well as in the proof of other statements in this section, we have to consider the point $\mathbf{j}$ of H_n with several different n's. To simplify our notation, however, we denote each one by just $\mathbf{j}$ since the dimensionality will be clear from the context. Let $x = \begin{bmatrix} v & y \\ {}^t y & u \end{bmatrix} \in X_n$ with $v \in X_q$. Put $w = \mathbf{j} + v$ and $z = \mathbf{j} + u$. Since

$$\begin{bmatrix} w & y \\ {}^t y & z \end{bmatrix} \begin{bmatrix} 1 & -w^{-1}y \\ 0 & 1 \end{bmatrix} = \begin{bmatrix} w & 0 \\ {}^t y & z - {}^t y w^{-1} y \end{bmatrix},$$

we have $\delta(\mathbf{j} + x) = \delta(\mathbf{j} + v)\delta(\mathbf{j} + u - {}^t y w^{-1} y)$. Therefore, for $\mathrm{Re}(s) > \lambda$,

$$(*) \quad \xi_n(1_n, h; s)$$
$$= \int_{\mathbf{C}_r^q} \int_{X_q} \int_{X_r} \mathbf{e}(-\tau(kv^* + \ell u^*))\delta(\mathbf{j} + v)^{-2s}\delta(\mathbf{j} + u - {}^t y w^{-1} y)^{-2s} du\, dv\, dy.$$

For $a \in \mathbf{H}_r^r$ put $a_* = \mathbf{j}a^*\mathbf{j}^{-1}$. Then for $b + c\mathbf{j} \in Z_r$ with $b \in X_r$ and $c \in Y_r$ we have $(b + c\mathbf{j})_* = b - c\mathbf{j}$. By Lemma 4.2, $w^{-1} \in Z_q$; then we easily see that ${}^t y w^{-1} y \in Z_r$, and hence we can put ${}^t y w^{-1} y = f - g\mathbf{j}$ with $f \in X_r$ and $g \in Y_r$. Then $2g\mathbf{j} = ({}^t y w^{-1} y)_* - {}^t y w^{-1} y = {}^t y(w_*^{-1} - w^{-1})y = {}^t y w^{-1}(w - w_*)w_*^{-1} y = {}^t y w^{-1}(2\mathbf{j})w_*^{-1} y$, so that $g = {}^t y(w^* w)^{-1}\bar{y}$. Similarly $2f = {}^t y w^{-1} y + ({}^t y w^{-1} y)_* = {}^t y w^{-1}(w + w_*)w_*^{-1} y = {}^t y w^{-1}(2v)w_*^{-1} y$. Put $1 + v^* v = a^2$ with $a \in Q_q$. Then $w^* w = a^2$, so that $w^{-1} = a^{-2}w^*$ and $w_*^{-1} = \mathbf{j}w\mathbf{j}^{-1} \cdot {}^t a^{-2} = (\mathbf{j} + \bar{v}) \cdot {}^t a^{-2}$. Thus $w^{-1}(2v)w_*^{-1} = 2a^{-2}(\bar{v} - \mathbf{j})v(\mathbf{j} + \bar{v}) \cdot {}^t a^{-2} = 2a^{-2}(\bar{v} + \bar{v}v\bar{v}) \cdot {}^t a^{-2} = 2a^{-2}\bar{v}$. Now, changing the variable u for $u + f$, we find that

$$\xi_n(1_n, h; s)$$
$$= \int_{\mathbf{C}_r^q} \int_{X_q} \int_{X_r} \mathbf{e}(-\tau(kv^* + \ell u^* + \ell f^*))\delta(\mathbf{j} + v)^{-2s}\delta((1 + g)\mathbf{j} + u)^{-2s} du\, dv\, dy$$
$$= \int_{\mathbf{C}_r^q} \int_{X_q} \mathbf{e}(-\tau(kv^* + \ell f^*))\delta(\mathbf{j} + v)^{-2s}\, \xi_r(1 + g, \ell; s)\, dv\, dy.$$

Since $a^2\bar{v} = \bar{v} \cdot {}^t a^2$, we have $a\bar{v} = \bar{v} \cdot {}^t a$, and hence $f = {}^t y a^{-1}\bar{v} \cdot {}^t a^{-1} y$. Put $y_1 = a^{-1}\bar{y}$. Then $g = y_1^* y_1$ and $f = y_1^* \overline{v} y_1$. Since $\delta(a) = \delta(\mathbf{j} + v)$ and $\tau(\ell f^*) = \tau(\bar{y}_1 \bar{\ell} y_1^* v^*)$, we obtain

$$\xi_n(1_n, h; s)$$
$$= \int_{\mathbf{C}_r^q} \int_{X_q} \mathbf{e}(-\tau((k + \bar{y}_1\bar{\ell}y_1^*)v^*))\delta(\mathbf{j} + v)^{2r - 2s}\, dv\, \xi_r(1 + y_1^* y_1, \ell; s)\, dy_1,$$

which gives the desired formula in Case IV.

We next treat Case III. Let $x = \begin{bmatrix} x_1 & x_2 \\ -{}^t x_2 & x_4 \end{bmatrix} \in \mathbf{C}_n^n$ with $x_4 \in \mathbf{C}_r^r$. Put $v = \varphi(x_1)E_q$, $u = \varphi(x_4)E_r$, and $y = \varphi(x_2)E_r$. Then there exists a matrix T in

$GL_{2n}(\mathbf{R})$ such that

$$T(1+\varphi(x)E_n)T^{-1} = \begin{bmatrix} w & y \\ -{}^t y & z \end{bmatrix}$$

with $w = 1+v$ and $z = 1+u$. Since

$$\begin{bmatrix} w & y \\ -{}^t y & z \end{bmatrix}\begin{bmatrix} 1 & -w^{-1}y \\ 0 & 1 \end{bmatrix} = \begin{bmatrix} w & 0 \\ -{}^t y & z+{}^t yw^{-1}y \end{bmatrix},$$

we have $\delta(1+\varphi(x)E_n) = \delta(1+v)\delta(1+u+{}^t yw^{-1}y)$. We easily see that $\tau\big(h\cdot(\varphi(x)E_n)^*\big) = \tau(kv^*)+\tau(\ell u^*)$. Thus we have a formula similar to (*) in the present case. Put ${}^t yw^{-1}y = g-f$ with ${}^t f = -f$ and ${}^t g = g$. Then $2f = {}^t y({}^t w^{-1} - w^{-1})y$ and ${}^t w^{-1} - w^{-1} = {}^t w^{-1}(w - {}^t w)w^{-1} = 2\cdot{}^t w^{-1}vw^{-1} = 2v\cdot{}^t w^{-1}w^{-1}$, as v commutes with both w and ${}^t w$. Since $w\cdot{}^t w \in Q$, we can put $w\cdot{}^t w = \varphi(b^{-2})$ with $0 < b = b^* \in \mathbf{C}_r^r$. Then $\varphi(b^2)v = v\varphi(b^2)$, and hence $\varphi(b)v = v\varphi(b)$. Thus $f = {}^t y\varphi(b)v\varphi(b)y = \varphi({}^t x_2\bar{b}\bar{x}_1 bx_2)E_r \in X_r$. Similarly $2g = {}^t y(w^{-1}+{}^t w^{-1})y = {}^t y\cdot{}^t w^{-1}(w+{}^t w)w^{-1}y = 2\cdot{}^t y\varphi(b)^2 y = 2\varphi({}^t x_2\bar{b}^2\bar{x}_2) \in Q$. Therefore, changing the variable u for $u+f$, we obtain

$$\xi_n(1_n, h;\, s) = \int_S\int_{X_q} \mathbf{e}\big(-\tau(kv^*+\ell f^*)\big)\delta(1+v)^{-2s}\xi_r(1+g, \ell;\, s)\,dv\,dy.$$

Next, put $y_1 = \varphi(b)y$. Then $\tau(\ell f^*) = \tau(\ell y_1^* v^* y_1) = \tau(y_1\ell y_1^* v^*)$ and $g = y_1^* y_1$. Since $\delta\big(\varphi(b)\big)^{-1} = \delta(1+v)$, we find the formula as stated in our proposition.

Cases I and II can be handled in a similar and much simpler way.

Proposition 4.4. *Let $n = q+r$ as above. Let $h = \mathrm{diag}[k, 0_r] \in X_n$ with $k \in X_q$ in Cases I, II, and IV; let $h = \varphi(\mathrm{diag}[c, 0_r])E_n \in X_n$ and $k = \varphi(c)E_q \in X_q$ with $c \in \mathbf{C}_q^q$ in Case III. Then*

$$\xi_n(1_{n'}, h;\, s) = 2^{r\alpha}\pi^{r\beta}\frac{\Gamma_r^\iota(2s-\mu)}{\Gamma_r^\iota\big(s+(\nu/2)\big)\Gamma_r^\iota(s)}\cdot\xi_q\big(1_{q'}, k;\, s-(\iota r/2)\big),$$

where $\mu = \mu(X_n)$, $\alpha = 2+(\iota/2)(r-1)-\nu-2s$, and $\beta = \mu+(\nu/2)$.

PROOF. Put $\beta_r(s) = \xi_r(1_{r'}, 0;\, s)$. Take $\ell = 0$ in Proposition 4.1. By (2.3) we have $\xi_r(1+y^*y, 0;\, s) = \beta_r(s)\delta(1+y^*y)^{\mu'-2s}$ with $\mu' = \mu(X_r)$. Therefore the integral over S of Proposition 4.3 becomes

$$\beta_r(s)\xi_q\big(1_{q'}, k;\, s-(\iota r/2)\big)\int_S \delta(1+y^*y)^{\mu'-2s}dy.$$

This together with Lemma 3.3 and (2.13) proves the desired formula.

Theorem 4.5. *Let h be an element of X_n in Case I, and let $m = \mathrm{rank}(h)/2$. Then there exists a real analytic function of $(g, s) \in Y\times\mathbf{C}$ holomorphic in s that coincides with*

$$\xi_n(g, h;\, s)\Gamma_{n-2m}^1\big(2s-(n-1)/2\big)^{-1}\prod_{k=1}^{n-m}\Gamma(2s+1-k) \tag{4.1}$$

for $\mathrm{Re}(s) > \lambda$, where we understand that $\Gamma_0^1 = 1$. In particular, if $n-2m \le 1$, then $\xi_n(g, h;\, s)\prod_{k=1}^m \Gamma(2s+1-k)$ is an entire function of s.

PROOF. If $h = 0$, this follows from (2.13). Now at the end of [S4] we showed that for every $h \in X, \neq 0$, in Case I there is an element $a \in \mathfrak{g}$ such that the product

$$(4.2)\qquad \xi_n(g,\, R_a^X(h);\, s)\Lambda_{h,\mathfrak{c}}(s)\Gamma_n^1(s+(1/2))\Gamma_3^1(s+1) \cdot \prod_{i=2}^{[n/2]} \Gamma(2s+1-i) \prod_{i=0}^{n} L(2s+1-i,\, \chi)$$

is an entire function. Here χ is a Hecke character of $\mathbf{Q}$ such that $\chi^2 \neq 1$ and $\chi_{\mathbf{a}}(x) = x/|x|$, $L(s,\, \chi)$ is the L-function of χ, and $\Lambda_{h,\mathfrak{c}}(s)$ is a finite product of certain L-functions. (Notice that $g(s)$ of [S4], (11.6.5), is $\gamma(s,\, e)$ of [S4], (11.6.3), with $e = 1$, so that $g(s) = \Gamma_3^1(s+1)\prod_{i=2}^{m}\Gamma(2s+1-i))$.) Moreover, the product of (4.2) must be real analytic in $(g,\, s)$, since it appears as a Fourier coefficient of an Eisenstein series $E_1^*(w,\, s)$ of [S4], (6.4.8), or rather its product with some gamma factors and L-functions which is real analytic in $(w,\, s)$ where it is finite, and which can be obtained by replacing E_1 of [S4], (11.6.2) by E_1^*. (We can also apply Lemma 4.7 below to these functions.) This shows at least meromorphic continuation of ξ_n to the whole $\mathbf{C}$. Put

$$\eta(s) = \xi_n(g,\, R_a^X(h);\, s)\Lambda_{h,\mathfrak{c}}(s) \prod_{k=1}^{[n/2]} \Gamma(2s+1-k),$$

$$R(s,\, \chi) = \Gamma\big((s+1)/2\big)L(s,\, \chi).$$

Arranging the gamma factors of (4.2) suitably, we see that

$$\eta(s) \prod_{k=0}^{n} R(2s+1-k,\, \chi)$$

is entire. Put $D_k = \{\, s \in \mathbf{C} \mid 0 < \mathrm{Re}(2s+1-k) < 1 \,\}$. Then $R(2s+1-k,\, \chi) \neq 0$ for $s \notin D_k$. Therefore $\eta(s)$ is holomorphic at s if $s \notin \bigcup_{k=0}^{n} D_k$. Let $s_0 \in \bigcup_{k=0}^{n} D_k$. Then $s_0 \in D_k$ for a unique k, and hence $\eta(s)R(2s+1-k,\, \chi)$ is holomorphic at s_0. This is so for any character χ of the above type. Now by a result of Rohrlich [R], we can find a Hecke character ψ of $\mathbf{Q}$ such that $\psi_{\mathbf{a}} = 1$, $R(2s_0+1-k,\, \chi\psi) \neq 0$, and the conductor of ψ is prime to that of χ. Replacing χ by $\chi\psi$, we find that η is holomorphic at s_0. Thus η is an entire function of s, and is real analytic in $(g,\, s)$ on $Q \times \mathbf{C}$.

Now suppose that h is invertible; then the function $\Lambda_{h,\mathfrak{c}}(s)$, which is given by [S4], (11.6.5), is 1. Therefore $\xi_n(g,\, R_a^X(h);\, s)\prod_{k=1}^{m}\Gamma(2s+1-k)$ is an entire function of s.

Next assume that $2m < n$. Employing Proposition 4.4 and applying the above result to ξ_q there, we can easily verify that the product of (4.1) with $g = 1_n$ and with $R_a^X(h)$ in place of h is entire. This conclusion is true for every $g \in Q$ and every h of rank $2m$ because of (2.2) and Lemma 3.2. Since (4.2) is real analytic in $(g,\, s)$, clearly (4.1), which is finite, must be real analytic in $(g,\, s)$. Finally the last assertion of our theorem can easily be derived from the first part.

In Cases II, III, and IV, we have weaker results than those in Case I; namely we have

Theorem 4.6. *The function* $\xi_n(g, h; s)$ *in all cases can be continued as a meromorphic function of* s *to the whole complex plane. More precisely, let* $q = \mathrm{rank}(h)/2$ *in Case III and let* $q = \mathrm{rank}(h)$ *in Cases II and IV. Then there exists a real analytic function of* $(g, s) \in Y \times \mathbf{C}$ *holomorphic in* s *that coincides with*

$$\text{(4.3)} \qquad \frac{\Gamma_r^\iota(s + (\nu/2))\Gamma_r^\iota(s)}{\Gamma_r^\iota(2s - \mu(X_n))} \cdot \xi_n(g, h; s)$$

for $\mathrm{Re}(s) > \lambda$, *where* $r = n - q$ *and we understand that* $\Gamma_0^\iota = 1$.

PROOF. If $h = 0$, this follows from (2.13). For invertible h we can show that $\xi_n(1_{n'}, h; s)$ belongs to the class of functions investigated by Wallach [W]. Then [W], Theorem 7.2, shows that it can be continued to an entire function of s. For this we have to verify that our function satisfies all the conditions imposed in [W]. Since this is nontrivial and requires a detailed discussion within the semisimple group acting on H, we will explain about it at the end of the next section. Once this is established, by combining this result for invertible h with (2.2) and Proposition 4.4, we see that (4.3) is an entire function of s. As for the real analyticity, we showed it for $\mathrm{Re}(s) > \lambda$ in Proposition 2.4. Then the real analyticity on the whole $Q \times \mathbf{C}$ follows from Lemma 4.7 below due to Fefferman and Narasimhan.

Combining Lemma 2.1 with Theorems 4.5 and 4.6, we see that the integral of (2.9) is convergent at least for $\mathrm{Re}(s) > \sigma_h$ with

$$\text{(4.4)} \qquad \sigma_h = \begin{cases} (n-1)/2 & \text{(Case I)}, \\ \mathrm{Max}(2n-2,\, 2n-q-(1/2)) & \text{(Case II)}, \\ n-1 & \text{(Case III)}, \\ \mathrm{Max}(n-1,\, n-(q/2)) & \text{(Case IV)}, \end{cases}$$

with q as in Theorem 4.6; $q = \mathrm{rank}(h)$ in Case I. Notice that $\sigma_h < \lambda$ only in Cases II and IV.

Lemma 4.7 (Fefferman and Narasimhan). *Let* M *be a real analytic manifold and let* f *be a* $\mathbf{C}$*-valued function on* $M \times \mathbf{C}$. *Suppose that for each fixed* $x \in M$ *the function* $f(x, s)$ *of* $s \in \mathbf{C}$ *is entire and that* f *is real analytic in* (x, s) *on* $M \times \{s \in \mathbf{C} \mid \mathrm{Re}(s) > 0\}$. *Then* f *is real analytic on the whole* $M \times \mathbf{C}$.

For the proof the reader is referred to [FN].

We end this section by proving a higher-dimensional analogue of (3.5) when $\mathrm{Re}(s) > \lambda$.

Proposition 4.8. *Suppose* h *is an invertible element of* X_n. *Then for any compact subset* C *of the domain* $\mathrm{Re}(s) > \lambda$ *and a positive real number* b_0, *there exists a positive constant* A *depending only on* b_0, C, *and* n *such that*

$$|\delta(g)^{2s-\mu}\xi_n(g, h; s)| \le A \cdot \exp\big(-\delta(gh)^{1/n}\big)$$

if $s \in C$ *and* $\sigma(R_a(h)) > b_0$, *where* $\mu = \mu(X_n)$ *and* $a = g^{1/2}$.

PROOF. We give a detailed proof only in Case IV. The other three cases can be handled similarly with some modifications. After the proof in Case IV we shall explain what types of modifications should be made in each case. Thus take ξ_n in Case IV. We have $\lambda = n$ in this case. We prove our assertion by induction on n. If $n = 1$, our assertion follows from (3.8.IV) and (3.5). Suppose $n = r + 1$ with $r > 0$ and the inequality holds for ξ_r. By (2.3) we can reduce our problem to the case $g = 1_n$. Given h, by Lemma 3.1 (2) we may assume that $h = \mathrm{diag}[k, \ell]$ with $0 < k \in \mathbf{R}$ and $\ell \in X_r$. Moreover, we may assume that $k^2 = \sigma(h)$, so that $k^2 \le \sigma(\ell)$ and $k^r \le \delta(\ell)$. In the following we fix b_0 and C, and we assume that $\sigma(h) \ge b_0$ and $s \in C$. We denote by $A_1, A_2, \ldots$ positive constants depending only on b_0, C, and n. Observe that $\delta(h) = k\delta(\ell) \le \delta(\ell)^{n/r}$, and hence $\delta(h)^{1/n} \le \delta(\ell)^{1/r}$. Now by Proposition 4.3 we have

$$\xi_n(1_n, h; s) = \int_S \xi_1(1, k + \ell\{z\}; s - (\iota r/2))\xi_r(1_r + z^*z, \ell; s)dz,$$

where $S = \mathbf{C}_r^1$. Put $w = k + \ell\{z\}$. Then by (3.8.IV) and Proposition 4.3 we obtain

$$\Gamma(s-r)\xi_n(1_n, h; s) = \pi \int_{S'} |\pi w|^{s-r-1} K(s - r - 1, \pi|w|)\xi_r(1_r + z^*z, \ell; s)dz,$$

where $S' = \{ z \in S \mid k + \ell\{z\} \neq 0 \}$. If $s \in C$, then $\mathrm{Re}(s - r - 1) > 0$, and hence by (3.6) the last integral is majorized by

$$A_1 \int_S |\xi_r(1_r + z^*z, \ell; s)|dz. \tag{4.5}$$

Put $g = 1 + z^*z$. Then $\delta(g) = 1 + zz^*$ by (3.1), and $\sigma(g^{1/2}\ell \cdot {}^t g^{1/2}) \ge \sigma(\ell)$ by Lemma 4.1. Therefore our induction assumption implies that

$$|\xi_r(1_r + z^*z, \ell; s)| \le A_2 |(1 + zz^*)^{\mu' - 2s}| \exp\big(-\eta(1 + zz^*)^\beta\big)$$

if $s \in C$ and $\sigma(\ell) \ge b_0$, where $\mu' = \mu(X_r)$, $\beta = 1/r$, and $\eta = \delta(\ell)^\beta$. If $\mathrm{Re}(s) > \lambda$, we easily see that $\mathrm{Re}(2s - \mu') > 0$, and so the integral of (4.5) is less than $A_3 \int_S \exp\big(-\eta(1 + zz^*)^\beta\big)dz$. By Lemma 3.5 this is less than $A_4 e^{-\eta}$ if $\sigma(\ell) \ge b_0$. Since $\eta^n \ge \delta(h)$, we obtain the desired estimate for ξ_n.

Let us now explain what kind of modifications are necessary in Cases I–III. The assertion in the lowest-dimensional cases can be obtained from (3.8.I–III) and (3.5). Therefore our explanation concerns the induction process. We use C as in our Proposition.

Case I: Put $n = 2 + r$ with $r = 2m > 0$. We have $\lambda = (2n - 3)/4$. By Lemma 3.2 we may assume that $h = \mathrm{diag}[k, \ell]$ with $k = b\zeta$ and $\ell = \mathrm{diag}[c_1\zeta, \ldots, c_m\zeta]$, where $\zeta = \begin{bmatrix} 0 & -1 \\ 1 & 0 \end{bmatrix}$ and $b, c_1, \ldots, c_m \in \mathbf{R}^\times$; we may assume that $|b| \le |c_i|$. Then $\delta(h) \le \delta(\ell)^{n/r}$. By Proposition 4.3 we have

$$\xi_n(1_n, h; s) = \int_S \xi_2(1_2, k + y\ell \cdot {}^t y; s - m)\xi_r(1_r + {}^t yy, \ell; s)dy$$

with $S = \mathbf{R}_r^2$. Put $y = [x_1 \ \cdots \ x_m]$ with $x_\nu \in \mathbf{R}_2^2$ and $q(y) = \sum_{\nu=1}^m c_\nu \det(x_\nu)$. Then $k + y\ell \cdot {}^t y = (q(y) + b)\zeta$. Putting $w = q(y) + b$, by (3.8.I) and (3.4) we have

$$\pi^{-1/2}\Gamma(2s - r)\xi_n(1_n, h; s)$$
$$= \int_S |2\pi w|^{2s-r-(1/2)} K(2s - r - (1/2), 2\pi|w|)\xi_r(1_r + {}^t yy, \ell; s)dy.$$

Let $s \in C$. Since $\lambda = (2n - 3)/4$, we have $\mathrm{Re}(2s) - r - (1/2) > 0$. Therefore by (3.6) the proof of Proposition 4.8 can be reduced to the estimate of (4.5). Now $\delta(1_r + {}^t yy) \geq 1 + \tau({}^t yy)$ by Lemma 3.4. Therefore we can reduce our problem to an integral of Lemma 3.5.

Case II: Put $n = r + 1$ with $r > 0$. In this case $\lambda = 2n - (1/2)$. By Lemma 3.2 we may assume that $h = \mathrm{diag}[k, \ell]$ with $k = 2aj$ and $\ell = pj$ with $0 < a \in \mathbf{R}$ and a real diagonal p. By Proposition 4.3 we have

$$\xi_n(1_n, h; s) = \int_S \xi_1(1, k + z\ell z^*; s - 2r)\xi_r(1_r + z^* z, \ell; s)dz$$

with $S = \mathbf{H}_r^1$. Then, using (3.8.II), we obtain the desired estimate.

Case III: Put $n = 2 + r$ with $r = 2m > 0$. In this case $\lambda = n - (3/2)$. By Lemma 3.2 we may assume that $h = \varphi(\mathrm{diag}[b\zeta, c_1\zeta, \ldots, c_m\zeta])E_n$ with ζ as in Case I, and $b, c_1, \ldots, c_m \in \mathbf{C}^\times$. Put $\ell = \varphi(\mathrm{diag}[c_1\zeta, \ldots, c_m\zeta])E_r$ and $k = \varphi(b\zeta)E_2$. By Proposition 4.3 we have

$$\xi_n(1_{2n}, h; s) = \int_S \xi_2(1_4, k + y\ell \cdot {}^t y; s - r)\xi_r(1_{2r} + {}^t yy, \ell; s)dy$$

with $S = \varphi(\mathbf{C}_r^2)E_r$. Put $y = \varphi(z)E_r$, $z = [z_1 \ \cdots \ z_m] \in \mathbf{C}_r^2$, with $z_\nu \in \mathbf{C}_2^2$ and $q(z) = \sum_{\nu=1}^m \overline{c}_\nu \det(z_\nu)$. Then $k + y\ell \cdot {}^t y = \varphi([q(z) + b]\zeta)E_2$ and ${}^t yy = \varphi({}^t z\overline{z})$. Putting $w = q(z) + b$, by (3.8.III) and (3.4) we have

$$\pi^{-1}\Gamma(2s - 2r)\xi_n(1_{2n}, h; s)$$
$$= \int_{\mathbf{C}_r^2} |2\pi w|^{2s-2r-1} K(2s - 2r - 1, 2\pi|w|)\xi_r(1_{2r} + \varphi({}^t z\overline{z}), \ell; s)dz.$$

Therefore this case can be handled in the same manner as in other cases.

5. *H* as a homogeneous space

Let us now define a group G in each case as follows:

(5.0) $$I_n = \begin{bmatrix} 0 & 1_n \\ 1_n & 0 \end{bmatrix}, \quad J_n = \begin{bmatrix} 0 & -1_n \\ 1_n & 0 \end{bmatrix}, \quad L = \begin{bmatrix} 0 & J_n \\ -J_n & 0 \end{bmatrix},$$

(5.1.I) $G = \{\alpha \in GL_{2n}(\mathbf{R}) \mid {}^t\alpha I_n \alpha = I_n\}$,

(5.1.II) $G = \{\alpha \in GL_{2n}(\mathbf{H}) \mid \alpha^* I_n \alpha = I_n\}$,

(5.1.III) $G = \{\alpha \in SL_{4n}(\mathbf{R}) \mid {}^t\alpha I_{2n} \alpha = I_{2n}, \ {}^t\alpha L \alpha = L\}$,

(5.1.IV) $G = Sp(n, \mathbf{C}) = \{\alpha \in GL_{2n}(\mathbf{C}) \mid {}^t\alpha J_n \alpha = J_n\}$.

We note that G in Case III can be given as $\Psi(G')$ with

$$\text{(5.2.III)} \qquad G' = \{\, \alpha \in GL_{2n}(\mathbf{C}) \mid {}^t\alpha I_n \alpha = I_n \,\},$$

$$\Psi\left(\begin{bmatrix} a & b \\ c & d \end{bmatrix}\right) = \begin{bmatrix} \varphi(a) & \varphi(b)E \\ E\varphi(c) & E\varphi(d)E \end{bmatrix} \qquad (a,\, b,\, c,\, d \in \mathbf{C}_n^n),$$

where E and φ are as in (1.5.III).

The group G is not connected in Cases I and III, but connected in the other two cases. Now in all cases define a bijection $z \mapsto z^\rho$ of Z onto itself by $(x + y\mathbf{j})^\rho = x - y\mathbf{j}$ for $x \in X$ and $y \in Y$. Then $z \mapsto -z^\rho$ is a bijection of H onto itself. Clearly $z^\rho = izi^{-1}$ in Case IV and $z^\rho = -z^*$ otherwise. We can now let G act on both H and H^ρ as follows: Given $z \in H$ or $z \in H^\rho$ and $\gamma = \begin{bmatrix} a & b \\ c & d \end{bmatrix} \in G$ with a, b, c, d of the same size, we can define $\gamma z = \gamma(z) \in H$ or $\in H^\rho$ by

$$\text{(5.3)} \qquad \gamma(z) = (az + b)(cz + d)^{-1}.$$

Moreover put $B(z) = \begin{bmatrix} z^\rho & z \\ 1 & 1 \end{bmatrix}$ for $z \in H$. Then we have two factors of automorphy $\lambda(\gamma,\, z)$ and $\mu(\gamma,\, z)$ given by

$$\text{(5.4)} \qquad \gamma B(z) = B(\gamma z)\mathrm{diag}[\lambda(\gamma,\, z),\, \mu(\gamma,\, z)] \qquad (z \in H).$$

In particular, we have $\gamma(z)^\rho = \gamma(z^\rho)$. Both λ and μ have values in $GL_n(\mathbf{R})$, $GL_n(\mathbf{H})$, $GL_{2n}(\mathbf{R})$, and $GL_n(\mathbf{H})$ in Cases I, II, III, and IV, respectively. These facts are elementary; Case IV was discussed in the Appendix of [S2]. We shall justify (5.3) and (5.4) in the other three cases after the proof of Theorem 5.3.

Define a parabolic subgroup P of G and a compact subgroup K of G by

$$\text{(5.5)} \qquad P = \{\, \gamma \in G \mid c_\gamma = 0 \,\}, \quad K = \{\, \alpha \in G \mid \alpha(\mathbf{j}) = \mathbf{j} \,\},$$

where c_γ is the lower left square block of γ which we wrote c in (5.3). Since $P(\mathbf{j}) = H$ as can easily be verified, we obtain $G = PK$. Thus the map $\alpha K \mapsto \alpha(\mathbf{j})$ gives a bijection of G/K onto H. We also define functions η and ε on H and a scalar factor of automorphy $m(\gamma,\, z)$ by

$$\text{(5.6)} \qquad \eta(x + y\mathbf{j}) = y, \quad \varepsilon(x + y\mathbf{j}) = \delta(y), \quad m(\gamma,\, z) = \delta\big(\mu(\gamma,\, z)\big).$$

From (5.4) we can easily derive, for every $\gamma \in G$,

$$\text{(5.7)} \qquad \eta(z) = \mu(\gamma,\, z)^*\eta(\gamma z)\mu(\gamma,\, z), \quad \varepsilon(\gamma z) = m(\gamma,\, z)^{-2}\varepsilon(z).$$

Lemma 5.1. *Given $q \in Q$ and a $\mathbf{C}$-valued polynomial function F on Z of degree e, there exists a positive constant A which depends only on q and F such that $|F(x + y\mathbf{j})| \le A \cdot \delta(x + y\mathbf{j})^e$ for every $x \in X$ and every $y \in Q,> q$.*

Proof. Put $z = x + y\mathbf{j}$. Let μ be the minimum eigenvalue of y. It is sufficient to prove that if $\mu \ge \mu_0 > 0$, then each entry of x or y has absolute value at most $A\delta(z)$ with a constant A depending only on n and μ_0. To prove this we may clearly assume that y is a real diagonal matrix whose diagonal entries are greater than μ_0. Put $g = y^{-1/2}$ and $u = gxg$ in Cases I–III and $u = {}^t gxg$ in Case IV. Then $(u + \mathbf{j})(u + \mathbf{j})^* = 1 + uu^*$, and so $\delta(g^2 z)^2 = \delta(1 + uu^*) \ge 1 + \tau(uu^*)$ by Lemma 3.4. Therefore

$$\delta(z)^2 \geq \delta(y)^2(1+\tau(uu^*)). \tag{5.8}$$

In particular $\delta(y) \leq \delta(z)$. Then every eigenvalue of y is at most $\mu_0^{1-n}\delta(z)$, which is the desired fact for the entries of y. As for x, we see from (5.8) that each entry of $\delta(y)gxg$ has absolute value at most $\delta(z)$. Then μ_0^{1-n} times each entry of x has the same property. This completes the proof.

Lemma 5.2. *Let $\zeta_1, \ldots, \zeta_m$ denote real coordinate functions on Z defined by $z = \sum_{i=1}^m \zeta_i(z)e_i$ for $z \in Z$ with a basis $\{e_1, \ldots, e_m\}$ of Z over $\mathbf{R}$. Then, given every complex polynomial ψ in m variables and $q \in Q$, there exist a real constant σ and a polynomial $A(s)$ in s such that*

$$\int_X \left|\psi(\partial/\partial\zeta_1, \ldots, \partial/\partial\zeta_m)\left[\varepsilon(z)^s\delta(z+v)^{-2s}\right]\right| dv < |A(s)|$$

for $\mathrm{Re}(s) > \sigma$ and every $z = x + y\mathbf{j} \in Z$ such that $y > q$.

PROOF. We can easily see that

$$\psi(\partial/\partial\zeta_1, \ldots, \partial/\partial\zeta_m)\left[\varepsilon(z)^s\delta(z+v)^{-2s}\right] = \sum_{a\in A,\, b\in B} F_{ab}(y,\, z+v,\, s)\delta(y)^{s-a}\delta(z+v)^{-2s-b}$$

with a polynomial F_{ab} in $(y,\, z+v,\, s)$, where A and B are finite sets of non-negative integers. Thus the problem can be reduced to integrals of the form $\int_X F(z+v)\delta(z+v)^{-2s-b}dv$ with a polynomial F. Then Lemma 5.1 reduces the problem to the case in which $F = 1$. Since we know the absolute convergence of (2.1) for $\mathrm{Re}(s) > \lambda$, we obtain our lemma.

Theorem 5.3. *For every $s \in \mathbf{C}$ and every $h \in X$ the functions $\varepsilon(z)^s$ and $\varepsilon(z)^s\delta(z+h)^{-2s}$, of $z = x + y\mathbf{j}$ are eigenfunctions of all G-invariant differential operators on H. The same is true for $\varepsilon(z)^s\mathbf{e}(\tau(hx^*))$ times the function of (4.1) (resp. of (4.3)) with y as g in Case I (resp. Cases II–IV).*

PROOF. For $g = pk \in G$ with $p \in P$ and $k \in K$ put $f(g) = \delta(d_p)^{-2s}$, where d_p is "the d-block" of p. By a well known principle (see [H], p. 303, line 9 from the bottom, for example) f as a function on G/K is an eigenfunction of all G-invariant differential operators. For $z = g(\mathbf{j})$ we have $z = p(\mathbf{j})$ so that $\varepsilon(z)^s = \delta(d_p)^{-2s} = f(g)$. Thus ε^s is an eigenfunction. Put $\gamma = \alpha\begin{bmatrix} 1 & h \\ 0 & 1 \end{bmatrix}$ with a fixed $h \in X$, where $\alpha = I_n$ in Cases I and II, $\alpha = \Psi(I_n)$ in Case III, and $\alpha = J_n$ in Case IV. Then $\varepsilon(z)^s\delta(z+h)^{-2s} = \varepsilon(\gamma z)^s$ by (5.7), and hence $\varepsilon(z)^s\delta(z+h)^{-2s}$ is also an eigenfunction. Now we have

$$\mathbf{e}(\tau(hx^*))\xi_X(y,\, h;\, s) = \int_X \mathbf{e}(\tau[h(x^*-v^*)])\delta(v+y\mathbf{j})^{-2s}dv \tag{5.9}$$

$$= \int_X \mathbf{e}(-\tau(hu^*))\delta(z+u)^{-2s}du,$$

where $z = x + y\mathbf{j}$. By Lemma 5.2 we know that

$$\int_X \mathbf{e}(-\tau(hu^*))\psi(\partial/\partial\zeta_1, \ldots, \partial/\partial\zeta_m)\left[\varepsilon(z)^s\delta(z+u)^{-2s}\right]du$$

is convergent locally uniformly on H, if $\mathrm{Re}(s)$ is sufficiently large. Therefore, since $\varepsilon(z)^s\delta(z+u)^{-2s}$ is an eigenfunction, differentiation of $\varepsilon(z)^s$ times (5.9) gives the expected fact on $\varepsilon(z)^s\mathbf{e}(\tau(hx^*))\xi_X(y,\, h;\, s)$, at least for large $\mathrm{Re}(s)$. We also see that each eigenvalue is the same as that for $\varepsilon(z)^s$, which is clearly a polynomial in s. Since our function in each case is real analytic in $(x,\, y,\, s)$, we obtain the desired result for every $s \in \mathbf{C}$.

It should be noted that a statement similar to the above theorem can be made for the functions $\xi(g,\, h;\, \alpha,\, \beta)$ on hermitian symmetric spaces we studied in [S1] when $\alpha = \beta$. If $\alpha \neq \beta$, we have to consider them relative to a suitable representation of the maximal compact subgroup of the group of transformations. In these hermitian cases, we established in [S1], Theorem 5.9, the real analyticity of some gamma factors times $\xi(g,\, h;\, \alpha,\, \beta)$ in $(g,\, \alpha,\, \beta)$ on the whole domain of definition, and so the function is an eigenfunction with no restriction on $(\alpha,\, \beta)$.

Let us now justify (5.3) and (5.4) in Cases I–III. To define the action of G on H in Case III, we first put

$$\mathfrak{X} = \left\{ y \in \mathbf{R}_{4n}^{4n} \;\middle|\; {}^t y I_{2n} y = \begin{bmatrix} -p & 0 \\ 0 & q \end{bmatrix} \text{ with } p,\, q \in S_{2n}^+, \right. \left. {}^t y L y = \begin{bmatrix} 0 & r \\ {}^t r & 0 \end{bmatrix} \text{ with } r \in \mathbf{R}_{2n}^{2n} \right\},$$

where S_{2n}^+ is the set of all positive definite symmetric matrices of size $2n$. Then we can easily show that the map $(z,\, \lambda,\, \mu) \mapsto B(z)\mathrm{diag}[\lambda,\, \mu]$ gives a bijection of $H \times GL_{2n}(\mathbf{R}) \times GL_{2n}(\mathbf{R})$ onto $\mathfrak{X}$. (The reasoning is similar to and simpler than the proof of [S3], Lemmas 6.2 and 7.10.) Clearly if $y \in \mathfrak{X}$ and $\gamma \in G$, then $\gamma y \in \mathfrak{X}$. Therefore, given $z \in H$ and $\gamma \in G$, we have $\gamma B(z) \in \mathfrak{X}$, so that $\gamma B(z) = B(w)\mathrm{diag}[\lambda,\, \mu]$ with some $w \in H$ and $\lambda,\, \mu \in GL_{2n}(\mathbf{R})$. We then put $w = \gamma z$, $\lambda = \lambda(\gamma,\, z)$, and $\mu = \mu(\gamma,\, z)$. Therefore we obtain (5.3) and (5.4) in Case III. We can also show that $\lambda(\gamma,\, z) = J\mu(\gamma,\, z)J^{-1}$.

In Case II we put

$$\mathfrak{X} = \left\{ y \in \mathbf{H}_{2n}^{2n} \;\middle|\; y^* I_n y = \begin{bmatrix} -p & 0 \\ 0 & q \end{bmatrix} \text{ with } p,\, q \in Q_n \right\}, \tag{5.10}$$

and observe that the map $(z,\, \lambda,\, \mu) \mapsto B(z)\mathrm{diag}[\lambda,\, \mu]$ gives a bijection of $H \times GL_n(\mathbf{H}) \times GL_n(\mathbf{H})$ onto $\mathfrak{X}$. Then we can justify (5.3) and (5.4) by the same technique as in Case III. Case I can be handled by taking $\mathbf{R}$ in place of $\mathbf{H}$.

We end this section by showing how the result of [W] is applicable to our functions $\xi_n(1_{n'},\, h;\, s)$ with invertible h. For this we have to verify that it belongs to the class of functions defined in [W] satisfying the conditions imposed there. This is so for *every* invertible h, which is nontrivial. First of all, we take our G and P to be G and P of [W], §6. Let N be the unipotent radical of P. This consists of the matrices $\begin{bmatrix} 1 & x \\ 0 & 1 \end{bmatrix}$ with $x \in X$. Given an invertible $h \in X$, we put $\eta\left(\begin{bmatrix} 1 & x \\ 0 & 1 \end{bmatrix}\right) = \mathbf{e}(\tau(hx^*))$ for $x \in X$. Then η corresponds to a nondegenerate map ψ of [W], §2. (We can take $\mathrm{diag}[1_n, -1_n]$ to be H there.) The groups M and M_ψ of [W], §6, are given by

$$M = \{\,\mathrm{diag}[a, \tilde{a}] \mid a \in \mathfrak{g}\,\}, \quad M_\psi = \{\,\mathrm{diag}[a, \tilde{a}] \in M \mid a^*ha = h\,\} \quad \text{in Case II},$$
$$M = \{\,\mathrm{diag}[a, \tilde{a}] \mid a \in \mathfrak{g}\,\}, \quad M_\psi = \{\,\mathrm{diag}[a, \tilde{a}] \in M \mid {}^t\bar{a}h\bar{a} = h\,\} \quad \text{otherwise},$$

where $\tilde{a} = (a^*)^{-1}$ in Case II and $\tilde{a} = {}^ta^{-1}$ otherwise, and $\mathfrak{g}$ is the group of (1.10.I–IV). Now for $s \in \mathbf{C}$ define a function f_s on G by $f_s(g) = \delta(\ell_g\ell_g^*)^{-s}$, where ℓ_g is the lower half of the matrix g. Then $f_s(pk) = \delta(d_p)^{-2s}$ for $p \in P$ and $k \in K$, where d_p is the d-block of p. Put $w_0 = \begin{bmatrix} 0 & \varepsilon b \\ \tilde{b} & 0 \end{bmatrix}$ with any fixed $b \in \mathfrak{g}$, where $\varepsilon = -1$ in Case IV and $\varepsilon = 1$ otherwise. Then we see that

$$\int_N \eta(n)^{-1} f_s(w_0 n)dn = \delta(b)^{2s} \int_X \mathbf{e}(-\tau(hx^*))\delta(1 + xx^*)^{-s}dx \tag{5.11}$$

The left-hand side is the integral of [W], Proposition 7.1, and the integral on the right hand side is $\xi_n(1_{n'}, h; s)$, because of (2.8). (Notice that $\mathfrak{a}$ of [W] is one-dimensional in the present case.) Now [W], Theorem 7.2, asserts that (5.11) can be continued to an entire function of s provided G has a finite subset W satisfying the following three conditions:

(1) $G = \bigsqcup_{w \in W} PwM_\psi N$.

(2) The element w_0 of (5.11) is a unique element of W such that $Pw_0M_\psi N$ is open in G. Moreover $Pw_0M_\psi N = Pw_0N$ and $w_0M_\psi w_0^{-1} = M_\psi$.

(3) If $w_0 \neq w \in W$, then η is nontrivial on $N \cap w^{-1}Pw$.

To verify these, we first put

$$\zeta_0 = 1_{2n}, \quad \zeta_r = \begin{bmatrix} 0 & 0 & \varepsilon 1_r & 0 \\ 0 & 1_{n-r} & 0 & 0 \\ 1_r & 0 & 0 & 0 \\ 0 & 0 & 0 & 1_{n-r} \end{bmatrix} \quad \text{if } r > 0.$$

(In Case III, it is simpler to take G' of (5.2.III) instead of G, and so for the purpose of verification of the above three conditions, we can take P, ζ_r, M_ψ, etc. within G'. Then M and M_ψ are defined with $GL_n(\mathbf{C})$ in place of $\mathfrak{g}$.) It is well known that $G = \bigsqcup_{r=0}^n P\zeta_r P$ and $P\zeta_r P = \{\,\alpha \in G \mid \mathrm{rank}(c_\alpha) = r\,\}$ (see [S3], Lemma 2.12).

Let $\mathbf{K} = \mathbf{R}$ in Case I, $\mathbf{K} = \mathbf{H}$ in Case II, and $\mathbf{K} = \mathbf{C}$ in Cases III, IV. Let Q'_r denote the subgroup of $GL_n(\mathbf{K})$ consisting of the matrices whose upper right $r \times (n-r)$-blocks are 0, and let $Q_r = \{\,\mathrm{diag}[a, \tilde{a}] \mid a \in Q'_r\,\}$. Now, by a result of Matsuki [M], Theorem 3, there exists a finite subset C_r of M that represents $Q_r\backslash M/M_\psi$. (Though [M], Theorem 3, concerns a semisimple group and our group M is not semisimple, the result is applicable, since the problem can be reduced to the simple factor of M.) Then we have $P = MN = Q_rC_rM_\psi N$. Since $Q_r \subset \zeta_r^{-1}P\zeta_r$, we have $P\zeta_rP = P\zeta_rC_rM_\psi N$. Thus taking a suitable finite subset W of $\bigcup_{r=0}^n \zeta_rC_r$, we obtain (1). Here ζ_r can be replaced by any element of $P\zeta_r$. Since $P\zeta_rP = \{\,\alpha \in G \mid \mathrm{rank}(c_\alpha) = r\,\}$, $P\zeta_rP$ is open in G if and only if $r = n$, in which case $Q_r = M$ and we can take $C_r = \{1\}$. Take w_0 of (5.11). Since $w_0 \in P\zeta_n$, we have $Pw_0P = P\zeta_nMN = PM\zeta_nN = Pw_0N$.

Now we can find an element b of $\mathfrak{g}$ so that $b^*hb = h^{-1}$ in Case II (cf. Lemma 3.2) and $b^*h\bar{b} = h^{-1}$ in the other three cases. (In Case III, if $h = \varphi(s)E$ with ${}^ts = -s \in \mathbf{C}_n^n$, then take $a \in GL_n(\mathbf{C})$ so that ${}^tasa = \bar{s}^{-1}$ and put $b = \varphi(\bar{a})$.) Then use this b in the above definition of w_0. Then we see that $w_0 M_\psi w_0^{-1} = M_\psi$, which completes the verification of (2).

Finally let $w = p\zeta_r m$ with $p \in P$ and $m \in M$. Then $N \cap w^{-1}Pw = m^{-1}(N \cap \zeta_r^{-1}P\zeta_r)m$. Let us now show that η is nontrivial on $N \cap w^{-1}Pw$ if $r < n$. We first observe that $N \cap \zeta_r^{-1}P\zeta_r$ consists of all the matrices $\begin{bmatrix} 1 & x \\ 0 & 1 \end{bmatrix}$ with $x \in X_n$ such that the upper left $(r \times r)$-block of x is 0. Take Case IV, for example. Put

$$m = \begin{bmatrix} c^{-1} & 0 \\ 0 & {}^tc \end{bmatrix}, \quad c^*h\bar{c} = \begin{bmatrix} a_1 & a_2 \\ {}^ta_2 & a_3 \end{bmatrix}, \quad x^* = \begin{bmatrix} 0 & x_2 \\ {}^tx_2 & x_3 \end{bmatrix},$$

where a_3 and x_3 are of size $n - r$. Then

$$\eta\left(m^{-1}\begin{bmatrix} 1 & x \\ 0 & 1 \end{bmatrix} m\right) = \mathbf{e}(\tau(h(cx \cdot {}^tc)^*)) = \mathbf{e}\Big(\mathrm{Re}(\mathrm{tr}(2 \cdot {}^ta_2x_2 + a_3x_3))\Big).$$

If h is invertible, then so is $c^*h\bar{c}$, and hence $2 \cdot {}^ta_2x_2 + a_3x_3$ can take any element of $\mathbf{C}_{n-r}^{n-r}$ for some x_2 and x_3. Thus condition (3) is satisfied in Case IV. The other three cases can be handled in the same manner.

6. Case V: $G = O(n + 1, 1)$

As mentioned at the beginning, there is one more case in which the group is isomorphic to $O(n+1, 1)$. The spaces X, Y, Z, H, and the group G are given by

(6.1) $$X = \mathbf{R}^n, \quad Y = \mathbf{R}, \quad Q = \{y \in \mathbf{R} \mid y > 0\},$$

(6.2) $$Z = \mathbf{R}^{n+1} = X + Y = \left\{ \begin{bmatrix} y \\ x \end{bmatrix} \,\middle|\, y \in Y, x \in X \right\},$$

(6.3) $$H = \left\{ \begin{bmatrix} y \\ x \end{bmatrix} \,\middle|\, y \in Q, x \in X \right\} = X + Q,$$

(6.4) $$G = \{\alpha \in GL_{n+2}(\mathbf{R}) \mid {}^t\alpha S\alpha = S\}, \quad S = \begin{bmatrix} 0 & 0 & -1/2 \\ 0 & 1_n & 0 \\ -1/2 & 0 & 0 \end{bmatrix}.$$

We put $\begin{bmatrix} y \\ x \end{bmatrix} = x + y\mathbf{j}$ in accordance with the notation in Cases I–IV.

To define the action of G on H, put

(6.5) $$\mathcal{Y} = \{w \in \mathbf{R}^{n+2} \mid {}^twSw < 0\},$$

(6.6) $$p(z) = \begin{bmatrix} y + {}^txx \\ x \\ 1 \end{bmatrix} \ (\in \mathbf{R}^{n+2}) \quad \textit{for} \quad z = x + y\mathbf{j},\ x \in X,\ y \in \mathbf{R}.$$

Then we easily see that the map $(z, \mu) \mapsto \mu \cdot p(z)$ for $z \in H$ and $\mu \in \mathbf{R}^\times$ gives a bijection of $H \times \mathbf{R}^\times$ onto $\mathcal{Y}$. Now let $\alpha \in G$ and $z = x + y\mathbf{j} \in H$ with $x \in X$ and $y \in Q$. Since $\alpha\mathcal{Y} = \mathcal{Y}$, we have $\alpha p(z) = \mu \cdot p(w)$ with $(w, \mu) \in H \times \mathbf{R}^\times$ uniquely determined by α and z. Putting $w = \alpha z$ and $\mu = \mu_\alpha(z)$, we have

(6.7) $$\alpha p(z) = \mu_\alpha(z)p(\alpha z).$$

For such $x+y\mathbf{j}$ we put

$$\delta(x+y\mathbf{j}) = y + {}^t xx, \qquad \varepsilon(x+y\mathbf{j}) = y. \tag{6.8}$$

Then we have $\varepsilon(\alpha z) = \mu_\alpha(z)^{-2}\varepsilon(z)$ for every $\alpha \in G$. The parabolic subgroup P in this case consists of the elements α of G such that $\alpha\mathbf{R}\mathbf{j} = \mathbf{R}\mathbf{j}$. Then $G = PK$ with $K = \{\,\alpha \in G \mid \alpha\mathbf{j} = \mathbf{j}\,\}$.

We now define ξ_X by

$$\xi_X(g,\, h;\, s) = \int_X \mathbf{e}(-{}^t xh)\delta(x+g\mathbf{j})^{-2s}dx \qquad (g \in Q,\, h \in X,\, s \in \mathbf{C}). \tag{6.9}$$

For $\mathrm{Re}(s) > 0$ we have

$$\pi^{-2s}\Gamma(2s)\delta(x+g\mathbf{j})^{-2s} = \int_0^\infty \exp\big(-\pi p(g + {}^t xx)\big)p^{2s-1}dp,$$

and therefore

$$\begin{aligned}\pi^{-2s}\Gamma(2s)\xi_X(g,\, h;\, s) &= \int_0^\infty \exp(-\pi pg)p^{2s-1}\int_X \exp(-\pi p\cdot {}^t xx)\mathbf{e}(-{}^t xh)dx\,dp\\ &= \int_0^\infty \exp\big(-\pi pg - \pi p^{-1}\cdot {}^t hh\big)p^{2s-(n/2)-1}dp.\end{aligned}$$

For $h = 0$ this is a gamma integral, and so we obtain

$$\xi_X(g,\, 0;\, s) = \pi^{n/2}g^{(n/2)-2s}\Gamma\big(2s-(n/2)\big)/\Gamma(2s). \tag{6.10}$$

Put $\|h\| = \big({}^t hh\big)^{1/2}$. Then from (3.4) we obtain

$$\begin{aligned}&\pi^{-2s}\Gamma(2s)\xi_X(g,\, h;\, s)\\ &= \big(\|h\|^2/g\big)^{s-(n/4)}K\big(2s-(n/2),\, \pi g^{1/2}\|h\|\big) \quad \text{if} \quad h \neq 0.\end{aligned} \tag{6.11}$$

Thus, in this case no further investigation of ξ_X is necessary. Every G-invariant differential operator on H is a polynomial of the Laplace-Beltrami operator

$$L = y\sum_{k=1}^n \frac{\partial^2}{\partial x_k^2} + 4y^2\frac{\partial^2}{\partial y^2} + (4-2n)y\frac{\partial}{\partial y}, \tag{6.12}$$

and we can easily verify that the functions $\varepsilon(z)^s$, $\varepsilon(z)^s\delta(z+h)^{-2s}$, and $\varepsilon(z)^s \cdot\mathbf{e}({}^t xh)\xi_X(y,\, h;\, s)$ are eigenfunctions of L. Putting $y = v^2$ with $0 < v \in \mathbf{R}$, we can take $(v,\, x)$ to be another natural set of parameters on H. Though the action of G on H is not rational in $(v,\, x)$, the operator L has a simpler expression; indeed, we have

$$L = v^2\sum_{k=1}^n \frac{\partial^2}{\partial x_k^2} + v^2\frac{\partial^2}{\partial v^2} + (1-n)v\frac{\partial}{\partial v}. \tag{6.13}$$

7. Some integral formulas

In this section we give a few results which will become necessary for the investigation of Eisenstein series and the mass formula on a quaternion unitary group mentioned in the introduction.

Now, we can define a G-invariant measure $d_H z$ on H in all five cases by

$$d_H(x + y\mathbf{j}) = \delta(y)^{-(\kappa+\mu)} dx dy, \tag{7.1}$$

where $\mu = \mu(X)$; we understand that $\kappa + \mu = (n+2)/2$ in Case V.

7.1. Lemma. *For* $z \in H$ *and* $\mathrm{Re}(s) > \kappa + \mu - 1$ *we have*

$$\int_H \varepsilon(w)^s \delta(w - z^\rho)^{-2s} d_H w = c(s)\varepsilon(z)^{-s}$$

with $c(s) = \begin{cases} 2^{n(1+\mu-2s)}\pi^{n\kappa-(n\nu/2)}\Gamma_n^\iota(s-\mu)/\Gamma_n^\iota(s+(\nu/2)) & \text{(Cases I–IV)}, \\ 2^{1-2s}\pi^{(n+1)/2}\Gamma(s-(n/2))/\Gamma(s-(1/2)) & \text{(Case V)}, \end{cases}$

where $(x + y\mathbf{j})^\rho = x - y\mathbf{j}$.

PROOF. We first prove, if $\mathrm{Re}(\beta - \alpha) > \kappa - 1$ and $\mathrm{Re}(\alpha) > \kappa - 1$, then

$$\int_Q \delta(y)^{\alpha-\kappa}\delta(1_n + y)^{-\beta} dy = \frac{\Gamma_n^\iota(\beta-\alpha)\Gamma_n^\iota(\alpha)}{\Gamma_n^\iota(\beta)}. \tag{7.2}$$

Indeed, by (2.5), $\Gamma_n^\iota(\beta)$ times the left-hand side equals

$$\int_Q\int_Q e^{-\tau(p(1+y))}\delta(p)^{\beta-\kappa}dp\,\delta(y)^{\alpha-\kappa}dy$$
$$= \int_Q e^{-\tau(p)}\delta(p)^{\beta-\kappa}\int_Q e^{-\tau(py)}\delta(y)^{\alpha-\kappa}dy\,dp$$
$$= \Gamma_n^\iota(\alpha)\int_Q e^{-\tau(p)}\delta(p)^{\beta-\alpha-\kappa}dp = \Gamma_n^\iota(\beta-\alpha)\Gamma_n^\iota(\alpha),$$

which proves (7.2). Now the integral of our lemma is invariant under the map $z \mapsto z + u$ with any $u \in X$, and so we may assume that $z = p\mathbf{j}$ with $p \in Q$. Then we can easily reduce the problem to the case $p = 1_n$. Then, in Cases I–IV, our integral equals

$$\int_Q\int_X \delta(y)^{s-\kappa-\mu}\delta\bigl(x + (1+y)\mathbf{j}\bigr)^{-2s} dx dy$$
$$= \xi_n(1_n, 0;\, s)\int_Q \delta(y)^{s-\kappa-\mu}\delta(1+y)^{\mu-2s}dy.$$

Therefore we obtain $c(s)$ by combining (7.2) with (2.13). Case V can be handled in the same manner.

Let $\mathbf{K}$ denote either $\mathbf{R}$, $\mathbf{C}$, or $\mathbf{H}$. Let $n = q + r$ with positive integers q and r. Define a group G by

$$G = \bigl\{\alpha \in GL_n(\mathbf{K}) \mid \alpha^* I_{q,r}\alpha = I_{q,r}\bigr\}, \quad I_{q,r} = \mathrm{diag}[1_q, -1_r]. \tag{7.3}$$

For $w, z \in \mathbf{K}_r^q$ we put

$$B(z) = \begin{bmatrix} 1_q & z \\ z^* & 1_r \end{bmatrix}, \quad \xi(w, z) = 1_r - w^*z, \quad \xi'(w, z) = 1_q - wz^*, \tag{7.4}$$

$$\xi(z) = \xi(z, z), \qquad \xi'(z) = \xi'(z, z), \tag{7.5}$$

$$\varepsilon(w, z) = \delta\bigl[\xi(w, z)\bigr], \qquad \varepsilon(z) = \varepsilon(z, z). \tag{7.6}$$

Then we easily see that $\delta[B(z)] = \delta[\xi'(z)] = \varepsilon(z)$, and

$$B(z)^* I_{q,r} B(z) = \mathrm{diag}[\xi'(z),\ -\xi(z)]. \tag{7.7}$$

Therefore $\xi(z) > 0$ if and only if $\xi'(z) > 0$. We now put

$$\mathfrak{B} = \{\, z \in \mathbf{K}_r^q \mid \xi(z) > 0 \,\}. \tag{7.8}$$

We can let G act on $\mathfrak{B}$. Namely, for $\alpha \in G$ and $z \in \mathfrak{B}$ we can define $\alpha z \in \mathfrak{B}$ and two factors of automorphy $\lambda_\alpha(z)$ and $\mu_\alpha(z)$ with values in $GL_n(\mathbf{K})$ by

$$\alpha B(z) = B(\alpha z)\mathrm{diag}[\lambda_\alpha(z),\ \mu_\alpha(z)]. \tag{7.9}$$

This can be shown by the same technique as in Section 5. If $\alpha = \begin{bmatrix} a & b \\ c & d \end{bmatrix} \in G(q, r)$ with a of size q, then $\alpha z = (az+b)(cz+d)^{-1}$ and $\mu_\alpha(z) = cz+d$. We then put

$$m_\alpha(z) = \delta(\mu_\alpha(z)). \tag{7.10}$$

We can easily verify that $m_\alpha(z) = \delta(\lambda_\alpha(z))$ and

$$\mu_\alpha(w)^*\xi(\alpha w,\ \alpha z)\mu_\alpha(z) = \xi(w,\ z), \quad \varepsilon(\alpha w,\ \alpha z)m_\alpha(w)m_\alpha(z) = \varepsilon(w,\ z) \tag{7.11}$$

for every $\alpha \in G$.

To distinguish the present G from the group of Section 5, let us denote the group G of (5.1.I, II) by $\mathcal{G}_n$. (We defined the group only for $\mathbf{K} = \mathbf{R}$ or $\mathbf{K} = \mathbf{H}$, but the case $\mathbf{K} = \mathbf{C}$ can be defined similarly.) We are going to define an injection $G \times G \to \mathcal{G}_n$ and another injection $\mathfrak{B} \times \mathfrak{B} \to H_n$. We first put

$$R = \begin{bmatrix} -2^{-1}1_n & 2^{-1}1_n \\ I_{q,r} & I_{q,r} \end{bmatrix}, \tag{7.12}$$

and observe that $R^* I_n R = \mathrm{diag}[-I_{q,r},\ I_{q,r}]$. Therefore puting

$$[\alpha,\ \beta]_R = R \cdot \mathrm{diag}[\alpha,\ \beta]R^{-1} \quad (\alpha,\ \beta \in G), \tag{7.13}$$

we see that $(\alpha,\ \beta) \mapsto [\alpha,\ \beta]_R$ gives an injection of $G \times G$ into $\mathcal{G}_n$. Put

$$L = \begin{bmatrix} 1_q & 0 & 0 & 0 \\ 0 & 0 & 0 & 1_r \\ 0 & 0 & 1_q & 0 \\ 0 & 1_r & 0 & 0 \end{bmatrix}. \tag{7.14}$$

Observe that $L \cdot \mathrm{diag}[a,\ b,\ c,\ d] = \mathrm{diag}[a,\ d,\ c,\ b]L$ for square matrices a, b, c, and d of size q, r, q, and r, respectively. Then given $(z,\ w) \in \mathfrak{B} \times \mathfrak{B}$, we can find an element $Y \in H_n$ and $M,\ N \in GL_n(\mathbf{K})$ such that

$$R \cdot \mathrm{diag}\big[B(z),\ B(w)\big]L = B(Y) \cdot \mathrm{diag}[M,\ N], \tag{7.15}$$

where $B(Y)$ is the element of $GL_{2n}(\mathbf{K})$ as in (5.4), since the left-hand side belongs to the set of (5.10). Put now $Y = f(z,\ w)$. Then f is an injection of $\mathfrak{B} \times \mathfrak{B}$ into H_n and

$$[\alpha,\ \beta]_R f(z,\ w) = f(\alpha z,\ \beta w), \tag{7.16}$$

$$\varepsilon\big(f(z,\ w)\big) = 2^{-n}\varepsilon(z)\varepsilon(w)\varepsilon(w,\ z)^{-2}, \tag{7.17}$$

where ε on the left-hand side is defined by (5.6). These are similar to [S3], Proposition 6.11, and can be proved in the same manner. A direct caculation shows

$$f(z, w) = \tfrac{1}{2}\begin{bmatrix} 1_q & z \\ w^* & 1_r \end{bmatrix}\begin{bmatrix} 1_q & -z \\ -w^* & 1_r \end{bmatrix}^{-1}. \tag{7.18}$$

7.2. Lemma. *For* $z \in \mathfrak{B}$ *and* $\mathrm{Re}(s) > (\iota n/2) - 1$ *we have*

$$\int_{\mathfrak{B}} \varepsilon(w)^s \varepsilon(w, z)^{-2s} d_{\mathfrak{B}} w = \varepsilon(z)^{-s} \pi^{\iota qr/2} \frac{\Gamma_q^\iota(s + 1 - (\iota/2)(r+1))}{\Gamma_q^\iota(s + 1 - (\iota/2))},$$

where $\iota = [\mathbf{K} : \mathbf{R}]$ *and* $d_{\mathfrak{B}} w = \varepsilon(w)^{-\iota n/2} dw$,

PROOF. By (7.11) we can reduce the problem to the case where $z = 0$. Then the integral in question is essentially the same as [S3], (A2.7.1), if $\mathbf{K} \neq \mathbf{H}$, and so the formula there gives the desired result in such cases. The case $\mathbf{K} = \mathbf{H}$ can be handled in the same manner as noted in [S3], p. 219, lines 4–7 from the bottom.

We assumed $qr > 0$ in the above. We can in fact include the cases $q = 0$ and $r = 0$ by understanding that $I_{q,r} = 1_n$ if $r = 0$ and $I_{q,r} = -1_n$ if $q = 0$. We take $\mathfrak{B}$ to be the set consisting of a single element $\mathbf{1}$ and we let G act on it trivially. Then (7.16) is valid if we put

$$f(\mathbf{1}, \mathbf{1}) = 2^{-1} 1_n. \tag{7.19}$$

References

[FN] C. L. Fefferman and R. Narasimhan, A note on joint analyticity, J. reine angew. Math. **509** (1999), 103–116

[H] S. Helgason, Groups and Geometric Analysis, Academic Press, 1984.

[M] T. Matsuki, The orbits of affine symmetric spaces under the action of minimal parabolic subgroups, J. Math. Soc. Japan, **31** (1979), 331–357.

[R] D. E. Rohrlich, Nonvanishing of L-functions for $GL(2)$, Inv. math. **97** (1989), 381–403.

[S1] G. Shimura, Confluent hypergeometric functions on tube domains, Math. Ann. **260** (1982), 269–302.

[S2] G. Shimura, On the transformation formulas of theta series, Amer. J. Math. **115** (1993), 1011–1052.

[S3] G. Shimura, Euler products and Eisenstein series, CBMS Regional Conference Series in Mathematics, No. 93, Amer. Math. Soc., 1997.

[S4] G. Shimura, An exact mass formula for orthogonal groups, Duke Math. J. **97** (1999), 1–66.

[S5] G. Shimura, Some exact formulas on quaternion unitary groups, J. reine angew. Math. **509** (1999), 67–102

[W] N. R. Wallach, Lie algebra cohomology and holomorphic continuation of generalized Jacquet integrals, in Representations of Lie groups, Adv. Stud. Pure Math. **14**, (1988), 123–151.

99d

Some exact formulas on quaternion unitary groups

Journal für die reine und angewandte Mathematik, 509 (1999), 67-102*

Introduction

The purpose of the present paper is to develop a theory in the symplectic case parallel, to a certain extent, to those we did in the unitary and orthogonal cases in [S5] and [S6]. Since we are mainly interested in the nonsplit case, our group can be given as the unitary group G^φ of a quaternion hermitian form φ over a division quaternion algebra B whose center is an algebraic number field F. To be precise, we take a free left B-module V of rank n, and a B-valued hermitian form $\varphi : V \times V \to B$ such that $\varphi(x, y)^\rho = \varphi(y, x)$ and $\varphi(ax, by) = a\varphi(x, y)b^\rho$ for $a, b \in B$ and $x, y \in V$, where ρ is the main involution of B. Then G^φ consists of all the B-linear automorphisms α of V such that $\varphi(x\alpha, y\alpha) = \varphi(x, y)$.

Though our primary result is an exact mass formula for such a G^φ, we include also some related results such as: (1) the explicit rational form for an Euler factor of a zeta function associated to an automorphic form on G^φ; (2) the explicit rational form for a nonarchimedean Euler factor of the constant term of an Eisenstein series on a group

$$(1) \qquad G^\eta = \{\alpha \in GL_{2n}(B) \mid \alpha\eta\alpha^* = \eta\}, \quad \eta = \eta_n = \begin{bmatrix} 0 & 1_n \\ 1_n & 0 \end{bmatrix}$$

with respect to the standard maximal parabolic subgroup, where $\alpha^* = {}^t\alpha^\rho$.

Let us now describe our exact mass formula. Fixing a maximal order $\mathfrak{o}$ in B, we take an $\mathfrak{o}$-lattice L in V which is maximal among the lattices on which φ takes its values in $\mathfrak{o}$. We then take the genus Λ of L, which consists of all the lattices in V that are equivalent to L with respect to the localizations of G^φ at all nonarchimedean primes. It is well known that there exists a finite set of lattices $\{L_i\}_{i=1}^h$ such that Λ is the disjoint union $\bigsqcup_{i=1}^h \{L_i\alpha \mid \alpha \in G^\varphi\}$. Let $G^\varphi_{\mathbf{a}}$ denote the product of all the archimedean localizations of G^φ, and $\mathcal{Z}^\varphi$ the quotient of $G^\varphi_{\mathbf{a}}$ by its maximal compact subgroup. We can show that $h = 1$ if $G^\varphi_{\mathbf{a}}$ is not compact. Now we define *the mass of the genus* Λ to be the real number $\mathfrak{m}(\varphi, \Lambda)$ given by

$$\mathfrak{m}(\varphi, \Lambda) = \begin{cases} \sum_{a \in \mathcal{B}} [\Gamma^a : 1]^{-1} & \text{if } G^\varphi_{\mathbf{a}} \text{ is compact}, \\ [\Gamma^1 \cap \{\pm 1\} : 1]^{-1} \mathrm{vol}(\Gamma^1 \backslash \mathcal{Z}^\varphi) & \text{otherwise}, \end{cases}$$

where $\Gamma^i = \{\gamma \in G^\varphi \mid L_i\gamma = L_i\}$. Assuming that φ is anisotropic, we give a formula for $\mathfrak{m}(\varphi, \Lambda)$ in an exact form (Theorem 4.7). We prove this for an arbitrary algebraic number field F. To give an idea about its nature, let us assume here that F is totally real and that G^φ_v is compact for every archimedean prime v of F ramified in B. Then $\mathcal{Z}^\varphi$ is isomorphic to the product of b copies of the

Siegel upper space of degree n with a nonnegative integer b; we have $b = 0$ if $G^{\varphi}_{\mathbf{a}}$ is compact. Then

$$(2) \quad \mathfrak{m}(\varphi, \Lambda) = |D_F|^{n^2} \prod_{k=1}^{n} \left\{ D_F^{1/2} \left[(2k-1)!(2\pi)^{-2k} \right]^a \zeta_F(2k) \right\}$$
$$\cdot \left\{ 2^{n^2+n} \pi^{(n^2+n)/2} \prod_{k=1}^{n} \frac{(k-1)!}{(2k-1)!} \right\}^b \cdot \prod_{\mathfrak{p} | \mathfrak{e}} \prod_{k=1}^{n} \{ N(\mathfrak{p})^k + (-1)^k \}.$$

Here D_F denotes the discriminant of F, $a = [F : \mathbf{Q}]$, ζ_F the Dedekind zeta function of F, and $\mathfrak{e}$ the product of all the prime ideals in F ramified in B. If $F = \mathbf{Q}$ and $b = 0$, the formula can be written

$$(3) \qquad \mathfrak{m}(\varphi, \Lambda) = \prod_{k=1}^{n} \frac{|B_{2k}|}{4k} \cdot \prod_{p | \mathfrak{e}} \prod_{k=1}^{n} \{ p^k + (-1)^k \},$$

where B_{2k} denotes the Bernoulli number.

The group Γ^1 when $b = 1$ was introduced by Siegel in [Si]. In that paper he computed the volume of the fundamental domain for $Sp(n, \mathbf{Z})$ and also showed that it was closely related to his product formula for the representation densities formulated for alternating matrices (instead of quadratic forms as he had investigated in his earlier papers). Then he mentioned the possibility of a similar product formula for our group G^{φ} but he said nothing concrete. In fact, he said "But we do not go into detail since the proof of this statement depends upon the analytic theory of hermitian forms, of which no complete account has hitherto been given." Curiously he was silent about the possibility of the volume formula in that case similar to his formula for $Sp(n, \mathbf{Z})$.

Anyway our theorem specialized as above gives $\mathrm{vol}(\Gamma^1 \backslash \mathcal{Z}^{\varphi})$ in an exact form. We also show that the $\mathfrak{p}$-Euler factor of the formula is the representation density at $\mathfrak{p}$ (Proposition 2.12 and the remark after its proof). We do not state, however, the product formula in Siegel's style, since we think that such is unnecessary once the formula for $\mathfrak{m}$ is given in an exact form.

It may also be noted that formula (2) for $n = 1$ is closely related to the result of Eichler in [E] concerning the class number of a totally definite quaternion algebra, which may be called the mass formula for the group $B^{\times}$ for such a B. Since our formula for $n = 1$ concerns the group $\{ x \in B \mid xx^{\rho} = 1 \}$, it is different from Eichler's formula; for one thing, his formula involves the class number of F, but ours does not. If $F = \mathbf{Q}$, the difference disappears. In fact, putting $n = 1$ in (3), we obtain the well-known equality $2\mathfrak{m}(\varphi, \Lambda) = (1/12) \prod_{p | \mathfrak{e}} (p - 1)$; we can remove the factor 2 if we define $\mathfrak{m}$ with $[\Gamma^a : \{\pm 1\}]$ instead of $[\Gamma^a : 1]$, as done in [E]. Thus (3) may be viewed as its generalization to the higher-dimensional case.

Our methods are the same as in the unitary and orthogonal cases in [S5] and [S6]. In [S5] we investigated zeta functions on a unitary group G associated with cusp forms and also an Eisenstein series on G associated with a cusp form on a subgroup of G. The same can be done for the present G^{φ} without great difficulties. For instance, we can associate an Euler product to a holomorphic cusp form on $\mathcal{Z}^\varphi$ in the setting of formula (3) and prove its meromorphic continuation to the whole complex plane. We do not develop such a theory in the present paper, however, except that we determine the rational expression for an Euler factor of the zeta function (Theorem 3.12) and that for an Euler factor of the constant term of an Eisenstein series on G^η (Proposition 3.5) as mentioned at the beginning; we also give the residue of the series at the largest pole in an exact form (Theorem 5.11). There is a new point by which the present paper can be distinguished from [S5] and [S6]: that we take an *arbitrary* number field to be our basic field F, which was totally real both in [S5] and in the essential part of [S6].

1. The basic setting

1.0. Notation. For an associative ring R with identity element and an R-module M we denote by $R^\times$ the group of all invertible elements of R and by M_n^m the R-module of all $m \times n$-matrices with entries in M; we put $M^m = M_1^m$ for simplicity. The identity and zero elements of the ring R_n^n are denoted by 1_n and 0_n (when n needs to be stressed). The transpose of a matrix x is denoted by ${}^t x$. If $x_1, \ldots, x_r$ are square matrices, $\mathrm{diag}[x_1, \ldots, x_r]$ denotes the matrix with $x_1, \ldots, x_r$ in the diagonal blocks and 0 in all other blocks. For a finite set X we denote by $\#X$ or $\#(X)$ the number of elements in X. We denote by $\mathbf{H}$ the Hamilton quaternion algebra, and put

$$\mathbf{T} = \{ z \in \mathbf{C} \mid |z| = 1 \}.$$

1.1. Let B be a quaternion algebra over a field F, by which we understand a central simple algebra over F such that $[B : F] = 4$. We include the case in which B is the matrix algebra F_2^2. Denoting by ρ the main involution of B, we put, for every $x \in B_n^m$ (assumed $m = n$ and invertible if necessary),

$$x^* = {}^t x^\rho, \quad x^{-\rho} = (x^\rho)^{-1}, \quad \widehat{x} = {}^t x^{-\rho}. \tag{1.1}$$

The algebra B can be embedded F-linearly into K_2^2 with a suitable algebraic extension K of F, and so there is an F-linear ring-injection $\psi : B_n^n \to K_{2n}^{2n}$. We then put

$$\lambda(\alpha) = \det[\psi(\alpha)], \qquad \tau(\alpha) = \mathrm{tr}[\psi(\alpha)] \quad \text{for} \quad \alpha \in B_n^n. \tag{1.2}$$

These are the reduced trace and norm maps of B_n^n into F and independent of the choice of K. In particular, if $n = 1$, we have $\lambda(a) = aa^\rho$ and $\tau(a) = a + a^\rho$ for $a \in B$. If $B = \mathbf{H}$, ρ is of course quaternion conjugation. Then, for $y = y^* \in \mathbf{H}_n^n$ we write $y > 0$ if $x^* y x > 0$ for every $x \in \mathbf{H}_1^n, \neq 0$, and $y > z$ if $y - z > 0$.

Let V be a free left B-module of rank n; we write then $n = \dim(V)$, denote by $\mathrm{End}(V, B)$ the ring of all B-linear endomorphisms of V, and put $GL(V, B) = \mathrm{End}(V, B)^\times$. Let $\varphi : V \times V \to B$ be an F-bilinear map such that

$$\varphi(x, y)^\rho = \varphi(y, x), \quad \varphi(ax, by) = a\varphi(x, y)b^\rho \quad (x, y \in V; a, b \in B).$$

Assuming that φ is nondegenerate, we put

$$G^\varphi = G(\varphi) = \{ \alpha \in GL(V, B) \mid \varphi(x\alpha, y\alpha) = \varphi(x, y) \}.$$

To make our formulas short, we put $\varphi[x] = \varphi(x, x)$.

Given another form $\varphi' : V' \times V' \to B$ of the same type, we write $(W, \psi) = (V, \varphi) \oplus (V', \varphi')$ if $W = V \oplus V'$ and $\psi(x + x', y + y') = \varphi(x, y) + \varphi'(x', y')$ for $x, y \in V$ and $x', y' \in V'$. We often take (V, φ) in the form $V = B_n^1$ and $\varphi(x, y) = x\varphi_0 y^*$ for $x, y \in B_n^1$ with $\varphi_0 = \varphi_0^* \in GL_n(B)$. To simplify our notation, we use the same symbol φ for the matrix φ_0, so that $\varphi(x, y) = x\varphi y^*$. In particular, by (H_r, η_r) we shall mean the structure given by

(1.3) $$H_r = B_{2r}^1, \quad \eta_r(x, y) = x\eta_r y^*, \quad \eta_r = \begin{bmatrix} 0 & 1_r \\ 1_r & 0 \end{bmatrix}.$$

If (V, φ) is isomorphic to $(H_r, \eta_r) \oplus (Z, \zeta)$ with some $\zeta : Z \times Z \to B$, then H_r has a B-basis consisting of $2r$ elements e_i, f_i such that

(1.4) $$V = \textstyle\sum_{i=1}^r (Be_i + Bf_i) + Z,$$

(1.5) $$\varphi(e_i, e_j) = \varphi(f_i, f_j) = 0, \quad \varphi(e_i, f_j) = \delta_{ij} \quad \textit{for every } i \textit{ and } j,$$

(1.6) $$Z = \{ v \in V \,|\, \varphi(e_i, v) = \varphi(f_i, v) = 0 \quad \textit{for every } i \}.$$

We call the expression of (1.4) a *weak Witt decomposition;* we call it a *Witt decomposition* if ζ is anisotropic. Here we include the case $Z = \{0\}$. Also we shall often consider the case in which $\dim(Z) = 1$. We then take a nonzero element g of Z and write (1.4) in the form $V = \sum_{i=1}^r (Be_i + Bf_i) + Bg$.

1.2. Let F be an algebraic number field or its completion at a nonarchimedean prime. We denote by $\mathfrak{g}$ the maximal order of F, which is a valuation ring in the local case. Let $\mathfrak{o}$ be a maximal order in B. By an *$\mathfrak{o}$-lattice* (resp. a *$\mathfrak{g}$-lattice*) in V we mean a finitely generated $\mathfrak{o}$-submodule (resp. $\mathfrak{g}$-submodule) of V which spans V over F; by a *$\mathfrak{g}$-ideal* we mean a $\mathfrak{g}$-lattice in F, that is, a fractional ideal in F. Also, by a *two-sided $\mathfrak{o}$-ideal* we mean a $\mathfrak{g}$-lattice $\mathfrak{a}$ in B such that $\mathfrak{o}\mathfrak{a}\mathfrak{o} \subset \mathfrak{a}$. For such an $\mathfrak{a}$ we can show that $x\mathfrak{a} \subset \mathfrak{o}$ if and only if $\mathfrak{a}x \subset \mathfrak{o}$; we denote by $\mathfrak{a}^{-1}$ the set of all such elements x. There is a two-sided $\mathfrak{o}$-ideal $\mathfrak{d}$ determined by

(1.7) $$\mathfrak{d}^{-1} = \{ x \in B \,|\, \tau(x\mathfrak{o}) \subset \mathfrak{g} \}.$$

We note here a basic fact:

(1.8) $$\tau(\mathfrak{o}) = \mathfrak{g}.$$

To prove this, put $\tau(\mathfrak{o}) = \mathfrak{b}$. Then $\mathfrak{b}$ is an integral $\mathfrak{g}$-ideal and $\mathfrak{b}^{-1}\mathfrak{o} \subset \mathfrak{d}^{-1}$, so that $\mathfrak{d} \subset \mathfrak{b}\mathfrak{o}$. It is well known that $\mathfrak{d}^2 = \mathfrak{e}\mathfrak{o}$ with a square free integral $\mathfrak{g}$-ideal $\mathfrak{e}$, and so we must have $\mathfrak{b} = \mathfrak{g}$.

Given an $\mathfrak{o}$-lattice L in V, we denote by $\mu(L)$ the $\mathfrak{g}$-ideal generated by the elements $\varphi[x]$ for all $x \in L$, and by $\mu_0(L)$ the two-sided $\mathfrak{o}$-ideal generated by the elements $\varphi(x, y)$ for all $x, y \in L$. We call L *μ-maximal* (resp. *μ_0-maximal*) (with respect to φ) if there is no $\mathfrak{o}$-lattice L', other than L itself, which contains L and such that $\mu(L') = \mu(L)$ (resp. $\mu_0(L') = \mu_0(L)$). The maximal lattices of these two types were investigated in [S1] and [S2], Section 6. In particular, we showed in [S2], Proposition 6.17, that if $n > 1$, an $\mathfrak{o}$-lattice L is μ-maximal if and only if L is μ_0-maximal and $\mu_0(L) = \mu(L)\mathfrak{d}^{-1}$; these two types of maximality are the same if $n = 1$. In the present paper, however, we consider only μ_0-maximal lattices. Therefore, hereafter we call them simply *maximal.*

1.3. Lemma. *Let L be an $\mathfrak{o}$-lattice in V and let $\mathfrak{a} = \mu_0(L)$. Suppose $x_i \in L$ and $y_i \in \mathfrak{a}^{-1}L$ for $1 \le i \le r$, $\varphi(x_i, x_j) = \varphi(y_i, y_j) = 0$ and $\varphi(x_i, y_j) = \delta_{ij}$ for every i and j. Then $L = \sum_{i=1}^{r}(\mathfrak{o}x_i + \mathfrak{a}y_i) + N$ with $N = L \cap W$, where*

$$W = \{ v \in V \mid \varphi(x_i, v) = \varphi(y_i, v) = 0 \text{ for every } i \}.$$

Moreover, L is maximal if and only if N is maximal among the $\mathfrak{o}$-lattices X in W such that $\mu_0(X) \subset \mathfrak{a}$.

This lemma can be proved easily in the same manner as in [S5], Lemma 4.9.

1.4. Let us now take F to be an algebraic number field. We denote by $\mathbf{a}$ and $\mathbf{h}$ the sets of archimedean and nonarchimedean primes of F; we put $\mathbf{v} = \mathbf{a} \cup \mathbf{h}$; further we denote by $\mathbf{r}$ and $\mathbf{c}$ the sets of real and imaginary archimedean primes of F, so that $\mathbf{a} = \mathbf{r} \cup \mathbf{c}$. For a subset $\mathbf{x}$ of $\mathbf{a}$ we put $|\mathbf{x}| = \#\mathbf{x}$.

Given an algebraic group G defined over F, we define G_v for each $v \in \mathbf{v}$ and the adelization $G_{\mathbf{A}}$ as usual, and view G and G_v as subgroups of $G_{\mathbf{A}}$. We then denote by $G_{\mathbf{a}}$ and $G_{\mathbf{h}}$ the archimedean and nonarchimedean factors of $G_{\mathbf{A}}$, respectively. For $x \in G$ we denote by x_v, $x_{\mathbf{a}}$, and $x_{\mathbf{h}}$ its projections to G_v, $G_{\mathbf{a}}$, and $G_{\mathbf{h}}$. In particular, F_v is the v-completion of F. We denote by $\mathfrak{g}_v$ the closure of $\mathfrak{g}$ in F_v. For $x \in F_{\mathbf{A}}^{\times}$ we put $|x|_{\mathbf{A}} = \prod_{v \in \mathbf{v}} |x_v|_v$ with the normalized valuation $|\ |_v$ at each $v \in \mathbf{v}$.

Given B, V, and a nondegenerate φ as above over the present F, we have naturally B_v, V_v, and G_v^{φ} for each $v \in \mathbf{v}$. Let $\mathbf{b}$ (resp. $\mathbf{d}$) be the set of all $v \in \mathbf{r}$ such that B_v is isomorphic to $(F_v)_2^2$ (resp. B_v is a division algebra). Then $B_{\mathbf{a}}$ is isomorphic to $(\mathbf{R}_2^2)^{\mathbf{b}} \times (\mathbf{C}_2^2)^{\mathbf{c}} \times \mathbf{H}^{\mathbf{d}}$. Similarly we let $\mathbf{h}_u$ (resp. $\mathbf{h}_r$) be the set of all $v \in \mathbf{h}$ such that B_v is isomorphic to $(F_v)_2^2$ (resp. B_v is a division algebra). For a matrix ξ and a ring (or a module) X we write $\xi \prec X$ if all the entries of ξ belong to X. There is a variation of this notation: Given an F-algebra R, a $\mathfrak{g}$-lattice $\mathfrak{a}$ in R, and $\alpha \in R_{\mathbf{A}}$, we write $\alpha \prec \mathfrak{a}$ if α_v has entries in $\mathfrak{a}_v$ for every $v \in \mathbf{h}$.

Let $\mathfrak{o}$ be a maximal order in B and L an $\mathfrak{o}$-lattice in V. For $v \in \mathbf{h}$ we define $\mathfrak{o}_v$ and L_v as usual. The following facts are either well known, or can easily be verified: $\mathfrak{o}_v$ is a maximal order of B_v; $\mu_0(L_v) = \mu_0(L)_v$; L is maximal if and only if L_v is maximal for every $v \in \mathbf{h}$.

Given $\xi \in G_{\mathbf{A}}^{\varphi}$, we denote by $L\xi$ the $\mathfrak{o}$-lattice in V such that $(L\xi)_v = L_v\xi_v$ for every $v \in \mathbf{h}$.

1.5. Let us now take $B = F_2^2$ in the setting of §1.1. Let $\{\varepsilon_{ij}\}$ be the standard matrix units in F_2^2. Given (V, φ) as above, let $W = \varepsilon_{11}V$. Then we can easily show that $\varphi(x, y) \in F\varepsilon_{12}$ if $x, y \in W$. Define an F-bilinear form $f : W \times W \to F$ by

$$\varphi(x, y) = f(x, y)\varepsilon_{12} \qquad (x, y \in W). \tag{1.9}$$

We see that $W\alpha \subset W$ for every $\alpha \in \mathrm{End}(V, B)$. Let α_1 denote the restriction of α to W. Then we can easily verify the following facts: (i) the map $\alpha \mapsto \alpha_1$

gives an F-linear ring-isomorphism of End(V, B) onto End(W, F), which maps $GL(V, B)$ onto $GL(W, F)$. (ii) If φ is nondegenerate, then f is nondegenerate and G^φ is mapped onto the group

$$(1.10)\qquad G^f = \{\alpha \in GL(W, F) \mid f(x\alpha, y\alpha) = f(x, y)\}.$$

Suppose now that F and $\mathfrak{g}$ are as in §1.2, still with $B = F_2^2$; take $\mathfrak{o} = \mathfrak{g}_2^2$. Then $\mathfrak{d} = \mathfrak{o}$. Given an $\mathfrak{o}$-lattice L in V, put $L_1 = \varepsilon_{11}L$. Then we easily see that $\mu_0(L) = \mu(L)\mathfrak{o}$, $L_1 = L \cap W$, and $\mu(L)$ is the $\mathfrak{g}$-ideal generated by $f(x, y)$ for all $x, y \in L_1$. Therefore if L is maximal, then L_1 is maximal with respect to f in the sense that if $L_1 \subset N$ with a $\mathfrak{g}$-lattice N in W and $f(N, N) \subset \mu(L)$, then $L_1 = N$. Suppose L is maximal; put $\mathfrak{b} = \mu(L)$. Then by [S1], Proposition 1.3, or [S5], Lemma 7.6, we can find an F-basis $\{e_1, \ldots, e_n, e'_1, \ldots, e'_n\}$ of W such that

$$(1.11)\qquad f(e_i, e_j) = f(e'_i, e'_j) = 0 \ \text{ and } \ f(e_i, e'_j) = 1 \ \text{ for every } i, j,$$

$$(1.12)\qquad L_1 = \textstyle\sum_{i=1}^n (\mathfrak{g}e_i + \mathfrak{b}e'_i).$$

Next suppose that $\mathfrak{g}$ is a valuation ring; let Y be a totally φ-isotropic B-submodule of V isomorphic to B_r^1 with some r. Put $Y_1 = \varepsilon_{11}Y$ and

$$(1.13)\qquad C = \{\alpha \in G^\varphi \mid L\alpha = L\},\quad C_1 = \{\beta \in G^f \mid L_1\beta = L_1\},$$

$$(1.14)\qquad P_J^\varphi = \{\alpha \in G^\varphi \mid Y\alpha = Y\},\quad P_1 = \{\beta \in G^f \mid Y_1\beta = Y_1\}.$$

Then we easily see that $f(Y_1, Y_1) = 0$, and

$$C_1 = \{\alpha_1 \mid \alpha \in C\},\quad P_1 = \{\alpha_1 \mid \alpha \in P_J^\varphi\}.$$

Now in [S5], Proposition 7.2, we showed that $G^f = P'C'$ with P' and C' which are conjugate to P_1 and C_1. Therefore we have $G^f = P_1C_1$, and consequently

$$(1.15)\qquad G^\varphi = P_J^\varphi C.$$

1.6. We consider B, V, φ, and G^φ still with $B = F_2^2$ but with an arbitrary field F. It is often convenient to take matrix representation of various objects. We first identify B_n^n with F_{2n}^{2n} by viewing an element $[b_{ij}]$ of B_n^n as a matrix of size $2n$ whose (i, j)-block of size 2 is b_{ij}. Put

$$(1.16)\qquad J_n = \begin{bmatrix} 0 & -1_n \\ 1_n & 0 \end{bmatrix},\quad K_n = \mathrm{diag}[J_1, \ldots, J_1]\ (\in F_{2n}^{2n}),$$

$$(1.17)\qquad Sp(n, F) = \{\alpha \in GL_{2n}(F) \mid \alpha J_n \cdot {}^t\alpha = J_n\}.$$

Then we easily see that $\alpha^* = K_n^{-1} \cdot {}^t\alpha K_n$ for $\alpha \in B_n^n$, where ${}^t\alpha$ denotes the transpose of α as a matrix of size $2n$. Thus, for $\varphi \in B_n^n$, we have $\varphi = \varphi^*$ if and only if φK_n is alternating. Therefore, given $\varphi = \varphi^* \in GL_n(B) = GL_{2n}(F)$, we can find an element σ of $GL_n(B)$ such that $\sigma\varphi\sigma^* = J_nK_n$. Then we easily see that

$$(1.18)\qquad \sigma G^\varphi \sigma^{-1} = G(J_nK_n) = Sp(n, F).$$

1.7. Lemma. *Let F, B, $\mathfrak{o}$, and L be as in §1.2; suppose that F is an algebraic number field. Then there exist a B-basis $\{g_i\}$ of V and left $\mathfrak{o}$-ideals $\mathfrak{a}_i$ such that $L = \sum_{i=1}^{n} \mathfrak{a}_i g_i$.*

PROOF. This is obvious if $n = 1$. Suppose $n > 1$; identify V with B_n^1 and put $M = \mathfrak{o}_n^1$. Then $L = M\alpha$ with $\alpha \in GL_n(B)_{\mathbf{A}}$. Take $c \in B_{\mathbf{A}}^{\times}$ so that $\lambda(c) = \lambda(\alpha)$ and put $\beta = \mathrm{diag}[1_{n-1}, c]$; put also $U = \{ y \in GL_n(B)_{\mathbf{A}} \mid M\beta y = M\beta \}$. Since strong approximation holds for $\{ x \in GL_n(B) \mid \lambda(x) = 1 \}$, we have $\beta^{-1}\alpha \in U\gamma$ with some $\gamma \in GL_n(B)$. Then $L = M\alpha = M\beta\gamma = \mathfrak{o}e_1\gamma + \cdots + \mathfrak{o}e_{n-1}\gamma + \mathfrak{o}ce_n\gamma$, where $\{e_i\}$ is the standard basis of B_n^1. Putting $\mathfrak{a}_i = \mathfrak{o}$ for $i < n$, $\mathfrak{a}_n = \mathfrak{o}c$, and $g_i = e_i\gamma$, we obtain our lemma.

2. Local theory

2.1. The objects in this section are nonarchimedean localizations of the global objects in the previous section. To simplify our notation, we fix v in $\mathbf{h}$, and drop the subscript v. Thus F denotes a finite algebraic extension of $\mathbf{Q}_p$ with a prime number p, $\mathfrak{g}$ its valuation ring, and $\mathfrak{p}$ the maximal ideal of $\mathfrak{g}$. We put $q = [\mathfrak{g} : \mathfrak{p}]$, denote by π any unspecified prime element of F, and define a valuation $x \mapsto |x|$ of F by $|\pi| = q^{-1}$. We consider a *division* quaternion algebra B over F and its unique maximal order $\mathfrak{o}$. Then $\mathfrak{d}$ of (1.7) is the unique maximal ideal of $\mathfrak{o}$. It is well known that $\mathfrak{d}^2 = \pi\mathfrak{o}$ and $[\mathfrak{o}/\mathfrak{d} : \mathfrak{g}/\mathfrak{p}] = 2$.

We now consider V and G^{φ} defined with respect to this B and a fixed nondegenerate φ. Since $F^{\times} = \lambda(B^{\times})$, we easily see that there is only one isomorphism class of (V, φ) when $\dim(V)$ is fixed. Thus, given (V, φ), we have (1.4) with $\dim(Z) \le 1$. If $\dim(Z) = 1$, we can put $Z = B$ and $\zeta(x, y) = xy^{\rho}$.

2.2. Lemma. *Let L be a maximal $\mathfrak{o}$-lattice in V. Put $\mathfrak{a} = \mu_0(L)$. Then there exists a Witt decomposition $V = \sum_{i=1}^{m}(Bx_i + By_i) + Bz$ such that $L = \sum_{i=1}^{m}(\mathfrak{o}x_i + \mathfrak{a}y_i) + \mathfrak{o}z$ and $\varphi[z]\mathfrak{g} = \mathfrak{a} \cap F$, where Bz and $\mathfrak{o}z$ must be ignored if $n = 2m$. Conversely, given a Witt decomposition $V = \sum_{i=1}^{m}(Be_i + Bf_i) + Bg$ and a two-sided $\mathfrak{o}$-ideal $\mathfrak{a}$ such that $\varphi[g]\mathfrak{g} = \mathfrak{a} \cap F$, put $L = \sum_{i=1}^{m}(\mathfrak{o}e_i + \mathfrak{a}f_i) + \mathfrak{o}g$; ignore Bg and $\mathfrak{o}g$ if $n = 2m$. Then L is maximal; $\mu_0(L) = \mathfrak{a}$ if $m > 0$, or if $\mathfrak{a} = (\mathfrak{a} \cap F)\mathfrak{o}$.*

This was given in [S1], Proposition 3.5. As an easy consequence of this lemma we observe that if L and L' are maximal $\mathfrak{o}$-lattices in V and $\mu_0(L) = \mu_0(L')$, then $L' = L\alpha$ for some $\alpha \in G^{\varphi}$.

2.3. Proposition. *Let L be a maximal $\mathfrak{o}$-lattice in V given in the form $L = \sum_{i=1}^{m}(\mathfrak{o}e_i + \mathfrak{a}f_i) + \mathfrak{o}g$ with a two-sided $\mathfrak{o}$-ideal $\mathfrak{a}$ and a Witt decomposition $V = \sum_{i=1}^{m}(Be_i + Bf_i) + Bg$ such that $\varphi[g]\mathfrak{g} = \mathfrak{a} \cap F$, where Bg and $\mathfrak{o}g$ must be ignored if $n = 2m$. Let $C = \{ \gamma \in G^{\varphi} \mid L\gamma = L \}$ and let Δ denote the set of elements $\beta \in G^{\varphi}$ such that $e_i\beta = b_i^{-\rho}e_i$, $f_i\beta = b_if_i$ for $1 \le i \le m$ and $g\beta = g$ with elemensts b_i of $B^{\times}$ such that $\mathfrak{o}b_1 \supset \cdots \supset \mathfrak{o}b_m \supset \mathfrak{o}$. Then $G^{\varphi} = C\Delta C$. Moreover $C\beta C$ for such a β is completely determined by the ideals $\mathfrak{o}b_i$, and vice versa. Furthermore $C\alpha C = C\alpha^{-1}C$ for every $\alpha \in G^{\varphi}$.*

The last assertion follows easily from the rest of the proposition, which was given in [S1], Proposition 3.10.

2.4. Throughout this and next sections we put

$$E = E^m = GL_m(\mathfrak{o}). \tag{2.1}$$

We call an element x of B_n^m, $m \geq n$, *primitive* if $[x \;\; y] \in E^m$ with a suitable $y \in B_{m-n}^m$; if $m < n$ we call x *primitive* if x^* is primitive. Given (V, φ) as above, we put $n = \dim(V)$ and fix a maximal $\mathfrak{o}$-lattice L in V such that $\mu_0(L) = \mathfrak{o}$. By Lemma 2.2 we can find a Witt decomposition $V = \sum_{i=1}^m (Be_i + Bf_i) + Bg$ such that $L = \sum_{i=1}^m (\mathfrak{o}e_i + \mathfrak{o}f_i) + \mathfrak{o}g$ and $\varphi[g] = 1$. Here and in the following we ignore g if $n = 2m$. Now, given a positive integer $r \leq m$, we put $t = n - 2r$ and

$$T = \textstyle\sum_{i=r+1}^m (Be_i + Bf_i) + Bg, \quad M = \sum_{i=r+1}^m (\mathfrak{o}e_i + \mathfrak{o}f_i) + \mathfrak{o}g, \tag{2.2}$$

$$C = C^\varphi = \{\gamma \in G^\varphi \,|\, L\gamma = L\}, \tag{2.3}$$

$$P = P_J^\varphi = \{\alpha \in G^\varphi \,|\, J\alpha = J\}, \quad J = \textstyle\sum_{i=1}^r Bf_i. \tag{2.4}$$

Then we have a weak Witt decomposition $V = \sum_{i=1}^r (Be_i + Bf_i) + T$ and $L = \sum_{i=1}^r (\mathfrak{o}e_i + \mathfrak{o}f_i) + M$. We denote by θ the matrix that represents the restriction of φ to $T \times T$ with respect to $\{e_{r+1}, \ldots, e_m, g, f_{r+1}, \ldots, f_m\}$. Taking the union of this basis and $\{e_i, f_i\}_{i=1}^r$ to be a basis of V, we can represent φ by a matrix

$$\varphi_0 = \begin{bmatrix} 0 & 0 & 1_r \\ 0 & \theta & 0 \\ 1_r & 0 & 0 \end{bmatrix}, \quad \theta = \begin{bmatrix} 0 & 0 & 1_{m-r} \\ 0 & 1 & 0 \\ 1_{m-r} & 0 & 0 \end{bmatrix}, \tag{2.5}$$

with respect to the decompositions $V = \sum_{i=1}^r Be_i + T + \sum_{i=1}^r Bf_i$ and $T = \sum_{i=r+1}^m Be_i + Bg + \sum_{i=r+1}^m Bf_i$. If $n = 2m$, we understand that $\theta = \eta_{m-r}$ with η of (1.3). We also identify V and $\mathrm{End}(V, B)$ with B_n^1 and B_n^n with respect to that basis. Then identifying φ with φ_0 of (2.5), we have $\varphi(x, y) = x\varphi y^*$. Clearly $C = G^\varphi \cap E^n$, $\varphi \in E^n$, and $\theta \in E^t$.

2.5. Lemma. (1) *Put* $x_0 = [0_{r+t}^r \;\; 1_r]$. *If x is a primitive element of $\mathfrak{o}_n^r$ such that $x\varphi x^* = 0$, then there exists an element $\gamma \in C$ such that $x = x_0\gamma$.*

(2) $G^\varphi = PC$.

PROOF. For $w \in B_n^r$ we let w_i denote the i-th row of w. Then $f_i = (x_0)_i$. Let x be as in (1). Then $x\varphi$ is primitive, and so we can find $y \in \mathfrak{o}_n^r$ such that $x\varphi y^* = 1_r$. Define $g = [g_{ij}] \in \mathfrak{o}_r^r$ by taking $g_{ij} = (y\varphi y^*)_{ij}$ for $i < j$, $g_{ij} = 0$ for $i > j$, and $g_{ii} \in \mathfrak{o}$ so that $\tau(g_{ii}) = (y\varphi y^*)_{ii}$, which is possible by (1.8). Put $z = y - gx$. Then $x\varphi z^* = 1_r$ and $z\varphi z^* = y\varphi y^* - g - g^* = 0$. By Lemma 1.3, $L = \sum_{i=1}^r (\mathfrak{o}z_i + \mathfrak{o}x_i) + N$ with $N = L \cap W$, where $W = \{w \in V \,|\, w\varphi x^* = w\varphi z^* = 0\}$. Since L is maximal and W has a maximal $\mathfrak{o}$-laltice N' such that $\mu(N') = \mathfrak{o}$, we see that N is maximal and $\mu_0(N) = \mathfrak{o}$. Then we can find a B-linear isomorphism β of T onto W such that $\varphi[v\beta] = \varphi[v]$ for every $v \in T$ and $M\beta = N$. We can extend β to an element γ of G^φ so that $e_i\gamma = z_i$ and $f_i\gamma = x_i$. Then we see that $\gamma \in C$ and $x_0\gamma = x$, which proves (1). To prove (2), let $\alpha \in G^\varphi$. Then $x_0\alpha = dx$ with a primitive x and $d \in GL_r(B)$. Observe that $x\varphi x^* = 0$. Therefore, by (1), $x = x_0\gamma$ with $\gamma \in C$. Then $x_0\alpha\gamma^{-1} = dx_0$, which shows that $\alpha\gamma^{-1} \in P$. Thus $\alpha \in P\gamma \subset PC$. This proves (2).

2.6. For $0 \le m \in \mathbf{Z}$ we define subgroups C^φ_m of C^φ by

$$C_m = C^\varphi_m = \{\gamma \in C^\varphi \mid L(\gamma - 1) \subset \mathfrak{p}^m L\}. \tag{2.6}$$

If $n = 1$, we have clearly

$$C = \{a \in \mathfrak{o}^\times \mid aa^\rho = 1\}, \qquad C_m = C \cap (1 + \mathfrak{p}^m\mathfrak{o}).$$

2.7. Lemma. *Let $W_m = \{\alpha \in \mathfrak{o}^n_n \mid \alpha\varphi\alpha^* - \varphi \prec \mathfrak{p}^m\mathfrak{o}\}$ for $0 < m \in \mathbf{Z}$. Then, for every $\alpha \in W_m$ there exists an element γ of C such that $\alpha - \gamma \prec \mathfrak{p}^m\mathfrak{o}$.*

PROOF. Clearly W_m is a subgroup of E. Given $\alpha \in W_m$, put $\xi = \alpha\varphi\alpha^* - \varphi$ and take $b_{ii} \in \mathfrak{p}^m\mathfrak{o}$ so that $\tau(b_{ii}) = \xi_{ii}$. Define $\beta = [b_{ij}] \in \mathfrak{p}^m\mathfrak{o}^n_n$ with these b_{ii}, $b_{ij} = \xi_{ij}$ for $i < j$, and $b_{ij} = 0$ for $i > j$. Then $\xi = \beta + \beta^*$. Put $\alpha_1 = \alpha - \beta\widehat{\alpha}\varphi^{-1}$. Then $\alpha - \alpha_1 \prec \mathfrak{p}^m\mathfrak{o}$ and $\alpha_1\varphi\alpha_1^* - \varphi \prec \mathfrak{p}^{m+1}\mathfrak{o}$. Repeating this procedure, we can find a sequence $\{\alpha_k\}_{k=0}^\infty$ in E such that $\alpha = \alpha_0$, $\alpha_k - \alpha_{k+1} \prec \mathfrak{p}^{m+k}\mathfrak{o}$, and $\alpha_k\varphi\alpha_k^* - \varphi \prec \mathfrak{p}^{m+k}\mathfrak{o}$. This sequence converges to the desired element γ.

2.8. Lemma. $[C : C_1] = q^2(q+1)$ *if* $n = 1$.

PROOF. In this case $W_1 = \{a \in \mathfrak{o}^\times \mid aa^\rho - 1 \in \mathfrak{p}\}$. Since the map $a \mapsto aa^\rho$ gives a surjective homomorphism of $(\mathfrak{o}/\mathfrak{p}\mathfrak{o})^\times$ onto $(\mathfrak{g}/\mathfrak{p})^\times$, we easily see that $[W_1 : 1 + \mathfrak{p}\mathfrak{o}] = q^2(q+1)$. Now, by the above lemma we have $W_1 = (1 + \mathfrak{p}\mathfrak{o})C$. Since $C_1 = C \cap (1 + \mathfrak{p}\mathfrak{o})$, we obtain our lemma.

2.9. Lemma. *Let χ be a $\mathbf{T}$-valued character of the additive group F such that $\mathfrak{g} = \{a \in F \mid \chi(a\mathfrak{g}) = 1\}$. Put $I(\varphi) = \sum_{s \in \mathfrak{g}/\mathfrak{p}} B_s(\varphi)$, where $B_s(\varphi) = \sum_{x \in L/\mathfrak{p}L} \chi(s\varphi[x]/\pi)$; put $r = [n/2]$ and $t = n - 2r$. Then*

$$I(\varphi) = \begin{cases} q^{4n} + q^{3n+1} - q^{3n} & \text{if } t = 0, \\ q^{4n} - q^{3n+1} + q^{3n} & \text{if } t = 1. \end{cases}$$

PROOF. We can take φ in the form (2.5) with $\theta = 1$ if $t = 1$; we ignore θ if $t = 0$. Clearly $B_s(\varphi) = B_s(\eta_r)B_s(\theta)$, $B_s(\eta_r) = B_s(\eta_1)^r$, and $B_s(\eta_1) = \sum_{a,b \in \mathfrak{o}/\mathfrak{p}\mathfrak{o}} \chi(s\tau(ab^\rho)/\pi)$. Since $\tau(\mathfrak{o}) = \mathfrak{g}$ and $\tau(\mathfrak{d}) = \mathfrak{p}$, we see that $\sum_{a \in \mathfrak{o}/\mathfrak{p}\mathfrak{o}} \chi(s\tau(ac)/\pi)$ for $s \in \mathfrak{g}^\times$ is q^4 or 0 according as $c \in \mathfrak{d}$ or $c \notin \mathfrak{d}$. Therefore $B_s(\eta_1) = q^6$ if $s \notin \mathfrak{p}$. Since $B_0(\varphi) = q^{4n}$, we have

$$I(\varphi) = q^{4n} + \sum_{s \in (\mathfrak{g}/\mathfrak{p})^\times} B_s(\varphi) = q^{4n} + q^{6r} \sum_{s \in (\mathfrak{g}/\mathfrak{p})^\times} B_s(\theta).$$

If $t = 0$, we have $B_s(\theta) = 1$, so that $I(\varphi) = q^{4n} + q^{6r+1} - q^{6r}$. Suppose $t = 1$. Then

$$I(\varphi) = q^{4n} + q^{6r}\big(I(\theta) - q^4\big) = q^{4n} + q^{6r}I(\theta) - q^{3n+1},$$

and $I(\theta) = \sum_{x \in \mathfrak{o}/\mathfrak{p}\mathfrak{o}} \sum_{s \in \mathfrak{g}/\mathfrak{p}} \chi(sxx^\rho/\pi)$. The last sum over $s \in \mathfrak{g}/\mathfrak{p}$ is 0 or q according as $x \notin \mathfrak{d}$ or $x \in \mathfrak{d}$, and hence $I(\theta) = q^3$. Therefore we obtain $I(\varphi)$ as stated in our lemma.

2.10. To state the next lemma, we need some symbols defined as follows:

$$X = \{\, x \in L \,|\, \varphi[x] \in \mathfrak{p} \,\}, \quad X_1 = \{\, x \in X \,|\, x \notin \mathfrak{d}L \,\},$$
$$X_0 = \{\, x \in X_1 \,|\, \varphi[x] = 0 \,\}.$$

For $\xi \in L$ or $\xi \in L \times L$ we denote by $\mu(\xi)$ the element of $L/\mathfrak{p}L$ or $(L/\mathfrak{p}L) \times (L/\mathfrak{p}L)$ represented by ξ; we then put $Y = \mu(X)$, $Y_1 = \mu(X_1)$, $Y_0 = \mu(X_0)$, $A = \#Y$, and $A_0 = \#Y_0$.

2.11. Lemma. *Let r and t be as in Lemma 2.9. If $r > 0$, then*

(2.7) $$\#\{\, \mu(x, w) \,|\, x \in X_0,\, w \in X_0,\, \varphi(x, w) = 1 \,\} = q^{4n-5} A_0,$$

$$A_0 = \begin{cases} q^{2n}(q^{n-1}+1)(q^n-1) & \text{if } t = 0, \\ q^{2n}(q^{n-1}-1)(q^n+1) & \text{if } t = 1. \end{cases}$$

PROOF. In this proof we understand that $M = \{0\}$ and $\theta = 0$ if $t = 0$. For a fixed $x \in X_0$ we want to determine the number of $\mu(w)$ with $w \in X_0$ such that $\varphi(x, w) = 1$. By Lemma 2.5(1) there exists an element $\gamma \in C$ such that $x = e_1\gamma$. Thus we may assume that $x = e_1$. Suppose $w \in X_0$ and $\varphi(e_1, w) = 1$; put $w = \sum_{i=1}^{r}(c_i e_i + d_i f_i) + z$ with $c_i, d_i \in \mathfrak{o}$ and $z \in M$. Then $d_1 = 1$, and so $\tau(c_1) = -\sum_{i=2}^{r}\tau(c_i d_i^{\rho}) - \theta[z]$. Since τ gives a surjective map of $\mathfrak{o}/\mathfrak{p}\mathfrak{o}$ onto $\mathfrak{g}/\mathfrak{p}$, we easily see that the number of $\mu(w)$ is exactly q^{4n-5}. Thus our task is to determine A_0. Put $L' = \sum_{i=1}^{r}(\mathfrak{o}e_i + \mathfrak{o}f_i)$. We first prove that

(*) $$\mu(y+z) \in Y_0 \quad \text{if} \quad y \in L',\, y \notin \mathfrak{d}L',\, z \in M, \text{ and } y + z \in X.$$

In fact, put $y = \sum_{i=1}^{r}(a_i e_i + b_i f_i)$ with $a_i, b_i \in \mathfrak{o}$. Suppose $y + z \in X$ and $y \notin \mathfrak{d}L'$; then we may assume that $a_1 \notin \mathfrak{d}$. We have $\sum_{i=1}^{r}(a_i b_i^{\rho} + b_i a_i^{\rho}) + \theta[z] = c$ with $c \in \mathfrak{p}$. Take $d \in \mathfrak{p}\mathfrak{o}$ so that $\tau(d) = c$. Changing b_1 for $b_1 - d a_1^{-\rho}$, we obtain an element $y' \in L'$ such that $\mu(y+z) = \mu(y'+z)$ and $y' + z \in X_0$. This proves (*). If $t = 0$, we obtain $Y_1 = Y_0$ from (*). Suppose $t = 1$; let $x = y + z \in X_1$ with $y \in L'$ and $z \in M$. If $y \in \mathfrak{d}L'$, then $\theta[z] = x\varphi x^* - y\varphi y^* \in \mathfrak{p}$, and hence $z \in \mathfrak{d}$. Thus $x \in \mathfrak{d}L$, a contradiction. Therefore (*) shows that $\mu(x) \in Y_0$. Thus $Y_1 = Y_0$ in this case too. Consequently $A - A_0 = q^{2n}$. Now observe that for $x \in L$ the sum $\sum_{s \in \mathfrak{g}/\mathfrak{p}} \chi(s\varphi[x]/\pi)$ is q or 0 according as $x \in X$ or $x \notin X$. Therefore qA coincides with the sum $I(\varphi)$ of Lemma 2.9. Combining the value of $I(\varphi)$ given there with the above result on $A - A_0$, we obtain A_0 as stated in our lemma.

2.12. Proposition. $[C^{\varphi} : C^{\varphi}_m] = q^{m(2n^2+n)} \prod_{k=1}^{n} \left[1 - (-q)^{-k}\right]$.

PROOF. By Lemma 2.7, $W_m = C^{\varphi}(1 + \mathfrak{p}^m \mathfrak{o}_n^n)$. Since $C^{\varphi}_m = C^{\varphi} \cap (1 + \mathfrak{p}^m \mathfrak{o}_n^n)$, we have

(2.8) $$[C^{\varphi} : C^{\varphi}_m] = [W_m : 1 + \mathfrak{p}^m \mathfrak{o}_n^n],$$

and hence

(2.9) $$[C^{\varphi}_m : C^{\varphi}_{m+1}] = [W_{m+1} : 1 + \mathfrak{p}^{m+1}\mathfrak{o}_n^n]/[W_m : 1 + \mathfrak{p}^m \mathfrak{o}_n^n].$$

Since $C^{\varphi} \subset W_{m+1}$, we have $W_m = W_{m+1}(1 + \mathfrak{p}^m \mathfrak{o}_n^n)$. Thus the right-hand side of (2.9) equals $[W_{m+1} \cap (1 + \mathfrak{p}^m \mathfrak{o}_n^n) : 1 + \mathfrak{p}^{m+1}\mathfrak{o}_n^n]$. Now for $x \in \mathfrak{p}^m \mathfrak{o}_n^n$ we have $1 + x \in W_{m+1}$ if and only if $x\varphi + \varphi x^* \prec \mathfrak{p}^{m+1}$. Then we easily see that the last group index is q^{2n^2+n}, which gives the value of (2.9). Therefore it is sufficient to prove our proposition when $m = 1$. If $n = 1$, this is given in Lemma 2.8. Thus let us assume $n > 1$, so that $r > 0$. Let $\mu(x, w)$ be an element of the set of (2.7). As shown in the proof of Lemma 2.5, there exists an element γ of C such that $x = e_1\gamma$ and $w = f_1\gamma$. Put

$$C' = \{\alpha \in C \mid e_1\alpha - e_1 \in \mathfrak{p}L,\ f_1\alpha - f_1 \in \mathfrak{p}L\}.$$

Then $C_1 \subset C' \subset C$, and Lemma 2.11 shows that $[C : C'] = q^{4n-5}A_0$. Put $U = \sum_{i>1}(Be_i + Bf_i) + T$. Then we can put $(V, \varphi) = (H_1, \eta_1) \oplus (U, \sigma)$, where σ is the restriction of φ to U. Then φ is of the form (2.5) with 1 and σ in place of 1_r and θ. Write $\alpha \in C'$ in the form

$$\alpha = \begin{bmatrix} a & b & c \\ g & e & f \\ h & \ell & d \end{bmatrix}$$

with a, c, d of size 1, $n-2$, 1 with respect to the decomposition $V = Be_1 \oplus U \oplus Bf_1$. Then we see that the blocks b, c, h, and ℓ have coefficients in $\mathfrak{po}$. Now we have

$$\alpha^{-1} = \varphi\alpha^*\varphi^{-1} = \begin{bmatrix} d^* & f^*\sigma^{-1} & c^* \\ \sigma\ell^* & \sigma e^*\sigma^{-1} & \sigma b^* \\ h^* & g^*\sigma^{-1} & a^* \end{bmatrix}.$$

Since $\sigma \in E^{n-2}$ and $\alpha^{-1} \in C'$, we see that $f \prec \mathfrak{po}$ and $g \prec \mathfrak{po}$. Now $a - 1 \in \mathfrak{po}$ and $d - 1 \in \mathfrak{po}$, and hence $\alpha \in C_1$ if and only if $e - 1 \prec \mathfrak{po}$. If $n = 2$, then $C' = C_1$ and so $[C : C_1] = q^{4n-5}A_0$ with $n = 2$. Suppose $n > 2$; then from the equality $\alpha\varphi\alpha^* = \varphi$ we can derive that $e\sigma e^* - \sigma = gf^* + fg^* \prec \mathfrak{po}$. Therefore by Lemma 2.7 there exists an element $k \in C^\sigma$ such that $k - e \prec \mathfrak{po}$. Clearly $\alpha \in C_1$ if and only if $k \in C_1^\sigma$. Thus, assigning k to α, we obtain an isomorphism of C'/C_1 onto C^σ/C_1^σ. In this way we find that

$$[C : C_1] = q^{4n-5}A_0[C^\sigma : C_1^\sigma].$$

Employing the value of A_0 given in Lemma 2.11, by induction we obtain $[C : C_1]$ as stated in our proposition, since the case $n \le 2$ is already proved.

In view of (2.8), we see that $[C^\varphi : C_1^\varphi]$ is essentially the representation density of φ by itself in Siegel's sense.

3. The series α and local Euler factors

3.1. Our setting is the same as in Section 2. Let K be an unramified quadratic extension of F contained in B and $\mathfrak{r}$ the maximal order of K. Then there exists an element δ of B such that $\delta^2 = \pi$ and $\delta a = a^\rho\delta$ for every $a \in K$. For such a δ we have $B = K + K\delta$, $\mathfrak{o} = \mathfrak{r} + \mathfrak{r}\delta$, and $\mathfrak{d} = \mathfrak{o}\delta$. Then we can define an F-linear injection $\psi : B \to K_2^2$ by

$$\psi(a + b\delta) = \begin{bmatrix} a & b \\ \pi b^\rho & a^\rho \end{bmatrix}. \tag{3.1}$$

Clearly $\psi(\mathfrak{o}) \subset \mathfrak{r}_2^2$. Then we define $\psi_n^m : B_n^m \to K_{2n}^{2m}$ by

$$(3.2)\quad \psi_n^m\left(\begin{bmatrix} c_{11} & \cdots & c_{1n} \\ \cdots & \cdots & \cdots \\ c_{m1} & \cdots & c_{mn} \end{bmatrix}\right) = \begin{bmatrix} \psi(c_{11}) & \cdots & \psi(c_{1n}) \\ \cdots & \cdots & \cdots \\ \psi(c_{m1}) & \cdots & \psi(c_{mn}) \end{bmatrix} \qquad (c_{ij} \in B).$$

Let us now put

$$(3.3)\qquad E = E^m = GL_m(\mathfrak{o}), \quad A = A^m = \mathfrak{o}_m^m \cap GL_m(B).$$

Notice that $\psi_n^n(E^n) \subset GL_{2n}(\mathfrak{r})$. Since $a\mathfrak{o} = \mathfrak{o}a$ for every $a \in B$, we have

$$(3.4)\qquad \xi \prec \mathfrak{o} \Rightarrow a\xi a^{-1} \prec \mathfrak{o} \ \text{ and } \ aEa^{-1} = E \quad \text{for every } a \in B^\times.$$

Now, for $x \in B_n^m$ we can always find elements $c \in \mathfrak{o}_n^m$ and $d \in A^m$ such that $x = d^{-1}c$ and the matrix $[c \ \ d]$ is primitive. Then we call the equality $x = d^{-1}c$ a *left reduced expression for* x, and put

$$(3.5)\qquad \nu_0(x) = \lambda(d)\mathfrak{g}, \qquad \nu(x) = |\lambda(d)|^{-1}.$$

Similarly we call the equality $x = ef^{-1}$ a *right reduced expression for* x if $e \in \mathfrak{o}_n^m$, $f \in A^n$, and $\begin{bmatrix} e \\ f \end{bmatrix}$ is primitive. These are similar to what was done in [S5], Section 3. The present case, though noncommutative, can be treated by the same methods with no new ideas. The easiest, if not the best, way to deal with these is as follows: First define $\nu_0(y)$ for a matrix y with entries in K as in [S5], Section 3, by taking K and $\mathfrak{r}$ there to be the present ones. Then we can easily show that, for $x \in B_n^m$,

$$(3.6)\qquad \nu_0(\psi_n^m(x)) = \nu_0(x)\mathfrak{r}.$$

Therefore we can often reduce our problems to the commutative case.

3.2. Lemma. *Define two formal Dirichlet series $\gamma_n(s)$ and $\beta_{mn}(s)$ by*

$$(3.7)\qquad \gamma_n(s) = \sum_{\alpha \in A/E} |\lambda(\alpha)|^s, \quad \beta_{mn}(s) = \sum_{b \in B_n^m/\mathfrak{o}_n^m} \nu(b)^{-s},$$

where A/E is considered multiplicatively and $B_n^m/\mathfrak{o}_n^m$ additively. Then we have

$$(3.8)\qquad \gamma_n(s) = \prod_{i=1}^n (1 - q^{2i-2-s})^{-1},$$

$$(3.9)\qquad \beta_{mn}(s) = \prod_{i=1}^m \frac{1 - q^{2i-2-s}}{1 - q^{2n+2i-2-s}} = \prod_{i=1}^n \frac{1 - q^{2i-2-s}}{1 - q^{2m+2i-2-s}}.$$

These can be proved in the same manner as in [S5], Lemmas 3.13 and 3.14. Notice that $|\lambda(\alpha)|^{-2} = [\mathfrak{o}_n^1 : \mathfrak{o}_n^1\alpha]$ for $\alpha \in A^n$ and $\gamma_n(2s)$ is the (local) zeta function of B.

3.3. Lemma. (1) *Let $\xi = \begin{bmatrix} p & r \\ 0 & q \end{bmatrix}$ with $p \in B_m^m$, $q \in B_n^n$, and $r \in B_n^m$. Let $p = a^{-1}b$ be a left reduced expression for p and $q = cd^{-1}$ a right reduced expression for q. Then $\nu_0(\xi) = \lambda(a)\lambda(d)\nu_0(ard)$.*

(2) *Let $\eta = [p \ \ r]$ and $\zeta = \begin{bmatrix} r \\ q \end{bmatrix}$ with $p = a^{-1}b$, $q = cd^{-1}$, and r as above. Then $\nu_0(\eta) = \lambda(a)\nu_0(ar)$ and $\nu_0(\zeta) = \lambda(d)\nu_0(rd)$.*

(3) *Let* $\alpha = [a_{ij}]$ *be an upper triangular element of* B_n^n. *Then* $\nu_0(\alpha) \subset \prod_{i=1}^n \nu_0(a_{ii})$.

PROOF. The first two assertions follow from [S5], Propositions 3.9 and 3.10, by means of (3.6). Then (3) follows from (1) by induction on n.

3.4. For $\varepsilon = \pm 1$, or more simply for $\varepsilon = \pm$, let us now put

(3.10) $$S = S_\varepsilon^n = \{ h \in B_n^n \mid h^\rho = \varepsilon h \}, \qquad R = R_\varepsilon^n = \mathfrak{o}_n^n \cap S_\varepsilon^n,$$

We shall often suppress the superscript n and the subscript ε when they are clear from the context.

3.5. Proposition. *Define a formal Dirichlet series* $\alpha(s)$ *by*

(3.11) $$\alpha(s) = \sum_{h \in S/R} \nu(h)^{-s}.$$

Then

$$\alpha(s) = \frac{1 - q^{-s}}{1 - q^{2n-s}} \prod_{i=1}^n \frac{1 - q^{4i-2s}}{1 - q^{2n+2i-1-2s}} \quad \textit{if } \varepsilon = -1,$$

$$\alpha(s) = \prod_{i=1}^n \frac{1 - q^{4i-4-2s}}{1 - q^{2n+2i-3-2s}} \quad \textit{if } \varepsilon = 1.$$

Notice that α for $\varepsilon = -1$ can be written

$$\alpha(s) = \prod_{i=1}^n \frac{(1 - q^{2i-2-s})(1 + q^{2i-s})}{1 - q^{2n+2i-1-2s}}.$$

PROOF. This can be proved in the same manner as for [S5], Proposition 15.4. We indicate here only how we must modify the arguments. Given $c = (c_1, \ldots, c_n)$ with $c_i \in \mathbf{Z}$ such that $0 \le c_1 \le \cdots \le c_n$, put

$$\sigma(c, s) = \begin{cases} \sum_{i=1}^n \{(2n - 2i - s)c_i + [(3c_i + 1)/2]\} & (\varepsilon = -1), \\ \sum_{i=1}^n \{(2n - 2i - s)c_i + [c_i/2]\} & (\varepsilon = 1). \end{cases}$$

Then by the same technique as in [S5], §§15.5–7, we can show that

$$\gamma_n(s)\alpha(s) = \sum_{0 \le c_1 \le \cdots \le c_n} \delta(c_1, \ldots, c_n) q^{\sigma(c, s)},$$

where $\delta(c_1, \ldots, c_n)$ is the number given in [S5], Lemma 15.2, with q^2 in place of q. Putting $t = q^{-s}$, call the last sum $Y_n(t)$. Then the technique of [S5], §15.7, with obvious modifications shows that

$$Y_n(t) = \sum_{i=0}^n g_i^n q^{(2n-\varepsilon)i} t^{2i} Y_i(q^{2n-2i}t) \sum_{k=0}^{n-i} g_k^{n-i} t^k q^{k^2 - k\varepsilon},$$

where $g_r^n = \prod_{i=0}^{r-1} (q^{2n} - q^{2i})/(q^{2r} - q^{2i})$, that is, g_r^n is the number h_r^n of [S5], (15.1.2), defined with q^2 in place of q. Put $W_n(t) = \prod_{i=0}^{n-1} (1 - q^{2i}t)^{-1}$. Then the same type of reasoning as in [S5], pp.124–125, shows that $Y_n(t) = W_n(q^{2n-\varepsilon}t^2)/W_n(-q^{1-\varepsilon}t)$, from which we obtain the desired fromulas for α.

3.6. We now consider the Hecke algebra $\mathfrak{R}(E, GL_m(B))$ consisting of all formal finite sums $\sum c_x ExE$ with $c_x \in \mathbf{Q}$ and $x \in GL_m(B)$, with the law of multiplication defined as in [S5], §11.1. Taking m indeterminates $t_1, \ldots, t_m$, we define a $\mathbf{Q}$-linear map

$$\omega_0 : \mathfrak{R}(E, GL_m(B)) \to \mathbf{Q}[t_1, \ldots, t_m, t_1^{-1}, \ldots, t_m^{-1}] \tag{3.12}$$

as follows: Given ExE with $x \in GL_m(K)$, put $ExE = \bigsqcup_y Ey$ with some y's. For each y we can find an upper triangular z such that $Ey = Ez$, which can be proved by the same technique as in [S5], Proposition 3.5. We may assume that the diagonal entries of z are $\delta^{e_1}, \ldots, \delta^{e_m}$ with $e_i \in \mathbf{Z}$. Then we put

$$\omega_0(ExE) = \sum_y \omega_0(Ey), \quad \omega_0(Ey) = \prod_{i=1}^{m} (q^{-2i}t_i)^{e_i}. \tag{3.13}$$

This is a ring-homomorphism.

3.7. Lemma. *The map ω_0 is injective. Consequently $\mathfrak{R}(E, GL_m(B))$ is commutative.*

PROOF. Given two matrices $x = \mathrm{diag}[\delta^{a_1}, \ldots, \delta^{a_m}]$ and $y = \mathrm{diag}[\delta^{b_1}, \ldots, \delta^{b_m}]$ such that $a_1 \leq \ldots \leq a_m$ and $b_1 \leq \ldots \leq b_m$, let us write $x < y$ if there exists an i such that $a_i < b_i$ and $a_k = b_k$ for $k < i$. Given a finite set X of such diagonal matrices, our task is to show that the $\omega_0(ExE)$ for $x \in X$ are linearly independent. For that purpose, take $x \in X$ so that $x < u$ for every $u \in X, \neq x$. Suppose the monomial $\omega_0(Ex)$ appears in $\omega_0(EyE)$ for some $y \in X$. Put $x = \mathrm{diag}[\delta^{a_1}, \ldots, \delta^{a_m}]$ and $y = \mathrm{diag}[\delta^{b_1}, \ldots, \delta^{b_m}]$. Then EyE contains an upper triangular matrix z whose diagonal entries are $\delta^{a_1}, \ldots, \delta^{a_m}$. Suppose $x \neq y$; then there exists an i such that $a_i < b_i$ and $a_k = b_k$ for $k < i$. We are going to show that this leads to a contradiction. For that purpose put $y' = \delta^{-b_i}y$, $c_k = a_k - b_i$, $d_k = b_k - b_i$, and $z' = \delta^{-b_i}z$. Then, by the principle of (3.4) and Lemma 3.3 (1), (3),

$$\mathfrak{p}^{-d_1-\cdots-d_i} = \nu_0(y') = \nu_0(z') \subset \mathfrak{p}^{-c_1-\cdots-c_i},$$

a contradiction. This fact, together with induction on the number of elements of X, proves the desired injectivity.

3.8. We now consider G^φ and C as in §2.4 and define P_J^φ by (2.4) with a fixed integer $r \leq n/2$. With respect to the basis of V chosen there, we can put

$$G^\varphi = \{\alpha \in GL_n(B) \mid \alpha\varphi\alpha^* = \varphi\}, \quad C = G^\varphi \cap E^n, \tag{3.14}$$

where we write φ for the matrix φ_0 of (2.5). Given θ as in (2.5), we can find an element $\mu \in \mathfrak{o}_t^\dagger$ such that

$$\theta = \mu + \mu^*. \tag{3.15}$$

Indeed, take an element w of $\mathfrak{o}$ so that $w + w^\rho = 1$, which is possible by virtue of (1.8). Then (3.15) is satisfied with

$$\mu = \begin{bmatrix} 0 & 0 & 1_{m-r} \\ 0 & w & 0 \\ 0 & 0 & 0 \end{bmatrix}, \tag{3.16}$$

if n is odd. The case of even n is easier.

3.9. Lemma. (1) *P_J^φ consists of the elements of the form*

$$\xi = \begin{bmatrix} a & b & c \\ 0 & e & f \\ 0 & 0 & d \end{bmatrix} \tag{3.17}$$

with $\widehat{a} = d \in GL_r(B)$, $e \in G^\theta$, $b \in B_t^r$, $f = -e\theta b^* d$, and $c = (s - b\mu b^*)d$, $s \in S_-^r$.

(2) *Given ξ as above, suppose that $r > 0$; suppose also that $e \prec \mathfrak{o}$ if $t > 0$. (If $t = 0$, ignore θ, μ, b, e, and f, and put $c = sd$.) Let $a = g^{-1}h$ be a left reduced expression for a, and $gb = j^{-1}k$ a left reduced expression for gb. Then*

$$\nu_0(\xi) = \lambda(ghj^2)\nu_0(jgsg^*j^*),$$

where we take $j = 1_r$ if $t = 0$.

This is an analogue of [S5], Lemma 2.10 and Proposition 16.10, and can be proved in the same manner by means of Lemma 3.3.

3.10. Putting $m = [n/2]$ and $n = 2m + t$, we take $r = m$ in the setting of §3.8. Thus $t \leq 1$. We then consider the Hecke algebra $\mathfrak{R}(C, G^\varphi)$ consisting of all formal finite sums $\sum c_\sigma C\sigma C$ with $c_\sigma \in \mathbf{Q}$ and $\sigma \in G^\varphi$, with the usual law of multiplication. Taking m indeterminates $t_1, \ldots, t_m$, we define a $\mathbf{Q}$-linear map

$$\omega : \mathfrak{R}(C, G^\varphi) \to \mathbf{Q}[t_1, \ldots, t_m, t_1^{-1}, \ldots, t_m^{-1}] \tag{3.18}$$

as follows: Given $C\sigma C$ with $\sigma \in G^\varphi$, we can put $C\sigma C = \bigsqcup_\xi C\xi$ with $\xi \in P$ of the form (3.17), by Lemma 2.5 (2). We then put

$$\omega(C\sigma C) = \textstyle\sum_\xi \omega(C\xi), \qquad \omega(C\xi) = \omega_0(E^m d_\xi), \tag{3.19}$$

where ω_0 is given by (3.12) and d_ξ is the d-block of ξ in (3.17). It can easily be verified that this is well-defined, and gives a ring-homomorphism. We observe that the formal sum $\sum_\xi E^m d_\xi$ is invariant under right multiplication by the elements of E^m. Therefore we can find a subset $\{\alpha\}$ of $\{\xi\}$ such that $\sum_\xi E^m d_\xi$ can be identified with $\sum_\alpha E^m d_\alpha E^m$ in an obvious sense. Putting $\eta(C\sigma C) = \sum_\alpha E^m d_\alpha E^m$, we thus obtain a $\mathbf{Q}$-linear map η of $\mathfrak{R}(C, G^\varphi)$ into $\mathfrak{R}(E^m, GL_m(B))$ with the property that $\omega = \omega_0 \circ \eta$.

By Proposition 2.3, $C\backslash G^\varphi/C$ can be given by the matrices $\sigma_z = \mathrm{diag}[\widehat{z}, 1_t, z]$ with $\widehat{z}$ in $E^m\backslash A^m/E^m$, and $C\alpha C = C\alpha^{-1}C$ for every $\alpha \in G^\varphi$.

3.11. Proposition. *The map ω is injective. Consequently $\mathfrak{R}(C, G^\varphi)$ is commutative and the map η is injective.*

Proof. We use the same notation as in the proof of Lemma 3.7. Given x and y in that proof with $a_m \leq 0$ and $b_m \leq 0$, suppose that $a_i < b_i$ for some i and $a_k = b_k$ for $k < i$; suppose also that the monomial $\omega(C\sigma_x)$ appears in $\omega(C\sigma_y C)$. Then $C\sigma_y C$ contains an element ξ of P such that $\omega_0(E^m d_\xi) = \omega_0(E^m x)$. Put

$\alpha = \delta^{-b_i}\sigma_y$, $\beta = \delta^{-b_i}\xi$, $c_\nu = a_\nu - b_i$ and $d_\nu = b_\nu - b_i$. Then $\beta \in E^n\alpha E^n$, and so by Lemma 3.3(3) we have

$$\mathfrak{p}^{-d_1-\cdots-d_i} = \nu_0(\alpha) = \nu_0(\beta) \subset \mathfrak{p}^{-c_1-\cdots-c_i},$$

a contradiction. Therefore we obtain the desired injectivity for the same reason as in Lemma 3.7.

3.12. Theorem. *Let $n = 2m + t$ with $m = [n/2]$; define a formal Dirichlet series $\mathfrak{T}(s)$ with coefficients in $\mathbf{Q}[t_1, \ldots, t_m, t_1^{-1}, \ldots, t_m^{-1}]$ by*

$$\mathfrak{T}(s) = \sum_{\tau \in A} \omega(C\sigma C)\nu(\sigma)^{-s}, \qquad A = C\backslash G^\varphi/C. \tag{3.20}$$

Then

$$\mathfrak{T}(s) = \frac{1 - q^{-s}}{1 - q^{2m-s}} \prod_{i=1}^{m} \frac{1 - q^{4i-2s}}{(1 - q^{n+t-1-s}t_i)(1 - q^{2m-s}t_i^{-1})}.$$

Proof. This can be proved in the same manner as in [S5], Theorem 16.16. Indeed, the proof there modified in an obvious way shows that

$$\mathfrak{T}(s) = \alpha(s)\beta_{mt}(2s - 2m - 1)\mathfrak{A}(s - n - 1 - t, s),$$

where α is the series of (3.11) with $\varepsilon = -1$, β_{mt} is defined by (3.7) (understood to be 1 if $t = 0$), and

$$\mathfrak{A}(s', s) = \sum_{x \in T} \omega_0(E^m x)\nu(x^{-1})^{-s'}\nu(x)^{-s}, \quad T = E^m\backslash GL_m(B).$$

By the same technique as in the proof of [S5], Lemma 16.5, we can show that

$$\mathfrak{A}(s', s) = \prod_{i=1}^{m} \frac{1 - q^{2i-2-s-s'}}{(1 - q^{-2-s'}t_i)(1 - q^{2m-s}t_i^{-1})}.$$

Combining this with the formulas of (3.9) and Proposition 3.5, we obtain the desired expression for $\mathfrak{T}$.

4. Main results

4.1. We take B and(V, φ) over an algebraic number field F and define $\mathbf{b}$, $\mathbf{c}$, $\mathbf{d}$, $\mathbf{h}_u$, and $\mathbf{h}_r$ as in §1.4. We denote by $\mathfrak{e}$ the product of all prime ideals for which $v \in \mathbf{h}_r$. We fix a maximal order $\mathfrak{o}$ in B and also a B-basis $\{e_i\}$ of V, and represent φ and the elements of $\mathrm{End}(V, B)$ by matrices of B_n^n. For $v \in \mathbf{b}\cup\mathbf{c}\cup\mathbf{h}_u$ we identify B_v with $(F_v)_2^2$ once for all. In particular, for $v \in \mathbf{h}_u$ we take the identification so that $\mathfrak{o}_v = (\mathfrak{g}_v)_2^2$. For $\alpha \in B_n^n$ and $v \in \mathbf{v}$ we have a well-defined element α_v of $(B_v)_n^n$. If $v \in \mathbf{b} \cup \mathbf{c} \cup \mathbf{h}_u$, we view it as an element of $(F_v)_{2n}^{2n}$ in the manner explained in §1.6. We put

$$\delta_v(z) = \begin{cases} |\det(z)| & (z \in \mathbf{C}_m^m,\ v \in \mathbf{b}), \\ \lambda_m(z) & (z \in \mathbf{H}_m^m,\ v \in \mathbf{c} \cup \mathbf{d}), \end{cases} \tag{4.1}$$

where λ_m denotes the reduced norm map $\mathbf{H}_m^m \to \mathbf{R}$. Thus $\delta_v(a1_m) = a^{2m}$ for $a \in \mathbf{R}$ if $v \in \mathbf{c} \cup \mathbf{d}$. For $y = y^* \in \mathbf{H}_n^n$ we write $y > 0$ if $x^*yx > 0$ for every $x \in \mathbf{H}_1^n$, $\neq 0$, and $y > z$ if $y - z > 0$.

To define the action of G^φ on a symmetric space, we first make a preliminary discussion when $v \in \mathbf{d}$. Given $n = q + r$ with nonnegative integers q and r, we define a group $G(q, r)$ and a space $\mathfrak{B}(q, r)$ by

$$G(q, r) = \{ \alpha \in GL_n(\mathbf{H}) \mid \alpha^* I_{q,r} \alpha = I_{q,r} \}, \quad I_{q,r} = \mathrm{diag}[1_q, -1_r], \tag{4.2}$$

$$\mathfrak{B}(q, r) = \{ z \in \mathbf{H}_r^q \mid 1_q - zz^* > 0 \}. \tag{4.3}$$

Assuming $qr > 0$ for the moment, for $z, w \in \mathfrak{B}(q, r)$ and $\alpha = \begin{bmatrix} a & b \\ c & d \end{bmatrix} \in G(q, r)$ with a of size q, we put

$$\Delta(w, z) = \delta_v(1_q - wz^*), \qquad \Delta(z) = \Delta(z, z), \tag{4.4}$$

$$\alpha z = (az + b)(cz + d)^{-1}, \qquad m(\alpha, z) = \delta_v(cz + d). \tag{4.5}$$

We understand that $I_{q,r} = 1_n$ if $r = 0$ and $I_{q,r} = -1_n$ if $q = 0$; in these cases we understand that $\mathfrak{B}(q, r)$ consists of a single element $\mathbf{1}$ and we let $G(q, r)$ act on it trivially; we can define m and Δ in an appropriate manner even when $qr = 0$, but we don't need them.

For each $v \in \mathbf{a}$ we fix an element $\sigma_v \in GL_n(B_v)$ so that

$$\sigma_v \varphi_v \sigma_v^* = \begin{cases} J_n K_n & (v \in \mathbf{b} \cup \mathbf{c}), \\ \mathrm{diag}[1_{q_v}, -1_{r_v}] & (v \in \mathbf{d}) \end{cases} \tag{4.6}$$

with J_n, K_n of (1.16) and nonnegative integers q_v, r_v. The existence of such a σ_v for $v \in \mathbf{d}$ is obvious. For $v \in \mathbf{b} \cup \mathbf{c}$ it is explained in §1.6. Then we have

$$\sigma_v G_v^\varphi \sigma_v^{-1} = \begin{cases} Sp(n, F_v) & (v \in \mathbf{b} \cup \mathbf{c}), \\ G(q_v, r_v) & (v \in \mathbf{d}). \end{cases}$$

We then put

$$\mathcal{Z}^\varphi = \prod\nolimits_{v \in \mathbf{a}} \mathcal{Z}_v^\varphi,$$

$$\mathcal{Z}_v^\varphi = \{ z \in \mathbf{C}_n^n \mid {}^t z = z,\ i(z^* - z) > 0 \} \qquad (v \in \mathbf{b}),$$

$$\mathcal{Z}_v^\varphi = \{ x + yj \in \mathbf{H}_n^n \mid {}^t x = x \in \mathbf{C}_n^n,\ 0 < y = y^* \in \mathbf{C}_n^n \} \qquad (v \in \mathbf{c}),$$

$$\mathcal{Z}_v^\varphi = \mathfrak{B}(q_v, r_v) \qquad (v \in \mathbf{d}),$$

where j is the standard quaternion unit such that $j^2 = -1$ and $ji = -ij$. For $v \in \mathbf{b} \cup \mathbf{c}$ we can let $Sp(n, F_v)$ act on $\mathcal{Z}_v^\varphi$ as usual (see [S4], Appendix, for details in the case $F_v = \mathbf{C}$). Namely, for $z \in \mathcal{Z}_v^\varphi$ and $\alpha = \begin{bmatrix} a & b \\ c & d \end{bmatrix} \in Sp(n, F_v)$ with a of size n, we put $\alpha z = (az + b)(cz + d)^{-1}$. Now defining z' for $z \in \mathcal{Z}_v^\varphi$, $v \in \mathbf{b} \cup \mathbf{c}$, by $(x + yi)' = x - yi$ if $v \in \mathbf{b}$ and $(x + yj)' = x - yj$ if $v \in \mathbf{c}$, we put, for $z, w \in \mathcal{Z}_v^\varphi$, $v \in \mathbf{b} \cup \mathbf{c}$, and α as above,

$$\Delta(w, z) = \delta_v(w - z'), \quad \Delta(z) = \delta_v(2^{-1}(z - z')), \tag{4.7}$$

$$m(\alpha, z) = \delta_v(cz + d). \tag{4.8}$$

Thus the action of $\sigma_v G_v^\varphi \sigma_v^{-1}$ on $\mathcal{Z}_v^\varphi$ is defined for every $v \in \mathbf{a}$. We then let $G_\mathbf{A}^\varphi$ act on $\mathcal{Z}^\varphi$ via the homomorphism $\alpha \mapsto (\sigma_v \alpha_v \sigma_v^{-1})_{v \in \mathbf{a}}$ of $G_\mathbf{A}^\varphi$ onto $\prod_{v \in \mathbf{a}} \sigma_v G_v^\varphi \sigma_v^{-1}$, ignoring $\alpha_\mathbf{h}$. Moreover, for $z, w \in \mathcal{Z}^\varphi$ and $\alpha \in G_\mathbf{A}^\varphi$ we put

$$(4.9)\qquad \Delta(z) = \prod\nolimits_{v\in\mathbf{a}'}\Delta(z_v), \qquad \Delta(z, w) = \prod\nolimits_{v\in\mathbf{a}'}\Delta(z_v, w_v),$$

$$(4.10)\qquad m(\alpha, z) = \prod\nolimits_{v\in\mathbf{a}'} m(\sigma_v\alpha_v\sigma_v^{-1}, z_v),$$

$$(4.10\text{a})\qquad \mathbf{a}' = \mathbf{b}\cup\mathbf{c}\cup\mathbf{d}', \qquad \mathbf{d}' = \{\, v\in\mathbf{d} \mid q_v r_v > 0 \,\}.$$

Thus $\Delta(z) = \Delta(z, w) = m(\alpha, z) = 1$ if $\mathbf{a}' = \varnothing$, that is, if $G^{\varphi}_{\mathbf{a}}$ is compact.

4.2. Hereafter we fix a maximal $\mathfrak{o}$-lattice L in V such that $\mu_0(L) = \mathfrak{o}$. We also take an arbitrary integral $\mathfrak{g}$-ideal $\mathfrak{c}$. With the basis $\{e_i\}$ as in §4.1, we put $L' = \sum_{i=1}^{n}\mathfrak{o}e_i$. We take and fix an element $\sigma\in GL_n(B)_{\mathbf{A}}$ so that $L'\sigma = L$ and σ_v for $v\in\mathbf{a}$ coincides with that of (4.5). We then put

$$(4.11)\qquad \varphi' = \sigma\varphi\sigma^*,$$

$$(4.12)\qquad C^{\varphi} = \{\, \alpha\in G^{\varphi}_{\mathbf{A}} \mid L\alpha = L \,\},$$

$$(4.13)\qquad D^{\varphi} = \{\, \gamma\in C^{\varphi} \mid L_v(\gamma_v - 1)\subset \mathfrak{c}_v L_v \ \text{ for every } v|\mathfrak{c} \,\}.$$

For a subgroup X of $GL_n(B)_{\mathbf{A}}$, we denote by X_v its projection to $GL_n(B_v)$. In fact, in all cases in which this notation is used we have $X_v = X\cap GL_n(B_v)$. Observe that

$$C^{\varphi}_v = G^{\varphi}_v \cap \sigma_v^{-1}GL_n(\mathfrak{o}_v)\sigma_v, \qquad D^{\varphi}_v = \{\, \gamma\in C^{\varphi}_v \mid (\sigma\gamma\sigma^{-1} - 1)_v \prec \mathfrak{c}_v \,\}.$$

With σ as above, for $x\in G^{\varphi}_{\mathbf{A}}$ we define an integral $\mathfrak{g}$-ideal $\nu^{\sigma}(x)$ by

$$(4.14)\qquad \nu^{\sigma}(x)_v = \nu_0\big((\sigma x\sigma^{-1})_v\big) \quad \text{for every } \ v\in\mathbf{h}.$$

Here $\nu_0(y)$ for $y\in(B_v)^n_n$ is defined by (3.5) if $v|\mathfrak{e}$. If $v\nmid\mathfrak{e}$ then we consider y an element of $(F_v)^{2n}_{2n}$ as explained in §4.1, and $\nu_0(y)$ is the denominator ideal defined in [S5], (3.7.1). Clearly $\nu^{\sigma}(x)$ depends only on $C^{\varphi}xC^{\varphi}$.

4.3. Lemma. *Define a subgroup $\mathfrak{X}$ of $G^{\varphi}_{\mathbf{A}}$ and a subgroup Ξ of G^{φ} by*

$$(4.15)\qquad \mathfrak{X} = \{\, y\in G^{\varphi}_{\mathbf{A}} \mid y_v\in D^{\varphi}_v \ \text{ for every } v|\mathfrak{c} \,\}, \qquad \Xi = \mathfrak{X}\cap G^{\varphi}.$$

For $a\in G^{\varphi}_{\mathbf{A}}$ put $\Gamma^a = G^{\varphi}\cap aD^{\varphi}a^{-1}$; let R^a be a complete set of representatives for $\Gamma^a\backslash\Xi$. Then the following assertions hold:

(1) There exists a finite subset $\mathcal{B}$ of $G^{\varphi}_{\mathbf{h}}$ such that $G^{\varphi}_{\mathbf{A}} = \bigsqcup_{a\in\mathcal{B}} G^{\varphi}aD^{\varphi}$ and $a_v = 1$ for every $a\in\mathcal{B}$ and every $v|\mathfrak{c}$.

(2) $D^{\varphi}\backslash\mathfrak{X}$ can be given by $\bigsqcup_{a\in\mathcal{B}} a^{-1}R^a$.

PROOF. Assertion (1) can be proved in the same manner as in [S5], Lemma 8.12. The key point is that weak approximation holds for G^{φ}, as proved in [S2], Proposition 6.3. Assertion (2) can easily be verified by observing that $\mathfrak{X} = \bigsqcup_{a\in\mathcal{B}} D^{\varphi}a^{-1}\Xi$.

4.4. Lemma. *If $G^{\varphi}_{\mathbf{a}}$ is not compact, then strong approximation holds for G^{φ}, that is, $G^{\varphi}_{\mathbf{A}} = G^{\varphi}Y$ for every open subgroup Y of $G^{\varphi}_{\mathbf{A}}$.*

This was given in [S2], Theorem 6.7.

Given any Hecke character χ of F, we put

$$(4.16)\qquad L_{\mathfrak{c}}(s, \chi) = \prod\nolimits_{\mathfrak{p}\nmid\mathfrak{c}}\big(1 - \chi^*(\mathfrak{p})N(\mathfrak{p})^{-s}\big)^{-1},$$

$$(4.17)\qquad \zeta_{F,\mathfrak{c}}(s) = \prod\nolimits_{\mathfrak{p}\nmid\mathfrak{c}}\big(1 - N(\mathfrak{p})^{-s}\big)^{-1},$$

where $\mathfrak{p}$ runs over all the prime ideals in F prime to $\mathfrak{c}$, and χ^* denotes the ideal character attached to χ. We denote these by $L(s, \chi)$ and ζ_F if $\mathfrak{c} = \mathfrak{g}$.

4.5. Proposition. *Put* $m = [n/2]$ *and*

$$\mathcal{T}(s) = \sum_{\tau \in D^\varphi \backslash \mathfrak{X}} N(\nu^\sigma(\tau))^{-s},$$

$$\Lambda'_\mathfrak{c}(s) = \zeta_{F,\mathfrak{c}}(s) \prod_{j=1}^{n} \zeta_{F,\mathfrak{c}}(2s - 2j) \cdot \prod_{\mathfrak{p}|\mathfrak{e},\, \mathfrak{p}\nmid\mathfrak{c}} \prod_{i=1}^{[(n+1)/2]} (1 - N(\mathfrak{p})^{4i-2-2s}).$$

Then

$$\Lambda'_\mathfrak{c}(s)\mathcal{T}(s) = \prod_{j=0}^{2n} \zeta_{F,\mathfrak{c}}(s - j) \prod_{\mathfrak{p}|\mathfrak{e},\, \mathfrak{p}\nmid\mathfrak{c}} \prod_{i=1}^{[(n+1)/2]} (1 - N(\mathfrak{p})^{2m+2i-s})(1 - N(\mathfrak{p})^{2i-1-s}).$$

PROOF. Clearly $\mathcal{T}(s) = \prod_{v\nmid\mathfrak{c}} \mathcal{T}_v(s)$ with $\mathcal{T}_v(s) = \sum_{\xi \in C_v^\varphi \backslash \mathfrak{X}_v} N(\nu^\sigma(\xi))^{-s}$. Fix a prime v that does not divide $\mathfrak{c}$, and put $C' = \sigma_v C_v^\varphi \sigma_v^{-1}$ and $\mathfrak{X}' = \sigma_v \mathfrak{X}_v \sigma_v^{-1}$. Then $C' = GL_n(\mathfrak{o}_v) \cap G(\varphi'_v)$, $\mathfrak{X}' = G(\varphi'_v)$, and $\mathcal{T}_v(s) = \sum_{\tau \in C' \backslash \mathfrak{X}'} \nu(\tau)^{-s}$, where $\nu(\tau) = N(\nu_0(\tau))$. Clearly $\mathcal{T}_v(s) = \sum_{\tau \in C' \backslash \mathfrak{X}'/C'} \#(C' \backslash C'\tau C')\nu(\tau)^{-s}$. Suppose $v \in \mathbf{h}_r$. Recall that $L'_v = \sum_i \mathfrak{o}_v e_i = L_v \sigma_v^{-1}$. Since L'_v is maximal with respect to φ'_v and $\varphi'_v(L'_v, L'_v)$ generates $\mathfrak{o}_v$, our groups $G(\varphi'_v)$ and C' are isomorphic to G^φ and C of §3.8. Therefore $\mathcal{T}_v(s)$ can be obtained from the series $\mathfrak{T}(s)$ of (3.20) by replacing $\omega(C\sigma C)$ there by $\#(C \backslash C\sigma C)$. This replacement can be achieved by substituting q^{2i} for the parameter t_i employed in the definition of $\omega(C\sigma C)$, where $q = N(\mathfrak{p})$ with the prime ideal $\mathfrak{p}$ at v. Then the formula of Theorem 3.12 gives

$$\mathcal{T}_v(s) = \frac{1 - q^{-s}}{1 - q^{2m-s}} \prod_{i=1}^{m} \frac{1 - q^{4i-2s}}{(1 - q^{n+t-1+2i-s})(1 - q^{2m-2i-s})}.$$

If $v \in \mathbf{h}_u$, then $G(\varphi'_v)$ and C' are isomorphic to $Sp(n, F_v)$ and $Sp(n, \mathfrak{g}_v)$, as explained in §1.4. Therefore $\mathcal{T}_v(s)$ can be obtained from the series $\mathcal{T}(s)$ of [S5], (16.15.3), by replacing $\omega(C\tau C)$ there by $\#(C \backslash C\tau C)$. This replacement can be achieved by substituting q^i for t_i there. Then the first formula of [S5], Theorem 16.16, gives

$$\mathcal{T}_v(s) = \frac{1 - q^{-s}}{1 - q^{n-s}} \prod_{i=1}^{n} \frac{1 - q^{2i-2s}}{(1 - q^{n+i-s})(1 - q^{n-i-s})}.$$

Taking the infinite product $\prod_{v\nmid\mathfrak{c}} \mathcal{T}_v$ and arranging the factors suitably, we obtain our proposition.

4.6. We now define a $G_\mathbf{a}^\varphi$-invariant measure $d_\varphi^* z$ on $\mathcal{Z}^\varphi$ by

$$(4.18) \qquad d_\varphi^* z = \prod_{v \in \mathbf{b}\cup\mathbf{c}\cup\mathbf{d}'} d_v^* z_v, \quad d_v^* z_v = \begin{cases} \Delta(z_v)^{-n-1} dz_v & (v \in \mathbf{b}), \\ \Delta(z_v)^{-n-(1/2)} dz_v & (v \in \mathbf{c}), \\ \Delta(z_v)^{-n} dz_v & (v \in \mathbf{d}'), \end{cases}$$

where $\mathbf{d}'$ is as in (4.10a) and dz_v is the standard measure on the $\mathbf{R}$-linear span of $\mathcal{Z}_v^\varphi$, which is identified with $\mathbf{R}^{n(n+1)}$, $\mathbf{R}^{2n^2+n}$, or $\mathbf{R}^{4q_v r_v}$ in an obvious way;

in particular, if $v \in \mathbf{c}$, for $z = x + yj \in \mathcal{Z}_v^\varphi$ we take $\left(\mathrm{Re}(x_{aa}), \mathrm{Im}(x_{aa}), y_{aa}\right)_a$ and $\left(\mathrm{Re}(x_{ab}), \mathrm{Im}(x_{ab}), \mathrm{Re}(y_{ab}), \mathrm{Im}(x_{ab})\right)_{a<b}$ to be the coordinates.

Now let D be an open subgroup of $G_{\mathbf{A}}^\varphi$ containing $G_{\mathbf{a}}^\varphi$ such that $D \cap G_{\mathbf{h}}^\varphi$ is compact. For each $x \in G_{\mathbf{A}}^\varphi$ put $\Gamma^x = G^\varphi \cap xDx^{-1}$. Take $\mathcal{B} \subset G_{\mathbf{A}}^\varphi$ so that $G_{\mathbf{A}}^\varphi = \bigsqcup_{a \in \mathcal{B}} G^\varphi a D$. We then define a real number $\mathfrak{m}(G^\varphi, D)$ by

$$\mathfrak{m}(G^\varphi, D) = \begin{cases} \sum_{a \in \mathcal{B}} [\Gamma^a : 1]^{-1} & \text{if } G_{\mathbf{a}}^\varphi \text{ is compact,} \\ [\Gamma^1 \cap \{\pm 1\} : 1]^{-1} \mathrm{vol}(\Gamma^1 \backslash \mathcal{Z}^\varphi) & \text{otherwise,} \end{cases} \tag{4.19}$$

where $\mathrm{vol}(\Gamma^1 \backslash \mathcal{Z}^\varphi)$ is defined with respect to the measure defined above. Here notice that if $G_{\mathbf{a}}^\varphi$ is not compact, then $G_{\mathbf{A}}^\varphi = G^\varphi D$ by Lemma 4.4, so that we can take $\{1\}$ to be $\mathcal{B}$. The reader is referred to [S5], §24.1, for the definition of the quantity $\mathfrak{m}$ for a more general type of group.

4.7. Theorem. *Suppose that φ is anisotropic. Let $\mathfrak{e}$ be the product of all the prime ideals ramified in B. Define an open subgroup D^φ of $G_{\mathbf{A}}^\varphi$ by (4.13) with an integral ideal $\mathfrak{c}$ in F. Then*

$$\mathfrak{m}(G^\varphi, D^\varphi) = |D_F|^{n^2+(n/2)} N(\mathfrak{c})^{2n^2+n} N(\mathfrak{e})^{n(n+1)/2} \prod_{k=1}^{n} \zeta_{F,\mathfrak{c}}(2k)$$
$$\cdot \prod_{\mathfrak{p}|\mathfrak{e}, \mathfrak{p}\nmid\mathfrak{c}} \prod_{k=1}^{n} \{1 + (-N(\mathfrak{p}))^{-k}\}$$
$$\cdot \left\{\pi^{(-n^2-n)/2} \prod_{k=1}^{n} (k-1)!\right\}^{|\mathbf{b}|} \left\{2^{-2n^2-n} \pi^{-n-n^2} \prod_{k=1}^{n} (2k-1)!\right\}^{|\mathbf{c}|}$$
$$\cdot \prod_{v \in \mathbf{d}} \left\{2^{-n^2-n} \pi^{-n-q_v^2-r_v^2} \prod_{i=1}^{q_v} (2i-1)! \prod_{k=1}^{r_v} (2k-1)!\right\},$$

where D_F is the discriminant of F, and (q_v, r_v) for $v \in \mathbf{d}$ is defined by (4.6).

The proof will be completed in Section 6.

The above expression for $\mathfrak{m}(G^\varphi, D^\varphi)$ leads us naturally to the comparison of the quantity with $\prod_{k=1}^{n} R_F(2k)$, where R_F is the function

$$R_F(s) = |D_F|^{s/2} \left\{\pi^{-s/2}\Gamma(s/2)\right\}^{|\mathbf{b}|+|\mathbf{d}|} \left\{(2\pi)^{-s}\Gamma(s)\right\}^{|\mathbf{c}|} \zeta_F(s), \tag{4.20}$$

for which we have $R_F(1-s) = R_F(s)$. A straightforward calculation shows that

$$\mathfrak{m}(G^\varphi, D^\varphi) = \varepsilon \cdot \pi^\gamma |D_F|^{n^2/2} \prod_{k=1}^{n} R_F(2k), \tag{4.21}$$

where $\gamma = -(n^2+n)|\mathbf{d}|/2 + \sum_{v \in \mathbf{d}'} 2q_v r_v$ and ε is a rational number whose explicit form can easily be obtained. For instance, we have

4.8. Corollary. *Suppose that φ is anisotropic and F is totally real. Then $\mathfrak{m}(G^\varphi, D^\varphi)$ is a rational number times π^α, where*

$$\alpha = (n^2+n)|\mathbf{b}|/2 + \textstyle\sum_{v \in \mathbf{d}'} 2q_v r_v.$$

In particular, if $\mathfrak{g} = \mathfrak{c}$ *and* $\mathbf{d}' = \varnothing$ *(so that* D^φ *is* C^φ *of (4.12)), then we have*

$$\mathfrak{m}(G^\varphi, C^\varphi) = |D_F|^{n^2} \prod_{k=1}^{n} \left\{ D_F^{1/2} \left[(2k-1)!(2\pi)^{-2k} \right]^{|\mathbf{a}|} \zeta_F(2k) \right\}$$

$$\cdot \left\{ 2^{n^2+n} \pi^{(n^2+n)/2} \prod_{k=1}^{n} \frac{(k-1)!}{(2k-1)!} \right\}^{|\mathbf{b}|} \cdot \prod_{\mathfrak{p}|\mathfrak{e}} \prod_{k=1}^{n} \{ N(\mathfrak{p})^k + (-1)^k \}.$$

The last formula can be obtained by arranging various factors of the formula of Theorem 4.7 suitably. The first assertion follows easily from the well known fact that $|D_F|^{1/2}\pi^{-2k|\mathbf{a}|}\zeta_F(2k) \in \mathbf{Q}$ for every $k \in \mathbf{Z}, > 0$, provided F is totally real. Thus the first line on the right-hand side of the last formula is rational. Notice that 2α is the dimension of $\mathcal{Z}^\varphi$ as a real manifold; $\mathcal{Z}^\varphi$ is hermitian if and only if $\mathbf{c} = \mathbf{d}' = \varnothing$.

4.9. Let us now discuss some special cases. First of all, if $n = 1$ and B is totally definite, then the above formula is closely related to the result of Eichler in [E], which has the following form:

(4.22) $$\sum_{i=1}^{t} [\mathfrak{o}_i^\times : \mathfrak{g}^\times]^{-1} = 2h_F |D_F|^{3/2} (2\pi)^{-2n} \zeta_F(2) \prod_{\mathfrak{p}|\mathfrak{e}} \{ N(\mathfrak{p}) - 1 \}.$$

Here we take a decomposition $B_\mathbf{A}^\times = \bigsqcup_{i=1}^t B^\times a_i E$ with $E = B_\mathbf{a}^\times \prod_{v \in \mathbf{h}} \mathfrak{o}_v^\times$ and put $\mathfrak{o}_i = a_i \mathfrak{o} a_i^{-1}$; h_F is the class number of F. Comparing (4.22) with the formula of the above corollary for $n = 1$, we obtain

(4.23) $$\sum_{i=1}^{t} [\mathfrak{o}_i^\times : \mathfrak{g}^\times]^{-1} = 2h_F \cdot \mathfrak{m}(G^\varphi, C^\varphi).$$

Notice that $G^\varphi = \{ x \in B \mid xx^\rho = 1 \}$ and $C^\varphi = G^\varphi \cap E$.

There is one more type of group associated to B: the orthogonal group G^ψ of the quadratic form $\psi[x] = xx^\rho$ defined on the space $W = \{ x \in B \mid x^\rho = -x \}$ which has dimension 3 over F. Putting $M = \mathfrak{o} \cap W$ and $C^\psi = \{ \alpha \in G^\psi \mid M\alpha = M \}$, we can define the mass $\mathfrak{m}(G^\psi, C^\psi)$ by the general principle of [S5], §24.1. Specializing the formula of [S6], Theorem 5.8, to the three-dimensional case (see also [S8], Theorem 6.4), and comparing it with $\mathfrak{m}(G^\varphi, C^\varphi)$, we find

(4.23) $$2^f \mathfrak{m}(G^\psi, C^\psi) = \mathfrak{m}(G^\varphi, C^\varphi),$$

where f is the number of prime ideals ramified in B.

4.10. Returning to an arbitrary n, if $|\mathbf{b}| = 1$ and $\mathbf{d}' = \varnothing$, the formula of Corollary 4.8 gives $2^{-1}\mathrm{vol}(\Gamma \backslash \mathcal{Z}^\varphi)$ for $\Gamma = \{ \gamma \in G^\varphi \mid L\gamma = L \}$. This Γ is essentially the same as the group treated by Siegel in [Si].

Next suppose that $F = \mathbf{Q}$ and φ is totally definite. By the Hasse principle, we may assume that $\varphi = 1_n$. Then $\mathfrak{o}_n^1$ is a maximal lattice. Recalling the formula $\zeta(2k) = (2\pi)^{2k}|B_{2k}|/[2 \cdot (2k)!]$, where B_{2k} denotes the Bernoulli number, we obtain

(4.25) $$\mathfrak{m}(G^\varphi, C^\varphi) = \prod_{k=1}^{n} \frac{|B_{2k}|}{4k} \cdot \prod_{p|e(B)} \prod_{k=1}^{n} \{ p^k + (-1)^k \},$$

where $e(B)$ is the product of all the prime numbers ramified in B. For $n = 1$ and 2, for example, the right-hand side takes the forms

$$\frac{1}{24}\prod_{p|e(B)}(p-1) \quad \text{and} \quad (2^7\cdot 3^2\cdot 5)^{-1}\prod_{p|e(B)}(p-1)(p^2+1). \tag{4.26}$$

The first one is essentially (4.22) for $F = \mathbf{Q}$. If $e(B) = 2$, for example, we can easily verify that the class number of G^φ is 1 for $n \le 3$, and > 1 for $n = 4$.

5. Eisenstein series

5.1. With B as in §1.4 we consider the group G^η in the matrix form as follows:

$$G^\eta = \{\alpha \in GL_{2n}(B) \mid \alpha\eta\alpha^* = \eta\}, \quad \eta = \eta_n = \begin{bmatrix} 0 & 1_n \\ 1_n & 0 \end{bmatrix}, \tag{5.1}$$

For $x = \begin{bmatrix} a & b \\ c & d \end{bmatrix} \in GL_{2n}(B)_\mathbf{A}$ with $a \in (B_\mathbf{A})_n^n$, we write $a = a_x$, $b = b_x$, $c = c_x$, and $d = d_x$. Let us now put

$$P = \{\xi \in G^\eta \mid c_\xi = 0\}, \quad Q = \{\xi \in P \mid b_\xi = 0\}, \tag{5.2}$$

$$S = S^n = \{h \in B_n^n \mid h^* = -h\}, \quad S(\mathfrak{a}) = S \cap (\mathfrak{a}\mathfrak{o})_n^n, \tag{5.3}$$

where $\mathfrak{a}$ is a fractional ideal in F.

We put $\mathfrak{H}^\eta = \prod_{v\in\mathbf{a}} \mathfrak{H}_v^\eta$ with $\mathfrak{H}_v^\eta$ defined by

$$\mathfrak{H}_v^\eta = \{z \in \mathbf{C}_{2n}^{2n} \mid {}^t z = z,\ \mathrm{Im}(z) > 0\} \qquad (v \in \mathbf{b}),$$

$$\mathfrak{H}_v^\eta = \{x + jy \in \mathbf{H}_{2n}^{2n} \mid {}^t x = x \in \mathbf{C}_{2n}^{2n},\ 0 < y = y^* \in \mathbf{C}_{2n}^{2n}\} \qquad (v \in \mathbf{c}),$$

$$\mathfrak{H}_v^\eta = \{z \in \mathbf{H}_n^n \mid z + z^* > 0\} \qquad (v \in \mathbf{d}).$$

Notice that $\mathfrak{H}_v^\eta$ coincides with $\mathcal{Z}_v^\eta$ for $v \in \mathbf{b} \cup \mathbf{c}$, but is merely diffeomorphic to $\mathfrak{B}(n, n)$ for $v \in \mathbf{d}$. We define "the origin" $\mathbf{i}$ of $\mathfrak{H}^\eta$ by $\mathbf{i} = (\mathbf{i}_v)_{v\in\mathbf{a}}$, $\mathbf{i}_v = i1_{2n}$ if $v \in \mathbf{b}$, $\mathbf{i}_v = j1_{2n}$ if $v \in \mathbf{c}$, and $\mathbf{i}_v = 1_n$ if $v \in \mathbf{d}$. We can let G_v^η act on $\mathfrak{H}_v^\eta$ as follows. For $v \in \mathbf{b} \cup \mathbf{c}$ we identify $(B_v)_{2n}^{2n}$ with $(F_v)_{4n}^{4n}$, put $T_n = \mathrm{diag}[1_{2n}, K_n]$ with K_n of (1.16), and observe that $T_n\eta_nT_n^* = J_{2n}K_{2n}$. As explained in §1.6, we have $T_nG(\eta_v)T_n^{-1} = Sp(2n, F_v)$. Then we let G_v^η act on $\mathfrak{H}_v^\eta$ via the isomorphism $\alpha \mapsto T_n\alpha T_n^{-1}$. For $v \in \mathbf{d}$ we observe that G_v^η and $\mathfrak{H}_v^\eta$ are the group and the space in Case II studied in [S7], and so the action is defined in [S7], Section 5.

Given $z \in \mathfrak{H}^\eta$, we put $z_v = x_v + y_vi$ with real x_v and y_v if $v \in \mathbf{b}$, $z_v = x_v + y_vj$ with complex x_v and y_v if $v \in \mathbf{c}$, and $z_v = x_v + y_v$ with x_v and y_v such that $x_v^* = -x_v$ and $y_v^* = y_v$ if $v \in \mathbf{d}$. Then we write $z = x + y\mathbf{i}$ with $x = (x_v)_{v\in\mathbf{a}}$ and $y = (y_v)_{v\in\mathbf{a}}$. Since $\mathfrak{H}_v^\eta = \mathcal{Z}_v^\eta$ for $v \in \mathbf{b}\cup\mathbf{c}$, we have the functions $\Delta(z)$ and $m(\alpha, z)$ for $z \in \mathfrak{H}_v^\eta$ and $\alpha \in Sp(2n, F_v)$. If $v \in \mathbf{d}$, we put

$$\Delta(z) = \delta_v(2^{-1}(z + z^*)), \quad m(\alpha, z) = \delta_v(c_\alpha z + d_\alpha) \tag{5.4}$$

for $z \in \mathfrak{H}_v^\eta$ and $\alpha \in G_v^\eta$. Then for $z \in \mathfrak{H}^\eta$ and $\xi \in G_\mathbf{A}^\eta$ we put

$$\Delta(z) = \prod_{v\in\mathbf{a}} \Delta(z_v), \qquad m(\xi, z) = \prod_{v\in\mathbf{a}} m(\tau_v\xi_v\tau_v^{-1}, z_v), \tag{5.5}$$

where $\tau_v = T_n$ if $v \in \mathbf{b} \cup \mathbf{c}$ and $\tau_v = 1$ if $v \in \mathbf{d}$. Clearly $\Delta(x+y\mathbf{i}) = \big(\delta_v(y_v)\big)_{v\in\mathbf{a}}$.

Given $\mathfrak{g}$-ideals $\mathfrak{y}$ and $\mathfrak{z}$ such that $\mathfrak{y}\mathfrak{z} \subset \mathfrak{g}$, we define subgroups $D[\mathfrak{y}, \mathfrak{z}]$ and $D_0[\mathfrak{y}, \mathfrak{z}]$ of $G^\eta_\mathbf{A}$ by

$$(5.6) \qquad D[\mathfrak{y}, \mathfrak{z}] = \{\, x \in G^\eta_\mathbf{A} \mid a_x \prec \mathfrak{o},\, b_x \prec \mathfrak{y}\mathfrak{o}, c_x \prec \mathfrak{z}\mathfrak{o},\, d_x \prec \mathfrak{o} \,\},$$

$$(5.7) \qquad D_0[\mathfrak{y}, \mathfrak{z}] = \{\, x \in D[\mathfrak{y}, \mathfrak{z}] \mid x(\mathbf{i}) = \mathbf{i} \,\}.$$

5.2. Lemma. *For a prime ideal* $\mathfrak{p}$ *in* F *we have*

$$[D[\mathfrak{z}^{-1}, \mathfrak{z}] : D[\mathfrak{z}^{-1}, \mathfrak{z}\mathfrak{p}]] = \begin{cases} q^{2n^2+n}\prod_{i=1}^{n}(1+q^{1-2i}) & \text{if } \mathfrak{p}|\mathfrak{e}, \\ q^{2n^2+n}\prod_{i=1}^{2n}(1+q^{-i}) & \text{if } \mathfrak{p}\nmid\mathfrak{e}, \end{cases}$$

where $q = N(\mathfrak{p})$.

PROOF. Clearly the index does not depend on $\mathfrak{z}$; so we may put $\mathfrak{z} = \mathfrak{g}$. Put $C = G^\eta_\mathfrak{p} \cap D[\mathfrak{g}, \mathfrak{g}]$, $C' = G^\eta_\mathfrak{p} \cap D[\mathfrak{g}, \mathfrak{p}]$, and $C_1 = \{\, \gamma \in C \mid \gamma - 1 \prec \mathfrak{p}\mathfrak{o} \,\}$. Then $[C : C']$ is the index in question. If $\mathfrak{p} \nmid \mathfrak{e}$, then $T_n C T_n^{-1} = Sp(2n, \mathfrak{g}_\mathfrak{p})$, and so $[C : C']$ is given in [S5], Proposition 17.15. To be precise, it is the index $[C : C^0]$ given there for $K = F$ and $\varepsilon = -1$ with $2n$ as n there. Therefore it is sufficient to prove the case $\mathfrak{p}|\mathfrak{e}$. In this case observe that $x \mapsto (d_x,\, b_x d_x^{-1})$ is an injection of C'/C_1 into $GL_n(\mathfrak{o}/\mathfrak{p}\mathfrak{o}) \times S(\mathfrak{g})/S(\mathfrak{p})$. By Lemma 2.7 this map is surjective. Now we can easily verify that

$$[GL_n(\mathfrak{o}/\mathfrak{p}\mathfrak{o}) : 1] = q^{3n^2-n}\prod_{i=1}^{n}(q^{2i}-1), \qquad [S(\mathfrak{g}) : S(\mathfrak{p})] = q^{2n^2+n}.$$

Since $[C : C_1]$ is given by Proposition 2.12 with $2n$ as n there, we obtain $[C : C_1]$ as stated in our lemma.

5.3. We now put

$$\mathbf{e}(z) = e^{2\pi i z} \qquad (z \in \mathbf{C}),$$

and define characters $\mathbf{e}_\mathbf{A} : F_\mathbf{A} \to \mathbf{T}$ and $\mathbf{e}_v : F_v \to \mathbf{T}$ for each $v \in \mathbf{v}$ as usual (see [S5], §18.2); note that for $v \in \mathbf{c}$ we have $\mathbf{e}_v(z) = \mathbf{e}(z + \overline{z})$. We then put $\mathbf{e}_\mathbf{h}(x) = \mathbf{e}_\mathbf{A}(x_\mathbf{h})$ and $\mathbf{e}_\mathbf{a}(x) = \mathbf{e}_\mathbf{A}(x_\mathbf{a})$ for $x \in F_\mathbf{A}$.

Whenever we take a finite-dimensional vector space Y over F, we normalize the Haar measure dy on $Y_\mathbf{A}$ so that the measure of $Y_\mathbf{A}/Y$ is 1, where the measure on Y is the standard discrete measure. We call it the *normalized measure* on $Y_\mathbf{A}$. Taking Y to be B, we have the normalized measure dx on $B_\mathbf{A}$. Now for each $v \in \mathbf{v}$ we define a Haar measure dx_v on B_v as follows. If $v \in \mathbf{h}$ we take dx_v so that the measure of $\mathfrak{o}_v$ is 1. This is independent of the choice of $\mathfrak{o}$. If $v \in \mathbf{b}$ or $v \in \mathbf{c}$, then B_v is ring-isomorphic to $\mathbf{R}^2_2$ or $\mathbf{C}^2_2$. Then we take dx_v to be the measure obtained by identifying these with $\mathbf{R}^4$ or $\mathbf{C}^4$ with respect to the standard matrix units $\{\varepsilon_{ij}\}$; here we take $d(x + iy) = dxdy$ on $\mathbf{C}$. If $v \in \mathbf{d}$, we identify B_v with $\mathbf{H}$, and define dx_v on B_v by identifying $\mathbf{H}$ with $\mathbf{R}^4$ with respect to the standard quaternion units $\{1, i, j, k\}$.

Next let dw denote the normalized measure on $W_{\mathbf{A}}$, where

(5.9) $$W = \{ w \in B \mid w^\rho = -w \}.$$

If $v \in \mathbf{h}$ we take a measure dw_v on W_v so that the measure of $\mathfrak{o}_v \cap W_v$ is 1. If $v \in \mathbf{b}$ or $v \in \mathbf{c}$, then we define dw_v on W_v by identifying W_v with F_v^3 with respect to the basis $\{ \varepsilon_{11} - \varepsilon_{22}, \varepsilon_{12}, \varepsilon_{21} \}$. If $v \in \mathbf{d}$, we define dw_v on W_v by identifying W_v with $\mathbf{R}^3$ with respect to the basis $\{ i, j, k \}$.

Finally we consider S and let $d\sigma$ denote the normalized measure on $S_{\mathbf{A}}$. Now S can be identified with $W^n \times B^{n(n-1)/2}$ in an obvious way. Therefore for each $v \in \mathbf{v}$ we can define a measure $d\sigma_v$ on S_v by means of the above measures dw_v on W_v and dx_v on B_v. Clearly $S(\mathfrak{g})_v$ has measure 1 for every $v \in \mathbf{h}$.

5.4. Lemma. *We have* $d\sigma = c(S) \prod_{v \in \mathbf{v}} d\sigma_v$ *with*

$$c(S) = 2^{(2n^2+n)|\mathbf{c}|+n^2|\mathbf{d}|} |D_F|^{-n^2-(n/2)} N(\mathfrak{e})^{-n(n+1)/2},$$

where $|\mathbf{c}| = \#\mathbf{c}$, $|\mathbf{d}| = \#\mathbf{d}$, D_F *is the discriminant of* F, *and* $\mathfrak{e}$ *is the product of all the prime ideals ramified in* B.

PROOF. We have $dx = c(B) \prod_{v \in \mathbf{v}} dx_v$ on $B_{\mathbf{A}}$ and $dw = c(W) \prod_{v \in \mathbf{v}} dw_v$ on $W_{\mathbf{A}}$ with constants $c(B)$ and $c(W)$, and $c(S) = c(W)^n c(B)^{n(n-1)/2}$. Thus our task is to compute $c(B)$ and $c(W)$. Take an F-basis $\{e_i\}$ of B and $\mathfrak{g}$-ideals $\mathfrak{a}_i$ so that $\mathfrak{o} = \sum_{i=1}^4 \mathfrak{a}_i e_i$. For each $v \in \mathbf{a}$ define a measure $d'x_v$ on B_v by identifying B_v with F_v^4 with respect to $\{e_i\}$. Take $d'x_v = dx_v$ for $v \in \mathbf{h}$. Then $dx = c'(B) \prod_{v \in \mathbf{v}} d'x_v$ with

$$c'(B) = \prod_{i=1}^4 \mathrm{vol}(F_{\mathbf{a}}/\mathfrak{a}_i)^{-1} = 2^{4|\mathbf{c}|} |D_F|^{-2} \prod_{i=1}^4 N(\mathfrak{a}_i)^{-1}$$

for the reason explained in [S5], (A6.2.1) and §A6.6. Now $\mathfrak{d}^{-1} = \sum_{i=1}^4 \mathfrak{a}_i^{-1} f_i$ with $f_i \in B$ such that $\tau(e_i f_j) = \delta_{ij}$. Thus

$$N(\mathfrak{e})^2 = [\mathfrak{o} : \mathfrak{d}] = |N_{F/\mathbf{Q}}(T')| \textstyle\prod_{i=1}^4 N(\mathfrak{a}_i)^2,$$

where $T' = \det\left[\left(\tau(e_i e_j) \right)_{i,j=1}^4 \right]$. Fix one $v \in \mathbf{a}$ and let $\{\varepsilon_i\}_{i=1}^4$ be an F_v-basis of B_v such that $d(\sum_i x_i \varepsilon_i) = \prod_i dx_i$ for $x_i \in F_v$. Then we easily see that $d'x_v = |T_v/T'|_v^{1/2} dx_v$ with $T_v = \det\left[\left(\tau(\varepsilon_i \varepsilon_j) \right)_{i,j=1}^4 \right]$. Now $T_v = -1$ for $v \in \mathbf{b} \cup \mathbf{c}$ and $T_v = -2^4$ for $v \in \mathbf{d}$. Therefore we obtain

(5.10) $$c(B) = c'(B) \prod_{v \in \mathbf{a}} |T_v/T'|_v^{1/2} = 2^{4|\mathbf{c}|+2|\mathbf{d}|} |D_F|^{-2} N(\mathfrak{e})^{-1}.$$

The same type of argument for W gives $c(W) = 2^{3|\mathbf{c}|+|\mathbf{d}|} |D_F|^{-3/2} N(\mathfrak{e})^{-1}$. Combining these we obtain the value of $c(S)$.

5.5. Lemma. *Let* D *be an open subgroup of* $G_{\mathbf{A}}^\eta$ *such that* $D \cap G_{\mathbf{h}}^\eta$ *is compact; put* $U = \{ \lambda(d_x) \mid x \in P_{\mathbf{A}} \cap D \}$ *and* $\Gamma = D \cap G^\eta$. *Then the following assertions hold:*

(1) *There exists a finite subset* T *of* $Q_{\mathbf{h}}$ *such that* $P_{\mathbf{A}} D = \bigsqcup_{t \in T} PtD$.

(2) *There exists a finite subset* $\mathcal{A}$ *of* G^η *such that* $G^\eta \cap P_{\mathbf{A}}D = \bigsqcup_{\alpha\in\mathcal{A}} P\alpha\Gamma$.

(3) *If* $n > 1$ *or* $\mathbf{a} \neq \mathbf{d}$, *then the map* $PxD \mapsto \lambda(d_x)$ *for* $x \in P_{\mathbf{A}}$ *gives a bijection of* $P\backslash P_{\mathbf{A}}D/D$ *onto* $F_{\mathbf{A}}^\times/F^\times U$.

PROOF. These can be proved in the same manner as in [S5], §9, which concerns the symplectic and unitary cases. In particular, see Lemmas 9.6, 9.8, and 9.10 of the volume. By Lemma 4.4 strong approximation holds for G^η, so that $xD \cap G^\eta \neq \varnothing$ for every $x \in G_{\mathbf{A}}^\eta$. Also, strong approximation holds for the simple factor of $GL_n(B)$ if and only if $n > 1$ or $\mathbf{a} \neq \mathbf{d}$, which is why we have that condition in (3).

5.6. Hereafter in this section we fix a $\mathfrak{g}$-ideal $\mathfrak{b}$. Now $G_{\mathbf{A}}^\eta = P_{\mathbf{A}}D_0[\mathfrak{b}^{-1}, \mathfrak{b}]$, since the product expression holds locally for every $v \in \mathbf{v}$. (See Lemma 2.5(2) for $v \in \mathbf{h}_r$, and (1.15) for $v \in \mathbf{h}_u$; as for $v \in \mathbf{a}$, it can be derived from the fact that $P_v(\mathbf{i}_v) = \mathfrak{H}_v^\eta$.) Therefore every element x of $G_{\mathbf{A}}^\eta$ belongs to $pD_0[\mathfrak{b}^{-1}, \mathfrak{b}]$ for some $p \in P_{\mathbf{A}}$. With such a p we define a positive real number $\varepsilon(x)$ and an ideal $\mathrm{il}_{\mathfrak{b}}(x)$ by

$$(5.11)\qquad \varepsilon(x) = |\lambda(d_p)|_{\mathbf{A}}, \qquad \mathrm{il}_{\mathfrak{b}}(x) = \lambda(d_p)\mathfrak{g}$$

with $|\ |_{\mathbf{A}}$ defined in §1.4. We can easily verify that

$$(5.12)\qquad \varepsilon(x_{\mathbf{a}}) = m(x, \mathbf{i}), \qquad \varepsilon(x_{\mathbf{h}}) = N\big(\mathrm{il}_{\mathfrak{b}}(x)\big)^{-1}.$$

To define our Eisenstein series, we take, in addition to $\mathfrak{b}$, an integral $\mathfrak{g}$-ideal $\mathfrak{c}$, and put, hereafter until the end of this section,

$$(5.13)\qquad D = D[\mathfrak{b}^{-1}, \mathfrak{bc}], \qquad D_0 = \{\, x \in D \mid x(\mathbf{i}) = \mathbf{i} \,\}.$$

We consider also the restriction of ε^{-s} to $P_{\mathbf{A}}D$; to be precise, we put

$$(5.14)\qquad \varepsilon_{\mathfrak{c}}(x, s) = \begin{cases} \varepsilon(x)^{-s} & \text{if } x \in P_{\mathbf{A}}D, \\ 0 & \text{if } x \notin P_{\mathbf{A}}D. \end{cases}$$

We then define an Eisenstein series $E(x, s)$ and its tranform $E^*(x, s)$ for $(x, s) \in G_{\mathbf{A}}^\eta \times \mathbf{C}$ by

$$(5.15)\qquad E(x, s) = E(x, s; D) = \sum_{\alpha\in A} \varepsilon_{\mathfrak{c}}(\alpha x, s), \quad A = P\backslash G^\eta,$$

$$(5.16)\qquad E^*(x, s) = E(x\eta_{\mathbf{h}}^{-1}, s; D),$$

where η is the matrix of (5.1). By virtue of the standard principle explained in [S5], (10.7.5), for every $r \in G_{\mathbf{h}}^\eta$ we can define functions $E_r(z, s)$ and $E_r^*(z, s)$ of $(z, s) \in \mathfrak{H}^\eta \times \mathbf{C}$ by

$$(5.17)\qquad E_r(\xi(\mathbf{i}), s) = E(r\xi, s), \quad E_r^*(\xi(\mathbf{i}), s) = E^*(r\xi, s)$$

for $\xi \in G_{\mathbf{a}}^\eta$. The series of (5.15), or rather the corresponding series for $E_r(z, s)$, is convergent for sufficiently large $\mathrm{Re}(s)$ (in fact, for $\mathrm{Re}(s) > 2n + 1$), which can be proved by the same method as in [S5], §A.3.

Now we have a Fourier expansion

$$E^*\left(\begin{bmatrix} q & \sigma\widehat{q} \\ 0 & \widehat{q} \end{bmatrix}, s\right) = \sum_{h\in S} c(h, q, s)\mathbf{e_A}(\tau(h\sigma)) \tag{5.18}$$

for $q \in GL_n(B)_\mathbf{A}$ and $\sigma \in S_\mathbf{A}$ with $c(h, q, s) \in \mathbf{C}$, where τ is the trace map of (1.2).

5.7. Lemma. *(1) If $r\eta_\mathbf{h} = \alpha^{-1}tw$ with $r, t \in G^\eta_\mathbf{h}$, $\alpha \in G^\eta$, and $w \in D[\mathfrak{bc}, \mathfrak{b}^{-1}]$, then $E_r(z, s) = E^*_t(\alpha z, s)$.*

(2) For $t = \operatorname{diag}[q_1, \widehat{q}_1]$ with $q_1 \in GL_n(B)_\mathbf{h}$ and $z = x + \mathbf{i}y \in \mathfrak{H}^\eta$ put $c_t(h, y, s) = c(h, q, s)$ with $q \in GL_n(B)_\mathbf{A}$ such that $q_\mathbf{h} = q_1$ and $q_\mathbf{a} = y^{1/2}$. Then

$$E^*_t(z, s) = \sum_{h\in S} c_t(h, y, s)\mathbf{e_a}(\tau(hx\kappa_n)),$$

where κ_n is the element of $(B^n_n)_\mathbf{a}$ such that $(\kappa_n)_v = K_n$ for $v \in \mathbf{b}\cup\mathbf{c}$ and $(\kappa_n)_v = 1$ for $v \in \mathbf{d}$.

(3) Put $D' = D[\mathfrak{b}^{-1}, \mathfrak{bc}']$ with a $\mathfrak{g}$-ideal $\mathfrak{c}'$ divisible by $\mathfrak{c}$; define $E'_r(z, s)$ by the principle of (5.17) with D' in place of D, that is, $E'_r(\xi(\mathbf{i}), s) = E(r\xi, s; D')$ for $\xi \in G^\eta_\mathbf{a}$. Let $\mathcal{B}$ be a subset of $G^\eta_\mathbf{h} \cap D$ such that $D = \bigsqcup_{b\in\mathcal{B}} D'b^{-1}$. Then

$$E_r(z, s) = \sum_{b\in\mathcal{B}} E'_{rb}(z, s).$$

These are similar to [S5], Lemmas 18.7, 12.7(2), and can easily be verified. Notice that

$$S_v(\kappa_n)_v = (\kappa_n)_v S_v = \begin{cases} \{x \in (F_v)^{2n}_{2n} \mid {}^t x = x\} & (v \in \mathbf{b}\cup\mathbf{c}), \\ \{x \in \mathbf{H}^n_n \mid x^* = -x\} & (v \in \mathbf{d}). \end{cases} \tag{5.19}$$

5.8. To describe the archimedean factors of $c(h, q, s)$, we need generalized Bessel functions on $\mathfrak{H}^\eta_v$ defined by

$$\xi_v(y, h; s) = \int_{S_v} \mathbf{e}_v(-\tau(h\sigma))\delta_v(\sigma(\kappa_n)_v^{-1} + y\mathbf{i}_v)^{-s}d_v\sigma \tag{5.20}$$

$$(s \in \mathbf{C},\ h \in S_v,\ y\mathbf{i} \in \mathfrak{H}^\eta),$$

where δ_v is defined by (4.1). Then we put

$$\Xi(y, h; s) = \textstyle\prod_{v\in\mathbf{a}} \xi_v(y_v, h_v; s). \tag{5.21}$$

Next, to obtain the nonarchimedean factors of $c(h, q, s)$, we first put

$$Z = Z^n = \{k \in S \mid \tau(kS(\mathfrak{g})) \subset \mathfrak{g}\}. \tag{5.22}$$

Also, we let $\mathfrak{d}(F/\mathbf{Q})$ denote the different of F relative to $\mathbf{Q}$, and for each $v \in \mathbf{h}$ take an element ε_v of F_v such that $\varepsilon_v\mathfrak{g}_v = \mathfrak{d}(F/\mathbf{Q})_v$. Given $\zeta \in S_\mathbf{A}$ such that $\zeta_v \in Z_v$ for every $v \in \mathbf{h}$, we put

$$\alpha_\mathfrak{c}(\zeta, s) = \prod_{v\nmid\mathfrak{c}} \alpha(\zeta_v, s), \tag{5.23}$$

$$\alpha(\zeta_v, s) = \sum_{\sigma\in S_v/\Lambda_v} \mathbf{e}_v(\varepsilon_v^{-1}\tau(\zeta\sigma))\nu(\sigma)^{-s}, \quad \Lambda_v = S(\mathfrak{g})_v,$$

where ν is the norm of the denominator ideal defined on $(F_v)^{2n}_{2n}$ by [S5], (3.11.2).

5.9. Proposition. (1) *Suppose that* $\mathfrak{c} \neq \mathfrak{g}$; *then* $c(h, q, s) \neq 0$ *only if* $(q^*hq)_v \in \left(\mathfrak{b}\mathfrak{c} \cdot \mathfrak{d}(F/\mathbf{Q})\right)_v^{-1} Z_v$ *for every* $v \in \mathbf{h}$, *in which case*

$$c(h, q, s) = c(S)N(\mathfrak{b}\mathfrak{c})^{-2n^2-n}|\lambda(q)_{\mathbf{h}}|_{\mathbf{A}}^{2n+1-s}|\lambda(q)_{\mathbf{a}}|_{\mathbf{A}}^{s} \cdot \alpha_{\mathfrak{c}}\left(\beta q^*hq, s\right)\Xi(qq^*, h; s),$$

where β *is an element of* $F_{\mathbf{h}}^{\times}$ *such that* $\beta\mathfrak{g} = \mathfrak{b}\mathfrak{d}(F/\mathbf{Q})$.

(2) *Suppose that* $\mathfrak{e} \supset \mathfrak{c} \neq \mathfrak{g}$ *and* $(q^*hq)_v \in \left(\mathfrak{b}\mathfrak{c} \cdot \mathfrak{d}(F/\mathbf{Q})\right)_v^{-1} Z_v$ *for every* $v \in \mathbf{h}$. *Put* $g^*hg = \mathrm{diag}[h', 0]$ *with* $g \in GL_n(B)$ *and* $h' \in S^r \cap GL_r(B)$, $0 \leq r \leq n$. *Let* ρ_h *be the Hecke character of* F *corresponding to* $F(c^{1/2})/F$, *where* $c = (-1)^r\lambda(h')$ *if* $r > 0$, *and* $c = 1$ *(and hence* $\rho_h = 1$*) if* $r = 0$; *let* π_v *be a prime element of* F_v. *Then*

$$\alpha_{\mathfrak{c}}(\beta q^*hq, s) = \Lambda_{\mathfrak{c}}(s)^{-1}\Lambda_{h,\mathfrak{c}}(s)\prod_{v\in\mathbf{x}} f_{h,q,v}\left(|\pi_v|^s\right)$$

with a finite subset $\mathbf{x}$ *of* $\mathbf{h}$, *polynomials* $f_{h,q,v}$ *with constant term 1 and coefficients in* $\mathbf{Z}$, *and functions* $\Lambda_{\mathfrak{c}}$ *and* $\Lambda_{h,\mathfrak{c}}$ *given as follows:*

$$\Lambda_{\mathfrak{c}}(s) = \zeta_{F,\mathfrak{c}}(s)\prod_{i=1}^{n}\zeta_{F,\mathfrak{c}}(2s-2i),$$

$$\Lambda_{h,\mathfrak{c}}(s) = L_{\mathfrak{c}}(s-2n+r, \rho_h)\prod_{i=1}^{n-r}\zeta_{F,\mathfrak{c}}(2s-2n-2i+1).$$

The set $\mathbf{x}$ *is determined as follows:* $\mathbf{x} = \varnothing$ *if* $r = 0$. *If* $r > 0$, *take* $g_v \in GL_n(\mathfrak{o}_v)$ *for each* $v \nmid \mathfrak{c}$ *so that* $(\beta q^*hq)_v = g_v^*\mathrm{diag}[\xi_v, 0]g_v$ *with* $\xi_v \subset Z_v^r$. *Then* $\mathbf{x}$ *consists of all the* v *prime to* $\mathfrak{c}$ *such that* $\lambda(2\xi_v) \notin \mathfrak{g}_v^{\times}$.

PROOF. Under the condition $\mathfrak{c} \neq \mathfrak{g}$, we can apply the methods of [S5], pp.151–156, to the present case to obtain (1). As for (2), since we assume that $\mathfrak{e} \supset \mathfrak{c}$, we have $B_v = (F_v)_2^2$ for every $v \nmid \mathfrak{c}$. Therefore $\alpha(\zeta_v, s)$ for such a v coincides with $\alpha(\zeta_v K_n, s)$ defined in [S5], (13.4.3), in the case $K = F$ and $\varepsilon' = 1$. (Here K, F, and ε' are the symbols of [S5], §13.1; also we have to take $2n$ to be n there.) Its rational expression in $|\pi_v|$ is given in [S5], Theorem 13.6; see also [S5], Proposition 19.2. Therefore we obtain (2).

5.10. Lemma. *The integral of (5.20) is convergent for* $\mathrm{Re}(s) > 2n$, *and so defines a holomorphic function there.*

PROOF. If $v \in \mathbf{b}$, then $\xi_v(y_v, h_v; s)$ coincides with $\xi(y_v, K_n h_v; s/2, s/2)$ of [S3], (1.25) in Case I, with $2n$ in place of n there, and so the integral of (5.20) for $v \in \mathbf{b}$ is convergent for $\mathrm{Re}(s) > 2n$, as noted in [S3], the last two lines on page 274. If $v \in \mathbf{c}$ (resp. $v \in \mathbf{d}$), then $\xi_v(y_v, h_v; s)$ coincides with $\xi_{2n}(y_v, 2K_n\overline{h}_v; s)$ in Case IV (resp. $\xi_n(y_v, 2h_v; s)$ in Case II) of [S7], (2.1), and so the convergence holds for $\mathrm{Re}(s) > 2n$ (resp. $\mathrm{Re}(s) > 2n - (1/2)$) by [S7], Proposition 2.2. Notice that δ_v of (4.1) is the square of δ of [S7], (1.8.II, IV) and the present τ is twice the function τ of [S7], (1.9.II, IV). If $h \neq 0$, then the domain of holomorphy of ξ_v can be extended beyond the line $\mathrm{Re}(s) = 2n$; for details, see [S3] and [S7]. However, we do not need that result in the present paper.

5.11. Theorem. *Let $\mu = 2n+1$ and let $r(\mathfrak{c})$ be the residue of $\zeta_{F,\mathfrak{c}}(s)$ at $s = 1$. Then the function $E_r(z, s)$ of (5.17) is holomorphic in s for $\mathrm{Re}(s) > \mu - 1$, except for a simple pole at $s = \mu$. The residue of E_r at $s = \mu$ is the number $R(\mathfrak{c})$ given by*

$$R(\mathfrak{c}) = r(\mathfrak{c})c(S)N(\mathfrak{b}\mathfrak{c})^{-n\mu}\Lambda_{\mathfrak{c}}(\mu)^{-1} \cdot 2^{\alpha}\pi^{\beta}\Gamma_{2n}^{1}(n + (1/2))^{-|\mathbf{b}|}\Gamma_{2n}^{2}(2n + (1/2))^{-|\mathbf{c}|}\Gamma_{n}^{4}(2n)^{-|\mathbf{d}|} \cdot \prod_{i=1}^{n}\zeta_{F,\mathfrak{c}}(2i + 1) \cdot \prod_{\mathfrak{p}|\mathfrak{e},\mathfrak{p}\nmid\mathfrak{c}}\prod_{i=1}^{n}\left(1 + N(\mathfrak{p})^{-2i}\right)^{-1},$$

where $c(S)$ is the quantity given in Lemma 5.4, $\alpha = n|\mathbf{b}| - 2n^2[F : \mathbf{Q}]$, $\beta = 2n^2[F : \mathbf{Q}] + n|\mathbf{b}| + n|\mathbf{c}|$, and Γ_m^{ι} is defined by (5.26) below.

PROOF. We first prove this when $\mathfrak{e} \supset \mathfrak{c} \neq \mathfrak{g}$. Now the analytic nature of $E_r(z, s)$ and $E_t^*(z, s)$ at a given point s is completely determined by $c(h, q, s)$, as explained in [S5], Proposition 19.1. Put $D' = D[\mathfrak{b}\mathfrak{c}, \mathfrak{b}^{-1}]$. Given $r \in G_{\mathbf{h}}^{\eta}$, we can find $\alpha \in G^{\eta}$ and $w \in D'$ so that $r\eta_{\mathbf{h}} = \alpha^{-1}w$, since strong approximation holds for G^{η} (see Lemma 4.4). Then $E_r(z, s) = E_1^*(\alpha z, s)$ by Lemma 5.7 (1). By Lemma 5.10 and Proposition 5.9, we easily see that $c(h, q; s)$ is holomorphic for $\mathrm{Re}(s) > 2n$ if $h \neq 0$. Now by Proposition 5.9 we have

$$c_1(0, y, s) = c(S)N(\mathfrak{b}\mathfrak{c})^{-n\mu}\Lambda_{\mathfrak{c}}(s)^{-1}\Lambda_{0,\mathfrak{c}}(s)\Delta(y\mathbf{i})^{s/2}\Xi(y, 0; s), \tag{5.24}$$

$$\Lambda_{0,\mathfrak{c}}(s) = \zeta_{F,\mathfrak{c}}(s - 2n)\prod_{i=1}^{n}\zeta_{F,\mathfrak{c}}(2s - 2n - 2i + 1).$$

Therefore $c_1(0, y, s)$ is meromorphic for $\mathrm{Re}(s) > 2n$, and the only possible pole is at $s = \mu$, which is at most simple, and the residue is

$$r(\mathfrak{c})c(S)N(\mathfrak{b}\mathfrak{c})^{-n\mu}\Lambda_{\mathfrak{c}}(\mu)^{-1}\Delta(y\mathbf{i})^{\mu/2}\Xi(y, 0; \mu)\prod_{i=1}^{n}\zeta_{F,\mathfrak{c}}(2i + 1).$$

Now from [S3], (1.31), and [S7], (2.13), we obtain

$$\delta_v(y_v)^{\mu/2}\xi_v(y, 0; \mu) = \begin{cases} 2^{n-2n^2}\pi^{2n^2+n}\Gamma_{2n}^{1}(n + (1/2))^{-1} & (v \in \mathbf{b}), \\ 2^{-4n^2}\pi^{4n^2+n}\Gamma_{2n}^{2}(2n + (1/2))^{-1} & (v \in \mathbf{c}), \\ 2^{-2n^2}\pi^{2n^2}\Gamma_{n}^{4}(2n)^{-1} & (v \in \mathbf{d}), \end{cases} \tag{5.25}$$

$$\Gamma_m^{\iota}(s) = \pi^{\iota m(m-1)/4}\prod_{k=0}^{m-1}\Gamma(s - (\iota k/2)). \tag{5.26}$$

Combining these, we obtain $R(\mathfrak{c})$ and also the first assertion of our theorem when $\mathfrak{e} \supset \mathfrak{c} \neq \mathfrak{g}$. Next, take an arbitrary $\mathfrak{c}$ and take a multiple $\mathfrak{c}'$ of $\mathfrak{c}$ so that $\mathfrak{e} \supset \mathfrak{c}' \neq \mathfrak{g}$ and $\mathfrak{c}^{-1}\mathfrak{c}'$ is a squarefree ideal prime to $\mathfrak{c}$. The above result is applicable to the series defined with $\mathfrak{c}'$ in place of $\mathfrak{c}$. By Lemma 5.7(3) we obtain the first assertion of our theorem; at the same time, we have $R(\mathfrak{c}) = [D : D']R(\mathfrak{c}')$ with D' defined there, since $R(\mathfrak{c}')$ is independent of r. Clearly $[D : D'] = \prod_{\mathfrak{p}}[D : D[\mathfrak{b}^{-1}, \mathfrak{b}\mathfrak{c}\mathfrak{p}]]$ where $\mathfrak{p}$ runs over the primes dividing $\mathfrak{c}^{-1}\mathfrak{c}'$. Employing the formula of Lemma 5.2, we obtain $R(\mathfrak{c})$ with no condition on $\mathfrak{c}$.

6. Proof of Theorem 4.7

6.1. With (V, φ) as in §4.1 we put

$$(X, \omega) = (V, \varphi) \oplus (V, -\varphi). \tag{6.1}$$

We take a B-basis of V as in §4.1; we then take the union of this basis and its copy to be the basis of X and consider matrix representation with respect to these bases. In particular we can put

$$\omega = \mathrm{diag}[\varphi, -\varphi]. \tag{6.2}$$

An element (β, γ) of $G^\varphi \times G^\varphi$ can be identified with the element $\mathrm{diag}[\beta, \gamma]$ of G^ω. Expressing an element of X in the form (u, v) with $u, v \in V$ with respect to the decomposition of (6.1), we put

$$U = \{ (v, v) \mid v \in V \}, \quad P^\omega = \{ \gamma \in G^\omega \mid U\gamma = U \}. \tag{6.3}$$

Since U is totally ω-isotropic, P^ω is the parabolic subgroup of G^ω relative to U.

Now with L as in §4.2, we can take $\kappa \in B_n^n$ so that

$$\varphi = \kappa + \kappa^* \quad \text{and} \quad x\kappa y^* \in \mathfrak{o} \quad \textit{for every} \quad x, y \in L. \tag{6.4}$$

Indeed, by Lemma 1.7 we can take left $\mathfrak{o}$-ideals $\mathfrak{a}_1, \ldots, \mathfrak{a}_n$ and a B-basis $\{g_i\}$ of V so that $L = \sum_{i=1}^n \mathfrak{a}_i g_i$. Then $\varphi(g_i, g_j) \in (\mathfrak{a}_j^\rho \mathfrak{a}_i)^{-1}$. Let g be the $(n \times n)$-matrix whose i-th row is g_i. By (1.8) we can find c_i in the right order of $\mathfrak{a}_i$ so that $\tau(c_i) = 1$. Define $\kappa \in B_n^n$ so that $(g\kappa g^*)_{ii} = c_i\varphi(g_i, g_i)$, $(g\kappa g^*)_{ij} = \varphi(g_i, g_j)$ for $i < j$, and $(g\kappa g^*)_{ij} = 0$ for $i > j$. Then (6.4) is satisfied. With this κ we put

$$\eta = \begin{bmatrix} 0 & 1_n \\ 1_n & 0 \end{bmatrix}, \qquad S = \begin{bmatrix} 1_n & -\kappa \\ -1_n & -\kappa^* \end{bmatrix}. \tag{6.5}$$

We easily see that

$$S\eta S^* = -\omega, \qquad US = \{ [0_n^1 \;\; x] \in B_{2n}^1 \mid x \in B_n^1 \}. \tag{6.6}$$

Consider G^η and define the standard parabolic subgroup P^η by (5.2). Then

$$S^{-1}G^\omega S = G^\eta, \qquad S^{-1}P^\omega S = P^\eta = \{ \pi \in G^\eta \mid US\pi = US \}. \tag{6.7}$$

6.2. Lemma. *Let $D^\eta = D[\mathfrak{b}, \mathfrak{c}]$ with the notation of (5.6) and with the integral ideal $\mathfrak{c}$ we took in §4.2; define C^φ, D^φ, and Ξ by (4.12), (4.13), and (4.15). Let q be an element of $G_{\mathbf{A}}^\eta$ such that $q_v = 1$ for every $v|\mathfrak{c}$. Put*

$$\Sigma = \mathrm{diag}[\sigma, \widehat{\sigma}] \quad (\in G_{\mathbf{A}}^\eta)$$

with σ of §4.2. Then the following assertions hold:

(1) $\Xi = \{ \xi \in G^\varphi \mid S^{-1}(\xi, 1)Sq \in P_{\mathbf{A}}^\eta D^\eta \Sigma \}$.

(2) $D^\varphi = \{ \alpha \in C^\varphi \mid (\Sigma S^{-1}(\alpha, 1)S\Sigma^{-1})_v \in D_v^\eta \;\; \textit{for every} \;\; v|\mathfrak{c} \}$
$= \{ \alpha \in C^\varphi \mid \Sigma S^{-1}(\alpha, 1)S\Sigma^{-1} \in D^\eta \}$.

6.3. Lemma. *Let* $\alpha = (\xi, 1)$ *with* $\xi \in \Xi$ *and* $q = S^{-1}(b, a)S$ *with* $a, b \in G^{\varphi}_{\mathbf{h}}$ *such that* $a_v = b_v = 1$ *for every* $v|\mathfrak{c}$; *let* $x = S^{-1}\alpha Sq\Sigma^{-1}$. *Then the following assertions hold:*

(1) $\nu^{\sigma}(a^{-1}\xi b)$ *is prime to* $\mathfrak{c}$.

(2) $\mathrm{il}_{\mathfrak{g}}(x) = \lambda(\sigma)\nu^{\sigma}(a^{-1}\xi b)^{-1}$.

(3) $(\widehat{\sigma} d_x - 1)_v \prec \mathfrak{c}_v$ *for every* $v|\mathfrak{c}$, *where* d_x *is the d-block of* x *in the sense of* §5.1.

These two lemmas are analogues of [S5], Lemmas 21.4 and 21.6, and [S6], Lemmas 7.2 and 7.3, and can be proved in the same manner. In fact, the present case is much easier than the cases in those articles, since $\sigma\kappa\sigma^* \prec \mathfrak{o}$ and $\varphi'_v \in GL_n(\mathfrak{o}_v)$ for every $v \in \mathbf{h}$. Notice also that $G^{\eta}_v \subset P^{\eta}_{\mathbf{A}} D^{\eta}$ if $v \nmid \mathfrak{c}$.

6.4. For each $v \in \mathbf{a}$ take φ'_v of (4.11) (see also (4.6)) and put

$$U_v = \begin{cases} TX_v^{-1}Q_v^{-1}R & (v \in \mathbf{b} \cup \mathbf{c}), \\ X_v^{-1} & (v \in \mathbf{d}), \end{cases}$$

$$T = \mathrm{diag}[1_{2n}, K_n], \quad R = \mathrm{diag}[1_n, J_n, 1_n], \quad X_v = Z_v\Sigma_v,$$

$$Q_v = \begin{bmatrix} 1_n & -2^{-1}\varphi'_v \\ -1_n & -2^{-1}\varphi'_v \end{bmatrix}, \quad Z_v = \begin{bmatrix} 1_n & \tau_v \\ 0 & 1_n \end{bmatrix}, \quad \tau = 2^{-1}\sigma(\kappa^* - \kappa)\sigma^*.$$

We can easily verify that $Q_v\eta Q_v^* = \mathrm{diag}[-\varphi'_v, \varphi'_v]$ and $\mathrm{diag}[-J_n, J_n] = RJ_{2n} \cdot {}^tR$. Clearly $X_v \in G^{\eta}_v$. Recall (§1.6) that $K_m A^* = {}^tAK_m$ for $A \in (B_v)^m_m = (F_v)^{2m}_{2m}$, $v \in \mathbf{b} \cup \mathbf{c}$. Since $T_n\eta_n T_n^* = J_{2n}K_{2n}$ and $\varphi'_v = J_nK_n$ if $v \in \mathbf{b} \cup \mathbf{c}$, we can verify that $U_v \in Sp(2n, F_v)$ for such a v.

Define $f_v : \mathcal{Z}^{\varphi}_v \times \mathcal{Z}^{\varphi}_v \to \mathfrak{H}^{\eta}_v$ by

$$f_v(z, w) = \begin{cases} \mathrm{diag}[-z', w] & (v \in \mathbf{b} \cup \mathbf{c}), \\ \dfrac{1}{2}\begin{bmatrix} 1_{q_v} & z \\ w^* & 1_{r_v} \end{bmatrix}\begin{bmatrix} 1_{q_v} & -z \\ -w^* & 1_{r_v} \end{bmatrix}^{-1} & (v \in \mathbf{d}), \end{cases}$$

where $(x + yi)' = x - yi$ for $v \in \mathbf{b}$ and $(x + yj)' = x - yj$ for $v \in \mathbf{c}$. We understand that $f_v(z, w) = 2^{-1}1_n$ if $q_vr_v = 0$. We then define two maps f and f_U of $\mathcal{Z}^{\varphi} \times \mathcal{Z}^{\varphi}$ into $\mathfrak{H}^{\eta}$ by

$$f(z, w) = \big(f_v(z_v, w_v)\big)_{v\in\mathbf{a}}, \quad f_U(z, w) = (U_v f(z_v, w_v))_{v\in\mathbf{a}} \qquad (z, w \in \mathcal{Z}^{\varphi}).$$

Put also

$$[\beta, \gamma]_S = S^{-1}(\beta, \gamma)S \quad \text{for} \quad (\beta, \gamma) \in G^{\varphi} \times G^{\varphi}.$$

Then $[\beta, \gamma]_S \in G^{\eta}$.

6.5. Lemma. *For* $(\beta, \gamma) \in G^{\varphi} \times G^{\varphi}$ *and* $(z, w) \in \mathcal{Z}^{\varphi} \times \mathcal{Z}^{\varphi}$ *we have*

$$[\beta, \gamma]_S f_U(z, w) = f_U(\beta z, \gamma w), \tag{6.8}$$

$$\Delta\big(f_U(z, w)\big) = 2^{-2n|\mathbf{d}|}|\lambda(\sigma)_{\mathbf{a}}|_{\mathbf{A}}^{-2}\Delta(w, z)^{-2}\Delta(z)\Delta(w). \tag{6.9}$$

PROOF. First we note that $S_v = \mathrm{diag}[\sigma, \sigma]_v^{-1}Q_vX_v$. Put $\beta' = \sigma\beta\sigma^{-1}$ and $\gamma' = \sigma\gamma\sigma^{-1}$; recall our convention of §4.1 that the action of β on G^{φ}_v is that of β'_v. Suppose $v \in \mathbf{d}$; observe that Q_v^{-1} coincides with $-R$ of [S7], (7.12), and hence, by [S7], (7.13) and (7.16), we have

$$([\beta,\gamma]s)_v f_U(z,w)_v = \{UQ^{-1}(\beta',\gamma')Q\}_v f(z,w)_v = U_v f(\beta z,\gamma w)_v.$$

Next suppose $v \in \mathbf{b} \cup \mathbf{c}$; put $[\beta'_v, \gamma'_v]_R = R^{-1}(\beta'_v, \gamma'_v)R$; observe that $[\beta'_v, \gamma'_v]_R \in Sp(2n, F_v)$ and $[\beta'_v, \gamma'_v]_R f_v(z,w) = f_v(\beta'_v z, \gamma'_v w)$. Therefore, in view of our convention of §5.1, we have

$$\begin{aligned}([\beta,\gamma]s)_v f_U(z,w)_v &= \{TS^{-1}(\beta,\gamma)ST^{-1}U\}_v f(z,w)_v \\ &= \{UR^{-1}(\beta',\gamma')R\}_v f(z,w)_v = U_v f(\beta z,\gamma w)_v.\end{aligned}$$

Thus we obtain (6.8). Now direct calculations show that

$$m(U_v, f(z,w)_v) = \begin{cases} |\lambda(\sigma)_v|_v \Delta(w_v, z_v) & (v \in \mathbf{b} \cup \mathbf{c}), \\ |\lambda(\sigma)_v|_v & (v \in \mathbf{d}). \end{cases}$$

In [S7], (7.17), we showed that $\Delta\big(f(z,w)_v\big) = 2^{-2n}\Delta(z_v)\Delta(w_v)\Delta(w_v,z_v)^{-2}$ for $v \in \mathbf{d}'$. (Notice that $\Delta = \varepsilon^2$ with the symbol ε there.) If $v \in \mathbf{d}$ and $v \in \mathbf{d}'$, then $f(z,w)_v = 2^{-1}1_n$, so that $\Delta\big(f(z,w)_v\big) = 2^{-2n}$. Clearly $\Delta\big(f(z,w)_v\big) = \Delta(z_v)\Delta(w_v)$ if $v \in \mathbf{b} \cup \mathbf{c}$. Since $\Delta\big(f_U(z,w)\big) = m\big(U, f(z,w)\big)^{-2}\Delta\big(f(z,w)\big)$, we obtain (6.9).

6.6. Let us now assume that φ is anisotropic. Our argument is the same as in [S5], §22.4ff., and [S6], §7.6, and so we only sketch the proof. First by [S5], (2.7.1), we have $G^\omega = \bigsqcup_{\xi \in G^\varphi} P^\omega(\xi, 1)$, and hence

$$G^\eta = \bigsqcup_{\xi \in G^\varphi} P^\eta S^{-1}(\xi, 1)S. \tag{6.10}$$

We consider E of (5.15) with $\mathfrak{b} = \mathfrak{g}$, and take $\mathcal{B}$ as in Lemma 4.3(1). Pick an elements $a \in \mathcal{B}$. We are interested in $E(xq\Sigma_{\mathbf{h}}^{-1}, s)$ with the elements Σ and $q = S^{-1}(1, a)S$ of $G^\eta_{\mathbf{h}}$ given in Lemmas 6.2 and 6.3. To be explicit, we have

$$E(xq\Sigma_{\mathbf{h}}^{-1}, s) = \sum_{\alpha \in A} \varepsilon_{\mathfrak{c}}(\alpha x q \Sigma_{\mathbf{h}}^{-1}, s) \qquad (x \in G^\eta_{\mathbf{A}},\ A = P^\eta \backslash G^\eta).$$

By (6.10) we can take the elements $S^{-1}(\xi, 1)S$ with $\xi \in G^\varphi$ to be A. From $E(xq\Sigma_{\mathbf{h}}^{-1}, s)$ we obtain a function $\mathcal{E}_a(\mathfrak{z}, s)$ of $(\mathfrak{z}, s) \in \mathfrak{H}^\eta \times \mathbf{C}$ by the same principle as in (5.17), that is, $\mathcal{E}_a\big(y(\mathbf{i}), s\big) = E(q\Sigma_{\mathbf{h}}^{-1}y, s)$ for every $y \in G^\eta_{\mathbf{a}}$. Then putting $\mathfrak{z} = y(\mathbf{i})$, we have $\mathcal{E}_a(\mathfrak{z}, s) = \sum_{\xi \in G^\varphi} p_\xi(\mathfrak{z})$ with

$$p_\xi(\mathfrak{z}) = \varepsilon_{\mathfrak{c}}(\alpha q \Sigma_{\mathbf{h}}^{-1} y, s) \quad \text{for} \quad \alpha = S^{-1}(\xi, 1)S.$$

By Lemma 6.2 (1) and by our definition of $\varepsilon_{\mathfrak{c}}$ we have $p_\xi \neq 0$ if and only if $\xi \in \Xi$. Also, by (5.12) and Lemma 6.3 (2),

$$p_\xi(\mathfrak{z}) = \varepsilon(\alpha_{\mathbf{h}} q \Sigma_{\mathbf{h}}^{-1})^{-s} \varepsilon(\alpha_{\mathbf{a}} y)^{-s} = |\lambda(\sigma)_{\mathbf{h}}|_{\mathbf{A}}^{-s} N\big(\nu^\sigma(a^{-1}\xi)\big)^{-s} \Delta(\alpha\mathfrak{z})^{s/2}.$$

To simplify our notation, put $N^\sigma_a(\xi) = N\big(\nu^\sigma(a^{-1}\xi)\big)$. Thus

$$\mathcal{E}_a(\mathfrak{z}, s) = |\lambda(\sigma)_{\mathbf{h}}|_{\mathbf{A}}^{-s} \sum_{\xi \in \Xi} N^\sigma_a(\xi)^{-s} \Delta(\alpha\mathfrak{z})^{s/2}.$$

For $(z, w) \in \mathcal{Z}^\varphi \times \mathcal{Z}^\varphi$ put $H_a(z, w; s) = \mathcal{E}_a\big(f_U(z, w), s\big)$. Then

$$H_a(z, w; s) = \sum_{\xi \in \Xi} p_\xi(f_U(z, w))$$
$$= 2^{-n|\mathbf{d}|s} |\lambda(\sigma)|_{\mathbf{A}}^{-s} \sum_{\xi \in \Xi} N_a^\sigma(\xi)^{-s} \Delta(\xi z)^{s/2} \Delta(w)^{s/2} \Delta(w, \xi z)^{-s}$$

by Lemma 6.5. Since $\varphi'_v \in GL_n(\mathfrak{g}_v)$ for every $v \in \mathbf{h}$, we see from (4.6) that $|\lambda(\sigma)|_{\mathbf{A}}^2 = |\lambda(\varphi)|_{\mathbf{A}}^{-1} = 1$.

Now we need a formula

$$\text{(6.11)} \qquad \int_{\mathcal{Z}^\varphi} \Delta(z)^{s/2} \Delta(w)^{s/2} \Delta(w, z)^{-s} d_\varphi^* w$$
$$= \left\{ 2^{n(n+3-2s)/2} \pi^{n(n+1)/2} \frac{\Gamma_n^1((s-n-1)/2)}{\Gamma_n^1(s/2)} \right\}^{|\mathbf{b}|}$$
$$\cdot \left\{ 2^{n(n+2-2s)} \pi^{n^2+(n/2)} \frac{\Gamma_n^2(s-n-1)}{\Gamma_n^2(s-(1/2))} \right\}^{|\mathbf{c}|} \prod_{v \in \mathbf{d}'} \left\{ \pi^{2q_v r_v} \frac{\Gamma_{q_v}^4(s-1-2r_v)}{\Gamma_{q_v}^4(s-1)} \right\},$$

where $d_\varphi^* w$ is defined by (4.18). The integral on the left-hand side can be decomposed into the product of the integrals over $\mathcal{Z}_v^\varphi$. For $v \in \mathbf{c}$ and $v \in \mathbf{d}'$ the integrals are given in [S7], Lemmas 7.1 and 7.2; for $v \in \mathbf{b}$ the value of the integral is given by [S5], Proposition A2.11. Thus we obtain (6.11).

6.7. Put $\Gamma^a = G^\varphi \cap aD^\varphi a^{-1}$, $\Phi_a = \Gamma^a \backslash \mathcal{Z}^\varphi$, $R^a = \Gamma^a \backslash \Xi$, and $H'_a(z, w) = \sum_{\xi \in R^a} p_\xi(f_U(z, w))$. Assuming that $\mathcal{Z}^\varphi$ is nontrivial, we consider

$$\text{(6.12)} \qquad \int_{\Phi_a} H_a(z, w; s) d_\varphi^* w \qquad (\Phi_a = \Gamma^a \backslash \mathcal{Z}^\varphi).$$

Then we can easily verify that

$$\text{(6.13)} \qquad H_a(z, w; s) = \sum_{\gamma \in \Gamma^a} H'_a(z, \gamma w),$$

so that (6.12) equals

$$\ell \int_{\mathcal{Z}^\varphi} H'_a(z, w) d_\varphi^* w,$$

where $\ell = [\Gamma^a \cap \{\pm 1\} : 1]$. By (6.11) this gives

$$\ell \cdot 2^{-n|\mathbf{d}|s} c(s) \sum_{\xi \in R^a} N_a^\sigma(\xi)^{-s},$$

where $c(s)$ is the product of (6.11). Notice that by Lemma 4.3(2) we have

$$\mathcal{T}(s) = \sum_{a \in \mathcal{B}} \sum_{\xi \in R^a} N(\nu^\sigma(a^{-1}\xi))^{-s}.$$

Thus we obtain

$$\text{(6.14)} \qquad 2^{-n|\mathbf{d}|s} c(s) \mathcal{T}(s) = \ell^{-1} \sum_{a \in \mathcal{B}} \int_{\Phi_a} H_a(z, w; s) d_\varphi^* w.$$

(In fact, $\mathcal{B}$ consists of a single element if $\mathcal{Z}^\varphi$ is nontrivial, but we use $\sum_{a \in \mathcal{B}}$ with our later discussion for trivial $\mathcal{Z}^\varphi$ in view.) Put $\mu = 2n+1$. Now the residue of $H_a(z, w; s)$ at $s = \mu$ is the residue of E_r at μ, where $r = q\Sigma_{\mathbf{h}}^{-1}$. Its value is $R(\mathfrak{c})$ given in Theorem 5.11. Thus the residue of the right-hand side of (6.14) at μ is exactly $R(\mathfrak{c})\mathfrak{m}(G^\varphi, D^\varphi)$. The residue of the left-hand side of (6.14) at μ can be obtained from the formula for $\mathcal{T}$ in Proposition 4.5. Arranging various factors suitably, we obtain $\mathfrak{m}(G^\varphi, D^\varphi)$ as stated in Theorem 4.7.

6.8. In the above computation we assumed that $\mathcal{Z}^\varphi$ is nontrivial. Suppose now that $G^\varphi_{\mathbf{a}}$ is compact, so that $\mathcal{Z}^\varphi$ consists of a single point $\mathbf{1}^\varphi$. Then (6.13) can be written

$$H_a(\mathbf{1}^\varphi, \mathbf{1}^\varphi;\, s) = 2^{-n|\mathfrak{d}|s}[\Gamma^a : 1]\sum_{\xi\in R^a} N^\sigma_a(\xi)^{-s},$$

and hence, instead of (6.14) we have

$$2^{-n|\mathfrak{d}|s}\mathcal{T}(s) = \sum_{a\in\mathcal{B}}[\Gamma^a : 1]^{-1}H_a(\mathbf{1}^\varphi, \mathbf{1}^\varphi;\, s).$$

Taking the residue at $s = \mu$, we obtain Theorem 4.7 when $G^\varphi_{\mathbf{a}}$ is compact. The formula in this case can be obtained from the noncompact case by taking $\mathbf{b} = \mathbf{c} = \mathbf{d}' = \varnothing$.

References

[E] M. Eichler, Über die Idealklassenzahl total definiter Quaternionen-Algebren, Math. Z. **43** (1937), 102–109.

[S1] G. Shimura, Arithmetic of alternating forms and quaternion hermitian forms, J. Math. Soc. Japan, **15** (1963), 33–65.

[S2] G. Shimura, Arithmetic of unitary groups, Ann. of Math. **79** (1964), 369–409.

[S3] G. Shimura, Confluent hypergeometric functions on tube domains, Math. Ann. **260** (1982), 269–302.

[S4] G. Shimura, On the transformation formulas of theta series, Amer. J. Math. **115** (1993), 1011–1052.

[S5] G. Shimura, Euler products and Eisenstein series, CBMS Regional Conference Series in Mathematics, No. 93, Amer. Math. Soc., 1997.

[S6] G. Shimura, An exact mass formula for orthogonal groups, Duke Math. J. **97** (1999), 1–66.

[S7] G. Shimura, Generalized Bessel functions on symmetric spaces, J. reine angew. Math. **509** (1999), 35–66.

[S8] G. Shimura, The number of representations of an integer by a quadratic form, Duke Math. J. **100** (1999), 59–92.

[Si] C. L. Siegel, Symplectic Geometry, Amer. J. M. **65** (1943), 1–86.

99e

André Weil as I knew him

Notices of the American Mathematical Society, vol. 46, No. 4 (April 1999), 428-433*

Bathed in the sunlight of late summer, I was walking a quiet street of Takanawa, a relatively fashionable district in southern Tokyo, toward the Prince Hotel Annex, where André Weil was staying. It was the afternoon on a warm day of early September in 1955. He was among the nine foreign participants of the International Symposium on Algebraic Number Theory, to be held in Tokyo and Nikko that month. The Korean war had ended two years earlier, and in the United States, Eisenhower's first term had begun in the same year. Five years later, in 1960, his planned visit to Japan would be hindered by the almost riotous demonstrations of labor unions and students in the city, but nobody foresaw it in the peaceful atmosphere of the mid 1950s. While walking, I had a mildly uplifted feeling of expectation and curiosity about what would happen, the first of those I would experience many times later, whenever I was going to see Weil.

My acquaintance with him began in 1953, when I sent my manuscript on "Reduction of algebraic varieties with respect to a discrete valuation of the basic field" to him in Chicago, asking his opinion. I told him my intention of applying the theory eventually to complex multiplication of abelian varieties. In his answer, dated December 23, 1953, he was quite favorable to the work, and encouraged me to proceed in that direction; he also advised me to send the paper to the *American Journal of Mathematics*, which I did. By that time I had read his trilogy *Foundations, Courbes algébriques,* and *Variétés abéliennes,* as well as his 1950 Congress lecture [50b][1)] and a few more papers of his. I was also aware of the existence of many of his other papers, or had some vague ideas about them, [28], [35b][2)], [49b], [51a], for example. But I don't think I had read all those before 1955. The article "L'avenir des mathématiques" [47a] and his review [51c] of Chevalley's book on algebraic functions were topics of conversation among young mathematicians in Tokyo. Later, while in Japan, when he was asked to offer his opinion on various things, he jokingly complained that he was being treated like a prophet, not a professor. But to some extent that was so even before his arrival.

In any case, when he accepted the invitation to the Tokyo-Nikko conference, we young mathematicians in Japan, expected him with a sense of keen anticipation. I shook hands with him for the first time on August 18, in a room in the Mathematics Department, University of Tokyo. He looked gentler than the photo I had seen somewhere. He was forty-nine at that time. Our meeting was short, and there was not much mathematical discussion, nor did he make any strong impression on me that day. He was given, perhaps a few days later, a set of mimeographed preliminary drafts of papers of most Japanese participants, including my 49-page manuscript titled "On complex multiplications", which was never published in its original form.[3)]

About two weeks later a message was forwarded to me: Professor Weil wishes to see me at his hotel. So I brought myself there at the appointed time. He appeared in the lobby wearing beige trousers with no jacket, or tie. He had read my manuscipt by then, and sitting on a patio chair in a small courtyard of the hotel, he asked many questions and made some comments. Then he started to talk about his ideas on polarization of an abelian variety and a Kummer variety. He scribbled various formulas on some hotel stationery, which I still keep in my possession. At some point he left his chair; pacing the courtyard from one end to the other, he impatiently tried to pour his ideas into my head. He treated me as if I was an expert who knew everything. I knew of course what a divisor meant and even the notion of linear and algebraic equivalence, as I had read his 1954 *Annalen* paper on that topic, but I lacked the true feeling of the matter, not to speak of the historical perspective. Therefore, though I tried hard to follow him, it is fair to say that I understood little of what he said. At the end we had tea, and he ate a rather large piece of cake, but I declined his offer of the same, perhaps because his grilling lessened my appetite.

During the conference and his stay in Tokyo afterward, I saw him many times. On each occasion he behaved very naturally, if in a stimulating way. It was as if that hotel encounter had the effect of immunization for me, and possibly for him too. I remember that I asked him about the nature of the periods of a differential form of the first kind on an abelian variety with complex multiplication. He said, "They are highly transcendental," which was not a satisfactory answer, but as good as anything under the circumstances. At least, and at long last, I found someone to whom I could ask such a question. Those several weeks were truly a memorable and exciting period. To make it more exciting, one of my colleagues would make a telephone call to the other, imitating Weil's voice and accent: "Hello, this is Weil. I didn't understand what you said the other day; so I'd like to discuss with you . . . ." Sometimes the prank worked. A few days before his leaving Tokyo for Chicago, I, together with three such naughty boys, visited him in another hotel in the same area. Taniyama promised to come, but didn't; apparently he overslept as usual. During our conversation, Weil advised us not to stick to a wrong idea too long. "At some point you must be able to tell whether your idea is right or wrong; then you must have the guts to throw away your wrong idea."

As he said in his *Collected Papers*, his stay in Japan was one of his most enjoyable and gratifying periods. He found an audience of young people who were not afraid of him and was sophisticated enough to understand, or at least willing enough to try to understand, his mathematics; he certainly had an audience in the United States then, but apparently of a different kind.

More than two years passed before I saw him again, which was in Paris in November 1957. Henri Cartan, accepting his suggestion, had secured a position of chargé de recherches at CNRS for me. Weil was on leave from Chicago for one year, and sharing an office with Roger Godement at the Institut Henri Poincaré, but he occupied it alone for most of the time. In that period he was working on various problems on algebraic groups, the topics which can be seen from [57c], [58d], and [60b], for example. He was giving lectures on one such subject at the École Normale Supérieure and regularly attended the Cartan seminar. He lived in an apartment at the southeast corner of the Luxembourg garden with a fantastic view of Sacré Coeur to the far north and the Eiffel tower to the west. One of his favorite restaurants was Au vieux Paris, in the back of the Panthéon. A few days after my arrival, he invited me to have lunch there. I remember that he had radis au beurre (radish with butter) and lapin (rabbit) sauté, a fairly common affair in those days, but perhaps somewhat old-fashioned nowadays. I don't remember his choice of wine, but most likely a fullsized glass of red wine for each of us. To tell the truth, it was not rare to find him snoozing during the seminar. From his apartment, the institute and the Panthéon could be reached in less than ten minutes on foot. Paris in the 1950s retained its legendary charm of an old city, which had not changed much — he once told me — since the days of his childhood. It is sad to note that the city went through an inevitable and drastic transformation in the 1970s.

Though I was working on a topic different from his, he was earnestly interested in my progress, and so I would drop into his office whenever I had something to talk about. For instance, one day I showed him some of my latest results for which I employed Poincaré's theorem on the number of common zeros of theta functions. He smiled and said, "Oh, you use it, but it is not a rigorously proved theorem." Then he advised me to take a different route, or to find a better proof; later he told me a recently proved result concerning divisors on an abelian variety, by which I was able to save my result, as well as Poincaré's theorem.

On another occasion, I heard some shouting in his office. As I had only a brief message to him, I knocked on the door. He opened it and introduced me to Friederich Mautner[4)], professor at Johns Hopkins, who was his shouting partner. After a minute or so I left. As soon as I closed the door, they started their shouting again. When I was walking through the corridor after spending half an hour in the library, the shouting match was still going on; I never knew when and how it began and ended, nor who won.

From time to time he fetched me for a walk in the city. The topics of our conversation during those walks were varied; he would suggest to me, for example, that I go to churches to listen to religious music; he said it was necessary for me only to stand up and sit down when others did. When asked about his faith, he said, "Pas du tout" ("Not at all"). According to him, one of the best way to learn French or any foreign language was to see the same movie in that language again and again, staying in the same seat in the same movie theatre, a piece of advice I followed perhaps too faithfully. It was the time when Brigitte Bardot and Zizi Jeanmaire were at their zenith. Another method he suggested was to read newspapers, but I was not so dilligent in this task. Perhaps as he became impatient with my slow progress in French, he asked me whether I was doing my homework in that respect. I dodged the issue by mentioning an old Oriental saying, "He who runs after two rabbits will catch neither." Maybe I subconsciously remembered the rabbit for his meal. "What's your rabbit? Hecke operators?" he asked. Then we discussed about the possible method or philosophy of how a Frobenius of a reduced variety can be lifted. A few days later he caught me in the library and asked again, "What about your rabbit?" He was an extremely sharp man, and clearly he sensed that I was up to something, which was true. In this article, however, I should leave the rabbit at large, merely mentioning that he would later say, "How is your rabbitry doing?"

Starting in the fall of 1958 he was at the Institute for Advanced Study permanently, and I was a member there for that academic year. So I practically followed him, and I had the same daily routine with him for another several months. Looking back on those days, I am filled with a sense of deep gratitude to him for paying such an unusual and personal attention to me; also, I must note, to my regret, that unaware of the real meaning of my situation at that time, I did not take full advantage of my fortunate privilege of being constantly with such an extraordinary man in his prime.

In the spring of 1961, he spent a few months in Japan with his wife, Eveline. Though they undoubtedly enjoyed their stay, and I was happy to have a person at hand who really understood me — perhaps the only one at that time — I may be excused for saying that overall his presence was less than a pale revival of his former visit.[5)] As for myself, after spending three years in Japan, I came back to Princeton in September 1962, when I began a new and long chapter of my relationship with him. To continue my narrative, I will now present some interesting aspects of his words and deeds in this period irrespective of the chronological order of the events.

As already mentioned above, he liked to walk, partly for the purpose of physical exercise. In Princeton every Sunday he would walk one and half miles from his home to buy the Sunday *New York Times*, and so, according to his daughters, his church denomination was pedestrian. At the Institute he would occasionally pick a walking partner among the members. He was not a good walker, however. Though he was physically fit and walked briskly, he often fell on his face by tripping on something on the ground. That happened when I was with him in the Institute wood, but I pretended to have seen nothing, as he hated being helped on such an occasion. Though he was not injured then, he was not so lucky other times. During such a walk, he would answer my questions, or would tell his stories. Here are some samples:

When he was twelve or thirteen,[6)] there was a magazine for elementary mathematics asking the reader to send in solutions to the problems; then they would print best solutions. He contributed many as he found great pleasure in seeing his name in the magazine, but he graduated from that level after about two years. Then he said: "Maybe I should have included some of the solutions in my oeuvres, he! he! he!"

Around the time when he was at Haverford, he asked Hermann Weyl to lend him some money. "How much?" asked Weyl. "Well, four or five hundred dollars."

Then Weyl brought out his checkbook, and after thinking awhile he signed a check for four hundred and fifty dollars.

When he was teaching at Lehigh, a student asked him for help in calculus. After they spent a lot of time struggling to find out what his problem was, the student finally said, "I don't seem to understand this symbol x."[7] He referred to his Lehigh days as his period of "overemployment".

A French gentleman's ideal is to have three concurrent loves: the first one, whom he cares about at present; the second, a potential one, whom he has his eye on with the hope that she will eventually be his principal love; the third, the past one, with whom he hasn't completely cut off his relations. Then he observed: "It's a good idea for a mathematician to have three mathematical loves in the same sense."

He would talk about Baudelaire, Proust, and Gide, their homosexuality in particular, Paul Claudel's treatment of his sister Camille, and also about the letter exchange between Paul Claudel and Madeleine Gide. He amused himself by twisting each story in his own fashion to make it funny, often with a piquant effect.

I asked, for what reason I don't remember, whether he read detective stories. "Yes, but only when I have a cold," he said, and added, "You know, when you have a cold, there is nothing else to do but read detective stories." He was rather apologetic, and so I asked, "How often do you catch cold?" "Very often" was his answer.

As to Fields medals, he said: "It's a kind of lottery. There are so many eligible candidates, and the whole selection process is a matter of chance. Therefore the prize could be given to *any of them* as in a lottery."[8]

He used to say that a good mathematician must have two good ideas. "It is possible for someone to have a really good idea, but it may be just a fluke. Once the person has a second good idea, then there is a good chance for him to develop into a better mathematician." He mentioned a well-known American as a prolific mathematician with a single idea. He also noted Mordell as a counterexample to his principle.

He could say something even harsher, but that was rare. In the summer of 1970 after the Nice Congress, I was talking with him somewhere in the Institute about French mathematicians. He observed that there were three young mathematicians in Paris who started brilliantly, and so there were high expectations for them. He mentioned three well-known names and said, "What happened to them? They utterly failed to produce anything great." That was more than a quarter century ago, but I cannot tell whether or not he changed his opinion, as we never talked on that matter again. After around 1975 he expressed, more than once, his pessimistic view that French mathematics had been declining for some time. Therefore we should perhaps take his criticism in that context.

He held Riemann and Poincaré in high esteem, which was more than natural; Hecke was also his favorite. He rarely talked about Hilbert in our conversation. He didn't think much of Klein, which is not surprising. Picard was depicted by him as formal and stiff. Among his contemporaries, he thought highly of Siegel, and spoke of Chevalley in amicable terms, but not so with Weyl, about whom he seemed to have a kind of ambivalence. He recognized the unusual talent of Eichler.[9] Hadamard was his teacher, and their relationship is well documented in his autobiography. He paid due respect to Hasse, though he remembered the fact that Hasse wore a Nazi uniform at some point.[10] He told me several anecdotes about Hardy, but he presented each story in a sarcastic tone. "Hardy's opinion that mathematics is a young man's game is nonsense," he said.

It may be too optimistic a view to say that most people mellow with advancing age. At least many do, and there are those who don't. It is told, for example, that Saint-Saëns achieved an ever-increasing reputation as a man of bad temper through his long life of eighty-six years. Weil did mellow, but even after the age of seventy he was capable, if rarely, of being childishly irritable, as can be seen from the following episode. But first let me note: Around 1976 or 1977 he declared, "I am no longer a mathematician; I am a mathematical historian." Apparently he realized that there were no more subjects he could handle better than the younger generation. Coming to my story: In my teens I somehow got hold of a copy of a pirate edition, which was being called the Shanghai edition, of *Eindeutige Analytische Funktionen* by Rolf Nevanlinna. I enjoyed reading the first one-third of the book, but gave up on the rest. Still, my reading of the book remains as one of my fond memories. When I recognized Nevanlinna in a lecture hall at the 1978 Helsinki Congress, I introduced myself and shook hands with him, an incident which, in my youth I never imagined, would happen. He was eighty-three then. Weil gave a lecture titled "History of mathematics: Why and how" there.

After the congress, I spent a week in Paris, and one day I was sipping coffee with Weil in a café near his apartment. I told him about that happy experience of mine at Helsinki. But he was much displeased with my story. He said with a grimace that Nevanlinna was not such a good mathematician who was worthy of my esteem, and so on. I was dumbfounded; I never idolized Nevanlinna, whose name I knew before acquainting myself with any of Weil's works, simply because the book was accidentally available. That must have been clear to him. After all, it was none other than Nevanlinna who saved him from being executed by the Finnish police, a fact he told me some years earlier, and narrated in his autobiography, which also includes a passage of the Weil couple's happy stay in Nevanlinna's villa in 1939.

I should add, however, that he could be found on the other side of the world. When there was a discussion of a new appointment at the Institute, Morton White, professor of the school of history, was fiercely against the proposition, and at the faculty meeting he expressed his opinion in a heated fashion. Then Weil, sitting next to him, said, "Calm down, please, calm down." White later told me that he thought the scene rather funny in view of the normal temperament of Weil.

After Eveline's passing away in May 1986 at the age of seventy-five, his daughter Nicolette bought a microwave oven for him. However, saying that he didn't like to "push the button," he never touched it, and so the oven was returned to the dealer. The Weils had been our regular dinner guests, but since then naturally he alone was with us, which happened not infrequently. It was sometime in December 1987. Weil, Hervé Jacquet, Karl Rubin, Alice Silverberg, my wife Chikako, and I had dinner at a Chinese restaurant, and were having dessert at our place. When I prodded the guests to tell their ambitions in their next lives, Jacquet said he would like to be an opera singer, and that was not a joke for him. In fact, opera singing was his first love, mathematics being merely the second. Next, "I want to be a Chinese scholar studying Chinese poems," said Weil. After visiting China twice, he had been reading English translations of Chinese standard literature like *The Dream of the Red Chamber.* "That may be a rather dull life, and I don't think a person like you can stand it," said I. "All right then, I will be a house cat. The life of a house cat is very comfortable." Pointing to our neighbor's female white cat who was also a guest, he said, "Maybe she will be my mother." Then Rubin said, "Perhaps a Chinese cat is a good solution." With laughter everybody accepted it. That was about a week or two before Christmas, and so after a few days, Chikako brought him a stuffed cat as a Christmas present, which pleased him greatly. In fact, the Weil family used to have a cat, and once he defended himself for having a Christmas tree in his house by saying that they had it because their cat loved it.

He was conscious of his old age, particularly after he became a widower. According to what he said: Eveline was afraid of becoming senile. But she was not at all senile when she died. A famous French mathematician, who lived beyond eighty, was senile in his last two years, but he knew it himself. So when he had visitors, he held a newspaper to show that he was at least able to read, but the paper was often upside down. Another, who lived longer, was not like that; even so, when Weil visited him, he brought out and showed him, one after another, the diploma of each of the many honorary degrees he had received.

As for Weil himself, he showed no such sign, as far as I remember. I talked with him sometime in November 1995 for half an hour or so in his office. He was alert and able to make a reasonable judgment on the matter for which I went to see him. There was a lunch party for his ninetieth birthday in May 1996 at a restaurant in Princeton; though he did not talk much, he was in a good mood. Before and after that Chikako had lunch at the Institute cafeteria several times; she would find him eating mostly alone, sometimes with his daughters; she would say hello to him, to which he would reply, "Is Goro here?" So she was relieved to find that he at least remembered her as someone related to me.

I saw him for the last time on December 19, 1996. For some reason he phoned me the day before. Since he had hearing difficulties, he finally suggested that I see him at the Institute. I proposed some date, but he said, "No, why don't you come tomorrow; otherwise I won't remember." So I had lunch with him there that day. From the previous night it had been drizzling endlessly. When I met him in the common room of Fuld Hall, he did not have his hearing aid, and he asked me to drive him home to get it. After getting it we went into the dining hall. He used to eat well, and almost twice as much as myself. Around 1980, André, Eveline, Chikako, and I had lunch together at a restaurant in New Hope, Pennsylvania. That was a buffet style affair, and he was in high spirits. I remember that his appetite impressed the remaining three. Incidentally, he was not fussy about wine. Not that he did not care, but it is my impression that Eveline cared more.

I was curious how he would eat this time. Not surprisingly, compared with what he ate sixteen years ago, the quantity he took was modest, less than half of the previous meal. Since he had hearing problems, it was difficult to conduct our conversation smoothly, and I often had to write words and sentences on a piece of paper. Unlike the occasion forty-one years ago, this time it was I who was writing. I was working on the Siegel mass formula[11] with a new idea at that time, and that was one of his favorite topics. So I asked him about the history of that subject. For example, I asked him whether or how he studied the works of Eisenstein, Minkowski, and Hardy. He said he didn't remember about Eisenstein,[12] but he had studied a little, but not much, of Minkowski's work; he never studied Hardy. He kept saying that it was a long time ago, and so he didn't remember, which must be true, and so we should not accept what he said at face value. In fact, to check that point, I asked him whether Minkowski was reliable. He said, "I think so." At that point I realized that his recollection was faulty, since Minkowski gave an incorrect formula, as Siegel pointed out, and that was known to most experts. If I was asking questions on what he did in his twenties or thirties, he might have remembered things better, but at that time I did not take into account the fact that he worked on the Siegel formula in his fifties.

I asked him whether he was writing something on a historical topic. He said, "I cannot write any more." To cheer him up I then said, "That's why I told you long ago to get a computer." He also said he was half blind. Toward the end of the meal he said, "I'd like to see the Riemann hypothesis settled before I die, but that is unlikely."

That reminded me of a party at Borel's place in the 1970s. Wei-Liang Chow was the guest of honor. I was talking with Chow and Borel about a passage in Charlie Chaplin's autobiography. In it Chaplin in his twenties met a fortuneteller in San Francisco, who told him that he would make a tremendous fortune, would be married so many times with so many children, and would die of bronchial pneumonia at the age of eighty-two. Hearing this story, Weil said, "Well, in my autobiography I might write that in my youth I was told by a fortuneteller that I would never be able to solve the Riemann hypothesis."

When we left the dining hall and were walking to the parking lot, he said, "You are certainly disappointed, but I am disappointed too," and added, after a few seconds, "with myself." He knew that I was expecting him to say something about Siegel's work. He again said, "I cannot write any more." I drove him home and left. He was able to walk slowly but I could not say he was in good shape; still he was not in terrible shape, and so I had a sense of relief. While driving home alone under still drizzling rain, I could not but recall our hotel encounter in 1955 and the lunch in 1957, though I did not think much about the possibility that I would never see him again.

André Weil as a mathematician will of course be remembered by his colossal accomplishments witnessed by the three volumes of his *Collected Papers* and several books, the trilogy mentioned at the beginning in particular. In my mind, however, he will remain chiefly as the figure with two mutually related characteristics: First, he was flexible and receptive of new ideas of others and new directions, quite unlike many of the younger people these days who can work only within a well-established framework. Second, more importantly and in a similar vein, he had deep and penetrating understanding of mathematics, or rather, he strived tirelessly to understand the real meaning of every basic mathematical phenomenon, and to present it in a clearer form and in a better perspective. He did so by endowing each subject with new concepts and setting up new frameworks, always in a fresh and fundamental way. In other words, he was not a mere problem-solver. Clearly his death marked the end of an era, and at the same time left a large vacuum which will not easily be filled for a long time to come.

Endnotes

1) Each number in brackets refers to the article designated by that number in his *Collected Papers* with "19" omitted.

2) It seems that [35b] is the first paper which mentions the fact that the coordinate ring of a variety is integral over a subring obtained by considering suitable hyperplanes (see *Collected Papers*, vol. I, p. 89). Zariski attributed it to E. Noether. It is my impression that she considered generic hyperplane sections, but not the fact of elements being integral. Weil agreed with me on this and said: "Perhaps Zariski didn't like to refer to the work of a younger colleague, a common psychological phenomenon." On the other hand, though he must have had his own citation policy, frankly I had difficulty in accepting it occasionally. See 9) below.

3) As to my paper on "Reduction of algebraic varieties etc." he said "il (Shimura) me dit, il eût plutôt eu en vue d'autres applications." (*Collected Papers*, vol. II, p. 542.) This is not correct. Probably he misunderstood me when I told him that I was interested in Brauer's modular representations at one time. Brauer was also a participant of the conference.

4) Mautner was responsible for introducing Weil to Tamagawa's idea; see Weil's comments on [59a].

5) In his *Collected Papers* he says practically nothing about his second visit, though he mentions it; see vol. II, p. 551.

6) This is what he told me. In his autobiography, however, the story is assigned to an earlier period, which may be true.

7) This is also what he told me. A somewhat different version is given in his autobiography.

8) There is a big difference. In order to win a lottery, we have to buy a ticket, but by doing so, we put our trust in the fairness of the system.

9) Whenever he spoke of strong approximation in algebraic groups, he always referred to Kneser's theorem. That is so in [65], for example, which is understandable. But that was always so even in his lectures in the 1960s, though in [62b] Eichler is mentioned in connection with the fact that the spinor genus of an indefinite quadratic form consists of a single class. However strange it may sound, it is possible, and even likely, that he was unable to recognize Eichler's fundamental idea and decisive result on strong approximation for simple algebras and orthogonal groups, and he knew only its consequence about the spinor genus. In his *Collected Papers* he candidly admits his ignorance in his youth. Though he had wide knowledge, his ignorance of certain well-known facts, even in his late years, surprised me occasionally. He knew Hecke's papers to the extent he quoted them in his own papers. It would be wrong, however, to assume that he was familiar with most of Hecke's papers. Besides, his comments in his *Collected Papers* include many insignificant references. For these reasons, the reader of those comments may be warned of their incompleteness and partiality.

10) According to Weil, once Hasse in such a uniform visited Julia, who became anxious about the possibility that he would be viewed as a collaborator.

11) In [65] he says, "On a ainsi retrouvé, quelque peu généralisées, tous les résultats démontrés par Siegel au cours de ses travaux sur les formes quadratiques, ainsi que ceux énoncés à la fin de [12] (Siegel's Annalen paper in 1952) á l'exception des suivants. Tout d'abord, ..." (Collected Papers, vol. III, p. 154). I think this is misleading, since the list of exceptions does not include the case of inhomogeneous forms, which Siegel investigated. It is true that Siegel's product formula for an inhomogeneous form in general can be obtained from the "formule de Siegel" (in Weil's generalized form, combined with some nontrivial calculations of the Fourier coefficients of Eisenstein series); and, one might say, that is not so important. Still, it should be mentioned at least that the inhomogeneous case is not just the matter of the Tamagawa number, and that nobody has ever made such explicit calculations in general, even in the orthogonal case. In the mid 1980s I asked Weil about this point, but he just said, "I don't remember."

12) In [76c] he reviews the complete works of Eisenstein; also the title of [76a] is *Elliptic functions according to Eisenstein and Kronecker.* It is believable, however, that he did not study Eisenstein's papers on quadratic forms in detail, though he must have been aware of them.

01b

Arithmeticity of Dirichlet series and automorphic forms on unitary groups

unpublished

Introduction

Our notation is basically the same as in [97] and [00]. For example, for an associative ring R and positive integers m and n we denote by R_n^m the module of $(m \times n)$-matrices with entries in R. Also, $\overline{\mathbf{Q}}$ denotes the algebraic closure of $\mathbf{Q}$ in $\mathbf{C}$. For a finite-dimensional vector space W over $\mathbf{Q}$ we denote by $\mathcal{S}(W)$ the set of all finite $\mathbf{C}$-linear combinations of functions on W each of which is the characteristic function of a coset of W modulo a $\mathbf{Z}$-lattice in W.

Now this paper can be divided into two parts, each part being composed of four sections. The problems in the first part are formulated in terms of automorphic forms on a unitary group

$$(0.1)\qquad G = \{\alpha \in GL_{2n}(K) \mid \alpha\eta_n \cdot {}^t\alpha^\rho = \eta_n\}, \qquad \eta_n = \begin{bmatrix} 0 & -1_n \\ 1_n & 0 \end{bmatrix},$$

where K is a totally imaginary quadratic extension of a totally real algebraic number field F, and ρ is the Galois involution of K/F. Our first principal problem concerns a function

$$(0.2)\qquad g(z) = \sum_{\xi \in K_n^n} \lambda(\xi)\det(\xi)^\kappa \exp\Bigl(2\pi i \sum_{v \in \mathbf{a}} \mathrm{tr}\bigl(({}^t\xi^\rho \omega \xi)^{\tau_v} z_v\bigr)\Bigr).$$

Here we let $\mathbf{a}$ denote the set of all archimedean primes of F, and fix, for each $v \in \mathbf{a}$, an isomorphic embedding τ_v of K into $\mathbf{C}$ that extends v; $z = (z_v)_{v \in \mathbf{a}}$ is the variable on $\mathcal{H}_n^{\mathbf{a}}$, the product of $\mathbf{a}$ copies of the space $\mathcal{H}_n$ defined by

$$(0.3)\qquad \mathcal{H}_n = \{z \in \mathbf{C}_n^n \mid i(z^* - z) > 0\};$$

$\lambda \in \mathcal{S}(K_n^n)$, $\kappa = \sum_{v \in \mathbf{a}} \kappa_v \tau_v$ with $0 < \kappa_v \in \mathbf{Z}$ and $\omega = {}^t\omega^\rho \in GL_n(K)$; we assume that $\omega^{\tau_v} > 0$ for every $v \in \mathbf{a}$. We can show that g is an automorphic form with respect to a congruence subgroup of G. As our first main result we shall prove (Theorem 3.5) that

$$(0.4)\qquad \langle g, g\rangle \in \pi^{-dn^2} p_K\bigl(\textstyle\sum_{v \in \mathbf{a}} \kappa_v \tau_v,\ 2n\tau\bigr)\overline{\mathbf{Q}},$$

provided $\kappa_v \geq n$ for every $v \in \mathbf{a}$ and λ is $\overline{\mathbf{Q}}$-valued, where $\langle g, g\rangle$ is the inner product whose precise definition is given in (1.6) below, $d = [F : \mathbf{Q}]$, and p_K is the period symbol introduced in [80] and [98]. In fact, $p_K\bigl(\sum_{v \in \mathbf{a}} \kappa_v \tau_v,\ 2n\tau\bigr)$ can be represented modulo $\overline{\mathbf{Q}}^\times$ by the value of a $\overline{\mathbf{Q}}$-rational Hilbert modular form h of weight $(2n\kappa_v)_{v \in \mathbf{a}}$ at a point $(w^{\tau_v})_{v \in \mathbf{a}}$ of $\mathcal{H}_1^{\mathbf{a}}$ with a suitable $w \in K$.

In [97] and [00] we associated an Euler product $\mathcal{Z}(s)$ to a Hecke eigenform $\mathbf{f}$ on $G_{\mathbf{A}}$ and investigated its critical values. In general $\mathbf{f}$ consists of finitely many automorphic forms on $\mathcal{H}_n^{\mathbf{a}}$, which may be called the components of $\mathbf{f}$. Suppose that the components of $\mathbf{f}$ are of type (0.2) with a fixed κ. Then our next main result (Corollary 3.6) in a simplified form can be stated as follows: Let σ_0 be an integer such that $2n-\kappa_v \le \sigma_0 \le n$; suppose that $\kappa_v > 2n$ for every $v \in \mathbf{a}$, and at least one Euler factor of $\mathcal{Z}$ is trivial. (If the last condition is not satisfied, then we only have to eliminate any Euler factor.) Then

$$(0.5)\qquad \mathcal{Z}(\sigma_0) \in \pi^{-bn} p_K\bigl(\textstyle\sum_{v\in\mathbf{a}} \kappa_v \tau_v,\ 2n\tau\bigr)\overline{\mathbf{Q}},\ \textit{where}\quad b = n\textstyle\sum_{v\in\mathbf{a}}(\kappa_v + 2\sigma_0 - 2n + 1).$$

The second part of the paper concerns a certain Dirichlet series $W(s)$ and a function $\mathcal{F}(\mathfrak{z}, s)$ of $(\mathfrak{z}, s) \in \mathfrak{B} \times \mathbf{C}$, where $\mathfrak{B}$ is a hermitian symmetric space on which another type of unitary group, $G(H)$ of (0.8) below, acts. To define $W(s)$, we take a set of data (m, n, Y, f, ζ), where $0 \le m \in \mathbf{Z}$, $0 < n \in \mathbf{Z}$, Y is a direct sum of finitely many CM-fields which contain K such that $[Y : K] = m+n$, and f is an automorphic form on $\mathcal{H}_n^{\mathbf{a}}$ of weight $p = (p_v)_{v\in\mathbf{a}}$ with respect to a congruence subgroup of G. We take its Fourier expansion

$$f(z) = \sum_{\sigma\in S} c_f(\sigma) \exp\Bigl(2\pi i \sum_{v\in\mathbf{a}} \mathrm{tr}(\sigma_v z_v)\Bigr) \qquad (z \in \mathcal{H}_n^{\mathbf{a}}),$$

where $S = \{\, \sigma \in K_n^n \mid \sigma = {}^t\sigma^\rho \,\}$. Then, in an easier case W is given by

$$(0.6)\qquad W(s) = \sum_{y\in U\backslash \mathcal{Y}} \alpha(y) \det(y^\varepsilon)^{-\kappa} |\det(y^\varepsilon)|^{\kappa-p-2s\mathbf{a}} c_f\bigl(\mathrm{Tr}_{Y/K}(y\zeta y^*)\bigr).$$

Here U is a subgroup of $GL_n(\mathfrak{r})$ of finite index, where $\mathfrak{r}$ is the maximal order of K, and $\mathcal{Y}$ is the set of all $y = (y_i)_{i=1}^n \in Y^n$ such that the y_i are linearly independent over K; $\alpha \in \mathcal{S}(Y^n)$, κ is the same as in (0.2), and $\zeta = \zeta^\rho \in Y^\times$; $\mathrm{Tr}_{Y/K}(y\zeta y^*)$ denotes the $(n\times n)$-matrix whose (i, j)-entry is $\mathrm{Tr}_{Y/K}(\zeta y_i y_j^\rho)$ for $y = (y_i)_{i=1}^n \in Y^n$, where ρ is extended to Y so that its restriction to every simple factor of Y is complex conjugation; $y^\varepsilon = (y_v^\varepsilon)_{v\in\mathbf{a}}$ and y_v^ε is the $n\times n$-matrix whose components are $y_i^{\varepsilon_{vj}}$ with some homomorphisms ε_{vj} of Y into $\mathbf{C}$.

Under some natural conditions on ζ and $\{\varepsilon_{vj}\}$, and with a suitable choice of U, we can show that the sum of (0.6) is meaningful and convergent for sufficiently large $\mathrm{Re}(s)$; moreover, it can be continued as a meromorphic function to the whole complex plane. Furthermore, suppose that α is $\overline{\mathbf{Q}}$-valued and $c_f(\sigma) \in \overline{\mathbf{Q}}$ for every σ; suppose also that $\kappa_v > 2n$ for every $v \in \mathbf{a}$. Let s_0 be an integer such that $\kappa_v - p_v - (2/n)(\kappa_v - 1) \le s_0 \le \kappa_v - p_v$ and $s_0 - \kappa_v + p_v \in 2\mathbf{Z}$ for every $v \in \mathbf{a}$; $s_0 \ge m$; $s_0 \ne m+1$ if $F = \mathbf{Q}$. Then $W(s)$ is finite at $s = s_0/2$, and

$$(0.7)\qquad W(s_0/2) \in \pi^c p_K\bigl(\textstyle\sum_{v\in\mathbf{a}} \kappa_v \tau_v,\ n\tau\bigr) p_Y\bigl(\textstyle\sum_{v\in\mathbf{a}}\sum_{i=1}^n \kappa_v \varepsilon_{vi},\ \Phi\bigr)\overline{\mathbf{Q}},$$

where $c = n\sum_{v\in\mathbf{a}} \kappa_v - [F:\mathbf{Q}]n(n-1)/2$ and Φ is a certain CM-type of Y (Theorem 5.3).

The proof of (0.7) requires the arithmeticity of the function $\mathcal{F}(\mathfrak{z}, s)$ mentioned above at $s = s_0/2$. As a function of $\mathfrak{z}$, $\mathcal{F}$ is a nonholomorphic automorphic form on $\mathfrak{B}$ with respect to a congruence subgroup of a unitary group

$$(0.8) \qquad G(H) = \{\, \alpha \in GL_{m+n}(K) \,|\, \alpha H \cdot {}^t\alpha^\rho = H \,\},$$

where H is a hermitian element of $GL_{m+n}(K)$ whose signature at an archimedean prime of K is (m, n) or $(m+n, 0)$. Since the explicit form of $\mathcal{F}$ as an infinite series, (6.13), requires many more symbols, we content ourselves here with noting several essential points as follows:

(i) $\mathcal{F}(\mathfrak{z}, s)$ can be viewed as a generalization of Eisenstein series.

(ii) $\mathcal{F}(\mathfrak{z}, s_0/2)$, for s_0 under some conditions similar to those for s_0 as in (0.7), is a nearly holomorphic function of $\mathfrak{z}$, and even arithmetic (Theorems 6.5 and 6.7).

(iii) The series $W(s)$ appears as $\mathcal{F}(w, s)$ with a suitable $\mathcal{F}$ and a suitable CM-point w of $\mathfrak{B}$.

(iv) The arithmeticity of $\mathcal{F}(\mathfrak{z}, s_0/2)$ follows from (0.7), and (0.7) follows from the arithmeticity of $\mathcal{F}(\mathfrak{z}, s_0/2)$ in the case where H is totally indefinite. In that special case the desired arithmeticity follows from the special case of (0.7) in which $[Y : K] = n$. That special case of (0.7), Theorem 5.5, can be derived from (0.4). Finally, (0.4) can be proved by expressing the inner product in terms of the special value of a certain Eisenstein series at a CM-point.

(v) In the course of our exposition we prove some more results on related problems. Without describing them in detail, we merely mention that they are Theorems 2.2, 2.3, 3.4, 8.2, and 8.3.

Let us conclude the introduction by adding a few more remarks. First of all it should be noted that the notion of near holomorphy is indispensable for the proof of (0.7), because we obtain $W(\sigma_0/2)$ as $\mathcal{F}(w, s_0/2)$ with a CM-point w and $\mathcal{F}(\mathfrak{z}, s_0/2)$ is nearly holomorphic in $\mathfrak{z}$, but not holomorphic in $\mathfrak{z}$ in general.

Next, if $n = 1$, then $SL_2(F) \subset G \subset K^\times \cdot SL_2(F)$, so that g of (0.2) is a Hilbert modular form. In this case (0.4) was proved in [76] when $F = \mathbf{Q}$, and in [78] in the general case. Also (0.7) when $n = 1$ was given in [88]. The series $\mathcal{F}$ was essentially introduced in [86], but its arithmeticity was proved in [88] only for $n = 1$. In fact, the present paper grew out of our desire to obtain higher-dimensional generalizations of all these results.

Results of type (0.7) in the case $n = 1$ under weaker conditions were obtained by the author in some earlier papers, [80] in particular, in connection with the problem of arithmeticity on orthogonal groups. The analogues of $\mathcal{F}$ were defined on orthogonal and other groups, and their near holomorphy and arithmeticity were investigated in [86] and [88]. In [B90] and [B94] A. Bluher obtained improvements on these, which deal with more values of s and also with a series defined adelically.

1. Preliminaries

1.1. We use the symbols $\mathcal{S}(W)$, K, F, ρ, $\mathbf{a}$, and $(\tau_v)_{v\in\mathbf{a}}$ in the same sense as in the introduction. In addition, we denote by $\mathbf{h}$ the set of all nonarchimedean primes of F, by $\mathfrak{g}$ resp. $\mathfrak{r}$ the maximal order of F resp. K, and by D_F resp. D_K the discriminatnt of F resp. K. We put $\tau = (\tau_v)_{v\in\mathbf{a}}$, which is a CM-type of K. We view $\mathbf{a}$ also as the set of all archimedean primes of K. Throughout the paper we fix one τ; then for $v \in \mathbf{a}$ and $\xi \in K$ we shall often write ξ_v for ξ^{τ_v}. For a square matrix x of size $2n$ we denote by a_x, b_x, c_x, and d_x the standard a-, b-, c-, and d-blocks of x. We put $\alpha^* = {}^t\alpha^\rho$ for a matrix α with entries in K, and $z^* = {}^t\overline{z}$ for a complex matrix z; we put also $\widehat{\alpha} = \alpha^\wedge = (\alpha^*)^{-1}$ if α is invertible. The adelization of an algebraic group X over F is denoted by $X_\mathbf{A}$; its archimedean and nonarchimedean factors are denoted by $X_\mathbf{a}$ and $X_\mathbf{h}$. For $a \in F_\mathbf{A}^\times$ we write $a \gg 0$ if $a_v > 0$ for every $v \in \mathbf{a}$.

In addition to G and $\mathcal{H}_n$ of the introduction we define various symbols as follows:

(1.1a) $$G_1 = G \cap SL_{2n}(K), \qquad P = \{\, \gamma \in G \mid c_\gamma = 0 \,\},$$

(1.1b) $$V = K_n^n, \quad S = \{\, \sigma \in V \mid \sigma^* = \sigma \,\},$$

(1.1c) $$S^+ = \{\, \sigma \in S \mid \sigma_v > 0 \ \ \text{for every} \ \ v \in \mathbf{a} \,\},$$

(1.1d) $$B(z) = \begin{bmatrix} z^* & z \\ 1_n & 1_n \end{bmatrix} \qquad (z \in \mathcal{H}_n).$$

For $\alpha \in G_\mathbf{A}$ and $z = (z_v)_{v \in \mathbf{a}} \in \mathcal{H}_n^\mathbf{a}$ we define $\alpha z \in \mathcal{H}_n^\mathbf{a}$ and $GL_n(\mathbf{C})$-valued factors of automorphy $\lambda_v(\alpha, z)$, $\mu_v(\alpha, z)$ by the relation (see [00, §3.3])

(1.2) $$\alpha_v B(z_v) = B\big((\alpha z)_v\big)\mathrm{diag}\big[\overline{\lambda_v(\alpha, z)}, \, \mu_v(\alpha, z)\big] \qquad (v \in \mathbf{a}).$$

For some reasons, we have to consider, in addition to $\mathbf{a}$, the set of all isomorphic embeddings of K into $\mathbf{C}$, which we denote by $\mathbf{b}$ or J_K. These two symbols represent the same thing, but are employed in different contexts. For the moment we use $\mathbf{b}$, which contains $\mathbf{a}$ as a subset, since we identify $v \in \mathbf{a}$ with τ_v. For $v \in \mathbf{a}$ we also denote by $v\rho$ the element $\rho\tau_v$ of $\mathbf{b}$. We put

(1.3a) $$\mu_{v\rho}(\alpha, z) = \lambda_v(\alpha, z) \qquad (v \in \mathbf{a}),$$

(1.3b) $$j_\alpha(z) = j(\alpha, z) = \big(j_v(\alpha, z)\big)_{v \in \mathbf{b}}, \quad j_v(\alpha, z) = \det\,[\mu_v(\alpha, z)] \quad (v \in \mathbf{b}),$$

(1.3c) $$\delta(z) = \big(\det((i/2)(z_v^* - z_v))\big)_{v \in \mathbf{a}} \qquad (z \in \mathcal{H}_n^\mathbf{a}).$$

Here we recall a basic formula (see [00, (3.23)])

(1.4) $$\det(\alpha)_v j_{v\rho}(\alpha, z) = j_v(\alpha, z) \qquad (v \in \mathbf{a}).$$

For $x, y \in \mathbf{C}^\mathbf{b}$ we put

(1.5) $$x^y = \prod_{v \in \mathbf{b}} x_v^{y_v},$$

whenever $x_v^{y_v}$ is meaningful. For $s \in \mathbf{C}$ we denote by $s\mathbf{a}$ the element of $\mathbf{Z}^\mathbf{b}$ such that $(s\mathbf{a})_v = s$ and $(s\mathbf{a})_{v\rho} = 0$ for every $v \in \mathbf{a}$. We view $\mathbf{C}^\mathbf{a}$ as the subspace of $\mathbf{C}^\mathbf{b}$ consisting of the elements whose $v\rho$-components are 0 for every $v \in \mathbf{a}$. For $a \in K$ and $v \in \mathbf{b}$ we put $a_v = a^v$, and view K as a subset of $\mathbf{C}^\mathbf{b}$ by identifying $a \in K$ with $(a_v)_{v \in \mathbf{b}} \in \mathbf{C}^\mathbf{b}$. Also, for $x \in \mathbf{C}^\mathbf{b}$ we put $|x| = (|x_v|)_{v \in \mathbf{b}}$, where $|y|$ for $y \in \mathbf{C}$ is the standard absolute value of y, so that $|x_v|^2$ is the normalized valuation at v. Under these conventions, we have $|x|^{2s\mathbf{a}} = \prod_{v \in \mathbf{a}} |x_v|^{2s} = N_{K/\mathbf{Q}}(x)^s$ for $x \in K$ and $s \in \mathbf{C}$, for example.

1.2. Hereafter we write simply $\mathcal{H}$ for the space $\mathcal{H}_n^{\mathbf{a}}$. For $k \in \mathbf{Z}^{\mathbf{b}}$, $\gamma \in G$, and $f : \mathcal{H} \to \mathbf{C}$ we define $f\|_k\gamma : \mathcal{H} \to \mathbf{C}$ by $(f\|_k\gamma)(z) = j_\gamma(z)^{-k}f(\gamma z)$, where $j_\gamma(z)^{-k}$ should be understood in the sense of (1.5). For a congruence subgroup Γ of G or G_1 we denote by $\mathcal{M}_k(\Gamma)$ the space of all holomorphic automorphic forms on $\mathcal{H}$ of weight k with respect to Γ, and by $\mathcal{M}_k$ the union of $\mathcal{M}_k(\Gamma)$ for all possible Γ; the subspaces of these consisting of cusp forms are denoted by $\mathcal{S}_k(\Gamma)$ and $\mathcal{S}_k$; see [00, §§5.2 and 5.8].

Given C^∞-functions f, g on $\mathcal{H}$ such that $f\|_k\gamma = f$ and $g\|_k\gamma = g$ for every $\gamma \in \Gamma$, we define the inner product $\langle g, f \rangle$ by

$$\text{(1.6)}\quad \langle g, f \rangle = \mathrm{vol}(\mathfrak{D})^{-1}\int_{\mathfrak{D}} \overline{g(z)}f(z)\delta(z)^m\mathbf{d}z, \quad \mathrm{vol}(\mathfrak{D}) = \int_{\mathfrak{D}} \mathbf{d}z, \quad \mathfrak{D} = \Gamma\backslash\mathcal{H},$$

whenever the integral is convergent, where $m = (k_v + k_{v\rho})_{v\in\mathbf{a}}$ and $\mathbf{d}z = \prod_{v\in\mathbf{a}} \mathbf{d}z_v$ with $\mathbf{d}z_v$ of [00, Lemma 3.4].

For $f \in \mathcal{M}_k$ we define its Fourier coefficients $c_f(\sigma)$ by

$$\text{(1.7)}\qquad f(z) = \sum_{\sigma\in S} c_f(\sigma)\mathbf{e}_{\mathbf{a}}^n(\sigma z) \qquad (z \in \mathcal{H}),$$

where

$$\text{(1.7a)}\qquad \mathbf{e}_{\mathbf{a}}^n(x) = \exp\left(2\pi i \textstyle\sum_{v\in\mathbf{a}} \mathrm{tr}(x_v)\right) \qquad (x \in (\mathbf{C}_n^n)^{\mathbf{a}}).$$

Given $f \in \mathcal{M}_k$ and $g \in \mathcal{M}_l$, we can find a subgroup U of $GL_n(\mathfrak{r})$ of finite index such that $c_f(a^*\sigma a) = \det(a)^{k\rho}c_f(\sigma)$ and $c_g(a^*\sigma a) = \det(a)^{l\rho}c_g(\sigma)$ for every $a \in U$; see [00, (5.21)]. Changing U suitably, we may assume that $0 \ll \det(a) \in \mathfrak{g}^\times$ for every $a \in U$. Fixing such a subgroup U, we define an equivalence relation $\sim$ in S^+ with respect to U by: $\sigma \sim \sigma'$ if and only if $\sigma' = a^*\sigma a$ with $a \in U$. We then put

$$\text{(1.8)}\qquad D(s;\, f,\, g) = \sum_{\sigma\in S^+/U} \nu_\sigma c_f(\sigma)\overline{c_g(\sigma)}\det(\sigma)^{-s\mathbf{a}-h} \qquad (s \in \mathbf{C}),$$

where

$$\text{(1.9a)}\qquad \nu_\sigma = [U_\sigma : 1]^{-1}, \quad U_\sigma = \{\, a \in U \mid a^*\sigma a = \sigma \,\},$$

$$\text{(1.9b)}\qquad h = (1/2)(k_v + k_{v\rho} + l_v + l_{v\rho})_{v\in\mathbf{a}}.$$

This series is essentially the same as that of [00, (22.4)], Thus it is convergent for sufficiently large $\mathrm{Re}(s)$; see [00, §A6.7]. Strictly speaking, $D(s; f, g)$ depends on U, and if U is changed, then the series is multiplied by a positive rational number. Since we are interested only in the properties of $D(s; f, g)$ not affected by such multiplication, we will not be concerned about this point.

Let us now assume that either f or g is a cusp form. Then, as shown in [00, (22.9)], we have

$$\text{(1.10)}\qquad \Gamma((s))D(s;\, f,\, g) = A\int_{\mathfrak{D}} f(z)\overline{g(z)}\,\overline{E(z,\, \overline{s}+n;\, m-m',\, \Gamma)}\delta(z)^m\,\mathbf{d}z.$$

Here $\Gamma((s)) = \prod_{v\in\mathbf{a}}\Gamma_n(s+h_v)(4\pi)^{-n(s+h_v)}$, $\Gamma_n(s) = \pi^{n(n-1)/2}\prod_{i=0}^{n-1}\Gamma(s-i)$, A is a positive rational number times $\mathrm{vol}(S_{\mathbf{a}}/L)^{-1}$ with a $\mathfrak{g}$-lattice L in S, $m = (k_v + k_{v\rho})_{v\in\mathbf{a}}$, $m' = (l_v + l_{v\rho})_{v\in\mathbf{a}}$, $\mathfrak{D} = \Gamma\backslash\mathcal{H}$, Γ is a suitable congruence subgroup of G_1, and

$$(1.11)\qquad E(z,\, s;\, t,\, \Gamma) = \sum_{\gamma\in(P\cap\Gamma)\backslash\Gamma} \delta(z)^{s\mathbf{a}-t/2}\|_t\gamma \qquad (t \in \mathbf{Z}^{\mathbf{a}}).$$

We easily see that $\overline{E(z,\, \overline{s};\, t,\, \Gamma)} = \delta(z)^{-t}E(z,\, s;\, -t,\, \Gamma)$, and hence

$$(1.12)\qquad \Gamma((s))D(s;\, f,\, g) = A\int_{\mathfrak{D}} \overline{g(z)}f(z)E(z,\, s+n;\, m'-m,\, \Gamma)\delta(z)^{m'}\,\mathbf{d}z.$$

1.3. Lemma. *Let Γ be a congruence subgroup of G_1 and let $t \in \mathbf{Z}^{\mathbf{a}}$. Then $E(z,\, s;\, t,\, \Gamma)$ can be continued as a meromorphic function of s to the whole plane, which is holomorphic for* $\mathrm{Re}(s) > n$. *Moreover, it is holomorphic at $s = n$ except when $t = 0$, in which case it has a simple pole at $s = n$, and the residue is a positive rational number times $\pi^{-n^2d}r_0^n(F)$ with a positive constant $r_0^n(F)$ defined by*

$$r_0^n(F) = \mathrm{Res}_{s=1}\zeta_F(s)\cdot\prod_{j=2}^{n} L(j,\, \theta^{j-1}),$$

where $d = [F : \mathbf{Q}]$, ζ_F is the Dedekind zeta function of F, θ is the Hecke character of F corresponding to K/F, and $L(s,\, \theta^{j-1})$ is the L-function of θ^{j-1}.

PROOF. By [00, Lemma 17.2], $E(z,\, s;\, t,\, \Gamma)$ is a positive rational number times a finite sum $\sum_{\chi,\alpha} E(z,\, s;\, t,\, \chi,\, \mathfrak{c})\|_t\alpha$ with some series $E(z,\, s;\, t,\, \chi,\, \mathfrak{c})$ defined in [00, (16.48)] in the special unitary case, where $\alpha \in G_1$ and χ is a Hecke character of F such that $\chi(x) = x^t|x|^{-t}$ for $x \in F_{\mathbf{a}}^{\times}$. This $E(z,\, s;\, t,\, \chi,\, \mathfrak{c})$ is the same as E_r of [97, Lemma 18.7 (3)] with $r = 1$. Therefore [97, Proposition 19.3] guarantees meromorphic continuation of $E(z,\, s;\, t,\, \Gamma)$ to the whole $\mathbf{C}$. Now, by [97, Theorem 19.7] it is holomorphic at $s = n$ except when $t = 0$ and $\chi = 1$, in which case it has a simple pole at $s = n$. The residue is written $R(\mathfrak{c})$ in that theorem, and depends on K, n, and two ideals $\mathfrak{b}$ and $\mathfrak{c}$ of F. Let R_0 denote the number $R(\mathfrak{c})$ in the case $\mathfrak{b} = \mathfrak{c} = \mathfrak{g}$. Since the residue in the general case is a positive rational number times R_0, as can be seen from the formula given there, our task is to calculate R_0. (The residue must be positive, since $E(z,\, s;\, 0, \Gamma)$ for $n < s \in \mathbf{R}$ is positive.) The formula for $R(\mathfrak{c})$ in [97, Theorem 19.7] involves several types of factors, of which two are essential; $\prod_{j=2}^{n} L(j,\, \theta^{j-1})$ and the quantity written $\Lambda_{\mathfrak{c}}(n)^{-1}$; other factors are easy. By [97, Proposition 19.2], $\Lambda_{\mathfrak{g}}(n) = \prod_{j=1}^{n} L(n+j,\, \theta^{n+j})$. Now, by [00, Theorem 18.12], $L(k, \theta^k) \in |D_F|^{1/2}\pi^{kd}\mathbf{Q}^{\times}$ if $0 < k \in 2\mathbf{Z}$, and $L(k, \theta^k) \in |D_F|^{1/2}(\pi i)^{kd}\mathbf{g}(\theta)\mathbf{Q}^{\times}$ if $0 \leq k-1 \in 2\mathbf{Z}$, where $\mathbf{g}(\theta)$ is the Gauss sum of θ defined by [00, (18.10)]. By [97, (A6.3.4)], $\mathbf{g}(\theta) = i^d|D_K|^{1/2}|D_F|^{-1}$. From these and other factors of the formula for the residue we obtain a factor of the form $|D_F^aD_K^b|$, but we easily see that $a,\, b \in \mathbf{Z}$. Notice also that $L(s,\, \theta) = \zeta_K(s)/\zeta_F(s) > 0$ for $1 \leq s \in \mathbf{R}$. Combining all these, we obtain the desired formula for the residue.

1.4. Lemma. *Put* $d = [F : \mathbf{Q}]$, $m = (k_v + k_{v\rho})_{v\in\mathbf{a}}$, *and* $m' = (l_v + l_{v\rho})_{v\in\mathbf{a}}$. *Suppose that either f or g is a cusp form. Then the following assertions hold:*

(1) *$D(s;\, f,\, g)$ can be continued as a meromorphic function of s to the whole plane, which is holomorphic for* $\mathrm{Re}(s) > 0$. *Moreover it is holomorphic at $s = 0$ if $m \neq m'$.*

(2) *The right-hand side of (22.4) is absolutely convergent for* $\mathrm{Re}(s) > 0$ *if both* f *and* g *are cusp forms.*

(3) *If* $m = m'$, *then* $D(s;\, f,\, g)$ *has at most a simple pole at* $s = 0$ *whose residue is* $|D_F^{n/2} D_K^{n(n-1)/4}|\pi^a r_0^n(F)\langle g,\, f\rangle$ *times a positive rational number, where* $a = n\sum_{v\in\mathbf{a}} m_v - dn(n-1)/2$ *and* $r_0^n(F)$ *is the constant of Lemma 1.3.*

PROOF. This is essentially included in [96, Lemma 3.2] or [00, Proposition 22.2], though the present formulation is somewhat different. The only new point is the precise nature of the residue, which was not given in those articles, but follows easily from (1.10). Indeed, from (1.10) we easily see that the residue is a positive rational number times $A\Gamma((0))^{-1}\mathrm{vol}(\mathfrak{D})\pi^{-dn^2} r_0^n(F)\langle g,\, f\rangle$. Now, by [97, (18.9.3) and (18.9.4)], A is a positive rational number times $|D_F^{n/2} D_K^{n(n-1)/4}|$; $\mathrm{vol}(\mathfrak{D})$ is π^{dn^2} times a positive rational number as explained in [00, p. 232]. Since f or g is a cusp form, we have $m_v \geq n$ for every $v \in \mathbf{a}$ by [94, Theorem 5.6], so that we easily find that $\Gamma((0))$ is a positive rational number times π^{-a} with the integer a given in our lemma. Thus we obtain the desired fact.

2. Two theorems on certain Dirichlet series associated with $GL_n(K)$

2.1. By a CM-algebra we understand a finite direct sum of CM-fields. For a CM-algebra Y we denote by J_Y the set of all nontrivial ring-homomorphisms of Y into $\mathbf{C}$, and I_Y the free $\mathbf{Z}$-module generated by the elements of J_Y. If Y is the direct sum of the CM-fields K_ν for ν in a finite set of indices N, then J_Y is clearly identified with the disjoint union of J_{K_ν} for all $\nu \in N$ and I_Y with the direct sum of I_{K_ν} for all $\nu \in N$. We recall that there is a $\mathbf{Z}$-bilinear map $p_Y : I_Y \times I_Y \to \mathbf{C}^\times/\overline{\mathbf{Q}}^\times$, called *the period symbol*, which gives the values of automorphic forms at CM-points, as well as the values of certain abelian integrals; for its definition and properties the reader is referred to [80, Section 1], [98, Section 32], and [00, §11.3].

Given $\varepsilon \in \mathcal{S}(V)$ and $r \in \mathbf{Z}^{\mathbf{b}}$, we put $q = (r_v + r_{v\rho})_{v\in\mathbf{a}}$, and consider a series

$$M(s) = \sum_{\xi\in GL_n(K)/U} \varepsilon(\xi)\det(\xi)^r|\det(\xi)|^{-q-2s\mathbf{a}}, \tag{2.1}$$

where U is a subgroup of $GL_n(\mathfrak{r})$ of finite index such that $0 \ll \det(\gamma) \in \mathfrak{g}^\times$ and $\varepsilon(\xi\gamma) = \varepsilon(\xi)$ for every $\gamma \in U$. Then the series is well-defined and convergent for sufficiently large $\mathrm{Re}(s)$, since the problem can be reduced to the convergence of the zeta function of the matrix algebra V; also M can be continued as a meromorphic function of s to the whole complex plane, as shown by (2.2) below.

2.2. Theorem. *Put* $e_v = |r_v - r_{v\rho}|$ *for* $v \in \mathbf{a}$; *define* $\sigma_v \in J_K$ *by* $\sigma_v = \tau_v$ *if* $r_v \leq r_{v\rho}$ *and* $\sigma_v = \tau_v\rho$ *if* $r_v > r_{v\rho}$. *Then the following assertions hold:*

(1) *We have*

$$M(s) = \sum_{i\in I} a_i b_i^s \prod_{j=0}^{n-1} L(s-j,\, \chi_{ij}) \tag{2.2}$$

with a finite set I, $a_i \in \mathbf{C}$, $0 < b_i \in \mathbf{R}$, *and Hecke characters* χ_{ij} *of* K *such that* $(\chi_{ij})_{\mathbf{a}}(x) = \prod_{v\in\mathbf{a}} \left(x^{\sigma_v}/|x^{\sigma_v}|\right)^{e_v}$ *for every* $x \in K^\times$.

(2) *Suppose that* ε *is* $\overline{\mathbf{Q}}$*-valued. Then we can take* a_i *and* b_i *of (2.2) to be algebraic. Moreover, if* s_0 *is an integer such that* $2n - e_v \le s_0 \le e_v$ *and* $s_0 - e_v \in 2\mathbf{Z}$ *for every* $v \in \mathbf{a}$*, then* M *is finite at* $s = s_0/2$*, and*

$$M(s_0/2) \in \pi^{nc/2} p_K\bigl(\textstyle\sum_{v\in\mathbf{a}}(r_{v\rho} - r_v)\tau_v,\ n\sum_{v\in\mathbf{a}}\sigma_v\bigr)\overline{\mathbf{Q}},$$

where $c = \sum_{v\in\mathbf{a}}(e_v + s_0 - n + 1)$.

PROOF. With σ_v as above, we have $x^{-r}|x|^q = \prod_{v\in\mathbf{a}}\bigl(x^{\sigma_v}/|x^{\sigma_v}|\bigr)^{e_v}$ for $x \in K^\times$, and

$$M(s) = \sum_{\xi\in GL_n(K)/U} \varepsilon(\xi)\det(\xi)^{-t}|\det(\xi)|^{e-2s\mathbf{a}},$$

where $t = \sum_{v\in\mathbf{a}} e_v\sigma_v$. By [97, Lemma 11.14 (3)] there exists a Hecke character ω of K such that $\omega_\mathbf{a}(x) = x^t|x|^{-e}$. Therefore we obtain (2.2) and the algebraicity of a_i and b_i in (2) from [84, Proposition 9.2]. (Take $f(\xi) = \det(\xi)^{-t}|\det(\xi)|^e$ and $\chi_i = \omega$ in [84, (9.5a, b)].) By [00, Theorem 18.16] we see that $L(\nu/2, \chi_{ij}) \in \pi^\gamma p_K\bigl(t, \sum_{v\in\mathbf{a}}\sigma_v\bigr)\overline{\mathbf{Q}}$ for every integer ν such that $-e_v < \nu \le e_v$ and $e_v - \nu \in 2\mathbf{Z}$ for every $v \in \mathbf{a}$, where $\gamma = (1/2)\sum_{v\in\mathbf{a}}(e_v + \nu)$. Recalling that $p_K(\rho\alpha, \beta) = p_K(-\alpha, \beta)$, we obtain the main part of (2).

2.3. Theorem. *Every* M *of type (2.1) with* $q = r = 0$ *has at most a simple pole at* $s = n$*. Moreover, the residue belongs to* $\pi^{dn(n+1)/2} r_0^n(F)\overline{\mathbf{Q}}$*, provided* ε *is* $\overline{\mathbf{Q}}$*-valued, where* $r_0^n(F)$ *is the constant of Lemma 1.3.*

PROOF. The first part follows immediately from expression (2.2). The second part will be proven in Section 4. It should be noted that we cannot derive the second part from expression (2.2).

We add two remarks:

(2.3) *The product* $\prod_{i=0}^{n-1}\zeta_K(s-j)$ *has a simple pole at* $s = n$*, and the residue is a positive rational number times* $|D_F^{n/2} D_K^{n(n-1)/4}|\pi^{dn(n+1)/2} r_0^n(F)$.

(2.4) *Such a residue can be obtained as the residue of a suitable* M *of type (2.1) with* $q = r = 0$ *and* $\overline{\mathbf{Q}}$*-valued* ε.

Here ζ_K is the zeta function of K and $d = [F : \mathbf{Q}]$. The first of these, (2.3), can easily be verified by means of the known facts on $L(j, \theta^j)$ already mentioned in the proof of Lemma 1.3. To prove (2.4), we consider

$$M_0(s) = \sum_{x\in X}\lambda(x)|\det(x_\mathbf{h})|_K^s, \quad X = GL_n(K)_\mathbf{A}/E,$$

where λ is the characteristic function of $\prod_{v\in\mathbf{h}}(\mathfrak{r}_v)_n^n$, $|a|_K$ is the idele norm of $a \in K_\mathbf{A}^\times$, and $E = GL_n(K)_\mathbf{a}\prod_{v\in\mathbf{h}}GL_n(\mathfrak{r}_v)$. We easily see that $M_0(s) = \prod_{j=0}^{n-1}\zeta_K(s-j)$. To show that M_0 is essentially (but not exactly) of type (2.1), take a disjoint decomposition $GL_n(K)_\mathbf{A} = \bigsqcup_{q\in Q} GL_n(K)qE$ with a finite subset Q of $GL_n(K)_\mathbf{h}$, and let $U_q = GL_n(K)\cap qEq^{-1}$ and $Y_q = GL_n(K)/U_q$ for $q \in Q$. Then we easily see that X can be given by $\{\alpha q \mid \alpha \in Y_q,\ q \in Q\}$. Thus

$$M_0(s) = \sum_{q\in Q} |\det(q)|_K^s \sum_{\alpha\in Y_q} \lambda(\alpha q)|\det(\alpha)|^{-2s\mathbf{a}}.$$

Put $U = \bigcap_{q\in Q} U_q \cap GL_n(\mathfrak{r})$ and $\varepsilon(\alpha) = \sum_{q\in Q}[Y_q : U]^{-1}|\det(q)|_K^n \lambda(\alpha q)$. Then M defined with these ε and U have the same residue as M_0 at $s = n$ as desired.

3. Inner products as period symbols

3.1. For two fractional ideals $\mathfrak{x}$ and $\mathfrak{y}$ in F such that $\mathfrak{x}\mathfrak{y} \subset \mathfrak{g}$ we put

$$D[\mathfrak{x}, \mathfrak{y}] = \{ x \in G_{\mathbf{A}} \mid a_x \prec \mathfrak{r},\ b_x \prec \mathfrak{r}\mathfrak{x},\ c_x \prec \mathfrak{r}\mathfrak{y},\ d_x \prec \mathfrak{r} \}, \tag{3.1}$$

where we write $\alpha \prec \mathfrak{x}$ if all the entries of α_v belong to $\mathfrak{x}_v$ for every $v \in \mathbf{h}$.

We now take a fractional ideal $\mathfrak{b}$ and an integral ideal $\mathfrak{c}$ in F, and consider the subgroup $D[\mathfrak{b}^{-1}\mathfrak{c}, \mathfrak{b}\mathfrak{c}]$ of $G_{\mathbf{A}}$. We put then

$$C = \{ x \in D[\mathfrak{b}^{-1}\mathfrak{c}, \mathfrak{b}\mathfrak{c}] \mid a_x - 1 \prec \mathfrak{r}\mathfrak{c} \}, \tag{3.2}$$

$$\mathfrak{X} = \{ x \in G_{\mathbf{A}} \mid x_v \in G_v \cap C \ \textit{for every}\ v|\mathfrak{c} \}. \tag{3.3}$$

Given $k \in \mathbf{Z}^{\mathbf{b}}$, we denote by $\mathcal{M}_k(C)$ the set of all functions $\mathbf{f} : G_{\mathbf{A}} \to \mathbf{C}$ satisfying the following two conditions:

(3.4a) $\mathbf{f}(\alpha x w) = j_w(\mathbf{i})^{-k}\mathbf{f}(x)$ *if* $\alpha \in G$, $w \in C$, *and* $w(\mathbf{i}) = \mathbf{i}$.

(3.4b) *For every* $p \in G_{\mathbf{h}}$ *there exists an element* f_p *of* $\mathcal{M}_k$, *called the p-component of* $\mathbf{f}$, *such that* $\mathbf{f}(py) = (f_p\|_k y)(\mathbf{i})$ *for every* $y \in G_{\mathbf{a}}$.

Here $\mathbf{i}$ is the point of $\mathcal{H}_n^{\mathbf{a}}$ whose v-component is $i1_n$ for every $v \in \mathbf{a}$. We denote by $\mathcal{S}_k(C)$ the set of all $\mathbf{f} \in \mathcal{M}_k(C)$ such that $f_p \in \mathcal{S}_k$ for every $p \in G_{\mathbf{h}}$. We consider the Hecke algebra $\mathcal{R}(C, \mathfrak{X})$ and let it act on $\mathcal{S}_k(C)$. To each Hecke eigenform $\mathbf{f}$ in $\mathcal{S}_k(C)$ and a Hecke character χ of K we can associate an Euler product $\mathcal{Z}(s, \mathbf{f}, \chi)$; for details, see [00, §§20.6 and 20.11].

3.2. We take $\mu \in \mathbf{Z}^{\mathbf{b}}$ such that $\mu_v \geq 0$ for every $v \in \mathbf{b}$ and $\mu_v\mu_{v\rho} = 0$ for every $v \in \mathbf{a}$, and take also $\omega \in S^+$. We then put $l = \mu + n\mathbf{a}$ and

$$\theta_\mu(z; \lambda, \omega) = \sum_{\xi\in V} \lambda(\xi)\det(\xi)^{\mu\rho}\mathbf{e}_{\mathbf{a}}^n(\xi^*\omega\xi z) \qquad (\lambda \in \mathcal{S}(V),\ z \in \mathcal{H}). \tag{3.5}$$

Here we understand that $\det(\xi)^{\mu\rho} = 1$ for every ξ if $\mu = 0$. In [00, §A5.5] we showed that $\theta_\mu(z; \lambda, \omega)$ belongs to $\mathcal{M}_l$ and is a cusp form if $\mu \neq 0$. (Correction to [00]: τ on page 277, line 13 should read 2τ.)

Fixing μ, λ, and ω, put $g(z) = \theta_\mu(z; \lambda, \omega)$. Then clearly $c_g(\sigma)$ is the sum of $\lambda(\xi)\det(\xi)^{\mu\rho}$ for all $\xi \in V$ such that $\xi^*\omega\xi = \sigma$. Take this g to be g of (1.8); take also U so that $\lambda(\xi a) = \lambda(\xi)$ for every $a \in U$. We are going to show that

$$D(s; f, g) = \det(\omega)^{-s\mathbf{a}-h} \sum_{\xi\in GL_n(K)/U} c_f(\xi^*\omega\xi)\det(\xi)^{\mu}\overline{\lambda(\xi)}|\det(\xi)|^{-2h-2s\mathbf{a}}. \tag{3.6}$$

Let R be a complete set of representatives for S^+/U. Put $Y_\sigma = \{\xi \in GL_n(K) \mid \xi^*\omega\xi = \sigma\}$ and observe that $GL_n(K)/U$ can be given by $\bigsqcup_{\sigma\in R} Y_\sigma/U_\sigma$. Our assumption on U implies that $\det(U_\sigma) = 1$. Therefore the right-hand side of (3.6) equals

$$(3.7)\qquad \sum_{\sigma\in R} |\det(\sigma)|^{-2h-2s\mathbf{a}} c_f(\sigma) \sum_{\xi\in Y_\sigma/U_\sigma} \det(\xi)^{\mu}\overline{\lambda(\xi)}$$
$$= \sum_{\sigma\in R} \det(\sigma)^{-h-s\mathbf{a}} c_f(\sigma)\nu_\sigma \overline{c_g(\sigma)},$$

which is exactly $D(s; f, g)$ as expected.

3.3. We denote by $\mathcal{T}_k$ the space spanned by the forms of type (3.5) over $\mathbf{C}$ with all choices of λ and ω under the condition that $k = \mu + n\mathbf{a}$. We assume that $\mu \neq 0$, so that $\mathcal{T}_k \subset \mathcal{S}_k$. We then denote by $\mathcal{T}_k(\overline{\mathbf{Q}})$ the set of all the $\overline{\mathbf{Q}}$-rational elements of $\mathcal{T}_k$, and by $\mathcal{T}_k(C)$ the set of all $\mathbf{f} \in \mathcal{S}_k(C)$ such that the p-component of $\mathbf{f}$ belongs to $\mathcal{T}_k$ for every $p \in G_{\mathbf{h}}$. By [00, (A5.4)], $\mathcal{T}_k(\overline{\mathbf{Q}})$ is stable under $\|_k G$. We call an element $\mathbf{f}$ of $\mathcal{M}_k(C)$ $\overline{\mathbf{Q}}$*-rational* if the p-component of $\mathbf{f}$ is $\overline{\mathbf{Q}}$-rational for every $p \in G_{\mathbf{h}}$. We denoe by $\mathcal{S}_k(C, \overline{\mathbf{Q}})$ the set of all $\overline{\mathbf{Q}}$-rational elements of $\mathcal{S}_k(C)$; we then put $\mathcal{T}_k(C, \overline{\mathbf{Q}}) = \mathcal{T}_k(C) \cap \mathcal{S}_k(C, \overline{\mathbf{Q}})$. Now $\mathcal{S}_k(C) = \mathcal{S}_k(C, \overline{\mathbf{Q}}) \otimes_{\overline{\mathbf{Q}}} \mathbf{C}$ by [00, Theorem 10.8]. Also, from the definition of $\mathcal{T}_k(C)$ we easily see that $\mathcal{T}_k(C) = \mathcal{T}_k(C, \overline{\mathbf{Q}}) \otimes_{\overline{\mathbf{Q}}} \mathbf{C}$. Also $\mathcal{T}_k(C, \overline{\mathbf{Q}})$ is stable under Hecke operators.

Every element of $\mathcal{S}(V)$ is a finite $\mathbf{C}$-linear combination of $\overline{\mathbf{Q}}$-valued elements. Therefore $\mathcal{T}_k$ is spanned by $\theta_\mu(z; \lambda, \omega)$ with $\overline{\mathbf{Q}}$-valued λ. Given $f \in \mathcal{T}_k$, we can thus find finitely many $g_i = \theta_\mu(z; \lambda_i, \omega_i)$ with $\overline{\mathbf{Q}}$-valued λ_i such that $f = \sum_i c_i g_i$ with $c_i \in \mathbf{C}$. Changing $\{g_i\}$ for its suitable subset, we may assume that the g_i are linearly independent over $\mathbf{C}$. If $f \in \mathcal{T}_k(\overline{\mathbf{Q}})$, then for $\sigma \in \mathrm{Aut}(\mathbf{C}/\overline{\mathbf{Q}})$ we have $\sum_i c_i g_i = f = f^\sigma = \sum_i c_i^\sigma g_i$, and hence $c_i^\sigma = c_i$, that is, $c_i \in \overline{\mathbf{Q}}$. This shows that

(3.8) $\mathcal{T}_k(\overline{\mathbf{Q}})$ *is spanned over* $\overline{\mathbf{Q}}$ *by* $\theta_\mu(z; \lambda, \omega)$ *with* $\overline{\mathbf{Q}}$*-valued* λ.

Also we easily see that

(3.9) $\theta_\mu(z; \lambda, \omega) = \theta_\mu(z; \lambda_1, \alpha^*\omega\alpha)$ *if* $\lambda_1(\xi) = \det(\alpha)^{\mu\rho}\lambda(\alpha\xi)$, $\alpha \in GL_n(K)$.

Therefore, we can always assume that ω is diagonal by changing λ suitably, without changing the function. If λ is $\overline{\mathbf{Q}}$-valued, then so is λ_1. We note an easy fact:

(3.10) *Given* $\omega, \omega' \in S^+$, *there exists* $\alpha \in GL_n(K)$ *such that* $\alpha^*\omega\alpha = \omega'$ *if and only if* $\det(\omega\omega') \in N_{K/F}(K^\times)$.

The "only if"-part is obvious. The "if"-part follows from the Hasse principle combined with the nonarchimedean local version of (3.10), which is given in [97, Proposition 5.3].

3.4. Theorem. *Let* $g(z) = \theta_\mu(z; \lambda, \omega)$ *and* $f(z) = \theta_\nu(z; \lambda', \omega')$. *Suppose* $\mu_v + \mu_{v\rho} = \nu_v + \nu_{v\rho}$ *for every* $v \in \mathbf{a}$; *suppose also that* f *is a cusp form. Then* $\langle g, f\rangle \neq 0$ *only if the following two conditions are satisfied:*

(i) $\det(\omega\omega') \in N_{K/F}(K^\times)$;

(ii) $\mu = \nu$ or $\mu = \nu\rho$.

PROOF. Suppose $\det(\omega\omega') \notin N_{K/F}(K^\times)$. For $\sigma \in S^+$ we have $c_f(\sigma)\overline{c_g(\sigma)} \neq 0$ only if there exist $\xi, \xi' \in GL_n(K)$ such that $\xi^*\omega\xi = (\xi')^*\omega'\xi' = \sigma$, which is impossible in view of (3.10). Thus $D(s; f, g) = 0$, so that $\langle g, f \rangle = 0$ by Lemma 1.4 (3). Suppose $\langle g, f \rangle \neq 0$. Then there exists $\alpha \in GL_n(K)$ such that $\alpha^*\omega\alpha = \omega'$. By (3.2), $f(z) = \theta_\nu(z; \lambda'', \omega)$ with some λ''. Thus, to show that (ii) is necesssary, we may assume that $\omega = \omega'$. The proof will be completed in §3.10.

3.5. Theorem. *Let $k = \nu + n\mathbf{a}$ with $\nu \in \mathbf{Z}^\mathbf{b}$ such that $\nu_v \geq 0, \nu_{v\rho} \geq 0$, and $\nu_v\nu_{v\rho} = 0$ for every $v \in \mathbf{a}$. Put $\sigma_v = \tau_v$ if $\nu_v = 0$, and $\sigma_v = \tau_v\rho$ if $\nu_v > 0$; put also $d = [F : \mathbf{Q}]$ and $\mathfrak{q} = p_K\big(\sum_{v\in\mathbf{a}}(\nu_{v\rho} - \nu_v)\tau_v,\ n\sum_{v\in\mathbf{a}}\sigma_v\big)$. Then the following assertions hold:*

(1) *If $\nu_v + \nu_{v\rho} \geq n$ for every $v \in \mathbf{a}$, then $\langle g, f \rangle \in \pi^{-dn^2}\mathfrak{q}^2\overline{\mathbf{Q}}$ for every $f, g \in \mathcal{T}_k(\overline{\mathbf{Q}})$, and $\langle \mathbf{f}, \mathbf{f} \rangle \in \pi^{-dn^2}\mathfrak{q}^2\overline{\mathbf{Q}}$ for every Hecke eigenform $\mathbf{f}$ belonging to $\mathcal{T}_k(C, \overline{\mathbf{Q}})$.*

(2) *If $\nu_v + \nu_{v\rho} > 2n$ for every $v \in \mathbf{a}$, then $\langle g, f \rangle \in \pi^{-dn^2}\mathfrak{q}^2\overline{\mathbf{Q}}$ for every $f \in \mathcal{T}_k(\overline{\mathbf{Q}})$ and $g \in \mathcal{M}_k(\overline{\mathbf{Q}})$.*

The proof will be completed in §4.5. If $n = 1$, our assertion was proven in [76, Proposition 5], [78, Theorem 5.5], and [80, Theorem 9.8].

3.6. Corollary. *Let $\mathbf{f}$ be as in Theorem 3.5 (1) and let χ be a Hecke character of K such that $\chi_\mathbf{a}(x) = x_\mathbf{a}^\ell|x_\mathbf{a}|^{-\ell}$ with $\ell \in \mathbf{Z}^\mathbf{a}$; let $\sigma_0 \in 2^{-1}\mathbf{Z}$. Put $m_v = k_v + k_{v\rho}$. Suppose that $m_v > 3n$, $4n - (2k_{v\rho} + \ell_v) \leq 2\sigma_0 \leq m_v - |k_v - k_{v\rho} - \ell_v|$ and $2\sigma_0 - \ell_v \in 2\mathbf{Z}$ for every $v \in \mathbf{a}$. Further exclude the following two cases:*

(I) *$2\sigma_0 = 2n + 1$, $F = \mathbf{Q}$, $\chi_1 = \theta$, and $k_v - k_{v\rho} = \ell_v$;*
(II) *$0 \leq 2\sigma_0 < 2n$, $\mathfrak{c} = \mathfrak{g}$, $\chi_1 = \theta^{2\sigma_0}$, and the conductor of χ is $\mathfrak{r}$.*

Here χ_1 is the restriction of χ to $F_\mathbf{A}^\times$ and θ is the Hecke character of F corresponding to K/F. Then $\mathcal{Z}(\sigma_0, \mathbf{f}, \chi) \in \pi^{n|m|+d\alpha}\mathfrak{q}^2\overline{\mathbf{Q}}$, where $\mathfrak{q}$ is as in Theorem 3.5, $|m| = \sum_{v\in\mathbf{a}} m_v$, and $\alpha = 2n\sigma_0 - 3n^2 + n$.

PROOF. In [00, (28.10)] we showed that $\mathcal{Z}(\sigma_0, \mathbf{f}, \chi)/\langle \mathbf{f}, \mathbf{f} \rangle \in \pi^{n|m|+d(\alpha+n^2)}\overline{\mathbf{Q}}$. Combining this result with Theorem 3.5 (1), we obtain our assertion.

Theorem 3.5 (2) and Corollary 3.6 are probably true under the weaker condition that $\nu_v + \nu_{v\rho} \geq n$, that is, $m_v \geq 2n$, for every $v \in \mathbf{a}$. We can indeed prove these if $n = 1$, in which case $G_1 = SL_2(F)$ The essential point of the proof of Corollary 3.6, or rather that of [00, (28.10)], is (2) of Theorem 3.5, which is proved in [80, Theorem 9.8] for $n = 1$ under such a condition on ν.

3.7. We consider $D(s; f, g)$ by taking $g(z) = \theta_\mu(z; \lambda, \omega)$ and $f(z) = \theta_\nu(z; \lambda', \omega')$ with $\mu, \nu \in \mathbf{Z}^\mathbf{b}$, $\overline{\mathbf{Q}}$-valued λ and λ', and $\omega, \omega' \in S^+$. Put $k = \nu + n\mathbf{a}$ and $l = \mu + n\mathbf{a}$; then $f \in \mathcal{M}_k$ and $g \in \mathcal{M}_l$. The purpose of this section is to express $D(s; f, g)$ in terms of the value of a certain Eisenstein series at a CM-point. We assume that ω and ω' are diagonal. Now take $\zeta \in K$ so that $\zeta^\rho = -\zeta$ and $\mathrm{Im}(\zeta_v) > 0$ for every $v \in \mathbf{a}$. Since $\zeta \cdot \mathrm{diag}[\omega', -\omega]$ is equivalent to

η_n, we can find $q \in GL_{2n}(K)$ such that $q\eta_n q^* = \zeta \cdot \mathrm{diag}[\omega', -\omega]$. Put $Y = K^{2n}$ and $Y^u = \{a \in Y \mid aa^\rho = 1\}$, where we extend ρ to Y in an obvious way. Define $h : Y \to K^{2n}_{2n}$ by $h(a) = q^{-1}\mathrm{diag}[a_1, \ldots, a_{2n}]q$ for $a = (a_i)_{i=1}^{2n} \in Y$ with $a_i \in K$. Then $h(a^\rho)\eta_n = \eta_n h(a)^*$ for every $a \in Y$, so that $h(Y^u)$ has a fixed point w on $\mathcal{H}$, which is a CM-point; see [00, §4.11]. We are going to evaluate a certain Eisenstein series at this CM-point.

Before proceeding further, we introduce some new symbols. For $v \in \mathbf{a}$, $z \in \mathcal{H}$, and $x \in K^r_{2n}$ with $0 < r \in \mathbf{Z}$ we define $\lambda_v[x, z] \in \mathbf{C}^r_n$ and $\mu_v[x, z] \in \mathbf{C}^r_n$ by

$$[\overline{\lambda_v[x, z]} \quad \mu_v[x, z]] = x_v B(z_v) \tag{3.11}$$

with $B(z)$ of (1.1d). Employing (1.2), we can easily verify that

$$\lambda_v[x\gamma, z] = \lambda_v[x, \gamma z]\lambda_v(\gamma, z), \qquad \mu_v[x\gamma, z] = \mu_v[x, \gamma z]\mu_v(\gamma, z) \qquad (\gamma \in G).$$

For $a \in Y^u$ we can diagonalize $\lambda_v(h(a), w)$ and $\mu_v(h(a), w)$, and the diagonal elements determine a "CM-type" Φ of Y according to the principle of [00, (4.30a), (4.37), (4.40)]. Namely, for $a \in Y$ we can put

$$A_v^{-1}\lambda_v(h(a), w)A_v = \mathrm{diag}[a^{\gamma_{v1}}, \ldots, a^{\gamma_{vn}}], \tag{3.12a}$$

$$B_v^{-1}\mu_v(h(a), w)B_v = \mathrm{diag}[a^{\varepsilon_{v1}}, \ldots, a^{\varepsilon_{vn}}] \tag{3.12b}$$

with $\gamma_{vt}, \varepsilon_{vt} \in J_Y$ and $A_v, B_v \in GL_n(\mathbf{C})$. Strictly speaking, the left-hand sides of these are meaningful only when $a \in Y^u$, but by linearity we can make them meaningful for every $a \in Y$; see [00, p. 28]. Then Φ is an element of I_Y given by $\Phi = \sum_{v \in \mathbf{a}} \sum_{i=1}^n (\gamma_{vi} + \varepsilon_{vi})$. We may assume that A_v and B_v have algebraic entries for the reason explained in [00, Lemma 4.13].

From [00, (3.11)] we obtain $i\eta_n = B(z)\mathrm{diag}[i(z^* - z), i(z - z^*)]^{-1}B(z)^*$ for every $z \in \mathcal{H}_n$, which combined with (3.11) gives

$$i(x\eta_n y^*)_v = \overline{\lambda_v[x, z]}Z_v^{-1}\cdot{}^t\lambda_v[y, z] - \mu_v[x, z]Z_v^{-1}\mu_v[y, z]^* \qquad (x, y \in K^1_{2n}), \tag{3.13}$$

where $Z_v = i(z_v^* - z_v)$. Let $\mathbf{1}$ be the identity element of Y. For $x \in Y = K^1_{2n}$ our definition of h shows that $xq = \mathbf{1}qh(x)$, and hence if $x \in Y^u$, then

$$\begin{aligned}\lambda_v[xq, w]A_v &= \lambda_v[\mathbf{1}q \cdot h(x), w]A_v = \lambda_v[\mathbf{1}q, w]\lambda_v(h(x), w)A_v \\ &= \lambda_v[\mathbf{1}q, w]A_v\mathrm{diag}[x^{\gamma_{v1}}, \ldots, x^{\gamma_{vn}}].\end{aligned}$$

Put $\lambda_v[\mathbf{1}q, w]A_v = (p_1^v, \ldots, p_n^v)$, $\mu_v[\mathbf{1}q, w]B_v = (r_1^v, \cdots, r_n^v)$, and

$$C_v = \mathrm{diag}[p_1^v, \ldots, p_n^v]A_v^{-1}, \qquad D_v = \mathrm{diag}[r_1^v, \ldots, r_n^v]B_v^{-1};$$

write simply x^{γ_v} and x^{ε_v} for the row vectors $(x^{\gamma_{v1}}, \ldots, x^{\gamma_{vn}})$ and $(x^{\varepsilon_{v1}}, \ldots, x^{\varepsilon_{vn}})$; then we obtain

$$\lambda_v[xq, w] = x^{\gamma_v}C_v, \qquad \mu_v[xq, w] = x^{\varepsilon_v}D_v \tag{3.14}$$

at least for $x \in Y^u$, but the equalities hold for every $x \in Y$, since Y^u spans Y over $\mathbf{Q}$. Substituting (xq, yq, w) for (x, y, z) in (3.13) and employing (3.14), we obtain, for $x = (x_t)_{t=1}^{2n}$ and $y = (y_t)_{t=1}^{2n}$ in K^1_{2n},

$$i\zeta_v \sum_{t=1}^{n}(x_t\omega'_t y_t^\rho)_v - i\zeta_v \sum_{t=n+1}^{2n}(x_t\omega_t y_t^\rho)_v = i\big(\zeta x \cdot \mathrm{diag}[\omega',\, -\omega]y^*\big)_v$$
$$= i\big(xq\eta_n(yq)^*\big)_v = (x^{\gamma_v})^\rho T'_v \cdot {}^t y^{\gamma_v} - x^{\varepsilon_v} T_v (y^{\varepsilon_v})^*$$

with some positive definite hermitian matrices T'_v and T_v. Now $x^{\gamma_{vj}}$ coincides with x_t^α with some t and some $\alpha \in J_K$. Since $\mathrm{Im}(\zeta_v) > 0$ for every $v \in \mathbf{a}$ and ω_t and ω'_t are totally positive, from the last equality we can conclude, after rearranging the γ_{vj} and ε_{vj}, that $x^{\varepsilon_{vt}} = x_t^{\tau_v}$ and $x^{\gamma_{vt}} = x_{n+t}^{\tau_v\rho}$ for every $t \le n$. Thus we can put

$$\text{(3.15)} \qquad \lambda_v[xq,\, w] = (x_{n+1},\, \ldots,\, x_{2n})^{\tau_v\rho} C_v, \qquad \mu_v[xq,\, w] = (x_1,\, \ldots,\, x_n)^{\tau_v} D_v.$$

Also, the CM-type $(Y,\, \Phi)$ is the direct sum of n copies of $(K,\, \tau)$ and n copies of $(K,\, \tau\rho)$. By [00, Lemma 4.13], the entries of w_v are algebraic, and hence so are the entries of $\lambda_v[x,\, w]$ and $\mu_v[x,\, w]$ for every $x \in K_n^\tau$; thus p_i^v and r_i^v, as well as the entries of C_v and D_v, are algebraic. Now, for $\xi \in K_{2n}^n$ let ξ_1 and ξ_2 be the left and right halves of the matrix ξ. Then from (3.15) we obtain

$$\text{(3.16)} \qquad \lambda_v[\xi q,\, w] = \xi_2^{\tau_v\rho} C_v, \qquad \mu_v[\xi q,\, w] = \xi_1^{\tau_v} D_v.$$

3.8. Since $c_f(\sigma)$ is the sum of $\lambda'(\xi_1)\det(\xi_1)^{\nu\rho}$ for all ξ_1 such that $\xi_1^*\omega'\xi_1 = \sigma$, by (3.6) we have

$$\text{(3.17)} \qquad \det(\omega')^{s\mathbf{a}+h} D(s;\, f,\, g) = \sum_{\xi_2 \in GL_n(K)/U} \sum_{\xi_1} \lambda'(\xi_1)\overline{\lambda(\xi_2)}\det(\xi_2)^\mu \det(\xi_1)^{\nu\rho}|\det(\xi_1)|^{-2h-2s\mathbf{a}},$$

where the second sum is taken over all ξ_1 such that $\xi_1^*\omega'\xi_1 = \xi_2^*\omega\xi_2$. (Notice that $\det(\omega)|\det(\xi_2)|^2 = \det(\omega')|\det(\xi_1)|^2$.) On the other hand, put

$$\text{(3.18a)} \qquad W = \{\, x \in K_{2n}^n \mid \mathrm{rank}(x) = n,\ x\eta_n x^* = 0 \,\},$$
$$\text{(3.18b)} \qquad W_0 = \{\, x \in K_{2n}^n \mid \mathrm{rank}(x) = n,\ x \cdot \mathrm{diag}[\omega',\, -\omega]x^* = 0 \,\}.$$

Clearly $W = W_0 q$. Changing ξ_1 and ξ_2 for ξ_1^* and ξ_2^*, and also changing U for a suitable group of the same type, we find that the right-hand side of (3.17) can be written

$$\text{(3.19)} \qquad \sum_{\xi = [\xi_1\ \xi_2] \in U\backslash W_0} \Lambda_0(\xi)\det(\xi_1)^\nu \det(\xi_2)^{\mu\rho}|\det(\xi_1)|^{-2h-2s\mathbf{a}}$$

with some $\overline{\mathbf{Q}}$-valued $\Lambda_0 \in \mathcal{S}(K_{2n}^n)$. Now, for $x \in K_{2n}^n$ we put

$$j[x,\, z] = \big(j_v[x,\, z]\big)_{v\in\mathbf{b}},$$
$$j_{v\rho}[x,\, z] = \det(\lambda_v[x,\, z]), \quad j_v[x,\, z] = \det(\mu_v[x,\, z]) \qquad (v \in \mathbf{a}).$$

From (3.16) we obtain $\det(\xi_1)^{\tau_v} = c_v j_v[\xi q,\, w]$ and $\det(\xi_2)^{\tau_v\rho} = c'_v j_{v\rho}[\xi q,\, w]$ with some $c_v,\, c'_v \in \overline{\mathbf{Q}}^\times$. Since $W = W_0 q$, (3.19) equals

$$\text{(3.20)} \qquad c_0^s \sum_{\xi \in U\backslash W} \Lambda(\xi) j[\xi,\, w]^\varphi \overline{j[\xi,\, w]}^{\varphi'} |j[\xi,\, w]|^{-2h-2s\mathbf{a}}$$

with some $c_0 \in \mathbf{R} \cap \overline{\mathbf{Q}}, > 0$, some $\overline{\mathbf{Q}}$-valued $\Lambda \in \mathcal{S}(K_{2n}^n)$ such that $\Lambda(a\xi) = \Lambda(\xi)$ for every $a \in U$, and $\varphi, \varphi' \in \mathbf{Z}^{\mathbf{b}}$ given by $\varphi_v = \nu_v$, $\varphi_{v\rho} = \mu_v$, $\varphi'_v = \nu_{v\rho}$, $\varphi'_{v\rho} = \mu_{v\rho}$ for $v \in \mathbf{a}$; also, $|j[\xi, w]| = \bigl(|j_v[\xi, w]|\bigr)_{v\in\mathbf{a}}$. Changing w for the variable z on $\mathcal{H}$, we put

$$T(z, s; \Lambda) = \sum_{x\in U\backslash W} \Lambda(x) j[x, z]^{\varphi} \overline{j[x, z]}^{\varphi'} |j[x, z]|^{-2h-2s\mathbf{a}}. \tag{3.21}$$

Going back to (3.17), we thus obtain

$$D(s; f, g) = a_0 b_0^s T(w, s; \Lambda) \tag{3.22}$$

with some $a_0 \in \overline{\mathbf{Q}}^{\times}$ and $b_0 \in \mathbf{R} \cap \overline{\mathbf{Q}}, > 0$.

3.9. Let us now connect the series T with Eisenstein series of type (1.11). We first put

$$\Gamma = \{\, \gamma \in G \cap SL_{2n}(\mathfrak{r}) \mid \Lambda(x\gamma) = \Lambda(x) \,\}. \tag{3.23}$$

This is a congruence subgroup of G_1. Since $G = PG_1$, we have $G = \bigsqcup_{\beta\in B} P\beta\Gamma$ with a finite subset B of G_1. Put $\Gamma_\beta = \beta\Gamma\beta^{-1}$ and take a complete set of representatives S_β for $(P \cap \Gamma_\beta)\backslash\Gamma_\beta$; then $G = \bigsqcup_{\beta\in B} PS_\beta\beta$. For $\alpha \in K_{2n}^{2n}$ let $\mathfrak{l}(\alpha)$ denote the last n rows of α. Then $\mathfrak{l}(\pi\alpha\beta) = d_\pi \mathfrak{l}(\alpha)\beta$ for $\pi \in P$ and $\beta \in G$. By [97, Lemma 2.1 or (2.14.4)] we have $W = \mathfrak{l}(G)$, and hence $W = \bigcup_{\beta\in B} \mathfrak{l}(PS_\beta\beta) = \bigcup_\beta GL_n(K)\mathfrak{l}(S_\beta)\beta$. Moreover we easily see that the map $(d, \alpha, \beta) \mapsto d \cdot \mathfrak{l}(\alpha\beta)$ gives a bijection of $\bigsqcup_{\beta\in B} \bigl(GL_n(K) \times S_\beta \times \{\beta\}\bigr)$ onto W. Put $\mu_\beta(d) = \Lambda(d \cdot \mathfrak{l}(\beta))$ for $d \in V$. Then $\mu_\beta \in \mathcal{S}(K_n^n)$, $\mu_\beta(ad) = \mu_\beta(d)$ for every $a \in U$, and $\Lambda(d \cdot \mathfrak{l}(\alpha\beta)) = \mu_\beta(d)$ for every $\alpha \in \Gamma_\beta$, since $\mathfrak{l}(\alpha\beta) = \mathfrak{l}(\beta)\beta^{-1}\alpha\beta$ and $\beta^{-1}\alpha\beta \in \Gamma$. Let Δ be a complete set of representatives for $U\backslash GL_n(K)$. Then the elements $d \cdot \mathfrak{l}(\alpha\beta)$ for $d \in \Delta$, $\alpha \in S_\beta$, and $\beta \in B$ represent $U\backslash W$ without repetition. Since $\alpha\beta \in G_1$, we have, by (1.4), $j_{v\rho}(\alpha\beta, z) = j_v(\alpha\beta, z)$ for every $v \in \mathbf{a}$. Observe that $\det(d)^{\varphi}\det(d)^{\rho\varphi'} = \det(d)^r$ with an element r of $\mathbf{Z}^{\mathbf{b}}$ given by

$$r_v = \nu_v + \mu_{v\rho} \quad\text{and}\quad r_{v\rho} = \mu_v + \nu_{v\rho} \qquad (v \in \mathbf{a}). \tag{3.24}$$

Also, we have $j_v[\mathfrak{l}(\gamma), z] = j_v(\gamma, z)$ for every $v \in \mathbf{b}$ and

$$j_v[d \cdot \mathfrak{l}(\gamma), z] = \det(d)_v j_v(\gamma, z), \qquad j_{v\rho}[d \cdot \mathfrak{l}(\gamma), z] = \det(d)_v^{\rho} j_v(\gamma, z),$$

for every $\gamma \in G_1$ and $v \in \mathbf{a}$. Since $k = \nu + n\mathbf{a}$ and $l = \mu + n\mathbf{a}$, by (1.9b) we have $2h = (\mu_v + \mu_{v\rho} + \nu_v + \nu_{v\rho} + 2n)_{v\in\mathbf{a}}$, and hence

$$j(\gamma, z)^{\varphi}\overline{j(\gamma, z)}^{\varphi'} |j(\gamma, z)|^{-2h} = j(\gamma, z)^{-t}|j(\gamma, z)|^{t-2n\mathbf{a}}$$

for every $\gamma \in G_1$, where

$$t = (\mu_{v\rho} + \nu_{v\rho} - \mu_v - \nu_v)_{v\in\mathbf{a}}. \tag{3.25}$$

Therefore, substituting $d\mathfrak{l}(\alpha\beta)$ for x in (3.21), we obtain

$$\begin{aligned} &\delta(z)^{s\mathbf{a}+n\mathbf{a}-t/2} T(z, s; \Lambda) \\ &= \sum_{\beta\in B}\sum_{d\in\Delta} \mu_\beta(d)\det(d)^r |\det(d)|^{-2h-2s\mathbf{a}} E(z, s+n; t, \Gamma_\beta)\|_t\beta. \end{aligned} \tag{3.26}$$

Observe that $r_v + r_{v\rho} = 2h_v - 2n$ for every $v \in \mathbf{a}$. Putting

(3.27) $$M_\beta(s) = \sum_{d\in\Delta} \mu_\beta(d)\det(d)^r|\det(d)|^{-2h+2n\mathbf{a}-2s\mathbf{a}}$$

and combining (3.26) with (3.22), we obtain

(3.28) $$D(s; f, g) = a_1 b_1^s \sum_{\beta\in B} M_\beta(s+n)\big[E(z, s+n; t, \Gamma_\beta)\|_t\beta\big]_{z=w}$$

with some $a_1 \in \overline{\mathbf{Q}}^\times$ and $b_1 \in \mathbf{R}\cap\overline{\mathbf{Q}}, > 0$, since $\delta(w)$ is algebraic.

3.10. To complete the proof of Theorem 3.4, we take f and g of that theorem in (3.28). We disregard the $\overline{\mathbf{Q}}$-rationality of λ and λ'. We still have (3.28) with some constants a_1 and b_1 which may not be algebraic. Suppose $\mu \neq \nu$ and $\mu \neq \nu\rho$. Since $\mu_v + \mu_{v\rho} = \nu_v + \nu_{v\rho} = m_v - n$, we have $r_{v\rho} - r_v = 2(\mu_v - \nu_v)$ and $t_v = 2(\nu_{v\rho} - \mu_v)$ for every $v \in \mathbf{a}$. Thus $(r_{v\rho} - r_v)_{v\in\mathbf{a}} \neq 0$, so that the character χ_{ij} of (2.2) determined with M_β as M there are all nontrivial. Thus M_β is entire. Since $t \neq 0$, $E(z, s; t, \Gamma_\beta)$ is finite at $s = n$ by Lemma 1.3. Therefore, from (3.28) we see that $D(s; f, g)$ is finite at $s = 0$, and hence $\langle g, f\rangle = 0$ by Lemma 1.4 (3). This completes the proof of Theorem 3.4.

3.11. We now evaluate (3.28) at a special value of s. Let $\mathbf{f} = (f_p) \in \mathcal{T}_k(C)$. Given a Hecke character χ of K, we can define $\mathcal{Z}(s, \mathbf{f}, \chi)$ as in [00, §20.11]. Combining (22.19) and (22.18b) of [00] together, we can find $\mathbf{g} = (g_p) \in \mathcal{T}_l$ (whose description is given in [00, §22.6]) with which the following equality holds:

(3.29) $$\mathcal{Z}(s, \mathbf{f}, \chi) = P(s)\Lambda(s)\sum_{p\in\mathcal{A}} a_p b_p^s D\big(s-(3n/2), f_p, g_p\big).$$

Here $\mathcal{A}$ is a finite subset of $G_\mathbf{h}$; $a_p \in \overline{\mathbf{Q}}$, $0 < b_p \in \mathbf{Q}$; P is a finite Dirichlet series with constant term 1 and algebraic coefficients;

(3.30) $$\Lambda(s) = \prod_{i=1}^{n} L_\mathfrak{c}(2s-n+1-i, \chi_1\theta^{n+i-1}),$$

where χ_1 is the restriction of χ to $F_\mathbf{A}^\times$ and θ is the Hecke character of F corresponding to K/F. These are explained in [00, §§22.9 and 28.3].

We first prove Theorem 3.5 under the following condition:

(3.31) $$\nu_v = 0 \text{ for every } v \in \mathbf{a}.$$

This means that $k_v = n$ and $k_{v\rho} = m_v - n$ for every $v \in \mathbf{a}$; the condition $\nu_v + \nu_{v\rho} \geq n$ implies that $\nu_{v\rho} \geq n$ for every $v \in \mathbf{a}$, so that $k_{v\rho} \geq k_v$ for every $v \in \mathbf{a}$.

We make the following special choice of μ and χ; we put $m_0 = \mathrm{Min}_{v\in\mathbf{a}} m_v$, $\mu_v = m_v - m_0$, $\mu_{v\rho} = 0$, and $\ell_v = m_0 - 2k_{v\rho}$ for every $v \in \mathbf{a}$; then we take a Hecke character χ of K such that $\chi_\mathbf{a}(x) = x_\mathbf{a}^\ell |x_\mathbf{a}|^{-\ell}$. Such a χ exists by virtue of [97, Lemma 11.14 (3)]. Then [00, (22.15a, b)] are satisfied with $t' = -\ell$, and we obtain (3.29); we take ψ of [00, Section 22] to be trivial. Now we put $\sigma_0 = m_0/2$, and evaluate (3.29) at $s = \sigma_0$ and (3.20) at $s = \sigma_0 - (3n/2)$. By [00, Lemma 17.5 (2)] we see that $\Lambda(m_0/2) \in \pi^{dnp}\overline{\mathbf{Q}}$, where $d = [F:\mathbf{Q}]$ and $p = m_0 - (3n-1)/2$. Next, assuming that $m_0 > 3n$, we apply [00, Theorems 28.5 and 28.8] to the present $\mathbf{f}$. Since $k_v - k_{v\rho} - \ell_v = m_v - m_0$ and $2k_{v\rho} + \ell_v = m_0$, the condition on σ_0 in [00, Theorem 28.8] can be written

$$4n - m_0 \leq 2\sigma_0 \leq m_0 \text{ and } 2\sigma_0 - m_0 \in 2\mathbf{Z},$$

which is clearly satisfied by $\sigma_0 = m_0/2$. Therefore that theorem shows that

(3.32) $$\mathcal{Z}(\sigma_0, \mathbf{f}, \chi)/\langle \mathbf{f}, \mathbf{f} \rangle \in \pi^{nq}\overline{\mathbf{Q}}$$

with $q = \sum_{v\in\mathbf{a}}(m_v + m_0 - 2n + 1)$. To evaluate (3.28) at $s = \sigma_0 - (3n/2)$, we see that $r_{v\rho} - r_v = 2m_v - m_0 - n \geq m_0 - n$ for every $v \in \mathbf{a}$, and hence, taking $s_0 = 2\sigma_0 - n$ in Lemma 2.2, we obtain

(3.33) $$M(\sigma_0 - (n/2)) \in \pi^{cn/2} p_K\bigl(\textstyle\sum_{v\in\mathbf{a}}(2m_v - m_0 - n)\tau_v, n\tau\bigr)\overline{\mathbf{Q}}$$

with $c = \sum_{v\in\mathbf{a}}(2m_v - 3n + 1)$.

Finally we apply [87a, Proposition 4.1 (i)] or [00, Theorem 17.7 (i)] to $E(z, \sigma_0 - n/2;\, t, \Gamma_\beta)$. Since $t_v = \nu_{v\rho} + \mu_{v\rho} - \nu_v - \mu_v = m_0 - n$ for every $v \in \mathbf{a}$, we find that $E(z, \sigma_0 - n/2;\, t, \Gamma_\beta) \in \mathcal{M}_t(\overline{\mathbf{Q}})$. Define the symbols $\mathfrak{p}_v(w)$ and $\mathfrak{P}_k(w)$ by [00, (11.4b) and (11.17a)]. Then the structure of the CM-type (Y, Φ) determined in §3.7 together with (3.12b) shows that $\det\bigl(\mathfrak{p}_v(w)\bigr) = p_K(\tau_v, n\tau)$, and hence $\mathfrak{P}_t(w) = p_K\bigl((m_0 - n)\tau, n\tau\bigr)$. Thus, by [00, Proposition 11.11],

$$\bigl[E(z, \sigma_0 - n/2;\, t, \Gamma_\beta)\|_t\beta\bigr]_{z=w} \in p_K\bigl((m_0 - n)\tau, n\tau\bigr)\overline{\mathbf{Q}}.$$

Therefore, by (3.28) and (3.29) we obtain

(3.34) $$\mathcal{Z}(\sigma_0, \mathbf{f}, \chi) \in \pi^{nc/2+dnp} p_K\bigl(\textstyle\sum_{v\in\mathbf{a}} 2(m_v - n)\tau_v, n\tau\bigr)\overline{\mathbf{Q}}.$$

By [00, Theorem 20.13], $\mathcal{Z}(\sigma_0, \mathbf{f}, \chi) \neq 0$, and hence, comparing (3.32) with (3.34), we obtain

(3.35) $$\langle \mathbf{f}, \mathbf{f} \rangle \in \pi^{-dn^2} p_K\bigl(\textstyle\sum_{v\in\mathbf{a}} 2(m_v - n)\tau_v, n\tau\bigr)\overline{\mathbf{Q}}.$$

Once this is established, the assertion concerning $\langle g, f \rangle$ of Theorem 3.5 (2) follows from [00, Corollary 28.6]. This proves Theorem 3.5 (2) under (3.31).

4. Proof of Theorem 2.3 and the remaining part of Theorem 3.5

We start with two lemmas, one elementary and the other quite nontrivial. We put $d = [F : \mathbf{Q}]$ throughout this section.

4.1. Lemma. *Let $p_1, \ldots, p_m$ be $\mathbf{C}$-valued continuous functions on $\mathcal{H}$ linearly independent over $\mathbf{C}$, and $\mathcal{W}$ a dense subset of $\mathcal{H}$. Then there exist m points $w_1, \ldots, w_m$ of $\mathcal{W}$ such that $\det\bigl(p_i(w_j)\bigr)_{i,j=1}^{m} \neq 0$.*

PROOF. Let $X_w = \bigl\{ y \in \mathbf{C}^m \bigm| \sum_{i=1}^m y_i p_i(w) = 0 \bigr\}$. Then $\bigcap_{w\in\mathcal{W}} X_w = \{0\}$, and hence we can find m points $w_1, \ldots, w_m$ of $\mathcal{W}$ such that $\bigcap_{i=1}^m X_{w_i} = \{0\}$. Then we obtain the desired conclusion.

4.2. Lemma. *Let Γ be a congruence subgroup of G_1 and let B be a finite subset of G_1 such that $G = \bigsqcup_{\beta\in B} P\beta\Gamma$; put $\Gamma_\beta = \beta\Gamma\beta^{-1}$ and $E_\beta(z, s;\, t) = E(z, s;\, t, \Gamma_\beta)\|_t\beta$ for $\beta \in B$. Let κ be an even integer $\geq 2n$; suppose $n > 1$*

or $F \neq \mathbf{Q}$. Then the functions $E_\beta(z, n; \kappa\mathbf{a})$ for $\beta \in B$ are linearly independent over $\mathbf{C}$.

PROOF. If $\kappa = 2n$, the fact is included in [00, Theorem 27.15] as a special case. Indeed, take $r = 0$ in that theorem. Then the space $\mathcal{E}_{\kappa\mathbf{a}}^{\psi,r}(\Gamma)$ of that theorem has dimension $\#(B)$ over $\mathbf{C}$, and generated by the functions $E_\beta(z, n; 2n\mathbf{a})$, which proves our lemma for $\kappa = 2n$. (Notice that $\mathcal{S}_k^{\varphi_0}(\cdots)$ there is $\mathbf{C}$ and X there is B here. We cannot apply [00, Theorem 27.15] to the present situation if $n = 1$ and $F = \mathbf{Q}$, though that case can be handled by a different method.) To prove the case $\kappa > 2n$, we introduce differential operators $\mathcal{D}^q$ and Δ_r^q defined for $q, r \in \mathbf{Z}^{\mathbf{a}}$ by

$$(4.1)\quad \mathcal{D}^q = \prod_{v\in\mathbf{a}} \left[\det(\partial/\partial z_{vij})_{i,j=1}^n\right]^{q_v}, \quad \Delta_r^q f = \delta(z)^{n\mathbf{a}-q-r}\mathcal{D}^q\left[\delta(z)^{q+r-n\mathbf{a}} f\right],$$

where f is a C^∞ function on $\mathcal{H}$ and z_{vij} is the (i,j)-entry of the matrix z_v. Then

$$(4.2)\quad \Delta_r^q(f\|_r\gamma) = (\Delta_r^q f)\|_{r+2q}\gamma \text{ for every } \gamma \in G_1,$$

$$(4.3)\quad \Delta_r^q E(z, s; r, \Gamma) = b_r^q(s)E(z, s; r + 2q, \Gamma)$$

with a polynomial function b_r^q which depends only on q, r, and n. Indeed, (4.2) was given in [81, p.842]; as for (4.3), see [87a,(4.12)] or [00, (17.20)]. (The operator Δ_r^q is essentially an operator of type Δ_ρ^Z of [94] and [00, Sections 12 and 13], and (4.2) is a special case of [00, (12.24c)]; see also [94, Lemma 4.3] and its proof.) Now suppose $\kappa > 2n$; put $p = (\kappa - 2n)/2$. Taking $s = n$, $r = 2n\mathbf{a}$, and $q = p\mathbf{a}$ and combining (4.2) and (4.3), we obtain

$$\Delta_r^q E_\beta(z, n; 2n\mathbf{a}) = b_r^q(n)E_\beta(z, n; \kappa\mathbf{a}).$$

As noted in [87a, p.21] and [00, p.146], employing the explicit form for b_r^q, we see that $b_r^q(n) \neq 0$. Therefore, to prove the linear independence of the $E_\beta(z, n; \kappa\mathbf{a})$, it is sufficient to show that if $\Delta_r^q f = 0$ for $f \in \mathcal{M}_{2n\mathbf{a}}$, then $f = 0$.

Suppose $0 \neq f \in \mathcal{M}_{2n\mathbf{a}}$ and $\Delta_r^q f = 0$; take the Fourier expansion (1.7) of f. By [94, Theorem 5.6], there exists an element σ_0 of S^+ such that $c_f(\sigma_0) \neq 0$. (Indeed, if there is no such σ_0, then $\operatorname{rank}(f) < n$, and hence $2n < n$ by that theorem, a contradiction.) Now we have $\mathcal{D}^{p\mathbf{a}}\left[\delta(z)^{p\mathbf{a}+n\mathbf{a}} f(z)\right] = 0$, which produces an equality

$$(4.4)\quad \delta(z)^{p\mathbf{a}+n\mathbf{a}} \cdot \mathcal{D}^{p\mathbf{a}} f + \sum_{i\in I} g_i(z)P_i(z^* - z) = 0,$$

where I is a finite set of indices, the g_i are holomorphic functions on $\mathcal{H}$, and the $P_i(z^* - z)$ are polynomials of $\overline{z}_{vji} - z_{vij}$ for all (v, i, j) of total degree $< dn(p+n)$. Now

$$\mathcal{D}^{p\mathbf{a}} f = (2\pi i)^{dnp} \sum_{\sigma\in S} c_f(\sigma)\det(\sigma)^{p\mathbf{a}}\mathbf{e}_{\mathbf{a}}^n(\sigma z) \neq 0,$$

since $c_f(\sigma_0) \neq 0$. Now the functions $\overline{z}_{vji} - z_{vij}$ for all (v, i, j) are algebraically independent over the field of meromorphic functions on $\mathcal{H}$, so that (4.4) is impossible. This completes the proof.

4.3. To prove the last part of Theorem 2.3, we may assume that $n > 1$, since the desired fact follows from (2.2) and (2.3) if $n = 1$. Given $\varepsilon \in \mathcal{S}(V)$, define $\Lambda \in \mathcal{S}(K^n_{2n})$ by $\Lambda(\xi_1, \xi_2) = \alpha'(\xi_1)\varepsilon(\xi_2)$ with any $\mathbf{Q}$-valued $\alpha' \in \mathcal{S}(V)$ such that $\alpha'(0) = 1$. Define also Γ by (3.23) for this Λ and take B so that $G = \bigsqcup_{\beta \in B} P\beta\Gamma$. We repeat what we did in §3.9 with the present Λ. Then we have (3.26) in the present case, that is,

$$\delta(z)^{s\mathbf{a}+n\mathbf{a}-t/2}T(z, s; \Lambda) = \sum_{\beta \in B} M_\beta(s+n)E_\beta(z, s+n; t), \tag{4.5}$$

where M_β is defined by (3.27) with $\mu_\beta(d) = \Lambda(d \cdot \iota(\beta))$.

We now take $\mu_v = \nu_v = 0$ and $\mu_{v\rho} = \nu_{v\rho} = e$ for every $v \in \mathbf{a}$ with an integer $e > n+1$. Then $r_v = r_{v\rho} = e$ and $t_v = 2e$ for every $v \in \mathbf{a}$, and hence

$$M_\beta(s) = \textstyle\sum_{d \in \Delta} \mu_\beta(d)|\det(d)|^{-2s\mathbf{a}}. \tag{4.6}$$

If $\beta = 1$, we have $d \cdot \iota(1) = (0, d)$, and hence $\mu_\beta(d) = \varepsilon(d)$. Thus $M_1(s)$ coincides with $M(s)$ of Theorem 2.3 times a positive rational number depending on U.

Let the symbols q, h, and w be as in §3.7. Put $q_1 = q\alpha^{-1}$ with any fixed $\alpha \in G$. Then we have $q_1\eta_n q_1^* = q\eta_n q^*$, and so we can repeat our discussion of §§3.7 and 3.8 with q_1 in place of q. The map h must be changed for h_1 defined by $h_1(a) = \alpha h(a)\alpha^{-1}$; then $h_1(Y^u)$ has αw as its fixed point. Let $m = \#(B)$. Observe that the set Gw is dense in $\mathcal{H}$. Therefore, combining Lemma 4.1 with Lemma 4.2, we can find m elements w_i of Gw such that $\det\left[E_\beta(w_i, n; t)\right]_{\beta,i} \neq 0$. Put $w_i = \alpha_i w$ with $\alpha_i \in G$; put also $q_i = q\alpha_i^{-1}$ and $\Lambda_i(\xi) = \Lambda(\xi q_i)$. Since (3.19) equals (3.20), we have

$$a_i b_i^s T(w_i, s; \Lambda) = \sum_{\xi = [\xi_1\ \xi_2] \in U\backslash W_0} \Lambda_i(\xi)\det(\xi_1)^\nu \det(\xi_2)^{\mu\rho}|\det(\xi_1)|^{-2h-2s\mathbf{a}} \tag{4.7}$$

with some $a_i \in \overline{\mathbf{Q}}^\times$ and $0 < b_i \in \mathbf{R} \cap \overline{\mathbf{Q}}$. We obtained (3.19) from (3.17). Instead we go backward; namely we start from the right-hand side of (4.7), and observe that it can be written as a finite sum

$$\sum_j \det(\beta)^{s\mathbf{a}+h} D(s; f_j, g_j) \tag{4.8}$$

with $f_j, g_j \in \mathcal{T}_k(\overline{\mathbf{Q}})$, since $\Lambda_i(\xi_1, \xi_2)$ is a finite sum of the functions of the form $\lambda_1(\xi_1)\lambda_2(\xi_2)$ with $\overline{\mathbf{Q}}$-valued $\lambda_1, \lambda_2 \in \mathcal{S}(V)$. Now we take the residues of (4.7) and (4.8) at $s = 0$. We take $e > 2n$. Then our result of §3.11 shows that $\langle f_j, g_j \rangle \in \pi^{-dn^2}\mathfrak{q}^2\overline{\mathbf{Q}}$, where $\mathfrak{q} = p_K(e\tau, n\tau)$. Therefore, by Lemma 1.4 (3), the residue of (4.8) at $s = 0$ belongs to $\pi^{dn\gamma}R\mathfrak{q}^2\overline{\mathbf{Q}}$ where $\gamma = e - (n-1)/2$. On the other hand, by [00, Theorem 17.9], $E_\beta(z, n; t)$ is $\pi^{dn\gamma'}$ times a $\overline{\mathbf{Q}}$-rational nearly holomorphic automorphic form of weight t, where $\gamma' = e - n$, and hence, by the characterization of near holomorphy in [00, §14.4], $E_\beta(w_i, n; t) \in \pi^{dn\gamma'}\mathfrak{q}^2\overline{\mathbf{Q}}$. Therefore, solving the linear equations (4.5) with $(z, s) = (w_i, 0)$ for $1 \le i \le m$, we find that each M_β has at most a simple pole at $s = n$, and the residue is an algebraic number times $\pi^{cdn}r_0^n(F)$ with $c = (n+1)/2$. This completes the proof of Theorem 2.3.

4.4. Let us now prove Theorem 3.5 (1) under (3.31). Take $\mu = \nu$ in (3.28); then $r_v = r_{v\rho}$ and $t = 2(\nu_{v\rho})_{v\in\mathbf{a}}$. Our assumption on ν together with (3.31) implies that $\nu_{v\rho} \geq n$ for every $v \in \mathbf{a}$. We consider the residue of (3.28) at $s = 0$. For the same reason as above, we have $E_\beta(w, n; t) \in \pi^{nb}\mathfrak{q}_t\overline{\mathbf{Q}}$, where $b = \sum_{v\in\mathbf{a}}(\nu_{v\rho} - n)$ and $\mathfrak{q}_t = p_K\big(\sum_{v\in\mathbf{a}} 2\nu_{v\rho}\tau_v, n\tau\big)$. Also, by Theorem 2.3 the residue of M_β at $s = n$ belongs to $\pi^{dn(n+1)/2}r_0^n(F)\overline{\mathbf{Q}}$. On the other hand, by Lemma 1.4 (3) the residue of $D(s; f, g)$ at $s = 0$ is a nonzero algebraic number times $\langle g, f\rangle\pi^{na}r_0^n(F)$ with $a = \sum_{v\in\mathbf{a}}(\nu_{v\rho}+n) - d(n-1)/2$. Therefore $\langle g, f\rangle \in \pi^{nc}\mathfrak{q}_t\overline{\mathbf{Q}}$ with $c = b - a + d(n+1)/2 = -dn$. This proves Theorem 3.5 (1) under (3.31).

4.5. We now prove Theorem 3.5 without (3.31), by showing that the general case can be reduced to the case with (3.31). We first recall that we have fixed a CM-type (K, τ) and our action of G on $\mathcal{H}$ depends on this CM-type. Namely, for $\alpha \in G$ and $z \in \mathcal{H}$ we have $(\alpha z)_v = \alpha^{\tau_v}(z_v)$. Now take another CM-type σ of K and define a map $z \mapsto z^\circ$ of $\mathcal{H}$ onto itself by $z_v^\circ = z_v$ if $\sigma_v = \tau_v$ and $z_v^\circ = {}^t z_v$ if $\sigma_v = \tau_v\rho$. Define also $\alpha\{z\} \in \mathcal{H}$, $\mu_v\{\alpha, z\}$ and $\lambda_v\{\alpha, z\}$ by $\alpha\{z\}_v = \alpha^{\sigma_v}(z_v)$, and $\alpha^{\sigma_v}B(z_v) = B(\alpha\{z\}_v)\mathrm{diag}\big[\overline{\lambda_v\{\alpha, z\}}, \mu_v\{\alpha, z\}\big]$. Then we can easily verify that $\alpha\{z^\circ\} = (\alpha z)^\circ$; $\lambda_v\{\alpha, z^\circ\} = \lambda_v(\alpha, z)$ and $\mu_v\{\alpha, z^\circ\} = \mu_v(\alpha, z)$ if $\sigma_v = \tau_v$; $\lambda_v\{\alpha, z^\circ\} = \mu_v(\alpha, z)$ and $\mu_v\{\alpha, z^\circ\} = \lambda_v(\alpha, z)$ if $\sigma_v = \tau_v\rho$; see [00, (5.31)].

For $f : \mathcal{H} \to \mathbf{C}$ define $f^\circ : \mathcal{H} \to \mathbf{C}$ by $f^\circ(z) = f(z^\circ)$. Then, for $\alpha \in G$ and $k \in \mathbf{Z}^{\mathbf{b}}$ we have

$$(f\|_k\alpha)^\circ(z) = j\{\alpha, z\}^{-k'} f^\circ(\alpha\{z\}). \tag{4.9}$$

Here $j\{\alpha, z\} = \big(j_v\{\alpha, z\}\big)_{v\in\mathbf{b}}$ with $j_v\{\alpha, z\} = \det\big(\mu_v\{\alpha, z\}\big)$ and $j_{v\rho}\{\alpha, z\} = \det\big(\lambda_v\{\alpha, z\}\big)$ for $v \in \mathbf{a}$; $(k'_{v\rho}, k'_v) = (k_{v\rho}, k_v)$ if $\sigma_v = \tau_v$ and $(k'_{v\rho}, k'_v) = (k_v, k_{v\rho})$ if $\sigma_v = \tau_v\rho$. These were essentially observed in [00, §5.11]. If $f(z) = \sum_{h\in S} c_f(h)\mathbf{e}_{\mathbf{a}}^n(hz)$, then $f^\circ(z) = \sum_{h\in S} c_f(h)\mathbf{e}\big(\sum_{v\in\mathbf{a}} \mathrm{tr}(h^{\sigma_v}z_v)\big)$, so that the $c_f(h)$ are the Fourier coefficients of f° defined with respect to $\{\sigma_v\}$ and the action $z \mapsto \alpha\{z\}$.

We can naturally define automorphic forms with respect to the action $z \mapsto \alpha\{z\}$ and $j\{\alpha, z\}$. Then $f \mapsto f^\circ$ sends the space $\mathcal{M}_k$ previously defined to such a new space of automorphic forms of weight k'. Clearly we have

$$\langle f, g\rangle = \langle f^\circ, g^\circ\rangle. \tag{4.10}$$

We next consider $f(z) = \theta_\nu(z; \lambda, \omega)$. Define $\nu' \in \mathbf{Z}^{\mathbf{b}}$ by $(\nu'_{v\rho}, \nu'_v) = (\nu_{v\rho}, \nu_v)$ if $\sigma_v = \tau_v$ and $(\nu'_{v\rho}, \nu'_v) = (\nu_v, \nu_{v\rho})$ if $\sigma_v = \tau_v\rho$. For $\xi \in V$ observe that $\mathrm{tr}\big((\xi^*\omega\xi)^{\tau_v}z_v^\circ\big) = \mathrm{tr}\big((\xi^*\omega\xi)^{\sigma_v}z_v\big)$ for every $v \in \mathbf{a}$. Also $\sum_{v\in\mathbf{a}}(\nu_v\tau_v + \nu_{v\rho}\tau_v\rho) = \sum_{v\in\mathbf{a}}(\nu'_v\sigma_v + \nu'_{v\rho}\sigma_v\rho)$. Therefore $\theta_\nu(z^\circ; \lambda, \omega)$ is the function of type (3.5) defined with $\big(\sigma, \sum_{v\in\mathbf{a}}(\nu'_v\sigma_v + \nu'_{v\rho}\sigma_v\rho)\big)$ in place of $\big(\tau, \sum_{v\in\mathbf{a}}(\mu_v\tau_v + \mu_{v\rho}\tau_v\rho)\big)$.

To complete the proof of Theorem 3.5, given $f(z) = \theta_\nu(z; \lambda', \omega)$, define $\sigma = \sum_{v\in\mathbf{a}}\sigma_v$ so that $\sigma_v = \tau_v$ if $\nu_v = 0$ and $\sigma_v = \tau_v\rho$ if $\nu_v > 0$. Then $\nu'_v = 0$ for every $v \in \mathbf{a}$, and hence we can apply what we proved in §§3.11 and 4.4 to f° to find that

$$\langle f^\circ, f^\circ\rangle \in \pi^{-dn^2}p_K\big(\textstyle\sum_{v\in\mathbf{a}} 2(m_v - n)\sigma_v, n\sigma\big)\overline{\mathbf{Q}}. \tag{4.11}$$

Recalling that $m_v - n = \nu_v + \nu_{v\rho}$, $\nu_v\nu_{v\rho} = 0$, and the basic equality $p_K(\rho\alpha, \beta) = p_K(-\alpha, \beta)$, we can replace the quantity $p_K(\cdots)$ of (4.11) by $p_K\bigl(\sum_{v\in\mathbf{a}} 2(\nu_{v\rho} - \nu_v)\tau_v,\ n\sigma\bigr)$. This combined with (4.10) completes the proof of Theorem 3.5.

5. Dirichlet series whose critical values are arithmetic

5.1. In addition to the unitary group G of (0.1) we need the unitary group $G(H)$ of (0.8) and also a CM-field L contained in K. To avoid possible confusion, hereafter we write G_K^n for the group G of (0.1), and $\mathcal{M}_{K,\omega}$ for the space of modular forms $\mathcal{M}_\omega$ of weight ω on $\mathcal{H}_n^{\mathbf{a}}$ with respect to the congruence subgroups of G_K^n.

Let Y be a CM-algebra which is a direct sum of CM-fields containing K such that $[Y : K] = m + n$ with a nonnegative integer m and a positive integer n. We denote by ρ the automorphism of Y that gives complex conjugation on each simple factor of Y. We define $\mathbf{Z}$-linear maps $\mathrm{Res}_{Y/K} : I_Y \to I_K$ and $\mathrm{Inf}_{Y/K} : I_K \to I_Y$ so that $\mathrm{Res}_{Y/K}(\alpha)$ for $\alpha \in J_Y$ is the restriction of α to K and $\mathrm{Inf}_{Y/K}(\beta)$ for $\beta \in J_K$ is the sum of all $\alpha \in J_Y$ such that $\mathrm{Res}_{Y/K}(\alpha) = \beta$.

We are going to define a certain Dirichlet series W which depends on two sets of data $\{\zeta, \gamma_{vi}, \varepsilon_{vj}\}$ and $\{Q, f, \alpha, k\}$. We impose the following conditions on the first set of data:

(5.1a) $\gamma_{vi}, \varepsilon_{vj} \in J_Y$ given for $v \in \mathbf{a}$, $1 \le i \le m$, $1 \le j \le n$, and
$$\{\rho\gamma_{v1}, \ldots, \rho\gamma_{vm}, \varepsilon_{v1}, \ldots, \varepsilon_{vn}\} = \{\sigma \in J_Y \mid \mathrm{Res}_{Y/K}(\sigma) = \tau_v\};$$

(5.1b) $\zeta \in Y^\times$, $\zeta^\rho = \zeta$; $\zeta^{\gamma_{vi}} < 0$ and $\zeta^{\varepsilon_{vi}} > 0$.

We next take a polynomial representation

$$R : GL_n(\mathbf{C})^{\mathbf{a}} \to GL(X) \tag{5.2a}$$

with a finite-dimensional complex vector space X and a polynomial map

$$Q : (\mathbf{C}_n^m)^{\mathbf{a}} \to \mathrm{Hom}(X, \mathbf{C}) \tag{5.2b}$$

such that $Q(xb) = Q(x)R(b)$ for $b \in GL_n(\mathbf{C})^{\mathbf{a}}$. We assume that X has a $\overline{\mathbf{Q}}$-structure and R is $\overline{\mathbf{Q}}$-rational. Taking $p \in \mathbf{Z}^{\mathbf{a}}$ such that $p_v \ge 0$ for every $v \in \mathbf{a}$, we consider a representation $\omega : GL_n(\mathbf{C})^{\mathbf{a}} \times GL_n(\mathbf{C})^{\mathbf{a}} \to GL(X)$ by $\omega(a, b) = \det(b)^p R({}^t a^{-1})$ for $(a, b) \in GL_n(\mathbf{C})^{\mathbf{a}} \times GL_n(\mathbf{C})^{\mathbf{a}}$. If $m = 0$, we take $X = \mathbf{C}$, R to be trivial, and $Q = 1$; thus $\omega(a, b) = \det(b)^p$.

We take an element f of $\mathcal{M}_{K,\omega}$. Then we have a Fourier expansion

$$f(z) = \sum_{\sigma\in S} c_f(\sigma)\mathbf{e}_{\mathbf{a}}^n(\sigma z) \qquad (z \in \mathcal{H}_n^{\mathbf{a}}) \tag{5.3}$$

with elements $c_f(\sigma)$ of X in the same sense as in (1.7). Then there exists (see [00, (5.21)]) a subgroup U of $GL_n(\mathfrak{r})$ of finite index such that

$$c_f(a\sigma a^*) = \omega(\bar{a}, a)c_f(\sigma) \quad \textit{for every} \quad a \in U. \tag{5.4}$$

The last two items α and k in our sets of data are as follows: α is an element of $\mathcal{S}(Y^n)$; $k = \sum_{v\in\mathbf{a}}(k_v\tau_v + k_{v\rho}\tau_v\rho) \in J_K$ with $0 \le k_v, k_{v\rho} \in \mathbf{Z}$; we define $g \in \mathbf{Z}^{\mathbf{a}}$ by

$$g_v = k_v + k_{v\rho} - p_v \qquad (v \in \mathbf{a}). \tag{5.5}$$

We now define a $\mathbf{C}$-valued series $W(s, Q)$ by

$$(5.6)\quad W(s, Q) = \sum_{y \in U \backslash \mathcal{Y}} \alpha(y) \det(y^{\varepsilon})^{-k} |\det(y^{\varepsilon})|^{g-2s\mathbf{a}} Q({}^t y^{\gamma}) c_f\bigl(\mathrm{Tr}_{Y/K}(y\zeta y^*)\bigr),$$

where U is a subgroup of $GL_n(\mathfrak{r})$ of finite index whose nature will be discussed below; $\mathcal{Y}$ is the set of all $y = (y_i)_{i=1}^n \in Y^n$ such that the y_i are linearly independent over K; $\mathrm{Tr}_{Y/K}(y\zeta y^*)$ denotes the $(n \times n)$-matrix whose (i, j)-entry is $\mathrm{Tr}_{Y/K}(\zeta y_i y_j^{\rho})$ for $y = (y_i)_{i=1}^n \in Y^n$;

$$(5.7\mathrm{a})\qquad y^{\varepsilon} = (y_v^{\varepsilon}, y_{v\rho}^{\varepsilon})_{v \in \mathbf{a}}, \quad y_v^{\varepsilon} = \begin{bmatrix} y_1^{\varepsilon_{v1}} & \cdots & y_1^{\varepsilon_{vn}} \\ \cdots & \cdots & \cdots \\ y_n^{\varepsilon_{v1}} & \cdots & y_n^{\varepsilon_{vn}} \end{bmatrix}, \quad y_{v\rho}^{\varepsilon} = (y_v^{\varepsilon})^{\rho},$$

$$(5.7\mathrm{b})\qquad y^{\gamma} = (y_v^{\gamma})_{v \in \mathbf{a}}, \quad y_v^{\gamma} = \begin{bmatrix} y_1^{\gamma_{v1}} & \cdots & y_1^{\gamma_{vm}} \\ \cdots & \cdots & \cdots \\ y_n^{\gamma_{v1}} & \cdots & y_n^{\gamma_{vm}} \end{bmatrix};$$

$Q({}^t y^{\gamma})$ should be ignored if $m = 0$. We view $\det(y^{\varepsilon})$ as an element of $\mathbf{C}^{\mathbf{b}}$, and $\det(y^{\varepsilon})^{-k}$ is defined acording to the notation of (1.5). Since $c_f(h) \neq 0$ only if h is totally nennegative, the sum of (5.6) is taken over all $y \in \mathcal{Y}$ such that $\mathrm{Tr}_{Y/K}(y\zeta y^*)$ is totally nonnegative. We shall show in §7.1 that y_v^{ε} is invertible for every such y and every $v \in \mathbf{a}$. Thus each term of (5.6) is meaningful. To make it dependent only on Uy, we take U so that $0 \ll \det(a) \in \mathfrak{g}^{\times}$ and $\alpha(ay) = \alpha(y)$ for every $a \in U$, and (5.4) holds. We also assume

(5.8) $X = \mathbf{C}$, R is trivial, and $Q = 1$ if $k_v < k_{v\rho}$ for some $v \in \mathbf{a}$.

Then, in view of (5.5), the sum of (5.6) is at least formally meaningful.

In order to ensure the convergence, we impose the following condition on f:

(5.9) $\|\omega(\overline{\sigma}, \sigma)^{-1/2} c_f(\sigma)\| \leq C \cdot \det(\sigma)^{a\mathbf{a}}$ *for every* $\sigma \in S^+$ *with constants* $C \in \mathbf{R}, > 0$, *and* $a \in \mathbf{R}, \geq 0$.

Here we assume that $X = \mathbf{C}^e$, $\|x\|^2 = \sum_{\nu=1}^{e} |x_\nu|^2$ for $x \in \mathbf{C}^e$, and ω has the property that $\omega(a^*, b^*) = \omega(a, b)^*$, which is the case if we change, if necessary, ω for a suitable representation equivalent to ω. It is conjecturable that (5.9) is always true. We can at least show that (5.9) is satisfied if ω is one-dimensional or f is a cusp form. Indeed, the case of one-dimensional ω is proven in [00, Proposition A6.4]. If f is a cusp form, then standard arguments show that $\omega({}^t y, y)^{1/2} f(z)$ is bounded on $\mathcal{H}_n^{\mathbf{a}}$, where $y = \bigl((i/2)(z_v^* - z_v)\bigr)_{v \in \mathbf{a}}$, from which we can derive the desired inequality with $a = 0$ by the same technique as in the scalar-valued case (see [00, Proposition A6.4]).

5.2. In order to introduce another type of series, we recall some symbols. This time we assume both m and n to be positive. We put $T_v = \mathbf{C}_n^m$ for each $v \in \mathbf{a}$ and $T = \prod_{v \in \mathbf{a}} T_v$. For $e \in \mathbf{Z}^{\mathbf{a}}$ with $e_v \geq 0$ for every $v \in \mathbf{a}$ we denote by $S_e(T)$ the set of all polynomial functions on T which are homogeneous of degree e_v with respect to the variable on T_v for each v. We define a representation $\tau^e : GL_m(\mathbf{C})^{\mathbf{a}} \times GL_n(\mathbf{C})^{\mathbf{a}} \to GL\bigl(S_e(T)\bigr)$ by

(5.10) $(\tau^e(a,b)h)(u) = h\big(({}^t a_v u_v b_v)_{v\in\mathbf{a}}\big)$ *for* $a \in GL_m(\mathbf{C})^{\mathbf{a}}$, $b \in GL_n(\mathbf{C})^{\mathbf{a}}$, $h \in S_e(T)$, *and* $u \in T$.

Then $S_e(T)$ can be decomposed into irreducible subspaces, whose description is unnecessary for the moment. Taking an element φ of $S_e(T)$, we define a scalar-valued series $W(s, \varphi)$ by

(5.11a) $$W(s,\varphi) = \sum_{y\in U\backslash \mathcal{Y}} \alpha(y)\det(y^{\varepsilon})^{-k}|\det(y^{\varepsilon})|^{g-2s\mathbf{a}}\varphi(\mathfrak{x}(y))c_f\big(\mathrm{Tr}_{Y/K}(y\zeta y^*)\big),$$

(5.11b) $$\mathfrak{x}(y) = (\mathfrak{x}_v)_{v\in\mathbf{a}},\quad \mathfrak{x}_v = \begin{cases} (y_v^{\gamma})^* \cdot {}^t(y_v^{\varepsilon})^{-1} & \text{if } k_v \ge k_{v\rho}, \\ {}^t(y_v^{\gamma})(y_v^{\varepsilon})^{\wedge} & \text{if } k_v < k_{v\rho}. \end{cases}$$

Here the symbols are the same as in (5.6) except that we have $\varphi(\mathfrak{x}(y))$ instead of $Q({}^t y^{\gamma})$ and we take $\omega(a,b) = \det(b)^p$. Notice that $\mathfrak{x}(uy) = \mathfrak{x}(y)$ for every $u \in GL_n(\mathfrak{r})$, so that the sum is formally well-defined.

To state our theorem concerning the critical values of our series, we need a few more symbols. Namely we put

(5.12a) $$\Phi = \textstyle\sum_{v\in\mathbf{a}}\big(\sum_{i=1}^m \gamma_{vi} + \sum_{j=1}^n \varepsilon_{vj}\big)\rho_v,$$

(5.12b) $$\mathfrak{p}' = (\mathfrak{p}'_v)_{v\in\mathbf{a}},\quad \mathfrak{p}'_v = \mathrm{diag}\big[p_Y(\gamma_{v1}\rho_v, \Phi), \ldots, p_Y(\gamma_{vm}\rho_v, \Phi)\big],$$

(5.12c) $$\mathfrak{p} = (\mathfrak{p}_v)_{v\in\mathbf{a}},\quad \mathfrak{p}_v = \mathrm{diag}\big[p_Y(\varepsilon_{v1}\rho_v, \Phi), \ldots, p_Y(\varepsilon_{vn}\rho_v, \Phi)\big],$$

(5.12d) $$\mathbf{p} = (\mathbf{p}_v)_{v\in\mathbf{a}},\quad \mathbf{p}_v = p_K\big(\tau_v\rho_v, \textstyle\sum_{u\in\mathbf{a}}\tau_u\rho_u\big),$$

where $\rho_v = 1$ if $k_v \ge k_{v\rho}$ and $\rho_v = \rho$ if $k_v < k_{v\rho}$.

5.3. Theorem. (1) *The sum of (5.11a) is convergent for sufficiently large* $\mathrm{Re}(s)$, *and can be continued as a meromorphic function to the whole complex plane. The same is true for* $W(s, Q)$ *under (5.9).*

(2) *Suppose that* α *is* $\overline{\mathbf{Q}}$*-valued and* Q, f *are* $\overline{\mathbf{Q}}$*-rational, and also that (5.9) is satisfied; put* $\kappa_v = |k_v - k_{v\rho}|$ *and suppose that*

(5.13a) $\kappa_v > 2n$ *for every* $v \in \mathbf{a}$.

Let s_0 *be an integer such that*

(5.13b) $\kappa_v - p_v - (2/n)(\kappa_v - 1) \le s_0 \le \kappa_v - p_v$ *and* $s_0 - \kappa_v + p_v \in 2\mathbf{Z}$ *for every* $v \in \mathbf{a}$; $s_0 \ge m$; $s_0 \ne m+1$ *if* $F = \mathbf{Q}$.

Then $W(s, Q)$ *is finite at* $s = s_0/2$, *and* $W(s_0/2, Q') \in \pi^c\mathbf{q}\mathfrak{q}\overline{\mathbf{Q}}$, *where*

(5.14a) $$c = n\textstyle\sum_{v\in\mathbf{a}}|k_v - k_{v\rho}| - [F:\mathbf{Q}]n(n-1)/2,$$

(5.14b) $$\mathbf{q} = p_K\big(\textstyle\sum_{v\in\mathbf{a}}(k_v - k_{v\rho})\tau_v,\ n\sum_{v\in\mathbf{a}}\tau_v\rho_v\big),$$

(5.14c) $$\mathfrak{q} = p_Y\big(\textstyle\sum_{v\in\mathbf{a}}\sum_{i=1}^n(k_v - k_{v\rho})\varepsilon_{vi},\ \Phi\big),$$

and Q' is defined by $Q'(x) = Q(\mathbf{p}^{-1}\mathfrak{p}'x)$.

(3) *Let* s_0 *be as above, and* $S_e(T)$ *be as in* §5.2. *Then* $W(s, \varphi)$ *is finite at* $s = s_0/2$ *for every* $\varphi \in S_e(T)$. *Moreover, if* α *is* $\overline{\mathbf{Q}}$*-valued and* f, φ *are* $\overline{\mathbf{Q}}$*-rational, then* $W(s_0/2, \varphi') \in \pi^b\mathbf{q}\mathfrak{q}\overline{\mathbf{Q}}$, *where* $b = c + \sum_{v\in\mathbf{a}} e_v$ *and* $\varphi' = \tau^e(\mathfrak{p}', \mathfrak{p})^{-1}\varphi$.

The proof will be given in §7.2. Here let us show that we may assume that $k_v \ge k_{v\rho}$ for every $v \in \mathbf{a}$, since the general case can be reduced to the case in which that condition is satisfied. Indeed, define $\{\tau'_v\}_{v\in\mathbf{a}}$ and $k' \in \mathbf{Z}^{\mathbf{b}}$ as follows: $\tau'_v = \tau_v$ and $(k'_v, k'_{v\rho}) = (k_v, k_{v\rho})$ if $k_v \geq k_{v\rho}$; $\tau'_v = \tau_v\rho$ and $(k'_v, k'_{v\rho}) = (k_{v\rho}, k_v)$ if $k_v < k_{v\rho}$. Change γ_{vi} and ε_{vj} for $\gamma_{vi}\rho$ and $\varepsilon_{vj}\rho$ if $\tau'_v = \tau_v\rho$. Let $W'(s, Q)$ and $W'(s, \varphi)$ denote the series defined with $\{\tau'_v\}_{v\in\mathbf{a}}$ in place of $\{\tau_v\}_{v\in\mathbf{a}}$ and with new γ_{vi} and ε_{vj}. Then we easily see that these coincide with the original series $W(s, Q)$ and $W(s, \varphi)$; also, the quantities of (5.12b, c, d) and (5.14b, c) do not change. Thus it is sufficient to prove the case in which $k_v \geq k_{v\rho}$ for every $v \in \mathbf{a}$.

5.4. If $m = 0$, our series is essentially the same as a series of the following type:

$$(5.15) \qquad \mathcal{D}(s) = \sum_{x\in U\backslash GL_n(K)} \alpha(x)\det(x)^{-k}|\det(x)|^{g-2s\mathbf{a}}c_f(x\beta x^*).$$

Here α is an element of $\mathcal{S}(K^n_n)$; $k \in \mathbf{Z}^{\mathbf{b}}$ and g is defined by (5.5); $f \in \mathcal{M}_{K,\omega}(\overline{\mathbf{Q}})$ with $\omega(a, b) = \det(b)^p$, $p \in \mathbf{Z}^{\mathbf{a}}$; β is an element of the set S^+ of (1.1c); U is a subgroup of $GL_n(\mathfrak{r})$ of finite index such that $0 \ll \det(a) \in \mathfrak{g}^\times$ and $\alpha(ax) = \alpha(x)$ for every $a \in U$, and (5.4) holds.

Let us now show that if $m = 0$, then $W(s, Q)$ of (5.6) can be written $ab^s\mathcal{D}(s)$ with $a \in \overline{\mathbf{Q}}^\times$, $0 < b \in \overline{\mathbf{Q}} \cap \mathbf{R}$, and a suitable $\mathcal{D}$ of type (5.15). We have $[Y : K] = n$ if $m = 0$. Let $\{e_i\}_{i=1}^n$ be a K-basis of Y. Define $r : K^1_n \to Y$ by $r(x) = \sum_{j=1}^n x_je_j$ for $x = (x_j)_{j=1}^n \in K^1_n$. Then we have an element $\beta = \beta^* \in K^n_n$ such that $a\beta b^* = \mathrm{Tr}_{Y/K}\bigl(r(a)\zeta r(b)^\rho\bigr)$ for $a, b \in K^1_n$. Next define $t : K^n_n \to Y^n$ by $t(x) = \bigl(\sum_{j=1}^n x_{ij}e_j\bigr)_{i=1}^n$ for $x = (x_{ij}) \in K^n_n$. Then $x\beta x^* = \mathrm{Tr}_{Y/K}\bigl(t(x)\zeta t(x)^*\bigr)$. From (5.1b) we see that $\beta \in S^+$. For $y = t(x)$ we have $y \in \mathcal{Y}$ if and only if $\det(x) \neq 0$; also $y_i^{\varepsilon_{vh}} = \sum_{j=1}^n x_{ij}^{\tau_v}e_j^{\varepsilon_{vh}}$, so that $y_v^\varepsilon = x^{\tau_v}E_v$, where $E_v = (e_j^{\varepsilon_{vh}})_{j,h}$. Therefore, substituting $t(x)$ for y in (5.6), we obtain the desired expression $W(s, Q) = ab^s\mathcal{D}(s)$.

In this special case we can derive the assertions of Theorem 5.3 from the results of Section 3; namely:

5.5. Theorem. *The notation being as above, the series of (5.15) is convergent for sufficiently large* $\mathrm{Re}(s)$, *and can be continued as a meromorphic function to the whole plane. Moreover, suppose that* α *is* $\overline{\mathbf{Q}}$*-valued and* f *is* $\overline{\mathbf{Q}}$*-rational. Let* s_0 *be an integer such that* $0 \leq s_0 \leq |k_v - k_{v\rho}| - p_v$ *and* $|k_v - k_{v\rho}| - p_v - s_0 \in 2\mathbf{Z}$ *for every* $v \in \mathbf{a}$. *Suppose that* $|k_v - k_{v\rho}| > 2n$ *for every* $v \in \mathbf{a}$, *and also that* $s_0 \neq 1$ *if* $F = \mathbf{Q}$. *Put* $\sigma_v = \tau_v$ *if* $k_v < k_{v\rho}$ *and* $\sigma_v = \tau_v\rho$ *if* $k_v \geq k_{v\rho}$. *Then* $\mathcal{D}$ *is finite at* $s_0/2$, *and* $\mathcal{D}(s_0/2) \in \pi^c\mathfrak{q}^2\overline{\mathbf{Q}}$, *where* $c = n\sum_{v\in\mathbf{a}}|k_v - k_{v\rho}| - [F : \mathbf{Q}]n(n-1)/2$ *and* $\mathfrak{q} = p_K\bigl(\sum_{v\in\mathbf{a}}(k_{v\rho} - k_v)\tau_v,\ n\sum_{v\in\mathbf{a}}\sigma_v\bigr)$.

PROOF. Define $\nu \in \mathbf{Z}^{\mathbf{b}}$ as follows: $\nu_{v\rho} = 0$ and $\nu_v = k_v - k_{v\rho}$ if $k_v \geq k_{v\rho}$; $\nu_{v\rho} = k_{v\rho} - k_v$ and $\nu_v = 0$ if $k_v < k_{v\rho}$. Then $\nu_v + \nu_{v\rho} = |k_v - k_{v\rho}|$ and $\nu_{v\rho} - \nu_v = k_{v\rho} - k_v$ for every $v \in \mathbf{a}$. Put

$$q(z) = \sum_{x\in V} \overline{\alpha}(x^*)\det(x)^{\nu\rho}\mathbf{e}^n_{\mathbf{a}}(x^*\beta xz) \qquad (z \in \mathcal{H}).$$

Then $q \in \mathcal{S}_{m'}$ with $m' = (\nu_v + \nu_{v\rho} + n)_{v\in\mathbf{a}}$. We easily see that $a^{-k}|a|^{-g} = a^{\nu\rho}|a|^{n\mathbf{a}-2h}$ for $a \in K^\times$ with $h = \bigl((p_v + m'_v)/2\bigr)_{v\in\mathbf{a}}$. Therefore, changing ξ for

ξ^* in (3.6), we find that $\mathcal{D}(s+n/2) = \det(\beta)^{s\mathbf{a}+h}D(s;\, f,\, q)$. Thus the convergence and meromorphic continuation are guaranteed. Put $s_1 = s_0 - n$, $\kappa = s_0 + n$, and $r = (m' - p - \kappa\mathbf{a})/2$ with s_0 as in our theorem. We now evaluate (1.12) at $s = s_1/2$. By (4.3) we have

$$(5.16) \qquad b^r_{\kappa\mathbf{a}}(\kappa/2)E(z,\, \kappa/2;\, m' - p,\, \Gamma) = \Delta^r_{\kappa\mathbf{a}}E(z,\, \kappa/2;\, \kappa\mathbf{a},\, \Gamma)$$

and $b^r_{\kappa\mathbf{a}}(\kappa/2) \neq 0$ as noted in [87a, p.21] and in [00, p.146]. From (1.12) and (5.16) we obtain $\Gamma((s_1/2))\mathcal{D}(s_0/2) = a \cdot \pi^{dn^2}\langle q,\, f\Delta^r_{\kappa\mathbf{a}}h\rangle$ with $h(z) = E(z,\, \kappa/2;\, \kappa\mathbf{a},\, \Gamma)$ and $a \in \overline{\mathbf{Q}}^\times$. By [00, Theorem 17.7 (i)], $h \in \mathcal{M}_{\kappa\mathbf{a}}(\overline{\mathbf{Q}})$ under our assumption on s_0, and hence, by [00, Lemma 15.8], there exists an element t of $\mathcal{M}_{m'}(\overline{\mathbf{Q}})$ such that $\langle q,\, f\Delta^r_{\kappa\mathbf{a}}h\rangle = \pi^{n|r|}\langle q,\, t\rangle$. By Theorem 3.5 (2), $\langle q,\, t\rangle \in \pi^{-dn^2}\mathfrak{q}^2\overline{\mathbf{Q}}$. Since $\Gamma((s_1/2)) \in \pi^b\overline{\mathbf{Q}}^\times$ with $b = dn(n-1)/2 - dns_1/2 - n\sum_{v\in\mathbf{a}} h_v$, we obtain our assertion concerning $\mathcal{D}(s_0/2)$.

If $n = 1$, Theorem 5.5 is true under a weaker condition that $|k_v - k_{v\rho}| > 0$ for every $v \in \mathbf{a}$. Indeed, the conclusion of Theorem 3.5 (2) is true if $n = 1$ and $\nu_v + \nu_{v\rho} > 0$ for every $v \in \mathbf{a}$, as remarked after the proof of Corollary 3.6, so that the above proof is valid under that condition. It is also conjecturable that Theorem 5.5 is true even when $s_0 = 1$ and $F = \mathbf{Q}$. We can verify it at least when f is a cusp form or $n = 1$. If $s_0 = 1$ and $F = \mathbf{Q}$, then $\kappa = n + 1$ and $E(z,\, \kappa/2;\, \kappa\mathbf{a},\, \Gamma)$ is nearly holomorphic and arithmetic by [87a, Proposition 4.1 (ii)] or by [00, Theorem 17.7]. Therefore, if f is a cusp form, we can employ [00, Proposition 15.6 (3)] instead of [00, Lemma 15.8] to obtain the form t in the above proof. If $n = 1$, we obtain t by virtue of [87b, Proposition 9.4].

Notice that $\mathbf{q} = \mathfrak{q}$ if $m = 0$ in the setting of Theorem 5.3. This can be verified by means of the equality $p_Y(\alpha,\, \mathrm{Inf}_{Y/K}(\beta)) = p_K(\mathrm{Res}_{Y/K}(\alpha),\, \beta)$, which follows easily from [98, Theorem 32.5 (4)].

6. Near holomorphy and arithmeticity of certain automorphic forms given as infinite series

6.1. Before introducing $G(H)$, we first define several symbols as follows:

$$(6.1\mathrm{a}) \quad U(m,\, n) = \{\alpha \in GL_{m+n}(\mathbf{C}) \mid \alpha^* I_{m,n}\alpha = I_{m,n}\},\quad I_{m,n} = \mathrm{diag}[1_m,\, -1_n],$$

$$(6.1\mathrm{b}) \qquad \mathfrak{B}_{m,n} = \mathfrak{B}(m,\, n) = \{\mathfrak{z} \in \mathbf{C}^m_n \mid 1_n - \mathfrak{z}^*\mathfrak{z} > 0\} \qquad (mn > 0).$$

$$(6.1\mathrm{c}) \quad B(\mathfrak{z}) = \begin{bmatrix} 1_m & \mathfrak{z} \\ \mathfrak{z}^* & 1_n \end{bmatrix},\quad \xi(\mathfrak{z}) = 1_m - \bar{\mathfrak{z}} \cdot {}^t\mathfrak{z},\quad \eta(\mathfrak{z}) = 1_n - \mathfrak{z}^*\mathfrak{z} \quad (\mathfrak{z} \in \mathbf{C}^m_n).$$

Then for $\alpha \in U(m,\, n)$ and $\mathfrak{z} \in \mathfrak{B}_{m,n}$ we can define $\alpha\mathfrak{z} \in \mathfrak{B}_{m,n}$, $\lambda(\alpha,\, \mathfrak{z}) \in GL_m(\mathbf{C})$, and $\mu(\alpha,\, \mathfrak{z}) \in GL_n(\mathbf{C})$ by the formula

$$(6.2) \qquad \alpha B(\mathfrak{z}) = B(\alpha\mathfrak{z})\mathrm{diag}\big[\overline{\lambda(\alpha,\, \mathfrak{z})},\, \mu(\alpha,\, \mathfrak{z})\big].$$

Here m and n are nonnegative integers such that $m+n > 0$. If $mn = 0$, then we understand that $\mathfrak{B}_{m,n}$ consists of a single element written 0 and $B(0) = 1_{m+n}$, and we let the group $U(m,\, n)$ act on it trivially. As for the definition of other symbols in the case $mn = 0$, see [00, (3.24a, b, c)].

Let F, K, τ, and other symbols be as in §1.1. For every matrix α with entries in K we denote by α_v its image under τ_v. Given $H = H^* \in GL_r(K)$ with $0 < r \in \mathbf{Z}$, we put

$$G(H) = \{\, \alpha \in GL_r(\mathbf{C}) \mid \alpha H \alpha^* = H \,\}. \tag{6.3}$$

For each $v \in \mathbf{a}$ let (m_v, n_v) be the signature of H_v. Take an element $Q_v \in GL_r(\overline{\mathbf{Q}})$ so that

$$H_v = Q_v I_{m_v, n_v} Q_v^*. \tag{6.4}$$

Clearly the map $\alpha \mapsto (Q_v^{-1}\alpha_v Q_v)_{v\in\mathbf{a}}$ for $\alpha \in G(H)$ gives an injection of $G(H)$ into $\prod_{v\in\mathbf{a}} U(m_v, n_v)$. We put then

$$\mathfrak{B} = \prod_{v\in\mathbf{a}} \mathfrak{B}(m_v, n_v), \tag{6.5a}$$

$$\lambda_v(\alpha, \mathfrak{z}) = \lambda(Q_v^{-1}\alpha_v Q_v, \mathfrak{z}_v), \quad \mu_v(\alpha, \mathfrak{z}) = \mu(Q_v^{-1}\alpha_v Q_v, \mathfrak{z}_v), \tag{6.5b}$$

$$j_\alpha(\mathfrak{z}) = j(\alpha, \mathfrak{z}) = \bigl(j_v(\alpha, \mathfrak{z})\bigr)_{v\in\mathbf{a}}, \quad j_v(\alpha, \mathfrak{z}) = \det\bigl[\mu_v(\alpha, \mathfrak{z})\bigr]. \tag{6.5c}$$

We let $\prod_{v\in\mathbf{a}} U(m_v, n_v)$ act on $\prod_{v\in\mathbf{a}} \mathfrak{B}(m_v, n_v)$ component-wise, and let $G(H)$ act on $\prod_{v\in\mathbf{a}} \mathfrak{B}(m_v, n_v)$ via the above injection $G(H) \to \prod_{v\in\mathbf{a}} U(m_v, n_v)$. Then we have

$$\alpha_v Q_v B(\mathfrak{z}_v) = Q_v B\bigl((\alpha\mathfrak{z})_v\bigr) \cdot \mathrm{diag}\bigl[\overline{\lambda_v(\alpha, \mathfrak{z})}, \mu_v(\alpha, \mathfrak{z})\bigr] \qquad (v \in \mathbf{a}). \tag{6.6}$$

We can define automorphic forms on $\mathfrak{B}$ with respect to a congruence subgroup in a natural way and their $\overline{\mathbf{Q}}$-rationality; for details, see [00, Sections 5 and 11].

6.2. For $\mathfrak{z} \in \mathfrak{B}$ and $x \in K_r^q$, $0 < q \in \mathbf{Z}$, we define matrices $\lambda_v[x, \mathfrak{z}] \in \mathbf{C}_{m_v}^q$ and $\mu_v[x, \mathfrak{z}] \in \mathbf{C}_{n_v}^q$ by

$$x_v Q_v B(\mathfrak{z}_v) = \bigl[\,\overline{\lambda_v[x, \mathfrak{z}]} \quad \mu_v[x, \mathfrak{z}]\,\bigr] \qquad (v \in \mathbf{a}). \tag{6.7}$$

If $m_v = 0$ (resp. $n_v = 0$), then we do not define the symbol $\lambda_v[x, \mathfrak{z}]$ (resp. $\mu_v[x, \mathfrak{z}]$), and we ignore it when it appears in formulas. We can easily verify that

$$\lambda_v[x\gamma, \mathfrak{z}] = \lambda_v[x, \gamma\mathfrak{z}]\lambda_v(\gamma, \mathfrak{z}), \quad \mu_v[x\gamma, \mathfrak{z}] = \mu_v[x, \gamma\mathfrak{z}]\mu_v(\gamma, \mathfrak{z}) \qquad \bigl(\gamma \in G(H)\bigr).$$

Observe that $B(\mathfrak{z})^* I_{m,n} B(\mathfrak{z}) = \mathrm{diag}[{}^t\xi(\mathfrak{z}), -\eta(\mathfrak{z})]$. Taking the inverse, we obtain $I_{m,n} = B(\mathfrak{z})\mathrm{diag}[{}^t\xi(\mathfrak{z})^{-1}, -\eta(\mathfrak{z})^{-1}]B(\mathfrak{z})^*$. Combining this with (6.4), we obtain

$$(xHy^*)_v = \overline{\lambda_v[x, \mathfrak{z}]} \cdot {}^t\xi(\mathfrak{z}_v)^{-1} \cdot {}^t\lambda_v[y, \mathfrak{z}] - \mu_v[x, \mathfrak{z}]\eta(\mathfrak{z}_v)^{-1}\mu_v[y, \mathfrak{z}]^* \tag{6.8}$$

for $x, y \in K_r^q$ and $v \in \mathbf{a}$.

6.3. We take a CM-field L contained in K and put $E = F \cap L$; we also take a subset $\mathbf{c}$ of $\mathbf{a}$ and put $\psi_v = \mathrm{Res}_{K/L}(\tau_v)$ for $v \in \mathbf{c}$ and $\psi = \sum_{v\in\mathbf{c}} \psi_v$. We assume the following conditions:

(6.9a) $\mathrm{Res}_{K/L}(\tau) = [K : L]\psi$.

(6.9b) *There exist two positive integers m and n such that $m + n = r$, $(m_v, n_v) = (m, n)$ if $v \in \mathbf{c}$, and $(m_v, n_v) = (r, 0)$ if $v \notin \mathbf{c}$.*

Then $\mathfrak{B} = \mathfrak{B}(m, n)^{\mathbf{c}}$. From (6.9a) we easily see that

(6.10) *The restriction of the members of* $\mathbf{c}$ *to* E *gives a bijection of* $\mathbf{c}$ *onto the set of all archimedean primes of* E, *and* ψ *is a CM-type of* L.

Thus we shall often identify $\mathbf{c}$ with the set of all archimedean primes of E.

We next take polynomial representations

$$R : GL_n(\mathbf{C})^{\mathbf{a}} \to GL(X) \quad \text{and} \quad R' : \prod_{v\in\mathbf{a}} GL_{m_v}(\mathbf{C}) \to GL(X')$$

with finite-dimensional complex vector spaces X and X' and a polynomial map

$$P : \prod_{v\in\mathbf{a}} \mathbf{C}_n^{m_v} \to \mathrm{Hom}(X, X')$$

such that $P(axb) = R'(a)P(x)R(b)$ for $a \in \prod_{v\in\mathbf{a}} GL_{m_v}(\mathbf{C})$ and $b \in GL_n(\mathbf{C})^{\mathbf{a}}$. We assume that X and X' have $\overline{\mathbf{Q}}$-structures, and that R, R', are $\overline{\mathbf{Q}}$-rational. We define a representation $R_0 : GL_n(\mathbf{C})^{\mathbf{c}} \to GL(X)$ by $R_0(c) = R(b)$, where $b_{v'} = c_v$ whenever $v = v'$ on E. Taking $p \in \mathbf{Z}^{\mathbf{c}}$, we consider a representation $\omega : GL_n(\mathbf{C})^{\mathbf{c}} \times GL_n(\mathbf{C})^{\mathbf{c}} \to GL(X)$ by $\omega(a, b) = \det(b)^p R_0({}^t a^{-1})$ for $(a, b) \in GL_n(\mathbf{C})^{\mathbf{c}} \times GL_n(\mathbf{C})^{\mathbf{c}}$.

We take an element f of $\mathcal{M}_{L,\omega}$. Then we have a Fourier expansion

$$f(z) = \sum_{\sigma\in S\cap L_n^n} c_f(\sigma)e_{\mathbf{c}}^n(\sigma z) \qquad (z \in \mathcal{H}_n^{\mathbf{c}}) \tag{6.11}$$

with elements $c_f(\sigma)$ of X in the same sense as in (1.7). Then there exists (see [00, (5.21)]) a subgroup U of $GL_n(\mathfrak{r} \cap L)$ of finite index such that

$$c_f(a\sigma a^*) = \omega(\overline{a}, a)c_f(\sigma) \quad \textit{for every} \quad a \in U. \tag{6.12}$$

Given $\alpha \in \mathcal{S}(K_r^n)$, such an f, and $\kappa \in \mathbf{Z}^{\mathbf{c}}$ such that $\kappa_v \geq 0$ for every $v \in \mathbf{c}$, we define an X'-valued infinite series $\mathcal{F}(\mathfrak{z}, s; \alpha)$ for $(\mathfrak{z}, s) \in \mathfrak{B} \times \mathbf{C}$ by

$$\mathcal{F}(\mathfrak{z}, s) = \mathcal{F}(\mathfrak{z}, s; \alpha) = \sum_{x\in U\backslash\mathcal{X}} \alpha(x)\delta(\mathfrak{z})^{s\mathbf{c}-q}|j[x, \mathfrak{z}]|^{2q-2s\mathbf{c}} j[x, \mathfrak{z}]^{-\kappa} \cdot P({}^t\lambda[x, \mathfrak{z}])c_f(-\mathrm{Tr}_{K/L}(xHx^*)). \tag{6.13}$$

Here $\mathcal{X} = \{ x \in K_r^n \mid \mathrm{rank}(x) = n \}$ and $q = (\kappa - p)/2$;

$$\lambda[x, \mathfrak{z}] = (\lambda_v[x, \mathfrak{z}])_{v\in\mathbf{a}} \Big(\in \prod_{v\in\mathbf{a}} \mathbf{C}_n^{m_v}\Big), \quad \mu[x, \mathfrak{z}] = (\mu_v[x, \mathfrak{z}])_{v\in\mathbf{c}} \ (\in (\mathbf{C}_n^n)^{\mathbf{c}}), \tag{6.14a}$$

$$j[x, \mathfrak{z}] = (j_v[x, \mathfrak{z}])_{v\in\mathbf{c}} \ (\in \mathbf{C}^{\mathbf{c}}), \quad j_v[x, \mathfrak{z}] = \det(\mu_v[x, \mathfrak{z}]), \tag{6.14b}$$

$$\delta(\mathfrak{z}) = (\delta(\mathfrak{z}_v))_{v\in\mathbf{c}} \ (\in \mathbf{C}^{\mathbf{c}}), \quad \delta(\mathfrak{z}_v) = \det(\eta(\mathfrak{z}_v)); \tag{6.14c}$$

for $h \in K_n^n$ we denote by $\mathrm{Tr}_{K/L}(h)$ the element of L_n^n such that $\mathrm{Tr}_{K/L}(h)_{ij} = \mathrm{Tr}_{K/L}(h_{ij})$. Since $c_f(h) \neq 0$ only if h is totally nonnegative, the above sum is extended over all x such that $-\mathrm{Tr}_{K/L}(xHx^*)$ is totally nonnegative. For such an x we can show that $j_v[x, \mathfrak{z}] \neq 0$ for every $v \in \mathbf{a}$; see [86, (5.12)]. Thus each term of (6.13) is meaningful. To make it dependent only on Ux, we take U to be a subgroup of $GL_n(\mathfrak{r} \cap L)$ of finite index such that $\alpha(ax) = \alpha(x)$ and $0 \ll \det(a) \in \mathfrak{g}^\times \cap E$ for every $a \in U$. We easily see that the sum of (6.13) is formally meaningful, and

$$\mathcal{F}(\gamma(\mathfrak{z}), s; \alpha) = j_\gamma(\mathfrak{z})^\kappa R'({}^t\lambda(\gamma, \mathfrak{z})^{-1})\mathcal{F}(\mathfrak{z}, s; \alpha^\gamma) \tag{6.15}$$

for every $\gamma \in G(H)$, where $\alpha^\gamma(x) = \alpha(x\gamma^{-1})$. Therefore $\mathcal{F}(\mathfrak{z}, s; \alpha)$ has the property of automorphy with respect to a congruence subgroup of $G(H)$.

6.4. To introduce another type of series, we consider $S_e(T)$ with $T = (\mathbf{C}_n^m)^{\mathbf{c}}$ as in §5.2 (with $\mathbf{c}$ instead of $\mathbf{a}$), and take a τ^e-irreducible subspace Z of $S_e(T)$. Then we define a series $\mathcal{F}^Z(\mathfrak{z}, s)$ with values in Hom$(Z, \mathbf{C})$ by

$$(6.16)\qquad \mathcal{F}^Z(\mathfrak{z}, s)(\varphi) = \mathcal{F}^Z(\mathfrak{z}, s; \alpha)(\varphi) = \sum_{x \in U\backslash \mathcal{X}} \alpha(x)\delta(\mathfrak{z})^{s\mathbf{c}-q}|j[x, \mathfrak{z}]|^{2q-2s\mathbf{c}}$$
$$\cdot j[x, \mathfrak{z}]^{-\kappa}\varphi\bigl(\xi(\mathfrak{z})^{-1}\lambda[x, \mathfrak{z}]^* \cdot {}^t\mu[x, \mathfrak{z}]^{-1}\bigr)c_f\bigl(-\mathrm{Tr}_{K/L}(xHx^*)\bigr),$$

for $\varphi \in Z$. Here the symbols are the same as in (6.13) except that: we take $\omega(a, b) = \det(b)^p$, so that f is $\mathbf{C}$-valued; we ignore $\lambda_v[x, \mathfrak{z}]$ for $v \notin \mathbf{c}$, so that $\lambda[x, \mathfrak{z}] \in (\mathbf{C}_m^n)^{\mathbf{c}}$; $\xi(\mathfrak{z}) = \bigl(\xi(\mathfrak{z}_v)\bigr)_{v\in\mathbf{a}}$ with ξ of (6.1c). This time we have, instead of (6.15),

$$(6.17)\qquad j_\gamma(\mathfrak{z})^{-\kappa}\mathcal{F}^Z\bigl(\gamma(\mathfrak{z}), s; \alpha\bigr)\bigl(\tau^e({}^t\lambda_\gamma(\mathfrak{z}), {}^t\mu_\gamma(\mathfrak{z}))^{-1}\varphi\bigr) = \mathcal{F}^Z(\mathfrak{z}, s; \alpha^\gamma)(\varphi)$$

for every $\gamma \in G(H)$, where τ^e is defined in §5.2. Using the notation of [00, §12.11], we can express the left-hand side of (6.17) as $\mathcal{F}^Z\|_{\omega\otimes\tau^e}\gamma$; see [00, (12.24a)].

In order to state our theorem concerning the critical values of $\mathcal{F}(\mathfrak{z}, s)$, we recall the definition of nearly holomorphic functions on $\mathfrak{B}$. Put $r_v(\mathfrak{z}) = -\xi(\mathfrak{z}_v)^{-1}\bar{\mathfrak{z}}_v$ for $\mathfrak{z} = (\mathfrak{z}_v)_{v\in\mathbf{c}} \in \mathfrak{B}$ and $v \in \mathbf{c}$; then we have $r_v(\mathfrak{z}) = -\bar{\mathfrak{z}}_v \cdot {}^t\eta(\mathfrak{z}_v)^{-1}$. Here ξ and η are defined by (6.1c). By a *nearly holomorphic function* on $\mathfrak{B}$ we mean a polynomial of the entries of $r_v(\mathfrak{z})$ for all $v \in \mathbf{c}$ with holomorphic functions on $\mathfrak{B}$ as coefficients. For $t = (t_v)_{v\in\mathbf{c}} \in \mathbf{Z}^{\mathbf{c}}$ with $t_v \geq 0$, we say that such a function is of degree $\leq t$ if it is of degree $\leq t_v$ as a polynomial of the entries of $r_v(\mathfrak{z})$ for each $v \in \mathbf{c}$. We can also define the arithmeticity of a nearly holomorphic function. For a detailed treatment, the reader is referred to [94] or [00, Sections 13 and 14].

6.5. Theorem. (1) *The series of (6.16) is locally uniformly convergent on* $\mathfrak{B} \times \{s \in \mathbf{C} \mid \mathrm{Re}(s) > c_0\}$ *for sufficiently large* c_0. *The same is true for the series of (6.13) if* f *satisfies (5.9) with* $(S^+ \cap L_n^n, \mathbf{c})$ *in place of* $(S^+, \mathbf{a})$.

(2) *Suppose that the series of (6.13) is locally uniformly convergent on* $\mathfrak{B} \times \{s \in \mathbf{C} \mid \mathrm{Re}(s) > c_0\}$ *for sufficiently large* c_0. *Then there exists a nonzero entire function* $A(s)$ *and a real analytic function* $B(\mathfrak{z}, s)$ *on* $\mathfrak{B} \times \mathbf{C}$ *such that* $A(s)\mathcal{F}(\mathfrak{z}, s) = B(\mathfrak{z}, s)$ *for* $\mathrm{Re}(s) > c_0$; *moreover,* $\mathcal{F}(\mathfrak{z}, s)$ *is holomorphic in* s *for* $\mathrm{Re}(s) > (m+n)[K : L]/2$. *The same assertions are true for* $\mathcal{F}^Z(\mathfrak{z}, s)$.

(3) *Let* s_0 *be an integer such that*

$$(6.18)\qquad \kappa_v - p_v - (2/n)(\kappa_v - 1) \leq s_0 \leq \kappa_v - p_v \text{ and } s_0 - \kappa_v + p_v \in 2\mathbf{Z} \text{ for every } v \in \mathbf{c};\ s_0 \geq (m+n)[K : L] - n;\ s_0 \neq (m+n)[F : \mathbf{Q}] - n + 1 \text{ if } E = \mathbf{Q}.$$

Then $\mathcal{F}(\mathfrak{z}, s)$ *is finite at* $s = s_0/2$, *and* $\mathcal{F}(\mathfrak{z}, s_0/2)$ *is a nearly holomorphic function on* $\mathfrak{B}$ *of degree* $\leq t$, *where*

$$(6.19)\qquad t = (n/2)(\kappa_v - p_v - s_0)_{v\in\mathbf{c}}.$$

(4) *For $Z \subset S_e(T)$ as in §6.4 and s_0, t as in (3) above, $\mathcal{F}^Z(\mathfrak{z}, s_0/2)$ is a nearly holomorphic function of degree $\le t+e$ provided that we assume*

(6.20) $$s_0 + \kappa_v + p_v \ge 2 \cdot \mathrm{Min}(m,\, n) \text{ for every } v \in \mathbf{c}.$$

PROOF. This is an improved version of [86, Proposition 5.1 and Theorem 5.2]. As will be explained in §7.5, our $\mathcal{F}$ is essentially the same as the series of [86, (5.11)], once we change the exponents $s\psi$ and $-2s\psi$ there for $s\mathbf{c}-q$ and $2q-2s\mathbf{c}$. These changes allow us to dispense with condition (5.13) of [86]. Now, as to the convergence, once we assume (5.9), then the proof in [86, Section 9] is valid with easy modifications. Strictly speaking, however, the results and the proof there are stated under the condition $m \ge n$. In [86] we employed l and m instead of m and n here, so that the condition was $l \ge m$. This condition was employed in formula (8.2) and Lemma 8.1 of [86], but we can include the case $l < m$ for the following reason. Formula (8.2) is true for any $(l,\, m)$ if we take $\mathrm{Min}\{l,\, m\}$ in place of m in (8.2). Also, the proof of Lemma 8.1 is valid for $l < m$ if we take w in the proof to be the $(m \times m)$-matrix whose first l rows form $a \cdot {}^t c$ and the last $m-l$ rows are 0. Then the proof in [86, Section 8] is applicable to the case $l < m$. It should also be noted that once we include the case $l < m$, then we have to make the following changes in [86]: Read $l \ne m$ for $l > m$ on page 378, line 6 and (8.16).

To prove meromorphic continuation of $\mathcal{F}$, we repeat the proof in [86, Section 8] with the above changes and with some modifications. We fix one $v \in \mathbf{a}$ (which corresponds to the symbol τ there), and consider an irreducible subspace Z of $S_e(T)$, $T = \mathbf{C}_n^m$ with $0 \le e \le \kappa_v$. (We use $(m,\, n)$ instead of $(l,\, m)$ of [86]. See also "Corrections to [86]" on page 259 of [88].) Then $-s$ in (8.1), (8.13), and (8.14) should be $q_v - s$. We take also $(\kappa + p)/2$ in place of k appearing in the exponents in (8.6), (8.7), and (8.9). Then we eventually obtain, instead of (8.15),

(6.21) $$(2\pi)^{-b} \prod_{v \in \mathbf{c}} \Gamma_n(s + p_v + q_v) \cdot \big[E_Z f(w,\, s)\big]\zeta = C \sum_{i \in I} x_i \int_{\Phi} \big[E_{Z'} \mathcal{E}(z,\, s)\big] B_i'(z,\, w;\, \zeta)\mathbf{d}z,$$

where $b = \sum_{v \in \mathbf{c}} n(s + p_v + q_v)$, C is a nonzero constant, and $\mathcal{E}(z,\, s) = E(z,\, s + \nu/2;\, \kappa - p + \nu\mathbf{c},\, \Gamma)$ with $E(\dots)$ of (1.11) defined for $z \in \mathcal{H}_n^{\mathbf{c}}$ with a congruence subgroup Γ of G_L^n and $\nu = 2n-(m+n)[K : L]$. Here $f(w,\, s)$ (with $w \in \mathfrak{B}$) is the function of [86] corresponding to $\mathcal{F}(\mathfrak{z},\, s)$ with the exponents changed as above, and E_Z is the operator of [86, (1.24)], which is the same as that of [00, (12.20) and (13.22)]; Z is an irreducible subspace of $S_e(\mathbf{C}_n^m)$, and $E_{Z'}$ is an operator of the same type defined on $\mathcal{H}_n^{\mathbf{c}}$ with an irreducible subspace Z' of $S_e(\mathbf{C}_n^n)$ of the same highest weight as Z; B_i' is explained in [86, p. 394]. Taking $e = 0$ (which means that E_Z and $E_{Z'}$ are the identity operators), we obtain meromorphic continuation of $f(w,\, s)$. By Lemma 1.3, $\mathcal{E}$ is holomorphic for $\mathrm{Re}(s + \nu/2) > n$. Thus we obtain assertion (2) for $\mathcal{F}$.

Next let s_0 be as in (3). By [87a, Theorem 4.2 (i)] or [00, Theorem 17.9 (ia)], $\mathcal{E}(z,\, s_0/2)$ is a nearly holomorphic function on $\mathcal{H}_n^{\mathbf{c}}$ of degree $\le t$ with t of

(6.19). As explained in [00, §13.11], such a function, say g, has the property that $E'_Z g = 0$ for every $Z' \subset S_{t_v+1}(\mathbf{C}^n_n)$ and every $v \in \mathbf{c}$, and vice versa. Therefore (6.19) shows that $E_Z f(w, s_0/2) = 0$ for every irreducible $Z \subset S_{t_v+1}(\mathbf{C}^m_n)$, since there is always a unique irreducible $Z' \subset S_e(\mathbf{C}^n_n)$ with the same highest weight as Z. We have to assume that $\kappa_v \geq t_v + 1$, which holds under (6.18). Thus we obtain (3). We prove the assertions concerning $\mathcal{F}^Z$ in §6.8.

6.6. Let us now recall the definition of arithmeticity. Put

$$\mathfrak{G} = \prod_{v\in\mathbf{a}} [GL_{m_v}(\mathbf{C}) \times GL_{n_v}(\mathbf{C})]$$

and take a $\overline{\mathbf{Q}}$-rational representation $\Omega : \mathfrak{G} \to GL(Y)$, where Y is a finite-dimensional complex vector space with a $\overline{\mathbf{Q}}$-structure. For each CM-point w of $\mathfrak{B}$, we define an element $\mathfrak{p}(w)$ of $\mathfrak{G}$ as in [00, §11.4] and put $\mathfrak{P}_\Omega(w) = \Omega(\mathfrak{p}(w))$. Let g be a Y-valued nearly holomorphic function on $\mathfrak{B}$ such that $g(\gamma(\mathfrak{z})) = \Omega(\lambda(\gamma, \mathfrak{z}), \mu(\gamma, \mathfrak{z}))g(\mathfrak{z})$ for every γ in a congruence subgroup of $G(H)$. Then we say that g is Ω- *arithmetic* (or simply, *arithmetic*) if $\mathfrak{P}_\Omega(w)^{-1}g(w)$ is $\overline{\mathbf{Q}}$-rational for every CM-point w of $\mathfrak{B}$ (see [00, §14.4]). In the following theorem $g(\mathfrak{z}) = \mathcal{F}(\mathfrak{z}, s_0/2)$, for which we take $\Omega(a, b) = \det(b)^\kappa R'({}^t a^{-1})$ in view of (6.15).

6.7. Theorem. Let s_0 be as in (3) of Theorem 6.5. Suppose that α is $\overline{\mathbf{Q}}$-valued and P, f are $\overline{\mathbf{Q}}$-rational; suppose also that $\kappa_u > 2n$ for every $u \in \mathbf{c}$. Then $\pi^{-c}\mathbf{q}^{-1}\mathfrak{Q}^{-1}\mathcal{F}(\mathfrak{z}, s_0/2)$ is arithmetic, where $c = n\sum_{u\in\mathbf{c}}\kappa_u - [E:\mathbf{Q}]n(n-1)/2$, $\mathbf{q} = p_L\big(\sum_{u\in\mathbf{c}}\kappa_u\psi_u, n\psi\big)$, and $\mathfrak{Q} = R'(\mathbf{p}_0)$, $\mathbf{p}_0 = (\mathbf{p}_{0v})_{v\in\mathbf{a}}$ with $\mathbf{p}_{0v} = p_K(\tau_v, \tau)1_m$ for $v \in \mathbf{c}$ and $\mathbf{p}_{0v} = p_K(\tau_v, \tau)1_r$ for $v \notin \mathbf{c}$. Moreover, for such an s_0 satisfying (6.20), $\pi^{-b}\mathbf{q}^{-1}\mathcal{F}^Z(\mathfrak{z}, s_0/2)$ is arithmetic, where $b = c + \sum_{v\in\mathbf{c}} e_v$.

The proof will be given in Section 7. Notice that (6.20) is satisfied as we assume $\kappa_v > 2n$ for every $v \in \mathbf{c}$.

6.8. To prove the convergence and meromorphic continuation of (6.16), we may assume that α takes only 0 and 1 as its values, since α of a general type is a linear combination of such α's. Next, put

$$g(\mathfrak{z}) = \delta(\mathfrak{z})^{sc-q}|\mu[x, \mathfrak{z}]|^{2q-2sc}\mu[x, \mathfrak{z}]^{-\kappa}$$

with a fixed $x \in K^n_r$ such that $\det\big(\mu[x, \mathfrak{z}]\big) \neq 0$. Take an irreducible subspace Z of $S_e(T)$ as in §6.4 and define the operator D^Z_κ as in [00, §12.11, (13.22)]; here we denote by D^Z_κ the operator D^Z_ρ for the representation ρ such that $\rho(a, b) = \det(b)^\kappa$. We can put $Z = \bigotimes_{v\in\mathbf{c}} Z_v$ with an irreducible subspace Z_v of $S_{e_v}(T_v)$ for each v. Then for every $\varphi \in Z$, we have

$$(D^Z_\kappa g)\varphi = \psi(s)\varphi(\xi(\mathfrak{z})^{-1}\lambda[x, \mathfrak{z}]^* \cdot {}^t\mu[x, \mathfrak{z}]^{-1})g(\mathfrak{z}), \tag{6.22}$$

$$\psi(s) = \psi_v(q_v - \kappa_v - s), \quad \psi_v(s) = \prod_{j=1}^{\nu}\prod_{i=1}^{r_j}(s - i + j), \tag{6.23}$$

where $\nu = \mathrm{Min}(m, n)$ and $(r_j)_{j=1}^{\nu}$ is the weight of Z_v in the sense of [00, Theorem 12.7]. Formula (6.22) is essentially proved in [00, Lemma 13.9], which concerns the case where x is the lower n rows of an element of $U(m, n)$, but the proof there (for Type AB) is valid for the present g.

Let $\mathcal{F}_0(\mathfrak{z}, s)$ denote the series $\mathcal{F}(\mathfrak{z}, s)$ of (6.13) with $X = X' = \mathbf{C}$, trivial R and R', and with the constant 1 as P. Then, by (6.22) and termwise application of D_κ^Z we obtain, at least formally,

$$D_\kappa^Z \mathcal{F}_0(\mathfrak{z}, s) = \psi(s)\mathcal{F}^Z(\mathfrak{z}, s). \tag{6.24}$$

To prove this in the strict sense, we introduce another variable $\mathfrak{w}$ on $\mathfrak{B}$, and consider, for $\sigma \in \mathbf{R}$,

$$\mathcal{G}(\mathfrak{z}, \mathfrak{w}; \sigma) = \sum_{x \in U \backslash \mathcal{X}} \alpha(x) j[x, \mathfrak{z}]^{q-\sigma c-\kappa} \overline{j[x, \mathfrak{w}]}^{q-\sigma c} c_f\bigl(-\mathrm{Tr}_{K/L}(xHx^*)\bigr). \tag{6.25}$$

Here for each x we choose branches of $j[x, \mathfrak{z}]^{q-\sigma c-\kappa}$ and $\overline{j[x, \mathfrak{w}]}^{q-\sigma c}$ so that their product, when $\mathfrak{w} = \mathfrak{z}$, becomes $|j[x, \mathfrak{z}]|^{2q-2\sigma c} j[x, \mathfrak{z}]^{-\kappa}$. By the Cauchy-Schwarz inequality, the last sum can be majorized by the square root of

$$\sum_{x \in U \backslash \mathcal{X}} \alpha(x) |j[x, \mathfrak{z}]^{q-\sigma c-\kappa}|^2 |c_f\bigl(-\mathrm{Tr}_{K/L}(xHx^*)\bigr)| \det(xTx^*)^{\kappa/2} \tag{6.26}$$
$$\cdot \sum_{x \in U \backslash \mathcal{X}} \alpha(x) |j[x, \mathfrak{w}]^{q-\sigma c}|^2 |c_f\bigl(-\mathrm{Tr}_{K/L}(xHx^*)\bigr)| \det(xTx^*)^{-\kappa/2},$$

where $T = (T_v)_{v \in \mathbf{a}}$ is defined by [86, (9.6)]. (Notice that $\alpha^2 = \alpha$ because of our choice of α.) Let us now prove the convergence of the two series of (6.26), which will establish $\mathcal{G}$ as a holomorphic function of $(\mathfrak{z}, \overline{\mathfrak{w}})$. For $h = -\mathrm{Tr}_{K/L}(xHx^*)$ and $v \in \mathbf{c}$ we have $\det(h)_v \le 2^n \delta(\mathfrak{z}_v)^{-1} |j_v[x, \mathfrak{z}]|^2$ by [86, p. 399, lines 12-13]. Also, by [00, Proposition A6.4], $|c_f(h)| \le C|\det(h)|^{p/2+ac}$ for every $h \in S^+ \cap L_n^n$ with constants C and $a \in \mathbf{R}, \ge 0$. Therefore the first sum of (6.26) is majorized by

$$C_1 \delta(\mathfrak{z})^{-ac-p/2} \sum_x |\alpha(x)| |j[x, \mathfrak{z}]|^{2ac-2\sigma c-\kappa} \det(xTx^*)^{\kappa/2},$$

with a constant C_1. By means of the inequality of [86, p. 399, line 12 from the bottom] we can reduce our series to that of [86, p. 396, line 7 from the bottom], and hence the desired convergence for $\sigma > \sigma_0$ with some σ_0 follows from Lemma 9.1 on the same page. The same method applies to the second sum of (6.26). Thus (6.25) defines a holomorphic function of $(\mathfrak{z}, \overline{\mathfrak{w}})$ for $\sigma > \sigma_0$.

Now the effect of D_κ^Z is essentially successive applications of two types of operations: (1) multiplication by a polynomial of the entries of $1_m - \bar{\mathfrak{z}}_v \cdot {}^t\mathfrak{z}_v$, $1_n - \mathfrak{z}_v^* \mathfrak{z}_v$, and their inverses for $v \in \mathbf{c}$; (2) holomorphic differentiation with respect to $\mathfrak{z}$. Substituting $\overline{\mathfrak{w}}$ for $\bar{\mathfrak{z}}$, we can apply each of these two types of operations to $\mathcal{G}(\mathfrak{z}, \mathfrak{w}; \sigma)$. Here we have to restrict $(\mathfrak{z}, \mathfrak{w})$ to the domain determined by

$$\textstyle\prod_{v \in \mathbf{c}} \det(1_m - \overline{\mathfrak{w}}_v \cdot {}^t\mathfrak{z}_v) \det(1_n - \mathfrak{w}_v^* \mathfrak{z}_v) \ne 0.$$

The holomorphy in $(\mathfrak{z}, \overline{\mathfrak{w}})$ of each term justifies the termwise application of each such operation, and eventually find that $D_\kappa^Z \mathcal{F}(\mathfrak{z}, \sigma)(\varphi)$ equals $\mathcal{D}\mathcal{G}(\mathfrak{z}, \mathfrak{w}; \sigma)$ at $\mathfrak{w} = \mathfrak{z}$ for $\sigma > \sigma_0$ with a certain differential operator $\mathcal{D}$. The termwize application of $\mathcal{D}$ shows that $\mathcal{D}\mathcal{G}(\mathfrak{z}, \mathfrak{w}; \sigma)$ at $\mathfrak{w} = \mathfrak{z}$ is $\mathcal{F}^Z(\mathfrak{z}, \sigma)(\varphi)$. This proves the convergence of (6.16) for $\sigma_0 < s \in \mathbf{R}$. Clearly the sum expression of (6.16) shows that it is convergent for $\mathrm{Re}(s) > \sigma_o$, and so (6.16) defines a holomorphic function of s for $\mathrm{Re}(s) > \sigma_o$. Now we have (6.24) for $\sigma_0 < s \in \mathbf{R}$. Since both sides are holomorphic in s, we have (6.24) for $\mathrm{Re}(s) > \sigma_0$. We already know meromorphic continuation of $\mathcal{F}_0(\mathfrak{z}, s)$ to the whole $\mathbf{C}$ in the sense of Theorem 6.5 (2), and hence (6.24) gives the same for $\mathcal{F}^Z(\mathfrak{z}, s)$. Since (6.23) shows that $\psi(s) \neq 0$ for $\mathrm{Re}(s) > (m+n)[K : L]/2$, we obtain the last assertion of Theorem 6.5 (2) concerning $\mathcal{F}^Z$. Similarly from (6.23) we see that $\psi(s_0/2) \neq 0$ under (6.20). Therefore, by [00, Proposition 13.15 (1)], (6.24) shows that $\mathcal{F}^Z(\mathfrak{z}, s_0/2)$ is nearly holomorphic of degree $\leq t+e$. This completes the proof of Theorem 6.5.

7. Proof of Theorems 5.3 and 6.7

7.1. Our proof will be given roughly in the following sequence of implications:

(7.0) Theorem 5.5 $\Longrightarrow$ Theorem 6.7 with $K = L$ $\Longrightarrow$

Theorem 5.3 with $m > 0$ $\Longrightarrow$ Theorem 6.7 in the general case.

We remind the reader that if $m = 0$, Theorem 5.3 can be reduced to Theorem 5.5, which was proved. In this and next subsections we prove the second implication. To be precise, we derive Theorem 5.3 with $m > 0$ from the special cases of Theorems 6.5 and 6.7 with $K = L$. The notation being as in §5.1, put $r = m+n$ and take a K-linear bijection $\ell : Y \to K_r^1$. Since $(x, y) \mapsto \mathrm{Tr}_{Y/K}(\zeta x y^\rho)$ defines a hermitian form on Y, we can find a hermitian element H of K_r^r such that $\ell(x) H \ell(y)^* = -\mathrm{Tr}_{Y/K}(\zeta x y^\rho)$. By (5.1a) we have $\mathrm{Tr}_{Y/K}(\zeta x y^\rho)_v = \sum_{i=1}^m (\zeta x^\rho y)^{\gamma_{vi}} + \sum_{j=1}^n (\zeta x y^\rho)^{\varepsilon_{vj}}$. Also, $y \mapsto \{y^\sigma\}_{\sigma \in J_Y}$ can be extended to an $\mathbf{R}$-linear bijection of $Y \otimes_{\mathbf{Q}} \mathbf{R}$ to $\mathbf{C}^{2d}$, $d = [Y : \mathbf{Q}]$. Therefore we see from (5.1b) that H_v has signature (m, n) for every $v \in \mathbf{a}$. Define $G(H)$ and other symbols with respect to this H; put $Y^u = \{\, y \in Y \mid y y^\rho = 1 \,\}$. Define a K-linear ring-injection $h : Y \to K_r^r$ by $\ell(xa) = \ell(x) h(a)$ for $x, a \in Y$. Then $h(a^\rho) H = H h(a)^*$ and $h(Y^u)$ has a unique fixed point w on $\mathfrak{B}$, which is a CM-point on $\mathfrak{B}$; moreover we have a set Φ of elements of J_Y with which we may call (Y, Φ) a generalized CM-type; for details, see [00, §4.11].

For $a \in Y^u$ we can diagonalize $\lambda_v(h(a), w)$ and $\mu_v(h(a), w)$, and the diagonal elements determine Φ according to the principle of [00, (4.36a), (4.37), (4.38), (4.40)]. Namely, for $a \in Y^u$ we can put

$$A_v^{-1} \lambda_v(h(a), w) A_v = \mathrm{diag}[a^{\alpha_{v1}}, \ldots, a^{\alpha_{vm}}], \tag{7.1a}$$

$$B_v^{-1} \mu_v(h(a), w) B_v = \mathrm{diag}[a^{\beta_{v1}}, \ldots, a^{\beta_{vn}}] \tag{7.1b}$$

with $A_v \in GL_m(\overline{\mathbf{Q}})$, $B_v \in GL_n(\overline{\mathbf{Q}})$, and $\alpha_{vi}, \beta_{vj} \in J_Y$. Then $\Phi = \sum_{v \in \mathbf{a}} \bigl(\sum_{i=1}^m \alpha_{vi} + \sum_{j=1}^n \beta_{vj} \bigr)$. This is similar to what was done in §3.7. Taking $a \in K$, we find that

$$\{\rho\alpha_{v1}, \ldots, \rho\alpha_{vm}, \beta_{v1}, \ldots, \beta_{vn}\} = \{\, \sigma \in J_Y \mid \sigma = \tau_v \text{ on } K \,\}. \tag{7.2}$$

Now we can find $C_v \in GL_m(\overline{\mathbf{Q}})$ and $D_v \in GL_n(\overline{\mathbf{Q}})$ such that

(7.3) $C_v A_v$ and $D_v B_v$ are diagonal matrices,

(7.4) $\lambda_v[\ell(y), w] = [y^{\alpha_{v1}} \ \dots \ y^{\alpha_{vm}}]C_v, \quad \mu_v[\ell(y), w] = [y^{\beta_{v1}} \ \dots \ y^{\beta_{vn}}]D_v$

for every $y \in Y$. This is similar to (3.14), and can be obtained by the same argument by taking $\ell(x)$ in place of xq there. Substituting $(\ell(x), \ell(y), w)$ for $(x, y, \mathfrak{z})$ in (6.8) with $x, y \in Y$, we obtain, in view of (7.2) and (7.4),

$$-\sum_{i=1}^{m}(\zeta x y^{\rho})^{\rho\alpha_{vi}} - \sum_{j=1}^{n}(\zeta x y^{\rho})^{\beta_{vj}} = -\mathrm{Tr}_{Y/K}(\zeta x y^{\rho})_v = (\ell(x)H\ell(y)^*)_v$$
$$= [x^{\rho\alpha_{v1}} \ \dots \ x^{\rho\alpha_{vm}}]\overline{C}_v \cdot {}^t\xi(w_v)^{-1} \cdot {}^tC_v[y^{\rho\alpha_{v1}} \ \dots \ y^{\rho\alpha_{vm}}]^*$$
$$- [x^{\beta_{v1}} \ \dots \ x^{\beta_{vn}}]D_v\eta(w_v)^{-1}D_v^*[y^{\beta_{v1}} \ \dots \ y^{\beta_{vn}}]^*.$$

Since $\xi(w_v)$ and $\eta(w_v)$ are positive definite, we see that

(7.5) $\zeta^{\alpha_{vi}} < 0$ for every (v, i) and $\zeta^{\beta_{vj}} > 0$ for every (v, j).

Comparing (7.2) and (7.5) with (5.1a, b), we see that $\{\alpha_{vi}\}$ coincides with $\{\gamma_{vi}\}$ as a whole, and $\{\beta_{vj}\}$ with $\{\varepsilon_{vj}\}$ as a whole. Therefore, changing A_v and B_v suitably, we may assume that $\alpha_{vi} = \gamma_{vi}$ and $\beta_{vj} = \varepsilon_{vj}$.

Define $\ell_n : Y^n \to K_r^n$ by

$$(7.6) \qquad \ell_n \begin{bmatrix} y_1 \\ \cdots \\ y_n \end{bmatrix} = \begin{bmatrix} \ell(y_1) \\ \cdots \\ \ell(y_n) \end{bmatrix} \qquad (y_i \in Y).$$

Then, from (5.7a, b) and (7.4) we obtain

$$(7.7) \qquad \lambda_v[\ell_n(y), w] = y_v^{\gamma} C_v, \qquad \mu_v[\ell_n(a), w] = y_v^{\varepsilon} D_v \qquad (y \in Y^n).$$

Notice also that $\mathrm{Tr}_{Y/K}(y\zeta y^*)$ of (5.6) for $y \in Y^n$ coincides with $-\ell_n(y)H\ell_n(y)^*$.

Let T be the vector space of all polynomial maps $Q : (\mathbf{C}_n^m)^{\mathbf{a}} \to \mathrm{Hom}(X, \mathbf{C})$ of a fixed degree such that $Q(xb) = Q(x)R(b)$ for $b \in GL_n(\mathbf{C})^{\mathbf{a}}$. Take a $\overline{\mathbf{Q}}$-rational basis $\{Q_i\}_{i=1}^t$ of T over $\mathbf{C}$, and define $P : (\mathbf{C}_n^m)^{\mathbf{a}} \to \mathrm{Hom}(X, \mathbf{C}^t)$ by $P(x)\xi = \sum_{i=1}^t (Q_i(x)\xi)e_i$ for $\xi \in X$ and $x \in (\mathbf{C}_n^m)^{\mathbf{a}}$, where $\{e_i\}_{i=1}^t$ is the standard basis of $\mathbf{C}^t$. Then we easily find a $\overline{\mathbf{Q}}$-rational polynomial representation $R' : GL_m(\mathbf{C})^{\mathbf{a}} \to GL_t(\mathbf{C})$ such that $P(ax) = R'(a)P(x)$. Notice that $P(xb) = P(x)R(b)$ for $b \in GL_n(\mathbf{C})^{\mathbf{a}}$.

We now consider $W(s, Q)$ and $W(s, \varphi)$ of (5.6) and (5.11a). By [86, (5.12)], $j_v[x, \mathfrak{z}] \neq 0$ for every $v \in \mathbf{a}$ if $-xHx^*$ is totally nonnegative, and hence, by (7.7), $\det(y_v^{\varepsilon}) \neq 0$ for every $v \in \mathbf{a}$ if $\mathrm{Tr}_{Y/K}(y\zeta y^*)$ is totally nonnegative. This shows that the sum of (5.11a) is meaningful.

As explained earlier, to prove Theorem 5.3, we may assume that $k_v \geq k_{v\rho}$. Put $\kappa = (k_v - k_{v\rho})_{v \in \mathbf{a}}$ and define $\mathcal{F}(\mathfrak{z}, s)$ by (6.13) with $K = L$ and with these H, P, and κ. Then we easily see that $c^{-k}|c|^g = c^{-\kappa}|c|^{2q}$ for $c \in K^\times$, and hence

$$(7.8a) \qquad \mathcal{F}(w, s; \alpha \circ \ell_n^{-1}) = ab^s R'({}^tC) \sum_{i=1}^{t} W(s, Q_i)e_i$$

with $a \in \overline{\mathbf{Q}}^\times$, $0 < b \in \overline{\mathbf{Q}} \cap \mathbf{R}$ and $C = (C_v)_{v \in \mathbf{a}}$. Similarly

(7.8b) $\quad \mathcal{F}^Z(w, s)(\varphi) = a_1 b_1^s W(s, \varphi')$ with $\varphi' = \tau^e\left(\overline{C\xi(w)}^{-1}, {}^tD^{-1}\right)\varphi$

and $a_1 \in \overline{\mathbf{Q}}^\times$, $0 < b_1 \in \overline{\mathbf{Q}} \cap \mathbf{R}$. Therefore the convergence and meromorphic continuation of $W(s, Q)$ and $W(s, \varphi)$ follow from those of $\mathcal{F}$ and $\mathcal{F}^Z$.

7.2. Our next task is to prove the arithmeticity of W at $s_0/2$. Since w is fixed by $h(Y^u)$, from [00, (3.19)] we obtain

(7.9) $$\lambda_v(h(a), w)^* \xi(w_v) \lambda_v(h(a), w) = \xi(w_v)$$

for every $a \in Y^u$. Combining this with (7.1a) and (7.3), we see that

$$\xi(w_v) C_v^{-1} \mathrm{diag}[a^{\alpha_{v1}}, \ldots, a^{\alpha_{vm}}] C_v = C_v^* \mathrm{diag}[a^{\alpha_{v1}}, \ldots, a^{\alpha_{vm}}]^\wedge C_v^\wedge \xi(w_v)$$

for every $a \in Y^u$. Since $aa^\rho = 1$, we have $\mathrm{diag}[a^{\alpha_{vi}}]^\wedge = \mathrm{diag}[a^{\alpha_{vi}}]$, and hence $C_v \xi(w_v)^{-1} C_v^*$ commutes with $\mathrm{diag}[a^{\alpha_{vi}}]$ for every $a \in Y$, from which we obtain

(7.10) $\quad C_v \xi(w_v)^{-1} C_v^*$ is diagonal for every $v \in \mathbf{a}$.

Let s_0 be as (5.13b). Define $\mathfrak{p}'$, $\mathfrak{p}'_v$, $\mathfrak{p}$, $\mathfrak{p}_v$ and $\mathbf{p}$ by (5.12b, c, d); put $\mathfrak{Q} = R'(\mathbf{p}1_m)$. Then the definition of $\mathfrak{p}_{v\rho}(w)$ and $\mathfrak{p}_v(w)$ in [00, (11.4a, b)] shows that $\mathfrak{p}_{v\rho}(w) = A_v \mathfrak{p}'_v A_v^{-1}$ and $\mathfrak{p}_v(w) = B_v \mathfrak{p}_v B_v^{-1}$, so that for the representation Ω defined by $\Omega(a, b) = \det(b)^\kappa R'({}^ta^{-1})$ we have $\mathfrak{P}_\Omega(w) = \mathfrak{q}\mathfrak{P}$, where $\mathfrak{q} = \prod_{v \in \mathbf{a}} \det(\mathfrak{p}_v)^{\kappa_v}$ and $\mathfrak{P} = R'\left(({}^t\mathfrak{p}_{v\rho}(w)^{-1})_{v \in \mathbf{a}}\right)$. Thus the arithmeticity stated in Theorem 6.7 implies that $\pi^{-c}\mathfrak{q}^{-1}\mathfrak{P}^{-1}\mathbf{q}^{-1}\mathfrak{Q}^{-1}\mathcal{F}(w, s_0/2)$ is a $\overline{\mathbf{Q}}$-rational element of X. By (5.12b) we have $\mathbf{q} = p_Y\left(\sum_{v \in \mathbf{a}} \sum_{i=1}^n \kappa_v \varepsilon_{vi}, \Phi\right)$. By (7.3) we have

(7.11) $$\mathfrak{p}_{v\rho}(w) = C_v^{-1}\mathfrak{p}'_v C_v \quad \text{and} \quad \mathfrak{p}_v(w) = D_v^{-1}\mathfrak{p}_v D_v,$$

so that $\mathfrak{P}^{-1} = R'({}^tC\mathfrak{p}' \cdot {}^tC^{-1})$. Therefore

$$\pi^{-c}\mathfrak{q}^{-1}\mathfrak{P}^{-1}\mathbf{q}^{-1}\mathfrak{Q}^{-1}\mathcal{F}(w, s_0/2) = a_2 \pi^{-c}\mathfrak{q}^{-1}\mathbf{q}^{-1}R'({}^tC)\sum_{i=1}^t W(s_0/2, Q'_i)e_i$$

with $a_2 \in \overline{\mathbf{Q}}^\times$ and $Q'_i(x) = Q_i(\mathbf{p}^{-1}\mathfrak{p}'x)$. Since $R'({}^tC)$ is algebraic, we thus obtain the algebraicity of $\pi^{-c}\mathfrak{q}^{-1}\mathbf{q}^{-1}W(s_0/2, Q'_i)$ for every i, which proves Theorem 5.3 (2).

As for $W(s_0/2, \varphi)$, we first take a nearly holomorphic form g on $\mathfrak{B}$ with values in $\mathrm{Hom}(Z, \mathbf{C})$ of weight $\omega \otimes \tau^e$, where $\omega(a, b) = \det(b)^\kappa$. Then, by [00, (12.24a)],

$$\left[(\omega \otimes \tau^e)\left(\mathfrak{p}(w)\right)^{-1} g(w)\right](\varphi) = \mathfrak{q}^{-1} g(w)\left(\tau^e\left({}^t\mathfrak{p}(w)\right)^{-1}\varphi\right)$$

for every $\varphi \in Z$ with $\mathfrak{q}$ as above. This quantity must be algebraic if g is arithmetic and φ is $\overline{\mathbf{Q}}$-rational. Now we take g to be $\pi^{-b}\mathbf{q}^{-1}\mathcal{F}^Z(\mathfrak{z}, s_0/2)$ with b and s_0 as in Theorem 5.3 (3). Then we find that

$$\pi^{-b}\mathbf{q}^{-1}\mathfrak{q}^{-1}\mathcal{F}^Z(w, s_0/2)\left[\tau^e({}^t\mathfrak{p}(w)^{-1})\varphi\right] \in \overline{\mathbf{Q}}$$

for every $\overline{\mathbf{Q}}$-rational φ. Put $\varphi_1 = \tau^e\left(\overline{C\xi(w)}^{-1}, {}^tD^{-1}\right)\tau^e({}^t\mathfrak{p}(w)^{-1})\varphi$. By (7.8b),

$$\pi^{-b}\mathbf{q}^{-1}\mathfrak{q}^{-1}W(s_0/2, \varphi_1) = a_3\pi^{-b}\mathbf{q}^{-1}\mathfrak{q}^{-1}\mathcal{F}^Z(w, s_0/2)\left(\tau^e({}^t\mathfrak{p}(w)^{-1})\varphi\right)$$

with $a_3 \in \overline{\mathbf{Q}}^\times$. By (7.11) and (7.10) we have ${}^tD_v^{-1} \cdot {}^t\mathfrak{p}_v(w)^{-1} = \mathfrak{p}_v^{-1} \cdot {}^tD_v^{-1}$ and

$$\overline{C}_v\overline{\xi(w_v)}^{-1} \cdot {}^t\mathfrak{p}_{v\rho}(w)^{-1} = (\mathfrak{p}_v')^{-1}\overline{C}_v\overline{\xi(w_v)}^{-1}.$$

Thus $\varphi_1 = \tau^e(\mathfrak{p}', \mathfrak{p})^{-1}\varphi_2$ with $\varphi_2 = \tau^e(\overline{C\xi(w)}^{-1}, {}^tD^{-1})\varphi$. Since C, D, and $\xi(w)$ are algebraic, φ is $\overline{\mathbf{Q}}$-rational if and only if φ_2 is $\overline{\mathbf{Q}}$-rational. Therefore we can conclude that $\pi^{-b}\mathfrak{q}^{-1}\mathfrak{q}^{-1}W(s_0/2, \varphi_1) \in \overline{\mathbf{Q}}$ if $\varphi_1 = \tau^e(\mathfrak{p}', \mathfrak{p})^{-1}\varphi_2$ with any $\overline{\mathbf{Q}}$-rational φ_2, which is exactly Theorem 5.3 (3).

7.3. Let us next prove the last implication of (7.0); namely we derive Theorem 6.7 from Theorem 5.3. Let the notation be as in (6.13). Since K is dense in $K \otimes_{\mathbf{Q}} \mathbf{R}$, we can easily find an element B of K_r^m such that BHB^* is totally positive. Put $H_1 = BHB^*$. Then H is equivalent to $\mathrm{diag}[H_1, -H_2]$ with $H_2 = H_2^* \in GL_n(K)$ such that H_{2v} is positive (resp. negative) definite if $v \in \mathbf{c}$ (resp. $v \notin \mathbf{c}$). Thus, we may assume that $H = \mathrm{diag}[H_1, -H_2]$ with such H_1 and H_2; we may even assume that H is diagonal. Then (6.4) is satisfied with $Q_v = \mathrm{diag}[H_1, H_2]_v^{1/2}$ for $v \in \mathbf{c}$ and $Q_v = H_v^{1/2}$ for $v \notin \mathbf{c}$.

Let $Y = K^r$ and define $h : Y \to H_r^r$ by $h(a) = \mathrm{diag}[a_1, \ldots, a_r]$. Also let $\mathbf{o}$ denote the point of $\mathfrak{B}$ whose v-component is 0 for every $v \in \mathbf{a}$. (Recall that $\mathfrak{B}(r, 0)$ consists of a single element written 0.) Then $h(a^\rho)H = Hh(a)^*$, and we find that $\mathbf{o}$ is the fixed point of $h(Y^u)$, and hence, is a CM-point. We easily see that

$$\lambda_v(h(a), \mathbf{o}) = \mathrm{diag}[a_1, \ldots, a_m]^{\tau_v\rho} \quad \text{and} \tag{7.12}$$

$$\mu_v(h(a), \mathbf{o}) = \mathrm{diag}[a_{m+1}, \ldots, a_r]^{\tau_v} \quad \text{if } v \in \mathbf{c},$$

$$\lambda_v(h(a), \mathbf{o}) = \mathrm{diag}[a_1, \ldots, a_r]^{\tau_v\rho} \quad \text{if } v \notin \mathbf{c}, \tag{7.13}$$

for every $a \in Y^u$. Thus the CM-type of Y determined at this point can be written $\sum_{i=1}^r(K_i, \tau^i)$, where K_i is the i-th factor of $Y = K^r$, identified with K; $\tau^i = \tau\rho$ for $i \le m$, and $\tau^i = \sum_{v\in\mathbf{c}} \tau_v + \sum_{v\notin\mathbf{c}} \tau_v\rho$ for $i > m$. Recall the property $p_K(\tau_v\rho, \tau\rho) = p_K(\tau_v, \tau)$. Therefore, putting $\sigma = \sum_{v\in\mathbf{c}} \tau_v + \sum_{v\notin\mathbf{c}} \tau_v\rho$, we have

$$\mathfrak{p}_{v\rho}(\mathbf{o}) = p_K(\tau_v, \tau)1_m \text{ and } \mathfrak{p}_v(\mathbf{o}) = p_K(\tau_v, \sigma)1_n \text{ if } v \in \mathbf{c}, \tag{7.14a}$$

$$\mathfrak{p}_{v\rho}(\mathbf{o}) = \mathrm{diag}\big[p_K(\tau_v, \tau)1_m, p_K(\tau_v\rho, \sigma)1_n\big] \text{ if } v \notin \mathbf{c}. \tag{7.14b}$$

Let $x = [x_1 \ x_2] \in K_r^n$ with $x_1 \in K_m^n$ and $x_2 \in K_n^n$. Clearly

$$\lambda_v[x, \mathbf{o}] = x_1^{\tau_v\rho}H_{1v}^{1/2} \text{ and } \mu_v[x, \mathbf{o}] = x_2^{\tau_v}H_{2v}^{1/2} \text{ if } v \in \mathbf{c}; \tag{7.15}$$

$$\lambda_v[x, \mathbf{o}] = x^{\tau_v\rho}H_v^{1/2} \text{ if } v \notin \mathbf{c}.$$

Therefore, for $\mathcal{F}$ of (6.13), we have, with suitable $a \in \overline{\mathbf{Q}}^\times$ and $0 < b \in \mathbf{R} \cap \overline{\mathbf{Q}}$,

$$\mathcal{F}(\mathbf{o}, s) = ab^sR'(H')\sum_x \alpha(x)|\det(x_2)|^{2q-2s\mathbf{c}}\det(x_2)^{-\kappa} \tag{7.16}$$

$$\cdot P(x')c_f\big(-\mathrm{Tr}_{K/L}(xHx^*)\big),$$

where $H' = \big((H_{1v}^{1/2})_{v\in\mathbf{c}},\, (H_v^{1/2})_{v\in\mathbf{a}-\mathbf{c}}\big)$ and $x' = \big(({}^t x_1^{\tau_v\rho})_{v\in\mathbf{c}},\, ({}^t x^{\tau_v\rho})_{v\in\mathbf{a}-\mathbf{c}}\big)$.

Let ζ be the element of $Y = K^r$ whose components are the diagonal entries of $-H$, that is, $-\zeta = (h_{ii})_{i=1}^r$. Now, for $y = (y_1, \ldots, y_n) \in Y^n$ with $y_i \in Y = K^r$ let $\ell(y)$ denote the element of K_r^n whose (i, j)-entry is the j-th component of y_i. Then we easily see that $\mathrm{Tr}_{K/L}\big(\ell(y)H\ell(y)^*\big)$ coincides with $-\mathrm{Tr}_{Y/L}(y\zeta y^*)$. Put $[K : L] = d$ and $\mu = m + r(d-1)$. Then $[Y : L] = rd = \mu + n$. We identify $\mathbf{c}$ with the set of archimedean primes of E.

We now consider W of (5.6) with $\{L, \psi, \mu, n; \alpha\circ\ell, R_0, f\}$ here as $\{K, \tau, m, n; \alpha, R, f\}$ there, defining $\{k, \varepsilon_{vi}, \gamma_{vj}, Q\}$ as follows: We take $k = \sum_{u\in\mathbf{c}} \kappa_u\psi_u$. Next, for $u \in \mathbf{c}$ and $1 \le i \le n$ we define $\varepsilon_{ui} \in J_Y$ by $y^{\varepsilon_{ui}} = x_{m+i}^{\tau_u}$, where $y = (x_1, \ldots, x_r) \in Y$ with $x_i \in K$. Then we define $\gamma_{uj} \in J_Y$ for $1 \le j \le \mu$ by the condition

$$(7.17) \qquad \{\rho\gamma_{u1}, \ldots, \rho\gamma_{u\mu}, \varepsilon_{u1}, \ldots, \varepsilon_{un}\} = \{\,\beta \in J_Y \,|\, \mathrm{Res}_{Y/L} = \psi_u\,\}.$$

Then we easily see that $\zeta^{\gamma_{uj}} < 0$, $\zeta^{\varepsilon_{ui}} > 0$, and y_u^ε defined by (5.7a) for $u \in \mathbf{c}$ coincides with $x_2^{\tau_u}$ if $x = \ell(y)$, $y \in Y^n$. We also see that the set $\{\gamma_{uj} \,|\, 1 \le j \le \mu\}$ for a fixed $u \in \mathbf{c}$ consists of all the $\gamma \in J_Y$ of the following two types: (i) $y^\gamma = x_i^{\rho\tau_u}$ for $1 \le i \le m$; (ii) $y^\gamma = x_i^{\rho\tau_v}$ for $1 \le i \le r$ and $v \notin \mathbf{c}$ such that $\tau_v = \tau_u$ on L. Put $\Phi = \sum_{u\in\mathbf{c}}\big(\sum_{i=1}^{\mu}\gamma_{ui} + \sum_{j=1}^{n}\varepsilon_{vj}\big)$. Then (Y, Φ) coincides with the CM-type $\sum_{i=1}^{r}(K_i, \tau^i)$ given above.

Next define a bijection $T : (\mathbf{C}_n^m)^{\mathbf{c}} \times (\mathbf{C}_n^r)^{\mathbf{a}-\mathbf{c}} \to (\mathbf{C}_n^\mu)^{\mathbf{c}}$ by

$$T\big((x_u)_{u\in\mathbf{c}},\, (x_v)_{v\in\mathbf{a}-\mathbf{c}}\big) = \big({}^t[{}^t x_u \;\; {}^t x_{v_2} \;\; \cdots \;\; {}^t x_{v_d}]\big)_{u\in\mathbf{c}},$$

where $\{u, v_2, \ldots, v_d\} = \{v \in \mathbf{a} \,|\, v = u \text{ on } E\}$. Then, with a natural arrangement of γ_{ui} we see that ${}^t y^\gamma$ with y^γ defined by (5.7b) for $y \in Y^n$ coincides with $T(x')$ with x' as in (7.16) if $x = \ell(y)$. Also, $\big(p_Y(\gamma_{ui}, \Phi)\big)_{u,i}$ with the same arrangement of the γ_{ui} coincides with $\big(\mathfrak{p}_{v\rho}(\mathbf{o})\big)_{v\in\mathbf{a}}$ given by (7.14a, b). Define $\mathfrak{p}'$, $\mathfrak{p}$, $\mathbf{p}$, c, $\mathbf{q}$, $\mathfrak{q}$ by (5.12b, c, d) and (5.14a, b, c) in the present case with E, L, $\mathbf{c}$, κ in place of F, K, $\mathbf{a}$, k there. Then we see that $p_Y(\varepsilon_{vi}, \Phi) = p_K(\tau_v, \sigma)$, so that $\mathfrak{q} = p_K\big(\sum_{u\in\mathbf{c}}\kappa_u\tau_u, n\sigma\big) = \prod_{u\in\mathbf{c}} \det\big(\mathfrak{p}_u(\mathbf{o})\big)^{\kappa_u}$. Put $\mathfrak{p}'(\mathbf{o}) = \big(\mathfrak{p}_{v\rho}(\mathbf{o})\big)_{v\in\mathbf{a}}$ and define $\mathbf{q}$, $\mathbf{p}_0$, and $\mathfrak{Q}$ as in Theorem 6.7. From what we said about $p_Y(\gamma_{ui}, \Phi)$ we obtain the first of the following equalities:

$$(7.18) \qquad \mathfrak{p}'T = T\mathfrak{p}'(\mathbf{o}) \quad and \quad \mathbf{p}T = T\mathbf{p}_0.$$

As for the second one, if $\mathrm{Res}_{K/L}(\tau_v) = \psi_u$, then $p_L(\psi_u, \psi) = p_K\big(\tau_v, \mathrm{Inf}_{K/L}(\psi)\big) = p_K(\tau_v, \tau)$, and hence $\mathbf{p}T = T\mathbf{p}_0$.

Now take $Q = P\circ T^{-1}$. Then (7.16) shows that $\mathcal{F}(\mathbf{o}, s) = ab^s R'(H')W(s, Q)$. (The present $W(s, Q)$ is X'-valued, and so, to obtain a $\mathbf{C}$-valued series, we have to consider $\chi\big(W(s, Q)\big)$ with a $\overline{\mathbf{Q}}$-rational element χ of $\mathrm{Hom}(X', \mathbf{C})$. Also, the sum for $W(s, Q)$ is extended over all $y = (y_i)_{i=1}^n \in Y^n$ such that the y_i are linearly independent over L and $\mathrm{Tr}_{Y/L}(y\zeta y^*)$ is totally nonnegative. As already noted, $\det(y_u^\varepsilon) \ne 0$ for every such y and every $u \in \mathbf{c}$, so that $\det(x_2) \ne 0$ if $x = \ell(y)$; then we see that $\mathrm{rank}(x) = n$, and hence the sum for $\mathcal{F}$ is consistent with that for W.) Let $\Omega(a, b) = \det(b)^\kappa R'({}^t a^{-1})$. Then $\mathfrak{P}_\Omega(\mathbf{o}) = \mathfrak{q}R'\big(\mathfrak{p}'(\mathbf{o})^{-1}\big)$. Let s_0 be as in (6.18). Since H' is diagonal, we have

$$\mathfrak{P}_\Omega(\mathbf{o})^{-1}\mathbf{q}^{-1}\mathfrak{Q}^{-1}\mathcal{F}(\mathbf{o}, s_0/2) = a_1R'(H')\mathbf{q}^{-1}\mathbf{q}^{-1}R'(\mathfrak{p}'(\mathbf{o})\mathfrak{p}_0^{-1})W(s_0/2, Q)$$
$$= a_1R'(H')\mathbf{q}^{-1}\mathbf{q}^{-1}W(s_0/2, Q_1)$$

with $a_1 \in \overline{\mathbf{Q}}^\times$ and $Q_1(x) = P\big(\mathfrak{p}'(\mathbf{o})\mathfrak{p}_0^{-1}T^{-1}(x)\big)$. By (7.18) we have $Q_1(x) = Q(\mathfrak{p}^{-1}\mathfrak{p}'x)$. Therefore, by Theorem 5.3, $\pi^{-c}\mathbf{q}^{-1}\mathbf{q}^{-1}W(s_0/2, Q_1)$ is $\overline{\mathbf{Q}}$-rational, and hence $\pi^{-c}\mathfrak{P}_\Omega(\mathbf{o})^{-1}\mathbf{q}^{-1}\mathfrak{Q}^{-1}\mathcal{F}(\mathbf{o}, s_0/2)$ is $\overline{\mathbf{Q}}$-rational. This is the desired arithmeticity at the point $\mathbf{o}$. Take any $\gamma \in G(H)$. From (6.15) we obtain $\mathcal{F}(\gamma(\mathbf{o}), s; \alpha) = j_\gamma(\mathbf{o})^\kappa R'({}^t\lambda(\gamma, \mathbf{o})^{-1})\mathcal{F}(\mathbf{o}, s; \alpha^\gamma)$. By [00, Proposition 11.5 (2)] we have $\mathfrak{P}_\Omega(\gamma(\mathbf{o})) = R'({}^t\lambda(\gamma, \mathbf{o})^{-1})j_\gamma(\mathbf{o})^\kappa\mathfrak{P}_\Omega(\mathbf{o})A$ with an algebraic matrix A. Since $\mathfrak{Q}$ commutes with $R'({}^t\lambda(\gamma, \mathbf{o}))$, we obtain

$$\text{(7.19)}\qquad \pi^{-c}\,\mathfrak{P}_\Omega\big(\gamma(\mathbf{o})\big)^{-1}\mathbf{q}^{-1}\mathfrak{Q}^{-1}\mathcal{F}(\gamma(\mathbf{o}), s_0/2; \alpha)$$
$$= \pi^{-c}A^{-1}\mathfrak{P}_\Omega(\mathbf{o})^{-1}\mathbf{q}^{-1}\mathfrak{Q}^{-1}\mathcal{F}(\mathbf{o}, s_0/2; \alpha^\gamma),$$

which is $\overline{\mathbf{Q}}$-rational, since the above result is applicable to any α. Thus $\pi^{-c}\mathbf{q}^{-1}\cdot\mathfrak{Q}^{-1}\mathcal{F}(\mathfrak{z}, s_0/2; \alpha)$ is arithmetic at $\mathfrak{z} = \gamma(\mathbf{o})$ in the sense of [00, §14.4] for every $\gamma \in G(H)$. Since all such points $\gamma(\mathbf{o})$ form a dense subset of $\mathfrak{B}$, $\pi^{-c}\mathbf{q}^{-1}\mathfrak{Q}^{-1}\cdot\mathcal{F}(\mathfrak{z}, s_0/2; \alpha)$ is arithmetic by virtue of [00, Theorem 14.9 (1)]. Once this is established, our assertion of Theorem 6.7 concerning $\mathcal{F}^Z$ follows immediately from [00, Theorem 14.9 (2)] combined with (6.24), since $\psi(s_0/2) \neq 0$ for s_0 satisfying (6.20) as noted at the end of §6.8. This completes the proof of the last implication of (7.0).

7.4. We finally prove the first implication of (7.0). Suppose that $K = L$ in the setting of §7.3. We may assume that $\alpha([x_1\ \ x_2]) = \alpha_1(x_1)\alpha_2(x_2)$ with $\alpha_1 \in \mathcal{S}(K_m^n)$ and $\alpha_2 \in \mathcal{S}(K_n^n)$. By (7.16) we obtain

$$\mathcal{F}(\mathbf{o}, s) = ab^sR'(H_1^{1/2})\sum_{x\in U\backslash\mathcal{X}}\alpha_1(x_1)\alpha_2(x_2)|\det(x_2)|^{2q-2s\mathbf{a}}\det(x_2)^{-\kappa}$$
$$\cdot P({}^tx_1^\rho)c_f(x_2H_2x_2^* - x_1H_1x_1^*),$$

where $x = [x_1\ \ x_2]$, $a \in \overline{\mathbf{Q}}^\times$, and $0 < b \in \mathbf{R} \cap \overline{\mathbf{Q}}$. Put

$$t(z) = g(z)f(z),\quad g(z) = \sum_{y\in K_m^n}\alpha_1(y)P({}^ty^\rho)\mathbf{e}_\mathbf{a}^n(yH_1y^*z)\qquad (z \in \mathcal{H}_n^\mathbf{a}).$$

By [97, Theorem A7.11] or by [86, Proposition 7.7] we have

$$\text{(7.20)}\qquad g\big(\gamma(z)\big) = j_\gamma(z)^{m\mathbf{a}}g(z)R\big({}^t\lambda(\gamma, z)\big)\ \text{ for every } \gamma \in \Gamma$$

with a congruence subgroup Γ of G_K^n. Therefore $t\big(\gamma(z)\big) = j_\gamma^{m\mathbf{a}+p}t(z)$ for γ in a suitable congruence subgroup of G_K^n. Clearly $c_t(\sigma) = \sum_y \alpha_1(y)P({}^ty^\rho)c_f(\sigma - yH_1y^*)$, and hence

$$c_t(x_2H_2x_2^*) = \sum_{x_1\in K_m^n}\alpha_1(x_1)P({}^tx_1^\rho)c_f\big(x_2H_2x_2^* - x_1H_1x_1^*\big).$$

Therefore we can easily verify that

$$\mathcal{F}(\mathbf{o}, s) = ab^sR'(H_1^{1/2})\sum_{\xi\in U\backslash GL_n(K)}\alpha_2(\xi)\det(\xi)^{-\kappa}|\det(\xi)|^{2q-2s\mathbf{a}}c_t(\xi H_2\xi^*).$$

Define $\mathcal{D}$ by (5.15) with $(\alpha_2,\ \kappa,\ t,\ p+m\mathbf{a},\ H_2)$ as $(\alpha,\ k,\ f,\ p,\ \beta)$ there. Then the last sum equals $\mathcal{D}(s-m/2)$. Let $s_1 = s_0 - m$ with s_0 as in (6.18). By Theorem 5.5, $\pi^{-c}\mathfrak{q}^{-2}\mathcal{D}(s_1/2)$ is $\overline{\mathbf{Q}}$-rational, where $c = n\sum_{v\in\mathbf{a}}\kappa_v - [F:\mathbf{Q}]n(n-1)/2$ and $\mathfrak{q} = p_K\big(\sum_{v\in\mathbf{a}}\kappa_v\tau_v,\ n\tau\big)$. (Since t is X'-valued, we are applying Theorem 5.5 to each component of $\mathcal{D}$. Notice also that t is $\overline{\mathbf{Q}}$-rational if P and f are $\overline{\mathbf{Q}}$-rational.)

Define $\mathbf{q}$ and $\mathfrak{Q}$ as in Theorem 6.7 with $K = L$; then $\mathbf{q} = \mathfrak{q}$ and $\mathfrak{Q} = R'\big((p_K(\tau_v,\ \tau)1_m)_{v\in\mathbf{a}}\big) = R'\big(\mathfrak{p}'(\mathbf{o})\big)$. Therefore

$$\mathfrak{P}_\Omega(\mathbf{o})^{-1}\pi^{-c}\mathbf{q}^{-1}\mathfrak{Q}^{-1}\mathcal{F}(\mathbf{o},\ s_0/2) = a_0R'(H_1^{1/2})\pi^{-c}\mathfrak{q}^{-2}\mathcal{D}(s_1/2)$$

with $a_0 \in \overline{\mathbf{Q}}$, and this quantity is $\overline{\mathbf{Q}}$-rational. Thus $\pi^{-c}\mathbf{q}^{-1}\mathfrak{Q}^{-1}\mathcal{F}(\mathfrak{z},\ s_0/2)$ is arithmetic at $\mathfrak{z} = \mathbf{o}$. Repeating the argument at the end of §7.3, we can show that $\pi^{-c}\mathbf{q}^{-1}\mathfrak{Q}^{-1}\mathcal{F}(\mathfrak{z},\ s_0/2)$ is arithmetic at $\mathfrak{z} = \gamma(\mathbf{o})$ for every $\gamma \in G(H)$, which proves the desired arithmeticity of $\pi^{-c}\mathbf{q}^{-1}\mathfrak{Q}^{-1}\mathcal{F}(\mathfrak{z},\ s_0/2)$. The assertion concerning $\mathcal{F}^Z$ in this case follows from this result for the reason explained at the end of §7.3. This establishes the first implication of (7.0). Since Theorem 5.5 is already proved, our sequence (7.0) is now completely proved.

7.5. Remark. In our previous articles [84], [86], and [98] the action of $G(H)$ on $\mathfrak{B}$ and the factors of automorphy were defined in a manner somewhat different from the present paper. Let us now show that the difference causes no problem.

We first observe that if $\alpha \in G(H)$, then $\alpha^\rho \in G(H^\rho)$. Now for $\beta \in G(H^\rho)$ and $\mathfrak{z} \in \mathfrak{B}$ let $\lambda'_v(\beta,\ \mathfrak{z})$, $\mu'_v(\beta,\ \mathfrak{z})$ and $\beta(\mathfrak{z})'$ denote the factors of automorphy and the image of $\mathfrak{z}$ under β defined in [86, Section 5]. This requires elements $Q_v \in GL_r(\mathbf{C})$ such that $\overline{H}_v = \overline{Q}_v I_{m_v,\,n_v}\cdot {}^tQ_v$. Clearly the Q_v of (5.4) satisfy that condition. Then we easily see that $\lambda'_v(\alpha^\rho,\ \mathfrak{z}) = \lambda_v(\alpha,\ \mathfrak{z})$, $\mu'_v(\alpha^\rho,\ \mathfrak{z}) = \mu_v(\alpha,\ \mathfrak{z})$, and $\alpha^\rho(\mathfrak{z})' = \alpha(\mathfrak{z})$. Therefore our automorphic forms are exactly those defined in [84] and [86] with respect to $G(H^\rho)$, and we can also easily see that the arithmeticity is consistent.

Also, for $x \in K_r^n$ define $\lambda'_v[x,\ \mathfrak{z}] \in \mathbf{C}^n_{m_v}$ and $\mu'_v[x,\ \mathfrak{z}] \in \mathbf{C}^n_{n_v}$ by

(7.21) $$x^{\tau_v\rho}Q_vB(\mathfrak{z}_v) = \big[\overline{\lambda'_v[x,\ \mathfrak{z}]}\quad \mu'_v[x,\ \mathfrak{z}]\big].$$

These λ'_v and μ'_v are the quantities defined in [86, (5.5)]. In [86] the series corresponding to the present $\mathcal{F}$ was defined with λ'_v and μ'_v and also with an automorphic form $A \in \mathcal{M}_{L,\chi}$, where $\chi(a,\ b) = \det(b)^\kappa R_0({}^tb^{-1})$. However, if we put $f'(z) = f({}^tz)$ and substitute x^ρ for x in (6.13), then the resulting series is that of [86, (5.11)] defined with $(f',\ P,\ H^\rho)$ as $(A,\ S,\ H)$ there, provided we change the exponents $s\psi$ and $-2s\psi$ there for $s\mathbf{c}-q$ and $2q-2s\mathbf{c}$; see [00, §5.11] for the relationship between the weight of f and that of f'.

8. The residues of W and $\mathcal{F}$

Instead of taking the special values of W and $\mathcal{F}$, we can examine their residues at some points, and prove their arithmeticity. We begin with preliminary results:

8.1. Proposition. (1) *The notation being as in Lemma 1.3, let* $s_e = n-(e/2)$ *with an integer* e *such that* $0 \le e < n$. *Then* $E(z,\ s;\ e\mathbf{a},\ \Gamma)$ *has at most a simple*

pole at s_e *and the residue is* $\pi^{dn(e-n)}r_e^n(F)$ *times an element of* $\mathcal{M}_{\mathbf{ea}}(\mathbf{Q}_{\mathrm{ab}})$, where $d = [F : \mathbf{Q}]$ and $r_e^n(F) = \mathrm{Res}_{s=1}\zeta_F(s) \cdot \prod_{j=2}^{n-e} L(j, \theta^{j-1})$.

(2) *The notation being as in Theorem 5.5, suppose that* $|k_v - k_{v\rho}| + n - p_v = e$ *for every* $v \in \mathbf{a}$ *with an integer* e *such that* $0 \le e < n$. *Then* $\mathcal{D}(s)$ *has at most a simple pole at* $(n-e)/2$. *Moreover, suppose that* α *is* $\overline{\mathbf{Q}}$*-valued,* f *is* $\overline{\mathbf{Q}}$*-rational, and* $|k_v - k_{v\rho}| > 2n$ *for every* $v \in \mathbf{a}$. *Then the residue at* $(n-e)/2$ *belongs to* $\pi^c r_e^n(F)\mathfrak{q}^2\overline{\mathbf{Q}}$ *with* c *and* $\mathfrak{q}$ *as in Theorem 5.5.*

PROOF. The first assertion for $e = 0$ is included in Lemma 1.3. If $e > 0$, it follows from [00, Theorems 7.11 and 17.8] for the reason explained at the beginning of the proof of Lemma 1.3. As for (2), define q as in the proof of Theorem 5.5. Then our question is the behavior of $D(s;\, f,\, q)$ at $s = -e/2$. We apply (1.12) to this D with $m = p$. Notice that $m' - p = e\mathbf{a}$. Let $t(z)$ be the residue of $E(z,\, s;\, m' - p,\, \Gamma)$ at s_e. Then we see that $D(s;\, f,\, q)$ has at most a simple pole at $s = -e/2$, and the residue is $A\Gamma((-e/2))^{-1}\mathrm{vol}(\mathfrak{D})\langle q,\, ft\rangle$. Therefore we obtain (2) by combining (1) with Theorem 3.5 (2).

8.2. Theorem. (1) *The notation being as in Theorem 5.3, suppose that (5.9) is satisfied and* $\kappa_v - p_v = m - n + e$ *for every* $v \in \mathbf{a}$ *with an integer* e *such that* $0 \le e < n$; *Then* $W(s,\, Q)$ *has at most a simple pole at* $s = (m+n-e)/2$.

(2) *Suppose moreover that* α *is* $\overline{\mathbf{Q}}$*-valued,* Q*,* f *are* $\overline{\mathbf{Q}}$*-rational, and* $\kappa_v > 2n$ *for every* $v \in \mathbf{a}$. *Then the residue of* $W(s,\, Q')$ *at* $(m+n-e)/2$ *belongs to* $\pi^c r_e^n(F)\mathbf{q}\mathfrak{q}\overline{\mathbf{Q}}$ *with* c*,* $\mathbf{q}$*,* $\mathfrak{q}$ *and* Q' *as in Theorem 5.3.*

This will be proven together with the following theorem.

8.3. Theorem. (1) *The notation being as in Theorem 6.5, suppose that (5.9) is satisfied and* $\kappa_u - p_u = (m+n)[K : L] - 2n + e$ *for every* $u \in \mathbf{c}$ *with an integer* e *such that* $0 \le e < n$; *put* $t_e = (1/2)\{(m+n)[K : L] - e\}$. *Then* $\mathcal{F}(\mathfrak{z},\, s)$ *has at most a simple pole at* $s = t_e$; *moreover the residue is a holomorphic automorphic form of weight* Ω*, provided* $\kappa_u \ge 1$ *for every* $u \in \mathbf{c}$*, where* $\Omega(a,\, b) = \det(b)^{\kappa} R'({}^t a^{-1})$.

(2) *Suppose moreover that* α *is* $\overline{\mathbf{Q}}$*-valued,* P*,* f *are* $\overline{\mathbf{Q}}$*-rational, and* $\kappa_u > 2n$ *for every* $u \in \mathbf{c}$. *Then the residue of* $\pi^{-c}r_e^n(E)^{-1}\mathbf{q}^{-1}\mathfrak{Q}^{-1}\mathcal{F}(\mathfrak{z},\, s)$ *at* $s = t_e$ *is a* $\overline{\mathbf{Q}}$*-rational holomorphic automorphic form, where* c*,* $\mathbf{q}$*, and* $\mathfrak{Q}$ *are as in Theorem 6.7.*

PROOF. We specialize (6.21) to the situation of (1) with $Z = \mathbf{C} = S_0(T)$ and $Z' = \mathbf{C} = S_0(\mathbf{C}_n^n)$, in which case E_Z and $E_{Z'}$ are the identity operators. Then $\mathcal{E}(z,\, s) = E(z,\, s + \nu/2;\, e\mathbf{c},\, \Gamma)$. (The present e is different from e in the proof of Theorem 6.5, which we do not use here.) From (1) of Proposition 8.1 we see that $\mathcal{F}(\mathfrak{z},\, s)$ (which corresponds to $f(w,\, s)$ of (6.21)) has at most a simple pole at $s = t_e$. Next we consider (6.21) with $Z = S_1(T)$ and $Z' = S_1(\mathbf{C}_n^n)$. Since the residue of $\mathcal{E}(z,\, s)$ at $s = t_e$ is holomorphic in view of Proposition 8.1 (1), it is annihilated by E'_Z. Therefore (6.21) shows that the residue of $f(w,\, s)$ is annihilated by E_Z, which means that it is holomorphic. The residue of $\mathcal{F}$ must be an automorphic form as can be seen from (6.15). This proves assertion (1).

Notice that we have to assume that $\kappa_v \geq 1$ in order to guarantee (6.21) for such Z and Z'.

Now Theorem 8.3 (2) and Theorem 8.2 can be proved by the same technique as in Section 7. We take the residues of the functions in place of the values at $s_0/2$, and consider a sequence of implications similar to (7.0). We first note that Theorem 8.2, if $m = 0$, is included in Proposition 8.1 (2) for the same reason as in §5.4. From that result we derive Theorem 8.3 (2) with $K = L$, which in turn implies Theorem 8.2 with $m > 0$. Finally we derive Theorem 8.3 (2) in the general case from Theorem 8.2 (2). We do not give details here, as the whole procedure consists of obvious modifications of the arguments in Section 7.

8.4. Remark. The residue of $\mathcal{F}(\mathfrak{z}, s)$ at $s = t_0$ can be given as a theta integral, as can be seen from (6.21). Let us now present its explicit form. First, for $(z, \mathfrak{z}) \in \mathcal{H}_n^{\mathbf{a}} \times \mathfrak{B}$ we define a $\mathrm{Hom}(X, X')$-valued function Θ by

$$(8.1)\qquad \Theta(z, \mathfrak{z}) = \sum_{x \in K_r^n} \alpha(x)\delta(\mathfrak{z})^{-\kappa}\overline{j[x, \mathfrak{z}]}^{\kappa}\mathbf{e}_{\mathbf{a}}^n(2^{-1}S[x;\, z, \mathfrak{z}])P({}^t\lambda[x, \mathfrak{z}]),$$

$$S[x;\, z, \mathfrak{z}]_v = (xHx^*)_v z_v + \begin{cases} \mu_v[x, \mathfrak{z}]\eta(\mathfrak{z}_v)^{-1}\mu_v[x, \mathfrak{z}]^*(z - z^*)_v & (v \in \mathbf{c}), \\ 0 & (v \notin \mathbf{c}). \end{cases}$$

Here the notation is the same as in (6.13) and η is defined by (6.1c). This is a special case of [97, (A7.10.3)] and is essentially the same as the series Θ of [86, p. 391] if we take e there to be 0. It can easily be seen that Θ satisfies the same type of formula as (6.15) under the transformation by the elements of $G(H)$. Also, by [97, Theorem A7.11] or by [86, Proposition 7.7] we have

$$(8.2)\qquad \Theta(\beta z, \mathfrak{z}) = j_\beta(z)^{(m+n)\mathbf{a}-2n\mathbf{c}-\kappa}|j_\beta(z)|^{2n\mathbf{c}+2\kappa}\Theta(z, \mathfrak{z})R({}^t\lambda(\beta, z))$$

for every β in a congruence subgroup of G_K^n. For $z \in \mathcal{H}_n^{\mathbf{c}}$ define $z' \in \mathcal{H}_n^{\mathbf{a}}$ by $z'_v = z_u$ if $v = u$ on E, and put $\theta(z, \mathfrak{z}) = \Theta(z', \mathfrak{z})$ for $(z, \mathfrak{z}) \in \mathcal{H}_n^{\mathbf{c}} \times \mathfrak{B}$. From (8.2) we easily obtain

$$(8.3)\qquad \theta(\gamma z, \mathfrak{z}) = j_\gamma(z)^{-\nu\mathbf{c}-\kappa}|j_\gamma(z)|^{2n\mathbf{c}+2\kappa}\theta(z, \mathfrak{z})R_0({}^t\lambda(\gamma, z))$$

for every γ in a congruence subgroup of G_L^n, where $\nu = 2n - (m+n)[K : L]$. Now we have

$$(8.4)\qquad (2\pi)^{-b}\prod_{v \in \mathbf{c}} \Gamma_n(s + p_v + q_v) \cdot \mathcal{F}(\mathfrak{z}, s)$$
$$= a_0 \int_{\mathfrak{D}} \theta(z, \mathfrak{z})f(z/2)E(z,\, s + \nu/2;\, \kappa - p + \nu\mathbf{c}, \Gamma)\delta(z)^{n\mathbf{c}+\kappa}\mathbf{d}z,$$

where $0 < a_0 \in \overline{\mathbf{Q}} \cap \mathbf{R}$, $b = \sum_{v \in \mathbf{c}} n(s + p_v + q_v)$, f is as in (6.11), and $\mathfrak{D} = \Gamma\backslash\mathcal{H}_n^{\mathbf{c}}$ with a suitable congruence subgroup Γ of G_L^n. This can be verified by the standard technique, and is essentially shown in [86, pp. 392-394]. Now assume that

$$(8.5)\qquad \kappa_u - p_u = (m+n)[K : L] - 2n \quad \textit{for every} \quad u \in \mathbf{c}.$$

This means that $\kappa - p + \nu\mathbf{c} = 0$. Take the residue of (8.4) at t_0. Then, by Lemma 1.3 we find that

$$(8.6) \qquad \lim_{s\to t_0}(s-t_0)\mathcal{F}(\mathfrak{z},\, s) = a_1\pi^c r_0^n(E)\int_{\mathfrak{D}}\theta(z,\, \mathfrak{z})f(z/2)\delta(z)^{nc+\kappa}\mathbf{d}z,$$

where $0 < a_1 \in \overline{\mathbf{Q}} \cap \mathbf{R}$ and $c = \sum_{u\in\mathbf{c}} \kappa_u - [E:\mathbf{Q}]n(n-1)/2$. Therefore Theorem 8.3 gives the nature of the integral of (8.6). In the simplest case where $K = L$, we have $\Theta(z,\, \mathfrak{z}) = \theta(z,\, \mathfrak{z})$, $\mathbf{c} = \mathbf{a}$, $E = F$, $t_0 = (m+n)/2$, and condition (8.5) is $\kappa_v - p_v = m - n$ for every $v \in \mathbf{a}$.

References

[B90] A. Bluher, Near holomorphy of some automorphic forms at critical points, Invent. math. **102** (1990), 335–376.

[B94] A. Bluher, Arithmeticity of some automorphic forms at critical points, Amer. J. Math. **116** (1994), 1283–1335.

[76] G. Shimura, The special values of the zeta functions associated with cusp forms, Comm. pure appl. Math. **29** (1976), 783–804.

[78] G. Shimura, The special values of the zeta functions associated with Hilbert modular forms, Duke M. J. **45** (1978), 637–679.

[80] G. Shimura, The arithmetic of certain zeta functions and automorphic forms on orthogonal groups, Ann. of Math. **111** (1980), 313–375.

[81] G. Shimura, Arithmetic of differential operators on symmetric domains, Duke M. J. **48** (1981), 813–843.

[84] G. Shimura, Differential operators and the singular values of Eisenstein series, Duke M. J. **51** (1984), 261–329.

[86] G. Shimura, On a class of nearly holomorphic automorphic forms, Ann. of Math. **123** (1986), 347–406.

[87a] G. Shimura, Nearly holomorphic functions on hermitian symmetric spaces, Math. Ann. **278** (1987), 1–28.

[87b] G. Shimura, On Hilbert modular forms of half-integral weight, Duke M. J. **55** (1987), 765–838.

[88] G. Shimura, On the critical values of certain Dirichlet series and the periods of automorphic forms, Inv. mat. **94** (1988), 245–305.

[94] G. Shimura, Differential operators, holomorphic projection, and singular forms, Duke M. J. **76** (1994), 141–173.

[96] G. Shimura, Convergence of zeta functions on symplectic and metaplectic groups, Duke Math. J. **82** (1996), 327–347.

[97] G. Shimura, Euler Products and Eisenstein series, CBMS Regional Conference Series in Mathematics, No. 93, Amer. Math. Soc., 1997.

[98] G. Shimura, Abelian varieties with complex multiplication and modular functions, Princeton University Press, 1998.

[00] G. Shimura, Arithmeticity in the theory of automorphic forms, Math. Surveys Monogr., vol. 82, Amer. Math. Soc., 2000.

01c

The relative regulator of an algebraic number field

unpublished

Introduction

Given an algebraic number field F, we let h_F denote the class number of F, R_F the regulator of F, E_F the group of units of F, W_F the group of all roots of unity contained in F, D_F the discriminant of F, $\mathbf{a}(F)$ the set of archimedean primes of F, r_F the number of real archimedean primes of F, c_F the number of complex archimedean primes of F, and ζ_F the Dedekind zeta function of F. We put

$$t_F = r_F + c_F - 1. \tag{0.1}$$

We recall a well-known formula due to Dedekind:

$$\mathrm{Res}_{s=1}\zeta_F(s) = 2^{r_F}(2\pi)^{c_F}[W_F : 1]^{-1}|D_F|^{-1/2}h_F R_F. \tag{0.2}$$

Let K be an algebraic extension of F of degree $m > 1$. We put

$$U = U(K/F) = \{x \in E_K \mid N_{K/F}(x) = 1\}. \tag{0.3}$$

If $m = 2$, we have $\zeta_K(s) = \zeta_F(s)L(s, \lambda)$ with the Hecke L-function $L(s, \lambda)$ of the quadratic character λ of F corresponding to the extension K/F. Then the value $L(1, \lambda)$ can be obtained as the quotient $\mathrm{Res}_{s=1}\zeta_K(s)/\mathrm{Res}_{s=1}\zeta_F(s)$, which involves R_K/R_F. Thus it is natural to ask whether there is a simpler expression for R_K/R_F. In fact, we can prove, as the first principal result of this paper, that R_K/R_F can be described in terms of U for an arbitrary m, as follows:

Let $b = t_K - t_F$ and let $\mathbf{b}$ be a subset of $\mathbf{a}(K)$ such that the complement of $\mathbf{b}$ in $\mathbf{a}(K)$ is mapped bijectively onto $\mathbf{a}(F)$ by the restriction of the elements of $\mathbf{a}(K)$ to F; Then $\#(\mathbf{b}) = b$ and $U/(U \cap W_K)$ is isomorphic to $\mathbf{Z}^b$. Moreover, put $R(K/F) = \left|\det\left(\log|\eta_i|_v\right)_{1\le i\le b,\, v\in\mathbf{b}}\right|$ with a set of independent generators $\{\eta_i\}_{i=1}^b$ of U modulo $U \cap W_K$; then

$$R_K/R_F = [W_F : N_{K/F}(W_K)]^{-1}[E_F : N_{K/F}(E_K)]R(K/F). \tag{0.4}$$

Here $|\ |_v$ is the normalized valuation of K at v (Theorem 1.3).

The number $R(K/F)$ may be called *the relative regulator* of the extension K/F. If $m = 2$ in particular, $L(1, \lambda)$ can be explicitly given in terms of $R(K/F)$. It may be emphasized here that $R(K/F)$ is defined by a set of generators for U in a canonical way, not by "suitably chosen" units of K, and that K is not necessarily normal over F.

To explain our next point, for any K/F such that $[K : F] = m$, put

$$I(K/F) = [W_K : N_{K/F}(W_K)]^{-1}[E_F : N_{K/F}(E_K)] \cdot (h_K/h_F) \cdot R(K/F). \tag{0.5}$$

Looking at linear relations among the group characters of a Galois group, we can find multiplicative relations between several functions of the form ζ_K/ζ_F, from which we can derive multiplicative relations between $I(K/F)$ for various extensions K/F. In particular, we can find nonisomorphic extensions K/F and K'/F' such that

$$(0.6) \qquad I(K/F) = I(K'/F').$$

We shall also prove the equalities between various kinds of invariants associated to K/F such as $|D_K/D_F|$ (Theorems 2.3 and 3.3). As an application of our methods, we obtain, in a setting which may be roughly called Kummer's type, a "canonical expression"

$$(0.7) \qquad L(1, \mu) = 2^t\pi^c d^{1/2} I(K/F)$$

for a Hecke character μ of an algebraic number field P whose order is an odd prime power m, where K/F is an extension of degree m, and t, c, d are positive integers determined by the functional equation for $L(s, \mu)$ (Corollary 3.4). Thus the essential factor of $L(1, \mu)$ can be given in terms of the units of $U(K/F)$, even when $[K : F] > 2$. It should be noted however that $R(K/F)$ depends on the choice of K/F, which is not unique for $L(s, \mu)$ in general, though the quantity $I(K/F)$ is completely determined by the L-function.

Once we have formula (0.4), we can derive (0.6) and its generalization for various K/F, and also (0.7), from the coincidence of group characters. In that sense we need no drastically new methods. Still, we think that (0.6) and (0.7) are noteworthy and were not explicitly stated before, though (0.6) in the case of CM-extensions was discussed in our paper [3]. Indeed, take a special case of (0.6) with $m = 2$ and $b = 1$. Then we obtain units $\eta \in K$ and $\eta' \in K'$ such that $|\eta|^e = |\eta'|^{e'}$, where $e = [E_F : N_{K/F}(E_K)] \cdot (h_K/h_F)$ and e' is defined similarly for K'/F'. This is reminiscent of the classical formula concerning cyclotomic units.

In the final section we shall add a few remarks about the connection of all these with the mass formula for an orthogonal group. In particular, we shall present a viewpoint that $R(K/F)$ or $I(K/F)$ is the mass of the algebrtaic group $\mathcal{G}(K/F)$ defined in (1.1) below.

1. The quotient of the regulators of an algebraic extension

1.1. For an algebraic extension K of F of degree $m > 1$ as in the introduction, we put

$$(1.1) \qquad \mathcal{G}(K/F) = \{\, x \in K \mid N_{K/F}(x) = 1 \,\}, \quad U = U(K/F) = E_K \cap \mathcal{G}(K/F),$$

$$(1.2) \qquad E_F^m = \{\, x^m \mid x \in E_F \,\}, \qquad W_F^m = \{\, x^m \mid x \in W_F \,\}.$$

Since E_F is isomorphic to the direct product of $\mathbf{Z}^{t_F}$ and W_F, we have

$$(1.3) \qquad [E_F : E_F^m] = m^{t_F}[W_F : W_F^m] \quad \text{in general,}$$
$$= 2^{t_F+1} \quad \text{if } m = 2.$$

1.2. Lemma. *The notation being as above, the following assertions hold:*

(1) $U/(U \cap W_K)$ *is isomorphic to* $\mathbf{Z}^b$, *where* $b = t_K - t_F$.

(2) $UW_K \cap E_F W_K = W_K$.

(3) $UE_F W_K / W_K$ *is isomorphic to* $\mathbf{Z}^{t_K}$, *and* $E_K / UE_F W_K$ *is isomorphic to* $N_{K/F}(E_K)/E_F^m N_{K/F}(W_K)$, *whose order divides* $[E_F : E_F^m]$.

(4) $$m^{t_F} / [N_{K/F}(E_K) : E_F^m N_{K/F}(W_K)] = [E_F : N_{K/F}(E_K)]/[W_F : N_{K/F}(W_K)].$$

(5) $[E_F : N_{K/F}(E_K)] = 2^{t_F+1}\left[N_{K/F}(E_K) : E_F^2\right]^{-1}$ *if* $m = 2$.

(6) $N_{K/F}(W_K) = \{1\}$ *if* $m = 2$ *and* $r_F > 0$.

PROOF. If $xW_K = yW_K$ with $x \in U$ and $y \in E_F$, then $y^m = N_{K/F}(x^{-1}y) \in N_{K/F}(W_K) \subset W_F$, and hence $y \in W_F$, which proves (2). We easily see that the map $N_{K/F} : E_K \to E_F$ gives the second isomorphism of (3); the assertion concerning the order is obvious. Consequently $[E_K : UE_F W_K]$ is finite, and so we obtain the first part of (3), which combined with (2) proves (1), since $E_F W_K / W_K$ is isomorphic to $\mathbf{Z}^{t_F}$. Suppose $m = 2$; then the equality of (5) follows from (1.3). Suppose moreover that F has a real archimedean prime v. If v splits in K, then $W_K = \{\pm 1\}$; if v is ramified in K, then $N_{K/F}(a) > 0$ at v for every $a \in K^\times$. Therefore $N_{K/F}(W_K) = 1$ in either case. As for (4), we have, by (1.3),

$$\frac{m^{t_F}}{[N_{K/F}(E_K) : E_F^m N_{K/F}(W_K)]} = \frac{[E_F : N_{K/F}(E_K)][N_{K/F}(E_K) : E_F^m]}{[W_F : W_F^m][N_{K/F}(E_K) : E_F^m N_{K/F}(W_K)]}$$
$$= \frac{[E_F : N_{K/F}(E_K)][E_F^m N_{K/F}(W_K) : E_F^m]}{[W_F : W_F^m]}.$$

Since $N_{K/F}(W_K) \cap E_F^m = W_F^m$, we see that the second factor of the numerator equals $[N_{K/F}(W_K) : W_F^m]$. Therefore we obtain (4).

1.3. Theorem. *The notation being as in the above lemma, take a disjoint decomposition* $\mathbf{a}(K) = \mathbf{b} \sqcup \mathbf{e}$ *such that the restriction of the members of* $\mathbf{e}$ *to* F *gives a bijection of* $\mathbf{e}$ *onto* $\mathbf{a}(F)$. *Then* $\#(\mathbf{b})$ *equals* $t_K - t_F$, *that is, the integer* b *of Lemma 1.2 (1). Taking a set of independent generators* $\{\eta_i\}_{i=1}^b$ *of* U *modulo* $U \cap W_K$, *put*

$$R(K/F) = \left|\det\left(\log|\eta_i|_v\right)_{1 \le i \le b,\, v \in \mathbf{b}}\right|, \tag{1.4}$$

where $|\ |_v$ *is the normalized valuation of* K *at* v. *(We understand that* $R(K/F) = 1$ *if* $b = 0$.*) Then* $R(K/F)$ *is well-defined independently of the choice of* $\mathbf{b}$ *and* $\{\eta_i\}$, *and*

$$R_K/R_F = [W_F : N_{K/F}(W_K)]^{-1}[E_F : N_{K/F}(E_K)]R(K/F). \tag{1.5}$$

Moreover, define $I(K/F)$ *by (0.5). Then*

$$(\zeta_K/\zeta_F)(1) = 2^{t_K - t_F}\pi^{c_K - c_F}|D_F/D_K|^{1/2} I(K/F). \tag{1.6}$$

In particular if $m = 2$, *then* b *is the number of primes in* $\mathbf{a}(F)$ *unramified in* K. *Furthermore, if* $t_K = t_F + 1$ *and* $m = 2$, *then*

$$(1.7)\quad L(1,\lambda)=\frac{2\pi^{c_K-c_F}|D_F/D_K|^{1/2}[E_F:N_{K/F}(E_K)]\cdot(h_K/h_F)\cdot|\log|\eta|_v|}{[W_K:N_{K/F}(W_K)]},$$

where λ is the Hecke character of F corresponding to K/F, η is a generator of U modulo $U\cap W_K$, and $|\ |_v$ is the normalized archimedean valuation of K at a place unramified over F.

In (1.7) we may assume, changing η for η^{-1} if necessary, that $|\eta|_v>1$.

PROOF. Since $\#(\mathbf{a}(K))=t_K+1$ and $\#(\mathbf{e})=t_F+1$, we have $\#(\mathbf{b})=t_K-t_F=b$. For simplicity put $t=t_K$ and $q=t_F$. Given any set of t elements $\{\xi_i\}_{i=1}^t$ of E_K, we put

$$(*)\quad R[\xi_1,\ldots,\xi_t]=\frac{1}{t+1}\left|\det\begin{bmatrix}\log|\xi_1|_{w_1} & \cdots & \log|\xi_t|_{w_1} & 1\\ \cdots & \cdots & \cdots & \cdots\\ \log|\xi_1|_{w_{t+1}} & \cdots & \log|\xi_t|_{w_{t+1}} & 1\end{bmatrix}\right|,$$

where $\{w_i\}_{i=1}^{t+1}=\mathbf{a}(K)$. This is R_K if the ξ_i generate E_K modulo W_K. Let $\{\varepsilon_i\}_{i=1}^q$ be a set of independent generators of E_F modulo W_F. Then by (2) and (3) of Lemma 1.2, $\{\eta_i\}_{i=1}^b\cup\{\varepsilon_i\}_{i=1}^q$ is a set of independent generators of UE_FW_K modulo W_K. For each fixed $v\in\mathbf{a}(F)$ denote by c_v the number of elements of $\mathbf{a}(K)$ lying above v; take $u_1,\ldots,u_a\in\mathbf{b}$ and $w\in\mathbf{e}$ so that $\{u_1,\ldots,u_a,w\}$ is exactly the set of such elements of $\mathbf{a}(K)$; thus $a=c_v-1$. Let us first assume that $b>0$. Then we consider the matrix on the right-hand side appearing in the expression (*) for $R[\eta_1,\ldots,\eta_b,\varepsilon_1,\ldots,\varepsilon_q]$, and look at the u_i-th rows and the w-th row, which form $(c_v\times(t+1))$-matrix. For every $\eta\in K$ we have $|\eta|_w\prod_{i=1}^a|\eta|_{u_i}=\left|N_{K/F}(\eta)\right|_v$, which is 1 if $\eta\in U$. Therefore, adding the u_i-th row of the matrix for all i to the w-th row, (which does not change the determinant), we obtain, for each fixed v, a matrix of the form

$$\begin{bmatrix}\log|\eta_1|_{u_1} & \cdots & \log|\eta_b|_{u_1} & \log|\varepsilon_1|_{u_1} & \cdots & \log|\varepsilon_q|_{u_1} & 1\\ \cdots & \cdots & \cdots & \cdots & \cdots & \cdots & \cdots\\ \log|\eta_1|_{u_a} & \cdots & \log|\eta_b|_{u_a} & \log|\varepsilon_1|_{u_a} & \cdots & \log|\varepsilon_q|_{u_a} & 1\\ 0 & \cdots & 0 & m\log|\varepsilon_1|_v & \cdots & m\log|\varepsilon_q|_v & c_v\end{bmatrix}.$$

(If $m=2$, v is real, and w is complex, then $a=0$, and the last matrix consists of the last single row.) Combining such matrices for all $v\in\mathbf{a}(F)$, we find that

$$\begin{aligned}&(t+1)R[\eta_1,\ldots,\eta_b,\varepsilon_1,\ldots,\varepsilon_q]\\ &\quad=\pm R(K/F)\cdot m^q\cdot\det\begin{bmatrix}\log|\varepsilon_1|_v & \cdots & \log|\varepsilon_q|_v & c_v\\ \cdots & \cdots & \cdots & \cdots\\ \log|\varepsilon_1|_{v'} & \cdots & \log|\varepsilon_q|_{v'} & c_{v'}\end{bmatrix},\end{aligned}$$

where $\{v,\ldots,v'\}=\mathbf{a}(F)$. Adding the first q rows to the last row, we find that

$$(t+1)R[\eta_1,\ldots,\eta_b,\varepsilon_1,\ldots,\varepsilon_q]=(t+1)m^qR(K/F)R_F.$$

By (3) of Lemma 1.2, the left-hand side is $[N_{K/F}(E_K):E_F^mN_{K/F}(W_K)]$ times $(t+1)R_K$, and hence we obtain

$$(1.8)\qquad R_K/R_F=m^{t_F}[N_{K/F}(E_K):E_F^mN_{K/F}(W_K)]^{-1}R(K/F)$$

when $b>0$. Suppose $b=0$. Then $m=2$, $c_F=0$, $r_K=0$, and $U=W_K$, and hence, ignoring the η_i in the above, we easily find that (1.8) is valid even when

$b = 0$, with the convention that $R(K/F) = 1$. Then (1.5) follows from (1.8) and Lemma 1.2 (4). Combining (1.5) with (0.2), we obtain (1.6). The next assertion concerning b when $m = 2$ can easily be verified. The last formula (1.7) follows immediately from (1.6).

To simplify our formulas, we hereafter put

$$c(K/F) = c_K - c_F, \qquad t(K/F) = t_K - t_F, \tag{1.9}$$

$$I(F) = h_F R_F / [W_F : 1]. \tag{1.10}$$

From (1.5) and (0.5) we obtain

$$I(K/F) = I(K)/I(F). \tag{1.11}$$

1.4. Formula (1.7) concerns the case where $t_K = t_F + 1$ and $m = 2$, which can happen exactly for the following two types of quadratic extensions K/F :

(I) $c_F = 0$ and one and exactly one archimedean prime of F splits in K.

(II) $c_F = 1$ and all real archimedean primes of F are ramified in K.

If the extension K/F is of type (I), then $W_K = W_F = \{\pm 1\}$ and $N_{K/F}(W_K) = \{1\}$, and hence (1.7) takes the form

$$L(1, \lambda) = \pi^{g-1} |D_F/D_K|^{1/2} [E_F : N_{K/F}(E_K)] \cdot (h_K/h_F) \cdot \big| \log |\eta|_v \big|, \tag{1.12}$$

where $g = [F : \mathbf{Q}]$. A formula essentially the same as (1.12) was proved by Stark in [7], Theorem 2, for the extensions of type (I), but there is a subtle difference between his formula and ours, since the factor $[E_F : N_{K/F}(E_K)] |\log |\eta|_v|$ was given there in the form $(2^{g-1}/u) \log \varepsilon$ with a unit ε in K and $u = 1$ or 2. His proof shows that ε and ± 1 generate the subgroup $\{\pm N_{K/F}(e)/e^2 \mid e \in E_K\}$ of U. Thus the point of (1.7) or (1.12) is that η generates $U/\{\pm 1\}$, the fact we already emphasized in the introduction.

1.5. Lemma. (1) *Given algebraic number fields F_i for $1 \le i \le n$, suppose $\prod_{i=1}^n \zeta_{F_i}(s)^{a_i} = 1$ with $a_i \in \mathbf{Z}$. Then*

$$\sum_{i=1}^n a_i = 0, \quad \sum_{i=1}^n a_i r_{F_i} = 0, \quad \sum_{i=1}^n a_i c_{F_i} = 0, \quad \sum_{i=1}^n a_i t_{F_i} = 0,$$

$$\sum_{i=1}^n a_i [F_i : \mathbf{Q}] = 0, \quad \prod_{i=1}^n |D_{F_i}|^{a_i} = 1, \quad \prod_{i=1}^n I(F_i)^{a_i} = 1.$$

(2) *Given algebraic extensions K_i/F_i of algebraic number fields for $1 \le i \le n$, suppose $\prod_{i=1}^n (\zeta_{K_i}/\zeta_{F_i})^{b_i} = 1$ with $b_i \in \mathbf{Z}$. Then*

$$\sum_{i=1}^n b_i c(K_i/F_i) = 0, \quad \sum_{i=1}^n b_i t(K_i/F_i) = 0,$$

$$\prod_{i=1}^n |D_{K_i}/D_{F_i}|^{b_i} = 1, \quad \prod_{i=1}^n I(K_i/F_i)^{b_i} = 1.$$

PROOF. Though (1) is already proven in Brauer [1], we give an easier proof here for the reader's convenience. First, looking at the order of the pole of $\prod_{i=1}^n \zeta_{F_i}(s)^{a_i}$ at $s = 1$, we obtain $\sum_{i=1}^n a_i = 0$. Now put

$$\Gamma_F(s) = |D_F|^{s/2}\pi^{-s[F:\mathbf{Q}]/2}\Gamma(s/2)^{r_F+c_F}\Gamma((s+1)/2)^{c_F}$$

and $Z_F(s) = \Gamma_F(s)\zeta_F(s)$. Then $Z_F(1-s) = Z_F(s)$, and we easily see that ζ_F has a zero of order t_F at $s = 0$ and a zero of order c_F at $s = -1$. Therefore we obtain the next four equalities of (1). Put $q(s) = \prod_{i=1}^{n} Z_{F_i}(s)^{a_i}$. Then $q(s) = A^{s/2}$ with $A = \prod_{i=1}^{n} |D_{F_i}|^{a_i}$. Since $q(1-s) = q(s)$, we obtain $A = 1$. Finally, looking at the value of $\prod_{i=1}^{n} \zeta_{F_i}(s)^{a_i}$ at $s = 1$, from (0.2) we obtain the last equality of (1). Assertion (2) follows immediately from (1) except for the last equality concerning $I(K/F)$, which follows from (1.6) or (1.11).

We note here two basic formulas:

(1.13a) $$-I(F) = \lim_{s\to 0} s^{-t_F}\zeta_F(s),$$

(1.13b) $$I(K/F) = \lim_{s\to 0} s^{-t(K/F)}(\zeta_K/\zeta_F)(s).$$

The first formula follows immediately from (0.2) and the functional equation for ζ_F. Then the second one follows from this and (1.11). Thus $I(K/F)$ is a quantity completely determined by the function ζ_K/ζ_F. This is trivial if we define $I(K/F)$ by (1.11), but not so with (0.5) as its definition. We remind the reader that the essential factor $R(K/F)$ of $I(K/F)$ is defined as the determinant of a matrix of size $t(K/F)$.

The following lemma is essentially included in the above proof, but given here for the sake of later references.

1.6. Lemma. *For K/F as above, the quantities $t(K/F)$, $c(K/F)$, $|D_K/D_F|$, and $I(K/F)$ are completely determined by the function ζ_K/ζ_F. Consequently, expression (1.6) for the value $(\zeta_K/\zeta_F)(1)$ is canonical in the sense that it depends only on the function ζ_K/ζ_F, and is independent of the choice of K/F.*

PROOF. The proof of Lemma 1.5 shows that ζ_K/ζ_F has a zero of order $t(K/F)$ at $s = 0$ and a zero of order $c(K/F)$ at $s = -1$. Put $d = |D_K/D_F|^{1/2}$ and $Z'(s) = d^{-s}(Z_K/Z_F)(s)$. Since $[K:\mathbf{Q}]-[F:\mathbf{Q}] = t(K/F)+c(K/F)$, we see that $d^{-s}(\Gamma_K/\Gamma_F)(s)$ is completely determined by ζ_K/ζ_F, and hence the same applies to $Z'(s)$. Now $d^{1-s}Z'(1-s) = d^sZ'(s)$, so that d is completely determined by ζ_K/ζ_F.

In the following sections we shall show that $\zeta_K/\zeta_F = \zeta_{K'}/\zeta_{F'}$ can often happen for nonisomorphic K/F and K'/F'.

2. Some identities between regulators and class numbers

2.1. Throughout this and the next sections, B denotes a (basic) algebraic number field, and M a Galois extension of B; we put $G = \mathrm{Gal}(M/B)$. In this section we assume that G is a dihedral group of order $4d$ with an integer $d > 1$, generated by two elements σ and τ such that $\tau^{2d} = \sigma^2 = \tau\sigma\tau\sigma = 1$. For $\nu \in \mathbf{Z}$ put $H_\nu = \{1, \sigma\tau^\nu\}$ and $H'_\nu = H_\nu \cup H_\nu\tau^d$. Let F_ν and K_ν be the subfields of M corresponding to H'_ν and H_ν, respectively; also let P be the subfield of

M corresponding to $\{1, \tau^d\}$. We take these extensions K_ν/F_ν to be K/F of Theorem 1.3, and so $m = 2$. Notice that $[F_\nu : B] = d$.

2.2. Lemma. *In the above setting let χ_ν be the character of G induced from the character ψ_ν of H'_ν such that $\psi_\nu(H_\nu) = 1$ and $\psi_\nu(\tau^d) = -1$. Then the following assertions hold:*

(1) *There exists an element α of G such that $\alpha H_\nu \alpha^{-1} = H_\mu$ if and only if $\nu - \mu \in 2\mathbf{Z}$, in which case $\alpha H'_\nu \alpha^{-1} = H'_\mu$.*

(2) *$\chi_\nu = \chi_\mu$ if $\nu - \mu \in 2\mathbf{Z}$; moreover, $\chi_0 = \chi_1$ if d is even, and $\chi_0 \neq \chi_1$ if d is odd.*

(3) *$\chi_0 + \chi_1$ coincides with the character of G induced from the nontrivial character of $\{1, \tau^d\}$.*

(4) *If d is even, then $W_{K_\nu} = W_{F_\nu}$ for every ν and $W_M = W_P$.*

PROOF. The first three assertions can be verified in a straightforward way. Let us just note here that $\chi_\nu(\sigma\tau^\mu) = 0$ for every ν and μ if d is even; $\chi_\nu(\sigma\tau^\mu) = (-1)^{\mu-\nu}$ if d is odd. Assertion (4) is obvious if $W_M = \{\pm 1\}$. Suppose W_M contains an element γ other than ± 1; let J be the subgroup of G corresponding to $B(\gamma)$. Then J is normal in G, and G/J, being isomorphic to $\mathrm{Gal}(B(\gamma)/B)$, is abelian. Therefore $\tau^2 = \sigma^{-1}\tau^{-1}\sigma\tau \in J$, and hence $[G : J]|4$. Suppose d is even; then $\tau^d \in J$, so that $\gamma \in P$. Thus $W_M = W_P$. If $\gamma \in K_\nu$, then $\gamma \in K_\nu \cap P = F_\nu$, and hence $W_{K_\nu} = W_{F_\nu}$. This proves (4).

2.3. Theorem. (1) *If d is even, then the following quantities do not depend on ν:*

$$\zeta_{K_\nu}/\zeta_{F_\nu}, \quad |D_{K_\nu}/D_{F_\nu}|, \quad c(K_\nu/F_\nu), \quad t(K_\nu/F_\nu),$$
$$\big[E_{F_\nu} : N_{K_\nu/F_\nu}(E_{K_\nu})\big] \cdot (h_{K_\nu}/h_{F_\nu}) \cdot R(K_\nu/F_\nu).$$

However, t_{F_1} is not necessarily equal to t_{F_0}.

(2) *For both even and odd d we have*

$$|D_M/D_P| = |D_{K_0}/D_{F_0}| \cdot |D_{K_1}/D_{F_1}| \quad \text{and} \quad I(M/P) = I(K_0/F_0)I(K_1/F_1).$$

PROOF. With χ_ν and ψ_ν as in Lemma 2.2, the Artin L-function $L(s, M/B, \chi_\nu)$ equals $L(s, M/F_\nu, \psi_\nu)$, which equals the Hecke L-function $L(s, \lambda_\nu)$ of the Hecke character λ_ν of F_ν associated with the quadratic extension K_ν/F_ν. By Lemma 2.2 (2) we have $\chi_\nu = \chi_0$, so that $\zeta_{K_\nu}/\zeta_{F_\nu} = L(s, \lambda_\nu) = L(s, \lambda_0) = \zeta_{K_0}/\zeta_{F_0}$. We have $W_{K_\nu} = W_{F_\nu}$ by Lemma 2.2 (4), and hence

$$2I(K_\nu/F_\nu) = \big[E_{F_\nu} : N_{K_\nu/F_\nu}(E_{K_\nu})\big] \cdot (h_{K_\nu}/h_{F_\nu}) \cdot R(K_\nu/F_\nu). \tag{2.1}$$

Therefore, applying Lemma 1.5 (2) to the present case, we find that the quantities of (1) are independent of ν. The last fact about t_F will be discussed in §2.5.

To prove (2), let Λ be the Hecke character of P such that $\zeta_M(s)/\zeta_P(s) = L(s, \Lambda)$. Then, by (3) of Lemma 2.2, the Artin L-function $L(s, M/B, \chi_0 + \chi_1)$ equals $L(s, \Lambda)$, so that $L(s, \Lambda) = L(s, \lambda_0)L(s, \lambda_1) = (\zeta_{K_0}/\zeta_{F_0})(\zeta_{K_1}/\zeta_{F_1})$. Therefore we obtain the equalities of (2) by Lemma 1.5 (2).

2.4. An essential point of Theorem 2.3 is that K_1/F_1 cannot be obtained from K_0/F_0 by an isomorphism, if d is even and $B = \mathbf{Q}$. (The matter is not so clear-cut if $B \neq \mathbf{Q}$. That $\zeta_{K_1}/\zeta_{F_1} = \zeta_{K_0}/\zeta_{F_0}$ can happen for nonisomorphic K_1/F_1 and K_0/F_0 was already noted in [3], p. 84.) In order to gain a perspective, let us put, using the notation of (1.1),

$$\text{(2.2)} \qquad \mathcal{G}_\nu = \mathcal{G}(K_\nu/F_\nu) \quad\text{and}\quad U_\nu = U(K_\nu/F_\nu).$$

Theorem 2.3 shows that if d is even, then $R(K_0/F_0)/R(K_1/F_1)$ is a rational number which can be explicitly given in terms of class numbers and the indices of certain unit groups. In the simplest case we can have $t_{K_0} - t_{F_0} = t_{K_1} - t_{F_1} = 1$. Let η_ν be a generator of U_ν modulo $U_\nu \cap W_{K_\nu}$ in such a case. Then we have

$$\text{(2.3)} \qquad |\eta_0|^{e_0} = |\eta_1|^{e_1} \text{ with } e_\nu = \left[E_{F_\nu} : N_{K_\nu/F_\nu}(E_{K_\nu})\right] \cdot h_{K_\nu}/h_{F_\nu}.$$

Here we have to choose η_ν and a normalized archimedean valuation $|\ |$ of K_ν unramified over F_ν so that $|\eta_\nu| > 1$. This equality is reminiscent of the well-known fact about cyclotomic units, though the nature is different. If the prime taken for η_0 is real, then $|\eta_0|$ belongs to a conjugate of K_0. If the prime is complex, however, $|\eta_0|$ may not belong to any conjugate of K_0.

Let us now show that there is an "isogeny" of $\mathcal{G}_0$ into $\mathcal{G}_1$ if d is even. Put $f(x) = xx^\tau$ for $x \in \mathcal{G}_0$. Since $x^\sigma = x$ for $x \in K_0$, we easily see that $f(x)^{\sigma\tau} = f(x)$, and hence $f(x) \in K_1$. Put $\rho = \tau^d$. Then $N_{K_\nu/F_\nu}(y) = yy^\rho$ for $y \in K_\nu$, and $f(x)f(x)^\rho = 1$ if $xx^\rho = 1$. Thus f gives a homomorphism of $\mathcal{G}_0$ into $\mathcal{G}_1$. If $x \in \mathcal{G}_0$ and $f(x) = 1$, then $x^\tau = x^\rho$, so that $x^{\tau^{d-1}} = 1$. Since $\tau^{2d} = 1$ and $d-1$ is odd, we have $x^\tau = x$, so that $x = \pm 1$. Thus f gives an injection of $\mathcal{G}_0/\{\pm 1\}$ into $\mathcal{G}_1$. Clearly $f(U_0) \subset U_1$. Since we know that U_0 and U_1 have the same rank, f maps $U_0/\{\pm 1\}$ isomorphically onto a subgroup of U_1 of finite index. This explains at least the commensurability of $R(K_0/F_0)$ with $R(K_1/F_1)$.

As for $I(M/P)$, we have clearly $U_0 U_1 \subset U(M/P)$. Since $K_0 \cap K_1 = B$, we see that $U_0 \cap U_1 = \{\pm 1\}$. Since $\zeta_M/\zeta_P = (\zeta_{K_0}/\zeta_{F_0})(\zeta_{K_1}/\zeta_{F_1})$, Lemma 1.5 (2) shows that $U(M/P)$ has the same rank as $U_0 \times U_1$. This partly explains why $R(M/P)$ is related to $R(K_0/F_0)$ and $R(K_1/F_1)$.

2.5. *Examples and counterexamples.* To show that the inequality $t_{F_1} \neq t_{F_0}$ can happen, take F_0 to be a real quadratic field and take $K_0 = F_0(\xi)$, $\xi^2 = \alpha$, with $\alpha \in F_0$ such that $N_{F_0/\mathbf{Q}}(\alpha) < 0$. Let M be the Galois closure of K_0 over $\mathbf{Q}$. Then $M = F_0(\xi, \xi')$ with an element $\xi'^2 = N_{F_0/\mathbf{Q}}(\alpha)/\alpha$, and the Galois group is a dihedral group of order 8. In this case we can define $\sigma, \tau \in G$ by $(\xi, \xi')^\sigma = (\xi, -\xi')$ and $(\xi, \xi')^\tau = (\xi', -\xi)$. Thus $\xi' = \xi^\tau$. Put $\beta = \xi\xi^\tau$ and $\omega = \xi + \xi^\tau$. Then $\beta^2 = N_{F_0/\mathbf{Q}}(\alpha) < 0$, $\beta^{\sigma\tau} = \beta$, $\omega^{\sigma\tau} = \omega$, and hence $F_1 = \mathbf{Q}(\beta)$ and $K_1 = F_1(\omega)$. Thus F_1 is imaginary, $t_{F_1} = 0$, $t_{K_1} = 1$, $t_{F_0} = 1$, and $t_{K_0} = 2$.

3. The case $[K : F] > 2$

3.1. Let us now present some identities similar to those of Section 2 in the case $[K : F] = m$ with an odd $m > 1$. We take six integers p, m, d, e, r, and ν satisfying the following conditions:

(3.1a) m *is an odd number* > 1, $d > 0$, *and* $e > 1$;

(3.1b) $r \pmod{m}$ *is an element of order* e *in* $(\mathbf{Z}/m\mathbf{Z})^\times$ *and* $r^e - 1 \in dm\mathbf{Z}$;

(3.1c) m *is a power of a prime* p *if* $r \neq -1$;

(3.1d) $\nu \sum_{i=0}^{e-1} r^i \in dm\mathbf{Z}$.

Let C be the subgroup of $(\mathbf{Z}/dm\mathbf{Z})^\times$ generated by $r \pmod{dm}$. Then we put

$$\text{(3.2a)} \qquad G = \left\{ \begin{bmatrix} x & y \\ 0 & 1 \end{bmatrix} \middle| \; x \in C,\ y \in \mathbf{Z}/dm\mathbf{Z} \right\},$$

$$\text{(3.2b)} \quad \tau = \begin{bmatrix} 1 & 1 \\ 0 & 1 \end{bmatrix}, \quad \delta = \tau^d, \quad \sigma = \begin{bmatrix} r & 0 \\ 0 & 1 \end{bmatrix}, \quad \sigma_\nu = \tau^\nu \sigma \qquad (0 \le \nu < dm).$$

If $r = -1$, then G is a dihedral group of order $2dm$. Notice that $\sigma_\nu \tau \sigma_\nu^{-1} = \tau^r$. Let N, T, and H_ν denote the subgroups of G generated by δ, $\delta^{m/p}$, and σ_ν respectively; let $H'_\nu = H_\nu T$ and $J_\nu = H_\nu N$. (T and H'_ν are defined only when m is a power of p.) We easily see that N has order m and H_ν has order e in view of (3.1d). (Clearly (3.1d) is satisfied if $r = -1$ or $\nu = 0$.)

We take this group G to be $\mathrm{Gal}(M/B)$. Then we let P, S, F_ν, K_ν, and Z_ν denote the subfields of M corresponding to N, T, H'_ν, H_ν, and J_ν, respectively. (S and F_ν are defined only when m is a power of p.) Then $[M : S] = [K_\nu : F_\nu] = p$, $[M : P] = [K_\nu : Z_\nu] = m$, and $[P : Z_\nu] = [S : F_\nu] = [M : K_\nu] = e$.

3.2. Lemma. *In the above setting let* ψ *be a character of* N *of order* m *and for* $k \in \mathbf{Z}$ *let* χ_k *be the character of* J_ν *induced from the character* ψ^k *of* N. *Also, let* φ *be a character of* H_ν *of order* e, *and* φ' *the character of* H'_ν *which coincides with* φ *on* H_ν *and is trivial on* T; *for* $i \in \mathbf{Z}$ *let* ω_i *(resp.* ξ_i*) be the character of* J_ν *induced from the character* φ^i *of* H_ν *(resp.* $(\varphi')^i$ *of* H'_ν*). Then the following assertions hold:*

(1) *There exists an element* α *of* G *such that* $\alpha H_\nu \alpha^{-1} = H_\mu$ *if and only if* $\nu - \mu$ *is divisible by the greatest common divisor of* $r - 1$ *and* dm, *in which case* $\alpha H'_\nu \alpha^{-1} = H'_\mu$ *and* $\alpha J_\nu \alpha^{-1} = J_\mu$.

(2) *Suppose* m *is a prime power; let* A *be a complete set of representatives for* $(\mathbf{Z}/m\mathbf{Z})^\times$ *modulo the subgroup generated by* $r \pmod{m}$. *Then* $\omega_i = \xi_i + \sum_{a \in A} \chi_a$.

(3) *Suppose* $r = -1$; *let* η_i *be the character of* J_ν *which coincides with* φ^i *on* H_ν *and is trivial on* N. *Then* $\omega_i = \eta_i + \sum_{a=1}^{(m-1)/2} \chi_a$.

(4) *If* $r - 1$ *is prime to* m, *then* $W_M = W_S = W_P$ *and* $W_{K_\nu} = W_{F_\nu} = W_{Z_\nu}$ *for every* ν *satisfying (3.1d).*

(5) *Suppose that* m *is a power of an odd prime* p, $r \pmod{m}$ *generates* $(\mathbf{Z}/m\mathbf{Z})^\times$, *and* d *divides both* $r - 1$ *and* $p - 1$. *Then* $\sum_{i=0}^{e-1} r^i \in dm\mathbf{Z}$, *where* $e = m(1 - p^{-1})$, *and consequently (3.1b), as well as (3.1d) for every* $\nu \in \mathbf{Z}$, *is satisfied.*

PROOF. To prove (1), fix ν, and observe that every element α of G is of the form $\alpha = \tau^x \sigma_\nu^y$ with integers x and y. Then $\alpha \sigma_\nu \alpha^{-1} = \tau^{(1-r)x} \sigma_\nu$, which belongs to H_μ if and only if $(1 - r)x + \nu - \mu \in dm\mathbf{Z}$. This proves the first part of

(1). The remaining part of (1) is then obvious. As for (2), clearly χ_a is 0 outside N. Putting $\zeta = \psi(\delta)$, we have $\chi_a(\delta^b) = \sum_{x\in X}\zeta^{abx}$, where X is the subgroup of $(\mathbf{Z}/m\mathbf{Z})^\times$ generated by $r \pmod m$, and hence $\sum_{a\in A}\chi_a(\delta^b) = \sum_{c\in Y}\zeta^{bc}$, where $Y = (\mathbf{Z}/m\mathbf{Z})^\times$. Assuming that m is a power of p, put $n = m/p$. Then the last sum over Y equals $m-n$, 0, or $-n$ according as $m|b$, $n\nmid b$, or $p\nmid b/n \in \mathbf{Z}$, that is, $\delta^b = 1$, $\delta^b \notin T$, or $1 \neq \delta^b \in T$. Next, for $\gamma = \delta^b\sigma_\nu^a \in J_\nu$ with integers a and b, we have

$$\xi_i(\gamma) = \textstyle\sum_{x=0}^{n-1}(\varphi')^i(\delta^{-x}\gamma\delta^x) = \sum_x(\varphi')^i(\delta^{b+c_ax}\sigma_\nu^a),$$

where $c_a = r^a - 1$. For $a = 0$ the sum is nonzero only if $\gamma \in T$, in which case $\xi_i(\gamma) = n$. Suppose $0 < a < e$; put $m = p^s$ and $r^a - 1 = p^tq$ with $0 \le t \in \mathbf{Z}$ and an integer q prime to p. Then $t < s$ and the last sum is extended over x such that $0 \le x < n$ and $b + p^tqx \equiv 0 \pmod n$. Such an x exists if and only if $b \in p^t\mathbf{Z}$, and the number of such x's is $\#(p^{s-1-t}\mathbf{Z}/p^{s-1}\mathbf{Z})$, which is p^t. Thus, for $0 < a < e$ we have $\xi_i(\gamma) \neq 0$ only if $b \in p^t\mathbf{Z}$, in which case $\xi_i(\gamma) = p^t\varphi^i(\sigma_\nu^a)$. The same type of reasoning applies to ω_i. Namely, for $a = 0$ we have $\omega_i(\gamma) \neq 0$ only if $\gamma = 1$, in which case $\xi_i(\gamma) = m$. For $0 < a < e$ we have $\omega_i(\gamma) \neq 0$ only if $b \in p^t\mathbf{Z}$, in which case $\omega_i(\gamma) = \#(p^{s-t}\mathbf{Z}/p^s\mathbf{Z})\varphi^i(\sigma_\nu^a) = p^t\varphi^i(\sigma_\nu^a)$. Comparing $\omega_i - \xi_i$ with $\sum_{a\in A}\chi_a$, we obtain the desired equality of (2). Assertion (3) can be verified in a similar and easier way.

To prove (4), given $\varepsilon \in W_M$, let V be the subgroup of G corresponding to $B(\varepsilon)$. Then G/V, being isomorphic to $\mathrm{Gal}(B(\varepsilon)/B)$, is abelian, and hence $\delta^{r-1} = \sigma\delta\sigma^{-1}\delta^{-1} \in V$. If $r-1$ is prime to m, then $\delta \in V$, so that $\varepsilon \in P$. This shows that $W_M = W_S = W_P$. Since $Z_\nu = K_\nu \cap P$, we obtain $W_{K_\nu} = W_{F_\nu} = W_{Z_\nu}$. Finally, as for (5), suppose $d|(r-1)$ and $d|(p-1)$. Then $d|e$, since $p-1$ divides e. We have $\sum_{i=0}^{e-1} r^i \equiv e \pmod d$, and hence $d|\sum_{i=0}^{e-1} r^i$. Also, $(r-1)\sum_{i=0}^{e-1} r^i = r^e - 1 \in m\mathbf{Z}$, so that $\sum_{i=0}^{e-1} r^i \in m\mathbf{Z}$, as $r-1$ is prime to p. Now d, being a divisor of $r-1$, is prime to p, and hence $\sum_{i=0}^{e-1} r^i \in dm\mathbf{Z}$. This proves (5).

3.3. Theorem. *In the setting of §3.1 let μ be the Hecke character of P of order m corresponding to the character ψ viewed as a character of* Gal(M/P). *Then the following assertions hold.*

(1) *If m is a prime power, then $I(K_\nu/F_\nu)$ is independent of ν, $I(M/S) = I(K_\nu/F_\nu)^e$, and*

$$I(K_\nu/F_\nu) = \{I(M/K_\nu)/I(S/F_\nu)\}^{1/(e-1)}$$
$$= 2^{-t(K_\nu/F_\nu)}\pi^{-c(K_\nu/F_\nu)}|D_{K_\nu}/D_{F_\nu}|^{1/2}\prod_{a\in A} L(1, \mu^a)$$

for every ν, where A is as in Lemma 3.2 (2).

(2) *If $r = -1$, then $I(K_\nu/Z_\nu)$ is independent of ν, $I(M/P) = I(K_\nu/Z_\nu)^2$, and*

$$I(K_\nu/Z_\nu) = I(M/K_\nu)/I(P/Z_\nu)$$
$$= 2^{-t(K_\nu/Z_\nu)}\pi^{-c(K_\nu/Z_\nu)}|D_{K_\nu}/D_{Z_\nu}|^{1/2}\prod_{a=1}^{(m-1)/2} L(1, \mu^a)$$

PROOF. Let λ (resp. Λ) be the Hecke character of F_ν (resp. K_ν) corresponding to the character φ' (resp. φ) viewed as a character of $\mathrm{Gal}(S/F_\nu)$ (resp. $\mathrm{Gal}(M/K_\nu)$). From Lemma 3.2 (2) we obtain

$$\begin{aligned}(3.3)\qquad L(s,\Lambda^i) &= L(s, M/Z_\nu, \omega_i)\\ &= L(s, M/Z_\nu, \xi_i)\prod_{a\in A} L(s, M/Z_\nu, \chi_a) = L(s,\lambda^i)\prod_{a\in A} L(s,\mu^a).\end{aligned}$$

Taking $i = 0$, we obtain

$$(3.4)\qquad \zeta_{K_\nu}(s) = \zeta_{F_\nu}(s)\prod_{a\in A} L(s,\mu^a).$$

Taking also the product of (3.3) for $0 < i < e$ and also for $0 \le i < e$, we obtain

$$(3.5)\qquad (\zeta_M/\zeta_{K_\nu})(s) = (\zeta_S/\zeta_{F_\nu})(s)\prod_{a\in A} L(s,\mu^a)^{e-1},$$

$$(3.6)\qquad \zeta_M(s)/\zeta_S(s) = \prod_{a\in A} L(s,\mu^a)^e = \{\zeta_{K_\nu}(s)/\zeta_{F_\nu}(s)\}^e.$$

Therefore we obtain the equalities of (1) by applying Lemma 1.5 (2) to (3.4), (3.5), and (3.6), and employing (1.6).

The proof of (2) is similar. Let κ be the Hecke character of Z_ν corresponding to the quadratic extension P/Z_ν. Then the equality of Lemma 3.2 (3) shows that

$$(3.7)\qquad L(s,\Lambda^i) = L(s,\kappa^i)\prod_{a=1}^{(m-1)/2} L(s,\mu^a).$$

Then we have

$$\begin{aligned}(3.8)\qquad \zeta_{K_\nu}(s)/\zeta_{Z_\nu}(s) = \prod_{a=1}^{(m-1)/2} L(s,\mu^a) &= L(s,\Lambda)/L(s,\kappa)\\ &= (\zeta_M/\zeta_{K_\nu})/(\zeta_P/\zeta_{Z_\nu})\end{aligned}$$

and $\zeta_M/\zeta_P = \prod_{a=1}^{(m-1)/2} L(s,\mu^a)^2$. Therefore we obtain (3) by Lemma 1.5 (2).

3.4. Corollary. *In the setting of Theorem 3.3 (1) the following assertions hold:*

(1) *If $r \pmod m$ generates $(\mathbf{Z}/m\mathbf{Z})^\times$, then*

$$(3.9)\qquad L(1,\mu) = 2^{t(K_\nu/F_\nu)}\pi^{c(K_\nu/F_\nu)}\left|D_{F_\nu}/D_{K_\nu}\right|^{1/2} I(K_\nu/F_\nu).$$

(2) *Suppose that e is divisible by $p-1$; let μ' be the Hecke character of P corresponding to a character of $\mathrm{Gal}(M/P)$ of order p; also let X_ν (resp. Y_ν) be the subfield of M corresponding the subgroup of G generated by $\{\delta^p, \tau^\nu\sigma^q\}$ (resp. $\{\delta, \tau^\nu\sigma^q\}$), where $q = e/(p-1)$ and ν is an integer such that $\nu\sum_{i=0}^{p-2} r^{qi} \in dp\mathbf{Z}$. Then*

$$(3.10)\qquad L(1,\mu') = 2^{t(X_\nu/Y_\nu)}\pi^{c(X_\nu/Y_\nu)}\left|D_{Y_\nu}/D_{X_\nu}\right|^{1/2} I(X_\nu/Y_\nu).$$

PROOF. Assertion (1) follows immediately from Theorem 3.3 (1), since we can take $A = \{1\}$ under our assumption. To prove (2), Let D (resp. E) be the subgroup of G generated by $\{\tau, \sigma^q\}$ (resp. δ^p). Then E is a normal subgroup of D, and D/E is a group of type (3.2a) with $(p-1, d, p, r^q)$ in place of (e, d, m, r); X_ν and Y_ν take the place of K_ν and F_ν. Applying the result of (1) to this setting, we obtain (3.10). If $m = p$ in the setting of (1) and (2), then (3.9) is the same as (3.10).

3.5. Let us now add a few remarks. First of all, in the setting of Corollary 3.4, formulas (3.9) and (3.10) are canonical expressions for $L(1, \mu)$ and $L(1, \mu')$ in the sense of Lemma 1.6, though there are nonisomorphic K_ν/F_ν and also nonisomorphic X_ν/Y_ν. We can also state several equalities between the invariants $t(K/F), c(K/F)$, and $|D_K/D_F|$ for various K/F similar to those of Theorem 2.3, which we do not state here, since they follow directly from identities such as (3.8) combined with Lemma 1.5 (2).

We can of course treat G with a more general m and with some conditions weaker than (3.1a, b, c, d). We imposed those conditions simply in order to obtain clear-cut formulas without complicated calculations. Indeed, it is not an impossible task to consider the case where m is not a prime power, but the proof becomes much longer, and therefore we decided not to treat such a general case.

Let us next present some examples.

(i) First, we can take $B = \mathbf{Q}$ and $M = \mathbf{Q}(\beta, \varepsilon)$ with a primitive $2m$-th root of unity ε and β such that $\beta^{2m} \in \mathbf{Q}^\times$, where m is a power of an odd prime p. With a suitable choice of β^{2m}, (a square-free integer $\neq \pm 1$, for example), $\mathrm{Gal}(M/\mathbf{Q})$ is isomorphic to G of (3.2a) with $e = m(1-p^{-1})$ and $d = 2$, if we take r to be an odd integer which gives a primitive root modulo m. By Lemma 3.2 (5) all the necessary conditions are satisfied; σ and τ are given by $\varepsilon^\sigma = \varepsilon^g$, $\beta^\sigma = \beta$, $\varepsilon^\tau = \varepsilon$, $\beta^\tau = \varepsilon\beta$, where g is an integer such that $rg - 1 \in 2m\mathbf{Z}$. It is well-known that $\mathbf{Q}(\varepsilon)$ contains a square root γ of $\pm p$, where the sign is determined by $\pm p \equiv 1 \pmod 4$. Then we can easily verify that $K_0 = \mathbf{Q}(\beta)$, $F_0 = \mathbf{Q}(\beta^p)$, $K_m = \mathbf{Q}(\beta\gamma)$, $F_m = \mathbf{Q}(\beta^p\gamma^p)$; K_ν/F_ν is isomorphic to K_0/F_0 or K_m/F_m according as ν is even or odd; K_0/F_0 is not isomorphic to K_m/F_m. Now $P = \mathbf{Q}(\varepsilon, \beta^m)$, and formulas (3.9) and (3.10) are applicable to the characters of $\mathrm{Gal}(M/P)$ of order m and p, respectively. We leave the easy task of examining X_ν and Y_ν to the reader.

(ii) An easier example with $d = 1$ can be obtained by taking $B = \mathbf{Q}$ and $M = \mathbf{Q}(\varepsilon, \beta^2)$. In this case $P = \mathbf{Q}(\varepsilon)$ and the K_ν/F_ν are the conjugates of $\mathbf{Q}(\beta^2)/\mathbf{Q}(\beta^{2p})$; (3.9) and (3.10) are applicable.

(iii) To find an example of G with $d > 2$, take a power m of p as above and a primitive root r modulo m. We take these so that $p \equiv r \equiv 1 \pmod 6$. (Take $r = 7$ and $m = 13$, for example.) By Lemma 3.2 (5) we can take $d = 6$. Let $M = \mathbf{Q}(\beta, \varepsilon)$ with a primitive $6m$-th root of unity ε and β such that $\beta^{6m} \in \mathbf{Q}^\times$. With a suitable choice of β^{2m}, $\mathrm{Gal}(M/\mathbf{Q})$ is isomorphic to G of (3.2a) with $C = (\mathbf{Z}/6m\mathbf{Z})^\times$ and $d = 6$. Let $B = \mathbf{Q}(\sqrt{-3})$. Then $\mathrm{Gal}(M/B)$ is isomorphic to G of (3.2a) with $e = m(1 - p^{-1})$ and $d = 6$; σ and τ can be chosen in the same manner as in example (i); this time we assume $gr - 1 \in 6m\mathbf{Z}$. Since $g - 1 \in 6\mathbf{Z}$, we see that $(\varepsilon^m)^\sigma = \varepsilon^m$, that is, σ is indeed the identity map on B. We have $P = \mathbf{Q}(\varepsilon, \beta^m)$, $K_0 = B(\beta)$, and $F_0 = B(\beta^p)$.

By Lemma 3.2 (1), K_ν/F_ν is an isomorphic image of K_μ/F_μ over B if and only if $\mu - \nu \in 6\mathbf{Z}$. Thus there exist 6 nonisomorphic classes of K_ν/F_ν under Gal(M/B), but $I(K_\nu/F_\nu)$ is independent of ν. Now Gal$(M/\mathbf{Q})$ is generated by Gal(M/B) and an element ρ such that $\varepsilon^\rho = \varepsilon^{-1}$ and $\beta^\rho = \beta$. We have $\rho H_\nu \rho^{-1} = H_{-\nu}$ and $\rho H'_\nu \rho^{-1} = H'_{-\nu}$, and therefore those 6 classes can be reduced to 4 nonisomorphic classes over $\mathbf{Q}$, represented by K_ν/F_ν for $0 \le \nu \le 3$. The condition $\nu \sum_{i=0}^{p-2} r^{qi} \in 6p\mathbf{Z}$ of Corollary 3.4 (2) is satisfied for every $\nu \in \mathbf{Z}$, and nonisomorphic classes of the extensions X_ν/Y_ν over $\mathbf{Q}$ can be represented exactly by those for $0 \le \nu \le 3$. Formulas (3.9) and (3.10) are applicable.

3.6. Now there are some previous papers on the regulators of number fields contained in a Galois extension (Brauer [1] and Stark [7]). The starting point of [1] is a relation of the form

$$\prod_A \zeta_A(s)^{n_A} = 1, \tag{3.11}$$

where A runs over the subfields of a Galois extension M of $\mathbf{Q}$ and $n_A \in \mathbf{Z}$, which was already mentioned in Lemma 1.5. Then, as stated there, we have $\prod_A h_A^{n_A} = \prod_A [W_A : 1]^{n_A} \prod_A R_A^{-n_A}$. The main purpose of [1] is to find an expression for $\prod_A h_A^{n_A}$ that does not involve the regulators. This can be achieved by expressing each R_A in terms of the Minkowski units of M. On the other hand, in [7], a certain invariant $R(\chi)$ is associated to each rational character χ of Gal(M/B) for a Galois extension M of B. Then it is shown that $L(1, M/B, \chi)$ is $R(\chi)$ times elementary factors in a manner similar to (1.6) or (1.12). The proof employs a relation $L(s, K/B, \chi) = \prod_A \zeta_A(s)^{m_A}$ with $m_A \in \mathbf{Q}$, and the expression for R_A of Brauer's type, or its generalization. See also §1.4.

Those investigations are natural, and the relation $\prod_{i=1}^n (\zeta_{K_i}/\zeta_{F_i})^{b_i} = 1$ of Lemma 1.5 (2) we have been discussing is equivalent to (3.11). However, in the present paper we have taken a different viewpoint. Namely, we put the group $\mathcal{G}(K/F)$ on the center stage, and we are interested in the algebraic relations among $R(K/F)$, or rather, the quantities $I(K/F)$ for various different K/F. It seems that a relation of type (2.3) has not been mentioned before, and we are presenting it in a new framework. Also, $R(K/F)$ has a smaller size than $R(K)$, and therefore a relation between various $I(K/F)$ involves "fewer units" than a relation between $h_K R_K$ in general. Of course we cannot expect that relations of types (3.9) and (3.10) hold in general, since our method requires the L-function in question to be of the form ζ_K/ζ_F. Still, we believe that the expression for $L(1, \mu)$ in terms of $R(K/F)$ is a natural thing to do, when that can be done.

There is another reason why $R(K/F)$ is a natural object of study, as will be explained in Section 5.

We end this section by one more historical remark. The equality $\zeta_K(s) = \zeta_F(s)L(s, \mu)$ of (3.4) when $m = 3$ was first discovered by Dedekind. In fact, he showed it when $F = \mathbf{Q}$, $K = \mathbf{Q}(\alpha^{1/3})$ with $\alpha \in \mathbf{Q}^\times$, and $P = \mathbf{Q}(\sqrt{-3})$; see [2], p. 66, line 11. He also mentioned that as a consequence of the equality, $L(1, \mu)$ can be obtained in terms of $\log \varepsilon$ with a generator ε of $E_K/\{\pm 1\}$, and gave some numerical examples of ε; see [2], p. 114, line 6 from the bottom and p. 122. Thus these formulas are at least one hundred years old and actually older, since he said in the introduction that the work had been done several years earlier.

4. The case of CM-fields

4.1. In the general setting of Section 1, let us now assume that $m = 2$, F is totally real, and K is totally imaginary, which is the case if and only if $t_F = t_K$. In such a case we call K a *CM-field*. The generator ρ of $\mathrm{Gal}(K/F)$ has the property that for every field-injection β of K into $\mathbf{C}$ and every $z \in K$, $z^{\rho\beta}$ is the complex conjugate of z^β (see Lemma 18.2 of [4]). Therefore we call ρ *the complex conjugation in K* and F *the maximal real subfield of K*. Thus F is uniquely determined by K; also $U = W_K$, $W_F = \{\pm 1\}$, and $R(K/F) = 1$. We have $[E_K : E_F] = [E_K : W_K E_F][W_K E_F : E_F]$, $E_K/W_K E_F$ is isomorphic to $N_{K/F}(E_K)/E_F^2$, and $W_K E_F/E_F$ is isomorphic to W_K/W_F. Therefore, putting $g = [F : \mathbf{Q}]$, we have

$$\frac{2^g}{[E_K : E_F]} = \frac{[E_F : E_F^2]}{[N_{K/F}(E_K) : E_F^2][W_K : W_F]} = \frac{2[E_F : N_{K/F}(E_K)]}{[W_K : N_{K/F}(W_K)]}, \tag{4.1}$$

and hence

$$I(K/F) = 2^{g-1}[E_K : E_F]^{-1} \cdot h_K/h_F. \tag{4.2}$$

Thus (1.6) for a CM-field K can be written

$$L(1, \lambda) = 2^{g-1}\pi^g [E_K : E_F]^{-1} |D_F/D_K|^{1/2} \cdot h_K/h_F. \tag{4.3}$$

We now consider the case where $\mathrm{Gal}(M/B)$ is a dihedral group of order $4d$ as in §2.1.

4.2. Theorem. *In the setting of §2.1 suppose that K_0 is a CM-field. Then M and K_ν for every ν are CM-fields, P and F_ν are their maximal real subfields, respectively, and (4.2) and (4.3) are applicable to K_ν/F_ν and M/P. Also, the conclusions of Theorem 2.3 are of course valid with $R(M/P) = R(K_\nu/F_\nu) = 1$.*

Proof. By Lemma 18.2 of [4], the Galois closure of K_0 over $\mathbf{Q}$, say X, is a CM-field and a subfield of X is either a CM-field or totally real. Thus M is a CM-field. Let ρ be the element of G that gives the complex conjugation in M. Now the complex conjugation in X is an element of order 2 belonging to the center of $\mathrm{Gal}(X/\mathbf{Q})$ (as remarked after that lemma), and therefore ρ is an element of order 2 of the center of G. Therefore ρ must coincide with τ^d. Thus we obtain the first assertion concerning M, K_ν, P, and F_ν. The remaining statements of our theorem are obvious.

4.3. The notation being as in Theorem 4.2, suppose that $B = \mathbf{Q}$; put $K = K_0$ and let Φ be a CM-type of K in the sense of [5]. Clearly Φ can be given in the form $\Phi = \sum_{a \in A} \tau^a$ with a subset A of $\mathbf{Z}/2d\mathbf{Z}$ such that $\mathbf{Z}/2d\mathbf{Z}$ is the disjoint union of A and $\{a + d \,|\, a \in A\}$. Let (K', Φ') be the reflex of (K, Φ) in the sense of [5]. Since the discussion of (K', Φ') in general is not so simple, let us treat here just one case by taking $\Phi = \sum_{a=0}^{d-1} \tau^a$. Then we can easily verify that $K' = K_{d-1}$ and $\Phi' = \sum_{a=0}^{d-1} \tau^{-a}$. In this case Theorem 4.2 says that

$$[E_K : E_F]^{-1} \cdot h_K/h_F = [E_{K'} : E_{F'}]^{-1} \cdot h_{K'}/h_{F'}, \tag{4.4a}$$

$$|D_K/D_F| = |D_{K'}/D_{F'}|. \tag{4.4b}$$

This was already noted in [3], Proposition A.7. In general the degree of the reflex field may be different from the degree of the original CM-field, and the relation between $I(K/F)$ and $I(K'/F')$ is not so simple as (4.3).

5. Relation with an exact mass formula

5.1. In our recent paper [6] we proved an exact mass formula for orthogonal groups. Let us now show that the formula is closely connected with $R(K/F)$.

Let V be an n-dimensional vector space over an algebraic number field F, and φ a symmetric F-bilinear form $V \times V \to F$. We then put

$$\varphi[x] = \varphi(x, x) \qquad (x \in V), \tag{5.1}$$

$$G = \{\, \alpha \in GL(V) \,|\, \varphi[x\alpha] = \varphi[x] \,\}, \tag{5.2}$$

$$G^+ = \{\, \alpha \in G \,|\, \det(\alpha) = 1 \,\}. \tag{5.3}$$

Hereafter we assume

(5.4) F is totally real and n is even.

Let $\mathfrak{g}$ be the maximal order of F, and $\mathbf{h}$ the set of all nonarchimedean primes of F. Define the adelization $G^+_{\mathbf{A}}$ and its archimedean and nonarchimedean factors $G^+_{\mathbf{a}}$ and $G^+_{\mathbf{h}}$ as usual. Take an open subgroup D of $G^+_{\mathbf{A}}$ containing $G^+_{\mathbf{a}}$ and such that $D \cap G^+_{\mathbf{h}}$ is compact. For each $x \in G^+_{\mathbf{A}}$ put $\Gamma^x = G^+ \cap xDx^{-1}$. Take $\mathcal{B} \subset G^+_{\mathbf{A}}$ so that $G^+_{\mathbf{A}} = \bigsqcup_{a \in \mathcal{B}} G^+ aD$. We then define real numbers $\mathfrak{m}(G^+, D)$ and $\nu(\Gamma^a)$ (cf. [4, §24.1] or [6, (5.5.1)) by

$$\mathfrak{m}(G^+, D) = \sum_{a \in \mathcal{B}} \nu(\Gamma^a), \tag{5.5}$$

$$\nu(\Gamma^a) = \begin{cases} [\Gamma^a : 1]^{-1} & \text{if } G^+_{\mathbf{a}} \text{ is compact}, \\ [\Gamma^a \cap \{\pm 1\} : 1]^{-1} \mathrm{vol}(\Gamma^a \backslash \mathcal{Z}^{\varphi}) & \text{otherwise}, \end{cases} \tag{5.6}$$

where $\mathcal{Z}^{\varphi}$ is a space diffeomorphic to $G^+_{\mathbf{a}}$ modulo its maximal compact subgroup; we let G^+ act on $\mathcal{Z}^{\varphi}$ naturally. For the explicit forms of $\mathcal{Z}^{\varphi}$ and its invariant measure, see Sections 4 and 5 of [6].

Let us now put $K = F\big([(-1)^{n/2} \det(\varphi)]^{1/2}\big)$ and denote by ψ the Hecke character of F corresponding to the extension K/F. We fix a $\mathfrak{g}$-lattice L in V which is $\mathfrak{g}$-maximal in the sense that $\varphi[x] \in \mathfrak{g}$ for every $x \in L$ and that if L' is a $\mathfrak{g}$-lattice in V containing L and $L' \neq L$, then $\varphi[y] \notin \mathfrak{g}$ for some $y \in L'$. We then take our D to be defined by

$$D = \{\, \xi \in G^+_{\mathbf{A}} \,|\, L_v \xi_v = L_v \quad \text{for every} \quad v \in \mathbf{h} \,\}. \tag{5.7}$$

Then in [6, Theorem 5.8] we proved an exact mass formula

$$\mathfrak{m}(G^+, D) = B_{\varphi} C(F, n) |D_K|^{n-2} |D_K/D_F|^{1/2} \pi^{-mg} L(m, \psi) \textstyle\prod_{\mathfrak{p}} N(\mathfrak{p})^{n-1}. \tag{5.8}$$

Here $m = n/2$ and $g = [F : \mathbf{Q}]$; $C(F, n)$ is a positive rational number which depends only on F and n; B_{φ} is a positive constant which depends only on the signatures of φ at the archimedean primes of F; the last product $\prod_{\mathfrak{p}}$ is extended over a certain finite set of prime ideals of F (see [6], p. 29 (4)).

5.2. We now consider the following case: $n = 2$, V is a quadratic extension of F, and $\varphi[x] = N_{V/F}(x)$ for $x \in V$. Then V coincides with the field K defined as above for this φ, so that we hereafter write K for V. We denote by ρ the generator of $\mathrm{Gal}(K/F)$, and by $\mathfrak{r}$ the maximal order of K. In this setting we have

$$G^+ = \mathcal{G}(K/F) = \{x \in K \mid N_{K/F}(x) = 1\}, \tag{5.9}$$

and the action of an element of $\mathcal{G}(K/F)$ on $V = K$ is by multiplication (see [6, Lemma 1.3]). We can take $L = \mathfrak{r}$, and hence (5.7) takes the form

$$D = \{\xi \in G^+_{\mathbf{A}} \mid x_v \in \mathfrak{r}_v^\times \quad \textit{for every} \quad v \in \mathbf{h}\}. \tag{5.10}$$

Since $G^+_{\mathbf{A}}$ is commutative, the decomposition $G^+_{\mathbf{A}} = \bigsqcup_{a\in\mathcal{B}} G^+aD$ is merely the coset decomposition of $G^+_{\mathbf{A}}$ modulo G^+D, and $\Gamma^a = U(K/F)$ for every $a \in G^+_{\mathbf{A}}$. Also, $G^+_{\mathbf{a}}$ is compact if and only if K is totally imaginary. Thus (5.5) can be written

$$\mathfrak{m}(G^+, D) = \begin{cases} [W_K : 1]^{-1}[G^+_{\mathbf{A}} : G^+D] & \textit{if } r_K = 0, \\ 2^{-1}[G^+_{\mathbf{A}} : G^+D]\mathrm{vol}(U\backslash\mathcal{Z}^\varphi) & \textit{if } r_K > 0. \end{cases} \tag{5.11}$$

The factors of formula (5.8) given in [6], Theorem 5.8 take simpler forms in this case, and we have

$$\mathfrak{m}(G^+, D) = 2^{1-\kappa}(2\pi)^{-c_K}|D_K/D_F|^{1/2}L(1, \psi), \tag{5.12}$$

where κ is the number of prime ideals of F ramified in K.

Let $C_{K/F}$ denote the subgroup of the ideal class group of K consisting of the classes represented by the ideals $\mathfrak{a}$ of K such that $\mathfrak{a}^\rho = \mathfrak{a}$. Then

$$[G^+_{\mathbf{A}} : G^+D] = h_K[C_{K/F} : 1]^{-1}, \tag{5.13}$$

$$[C_{K/F} : 1] = 2^{\kappa+c_K-1}[E_F : N_{K/F}(E_K)]^{-1}h_F. \tag{5.14}$$

The first equality is given in [4, (24.5.5)]. The second one is a special case of a well-known formula in class-field theory; see [3] p. 81, (a.2). Combining these with (5.11) and (5.12), we obtain

$$L(1, \psi) = \begin{cases} \pi^g|D_F/D_K|^{1/2}[W_K : 1]^{-1}[E_F : N_{K/F}(E_K)] \cdot (h_K/h_F) \\ \qquad\qquad\qquad\qquad\qquad\qquad \textit{if } r_K = 0, \\ 2^{-1}\pi^{c_K}|D_F/D_K|^{1/2}[E_F : N_{K/F}(E_K)] \\ \quad \cdot (h_K/h_F) \cdot \mathrm{vol}(U\backslash\mathcal{Z}^\varphi) \qquad\qquad \textit{if } r_K > 0, \end{cases} \tag{5.15}$$

where $g = [F : \mathbf{Q}]$. If $r_K = 0$, then K is a CM-field, and our formula for $L(1, \psi)$ is the same as (4.3), in view of (4.1).

Suppose now $r_K > 0$; put $\mathbf{R}_+ = \{x \in \mathbf{R} \mid x > 0\}$. The definition of $\mathcal{Z}^\varphi$ and the action of G^+ on $\mathcal{Z}^\varphi$ are explained in [6], Section 4. In the present case we have $\mathcal{Z}^\varphi = \mathbf{R}_+^{\mathbf{b}}$ with $\mathbf{b}$ as in Theorem 1.3, and an element a of G^+ acts on $\mathcal{Z}^\varphi$ by $(x_v)_{v\in\mathbf{b}} \mapsto (a_v^2 x_v)_{v\in\mathbf{b}}$. Also, the measure on $\mathcal{Z}^\varphi$ is given by $\prod_{v\in\mathbf{b}} x_v^{-1}dx_v$. Therefore we easily see that $\mathrm{vol}(U\backslash\mathcal{Z}^\varphi) = 2^b R(K/F)$, where $b = \#(\mathbf{b}) = t(K/F)$ as before. Substituting this into (5.15), we obtain

$$L(1,\,\psi) = 2^{b-1}\pi^{c_K}|D_F/D_K|^{1/2}[E_F : N_{K/F}(E_K)]\cdot(h_K/h_F)\cdot R(K/F),$$

which is a special case of (1.6). Thus we obtain the formula for $L(1,\,\psi)$ in terms of $R(K/F)$ from the exact mass formula for $n = 2$. Though F is assumed totally real here, the same method should work for a more general F. For this reason, for an extension K/F of an arbitrary degree we are tempted to call $I(K/F)$ the mass of the group $\mathcal{G}(K/F)$, and formula (1.6) the mass formula for that group.

It should also be pointed out that we obtain $R(K/F)$ directly as $\mathrm{vol}(U\backslash\mathcal{Z}^{\varphi})$, not as R_K/R_F, and also (5.15) without employing Dedekind's formula (0.2) for the residues of ζ_K and ζ_F at $s = 1$. This fact is actually the genesis of our investigation of $R(K/F)$.

5.3. Returning to formula (5.8), let us consider the case in which $n > 2$ and $G^+_{\mathbf{a}}$ is not compact. Then, by [6], Theorem 5.10, $\nu(\Gamma^a)$ does not depend on a, and $\mathfrak{m}(G^+,\,D)$ is $\nu(\Gamma^a)$ times a certain group index in $F^\times_{\mathbf{A}}$, which equals the number of spinor genera contained in the genus of L. Therefore (5.8) gives a formula for $\mathrm{vol}(\Gamma^a\backslash\mathcal{Z}^{\varphi})$ in terms of $L(m,\,\psi)$. We have $|D_K/D_F|^{1/2}\pi^{-mg}L(m,\,\psi) \in \mathbf{Q}$ if $\psi_v(x) = x^{-m}|x|^m$ for every $x \in F_v^\times$ and every $v \in \mathbf{a}(F)$, which can happen if and only if φ has an even number of negative eigenvalues at every $v \in \mathbf{a}(F)$. In such a case $\mathrm{vol}(\Gamma^a\backslash\mathcal{Z}^{\varphi})$ is a power of π times a rational number. In other cases, however, there is no such result about $L(m,\,\psi)$. Therefore, instead of looking for such a result, we can view (5.8) as a formula for $L(m,\,\psi)$ in terms of $\mathrm{vol}(\Gamma^a\backslash\mathcal{Z}^{\varphi})$ in exactly the same sense as in (1.6) and (5.15). Of course the group G cannot be uniquely associated with the value $L(m,\,\psi)$. Even in the case $n = 2$, the isomorphism class of K/F is not uniquely determined by the function $L(s,\,\lambda)$ as we showed in Section 2. Thus we must accept the plurality of the interpretation of $L(1,\,\lambda)$ or $L(m,\,\psi)$ in terms of such a volume as natural phenomena.

References

[1] R. Brauer, Beziehungen zwischen Klassenzahlen von Teilkörpern eines galoisschen Körpers, Math. Nachr. **4** (1950/51), 158–174.

[2] R. Dedekind, Über die Anzahl der Idealklassen in reinen kubischen Zahlkörpern, J. reine und angew. Math. **121** (1900), 40–123.

[3] G. Shimura, On abelian varieties with complex multiplication, Proc. London Math. Soc. 3rd ser. **34** (1977), 65–86.

[4] G. Shimura, Euler products and Eisenstein series, CBMS Regional Conference Series in Mathematics, No. 93, Amer. Math. Soc. 1997.

[5] G. Shimura, Abelian varieties with complex multiplication and modular functions, Princeton University Press, 1998.

[6] G. Shimura, An exact mass formula for orthogonal groups, Duke Math. J. **97** (1999), 1–66.

[7] H. M. Stark, L-functions at $s = 1$. II. Artin L-functions with rational characters, Adv. in Math. **17** (1975), 60–92.

Notes IV

Numbers in brackets set in boldface such as **[79a]** mean items in the list of articles; those in roman such as [17] and [94b] are references originally cited in the article to which notes are given. Some of the corrections were already made in the original articles, but they are superseded by the corrections made in these notes.

89a. Yutaka Taniyama and his time

The reprints of this article were sent along with the following letter.

May, 1989

Dear Friend and Colleague:

I am sending you herewith a copy of "Yutaka Taniyama and His Time." In December, 1986 I wrote its first draft impulsively for some inexplicable reasons. I showed it to some of my friends, whose encouragement eventually led me to publish it in the present form.

I expended some effort to make it readable for nonmathematicians. Therefore I would be glad if you could show it to whomever you think appropriate, to your spouse, for example.

In the article I did not refrain from writing about myself to the extent relevant to its principal theme, as I thought the complete suppression of my relationship with him would make the writing artificial and uninteresting; this way I could write more freely and truthfully. I hope you understand this approach.

Let me finally note one thing: After the incident mentioned on the first several lines of page 190, I told Taniyama and other people present in the room that someday I would write it in my memoir as one of his anecdotes. Everybody laughed. My prediction was fulfilled in an unexpected fashion.

Sincerely yours,
Goro Shimura

As to when and how I stated the conjecture that every **Q**-rational elliptic curve is modular, see notes to **[64e]**. Here I add some more comments on it in connection with what I was doing during the period covered by the present article. (See also **[96b]** for some more relevant comments.)

The paper **[55]** was written in 1952-53 with the intention of applying it to complex multiplication of abelian varieties. After finishing it, early in 1954, I started to read Kronecker's paper "Zur Theorie der elliptischen Funktionen" (1883-1890). My idea was to study what happens to isogenies of an elliptic curve when they are reduced modulo a prime. In the case of a curve with complex multiplication, their behavior leads to the reciprocity-law of an abelian extension of an imaginary quadratic field. Soon I was able to treat the case of higher-dimensional abelian varieties at least to the extent explained in **[56]**. Eventually the work in collaboration with Taniyama was published in **[57b]** in Japanese, as described in the present article.

At the same time, I was attempting to develop the theory of elliptic modular functions, especially that of modular correspondences, in connection with points of finite order and such isogenies of generic elliptic curves. I found one crucial idea by reformulating a result in the paper of Kronecker. By June 1956 I was able to determine the zeta function of a modular curve in terms of the Mellin transforms of cusp forms of weight 2. All these were published in **[57a]** and **[59a]**; see also **[64e]**.

Taniyama was of course aware of, or rather, much interested in my work. In fact, he mentioned my work on this subject in two of his articles written in Japanese (October, 1955 and July, 1956). However, he himself never worked on the zeta functions of curves or varieties over number fields, other than abelian varieties with complex multiplication, though he was aware of the results known at that time. I don't think he ever made any numerical calculations on the zeta function of an elliptic curve. Also, though I understood what he meant by his problems, we never discussed the topic beyond the level of casual conversations, except that I showed him my results mentioned above, and I wrote letters to him from Paris explaining the results of **[58b]** and **[59a]**. Taniyama said "other special types of automorphic functions are necessary," and he meant Hecke's nonmodular triangle functions, but I was never interested in those triangle groups, though I discussed *arithmetic* triangle groups in **[67b]**.

There was an informal conference at Kyoto University in July, 1955, which was a kind of rehearsal of the Tokyo-Nikko Symposium in September, 1955. He did not attend this Kyoto conference, and he asked me to read his paper, which I did. In his letter to me, dated July 18, 1955, he naturally thanked me for doing the job, and also listed the wrong points of the paper. At the end he thanked me also for pointing out his earlier mistakes.

These together with what I wrote in the present article exhaust all essential aspects of my mathematical interaction with Taniyama.

90a. Invariant differential operators on hermitian symmetric spaces

p. 255, (4.14a, b); p. 257, the second paragraph: Here the functions are assumed to have compact support or C^∞ functions in a L^2 space. In applications we have to deal with functions of more general types. A sufficient condition for the validity of the formulas is given in [**94c**, Theorem 2.1 and the subsequent paragraph]; see also [**94c**, Lemma A.2], which is reproduced in [**00**, Lemma A8.3].

p. 271, Proposition 7.3. We can similarly prove two formulas about $\iota_+(u_1)\cdots\iota_+(u_e)f^\rho$ and $\iota_-(u_1)\cdots\iota_-(u_e)f^\rho$ for $u_i \in T$; see [**94c**, Proposition 2.2; p. 151, lines 1 and 2].

We present here two types of results supplementary to the present article: one concerning differential operators on G/K, and the other on the questions about compact G. See also Theorem 1 in notes to [**99c**].

1. Differential operators on G/K. We defined $\mathcal{D}(\rho)$ in terms of $U(\mathfrak{g})$, but did not study its relation with differential operators on G/K except proving some results in Section 7, Proposition 7.3 in particular. If ρ is trivial, it is well-known that $\mathcal{D}(\rho)$ is isomorphic to the ring of all G-invariant differential operators on G/K. Let us now generalize this to the case of arbitrary irreducible ρ.

We first take our setting to be that of Section 7; we define symbols $\mathcal{K}$, $\Lambda(g, z)$, and $\mathbf{o}$ as in that section, and consider $\rho : \mathcal{K} \to GL(V)$. For simplicity we take $V = \mathbf{C}^m$ so that ρ has values in $GL_m(\mathbf{C})$. For $f \in C^\infty(\mathcal{H}, \mathbf{C}^m)$ and $\alpha \in G$ we define $f\|_\rho\alpha \in C^\infty(\mathcal{H}, \mathbf{C}^m)$ and $f^\rho \in C^\infty(G, \mathbf{C}^m)$ by

$$(f\|_\rho\alpha)(z) = \rho\big(\Lambda(\alpha, z)^{-1}\big) f(\alpha z), \quad f^\rho(g) = (f\|_\rho g)(\mathbf{o}) \qquad (g \in G).$$

The latter is the same as what is defined in Proposition 7.3. Define $\rho_0 : K \to GL_m(\mathbf{C})$ by $\rho_0(k) = \rho\big(\Lambda(k, \mathbf{o})\big)$ for $k \in K$. Then we easily see that $f \mapsto f^\rho$ gives a bijection of $C^\infty(\mathcal{H}, \mathbf{C}^m)$ onto the set $C^\infty(\rho_0)$ defined in Section 2.

This much can be done also on G/K of general type considered in Section 2. We simply put $\mathcal{H} = G/K$, denote by z the variable on G/K, and assume that a matrix-valued factor of automorphy $\Lambda(g, z)$ is given; we take $\mathbf{o}$ to be the origin of G/K represented by 1. We assume that Λ is C^∞ in $(g, z) \in G \times \mathcal{H}$ and ρ is a real nalytic representation with which $\rho\big(\Lambda(g, z)\big)$ is meaningful.

For either hermitian or nonhermitian G/K we consider a matrix differential operator $W = (W_{ij})_{i,j=1}^m$ whose entries W_{ij} are ordinary scalar-valued C^∞ differential operators on $\mathcal{H}$. For $f = (f_i)_{i=1}^m \in C^\infty(\mathcal{H}, \mathbf{C}^m)$ with $f_i \in C^\infty(\mathcal{H})$ we define $Wf \in C^\infty(\mathcal{H}, \mathbf{C}^m)$ by $Wf = (\sum_{j=1}^m W_{ij} f_j)_{i=1}^m$. Then we let $\mathfrak{D}(\rho)$ denote the set of all W of that type such that $W(f\|_\rho\alpha) = (Wf)\|_\rho\alpha$ for every $f \in C^\infty(\mathcal{H}, \mathbf{C}^m)$ and every $\alpha \in G$. Define $\mathcal{D}(\rho_0)$ as in Section 2.

Theorem D. (1) *If ρ_0 is irreducible, then there is a ring-isomorphism $W \mapsto W^\rho$ of $\mathfrak{D}(\rho)$ onto $\mathcal{D}(\rho_0)$ such that $W^\rho(f^\rho) = (Wf)^\rho$ for every $f \in C^\infty(\mathcal{H}, \mathbf{C}^m)$.*
(2) *If ρ is one-dimensional, then $\mathfrak{D}(\rho)$ is commutative.*

PROOF. By Proposition 2.3 (1), $\mathcal{D}(\rho_0)$ is commutative if ρ_0 is one-dimensional. Therefore (2) follows from (1). To prove (1), take a Cartan decomposition $\mathfrak{g}_0 = \mathfrak{k}_0 + \mathfrak{p}_0$ and take an $\mathbf{R}$-basis $\{X_i\}_{i=1}^n$ of $\mathfrak{p}_0$; put $M(g) = \rho\big(\Lambda(g, \mathbf{o})\big)$ for $g \in G$ and $e(t) = \exp(\sum_{i=1}^n t_i X_i)$ for $t = (t_i)_{i=1}^n \in \mathbf{R}^n$. Since $G = \exp(\mathfrak{p}_0)K$, the map $t \mapsto e(t)\mathbf{o}$ gives a diffeomorphism of $\mathbf{R}^n$ onto $\mathcal{H}$. Therefore, given $W = (W_{ij}) \in \mathfrak{D}(\rho)$, we can find a matrix $P = (P_{ij})_{i,j=1}^m$ with polynomials $P_{ij}(\partial/\partial t_1, \ldots, \partial/\partial t_n)$ such that

$$(*) \qquad (Wf)\big(e(t)\mathbf{o}\big) = P(\partial/\partial t_1, \ldots, \partial/\partial t_n) f\big(e(t)\mathbf{o}\big) \text{ at } t = 0$$

for every $f \in C^\infty(\mathcal{H}, \mathbf{C}^m)$. Clearly there is another polynomial matrix $Q = (Q_{ij})_{i,j=1}^m$ with polynomials $Q_{ij}(\partial/\partial t_1, \ldots, \partial/\partial t_n)$ such that

$$(**) \qquad \big[P(\partial/\partial t_1, \ldots, \partial/\partial t_n)\big((M \circ e)\varphi\big)\big](0) = \big[Q(\partial/\partial t_1, \ldots, \partial/\partial t_n)\varphi\big](0)$$

for every $\varphi \in C^\infty(\mathbf{R}^n, \mathbf{C}^m)$. Let $\mathfrak{S}(\mathfrak{p})$ and $\psi : \mathfrak{S}(\mathfrak{p}) \to U(\mathfrak{g})$ be as in Section 2. Put $D = (D_{ij})_{i,j=1}^m$ with $D_{ij} = \psi\big(Q_{ij}(X_1, \ldots, X_n)\big)$ and $h(t) = M\big(e(t)\big)^{-1} f\big(e(t)\mathbf{o}\big)$ with $f \in C^\infty(\mathcal{H}, \mathbf{C}^m)$. Then $f^\rho\big(e(t)k\big) = \rho_0(k^{-1})h(t)$ for $k \in K$, and hence

$$(Df^\rho)(1) = \big[Q(\partial/\partial t_1, \ldots, \partial/\partial t_n)h\big](0).$$

By (**) and (*) this equals

$$P(\partial/\partial t_1, \ldots, \partial/\partial t_n) f\big(e(t)\mathbf{o}\big)_{t=0} = (Wf)(\mathbf{o}).$$

Thus $(Df^\rho)(1) = (Wf)(\mathbf{o})$, which is valid for every $f \in C^\infty(\mathcal{H}, \mathbf{C}^m)$. Take $f\|_\rho\gamma$ in place of f with any $\gamma \in G$. Observe that $(f\|_\rho\gamma)^\rho = f^\rho \circ \gamma$, where we define $\varphi \circ \gamma$ by $(\varphi \circ \gamma)(g) = \varphi(\gamma g)$ for $g \in G$; hence

$$(Df^\rho)(\gamma) = \big(D(f^\rho \circ \gamma)\big)(1) = W(f\|_\rho\gamma)(\mathbf{o}) = \big((Wf)\|_\rho\gamma\big)(\mathbf{o}) = (Wf)^\rho(\gamma),$$

that is, $Df^\rho = (Wf)^\rho$. Since ρ_0 is irreducible, there is an element E in $U(\mathfrak{g})$ such that $Ef^\rho = Df^\rho$, as explained in the paragraph including (2.1) and (2.2). Thus $Ef^\rho = (Wf)^\rho$. This shows that E maps $C^\infty(\rho_0)$ into itself, and hence E defines an element of $\mathcal{D}(\rho_0)$, which is determined by W, since $f \mapsto f^\rho$ is a bijection of $C^\infty(\mathcal{H}, \mathbf{C}^m)$ onto $C^\infty(\rho_0)$. Writing it W^ρ, we see that $W \mapsto W^\rho$ is a ring-injection.

To prove that this is surjective, let $Y \in \mathcal{D}(\rho_0)$. We can define a $\mathbf{C}$-linear endomorphism Z of $C^\infty(\mathcal{H}, \mathbf{C}^m)$ by $(Zf)^\rho = Yf^\rho$ for $f \in C^\infty(\mathcal{H}, \mathbf{C}^m)$. We can then put $Z = (Z_{ij})$ with endomorphisms Z_{ij} of $C^\infty(\mathcal{H})$ so that $Zf = \big(\sum_{j=1}^m Z_{ij} f_j\big)_{i=1}^m$ for $f = (f_i)_{i=1}^m \in C^\infty(\mathcal{H}, \mathbf{C}^m)$. Now, for $g \in G$ we have

$$(\#) \qquad (Zf)(g\mathbf{o}) = M(g)Y\big(M(g)^{-1} f(g\mathbf{o})\big).$$

Take $f = (f_i)$ with $f_2 = \cdots = f_m = 0$; suppose $f_1 = 0$ in a neighborhood of a point $\xi\mathbf{o}$ with $\xi \in G$. Then the right-hand side of (#) is 0 for g in a neighborhood of ξ. Therefore $Z_{i1} f_1 = 0$ in a neighborhood of $\xi\mathbf{o}$ for every i. This shows that the support of $Z_{i1} f_1$ is contained in the support of f_1. Consequently, by a well-known principle (see [H2, p. 236, Theorem 1.4], for example), Z_{i1} is a differential operator on $\mathcal{H}$. Clearly the same is true for Z_{ij} for all (i, j). For $\gamma \in G$ and an arbitrary f we have

$$\big(Z(f\|_\rho\gamma)\big)^\rho = Y(f\|_\rho\gamma)^\rho = Y(f^\rho \circ \gamma) = (Yf^\rho) \circ \gamma = (Zf)^\rho \circ \gamma = \big((Zf)\|_\rho\gamma\big)^\rho,$$

so that $Z(f\|_\rho\gamma) = (Zf)\|_\rho\gamma$, that is, $Z \in \mathfrak{D}(\rho)$. This proves the surjectivity, and completes the proof.

Now suppose ρ is one-dimensional, $\mathcal{H}$ is hermitian, and G is classical. Define L_Z and M_Z by (3.15); define also operators L_ρ^Z and M_ρ^Z by [**94c**, (2.20)], which belong to $\mathfrak{D}(\rho)$ in view of [**94c**, (2.21)]. By Proposition 4.1, $L_Z = \theta E^W D^Z$, and hence by [**94c**, Proposition 2.2] we have $L_Z f^\rho = \pm(M_\rho^Z f)^\rho$ for $f \in C^\infty(\mathcal{H}, \mathbf{C}^m)$. (Unfortunately, the use of the letters L and M in [**94c**] is not consistent with that in [**90a**].) By Theorem 3.6, $\mathcal{D}(\rho_0)$ is spanned by the operators obtained from these L_Z. Therefore $\mathfrak{D}(\rho)$ is spanned by the operators M_ρ^Z, and similarly by the L_ρ^Z. Moreover, we can find a canonical set of generators of $\mathfrak{D}(\rho)$ formed by L_ρ^Z (or M_ρ^Z) with the basic irreducible Z, by virtue of Theorem 3.6 (3), (4).

For example, if $G = SL_2(\mathbf{R})$, $\mathcal{H} = \{z \in \mathbf{C} | \mathrm{Im}(z) > 0\}$, and $\rho(\Lambda(g, z)) = (cz + d)^\kappa$ with $\kappa \in \mathbf{Z}$ for $g = \begin{bmatrix} a & b \\ c & d \end{bmatrix} \in G$, then $L_\rho^Z = -4y^2\partial^2/\partial z\partial\bar{z} + 2i\kappa y\partial/\partial\bar{z}$ with $y = \mathrm{Im}(z)$ for the simplest Z, and $\mathfrak{D}(\rho)$ consists of the polynomials of this operator.

More generally, define L_r as in [**84a**, Proposition 11.13]. Comparing [**84a**, (2.9a, b)] with [**94c**, (2.20), (4.9c, d)], we see that L_r coincides with M_ρ^Z with $Z = Z_r$ defined in [**94c**, p. 159]. Such Z_r for $0 < r \leq m$ form the set of basic Z. Consequently $\mathfrak{D}(\rho)$ is generated by the L_r, which gives an affirmative answer to the question in [**84a**, p. 325, line 5 from the bottom].

2. The case of compact G. In this article we treated noncompact G, but most of the results are valid for compact G with suitable modifications, which we will now explain.

Let (G, K) be a Riemannian symmetric pair in the sense of [H3] with a compact semisimple connected Lie group G and a compact subgroup K of G. We naturally assume that there is an analytic automorphism θ of G such that $(K_\theta)^\circ \subset K \subset K_\theta$ where $K_\theta = \{x \in G | \theta(x) = x\}$ and $(K_\theta)^\circ$ is the identity component of K_θ. Define the symbols $\mathfrak{g}_0$, $\mathfrak{k}_0$, $\mathfrak{p}_0$, $\mathfrak{g}$, $\mathfrak{k}$, $\mathfrak{p}$, and $U(\mathfrak{g})$ as in Section 2 for the present G and K; we have $\mathfrak{k}_0 = \{X \in \mathfrak{g}_0 | d\theta X = X\}$ and $\mathfrak{p}_0 = \{X \in \mathfrak{g}_0 | d\theta X = -X\}$. Taking also a continuous representation $\rho : K \to GL(V)$, we define $D(\rho)$, $\mathcal{D}(\rho)$, and ρ_u in the same way as in Section 2. The point we must remember is that K is not necessarily connected. (If G/K is hermitian or every simple factor of G is noncompact, then K is connected.) Therefore we define the symbol $\mathfrak{X}^K$ to be the set of $Ad(K)$-invarinat elements of $\mathfrak{X}$, without considering K^c. Then:

Fact 1. *Proposition 2.1 is true for compact G as above.*

PROOF. Take an $\mathbf{R}$-basis $\{X_i\}_{i=1}^n$ of $\mathfrak{p}_0$ and put $\varepsilon(t) = \exp\left(\sum_{i=1}^n t_i X_i\right)$ for $t = (t_i)_{i=1}^n \in \mathbf{R}^n$; take a bounded open neighborhood H of 0 in $\mathbf{R}^n$ so that the map $t \mapsto \varepsilon(t)K$ gives a diffeomorphism of H onto an open subset of G/K. Define a function g on $\varepsilon(H)K$ by $g(\varepsilon(t)k) = \rho(k)^{-1}h(t)$ for $h \in C^\infty(H, V)$. Take open neighborhoods H_1 and H_2 of 0 in $\mathbf{R}^n$ so that $\overline{H_1} \subset H_2 \subset \overline{H_2} \subset H$, and take $\varphi \in C^\infty(G)$ so that $\varphi = 1$ on $\varepsilon(H_1)K$ and $\varphi = 0$ outside of $\varepsilon(H_2)K$. Put $f_h(x) = \varphi(x)g(x)$ for $x \in \varepsilon(H)K$ and $f_h(x) = 0$ for $x \notin \varepsilon(H)K$. Then $f_h \in \mathbf{C}^\infty(\rho)$ and $f_h(\varepsilon(t)k) = \rho(k)^{-1}h(t)$ for $t \in H_1$. Since we can take any C^∞ function h on H, we can now repeat the original proof of Proposition 2.1 using this f_h.

Fact 2. *Proposition 2.2 is true for compact G as above if $d\rho$ is irreducible.*

PROOF. Since K is not necessarily connected, the irreducibility of ρ does not necessarily mean the irreducibility of $d\rho$. However, once we assume that $d\rho$ is irreducible, the original proof applies to the compact case, provided we take f_h in the proof to be that defined in the above modified proof of Proposition 2.1.

Fact 3. *Proposition 2.3 is true for compact G as above.*

PROOF. There is no problem about assertions (2) and (3). As for (1), the paper [L] deals only with noncompact G, and so is not applicable directly. However, we can use the result of [L] by considering the subalgebra $\mathfrak{g}_1 = \mathfrak{k}_1 + \mathfrak{p}_1$ of $\mathfrak{g}$ with $\mathfrak{k}_1 = \mathfrak{k}_0$ and $\mathfrak{p}_1 = i\mathfrak{p}_0$. Let G_1 and K_1 be the connected subgroups of G^c corresponding to $\mathfrak{g}_1$ and $\mathfrak{k}_1$. Then G_1 is noncompact, and [L] is applicable to G_1/K_1. Since K_1 is connected, we have $Ad(K_1) \subset Ad(K)$, and hence $U(\mathfrak{g})^K \subset U(\mathfrak{g})^{K_1}$. Therefore $U(\mathfrak{g})^K/\left[U(\mathfrak{g})^K \cap U(\mathfrak{g})\mathfrak{N}_\rho\right]$ is isomorphic to a subring of $U(\mathfrak{g})^{K_1}/\left[U(\mathfrak{g})^{K_1} \cap U(\mathfrak{g})\mathfrak{N}_\rho\right]$. Since the last group is commutative as explained in the original proof, we obtain (1) for the present G/K in view of (3) of Proposition 2.1.

Fact 4. *If G is a compact group as above and classical, then the natural map of $C(\mathfrak{g})$ into $\mathcal{D}(\rho)$ is surjective for every one-dimensional ρ.*

Proof. Using the same notation as in the proof of Fact 3, we have $\mathfrak{S}^r(\mathfrak{p})^K \subset \mathfrak{S}^r(\mathfrak{p})^{K_1}$, since $Ad(K_1) \subset Ad(K)$. Also, an element of $\mathfrak{S}^r(\mathfrak{g})$ is $Ad(G)$-invariant if and only if $Ad(G_1)$-invariant, since both $Ad(G)$ and $Ad(G_1)$ are Zariski-dense in $Ad(G^c)$. Now (2.4) is true with G_1 and K_1 in place of G and K, since G_1 is noncompact, and hence it is true with the present G and K. Then we can repeat the proof of Theorem 2.4 to obtain the desired fact.

Fact 5. *Suppose that G/K with compact G as above is classical and hermitian. Then all the results of Sections 3, 4, and 6 are valid.*

Proof. We know that K is connected if G/K is hermitian. Therefore, using the same notation as in the proof of Fact 3, we have $Ad(K_1) = Ad(K)$. Consequently $U^r(\mathfrak{g})^K = U^r(\mathfrak{g})^{K_1}$ and $\mathfrak{S}^r(\mathfrak{p})^K = \mathfrak{S}^r(\mathfrak{p})^{K_1}$; also, $I(\mathfrak{p})$, $I_r(\mathfrak{p})$, and $I^r(\mathfrak{p})$ of (3.10) are the same for both K and K_1. We take an element H_0 of $\mathfrak{k}_0$ as usual, by which we define the complex structure of G/K and also that of G_1/K_1; we put $\mathfrak{p}_\pm = \left\{X \in \mathfrak{p} \,\middle|\, [H_0, X] = \pm X\right\}$; then $\mathfrak{p} = \mathfrak{p}_+ \oplus \mathfrak{p}_-$. In view of Propositions 2.1 and 2.3, which are true in the compact case, the structure of $\mathcal{D}(\rho)$ for G/K is the same as that for G_1/K_1. Thus we obtain the desired fact.

We can naturally ask whether there exist some results in the compact case parallel to those of Sections 5 and 7. In order to give an affirmative answer, let us start with explicit forms of G/K and "quasi-factors of automorphy" according to the standard classification of simple groups which give rise to irreducible hermitian symmetric spaces. For each type we present G/K as $L/GL_m(\mathbf{C})$ with a certain subset of L of $\mathbf{C}_m^N$ on which we have natural right action of $GL_m(\mathbf{C})$. There is a parabolic subgroup P of G^c with which we have

$$G^c = PG \subset GL_N(\mathbf{C}), \quad G = U(N) \cap G^c, \quad K = G \cap P = U(N) \cap P.$$

Clearly G/K can be identified with G^c/P.

There is also a complex vector space T and a map $p : T \to L$ which gives a bijection of T onto a dense open subset of G/K. Thus T can be viewed as a (large) coordinate neighborhood on G/K. Also, for each $\alpha \in G$ there is an open dense subset T_α and the "quasi-factors of automorphy" $\lambda_\alpha(z)$, $\mu_\alpha(z)$ defined for $z \in T_\alpha$. These λ_α and μ_α, as well as the action of α, are holomorphic on G/K. We shall describe T_α explicitly only for Type A. We have

$$(1) \qquad d(\alpha z) = {}^t\lambda_\alpha(z)^{-1} \cdot dz \cdot \mu_\alpha(z)^{-1} \qquad (z \in T_\alpha, \alpha \in G).$$

Moreover we have positive definite hermitian matrices $\xi(z)$ and $\eta(z)$ with which we can give explicit forms of differential operators in the same manner as in Proposition 7.3. In addition to the symbols about classical groups on page 258, we put $SO(n) = SO(1_n, \mathbf{R})$; for a complex matrix z we put $z^* = {}^t\bar{z}$. We now give explicit forms of all these objects. (The explicit form of each G/K in terms of $L/GL_m(\mathbf{C})$ and its kähler structure are given in [**87a**, Section 2], in which the map p is denoted by λ.)

Type A. $G^c = SL_N(\mathbf{C})$, $N = n + m$, $P = \left\{ \begin{bmatrix} a & 0 \\ c & d \end{bmatrix} \in G^c \,\middle|\, d \in GL_m(\mathbf{C}) \right\}$.

$$G = SU(N),\ K = \{\mathrm{diag}[u, v] \in G^c \,|\, u \in U(n),\ v \in U(m)\},$$

$$L = \{w \in \mathbf{C}_m^N \,|\, \mathrm{rank}(w) = m\},\ T = \mathbf{C}_m^n,\ p(z) = \begin{bmatrix} z \\ 1_m \end{bmatrix},$$

$$B(z) = \begin{bmatrix} 1_n & z \\ -z^* & 1_m \end{bmatrix},\quad \xi(z) = 1_n + \bar{z} \cdot {}^t x,\quad \eta(z) = 1_m + z^* z \quad (z \in T).$$

Let $\mathfrak{X}$ be the set of all $X \in GL_N(\mathbf{C})$ such that $X^*X = \mathrm{diag}[a, b]$ with a and b of size n and m. Let $\mathfrak{X}_0$ be the subset of $\mathfrak{X}$ consisting of all X such that the determinant of the upper left n^2 entries of X is nonzero. Then we easily see that the map $(z, \lambda, \mu) \mapsto B(z)\mathrm{diag}[\lambda, \mu]$ gives a bijection of $T \times GL_n(\mathbf{C}) \times GL_m(\mathbf{C})$ onto $\mathfrak{X}_0$. For $\alpha \in G$ put $T_\alpha = \{z \in T \,|\, \alpha B(z) \in \mathfrak{X}_0\}$. Then T_α is open and dense in T. Now for $z \in T_\alpha$ we can define $\alpha z \in T$, $\lambda_\alpha(z) \in GL_n(\mathbf{C})$ and $\mu_\alpha(z) \in GL_m(\mathbf{C})$ by the relation

$$\alpha B(z) = B(\alpha z)\mathrm{diag}[\overline{\lambda_\alpha(z)}, \mu_\alpha(z)] \qquad (z \in T_\alpha).$$

Type B. $G^c = SO(R, \mathbf{C})$, $\quad N = n + 2$, $\quad R = \mathrm{diag}\left[1_n, \begin{bmatrix} 0 & -1 \\ -1 & 0 \end{bmatrix}\right]$,

$$G \cong SO(N),\quad P = \left\{ \begin{bmatrix} a & b & 0 \\ 0 & d^{-1} & 0 \\ d \cdot {}^t ba & d \cdot {}^t bb/2 & d \end{bmatrix} \,\middle|\, a \in SO(n, \mathbf{C}),\ b \in \mathbf{C}^n,\ d \in \mathbf{C}^\times \right\},$$

$$K = \{\mathrm{diag}[a, d^{-1}, d] \,|\, a \in SO(n),\ |d| = 1\},\quad L = \{w \in \mathbf{C}^N \,|\, {}^t wRw = 0,\ w \neq 0\},$$

$$T = \mathbf{C}^n,\quad p(z) = \begin{bmatrix} z \\ w \\ 1 \end{bmatrix},\quad w = {}^t zz/2,\quad \eta(z) = p(z)^* p(z) = z^* z + |w|^2 + 1 \quad (z \in T),$$

$$B(z) = \begin{bmatrix} 1 & -\bar{z} & z \\ {}^t z & 1 & w \\ 0 & \overline{w} & 1 \end{bmatrix} \begin{bmatrix} 1_n & 0 & 0 \\ 0 & 1 & 0 \\ -\eta^{-1} \cdot {}^t u & 0 & 1 \end{bmatrix} = \begin{bmatrix} 1 - \eta^{-1} z \cdot {}^t u & -\bar{z} & z \\ {}^t z - \eta^{-1} w \cdot {}^t u & 1 & w \\ -\eta^{-1} \cdot {}^t u & \overline{w} & 1 \end{bmatrix},$$

$$\xi(z) = 1_n + \bar{z} \cdot {}^t z - \eta^{-1} \bar{u} \cdot {}^t u,\quad \text{where } \eta = \eta(z) \text{ and } u = \bar{z} + \overline{w} z,$$

$$\alpha B(z) = B(\alpha z)\mathrm{diag}[{}^t\lambda_\alpha(z)^{-1}, \overline{\mu_\alpha(z)}, \mu_\alpha(z)] \qquad (z \in T_\alpha).$$

Type C. $G^c = Sp(n, \mathbf{C})$, $\quad N = 2n$, $\quad K = \{\mathrm{diag}[\bar{d}, d] \,|\, d \in U(n)\} \cong U(n)$,

$$P = \left\{ \begin{bmatrix} {}^t d^{-1} & 0 \\ dz & d \end{bmatrix} \,\middle|\, d \in GL_n(\mathbf{C}),\ z \in T \right\},\quad T = \{z \in \mathbf{C}_n^n \,|\, {}^t z = z\},$$

$$L = \{w \in \mathbf{C}_n^{2n} \,|\, {}^t wJw = 0,\ \mathrm{rank}(w) = n\},\quad J = \begin{bmatrix} 0 & -1_n \\ 1_n & 0 \end{bmatrix},\quad p(z) = \begin{bmatrix} z \\ 1_n \end{bmatrix},$$

$$B(z) = \begin{bmatrix} 1_n & z \\ -\bar{z} & 1_n \end{bmatrix},\quad \alpha B(z) = B(\alpha z)\mathrm{diag}[\overline{\mu_\alpha(z)}, \mu_\alpha(z)],$$

$\xi(z) = \eta(z) = 1_n + z^*z, \quad \lambda_\alpha(z) = \mu_\alpha(z).$

All the formulas can be obtained from those for Type A by taking $n = m$ and $z = {}^t z$.

Type D. $G^c = SO(Q, \mathbf{C}), \quad Q = \begin{bmatrix} 0 & 1_n \\ 1_n & 0 \end{bmatrix}, \quad N = 2n > 2, \quad G \cong SO(2n),$

$$K = \{\mathrm{diag}[\overline{d}, d] \mid d \in U(n)\} \cong U(n), \quad \widetilde{L} = \{w \in \mathbf{C}_n^{2n} \mid {}^t w Q w = 0, \mathrm{rank}(w) = n\},$$

$$L = \text{the connected component of } \widetilde{L} \text{ containing } p(0), \quad p(z) = \begin{bmatrix} z \\ 1_n \end{bmatrix},$$

$$P = \left\{ \begin{bmatrix} {}^t d^{-1} & 0 \\ dz & d \end{bmatrix} \middle| d \in GL_n(\mathbf{C}), z \in T \right\}, \quad T = \{z \in \mathbf{C}_n^n \mid {}^t z = -z\},$$

$$B(z) = \begin{bmatrix} 1_n & z \\ -\overline{z} & 1_n \end{bmatrix}, \quad \alpha B(z) = B(\alpha z)\mathrm{diag}[\overline{\mu_\alpha(z)}, \mu_\alpha(z)].$$

$\xi(z) = \eta(z) = 1_n + z^*z, \quad \lambda_\alpha(z) = \mu_\alpha(z).$

All the formulas can be obtained from those for Type A by taking $n = m$ and $z = -{}^t z$.

Now define $\varphi : \mathcal{K} \to K^c$ and $\iota_\pm : T \to \mathfrak{p}_\pm$ for all four types as in Section 5, pp. 259-261. G^c and T are the same, but the present $\mathfrak{p}_0$ is contained in the Lie algebra of compact G; thus formula (5.3) becomes $i\mathfrak{p}_0 = \{\iota_+(z) + \iota_-(z) \mid z \in T\}$, so that

$$\text{(2)} \qquad \mathfrak{p}_0 = \{\iota_+(u) - \iota_-(u) \mid u \in T\}$$

with the present $\mathfrak{p}_0$. Define the origin $\mathbf{o}$ of $\mathcal{H} = G/K$ by $\mathbf{o} = p(0)$. Then $K = \{\alpha \in G \mid \alpha\mathbf{o} = \mathbf{o}\}$, and all the formulas of Section 7 can be stated and proved in the compact case with easy modifications. For example $\Lambda(g, z)$ of (7.1) is meaningful for $z \in T_g$ and $g \in G$; $\Lambda(g, z)$ is $(\lambda_g(z), \mu_g(z))$ and $\Lambda(g, z)dz = {}^t\lambda_g(z)^{-1} dz \mu_g(z)^{-1}$; also, D and $\overline{D}$ can be defined by (7.5). Given $f \in C^\infty(\mathcal{H}, V)$ with a finite-dimensional complex vector space V, define $F \in C^\infty(G, V)$ by $F(g) = f(g\mathbf{o})$ for $g \in G$. Then we have

$$\text{(3)} \quad [\iota_+(u)F](g) = (Df)(g\mathbf{o})[\Lambda(g, \mathbf{o})u], \quad [-\iota_-(u)F](g) = (\overline{D}f)(g\mathbf{o})[\Lambda(g, \mathbf{o})u]$$

for every $u \in T$. The main difference of this from Lemma 7.1 is that here we have $-\iota_-(u)F$ instead of $\iota_-(u)F$. Lemma 7.2 holds also in the present case. These can be proved in the same manner. Formulas (7.7) and (7.8) have obvious analogues; we have $\Xi(z) = (\xi(z), \eta(z))$. Operators D_ρ and E can be defined, and we can prove

$$\text{(4)} \qquad \iota_+(u)f^\rho = (D_\rho f)^{\rho\otimes\tau}(u), \quad -\iota_-(u)f^\rho = (Ef)^{\rho\otimes\sigma}(u) \qquad (u \in T)$$

in the same way as in Proposition 7.3.

90b. On the fundamental periods of automorphic forms of arithmetic type

p. 406, line 11: For "$\mathbf{f}'_\lambda$" read "f'_λ".
p. 408, line 4: For "$\mathbf{a}_\alpha$" read "a_α".
p. 413, line 5 from the bottom: For "(4.2)" read "(4.3)".
p. 422, line 16: For "$\mathbf{w}^0$" read "w^0".
p. 422, line 17: For "$\mathbf{s}$" read "s".

90c. Some old and recent arithmetical results concerning modular forms and related zeta functions

This appeared in the special issue *International Symposium on Algebra and Number Theory in honour of Cahit Arf*. The symposium was held September 3–7, 1990 in Silivri, Turkey, and this is the text of my lecture given there. The editor did not let me read the proofs, and consequently there are many typographical errors. Besides, no new results were included, and so the article was not reproduced.

91. The critical values of certain Dirichlet series attached to Hilbert modular forms

p. 570, Proposition 2.3: The formula holds under a weaker condition that $\tau\mathfrak{b}^{-1}\mathfrak{d} = \mathfrak{q}^2\mathfrak{r}$, where $0 \ll \tau \in F$, $\mathfrak{q}$ is a fractional ideal, and $\mathfrak{r}$ is a square-free integral ideal.

p. 579, line 4: For "D" read "$D[\mathfrak{e}^{-1}, \mathfrak{e}]$".

p. 579, (5.6): This can be given also in the form $\Gamma_v(s) = \Gamma(s + |q_v|/2)\Gamma(s + \iota_v/4)$ with $\iota_v = (-1)^{|q_v|+1/2}$.

p. 588, line 12: For "$\mathcal{S}_k($" read "$\mathcal{S}_k(($".

p. 602, line 11 from the bottom: The product on this line can be written

$$\prod_{v \in \mathbf{a}} \Gamma\big((s - \tau_v + i\kappa_v)/2\big)\Gamma\big((s + k_v - 1 + i\kappa_v)/2\big)\Gamma\big((s + k_v - 2 + i\kappa_v)/2\big).$$

p. 609, last line: For "$m - k_0 u/2$" read "$(m - k_0 u)/2$".

94a. Fractional and trigonometric expressions for matrices

This article appeared along with the following biographical sketch on page 790 of the same issue.

GORO SHIMURA is self-taught to a good extent, since he attended neither graduate school nor kindergarten, though he managed to graduate from the University of Tokyo in 1952. Princeton University has been liberal enough to permit him to teach since 1962. His youthful aspiration was to become a fashion designer, but he reconciled himself to living in a less competitive world of mathematics and testing his meager knowledge of low-dimensional topology by designing his wife's dress occasionally.

94b. Euler products and Fourier coefficients of automorphic forms on symplectic groups

p. 547, line 8: For "$\varepsilon(\Phi)$" read "$\varepsilon(\Psi)$".

p. 551, line 9: For "4.2" read "4.1".

p. 551, line 14: For "$\lambda(K)$" read "$\gamma(K)$".

p. 556, line 7 from the bottom: Multiply the left-hand side of the first equality by $\psi_{\mathfrak{c}}\big(\det(g)\big)$.

p. 560, (6.9): For "$(p^{-1}\xi)\mathfrak{g}$" read "$(p^{-1}\xi)\mathfrak{g})$".
p. 563, line 6: For "Theorem" read " Proposition".
p. 563, line 10: For "$|\pi_v|^{2s}$" read "$|\pi_v|^{2s+\lambda/2}$".
p. 564, line 10 from the bottom: For "the first set" read "the set of possible poles described in the case in which $\chi^2 = 1$ and $\mathfrak{c} \neq \mathfrak{g}$".
p. 565, line 9; p. 567, (7.14): For "±1" read "0 or 1".

For a more detailed treatment of the series $\alpha(\zeta)$ of §3, see [**97**, Sections 13, 14, and 15] and [**00**, Section A1].

94c. Differential operators, holomorphic projection, and singular forms

p. 150, line 2 from the bottom: For "$(D_\rho^{p+1} f)$" read "$(D_\rho^{p+1} f)^{\rho\otimes\tau^{p+1}}$".
p. 156, line 16: For "$\sum_{i=1}^{q}$" read "$\sum_{i=0}^{q}$".

95a. Eisenstein series and zeta functions on symplectic groups

p. 539, line 10 from the bottom: For "Im(z)" read "Im(z))".
p. 543, (1.8): For "**H**" read "$\mathcal{H}$".
p. 547, line 3: For "$X_k^r\|$" read "$X_k^r|$".
p. 549, line 16 from the bottom: For "$> n+1$" read "$> n$".
p. 551, (3.1): A detailed proof is given in [**97a**, Lemma 9.2].

p. 563, Proof of Lemma 5.1, line 6 from the bottom: The last sum on this line vanishes for $W = \mathrm{diag}[z, w]$, which leads to (5.6) on the next page. However, the proof of this vanishing given on the last five lines of p. 563 is not adequate. It must be given as follows. We fix our attention on one $v \in \varepsilon$, and drop the subscript v; for example we write S^{2n} for S_v^{2n}. For a complex vector space $\mathfrak{X}$, denote by $\mathfrak{S}_d(\mathfrak{X})$ the space of all **C**-valued homogeneous polynomial functions on $\mathfrak{X}$ of degree d. Then $GL_{2n}(\mathbf{C})$ (resp. $GL_n(\mathbf{C}) \times GL_n(\mathbf{C})$) acts on $\mathfrak{S}_d(S^{2n})$ (resp. $\mathfrak{S}_d(\mathbf{C}_n^n)$). (See [**84b**] or [**94c**].) Let $\det_i(x)$ denote the determinant of the upper left i^2 entries of a matrix x of size $\geq i$. For $1 \leq i \leq n$ let Y_i (resp. Z_i) be the irreducible subspace of $\mathfrak{S}_i(\mathbf{C}_n^n)$ (resp. $\mathfrak{S}_i(S^{2n})$) whose highest weight vector is $\det_i(x)$. For $h \in \mathfrak{S}_i(\mathbf{C}_n^n)$ define $\xi_h \in \mathfrak{S}_i(S^{2n})$ by $\xi_h(W) = h\big(q(W)\big)$, where $q = q(W)$ is defined by (5.1). Then the map $h \mapsto \xi_h$ sends Y_i into Z_i. Indeed, let $h(x) = \det_i(x)$ and let Z' be the irreducible subspace of $\mathfrak{S}_i(S^{2n})$ containing ξ_h. Then Z' has a highest weight vector of the form $p(W) = \prod_{v=1}^{2n} \det_v(W)^{c_v}$, $\sum v c_v = i$. Suppose $Z' \neq Z_i$; then $i > 1$ and $p(W)$ is of degree > 1 in z_{11}, where z is determined for W as in (5.1). On the other hand we can easily verify that $\xi_h({}^tAWA)$ is of degree ≤ 1 in z_{11} for every $A \in GL_{2n}(\mathbf{C})$, a contradiction. Thus $Z' = Z_i$. Now if $h_0(x) = h({}^taxb)$ with $a, b \in GL_n(\mathbf{C})$, then $\xi_{h_0}(W) = \xi_h({}^tAWA)$ with $A = \mathrm{diag}[a, b]$. Thus Y_i is sent into Z_i. Let $\partial/\partial W$ denote the matrix of (5.3) and let $\partial/\partial q = (\partial/\partial q_{ij})$. For $\eta \in \mathfrak{S}_n(S^{2n})$ defined by $\eta(W) = \det(q)$, we have $\eta(\partial/\partial W) = \det(\partial/\partial q)$, and hence, by [**94c**, Lemma 4.6]

([**94c**] is [**94b**] of the present article),

$$\eta(\partial/\partial W)[\delta(W)^s g] = \sum_{i=0}^{n} \binom{n}{i} \sum_{\nu} \{\omega_\nu^i(\partial/\partial q)\delta(W)^s\}\zeta_\nu^i(\partial/\partial q)g,$$

where $\{\omega_\nu^i\}$ resp. $\{\zeta_\nu^i\}$ is a basis of Y_{n-i} resp. Y_i. Now. for $\omega \in Y_{n-i}$ with $n > i$ we have $\xi_\omega \in Z_{n-i}$ as shown above; hence

$$\omega(\partial/\partial q)\delta(W)^s = \xi_\omega(\partial/\partial W)\delta(W)^s = c(s)\delta(W)^s\xi_\omega(\mathrm{Im}(W)^{-1})$$

by [**84b**, Theorem 4.3], where $c(s)$ is a constant depending only on s and Z_{n-i}. Clearly $\xi_\omega(\mathrm{Im}(W)^{-1}) = 0$ for $W = \mathrm{diag}[z, w]$. Thus $\eta(\partial/\partial W)[\delta(W)^s g] = \delta(W)^s \cdot \det(\partial/\partial q)g$ at $W = \mathrm{diag}[z, w]$. This proves (5.6). After obtaining (5.6), we can continue the original proof on page 564.

For a better and less computational approach to formulas of the same type as Lemma 5.1, see [**00**, Section 25, (25.5a, b)].

p. 565, line 6 For "$\|_{k'}^{-1}$" read "$\|_{k'}\iota^{-1}$".
p. 573, Theorem 7.3, line 2: For "Γ'''" read "$\Gamma' \cap \mathcal{P}^{m,n}$".
p. 573, Theorem 7.3, line 3: For "G^n" read "G^m".
p. 576, lines 6, 11: For "$\delta \circ \omega_r$" read "$\delta \circ \omega_r \circ \alpha$".
p. 578, (8.5): For "f" read "p_ξ".
p. 581, line 13: For "7.10" read "8.10".
p. 581, line 19: For "6.3" read "7.3".

The functions δ of (1.32) and $E_k^{n,r}$ of (1.34) are eigenfunctions of all differential operators W on $\mathcal{H}_n$ such that $W(f\|_k\gamma) = (Wf)\|_k\gamma$. See Theorem 3 of notes to [**99c**]. Let us now prove

Theorem E. *There exists a function $T(z, w; s)$ of $(z, w) \in \mathcal{H} \times \mathcal{H}$ and* $\mathrm{Re}(s) > (n + r + 1)/2$ *holomorphic in $(z, \overline{w}, s)$ such that $T(\gamma z, \gamma w; s) = j_\gamma^k(z) \cdot T(z, w; s)$ for every $\gamma \in \Gamma$ and $T(z, z; s) = E_k^{n,r}(z, s; f, \Gamma)$ for $z \in \mathcal{H}$ and* $\mathrm{Re}(s) > (n + r + 1)/2$.

PROOF. We first put $\varepsilon(z, w) = (\det[(i/2)(w_v^* - z_v)])_{v\in\mathbf{a}}$ and observe that

(1) $$\varepsilon(z, w) = j_\alpha(z)\overline{j_\alpha(w)}\varepsilon(\alpha z, \alpha w) \text{ for every } \alpha \in G_{\mathbf{A}}.$$

Put also $h_\alpha(z, w) = \varepsilon(\alpha z, \alpha w)\varepsilon(\omega_r(\alpha z), \omega_r(\alpha w))^{-1}$ for $(z, w) \in \mathcal{H} \times \mathcal{H}$ and $\alpha \in G_{\mathbf{A}}$. Define $\det(-iu)^s = \exp(s \cdot \log[\det(-iu)])$ for $u \in H_n$ and $s \in \mathbf{C}$ with the branch of $\log[\det(-iu)]$ that is real for $u = i1_n$. Then we consider

(2) $$T(z, w; s) = \sum_{\gamma \in A} f(\omega_r(\gamma z)) j_\gamma^k(z)^{-1} h_\gamma(z, w)^{su-k/2}, \quad A = (\mathcal{P}^{n,r} \cap \Gamma)\backslash\Gamma.$$

This is well-defined. To prove the convergence, fix a member v of $\mathbf{a}$, and we consider $h_\alpha(z, w)_v$. Let C be a compact subset of H_n. Then there exist two positive constants B_1 and B_2 such that

(3) $$B_1 \le |h_\alpha(z, w)_v / h_\alpha(i1_n, i1_n)_v| \le B_2 \text{ for every } (\alpha, z, w) \in G_v \times C \times C.$$

To prove this, put $\mathfrak{K} = \{\alpha \in G_v | \alpha(i1_n) = i1_n\}$. Clearly we can find B_1 and B_2 such that the inequalities of (2) hold for $(\alpha, z, w) \in \mathfrak{K} \times C \times C$. If $\pi \in P_v^{n,r}$, then $h_{\pi\alpha}(z, w)_v = h_\alpha(z, w)_v|\lambda_r(\pi)|^{-2}$ with λ_r of (1.22), and hence $h_\alpha(z, w)_v / h_\alpha(i1_n, i1_n)_v$ depends only on $P_v^{n,r}\alpha$. Since $G_v = P_v^{n,r}\mathfrak{K}$, we obtain (3). Now we have $|f(\mathfrak{z})| \le M_1\varepsilon(\mathfrak{z}, \mathfrak{z})^{-k/2}$ for every $\mathfrak{z} \in \mathcal{H}_r$ with a constant M_1, since f is a cusp form. Let Z be a compact subset of $\mathcal{H}_n$ and Σ a compact subset of $\{s \in \mathbf{C} | \mathrm{Re}(s) > (n+r+1)/2\}$. By (1) we have $\varepsilon(\alpha z, \alpha z) = |j_\alpha(z)|^{-2}\varepsilon(z, z)$, and hence

$$|f(\omega_r(\gamma z)) j_\gamma^k(z)^{-1}| \leq M_2 \varepsilon(\omega_r(\gamma z), \omega_r(\gamma z))^{-k/2} \varepsilon(\gamma z, \gamma z)^{k/2} = M_2 h_\gamma(z, z)^{k/2}$$

for $z \in Z$ with a constant M_2. For $s \in \Sigma$ and $u \in H_n$ we have

$$N_1 |\det(-iu)|^{\mathrm{Re}(s)} \leq |\det(-iu)^s| \leq N_2 |\det(-iu)|^{\mathrm{Re}(s)}$$

with positive constants N_1 and N_2; see [**82c**, Lemma 1.1]. Therefore, for $s \in \Sigma$ and $z, w \in Z$ we have

$$|f(\omega_r(\gamma z)) j_\gamma^k(z)^{-1} h_\gamma(z, w)^{su-k/2}| \leq M_3 h_\gamma(\mathbf{i}, \mathbf{i})^{\sigma u}$$

with a constant M_3, where $\sigma = \mathrm{Re}(s)$ and $\mathbf{i}$ is defined by (1.6). Now $h_\gamma(z, z)^u$ coincides with $\Delta(\gamma z)$ with Δ of [**97a**, (A3.1.2)], and $\sum_{\gamma \in A} \Delta(\gamma \mathbf{i})^\sigma$ is convergent for $\sigma > (n+r+1)/2$ as noted in [**97a**, §A3.9]. This proves the local uniform convergence of the series of (2). Since each term is holomorphic in $(z, \overline{w}, s)$, so is the sum. Putting $z = w$, we obtain the series of (1.34). The equality $T(\gamma z, \gamma w; s) = j_\gamma^k(z) T(z, w; s)$ follows clearly from (2). This completes the proof.

As a consequence of the above theorem, we see that $E_k^{n,r}(z, s; f, \Gamma)$ is real analytic in (z, s) at least for $\mathrm{Re}(s) > (n+r+1)/2$. Then meromorphic continuation with respect to s, together with a general principle (Lemma 4.7 of [**99c**] due to Fefferman and Narasimhan, which is a special case of a result of Shiffman as noted in notes to [**99c**]), guarantees the real analyticity at the point $(z, s) \in \mathcal{H}_n \times \mathbf{C}$ where the function is finite. (To be precise, we must invoke Lemma 3.7 and Theorem 7.1.) Also we can apply any C^∞ differential operator to the series of (1.34) termwise, because of expression (2). See also notes to [**99c**].

Finally we note that a result parallel to the above theorem can be proved for the Eisenstein series on unitary groups studied in [**97a**] in the same manner. Indeed, we take $\delta(w, z)$ of [**97a**, (6.6.8)] in place of $\varepsilon(z, w)$; equality (1) in this case is [**97a**, (6.6.9)]. Notice that $\delta(w, z) \neq 0$ for every z, w by [**97a**, (6.6.10)].

95b. Zeta functions and Eisenstein series on metaplectic groups

p. 22, line 17: For "G^n" read "G^r".
p. 25, (1.6a): For "x" (two places) read "$\mathfrak{x}$".
p. 25, (1.6b): For "(x_v)" read "$(\mathfrak{x}_v)$".
p. 40, line 18 from the bottom: Insert "that" after "that".

96a. Convergence of zeta functions on symplectic and metaplectic groups

p. 331, (1.14): For "$\mathfrak{h}$" read "$\mathbf{h}$".

p. 332, (1.18): For "$\mathfrak{p}_{\mathfrak{c}}$" read "$\mathfrak{p}'$".

p. 336: Proposition 2.1: The conditions on θ and β must be as follows: $\theta \in \mathcal{M}_{\ell}(C, \eta')$ and $\beta \in G \cap \mathrm{diag}[r, \widetilde{r}]C$, where $C = \{x \in D[\mathfrak{b}^{-1}, \mathfrak{bc}] \,|\, \det(a_x) - 1 \in \mathfrak{e}\}$.

p. 336, line 12 from the bottom: For "$y \in E'_v, v \nmid \mathfrak{f}$" read "$y \in E'_v, v|\mathfrak{f}$".

p. 339, line 17: Insert "of" after "series".

p. 343, line 2: For "g" read "θ".

96b. Response

This is my response when I received Leroy P. Steele Prize for Lifetime Achievement from the American Mathematical Society at the Summer Mathfest held at the University of Washington, August 1996.

I mentioned "a lecture" by Eichler in Tokyo and "a hexagon" he drew on the blackboard. Actually he gave several lectures from April 21 through 24, 1958 at the University of Tokyo. Notes were taken by Y. Taniyama, and published (in Japanese) in Sugaku, 10 (1959), 182–190. I did not attend the lectures, as I was in Paris at that time, but found the hexagon in the notes. There was a colloquium in honor of Eichler's 70th birthday at the University of Basel, June 4 and 5, 1982, and I was asked to talk on his work, which I gladly did. At the beginning of the talk I drew on the blackboard a hexagon whose vertices were named as in the present article, saying that it was the same as what he drew 24 years ago. He was much amused by it and said that he had completely forgotten about it.

97b. Zeta functions and Eisenstein series on classical groups

This is the text of my lecture at the colloquium "Elliptic curves and modular forms" March 15–17, 1996 at the National Academy of Sciences, Washington, DC. This is another case of incompetent copyeditor and typesetter. Therefore, instead of reproducing the published version, we present the original manuscript version.

99a. An exact mass formula for orthogonal groups

p. 27, line 8 from the bottom: For "a suitable $\mathfrak{g}_v$-basis of L'_v" read "$\{e_i\}$".

p. 61, line 2 from the bottom: The symbol in brackets should be "$\begin{bmatrix} q & s\widehat{q} \\ 0 & \widehat{q} \end{bmatrix}$".

99c. Generalized Bessel functions on symmetric spaces

Lemma 4.7, due to Fefferman and Narasimhan, is actually contained in a much stronger result by B. Shiffman: Separate analyticity and Hartogs theorems, Indiana University Mathematics Journal, 38 (1989), 943–957, Theorem 1.

We can prove Theorem 5.3 without Lemmas 5.1 and 5.2, and even without knowing the real analyticity of $\xi_X(y, h; s)$ in (y, s). First we take the Laplace-Beltrami operator L on H. Then the integrand of the last integral over X of (5.9) is a solution f of an equation of the form $\big(L + \partial^2/\partial s\partial\bar{s} - c(s)\big)f = 0$ with a polynomial $c(s)$. Therefore we can apply Theorem A (ii) of notes to **[87a]** to (5.9), and find that $\xi_X(y, h; s)$ is real analytic in (y, s) for $\mathrm{Re}(s) > \lambda$. By Theorem A (i), we can apply every G-invariant differential operator to (5.9) under the integral sign, and obtain Theorem 5.3.

In the paragraph after the proof of Theorem 5.3 it is stated that the result of the same type holds for the generalized confluent hepergeometric functions studied in **[82c]**. Since the proof requires a nontrivial modification of a standard fact in [H] cited on lines 2–3 of the proof of the theorem, we present here a self-contained treatment of the subject, by first proving a principle which generalizes the fact just mentioned.

We consider a noncompact connected semisimple Lie group G with finite center and a maximal compact subgroup K of G, so that G/K is a symmetric space of noncompact type; this may or may not be hermitian. Given a one-dimensional analytic representation $\rho : K \to \mathbf{C}^\times$, we denote by $C^\infty(G, \rho)$ the set of all C^∞ functions f on G such that $f(gk) = \rho(k^{-1})f(g)$ for every $k \in K$ and $g \in G$. We assume that $G = RBK$ with closed Lie subgroups R and B such that

$$(\#) \qquad RB \cap K = B \cap K \text{ and } bRb^{-1} \subset R \text{ for every } b \in B.$$

For example, we can take a parabolic subgroup P of G such that $G = PK$, and consider a decomposition $P = RB$ with the unipotent radical R of P and a reductive factor B of P. An Iwasawa decomposition $G = NAK$ is another example. Let $\mathfrak{g}_0$, $\mathfrak{r}_0$, $\mathfrak{b}_0$, and $\mathfrak{k}_0$ be the Lie algebras of G, R, B, and K. We denote by $\mathfrak{g}$, $\mathfrak{r}$, $\mathfrak{b}$, and $\mathfrak{k}$ the complexifications of these, by $U(\mathfrak{x})$ the universal enveloping algebra of $\mathfrak{x}$, and by $D(\rho)$ the set of all elements of $U(\mathfrak{g})$ that map $C^\infty(G, \rho)$ into itself. We then put

$$\mathcal{E}_R(\rho) = \big\{f \in C^\infty(G, \rho) \big| f(rg) = f(g) \text{ for every } r \in R \text{ and } g \in G\big\}.$$

Let ρ_1 be the restriction of ρ to $B \cap K$. Taking $\{B, B \cap K, B \cap R, \rho_1\}$ in place of $\{G, K, R, \rho\}$, we can similarly define a subalgebra $D(\rho_1)$ of $U(\mathfrak{b})$ and a subset $\mathcal{E}_{B\cap R}(\rho_1)$ of $C^\infty(B, \rho_1)$. For every $f \in \mathcal{E}_R(\rho)$ the restriction of f to B belongs to $\mathcal{E}_{B\cap R}(\rho_1)$, and we easily see that the restriction gives a bijection of $\mathcal{E}_R(\rho)$ onto $\mathcal{E}_{B\cap R}(\rho_1)$. Indeed, if $rbk = r_1b_1k_1$ with $r, r_1 \in R$, $b, b_1 \in B$, and $k, k_1 \in K$, then $kk_1^{-1} \in RB \cap K = B \cap K$ by (#), and hence $r_1^{-1}r \in R \cap B$.

Theorem 1. *Given $f \in \mathcal{E}_R(\rho)$, let φ be the restriction of f to B. Suppose φ is an eigenfunction of $D(\rho_1)$ in the sense that $T\varphi = \beta(T)\varphi$ with $\beta(T) \in \mathbf{C}$ for every $T \in D(\rho_1)$. Then $Yf = \gamma(Y)f$ with $\gamma(Y) \in \mathbf{C}$ for every $Y \in D(\rho)$. Moreover, $\gamma(Y)$ is completely determined by ρ, Y, and β.*

PROOF. We first prove:

(∗) *If $h \in C^\infty(G)$ and $h(rg) = h(g)$ for every $r \in R$ and $g \in G$, then $(Xh)(b) = 0$ for every $X \in \mathfrak{r}$ and every $b \in B$.*

To prove this, we may assume that $X \in \mathfrak{r}_0$. Then, for $b \in B$ we have

$$(Xh)(b) = (d/dt)h\big(b \cdot \exp(tX)\big)_{t=0} = (d/dt)h\big(b \cdot \exp(tX)b^{-1}b\big) = (d/dt)h(b) = 0,$$

since $b\cdot\exp(tX)b^{-1} \in R$. This proves (*). Now extend $-d\rho : \mathfrak{k} \to \mathbf{C}$ to a homomorphism $\alpha : U(\mathfrak{k}) \to \mathbf{C}$. Then $Wf = \alpha(W)f$ if $W \in U(\mathfrak{k})$ and $f \in C^\infty(G, \rho)$. Let $Y \in D(\rho)$. Since $\mathfrak{g} = \mathfrak{r} + \mathfrak{b} + \mathfrak{k}$, by a well-known principle, Y can be expressed as a finite sum of elements of the form TZW with $T \in U(\mathfrak{r})$, $Z \in U(\mathfrak{b})$, and $W \in U(\mathfrak{k})$. (The expression $\mathfrak{g} = \mathfrak{r}+\mathfrak{b}+\mathfrak{k}$ may not be a direct sum, but that causes no problem.) Therefore we can find a finite sum $E = \sum ZW$ with such Z and W, such that $Y - E \in \mathfrak{r}U(\mathfrak{g})$. Let $f \in \mathcal{E}_R(\rho)$. Then for any $Q \in U(\mathfrak{g})$ we can take Qf to be h in (*). Therefore $\big[(Y - E)f\big](b) = 0$ for every $b \in B$. Thus

$$(Yf)(b) = (Ef)(b) = \sum (ZWf)(b) = \sum \big(\alpha(W)Zf\big)(b) = (T\varphi)(b),$$

where $T = \sum \alpha(W)Z$, which is an element of $U(\mathfrak{b})$. Since $Y \in D(\rho)$ and $f \in \mathcal{E}_R(\rho)$, we easily see that $Yf \in \mathcal{E}_R(\rho)$, and hence $T\varphi \in \mathcal{E}_{B\cap R}(\rho_1)$. Now $f \mapsto \varphi$ is a bijection of $\mathcal{E}_R(\rho)$ onto $\mathcal{E}_{B\cap R}(\rho_1)$. Therefore $T\varphi \in \mathcal{E}_{B\cap R}(\rho_1)$ for every $\varphi \in \mathcal{E}_{B\cap R}(\rho_1)$, so that $T \in D(\rho_1)$. Thus $T\varphi = \beta(T)\varphi$ by our assumption, and hence Yf coincides with $\beta(T)f$ on B. Since both Yf and $\beta(T)f$ belong to $\mathcal{E}_R(\rho)$ and they coincide on B, we obtain $Yf = \beta(T)f$, which proves our theorem.

Now we consider the tube domains considered in **[82c]**. To be explicit, we put

$$J_m = \begin{bmatrix} 0 & -1_m \\ 1_m & 0 \end{bmatrix}, \qquad J_m^{(2)} = \mathrm{diag}[J_m, J_m],$$

and define a group G, a vector space V, a domain of positivity P, and a space $H = V + iP$ of tube type as follows:

Type A. $\quad G = \{\alpha \in SL_{2m}(\mathbf{C}) \,|\, \alpha^* J_m \alpha = J_m\}$, $V = \{x \in \mathbf{C}_m^m \,|\, x = x^*\}$,

$$P = \{x \in V \,|\, x > 0\}, \qquad H = \{z \in \mathbf{C}_m^m \,|\, i(z^* - z) > 0\}.$$

Type B. $\quad G =$ the identity component of $\{\alpha \in SL_{m+2}(\mathbf{R}) \,|\, {}^t\alpha S \alpha = S\}$,

$$S = \mathrm{diag}\left[Q, -\begin{bmatrix} 0 & 1 \\ 1 & 0 \end{bmatrix}\right], \quad V = \mathbf{R}^m, \quad P = \{x \in V \,|\, {}^t x Q x < 0,\ {}^t x Q e < 0\},$$

$$H = \{z \in \mathbf{C}^m \,|\, i(\bar{z} - z) \in P\}.$$

Here Q is a symmetric element of $GL_m(\mathbf{R})$ of signature $(m-1, 1)$, and e is a fixed element of $\mathbf{R}^m$ such that ${}^t e Q e < 0$.

Type C. $\quad G = \{\alpha \in GL_{2m}(\mathbf{R}) \,|\, {}^t\alpha J_m \alpha = J_m\}$, $\quad V = \{x \in \mathbf{R}_m^m \,|\, {}^t x = x\}$,

$$P = \{x \in V \,|\, x > 0\}, \quad H = \{z \in \mathbf{C}_m^m \,|\, {}^t z = z,\ i(z^* - z) > 0\}.$$

Type D. $\quad G = \{\alpha \in SL_{4m}(\mathbf{C}) \,|\, \alpha^* J_{2m} \alpha = J_{2m},\ \bar{\alpha} J_m^{(2)} = J_m^{(2)} \alpha\}$,

$$V = \{x \in \mathbf{C}_{2m}^{2m} \,|\, {}^t x J_m = J_m x,\ x^* = x\}, \quad P = \{x \in V \,|\, x > 0\},$$

$$H = \{z \in \mathbf{C}_{2m}^{2m} \,|\, {}^t z J_m = J_m z,\ i(z^* - z) > 0\}.$$

Types A, B, C, and D correspond to Cases II, IV, I, and III of **[82c]**; the symbols V, P, and H are the same as V_m, P_m, and H_m in that paper; for Type B we put $\sigma(x, y) = -{}^t x Q y$, and take this to be σ of **[82c**, (1.5.IV)**]**. However, the present G is a semisimple group acting on H, and not the group G_m of **[82c]**. These P, V, and H are essentially the same as those defined in **[84b**, Section 3] and **[94c**, Section 1]. In each case we have a point $\mathbf{o}$ of H, which we call the *origin* of H, and with which we define K by

$$K = \{\alpha \in G \,|\, \alpha\mathbf{o} = \mathbf{o}\}.$$

Then H can be identified with G/K. Also we have a function $\varepsilon(z)$ on H and a $\mathbf{C}^\times$-valued factor of automorphy $j(\alpha, z)$ defined for $\alpha \in G$ and $z \in H$ by

$$\varepsilon(z) = \begin{cases} \det\big((i/2)(z^* - z)\big) & \text{(Types A and C)}, \\ (1/2)\cdot {}^t(z - \bar{z})Q(z - \bar{z}) & \text{(Type B)}, \\ \det\big((i/2)(z^* - z)\big)^{1/2} & \text{(Type D)}, \end{cases}$$

$$j(\alpha, z) = \begin{cases} \det(cz + d) & \text{(Types A, C, and D)}, \\ cz + (p/2)\cdot {}^t zQz + d & \text{(Type B)}, \end{cases}$$

Here $[c \;\; p \;\; d]$ with an m-dimensional row vector c is the last row of α for Type B; $[c \;\; d]$ is the lower half of α otherwise.

With a fixed $\kappa \in \mathbf{Z}$ we consider $\rho : K \to \mathbf{C}^\times$ defined by $\rho(k) = j(k, \mathbf{o})^\kappa$ for $k \in K$. Given $f \in C^\infty(H)$ and $\gamma \in G$ we define $f\|_\kappa\gamma \in C^\infty(H)$ by $(f\|_\kappa\gamma)(z) = j(\gamma, z)^{-\kappa} f(\gamma z)$. Let $\mathfrak{D}_\kappa$ be the set of all C^∞ differential operators W on H such that $W(f\|_\kappa\gamma) = (Wf)\|_\kappa\gamma$ for every $f \in C^\infty(H)$ and $\gamma \in G$. In Theorem D of notes to **[90a]** we proved that there is an isomorphism $W \mapsto W^\rho$ of $\mathfrak{D}_\kappa$ onto the ring of differential operators on $C^\infty(G, \rho)$ obtained from $D(\rho)$ such that $(Wf)^\rho = W^\rho f^\rho$ for every $f \in C^\infty(H)$, where $f^\rho(g) = (f\|_\kappa g)(\mathbf{o})$ for $g \in G$.

We now consider the functions defined in **[82c]**. Define $\delta(z)^s$ and $\sigma(x, y)$ as in **[82c**, pp. 272–273], and $\xi(g, h; \alpha, \beta)$ for $g \in P$, $h \in V$, and $\alpha, \beta \in \mathbf{C}$ by **[82c**, (1.29)]. Put $\mathbf{e}(x) = \exp(2\pi i x)$ for $x \in \mathbf{C}$.

Theorem 2. *For every $s \in \mathbf{C}$ and every $h \in V$ the functions $\varepsilon(z)^s$ and $\varepsilon(z)^s\delta(z + h)^{-\iota\kappa - s}\delta(\bar{z} + \bar{h})^{-s}$ are eigenfunctions of all elements of $\mathfrak{D}_\kappa$, where $\iota = 2$ for Type D and $\iota = 1$ otherwise. Moreover,*

$$(**) \qquad \varepsilon(z)^s\mathbf{e}\big(\sigma(h, x)\big)\xi(y, h; \iota\kappa + s, s)$$

as a function of $z = x + iy$ with $x \in V$ and $y \in P$, when finite, is an eigenfunction of all elements of $\mathfrak{D}_\kappa$. The eigenvalues of an element of $\mathfrak{D}_\kappa$ for these three types of functions are the same, and depend only on s, κ, and H.

PROOF. We can easily find a natural Iwasawa decomposition $G = NAK$ such that $j(n, z) = 1$ for every $n \in N$ and $j(a, z)$ for $a \in A$ is a product of "real coordinates" of a. (For Type B, each $a \in A$ has two coordinates, but only one of them gives $j(a, z)$.) We consider f^ρ with $f(z) = \varepsilon(z)^s$ and observe that Theorem 1 (with N and A as R and B there) is applicable to f^ρ. Thus f^ρ is an eigenfunction of $D(\rho)$, and hence $\varepsilon(z)^s$ is an eigenfunction of $\mathfrak{D}_\kappa$. Given $h \in V$, we take an element α of G as follows:

$$\alpha = \begin{bmatrix} 0 & -1_{\iota m} \\ 1_{\iota m} & h \end{bmatrix} \text{ (Types A, C, D);} \quad \alpha = \begin{bmatrix} 1 & 0 & h \\ 0 & 0 & -1 \\ {}^t h\sigma & -1 & {}^t h\sigma h/2 \end{bmatrix} \text{ (Type B).}$$

Then $j(\alpha, z) = 2^{-1}\delta(z + h)$ for Type B and $j(\alpha, z) = \delta(z + h)^\iota$ otherwise. Thus

$$(\varepsilon^s\|_\kappa\alpha)(z) = c^{\kappa+2s}\varepsilon(z)^s\delta(z + h)^{-\iota\kappa - s}\delta(\bar{z} + \bar{h})^{-s},$$

where $c = 2$ for Type B and $c = 1$ otherwise. This shows that the second function of our theorem is an eigenfunction of $\mathfrak{D}_\kappa$. Next, from the definition of $\xi(y, h; \alpha, \beta)$ in [**82c**, (1.25)], for $z = x + iy$ with $x \in V$ and $y \in P$ we have

$$\mathbf{e}(\sigma(h, x))\xi(y, h; \alpha, \beta) = \int_V \mathbf{e}(\sigma(h, x - v))\delta(v + iy)^{-\alpha}\delta(v - iy)^{-\beta}dv$$
$$= \int_V \mathbf{e}(-\sigma(h, u))\delta(z + u)^{-\alpha}\delta(\overline{z} + \overline{u})^{-\beta}du,$$

when the last integral is convergent. Here we used the fact that $\delta(x - iy + u) = \delta(\overline{z} + \overline{u})$, which is obvious for Types B and C; for Types A and D we have $z^* + u^* = x - iy + u$ and $\delta(w) = \delta({}^t w)$, so that $\delta(x - iy + u) = \delta(z^* + u^*) = \delta(\overline{z} + \overline{u})$. Therefore, applying an element of $\mathfrak{D}_\kappa$ to $\varepsilon(z)^s$ times the last integral, we find that the function of (**) is an eigenfunction, provided differentiation under the integral sign can be justified. The last integral over V is locally uniformly convergent with respect to z, provided $\mathrm{Re}(\alpha)$ and $\mathrm{Re}(\beta)$ are sufficiently large; see [**82c**, (1.23) and (1.28)]. Now the real analyticity of $\xi(y, h; \alpha, \beta)$ times certain gamma factors in (y, α, β) was established in [**82c**, Theorem 5.9]. Therefore, by Lemma A (i) of notes to [**87a**], differentiation under the integral sign can be justified, and we know that (**) is an eigenfunction for sufficiently large $\mathrm{Re}(s)$. By analyticity, the same is true for arbitrary s, provided the function is finite, since the eigenvalue is clearly a polynomial of s. This completes the proof.

Explicit expressions for the eigenvalues of certain elements of $\mathfrak{D}_\kappa$ as polynomials of s are given in [**84a**, Proposition 11.12].

Next we consider the Eisenstein series studied in [**95a**] and [**95b**]. Though we can treat more general cases including the case of unitary groups and also forms of half-integral weight, here we discuss only the case of forms of integral weight on symplectic groups. Let the notation be as in [**95a**, Section 1]; in particular, we define $G^n = Sp(n, F)$, a space $\mathcal{H}_n$, and a parabolic subgroup $P^{n,r}$ as in [**95a**, (1.2a, c) and (1.21a)]. We fix a weight $k \in \mathbf{Z}^{\mathbf{a}}$ and define the ring $\mathfrak{D}_k^n$ of differential operators on $\mathcal{H}_n$ with respect to the factor of automorphy $j_\alpha(z)^k$ of [**95a**, (1.10a)], in the same manner as for $\mathfrak{D}_\kappa$. Define also $\delta(z, s; f)$ and $E_k^{n,r}(z, s; f, \Gamma)$ by [**95a**, (1.32) and (1.34)] with a C^∞ function f on $\mathcal{H}_r$ such that $f\|_k\gamma = f$ for every γ in a congruence subgroup of G^r and $|f(z)\delta(z)^{k/2}|$ is bounded on $\mathcal{H}$; we do not assume that f is holomorphic. The proof of convergence in [**97a**, Proposition A3.7 and §A3.9] is applicable to this case, and so $E_k^{n,r}(z, s; f, \Gamma)$ is meaningful for $\mathrm{Re}(s) > (n + r + 1)/2$.

Theorem 3. *Suppose f is an eigenfunction of every element of $\mathfrak{D}_k^r$; then $\delta(z, s; f)$, for every $s \in \mathbf{C}$, is an eigenfunction of every element of $\mathfrak{D}_k^n$. The same is true for $E_k^{n,r}(z, s; f, \Gamma)$ if $\mathrm{Re}(s) > (n+r+1)/2$. Moreover, every holomorphic cusp form on $\mathcal{H}_r$ is an eigenfunction of every element of $\mathfrak{D}_k^r$, and $E_k^{n,r}(z, s; f, \Gamma)$ for such a holomorphic f is an eigenfunction of every element of $\mathfrak{D}_k^n$ at the point (z, s) where the function is finite.*

PROOF. Let R' be the unipotent radical of $P^{n,r}$, and T (resp. Δ) the group of all elements of the form $\mathrm{diag}[{}^t d^{-1}, d]$ with upper triangular unipotent matrices d in $GL_{n-r}(\mathbf{R})$ (resp. diagonal elements d of $GL_{n-r}(\mathbf{R})$ with positive diagonal entries). Using the notation $\beta \times \gamma$ of [**95a**, (1.25)], we take R and B to be given by $R = R'_{\mathbf{a}}(1 \times T^{\mathbf{a}})$ and $B =$

$G'_{\mathbf{a}} \times \Delta^{\mathbf{a}}$. We have then $G^n_{\mathbf{a}} = RBK$ with $K = \{\alpha \in G^n_{\mathbf{a}} | \alpha(\mathbf{i}_n) = \mathbf{i}_n\}$ with $\mathbf{i}_n$ of [**95a**, (1.6)]. Define $\rho : K \to \mathbf{C}^\times$ by $\rho(t) = j_t(\mathbf{i}_n)^k$ for $t \in K$. Given f as in our theorem, define $h \in C^\infty(G'_{\mathbf{a}})$ by $h(\gamma) = (f\|_k\gamma)(\mathbf{i}_r)$ for $\gamma \in G'_{\mathbf{a}}$, and define similarly $q \in C^\infty(G^n_{\mathbf{a}})$ by taking $\delta(z, s; f)$ in place of f. Define also a function φ on B by $\varphi(\gamma \times d) = h(\gamma)\det(d)^{-2su}$ for $\gamma \in G'_{\mathbf{a}}$ and $d \in \Delta^{\mathbf{a}}$. A simple calculation shows that $q \in \mathcal{E}_R(\rho)$ and φ is the restriction of q to B. By Theorem D in notes to [**90a**], h is an eigenfunction on $G'_{\mathbf{a}}$; then we can easily verify that φ is an eigenfunction of $D(\rho_1)$, so that by Theorem 1, q is an eigenfunction of $D(\rho)$. Again by Theorem D, $\delta(z, s; f)$ is an eigenfunction of $\mathfrak{D}^n_k$. Now $E^{n,r}_k(z, s; f, \Gamma) = \sum_{\gamma \in A} \delta(z, s; f)\|_k\gamma$, and the effect of an element of $\mathfrak{D}^n_k$ can be achieved by termwise application, by virtue of Lemma A (i) in notes to [**87a**]. This shows that $E^{n,r}_k$ is an eigenfunction for $\mathrm{Re}(s) > (n + r + 1)/2$.

Here we need to know that $E^{n,r}_k(z, s; f, \Gamma)$ is at least C^∞, which was proved in notes to [**95a**] when f is a holomorphic cusp form. This point can be proved, for both holomorphic and nonholomorphic f, more directly in the following way. The Casimir element of $U(\mathfrak{g})$ produces an element Z of $\mathfrak{D}^n_k$. An eigenfunction h of Z is a solution of an equation $Zh = c(s)h$ with a polynomial $c(s)$. Put $Q = Z - c(s) + \partial^2/\partial s\partial\bar{s}$. This is real analytic and elliptic on $\mathcal{H} \times \mathbf{C}$. By (ii) of Lemma A to notes to [**87a**], applying Q to $\sum_\gamma \delta(z, s; f)\|_k\gamma$, we find that $E^{n,r}_k(z, s; f, \Gamma)$ is real analytic in (z, s) at least for $\mathrm{Re}(s) > (n + r + 1)/2$.

Now suppose f is a holomorphic cusp form. As remarked in notes to [**90a**], $\mathfrak{D}^r_k$ is generated by the operators $L^Z_\rho = (-1)^p\theta D^Z_{\rho\otimes\sigma_Z} E^Z$ of [**94c**, (2.20)]. Since E^Z annihilates holomorphic functions, we have $L^Z_\rho f = 0$, and hence f is an eigenfunction of $\mathfrak{D}^r_k$. Thus the above result is applicable. Also [**95a**, Lemma 3.7 and Theorem 7.1] guarantee meromorphic continuation of $E^{n,r}_k(z, s; f, \Gamma)$ in s to the whole $\mathbf{C}$. Then Lemma 4.7 establishes the real analyticity in (z, s), so that we obtain the last assertion of our theorem.

99e. André Weil as I knew him

One of the anecdotes Weil told about Hardy is as follows: Once Hardy, in the middle of his lecture, said that a certain mathematical statement was trivial, and then became unsure of it. He left the lecture hall and came back after fifteen minutes. Then he said, "Yes, it is trivial," and continued his lecture.

His comment "Hardy's opinion that mathematics is a young man's game is nonsense" was actually made as a response to my question. In 1962 I told him my opinion that if someone wishes to be a first-rate mathematician, then he gradually elevates himself and reaches a certain height at the age of perhaps 40 or 45; after that he should try (or he must be able) to keep that level as long as possible. I told him that that was my ideal, and asked him what he thought. He merely said, "You will find out." At that time I took it to be negative. Then, around 1980 I reminded him of this conversation, which he remembered. He said that he didn't say it negatively, and added that "Hardy's opinion that mathematics is a young man's game is nonsense." I am inclined to think, however, that he was not completely sure in 1962; maybe he found out himself.